BIOLOGY

LIFE ON EARTH WITH PHYSIOLOGY 12E

TERESA AUDESIRK
UNIVERSITY OF COLORADO DENVER

GERALD AUDESIRK
UNIVERSITY OF COLORADO DENVER

BRUCE E. BYERS
UNIVERSITY OF MASSACHUSETTS AMHERST

Courseware Portfolio Manager: Cady Owens
Director of Portfolio Management: Beth Wilbur
Content Producer: Anastasia Slesareva
Managing Producer: Michael Early
Courseware Director, Content Development: Ginnie Simione Jutson
Senior Development Editor: Jennifer Angel
Courseware Editorial Assistant: Chelsea Noack
Rich Media Content Producer: Robert Johnson
Full-Service Vendor: Mary Tindle, Pearson CSC
Copyeditor: Lucy Mullins
Compositor: Pearson CSC
Illustrators: ImagineeringArt.com, Inc.
Design Manager: Mark Ong
Interior and Cover Designer: Jeff Puda
Rights & Permissions Project Manager: Eric Schrader
Rights & Permissions Management: Matthew Perry, Pearson CSC
Photo Researcher: Kristin Piljay
Manufacturing Buyer: Stacey Weinberger
Director of Product Marketing: Allison Rona
Cover Photo Credit: Jay Fleming/Getty Images

Library of Congress Cataloging-in-Publication Data
Names: Audesirk, Teresa, author. | Audesirk, Gerald, author. | Byers, Bruce E., author.
Title: Biology : life on earth with physiology / Teresa Audesirk, University of Colorado Denver, Gerald Audesirk, University of Colorado Denver, Bruce E. Byers, University of Massachusetts Amherst.
Description: Twelfth edition. | New York: Pearson, [2020] | Includes index.
Identifiers: LCCN 2018050171 | ISBN 9780134813448 (loose leaf) | ISBN 0134813448 (loose leaf)
Subjects: LCSH: Biology.
Classification: LCC QH308.2 .A93 2020 | DDC 570—dc23 LC record available at https://lccn.loc.gov/2018050171

ISBN 10: 0-134-81344-8; ISBN 13: 978-0-134-81344-8 (Student edition)
ISBN 10: 0-135-44398-9; ISBN 13: 978-0-135-44398-9 (Instructor's Review Copy)
www.pearson.com

About the Authors

TERRY AND GERRY AUDESIRK grew up in New Jersey, where they met as undergraduates, Gerry at Rutgers University and Terry at Bucknell University. After marrying in 1970, they moved to California, where Terry earned her doctorate in marine ecology at the University of Southern California and Gerry earned his doctorate in neurobiology at the California Institute of Technology. As postdoctoral students at the University of Washington's marine laboratories, they worked together on the neural bases of behavior, using a marine mollusk as a model system.

They are now emeritus professors of biology at the University of Colorado Denver, where they taught introductory biology and neurobiology from 1982 through 2006. In their research, funded primarily by the National Institutes of Health, they investigated the mechanisms by which neurons are harmed by low levels of environmental pollutants and protected by estrogen.

Terry and Gerry are long-time members of many conservation organizations and share a deep appreciation of nature and of the outdoors. They enjoy hiking in the Rockies, walking and horseback riding near their home outside Steamboat Springs, and singing in the community chorus. Keeping up with the amazing and endless stream of new discoveries in biology provides them with a continuing source of fascination and stimulation. They are delighted that their daughter Heather has become a teacher and is inspiring a new generation of students with her love of chemistry.

BRUCE E. BYERS is a Midwesterner transplanted to the hills of western Massachusetts, where he is a professor in the biology department at the University of Massachusetts Amherst. He has been a member of the faculty at UMass (where he also completed his doctoral degree) since 1993. Bruce teaches courses in evolution, ornithology, and animal behavior, and does research on the function and evolution of bird vocalizations.

With love to Jack, Lori, and Heather and in loving memory of Eve and Joe.

— T. A. & G. A.

In memory of Bob Byers, a biologist at heart.

— B. E. B.

ABOUT THE COVER A green sea turtle (***Chelonia mydas***) swims above a bed of seagrass in the Caribbean Sea. Green sea turtles are found in tropical and subtropical marine waters worldwide, where adults feed mainly on algae and marine plants. Immature green sea turtles have a more omnivorous diet, consuming mollusks, jellyfish, and other invertebrates. Like other sea turtle species, green sea turtles migrate long distances from their feeding areas to their mating areas. After mating, a female emerges from the water onto a sandy beach, where she buries her eggs. A couple of months later, the eggs hatch and the hatchling turtles make their way back to the sea. The International Union for Conservation of Nature classifies green sea turtles as endangered. Many turtles drown when they become entangled in fishing nets, and people in many places harvest sea turtles and their eggs for food. Housing developments on sea turtle nesting beaches pose an additional threat.

Table of Contents

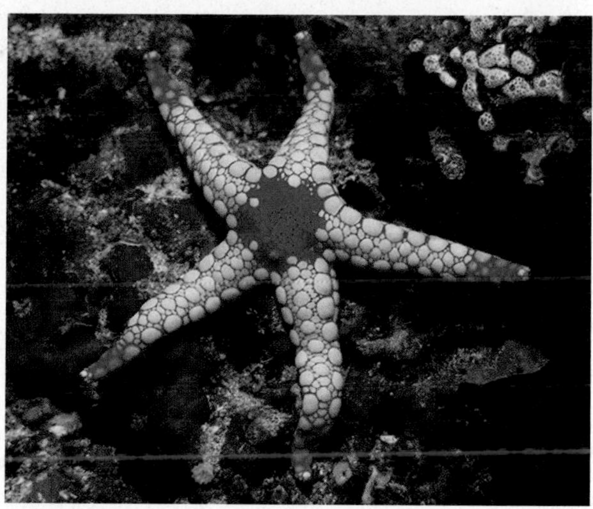

UNIT 4

Behavior and Ecology 429

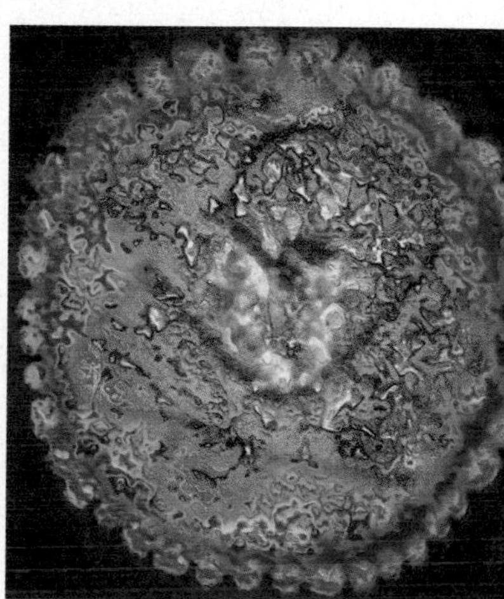

Case Studies

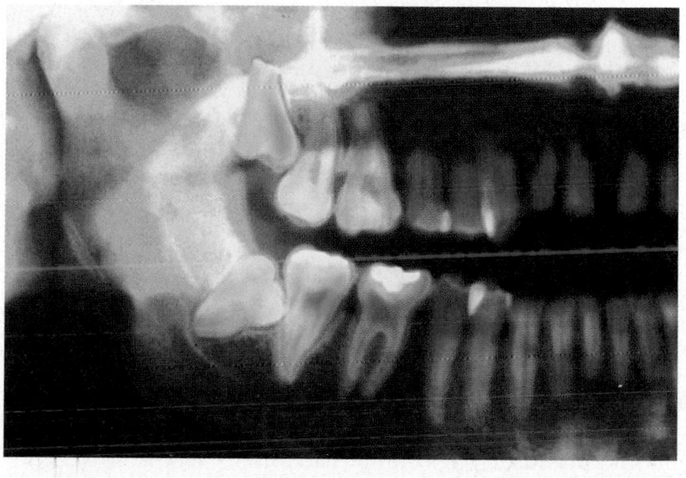

Essays

Earth Watch

Health Watch

Doing Science

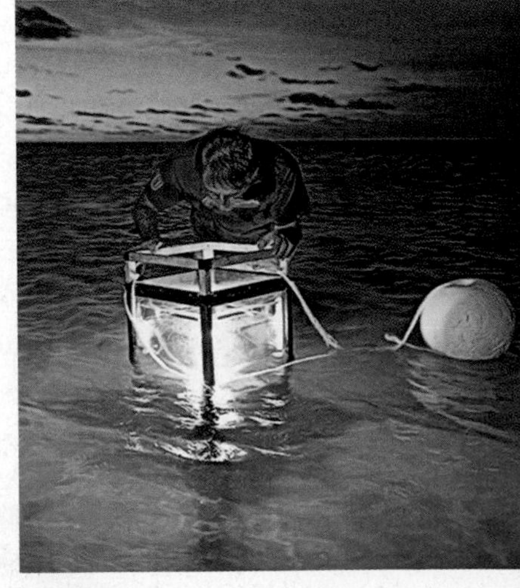

Have You Ever Wondered . . .

Preface

THE CASE FOR SCIENTIFIC LITERACY

Climate change, biofuels versus food and forests, bioengineering, stem cells in medicine, potential flu pandemics, the plight of polar bears and pandas, human population growth and sustainability: these are just some of the very real, urgent, and interrelated concerns facing our increasingly connected human societies. The Internet places a wealth of information—and a flood of misinformation—at our fingertips. Never have scientifically literate students been more important to humanity's future. As educators, we feel humbled before this massive challenge. As authors, we feel hopeful that the Twelfth Edition of *Biology: Life on Earth* will help lead introductory biology students along paths to understanding.

Scientific literacy requires a foundation of factual knowledge that provides a solid and accurate cognitive framework into which new information can be integrated. But more importantly, it endows people with the mental tools to separate the wealth of data from the morass of misinformation. Scientifically literate citizens are better able to evaluate facts and make informed choices in their personal lives and in the political arena.

This Twelfth Edition of *Biology: Life on Earth* continues our tradition of:

- Helping instructors present biological information in a way that will foster scientific literacy among their students.
- Helping to inspire students with a sense of wonder about the natural world, fostering an attitude of inquiry and a keen appreciation for knowledge gained through science.
- Helping students to recognize the importance of what they are learning to their future roles in our rapidly changing world.

BIOLOGY: LIFE ON EARTH, TWELFTH EDITION

. . . Is Organized Clearly and Uniformly

Navigational aids help students explore each chapter. An important goal of this organization is to present biology as a hierarchy of closely interrelated concepts rather than as a compendium of independent topics.

- Major sections are introduced as broad questions that stimulate students to think about the material to follow; subheadings are statements that summarize their specific content.
- Each major section concludes with "Check Your Learning" questions that remind students of the key concepts and content in the preceding passage.
- "Case Study Continued" segments end with probing questions to help students anticipate what they will learn next.
- A "Summary of Key Concepts" section ends each chapter, providing a concise, efficient review of the chapter's major topics.

. . . Engages and Motivates Students

Scientific literacy cannot be imposed on students—they must actively participate in acquiring the necessary information and skills. To be inspired to accomplish this, they must first recognize that biology is about their own lives. For example, we help students acquire a basic understanding and appreciation of how their own bodies function by including information about diet and weight, cancer, and lower back pain.

We fervently hope that students who use this text will come to see their world through keener eyes. For example, they will perceive forests, fields, and ponds as vibrant and interconnected ecosystems brimming with diverse life-forms rather than as mundane features of their everyday surroundings. If we have done our job, students will also gain the interest, insight, and information they need to look at how humanity has intervened in the natural world. If they ask the question, "Is this activity sustainable?" and then use their new knowledge and critical thinking skills to seek some answers, we can be optimistic about the future.

In support of these goals, the Twelfth Edition has updated features that make Biology more engaging and accessible.

- **Case Studies** Each chapter opens with an attention-grabbing "Case Study" that highlights topics of emerging relevance in today's world. Case Studies, including "Unstable Atoms Unleashed" (Chapter 2), "New Parts for Human Bodies" (Chapter 4), and "Unwelcome Dinner Guests" (Chapter 20), are based on news events, personal interest stories, or particularly fascinating biological topics. "Case Study Continued" segments weave the topic throughout the chapter, and "Case Study Revisited" completes the chapter, exploring the topic further in light of the chapter's lessons.
- **Boxed Essays** Three categories of essays enliven the text. "Earth Watch" essays explore pressing environmental issues; "Health Watch" essays cover important or intriguing medical topics; and "Doing Science" essays explain how scientific knowledge is acquired. In addition, "In Greater Depth" essays that provide more detailed information on some topics are available on Mastering Biology.
- **"Have You Ever Wondered" Questions** These popular features demystify common and intriguing questions, showing biology in the real world.

- **End-of-Chapter Questions** The questions that conclude each chapter allow students to review the material in different formats—multiple choice, fill-in-the-blank, short-answer, and essay—that help them to study and assess what they have learned. Answers to the multiple choice, fill-in-the-blank, and short-answer questions are included in the back of the book. Hints for the essay questions are included on Mastering Biology.
- **Key Terms and a Complete Glossary** Boldfaced key terms are defined clearly within the text as they are introduced. These terms are also available on Mastering Biology as flashcards, providing students with an opportunity to quiz themselves on important definitions. The glossary, carefully written by the authors, provides complete definitions for all key terms, as well as for other important biological terms.

. . . Encourages Critical Thinking

Throughout the text, we aim to help students develop and improve their ability to think critically about scientific information and concepts. Aspects of the text that foster critical thinking include:

- **Two Question Types in Essays and Figure Captions** In each chapter, students encounter a number of questions designed to encourage them to think critically about the content. "Think Critically" questions focus on solving problems, thinking about scientific data, or evaluating a hypothesis. "Consider This" questions invite students to form an opinion or pose an argument for or against an issue, based on valid scientific information. Answers to "Think Critically" questions are included in the back of the book; hints for "Consider This" questions are included on Mastering Biology.
- **In-depth Questions for Classroom Discussion** A number of new "Think Deeper" questions, available in the Instructor Resource area, follow up and extend "Think Critically" questions by asking additional, related questions that require more extensive thought and analysis.
- **Chapter-Ending Thought Questions** The end-of-chapter material for each chapter concludes with "Applying the Concepts" questions that challenge students to apply knowledge gained in the chapter to novel problems.
- **"Doing Science" Essays** These essays illustrate the process of science in a methodical way, emphasizing the process of what scientists do. Essays describe the details of experiments, highlighting exciting technology and data. They are structured to emphasize the process of inquiry, with sections titled "What Question Was Asked?", "How Was Evidence Gathered?", and "What Was Learned?" All "Doing Science" essays conclude with a "Think Critically" or "Consider This" question, encouraging students to analyze data or engage with the topics presented in the essay.
- **Data in "Earth Watch" Essays** Students will find examples of real scientific data in the form of graphs and tables. The essays are accompanied by "Think Critically" questions that challenge students to interpret the data, fostering increased understanding of how science is communicated.

. . . Is a Comprehensive Learning Package

The Twelfth Edition of *Biology: Life on Earth* is a complete learning package, providing updated and innovative teaching aids for instructors and learning aids for students.

ACKNOWLEDGMENTS

Biology: Life on Earth enters its Twelfth Edition guided by the excellent team at Pearson. Beth Wilbur, Director of Portfolio Management, continues to oversee the enterprise with warmth, competence, and first-rate leadership. Courseware Portfolio Manager Cady Owens, Content Development Director Ginnie Simione Jutson, and Content Producer Anastasia Slesareva coordinated this complex and multifaceted endeavor. Ginnie and Cady did a wonderful job of working with us to develop a revision plan to further extend the text's appeal and reach. We also appreciate Cady's extensive travel to share her enthusiasm for the text and its extensive ancillary resources with educators across the country. Mary Tindle, Content Producer at SPi Global, did a marvelous job of keeping everything on track and on schedule. Senior Development Editor Jennifer Angel carefully reviewed every word and figure in the manuscript, making sure the prose was clear, the images informative, and the organization logical. In addition, Jennifer worked diligently and with ingenuity to ensure that page layouts were attractive and effective. We very much appreciate her attention to detail and thoughtful suggestions. Our outstanding copyeditor, Lucy Mullins, improved the text and art and caught errors that we had overlooked. The book boasts a large number of excellent new photos, tracked down with skill and persistence by Kristin Piljay. Matthew Perry made sure we had proper permission to use the photos we chose.

We are grateful to Imagineeringart.com, Inc., under the direction of Project Manager Stephanie Marquez, for deciphering our art instructions and patiently making new adjustments to already outstanding figures. We owe our beautifully redesigned text and delightful new cover to Jeff Puda and Design Manager Mark Ong.

We thank Allison Rona, Director of Product Marketing, for making sure the finished product reached your desk. In her role as Manufacturing Buyer, Stacey Weinberger's expertise has served us well. The ancillaries are an endeavor fully as important as the text itself. Thanks also to Robert Johnson and Ashley Gordon for developing the outstanding Mastering Biology Web site that accompanies this text.

We are extremely fortunate to be working with the Pearson team. This Twelfth Edition of *Biology: Life on Earth* reflects their exceptional abilities and dedication.

Finally, we appreciate the 400+ faculty members from around the country who have checked and rechecked the text for accuracy and clarity over the past several years. Reviewers specific to the past two editions are listed below.

With gratitude,

TERRY AUDESIRK, GERRY AUDESIRK, AND BRUCE BYERS

Twelfth Edition Reviewers

Dana Almengor
Lenoir Community College
Erin Baumgartner
Western Oregon University
Molly Baxter
Southeastern Community College
Joel Bergh
Texas State San Marcos
Carol Chaffee
Cal State University Fullerton
Reggie Cobb
Nash Community College
Karen Dunbar
Ivy Tech Community College
Cindy Malone
Cal State University Northridge
Melissa Meador
Arkansas State University Beebe
Pam Oliver
Coffeyville Community College
Justin Rosemier
Lakeland Community College
Mike Vieth
Quinnipiac University
Alan Wasmoen
Metropolitan Community College
Taek You
Campbell University

Eleventh Edition Reviewers

Aekam Barot
Lake Michigan College
Mark Belk
Brigham Young University
Karen Bledsoe
Western Oregon University
Christine Bozarth
Northern Virginia Community College
Britt Canada
Western Texas College
Reggie Cobb
Nash Community College
Rachel Davenport
Texas State University, San Marcos
Diane Day
Clayton State University
Lewis Deaton
University of Louisiana at Lafayette
Peter Ekechukwu
Horry-Georgetown Technical College
Janet Gaston
Troy University
Mijitaba Hamissou
Jacksonville State University
Karen Hanson
Carroll Community College
Brian Ingram
Jacksonville State University
Karen Kendall-Fite
Columbia State Community College
Neil Kirkpatrick
Moraine Valley Community College
Damaris-Lois Lang
Hostos Community College
Tiffany McFalls-Smith
Elizabethtown Community and Technical College
Mark Meade
Jacksonville State University
Samantha Parks
Georgia State University
Indiren Pillay
Georgia College
John Plunket
Horry-Georgetown Technical College
Cameron Russell
Tidewater Community College
Roger Sauterer
Jacksonville State University
Terry Sellers
Spartanburg Methodist College
David Serrano
Broward College
Philip Snider
Gadsden State Community College
Judy Staveley
Carroll Community College
Katelynn Woodhams
Lake Michigan College
Min Zhong
Auburn University
Deborah Zies
University of Mary Washington

1 An Introduction to Life on Earth

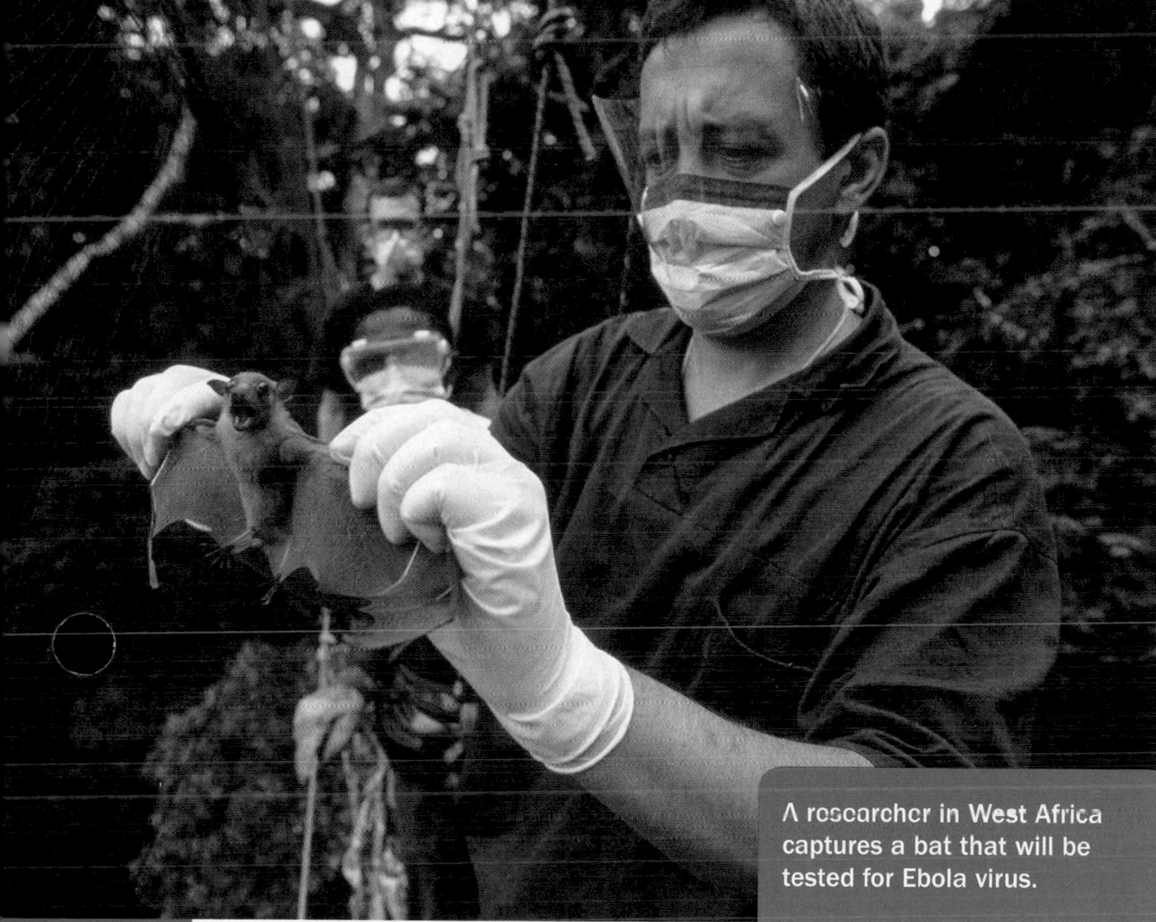

A researcher in West Africa captures a bat that will be tested for Ebola virus.

CASE STUDY

The Boundaries of Life

IN LATE 2013, two-year-old Emile Ouamouno became the first victim of a disease outbreak that would ultimately kill thousands. Emile lived in a small village in the country of Guinea, West Africa, where he became infected with the Ebola virus, most likely through contact with bats that roosted in a huge, hollow tree near the village. The infection soon killed Emile. Within weeks of his death, Emile's mother, sister, and grandmother had also died of Ebola. Other villagers who caught the disease from Emile and his family carried the virus to nearby settlements before they, too, died. Thus began a chain of Ebola virus transmission that eventually sickened more than 28,000 people, killing 11,300 of them by the time the epidemic subsided in mid-2016.

Victims of Ebola experience symptoms that include fever, headache, joint and muscle aches, and stomach pains, followed by vomiting, bloody diarrhea, and organ failure. Internal hemorrhaging can leave victims bleeding from nearly every orifice. Death usually occurs within 7 to 16 days after the onset of symptoms. Ebola is transmitted by contact with the body fluids of infected individuals, so caregivers and undertakers wear "moon suits" to protect themselves. There is no cure for Ebola. However, the recent epidemic spurred a successful effort to develop an effective vaccine, raising the hope that future Ebola outbreaks can be prevented.

Ebola is one of many diseases caused by viruses. Although some viral diseases, such as smallpox and polio, have been largely eradicated, others, like the common cold and influenza (flu), continue to make us miserable. Perhaps most alarming are dangerous viral diseases that seem to emerge suddenly. AIDS was unknown until 1981, Ebola emerged in 1975, and Marburg virus disease appeared for the first time in 1967. New variants of the flu virus emerge regularly; some of these previously unknown variants have sparked deadly epidemics.

Although the viruses that infect humans are of special interest to us, viruses also infect every other form of life. Viruses have some things in common with the organisms they infect; for example, they reproduce and they evolve. But despite these lifelike qualities, scientists disagree about whether viruses are living organisms. The reason for this disagreement may surprise you: There is no universally accepted scientific definition of life. What is life, anyway?

AT A GLANCE

1.1 WHAT IS LIFE?

Biology is the study of life. But what is life? This simple question does not have a simple answer. Although each of us has an intuitive understanding of what it means to say that something is alive, it has proved impossible to devise a precise scientific definition that neatly divides the living from the nonliving. Dictionary definitions typically fall back on phrases like, "the quality that distinguishes living organisms from inorganic objects and dead organisms," without providing much insight as to what that "quality" might be.

Because we don't have a precise definition of life, we must build our definition bit by bit, by describing a set of features. All living things, or **organisms**, share certain characteristics that, taken together, define life:

- Organisms actively maintain organized complexity.
- Organisms acquire and use energy and materials.
- Organisms sense and respond to stimuli.
- Organisms grow.
- Organisms reproduce.
- Organisms evolve.

Nonliving objects may possess some of these attributes, but only living things possess them all. In the sections below, we introduce the characteristics of life.

Organisms Actively Maintain Organized Complexity

Compared with nonliving matter, living things are highly complex and organized. A nonliving crystal of table salt consists of just two chemical elements, sodium and chlorine, arranged in a precise way; the salt crystal is organized but simple. The nonliving water of an ocean contains atoms of all the naturally occurring elements, but these atoms are randomly distributed; the oceans are complex but not organized. In contrast, even the simplest organisms contain the atoms of dozens of different elements linked together in thousands of specific combinations to form a cell, the basic unit of life (**FIG. 1-1**). Cells are both complex and organized. Each cell contains a huge variety of structures and chemicals enclosed by a thin sheet called the **plasma membrane**.

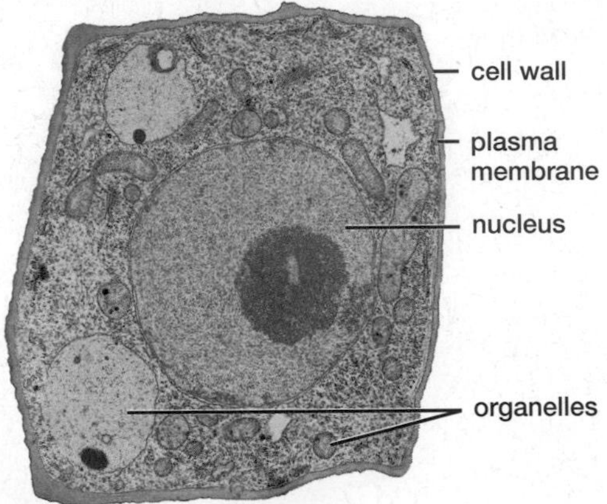
cell wall
plasma membrane
nucleus
organelles

▲ **FIGURE 1-1 The cell is the smallest unit of life** This artificially colored micrograph of a plant cell (a eukaryotic cell) shows a supporting cell wall (blue) that surrounds plant cells. Just inside the cell wall, the plasma membrane (found in all cells) has control over which substances enter and leave. Cells also contain several types of specialized organelles, including the nucleus, suspended within a fluid environment (orange).

Cells fall into two main groups: eukaryotic cells and prokaryotic cells. **Eukaryotic** cells contain a variety of **organelles**, which are structures (often surrounded by membrane) that carry out functions such as synthesizing large molecules, digesting food molecules, or obtaining energy. The term "eukaryotic" comes from Greek words meaning "true nucleus." As the name suggests, the **nucleus**, a membrane-enclosed organelle that contains the cell's DNA, is a prominent feature of eukaryotic cells (see Fig. 1-1). **Prokaryotic** cells are simpler than eukaryotic cells; they lack organelles enclosed by membranes. As their name—meaning "before the nucleus"—suggests, a prokaryotic cell's DNA is not confined within a nucleus. All organisms with prokaryotic cells are **unicellular** (exist as single cells); organisms with eukaryotic cells may be unicellular or **multicellular** like the water flea in **FIGURE 1-2**.

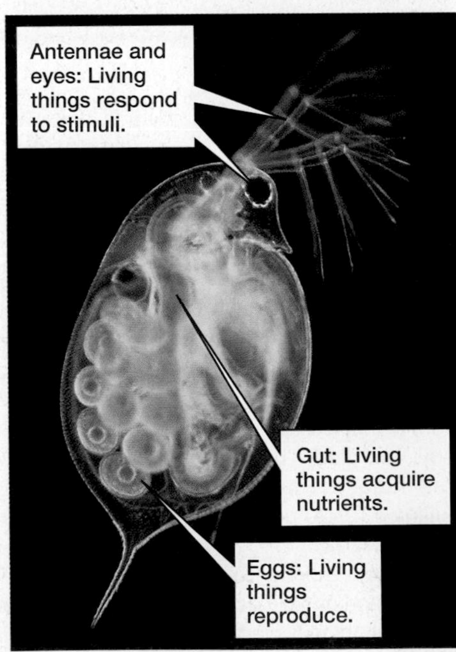
Antennae and eyes: Living things respond to stimuli.
Gut: Living things acquire nutrients.
Eggs: Living things reproduce.

▲ **FIGURE 1-2 Properties of life** Evolution has molded the adaptations that allow this water flea to respond to stimuli, acquire nutrients, grow, and reproduce.

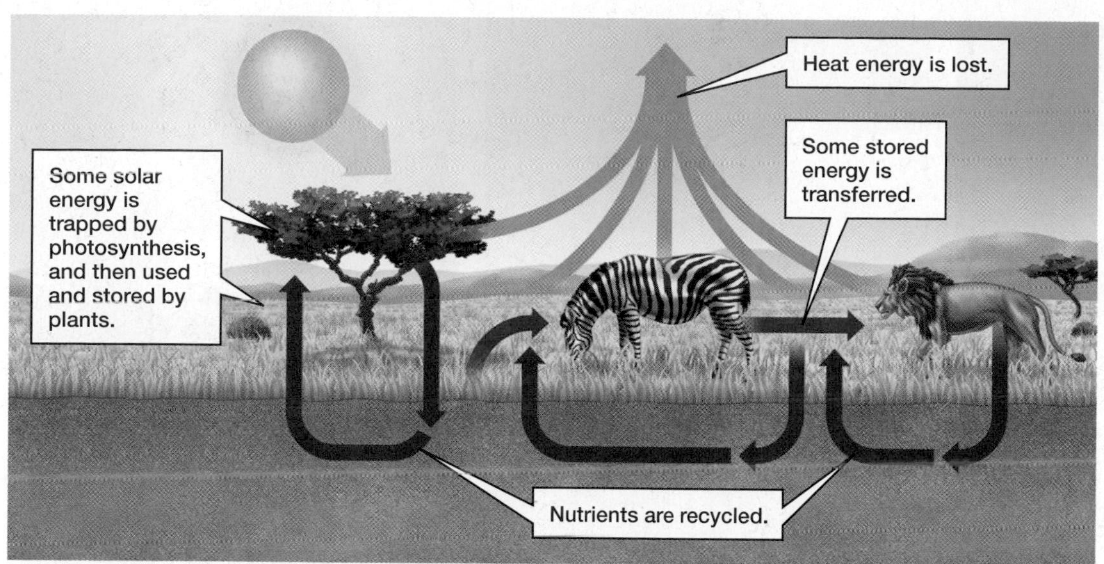

Heat energy is lost.
Some stored energy is transferred.
Some solar energy is trapped by photosynthesis, and then used and stored by plants.
Nutrients are recycled.

FIGURE 1-3 The flow of energy and the recycling of nonliving nutrients

THINK CRITICALLY Describe the source of the energy stored in the meat and the bun of a hamburger, and explain how the energy got from the source to the two foodstuffs.

Organisms Acquire and Use Energy and Materials

Organization and complexity tend to break down unless energy is used to maintain them. Living things, representing the ultimate in organized complexity, continuously use energy to maintain themselves. Almost all the energy that sustains life comes from sunlight. Some organisms capture solar energy directly through a process called **photosynthesis**. Photosynthetic organisms (plants and many single-celled organisms) trap and store the sun's energy for their own use. When these organisms are consumed by nonphotosynthetic organisms, this stored energy also powers the consumers. So, energy flows in a one-way path from the sun to photosynthetic organisms to all other forms of life (**FIG. 1-3**). However, some energy is lost as heat at each transfer from one organism to another, making less energy available with each transfer.

The energy that organisms acquire is continuously expended to maintain very specific internal conditions. The ability of an organism to maintain its internal environment within the limits required to sustain life is called **homeostasis**. To maintain homeostasis, cell membranes constantly pump certain substances in and others out. Organisms use both physiological and behavioral mechanisms to maintain the narrow temperature range that allows life-sustaining reactions to occur in their cells (**FIG. 1-4**).

FIGURE 1-4 Organisms maintain relatively constant internal conditions Evaporative cooling by water, both from sweat and from a bottle, helps this athlete maintain his body temperature during vigorous exercise.

Organisms obtain the materials that make up their bodies—such as minerals, water, and other simple chemical building blocks—from the air, water, soil, and bodies of other living things. Because life neither creates nor destroys matter, materials are continuously exchanged and recycled among organisms and their surroundings (see Fig. 1-3).

Organisms Sense and Respond to Stimuli

To obtain energy and nutrients, organisms must sense and respond to stimuli in their environments. Animals use specialized cells to detect light, temperature, sound, gravity, touch, chemicals, and many other stimuli from their external and internal surroundings. For example, when your brain detects both a low level of sugar in your blood (an internal stimulus) and the smell of food (an external stimulus), it causes your mouth to water, or *salivate* (a response) in anticipation of a meal. Plants, fungi, and unicellular organisms respond to stimuli using mechanisms that are effective for their needs (**FIG. 1-5**).

FIGURE 1-5 Bending toward the light Plants perceive and often bend toward light, which provides them with the energy they need to survive.

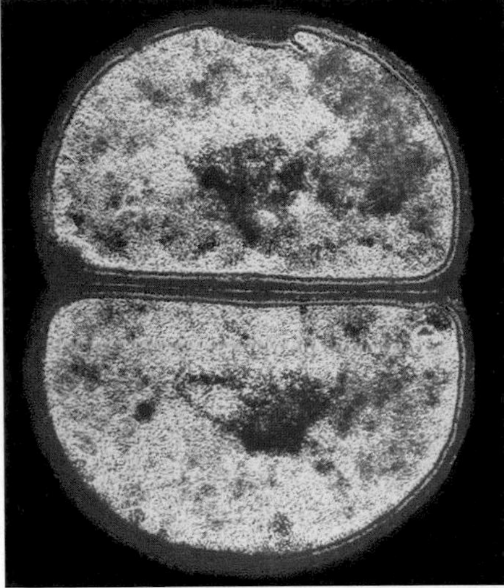

(a) Dividing *Streptococcus* bacterium (b) Dandelion producing seeds (c) Panda with its baby

▲ **FIGURE 1-6 Organisms reproduce**

Organisms Grow

At some time in its life, every organism grows. The water flea in Figure 1-2 was once much smaller—the size of one of the eggs you see in its body. Unicellular organisms such as bacteria grow to about double their original size before dividing in half to reproduce. Animals and plants grow by increasing the number of cells in their bodies via cell division, and in some cases by increasing the size of individual cells, as occurs in muscle and fat cells in animals and in food storage cells in plants. In multicellular organisms, growth may be accompanied by *development*, in which a growing organism becomes increasingly complex. For example, a fertilized egg develops into an adult with many complex structures.

Organisms Reproduce

Organisms reproduce in a variety of ways (**FIG. 1-6**); for example, by dividing in half, producing seeds, or bearing live young. The end result is always the same: new individuals of the same type of organism as the parent.

When organisms reproduce, the offspring inherit characteristics of the parents. The instructions for producing these characteristics are carried in the molecule **deoxyribonucleic acid (DNA)**, which is present in every cell and passed on to descendants (**FIG. 1-7**). Specific segments of DNA called **genes** are the basic units of heredity. The complete set of genes contained in each cell provides detailed instructions for building and maintaining an organism, much as a recipe provides instructions for baking a cake.

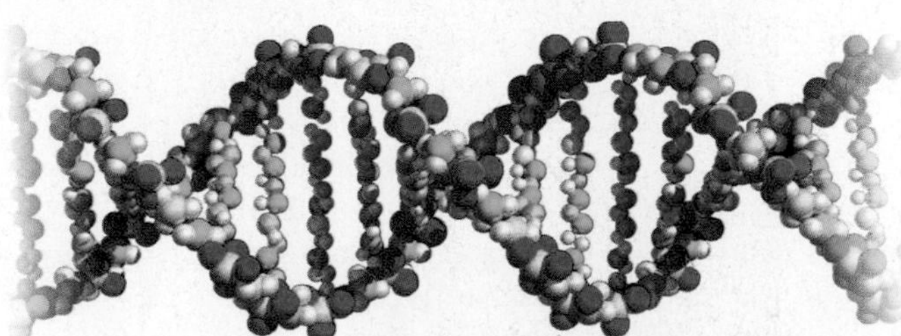

▲ **FIGURE 1-7 DNA** According to James Watson, codiscoverer of the structure of DNA, "A structure this pretty just had to exist."

Organisms Evolve

All living organisms descended from an ancient common ancestor. Today's diverse forms of life have arisen though a process of descent with modification known as **evolution**. All populations evolve (a population is a group of the same type of organism inhabiting the same area), and every biological structure and process arose through evolution. In the words of biologist Theodosius Dobzhansky, "Nothing in biology makes sense except in the light of evolution."

CHECK YOUR LEARNING

Can you . . .

- explain the characteristics that define life?

CASE STUDY CONTINUED

The Boundaries of Life

Are viruses alive? Viruses do not obtain or use energy or materials, maintain themselves, or grow. Therefore, viruses do not meet the criteria for life. They do, however, possess some lifelike characteristics. For example, viruses respond to stimuli by binding to cells in response to the presence of particular proteins on the cell surface. In addition, viruses reproduce by releasing viral genetic material inside a cell and inducing the infected cell to use its own energy supplies and biochemical machinery to churn out many copies of viral parts, which then assemble into new viruses that emerge from the host cell. Viruses also evolve, often rapidly. How does evolution occur in viruses and organisms?

1.2 WHAT IS EVOLUTION?

The scientific theory of evolution states that all organisms are related by common ancestry and have changed over time. The theory was formulated in the mid-1800s by two English naturalists, Charles Darwin and Alfred Russel Wallace. Since that time, the theory has been supported by a vast amount of evidence from fossils, geology, genetics, molecular biology, biochemistry, and more. Evolution not only accounts for the enormous diversity of life, but also accounts for the remarkable similarities among different types of organisms. For example, people share many features with chimpanzees, and the sequence of our DNA is nearly identical to that of chimpanzees. The similarities provide strong evidence that people and chimps descended from a common ancestor; the differences (**FIG. 1-8**) reflect changes since the evolutionary paths of chimps and humans diverged.

Natural Selection Causes Evolution

The most important process by which evolution occurs is **natural selection**. Natural selection occurs because the characteristics of the different individuals in a population vary, with some individuals possessing traits that help them survive and reproduce more successfully than do others that lack those traits. The individuals with these favorable traits tend to have a greater number of offspring, which inherit those traits. The favorable traits, and the genes that encode them, thus become more common in the population.

For example, consider how natural selection might have influenced the evolution of beaver teeth. In a population of beavers, tooth size varies. Why? Because different beavers may have different versions of the genes that influence tooth size, as a result of past mutations. **Mutations** are changes in genes caused by DNA-copying errors or by damage to DNA. Past beavers with genes for larger teeth might have been able to chew down trees more efficiently, build bigger dams and lodges, and eat more bark than "ordinary" beavers. Because these big-toothed beavers obtained more food and better shelter than their smaller-toothed relatives, they raised more offspring. The offspring inherited their parents' genes for larger teeth. Over time, less-successful, smaller-toothed beavers became increasingly scarce. After many generations, all beavers had large teeth.

▲ **FIGURE 1-8 Chimpanzees and people are closely related**

Natural Selection Results in Adaptation

Structures, physiological processes, or behaviors that arise through natural selection are called **adaptations**. Adaptations help an organism survive and reproduce in a particular environment. Most of the features that we admire so much in other life-forms, such as the fleet, agile limbs of deer, the broad wings of eagles, and the mighty trunks of redwood trees, are adaptations. Adaptations help organisms escape predators, capture prey, reach the sunlight, or accomplish other feats that help ensure their survival and reproduction. The huge array of adaptations found in living things today was molded by natural selection acting on random mutations.

Evolution Can Produce New Species

Although natural selection is responsible for adaptations, it cannot by itself explain how life has diversified to include so many different kinds of organisms. How did deer, eagles, redwoods, people, and the rest of Earth's varied inhabitants all arise from the first single-celled life that appeared billions of years ago? The evolutionary process of diversification begins when a population becomes fragmented. For example, a violent storm carries some members of a population from the mainland to an offshore island. The mainland population and the newly arrived island population will initially consist of individuals of the same **species** (organisms of the same type that can interbreed). But if the island's environment differs from that of the mainland, natural selection will favor different adaptations on the island than on the mainland. These differences may eventually become great enough that the two populations can no longer interbreed. At that point, a new species will have evolved.

Extinction Eliminates Species

What helps an organism survive today may become a liability in the future. If environments change—for example, as global climate change occurs—the traits that are adaptive will change as well. For example, in a location where temperatures are rising, if a random mutation helps an organism survive and reproduce in a warmer climate, the mutation will be favored by natural selection and will become more common in the population with each new generation.

If mutations that are adaptive do not occur, a changing environment may doom a species to **extinction**—complete elimination. Dinosaurs flourished for 100 million years, but

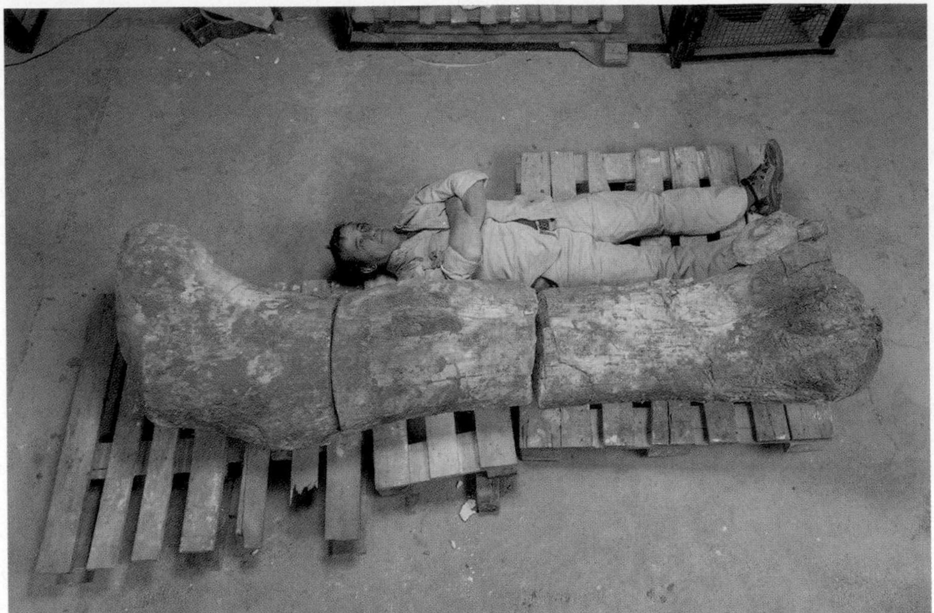

FIGURE 1-9 A fossil from a newly discovered dinosaur, *Titanosaurus* The most widely accepted hypothesis for the extinction of dinosaurs about 66 million years ago is a massive meteorite strike that rapidly and radically altered their environment. This thigh bone, estimated to be 95 million years old, is from a plant-eating giant with an estimated length of 130 feet (40 meters) and a weight of about 176,000 pounds (80 metric tons).

THINK CRITICALLY The largest dinosaurs were plant-eaters. Based on Figure 1-3, can you suggest a reason why?

because they did not adapt to changing conditions, they became extinct (**FIG. 1-9**). In recent decades, human activities such as burning fossil fuels and converting tropical forests to farmland have drastically accelerated the rate of environmental change, and consequently the rate of extinction has increased dramatically.

CHECK YOUR LEARNING

Can you . . .

- explain how natural processes lead inevitably to evolution by natural selection?
- explain what mutations are and the role they play in evolution?
- describe how species arise and how they become extinct?

CASE STUDY CONTINUED

The Boundaries of Life

One lifelike property of viruses is their capacity to evolve. Viruses may evolve by natural selection to become more deadly, become more easily transmitted, or gain the ability to infect new hosts. The genetic material of many viruses, including Ebola, HIV, and flu, is copied very inefficiently, resulting in a mutation rate that is about 1,000 times higher than that of the average animal cell. One consequence of this high mutation rate is that viruses evolve rapidly, quickly acquiring adaptations that help them resist both our immune system and antiviral drugs. Rapid viral evolution explains why flu shots must be updated every year and why HIV has become resistant to the drugs used to treat it. Because natural selection in a population exposed to an antiviral drug will inevitably favor viruses with mutations that make them resistant to the drug, HIV patients are given "cocktails" of three or four different drugs. A virus would be resistant to all the drugs only if it carried multiple different resistance mutations, which is very unlikely.

1.3 HOW DO SCIENTISTS STUDY LIFE?

The science of biology encompasses many different areas of inquiry. In fact, biology is not a single scientific discipline, but many—linked by the amazing complexity of life.

Life Can Be Studied at Different Levels

Biologists often view the living world as a series of levels of organization, with each level providing the building blocks for the one above it (**FIG. 1-10**). At the lowest level of organization are atoms. An **atom** is the smallest particle of an element that retains all the properties of the element (an element is a substance that cannot be broken down or converted into a simpler substance). For example, the smallest possible unit of a piece of gold is an individual gold atom. Atoms may combine to form **molecules**; for example, one oxygen atom can combine with two hydrogen atoms to form a molecule of water. Complex biological molecules containing carbon atoms form the building blocks of **cells**, which are the basic units of life. In multicellular organisms, cells of a similar type may combine to form **tissues**, such as the epithelial tissue that lines the stomach. Different types of tissues, in turn, unite to form functional units called **organs**, such as the entire stomach. The grouping of two or more organs that work together to perform a specific body function is called an **organ system**; for example, the stomach is part of the digestive system. Multiple organ systems work together within multicellular organisms. Organisms of the same species that live in a defined area form a **population**. A collection of all the populations of organisms that are similar enough to interbreed forms a species. The different species that live in an area and interact with one another constitute a **community**. A community and the nonliving environment that surrounds it make up an **ecosystem**. Finally, all the ecosystems on Earth together compose the **biosphere**.

One of the first choices a biologist makes when designing an experiment is the appropriate level of organization at

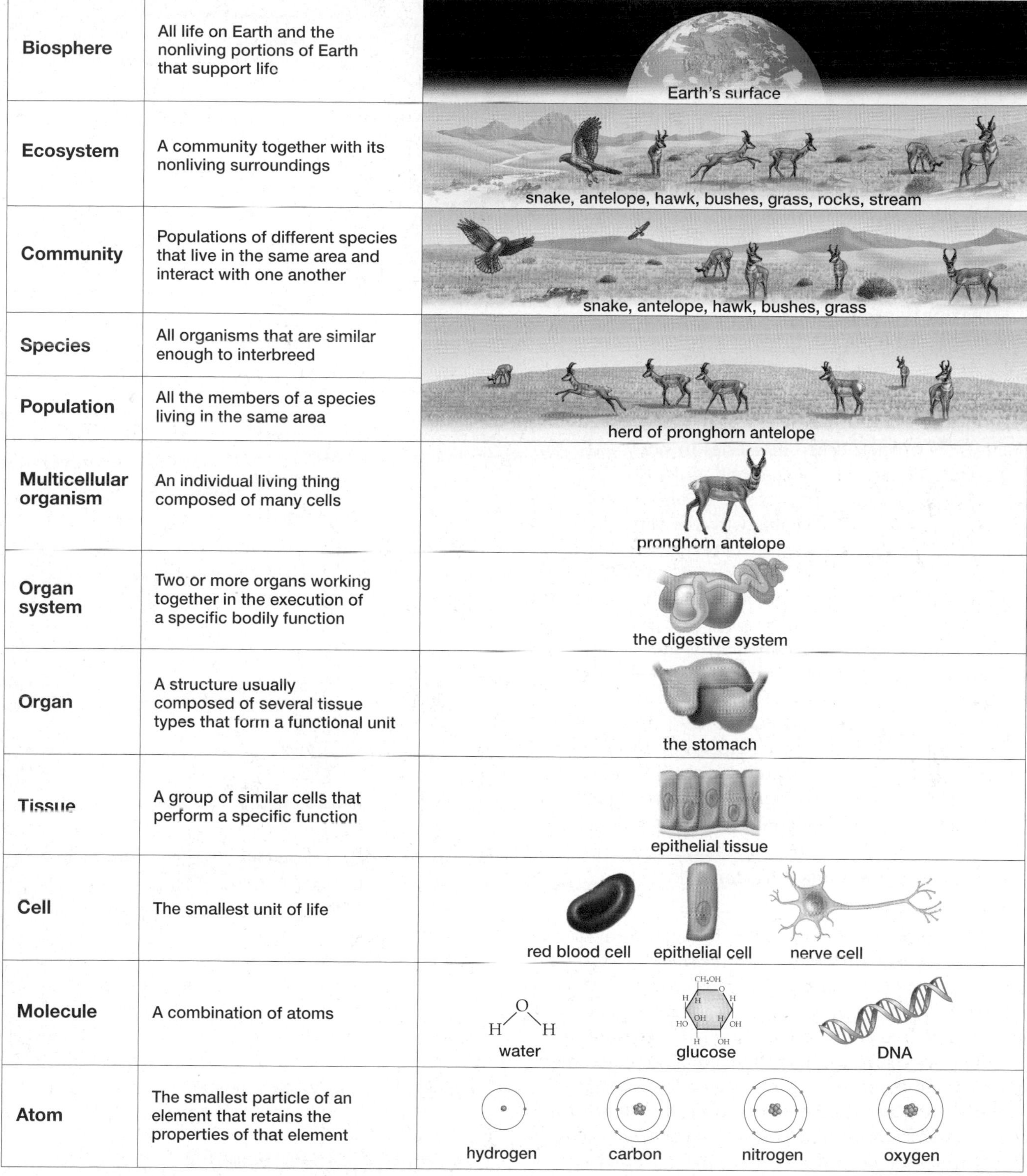

Level	Definition	Example
Biosphere	All life on Earth and the nonliving portions of Earth that support life	Earth's surface
Ecosystem	A community together with its nonliving surroundings	snake, antelope, hawk, bushes, grass, rocks, stream
Community	Populations of different species that live in the same area and interact with one another	snake, antelope, hawk, bushes, grass
Species	All organisms that are similar enough to interbreed	herd of pronghorn antelope
Population	All the members of a species living in the same area	
Multicellular organism	An individual living thing composed of many cells	pronghorn antelope
Organ system	Two or more organs working together in the execution of a specific bodily function	the digestive system
Organ	A structure usually composed of several tissue types that form a functional unit	the stomach
Tissue	A group of similar cells that perform a specific function	epithelial tissue
Cell	The smallest unit of life	red blood cell, epithelial cell, nerve cell
Molecule	A combination of atoms	water, glucose, DNA
Atom	The smallest particle of an element that retains the properties of that element	hydrogen, carbon, nitrogen, oxygen

▲ **FIGURE 1-10 Levels of biological organization** Each level provides building blocks for the one above it, which has new properties that emerge from the interplay of the levels below.

THINK CRITICALLY Which level of organization would be most appropriate for investigating how photosynthesis converts solar energy to stored energy?

which to study a problem. This decision is ordinarily based on the question to be answered. For example, if you wanted to know how frogs make croaking sounds, you would study organs within the frog's body. The question of how frogs croak would be impossible to answer if you focused on frog cells or frog communities. On the other hand, if you wanted to know whether global climate change is reducing the number of frogs in the world, it would do you no good to study frog organs. To answer that question, you would have to study frog populations. It is important for scientists to recognize and choose the level of organization that is most appropriate to the question at hand.

Biologists Classify Organisms Based on Their Evolutionary Relationships

Scientists classify Earth's diverse species on the basis of their evolutionary relatedness, placing them into three major **domains**: Bacteria, Archaea, and Eukarya (**FIG. 1-11**).

Cell Structure Distinguishes the Bacteria, Archaea, and Eukarya

Members of the three domains can be distinguished by the characteristics of their cells. All species in both Bacteria and Archaea are prokaryotic and unicellular, although some form loosely organized strands, mats, or films. In keeping with the very distant evolutionary relationship between the two domains, bacterial and archaeal cells differ significantly in structure and chemical composition. Organisms in Eukarya consist of one or more eukaryotic cells. This domain includes fungi, plants, animals, and a diverse collection of organisms collectively known as *protists*. Although most protists are unicellular, all plants and animals and nearly all fungi are multicellular; their lives depend on intimate communication and cooperation among numerous specialized cells. You will learn more about life's incredible diversity and how it evolved in Unit 3.

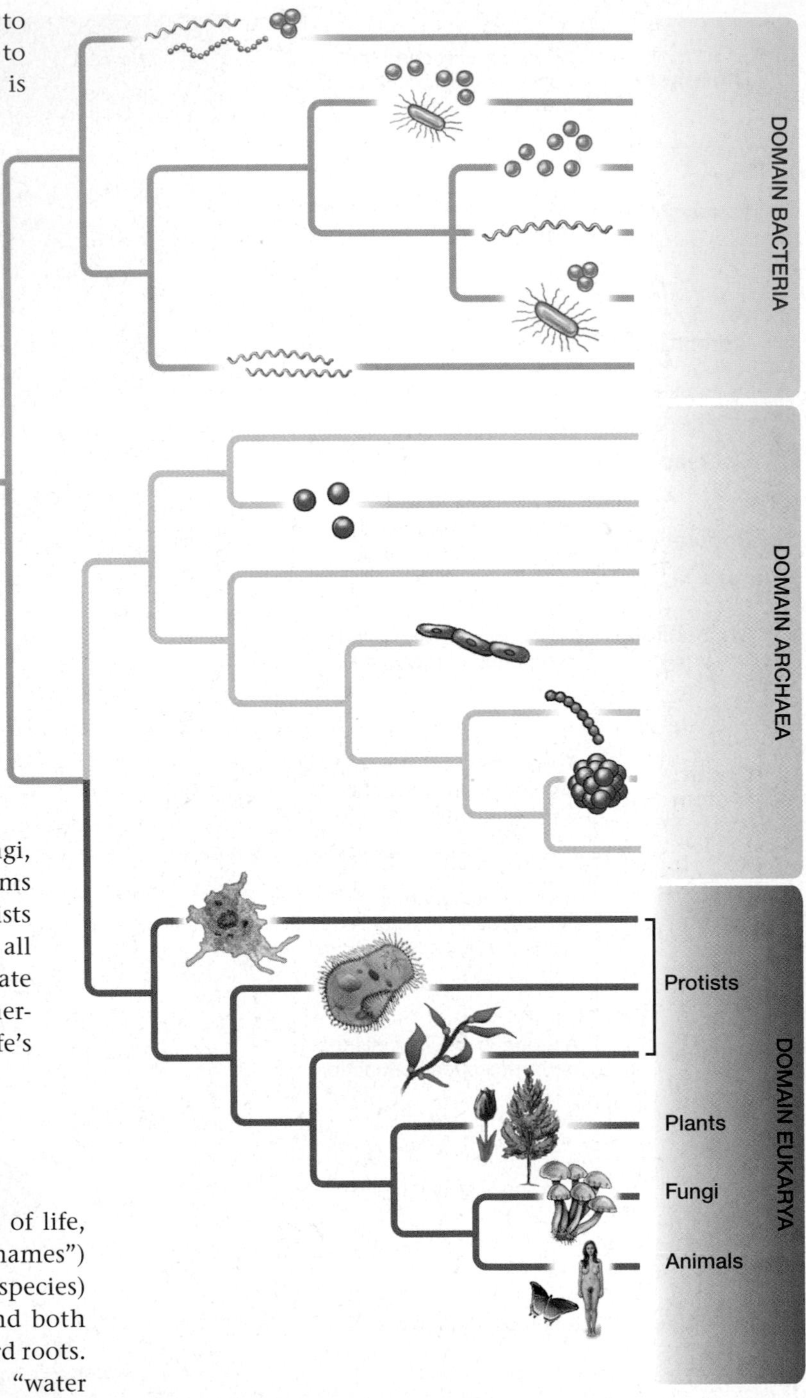

▲ **FIGURE 1-11 The domains of life**

Biologists Use the Binomial System to Name Organisms

To provide a unique scientific name for each form of life, biologists use a **binomial system** (literally "two names") consisting of the genus (a group of closely related species) and the species. The genus name is capitalized, and both names are italicized and based on Latin or Greek word roots. The animal in Figure 1-2 has the common name "water flea," but there are many types of water fleas, and people who study them need to be precise. So this water flea has been given the scientific name *Daphnia longispina*, placing it in the genus *Daphnia* (which includes many similar species of water fleas) and the species *longispina* (referring to its long spine, visible in Figure 1-2). People are classified as *Homo sapiens*; we are the only surviving members of our genus.

CHECK YOUR LEARNING

Can you . . .

- describe the levels of biological organization?
- explain how scientists categorize and name diverse forms of life?
- describe some differences between the three domains of life?

1.4 WHAT IS SCIENCE?

Science can be defined as systematic inquiry, through observation and experiment, into all aspects of the physical universe.

Science Is Based on General Underlying Principles

Three basic principles provide the foundation for scientific inquiry. The first is that all events can be traced to natural causes. In ancient times, it was common to believe that supernatural forces were responsible for natural events. For example, ancient Greeks believed that lightning bolts were weapons hurled by the god Zeus and that epileptic seizures were the result of a visit from the gods. Today, we understand from scientific research that lightning is a massive electrical discharge, and epilepsy is a brain disorder caused by uncontrolled firing of nerve cells. Science is an unending quest to discover the causes of phenomena that we don't yet understand.

The second principle of science is that natural laws do not change over time or distance. The laws of gravity, for example, are the same today as they were 10 billion years ago, and they apply everywhere in our universe.

The third principle is that scientific findings are "value neutral." Science, in its ideal form, provides us with facts that are not influenced by subjective values; scientific data are independent of any belief system. For example, science can objectively describe the events that occur when a human egg is fertilized, but cannot answer the subjective, value-based question of whether a fertilized egg is a person.

The Scientific Method Is an Important Tool of Scientific Inquiry

To learn about the world, scientists in many disciplines, including biology, use the **scientific method**. This method consists of six interrelated elements: observation, question, hypothesis, prediction, experiment, and conclusion. Scientific inquiry begins with an **observation** of a phenomenon. The observation, in turn, leads to a **question** about what was observed. After carefully studying earlier investigations, thinking, and often conversing with colleagues, the investigator forms a hypothesis. A **hypothesis** is a proposed explanation for the observation, an answer to the question. To be useful, a hypothesis must lead to a **prediction**, which is an outcome expected after testing if the hypothesis is correct. The prediction is tested by carefully designed additional observations or carefully controlled manipulations called **experiments**. Experiments produce results that either support or refute the hypothesis, allowing the scientist to reach a **conclusion** about whether the hypothesis is valid or not. For the conclusion to be valid, the experiment and its results must be repeatable not only by the original researcher but also by others.

We use less-formal versions of the scientific method in our daily lives. For example, suppose you are late for an important date, so you rush to your car, turn the ignition key, and make the *observation* that the car won't start. Your *question*, "Why won't the car start?" leads to a *hypothesis*: The battery is dead. This hypothesis leads to the *prediction* that a jump start will solve the problem. You *experiment* by attaching jumper cables from your roommate's car battery to your own. The result? Your car starts immediately, leading to the *conclusion* that your experiment supported your hypothesis about the dead car battery.

Experiments Incorporate Controls

Many experiments test the hypothesis that a single factor, or **variable**, is the cause of an observed phenomenon. The most effective test of such a hypothesis is usually an experiment in which only a single variable is manipulated. In most experiments, however, it is difficult to be certain that the manipulation did not inadvertently change more than one factor. For example, in the car battery experiment, jump-starting the car might have both delivered a charge to the battery and knocked some corrosion off the battery terminal that was preventing the battery from delivering power—the battery might actually have been fully charged. To guard against the effects of unnoticed variables, experiments usually also include **controls**, sections of the experiment in which no variable is changed. Results from the control condition can then be compared with those from the experimental condition (see "Doing Science: Controlled Experiments Provide Reliable Data" on page 10). In real experiments, scientists must attempt to control for all the possible effects of any manipulation they perform, so frequently more than one control is needed.

Experiments Are Not Always Possible

A well-designed experiment is usually the most convincing way to test a hypothesis, but biology includes many hypotheses that are not suited to experimental tests. For example, evolutionary biologists often ask questions about events from the historical past. Consider, for example, the hypothesis that the ancestors of today's birds were dinosaurs. These hypothesized ancestors went extinct long ago, of course, and there is no experiment that can demonstrate how they evolved millions of years ago. Nonetheless, the biologists who study this question do apply the other parts of the scientific method, using their hypotheses to make testable predictions. For example, if dinosaurs were the ancestors of birds, then we predict the discovery of fossils of dinosaurs with feathers. Such fossils have indeed been found, providing evidence that supports the hypothesis.

In some cases, an experiment would be theoretically possible but is impractical or unethical. For example, consider the hypothesis that smoking causes lung cancer in people. In principle, we could test this hypothesis with an experiment that divided a large sample of people who had never smoked into two groups. The members of one group would be required to smoke a pack of cigarettes each day, and the members of the other group would serve as controls and would not be allowed to smoke. After, say, 20 years, we could count the number of cases of lung cancer in each group. Such an experiment would provide a powerful test of the hypothesis but would, needless to say, be highly unethical. Again, though, an inability to experiment does not mean that the scientific method must be abandoned, because predictions can be tested by careful observation. If smoking causes lung cancer, then we predict that a random sample of smokers should contain more lung cancer victims than a comparable sample of nonsmokers. Many studies have indeed found an association between smoking and lung cancer, evidence that supports the hypothesis that smoking causes cancer.

DOING Science: Controlled Experiments Provide Reliable Data

A classic experiment by the Italian physician Francesco Redi (1626–1697) beautifully demonstrates the scientific method and helps to illustrate the scientific principle that all events can be traced to natural causes. Redi investigated why maggots (fly larvae) appear on spoiled meat. In Redi's time, refrigeration was unknown, and meat was stored in the open. Many people of that time believed that the appearance of maggots on meat was evidence of **spontaneous generation**, the emergence of life from nonliving matter.

Redi *observed* that flies swarm around fresh meat and that maggots appear on meat left out for a few days. On the basis of this observation, he posed the *question* of whether there was a connection between the presence of flies and the appearance of maggots. He then formed a testable *hypothesis*: Flies produce maggots. This led to the *prediction* that keeping flies off the meat would prevent maggots from appearing. In his *experiment*, Redi wanted to test one variable—the access of flies to the meat. Therefore, he placed similar pieces of meat in each of two clean jars. He left one jar open (the control jar) and covered the other with gauze to keep out flies (the experimental jar). He did his best to keep all the other conditions the same (for example, the type of jar, the type of meat, and the temperature). After a few days, he observed maggots on the meat in the open jar but saw none on the meat in the covered jar. Redi *concluded* that his hypothesis was correct and that maggots are produced by flies, not by the nonliving meat (**FIG. E1-1**). Only through this and other controlled experiments could the age-old belief in spontaneous generation be laid to rest.

Today, more than 300 years later, scientists still perform controlled experiments. Consider the experiments of Malte Andersson, who investigated the long tails of male widowbirds. Andersson *observed* that male, but not female, widowbirds have extravagantly long tails, which they display while flying across African grasslands. Andersson asked the *question*, "Why do male birds have such long tails?" His *hypothesis* was that females prefer to mate with long-tailed males, and so these males have more offspring, who inherit the genes for long tails. Andersson *predicted* that if his hypothesis were true, more females would build nests on the territories of males with artificially lengthened tails than on the territories of males with artificially shortened tails. To test this prediction, he captured some males, trimmed their tails to about half their original length, and released them (*experimental* group 1). He took another group of males and glued on the tail feathers that he had removed from the first group, creating exceptionally long tails (*experimental* group 2). Then, in *control* group 1, he cut the tail feathers but then glued them back in place (to control for the effects of capturing the birds and manipulating their feathers). In *control* group 2, he simply captured and released a group of male birds to control for behavioral changes caused by the stress of being caught and handled.

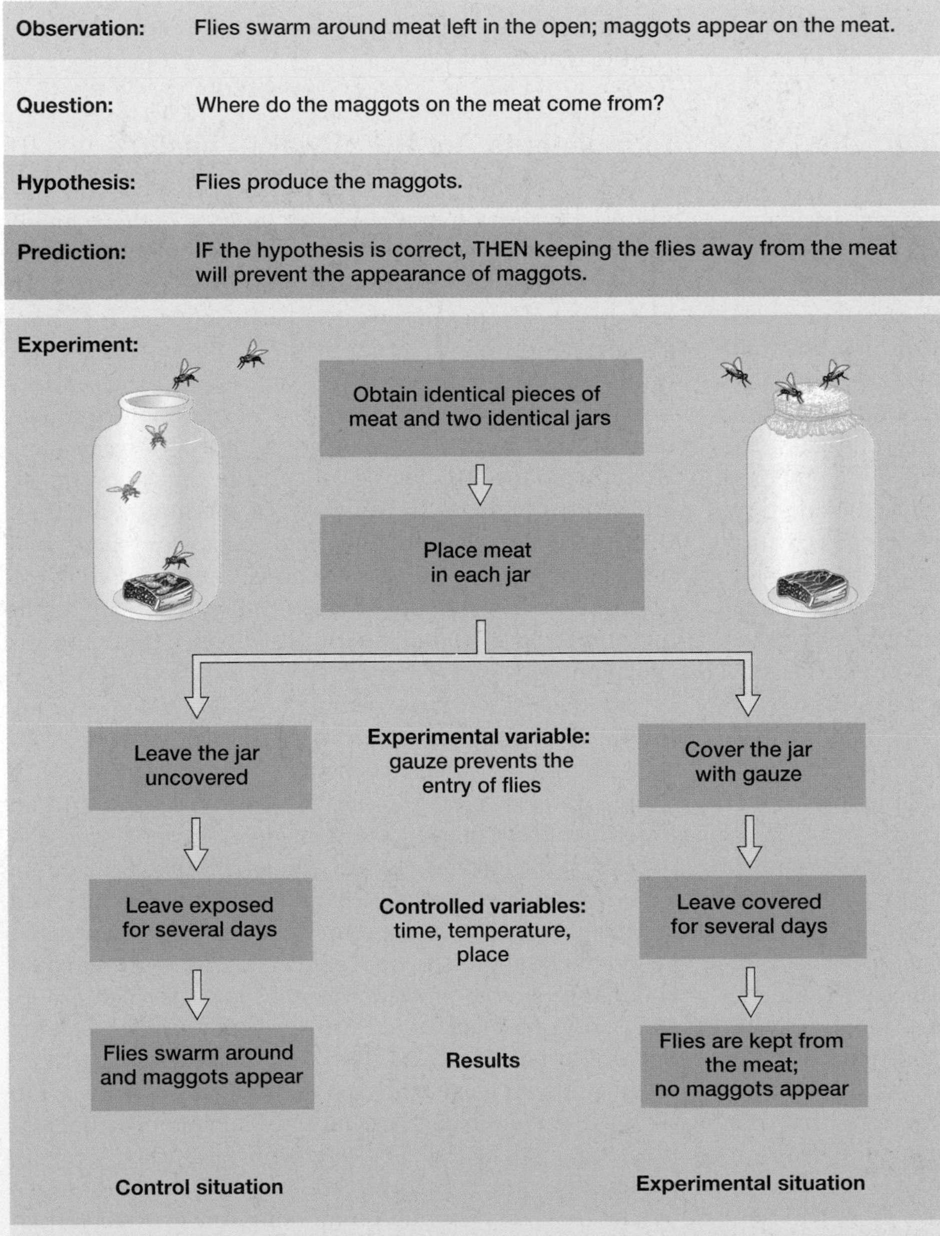
Observation: Flies swarm around meat left in the open; maggots appear on the meat.
Question: Where do the maggots on the meat come from?
Hypothesis: Flies produce the maggots.
Prediction: IF the hypothesis is correct, THEN keeping the flies away from the meat will prevent the appearance of maggots.
Experiment:
Obtain identical pieces of meat and two identical jars
Place meat in each jar
Leave the jar uncovered
Experimental variable: gauze prevents the entry of flies
Cover the jar with gauze
Leave exposed for several days
Controlled variables: time, temperature, place
Leave covered for several days
Flies swarm around and maggots appear
Results
Flies are kept from the meat; no maggots appear
Control situation
Experimental situation
Conclusion: The experiment supports the hypothesis that flies are the source of maggots and that spontaneous generation of maggots does not occur.

FIGURE E1-1 The experiment of Francesco Redi illustrates the scientific method

Later, Andersson counted the number of nests that females had built on each male's territory, which indicated how many females had mated with that male. He found that males with lengthened tails had the most nests on their territories, males with shortened tails had the fewest, and control males (with normal-length tails, either untouched or cut and glued together) had an intermediate number (**FIG. E1-2**). Andersson *concluded* that his results supported the hypothesis that female widowbirds prefer to mate with long-tailed males.

Although controlled experiments like those of Redi and Andersson are the gold standard of scientific evidence, scientific research does not always include all the elements we have described here. For example, the early stages of a research project might consist of exploratory observations, conducted without a specific hypothesis or prediction. In some scientific fields, such as paleontology (the study of fossils), experiments are not even possible. As we explore examples of scientific investigation in "Doing Science" boxes later in this text, you will see different variants of the scientific method. In most cases, however, the examples will feature a scientific question (indicated by the heading "What Question Was Asked?"), the methods used to collect evidence ("How Was Evidence Gathered?"), and the results and conclusions ("What Was Learned?").

THINK CRITICALLY Did Redi's experiment convincingly demonstrate that flies produce maggots? What kind of follow-up experiment would help confirm the source of maggots?

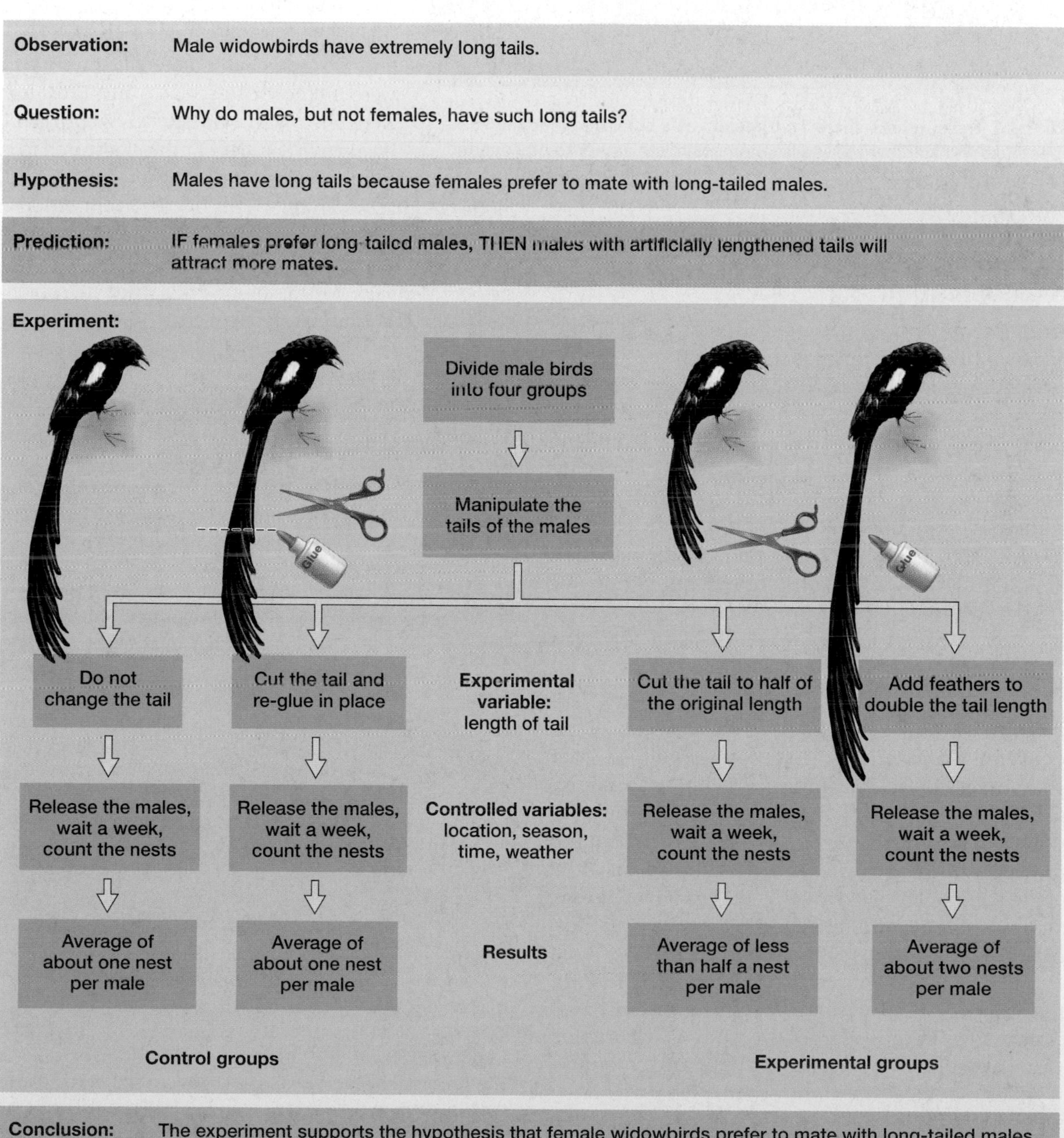

Observation: Male widowbirds have extremely long tails.
Question: Why do males, but not females, have such long tails?
Hypothesis: Males have long tails because females prefer to mate with long-tailed males.
Prediction: IF females prefer long-tailed males, THEN males with artificially lengthened tails will attract more mates.
Experiment:
Divide male birds into four groups
Manipulate the tails of the males
Glue
Do not change the tail
Cut the tail and re-glue in place
Experimental variable: length of tail
Cut the tail to half of the original length
Add feathers to double the tail length
Release the males, wait a week, count the nests
Release the males, wait a week, count the nests
Controlled variables: location, season, time, weather
Release the males, wait a week, count the nests
Release the males, wait a week, count the nests
Average of about one nest per male
Average of about one nest per male
Results
Average of less than half a nest per male
Average of about two nests per male
Control groups
Experimental groups
Conclusion: The experiment supports the hypothesis that female widowbirds prefer to mate with long-tailed males (and are less likely to mate with short-tailed males).

▲ **FIGURE E1-2 The experiment of Malte Andersson**

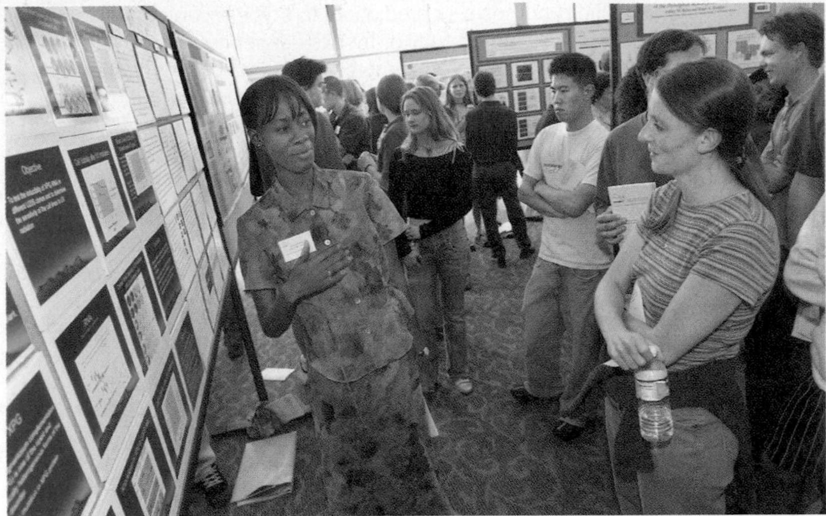

▲ **FIGURE 1-12 Researchers share their results at a scientific meeting** Posters are often used used to summarize results when scientists gather to present and discuss their findings.

Science Requires Repeatability and Communication

If an experiment's result is valid, the experiment will yield the same result each time it is performed. Thus, to ensure validity, researchers perform multiple repetitions of an experiment, setting up several replications of each control condition and each experimental condition. Scientists are most confident in an experimental result when it has also been replicated by researchers other than the ones who made the initial finding.

Even the best experiment is useless if it is not communicated (**FIG. 1-12**). Good scientists publish their results, explaining their methods in detail so others can repeat and build on their experiments.

Scientific Theories Have Been Thoroughly Tested

Scientists use the word "theory" in a way that differs from its everyday usage. If Dr. Watson asked Sherlock Holmes, "Do you have a theory as to the perpetrator of this foul deed?" then in scientific terms, he would be asking Holmes for a hypothesis—a proposed explanation based on clues that provide incomplete evidence. A **scientific theory**, in contrast, is a general and reliable explanation of important natural phenomena that has been developed through extensive and reproducible observations and experiments. In short, a scientific theory is best described as a **natural law**, a basic principle derived from the study of nature that has never been disproven by scientific inquiry. For example, scientific theories such as the atomic theory (that all matter is composed of atoms) and the theory of gravitation (that objects exert attraction for one another) are fundamental to the science of physics. Likewise, the **cell theory** (that all living organisms are composed of cells) and the theory of evolution are fundamental to the study of biology.

Scientists describe fundamental principles as "theories" rather than "facts" because even scientific theories can potentially be disproved, or falsified. If compelling evidence arises that renders a scientific theory invalid, that theory must be modified or discarded. A modern example of the need to modify basic principles in the light of new scientific evidence is the discovery of prions, which are infectious proteins (see Chapter 3). Before the early 1980s, all known infectious disease agents replicated using instructions from genetic material. Then in 1982, neurologist Stanley Prusiner published evidence that scrapie (an infectious disease of sheep that causes brain degeneration) is actually triggered and transmitted by a protein and has no genetic material. Infectious proteins were unknown to science, and Prusiner's results were met with widespread skepticism. It took nearly two decades of further research to convince most of the scientific community that a protein alone could act as an infectious disease agent. Prions are now known to cause mad cow disease and two fatal human brain disorders. Prusiner was awarded the Nobel Prize in Physiology or Medicine for his pioneering work. By accepting, on the basis of accumulated scientific evidence, the conclusion

Have You Ever Wondered ... Why Scientists Study Obscure Organisms?

Some people are puzzled by the willingness of governments to fund research that seems obscure or pointless. One reason to fund such research is that no one can say where a research idea might lead, and allowing scientists to follow their curiosity may lead to unexpected and valuable findings. Research on seemingly unimportant organisms like fruit flies, bacteria from hot springs, sea jellies, Gila monsters, and burdock burrs has improved people's lives.

Fruit flies have been used for over 100 years to study how genes influence traits. Their genes are similar enough to ours that studying fruit flies has helped us to better understand many human genetic diseases. Research on an obscure bacterium found in hot springs revealed a protein that is now an essential component of a laboratory process that rapidly copies DNA. Thanks to this discovery, the DNA in a few skin cells left at a crime scene can generate a sample large enough to be compared to the DNA of a suspect. Scientific investigation of a sea jelly uncovered a fluorescent protein that can be attached to a gene, protein, or virus, making it glow and allowing researchers to monitor its activity. A protein found in a Gila monster's venomous saliva is now used as a drug that helps diabetics maintain more constant blood sugar levels. And what did microscopic examination of a burr lead to? The inspiration for Velcro.

Gila monster

that prions can act as infectious proteins, scientists maintained the integrity of the scientific process while expanding our understanding of how diseases can occur. Ongoing scientific inquiry continuously tests scientific theories.

Science Is a Human Endeavor

Scientists are people, driven by the pride, fears, and ambition common to humanity. Accidents, lucky guesses, competition with other scientists, and, of course, the intellectual curiosity of individual scientists all contribute to scientific advances. Even mistakes can play a role. Let's consider an actual case.

Microbiologists often study pure cultures—a single type of bacterium grown in sterile, covered dishes free from contamination by other bacteria and molds. At the first sign of contamination, a culture is usually thrown out, often with mutterings about sloppy technique. In the late 1920s, however, Scottish bacteriologist Alexander Fleming turned a ruined bacterial culture into one of the greatest medical advances in history.

One of Fleming's cultures became contaminated with a mold (a type of fungus) called *Penicillium*. But instead of discarding the dish, Fleming observed that no bacteria were growing near the mold (**FIG. 1-13**). He asked the question, "Why aren't bacteria growing in this region?" Fleming then formulated the hypothesis that *Penicillium* releases a substance that kills bacteria, and he predicted that a solution in which the mold had grown would contain this substance and kill bacteria. To test this hypothesis, Fleming performed an experiment. He grew *Penicillium* in a liquid nutrient broth and then filtered out the mold and poured some of the mold free broth on a plate with a pure bacterial culture. Sure enough, something in the liquid killed the bacteria, supporting his hypothesis. This (and more experiments that confirmed his results) led to the conclusion that *Penicillium* secretes a substance that kills bacteria. Further research into these mold extracts resulted in the production of the first antibiotic—penicillin.

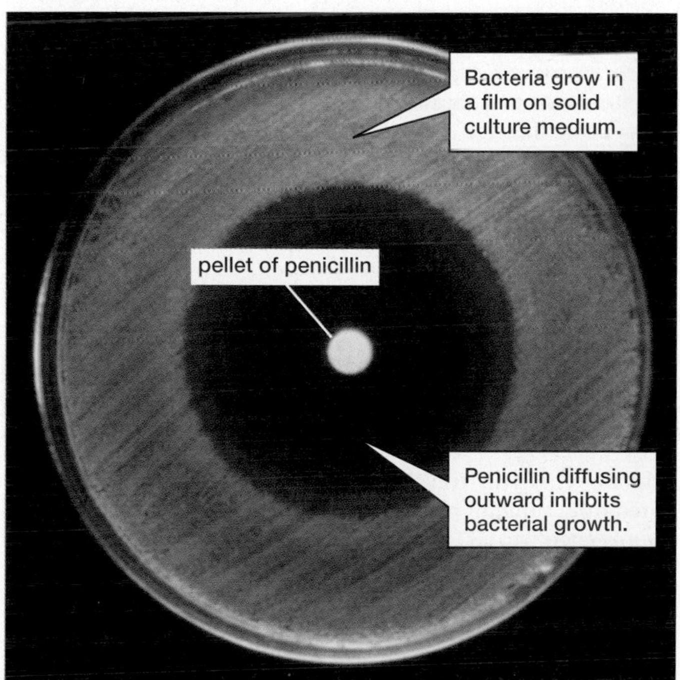
Bacteria grow in a film on solid culture medium.
pellet of penicillin
Penicillin diffusing outward inhibits bacterial growth.

▲ **FIGURE 1-13 Penicillin kills bacteria** Alexander Fleming observed similar inhibition of bacterial growth around colonies of *Penicillium* mold.

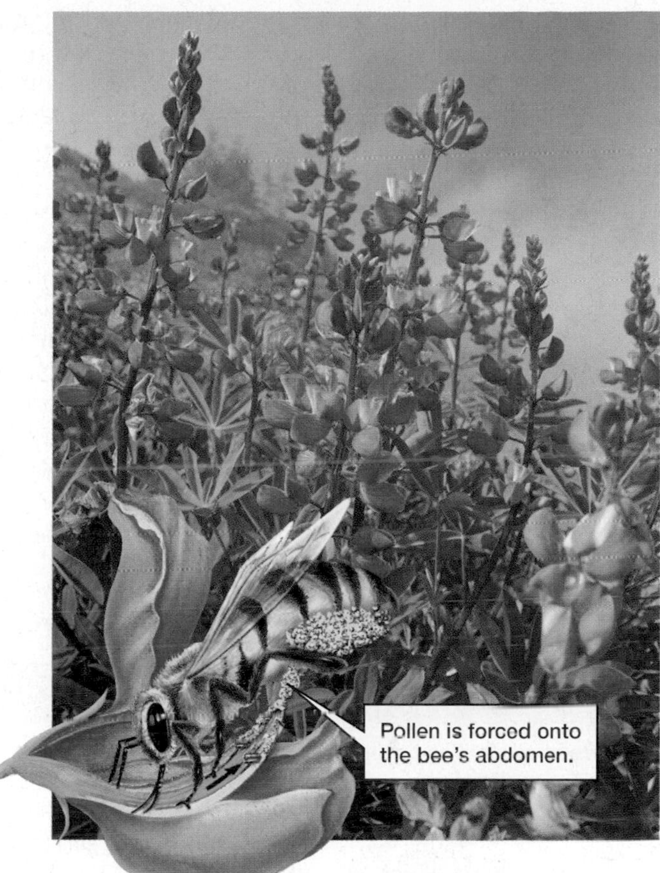
Pollen is forced onto the bee's abdomen.

▲ **FIGURE 1-14 Adaptations in lupine flowers** Understanding life helps people notice and appreciate the small wonders at their feet. (Inset) A lupine flower deposits pollen on a foraging bee's abdomen.

Fleming's experiments are a classic example of the scientific method, but they would never have happened without the combination of a mistake, a chance observation, and the curiosity to explore it. The outcome has saved millions of lives. As French microbiologist Louis Pasteur said, "Chance favors the prepared mind."

Biology Illuminates Life

Some people feel that science promotes a cold, clinical view of life and that scientific explanations of the natural world rob us of wonder and awe. Nothing could be further from the truth. Biological knowledge only deepens our appreciation of nature's majesty.

Let's look closely at lupine flowers. Their two lower petals form a tube surrounding both male and female reproductive parts (**FIG. 1-14**). In young flowers, the weight of a bee on this tube forces pollen grains (which carry sperm) out of the tube and onto the bee's abdomen. In older lupine flowers that are ready to be fertilized, the female part grows and emerges through the end of the tube. When a pollen-dusted bee visits, it deposits some pollen on the female organ, allowing the lupine to produce the seeds of the next generation.

Do these insights detract from our appreciation of lupines? Far from it. There is added delight in watching and understanding the intertwined form and function of bee and flower that resulted as these organisms evolved together. Soon after learning the lupine's pollination mechanism, two of the authors of this text crouched beside a wild lupine to watch it happen. An elderly man passing by stopped to ask what they were looking at so intently. He listened with interest as they explained about what happened when a bee landed on the lupine's petals, and he immediately went to observe another patch of lupines where bees were foraging. He, too, felt the heightened sense of appreciation and wonder that comes with understanding.

Throughout this text, we try to convey that biology is not just another set of facts to memorize. It is a pathway to understanding yourself and the life around you. It is also important to recognize that biology is not a completed work, but an ongoing exploration. As Alan Alda, best known for playing "Hawkeye" in the TV show *M*A*S*H*, stated: "With every door into nature we nudge open, 100 new doors become visible."

CHECK YOUR LEARNING

Can you . . .

- describe the principles underlying science?
- outline the scientific method?
- explain why controls are crucial in biological studies?
- explain why fundamental scientific principles are called theories?

CASE STUDY REVISITED

The Boundaries of Life

If viruses aren't living organisms, what are they? A virus by itself is an inert particle far simpler than even the least complex cell. The simplest viruses, such as the ones that cause smallpox (**FIG. 1-15**), consist of nothing more than a protein coat surrounding a small amount of genetic material. The uncomplicated structure of viruses has made it possible for researchers to synthesize viruses in the laboratory, using off-the-shelf chemicals and the instructions contained in viral genetic material. The first virus to be synthesized was the small, simple poliovirus. This feat was accomplished in 2002 by Eckard Wimmer and co-workers at Stony Brook University, who titled their work, "The Test-Tube Synthesis of a Chemical Called Poliovirus."

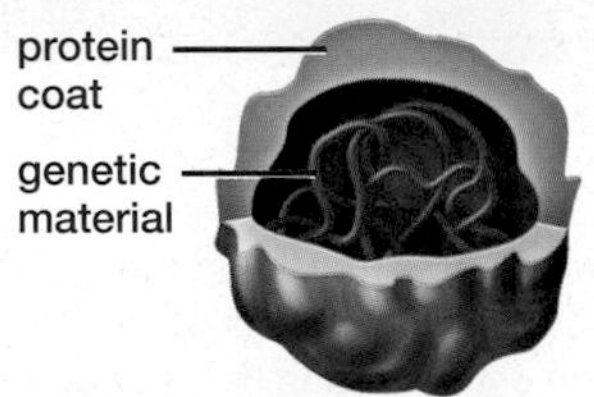

▲ **FIGURE 1-15 A smallpox virus**

Did these researchers create life in the laboratory? A few scientists would say, "yes," defining life by its ability to reproduce and to evolve. Wimmer himself describes viruses as entities that switch between a nonliving phase outside the cell and a living phase inside. Although most scientists agree that viruses aren't alive and support the definition of life presented in this text, the controversy continues. As virologist Luis Villarreal puts it, "Viruses are parasites that skirt the boundaries between life and inert matter."

CONSIDER THIS When Wimmer and coworkers announced that they had synthesized the poliovirus, controversy erupted. Some people feared that the newly developed methods could be used by terrorists to synthesize deadly and highly contagious viruses. The researchers responded that they were merely applying current knowledge and techniques to demonstrate that viruses are basically chemical entities that can be synthesized in the laboratory. Do you think scientists should synthesize viruses or other agents that can cause infectious disease? What are the implications of forbidding such research?

CHAPTER REVIEW

Go to **Mastering Biology** to access the Pearson eText, vocabulary review, practice quizzes, activities, videos, current events, and more.

*Answers to **Think Critically and Thinking Through the Concepts** questions can be found in the **Answers** section at the back of the book.*

Summary of Key Concepts

1.1 What Is Life?

Living organisms actively maintain organized complexity, perceive and respond to stimuli, grow, reproduce, and evolve. Organisms also acquire and use materials and energy. Materials are obtained from other organisms or the nonliving environment and are repeatedly recycled. Energy must be continuously captured from sunlight by photosynthetic organisms, whose bodies supply energy to all other organisms.

1.2 What Is Evolution?

Evolution is the scientific theory that modern organisms descended, with changes, from earlier organisms. Evolution occurs as a consequence of (1) genetic differences among members of a population; (2) inheritance of these differences by offspring; and

(3) natural selection favoring individuals with the characteristics that are the best adaptations to the organism's environment.

1.3 How Do Scientists Study Life?

Scientists identify a hierarchy of levels of organization, each more encompassing than those beneath (see Fig. 1-10). Biologists categorize organisms into three domains: Archaea, Bacteria, and Eukarya. Members of Archaea and Bacteria consist of single prokaryotic cells, but fundamental structural and chemical differences distinguish them. Members of Eukarya are composed of one or more eukaryotic cells. Organisms are assigned scientific names that identify them as a unique species within a specific genus.

1.4 What Is Science?

Science is based on three principles: (1) all events can be traced to natural causes that can be investigated; (2) the laws of nature are unchanging; and (3) scientific findings are independent of values except honesty in reporting data. Knowledge in biology is acquired through the scientific method, in which an observation leads to a question that leads to a hypothesis. The hypothesis generates a prediction that is then tested by controlled experiments or precise observation. The experimental results, which must be repeatable, lead to a conclusion that either supports or refutes the hypothesis. A scientific theory is a general explanation of natural phenomena developed through extensive and reproducible experiments and observations.

Thinking Through the Concepts

Bloom's: Remembering, Understanding

Multiple Choice

1. Evolution is
   a. a belief.
   b. a scientific theory.
   c. a hypothesis.
   d. never observed in the modern world.
2. Which of the following is *not* true of science?
   a. Science is based on the premise that all events can be traced to natural causes.
   b. Important science can be based on chance observations.
   c. A hypothesis is basically a wild guess.
   d. Scientific theories can potentially be disproved.
3. Which of the following statements about natural selection is *not* true?
   a. Natural selection favors the same traits in all environments.
   b. Natural selection produces adaptations.
   c. Natural selection affects only traits that are inherited.
   d. Natural selection acts on variation that is caused by mutations.
4. Viruses
   a. have DNA confined in a nucleus.
   b. are relatively rare compared to living organisms.
   c. do not evolve.
   d. require a host cell to reproduce.
5. Which one of the following is true?
   a. The presence of a cell nucleus distinguishes Bacteria from Archaea.
   b. All cells are surrounded by a plasma membrane.
   c. All members of Eukarya are multicellular.
   d. Viruses are the simplest cells.

Fill-in-the-Blank

1. Organisms respond to ____________. Organisms acquire and use ____________ and ____________ from the environment. Organisms are composed of cells whose structure is both ____________ and ____________. Populations of organisms ____________ over time.
2. The smallest particle of an element that retains all the properties of that element is a(n) ____________. The smallest unit of life is the ____________. Cells of a specific type within multicellular organisms combine to form ____________. A(n) ____________ consists of all the members of a species within a defined area. A(n) ____________ consists of all the interacting populations within the same area. A(n) ____________ consists of the community and its nonliving surroundings.
3. A(n) ____________ is a general explanation of natural phenomena supported by extensive, reproducible tests and observations. In contrast, a(n) ____________ is a proposed explanation for observed events. To answer specific questions about life, biologists use a general process called the ____________.
4. An important scientific theory that explains why organisms are at once so similar and so diverse is the theory of ____________. This theory explains life's diversity as having originated primarily through the process of ____________.
5. The molecule that guides the construction and operation of an organism's body is called (complete term) ____________, abbreviated as ____________. This large molecule contains discrete segments with specific instructions; these segments are called ____________.

Review Questions

1. What properties are shared by all forms of life?
2. Why do organisms require energy? Where does the energy come from?
3. Define *evolution*, and explain the process of natural selection.
4. What are the three domains of life?
5. What are some differences between prokaryotic and eukaryotic cells? In which domain(s) is each found?
6. What basic principles underlie scientific inquiry?
7. What is the difference between a scientific theory and a hypothesis? Why do scientists refer to basic scientific principles as "theories" rather than "facts"?
8. What factors did Redi control for in his open jar of meat? What factors did Andersson control for?
9. List the steps in the scientific method with a brief description of each step.

Applying the Concepts

Bloom's: Applying, Analyzing, Evaluating

1. What misunderstanding causes some people to dismiss evolution as "just a theory"?
2. How would this textbook's definition of life need to be changed to allow viruses to qualify as life-forms? For prions to be considered alive?
3. Review Alexander Fleming's experiment that led to the discovery of penicillin. What would be an appropriate control for the experiment in which Fleming applied filtered medium from a *Penicillium* culture to plates of bacteria?
4. Explain an instance in which your own understanding of a phenomenon enhances your appreciation of it.
5. In using the scientific method to help start your car, if jump-starting didn't work, what hypothesis would you test next?

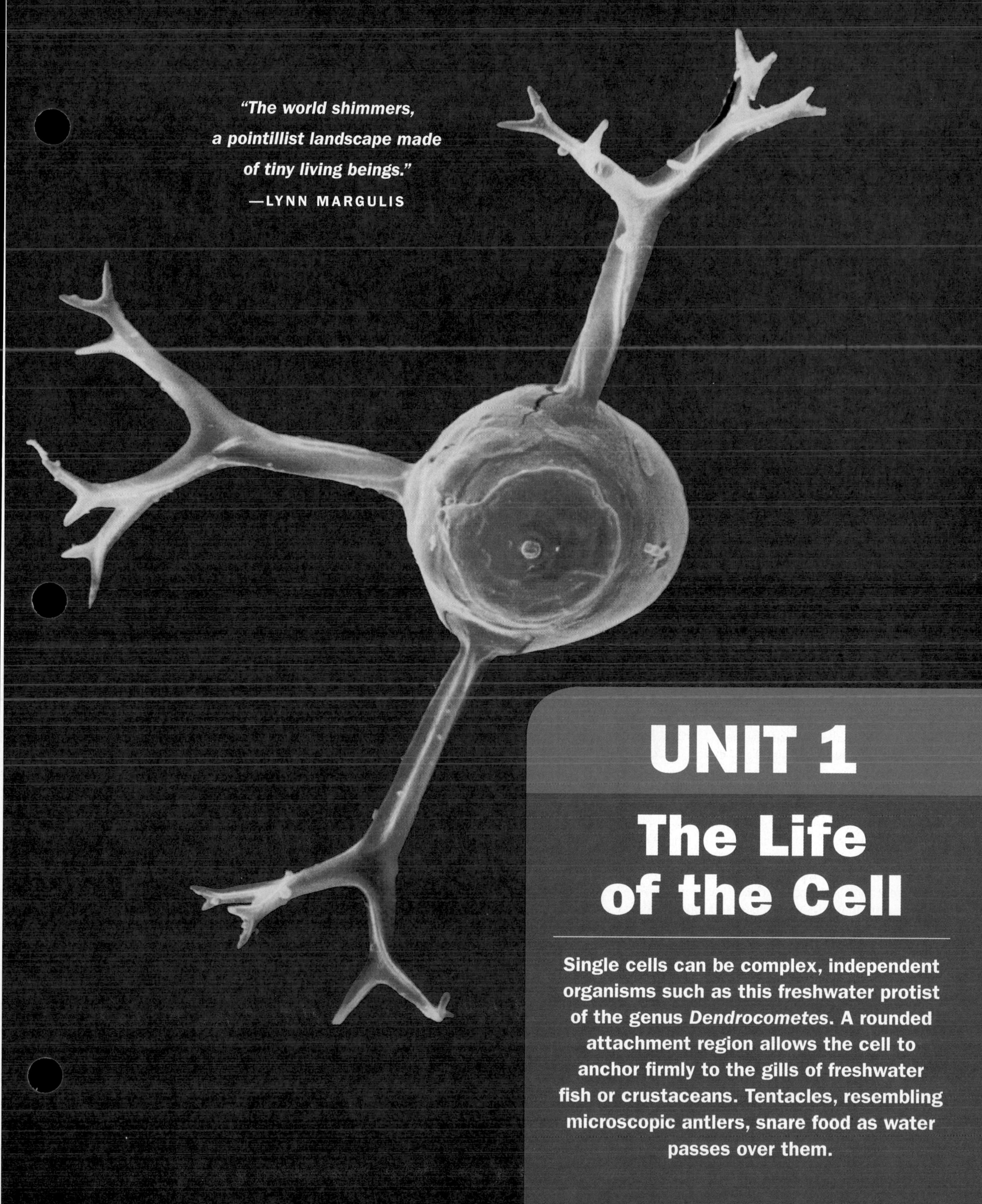

"The world shimmers, a pointillist landscape made of tiny living beings."

—LYNN MARGULIS

UNIT 1
The Life of the Cell

Single cells can be complex, independent organisms such as this freshwater protist of the genus *Dendrocometes*. A rounded attachment region allows the cell to anchor firmly to the gills of freshwater fish or crustaceans. Tentacles, resembling microscopic antlers, snare food as water passes over them.

2 Atoms, Molecules, and Life

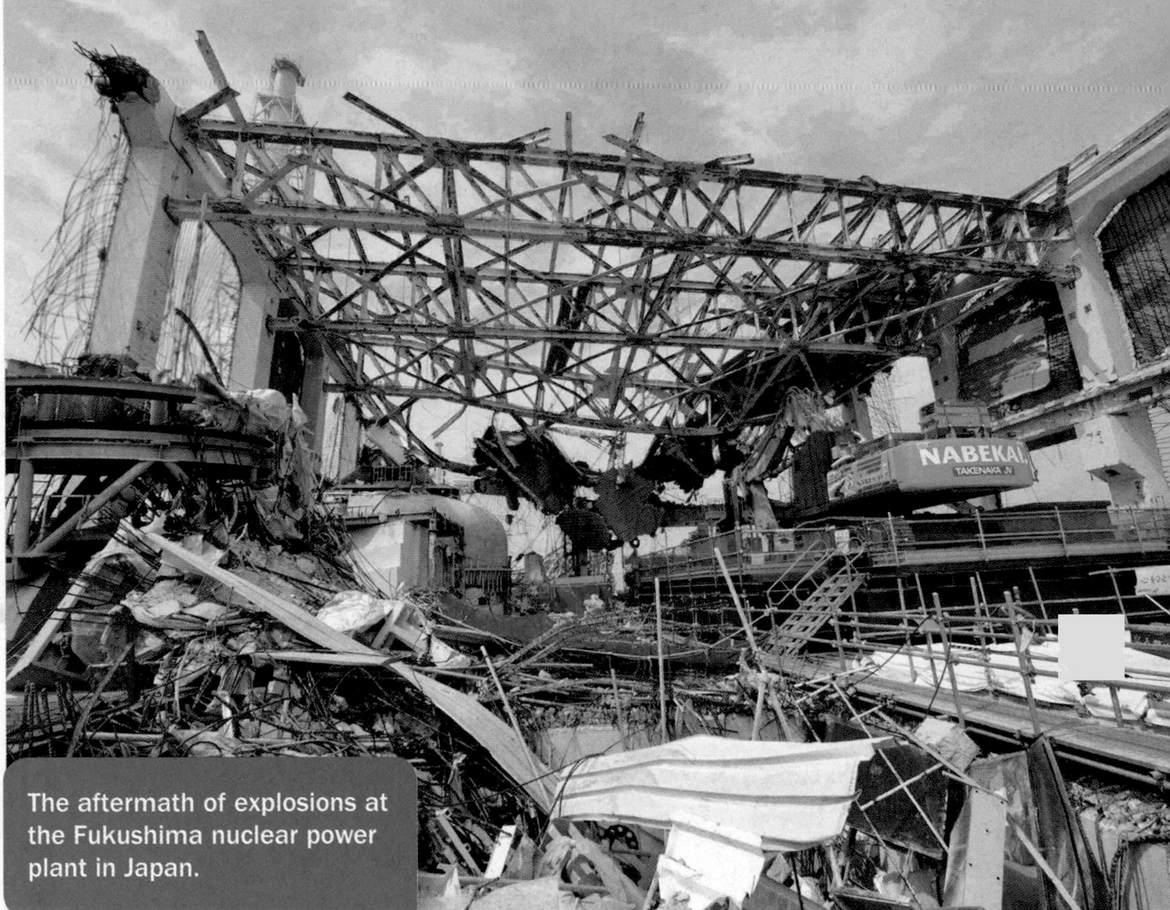

The aftermath of explosions at the Fukushima nuclear power plant in Japan.

CASE STUDY

Unstable Atoms Unleashed

ON MARCH 11, 2011, an earthquake of epic magnitude—9.0 on the Richter scale—shook the northeast coast of Japan. Soon after, a tsunami caused by the massive quake slammed into the Fukushima Daiichi nuclear power plant on Japan's eastern coast. Towering waves nearly 50 feet high flooded the plant and knocked out its main electrical power supply and backup generators, with disastrous consequences.

The core of a nuclear reactor, including the six that were at the Fukushima plant, contains thousands of fuel rods consisting of zirconium metal tubes filled with uranium fuel. Nuclear reactions in the core produce heat that converts water to steam that drives electricity-producing turbines. Each core is surrounded by two thick steel containment vessels. Water is pumped continuously around the vessels to absorb the intense heat generated by the nuclear reactions taking place inside them. This cooling system failed when the Fukushima plant lost power, and the temperature of the cores began to increase.

In a desperate attempt to cool the overheating cores, plant operators injected seawater into the inner containment vessels. But their efforts failed; the core temperature rose to over 1,800°F (about 1,000°C), melting the zirconium tubes and releasing radioactive fuel into the inner vessels. Stressed by the heat, the vessels cracked, allowing water and steam to escape. The steam reacted with the melted zirconium to generate hydrogen gas. As the pressure of the steam and hydrogen gas increased, it threatened to rupture the outer containment vessels.

In an attempt to prevent a rupture, plant operators vented the gas mixture—which also contained radioactive elements from the melted fuel rods—into the atmosphere. As the hot hydrogen gas encountered oxygen in the atmosphere, the two combined explosively, destroying parts of the buildings housing the containment vessels. Despite the venting, the outer containment vessels eventually gave way, unable to withstand the intense heat and pressure. Contaminated water flowed into the ocean from the compromised vessels for months following the disaster.

In the aftermath of the disaster, tens of thousands of people were evacuated from all habitations within 12 miles of the plant; the evacuees were not allowed to return until more than 6 years later. Even today, the melted core continues to generate contaminated water.

Why were people evacuated from their homes when radioactive gases were released into the atmosphere? How do the atoms of radioactive elements differ from those of nonradioactive elements?

AT A GLANCE

2.1 WHAT ARE ATOMS?

When you write "atom" with a pencil, the letters on the page are made of carbon. Now imagine cutting up that carbon into smaller and smaller particles, until all you have left is a pile of the smallest possible units of carbon: individual carbon atoms. A carbon atom is so small that 100 million of them placed in a row would span less than half an inch (1 centimeter).

Atoms Are the Basic Structural Units of Elements

Carbon is an example of an **element**—a substance that can neither be separated into simpler substances nor converted into a different substance by an ordinary **chemical reaction** (a process that forms or breaks bonds between atoms). Elements, alone or combined with other elements, form all matter. An **atom** is the smallest unit of an element, and each atom retains all the chemical properties of that element.

Ninety-two different elements occur in nature. Each is given an abbreviation, its *atomic symbol*. Most elements are present in only small quantities in the biosphere, and relatively few are essential to life on Earth. **TABLE 2-1** lists the elements most common in living things.

TABLE 2-1 Common Elements in Living Organisms

Element	Atomic Number[1]	Mass Number[2]	% by Weight in the Human Body
Oxygen (O)	8	16	65
Carbon (C)	6	12	18.5
Hydrogen (H)	1	1	9.5
Nitrogen (N)	7	14	3.0
Calcium (Ca)	20	40	1.5
Phosphorus (P)	15	31	1.0
Potassium (K)	19	39	0.35
Sulfur (S)	16	32	0.25
Sodium (Na)	11	23	0.15
Chlorine (Cl)	17	35	0.15
Magnesium (Mg)	12	24	0.05
Iron (Fe)	26	56	Trace
Fluorine (F)	9	19	Trace
Zinc (Zn)	30	65	Trace

[1]Atomic number: number of protons in the atomic nucleus.
[2]Mass number: total number of protons and neutrons.

TABLE 2-2 Mass and Charge of Subatomic Particles

Subatomic Particle	Mass (in atomic mass units)	Charge
Neutron (*n*)	1	0
Proton (p^+)	1	+1
Electron (e^-)	0.00055	−1

Atoms Are Composed of Still Smaller Particles

Atoms are composed of *subatomic particles*: **neutrons** (*n*), which have no charge; **protons** (p^+), each of which carries a single positive charge; and **electrons** (e^-), each of which carries a single negative charge. An atom as a whole is uncharged, or *neutral*, because it contains equal numbers of protons and electrons, whose positive and negative charges electrically balance each other. The mass of a subatomic particle is measured in *atomic mass units*. As you can see in **TABLE 2-2**, each proton and neutron has a mass of 1 atomic mass unit, whereas an electron's mass is negligible compared to that of the larger particles. The total number of protons and neutrons in the nucleus of an atom is known as its **mass number**.

Protons and neutrons cluster together in the center of each atom, forming its **atomic nucleus**. An atom's tiny electrons are in continuous rapid motion around its nucleus within a defined three-dimensional space, as illustrated in **FIGURE 2-1**, which shows the two simplest atoms, hydrogen

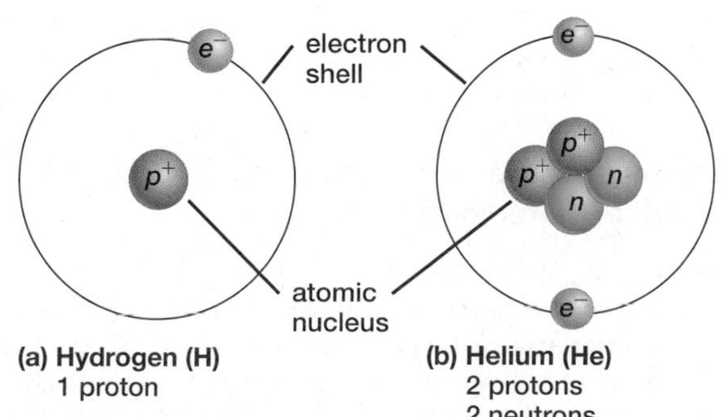

▲ **FIGURE 2-1 Atomic models** Orbital models of **(a)** hydrogen (the only atom with no neutrons) and **(b)** helium. In these simplified models, the electrons (pale blue) are represented as miniature planets, orbiting around a nucleus that contains protons and neutrons.

THINK CRITICALLY What is the mass number of hydrogen? Of helium?

and helium. These *orbital models* of atomic structure are extremely simplified to make atoms easier to visualize. Atoms are never drawn to scale; if they were, and if this dot · were the nucleus, the electrons would be somewhere in the next room (or outside)—roughly 30 feet away.

Elements Are Defined by Their Atomic Numbers

The characteristic that defines each element, making it distinct from all others, is its **atomic number**—the number of protons in its nucleus. For example, a hydrogen atom has one proton, a carbon atom has six, and an oxygen atom has eight, giving these atoms atomic numbers of 1, 6, and 8, respectively. The **periodic table** in Appendix II organizes the elements according to their atomic numbers (rows) and their general chemical properties (columns).

Isotopes Are Atoms of an Element with Different Numbers of Neutrons

Although every atom of an element has the same number of protons, different atoms of an element may have different numbers of neutrons. Atoms of a given element with different numbers of neutrons are called **isotopes**. Each isotope of an element has a different mass number. An isotope's mass number is shown as a superscript preceding the atomic symbol.

Some Isotopes Are Radioactive

Most isotopes are stable; their nuclei do not change spontaneously. A few, however, are **radioactive**, meaning that their nuclei spontaneously break apart, or decay. Radioactive decay always emits energy and often emits subatomic particles as well. Radioactive decay of nuclei may convert an element to a different element. For example, nearly all carbon exists as stable ^{12}C. But a radioactive isotope called carbon-14 (^{14}C; 6 protons + 8 neutrons) is produced continuously by reactions in the atmosphere. When a radioactive ^{14}C molecule decays, energy is released and a neutron is converted to a proton, producing a stable nitrogen atom (^{14}N; 7 protons + 7 neutrons).

Radioactive Isotopes Are Important in Scientific Research and Medicine

Scientists often make use of radioactive isotopes. For example, archeologists can determine the age of artifacts because they know that after an organism dies, the ratio of ^{14}C to ^{12}C in its body declines at a predictable rate as the ^{14}C decays. By measuring this ratio in artifacts such as mummies, ancient trees, skeletons, or tools made of wood or bone, researchers can accurately assess the age of artifacts up to about 50,000 years old.

In laboratory research, scientists often insert radioactive isotopes into molecules within organisms. The radioactivity "labels" the molecules, making them easier to locate (and trace if they move). For example, experiments with radioactively labeled DNA and protein allowed scientists to conclude that DNA is the genetic material of cells (described in Chapter 12).

Modern medicine also makes extensive use of radioactive isotopes. For example, radiation therapy is frequently used to treat cancer. A radioactive isotope may be introduced into the bloodstream or implanted in the body near the cancer, or radiation may be directed into the tumor by an external device. Radiation damages DNA, so rapidly dividing cancer cells (which require intact DNA to divide) are particularly vulnerable. The radiation that kills cancer cells can also cause mutations in the DNA of healthy cells. These mutations slightly increase the chance that the patient will develop cancer again in the future, but most patients consider this a risk worth taking. For more information about uses for radioactive isotopes, see "Doing Science: Radioactive Revelations."

CASE STUDY CONTINUED

Unstable Atoms Unleashed

Because exposure to radioactivity can cause cancer, Japanese authorities have performed regular cancer screenings on hundreds of thousands of children exposed to radioactivity by the Fukushima power plant disaster. Fortunately, recent surveys have found no evidence of increased cancer rates.

But years after the meltdown, engineers at the Fukushima power plant—using remote-controlled robotic instruments—discovered hot spots of radiation so intense that a person exposed for an hour would be dead within a few weeks. Why would death come so fast? Extremely high doses of radiation damage DNA and other biological molecules so badly that cells can no longer function. Skin cells are destroyed. Cells lining the stomach and intestine break down, causing nausea and vomiting. Bone marrow, where blood cells and platelets are produced, is destroyed. The loss of white blood cells allows infections to flourish, and the loss of platelets crucial for blood clotting leads to internal bleeding.

Fortunately, radioactive elements such as those released by the Fukushima disaster are rare in nature. Why do most elements remain stable?

Electrons Are Responsible for the Interactions Among Atoms

Nuclei and electrons play complementary roles in atoms. Nuclei (unless they are radioactive) provide stability; they remain unchanged during ordinary chemical reactions. Electrons, in contrast, are dynamic; they can capture and release energy, and as we describe later, they form the bonds that link atoms together into molecules.

Electrons Occupy Shells of Increasing Energy

Electrons occupy **electron shells**, complex three-dimensional regions around the nucleus. For simplicity, we will depict these shells as a series of increasingly large, concentric rings around

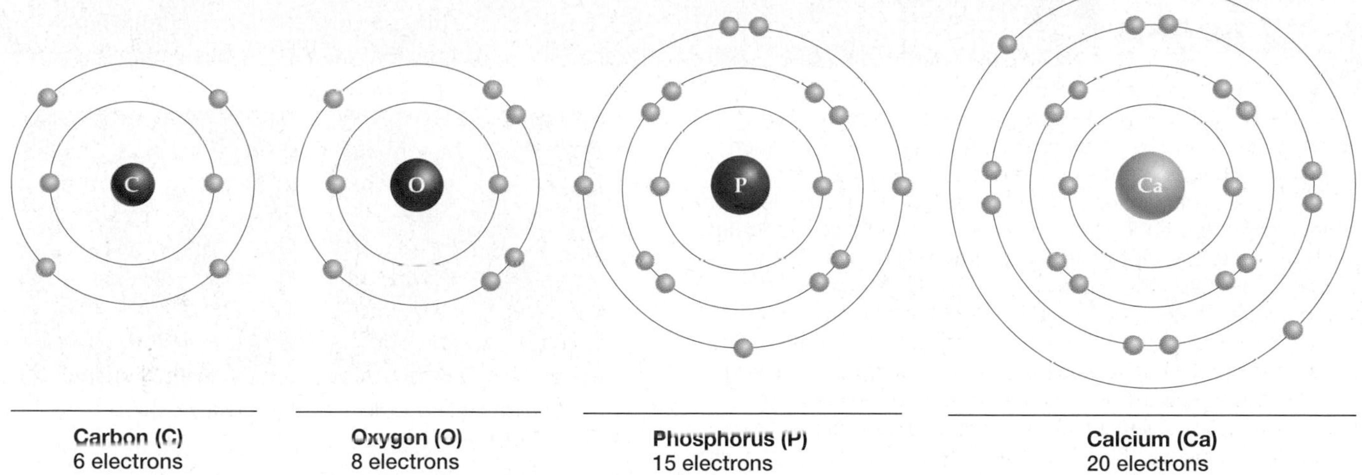

▲ **FIGURE 2-2 Electron shells in atoms** Most biologically important atoms have two or more shells of electrons. The shell closest to the nucleus can hold two electrons; the next three shells can each contain eight electrons.

THINK CRITICALLY Why do atoms with unfilled outer electron shells tend to react with one another?

the nucleus where electrons travel like planets orbiting the sun (**FIG. 2-2**). Each shell has a specific level of energy associated with it. The farther a shell extends from the nucleus, the greater the amount of energy stored in the electrons occupying the shell.

Electrons Can Capture and Release Energy

When an atom absorbs energy, the energy can cause an electron to jump from a lower-energy electron shell to a higher-energy shell. Soon afterward, the electron spontaneously falls back into its original electron shell and releases its extra energy as heat and often also as light (**FIG. 2-3**).

We make use of the ability of electrons to capture and release energy every time we switch on a light bulb. Although incandescent bulbs are rapidly becoming obsolete, they are the easiest type to understand. Electricity flows through a thin wire, heating it to around 4,500°F (about 2,500°C) for a 100-watt bulb. The heat energy bumps some electrons in the wire into higher-energy electron shells. As the electrons drop back down into their original shells, they emit some of the energy as light. Unfortunately, more than 90% of the energy absorbed by the wire is re-emitted as heat rather than light, making an incandescent bulb an extremely inefficient light source.

As Atomic Number Increases, Electrons Fill Shells Increasingly Distant from the Nucleus

Each electron shell can hold a specific number of electrons; the shell nearest the nucleus can hold only two, and more distant shells can hold eight or more. Electrons always fill the

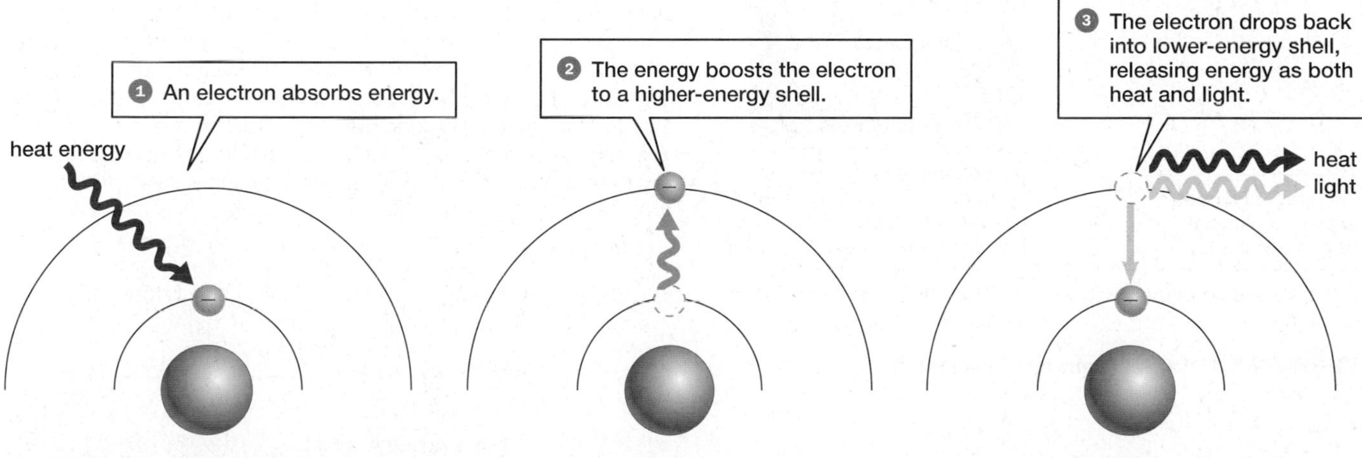

1 An electron absorbs energy.
heat energy
2 The energy boosts the electron to a higher-energy shell.
3 The electron drops back into lower-energy shell, releasing energy as both heat and light.
heat
light

▲ **FIGURE 2-3 Energy capture and release**

THINK CRITICALLY What causes the coals of a campfire to glow?

DOING Science Radioactive Revelations

Radioactivity plays a key role in imaging technologies that are widely used by researchers and physicians. One such technology is positron emission tomography (PET). To perform a PET scan, sugar molecules are tagged with a radioactive isotope and injected into a patient's bloodstream. The tagged molecules tend to move to the more metabolically active regions of the body, which require more sugar for energy. To discover where the radioactive isotope has accumulated, the patient's body is moved through a ring of detectors that respond to the energetic particles (positrons) emitted as the isotope decays. A powerful computer then uses these data to calculate precisely where the decays occurred and generates a color-coded map of the frequency of decays within each "slice" of body passing through the detector ring (**FIG. E2-1a**).

Among the uses of PET scans is medical diagnosis. A PET scan may reveal the location and size of a cancerous tumor or help determine if a patient has Alzheimer's disease. But how can physicians be confident that a PET scan will improve their diagnostic capabilities?

What Question Was Asked?

Can a PET brain scan reliably detect Alzheimer's disease and distinguish it from other causes of declining cognitive function? Before a diagnostic procedure can be widely adopted by physicians, its usefulness must be convincingly and repeatedly demonstrated by research. In one example of such research, investigators asked whether PET scans were any more effective than a physician's exam in identifying patients with Alzheimer's disease.

How Was Evidence Gathered?

To answer their question, the researchers focused on 45 people who had been examined by a medical team, had undergone a PET scan a few months after the exam, and were autopsied years later after death. About half of patients in the study had been diagnosed with Alzheimer's on the basis of the initial medical exam alone, and the other half had been judged Alzheimer's-free. The researchers then compared autopsy results (which confirm an Alzheimer's diagnosis with high reliability) with both the initial diagnosis by the medical team and a new diagnosis based on the PET scan. The researchers were careful to ensure that the diagnosticians analyzing the PET scans did not know the outcome of the autopsies.

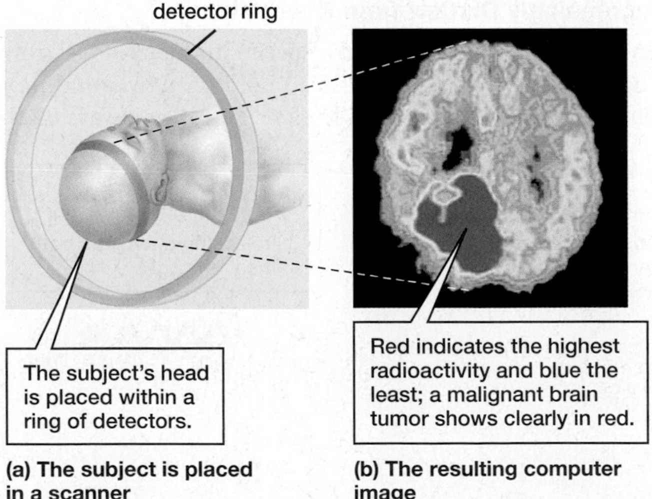

detector ring
The subject's head is placed within a ring of detectors.
Red indicates the highest radioactivity and blue the least; a malignant brain tumor shows clearly in red.

(a) The subject is placed in a scanner

(b) The resulting computer image

▲ **FIGURE E2-1 Positron emission tomography**

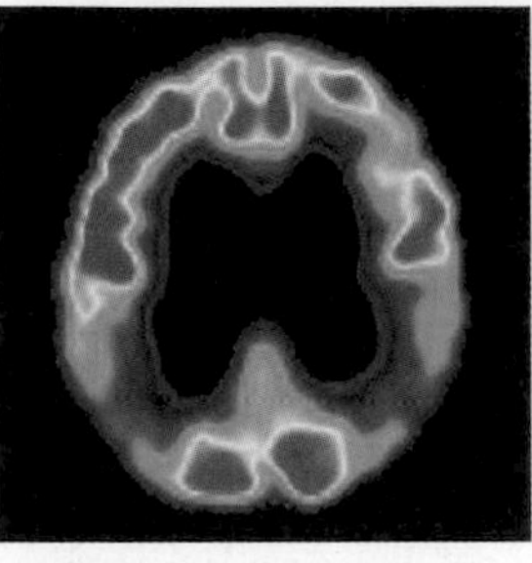

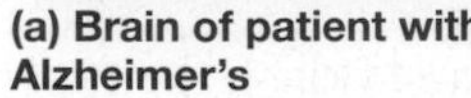

(a) Brain of patient with Alzheimer's

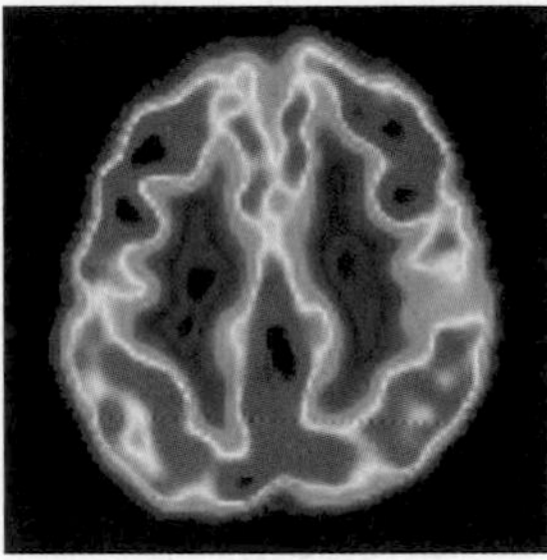

(b) Healthy brain

▲ **FIGURE E2-2 PET reveals differences in brain function** Brain activity is rainbow color-coded, with red indicating the highest activity and blue the lowest; black areas are fluid-filled.

What Was Learned?

The results of the Alzheimer's diagnosis research showed that PET scans were more likely than physicians' exams to accurately diagnose Alzheimer's. This finding suggests that the effort and expense of a PET scan is worthwhile for patients who might have Alzheimer's. Subsequent PET scan research has identified subtle changes in brain activity that indicate very early stages of the disease, raising hope that early detection will help development of effective treatments.

PET scans provide effective diagnosis of Alzheimer's because the brain of a patient with Alzheimer's is far less active than that of a healthy individual (**FIG. E2-2**). In contrast, cancerous brain tumors show up in PET scans as "hot spots" of high activity, because their rapid cell division uses large amounts of sugar (**FIG. E2-1b**). In addition to its medical uses, PET is also used by researchers to learn which brain regions are active when a person performs a particular mental task, such as solving a math problem or recalling a past event.

THINK CRITICALLY In addition to lower brain activity, what other problem has occurred in the brain of the Alzheimer's patient as shown by the images in Fig. E2-2?

lowest-energy shell (the shell nearest the nucleus) first, and then fill higher-energy shells. Elements with increasingly large numbers of protons in their nuclei require more electrons to balance these protons, so their electrons will occupy shells at increasing distances from the nucleus. For example, the two electrons in helium (He) occupy the first electron shell (see Fig. 2-1b). A carbon atom (C) with six electrons will have two electrons filling its first shell and four occupying its second shell, which can contain up to eight electrons (see Fig. 2-2).

CHECK YOUR LEARNING

Can you . . .

- define *element* and *atom*?
- name and describe the subatomic particles that make up the atom?
- explain atomic number and mass number?
- explain radioactivity and its dangers and benefits?
- describe electron shells?

2.2 HOW DO ATOMS INTERACT TO FORM MOLECULES?

Most matter that we encounter in our daily lives consists of atoms linked together to form **molecules**. Simple examples are oxygen gas (O_2; two oxygen atoms) and water (H_2O; two hydrogen atoms and one oxygen atom). How and why do molecules form?

Atoms Form Molecules to Fill Vacancies in Their Outer Electron Shells

In the atoms of some elements, the electrons needed to balance the protons completely fill one or more electron shells. An atom whose outermost electron shell is completely full will not react with other atoms. Such an atom (e.g., helium in Fig. 2-1b) is extremely stable and is described as *inert*.

In the atoms of most elements, electrons fill the inner shells, but do not completely fill the outermost shell. An atom whose outermost electron shell is only partially full will react readily with other atoms. Such an atom (e.g., hydrogen in Fig. 2-1a) is described as *reactive*.

Among atoms and molecules with unfilled outer shells, some—called **free radicals**—are so reactive that they can tear other molecules apart. Free radicals are produced in large numbers in the body by reactions that make energy available to cells. Although these reactions are essential to life, the damage to cells caused by the resulting free radicals may contribute to aging and ultimately death. Learn more in "Health Watch: Free Radicals—Friends and Foes?" on page 25.

Chemical Bonds Hold Atoms Together in Molecules

Reactive atoms form **chemical bonds**, which are attractive forces that hold atoms together in molecules. Bonds are formed when atoms acquire, lose, or share electrons to gain stability. There are three major types of bonds: *ionic bonds*, *covalent bonds*, and *hydrogen bonds* (**TABLE 2-3**).

Ionic Bonds Form Between Ions

Because the number of protons in an atom is equal to the number of electrons, an atom's overall electrical charge is neutral. Electrical neutrality, however, does not necessarily make an atom as stable as possible. For example, an atom whose outermost electron shell is nearly empty can become more stable by losing electrons to completely empty the outer shell, giving the atom a positive charge. Conversely, an atom with a nearly full outer shell can become more stable by gaining electrons to completely fill the shell, giving the atom a negative charge. When an atom becomes stable by losing or gaining one (or a few) electrons and thus acquiring an overall positive or negative charge, it becomes an **ion**.

Ions with opposite charges attract one another, and the electrical attraction between positively and negatively charged ions forms **ionic bonds**. For example, the white crystals in your salt shaker are sodium and chloride ions linked by

TABLE 2-3 Common Types of Bonds in Biological Molecules

Type	Type of Interaction	Example
Ionic bond	An electron is transferred between atoms, creating positive and negative ions that attract one another.	Occurs between the sodium (Na^+) and chloride (Cl^-) ions of table salt (NaCl)
Covalent bond	Electrons are shared between atoms.	
Nonpolar	Electrons are shared equally between atoms.	Occurs between the two hydrogen atoms in hydrogen gas (H_2)
Polar	Electrons are shared unequally between atoms.	Occurs between hydrogen and oxygen atoms of a water molecule (H_2O)
Hydrogen bond	Attractions occur between polar molecules in which hydrogen is bonded to oxygen or nitrogen. The slightly positive hydrogen attracts the slightly negative oxygen or nitrogen of a nearby polar molecule.	Occurs between water molecules, where slightly positive charges on hydrogen atoms attract slightly negative charges on oxygen atoms of nearby molecules

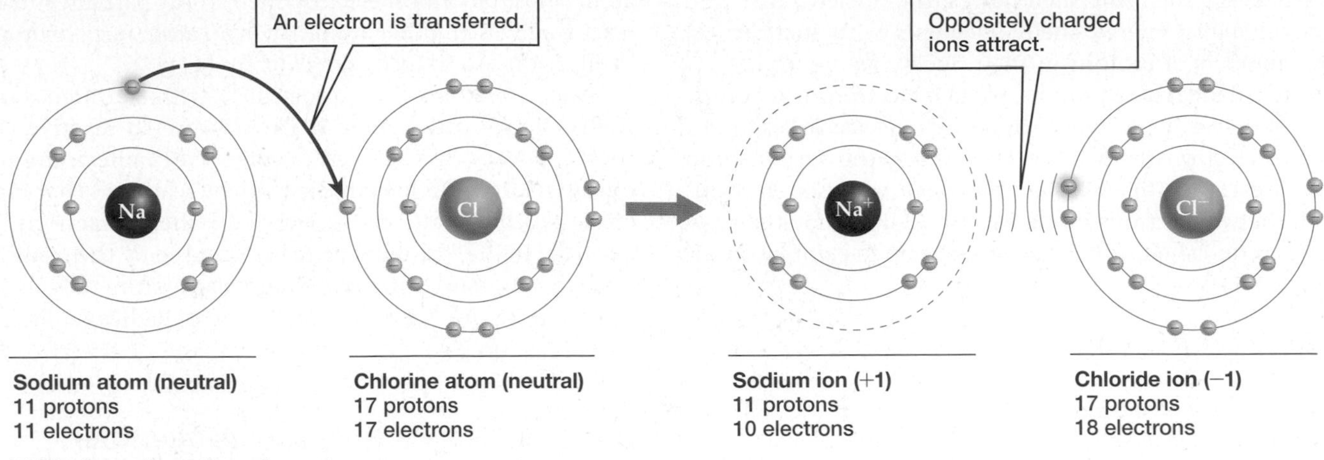

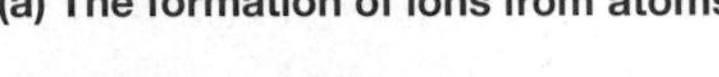

(a) The formation of ions from atoms

(b) An ionic molecule: NaCl

◀ **FIGURE 2-4 Ions and ionic bonds (a)** Stable ions form when sodium loses an electron (Na^+) and chlorine gains an electron (Cl^-). **(b)** Sodium and chloride ions nestle closely together in cubic crystals of table salt (NaCl).

ionic bonds. Sodium (Na) has only one electron in its outermost electron shell, so it can become stable by losing this electron, forming the ion Na^+. Chlorine (Cl) has seven electrons in its outer shell, which can hold eight electrons. So chlorine can become stable by gaining an electron (in this case from sodium), forming the ion Cl^- (**FIG. 2-4a**) and producing an ionic bond between Na^+ and Cl^-. These ionic bonds result in crystals composed of a repeating, orderly array of Na^+ and Cl^- (**FIG. 2-4b**). As we describe later, water is attracted to ions and can break ionic bonds, as occurs when water dissolves salt.

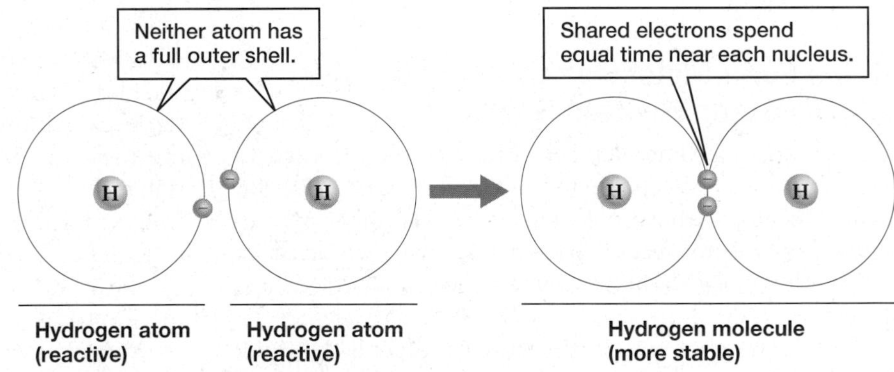

▲ **FIGURE 2-5 Nonpolar covalent bonds** A hydrogen molecule (H_2) is formed when an electron from each of two hydrogen atoms is shared equally, forming a single nonpolar covalent bond.

Covalent Bonds Form by Sharing Electrons

Atoms with partially full outermost electron shells can also become stable by forming **covalent bonds**, in which two atoms share electrons, filling the outer shells of both. For example, two hydrogen atoms can become more stable if they share their single outer electrons, allowing each to behave almost as if it had two electrons in its outer shell (**FIG. 2-5**). This covalent bond forms hydrogen gas (H_2). Covalent bonds join the atoms in most biological molecules, such as proteins, sugars, and fats (**TABLE 2-4**).

TABLE 2-4 Electrons and Bonds in Atoms Common in Biological Molecules

Atom	Capacity of Outer Electron Shell	Electrons in Outer Shell	Number of Covalent Bonds Usually Formed
Hydrogen (H)	2	1	1
Carbon (C)	8	4	4
Nitrogen (N)	8	5	3
Oxygen (O)	8	6	2
Sulfur (S)	8	6	2

Health WATCH Free Radicals—Friends and Foes?

Any atom with a partially full outer electron shell will be reactive, but reactivity increases dramatically if the unfilled shell also contains an uneven number of electrons. Atoms with unpaired electrons in their outer shells are called free radicals. Free radicals react vigorously with other molecules, capturing or releasing electrons to achieve a more stable configuration that lacks unpaired electrons. Such reactions may result in damage to crucial biological molecules, including proteins and DNA.

Our bodies continuously produce free radicals as a by-product of reactions that generate cellular energy. Free radicals are also formed when our cells are bombarded by sunlight's ultraviolet (UV) radiation, X-rays, radioactive isotopes, and toxic chemicals in the environment. Our bodies counteract free radicals by generating **antioxidants**, molecules that react with free radicals and render them harmless. We also obtain antioxidants in our diets, since they occur naturally in many plant-derived foods. But when free radicals overwhelm the body's ability to counteract them, the resulting *oxidative stress* can injure cells. Oxidative stress contributes to cardiovascular disease, lung disorders such as asthma, some forms of cancer, and neurological diseases such as Alzheimer's. The most visible signs of aging—graying hair and wrinkled skin—are a result of oxidative stress. Much of the oxidative stress that causes wrinkles, even in younger adults, is caused by exposure to UV radiation from the sun (**FIG. E2-3**).

Strong evidence suggests that diets high in antioxidant-containing fruits and vegetables are associated with a lower incidence of cardiovascular disease and some cancers. As a result, some people have concluded that consuming antioxidant supplements such as vitamins C and E and beta-carotene might provide a shortcut to health. But an analysis that combined numerous studies of large groups of people showed no health benefits from taking such supplements, and suggested that beta-carotene and vitamin E supplements may in some cases have an adverse effect on health. Why might this be?

During the course of evolution, organisms have been continuously exposed to free radicals and have evolved ways to either use them constructively or deactivate them. Free radicals are involved in regulating blood pressure, healing wounds, and defending against disease-causing microbes. Growing evidence supports the hypothesis that health requires a complex balance of free radicals and antioxidants and that the high doses of purified antioxidants in supplements can upset this delicate equilibrium. Fruits and vegetables contain natural antioxidants at levels generally well below those found in supplements, so it is most prudent to obtain antioxidants from a diet rich in fruits and vegetables (unless you have a medical condition that requires using supplements).

FIGURE E2-3 Free-radical damage Free radicals, caused by UV radiation and other factors, damage proteins that give skin its elasticity, causing wrinkles.

Which types of fruits and vegetables should you eat? Would you believe dark chocolate makes the list? Although it seems a stretch to put chocolate in the fruit or vegetable category, chocolate is made from cocoa beans, which are the seeds found in the fruit of the cacao tree. Dark chocolate is especially rich in antioxidant molecules called *flavonols* produced by the cacao plant. Although it sounds almost too good to be true, controlled studies have reported that consuming dark chocolate (**FIG. E2-4**) has a beneficial effect on high blood pressure and other risk factors for cardiovascular disease. Several large studies on human populations have also found a correlation between higher consumption of chocolate and a reduced incidence of cardiovascular disease, including strokes and heart attacks (correlations provide suggestive evidence, but are less conclusive than the results of controlled studies). Don't like chocolate? Flavonols are also found in green tea, cranberries, apples, onions, kale, and other plant foods.

THINK CRITICALLY At a physical exam, Thomas, a sedentary individual, was warned of his dangerously high blood pressure. After reading that chocolate is good for cardiovascular health, he immediately stocked up on 6-ounce dark chocolate bars and resolved to add one daily to his regular diet. Predict some results of his next annual physical exam, and provide him with some good health advice.

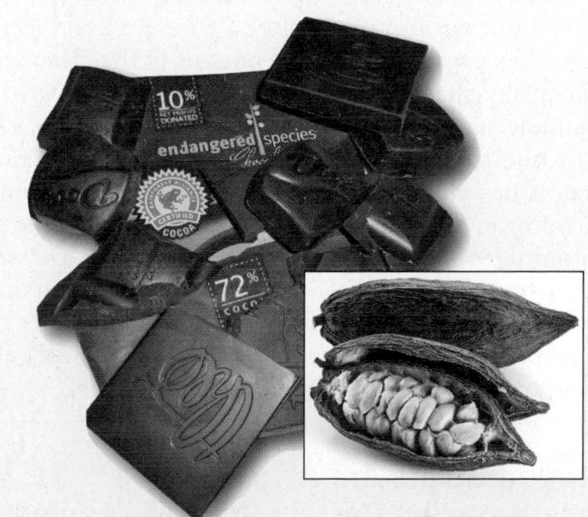

FIGURE E2-4 Chocolate Chocolate comes from cacao beans found inside pods (inset) that grow on trees native to South America.

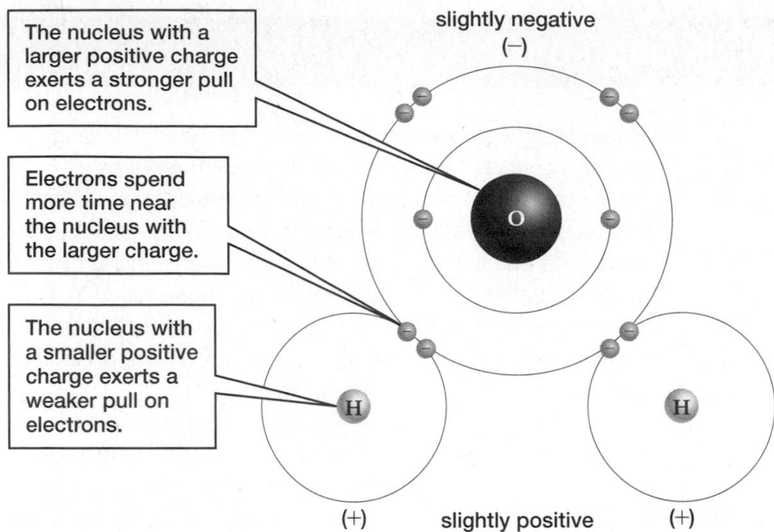
The nucleus with a larger positive charge exerts a stronger pull on electrons.
Electrons spend more time near the nucleus with the larger charge.
The nucleus with a smaller positive charge exerts a weaker pull on electrons.
slightly negative
(−)
O
H
H
(+)
slightly positive
(+)

▲ **FIGURE 2-6 Polar covalent bonds** Oxygen (O) needs two electrons to fill its outer shell, allowing it to form covalent bonds with two hydrogen atoms (H), which produces water (H_2O). The oxygen atom exerts a greater pull on the electrons than do the hydrogen atoms, producing a slight negative charge near the oxygen and a slight positive charge near the two hydrogens.

Covalent Bonds May Produce Nonpolar or Polar Molecules

If the atoms participating in a covalent bond share electrons equally or nearly equally, the result will be a **nonpolar covalent bond** in which there is no charge on any part of the molecule. All covalent bonds between atoms of the same element are nonpolar. For example, because the two nuclei in H_2, are identical, the electrons in the molecule spend equal time near each nucleus, so neither end, or *pole*, of the molecule is charged. The covalent bonds in oxygen gas (O_2) and nitrogen gas (N_2) are also nonpolar. In some cases, covalent bonds between atoms of different elements are also nonpolar, for example in carbon dioxide (CO_2) and certain biological molecules such as oils and fats (described in Chapter 3).

If the nucleus of one atom in a covalent bond attracts electrons more strongly than the nucleus of the other atom, the unequally shared electrons produce a **polar covalent bond**. Molecules containing such bonds are called *polar molecules*. Although a polar molecule as a whole is electrically neutral, it has charged poles. In water (H_2O), for example, each hydrogen atom shares an electron with the single oxygen atom (**FIG. 2-6**). The oxygen nucleus exerts a stronger attraction on the electrons than does either hydrogen nucleus. By attracting electrons, the oxygen pole of a water molecule becomes slightly negative, leaving each hydrogen atom slightly positive.

Hydrogen Bonds Are Attractive Forces Between Certain Polar Molecules

Biological molecules, including sugars, proteins, and nucleic acids, often contain many polar covalent bonds between hydrogen and oxygen or between hydrogen and nitrogen. In these cases, the hydrogen is slightly positive and the oxygen or nitrogen is slightly negative. As you know, opposite charges attract. A **hydrogen bond** is the attraction between molecules that occurs when the slightly positive region of one molecule attracts the slightly negative region of another. Hydrogen bonds between water molecules link the molecules in a loosely connected, ever-changing network (**FIG. 2-7**). As you will see shortly, these hydrogen bonds give water several unusual properties that are essential to life as we know it.

CHECK YOUR LEARNING

Can you . . .

- explain what makes an atom reactive?
- define *molecules* and *chemical bonds*?
- describe and provide examples of ionic, covalent, and hydrogen bonds?

2.3 WHY IS WATER SO IMPORTANT TO LIFE?

As naturalist Loren Eiseley eloquently stated, "If there is magic on this planet, it is contained in water." What makes water so special?

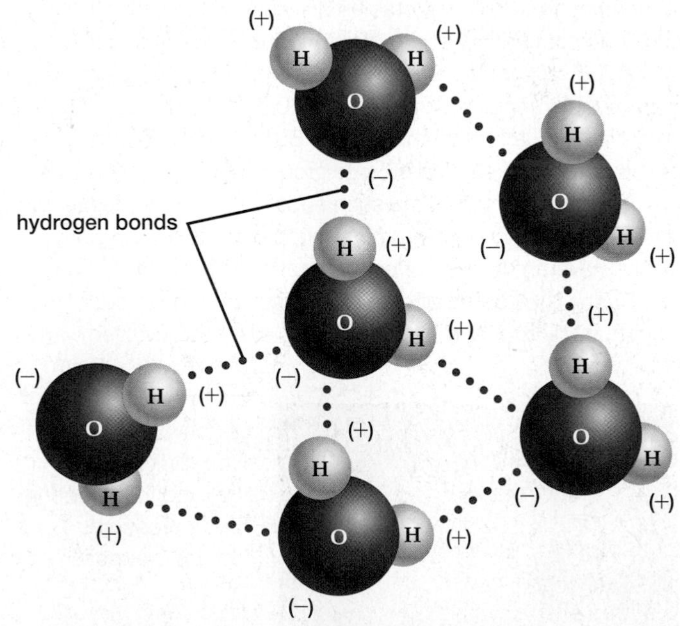
hydrogen bonds
H
O
(+)
(−)

▲ **FIGURE 2-7 Hydrogen bonds in water** The slight charges on opposite poles of water molecules (shown in parentheses) produce hydrogen bonds (dotted lines) between the oxygen and hydrogen atoms in adjacent water molecules. Each water molecule can form up to four hydrogen bonds. In liquid water, these bonds constantly break as new ones form.

(a) Cohesion and adhesion

(b) Cohesion causes surface tension

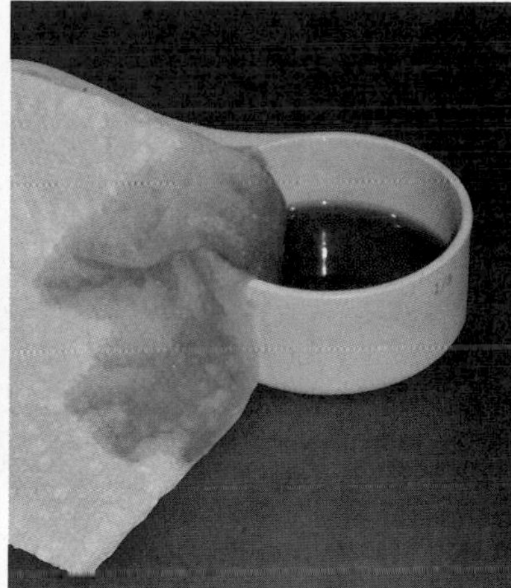

(c) Capillary action

▲ **FIGURE 2-8 Water molecules have cohesive and adhesive properties (a)** Cohesion causes water to form droplets; adhesion sticks them to spiderweb silk. **(b)** This water strider uses surface tension as it moves across the water's surface. **(c)** Cohesion and adhesion work together in capillary action, which draws water into narrow spaces among charged surfaces.

Water Molecules Attract One Another

Many of water's distinctive properties result from the hydrogen bonds that interconnect water molecules. The bonds result in **cohesion**, the tendency of water molecules to stick together; cohesion is what causes water to form droplets (**FIG. 2-8a**). Cohesion also produces **surface tension**, the tendency for a water surface to resist being broken. Surface tension arises because the water molecules at a surface are attracted more strongly to one another and to the water molecules beneath them than to the molecules of the air above. Surface tension is what causes water to brim slightly above the top of a glass just before it overflows and what allows the surface of a pond to support water insects (**FIG. 2-8b**).

Hydrogen bonds also play a role in **adhesion**, the tendency of water molecules to cling to other substances. Water adheres to materials whose molecules contain charged regions, such as glass, wood, paper, and the silk of spiderwebs (see Fig. 2-8a). When adhesion attracts water to a surface and cohesion then drags more water molecules along, the result is **capillary action**. During capillary action, water moves spontaneously into very narrow spaces, such as those between the cellulose fibers of a paper towel (**FIG. 2-8c**).

The tendency of water molecules to cohere and adhere plays a crucial role in land plants, enabling water absorbed by a plant's roots to reach the plant's leaves, even if the plant is a 300-foot-tall redwood tree. A plant's roots, stem, and leaves are connected by water-filled tubes. As water evaporates continuously from leaves, each evaporating molecule pulls the one below it to the leaf's surface, much like a chain being dragged upward. The hydrogen bonds that link water molecules are stronger than the downward pull of gravity, so the water chain doesn't break. In addition, water adheres to the walls of the conducting tubes, which are composed of cellulose and are microscopically narrow. These properties allow capillary action to contribute to the transport of water from roots to leaves. Without the cohesion and adhesion of water, there could be no large land plants, and life on Earth would be radically different.

Given that water molecules are attracted to one another by hydrogen bonds, why is water still able to flow freely? The hydrogen bonds in liquid water constantly break and re-form, like square dancers continually moving from one partner to the next, releasing hands and clasping new ones as they go. This dynamic change in hydrogen bonding allows water to flow.

Water Interacts with Many Other Molecules

Water is an excellent **solvent**, a substance that completely surrounds and disperses the atoms or molecules of another substance. When this occurs, the solvent is said to **dissolve** the substance it disperses. A solvent that contains one or

Have You Ever Wondered . . .

Why It Hurts So Much to Do a Belly Flop?

The slap of a belly flop provides first-hand experience of the power of cohesion among water molecules. Because of the hydrogen bonds that interconnect water molecules, the water surface resists being broken. When you suddenly force a large number of water molecules apart with your belly, the result can be a bit painful. Do you think that belly-flopping into a pool of (nonpolar) vegetable oil would hurt as much? If you encountered a deep pool filled with vegetable oil, could you float or swim in it?

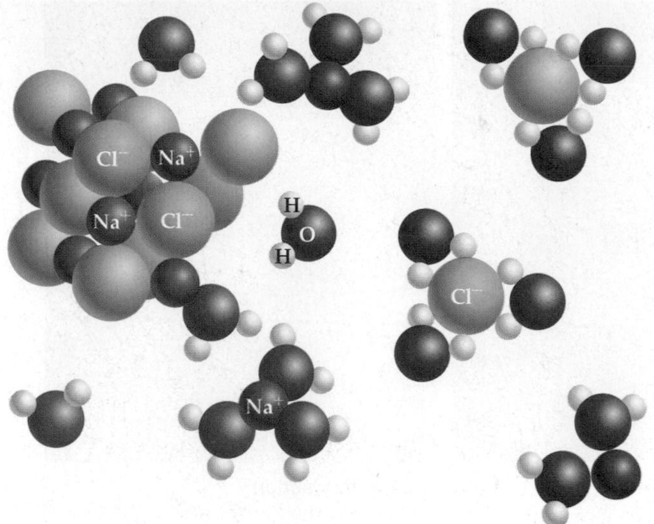

▲ **FIGURE 2-9 Water as a solvent** When a salt crystal (NaCl) is dropped into water, the water surrounds the sodium and chloride ions, with the positive poles of water molecules facing the Cl^-, and the negative poles facing the Na^+. The ions disperse as the surrounding water molecules isolate them from one another, dissolving the salt crystal.

THINK CRITICALLY If you placed a salt crystal in a nonpolar liquid (like oil), would it dissolve? Explain.

◀ **FIGURE 2-10 Oil and water don't mix** Yellow oil poured into water remains in discrete droplets as it rises to the surface. Oil floats because it is less dense than water.

THINK CRITICALLY Predict how a drop of water on an oil-coated surface would differ in shape from a drop on a clean glass surface.

more dissolved substances is called a **solution**. Water is an especially good solvent because the positive and negative poles of water molecules are attracted to charges on other polar molecules and ions. For example, a crystal of table salt is held together by ionic bonds between positively charged sodium ions (Na^+) and negatively charged chloride ions (Cl^-; see Fig. 2-4b). When a salt crystal is dropped into water, the positively charged hydrogen poles of water molecules are attracted to the Cl^-, and the negatively charged oxygen poles are attracted to the Na^+. As water molecules surround the ions, shielding them from interacting with each other, the ions separate from the crystal and drift away in the water—the salt dissolves (**FIG. 2-9**). Because ions and polar molecules are attracted to (and dissolve in) water, they are described as **hydrophilic** (Gk. *hydro*, water, and *phylos*, loving).

Gases such as oxygen and carbon dioxide also dissolve in water even though they are nonpolar. How? These molecules are so small that they fit into the spaces between water molecules without disrupting the hydrogen bonds. Water's ability to dissolve oxygen allows fish and other aquatic organisms to flourish.

Larger nonpolar molecules, such as fats and oils, are **hydrophobic** (Gk. *phobic*, fearing) and do not dissolve in water. Nevertheless, water has an important effect on such molecules. By sticking together, water molecules exclude hydrophobic molecules. So, for example, nonpolar oil molecules are forced together into drops, surrounded by water molecules that form hydrogen bonds with one another but not with the oil (**FIG. 2-10**).

Water Moderates the Effects of Temperature Changes

The hydrogen bonds linking water molecules (see Fig. 2-7) allow water to moderate temperature changes.

It Takes a Lot of Energy to Heat Water

The **specific heat** of a substance is the amount of energy required to heat 1 gram of the substance by 1°C. Water's specific heat is far higher than that of any other common substance. Because of its high specific heat, water moderates temperature changes. One reason you can sit on hot sand in the hot sun without instantly overheating is that your body is about 60% water, which can absorb considerable heat before its temperature changes.

Why is water's specific heat so high? At any temperature above absolute zero (−459°F or −273°C), all molecules are in constant motion. If energy is added to a substance to make its molecules move faster, the substance becomes warmer. But with water, any added energy must first break the water's hydrogen bonds in order to make the molecules move faster. Breaking the bonds consumes a considerable amount of energy, and so less energy is available to raise the water's temperature. In contrast, for substances that lack hydrogen bonds, a greater proportion of added energy is available to raise their temperatures. Thus, for example, a given amount of energy would increase the temperature of granite rock (which lacks hydrogen bonds) about five times as much as it would the same weight of water.

It Takes a Lot of Energy to Evaporate Water

While the high specific heat of water helps slow the rate at which a person's body temperature rises in warm conditions, another property of water helps the body to offload excess

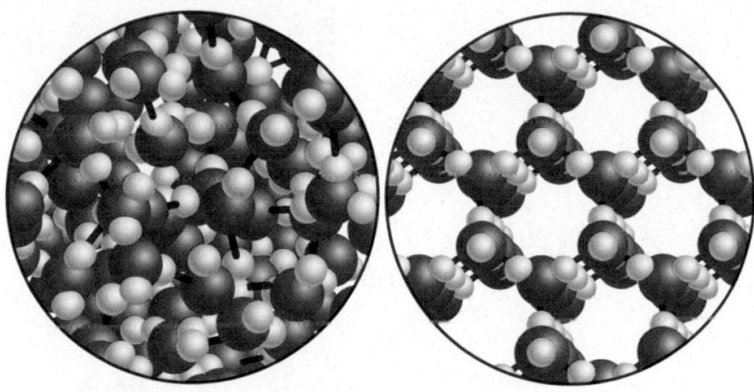

▲ **FIGURE 2-11 Liquid water (left) and ice (right)**

THINK CRITICALLY How do these configurations explain why ice floats?

heat. When we perspire to maintain our body temperature in hot conditions, we are making use of water's extremely high **heat of vaporization**, which is the amount of heat needed to cause a substance to evaporate (change from a liquid to a vapor). In order for water molecules to move fast enough to escape and evaporate into the air, they must absorb enough energy to break the hydrogen bonds that interconnect them. Thus, when we perspire, the water in our sweat absorbs a great deal of body heat as it evaporates, cooling us as its fastest-moving molecules vaporize.

CASE STUDY CONTINUED

Unstable Atoms Unleashed

The high specific heat of water makes it an ideal coolant for nuclear power plants. Compared to nonpolar liquids like alcohol, a great deal of heat is required to raise the temperature of water. The tsunami that hit the Fukushima power plant disrupted the electrical supply running the pumps that kept water circulating over the fuel rods. Without enough water to absorb the excess heat, the metal tubes surrounding the fuel rods melted.

Water Forms an Unusual Solid: Ice

Even solid water is unusual. Most liquids become denser when they solidify, but ice is actually less dense than liquid water. When water freezes, each molecule forms stable hydrogen bonds with four other water molecules, creating an open, hexagonal (six-sided) arrangement (**FIG. 2-11**). This arrangement keeps the water molecules farther apart than they are in liquid water. Thus, ice is less dense than liquid water, which is why icebergs and ice cubes float.

This property of water is crucial to the distribution of aquatic life. When a pond or lake starts to freeze in winter, the floating ice forms an insulating layer that delays the freezing of the water below. This insulation allows fish and other aquatic organisms to survive in the liquid water below the ice (**FIG. 2-12**). If ice did not float, many ponds and lakes around the world would freeze solid from the bottom up during the winter, killing most of their inhabitants.

▲ **FIGURE 2-12 Ice floats** The floating ice insulates the water beneath it, helping to protect bodies of water from freezing solid and allowing aquatic organisms such as this freshwater seal in Lake Baikal, Siberia, to survive beneath it.

Water-Based Solutions Can Be Acidic, Basic, or Neutral

At any given moment, a tiny fraction of the molecules in a quantity of water (H_2O) will have split into hydroxide ions (OH^-) and hydrogen ions (H^+) (**FIG. 2-13**). Pure water contains equal concentrations of each. But when a substance that releases OH^- or H^+ is dissolved in water, the resulting solution no longer has equal concentrations of OH^- and H^+.

If the concentration of H^+ in a solution exceeds the concentration of OH^-, the solution is **acidic**. An **acid** is a substance that increases the concentration of hydrogen ions when it dissolves in water. For example, when hydrochloric acid (HCl) is added to pure water, almost all of the HCl molecules separate into H^+ and Cl^-. As a result, the

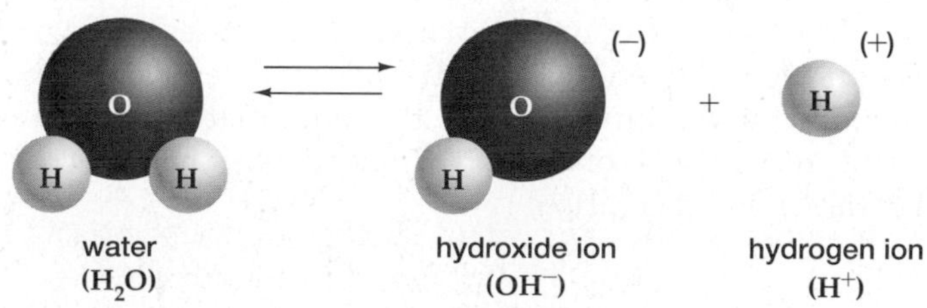

▲ **FIGURE 2-13 Some water is always ionized**

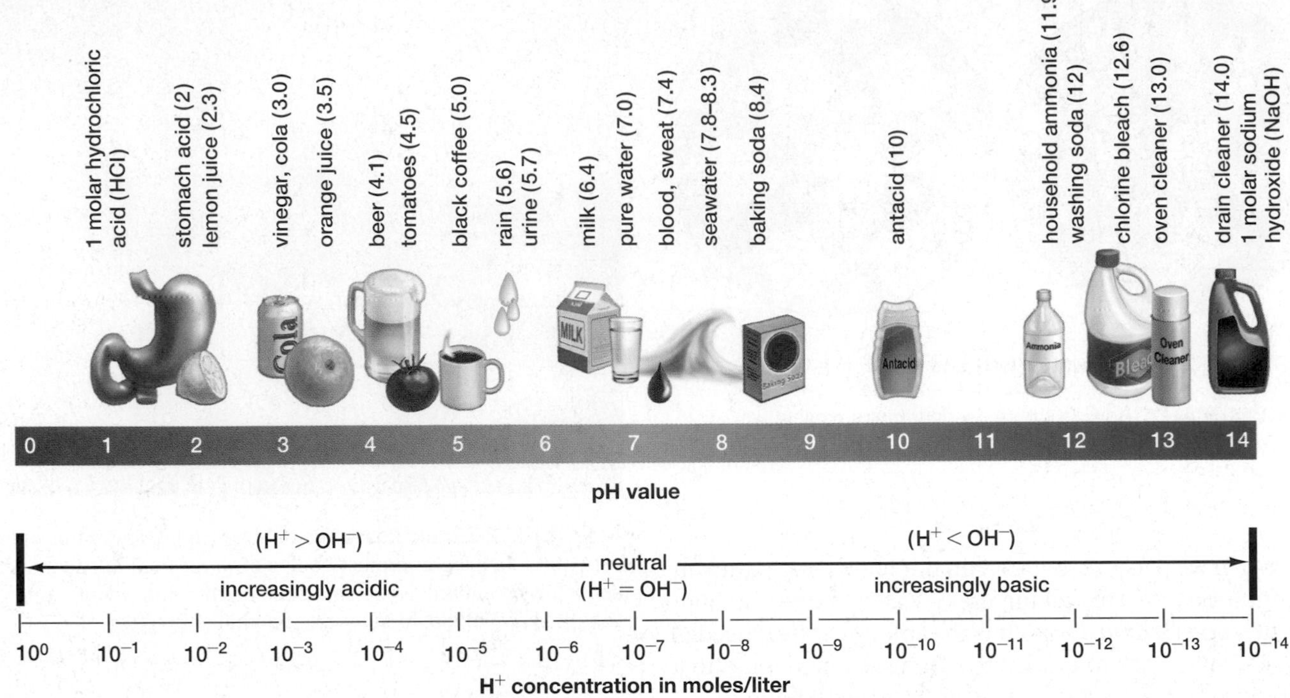

▲ **FIGURE 2-14 The pH scale** The pH scale reflects the concentration of hydrogen ions in a solution. Notice that pH (upper scale; 0 to 14) is the negative log of the H^+ concentration (lower scale). Each unit on the scale represents a tenfold change. Lemon juice, for example, is about 10 times more acidic than orange juice.

concentration of H^+ exceeds the concentration of OH^-, and the resulting solution is acidic. Acidic substances—think lemon juice (containing citric acid) or vinegar (acetic acid)—taste sour because sour receptors on your tongue respond to excess H^+.

If the concentration of OH^- in a solution is greater than the concentration of H^+, the solution is **basic**. A **base** is a substance that increases the concentration of hydroxide ions when it dissolves in water. If, for instance, sodium hydroxide (NaOH) is added to water, the NaOH molecules separate into Na^+ and OH^-. As a result, the concentration of OH^- exceeds the concentration of H^+, and the resulting solution is basic. Bases are used in many cleaning solutions, and also in many antacids like Tums to neutralize heartburn caused by excess hydrochloric acid in the stomach.

The **pH scale** measures how acidic or basic a solution is (**FIG. 2-14**). Values measured on the pH scale range from 0 to 14. Pure water (equal concentrations of H^+ and OH^-) has a neutral pH of 7. Acids have a pH below 7, and bases have a pH above 7. Each unit on the pH scale represents a tenfold change in the concentration of H^+. Thus, the concentration of H^+ in a soft drink with a pH of 3 is 10,000 times greater than the concentration in water (pH 7).

A **buffer** is a molecule that tends to maintain a solution at a nearly constant pH by accepting or releasing H^+ in response to small changes in H^+ concentration. In the presence of excess H^+, a buffer combines with the H^+, reducing its concentration. In the presence of excess OH^-, buffers release H^+, which combines with the OH^- to form H_2O. Buffers play an important role in maintaining constant conditions within organisms. For example, various buffers allow humans and other mammals to maintain a pH in body fluids that is just slightly basic (about 7.4), even though the numerous chemical reactions that take place in cells would otherwise alter pH. This buffering is essential, because if your blood became as acidic as 7.0 or as basic as 7.8, you would likely die; even small changes in pH cause drastic changes in biological molecules.

CHECK YOUR LEARNING

Can you . . .

- describe the unique properties of water and the importance of these properties to life?
- explain how polar covalent and hydrogen bonds contribute to the unique properties of water?
- explain the concept of pH and how acids, bases, and buffers affect solutions?

CASE STUDY REVISITED

Unstable Atoms Unleashed

The chain reaction that generates heat in nuclear power plants begins when radioactive uranium atoms release neutrons. These neutrons bombard other uranium atoms and cause them to split, releasing energy and more neutrons that repeat the bombardment in a self-sustaining chain reaction.

When the tsunami struck the Fukushima plant, neutron-absorbing rods were immediately lowered around the fuel, halting the chain reaction. But the uranium's decay generated additional radioactive isotopes that continued to spontaneously decay and generate heat. This heat contributed to the disastrous breach that released the isotopes into the surrounding environment. One isotope of particular concern is radioactive iodine.

Iodine enters the body in food and water. It becomes concentrated in the thyroid gland, which uses iodine to synthesize thyroid hormone. Unfortunately, the thyroid gland does not distinguish between radioactive and non-radioactive iodine. Children exposed to radioactive iodine are at increased risk for thyroid cancer, which may occur decades after exposure. To help protect them, Japanese authorities distributed iodine tablets to children near the failed reactor. This non-radioactive iodine saturates the thyroid, which is then less likely to take up the radioactive form. Only time will reveal the full health effects of the fallout from Fukushima.

CONSIDER THIS How do the risks associated with nuclear power compare to the risks of burning fossil fuels? Should societies subsidize the cost of improving renewable energy technologies such as wind and solar power?

CHAPTER REVIEW

Go to **Mastering Biology** to access the Pearson eText, vocabulary review, practice quizzes, activities, videos, current events, and more.

Answers to ***Think Critically*** *and* ***Thinking Through the Concepts*** *questions can be found in the* ***Answers*** *section at the back of the book.*

Summary of Key Concepts

2.1 What Are Atoms?

An element is a substance that can neither be broken down nor converted to different substances by ordinary chemical reactions. The smallest possible particle of an element is the atom, which is composed of positively charged protons, uncharged neutrons, and negatively charged electrons. All atoms of a given element have the same unique number of protons. Neutrons and protons cluster to form atomic nuclei. Electrons orbit the nucleus within specific regions called electron shells. Shells at increasing distances from the nucleus contain electrons with increasing amounts of energy. Each shell can contain a fixed maximum number of electrons. An atom is most stable when its outermost shell is full. Isotopes are atoms of the same element with different numbers of neutrons. Some isotopes are radioactive; their nuclei spontaneously break down, forming new elements and releasing energy and often subatomic particles.

2.2 How Do Atoms Interact to Form Molecules?

Atoms gain stability by filling or emptying their outer electron shells. They do this by acquiring, losing, or sharing electrons during chemical reactions. This produces attractive forces called chemical bonds, which link atoms to form molecules.

There are three types of bonds: ionic, covalent, and hydrogen. Atoms that have lost or gained electrons are called ions. Ionic bonds are electrical attractions between negatively and positively charged ions that hold them together in crystals. Covalent bonds form when atoms fill their outer electron shells by sharing electrons. In a nonpolar covalent bond, the two atoms share electrons equally. In a polar covalent bond, one atom attracts electrons more strongly than the other atom does, giving the molecule slightly positive and negative poles. Polar covalent bonds allow hydrogen bonding, the attraction between the slightly positive regions of one polar molecule and slightly negative regions of another polar molecule.

2.3 Why Is Water So Important to Life?

Water's unique properties allowed life as we know it to evolve. Water is polar and dissolves polar substances and ions. Water forces nonpolar substances, such as oil, to form clumps. Water molecules cohere to each other using hydrogen bonds, producing surface tension. Water also adheres to other polar surfaces. Water's extremely high specific heat and high heat of vaporization function in some animals to maintain relatively stable body temperatures despite large outside temperature fluctuations. Water is unusual in being less dense in its frozen state than in its liquid state.

Pure water contains equal numbers of H^+ and OH^- (pH 7), but dissolved substances can make solutions acidic (more H^+ than OH^-) or basic (more OH^- than H^+). Buffers help maintain a constant pH.

Thinking Through The Concepts

Bloom's: Remembering, Understanding

Multiple Choice

1. Which of the following is false?
   a. An element is defined by its atomic number.
   b. Ninety-two elements occur naturally.
   c. An atom consists of subatomic particles.
   d. Electron shells increase in energy closer to the nucleus.
2. The mass number of an element is equal to
   a. the mass of its atom's protons and neutrons.
   b. the mass of its atom's neutrons.
   c. the mass of its atom's protons.
   d. its atomic number.

3. Isotopes are defined as
   a. atoms of the same element with different numbers of protons.
   b. radioactive elements.
   c. stable elements.
   d. atoms of the same element with different numbers of neutrons.
4. Molecules
   a. always consist of different elements bonded together.
   b. may be polar or nonpolar.
   c. may be held together entirely by hydrogen bonds.
   d. cannot be held together by ionic bonds.
5. Covalent bonds
   a. link water molecules in ice.
   b. create only nonpolar molecules.
   c. bind sodium and chlorine in table salt.
   d. link hydrogen to oxygen in a water molecule.

Fill-in-the-Blank

1. An atom consists of an atomic nucleus composed of positively charged ____________ and uncharged ____________. Orbiting around the nucleus are ____________ that occupy discrete spaces called ____________.
2. An atom that has lost or gained one or more electrons is called a(n) ____________. If an atom loses an electron, it takes on a(n) ____________ charge. Atoms with opposite charges attract one another, forming ____________ bonds.
3. Atoms of the same element that differ in the number of neutrons in their nuclei are called ____________. Some of these atoms spontaneously break apart, and in this process, they sometimes become different ____________. Atoms that behave this way are described as being ____________.
4. An atom with an outermost electron shell that is either completely full or empty is described as ____________. Atoms with partially full outer electron shells are ____________. Covalent bonds are formed when atoms ____________ electrons, filling their outer shells.
5. Water is described as ____________ because each water molecule has slightly negative and positive poles. This property allows water molecules to form ____________ bonds with one another. The bonds between water molecules give water ____________ that produces surface tension.

Review Questions

1. Based on Table 2-1, how many neutrons are there in oxygen? In hydrogen? In nitrogen?
2. Distinguish between atoms and molecules and among protons, neutrons, and electrons.
3. Compare and contrast covalent bonds and ionic bonds.
4. Explain how polar covalent bonds allow hydrogen bonds to form, and provide an example.
5. Why can water absorb a great amount of heat with little increase in its temperature? What is this property called?
6. Describe how water dissolves a salt.
7. Define *pH scale, acid, base,* and *buffer.* How do buffers reduce changes in pH when hydrogen ions or hydroxide ions are added to a solution? Why is this phenomenon important in organisms?

Applying the Concepts

Bloom's: Applying, Analyzing, Evaluating

1. Detergents help clean by dispersing fats and oils in water so that they can be rinsed away. What general chemical structures (for example, polar or nonpolar parts) must a soap or detergent have, and why?
2. What do people mean when they say, "It's not the heat, it's the humidity"? Why does high humidity make a hot day less bearable?
3. You can now buy "whiskey stones" made of granite, which you place in your freezer and use instead of ice cubes to cool drinks. Would whiskey stones cool drinks more or less effectively than an equal weight of ice cubes from the same freezer? Explain your answer.

Mad cow disease may have emerged as a result of cows eating feed containing protein from the remains of sheep infected with scrapie.

3 Biological Molecules

CASE STUDY

Puzzling Proteins

"YOU KNOW, LISA, I think something is wrong with me," Charlene Singh told her sister. The vibrant 22-year-old scholarship winner had difficulty remembering things, and she suffered from mood swings. During the next 3 years, her symptoms worsened. Her hands shook, she experienced uncontrollable episodes of biting and striking people, and she became unable to walk or swallow. Ultimately, Charlene Singh became the first U.S. resident to die of the human form of mad cow disease, which she had almost certainly contracted more than 10 years earlier while living in England.

It was not until the mid-1990s that health officials recognized that mad cow disease, or bovine spongiform encephalitis (BSE), could spread to people who ate meat from infected cattle. Although millions of people may have eaten tainted beef, fewer than 250 people worldwide have contracted the human version of BSE, called variant Creutzfeldt-Jakob disease (vCJD). But for those infected, the disease is always fatal, riddling the brain with microscopic holes that give it a spongy appearance.

Where did mad cow disease come from? One hypothesis is that it was derived from a sheep disease called scrapie, whose symptoms are almost identical to those of BSE. The brains of sheep suffering from scrapie become spongy, and the animals lose weight and become uncoordinated, nervous, or aggressive. A mutated form of scrapie may have become capable of infecting cattle, perhaps in the early 1980s. At that time, cattle feed often included body parts from sheep, some of which may have harbored scrapie. The use of sheep, cow, and goat parts in cattle feed was banned in 1988, two years after BSE was first identified in British cattle.

British beef exports were temporarily halted in 1996, after experts confirmed that BSE could spread to people who ate infected meat. As a precautionary measure at that time, more than 4.5 million cattle in Britain were slaughtered and their bodies were burned—a tragedy for both the cows and the British farmers.

Mad cow disease is particularly fascinating to scientists because unlike other contagious diseases, it is not caused by a microorganism or a virus. In the early 1980s, researcher Stanley Prusiner found evidence that scrapie is caused by a protein, and that the protein can transmit the disease to other animals. He dubbed the infectious proteins "prions" (pronounced PREE-ons).

What are proteins? How do they differ from DNA and RNA? How can a protein infect another organism, increase in number, and produce a fatal disease? Are BSE and vCJD still a threat?

AT A GLANCE

3.1 WHY IS CARBON SO IMPORTANT IN BIOLOGICAL MOLECULES?

You have probably seen fruits and vegetables in the supermarket labeled as "organic," meaning that they were grown without synthetic fertilizers or pesticides. But in chemistry, the word **organic** describes molecules that contain carbon and usually oxygen and hydrogen. Many organic molecules are synthesized by organisms, hence the name organic. In contrast, **inorganic** molecules generally lack carbon atoms (examples are water and salt). Inorganic molecules, such as those that make up Earth's rocks and metal deposits, are far less diverse and generally much simpler than organic molecules.

Life is characterized by an amazing array of **biological molecules**, which we define as all molecules produced by living things. Nearly all of these are organic, based on the carbon atom. Biological molecules interact in dazzlingly complex ways, and as they interact, their structures and chemical properties change. These precisely orchestrated changes allow cells to acquire and use nutrients, eliminate wastes, move, grow, and reproduce. All of this coordinated chemical activity is made possible by the versatility of the carbon atom.

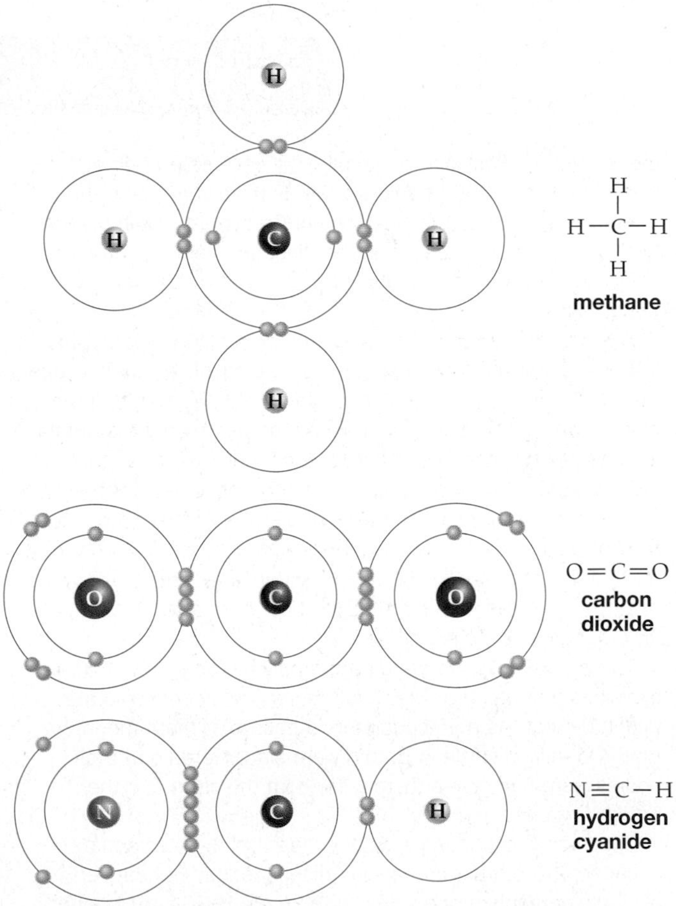

▲ **FIGURE 3-1 Covalent bonding by carbon atoms** Carbon must form four covalent bonds to fill its outer electron shell and become stable. It can do this by forming single, double, or triple covalent bonds. In these examples, carbon forms methane (CH_4), carbon dioxide (CO_2), and hydrogen cyanide (HCN).

THINK CRITICALLY Which of these molecules is/are polar? (You may need to refer back to Chapter 2.)

The Bonding Properties of Carbon Are Key to the Complexity of Organic Molecules

As described in Chapter 2, atoms whose outermost electron shells are only partially filled tend to react with one another, gaining stability by filling their shells and forming covalent bonds. Depending on the number of vacancies in their outer shells, two bonded atoms can share two, four, or six electrons—forming a single, double, or triple covalent bond (**FIG. 3-1**). The bonding patterns of the four elements most commonly found in biological molecules are shown in **FIGURE 3-2**. Covalent bonds are represented by solid lines drawn between atomic symbols.

Life on Earth depends on a tremendous variety of biological molecules, a variety made possible by the bonding flexibility of the carbon atom. With four electrons in an outermost shell that can accommodate eight, a carbon atom (C) readily forms single or double bonds with other carbon atoms and can bond with up to four other atoms of carbon or other elements (see Fig. 3-1). Additional diversity arises from the range of complex shapes that organic molecules can assume.

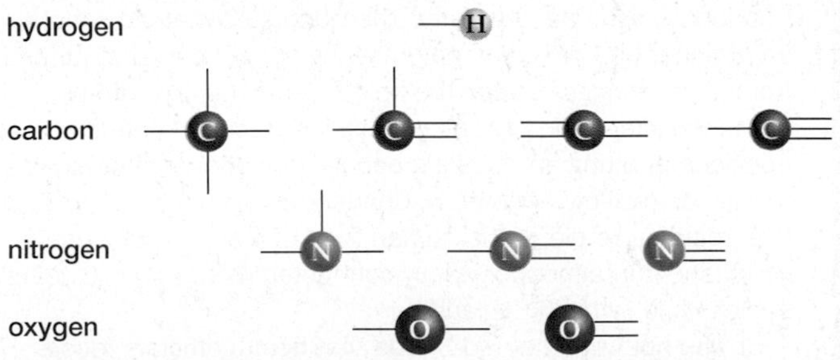

▲ **FIGURE 3-2 Bonding patterns** The bonding patterns of the four most common atoms in biological molecules. Each line indicates a covalent bond.

Functional Groups Attach to the Carbon Backbone of Organic Molecules

Functional groups are common atoms or groups of atoms that bond to the carbon backbone of organic molecules. The functional groups in a biological molecule determine its distinctive properties and tendency to react with other molecules. In general, functional groups are less stable than a carbon backbone and more likely to participate in chemical reactions. **TABLE 3-1** describes seven functional groups that are important in biological molecules.

CHECK YOUR LEARNING

Can you . . .

- define organic molecules and explain why carbon is so important to life?
- explain why functional groups are important in biological molecules?
- name and describe the properties of seven functional groups?

3.2 HOW ARE LARGE BIOLOGICAL MOLECULES SYNTHESIZED?

Although a complex molecule could in principle be constructed by painstakingly attaching one atom at a time, the machinery of life works far more efficiently by preassembling molecular subunits and hooking them together. Just as numerous train cars are joined to make a train, small organic molecules (for example, sugars or amino acids) are joined to form large biological molecules (for example, starches or proteins). The individual subunits are called **monomers** (Gk. *mono*, one); chains of monomers are called **polymers** (Gk. *poly*, many).

Biological Polymers Are Formed by the Removal of Water and Broken Down by the Addition of Water

The subunits of large biological molecules are usually joined by a chemical reaction called **dehydration synthesis**, literally meaning "removing water to put together." In

TABLE 3-1 Important Functional Groups in Biological Molecules

Group	Structure	Properties	Found In
Hydroxyl	–O–H	Polar; involved in dehydration and hydrolysis reactions; forms hydrogen bonds	Sugars, polysaccharides, nucleic acids, alcohols, some amino acids, steroids
Carbonyl	C=O	Polar; makes parts of molecules hydrophilic (water soluble)	Sugars (linear forms), steroid hormones, peptides and proteins, some vitamins
Carboxyl (ionized form)	–C(=O)O⁻	Polar and acidic; the negatively charged oxygen may bond H^+, forming carboxylic acid (–COOH); involved in peptide bonds	Amino acids, fatty acids, carboxylic acids (such as acetic and citric acids)
Amino	–N(H)H	Polar and basic; may become ionized by binding a third H^+; involved in peptide bonds	Amino acids, nucleic acids, some hormones
Sulfhydryl	–S–H	Nonpolar; forms disulfide bonds in proteins	Cysteine (an amino acid), many proteins
Phosphate (ionized form)	–O–P(=O)(O⁻)–O⁻	Polar and acidic; links nucleotides in nucleic acids; forms high-energy bonds in ATP (ionized form occurs in cells)	Phospholipids, nucleotides, nucleic acids
Methyl	–C(H)(H)H	Nonpolar; may be attached to nucleotides in DNA (methylation), changing gene expression	Steroids, methylated nucleotides in DNA

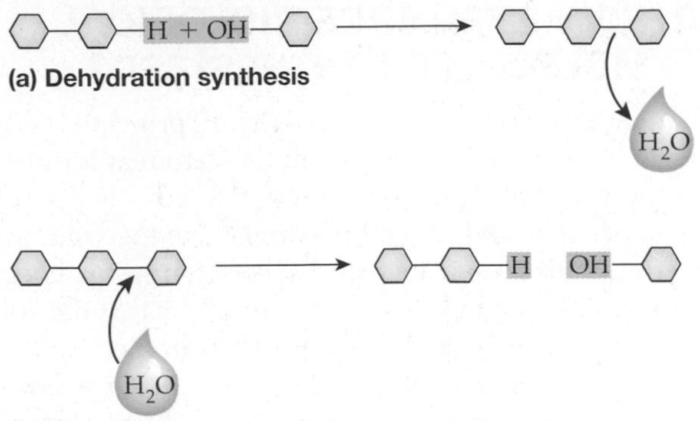

(a) Dehydration synthesis

(b) Hydrolysis

▲ **FIGURE 3-3 Dehydration synthesis and hydrolysis** Biological polymers are formed by **(a)** linking monomer subunits in a reaction that removes H_2O from polar functional groups. These polymers may be broken apart by adding the atoms in H_2O **(b)**, which re-creates the subunits.

dehydration synthesis, a hydrogen ion (H^+) is removed from one subunit and a hydroxyl ion (OH^-) is removed from a second subunit, leaving vacancies in the outer electron shells of atoms in the two subunits. These vacancies are filled when the subunits share electrons to form a covalent bond. The hydrogen ion and the hydroxyl ion combine to form a molecule of water (H_2O), as shown in **FIGURE 3-3a**.

The reverse reaction that breaks down a large biological molecule into its subunits is **hydrolysis** (Gk. *hydro,* water, and *lysis*, break apart). To break a bond between two subunits, hydrolysis consumes a molecule of water, which donates a hydrogen ion to one subunit and a hydroxyl ion to the other (**FIG. 3-3b**). Hydrolysis is important in the reactions that break down food. For example, enzymes in our saliva and small intestines promote hydrolysis of starch, which consists of a chain of sugar molecules, into individual sugar molecules that can be absorbed into the body.

Although biological molecules are tremendously diverse, nearly all fall into one of four general categories: carbohydrates, proteins, nucleic acids, and lipids (**TABLE 3-2**).

CHECK YOUR LEARNING

Can you . . .

- name and describe the reactions that build and break apart biological polymers?

TABLE 3-2 Four Principal Classes of Biological Molecules

Name and General Structure	Types	Example(s)	Typical Function
Carbohydrates: Molecules composed primarily of C, H, and O in the ratio $(CH_2O)_n$, where *n* is the number of C's in the molecule's backbone. ***Subunit***: Monosaccharide	**Monosaccharides**: Simple sugars	Glucose, fructose	Short-term energy storage in plants
	Disaccharides: Two linked monosaccharides	Sucrose	
	Polysaccharides: Polymers of monosaccharides	Starch, glycogen	Long-term energy storage in plants and animals, respectively
		Cellulose, chitin	Structural support in plants and arthropods, respectively
Proteins: Molecules with one or more chains of amino acids. Proteins have up to four levels of structure. ***Subunit***: Amino acid	**Peptides**: Short chains of amino acids	Insulin, oxytocin	Hormones involved in blood sugar regulation and reproduction, respectively
	Polypeptides: Long chains (polymers) of amino acids	Hemoglobin	Oxygen transport
		Keratin	Structural component of hair
Nucleic acids: Molecules composed of polymers of nucleotides, each consisting of a simple sugar, an N-containing base, and a phosphate group. ***Subunit***: Nucleotide	**Deoxyribonucleic acid (DNA)**: A polymer of nucleotides whose simple sugar is deoxyribose	DNA	Codes for genetic information
	Ribonucleic acids (RNA): Polymers of nucleotides whose simple sugar is ribose	Messenger RNA, transfer RNA, ribosomal RNA	Work together to form proteins from amino acids based on nucleotide sequences in DNA
Lipids: Diverse group of molecules containing nonpolar (hydrophobic) regions that make them insoluble in water. ***Subunit***: No consistent subunit; not polymers	**Fats, oils, and waxes**: Contain one or more fatty acids, hydrophobic chains of carbon atoms that terminate in a carboxylic acid functional group	Animal fats, vegetable oils	Long-term energy storage in animals and plants, respectively
		Beeswax	Structural component of beehives
	Phospholipids: Contain two fatty acids (hydrophobic) and two hydrophilic functional groups, one of which is phosphate	Lecithin	Structural component of cell membranes
	Steroids: Contain four rings of carbon atoms, with different functional groups attached	Cholesterol	Component of cell membranes
		Testosterone, estrogen	Male and female sex hormones, respectively

3.3 WHAT ARE CARBOHYDRATES?

Carbohydrate molecules are composed of carbon, hydrogen, and oxygen in the approximate ratio of 1C:2H:1O. This ratio explains the origin of the word "carbohydrate," which literally means "carbon plus water" (C + H_2O). All carbohydrates are either small, water-soluble **sugars** or polymers of sugar, such as starch. If a carbohydrate consists of just one sugar molecule, it is called a **monosaccharide** (Gk. *mono*, one, and *saccharum*, sugar). When two monosaccharides are linked, they form a **disaccharide**. A polymer of many monosaccharides is called a **polysaccharide**.

▼ **FIGURE 3-4 Sugar dissolving in water** Glucose dissolves as the polar hydroxyl groups of each sugar molecule form hydrogen bonds with nearby water molecules.

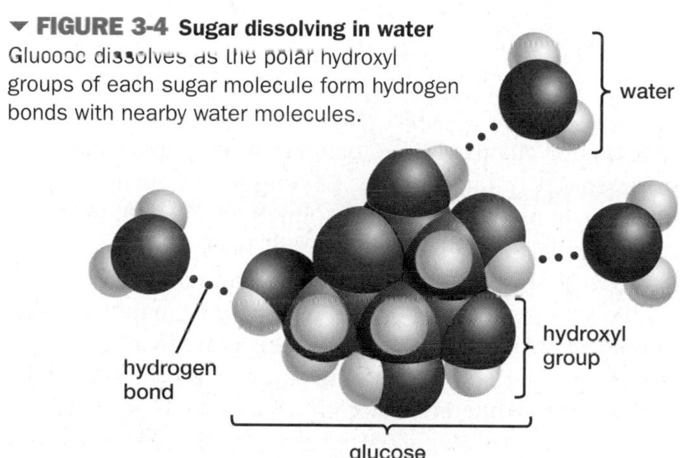

Different Monosaccharides Have Slightly Different Structures

Monosaccharides have a backbone of three to seven carbon atoms. Most of these carbon atoms have both a hydrogen (—H) and a hydroxyl group (—OH) attached to them; therefore, carbohydrates generally have the chemical formula $(CH_2O)_n$, where *n* is the number of carbons in the backbone. Monosaccharide molecules are hydrophilic; their hydroxyl functional groups are polar and form hydrogen bonds with polar water molecules (**FIG. 3-4**). When a monosaccharide is dissolved in water, such as inside a cell, its carbon backbone usually forms a ring. Similarly, each monomer in a polysaccharide forms a ring.

Glucose is the most common monosaccharide in organisms and the primary energy source of cells. Glucose has six carbons, so its chemical formula is $C_6H_{12}O_6$. Figure 3-4 and **FIGURE 3-5** show various ways of depicting the chemical structure of glucose; keep in mind that any unlabeled "joint" in a ring or chain actually represents a carbon atom.

Many organisms synthesize monosaccharides that have the same chemical formula as glucose but slightly different structures. For example, some plants store energy in *fructose* (L. *fruct*, fruit), which we consume in fruits, juices, honey, corn syrup, and soft drinks. *Galactose* is secreted by mammals in their milk as an energy source for their young (**FIG. 3-6**). Fructose and galactose must be converted to glucose before cells can use them as a source of energy.

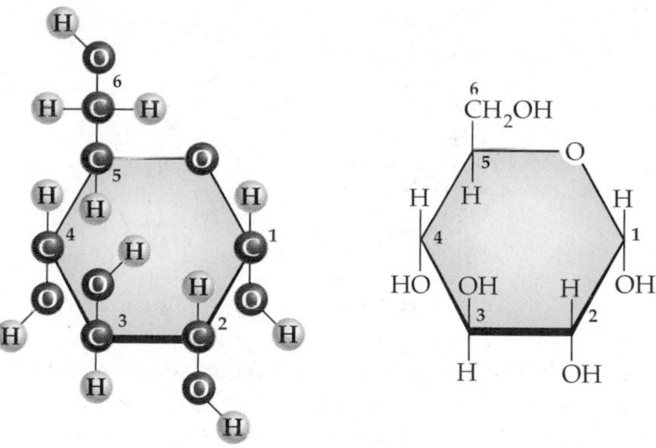

▲ **FIGURE 3-5 Depictions of chemical structures** The molecule glucose ($C_6H_{12}O_6$) can be drawn as (left) a ball-and-stick model showing each atom or (right) a simplified version in which each unlabeled joint is a carbon atom. The carbon atoms are numbered for reference. The space-filling structure of glucose is shown in Figure 3-4.

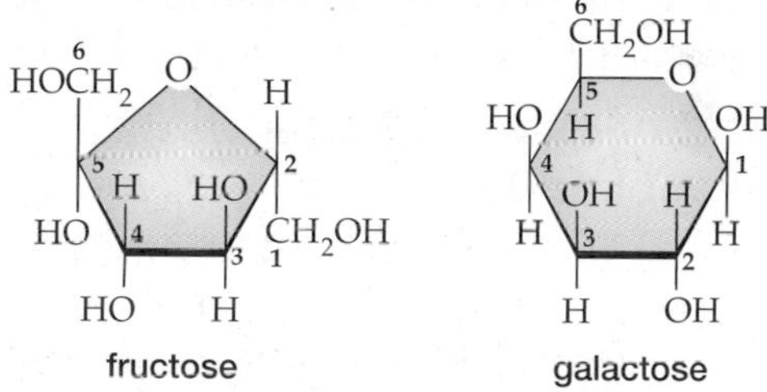

▲ **FIGURE 3-6 Some six-carbon monosaccharides**

Other common monosaccharides, such as ribose and deoxyribose (found in the nucleic acids RNA and DNA, respectively), have five carbons. Notice in **FIGURE 3-7** that deoxyribose has one fewer oxygen atom than ribose because one of the hydroxyl groups in ribose is replaced by a hydrogen atom in deoxyribose.

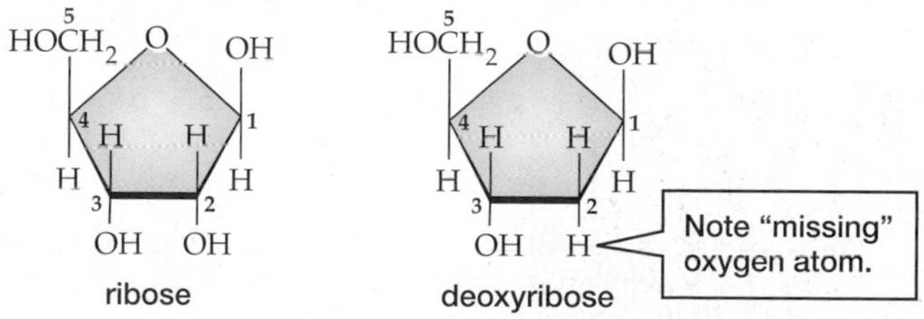

ribose

deoxyribose

Note "missing" oxygen atom.

▲ **FIGURE 3-7 Some five-carbon monosaccharides**

Disaccharides Consist of Two Monosaccharides Linked by Dehydration Synthesis

Two monosaccharides can be linked by dehydration synthesis to form a disaccharide (**FIG. 3-8**). Disaccharides are often used for short-term energy storage in plants. When energy is required, the disaccharides are broken apart by hydrolysis into their monosaccharide subunits (see Fig. 3-3), which are, if necessary, converted to glucose that is further broken down to release energy.

Perhaps you had toast and coffee with cream and sugar for breakfast. If so, you stirred *sucrose* (glucose plus fructose)

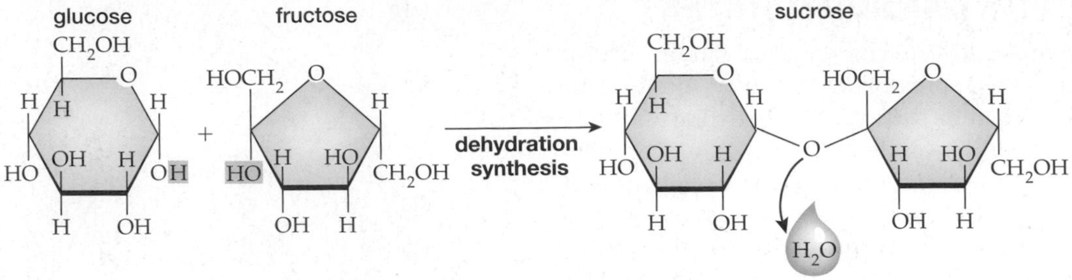

glucose fructose sucrose

dehydration synthesis

▲ **FIGURE 3-8 Synthesis of a disaccharide** Sucrose is synthesized by a dehydration reaction in which a hydrogen is removed from glucose and a hydroxyl group is removed from fructose. This forms water and leaves the two monosaccharide rings joined by single bonds to the remaining oxygen atom.

THINK CRITICALLY Describe hydrolysis of this molecule.

into your coffee and then added cream containing *lactose* (glucose plus galactose). The disaccharide *maltose* (glucose plus glucose) was formed when enzymes in your digestive tract hydrolyzed the wheat starch in your toast.

Polysaccharides Are Chains of Monosaccharides

Most polysaccharides do not dissolve in water at body temperatures. They are insoluble because their polar hydroxyl groups were lost during dehydration synthesis, which linked the monosaccharides together in long chains and released water (see Fig. 3-8). Despite their lack of solubility, polysaccharides can be hydrolyzed under the right conditions. For example, if you take a bite of a bagel and chew it for a minute or so, you may notice that it gradually tastes sweeter. This is because enzymes in saliva cause hydrolysis of the starch in the bagel into its component glucose molecules, which dissolve in your saliva and stimulate receptors on your tongue that respond to sweetness.

Polysaccharides often serve as energy-storage molecules in organisms. For example, **starch** (**FIG. 3-9**), which consists of branched chains of up to half a million glucose subunits, stores energy in many plants. **Glycogen**, an energy-storage molecule in humans and other animals, is also a chain of glucose subunits but is much more highly branched than starch. Glycogen is stored primarily in the liver and muscles.

Polysaccharides also serve as structural materials. One important structural polysaccharide is **cellulose**, the main component of the walls of living plant cells; cellulose makes up the fluffy white bolls of cotton plants and accounts for about half the dry weight of tree trunks (**FIG. 3-10**). Scientists estimate that plants synthesize about a trillion tons of cellulose each year, making it the most abundant organic molecule on Earth.

Cellulose, like starch, is a polymer of glucose. But in cellulose, every other glucose is "upside down," as you will see when you compare Figure 3-9c with Figure 3-10d. Although most animals easily digest starch, no vertebrate can digest the bonds that link the glucose molecules in cellulose.

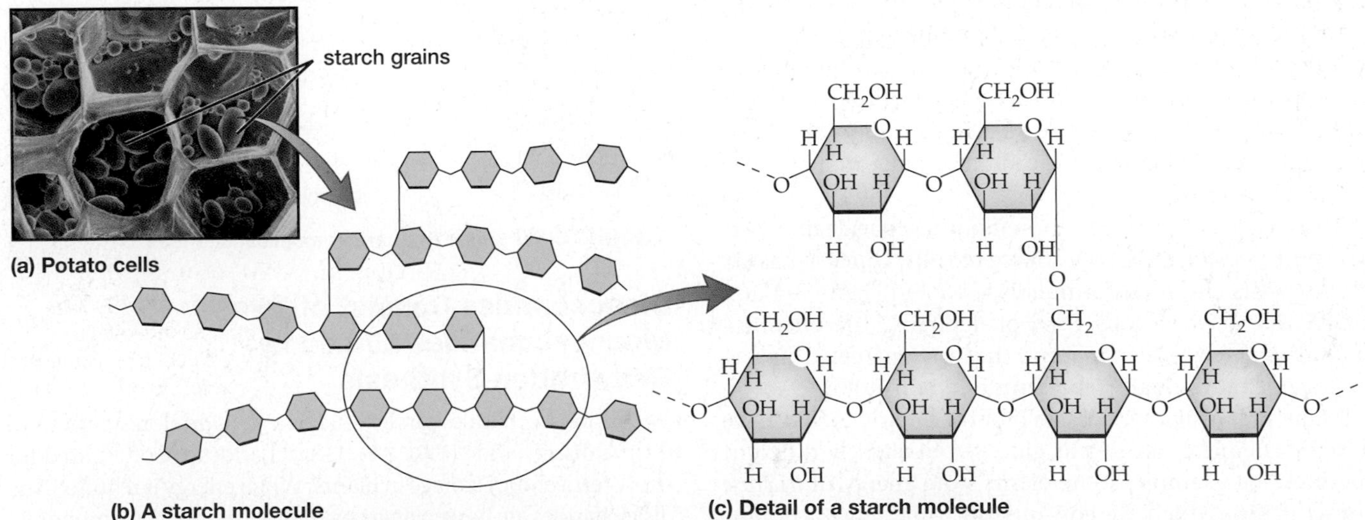

▲ **FIGURE 3-9 Starch structure and function (a)** Starch grains inside potato cells store energy that will allow the potato to generate new plants in the spring. **(b)** A section of a single starch molecule. Starches consist of branched chains of up to half a million glucose subunits. **(c)** The precise structure of the circled portion of the starch molecule in (b). Notice the linkage between the individual glucose subunits for comparison with cellulose (**Fig. 3-10**).

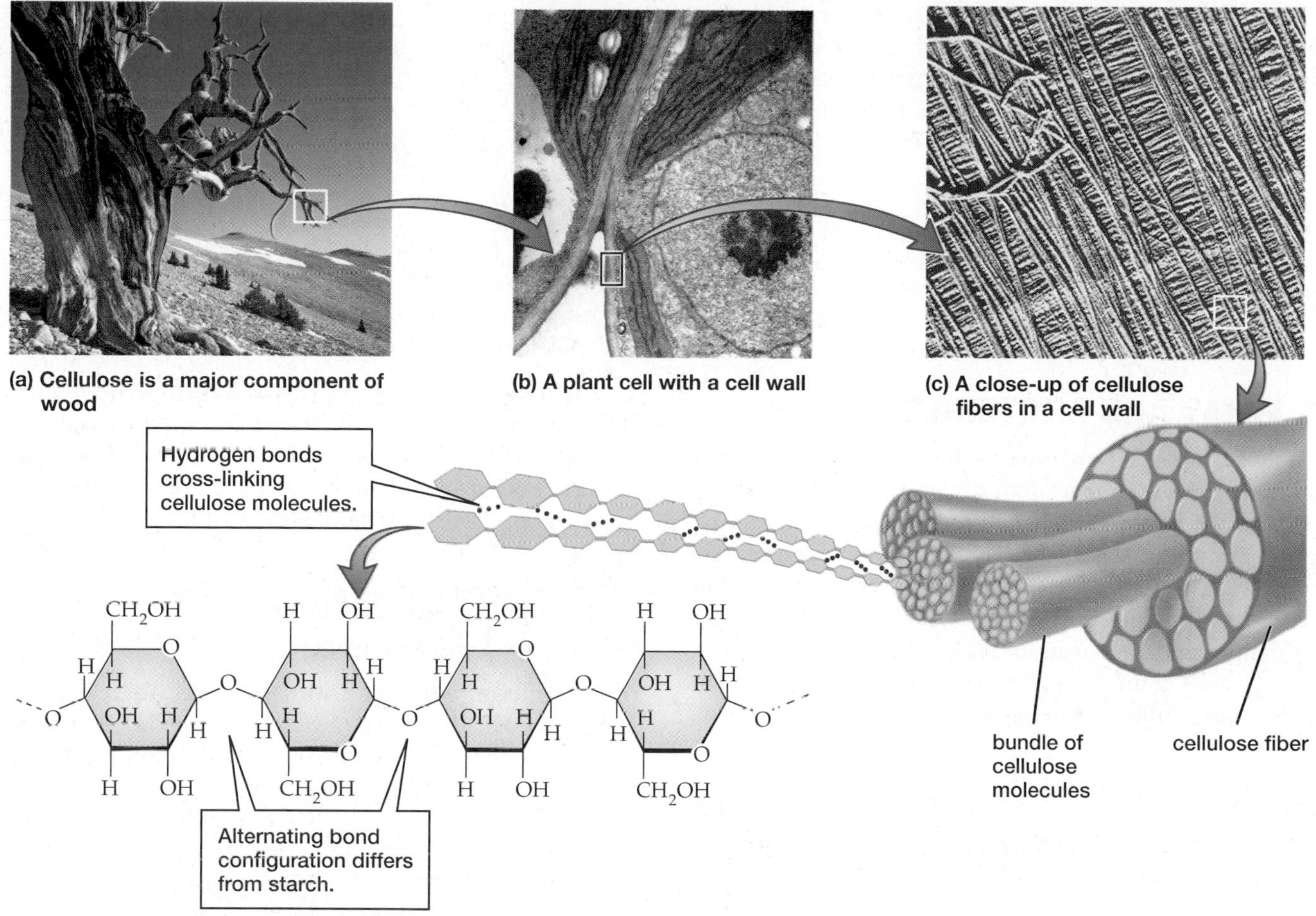
(a) Cellulose is a major component of wood
(b) A plant cell with a cell wall
(c) A close-up of cellulose fibers in a cell wall
Hydrogen bonds cross-linking cellulose molecules.
bundle of cellulose molecules
cellulose fiber
Alternating bond configuration differs from starch.
(d) Detail of a cellulose molecule

▲ **FIGURE 3-10 Cellulose structure and function (a)** Wood in this 3,000 year-old bristlecone pine is primarily cellulose. **(b)** Cellulose forms the cell wall that surrounds each plant cell. **(c)** Plant cell walls often consist of cellulose fibers in layers that run at right angles to each other to resist tearing in both directions. **(d)** Cellulose is composed of up to 10,000 glucose subunits. Compare this structure with Figure 3-9c, and notice that every other glucose molecule in cellulose is "upside down."

A few vertebrates, such as cows and rabbits, harbor cellulose-digesting microbes in their digestive tracts and benefit from the glucose subunits that the microbes release. In humans, cellulose fibers pass intact through the digestive system; cellulose supplies no nutrients, but it adds bulk that has digestive benefits.

Another structural polysaccharide is **chitin**, which makes up the outer coverings (exoskeletons) of insects, crabs, and spiders. Chitin also stiffens the cell walls of many fungi, including mushrooms. Chitin is similar to cellulose, except the glucose subunits bear a nitrogen-containing functional group (**FIG. 3-11**).

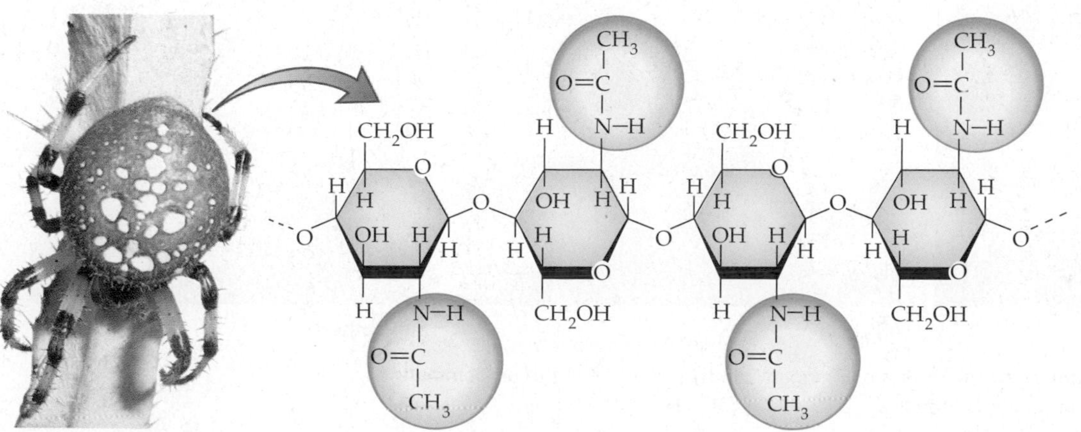

◀ **FIGURE 3-11 Chitin structure and function** Chitin has the same glucose bonding configuration as cellulose, but the glucose subunits have a nitrogen-containing functional group replacing one of the hydroxyls. Chitin supports the otherwise soft bodies of arthropods (including spiders such as this one, insects, and crabs and their relatives) as well as most fungi.

Carbohydrates may also form parts of larger molecules; for example, the plasma membrane that surrounds each cell is studded with proteins to which carbohydrates are attached. Nucleic acids (discussed later) also contain sugar molecules.

CHECK YOUR LEARNING

Can you . . .

- describe the major types of carbohydrates?
- provide examples of each type of carbohydrate and explain how organisms use them?

variable group (R)

amino group

carboxylic acid group

hydrogen

▲ **FIGURE 3-13 Amino acid structure**

3.4 WHAT ARE PROTEINS?

Like polysaccharides, **proteins** are biological polymers synthesized by linking simple subunits. Scientists estimate that the human body contains between 250,000 and 1 million different proteins (**TABLE 3-3**). Some of these are **enzymes**, proteins that promote specific chemical reactions. Other proteins are structural, including *keratin*, which forms hair, horns, nails, scales, and feathers (**FIG. 3-12**). Nutritional proteins, such as albumin in egg white and casein in milk, nourish developing animals. The protein hemoglobin transports oxygen in the blood. Actin and myosin are proteins in muscle that allow animal bodies to move. Some proteins are hormones (insulin and growth hormone, for example), others are antibodies (which help fight disease and infection), and a few are toxins (such as rattlesnake venom).

TABLE 3-3 Functions of Proteins

Function	Example(s)
Structural	Keratin (forms hair, nails, scales, feathers, and horns); silk (forms webs and cocoons)
Movement	Actin and myosin (found in muscle cells; allow contraction)
Defense	Antibodies (found in the bloodstream; fight disease organisms; some neutralize venoms); venoms (found in venomous animals; deter predators and disable prey)
Storage	Albumin (in egg white; provides nutrition for an embryo)
Signaling	Insulin (secreted by the pancreas; promotes glucose uptake into cells)
Catalyzing reactions	Amylase (found in saliva and the small intestine; digests carbohydrates)

Proteins Are Formed from Chains of Amino Acids

The subunits of proteins are **amino acids**. There are 20 different amino acids commonly found in proteins, all of which have the same basic structure. A central carbon is bonded to a hydrogen atom and to three functional groups: a nitrogen-containing amino group ($-NH_2$), a carboxylic acid group ($-COOH$), and an "R" group that varies among different amino acids (**FIG. 3-13**). The R group gives each amino acid distinctive properties (**FIG. 3-14**). Some amino acids are hydrophilic and water soluble because their R groups are polar. Others are hydrophobic, with nonpolar R groups that are insoluble in water. The amino acid cysteine (Fig. 3-14c) is unique in having a sulfur-containing R group that can form covalent **disulfide bonds** with the sulfur of another cysteine molecule. These disulfide bonds play important roles in proteins, as described later.

Like polysaccharides, proteins are formed by dehydration synthesis. The nitrogen in the amino group of one amino acid is joined by a single covalent bond to the carbon in the carboxylic acid group of another amino acid, and

(a) Hair

(b) Horn

(c) Feathers

▲ **FIGURE 3-12 Structural proteins** Keratin is a common structural protein. It is the predominant protein found in **(a)** hair, **(b)** horn, and **(c)** feathers.

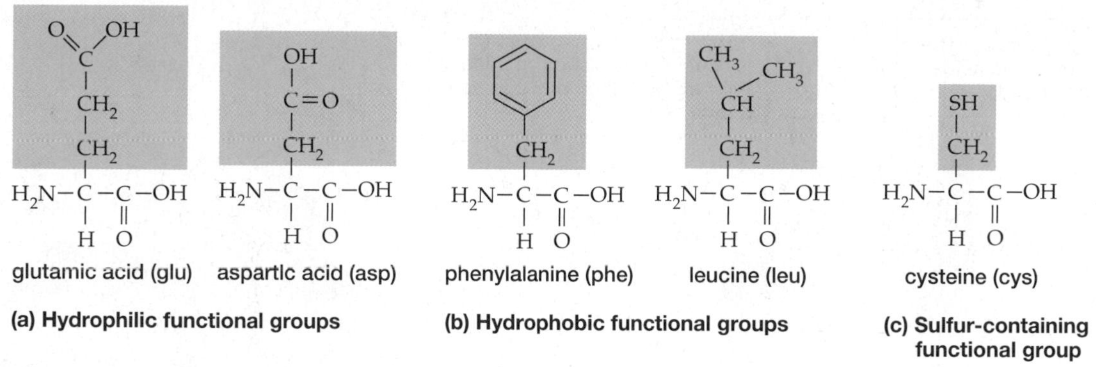

glutamic acid (glu) aspartic acid (asp)

(a) Hydrophilic functional groups

phenylalanine (phe) leucine (leu)

(b) Hydrophobic functional groups

cysteine (cys)

(c) Sulfur-containing functional group

▲ **FIGURE 3-14 Amino acid diversity** The diversity of amino acids is caused by the different R functional groups (green backgrounds), which may be **(a)** hydrophilic or **(b)** hydrophobic. **(c)** The R group of cysteine has a sulfur atom that can form covalent bonds with the sulfur in other cysteines.

THINK CRITICALLY Look up the rest of the amino acids and, based on their structures, identify three others that have hydrophobic functional groups.

water is liberated (**FIG. 3-15**). This process forms a **peptide bond**, and the resulting chain of two amino acids is called a **peptide**, a term used to describe any relatively short chain of amino acids (up to 50 or so). Additional amino acids are added to the growing peptide until a long *polypeptide* chain is completed. Polypeptide chains can be up to thousands of amino acids in length. A protein consists of one or more polypeptide chains.

A Protein Can Have up to Four Levels of Structure

Interactions among the amino acids in polypeptide chains cause the chains to twist, fold, and connect in ways that give proteins their three-dimensional structure. Up to four levels of protein structure are possible: primary, secondary, tertiary, and quaternary (**FIG. 3-16**). A protein's **primary structure** is simply the sequence of its amino acids, which is specified by instructions encoded in DNA (see Fig. 3-16a).

The **secondary structure** of a protein is the shape it folds into as a result of hydrogen bonds between different amino acids in the polypeptide sequence. These hydrogen bonds do not involve the R groups, but form between the slightly negative carbonyl (C=O) group in one amino acid and the slightly positive N–H group from an amino acid farther along the chain (see the carbonyl group in Table 3-1). When such hydrogen bonds form between every fourth amino acid, they create the coils of a spring-like **helix**. Such helices are found in the keratin protein of hair and in the polypeptide subunits of hemoglobin (see Fig. 3-16b). Another common secondary structure, the **pleated sheet**, occurs when polypeptide chains repeatedly fold back upon themselves, anchored by hydrogen bonds (**FIG. 3-17**).

CASE STUDY CONTINUED

Puzzling Proteins

The infectious prion that causes mad cow disease is an abnormally folded version of a protein. The normal prion protein has a secondary structure that is primarily helical, but the infectious version is folded into pleated sheets. The pleated sheets are so stable they are unaffected by the enzymes that break down normal prion protein. As a result, infectious prions accumulate destructively in the brain.

Helices and pleated sheets are the two major secondary structures of proteins. What do tertiary and quaternary structures look like?

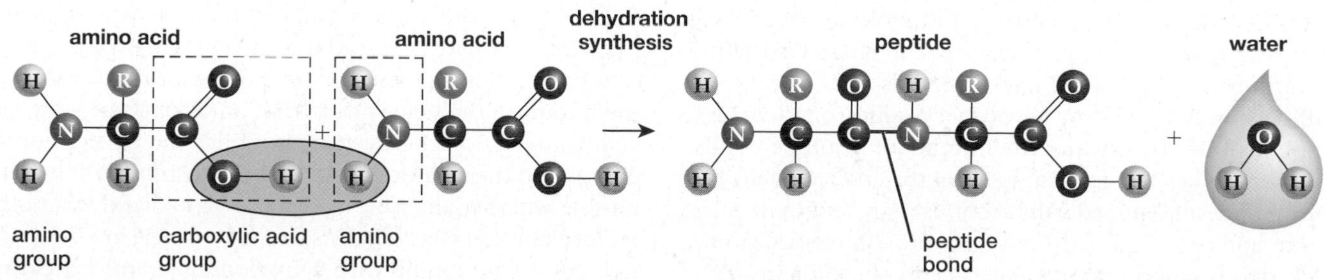

▲ **FIGURE 3-15 Protein synthesis** A dehydration reaction forms a peptide bond between the carbon of the carboxylic acid group of one amino acid and the nitrogen of the amino group of a second amino acid.

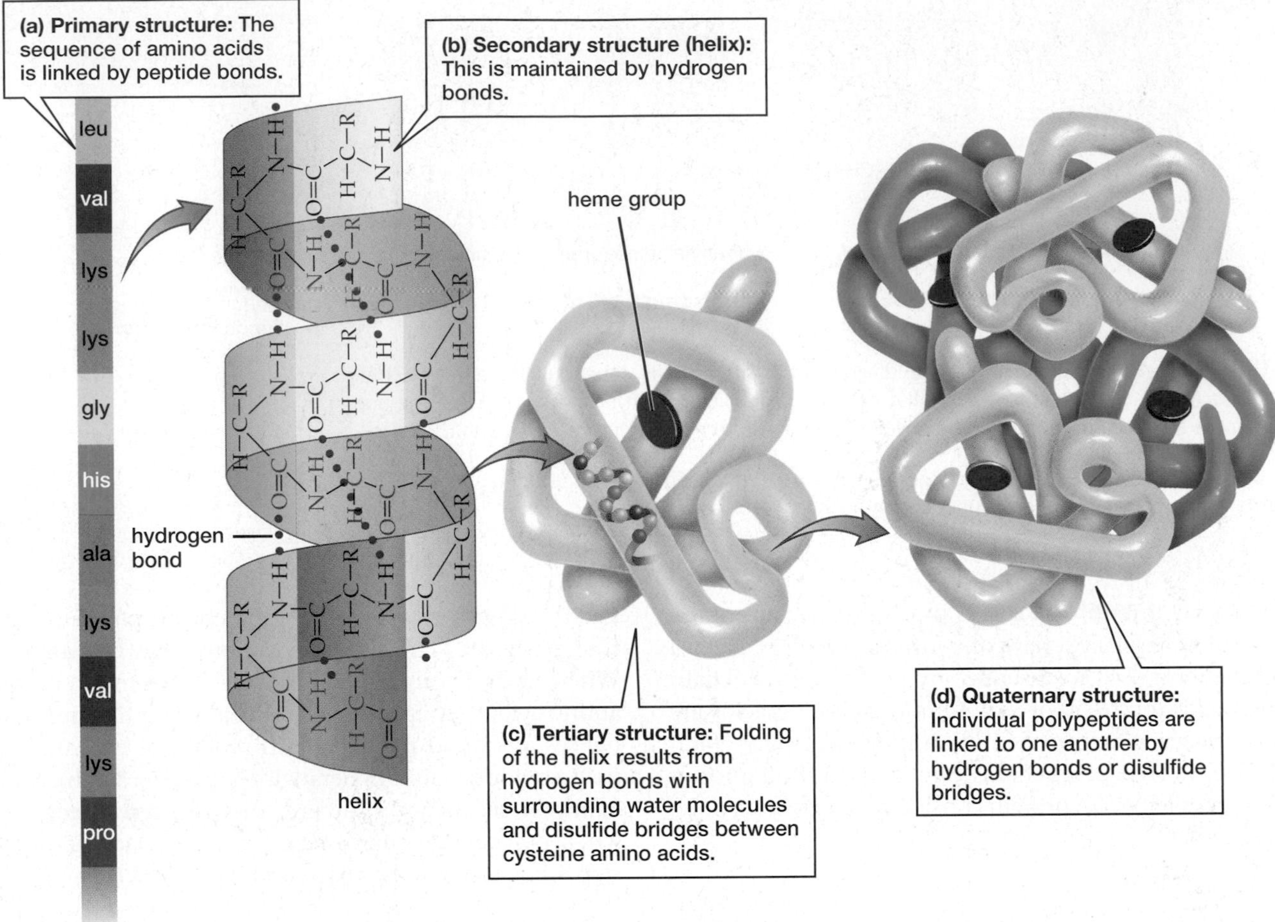

(a) Primary structure: The sequence of amino acids is linked by peptide bonds.
(b) Secondary structure (helix): This is maintained by hydrogen bonds.
heme group
leu
val
lys
lys
gly
his
ala
lys
val
lys
pro
hydrogen bond
helix
(c) Tertiary structure: Folding of the helix results from hydrogen bonds with surrounding water molecules and disulfide bridges between cysteine amino acids.
(d) Quaternary structure: Individual polypeptides are linked to one another by hydrogen bonds or disulfide bridges.

▲ **FIGURE 3-16 The four levels of protein structure** Hemoglobin is the oxygen-carrying protein in red blood cells. Red disks represent the iron-containing heme group that binds oxygen.

THINK CRITICALLY Why do many proteins, when heated excessively, lose their ability to function?

Each protein also contorts into a **tertiary structure** (see Fig. 3-16c) determined by the protein's primary and secondary structure and by its environment. For example, a protein in a watery environment such as the interior of a cell may fold into a roughly spherical shape that exposes the protein's hydrophilic amino acids to the water and buries its hydrophobic amino acids deeper within the molecule. These somewhat water-soluble *globular proteins* include hemoglobin, enzymes, and albumin. The tertiary structure of other proteins results in long, thin strands. These *fibrous proteins* contain many hydrophobic amino acids and are insoluble in water. One widespread fibrous protein is keratin. The tertiary structure of keratin consists of paired helical strands held together by disulfide bonds that form between cysteines of each helical polypeptide. The greater the number of disulfide bonds, the stiffer the keratin; for example, keratin in fingernails has more cysteine and more disulfide bonds than keratin in hair.

An additional, fourth level of protein organization, called **quaternary structure**, occurs in proteins that contain multiple polypeptides linked by hydrogen bonds, disulfide bonds, or attractions between oppositely charged portions of different amino acids. Hemoglobin, for example, consists of four polypeptide chains held together by hydrogen bonds (see Fig. 3-16d). Each of the four polypeptides has an iron-containing region called a heme group that can bind one molecule of oxygen.

Protein Function Is Determined by Protein Structure

The structure of a protein is closely linked to its biological function. For example, an enzyme's shape may allow it to interact only with certain other molecules, as described in Chapter 6. Similarly, the structure of hemoglobin ensures that certain R groups are present at the precise location that allows them to form the heme group that binds oxygen. More generally, interactions between hydrophilic and hydrophobic R groups and their watery environment are important in determining whether and how a protein will fold and what other molecules it can react with. As a result, a single mutation that replaces a hydrophilic with a hydrophobic amino acid can have a significant effect on protein function. For example, this type of mutation in hemoglobin causes the genetic disorder sickle-cell anemia (described in Chapter 11).

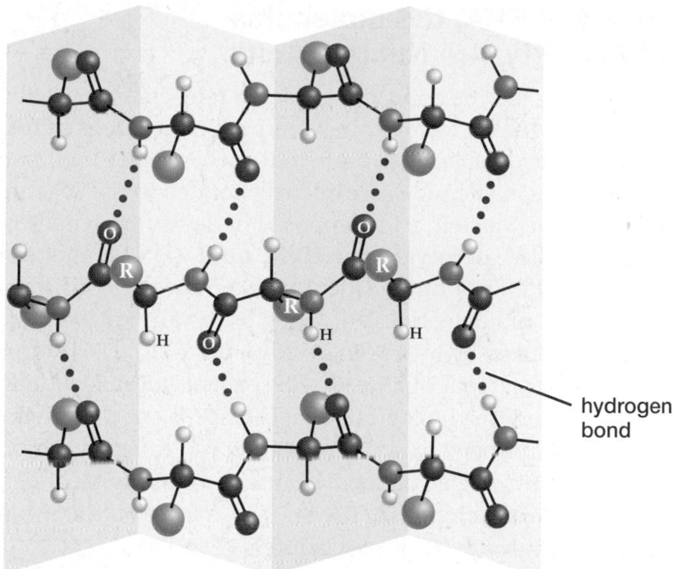

▲ **FIGURE 3-17 The pleated sheet and the structure of silk protein** In a pleated sheet, a single polypeptide chain is folded back upon itself repeatedly (the loops formed by these folds are not shown). Adjacent segments of the folded polypeptide are linked by hydrogen bonds (dotted lines). The R groups (green) project alternately above and below the sheet.

A protein is described as **denatured** when its normal three-dimensional structure is destroyed, leaving its primary structure intact. For example, egg white consists of albumin protein, which is normally transparent and relatively fluid. But the heat of a frying pan rips its hydrogen bonds apart, destroying the albumin's secondary and tertiary structure and causing it to become opaque, white, and solid (**FIG. 3-18**). Keratin in hair is denatured by a permanent wave. Bacteria and viruses can be destroyed by using heat or ultraviolet light to denature their proteins. Proteins can also be denatured by solutions that are highly salty or acidic. For example, dill pickles are protected from spoiling by their salty, acidic brine, which denatures bacterial proteins.

▲ **FIGURE 3-18 Heat denatures albumin**

Have You Ever Wondered ...

Why a Perm Is (Temporarily) Permanent?

Whether your hair is naturally straight or curly is determined by the shape of your hair follicles and the hair shafts they produce. The follicles of straight hair are round in cross section, whereas the follicles of curly hair are flatter. But genes are not destiny! Even if your follicles are round, you can create curls chemically, by altering the secondary structure of keratin.

A strand of hair consists of bundles within bundles of keratin. Keratin's helical, spring-like secondary structure is created by hydrogen bonds, which are easily disrupted by the attraction of polar water molecules. If you break hydrogen bonds in straight hair by wetting it, and then let the hair dry while it's wrapped around curlers, hydrogen bonds will re-form as the water disappears, but in new places because of the distortion caused by the curler. Voila, curls. But if it's rainy, you can say goodbye to your hydrogen-bonded curls; water disrupts the bonds and causes your hair to revert to its natural straightness.

How is a permanent wave created? Keratin contains a lot of cysteine amino acids, and a perm alters the locations of the strong covalent disulfide bonds between cysteines. First, the hair is soaked in a solution that breaks the natural disulfide bonds linking adjacent keratin molecules. The hair is then set on curlers and saturated with a different solution that causes disulfide bonds to re-form. The curlers force these bonds to form in new locations, and the strong disulfide bonds permanently maintain the curl. Genetically straight hair has been transformed into artificially curly hair—at least until new hair grows in.

A perm changes the location of bonds between keratin molecules

CHECK YOUR LEARNING

Can you . . .

- describe protein subunits and how proteins are synthesized?
- explain the four levels of protein structure and why a protein's three-dimensional structure is important?
- list several functions of proteins and provide examples of proteins that perform each function?
- explain what occurs when a protein is denatured?

3.5 WHAT ARE NUCLEOTIDES AND NUCLEIC ACIDS?

A **nucleotide** is a molecule with three parts: a five-carbon sugar, one or more phosphate functional groups, and a nitrogen-containing **base**. The sugar may be either ribose or deoxyribose (see Fig. 3-7). The bases are composed of carbon and nitrogen atoms linked together in either a single ring (in the bases thymine, uracil, and cytosine) or double rings (in the bases adenine and guanine). A deoxyribose nucleotide with an adenine base is illustrated in **FIGURE 3-19**. Nucleotides may function as energy-carrier molecules or intracellular messenger molecules, or as subunits of long polymers called nucleic acids, which encode genetic information.

Some Nucleotides Act As Energy Carriers

Adenosine triphosphate (ATP) is a ribose nucleotide with an adenine base and three phosphate functional groups (**FIG. 3-20**). This molecule is formed in cells by reactions that release energy, such as the reaction that breaks down a sugar molecule. ATP stores energy in bonds between its phosphate groups and releases energy when the bond linking the last phosphate is broken. This energy is then available to drive energy-demanding reactions, such as the reaction that links amino acids to form proteins.

Other nucleotides are *electron carriers*, so called because they transport energy in the form of high-energy electrons. The energy and electrons carried by these molecules are used in ATP synthesis when cells break down sugar (see Chapter 8).

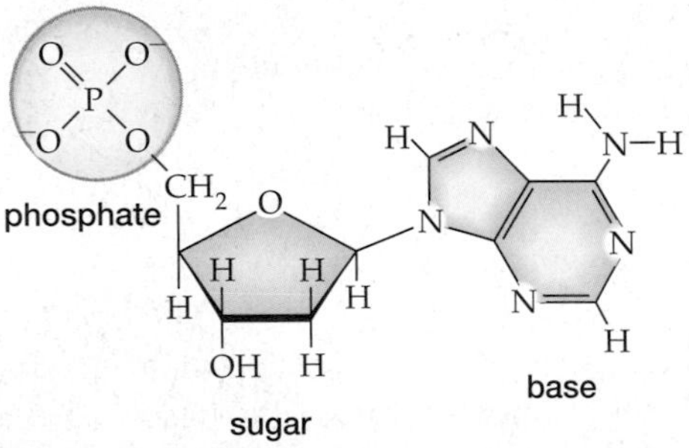

▲ **FIGURE 3-19 A deoxyribose nucleotide**

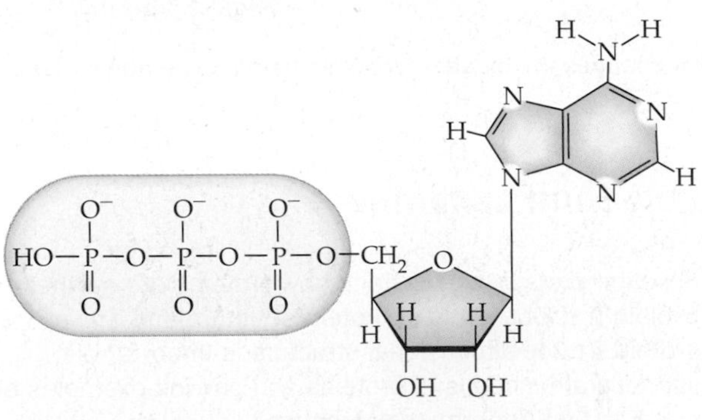

▲ **FIGURE 3-20 The energy-carrier molecule adenosine triphosphate (ATP)**

DNA and RNA, the Molecules of Heredity, Are Nucleic Acids

Nucleotides may be bonded together in long chains by dehydration synthesis, forming polymers called **nucleic acids**. In nucleic acids, an oxygen atom in the phosphate functional group of one nucleotide is covalently bonded to the sugar of the next. The nucleic acid composed of deoxyribose nucleotides is called **deoxyribonucleic acid (DNA)** and can contain millions of nucleotides. A DNA molecule consists of two strands of nucleotides entwined in the form of a double helix and linked by hydrogen bonds (**FIG. 3-21**). DNA forms the genetic material of all cells. Its sequence of nucleotides, like the letters of a biological alphabet, spells out the genetic information needed to construct proteins. Protein synthesis is directed by single-stranded chains of ribose nucleotides, called **ribonucleic acid (RNA)**, that transcribe the information encoded in DNA (see Chapters 11 and 12).

CHECK YOUR LEARNING

Can you . . .

- describe the general structure of nucleotides?
- list three different functions of nucleotides?
- explain how nucleic acids are synthesized?
- give two examples of nucleic acids and their functions?

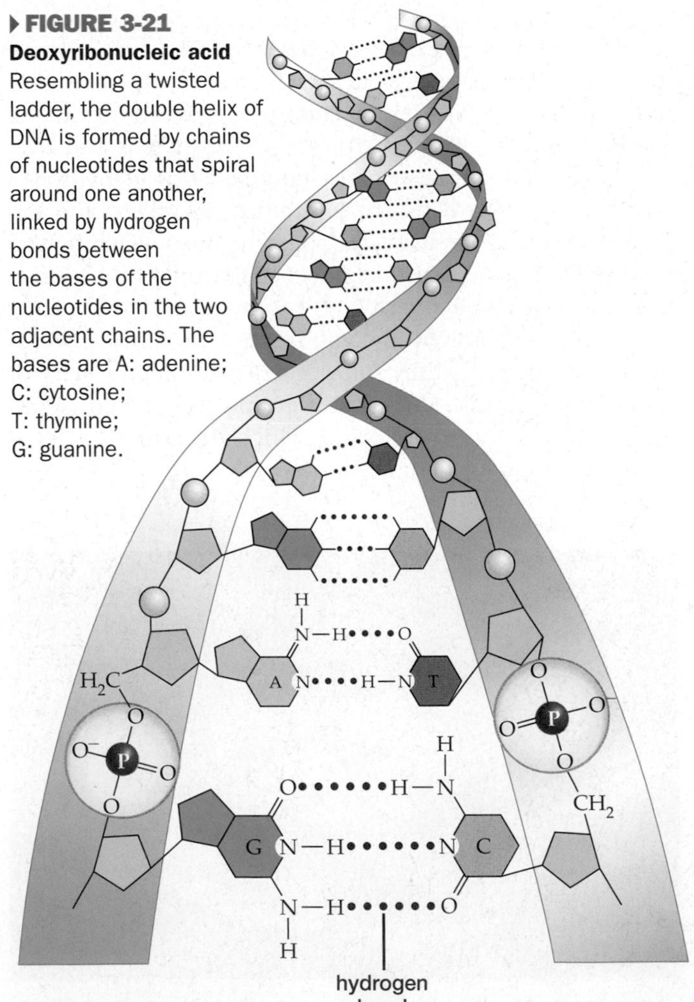

▶ **FIGURE 3-21 Deoxyribonucleic acid** Resembling a twisted ladder, the double helix of DNA is formed by chains of nucleotides that spiral around one another, linked by hydrogen bonds between the bases of the nucleotides in the two adjacent chains. The bases are A: adenine; C: cytosine; T: thymine; G: guanine.

CASE STUDY **CONTINUED**

Puzzling Proteins

All disease-causing organisms use DNA to encode the instructions for producing more organisms, and viruses use either DNA or RNA to encode the instructions for producing more viruses. Before the discovery of prions, no known infectious agent lacked genetic material (DNA or RNA). Scientists were extremely skeptical of the hypothesis that prion proteins could reproduce themselves without genetic material, but repeated studies of prions have not found any trace of nucleic acids.

In addition to lacking genetic material, prions also lack another component that all other infectious agents possess: a surrounding membrane. What kinds of molecules are involved in the construction of membranes?

3.6 WHAT ARE LIPIDS?

Lipids are a diverse group of molecules that contain regions composed almost entirely of hydrogen and carbon, with nonpolar carbon–carbon and carbon–hydrogen bonds. These regions are hydrophobic, which makes lipids insoluble in water. Unlike carbohydrates, proteins, and nucleic acids, lipids are not formed by linking monomer subunits into polymers. Lipids can store energy, provide waterproof coatings on plants, form major components of cell membranes, or function as hormones. Lipids fall into three major groups: (1) oils, fats, and waxes, (2) phospholipids, and (3) steroids.

Oils, Fats, and Waxes Contain Only Carbon, Hydrogen, and Oxygen

Oils, fats, and waxes are built from only three types of atoms: carbon, hydrogen, and oxygen. Each contains one or more **fatty acids**, long chains of carbon and hydrogen with a carboxylic acid functional group (–COOH) at one end.

Fats and **oils** are formed when dehydration synthesis links three fatty acid subunits to one molecule of *glycerol,* a three-carbon molecule (**FIG. 3-22**). This structure gives fats and oils their chemical name: **triglycerides**. Fats and oils are used primarily as energy storage molecules; they contain more than twice as many calories per gram as do carbohydrates and proteins. Fats (such as butter and lard) are produced primarily by animals, whereas oils (found in corn, canola, olives, and avocados) are produced primarily by plants (**FIG. 3-23**).

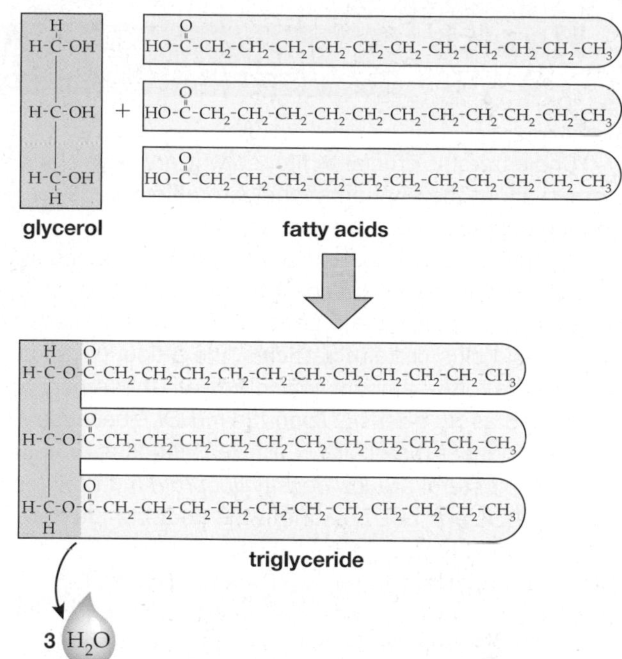

▲ **FIGURE 3-22 Synthesis of a triglyceride** Dehydration synthesis links a single glycerol molecule with three fatty acids to form a triglyceride and three water molecules.

THINK CRITICALLY What kind of reaction breaks this molecule apart?

The difference between fats, which are solid at room temperature, and oils, which are liquid at room temperature, lies in the structure of their fatty acid subunits. In fats, the carbons of the fatty acid subunits are linked entirely by single bonds, and the fatty acids are described as **saturated**, because they contain as many hydrogen atoms as possible. Saturated fatty acid chains are straight and can

(a) Fat is stored prior to hibernation.

(b) Avocado flesh is rich in oil.

◀ **FIGURE 3-23 Energy storage (a)** A grizzly bear stores fat to provide both insulation and energy as he prepares to hibernate. If he stored the same amount of energy in carbohydrates, he would probably be unable to walk. **(b)** Oily avocado flesh likely originally evolved to entice enormous seed-dispersing mammals (such as giant ground sloths, extinct for about 10,000 years), which would swallow the seeds and excrete them intact.

Cholesterol, Trans Fats, and Your Heart

Cholesterol is crucial to life, so why are physicians so concerned about their patients' cholesterol levels? Is there something dangerous about cholesterol?

Cholesterol molecules are insoluble and are transported in the bloodstream as microscopic particles surrounded by hydrophilic phospholipids and proteins. In some of these *lipoprotein* (lipid plus protein) particles, the amount of protein is high relative to the amount of cholesterol; these particles are described as *high-density lipoprotein* (*HDL*), because proteins are denser than lipids. Particles with less protein and more cholesterol are *low-density lipoprotein* (*LDL*). The amounts of HDL and LDL present in the body can be determined by a blood test.

Why is it important to know HDL and LDL levels? Because elevated HDL is associated with a reduced risk of heart disease and stroke, whereas elevated LDL is a risk factor for cardiovascular disease. LDL deposits cholesterol in artery walls, where it participates in the formation of fatty deposits called *plaques* (**FIG. E3-1**). Blood clots may form around the plaques. If a clot breaks loose and blocks an artery supplying blood to the heart or brain, it can cause a heart attack or a stroke. In contrast to the damaging effects of LDL, HDL particles can absorb cholesterol from plaques accumulating in artery walls and transport it to the liver, which uses cholesterol to synthesize bile. The bile is secreted into the small intestine to aid in fat digestion.

Cholesterol-containing foods, such as egg yolks, sausages, bacon, whole milk, and butter, account for a relatively modest portion of the cholesterol in our blood—typically about 15% to 20%. Although most cholesterol is synthesized by the body, the amount produced can be affected by diet. Saturated fats (such as those in dairy and red meat) stimulate the liver to churn out more LDL cholesterol, but diets rich in unsaturated fats (found in fish, nuts, and most vegetable oils) reduce LDL and are associated with a decreased risk of heart disease. Lifestyle also affects cholesterol levels: Exercise tends to increase HDL, whereas obesity and smoking increase LDL levels.

The worst dietary fat is **trans fat**, made artificially when hydrogen atoms are added to oil in a configuration that causes the kinked fatty acid tails to straighten such that the oil is solid at room temperature. Trans fats—also called "partially hydrogenated oils"—are very stable, so they extend the shelf life of foods such as margarine, cookies, crackers, and fried foods and help retain their flavor. Trans fat simultaneously decreases HDL and increases LDL and so puts people who consume it at a higher risk of heart disease. Fortunately, the FDA recently declared that trans fats are "not generally recognized as safe," and as of 2018, manufacturers and retailers are no longer allowed to sell foods containing trans fats.

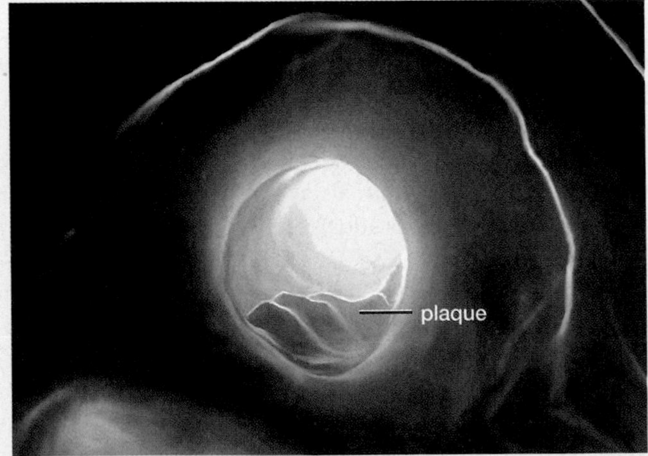

▲ **FIGURE E3-1 Plaque** A plaque deposit (rippled structure) partially blocks this carotid artery.

THINK CRITICALLY An obese 55-year-old woman consults her physician about minor chest pains during exercise. Explain the physician's preliminary diagnosis, list the questions she should ask her patient, describe the tests she would perform, and provide the advice she should give.

pack closely together, thus forming a solid at room temperature (**FIG. 3-24a**). In oils, however, some of the carbons in fatty acids are linked by double bonds, which produce kinks in the fatty acid chains that prevent the molecules from packing closely together (**FIG. 3-24b**). The fatty acids in oil are described as **unsaturated**, because they contain fewer than the maximum possible number of hydrogens. The commercial process of hydrogenation—which breaks some of the double bonds and adds hydrogens to the carbons—can convert liquid oils to solids, but with health consequences (see "Health Watch: Cholesterol, Trans Fats, and Your Heart").

Although **waxes** are chemically similar to fats, humans and most other animals do not have the appropriate enzymes

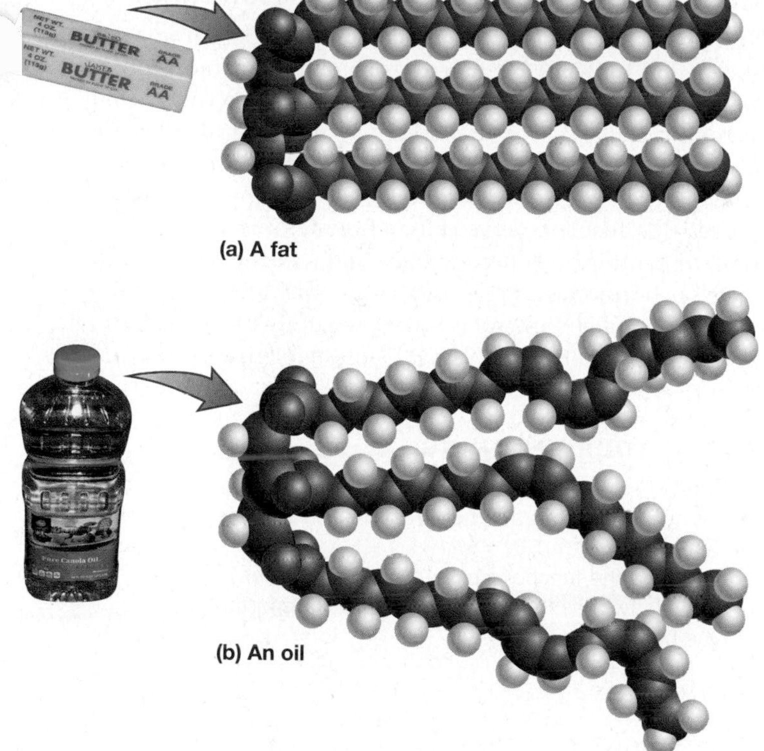

▲ **FIGURE 3-24 Fats and oils (a)** Fats have straight chains of carbon atoms in their fatty acid tails. **(b)** The fatty acid tails of oils have double bonds between some of their carbon atoms, creating kinks in the chains. Oils are liquid at room temperature because their kinky tails keep the molecules farther apart.

▲ **FIGURE 3-25 Waxes** Waxes are highly saturated lipids that remain solid at outdoor temperatures. Bees form wax into the hexagons of this honeycomb.

to break them down. Waxes are highly saturated and are solid at outdoor temperatures. They form a water-repellent coating on the leaves and stems of land plants, and birds distribute waxy secretions over their feathers, causing them to shed water. Honey bees use waxes to build intricate honeycomb structures, where they store honey and lay their eggs (**FIG. 3-25**).

Phospholipids Have Water-Soluble Heads and Water-Insoluble Tails

All cells are surrounded by a plasma membrane that contains several types of **phospholipids**. A phospholipid resembles an oil, but with one of its three fatty acids replaced by a phosphate group. The phosphate is linked to one of several polar functional groups that typically contain nitrogen (**FIG. 3-26**). A phospholipid molecule has two dissimilar ends. At one end is the phosphate–nitrogen "head," which is polar and water soluble. At the other end are the two fatty acid "tails," which are nonpolar and insoluble in water. The difference between the polar heads and nonpolar tails of phospholipids is crucial to the structure and function of cell membranes (see Chapter 5).

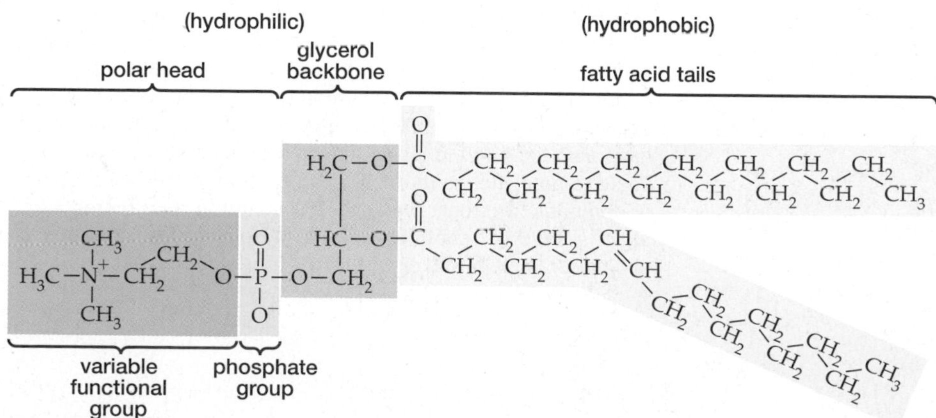

◀ **FIGURE 3-26 Phospholipids** Phospholipids have two hydrophobic fatty acid tails attached to the three-carbon glycerol backbone. The third position on glycerol is occupied by a polar head (left) consisting of a phosphate group to which a second (usually nitrogen-containing) variable functional group is attached. The phosphate group bears a negative charge, and the nitrogen-containing functional group bears a positive charge, making the head hydrophilic.

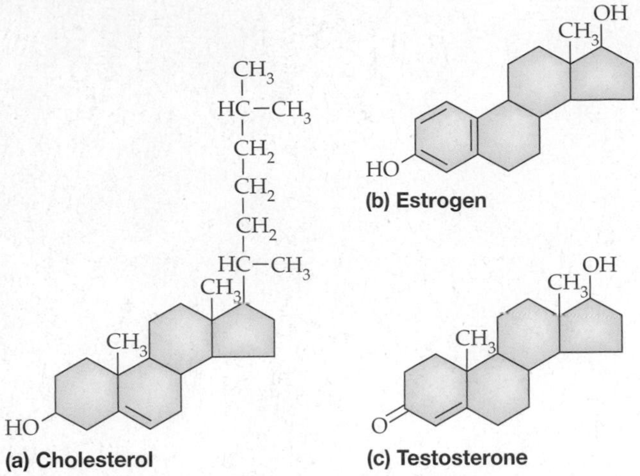

(a) Cholesterol

(b) Estrogen

(c) Testosterone

▲ **FIGURE 3-27 Steroids** All steroids have a similar, nonpolar molecular structure with four fused carbon rings. Differences in steroid function result from differences in functional groups attached to the rings. **(a)** Cholesterol, the molecule from which other steroids are synthesized; **(b)** the female sex hormone estrogen (estradiol); **(c)** the male sex hormone testosterone. Note the similarities in structure between the sex hormones.

THINK CRITICALLY Why are steroid hormones able to diffuse through cell membranes to exert their effects?

Steroids Contain Four Fused Carbon Rings

Steroids are composed of four rings of carbon atoms. As shown in **FIGURE 3-27**, the rings share one or more sides, with various functional groups protruding from them. One steroid, *cholesterol*, is a vital component of the membranes of animal cells. It makes up about 2% of the human brain, where it is an important component of the lipid-rich membranes that insulate nerve cells. Cholesterol is also used by cells to synthesize other steroids, such as the female and male sex hormones *estrogen* and *testosterone*. Too much of the wrong form of cholesterol is linked to cardiovascular disease, as explained in "Health Watch: Cholesterol, Trans Fats, and Your Heart."

CHECK YOUR LEARNING

Can you . . .

- compare and contrast the structure and synthesis of fats and oils?
- describe the functions of fats, oils, and waxes?
- provide two reasons why cholesterol is important in the body?

CASE STUDY REVISITED

Puzzling Proteins

Stanley Prusiner and his associates coined the term "prion" to refer to the misfolded, infectious version of a normal protein called PrP^c, which is found on cell membranes. How do prions replicate themselves? Prions interact with normal helical PrP^c proteins, forcing them to change into the pleated sheet configuration of the infectious form. These new "prion converts" then go on to transform other normal PrP^c proteins in an ever-expanding chain reaction. The chain reaction apparently occurs slowly enough that, as in Charlene Singh's case, it can be a decade or more after infection before disease symptoms occur. Fortunately, both vCJD and BSE have been nearly eradicated worldwide. However, there is evidence that a rare mutation in the gene coding for PrP^c can cause BSE in cows that have not been exposed to prions. Careful surveillance of cattle continues.

The discovery that prions are the disease agent in vCJD has focused attention on the role of the normal PrP^c protein. It is found in cell membranes throughout the body and in relatively high levels in the brain. Preliminary evidence suggests that PrP^c has diverse functions, which include protecting cells from oxidative stress, contributing to the growth of neurons, and helping to maintain the new connections between neurons that form when an animal learns.

When Prusiner accepted the 1997 Nobel Prize for his research, he predicted that further research into how prions cause vCJD could lead to a better understanding of other neurodegenerative disorders. Researchers are now exploring the hypothesis that diseases including Alzheimer's, Parkinson's, and amyotrophic lateral sclerosis (ALS) may arise from misfolded proteins that propagate and accumulate in the nervous system.

CONSIDER THIS A disorder called chronic wasting disease (CWD) affects deer and elk in at least 20 U.S. states. Like scrapie and BSE, CWD is a fatal brain disorder caused by prions. The disease spreads among animals by contact with saliva, urine, and feces, which may contain prions. Prions have also been found in the muscles of deer with CWD. There is no evidence that CWD can be transmitted to people or domestic livestock. If you were a hunter in an affected region, would you continue to hunt? Would you eat deer or elk meat? Explain why or why not.

CHAPTER REVIEW

Go to **Mastering Biology** to access the Pearson eText, vocabulary review, practice quizzes, activities, videos, current events, and more.

*Answers to **Think Critically** and **Thinking Through the Concepts** questions can be found in the **Answers** section at the back of the book.*

Summary of Key Concepts

3.1 Why Is Carbon So Important in Biological Molecules?

Organic molecules have a carbon backbone. They are very diverse because the carbon atom is able to form bonds with up to four other molecules. This allows organic molecules to form complex shapes. The presence of functional groups produces further diversity among biological molecules (see Table 3-1).

3.2 How Are Large Biological Molecules Synthesized?

Most large biological molecules are polymers synthesized by linking many smaller monomer subunits using dehydration synthesis. Hydrolysis reactions break these polymers apart. The most important organic molecules fall into four classes: carbohydrates, lipids, proteins, and nucleotides/nucleic acids (see Table 3-2).

3.3 What Are Carbohydrates?

Carbohydrates include sugars, starches, cellulose, and chitin. Sugars include monosaccharides and disaccharides; they are used for temporary energy storage and the construction of other molecules. Starches and glycogen are polysaccharides that provide longer-term energy storage in plants and animals, respectively. Cellulose forms the cell walls of plants, and chitin strengthens the exoskeletons of many invertebrates and the cell walls of fungi.

3.4 What Are Proteins?

Proteins consist of one or more amino acid chains called polypeptides with up to four levels of structure. Primary structure is the sequence of amino acids; secondary structure consists of helices or pleated sheets. These may fold to produce tertiary structure. Proteins with two or more linked polypeptides have quaternary structure. Some proteins or parts of proteins are disordered and lack a stable secondary or tertiary structure. The function of a protein is determined by its shape and by how its amino acids interact with their surroundings and with each other. See Table 3-3 for protein functions and examples.

3.5 What Are Nucleotides and Nucleic Acids?

A nucleotide is composed of a phosphate group, a five-carbon sugar (ribose or deoxyribose), and a nitrogen-containing base. Molecules formed from single nucleotides include energy-carrier molecules such as ATP. Nucleic acids are chains of nucleotides. DNA carries the hereditary instructions, and RNA directs the synthesis of proteins.

3.6 What Are Lipids?

Lipids are nonpolar, water-insoluble molecules. Oils, fats, waxes, and phospholipids all contain fatty acids, which are chains of carbon and hydrogen atoms with a carboxylic acid group at the end. Steroids all have four fused rings of carbon atoms with functional groups attached. Lipids are used for energy storage (oils and fats), as waterproofing for the outside of many plants and animals (waxes), as the principal component of cellular membranes (phospholipids and cholesterol), and as hormones (steroids).

Thinking Through the Concepts

Bloom's: Remembering, Understanding

Multiple Choice

1. Polar molecules
   a. dissolve in lipids.
   b. are hydrophobic.
   c. form covalent bonds.
   d. form ionic bonds.
2. Which match is correct?
   a. monosaccharide–sucrose
   b. polysaccharide–maltose
   c. disaccharide–lactose
   d. disaccharide–glycogen
3. Which of the following statements is false?
   a. In starch breakdown, water is formed.
   b. In chitin, glucoses are linked as in cellulose.
   c. Peptide bonds determine primary peptide structure.
   d. Disulfide bridges are formed by covalent bonds.
4. Which of the following is *not* composed of repeating subunits?
   a. starch
   b. protein
   c. nucleic acid
   d. lipid
5. Which of the following statements is false?
   a. Carbohydrates are the most efficient energy-storage molecules by weight.
   b. Lipids are not soluble in water.
   c. Nucleotides may act as energy-carrier molecules.
   d. Very acidic or salty solutions may denature proteins.

Fill-in-the-Blank

1. In organic molecules made of chains of subunits, each subunit is called a(n) ____________, and the chains are called ____________. Carbohydrates consisting of long chains of sugars are called ____________. These sugar chains can be broken down by ____________ reactions. Three types of carbohydrates consisting of long glucose chains are ____________, ____________, and ____________.
2. Fill in the following with the specific bond(s): Maintain(s) the helical structure of many proteins: ____________; link(s) polypeptide chains and can cause proteins to bend: ________ and ________; join(s) the two strands of the double helix of DNA: ____________; link(s) amino acids to form the primary structure of proteins: ____________.

3. Proteins are synthesized by a reaction called ________ synthesis, which releases ________. Subunits of proteins are called ____________. The sequence of protein subunits is called the ______________ structure of the protein. Two regular configurations of secondary protein structure are ______________ and ______________. When a protein's secondary or higher-order structure is destroyed, the protein is said to be ______________.
4. A nucleotide consists of three parts: ______________, ______________, and ______________. A nucleotide that acts as an energy carrier is ______________. The four bases found in deoxyribose nucleotides are __________, __________, __________, and __________. Two important nucleic acids are ______________ and ______________. The functional group that joins nucleotides in nucleic acids is __________.
5. Fill in the following with the appropriate type of lipid: Unsaturated, liquid at room temperature: __________; bees use to make honeycombs: __________; stores energy in animals: ______________; sex hormones are synthesized from these: ______________; the LDL form of this contributes to heart disease: ______________; a major component of cell membranes that has polar heads: ______________.

Review Questions

1. What does the term "organic" mean to a chemist?
2. List the four principal classes of biological molecules, and give an example of each.
3. What roles do nucleotides play in living organisms?
4. How are fats and oils similar? How do they differ, and how do their differences explain whether they are solid or liquid at room temperature?
5. Describe and compare dehydration synthesis and hydrolysis. Give an example of a substance formed by each chemical reaction, and describe the specific reaction in each instance.
6. Describe the synthesis of a protein from amino acids. Then describe the primary, secondary, tertiary, and quaternary structures of a protein.
7. Where in nature do we find cellulose? Where do we find chitin? In what way(s) are these two polymers similar? How are they different?

Applying the Concepts

Bloom's: Applying, Analyzing, Evaluating

1. Based on their structure, sketch and explain how phospholipids would organize themselves in water.
2. Compare the way fat and carbohydrates interact with water, and explain why this interaction gives fat an extra advantage for weight-efficient energy storage.
3. In an alternate universe where people could digest cellulose molecules, how might this affect our way of life?

4 Cell Structure and Function

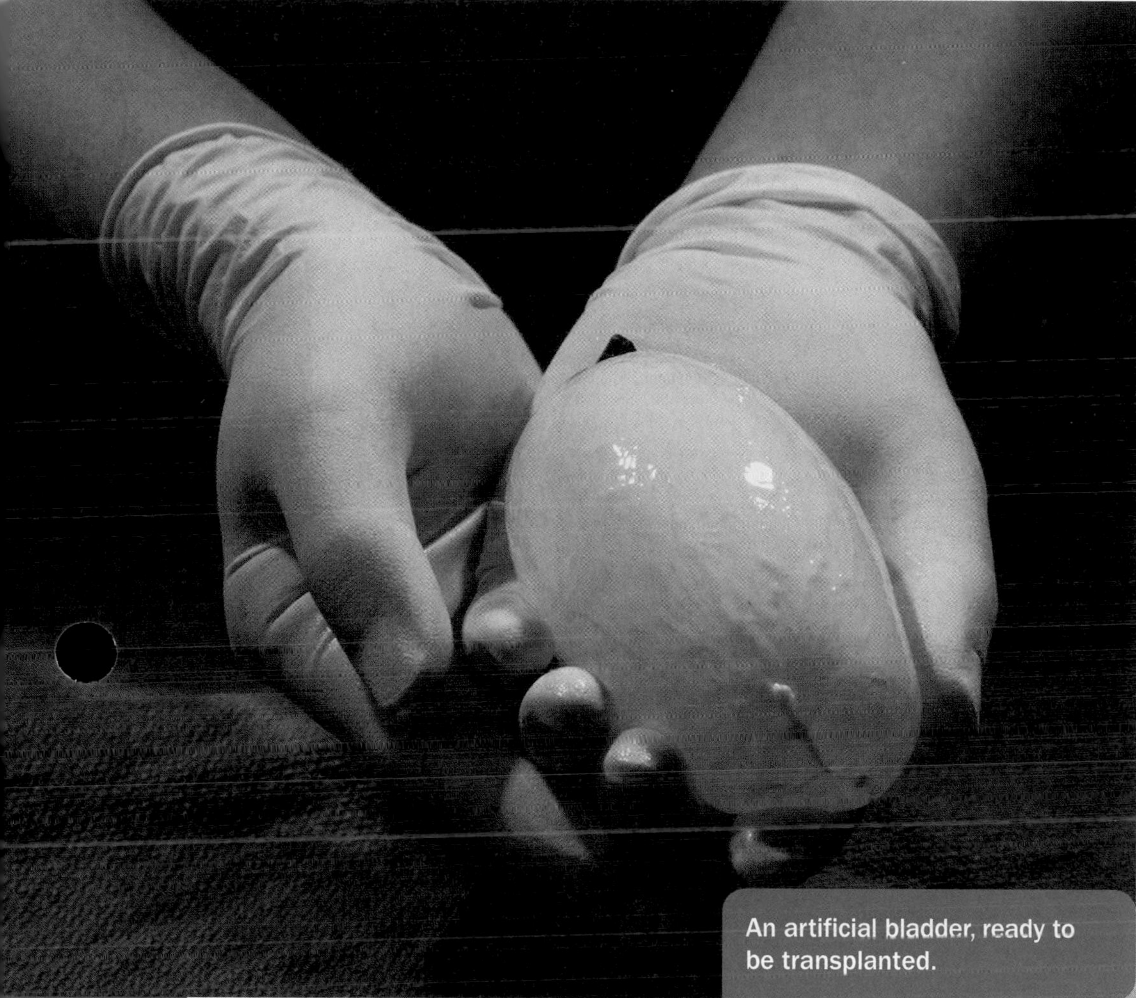

An artificial bladder, ready to be transplanted.

CASE STUDY

New Parts for Human Bodies

WHEN LUKE MASSELLA WAS 10 YEARS OLD, he faced a very serious health crisis. Complications from a birth defect had left Luke with a bladder that could not fully function, and his dysfunctional bladder was inflicting damage on his kidneys. Without some kind of intervention, Luke's kidneys would fail, with potentially fatal consequences. With Luke's health declining rapidly, he underwent an experimental surgery in which his bladder was replaced with a new one. But the new bladder was not a natural organ taken from a cadaver. Instead, Luke's new bladder was grown in a laboratory, bioengineered just for him.

Luke's bioengineered bladder was built from his own cells. A team led by Dr. Anthony Atala removed a very small piece of tissue from Luke's urinary tract. The scientists induced the cells to multiply and then spread the resulting cells over a biodegradable, bladder-shaped mold made largely of the protein collagen. With a thin layer of muscle cells on the outside, and a thin layer of epithelial cells on the inside, the mold was placed in a nutrient broth inside a temperature-controlled "bioreactor," where further cell proliferation ultimately formed a complete bladder. When the artificial bladder was ready, surgeons used it to replace Luke's damaged organ.

In the years after his pioneering surgery, Luke went on to become the captain of his high school wrestling team, graduate from college, and begin his career. Meanwhile, Dr. Atala and his team, along with scientists in numerous other labs around the world, have made further progress toward the goal of producing bioengineered tissues and organs to help people who need theirs repaired or replaced.

Bioengineering human organs demonstrates our expanding ability to manipulate cells, the fundamental units of life. What structures make up cells? What new bioengineering techniques involving human or animal cells are being developed and tested?

AT A GLANCE

4.1 WHAT IS THE CELL THEORY?

The discovery of cells in the 1600s was the first step toward understanding their importance. In 1838, the German botanist Matthias Schleiden concluded that cells and substances produced by cells form the basic structure of plants and that plant growth occurs by adding new cells. In 1839, German biologist Theodor Schwann (Schleiden's friend and collaborator) drew similar conclusions about animal cells. The work of Schleiden and Schwann provided a unifying theory of cells as the fundamental units of life. In 1855, the German physician Rudolf Virchow completed the **cell theory**—a fundamental concept of biology—by concluding that all cells come from previously existing cells.

The cell theory consists of three principles:

1. Every organism is made up of one or more cells.
2. The smallest organisms are single cells, and cells are the functional units of multicellular organisms.
3. All cells arise from preexisting cells.

CHECK YOUR LEARNING

Can you . . .

- trace the historical development of the cell theory?
- list the three principles of the cell theory?

4.2 HOW DO SCIENTISTS VISUALIZE CELLS?

Although cells form the basis of life, they're so small that people didn't realize cells existed until they could actually be seen. In 1665, the English scientist and inventor Robert Hooke aimed his primitive light microscope at an "exceeding thin . . . piece of Cork" and saw "a great many little Boxes," which he drew with great skill (**FIG. 4-1a**). Hooke called the boxes "cells," because he thought they resembled the tiny rooms (called cells) occupied by monks in a monastery. Cork comes from the dry outer bark of the cork oak tree, and we

(a) Hooke's light microscope and his drawing of cork cells

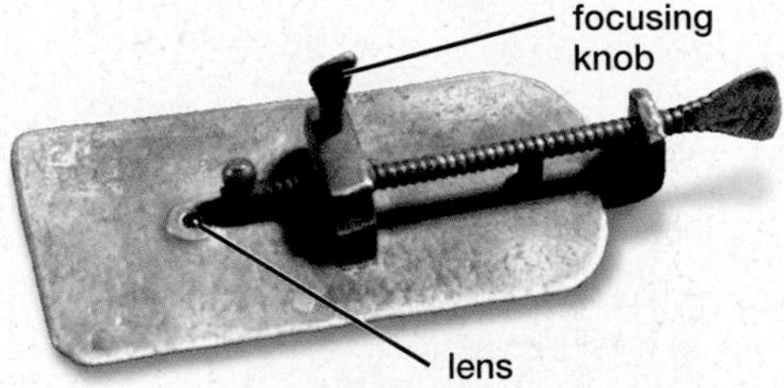

(b) van Leeuwenhoek's light microscope

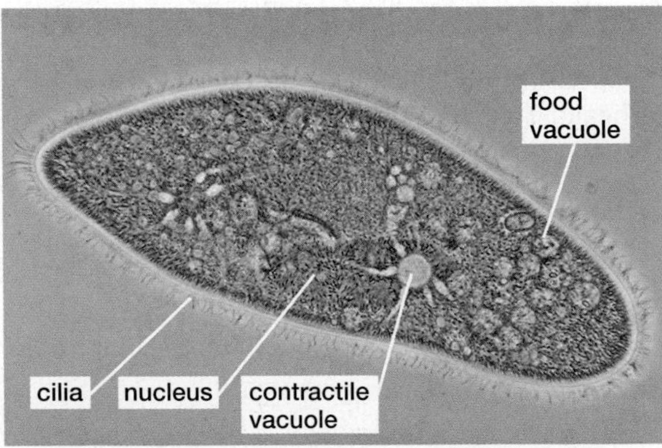

(c) Modern light micrograph *(Paramecium)*

▲ **FIGURE 4-1 Light microscopy yesterday and today (a)** Hooke observed the walls of cork cells through his elegant microscope. **(b)** Hooke's contemporary van Leeuwenhoek built a microscope that produced images superior to those of Hooke's. **(c)** Modern light microscopes can reveal structures within cells, as seen in this light micrograph of a living, single-celled protist of the genus *Paramecium*.

now know that he was looking at the nonliving cell walls that surround all plant cells. Hooke wrote that in the living oak and other plants, "These cells [are] fill'd with juices."

Light Microscopes Can View Living Cells

In the 1670s, Dutch microscopist Anton van Leeuwenhoek constructed his own simple microscopes (**FIG. 4-1b**) and observed a previously unknown living world. Although Hooke described Van Leeuwenhoek's microscopes as "offensive to my eye" because of their primitive appearance, their superior lenses provided clearer images and higher magnification than did Hooke's microscope. Van Leeuwenhoek's descriptions of myriad "animalcules" (mostly single-celled organisms) in rain, pond, and well water were greeted with amazement. Over the years, he described an enormous range of microscopic specimens, including blood cells, sperm cells, and the eggs of aphids and fleas. Observing plaque scraped from his teeth, van Leeuwenhoek saw swarms of cells that we now recognize as bacteria. Disturbed by these animalcules in his mouth, he tried to kill them with vinegar and hot coffee—but with little success.

Since the pioneering efforts of early microscopists, biologists, physicists, and engineers have collaborated to develop a variety of advanced microscopes to view cells and their components. *Light microscopes* use lenses made of glass or quartz to bend, focus, and transmit light rays that have passed through or bounced off a specimen. The resolving power (the smallest structure distinguishable under ideal conditions) of modern light microscopes is generally about 200 nanometers (see Fig. 4-3). This is sufficient to see most prokaryotic cells, some structures inside eukaryotic cells, and living cells such as a swimming *Paramecium* (**FIG. 4-1c**). In recent years, researchers have developed several types of "super resolution" light microscopes that use lasers and fluorescent molecules to produce images that can resolve even smaller structures within living cells.

Electron Microscopes Provide High Resolution

Instead of using light focused by lenses, electron microscopes (**FIG. 4-2**) use beams of electrons focused by magnetic fields. *Transmission electron microscopes* pass electrons through a thin specimen and can reveal the details of interior cell structure. Some modern transmission electron microscopes can resolve structures as small as 0.05 nanometer, allowing scientists to see molecules such as DNA and even individual carbon atoms.

Scanning electron microscopes bounce electrons off specimens that are dry and hard or that have been covered with an ultrathin coating of metal such as gold. Scanning electron microscopes can be used to view the three-dimensional surface details of structures that range in size from entire small insects down to cells and their components, with a maximum resolution of about 1.5 nanometers.

CHECK YOUR LEARNING

Can you . . .

- explain how light microscopes and electron microscopes differ?
- describe the difference between transmission electron microscopes and scanning electron microscopes?

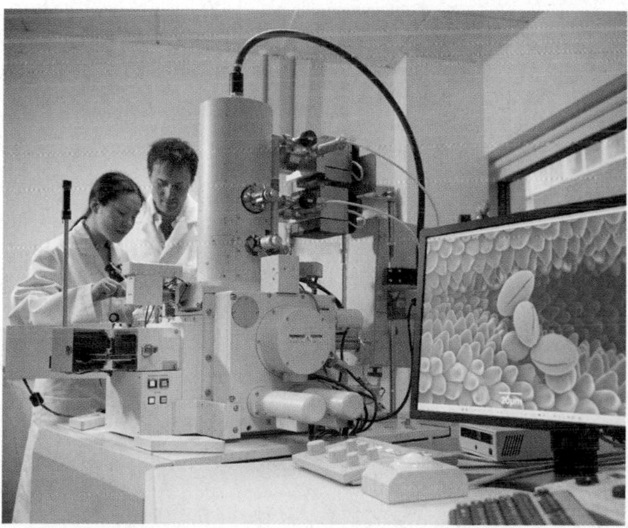

Electron microscope

TEM of carbon atoms

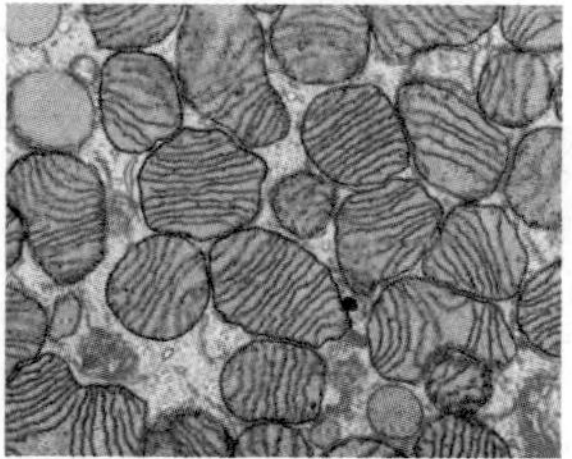

TEM of mitochondria

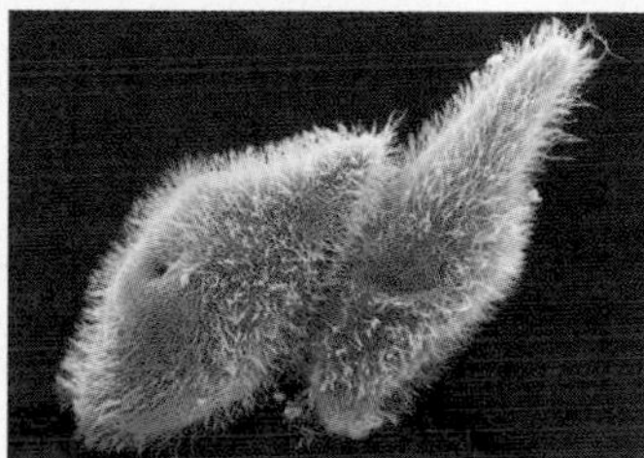

SEM of two *Paramecium*

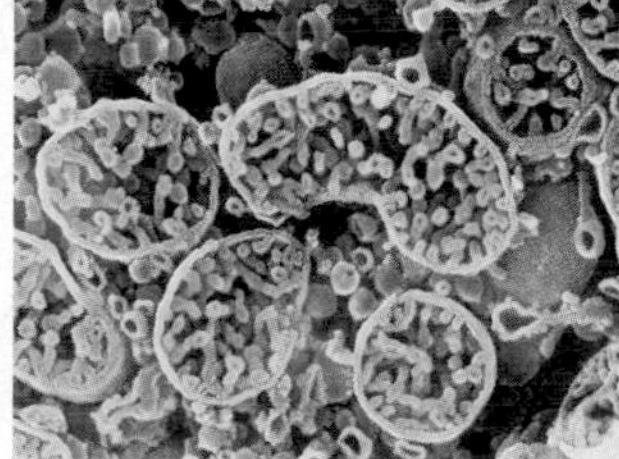

SEM of mitochondria

▲ **FIGURE 4-2 Electron microscopy** Many of today's electron microscopes produce both transmission electron micrographs (TEMs) and scanning electron micrographs (SEMs). An SEM image of pollen grains is visible on the computer screen of the electron microscope pictured. All colors in the electron micrographs (SEMs or TEMs) have been added artificially.

THINK CRITICALLY Compare the micrograph in Figure 4-1 with those in Figure 4-2. Based on these images, what are the advantages of each kind of micrograph (light, TEM, and SEM) for visualizing microscopic structures?

4.3 WHAT ARE THE BASIC ATTRIBUTES OF CELLS?

All living things are composed of cells, which fall into two types: **prokaryotic** and **eukaryotic**. In eukaryotic cells, the genetic material is contained within a membrane-enclosed **nucleus**, but in prokaryotic cells it is not. The single cells of bacteria and archaea, the simplest forms of life, are prokaryotic. Eukaryotic cells are more complex and make up the bodies of animals, plants, fungi, and protists.

Cells Are Small

Most cells are very small, ranging from about 1 to 100 micrometers (μm; millionths of a meter) in diameter (**FIG. 4-3**). Why are most cells so small? Because they exchange nutrients and wastes with their external environment via diffusion (see Chapter 5). Diffusion—a key process by which molecules dissolved in fluids move—is relatively slow, so all parts of the cell must remain close to the external environment to have ready access to necessary materials and the ability to get rid of wastes. Thus, cells have a very small diameter.

All Cells Share Common Features

All cells are descended from an ancestor that arose about 3.5 billion years ago. As a result of this common ancestry, all cells share some important features.

The Plasma Membrane Encloses the Cell and Allows Interactions Between the Cell and Its Environment

Each cell is surrounded by an extremely thin membrane called the **plasma membrane** (**FIG. 4-4**). The plasma membrane consists of proteins embedded in a double layer, or *bilayer*, of phospholipids.

The phospholipid and protein components of plasma membranes play very different roles. The phospholipid bilayer helps isolate the cell from its surroundings, allowing the cell to maintain essential differences in the concentrations of materials inside and out. In contrast, the huge variety of proteins within the bilayer facilitate communication between the cell and its environment. For example, *channel proteins* allow specific molecules or ions to pass into or out of the cell (see Fig. 4-4). Other proteins embedded in the membrane act as *receptor proteins* that bind to messenger molecules such as hormones or neurotransmitters, and initiate a cell's response to the message.

All Cells Contain Cytoplasm

The **cytoplasm** consists of all the fluid and structures that lie inside the plasma membrane but outside of the nucleus (see Figs. 4-6 and 4-7). The fluid portion of the cytoplasm, called the **cytosol**, contains water, salts, and an assortment of organic molecules, including proteins, lipids, carbohydrates, sugars, amino acids, and nucleotides. Most of a cell's

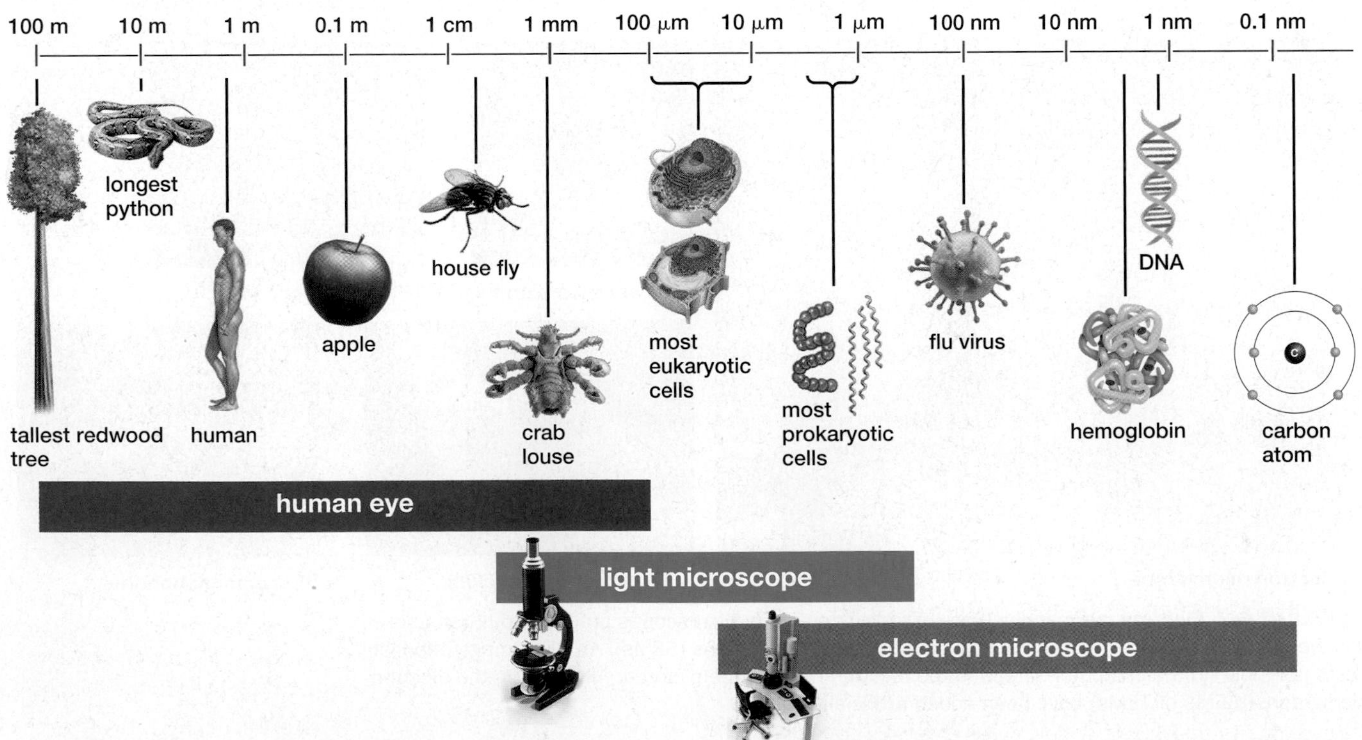

▲ **FIGURE 4-3 Relative sizes** Dimensions encountered in biology range from about 100 meters (the height of the tallest redwood trees) to a few nanometers (nm; the diameter of many large molecules).

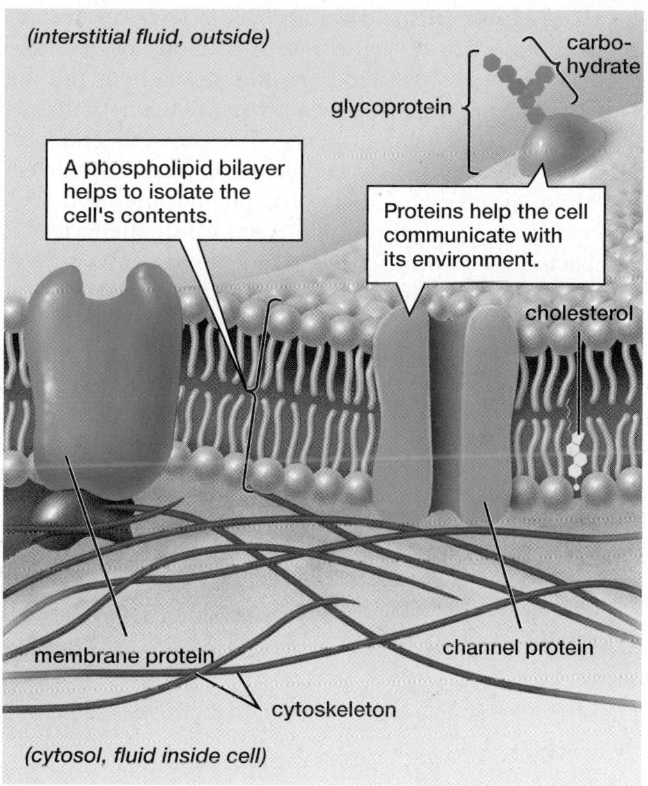

▲ **FIGURE 4-4 The plasma membrane** The plasma membrane encloses the cell in a double layer of phospholipids associated with a variety of proteins. The membrane is supported by the cytoskeleton.

metabolic activities—the biochemical reactions that support life—occur in the cytoplasm.

The **cytoskeleton** consists of a variety of protein filaments within the cytoplasm. These filaments provide support, transport structures within the cell, and allow cells to move and change shape. The cytoskeleton also plays a key role in cell division (see chapter 9).

All Cells Use DNA As Hereditary Instructions and RNA to Guide Construction of Cell Parts

The genetic material in all cells consists of **deoxyribonucleic acid (DNA)** that encodes an inherited set of instructions in segments called genes. Genes store the instructions for making all the parts of a cell and for producing new cells (see Chapter 12). DNA genes are copied to **ribonucleic acid (RNA)**, which is chemically similar to DNA and helps construct proteins based on the genetic instructions. The proteins are constructed on **ribosomes**, cellular "workbenches" composed of a specialized type of RNA called *ribosomal RNA*.

CHECK YOUR LEARNING

Can you . . .

- describe the structure and features shared by all cells?
- distinguish prokaryotic from eukaryotic cells?
- explain why cells are small?

CASE STUDY CONTINUED

New Parts for Human Bodies

Why was Luke Massella's artificial bladder considered a scientific breakthrough? One reason is that the patient's own cells were used to grow the new body part, so his immune system was unlikely to reject the cells. The plasma membranes of the cells in a body bear glycoprotein molecules that are unique to each individual. These distinctive glycoproteins allow a person's immune system to recognize his or her own cells as "self." But cells from any other person (except an identical twin) bear different glycoproteins, so the cells of an organ transplanted from one person to another will be identified by the immune system as foreign and attacked, which can cause rejection of the organ. To prevent organ rejection, transplant patients must take drugs that suppress the immune system, which increases vulnerability to cancers and infections that the immune system would normally target and destroy. Transplanting a bioengineered organ built from the patient's own cells avoids this problem.

The cells most likely to cause problems for immune-suppressed patients are prokaryotic. What are the features of these simple cells?

4.4 WHAT ARE THE MAJOR FEATURES OF PROKARYOTIC CELLS?

Prokaryotic cells have a relatively simple internal structure and are generally less than 5 micrometers in diameter (in comparison, eukaryotic cells range from 10 to 100 micrometers in diameter). Prokaryotes also lack the complex internal membrane-enclosed structures that are the most prominent features of eukaryotic cells.

Prokaryotes are unicellular (consist of a single prokaryotic cell) and make up two of life's domains: Archaea and Bacteria. Many archaea inhabit extreme environments, such as hot springs and cow stomachs, but they are also found in more familiar locales, such as soil and oceans. In this chapter, we focus mainly on the more well-known bacteria as representative prokaryotic cells (**FIG. 4-5**).

Prokaryotic Cells Have Specialized Cytoplasmic Structures

The cytoplasm of a typical prokaryotic cell contains several specialized structures. A distinct region called the **nucleoid** (meaning "like a nucleus"; see Fig. 4-5a) contains a single circular chromosome that consists of a long, coiled strand of DNA. Unlike the nucleus of a eukaryotic cell, the nucleoid is not separated from the cytoplasm by a membrane. In addition to the DNA in the nucleoid, most prokaryotic cells also contain small rings of DNA called **plasmids**. Plasmids usually carry genes that give the cell particular properties;for example, some disease-causing bacteria have plasmids that encode proteins that deactivate antibiotics. Bacterial cytoplasm also includes ribosomes, where proteins are synthesized, as well as food granules that store energy-rich molecules such as glycogen. Prokaryotes also contain an extensive cytoskeleton that functions in cell division and helps regulate the shape of the cell.

Prokaryotic Cells Have Distinctive Surface Features

Nearly all prokaryotic cells are surrounded by a **cell wall**, which is a relatively stiff covering that the cell secretes around itself. The cell wall provides protection and helps the prokaryotic cell maintain its shape, which may be rod-like, spiral, or spherical (see Figs. 4-5a, b, c). Bacterial cell walls are composed of *peptidoglycan*, a polymer in which short peptides link chains of sugar molecules that have amino functional groups.

Many bacteria secrete polysaccharide coatings called *capsules* and *slime layers* outside their cell walls. (The two coatings are similar, but slime layers are somewhat more loosely attached to the cell.) In bacteria such as those that cause tooth decay, diarrhea, pneumonia, or urinary tract infections, capsules and slime layers help the cells adhere to tissues such as the surface of a tooth or the lining of the small intestine, lungs, or bladder. Capsules and slime layers allow some bacteria to form films, such as those that may coat unbrushed teeth or unwashed toilet bowls.

Pili (singular, **pilus**; meaning "hairs") are surface proteins that project from the cell walls of many bacteria (see Fig. 4-5a). There are two types of pili: attachment pili and sex pili. Short, abundant *attachment pili* act on their own or with capsules and slime layers to help bacteria adhere to structures. For example, *Streptococcus* bacteria (which can cause strep throat, skin infections, pneumonia, and toxic shock syndrome) use pili to help them infect their victims. Many bacteria also form *sex pili*, which are fewer in number and longer than attachment pili. A sex pilus from one bacterium binds to a nearby bacterium of the same type and draws the two cells together. The two bacteria then form a short bridge that links their cytoplasm and allows them to transfer plasmids.

Some bacteria and archaea possess **flagella** (singular, **flagellum**; meaning "whip"), which extend from the cell surface and rotate to propel the cell through a fluid environment (see Fig. 4-5a). Prokaryotic flagella differ from eukaryotic flagella, which are described later in this chapter.

CHECK YOUR LEARNING

Can you . . .

- describe the structure and function of the major features of prokaryotic cells?
- describe some cytoplasmic and surface features of bacteria and explain their functions?

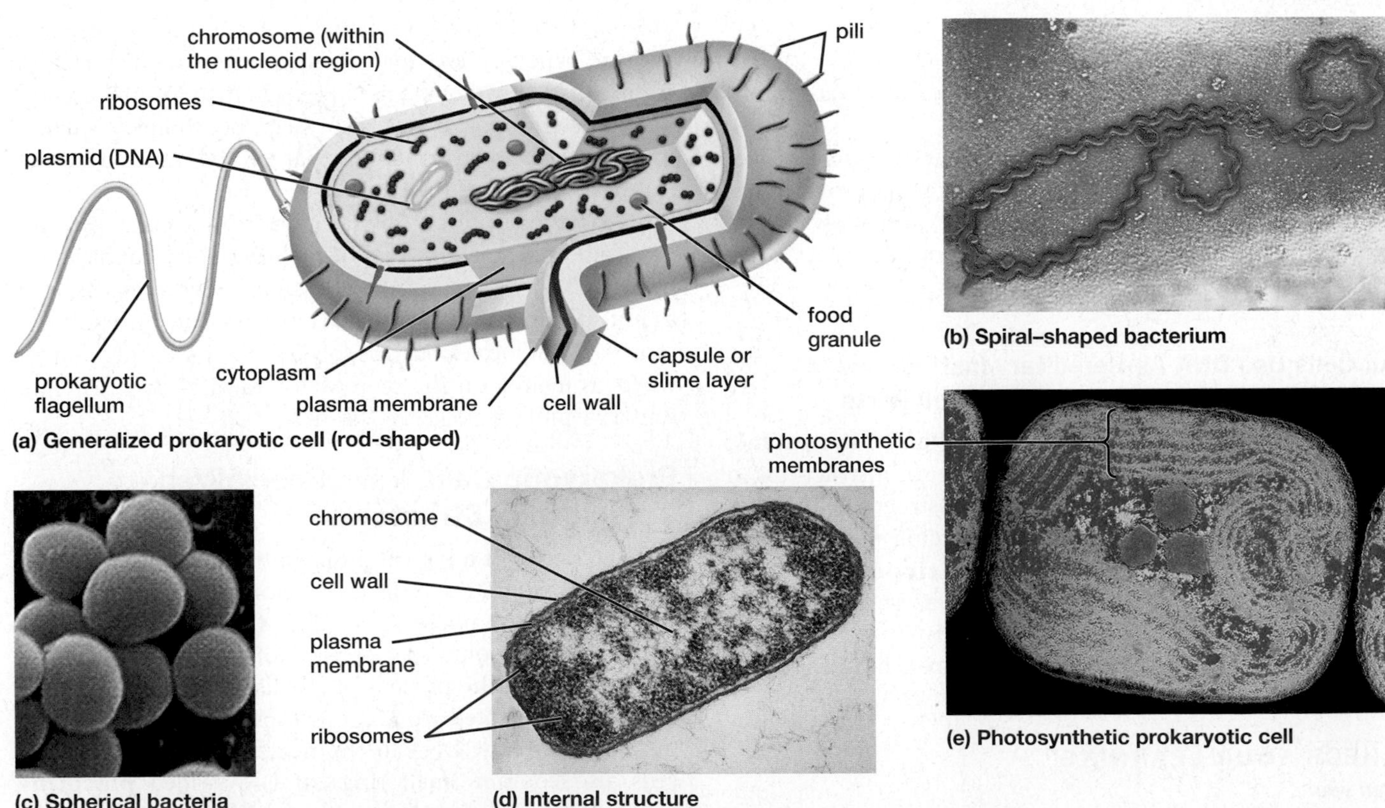

▲ **FIGURE 4-5 Prokaryotic cells** Prokaryotes come in different shapes, including **(a)** rod-shaped, **(b)** spiral-shaped, and **(c)** spherical. Internal structures are revealed in the TEMs in **(d)** and **(e)**. Some photosynthetic bacteria have internal membranes where photosynthesis occurs, as shown in (e).

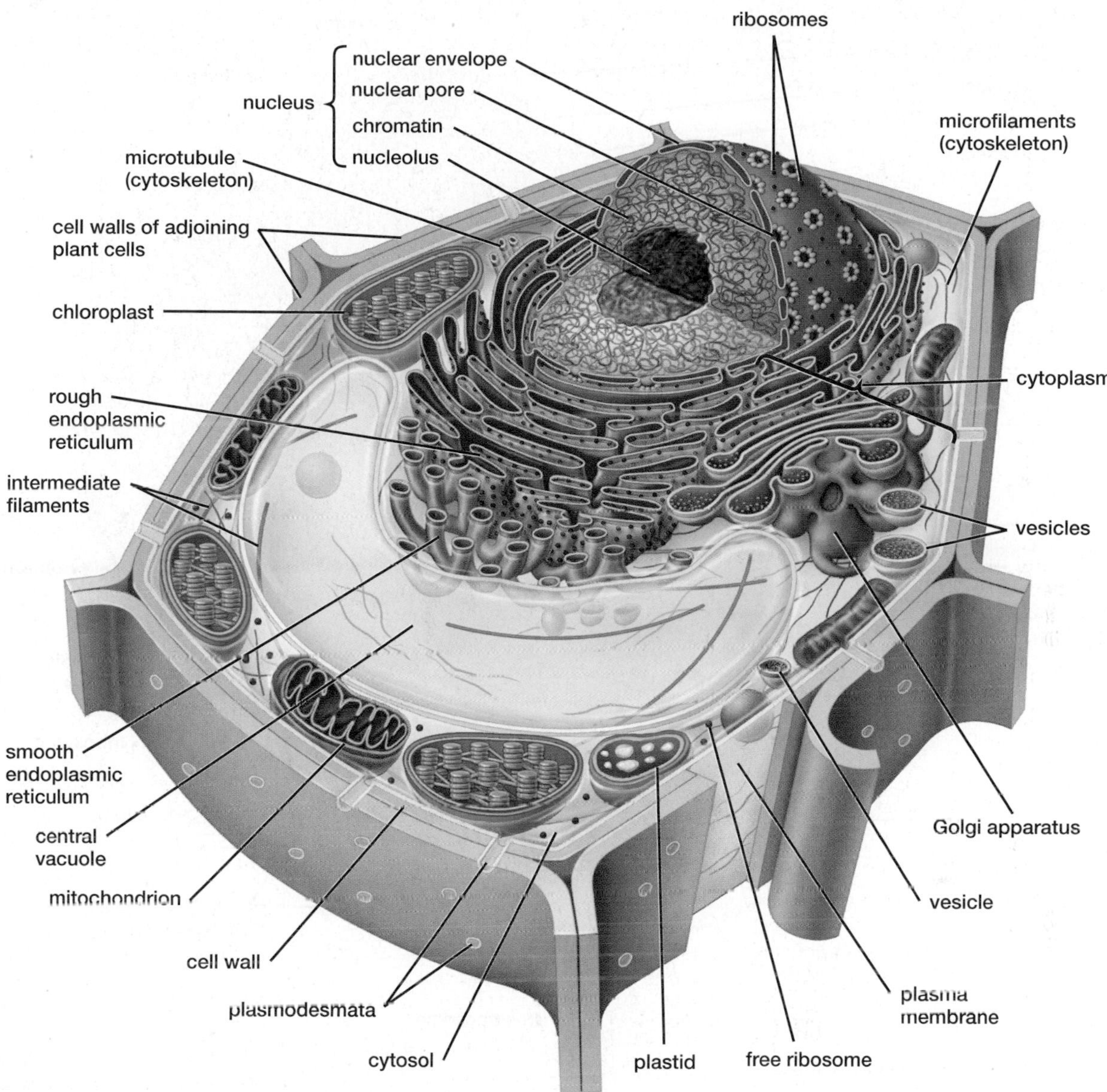

▲ **FIGURE 4-6 A generalized plant cell**

4.5 WHAT ARE THE MAJOR FEATURES OF EUKARYOTIC CELLS?

Eukaryotic cells make up the bodies of organisms in the domain Eukarya: animals, plants, protists, and fungi. As you might imagine, these cells are extremely diverse. The cells that form the bodies of unicellular protists can perform all the activities necessary for independent life. In contrast, each cell in the body of a multicellular organism is specialized to perform a specific function and generally cannot live independently. Here, we focus on plant and animal cells.

Unlike prokaryotic cells, eukaryotic cells contain **organelles** ("little organs"), membrane-enclosed structures specialized for a particular function. Organelles contribute to the complexity of eukaryotic cells. **FIGURE 4-6** illustrates a generalized plant cell, and **FIGURE 4-7** illustrates a generalized animal cell, each with some distinctive structures. Plant cells have cell walls, central vacuoles, and plastids (including chloroplasts), which are absent in animal cells, and animal cells have centrioles, lysosomes, cilia, and flagella, which are not found in the most common plant cells. **TABLE 4-1** summarizes the principal features of plant, animal, and prokaryotic cells.

Extracellular Structures Surround Animal and Plant Cells

The plasma membrane, which is only about two molecules thick and has the consistency of viscous oil, would be torn apart in the absence of reinforcing structures. For animal cells, the reinforcing structure is a complex **extracellular matrix (ECM)**, secreted by the cell. The ECM includes an array of supporting and adhesive proteins embedded in a

▲ **FIGURE 4-7 A generalized animal cell**

gel composed of polysaccharides that are linked together by proteins (**FIG. 4-8**). In addition to structural support, the ECM provides biochemical support functions, such as those provided by proteins called *growth factors*, which promote cell survival and growth. The ECM also binds adjacent cells to one another, transmits molecular signals between cells, and guides cells as they migrate and differentiate (become a particular type of cell) during development. It anchors cells and provides a supporting framework within tissues; for example, a stiff ECM forms the scaffolding for bone and cartilage.

For plant cells, the ECM is the cell wall, which is composed mainly of overlapping cellulose fibers. Plant cell walls protect and support each cell and attach adjacent cells to one another. The walls are porous and allow oxygen, carbon dioxide, and water (containing dissolved substances) to flow through them.

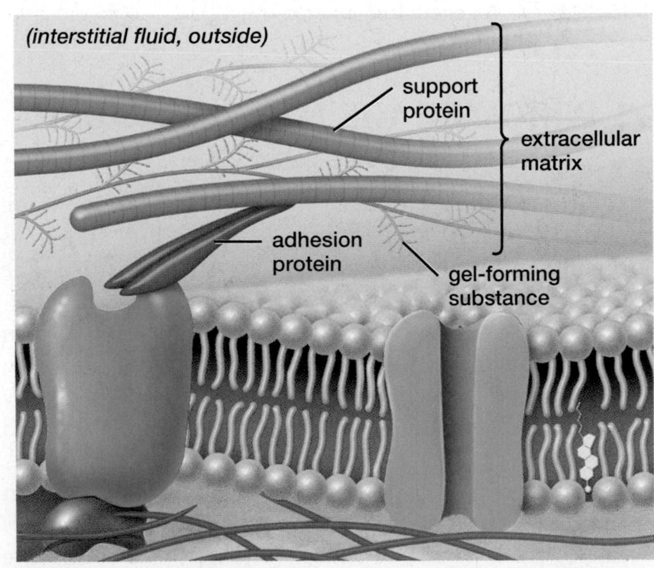

▲ **FIGURE 4-8 The extracellular matrix** Extracellular proteins perform a variety of functions.

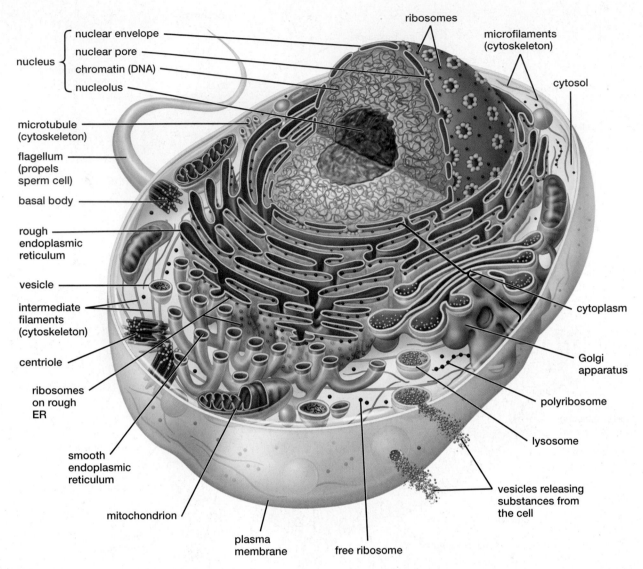

TABLE 4-1 Functions and Distribution of Cell Structures

Structure	Function	Prokaryotes	Eukaryotes: Plants	Eukaryotes: Animals
Cell Surface				
Extracellular matrix	Surrounds cells, providing biochemical and structural support	Absent	Present	Present
Cilia	Move the cell through fluid or move fluid past the cell surface	Absent	Absent (in most)	Present
Flagella	Move the cell through fluid	Present[1]	Absent (in most)	Present
Plasma membrane	Isolates the cell contents from the environment; regulates movement of materials into and out of the cell; allows communication with other cells	Present	Present	Present
Organization of Genetic Material				
Genetic material	Encodes the information needed to construct the cell and to control cellular activity	DNA	DNA	DNA
Chromosomes	Contain and control the use of DNA	Single, circular	Many, linear	Many, linear
Nucleus[2]	Contains chromosomes and nucleoli	Absent	Present	Present
Nuclear envelope	Encloses the nucleus; regulates movement of materials into and out of the nucleus	Absent	Present	Present
Nucleolus	Synthesizes ribosomes	Absent	Present	Present
Cytoplasmic Structures				
Ribosomes	Provide sites for protein synthesis	Present	Present	Present
Mitochondria[2]	Produce energy by aerobic metabolism	Absent	Present	Present
Chloroplasts[2]	Perform photosynthesis	Absent	Present	Absent
Endoplasmic reticulum[2]	Synthesizes membrane components, proteins, and lipids	Absent	Present	Present
Golgi apparatus[2]	Modifies, sorts, and packages proteins and lipids	Absent	Present	Present
Lysosomes[2]	Contain digestive enzymes; digest food and worn-out organelles	Absent	Absent (in most)	Present
Plastids[2]	Store food, pigments	Absent	Present	Absent
Central vacuole[2]	Contains water and wastes; provides turgor pressure to support the cell	Absent	Present	Absent
Other vesicles and vacuoles[2]	Transport secretory products; contain food obtained through phagocytosis	Absent	Present	Present
Cytoskeleton	Gives shape and support to the cell; positions and moves cell parts	Present	Present	Present
Centrioles	Produce the basal bodies of cilia and flagella	Absent	Absent (in most)	Present

[1]Some prokaryotes have structures called flagella, which lack microtubules and move in a fundamentally different way than do eukaryotic flagella.
[2]Indicates organelles, which are surrounded by membranes and found only in eukaryotic cells.

CASE STUDY CONTINUED

New Parts for Human Bodies

In addition to growing hollow organs like Luke Massena's replacement bladder, bioengineers are growing muscles as well. In the past, a major muscle injury could mean amputation and a prosthetic limb, because muscles have limited ability to regenerate, and the scar tissue that forms after an injury interferes with muscle function. But Stephen Badylak and colleagues are developing ways to use the ECM to help muscles heal and even regenerate.

After 28-year-old Marine Ron Strang's quadriceps muscle was almost ripped from his leg by a roadside bomb in Afghanistan, he volunteered for a new bioartificial muscle treatment developed by Badylak. The treatment used ECM from pig bladders. First, all cells were removed from the pig tissue (to ensure that the replacement muscle would not be rejected); only the ECM remained. Badylak's team then cut away scar tissue from Strang's thigh muscle and placed the pig matrix in the resulting cavity. There, the ECM supplied both scaffolding proteins and growth factors that recruited muscle stem cells to the site. The treatment worked a major transformation. After 6 months, the pig matrix had been broken down and replaced by healthy human tissue. Strang went from hobbling to hiking and riding a bike.

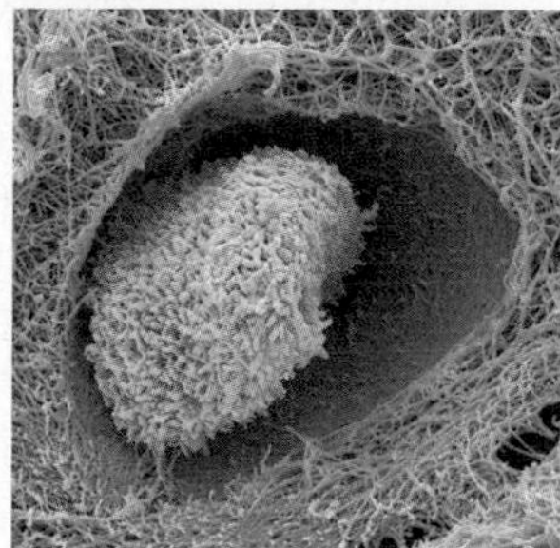

EMC (blue) surrounds a human cartilage cell

Strang's treatment worked in part because the ECM contains growth factors that cause cells to respond in particular ways. Which cell structures direct and enact those responses?

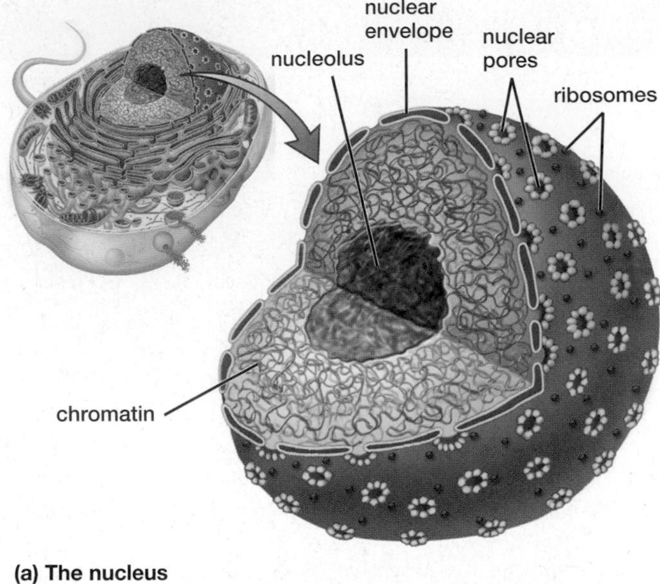

(a) The nucleus

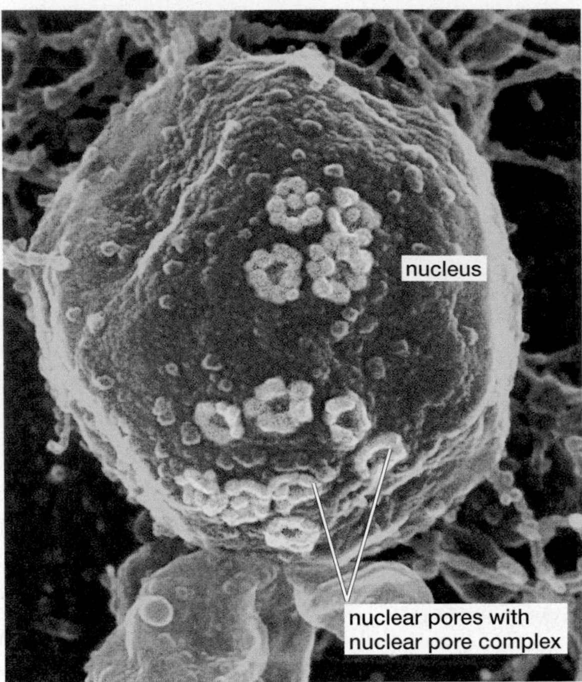

(b) Nucleus of a yeast cell

▲ **FIGURE 4-9 The nucleus** The nucleus is bounded by a double outer membrane perforated by pores. **(b)** SEM of the nucleus of a yeast cell.

The Nucleus Is the Control Center of the Eukaryotic Cell

A eukaryotic cell's DNA is housed within a membrane-enclosed nucleus. The DNA in the nucleus stores all the information needed to construct the cell and direct the countless chemical reactions necessary for life. Only a portion of the instructions in a eukaryotic cell's DNA is used at any given time; the portion used depends on the cell's stage of development, its environment, and its function in a body composed of diverse cell types. The nucleus has three major parts: the nuclear envelope, chromatin, and the nucleolus (**FIG. 4-9**).

The Nuclear Envelope Allows Selective Exchange of Materials

The nucleus is isolated from the rest of the cell by a double membrane, the **nuclear envelope**, which is perforated by protein-lined nuclear pores. Water, ions, and small molecules can pass freely through the pores, but the passage of large molecules—particularly proteins, parts of ribosomes, and RNA—is regulated by gatekeeper proteins that line each pore (see Fig. 4-9). Ribosomes stud the outer nuclear membrane, which is continuous with membranes of the rough endoplasmic reticulum, described later in the chapter.

Chromatin Consists of Strands of DNA Associated with Proteins

Early observers of the nucleus named the nuclear material **chromatin** (meaning "colored substance") because the stains used in light microscopy gave it a dark color. Biologists have since learned that chromatin consists of **chromosomes** (literally, "colored bodies") made of DNA molecules and their associated proteins. When a cell is not dividing, its chromosomes take the form of extremely long strands that are so thin they cannot be distinguished from one another when viewed with a light microscope. During cell division, the chromosomes become condensed and are easily visible with a light microscope (**FIG. 4-10**).

The DNA in chromosomes includes genes—sequences of nucleotides that provide instructions for the synthesis of proteins and ribosomes. Because DNA is confined to the nucleus but proteins are synthesized in the cytoplasm, copies of the genetic code for proteins must be ferried from the nucleus into the cytoplasm. To accomplish this, the genetic information is transcribed from DNA into molecules of *messenger RNA*

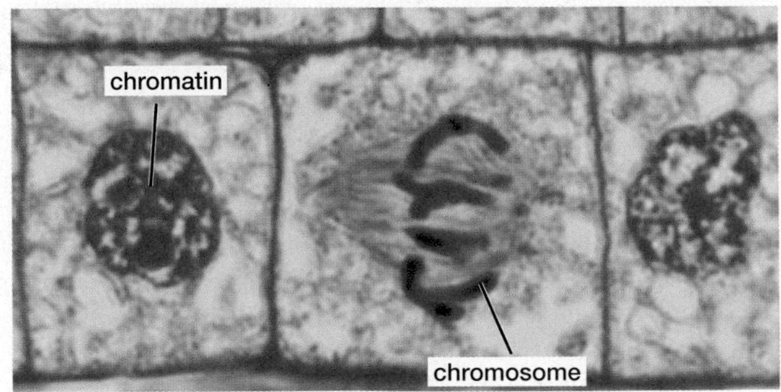

▲ **FIGURE 4-10 Chromosomes** Chromosomes, seen in a light micrograph of a dividing cell (center) in an onion root tip. Chromatin is visible in adjacent cells.

THINK CRITICALLY Why do the chromosomes in chromatin condense in dividing cells?

(mRNA). The mRNA then moves through the nuclear pores into the cytosol. In the cytosol, the sequence of nucleotides in mRNA is used to direct protein synthesis, which occurs on ribosomes. (Protein synthesis is described in Chapter 13.)

The Nucleolus Is the Site of Ribosome Assembly

A nucleus contains at least one **nucleolus** (plural, **nucleoli**; see Fig. 4-9), which is the site of ribosome synthesis. A nucleolus consists of proteins, *ribosomal RNA* (rRNA), parts of chromosomes that carry genes coding for rRNA, and ribosomes in various stages of construction.

Ribosomes, which are composed of rRNA and proteins, act as workbenches for the synthesis of proteins. Just as a carpenter's workbench can be used to construct many different objects, a ribosome can be used to synthesize a multitude of different proteins (depending on the sequence of the mRNA to which it attaches). In electron micrographs of cells, ribosomes appear as dark granules. They may be dispersed singly in the cytoplasm, or they may be embedded in the nuclear envelope and endoplasmic reticulum. Some are strung along strands of mRNA in the cytoplasm, forming *polyribosomes* (**FIG. 4-11**).

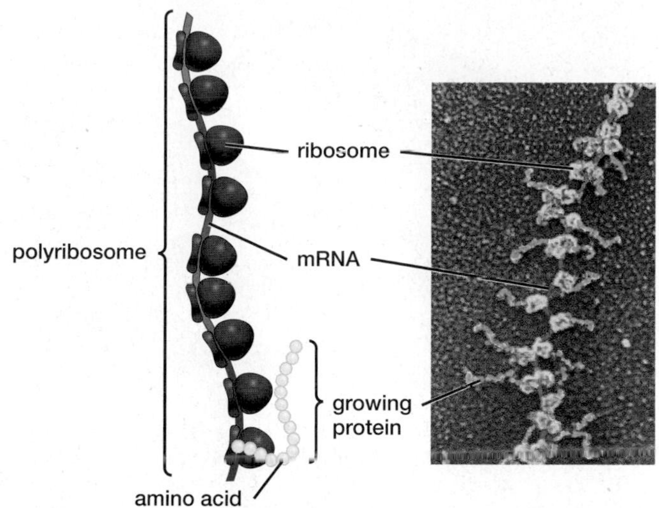

▲ **FIGURE 4-11 A polyribosome** Ribosomes strung along a messenger RNA molecule form a polyribosome. In the TEM (right), individual ribosomes are synthesizing multiple copies of a protein, visible as strands projecting from some of the ribosomes.

Mitochondria Extract Energy from Food Molecules, and Chloroplasts Capture Solar Energy

Cells need energy to survive, grow, and reproduce. Organelles called mitochondria play a key role in making energy available to meet a cell's needs. In eukaryotes that can harvest energy from sunlight, the chemical reactions that do the job take place in chloroplasts.

Mitochondria Use Energy Stored in Food Molecules to Produce ATP

All eukaryotic cells contain **mitochondria** (singular, **mitochondrion**), organelles that extract energy from food molecules and store it in the high-energy bonds of ATP. Because of their role in making energy available, mitochondria are sometimes called the powerhouses of the cell.

A mitochondrion is surrounded by a double membrane (**FIG. 4-12**). The outer membrane is smooth, whereas the inner membrane forms deep folds called cristae (singular, *crista*; meaning "crest"). The mitochondrial membranes enclose

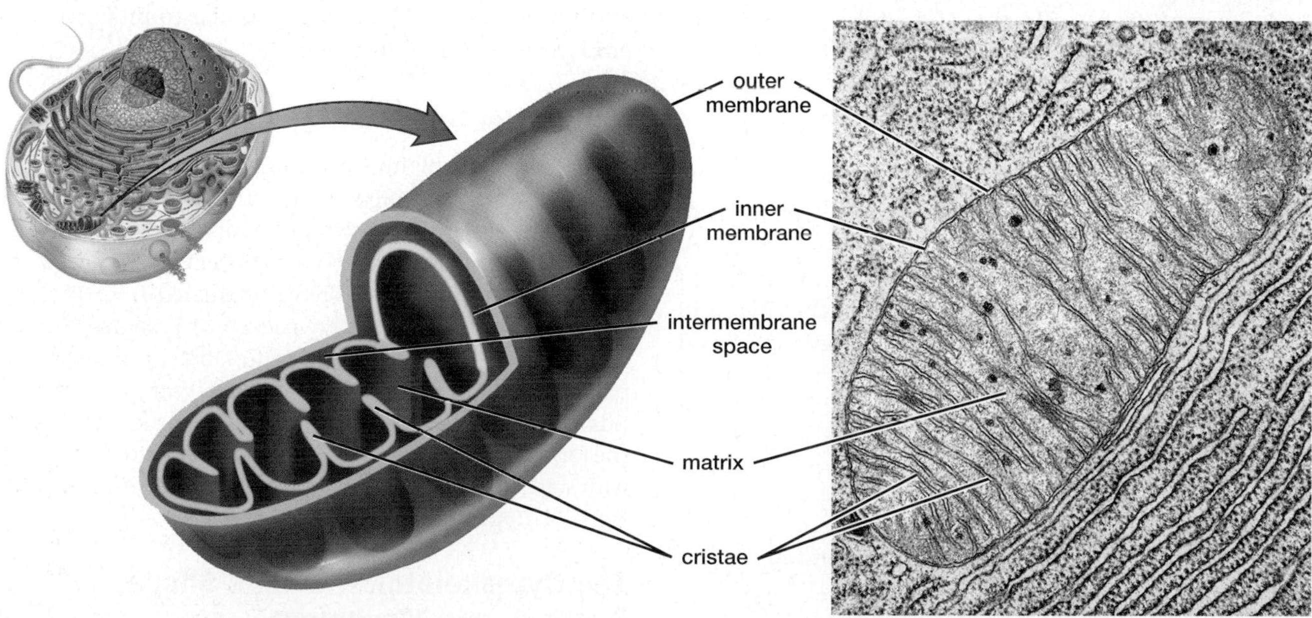

▲ **FIGURE 4-12 A mitochondrion** (Left) Mitochondrial membranes enclose two fluid compartments. The inner membrane forms deep folds called cristae. (Right) A TEM showing mitochondrial structures.

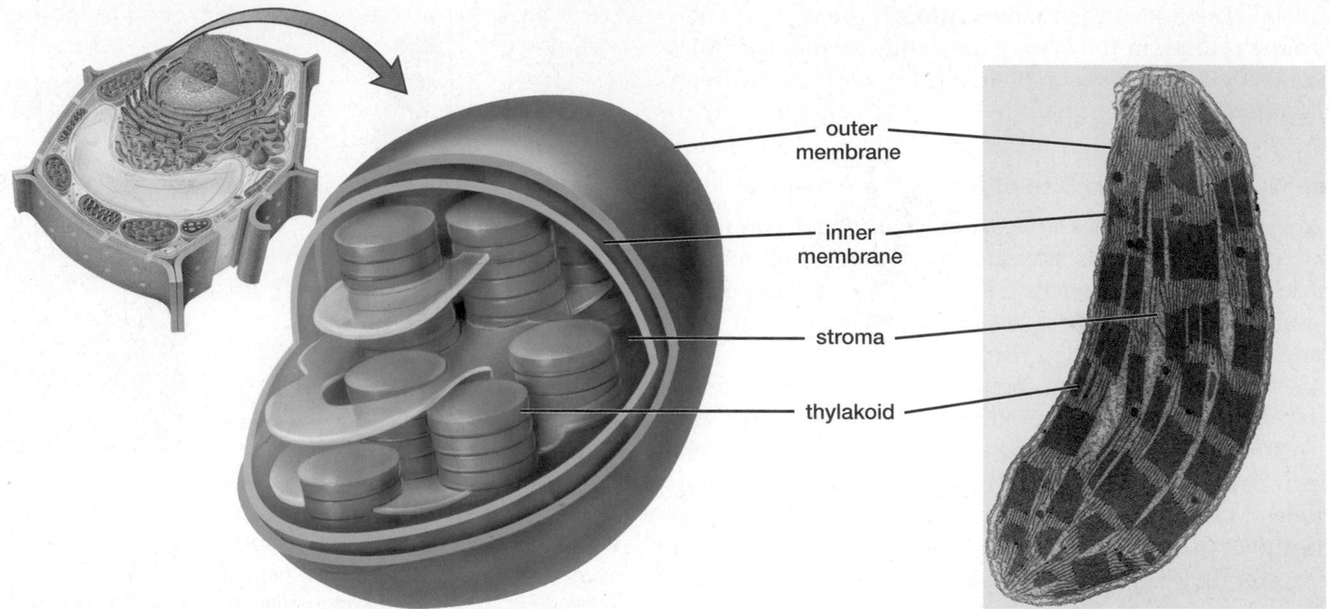

▲ **FIGURE 4-13 The chloroplast is a complex plastid** (Left) A chloroplast is surrounded by a double membrane that encloses the fluid stroma. (Right) A TEM showing chloroplast structures.

two fluid-filled spaces: (1) the *intermembrane compartment* lies between the two membranes, and (2) the *matrix* fills the space within the inner membrane. Some of the reactions that break down food molecules occur in the fluid of the matrix; the rest are conducted by enzymes attached to the membranes of the cristae. (The role of mitochondria in energy production is described in Chapter 8.)

Chloroplasts Are the Sites of Photosynthesis

The cells of plants and some protists contain **chloroplasts**, organelles in which photosynthesis occurs. Photosynthesis captures energy from sunlight and stores it in sugar molecules. Like a mitochondrion, a chloroplast is surrounded by a double membrane (**FIG. 4-13**). The inner membrane of the chloroplast encloses a fluid called the *stroma*. Within the stroma are interconnected stacks of hollow, membranous sacs. An individual sac is called a *thylakoid*.

The thylakoid membranes contain the pigment molecule **chlorophyll** (which gives plants their green color). During photosynthesis, chlorophyll captures the energy of sunlight and transfers it to other molecules in the thylakoid membranes. These molecules transfer the energy to ATP and other energy carriers. The energy carriers diffuse into the stroma, where their energy is used to drive the synthesis of sugar from carbon dioxide and water. The sugar stores energy that powers nearly all life.

Mitochondria and Chloroplasts Are Descended from Prokaryotes

Mitochondria and chloroplasts are descendants of ancient prokaryotic species that once lived independently. According to the widely accepted **endosymbiont hypothesis** (see Chapter 18), both mitochondria and chloroplasts evolved from bacteria. Roughly 1.7 billion years ago, these bacteria took up residence within other prokaryotic cells, a process called endosymbiosis (Gk. *symbiosis*, "living together"). This history is reflected in the double membranes of mitochondria and chloroplasts; the outer membrane came from the original host cell, and the inner membrane from the guest cell. The prokaryotic heritage of these organelles is revealed by several shared characteristics. Mitochondria and chloroplasts are roughly the same size as bacteria (1 to 5 micrometers in diameter), and both contain enzymes that synthesize ATP (as would have been needed by an independent cell). In addition, both mitochondria and chloroplasts contain their own DNA and ribosomes, which are more similar to prokaryotic than eukaryotic DNA and ribosomes.

Plants Use Plastids for Storage

Chloroplasts are highly specialized **plastids**, organelles surrounded by double membranes and used for the synthesis and/or storage of pigments or food molecules. Plastids are found only in plants and photosynthetic protists (**FIG. 4-14**); all plastids are thought to have originated from prokaryotic cells. Some storage plastids are packed with pigments that give ripe fruits or flower petals their yellow, orange, or red colors. In plants that continue growing from one year to the next, plastids store food produced during the growing season, usually in the form of starch granules. For example, potato cells are stuffed with starch-filled plastids, food for the next spring's growth (see Fig. 4-14, upper right).

The Cytoskeleton Provides Shape, Support, and Movement

The cytoskeleton is a dynamic network of protein fibers within the cytoplasm (**FIG. 4-15**). Cytoskeletal proteins come in three major types: thin **microfilaments** (composed of

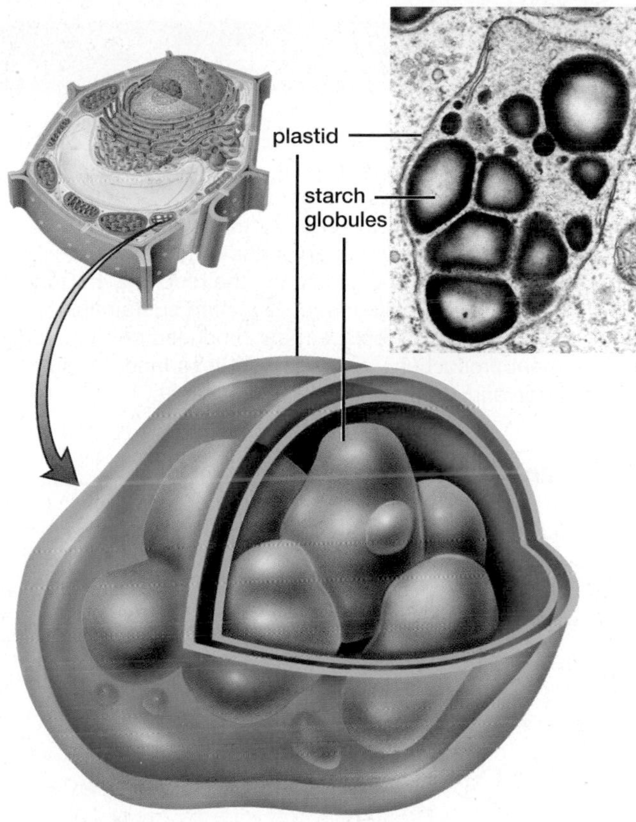

▲ **FIGURE 4-14 A simple storage plastid** Plastids are surrounded by a double outer membrane. This potato plastid stores starch, also seen in the upper right TEM.

actin protein), medium-sized **intermediate filaments** (composed of various proteins), and thick **microtubules** (composed of tubulin protein). Cytoskeletal proteins contribute to several important functions:

- ***Support and Connection*** In animal cells, a scaffolding of intermediate filaments supports the cell and links cells to one another and to the ECM. An array of microfilaments concentrated just inside the plasma membrane provides additional support and also connects with the ECM.
- ***Shape*** Cytoskeletal proteins help maintain a cell's shape. They can also use energy from ATP to alter the shape, either by changing protein length (by adding or removing subunits) or by sliding past one another.
- ***Movement*** Many animal cells can move as microtubules and microfilaments extend by adding subunits at one end and releasing subunits at the other end. Cytoskeletal proteins may be associated with *motor proteins*, which use energy from ATP to generate movement. For example, motor proteins transport organelles within the cell, using microfilaments and microtubules as "railroad tracks." Motor proteins can also cause actin microfilaments to slide past one another, as occurs during the contraction of muscle cells. Muscle cells must contract to increase in size, as scientists discovered when they attempted to grow muscle protein in the laboratory for possible human consumption. Learn more in "Earth Watch: Would You Like Fries with Your Cultured Cow Cells?" on page 64.
- ***Cell Division*** During cell division, microtubules guide chromosome movements; in animal cells, microfilaments pinch the dividing cell into two daughter cells. (Cell division is covered in Chapter 9.)

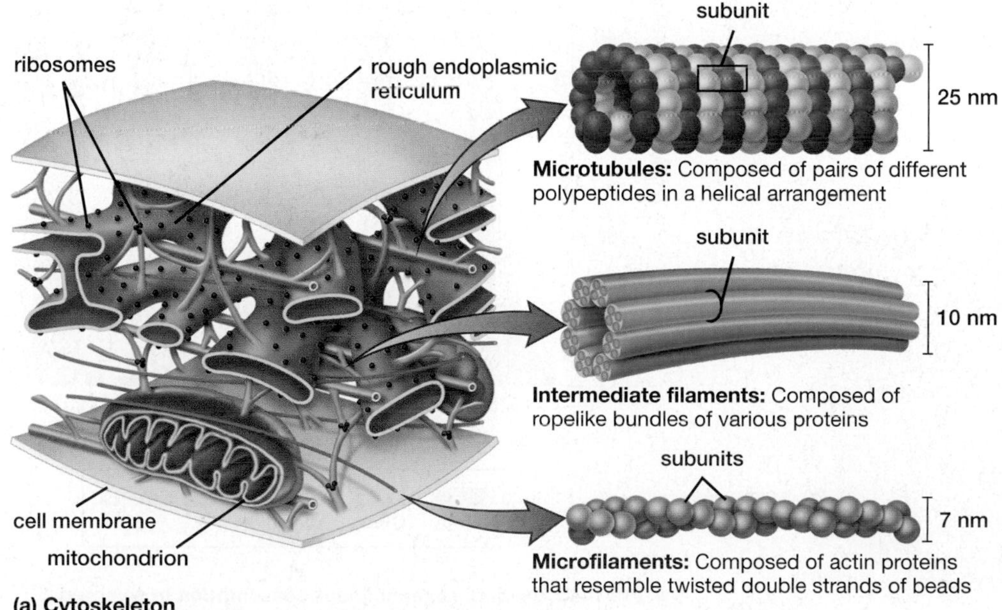

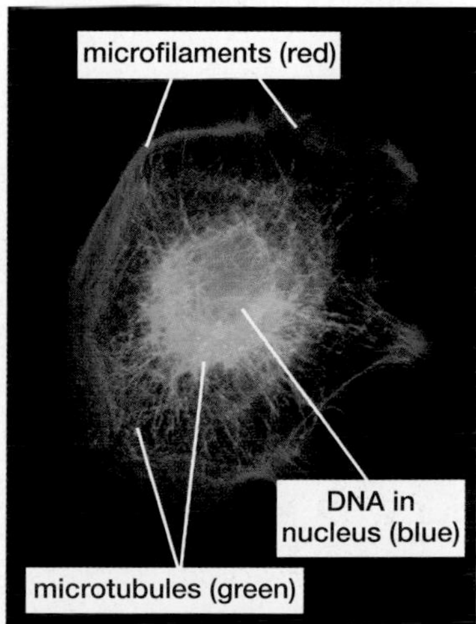

(b) Light micrograph showing the cytoskeleton

▲ **FIGURE 4-15 The eukaryotic cytoskeleton (a)** Three types of protein strands form the cytoskeleton. **(b)** In this light micrograph, cells treated with fluorescent stains reveal microtubules, microfilaments, and nuclei.

Earth WATCH Would You Like Fries with Your Cultured Cow Cells?

What do you get when you combine 20,000 paper-thin yellowish-pink strips of cow muscle stem cells, some lab-grown fat cells, beet coloring, egg powder, bread crumbs, and a dash of salt? These are the unlikely ingredients of the world's first lab-grown hamburger (**FIG. E4-1**). To create it, the stem cells were allowed to multiply in a nutrient broth and then seeded onto strips of gel and stimulated repeatedly by pulses of electricity. In response to the stimulation, the cells' actin-based filaments contracted repeatedly, causing the cells to grow and "bulk up," much as human muscle cells do when exercised. Finally, the other ingredients were added. The resulting artificial burger cost roughly $425,000 to produce, and its flavor was found to be somewhat lacking by fast-food aficionados. Another group of food scientists has grown chicken meat in the lab that reportedly tastes pretty good and can be produced for "only" $9,000 per pound. Still, lab-grown meat remains scarce and super-expensive, so what's the point of producing it?

Worldwide demand for meat is growing (**FIG. E4-2**). The increased demand is fueled partly by an expanding population, but also by increasing incomes and appetite for meat in developing countries, especially China. The Food and Agriculture Organization of the United Nations estimates that world meat production in 2050 will be 500 million tons (compared to about 360 million tons in 2018).

To accommodate our increasing demand for meat, we are burdening Earth's natural ecosystems and altering its climate. Grazing and growing food for livestock already require about 30% of Earth's total land (compared to about 6% used for growing crops directly for human consumption), and meat production accounts for roughly 18% of human-caused greenhouse gas emissions. Cattle have by far the greatest environmental impact among meat-producing livestock; raising beef cattle requires about three times as much land per pound of protein as does raising chicken or pork. Increasing beef production occurs primarily at the expense of rain forest, which is cleared to provide low-quality land for cattle grazing.

Clearly, current methods of meat production are unsustainable in light of the steadily increasing demand for meat. Scientists growing "test-tube beef" argue that if their techniques could be refined and scaled up, the resulting environmental benefits would make meat production sustainable. If lab-made meat were to replace meat produced on farms and ranches, meat production would use 99% less land, consume far less water, and emit fewer greenhouse gases.

THINK CRITICALLY Based on your analysis of the graph in Figure E4-2, in which year was per-person meat consumption in China roughly half that in the United States? By roughly how many pounds did total worldwide meat consumption increase between 1973 and 2013? (Note: In 1973, Earth's human population was about 3.92 billion, and in 2013 it was about 7.21 billion.)

▲ **FIGURE E4-1 Will hamburger of the future be grown in a lab?**

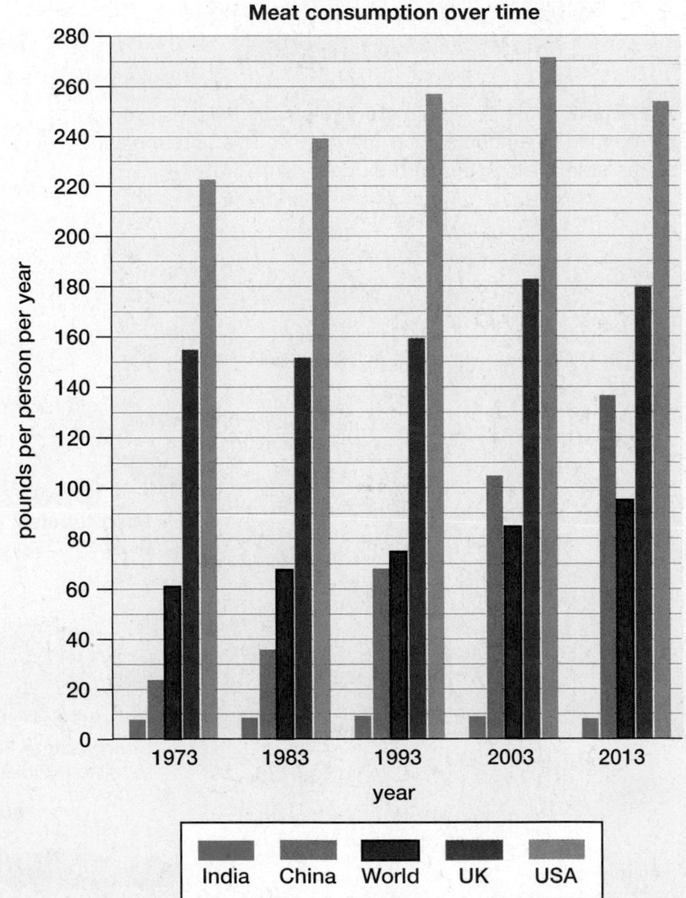

▲ **FIGURE E4-2 Changes in meat consumption in selected countries**

Source: FAOSTAT (Food and Agricultural Organization of the United Nations), Food Supply

Eukaryotic Cytoplasm Contains Membranes That Compartmentalize the Cell

All eukaryotic cells contain internal membranes that create loosely connected compartments within the cytoplasm. These membranes, collectively called the **endomembrane system**, segregate molecules from the surrounding cytosol and ensure that biochemical processes occur in an orderly fashion. The endomembrane system encloses regions within which an enormous variety of molecules are synthesized, broken down, and transported for use inside the cell or export outside the cell. This system of intracellular membranes includes the nuclear envelope (described earlier), vesicles, the endoplasmic reticulum, the Golgi apparatus, and lysosomes.

Vesicles Bud from the Endomembrane System and the Plasma Membrane

Vesicles are temporary sacs that ferry biological molecules throughout the cell. They bud from the endomembrane system or the plasma membrane. Vesicles that bud from the plasma membrane are formed by a process called *endocytosis* (Gk. *endo*, "inside"), in which the membrane extends to surround material just outside the cell and then fuses and pinches off to form a vesicle inside the cell. At its destination, a vesicle fuses with an endomembrane compartment to release its contents into the compartment, or it fuses with the plasma membrane to release its contents outside the cell, a process called *exocytosis* (Gk. *exo*, "outside").

Vesicles are transported within the cell by motor proteins running along tracks of microtubules. How do the vesicles know where to go? Proteins embedded in vesicle membranes contain specific sequences of amino acids that serve as "mailing labels," providing the address for delivery of the vesicle and its payload. Membrane proteins and proteins exported from the cell are synthesized in the rough endoplasmic reticulum.

The Endoplasmic Reticulum Forms Membrane-Enclosed Channels Within the Cytoplasm

The **endoplasmic reticulum (ER)** is a labyrinth of tubular membrane that forms interconnected sacs and channels throughout the cytosol (endoplasmic, "inside the cytoplasm," and reticulum, "network"). The ER typically makes up at least 50% of the total cellular membrane (**FIG. 4-16**). This organelle plays a major role in synthesizing, modifying, and transporting biological molecules throughout the cell. The ER also produces new membranes that are distributed throughout the endomembrane system. The ER has both rough and smooth membranes, which are continuous with one another.

Rough Endoplasmic Reticulum

Rough ER extends from the outer nuclear membrane (see Fig. 4-7), and ribosomes studding its outer surface make it appear rough under the electron microscope. These ribosomes are important sites of protein synthesis. As proteins are synthesized on ER ribosomes, some are inserted into the ER membrane. These proteins may remain in the ER membrane or become part of vesicle membranes that bud from the ER. Other proteins synthesized on ER ribosomes are destined to be secreted from the cell or used within the cell; these proteins move to the interior of the ER, where they are chemically modified and folded. Eventually, these proteins accumulate in pockets of ER membrane that pinch off as vesicles and travel to the Golgi apparatus. Proteins produced by the rough ER for export differ with cell type; they include enzymes, infection-fighting antibodies, and proteins that form the extracellular matrix. Proteins that remain in the cell include the digestive enzymes within lysosomes (described later in the chapter) and plasma membrane proteins.

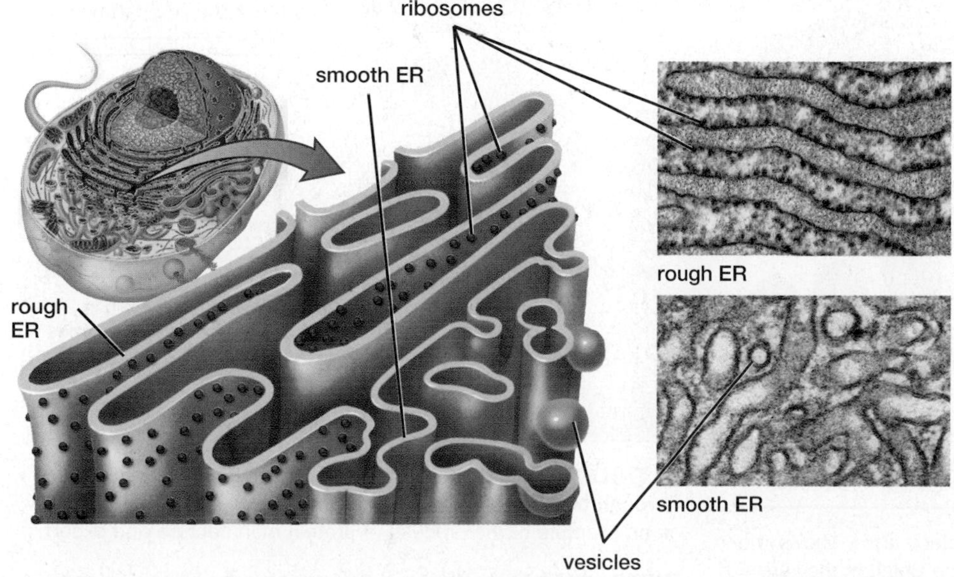

(a) Endoplasmic reticulum may be rough or smooth **(b) Rough and smooth ER**

FIGURE 4-16 Endoplasmic reticulum **(a)** Ribosomes (black dots) stud the outside of the rough ER membrane. Rough ER is continuous with the outer nuclear envelope. Smooth ER is less flattened and more cylindrical than rough ER and may be continuous with rough ER. **(b)** TEMs of rough and smooth ER.

Smooth Endoplasmic Reticulum

Smooth ER, which lacks ribosomes, functions in the synthesis of cell membrane phospholipids. It also has other specialized functions in certain cell types. For example, smooth ER is abundant in the cells of vertebrate reproductive organs that synthesize steroid sex hormones. In muscle cells, smooth ER is specialized to store calcium ions, which play a central role in muscle contraction. Membranes of smooth ER in liver cells contain a variety of embedded enzymes. Some of these enzymes help convert stored glycogen into glucose to provide energy. Others promote synthesis of the lipid portion of lipoproteins. Still others break down metabolic wastes such as ammonia, drugs such as alcohol, and poisons such as certain pesticides.

The Golgi Apparatus Modifies, Sorts, and Packages Important Molecules

Named for the Italian physician and cell biologist Camillo Golgi, who discovered it in 1898, the **Golgi apparatus** (or simply **Golgi**) is a specialized set of membranes resembling a stack of flattened and interconnected sacs (**FIG. 4-17**). The compartments of the Golgi are like the finishing rooms of a factory, where final touches are added to products in preparation for packaging and export. At the receiving side of the Golgi apparatus, vesicles arrive from the rough ER and fuse with the Golgi membrane, emptying their contents into the Golgi compartments. Within the Golgi compartments, proteins delivered from the rough ER may be further modified. Finally, vesicles bud off from the "shipping" side of the Golgi, carrying away finished products for use in the cell or export out of the cell.

The Endomembrane System Synthesizes, Modifies, and Transports Proteins to Be Secreted

To understand how some of the components of the endomembrane system work together, let's look at the manufacture and export of antibodies (**FIG. 4-18**). Antibodies, produced by white blood cells, bind to foreign invaders (such as disease-causing bacteria) and help destroy them. Antibody proteins are synthesized on ribosomes of the rough ER and released into the ER channels ❶, where they are packaged

Protein-carrying vesicles from the ER merge with the Golgi apparatus.

Golgi apparatus

Vesicles carrying modified protein leave the Golgi apparatus.

▲ **FIGURE 4-17 The Golgi apparatus** The black arrow shows the direction of movement of materials through the Golgi as they are modified and sorted. Vesicles bud from the face of the Golgi opposite the ER.

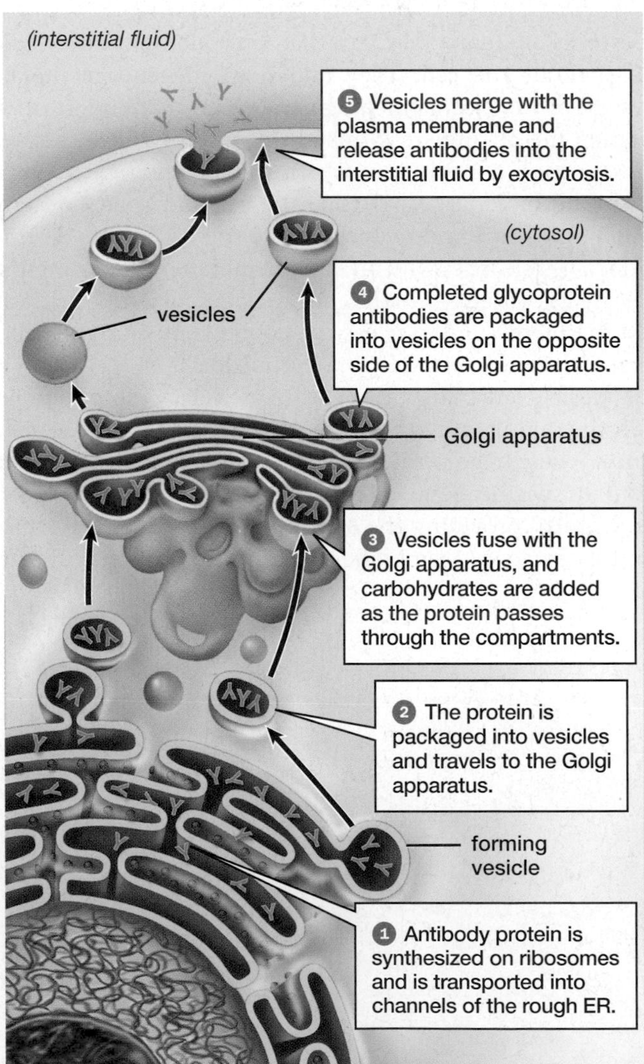

(interstitial fluid)
5 Vesicles merge with the plasma membrane and release antibodies into the interstitial fluid by exocytosis.
(cytosol)
vesicles
4 Completed glycoprotein antibodies are packaged into vesicles on the opposite side of the Golgi apparatus.
Golgi apparatus
3 Vesicles fuse with the Golgi apparatus, and carbohydrates are added as the protein passes through the compartments.
2 The protein is packaged into vesicles and travels to the Golgi apparatus.
forming vesicle
1 Antibody protein is synthesized on ribosomes and is transported into channels of the rough ER.

▲ **FIGURE 4-18 A protein is manufactured and exported through the endomembrane system** The formation of an antibody is an example of the process of protein manufacture and export.

THINK CRITICALLY Why is it advantageous for all cellular membranes to have a fundamentally similar composition?

into vesicles formed from ER membrane. These vesicles travel to the Golgi ❷, where their membranes fuse with the Golgi membranes and release the antibodies inside. Within the Golgi, carbohydrates are attached to the antibodies (transforming them into glycoproteins) ❸, which are then repackaged into vesicles formed from Golgi membrane ❹. The vesicle containing the completed antibodies travels to the plasma membrane and fuses with it, releasing the antibodies outside the cell by exocytosis ❺. From there, they will make their way into the bloodstream to help defend the body against infection.

Lysosomes Serve as the Cell's Digestive System

Lysosomes are membrane-bound sacs that digest food particles ranging from individual proteins to microorganisms such as bacteria (**FIG. 4-19**). Lysosomes contain dozens of different enzymes. These enzymes use hydrolysis to break down almost all large biological molecules, including carbohydrates, lipids, proteins, and nucleic acids. The enzymes of lysosomes require an acidic environment to function effectively, so they are almost nonfunctional in the ER compartment where they are manufactured, where the pH is about 7.2 ❶. Lysosomal enzymes are transported to the Golgi in vesicles that bud from the ER ❷. In the Golgi, a carbohydrate "mailing label" is added to the enzymes ❸; this tag directs them into specific Golgi vesicles that will travel to lysosomes ❹. Lysosomal membranes expend energy to pump hydrogen ions inside, creating an acidic environment (about pH 5) that allows the enzymes to perform optimally.

Many cells of animals and protists "eat" by endocytosis—that is, by engulfing particles from just outside the cell ❺. The plasma membrane with its enclosed food then pinches off inside the cytosol and forms a large vesicle called a **food vacuole**. Lysosomes merge with these food vacuoles ❻, and the lysosomal enzymes digest the food into small molecules such as monosaccharides, fatty acids, and amino acids. Lysosomes also digest worn-out or defective organelles within the cell, breaking them down into their component molecules. All of these small molecules are released into the cytosol through the lysosomal membrane, where they are used in the cell's metabolic processes.

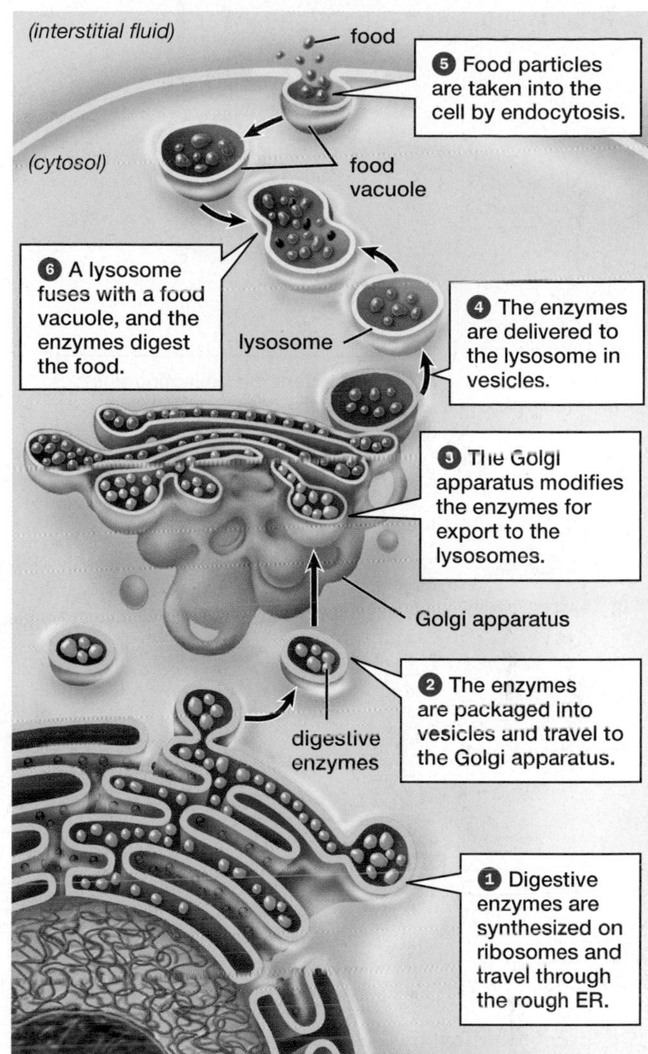

▲ **FIGURE 4-19 Lysosomes and food vacuoles are formed by the endomembrane system**

THINK CRITICALLY Why is it important for lysosomal enzymes to be inactive at pH 7.2?

Vacuoles Serve Many Functions, Including Water Regulation, Storage, and Support

Some types of vacuoles, such as food vacuoles, are temporary structures. Other vacuoles, however, persist for the lifetime of a cell. In the following sections, we describe the permanent vacuoles found in some freshwater protists and in plant cells.

Freshwater Protists Have Contractile Vacuoles

Freshwater protists such as *Amoeba* and *Paramecium* possess **contractile vacuoles** composed of collecting ducts, a central reservoir, and a tube leading to a pore in the plasma membrane (**FIG. 4-20**). Fresh water constantly leaks into the cell through the plasma membrane and then into contractile vacuoles. Why doesn't this continuous influx of water eventually burst the fragile organism? *Paramecium* expends cellular energy to draw water from its cytosol into collecting ducts. The water then drains into the central reservoir. When the reservoir is full, the contractile vacuole contracts, squirting the water out through a pore in the plasma membrane.

Plant Cells Have Central Vacuoles

A large **central vacuole** occupies three-quarters or more of the volume of most mature plant cells (see Fig. 4-6) and serves several functions. It stores wastes and toxins extracted from the cytosol; in some plants, the stored toxins deter animals from munching on the plant's leaves. Vacuoles may also store

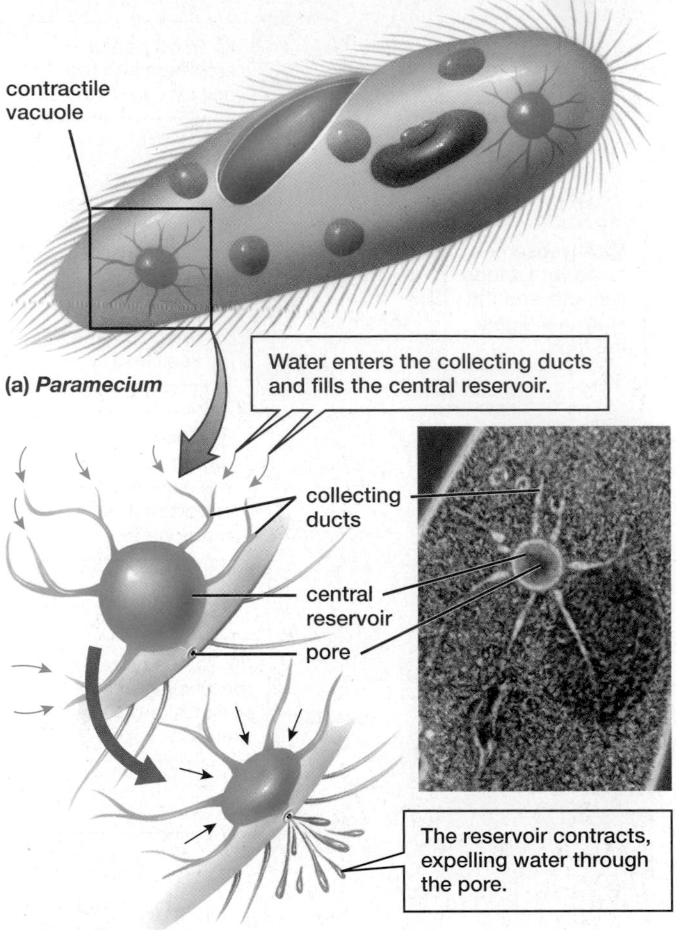
contractile
vacuole
(a) Paramecium
Water enters the collecting ducts and fills the central reservoir.
collecting ducts
central reservoir
pore
The reservoir contracts, expelling water through the pore.
(b) Contractile vacuole

▲ **FIGURE 4-20 A contractile vacuole (a)** *Paramecium* lives in freshwater ponds and lakes. **(b)** (Left) The vacuole collects and expels water. (Right) The contractile vacuole seen under a light microscope using fluorescent dyes.

sugars and amino acids that are not immediately needed by the cell. In addition, the central vacuole's membrane helps regulate the ion content of the cytosol. Pigments stored in central vacuoles are responsible for the colors of many flowers.

Central vacuoles also provide support for plant cells. Dissolved substances cause water to move by osmosis into the vacuole. The resulting water pressure, called *turgor pressure,* within the vacuole pushes the fluid portion of the cytoplasm up against the cell wall with considerable force. Cell walls are usually somewhat flexible, so both the overall shape and the rigidity of the cell depend on turgor pressure within the cell. Turgor pressure thus provides support for the non-woody parts of plants.

Cilia and Flagella May Move Cells Through Fluid or Move Fluid Past Cells

Both **cilia** (singular, **cilium**; "eyelash") and flagella are beating structures, covered by plasma membrane, that extend outward from some cell surfaces. (Here we describe the cilia and flagella of eukaryotes. The structure of prokaryotic flagella differs from that of eukaryotic flagella; cilia are not present in prokaryotes.) In eukaryotes, cilia and flagella are supported and moved by microtubules of the cytoskeleton. Each cilium or flagellum contains a ring of nine fused pairs of microtubules surrounding an unfused pair (**FIG. 4-21**). Cilia and flagella beat almost continuously, powered by motor proteins that extend like tiny arms and attach neighboring pairs of microtubules (see Fig. 4-21a). These sidearms use ATP energy to slide the microtubules past one another, causing the cilium or flagellum to bend.

In general, cilia are shorter and more numerous than flagella. Cilia beat in unison to produce a force on the surrounding fluid that is similar to that created by oars on a rowboat. A flagellum, in contrast, rotates in a corkscrew motion that propels a cell through fluid, acting somewhat like the propeller on a motorboat. Cells with flagella usually have only one or two of them.

Many small protists, such as *Paramecium,* use cilia or flagella to swim through water. Animals, in contrast, use cilia to move fluids over cell surfaces. Ciliated cells line such diverse structures as the gills of oysters (where they circulate water rich in food and oxygen), the female reproductive tract of vertebrates (where cilia transport egg cells to the uterus), and the respiratory tracts of most land vertebrates (where cilia convey mucus that carries debris and microorganisms out of the air passages; see Fig. 4-21b). Flagella propel the sperm cells of nearly all animals (see Fig. 4-21c).

Have You Ever Wondered ...

How Many Cells Form The Human Body?

How many cells are in your body? According to the best estimate of scientists, about 30 trillion. The 30 trillion human cells in the body are joined by a roughly equal (or perhaps slightly larger) number of prokaryotic cells residing in a community called the microbiome. Each of us is host to a community consisting of about a half pound (200 grams) of bacteria and archaea. These organisms colonize the nose, skin, vagina, and digestive tract from mouth to anus, where they perform a host of functions essential to our health. (You may read elsewhere that our bodies contain 10 or more times as many prokaryotic cells as human cells, but this older estimate is not consistent with the latest data about our microbiome.)

CHECK YOUR LEARNING

Can you . . .

- define organelles?
- list the structures found in animal but not plant cells, and vice versa?
- describe the structure and function of each major structure found in eukaryotic cells?

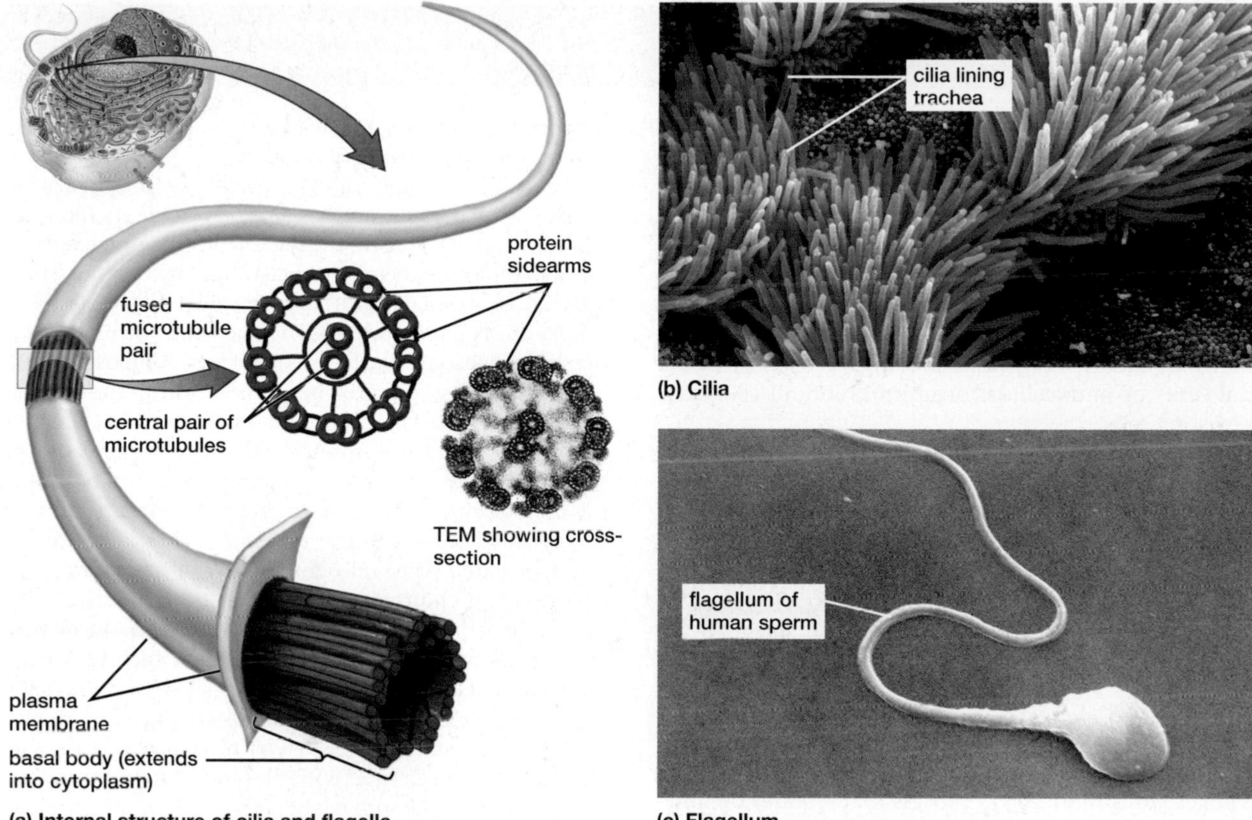

▲ **FIGURE 4-21 Cilia and flagella (a)** These structures are filled with microtubules produced by the basal body. **(b)** Cilia, shown in this SEM, line the trachea and sweep out debris. **(c)** A human sperm cell, shown in this SEM, uses its flagellum to swim to the egg.

THINK CRITICALLY What problems would arise if the trachea were lined with flagella instead of cilia?

CASE STUDY REVISITED

New Parts for Human Bodies

Recent progress in understanding how cells interact with the extracellular matrix has set the stage for constructing complex bioengineered organs. Scientists now recognize that the ECM is in a dynamic partnership with the cells that secrete it; different cell types secrete distinctive matrices that support their own specific needs. The matrix of an organ, even when isolated from its cells, retains molecular cues that attract appropriate stem cells and stimulate them to differentiate into functioning cells typical of the original organ. This phenomenon has generated hope that the ECM of organs from animals such as pigs might serve as scaffolds for growing human organs. In this scenario, an animal organ is stripped of cells, leaving the matrix intact, to be seeded with stem cells taken from the person needing a transplant.

Scientists must overcome a number of challenges before they can start cranking out artificial organs for transplanting. For example, to produce a complex organ such as a liver, researchers must induce stem cells to produce not only the cells of the organ, but also an extensive network of blood vessels to keep the organ alive and functioning. In a recent effort to help solve this problem, researchers first flushed detergent solution through the natural blood supply of animal livers. This procedure washed away all the liver and blood vessel cells, but left the three-dimensional channels of the blood vessel network intact within the ECM. The researchers then injected suspensions of immature human liver and blood vessel cells into the entry vessel, which conducted the cells throughout the vessel network. After a week in a nutrient-filled bioreactor, the human blood vessel cells had formed a lining within the blood vessel channels, and the liver cells had multiplied within the rest of the scaffold. The resulting 1-inch-diameter liver-like tissues represent only one step along a long path to fully functional artificial livers, but the rapid progress of bioengineering research raises hope that bioartificial body parts will help people live longer, healthier lives in the coming decades.

CONSIDER THIS What advantages do bioengineered organs have over transplants from human donors? If you had a failing organ and a human trial of an experimental bioengineered organ was just starting, would you volunteer to be a recipient? Why or why not?

CHAPTER REVIEW

Go to **Mastering Biology** to access the Pearson eText, vocabulary review, practice quizzes, activities, videos, current events, and more.

*Answers to **Think Critically** and **Thinking Through the Concepts** questions can be found in the **Answers** section at the back of the book.*

Summary of Key Concepts

4.1 What Is the Cell Theory?

The cell theory states that every living organism consists of one or more cells, the smallest organisms are single cells, cells are the functional units of multicellular organisms, and all cells arise from preexisting cells.

4.2 How Do Scientists Visualize Cells?

Scientist were unaware of the existence of cells until the invention of the microscope in the 1700s. Today, scientists view cells through light microscopes that use lenses to focus visible light and through electron microscopes that use magnets to focus beams of electrons. Electron microscopes can resolve smaller structures than can light microscopes.

4.3 What Are the Basic Attributes of Cells?

Cells are small because they must exchange materials with their surroundings by diffusion, a slow process that requires the interior of the cell to be close to the plasma membrane. All cells are surrounded by a plasma membrane that regulates the interchange of materials between the cell and its environment. All cells use DNA as genetic instructions and RNA to direct protein synthesis. There are two fundamentally different types of cells. Prokaryotic cells lack membrane-enclosed organelles. Eukaryotic cells are generally larger than prokaryotic cells and have a variety of organelles, including a nucleus. See Table 4-1 for a comparison of prokaryotic and eukaryotic cells.

4.4 What Are the Major Features of Prokaryotic Cells?

All members of the domains Archaea and Bacteria are unicellular and prokaryotic. Prokaryotic cells are generally smaller than eukaryotic cells and have a simpler internal structure that lacks membrane-enclosed organelles. Some prokaryotes have flagella. Most bacteria are surrounded by a cell wall made of peptidoglycan. Some bacteria, including many that cause disease, attach to surfaces using external capsules or slime layers and/or protein strands called attachment pili. Sex pili draw bacteria together to allow transfer of plasmids, small rings of DNA that confer features such as antibiotic resistance. Most bacterial DNA is in a single chromosome in the nucleoid region. Bacterial cytoplasm includes ribosomes and a cytoskeleton. Photosynthetic bacteria may have internal membranes where the reactions of photosynthesis occur.

4.5 What Are the Major Features of Eukaryotic Cells?

Eukaryotic cells have a variety of membrane-enclosed structures called organelles, some of which differ between plant and animal cells (see Table 4-1). Both plant and animal cells secrete an extracellular matrix. In animal cells, the ECM consists of proteins and polysaccharides that provide structural and biochemical support. In plant cells, the ECM is the supportive and porous cell wall composed primarily of cellulose.

Eukaryotic genetic material (DNA) is contained within the nucleus, which is surrounded by the double membrane of the nuclear envelope. Pores in the nuclear envelope regulate the movement of molecules between nucleus and cytoplasm. The genetic material is organized into strands called chromosomes, which consist of DNA and proteins. The nucleolus within the nucleus is the site of ribosome synthesis. Ribosomes, composed of rRNA and protein, are the sites of protein synthesis within the cytoplasm.

All eukaryotic cells contain mitochondria, which extract energy from food and store it in the high-energy bonds of ATP. Cells of plants and photosynthetic protists contain plastids. Plastids include chloroplasts, in which photosynthesis captures solar energy and stores it in sugar molecules. Other plastids store pigments or starch. The endosymbiont hypothesis states that mitochondria and chloroplasts (as well as other plastids) originated from independently living prokaryotic cells. Eukaryotic cells have an internal cytoskeleton of protein filaments that transports and anchors organelles; in animal cells, it also shapes the cells, aids in cell division, and allows certain cells to move.

The endomembrane system within eukaryotic cells includes the nuclear envelope, endoplasmic reticulum (ER), Golgi apparatus, and lysosomes and other vesicles. The ER forms a series of interconnected membranous compartments and is a major site of new membrane production. Rough ER, continuous with the outer nuclear envelope, bears ribosomes where proteins are manufactured. These proteins are modified, folded, and transported within the ER channels. Smooth ER, which lacks ribosomes, manufactures lipids such as steroid hormones, detoxifies drugs and metabolic wastes, and stores calcium. The Golgi apparatus is a series of flattened membranous sacs. The Golgi processes and modifies materials that are synthesized in the rough ER. Substances modified in the Golgi are sorted and packaged into vesicles for transport elsewhere in the cell. Lysosomes are specialized sacs of membrane containing digestive enzymes. These merge with food vacuoles and break down food particles. Lysosomes also digest defective organelles. Some freshwater protists have contractile vacuoles that collect and expel excess water. Plants use central vacuoles to support their cells and to store nutrients, pigments, wastes, and toxic materials. Some eukaryotic cells have cilia or flagella, extensions of the plasma membrane that contain microtubules arranged in a characteristic pattern. These structures move fluids past the cell or move the cell through a fluid.

Thinking Through the Concepts

Bloom's: Remembering, Understanding

Multiple Choice

1. Which of the following is/are found only in prokaryotic cells?
   a. plasmids
   b. a cytoskeleton
   c. mitochondria
   d. ribosomes

2. Which of the following is *not* a function of the cytoskeleton?
   a. organelle movement
   b. extracellular support
   c. cell movement
   d. maintenance of cell shape
3. Which of the following statements is false?
   a. Cilia and flagella can move cells through fluids.
   b. Cilia and flagella are supported and moved by microfilaments.
   c. Cilia are shorter and more numerous than flagella.
   d. Flagella propel human sperm.
4. Which of the following is *not* a location of ribosomes?
   a. on the nuclear membrane
   b. free in the cytoplasm
   c. strung along messenger RNA
   d. inside the rough ER
5. Which of the following statements is false?
   a. Photosynthesis occurs in plastids.
   b. Chloroplasts extract energy from food storage molecules.
   c. Mitochondria likely originated from prokaryotic cells.
   d. Plant cell walls are a type of extracellular matrix.

Fill-in-the-Blank

1. The plasma membrane is composed of two major types of molecules, ____________ and ____________. Which type of molecule is responsible for each of the following functions? Isolation from the surroundings: ____________; interactions with other cells: ____________.
2. The three types of cytoskeleton fibers are ____________, ____________, and ____________. Which of these supports cilia? ____________. Moves organelles? ____________. Allows muscle contraction? ____________. Provides a supporting internal framework for the cell? ____________.
3. After each description, fill in the appropriate term: "workbenches" of the cell: ____________; comes in rough and smooth forms: ____________; site of ribosome production: ____________; a stack of flattened membranous sacs: ____________; outermost layer of plant cells: ____________; ferries instructions for protein production from the nucleus to the cytoplasm: ____________.
4. Antibody proteins are synthesized on ribosomes associated with the ____________. The antibody proteins are packaged into membranous sacs called ____________ and are then transported to the ____________. There, what type of molecule is added to the protein? ____________ After the antibody is completed, it is packaged into vesicles that fuse with the ____________.
5. After each description, fill in the appropriate structure: "powerhouses" of the cell: ____________; capture solar energy: ____________; structure outside of animal cells: ____________; region of prokaryotic cell containing DNA: ____________; propel fluid past cells: ____________; consists of the cytosol and the organelles within it: ____________.
6. Two organelles that are believed to have evolved from prokaryotic cells are ____________ and ____________. Evidence for this hypothesis is that both have ____________ membranes, both have groups of enzymes that synthesize ____________, and their ____________ is similar to that of prokaryotic cells.
7. What structure in bacterial cells is composed of peptidoglycan? ____________ What structure in prokaryotic cells serves a similar function to the nucleus in eukaryotic cells? ____________ Short segments of bacterial DNA that confer special features such as antibiotic resistance are called ____________. Bacterial structures called ____________ pull bacterial cells together so they can transfer DNA.

Review Questions

1. What are the three principles of the cell theory?
2. Which cytoplasmic structures are common to both plant and animal cells, and which are found in one type but not the other?
3. Name the proteins of the eukaryotic cytoskeleton; describe their relative sizes and major functions.
4. Describe the nucleus and the function of each of its components, including the nuclear envelope, chromatin, chromosomes, DNA, and the nucleolus.
5. What are the functions of mitochondria and chloroplasts? Why do scientists believe that these organelles arose from prokaryotic cells? What is this hypothesis called?
6. What is the function of ribosomes? Where in the cell are they found? Are they limited to eukaryotic cells?
7. Describe the structure and function of the endoplasmic reticulum (smooth and rough) and the Golgi apparatus and how they work together.
8. How are lysosomes formed? What is their function?
9. Diagram the structure of eukaryotic cilia and flagella. Describe how each moves and what its movement accomplishes.
10. List the structures of bacterial cells that have the same name and function as some eukaryotic structures, but a different molecular composition.

Applying the Concepts

Bloom's: Applying, Analyzing, Evaluating

1. If samples of muscle tissue were taken from the legs of a world-class marathon runner and a sedentary individual, which would you expect to have a higher density of mitochondria? Why?
2. One of the functions of the cytoskeleton in animal cells is to give shape to the cell. Plant cells have a fairly rigid cell wall surrounding the plasma membrane. Does this mean that a cytoskeleton is unnecessary for a plant cell? Explain.
3. What problems would an enormous spherical cell encounter? What adaptations might help a very large cell survive?

5 Cell Membrane Structure and Function

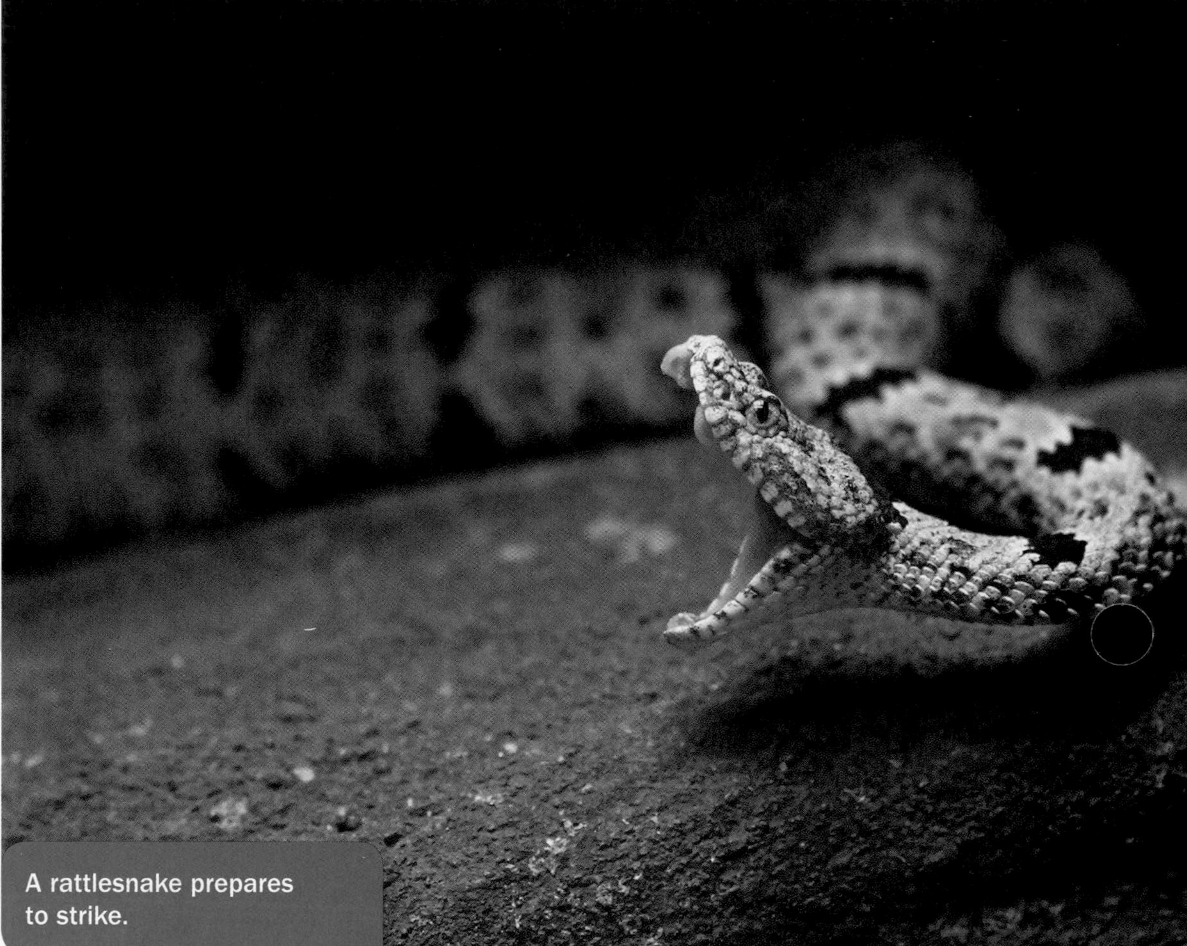

A rattlesnake prepares to strike.

CASE STUDY

Vicious Venoms

THIRTEEN-YEAR-OLD JUSTIN SCHWARTZ was enjoying his stay at a summer camp near Yosemite National Park. But that changed in a hurry when, during a long hike, Justin stopped for a rest on some sunny rocks. Suddenly, he felt a piercing pain in his left palm. A 5-foot rattlesnake—probably feeling threatened by Justin's dangling arm—had struck without warning.

Justin's campmates stared in alarm as the snake slithered into the undergrowth, but Justin focused on his hand, which was swelling rapidly. He suddenly felt weak and dizzy, and the pain was agonizing. The symptoms worsened as Justin's counselors and campmates carried him four miles down the trail; the pain spread up his now-discolored arm, and his hand felt as if it were going to burst. At the trailhead, a helicopter whisked Justin to a hospital, where he fell unconscious. A day later, he regained consciousness at the University of California Davis Medical Center. He spent more than a month there, undergoing 10 surgeries. The surgeries relieved the enormous pressure from swelling in his arm, removed dead muscle tissue, and began the long process of repairing the extensive damage to his hand and arm.

Diane Kiehl's ordeal began as she dressed for an informal Memorial Day celebration with her family, pulling on blue jeans that she had tossed on the bathroom floor the previous night. Feeling a sting on her right thigh, she ripped off the jeans and watched with irritation as a long-legged spider crawled out. Spiders are usually harmless, so Diane at first paid little attention to the two small puncture wounds on her leg. But this bite was not from a harmless spider, it was from a brown recluse. The next day, an extensive, itchy rash appeared at the site of the bite. By the third day, intermittent pain pierced like a knife through Diane's thigh. A physician gave her painkillers, steroids to reduce the swelling, and antibiotics to combat bacteria introduced by the spider's mouthparts. Nonetheless, the next 10 days were a nightmare of pain from the growing sore, now covered with oozing blisters and underlain with clotting blood. It took 4 months for the lesion to heal. Even a year later, Diane sometimes felt pain in the large scar that remained.

How do rattlesnake and brown recluse spider venoms cause leaky blood vessels, disintegrating skin and tissue, and sometimes life-threatening symptoms throughout the body? What do venoms have to do with cell membranes?

AT A GLANCE

5.1 HOW IS THE STRUCTURE OF THE CELL MEMBRANE RELATED TO ITS FUNCTION?

All cells, as well as organelles within eukaryotic cells, are surrounded by membranes. The membranes of a cell all have the same basic structure: proteins suspended in a double layer of phospholipids (**FIG. 5-1**). The structures of a membrane's proteins and phospholipids can change dynamically in response to the environment and the cell's changing needs.

Cell membranes perform several crucial functions:

- They isolate the contents of cells from the surrounding fluid, and isolate the contents of membrane-enclosed organelles from the surrounding cytosol.
- They regulate the exchange of substances between cells and their surroundings, and between membrane-enclosed organelles and the cytosol.
- They allow communication between the cells of multicellular organisms.
- They create attachments within and between cells.
- They regulate many biochemical reactions.

These are formidable tasks for a structure so thin that 10,000 membranes stacked atop one another would scarcely equal the thickness of a piece of paper.

Membranes Are "Fluid Mosaics" in Which Proteins Move Within Layers of Lipids

In 1972, S. J. Singer and G. L. Nicolson proposed the **fluid mosaic model** of cell membranes, which forms the basis

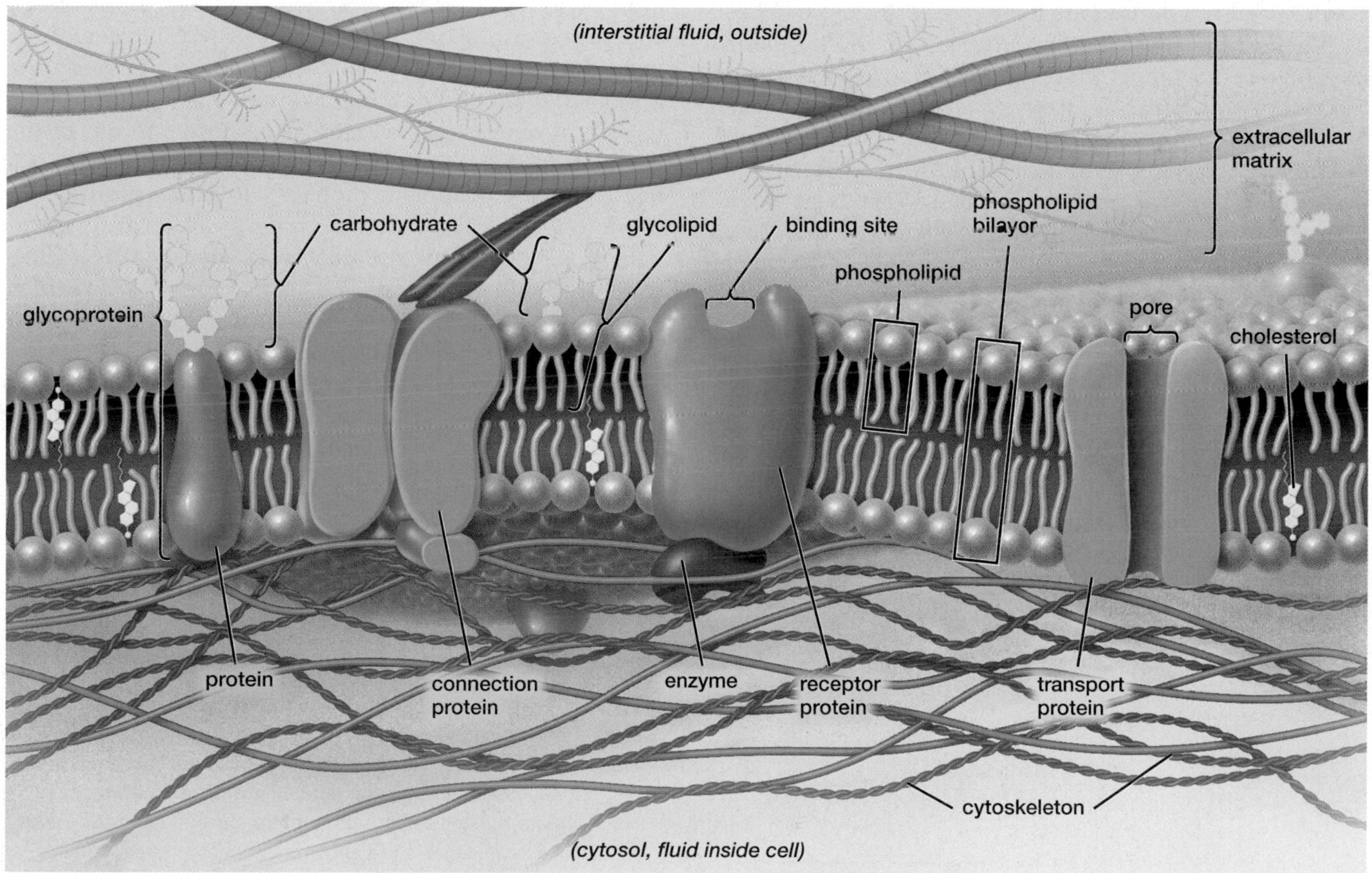

▲ **FIGURE 5-1 The plasma membrane** The plasma membrane is a bilayer of phospholipids interspersed with cholesterol molecules and embedded with proteins (blue). Membrane proteins include recognition, connection, receptor, and transport proteins, as well as enzymes. There are also many glycoproteins and glycolipids with attached carbohydrates.

for our understanding of membrane structure and function. A **fluid** is any substance whose molecules can flow past one another; fluids include gases, liquids, and cell membranes. The word *mosaic* refers to anything that resembles a mosaic artwork made by setting small colored pieces, such as pieces of broken tile, in cement-like mortar. A membrane viewed at high magnification looks something like a lumpy, constantly shifting mosaic of tiles. A double layer of phospholipids forms a thick but still liquid "mortar" for the mosaic, and various proteins are the embedded "tiles," which can move about within the phospholipid layers (see Fig. 5-1).

The Fluid Phospholipid Bilayer Helps to Isolate the Cell's Contents

As shown in **FIGURE 5-2**, a phospholipid molecule consists of two very different parts: a "head" that is polar and hydrophilic (attracted to water) and a pair of fatty acid "tails" that are nonpolar and hydrophobic (not attracted to water). In water, phospholipids spontaneously arrange themselves into a double layer called a **phospholipid bilayer**; hydrogen bonds between water and the hydrophilic phospholipid heads cause the heads to face the water, while the hydrophobic phospholipid tails cluster together within the bilayer.

This bilayer arrangement also occurs in membranes, because watery solutions are found on both sides of all cell membranes. In all cells, the inner surface of the plasma membrane is in contact with the *cytosol* (the fluid portion of the cytoplasm), which is mostly water. The watery solution in contact with the outside surface of the membrane differs among cell types. In animal cells, the outer surfaces are bathed in watery **interstitial fluid**, a weakly salty liquid. Plant cells are surrounded by cellulose walls that are saturated with water. And single-celled organisms live in water or in moist places such as soil or inside multicellular organisms.

Membranes Are Flexible and Dynamic

The components of cell membranes are in constant motion. To understand why, it is important to be aware that, at any temperature above absolute zero (−459.67°F or −273.15°C), atoms, molecules, and ions are in constant random motion. As the temperature increases, their rate of motion increases; at temperatures that support life, these particles move rapidly indeed. The phospholipid molecules in a membrane are not chemically bonded to one another, so their constant random motion causes them to shift about freely, although they rarely flip between the two layers of membrane.

The flexibility of the bilayer is crucial for membrane function. If plasma membranes were stiff, your cells would break open and die as you moved about. The fluid nature of membranes also allows membranes to merge with one another. For example, vesicles from the Golgi expel their contents to the outside of the cell by merging with the plasma membrane (see Chapter 4). To maintain flexibility, most membranes have a consistency similar to room-temperature olive oil. But olive oil becomes a solid if you refrigerate it, so what happens to membranes when an organism gets cold? To find out, see "Health Watch: Membrane Fluidity, Phospholipids, and Fumbling Fingers."

The Phospholipid Bilayer Blocks the Passage of Most Molecules

Most molecules used by cells, including salts, amino acids, and sugars, are hydrophilic. Because these molecules are polar and water soluble, they cannot move through the nonpolar, hydrophobic fatty acid tails of the phospholipid bilayer. Movement of polar molecules into and out of the cell is controlled by the mosaic of proteins within the membrane. Because membrane proteins allow only specific ions or molecules to pass through, or permeate, plasma membranes are described as **selectively permeable**. Selective permeability gives the cell greater control over which substances enter and leave it. However, some biological molecules, including fat-soluble vitamins and steroid hormones such as estrogen and testosterone (see Chapter 3), are hydrophobic and can diffuse directly through the phospholipid bilayer.

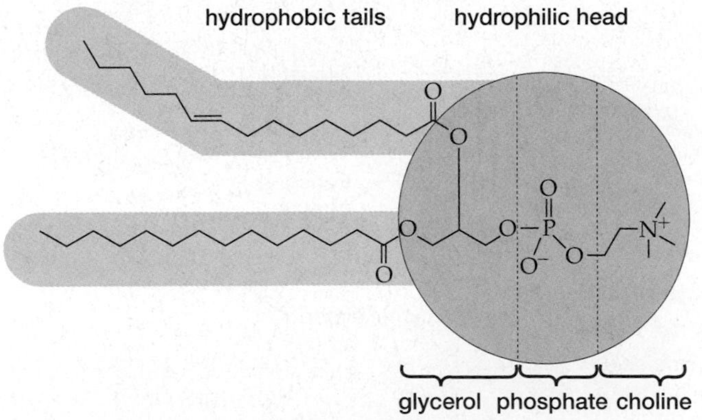

hydrophobic tails
hydrophilic head
glycerol phosphate choline

▲ **FIGURE 5-2 A phospholipid** Phosphatidylcholine (shown here) is abundant in cell membranes. The double bond in one of the fatty acid tails causes the tail to bend.

CASE STUDY CONTINUED

Vicious Venoms

Some of the most devastating effects of certain snake and spider venoms occur because the venoms contain *phospholipases*, enzymes that break down phospholipids. You now know that phospholipids are a major component of cell membranes, which isolate the cell's contents from its surroundings. As the phospholipids degrade, the membranes become leaky, causing the cells to die.

As if phospholipases aren't deadly enough, snake venoms also contain enzymes that break down proteins. What functions do membrane proteins serve?

Health WATCH Membrane Fluidity, Phospholipids, and Fumbling Fingers

To function properly, cell membranes must remain fluid. But can they maintain optimal fluidity in cold temperatures? As temperatures cool, the phospholipid molecules in membranes move about less vigorously and pack together more tightly, causing the membrane to stiffen. If a membrane loses too much fluidity, the proteins embedded in the membrane may lose their ability to function effectively.

Organisms that experience seasonal changes in temperature often possess mechanisms for maintaining optimal membrane fluidity in the face of such changes. Scientists studying plants, fish, frogs, snails, and mammals have found that these organisms can modify the composition of their cell membranes. As the temperature falls, they incorporate phospholipids containing more unsaturated fatty acids into their membranes. When the temperature rises, the organisms restore the saturated phospholipids. Why? The fluidity of a membrane at a given temperature is strongly influenced by the relative amounts of saturated and unsaturated fatty acid tails in membrane phospholipids. Saturated fatty acids (with no double-bonded carbons) are straight and can pack tightly together, forming a relatively stiff membrane. Unsaturated fatty acids have one or more double-bonded carbons, each introducing a kink into the tail (see Fig. 5-2). This structure forces the phospholipids farther apart, making the membrane more fluid (**FIG. E5-1**). Adjusting the fatty acids in the phospholipids helps organisms maintain optimal membrane fluidity during seasonal temperature changes.

But when you get caught out in the cold, some of your cell membranes may stiffen, because they won't have time to adjust their phospholipids. If your core temperature suddenly begins to fall, your body will reduce blood flow to your hands and feet, conserving warmth for vital organs such as your heart and brain. As your hands get colder, it becomes harder to control your fingers, and your sense of touch diminishes (**FIG. E5-2**).

Is this loss of sensation and coordination connected to stiff cell membranes? Perhaps so. Researchers have found evidence supporting the hypothesis that as membranes stiffen in the cold, the stiffness hampers the functioning of the ion channel proteins responsible for transmitting nerve impulses. As a result, impulses are slowed in the nerves that carry sensations and control muscles, which makes it difficult to coordinate delicate hand movements like zipping a coat or lighting a fire.

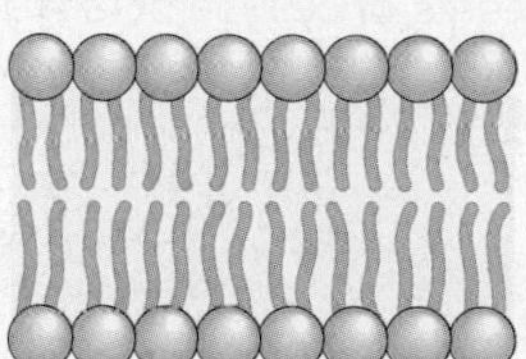

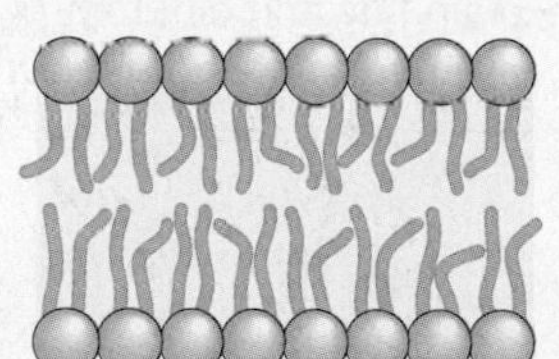

▲ **FIGURE E5-1 Tail kinks in phospholipids increase membrane fluidity**

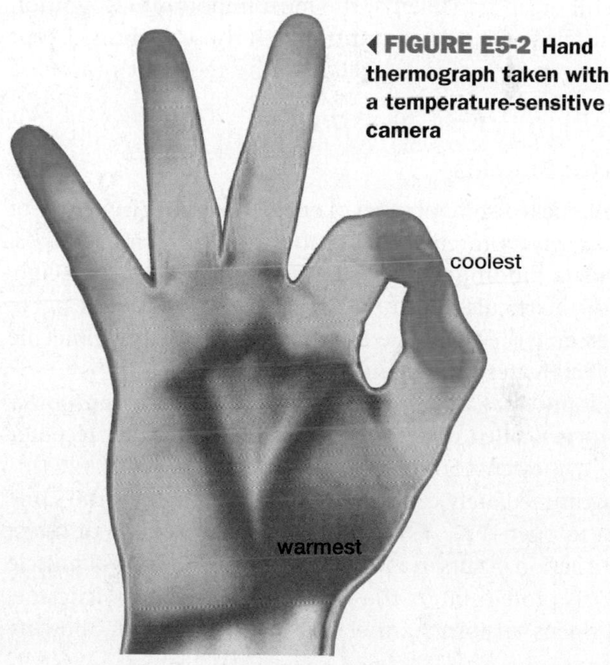

◀ **FIGURE E5-2 Hand thermograph taken with a temperature-sensitive camera**

THINK CRITICALLY Researchers have recently discovered ion channels in pain-generating skin nerve cells whose ability to conduct ions is actually enhanced by cold, making the nerve cell more active. Develop a hypothesis to explain why these receptors evolved.

A Variety of Proteins Form a Mosaic Within the Membrane

Thousands of different membrane proteins are embedded within or attached to the surfaces of the phospholipid bilayer of cell membranes. Many membrane proteins bear carbohydrate groups that project from the outer membrane surface (see Fig. 5-1). These proteins are called **glycoproteins** (Gk. *glyco*, sweet; referring to the carbohydrate sugar subunits). Membrane proteins may be grouped into five major categories based on their function: enzymes, recognition proteins, receptor proteins, connection proteins, and transport proteins.

Enzymes

Proteins called **enzymes** promote chemical reactions that synthesize or break apart biological molecules (see Chapter 6). Enzymes perform a variety of functions in cell membranes. For example, enzymes involved in ATP synthesis are embedded in the inner mitochondrial membrane. Enzymes in the plasma membrane help synthesize the supportive extracellular matrix that fills spaces between animal cells. In cells lining the small intestine, plasma membrane enzymes complete the breakdown of carbohydrates and proteins as these nutrients are taken into the cells.

Recognition Proteins

Recognition proteins are glycoproteins that serve as identification tags. Animal cells bear distinctive glycoproteins that identify an individual's cells as "self." Immune cells ignore self-cells and attack invading cells, such as bacteria, that have different recognition proteins on their membranes. For successful organ transplants, the most important recognition glycoproteins of the donor must match those of the recipient so the organ won't be attacked by the recipient's immune system.

Receptor Proteins

Most cells bear dozens of types of **receptor proteins** (some of which are glycoproteins) that span their plasma membranes. Each has a binding site specific for a messenger molecule, such as a particular hormone or neurotransmitter (a nerve cell messenger). When the appropriate messenger molecule binds, it activates the receptor protein.

In some cases, a receptor protein also acts as a transporter in a process called direct receptor action. In direct receptor action, a messenger molecule binds the receptor, and this binding immediately causes an ion channel within the same protein to open (**FIG. 5-3a**). A well-known example of direct receptor action occurs in a protein in the membranes of muscle cells. This protein binds the neurotransmitter acetylcholine, which opens an ion channel in the same protein, allowing ions to flow and stimulate the muscle cell to contract. Other receptor proteins produce their effects through indirect action. In indirect receptor action, a messenger molecule—which can be a neurotransmitter or a hormone—binds to a receptor protein; in response, the protein changes its shape and biochemical activity. This change starts a series of reactions within the cell that produce effects (such as opening many ion channels) at different sites (**FIG. 5-3b**). Most neurotransmitters and hormones stimulate indirect receptor action.

Connection Proteins

A diverse group of **connection proteins** anchors cell membranes in various ways. Some connection proteins help maintain cell shape by linking the plasma membrane to the cell's cytoskeleton. Other connection proteins span the plasma membrane, linking the cytoskeleton inside the cell with the extracellular matrix outside, which helps anchor the cell in place within a tissue (see Fig. 5-1). Connection proteins also link adjacent cells.

Transport Proteins

Transport proteins span the phospholipid bilayer and regulate the movement of hydrophilic molecules across the membrane. Some transport proteins form pores (channels) that can be opened or closed to allow specific substances to pass across the membrane. Other transport proteins bind substances and conduct them through the membrane, sometimes using cellular energy. Transport proteins are described in more detail later in this chapter.

CHECK YOUR LEARNING

Can you . . .

- describe the components, structure, and functions of cell membranes?
- diagram and describe the fluid mosaic model of cell membranes?
- explain how the different components of cell membranes contribute to membrane functions?

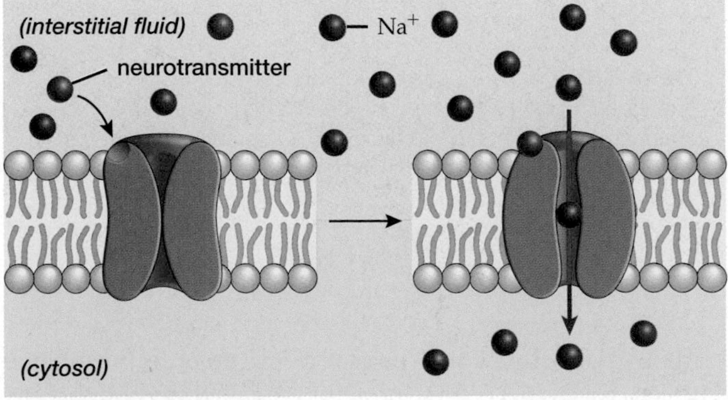

(a) Direct receptor action

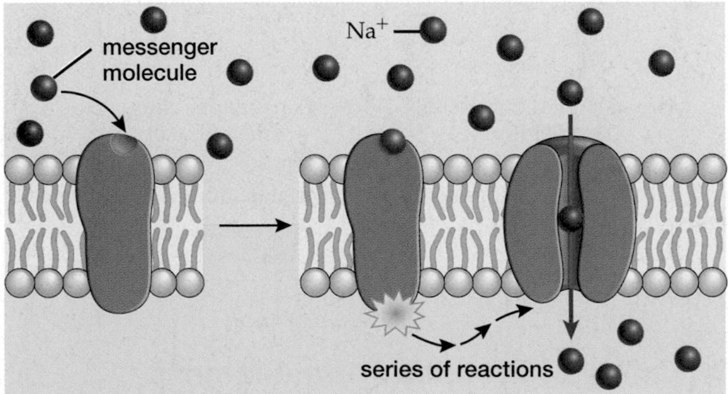

(b) Indirect receptor action

▲ **FIGURE 5-3 Receptor protein activation (a)** A neurotransmitter binds a receptor site on a membrane channel protein, causing the channel to open and allow a flow of ions. **(b)** A messenger molecule (such as a hormone or neurotransmitter) binds a membrane receptor, which stimulates a series of reactions inside the cell that cause channel proteins to open.

CASE STUDY CONTINUED

Vicious Venoms

Most snake venoms are nasty cocktails of toxins. In addition to toxins that break down phospholipids, some snake venoms contain toxins that bind to and inhibit the activity of receptor proteins for the neurotransmitter acetylcholine. For example, krait snakes, native to Asia, produce a venom protein that binds to the direct-acting membrane channel receptors for acetylcholine, including those on skeletal muscle cells. Once bound, the protein remains firmly attached, blocking acetylcholine from binding. This prevents muscles from contracting; an untreated krait bite victim often dies when the skeletal muscles that control breathing become paralyzed.

The ability of substances to move across membranes is crucial not only to breathing, but also to all other aspects of staying alive. What are the physical processes that underlie the movement of substances across membranes?

5.2 WHICH PHYSICAL PROCESSES MOVE MOLECULES IN FLUIDS?

Before exploring how substances are transported across membranes, let's look at two physical processes that play important roles in membrane transport: diffusion and osmosis. We begin with a few definitions:

- A **solute** is a substance that can be dissolved (dispersed into individual atoms, molecules, or ions) in a **solvent**, which is a fluid (usually a liquid) capable of dissolving the solute. Water, in which all of a cell's chemical processes occur, dissolves so many different solutes that it is sometimes called the "universal solvent."
- **Concentration** is the amount of solute in a given volume of solvent.
- A **gradient** is a difference in certain properties—such as temperature, pressure, electrical charge, or concentration—between two adjacent regions. Creating or maintaining a gradient requires the expenditure of energy. Thus, a gradient decreases over time unless energy is supplied to maintain it (or unless an impenetrable barrier separates the adjacent regions).

Molecules in Fluids Diffuse in Response to Gradients

Because atoms, molecules, and ions are in constant motion, over time random movement of solute molecules produces a net movement from regions of high concentration to regions of low concentration, a process called **diffusion**. The greater the concentration gradient, the more rapidly diffusion occurs. If nothing opposes this diffusion, then the random movement of molecules will eventually cause the substance to become evenly dispersed throughout the fluid. Molecules moving from regions of high concentration to regions of low concentration are described as moving "down" their concentration gradients.

To watch diffusion in action, place a drop of food coloring in a glass of water (**FIG. 5-4** ❶). Random motion propels dye molecules both out of and back into the dye droplet, but there is a net transfer of dye into the water and of water into the dye, down their respective concentration gradients ❷. The net movement of dye will continue until it is uniformly dispersed in the water ❸. If you compare diffusion of a dye in hot water to diffusion in cold water, you will see that heat increases the diffusion rate by causing molecules to move faster.

SUMMING UP: Principles of Diffusion

- Diffusion is the net movement of molecules down a gradient from high to low concentration.
- The greater the concentration gradient, the faster the rate of diffusion.
- The higher the temperature, the faster the rate of diffusion.
- If no other processes intervene, diffusion will continue until the concentrations become equal throughout the solution, that is, until the concentration gradient is eliminated.

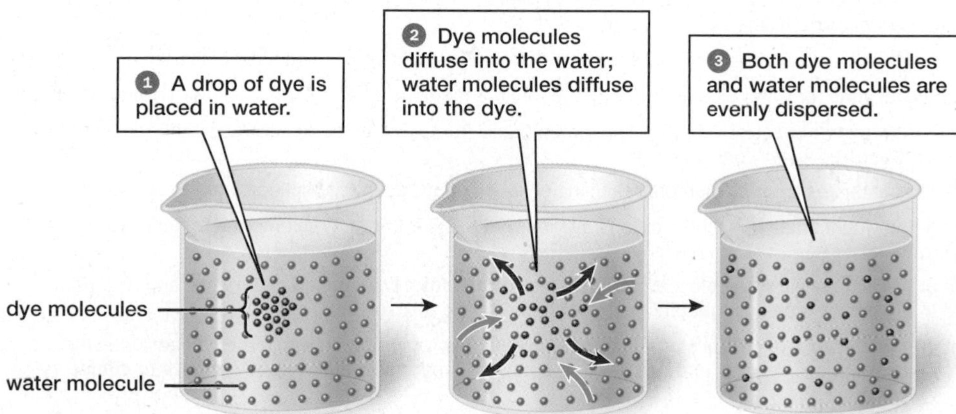

❶ A drop of dye is placed in water.
❷ Dye molecules diffuse into the water; water molecules diffuse into the dye.
❸ Both dye molecules and water molecules are evenly dispersed.
dye molecules
water molecule

FIGURE 5-4 Diffusion of a dye in water

Osmosis Is the Diffusion of Water Across Selectively Permeable Membranes

Osmosis is the diffusion of water across a selectively permeable membrane in response to gradients of concentration, pressure, or temperature. Here, we will focus on osmosis from a region of higher water concentration to a region of lower water concentration.

What do we mean by a "high water concentration" or a "low water concentration"? Water containing no solutes has the highest possible water concentration, which causes more water molecules to collide with—and so move through—a water-permeable membrane. Adding any solute to water reduces the water concentration by replacing some of the water molecules in a given volume of the solution. Thus, the higher the concentration of solute, the lower the concentration of water. Osmosis (like other forms of diffusion) will cause a net movement of water molecules from the solution with a higher water concentration (a lower solute concentration) into the solution with a lower water concentration (a higher solute concentration). For example, water will move by osmosis from a solution with less dissolved sugar into a solution with more dissolved sugar.

Solutions with equal concentrations of solute—and thus equal concentrations of water—are described as being **isotonic** to one another (Gk. *iso*, same). When isotonic solutions are separated by a water-permeable membrane, water moves equally through the membrane in both directions, so there is no net movement of water. When comparing two solutions with different concentrations of a solute, the solution with the greater concentration of solute is described as being **hypertonic** (Gk. *hyper*, greater than) to the less concentrated solution. The more dilute solution is described as **hypotonic** (Gk. *hypo*, below). Water tends to move through water-permeable membranes from hypotonic solutions into hypertonic solutions. This movement will continue until the water concentrations (and thus the solute concentrations) on both sides of the membrane are equal.

SUMMING UP: Principles of Osmosis

- Osmosis is the movement of water through a selectively water-permeable membrane.
- Water moves down its concentration gradient from a higher concentration of water molecules to a lower concentration of free water molecules.
- Dissolved substances, called solutes, reduce the concentration of free water molecules in a solution.
- When comparing two solutions, the one with a higher solute concentration is hypertonic, and the solution with the lower solute concentration is hypotonic.

CHECK YOUR LEARNING

Can you . . .

- explain the process of diffusion?
- explain the process of osmosis?

5.3 HOW DO SUBSTANCES MOVE ACROSS MEMBRANES?

The plasma membrane permits substances to move through it in two different ways: passive transport and energy-requiring transport (**TABLE 5-1**). **Passive transport** involves diffusion of substances across cell membranes down their concentration gradients, whereas **energy-requiring transport** requires that the cell expend energy to move substances across membranes. Energy is required to transport substances against concentration gradients or to move particles or fluid droplets into or out of the cell.

Passive Transport Includes Simple Diffusion, Facilitated Diffusion, and Osmosis

Substances can diffuse across a membrane that is permeable to the diffusing substance. Many molecules cross plasma membranes by diffusion, driven by concentration differences between the cytosol and the surrounding fluid.

TABLE 5-1	Transport Across Membranes
Passive Transport	Diffusion of substances across a membrane down a gradient of concentration, pressure, or electrical charge; does not require cellular energy
Simple diffusion	Diffusion of water, dissolved gases, or lipid-soluble molecules through the phospholipid bilayer of a membrane
Facilitated diffusion	Diffusion of water, ions, or water-soluble molecules through a membrane via a channel or carrier protein
Osmosis	Diffusion of water across a selectively permeable membrane from a region of higher free water concentration to a region of lower free water concentration
Energy-Requiring Transport	Movement of substances through membranes using cellular energy, usually supplied by ATP
Active transport	Movement of individual small molecules or ions against their concentration gradients through membrane-spanning proteins
Endocytosis	Movement of fluids, specific molecules, or particles into a cell; occurs as the plasma membrane engulfs the substance in a membranous sac that pinches off and enters the cytosol
Exocytosis	Movement of particles or large molecules out of a cell; occurs as a membrane within the cell encloses the material, moves to the cell surface, and fuses with the plasma membrane, allowing its contents to diffuse out

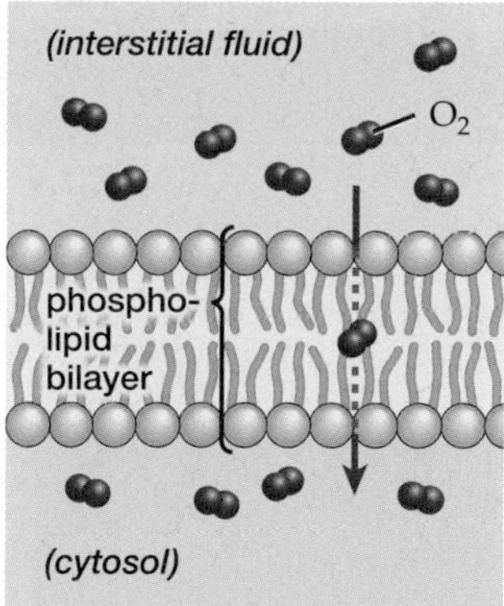

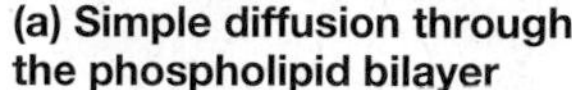

(a) Simple diffusion through the phospholipid bilayer

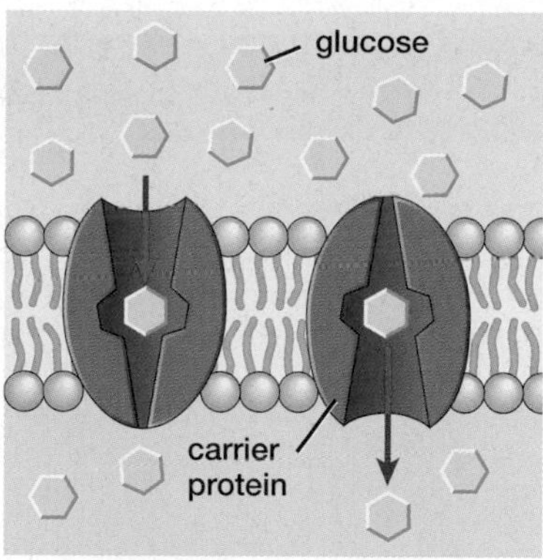

(b) Facilitated diffusion through carrier proteins

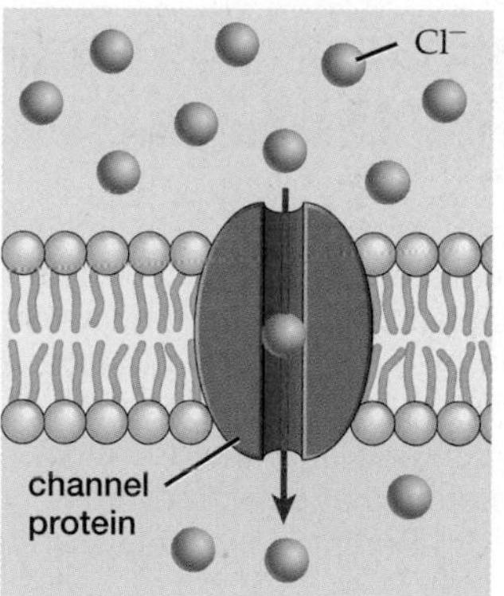

(c) Facilitated diffusion through channel proteins

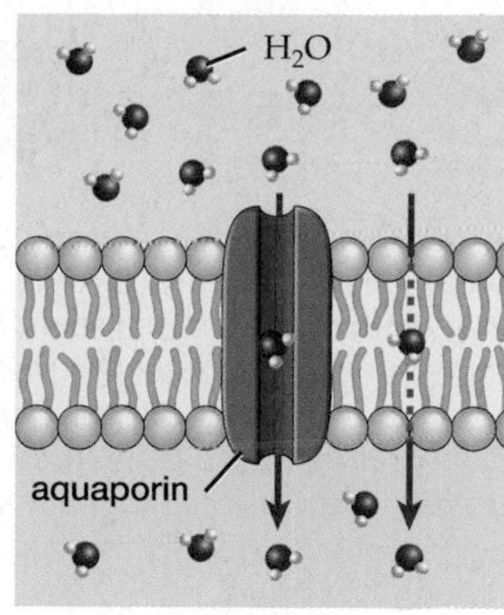

(d) Osmosis through aquaporins or the phospholipid bilayer

▲ **FIGURE 5-5 Types of diffusion through the plasma membrane (a)** Small, uncharged, or lipid-soluble molecules diffuse directly through the phospholipid bilayer. Here, oxygen molecules diffuse down their concentration gradient (red arrow). **(b)** Carrier proteins have binding sites for specific molecules, such as glucose. Binding causes the carrier to change shape and shuttle the molecule across the membrane down its concentration gradient. **(c)** Facilitated diffusion through specific channel proteins allows ions, such as chloride, to cross membranes. **(d)** Osmosis is the diffusion of water. Water molecules can pass through the phospholipid bilayer by simple diffusion, or they move far more rapidly by facilitated diffusion through aquaporins.

Some Molecules Move Across Membranes by Simple Diffusion

Some molecules diffuse directly through the phospholipid bilayer, a process called **simple diffusion** (**FIG. 5-5a**). Molecules that can cross cell membranes by simple diffusion include very small molecules with no net charge, such as oxygen and carbon dioxide, as well as lipid-soluble molecules, including alcohol, certain vitamins, and steroid hormones. The rate of simple diffusion is increased by larger concentration gradients, higher temperatures, smaller molecular sizes, and greater solubility in lipids.

Water can also cross cell membranes by simple diffusion, even though it is a polar molecule. How can water diffuse directly through the hydrophobic (literally "water-fearing") phospholipid bilayer? Water molecules are so small that some stray into the thicket of phospholipid tails, and the random movement of these water molecules carries them through the membrane. But because simple diffusion of water through the phospholipid bilayer is relatively slow, many cell types have specific transport proteins for water, as described later in this chapter.

Some Molecules Cross Membranes by Facilitated Diffusion Using Membrane Transport Proteins

The phospholipid bilayers of cell membranes are impermeable to many molecules. For example, most polar molecules, such as sugars, cannot pass through the bilayer because they are too large and are not lipid soluble. Ions, such as K^+, Na^+, Cl^-, and Ca^{2+}, are also excluded even though they are small, because their charges cause polar water molecules to cluster around them, forming an aggregation that is too large to move directly through the phospholipid bilayer. Therefore, ions and polar molecules must use transport proteins to move through cell membranes. The transport proteins help these molecules move down a concentration gradient in a process called **facilitated diffusion**. Facilitated diffusion does not require the cell to expend energy. Two types of proteins allow facilitated diffusion: carrier proteins and channel proteins.

Carrier proteins span the cell membrane and have regions that loosely bind certain ions or specific molecules such as sugars or small proteins. This binding of ions or molecules causes the carrier proteins to change shape and transfer the bound molecules across the membrane. For example, glucose carrier proteins in plasma membranes allow this sugar to diffuse down its concentration gradient from the interstitial fluid into cells, which continuously use up glucose to meet their energy needs (**FIG. 5-5b**).

Channel proteins form pores through a cell membrane. For example, pores in the membranes of mitochondria and chloroplasts allow passage of many different water-soluble substances. In contrast, *ion channel proteins* (ion channels) are highly selective and remain closed unless they are opened by specific stimuli, such as a neurotransmitter binding to a receptor protein (**FIG. 5-5c**). Ion channel proteins are selective because their interior diameters limit the size of the ions that can pass through and because specific amino acids that line the pores have weak electrical charges that attract specific ions and repel others. For example, slight negative charges inside Na^+ channels attract Na^+ but repel Cl^-.

Many cells have specialized channel proteins called **aquaporins** (literally, "water pores") that are selective for water molecules (**FIG. 5-5d**). To learn more about aquaporins, see "Doing Science: Discovering Aquaporins."

DOING Science Discovering Aquaporins

Sometimes a chance observation leads to a scientific breakthrough. By the 1980s, scientists had long understood that osmosis directly through the phospholipid bilayer is much too slow to account for the rapid rate at which water was known to move across certain cell membranes, such as those of kidney cells and red blood cells. It thus seemed likely that the membranes of these cells contain a protein that selectively transports water. Nonetheless, repeated attempts to identify such a protein had failed.

What Question Was Asked?

In the mid-1980s, Peter Agre was attempting to determine the structure of a glycoprotein found on red blood cells. He isolated the glycoprotein but found that it was contaminated with large quantities of an unknown protein. Intrigued by the mystery protein, Agre and his coworkers wondered whether it was involved in water transport. Their interest grew when they discovered that the same protein occurs in kidney cells.

How Was Evidence Gathered?

To test their hypothesis that the unidentified protein helped transport water across membranes, Agre's team performed an experiment using frog egg cells, whose membranes are nearly impermeable to water. The team predicted that if the protein were a water channel, inserting it into the egg cells would cause them to swell when they were placed in a hypotonic solution. The researchers injected frog eggs with messenger RNA that coded for the unidentified protein, causing the eggs to synthesize the protein and insert it into their plasma membranes. Control eggs were injected with an equal quantity of water.

What Was Learned?

Three days later, the control and experimental eggs appeared identical—until they were placed in a hypotonic solution. The control eggs swelled only slightly, whereas those with the inserted protein swelled rapidly and burst (**FIG. E5-3**). Further studies revealed that only water could move through this channel protein. Agre concluded that the protein he had isolated from red blood cells was a water channel. He named it aquaporin (literally "water pore"). Billions of water molecules can move through an aquaporin in single file every second, while larger molecules and small positively charged ions (such as hydrogen ions) are excluded.

Many subtypes of aquaporin proteins have now been identified, and these water channels have been found in all forms of life that have been investigated. For example, the membrane of the central vacuole of plant cells is rich in aquaporins, allowing the vacuole to fill rapidly when water is available (see Fig. 5-7). Kidney cells insert aquaporins into their plasma membranes when the body becomes dehydrated and water must be conserved. Aquaporins have also been implicated in many pathological conditions including brain swelling, glaucoma, and cancer; researchers are working to design therapeutic drugs for these disorders that will either block or facilitate water movement through these channels, depending on which function has gone awry.

In 2003, Peter Agre shared the Nobel Prize in Chemistry for his discovery. In his Nobel lecture, he shared an insight that is fundamental to scientific advances today: "In science, one should use all available resources to solve difficult problems. One of our most powerful resources is the insight of our colleagues." As a result of collaboration, careful observation, persistence, and perhaps a bit of what he described modestly as the "scientific approach known as sheer blind luck," Agre and his team identified the elusive transport protein for water.

(a) Frog eggs

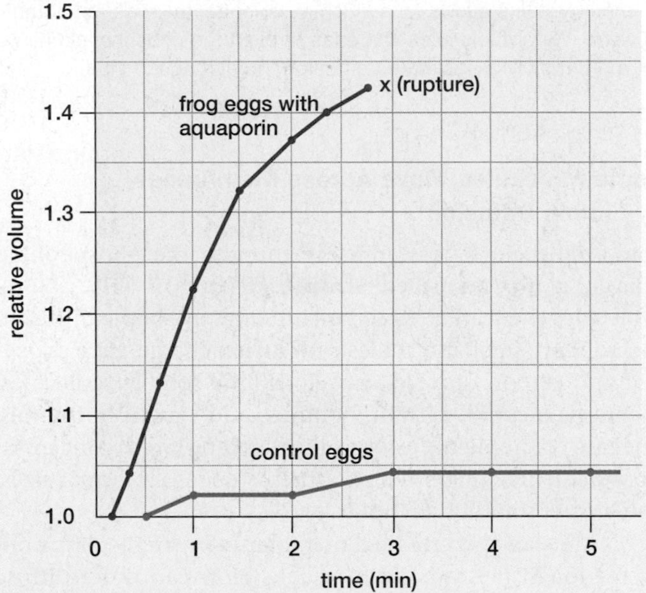

(b) Comparison of swelling in hypotonic medium

▲ **FIGURE E5-3 Investigating aquaporins (a)** The frog egg on the right with aquaporins inserted into its plasma membrane burst after immersion in a hypotonic solution. The normal frog egg on the left swelled only slightly. **(b)** When transferred from a normal to a hypotonic solution, the average volume of frog eggs increased rapidly until they burst, while control eggs swelled very slightly (the relative volume of 1 is the size of the egg just prior to transfer).

THINK CRITICALLY What is likely to have happened if, instead of being immersed in a hypotonic solution, eggs with and without aquaporins had been immersed in a concentrated salt solution? Explain your reasoning.

Osmosis Across the Plasma Membrane Plays an Important Role in the Lives of Cells

Osmosis across plasma membranes is crucial to many biological processes, including water uptake by the roots of plants, absorption of dietary water from the intestine, and the reabsorption of water into the bloodstream that occurs in kidneys.

In animals, the interstitial fluid that surrounds cells is isotonic to the cell cytosol. Although the concentrations of specific solutes are rarely the same both inside and outside of cells, the total concentrations of water and solutes inside and outside are equal. As a result, there is no overall tendency for water to enter or leave animal cells. To demonstrate the importance of maintaining isotonic conditions between cells and their surrounding interstitial fluid, we can observe red blood cells placed in solutions with different solute concentrations, as illustrated in **FIGURE 5-6**. The membranes of red blood cells are rich in aquaporins and therefore very water permeable; cells placed in an isotonic salt solution retain their normal size. But if red blood cells are placed in a salt solution that is hypertonic to their cytosol, water leaves by osmosis, causing the cells to shrivel. In contrast, immersion in a hypotonic salt solution causes the cells to swell (and eventually burst) as water diffuses in.

Have You Ever Wondered ... Why Bacteria Die When You Take Antibiotics?

Penicillin and related antibiotics fight bacterial infections by interfering with cell wall synthesis in bacterial cells. Under normal conditions, a bacterium uses active transport to maintain an internal environment that is hypertonic to its surroundings. This allows the bacterial cytoplasm to maintain turgor pressure against its tough cell wall, much as in a plant cell. If an antibiotic disrupts construction of the confining cell wall, osmosis into the hypertonic cytosol of the bacterium will cause the bacterial cell to swell and rupture its fragile plasma membrane, with fatal consequences.

Antibiotics help us fight bacterial infections; unfortunately, bacteria rapidly evolve resistance to antibiotics. Adaptations that help bacteria resist antibiotics include enzymes that break down the antibiotic, membranes that block entry of the antibiotic, and membrane pumps that expel the antibiotic. As a result, hospitals are now battling bacterial strains that cause serious infections but resist nearly all antibiotics.

Nearly every living plant cell is supported by water that enters through osmosis. Most plant cells have a large central vacuole enclosed by a membrane that is rich in aquaporins.

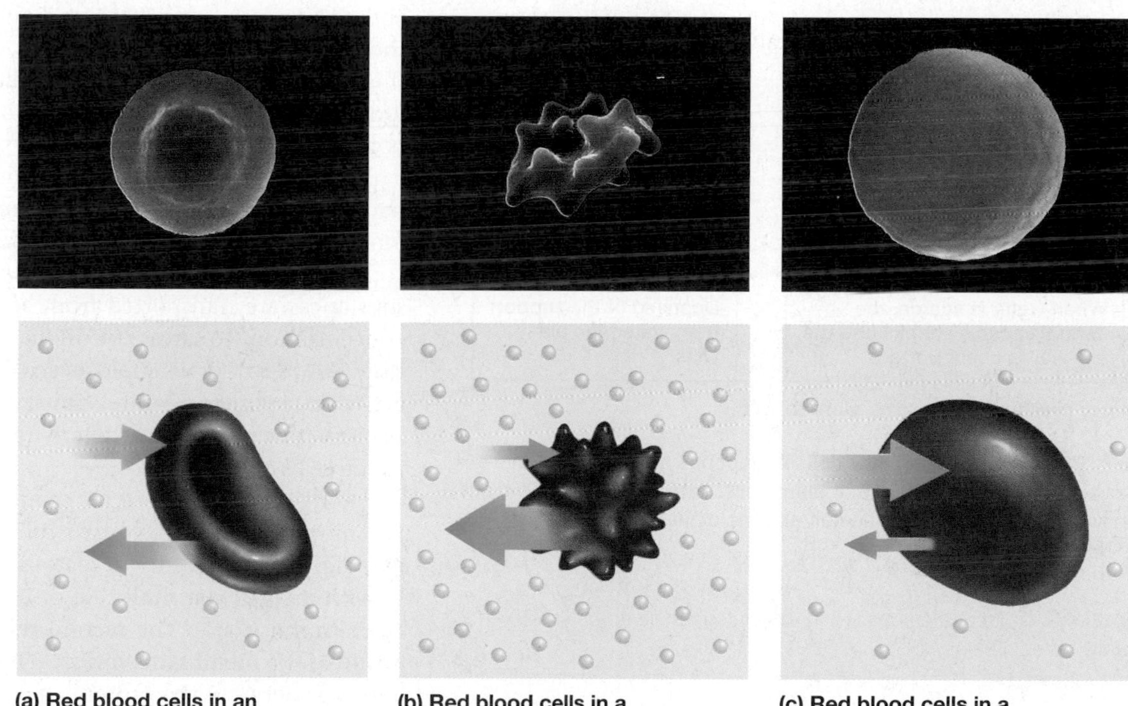

▲ **FIGURE 5-6 The effects of osmosis on red blood cells** The plasma membrane of a red blood cell is rich in aquaporins, so water flows readily in or out along its concentration gradient. **(a)** Cells immersed in an isotonic solution retain their normal dimpled shape. **(b)** Cells in a hypertonic solution shrivel as more water moves out than flows in. **(c)** Cells in a hypotonic solution expand.

THINK CRITICALLY A student pours some distilled water into a sample of blood. Later, she looks at the blood under a microscope and sees no blood cells at all. What happened?

cytoplasm
central vacuole
When water is plentiful, it fills the central vacuole, pushes the cytoplasm against the cell wall, and helps maintain the cell's shape.
Water pressure supports the leaves of this impatiens plant.

(a) Turgor pressure provides support

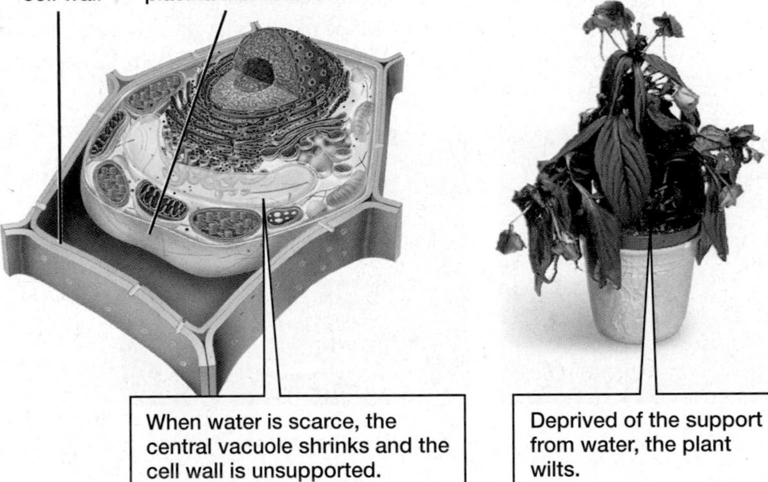

cell wall
plasma membrane
When water is scarce, the central vacuole shrinks and the cell wall is unsupported.
Deprived of the support from water, the plant wilts.

(b) Loss of turgor pressure causes the plant to wilt

▲ **FIGURE 5-7 Turgor pressure in plant cells** Aquaporins allow water to move rapidly in and out of the central vacuoles of plant cells. **(a)** The cell and the plant are supported by turgor pressure. **(b)** The cell and the plant have lost turgor pressure and support due to dehydration.

THINK CRITICALLY If a plant cell is placed in water containing no solutes, will the cell eventually burst? Explain.

Dissolved substances stored in the vacuole make its contents hypertonic to the surrounding cytosol, which in turn is usually hypertonic to the interstitial fluid that bathes the cells. Water therefore flows through the cell wall into the cytosol and then into the central vacuole by osmosis. This produces **turgor pressure**, which causes the plasma membrane to press against the cell wall (**FIG. 5-7a**). If you forget to water a houseplant, the cytosol and cell vacuole lose water, causing the plant's cells to shrink away from their cell walls. The plant droops as its cells lose turgor pressure (**FIG. 5-7b**). Now you know why grocery stores are always spraying their leafy produce: to keep it looking perky and fresh with full central vacuoles.

Energy-Requiring Transport Includes Active Transport, Endocytosis, and Exocytosis

Many cellular activities rely on energy-requiring transport. Active transport, endocytosis, and exocytosis are crucial to maintaining concentration gradients, acquiring food, excreting wastes, and (in multicellular organisms) communicating with other cells.

Cells Maintain Concentration Gradients Using Active Transport

By building gradients and then allowing the gradients to run down under specific circumstances, cells generate ATP and respond to stimuli. For example, concentration gradients of various ions provide the energy to form ATP in mitochondria and chloroplasts (see Chapters 7 and 8), power the electrical signals of nerve cells, and trigger the contraction of muscles. But gradients cannot form spontaneously—they require active transport across a membrane.

During **active transport**, membrane proteins use cellular energy to move molecules or ions across a plasma membrane *against* their concentration gradients, which means that the substances are transported from areas of lower concentration to areas of higher concentration (**FIG. 5-8**). For example, every cell must use active transport to acquire some nutrients that are less concentrated in the environment than in the cell's cytoplasm.

Active transport proteins span the width of the membrane and have two binding regions (see Fig. 5-8). One of these regions binds loosely with a particular molecule or ion, such as a calcium ion (Ca^{2+}). The second region, on the inside of the membrane, binds ATP ❶. The ATP donates energy to the protein, causing the protein to change shape and move the calcium ion across the membrane ❷. The energy for active transport comes from breaking the high-energy bond that links the last of the three phosphate groups in ATP. As it loses a phosphate group, releasing its stored energy, ATP becomes ADP (adenosine diphosphate) plus a free phosphate ❸. Active transport proteins are often called *pumps* because, just as pumping water into an elevated storage tank requires energy, active transport proteins use energy to move ions or molecules "uphill" against a concentration gradient.

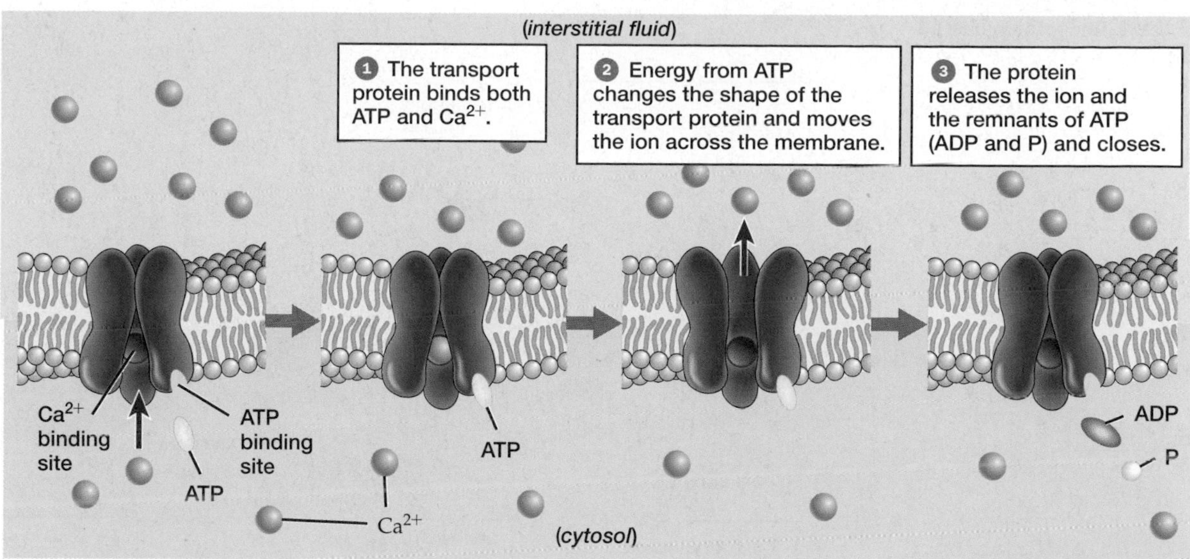
(interstitial fluid)
1 The transport protein binds both ATP and Ca^{2+}.
2 Energy from ATP changes the shape of the transport protein and moves the ion across the membrane.
3 The protein releases the ion and the remnants of ATP (ADP and P) and closes.
Ca^{2+} binding site
ATP binding site
ATP
ATP
Ca^{2+}
ADP
P
(cytosol)

▲ **FIGURE 5-8 Active transport** Cellular energy is used to move molecules across the plasma membrane against a concentration gradient. The active transport protein has an ATP binding site and a binding site for the transported substance, such as these calcium ions. When ATP donates its energy, it loses its third phosphate group and becomes ADP plus a free phosphate.

THINK CRITICALLY Would a cell ever use active transport to move water across its membrane? Explain.

Endocytosis Allows Cells to Engulf Particles or Fluids

A cell may need to acquire materials from its environment that are too large to move directly through the membrane. Such materials are engulfed by the plasma membrane and transported within the cell inside vesicles. This energy-requiring process is called **endocytosis** (Gk. *endo*, inside). Here, we describe three forms of endocytosis that differ in the size and type of material acquired and the method of acquisition: pinocytosis, receptor-mediated endocytosis, and phagocytosis.

Pinocytosis (Gk. *pino*, drink) moves a droplet of interstitial fluid into the cell (**FIG. 5-9**). As a result, the cell acquires materials in the same concentration as in the interstitial fluid. Pinocytosis begins when a very small patch of plasma membrane dimples inward as it surrounds interstitial fluid ❶. The dimple deepens as it encloses fluid from outside the cell ❷. Then the membrane buds off into the cytosol as a tiny vesicle ❸.

Cells use **receptor-mediated endocytosis** to selectively take up specific molecules or complexes of molecules that cannot move through channels or diffuse through the plasma membrane (**FIG. 5-10**). Receptor-mediated endocytosis occurs in thickened depressions called *coated pits*. In the pit, receptor proteins for a specific substance project from the plasma membrane. These receptors bind the molecules to be transported. Then the depression deepens into a pocket that pinches off, forming a coated vesicle that carries the molecules into the cytosol. Molecules moved by receptor-mediated endocytosis include most protein hormones and packets of lipoprotein that contain cholesterol.

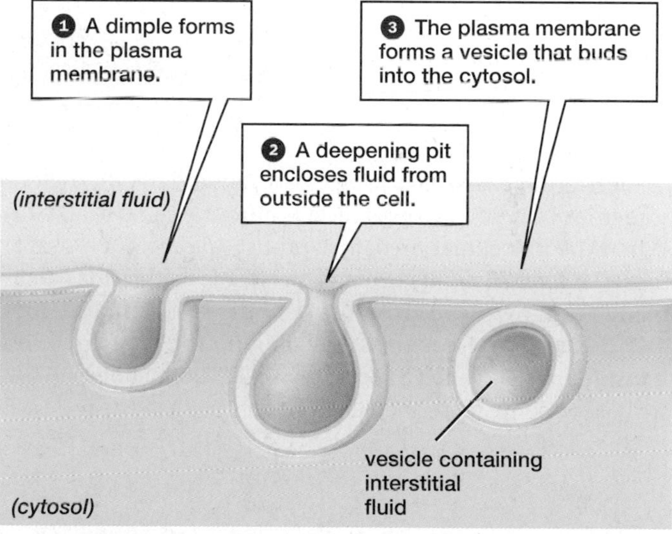
1 A dimple forms in the plasma membrane.
3 The plasma membrane forms a vesicle that buds into the cytosol.
2 A deepening pit encloses fluid from outside the cell.
(interstitial fluid)
vesicle containing interstitial fluid
(cytosol)

(a) Pinocytosis

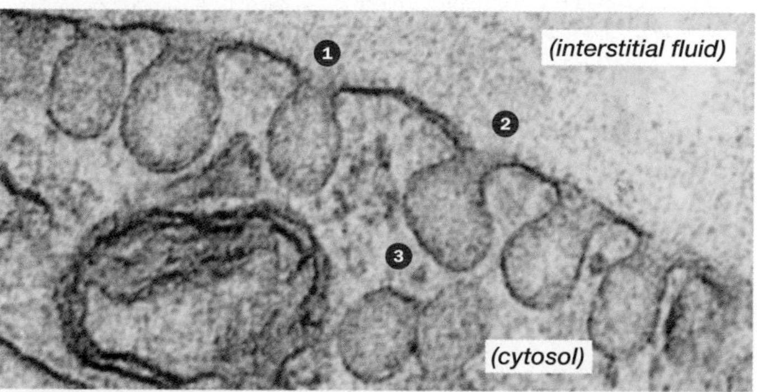
1
(interstitial fluid)
2
3
(cytosol)

(b) TEM of pinocytosis

▲ **FIGURE 5-9 Pinocytosis**

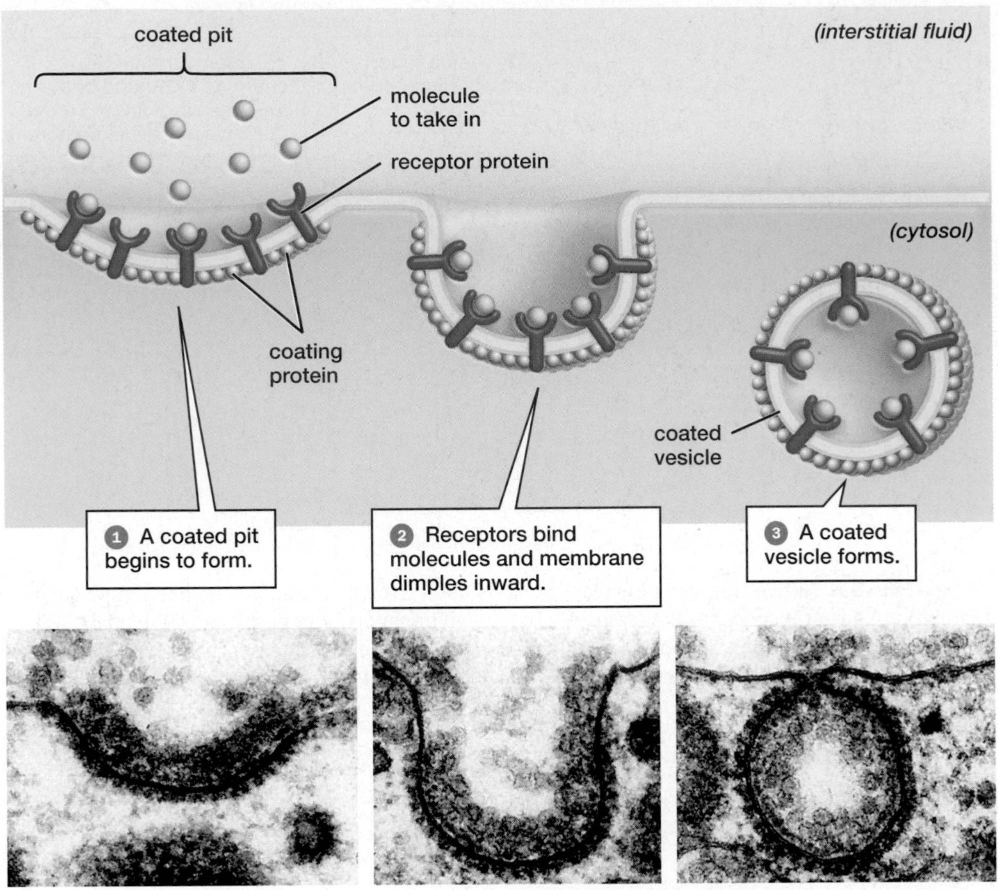

▶ **FIGURE 5-10 Receptor-mediated endocytosis**

Phagocytosis (Gk. *phago*, eat) moves large particles—sometimes whole microorganisms—into the cell (**FIG. 5-11a**). When the unicellular predatory protist *Amoeba,* for example, senses *Paramecium,* it extends parts of its plasma membrane, forming pseudopods (Gk. *pseudo*, false, and *pod*, foot; **FIG. 5-11b**). The pseudopods fuse around the prey, enclosing it inside a vesicle called a **food vacuole**. The vacuole will fuse with a lysosome (described in Chapter 4) where the food will be digested. White blood cells use phagocytosis, followed by digestion, to engulf and destroy invading bacteria (**FIG. 5-11c**).

Exocytosis Moves Material Out of the Cell

Cells also use energy to dispose of undigested particles or to secrete substances—such as enzymes, hormones, or antibodies—into the interstitial fluid, a process called **exocytosis**

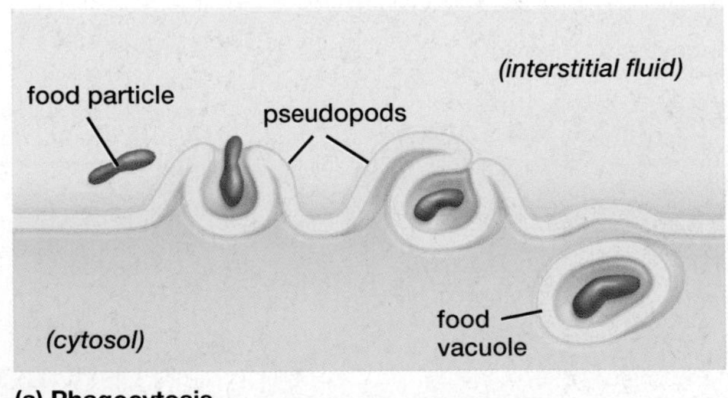

(a) Phagocytosis

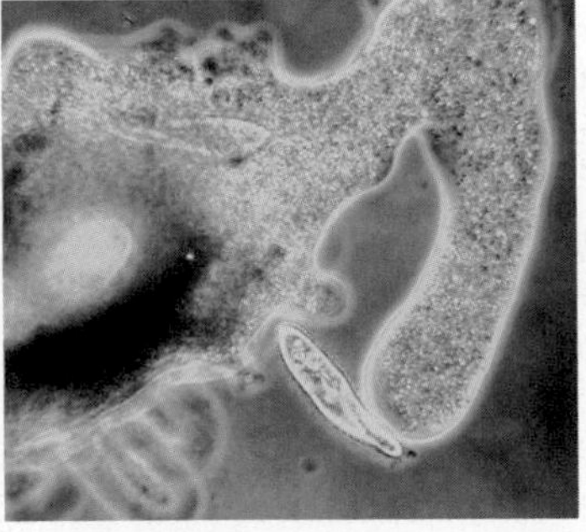

(b) *Amoeba* engulfs *Paramecium*

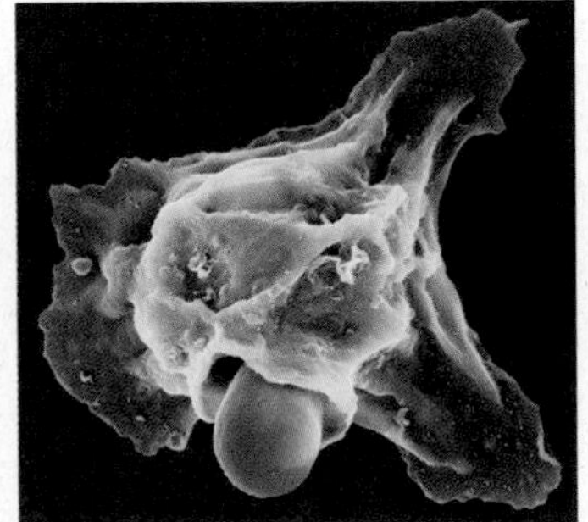

(c) A white blood cell engulfs a disease-causing fungal cell

▲ **FIGURE 5-11 Phagocytosis (a)** The mechanism of phagocytosis. **(b)** Amoebas use phagocytosis to feed, and **(c)** white blood cells use phagocytosis to engulf disease-causing microorganisms.

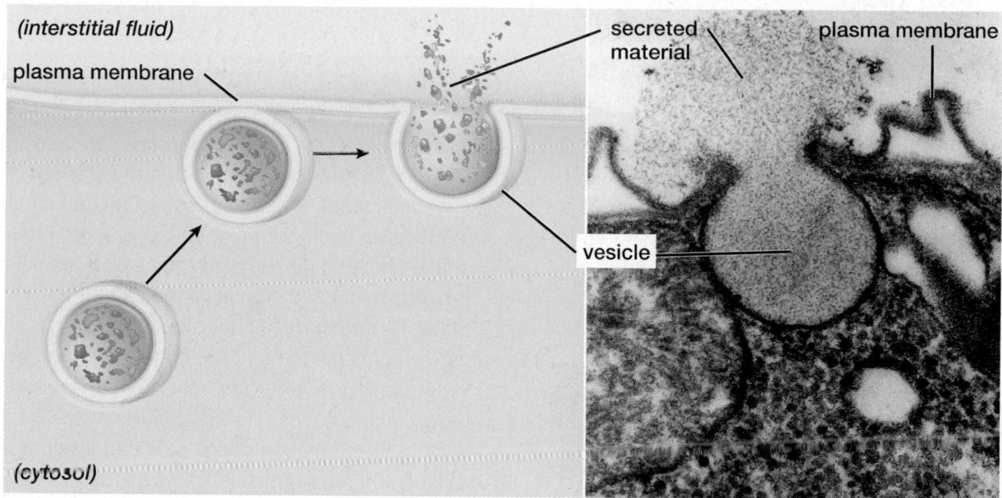

▲ **FIGURE 5-12 Exocytosis** Exocytosis is functionally the reverse of endocytosis. Material enclosed in a vesicle from inside the cell is transported to the cell surface. The vesicle then fuses with the plasma membrane, releasing its contents into the surrounding fluid.

THINK CRITICALLY How does exocytosis differ from diffusion of materials out of a cell?

(Gk. *exo*, outside; **FIG. 5-12**). During exocytosis, a membrane-enclosed vesicle carrying material to be expelled moves to the cell surface, where the vesicle's membrane fuses with the cell's plasma membrane. The vesicle's contents then diffuse into the fluid outside the cell.

Exchange of Materials Across Membranes Influences Cell Size and Shape

Why are cells so small? Cells rely on the slow process of diffusion to acquire nutrients and eliminate wastes, so a cell must be small enough that no part of it is too far removed from the surrounding fluid. The larger a cell's diameter, the farther its innermost contents are from the plasma membrane. In a hypothetical spherical giant cell 8.5 inches (20 centimeters) in diameter, oxygen molecules entering across the plasma membrane would take more than 200 days to diffuse to the center of the cell. A waste molecule produced in the center of the cell would take a similar amount of time to reach the membrane and be expelled. No actual cell could survive such a long wait to acquire necessary molecules and expel undesirable ones.

More generally, as cell size increases, the volume of cytoplasm (where all metabolic reactions occur) increases more rapidly than does surface area (through which nutrients and wastes must be exchanged; **FIG. 5-13**). This constraint limits the size of most cells. However, some cells, such as nerve and muscle cells, can be relatively large, due to an extremely elongated shape that increases membrane surface area, keeping the ratio of surface area to volume relatively high.

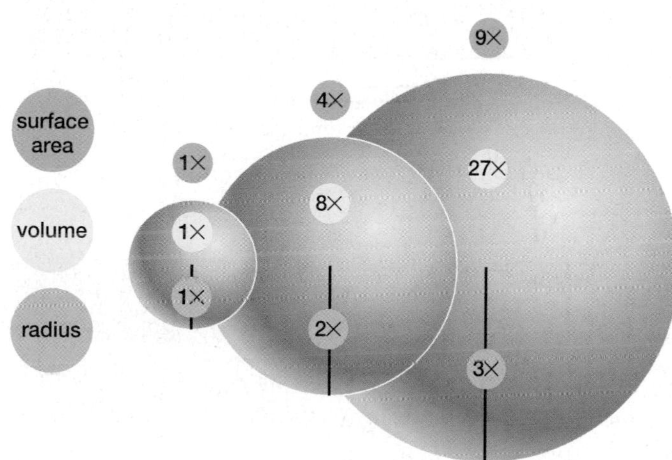

▲ **FIGURE 5-13 Surface area and volume relationships** If the radius of a sphere increases by a factor of 3, then the volume increases by a factor of 27 but the surface area only increases by a factor of 9.

CHECK YOUR LEARNING

Can you . . .

- explain simple diffusion, facilitated diffusion, and osmosis?
- describe active transport, endocytosis, and exocytosis?
- explain how the need to exchange materials across membranes influences cell size and shape?

CASE STUDY REVISITED

Vicious Venoms

The "witches' brews" of rattlesnake and brown recluse spider venoms contain phospholipases that destroy the tissue around the bite (**FIG. 5-14**). When phospholipases attack the membranes of cells that make up capillaries, these tiny blood vessels rupture and release blood into the tissue surrounding the wound.

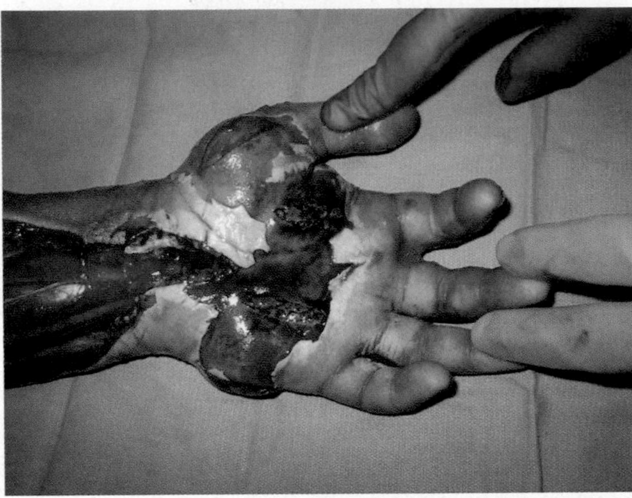

▲ **FIGURE 5-14 Phospholipases in venoms can destroy cells** Justin Schwartz's hand 36 hours after the rattlesnake bite.

In extreme cases, capillary damage can lead to internal bleeding. By attacking the membranes of red blood cells, the phospholipases in rattlesnake venom can cause anemia (an inadequate number of oxygen-carrying red blood cells). Rattlesnake phospholipases also attack muscle cell membranes; this attack caused extensive damage to muscles in Justin Schwartz's forearm. Justin required large quantities of antivenin, which contains proteins that bind and neutralize snake venom proteins. Unfortunately, no antivenin is available for brown recluse bites, and treatment generally consists of preventing infection, controlling pain and swelling, and waiting—sometimes for months—for the wound to heal.

Although both snake and spider bites can have serious consequences, very few of the spiders and snakes found in North America are dangerous to people. The best defense is to learn which venomous animals live in your area and where they prefer to hang out. If your activities bring you to such places, wear protective clothing—and always look before you reach! Knowledge can help us coexist with spiders and snakes, avoid their bites, and keep our cell membranes intact.

THINK CRITICALLY Phospholipases are found in animal digestive tracts as well as in snake venom. Formulate a hypothesis about how the role of phospholipases in digestion might differ from their role in venom.

CHAPTER REVIEW

Go to **Mastering Biology** to access the Pearson eText, vocabulary review, practice quizzes, activities, videos, current events, and more.

*Answers to **Think Critically** and **Thinking Through the Concepts** questions can be found in the **Answers** section at the back of the book.*

Summary of Key Concepts

5.1 How Is the Structure of the Cell Membrane Related to Its Function?

The cell membrane consists of a bilayer of phospholipids in which a variety of proteins are embedded, often described as a fluid mosaic. The plasma membrane isolates the cytoplasm from the external environment, regulates the flow of materials into and out of the cell, allows communication between cells, allows attachments within and between cells, and regulates many biochemical reactions. There are five major types of membrane proteins: (1) enzymes, which promote chemical reactions; (2) recognition proteins, which label the cell; (3) transport proteins, which regulate the movement of most water-soluble substances through the membrane; (4) receptor proteins, which bind molecules and trigger changes within the cell; and (5) connection proteins, which anchor the plasma membrane to the cytoskeleton and extracellular matrix or bind cells to one another.

5.2 Which Physical Processes Move Molecules in Fluids?

Diffusion is the movement of substances from regions of higher concentration to regions of lower concentration. Osmosis is the diffusion of water across a selectively permeable membrane down its concentration gradient through the phospholipid bilayer or through aquaporins.

5.3 How Do Substances Move Across Membranes?

In simple diffusion, water, dissolved gases, and lipid-soluble molecules diffuse through the phospholipid bilayer. During facilitated diffusion, carrier proteins or channel proteins allow water and water-soluble molecules to cross the membrane down their concentration gradients without expending cellular energy.

Energy-requiring transport includes active transport, in which carrier proteins use cellular energy (ATP) to drive the movement of molecules across the plasma membrane against concentration gradients. Interstitial fluid, large molecules, and food particles may be acquired by endocytosis, which includes pinocytosis, receptor-mediated endocytosis, and phagocytosis.

The secretion of substances and the excretion of particulate cellular wastes are accomplished by exocytosis.

Cells exchange materials between the cytoplasm and the external environment principally by the slow process of diffusion through the plasma membrane. This requires that no part of the cell be too far from the plasma membrane, limiting the diameter of cells.

Thinking Through the Concepts

Bloom's: Remembering, Understanding

Multiple Choice

1. Animal cells are surrounded by ________ fluid. This fluid is ________ to the cytosol.
   a. phospholipid; isotonic
   b. plasma; hypertonic
   c. interstitial; isotonic
   d. interstitial; hypotonic
2. Which of the following *cannot* enter a cell by simple diffusion?
   a. water
   b. sugar
   c. estrogen
   d. oxygen
3. Glycoproteins are important for
   a. binding hormones.
   b. cell recognition by the immune system.
   c. forming ion channels.
   d. active transport through the membrane.
4. Diffusion
   a. is always facilitated.
   b. occurs by active transport.
   c. is increased when temperature increases.
   d. requires aquaporin proteins.
5. Which of the following is *not* true of endocytosis?
   a. It is a form of passive transport.
   b. It includes pinocytosis.
   c. It can occur in coated pits.
   d. It is used by *Amoeba* to feed on *Paramecium*.

Fill-in-the-Blank

1. Membranes consist of a bilayer of ____________. The five major categories of protein within the bilayer are ____________, ____________, ____________, ____________, and ____________ proteins.
2. A membrane that is permeable to some substances but not to others is described as being ____________. The movement of a substance through a membrane down its concentration gradient is called ____________. When applied to water, this process is called ____________. Channels that are specific for water are called ____________. The process that moves substances through a membrane against their concentration gradient is called ____________.
3. Facilitated diffusion involves either ____________ proteins or ____________ proteins. Diffusion directly through the phospholipid bilayer is called ____________ diffusion, and molecules that take this route must be soluble in ____________ or be very small and have no net electrical charge.
4. After each molecule, place the two-word term that most specifically describes the *process* by which it moves through a plasma membrane. Carbon dioxide: ____________; ethyl alcohol: ____________; a sodium ion: ____________; glucose: ____________.
5. The general process by which fluids or particles are transported out of cells is called ____________. Does this process require energy? ____________ The substances to be expelled are transported within the cell in membrane-enclosed sacs called ____________.

Review Questions

1. Describe and diagram the structure of a plasma membrane. What are the two principal types of molecules in plasma membranes, and what is the general function of each?
2. Sketch the configuration that 10 phospholipid molecules would assume if placed in water. Explain why they arrange themselves this way.
3. What are the five categories of proteins commonly found in plasma membranes, and what is the function of each one?
4. Define *diffusion* and *osmosis*. Explain how osmosis helps plant leaves remain firm. What is the term for water pressure inside plant cells?
5. Define *hypotonic*, *hypertonic*, and *isotonic*. What would be the fate of an animal cell immersed in each of these three types of solution?
6. Describe the following types of transport processes in cells: simple diffusion, facilitated diffusion, active transport, pinocytosis, receptor-mediated endocytosis, phagocytosis, and exocytosis.
7. Name the protein that allows facilitated diffusion of water. What experiment demonstrated the function of this protein?
8. Imagine a container of glucose solution, divided into two compartments (A and B) by a membrane that is permeable to water and glucose but not to sucrose. If some sucrose is added to compartment A, how will the contents of compartment B change? Explain.

Applying the Concepts

Bloom's: Applying, Analyzing, Evaluating

1. Different cells have different plasma membranes. The plasma membrane of *Paramecium*, for example, is only about 1% as permeable to water as the plasma membrane of a human red blood cell. Hypothesize why this is the case. Is *Paramecium* likely to have aquaporins in its plasma membrane? Explain your answer.
2. Predict and sketch the configuration that 10 phospholipid molecules would assume if they were completely submerged in vegetable oil. Explain your prediction.

6

Energy Flow in the Life of a Cell

Runners in the New York Marathon convert energy stored in their bodies into the energy of movement and heat. Their pounding footsteps shake the Verrazano Narrows Bridge.

CASE STUDY

Energy Unleashed

PICTURE THE NEW YORK CITY MARATHON, in which well over 50,000 people from countries throughout the world gather to run 26.2 miles. The race is a testimony to persistence, endurance, and the ability of the human body to use energy. On average, each runner expends roughly 3,000 Calories before reaching the finish line. Once finished, the runners douse their overheated bodies with water and replenish their energy stores with celebratory meals.

Training for a marathon takes months and is especially grueling for people unaccustomed to running long distances. Training triggers important physiological changes that prepare the body to expend the tremendous amount of energy necessary for the race. In muscle cells, the number of mitochondria increases, improving the cells' ability to metabolize glucose. Muscle cells also increase their ability to store glycogen, a polymer of glucose. Near the muscles, capillaries proliferate to supply the extra oxygen that muscle mitochondria need to break down glucose.

What exactly is energy? Do our bodies use energy according to the same principles that govern its use in the engines of cars and airplanes? Why do our bodies generate heat, and why do we give off more heat when exercising than when studying or watching TV?

AT A GLANCE

6.1 WHAT IS ENERGY?

Energy is the capacity to do work. **Work**, in turn, is the transfer of energy to an object that causes the object to move. It is obvious that marathoners are working; their chests heave, their arms pump, and their legs stride, moving their bodies relentlessly forward. This muscular work is powered by the energy available in the bonds of molecules. The molecules that provide this energy—including glucose, glycogen, and fat—are stored in the cells of the runners' bodies. Within each cell, specialized molecules such as ATP accept, briefly store, and then transfer energy from reactions that release energy to reactions that demand it, such as muscle contraction.

There are two fundamental types of energy: potential energy and kinetic energy, each of which takes several forms. **Potential energy** is stored energy. It includes the elastic energy stored in a compressed spring or a drawn bow, and the gravitational energy stored in water behind a dam or a roller-coaster car about to begin its downward plunge (**FIG. 6-1**). Potential energy also includes **chemical energy**, which is energy stored in, for example, the battery that powers your phone, the biological molecules that power marathon runners, and the fossil fuels that power motorized vehicles. **Kinetic energy** is the energy of movement. It includes radiant energy (such as light, X-rays, and other forms of *electromagnetic radiation*), heat or thermal energy (the motion of molecules or atoms), and electrical energy (electricity; the flow of charged particles). Kinetic energy also includes any motion of larger objects, such as plummeting roller-coaster cars or running marathoners.

Under the right conditions, potential energy can be transformed into kinetic energy, and vice versa. For example, as a roller-coaster car plunges downward from atop a peak its gravitational potential energy is converted to the kinetic energy of movement. As the car coasts to the top of the next rise, its kinetic energy is converted back to potential energy. At a molecular level, during photosynthesis, the kinetic energy of light is captured and transformed into the potential energy of chemical bonds (see Chapter 7). When a marathoner's muscle cells break down glucose, the potential energy of chemical bonds is converted to the kinetic energy of movement and heat.

The Laws of Thermodynamics Describe the Basic Properties of Energy

To better understand how energy flows and changes, we need to know more about its properties; the laws of thermodynamics describe some of these properties.

▲ **FIGURE 6-1 Converting potential energy to kinetic energy** Roller coaster cars convert gravitational potential energy to kinetic energy as they plummet downhill.

THINK CRITICALLY Could one design a roller coaster that didn't use any motors to pull the cars uphill after they were released from a high point? Explain your answer.

Energy Is Neither Created nor Destroyed

The **first law of thermodynamics** states that energy can be neither created nor destroyed by ordinary processes. (Nuclear reactions are the exception.) This means that within an **isolated system**—a space where neither mass nor energy can enter or leave—the total amount of energy before and after any process will be unchanged. For this reason, the first law of thermodynamics is often called the **law of conservation of energy**.

To illustrate the law of conservation of energy, consider a gasoline powered car. Before you turn the ignition key, the energy in the car is all potential energy, stored in the chemical

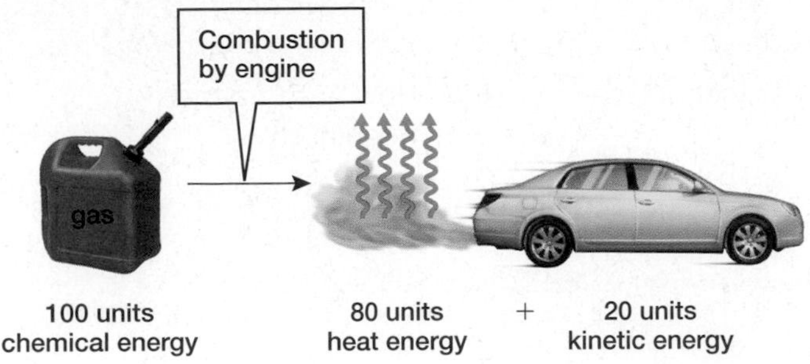

▲ **FIGURE 6-2 All energy conversions result in a loss of useful energy**

bonds of its fuel. As you drive, only about 20% of this potential energy is converted into the kinetic energy of motion. But since energy is neither created nor destroyed, what happens to the other 80% of the energy? The burning fuel also heats up the engine, the exhaust system, and the air around the car, while friction from the tires heats the road. So, as the first law dictates, the total amount of energy remains the same, although it has changed in form—with about 20% of it converted to kinetic energy and 80% of it to heat (**FIG. 6-2**).

Energy Conversion Reduces the Amount of Useful Energy

The **second law of thermodynamics** states that when energy is converted from one form to another, the amount of useful energy decreases. In other words, all ordinary (nonnuclear) processes cause energy to be converted from more useful into less useful forms. In our car example, combustion converted useful chemical energy to less-useful heat, which merely increased the random movement of molecules in the car, the air, and the road.

Now consider the human body. Whether running or reading, your body "burns" food to release the chemical energy stored in the food's molecules. Some of the energy is released as heat, which warms your body and is radiated to your surroundings. The energy lost as heat is not available to power muscle contraction or to help brain cells interpret written words. The second law tells us that no energy conversion process in the body is 100% efficient. Some energy is lost to the environment—almost always in the form of heat—and cannot be used to power useful activity.

Disorder Tends to Increase

The second law of thermodynamics also tells us something about the organization of matter. Useful energy tends to be stored in highly ordered matter, such as in the bonds of complex molecules. As a result, whenever energy is used within an isolated system, there is an overall loss of organization as complex molecules are broken apart into simpler ones. This tendency toward the loss of complexity, orderliness, and useful energy is called **entropy**. The effects of entropy are apparent as we perform the activities of daily life: Dirty dishes accumulate, clothes collect in confusion, the bed gets rumpled, and books and papers pile up (**FIG. 6-3**). This randomness and disorder can only be reversed by adding energy to the system through energy-demanding cleaning and organizing efforts.

▲ **FIGURE 6-3 Entropy at work**

At the molecular level, we see entropy at work. For example, consider what happens when we burn glucose sugar:

$$\underset{\text{glucose}}{C_6H_{12}O_6} + \underset{\text{oxygen}}{6\,O_2} \rightarrow \underset{\substack{\text{carbon}\\\text{dioxide}}}{6\,CO_2} + \underset{\text{water}}{6\,H_2O} + \text{heat energy}$$

The number and types of atoms is unchanged by the reaction, but if you count the molecules on each side of the equation, you will see that there is an overall increase in the number of molecules as a complex molecule of sugar is combined with oxygen and broken down to form simpler product molecules (carbon dioxide and water). The heat energy that is released causes the product molecules to move about randomly and more rapidly. To counteract entropy—for example, to synthesize glucose from carbon dioxide and water—energy must be added to the system from an outside source, ultimately the sun. When the eminent Yale scientist G. Evelyn Hutchinson stated, "Disorder spreads through the universe, and life alone battles against it," he was making an eloquent reference to entropy and the second law of thermodynamics.

CASE STUDY **CONTINUED**

Energy Unleashed

Like a car's engine, a marathoner's muscles are only about 20% efficient in converting chemical energy into movement. Much of the wasted 80% is lost as heat, which can overheat the runner. Sweating helps to prevent overheating because the water in sweat absorbs large amounts of heat as it evaporates. A sweat-drenched marathoner is burning a lot of fuel, but even when we're doing something non-sweaty like sitting at a computer or listening to a lecture, we still use energy, just to stay alive. Where does this energy come from?

Living Things Use Solar Energy to Maintain Life

If you think about the second law of thermodynamics, you may wonder how life can exist at all. If chemical reactions, including those inside living cells, cause the amount of usable energy to decrease, and if matter tends toward increasing randomness and disorder, how can organisms maintain their amazingly organized complexity? Where does useful energy originate, and where does all the waste heat go? The answer is that cells, organisms, and Earth itself are not isolated systems; they receive useful solar energy released by nuclear reactions in the sun, 93 million miles away.

Living things "battle against disorder" by using a continuous inflow of sunlight to synthesize complex molecules and maintain their intricate bodies. Energy from sunlight enters the biosphere through photosynthetic organisms such as plants and algae. The chemical equation for photosynthesis is:

$$6\,CO_2 + 6\,H_2O + \text{light energy} \rightarrow C_6H_{12}O_6 + 6\,O_2$$

Notice that photosynthesis reverses the reaction that breaks down glucose; the energy of sunlight is captured and stored in the chemical bonds of glucose. Thus, the highly ordered (and therefore low-entropy) systems that characterize life do not violate the second law of thermodynamics; such systems are maintained by a continuous inflow of useful energy from the sun. When you organize your desk or make your bed, your muscles are (indirectly) using solar energy that was originally trapped by photosynthesis.

CHECK YOUR LEARNING

Can you . . .

- define energy and work?
- define potential energy and kinetic energy and provide three examples of each?
- state and explain the first and second laws of thermodynamics?

6.2 HOW IS ENERGY TRANSFORMED DURING CHEMICAL REACTIONS?

A **chemical reaction** is a process that breaks and forms chemical bonds. Chemical reactions convert one combination of molecules, the **reactants**, into different molecules, the **products**, that contain the same number and types of atoms as the reactants. A reaction is **exergonic** if there is an overall release of heat, that is, if the products contain less energy than the original reactants (**FIG. 6-4a**). A reaction is **endergonic** if it requires a net input of energy, that is, if the products contain more energy than the reactants. Endergonic reactions require an overall inflow of energy from an outside source (**FIG. 6-4b**).

Exergonic Reactions Release Energy

Sugar can be set afire, as any cook can tell you. As it burns, sugar combines with oxygen to produce carbon dioxide and water, while generating heat. In organisms, sugar undergoes the same overall reaction, but some of the energy released is stored as chemical energy in ATP (the rest is released as heat). The total energy in the reactant molecules (glucose and oxygen) is much higher than in the product molecules (carbon dioxide and water), so burning sugar is an exergonic reaction. It may be helpful to think of exergonic reactions as running "downhill," from a higher energy state to a lower energy state, like a rock rolling down a hill to rest at the bottom.

Endergonic Reactions Require a Net Input of Energy

Many reactions in living things are endergonic, requiring a net input of energy and yielding products that contain more energy than the reactants. The synthesis of large biological molecules is endergonic. For example, the proteins in a muscle cell contain more energy than the individual amino acids that were linked together to synthesize them. How do organisms power endergonic reactions? They use high-energy molecules synthesized using solar energy that was captured during photosynthesis. We can think of endergonic reactions as "uphill" reactions because they require a net input of energy, just as pushing a rock up a hill requires effort.

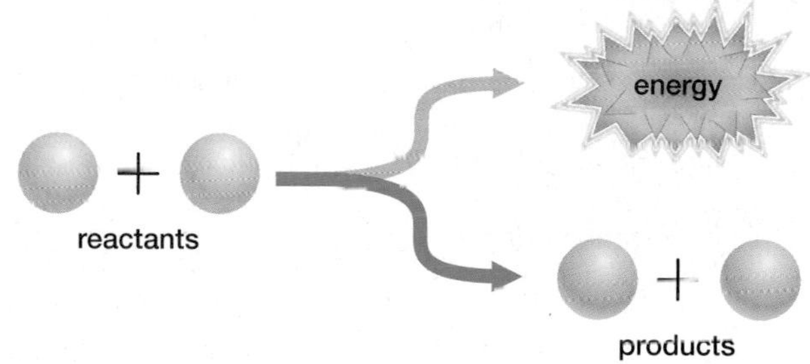

(a) An exergonic reaction

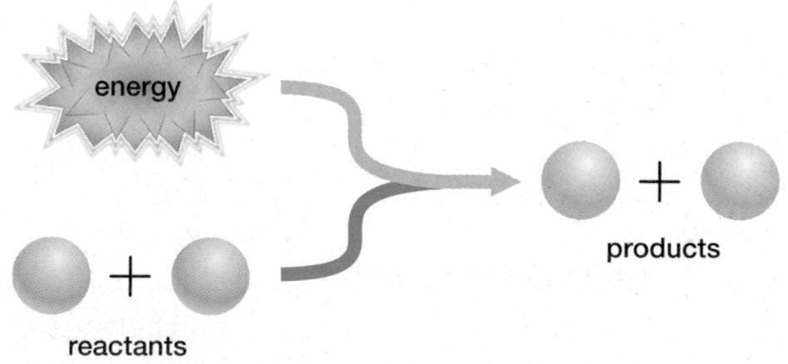

(b) An endergonic reaction

▲ **FIGURE 6-4 Exergonic and endergonic reaction (a)** In an exergonic reaction, the products contain less energy than the reactants. **(b)** In an endergonic reaction, the products contain more energy than the reactants.

THINK CRITICALLY Is glucose breakdown endergonic or exergonic? What about photosynthesis?

All Chemical Reactions Require Activation Energy to Begin

All chemical reactions require **activation energy** to get started (**FIG. 6-5a**). Think of a rock sitting at the top of a hill; it will remain there indefinitely unless a push starts it rolling down. Like a rolling rock, many chemical reactions continue spontaneously if enough activation energy is supplied to start them. We see this in wood burning in a campfire or in a marshmallow ignited by the fire's flames (**FIG. 6-5b**). After an input of energy ignites the sugar in a marshmallow, the sugar reacts with oxygen from the air, and the marshmallow burns spontaneously.

Why is activation energy required for chemical reactions? Shells of negatively charged electrons surround all atoms (see Chapter 2). These negative charges repel one another and tend to keep the atoms separated. Activation energy is required to overcome this repulsion and force the atoms close enough together to react and form new chemical bonds.

Activation energy can be provided by the kinetic energy of moving molecules. Atoms and molecules are in constant motion. If they are moving fast enough, collisions between reactive molecules force their electron shells to mingle and react. Because molecules move faster as the temperature increases, most chemical reactions occur more readily at high temperatures; this is why a flame's heat can ignite a marshmallow.

CHECK YOUR LEARNING

Can you . . .

- describe how energy is captured and released by chemical reactions?
- explain exergonic and endergonic reactions and provide examples of each?
- explain activation energy?

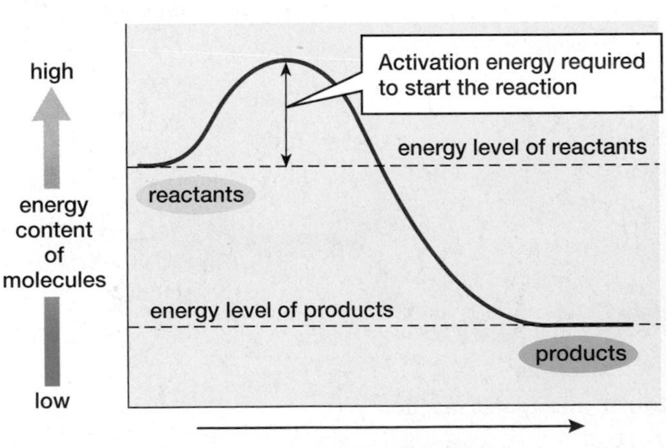

(a) An exergonic reaction

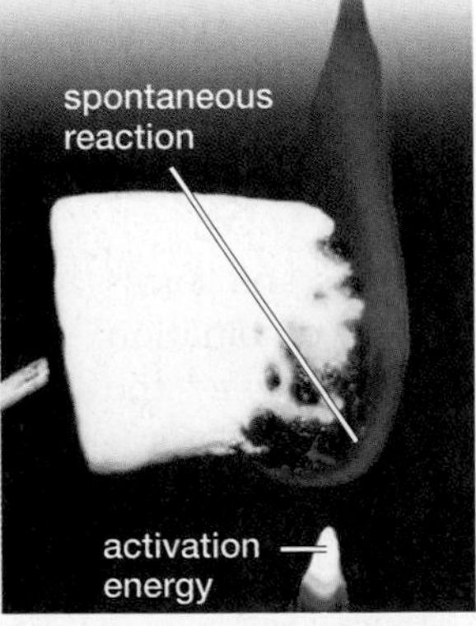

(b) A flame ignites the sugar in a marshmallow

▲ **FIGURE 6-5 Activation energy in an exergonic reaction (a)** After the activation hump is overcome, there will be a net release of energy. **(b)** Heat released by the burning sugar will allow the reaction to continue spontaneously.

CASE STUDY CONTINUED

Energy Unleashed

Marathoners rely on glycogen stored in their muscles and liver for much of the energy to power their run. Glycogen consists of long, branched chains of glucose molecules. When energy is needed, glucose molecules are cut from the chain and then broken down into carbon dioxide and water. This exergonic reaction releases the chemical energy that will power muscle contraction. The carbon dioxide produced by the reaction is exhaled as the runners breathe rapidly to supply their muscles with adequate oxygen. The water generated by the reaction (as well as the water that the runners drink during the race) is lost as cooling sweat.

The glucose molecules in muscle and other cells are not ignited and do not literally burn as they are broken down. So how do cells provide activation energy and control the release of chemical energy to allow the released energy to do work?

6.3 HOW IS ENERGY TRANSPORTED WITHIN CELLS?

Most organisms are powered by the chemical energy supplied by the exergonic breakdown of glucose. But to be used, the chemical energy stored in glucose must first be transferred to energy-carrier molecules, such as ATP. **Energy-carrier molecules** are high-energy molecules that are synthesized at the site of an exergonic reaction, where they capture and temporarily store some of the released chemical energy. Just as rechargeable flashlight batteries can store electrical energy that is later released as light, energy-carrier molecules are charged up by exergonic reactions and then release their energy to drive endergonic reactions. They can then be recharged, as described later. Energy-carrier molecules capture and transfer energy only within cells; they cannot ferry energy through cell membranes, nor are they used for long-term energy storage.

ATP and Electron Carriers Transport Energy Within Cells

Many exergonic reactions in cells, such as breaking down sugars and fats, produce ATP, the most common energy-carrier molecule in the body. ATP (adenosine triphosphate) is a nucleotide composed of the nitrogen-containing base adenine, the sugar ribose, and three phosphate groups. ATP is produced by an endergonic

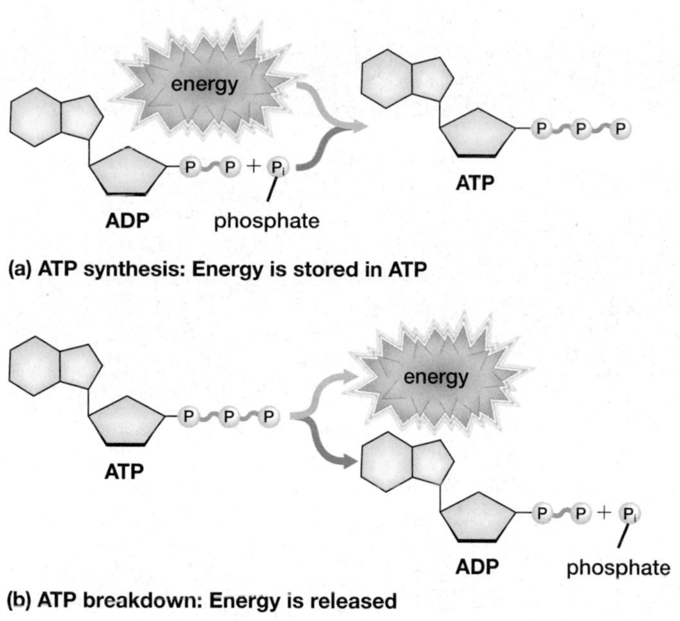

(a) ATP synthesis: Energy is stored in ATP

(b) ATP breakdown: Energy is released

▲ **FIGURE 6-6 The interconversion of ADP and ATP (a)** Energy is captured when a phosphate group (P_i) is added to adenosine diphosphate (ADP) to make adenosine triphosphate (ATP). **(b)** Energy to power cellular work is released when ATP is broken down into ADP and P_i.

reaction (**FIG. 6-6a**) in which energy from exergonic reactions is used to combine inorganic phosphate (HPO_4^{2-}, also designated P_i) with ADP (adenosine diphosphate).

ATP diffuses throughout the cell, carrying energy to sites where endergonic reactions occur. There its energy is liberated as it breaks down, regenerating ADP and P_i (**FIG. 6-6b**). The life span of an ATP molecule in a cell is very short; each molecule is recycled roughly 1,400 times every day. A marathon runner may use a pound (0.45 kilogram) of ATP molecules every minute, so if ATP were not almost instantly recycled, marathons would never happen.

ATP is not the only energy-carrier molecule within cells. In some exergonic reactions, such as glucose breakdown, some energy is transferred to electrons. These energetic electrons, along with hydrogen ions (H^+; present in the cytosol of cells), are captured by molecules called **electron carriers**. The loaded electron carriers donate their high-energy electrons to other molecules, which are often involved in reaction pathways that generate ATP. Common electron carriers include NADH (nicotinamide adenine dinucleotide) and its relative, $FADH_2$ (flavin adenine dinucleotide). You will learn more about electron carriers in Chapters 7 and 8.

Coupled Reactions Link Exergonic with Endergonic Reactions

In a **coupled reaction**, an exergonic reaction provides the energy needed to drive an endergonic reaction (**FIG. 6-7**), using ATP or electron carriers as intermediaries. During photosynthesis, for example, plants use sunlight (from exergonic reactions in the sun's core) to drive the endergonic synthesis of high-energy glucose molecules from lower-energy reactants. Nearly all organisms use the energy released by exergonic reactions (such as the breakdown of glucose) to drive endergonic reactions (such as the synthesis of proteins from amino acids). Because some energy is lost every time it is transformed, in coupled reactions the energy released by exergonic reactions always exceeds the energy needed to drive the endergonic reactions. Thus, the coupled reaction overall is exergonic.

CHECK YOUR LEARNING

Can you . . .

- name and describe two important energy-carrier molecules in cells?
- explain coupled reactions?

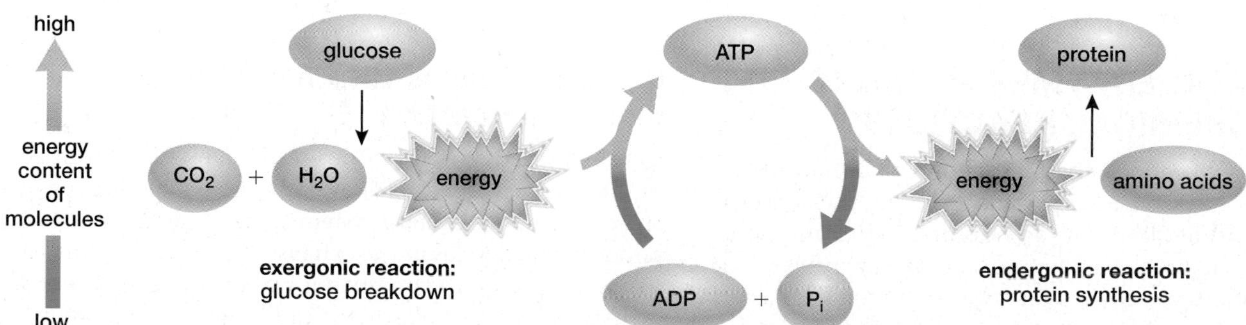

▲ **FIGURE 6-7 Coupled reactions within cells** Exergonic reactions, such as glucose breakdown, drive the endergonic reaction that synthesizes ATP from ADP and P_i. The ATP molecule carries its chemical energy to a part of the cell where energy is needed to drive an endergonic reaction, such as protein synthesis.

THINK CRITICALLY Why is the overall coupled reaction exergonic?

Earth WATCH Enzymes Versus Plastic

The world is awash in plastic. In the 70 or so years since plastic came into widespread use, people worldwide have manufactured 18 billion tons of it. Most types of plastic are made from petroleum products; plastic molecules consist of tens of thousands of carbon and hydrogen atoms, plus some oxygen, nitrogen, sulfur, or chlorine atoms. Plastic breaks down very slowly, so virtually all of the plastic ever manufactured is still with us. Some of it is still in use, but most of it, almost 14 billion tons, ended up as waste. Of this waste, a small portion (9%) was recycled, and some (12%) was burned. But the rest was simply discarded, dumped in landfills and the environment. Today, plastic waste is found almost everywhere, including in all of Earth's oceans (**Fig. E6-1**). Waste plastic and the chemicals that seep out of it pose a significant threat to wildlife and human health. And the problem is likely to worsen; if current trends continue, we will generate an additional 43 billion tons of plastic waste by 2050.

Finding solutions to the problem of plastic waste is a huge challenge, but help might come from an unexpected source: enzymes. Researchers recently discovered a bacterium that digests polyethylene terephthalate (PET), a widely used type of plastic. The researchers found that the bacterium completely breaks down and metabolizes PET, using two previously unknown enzymes. Another research group discovered that larvae (caterpillars) of the greater wax moth eat and degrade polyethylene, the kind of plastic used in shopping bags and many other goods. The caterpillars' normal diet includes beeswax, and the enzymes that digest wax apparently work on plastic as well. However, researchers have not yet identified the caterpillar's plastic-digesting enzymes.

Can plastic-eating organisms help reduce plastic waste? A mass release of hordes of caterpillars is unlikely, but seeding an area with plastic-eating bacteria might one day be feasible. Still, because widespread release of organisms into the environment would have ecological consequences that are difficult to predict, serious consideration of this approach will be possible only after extensive additional research. A more likely scenario is that the enzymes used by plastic-eating organisms will be characterized, extracted or synthesized, and used to develop processes to degrade or recycle plastic waste on a large scale. Nature's consumers of plastic may help point the way to dealing with our self-induced onslaught of the stuff.

▲ **FIGURE E6-1 Plastic is everywhere** This beach on Henderson Island in the South Pacific is covered with plastic debris, even though the island is uninhabited, 3,500 miles from the nearest continent, and one of the most remote places on Earth.

CONSIDER THIS In hope of reducing plastic waste, some towns and cities have banned the use of plastic shopping bags by stores. And in a few places, the sale of water in plastic bottles is prohibited. Do you think such laws are a good idea? Why or why not?

6.4 HOW DO ENZYMES PROMOTE BIOCHEMICAL REACTIONS?

In general, the likelihood that a reaction will occur is determined by its activation energy. Some reactions, such as sugar dissolving in water, have low activation energies and occur rapidly at human body temperature (approximately 98.6°F, or 37°C). In contrast, you could store sugar at this temperature in the presence of oxygen for decades and it would remain virtually unchanged. Why? Because the reaction of sugar with oxygen to yield carbon dioxide and water has a high activation energy. Nonetheless, sugar is one of the most important energy sources for living cells. If sugar molecules in cells do not break down on their own, how is their energy released? With the help of **catalysts**—substances that lower the activation energy of a reaction.

Catalysts Reduce the Energy Required to Start a Reaction

A catalyst speeds up the rate of a reaction by reducing its activation energy (**FIG. 6-8**). For example, in a catalytic converter on an automobile exhaust system, a metal catalyst causes the carbon monoxide gas in the exhaust to combine rapidly with atmospheric oxygen, producing carbon dioxide. If not for catalytic converters, poisonous carbon monoxide, produced by incomplete combustion of gasoline, could reach dangerous levels.

Enzymes Are Biological Catalysts

Most inorganic catalysts speed up a number of different chemical reactions. But in cells, indiscriminately speeding up dozens of different reactions would almost certainly be deadly.

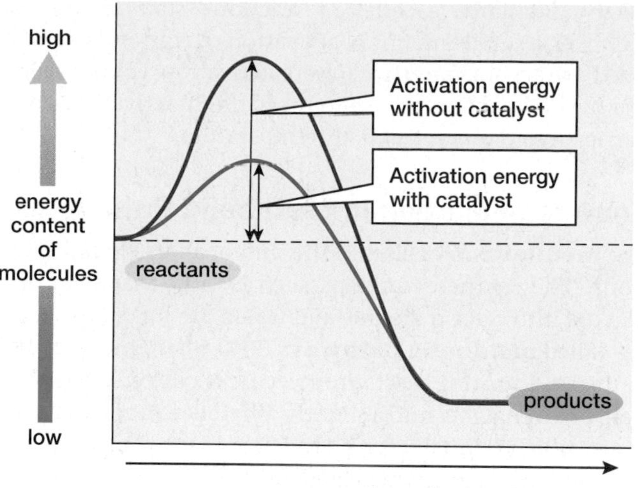
high
Activation energy without catalyst
Activation energy with catalyst
energy content of molecules
reactants
products
low
progress of reaction

▲ **FIGURE 6-8 Catalysts lower activation energy** At any given temperature, a reaction is much more likely to proceed in the presence of a catalyst.

THINK CRITICALLY Can an enzyme catalyst make an endergonic reaction occur spontaneously at body temperature? Explain your answer.

Instead, cells employ highly specific biological catalysts called **enzymes**, nearly all of which are proteins. Most enzymes catalyze only one or a few types of chemical reactions involving particular molecules, leaving very similar molecules unchanged. (Enzymes that catalyze the breakdown of plastic molecules may help reduce the amount of plastic in the environment, as we explore in "Earth Watch: Enzymes Versus Plastic.")

Both exergonic and endergonic reactions are catalyzed by enzymes. The synthesis of ATP from ADP and P_i, for example, is catalyzed by the enzyme *ATP synthase*. The breakdown of ATP to power an endergonic reaction is catalyzed by *ATPase*, another enzyme. As you read about enzymes, be aware that enzyme names are not consistent. In some cases, the suffix "ase" is added to what the enzyme does (ATP synthase); in other cases, "ase" is added to the molecule upon which the enzyme acts (ATPase). Furthermore, some enzymes have names with no "ase" at all (pepsin).

The Structure of Enzymes Allows Them to Catalyze Specific Reactions

The function of an enzyme, like the function of any protein, is determined by its structure (see Chapter 3). Each enzyme's distinctive shape is determined by its amino acid sequence and the precise way in which the amino acid chain is twisted and folded. The three-dimensional structure of enzymes allows them to orient, distort, and reconfigure other molecules, inducing these molecules to react, while the enzyme emerges unchanged. When a catalyst acts to speed up a reaction, it is neither used up nor permanently altered in the process.

Each enzyme has a pocket, called the **active site**, into which reactant molecules, called **substrates**, can enter. The shape of the active site, as well as the charges on the amino acids that form the active site, determine which molecules can enter. Consider the enzyme amylase, for example. Amylase catalyzes the breakdown of starch molecules by hydrolysis, but leaves cellulose molecules intact, even though both starch and cellulose consist of chains of glucose molecules. Why? Because the bonding pattern between glucose molecules in starch allows the glucose chain to fit into the active site of amylase, but the bonding pattern in cellulose does not.

How does an enzyme catalyze a reaction? You can follow the events in **FIGURE 6-9**, which illustrates two substrate molecules combining into a single product. The shape and charge of the active site allow substrates to enter the enzyme only in specific orientations ❶. When appropriate substrates enter the active site, both the substrates and active site change shape slightly as weak chemical bonds form between specific amino acids in the active site and specific parts of the substrate ❷. This shape change distorts the original bonds within the substrate, making these bonds easier to break. The combination of substrate selectivity, substrate orientation, temporary bonds, and the distortion of existing bonds promotes the specific chemical reaction catalyzed by a particular enzyme. This holds true for enzymes that cause two molecules to react with one another and for enzymes that cause a single molecule to split into smaller products. When the reaction is complete, the product no longer fits properly into the active site and drifts away ❸. The enzyme reverts to its original configuration, and it is then ready to accept another set of the same substrates. When substrate

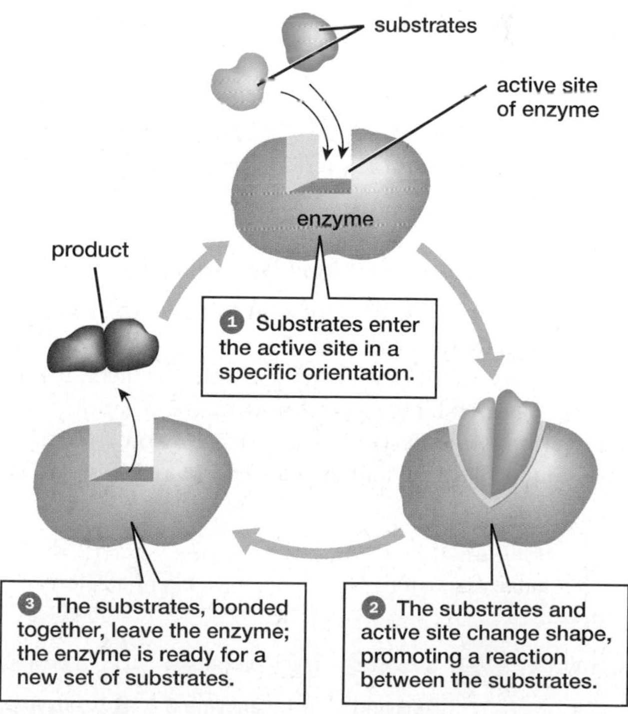
substrates
active site of enzyme
enzyme
product
❶ Substrates enter the active site in a specific orientation.
❷ The substrates and active site change shape, promoting a reaction between the substrates.
❸ The substrates, bonded together, leave the enzyme; the enzyme is ready for a new set of substrates.

▲ **FIGURE 6-9 The cycle of enzyme–substrate interactions** This diagram shows two reactant substrate molecules combining to form a single product. Enzymes can also catalyze the breakdown of a single substrate into two product molecules.

Have You Ever Wondered... If Plants Can Glow in the Dark?

You may have seen the almost magical glow of fireflies, but did you know that plants can be bio-engineered to glow in the dark, too? Fireflies' natural glow comes from the fluorescent chemical luciferin. Luciferin serves as the substrate for the enzyme luciferase. Luciferase catalyzes a reaction that modifies luciferin, using energy from ATP to boost electrons briefly into a higher-energy electron shell. As the electrons fall back into their original shell, they emit their excess energy as light.

Plants don't glow naturally, but bioluminescence has been engineered into watercress plants by researchers at the Massachusetts Institute of Technology. To produce glowing plants, the researchers first manufactured nanoparticles—incredibly tiny particles, made of silica or polymers, that are far smaller than a flu virus. Some of the nanoparticles were engineered to carry luciferin, and others to carry luciferase. The researchers than injected the particles into watercress plants. The luciferase-carrying particles had been designed to cross cell walls and enter leaf cells, whereas the luciferin-carrying particles remained in the extracellular space around the cells. Gradually, the particles outside the cells released their luciferin, which entered the cells and reacted with the luciferase enzyme there. The result: plants that glowed for about 3½ hours. The researcher hope to eventually develop a plant that will glow throughout its life. One day, we might see trees that are also streetlights.

Reading by "plant-light"

molecules are abundant, some fast-acting enzymes can catalyze tens of thousands of reactions per second, while others act far more slowly.

Enzyme-Catalyzed Reactions Occur in Steps

The breakdown or synthesis of a molecule within a cell usually occurs in many small steps, each catalyzed by a different enzyme. Each of these enzymes lowers the activation energy for its particular reaction, allowing the reaction to occur readily at body temperature. Consider how much easier it is to walk up a flight of stairs than to scale a cliff of the same height. Similarly, a series of reaction "stair steps"—each requiring a small amount of activation energy and each catalyzed by an enzyme that lowers activation energy—allows the overall reaction to surmount its high activation energy barrier and to proceed at body temperature.

Enzymes Function in Metabolic Pathways

The **metabolism** of a cell is the sum of all its chemical reactions. Many of these reactions, such as those that break down glucose into carbon dioxide and water, are linked in sequences called **metabolic pathways** (**FIG. 6-10**). In a metabolic pathway, a starting reactant molecule is converted, with the help of an enzyme, into a slightly different intermediate molecule, which is modified with the help of yet another enzyme to form a second intermediate, and so on, until an end product is produced. Photosynthesis (see Chapter 7), for example, is a metabolic pathway, as is the breakdown of glucose (Chapter 8). Different metabolic pathways often use some of the same molecules; as a result, all the thousands of metabolic pathways within a cell are interconnected.

CHECK YOUR LEARNING

Can you . . .

- explain how catalysts reduce activation energy?
- explain how enzymes function as biological catalysts?

6.5 HOW ARE ENZYMES REGULATED?

In a test tube, the rate of an enzyme-catalyzed reaction will depend mainly on the concentrations of the enzyme and substrate molecules. In general, increasing the concentration of either the enzyme or the substrate (or both) will increase the reaction rate, because it will boost the chances that the two types of molecules will meet. Living cells, however, cannot rely on this simple mechanism for determining how rapidly reactions occur. Cells require more complex methods of control.

Cells Regulate Metabolic Pathways by Controlling Enzyme Synthesis and Activity

Cells must keep the amounts of end products within narrow limits, even when the amounts of reactants (enzyme substrates) fluctuate considerably. For example, when glucose

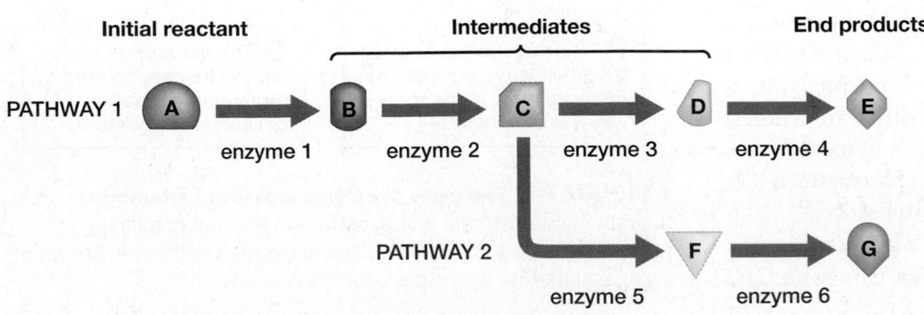

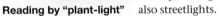

FIGURE 6-10 Simplified metabolic pathways The initial reactant molecule (A) undergoes a series of reactions, each catalyzed by a specific enzyme. The product of each reaction serves as the reactant for the next reaction in the pathway. Metabolic pathways are commonly interconnected, so the product of a step in one pathway (C in pathway 1) often serves as a substrate for an enzyme in a different pathway (enzyme 5 in pathway 2).

Lack of an Enzyme Leads to Lactose Intolerance

If you enjoy ice cream and pizza, it might be hard to imagine life without these treats. However, much of the world's population cannot enjoy dairy-containing foods. Although all young children normally produce lactase (the enzyme that breaks down lactose, or "milk sugar"), about 65% of people worldwide, including 30 to 50 million people in the United States, produce very little of this enzyme after early childhood, a condition called *lactose intolerance*. In the worst cases, lactose-intolerant people may experience abdominal pain, flatulence, nausea, and diarrhea after consuming milk products (**FIG. E6-2**).

Why do people stop synthesizing the enzyme needed to digest a nutritious food? In our early ancestors, natural selection favored individuals who lost the ability to digest lactose after childhood. Early humans had not yet domesticated livestock, so after weaning they had no access to milk. With no access to milk, it was disadvantageous to expend energy to produce an enzyme that had no use. As a result, humans evolved a regulatory mechanism that turned off the lactase gene after weaning. This adaptation to our early environment persists in most modern humans. Lactose intolerance is particularly prevalent in people of East Asian, West African, and Native American descent.

Why isn't lactose intolerance universal among modern humans? Between 10,000 and 6,000 years ago, some people in northern Europe and the Middle East acquired mutations that allowed them to digest lactose throughout their lives. These mutations were advantageous and gradually spread because they provided better nutrition for members of agricultural societies, who could obtain milk as well as meat from their livestock. Their descendants today continue to enjoy milk, ice cream, and extra-cheese pizzas.

▲ **FIGURE E6-2 Risky behavior?** For the majority of the world's adults, drinking milk invites unpleasant consequences.

THINK CRITICALLY A family brings their 8-year-old adopted child to a pediatric clinic because she has begun to suffer from diarrhea and stomach cramps after drinking milk. What would the pediatrician suspect was the cause? If tests confirm his suspicions, what approaches would he recommend to deal with the issue? Are there dairy products that would not cause the reaction? How might they work?

molecules flood into the bloodstream after a meal, it would not be desirable to metabolize them all at once, producing far more ATP than the cell needs. Instead, some of the glucose molecules are stored as glycogen or fat for later use. To be effective, then, metabolic reactions within cells must be precisely regulated; they must occur at the proper times and proceed at the proper rates. Cells regulate their metabolic pathways by controlling the type, quantity, and activity levels of the enzymes they produce.

Genes That Code for Enzymes May Be Turned On or Off

A very effective way for cells to regulate enzymes is to turn the genes that code for specific enzymes on or off depending on the cell's changing needs. For example, when glucose enters the bloodstream after a starchy meal, it stimulates a response that ultimately causes many body cells to turn on the gene that codes for the enzyme that catalyzes the first step in the metabolic pathway that breaks down glucose. Turning on this gene causes the cells to synthesize larger amounts of the enzyme, increasing glucose metabolism. Some enzymes are synthesized only during specific stages in an organism's life, as described in "Health Watch: Lack of an Enzyme Leads to Lactose Intolerance."

Some Enzymes Are Synthesized in Inactive Forms

Some enzymes are synthesized in an inactive form that is activated only under the conditions found where the enzyme is needed. For example, the protein-digesting enzyme pepsin is initially produced with its active site covered, preventing it from digesting and killing the cell that manufactures it. Exposure to acidic conditions in the stomach transforms the inactive pepsin to expose its active site and allow it to begin catalyzing the breakdown of proteins from a meal.

Enzyme Activity May Be Controlled by Competitive or Noncompetitive Inhibition

After an enzyme has been synthesized and is in its active state, there are two additional ways in which the enzyme can be inhibited to control metabolic pathways: competitive inhibition and noncompetitive inhibition. In both cases, an inhibitor molecule binds temporarily to the enzyme. The higher the concentration of inhibitor molecules, the more likely they are to bind to enzymes.

In **competitive inhibition**, a substance that is not the enzyme's normal substrate can bind to the active site of the enzyme, competing directly with the substrate for the active site (**FIG. 6-11a, b**). Usually, a competitive inhibitor molecule has structural similarities to the usual substrate that allow it to occupy the active site.

In **noncompetitive inhibition**, a molecule binds to a site on the enzyme that is distinct from the active site. This causes the active site to change shape and become unavailable, making the enzyme unable to catalyze the reaction (**FIG. 6-11c**).

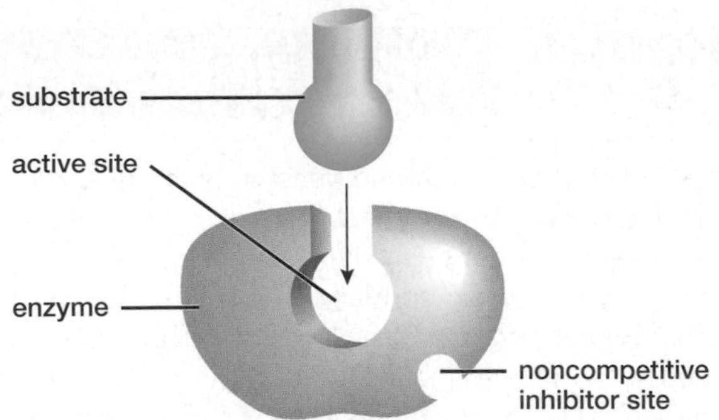

(a) A substrate binding to an enzyme

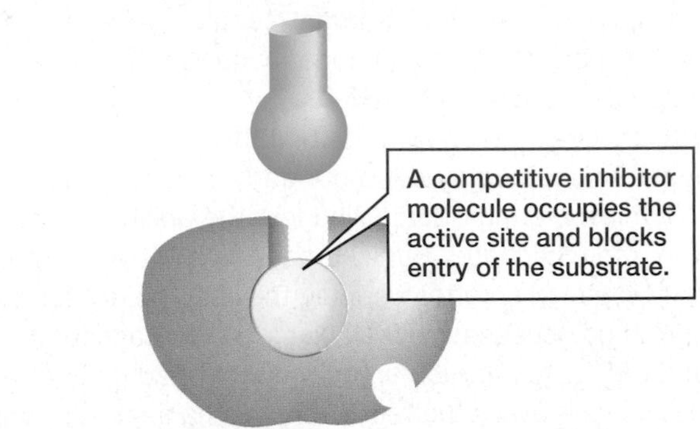

(b) Competitive inhibition

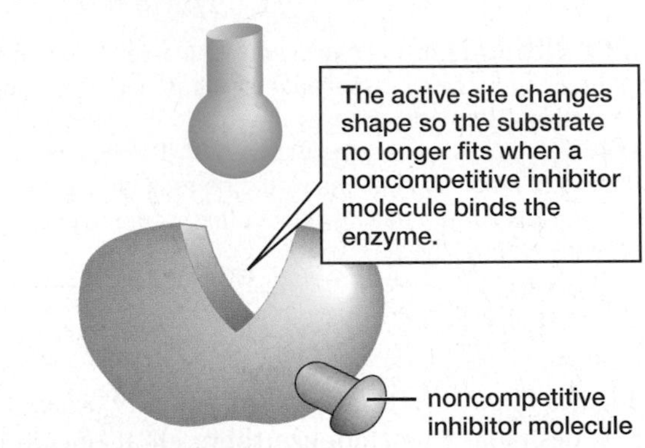

(c) Noncompetitive inhibition

▲ **FIGURE 6-11 Competitive and noncompetitive inhibition**
(a) The normal substrate fits into the enzyme's active site.
(b) In competitive inhibition, a competitive inhibitor molecule that resembles the substrate enters and blocks the active site.
(c) In noncompetitive inhibition, a molecule binds to a different site on the enzyme, distorting the active site so that the enzyme's substrate no longer fits.

Poisons, Drugs, and Environmental Conditions Influence Enzyme Activity

Poisons and drugs that act on enzymes usually inhibit them, either competitively or noncompetitively. In addition, environmental conditions can denature enzymes, distorting the three-dimensional structure that is required for their function.

Some Poisons and Drugs Are Competitive or Noncompetitive Inhibitors of Enzymes

Some poisons competitively inhibit enzymes. For example, nerve gases such as sarin and insecticides such as malathion permanently block the active site of the enzyme acetylcholinesterase, which catalyzes the breakdown of acetylcholine (a substance that nerve cells release to activate muscles). As a result, acetylcholine builds up and overstimulates muscles, causing paralysis. Death may result because victims become unable to breathe. Other poisons are noncompetitive inhibitors of enzymes. For example, potassium cyanide causes rapid death by noncompetitively inhibiting an enzyme that is crucial for the production of ATP.

Many drugs act by competitively inhibiting enzymes. For example, the antibiotic penicillin destroys bacteria by competitively inhibiting an enzyme that is needed to synthesize bacterial cell walls. Both aspirin and ibuprofen (Advil) are competitive inhibitors of an enzyme that catalyzes the synthesis of molecules that contribute to swelling, pain, and fever. Statin drugs (such as Lipitor) reduce blood cholesterol levels by competitively inhibiting an enzyme in the pathway that synthesizes cholesterol. Many anticancer drugs block cancer cell proliferation by interfering with one or more of the enzymes required to copy DNA during cell division. Unfortunately, these anticancer drugs also interfere with the growth of other rapidly dividing cells, such as those in hair follicles and the lining of the digestive tract. This effect explains why cancer chemotherapy may cause hair loss and nausea.

The Activity of Enzymes Is Influenced by Their Environment

The complex three-dimensional structures of enzymes are sensitive to environmental conditions. The hydrogen bonds that are important in determining the three-dimensional structure of proteins (see Chapter 3) are stable only within a narrow range of conditions, including the proper pH, temperature, and salt concentration. Thus, most enzymes function optimally only within a narrow range of conditions. When conditions fall outside this range, the enzyme becomes **denatured**, which means that it loses the exact three-dimensional structure required for it to function properly.

In humans, cellular enzymes generally work best at a pH around 7.4, the level maintained in and around our cells (**FIG. 6-12a**). For these enzymes, an acidic pH alters the charges on amino acids by adding hydrogen ions to them, which in turn will change the enzyme's shape and compromise its ability to function. Enzymes that operate in the human digestive tract, however, may function outside of the pH range maintained within cells. The protein-digesting

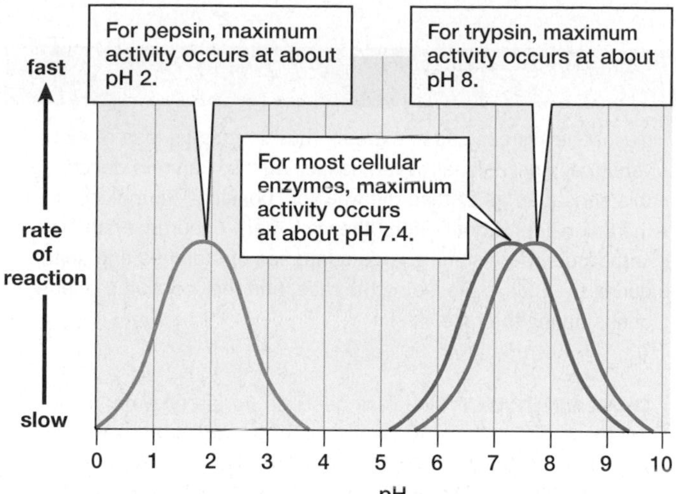
For pepsin, maximum activity occurs at about pH 2.
For trypsin, maximum activity occurs at about pH 8.
For most cellular enzymes, maximum activity occurs at about pH 7.4.
fast
rate of reaction
slow
0 1 2 3 4 5 6 7 8 9 10
pH
(a) Effect of pH on enzyme activity

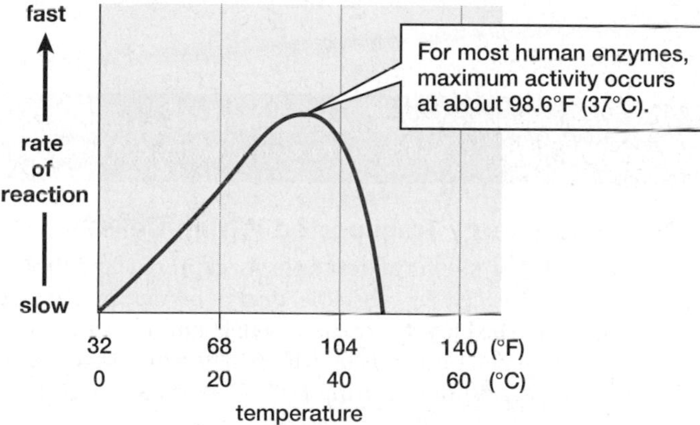
For most human enzymes, maximum activity occurs at about 98.6°F (37°C).
fast
rate of reaction
slow
32 68 104 140 (°F)
0 20 40 60 (°C)
temperature
(b) Effect of temperature on enzyme activity

▲ **FIGURE 6-12 Human enzymes function best within narrow ranges of pH and temperature (a)** The digestive enzyme pepsin, released into the stomach, works best at an acidic pH. Trypsin, released into the small intestine, works best at a basic pH. Most enzymes within cells work best at the pH found in the blood, interstitial fluid, and cytosol (about 7.4). **(b)** The maximum activity of most human enzymes occurs at human body temperature.

enzyme pepsin, for example, requires the acidic conditions of the stomach (pH around 2). In contrast, the protein-digesting enzyme trypsin, found in the small intestine where alkaline conditions prevail, works best at a pH close to 8.

Enzyme-catalyzed reactions are slowed by lower temperatures and accelerated by moderately higher temperatures (**FIG. 6-12b**). Why? Recall that molecular motion increases as temperature rises and decreases as the temperature falls. The rate of movement of molecules, in turn, influences how likely they are to encounter the active site of an enzyme. Cooling the body can drastically slow human metabolic reactions. Consider the real-life example of a young boy who fell through the ice on a lake, where he remained submerged for about 20 minutes before being rescued. At normal body temperature, the brain dies from lack of ATP after about 4 minutes without oxygen. But, fortunately, this child recovered because the icy water drastically reduced his need for oxygen by lowering his body temperature and thus slowing his metabolic rate.

In contrast, when temperatures rise too high, the hydrogen bonds that determine protein shape may be broken by excessive molecular motion, denaturing the protein. Think of the protein in egg white and how its appearance and texture are completely altered by cooking. Temperatures far lower than those required to fry an egg can still be too hot to allow enzymes to function properly. Excessive heat can even be fatal; every summer, children die when left unattended in overheated cars.

Salt concentration also affects enzyme activity. Salts dissociate into ions, which form bonds with amino acids in enzymes. Too much salt interferes with the three-dimensional structure of enzymes, destroying their activity. For example, a salty solution can slow the enzyme-catalyzed reactions that allow bacteria and fungi (which can spoil food) to grow and reproduce. Before the advent of refrigeration, meat was commonly preserved by using concentrated salt solutions; think of bacon or salt pork. Dill pickles are very well preserved in a vinegar-salt solution, which combines both highly salty and acidic conditions (**FIG. 6-13**).

CHECK YOUR LEARNING

Can you . . .

- describe how cells regulate the rate at which metabolic reactions proceed?
- explain how poisons, drugs, and environmental conditions influence enzyme activity, and provide examples?

"CRUNCHY!"
▲ **FIGURE 6-13 Preservation** Only the pickled cucumbers will be edible months from now.

CASE STUDY REVISITED

Energy Unleashed

During the course of a marathon, a runner burns a great deal of glucose to provide enough ATP to power her muscles through roughly 34,000 running steps. People store glucose molecules linked together in long branched chains of glycogen. Glycogen is most abundant in the muscles and liver. Adults typically store about 3.5 ounces (100 grams) of glycogen in the liver and about 10 ounces (280 grams) in muscles. Compared to an average person, a highly trained distance runner may store 50% more glycogen in her liver and twice as much in her muscles.

Glycogen storage is crucial for marathon runners. During a marathon, a runner depletes essentially all of her stored glycogen, often by about 90 minutes into the race. The body then begins converting fat into glucose. This is a much slower process, which can leave muscles and the brain starved for glucose. Low blood glucose levels can cause extreme muscle fatigue, loss of motivation, and occasionally even hallucinations. Runners describe this sensation as "hitting the wall" or "bonking." To make it to the finish line without hitting the wall, many endurance athletes carbo-load by consuming large quantities of starches and sugars during the 3 days preceding the race, and they consume energy drinks during the race.

THINK CRITICALLY Sketch or describe, for an enzyme in the metabolic pathway for synthesis of glycogen, the line you would predict on a graph, with enzyme activity (rate of the reaction catalyzed by the enzyme) on the *y*-axis and temperature (ranging from 0 to 50°C) on the *x*-axis. Explain your prediction.

CHAPTER REVIEW

Go to **Mastering Biology** to access the Pearson eText, vocabulary review, practice quizzes, activities, videos, current events, and more.

Answers to ***Think Critically*** *and* ***Thinking Through the Concepts*** *questions can be found in the* ***Answers*** *section at the back of the book.*

Summary of Key Concepts

6.1 What Is Energy?

Energy is the capacity to do work. Potential energy is energy that has been stored, and kinetic energy is the energy of movement. Potential energy and kinetic energy can be interconverted. The first law of thermodynamics states that in an isolated system, the total amount of energy remains constant, although the energy may change form. The second law of thermodynamics states that any use of energy causes a decrease in the quantity of useful energy and an increase in entropy (disorder or less useful forms of energy such as heat). The highly organized, low-entropy systems that characterize life do not violate the second law of thermodynamics because they are achieved through a continuous influx of usable energy from the sun.

6.2 How Is Energy Transformed During Chemical Reactions?

All chemical reactions involve making and breaking bonds. In exergonic reactions, the reactant molecules have more energy than do the product molecules, so the overall reaction releases energy. In endergonic reactions, the reactants have less energy than do the products, so the reaction requires a net input of energy. Exergonic reactions can occur spontaneously, but all reactions, including exergonic ones, require an initial input of activation energy to overcome electrical repulsions between reactant molecules. Exergonic and endergonic reactions may be coupled such that the energy liberated by an exergonic reaction is stored in ATP, which can then drive an endergonic reaction.

6.3 How Is Energy Transported Within Cells?

Energy released by chemical reactions is captured and transported within the cell by unstable energy-carrier molecules, such as ATP and the electron carriers NADH and $FADH_2$. These molecules are the major means by which cells couple exergonic and endergonic reactions occurring at different places in the cell.

6.4 How Do Enzymes Promote Biochemical Reactions?

Enzymes are proteins that act as biological catalysts by lowering activation energy and allowing biochemical reactions to occur without a permanent change of the enzyme. Enzymes usually promote one or a few specific reactions. The reactants temporarily bind to the active site of the enzyme, reducing the activation energy needed to initiate the reaction. Cells control their reactions by regulating the synthesis and use of enzymes. Enzymes allow the breakdown of energy-rich molecules such as glucose in a series of small steps so that energy is released gradually and can be captured in ATP for use in endergonic reactions. Cellular metabolism involves complex, interconnected sequences of reactions called metabolic pathways.

6.5 How Are Enzymes Regulated?

Cellular reactions are catalyzed by specific enzymes. Cells precisely control the amounts and activities of these enzymes by altering the rate of enzyme synthesis, activating previously inactive enzymes, or competitive and noncompetitive inhibition. Many poisons and drugs act as enzyme inhibitors. Environmental conditions—including pH, salt concentration, and temperature—can promote or inhibit enzyme function by altering or preserving the enzyme's three-dimensional structure.

Thinking Through the Concepts

Bloom's: Remembering, Understanding

Multiple Choice

1. Which of the following is true?
   a. Enzymes increase activation energy requirements.
   b. Activation energy is required to initiate exergonic reactions.
   c. Heat cannot supply activation energy.
   d. Stomach acid inactivates pepsin.
2. Which is *not* an example of an exergonic reaction?
   a. photosynthesis
   b. a nuclear reaction in the sun
   c. $ATP \rightarrow ADP + P_i$
   d. glucose breakdown
3. Which of the following is true?
   a. ATP is a long-term energy storage molecule.
   b. ATP can carry energy from one cell to another.
   c. ADP inhibits glucose breakdown in cells.
   d. ATP is produced by exergonic reactions.
4. Coupled reactions
   a. are endergonic overall.
   b. both synthesize and break down ATP.
   c. are catalyzed by the same enzyme.
   d. end with reactants that contain more energy than their products.
5. Enzymes
   a. increase the rate of a reaction.
   b. are active across a wide range of temperature and pH.
   c. are not very specific in the substrates they bind.
   d. are used up during catalysis.

Fill-in-the-Blank

1. According to the first law of thermodynamics, energy can be neither ____________ nor ____________. Energy occurs in two major forms: ____________, the energy of movement, and ____________, stored energy
2. According to the second law of thermodynamics, when energy changes forms some is always converted into ____________ useful forms. This tendency is called ____________.
3. Once started, some reactions release energy and are called ____________ reactions. Others require a net input of energy and are called ____________ reactions. Which type of reaction will continue spontaneously once it starts? ____________. Which type of reaction allows the formation of complex biological molecules from simpler molecules? ____________.
4. The abbreviation ATP stands for ____________. The molecule is synthesized by cells from ____________ and ____________. This synthesis requires an input of ____________, which is temporarily stored in ATP.
5. Enzymes are what type of biological molecule? ____________ Enzymes promote reactions in cells by acting as biological ____________ that lower the ____________. Each enzyme possesses a region called a(n) ____________ that binds specific biological molecules.
6. Some poisons and drugs act by ____________ enzymes. When a drug is similar to the enzyme's substrate, it acts as a(n) ____________ inhibitor.

Review Questions

1. Explain why organisms do not violate the second law of thermodynamics. What is the ultimate energy source for most forms of life on Earth?
2. Define *potential energy* and *kinetic energy*, and provide two specific examples of each. Explain how one form of energy can be converted into another. Will some energy be lost during this conversion? If so, what form will it take?
3. Define *metabolism*, and explain how reactions can be coupled to one another.
4. What is activation energy? How do catalysts affect activation energy? How do catalysts affect the reaction rate?
5. Compare breaking down glucose in a cell to setting it on fire with a match. What is the source of activation energy in each case?
6. Compare the mechanisms of competitive and noncompetitive inhibition of enzymes.
7. Describe the structure and function of enzymes. How is enzyme activity regulated?

Applying the Concepts

Bloom's: Applying, Analyzing, Evaluating

1. While vacuuming, you show off by telling a friend that you are using electrical energy to create a lower-entropy state. She replies that you are taking advantage of increased entropy in the sun. Explain this conversation.
2. Refute the following: "According to evolutionary theory, organisms have increased in complexity through time. However, an increase in complexity contradicts the second law of thermodynamics. Therefore, evolution is impossible."
3. Can a bear use all the energy contained in the body of the fish it eats? Explain, and based on your explanation, predict and further explain whether a forest would likely have more predators or more prey animals (by weight).

7 Capturing Solar Energy: Photosynthesis

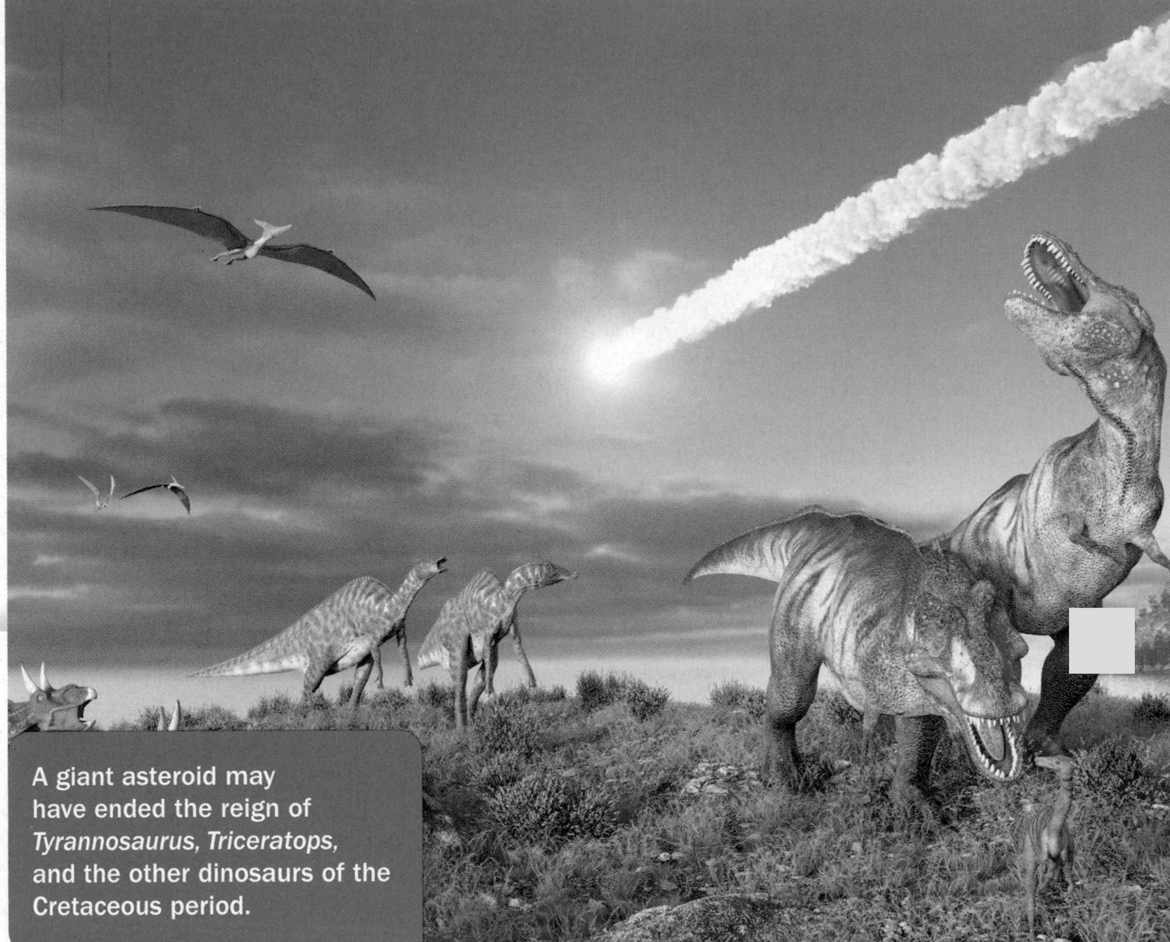

A giant asteroid may have ended the reign of *Tyrannosaurus, Triceratops,* and the other dinosaurs of the Cretaceous period.

CASE STUDY

Did the Dinosaurs Die from Lack of Sunlight?

AS THE CRETACEOUS PERIOD ENDED about 66 million years ago, life on Earth suffered a catastrophic blow. A devastating mass extinction swept the planet, eliminating at least 70% of Earth's species. The 160-million-year reign of the dinosaurs came to an abrupt end. It would be many millions of years before Earth became repopulated with a diversity of species even approaching that of the late Cretaceous.

What could have caused a calamity so vast that it wiped out the dinosaurs? In 1980, a team of scientists proposed what was then a very controversial hypothesis: A massive asteroid had brought the Cretaceous to an abrupt and violent end. The scientists—physicist Luis Alvarez, his geologist son Walter Alvarez, and nuclear chemists Helen Michel and Frank Asaro—based their claim on evidence drawn from a thin layer of clay deposited 66 million years ago and found at sites throughout the world. Known as the "K-T boundary layer," this deposit contains strangely high levels of the silvery-white metal iridium—30 to 160 times the amount typically found in Earth's crust. Although iridium is extremely rare on Earth, it is abundant in certain types of asteroids. So Alvarez and his colleagues concluded that the most likely explanation for a 66-million-year-old layer of iridium-rich clay is that a large asteroid collided with Earth.

How large would this asteroid need to have been to blanket Earth with iridium? The Alvarez team calculated that the incoming space rock must have been at least 6 miles (10 kilometers) in diameter. As it crashed into Earth, its impact released energy equivalent to 8 billion of the atomic bombs that destroyed Hiroshima and Nagasaki. The asteroid's impact blasted out a plume of pulverized rock, some of which reached the moon and beyond. Most of the debris burned as it fell back to Earth, plummeting down in a fiery shower that ignited wildfires over much of Earth's surface. A shroud of dust and soot blocked the sun's rays, and the broiling heat gave way to a cooling darkness that enveloped Earth.

Why did the asteroid impact extinguish so many species? In the months following the firestorms, one of the most damaging effects would have been darkness that disrupted photosynthesis, the most important biochemical pathway on Earth. What occurs during photosynthesis? What makes this process so important that interrupting it would wipe out much of Earth's biodiversity?

AT A GLANCE

7.1 WHAT IS PHOTOSYNTHESIS?

Roughly 3.5 billion years ago, some bacteria evolved the ability to harness the energy of sunlight. That is, they evolved **photosynthesis**—the process by which light energy is captured and then stored as chemical energy in the bonds of sugar molecules (**FIG. 7-1**). The evolution of photosynthesis made life as we know it possible. This amazing process provides the fuel for life; the energy-rich molecules synthesized by photosynthetic organisms eventually become available to feed all other organisms. In lakes and oceans, photosynthesis is performed primarily by photosynthetic bacteria and protists, and on land mostly by plants. Fundamentally similar reactions occur in all photosynthetic organisms; here we will concentrate on the most familiar of these: land plants.

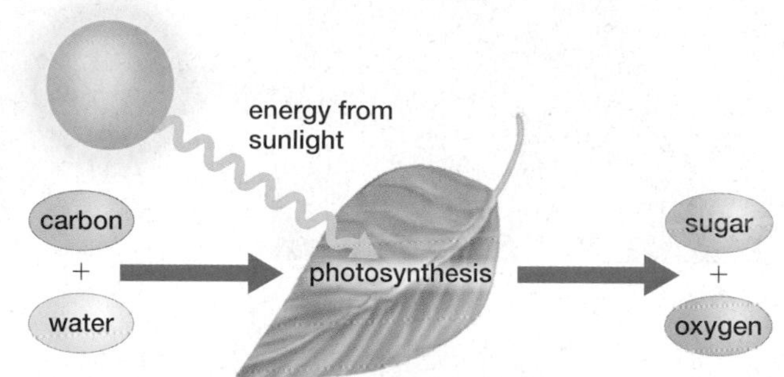

▲ **FIGURE 7-1 An overview of photosynthesis**

Chloroplasts and Leaves Are Adaptations for Photosynthesis

Photosynthesis takes place in **chloroplasts**, organelles that consist of a double outer membrane enclosing a fluid, the **stroma**. Embedded in the stroma are interconnected membrane-enclosed sacs called **thylakoids**. Each of these sacs encloses a fluid-filled region called the thylakoid space.

The leaves of plants are beautifully adapted to the demands of photosynthesis. A leaf's flattened shape exposes a large surface area to the sun, and its thinness ensures that sunlight can penetrate to reach the chloroplasts. Inside the leaf are layers of cells collectively called **mesophyll** (Gk. *meso*, middle). Mesophyll cells contain most of a leaf's chloroplasts; each cell may contain 40 to 50 chloroplasts. In the leaf's center, mesophyll cells are loosely packed, allowing air to circulate around them and CO_2 and O_2 to be exchanged through their moist membranes. Air enters the leaf through pores, called **stomata** (singular, stoma), that can be opened and closed (**FIG. 7-2**). *Vascular bundles*, which form veins in the leaf, supply water and minerals to the leaf's cells and carry the sugar molecules produced during photosynthesis to other parts of the plant. Both the upper and lower surfaces of a leaf consist of a layer of transparent cells that form the **epidermis**, which protects the inner parts of the leaf while allowing light to penetrate. The outer surface of the epidermis is covered by the **cuticle**, a transparent, waxy, waterproof covering that reduces the evaporation of water from the leaf. The structure of a leaf is illustrated in **FIGURE 7-3**.

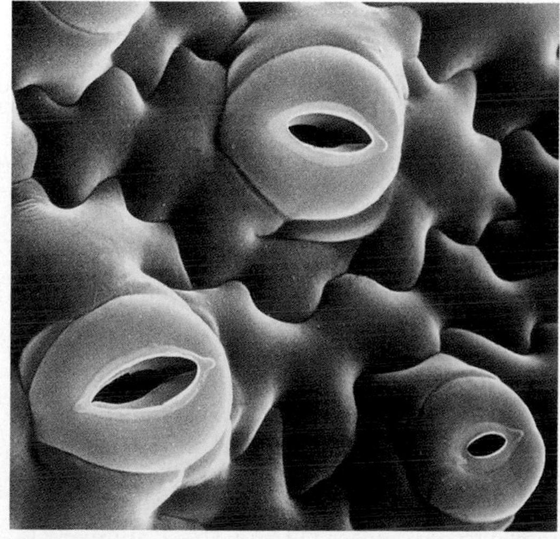

(a) Stomata open

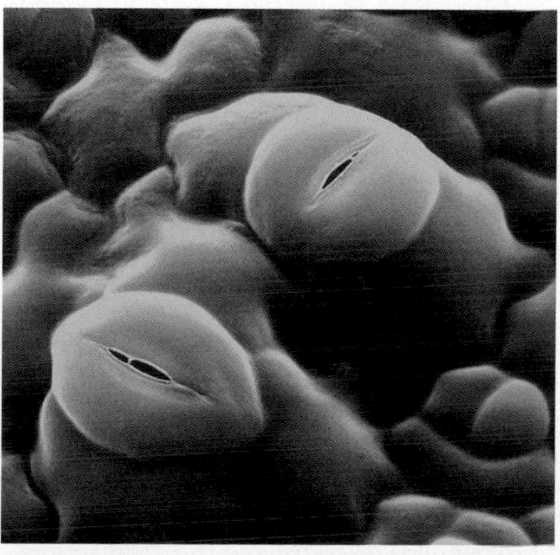

(b) Stomata closed

◀ **FIGURE 7-2 Stomata** **(a)** Open stomata allow CO_2 to diffuse in and O_2 to diffuse out. **(b)** Closed stomata reduce water loss by evaporation but prevent CO_2 from entering and O_2 from leaving.

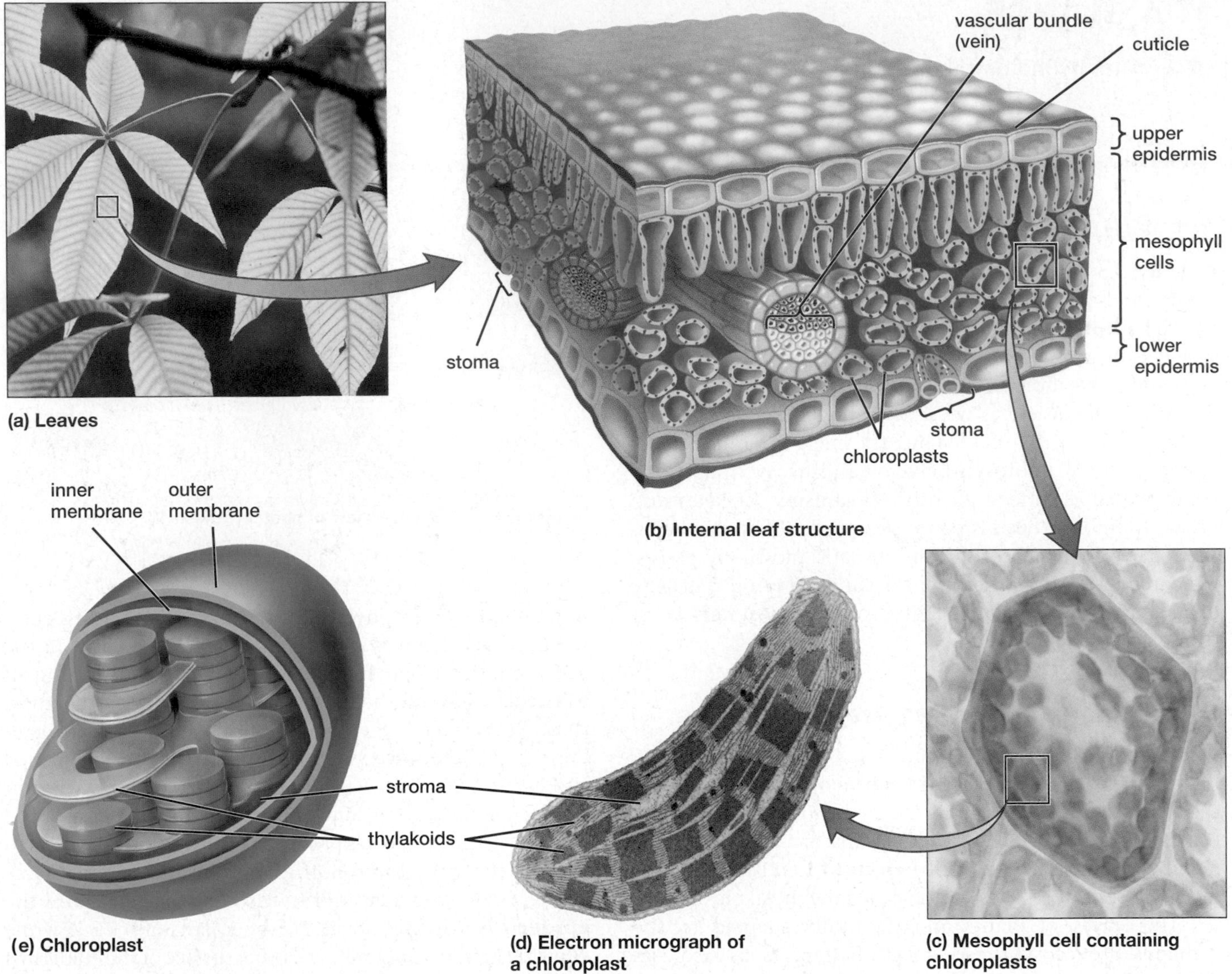

▲ **FIGURE 7-3 Photosynthetic structures (a)** Photosynthesis occurs primarily in the leaves. **(b)** A section of a leaf. **(c)** A light micrograph of a single mesophyll cell, packed with chloroplasts. **(d)** A TEM of a single chloroplast, showing the stroma and thylakoids where photosynthesis occurs. **(e)** An illustrated chloroplast.

Photosynthesis Consists of the Light Reactions and the Calvin Cycle

Starting with the simple molecules of carbon dioxide and water, photosynthesis converts the energy of sunlight into chemical energy stored in the bonds of glucose and releases oxygen as a by-product (**FIG. 7-4**). The overall chemical reaction for photosynthesis is:

$$6\,CO_2 + 6\,H_2O + \text{light energy} \rightarrow C_6H_{12}O_6\ (\text{sugar}) + 6\,O_2$$

This straightforward equation disguises the fact that photosynthesis actually involves dozens of individual reactions, each catalyzed by a different enzyme. These reactions occur in two distinct stages: the light-dependent reactions and the Calvin cycle. Each stage takes place in a different region of the chloroplast, but the two are connected by an important link: energy-carrier molecules.

In the **light reactions** (the "photo" part of photosynthesis), chlorophyll and other molecules embedded in the thylakoid membranes of the chloroplast capture sunlight energy and convert it into chemical energy. This chemical energy is stored in the energy-carrier molecules ATP (adenosine triphosphate) and **NADPH (nicotinamide adenine dinucleotide phosphate)**. Water is split apart, and oxygen gas is released as a by-product.

The reactions of the **Calvin cycle** (the "synthesis" part of photosynthesis) use the energy captured by the light reactions to manufacture sugar. During the Calvin cycle,

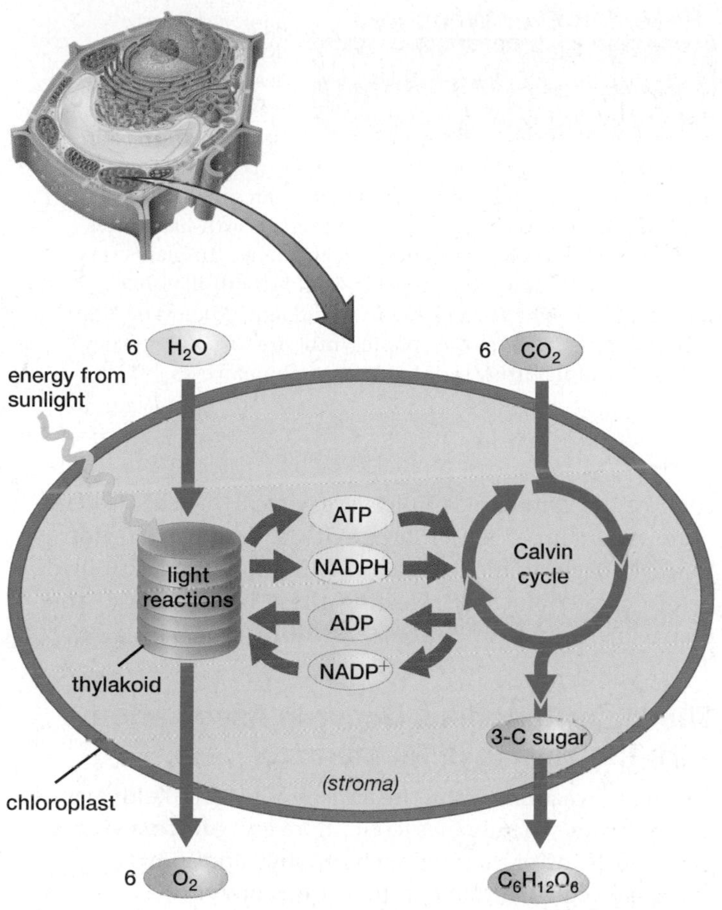

▲ **FIGURE 7-4 The relationship between the light reactions and the Calvin cycle** Notice that H_2O and CO_2, the raw materials for photosynthesis, enter at different stages and are used in different parts of the chloroplast. The O_2 liberated by photosynthesis is derived from H_2O, while the carbon used in the synthesis of sugar is obtained from CO_2. (Look for smaller versions of this figure in later figures.)

enzymes in the stroma surrounding the thylakoids catalyze reactions that use CO_2 and chemical energy from ATP and NADPH to make a three-carbon sugar that is then used to make glucose.

Figure 7-4 shows the locations at which the light reactions and the Calvin cycle occur and illustrates the interdependence of the two processes. In the following sections, we examine each stage of photosynthesis.

CHECK YOUR LEARNING

Can you . . .

- explain why photosynthesis is important?
- diagram the structure of leaves and chloroplasts and explain how these structures function in photosynthesis?
- write out and explain the basic equation for photosynthesis?
- summarize the main events of the light reactions and the Calvin cycle and explain the relationship between these two processes?

CASE STUDY CONTINUED

Did The Dinosaurs Die from Lack of Sunlight?

More than 2 billion years before the K-T extinction event, the photosynthetic organisms that filled the seas became abundant enough to release an appreciable amount of what was then a deadly gas: oxygen. Oxygen accumulated in what had originally been an oxygen-free atmosphere, radically altering Earth's environment. This "Great Oxygenation Event" triggered a massive extinction among the single-celled organisms that populated the planet (multicellular organisms had not yet evolved). Fortunately, some organisms escaped extinction, and a serendipitous combination of random mutations allowed some of them to not only survive exposure to oxygen, but also use it to their advantage. These organisms became the ancestors of nearly all forms of life, including the plants that later invaded the land and provided sustenance for herbivorous giants such as the 12-ton *Triceratops*.

What chemical reactions allow plants to capture solar energy and store it in chemical bonds, releasing oxygen in the process?

7.2 THE LIGHT REACTIONS: HOW IS LIGHT ENERGY CONVERTED TO CHEMICAL ENERGY?

The molecules that make the light reactions possible, including light-capturing pigments and enzymes, are anchored in the membranes of the thylakoids. As you read this section, notice how the thylakoid membranes and the spaces they enclose support the light reactions.

Light Is Captured by Pigments in Chloroplasts

The sun emits energy that spans a broad spectrum of electromagnetic radiation. The **electromagnetic spectrum** ranges from short-wavelength gamma rays, through ultraviolet, visible, and infrared light, to very-long-wavelength radio waves (**FIG. 7-5**). Light and all other electromagnetic waves are composed of individual packets of energy called **photons**. The energy of a photon corresponds to its wavelength: Short-wavelength photons, such as gamma and X-rays, are very energetic, whereas long-wavelength photons, such as microwaves and radio waves, carry lower energies. Visible light consists of wavelengths with energies that are high enough to alter biological *pigment molecules* (light-absorbing molecules) such as chlorophyll, but not high enough to break the bonds of crucial molecules such as DNA. (These wavelengths, with just the right amount of energy, also stimulate the pigments in our eyes, allowing us to see.)

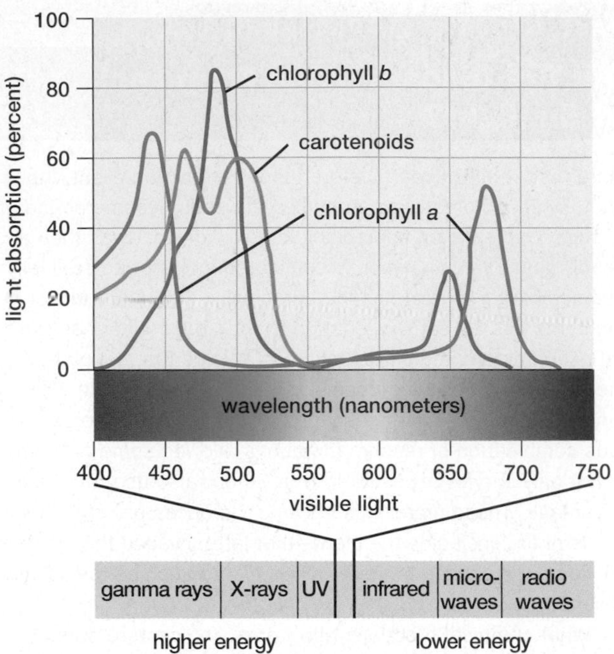

▲ **FIGURE 7-5 Light and chloroplast pigments** The rainbow colors that we perceive are a small part of the electromagnetic spectrum. Chlorophyll *a* and *b* (green and blue curves, respectively) strongly absorb violet, blue, and red light, reflecting a green or yellowish-green color to our eyes. Carotenoids (orange curve) absorb blue and green wavelengths.

THINK CRITICALLY Imagine that you continuously monitor the photosynthetic oxygen production from the leaf of a plant illuminated by white light. How and why would oxygen production change if you placed filters in front of the light source that transmit (a) only red, (b) only infrared, and (c) only green light onto the leaf?

When a specific wavelength of light strikes an object such as a leaf, one of three events occurs: The light may be reflected (bounced back), transmitted (passed through), or absorbed (captured). Wavelengths of light that are reflected or transmitted can reach the eyes of an observer; these wavelengths are seen as the color of the object. Light energy that is absorbed can drive biological processes such as photosynthesis.

Chloroplasts contain a variety of pigment molecules that absorb different wavelengths of light. **Chlorophyll *a*,** the key light-capturing pigment molecule in chloroplasts, strongly absorbs violet, blue, and red light, but reflects green, thus giving green leaves their color (see Fig. 7-5). Chloroplasts also contain other molecules, collectively called **accessory pigments**, which absorb additional wavelengths of light energy and transfer their energy to chlorophyll *a*. Accessory pigments include chlorophyll *b*, a form of chlorophyll that reflects yellow-green light and absorbs some of the blue and red-orange wavelengths of light that are missed by chlorophyll *a*. **Carotenoids** are accessory pigments that absorb blue and green light and therefore appear mostly yellow or orange (see Fig. 7-5). The color of the carotenoids in leaves is usually masked by the more abundant green chlorophyll. In temperate regions, as leaves begin to die in autumn, chlorophyll breaks down before carotenoids do, revealing these bright yellow and orange pigments as fall colors (**FIG. 7-6**).

Have You Ever Wondered ...

What Color Plants Might Be on Other Planets?

Biologist Nancy Kiang and her colleagues at NASA have developed hypotheses about alien plant colors. M-type stars, the most abundant type in our galaxy, emit light that is redder and dimmer than that of our sun. If photosynthetic organisms happened to evolve on an Earth-like planet circling an M-type star, to capture enough energy, the plants very possibly would require pigments that would absorb all visible wavelengths of light. Such pigments would reflect almost no light back to our eyes, so these alien photosynthesizers would probably be black, creating a truly eerie landscape to human eyes.

The Light Reactions Occur in Association with the Thylakoid Membranes

The light reactions occur in and on the thylakoid membranes. These membranes contain many **photosystems**, each consisting of a cluster of chlorophyll and accessory pigment molecules surrounded by various proteins. Two types of photosystems—named photosystem I and photosystem II—work together during the light reactions. The photosystems are named according to the order in which they were discovered, but the light reactions start with photosystem II and then proceed to photosystem I.

Adjacent to each photosystem is an **electron transport chain (ETC)** consisting of a series of electron-carrier molecules embedded in the thylakoid membrane. Electrons flow through the following pathway in the light reactions:

▲ **FIGURE 7-6 Loss of chlorophyll reveals carotenoid pigments** As winter approaches, chlorophyll in these maple leaves breaks down, revealing yellow and orange carotenoid pigments.

photosystem II → electron transport chain → photosystem I → electron transport chain → NADPH (**FIG. 7-7**).

You can think of the light reactions as a sort of arcade pinball game: Energy (sunlight) is introduced when a player pulls back and releases a knob. The energy is transferred from spring-driven pistons (chlorophyll molecules) to a ball (an electron), propelling it upward (into a higher-energy level). As the ball bounces back downhill, the energy it releases can be used to turn a wheel (generate ATP) and ring a bell (generate NADPH). With this overall scheme in mind, let's look more closely at the sequence of events in the light reactions.

Photosystem II and Its Electron Transport Chain Capture Light Energy, Create a Hydrogen Ion Gradient, and Split Water

As you read the following descriptions, refer to the numbered steps in Figure 7-7. The light reactions begin when photons of light are absorbed by pigment molecules clustered in photosystem II ❶. The energy from light hops from one pigment molecule to the next until it is funneled into the photosystem II reaction center ❷. The **reaction center** of each photosystem consists of a pair of chlorophyll *a* molecules and a *primary electron acceptor* molecule embedded in a complex of proteins. When the energy from light reaches the reaction center, it boosts an electron from one of the reaction center chlorophylls to the primary electron acceptor, which captures the energized electron ❸.

For photosynthesis to continue, the electrons that were boosted out of the reaction center chlorophylls in photosystem II must be replaced. The replacement electrons come from water (see ❷). Water molecules are split by an enzyme associated with photosystem II, liberating electrons that will replace those lost by the reaction center chlorophyll molecules. Splitting water also releases hydrogen ions and oxygen. For every two water molecules split, one molecule of O_2 is produced.

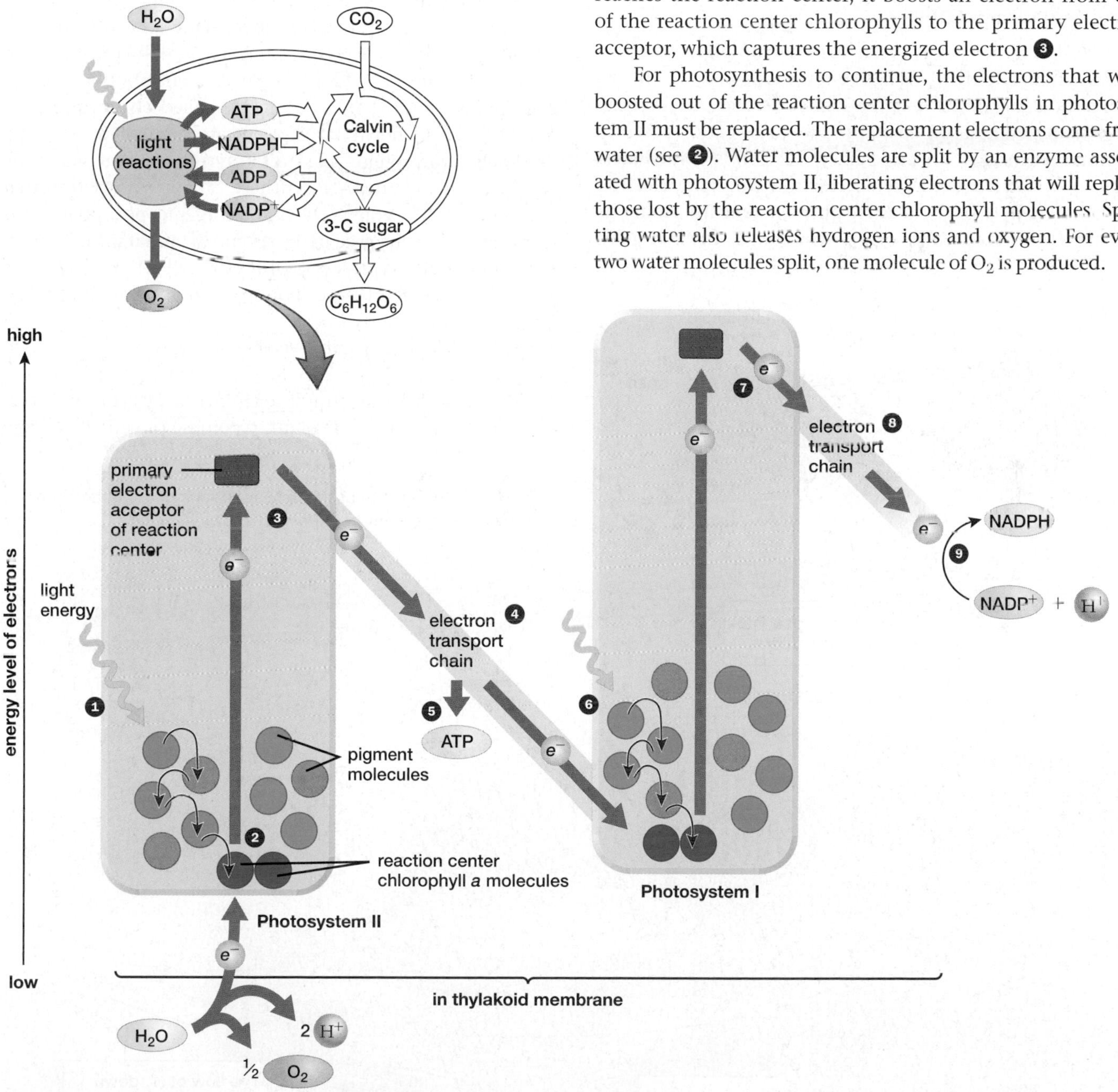

▲ **FIGURE 7-7 Energy transfer and the light reactions of photosynthesis** Light reactions occur in and adjacent to the thylakoid membrane. The vertical axis indicates the relative energy levels of the molecules involved.

Once the primary electron acceptor in photosystem II captures the electron, it passes the electron to the first molecule of the adjacent electron transport chain in the thylakoid membrane ❹. The electron then travels from one electron-carrier molecule to the next, releasing energy as it goes. Some of this energy is harnessed to pump hydrogen ions (H^+) across the thylakoid membrane and into the thylakoid space, contributing to the H^+ gradient that generates ATP (❺; to be discussed shortly). Finally, the energy-depleted electron leaves the ETC and enters the reaction center of photosystem I.

Photosystem I and Its Electron Transport Chain Generate NADPH

Meanwhile, light has also been striking the pigment molecules of photosystem I. This light energy is passed to a chlorophyll *a* molecule in the reaction center ❻. Here, it energizes an electron that is absorbed by the primary electron acceptor of photosystem I ❼. (This energized electron is immediately replaced by an energy-depleted electron from the electron transport chain associated with photosystem II.) From the primary electron acceptor of photosystem I, the energized electron is passed to a second electron transport chain adjacent to photosystem I in the thylakoid membrane ❽. Here, the final electron carrier is an enzyme that catalyzes the synthesis of NADPH. The reaction that forms NADPH combines $NADP^+$ and H^+ (both dissolved in the stroma) with two energized electrons from the electron transport chain ❾.

The Hydrogen Ion Gradient Generates ATP by Chemiosmosis

The light reactions in photosystem II create an H^+ gradient that drives ATP synthesis. As an energized electron travels along the electron transport chain associated with photosystem II, it releases energy in steps. Some of this energy is harnessed to pump H^+ across the thylakoid membrane and into the thylakoid space (**FIG. 7-8**, ❶). This creates a high concentration of H^+ inside the space ❷ and a low concentration in the surrounding stroma. Then, in a process called **chemiosmosis**, H^+ flows back down its concentration gradient through channels in a protein called **ATP synthase** that spans the thylakoid membrane. ATP synthase catalyzes a reaction that synthesizes ATP from ADP and inorganic phosphate ions (P_i in Figure 7-8) dissolved in the stroma ❸. About three H^+ must pass through ATP synthase to synthesize one ATP molecule.

The H^+ gradient stores potential energy, much as water behind a dam at a hydroelectric plant stores energy. When water is released at the hydroelectric plant, it is channeled downward through turbines. Similarly, the hydrogen ions in the thylakoid space are funneled through ATP synthase channels. In the hydroelectric plant, turbines convert the energy

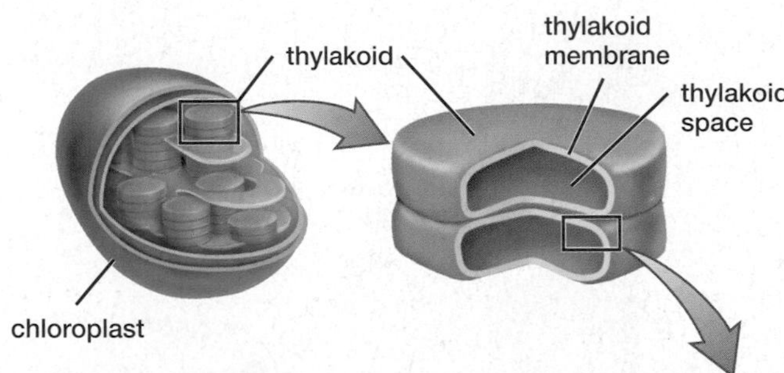

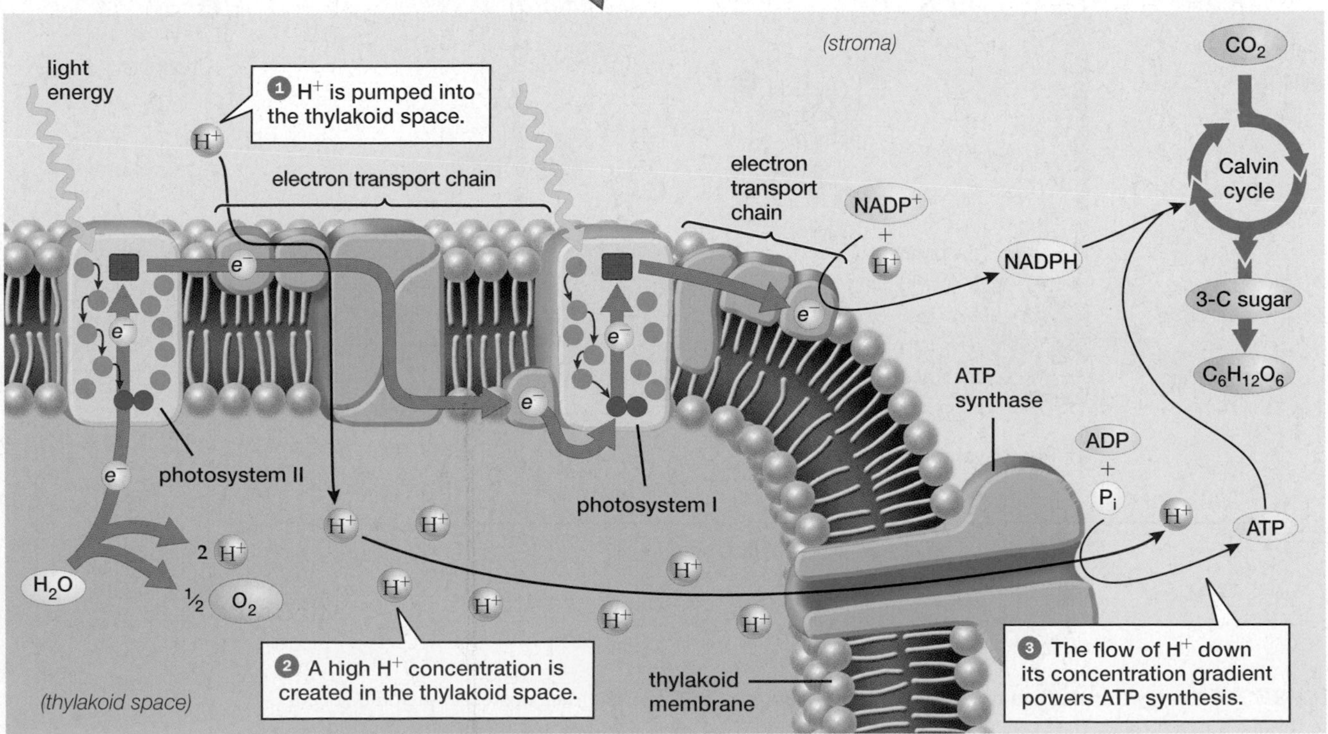

◀ **FIGURE 7-8 Events of the light reactions occur in and near the thylakoid membrane**

of moving water into electrical energy. In similar fashion, ATP synthase converts the energy liberated by the flow of H^+ into chemical energy stored in the bonds of ATP.

SUMMING UP: Light Reactions

- Chlorophyll and carotenoid pigments of photosystem II absorb light that energizes and ejects an electron from a reaction center chlorophyll *a* molecule. The energized electron is captured by the primary electron acceptor molecule.
- The electron is passed from the primary electron acceptor to the adjacent ETC, where it moves from molecule to molecule, releasing energy with each transfer. Some of the energy is used to create a hydrogen ion gradient across the thylakoid membrane. This gradient is used to drive ATP synthesis by chemiosmosis.
- Enzymes associated with photosystem II split water. This process releases electrons that replace those ejected from the reaction center chlorophylls, supplies H^+ that enhances the H^+ gradient for ATP production, and liberates O_2.
- Chlorophyll and carotenoid pigments in photosystem I absorb light that energizes and ejects an electron from a reaction center chlorophyll *a* molecule into the primary electron acceptor molecule. This electron is replaced by an energy-depleted electron from the ETC associated with photosystem II.
- The energized electron passes from the primary electron acceptor into the adjacent ETC, where it moves from molecule to molecule, releasing energy.
- The final molecule in this second ETC is an enzyme that synthesizes the energy-carrier NADPH.
- The overall products of the light reactions are the energy carriers ATP and NADPH; O_2 is released as a by-product.

CASE STUDY CONTINUED

Did the Dinosaurs Die from Lack of Sunlight?

Air bubbles trapped in amber from the Cretaceous period have revealed that oxygen made up nearly 35% of the atmosphere at that time, compared to 21% today. After the giant asteroid hit the Earth, this abundant atmospheric oxygen would have intensified the fires caused by the flaming re-entry of debris from the impact.

A large portion of Earth's terrestrial vegetation was likely consumed by fire, and many of the land plants that survived the fires would have succumbed during the cold, dark "global winter" that began as Earth was encompassed by soot and dust. Most plant-eating animals that survived the initial blast would have soon starved, especially enormous ones like the 12-ton *Triceratops,* which needed to consume hundreds of pounds of vegetation daily. Predators such as *Tyrannosaurus,* which relied on plant-eaters for food, would have died soon afterward.

Which reactions allow plants to store the high-energy molecules that they and most other forms of life rely on?

CHECK YOUR LEARNING

Can you . . .

- list the light-capturing molecules in chloroplasts and describe their functions?
- diagram and describe the molecules within the thylakoid membranes and explain how they capture and transfer light energy?
- explain how ATP and NADPH are generated?

7.3 THE CALVIN CYCLE: HOW IS CHEMICAL ENERGY STORED IN SUGAR MOLECULES?

Although our cells require sugar to burn for energy, they are not able to make the sugar molecules they need by capturing (or *fixing*) carbon atoms from their surroundings. Photosynthesis, however, can accomplish this feat. In the course of photosynthesis, carbon is captured from atmospheric CO_2 during the reactions of the Calvin cycle, which are powered by energy from sunlight harnessed during the light reactions.

The Calvin Cycle Captures Carbon Dioxide

The Calvin cycle takes place in the fluid stroma that surrounds the thylakoids. There, the ATP and NADPH synthesized during the light reactions power synthesis of the three-carbon sugar *glyceraldehyde-3-phosphate (G3P)* from CO_2. This metabolic pathway is described as a "cycle" because it begins and ends with the same five-carbon molecule, *ribulose bisphosphate (RuBP).*

For simplicity, we illustrate the cycle starting and ending with three molecules of RuBP. Each "turn" of the cycle captures three molecules of CO_2 and produces one molecule of the simple sugar end product: G3P. The Calvin cycle is best understood if we divide it into three parts: (1) carbon fixation, (2) the synthesis of G3P, and (3) the regeneration of RuBP that allows the cycle to continue (**FIG. 7-9**).

During **carbon fixation**, carbon from CO_2 is incorporated into organic molecules. The enzyme **rubisco** catalyzes the combination of three CO_2 molecules with three five-carbon RuBP molecules to produce six molecules of *phosphoglyceric acid* (*PGA*, a three-carbon molecule) ❶.

Synthesis of the three-carbon sugar G3P occurs via a series of reactions that use energy donated by ATP and NADPH generated during the light reactions. During these reactions, six three-carbon PGA molecules are rearranged to form six three-carbon G3P molecules ❷.

Five of the six G3P molecules are used to regenerate three five-carbon RuBP molecules, using ATP generated during the light reactions ❸. The single remaining G3P molecule exits the Calvin cycle ❹.

Carbon Fixed During the Calvin Cycle Is Used to Synthesize Glucose

In reactions that occur outside of the Calvin cycle, two three-carbon G3P sugar molecules can be combined to form

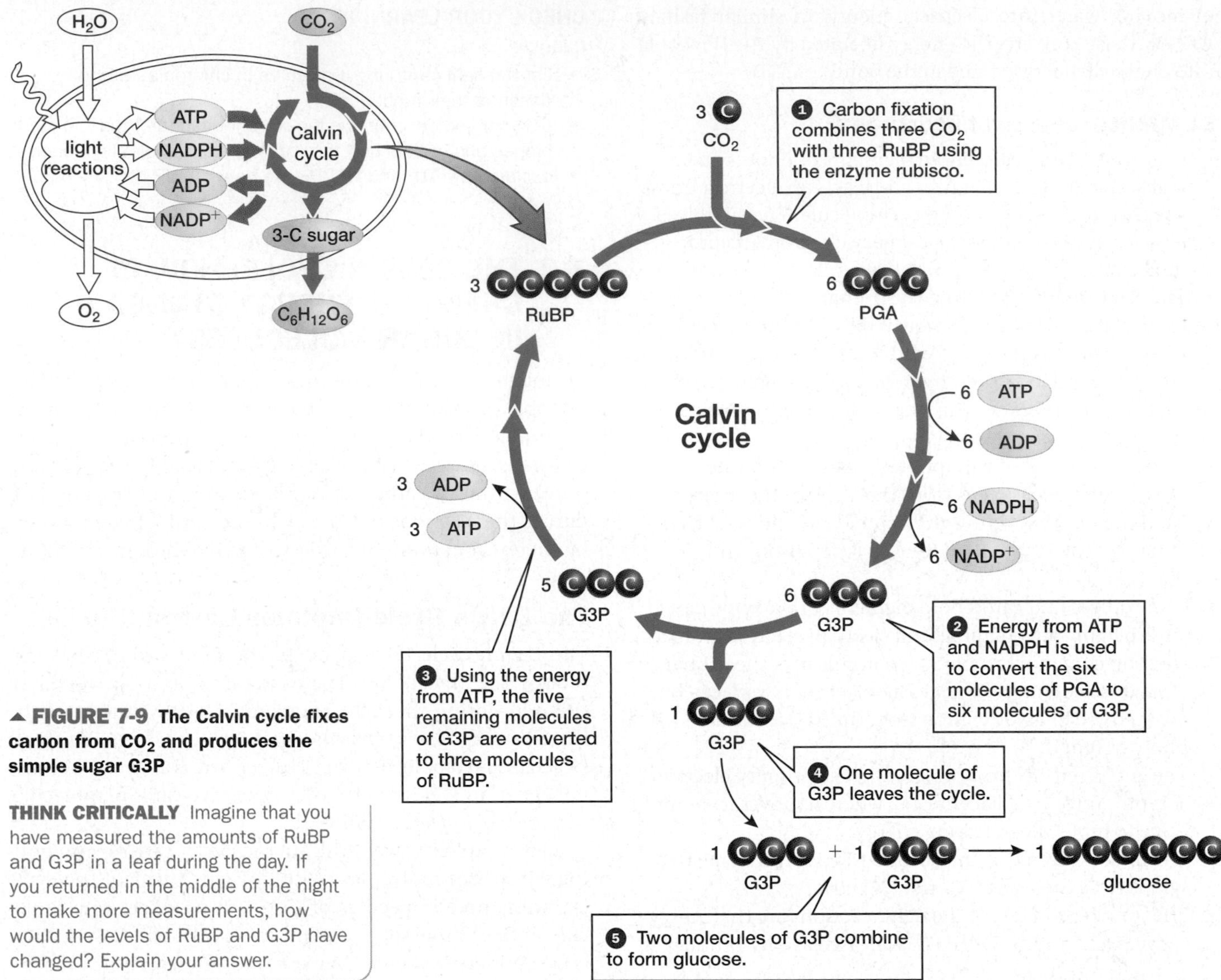

H_2O
CO_2
ATP
NADPH
ADP
$NADP^+$
light reactions
Calvin cycle
3-C sugar
O_2
$C_6H_{12}O_6$
3 C
CO_2
1 Carbon fixation combines three CO_2 with three RuBP using the enzyme rubisco.
3 RuBP
6 PGA
Calvin cycle
6 ATP
6 ADP
6 NADPH
6 $NADP^+$
3 ADP
3 ATP
5 G3P
6 G3P
2 Energy from ATP and NADPH is used to convert the six molecules of PGA to six molecules of G3P.
3 Using the energy from ATP, the five remaining molecules of G3P are converted to three molecules of RuBP.
1 G3P
4 One molecule of G3P leaves the cycle.
1 G3P + 1 G3P → 1 glucose
5 Two molecules of G3P combine to form glucose.

▲ **FIGURE 7-9 The Calvin cycle fixes carbon from CO_2 and produces the simple sugar G3P**

THINK CRITICALLY Imagine that you have measured the amounts of RuBP and G3P in a leaf during the day. If you returned in the middle of the night to make more measurements, how would the levels of RuBP and G3P have changed? Explain your answer.

one six-carbon glucose molecule (see Fig. 7-9 ❺). Glucose can then be broken down to provide energy for an organism's needs. Glucose can also be used to synthesize storage molecules such as the disaccharide sucrose or the polymer starch, or to synthesize the polymer cellulose, a major component of plant cell walls. In addition, some plants convert glucose into lipids for storage.

The storage products of photosynthesis are being eyed by Earth's growing and energy-hungry human population as a substitute for fossil fuels. These "biofuels" have the potential advantage of not adding additional CO_2 (a greenhouse gas that contributes to global climate change) to the atmosphere, but do they live up to their promise? We explore this question in "Earth Watch: Biofuels—Are Their Benefits Bogus?"

SUMMING UP: The Calvin Cycle

The Calvin cycle can be divided into three stages:

1. Carbon fixation: Three RuBP capture three CO_2, forming six PGA.
2. G3P synthesis: A series of reactions, driven by energy from ATP and NADPH (from the light reactions), produces six G3P. One G3P leaves the cycle and is available to form glucose.
3. RuBP regeneration: Three RuBP molecules are regenerated from the remaining five G3P using ATP energy, allowing the cycle to continue.

In a separate process outside the chloroplast, two G3P molecules produced by the Calvin cycle combine to form glucose.

CHECK YOUR LEARNING

Can you . . .

- describe the function of the Calvin cycle and where it occurs?
- list the three stages of the Calvin cycle, including the molecules that enter the cycle and those that are formed at each stage?
- describe the fate of the simple sugar G3P generated by the Calvin cycle?

Biofuels—Are Their Benefits Bogus?

When you drive a car, turn up the thermostat, or flick on your desk lamp, you are actually unleashing the energy of sunlight trapped by prehistoric photosynthetic organisms. Over hundreds of millions of years, the bodies of these organisms—with their stored solar energy and carbon captured from ancient atmospheric CO_2— have been converted by heat and pressure into coal, oil, and natural gas. Without human intervention, these fossil fuels would remain trapped deep underground.

A major contributor to global climate change is increased burning of fossil fuels by a growing human population. This combustion releases CO_2 into the atmosphere; the added carbon dioxide traps heat that would otherwise radiate into space. Since we began using fossil fuels during the industrial revolution in the mid-1800s, humans have increased the CO_2 content of the atmosphere by about 38%. As a result, Earth is growing warmer, and many experts fear that a hotter future climate will place extraordinary stresses on Earth's inhabitants, ourselves included (see Chapter 29).

To reduce CO_2 emissions and reliance on imported oil, many governments are subsidizing and promoting the use of biofuels, especially ethanol and biodiesel. Ethanol is produced by fermenting plants rich in sugars, such as sugarcane and corn, to produce alcohol (fermentation is described in Chapter 8). Biodiesel fuel is made primarily from oil derived from plants such as soybeans, canola, or palms. Because the carbon stored in biofuels was removed from the modern atmosphere by photosynthesis, burning them seems to simply restore CO_2 that was recently present in the atmosphere. Is this a solution to global climate change?

The environmental and social benefits of burning fuels derived from food crops are hotly debated. Over 35% of the U.S. corn crop is now feeding vehicles rather than animals and people. The increased diversion of corn to ethanol production resulted in a sharp increase in corn prices (**FIG. E7-1**). Although corn prices have subsided in recent years, higher food costs remain a likely long-term effect of ethanol production. Another concern is that growing corn and converting it into ethanol uses large quantities of fossil fuel, negating corn ethanol's advantages over burning gasoline.

The environmental costs of using food crops as an alternative fuel source are also enormous. For example, Indonesia's luxurious tropical rain forests—home to orangutans, Sumatran tigers, and clouded leopards—are being destroyed to make room for oil palms grown for biofuels; more than 25 million acres of forest have already been destroyed (an area the size of Kentucky), and an additional 2 million more acres are cleared each year (**FIG. E7-2**). In Brazil, soybean plantations for biofuels have replaced large expanses of rain forest. Ironically, clearing these forests for agriculture increases atmospheric CO_2 because rain forests trap far more carbon than the crops that replace them.

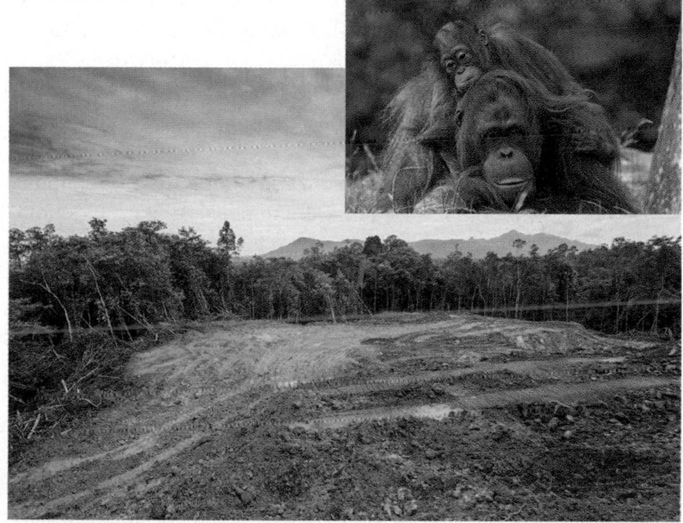

▲ **FIGURE E7-2 Cleared tropical forest** This aerial view shows the aftermath of clearing lush tropical forest, the former home of rare Sumatran tigers, elephants, leopards, orangutans, and a wealth of bird species. The cleared area will become a palm oil plantation for biofuels. When endangered orangutans (such as the ones pictured) lose their homes to deforestation, they are often killed as they are forced closer to human settlements.

Biofuels would have far less environmental and social impact if they were not produced from food crops or by destroying Earth's dwindling rain forests. Algae show great promise as an alternative. Some algae produce starch that can be fermented into ethanol; others produce oil that can become biodiesel. Some of these microscopic photosynthesizers can potentially produce 60 times as much oil per acre as soybeans and 5 times as much as oil palm. Researchers are also attempting to break cellulose into its component sugars, which would allow ethanol to be generated from corn stalks, wood chips, or grasses. Commercial scale cellulose biorefineries have recently opened in the United States; their long-term success remains uncertain. Although the benefits of most biofuels in wide use today may not justify their environmental costs, there is hope that this equation will change as we develop better technologies to harness the energy captured by photosynthesis.

THINK CRITICALLY What would be the advantages of producing ethanol from cellulose rather than from the sugars in crops such as corn?

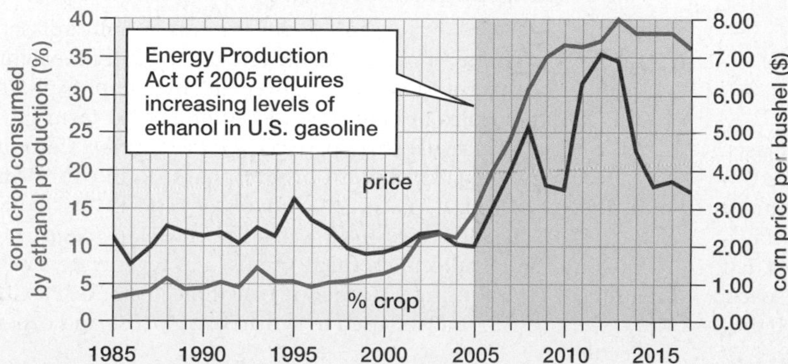

▲ **FIGURE E7-1 Corn prices increased dramatically after corn ethanol was required to be added to gasoline**

CASE STUDY REVISITED

Did the Dinosaurs Die from Lack of Sunlight?

Did an asteroid really end the reign of dinosaurs? The Alvarez hypothesis was initially met with skepticism. If such a cataclysmic event had occurred, where was the crater? In 1991, scientists finally located it near the coastal town of Chicxulub on Mexico's Yucatán Peninsula. The crater, estimated at over 110 miles in diameter and 10 miles deep, was filled with debris and sedimentary rock laid down during the 66 million years since the impact. Ocean and dense vegetation hid most remaining traces from satellite images. The final identification of the Chicxulub crater was based on rock core samples, unusual gravitational patterns, and faint surface features.

Some paleontologists argue that the asteroid impact may have done nothing more than add to existing stresses caused by more gradual changes in climate, to which the dinosaurs (with the exception of those ancestral to modern-day birds) could not adapt. Such changes might have been caused by the prolonged intense volcanic activity that is known to have occurred in the late Cretaceous. Volcanoes spew out soot, ash, and climate-changing gases, so furious volcanism could have significantly worsened conditions for plant growth.

Recently, alternative hypotheses to the asteroid impact were dealt a blow when an expert group of 41 researchers published a review article in the journal *Science*. This article analyzed the previous 20 years of research by paleontologists, geochemists, geophysicists, climatologists, and sedimentation experts dealing with the K-T extinction event. The conclusion: Land and ocean ecosystems were destroyed extremely rapidly, and evidence overwhelmingly supports the asteroid impact hypothesis first proposed by the Alvarez group 30 years earlier.

CONSIDER THIS The K-T extinction event is the most recent of five major mass extinctions documented in the fossil record. The ultimate cause of any such event is a massive environmental change occurring on a timescale too rapid to allow species to adapt.

Today, humans are having profound effects on the environment. We have modified roughly half of all Earth's land area, and we are changing Earth's climate at a rate at least 10 times that of previous natural periods of warming. As a result of these human impacts, Earth is currently undergoing a sixth mass extinction. Current extinction rates are estimated to be from 100 to 1,000 times the rate that would occur in the absence of human activity.

The impact of humanity is collective, but the human population consists of individuals. What global policy changes and what individual choices can help us sustain the planet that sustains us?

CHAPTER REVIEW

Go to **Mastering Biology** to access the Pearson eText, vocabulary review, practice quizzes, activities, videos, current events, and more.

Answers to ***Think Critically*** *and* ***Thinking Through the Concepts*** *questions can be found in the* ***Answers*** *section at the back of the book.*

Summary of Key Concepts

7.1 What Is Photosynthesis?

Photosynthesis is the process that captures the energy of sunlight and uses it to convert inorganic molecules of carbon dioxide and water into a high-energy sugar molecule, releasing oxygen as a by-product. In plants, photosynthesis takes place in the chloroplasts, using two major reaction sequences: the light reactions and the Calvin cycle.

7.2 The Light Reactions: How Is Light Energy Converted to Chemical Energy?

The light reactions occur in the thylakoids of chloroplasts. Light energizes electrons in chlorophyll molecules located in photosystems II and I. Energized electrons jump to a primary electron acceptor and then move into adjacent electron transport chains. Energy lost as the electrons travel through the first ETC is used to pump hydrogen ions into the thylakoid space, creating an H^+ gradient across the thylakoid membrane. Hydrogen ions flow down this concentration gradient through ATP synthase channels in the membrane, driving ATP synthesis by chemiosmosis. For every two energized electrons that pass through the second ETC, one molecule of the energy-carrier NADPH is formed from $NADP^+$ and H^+. Electrons lost from photosystem II are replaced by electrons liberated by splitting water, which also generates H^+ and O_2. Energized electrons lost from photosystem I are replaced by energy-depleted electrons from photosystem II.

7.3 The Calvin Cycle: How Is Chemical Energy Stored in Sugar Molecules?

The Calvin cycle, which occurs in the stroma of chloroplasts, uses energy from the ATP and NADPH generated during the light reactions to drive the synthesis of glyceraldehyde-3-phosphate (G3P). Two molecules of G3P are then combined to form glucose. The Calvin cycle has three parts: (1) *Carbon fixation*: Carbon dioxide combines with ribulose bisphosphate (RuBP) to form phosphoglyceric acid (PGA). (2) *Synthesis of G3P*: PGA is converted to G3P, using energy from ATP and NADPH. (3) *Regeneration of RuBP*: Five molecules of G3P are used to regenerate three molecules of RuBP, using ATP energy. One molecule of G3P exits the cycle; this G3P may be used to synthesize glucose and other molecules.

Thinking Through the Concepts

Bloom's: Remembering, Understanding

Multiple Choice

1. Which of the following is true?
   a. Photosynthesis evolved in an atmosphere with little or no oxygen.
   b. Photosynthesis occurs only in plants.
   c. Oxygen is necessary for photosynthesis.
   d. Carbon dioxide is necessary for photorespiration.

2. The Calvin cycle
   a. can only occur when light is present.
   b. is the part of photosynthesis where carbon is captured.
   c. produces ATP and NADPH.
   d. occurs in the thylakoids.

3. Which of the following is true?
   a. Chloroplasts are found primarily in leaf mesophyll cells.
   b. The region between the outer and inner membranes of chloroplasts is called the thylakoid space.
   c. Mesophyll cells in the leaf's center are packed tightly together.
   d. Vascular bundles of the leaf circulate CO_2 and O_2.

4. Carotenoids
   a. include chlorophylls *a* and *b*.
   b. serve as accessory pigments.
   c. are produced in the fall in temperate climates.
   d. absorb mostly yellow and orange light.

5. Which of the following is false?
   a. Photosystem II splits water.
   b. Photosystem II captures energy directly from light.
   c. Reaction centers contain chlorophyll *a* molecules.
   d. Photosystem I generates O_2.

Fill-in-the-Blank

1. Plant leaves contain pores called ____________ that allow the plant to release ____________ and take in ____________. Photosynthesis occurs in organelles called ____________ that are concentrated within the ____________ cells of most plant leaves.

2. Chlorophyll *a* captures wavelengths of light that correspond to the three colors ____________, ____________, and ____________. What color does chlorophyll reflect? ____________ Accessory pigments that reflect yellow and orange are called ____________. These pigments are located in clusters called ____________ in the ____________ membrane of the chloroplast.

3. During the first stage of photosynthesis, light is captured and funneled into chlorophyll *a* molecules in the ____________. The energized electron is then passed into a(n) ____________. From here, it is transferred through a series of molecules called the ____________. Energy lost during these transfers is used to create a gradient of ____________. The process that uses this gradient to generate ATP is called ____________.

4. The oxygen produced as a by-product of photosynthesis is derived from ____________, and the carbons used to make glucose are derived from ____________. The biochemical pathway that captures atmospheric carbon is called the ____________. The process of capturing carbon is called ____________.

5. Light reactions generate the energy-carrier molecules ____________ and ____________, which are then used in the ____________ cycle. Carbon fixation combines carbon dioxide with the five-carbon molecule ____________. Two molecules of ____________ can be combined to produce the six-carbon sugar ____________.

Review Questions

1. Explain what would happen to life if photosynthesis ceased. Why would this occur?
2. Write and then explain the equation for photosynthesis.
3. Draw a simplified diagram of a leaf cross-section and label it. Explain how a leaf's structure supports photosynthesis.
4. Draw a simplified diagram of a chloroplast and label it. Explain how the individual parts of the chloroplast support photosynthesis.
5. Trace the flow of energy in chloroplasts from sunlight to ATP, including an explanation of chemiosmosis.
6. Summarize the events of the Calvin cycle. Where does it occur? What molecule is fixed? What is the product of the cycle? Where does the energy come from to drive the cycle? What molecule is regenerated?

Applying the Concepts

Bloom's: Applying, Analyzing, Evaluating

1. Suppose an experiment is performed in which plant I is supplied with normal carbon dioxide but with water that contains radioactive oxygen atoms. Plant II is supplied with normal water but with carbon dioxide that contains radioactive oxygen atoms. Each plant is allowed to perform photosynthesis, and the oxygen gas and sugars produced are tested for radioactivity. Which plant would you expect to produce radioactive sugars, and which plant would you expect to produce radioactive oxygen gas? Explain why.

2. If you were to measure the pH in the space surrounded by the thylakoid membrane in an actively photosynthesizing plant, would you expect it to be acidic, basic, or neutral? Explain your answer.

3. Assume you want to add an accessory pigment to the photosystems of a plant chloroplast to help it photosynthesize more efficiently. What wavelengths of light would the new pigment absorb? Describe the color of your new pigment. Would this be a useful project for genetic engineers? Explain.

8 Harvesting Energy: Glycolysis and Cellular Respiration

King Richard III died in the Battle of Bosworth. More than 500 years later, researchers unearthed his skeleton.

CASE STUDY

Raising a King

"A horse! A horse! My kingdom for a horse!" shouts King Richard III in Shakespeare's *King Richard III*, moments after his horse is slain at the Battle of Bosworth. The 32-year-old monarch also perished in this battle, fought in 1485 in Leicester, England. He had ruled for only 2 years, and his short life was filled with conspiracy and political intrigue. Now, thanks to the DNA in mitochondria, his skeleton has been identified, providing details of his final moments, as well as some clues about his appearance. For example, although he did not have the hunched back portrayed by Shakespeare, his skeleton does reveal a severe spinal deformity.

No one knows what King Richard III was really shouting as he died, but we do know that his wounds were horrific. His eight head injuries included a sword wound that pierced his skull, penetrating entirely through his brain, and mutilation from an axe-like weapon that hacked out a large chunk of his skull. Wounds on the defeated king's ribs and pelvis suggest that his body was further mutilated after death. Franciscan friars apparently buried him hastily in the Greyfriars Church—naked, without a coffin, and in a shallow grave that was too short for his body.

A few decades after the king's death, the church was leveled and new buildings erected on the site; in time, the king's burial site was forgotten. Its location remained a mystery until a few years ago, when a historian and archaeologists (aided by ground-penetrating radar) did some detective work. On the basis of their findings, the researchers hypothesized that the ruins were likely to be found beneath a Leicester parking lot. Despite this compelling hypothesis, experts still believed there was only a very remote possibility that Richard III's remains would be found there. So observers were astonished when researchers discovered and excavated a skeleton disfigured by battle wounds and severe spinal curvature.

Painstaking investigations led by geneticist Turi King of the University of Leicester identified the body as Richard III with near certainty. The clinching evidence came from analysis of DNA recovered from one of the skeleton's teeth. The recovered DNA was mitochondrial DNA, which occurs in tiny circular loops inside mitochondria.

Mitochondria provide energy for every cell in the bodies of eukaryotic organisms. These complex organelles powered the muscles of the soldiers and their horses at the Battle of Bosworth, while simultaneously performing the mundane task of keeping their teeth alive. How do mitochondria work? Why do we begin dying in seconds if their function is blocked? And why do mitochondria have their own DNA?

AT A GLANCE

8.1 HOW DO CELLS OBTAIN ENERGY?

Cells require a continuous supply of energy to power the multitude of metabolic reactions that are essential just to stay alive. In this chapter, we describe the cellular reactions that transfer energy from energy-storage molecules, particularly glucose, to energy carrier molecules, such as ATP.

Photosynthesis Is the Ultimate Source of Cellular Energy

The energy used by life on Earth comes almost entirely from sunlight. Solar energy is captured during photosynthesis by plants and other photosynthetic organisms and stored in the chemical bonds of sugars and other organic molecules (see Chapter 7). To gain access to this stored energy, almost all organisms, including those that photosynthesize, break down molecules synthesized by photosynthesis and capture some of the released energy as ATP. Although most cells use a variety of organic molecules to produce ATP, in this chapter we focus on the breakdown of glucose ($C_6H_{12}O_6$), which all cells can use as an energy source. The relationship between photosynthesis and the breakdown of glucose is illustrated in **FIGURE 8-1**.

The chemical equation describing glucose formation by photosynthesis is the reverse of the equation describing complete glucose breakdown. The products of photosynthesis are the reactants of glucose breakdown, and vice versa. However, the forms of energy involved in the two reactions differ. The *light energy* captured by photosynthesis is converted during glucose breakdown to *chemical energy* stored in ATP, with some energy lost as heat during both reactions.

Photosynthesis

$$6\ CO_2 + 6\ H_2O + \text{light energy} \rightarrow C_6H_{12}O_6 + 6\ O_2$$

Complete Glucose Breakdown

$$C_6H_{12}O_6 + 6\ O_2 \rightarrow 6\ CO_2 + 6\ H_2O + \text{ATP chemical energy}$$

All Cells Can Use Glucose As a Source of Energy

The breakdown of glucose occurs via two major processes: It starts with *glycolysis* and proceeds to *cellular respiration* if oxygen is available. Glucose breakdown begins with glycolysis in the cell cytosol, liberating small quantities of ATP.

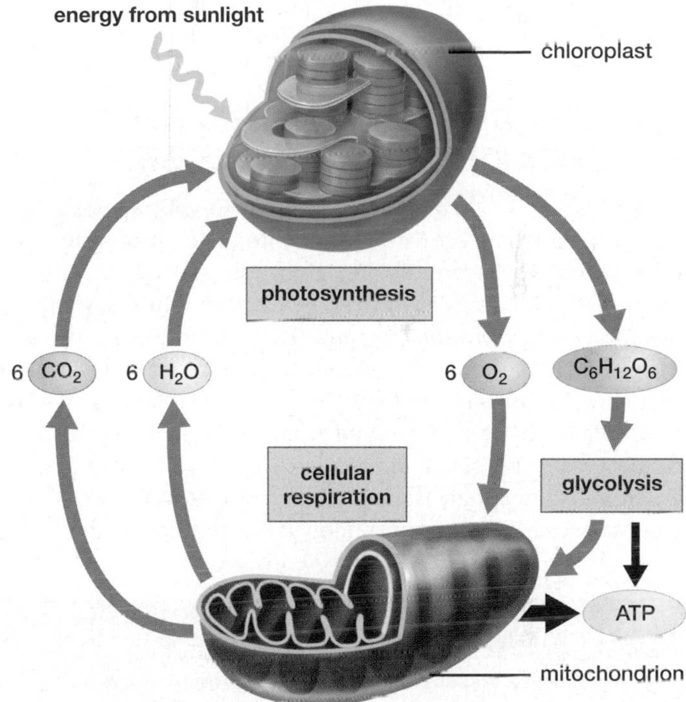

(a) Complementary processes

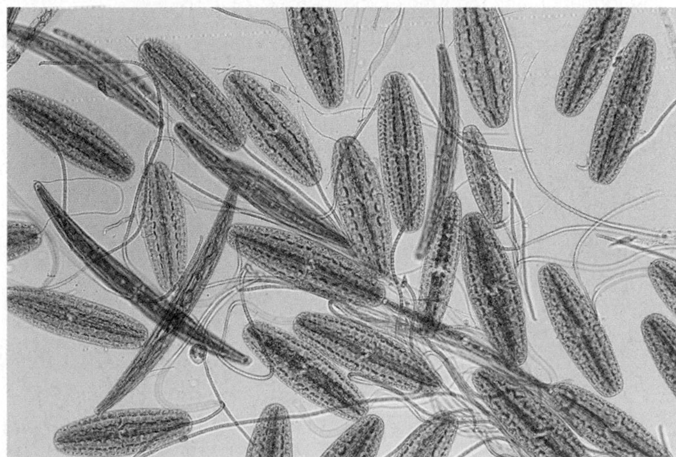

(b) Unicellular green algae

▲ **FIGURE 8-1 Photosynthesis and glucose breakdown (a)** The products of each process are used by the other. **(b)** The cells of photosynthetic organisms, such as these green algae, contain both chloroplasts that produce glucose and mitochondria that break it down.

Then the end product of glycolysis is further broken down during cellular respiration in mitochondria, supplying far greater amounts of energy in ATP (**FIG. 8-2**, **TABLE 8-1**). During cellular respiration, cells use oxygen (originally released by photosynthetic organisms) and liberate both water and carbon dioxide—the raw materials for photosynthesis.

CHECK YOUR LEARNING

Can you . . .

- explain how photosynthesis and glucose breakdown are related to one another using their overall chemical equations?
- summarize glucose breakdown in the presence and absence of oxygen?

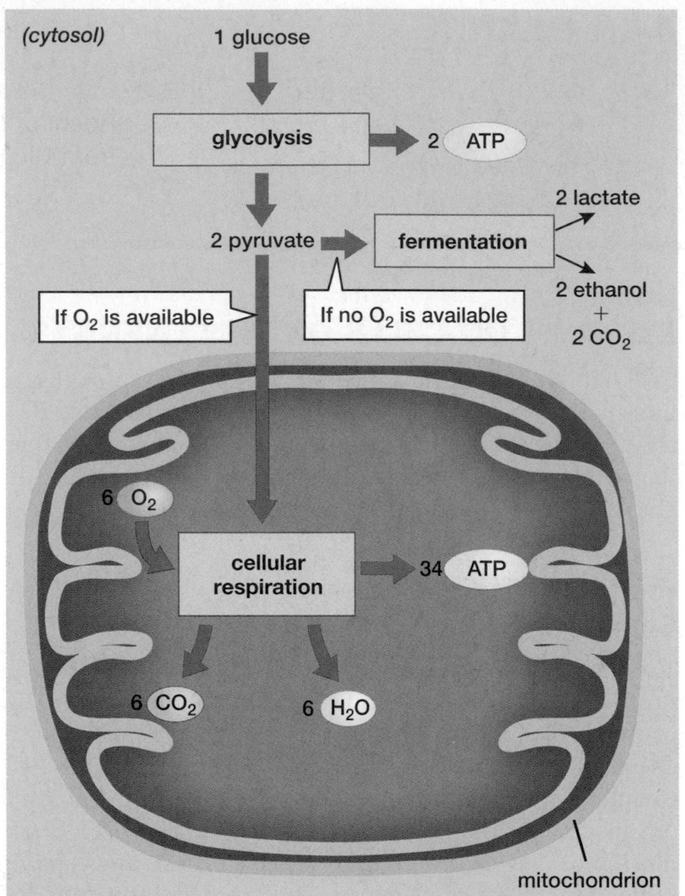

▲ **FIGURE 8-2 An overview of glucose breakdown**

8.2 HOW DOES GLYCOLYSIS BEGIN BREAKING DOWN GLUCOSE?

Glycolysis (Gk. *glyco*, sweet, and *lysis*, to split apart) splits a six-carbon glucose molecule into two molecules of three-carbon pyruvate. To get glycolysis started, an initial investment of energy from ATP is required. Thus, glycolysis has an *energy investment stage* that is followed by an *energy harvesting stage* (**FIG. 8-3**). During a series of reactions that constitutes the energy investment stage, each of two ATP molecules donates a phosphate group and energy to glucose, forming an "energized" molecule of fructose bisphosphate. Fructose bisphosphate is much more easily broken down than glucose because of the extra energy it has acquired from ATP.

TABLE 8-1 A Summary of Glucose Breakdown

Stage of Glucose Breakdown	Electron Carriers	Net ATP Produced	Location
Glycolysis	Energy captured in 2 NADH	2 ATP	Cytosol
Cellular respiration stage 1: Acetyl CoA formation and the Krebs cycle	Energy captured in 8 NADH and 2 $FADH_2$	2 ATP	Mitochondrial matrix
Cellular respiration stage 2: Electron transport chain and chemiosmosis	Energy released from 10 NADH and 2 $FADH_2$	32 ATP	Inner mitochondrial membrane and intermembrane space
Fermentation	2 NAD^+ regenerated	0 ATP	Cytosol
Total	0 NADH and 0 $FADH_2$	36 ATP	Cytosol and mitochondria

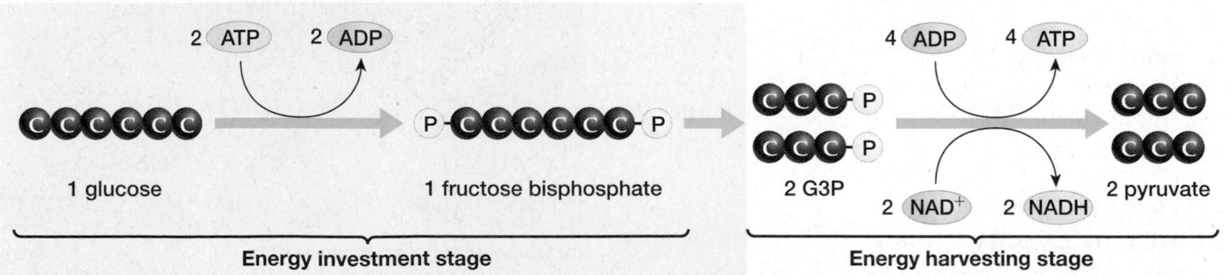

▲ **FIGURE 8-3 The essentials of glycolysis** In the energy investment stage, the energy of two ATP molecules is used to convert glucose into the fructose bisphosphate, which then breaks down into two molecules of G3P. In the energy-harvesting stage, the two G3P molecules undergo a series of reactions that capture energy in four ATP and two NADH molecules. (Only carbon skeletons are shown.)

THINK CRITICALLY What is the net energy yield in ATP and NADH produced?

Next, during the energy-harvesting stage, fructose bisphosphate is converted into two three-carbon molecules of glyceraldehyde-3-phosphate, or G3P. Each G3P molecule then undergoes a series of reactions that converts the G3P to pyruvate. During these reactions, energy is stored when two high-energy electrons and a hydrogen ion (H^+) are added to the electron acceptor **nicotinamide adenine dinucleotide (NAD^+)** to produce the electron carrier **NADH**. Two molecules of NADH are produced for every glucose molecule broken down. Additional energy is captured in ATP, with four molecules of ATP produced for each glucose molecule broken down. But because two ATP were used to form fructose bisphosphate during the energy investment stage, there is a net gain of only two ATP per glucose molecule during glycolysis.

CHECK YOUR LEARNING

Can you . . .

- explain the energy investment and energy-harvesting phases of glycolysis?
- describe the two types of high-energy molecule produced by glucose breakdown?

8.3 HOW DOES CELLULAR RESPIRATION EXTRACT ENERGY FROM GLUCOSE?

In most organisms, if oxygen is available, glycolysis is followed by the second process in glucose breakdown, called **cellular respiration**. Cellular respiration breaks down the two pyruvate molecules produced by glycolysis into six molecules of carbon dioxide and six molecules of water. During this process, the chemical energy from the two pyruvate molecules is used to produce 34 ATP.

In prokaryotic cells, cellular respiration takes place in the cytosol and plasma membrane. Here we focus on cellular respiration in eukaryotic cells, which occurs within **mitochondria**, organelles that are sometimes called the "powerhouses of the cell." A mitochondrion has two membranes. The inner membrane encloses a central compartment containing the fluid **matrix**, and the outer membrane surrounds the organelle, producing an **intermembrane space** between the two membranes. Each structure plays a crucial role in the process of cellular respiration (**FIG. 8-4**).

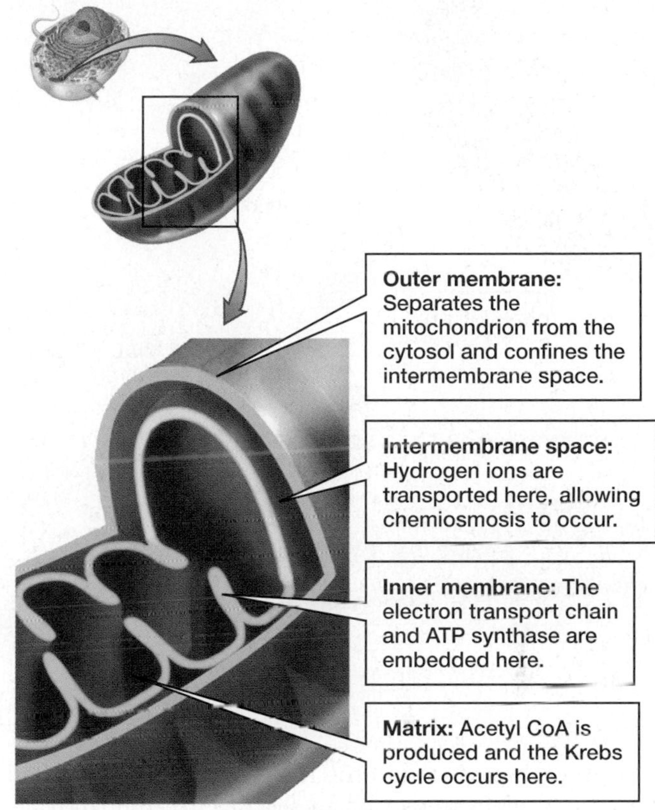
Outer membrane: Separates the mitochondrion from the cytosol and confines the intermembrane space.
Intermembrane space: Hydrogen ions are transported here, allowing chemiosmosis to occur.
Inner membrane: The electron transport chain and ATP synthase are embedded here.
Matrix: Acetyl CoA is produced and the Krebs cycle occurs here.

(a) Mitochondrial structures and their functions

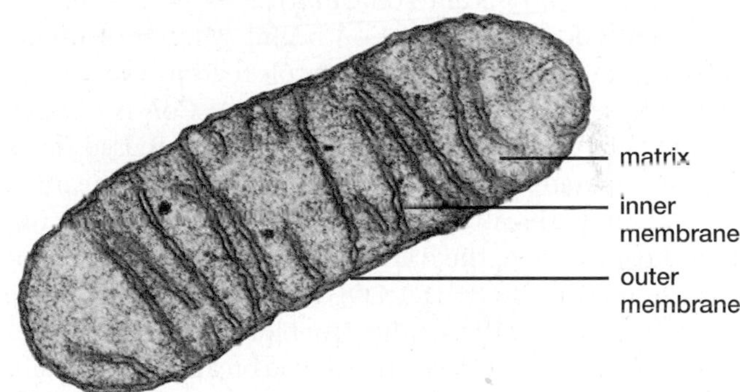
matrix
inner membrane
outer membrane

(b) TEM of a mitochondrion

▲ **FIGURE 8-4 The mitochondrion**

In the following sections, we discuss the two major stages of cellular respiration: first, the formation of acetyl CoA and its breakdown via the Krebs cycle, and second, the transfer of electrons along the electron transport chain and the generation of ATP by chemiosmosis.

Cellular Respiration Stage 1: Acetyl CoA Is Formed and Travels Through the Krebs Cycle

Pyruvate, the end product of glycolysis, is synthesized in the cytosol. Before cellular respiration can occur, the pyruvate diffuses from the cytosol through the porous outer mitochondrial membrane. It is then actively transported through the inner mitochondrial membrane and into the matrix, where cellular respiration begins.

Two sets of reactions occur within the mitochondrial matrix during stage 1 of cellular respiration: the formation of acetyl CoA and the Krebs cycle (**FIG. 8-5**). Acetyl CoA consists of a two-carbon functional (acetyl) group attached to a molecule called coenzyme A (CoA). To generate acetyl CoA, pyruvate from glycolysis is split, releasing CO_2 and leaving behind an acetyl group. The acetyl group reacts with CoA, forming acetyl CoA. This reaction liberates energy that is stored by transferring two high-energy electrons and a hydrogen ion to NAD^+, forming NADH.

The next set of reactions is known as the **Krebs cycle**, named after its discoverer, Hans Krebs, who won a Nobel Prize for this work in 1953. This metabolic pathway is called a cycle because it continuously regenerates the same substrate molecule with which it begins: oxaloacetate.

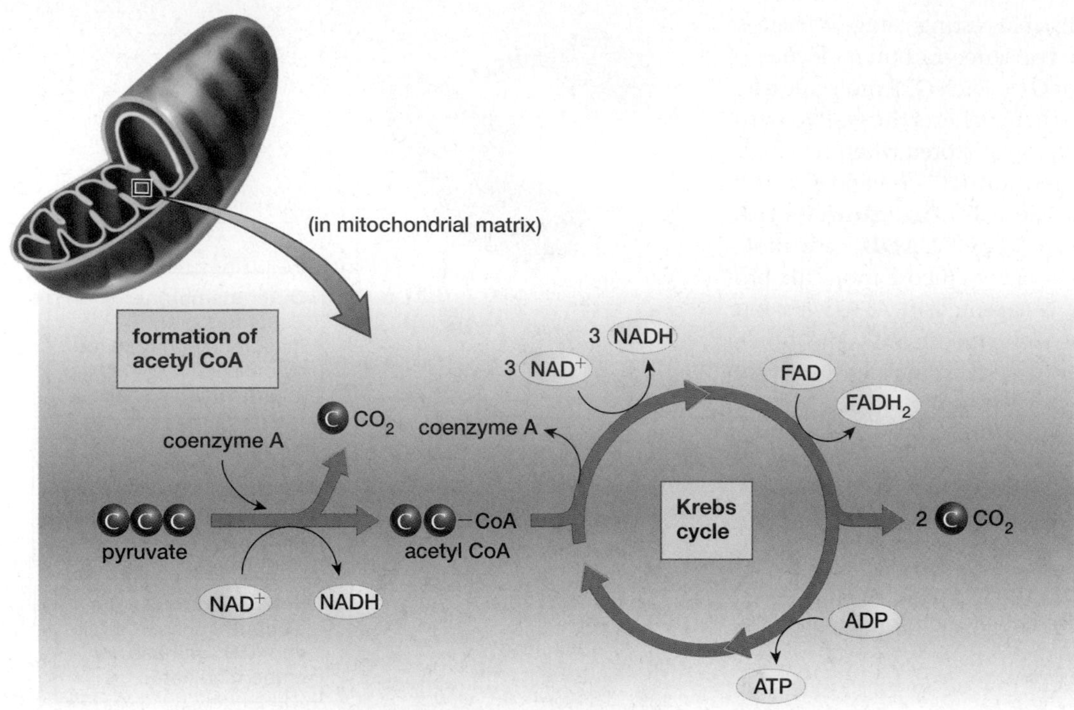

FIGURE 8-5 Reactions in the mitochondrial matrix: acetyl CoA formation and the Krebs cycle

With each pass around the Krebs cycle, the two carbon atoms that enter in acetyl-CoA are released in carbon dioxide, liberating energy. Some of this energy is captured in high-energy electron carriers and some in ATP.

The breakdown of acetyl CoA begins when it combines with a four-carbon oxaloacetate molecule to form a six-carbon citrate molecule, releasing the catalyst CoA. CoA is not permanently altered during these reactions and is reused many times. As the Krebs cycle proceeds, enzymes within the mitochondrial matrix break down the acetyl group, releasing two CO_2 and regenerating the oxaloacetate molecule to continue the cycle. In your body, the CO_2 generated in cells during the stage 1 reactions diffuses into your blood, which carries the CO_2 to your lungs. This is why the air you breathe out contains more CO_2 than the air you breathe in.

During the reactions of the Krebs cycle, chemical energy is captured in energy-carrier molecules. The breakdown of the acetyl group from each acetyl CoA produces one ATP and three NADH. It also produces one **flavin adenine dinucleotide ($FADH_2$)**, a high-energy electron carrier similar to NADH, when the electron acceptor **FAD** picks up two energetic electrons along with two H^+. Remember that two pyruvate molecules enter the Krebs cycle for each glucose molecule that undergoes glycolysis, so for each glucose molecule, the number of energy-carrier molecules produced by the cycle is twice the number produced per pyruvate.

Cellular Respiration Stage 2: High-Energy Electrons Traverse the Electron Transport Chain and Chemiosmosis Generates ATP

By the end of stage 1, the cell has gained only four ATP from the original glucose molecule (a net of two during glycolysis and two during the Krebs cycle). However, the cell has also captured many high-energy electrons in a total of 10 NADH and two $FADH_2$ molecules for each glucose molecule.

In the second stage of cellular respiration, the high-energy electron carriers each release two high-energy electrons into an **electron transport chain (ETC)**, a series of electron-transporting molecules, many copies of which are embedded in the inner mitochondrial membrane (**FIG. 8-6**). The depleted carriers are then available for recharging by glycolysis and the Krebs cycle.

The Electron Transport Chain Releases Energy in Steps

The electron transport chains in the mitochondrial membrane serve the same function as those embedded in the thylakoid membrane of chloroplasts (see Chapter 7). High-energy electrons jump from molecule to molecule along the ETC, releasing small amounts of energy at each step. The energy liberated in these stages is just the right amount to pump H^+ across the inner membrane, from the matrix into the intermembrane space (although some is always lost as heat). This ion pumping produces a concentration gradient of H^+, high in the intermembrane space and low in the matrix (see Fig. 8-6). Expending energy to create an H^+ gradient is similar to charging a battery. This H^+ battery will be discharged as ATP is generated by chemiosmosis, discussed later.

Oxygen Is the Final Electron Acceptor

Finally, at the end of the electron transport chain, the energy-depleted electrons are transferred to oxygen, which acts as an electron acceptor. The energy-depleted electrons combine with oxygen and hydrogen ions to form water. One

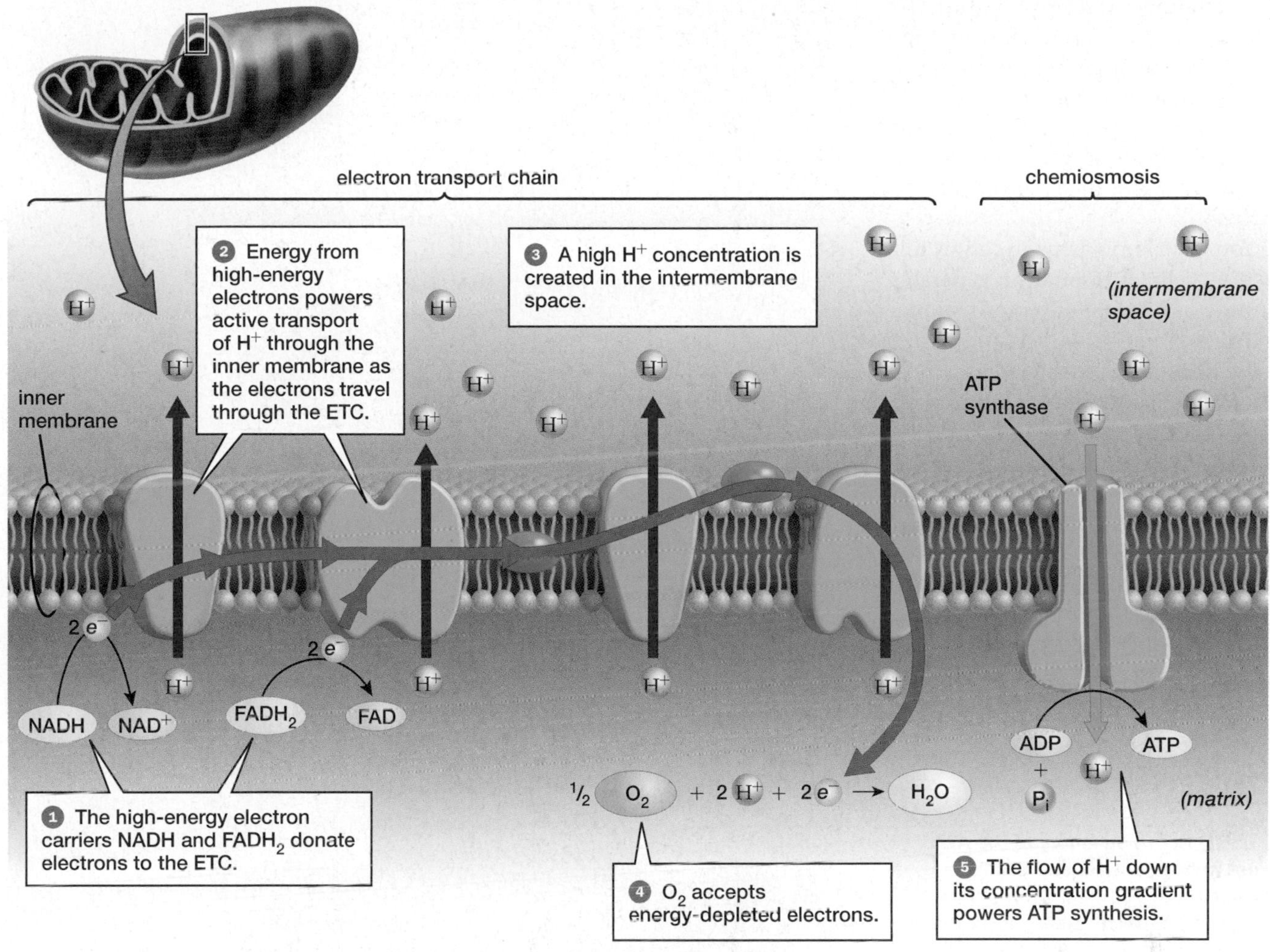
electron transport chain
chemiosmosis
2 Energy from high-energy electrons powers active transport of H+ through the inner membrane as the electrons travel through the ETC.
3 A high H+ concentration is created in the intermembrane space.
(intermembrane space)
inner membrane
ATP synthase
2 e−
2 e−
NADH
NAD+
FADH2
FAD
ADP
+
Pi
ATP
1/2 O2 + 2 H+ + 2 e− → H2O
(matrix)
1 The high-energy electron carriers NADH and FADH2 donate electrons to the ETC.
4 O2 accepts energy-depleted electrons.
5 The flow of H+ down its concentration gradient powers ATP synthesis.

▲ **FIGURE 8-6 The electron transport chain and chemiosmosis** Many copies of the electron transport chain and ATP synthase are embedded in the inner mitochondrial membrane.

THINK CRITICALLY How would the rate of ATP production be affected by the absence of oxygen?

water molecule is produced for every two electrons that traverse the electron transport chain (see Fig. 8-6).

If no oxygen were present to accept electrons, the electron transport chain would become saturated with electrons and could not acquire more from NADH and $FADH_2$. With electrons unable to move through the electron transport chain, H^+ could not be pumped across the inner membrane. The H^+ gradient would rapidly dissipate, and ATP synthesis by chemiosmosis would stop. Because of their high demand for energy from ATP, most eukaryotic cells die within minutes without a steady supply of oxygen to accept electrons. Your body obtains oxygen through the air you breathe, which enters your lungs and is transported in the bloodstream to every cell. Because of cellular respiration, the air you breathe out contains less oxygen than the air you breathe in.

Have You Ever Wondered ...

Why Cyanide is So Deadly?

Cyanide is a favorite poison in old murder mysteries, causing the hapless victim to die almost instantly. Cyanide exerts its lethal effects by blocking the last protein in the electron transport chain: an enzyme that combines energy-depleted electrons with oxygen. If these electrons are not carried away by oxygen, they act like a plug in a pipeline. Additional high-energy electrons cannot travel through the electron transport chain, so no more hydrogen can be pumped across the membrane, and ATP production by chemiosmosis stops abruptly. Because the energy demands of our cells are so great, blocking cellular respiration with cyanide can kill a person within a few minutes.

Chemiosmosis Captures Energy in ATP

Chemiosmosis is the process by which some of the energy stored in the concentration gradient of H^+ is captured in ATP as H^+ flows down its gradient. How is the energy captured? The inner membranes of mitochondria are permeable to H^+

only at channels that are part of an ATP synthase enzyme. As hydrogen ions flow from the intermembrane space into the matrix through these ATP-synthesizing enzymes, ATP is formed from ADP and inorganic phosphate ions (P_i in Fig. 8-6) dissolved in the matrix.

You now know why glycolysis followed by cellular respiration generates far more ATP than glycolysis alone. **FIGURE 8-7** and Table 8-1 summarize the breakdown of one glucose molecule in a eukaryotic cell with oxygen present, showing the energy produced during each stage and the general locations where the pathways occur. In summary, the two ATPs formed during glycolysis are supplemented by two more formed during the Krebs cycle and an additional 32 via chemiosmosis, for a total of 36 ATP per glucose molecule.

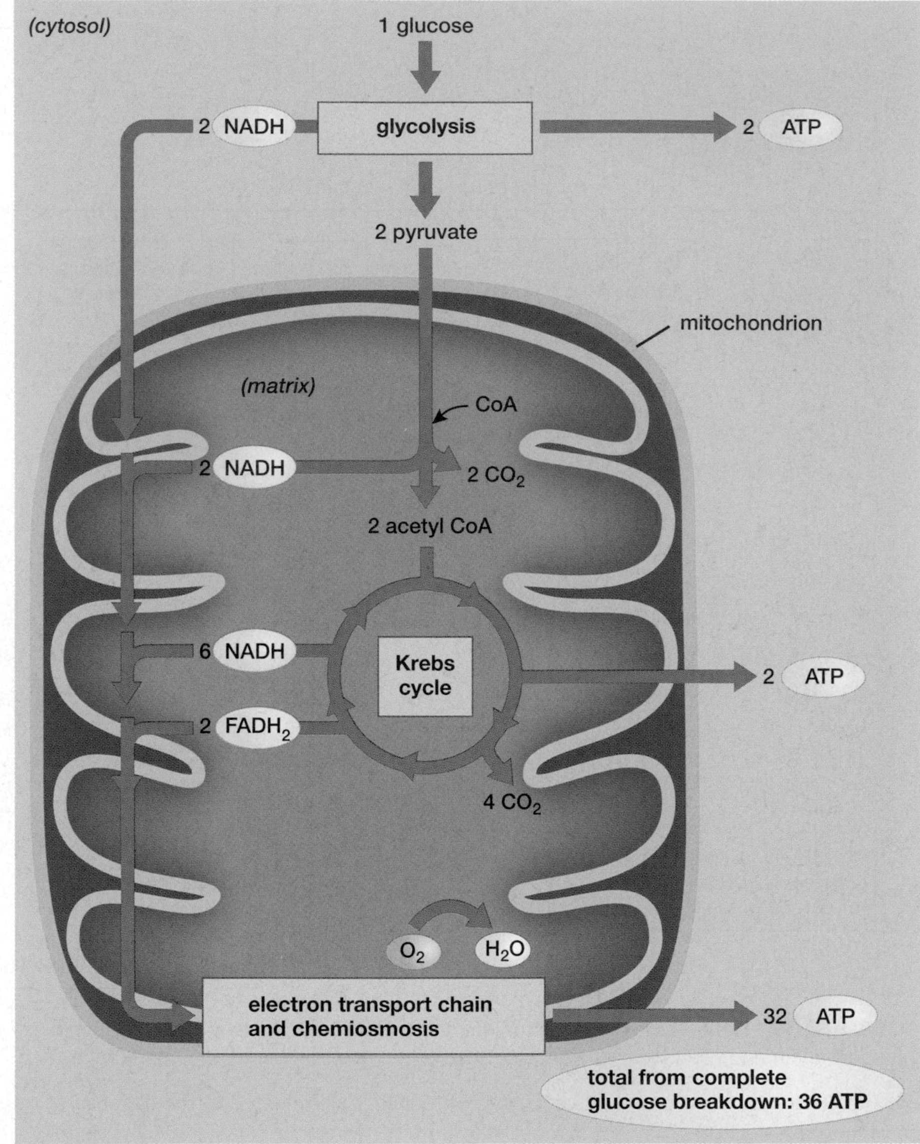

▸ **FIGURE 8-7 A summary of the ATP harvest from glycolysis and cellular respiration** Cellular respiration provides the biggest ATP payoff. Nearly all of the ATP comes from high-energy electrons donated by NADH and $FADH_2$. As the electrons flow through the electron transport chain, they generate the H^+ gradient, which allows chemiosmosis to occur.

CASE STUDY CONTINUED

Raising a King

Mitochondrial DNA (mtDNA), the distinctive molecule that researchers used to identify the remains of Richard III, plays a crucial role in the human body. Although human mtDNA contains only 37 genes (compared to about 20,000 in the nucleus), the genes are essential. For example, 13 mitochondrial genes code for parts of enzymes that participate in the electron transport chain and chemiosmosis. Some of these genes contribute to the enzymes that cause NADH and $FADH_2$ to release their high-energy electrons into the chain. Other mtDNA genes code for the final enzyme in the chain, which helps energy-depleted electrons combine with oxygen molecules and hydrogen ions to form water. Still other genes in mtDNA code for parts of the ATP synthase enzyme on the inner mitochondrial membrane. If mtDNA were to disappear, cellular respiration would come to a screeching halt.

We've seen how NADH and $FADH_2$ can carry high-energy electrons that originated in glucose. Can these electron carriers also obtain high-energy electrons from other molecules in our diets?

Cellular Respiration Can Extract Energy from a Variety of Foods

Animals, including people, often acquire glucose by consuming starch (a long chain of glucose molecules) or sucrose (table sugar; glucose linked to fructose). A typical human also acquires energy by consuming fat and protein. We can use fat and protein as sources of energy thanks to metabolic pathways that break down molecules other than glucose to produce various intermediate molecules of cellular respiration. These intermediates then enter the cellular respiration pathway at various stages and are broken down to produce ATP (**FIG. 8-8**). For example, some of the 20 amino acids found in protein can be directly converted into pyruvate, and the others can be transformed through complex pathways into molecules involved in the Krebs cycle. To release the energy stored in fats, the long fatty acid tails (which comprise most of each

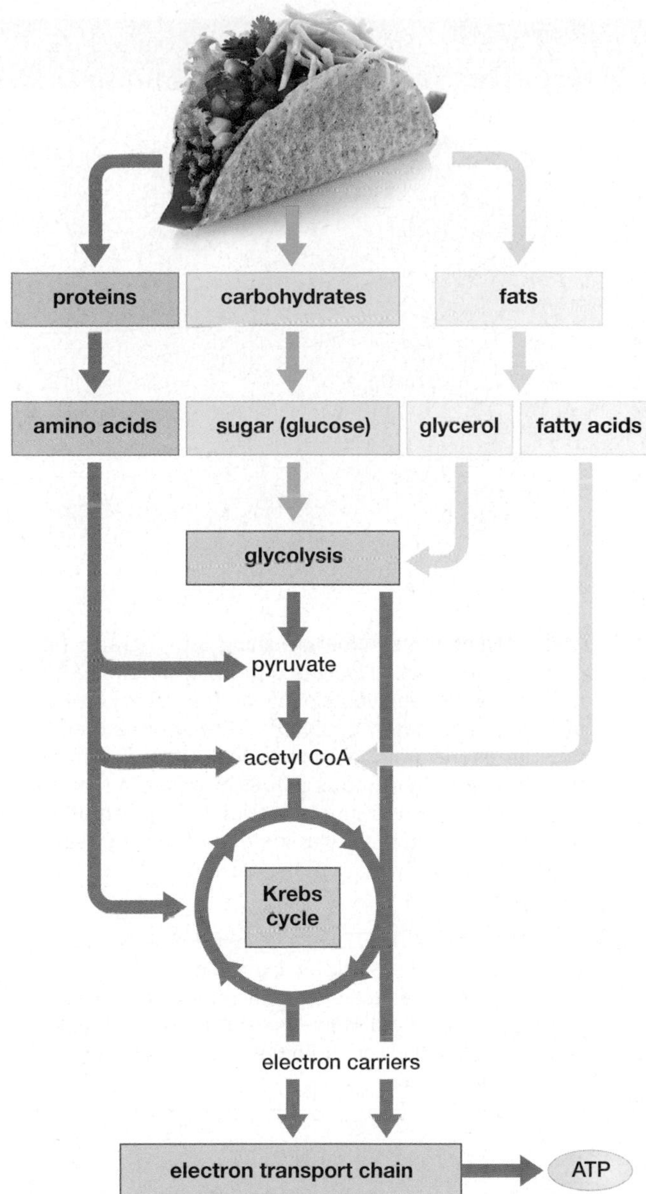

▲ **FIGURE 8-8 Proteins, carbohydrates, and fats are broken down and release ATP**

fat molecule; see Chapter 3) are broken into two-carbon fragments and combined with CoA to produce acetyl CoA, which enters the Krebs cycle.

When a person's diet includes too much sugar and starch, excess intermediate molecules produced during glucose breakdown can be converted to fat, as described in "Health Watch: How Can You Get Fat by Eating Sugar?"

CHECK YOUR LEARNING

Can you . . .

- summarize the two major stages of cellular respiration?
- explain how ATP is generated by chemiosmosis?
- describe the role of oxygen in cellular respiration?

8.4 HOW DOES FERMENTATION ALLOW GLYCOLYSIS TO CONTINUE WHEN OXYGEN IS LACKING?

Glycolysis is employed by virtually every organism on Earth, providing evidence that this is one of the most ancient of all biochemical pathways. Under **aerobic** conditions—that is, when oxygen is available—cellular respiration usually follows. But scientists have concluded that the earliest forms of life arose under the **anaerobic** (no oxygen) conditions that existed before photosynthesis evolved and enriched the air with oxygen. Today, many microorganisms still thrive in places where oxygen is rare or absent, such as in the stomach and intestines of animals (including humans), deep in soil, or in bogs and marshes.

Fermentation Takes Place in Anaerobic Conditions

Anaerobic organisms rely on **fermentation** for ATP production. In fermentation, the pyruvate produced by glycolysis enters a metabolic pathway that, in modern anaerobic organisms, converts the pyruvate to either lactate or to ethanol and CO_2. Some microorganisms are completely dependent on fermentation and lack the enzymes for cellular respiration. Others use fermentation when oxygen is absent, but switch to more efficient cellular respiration when oxygen is available.

Fermentation Produces Either Lactate or Alcohol and Carbon Dioxide

Lactate fermentation, which converts pyruvate to lactic acid, occurs in many microorganisms, but it may also occur in vertebrates, especially during intense muscular activity. If muscles contract so vigorously that blood cannot supply adequate oxygen for cellular respiration, muscle cells may briefly use glycolysis and lactate fermentation to generate ATP. This pathway regenerates NAD^+ by using the electrons and hydrogen ions from NADH to convert pyruvate into lactate (the dissolved form of lactic acid; **FIG. 8-9**). If you feel your muscles "burning" during vigorous exercise, they are probably fermenting pyruvate into lactic acid.

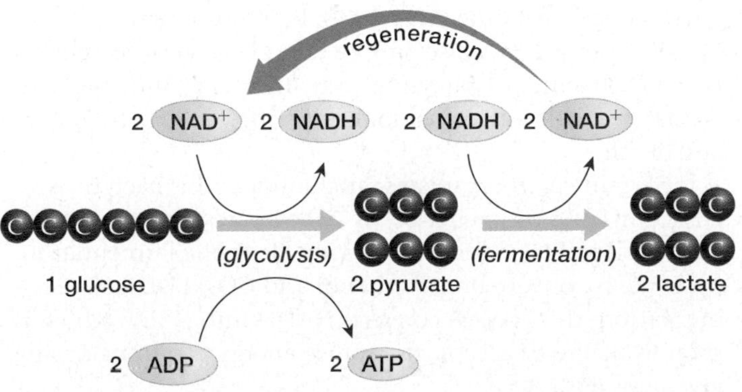

▲ **FIGURE 8-9 Glycolysis followed by lactic acid fermentation**

Health WATCH How Can You Get Fat by Eating Sugar?

From an evolutionary perspective, feeling hungry whenever food is available is highly adaptive. For most of the 300,000-year history of humans, sources of rich food were few and far between, so eating at every opportunity was a beneficial defense against starvation. But now that many people have continuous access to high-calorie food, obesity is an expanding health problem. Why do we accumulate fat? Because a gram of fat stores twice as much energy as a gram of carbohydrate, and stockpiling energy with minimum weight was important to our prehistoric ancestors, who needed to move quickly to catch prey or to avoid becoming prey themselves.

Acquiring fat by eating sugar and other carbohydrates is common among animals. How is fat made from sugar? As glucose is broken down during the Krebs cycle, acetyl CoA is formed. Excess acetyl CoA molecules are used as raw materials to synthesize the fatty acids that will be linked together to form a fat molecule (**FIG. E8-1**). Starches, such as those in bread, potatoes, or pasta, are actually long chains of glucose molecules, so you can see how eating excess starch can also make you fat.

To understand why storing fat rather than sugar can be advantageous, let's look at the ruby-throated hummingbird, which begins the summer weighing 3 to 4 grams (in comparison, a nickel weighs 5 grams). In late summer, hummingbirds feed voraciously on the sugary nectar of flowers and nearly double their weight in stored fat. The energy from fat powers their migration from the eastern United States across the Gulf of Mexico and into Mexico or Central America for the winter. If hummingbirds stored sugar instead of fat, they would be too heavy to fly.

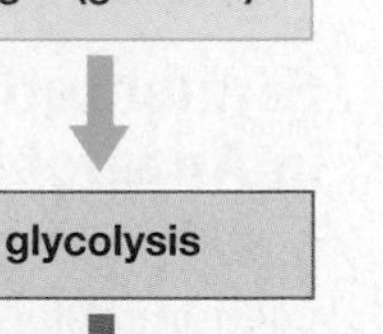

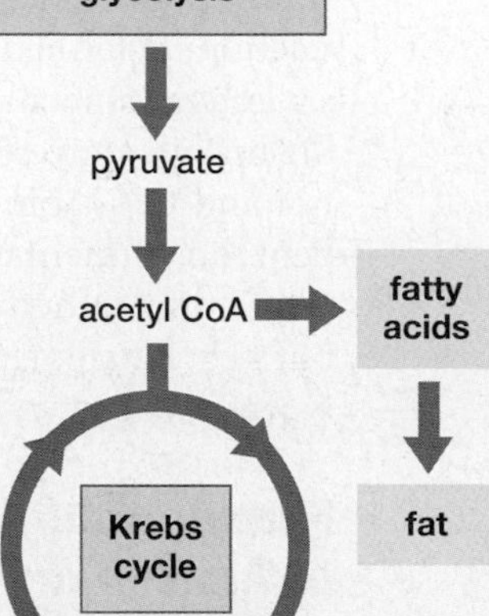

▲ **FIGURE E8-1 How sugar is converted to fat**

THINK CRITICALLY Colin, a 45-year-old obese man, comes to you, his physician, complaining that he has been on a fat-free diet for months without losing weight. What do you hypothesize about Colin's weight issue? To develop your hypothesis, what questions would you ask him?

Compared to cellular respiration, glycolysis followed by fermentation uses far more glucose to produce a given amount of ATP. But even though the simpler fermentation pathway is much less efficient that cellular respiration, it generates ATP much more quickly. This speedy production of ATP can provide the energy needed for a final, brief burst to the finish line (**FIG. 8-10**). Rapid production of ATP by fermentation can also supply energy for actions such as fighting, fleeing, or pursuing prey, in which the ability to persist just a bit longer can make the difference between life and death.

Certain microorganisms, including some bacteria and all forms of yeast, engage in **alcoholic fermentation** under anaerobic conditions. During alcoholic fermentation, pyruvate is converted into ethanol and CO_2. Like lactate fermentation, this process converts NADH into NAD^+, which is then available to accept more high-energy electrons during glycolysis (**FIG. 8-11**).

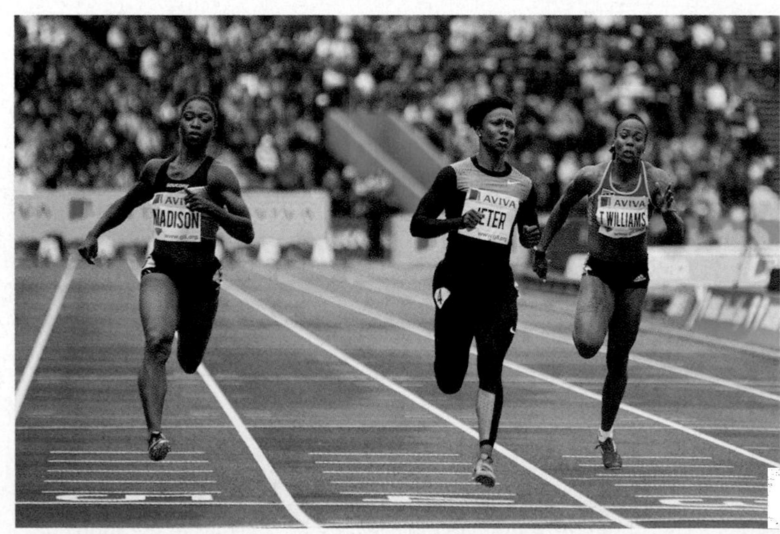

▲ **FIGURE 8-10 Lactic acid fermentation in action**

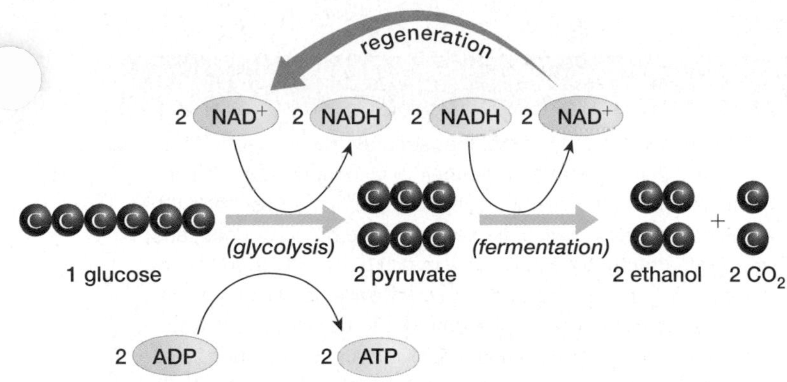

▲ **FIGURE 8-11 Glycolysis followed by alcoholic fermentation**

THINK CRITICALLY What would happen if cells were prevented from producing lactic acid or alcohol after glycolysis?

▲ **FIGURE 8-12 Some products of fermentation**

Fermentation Has Played a Long and Important Role in the Human Diet

The poet Omar Khayyam (1048–1122) described his vision of paradise on Earth as "A Jug of Wine, a Loaf of Bread—and Thou Beside Me" (**FIG. 8-12**). Khayyam's paradise (at least the part attributable to wine and bread) depends heavily on fermentation.

Historical evidence suggests that wine and beer, whose alcohol is produced by yeast, were being made roughly 7,000 years ago. Yeasts (single-celled fungi) are opportunists; they engage in efficient cellular respiration if oxygen is available, but switch to alcoholic fermentation (producing alcohol and CO_2) if they run out of oxygen. To make beer or wine, sugars from mashed grain (beer) or grapes (wine) are fermented by specialized strains of yeast. Fermentation is carried out in casks with valves that prevent air from entering (so cellular respiration can't occur) but that allow the CO_2 to escape (so the cask doesn't explode). To put the characteristic fizz in beer and champagne, fermentation is allowed to continue after the bottle is sealed, trapping CO_2 under pressure.

CASE STUDY **CONTINUED**

Raising a King

Richard III included plenty of fermented products in his diet. How do we know? Researchers can discover the types of food a person consumed by measuring isotopes in teeth and bones. Different foods contain characteristic ratios of isotopes of certain elements, and these isotope ratios are incorporated in the bones and teeth of people who eat the foods. Researchers have analyzed such isotopes to glean information about Richard III's lifestyle. Based on oxygen isotope ratios in his bones, it appears that during the last few years of his life, roughly one-quarter of his fluid intake consisted of wine. This beverage—fit for a king—was produced, of course, by alcoholic fermentation. What other ancient and modern staples result from fermentation?

In addition to its role in wine and beer production, fermentation also gives bread its airy texture. As bread-making begins, enzymes in yeast cells break the starch in flour into its component glucose molecules. The yeast cells rapidly grow and divide, releasing CO_2, first during cellular respiration and later during alcoholic fermentation after the O_2 dissolved in the water used to make the dough is used up. The dough, made stretchy and resilient by kneading, traps the CO_2 gas, which expands in the heat of the oven. The alcohol evaporates as the bread is baked.

A variety of microorganisms meet their energy needs primarily by glycolysis followed by lactate fermentation. These include the lactic acid bacteria that assist in transforming milk into yogurt, sour cream, and cheese. These bacteria first split lactose (milk sugar) into glucose and galactose; then these simple sugars enter glycolysis followed by lactate fermentation, which produces lactic acid. Lactic acid denatures milk protein, altering its three-dimensional structure and giving sour cream and yogurt their semisolid textures. Like all acids, lactic acid tastes sour and contributes to the distinctive tastes of these foods. Lactic acid bacteria are also used to begin the coagulation of milk during cheese production. In addition, lactate fermentation by salt-tolerant bacteria converts sugars in vegetables such as cucumbers and cabbage into lactic acid. The result: dill pickles and sauerkraut.

CHECK YOUR LEARNING

Can you . . .

- explain the function of fermentation and the conditions under which it occurs?
- compare the two types of fermentation?
- list some examples of human uses of each type of fermentation?

CASE STUDY REVISITED

Raising a King

Richard III's skeleton and the mtDNA it contained confirmed the king's identity because mtDNA is uniquely valuable for tracing the hereditary relationships of ancient remains. A person's mtDNA originates from the mitochondria in the cytoplasm of the egg cell from which that person develops; sperm mitochondria do not enter the egg when it is fertilized. As a result, mtDNA is passed directly from a mother to her children in an unbroken chain that can extend through thousands of generations on the mother's side of the family.

Over millennia, harmless mutations have accumulated in regions of the mtDNA molecule that do not code for functional proteins. Ancient mutations persist as new mutations are gradually added to the noncoding regions. Scientists can sequence the mtDNA of many people and divide people into distinct subgroups on the basis of their newer—and therefore less widespread—mutations. Each of these subgroups traces its ancestry back to a single woman, whose mutated mtDNA was passed to her descendants. These descendants were originally confined to a particular geographic area, and even today the unique mtDNA signature of each subgroup remains common in that location. As a result, people alive today who have particular mutations can trace their ancestry to a distinct subgroup with the same genetic signature and to the particular geographic location where the mutations first emerged.

Modern techniques can recover mtDNA even from severely decomposed remains, such as the skeleton under the Leicester parking lot. When Turi King sequenced the skeleton's mtDNA, she found that its sequence is relatively rare, shared by only 1% to 2% of the population of the United Kingdom.

Meanwhile, genealogists had identified two living descendants of an unbroken maternal line from Cecily Neville, mother of Richard III. One remains anonymous; the other is Michael Ibsen, a Canadian-born carpenter living in London (**FIG. 8-13**). The five centuries and 18 generations that separate Cecily Neville and Michael Ibsen have not altered their mitochondrial DNA, which matches that of the skeleton. The location of the skeleton in the ruins of Greyfriars Church, its deformity and battle wounds, and, most importantly, the remarkable match of its mtDNA to that of the only known descendants of Cecily Neville, led Leicester University's lead archaeologist to state: " . . . beyond reasonable doubt, the individual exhumed at Greyfriars in September 2012 is indeed Richard III."

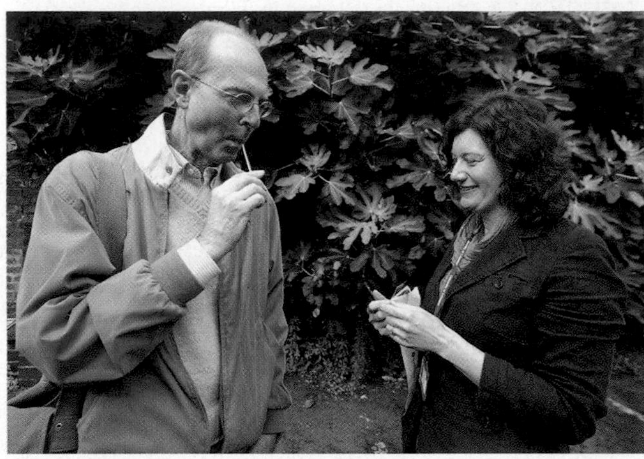

▲ **FIGURE 8-13 Michael Ibsen provides cheek cells to Turi King, who used them to sequence his mitochondrial DNA**

THINK CRITICALLY Jeremy has always had difficulty walking rapidly and for long distances. Shortly before Jeremy's wedding, genetic testing revealed that his problem was caused by an mtDNA mutation. Should Jeremy be concerned about a future daughter inheriting the faulty gene? What about a son?

CHAPTER REVIEW

Go to **Mastering Biology** to access the Pearson eText, vocabulary review, practice quizzes, activities, videos, current events, and more.

*Answers to **Think Critically** and **Thinking Through the Concepts** questions can be found in the **Answers** section at the back of the book.*

Summary of Key Concepts

8.1 How Do Cells Obtain Energy?

The ultimate source of energy for nearly all life is sunlight, captured during photosynthesis and stored in molecules such as glucose. Cells produce chemical energy by breaking down glucose and capturing some of the released energy as ATP. During glycolysis, glucose is broken down in the cytosol, forming pyruvate and generating two ATP and two NADH molecules, which are high-energy electron carriers. If oxygen is available, cellular respiration in the mitochondria breaks down pyruvate, generating 34 additional molecules of ATP.

8.2 How Does Glycolysis Begin Breaking Down Glucose?

Figure 8-3 and Table 8-1 summarize glycolysis. During the energy investment stage of glycolysis, glucose is energized by adding energy-carrying phosphate groups from two ATP molecules, forming fructose bisphosphate. Then, during the energy-harvesting stage, a series of reactions breaks down the fructose bisphosphate into two molecules of pyruvate. This produces a net energy yield of two ATP molecules and two NADH molecules.

8.3 How Does Cellular Respiration Extract Energy from Glucose?

Cellular respiration requires O_2; the process is summarized in Figures 8-5 and 8-6 and Table 8-1. The first stage of cellular respiration occurs in the mitochondrial matrix. Pyruvate from glycolysis is converted to acetyl CoA, releasing CO_2 and generating NADH. The acetyl CoA then enters the Krebs cycle, which generates ATP, NADH, and $FADH_2$, and releases additional CO_2. Overall, for each molecule of glucose that enters glycolysis, the first stage of cellular respiration produces two ATP, eight NADH, and two $FADH_2$.

During the second stage of cellular respiration, the NADH and $FADH_2$ generated in the first stage deliver high-energy electrons to the electron transport chain (ETC) within the inner mitochondrial membrane. As the electrons pass along the ETC, energy is released and used to pump hydrogen ions across the inner membrane from the matrix into the intermembrane space, creating a hydrogen ion gradient. At the end of the ETC, the depleted electrons combine with hydrogen ions and oxygen to form water. During chemiosmosis, the energy stored in the hydrogen ion gradient is used to produce ATP as the hydrogen ions diffuse down their concentration gradient through ATP synthase channels in the inner membrane.

Complete breakdown of a molecule of glucose from glycolysis followed by cellular respiration yields 2 ATP from glycolysis, 2 ATP from the Krebs cycle, and 32 ATP from chemiosmosis, for a total net yield of 36 ATP.

8.4 How Does Fermentation Allow Glycolysis to Continue When Oxygen Is Lacking?

Glycolysis uses NAD^+ to produce NADH as glucose is broken down into pyruvate. For these reactions to continue, NAD^+ must be continuously recycled. Under anaerobic conditions, NADH cannot release its high-energy electrons to the electron transport chain because there is no oxygen to accept them. Instead, fermentation regenerates NAD^+ from NADH by converting pyruvate to lactate (via lactate fermentation) or to ethanol and CO_2 (via alcoholic fermentation), allowing glycolysis to continue. People harness fermentation by microorganisms to produce beer, wine, bread, cheese, sauerkraut, and pickles.

Thinking Through the Concepts

Bloom's: Remembering, Understanding

Multiple Choice

1. Which of the following is true for one glucose molecule?
   a. Fermentation produces two ATP.
   b. Glycolysis followed by fermentation nets four ATP.
   c. Ethanol is one end product of glycolysis.
   d. The overall equation for photosynthesis is the reverse of that for aerobic glucose breakdown.
2. The portion of glucose breakdown that produces the most ATP is
   a. chemiosmosis.
   b. glycolysis.
   c. the Krebs cycle.
   d. fermentation.
3. ATP synthase enzymes are located in the
   a. cytosol.
   b. inner mitochondrial membrane.
   c. intermembrane space.
   d. mitochondrial matrix.
4. Fermentation
   a. regenerates NADH.
   b. follows cellular respiration when oxygen is lacking.
   c. generates additional ATP after glycolysis.
   d. uses pyruvate as its substrate.
5. Which of the following is produced in the intermembrane space of mitochondria?
   a. ATP
   b. a high concentration of H^+
   c. NADH and $FADH_2$
   d. acetyl CoA

Fill-in-the-Blank

1. The complete breakdown of glucose in the presence of oxygen occurs in two major stages: ____________ and ____________. The first of these stages occurs in the ____________ of the cell, and the second stage occurs in organelles called ____________. Conditions in which oxygen is present are described as ____________.
2. Conditions in which oxygen is absent are described as ____________. Some microorganisms break down glucose in the absence of oxygen using ____________, which generates only ____________ molecules of ATP. This process is followed by ____________, in which no more ATP is produced, but the electron-acceptor molecule ____________ is regenerated so it can be used in further glucose breakdown.
3. Yeasts in bread dough and alcoholic beverages use a type of fermentation that generates ____________ and ____________. Muscles pushed to their limit use ____________ fermentation. Which form of fermentation is used by microorganisms that produce yogurt, sour cream, and sauerkraut? ____________
4. During cellular respiration, the electron transport chain pumps H^+ out of the mitochondrial ____________ into the ____________, producing a ____________ of H^+. The ATP produced by cellular respiration is generated by a process called ____________. ATP is generated as H^+ travels through membrane channels within ____________.
5. The cyclic portion of cellular respiration is called the ____________ cycle. The molecule that enters this cycle is ____________. How many ATP molecules are generated by this cycle per molecule of glucose? ____________ What types of high-energy electron-carrier molecules are generated during the cycle? ____________ and ____________

Review Questions

1. Starting with glucose ($C_6H_{12}O_6$), write the overall equation for glucose breakdown in the presence of oxygen, compare this to the overall equation for photosynthesis, and explain how the energy components of the equations differ.
2. Draw and label a mitochondrion, and explain how each structure relates to its function.
3. What role do the following play in breaking down and harvesting energy from glucose: glycolysis, cellular respiration, chemiosmosis, fermentation, and the electron acceptors NAD^+ and FAD?
4. Outline the two major stages of glycolysis. How many ATP molecules (overall) are generated per glucose molecule during glycolysis? Where in the cell does glycolysis occur?

5. What molecule is the end product of glycolysis? How are the carbons of this molecule used in stage 1 of cellular respiration? In what form is most of the energy from the Krebs cycle captured?
6. Describe the electron transport chain and the process of chemiosmosis.
7. Why is oxygen necessary for cellular respiration to occur?
8. Compare the structure of chloroplasts (described in Chapter 7) to that of mitochondria, and describe how the similarities in structure relate to similarities in function.

Applying the Concepts

Bloom's: Applying, Analyzing, Evaluating

1. Some species of bacteria use aerobic respiration, and other species use fermentation. In an oxygen-rich environment, would either type be at a competitive advantage? What about in an oxygen-poor environment?
2. Many microorganisms in lakes use cellular respiration to generate energy. Dumping large amounts of raw sewage into rivers or lakes typically leads to massive fish kills, even if the sewage itself is not toxic to fish. What kills the fish? How might you reduce fish mortality after raw sewage is accidentally released into a small pond?
3. Imagine a hypothetical situation in which a starving cell reaches the stage where every bit of its ATP has been depleted and converted to ADP plus phosphate. If at this point you place the cell in a solution containing glucose, will it recover and survive? Explain your answer based on what you know about glucose breakdown.
4. Some species of bacteria that live at the surface of sediment on the bottom of lakes are capable of using either glycolysis plus fermentation or cellular respiration to generate ATP. There is very little circulation of water in lakes during the summer. Predict and explain what will happen to the bottom-most water of a deep lake as the summer progresses, and describe how this situation will affect the amount of energy production by bacteria.

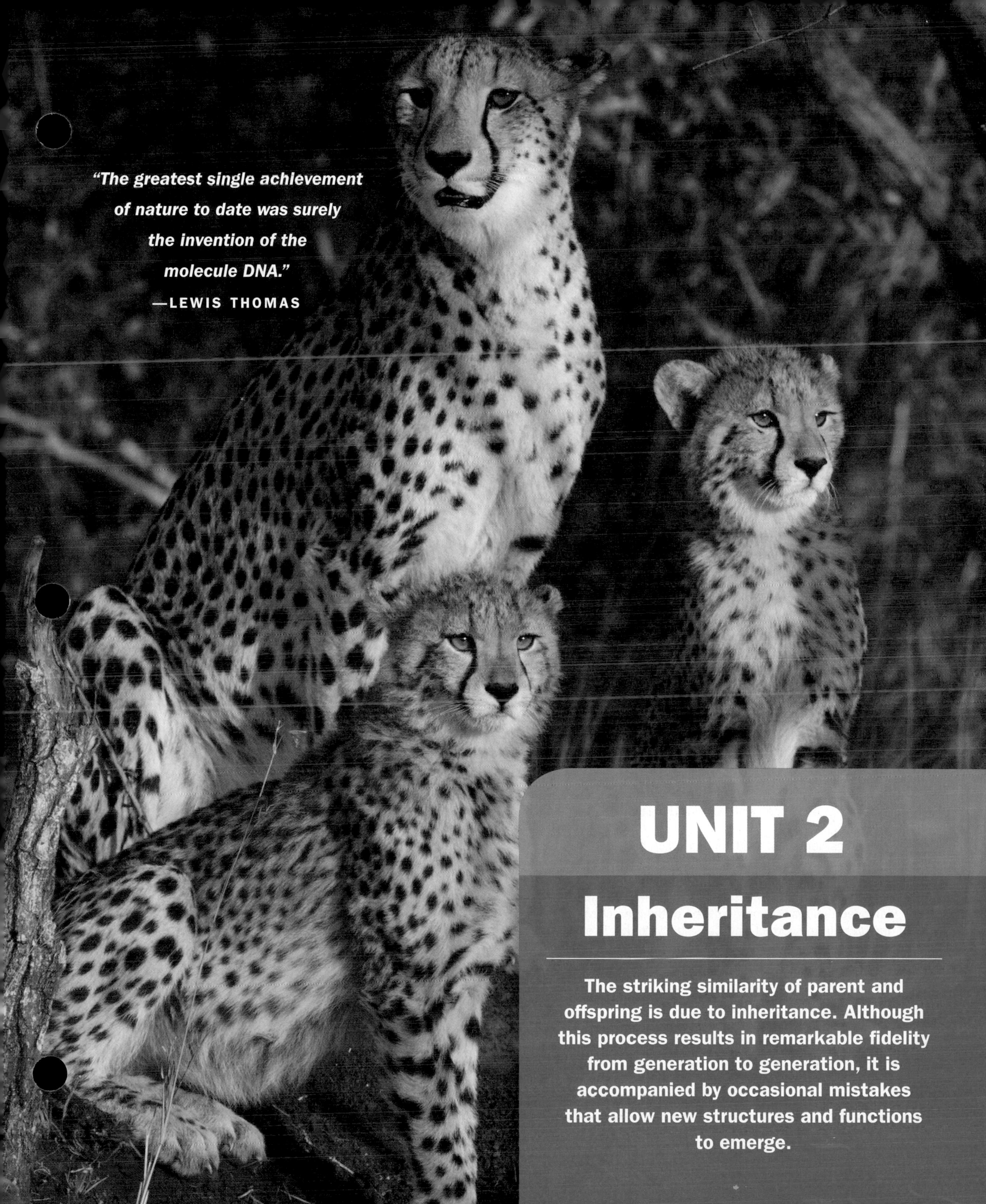

"The greatest single achievement of nature to date was surely the invention of the molecule DNA."

—LEWIS THOMAS

UNIT 2

Inheritance

The striking similarity of parent and offspring is due to inheritance. Although this process results in remarkable fidelity from generation to generation, it is accompanied by occasional mistakes that allow new structures and functions to emerge.

9 Cellular Reproduction

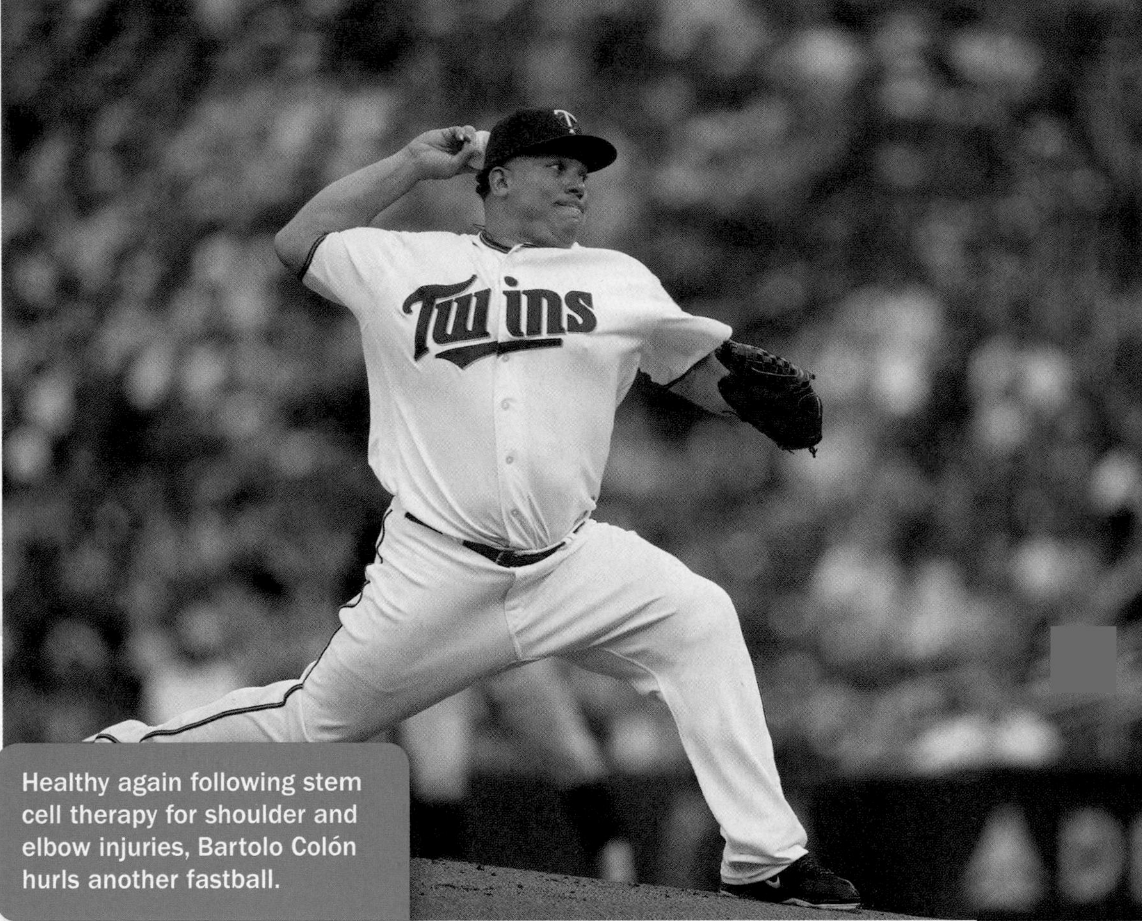

Healthy again following stem cell therapy for shoulder and elbow injuries, Bartolo Colón hurls another fastball.

CASE STUDY

Body, Heal Thyself

WITH A 95-MILES-PER-HOUR FASTBALL, Bartolo Colón was at the top of his game when he won the Cy Young Award as the best pitcher in the American League in 2005. But throwing that hard takes a toll, and Colón injured his arm near the end of his award-winning season. He had damaged ligaments and tendons in his throwing shoulder and elbow, and the injury kept him on the bench for much of the next 4 years. Why didn't his arm heal after all that time?

The ligaments and tendons that Colón had injured consist mostly of proteins organized in an arrangement that provides both strength and flexibility. For Colón to be able to throw as hard as he once did, his body would need to rebuild the damaged tissues with new proteins of the correct types, organized in the correct way. Can a body do that? Yes, the body's healing processes can eventually repair ligaments and tendons, so they return to their original size, strength, and flexibility. Unfortunately, though, the healing process is slow and isn't always completely successful. It didn't work very well for Colón.

In the spring of 2010, in an attempt to speed the healing process, Colón tried a new treatment. Physicians removed cells called mesenchymal stem cells from Colón's bone marrow and fat, and injected them into his shoulder and elbow. Mesenchymal stem cells can, with the right stimuli, multiply and produce populations of specialized cells that form cartilage, ligaments, tendon, bone, or other tissues. The hope was that the stem cells would repair Colón's damaged ligaments and tendons. Because Colón's own cells were used, there was no risk of rejection.

By late 2010, Colón was pitching again, in a Puerto Rican winter league. Meanwhile, the New York Yankees were looking for a good pitcher, and Colón was hoping to make a comeback in the major leagues. The Yankees worried that Colón might never return to top form but signed him anyway. The gamble paid off: In 2011, Colón's trademark fastballs were once again confounding hitters, and he helped the Yankees win a division title. Since then, Colón has pitched for the Athletics, the Mets, the Braves, and the Twins. He made the All-Star team in 2016 and was still going strong as of 2018. In the wake of Colón's career rejuvenation, stem cell treatments have become increasingly popular among major league pitchers with serious arm injuries.

In Bartolo Colón's repaired shoulder, the cells in newly grown tissue have exactly the same genes as the cells in the older, damaged tissue. When cells divide, why are the offspring cells genetically identical to the cells they came from?

AT A GLANCE

9.1 WHAT ARE THE FUNCTIONS OF CELL DIVISION?

"All cells come from cells." This insight, first stated by the German physician Rudolf Virchow in the mid-1800s, captures the critical importance of cellular reproduction for all living organisms. Cells reproduce by **cell division**, in which a parent cell divides into two **daughter cells**.

The Genetic Material Is Replicated During Cell Division

In a typical cell division, each daughter cell receives a complete set of hereditary information, identical to that of the parent cell, and about half the parent cell's cytoplasm. The hereditary information of all living cells is contained in **deoxyribonucleic acid (DNA)**. DNA is a polymer composed of subunits called **nucleotides** (**FIG. 9-1a**; see also Chapter 3). In all cells, DNA is packaged into **chromosomes**. The DNA in a chromosome consists of two long strands of nucleotides wound around each other, like a ladder twisted into a corkscrew shape. This structure is called a double helix (**FIG. 9-1b**). Each chromosome contains a double helix of DNA as well as proteins that organize the DNA's three-dimensional structure and regulate its use. The units of inheritance, called **genes**, are segments of chromosomal DNA that range from a few hundred to many thousands of nucleotides in length. Like the letters of an alphabet spelling out very long sentences, the specific sequences of nucleotides in genes spell out the instructions for making the proteins of a cell.

Cell Division Is Required for Growth, Development, and Repair of Multicellular Organisms

As you grew and developed from a fertilized egg, cell division produced all the cells in your body. This growth and development was the result of many rounds of **mitotic cell division**, in which each division produces two daughter cells that are genetically identical to the parent cell. Even now that you have reached your adult size, mitotic cell

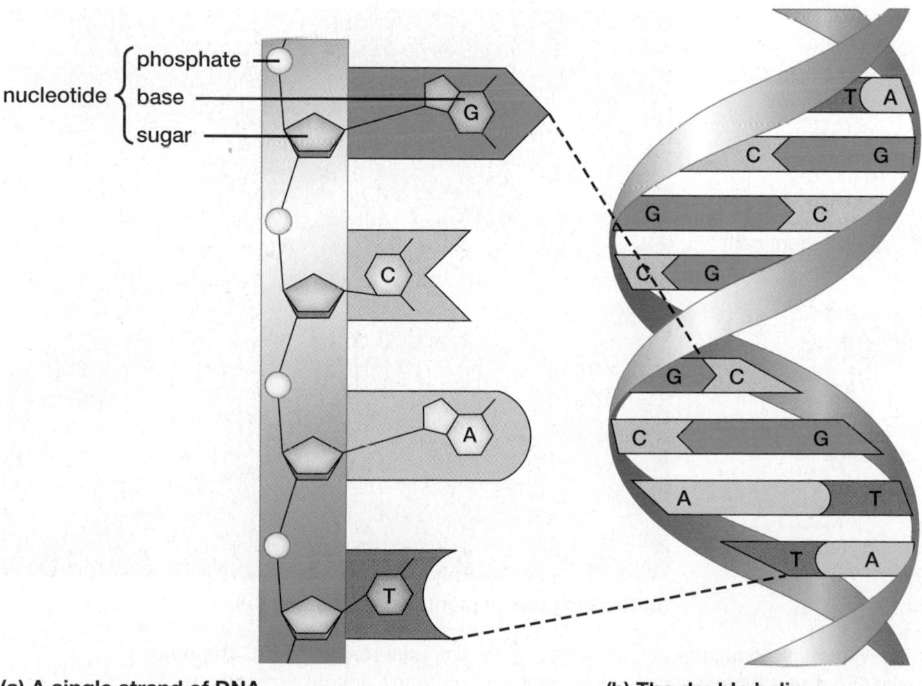

FIGURE 9-1 The structure of DNA **(a)** A nucleotide consists of a phosphate, a sugar, and one of four bases—adenine (A), thymine (T), guanine (G), or cytosine (C). A single strand of DNA consists of a long chain of nucleotides held together by chemical bonds between the phosphate of one nucleotide and the sugar of the next. **(b)** Two DNA strands twist around one another to form a double helix.

division continues to be essential, replacing cells that are killed by everyday life, such as cells in your digestive tract that are destroyed by stomach acid and digestive enzymes, or skin cells that are worn away by rubbing on your clothes. Mitotic cell division is also required to repair injuries, such as the damage that throwing thousands of fastballs inflicted on Bartolo Colón's arm.

The daughter cells formed by cell division may grow and divide again, in a repeating pattern called the **cell cycle**. Many of the daughter cells **differentiate**, becoming specialized for specific functions, such as contraction (muscle cells) or fighting infections (white blood cells). Most multicellular eukaryotic organisms typically have three categories of cells, based on their abilities to divide and differentiate:

- *Stem Cells* Most of the cells formed by the first few cell divisions of a fertilized egg, and some cells in adult animals, are **stem cells** (**FIG. 9-2a**). Stem cells have two important characteristics: self-renewal and potency. *Self-renewal* means that stem cells retain the capacity to divide, in some cases for the entire life of the organism. *Potency* means that dividing stem cells produce daughter cells that can differentiate into a variety of specialized cell types. Some stem cells in early embryos can produce any of the specialized cell types of the entire body. Stem cells in adults are usually more limited and produce daughter cells that can differentiate into only a few cell types.

  Plants also have stem cells, usually called *meristem cells*. The growing tips of roots, stems, and branches often contain clusters of meristem cells (**FIG. 9-2b**). Cell division and differentiation of some of the daughter cells produce the various structures of the plant body.
- *Differentiated Cells Capable of Dividing* Some differentiated cells can divide, but their daughter cells typically differentiate into only one or two cell types. For example, if part of your liver is seriously damaged, differentiated liver cells start dividing to replace the lost liver tissue; their daughter cells can only become more liver cells. Similarly, if xylem tissue—which conducts water and nutrients upward from a plant's roots—is damaged, certain differentiated cells in the tissue divide to repair the damage by forming new xylem cells.
- *Permanently Differentiated Cells* Some cells differentiate and never divide again. For example, as a human red blood cell differentiates, it loses its cell nucleus and its ability to divide. Most of the cells in your heart cannot divide, nor can most of the cells in the outermost layer of your skin. Likewise, almost all brain cells are permanently differentiated; the resulting inability to divide is one of the reasons that serious brain injuries do not heal.

Cell Division Is Required for Sexual and Asexual Reproduction

Reproduction requires cell division and may be sexual or asexual. **Sexual reproduction** occurs when offspring are produced by the fusion of **gametes** (sperm and eggs). To produce gametes, cells in the adult's reproductive system undergo a specialized type of cell division called meiotic cell division, which we will describe in Chapter 10.

Reproduction in which offspring are formed from a single parent, without having sperm fertilize eggs, is called

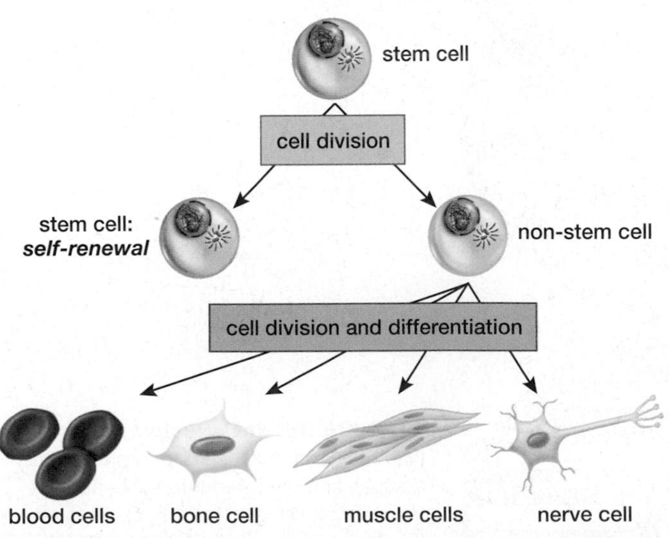

(a) Self-renewal and potency

(b) Growing tips of plants contain stem cells

▲ **FIGURE 9-2 Stem cells (a)** When a stem cell divides, one daughter cell remains a stem cell (self-renewal, left). The other daughter cell may divide a few times, but eventually differentiates into a specialized cell type (potency, animal examples at bottom). **(b)** The tips of plant shoots and roots, where growth occurs, contain self-renewing, potent stem cells (meristem cells).

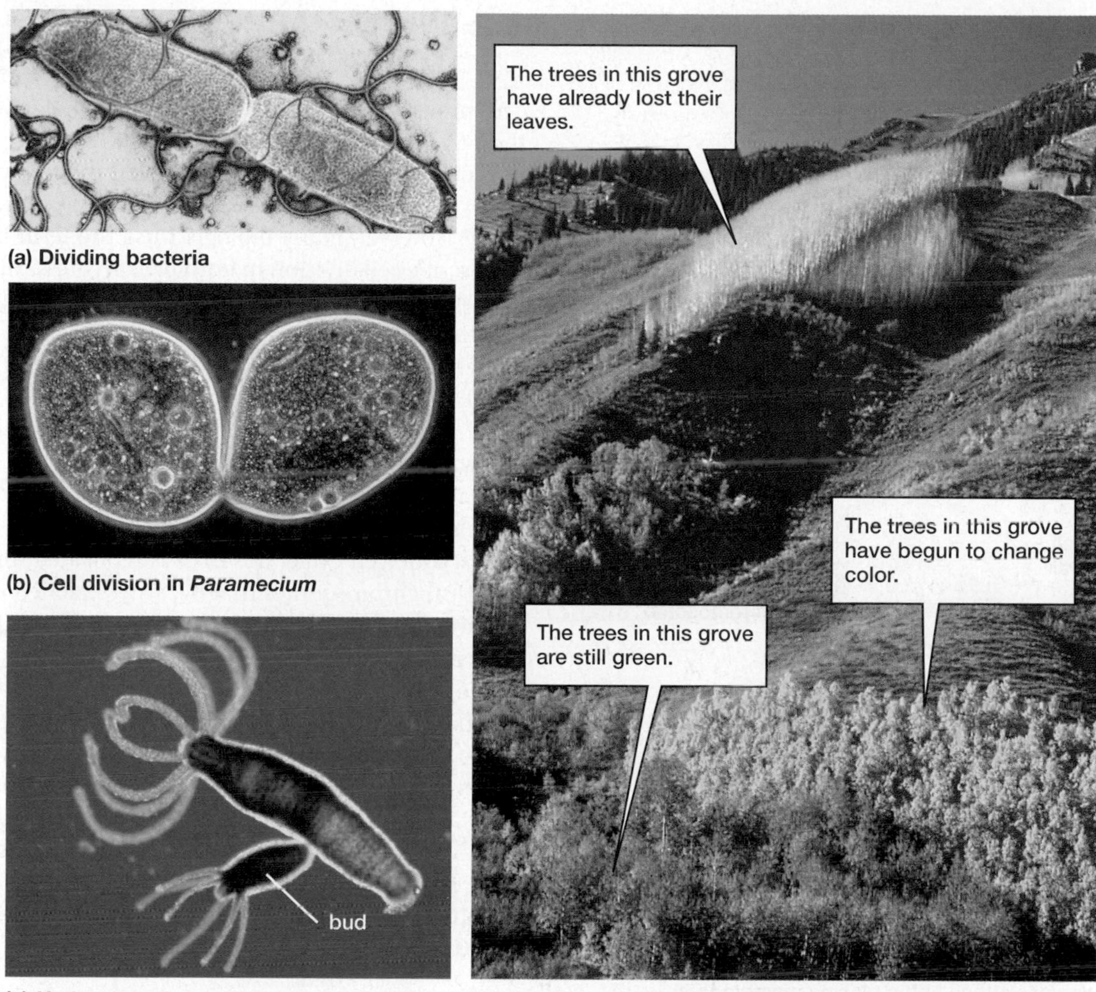

The trees in this grove have already lost their leaves.
The trees in this grove have begun to change color.
The trees in this grove are still green.
bud

(a) Dividing bacteria

(b) Cell division in *Paramecium*

(c) *Hydra* reproduces asexually by budding

(d) A grove of aspens often consists of genetically identical trees produced by asexual reproduction

▲ **FIGURE 9-3 Cell division enables asexual reproduction (a)** Prokaryotes, such as the bacteria shown here, reproduce asexually by dividing in two. **(b)** In single-celled eukaryotic microorganisms, such as the freshwater protist *Paramecium*, cell division produces two new, independent organisms. **(c)** *Hydra*, a freshwater relative of the sea anemone, grows a miniature replica of itself (a bud) on its side. When fully developed, the bud breaks off and assumes independent life. **(d)** The trees in aspen groves are often genetically identical. Here, the timing of fall colors and leaf drop shows the genetic identity within a grove and the genetic difference between three separate groves.

asexual reproduction. Asexual reproduction produces offspring that are genetically identical to the parent and to each other, but it is achieved by various methods across the domains of life. Bacteria (**FIG. 9-3a**) and archaea reproduce asexually by a type of cell division called prokaryotic fission (see Section 9.2). Many single-celled eukaryotic organisms, such as *Paramecium* (**FIG. 9-3b**), reproduce asexually by mitotic cell division. Some multicellular eukaryotes can reproduce asexually, using mitotic cell division to produce new, genetically identical, miniature versions of the adult. For example, *Hydra* reproduce by budding. First an individual grows a small replica of itself, called a bud, on its body (**FIG. 9-3c**). Eventually, the bud separates from its parent, forming a new individual.

Many plants and fungi can reproduce both asexually and sexually. For example, the fungus (yeast) cells in a batch of bread dough proliferate by asexual reproduction, their numbers increasing rapidly as they produce carbon dioxide that makes the dough rise. Among plants, aspen groves, develop asexually from shoots growing up from the root system of a single parent tree (**FIG. 9-3d**). Although a grove looks like a cluster of separate trees, it is often a single individual whose multiple trunks are connected by a common root system. Aspen can also reproduce by seeds, which result from sexual reproduction.

CHECK YOUR LEARNING

Can you . . .

- describe how the types of cells found in a multicellular organism are distinguished by their ability to divide and differentiate?
- describe the functions of cell division in single-celled and multicellular organisms?

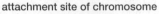

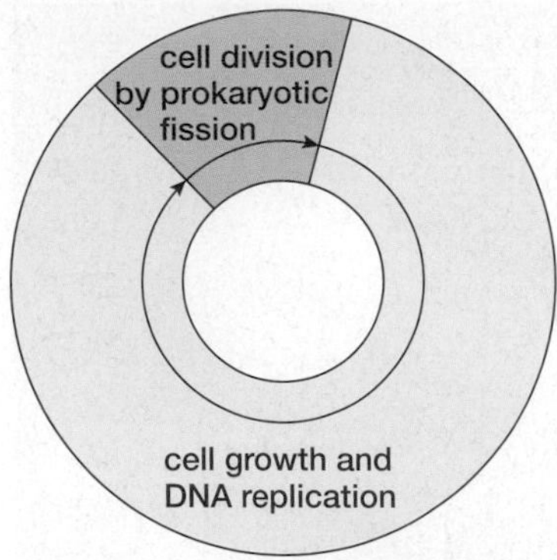

(a) The prokaryotic cell cycle

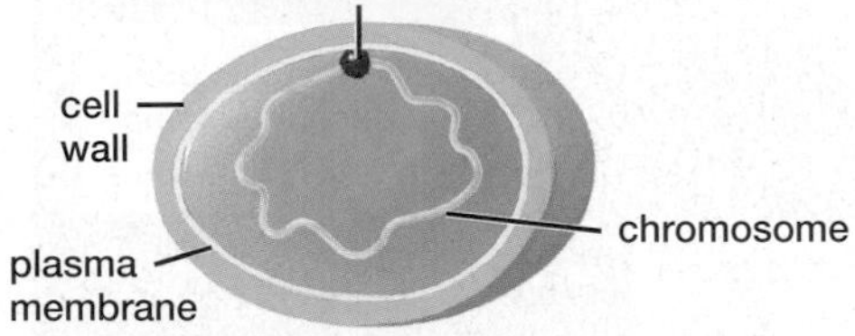

1 The prokaryotic chromosome, a circular DNA double helix, is attached to the plasma membrane at one point.

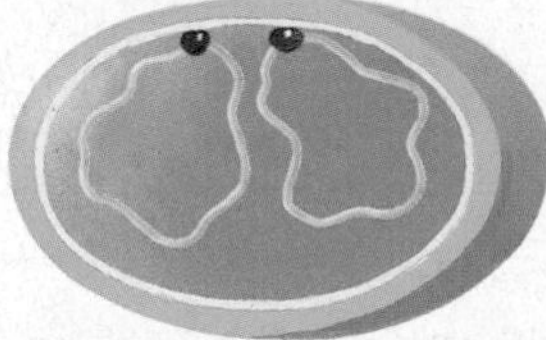

2 The DNA replicates and the resulting two chromosomes attach to the plasma membrane at nearby points.

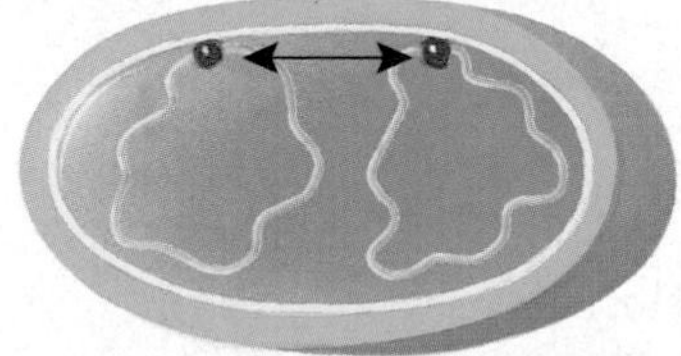

3 New plasma membrane is added between the attachment points, pushing the two chromosomes farther apart.

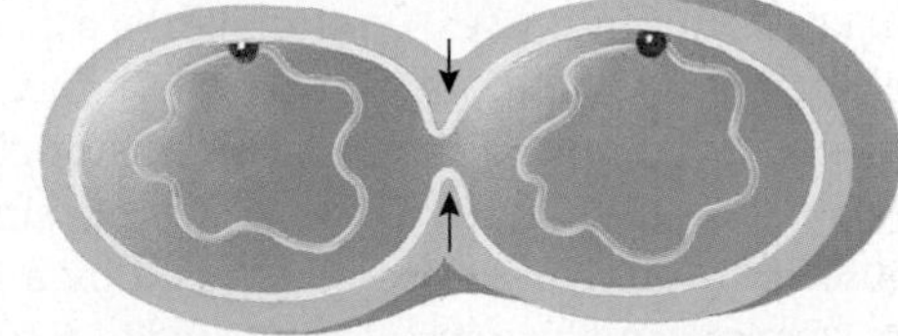

4 The plasma membrane grows inward at the middle of the cell.

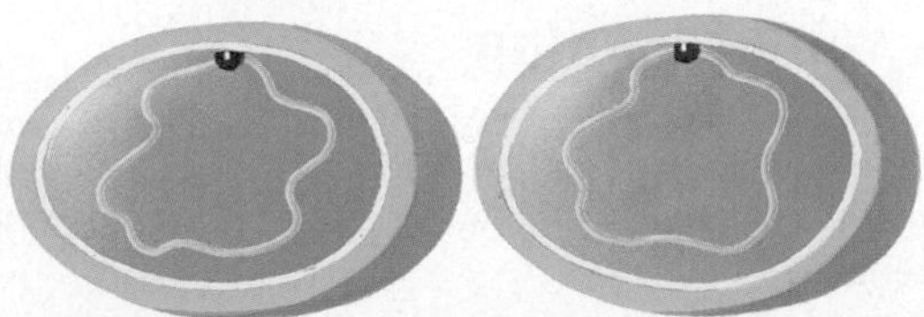

5 The parent cell divides into two daughter cells.

(b) Prokaryotic fission

9.2 WHAT HAPPENS DURING THE PROKARYOTIC CELL CYCLE?

The prokaryotic cell cycle consists of a relatively long period of growth, during which the cell replicates its DNA, followed by a type of cell division called **prokaryotic fission** (**FIG. 9-4a**). Prokaryotic fission is often called "binary fission." However, many biologists use the term binary fission to describe cell division in both prokaryotes and single-celled eukaryotes. To avoid confusion, we will use the term prokaryotic fission.

FIGURE 9-4b shows the process of prokaryotic fission. The DNA of a prokaryotic cell is contained in a single, circular chromosome. The chromosome is not contained in a membrane-bound nucleus (see Chapter 4). Instead, the chromosome is usually attached to the inside of the plasma membrane of the cell 1. During the growth phase of the prokaryotic cell cycle, the DNA is replicated, producing two identical chromosomes that become attached to the plasma membrane at nearby, but separate, sites 2. As the cell grows, new plasma membrane is added between the attachment sites of the chromosomes, pushing them apart 3. When the cell has approximately doubled in size, the plasma membrane around the middle of the cell grows inward between the two attachment sites 4. The plasma membrane then fuses along the equator of the cell, producing two daughter cells, each containing one of the chromosomes 5. Because DNA replication yields two identical DNA molecules, the two daughter cells are genetically identical to one another (and to the parent cell that produced them).

CHECK YOUR LEARNING

Can you . . .

- describe the prokaryotic cell cycle and the major events of prokaryotic fission?

9.3 HOW IS THE DNA IN EUKARYOTIC CHROMOSOMES ORGANIZED?

Eukaryotic chromosomes differ from prokaryotic chromosomes in several respects. They are separated from the cytoplasm within a membrane-bound nucleus, and they are linear instead of circular. Eukaryotic chromosomes also contain much more protein than prokaryotic chromosomes do, and their proteins are different. Finally, eukaryotic chromosomes usually contain far more DNA than prokaryotic chromosomes do. A human chromosome, for example, contains 10 to 50 times more DNA than the typical prokaryotic chromosome. If the DNA in a human cell were completely relaxed and extended, it would be about 6 feet (1.8 meters) long. The number of chromosomes in the cells of eukaryotic organisms varies tremendously—only one chromosome is found in the cells of male jack jumper ants,

◀ **FIGURE 9-4 The prokaryotic cell cycle (a)** The prokaryotic cell cycle consists of growth and DNA replication, followed by prokaryotic fission. **(b)** The process of prokaryotic fission.

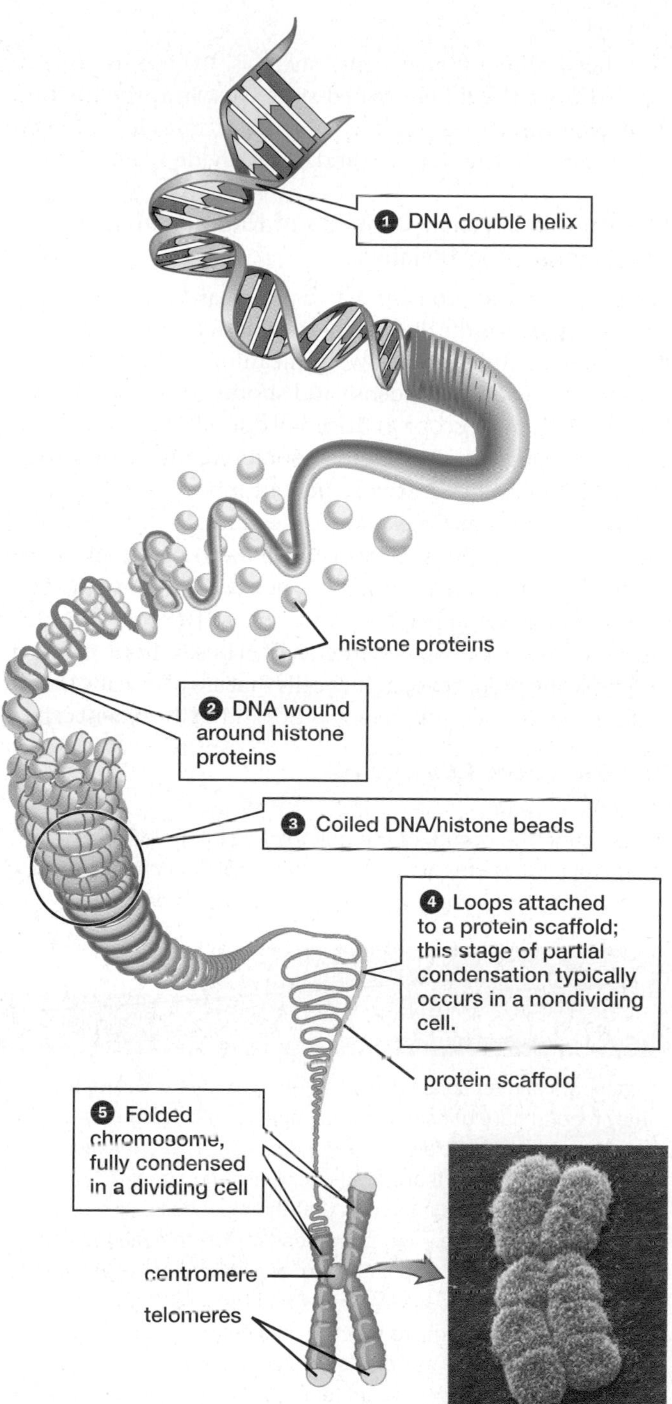

1 DNA double helix
histone proteins
2 DNA wound around histone proteins
3 Coiled DNA/histone beads
4 Loops attached to a protein scaffold; this stage of partial condensation typically occurs in a nondividing cell.
protein scaffold
5 Folded chromosome, fully condensed in a dividing cell
centromere
telomeres

▲ **FIGURE 9-5 Chromosome structure** Proteins in a eukaryotic chromosome wrap, coil, and fold the DNA into a compact structure. The ends of the chromosome are protected by telomeres. The centromere will be the site of attachment of microtubules that move the chromosome during mitotic cell division. (Inset) The fuzzy edges visible in the scanning electron micrograph are loops of folded chromosome.

but most animals have dozens, and some plants have more than 1,200!

The complex events of eukaryotic cell division are largely an evolutionary solution to the problem of duplicating and parceling out a large number of long chromosomes. To understand eukaryotic cell division, we will begin by taking a closer look at the structure of the eukaryotic chromosome.

The Eukaryotic Chromosome Consists of a Linear DNA Double Helix Bound to Proteins

Fitting a huge amount of DNA into a nucleus only a few ten-thousandths of an inch in diameter is no trivial task. The eukaryotic cell solves this problem by wrapping the DNA around protein supports, greatly reducing its length (**FIG. 9-5**). For most of a cell's life, the DNA double helix in a chromosome is wound around proteins called *histones* ❶, ❷. Other proteins coil up the DNA/histone beads, much like a spring or Slinky toy ❸. These coils are attached in loops to protein "scaffolding" to complete chromosome packaging as it occurs during most of the life of a cell. All of this winding, coiling, and looping condenses the DNA to about 1/1,000th of its extended length ❹, but even this enormous degree of compaction still leaves the chromosomes much too long to be sorted out and moved into daughter nuclei during cell division. However, as cell division begins, proteins fold up the chromosome, yielding about another 10-fold condensation ❺. The chromosome is now a compact structure less than 2 ten-thousandths of an inch long (about 4 micrometers).

Every chromosome has specialized regions that are crucial to its structure and function: two telomeres and one centromere. **Telomeres** ("end part" in Greek) are protective caps at each end of a chromosome (see Fig. 9-5). Without telomeres, genes located at the ends of the chromosomes would be lost during DNA replication. The **centromere**, as we will see, has two principal functions: (1) It temporarily holds two daughter DNA double helices together after DNA replication, and (2) it is the attachment site for microtubules that move the chromosomes during cell division.

CHECK YOUR LEARNING

Can you . . .

- describe the structure of a eukaryotic chromosome?
- describe the functions of telomeres and centromeres?

9.4 WHAT HAPPENS DURING THE EUKARYOTIC CELL CYCLE?

In the eukaryotic cell cycle, newly formed cells usually acquire nutrients from their environment, synthesize more cytoplasm and organelles, and grow larger. Some of these newly formed cells differentiate and never divide again. Other cells, however, may divide, and each resulting daughter cell may then enter another cell cycle and divide again. In general, cells divide only if they receive chemical signals that cause them to enter another cell cycle (see Section 9.6).

The Eukaryotic Cell Cycle Consists of Interphase and Mitotic Cell Division

The eukaryotic cell cycle is divided into two major phases: interphase and mitotic cell division (**FIG. 9-6**).

During Interphase, a Cell Grows in Size, Replicates Its DNA, and Often Differentiates

Most eukaryotic cells spend the majority of their time in **interphase**, the period between cell divisions. Interphase contains three subphases: G1 (the first growth phase and the first gap in DNA synthesis), S (when DNA synthesis occurs), and G_2 (the second growth phase and the second gap in DNA synthesis).

A newly formed daughter cell that enters the G_1 portion of interphase undergoes one or more of three processes. First, it almost always grows in size. Second, it often differentiates, developing the structures and biochemical pathways that allow it to perform a specialized function. For example, most nerve cells grow long strands, called axons, that allow them to connect with other cells, whereas liver cells produce a fluid that aids digestion, proteins that aid blood clotting, and enzymes that detoxify many poisonous materials. Third, the cell responds to internal and external signals that determine whether or not it will divide. If the cell is stimulated to divide, it must first duplicate its chromosomes, including making exact copies of the DNA of each chromosome. Duplicating the chromosomes occurs during the S phase. When the chromosomes have been duplicated, the cell proceeds to the G_2 phase, during which it may grow some more and synthesize the proteins needed for cell division.

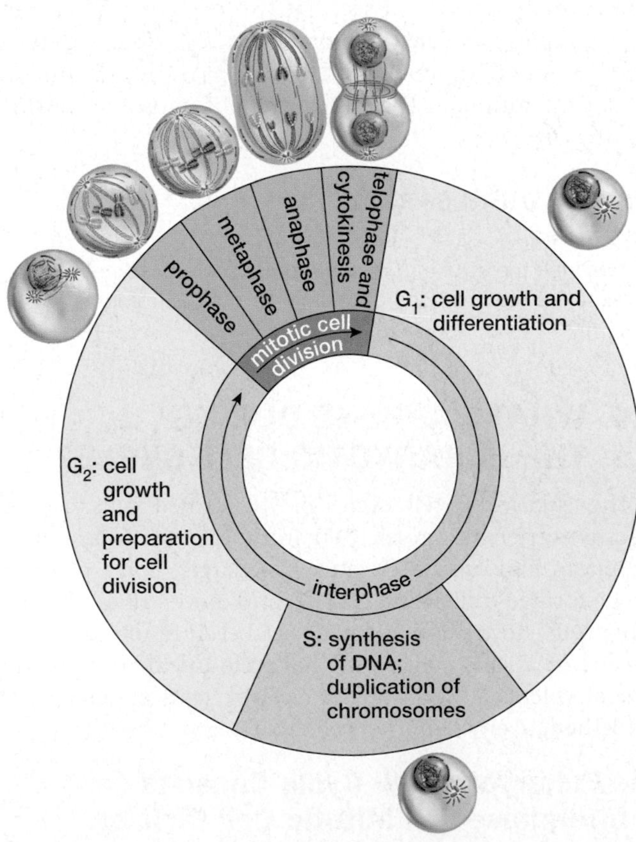

▲ **FIGURE 9-6 The eukaryotic cell cycle** The eukaryotic cell cycle consists of interphase and mitotic cell division.

Many differentiated cells, such as liver cells, can be recalled from the differentiated state back into the dividing state, whereas others, such as most heart muscle and nerve cells, remain in the G_1 phase and never divide again.

Mitotic Cell Division Consists of Nuclear Division and Cytoplasmic Division

Mitotic cell division consists of two processes: mitosis and cytokinesis. **Mitosis** is the division of the nucleus. The word "mitosis" is derived from a Greek word meaning "thread," because, as the chromosomes condense and shorten, they become visible in a light microscope as thread-like structures. Mitosis produces two daughter nuclei, each containing one copy of each of the chromosomes that were present in the parent nucleus. **Cytokinesis** (from Greek words meaning "cell movement") is the division of the cytoplasm. Cytokinesis places about half the cytoplasm, half the organelles (such as mitochondria, ribosomes, and Golgi apparatus), and one of the newly formed nuclei into each of two daughter cells. Thus, mitotic cell division typically produces daughter cells that are physically similar and genetically identical to each other and to the parent cell.

CHECK YOUR LEARNING

Can you . . .

- describe the events of the eukaryotic cell cycle?
- explain the difference between mitotic cell division and mitosis?

CASE STUDY CONTINUED

Body, Heal Thyself

Ligaments and tendons have a limited capacity for self-repair. They tend to have a meager blood supply and contain only a small number of specialized cells that produce proteins, such as collagen and elastin, that provide flexibility and strength. In Bartolo Colón's case, the hope was that the stem cells injected into his shoulder and elbow would progress rapidly through the cell cycle, producing large populations of specialized daughter cells that would regenerate his ligaments and tendons. How would mitotic cell division ensure that the daughter cells contained accurate copies of all of Colón's chromosomes, including the genes that specify all of the proteins needed to repair his arm?

9.5 HOW DOES MITOTIC CELL DIVISION PRODUCE GENETICALLY IDENTICAL DAUGHTER CELLS?

All of a cell's chromosomes (**FIG. 9-7a**) are copied during the S phase of interphase, before mitotic cell division starts. Each resulting **duplicated chromosome** consists of two identical DNA double helices (and their associated proteins), called sister **chromatids**, which are attached to each other at the centromere (**FIG. 9-7b**). During mitotic cell division, the two sister chromatids separate, each becoming an independent chromosome that is delivered to one of the two daughter cells (**FIG. 9-7c**).

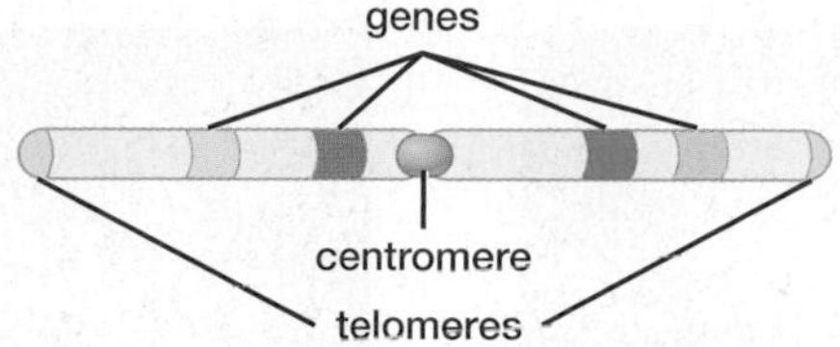

(a) A eukaryotic chromosome (one DNA double helix) before DNA replication

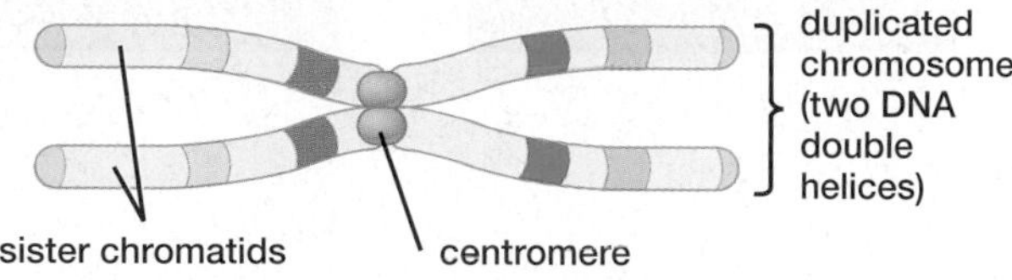

(b) A eukaryotic chromosome after DNA replication

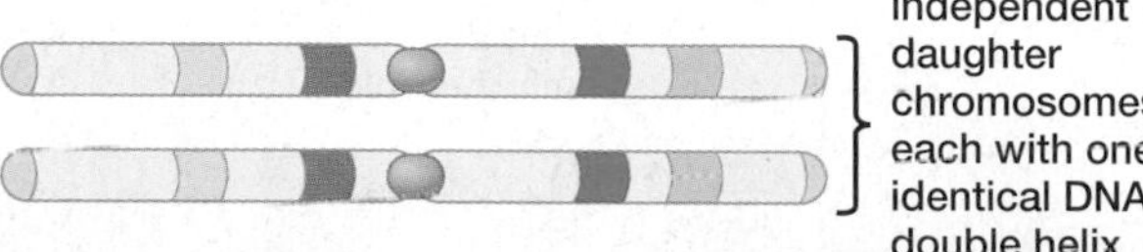

(c) Separated sister chromatids become independent chromosomes

▲ **FIGURE 9-7 A eukaryotic chromosome during cell division** **(a)** Before DNA replication. **(b)** After DNA replication, the two sister chromatids are held together at the centromere. **(c)** The sister chromatids separate during cell division to become two independent, genetically identical chromosomes.

For convenience, biologists divide mitosis into four phases, based on the appearance and behavior of the chromosomes: prophase, metaphase, anaphase, and telophase (**FIG. 9-8**). However, these phases are not really separate events; they instead form a continuum, with each phase merging into the next.

During Prophase, the Spindle Forms, the Nuclear Envelope Breaks Down, and Condensed Chromosomes Are Captured by Spindle Microtubules

The first phase of mitosis is called **prophase** (meaning "the stage before" in Greek). During prophase, four major events occur: (1) The duplicated chromosomes condense and the nucleolus disappears, (2) the spindle microtubules form, (3) the nuclear envelope breaks down, and (4) some of the spindle microtubules capture the chromosomes (**FIGS. 9-8b, c**).

Microtubules Form the Spindle

As the duplicated chromosomes condense, the **spindle** begins to form. The spindle is composed of microtubules, called **spindle microtubules** (see Fig. 9-8c). In animal cells, the spindle microtubules originate from a region that contains a pair of microtubule-containing structures called **centrioles**. The cells of plants, fungi, and many algae do not contain centrioles. Nevertheless, these cells form functional spindles during mitotic cell division, showing that centrioles are not required for spindle formation.

In animal cells, a new pair of centrioles forms during interphase near the previously existing pair. During prophase, the two centriole pairs migrate to opposite sides of the nucleus (see Fig. 9-8b). The area of cytoplasm around each centriole pair, called the spindle pole, controls the formation of the spindle microtubules. These microtubules radiate inward toward the nucleus and outward toward the plasma membrane (see Fig. 9-8c). To visualize this, picture the cell as a globe. The spindle poles are roughly where the north and south poles would be, and the spindle microtubules correspond to the lines of longitude. As on a globe, the equator of the cell cuts across the middle, halfway between the poles. Because one pair of centrioles is located at each spindle pole, each daughter cell will receive a pair of centrioles when the cell divides.

The Nuclear Envelope Disintegrates, and Chromosomes Attach to the Spindle

As the spindle microtubules form, the nuclear envelope disintegrates, releasing the duplicated chromosomes. Each sister chromatid in a duplicated chromosome has a **kinetochore**, a protein-containing structure located at the centromere. The kinetochores of the two sister chromatids are arranged back to back, facing away from one another. The kinetochore of one sister chromatid binds to the ends of spindle microtubules leading to one pole of the cell, while the kinetochore of the other sister chromatid binds to spindle microtubules leading to the opposite pole (see Fig. 9-8c). Thus, at the end of prophase, the two kinetochores of a duplicated chromosome are connected to spindle microtubules that lead to opposite poles of the cell. The spindle microtubules that attach to the kinetochore are called *kinetochore microtubules* to distinguish them from *polar microtubules*, whose role is described later.

During Metaphase, the Chromosomes Line Up Along the Equator of the Cell

During **metaphase** (the "middle stage"), each chromosome lines up along the equator of the cell, with one kinetochore facing each pole (**FIG. 9-8d**). The chromosomes move into this alignment as a result of a molecular "tug-of-war," in which microtubules lengthen or shorten, causing the two kinetochores on a duplicated chromosome to face opposite poles of the cell.

During Anaphase, Sister Chromatids Separate and Are Pulled to Opposite Poles of the Cell

At the beginning of **anaphase** (**FIG. 9-8e**), the sister chromatids separate, becoming independent daughter chromosomes. After this separation, each kinetochore moves its chromosome poleward, while simultaneously disassembling the end of the attached microtubule,

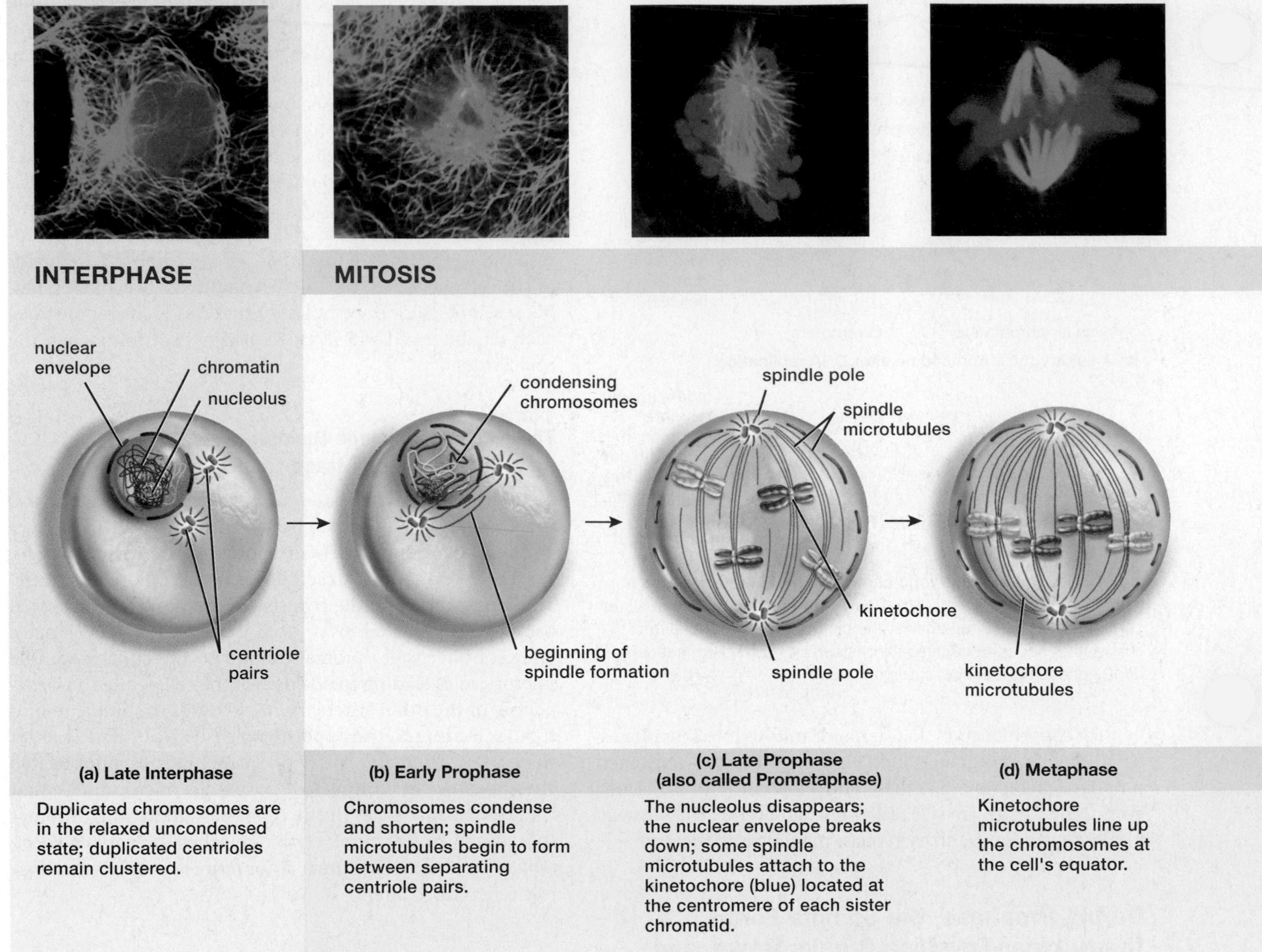
INTERPHASE
MITOSIS
nuclear envelope
chromatin
nucleolus
centriole pairs
condensing chromosomes
beginning of spindle formation
spindle pole
spindle microtubules
kinetochore
spindle pole
kinetochore microtubules
(a) Late Interphase
Duplicated chromosomes are in the relaxed uncondensed state; duplicated centrioles remain clustered.
(b) Early Prophase
Chromosomes condense and shorten; spindle microtubules begin to form between separating centriole pairs.
(c) Late Prophase (also called Prometaphase)
The nucleolus disappears; the nuclear envelope breaks down; some spindle microtubules attach to the kinetochore (blue) located at the centromere of each sister chromatid.
(d) Metaphase
Kinetochore microtubules line up the chromosomes at the cell's equator.

▲ **FIGURE 9-8 Mitotic cell division in an animal cell**

THINK CRITICALLY What would the consequences be if one set of sister chromatids failed to separate at anaphase?

thereby shortening it (a mechanism appropriately called "Pac-Man" movement, as the microtubule looks like it's being nibbled away). Each of the two daughter chromosomes derived from a parental chromosome moves to a different pole of the cell. Because the daughter chromosomes are identical copies of the parental chromosomes, each cluster of chromosomes that forms at opposite poles of the cell contains one copy of every chromosome that was in the parent cell.

At about the same time that the daughter chromosomes begin to move toward the poles, the free ends of polar microtubules radiating from each pole attach to one another where they overlap at the equator. These polar microtubules then simultaneously lengthen and push on one another, forcing the poles of the cell apart (see Fig. 9-8e).

During Telophase, a Nuclear Envelope Forms Around Each Group of Chromosomes

When the chromosomes reach the poles, **telophase** (the "end stage") begins (**FIG. 9-8f**). The spindle microtubules disintegrate, and a nuclear envelope forms around each group of chromosomes. The chromosomes revert to their extended state, and nucleoli begin to re-form. Cytokinesis occurs along with telophase, isolating each daughter nucleus in its own daughter cell (**FIG. 9-8g**).

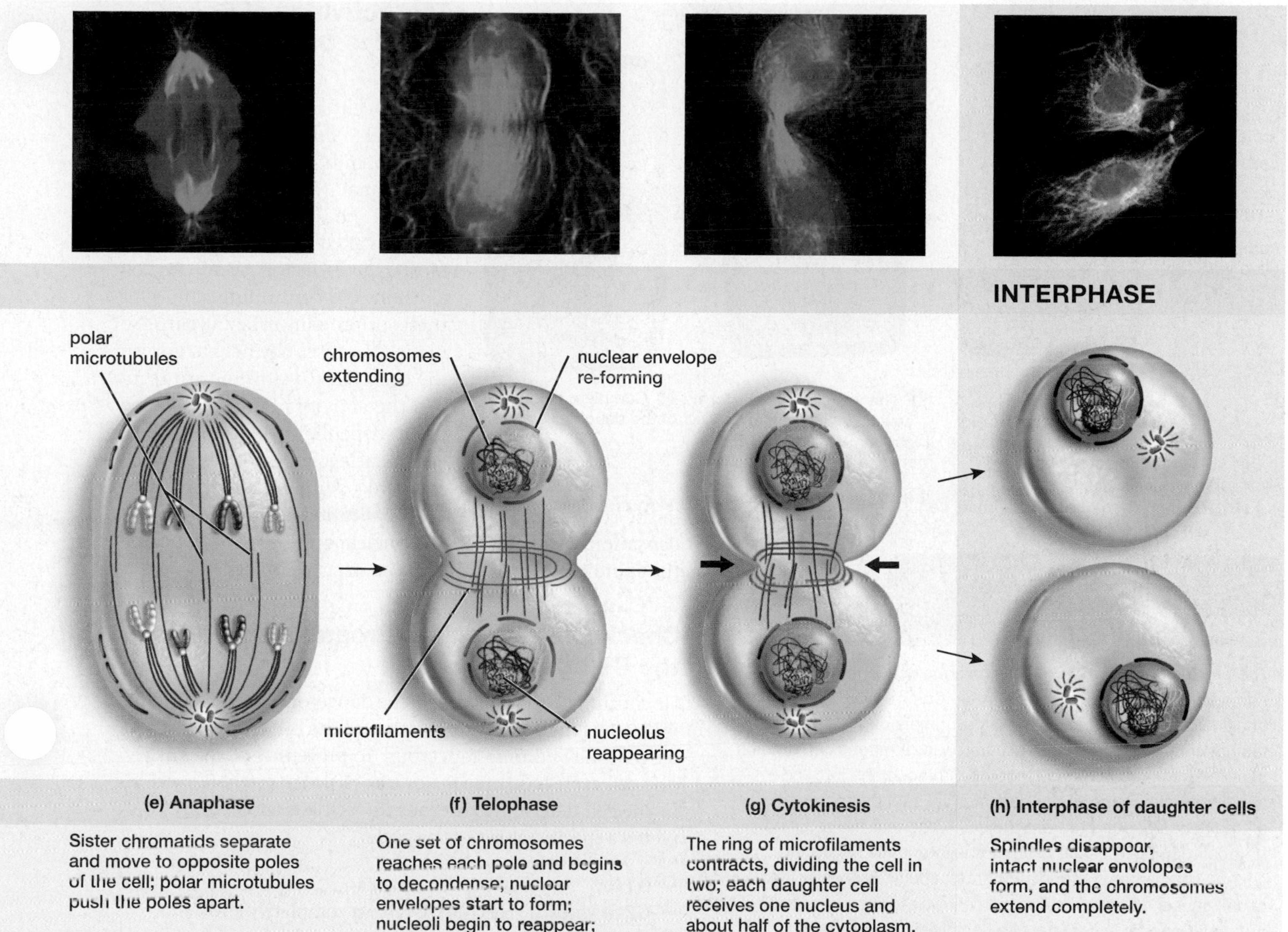

(e) Anaphase

Sister chromatids separate and move to opposite poles of the cell; polar microtubules push the poles apart.

(f) Telophase

One set of chromosomes reaches each pole and begins to decondense; nuclear envelopes start to form; nucleoli begin to reappear; spindle microtubules begin to disappear; microfilaments form rings around the equator.

(g) Cytokinesis

The ring of microfilaments contracts, dividing the cell in two; each daughter cell receives one nucleus and about half of the cytoplasm.

(h) Interphase of daughter cells

Spindles disappear, intact nuclear envelopes form, and the chromosomes extend completely.

During Cytokinesis, the Cytoplasm Is Divided Between Two Daughter Cells

Cytokinesis differs considerably between animal cells and plant cells. In animal cells, microfilaments attached to the plasma membrane assemble into a ring around the equator of the cell, usually late in anaphase or early in telophase (see Fig. 9-8f). The ring contracts and constricts the cell's equator, much like pulling the drawstring on sweatpants tightens the waist (see Fig. 9-8g). Eventually the "waist" of the parent cell constricts completely, dividing the cytoplasm into two new daughter cells (**FIG. 9-8h**).

Cytokinesis in plant cells is quite different, perhaps because their stiff cell walls make it impossible to divide one cell into two by pinching at the middle. Instead, carbohydrate-filled sacs called vesicles bud off the Golgi apparatus and line up along the equator of the cell between the two nuclei (**FIG. 9-9**). The vesicles fuse, producing a structure called a cell plate. When enough vesicles have fused, the edges of the cell plate merge with the plasma membrane around the circumference of the cell. The membranes on the two sides of the cell plate become new plasma membranes between the two daughter cells. The carbohydrates previously contained in the vesicles remain between the plasma membranes, forming the new cell wall.

CHECK YOUR LEARNING

Can you . . .

- describe the steps of mitotic cell division?
- describe the usual outcome of mitotic cell division?
- explain how cytokinesis differs in plant and animal cells?

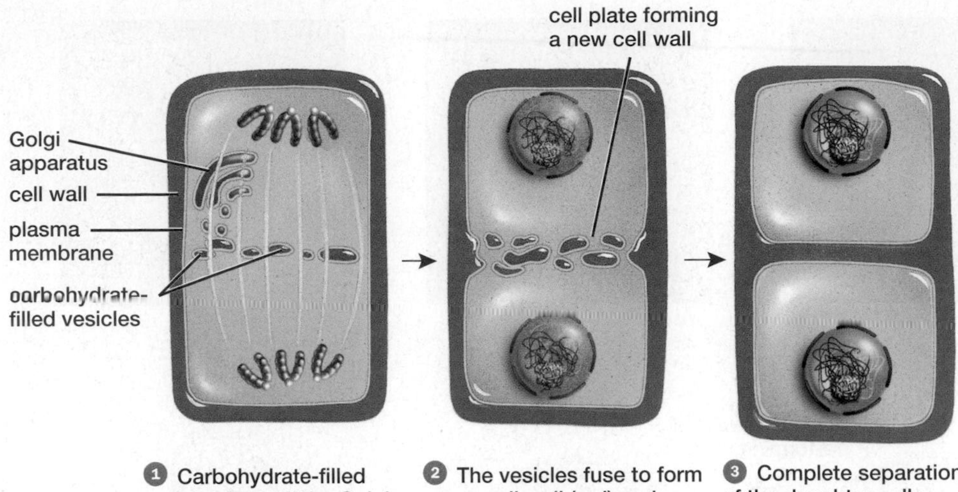

1 Carbohydrate-filled vesicles bud off the Golgi apparatus and move to the equator of the cell.

2 The vesicles fuse to form a new cell wall (red) and plasma membranes (yellow) between the daughter cells.

3 Complete separation of the daughter cells.

▲ **FIGURE 9-9 Cytokinesis in a plant cell**

CASE STUDY CONTINUED

Body, Heal Thyself

The precision of mitotic cell division is essential for repairing damaged tissues like those in Bartolo Colón's pitching arm. Imagine what might happen if DNA synthesis during interphase did not copy all of the genes accurately, or if mitotic cell division sent random numbers and types of chromosomes into the daughter cells. Some of the daughter cells might not contain all the genes needed to form the cell types that are required to repair damaged tissues. Other daughter cells might have genetic changes that stimulate unrestrained cell division and cause cancer. In cancer cells, the cell cycle spins out of control, but under normal circumstances cell division is precisely regulated. How does the body usually control the cell cycle?

9.6 HOW IS THE CELL CYCLE CONTROLLED?

Cell division is regulated by a diverse array of molecules, not all of which have been identified and studied. Nevertheless, some general principles apply to cell cycle control in most eukaryotic cells.

Have You Ever Wondered ...

Why Dogs Lick Their Wounds?

The saliva of dogs, like the saliva of most mammals (including humans), contains enzymes, antibacterial compounds, and growth factors. When a dog licks a wound, it not only cleans out some of the dirt and kills some of the bacteria that may have entered, but also leaves growth factors behind. The growth factors speed up the synthesis of cyclins, thereby stimulating the division of cells that regenerate the skin, helping to heal the wound more rapidly.

The Activities of Specific Proteins Drive the Cell Cycle

When cell division is needed, such as during development, after an injury, or to compensate for normal wear and tear, many cells in the body release molecules called **growth factors**. Most growth factors stimulate cell division by controlling the synthesis of proteins called cyclins, so named because they help to govern the cell cycle. Cyclins in turn regulate the activity of enzymes called cyclin-dependent kinases (Cdks). Particular cyclins and their associated Cdks stimulate different steps in the cell cycle, such as DNA replication, chromosome condensation, breakdown of the nuclear envelope, formation of the spindle, and separation of the sister chromatids.

Checkpoints Regulate Progress Through the Cell Cycle

Uncontrolled cell division can be dangerous. If a cell contains mutations that disrupt regulation of the cell cycle, its daughter cells may become cancerous. To prevent this, the eukaryotic cell cycle has three major **checkpoints**, where proteins in the cell determine whether the cell has successfully completed a specific phase of the cycle:

- ***G1 to S*** Is the cell's DNA intact and suitable for replication?
- ***G2 to Mitosis*** Has the DNA been completely and accurately replicated?
- ***Metaphase to Anaphase*** Are all the chromosomes attached to the spindle and aligned properly at the equator of the cell?

The checkpoint proteins usually regulate the production of cyclins or the activity of Cdks, or both, thereby regulating progression from one phase of the cell cycle to the next. In most cases, if the checkpoint proteins are activated, for example, by mutated DNA or misaligned chromosomes, they stop the cell cycle until the defect is repaired. If the defect is not repaired, the defective cells usually either destroy themselves or are killed by the immune system. When checkpoint control malfunctions, the result may be cancer, as we explore in "Health Watch: Cancer—Running the Stop Signs at Cell Cycle Checkpoints."

CHECK YOUR LEARNING

Can you . . .

- describe the interactions among growth factors, cyclins, and cyclin-dependent kinases (Cdks) that control the eukaryotic cell cycle?
- explain how a cell protects against producing defective daughter cells?

Cancer—Running the Stop Signs at Cell Cycle Checkpoints

A cancer is a cluster of cells that multiply without control and can invade other parts of the body. The ultimate causes of most cancers are **mutations**—damage to DNA. In most cases, a mutation is quickly fixed by enzymes that repair DNA. But if a mutation goes unrepaired, the cell containing it usually fails to divide and proliferate, because checkpoints in the cell cycle prevent division of cells in which problems such as mutations have occurred. Occasionally, however, a renegade mutated cell evades the checkpoints and multiplies uncontrollably, resulting in a cancer.

Responses to Growth Factors

Many cancerous cells contain mutated versions of genes collectively called *oncogenes* (literally, "to cause cancer"). Mutated oncogenes promote uncontrolled cell division, in many cases by influencing production of or response to growth factors. Most cells divide only when stimulated by growth factors. Cells containing mutated oncogenes, however, may overproduce growth factor receptors or produce receptors that are permanently activated, even in the absence of growth factors (**FIG. E9-1**). Alternatively, oncogene mutations may cause cyclins to be synthesized at a high rate, again independently of growth factors. The result: an abnormally large supply of activated Cdks and other molecules that stimulate cell division. Like a driver who hits the accelerator instead of the brake while approaching a stop sign, a cell with these mutations is likely to barge right through the checkpoints and multiply without control.

Evading the Checkpoint Stop Signs

Cells, however, have ways of enforcing the checkpoint stop signs. All cells contain a variety of proteins collectively called *tumor suppressors*. These proteins prevent uncontrolled cell division and block the production of daughter cells that have mutated DNA. For example, a tumor suppressor called p53 monitors the integrity of a cell's DNA. Healthy cells, with intact DNA, contain little p53. However, p53 levels rapidly increase in cells with damaged DNA. The p53 protein activates processes that inhibit Cdks and block DNA synthesis, halting the cell cycle at the checkpoint between the G_1 and S phases. The p53 protein also stimulates the synthesis of DNA repair enzymes. After the DNA has been repaired, p53 levels decline, Cdks become active, and the cell enters the S phase. If the DNA cannot be repaired, p53 triggers a special form of cell death called *apoptosis*, in which the cell cuts up its DNA and effectively commits suicide. Thus, p53 acts as a checkpoint enforcer, much like the tire-spiking strips that police sometimes use to prevent criminals from driving through roadblocks. Most cells with dangerous mutations cannot plow through the G_1 to S checkpoint, so they cannot continue through the cell cycle.

But what if the gene encoding the p53 protein is mutated, causing the production of defective p53? Then, even if a cell's DNA is damaged, the cell skips through the G_1 to S checkpoint. Not surprisingly, about half of all cancers—including many breast, lung, brain, pancreas, bladder, stomach, and colon cancers—have mutations in the p53 gene.

From Mutated Cells to Metastasizing Tumors

All of us probably have some cells with mutations in oncogenes or tumor suppressor genes, or both. Usually, a single mutation in one of these gene families will cause cells to multiply faster than usual and form a *benign tumor*—a cluster of cells that has multiplied independently of its surroundings, forming a distinct patch or lump. Benign tumors are common; moles, birthmarks, and some types of warts are benign tumors. "Benign" means that the tumor is not cancerous, at least not yet. It grows slowly, if at all, and it doesn't spread, or *metastasize*, to other parts of the body. Some benign tumors, however, can become cancerous, or malignant, over time. A *malignant tumor* is a lump of cells that grows rapidly and may metastasize.

Metastasis of a malignant tumor occurs in several steps, generally as cells in the tumor accumulate mutations over time. Some of these mutations promote the growth of blood vessels in the tumor, nourishing the cancerous cells and helping the tumor to grow larger. Other mutations allow some of the cells to break away, invade the blood vessels, and spread throughout the body. Typically, several mutations are required for a cell to progress from the simple growth of a benign tumor to stimulation of blood vessel growth, to loss of attachment to surrounding cells, and finally to metastasis. Once a cancer metastasizes and tumors begin to grow in multiple sites in the body, the cancer is extremely difficult to treat.

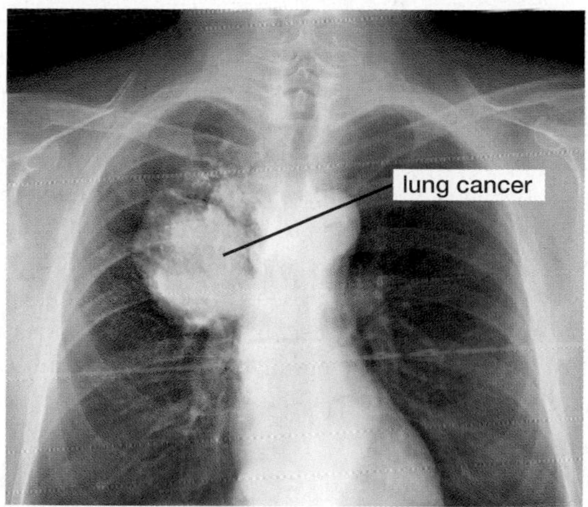

▲ **FIGURE E9-1 A colorized X-ray of advanced lung cancer** In women and in people who have never smoked, about 40% to 50% of lung cancers seem to be caused by too many receptors for growth factors or by mutated receptors that are active even in the absence of growth factors.

THINK CRITICALLY While Daniel was changing his shirt after a basketball game, one of his friends asked, "Have you always had that spot on your back?" Looking in the mirror, Daniel saw a dark brown, irregularly shaped mole. He checked in with a physician at the health center. She told him, "It's probably just a large mole, but we should do a biopsy to find out for sure." What genetic differences would you expect the pathology lab to find between a malignant tumor and an ordinary mole?

CASE STUDY REVISITED

Body, Heal Thyself

Bartolo Colón's saga sounds like a fairy tale come true: Injured, aging pitcher receives stem cell therapy and returns to stardom. But did stem cell therapy really help Colón? The truth is, no one really knows. Although there are several reports of spectacular results for individuals such as Colón, maybe he would have healed anyway. Or maybe he just happened to have an injury that stem cells were able to heal, but most other people would not be so lucky.

There have been very few clinical trials of stem cell therapies. Although stem cells taken from bone marrow are routinely used as treatments for certain cancers of the blood and immune system, clinical trials of other stem cell therapies are just beginning. Researchers can't be sure that they will work, but the range of possible applications is breathtaking—not just for joint injuries, but for multiple sclerosis, Parkinson's disease, amyotrophic lateral sclerosis (ALS, or Lou Gehrig's disease), and certain types of blindness.

THINK CRITICALLY Recently, a clinical trial testing the effectiveness of stem cells as a treatment for age-related blindness was halted when researchers discovered mutations in the retinal cells of treated patients (the retina is the part of the eye into which stem cells had been introduced). Why do you think the researchers were concerned by this discovery?

CHAPTER REVIEW

Go to **Mastering Biology** to access the Pearson eText, vocabulary review, practice quizzes, activities, videos, current events, and more.

Answers to ***Think Critically*** *and* ***Thinking Through the Concepts*** *questions can be found in the* ***Answers*** *section at the back of the book.*

Summary of Key Concepts

9.1 What Are the Functions of Cell Division?

Growth of multicellular eukaryotic organisms and replacement of cells that die during an organism's life occur through cell division and differentiation of the daughter cells. Asexual reproduction also occurs through cell division.

9.2 What Happens During the Prokaryotic Cell Cycle?

A prokaryotic cell contains a single, circular chromosome. The prokaryotic cell cycle consists of growth, replication of the DNA, and division of the cell by prokaryotic fission. The two resulting daughter cells are genetically identical to one another and to the parent cell.

9.3 How Is the DNA in Eukaryotic Chromosomes Organized?

Each chromosome in a eukaryotic cell consists of a single DNA double helix and proteins that organize the DNA and regulate its use. Genes are segments of DNA found at specific locations on a chromosome. During cell division, the chromosomes are duplicated and condense into short, thick structures.

9.4 What Happens During the Eukaryotic Cell Cycle?

The eukaryotic cell cycle consists of interphase and mitotic cell division. During interphase, the cell grows and duplicates its chromosomes. Interphase is divided into G_1 (growth phase 1), S (DNA synthesis), and G_2 (growth phase 2). During G_1, cells may differentiate to perform a specific function. Some differentiated cells can re-enter the dividing state; other cells remain differentiated for the life of the organism and never divide again.

9.5 How Does Mitotic Cell Division Produce Genetically Identical Daughter Cells?

A cell's chromosomes are duplicated during interphase, prior to mitotic cell division. A duplicated chromosome consists of two identical sister chromatids that remain attached to one another at the centromere during the early stages of mitotic cell division. Mitosis (nuclear division) consists of four phases, usually accompanied by cytokinesis (cytoplasmic division) during the last phase (see Fig. 9-8):

- ***Prophase*** The chromosomes condense, and their kinetochores attach to microtubules that form at this time.
- ***Metaphase*** Kinetochore microtubules move the chromosomes to the equator of the cell.
- ***Anaphase*** The two chromatids of each duplicated chromosome separate and become independent chromosomes. The kinetochore microtubules move the chromosomes to opposite poles of the cell. Meanwhile, polar microtubules force the cell to elongate.
- ***Telophase*** The chromosomes decondense, and nuclear envelopes re-form around each new daughter nucleus.
- ***Cytokinesis*** Cytokinesis usually occurs during telophase and divides the cytoplasm into approximately equal halves, each containing a nucleus. In animal cells, a ring of microfilaments pinches the plasma membrane in along the equator. In plant cells, new plasma membranes form on either side of the equator by the fusion of vesicles produced by the Golgi apparatus.

9.6 How Is the Cell Cycle Controlled?

Complex interactions among many proteins, particularly cyclins and cyclin-dependent protein kinases (Cdks), drive the cell cycle. There are three major checkpoints where progress through the cell cycle is regulated: between G_1 and S, between G_2 and mitosis, and between metaphase and anaphase. These checkpoints ensure that the DNA is intact and replicated accurately and that the chromosomes are properly arranged for mitosis before the cell divides.

Thinking Through the Concepts

Bloom's: Remembering, Understanding

Multiple Choice

1. A cell that remains capable of dividing throughout the life of an organism, and that produces daughter cells that can mature into any of several different cell types is a
   a. cancerous cell.
   b. differentiated cell.
   c. stem cell.
   d. gamete.
2. The chromosomes of a cell are lined up along the equator during
   a. prophase.
   b. metaphase.
   c. anaphase.
   d. telophase.
3. The chromosomes first attach to the spindle during
   a. prophase.
   b. metaphase.
   c. anaphase.
   d. telophase.
4. How does prokaryotic fission differ from eukaryotic cell division?
   a. Prokaryotic cells do not have chromosomes.
   b. Daughter cells are not genetically identical to the parent cells.
   c. Prokaryotic cell division does not require replication of DNA.
   d. Prokaryotic cells do not form spindles during cell division.
5. Which of the following is *not* true of mitotic cell division?
   a. The daughter cells are genetically identical.
   b. Chromosomes are moved to opposite poles of the cell.
   c. Mitotic cell division is required for asexual reproduction.
   d. Mitotic cell division is the mechanism by which bacterial cells divide.

Fill-in-the-Blank

1. The genetic material of all living organisms is ____________, which is contained in chromosomes.
2. Prokaryotic cells divide by a process called ____________.
3. Growth and development of eukaryotic organisms occur through ____________ cell division and ____________ of the resulting daughter cells. ____________ cells in multicellular eukaryotes remain capable of dividing throughout the life of the organism; their daughter cells can differentiate into a variety of cell types.
4. Eukaryotic cells are often stimulated to divide by hormone-like molecules called ____________. ____________ monitor progress through the cell cycle. Two categories of genes that, when mutated, often allow unregulated cell division are ____________ and ____________.
5. The four phases of mitosis are ____________, ____________, ____________, and ____________. Division of the cytoplasm into two cells, called ____________, occurs during which phase?____________
6. Chromosomes attach to spindle microtubules at structures called ____________. Some spindle microtubules, called ____________ microtubules, do not bind to chromosomes, but have free ends that overlap along the equator of the cell. These microtubules push the poles of the cell apart.

Review Questions

1. Diagram and describe the eukaryotic cell cycle. Name the phases, and briefly describe the events that occur during each.
2. Define *mitosis* and *cytokinesis*.
3. Diagram the stages of mitosis. How does mitosis ensure that each daughter nucleus receives a full set of chromosomes?
4. Define the following terms: *centromere, telomere, kinetochore, chromatid*, and *spindle*.
5. Describe and compare the process of cytokinesis in animal cells and in plant cells.
6. How is the cell cycle controlled? Why is it important to regulate progression through the cell cycle?
7. Diagram and describe the prokaryotic cell cycle.

Applying the Concepts

Bloom's: Applying, Analyzing, Evaluating

1. Most nerve cells in the adult human central nervous system, as well as heart muscle cells, do not divide. In contrast, cells lining the inside of the small intestine divide frequently. Discuss this difference in terms of why damage to the nervous system and heart muscle (for example, damage caused by a stroke or heart attack) is so dangerous.
2. Cancer cells divide out of control. Side effects of the cancer treatments chemotherapy and radiation therapy include loss of hair and of the intestinal lining, the latter producing severe nausea. What can you infer about the mechanisms of these treatments? What would you look for in an improved cancer therapy?

10 Meiosis: The Basis of Sexual Reproduction

The striking diversity of the Giddings family is the result of meiosis.

CASE STUDY

Diversity Runs in the Family

THE FOUR CHILDREN OF TESS AND CHRIS GIDDINGS look quite different from one another. Jacob, the oldest, has blue eyes like his mother's, but olive skin and curly brown hair. Savannah, the next oldest, looks a lot like Jacob, though her hair is lighter than his. Amiah, the next in line, has very pale skin, with straight, sandy-brown hair. Zion, the youngest, has dark skin, curly black hair, and brown eyes, much like his father. The Giddings clan is strikingly diverse.

The varied appearance of the Giddings children is a bit perplexing, sometimes even to Tess and Chris. When Amiah was born, she had low blood sugar and was whisked away to be checked out by a specialist, so quickly that the hospital staff didn't have time to put an ID wristband on her. When Amiah was returned to her parents a short while later, they were astounded at how white her skin was. They asked the inevitable question: Was she switched with another baby by mistake? Just to be sure, the Giddings agreed to a DNA test. The results showed that Tess and Chris were indeed Amiah's parents. When Zion was born a few years later, Chris burst out, "Oh my God, he's black!" To which the astounded midwife could only reply, "You do know you're a black man, don't you?"

Although they sometimes tire of questions and comments from curious strangers, Tess and Chris are pleased to have such diverse children. "They are a great joy to us and we couldn't be happier," says Tess. "It's so nice having children who are all a bit different." Or as Chris puts it, "We're like a box of chocolates—dark ones, brown ones, and white ones.

How could one couple have such a diverse family? As we will see in this chapter, sexual reproduction can combine characteristics inherited from parents in many ways, allowing a remarkable variety of different offspring. How does sexual reproduction produce such diversity?

AT A GLANCE

10.1 HOW DOES SEXUAL REPRODUCTION PRODUCE GENETIC VARIABILITY?

Individuals of a species tend to differ from one another. For example, you may have some classmates who are tall with straight blond hair, others who are short with curly black hair, still others who are tall with straight brown hair, and so on. This variability in appearance is due in large measure to underlying variability in genes. In this chapter, we will examine the processes that give rise to this genetic variability. To better understand these processes, we must first briefly review how genetic information is stored and organized.

Genetic Information Is Packaged in Chromosomes That In Eukaryotes Usually Occur in Pairs

In all organisms, hereditary information is encoded in molecules of deoxyribonucleic acid (DNA) that are packaged into chromosomes. In this chapter, we focus on chromosomes in eukaryotic organisms. The complete set of chromosomes from a single eukaryotic cell is its **karyotype** (**FIG. 10-1**).

The karyotype of most eukaryotic organisms consists of pairs of chromosomes. Humans have 23 pairs, for a total of 46 chromosomes per cell. The two chromosomes that make up a pair are called **homologous chromosomes**, or **homologues**, from Greek words that mean "to say the same thing." As the name suggests, the two chromosomes in a homologous pair contain the same genes. Cells with pairs of homologous chromosomes are called **diploid**, meaning "double." One homologue of each pair, called the maternal homologue, is inherited from the mother, and the other, called the paternal homologue, is inherited from the father.

In Some Cells, Chromosomes Are Not Paired

If a cell contains only one member of each pair of homologues, it is **haploid**. As we will see in Section 10.2, the reproductive cells produced by diploid organisms contain only one member of each pair of homologous chromosomes and so are

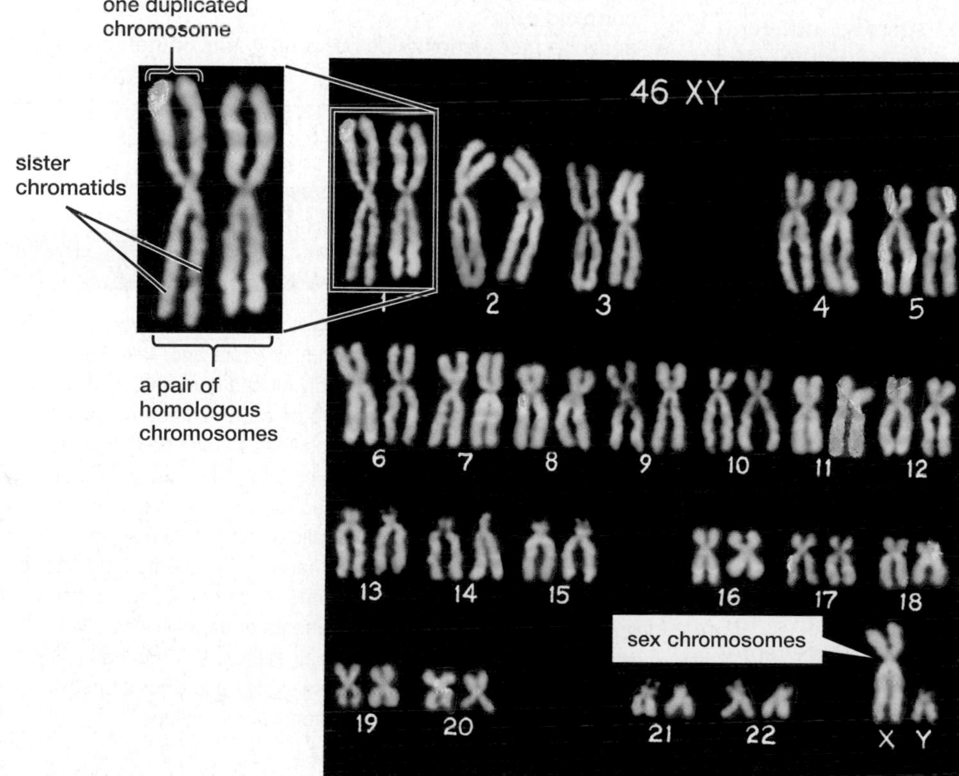

FIGURE 10-1 The karyotype of a human male To display a karyotype, technicians stain and photograph the entire set of duplicated, condensed chromosomes from a single cell. Pictures of the individual chromosomes are cut out and arranged in descending order of size. Chromosomes 1 through 22 are the autosomes; members of each pair of autosomes (homologues) are similar in size. The X and Y chromosomes are the sex chromosomes; they differ in size. If this were a female karyotype, it would have two X chromosomes and no Y chromosome.

haploid. In some organisms, nonreproductive cells may be haploid; for example, most fungi consist only of haploid cells for most of their life cycle.

Some organisms have more than two copies of each homologous chromosome in each cell and are **polyploid**. Many plants, for example, are polyploid, with four (tetraploid), six (hexaploid), or even more copies of each homologue per cell.

The number of different types of chromosomes in a species is called the haploid number and is designated n. For humans, $n = 23$ because we have 23 different chromosomes. Diploid cells contain $2n$ chromosomes.

Some Chromosomes Are Sex Chromosomes

Many organisms have two types of chromosome: autosomes and sex chromosomes. **Autosomes** are pairs of chromosomes with nearly identical DNA sequences that are found in diploid cells of both sexes. People have 22 pairs of autosomes. In addition to autosomes, animals and some plants have one pair of **sex chromosomes** that determine an individual's sex. In mammals, including humans, each individual's karyotype includes either two **X chromosomes** (in females) or an X and a **Y chromosome** (in males). X and Y chromosomes are quite different in size (see Fig. 10-1) and genetic composition.

A Gene May Have Multiple Variants

Each eukaryotic chromosome contains a few dozen to a few thousand genes. A **gene** is a sequence of nucleotides, found at a specific location on a chromosome, that encodes information for building functional molecules (mostly proteins). Genes are the basic units of inheritance, passed from parent to offspring.

Compared across members of a species, different instances of a given gene have extremely similar, but usually not identical, nucleotide sequences. The slightly different versions of a gene are called **alleles**. If we could survey all the members of a given species, we might find a few, several dozen, or even hundreds of alleles of each gene. Within an individual, a pair of homologous chromosomes may contain the same alleles of some genes and different alleles of other genes (**FIG. 10-2**).

Genetic Variability Originates Through Mutations

Where do alleles come from? Alleles arise as a result of **mutations**, changes to DNA that can occur when a cell makes a mistake copying its DNA or when DNA is damaged by factors such as ultraviolet light from the sun or harmful chemicals in the environment. If a mutation occurs in a gene, the result is a new allele. If a mutation that gives rise to an allele happens to occur in the cells that produce reproductive cells, the allele may be inherited by offspring and subsequently passed from one generation to the next until it eventually occurs in many members of a species.

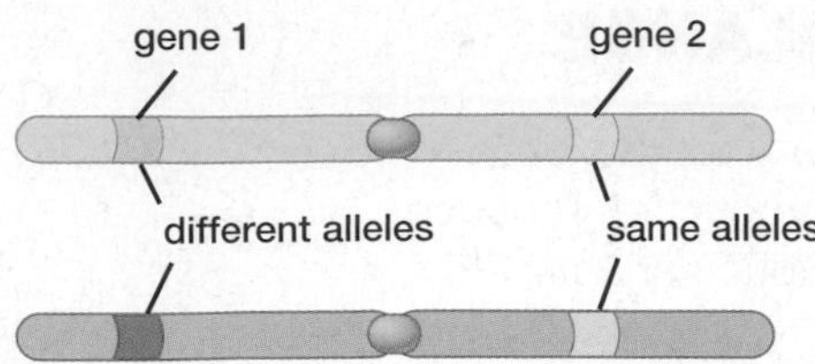

▲ **FIGURE 10-2 Homologous chromosomes are usually not identical** Homologous chromosomes have the same genes at the same locations. The homologues may have different alleles of some genes (left) and the same alleles of other genes (right).

Sexual Reproduction Generates Genetic Differences Between the Members of a Species

The diverse methods by which organisms reproduce fall into two categories: asexual and sexual. In asexual reproduction, mitotic cell division in a parent organism produces offspring that are genetically identical to one another and to the parent, apart from occasional differences due to mutations. In **sexual reproduction**, haploid reproductive cells called **gametes** from two parent organisms unite to produce offspring that are genetically different from one another and from either parent. As a result of these genetic differences between parents and offspring, sexual reproduction tends to increase the number of different allele combinations in a population.

CHECK YOUR LEARNING

Can you . . .

- define the terms *homologous chromosome*, *autosome*, and *sex chromosome*?
- explain the differences between diploid, haploid, and polyploid cells?
- describe the relationships between genes, mutations, and alleles?

CASE STUDY CONTINUED

Diversity Runs in the Family

The genetic variability of the Giddings children has its roots in mutations that occurred thousands of years ago. Take hair color: Our distant ancestors probably all had dark hair, its color controlled by multiple genes located on several different chromosomes. The alleles that produced Tess's blond hair originated as mutations in these genes. Tess probably inherited only "pale hair" alleles of all of these genes, so for any given hair color gene, she has the same pale hair allele on both homologous chromosomes. Chris, on the other hand, likely inherited both dark and pale hair alleles for at least some of the genes, so his homologues have different alleles. The effects of the dark hair alleles override those effects of the pale hair alleles, so Chris has black hair. Which alleles, packaged in Tess's eggs and Chris's sperm, might have combined to produce their diverse children?

10.2 HOW DOES MEIOTIC CELL DIVISION PRODUCE GENETICALLY VARIABLE, HAPLOID CELLS?

Sexual reproduction produces genetically diverse offspring in two steps (**FIG. 10-3**). First, haploid gametes are produced. Production of gametes begins with **meiotic cell division**, in which a diploid cell gives rise to haploid daughter cells that each contain one member of each pair of homologues ❶. The haploid cells, or their descendants produced by mitotic cell division, become gametes (in animals, sperm or eggs). Second, two gametes fuse at fertilization. Fusion of a haploid egg and a haploid sperm produces a diploid cell called a **zygote,** which grows by mitotic cell division into a diploid offspring ❷.

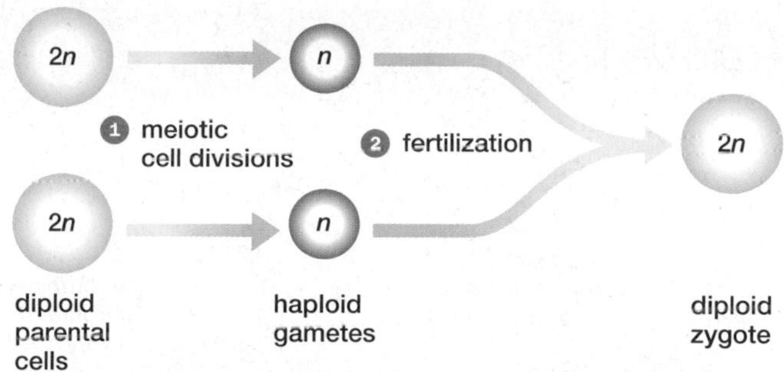
2n
n
❶ meiotic cell divisions
❷ fertilization
2n
2n
n
diploid parental cells
haploid gametes
diploid zygote

▲ **FIGURE 10-3 Meiotic cell division is essential for sexual reproduction** In sexual reproduction, diploid reproductive cells of the parents (2*n*) undergo meiosis to produce haploid cells (*n*). In animals, these cells become gametes (sperm or eggs). When an egg is fertilized by a sperm, the resulting zygote is diploid once again (2*n*).

Meiotic Division of a Diploid Cell Yields Four Haploid Daughter Cells

Meiotic cell division consists of meiosis and cytokinesis. **Meiosis** is a type of nuclear division in which a diploid nucleus divides twice to produce four haploid nuclei. Cytokinesis packages the four nuclei into separate cells.

The two nuclear divisions of meiosis are known as meiosis I and meiosis II. Before meiosis begins, the parent cell's DNA is replicated, so that each chromosome consists of duplicate sister chromatids (**FIG. 10-4** ❶). **Meiosis I** separates the homologues of each pair of homologous chromosomes into two daughter nuclei; each daughter nucleus receives one homologue from each pair ❷. The two daughter nuclei are haploid, though each chromosome still consists of two chromatids. **Meiosis II** then separates the chromatids into independent chromosomes and distributes each chromosome to a daughter nucleus ❸. Each daughter nucleus receives a single copy of each chromosome. Overall, each initial parental nucleus yields four haploid daughter nuclei—meiosis I yields two, each of which divides during meiosis II.

The phases of meiosis have the same names as similar phases in mitosis, followed by I or II to distinguish the two nuclear divisions.

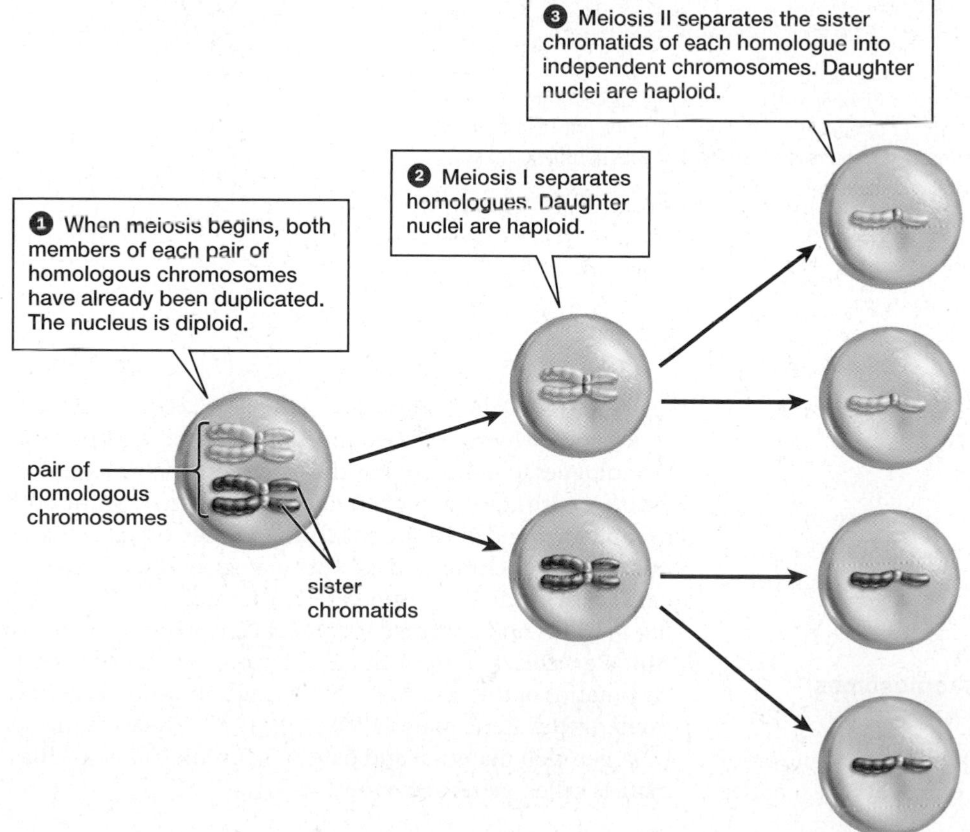
❸ Meiosis II separates the sister chromatids of each homologue into independent chromosomes. Daughter nuclei are haploid.
❷ Meiosis I separates homologues. Daughter nuclei are haploid.
❶ When meiosis begins, both members of each pair of homologous chromosomes have already been duplicated. The nucleus is diploid.
pair of homologous chromosomes
sister chromatids

◀ **FIGURE 10-4 Meiosis halves the number of chromosomes** For simplicity, only a single pair of homologous chromosomes is shown (maternal in purple and paternal in yellow).

MEIOSIS I

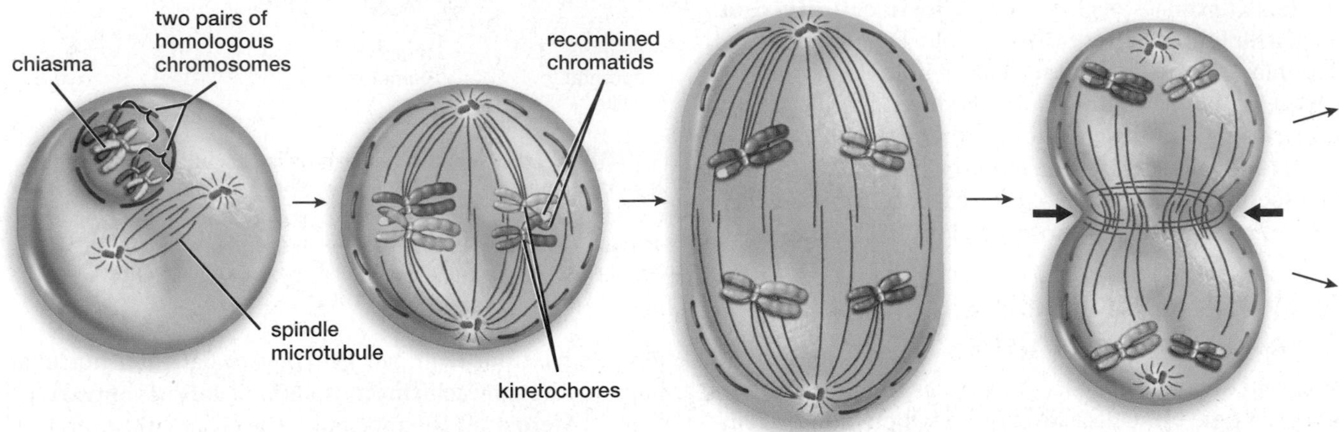

(a) Prophase I	(b) Metaphase I	(c) Anaphase I	(d) Telophase I
Duplicated chromosomes condense. Homologous chromosomes pair up and chiasmata occur as chromatids of homologues exchange parts by crossing over. The nuclear envelope disintegrates, and spindle microtubules form.	Paired homologous chromosomes line up along the equator of the cell. One homologue of each pair faces each pole of the cell and attaches to the spindle microtubules via the kinetochore (blue).	Homologues separate, one member of each pair going to each pole of the cell. Sister chromatids do not separate.	Spindle microtubules disappear. Two clusters of chromosomes have formed, each containing one member of each pair of homologues. The daughter nuclei are therefore haploid. Cytokinesis commonly occurs at this stage. There is little or no interphase between meiosis I and meiosis II.

▲ **FIGURE 10-5 Meiotic cell division** In meiotic cell division, the homologous chromosomes of a diploid cell are separated, producing four haploid daughter cells. For simplicity, only two pairs of homologous chromosomes are shown. In humans, 23 pairs are involved in this complex dance.

THINK CRITICALLY What would be the consequences for the resulting gametes and offspring if one pair of homologues failed to separate at anaphase I?

Meiosis I Separates Homologous Chromosomes into Two Haploid Daughter Nuclei

When meiosis I begins, the parent cell's chromosomes have already been duplicated during interphase, and the sister chromatids of each chromosome are attached to each other at the centromere.

During Prophase I, Homologous Chromosomes Pair Up and Exchange DNA

During prophase I, the duplicated homologous chromosomes line up side by side and their chromatids exchange segments of DNA (**FIG. 10-5a** and **FIG. 10-6**). This process begins when proteins bind the maternal and paternal homologues together so that they align precisely along their entire length. Enzymes then cut through the DNA of both homologues and graft the cut ends together, often exchanging part of a chromatid of the maternal homologue for part of a chromatid of the paternal homologue. The binding proteins and enzymes then depart, leaving **chiasmata** (singular, chiasma; from the Greek word for "cross") where chromatids of the maternal and paternal chromosomes have exchanged genetic material (see Fig. 10-6). The exchange of DNA between maternal and paternal chromosomes at chiasmata is called **crossing over**.

MEIOSIS II

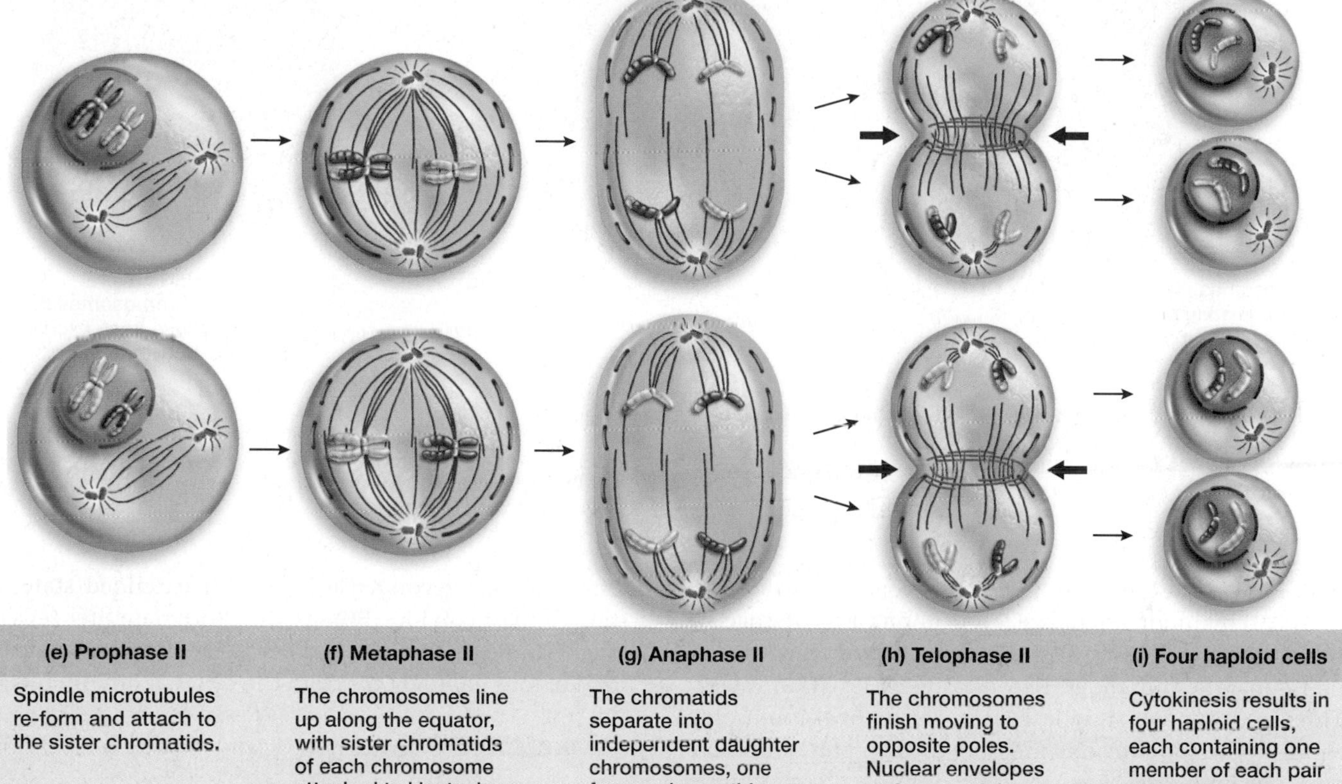

(e) Prophase II	(f) Metaphase II	(g) Anaphase II	(h) Telophase II	(i) Four haploid cells
Spindle microtubules re-form and attach to the sister chromatids.	The chromosomes line up along the equator, with sister chromatids of each chromosome attached to kinetochore microtubules that lead to opposite poles.	The chromatids separate into independent daughter chromosomes, one former chromatid moving toward each pole.	The chromosomes finish moving to opposite poles. Nuclear envelopes re-form, and the chromosomes decondense again (not shown here).	Cytokinesis results in four haploid cells, each containing one member of each pair of homologous chromosomes (shown here in the condensed state).

As in prophase of mitosis, spindle microtubules begin to assemble outside the nucleus during prophase I. Near the end of prophase I, the nuclear envelope breaks down and spindle microtubules invade the nuclear region, capturing the chromosomes by attaching to their kinetochores.

During Metaphase I, Paired Homologous Chromosomes Line Up at the Equator of the Cell

During metaphase I, interactions between kinetochores and spindle microtubules move the paired homologues to the equator of the cell (**FIG. 10-5b**). Unlike in metaphase of mitosis, in which *individual* duplicated chromosomes line up along the equator, in metaphase I of meiosis, *homologous pairs* of duplicated chromosomes, held together by chiasmata, line up along the equator. Which member of a pair of homologous chromosomes faces which pole of the cell is random—the maternal homologue may face "north" for some pairs and "south" for other pairs. This randomness, together with new genetic combinations caused by crossing over, ensures that the haploid cells produced by meiosis are genetically diverse.

During Anaphase I, Homologous Chromosomes Separate

Anaphase in meiosis I differs considerably from anaphase in mitosis. In anaphase of mitosis, sister chromatids of each duplicated homologue separate and move to opposite

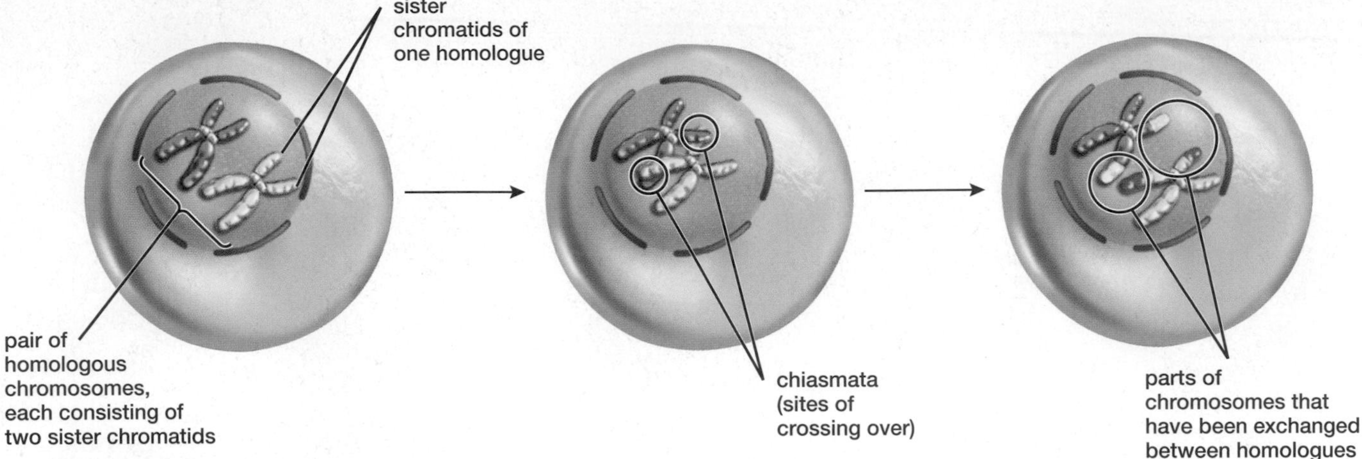

▲ **FIGURE 10-6 Crossing over** Paternal and maternal homologues exchange DNA at chiasmata.

THINK CRITICALLY What would be the genetic consequences for the gametes and offspring if crossing over occurred between two nonhomologous chromosomes?

poles. In contrast, in anaphase I of meiosis, the sister chromatids remain attached to each other and move together to the same pole. However, the two homologues in each pair separate and move to opposite poles (**FIG. 10-5c**). Thus, at the end of anaphase I, the cluster of chromosomes at each pole contains one member of each pair of homologous chromosomes. Each cluster contains a haploid number of chromosomes (although each chromosome is still duplicated and consists of sister chromatids attached at the centromere).

During Telophase I, Two Haploid Clusters of Duplicated Chromosomes Form

Telophase I in meiosis is similar to telophase in mitosis. In telophase I, the spindle microtubules disappear and cytokinesis commonly occurs (**FIG. 10-5d**). Nuclear envelopes may re-form. Telophase I is usually followed immediately by meiosis II, with little or no intervening interphase. The chromosomes do not replicate between meiosis I and meiosis II.

Meiosis II Separates Sister Chromatids into Four Daughter Nuclei

During meiosis II, the sister chromatids of each duplicated chromosome separate in a process that is virtually identical to mitosis in a haploid cell. During prophase II, the spindle microtubules re-form (**FIG. 10-5e**) and the kinetochores of the sister chromatids of each duplicated chromosome attach to spindle microtubules extending to opposite poles of the cell. During metaphase II, the duplicated chromosomes line up at the cell's equator (**FIG. 10-5f**). During anaphase II, the sister chromatids separate and move to opposite poles (**FIG. 10-5g**). Telophase II and cytokinesis conclude meiosis II as nuclear envelopes re-form, the chromosomes decondense into their extended state, and the cytoplasm divides (**FIG. 10-5h**). Both daughter cells produced in meiosis I usually undergo meiosis II, producing a total of four haploid cells from the original diploid parental cell (**FIG. 10-5i**).

TABLE 10-1 compares mitotic and meiotic cell division.

CHECK YOUR LEARNING

Can you . . .

- describe the steps and outcome of meiotic cell division?
- explain the results of crossing over?

CASE STUDY CONTINUED

Diversity Runs in the Family

When Tess and Chris Giddings produce eggs and sperm, respectively, meiosis separates their homologous chromosomes. If Tess has only "pale" alleles for all of the genes that might contribute to hair color, but Chris has alleles for both dark and pale hair color, then all of the eggs formed in Tess's ovaries would receive pale hair alleles. In Chris's sperm, however, the outcome would be more complicated. Some sperm might receive a dark allele for one hair color gene, but a pale allele for another. Other sperm would have different combinations of dark and pale alleles, including some sperm with all pale alleles and others with all dark alleles. As a result, depending on the alleles carried by the particular sperm that fertilizes Tess's egg, a Giddings child might end up with a combination of paternal and maternal alleles that yields pale, medium, or dark hair. But can the process of gamete formation and union explain the full range of diverse traits exhibited by the Giddings children?

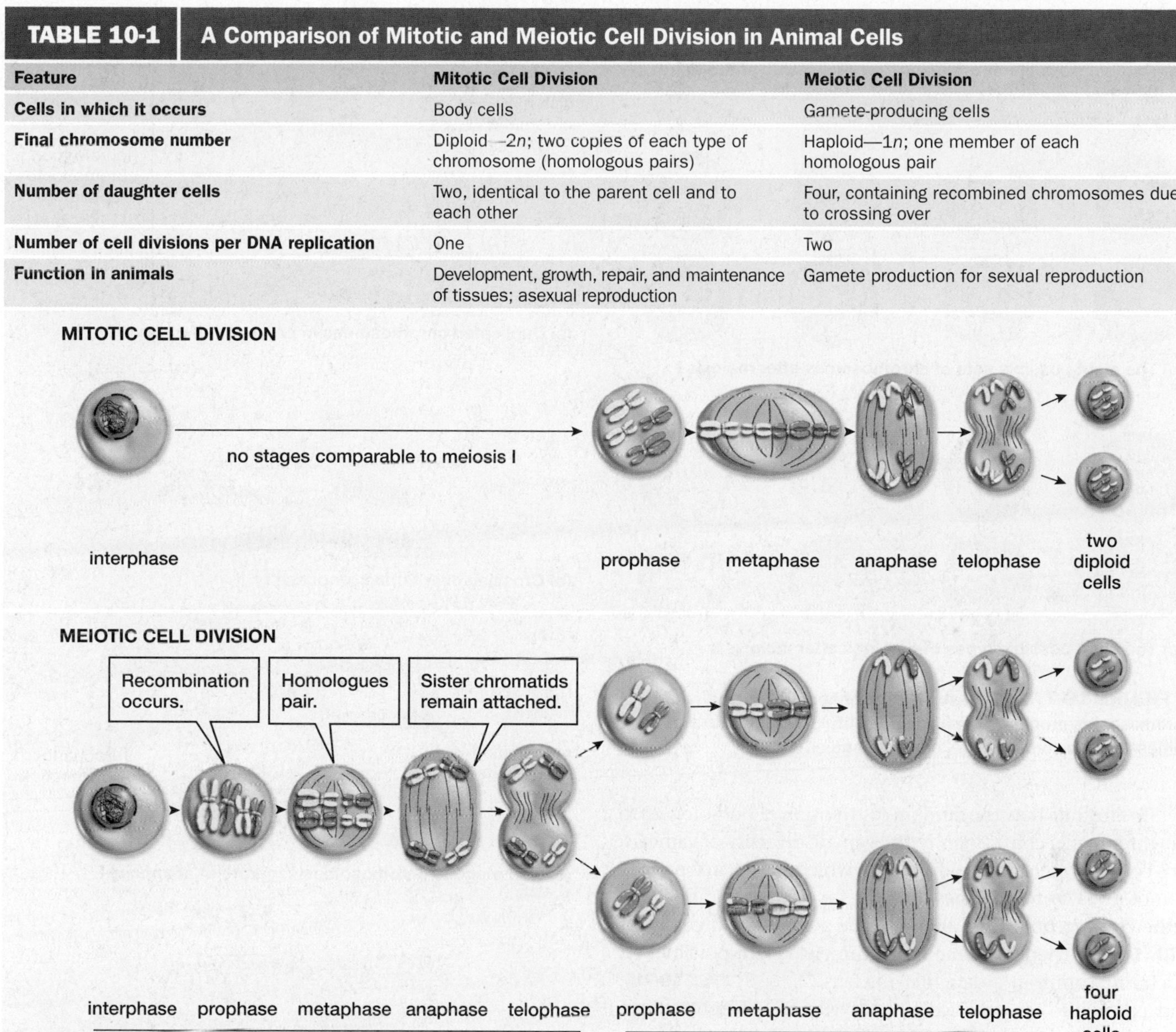

TABLE 10-1 A Comparison of Mitotic and Meiotic Cell Division in Animal Cells

Feature	Mitotic Cell Division	Meiotic Cell Division
Cells in which it occurs	Body cells	Gamete-producing cells
Final chromosome number	Diploid—2*n*; two copies of each type of chromosome (homologous pairs)	Haploid—1*n*; one member of each homologous pair
Number of daughter cells	Two, identical to the parent cell and to each other	Four, containing recombined chromosomes due to crossing over
Number of cell divisions per DNA replication	One	Two
Function in animals	Development, growth, repair, and maintenance of tissues; asexual reproduction	Gamete production for sexual reproduction

In these diagrams, comparable phases are aligned. In both mitosis and meiosis, chromosomes are duplicated during interphase. Meiosis I, with the pairing of homologous chromosomes, formation of chiasmata, exchange of chromosome parts, and separation of homologues to form haploid daughter nuclei, has no counterpart in mitosis. Meiosis II, however, is virtually identical to mitosis in a haploid cell.

10.3 HOW DO MEIOSIS AND UNION OF GAMETES PRODUCE GENETICALLY VARIABLE OFFSPRING?

Mutations are the source of new alleles, so mutations are responsible for the variety of alleles present in a population. Mutation alone, however, cannot explain why the individuals in a population carry so many different combinations of alleles. This aspect of genetic variability results almost entirely from meiosis and sexual reproduction.

Shuffling the Homologues Creates Novel Combinations of Chromosomes

One major source of genetic diversity is the random distribution of maternal and paternal homologues to daughter nuclei during meiosis I. Remember that at metaphase I the paired homologues line up at the cell's equator. In each pair of homologues, the maternal chromosome faces one pole and the paternal chromosome faces the opposite pole, but which homologue faces which pole is random and is not affected by the orientation of the homologues of other chromosome pairs.

(a) The four possible chromosome arrangements at metaphase of meiosis I

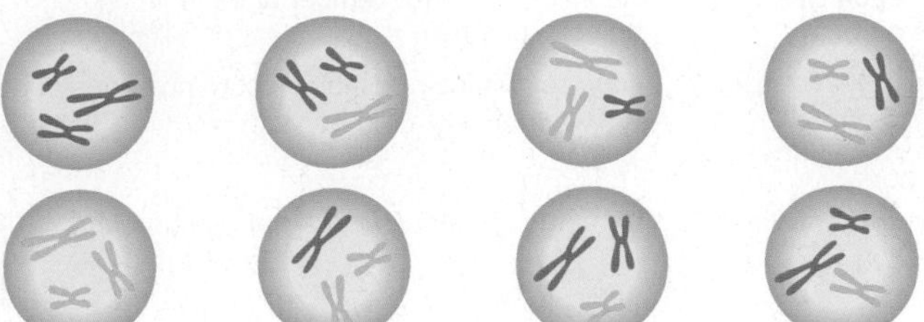

(b) The eight possible sets of chromosomes after meiosis I

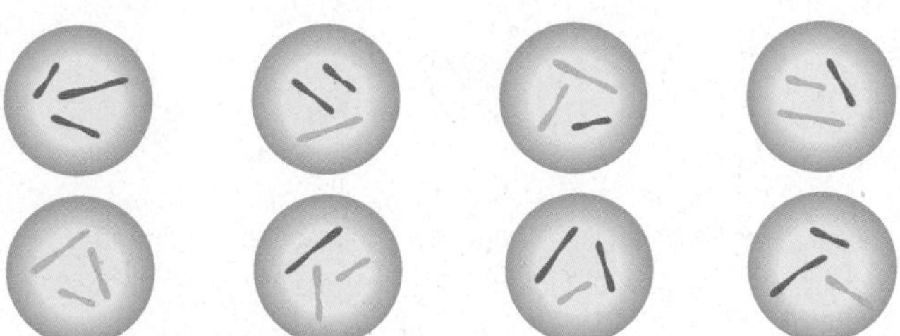

(c) The eight possible types of gametes after meiosis II

▲ **FIGURE 10-7 Random separation of homologous pairs of chromosomes produces genetic variability** For clarity, the chromosomes are depicted as large, medium, and small.

To illustrate how the random distribution of homologues to daughter nuclei contributes to the genetic diversity of gametes, let's consider meiosis in mosquitoes, which have three pairs of homologous chromosomes ($n = 3, 2n = 6$). At metaphase I, the chromosomes can align in four possible configurations (**FIG. 10-7a**). Therefore, anaphase I can yield eight possible sets of chromosomes in a daughter nucleus ($2^3 = 8$; **FIG. 10-7b**), and a mosquito can produce gametes with any of eight unique sets of chromosomes (**FIG. 10-7c**). In a human, meiosis randomly shuffles 23 pairs of homologous chromosomes, theoretically producing gametes with any of more than 8 million (2^{23}) different combinations of maternal and paternal chromosomes.

Crossing Over Creates Chromosomes with Novel Combinations of Genes

Recall that the two members of a pair of homologous chromosomes may have different alleles of some genes (see Fig. 10-2). If so, then crossing over may produce genetic **recombination**: formation of chromosomes with combinations of alleles that differ from those of either parent (**FIG. 10-8**). Chromosomes are very long—human chromosomes range from about 50 million to 250 million nucleotides in length—and crossing over can occur almost anywhere along the chromosome. Therefore, even in a single person, gamete production can yield a tremendously large number of genetically different recombined chromosomes.

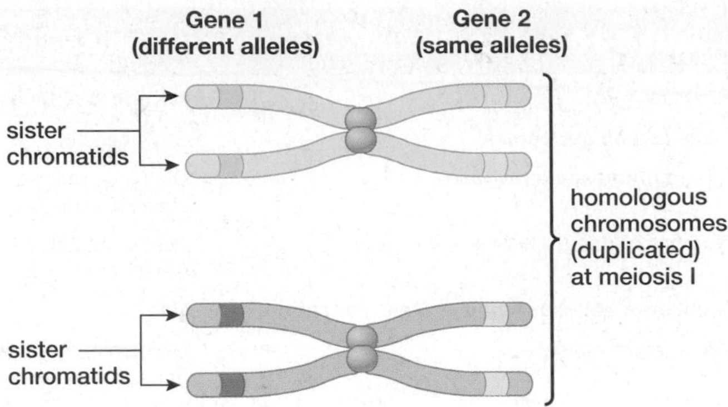

(a) Duplicated chromosomes in prophase of meiosis I

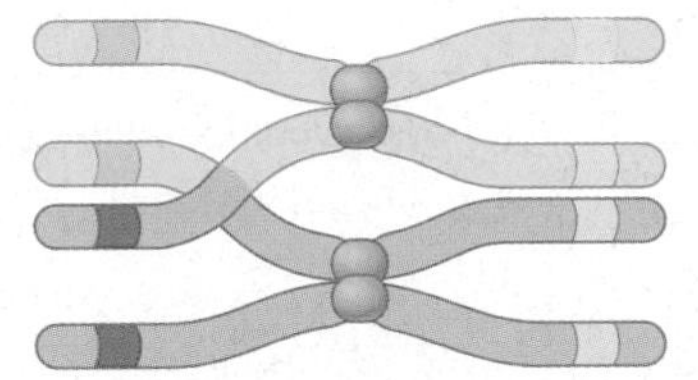

(b) Crossing over during prophase I

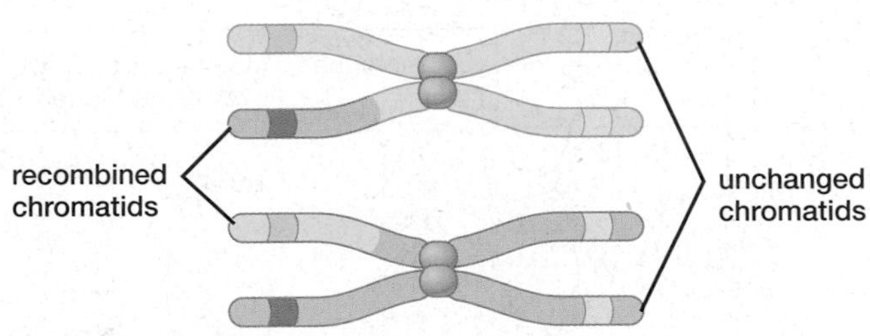

(c) Homologous chromosomes separate at anaphase I

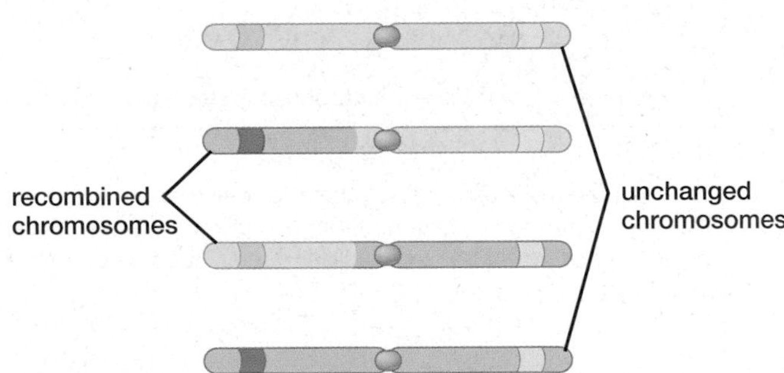

(d) Unchanged and recombined chromosomes after meiosis II

▲ **FIGURE 10-8 Crossing over recombines alleles on homologous chromosomes (a)** During prophase of meiosis I, duplicated homologous chromosomes pair up. **(b)** The two homologues exchange DNA by crossing over. **(c)** When the homologous chromosomes separate during anaphase of meiosis I, one chromatid of each of the homologues now contains a piece of DNA from a chromatid of the other homologue. **(d)** After meiosis II, two chromosomes are unchanged and two chromosomes show genetic recombination, with allele arrangements that did not occur in the parental chromosomes.

Fusion of Gametes Adds Further Genetic Variability to the Offspring

At fertilization, two gametes, each containing a unique combination of alleles, fuse to form a diploid zygote. As we have seen, if we ignore crossing over, a single person can produce gametes with any of 8 million chromosome combinations. Therefore, the chances that your parents could produce another child who is genetically identical to you are about 1/8,000,000 × 1/8,000,000, or about 1 in 64 trillion! When we factor in the almost endless variability produced by crossing over, we can confidently say that (unless you are an identical twin) there never has been, and never will be, anyone just like you.

CHECK YOUR LEARNING

Can you . . .

- explain how meiosis and sexual reproduction generate genetic variability in populations?

CASE STUDY CONTINUED

Diversity Runs in the Family

Tess Giddings probably produces eggs with "pale" alleles of the many genes that influence the color of hair, skin, and eyes, while Chris produces sperm with various pale and dark alleles of these genes. The coloring of the children was determined mainly by the genes in Chris's sperm. Crossing over and separation of homologues produced sperm containing a wide variety of combinations of color gene alleles.

10.4 HOW DO ERRORS IN MEIOSIS CAUSE INHERITED DISORDERS?

During meiotic cell division, there are occasional stumbles, resulting in gametes that have too many or too few chromosomes. Such errors in meiosis, called **nondisjunction**, can affect the number of chromosomes in a gamete (**FIG. 10-9**).

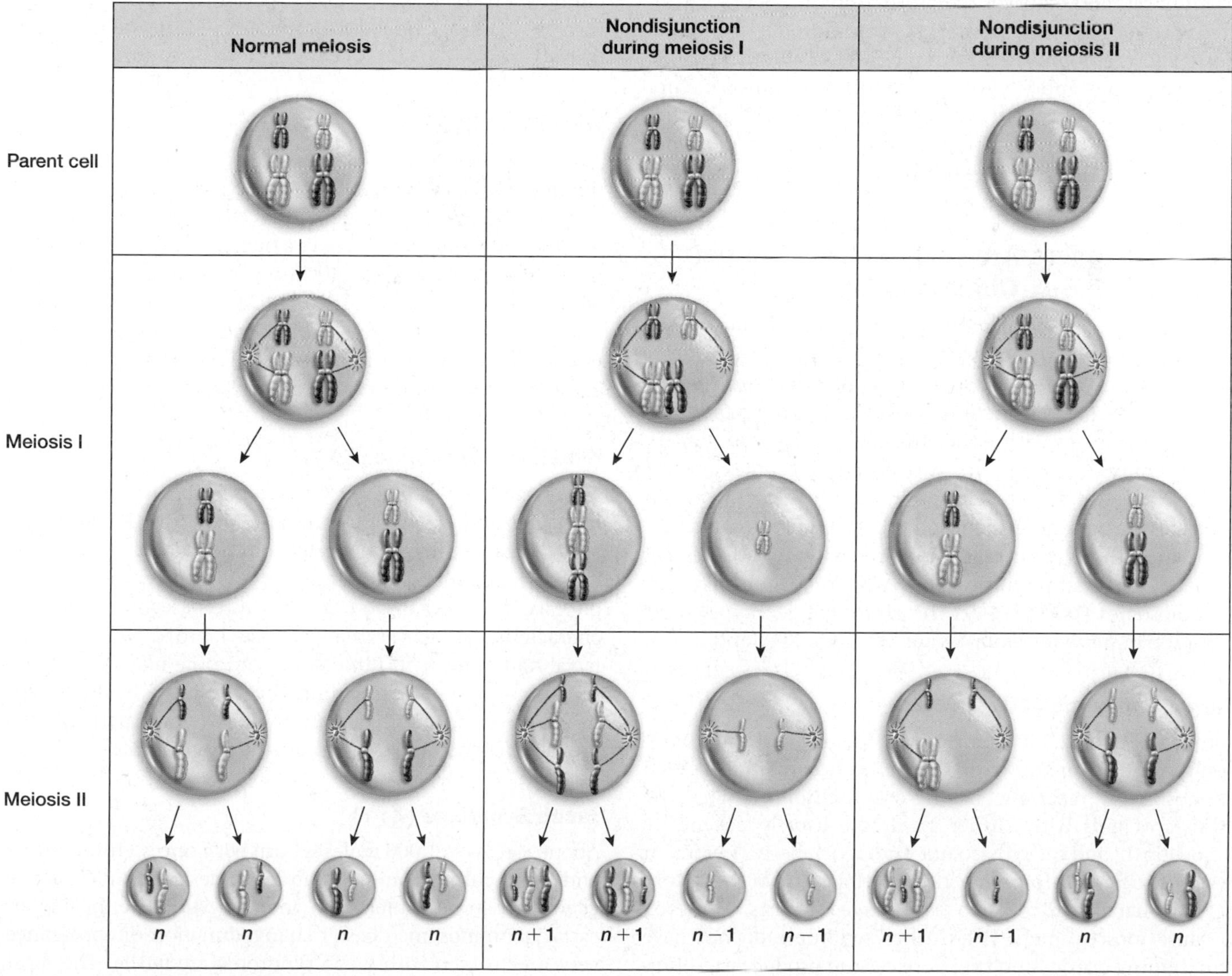

▲ **FIGURE 10-9 Nondisjunction during meiosis** Nondisjunction may occur either during meiosis I or meiosis II, resulting in gametes with too many ($n + 1$) or too few ($n - 1$) chromosomes.

TABLE 10-2 Effects of Nondisjunction of the Sex Chromosomes During Meiosis

Nondisjunction in the Father			
Sex Chromosomes of Defective Sperm	**Sex Chromosomes of Normal Egg**	**Sex Chromosomes of Offspring**	**Characteristics of Offspring**
O (none)	X	XO	Female—Turner syndrome
XX	X	XXX	Female—Trisomy X
XY	X	XXY	Male—Klinefelter syndrome
YY	X	XYY	Male—Jacob syndrome
Nondisjunction in the Mother			
Sex Chromosomes of Normal Sperm	**Sex Chromosomes of Defective Egg**	**Sex Chromosomes of Offspring**	**Characteristics of Offspring**
X	O (none)	XO	Female—Turner syndrome
Y	O (none)	YO	Dies as early embryo
X	XX	XXX	Female—Trisomy X
Y	XX	XXY	Male—Klinefelter syndrome

In humans, most embryos that arise from the fusion of gametes with an abnormal number of chromosomes spontaneously abort, accounting for 20% to 50% of all miscarriages. However, some embryos with an abnormal number of chromosomes survive to birth or beyond. The likelihood that an embryo will have an abnormal number of chromosomes increases with the age of the parents.

Some Disorders Are Caused by Abnormal Numbers of Sex Chromosomes

In humans and other mammals, each gamete normally contains a single sex chromosome. Normal eggs contain an X chromosome, and normal sperm contain either an X or a Y chromosome. Nondisjunction of sex chromosomes produces abnormal gametes with either two sex chromosomes or no sex chromosomes. Eggs with two sex chromosomes are XX; sperm with two sex chromosomes may be XX, YY, or XY. Gametes that lack sex chromosomes are often designated "O."

When these defective sperm or eggs fuse with normal gametes, the resulting embryos have abnormal numbers of sex chromosomes (**TABLE 10-2**). The abnormal genotypes most common in surviving embryos are XO, XXX, XXY, and XYY.

Turner Syndrome (XO)

About 1 in every 2,500 female babies has only one X chromosome, a condition known as **Turner syndrome** (also called monosomy X, meaning "having one X chromosome"). The ovaries of girls with Turner syndrome usually degenerate before birth, and the girls do not undergo puberty. Treatment with estrogen can promote the development of secondary sexual characteristics, such as enlarged breasts. However, because most women with Turner syndrome do not have functioning ovaries and therefore cannot produce eggs, hormone treatment does not make it possible for them to bear children. Women with Turner syndrome also experience increased risk of cardiovascular disease, kidney defects, and hearing loss.

Trisomy X (XXX)

About 1 in every 1,000 women has three X chromosomes, a condition known as **trisomy X,** or triple X. Most of these women have no detectable differences from XX women, except for a tendency to be taller and to have a higher incidence of learning disabilities. Unlike women with Turner syndrome, most women with trisomy X are fertile. Their children almost always have a normal complement of sex chromosomes (XX or XY), so some unknown mechanism must operate during meiosis to prevent an extra X chromosome from being included in their eggs.

Klinefelter Syndrome (XXY)

About 1 in every 500 to 1,000 males is born with two X chromosomes and one Y chromosome. Men with **Klinefelter syndrome** usually have small testes that do not produce as much testosterone as the testes of XY men typically do. At puberty, some XXY men develop mixed secondary sexual characteristics, such as enlarged breasts, broad hips, and thin facial hair. Men with Klinefelter syndrome may be infertile because of a low sperm count; the syndrome is often diagnosed when an affected man and his female partner seek medical help because they are unable to have children.

Jacob Syndrome (XYY)

About 1 in every 1,000 males is born with one X chromosome and two Y chromosomes. Y chromosomes contain few active genes, and in most men with **Jacob syndrome**, having an extra Y chromosome doesn't change function or appearance very much. Men with Jacob syndrome are fertile. The most common effect is that XYY males tend to be taller than average.

Have You Ever Wondered ...

Why Mules Are Sterile?

A mule is a cross between a female horse and a male donkey; like many hybrid animals, mules are sterile. Mules cannot produce viable gametes because they have an odd number of chromosomes. A mule's horse parent has 64 chromosomes ($n = 32$), and its donkey parent has 62 ($n = 31$), so a mule ends up with 63 chromosomes. When homologous pairs line up during meiosis I, one chromosome will not have a partner. In addition, many horse chromosomes are not homologous to donkey chromosomes, so many of a mule's chromosomes do not have a homologue to match up with. As a result, the normal mechanisms of meiosis fail, and mule gametes receive a nearly random mixture of horse and donkey chromosomes. Crucial genes are almost always missing, so whether a mule mates with a horse, a donkey, or another mule, the resulting fertilized eggs cannot develop. Very rarely, however, a mule does reproduce: There is one report of a female mule that apparently produced an egg with *all* horse chromosomes and *no* donkey chromosomes; she had a foal sired by a male donkey. Appropriately, the foal was named Blue Moon.

Some Disorders Are Caused by Abnormal Numbers of Autosomes

Nondisjunction of the autosomes produces eggs or sperm that are missing an autosome or that have two copies of an autosome. Fusion with a normal gamete bearing one copy of each autosome leads to an embryo with either one or three copies of the affected autosome. Embryos that have only one copy of any autosome almost always abort so early in development that the woman never knows she was pregnant. Embryos with three copies of an autosome (trisomy) also usually spontaneously abort. However, a small fraction of embryos with three copies of chromosomes 13, 18, or 21 survive to birth. In the case of trisomy 21, the child may live into adulthood.

Trisomy 21 (Down Syndrome)

An extra copy of chromosome 21 results in a condition called **trisomy 21**, or **Down syndrome**. Children with Down syndrome often show several distinctive physical characteristics, including weak muscle tone, a small mouth held partially open because it cannot accommodate the tongue, and distinctively shaped eyes (**FIG. 10-10**). More serious problems include varying degrees of mental impairment, low resistance to infectious diseases, and heart defects. Trisomy 21 can be diagnosed before birth by examining the chromosomes of fetal cells and, with less certainty, by biochemical tests and ultrasound examination of the fetus (see "Health Watch: Prenatal Genetic Screening" in Chapter 14).

Down syndrome occurs in about 1 of every 700 births overall, but the rate varies tremendously with the age of the parents, especially the mother. Among children born to 20-year-old women, only about 0.05% have Down syndrome, but in children born to women over 45 years of age, the rate is more than 3%. Nondisjunction in sperm accounts for about 10% of the cases of Down syndrome, and there is a small increase in risk of occurrence with increasing age of the father.

CHECK YOUR LEARNING

Can you . . .

- explain how nondisjunction causes offspring to have too many or too few chromosomes?
- describe some of the human genetic disorders that are caused by nondisjunction?

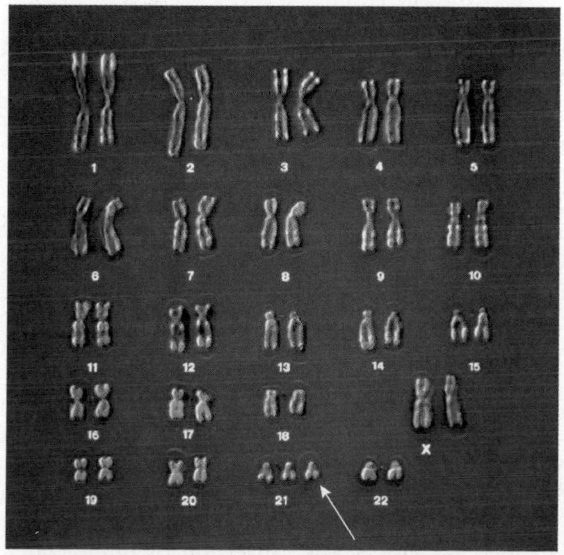

(a) Karyotype showing three copies of chromosome 21

(b) Girl with Down syndrome (right) and her older sister

▲ **FIGURE 10-10 Trisomy 21, or Down syndrome (a)** This karyotype of a child with Down syndrome reveals three copies of chromosome 21 (arrow). **(b)** Down syndrome is almost always caused by nondisjunction and seldom runs in families.

CASE STUDY REVISITED

Diversity Runs in the Family

Many people are astounded by the diversity of the Giddings children, but basic biology easily explains how such diversity arises. First, most genes have multiple alleles; second, meiotic cell division separates homologous chromosomes—and the alleles they carry—into different sperm and eggs; and third, the sperm and eggs unite at random. However, this straightforward explanation of how meiosis and sexual reproduction result in variable skin color leaves open a related question: Why does human skin color vary?

Evolutionary biologists hypothesize that natural selection favored different skin colors in different environments. In particular, skin color evolved in response to geographic variation in the amount of sunlight present. In terms of human health, sunlight provides both advantages and disadvantages. The ultraviolet rays in sunlight stimulate the synthesis of vitamin D, but they break down vitamin B_9 (folate); both vitamins are essential to human health. The pigments in dark skin absorb ultraviolet rays, minimizing the destruction of folate but also inhibiting synthesis of vitamin D. In the equatorial regions where the earliest humans lived, sunlight is so abundant that even dark-skinned individuals can produce sufficient vitamin D, and natural selection favored dark skin that protected against folate depletion. Eventually, some humans migrated to northern Europe, where sunlight is weak enough to minimize destruction of folate, but dark-skinned individuals cannot produce enough vitamin D. In European populations, natural selection therefore favored paler skin. As a result of this history, alleles for dark pigmentation occur most frequently in people whose ancestors lived in equatorial regions, and alleles for pale pigmentation predominate in people of northern European ancestry.

THINK CRITICALLY Ultraviolet rays in sunlight cause skin cancer, and pale-skinned people are more susceptible. Would the risk of skin cancer have selected against pale-skinned people in equatorial regions?

CHAPTER REVIEW

Go to **Mastering Biology** to access the Pearson eText, vocabulary review, practice quizzes, activities, videos, current events, and more.

*Answers to **Think Critically** and **Thinking Through the Concepts** questions can be found in the **Answers** section at the back of the book.*

Summary of Key Concepts

10.1 How Does Sexual Reproduction Produce Genetic Variability?

Eukaryotic cells typically contain pairs of chromosomes, called homologues, that carry the same genes with similar, although usually not identical, nucleotide sequences. These slightly different nucleotide sequences of a gene are called alleles. Cells containing paired homologous chromosomes are called diploid. Cells with only a single copy of each type of chromosome are called haploid. Cells with three or more copies of each type of chromosome are called polyploid.

10.2 How Does Meiotic Cell Division Produce Genetically Variable, Haploid Cells?

Meiotic cell division (meiosis followed by cytokinesis) separates homologous chromosomes and produces haploid cells with only one homologue from each pair. During interphase before meiosis, chromosomes are duplicated. The cell then undergoes two specialized divisions—meiosis I and meiosis II—to produce four haploid daughter cells (see **FIG. 10-5**).

Meiosis I

During prophase I, homologous duplicated chromosomes, each consisting of two chromatids, pair up and exchange DNA by crossing over. During metaphase I, homologues move together as pairs to the cell's equator, one member of each pair facing opposite poles of the cell. Homologous chromosomes separate during anaphase I, and two nuclei form during telophase I. Cytokinesis usually occurs during telophase I. Each daughter nucleus receives only one member of each pair of homologues and, therefore, is haploid. The sister chromatids of each chromosome remain attached to each other throughout meiosis I.

Meiosis II

Meiosis II resembles mitosis, but in a haploid cell. The duplicated chromosomes move to the cell's equator during metaphase II. The two chromatids of each chromosome separate and move to opposite poles of the cell during anaphase II. This second division produces four haploid nuclei. Cytokinesis normally occurs during or shortly after telophase II, producing four haploid cells.

10.3 How Do Meiosis and Union of Gametes Produce Genetically Variable Offspring?

The random shuffling of homologous maternal and paternal chromosomes during meiosis I creates new chromosome combinations. Crossing over creates chromosomes with allele combinations that may never have occurred before on single chromosomes. Because of the separation of homologues and crossing over, a parent probably never produces any gametes that are completely identical. The fusion of two genetically unique gametes adds further genetic variability to the offspring.

10.4 How Do Errors in Meiosis Cause Inherited Disorders?

Errors in meiosis can result in gametes with abnormal numbers of sex chromosomes or autosomes. Many people with abnormal numbers of sex chromosomes have distinguishing physical characteristics, and some have difficulty reproducing. An abnormal number of autosomes typically leads to spontaneous abortion early in pregnancy; in rare instances, the fetus may survive to birth, but mental or physical deficiencies always occur. The likelihood of producing gametes with abnormal numbers of chromosomes increases with increasing age of the mother and, to a lesser extent, the father.

Thinking Through the Concepts

Bloom's: Remembering, Understanding

Multiple Choice

1. Chromosomes that are found in every cell of both males and females are called
   a. sex chromosomes.
   b. autosomes.
   c. polyploid.
   d. chromatids.
2. A cell with three or more copies of each homologous chromosome is called
   a. a gamete.
   b. haploid.
   c. trisomy X.
   d. polyploid.
3. During crossing over,
   a. chromatids of homologous chromosomes exchange DNA.
   b. mutations occur with higher than average frequency.
   c. chromatids of nonhomologous chromosomes exchange DNA.
   d. nondisjunction occurs.
4. Which of the following does *not* contribute to genetic variability?
   a. accurate replication of DNA
   b. crossing over
   c. random alignment of homologous chromosomes during metaphase I of meiosis
   d. union of sperm and egg
5. Haploid nuclei are first formed at what stage of meiosis?
   a. metaphase I
   b. telophase I
   c. metaphase II
   d. telophase II

Fill-in-the-Blank

1. Meiotic cell division produces ____________ (how many) haploid daughter cells from each diploid parental cell. In animals, the haploid daughter cells produced by meiotic cell division become ____________.
2. During ____________ of meiosis I, paired homologous chromosomes form structures called ____________. These structures are the sites of what event? ____________
3. Sister chromatids remain attached and homologues separate during ________ of ________. Sister chromatids detach and separate during ________ of ________.
4. Three processes that promote genetic variability of offspring during sexual reproduction are ____________, ____________, and ____________.
5. Women with ____________ syndrome have a single X chromosome. They typically ________ (do/do not) undergo puberty and ________ (can/cannot) bear children. Men with ________ syndrome typically have reduced male secondary sexual characteristics. Their sex chromosomes are ____________ (list the number of X and Y chromosomes).

Review Questions

1. Diagram the events of meiosis. At which stage do homologous chromosomes separate?
2. Describe crossing over. At which stage of meiosis does it occur? Name two functions of chiasmata.
3. In what ways are mitosis and meiosis similar? In what ways are they different?
4. Describe how meiosis adds to genetic variability. If an animal had a haploid number of two (no sex chromosomes), how many genetically different gametes could it produce? (Assume no crossing over.) What if it had a haploid number of five?
5. Define *nondisjunction*, and describe common syndromes caused by nondisjunction of sex chromosomes and autosomes.

Applying the Concepts

Bloom's: Applying, Analyzing, Evaluating

1. Many plants can reproduce sexually or asexually. Strawberries, for example, can reproduce asexually by sending out horizontal stems called runners that root and form new plants, or they can reproduce sexually by flowering and producing fruit and seeds. Describe some advantages and disadvantages of each type of reproduction in wild plants. Include in your discussion the important aspects of the environments in which runners and seeds are likely to find themselves.

11 Patterns of Inheritance

CASE STUDY

Sudden Death on the Court

FLO HYMAN, 6 feet, 5 inches tall, graceful and athletic, was probably the best woman volleyball player of her time. Captain of the American women's volleyball team that won the silver medal in the 1984 Olympics, Hyman later joined a professional squad. In 1986, she came out of a game for a short breather and died while sitting quietly on the bench. Hyman was only 31 years old. How could this happen to someone so young and fit?

Hyman had a rare genetic disorder called Marfan syndrome. People with Marfan syndrome are typically tall and slender, with long limbs and large hands and feet. For some people with Marfan syndrome, these characteristics help bring fame and fortune. Unfortunately, Marfan syndrome can also be deadly.

Hyman died from a rupture of her aorta, the massive artery that carries blood from the heart to most of the body. How could the condition that made Hyman tall and large-handed also weaken her aorta? Because defects in a single protein—fibrillin—caused all of these effects.

Marfan syndrome is caused by a mutation in the gene that encodes a protein called fibrillin; people with the syndrome produce an abnormal version of the protein. Normal fibrillin strengthens certain kinds of tissues, including tendons that attach muscles to bones, ligaments that fasten bones to other bones in joints, and the walls of arteries. Fibrillin also traps growth factors, preventing them from stimulating excessive cell division in connective tissues such as bone, cartilage, ligaments, and tendons. Defective fibrillin cannot trap these growth factors, nor can it effectively strengthen tissues. Thus, people with Marfan syndrome have high concentrations of growth factors that cause their arms, legs, hands, and feet to become unusually long, and also have weak bones, ligaments, tendons, and artery walls.

Because we are diploid organisms, each of us carries two copies of the fibrillin gene. A single defective copy is enough to cause Marfan syndrome. What does this observation reveal about the inheritance of Marfan syndrome? Are all inherited diseases caused by a single defective copy of a gene? To find out, we must go back in time and visit the garden of Gregor Mendel.

Olympic volleyball silver medalist Flo Hyman was struck down by Marfan syndrome at the height of her career.

AT A GLANCE

11.1 WHAT IS THE PHYSICAL BASIS OF INHERITANCE?

Inheritance is the process by which the traits of organisms are passed to their offspring. In this chapter, our discussion of the process will be confined to inheritance in diploid, sexually reproducing organisms, a group that includes most plants and animals. We begin our exploration with a brief review of the structures that form the physical basis of inheritance.

Genes Are Sequences of Nucleotides at Specific Locations on Chromosomes

The units of inheritance are **genes**—segments of DNA that encode the information needed to construct an organism. Genes are parts of chromosomes, which consist of a double helix of DNA packaged with a variety of proteins (see Figs. 9-1 and 9-5). A gene's physical location on a chromosome is called its **locus** (plural, loci; **FIG. 11-1**).

The chromosomes of diploid organisms occur in pairs called homologues. A pair of homologues includes a paternal chromosome inherited from the male parent and a maternal chromosome inherited from the female parent. Each member of a pair of homologues carries the same genes at the same loci. However, the nucleotide sequences of a given gene may differ somewhat between the two homologues of an individual, and between different members of a species. These different versions of a gene are called **alleles** (see Fig. 11-1).

Mutations Are the Source of Alleles

Why do genes have multiple alleles? All alleles originally arose as **mutations**—changes in the sequence of nucleotides in the DNA of a gene. If a mutation occurs in a cell that becomes a sperm or egg, it can be passed on from parent to offspring. An offspring carrying the new version of the gene may pass it to its own offspring, and so on down the generations. Thus, most of the alleles present in an organism first appeared as mutations in the reproductive cells of the organism's ancestors.

An Organism's Two Alleles May Be the Same or Different

Because a diploid organism has pairs of homologous chromosomes, and both members of a pair contain the same gene loci, the organism has two copies of each gene. If both homologues have the same allele at a given locus, the organism is said to be **homozygous** at that locus. ("Homozygous" comes from Greek words meaning "same pair.") The chromosomes shown in Figure 11-1 are homozygous at two loci. If two homologous chromosomes have different alleles at a locus, the organism is **heterozygous** ("different pair") at that locus. The chromosomes in Figure 11-1 are heterozygous at one locus.

CHECK YOUR LEARNING

Can you . . .

- describe the relationships among chromosomes, genes, loci, mutations, and alleles?
- explain what it means for an organism to be heterozygous or homozygous for a gene?

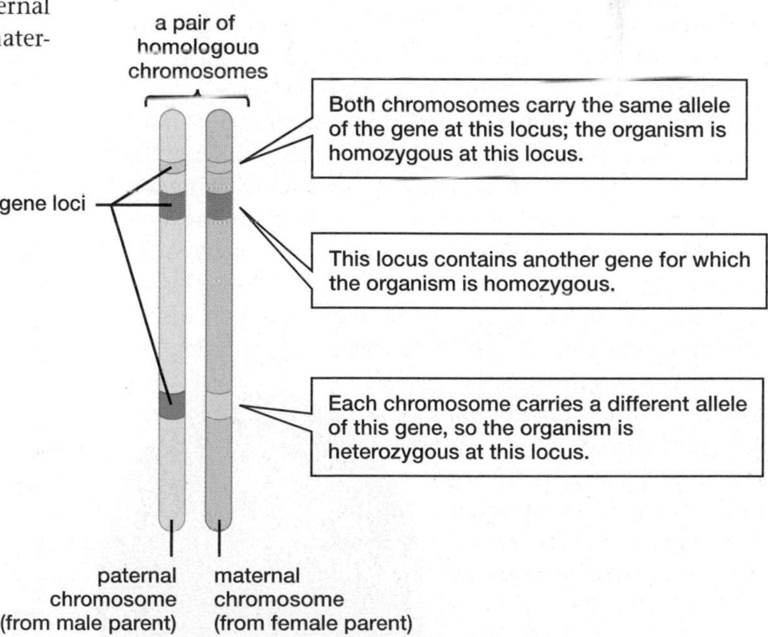

a pair of homologous chromosomes
gene loci
Both chromosomes carry the same allele of the gene at this locus; the organism is homozygous at this locus.
This locus contains another gene for which the organism is homozygous.
Each chromosome carries a different allele of this gene, so the organism is heterozygous at this locus.
paternal chromosome (from male parent)
maternal chromosome (from female parent)

▲ **FIGURE 11-1 The relationships among genes, alleles, and chromosomes** Each homologous chromosome carries the same set of genes. Each gene is located at the same position, or locus, on each homologue. Differences in nucleotide sequences at the same gene locus produce different alleles of the gene. Diploid organisms have two alleles of each gene, one on each homologue. The alleles on the two homologues may be the same or different.

11.2 HOW WERE THE PRINCIPLES OF INHERITANCE DISCOVERED?

In the mid-1800s, experiments by an Austrian monk, Gregor Mendel (**FIG. 11-2**), revealed many important principles of inheritance. Although Mendel worked long before DNA, chromosomes, or meiosis had been discovered, his research revealed essential facts about genes and alleles and how they are inherited during sexual reproduction.

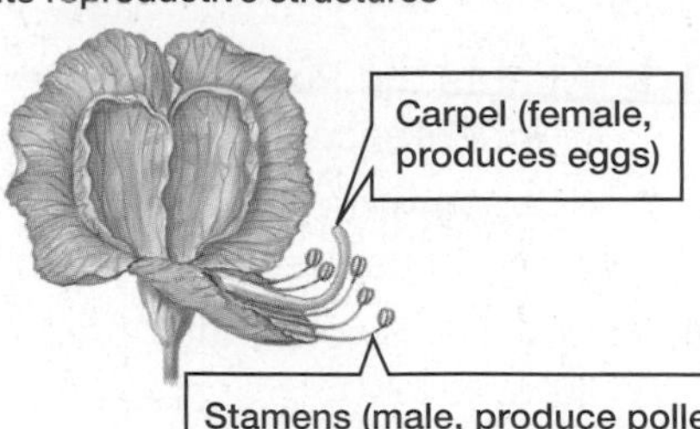

▲ **FIGURE 11-3 Flowers of the edible pea plant** In the intact pea flower (left), the lower petals enclose the reproductive structures—the stamens (male) and carpel (female). Pollen normally cannot enter the flower from outside, so peas usually self-pollinate and, hence, self-fertilize. If the flower is opened (right), it can be cross-pollinated by hand.

Doing It Right: The Secrets of Mendel's Success

Mendel experimented on edible pea plants (**FIG. 11-3**). In this species, male and female reproductive structures occur on the same flower. The male reproductive structures, called stamens, produce pollen grains, which contain sperm. The female reproductive structure, called the carpel, contains an ovary. A pea flower's petals enclose both reproductive structures, preventing another flower's pollen from entering. Therefore, the eggs in a pea flower must be fertilized by sperm from the pollen of the same flower. When an organism fertilizes its own eggs in this fashion, the process is called **self-fertilization**.

Self-fertilization did not always suit Mendel's goals, as he often wanted to mate two different pea plants to see what characteristics their offspring would inherit. To accomplish such a mating, he would open a pea flower and remove its stamens, preventing self-fertilization. Then he dusted the sticky tip of the carpel with pollen from the flower of another plant. When sperm from one individual fertilize the eggs of another individual, the process is called **cross-fertilization**. If the cross-fertilizing parents differ in at least one genetically determined trait, the offspring are called **hybrids**.

Mendel's experimental design was simple, but brilliant. Earlier researchers had generally tried to study inheritance by simultaneously considering all of the features of entire organisms, including traits that differed only slightly among individuals. Not surprisingly, these investigators were often confused rather than enlightened. Mendel, in contrast, began by studying only one trait at a time, and he studied traits with unmistakably different forms, such as white versus purple flowers.

In his experiments, Mendel followed the inheritance of traits for several generations, counting the numbers of offspring with each version of a trait. When he analyzed these numbers, the basic patterns of inheritance became clear. Today, quantifying experimental results and applying statistical analysis are essential tools in virtually every field of biology. But in Mendel's time, numerical analysis was an innovation.

▶ **FIGURE 11-2 Gregor Mendel** This portrait was painted in about 1888, after he had completed his pioneering genetics experiments.

CHECK YOUR LEARNING

Can you . . .

- distinguish between self-fertilization and cross-fertilization?
- explain the important features of Mendel's experimental design?

11.3 HOW ARE SINGLE TRAITS INHERITED?

Even today, one of the best ways to learn the basic principles of inheritance is to review the results of Mendel's experiments.

In his first set of experiments, Mendel cross-fertilized pea plants that were true-breeding for different variants of a single trait. (**True-breeding** organisms possess a trait that is inherited by all offspring produced by self-fertilization.) In one of these experiments, Mendel cross-fertilized true-breeding white-flowered plants with true-breeding purple-flowered plants. These plants were the parental generation, denoted by the letter P. To determine the traits of the offspring, Mendel saved the hybrid seeds and grew them the following year.

When Mendel grew the hybrid seeds, he found that all the first-generation offspring (the "first filial," or F_1 generation) produced purple flowers (**FIG. 11-4**). What had happened to the white color? The flowers of the F_1 hybrids were just as purple as their true-breeding purple parent. The white color of their true-breeding white parent seemed to have disappeared.

Mendel then allowed the F_1 plants to self-fertilize, collected the seeds, and planted them the following spring. In the second (F_2) generation, Mendel counted 705 plants with

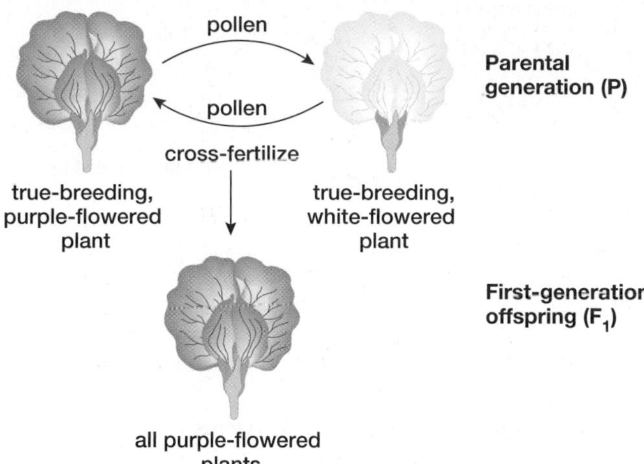

▲ **FIGURE 11-4 Cross of pea plants true-breeding for white or purple flowers** All of the offspring bear purple flowers.

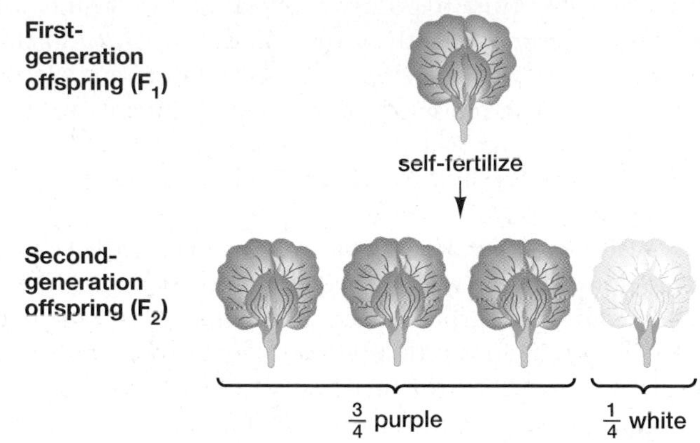

▲ **FIGURE 11-5 Self-fertilization of F_1 pea plants with purple flowers** Three-quarters of the offspring bear purple flowers and one-quarter bear white flowers.

purple flowers and 224 plants with white flowers. Approximately three-fourths of the plants had purple flowers, and one-fourth had white flowers, a ratio of about 3 purple to 1 white (**FIG. 11-5**). This result showed that the capacity to produce white flowers had not disappeared in the F_1 hybrids, but had only been hidden.

Mendel then allowed the F_2 plants to self-fertilize and produce a third (F_3) generation. He found that all the white-flowered F_2 plants produced white-flowered offspring; that is, they were true-breeding. In contrast, when purple-flowered F_2 plants self-fertilized, their offspring were of two types. One-third of the offspring were true-breeding for purple, but the other two-thirds produced both purple- and white-flowered offspring, again in the ratio of 3 purple to 1 white. Therefore, the F_2 generation overall included one-quarter true-breeding white plants, one-quarter true-breeding purple plants, and one-half hybrid purple plants.

The Inheritance of Dominant and Recessive Alleles on Homologous Chromosomes Explains the Results of Mendel's Crosses

Mendel's results, supplemented by modern knowledge of genes and chromosomes, allow us to develop a five-part hypothesis to explain the inheritance of single traits:

- Traits are determined by distinct physical units—genes—that come in pairs. Each organism has two alleles for each gene, one on each homologous chromosome.
- Homologous chromosomes separate, or segregate, from each other during meiosis, thus separating the alleles they carry. This is known as Mendel's **law of segregation**: Each gamete receives only one allele of each gene. When a sperm fertilizes an egg, the resulting offspring receives one allele from the father and one from the mother.
- Homologous chromosomes separate randomly during meiosis, so the distribution of alleles into gametes is also random.
- True-breeding individuals have two copies of the same allele for a given gene and are therefore homozygous for that gene. All of the gametes from a homozygous individual receive the same allele of the gene (**FIG. 11-6a**). Hybrid individuals have two different alleles for a given gene and so are heterozygous for that gene. Half of a heterozygous individual's gametes will contain one allele for that gene, and half will contain the other allele (**FIG. 11-6b**).
- When two different alleles are present in an organism, one—the **dominant** allele—may mask the expression of the other—the **recessive** allele. The recessive allele, however, is still present.

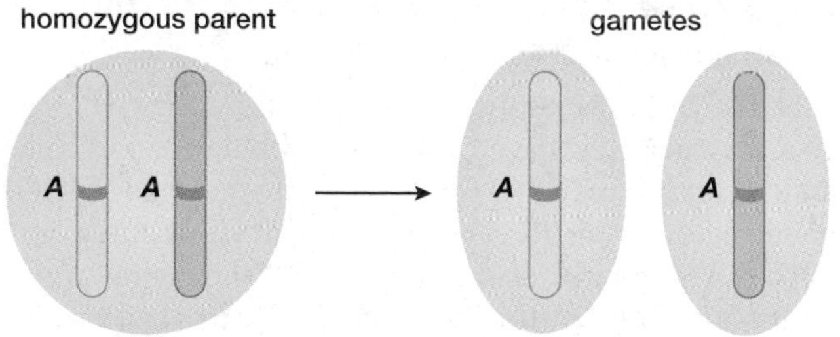

(a) Gametes produced by a homozygous parent

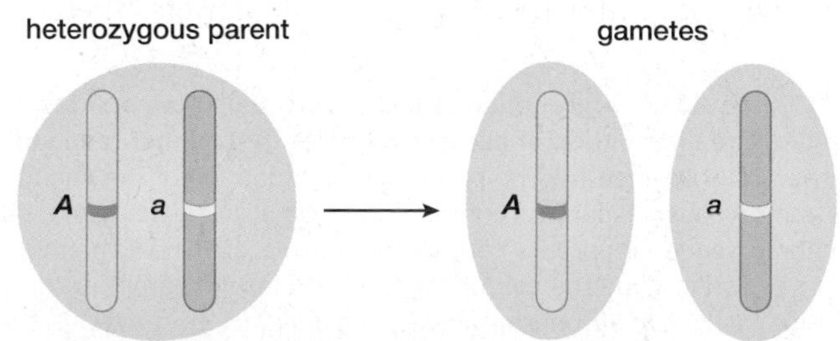

(b) Gametes produced by a heterozygous parent

▲ **FIGURE 11-6 The distribution of alleles in gametes (a)** All of the gametes produced by homozygous organisms contain the same allele. **(b)** Half of the gametes produced by heterozygous organisms contain one allele, and half of the gametes contain the other allele.

Let's see how this hypothesis explains the results of Mendel's experiments with flower color (**FIG. 11-7**). In the explanation, we will use letters to represent different alleles. Because the allele for purple flowers is dominant, we use the uppercase letter *P* to represent it. The allele for white flowers is recessive, and we assign the lowercase letter *p* to represent it.

Recall that Mendel began by crossing true-breeding purple-flowered and white-flowered plants. A plant that is true-breeding for purple flowers is homozygous, with two alleles for purple flower color (*PP*), so all of its sperm and eggs carry the *P* allele. A true-breeding white-flowered plant is also homozygous, but with two alleles for white flower color (*pp*); all of its sperm and eggs carry the *p* allele (**FIG. 11-7a**). Thus, all of the F_1 offspring from the cross were produced either when *P* sperm fertilized *p* eggs or when *p* sperm fertilized *P* eggs. In both cases, the offspring were heterozygous *Pp*. Because *P* is dominant over *p*, all of the offspring were purple (**FIG. 11-7b**).

To produce the F_2 generation, Mendel allowed the F_1 plants to self-fertilize. The heterozygous plants of the F_1 generation produced equal numbers of *P* and *p* sperm and equal numbers of *P* and *p* eggs. When a *Pp* plant self-fertilizes, each type of sperm has an equal chance of fertilizing each type of egg (**FIG. 11-7c**). Therefore, Mendel's F_2 generation contained three types of offspring: *PP*, *Pp*, and *pp*. The three types occurred in the approximate proportions of one-quarter *PP* (homozygous purple), one-half *Pp* (heterozygous purple), and one-quarter *pp* (homozygous white).

Observable Traits Do Not Always Reveal Underlying Alleles

Mendel's findings showed that two organisms that look alike may actually have different combinations of alleles. That is, organisms with the same *phenotype* may have different *genotypes*. An organism's **phenotype** consists of its traits, including its outward appearance, behavior, and any other observable feature. An organism's **genotype** consists of the combination of alleles it carries (for example, *PP* or *Pp*). As we have seen, plants with either the *PP* or the *Pp* genotype have the phenotype of purple flowers. Therefore,

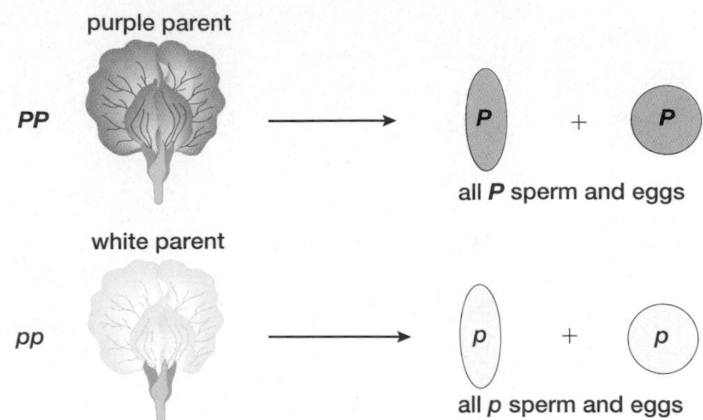

(a) Gametes produced by homozygous parents

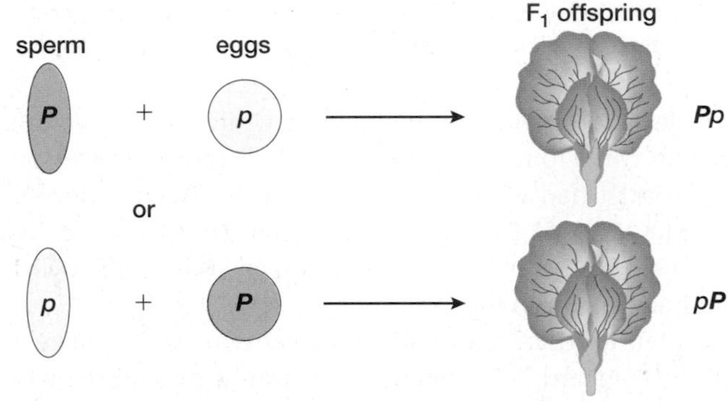

(b) Fusion of gametes produces F_1 offspring

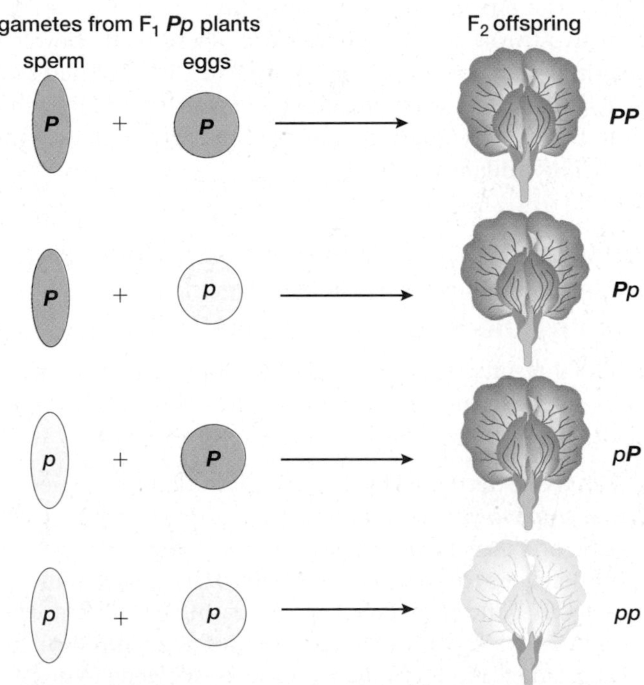

(c) Fusion of gametes from the F_1 generation produces F_2 offspring

▸ FIGURE 11-7 Segregation of alleles and fusion of gametes predict the distribution of alleles and traits in the inheritance of flower color in peas (a) The parental generation: All of the gametes of homozygous *PP* parents contain the *P* allele; all of the gametes of homozygous *pp* parents contain the *p* allele. **(b)** The F_1 generation: Fusion of gametes containing the *P* allele with gametes containing the *p* allele produces only *Pp* offspring. (Note that *Pp* is the same genotype as *pP*.) **(c)** The F_2 generation: Half of the gametes of heterozygous *Pp* parents contain the *P* allele and half contain the *p* allele. Fusion of these gametes produces *PP*, *Pp*, and *pp* offspring.

even though the F_2 generation of Mendel's peas included individuals with three different genotypes (one-quarter *PP*, one-half *Pp*, and one-quarter *pp*), it contained only two phenotypes (three-quarters purple and one-quarter white).

"Genetic Bookkeeping" Can Predict Genotypes and Phenotypes of Offspring

The **Punnett square method**, named after R. C. Punnett, a famous geneticist of the early 1900s, is a convenient way to predict the genotypes and phenotypes of offspring. **FIGURE 11-8a** shows how to use a Punnett square to determine the expected proportions of offspring that arise from breeding two organisms that are heterozygous for a single trait. **FIGURE 11-8b** shows how to calculate the proportions of offspring using the probabilities that each type of sperm will fertilize each type of egg.

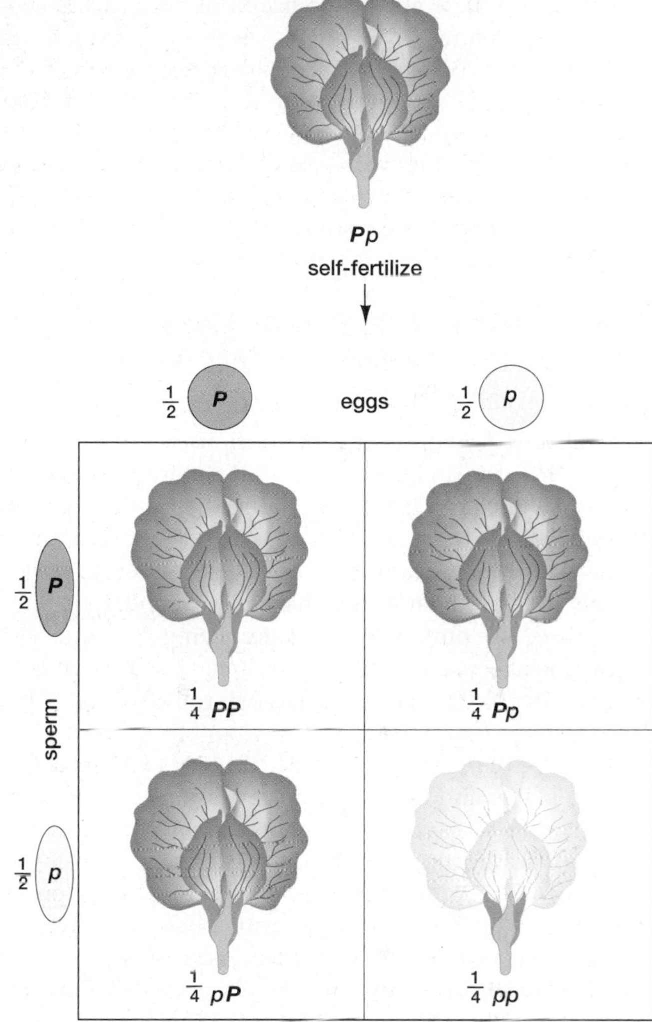

(a) Punnett square of a single-trait cross

sperm		eggs		offspring genotypes	genotypic ratio (1:2:1)	phenotypic ratio (3:1)
$\frac{1}{2}$ *P*	×	$\frac{1}{2}$ *P*	=	$\frac{1}{4}$ *PP*	$\frac{1}{4}$ *PP*	
$\frac{1}{2}$ *P*	×	$\frac{1}{2}$ *p*	=	$\frac{1}{4}$ *Pp*		
					$\frac{1}{2}$ *Pp*	$\frac{3}{4}$ purple
$\frac{1}{2}$ *p*	×	$\frac{1}{2}$ *P*	=	$\frac{1}{4}$ *pP*		
$\frac{1}{2}$ *p*	×	$\frac{1}{2}$ *p*	=	$\frac{1}{4}$ *pp*	$\frac{1}{4}$ *pp*	$\frac{1}{4}$ white

(b) Using probabilities to determine the offspring of a single-trait cross

▸ **FIGURE 11-8 Determining the outcome of a single-trait cross**
(a) The Punnett square allows you to predict both genotypes and phenotypes of specific crosses; here we use it for a cross between pea plants that are heterozygous for a single trait—flower color.

1. Assign letters to the different alleles; use uppercase for dominant alleles and lowercase for recessive alleles.
2. Determine all the types of genetically different gametes that can be produced by the male and female parents.
3. Draw the Punnett square, with the columns labeled with all possible genotypes of the eggs and the rows labeled with all possible genotypes of the sperm. (We also show the fractions of each genotype.)
4. Fill in the genotype of the offspring in each box by combining the genotype of the sperm in its row with the genotype of the egg in its column. (Multiply the fraction of sperm of each type in the row headers by the fraction of eggs of each type in the column headers.)
5. Count the number of offspring with each genotype. Note that *Pp* is the same genotype as *pP*.
6. Convert the number of offspring of each genotype to a fraction of the total number of offspring. In this example, out of four fertilizations, only one is predicted to produce the *pp* genotype, so one-quarter of the total number of offspring produced by this cross is predicted to be white. To determine phenotypic fractions, add the fractions of genotypes that would produce a given phenotype. For example, purple flowers are produced by $\frac{1}{4}PP + \frac{1}{4}Pp + \frac{1}{4}pP$, for a total of three-quarters of the offspring.

(b) Probabilities may also be used to predict the outcome of a single-trait cross. Determine the fractions of eggs and sperm of each genotype, and multiply these fractions together to calculate the fraction of offspring of each genotype. When two genotypes produce the same phenotype (e.g., *Pp* and *pP*), add the fractions of each genotype to determine the phenotypic fraction.

THINK CRITICALLY If you crossed a heterozygous ***Pp*** plant with a homozygous recessive ***pp*** plant, what would be the expected ratio of offspring? How does this differ from the offspring of a *PP* × *pp* cross? Try working this out before you read further in the text.

As you use these genetic bookkeeping techniques, keep in mind that in a real experiment, the actual offspring will usually not occur in exactly the predicted proportions. Why not? Let's consider a familiar example. Each time a baby is conceived, it has an equal chance of being a boy or a girl. However, many families with two children do not have one girl and one boy. We can expect to reliably observe a 1:1 ratio of girls to boys only if we combine the outcome from many families to create a very large sample.

Mendel's Hypothesis Can Be Used to Predict the Outcome of New Types of Single-Trait Crosses

You probably recognize that Mendel used the scientific method: He made an observation and used it to formulate a hypothesis. But did Mendel's hypothesis accurately predict the results of further experiments? Based on the hypothesis that heterozygous F_1 plants have one allele for purple flowers and one for white (that is, they have the *Pp* genotype), Mendel predicted the outcome of cross-fertilizing *Pp* plants with homozygous recessive white plants (*pp*): There should be equal numbers of *Pp* (purple) and *pp* (white) offspring. This is indeed what he found.

This type of experiment has practical uses for breeders of domestic plants and animals, who may want to know if an organism with a desirable dominant trait will pass that trait on to all of its offspring or only to some of them. Crossing an organism with a dominant phenotype (in this case, a purple flower) but an unknown genotype with a homozygous recessive organism (a white flower) is called a **test cross**, because it tests whether the organism with the dominant phenotype is homozygous or heterozygous (**FIG. 11-9**). When crossed with a homozygous recessive (*pp*), a homozygous dominant (*PP*) produces all phenotypically dominant offspring, whereas a heterozygous dominant (*Pp*) yields offspring with both dominant and recessive phenotypes in a 1:1 ratio.

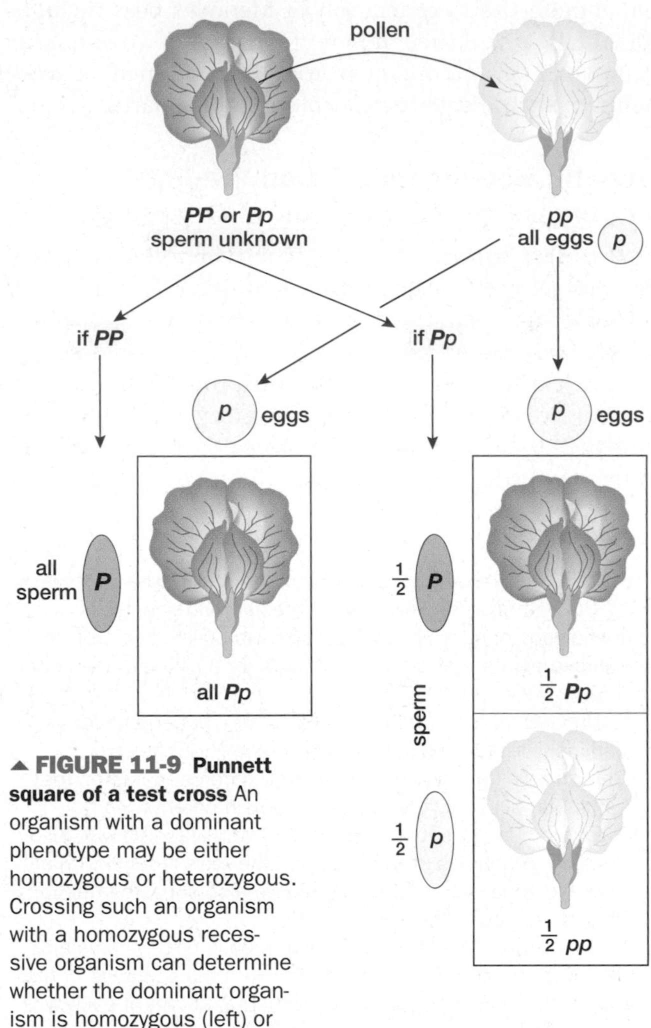

▲ **FIGURE 11-9 Punnett square of a test cross** An organism with a dominant phenotype may be either homozygous or heterozygous. Crossing such an organism with a homozygous recessive organism can determine whether the dominant organism is homozygous (left) or heterozygous (right).

CHECK YOUR LEARNING

Can you . . .

- describe the pattern of inheritance of a trait controlled by a single gene with two alleles, one dominant and one recessive?
- distinguish between genotype and phenotype?
- calculate the proportions of offspring with each genotype and phenotype that would be produced by mating parents with various combinations of the two alleles?

CASE STUDY CONTINUED

Sudden Death on the Court

Many traits are inherited in simple Mendelian fashion. Marfan syndrome, for example, is inherited as a dominant trait, which means that a single defective fibrillin allele is enough to cause the disorder. Flo Hyman inherited her defective allele from her father. Does the pattern of inheritance that Mendel discovered also apply when we consider several traits at once?

11.4 HOW ARE MULTIPLE TRAITS INHERITED?

Having determined how single traits are inherited, Mendel turned to the inheritance of multiple traits. In his next series of experiments, he again used edible pea plants and again studied traits that are expressed as two alternative phenotypes (**FIG. 11-10**).

▲ **FIGURE 11-10 Two of the traits studied by Mendel** Pea color may be yellow or green; pea shape may be smooth or wrinkled.

Mendel Extended His Experiments with More Complex Crosses

Mendel began his exploration of multiple-trait inheritance by cross-fertilizing plants that differed in two traits—seed color (yellow or green) and seed shape (smooth or wrinkled). From earlier crosses of plants with these traits, Mendel already knew that the smooth allele of the seed shape gene (*S*) is dominant to the wrinkled allele (*s*) and that the yellow allele of the seed color gene (*Y*) is dominant to the green allele (*y*). He crossed true-breeding plants grown from smooth, yellow seeds (genotype *SSYY*) with true-breeding plants grown from wrinkled, green seeds (*ssyy*). The *SSYY* plants produced only *SY* gametes, and the *ssyy* plants produced only *sy* gametes. Therefore, all the seeds produced by this cross were heterozygotes with genotype *SsYy* and a smooth, yellow phenotype.

Mendel allowed these heterozygous F_1 seeds to grow into mature plants and self-fertilize. Then he recorded the phenotypes of the F_2 generation: 315 smooth, yellow seeds; 101 wrinkled, yellow seeds; 108 smooth, green seeds; and 32 wrinkled, green seeds—a ratio of about 9:3:3:1. The offspring produced from other crosses of plants that were heterozygous for two traits also had phenotypic ratios of about 9:3:3:1.

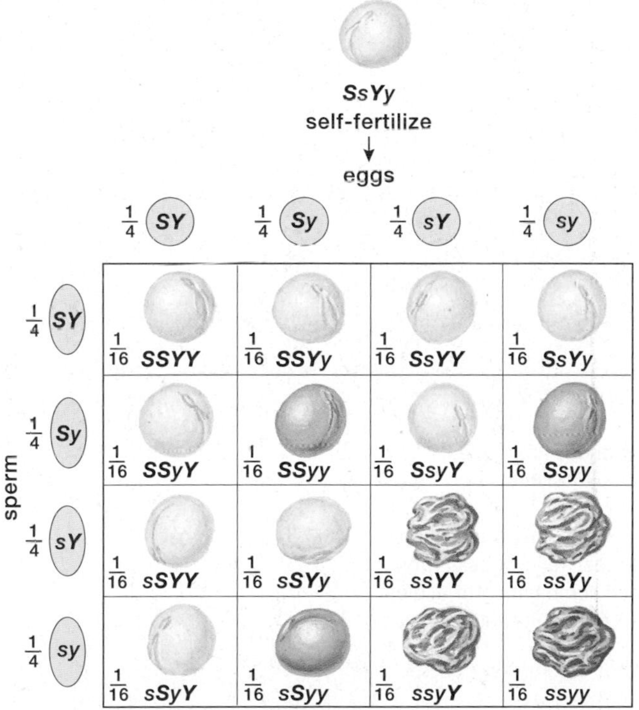

(a) Punnett square of a two-trait cross

seed shape		seed color		phenotypic ratio (9:3:3:1)
$\frac{3}{4}$ smooth	×	$\frac{3}{4}$ yellow	=	$\frac{9}{16}$ smooth yellow
$\frac{3}{4}$ smooth	×	$\frac{1}{4}$ green	=	$\frac{3}{16}$ smooth green
$\frac{1}{4}$ wrinkled	×	$\frac{3}{4}$ yellow	=	$\frac{3}{16}$ wrinkled yellow
$\frac{1}{4}$ wrinkled	×	$\frac{1}{4}$ green	=	$\frac{1}{16}$ wrinkled green

(b) Using probabilities to determine the offspring of a two-trait cross

◀ FIGURE 11-11 Predicting genotypes and phenotypes for a cross between parents that are heterozygous for two traits In pea seeds, yellow color (*Y*) is dominant to green (*y*), and smooth shape (*S*) is dominant to wrinkled (*s*). **(a)** In this cross, an individual heterozygous for both traits (*SsYy*) self-fertilizes. In a cross involving two independent genes, there will be equal numbers of gametes with all of the possible combinations of alleles of the two genes—*SY*, *Sy*, *sY*, and *sy*. Place these gamete combinations as the labels for the rows and columns in the Punnett square and then calculate the offspring as explained in Figure 11-8. Note that the Punnett square predicts both the frequencies of combinations of traits ($\frac{9}{16}$ smooth, yellow; $\frac{3}{16}$ smooth, green; $\frac{3}{16}$ wrinkled, yellow; and $\frac{1}{16}$ wrinkled, green) and the frequencies of individual traits ($\frac{3}{4}$ yellow, $\frac{1}{4}$ green, $\frac{3}{4}$ smooth, and $\frac{1}{4}$ wrinkled). **(b)** The probability of two independent events occurring together is the product (multiplication) of their individual probabilities. For example, to find the probability of tossing two coins and having both come up heads, multiply the probabilities of each coin coming up heads $\left(\frac{1}{2} \times \frac{1}{2} = \frac{1}{4}\right)$. Seed shape is independent of seed color. Therefore, multiplying the individual probabilities of the genotypes or phenotypes for each trait produces the predicted frequencies for the combined genotypes or phenotypes of the offspring. These frequencies are identical to those generated by the Punnett square.

THINK CRITICALLY Can the genotype of a plant grown from a smooth, yellow seed be revealed by a test cross with a plant grown from a wrinkled, green seed? Explain your answer.

Mendel Hypothesized That Traits Are Inherited Independently

Mendel realized that the 9:3:3:1 ratio he observed in the F_2 seeds was the ratio that would be expected if the genes for seed color and seed shape were inherited independently of each other and did not influence each other during gamete formation. If these traits were in fact inherited independently, then three-quarters of the seeds from the F_2 generation would be yellow and three-quarters would be smooth. That is, if we look just at the color of the seeds and ignore their shape, the dominant yellow color and the recessive green color should occur in the 3:1 ratio that we expect from self-fertilization of a parent heterozygous for seed color. Likewise, if we look just at the shape of the seeds and ignore their color, the dominant smooth shape and the recessive wrinkled shape should also occur in a 3:1 ratio. This result is just what Mendel observed. He found that among smooth and wrinkled seeds pooled together, 416 were yellow and 140 were green, about a 3:1 ratio. Similarly, among yellow and green seeds pooled together, 423 were smooth and 133 were wrinkled, a ratio of about 3:1. **FIGURE 11-11** shows how a Punnett square or probability calculation can be used to estimate the proportions of genotypes and phenotypes of the offspring of a cross between organisms that are heterozygous for two traits.

The independent inheritance of two or more traits is called the **law of independent assortment**. Multiple traits are inherited independently if the alleles of one gene

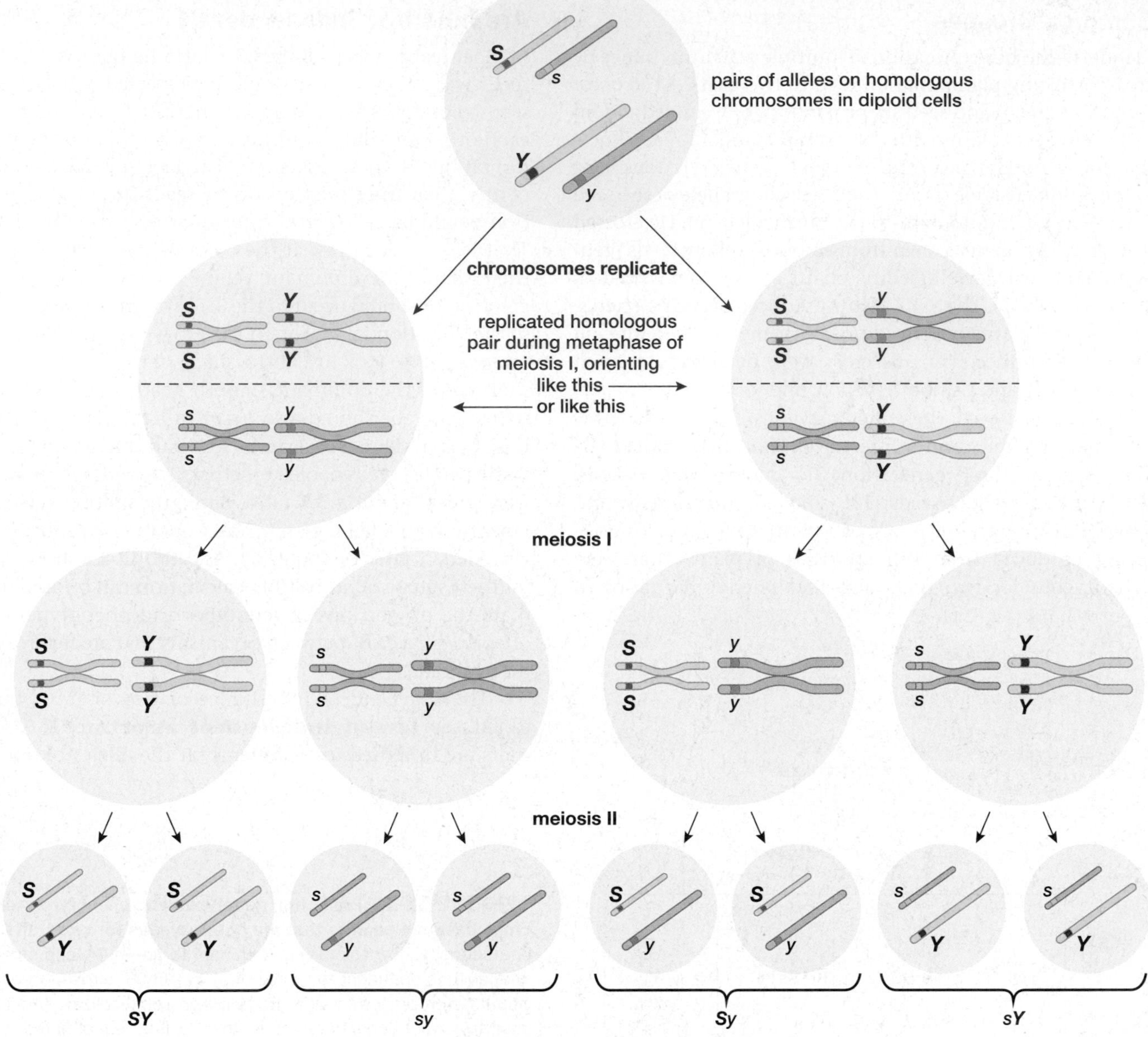

▲ **FIGURE 11-12 Independent assortment of alleles** Chromosome movements during meiosis produce independent assortment of alleles for two genes on different chromosomes. Yellow designates the paternal homologue and purple the maternal homologue of each pair. Each combination of alleles is equally likely to occur, producing gametes in the predicted proportions $\frac{1}{4}$ *SY*, $\frac{1}{4}$ *sy*, $\frac{1}{4}$ *Sy*, and $\frac{1}{4}$ *sY*.

THINK CRITICALLY If the genes for seed color and seed shape were on the same chromosome rather than on different chromosomes, would their alleles assort independently? Why or why not?

are distributed to gametes independently of the alleles of other genes. The physical basis for this independence is found in the events of meiosis. When paired homologous chromosomes line up during metaphase I, homologues face the poles of the cell at random, and the orientation of any one homologous pair does not influence the orientation of other pairs. Therefore, when the homologues separate during anaphase I, the alleles of genes on different chromosomes are distributed, or "assorted," independently of one another (**FIG. 11-12**).

CHECK YOUR LEARNING

Can you . . .

- describe the pattern of simultaneous inheritance of two traits if each of the traits is controlled by a separate gene with only two alleles, one dominant and one recessive?
- explain the law of independent assortment?
- calculate the frequencies of the genotypes and phenotypes of the offspring that would be produced by mating organisms with various combinations of the two alleles of each gene, assuming independent assortment of the two genes?

11.5 DO THE MENDELIAN RULES OF INHERITANCE APPLY TO ALL TRAITS?

In our discussion thus far, we have assumed that each trait is completely controlled by a single gene, that there are only two possible alleles of each gene, and that one allele is completely dominant to the other. Most traits, however, are influenced in more varied and subtle ways.

In Incomplete Dominance, the Phenotype of Heterozygotes Is Intermediate Between the Phenotypes of the Homozygotes

When one allele is completely dominant over a second allele, heterozygotes with one dominant allele have the same phenotype as homozygotes with two dominant alleles (see Figs. 11-8 and 11-9). However, in some cases the heterozygous phenotype is intermediate between the two homozygous phenotypes, a pattern of inheritance called **incomplete dominance**. For example, consider two alleles that influence coat color in horses: chestnut (C_1) and cremello (C_2). Horses that are homozygous for the C_1 allele have reddish-brown chestnut coats, and horses that are homozygous for the C_2 allele are cremellos, with pale creamy coats. Heterozygous C_1C_2 horses, however, have neither chestnut nor cremello coats. Instead, they have golden palomino coats that are intermediate between the colors of the two homozygous phenotypes. Because palominos are heterozygotes (C_1C_2), they do not breed true; a cross between palominos can produce chestnut, palomino, or cremello foals, with probabilities of one-quarter chestnut (C_1C_1), one-half palomino (C_1C_2), and one-quarter cremello (C_2C_2; **FIG. 11-13**).

A Single Gene May Have Multiple Alleles

Although an individual organism can have at most two different alleles of a gene (one on each of two homologous chromosomes), if we examined all the members of a species, we might find dozens, even hundreds, of different alleles for some genes. Researchers have discovered about 2,000 alleles of a gene known as *CFTR*. This gene has been closely studied because some of its alleles are associated with cystic fibrosis, a genetic disorder (see Section 11.8).

Human blood types are a familiar example of a trait influenced by multiple alleles of a single gene. The blood types A, B, AB, and O arise as a result of three different alleles of a gene (designated *A*, *B*, and *o*). This gene codes for an enzyme that adds sugar molecules to the ends of glycoproteins that protrude from the surfaces of red blood cells. Alleles *A* and *B* code for enzymes that add different sugars to the glycoproteins. Allele *o* codes for a nonfunctional enzyme that doesn't add any sugar molecules.

A person may have one of six genotypes at the blood type locus: *AA, BB, AB, Ao, Bo,* or *oo*. Alleles *A* and *B* are dominant to *o*. Therefore, people with genotypes *AA* or *Ao* make only type A glycoproteins and have type A blood. Those with genotypes *BB* or *Bo* synthesize only type B glycoproteins and have type B blood. Homozygous recessive *oo* individuals lack both types of glycoproteins and have type O blood. People with an *AB* genotype produce both A and B glycoproteins and have type AB blood. When a heterozygote expresses the phenotypes of both of the homozygotes (in this case, both A and B glycoproteins), the pattern of inheritance is called **codominance**.

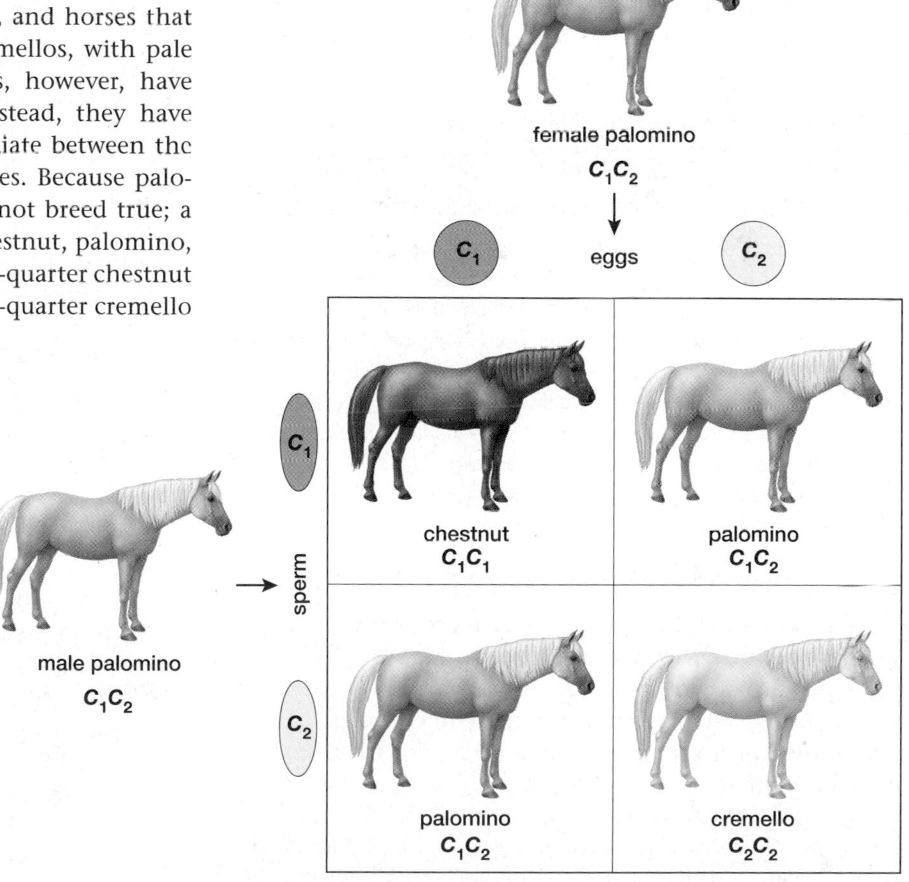

▲ **FIGURE 11-13 Incomplete dominance** The inheritance of palomino coat color in horses is an example of incomplete dominance. Palominos are heterozygotes with one chestnut allele (C_1) and one cremello allele (C_2). Foals produced by breeding palominos may have chestnut, palomino, or cremello coat colors, in the approximate ratio of $\frac{1}{4}$ chestnut: $\frac{1}{2}$ palomino: $\frac{1}{4}$ cremello.

THINK CRITICALLY What is the only breeding combination that will ensure a palomino foal?

▲ **FIGURE 11-14 Skin color in humans** Polygenic inheritance and variable amounts of suntan produce a continuous gradation of skin colors.

Single Genes Typically Have Multiple Effects on Phenotype

Single genes often have multiple phenotypic effects, a phenomenon called **pleiotropy**. For example, in lab mice, a mutation in a single gene results in "nude" mice. In addition to being hairless, these mice lack a thymus gland, have a defective immune system, and do not develop functional mammary glands. In another example of pleiotropy, house cats that are entirely white due to a dominant allele at a particular coat-color locus usually also have blue eyes and are deaf.

CASE STUDY | CONTINUED

Sudden Death on the Court

In people with Marfan syndrome, a single defective fibrillin allele causes increased height, long limbs, large hands and feet, weak walls in the aorta, and in many cases dislocated lenses in one or both eyes—a striking example of pleiotropy in humans. However, the types and severity of symptoms vary, even among family members who carry the same defective fibrillin allele. This variability suggests that environmental factors or the actions of other genes may affect the Marfan phenotype. How are other traits influenced by the environment and the alleles of other genes?

Many Traits Are Influenced by Several Genes

Many traits are influenced by two or more genes, a process called **polygenic inheritance**. As you might imagine, the more genes that contribute to a single trait, the greater the number of possible phenotypes and the finer the gradations among them. For example, human skin color is affected by at least 10 different genes (**FIG. 11-14**). Some of these genes have extremely large effects: People who are homozygous for a recessive allele of one particular gene lack pigmentation in skin, eyes, and hair. Other skin-color genes have smaller effects, with various alleles causing slightly darker or slightly lighter skin. At least 400 genes contribute to human height. Not surprisingly, variation in height is continuous—a person's height may fall anywhere within the range of human height.

The Environment Influences the Expression of Genes

An organism's phenotype is not determined solely by its genotype. The organism's environment also plays a large role, by influencing the effects of genes. Environmental effects on gene action are vividly illustrated by fur color in Siamese cats. All Siamese cats are born with pale fur, but within the first few weeks of life, the ears, nose, paws, and tail turn dark (**FIG. 11-15**). One of a Siamese cat's genes codes for an enzyme that produces dark fur. This enzyme is synthesized in pigment cells everywhere on the cat's body. So why aren't Siamese cats completely black? Because the enzyme that produces dark pigment is inactive at temperatures above about 93°F (34°C). Unborn kittens inside their mother's uterus are warm all over, so newborn Siamese kittens have pale fur on their entire bodies. After they are born, however, the ears, nose, paws, and tail become cooler than the rest of the body, so dark pigment is produced in those areas.

▲ **FIGURE 11-15 Environmental influence on phenotype** The distribution of dark fur in the Siamese cat is an interaction between genotype and environment, producing a particular phenotype. Newborn Siamese kittens have pale fur everywhere on their bodies. In an adult Siamese, the allele for dark fur is expressed only in the cooler areas (nose, ears, paws, and tail).

Have You Ever Wondered ...

Why Dogs Vary So Much in Size?

Dogs come in a wide variety of sizes, from huge Great Danes and Irish wolfhounds to minuscule toy breeds such as Chihuahuas and Pomeranians. Researchers have identified six genes that account for most of the size difference between breeds. Toy breeds are usually homozygous for "small" alleles of most of these genes. Most large breeds are homozygous for the "large" alleles of all six genes. Medium-sized dogs tend to be heterozygous for about half the genes. These patterns suggest that size in dogs is controlled by polygenic inheritance, with incomplete dominance between two or more alleles of each gene.

Although dogs evolved from wolves, wolves do not have any small alleles of the size genes. Why do dogs but not wolves have these alleles? Small alleles probably did arise in both species as a result of random mutations. But when small alleles arose in wolves, they would have been quickly eliminated by natural selection—just imagine the fate of a Chihuahua-sized wolf in the wild. In contrast, small alleles in dogs were preserved by people who liked small dogs and therefore kept only the smallest dogs of each litter as breeding stock. These early dog breeders ensured the preservation of the small alleles of the size genes.

Most environmental influences on phenotype are more complex than the effect of temperature on Siamese cat color. For example, human height is strongly influenced by nutrition, which varies quite a bit among individuals. When this environmental effect is combined with height's complex polygenic inheritance, the result is both continuous variation in height among individuals and change over time in the average height within populations. In many countries, average height has increased by about 4 inches over the last 150 years, as improved nutrition allowed more people to achieve their full genetic potential.

CHECK YOUR LEARNING

Can you . . .

- describe the patterns of inheritance of traits that show incomplete dominance, traits that show codominance, and traits affected by multiple alleles?
- explain how polygenic inheritance and environmental influences combine to produce nearly continuous variation in many traits?

11.6 HOW ARE GENES LOCATED ON THE SAME CHROMOSOME INHERITED?

Each chromosome contains many genes, up to several thousand in a really large chromosome. The presence of multiple genes per chromosome has important consequences for inheritance.

Genes on the Same Chromosome Tend to Be Inherited Together

Because chromosomes assort independently into gametes during meiosis I, genes located on *different chromosomes* also assort independently. However, genes located on the *same chromosome* do not assort independently. Instead, they tend to be inherited together, a phenomenon called **gene linkage**. One of the first pairs of linked genes to be discovered was found in the sweet pea, a different species from Mendel's edible pea. In sweet peas, the gene for flower color (purple versus red) and the gene for pollen grain shape (round versus long) are carried on the same chromosome (**FIG. 11-16**). Thus, the alleles for these genes usually assort together into gametes during meiosis and are inherited together.

Consider a heterozygous sweet pea plant with purple flowers and long pollen. Let's assume the paternal homologue has the dominant purple allele of the flower-color gene and the dominant long allele of the pollen-shape gene (Fig. 11-16, left) and the maternal homologue has the recessive red allele of the flower-color gene and the recessive round allele of the pollen-shape gene (Fig. 11-16, right). Therefore, the gametes

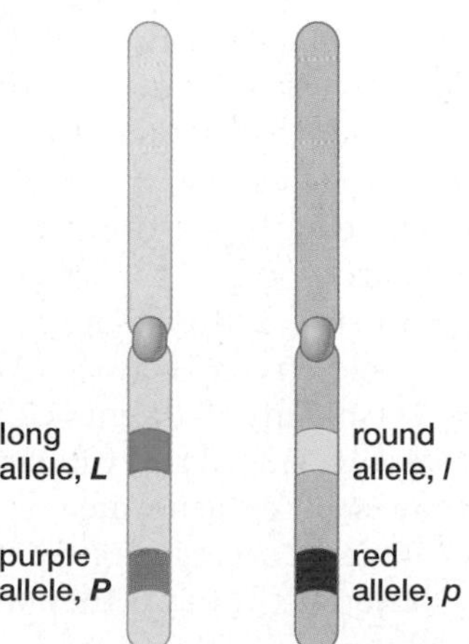

▲ **FIGURE 11-16 Linked genes on homologous chromosomes in the sweet pea** The genes for flower color and pollen shape are close together on the same chromosome, so they tend to be inherited together. The flower-color gene has alleles *P* and *p* (with purple dominant to red); the pollen-shape gene has alleles *L* and *l* (with long dominant to round).

produced by this plant are likely to have either purple and long alleles or red and round alleles. This pattern of inheritance does not conform to the law of independent assortment because the alleles for flower color and pollen shape do not segregate independently of one another, but tend to stay together during meiosis.

Crossing Over Creates New Combinations of Linked Alleles

Although genes on the same chromosome tend to be inherited together, they do not always stay together. If you cross-fertilized two sweet peas with the chromosomes shown in Figure 11-16, you might expect that all of the offspring would have either purple flowers with long pollen grains or red flowers with round pollen grains. (Try working this out with a Punnett square.) In reality, you would usually find a few offspring with purple flowers and round pollen and a few with red flowers and long pollen, as if, sometimes, the genes for flower color and pollen shape became unlinked. How can this happen?

Recall that during prophase I of meiosis, homologous chromosomes sometimes exchange genetic material, a process called crossing over (see Chapter 10, Fig. 10-8). In most chromosomes, at least one exchange between each homologous pair occurs during meiotic cell division. The exchange of corresponding segments of DNA during crossing over produces **genetic recombination**: new combinations of alleles on both homologous chromosomes. When the homologues separate at anaphase I, the haploid daughter cells will receive chromosomes with sets of alleles that are different from those of the parent cell.

The Strength of Linkage Between Two Genes Depends on the Distance Between Them

The likelihood that two genes on a chromosome will be inherited together depends on how far apart they are. To see why, imagine two long strings, each with a red stripe at one end, a blue stripe very close to the red one, and a yellow stripe at the opposite end. If you throw the strings on the floor so that one lands on top of the other, the strings will almost always cross between the far-apart blue and yellow stripes, but will very seldom cross between the close-together blue and red stripes. Similarly, two genes close together on a chromosome are strongly linked and will rarely be separated by a crossover. However, if two genes are very far apart, crossing over between the genes occurs so often that they seem to be independently assorted, just as if they were on different chromosomes.

When Gregor Mendel discovered independent assortment, he was not only clever and careful, he was also lucky. All the traits that he studied were controlled by genes on only four different chromosomes. He observed independent assortment because the genes that were on the same chromosomes happened to be far apart.

CHECK YOUR LEARNING

Can you . . .

- describe how the patterns of inheritance differ between traits controlled by genes on a single chromosome and traits controlled by genes on different chromosomes?

11.7 HOW ARE SEX AND SEX-LINKED TRAITS INHERITED?

In many animals, an individual's sex is determined by its **sex chromosomes**. In mammals, females have two sex chromosomes, both of which are **X chromosomes**, whereas males have one X chromosome and one **Y chromosome**. Although X and Y chromosomes differ considerably in size and genetic composition, they act as homologues, pairing up during prophase of meiosis I and separating during anaphase I. (However, unlike other homologues, they do not cross over during meiosis.) During sperm formation in males (XY), the sex chromosomes segregate, and each sperm receives either an X or a Y chromosome. The sex chromosomes also segregate during egg formation in females (XX), and every egg receives one X chromosome. Thus, a male offspring is produced if an egg is fertilized by a Y-bearing sperm, and a female offspring is produced if an egg is fertilized by an X-bearing sperm (**FIG. 11-17**).

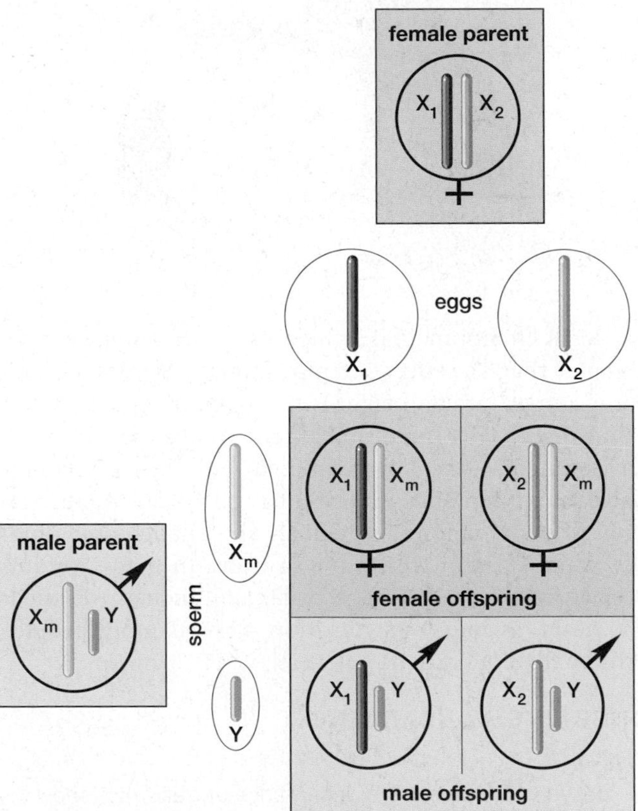

▲ **FIGURE 11-17 Sex determination in mammals** Male offspring receive their Y chromosome from their father; female offspring receive the father's X chromosome (labeled X_m). Both male and female offspring receive an X chromosome (either X_1 or X_2) from their mother.

Sex-Linked Genes Are Found Only on the X or Only on the Y Chromosome

Genes that are located only on sex chromosomes are referred to as **sex-linked**. In mammals, the Y chromosome carries relatively few genes. The human Y chromosome contains several dozen genes, many of which play a role in male reproduction. The most well-known Y-linked gene is the sex-determining gene, called *SRY*. During embryonic life, the action of *SRY* sets in motion the entire male developmental pathway.

In contrast to the small Y chromosome, the human X chromosome contains more than 1,000 genes, most of which have no counterpart on the Y chromosome. Most of the genes on the X chromosome affect traits that are important in both sexes, such as color vision, blood clotting, and the presence of certain structural proteins in muscles. Because females have two X chromosomes, they can be either homozygous or heterozygous for genes on the X chromosome, and dominant versus recessive relationships among alleles will be expressed. Males, in contrast, fully express all the alleles they have on their single X chromosome, regardless of whether those alleles would be dominant or recessive in females.

Inheritance of Sex-Linked Traits Differs for Males and Females

Let's look at an example sex-linked trait: red-green color vision deficiency (**FIG. 11-18**). Color vision deficiency is caused by certain defects in either of two genes located on the X chromosome. The normal, dominant alleles of these genes (we will call them both *N*) encode proteins that allow one set of color-vision cells in the eye, called cones, to be most sensitive to red light and another set to be most sensitive to green light. There are several defective recessive alleles of these genes (we will call them all *n*). Certain extremely defective alleles make an affected person truly red-green color-blind—unable to distinguish red from green. For people with the much more common condition of color vision deficiency, however, fire engines still look red and grass still looks green, but many "reddish" or "greenish" colors cannot be distinguished from one another (**FIG. 11-18a**).

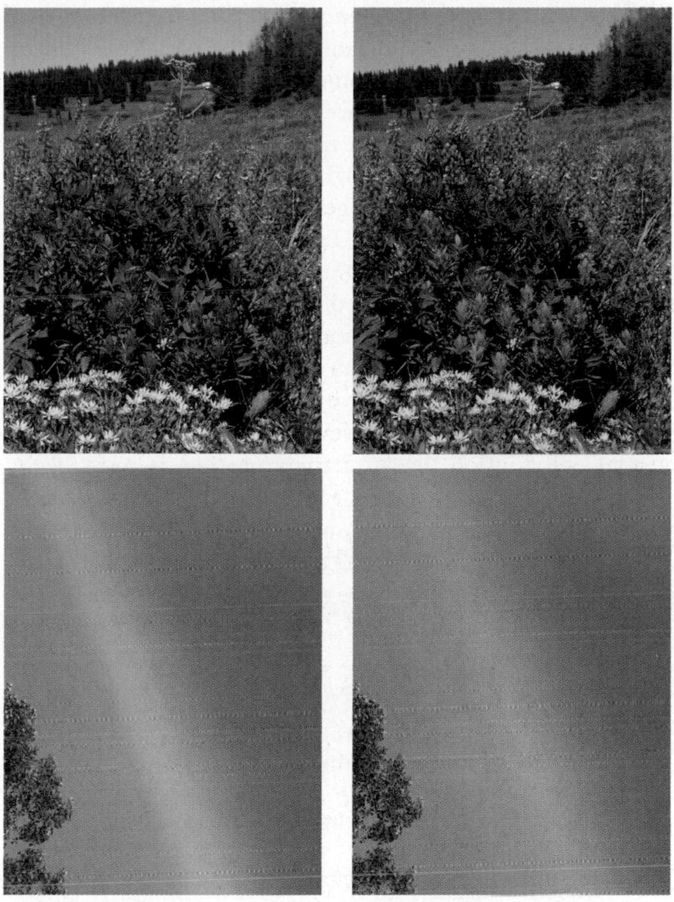

(a) Normal color vision (left); simulation of red-green color deficiency (right)

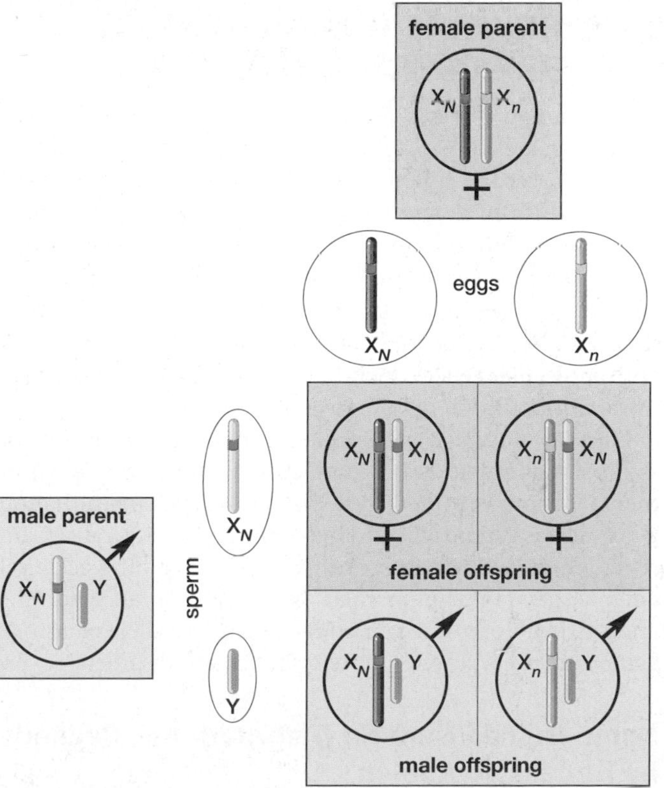

(b) Expected children of a man with normal color vision (X_NY), and a heterozygous woman (X_NX_n)

▶ **FIGURE 11-18 Sex-linked inheritance of red-green color deficiency (a)** These photographs show people with normal color vision what the world looks like through the eyes of a person with red-green color deficiency. **(b)** A Punnett square shows the inheritance of color deficiency from a heterozygous woman (X_NX_n) to her sons.

THINK CRITICALLY If a color-deficient woman and a man with normal color vision have a daughter, what are the chances that she will be color-deficient? If they have a son, what are the chances that he will be color-deficient? Explain your answers.

How is color-vision deficiency inherited? A man can have the genotype X_NY or X_nY, meaning that he has a color-vision allele *N* or *n* on his X chromosome and no color-vision gene on his Y chromosome. He will have normal color vision if his X chromosome bears the *N* allele or be color-deficient if it bears the *n* allele, as it does in the roughly 7% of men who have deficient color vision. A woman's genotype may be X_NX_N, X_NX_n, or X_nX_n. Women with X_NX_N or X_NX_n genotypes will have normal color vision. Only women with X_nX_n genotypes will be color-deficient; about 0.5% of women have this genotype. Most women (about 93%) are homozygous dominant, X_NX_N.

A color-deficient man (X_nY) can pass his defective *n* allele only to his daughters, because only his daughters inherit his X chromosome. Usually, however, his daughters will have normal color vision, because their mother is most likely homozygous dominant (X_NX_N) and will therefore pass a normal *N* allele to her children. Among the children of a father with normal color vision (X_NY) and a mother who is heterozygous (X_NX_n), half the sons are expected to be color-deficient (X_nY), but none of the daughters will be (**FIG. 11-18b**).

CHECK YOUR LEARNING

Can you . . .

- explain why the father's sperm determine the sex of offspring in mammals?
- explain why most sex-linked traits are controlled by genes on the X chromosome?
- describe the pattern of inheritance of sex-linked traits?

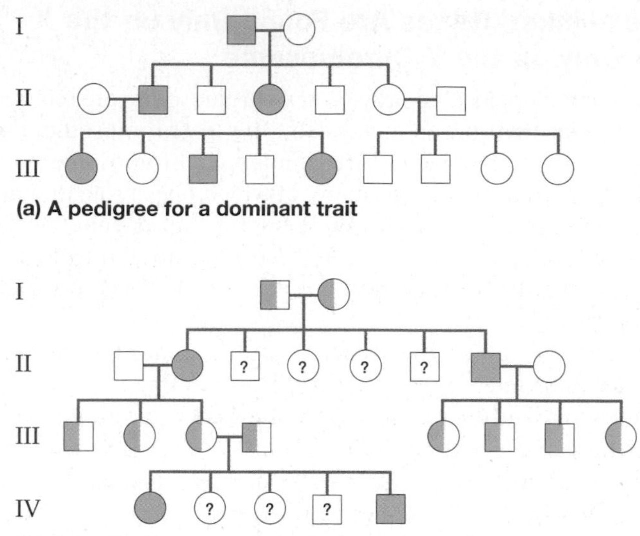

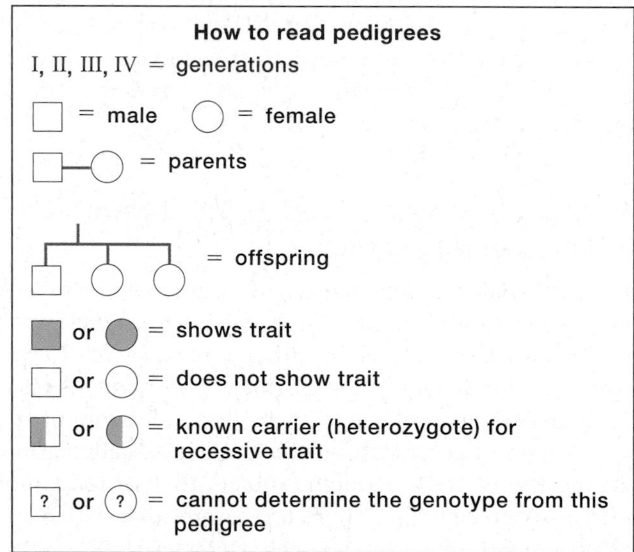

▲ **FIGURE 11-19 Family pedigrees (a)** A pedigree for a dominant trait. Note that any offspring showing a dominant trait must have at least one parent with the trait. **(b)** A pedigree for a recessive trait. Any individual showing a recessive trait must be homozygous recessive. If that person's parents did not show the trait, then both parents must be heterozygotes (carriers). Note that the genotype cannot be determined for some offspring, who may be either carriers or homozygous dominants.

11.8 HOW ARE HUMAN GENETIC DISORDERS INHERITED?

Many human diseases are influenced by genetics. Experimental crosses of people are of course out of the question, so to discover the genetic basis of disease, geneticists search medical, historical, and family records to study past crosses. Records extending across several generations can be arranged in the form of family **pedigrees**, diagrams that show the genetic relationships among a set of related individuals (**FIG. 11-19**). Careful analysis of human pedigrees, combined with molecular genetic technology, has allowed great strides toward understanding human genetic diseases.

Just as multiple genes influence traits such as height and skin color, multiple genes, interacting with complex environmental factors, may predispose people to develop health problems such as Parkinson's and Alzheimer's diseases, cancer, and schizophrenia. However, some disorders—including sickle-cell anemia, hemophilia, muscular dystrophy, Marfan syndrome, and cystic fibrosis—are caused by a defective allele of a single gene. Here, we will focus on such single-gene disorders.

Some Human Genetic Disorders Are Caused by Recessive Alleles

Some genetic disorders occur only in people who inherit two defective copies of a gene that normally codes for an essential protein. In these cases, a single normal allele generates a sufficient amount of functional protein, so a person who inherits one defective allele and one normal allele has the same phenotype as a person with two normal alleles. That is, the defective allele is recessive, and the normal allele is dominant.

A **carrier** for a genetic disorder is a person who is heterozygous at a disorder locus, with one dominant normal allele and one recessive defective allele. Carriers are phenotypically healthy but can pass on defective alleles to their offspring. In all likelihood, everyone carries some recessive alleles that would cause serious genetic disorders in homozygotes. For each locus at which these alleles occur, each of a carrier's children has a 50:50 chance of inheriting the defective allele. Fortunately, passing along a defective allele is usually harmless, because an unrelated man and woman typically carry defective alleles of different genes; it is unlikely that their children will inherit two defective alleles of the same gene. Nonetheless, some people do inherit two defective alleles of the same gene. The chances of this happening increase if the parents are related (for example, first cousins).

Sickle-Cell Anemia Is Caused by a Defective Allele for Hemoglobin Synthesis

Sickle-cell anemia is an inherited disorder that, as its name suggests, results in anemia—a deficiency of red blood cells and hemoglobin, a protein found in red blood cells that transports oxygen in the blood. Sickle-cell anemia is caused by an allele that encodes defective hemoglobin; the disorder appears in people who inherit two copies of the defective allele.

When people with sickle-cell anemia exercise or move to a high altitude, oxygen concentrations in their blood drop and defective hemoglobin proteins inside their red blood cells stick together. The resulting clumps of hemoglobin force red blood cells out of their usual flexible, disk shapes (**FIG. 11-20a**) into long, stiff sickle shapes (**FIG. 11-20b**). The sickled cells are fragile and easily damaged, so they are destroyed before their usual life span is completed, resulting in anemia. Sickled cells flow through capillaries less freely than do normal red blood cells, so they tend to jam up, causing blood clots. Tissues downstream of a clot do not receive enough oxygen, and a clot in a vessel leading to the brain can result in a stroke.

People who are heterozygous for the sickle-cell allele produce both normal and abnormal hemoglobin, but they have very few sickled red blood cells and seldom show any symptoms. However, during exceptionally strenuous exercise, some heterozygotes may experience life-threatening complications.

Some Human Genetic Disorders Are Caused by Dominant Alleles

Some serious genetic disorders are caused by dominant alleles. Because these disorders are present in everyone who carries a copy of a defective dominant allele, anyone who inherits a dominant genetic disorder must have at least one parent with the disease. The only exceptions are the rare cases in which a dominant allele that causes a disorder is the result of a mutation in the egg or sperm of a parent who is otherwise unaffected.

How can a defective allele be dominant to the normal, functional allele? Some defective dominant alleles encode an abnormal protein that interferes with the function of the normal one. Other dominant alleles may encode a protein that is overactive, performing its function at inappropriate times and places in the body. Still other dominant alleles may encode proteins that carry out new, toxic reactions.

Huntington Disease Is Caused by a Defective Protein That Kills Cells in Specific Brain Regions

One dominant disorder caused by a protein with toxic effects is **Huntington disease**. This disorder causes a slow, progressive deterioration of parts of the brain, resulting in loss of coordination, flailing movements, personality disturbances, and eventual death. The symptoms of Huntington disease typically do not appear until 30 to 50 years of age. Therefore, before they experience their first symptoms, many people who have the dominant Huntington allele have already become parents and passed the allele to some of their children. Geneticists isolated the Huntington gene in 1993 and, a few years later, identified the gene's protein product. The normal protein affects gene transcription, cytoskeleton function, and the movement of organelles within brain cells. The mutant protein is cut up into toxic fragments inside cells, ultimately killing them.

Some Human Genetic Disorders Are Sex-Linked

As you learned in Section 11.7, X-linked alleles have a distinctive inheritance pattern in which phenotypes caused

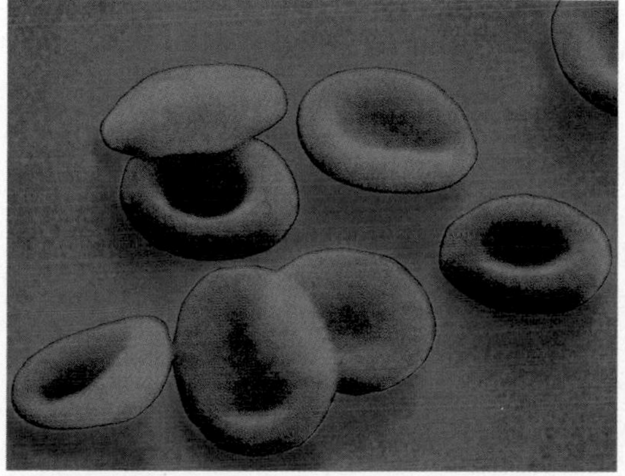

(a) Normal red blood cells

(b) Sickled red blood cells

FIGURE 11-20 Sickle-cell anemia (a) Normal red blood cells are disk shaped with indented centers. **(b)** When blood oxygen is low, the red blood cells in a person with sickle-cell anemia become long, slender, and curved, resembling a sickle.

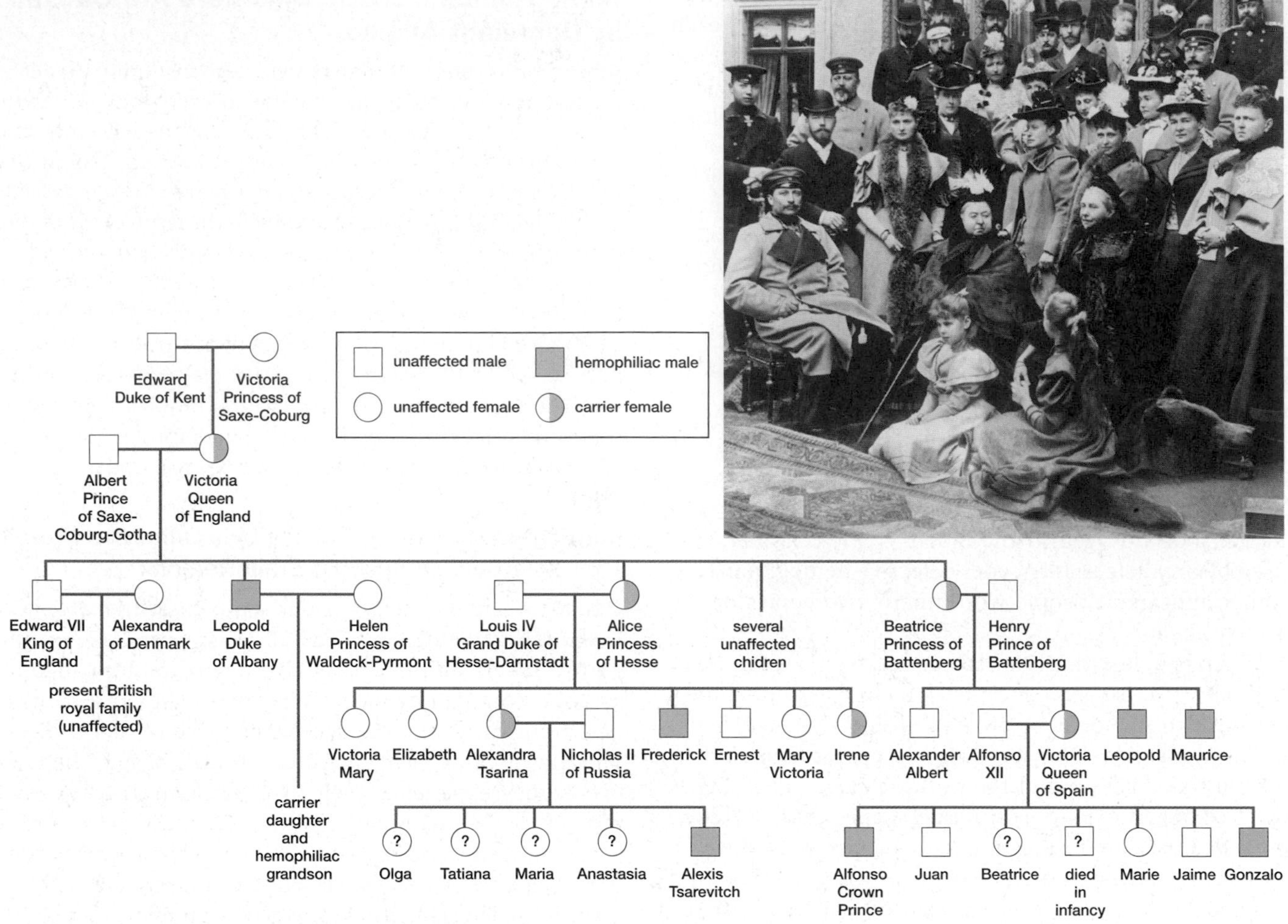

▲ **FIGURE 11-21 Hemophilia among the royal families of Europe** A famous genetic pedigree shows the transmission of sex-linked hemophilia from Queen Victoria of England (seated center front, with cane, in 1885) to her offspring and eventually to virtually every royal house in Europe, because of the extensive intermarriage of her children to the royalty of other European nations. Because Victoria's ancestors were free of hemophilia, the hemophilia allele must have arisen as a mutation either in Victoria herself or in one of her parents (or as a result of marital infidelity).

THINK CRITICALLY Why is it not possible that a mutation in Victoria's husband, Albert, was the original source of hemophilia in this family pedigree?

by recessive alleles appear far more frequently in males and typically skip generations. Most commonly, an affected male passes the trait to a phenotypically normal, carrier daughter, who in turn bears some affected sons. The most familiar genetic disorders due to recessive alleles of X-chromosome genes are red-green color-vision deficiency, hemophilia, and muscular dystrophy (see "Health Watch: The Genetics of Muscular Dystrophy").

Hemophilia is caused by a recessive allele on the X chromosome that results in a deficiency in one of the proteins needed for blood clotting. People with hemophilia bruise easily and may bleed extensively from minor injuries. They often have anemia due to blood loss. Nevertheless, even before modern treatment with clotting factors, some hemophiliac males survived to pass on their defective allele to their daughters, who in turn could pass it to their sons (**FIG. 11-21**).

CHECK YOUR LEARNING

Can you . . .

- use pedigrees to determine the pattern of inheritance of a trait?
- describe why some genetic disorders might be dominant or recessive and give examples of each?
- explain why sex-linked recessive disorders mainly affect males and tend to skip generations?

The Genetics of Muscular Dystrophy

When you lift something heavy, do some pull-ups, or ride a bike up a steep hill, you put considerable strain on your muscles. How do your muscles withstand the stress? Contracting muscles remain intact thanks to *dystrophin*, a very long protein that binds muscle cells together.

Unfortunately, about 1 in 3,500 boys makes faulty dystrophin proteins and suffers from **muscular dystrophy**, which literally means "degeneration of the muscles." The lack of functional dystrophin means that ordinary muscle contraction tears the muscle cells, which die and are replaced by fat and connective tissue (**FIG. E11-1**). Boys with Duchenne muscular dystrophy, the most severe form of the disorder, lose the ability to walk by age 7 or 8. Death usually occurs in the early 20s from heart and respiratory problems.

Girls almost never have Duchenne muscular dystrophy, because the dystrophin gene is on the X chromosome, and muscular dystrophy alleles are recessive. Although a boy will have muscular dystrophy if he has a defective dystrophin allele on his single X chromosome, a girl, with two X chromosomes, would have the disorder only if she inherited two defective copies. Such inheritance virtually never happens. For it to occur, a girl would have to inherit one of her defective dystrophin alleles from her father, but males with the allele almost never become fathers because they die young from Duchenne muscular dystrophy.

There is as yet no medical treatment that cures Duchenne muscular dystrophy, but treatments are available that slow muscle degeneration, prolong life, and make affected boys more comfortable. Improved treatments may come from drugs, currently in clinical trials, that stimulate synthesis of utrophin, a naturally occurring muscle protein that may be able to partially substitute for dystrophin. Researchers have high hopes for treatments based on gene editing, which has become easier and more accurate with the advent of new molecular tools, especially the one known as CRISPR-Cas9. In principle, CRISPR-Cas9 could one day be used to repair defective dystrophin genes in enough of a patient's cells to effectively cure the disorder.

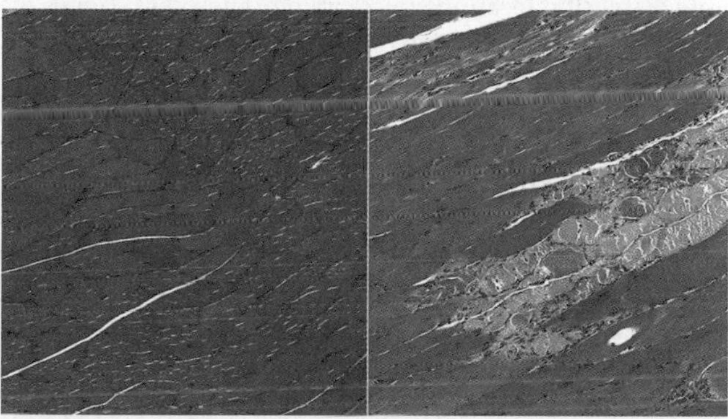

▲ **FIGURE E11-1 The effects of muscular dystrophy** The micrograph on the left shows a normal muscle, with little space between the cells. A dystrophic muscle (right) has fewer and more irregular muscle cells, with spaces between the cells filled with fat and connective tissue.

THINK CRITICALLY A mother of a young boy is devastated to find that her son has Duchenne muscular dystrophy. She takes a DNA test and discovers that she is a carrier for a defective dystrophin allele. If she decides to have another child, what is the likelihood that the second child will have the disorder? The woman has a sister. What is the likelihood that she is also a carrier?

CASE STUDY REVISITED

Sudden Death on the Court

Although Marfan syndrome caused Flo Hyman's death, it need not be fatal. In 2014, Baylor University basketball star Isaiah Austin (**FIG. 11-22**) decided to play professional ball after his sophomore year. Luckily for him, the NBA screens all players for health problems before they are eligible for the draft. NBA physicians diagnosed Austin with Marfan syndrome and found that he has an enlarged aorta with weak walls. Had he continued to play college basketball instead of trying to turn pro, he may well have suffered Flo Hyman's fate.

The NBA's physicians advised Austin to permanently abstain from competitive sports, because exercise could put too much stress on his aorta and cause it to rupture. But 3 years after his declaration for the draft, Austin announced that his cardiologist had cleared him to play. Shortly afterward, Austin returned to the court, signing with a professional team in Serbia. Isaiah Austin says that playing professionally is a risk worth taking, because "basketball brings me so much joy."

CONSIDER THIS It may soon be technically possible to repair the defective allele in the eggs or sperm of a prospective parent with Marfan syndrome, thereby ensuring that offspring do not inherit the condition. Do you think that this kind of genetic modification of sperm and eggs (or embryos) should be permitted? Why or why not?

◀ **FIGURE 11-22 Isaiah Austin dunks**

CHAPTER REVIEW

Go to **MasteringBiology** to access the Pearson eText, vocabulary review, practice quizzes, activities, videos, current events, and more.

*Answers to **Think Critically** and **Thinking Through the Concepts** questions can be found in the **Answers** section at the back of the book.*

Summary of Key Concepts

11.1 What Is the Physical Basis of Inheritance?

The units of inheritance are genes, which are segments of DNA found at specific locations (loci) on chromosomes. Genes may exist in two or more alternative forms, called alleles. When both homologous chromosomes carry the same allele at a given locus, the organism is homozygous for that gene. When two homologous chromosomes have different alleles at a given locus, the organism is heterozygous for that gene.

11.2 How Were the Principles of Inheritance Discovered?

Gregor Mendel deduced many principles of inheritance in the mid-1800s, before the discovery of DNA, genes, chromosomes, or meiosis. He did this by choosing an appropriate experimental subject, designing his experiments carefully, following progeny for several generations, and analyzing his data statistically.

11.3 How Are Single Traits Inherited?

A trait is an observable or measurable feature of an organism's phenotype, such as flower color or blood type. Each parent provides its offspring with one allele of every gene, so the offspring inherits a pair of alleles for every gene. The combination of alleles in the offspring determines its phenotype. Dominant alleles mask the expression of recessive alleles. The masking of recessive alleles can result in organisms with the same phenotype but different genotypes. Organisms with two dominant alleles (homozygous dominant) have the same phenotype as do organisms with one dominant and one recessive allele (heterozygous). Because each allele segregates randomly during meiosis, we can predict the relative proportions of offspring with a particular trait, using Punnett squares or the rules of probability.

11.4 How Are Multiple Traits Inherited?

If the genes for two traits are located on separate chromosomes, their alleles assort independently of one another into the egg or sperm; that is, the distribution of alleles of one gene into the gametes does not affect the distribution of the alleles of the other gene. Thus, breeding two organisms that are heterozygous at two loci on separate chromosomes produces offspring with nine different genotypes. For typical dominant and recessive alleles, the offspring will display only four different phenotypes.

11.5 Do the Mendelian Rules of Inheritance Apply to All Traits?

Not all inheritance follows the simple dominant-recessive pattern. In incomplete dominance, heterozygotes have a phenotype that is intermediate between the two homozygous phenotypes. If we examine the genes of many members of a given species, we find that many genes have more than two alleles. Codominance results when two alleles of a single gene independently contribute to the observed phenotype. Pleiotropy occurs when a single gene has effects on several, seemingly unrelated, aspects of an organism's phenotype. In polygenic inheritance, several different genes contribute to the phenotype. The environment influences the phenotypic expression of virtually all traits.

11.6 How Are Genes Located on the Same Chromosome Inherited?

Genes on the same chromosome tend to be inherited together. However, crossing over during meiosis results in some recombination between homologues, which may separate alleles that were formerly inherited together. Crossing over between gene loci will occur more often the farther apart on a chromosome the genes are located.

11.7 How Are Sex and Sex-Linked Traits Inherited?

In many animals, sex is determined by sex chromosomes, often designated X and Y. In mammals, females have two X chromosomes; males have one X and one Y chromosome. Male sperm contain either an X or a Y chromosome, whereas a female's egg cells always have an X chromosome. Therefore, sex is determined by the sex chromosome in the sperm that fertilizes an egg.

Sex-linked genes are found on the X or Y chromosome. In mammals, the Y chromosome has many fewer genes than the X chromosome, so most sex-linked genes are found on the X chromosome. Because males have only one copy of X chromosome genes, recessive alleles of genes on the X chromosome are more likely to be phenotypically expressed in males.

11.8 How Are Human Genetic Disorders Inherited?

Molecular genetic techniques and analysis of family pedigrees are used to determine the mode of inheritance of human traits. Some genetic disorders are inherited as recessive traits; therefore, only homozygous recessive persons show symptoms of the disease. Heterozygotes are called carriers; they carry the recessive allele but do not express the trait. Other disorders are inherited as simple dominant traits. In such cases, only one copy of the dominant allele is needed to cause full disease symptoms. Some human genetic disorders are sex-linked.

Thinking Through the Concepts

Bloom's: Remembering, Understanding

Multiple Choice

1. The physical position of a gene on a chromosome is its __________; slightly different forms of a gene are called __________.
   a. locus; alleles
   b. locus; polygenic
   c. chiasma; alleles
   d. trait; hybrids
2. If an organism has two different alleles (call the alleles *A* and *a*) of a gene,
   a. its phenotype will be the same as an organism with two identical alleles of this gene.
   b. all of its gametes will contain both the *A* allele and the *a* allele.
   c. it is homozygous for that gene.
   d. it is heterozygous for that gene.
3. Independent assortment means that
   a. two genes on the same chromosome tend to be inherited together.
   b. alleles are sorted into gametes after meiosis concludes.
   c. the particular allele of a gene does not determine the allele of a second gene in the same gamete.
   d. homologous chromosomes do not separate during meiosis.
4. If a gene is located on the X chromosome of a mammal, it is
   a. expressed only in females.
   b. expressed only in males.
   c. sex-linked, with females more likely to show recessive traits.
   d. sex-linked, with males more likely to show recessive traits.
5. A test cross is used to determine
   a. the genotype of an organism with a phenotypically dominant trait.
   b. the genotype of an organism with a phenotypically recessive trait.
   c. the genotype of an organism showing pleiotropic effects of a gene.
   d. if a trait is inherited polygenically.

Fill-in-the-Blank

1. An organism is described as *Rr*, with red coloring. *Rr* is the organism's __________, while red color is its __________. This organism would be __________ (homozygous/heterozygous) for this color gene.
2. The inheritance of multiple traits depends on the locations of the genes that control the traits. If the genes are on different chromosomes, then the traits are inherited __________ (as a group/independently). If the genes are located close together on a single chromosome, then the traits tend to be inherited __________ (as a group/independently). Genes on the same chromosome are said to be __________.
3. In mammals, males have __________ (XX/XY/YY) sex chromosomes and females have __________ (XX/XY/YY) sex chromosomes. The sex of offspring depends on which chromosome is present in the __________ (sperm/egg).
4. Genes that are present on one sex chromosome but not the other are called __________.
5. When the phenotype of heterozygotes is intermediate between the phenotypes of the two homozygotes, this pattern of inheritance is called __________. When heterozygotes express phenotypes of both homozygotes (not intermediate, but showing both traits), this is called __________. In __________, many genes, usually with similar effects on phenotype, control the inheritance of a trait.

Review Questions

1. Define the following terms: *gene, allele, dominant, recessive, true-breeding, homozygous, heterozygous, cross-fertilization,* and *self-fertilization.*
2. Explain why genes located on the same chromosome are said to be linked. Why do alleles of linked genes sometimes separate during meiosis?
3. Define *polygenic inheritance*. How does polygenic inheritance sometimes allow parents to produce offspring that are notably different in skin color than either parent?
4. What is sex linkage? In mammals, which sex would be most likely to show recessive sex-linked traits?
5. What is the difference between a phenotype and a genotype? Does knowledge of an organism's phenotype always allow you to determine the genotype? What type of experiment would you perform to determine the genotype of a phenotypically dominant individual?
6. In the pedigree of part (a) of Figure 11-19, do you think that the individuals showing the trait are homozygous or heterozygous? How can you tell from the pedigree?

Applying the Concepts

Bloom's: Applying, Analyzing, Evaluating

1. Sometimes the term *gene* is used rather casually. Compare the terms *allele* and *gene*.
2. Imagine an alternate universe in which all the genes in all species have only two alleles, one dominant and one recessive. Would every trait have only two phenotypes? Would all members of a species that are dominant for a given gene have exactly the same phenotype? Explain your reasoning.

Turn the page for more questions.

Genetics Problems

1. In certain cattle, hair color can be red (homozygous R_1R_1), white (homozygous R_2R_2), or roan (a mixture of red and white hairs, heterozygous R_1R_2).
   a. When a red bull is mated to a white cow, what genotypes and phenotypes of offspring could be obtained?
   b. If one of the offspring bulls in part (a) were mated to a white cow, what genotypes and phenotypes of offspring could be produced? In what proportion?
2. In the edible pea, tall (*T*) is dominant to short (*t*), and green pods (*G*) are dominant to yellow pods (*g*). List the types of gametes and offspring that would be produced in the following crosses, assuming the genes for these traits are located on separate chromosomes:
   a. *TtGg* × *TtGg*
   b. *TtGg* × *TTGG*
   c. *TtGg* × *Ttgg*
3. In tomatoes, round fruit (*R*) is dominant to long fruit (*r*), and smooth skin (*S*) is dominant to fuzzy skin (*s*). A true-breeding round, smooth tomato (*RRSS*) was crossbred with a true-breeding long, fuzzy tomato (*rrss*). All the F_1 offspring were round and smooth (*RrSs*). When these F_1 plants were bred, the following F_2 generation was obtained:
   Round, smooth: 43
   Long, fuzzy: 13
   Are the genes for skin texture and fruit shape likely to be on the same chromosome or on different chromosomes? Explain your answer.
4. In the tomatoes of Problem 3, an F_1 offspring (*RrSs*) was crossed with a homozygous recessive (*rrss*). The following offspring were obtained:
   Round, smooth: 583 Long, fuzzy: 602
   Round, fuzzy: 21 Long, smooth: 16
   What is the most likely explanation for this distribution of phenotypes?
5. In humans, hair color is controlled in part by two interacting genes located on separate chromosomes. The same pigment, melanin, is present in both brown-haired and blond-haired people, but brown hair has much more of it. Brown hair (*B*) is dominant to blond (*b*). Whether any melanin can be synthesized depends on another gene. The dominant form of this second gene (*M*) allows melanin synthesis; the recessive form (*m*) prevents melanin synthesis. Homozygous recessives (*mm*) have unpigmented hair. What will be the expected proportions of phenotypes in the children of the following parents?
   a. *BBMM* × *BbMm*
   b. *BbMm* × *BbMm*
   c. *BbMm* × *bbmm*

12 DNA: The Molecule of Heredity

What turns a bull into an incredible hulk? A tiny change in its DNA makes all the difference.

CASE STUDY

Muscles, Mutations, and Myostatin

NO, THIS BULL hasn't been pumping iron—he's a Belgian Blue. All Belgian Blues have big, bulging muscles that make them look like bodybuilders compared with ordinary bulls. Belgian Blues have more, and larger, muscle cells than do ordinary cattle because they carry abnormal versions of the gene that encodes a protein called myostatin. A Belgian Blue has a change, or mutation, in the DNA of its myostatin gene, making it slightly different from the DNA of the myostatin gene in most other cattle. As a result, Belgian Blues produce defective myostatin.

Why does defective myostatin result in abnormally large muscles? Because myostatin's normal function is to put the brakes on muscle development. When a mammal develops normally, its cells divide many times but eventually stop dividing and become specialized for a specific function. For example, some cells in a developing mammal are destined to form muscles. Normal myostatin slows down—and eventually stops—the multiplication of these pre-muscle cells. Myostatin also regulates the ultimate size of muscle cells. Because Belgian Blues lack functional myostatin, their pre-muscle cells multiply more than normal, and the cells become extra-large as they develop into muscle. The result: remarkably ripped cattle.

How does DNA encode the instructions for traits such as muscle size? How are these instructions passed from generation to generation? And why do the instructions sometimes change?

AT A GLANCE

12.1 What Is the Structure of DNA?

12.2 How Does DNA Encode Genetic Information?

12.3 How Does DNA Replication Ensure Genetic Constancy During Cell Division?

12.4 What Are Mutations, and How Do They Occur?

12.1 WHAT IS THE STRUCTURE OF DNA?

By the early 1950s, scientists had produced solid evidence that DNA is the molecule of inheritance (see "Doing Science: Discovering the Hereditary Molecule" on page 180). But knowing that genes are made of DNA does not answer other critical questions about inheritance: How does DNA encode genetic information? How is DNA replicated so that a cell can pass its hereditary information to its daughter cells? The secrets of DNA function and replication are found in the three-dimensional structure of the DNA molecule.

DNA Is Composed of Four Nucleotides

DNA consists of **nucleotides**. Each nucleotide has three parts: a phosphate group, a sugar called deoxyribose, and one of four nitrogen-containing **bases**. The four bases are **adenine (A)**, **guanine (G)**, **thymine (T)**, and **cytosine (C)** (**FIG. 12-1**). Adenine and guanine both consist of fused five- and six-member rings of carbon and nitrogen atoms, with different functional groups attached to the six-member ring. Thymine and cytosine consist of a single six-member ring of carbon and nitrogen atoms, again with different functional groups attached to the ring.

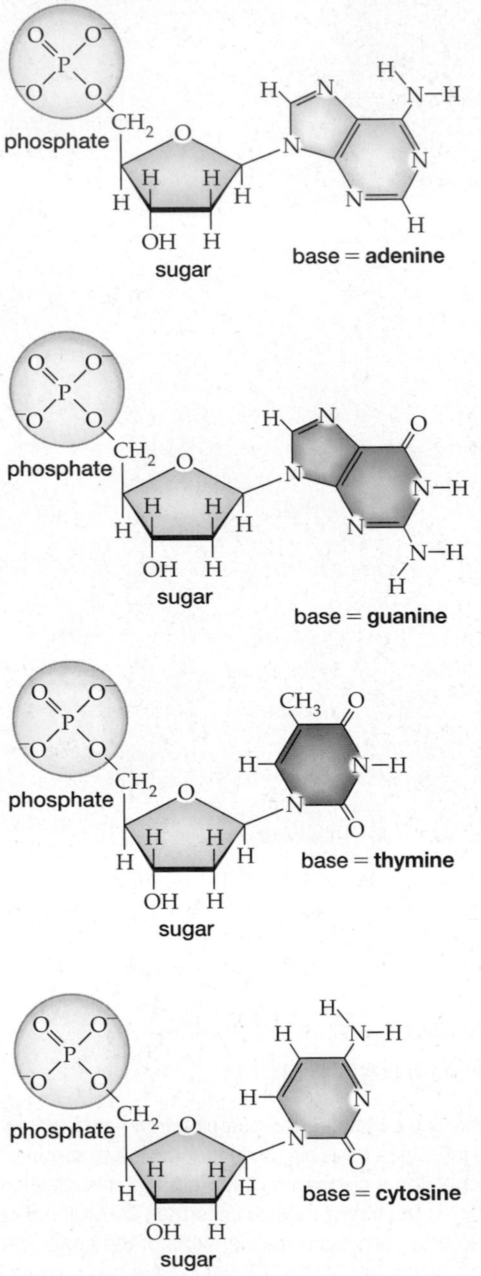

▲ **FIGURE 12-1 DNA nucleotides**

DNA Is a Double Helix of Two Nucleotide Strands

In the late 1940s, several scientists began to investigate the structure of DNA. British researchers Rosalind Franklin and Maurice Wilkins used a technique called X-ray diffraction to study the DNA molecule (**FIG. 12-2**). Although X-ray diffraction patterns do not provide a direct picture of molecules, they provide considerable information about molecular shape and structure. From their experiments, Franklin and Wilkins made several deductions. First, a molecule of DNA is long and thin, with a uniform width of 2 nanometers (2 billionths of a meter). Second, DNA is helical, twisted like a spiral staircase. Third, DNA is a double helix; that is, two strands coil around one another. Fourth, DNA consists of repeating subunits. And fifth, phosphates are on the outside of the helix.

Given enough time, Franklin and Wilkins would probably have deduced the correct structure of DNA. However, they were scooped by two young scientists, James Watson and Francis Crick (**FIG. 12-3**). Wilkins had shared the X-ray diffraction data with them, so they knew the general size and shape of a DNA molecule. Based in part on their intuition that "important biological objects come in pairs," as Watson put it, Watson and Crick offered a detailed molecular model for the structure of DNA.

Watson and Crick proposed that a single strand of DNA is a polymer consisting of many nucleotide subunits.

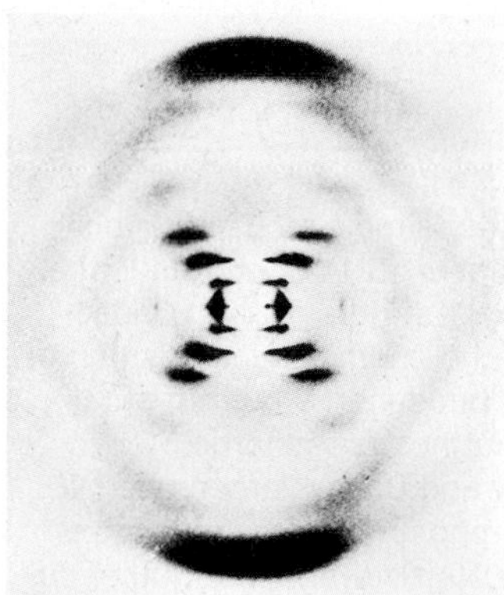

(a) X-ray diffraction image of DNA

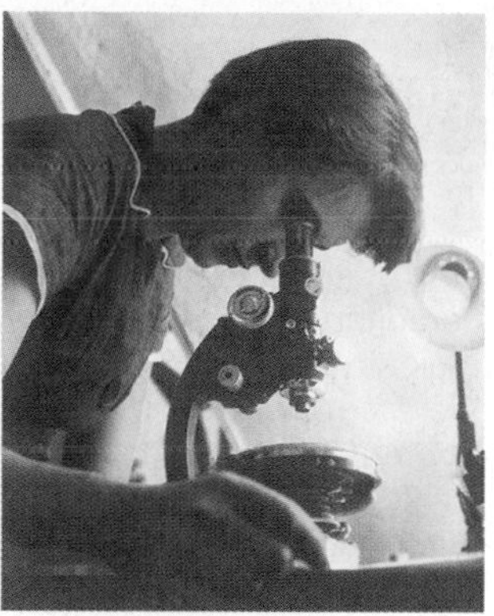

(b) Rosalind Franklin at work

▲ **FIGURE 12-2 Laying the groundwork for the discovery of DNA's structure (a)** In an X-ray diffraction image, a crossing pattern of dark spots is characteristic of helical molecules such as DNA. Measurements of various aspects of the pattern indicate the dimensions of the DNA helix. **(b)** The X-ray image was produced by Rosalind Franklin, working in collaboration with Maurice Wilkins.

▲ **FIGURE 12-3 James Watson (left) and Francis Crick with their model of DNA**

The phosphate group of one nucleotide is bonded to the sugar of the next nucleotide in the strand, thus producing a **sugar-phosphate backbone** of alternating, covalently bonded sugars and phosphates (**FIG. 12-4**). The bases of the nucleotides are attached to this sugar-phosphate backbone.

All of the nucleotides in a single DNA strand are oriented in the same direction. Therefore, the two ends of a DNA strand differ; one end has a "free" or unbonded sugar, and the other end has a "free" or unbonded phosphate (**FIG. 12-4a**). Picture a long line of cars stopped on a crowded one-way street at

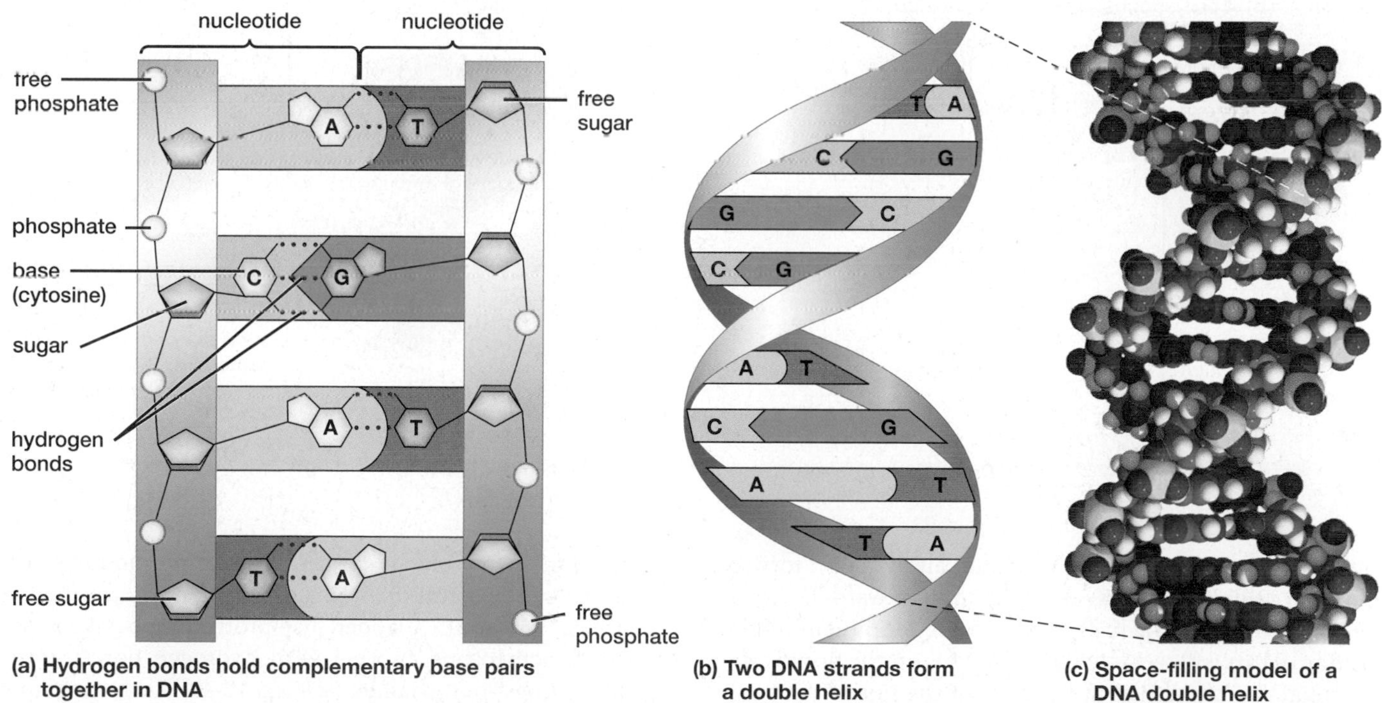

▲ **FIGURE 12-4 The Watson–Crick model of DNA structure (a)** Hydrogen bonding between complementary base pairs holds the two strands of DNA together. Three hydrogen bonds hold guanine to cytosine, and two hydrogen bonds hold adenine to thymine. Note that each strand has a free phosphate on one end and a free sugar on the opposite end, but the two strands run in opposite directions. **(b)** Strands of DNA wind about each other in a double helix, like a twisted ladder, with the sugar-phosphate backbone forming the uprights and the complementary base pairs forming the rungs. **(c)** A space-filling model of DNA structure.

THINK CRITICALLY Which do you think would be more difficult to break apart: an A–T base pair or a C–G base pair?

DOING Science Discovering the Hereditary Molecule

By the early 1900s, scientists had learned that inherited traits are carried in units they called genes and that genes are parts of chromosomes. But the identity of the molecule that makes up genes was not known with certainty until 1952, when Alfred Hershey and Martha Chase conducted a brilliant set of experiments.

What Question Was Asked?

When Hershey and Chase planned their experiments, they knew from others' earlier research that chromosomes are composed only of protein and DNA. So, one or the other of these molecules must be the molecule of heredity. But which one?

How Was Evidence Gathered?

Hershey and Chase studied a type of virus, called a **bacteriophage** ("phage" for short), that infects bacteria (**FIG. E12-1**). When a phage encounters a bacterium, it attaches to the bacterial cell wall and injects its genetic material into the bacterium ❶. The outer coat of the phage remains outside. The bacterium cannot distinguish phage genes from its own genes, so it "reads" the phage genes and uses that information to produce more phages ❷. Finally, the bacterium bursts, freeing the new phages ❸.

▼ **FIGURE E12-1 Bacteriophages (a)** Many bacteriophages have a complex structure, including a head containing genetic material, tail fibers that attach to the surface of a bacterium, and an elaborate apparatus for injecting their genetic material into the bacterium. **(b)** A bacteriophage reproduces inside a bacterium.

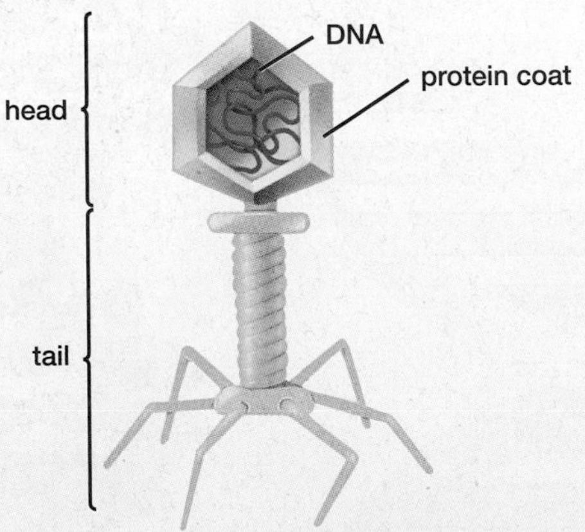
DNA
protein coat
head
tail

(a) Structure of a bacteriophage

Most phages are chemically very simple, consisting only of DNA and protein. To deduce whether DNA or protein is the hereditary molecule of bacteriophages, Hershey and Chase took advantage of a difference in the chemical composition of the two molecules. DNA contains phosphorus but not sulfur, whereas proteins contain sulfur but not phosphorus.

To conduct their test, Hershey and Chase forced one culture of phages to use radioactive phosphorus to synthesize DNA, thereby labeling the phage DNA. They forced another culture of phages to use radioactive sulfur to synthesize protein, thereby labeling the phage protein (**FIG. E12-2** ❶). Bacteria were infected by one of these two labeled phage cultures ❷. Then the bacteria were whirled in a blender to shake the phage coats off the bacteria ❸ and centrifuged to separate the phage coats from the bacteria ❹.

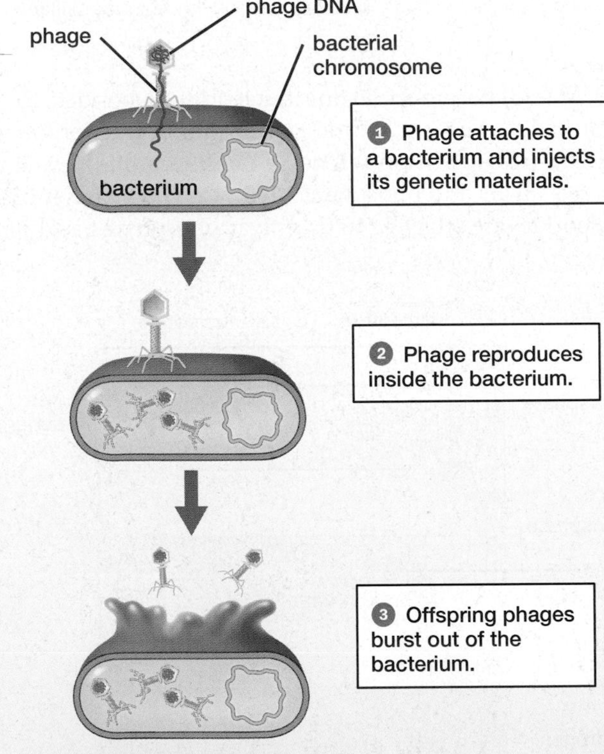
phage DNA
phage
bacterial chromosome
bacterium
❶ Phage attaches to a bacterium and injects its genetic materials.
❷ Phage reproduces inside the bacterium.
❸ Offspring phages burst out of the bacterium.

(b) Bacteriophage reproduction

night; the cars' headlights (phosphates) always point forward, and their taillights (sugars) always point backward. If the cars are jammed tightly together, a pedestrian standing in front of the line of cars will see only the headlights on the first car (free phosphate); a pedestrian at the back of the line will see only the taillights of the last car (free sugar).

Hydrogen Bonds Between Complementary Bases Hold Two DNA Strands Together

Watson and Crick's crucial insight was that the two strands of a DNA molecule are assembled like a ladder made out of many similar, but not identical, nucleotide modules. The sugar-phosphate backbones of the two strands form the two "uprights" of the DNA ladder. The protruding bases of each strand attach to one another with hydrogen bonds, forming the "rungs" of the ladder (see Fig. 12-4a). Each rung must span the same distance, given that the X-ray data showed that a DNA molecule has a uniform width. But the bases are of different sizes. Adenine and guanine each contain two fused rings, so they are larger. Thymine and cytosine, each with only a single ring, are smaller. Thus, the DNA ladder will have a uniform width only if each rung consists of one small and one large base.

What Was Learned?

Hershey and Chase found that, if bacteria were infected by phages containing radioactively labeled protein, the resulting phage coats were radioactive but the bacteria were not. If bacteria were infected by phages containing radioactive DNA, the bacteria became radioactive but the phage coats did not ❺. Therefore, the substance injected by the phages into the bacteria was DNA, not protein. Further, the infected bacteria produced new phages even after the protein coats were removed, showing that the injected DNA, not the protein in the coat, was the genetic material. In the words of James Watson, this experiment provided "powerful new proof that DNA is the primary genetic material."

THINK CRITICALLY If viral genetic material had the same structure as eukaryote chromosomes, would the Hershey–Chase experiment have succeeded? Why or why not?

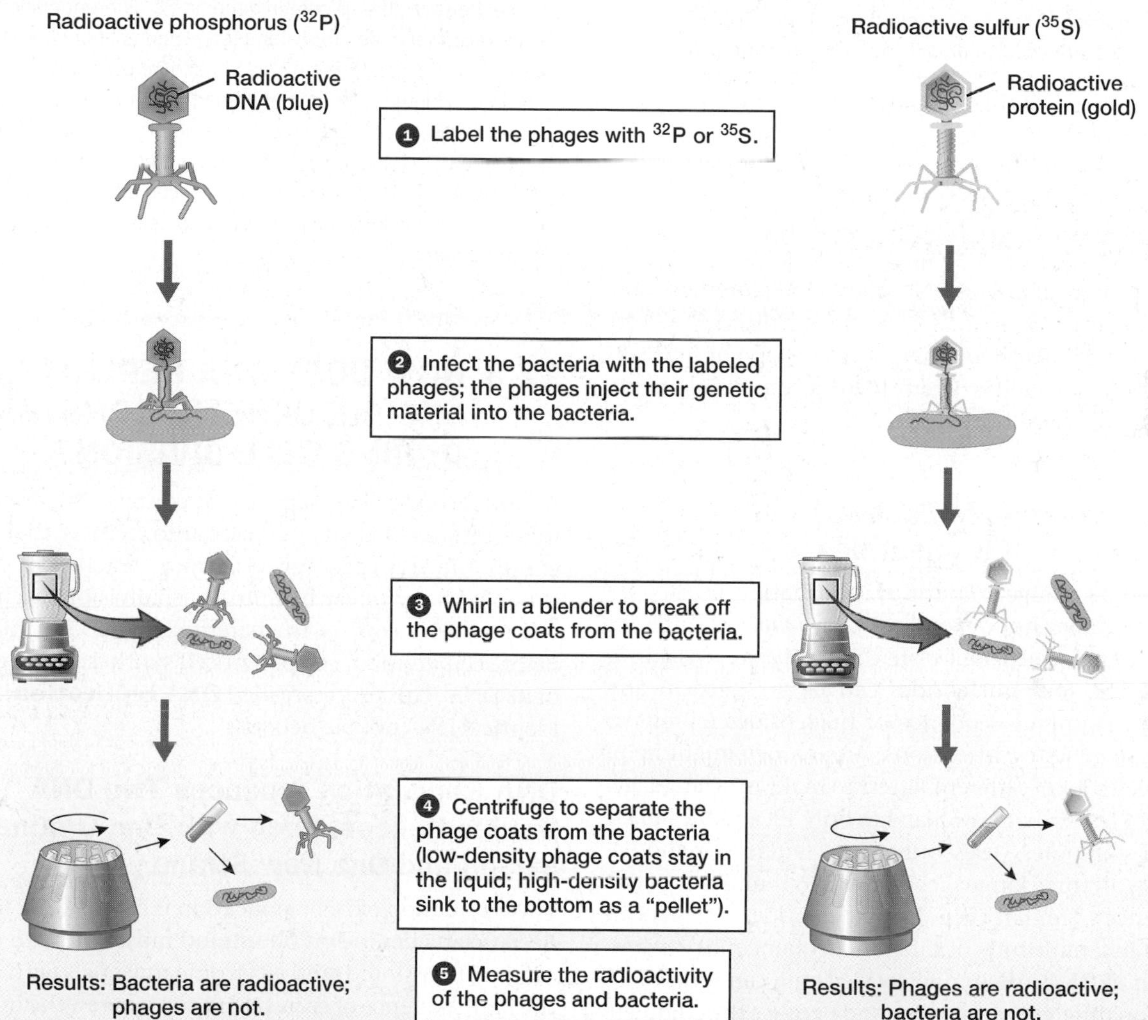

▲ **FIGURE E12-2 The Hershey–Chase experiment** By radioactively labeling either the DNA or the protein of bacteriophages, Hershey and Chase tested whether the genetic material of phages is DNA (left side) or protein (right side).

Which base pairs plug together to form a rung? Adenine can form hydrogen bonds only with thymine, and guanine can form hydrogen bonds only with cytosine. These A–T and G–C pairs are called **complementary base pairs**. Every rung of the DNA ladder is made of a complementary base pair. Therefore, if you know the base sequence of one DNA strand, you can deduce the base sequence of the other strand. For example, if one strand's sequence is A-T-T-C-C, the other strand's sequence must be T-A-A-G-G.

As the X-ray data showed, the DNA ladder isn't straight. The two strands are wound about each other to form a **double helix**, like a ladder twisted lengthwise into the shape of a spiral staircase (**FIG. 12-4b**). Further, the two strands in a DNA double helix are antiparallel to one another; that is, they are oriented in opposite directions. In Figure 12-4a, note that the left-hand DNA strand has a free phosphate group at the top and a free sugar on the bottom, but the ends are reversed on the right-hand DNA strand. Again imagine an evening traffic jam, this time on a crowded highway. A pedestrian on an overpass would see only the headlights of cars on one side of the highway and only the taillights of cars on the other side.

Watson and Crick's discovery was a watershed moment in the history of science. On March 7, 1953, at the Eagle Pub in

Cambridge, England, Francis Crick proclaimed to the lunchtime crowd, "We have discovered the secret of life." This claim was not far from the truth. The discovery of the double helix revolutionized much of biology, including genetics, evolutionary biology, and medicine. The revolution continues today.

CHECK YOUR LEARNING

Can you . . .

- name and compare the four nucleotides found in DNA?
- describe the components and three-dimensional structure of a DNA strand?

12.2 HOW DOES DNA ENCODE GENETIC INFORMATION?

Look again at the structure of DNA shown in Figure 12-4, and consider the many characteristics of an organism. How can the color of a bird's feathers, the size and shape of its beak, and its ability to sing all be determined by a molecule made from only four different nucleotides?

Genetic Information Is Encoded in the Sequence of Nucleotides

DNA can encode a huge amount of information because its coding method does not depend on the *number* of different nucleotides, but instead relies on their *sequence*. Within a DNA strand, the four nucleotides can be arranged in any order, and each unique sequence of nucleotides represents a unique set of genetic instructions. An analogy might help: You don't need a lot of different letters to make up a language. English has 26 letters, but Hawaiian has only 12, and the binary language of computers uses only two "letters" (0 and 1). Nevertheless, all three languages can spell out millions of different sentences. Similarly, the four nucleotides of DNA can be arranged in a multitude of different sequences. For example, a stretch of DNA just 10 nucleotides long can form more than a million different sequences. And because the complete sequence of DNA in an organism—its **genome**—contains millions or even billions of nucleotides, the amount of information that can be encoded is staggering. The number of possible human genome sequences is far larger than the number of atoms in the universe.

As we will describe in Chapter 13, the DNA sequence of most genes encodes the information needed to synthesize a protein. Different sequences encode different proteins, just as the words "leaf" and "flea" contain the same letters but in a different sequence, and mean different things.

CHECK YOUR LEARNING

Can you . . .

- explain how DNA encodes hereditary information?

CASE STUDY CONTINUED

Muscles, Mutations, and Myostatin

The sequence of nucleotides in a gene determines the function of the protein that the gene encodes. The myostatin gene of most cattle breeds has a nucleotide sequence that codes for normal myostatin protein that limits muscle size. In Belgian Blue cattle, however, the myostatin gene has a different nucleotide sequence that codes for a completely nonfunctional protein. As a result, a Belgian Blue's muscles become oversized. The gene for nonfunctional myostatin is present in every cell in a Belgian Blue's body. Why? Because as the animal's body grows by repeated cell division, its DNA is duplicated before each division by a process that almost always produces exact copies of the original nucleotide sequences. How do cells replicate their DNA so accurately?

12.3 HOW DOES DNA REPLICATION ENSURE GENETIC CONSTANCY DURING CELL DIVISION?

Almost every cell of your body contains identical genetic information—the same genetic information that was present in the fertilized egg cell from which you developed. When cells reproduce by mitotic cell division, each daughter cell receives a copy of the parent cell's genetic information. Before cell division, the parent cell synthesizes an exact copy of its DNA. This process, called **DNA replication**, yields two identical DNA double helices.

DNA Replication Produces Two DNA Double Helices, Each with One Original Strand and One New Strand

The foundation of DNA replication is base pairing (**FIG. 12-5**). Because an adenine on one strand must pair with a thymine on the other strand, and a cytosine must pair with a guanine, the base sequence of each strand contains all the information needed to replicate the other strand. This replication requires **free nucleotides** (nucleotides not yet part of a DNA strand) that were previously synthesized in the cytoplasm and imported into the nucleus. The other essential ingredients for replication are the parental DNA strands ❶ and a variety of enzymes that unwind the parental DNA double helix and synthesize new DNA strands.

In the first step of replication, enzymes pull apart the parental double helix, so that the bases of the two DNA strands are no longer bonded to one another ❷. Second, enzymes called **DNA polymerases** move along each separated parental DNA strand, matching bases on the parental strands with complementary free nucleotides ❸. For example, one type of DNA polymerase pairs an exposed adenine in the parental strand with a free thymine. DNA polymerase also connects

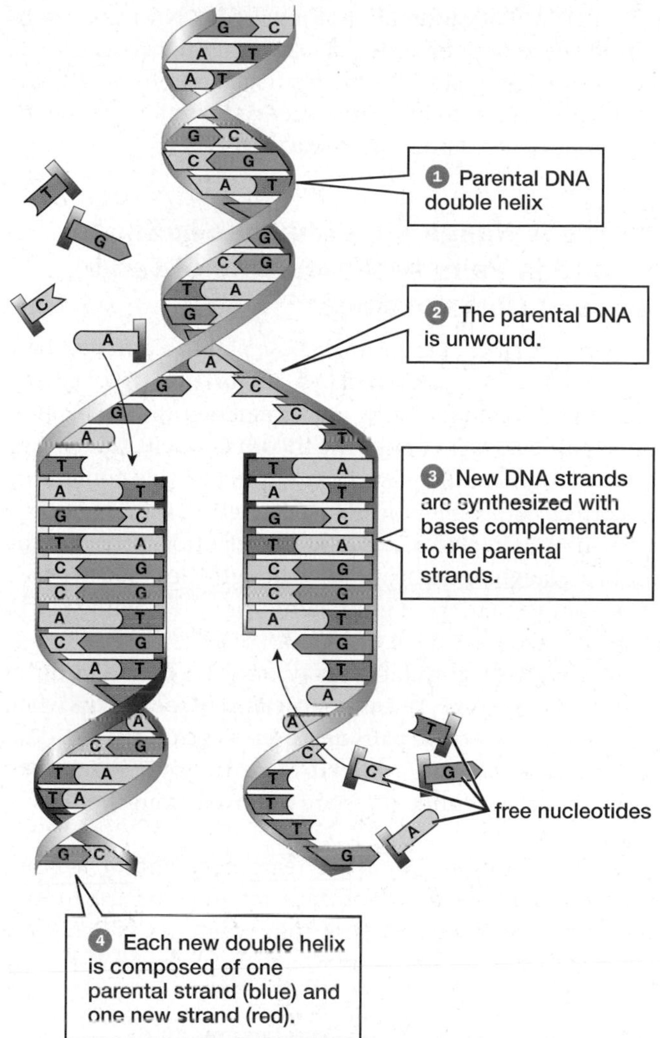

1 Parental DNA double helix
2 The parental DNA is unwound.
3 New DNA strands are synthesized with bases complementary to the parental strands.
free nucleotides
4 Each new double helix is composed of one parental strand (blue) and one new strand (red).

▲ **FIGURE 12-5 Basic features of DNA replication** During replication, the two strands of the parental DNA double helix separate. Free nucleotides that are complementary to those in each strand are joined to make new daughter strands. Each parental strand and its new daughter strand then form a new double helix.

these newly-paired free nucleotides with one another to form two new DNA strands, one new complementary strand for each parental strand. When the replication process is complete, each parental DNA strand and its newly synthesized, complementary daughter DNA strand wind together to form a new double helix ❹.

Replication of a double helix yields two double helices, each of which contains one parental DNA strand and one newly synthesized strand (**FIG. 12-6**). Thus, the replication process keeps, or *conserves*, one parental strand in each replicated double helix. The process is therefore known as **semiconservative replication** (*semi* means "half"). If no mistakes have been made during replication, the base sequences of both new DNA double helices are identical to the base sequence of the parental DNA double helix.

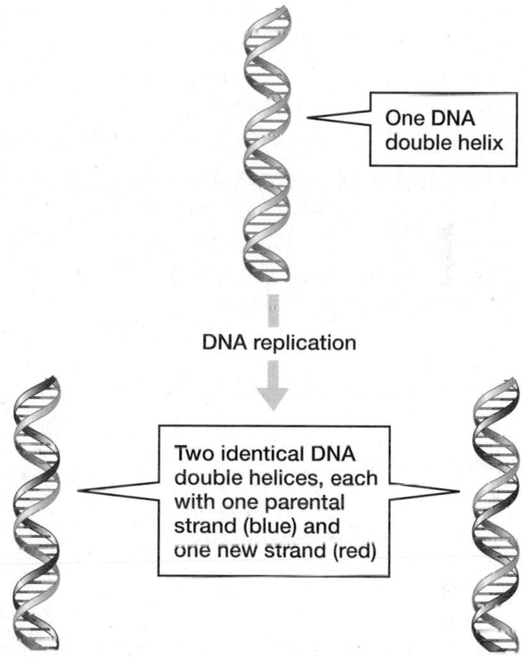

One DNA double helix
DNA replication
Two identical DNA double helices, each with one parental strand (blue) and one new strand (red)

▲ **FIGURE 12-6 Semiconservative replication of DNA**

CHECK YOUR LEARNING

Can you . . .

- describe the process of DNA replication, including the enzymes involved and the actions that they perform?
- explain why DNA replication is called "semiconservative"?

CASE STUDY CONTINUED

Muscles, Mutations, and Myostatin

"Double-muscled" cattle were first reported in the early 1800s by agriculturalists who were impressed by the cattle's potential to produce more meat from a given amount of feed. Sometime in the late 1700s or early 1800s, a mutation must have occurred in the myostatin gene in the reproductive cells of the Belgian Blue ancestor, changing the nucleotide sequence of the gene. If DNA replication is so accurate, how do such mutations happen?

12.4 WHAT ARE MUTATIONS, AND HOW DO THEY OCCUR?

Although the nucleotide sequence of DNA is preserved with a high rate of accuracy as cells divide and organisms reproduce, sequence changes sometimes do occur.

Accurate Replication, Proofreading, and DNA Repair Produce Almost Error-Free DNA

The error rate during DNA replication is very low. DNA polymerases incorporate incorrect bases only about once in every 10 thousand to 1 million base pairs. This small number of errors is reduced even further by DNA repair enzymes that proofread each daughter strand during and after its synthesis. For example, some forms of DNA polymerase recognize a base pairing mistake as it is made. These types of DNA polymerase pause, fix the mistake, and then continue synthesizing more DNA. As a result of replication's low error rate and effective repair mechanisms, a completed DNA strand contains only about one mistake in every 100 million to 10 billion base pairs, a phenomenally low error rate.

Toxic Chemicals, Radiation, or Occasional Mistakes During DNA Replication May Cause Mutations

Despite the amazing accuracy of DNA replication, no organism has error-free DNA. Occasionally, mistakes made during normal DNA replication are not repaired. DNA may also be damaged by toxic chemicals (such as components of cigarette smoke) or some types of radiation (such as ultraviolet rays in sunlight). Changes in DNA sequence that are not fixed by repair enzymes are known as **mutations**.

Mutations Range from Changes in Single Nucleotide Pairs to Movements of Large Pieces of Chromosomes

A mutation that changes a single base pair to a different base pair is known as a **substitution mutation** (**FIG. 12-7a**). Normally, if a pair of bases is mismatched during replication, repair enzymes cut out the incorrect nucleotide on the daughter strand and replace it with a nucleotide containing the correct complementary base. Sometimes, however, the repair enzymes replace the parental nucleotide instead of the incorrect daughter nucleotide. Although the resulting base pair is complementary, it is different from the original pair; a substitution mutation has occurred.

Alternatively, a mutation may result in the addition or removal of base pairs. An **insertion mutation** occurs when one or more nucleotide pairs are inserted into the DNA double helix (**FIG. 12-7b**). A **deletion mutation** occurs when one or more nucleotide pairs are removed from the double helix (**FIG. 12-7c**).

Some mutations result in rearrangement of a chromosome. Pieces of chromosomes ranging in size from a single nucleotide pair to very long stretches of DNA are

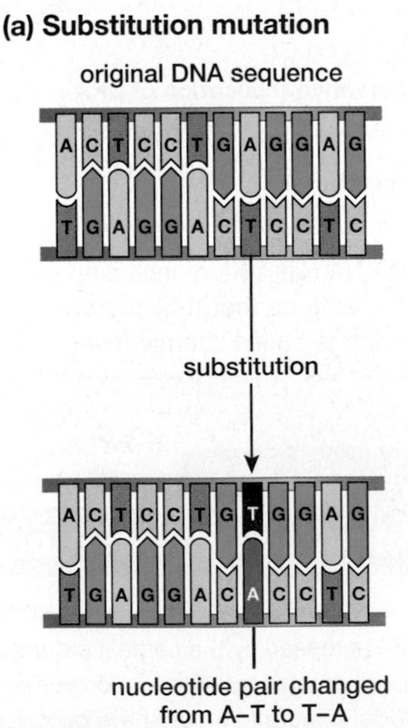

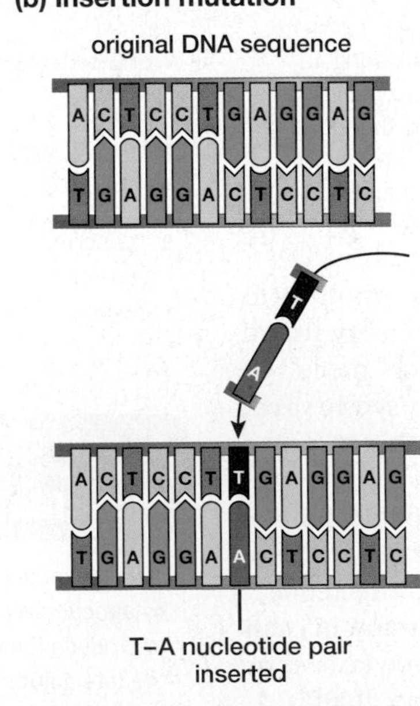

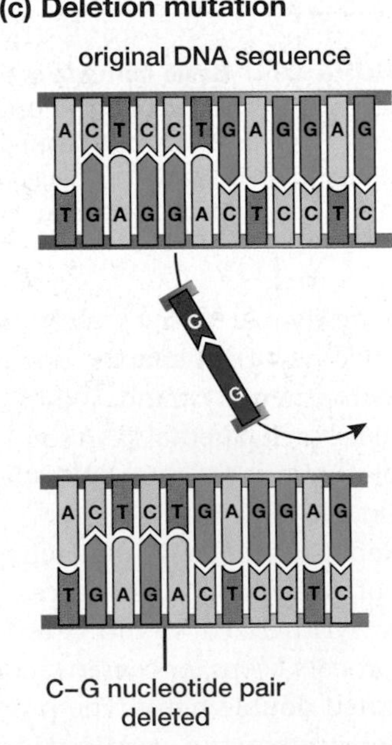

▲ **FIGURE 12-7 Mutations involving only one or a few pairs of nucleotides (a)** Substitution mutation. **(b)** Insertion mutation. **(c)** Deletion mutation. The original DNA bases are in pale colors with black letters; mutations are in dark colors with white letters.

(a) Inversion

original DNA sequence

breaks

DNA segment inverted

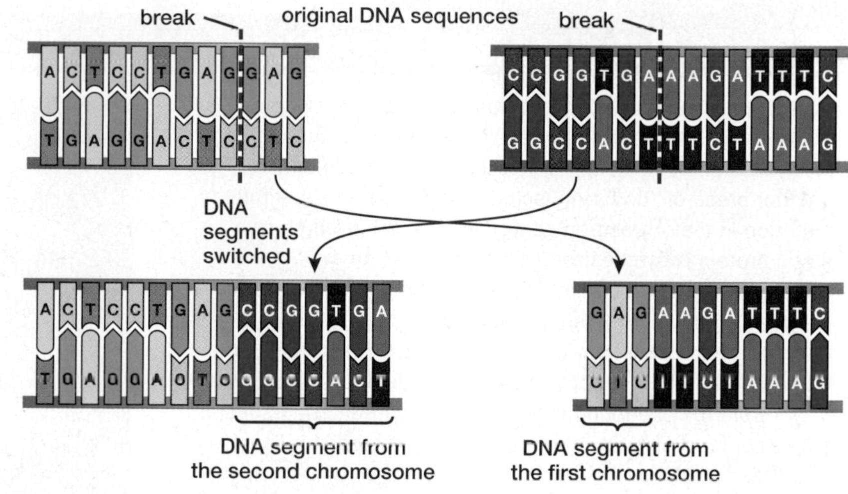

FIGURE 12-8 Mutations that rearrange pieces of chromosomes (a) Inversion mutation. **(b)** Translocation of pieces of DNA between two different chromosomes. In part **(a)**, bases in the unchanged part of the chromosome are in pale colors with black letters; bases in the part of the chromosome that is inverted are in dark colors with white letters. In part **(b)**, the DNA bases of one chromosome are in pale colors with black letters, and the DNA bases of the second chromosome are in dark colors with white letters.

occasionally rearranged. An **inversion** occurs when a piece of DNA is cut out of a chromosome, turned around, and reinserted into the gap (**FIG. 12-8a**). A **translocation** results when a chunk of DNA, sometimes very large, is removed from one chromosome and attached to a different one (**FIG. 12-8b**).

Note that in substitutions, insertions, deletions, inversions, and translocations, the mutated DNA still has complementary base pairs. As a result, the mutated DNA can be accurately replicated during future cell divisions, thereby perpetuating the mutation. The mutation becomes a permanent part of the chromosome and will be inherited by all the cell's descendants.

In most cases, the result of a mutation is harmful, much as random changes to the letters in a word would most likely make the word meaningless. Nonetheless, a mutation may have no effect on an organism or, in very rare instances, may even be beneficial. Mutations that benefit an organism will be favored by natural selection, and are the basis for the evolution of life on Earth. Note that in multicellular organisms, mutations can affect evolution only if they occur in reproductive cells, whose DNA is passed to subsequent generations.

CHECK YOUR LEARNING

Can you . . .

- explain what mutations are and how they occur?
- explain why mutations are rare?
- describe the different types of mutations?

Have You Ever Wondered . . .

How Much Genes Influence Athletic Prowess?

Face it—you'll never swim like Olympic champion Katie Ledecky. How much of her fantastic ability is genetic? A few genes are known to make significant contributions to athletic performance. For example, some alleles of the myostatin gene can boost strength and speed. One allele of the gene for a muscle fiber protein is associated with increased endurance. However, most of the individual genes that influence athletic performance have only small effects. At least 240 genes contribute to human athletic performance, so super-athletes like Ledecky most likely won the "genetic lottery" and inherited scores of alleles that each boost performance just a little but add up to unsurpassed athleticism.

CASE STUDY REVISITED

Muscles, Mutations, and Myostatin

Belgian Blue cattle are homozygous for a deletion mutation in their myostatin gene. As a result, their cells synthesize an incomplete, nonfunctional myostatin protein. Piedmontese, another breed of "double-muscled" cattle, have a substitution mutation in their myostatin gene. Although a full-length myostatin protein is synthesized, it doesn't fold into the correct three-dimensional structure and is completely inactive, so Piedmontese cattle have essentially the same phenotype as Belgian Blues. Other animals may also have mutated myostatin genes. For example, "bully" whippet dogs have a deletion mutation, different from the one found in Belgian Blue cattle, that results in short, nonfunctional myostatin and a huge increase in muscle size (**FIG. 12-9**).

Humans produce myostatin, too. A few people inherit defective myostatin alleles from one or both parents, resulting in a very rare condition called myostatin-related muscle hypertrophy. The functional and defective myostatin alleles are incompletely dominant to one another. Homozygotes for the defective allele have far greater muscle bulk and strength than people who are homozygous for the functional allele; heterozygotes have an intermediate increase in muscle size and strength.

Myostatin mutations reveal an important feature of the language of DNA: The nucleotides of genes must be ordered just right for the resulting proteins to function. In contrast, any one of an enormous number of possible mistakes will render the proteins useless.

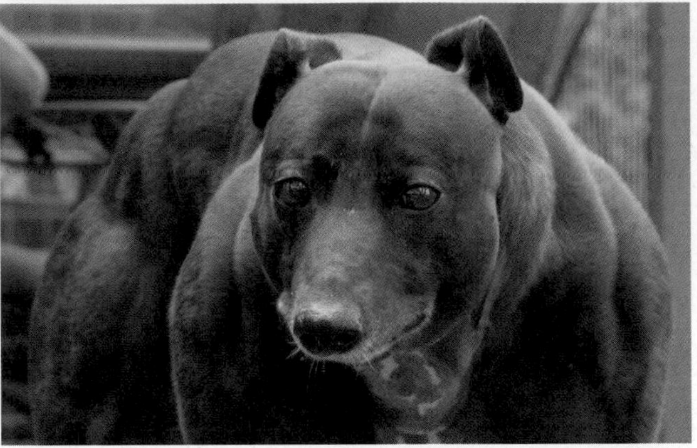

▲ **FIGURE 12-9 Myostatin mutation in whippets** "Bully" whippets have nonfunctional myostatin, resulting in enormous muscles.

THINK CRITICALLY Whippets that are homozygous for the defective "bully" allele have lots of muscle but often suffer from cramps in the shoulder and thigh, and they are not fast runners. Homozygous normal whippets are skinny and quite fast. However, the fastest whippets are heterozygous, with intermediate muscle size and phenomenal speed. If whippets were wild dogs that chased down their prey, what do you think would be the outcome of natural selection on myostatin alleles? Explain your answer.

CHAPTER REVIEW

Go to **Mastering Biology** to access the Pearson eText, vocabulary review, practice quizzes, activities, videos, current events, and more.

*Answers to **Think Critically** and **Thinking Through the Concepts** questions can be found in the **Answers** section at the back of the book.*

Summary of Key Concepts

12.1 What Is the Structure of DNA?

The structure of DNA was discovered by James Watson and Francis Crick. DNA consists of nucleotides that are linked into long strands. Each nucleotide consists of a phosphate group, the five-carbon sugar deoxyribose, and a nitrogen-containing base. Four types of bases occur in DNA: adenine, guanine, thymine, and cytosine. The sugar of one nucleotide is linked to the phosphate of the next nucleotide, forming a sugar-phosphate backbone for each strand. The bases are attached to this backbone.

Two nucleotide strands wind together to form a DNA double helix, which resembles a twisted ladder. The sugar-phosphate backbones form the sides of the ladder. The bases of each strand pair up in the middle of the helix, held together by hydrogen bonds and forming the rungs of the ladder. Only complementary base pairs can bond together in the helix: Adenine bonds with thymine, and guanine bonds with cytosine.

12.2 How Does DNA Encode Genetic Information?

Genetic information is encoded as the sequence of nucleotides in a DNA molecule. Just as a language can form thousands of words and complex sentences from a small number of letters, DNA can encode large amounts of information by varying the sequences and numbers of nucleotides in different genes. In genomes, which are millions to billions of nucleotides long, DNA encodes huge amounts of information.

12.3 How Does DNA Replication Ensure Genetic Constancy During Cell Division?

When cells reproduce, they must replicate their DNA so that each daughter cell receives all the original genetic information. During DNA replication, enzymes unwind and separate part of the two parental DNA strands. Then DNA polymerase enzymes bind to each parental DNA strand. Free nucleotides form hydrogen bonds with complementary bases on the parental strands, and DNA polymerases link the free nucleotides to form new DNA strands. Replication is semiconservative because both new DNA double helices consist of one parental DNA strand and one newly synthesized, complementary daughter strand. The two new DNA double helices are duplicates of the parental DNA double helix.

12.4 What Are Mutations, and How Do They Occur?

Mutations are changes in the base sequence of DNA. DNA polymerases and other repair enzymes proofread and repair the DNA, minimizing the number of mistakes made during replication. Mistakes that are not fixed become mutations. Other mutations occur as a result of radiation and damage from toxic chemicals. Mutations include substitutions, insertions, deletions, inversions, and translocations. Most mutations are harmful, but many are neutral and a few are beneficial and may be favored by natural selection.

Thinking Through the Concepts

Bloom's: Remembering, Understanding

Multiple Choice

1. If a parental DNA strand has the sequence A-T-T-G-C-A-C-T, DNA polymerases would synthesize a new strand with the sequence
   a. A-T-T-G-C-A-C-T.
   b. T-A-A-C-G-T-G-A.
   c. C-G-G-T-A-C-A-G.
   d. The sequence of the new strand cannot be determined from the information given.
2. At the conclusion of DNA replication, which of the following is true?
   a. Each daughter double helix consists of one original DNA strand and one new DNA strand.
   b. One daughter double helix consists of the two original DNA strands, and the other daughter double helix consists of two new DNA strands.
   c. Each resulting DNA strand consists of part of one of the original DNA strands and part of a new DNA strand.
   d. The resulting DNA daughter strands contain nucleotide sequences that were not present in the parental DNA strands.
3. An insertion mutation occurs when
   a. a nucleotide in a DNA sequence is replaced by a different nucleotide.
   b. one or more nucleotide pairs are added in the middle of a DNA sequence.
   c. one or more nucleotides are removed from the middle of a DNA sequence.
   d. a piece of DNA is removed from one chromosome and attached to a different chromosome.
4. The "rungs" of the DNA double helix consist of
   a. any combination of bases.
   b. any combination of one double-ring base and one single-ring base.
   c. specific combinations of double-ring bases.
   d. specific combinations of single-ring and double-ring bases.
5. The "rungs" of the DNA double helix are held together by
   a. ionic bonds.
   b. hydrogen bonds.
   c. covalent bonds.
   d. the force of the backbones on the outside of the helix pushing them together.

Fill-in-the-Blank

1. DNA consists of subunits called __________. Each subunit consists of three parts: __________, __________, and __________.
2. The subunits of DNA are assembled by linking the __________ of one nucleotide to the __________ of the next. In chromosomes, two DNA polymers are wound together into a structure called a(n) __________.
3. The "base pairing rule" in DNA is that adenine pairs with __________ and guanine pairs with __________. Bases that can form pairs in DNA are called __________.
4. When DNA is replicated, two new DNA double helices are formed, each consisting of one parental strand and one new daughter strand. For this reason, DNA replication is called __________.
5. In DNA replication, daughter DNA strands are synthesized by enzymes called __________. The daughter strands are built by adding __________ to the growing strand.
6. Sometimes mistakes are made during DNA replication. If uncorrected, these mistakes are called __________. When a single nucleotide is changed, this is called a(n) __________.

Review Questions

1. Describe the experimental evidence that DNA is the hereditary material of bacteriophages.
2. Draw the general structure of a DNA nucleotide. Which components are identical in all four DNA nucleotides, and which vary?
3. Describe the structure of DNA. Where are the bases, sugars, and phosphates in the structure? Which bases are complementary to one another? How are they held together in the double helix of DNA?
4. How is information encoded in the DNA molecule?
5. Describe the process of DNA replication.
6. How do mutations occur? Describe the principal types of mutations.

Applying the Concepts

Bloom's: Applying, Analyzing, Evaluating

1. Suppose that in an alternate universe, although proteins are still constructed of combinations of 20 different amino acids, DNA is constructed of six different nucleotides, not four as on Earth. Would you expect organisms in this universe to have more precise genetic instructions and/or more different genes than life on Earth? Would you expect the length of a typical gene to be the same, shorter, or longer than that of a typical gene on Earth? Explain your answers.
2. Genetic information is encoded in the sequence of nucleotides in DNA. Let's suppose that the nucleotide sequence on one strand of a double helix encodes the information needed to synthesize a hemoglobin molecule. Do you think that the sequence of nucleotides on the other strand of the double helix also encodes useful information? Why or why not? Why do you think DNA is double-stranded?

13 Gene Expression and Regulation

CASE STUDY

Cystic Fibrosis

Alice Martineau, shown here in a portrait painted by her brother Luke, hoped that " ... people will realize when they hear the music, I am a singer-songwriter who just happens to be ill."

IF YOU KNEW of Alice Martineau only through her music, you'd think she had it made—a young, talented singer-songwriter under contract with a major recording label. However, like about 70,000 other people worldwide, Martineau had cystic fibrosis, a devastating genetic disorder. Cystic fibrosis shortens lives; as recently as the 1950s, most people with cystic fibrosis died by age 4 or 5. Even now, with improved treatments, the average life span of a person with cystic fibrosis is only 35 to 40 years. Alice Martineau died when she was 30, less than a year after the release of her debut album. The story of how she pursued her calling despite her disease is told in ***A Song for Tomorrow***, a novel based on her life.

Cystic fibrosis occurs in people who are homozygous for a defective allele of the gene that encodes a crucially important protein called CFTR. This protein is found in many parts of the body, but probably its most essential role is in the cells that line the airways of the lungs. Normally, the action of the CFTR protein ensures that the airways are covered with a film of thin, watery mucus that traps bacteria and debris. The bacteria-laden mucus is then swept out of the lungs by cilia on the cells of the airways. However, people with cystic fibrosis produce defective CFTR proteins, resulting in thick mucus that the cilia can't move out of the lungs. Consequently, the airways become clogged, and bacteria multiply in the mucus, causing chronic lung infections.

In this chapter, we examine the processes by which the instructions in genes are translated into proteins. How do changes in those instructions—mutations—alter the structure and function of proteins such as CFTR?

AT A GLANCE

13.1 HOW IS THE INFORMATION IN DNA USED IN A CELL?

Information by itself doesn't do anything. For example, a recipe may provide all the information needed to bake a cake, but unless that information is translated into action by a baker, no cake will be baked. Likewise, although the nucleotide sequence of DNA, the molecular recipe of every cell, contains an incredible amount of information, DNA cannot carry out any actions on its own. So how does DNA determine whether you have dark or light hair or whether you have normal lung function or cystic fibrosis?

DNA Provides Instructions for Protein Synthesis via RNA Intermediaries

Although DNA is the hereditary molecule of all cells, proteins are a cell's "molecular workers." Protein enzymes catalyze the chemical reactions in a cell, and proteins form many cellular structures, such as the cytoskeleton and ion channels in the plasma membrane. Therefore, to build and operate a cell, information must flow from DNA to protein. DNA directs protein synthesis through intermediary molecules of **ribonucleic acid**, or **RNA**. RNA is structurally similar to DNA but differs in three respects: (1) Instead of the deoxyribose sugar found in DNA, the backbone of RNA contains the sugar ribose (the "R" in RNA); (2) RNA is usually single-stranded instead of double-stranded; and (3) RNA has nucleotides that contain the base uracil instead of the base thymine (**TABLE 13-1**). DNA codes for the synthesis of three types of RNA that play roles in protein synthesis: messenger RNA, transfer RNA, and ribosomal RNA (**FIG. 13-1**).

Messenger RNA Carries the Code for Protein Synthesis from DNA to Ribosomes

If you think of the information encoded in the DNA of a eukaryotic cell's nucleus as a valuable document stored in a library, then **messenger RNA (mRNA)** can be viewed as a kind of molecular photocopy of the information (**FIG. 13-1a**). Messenger RNA carries the copied information from the nucleus to the cytoplasm, where it will be used to direct protein synthesis.

TABLE 13-1 A Comparison of DNA and RNA

	DNA	RNA	
Strands	Two	One	
Sugar	Deoxyribose	Ribose	
Types of bases	Adenine (A), thymine (T) cytosine (C), guanine (G)	Adenine (A), uracil (U) cytosine (C), guanine (G)	
Base pairs	DNA–DNA	RNA–DNA	RNA–RNA
	A–T	A–T	A–U
	T–A	U–A	U–A
	C–G	C–G	C–G
	G–C	G–C	G–C
Function	Contains genes; the sequence of bases in most genes determines the amino acid sequence of a protein	Messenger RNA (mRNA): carries the code for a protein-coding gene from DNA to ribosomes Transfer RNA (tRNA): carries amino acids to the ribosomes Ribosomal RNA (rRNA): combines with proteins to form ribosomes, the structures that link amino acids to form a protein Noncoding RNA: regulates transcription and translation by binding to DNA or mRNA	

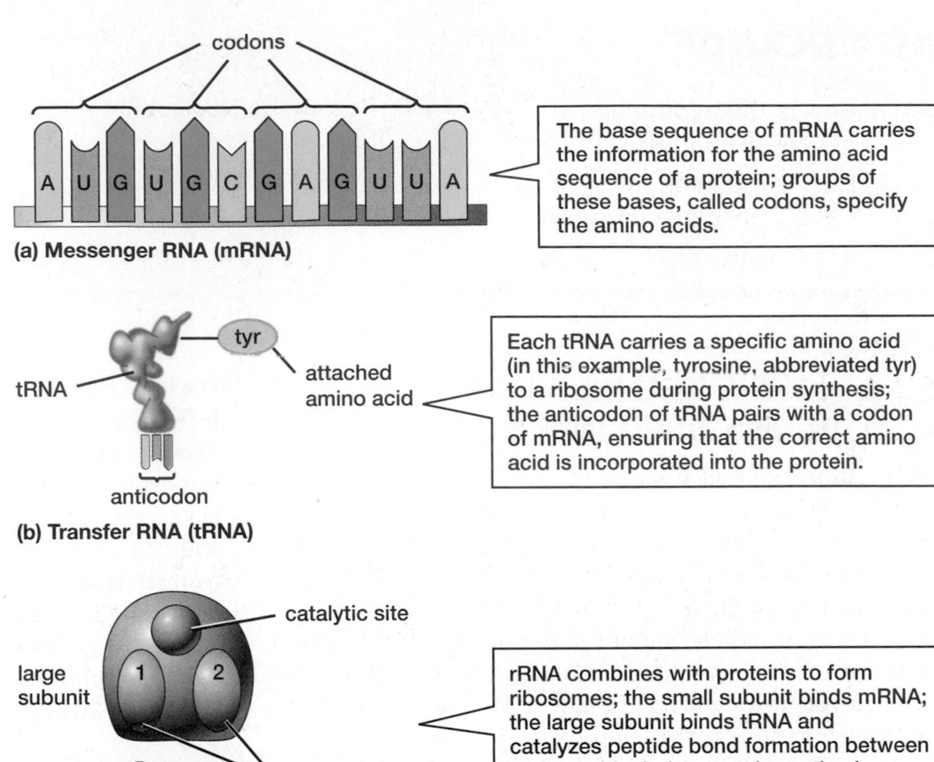
codons
A U G U G C G A G U U A
The base sequence of mRNA carries the information for the amino acid sequence of a protein; groups of these bases, called codons, specify the amino acids.
(a) Messenger RNA (mRNA)
tyr
tRNA
attached amino acid
anticodon
Each tRNA carries a specific amino acid (in this example, tyrosine, abbreviated tyr) to a ribosome during protein synthesis; the anticodon of tRNA pairs with a codon of mRNA, ensuring that the correct amino acid is incorporated into the protein.
(b) Transfer RNA (tRNA)
catalytic site
large subunit
1
2
rRNA combines with proteins to form ribosomes; the small subunit binds mRNA; the large subunit binds tRNA and catalyzes peptide bond formation between amino acids during protein synthesis.
tRNA/amino acid binding sites
small subunit
(c) Ribosome: contains ribosomal RNA (rRNA)

▸ FIGURE 13-1 Cells synthesize three major types of RNA that are required for protein synthesis

Transfer RNA Carries Amino Acids to the Ribosomes

Transfer RNA (tRNA) delivers amino acids to **ribosomes**, the cellular structures that synthesize proteins. Transfer RNA comes in different varieties, each of which binds to one of the 20 different amino acids used in proteins (**FIG. 13-1b**). Transfer RNA molecules carry their attached amino acids to a ribosome, where the amino acids are linked to form a protein.

Ribosomal RNA and Proteins Form Ribosomes

Ribosomes are composed of **ribosomal RNA (rRNA)** and dozens of proteins. Each ribosome consists of two subunits—one small and one large (**FIG. 13-1c**). The small subunit has binding sites for mRNA, a tRNA, and several proteins that are essential for assembling the ribosome and beginning protein synthesis. The large subunit has binding sites for two tRNA molecules and a site that catalyzes the formation of the peptide bonds between amino acids. During protein synthesis, the two subunits come together, clasping an mRNA molecule between them.

Overview: Genetic Information Is Transcribed into RNA and Then Translated into Proteins

Information in DNA is used to direct the synthesis of proteins in two steps, called *transcription* and *translation* (**FIG. 13-2** and **TABLE 13-2**).

1. In **transcription**, the information contained in the DNA of a gene is copied into mRNA. In eukaryotic cells, transcription occurs in the nucleus.
2. During **translation**, the mRNA nucleotide sequence is decoded and its information is converted to the sequence of amino acids in a protein. In eukaryotic cells, translation occurs in the cytoplasm, at ribosomes.

TABLE 13-2 Transcription and Translation

Process	Information for the Process	Product	Major Enzyme or Structure Involved in the Process	Type of Base Pairing Required
Transcription (synthesis of RNA)	A segment of one DNA strand	One RNA molecule (e.g., mRNA, tRNA, or rRNA)	RNA polymerase	RNA with DNA: RNA bases pair with DNA bases as an RNA molecule is synthesized
Translation (synthesis of a protein)	mRNA	One protein molecule	Ribosome (also requires tRNA)	mRNA with tRNA: A codon in mRNA forms base pairs with an anticodon in tRNA

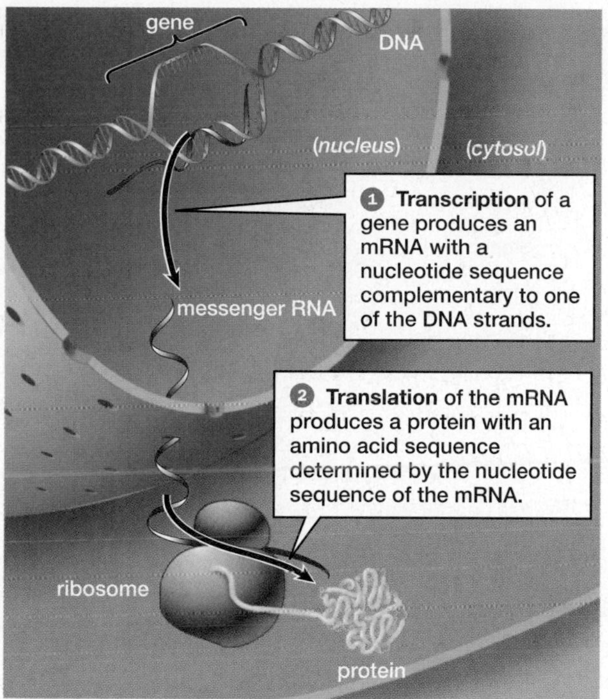
gene
DNA
(nucleus)
(cytosol)
1 Transcription of a gene produces an mRNA with a nucleotide sequence complementary to one of the DNA strands.
messenger RNA
2 Translation of the mRNA produces a protein with an amino acid sequence determined by the nucleotide sequence of the mRNA.
ribosome
protein

▲ **FIGURE 13-2 Genetic information flows from DNA to RNA to protein** Transcription takes place in the nucleus, translation in the cytoplasm.

It's easy to confuse transcription and translation. To help remember the difference, consider the meanings of the two terms in everyday English. In everyday usage, to "transcribe" means to make a written copy of something, usually in the same language. In an American courtroom, for example, verbal testimony is transcribed into a written copy, and both the testimony and the transcriptions are in English. In biology, transcription is the process of copying information from DNA to RNA using the common language of the bases found in their nucleotides. In contrast, the everyday meaning of "translate" is to convert words from one language to another language. In biology, translation means to convert information from the "nucleotide language" of RNA to the "amino acid language" of proteins.

The Genetic Code Uses Three Nucleotides to Specify an Amino Acid

Before we examine transcription and translation in detail, let's see how geneticists deciphered the **genetic code**—the rules that specify how the nucleotide sequences of DNA and mRNA are translated into the amino acid sequences of proteins.

DNA and RNA nucleotides each have four different bases (see Table 13-1), but proteins are made of 20 different amino acids. So, one base cannot directly translate into one amino acid. If a sequence of two bases coded for an amino acid, there would be 16 possible combinations (each of four possible first bases paired with each of four possible second bases, or $4 \times 4 = 16$). This still isn't enough to code for 20 amino acids. A three-base sequence, however, gives 64 possible combinations ($4 \times 4 \times 4 = 64$). Using this reasoning, physicist George Gamow hypothesized in 1954 that sets of three bases in mRNA, called **codons**, specify amino acids. In 1961, Francis Crick and three coworkers demonstrated that this hypothesis is correct.

To decipher the codons, Marshall Nirenberg and Heinrich Matthaei ruptured bacteria, producing a cytoplasmic mixture that could synthesize proteins if mRNA was added. To this mixture, they added lab-manufactured mRNA with a known sequence of nucleotides and analyzed the proteins that were formed. For example, they found that an mRNA strand composed entirely of uracil (UUUUUU . . .) directed the mixture to synthesize a protein composed solely of the amino acid phenylalanine. Therefore, the triplet UUU must be a codon that translates into phenylalanine. Because the genetic code was deciphered using mRNAs, it is usually written in terms of the base triplets in mRNA (rather than in DNA) that code for each amino acid (**TABLE 13-3**).

Certain Codons Start and Stop Translation

How does a cell recognize where the code for a protein starts and stops? Translation always begins with the codon AUG, appropriately known as the **start codon**. Because AUG also codes for the amino acid methionine, all proteins originally begin with methionine, although it may be removed after the protein is synthesized. Only the first AUG codon in an mRNA acts as a start codon; AUG codons that occur further on in the mRNA simply code for methionine. Three codons—UAG, UAA, and UGA—are **stop codons** and don't code for any amino acids. When the ribosome encounters a stop codon, it releases both the newly synthesized protein and the mRNA.

How does a cell recognize where individual codons start? Actually, there is no need for a special mechanism; "spaces" between codon "words" are unnecessary. To understand why, consider what would happen if English used only three-letter words. As long as you knew where to start reading, a sentence such as THEDOGSAWTHECAT would be perfectly understandable, even without spaces between the words. Similarly, because the beginning of a protein-coding sequence is specified by a start codon, and all codons consist of three bases, an mRNA sequence is readable even without a boundary marker between codons.

Because the genetic code has three stop codons, 61 triplets remain to specify only 20 amino acids. It turns out that several different codons may code for the same amino acid. For example, six codons—UUA, UUG, CUU, CUC, CUA, and CUG—code for leucine (see Table 13-3). However, each of the 61 non-stop codons specifies one, and only one, amino acid.

TABLE 13-3 The Genetic Code (Codons of mRNA)

First Base	U		C		A		G		Third Base
U	UUU	Phenylalanine (Phe)	UCU	Serine (Ser)	UAU	Tyrosine (Tyr)	UGU	Cysteine (Cys)	U
	UUC	Phenylalanine	UCC	Serine	UAC	Tyrosine	UGC	Cysteine	C
	UUA	Leucine (Leu)	UCA	Serine	UAA	Stop	UGA	Stop	A
	UUG	Leucine	UCG	Serine	UAG	Stop	UGG	Tryptophan (Trp)	G
C	CUU	Leucine	CCU	Proline (Pro)	CAU	Histidine (His)	CGU	Arginine (Arg)	U
	CUC	Leucine	CCC	Proline	CAC	Histidine	CGC	Arginine	C
	CUA	Leucine	CCA	Proline	CAA	Glutamine (Gln)	CGA	Arginine	A
	CUG	Leucine	CCG	Proline	CAG	Glutamine	CGG	Arginine	G
A	AUU	Isoleucine (Ile)	ACU	Threonine (Thr)	AAU	Asparagine (Asp)	AGU	Serine (Ser)	U
	AUC	Isoleucine	ACC	Threonine	AAC	Asparagine	AGC	Serine	C
	AUA	Isoleucine	ACA	Threonine	AAA	Lysine (Lys)	AGA	Arginine (Arg)	A
	AUG	Methionine (Met) Start	ACG	Threonine	AAG	Lysine	AGG	Arginine	G
G	GUU	Valine (Val)	GCU	Alanine (Ala)	GAU	Aspartic acid (Asp)	GGU	Glycine (Gly)	U
	GUC	Valine	GCC	Alanine	GAC	Aspartic acid	GGC	Glycine	C
	GUA	Valine	GCA	Alanine	GAA	Glutamic acid (Glu)	GGA	Glycine	A
	GUG	Valine	GCG	Alanine	GAG	Glutamic acid	GGG	Glycine	G

Second Base: U, C, A, G (columns)

CHECK YOUR LEARNING

Can you . . .

- describe three types of RNA that play roles in protein synthesis and explain the role of each type?
- describe how information is encoded in DNA and RNA and how this information flows from DNA to RNA to protein?
- explain why transcription and translation are appropriate terms for the processes they describe?

13.2 HOW IS THE INFORMATION IN A GENE TRANSCRIBED INTO RNA?

Transcription (**FIG. 13-3**) consists of three steps: (1) initiation, (2) elongation, and (3) termination. The three steps correspond to the three major parts of most genes: (1) a promoter region at the beginning of the gene, where transcription is initiated; (2) the "body" of the gene, where elongation of the RNA strand occurs; and (3) a termination signal at the end of the gene that sets in motion the process by which transcription stops.

Transcription Begins When RNA Polymerase Binds to the Promoter of a Gene

The enzyme **RNA polymerase** catalyzes the synthesis of RNA. Near the beginning of each gene are DNA sequences that make up the gene's **promoter**. When RNA polymerase binds to a gene's promoter, the DNA double helix at the beginning of the gene unwinds, and transcription begins (Fig. 13-3 ❶).

Elongation Generates a Growing Strand of RNA

After binding to the promoter, RNA polymerase travels down one of the DNA strands, called the **template strand**, synthesizing a single strand of RNA with nucleotides complementary to those in the DNA template ❷. Base pairing between DNA and RNA is the same as between two strands of DNA, except that uracil in RNA pairs with adenine in DNA (see Table 13-1).

After about 10 nucleotides have been added to the growing RNA chain, the first nucleotides of the RNA separate from the DNA template strand. This separation allows the two DNA strands to rewind into a double helix ❸. As the RNA molecule continues to elongate, one end drifts away from the DNA, while RNA polymerase keeps the other end attached to the template strand of the DNA. Sometimes multiple RNA polymerases land on the template strand, one after another, and transcribe a gene dozens of times in rapid succession (**FIG. 13-4**).

Transcription Stops After a Termination Signal Is Transcribed

RNA polymerase continues along the template strand of the gene, eventually reaching a sequence of DNA known as the *termination signal.* This termination sequence encodes an RNA sequence that attracts certain enzymes to the elongating strand. These enzymes cut the completed RNA molecule so that it detaches from RNA polymerase. The RNA polymerase then detaches from the DNA (Fig. 13-3 ❹).

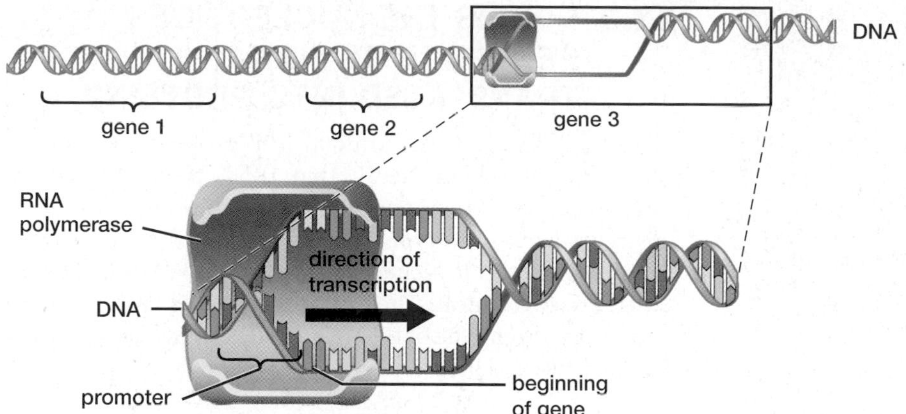

1 Initiation: RNA polymerase binds to the promoter region of DNA near the beginning of a gene, separating the double helix near the promoter.

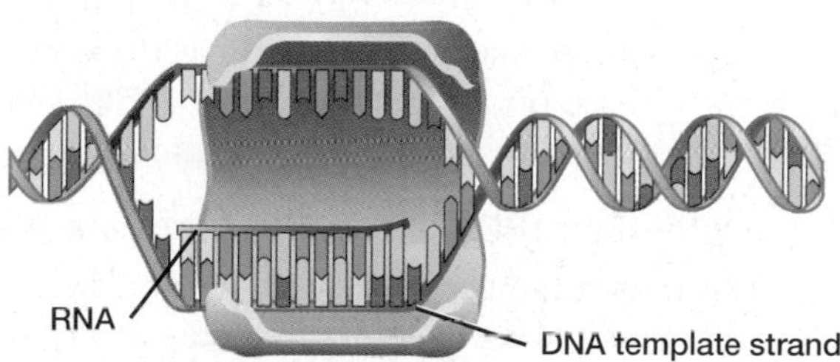

2 Elongation: RNA polymerase travels along the DNA template strand (blue), unwinding the DNA double helix and synthesizing RNA by catalyzing the addition of ribose nucleotides into an RNA molecule (red). The nucleotides in the RNA are complementary to the template strand of the DNA.

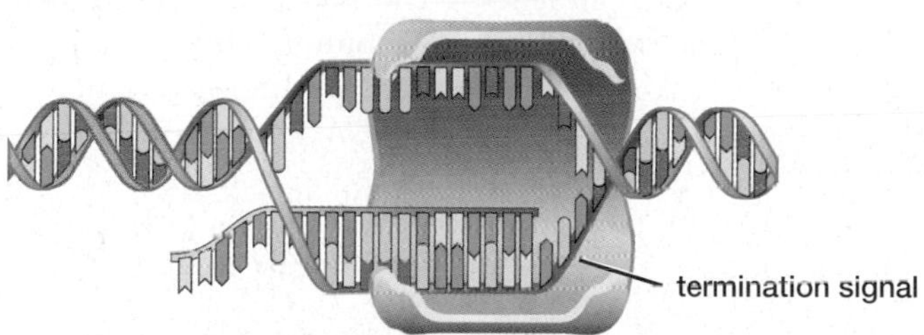

3 Termination: At the end of the gene, RNA polymerase encounters a DNA sequence called a termination signal, which is transcribed to a complementary sequence in the RNA. This RNA sequence binds to enzymes that release the RNA molecule and cause RNA polymerase to detach from the DNA.

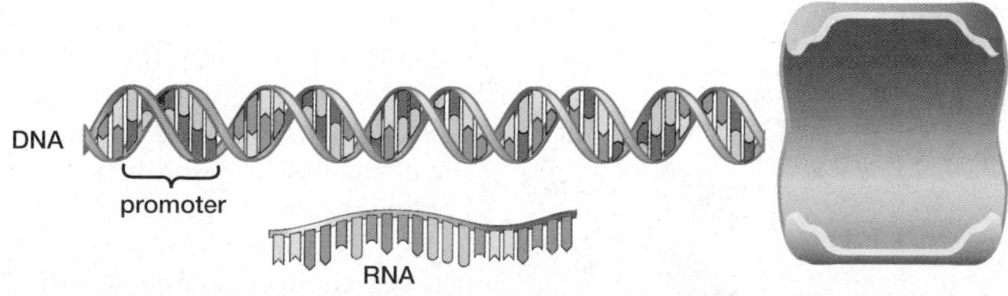

4 Conclusion of transcription: After termination, the DNA completely rewinds into a double helix. The RNA molecule is free to move from the nucleus to the cytoplasm for translation, and RNA polymerase may move to another gene (or back to the beginning of the same gene) and begin transcription once again.

▲ **FIGURE 13-3 Transcription is the synthesis of RNA from instructions in DNA** One of the DNA strands that make up the double helix of a gene serves as the template for the synthesis of an RNA molecule with bases complementary to the bases in the DNA strand.

In Eukaryotes, Precursor mRNA Is Processed to Form Finished mRNA

Although termination is the final step in transcription, most types of RNA molecules must be modified before they can carry out their functions. Prior to this modification, the mRNA strands are called *precursor* mRNA. Most eukaryotic genes contain two or more sequences called **introns**—meaning "within a gene"—that do not code for any part of a protein. Introns must be removed before the RNA is translated (**FIG. 13-5**). As introns are removed, the remaining coding segments of the gene—called **exons**, because they are *ex*pressed in protein—must be joined together, a process known as *splicing*. In humans, the average gene contains eight or nine exons.

Why do eukaryotic genes contain introns and exons? One advantage of this gene structure is that it allows a cell to produce several different proteins from a single gene by splicing exons together in different ways. For example, a gene called *CT/CGRP* is transcribed in both the thyroid gland and the brain. In the thyroid, one splicing arrangement results in the synthesis of the hormone calcitonin, which helps regulate calcium concentrations in the blood. In the brain, a different splicing arrangement results in the synthesis of a protein used as a neurotransmitter (a messenger for communication between nerve cells).

CHECK YOUR LEARNING

Can you . . .

- describe the process of transcription, explaining how DNA, RNA polymerase, and RNA nucleotides interact to produce a strand of RNA?
- describe how base pairing between DNA and RNA differs from base pairing between two strands of DNA?
- describe an example of post-transcription modification of RNA?

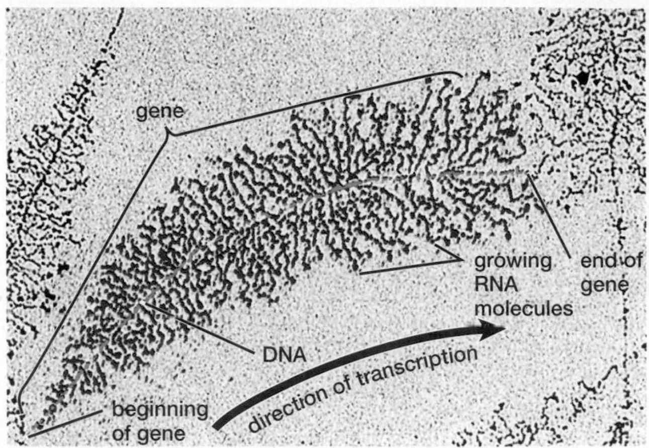

▲ **FIGURE 13-4 RNA transcription in action** In each tree-like structure in this colorized electron micrograph, the central "trunk" is DNA and the "branches" are RNA molecules. A series of RNA polymerase enzymes (too small to be seen here) is traveling down the DNA, with each enzyme synthesizing a strand of RNA. The short RNA molecules at the beginning of the gene on the left have just begun to be synthesized; the long RNA molecules on the right are almost finished.

THINK CRITICALLY Why do you think so many mRNA molecules are being transcribed from the same gene?

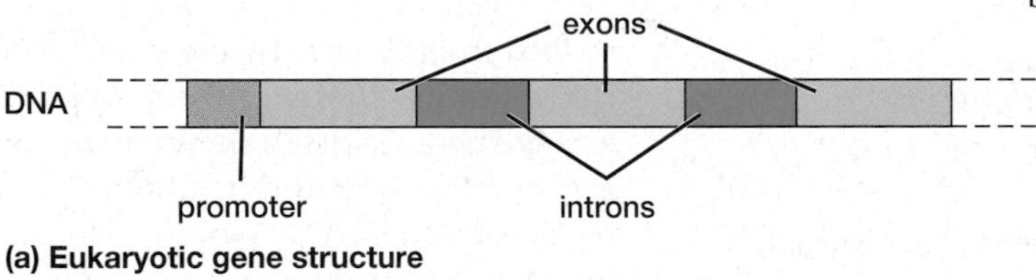

(a) Eukaryotic gene structure

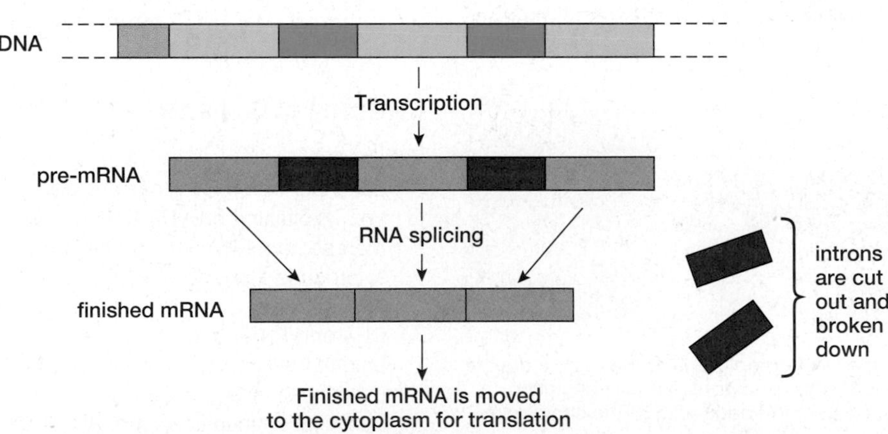

(b) RNA synthesis and processing in eukaryotes

▲ **FIGURE 13-5 Messenger RNA synthesis in eukaryotic cells (a)** Eukaryotic genes consist of exons (medium blue), which code for the amino acid sequence of a protein, and introns (dark blue), which do not. **(b)** Eukaryotic cells synthesize mRNA (red) in several steps.

13.3 HOW IS THE NUCLEOTIDE SEQUENCE OF mRNA TRANSLATED INTO PROTEIN?

Translating the codons of mRNA into protein is the job of tRNA and ribosomes. Recall that tRNA transports amino acids to the ribosomes, and that tRNA molecules are of various types, with each type carrying a particular amino acid. Each tRNA molecule includes a group of three exposed bases, called an **anticodon**. The anticodon's three-base sequence is unique to each type of tRNA, and each sequence is complementary to a specific codon in mRNA.

During Translation, mRNA, tRNA, and Ribosomes Interact to Synthesize Proteins

Translation, like transcription, has three steps: (1) initiation, (2) elongation, and (3) termination (**FIG. 13-6**).

Initiation: tRNA and mRNA Bind to a Ribosome

As translation begins, a "preinitiation complex"—composed of a small ribosomal subunit, a start (methionine) tRNA, and several other proteins ❶—binds to an mRNA molecule. The preinitiation complex moves along the mRNA until it finds a start (AUG) codon, which forms base pairs with the UAC anticodon of the methionine tRNA ❷. A large ribosomal subunit then attaches to the small subunit, sandwiching the mRNA between the two subunits. On the large subunit, the first tRNA binding site holds the methionine tRNA, and the second, empty tRNA binding site is aligned with the second codon in the mRNA molecule ❸. The ribosome is now ready to translate the mRNA.

Elongation: Amino Acids Are Added One at a Time to the Growing Protein Chain

Next, a second tRNA, with an anticodon complementary to the second codon of the mRNA, moves into the second of the two tRNA binding sites on the large subunit ❹. Ribosomal RNA at the catalytic site of the large subunit breaks the bond holding the first amino acid (methionine) to its tRNA and forms a peptide bond between the methionine and the amino acid attached to the second tRNA ❺.

After the peptide bond is formed, the first tRNA is no longer attached to an amino acid, and the second tRNA carries a two-amino-acid chain. The ribosome releases the empty tRNA and shifts to the next codon on the mRNA molecule ❻. The tRNA holding the chain of

Initiation:

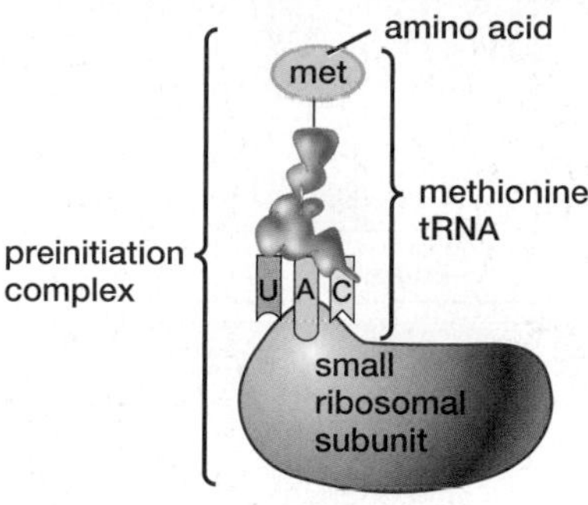

1 A tRNA with an attached methionine amino acid binds to a small ribosomal subunit, forming a preinitiation complex.

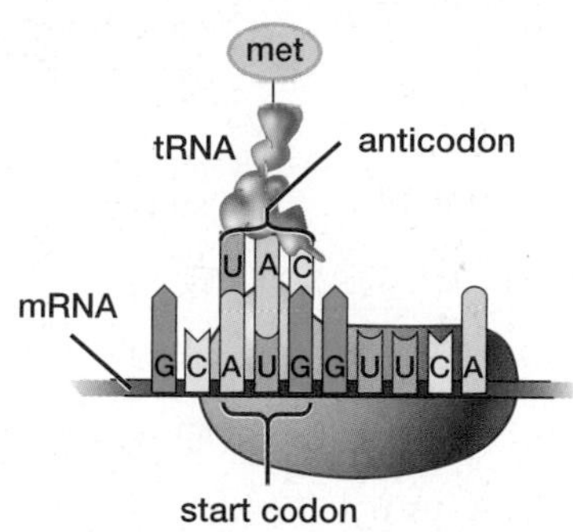

2 The preinitiation complex binds to an mRNA molecule. The methionine (met) tRNA anticodon (UAC) base-pairs with the start codon (AUG) of the mRNA.

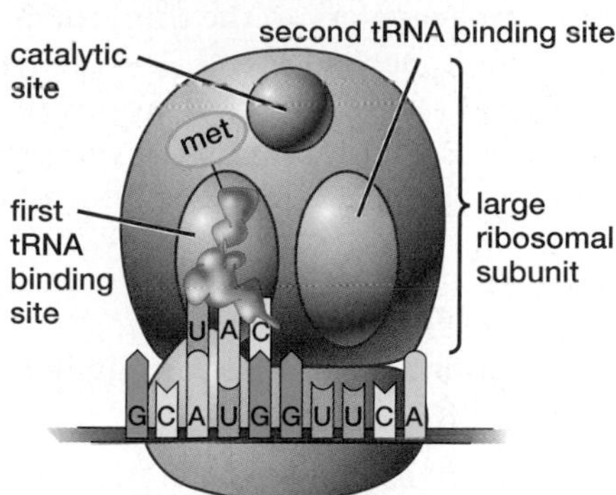

3 The large ribosomal subunit binds to the small subunit. The methionine tRNA binds to the first tRNA site on the large subunit.

Elongation:

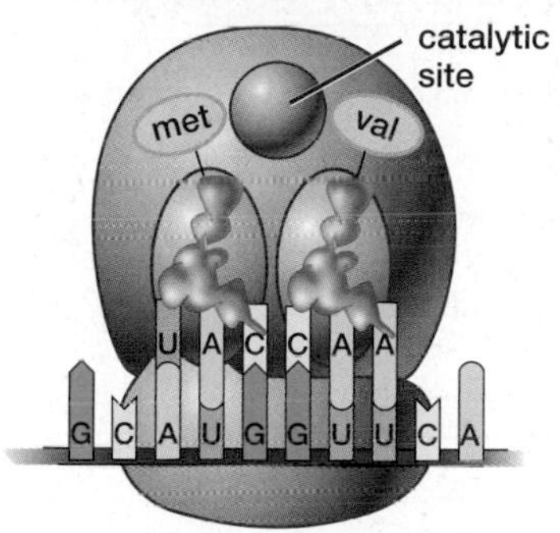

4 The second codon of mRNA (GUU) base-pairs with the anticodon (CAA) of a second tRNA carrying the amino acid valine (val). This tRNA binds to the second tRNA site on the large subunit.

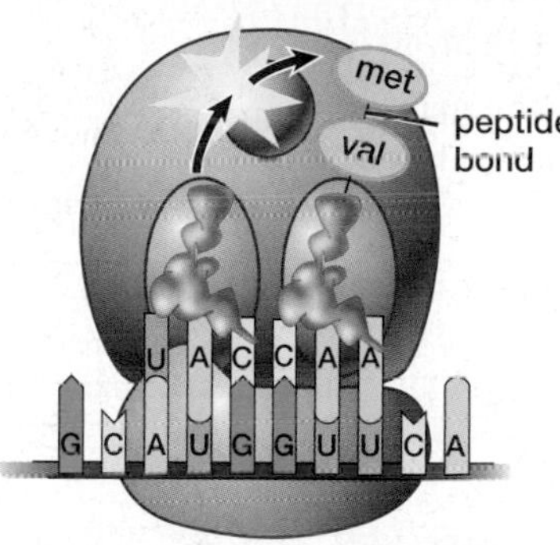

5 The catalytic site on the large subunit catalyzes the formation of a peptide bond linking the amino acids methionine and valine. The two amino acids are now attached to the tRNA in the second binding site.

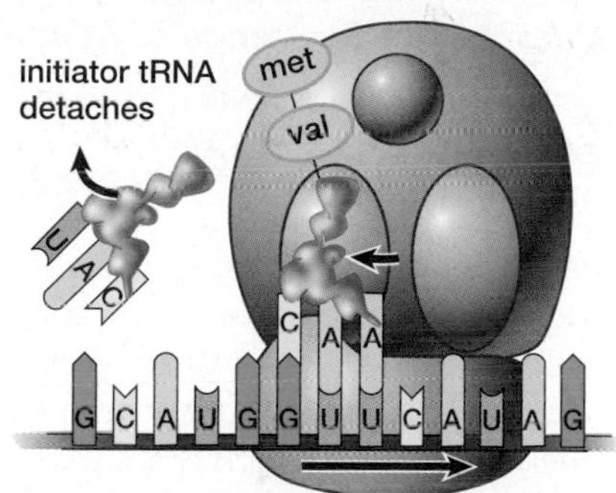

6 The "empty" tRNA is released and the ribosome moves down the mRNA, one codon to the right. The tRNA that is attached to the two amino acids is now in the first tRNA binding site and the second tRNA binding site is empty.

Termination:

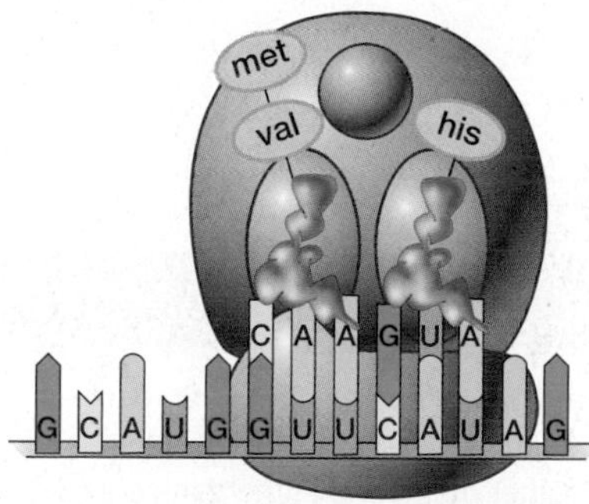

7 The third codon of mRNA (CAU) base-pairs with the anticodon (GUA) of a tRNA carrying the amino acid histidine (his). This tRNA enters the second tRNA binding site on the large subunit.

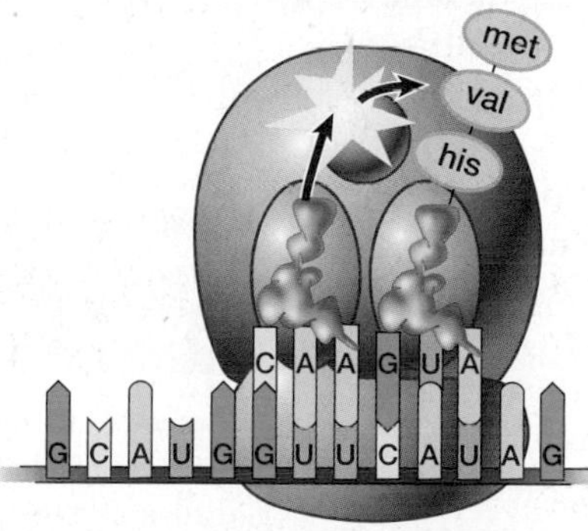

8 The catalytic site forms a peptide bond between valine and histidine, leaving the peptide attached to the tRNA in the second binding site. The tRNA in the first site leaves, and the ribosome moves one codon over on the mRNA.

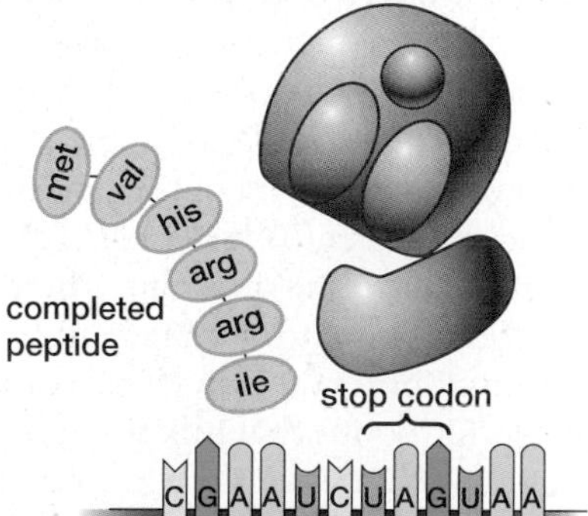

9 This process repeats until a stop codon is reached; the mRNA and the completed peptide are released from the ribosome, and the subunits separate.

▲ **FIGURE 13-6 Translation is the process of protein synthesis** Translation decodes the base sequence of an mRNA into the amino acid sequence of a protein.

amino acids also shifts, moving from the second to the first binding site of the ribosome. A new tRNA, with an anticodon complementary to the third codon of the mRNA, binds to the empty second site ❼. The catalytic site now joins the third amino acid to the growing protein chain ❽. The empty tRNA leaves the ribosome, the ribosome shifts to the next codon on the mRNA, and the process repeats, one codon at a time.

Termination: A Stop Codon Signals the End of Translation

When the ribosome reaches a stop codon in the mRNA, protein synthesis terminates. Stop codons do not bind to tRNA. Instead, the ribosome releases the finished protein chain and the mRNA ❾. The ribosome then disassembles into its large and small subunits.

SUMMING UP: Decoding the Sequence of Bases in DNA into the Sequence of Amino Acids in Protein

Let's summarize how a eukaryotic cell decodes the genetic information of DNA and uses the information to synthesize a protein (**FIG. 13-7**):

❶ With some exceptions, such as the genes for tRNA and rRNA, each gene codes for the amino acid sequence of a protein. The DNA of a gene consists of the template strand, which is transcribed into mRNA, and its complementary strand, which is not transcribed.

❷ In the nucleus, transcription produces an RNA molecule that is complementary to the template strand. This RNA molecule undergoes splicing to produce the final mRNA that will be translated. Sequences of three bases in mRNA, called codons, specify either the beginning of translation (the start codon, AUG), an amino acid, or the end of translation (a stop codon).

❸ Meanwhile, enzymes in the cytoplasm attach the specified amino acid to each tRNA. Each tRNA also includes a three-base sequence called an anticodon. An anticodon's sequence corresponds to the particular amino acid carried by a tRNA molecule.

❹ The mRNA moves out of the nucleus to a ribosome in the cytoplasm. Transfer RNAs carry their attached amino acids to the ribosome. There, the bases in tRNA anticodons bind to complementary bases in mRNA codons. Ribosomal RNA (rRNA) catalyzes the formation of peptide bonds that link the amino acids to form a protein with the amino acid sequence specified by the sequence of bases in the mRNA. When a stop codon is reached, the finished protein is released from the ribosome.

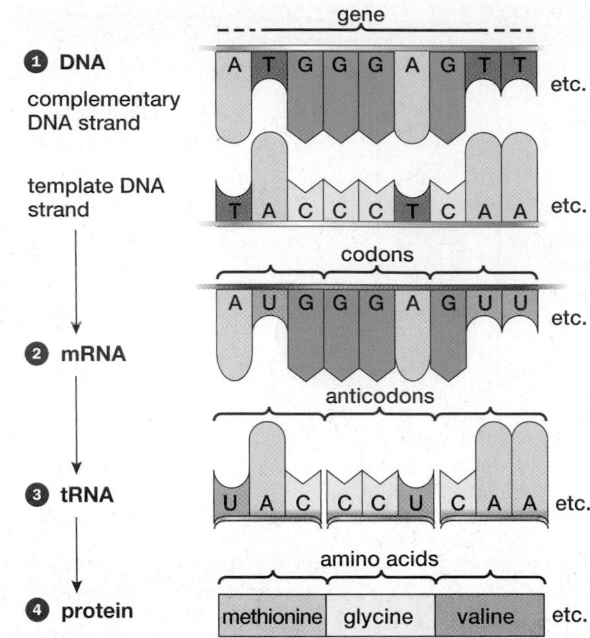

▲ **FIGURE 13-7 Complementary base pairing is required to decode genetic information**

THINK CRITICALLY Examine the mRNA sequence shown in step 2. If mutations changed all of the guanine bases visible in the sequence to uracil, how would the translated peptide differ from the one shown?

CHECK YOUR LEARNING

Can you . . .

- explain what an anticodon is?
- describe the role of the ribosome in protein synthesis?
- describe the process of translation, explaining how ribosomes, rRNA, mRNA, and tRNA interact to synthesize a peptide?
- summarize the steps by which the information in a DNA sequence directs the synthesis of a protein?

CASE STUDY CONTINUED

Cystic Fibrosis

Some of the many possible mutations in the *CFTR* gene result in a complete absence of correctly spliced mRNA molecules and cause severe cystic fibrosis. Other mutations in the gene seem to "confuse" the splicing machinery so that both correct and incorrect mRNA molecules are made. Many of the mutations that result in a defective *CFTR* gene involve changes to one or a few codons in an exon of the gene. How can relatively small changes in an exon's DNA sequence render a protein like CFTR nonfunctional?

13.4 HOW DO MUTATIONS AFFECT PROTEIN STRUCTURE AND FUNCTION?

Mistakes during DNA replication, ultraviolet rays in sunlight, chemicals in cigarette smoke, and a host of other environmental factors may cause mutations—changes in the sequence of bases in DNA. The consequences for an organism's structure and function depend on how the mutation affects the protein encoded by the mutated gene.

The Effects of Mutations Depend on How They Alter the Codons of mRNA

The types of mutations that may affect protein structure include substitutions, deletions, and insertions (see Fig. 12-7). Different types of mutations differ greatly in their likelihood of producing significant changes in protein structure and function.

Substitutions

In a **substitution mutation**, a single base pair in DNA is changed. A substitution within a protein-coding gene can produce one of four possible outcomes (**FIG. 13-8**). For example, consider four outcomes of substitution mutations in the gene that encodes the protein beta-globin, a subunit of hemoglobin that helps red blood cells carry oxygen to body tissues:

- ***The amino acid sequence of the protein may be unchanged.*** Because most amino acids are encoded by several different codons, some substitution mutations do not change the protein. Consider the sixth codon in the beta-globin mRNA, GAG, which specifies glutamic acid. If a substitution mutation changes this codon from GAG to GAA, the new codon still codes for glutamic acid. The protein synthesized from the mutated gene remains the same, even though the DNA sequence is different.
- ***The amino acid sequence may be altered, but protein function may be essentially unchanged.*** Many proteins have regions in which the exact amino acid sequence is relatively unimportant. For example, in beta-globin, the amino acids on the outside of the protein must be hydrophilic to keep the protein dissolved in the cytoplasm of red blood cells, but exactly *which* hydrophilic amino acids are on the outside doesn't matter much. Thus, a substitution mutation that changes the sixth codon in beta-globin mRNA from GAG to CAG replaces glutamic acid (hydrophilic) with glutamine (also hydrophilic) and does not affect the ability of beta-globin to function normally.

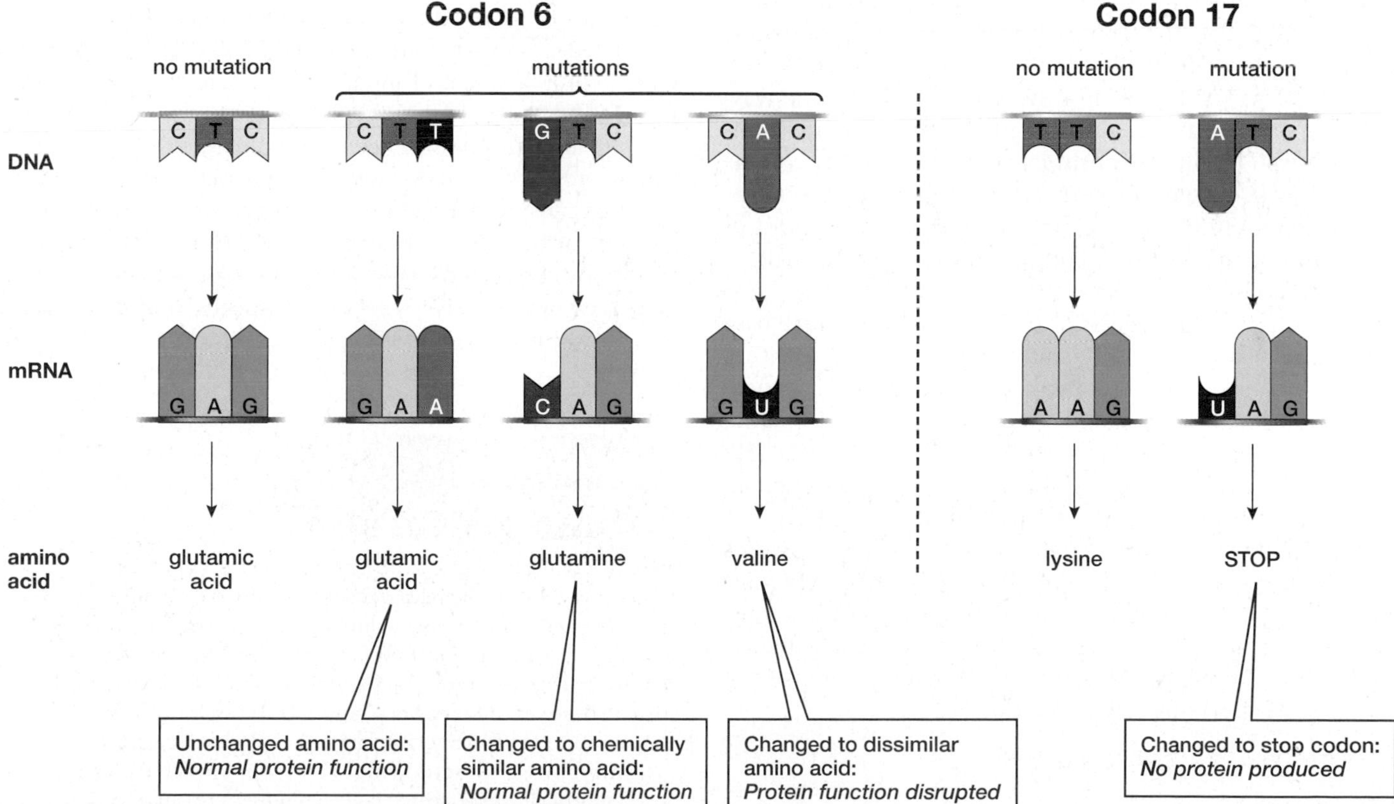

▲ **FIGURE 13-8 Substitution mutations can have four kinds of effect on protein function** The examples shown here are for mutations in the 6th and 17th codons of the human beta-globin gene.

- ***Protein function may be changed by an altered amino acid sequence.*** A substitution mutation that changes the sixth codon in beta-globin mRNA from GAG to GUG replaces glutamic acid (hydrophilic) with valine (hydrophobic). This substitution prevents beta-globin from functioning properly, and causes the genetic disease sickle-cell anemia (see Chapter 11).
- ***Protein function may be destroyed by a premature stop codon.*** The 17th codon in beta-globin mRNA, AAG, specifies lysine. A substitution mutation that changes the codon from AAG to UAG changes the lysine codon to a stop codon. This premature stop codon halts translation of beta-globin mRNA before the protein is completed; the protein is nonfunctional. Individuals who inherit this mutation from both parents have a condition called beta-thalassemia, which can be fatal unless treated with blood transfusions.

Deletions and Insertions

In a **deletion mutation**, one or more base pairs are removed from a gene. In an **insertion mutation**, one or more base pairs are inserted into a gene. If one or two pairs of nucleotides are deleted or inserted, protein function is usually completely ruined. To understand why, consider a sentence composed of three-letter words: THE DOG SAW THE CAT SIT AND THE FOX RUN. Deleting or inserting one letter changes all of the following words. If, for example, we delete the first E but still require three-letter words, all of the words in the sentence are changed: THD OGS AWT HEC ATS ITA NDT HEF OXR UN. In the same way, deleting or inserting one or two nucleotides, or any number that isn't a multiple of three, changes all of the codons that follow the deletion or insertion. In a protein synthesized from an mRNA containing such a mutation, most of the amino acids will be incorrect and the protein will be nonfunctional.

A mutation that deletes or inserts a group of three base pairs usually results in only small changes to the resulting protein, because only one or two codons are changed. Returning to our model sentence, let's suppose that we delete OGS. The sentence now reads: THE DAW THE CAT SIT AND THE FOX RUN, most of which is unchanged from the original. Similarly, if we add a new three-letter word, such as FAT, even in the middle of one of the original words, most of the words stay the same, as in THE DOG SAF ATW THE CAT SIT AND THE FOX RUN. As with substitution mutations that change an amino acid, the small changes in amino acid sequence caused by a three-base deletion or insertion may or may not have a large effect on protein function.

CHECK YOUR LEARNING

Can you . . .

- describe three different types of mutations?
- describe and explain the effects that each type of mutation can have on protein function?

CASE STUDY CONTINUED

Cystic Fibrosis

There are more than 1,900 different defective alleles of the *CFTR* gene. The most common one originated as a deletion mutation that removed three nucleotides—one codon. Losing this codon deletes a crucial amino acid from the CFTR protein, resulting in a misshapen protein that is quickly broken down by the cell. Other common defective *CFTR* alleles originated as substitutions that introduced a stop codon in the middle of the mRNA, terminating translation of the CFTR protein partway through. Still other alleles resulting from substitution mutations produce complete proteins, but with amino acid changes that render them unable to perform their normal function.

Some *CFTR* alleles can produce functional protein, but nevertheless cause cystic fibrosis. How can that be? These alleles affect when and how often the protein is produced. How can a gene affect the timing and rate of transcription and translation?

13.5 HOW IS GENE EXPRESSION REGULATED?

The complete human genome contains about 20,000 genes. All of these genes are present in the nucleus of each cell, but in any given cell only a small fraction of the genes is *expressed*—transcribed and, for genes that encode proteins, translated. Some genes are expressed in all cells because they encode proteins or RNA molecules that are essential for the life of any cell. For example, all cells must synthesize proteins, so they all transcribe the genes for tRNA, rRNA, and ribosomal proteins. Other genes are expressed only in certain types of cells, under certain conditions, or at certain times in an organism's life. For example, even though every cell in your body contains the gene for the milk protein casein, that gene is expressed only in certain breast cells, only in women, and only during a period of breast-feeding.

Have You Ever Wondered . . .

Why Bruises Turn Colors?

Bruises typically progress from purple to green to yellow. This sequence is visual evidence of the control of gene expression. If you bang your shin on a chair, blood vessels break and release red blood cells, which burst and spill their hemoglobin. Hemoglobin and its iron-containing heme group are dark bluish-purple in the deoxygenated state, so fresh bruises are purple. Heme, which if not broken down is toxic to the liver, kidneys, brain, and blood vessels, stimulates transcription of the heme oxygenase gene. Heme oxygenase is an enzyme that converts heme to biliverdin, which is green. Next, another enzyme, which is always present because its gene is always expressed, converts biliverdin to bilirubin, which is yellow. The bruise finally disappears as bilirubin moves to the liver, where it is secreted into the bile.

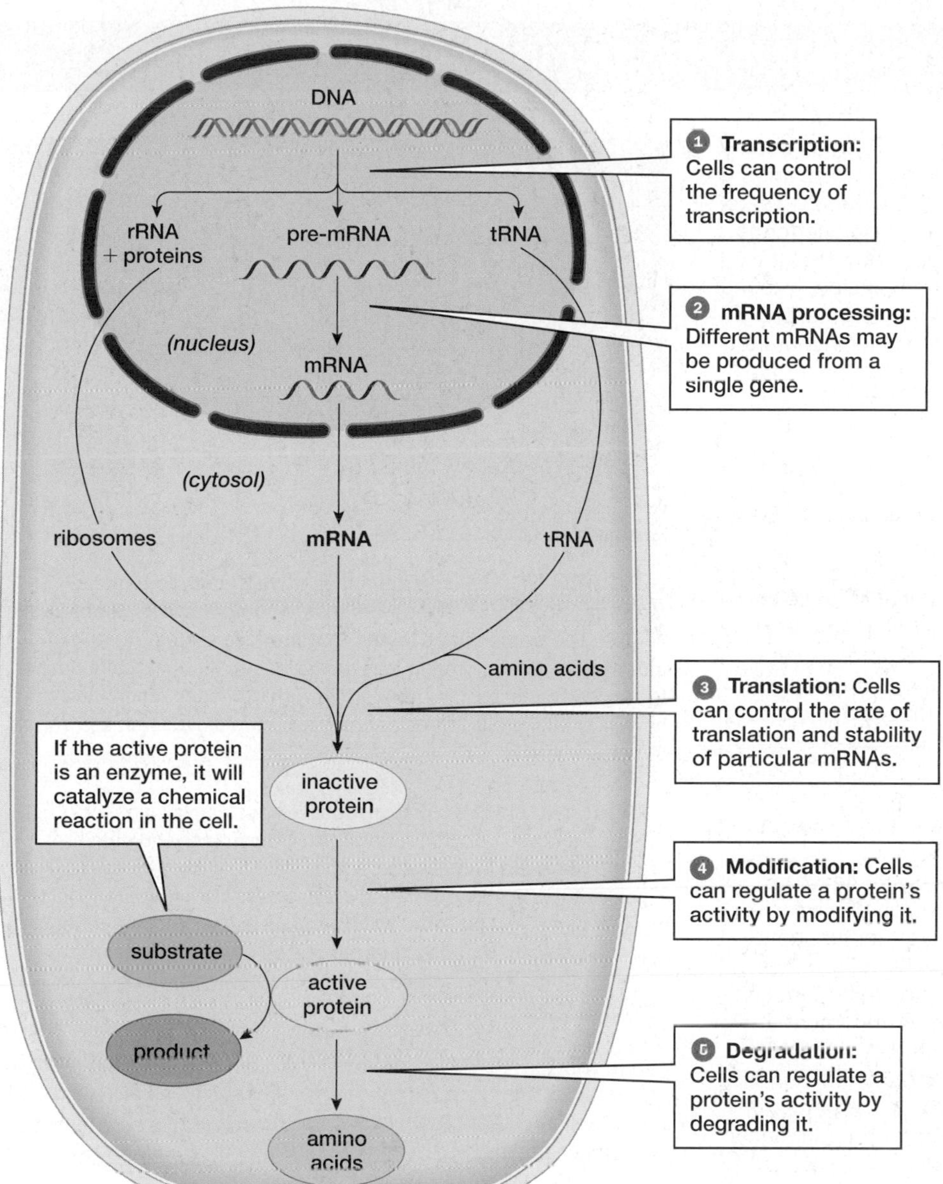

DNA
rRNA
+ proteins
pre-mRNA
tRNA
(nucleus)
mRNA
(cytosol)
ribosomes
mRNA
tRNA
amino acids
If the active protein is an enzyme, it will catalyze a chemical reaction in the cell.
inactive protein
substrate
active protein
product
amino acids
1 Transcription: Cells can control the frequency of transcription.
2 mRNA processing: Different mRNAs may be produced from a single gene.
3 Translation: Cells can control the rate of translation and stability of particular mRNAs.
4 Modification: Cells can regulate a protein's activity by modifying it.
5 Degradation: Cells can regulate a protein's activity by degrading it.

FIGURE 13-9 An overview of information flow in a eukaryotic cell Expression of a given gene may be regulated at all of these steps or just at some of them.

In Eukaryotes, Gene Expression Is Regulated at Many Levels

Gene expression in a eukaryotic cell is a multistep process, beginning with transcription of DNA and commonly ending with a protein performing a particular function. Regulation of gene expression can occur at any of these steps (**FIG. 13-9**):

1. Cells can control how frequently a gene is transcribed. The rate of transcription of a given gene differs among organisms, among cell types in a given organism, within a given cell type at different stages in the organism's life, and within a cell or organism depending on environmental conditions.
2. A single gene may produce different mRNAs and proteins. A single gene may produce more than one protein, depending on how the precursor mRNA is spliced to form the finished mRNA that is translated into protein.
3. Cells can control the stability and translation of mRNAs. Some mRNAs are long lasting and are translated into protein many times, whereas others are translated only a few times before they are broken down. In addition, cells may selectively block translation of some mRNAs or target some mRNAs for destruction.
4. Cells may modify proteins to regulate their activity. Many proteins, especially enzymes, may be modified after translation to change their function. Such changes may be temporary and reversible, thereby providing second-to-second control of a protein's activity. For example, some ion channels in a cell's plasma membrane open when a phosphate is added to a channel protein, and close when the phosphate is removed. Other modifications may be permanent, for example when an enzyme is converted from an inactive form to an active one.

Androgen Insensitivity Syndrome

Sometime before reaching 14 years of age, a girl usually goes through puberty: Her breasts swell, her hips widen, and she begins to menstruate. In rare instances, however, the years go by, but the girl never menstruates. If her physician performs a chromosome test, in some cases the results show that the girl's sex chromosomes are XY (the combination ordinarily found in males). The reason she has not begun to menstruate is that she lacks ovaries and a uterus but instead has immature testes inside her abdominal cavity. Further, she has about the same concentrations of androgens (male sex hormones such as testosterone) circulating in her blood, as would be found in most boys her age. In fact, androgens have been present in her body since early in her development. However, her cells cannot respond to them—a condition called androgen insensitivity syndrome.

Androgen insensitivity and other disruptions of male sexual development may be caused by any of 400 recessive alleles of the gene encoding a protein known as the androgen receptor. In typical males, androgens bind to this receptor protein, stimulating transcription of multiple genes that help produce male features, such as a penis and testes that descend into sacs outside the body cavity. But these features do not develop in individuals with androgen insensitivity, because their alleles for the receptor protein contain mutations encoding a premature stop codon that completely eliminates androgen receptor function.

The androgen receptor gene is on the X chromosome. A person who is genetically XY inherits a single allele for the androgen receptor. If this allele codes for nonfunctional androgen receptor proteins, then the person's cells will be unable to respond to testosterone, and male characteristics will not develop. In many respects, female development is the default option in humans; without functional androgen receptors, the body will develop female characteristics. Thus, a mutation that changes the nucleotide sequence of just one gene causes the production of a single type of nonfunctional protein, which results in a person who is genetically XY but female (**FIG. E13-1**).

▲ **FIGURE E13-1 Androgen insensitivity leads to female features** The cells of these women have both X and Y chromosomes. The women have testes that produce testosterone, but a mutation in their androgen receptor genes make their cells unable to respond to testosterone. Women with androgen insensitivity have normal life expectancies and lead normal lives but are unable to bear children.

THINK CRITICALLY Envision yourself as a physician. A mother, father, and their daughter come to you because the daughter is 16 years old and hasn't yet menstruated, whereas all of her girlfriends started menstruating years ago. You produce a karyotype and find that she is XY. Further molecular genetic testing reveals that she has a mutated androgen receptor allele on her X chromosome. The parents want to know how their daughter inherited the syndrome, why they don't have it, and, if any of their younger children might be androgen insensitive. How would you explain, in terms understandable to a layperson, the inheritance of androgen insensitivity and the likelihood that other children of these parents would have androgen insensitivity syndrome? Include diagrams to help them understand.

5. Cells can control the rate at which proteins are degraded. By preventing or speeding up a protein's degradation, a cell can rapidly adjust the amount of a particular protein it contains.

Next, let's examine a few of the mechanisms by which cells control transcription and translation.

Regulatory Proteins Binding to a Gene's Promoter Alter Its Rate of Transcription

The promoter regions of virtually all genes contain several sequences known as *response elements,* so named because they allow a cell to respond to changing conditions. Proteins called **transcription factors** may attach to a response element, enhancing or suppressing binding of RNA polymerase to the promoter and, consequently, enhancing or suppressing transcription of the gene.

In many cases, a transcription factor must be activated before it can affect gene transcription. For example, a transcription factor in birds that promotes transcription of albumin, the main protein in egg white, is activated by the hormone estrogen. The gene for albumin is not transcribed in winter when birds are not breeding and estrogen levels are low. During the breeding season, the ovaries of female birds release estrogen, which enters cells in the oviduct and binds to the transcription factor. The complex of estrogen and the transcription factor then attaches to an estrogen response element in the promoter of the albumin gene, making it easier for RNA polymerase to bind to the promoter and start transcribing mRNA. The mRNA is translated into large amounts of albumin for egg production. Similar activation of gene transcription by steroid hormones occurs in other animals, including humans. The importance of hormonal regulation of transcription during development is illustrated by genetic defects in which

Health WATCH The Strange World of Epigenetics

Most mechanisms for controlling gene expression work for time periods ranging from a few seconds to a few days and then fade away. Epigenetic controls, however, often remain in place for the life of an organism. For example, adding methyl groups to the promoter of the insulin gene turns off transcription unless and until the methyl groups are removed. In early embryos, all cells have methylated, silenced insulin genes. Later in development, methyl groups on the insulin gene are selectively removed in cells destined to become insulin-secreting cells of the pancreas. The other cells of the body retain methylated, silenced insulin genes.

Coat color in mice provides another example of epigenetic control of gene expression (**FIG. E13-2**). In certain strains of mice, the offspring in a single litter of genetically identical mice can have fur colors that range from yellow to brown. The color is controlled by methylation of a single gene; the greater the amount of methylation, the less the gene is expressed and the browner the fur. Mice with very little methylation of this gene have high gene expression and yellow fur (and they also become obese). Feeding pregnant mice a diet high in methionine, folate, soy protein, and vitamin B_{12} increases DNA methylation and produces litters in which all the pups are brown and not prone to obesity. Does this result show that epigenetic changes caused by a parent's experiences can be passed on to offspring?

In the vast majority of cases, methylation patterns on DNA are erased during meiotic cell division or gamete development, so epigenetic changes do not pass from generation to generation. However, there are exceptions. Methyl groups may be added to certain genes in the sperm or in the egg, resulting in genomic imprinting, in which one parent passes on a methylated version of a gene. In these cases, only the version of the gene inherited from the other parent is expressed. For example, Angelman syndrome, a rare genetic disease caused by a deletion mutation in chromosome 15, is inherited from the mother. The gene from the father is methylated (silenced) in his sperm and therefore cannot compensate for the mother's mutation.

Epigenetic changes that last for generations have been found in bacteria, protists, fungi, plants, and animals. These findings lead to an intriguing question: In people, can epigenetic changes caused by parents' life experiences or environment be inherited by their offspring?

▲ **FIGURE E13-2 Epigenetic differences can cause phenotypic differences in genetically identical mice** The obese yellow mouse has far fewer methyl groups on a gene that controls fur color than does the slim brown mouse.

For now, the answer seems to be "maybe." Evidence for multigenerational epigenetic inheritance in humans comes mainly from "natural experiments" in which some major event affected a fairly large number of people. For example, people whose fathers were conceived during the Dutch famine of 1944–1945 at the end of World War II are more likely to be obese than people whose fathers were conceived during times of plenty (and thus were not undernourished during their prenatal development). Another study in England found that when men start smoking at a very early age (before 11 years old), their sons tend to be overweight. These results are intriguing, but no one knows which genes were involved or whether epigenetic methylation of DNA caused the observed differences.

THINK CRITICALLY In some people with type 2 diabetes, the pancreas doesn't secrete enough insulin. If you could analyze the epigenetic control of the insulin gene and its promoter in such people, what would you expect to find? If you devised a treatment that could add or remove epigenetic controls on the gene, what would you do to attempt to normalize insulin secretion in these type 2 diabetics?

receptors for sex hormones are nonfunctional (see "Health Watch: Androgen Insensitivity Syndrome."

Epigenetic Controls Alter Gene Transcription and Translation

Epigenetics (which means "in addition to genetics") is the study of processes that change gene expression and function without changing the base sequence of DNA. In general, epigenetic control of gene expression involves (1) modification of DNA, (2) modification of chromosomal proteins, or (3) the action of particular kinds of RNA molecules collectively known as *noncoding RNA*. Some epigenetic modifications may be inherited from one generation to the next, as we explore in "Health Watch: The Strange World of Epigenetics."

Modification of DNA May Suppress Transcription

Certain enzymes in a cell add methyl groups ($-CH_3$) to cytosine bases in specific locations in the cell's DNA, a process called *methylation*. If a gene or its promoter has lots of methylated cytosines, the gene usually will not be transcribed into mRNA and its instructions will not be used to make proteins.

Modification of Chromosomal Proteins May Enhance Transcription

In eukaryotic chromosomes, DNA is wound around "spools" made of proteins called *histones* (see Chapter 9). When the DNA is tightly wound, RNA polymerase can't get to the promoters of genes, so transcription occurs slowly, if at all. However, when acetyl groups ($-COCH_3$) are added to histones, the DNA partially unwinds, giving RNA polymerase better access to the promoters and making gene transcription more likely.

Noncoding RNA May Regulate Transcription

Some DNA is transcribed into various types of noncoding RNA that help to control gene expression. Some types of noncoding RNA inhibit the binding of RNA polymerase to specific gene promoters, thereby blocking transcription. Others stimulate or inhibit epigenetic changes to DNA or histones, either enhancing or reducing transcription, depending on the particular modification involved.

Noncoding RNA May Regulate Translation

The quantity of any particular protein that a cell synthesizes depends in part on how much of the mRNA produced actually gets translated. Translation may be regulated by RNA interference, in which noncoding RNA is cut into very short strands that interfere with translation. In some cases, these short strands base-pair with specific segments of complementary mRNA, forming short sections of double-stranded RNA that cannot be translated. In other cases, noncoding RNA combines with enzymes to cut up complementary mRNA, which also prevents translation.

CHECK YOUR LEARNING

Can you . . .

- define *gene expression*?
- describe the general processes, at each step of the pathway from DNA to protein activity, by which cells may regulate gene expression?
- describe what transcription factors do, and explain how they do it?
- describe some epigenetic controls of gene expression?

CASE STUDY REVISITED

Cystic Fibrosis

All defective alleles of the *CFTR* gene are recessive to the functional *CFTR* allele. People who are heterozygous, with one normal *CFTR* allele and one copy of any of the defective alleles, produce enough functional CFTR protein to produce normal, watery secretions in their lungs; they do not develop cystic fibrosis. Someone with two defective alleles, however, will produce only proteins that don't work properly and will develop the disease.

Genetic diseases such as cystic fibrosis can't be "cured" in the way that an infection can be cured by killing offending bacteria or viruses. For many genetic diseases, treatment consists mainly of efforts to relieve the symptoms. For cystic fibrosis, such treatments include antibiotics, medicines that open the airways, and physical therapy to loosen thick secretions in the lungs.

The severity of the symptoms experienced by a person with cystic fibrosis depends on how defective the mutant alleles are. Canadian triathlete Lisa Bentley, for example, has a relatively mild case of cystic fibrosis (**FIG. 13-10**). Nonetheless, during a 9-hour triathlon, she produces copious amounts of very salty sweat. Why? One of the roles of the CFTR protein is to promote reabsorption of salt from sweat and transport the salt back into the blood. In cystic fibrosis, salt reabsorption fails, and people can lose life-threatening amounts of salt during exercise. Therefore, it's a constant challenge for Bentley to keep her body supplied with salt during a race. Despite this challenge, Bentley has won 11 Ironman triathlons, including the Australian Ironman Triathlon for 5 straight years. Bentley carefully controls her diet, especially her salt intake. Vigorous exercise helps to clear out her lungs. She also scrupulously avoids situations where she might be exposed to contagious diseases. Her cystic fibrosis hasn't kept her from becoming one of the finest athletes in the world.

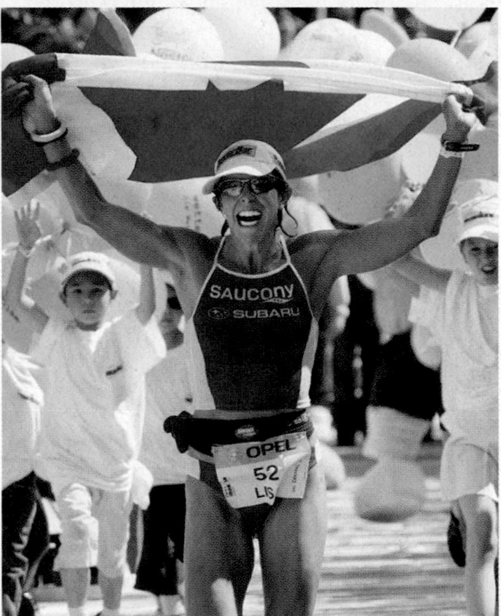

▲ **FIGURE 13-10 Lisa Bentley, sometimes called the Iron Queen, wins another triathlon**

CONSIDER THIS About 5% of cases of cystic fibrosis arise from a substitution mutation in which a full-length CFTR protein is synthesized but fails to function normally. In 2012, the U.S. Food and Drug Administration approved a drug called ivacaftor to treat this form of cystic fibrosis. Today, treatment with ivacaftor costs more than $300,000 per year. How do you think such treatments should be financed: by patients, by health insurance, or by governments?

CHAPTER REVIEW

Go to **Mastering Biology** to access the Pearson eText, vocabulary review, practice quizzes, activities, videos, current events, and more.

*Answers to **Think Critically** and **Thinking Through the Concepts** questions can be found in the **Answers** section at the back of the book.*

Summary of Key Concepts

13.1 How Is the Information in DNA Used in a Cell?

Genes are segments of DNA that can be transcribed into RNA and, for most genes, translated into protein. Transcription produces the three types of RNA needed for translation: mRNA, tRNA, and rRNA. Messenger RNA carries the genetic information of a gene from the nucleus to the cytoplasm, where ribosomes use the information to synthesize a protein. There are many different tRNAs. Each tRNA binds a specific amino acid and carries it to a ribosome for incorporation into a protein. Ribosomes are composed of rRNA and proteins, organized into large and small subunits. The genetic code of mRNA consists of codons, which are sequences of three bases that specify the start of translation (start codon), the amino acids in the protein chain, and the end of protein synthesis (stop codons).

13.2 How Is the Information in a Gene Transcribed into RNA?

When the cell requires the product of a gene, RNA polymerase binds to the promoter region of the gene and synthesizes a strand of RNA. This RNA is complementary to the template strand in the gene's DNA double helix. After a gene has been transcribed into mRNA, the mRNA molecule is usually modified—for example, by removing introns—prior to translation.

13.3 How Is the Nucleotide Sequence of mRNA Translated into Protein?

In eukaryotes, mRNA carries genetic information from the nucleus to the cytoplasm, where ribosomes use the information to synthesize a protein. The two ribosomal subunits come together at the start codon of the mRNA molecule to form a complete protein-synthesizing assembly. Transfer RNAs deliver the appropriate amino acids to the ribosome for incorporation into the growing protein, which depends on base pairing between the anticodon of the tRNA and the codon of the mRNA. Two tRNAs, each carrying an amino acid, bind simultaneously to the ribosome; rRNA in the large subunit catalyzes the formation of peptide bonds between the amino acids. As each new amino acid is attached, one tRNA detaches, and the ribosome moves over one codon, binding to another tRNA that carries the next amino acid specified by mRNA. Addition of amino acids to the growing protein continues until a stop codon is reached, causing the ribosome to disassemble and to release both the mRNA and the newly formed protein.

13.4 How Do Mutations Affect Protein Structure and Function?

A mutation is a change in the nucleotide sequence of a gene. Mutations can be caused by mistakes in base pairing during replication, by chemical agents, and by environmental factors such as radiation. Common types of mutations include substitutions, deletions, and insertions. Mutations vary in their effects on protein function. Some mutations produce codons that specify the same amino acid, or a very similar one, as the original codon; in these cases, protein function usually will not change significantly. Other mutations may encode a functionally different amino acid or a stop codon. These mutations may destroy protein function.

13.5 How Is Gene Expression Regulated?

Within a given cell, only certain genes are expressed. Expression of a gene requires that it be transcribed and, for most genes, also translated, and that the resulting RNA or protein perform some action within the cell. Which genes are expressed in a cell at any given time is determined by the function of the cell, the developmental stage of the organism, and the environment. Gene expression is regulated by a variety of mechanisms at various stages of the pathway from gene to protein activity. The amount of mRNA transcribed from a particular gene is often regulated in eukaryotes via transcription factors that bind to histones or DNA near the promoter. Transcription factors that add acetyl groups to histones enhance gene transcription, whereas those that add methyl groups to DNA suppress gene transcription. Noncoding RNA may suppress transcription or speed up mRNA degradation. The amount of protein translated from mRNA may be regulated by short strands of noncoding RNAs that inhibit translation by binding to mRNA. Some proteins, once synthesized, may be regulated by chemical modifications that cause them to either function or stop functioning.

Thinking Through the Concepts

Bloom's: Remembering, Understanding

Multiple Choice

1. The molecule that carries the genetic information of DNA from the nucleus to the cytoplasm to be used in protein synthesis is
   a. ribosomal RNA.
   b. messenger RNA.
   c. transfer RNA.
   d. noncoding RNA.
2. Which of the following is *not* true of RNA?
   a. It contains the base thymine.
   b. It contains the sugar ribose.
   c. It contains the base adenine.
   d. It is transcribed from DNA.
3. A stop codon
   a. signals the end of protein synthesis on a ribosome.
   b. codes for the amino acid methionine.
   c. signals the end of RNA synthesis.
   d. marks the boundary between an exon and an intron.
4. If the template strand of DNA reads ATTCGTAG,
   a. the complementary DNA strand reads UAAGCAUC.
   b. the base sequence of the RNA transcribed from this sequence is AUUCGUAG.
   c. the base sequence of the RNA transcribed from this sequence is UAAGCAUC.
   d. this segment of DNA contains the information for four codons.

5. Epigenetic modification of gene expression
   a. involves changes to a DNA sequence.
   b. always stimulates gene expression.
   c. is erased from the DNA following mitotic cell division.
   d. may sometimes be transmitted from generation to generation.

Fill-in-the-Blank

1. Synthesis of RNA from the instructions in DNA is called _________. Synthesis of a protein from the instructions in mRNA is called _________. Which structure in the cell is the site of protein synthesis? _________
2. The three types of RNA that are essential for protein synthesis are _________, _________, and _________. Another type of RNA, which can interfere with translation, is called _________.
3. The genetic code uses _________ (how many?) bases to code for a single amino acid. This sequence of bases in mRNA is called a(n) _________. The complementary sequence of bases in tRNA is called a(n) _________.
4. The enzyme _________ synthesizes RNA from the instructions in DNA. For any given gene, only one DNA strand, called the _________ strand, is transcribed. To begin transcribing a gene, this enzyme binds to a specific sequence of DNA bases located at the beginning of the gene. This DNA sequence is called the _________. Transcription ends when the enzyme encounters a DNA sequence at the end of the gene called the _________.
5. Translation begins with the _________ codon of mRNA and continues until a(n) _________ codon is reached. Individual amino acids are brought to the ribosome by _________. These amino acids are linked into protein by _________ bonds.
6. If one nucleotide is replaced by a different nucleotide, this is called a(n) _________ mutation. _________ mutations occur if one or more nucleotides are added in the middle of a gene. _________ mutations occur if one or more nucleotides are removed from the middle of a gene.

Review Questions

1. How does RNA differ from DNA?
2. Name the three types of RNA that are essential to protein synthesis. What is the function of each?
3. Define the following terms: *genetic code, codon,* and *anticodon*. What is the relationship among the bases in DNA, the codons of mRNA, and the anticodons of tRNA?
4. In eukaryotes, how is mRNA formed from a gene?
5. Diagram and describe protein synthesis.
6. Explain how complementary base pairing is involved in both transcription and translation.
7. Describe the principal mechanisms of regulating gene expression.
8. Define *mutation*. Describe four different possible effects of substitution mutations on protein sequence and function.

Applying the Concepts

Bloom's: Applying, Analyzing, Evaluating

1. If you were to compare an mRNA molecule ready for translation in a eukaryotic cell to the section of DNA (gene) from which the mRNA was transcribed, would the gene be longer, would the mRNA be longer, or would both be the same size? Explain your answer.
2. The defective *CFTR* alleles that cause cystic fibrosis vary tremendously in the severity of their effects. Some produce completely nonfunctional proteins, while others produce proteins that can function, but not as efficiently as the normal proteins do. Many people with cystic fibrosis are heterozygotes: They have two different defective *CFTR* alleles. How might the severity of cystic fibrosis symptoms depend on the allele combination in heterozygotes? As good as the better-functioning allele? As bad as the worse-functioning one? Somewhere in between? Why?

14 Biotechnology

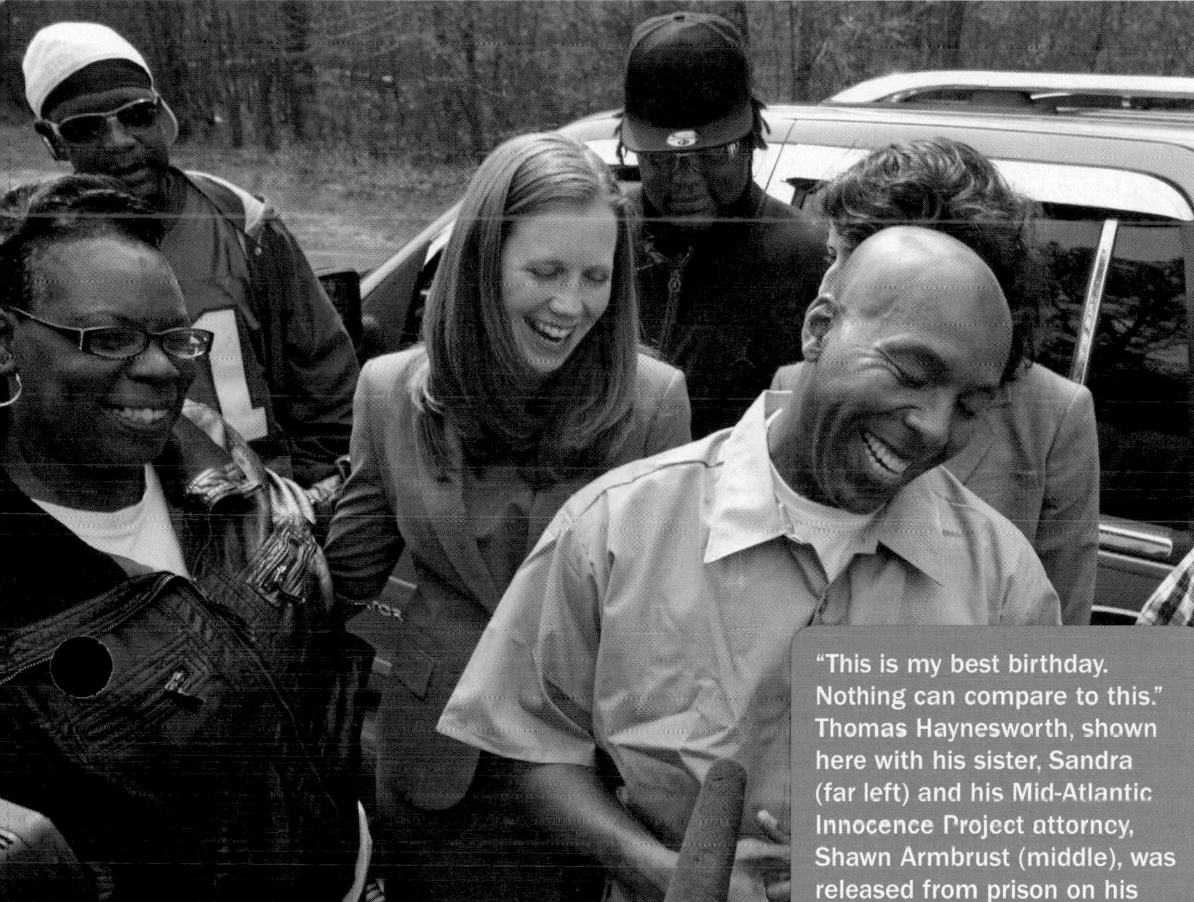

"This is my best birthday. Nothing can compare to this." Thomas Haynesworth, shown here with his sister, Sandra (far left) and his Mid-Atlantic Innocence Project attorney, Shawn Armbrust (middle), was released from prison on his 46th birthday.

CASE STUDY

Guilty or Innocent?

Thomas Haynesworth spent 27 years living a waking nightmare, locked up in prison for crimes he did not commit. His ordeal began in early 1984, when five women were sexually assaulted in the East End neighborhood of Richmond, Virginia. Shortly thereafter, Haynesworth, then 18 years old, was walking to the grocery store when he was spotted by one of the assault victims. She identified him as her assailant, and the other four women subsequently picked him out of a photo lineup. Haynesworth was swiftly convicted of rape, attempted robbery, and kidnapping.

But after Haynesworth was imprisoned, the barrage of rapes in the East End continued. Between April and December, 12 additional women were raped. In each case, the victim described her attacker as a man who, like Thomas Haynesworth, was young and black. Finally, in late December, police arrested Leon Davis, and the epidemic of sexual assaults ended. Davis was tried, convicted, and sentenced to multiple life terms.

Even though Davis's crimes were extremely similar to those for which Haynesworth had been convicted, the authorities did not revisit Haynesworth's case. Years later, though, his case file was one of thousands reviewed under a 2005 directive from then-Governor Mark Warner, who ordered that any biological evidence remaining in the files be reexamined. Finally, in 2009, a lab tested DNA that had been preserved in one of the Haynesworth files. The test results showed that Haynesworth was innocent of the crimes for which he had been convicted; the actual rapist was Leon Davis.

You might think that Haynesworth, now exonerated, would be set free immediately. Instead, he continued to languish in prison. He might still be there had his case not been taken up by the Mid-Atlantic Innocence Project, an organization based at the George Washington University Law School. With the help of Innocence Project staff, Haynesworth was released on parole on March 21, 2011. On December 6, he was declared innocent of all the charges against him.

What methods do crime scene investigators use to analyze and compare DNA samples? How can related methods of manipulating DNA be used to produce useful products and medical treatments? Should these kinds of technologies be used to change the genetic makeup of crops, livestock, or even people?

AT A GLANCE

14.1 WHAT IS BIOTECHNOLOGY?

Biotechnology is the manipulation or modification of living organisms or their components to produce useful products or services. Some kinds of biotechnology are ancient. For example, people have been using yeasts to produce beer, wine, and bread for thousands of years, and domestication of plants and animals also has a long history. People first domesticated wheat, grapes, dogs, pigs, and cattle between 6,000 to 15,000 years ago. In most cases, domestication was accompanied by *selective breeding*, in which people select and breed individual plants or animals with desirable traits, thereby enhancing these traits over many generations. For example, selective breeding transformed relatively slim wild boars, with long tusks and fierce temperaments, into much heavier, more placid domestic pigs. Because selective breeding is in essence a technique for changing the genetic makeup of populations of domestic organisms, it is a method of genetic manipulation.

Although selective breeding remains an important part of biotechnology, modern biotechnology also uses **genetic engineering** to manipulate cells or organisms by adding, deleting, or modifying genes. Genetic engineering can be used to improve plants and animals for agriculture, to combat disease, to produce valuable biological molecules, and to study how cells and genes work. Organisms containing DNA that has been modified through genetic engineering are called **genetically modified organisms (GMOs)**.

A key tool in modern biotechnology is **recombinant DNA**, which is DNA that contains genes or parts of genes from two or more organisms, usually of different species. Recombinant DNA can be produced in bacteria, viruses, or yeast and then transferred into other species. GMOs in which the modification includes the transfer of genes from another species are known as **transgenic** organisms.

Transgenic organisms and many of the tools of biotechnology originated as the result of basic biological research. These tools have, in turn played a huge role in the continuing advance of biological knowledge. Consider the Human Genome Project. Completed in 2003, the project discovered the entire 3-billion-base-pair sequence of a human genome, at a cost of about $2.7 billion. The researchers who worked on the project developed a host of new methods in order to complete their work, sparking a wave of innovation by biotechnology companies. Today, sequencing a human genome costs about $1,000, and will cost even less in the future. Whole-genome sequencing is now routine in both biotechnology applications and scientific research labs.

The rapid advance of gene sequencing and other DNA technologies has helped drive an improved understanding of human genetics that is having an enormous impact on medical practice. The World Health Organization estimates that more than 10,000 human diseases are caused or made more likely to occur by defective alleles, and researchers are identifying an ever-increasing number of such alleles. As a result, a growing number of defective alleles can be identified by genetic testing, and the results of such tests increasingly inform medical treatment decisions.

In this chapter, we will provide an overview of the methods and applications of biotechnology and discuss the impacts of biotechnology on society. We will organize our discussion around six major themes: (1) recombinant DNA mechanisms found in nature, (2) key techniques used in biotechnology, (3) DNA profiles and criminal forensics, (4) genetically modified plants and animals, (5) applications of biotechnology in medicine, and (6) ethical issues surrounding biotechnology.

CHECK YOUR LEARNING

Can you . . .

- define biotechnology?
- describe applications of genetic engineering?
- define GMO and transgenic organism?

14.2 WHAT NATURAL PROCESSES RECOMBINE DNA BETWEEN SPECIES?

Some recombinant DNA technologies are based on natural processes that transfer DNA from one organism to another. Two such processes are bacterial transformation and viral replication.

Transformation May Combine DNA from Different Bacterial Species

In **transformation**, bacteria acquire pieces of DNA from the environment (**FIG. 14-1**). The DNA may be part of the chromosome from another bacterium (**FIG. 14-1a**), sometimes from another species, or it may be tiny circular DNA molecules called **plasmids** (**FIG. 14-1b**). A single bacterium may contain dozens or even hundreds of copies of a plasmid.

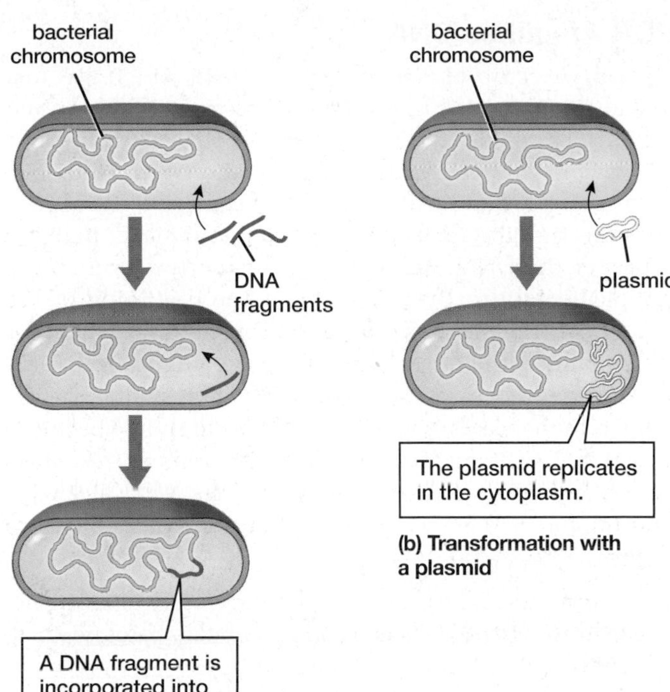

▲ **FIGURE 14-1 Transformation in bacteria** Bacterial transformation occurs when living bacteria take up **(a)** fragments of chromosomes or **(b)** plasmids.

When the bacterium dies, its plasmids are released into the environment, where they may be acquired by other bacteria of the same or different species. In addition, living bacteria can often pass plasmids directly to other living bacteria, or even to yeast or plant cells.

Viruses May Transfer DNA Between Species

Viruses sometimes transfer some of their host's genetic material to a different host species. Understanding this process requires some background on how viruses replicate.

Viruses can replicate only with the help of a cell (**FIG. 14-2**). The process begins when a virus attaches to specific molecules on the surface of a suitable host cell ❶. Usually the virus then enters the cytoplasm of the host ❷, where it releases its genetic material ❸. The host cell replicates the viral genetic material and synthesizes viral proteins ❹. The replicated viral genes and proteins assemble into new viruses inside the cell ❺. Eventually, the viruses are released and may infect other cells ❻.

In some cases, the cycle of viral replication may cause genes to be transferred from one organism to another. In these instances, viral DNA is inserted into one of the host cell's chromosomes (see Fig. 14-2 ❸), where it may remain for days, months, or even years. Every time the cell divides, it replicates the viral DNA along with its own DNA. When new viruses are finally produced, some of the host cell's genes may be attached to the DNA in the daughter viruses. If these now-recombinant viruses infect another cell and insert their DNA into the new host cell's chromosomes, pieces of the previous host's DNA may also be inserted. Most of the time, host DNA that is transferred in this manner will move between different individuals of a single species. However, some viruses can infect more than one species and may therefore transfer genes from one species to another.

CHECK YOUR LEARNING

Can you . . .

- describe natural processes that recombine DNA, including mechanisms that may combine DNA across species?

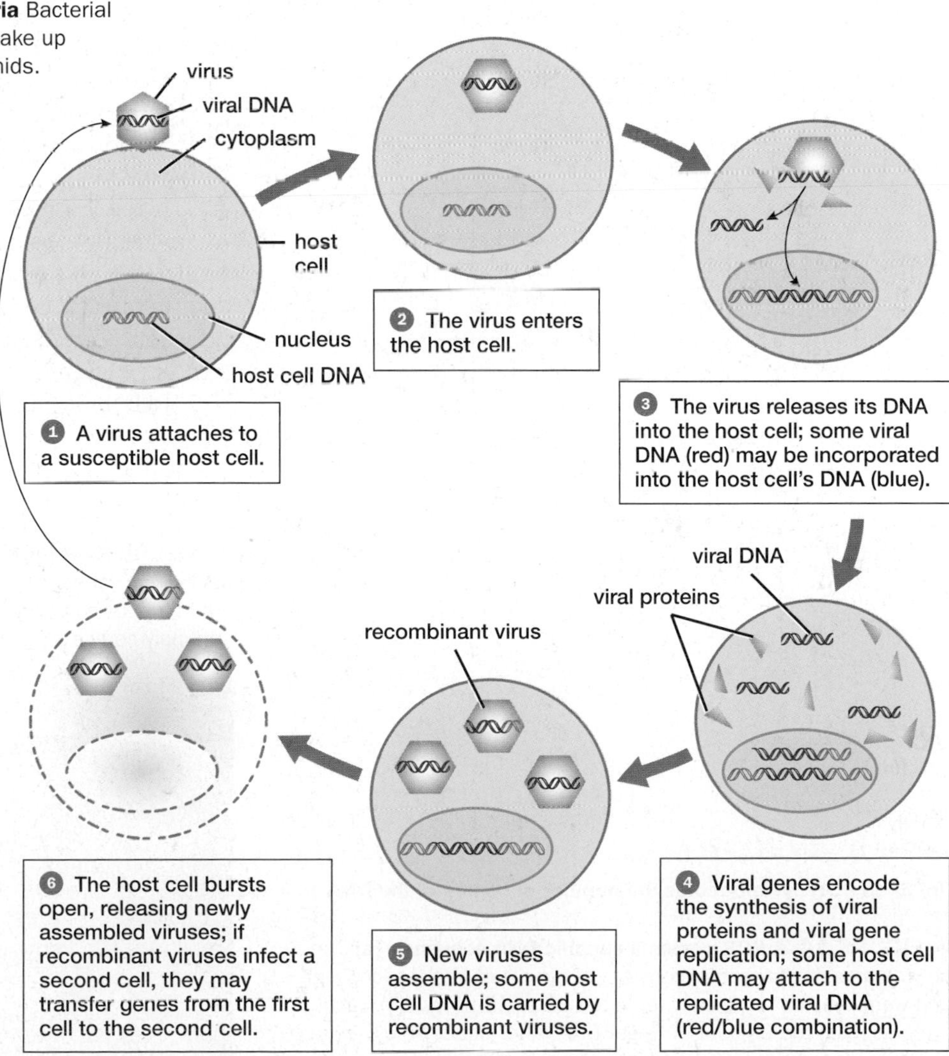

virus
viral DNA
cytoplasm
host cell
nucleus
host cell DNA
❶ A virus attaches to a susceptible host cell.
❷ The virus enters the host cell.
❸ The virus releases its DNA into the host cell; some viral DNA (red) may be incorporated into the host cell's DNA (blue).
viral DNA
viral proteins
❹ Viral genes encode the synthesis of viral proteins and viral gene replication; some host cell DNA may attach to the replicated viral DNA (red/blue combination).
recombinant virus
❺ New viruses assemble; some host cell DNA is carried by recombinant viruses.
❻ The host cell bursts open, releasing newly assembled viruses; if recombinant viruses infect a second cell, they may transfer genes from the first cell to the second cell.

▶ **FIGURE 14-2 The life cycle of a typical virus** In some cases, viral infections may transfer DNA from one host cell to another.

14.3 WHAT ARE SOME KEY METHODS FOR MANIPULATING DNA?

Before we turn to examples of how biotechnology is used, we will describe two techniques that play especially important roles in the field. These techniques are such key components of the biotech toolbox that researchers and bioengineers know them by their acronyms: PCR and CRISPR-Cas9.

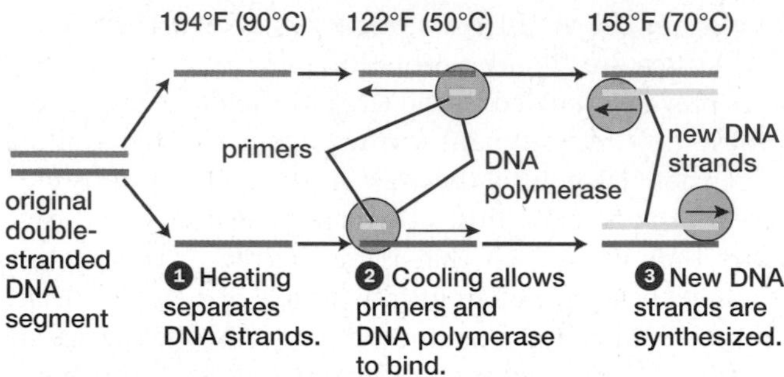

(a) One PCR cycle

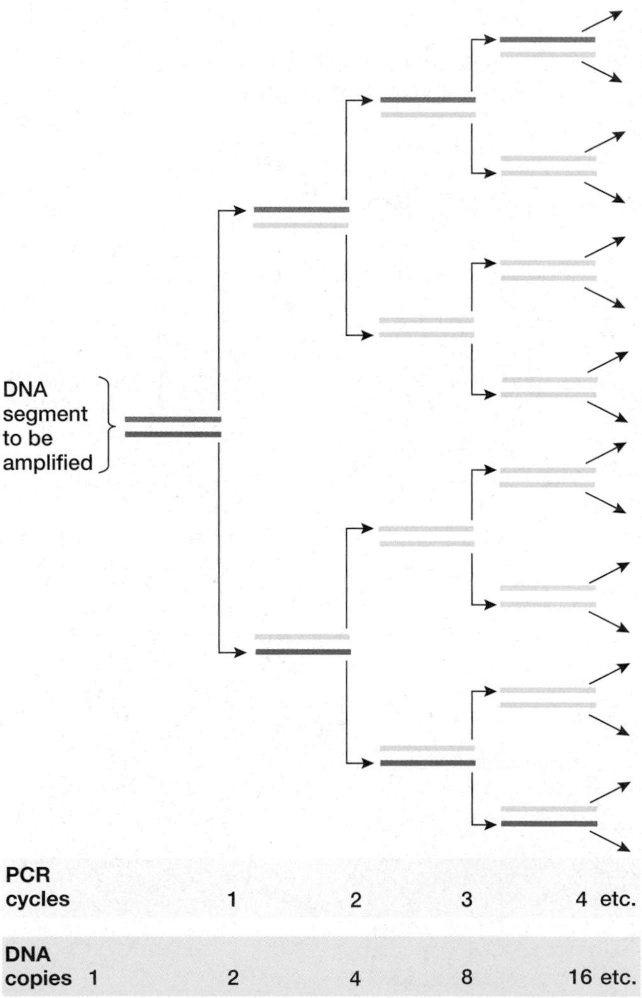

(b) Each PCR cycle doubles the number of copies of the DNA

▲ **FIGURE 14-3 PCR copies a specific DNA sequence (a)** The polymerase chain reaction consists of a cycle of heating, cooling, and warming that is typically repeated 30 to 40 times. **(b)** Each cycle doubles the amount of target DNA. After about 30 cycles, a billion copies of the target DNA have been synthesized.

PCR Amplifies DNA

The **polymerase chain reaction (PCR)**, which was first developed in the 1980s, can be used to make billions, even trillions, of copies of selected pieces of DNA. PCR involves two major steps: (1) synthesizing two short pieces of DNA, called *primers,* that identify the DNA segment to be copied, and (2) running repetitive reactions to make multiple copies of the DNA. The sequences of the two primers are complementary to the sequences at the beginning of the *target DNA* (the piece of DNA to be copied), one primer for each strand of the target. During the copying process, DNA polymerase recognizes the sequences complementary to the primers as the places where DNA replication should begin.

In PCR, the target DNA is mixed with primers, free nucleotides, and DNA polymerase in a small test tube. The reaction mixture is then cycled through a series of temperature changes (**FIG. 14-3a**):

1. The mixture is heated to a high temperature to break the hydrogen bonds between complementary bases, separating the DNA into single strands ❶.
2. The temperature is lowered to allow the two primers to form complementary base pairs with the beginning of the target DNA on each strand ❷.
3. The temperature is warmed, and DNA polymerase uses the free nucleotides to make copies of the DNA segment bounded by the primers ❸. PCR uses a DNA polymerase isolated from bacteria that live in hot springs (**FIG. 14-4**), because most other DNA polymerases do not function at the high temperatures required during the procedure.
4. This cycle is repeated, usually 30 to 40 times, until the free nucleotides have been used up. In PCR, the amount of DNA doubles with every temperature cycle (**FIG. 14-3b**). Twenty PCR cycles make about a million copies, and about 30 cycles make a billion copies. Each cycle takes only a few minutes, so PCR can produce billions of copies of a DNA segment in an afternoon. The DNA is then available for forensics, cloning, making transgenic organisms, or many other purposes.

▼ **FIGURE 14-4 PCR uses polymerase from a heat-loving bacterium** Hot springs like this one in Yellowstone National Park are home to the bacterium *Thermus aquaticus,* which produces the DNA polymerase used in PCR.

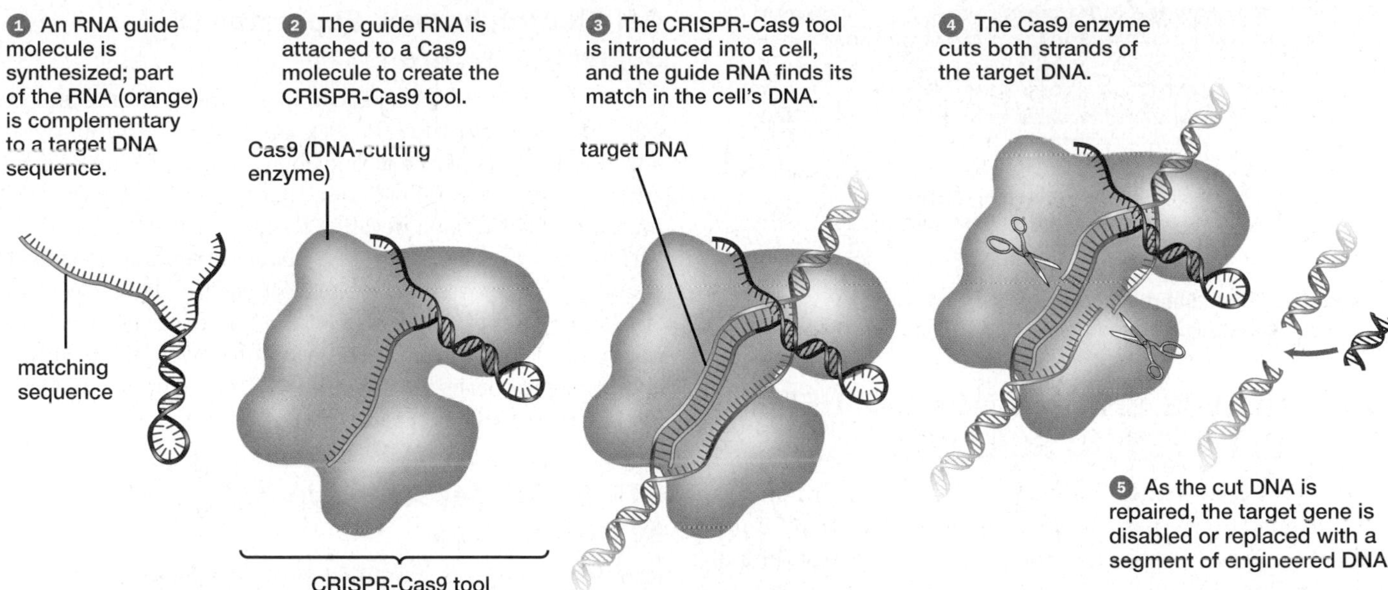

1 An RNA guide molecule is synthesized; part of the RNA (orange) is complementary to a target DNA sequence.
matching sequence
2 The guide RNA is attached to a Cas9 molecule to create the CRISPR-Cas9 tool.
Cas9 (DNA-cutting enzyme)
CRISPR-Cas9 tool
3 The CRISPR-Cas9 tool is introduced into a cell, and the guide RNA finds its match in the cell's DNA.
target DNA
4 The Cas9 enzyme cuts both strands of the target DNA.
5 As the cut DNA is repaired, the target gene is disabled or replaced with a segment of engineered DNA.

▲ **FIGURE 14-5 A CRISPR-Cas9 tool cuts DNA at any desired location**

CRISPR-Cas9 Allows Precise Editing of DNA

CRISPR-Cas9—often called just CRISPR, pronounced "crisper"—is a molecular tool for making precisely targeted changes to a cell's DNA. The key to its effectiveness is its ability to cut a DNA molecule at any desired sequence. Although biotechnologists had developed other methods for cutting a targeted DNA sequence, most of these methods required a uniquely engineered enzyme to cut each DNA sequence of interest. Using these methods to edit a target gene was slow, difficult, and expensive. CRISPR-Cas9, in contrast, uses an all-purpose enzyme—Cas9—that can cut any DNA sequence; CRISPR-Cas9 tools can be built far more quickly, easily, and cheaply than earlier gene editing tools.

CRISPR-Cas9 Is Derived from Natural Bacterial Defenses

The CRISPR-Cas9 tool is based on a natural defense mechanism that some bacteria use to chop up the genetic material of viruses that infect them. The RNA molecules that play a key role in this mechanism are encoded by bacterial DNA sequences called *clustered regularly interspaced short palindromic repeats*—CRISPR for short. In the early 2010s, researchers adapted this bacterial defense system into a powerful gene editing tool.

A CRISPR-Cas9 Tool Consists of RNA and Cas9 Enzyme

To build a CRISPR-Cas9 tool, a biologist first synthesizes an RNA molecule that includes a guide sequence that is complementary to the gene to be edited (**FIG. 14-5** ❶). The completed RNA molecule is then attached to a Cas9 enzyme molecule, creating a CRISPR-Cas9 tool ❷. The tool is introduced to a cell, where the guide RNA finds its complementary match in the cell's genome ❸. The Cas9 enzyme binds to the target gene, and the enzyme's active sites cut both strands of the DNA ❹. The cell's DNA repair mechanism reconnects the ends of the severed DNA molecule, but in so doing often introduces a small deletion or insertion that disables the gene. Or, if a biotechnologist has also introduced a segment of DNA (say, an alternative version of the cut gene), the cell's DNA repair mechanism usually inserts the new gene at the site of the cut ❺.

CHECK YOUR LEARNING

Can you . . .

- summarize how PCR works and explain why it is useful?
- summarize how CRISPR-Cas9 works and explain why it is useful?

CASE STUDY CONTINUED

Guilty or Innocent?

The evidence available to Innocence Project lawyers is often quite old. When a semen sample was found in one of the Haynesworth case files in 2009, the sample was 25 years old. Fortunately, DNA doesn't degrade very quickly. Forensic lab technicians used PCR to copy the DNA in the sample so that they had enough material to analyze. How did the lab use the copied DNA to determine that it came from Leon Davis, and not Haynesworth?

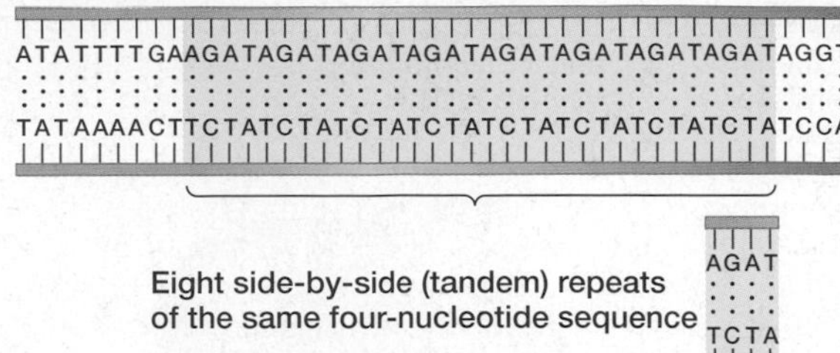

▲ **FIGURE 14-6 Short tandem repeats** This STR locus contains the sequence AGAT, repeated from 7 to 15 times in different alleles.

14.4 HOW IS BIOTECHNOLOGY USED IN FORENSIC SCIENCE?

In the few decades since the 1985 trial of the first criminal to be convicted on the basis of DNA evidence, DNA analysis has become one of the most commonly used tools in criminal investigations.

Differences in Short Tandem Repeats are Used to Identify Individuals by Their DNA

To analyze DNA left at a crime scene, forensics experts use segments of DNA called **short tandem repeats (STRs)**. STRs are ***short*** (about 20 to 250 nucleotides), ***repeating*** (consisting of the same sequence of 2 to 5 nucleotides repeated up to 50 times), and ***tandem*** (having all of the repetitions side by side). You can think of STRs as very small genes (**FIG. 14-6**). Different STRs occur at different loci in the genome, and, as with any gene, there may be alternative forms, or alleles, at each locus. The alleles of a given STR have different numbers of repeats of the same short nucleotide sequence. In forensic analysis, the STR alleles present in a suspect's DNA are compared to the alleles present in DNA from the crime scene.

A forensic lab usually begins its analysis of STR alleles by using PCR to amplify a DNA sample, increasing the size of the sample so that it is large enough to analyze. The PCR uses primers that amplify only the target STRs and the DNA directly adjacent to them. The primers also contain dye molecules, with a different color attached to each different STR primer.

Because STR alleles vary in how many repeats they contain, they vary in size; an STR allele with more repeats is larger than one with fewer repeats. Therefore, the forensic lab needs to identify the STRs in a DNA sample and determine the size of each one. In modern labs, sophisticated machines identify STRs by detecting the colors of their attached dye molecules, and determine the alleles of each STR by separating DNA segments by size.

Gel Electrophoresis Separates DNA Segments by Size

A mixture of DNA fragments can be separated by a technique called **gel electrophoresis** (**FIG. 14-7**). In gel electrophoresis, a DNA mixture is placed at one end of a slab of semisolid gel, and an electric field is applied so that the end where the DNA mixture is loaded has a negative charge and the far end of the gel has a positive charge. Because DNA fragments are negatively charged, they move toward the positively charged electrode. Smaller fragments slip through the gel more easily than larger fragments do, so they move more rapidly toward the positively charged electrode. Eventually, the DNA fragments are separated by size, forming distinct bands on the gel.

STR Genotypes Are Revealed by DNA Profiles

Running a sample of amplified STRs on a gel produces a pattern called a **DNA profile** (**FIG. 14-8**). The positions of the bands on the gel are determined by the numbers of repeats of the short nucleotide sequence of each STR allele. In the United States, a profile includes a standard set of 13 STR loci established by the Department of Justice. Most crime labs also include a gene that shows whether the DNA sample came from a man or a woman.

A DNA profile reveals a person's genotype at each STR locus. As with any gene, a person may be homozygous for a given STR (both alleles at the locus have the same number of repeats) or heterozygous (each allele has a different number of repeats). For example, in the D16 STR samples shown on the right side of Figure 14-8, the first person's DNA has a single band at 12 repeats (this person is homozygous for the D16 STR), but the second person's DNA has two bands—at 13 and 12 repeats (this person is heterozygous for the D16 STR). Also notice in Figure 14-8 that although some alleles are shared among people (for example, the second, fourth, and fifth samples for D16), no one's DNA had the same number of repeats for all four STRs.

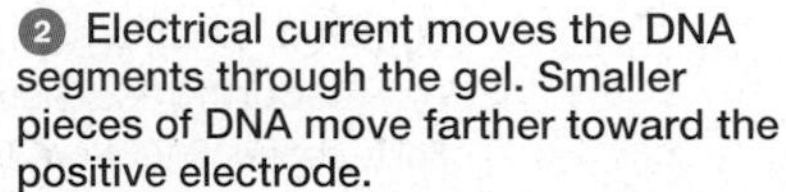
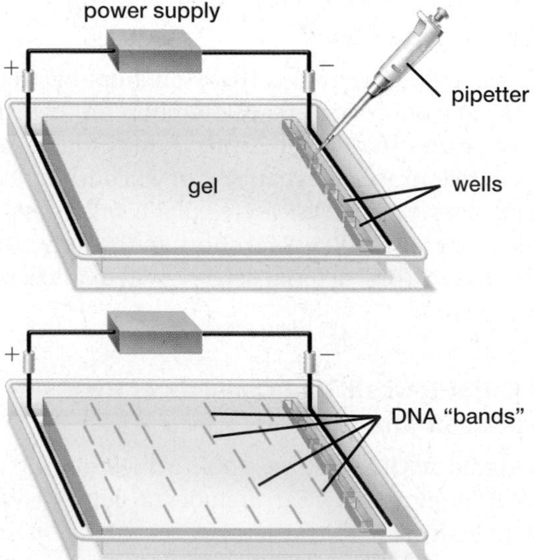

1 DNA samples are pipetted into wells (shallow slots) in the gel. Electrical current is sent through the gel (negative at the end with the wells and positive at the opposite end).
power supply
+
–
pipetter
gel
wells
2 Electrical current moves the DNA segments through the gel. Smaller pieces of DNA move farther toward the positive electrode.
DNA "bands"

▶ **FIGURE 14-7 Gel electrophoresis separates segments of DNA**

Unrelated People Almost Never Have Identical DNA Profiles

Are 13 STRs enough to uniquely identify a person, given the huge human population? To help answer this question, let's look at a simple hypothetical case. Imagine that a crime lab analyzes five STRs, each with 10 possible alleles (for example, all people have either 6, 7, 8, 9, 10, 11, 12, 13, 14, or 15 repeats). Let's also assume that all 10 alleles occur with equal probability in the human population (that is, a person has a 1 in 10, or 1/10, chance of having any one of the alleles). Finally, assume that the STRs are independently assorted (see page 164). In this scenario, the probability that two unrelated people will share the same alleles of all five STRs is the product (multiplication) of the separate probabilities, or $1/10 \times 1/10 \times 1/10 \times 1/10 \times 1/10 = 1$ chance in 100,000. (Note that close relatives would have a higher likelihood of sharing STR alleles, and identical twins would share all of their STR alleles.)

In actual DNA profiling, the chances of an erroneous match are much smaller than 1 in 100,000, because the number of STR loci tested and the number of alleles at each locus are higher than in our hypothetical example. Worldwide, different people may have as few as 3 to as many 50 repeats in a given STR. With 13 STRs, each with dozens of possible alleles, the chances of a random match are incredibly small. If two DNA samples match perfectly for both alleles at all 13 STR loci used in the United States, the probability is far less than one in a trillion that the samples match purely by chance. The odds of misidentifying a suspect in a criminal case are extremely low. Finally, a *mismatch* in DNA profiles is absolute proof that two samples did not come from the same person.

The United States Maintains a Database of DNA Profiles

In the United States, anyone convicted of certain crimes must give a blood sample. Crime lab technicians then determine the criminal's DNA profile and code the results as the number of repeats in each STR. The profile is stored in computer files at a state agency, at the FBI, or both. (Detectives on TV crime shows often refer to "CODIS," which stands for "Combined DNA Index System," a DNA profile database kept on FBI computers.) Because all U.S. forensic labs use the same 13 STRs, computers can easily determine if DNA left behind at a crime scene matches one of the profiles stored in the CODIS database. If the STRs match, then the odds are overwhelming that the crime scene DNA was left by the person with the matching profile. Whether or not there is a match, the crime scene DNA profile will remain in CODIS. Sometimes, years later, a new DNA profile will match an archived crime scene profile, and a "cold case" will be solved.

People aren't the only organisms that can be identified by their DNA sequences. An international group of government and private organizations is putting together the "Barcode of Life" to enable rapid DNA identification of all species. DNA barcodes have other applications as well, as we explain in "Earth Watch: What's Really in That Sushi?".

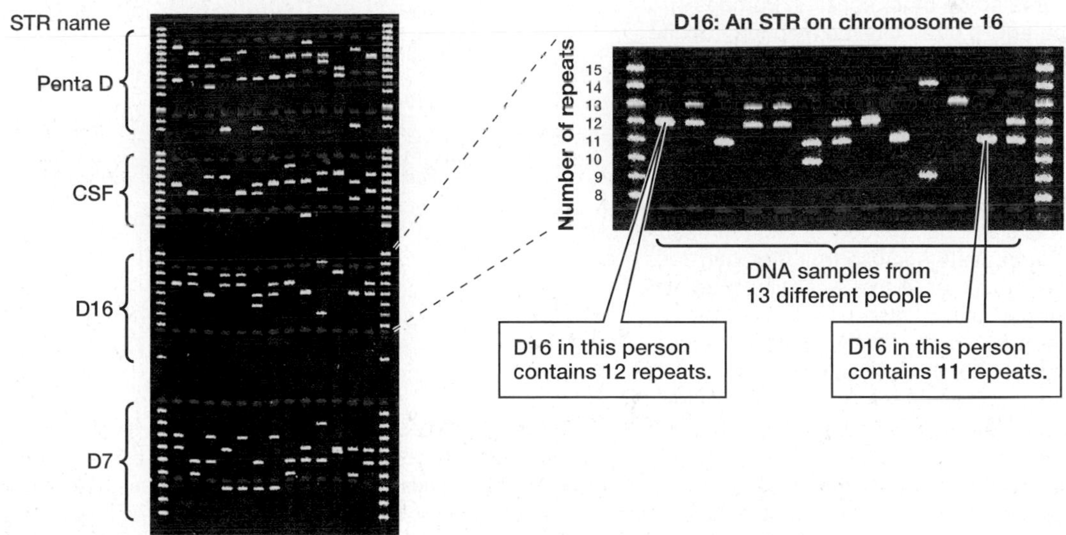

▲ **FIGURE 14-8 DNA profiling** The lengths of short tandem repeats of DNA form characteristic patterns on a gel. This gel displays four different STRs (Penta D, CSF, D16, and D7). The columns of evenly spaced yellow bands on the far left and far right sides of the gel show the number of repeats in the different STR alleles. DNA samples from 13 different people were run between these standards, resulting in one or two yellow bands in each vertical lane. The position of each band corresponds to the number of repeats in that STR allele (more repeats mean more nucleotides, so the allele is larger). (Photo courtesy of Dr. Margaret Kline, National Institute of Standards and Technology.)

THINK CRITICALLY For any single person, a given STR always has either one or two bands. Why? Further, single bands are always about twice as bright as each band of a pair. For example, in the D16 STR on the right, the single bands of the first and third DNA samples are twice as bright as the pairs of bands of the second, fourth, and fifth samples. Why?

What's Really in That Sushi?

Many of the students enrolled in the Introduction to Marine Science course at UCLA were surprised to receive a very unusual assignment: Go to a sushi restaurant and order some sushi (**FIG. E14-1**). Once the sushi arrived at their table, the students were to get out their forceps and scissors, slice off a tiny piece of fish, and place it in a vial of alcohol. Why? The students were helping with a research project that aimed to find out if sushi restaurant menus accurately describe what is served. If you order halibut sushi, do you really get halibut?

Students from the marine science course collected 364 fish samples from 26 restaurants. One would think it would be pretty difficult to identify the species in the samples. After all, by the time a diner sees the fish, it has been beheaded, cleaned, skinned, and cut into pieces. Fortunately, biotechnology makes the job of identification simple, through an analytical process called DNA barcoding.

The DNA barcode of an animal consists of the sequence of a 650-nucleotide fragment of DNA from a gene found in the mitochondria of virtually all eukaryotes (**FIG. E14-2**). For plants, a particular segment of chloroplast DNA serves as the standard barcode. Only extremely closely related species have the same nucleotide sequence in the barcode DNA. Thus, DNA barcoding is a simple, inexpensive way to identify species.

DNA barcoding of the sushi samples collected by the students revealed that about half of the fish were imposters. For some items on sushi menus, misidentification was especially widespread: 100% of sushi sold as halibut or red snapper across restaurants in the area was, in fact, some other species. Perhaps unsurprisingly, the errors often labeled a cheap, readily available fish as a more expensive species. In addition to swindling customers, mislabeling also thwarts their efforts to avoid consuming threatened or overfished species, or species known to have high mercury levels.

DNA barcoding is useful for more than just checking up on your local sushi bar. Barcoding is often used to identify agricultural pests such as fruit flies and public health threats such as disease-carrying mosquitoes. Authorities in several states have used DNA barcoding to determine whether herbal dietary supplements actually contain the plant species listed on their labels. The Federal Aviation Administration reads the DNA barcodes of feathers to find out what kinds of birds collide with planes.

DNA barcoding can also help to stop illegal trafficking of endangered species. For example, barcoding can determine whether leather marketed as alligator skin from farmed animals was in fact taken from an endangered crocodile species.

Identifying the species of origin of meat, skin, feathers, and other animal parts is often difficult, even for experts, but DNA barcodes are foolproof. The day may come when barcoding not only verifies your sushi but also puts a stop to the exploitation of endangered species.

▲ **FIGURE E14-1 Working on a research project**

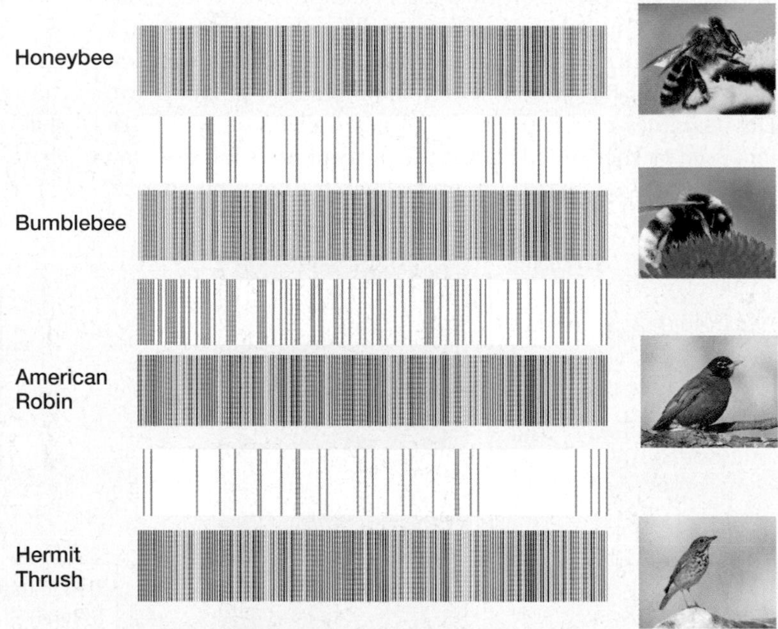

▲ **FIGURE E14-2 DNA barcoding** The different colors in the barcodes represent different bases in the DNA sequence of a fragment of a mitochondrial gene. The black lines between barcodes indicate base pairs that are different between the two barcodes. Closely related organisms have more-similar barcodes than distantly related organisms do, so the number of differences between the two bird species shown here is smaller than the number of differences between the bumblebee and the American robin. Every species, however, has a unique barcode.

THINK CRITICALLY How might ecologists use DNA barcoding to find out which species are present in a rain forest or which kinds of animals a predator eats?

CASE STUDY CONTINUED

Guilty or Innocent?

Thomas Haynesworth's DNA profile did not match the profile obtained from the crime scene semen sample, so Haynesworth could not have been the assailant. The DNA profile of Leon Davis, which had been stored in CODIS, was a match.

CHECK YOUR LEARNING

Can you . . .

- describe how STRs are used to create DNA profiles?
- explain why DNA profiles are unique for unrelated people?

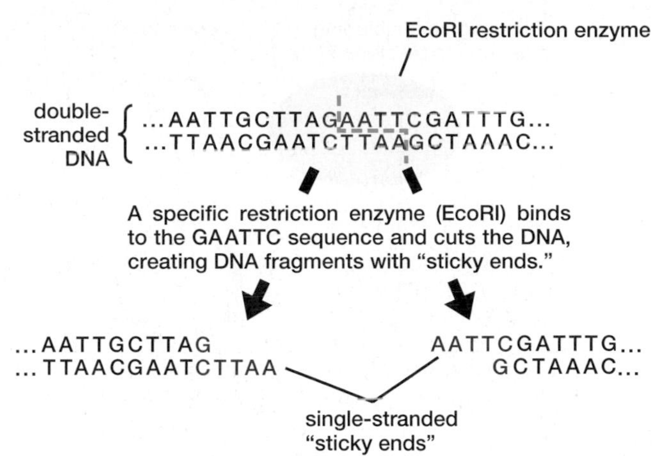

▲ **FIGURE 14-9 Restriction enzymes cut DNA at specific nucleotide sequences**

THINK CRITICALLY Restriction enzymes are isolated from bacteria. Why would bacteria synthesize enzymes that cut up DNA? (Hint: Bacteria can be infected by viruses called bacteriophages). Why wouldn't a bacterium's restriction enzymes destroy the DNA of its own chromosome?

14.5 HOW ARE TRANSGENIC ORGANISMS MADE?

Biotechnology can be used to modify genes or move genes from one species to another, processes that result in genetically modified organisms (GMOs). To produce a GMO containing genes from another species (a transgenic organism), a genetic engineer generally begins by obtaining the desired gene, cloning the gene, and then inserting it into the cells of a host organism.

The Desired Gene Is Isolated or Synthesized

To obtain a gene, biotechnologists either extract it from an organism or synthesize it. For much of the history of biotechnology, the only way to acquire a target gene was to isolate it from an organism that possessed it. For example, one could extract chromosomes from cells, cut them up with enzymes, and use gel electrophoresis to separate the DNA fragments containing the desired gene from the rest of the DNA. Today, however, biotechnologists can rapidly assemble nucleotides into DNA of any sequence to synthesize a desired gene.

The Gene Is Cloned

Once a gene has been obtained, it can be used for medical treatments or to make transgenic organisms. For such purposes, it is often necessary to have a huge number of copies of the gene, far more than are usually made by PCR. The simplest way to generate many copies of a gene is to let living organisms do it, by **DNA cloning**. In DNA cloning, a gene is inserted into single-celled organisms, such as bacteria or yeasts, that multiply very rapidly, manufacturing copies of the gene as they do.

In the most common method of DNA cloning, a gene is inserted into a bacterial plasmid and replicated when the bacteria containing the plasmid multiply. Genes are inserted into plasmids using **restriction enzymes**, which cut DNA. There are hundreds of different restriction enzymes, each of which cuts only at a particular DNA sequence. Many restriction enzymes cut straight across the double helix of DNA, but the ones used in DNA cloning make a staggered cut, snipping the DNA in a different location on each of the two strands so that single-stranded sections hang off the cut ends of the DNA. These single-stranded regions are commonly called "sticky ends," because they can base-pair with, and thus stick to, other single-stranded pieces of DNA with complementary bases (**FIG. 14-9**).

To insert a gene into a plasmid, a restriction enzyme is used to split open the circle of plasmid DNA and to cut the DNA on both ends of the gene (**FIG. 14-10** ❶). Because the opened-up plasmid and the cut-out gene were both cut with the same enzyme, they have complementary sticky ends and can base-pair with each other. When the cut genes and plasmids are mixed together, some copies of the gene will be inserted between the cut ends of the plasmids, held there by the complementary sticky ends ❷.

These recombinant plasmids are then introduced into bacteria (see Fig. 14-1b). Under the right conditions, when the bacteria multiply, they replicate the plasmids, too. Large colonies of the bacteria can rapidly produce as many copies of the cloned gene as needed. After being cloned, the gene may be isolated from the plasmids for use in making transgenic organisms, or whole plasmids may be used.

The Gene Is Inserted into a Host Organism

Next, the cloned gene is inserted into a host organism, a process known as **transfection**. In one method of transfection, recombinant plasmids or genes isolated from them are inserted into harmless bacteria or viruses, called *vectors,* and

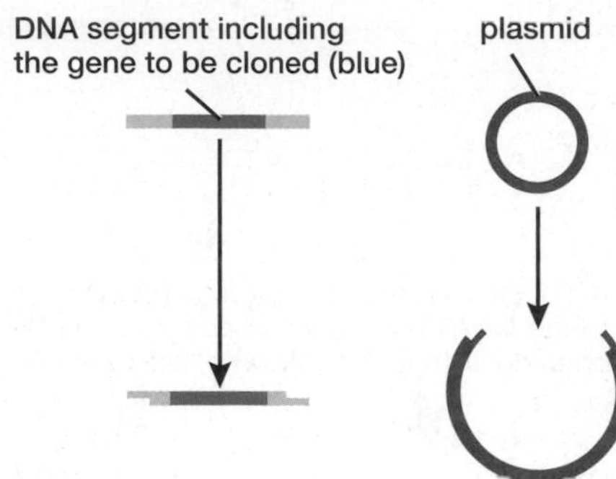

❶ The plasmid and the DNA segment containing the desired gene are cut with the same restriction enzyme.

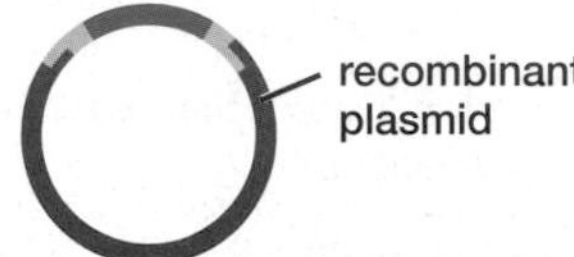

❷ The plasmids and the DNA segment containing the gene, both with the same complementary sticky ends, are mixed together; enzymes bond the genes into the plasmids.

▲ **FIGURE 14-10 Inserting a gene into a plasmid for DNA cloning**

the host organism is infected with them. Ideally, the bacteria or viruses insert the new gene into the chromosomes of the host organism's cells, where it becomes a permanent part of the host's genome.

Various chemical treatments also can be used to transfect cultured animal cells and plant or fungal cells that have had their cell walls removed or disrupted. Typically, the DNA is incorporated into tiny lipid vesicles, which fuse with the plasma membrane of the target cell and move the DNA to the target cell cytoplasm. Other transfection methods temporarily render the plasma membrane permeable so that DNA can enter the target cells.

Finally, plasmids or isolated genes can be directly injected into animal cells, usually fertilized eggs (**FIG. 14-11**). Tiny glass pipettes are loaded with a suitable solution containing the DNA. The pipettes have tips that are sharp enough to impale a cell without damaging it. Pressure applied to the back of the pipette pushes some of the DNA into the cell.

Regardless of the method of transfection, an introduced gene is most useful if it is inserted at an appropriate location in the host genome, so that the gene will be expressed appropriately and its insertion does not accidentally disrupt other genes. Controlling the insertion location of introduced genes has historically been among the most difficult challenges of genetic engineering, but the development of CRISPR-Cas9, with its ability to precisely direct and control genome modification, has greatly simplified this task. Today, genetic engineers are likely to include a CRISPR-Cas9 tool with the plasmids or genes used to transfect cells.

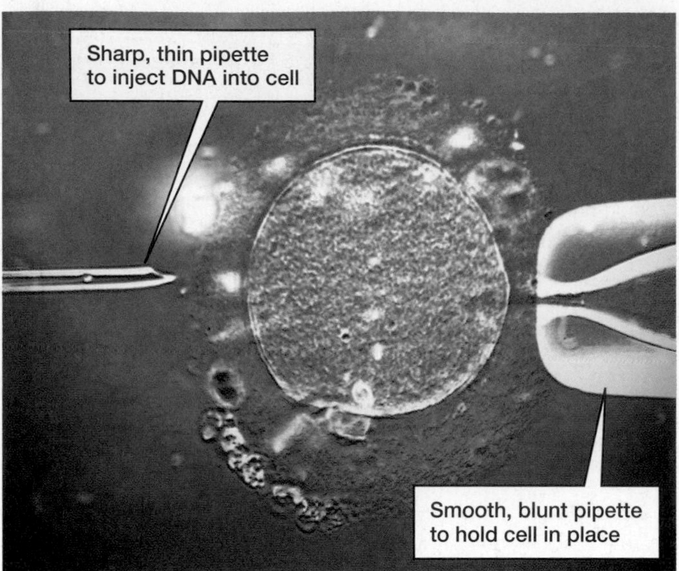

▲ **FIGURE 14-11 Transfecting a fertilized egg by injecting foreign DNA**

CHECK YOUR LEARNING

Can you . . .

- explain how genes are inserted into a plasmid and why that is useful in making a transgenic organism?
- describe the procedures used to transfect an organism with a foreign gene?

CASE STUDY **CONTINUED**

Guilty or Innocent?

Many countries regulate the use of genetically modified organisms, and enforcing the regulations requires methods of distinguishing modified and unmodified organisms whose outward appearances may be very similar. Some of the most effective methods are similar to the DNA profiling used in criminal investigations. For example, investigators in Europe routinely test crop samples by using PCR of target loci followed by gel electrophoresis to produce DNA profiles much like those used to identify criminals. These profiles are compared to those in a database of profiles from samples known to be either normal or genetically modified.

What kinds of genetically modified organisms are the European investigators trying to detect?

14.6 HOW ARE GENETICALLY MODIFIED ORGANISMS USED?

When you think about how GMOs are used, you might first think of crops. For example, you may have heard of corn that has been genetically modified to be more resistant to insect pests. Agricultural applications are indeed an important use

TABLE 14-1 Genetically Engineered Crops with USDA Approval

Genetically Engineered Trait	Potential Advantage	Examples
Resistance to herbicide	Application of herbicide kills weeds but not crop plants, producing higher crop yields	Beet, canola, corn, cotton, flax, potato, rice, soybean, tomato
Resistance to pests	Crop plants suffer less damage from insects, producing higher crop yields	Corn, cotton, potato, rice, soybean
Resistance to disease	Plants are less prone to infection by viruses, bacteria, or fungi, producing higher crop yields	Papaya, potato, squash
Sterile	Transgenic plants cannot cross with wild varieties, making them safer for the environment and more economically valuable for the seed companies that produce them	Chicory, corn
Altered oil content	Oils can be made healthier for human consumption or can be made similar to more expensive oils (such as palm or coconut)	Canola, soybean
Delayed browning or bruising	Produce is more appealing to consumers; waste is reduced if damaged produce is less likely to be discarded	Apple, potato

of GMOs, but GMOs are also used or under development for a wide variety of industrial, environmental, and medical applications. In the upcoming section, we explore some of these uses.

Many Crops Are Genetically Modified

In many countries, genetic engineering is rapidly replacing selective breeding techniques as the method of choice for producing improved crops. More than 90% of soybeans, corn, and cotton grown in the United States is transgenic, and U.S. farmers also grow genetically modified canola, alfalfa, sugar beets, papaya, apples, potatoes, and squash. Globally, farmers in 26 countries planted more than 470 million acres of transgenic crops in 2017.

Crops are most commonly modified to improve their resistance to pests or poisons (**TABLE 14-1**). For example, a number of transgenic crops have been transfected with bacterial genes that provide resistance to herbicides (weed killers). Herbicide-resistant crops allow farmers to kill weeds without harming their crops. Less competition from weeds means more water, nutrients, and light for the crops and, hence, larger harvests. However, growing herbicide-resistant crops allows greater use of herbicides, which has resulted in the evolution of resistant weeds.

Many crops have been made more resistant to insect pests by giving them a gene, called *Bt*, from the bacterium *Bacillus thuringiensis*. The protein encoded by the *Bt* gene damages the digestive tract of insects, but not mammals. Transgenic *Bt* crops often suffer far less damage from insects than regular crops do, so farmers can apply less pesticide to their fields (**FIG. 14-12**).

Although the most widely planted and commercially successful genetically modified crops are produced by inserting genes from other species, some have been produced by modifying a crop species' own genes. For example, apples that don't turn brown after slicing were produced by silencing a gene that encodes the enzyme that causes discoloration in damaged fruit. Crops that have been genetically modified without inserting foreign genes are becoming increasingly common, in part because CRISPR-Cas9 gene editing has made development of such crops much easier and faster. In addition, gene-edited crops receive less stringent regulatory scrutiny than do transgenic crops. (Scientists may use biotech tools to improve crops without genetic engineering; see "Doing Science: Using Genetic Markers to Breed Tastier Fruits and Veggies.")

◀ **FIGURE 14-12** ***Bt*** **plants resist insect attack** Transgenic cotton plants expressing the *Bt* gene (right) resist attack by bollworms, which eat cotton seeds. The transgenic plants therefore produce far more cotton than nontransgenic plants do (left).

DOING Science — Using Genetic Markers to Breed Tastier Fruits and Veggies

In the past, fruits such as tomatoes and strawberries were available only seasonally, but today you can buy these fruits at any time of year. However, this convenience comes with a price: loss of flavor. Fruits now found in supermarkets are much less tasty than old-fashioned fruits, as they have been bred to survive storage and shipping and as a result have lost much of their flavor.

Ideally, we would like to have fruits that are both durable and tasty. Plant breeders are hard at work realizing this desire, but progress can be slow. In general, breeders bring different properties together in one fruit by crossbreeding plants whose fruits show one of the preferred properties. The offspring of crossbred plants must grow to maturity and produce fruits that can be tested and selected, and this process must be repeated for many generations. Increasingly, however, the process of improving fruits is moving much faster, thanks to a technique known as "marker-assisted breeding."

What Question Was Asked?

Consider the cantaloupe. Slow, conventional breeding has produced a cantaloupe variety that is very flavorful and remains firm for several weeks after harvest. However, melons of this variety are very small. Plant breeders would like to develop a variety that combines good flavor, durability, and large fruit size. Can marker-assisted breeding help achieve this goal?

How Was Evidence Gathered?

Breeders began by performing genetic tests on both the tasty and durable (but tiny) cantaloupes, and on conventional cantaloupes that are large but bland. In these tests, a number of specific "marker" DNA sequences were tagged with identifiers, such as fluorescent molecules, that revealed which marker sequences were present in the DNA of each type of cantaloupe. The procedure revealed associations between particular DNA markers and particular desirable traits; such associations are usually due to the presence near the marker sequence of a linked gene that influences the trait (**FIG. E14-3**). The researchers then crossbred the two types of cantaloupes. But instead of planting all the seeds, waiting for new plants to mature, and then laboriously testing the fruits to find the best ones, the breeders screened the seeds for desirable combinations of genetic markers, using machines to rapidly test thousands of seeds. The machines harmlessly snipped tiny slivers from the seeds, extracted DNA, and tagged the marker sequences by attaching fluorescent molecules. Any seed that, by chance, had genetic markers for all the desired traits was set aside for later planting.

What Was Learned?

The presence of a marker sequence associated with a trait reliably indicates that the seed will yield plants with fruits that have the trait. Seeds with markers for multiple traits will produce fruits with all of those traits.

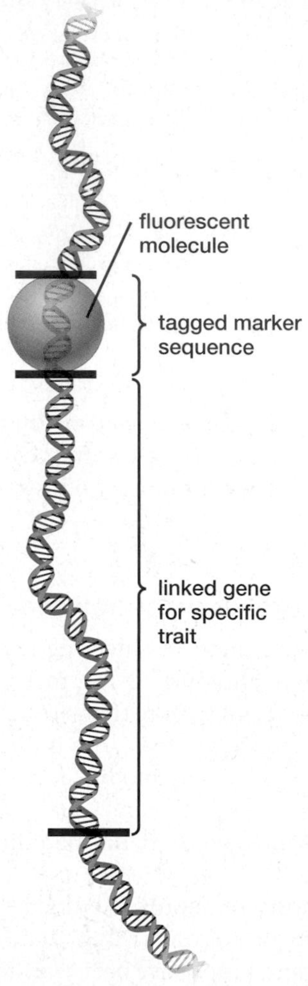

▲ **FIGURE E14-3 Marker-assisted selection**

Researchers worldwide are using genetic marker biotechnology to search for ideal combinations of features. They've produced firm, disease-resistant tomatoes with old-fashioned flavor, miniature bell peppers, and a crispy yet flavorful hybrid of iceberg and Romaine lettuce. These new crops may appeal to consumers who wish to avoid genetically modified foods, because the plants' genes have been recombined by traditional breeding techniques.

CONSIDER THIS Roughly 30% of harvested cantaloupes are wasted because they become overripe before reaching consumers. Agricultural scientists have used genetic engineering to insert an apple gene into the cantaloupe genome to produce a transgenic cantaloupe that ripens more slowly after harvesting. Is this approach more desirable than the marker-assisted breeding described above? Why or why not?

Have You Ever Wondered ...
If the Food You Eat Has Been Genetically Modified?

Corn and soy products are found in an amazing variety of foods, not just the obvious ones like tortilla chips, margarine, and soy sauce. For example, corn syrup is an ingredient in foods as diverse as soda, ketchup, salad dressing, and breakfast cereal. Soybean oil or protein is an important ingredient in cookies, cake mixes, and veggie burgers. Almost all of the corn and soy grown in the United States is genetically modified, with the result that about 80% of the packaged foods in American supermarkets contain ingredients made from GMOs. Many countries, including those in the European Union, require labeling of foods containing GMO ingredients, but the United States does not. Therefore, if you live in the United States, unless you are extremely motivated to avoid GMOs in your diet, you are probably eating them.

Genetically Modified Animals May Be Useful for Agriculture and Industry

To date, it has proven difficult to create commercially valuable genetically modified livestock. Only one transgenic animal has been approved for use as food in the United States: Atlantic salmon that contain extra genes for growth hormone and therefore grow faster than wild-type Atlantic salmon. But biotechnology companies are working on producing a variety of other genetically modified animals, including sheep that produce more wool, cattle that produce more protein in their milk, and pigs that produce leaner meat. Researchers in England and Scotland have developed transgenic chickens that cannot spread the H5N1 influenza virus, which causes avian flu. Worldwide, many millions of chickens have been killed to stop outbreaks of avian flu, so flu-resistant chickens could potentially be very valuable livestock. Transgenic goats housed at Utah State University secrete spider silk proteins in their milk. Spider silk is as strong as Kevlar, the fiber usually used in bulletproof vests, so the hope is that silk protein from these goats will one day be used to make flexible, lightweight body armor.

Genetically modified animals play an important role in biomedical research. For example, to investigate the causes of certain diseases and develop possible treatments, researchers often work with mice that have been genetically engineered to have a particular disease or physiological condition. These mice may have had their genomes edited to knock out particular genes, or they may be transgenic, carrying genes associated with human diseases such as Alzheimer's disease, Marfan syndrome, or cystic fibrosis.

Genetically Modified Organisms May Be Used for Environmental Bioengineering

Advances in biotechnology have raised hopes that some troubling environmental and public health problems might be solved by releasing genetically modified organisms into the environment. For example, GMOs may be able to help restore rare or endangered species, clean up soil and water pollution, and reduce the incidence of insect-borne diseases.

Rare Trees May Be Engineered for Disease Resistance

When Europeans first came to America, one of the most common trees in the deciduous forests was the majestic American chestnut. Today, however, American chestnuts are very rare. Most have been killed by the chestnut blight fungus, which arrived around 1900, accidentally imported with Chinese chestnut trees. For decades, in an effort to restore the chestnut to America's forests, horticulturists have been crossbreeding American and Chinese chestnuts, trying to get a mostly American variety that resists blight. Some of these hybrids are now being planted in restoration projects. Biotechnology now offers a second path to blight resistance. Researchers have developed transgenic American chestnut trees that contain a gene from wheat that prevents the blight fungus from harming the trees.

Bacteria May Be Engineered to Clean Up Pollution

At many old mine sites, the soil and water are badly contaminated with toxic heavy metals, including mercury, lead, and cadmium. Some bacteria thrive in high concentrations of some heavy metals by removing the metals from water and soil and storing them in their cytoplasm; the metals are retained in the bacteria and often converted to molecular forms that pose no threat to other organisms. Researchers have genetically modified some of these bacterial species so that they take up the metals faster and more efficiently. Environmental engineers hope that genetically engineered bacteria can provide an effective and affordable way to clean up contaminated soils and waterways.

Mosquitoes May Be Engineered to Halt the Spread of Diseases

Some of the world's most dangerous infectious diseases are spread by insects, particularly mosquitoes. When female mosquitoes feed on people to acquire the blood meal they need to produce eggs (**FIG. 14-13**), they may spread serious and sometimes fatal diseases, including malaria and dengue fever. These diseases affect hundreds of millions of people globally each year, principally in warmer regions. Biotechnologists hope that genetically modified mosquitoes can help reduce the number of infections.

Researchers have taken three general approaches to modifying mosquitoes. One approach involves inserting genes that kill the disease-causing microbial parasites carried by several species of mosquito. For example, researchers have modified malaria-transmitting *Anopheles* mosquitoes by inserting genes

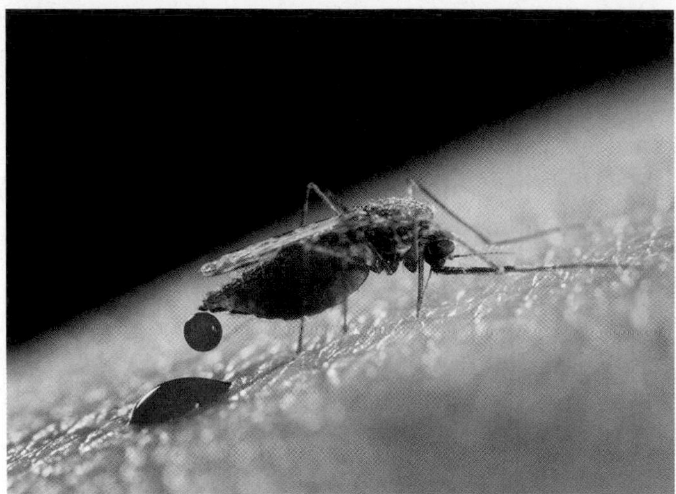

▲ **FIGURE 14-13 An *Anopheles* mosquito taking a meal of human blood** The World Health Organization estimates that *Anopheles* mosquitoes transmit malaria to more than 200 million people each year, causing about 450,000 deaths.

that encode antibodies that attack and kill the malaria parasite. A second approach seeks to distort the mosquito sex ratio so that it favors males, which do not bite people or transmit disease. For example, researchers have engineered male *Anopheles* mosquitoes to carry a CRISPR-Cas9 sequence that, when expressed during meiosis, targets and cuts up a DNA sequence found only on the X chromosome. As a result, the male's X chromosomes are shredded—all of his viable sperm carry only Y chromosomes, and all of his offspring are therefore male. If these engineered male mosquitoes were released into natural populations, few females would be born, and eventually the populations would crash.

A third approach is to insert a gene that kills the mosquitoes. For example, the biotech firm Oxitec has genetically modified male *Aedes aegypti* mosquitoes, the species that transmits dengue fever and Zika virus, to carry a lethal gene. When the genetically modified males mate with wild-type females, the offspring inherit the gene and die before they become adults. In field tests, releasing huge numbers of genetically modified male mosquitoes reduced the mosquito population by 80% to 96%. On the basis of these results, Brazil has released billions of Oxitec's genetically modified mosquitoes to fight dengue fever. In the United States, a field trial in the Florida Keys is underway, to see if the Oxitec mosquitoes can slow the spread of Zika virus.

CHECK YOUR LEARNING

Can you . . .

- describe the advantages of genetically modified crops and provide some examples?
- list some examples of how genetically modified animals might be used?
- describe how GMOs may be used to restore species, clean up pollution, and fight the spread of infectious diseases?

14.7 HOW IS BIOTECHNOLOGY USED IN MEDICINE?

Biotechnology plays an important role in medicine. It is used to diagnose inherited disorders and infectious diseases, and to produce therapeutic drugs and other treatments. A great deal of current research is aimed at developing additional treatments and cures, so the importance of biotechnology in medicine is likely to grow.

DNA Technology Can Be Used to Diagnose Inherited Disorders

Analysis of DNA provides accurate tests for many genetic diseases, including in fetuses (see "Health Watch: Prenatal Genetic Screening" on page 220). DNA-based tests are effective because they can determine if a person has inherited one or more defective alleles with nucleotide sequences that differ from those of normal, functional alleles. Public awareness of the impact of genetic testing got a boost in 2013, when actress Angelina Jolie underwent a preventive double mastectomy after learning that she has a defective allele of *BRCA1*, a tumor suppressor gene that is crucial to preventing breast and ovarian cancer (**FIG. 14-14**). Some defective *BRCA1* alleles increase the lifetime risk of developing breast cancer from 12% to about 65% to 85% and the lifetime risk of ovarian cancer from a little over 1% to about 30% to 60%. In 2015, Jolie also had her ovaries removed.

Disease Alleles Can Be Detected with PCR

If the DNA sequences of the defective alleles responsible for a genetic disorder are known, PCR alone can in some cases serve as a tool to diagnose the disease. Recall that PCR uses specific DNA primers to amplify particular DNA sequences. A medical testing company can detect the presence of a disease-causing allele by designing a PCR primer that is specific to that allele. So if PCR using that primer amplifies a patient's DNA sample, the patient carries the disease-causing allele.

▶ **FIGURE 14-14 Angelina Jolie has raised awareness of genetic testing**

Disease Alleles Can Be Detected with Restriction Enzymes

Sickle-cell anemia is a serious inherited disease caused by a single nucleotide substitution mutation in the gene for beta-globin. A common diagnostic test for sickle-cell anemia relies on restriction enzymes that cut DNA only at specific nucleotide sequences. To begin a test for the presence of the sickle-cell allele in a person's DNA, PCR is used to amplify a section of DNA that includes the mutation site. Then, the amplified DNA is mixed with a restriction enzyme called MstII that cuts the normal sequence (CCTG**A**GGAG), but not the sickle-cell sequence (CCTG**T**GGAG); normal alleles in the sample will be cut in half, whereas sickle-cell alleles will remain intact. Gel electrophoresis then reveals which alleles are present, separating the larger, intact sickle-cell allele from the smaller, cut pieces of the normal allele.

Disease Alleles Can Be Detected with DNA Probes

Another way to detect the presence of disease-causing alleles whose sequence is known is to determine whether a patient's DNA binds to synthetic DNA segments called **DNA probes**, which have sequences that are complementary to those of defective alleles. Such a test has been developed for cystic fibrosis, which is caused by defective alleles of the gene for a protein called CFTR. Any one of more than 1,900 different defective *CFTR* alleles can cause the disease, but just 32 alleles account for about 90% of the cases of cystic fibrosis. Although it is possible to sequence the entire *CFTR* gene and search for any defective allele, most screening for cystic fibrosis relies on more rapid and less expensive tests that use DNA probes to focus on the 32 common alleles.

Although there are many different technologies for detecting the cystic fibrosis alleles, all are based on complementary base pairing. The simplest screening consists of an array of different single-stranded DNA probes bound to a piece of specialized paper (**FIG. 14-15**). Each probe is complementary to one strand of a particular defective *CFTR* allele (**FIG. 14-15a**). A person's DNA is tested by cutting it into small pieces, separating the pieces into single strands, and labeling the strands with a colored molecule (**FIG. 14-15b**). The paper with the bound probes is then bathed in a solution containing the labeled DNA fragments. The person's DNA will bind only to a probe that has a perfectly complementary nucleotide sequence, thereby showing which *CFTR* alleles the person carries (**FIG. 14-15c**).

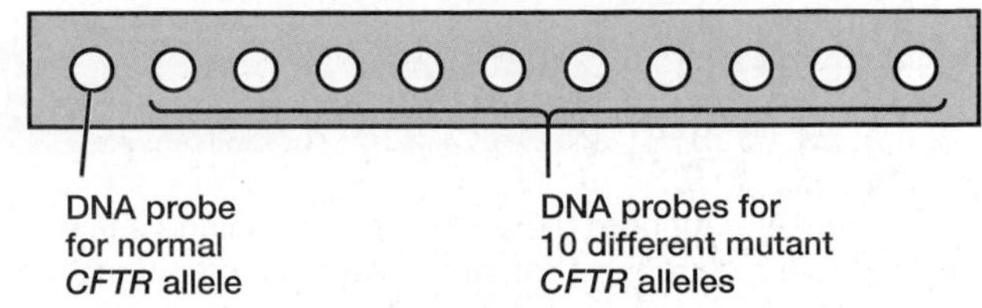

(a) Linear array of probes for cystic fibrosis

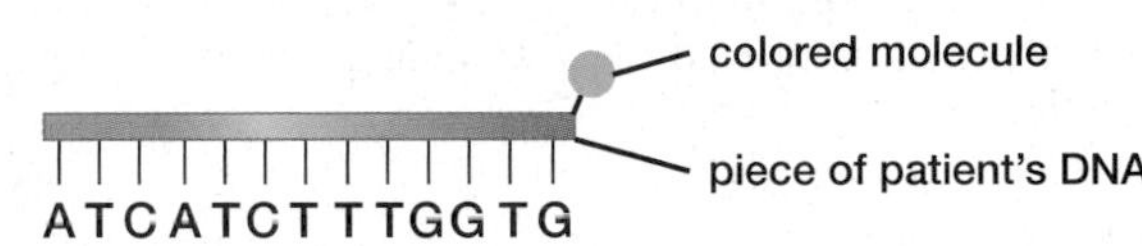

(b) *CFTR* allele labeled with a colored molecule

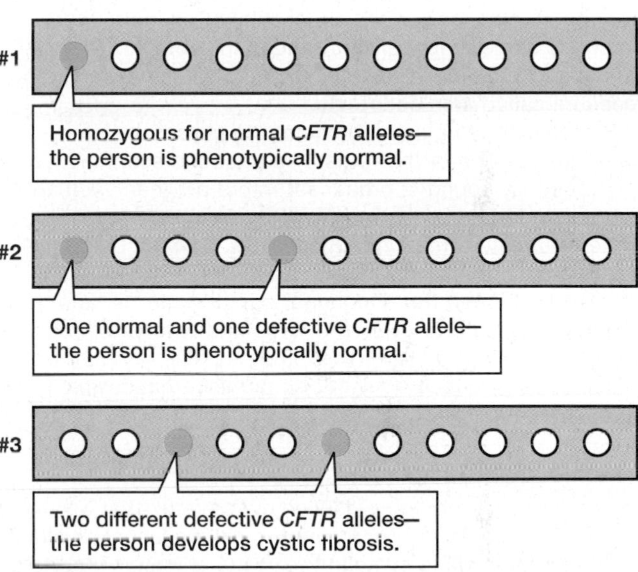

#1
Homozygous for normal *CFTR* alleles—the person is phenotypically normal.
#2
One normal and one defective *CFTR* allele—the person is phenotypically normal.
#3
Two different defective *CFTR* alleles—the person develops cystic fibrosis.

(c) Linear arrays with labeled DNA samples from three different people

▲ **FIGURE 14-15 A cystic fibrosis diagnostic array (a)** A typical diagnostic array for cystic fibrosis consists of specialized paper to which DNA segments complementary to the normal *CFTR* allele (far left spot) and several of the most common defective *CFTR* alleles (the other 10 spots) are attached. **(b)** DNA from a patient is cut into small pieces and separated into single strands, and the *CFTR* alleles are labeled with colored molecules. **(c)** The array is bathed in a solution of the patient's labeled DNA. The labeled DNA of a patient binds to one or two spots on the array, depending on which *CFTR* alleles the patient possesses.

DNA Technology Can Be Used to Diagnose Infectious Diseases

In addition to its role in detecting genetic diseases, biotechnology is also improving the speed and accuracy with which diseases caused by pathogenic microbes are diagnosed. Tests based on DNA probes or sequencing are increasingly important components of the medical toolkit.

An Array of Probes Can Test for Many Diseases Simultaneously

Diagnostic tests for infectious disease often employ a *microarray*. A microarray is a piece of plastic or glass that holds hundreds or even thousands of tiny dots of DNA in a grid pattern, with each dot containing a different probe sequence. Microarrays have many uses, but to diagnose infectious diseases, a microarray is loaded with DNA probes that are complementary to sequences from different types of

Health WATCH Prenatal Genetic Screening

Thanks to modern biotechnology, physicians can now perform prenatal screening for various genetic disorders, including cystic fibrosis, sickle-cell anemia, muscular dystrophy, and Down syndrome. Prenatal screening requires samples of fetal cells or chemicals produced by the fetus. Three techniques are commonly used to obtain samples for prenatal diagnosis: amniocentesis, chorionic villus sampling, and maternal blood collection.

Amniocentesis

The human fetus, like all mammal fetuses, develops in a watery environment. A waterproof membrane called the amnion surrounds the fetus and holds the fluid. As the fetus develops, it releases various chemicals (often in its urine) and sheds some of its cells into the amniotic fluid. When a fetus is 15 weeks or older, amniotic fluid can be collected by a procedure called **amniocentesis**.

Before extracting amniotic fluid, the physician determines the position of the fetus by ultrasound scanning. High-frequency sound is broadcast into a pregnant woman's abdomen, and instruments convert the echoes bouncing off the fetus into a real-time image (**FIG. E14-4**). Using the image as a guide, the physician carefully inserts a sterilized needle through the abdominal wall, the uterus, and the amnion (being sure to avoid the fetus and placenta), and withdraws 10 to 20 milliliters of amniotic fluid (**FIG. E14-5**). Amniocentesis causes a slight risk of miscarriage, which occurs in about 0.5% of cases.

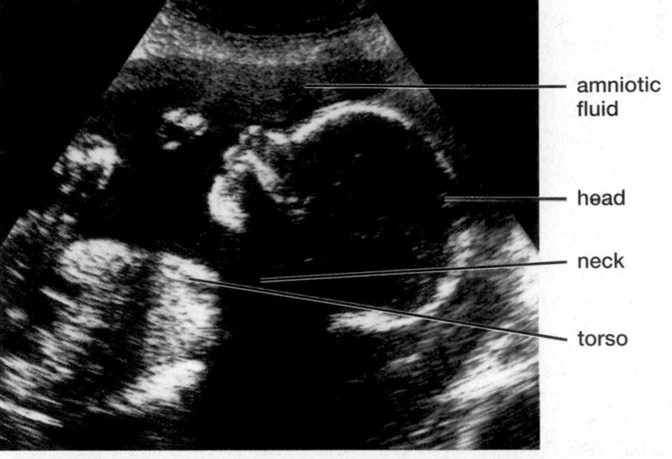

▲ **FIGURE E14-4 A human fetus imaged with ultrasound**

Chorionic Villus Sampling

The chorion, a membrane produced by the fetus that becomes part of the placenta, produces many small projections called villi. In **chorionic villus sampling (CVS)**, a physician inserts a small tube through the mother's vagina and into the uterus and suctions off a few villi for analysis (see Fig. E14-5). The loss of a few villi does not harm the fetus.

CVS has two major advantages over amniocentesis. First, it can be done much earlier in pregnancy—as early as the 8th week, but usually between the 10th and 12th weeks. This is especially important if the woman is contemplating an abortion if her fetus has a major defect. Second, the sample contains far more fetal cells than can be obtained by amniocentesis. However, CVS appears to have a slightly greater risk of causing a miscarriage than amniocentesis does. Also, because the chorion is outside of the amniotic sac, CVS does not obtain a sample of the amniotic fluid, which is needed to diagnose certain disorders. Finally, in some cases chorionic cells have chromosomal abnormalities that are not in fact present in the fetus. For these reasons, CVS is less commonly performed than amniocentesis.

Maternal Blood Collection

Because a tiny number of fetal cells cross the placenta and enter the mother's bloodstream, sampling maternal blood is another method used for prenatal screening. In addition to fetal cells, maternal blood also contains fragments of cell-free fetal DNA, as well as proteins and other chemicals produced by the fetus. Collecting a sample of the mother's blood is quick, easy, poses no risk to the fetus, and can be done earlier in pregnancy than either amniocentesis or CVS. Advances in biotechnology have helped overcome the technical challenge of separating tiny amounts of fetal cells and DNA from the large number of maternal cells in a blood sample.

disease-causing microbes. DNA is extracted from a patient's tissues and labeled; any binding of the extracted DNA to the sequences on the microarray reveals which microbes infect the patient.

Infectious Organisms Can Be Identified by Sequencing

Infections can also be diagnosed by sequencing the DNA of microbes present in a patient's tissues. For example, when physicians were unable to identify the cause of fevers, seizures, and brain inflammation that recently threatened the life of a Wisconsin teenager, they enlisted the help of researchers who sequenced DNA from the patient's cerebrospinal fluid. The researchers compared the sample sequence to known bacterial DNA sequences and identified the culprit, the bacterium *Leptospira santarosai*. As the cost of sequencing declines and this approach is commercialized, it is likely to become a common way to diagnose infections.

Analyzing the Samples

Some information about a fetus can be gleaned from chemicals in amniotic fluid, chorionic villi, or maternal plasma (the fluid portion of the mother's blood). For example, high concentrations of a protein called alpha-fetoprotein may indicate that the fetus has a nervous system disorder such as spina bifida, in which the spinal cord is incomplete, or anencephaly, in which major portions of the brain fail to develop. Particular combinations of alpha-fetoprotein, estrogen, and other chemicals in maternal plasma indicate the likelihood of Down syndrome and other disorders. However, these biochemical screening tests do not provide definitive diagnoses. Therefore, if initial tests indicate that a disorder may be present, then other tests such as karyotyping (examination of chromosomes), DNA analysis using methods such as those described in Section 14.7, or detailed ultrasound inspection of the fetus are employed to find out whether the fetus actually has one of these conditions.

Karyotyping requires fetal cells. Amniotic fluid and maternal blood contain very small numbers of fetal cells, so to obtain enough cells from these sources for karyotyping, the usual procedure is to grow the cells in culture for a week or two. CVS obtains a larger number of fetal cells, so karyotyping can usually be performed without culturing the cells first. Karyotyping fetal cells can reveal if the fetus has a condition caused by chromosomes with structural abnormalities or by having too many or too few copies of a chromosome. Down syndrome, for example, results from the presence of three copies of chromosome 21 (see page 153).

THINK CRITICALLY Explain how fetal DNA could be used to establish paternity of a fetus.

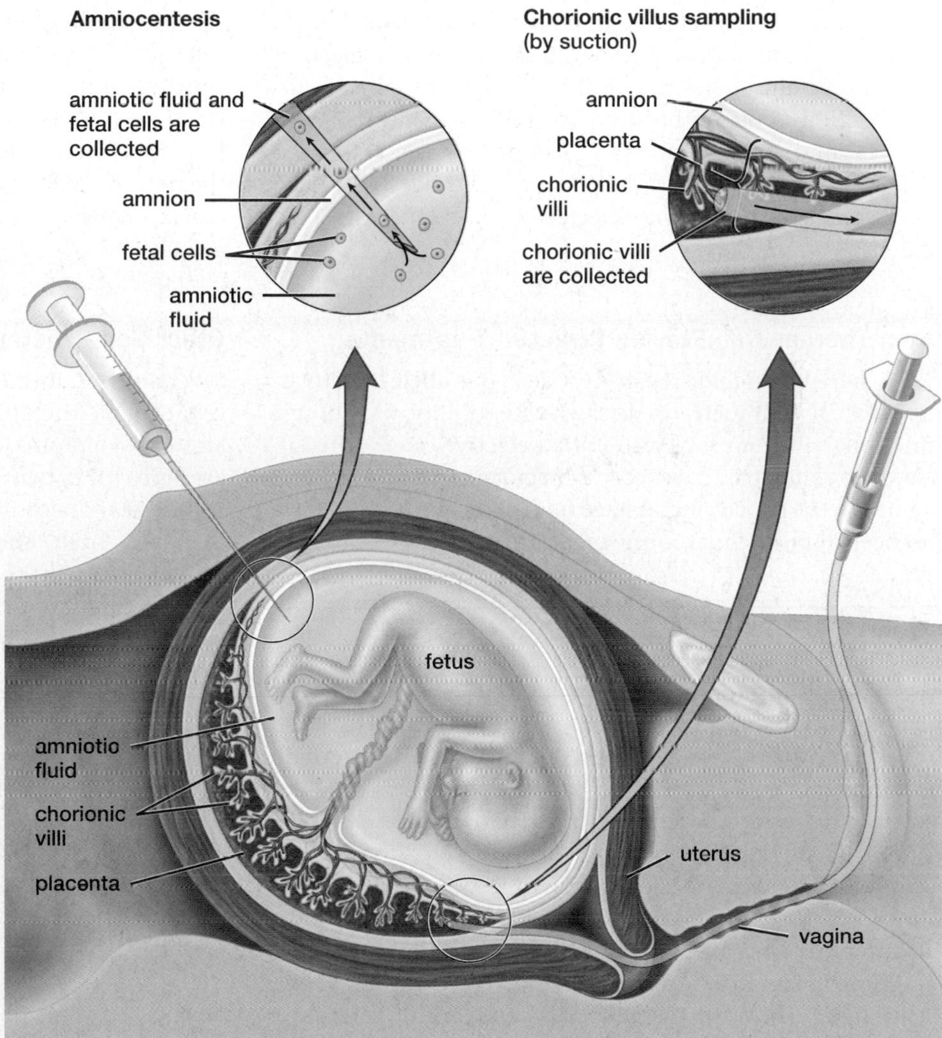

▶ **FIGURE E14-5 Prenatal sampling techniques** The two most common ways of obtaining samples for prenatal diagnosis are amniocentesis and chorionic villus sampling. (CVS is usually performed when the fetus is much younger than the one depicted in this illustration.)

DNA Technology Can Help to Treat Diseases

The two principal applications of DNA technology for treating disease are (1) medicines produced by transgenic organisms and (2) **gene therapy**, which seeks to cure diseases by inserting, deleting, or altering genes in a patient's cells.

Transgenic Organisms Can Produce Medicines

Genetically modified bacteria or eukaryotic cells are often used to produce lifesaving medicines. For example, yeasts have been engineered to produce the anti-malaria drug artemisinin, which otherwise must be extracted from the plants in which it occurs naturally. Transgenic microbes also produce some medically important human proteins, including growth hormone, clotting factors, antibodies, and the insulin needed by people with diabetes.

Although most biotechnological production of medicines relies on microbes or cultured cells, the tools of biotechnology can also be used to insert medically useful genes into animals or plants. Goats have been genetically modified to produce milk that contains the human protein antithrombin, which is used to prevent blood clots. Transgenic chickens produce eggs containing a protein used to treat a genetic disorder that causes liver damage. And transgenic plants produce an enzyme needed by people with Gaucher's disease, a genetic disorder that causes anemia, joint pain, and neurological symptoms.

Transgenic plants and animals can also produce human antibodies that combat infectious diseases. When a disease-causing microbe invades your body, it takes several days for your immune system to respond, whereas direct injection of the right antibodies might allow a faster response. During the 2014 Ebola outbreak in West Africa, an experimental cocktail of three antibodies made in transgenic tobacco plants is thought to have helped save the lives of health professionals who became infected while caring for Ebola patients.

Gene Therapy Can Provide Targeted Treatments

Gene therapies include repairing a defective allele, inactivating an allele that increases disease susceptibility, or adding a functional allele to substitute for a defective one. Although a few gene therapies have received regulatory approval, most are still in the experimental stage or in clinical trials and have not been approved for routine medical practice.

Gene Editing for AIDS

The human immunodeficiency virus (HIV) enters and kills several kinds of immune cells, including helper T cells that play a crucial role in responding to infection. When the body's supply of helper T cells becomes too low, the immune response falters, ordinarily trivial infections become life-threatening, and full-blown AIDS develops. To enter cells, HIV first binds to a receptor protein called CCR5, which is found on the surface of susceptible immune cells. But a tiny number of people have a mutated *CCR5* gene and don't make the receptors; they are resistant to infection by HIV.

Biotechnology offers the possibility of eliminating the CCR5 receptor in patients with AIDS and curing, or at least greatly alleviating, their disease. Researchers have used molecular gene editing tools, including CRISPR-Cas9, to target and disable the *CCR5* gene. In a treatment currently under development, immune cells are removed from a patient and their *CCR5* genes are disabled. These *CCR5*-deleted cells are put back into the patient, where they multiply. In two small clinical trials, this type of treatment greatly increased the number of functioning immune cells in AIDS patients.

Gene Replacement for Severe Combined Immune Deficiency

Severe combined immune deficiency (SCID) is a rare disorder in which a child's immune system fails to develop normally. In children with SCID, infections that would be trivial in a normal child become life threatening. In some cases, a bone marrow transplant from a compatible donor can give the child functioning stem cells so that he or she can develop a working immune system. Most children with SCID, however, die before their first birthday.

In one type of SCID, called ADA-SCID, affected children are homozygous for a recessive defective allele that normally codes for an enzyme called adenosine deaminase. Recent clinical trials have used gene therapy to cure ADA-SCID. Researchers removed bone marrow stem cells from children with the disorder, inserted a functional copy of the adenosine deaminase gene into the cells, and returned the repaired cells into the children. Because bone marrow stem cells continue to produce new white blood cells throughout life, the hope is that these children might be permanently cured. In clinical trials, more than 40 patients have received gene therapy for ADA-SCID, and 70% of them seem to be fully cured. The treatment is currently available as a commercial product.

Other Gene Therapies

Several other disorders have been treated with varying success by gene therapy. Gene therapy has been used to alleviate the symptoms of Parkinson's disease, partially restore eyesight in patients with a type of inherited blindness, restore blood clotting to hemophiliacs, cure sickle-cell anemia, and "train" the immune system to destroy some types of leukemia. Further development of these and other gene therapies has been slowed in part by problems with harmful side effects caused when the process of inserting a new gene inadvertently damages other genes. Researchers are hopeful that the availability of CRISPR-Cas9, with its ability to edit genomes with great precision, can help eliminate these "off target" effects and speed progress on developing successful therapies.

CHECK YOUR LEARNING

Can you . . .

- explain how biotechnology is used to diagnose both inherited and infectious diseases?
- describe how transgenic organisms are used to produce medicines and give examples?
- describe the procedures and advantages of gene therapy to treat inherited diseases?

14.8 WHAT ARE THE MAJOR ETHICAL ISSUES OF MODERN BIOTECHNOLOGY?

Modern biotechnology offers the promise—some would say the threat—of greatly changing our lives. The changes arising from the advance of biotechnology are accompanied by complex ethical and public policy challenges. Here we will explore two issues that raise such challenges: the presence of genetically modified organisms in our food supply and environment, and our rapidly advancing ability to genetically modify human beings.

Should Genetically Modified Crops and Livestock Be Permitted?

The transgenic crops produced by modern biotechnology have clear advantages for farmers. Herbicide-resistant crops allow farmers to use herbicides to rid their fields of weeds, increasing harvests by 10% or more. Insect-resistant crops decrease the need to apply pesticides, saving the cost of the pesticides themselves, as well as tractor fuel and labor. Therefore, transgenic crops may produce larger harvests at lower cost. These savings may be passed along to the consumer. Transgenic crops also have the potential to be more nutritious than standard crops (**FIG. 14-16**). But despite the current and potential benefits of transgenic crops and livestock, many people strenuously object to them, concerned that they may be harmful to human health or dangerous to the environment.

▲ **FIGURE 14-16 Golden Rice** Many people around the world suffer from vitamin A deficiency, and the developers of Golden Rice hope that this GMO will one day help remedy this problem. Golden Rice has been genetically engineered to contain beta-carotene, a pigment that the human body easily converts into vitamin A. The high beta-carotene content of Golden Rice gives it a bright yellow color. Normal rice lacks beta-carotene and is off-white.

Are Foods from GMOs Dangerous to Eat?

In most cases, there is no reason to think that GMOs are dangerous to eat. For example, tests have shown that the protein encoded by the *Bt* gene is not toxic to mammals, so it should not be dangerous to human health. If growth-enhanced livestock are ever marketed, they will simply have more meat, composed of the same proteins that exist in nontransgenic animals, so they shouldn't be dangerous either. The U.S. FDA has already declared that the flesh of transgenic salmon developed by a company called AquaBounty is "as safe as food from conventional Atlantic salmon." The genetically modified salmon contain extra genes for growth hormone and grow faster than wild-type Atlantic salmon, but their flesh contains the same proteins as are found in wild salmon.

On the other hand, some people might be allergic to certain genetically modified organisms. In the 1990s, a gene from Brazil nuts was inserted into soybeans in an attempt to improve the balance of amino acids in soybean protein. But concerns that people allergic to Brazil nuts would probably also be allergic to the transgenic soybeans meant that these transgenic soybean plants never made it to the farm. The FDA now requires all new transgenic crop plants to be tested for allergenic potential.

Overall, concerns that GMO foods are unsafe have not been supported by scientific evidence, despite extensive testing. In 2016, the National Academies of Sciences, Engineering, and Medicine conducted a detailed review of the scientific literature and reported that there was "no substantiated evidence that foods from GE [genetically engineered] crops were less safe than foods from non-GE crops." Over the past 15 years, similar statements have been issued by the World Health Organization, the American Association for the Advancement of Science, and many other health and scientific organizations.

Are GMOs Hazardous to the Environment?

The environmental effects of GMOs are more debatable. One clear positive effect of *Bt* crops is that farmers who plant them usually apply less pesticide to their fields. This should result in less pollution of the environment and less harm to the farmers. For example, in India, farmers growing *Bt* cotton use less than half as much pesticide as farmers growing conventional cotton. In the United States, increasing adoption of *Bt* crops has resulted in a corresponding decline in pesticide use, which fell by about 33% between 1995 and 2015.

An undesirable side effect of growing genetically modified crops is that *Bt* or herbicide-resistance genes might spread outside a farmer's fields. Because these genes are incorporated into the genome of the transgenic crop, they will be in its pollen, and a farmer cannot control where pollen from a transgenic crop will go.

Does it matter if genes escape from transgenic crops? It might, because many crops have wild relatives living nearby. Suppose these wild relatives are pollinated by transgenic crop pollen, resulting in hybrids that are resistant to herbicides or pests. There is already evidence this is happening: In 2006, researchers discovered herbicide-resistant grasses more than 2 miles away from a test plot of transgenic grass in Oregon and concluded that the herbicide-resistance genes had escaped in pollen and seeds. Since then, wild plants carrying herbicide-resistance genes have also been found in several other agricultural areas. Could such accidentally transgenic wild plants become significant weed problems? Could they displace other plants in the wild because they would be less likely to be eaten by insects? Even if transgenic crops have no close relatives in the wild, bacteria and viruses sometimes transfer genes among unrelated plant species. Could viruses spread unwanted genes into wild plant populations? No one knows the answers to these questions.

Should Biotechnology Be Used to Modify the Human Genome?

Some of the ethical dilemmas associated with medical biotechnology are the result of advances that simply add new dimensions to long-standing issues. For example, people have for decades passionately disagreed about the appropriate course of action when a major defect is detected in a fetus, and advances in prenatal genetic testing have intensified this controversy by making it possible to detect a growing number of disorders in fetuses. However, other ethical concerns that have arisen with advances in biotechnology are unprecedented. For example, we will soon face the question of whether people should be allowed to modify the genomes of their offspring.

People using in vitro fertilization (IVF) already have a limited way of selecting the genotype of their offspring. Shortly after an egg has been fertilized in vitro, but before it is implanted into the uterus, it divides a few times, forming an embryo. A cell can be removed from this very early embryo and its genome sequenced. Thus, parents can learn if an embryo has a genetic defect, and they may choose to implant only embryos that lack defects. Most countries regulate preimplantation genetic diagnosis and allow selection of embryos to keep based only on the absence of alleles for serious inherited disorders. In principle, however, the procedure could be used to select embryos on the basis of "elective" traits such as skin or eye color, or potential height and athletic ability.

The advent of gene editing tools such as CRISPR-Cas9 makes the issues surrounding embryo selection even more challenging. The same technologies used to edit genes in stem cells to cure SCID could be used to change the genes of human embryos (**FIG. 14-17**). In the not too distant future, it will likely become possible to, for example, insert functional *CFTR* alleles into human embryos, thereby preventing cystic fibrosis. Would this modification of the human genome be ethical? How about editing the genome to increase intelligence or reduce the likelihood of obesity? Or to make bigger weight lifters or faster runners? If and when gene editing technology becomes able to cure genetic diseases, it will be difficult to prevent it from being used for elective procedures. Who will determine which uses are allowed and which are not?

CHECK YOUR LEARNING

Can you . . .

- explain why people might be opposed to the use of genetically modified organisms in agriculture?
- envision circumstances in which it might be ethical to modify the genome of a human embryo?

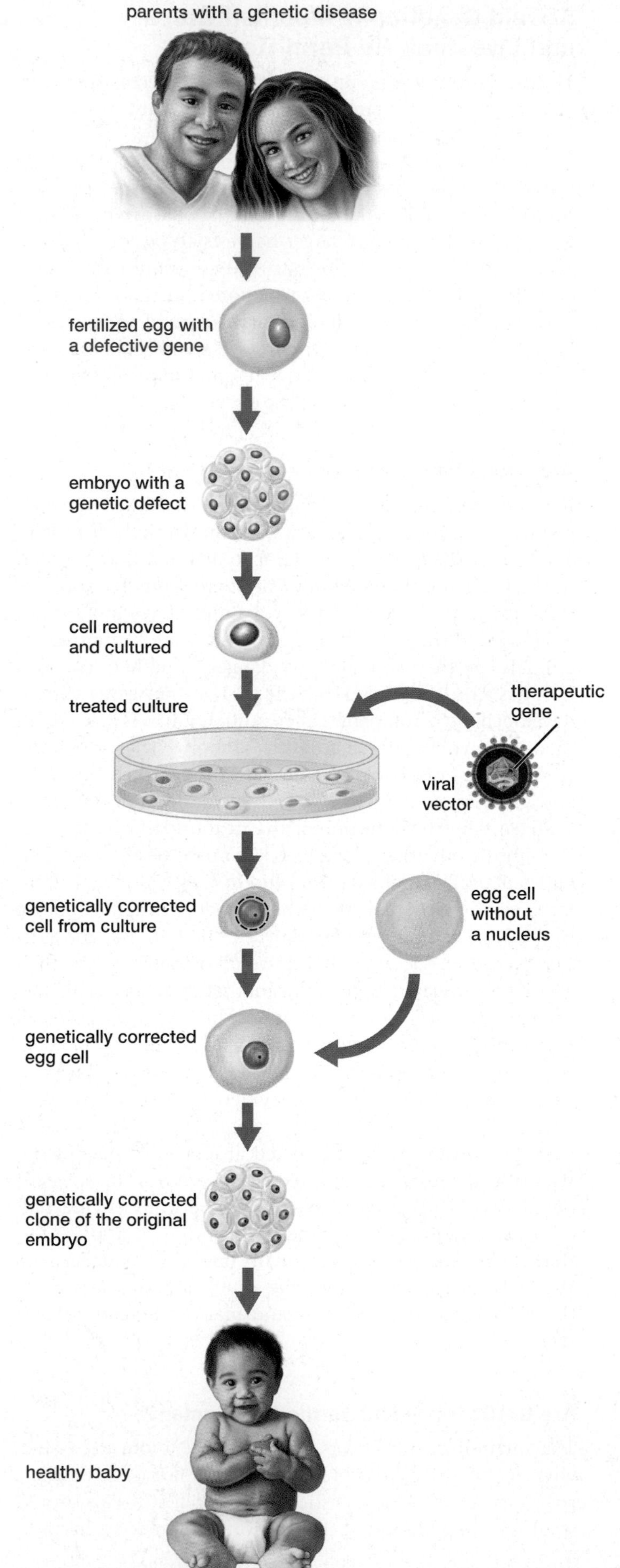

▶ **FIGURE 14-17 Using biotechnology to correct genetic defects in human embryos** In this hypothetical example, a couple who carry the alleles for a serious genetic disorder wish to have a child. The woman's eggs are fertilized in vitro by her partner's sperm. When an embryo containing a defective gene grows into a small cluster of cells, a single cell is removed from the embryo, and the defective allele in the cell is replaced using an appropriate vector, usually a disabled virus. The nucleus of another egg cell (taken from the same mother) is removed. The genetically repaired nucleus is then injected into the egg whose nucleus had been removed. The now repaired egg cell is allowed to divide a few times, and the resulting embryo is implanted in the woman's uterus for further development.

CASE STUDY REVISITED

Guilty or Innocent?

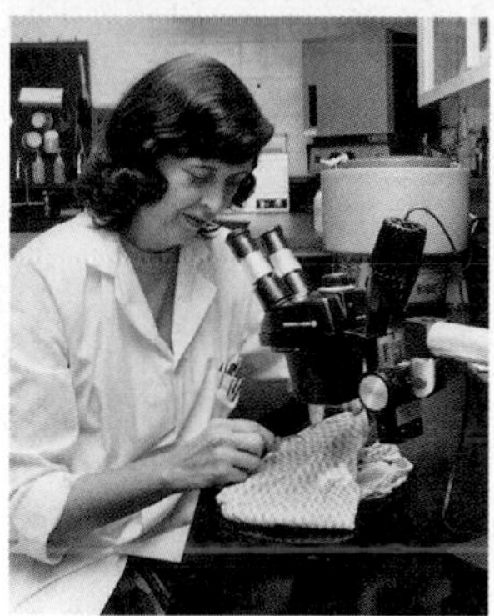

▲ **FIGURE 14-18 Mary Jane Burton**

Former Virginia Governor Mark Warner credits his state's ability to conduct a massive review of old cases mostly to one woman: lab technician Mary Jane Burton (**FIG. 14-18**). During Burton's time at the state forensic lab, it was standard practice to return evidence to local authorities, who usually destroyed it a few years after a case was closed. But Burton kept bits of evidence taped to her case files—no one is really sure why. Her efforts gave new life not only to Thomas Haynesworth, but also to Marvin Anderson, Julius Ruffin, Arthur Whitfield, Philip Thurman, Victor Burnette, and Willie Davidson. Burton, who died in 1999, did not live to see her legacy, but these men will never forget her.

The wrongly convicted cannot always be cleared by DNA evidence. For two of the rapes that Thomas Haynesworth was accused of, no biological evidence was saved. Nonetheless, Mid-Atlantic Innocence Project attorney Shawn Armbrust and former Virginia Attorney General Ken Cuccinelli stepped up and persuaded the courts to issue a "writ of actual innocence" for Haynesworth, officially clearing him of all crimes. Cuccinelli also hired Haynesworth to work in the attorney general's office mailroom, where he is now the supervisor. Haynesworth, a truly remarkable man, also volunteers for the Innocence Project, to help other people who may be accused and convicted of crimes they did not commit.

CONSIDER THIS Who are the heroes in these stories? There are the obvious ones, of course—Mary Jane Burton; the lawyers of the Innocence Project who have helped to free over 300 people wrongly convicted of crimes they did not commit; and, of course, unjustly imprisoned men who, like Thomas Haynesworth, have become gracious, productive members of society. But what about molecular biologist Kary Mullis, who discovered PCR? Or Thomas Brock, whose discovery of *Thermus aquaticus* in Yellowstone's hot springs provided the source of the heat-stable DNA polymerase that is so essential to PCR? Or the hundreds of biologists, chemists, and mathematicians who developed procedures for gel electrophoresis and statistical analysis of DNA profile matching? Scientists often say that science is worthwhile for its own sake, and that it is difficult to predict which discoveries will lead to the greatest benefits for humanity. Nonscientists, when asked to pay the costs of scientific projects, are sometimes skeptical of such claims. How do you think public funds that support science should be allocated? Fifty years ago, would you have voted to give Thomas Brock public funds to see what types of organisms lived in hot springs?

CHAPTER REVIEW

Go to **Mastering Biology** to access the Pearson eText, vocabulary review, practice quizzes, activities, videos, current events, and more.

*Answers to **Think Critically and Thinking Through the Concepts** questions can be found in the **Answers** section at the back of the book.*

Summary of Key Concepts

14.1 What Is Biotechnology?

Biotechnology is the manipulation or modification of living organisms or their components to produce useful products or services. Modern biotechnology uses genetic engineering to modify DNA. Organisms with such modifications are known as genetically modified organisms (GMOs). When the modification involves transfer of DNA from one species to another, the recipients are called transgenic. Applications of modern biotechnology include increasing our understanding of gene function, treating disease, improving agriculture, and solving crimes.

14.2 What Natural Processes Recombine DNA Between Species?

DNA recombination occurs naturally through processes such as bacterial transformation, in which bacteria acquire DNA from plasmids or other bacteria, and viral infection, in which viruses incorporate fragments of DNA from their hosts and transfer the fragments to members of the same or other species.

14.3 What Are Some Key Methods for Manipulating DNA?

The polymerase chain reaction (PCR) amplifies (makes many copies of) specific regions of very small quantities of DNA. CRISPR-Cas9 is a molecular tool for precisely editing DNA. It can be used to disable a gene or insert a new gene at a particular location in a genome.

14.4 How Is Biotechnology Used in Forensic Science?

The genome regions most commonly used in forensics are short tandem repeats (STRs). Gel electrophoresis separates the STRs of an individual by size to create a pattern of STRs, called a DNA profile. DNA profiles based on 13 STR regions of the genome can be used to match DNA found at a crime scene with DNA from suspects with extremely high accuracy.

14.5 How Are Transgenic Organisms Made?

There are three steps to making a transgenic organism. First, the desired gene is obtained from another organism or synthesized. Second, the gene is cloned, often into a bacterial plasmid, to provide multiple copies of the gene. Third, the gene is inserted into a host organism, often by using bacteria or viruses, by treatments that disrupt the plasma membrane of the target cell, or by injection into cells (especially fertilized eggs).

14.6 How Are Genetically Modified Organisms Used?

Many crop plants have been modified by the addition of genes that promote herbicide resistance or insect resistance, or by manipulating the genome to enhance desired characteristics. Genetically modified animals may be produced with properties such as faster growth, increased production of valuable products such as wool, or the ability to resist infection. Genetically modified organisms may be used more in the future to restore populations of rare species, clean up pollution, and fight the spread of mosquito-borne diseases.

14.7 How Is Biotechnology Used in Medicine?

Inherited diseases are caused by defective alleles of crucial genes. Tools of biotechnology—including PCR, restriction enzymes, gel electrophoresis, DNA probes, sequencing, and microarrays—may be used to diagnose genetic disorders such as sickle-cell anemia and cystic fibrosis or to diagnose which type of microbe is causing an infection. Transgenic organisms are used to produce medicines and human proteins—including hormones, antibodies, and enzymes—that can treat people with certain genetic disorders or acquired diseases. Gene therapy treats disorders by repairing or replacing damaged or defective DNA in patients' cells.

14.8 What Are the Major Ethical Issues of Modern Biotechnology?

The use of genetically modified organisms in agriculture is controversial for two major reasons: food safety and potentially harmful effects on the environment. In general, GMOs contain proteins that are harmless to mammals or are already found in nontransgenic foods. Environmental effects of GMOs are more difficult to predict. It is possible that foreign genes, such as those for pest or herbicide resistance, might be transferred to wild plants, with resulting damage to agriculture and/or disruption of ecosystems.

Genetically selecting or modifying human embryos is highly controversial. As technologies improve, society may be faced with decisions about the extent to which parents should be allowed to correct or "enhance" the genomes of their children.

Thinking Through the Concepts

Bloom's: Remembering, Understanding

Multiple Choice

1. The polymerase chain reaction (PCR)
   a. produces many copies of specific DNA sequences.
   b. is a method for sequencing DNA.
   c. is a technique for editing a genome.
   d. is used to determine whether a specific gene is expressed.
2. Imagine a DNA profile showing the STR patterns of a mother and her child. Will all of the child's DNA bands match the mother's?
   a. Yes, because the mother's DNA and her child's DNA are identical.
   b. Yes, because the child developed from her mother's egg.
   c. No, because half of the child's DNA is inherited from its father.
   d. No, because the child's DNA is a random sampling of its mother's.
3. The most widely planted transgenic crops have been genetically modified to
   a. taste better.
   b. be more nutritious.
   c. resist herbicides and insect pests.
   d. grow faster.
4. A restriction enzyme
   a. cuts DNA at a specific nucleotide sequence.
   b. cuts DNA at a random nucleotide sequence.
   c. splices DNA together at a specific nucleotide sequence.
   d. splices DNA together without regard to the nucleotide sequence.
5. DNA cloning is
   a. transcribing DNA to messenger RNA.
   b. making multiple copies of a piece of DNA.
   c. inserting DNA into a cell.
   d. changing the nucleotide sequence of a strand of DNA.

Fill-in-the-Blank

1. ____________ are organisms that contain DNA that has been modified (usually through use of recombinant DNA technology) or derived from other species.
2. ____________ is the process whereby bacteria acquire DNA from their environment. This DNA may be part of a chromosome, or it may be tiny circles of DNA called ____________.
3. ____________ is a technique for multiplying DNA in the laboratory.
4. Matching DNA samples in forensics uses a specific set of small "genes" called ____________. The alleles of these genes differ in ____________ between nonrelated people. The pattern of these alleles that a given person possesses is called his or her ____________.
5. Pieces of DNA can be separated according to size by a process known as ____________. The identity of a specific sample of DNA is usually determined by binding a synthetic piece of DNA to the sample DNA by ____________.

Review Questions

1. Describe two natural forms of genetic recombination, and discuss the similarities and differences between recombinant DNA technology and these natural forms of genetic recombination.
2. What is a plasmid? How are plasmids involved in bacterial transformation?
3. What is a restriction enzyme? How can restriction enzymes be used to splice a piece of human DNA into a plasmid?
4. Describe the polymerase chain reaction.
5. Describe how the CRISPR-Cas9 tool works.
6. What is a short tandem repeat (STR)? How are short tandem repeats used in forensics?
7. How does gel electrophoresis separate pieces of DNA?
8. Describe several uses of genetic engineering in agriculture.
9. Describe several uses of genetic engineering in human medicine.
10. Describe amniocentesis, chorionic villus sampling, and maternal blood sampling, including the advantages and disadvantages of each. What are their medical uses?

Applying the Concepts

Bloom's: Applying, Analyzing, Evaluating

1. Many insects have evolved resistance to common pesticides. Do you think that insects might evolve resistance to *Bt* crops? If this is a risk, do you think that *Bt* crops should be planted anyway? Why or why not?
2. One type of severe combined immune deficiency, called X-linked SCID, is caused by a defective recessive allele of a gene located on the X chromosome. All children born with X-linked SCID are boys. Explain why.

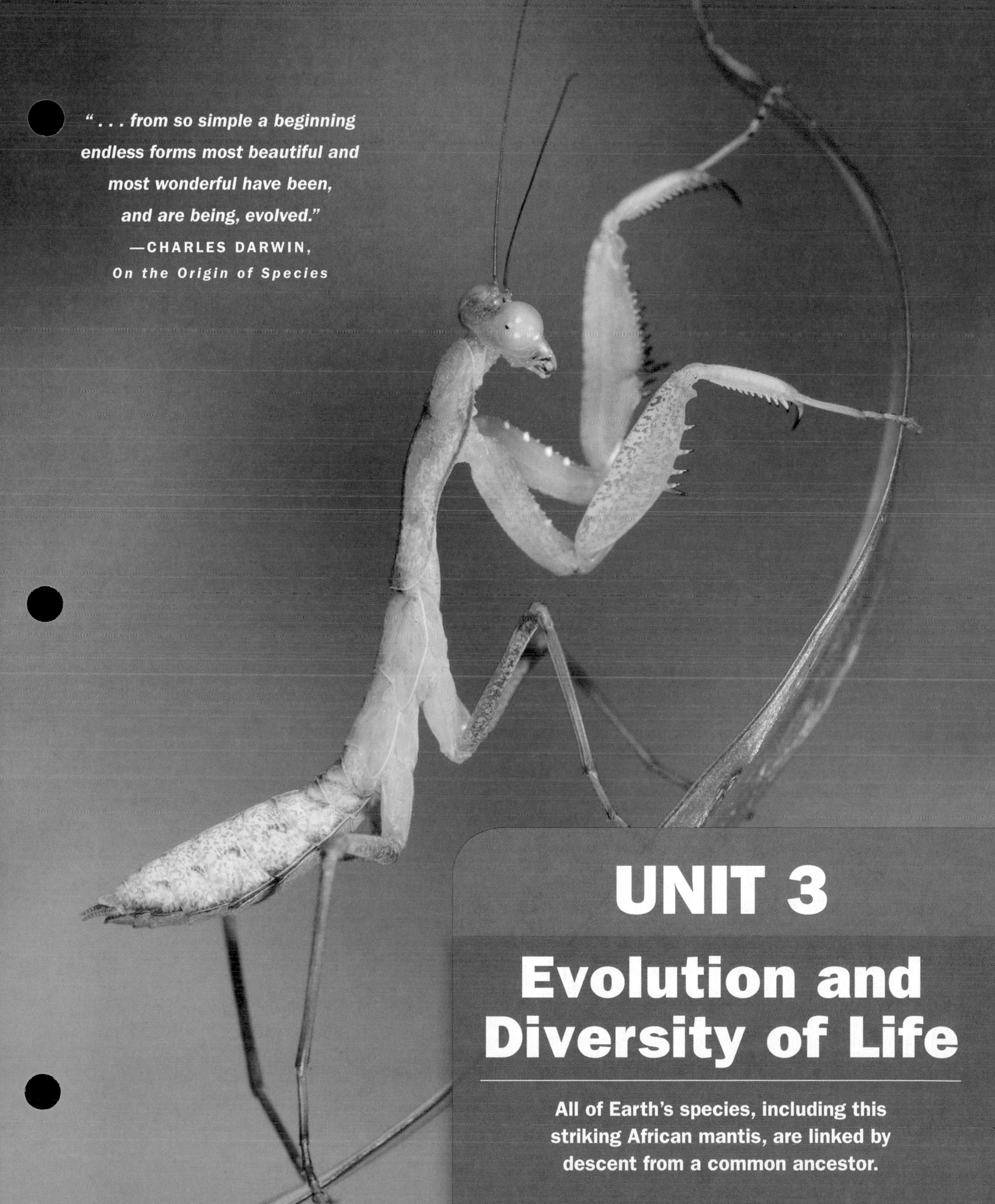
"... from so simple a beginning endless forms most beautiful and most wonderful have been, and are being, evolved."
—CHARLES DARWIN, On the Origin of Species
UNIT 3
Evolution and Diversity of Life
All of Earth's species, including this striking African mantis, are linked by descent from a common ancestor.

15 Principles of Evolution

This massive, earthbound ostrich has wings, a legacy of its evolutionary heritage.

CASE STUDY

What Good Are Wisdom Teeth and Ostrich Wings?

HAVE YOU HAD YOUR WISDOM TEETH REMOVED YET? If not, it's probably only a matter of time. Almost all of us will visit an oral surgeon to have our wisdom teeth extracted. There's just not enough room in our jaws for these rearmost molars, and removing them is the best way to prevent the pain, infections, and gum disease that can accompany the development of wisdom teeth. Removal is harmless because we don't really need wisdom teeth.

If you've already suffered through a wisdom tooth extraction, you may have found yourself wondering why we even have these extra molars. Biologists hypothesize that we have them because our apelike ancestors had them and we inherited them, even though we don't need them. Other living species, such as apes, also have these teeth, but with their ancestral function preserved. Even though in people these rearmost molars have lost their original function, the fact that they are present in apes as well reveals that we share an ancestor with apes.

Flightless birds also illustrate the connection between evolutionary ancestry and structures that do not perform their original function. Consider the ostrich, a bird that can grow to 8 feet tall and weigh 300 pounds (see the photo above). These massive creatures cannot fly. Nonetheless, they have wings, just as sparrows and ducks do. Why do ostriches have wings? Because the ancestor of all living birds had wings, and so do all of its descendants, even those that cannot fly. Many other organisms have, like ostriches and people, inherited hand-me-downs that no longer serve their original functions. What does this observation tell us about evolution? What other evidence shows us that evolution has occurred and reveals the mechanisms that cause evolution?

AT A GLANCE

15.1 How Did Evolutionary Thought Develop?

15.2 How Does Natural Selection Work?

15.3 How Do We Know That Evolution Has Occurred?

15.4 What Is the Evidence That Populations Evolve by Natural Selection?

15.1 HOW DID EVOLUTIONARY THOUGHT DEVELOP?

When you began studying biology, you may not have seen a connection between your wisdom teeth and an ostrich's wings. But the connection is there, provided by the concept that unites all of biology: **evolution**, or change over time in the characteristics of a population. (A **population** consists of all the individuals of one species in a particular area.)

Modern biology is based on our understanding that life has evolved, but early scientists did not recognize this fundamental principle. The main ideas of evolutionary biology became widely accepted only after the publication of Charles Darwin's work in the nineteenth century. Nonetheless, the intellectual foundation on which these ideas rest developed gradually over the centuries before Darwin's time.

Early Biological Thought Did Not Include the Concept of Evolution

Pre-Darwinian science, heavily influenced by theology, held that all organisms were created simultaneously by God and that each distinct life-form remained fixed and unchanging from the moment of its creation. This explanation of how life's diversity arose was elegantly expressed by the ancient Greek philosophers, especially Plato and Aristotle. Plato (427–347 B.C.) proposed that each object on Earth is merely a temporary reflection of its divinely inspired "ideal form." Plato's student Aristotle (384–322 B.C.) categorized all organisms into a linear hierarchy that he called the "Ladder of Nature" (**FIG. 15-1**).

These ideas formed the basis of the view that the form of each type of organism is permanently fixed. This view reigned unchallenged for more than 2,000 years. By the eighteenth century, however, several lines of newly emerging evidence began to undermine this static view of creation.

Exploration of New Lands Revealed a Staggering Diversity of Life

The Europeans who explored and colonized Africa, Asia, and the Americas were often accompanied by naturalists who observed and collected the plants and animals of these previously unknown (to Europeans) lands. By the 1700s, the accumulated observations and collections of the naturalists had begun to reveal the true scope of life's variety. The number of species, or different types of organisms, was much greater than anyone had suspected.

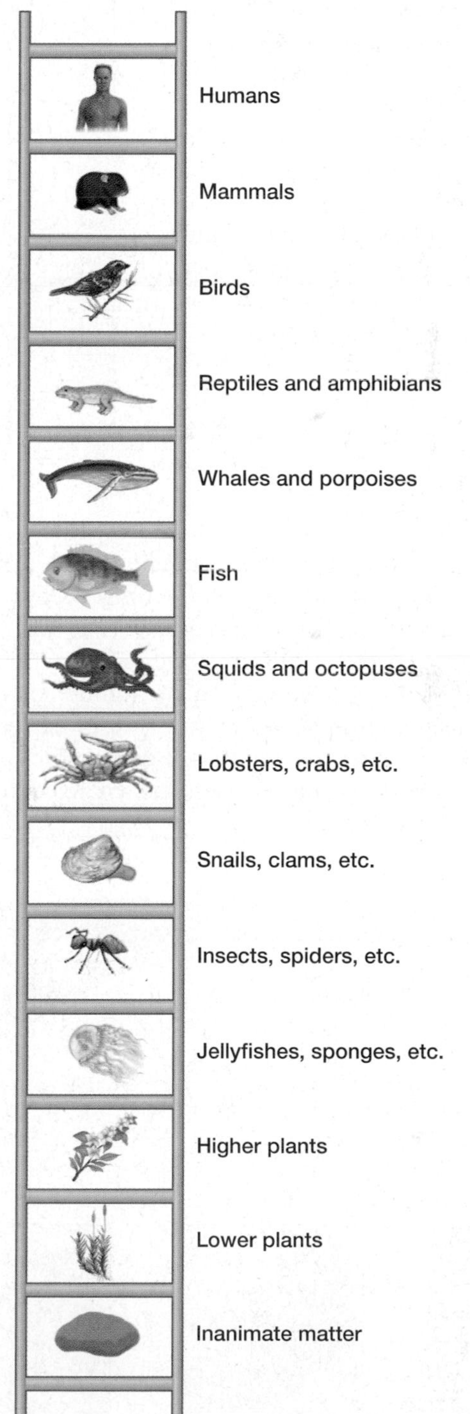

▲ **FIGURE 15-1 Aristotle's "Ladder of Nature"** In Aristotle's view, fixed, unchanging species can be arranged in order of increasing closeness to perfection.

Stimulated by the new evidence of life's incredible diversity, some eighteenth-century naturalists began to take note of fascinating patterns. They noticed, for example, that each geographic area had its own distinctive set of species. In addition, the naturalists saw that some of the species in a given location closely resembled one another, yet differed in some characteristics. To some scientists of the day, the differences between the species of different geographic areas and the existence of clusters of similar species within areas seemed inconsistent with the idea that species were fixed and unchanging. (You may wish to refer to the timeline in **FIGURE 15-2** as you read the following account.)

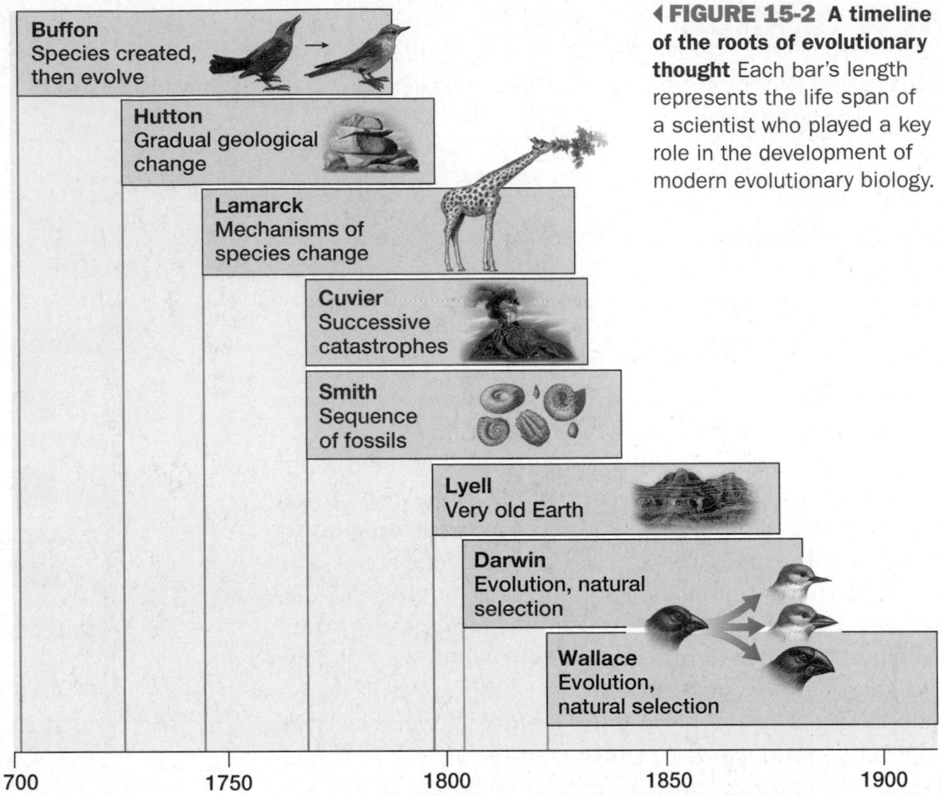

Buffon
Species created, then evolve
Hutton
Gradual geological change
Lamarck
Mechanisms of species change
Cuvier
Successive catastrophes
Smith
Sequence of fossils
Lyell
Very old Earth
Darwin
Evolution, natural selection
Wallace
Evolution, natural selection
1700 1750 1800 1850 1900

◀ **FIGURE 15-2 A timeline of the roots of evolutionary thought** Each bar's length represents the life span of a scientist who played a key role in the development of modern evolutionary biology.

A Few Scientists Speculated That Life Had Evolved

A few eighteenth-century scientists went so far as to speculate that species had, in fact, changed over time. For example, the French naturalist Georges Louis Leclerc (1707–1788), known by the title Comte de Buffon, suggested that the original creation provided a relatively small number of founding species, after which some might have "improved" or "degenerated," perhaps after moving to new geographic areas. That is, Buffon suggested that species had changed over time through natural processes.

Fossil Discoveries Showed That Life Has Changed over Time

As Buffon and his contemporaries pondered the implications of new biological discoveries, developments in geology cast further doubt on the idea of permanently fixed species. Especially important was the discovery, during excavations for roads, mines, and canals, of rock fragments that resembled parts of living organisms. People had known of such objects since the fifteenth century, but most thought they were ordinary rocks that wind, water, or people had worked into lifelike forms. As more and more organism-shaped rocks were discovered, however, it became obvious that they were **fossils**, the preserved remains or traces of organisms that had died long ago (**FIG. 15-3**). Many fossils are bones, wood, shells, or their

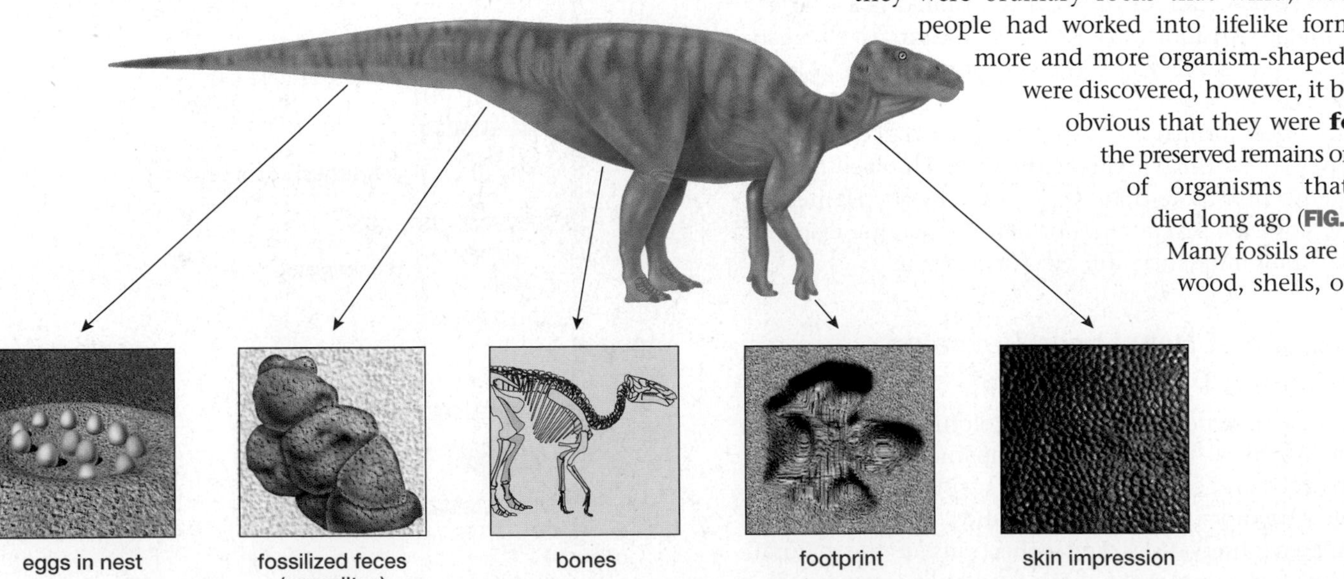

▲ **FIGURE 15-3 Types of fossils** Any preserved part or trace of an organism is a fossil.

impressions in mud that have been petrified, or converted to stone. Fossils also include other kinds of preserved traces, such as tracks, burrows, pollen grains, eggs, and feces.

By the beginning of the nineteenth century, some pioneering investigators realized that the distribution of fossils in rock was also significant. Many rocks occur in layers, with newer layers positioned over older layers. The British surveyor William Smith (1769–1839), who studied rock layers and the fossils embedded in them, recognized that certain fossils were always found in the same layers of rock. Further, the organization of fossils and rock layers was consistent across different areas: Fossil type A could always be found in a rock layer resting beneath a younger layer containing fossil type B, which in turn rested beneath a still-younger layer containing fossil type C, and so on.

Scientists of the period also discovered that fossil remains showed a remarkable progression. Most fossils found in the oldest layers were very different from modern organisms, and the resemblance to modern organisms gradually increased in progressively younger rocks (**FIG. 15-4**). Many of the fossils

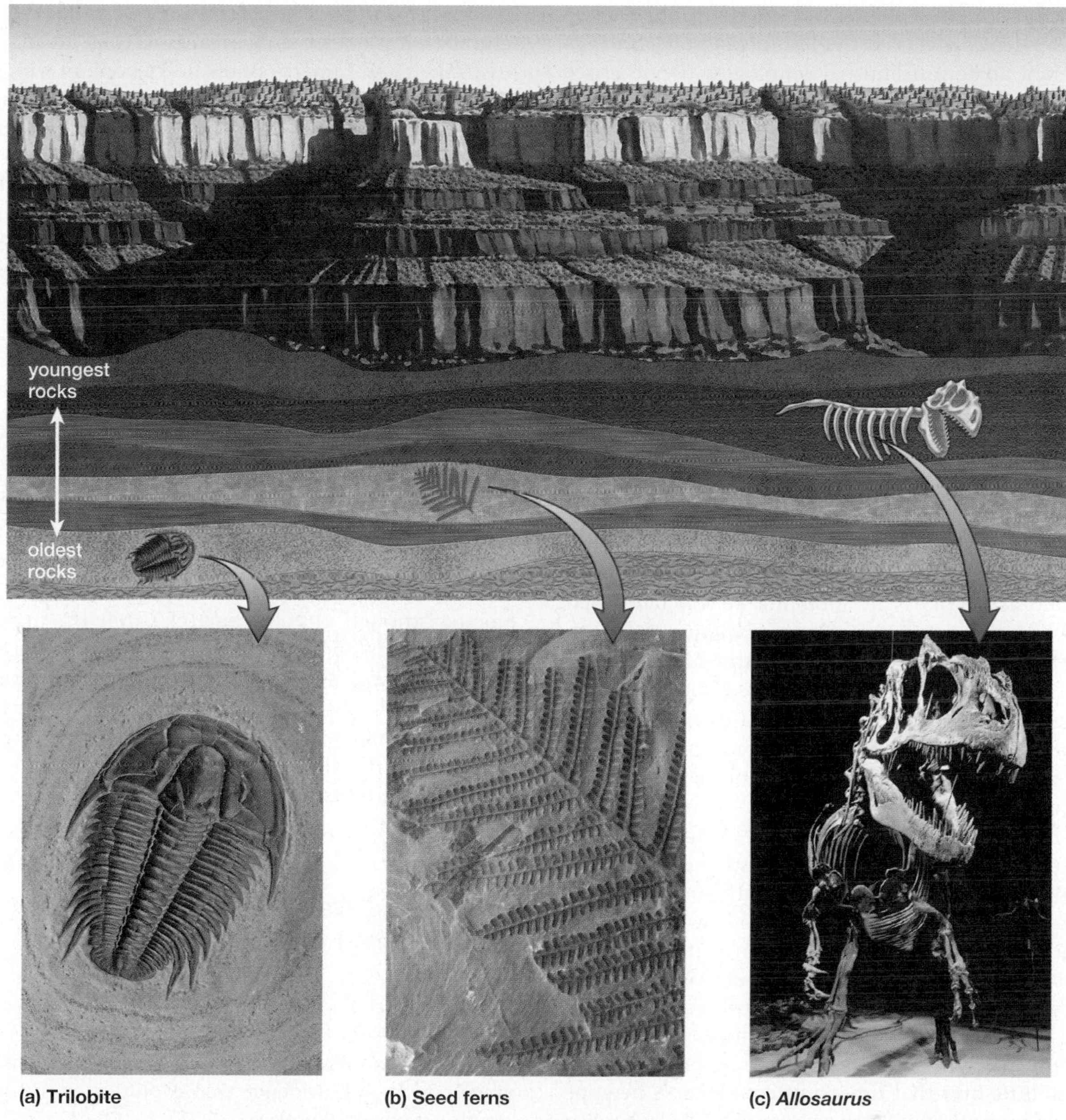

▲ **FIGURE 15-4 Different fossils are found in different rock layers** Fossils provide strong support for the idea that today's organisms were not created all at once but arose over time by the process of evolution. If all species had been created simultaneously, we would not expect **(a)** the earliest trilobites to be found in older rock layers than **(b)** the earliest seed ferns, which in turn would not be expected in older layers than **(c)** dinosaurs, such as *Allosaurus*. Trilobites first appeared about 520 million years ago, seed ferns (which were not actually ferns but had fern-like foliage) about 380 million years ago, and dinosaurs about 230 million years ago.

were from plant or animal species that had gone *extinct*; that is, no members of the species still lived on Earth.

Putting all of these facts together, some scientists came to an inescapable conclusion: Different types of organisms had lived at different times in the past.

Some Scientists Devised Nonevolutionary Explanations for Fossils

Despite the growing fossil evidence, many scientists of the period did not accept the proposition that species changed and new ones arose over time. To account for extinct species while preserving the notion of a single creation by God, Georges Cuvier (1769–1832) advanced the idea of *catastrophism*. Cuvier, a French anatomist and paleontologist, hypothesized that a vast supply of species was created initially. Successive catastrophes (such as the Great Flood described in the Bible) produced layers of rock and destroyed many species, fossilizing some of their remains in the process. The organisms of the modern world, he speculated, are the species that survived the catastrophes.

Geology Provided Evidence That Earth Is Exceedingly Old

Cuvier's hypothesis of a world shaped by successive catastrophes was challenged by the work of the geologist Charles Lyell (1797–1875). Lyell, building on the earlier thinking of James Hutton (1726–1797), considered the forces of wind, water, and volcanoes and concluded that there was no need to invoke catastrophes to explain the findings of geology. Don't flooding rivers lay down layers of sediment? Don't lava flows produce layers of basalt? Shouldn't we conclude, then, that layers of rock are evidence of ordinary natural processes, occurring repeatedly over long periods of time? This concept, that Earth's present landscape was produced by past action of the same gradual geological processes that we observe today, is called *uniformitarianism*. Acceptance of uniformitarianism by scientists of the time had a profound impact, because the idea implies that Earth is very old.

Before the 1830 publication of Lyell's evidence in support of uniformitarianism, few scientists suspected that Earth could be more than a few thousand years old. Counting generations in the Old Testament, for example, yields a maximum age of 4,000 to 6,000 years. An Earth this young poses problems for the idea that life has evolved. For example, ancient writers such as Aristotle described wolves, deer, lions, and other organisms that were identical to those present in Europe more than 2,000 years later. If organisms had changed so little over that time, how could whole new species possibly have arisen if Earth was created only a couple of thousand years before Aristotle's time?

But if, as Lyell suggested, rock layers thousands of feet thick were produced by slow, natural processes, then Earth must be old indeed, many millions of years old. Lyell, in fact, concluded that Earth was eternal. Modern geologists estimate that Earth is about 4.5 billion years old (see "Doing Science: Discovering the Age of a Fossil" in Chapter 18).

Lyell (and his intellectual predecessor Hutton) showed that there was enough time for evolution to occur. But what was the mechanism? What process could cause evolution?

Some Pre-Darwin Biologists Proposed Mechanisms for Evolution

One of the first scientists to propose a mechanism for evolution was the French biologist Jean Baptiste Lamarck (1744–1829). Lamarck was impressed by the sequences of organisms in rock layers. He observed that older fossils tend to be less like existing organisms than are more recent fossils. In 1809, Lamarck published a book in which he hypothesized that organisms evolved through the inheritance of acquired characteristics, a process in which the bodies of living organisms are modified through the use or disuse of parts, and these modifications are inherited by offspring. Why would bodies be modified? Lamarck proposed that all organisms possess an innate drive for perfection. For example, if ancestral giraffes tried to increase their feeding opportunities by stretching upward to reach leaves growing high up in trees, their necks became slightly longer as a result. Their offspring would inherit these longer necks and then stretch even farther to reach still higher leaves. Eventually, this process would produce modern giraffes with very long necks indeed.

Today, we understand how inheritance works and can see that Lamarck's proposed evolutionary process could not work. Acquired characteristics are not inherited. The fact that a prospective father pumps iron doesn't mean that his child will look like a champion bodybuilder. Remember, though, that in Lamarck's time the principles of inheritance had not yet been discovered. Gregor Mendel's pioneering work demonstrating inheritance in pea plants was not widely recognized until 1900 (see Chapter 11). In any case, Lamarck's insight that inheritance plays an important role in evolution had an important influence on the later biologists who discovered the key mechanism of evolution.

Darwin and Wallace Proposed a Mechanism of Evolution

By the mid-1800s, a growing number of biologists had concluded that present-day species had evolved from earlier ones. But how? In 1858, Charles Darwin (1809–1882) and Alfred Russel Wallace (1823–1913), working separately, provided convincing evidence that evolution was driven by a simple yet powerful process.

Although their social and educational backgrounds were very different, Darwin and Wallace were quite similar in some respects. Both had traveled extensively in the tropics and had studied the plants and animals living there. Both observed that some species differed in only a few features

(a) Large ground finch: beak suited to large seeds

(b) Small ground finch: beak suited to small seeds

(c) Warbler finch: beak suited to insects

(d) Vegetarian tree finch: beak suited to leaves

FIGURE 15-5 Darwin's finches, residents of the Galápagos Islands Darwin studied a group of closely related species of finches on the Galápagos Islands. Each species specializes in eating a different type of food and has a beak of characteristic size and shape because past individuals whose beaks were best suited to exploit each local food source produced more offspring than did individuals with less effective beaks.

(**FIG. 15-5**). Both were familiar with the fossils that had been discovered, many of which showed a trend through time of increasing similarity to modern organisms. Finally, both were aware of the studies of Hutton and Lyell, who had proposed that Earth is extremely ancient. These facts suggested to both Darwin and Wallace that species change over time. Both men sought a mechanism that might cause such evolutionary change.

Of the two, Darwin was the first to propose a mechanism for evolution, which he sketched out in 1842 and described more fully in an essay in 1844. He sent the essay to a few colleagues, but did not submit it for publication, perhaps because he was fearful of the controversy that publication would cause. Some historians wonder if Darwin would ever have published his ideas had he not received, some 16 years after his initial draft, a paper by Wallace that outlined ideas remarkably similar to Darwin's own. Darwin realized that he could delay no longer.

In separate but similar papers that were presented to the Linnaean Society in London in 1858, Darwin and Wallace each described the same mechanism for evolution. Initially, their papers had little impact. The secretary of the society, in fact, wrote in his annual report that nothing very interesting happened that year. Fortunately, the next year, Darwin published his monumental book, *On the Origin of Species by Means of Natural Selection,* which attracted a great deal of attention to the new ideas about how species evolve. (To learn more about Darwin's work, see "Doing Science: Charles Darwin and the Mockingbirds.")

CHECK YOUR LEARNING

Can you . . .

- identify some of the thinkers whose ideas set the stage for the development of the theory of evolution?
- describe the key ideas of those thinkers?
- define evolution?

15.2 HOW DOES NATURAL SELECTION WORK?

Darwin and Wallace proposed that life's huge variety arose by a process of descent with modification, in which individuals in each generation differ slightly from the members of the preceding generation. Over long stretches of time, these small differences accumulate to produce major transformations.

Darwin and Wallace's Theory Rests on Four Postulates

The chain of logic that led Darwin and Wallace to their proposed process of evolution turns out to be surprisingly

DOING **Science**

Charles Darwin and the Mockingbirds

Charles Darwin

How did mockingbirds provide key evidence for evolution? The story begins in 1831, when 22-year-old Charles Darwin secured a position as the official naturalist on the HMS *Beagle*. The *Beagle* soon embarked on a 5-year surveying expedition along the coastline of South America and then around the world.

Darwin's task was to observe and collect geological and biological specimens. The *Beagle* made many stops along the South American coast, but perhaps the most significant stopover of the voyage was the month spent on the Galápagos Islands.

What Question Was Asked?

Although Darwin's initial scientific mission was mainly to collect as many facts as possible about the landscapes and organisms he encountered, his observations led to questions. Darwin wondered, for example, how species had come to be distributed in the particular patterns that he observed. He was especially intrigued by the distribution of various mockingbirds among the Galápagos Islands (**FIG. E15-1**).

How Was Evidence Gathered?

When the *Beagle* reached the Galápagos, the first island that Darwin visited was then called Chatham Island. He noticed that the mockingbirds there seemed different than the ones he had observed on the South American mainland. A short while later, Darwin visited Charles Island and was surprised to find mockingbirds that differed in many respects from the ones on nearby Chatham. And, traveling onward, he found mockingbirds on James Island that, to his eye, were different from any he had seen on the other islands.

Darwin was astonished and impressed by his discovery that different islands had different mockingbirds. He was similarly struck by reports that the archipelago's gigantic tortoises also differed from island to island. Darwin began to wonder if the differences in the tortoises and in the mockingbirds arose after they had become isolated on separate islands.

What Was Learned?

A few weeks after leaving the Galápagos, Darwin, pondering his observations, wrote that "such facts undermine the stability of Species." This journal entry represents the first tentative indication that Darwin had concluded that species could change over time.

When the *Beagle* returned to London in 1836, Darwin asked various experts to examine the specimens he had collected during his journey. The ornithologist John Gould studied Darwin's bird specimens and judged that the roving naturalist had indeed collected three different species of mockingbird, each of which inhabited a different island or small set of islands. Gould's confirmation, made possible by Darwin's systematic collecting and careful documentation, became one of the clinching pieces of evidence in Darwin's conversion to evolutionary thinking. A few months later, Darwin drew in his journal a small tree-like diagram, a representation of his emerging idea that species are linked by descent from a common ancestor (**FIG. E15-2**).

THINK CRITICALLY A recent study found that Galápagos mockingbirds on a given island are more genetically similar to mockingbirds on nearby islands than to mockingbirds on more distant islands. From this information, what can you conclude about the evolutionary history of Galápagos mockingbirds?

▲ **FIGURE E15-1 Galápagos mockingbirds** Each island in the Galápagos contains a unique mockingbird species. Darwin and his contemporary John Gould described three species, but modern ornithologists recognize four. Clockwise from top left: the Galápagos, Hood, Charles, and Chatham mockingbirds.

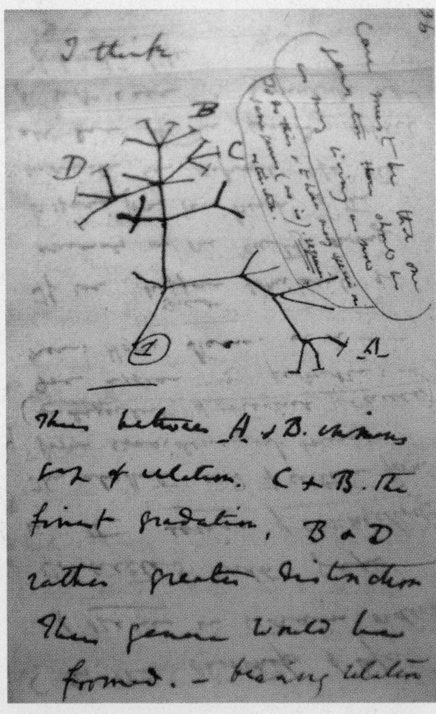

◀ **FIGURE E15-2 A sketch from Darwin's notebook, 1837.**

simple and straightforward. It is based on four postulates about populations:

Postulate 1: Individual members of a population differ from one another in many respects.
Postulate 2: At least some of the differences among members of a population are due to characteristics that may be passed from parent to offspring.
Postulate 3: In each generation, some individuals in a population survive and reproduce successfully but others do not.
Postulate 4: The fate of individuals is not determined entirely by chance or luck. Instead, an individual's likelihood of survival and reproduction depends on its characteristics. Individuals with advantageous traits survive longest and leave the most offspring, a process known as **natural selection**.

Darwin and Wallace understood that if all four postulates were true, populations would inevitably change over time. If members of a population have different traits, and if the individuals that are best suited to their environment leave more offspring, and if those individuals pass their favorable traits to their offspring, then the favorable traits will become more common in subsequent generations. The characteristics of the population will change slightly with each generation. This process is evolution by natural selection.

Are the four postulates true? Darwin thought so, and devoted much of *On the Origin of Species* to describing supporting evidence. Let's briefly examine each postulate, in some cases with the advantage of knowledge that had not yet come to light during the lifetimes of Darwin and Wallace.

Postulate 1: Individuals in a Population Vary

The accuracy of postulate 1 is apparent to anyone who has glanced around a crowded room. People differ in size, eye color, skin color, and many other physical features. Similar variability is present in populations of other organisms, although it may be less obvious to the casual observer (**FIG. 15-6**).

Postulate 2: Traits Are Passed from Parent to Offspring

The principles of genetics had not yet been discovered when Darwin published *On the Origin of Species*. Therefore, although observation of people, pets, and farm animals seemed to show that offspring generally resemble their parents, Darwin and Wallace did not have scientific evidence in support of postulate 2. Mendel's later work, however, demonstrated conclusively that particular traits can be passed to offspring. Since Mendel's time, genetics researchers have produced a detailed picture of how inheritance works.

Postulate 3: Some Individuals Fail to Survive and Reproduce

Darwin's formulation of postulate 3 was heavily influenced by Thomas Malthus's *Essay on the Principle of Population* (1798), which described the perils of unchecked growth of human populations. Darwin was keenly aware that organisms can produce far more offspring than are required merely to replace the parents. He calculated, for example, that a single pair of elephants would multiply to a population of 19 million in 750 years if each descendant had six offspring that lived to reproduce.

But we aren't overrun with elephants. The number of elephants, like the number of individuals in most natural populations, tends to remain relatively constant. Therefore, more organisms must be born than survive long enough to reproduce. In each generation, many individuals must die young. Even among those that survive, many must fail to reproduce, produce few offspring, or produce less-vigorous offspring that, in turn, fail to survive and reproduce. As you might expect, whenever biologists have measured reproduction in a population, they have found that some individuals have more offspring than others.

Postulate 4: Survival and Reproduction Are Not Determined by Chance

If unequal reproduction is the norm in populations, what determines which individuals leave the most offspring? A large amount of scientific evidence has shown that reproductive success depends on an individual's characteristics. For example, scientists found that larger male elephant seals in a California population have more offspring than smaller males

▲ **FIGURE 15-6 Variation in a population of snails** Although these snail shells are all from members of the same population, no two are exactly alike.

THINK CRITICALLY Is sexual reproduction required to generate the variability in structures and behaviors that is necessary for natural selection?

(because females are more likely to mate with large males). In a Colorado population of snapdragons, plants with white flowers have more offspring than plants with yellow flowers (because pollinators find white flowers more attractive). These results, and hundreds of other similar ones, show that in the competition to survive and reproduce, winners are for the most part determined not by chance but by the traits they possess.

Natural Selection Modifies Populations over Time

Observation and experiment suggest that the four postulates of Darwin and Wallace are sound. Logic suggests that the resulting consequence ought to be change over time in the characteristics of populations. In *On the Origin of Species*, Darwin proposed the following example: "Let us take the case of a wolf, which preys on various animals, securing [them] by . . . fleetness. . . . The swiftest and slimmest wolves would have the best chance of surviving, and so be preserved or selected. . . . Now if any slight innate change of habit or structure benefited an individual wolf, it would have the best chance of surviving and of leaving offspring. Some of its young would probably inherit the same habits or structure, and by the repetition of this process, a new variety might be formed." The same logic applies to the wolf's prey: The fastest or most alert or best camouflaged would be most likely to avoid predation and would pass these traits to its offspring.

Notice that natural selection acts on individuals. Eventually, however, the influence of natural selection on the fates of individuals has consequences for the population as a whole. Over generations, the population changes as the percentage of individuals inheriting favorable traits increases. An individual cannot evolve, but a population can.

CHECK YOUR LEARNING

Can you . . .

- explain how natural selection works and how it affects populations?
- describe the logic, based on four postulates, by which Darwin and Wallace deduced that populations must evolve by natural selection?

15.3 HOW DO WE KNOW THAT EVOLUTION HAS OCCURRED?

Today, evolution is an accepted scientific theory. (A scientific theory is a general explanation of important natural phenomena, developed through extensive, reproducible observations; see Chapter 1). An overwhelming body of evidence supports the conclusion that evolution has occurred. The key lines of evidence come from fossils, comparative anatomy (the study of how body structures differ among species), embryology (the study of developing organisms in the period from fertilization to birth or hatching), biochemistry, and genetics.

Fossils Provide Evidence of Evolutionary Change over Time

If many fossils are the remains of species ancestral to modern species, we might expect to find fossils in a progressive series that starts with an ancient organism, progresses through several intermediate stages, and culminates in a modern species. Such series have indeed been found. For example, fossils of the ancestors of modern whales illustrate stages in the evolution of an aquatic species from land-dwelling ancestors (**FIG. 15-7**). Series of fossil giraffes, elephants, horses, and mollusks also show the evolution of body structures over time. These fossil series suggest that new species evolved from, and replaced, previous species.

Comparative Anatomy Gives Evidence of Descent with Modification

Fossils provide snapshots of the past that allow biologists to trace evolutionary changes, but careful examination of today's organisms can also uncover evidence of evolution. Comparing the bodies of organisms of different species can reveal similarities that can be explained only by shared ancestry and differences that could result only from evolutionary change during descent from a common ancestor. In this way, the study of comparative anatomy has supplied strong evidence that different species are linked by a common evolutionary heritage.

Homologous Structures Provide Evidence of Common Ancestry

A body structure may be modified by evolution to serve different functions in different species. The forelimbs of birds and mammals, for example, are variously used for flying, swimming, running, and grasping objects. Despite this enormous diversity of function, the internal anatomy of all bird and mammal forelimbs is remarkably similar (**FIG. 15-8**). It seems inconceivable that the same bone arrangements would be used to serve such diverse functions if each animal had been created separately. Such similarity is exactly what we would expect, however, if bird and mammal forelimbs were derived from the forelimb of a common ancestor. Through natural selection, the ancestral forelimb has undergone different modifications in different kinds of animals. The resulting internally similar structures are called **homologous structures**, meaning that they have the same evolutionary origin despite any differences in current function or appearance.

Vestigial Structures Are Inherited from Ancestors

A **vestigial structure** no longer performs the function for which it evolved in a species' ancestors. Although vestigial structures are sometimes co-opted for new uses, they often seem to serve no function at all. Examples of functionless

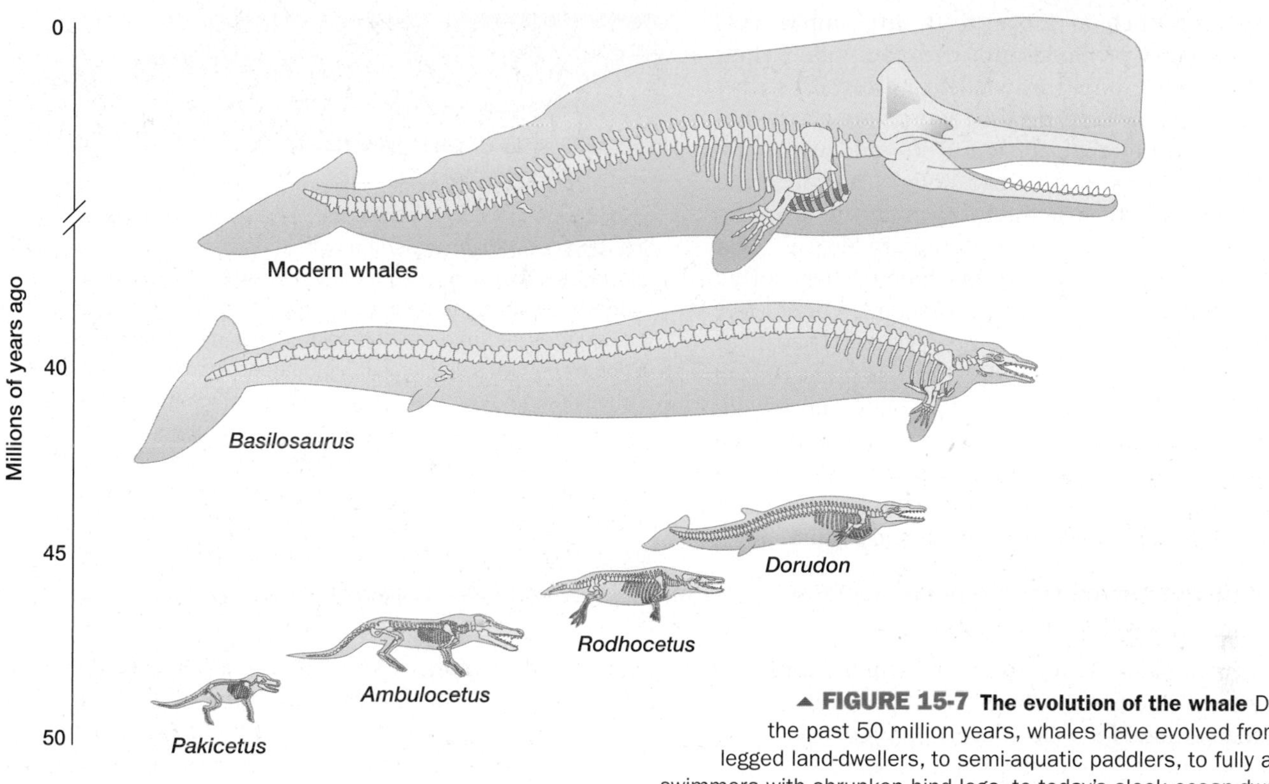

▲ **FIGURE 15-7 The evolution of the whale** During the past 50 million years, whales have evolved from four-legged land-dwellers, to semi-aquatic paddlers, to fully aquatic swimmers with shrunken hind legs, to today's sleek ocean-dwellers.

THINK CRITICALLY The fossil history of some kinds of modern organisms, such as sharks and crocodiles, shows that their structure and appearance have changed very little over hundreds of millions of years. Is this lack of change evidence that such organisms have not evolved during that time?

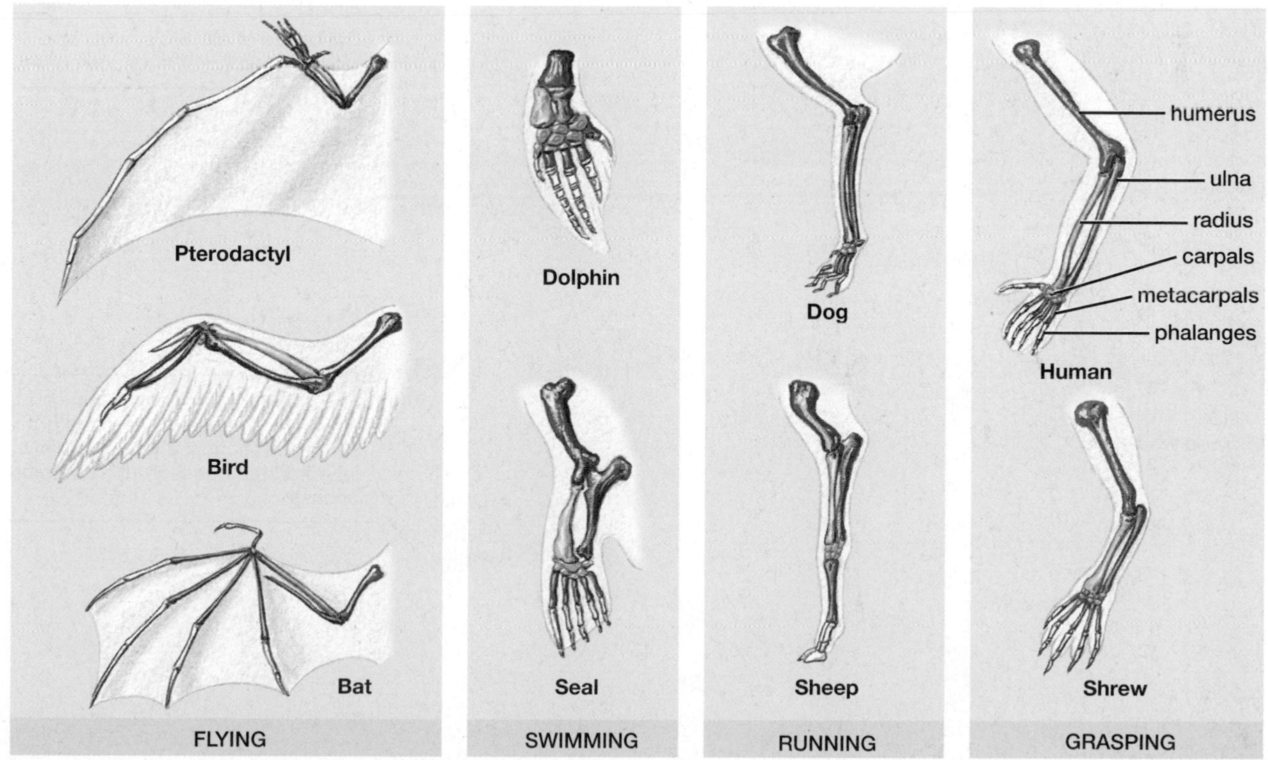

▲ **FIGURE 15-8 Homologous structures** Despite wide differences in function, the forelimbs of all of these animals contain the same set of bones, inherited from a common ancestor. The different colors of the bones highlight the correspondences among the various species.

vestigial structures include molar teeth in vampire bats (which live on a diet of blood and, therefore, don't chew their food) and pelvic bones in whales and certain snakes (**FIG. 15-9**). Both of these vestigial structures are clearly homologous to structures that are found in—and used by—other vertebrates (animals with a backbone). The continued existence in organisms of structures for which they have no use is best explained as a sort of "evolutionary baggage." For example, the ancestral mammals from which whales evolved had four legs and a well-developed set of pelvic bones (see Fig. 15-7). Whales do not have hind legs, yet they have small pelvic and leg bones embedded in their sides. During whale evolution, losing the hind legs provided an advantage, better streamlining the body for movement through water. The result is the modern whale with small, useless pelvic bones that persist because they have shrunk to the point that they no longer constitute a survival-reducing burden.

CASE STUDY CONTINUED

What Good Are Wisdom Teeth and Ostrich Wings?

Ostrich wings are vestigial because they are too rudimentary to perform the function for which they evolved in the species' flying ancestor. Nonetheless, the ostrich uses its wings for other purposes. For example, an ostrich may extend its wings to the side while running, to help maintain balance, and it may spread its wings as part of a threat display. These uses show that evolution by natural selection can sometimes repurpose vestigial structures that have lost the function for which they originally evolved. But whether a vestigial structure remains useless or acquires a new function, it is homologous to the version that retains its original function in other organisms and provides evidence of common ancestry. But are all similarities between different organisms the result of shared ancestry?

Some Anatomical Similarities Result from Evolution in Similar Environments

The study of comparative anatomy has demonstrated the shared ancestry of life by identifying a host of homologous structures that different species have inherited from common ancestors, but comparative anatomists have also identified many anatomical similarities that do not stem from common ancestry. Instead, these similarities arose through **convergent evolution**, in which natural selection causes non-homologous structures that serve similar functions to resemble one another. For example, both birds and insects have wings, but this similarity did not arise from evolutionary modification of a structure that both birds and insects inherited from a common ancestor. Instead, the similarity arose from parallel modification of two different, non-homologous structures. Because natural selection

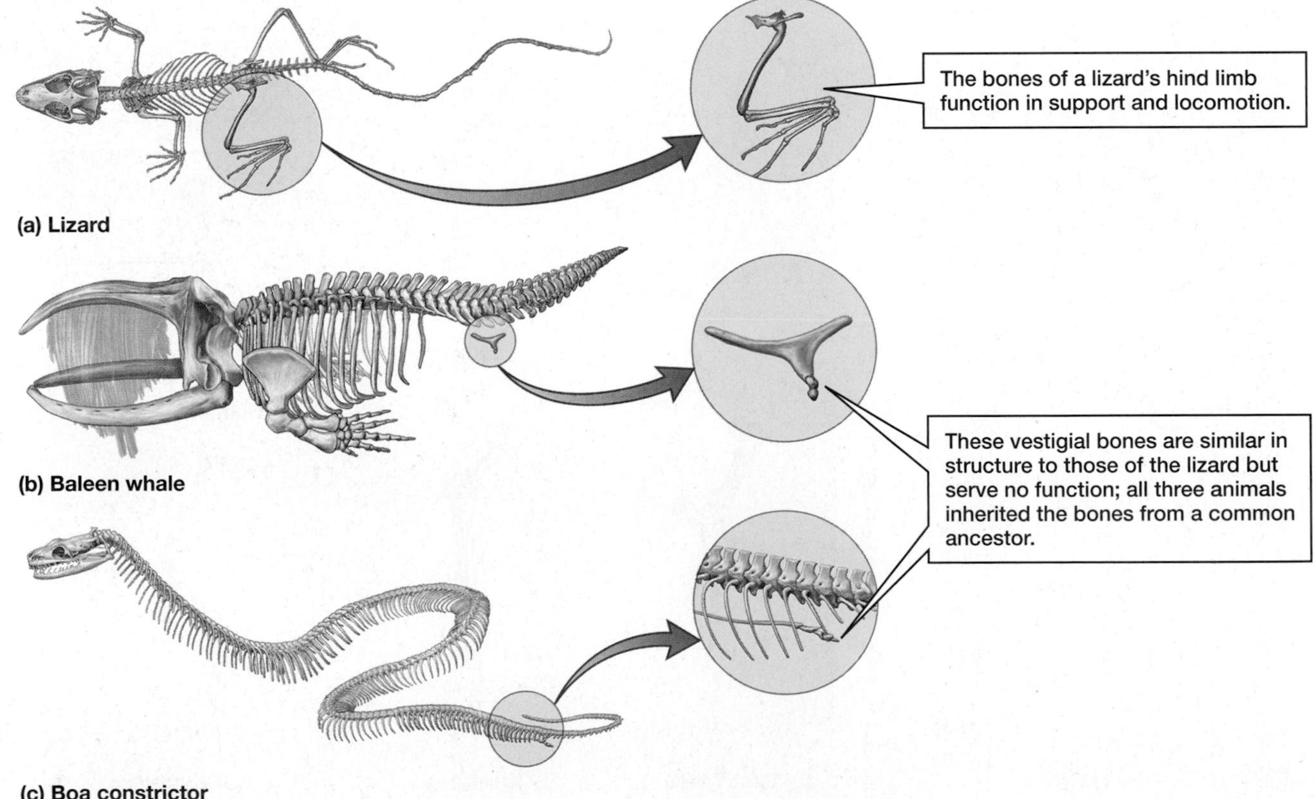

The bones of a lizard's hind limb function in support and locomotion.
(a) Lizard
(b) Baleen whale
These vestigial bones are similar in structure to those of the lizard but serve no function; all three animals inherited the bones from a common ancestor.
(c) Boa constrictor

▲ **FIGURE 15-9 Vestigial structures** Many organisms have vestigial structures that serve no apparent function. The **(a)** lizard, **(b)** baleen whale, and **(c)** boa constrictor all inherited hind limb bones from a common ancestor. These bones remain functional in the lizard but are vestigial in the whale and snake.

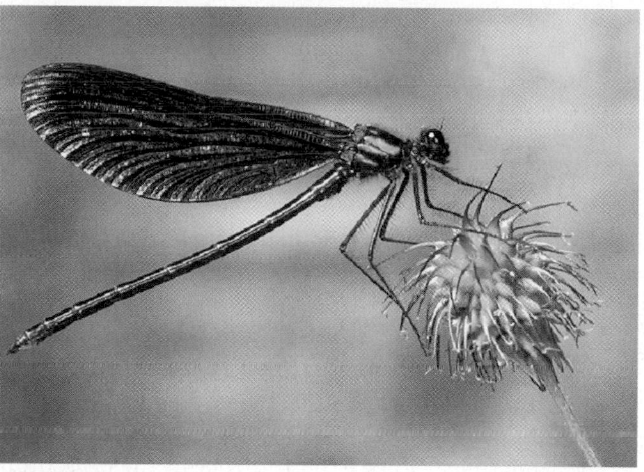

(a) Damselfly

(b) Swallow

▲ **FIGURE 15-10 Analogous structures** Convergent evolution can produce outwardly similar structures that differ anatomically, such as the wings of **(a)** insects and **(b)** birds.

THINK CRITICALLY Are a peacock's tail and a dog's tail homologous structures or analogous structures?

favored flight in both birds and insects, the two groups evolved wings of roughly similar appearance, but the similarity is superficial. Such outwardly similar but non-homologous structures are called **analogous structures** (**FIG. 15-10**). Analogous structures are typically very different in internal anatomy, because the parts are not derived from common ancestral structures.

Embryological Similarity Suggests Common Ancestry

Evidence of common ancestry is apparent in the striking similarity of embryos of different species (**FIG. 15-11**). For example, in their early embryonic stages, fish, turtles, chickens, mice, and humans all develop tails and gill slits (also called gill grooves).

Why are vertebrates that are so different as adults so similar at an early stage of development? The only plausible explanation is that all of these species descended from an ancestral vertebrate that possessed genes that directed the development of gills and tails. All of the descendants still have those genes. In fish, these genes are active throughout development, resulting in adults with fully developed tails and gills. In humans and chickens, these genes are active only during early developmental stages; the structures are lost or become inconspicuous before adulthood.

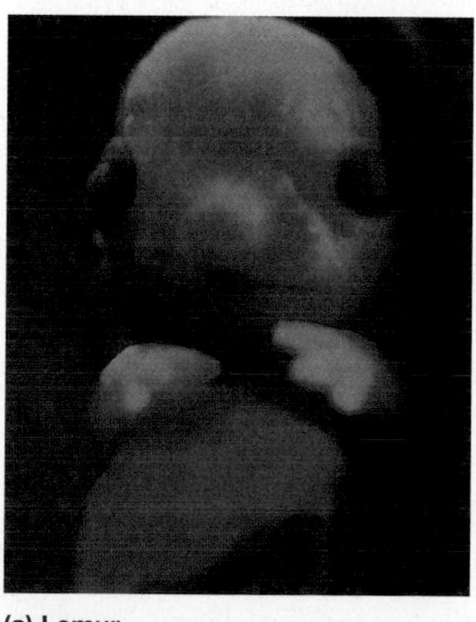

(a) Lemur

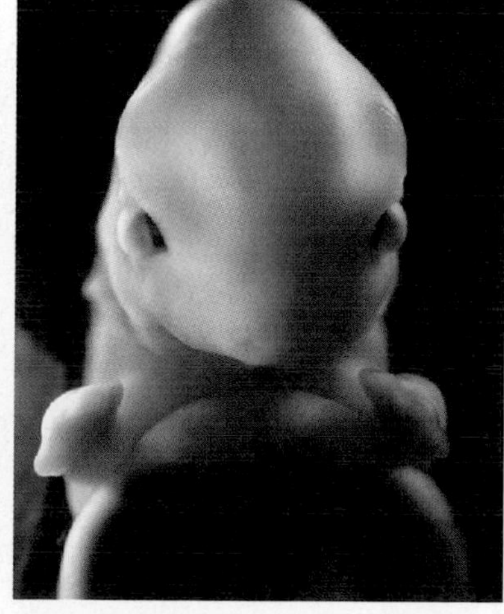

(b) Pig

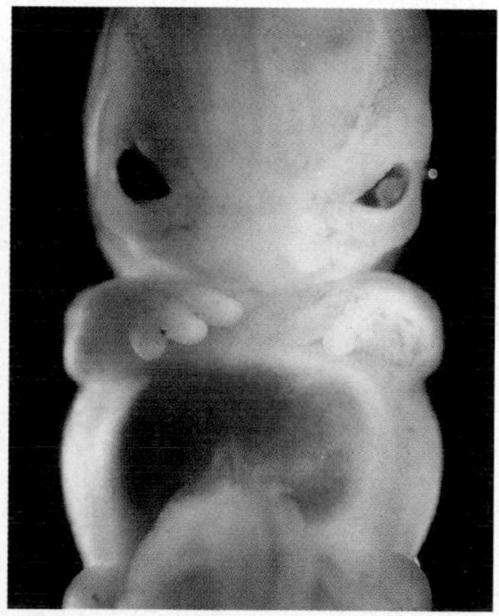

(c) Human

▲ **FIGURE 15-11 Embryonic stages reveal evolutionary relationships** Early embryonic stages of a **(a)** lemur, **(b)** pig, and **(c)** human, showing strikingly similar anatomical features.

Have You Ever Wondered ... Why Backaches Are So Common?

Between 70% and 85% of people will experience lower back pain at some point in life, and for many people, the condition is chronic. This state of affairs is an unfortunately painful consequence of the evolutionary process. We walk upright on two legs, but our distant ancestors walked on all fours. Thus, natural selection formed our vertically oriented spine by remodeling one whose normal orientation was parallel to the ground. Our spinal anatomy evolved some modifications in response to its new posture, but as is often the case with evolution, the changes involved some trade-offs. The arrangements of bone and muscle that permit our smooth, bipedal gait also generate vertical compression of the spine, and the resulting pressure can, and frequently does, cause painful damage to muscle and nerve tissues.

Modern Biochemical and Genetic Analyses Reveal Relatedness Among Diverse Organisms

Biologists have been aware of anatomical and embryological similarities among organisms for centuries, but it took the emergence of modern technology to reveal similarities at the molecular level. Biochemical similarities among organisms provide perhaps the most striking evidence of their evolutionary relatedness. Just as relatedness is revealed by homologous anatomical structures, it is also revealed by homologous molecules.

Today's scientists have access to a powerful tool for revealing molecular homologies: DNA sequencing. It is now possible to quickly determine the sequence of nucleotides in a DNA molecule and to compare the DNA of different organisms. For example, consider the gene that encodes the protein cytochrome *c* (see Chapters 12 and 13 for information on DNA and how it encodes proteins). Cytochrome *c* is present in all plants and animals (and many single-celled organisms) and performs the same function in all of them. The sequence of nucleotides in the gene for cytochrome *c* is similar in these diverse species (**FIG. 15-12**). The widespread presence of the same complex protein, encoded by the same gene and performing the same function, is evidence that the common ancestor of plants and animals had cytochrome *c* in its cells. At the same time, though, the sequence of the cytochrome *c* gene differs slightly in different species, showing that variations arose during the independent evolution of Earth's multitude of plant and animal species.

Some biochemical similarities are so fundamental that they extend to all living cells. For example:

- All cells have DNA as the carrier of genetic information.
- All cells use RNA, ribosomes, and approximately the same genetic code to translate that genetic information into proteins.
- All cells use roughly the same set of 20 amino acids to build proteins.
- All cells use ATP as a cellular energy carrier.

The most plausible explanation for such widespread sharing of complex and specific biochemical traits is that the

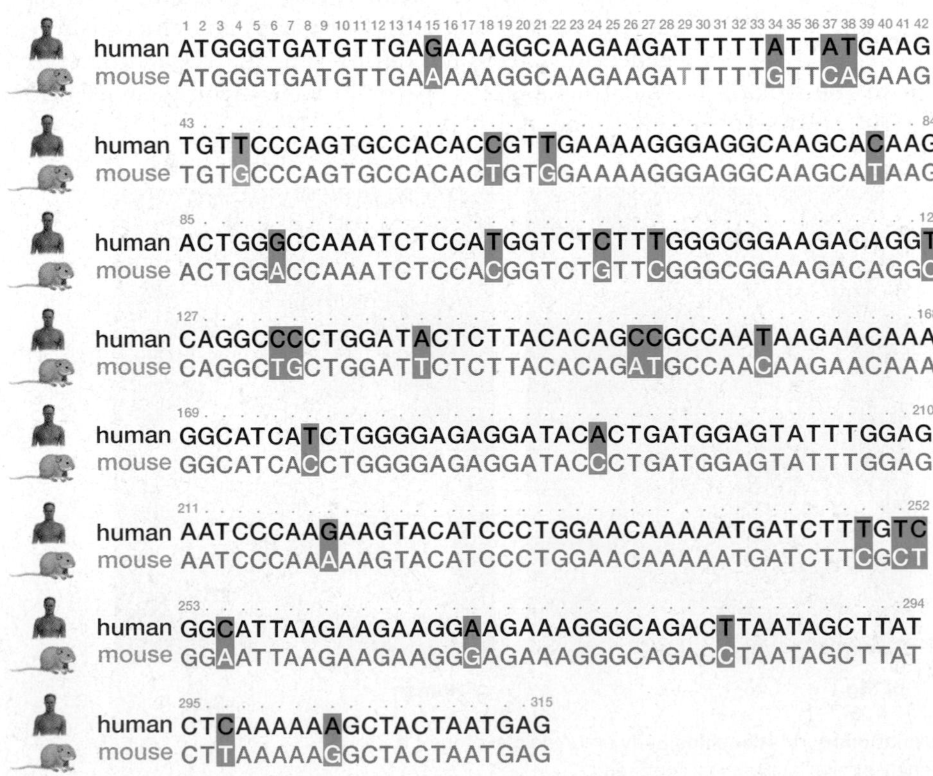

FIGURE 15-12 Molecular similarity shows evolutionary relationships The DNA sequences of the genes that code for cytochrome *c* in a human and a mouse. Of the 315 nucleotides in the gene, only 30 (shaded blue) differ between the two species.

traits are homologies. That is, they arose only once, in the common ancestor of all living things, from which all of today's organisms inherited them.

CHECK YOUR LEARNING

Can you . . .

- describe the evidence that evolution has occurred?
- explain the difference between similarity due to homology and similarity due to convergent evolution?

CASE STUDY CONTINUED

What Good Are Wisdom Teeth and Ostrich Wings?

Just as anatomical homology can lead to vestigial structures such as human wisdom teeth and the wings of flightless birds, genetic homology can lead to vestigial DNA sequences. For example, most mammal species produce an enzyme, L-gulonolactone oxidase, that catalyzes the last step in the production of vitamin C. The species that produce the enzyme are able to do so because they all inherited the gene that encodes it from a common ancestor. Humans, however, do not produce L-gulonolactone oxidase, so we can't produce vitamin C ourselves and must consume it in our diets. But even though we don't produce the enzyme, our cells do contain a stretch of DNA with a sequence very similar to that of the enzyme-producing gene present in rats and most other mammals. The human version, though, does not encode the enzyme (or any protein). We inherited this stretch of DNA from an ancestor that we share with other mammal species, but in us, the sequence has undergone a change that rendered it nonfunctional. (The change probably did not confer a strong disadvantage, because our ancestors got sufficient vitamin C in their diets.) The nonfunctional sequence remains as a vestigial trait, evidence of our shared ancestry.

Vestigial traits are evidence of both shared ancestry and change in traits over time. What kinds of observations and experiments show that natural selection contributes to evolutionary change?

(a) Gray wolf

(b) Diverse dogs

▲ **FIGURE 15-13 Dog diversity illustrates artificial selection** A comparison of **(a)** the ancestral dog (the gray wolf, *Canis lupus*) and **(b)** various breeds of dog. Artificial selection by humans has caused great divergence in the size and shape of dogs in only a few thousand years.

15.4 WHAT IS THE EVIDENCE THAT POPULATIONS EVOLVE BY NATURAL SELECTION?

We have seen that evidence of evolution comes from many sources. But what is the evidence that evolution occurs by the process of natural selection?

Controlled Breeding Modifies Organisms

One line of evidence supporting evolution by natural selection is **artificial selection**, the breeding of domestic plants and animals to produce specific desirable features. The various dog breeds provide a striking example of artificial selection (**FIG. 15-13**). Dogs descended from wolves, and even today, the two will readily crossbreed. With few exceptions, however, modern dogs do not closely resemble wolves. Some breeds are so different from one another that they would be considered separate species if they were found in the wild. Humans produced these radically different dogs in a few thousand years by doing nothing more than repeatedly selecting individuals with desirable traits for breeding. Therefore, it is quite plausible that natural selection could, by a comparable process acting over hundreds of millions of years, produce the spectrum of living organisms. Darwin was so impressed by the connection between artificial selection and natural selection that he devoted a chapter of *On the Origin of Species* to the topic.

Evolution by Natural Selection Occurs Today

Additional evidence of natural selection comes from scientific observation and experimentation. The logic of natural selection gives us no reason to believe that evolutionary change is limited to the past. After all, inherited variation and competition for access to resources are certainly not limited to the past. If Darwin and Wallace were correct that those conditions lead inevitably to evolution by natural selection, then researchers ought to be able to detect evolutionary change as it occurs. And they have. Next, we will consider some examples that give us a glimpse of natural selection at work.

Natural Selection Has Silenced Calling Crickets

Among the species found on the Hawaiian island of Kauai is the Polynesian field cricket (**FIG. 15-14a**). Researchers studying the crickets on Kauai in the 1990s found that males produced a loud call by rubbing together two structures on their wings: the scraper (a raised ridge) and the file (a modified vein studded with evenly spaced tiny teeth). The males called at night, and female crickets moved toward the calls to find and choose mates. Strangely, however, each year that the researchers returned to Kauai, they heard fewer and fewer cricket calls. By 2003, the silence was virtually complete; almost no crickets were calling. But the crickets had not disappeared; a nighttime searcher with a strong flashlight could still find a multitude of quietly active crickets. Why didn't the male crickets make any calls? Because the files on their wings were tiny and distorted, these "flatwing" males were not capable of calling (**FIG. 15-14b**). What had happened?

Some time during the 1990s, the crickets' environment had changed drastically with the arrival from North America of a fly species that had not previously been present on Kauai. The fly is a deadly parasite, with larvae that burrow into the bodies of crickets, eating them alive. How does a fly find a cricket to parasitize? By moving toward the sound of calling crickets. Thus, after the arrival of the flies, loudly calling crickets were less likely to survive than those that happened to call more softly or not at all. Natural selection favored quieter crickets, and within fewer than 20 generations, the silent males had almost completely replaced callers. As a result of natural selection, the structure of male cricket wings had changed significantly. There must have been a corresponding evolutionary change in female mating behavior, as females are now willing to mate with silent flatwing males.

Natural Selection Can Lead to Herbicide and Pesticide Resistance

Natural selection is also evident in weed species that have evolved resistance to the herbicides with which we try to control them. Successful agriculture depends on farmers' ability to kill weeds that compete with crop plants, but many members of weed species can no longer be killed by a formerly lethal dose of glyphosate, the active ingredient in Roundup, the world's most widely used herbicide. How did these glyphosate-resistant "superweeds" arise? They arose because the herbicide has acted as an agent of natural selection. Consider, for example, giant ragweed, one of the highly destructive weed species that are now resistant to glyphosate in some places. When a field is sprayed with Roundup, almost all of the giant ragweed plants there are killed, because glyphosate inactivates an enzyme that is essential to the plants' survival. A few ragweed plants, however, survive. Researchers have discovered that some of these survivors carry a mutation that causes them to produce a tremendous amount of the enzyme that glyphosate attacks, more enzyme than the usual dose of glyphosate can destroy. In the face of repeated applications of Roundup, the formerly rare protective mutation has become common in many giant ragweed populations. (For additional examples of how humans influence evolution, see "Earth Watch: People Promote High-Speed Evolution.")

The evolution of glyphosate-resistant superweeds was a direct result of changes in agricultural practice. In the 1990s, the biotechnology company Monsanto began selling seeds that had been genetically engineered to produce crops that are not harmed by glyphosate. These "Roundup Ready" crops, which now account for the vast majority of soybean, corn, and cotton plantings in the United States and other countries, allow farmers to freely apply glyphosate to their fields without fear of harming crop plants. As a result, use of glyphosate has skyrocketed. Today, large-scale agriculture is

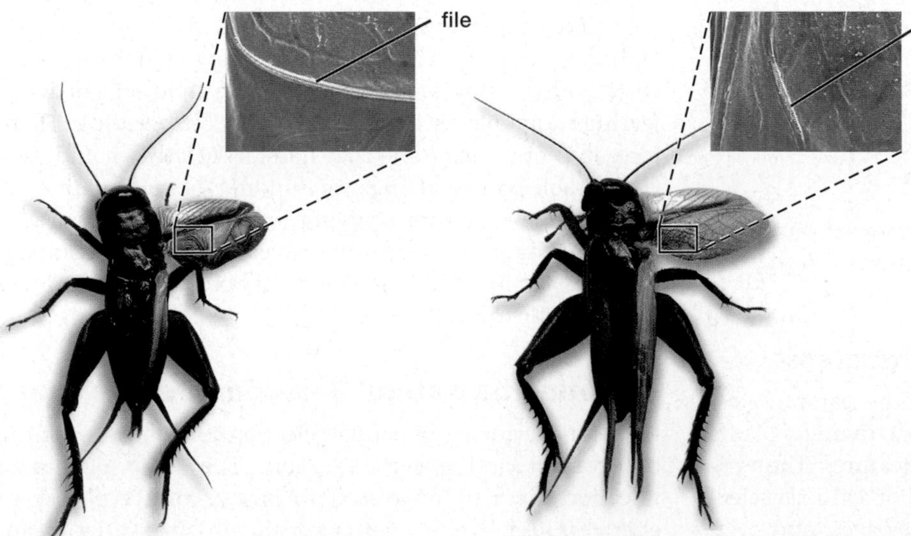

(a) Polynesian cricket, wing with file

(b) Polynesian cricket, wing with lost file

FIGURE 15-14 Crickets evolve to become silent when calls attract parasites A few decades ago, Polynesian field crickets on the island of Kauai used their wings to produce loud calls. But when the environment changed to include parasitic flies that are attracted to cricket calls, natural selection favored crickets with wings that cannot produce sound. **(a)** Noise-producing "files" were present on male cricket wings before the flies appeared in the environment, but **(b)** the files all but disappeared within a few years of the flies' arrival.

People Promote High-Speed Evolution

You probably don't think of yourself as a major engine of evolution. Nonetheless, as you go about the routines of your daily life, you are contributing to what is perhaps today's most significant cause of rapid evolutionary change. Human activity has changed Earth's environments tremendously, and the biological logic of natural selection, spelled out so clearly by Darwin, tells us that environmental change leads inevitably to evolutionary change. Thus, by changing the environment, humans have become a major agent of natural selection.

Unfortunately, many of the evolutionary changes we have caused have turned out to be bad news for us. Our liberal use of pesticides has selected for resistant pests that frustrate efforts to protect our food supply. By overmedicating ourselves with antibiotics and other drugs, we have selected for resistant "supergerms" and diseases that are ever more difficult to treat (see Chapter 16). Heavy fishing in the world's oceans has favored smaller fish that can slip through nets more easily, thereby selecting for slow-growing fish that remain small even as mature adults. As a result, fish of many commercially important species are now so small that our ability to extract food from the sea is compromised.

Our use of pesticides, antibiotics, and fishing technology has caused evolutionary changes that threaten our health and welfare, but the scope of these changes may be dwarfed by those that will arise from human-caused modification of Earth's climate. Human activities, especially activities that use energy derived from fossil fuels, modify the climate by contributing to global warming. In coming years, species' evolution will be increasingly influenced by environmental changes associated with a warming climate, such as reduced ice and snow, longer, hotter summers, and shifts in the life cycles of other species that provide food or shelter.

There is growing evidence that climate change is already causing evolutionary change. Warming-related evolution has been found in a number of plant and animal populations. For example, in Finland, tawny owls have changed color in response to a warmer climate (**FIG. E15-3**). Owls of this species come in two varieties, gray or brown. In the past, most owls were gray, and the brown variety was rare. Today, however, about 40% of Finnish owls are brown, and researchers have shown that the increase in brown owls was probably caused by climate change. In particular, the researchers found that in snowy winters, gray owls survive much better than brown owls, perhaps because they are better camouflaged against the snow and therefore suffer less predation by eagles. In winters with less snow, brown owls are better camouflaged and more likely to survive. As temperatures have risen in recent decades, snowy winters have become increasingly rare, and the resulting reduced snow cover has favored the survival of brown owls. Thus, a brown owl is more likely than a gray one to survive the winter and, because feather color is inherited, produce brown offspring. The proportion of brown owls in the population has grown in response to natural selection associated with warming temperatures.

▲ **FIGURE E15-3 Tawny owl populations have evolved in response to global climate change** The proportion of brown owls is increasing because brown owls survive better than gray ones over winters with less snow.

The available evidence suggests that global climate change will have an enormous evolutionary impact, potentially affecting the evolution of almost every species. How will these evolutionary changes affect us and the ecosystems on which we depend? This question is not readily answerable, because the path of evolution is not predictable. We can hope, however, that careful monitoring of evolving species and increased understanding of evolutionary processes will help us take appropriate steps to safeguard our health and well-being as Earth warms.

THINK CRITICALLY To reduce the incidence of pesticide resistance, farmers are advised to place fields that are free of pesticides or pesticide-containing crops beside fields in which pesticides are used as usual. Given your understanding of how evolution works, explain how this method would slow the evolution of pesticide resistance in insects.

extremely dependent on this single herbicide, even as its effectiveness declines steadily due to the evolution of weeds that are resistant.

Just as weeds have evolved to resist herbicides, many of the insects that attack crops have also evolved to resist the pesticides that farmers use to control them. Such resistance has been documented in more than 500 species of crop-damaging insects, and virtually every pesticide has fostered the evolution of resistance in at least one insect species. We pay a heavy price for this evolutionary phenomenon. The additional pesticides that farmers apply in their attempts to control resistant insects cost almost $2 billion each year in the United States alone and add millions of tons of poisons to Earth's soil and water.

Experiments Can Demonstrate Natural Selection

In addition to observing natural selection in the wild, scientists have also devised numerous experiments that confirm the action of natural selection. For example, one group of evolutionary biologists released small groups of *Anolis sagrei* lizards onto 14 small Bahamian islands that were previously uninhabited by lizards. The original lizards came from a population on Staniel Cay, an island with tall vegetation, including plenty of trees. In contrast, the islands to which the lizards were introduced had few or no trees and were covered mainly with small shrubs and other low-growing plants.

The biologists returned to those islands 14 years later and found that the original small groups of lizards had given rise to thriving populations of hundreds of individuals. On all 14 of the experimental islands, lizards had legs that were shorter and thinner than those of lizards from the original source population on Staniel Cay. In just over a decade, it appeared, the lizard populations had changed in response to new environments.

Why had the new lizard populations evolved shorter, thinner legs? The researchers found that lizards with short, thin legs were slower but more agile than lizards with long, thick legs. They hypothesized that speed was especially important for escaping predators on the thick-branched trees of Staniel Cay, but that agility was more important on the thin-branched bushes of the experimental islands. Therefore, in the new environment, agile individuals with shorter, thinner legs were better able to survive and produce a greater number of offspring, so members of subsequent generations had shorter, thinner legs on average.

Selection Acts on Random Variation to Favor Traits That Work Best in Particular Environments

Two important points underlie the evolutionary changes just described:

- ***The variations on which natural selection works are produced by chance mutations.*** Silent wings in Hawaiian crickets, extra enzyme in giant ragweed plants, and shorter legs in Bahamian lizards were not *produced* by the parasitic flies, Roundup herbicide, or thinner branches. The mutations that produced each of these beneficial traits arose spontaneously.
- ***Natural selection favors organisms that are best adapted to a particular environment.*** Natural selection is not a process for producing ever-greater degrees of perfection. Natural selection does not select for the "best" in any absolute sense, but only for what is best in the context of a particular environment, which varies from place to place and which may change over time. A trait that is advantageous under one set of conditions may become disadvantageous if conditions change. For example, lizards with longer legs were better at escaping predators in the forested environment of Staniel Cay, but were worse at escaping predators in the shrubby environments of other islands.

CHECK YOUR LEARNING

Can you . . .

- describe some observations and experiments that demonstrate that populations evolve by natural selection?

CASE STUDY REVISITED

What Good Are Wisdom Teeth and Ostrich Wings?

Wisdom teeth are but one of many human anatomical structures that appear to no longer serve an important function (**FIG. 15-15**). Darwin himself noted many of these "useless, or nearly useless" traits in the very first chapter of *On the Origin of Species* and declared them to be prime evidence that humans had evolved from earlier species.

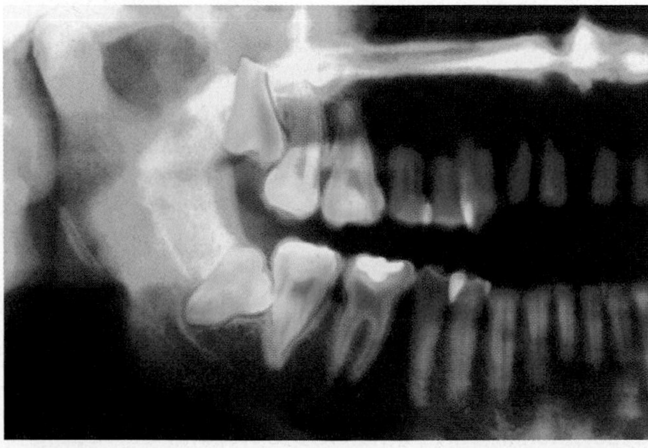

▲ **FIGURE 15-15 Wisdom teeth** Squeezed into a jaw that is too short to contain them, wisdom teeth often become impacted—unable to erupt through the surface of the gum. The leftmost upper and lower teeth in this X-ray image are impacted wisdom teeth.

Body hair is another vestigial human trait. It seems to be an evolutionary relic of the fur that kept our distant ancestors warm (and that still warms our closest evolutionary relatives, the great apes). Not only do we retain useless body hair, we also still have the muscles that allow other mammals to puff up their fur for better insulation. In humans, these vestigial structures just give us goose bumps.

Though humans don't have and don't need a tail, we nonetheless have a tailbone. The tailbone consists of a few tiny vertebrae fused into a small structure at the base of the backbone, where a tail would be if we had one. People born without a tailbone or who have theirs surgically removed suffer no ill effects.

THINK CRITICALLY Some advocates of the view that all organisms were created simultaneously by God argue that vestigial structures do not constitute evidence of evolution, because they show only that a divinely created structure can degenerate over time. According to this view, human tailbones are not evidence of evolution because they do not show that an adaptive improvement has occurred. Is this a valid argument?

CHAPTER REVIEW

Go to **Mastering Biology** to access the Pearson eText, vocabulary review, practice quizzes, activities, videos, current events, and more.

*Answers to **Think Critically, and Thinking Through the Concepts** questions can be found in the **Answers** section at the back of the book.*

Summary of Key Concepts

15.1 How Did Evolutionary Thought Develop?

Historically, the most common explanation for the origin of species was the divine creation of each species in its present form, and species were believed to remain unchanged after their creation. This view was challenged by evidence from fossils, geology, and biological exploration. Since the middle of the nineteenth century, scientists have realized that species originate and evolve by the operation of natural processes that change the genetic makeup of populations.

15.2 How Does Natural Selection Work?

Charles Darwin and Alfred Russel Wallace independently proposed the theory of evolution by natural selection. Their theory expresses the logical consequences of four postulates about populations: (1) populations are variable, (2) the variable traits can be inherited, (3) there is unequal reproduction, and (4) differences in reproductive success depend on the traits of individuals. If these four postulates are true, then the characteristics of successful individuals will be "naturally selected" and become more common over time.

15.3 How Do We Know That Evolution Has Occurred?

Many lines of evidence indicate that evolution has occurred, including the following:

- Fossils of ancient species tend to be simpler in form than modern species. Sequences of fossils have been discovered that show a graded series of changes in form. Both of these observations would be expected if modern species evolved from older species.
- Species thought to be related through evolution from a common ancestor possess many similar anatomical structures.
- Stages in early embryonic development are quite similar among very different types of vertebrates.
- Living cells share similarities in biochemical traits, such as the use of DNA as the carrier of genetic information.

15.4 What is the Evidence That Populations Evolve by Natural Selection?

Similarly, many lines of evidence indicate that natural selection is the chief mechanism driving changes in the characteristics of species over time, including the following:

- Inheritable traits have been changed rapidly in populations of domestic animals and plants by selectively breeding organisms with desired features (artificial selection). The immense variations in species produced in a few thousand years by artificial selection make it almost inevitable that much larger changes would be wrought by hundreds of millions of years of natural selection.
- Both natural and human activities can drastically change the environment over short periods of time. Inherited characteristics of species have been observed to change significantly in response to such environmental changes.

Thinking Through the Concepts

Bloom's: Remembering, Understanding

Multiple Choice

1. Whale skeletons contain nonfunctional pelvic bones
   a. as a result of convergent evolution.
   b. due to catastrophism.
   c. because whales evolved from ancestors that had hind legs.
   d. because the bones might be needed for a future adaptation.
2. Darwin was influenced by Malthus's thinking about __________ and by Lamarck's conclusion that __________
   a. the formation of fossils; Earth has a complex history
   b. the geographic distribution of species; individuals compete to survive and reproduce
   c. the principles of geology; all organisms are related by common descent
   d. limits on population growth; species change over time
3. Natural selection is
   a. a preference for natural traits.
   b. the method by which domestic dog breeds originated.
   c. increased reproduction due to particular traits.
   d. the reason that mutations occur.
4. Which of the following is *not* required for evolution by natural selection to occur?
   a. Individuals in a population vary.
   b. Offspring inherit traits from their parents.
   c. Thousands of generations must pass.
   d. Some individuals have more offspring than others.
5. Which of the following is *not* evidence of evolution (i.e., that species are linked by shared ancestry and have changed over time)?
   a. Different species share homologous structures.
   b. Different species share analogous structures.
   c. The older a fossil is, the less likely it is to be similar to a living organism.
   d. The cells of all species contain DNA.

Fill-in-the-Blank

1. The flipper of a seal is homologous with the __________ of a bird, and both of these are homologous with the __________ of a human. The wing of a bird and the wing of a butterfly are described as __________ structures that arose as a result of __________ evolution. Remnants of structures in animals that have no use for them, such as the small hind leg bones of whales, are described as __________ structures.

2. The finding that all organisms share the same genetic code provides evidence that all descended from a(n) ____________. Further evidence is provided by the fact that all cells use roughly the same set of ____________ to build proteins, and all cells use the molecule ____________ as an energy carrier.
3. Georges Cuvier supported a concept called ____________ to explain layers of rock with embedded fossils. Charles Lyell, building on the work of James Hutton, proposed an alternative explanation called ____________, which states that layers of rock and many other geological features can be explained by gradual processes that occurred in the past just as they do in the present. This concept provided important support for evolution because it required that Earth be extremely ____________.
4. The process by which inherited characteristics of populations change over time is called ____________. Variability among individuals is the result of chance changes called ____________ that occur in the hereditary molecule ____________.
5. The process by which individuals with traits that provide an advantage in their natural habitats are more successful at reproducing is called ____________. People who breed animals or plants can produce large changes in their characteristics in a relatively short time, a process called ____________.
6. Darwin's postulate 2 states that ____________. The work of ____________ provided the first experimental evidence for this postulate.

Review Questions

1. Natural selection acts on individuals, but only populations evolve. Explain why this statement is true.
2. Distinguish between catastrophism and uniformitarianism. How did these hypotheses contribute to the development of evolutionary theory?
3. Describe Lamarck's theory of inheritance of acquired characteristics. Why is it invalid?
4. What is natural selection? Describe how natural selection might have caused unequal reproduction among the ancestors of a fast-swimming predatory fish, such as the barracuda.
5. Describe how evolution occurs. In your description, include discussion of the reproductive potential of species, the stability of natural population sizes, variation among individuals of a species, inheritance, and natural selection.
6. What is convergent evolution? Give an example.
7. How do biochemistry and molecular genetics contribute to the evidence that evolution has occurred?
8. In what sense are humans currently acting as agents of selection on other species? Name some traits that are favored by the environmental changes humans cause.

Applying the Concepts

Bloom's: Applying, Analyzing, Evaluating

1. In discussions of untapped human potential, it is commonly said that the average person uses only 10% of his or her brain. Is this conclusion likely to be correct? Explain your answer in terms of natural selection.
2. Does evolution through natural selection produce "better" organisms in an absolute sense? Are we climbing the "Ladder of Nature"? Defend your answer.

16 How Populations Evolve

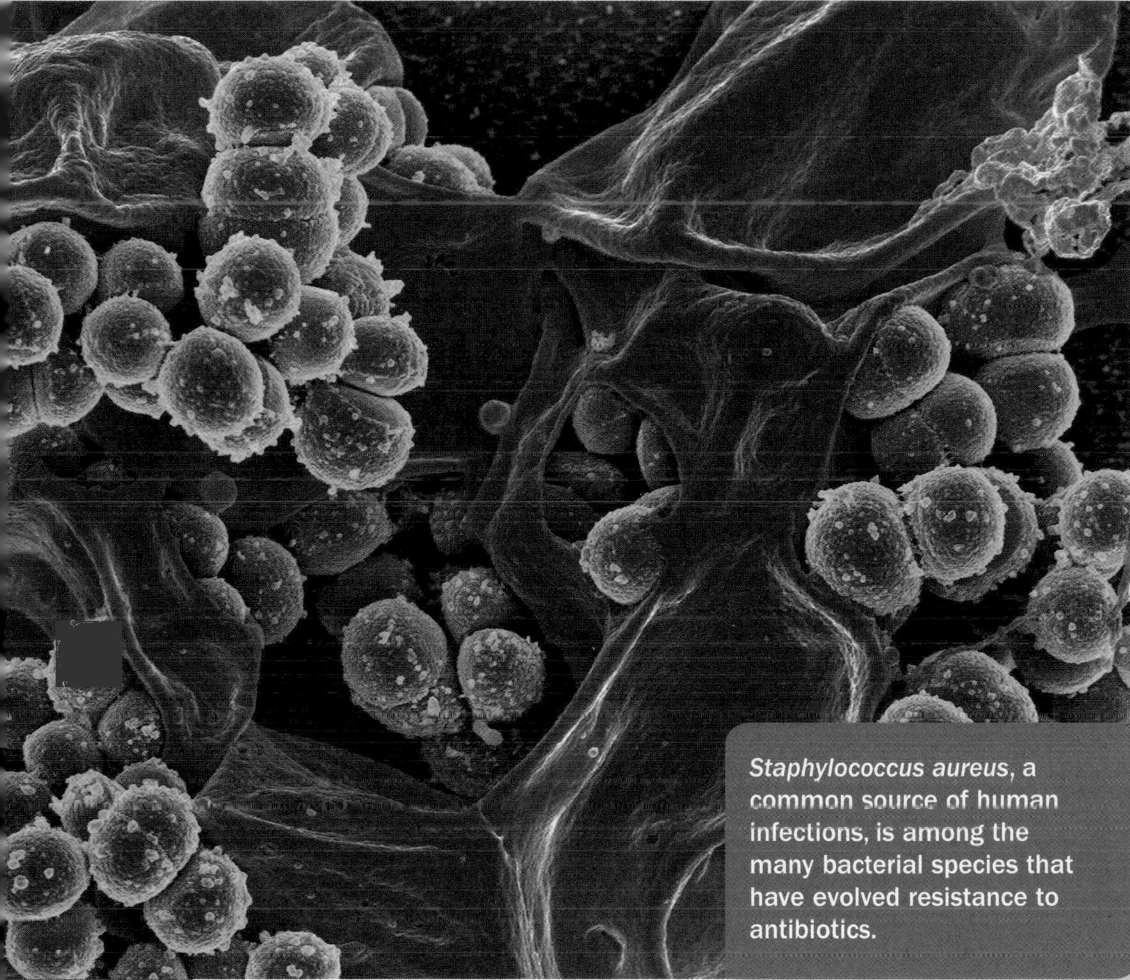

Staphylococcus aureus, a common source of human infections, is among the many bacterial species that have evolved resistance to antibiotics.

CASE STUDY

Evolution of a Menace

ON A FEBRUARY DAY NOT LONG AGO, a 20-year-old student arrived at the health center at Western Washington University. He had been bothered by a lingering cough for a couple of weeks, and when his symptoms worsened to include a fever and vomiting, he sought medical attention. The health center staff quickly determined that the student had pneumonia and began treatment. His condition deteriorated, however, and he was transferred to the local hospital. A few days later, he died.

Why couldn't doctors save a previously healthy young man from a normally curable disease? Because the victim's pneumonia was caused by methicillin-resistant *Staphylococcus aureus* (MRSA) bacteria. *Staphylococcus aureus*, sometimes referred to as "staph," is a common bacterium that can infect the skin, blood, or respiratory system. Many staph infections can be successfully treated with antibiotics, but MRSA bacteria are antibiotic resistant and cannot be killed by many of the most commonly used antibiotics. Until recently, MRSA infections occurred almost exclusively in hospitals. Today, however, resistant staph is widespread, and more than half of MRSA infections occur outside of hospitals, in homes, schools, and workplaces.

In the United States, MRSA infections kill at least 11,000 people each year. And, unfortunately, *Staphylococcus* is by no means the only disease-causing bacterium that is becoming less vulnerable to antibiotics. For example, antibiotic resistance has also appeared in the bacteria that cause tuberculosis, a disease that kills almost 2 million people each year. In an increasing number of tuberculosis cases, the disease does not respond to any of the drugs commonly used to treat it. Multidrug resistance is also becoming more prevalent in the bacteria responsible for the widespread sexually transmitted disease gonorrhea. In addition, drug resistance is common in the bacteria that cause food poisoning, blood poisoning, dysentery, pneumonia, meningitis, and urinary tract infections. We are experiencing a global onslaught of resistant "supergerms," and are facing the specter of diseases that cannot be cured.

Many physicians and scientists believe that the most effective way to combat the rise of resistant diseases is to reduce the use of antibiotics. Why might such a strategy be effective? Because the upsurge of antibiotic resistance is a consequence of evolutionary change in populations of bacteria, and the agent of this change is natural selection imposed by antibiotic drugs. Can understanding the mechanisms by which populations evolve help us understand how the crisis of antibiotic resistance arose, and how we might resolve it?

AT A GLANCE

16.1 HOW ARE POPULATIONS, GENES, AND EVOLUTION RELATED?

If you live in an area with a seasonal climate and you own a dog or cat, you have probably noticed that your pet's fur gets thicker and heavier as winter approaches. Has the animal evolved? No. The changes that we see in an individual organism over the course of its lifetime are not evolutionary changes. Instead, evolutionary changes occur from generation to generation, causing descendants to be different from their ancestors.

Furthermore, we can't detect evolutionary change across generations by looking at a single set of parents and offspring. For example, if you observed that a 6-foot-tall man had an adult son who stood 5 feet tall, could you conclude that humans were evolving to become shorter? Obviously not. Rather, if you wanted to learn about evolutionary change in human height, you would begin by measuring many humans over many generations to see if the average height is changing with time. Evolution is a property not of individuals but of populations. A **population** is a group that includes all the members of a species living in a given area.

Recognizing that evolution is a population-level phenomenon was one of Darwin's key insights. But populations are composed of individuals, and the actions and fates of individuals determine which characteristics will be passed on to descendant populations. In this fashion, inheritance provides the link between the lives of individual organisms and the evolution of populations. We will therefore begin our discussion of the processes of evolution by reviewing some principles of genetics as they apply to individuals. We will then extend those principles to the genetics of populations.

Genes and the Environment Interact to Determine Traits

Each cell of every organism contains genetic information encoded in the DNA of its chromosomes. The combined DNA in an organism's set of chromosomes is its *genome.* A *gene* is a segment of DNA located at a particular place on a chromosome (see Chapter 11). The sequence of nucleotides in a gene encodes the sequence of amino acids in a protein, usually an enzyme that catalyzes a particular reaction in the cell. At a given gene's location, different members of a species may have slightly different nucleotide sequences, called *alleles.* Different alleles encode different forms of the same enzyme. For example, various alleles of the genes that influence eye color in humans generate enzymes that help produce eyes that are brown, or blue, or green, and so on.

In any population of organisms, there are usually two or more alleles of each gene. An individual of a diploid species whose alleles of a particular gene are both the same is *homozygous* for that gene, and an individual with different alleles for that gene is *heterozygous*. The specific alleles borne on an organism's chromosomes (its *genotype*) influence the development of its physical and behavioral traits (its *phenotype*) (**FIG. 16-1**).

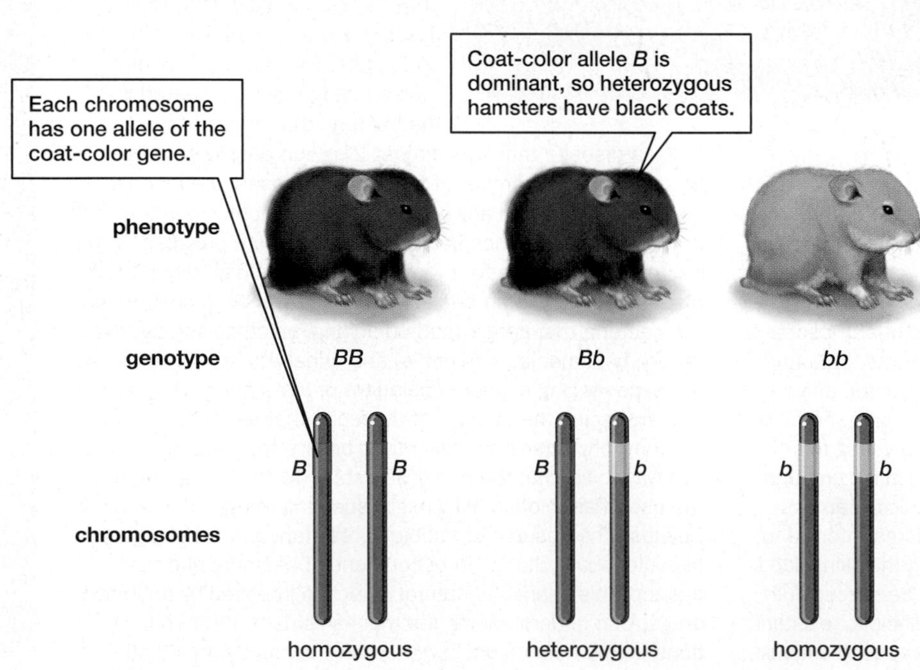

FIGURE 16-1 Alleles, genotype, and phenotype in individuals An individual's particular combination of alleles is its genotype. The word "genotype" can refer to the alleles of a single gene (as shown here), to a set of genes, or to all of an organism's genes. An individual's phenotype is determined by its genotype and environment. "Phenotype" can refer to a single trait, a set of traits, or all of an organism's traits.

Let's illustrate these principles with an example. A black hamster's coat is colored black because a chemical reaction in its hair follicles produces a black pigment. When we say that a hamster has the allele for a black coat, we mean that a particular stretch of DNA on one of the hamster's chromosomes contains a sequence of nucleotides that codes for the enzyme that catalyzes a pigment-producing reaction that results in a black coat. A hamster with the allele for a brown coat has a different sequence of nucleotides at the corresponding chromosomal position. That different sequence codes for an enzyme that cannot produce black pigment. If a hamster is homozygous for the black allele (two black alleles) or is heterozygous (one black allele and one brown allele), its fur contains the pigment and is black. But if a hamster is homozygous for the brown allele, its hair follicles produce no black pigment and its coat is brown. Because the hamster's coat is black even when only one copy of the black allele is present, the black allele is considered *dominant* and the brown allele *recessive*.

The Gene Pool Comprises All of the Alleles in a Population

Looking at evolution in terms of its effects on genes has proven to be a useful way to understand evolutionary processes. In particular, evolutionary biologists have made excellent use of the tools of a branch of genetics, called *population genetics*, that deals with the frequency, distribution, and inheritance of alleles in populations. To take advantage of this powerful aid to understanding evolution, you will need to learn a few of the basic concepts of population genetics.

Population genetics defines a **gene pool** as a set that contains all of the alleles of all of the genes from all of the individuals in a population. A gene pool is not an actual physical entity but is instead a concept that can help us understand the process of evolution. You can think of a gene pool as the contents of an imaginary bucket into which each member of a population has tossed one copy of its genotype. Thus, for any given allele, the gene pool receives one copy from each individual that is heterozygous for that allele (and therefore has only one copy of the allele in question), and two copies from each individual that is homozygous for that allele (and therefore has two copies).

Each particular gene can be considered to have its own gene pool, which comprises all of the alleles of that specific gene in a population (**FIG. 16-2**). If we count the number of copies of each allele present in a gene pool, we can determine the relative proportion of each allele in the gene pool. An allele's proportion in the gene pool is its **allele frequency**. For example, the gene pool of the population of 25 hamsters portrayed in Figure 16-2 contains 50 alleles of the gene that controls coat color (because hamsters are diploid and each hamster thus has two copies of each gene). Twenty of those 50 alleles are of the type that codes for black coats, so the frequency of that allele in the population is 20/50 = 0.40 (or 40%).

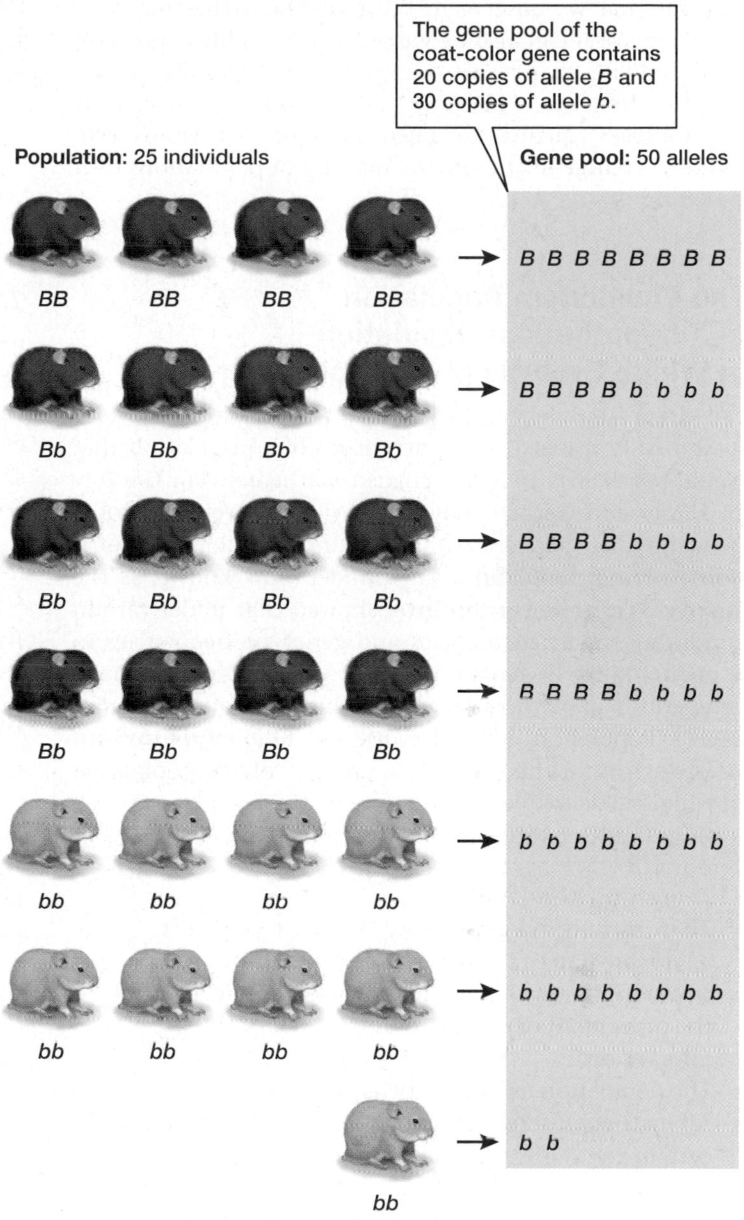

The gene pool of the coat-color gene contains 20 copies of allele *B* and 30 copies of allele *b*.

Population: 25 individuals

Gene pool: 50 alleles

BB BB BB BB → *B B B B B B B B*

Bb Bb Bb Bb → *B B B B b b b b*

Bb Bb Bb Bb → *B B B B b b b b*

Bb Bb Bb Bb → *B B B B b b b b*

bb bb bb bb → *b b b b b b b b*

bb bb bb bb → *b b b b b b b b*

bb → *b b*

▲ **FIGURE 16-2 A gene pool** In diploid organisms, each individual in a population contributes two alleles of each gene to the gene pool.

Evolution Is the Change of Allele Frequencies in a Population

A casual observer might define evolution on the basis of changes in the outward appearance or behaviors of the members of a population. However, many of the outward changes that we observe in the individuals that make up the population can also be viewed as the visible expression of underlying changes to the gene pool. A population geneticist, therefore, defines evolution as changes over time in the allele frequencies of a gene pool. In other words, evolution is change in the genetic makeup of populations over generations.

The Equilibrium Population Is a Hypothetical Population in Which Evolution Does Not Occur

It is easier to understand what causes populations to evolve if we first consider the characteristics of a population that would *not* evolve. In 1908, English mathematician Godfrey H. Hardy and German physician Wilhelm Weinberg independently developed a simple mathematical model of a non-evolving population. This model, now known as the **Hardy–Weinberg principle**, showed that under certain conditions, allele frequencies and genotype frequencies in a population will remain constant no matter how many generations pass. In other words, this population will not evolve. Population geneticists use the term **equilibrium population** for this hypothetical non-evolving population in which allele frequencies do not change as long as the following conditions are met:

- There must be no mutation.
- There must be no **gene flow**. That is, there must be no movement of alleles into or out of the population (as would be caused, for example, by the movement of organisms into or out of the population).
- The population must be very large.
- All mating must be random, with no tendency for certain genotypes to mate with specific other genotypes.
- There must be no natural selection. That is, all genotypes must reproduce with equal success.

Under these conditions, allele frequencies in a population will remain the same indefinitely. If one or more of these conditions is violated, then allele frequencies may change: The population will evolve.

As you might expect, few natural populations are truly in equilibrium. What, then, is the importance of the Hardy–Weinberg principle? The Hardy–Weinberg conditions are useful starting points for studying the mechanisms of evolution. In the following sections, we will examine some of these conditions, show that natural populations typically fail to meet them, and illustrate the consequences of such failures. In this way, we can better understand both the inevitability of evolution and the processes that drive evolutionary change.

CHECK YOUR LEARNING

Can you . . .

- define evolution in terms of concepts from population genetics?
- define equilibrium population and describe the conditions under which a population is expected to remain at equilibrium?

16.2 WHAT CAUSES EVOLUTION?

Population genetics predicts that the Hardy–Weinberg equilibrium can be disturbed by deviations from any of its five conditions. Therefore, we can predict five major causes of evolutionary change: mutation, gene flow, small population size, nonrandom mating, and natural selection.

Mutations Are the Original Source of Genetic Variability

A population remains in evolutionary equilibrium only if there are no **mutations** (changes in DNA sequence). Most mutations occur during cell division, when a cell makes a copy of its DNA. Sometimes, errors occur during the copying process and the copied DNA does not match the original. Most such errors are quickly corrected by cellular systems that identify and repair DNA copying mistakes, but some changes in nucleotide sequence slip past the repair systems. An unrepaired mutation in a cell that gives rise to gametes (eggs or sperm) may be passed to offspring and enter the gene pool of a population.

Inherited Mutations Are Rare But Important

How significant are mutations in changing the gene pool of a population? For any given gene, only a tiny proportion of a population inherits a new mutation from the previous generation. For example, the best estimates of human mutation rates suggest that a mutation at any particular site (base pair) in the genome will appear in only about 1 out of every 80 million newborns. Therefore, mutation by itself generally causes only very small changes in the frequency of any particular allele.

Despite the rarity of inherited mutations at any particular location in the genome, the cumulative effect of mutations is essential to evolution. The genomes of most organisms contain a large number of DNA base pairs, so although the rate of mutation is low for any particular base pair, the sheer number of possibilities means that each new generation of a population is likely to include some mutations. For example, the diploid human genome contains about 6 billion base pairs. Thus, even though each base pair has, on average, only a 1 in 80 million chance of mutation, most newborns will probably inherit 70 or 80 mutations. These mutations are new alleles—new variations on which other evolutionary processes can work. As such, they are the foundation of evolutionary change. Without mutations, there would be no evolution.

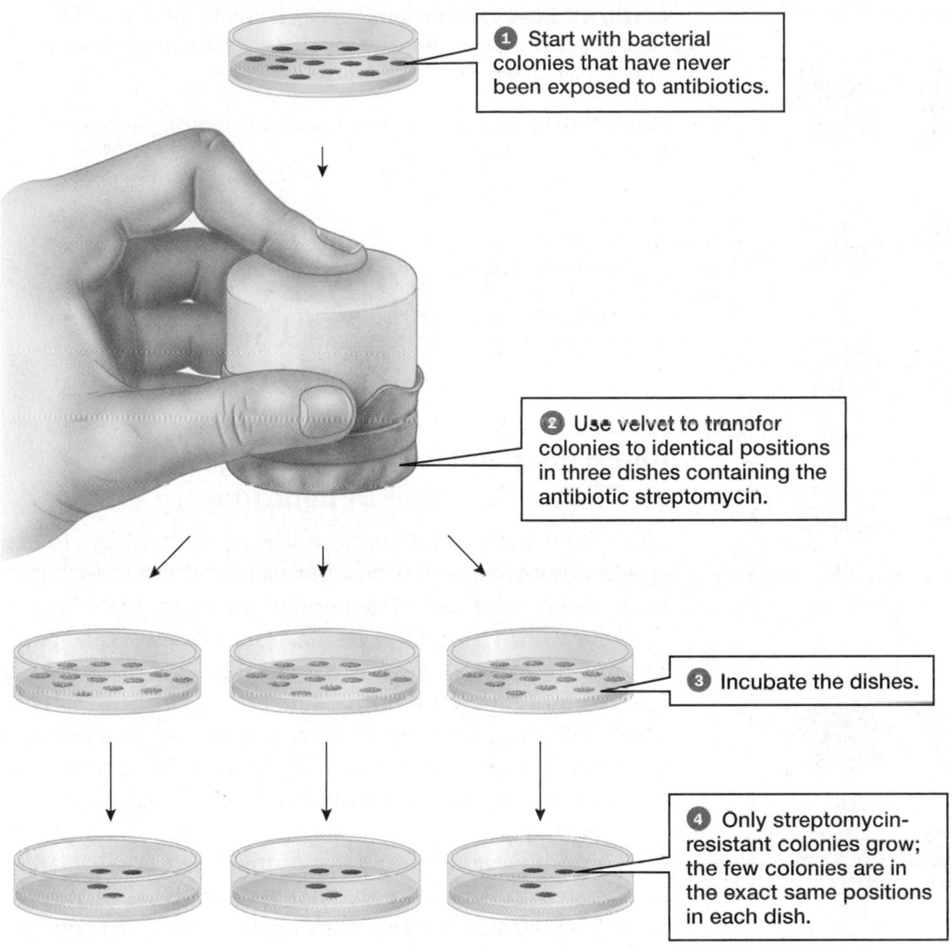
1 Start with bacterial colonies that have never been exposed to antibiotics.
2 Use velvet to transfer colonies to identical positions in three dishes containing the antibiotic streptomycin.
3 Incubate the dishes.
4 Only streptomycin-resistant colonies grow; the few colonies are in the exact same positions in each dish.

FIGURE 16-3 Mutations occur spontaneously This experiment demonstrates that mutations occur spontaneously and not in response to environmental conditions. When bacterial colonies that have never been exposed to antibiotics are exposed to the antibiotic streptomycin, only a few colonies grow. The observation that these surviving colonies grow in the exact same positions in all dishes shows that the mutations for resistance to streptomycin were present in the original dish before exposure to streptomycin.

THINK CRITICALLY If it were true that mutations do occur in response to the presence of antibiotics, how would the result of this experiment have differed from the actual result?

Mutations Are Not Goal Directed

A mutation does not arise as a result of, or in anticipation of, the needs of an organism. A mutation simply happens and may produce a change in a structure or function of the organism. Whether that change is helpful or harmful or neutral, now or in the future, depends on environmental conditions over which the organism has little or no control (**FIG. 16-3**). The mutation merely provides a potential for evolutionary change. Other processes, especially natural selection, may act to spread the mutation through the population or to eliminate it.

Gene Flow Between Populations Changes Allele Frequencies

The movement of alleles between populations, known as *gene flow*, changes how alleles are distributed among populations. When individuals move from one population to another and interbreed at the new location, alleles are transferred from one gene pool to another. For example, newly hatched clam larvae may be carried by ocean currents to distant locations, where they settle and mature, bringing their alleles to the local clam population. Similarly, baboons live in social groupings called troops, and some individuals—usually juvenile males—routinely leave their troop and move to new populations. If the departing baboons are fortunate, they join another troop and achieve sufficient social status to breed. In this way, the male offspring of one troop may carry alleles to the gene pools of other troops.

In some kinds of organisms, alleles move between populations only at certain stages of the life cycle. In flowering plants, for example, most gene flow is due to the movement of seeds and pollen. Pollen, which contains sperm cells, may be carried long distances by wind or by animal pollinators. If the pollen ultimately reaches the flowers of a different population of its species, it may fertilize eggs and add its collection of alleles to the local gene pool. Similarly, seeds may be borne by wind, water, or animals to distant locations where they can germinate to become part of a population far from their place of origin.

The main evolutionary effect of gene flow is to increase the genetic similarity of different populations of a species. Movement of alleles from one population to another tends to change the gene pool of the destination population so that it is more similar to the source population.

(a) Generation 1 frequency of $B = 50\%$
frequency of $b = 50\%$

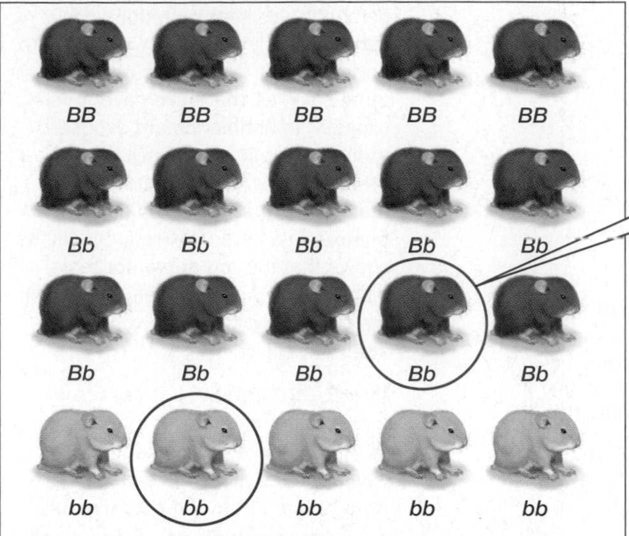

(b) Generation 2 frequency of $B = 25\%$
frequency of $b = 75\%$

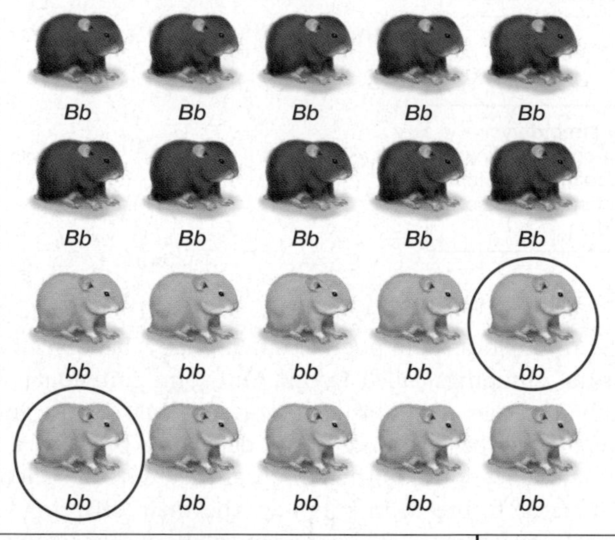

(c) Generation 3 frequency of $B = 0\%$
frequency of $b = 100\%$

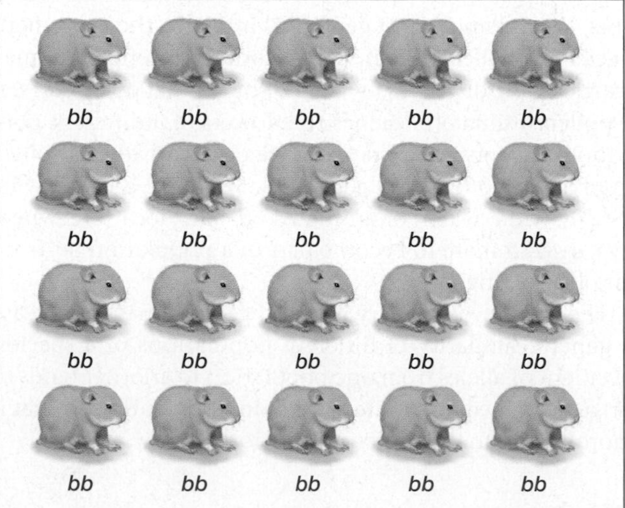

◀ **FIGURE 16-4 Genetic drift** If chance events prevent some members of a population from reproducing, allele frequencies can change randomly.

THINK CRITICALLY Explain how the distribution of genotypes in generation 2 is calculated.

Allele Frequencies May Change by Chance in Small Populations

Allele frequencies in populations can be changed by chance events other than mutations. For example, if bad luck prevents some members of a population from reproducing, their alleles will ultimately be removed from the gene pool, altering its makeup. Any unpredictable event, such as a flood or a fire, that arbitrarily cuts lives short or otherwise allows only a random subset of a population to reproduce can cause random changes in allele frequencies. The process by which chance events change allele frequencies is called **genetic drift**.

To see how genetic drift works, imagine a population of 20 hamsters in which the frequency of the black coat color allele *B* is 0.50 and the frequency of the brown coat color allele *b* is 0.50 (**FIG. 16-4a**). If all of the hamsters in the population were to interbreed to yield another population of 20 animals, the frequencies of the two alleles would not change in the next generation. But if we instead allow only two, randomly chosen hamsters (the ones circled in Figure 16-4a), to breed and become the parents of the next generation of 20 animals, allele frequencies might be quite different in generation 2 (**Figure 16-4b**; the frequency of *B* has decreased and the frequency of *b* has increased). And if breeding in the second generation were again restricted to two randomly chosen hamsters (circled in Figure 16-4b), allele frequencies

Have You Ever Wondered ... Why You Need to Get a Flu Shot Every Year?

A flu vaccination stimulates your immune system to recognize and attack the viruses that cause influenza. The immune system recognizes the viruses by the proteins they contain, but these proteins change from year to year. Why? Because of genetic drift, which is especially rapid in flu viruses due to their high mutation rate. After a year of genetic drift in the flu virus population, the immune system of a previously vaccinated person can no longer recognize the virus. The person needs a fresh vaccination designed to protect against the evolved version of the virus.

might change again in generation 3 (**FIG. 16-4c**). Allele frequencies will continue to change in random fashion as long as reproduction is restricted to a random subset of the population. Note that the changes caused by genetic drift can include the disappearance of an allele from the population, as illustrated by the disappearance of the *B* allele (and therefore the black coat phenotype) in generation 3 in Figure 16-4.

Population Size Matters

Genetic drift occurs more rapidly and has a greater effect in small populations than in large ones. If a population is large, chance events that randomly prevent some individuals from breeding are unlikely to significantly alter the population's genetic composition. The individuals that do breed will constitute a random sample large enough to ensure that its genetic composition is roughly the same as that of the source population. In a small population, however, a random sample of breeders may be quite small, and its genetic composition is therefore more likely to differ from that of the source population. An allele that occurs at low frequency in a small population (and is therefore present in only a few individuals) can be completely eliminated from the population if a chance event prevents its only carriers from breeding.

One way to test these predictions about genetic drift in large versus small populations is to write a computer program that simulates how the frequencies of the alleles would change over many generations in which only a random subset of the population breeds. **FIGURE 16-5** shows the results of simulations of three populations of the hypothetical hamsters we introduced in Figure 16-4. In these simulations, the program began with a population of a specified size in which the initial frequency of each allele was 50%. Then gametes were selected at random from the population's gene pool and combined to create a new generation with the same population size; the process was repeated for 100 generations.

FIGURE 16-5a shows the results of 10 runs in which the simulated population size was large (2,000 individuals). Notice that the frequency of allele *B* remains close to its initial frequency, but nonetheless changes over time. **FIGURE 16-5b** shows 10 simulations of a smaller population (40 individuals). In this case, the frequency of allele *B* is quite likely to diverge from its initial level, in some instances reaching a frequency of 100% (all hamsters are black) or 0% (all hamsters are brown). **FIGURE 16-5c** shows the fate of allele *B* in 10 runs of a simulation of a very small population (four individuals). Here, allele frequencies change very rapidly; in all 10 runs, allele *B* reaches a frequency of 100% or 0% after no more than 20 generations. Overall, the smaller the population, the more dramatic the effects of genetic drift. In populations of all sizes, however, each run of the simulation had a different outcome, because genetic drift is a random process.

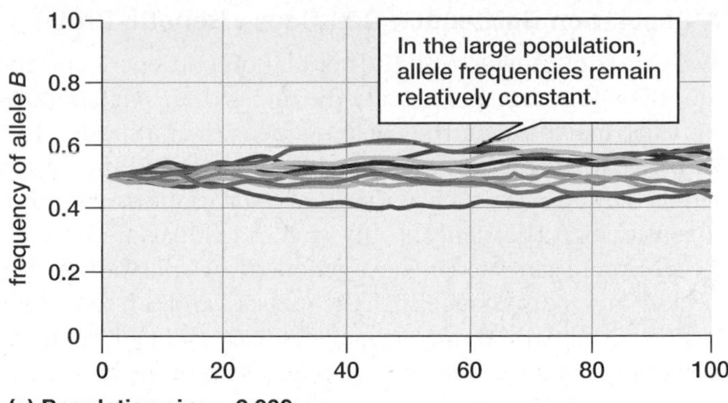

(a) Population size = 2,000

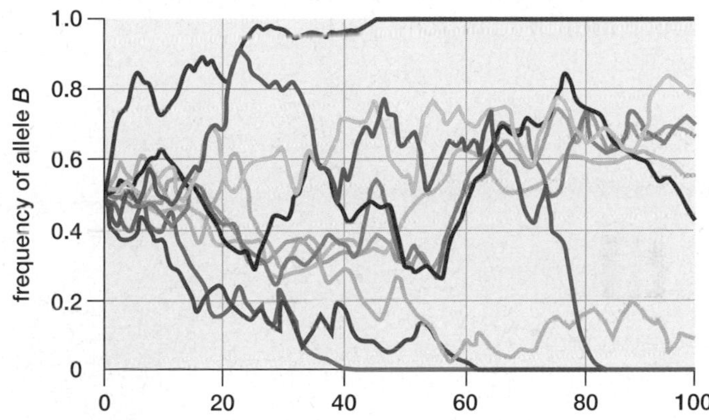

(b) Population size = 40

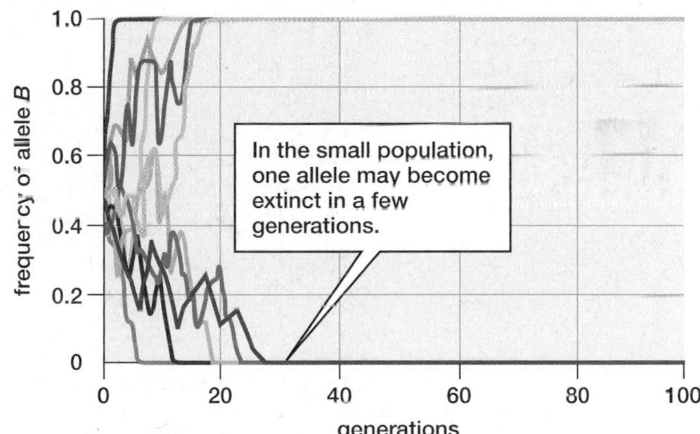

(c) Population size = 4

▲ **FIGURE 16-5 The effect of population size on genetic drift** Each colored line represents one computer simulation of the change over time in the frequency of allele *B* in (a) a large population, (b) a smaller population, and (c) a very small population. Half of the alleles in each starting population were *B* (50%) and, in each generation, randomly chosen individuals reproduced.

THINK CRITICALLY How would each plot change if the initial frequency of allele *B* were 0.7 instead of 0.5?

A Population Bottleneck Can Cause Genetic Drift

Two causes of genetic drift, the population bottleneck and the founder effect, further illustrate the impact that small population size may have on the allele frequencies of a species. In a **population bottleneck**, a population's size is drastically reduced by an event such as a natural catastrophe or overhunting. After such a bottleneck, only a few individuals are available to contribute genes to the next generation. Population bottlenecks can rapidly change allele frequencies and can reduce genetic variability by eliminating alleles (**FIG. 16-6a**). Even if the population later increases, the genetic effects of the bottleneck may remain for hundreds or thousands of generations.

Loss of genetic variability due to bottlenecks has been documented in numerous species, including the northern elephant seal (**FIG. 16-6b**). The elephant seal was hunted almost to extinction in the 1800s; by the 1890s, only about 20 individuals survived. Dominant male elephant seals typically monopolize breeding, so a single male may have fathered all the offspring at this extreme bottleneck point. Since the late nineteenth century, elephant seals have increased in number to more than 200,000 individuals, but biochemical analysis shows that all northern elephant seals are genetically almost identical. With so little genetic variation, the elephant seal has little potential to evolve in response to environmental changes (see "Earth Watch: The Perils of Shrinking Gene Pools"). Because of the limited genetic variation in this species, the species remains vulnerable to extinction regardless of how many elephant seals there are.

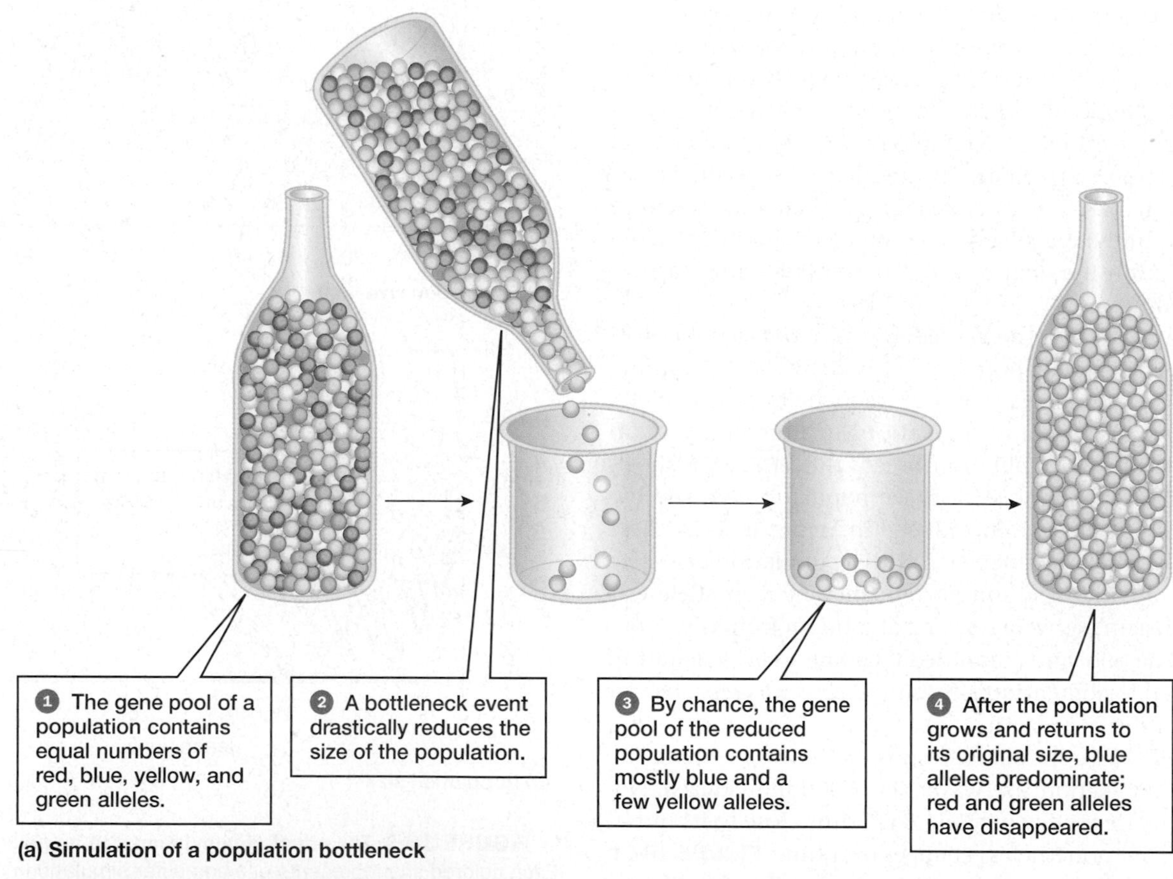

1 The gene pool of a population contains equal numbers of red, blue, yellow, and green alleles.
2 A bottleneck event drastically reduces the size of the population.
3 By chance, the gene pool of the reduced population contains mostly blue and a few yellow alleles.
4 After the population grows and returns to its original size, blue alleles predominate; red and green alleles have disappeared.

(a) Simulation of a population bottleneck

(b) Elephant seals

FIGURE 16-6 Population bottlenecks reduce variation (a) A population bottleneck may drastically reduce genotypic and phenotypic variation because the few organisms that survive may all carry similar sets of alleles. (b) The northern elephant seal passed through a population bottleneck in the recent past. As a result, the population's genetic diversity is extremely low.

THINK CRITICALLY If a population grows large again after a bottleneck, genetic diversity will eventually increase. Why?

The Perils of Shrinking Gene Pools

Many of Earth's species are in danger. According to the International Union for Conservation of Nature, more than 20,000 species of plants and animals are currently threatened with extinction. For most of these endangered species, the main threat is habitat destruction. When a species' habitat shrinks, its population size almost invariably follows suit. Many people, organizations, and governments are concerned about the plight of endangered species and are working to protect them and their habitats. Unfortunately, a population that has already become small enough to warrant endangered status is likely to undergo evolutionary changes that increase its chances of going extinct.

One problem is that, in small populations, mating choices are limited and a high proportion of matings may be between close relatives. This inbreeding increases the odds that offspring will be homozygous for harmful recessive alleles. These less-fit individuals may die before reproducing, further reducing the size of the population.

The greatest threat to small populations, however, stems from their inevitable loss of genetic diversity (**FIG. E16-1**). From our discussion of population bottlenecks, it is apparent that, when populations shrink to very small sizes, many of the alleles that were present in the original population will not be represented in the gene pool of the remnant population. Furthermore, we have seen that genetic drift in small populations will cause many of the surviving alleles to subsequently disappear permanently from the population (see FIG. 16-5c). Because genetic drift is a random process, many of the lost alleles will be advantageous ones that were previously favored by natural selection. Inevitably, the number of different alleles in the population grows ever smaller. Even if the size of an endangered population eventually begins to grow, it may take hundreds of generations to restore the lost genetic diversity.

Why does it matter if a population's genetic diversity is low? Low diversity creates two main risks. First, the fitness of the population as a whole is reduced by the loss of advantageous alleles that underlie adaptive traits. A less-fit population is unlikely to thrive. Second, a genetically impoverished population lacks the variation that will allow it to adapt when environmental conditions change. When the environment changes, as it inevitably will, a genetically uniform species is less likely to contain individuals well suited to survive and reproduce under the new conditions. A species unable to adapt to changing conditions is at very high risk of extinction.

▲ **FIGURE E16-1 Shrunken gene pools** The Ethiopian wolf is among the critically endangered species known to have extremely low genetic diversity.

What can be done to preserve the genetic diversity of endangered species? The best solution, of course, is to preserve plenty of diverse types of habitat so that species never become endangered in the first place. The human population, however, has grown so large and has thus appropriated so large a share of Earth's resources that this solution is impossible in many places. For many species, the only solution is to ensure that areas of preserved habitat are large enough to hold populations of sufficient size to contain most of a threatened species' total genetic diversity.

THINK CRITICALLY In many cases, circumstances prevent preservation of a large, unbroken area of a threatened species' habitat. Sometimes, however, several smaller areas can be protected. In such cases, conservation biologists stress that the small areas must be linked by corridors of the appropriate habitat. Can you explain why?

Isolated Founding Populations May Produce Bottlenecks

The **founder effect** occurs when isolated colonies are founded by a small number of organisms. A small flock of birds, for instance, that becomes lost during migration or is blown off course by a storm may settle on an isolated island. This founder group may, by chance, have allele frequencies that are very different from the frequencies of the parent population. If this is the case, the gene pool of the future population in the new location will be quite unlike that of the larger population from which it sprang. Consider, for example, the Amish inhabitants of Lancaster County, Pennsylvania, who are descended from only 200 or so eighteenth-century immigrants. Among today's Lancaster County Amish, a genetic disorder known as Ellis–van Creveld syndrome is far more common than it is among the general population (**FIG. 16-7**). The prevalence of the syndrome among the Amish stems from a single immigrant couple

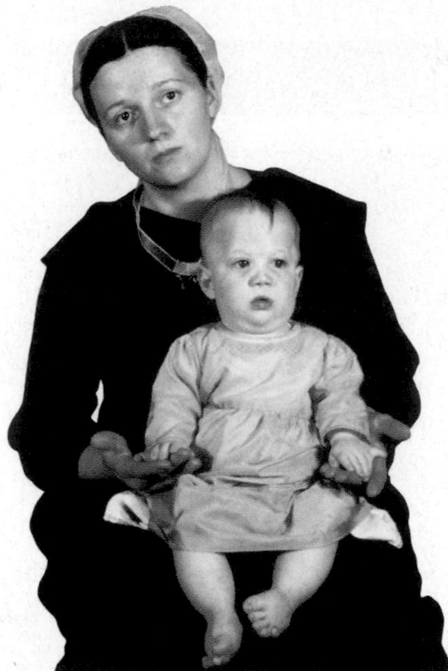

▲ **FIGURE 16-7 A human example of the founder effect** The child of this Amish woman suffers from a set of genetic defects known as Ellis–van Creveld syndrome. Symptoms of the syndrome include short arms and legs, extra fingers, and, in some cases, heart defects. The founder effect accounts for the prevalence of Ellis–van Creveld syndrome among the Amish residents of Lancaster County, Pennsylvania.

who carried the Ellis–van Creveld allele. Because the founder population was so small, this single occurrence meant that the allele was carried by a comparatively high proportion of the population (1 or 2 carriers out of 200 versus about 1 in 1,000 in the general population). This high initial allele frequency, a result of the founder effect, combined with subsequent genetic drift, has led to extraordinarily high levels of Ellis–van Creveld syndrome among this Amish group.

CASE STUDY CONTINUED

Evolution of a Menace

The mutant alleles that confer antibiotic resistance on members of a bacterial population can move to other populations by gene flow and increase by genetic drift. For example, MRSA living on a person's skin might be transferred to her roommate's skin via a shared towel or article of clothing. After joining the population of nonresistant bacteria on the roommate's skin, the newly arrived MRSA may reproduce and may even transfer resistance alleles directly to local bacteria through a process known as conjugation (see Chapter 20). The resistance alleles, introduced to the bacterial population on the roommate's skin by gene flow, may subsequently experience genetic drift that increases their frequency. But what will happen if antibiotics are introduced to the new environment (that is, the roommate's body)? Find out in Section 16.3.

Mating Within a Population Is Almost Never Random

The effects of *nonrandom mating* can play a significant role in evolution, because organisms seldom mate strictly randomly. For example, many organisms have limited mobility and tend to remain near their place of birth, hatching, or germination. In such species, most of the offspring of a given parent live in the same area; thus, when they reproduce, there is a good chance that they will be related to their reproductive partners. Such sexual reproduction between relatives is called *inbreeding*.

Because relatives are genetically similar, inbreeding tends to increase the number of individuals that inherit the same alleles from both parents and are therefore homozygous for many genes. This increase in homozygotes can have harmful effects, such as increased occurrence of genetic diseases or defects. Many gene pools include harmful recessive alleles that persist in the population because their negative effects are masked in heterozygous carriers (which have only a single copy of the harmful allele). Inbreeding, however, increases the odds of producing homozygous offspring with two copies of the harmful allele.

In animals, nonrandom mating can also arise if individuals have preferences or biases that influence their choice of mates. The snow goose is a case in point. Individuals of this species come in two "color phases"; some snow geese are white, and others are blue-gray (**FIG. 16-8**). Although both white and blue-gray geese belong to the same species, mate choice is not random with respect to color. The birds exhibit a strong tendency to mate with a partner of the same color. This preference for mates that are similar is known as *assortative mating*.

Neither inbreeding nor assortative mating by themselves will alter allele frequencies in a population. Nonetheless, they can have large effects on the distribution of different genotypes, and thus on the distribution of phenotypes, in the population.

▲ **FIGURE 16-8 Nonrandom mating among snow geese** Snow geese, which have either white plumage or blue-gray plumage, are most likely to mate with other birds of the same color.

TABLE 16-1 Causes of Evolution

Process	Consequence
Mutation	Creates new alleles; increases variability
Gene flow	Increases similarity of different populations
Genetic drift	Causes random change of allele frequencies; can eliminate alleles
Nonrandom mating	Changes genotype frequencies, but not allele frequencies
Natural and sexual selection	Increases frequency of favored alleles; produces adaptations

All Genotypes Are Not Equally Beneficial

In a hypothetical equilibrium population, individuals of all genotypes survive and reproduce equally well; no genotype has any advantage over the others. This condition, however, is probably met only rarely, if ever, in real populations. Even though many alleles are neutral, in the sense that organisms possessing any of several alleles are equally likely to survive and reproduce, some alleles confer an advantage on their possessor. Any time an allele provides, in Alfred Russel Wallace's words, "some little superiority," the individuals who carry it are favored by **natural selection**, the process in which individuals with traits that help them survive and reproduce leave more offspring than do individuals that lack those traits. We examine the impact of natural selection in greater depth in Section 16.3.

TABLE 16-1 summarizes the different causes of evolution.

CHECK YOUR LEARNING

Can you . . .

- describe how mutation, gene flow, genetic drift, nonrandom mating, and natural selection affect evolution?

16.3 HOW DOES NATURAL SELECTION WORK?

Unlike the other causes of evolution that we have discussed, natural selection leads to the evolution of traits that enhance an organism's ability to survive and reproduce. Studying the adaptive evolution that results from natural selection has been a major focus of evolutionary biology.

Natural Selection Stems from Unequal Reproduction

The British economist Herbert Spencer, writing in 1864, coined the phrase "survival of the fittest" to summarize the process that Darwin had named natural selection. But this formulation is not quite accurate: Natural selection favors traits that increase their possessors' survival only to the extent that improved survival leads to improved reproduction. A trait that improves survival may, for example, increase the likelihood that an individual survives long enough to reproduce, or might increase an organism's life span and, therefore, its number of opportunities to reproduce. But ultimately, it is reproductive success that determines the future of an individual's alleles and the prevalence in the next generation of the traits associated with those alleles. Thus, the main driver of natural selection is differences in reproduction: Individuals bearing certain alleles leave more offspring (who inherit those alleles) than do other individuals with different alleles. In the terminology of evolutionary biology, individuals with greater lifetime reproductive success are said to have greater **fitness** than do individuals with lower reproductive success. (For an example of how the concept of fitness might aid medical practice, see "Health Watch: Cancer and Darwinian Medicine" on page 258.)

Natural Selection Acts on Phenotypes

Although we have defined evolution as changes in the genetic composition of a population, it is important to recognize that natural selection does not act directly on the genotypes of individual organisms. Rather, natural selection acts on phenotypes, the structures and behaviors displayed by the members of a population. This selection of phenotypes, however, inevitably affects the genotypes present in a population, because phenotypes and genotypes are closely tied. For example, we know that a pea plant's height is strongly influenced by the plant's alleles of certain genes. If a population of pea plants were to encounter environmental conditions that favored taller plants, then taller plants would leave more offspring. These offspring would carry the alleles that contributed to their parents' height. Thus, if natural selection favors a particular phenotype, it will necessarily also favor the underlying genotype.

Some Phenotypes Reproduce More Successfully Than Others

As we have seen, natural selection simply means that some phenotypes reproduce more successfully than others do. This simple process is such a powerful agent of change because only the fittest phenotypes pass traits to subsequent generations. But what makes a phenotype fit? Successful phenotypes are those that have the best **adaptations**—characteristics that help an individual survive and reproduce in a particular environment.

An Environment Has Nonliving and Living Components

Individual organisms must cope with an environment that includes both nonliving physical factors and the other living organisms with which the individual interacts. The nonliving component of the environment includes such factors as climate, availability of water, and availability of nutrients.

Health WATCH Cancer and Darwinian Medicine

In recent years, an increasing number of physicians and biomedical researchers have come to the realization that medical practice should be informed by an appreciation of how evolution affects human health and disease. Consider, for example, cancer. Although a cancerous tumor is an abnormal, harmful growth of tissue, it is also an evolving population of cells. The population of cells that forms the tumor inhabits a particular habitat within a body, and the cells in the population tend to have a variety of different genotypes. This variety arises because cancer cells typically contain multiple mutations. When the cells replicate, which they do far more frequently than do normal cells, they are likely to acquire additional mutations. The cells in a tumor essentially compete with one another to survive and reproduce. The cells that are best adapted to the local environment leave more offspring, and over time the tumor evolves—its genetic makeup changes.

If you understand how natural selection works, you can probably predict what happens when a tumor is treated with a chemotherapy drug. The drug changes the tumor's environment such that many cancer cells cannot survive. Often, however, some of the cells have mutant alleles that allow them to resist the drug. These resistant cells survive the attack, and they or their descendants may move to new environments in different parts of the body. Eventually, the resistant cells proliferate, and the cancer patient now has multiple tumors, all of which are resistant to treatment by chemotherapy. This evolutionary scenario helps explain why a post-chemotherapy cancer patient sometimes seems to be free of cancer (because all but a few tumor cells have been killed, and the few remaining ones are undetectable), only to later suffer an even more threatening recurrence (because natural selection has favored the evolution of drug-resistant cancer cell populations that eventually spread and grow).

A better understanding of the details of tumor evolution could help improve cancer treatments (**FIG. E16-2**). For example, suppose that genetic analysis of a tumor revealed that most of its cells were susceptible to drug A, but that a few cells carried mutations known to confer resistance to drug A but not drug B. Physicians could initially treat the tumor with drug A, thereby altering the tumor's environment in a way that gave the resistant cells a competitive advantage and caused them to be favored by natural selection. As a result, the tumor would evolve such that its most abundant cells would be the formerly rare ones resistant to drug A but not drug B. At that point, drug B could be administered, killing cells of the type now most abundant in the evolved tumor. Finally, a subsequent re-treatment with drug A could kill any cells of the originally common type that had managed to survive the initial modification of their environment.

But what about tumors in which some cells contain mutations for which no treatment is known? Traditionally, such tumors are treated with a high dose of a drug that kills the susceptible cells, and the tumor shrinks rapidly. But, almost inevitably, the remaining resistant cells proliferate and a drug-resistant tumor regrows. In an alternative, evolutionarily-informed approach, the initial drug dosage is lower, so that many susceptible cells are killed but some survive. The now-smaller tumor contains a mix of susceptible and resistant cells that compete with one another; the competition keeps the population of resistant cells in check. Even though the "mixed" tumor may begin to regrow, it still contains many susceptible cells; therefore, it can be controlled, if not eliminated, by periodic repetition of the low-dose drug treatment.

THINK CRITICALLY A team of physicians treated four patients with breast cancer. Each patient received a course of chemotherapy (the same combination of drugs for each patient). In addition, researchers sequenced the genotype of cells in healthy skin tissue and in each patient's tumor at intervals during the treatment period. At the end of the course of treatment, all four patients were declared free of detectable cancer. But within 18 months, three of the patients had suffered a relapse. The researchers sequenced samples from the new tumors. If you compared the genetic information from the four patients, what kind of difference might you expect to find between the cured patient and the ones who suffered relapses? How might the theory of evolution by natural selection explain the difference?

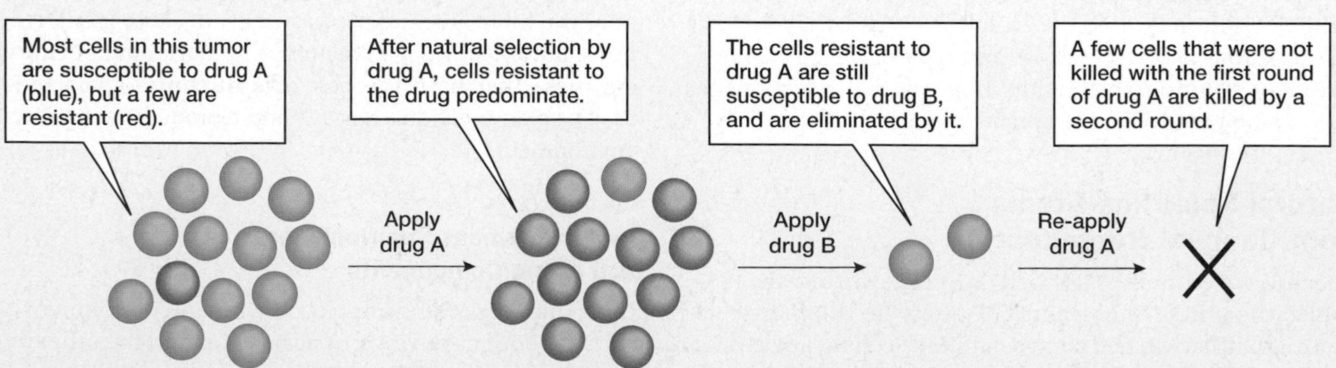

Most cells in this tumor are susceptible to drug A (blue), but a few are resistant (red).
After natural selection by drug A, cells resistant to the drug predominate.
The cells resistant to drug A are still susceptible to drug B, and are eliminated by it.
A few cells that were not killed with the first round of drug A are killed by a second round.
Apply drug A
Apply drug B
Re-apply drug A
X

▲ **FIGURE E16-2 Evolved resistance to one treatment may leave tumors vulnerable to another treatment**

These nonliving factors play a large role in determining the traits that help an organism to survive and reproduce. However, adaptations also arise because of interactions with the living component of the environment, namely, other organisms. A simple example illustrates this concept.

Consider a buffalo grass plant growing in a small patch of soil in the eastern Wyoming plains (**FIG. 16-9**). The plant's roots must be able to take up enough water and minerals for growth and reproduction, and to that extent, it must be adapted to its nonliving environment. But even in the dry prairies of Wyoming, this requirement is relatively trivial, provided that the plant is alone and protected in its square yard of soil. In reality, however, many other plants—other buffalo grass plants as well as other grasses, sagebrush bushes, and annual wildflowers—also sprout in that same patch of soil. If our buffalo grass is to survive, it must compete with the other plants for resources. Its long, deep roots and efficient methods of mineral uptake have evolved not only because the plains are dry but also because the buffalo grass must share the dry prairies with other plants. Further, buffalo grass must also coexist with animals that wish to eat it, such as the cattle and other plant-eating animals that graze the prairie. So over time, tougher, harder-to-eat buffalo grass plants survived better and reproduced more on the prairie than did less-tough buffalo grass plants. As a result, buffalo grass leaves are quite tough; embedded silica compounds reinforce them.

▲ **FIGURE 16-9 Buffalo grass**

Competition Acts As an Agent of Selection

As the example of buffalo grass shows, one of the major sources of natural selection is **competition** with other organisms for scarce resources. Competition for resources is most intense among members of the same species because, as Darwin wrote in *On the Origin of Species*, "they frequent the same districts, require the same food, and are exposed to the same dangers." In other words, no two competing organisms have such similar requirements for survival as do two members of the same species. Although different species may also compete for the same resources, they generally do so to a lesser extent than do individuals within a species.

Both Predators and Prey Act As Agents of Selection

When two species interact extensively, each exerts strong selection on the other. When one evolves a new feature or modifies an old one, the other typically evolves new adaptations in response. This process in which species mutually affect one another's evolution is called **coevolution**. Perhaps the most familiar form of coevolution is found in predator–prey relationships.

Predation describes any interaction in which one organism consumes another. In some instances, coevolution between predators (those that do the consuming) and prey (those that are consumed) is a sort of biological arms race, with each side evolving new adaptations in response to escalations by the other. Darwin used the example of wolves and deer: Wolf predation selects against slow or careless deer, thus leaving faster, more-alert deer to reproduce and pass on these traits. The resulting alert, swift deer select in turn against slow, clumsy wolves, because such predators cannot acquire enough food.

CASE STUDY **CONTINUED**

Evolution of a Menace

Antibiotic resistance evolves by natural selection. To see how, imagine a hospital patient with an infected wound. A doctor decides to treat the infection with an intravenous drip of penicillin. As the antibiotic courses through the patient's blood vessels, millions of bacteria die before they can reproduce. A few bacteria, however, carry a rare allele that codes for an enzyme that destroys penicillin. The bacteria carrying this rare allele are able to survive and reproduce, and their offspring inherit the penicillin-destroying allele. After a few generations, the frequency of the penicillin-destroying allele in the bacteria has soared to nearly 100%, and the frequency of the normal allele has declined to near zero. As a result of natural selection imposed by the antibiotic's killing power, the population of bacteria within the patient's body has evolved. The gene pool of the population has changed, and natural selection, in the form of bacterial destruction by penicillin, has caused the change.

Antibiotic Resistance Illustrates Key Points About Natural Selection

The example of antibiotic resistance highlights some important features of natural selection.

- **Natural selection does not cause genetic changes in individuals.** Alleles for antibiotic resistance arise spontaneously in some bacteria, long before the bacteria encounter an antibiotic. Antibiotics do not cause resistance to appear; their presence merely favors the survival of bacteria with antibiotic-destroying alleles over that of bacteria without such alleles.
- **Natural selection acts on individuals, but it is populations that are changed by evolution.** The agent of natural selection—in this example, antibiotics—acts on individual bacteria. As a result, some individuals reproduce and some do not. However, it is the population as a whole that evolves as its allele frequencies change.
- **Evolution by natural selection is not progressive; it does not make organisms "better."** The traits favored by natural selection change as the environment changes. Resistant bacteria are favored only when antibiotics are present. At a later time, when the environment no longer contains antibiotics, resistant bacteria may be at a disadvantage relative to other bacteria.

▲ **FIGURE 16-11 The peacock's showy tail has evolved through sexual selection** The ancestors of today's peahens were apparently picky when deciding on a male with which to mate, favoring males with longer and more colorful tails.

Sexual Selection Favors Traits That Help an Organism Mate

In many animal species, males have conspicuous features such as bright colors, long feathers or fins, or elaborate antlers. Males may also exhibit elaborate courtship behaviors. Although these extravagant features and behaviors typically play a role in mating, they also seem to be at odds with efficient survival and reproduction. Exaggerated ornaments and displays may help males gain access to females, but they may also make the males more conspicuous and thus vulnerable to predators. Darwin was intrigued by this apparent contradiction. He coined the term **sexual selection** to describe the special kind of selection that acts on traits that help an animal acquire a mate.

Darwin recognized that sexual selection could be driven either by sexual contests among males or by female preference for particular male phenotypes. Male–male competition for access to females can favor the evolution of features that provide an advantage in fights or ritual displays of aggression (**FIG. 16-10**). In animal species in which females actively choose their mates, females often seem to prefer males with the most elaborate ornaments or most extravagant displays (**FIG. 16-11**). Why?

One hypothesis is that male structures, colors, and displays that do not enhance survival might instead provide a female with an outward sign of a male's condition. Only a vigorous, energetic male can survive when burdened with conspicuous coloration or a large tail that might make him more vulnerable to predators. Conversely, males that are sick or under parasitic attack are dull and frumpy compared with healthy males. A female that chooses the brightest, most ornamented male is also choosing the healthiest male. By doing so, she gains fitness if, for example, the healthiest male provides superior parental care to offspring or if he carries alleles for disease resistance that will be inherited by offspring and help ensure their survival. Females thus gain a reproductive advantage by choosing the most highly ornamented

◀ **FIGURE 16-10 Competition between males favors the evolution, through sexual selection, of structures for ritual combat** Two male bighorn sheep spar during the fall mating season. In many species, the losers of such contests are unlikely to mate, while winners enjoy tremendous reproductive success.

THINK CRITICALLY If we studied a population of bighorn sheep and were able to identify the father and mother of each lamb born, would you predict that the difference in number of offspring between the most reproductively successful adult and the least successful adult would be greater for males or for females?

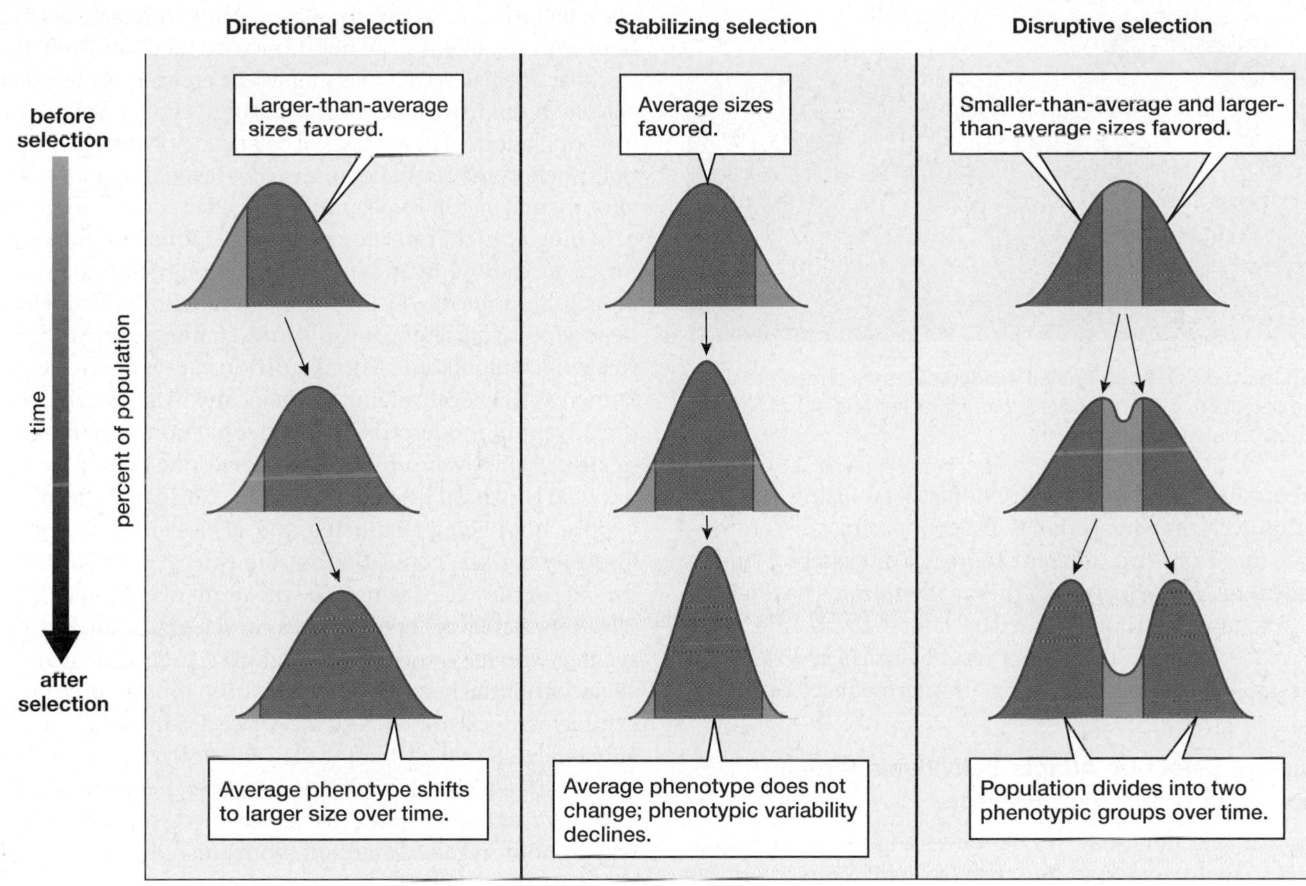

▲ **FIGURE 16-12 Three ways that selection affects a population over time** A graphical illustration of three ways natural and/or sexual selection, acting on a normal distribution of phenotypes, can affect a population over time. In all graphs, the blue areas represent individuals that are selected against—that is, the individuals that do not reproduce as successfully as do the individuals in the purple range.

THINK CRITICALLY When selection is directional, is there any limit to how extreme the trait under selection will become? Why or why not?

males, and the traits (including the exaggerated ornament) of these flashy males will be passed to subsequent generations.

Selection Can Influence Populations in Three Ways

Natural selection and sexual selection can lead to various patterns of evolutionary change. Evolutionary biologists group these patterns into three categories (**FIG. 16-12**):

- **Directional selection** favors individuals with an extreme value of a trait and selects against both average individuals and individuals at the opposite extreme. For example, directional selection might favor small size and select against both average and large individuals in a population.
- **Stabilizing selection** favors individuals with the average value of a trait (for example, intermediate body size) and selects against individuals with extreme values.
- **Disruptive selection** favors individuals at both extremes of a trait (for example, both large and small body sizes) and selects against individuals with intermediate values.

Directional Selection Shifts Character Traits in a Specific Direction

If environmental conditions change in a consistent way, a species may respond by evolving in a consistent direction. For example, during past ice age periods in which Earth's climate cooled considerably, many mammal species evolved thicker fur. The evolution of antibiotic resistance in bacteria is another example of directional selection: When antibiotics are present in a bacterial species' environment, individuals with greater resistance reproduce more prolifically than do individuals with less resistance.

Stabilizing Selection Acts Against Individuals Who Deviate Too Far from the Average

Directional selection can't go on forever. What happens once a species is well adapted to a particular environment? If the environment is unchanging, most new variations that appear will be harmful. Under these conditions, we expect species to be subject to stabilizing selection, which favors the survival

▲ **FIGURE 16-13 Black-bellied seedcrackers** As a result of disruptive selection, each black-bellied seedcracker has either a large beak (left) or a small beak (right).

and reproduction of average individuals. Stabilizing selection commonly occurs when a trait is under opposing environmental pressures from two different sources. For example, among lizards of the genus *Aristelliger*, the smallest lizards have a hard time defending territories, but the largest lizards are more likely to be eaten by owls. As a result, *Aristelliger* lizards are under stabilizing selection that favors intermediate body size.

Disruptive Selection Adapts Individuals Within a Population to Different Habitats

Disruptive selection may occur when a population inhabits an area with more than one type of useful resource. In this situation, the most adaptive characteristics may be different for each type of resource. For example, the food source of the black-bellied seedcracker (**FIG. 16-13**), a small, seed-eating bird found in the forests of Africa, includes both hard seeds and soft seeds. Cracking hard seeds requires a large, stout beak, but a smaller, pointier beak is a more efficient tool for processing soft seeds. Consequently, black-bellied seedcrackers have beaks in one of two sizes. A bird may have a large beak or small beak, but very few birds have a medium-sized beak; individuals with intermediate-sized beaks have a lower survival rate than individuals with either large or small beaks. Disruptive selection in black-bellied seedcrackers thus favors birds with large beaks and birds with small beaks, but not those with medium-sized beaks.

Black-bellied seedcrackers represent an example of *balanced polymorphism*, in which two or more phenotypes are maintained in a population. In many cases of balanced polymorphism, multiple phenotypes persist because each is favored by a separate environmental factor. For example, consider two different forms of hemoglobin that are present in some human populations in Africa. In these populations, the hemoglobin molecules of people who are homozygous for a particular allele produce defective hemoglobin that clumps up into long chains, which distort and weaken red blood cells. This distortion causes a serious illness known as sickle-cell anemia, which can kill its victims. Before the advent of modern medicine, people homozygous for the sickle-cell allele were unlikely to survive long enough to reproduce. So why hasn't natural selection eliminated the allele?

Far from being eliminated, the sickle-cell allele is present in nearly half the population in some areas of Africa. The persistence of the allele seems to be the result of counterbalancing selection that favors heterozygous carriers of the allele. Heterozygotes, who have one allele for defective hemoglobin and one allele for normal hemoglobin, suffer from mild anemia but also exhibit increased resistance to malaria, a deadly disease affecting red blood cells that is widespread in equatorial Africa. In areas of Africa with a high risk of malaria infection, heterozygotes must have survived and reproduced more successfully than either type of homozygote. As a result, both the normal hemoglobin allele and the sickle-cell allele have been preserved.

CHECK YOUR LEARNING

Can you . . .

- describe why selection of phenotypes can affect the evolution of genotypes?
- explain how competition and predation influence evolution?
- explain how sexual selection works and describe examples of its outcome?
- compare and contrast directional selection, stabilizing selection, and disruptive selection?

CASE STUDY REVISITED

Evolution of a Menace

Overuse of antibiotics has accelerated the evolution of antibiotic resistance. Each year, U.S. physicians write more than 100 million prescriptions for antibiotics; the Centers for Disease Control and Prevention estimates that about half of these prescriptions are unnecessary.

Although medical use and misuse of antibiotics are the most important sources of natural selection for antibiotic resistance, antibiotics also pervade the environment outside our bodies. More than 29 million pounds of antibiotics are fed to farm animals in the United States each year. In addition, Earth's soils and water are laced with antibiotics that enter the environment through human and livestock wastes, and from the antibacterial soaps and cleansers that are now routinely used in many households and workplaces. As a result of this massive alteration of the environment, resistant bacteria are now found not only in hospitals and the bodies of sick people but also in our food, water, and soil. Susceptible bacteria are under constant attack, and resistant strains have little competition. In our fight against disease, we have rashly overlooked some basic principles of evolutionary biology and are now paying a heavy price.

THINK CRITICALLY Microbiologists have discovered that alleles associated with antibiotic resistance are present in bacteria that live in soil, even in environments that are comparatively free of antibiotic pollution from human activities. Why are such alleles present (albeit at low levels) in bacterial populations? Conversely, if resistance alleles are beneficial, why are they rare in natural populations of bacteria?

CHAPTER REVIEW

Go to **Mastering Biology** to access the Pearson eText, vocabulary review, practice quizzes, activities, videos, current events, and more.

Answers to ***Think Critically,*** *and* ***Thinking Through the Concepts*** *questions can be found in the* ***Answers*** *section at the back of the book.*

Summary of Key Concepts

16.1 How Are Populations, Genes, and Evolution Related?

Evolution is change in the frequencies of alleles in a population's gene pool. Allele frequencies in a population will remain constant over generations only if the following conditions are met: (1) There is no mutation, (2) there is no gene flow, (3) the population is very large, (4) all mating is random, and (5) all genotypes reproduce equally well (that is, there is no natural selection). An equilibrium population in which all of these conditions are met will not evolve. Most natural populations, however, fail to meet at least one of the conditions; therefore, they evolve.

16.2 What Causes Evolution?

Evolutionary change is caused by mutation, gene flow, small population size, nonrandom mating, and natural selection.

Mutations are random, undirected changes in DNA composition. Although most mutations are neutral or harmful to the organism, some prove advantageous in certain environments. Mutations are rare and do not by themselves change allele frequencies very much, but they provide the raw material for evolution by other processes.

Gene flow is the movement of alleles between different populations of a species. It occurs when organisms or their gametes move from place to place. Gene flow tends to reduce differences in the genetic composition of different populations.

If a population is small, chance events may reduce the survival and reproduction of a disproportionate number of individuals that bear a particular allele, thereby greatly changing the allele's frequency in the population; this is genetic drift. Causes of genetic drift include population bottlenecks that reduce population size relatively suddenly, and founder effects that occur when a small portion of a population becomes isolated in a new location.

Nonrandom mating, such as assortative mating and inbreeding, can change the distribution of genotypes in a population, in particular by increasing the proportion of homozygotes.

The survival and reproduction of organisms are influenced by their phenotypes. Because phenotype depends at least partly on genotype, natural selection tends to favor the persistence of certain alleles at the expense of others.

16.3 How Does Natural Selection Work?

Natural selection is driven by differences in reproductive success among different genotypes. Natural selection stems from the interactions of organisms with both the living and nonliving parts of their environments. Between-organism interactions that result in natural selection include competition and predation. When two species interact intensively, both of them may evolve in response. Phenotypes that help organisms mate can evolve by sexual selection. The effects of natural selection and sexual selection may be directional, stabilizing, or disruptive.

Thinking Through the Concepts

Bloom's: Remembering, Understanding

Multiple Choice

1. The alleles responsible for antibiotic resistance in bacteria
   a. arise in response to the presence of antibiotics.
   b. are identical to the alleles responsible for pesticide resistance in insects.
   c. are present in bacterial populations that have never been exposed to antibiotics.
   d. are formed by interactions between antibiotic molecules and bacterial DNA.
2. Stabilizing selection on a trait tends to
   a. make the trait more extreme.
   b. reduce variability in the trait.
   c. decrease the frequency of alleles associated with the trait.
   d. result in elaborate male ornaments.
3. An adaptation is
   a. any trait that arises from a mutation.
   b. a trait that increases the reproductive success of its bearer.
   c. any trait that changes during the lifetime of an organism.
   d. a trait that arises due to gene flow or genetic drift.
4. Which of the following statements about mutations is false?
   a. Mutations at a given chromosomal site are rare.
   b. The genomes of most people contain some mutant alleles that are present in neither parent.
   c. Mutations are the ultimate source of genetic variability.
   d. Beneficial mutations are more likely to occur when an organism's needs change.
5. Genetic drift occurs
   a. when different phenotypes have different reproductive success.
   b. in small populations but not in large populations.
   c. when only a random subset of a population reproduces.
   d. in mammals but not in bacteria.

Fill-in-the-Blank

1. The ____________ provides a simple mathematical model for a non-evolving population, also called a(n) ____________ population, in which ____________ frequencies do not change over time. Are such populations likely to be found in nature? ____________
2. Different versions of the same gene are called ____________. These versions arise as a result of changes in the sequence of ____________ that form the gene. These changes are known as ____________. An individual with two identical copies of a given gene is described as being ____________ for that gene, while an individual with two different versions of that gene is described as ____________.

3. An organism's ____________ refers to the specific alleles found within its chromosomes, while the traits that these alleles produce are called its ____________. Which of these does natural selection act on? ____________
4. A random form of evolution is called ____________. This form of evolution occurs more rapidly and has a greater effect in populations that are ____________. Two important causes of this form of evolution are ____________ and ____________. Which of these would apply to a population started by a breeding pair that was stranded on an island? ____________
5. Competition is most intense between members of a(n) ____________. Predators and their prey act as agents of ____________ on one another, resulting in a form of evolution called ____________. This results in the evolution of characteristics called ____________ that help both predators and their prey survive and reproduce.
6. The evolutionary fitness of an organism is measured by its success at ____________. The fitness of an organism can change if its ____________ changes.

Review Questions

1. What is a gene pool? How would you determine the allele frequencies in a gene pool?
2. Define *equilibrium population*. Outline the conditions that must be met for a population to stay in genetic equilibrium.
3. How does population size affect the likelihood of changes in allele frequencies by chance alone? Can significant changes in allele frequencies (that is, evolution) occur as a result of genetic drift?
4. If you measured the allele frequencies of a gene and found large differences from those predicted by the Hardy–Weinberg principle, would that prove that natural selection is occurring in the population you are studying? Review the conditions that lead to an equilibrium population, and explain your answer.
5. People like to say that "you can't prove a negative." Study the experiment in Figure 16-3 again, and comment on what it demonstrates.
6. Describe the three ways in which natural selection can affect a population over time. Which way(s) is (are) most likely to occur in stable environments, and which way(s) might occur in rapidly changing environments?
7. What is sexual selection? How is sexual selection similar to and different from other forms of natural selection?

Applying the Concepts

Bloom's: Applying, Analyzing, Evaluating

1. In North America, the average height of adult humans has been increasing steadily for decades. Is directional selection occurring? What data would justify your answer?
2. By the 1940s, the whooping crane population had been reduced to only 15 individuals. Thanks to conservation measures, its numbers are now increasing. What special evolutionary problems do whooping cranes face now that they have passed through a population bottleneck?

17 The Origin of Species

Paedophryne amauensis is one of a number of previously undiscovered species recently found in the forests of New Guinea.

CASE STUDY

Discovering Diversity

AS DARKNESS FELL ON A WARM, HUMID EVENING in a rain forest on the island of New Guinea, a team of scientists carefully scrutinized the forest floor. They were seeking the source of a high-pitched call that sounded like an insect. But the scientists' careful search detected no animal that might have produced the sound. Finally, the frustrated searchers scooped up a bit of leaf litter from a spot that seemed to be the source of the calls, put it in a plastic bag, and headed back to the lab. There, concealed among the decaying leaves in the bag, they found what they had earlier missed: a tiny frog, smaller than a shirt button. The frog was smaller, in fact, than any other known animal with a backbone. And what's more, the frog was of a type previously unknown to science: It was a newly discovered species. The new frog species, which its discoverers dubbed *Paedophryne amauensis*, is one of many species recently discovered in New Guinea, including birds, mammals, butterflies, and flowering plants.

New Guinea is not the only location to yield interesting new species. One particularly surprising find was in the Annamite Mountains of Vietnam, where the saola, a hoofed, horned antelope, was discovered in the early 1990s. The discovery of a new species of large mammal at that late date was a complete shock—after centuries of human exploration and exploitation of nearly every corner of the world, scientists had been certain that no large mammal species could have escaped detection. More recent surprises include the discovery, reported in 2013, of the olinguito, a nocturnal relative of the raccoon that inhabits high-altitude cloud forests in the Andes Mountains of South America.

Our discussion of newly discovered species leads us to some important questions: What do we mean when we say that some organisms constitute a species? And how do such species originate?

AT A GLANCE

17.1 WHAT IS A SPECIES?

Although Darwin brilliantly explained how evolution shapes complex organisms, his ideas did not fully explain life's diversity. In particular, the process of natural selection cannot by itself explain how living things came to be divided into groups, with each group distinctly different from all other groups. When we look at big cats, we don't see a continuous array of different tiger phenotypes that gradually grades into the phenotype of a lion. We see lions and tigers as separate, distinct types with no overlap. Each distinct type is known as a species.

In everyday life, most of us make unthinking use of an informal, nonscientific conception of species. We perceive sparrows as clearly different from eagles, which are obviously different from ducks. But we sometimes run into trouble when we try to make finer distinctions. It is not easy, for example, to distinguish among different species of sparrows, especially if we don't have a precise idea of what constitutes a species. How, then, do scientists make these finer distinctions?

Have You Ever Wondered ... How Many Species Inhabit the Planet?

One way to determine the number of species on Earth might be to simply count them. You could comb the scientific literature to find all the species that scientists have discovered and named, and then tally up the total number. One attempt to do just that, the Catalogue of Life project, has compiled an online searchable database that listed 1,736,852 species as of 2018. But even the Catalogue of Life can't tell you how many species are on Earth.

Why doesn't counting work? Because most of the planet's species remain undiscovered. Relatively few scientists are engaged in the search for new species, and nearly all undiscovered species are small and inconspicuous, or live in poorly explored habitats such as the floor of the ocean or the topmost branches of tropical rain forests. So no one knows the actual number of species on Earth. But biologists agree that the number must be much higher than the number of named species. A commonly held view is that the actual number is close to 8.7 million, the number estimated by a recent analysis that used sophisticated statistical methods to extrapolate past trends in species discovery.

Each Species Evolves Independently

Today, biologists define a **species** as a group of populations that evolves independently. That is, even though different species may interact, each species' gene pool evolves separately from the gene pools of other species. This definition, however, does not clearly state the standard by which such evolutionary independence is judged. The most widely used standard defines species as "groups of actually or potentially interbreeding natural populations, which are reproductively isolated from other such groups." This definition, known as the *biological species concept*, is based on the observation that **reproductive isolation** (inability to successfully breed outside the group) ensures evolutionary independence.

The biological species concept has two major limitations. First, because the definition is based on patterns of sexual reproduction, it does not help us determine species boundaries among asexually reproducing organisms. Second, it is not always practical or even possible to directly observe whether members of two different groups interbreed. Thus, a biologist who wishes to determine if a group of organisms is a separate species must often make the determination without knowing for sure if group members breed with organisms outside the group.

Despite the limitations of the biological species concept, most biologists accept it for identifying species of sexually reproducing organisms. However, alternative definitions are required by scientists who study bacteria and other organisms that mainly reproduce asexually. Even some biologists who study sexually reproducing organisms prefer species definitions that do not depend on a property (reproductive isolation) that can be difficult to measure. Several such alternatives to the biological species concept have been proposed. (One alternative, the phylogenetic species concept, is described in Chapter 19.)

Appearance Can Be Misleading

Biologists have found that differences in appearance do not always mean that two populations belong to different species. For example, a northwestern garter snake may be brown, black, gray, green, or some shade in between, and it may be striped or unstriped (**FIG. 17-1**). If it is striped, the stripes may be broad or narrow, and they could be any of a variety of colors. Despite their diversity of appearance, though, all northwestern garter snakes are members of the same species.

(a) Green-striped northwestern garter snake

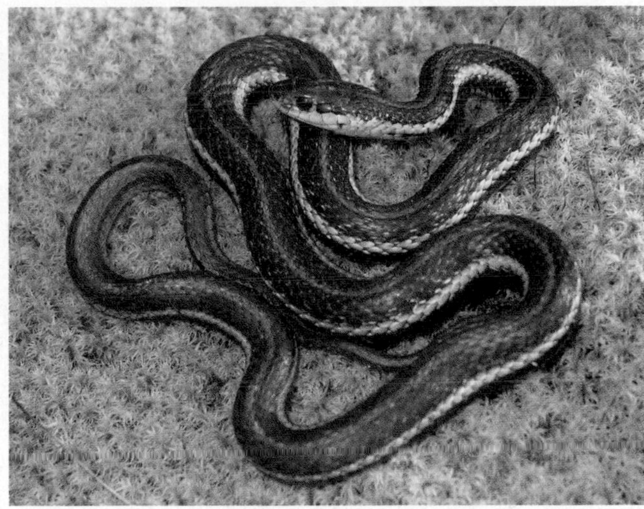

(b) Red-striped northwestern garter snake

▲ **FIGURE 17-1 Members of a species may differ in appearance (a)** This green-striped northwestern garter snake and **(b)** this red-striped northwestern garter snake are members of the same species.

THINK CRITICALLY Southern Wisconsin is home to several populations of squirrels with black fur. These squirrels are hypothesized to be members of the squirrel species *Sciurus carolinensis*, which usually has gray fur. How could you determine if the squirrels with black fur are in fact of the same species as the squirrels with gray fur?

Conversely, some organisms with very similar appearances belong to different species. For example, the cordilleran flycatcher and the Pacific-slope flycatcher are so similar that even experienced birdwatchers cannot tell them apart (**FIG. 17-2**). Likewise, there are virtually no visible differences between the Asian mosquito species *Anopheles dirus* and *Anopheles harrisoni*. This similarity in appearance, however, disguises a crucial difference: *A. dirus* spreads malaria from person to person, but *A. harrisoni* generally does not.

CHECK YOUR LEARNING

Can you . . .

- describe how biologists define species and explain why it is difficult to develop a criterion for distinguishing species?
- describe the biological species concept and discuss its limitations?
- list some reasons why it can be hard to tell different species apart?

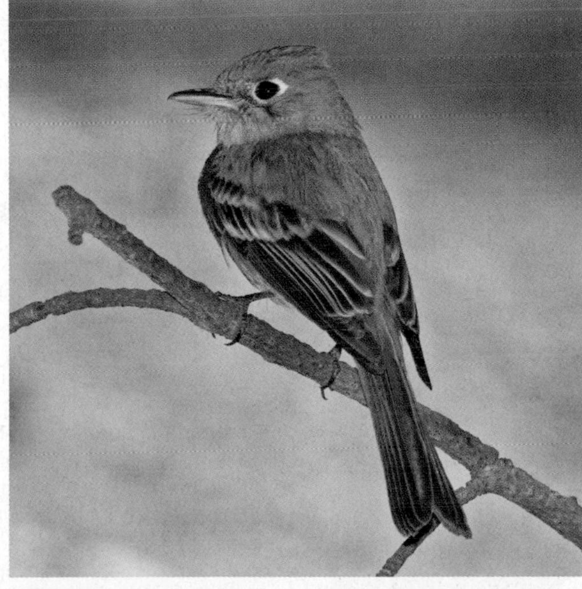

(a) Cordilleran flycatcher

(b) Pacific-slope flycatcher

▲ **FIGURE 17-2 Members of different species may be similar in appearance** The **(a)** cordilleran flycatcher and **(b)** Pacific-slope flycatcher are different species.

TABLE 17-1 Mechanisms of Reproductive Isolation

Premating Isolating Mechanisms	Factors that prevent organisms of two species from mating
Geographic isolation	The species do not interbreed because a physical barrier separates them.
Ecological isolation	The species do not interbreed even if they are within the same area because they occupy different habitats.
Temporal isolation	The species do not interbreed because they breed at different times.
Behavioral isolation	The species do not interbreed because they have different courtship and mating rituals.
Mechanical incompatibility	The species do not interbreed because their reproductive structures are incompatible.
Postmating Isolating Mechanisms	Factors that prevent organisms of two species from producing vigorous, fertile offspring after mating
Gametic incompatibility	Sperm from one species cannot fertilize eggs of another species.
Hybrid inviability	Hybrid offspring fail to survive.
Hybrid infertility	Hybrid offspring are sterile or have low fertility.

CASE STUDY CONTINUED

Discovering Diversity

The tiny frog ***Paedophryne amauensis*** was discovered by researchers purposefully searching for the source of an unusual sound, but sometimes similarity between two species can lead scientists to accidental discoveries. A previously unknown species of wild cat was inadvertently discovered by researchers who were sequencing the DNA of a house cat–sized South American wild cat called the tigrina. The DNA analysis revealed that many alleles found in the supposed tigrinas living in northeastern Brazil are not shared with other tigrinas. This finding suggested that the northeastern cats do not interbreed with other tigrinas and are therefore a different species. What prevents the two species from interbreeding?

17.2 HOW IS REPRODUCTIVE ISOLATION BETWEEN SPECIES MAINTAINED?

The traits that prevent interbreeding and maintain reproductive isolation are called **isolating mechanisms** (**TABLE 17-1**). Isolating mechanisms provide a clear benefit to individuals. An individual that breeds with a member of another species will probably produce no offspring (or offspring that are unfit or sterile), thereby wasting its reproductive effort and failing to contribute to future generations. Thus, natural selection favors traits that prevent reproduction across species boundaries.

Premating Isolating Mechanisms Prevent Mating Between Species

Reproductive isolation can be maintained by a variety of mechanisms, but those that prevent mating are especially effective. The mechanisms that prevent mating between species are collectively called **premating isolating mechanisms**.

Members of Different Species May Not Meet

Members of different species cannot mate if they never get near one another. **Geographic isolation** prevents interbreeding between populations that do not come into contact because they live in different, physically separated places (**FIG. 17-3**). However, we cannot determine if geographically separated populations are actually distinct species. Should the physical barrier separating the two populations disappear, the reunited populations might interbreed freely and not be separate species

(a) Kaibab squirrel

(b) Abert's squirrel

FIGURE 17-3 Geographic isolation To determine if these two squirrels are members of different species, we must know if they are "actually or potentially interbreeding." Unfortunately, it is hard to tell, because (a) the Kaibab squirrel lives only on the north rim of the Grand Canyon and (b) the Abert's squirrel lives on the south rim. The two populations are geographically isolated but still quite similar. Have they diverged enough since their separation to become reproductively isolated? Because they remain geographically isolated, we cannot say for sure.

after all. Geographic isolation, therefore, is usually not considered to be a mechanism that maintains reproductive isolation between species. Instead, it is a mechanism that allows new species to form. If populations cannot interbreed after geographic barriers have been eliminated, then other premating isolating mechanisms must have developed.

Different Species May Occupy Different Habitats

Two populations that use different resources may spend time in different habitats within the same general area and thus exhibit **ecological isolation**. White-crowned sparrows and white-throated sparrows, for example, have extensively overlapping geographic ranges. The white-throated sparrow, however, frequents dense thickets, whereas the white-crowned sparrow inhabits fields and meadows, seldom penetrating far into dense growth. The two species may coexist within a few hundred yards of one another and yet seldom meet during the breeding season. A more dramatic example is provided by the more than 300 species of fig wasp (**FIG. 17-4**). In most cases, fig wasps of a given species breed in (and pollinate) the fruits of one particular species of fig, and each fig species hosts only one or two species of pollinating wasp. Thus, fig wasps of different species only rarely encounter one another during breeding, and pollen from one fig species is not ordinarily carried to flowers of a different species.

▲ **FIGURE 17-4 Ecological isolation** This female fig wasp's eggs were fertilized by mating that took place within a fig. She will find another fig of the same species, enter it through a pore, lay eggs, and die. Her offspring will hatch, develop, and mate within the fig. Because each species of fig wasp reproduces only in its own particular fig species, each wasp species is reproductively isolated.

Different Species May Breed at Different Times

Even if two species occupy similar habitats, they cannot mate if they breed at different times, a phenomenon called **temporal isolation** (time-based isolation). For example, colonies of the Caribbean coral *Montastraea franksi* are found in close proximity to colonies of *Montastraea annularis*, but *M. franksi* spawns (releases its sperm and eggs) shortly after sunset, whereas *M. annularis* does not spawn until hours later (**FIG. 17-5**). As a result, the two species do not interbreed.

In plants, the reproductive structures of different species may mature at different times. For example, Bishop pines and Monterey pines grow in the same area near Monterey on the California coast, but the two species release their sperm-containing pollen (and have eggs ready to be fertilized) at different times: The Monterey pine releases pollen in early spring, the Bishop pine in summer. For this reason, the two species do not interbreed under natural conditions.

Different Species May Have Different Courtship Signals

Among animals, elaborate courtship colors and behaviors can prevent mating with members of other species. Signals and behaviors that differ from species to species create **behavioral isolation**. For example, the extravagant plumes and arresting pose of a courting male Raggiana bird of paradise are conspicuous indicators of his species, and there is little chance that females of another species will mate with him by mistake (**FIG. 17-6**). Among frogs, males are often impressively indiscriminate, jumping on every female in sight regardless of the species. Females, however, approach only male frogs that utter the call appropriate to their species. If they do find themselves in an unwanted embrace, they give the "release call," which causes the male to let go. As a result, few **hybrids**—offspring of parents of different species—are produced.

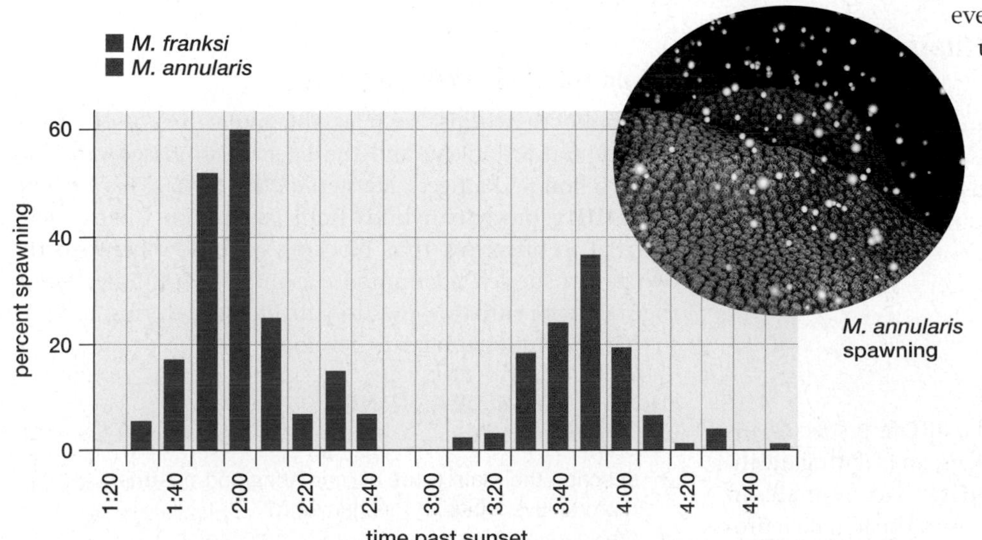

◀ **FIGURE 17-5 Temporal isolation** The coral species *Montastraea franksi* and *Montastraea annularis* inhabit the same Caribbean reefs. The species do not interbreed, however, because each species releases its sperm and eggs into the water at a different time of day.

▲ **FIGURE 17-6 Behavioral isolation** The mate-attraction display of a male Raggiana bird of paradise includes distinctive posture, movements, plumage, and vocalizations that do not resemble those of other bird of paradise species.

Differing Sexual Organs May Foil Mating Attempts

If a male and a female of different species attempt to mate, the attempt may be disrupted by physical differences between the species that are collectively known as **mechanical incompatibilities**. Among animal species with internal fertilization (in which the sperm is deposited inside the female's reproductive tract), the male's and female's sexual organs simply may not fit together. Incompatible body shapes may also make copulation between species impossible. For example, snails of species whose shells have left-handed spirals may be unable to successfully copulate with closely related snails whose shells have right-handed spirals, because the shell mismatch is accompanied by a body orientation mismatch that prevents the genitals of the two species from lining up properly during attempted copulations. Among plants, flowers with different structures may attract different pollinators, thereby preventing pollen transfer between species.

Postmating Isolating Mechanisms Limit Hybrid Offspring

When premating isolating mechanisms fail or have not yet evolved, members of different species may mate. However, **postmating isolating mechanisms** may prevent the formation of vigorous, fertile hybrid offspring, with the result that the two species remain separate, with little or no gene flow between them.

One Species' Sperm May Fail to Fertilize Another Species' Eggs

Even if a male inseminates a female of a different species, his sperm may not be able to fertilize her eggs, an isolating mechanism called **gametic incompatibility**. For example, in animals with internal fertilization, fluids in the female reproductive tract may weaken or kill sperm of other species.

Gametic incompatibility may be an especially important isolating mechanism in species that reproduce by scattering gametes in the water or in the air, such as marine invertebrate animals and wind-pollinated plants. For example, sea urchin sperm cells contain a protein that allows them to bind to eggs. The structure of the protein differs among species so that sperm of one sea urchin species cannot bind to the eggs of another species. In abalones (a type of mollusk), eggs are surrounded by a membrane that can be penetrated only by sperm containing a particular enzyme. Each abalone species has a distinctive version of the enzyme, so hybrids are rare, even though several species of abalones coexist in the same waters and spawn during the same period. Among plants, a similar chemical incompatibility may prevent the germination of pollen from one species that lands on the stigma (pollen-catching structure) of the flower of another species.

Hybrid Offspring May Fail to Survive or Reproduce

If cross-species fertilization does occur, the resulting hybrid may be unable to survive, a situation called **hybrid inviability**. The genetic instructions directing development of the two species may be so different that hybrids abort early in development. For example, captive leopard frogs can be induced to mate with wood frogs, and the matings generally yield fertilized eggs. The resulting embryos, however, inevitably fail to survive more than a few days.

In other animal species, a hybrid might survive to adulthood but fail to reproduce because it exhibits ineffective breeding behavior. Hybrids between certain species of lovebirds, for example, have great difficulty building nests. Members of each parental species inherit a particular behavior for carrying nest material; one species tucks the material under its rump feathers and the other carries it in its beak. Hybrids, however, use a nonfunctional mixture of the two behaviors. They repeatedly attempt to tuck nest material under their feathers, but are unable to do so because they don't release the material from their beaks. Hybrids with such ineffective nest-building behavior probably could not reproduce in the wild.

Hybrid Offspring May Be Infertile

Most animal hybrids, such as the mule (a cross between a horse and a donkey) and the liger (a zoo-based cross between a lion and a tiger), are sterile (**FIG. 17-7**). This **hybrid infertility** prevents hybrids from passing on their genetic material to offspring, thus blocking gene flow between the two parent species. A common reason for hybrid infertility is the failure of chromosomes to pair properly during meiosis, so that eggs and sperm fail to develop.

CHECK YOUR LEARNING

Can you . . .

- describe the main types of premating and postmating reproductive isolating mechanisms?
- provide examples of each type of mechanism?

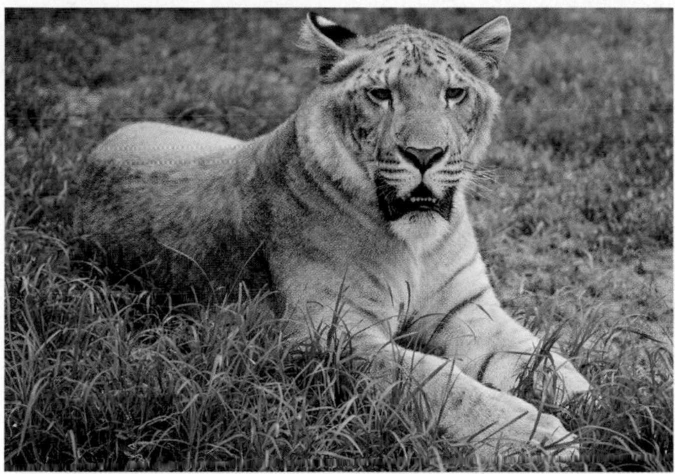

▲ **FIGURE 17-7 Hybrid infertility** This liger, the hybrid offspring of a lion and a tiger, is sterile. The gene pools of its parent species remain separate.

17.3 HOW DO NEW SPECIES FORM?

Despite his exhaustive exploration of the process of natural selection, Charles Darwin did not propose a complete mechanism of **speciation**, the process by which new species form. Today, however, biologists recognize that speciation depends on two processes: isolation and genetic divergence.

- ***Isolation of Populations*** If individuals (or their gametes) move freely between two populations, interbreeding and the resulting gene flow will cause changes in one population to become widespread in the other as well. Thus, two populations cannot grow increasingly different unless something happens to block interbreeding between them. Speciation depends on isolation.
- ***Genetic Divergence of Populations*** It is not sufficient for two populations simply to be isolated. They will become separate species only if, during the period of isolation, they evolve sufficiently large differences. The differences must be large enough that, if the isolated populations were reunited, they could no longer interbreed and produce vigorous, fertile offspring. That is, speciation is complete only if divergence results in evolution of an isolating mechanism. Such differences can arise by chance (genetic drift), especially if at least one of the isolated populations is small (see Chapter 16). Large genetic differences can also arise through natural selection if the isolated populations experience different environmental conditions.

Speciation always requires isolation followed by divergence, but these steps can take place in several different ways. Evolutionary biologists group the different pathways to speciation into two broad categories: **allopatric speciation**, in which two populations are geographically separated from one another, and **sympatric speciation**, in which two populations share the same geographic area. (To learn more about how scientists study the outcome of speciation, see "Doing Science: Seeking the Secrets of the Sea" on page 272.)

1 Part of a mainland population reaches an isolated island.
many generations pass
2 Over many generations, the isolated populations begin to diverge due to genetic drift and natural selection.
many generations pass
3 Divergence may eventually become sufficient to cause reproductive isolation.

▲ **FIGURE 17-8 Allopatric isolation and divergence** In allopatric speciation, some event causes a population to be divided by an impassable geographic barrier. One way the division can occur is by colonization of an isolated island. The two now-separated populations may diverge genetically. If the genetic differences between the two populations become large enough to prevent interbreeding, then the two populations constitute separate species.

THINK CRITICALLY Make a list of events or processes that could cause geographic subdivision of a population. Do you think the items on your list are sufficient to account for the formation of the millions of species that have inhabited Earth?

Geographic Separation of a Population Can Lead to Allopatric Speciation

New species can arise by allopatric speciation when an impassible barrier physically separates different parts of a population.

Organisms May Colonize Isolated Habitats

A small population can become isolated if it moves to a new location (**FIG. 17-8**). For example, some members of a

DOING Science Seeking the Secrets of the Sea

In addition to studying how new species arise, biologists also investigate the outcome of millennia of speciation: life's current diversity of species. One ambitious effort to increase our understanding of life's diversity is the Census of Marine Life.

What Question Was Asked?

The Census of Marine Life was a huge collaborative effort to systematically explore the least explored part of Earth: its oceans. The census project was not designed to test a specific hypothesis, but instead had a broad goal: to "assess and explain the diversity, distribution, and abundance of marine life." The project lasted 10 years and involved 2,700 scientists from more than 80 countries (**Fig. E17-1**).

▲ **FIGURE E17-1 Investigating marine biodiversity** A researcher with the Census of Marine Life uses a light box to examine organisms in a shallow bay.

How Was Evidence Gathered?

The scientists conducted 540 expeditions at a cost of $650 million. The expeditions spanned the globe from the tropics to polar regions, explored coastal waters and the open ocean, and examined life from the water's surface down to its deepest depths. The census scientists studied a huge range of organisms—from microbes to whales—counting them, tracking their movements (**Fig. E17-2**), mapping their locations, analyzing their DNA, and bringing many back to the lab for further study.

What Was Learned?

The census yielded a massive amount of information. The scientists found living organisms in every habitat that they explored, including ocean depths that lack oxygen. They documented previously unknown animal migration routes and compiled millions of records of organisms' locations and abundance that can serve as a baseline for tracking the effects of human activities on marine organisms. The census also identified "hotspots" in the ocean where life is especially abundant and discovered more than 6,000 new species. Overall, the Census of Marine Life dramatically increased our knowledge of life in the oceans and demonstrated the value of large-scale, coordinated scientific investigation of biodiversity.

THINK CRITICALLY How might conservation scientists use the map shown in FIG. E17-2 to help choose the best location for a proposed marine reserve?

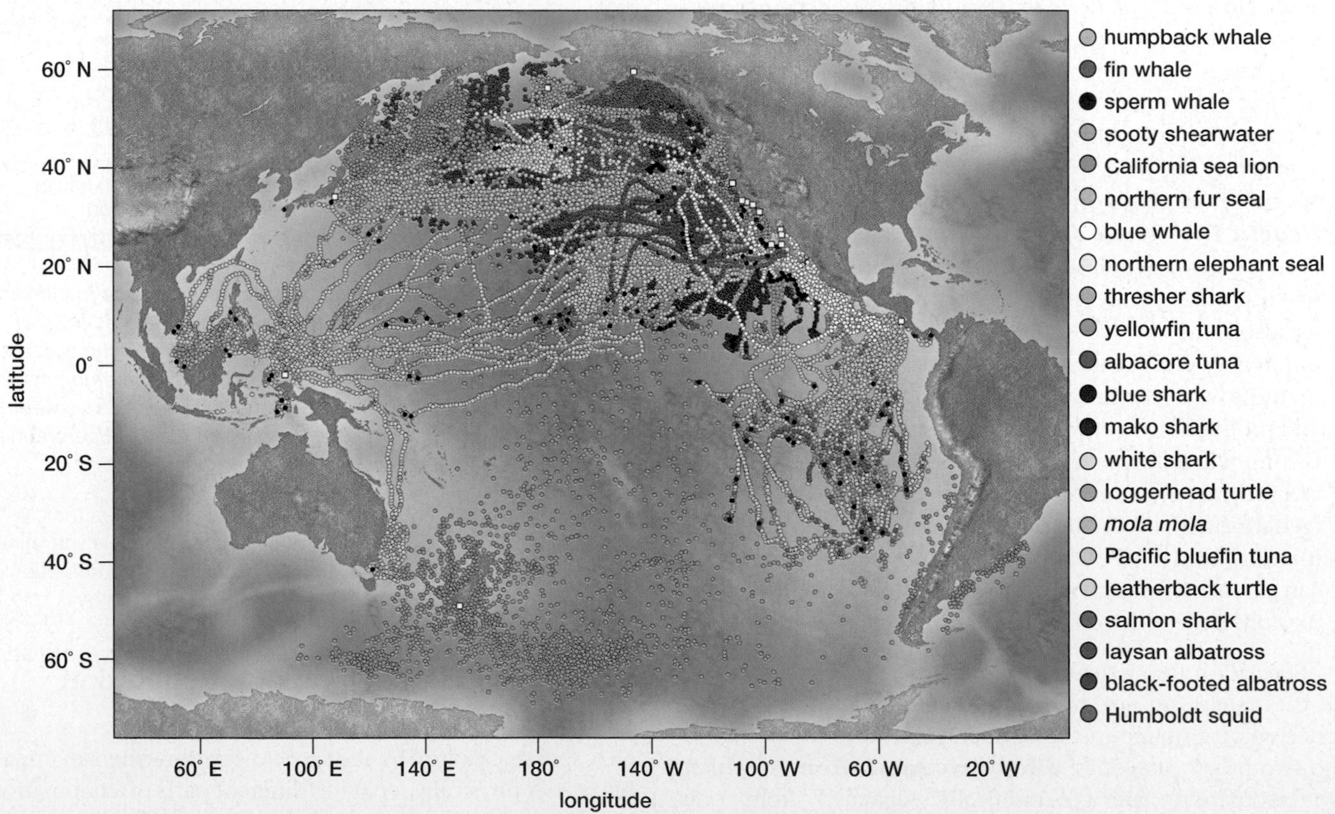

▲ **FIGURE E17-2 Tracking species** To understand the movements and geographic distribution of species, census researchers attached tracking tags to 22 species of top predators in the Pacific Ocean. The tags beamed signals to satellites, which relayed them to the scientists' computers. The resulting data were plotted on maps, revealing a wealth of information about where and when the animals move. Block et al. 2011. *Nature* 475:86–90.

population of land-dwelling organisms might colonize an oceanic island. The colonists might be birds, flying insects, fungal spores, or wind-borne seeds blown by a storm. More earthbound organisms might reach the island on a drifting raft of vegetation torn from the mainland coast. Whatever the means, we know that such colonization occurs regularly, given the presence of living things on even the most remote islands.

Isolation by colonization is not limited to islands. For example, different coral reefs may be separated by miles of open ocean, so any reef-dwelling sponges, fishes, or algae that were carried by ocean currents to a distant reef would be effectively isolated from their original populations. Any habitat that is bounded by a large expanse of a very different habitat can isolate arriving colonists.

CASE STUDY CONTINUED

Discovering Diversity

It is not surprising that the forests of New Guinea are home to a variety of distinctive species like the miniature frog *Paedophryne amauensis*. New Guinea is, after all, an island. It is likely that, in the past, populations colonized the island and became genetically isolated from mainland populations, thereby initiating the process of speciation. But what about non-island species, such as the saola, the olinguito (**FIG. 17-9**), and the other unique species of the Annamite and Andes Mountains? How might populations inhabiting mainland forests in Vietnam or Ecuador have become isolated from other populations?

(a) Saola

(b) Olinguito

▲ **FIGURE 17-9 Hidden mammals** Discoveries of previously unknown mammal species are uncommon, but remote forests can conceal species like the recently discovered **(a)** saola and **(b)** olinguito.

Geological and Climate Changes May Divide Populations

Isolation can also result from landscape changes that divide a population. For example, rising sea levels might transform a coastal hilltop into an island, isolating the residents. New rock from a volcanic eruption can divide a previously continuous sea or lake, splitting populations. A river that changes course can also divide populations, as can a newly formed mountain range. Climate shifts, such as those that happened in past ice ages, can change the distribution of vegetation and strand portions of populations in isolated patches of suitable habitat.

Throughout the history of Earth, many populations have been divided by continental drift. Earth's continents float on molten rock and slowly move about the surface of the planet. On a number of occasions during Earth's long history, continental landmasses have broken into pieces that subsequently moved apart (see Fig. 18-11). Each of these breakups must have split a multitude of populations.

Natural Selection and Genetic Drift May Cause Isolated Populations to Diverge

If two populations become geographically isolated for any reason, there will be no gene flow between them. If the environments of the locations differ, then natural selection may favor different traits in the different locations, and the populations may accumulate genetic differences. Alternatively, genetic differences may arise if one or more of the separated populations is small enough that substantial genetic drift occurs, which may be especially likely in the aftermath of a founder event in which a few individuals become isolated from the main body of the species. In either case, genetic differences between the separated populations may eventually become large enough to make interbreeding impossible. At that point, the two populations will have become separate species. Most evolutionary biologists believe that geographic isolation followed by allopatric speciation has been the most common source of new species, especially among animals.

Genetic Isolation Without Geographic Separation Can Lead to Sympatric Speciation

Genetic isolation—limited gene flow—is required for speciation, but populations can become genetically isolated without geographic separation. Thus, new species can arise by sympatric speciation.

Ecological Isolation Can Reduce Gene Flow

If a geographic area contains two distinct types of habitats (each with distinct food sources, places to raise young, and so on), different members of a single species may begin to specialize in one habitat or the other. If conditions are right, natural selection in the two different habitats may lead to the evolution of different traits in the two groups. Eventually, these differences may become large enough to prevent members of the two groups from interbreeding, and the formerly single species will have split into two species.

Such a split seems to be taking place right before biologists' eyes, so to speak, in the case of the fruit fly *Rhagoletis pomonella*. *Rhagoletis* is a parasite of the American hawthorn tree. This fly lays its eggs in the hawthorn's fruit; when the maggots hatch, they eat the fruit. About 150 years ago, scientists noticed that *Rhagoletis* had begun to infest apple trees, which were introduced into North America from Europe. Today, it appears that *Rhagoletis* is splitting into two species—one that breeds on apples, and one that breeds on hawthorns (**FIG. 17-10**). The two groups have evolved substantial genetic differences, some of which—such as those that affect the time of year at which adult flies emerge and begin to mate—are important for survival on a particular host plant.

The two kinds of flies will become two species only if they maintain reproductive separation. Apple trees and hawthorns typically grow in the same areas, and flies, after all, can fly. So why don't apple flies and hawthorn flies interbreed and cancel out any genetic differences between them? First, female flies usually lay their eggs in the same type of fruit in which they developed. Males also tend to prefer the same type of fruit in which they developed. Therefore, apple-liking males will encounter and mate with apple-liking females. Second, apples mature 2 to 3 weeks later than do hawthorn fruits, and the two types of flies emerge with timing appropriate for their chosen host fruit. Thus, the two varieties of flies have very little chance of meeting. Although some interbreeding between the two types of flies occurs, they seem to be well on their way to speciation. Will they make it? Entomologist Guy Bush suggests, "Check back with me in a few thousand years."

1 Part of a fly population that lives only on hawthorn trees moves to an apple tree.
many generations pass
2 If flies living on the apple tree rarely encounter flies living on the hawthorn tree, the populations may diverge over many generations.

▲ **FIGURE 17-10 Sympatric isolation and divergence** In sympatric speciation, some event blocks gene flow between two parts of a population that remains in a single geographic area. One way in which this genetic isolation can occur is if a portion of a population begins to use a previously unexploited resource, such as when some members of an insect population shift to a new host plant species (as has occurred in the fruit fly species *Rhagoletis pomonella*). The two now-isolated populations may diverge genetically. If the genetic differences between the two populations become large enough to prevent interbreeding, then the two populations constitute separate species.

THINK CRITICALLY How might future scientists test whether *R. pomonella* has become two species?

Mutations Can Lead to Genetic Isolation

In some instances, new species can arise nearly instantaneously as a result of mutations that change the number of chromosomes in an organism's cells. The acquisition of multiple copies of each chromosome is known as **polyploidy** and has been a frequent cause of sympatric speciation. In general, polyploid individuals cannot mate successfully with normal diploid individuals. Thus, a polyploid mutant is genetically isolated from its parent species. If, however, it somehow reproduces and leaves offspring, its descendants may form a new, reproductively isolated species.

Polyploid plants are more likely than polyploid animals to be able to reproduce, so speciation by polyploidy is more common in plants than in animals. Unlike animals, many plants can either self-fertilize or reproduce asexually, or both. Therefore, a polyploid plant is much more likely than a polyploid animal to become the founding member of a new, polyploid species.

Under Some Conditions, Many New Species May Arise

In the same way that the history of your family can be represented by a family tree, the history of life can be represented by an *evolutionary tree*. The base of the evolutionary tree of life represents Earth's earliest organisms, and each of the endmost branches represents one of today's living species. Each fork in the branches represents a speciation event, when one species split into two. Hypotheses and discoveries about the

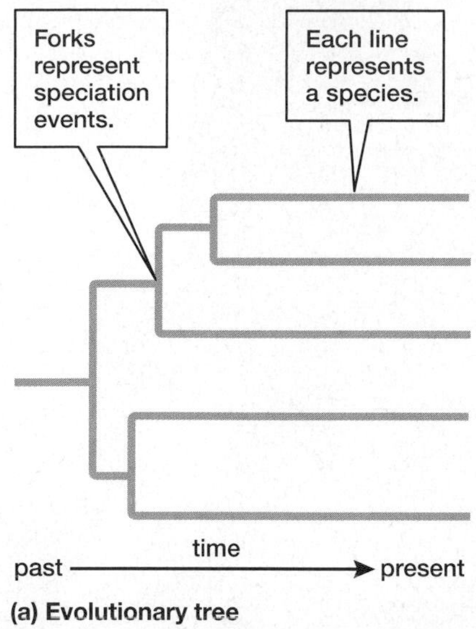

Forks represent speciation events.
Each line represents a species.
time
past
present

(a) Evolutionary tree

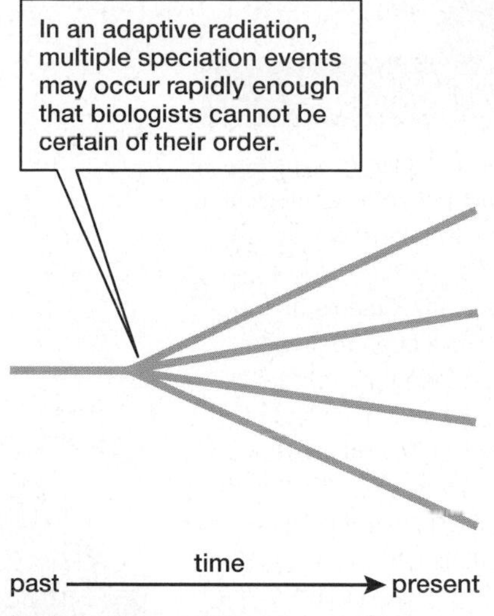

In an adaptive radiation, multiple speciation events may occur rapidly enough that biologists cannot be certain of their order.
time
past
present

(b) Evolutionary tree representing adaptive radiation

FIGURE 17-11 Interpreting evolutionary trees Evolutionary history is often represented by **(a)** an evolutionary tree, a graph in which the horizontal axis plots time. In **(b)**, an evolutionary tree representing an adaptive radiation, many lines may branch from a single point. This pattern reflects biologists' uncertainty about the order in which the multiple speciation events of the radiation took place. With more research, it may be possible to replace the "starburst" pattern with a more informative tree.

THINK CRITICALLY Imagine that for two neighboring branches on an evolutionary tree, one species is found only on an island distant from the mainland and the other species is found only on the mainland. How do you think the two species arose?

evolutionary relationships among species are often communicated by depictions of a portion of life's evolutionary tree (**FIG. 17-11a**).

In some cases, many new species have arisen in a relatively short time (**FIG. 17-11b**). This process, called **adaptive radiation**, can occur when populations of one species invade a variety of new habitats and evolve in response to the differing environmental pressures in those habitats.

Adaptive radiation has occurred many times and in many groups of organisms, typically when species encounter a wide variety of unoccupied habitats. For example, episodes of adaptive radiation took place when some wayward finches colonized the Galápagos Islands, when a population of cichlid fish reached isolated Lake Malawi in Africa, and when an ancestral silversword plant species arrived at the Hawaiian Islands (**FIG. 17-12**). These events gave rise to adaptive radiations of 13 species of Darwin's finches in the Galápagos, more than 300 species of cichlids in Lake Malawi, and 30 species of silversword plants in Hawaii. In these examples, the invading species faced no competitors except other members of their own species, and all the available habitats were rapidly exploited by new species that evolved from the original invaders.

CHECK YOUR LEARNING

Can you . . .

- describe the two general steps that are required for a new species to arise?
- explain the difference between allopatric and sympatric speciation and describe each process?
- explain adaptive radiation and describe the process by which it might arise?
- interpret an evolutionary tree diagram?

(a) Ahinahina

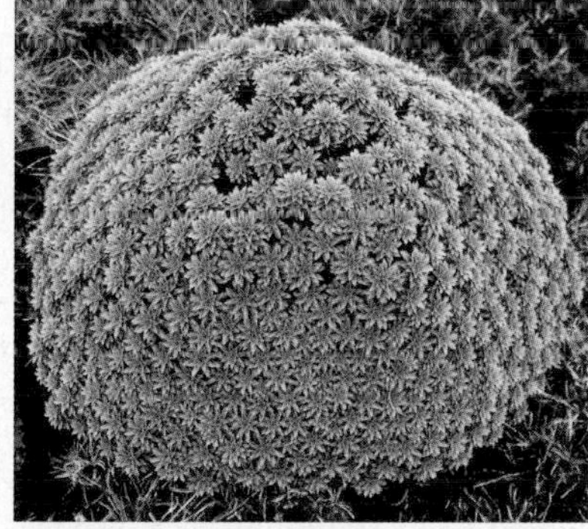

(b) Waialeale dubautia

(c) Kupaoa

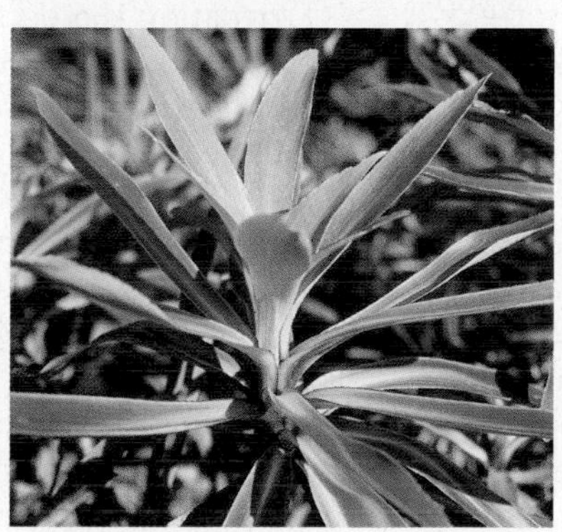

(d) Na'ena'e 'ula

FIGURE 17-12 Adaptive radiation About 30 species of silversword plants inhabit the Hawaiian Islands. These species are found nowhere else, and all of them descended from a single ancestral population within a few million years. This adaptive radiation has led to a collection of closely related species of diverse form and appearance, with an array of adaptations for exploiting the many different habitats in Hawaii, from warm, moist rain forests to cool, barren volcanic mountaintops.

THINK CRITICALLY Why do you suppose there are so many *endemic* species—that is, species found nowhere else—on islands? Why have the overwhelming majority of recent extinctions occurred on islands?

CASE STUDY CONTINUED

Discovering Diversity

One possible explanation for the distinctive collection of species found in the Annamite and Andes Mountains lies in the geological history of these regions. During the ice ages that have occurred repeatedly during the past million years or so, the area covered by tropical forests must have shrunk dramatically. Organisms that depended on the forests for survival would have been restricted to any remaining "islands" of forest, isolated from their fellows in other, distant patches of forest. As we have learned, this kind of isolation can set the stage for allopatric speciation and may have created the conditions that gave rise to the saola, the olinguito, and other unique denizens of tropical forests. Once a new species arises, can we expect it to persist indefinitely?

17.4 WHAT CAUSES EXTINCTION?

The ultimate fate of any species is **extinction**, the death of all of its members. In fact, at least 99.9% of all the species that have ever existed are now extinct. The natural course of evolution, as revealed by fossils, is continual turnover of species as new ones arise and old ones become extinct.

The immediate cause of extinction is probably always environmental change, in either the nonliving or the living parts of the environment. Environmental changes that can lead to extinction include habitat destruction and increased competition among species. In the face of such changes, species with small geographic ranges or highly specialized adaptations are especially susceptible to extinction.

Localized Distribution Makes Species Vulnerable

Species vary widely in their range of distribution and, hence, in their vulnerability to extinction. Some species, such as herring gulls, white-tailed deer, and humans, inhabit entire continents or even the whole Earth; others, such as the Devil's Hole pupfish, which is found in only one spring-fed water hole in the Nevada desert, have extremely limited ranges. Obviously, if a species inhabits only a very small area, any disturbance of that area could easily result in extinction. If Devil's Hole dries up due to a drought or well drilling nearby, its pupfish will immediately vanish. Conversely, wide-ranging species will not succumb to local environmental catastrophes.

Specialization Increases the Risk of Extinction

Another factor that may make a species vulnerable to extinction is extreme specialization. Each species evolves adaptations that help it survive and reproduce in its environment. In some cases, these adaptations include specializations that favor survival in a particular and limited set of environmental conditions. The Karner blue butterfly, for example, feeds only on the blue lupine plant (**FIG. 17-13**). The butterfly is therefore found only where the plant thrives. But the blue lupine has become quite rare because farms and development have largely replaced its habitat of sandy, open woods and clearings in northeast North America. If the lupine disappears, the Karner blue butterfly will surely become extinct along with it (see "Earth Watch: Why Preserve Biodiversity?").

▲ **FIGURE 17-13 Extreme specialization places species at risk** The Karner blue butterfly feeds exclusively on the blue lupine, found in dry forests and clearings in the northeastern United States. Such behavioral specialization renders the butterfly extremely vulnerable to any environmental change that may exterminate its single host plant species.

THINK CRITICALLY If specialization puts a species at risk for extinction, how could this hazardous trait have evolved?

Interactions with Other Species May Drive a Species to Extinction

Interactions such as predation and competition serve as agents of natural selection (see Chapter 16). In some cases, these same interactions can lead to extinction rather than to adaptation.

Predation is especially likely to contribute to extinction when species encounter predators to which they had not previously been exposed. For example, when the predatory brown tree snake was accidentally introduced to the Pacific island of Guam in the 1940s, bird populations began to decline. No predatory snakes had been previously present on Guam, so the island's birds had evolved no defenses against them. Within a few decades, almost all of Guam's native birds had disappeared, including two species that were found only on Guam, and are therefore now extinct.

Increased competition can also contribute to extinction. A possible example of extinction through competition began about 3 million years ago, when the isthmus of Panama rose

Why Preserve Biodiversity?

Today, most extinctions occur in the tropics, where the vast majority of species live. The main cause of these extinctions is environmental change, especially habitat destruction. Unfortunately, tropical habitats are being rapidly destroyed and disrupted by human activities (**FIG. E17-3**). For example, a recent study based on analysis of high-resolution satellite photos estimated that worldwide tropical rainforest cover decreased by about 50,000 square miles per year between 1990 and 2010. Although the rate of loss is high everywhere, it has recently slowed in some countries (**FIG. E17-4**). Most of the lost forest was destroyed by logging or clearing land for agriculture. Similarly, a worldwide survey of coral reefs revealed that about 20% of Earth's reef area has already been destroyed and an additional 20% is severely damaged, again, mostly as the result of human influences such as pollution.

The rapid destruction of habitats in the tropics is causing many species to go extinct, as their homes disappear. Recent ecological research suggests that the current rate of extinction is extremely high, perhaps higher than ever before in the history of life on Earth. Does it matter? Is there any reason for us to try to slow the loss of biodiversity?

One reason to protect Earth's biodiversity is that our ecological self-interest may be at stake. For example, Earth's species form communities, highly complex webs of interdependent life-forms whose interactions sustain one another. These communities play a crucial role in processes that purify the air we breathe and the water we drink, build the rich topsoil in which we grow our crops, provide the bounty of food that we harvest from the oceans, and decompose and detoxify our waste. We depend entirely on these "ecosystem services." When our activities cause species to disappear from communities, we take a big risk. If we remove too many species, or remove some especially crucial species, we may disrupt the finely tuned processes of the community and undermine its ability to sustain us.

▲ **FIGURE E17-3 Biodiversity threatened** Destruction of tropical rain forests by indiscriminate logging threatens Earth's greatest storehouse of biological diversity.

CONSIDER THIS The precarious state of rare species around the world poses profound ethical dilemmas. For example, in many cases, the habitat destruction that endangers some species also helps people by making space for farmland, housing, and workplaces needed by a growing human population. How can we reconcile the conflict between valid human needs and the needs of endangered species? Furthermore, it is becoming clear that, even with the best of intentions, we cannot save all of the species currently threatened with extinction. The resources available to preserve and manage protected habitats are limited, and we must make choices that will cause some species to survive and others to perish. Who should decide which species will live and which will die? On what criteria should such decisions be based?

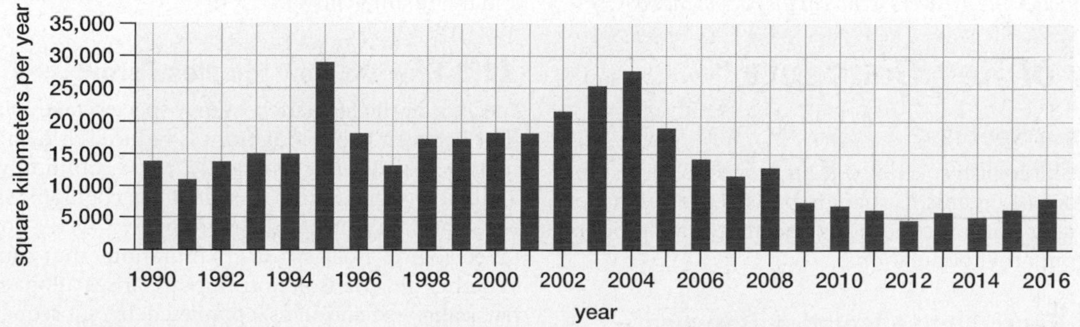

▲ **FIGURE E17-4 Yearly deforestation in the Brazilian Amazon** Earth and its forests are vast, which makes accurate measurement of deforestation difficult. One of the best tools for this task is satellite imagery; photos from space can be used to compare forest cover from year to year. Brazil's National Institute for Space Research has used this tool to track the country's loss of rain forest. The resulting data, displayed in this graph, show that, despite a recent period of slowing deforestation, the rate of forest destruction is now trending upward. An area in Brazil larger than the state of Delaware was cleared in 2016.

above sea level and formed a land bridge between North America and South America. After the previously separated continents were connected, the mammal species that had evolved in isolation on each continent were able to mix. Ul-timately, the North American species that moved south diversified and underwent an adaptive radiation that displaced the vast majority of the South American species, many of which went extinct. Although the reasons for the extinctions are not completely understood, it is likely that competition played a role; the species arriving from North America could exploit resources more efficiently than could their South American counterparts.

Habitat Change and Destruction Are the Leading Causes of Extinction

Habitat change, both contemporary and prehistoric, is the single greatest cause of extinctions. Present-day habitat destruction due to human activities is proceeding at a rapid pace. Many biologists believe that we are presently in the midst of the fastest-paced and most widespread episode of species extinction in the history of life. Loss of tropical forests is especially devastating to species diversity. As many as half the species presently on Earth may be lost during the next 50 years as the tropical forests that contain them are cut for timber or to clear land for cattle and crops. (In Chapter 18, we will discuss extinctions due to prehistoric habitat change.)

CHECK YOUR LEARNING

Can you . . .

- describe the main causes of extinction?
- describe some examples of living species that are at risk of extinction?

CASE STUDY REVISITED

Discovering Diversity

Ironically, recent discoveries of previously unknown species come at a time when the remote forests that host many of them are in danger of disappearing. Economic development has brought logging and mining to ever more remote regions, and forests in New Guinea (home of *Paedophryne amauensis*), Vietnam (home of the saola), and many other developing nations are being cleared at an unprecedented rate. As a result, newly discovered species are often very rare. For example, despite intensive searching by biologists, there have been only two verified observations of a live saola in the past 25 years: a photo taken by an unattended wildlife camera in 1999 and a second photo in 2013.

THINK CRITICALLY Given that genetic isolation is the first step in speciation, could human activities that reduce many species to small, isolated populations actually increase biodiversity by creating conditions that lead to the formation of new species? Why or why not?

CHAPTER REVIEW

Go to Mastering Biology for practice quizzes, activities, eText, videos, current events, and more.

*Answers to **Think Critically, and Thinking Through the Concepts** questions can be found in the **Answers** section at the back of the book.*

Summary of Key Concepts

17.1 What Is a Species?

According to the biological species concept, a species consists of all the populations of organisms that are potentially capable of interbreeding under natural conditions and that are reproductively isolated from other populations.

17.2 How Is Reproductive Isolation Between Species Maintained?

Reproductive isolation between species may be maintained by one or more of several mechanisms, collectively known as premating isolating mechanisms and postmating isolating mechanisms. Premating isolating mechanisms include geographic isolation, ecological isolation, temporal isolation, behavioral isolation, and mechanical incompatibility. Postmating isolating mechanisms include gametic incompatibility, hybrid inviability, and hybrid infertility.

17.3 How Do New Species Form?

Speciation, the formation of new species, takes place when gene flow between two populations is reduced or eliminated and the populations diverge genetically. Most commonly, speciation is allopatric—gene flow is restricted by geographic isolation. However, speciation can also be sympatric—gene flow is restricted by ecological isolation or by mutations that cause polyploidy. Whether genetic isolation initially arises allopatrically or sympatrically, speciation is completed by subsequent genetic divergence of the separated populations through genetic drift or natural selection.

17.4 What Causes Extinction?

Factors that cause extinctions include competition among species and habitat destruction. Localized distribution and extreme specialization increase a species' vulnerability to extinction.

Thinking Through the Concepts

Bloom's: Remembering, Understanding

Multiple Choice

1. The biological species concept is difficult or impossible to apply to
   a. asexually reproducing organisms.
   b. large organisms.
   c. rapidly evolving organisms.
   d. plants.
2. Which of the following does *not* describe a premating isolating mechanism?
   a. the courtship display of a bird of paradise
   b. the sterility of the offspring of a horse and a donkey
   c. the difference between the flowering periods of the Monterey pine and the Bishop pine
   d. the tendency of each species of fig wasp to breed only in the fruits of a particular species of fig
3. All instances of speciation require
   a. genetic isolation and divergence.
   b. genetic drift.
   c. geographic subdivision of a population.
   d. adaptive radiation.
4. Analysis of *Rhagoletis* fly populations in North America provides evidence that if some members of an insect population shift to a new host species,
   a. sympatric speciation may eventually result.
   b. adaptive radiation is likely.
   c. little or no subsequent genetic divergence is likely.
   d. sympatric speciation is probably impossible in insects.
5. In the initial phase of allopatric speciation, gene flow between parts of a population is inhibited by
   a. competition.
   b. hybrid inviability.
   c. geographic isolation.
   d. ecological isolation.

Fill-in-the-Blank

1. A species is a group of ____________ that evolves ____________. The biological species concept identifies species on the basis of their ____________. The biological species concept cannot be applied to species that reproduce ____________.
2. Fill in the following with the appropriate isolating mechanism: Occurs when members of two populations have different courtship behaviors: ____________; occurs when hybrid offspring fail to survive to reproduce: ____________; occurs when members of two populations have different breeding seasons: ____________; occurs when sperm from one species fails to fertilize the eggs of another species: ____________; occurs when the sexual organs of two species are incompatible: ____________.
3. Formation of a new species occurs when two populations of an existing species first become and then ____________. The process in which geographic separation of parts of a population leads to the formation of new species is called ____________. Isolated populations may diverge through the action of ____________ or ____________.
4. The process by which many new species arise in a relatively short period of time is known as ____________. This process often occurs when a species arrives in a previously unoccupied ____________.
5. A species may be at higher risk of extinction if its geographic range includes a(n) ____________ area, or if its food or habitat requirements are ____________. The leading direct cause of extinction is ____________.

Review Questions

1. Define the following terms: *species*, *speciation*, *allopatric speciation*, and *sympatric speciation*. Explain how allopatric and sympatric speciation might work, and give a hypothetical example of each.
2. Many of the oak tree species in central and eastern North America hybridize (interbreed). Are they "true species"?
3. Review the material on the possibility of sympatric speciation in *Rhagoletis* flies. What types of genotypic, phenotypic, or behavioral data would convince you that the two forms have become separate species?
4. A drug called colchicine prevents cell division after the chromosomes have doubled at the start of meiosis. Describe how you would use colchicine to produce a new polyploid plant species.
5. What are the two major types of reproductive isolating mechanisms? Give examples of each type, and describe how they work.

Applying the Concepts

Bloom's: Applying, Analyzing, Evaluating

1. It is difficult to perform experiments that test hypotheses about how new species form. But what if people lived for a really long time? Design an experiment lasting 100,000 years to test whether allopatric separation leads to speciation. What would your study organism be? Why? What would you measure, how often would you measure it, and what would you expect to find if the allopatric speciation hypothesis is correct?

18 The History of Life

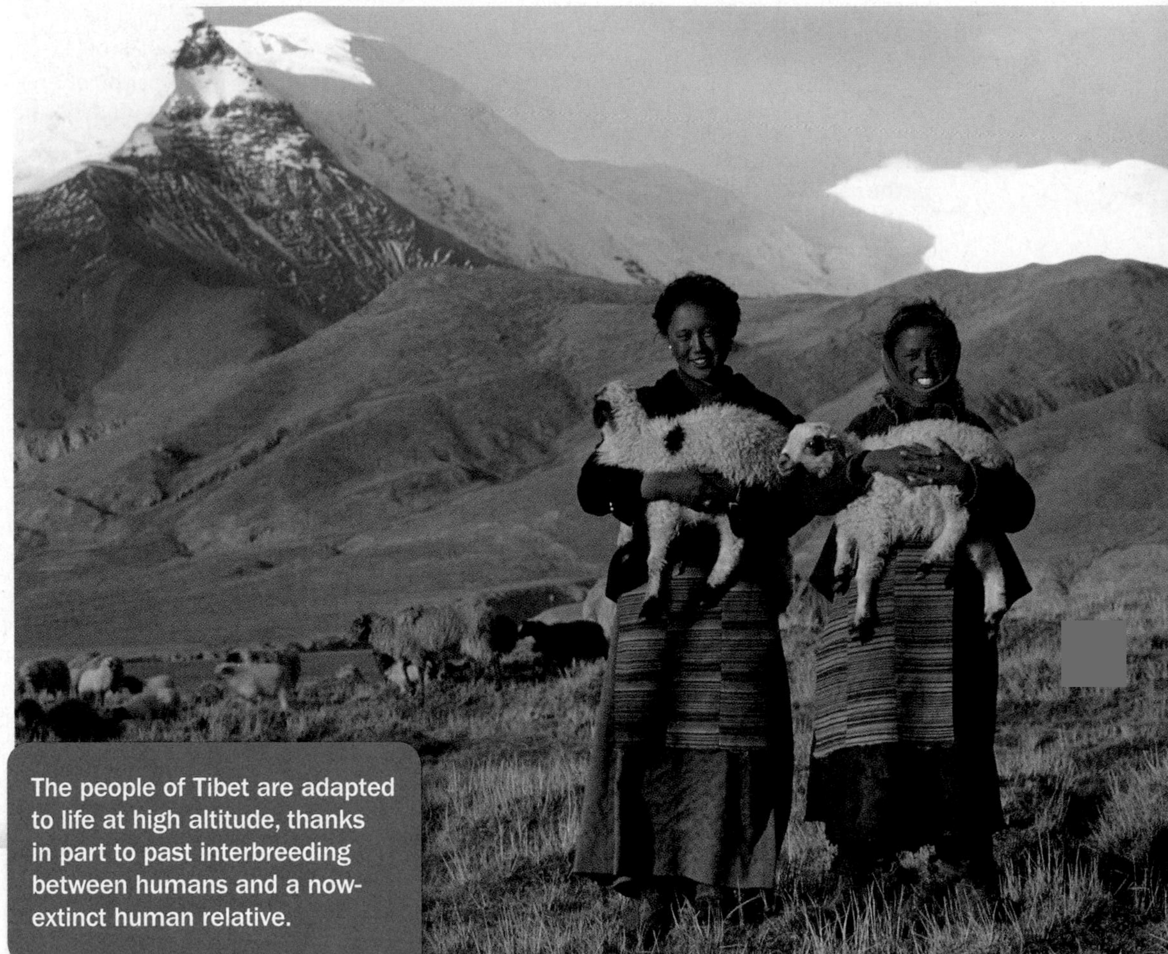

The people of Tibet are adapted to life at high altitude, thanks in part to past interbreeding between humans and a now-extinct human relative.

CASE STUDY

Ancient DNA Has Stories to Tell

THE PEOPLE OF TIBET LIVE AND WORK at altitudes higher than 13,000 feet, where there is much less oxygen in the air than at lower elevations. If you are not a Tibetan and you tried to live at that altitude, you would probably get sick, or at least tire easily and have a hard time catching your breath. How do the Tibetans manage? Their bodies have a number of adaptations to life at high altitude, including a special variant of a gene known as *EPAS1*. The Tibetan's version of *EPAS1*, which is not found in other human populations, improves their bodies' ability to function efficiently in low-oxygen conditions.

How did Tibetans come to have their special version of the gene? Did it originate as a lucky mutation in the early human inhabitants of Tibet? Apparently not. Researchers recently discovered that the gene variant has a surprising history: Its appearance in Tibetans is the result of past interbreeding with members of a hominin species, known as the Denisovans, that has been extinct for tens of thousands of years. (The word *hominin* describes the group that includes humans and the extinct species that are our closest relatives.)

How did researchers discover this fascinating bit of evolutionary history? Until recently, its discovery would have been impossible. But researchers have learned how to extract and sequence DNA from the ancient remains of extinct organisms. And when they examined the Denisovan genome, they found that it contained the same variant of *EPAS1* that today is found only in Tibetans.

Until recently, our knowledge of life's history came only from fossils and the characteristics of living organisms, including their DNA. But the fossil record can be spotty, and modern DNA provides only indirect inferences about past organisms, rather than direct evidence. These tools have nonetheless provided a tremendous amount of information about the past, but access to ancient DNA opens a new and exciting window to the later chapters of life's history, including the history of humans.

Aside from the surprising source of the Tibetans' adaptation, what else have we learned from ancient DNA? What other parts of life's story have been pieced together from DNA clues?

AT A GLANCE

18.1 HOW DID LIFE BEGIN?

Before Darwin, most people thought that all species were simultaneously created by God a few thousand years ago. Further, until the nineteenth century most people thought that new members of existing species sprang up all the time, through **spontaneous generation** from both nonliving matter and other, unrelated forms of life. Microorganisms were thought to arise spontaneously from broth, maggots from meat, and mice from mixtures of sweaty shirts and wheat.

In 1668, the Italian physician Francesco Redi disproved the maggots-from-meat hypothesis simply by keeping flies (whose eggs hatch into maggots) away from uncontaminated meat (see "Doing Science: Controlled Experiments Provide Reliable Data" in Chapter 1). In the mid-1800s, Louis Pasteur in France and John Tyndall in England disproved the broth-to-microorganism idea by showing that microorganisms did not appear in sterile broth unless the broth was first exposed to existing microorganisms in the surrounding environment (**FIG. 18-1**). Although Pasteur's and Tyndall's work effectively demolished the notion of spontaneous generation, it did not address the question of how life on Earth originated in the first place. Or, as the biochemist Stanley Miller put it, "Pasteur never proved it didn't happen once; he only showed that it doesn't happen all the time."

The First Living Things Arose from Nonliving Ones

Modern scientific ideas about the origin of life began to emerge in the 1920s, when Alexander Oparin in Russia and J. B. S. Haldane in England noted that today's oxygen-rich atmosphere would not have permitted the spontaneous formation of the complex organic molecules necessary for life. Oxygen reacts readily with other molecules, disrupting chemical bonds. Thus, an oxygen-rich environment tends to keep molecules simple.

Oparin and Haldane speculated that the atmosphere of the young Earth must have contained very little oxygen and that, under such atmospheric conditions, complex organic molecules could have arisen through ordinary chemical reactions. Some kinds of molecules could persist in the lifeless environment of early Earth better than others and would therefore become more common over time. This chemical version of the "survival of the fittest" is called *prebiotic* (meaning "before life") evolution. In the scenario envisioned by Oparin and Haldane, prebiotic chemical evolution gave rise to progressively more complex molecules and eventually to living organisms.

Organic Molecules Can Form Spontaneously Under Prebiotic Conditions

Inspired by the ideas of Oparin and Haldane, Stanley Miller and Harold Urey set out in 1953 to simulate prebiotic evolution

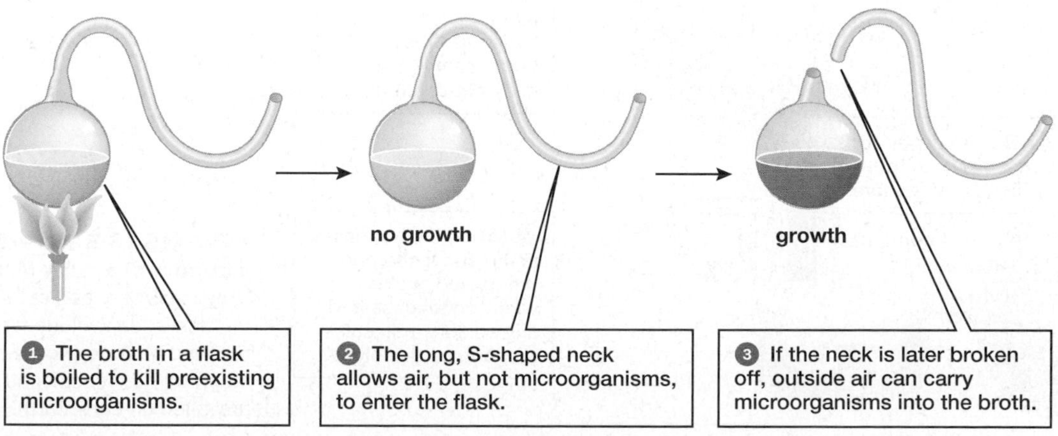

no growth
growth
❶ The broth in a flask is boiled to kill preexisting microorganisms.
❷ The long, S-shaped neck allows air, but not microorganisms, to enter the flask.
❸ If the neck is later broken off, outside air can carry microorganisms into the broth.

▲ **FIGURE 18-1 Spontaneous generation refuted** Louis Pasteur's experiment disproved the spontaneous generation of microorganisms in broth.

in the laboratory. They knew that, on the basis of the chemical composition of the rocks that formed early in Earth's history, geochemists had concluded that the early atmosphere probably contained virtually no oxygen gas, but did contain methane (CH_4), ammonia (NH_3), hydrogen (H_2), and water vapor (H_2O). Miller and Urey simulated the oxygen-free atmosphere of early Earth by mixing these components in a flask. Electrical sparks mimicked the intense energy of early Earth's lightning storms. In this experimental microcosm, the researchers found that simple organic molecules appeared after just a few days (**FIG. 18-2**). The experiment showed that small molecules likely present in the early atmosphere can combine to form larger organic molecules if electrical energy is present. (Recall from Chapter 6 that reactions that synthesize biological molecules from smaller ones are endergonic—they consume energy.) Similar experiments by Miller and others produced amino acids, peptides, nucleotides, adenosine triphosphate (ATP), and other molecules characteristic of living things.

In recent years, new evidence has convinced most geochemists that the actual composition of Earth's early atmosphere probably differed from the mixture of gases used in the pioneering Miller–Urey experiment. However, more recent experiments with simulated atmospheres that more closely resembled the probable atmosphere of early Earth have also yielded organic molecules. In addition, these experiments have shown that electricity is not the only suitable energy source. Other energy sources that were available on early Earth, such as heat or ultraviolet (UV) light, can also drive the formation of organic molecules in experimental simulations of prebiotic conditions. Thus, even though we may never know exactly what the earliest atmosphere was like, we can be confident that organic molecules formed on early Earth.

Additional organic molecules probably arrived from space when meteorites and comets crashed into Earth's surface. Analysis of present-day meteorites recovered from impact craters on Earth has revealed that some meteorites contain relatively high concentrations of amino acids and other simple organic molecules. Laboratory experiments suggest that these molecules could have formed in interstellar space before plummeting to Earth.

Organic Molecules Can Accumulate Under Prebiotic Conditions

Prebiotic synthesis was neither very efficient nor very fast. Nonetheless, large quantities of organic molecules eventually accumulated. Today, most organic molecules have a short life because they are either digested by living organisms or they react with atmospheric oxygen. Early Earth, however, lacked both life and free oxygen, so organic molecules would not have been exposed to these threats.

Still, prebiotic molecules could have been broken down by other chemical reactions or by the sun's high-energy UV radiation. Although UV light can provide energy for the formation of organic molecules, it can also break them apart. However, laboratory researchers have identified conditions under which molecules likely to have been present on prebiotic Earth are stable and can persist and even join together to form more complex molecules. Where on early Earth might such conditions have been found? Possibilities include the waters of mineral-rich hot springs, sheltered spots beneath rock ledges at the sea's edge, pores in the rock columns that form at hydrothermal vents on the ocean floor, and tiny crevices between ice crystals.

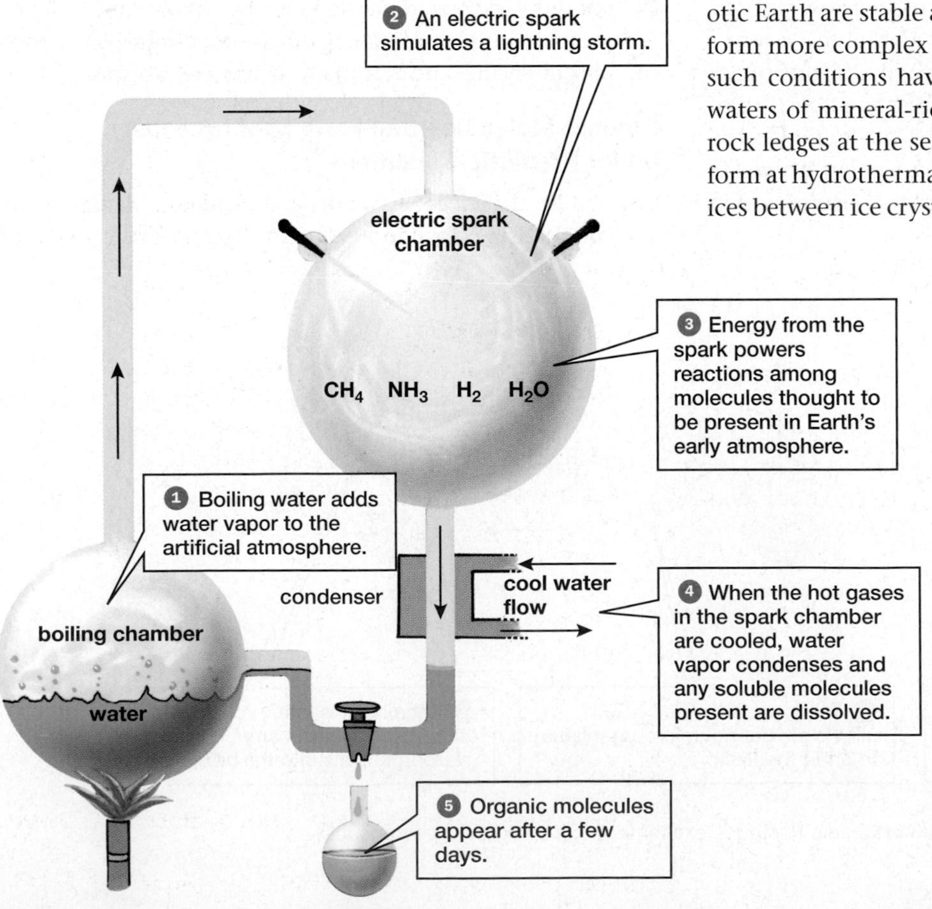
2 An electric spark simulates a lightning storm.
electric spark chamber
CH4 NH3 H2 H2O
3 Energy from the spark powers reactions among molecules thought to be present in Earth's early atmosphere.
1 Boiling water adds water vapor to the artificial atmosphere.
condenser
cool water flow
4 When the hot gases in the spark chamber are cooled, water vapor condenses and any soluble molecules present are dissolved.
boiling chamber
water
5 Organic molecules appear after a few days.

◀ **FIGURE 18-2 The experimental apparatus of Stanley Miller and Harold Urey** Life's very earliest stages left no fossils, so evolutionary scientists pursued a strategy of re-creating in the laboratory the conditions that may have prevailed on early Earth.

THINK CRITICALLY How would the experiment's result change if oxygen (O_2) were included in the spark chamber?

Clay May Have Catalyzed the Formation of Larger Organic Molecules

In the next stage of prebiotic evolution, wherever its location, simple molecules must have combined to form larger molecules. The chemical reactions that formed the larger molecules required that the reacting molecules be packed closely together. Scientists have proposed several processes by which the required high concentrations might have been achieved on early Earth. One possibility is that small molecules accumulated on the surfaces of clay particles, which often have a small electrical charge that attracts dissolved molecules with the opposite charge. Clustered on such a clay particle, small molecules would have been sufficiently close together to allow chemical reactions between them. Researchers have demonstrated the plausibility of this scenario with experiments in which adding clay to solutions of dissolved small organic molecules catalyzed the formation of larger, more complex molecules, including RNA. Such molecules might have gone on to become the building blocks of the first living organisms.

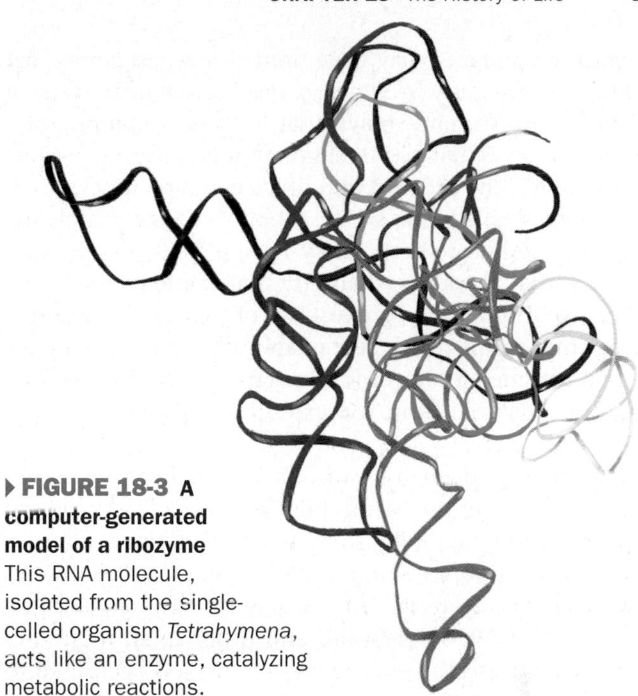

FIGURE 18-3 A computer-generated model of a ribozyme This RNA molecule, isolated from the single-celled organism *Tetrahymena*, acts like an enzyme, catalyzing metabolic reactions.

RNA May Have Been the First Self-Reproducing Molecule

Although all modern organisms use DNA to encode and store genetic information, it is unlikely that DNA was the earliest informational molecule. DNA can reproduce itself only with the help of large, complex protein enzymes, but the instructions for building these enzymes are encoded in DNA itself. For this reason, the origin of DNA's role as life's information storage molecule poses a "chicken and egg" puzzle: DNA requires proteins, but those proteins require DNA. It is thus difficult to construct a plausible scenario for the origin of self-replicating DNA unless we assume that the current DNA-based system of information storage evolved from an earlier system.

RNA Can Act As a Catalyst

A prime candidate for the first self-replicating informational molecule is RNA. In the 1980s, Thomas Cech and Sidney Altman, working with the single celled organism *Tetrahymena thermophila*, discovered a cellular reaction that was catalyzed not by a protein, but by a small RNA molecule. Because this special RNA molecule performed a function previously thought to be performed only by protein enzymes, Cech and Altman gave their catalytic RNA molecule the name **ribozyme** (**FIG. 18-3**).

In the years since the discovery of ribozymes, researchers have found dozens of naturally occurring ones that catalyze a variety of different reactions, including cutting other RNA molecules and splicing RNA fragments together. Ribozymes are also found in ribosomes, where they catalyze the attachment of amino acid molecules to growing proteins. In addition, researchers have been able to synthesize various ribozymes in the laboratory, including some that can catalyze the replication of small RNA molecules. The most effective replication ribozyme so far synthesized can copy RNA sequences up to 206 nucleotides long.

Earth May Once Have Been an RNA World

The discovery that RNA molecules can act as catalysts for diverse reactions, including RNA replication, provides support for the hypothesis that life arose in an "RNA world." According to this view, the current era of DNA-based life was preceded by one in which RNA served as both the information-carrying genetic molecule and the catalyst for its own replication. This RNA world may have emerged after hundreds of millions of years of prebiotic chemical synthesis, during which RNA nucleotides would have been among the molecules synthesized. After reaching a sufficiently high concentration, perhaps on clay particles, the nucleotides probably bonded together to form short RNA chains.

Let's suppose that, purely by chance, one of these RNA chains was a ribozyme that could catalyze the production of copies of itself. This first self-reproducing ribozyme probably wasn't very good at its job and likely produced copies with lots of errors. These mistakes were the first mutations. Like modern mutations, most undoubtedly ruined the catalytic abilities of the "daughter molecules," but a few may have been improvements. Such improvements set the stage for natural selection among RNA molecules, as variant ribozymes with increased speed and accuracy of replication copied themselves more rapidly than did less efficient RNA molecules, and thereby became increasingly common. Molecular evolution in the RNA world proceeded until, by some still-unknown chain of events, RNA gradually receded into its present role as an intermediary between DNA and protein enzymes.

Membrane-like Vesicles May Have Enclosed Ribozymes

Self-replicating molecules on their own do not constitute life; in all living cells such molecules are contained within some kind of enclosing membrane. The precursors of the earliest biological membranes may have been simple structures that formed spontaneously from purely mechanical processes. For example, chemists have shown that if water containing proteins and lipids is agitated to simulate waves beating against ancient shores, the proteins and lipids combine to form hollow structures called *vesicles*. These hollow balls resemble living cells in several respects. They have a well-defined outer boundary that separates their internal contents from the external solution. If the composition of the vesicle is right, a "membrane" forms that is remarkably similar in appearance to a real cell membrane. Under certain conditions, vesicles can absorb material from the external solution, grow, and even divide.

If a vesicle happened to surround the right ribozymes, it would form something resembling a living cell. We could call it a **protocell**, structurally similar to a cell but not alive. In the protocell, ribozymes and any other enclosed molecules would have been protected from degradation by free-roaming reactive molecules. Nucleotides and other small molecules might have diffused across the membrane and been used to synthesize new ribozymes and other complex molecules. After sufficient growth, the vesicle may have divided, with a few copies of the ribozymes becoming incorporated into each daughter vesicle. If this process occurred, the evolution of the first cells would be nearly complete.

Was there a particular moment when a nonliving protocell gave rise to a living organism? Probably not. Like most evolutionary transitions, the change from protocell to living cell was a gradual process, with no sharp boundary between one state and the next.

But Did All This Really Happen?

The scenario just described, although plausible and consistent with many research findings, is by no means certain. One of the most striking aspects of origin-of-life research is a great diversity of assumptions, experiments, and contradictory hypotheses. Researchers disagree about whether life arose in quiet terrestrial pools, at the sea's edge, in hot deep-sea vents, or in polar ice. A few researchers even argue that life arrived on Earth from space. Can we draw any firm conclusions from the research conducted so far? No, but we can make a few reasonable deductions.

First, the experiments of Miller and others show that amino acids, nucleotides, and other organic molecules, along with simple membrane-like structures, are likely to have formed in abundance on early Earth. Second, chemical evolution had long periods of time and huge areas of the Earth available to it. Given sufficient time and a sufficiently large pool of reactant molecules, even extremely rare events can occur many times. And given the vast expanses of time and space available, each small step on the path from primordial soup to living cell had ample opportunity to take place.

No particular account of life's origin can be tested definitively. The origin of life left no record, and researchers exploring this mystery can proceed only by developing a hypothetical scenario and then conducting laboratory investigations to determine if the scenario's steps are chemically and biologically plausible.

CHECK YOUR LEARNING

Can you . . .

- describe a likely scenario for the origin of life?
- describe, for each step in the scenario, some evidence that suggests the step is plausible?

18.2 WHAT WERE THE EARLIEST ORGANISMS LIKE?

When Earth first formed about 4.55 billion years ago, it was quite hot (**FIG. 18-4**). A multitude of meteorites smashed

▼ **FIGURE 18-4 Early Earth** In the immediate aftermath of Earth's formation 4.5 billion years ago, the planet was characterized by intense heat, abundant volcanic activity, and repeated meteorite strikes.

into the forming planet, and the kinetic energy of these extraterrestrial rocks was converted into heat on impact. Still more heat was released by the decay of radioactive atoms. Earth melted, and heavier elements such as iron and nickel sank to the center of the planet, where they remain molten even today. Nonetheless, geological evidence suggests that Earth had cooled enough for water to exist in liquid form by 4.3 billion years ago. Once liquid water was available, the prebiotic evolution that ultimately led to the first living organisms could begin.

The oldest fossil organisms found so far are in rocks that are about 3.4 billion years old. (Their age was determined using radiometric dating techniques; see "Doing Science: Discovering the Age of a Fossil" on page 287.) Chemical traces in older rocks have led some paleontologists to believe that life is even older, perhaps as old as 3.9 billion years.

The immense span of time in which life's origin and early history took place is known as the Precambrian. This name is among those assigned by geologists and paleontologists, who have devised a hierarchical naming system of eras, periods, and epochs to delineate geological time (**TABLE 18-1**).

The First Organisms Were Anaerobic Prokaryotes

The first cells to arise in Earth's oceans were **prokaryotes**, cells whose genetic material was not contained within a nucleus. These cells probably obtained nutrients and energy by absorbing organic molecules from their environment. There was no oxygen gas in the atmosphere, so the cells must have metabolized the organic molecules anaerobically. (You may recall from Chapter 8 that anaerobic metabolism yields only small amounts of energy.)

Thus, the earliest cells were primitive anaerobic bacteria. As these bacteria multiplied, they must have eventually used up the organic molecules produced by prebiotic chemical reactions. Simpler molecules, such as carbon dioxide and water, would still have been very abundant, as was energy in the form of sunlight. What was lacking, then, was not materials or energy but energetic molecules—molecules in which energy is stored in chemical bonds.

Some Organisms Evolved the Ability to Capture the Sun's Energy

Eventually, some cells evolved the ability to use the energy of sunlight to drive the synthesis of complex, high-energy molecules from simpler molecules; in other words, photosynthesis appeared. Photosynthesis requires a source of hydrogen, and the very earliest photosynthetic bacteria probably used hydrogen sulfide gas dissolved in water for this purpose (as purple photosynthetic bacteria do today). Eventually, however, Earth's supply of hydrogen sulfide (which is produced mainly by volcanoes) must have run low. The shortage of hydrogen sulfide set the stage for the evolution of photosynthetic bacteria that were able to use the planet's most abundant source of hydrogen—water (H_2O).

Water-based photosynthesis converts water and carbon dioxide to energy-containing molecules of sugar, releasing oxygen as a by-product. The emergence of this new method for capturing energy introduced significant amounts of free oxygen into the atmosphere for the first time. At first, the newly liberated oxygen was quickly consumed by reactions with other molecules in the atmosphere and in Earth's crust. One especially common reactive atom in the crust was iron, and much of the new oxygen combined with iron atoms to form huge deposits of iron oxide (rust). As a result, iron oxide is abundant in rocks formed during this period.

After most of the accessible iron had turned to rust, the concentration of oxygen gas in the atmosphere began to increase. Chemical analysis of rocks suggests that significant amounts of oxygen first appeared in the atmosphere about 2.4 billion years ago, produced by bacteria that were probably very similar to modern photosynthetic bacteria.

Aerobic Metabolism Arose in Response to Dangers Posed by Oxygen

Oxygen is potentially very dangerous to living things, because it can react with organic molecules, breaking them down. Many of today's anaerobic bacteria perish when exposed to oxygen, which is for them a deadly poison. The accumulation of oxygen in the atmosphere of early Earth probably exterminated many organisms and fostered the evolution of cellular mechanisms for detoxifying oxygen. This crisis for evolving life also provided the environmental pressure for the next great advance: the ability to use oxygen in metabolism. This ability not only provides a defense against the chemical action of oxygen, but actually channels oxygen's destructive power through aerobic respiration to generate useful energy for the cell (see Chapter 8). Because the amount of energy available to a cell is vastly increased when oxygen is used to metabolize food molecules, aerobic cells had a significant selective advantage.

Some Organisms Acquired Membrane-Enclosed Organelles

Hordes of bacteria would offer a rich food supply to any organism that could eat them. Paleobiologists speculate that, once this potential prey population appeared, predation would have evolved quickly. These early predators were probably prokaryotes that were larger than typical bacteria. In addition, they must have lost the rigid cell wall that surrounds most bacterial cells, so that their flexible plasma membrane was in contact with the surrounding environment. Thus, the predatory cells were able to envelop smaller bacteria in an infolded pouch of membrane and in this fashion engulf whole bacteria as prey.

These early predators were probably not photosynthetic, nor were they capable of using aerobic metabolism

TABLE 18-1 The History of Life on Earth

Era	Period	Epoch	Millions of Years Ago	Major Events
Cenozoic	Quaternary	Holocene Pleistocene	0.01–present 2.6–0.01	Evolution of genus *Homo*
	Neogene	Pliocene Miocene	5.3–2.6 23–5.3	First grasslands and kelp forests, earliest hominins
	Paleogene	Oligocene Eocene Paleocene	34–23 56–34 66–56	Widespread flourishing of birds, mammals, insects, and flowering plants
Mesozoic	Cretaceous		145–66	Flowering plants appear and become dominant Mass extinction of marine and terrestrial life, including dinosaurs
	Jurassic		201–145	Dominance of dinosaurs and conifers First birds
	Triassic		252–201	First mammals and dinosaurs Forests of gymnosperms and tree ferns
Paleozoic	Permian		299–252	Massive marine extinctions, including trilobites Flourishing of reptiles and the decline of amphibians
	Carboniferous		359–299	Forests of tree ferns and club mosses Dominance of amphibians and insects First reptiles and conifers
	Devonian		419–359	Fishes and trilobites flourish First amphibians, insects, seeds, and pollen
	Silurian		444–419	Many fishes, trilobites, and mollusks First vascular plants
	Ordovician		485–444	Dominance of arthropods and mollusks in the ocean Invasion of land by plants and arthropods First fungi
	Cambrian		541–485	Marine algae flourish Origin of most marine invertebrate phyla First fishes
Precambrian			630	First animals (soft-bodied marine invertebrates)
			1,200	First multicellular organisms
			1,700	First eukaryotes
			2,400	Accumulation of free oxygen in the atmosphere
			3,500	Origin of photosynthesis (in cyanobacteria)
			3,900–3,500	First living cells (prokaryotes)
			4,000–3,900	Appearance of the first rocks on Earth
			4,550	Origin of the solar system and Earth

DOING Science Discovering the Age of a Fossil

Scientists often determine a fossil's age by measuring the age of nearby rocks. The process for making such measurements is based on the outcome of radioactive decay, in which the nucleus of a radioactive isotope spontaneously breaks down to yield a more stable nucleus. Radioactive decay in many cases changes the number of protons in the nucleus, and hence changes the identity of the element. Each type of radioactive isotope decays at a characteristic rate, and the time it takes for half of its nuclei to decay is called its *half-life*.

If we know a radioactive isotope's rate of decay and can measure the amounts of decayed and undecayed nuclei in a rock, we can estimate how much time has passed since the element became trapped in the rock. This process is called *radiometric dating*. One radiometric dating technique measures the decay of potassium-40 (^{40}K) into argon-40 (^{40}Ar) gas. Suppose that a volcano erupts with a massive lava flow, covering the countryside. The lava will contain both ^{40}K and ^{40}Ar, but all the ^{40}Ar, being a gas, will bubble out. So, when the lava first cools and solidifies into rock, it will not contain any ^{40}Ar (**FIG. E18-1**). Over time, however, any ^{40}K present in the hardened lava will very slowly decay into ^{40}Ar, with half of the ^{40}K decaying every 1.25 billion years. This ^{40}Ar gas will be trapped in the rock. A geologist could take a sample of the rock and measure the ratio of ^{40}K to ^{40}Ar to determine the rock's age. For example, if the analysis finds equal amounts of the two elements, the geologist will conclude that the lava hardened 1.25 billion years ago.

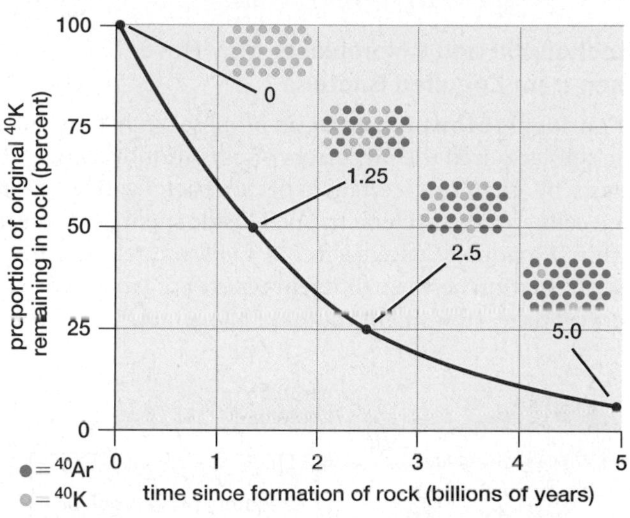

▲ **FIGURE E18-1 The relationship between time and the decay of radioactive ^{40}K to ^{40}Ar**

What Question Was Asked?

Knowing the age of a fossil is often very helpful to scientists trying to answer questions about evolutionary history. For example, researchers studying the evolution of hominins (humans and their extinct relatives) can better understand when key human adaptations arose if they know the ages of hominin fossils. So, when investigators discovered "Turkana Boy," a nearly complete *Homo erectus* fossil skeleton named for the Kenyan lake near which it was found, they wanted to know how old it is.

How Was Evidence Gathered?

To learn the age of Turkana Boy, researchers estimated the age of the volcanic rock layer about 18 inches below the sediments in which the fossil was found. The researchers ground several samples of the volcanic rock into fine dust and processed the dust to remove impurities. Then, using instruments that can detect and identify even tiny quantities of isotopes, they measured the amounts of radioactive ^{40}K and stable ^{40}Ar. After calculating the ratio of ^{40}K to ^{40}Ar, the researchers could estimate the age of the rock layer.

What Was Learned?

Potassium-argon radiometric dating revealed that the volcanic rocks below the fossil were 1.65 million years old. On the basis of this finding (and an estimate of the rate at which sediments were laid down on top of the rock layer), the researchers concluded that Turkana Boy's skeleton fossilized about 1.6 million years ago.

The Turkana Boy specimen is the best preserved and most complete fossil of *Homo erectus*. It is the source of much of our knowledge of the species' anatomy, which includes shoulders that allowed the arms to swing as *H. erectus* walked (just as our arms do), and long legs that helped the species spread across Africa and over much of Asia. Turkana Boy's skull confirms that hominin brains were large relative to body size more than a million years prior to the origin of *Homo sapiens*.

THINK CRITICALLY Uranium-235, with a half-life of 713 million years, decays to lead-207. If you analyze a rock and find that it contains uranium-235 and lead-207 in a ratio of 1:1, how old is the rock (assuming that decay of uranium-235 is the only source of lead-207)?

to efficiently extract energy from the prey they consumed. Although they could ingest smaller bacteria, they metabolized them inefficiently. By about 1.7 billion years ago, however, one predator probably gave rise to the first eukaryotic cell. Eukaryotic cells differ from prokaryotic cells in that they have an elaborate system of internal membranes, many of which enclose organelles such as a nucleus that contains the cell's genetic material. Organisms composed of one or more eukaryotic cells are known as **eukaryotes**.

The Internal Membranes of Eukaryotes May Have Arisen Through Infolding of the Plasma Membrane

The internal membranes of eukaryotic cells may have originally arisen through inward folding of the cell membrane of a single-celled predator. If, as in most of today's bacteria, the DNA of the eukaryotes' ancestor was attached to the inside of its cell membrane, an infolding of the membrane near the point of DNA attachment may have pinched off and become the precursor of the cell nucleus.

In addition to the nucleus, other key eukaryotic structures include the organelles used for energy metabolism: mitochondria (in all eukaryotes) and chloroplasts (in plants and algae). How did these organelles evolve?

Mitochondria and Chloroplasts May Have Arisen from Engulfed Bacteria

The **endosymbiont hypothesis** proposes that early eukaryotic cells acquired the precursors of mitochondria and chloroplasts by engulfing certain types of bacteria (**FIG. 18-5**). These cells and the bacteria trapped inside them (*endo* means "within") gradually entered into a *symbiotic* relationship, a close association between different types of organisms over an extended time. How might this have happened?

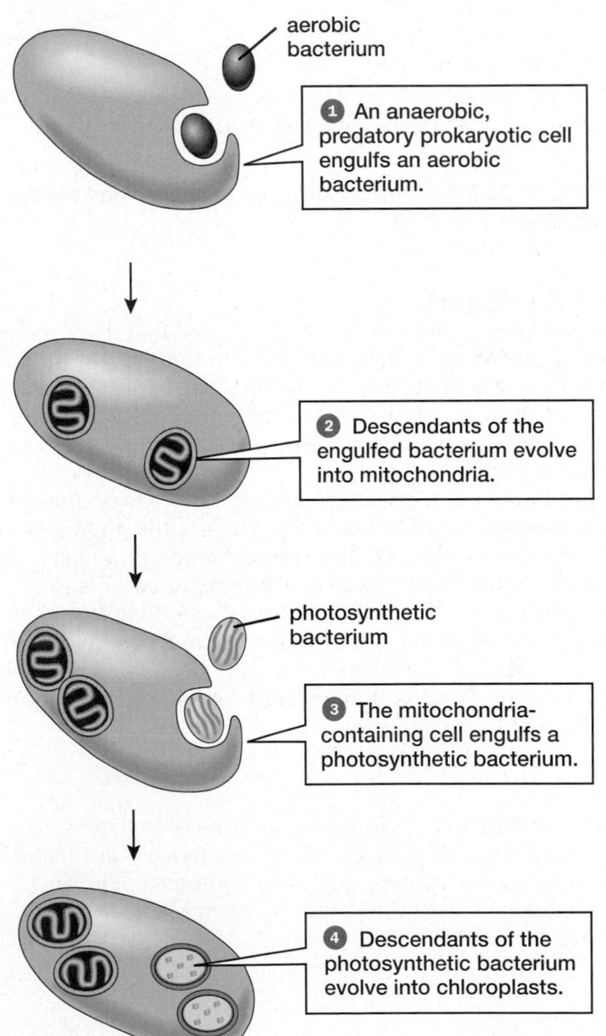
aerobic bacterium
1 An anaerobic, predatory prokaryotic cell engulfs an aerobic bacterium.
2 Descendants of the engulfed bacterium evolve into mitochondria.
photosynthetic bacterium
3 The mitochondria-containing cell engulfs a photosynthetic bacterium.
4 Descendants of the photosynthetic bacterium evolve into chloroplasts.

▲ **FIGURE 18-5 The probable origin of mitochondria and chloroplasts in eukaryotic cells**

THINK CRITICALLY Scientists have identified a free-living bacterium believed to be descended from the endosymbiont that gave rise to mitochondria. Would you expect the DNA sequence of this modern bacterium to be most similar to the sequence of DNA from a plant chloroplast, an animal cell nucleus, or a plant mitochondrion?

Let's suppose that an anaerobic predatory cell captured an aerobic bacterium for food, as it often did, but for some reason failed to digest this particular prey ❶. The aerobic bacterium remained alive and well, protected from other predatory cells. In fact, it was better off than ever, because the cytoplasm of its predator–host was chock-full of half-digested food molecules, the remnants of anaerobic metabolism. The aerobe absorbed these molecules and used oxygen to metabolize them, thereby gaining enormous amounts of energy. So abundant were the aerobes' food resources, and so bountiful their energy production, that the aerobes must have leaked energy, probably as ATP or similar molecules, back into their host's cytoplasm. The anaerobic predatory cell with its symbiotic bacteria could now metabolize food aerobically, gaining a great advantage over other anaerobic cells and leaving a greater number of offspring. Eventually, the endosymbiotic bacterium lost its ability to live independently of its host, and the mitochondrion was born ❷.

One of these successful new cellular partnerships managed a second feat: It captured a photosynthetic bacterium and again failed to digest its prey ❸. The bacterium flourished in its new host and gradually evolved into the first chloroplast ❹. Other eukaryotic organelles may have also originated through endosymbiosis. Cilia, flagella, centrioles, and microtubules may all have evolved from a symbiosis between a spirilla-like bacterium (a form of bacterium with an elongated corkscrew shape) and an early eukaryotic cell.

Evidence for the Endosymbiont Hypothesis Is Strong

Evidence that supports the endosymbiont hypothesis includes the many distinctive biochemical features shared by eukaryotic organelles and living bacteria. In addition, mitochondria and chloroplasts each contain their own minute supply of DNA, which many researchers interpret as remnants of the DNA originally contained within the engulfed bacteria.

Another kind of support comes from *living intermediates*, organisms alive today that are similar to hypothetical ancestors and thus help show that a proposed evolutionary pathway is plausible. For example, the amoeba *Pelomyxa palustris* lacks mitochondria but hosts a permanent population of aerobic bacteria that carry out much the same role. A variety of other protists also harbor symbiotic bacteria inside their cells (**FIG. 18-6**), as do many insect species.

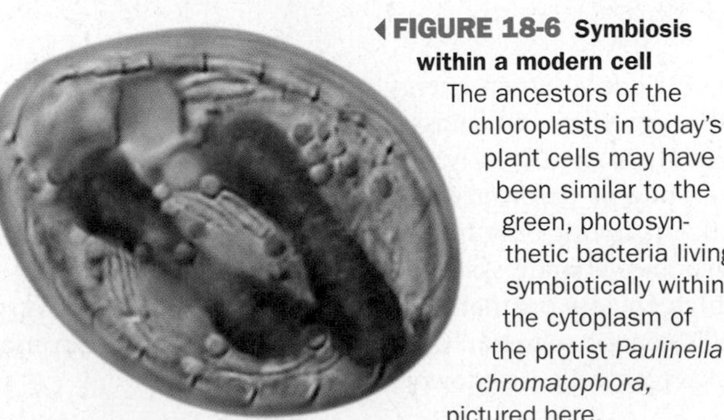

◀ **FIGURE 18-6 Symbiosis within a modern cell** The ancestors of the chloroplasts in today's plant cells may have been similar to the green, photosynthetic bacteria living symbiotically within the cytoplasm of the protist *Paulinella chromatophora*, pictured here.

These examples of modern cells that host bacterial endosymbionts suggest that similar symbiotic associations could have occurred almost 2 billion years ago and led to the first eukaryotic cells.

CHECK YOUR LEARNING

Can you . . .

- describe scenarios for the major evolutionary events and innovations that occurred during the period in which all organisms were single celled, including the origins of photosynthesis, atmospheric oxygen, aerobic respiration, and eukaryotic organelles?
- state the order in which these events occurred and list evidence that supports these scenarios?

18.3 WHAT WERE THE EARLIEST MULTICELLULAR ORGANISMS LIKE?

Once predation had evolved, increased size became an advantage. In the marine environments to which life was restricted, a larger cell could easily engulf a smaller cell and would also be difficult for other predatory cells to ingest. But enormous single cells have problems. The larger a cell becomes, the less surface membrane is available per unit volume of cytoplasm (see Fig. 5-13). Thus, as a cell grows larger, the process of diffusion across its plasma membrane becomes progressively less able to accommodate the oxygen and nutrients that must move into the cell and the waste products that must move out. One way for an organism to overcome this limit on cell size is to be multicellular; that is, it can consist of many small cells packaged into a larger, unified body.

Some Algae Became Multicellular

The oldest fossils of multicellular organisms are about 1.2 billion years old. They consist of impressions of multicellular algae that arose from single-celled eukaryotic organisms containing chloroplasts. Multicellularity would have provided at least two advantages for these organisms. First, large, many-celled algae would have been difficult for single-celled predators to engulf. Second, specialization of cells would have provided the potential for staying in one place in the brightly lit waters of the shoreline, as rootlike structures burrowed in sand or clutched onto rocks, while leaflike structures floated above in the sunlight. The green, brown, and red algae lining our shores today are the descendants of these early multicellular algae.

Animal Diversity Arose in the Precambrian

The oldest known unequivocal traces of animals include fossil embryos found in Precambrian deposits that are 630 million years old. Fossils of apparently adult animal bodies first appear in rocks laid down between 610 million and 541 million years ago. Some of these ancient invertebrate animals (animals lacking a backbone) are quite different in appearance from any animals that appear in later fossil layers and may represent types of animals that left no descendants. Other fossils in these rock layers, however, appear to be ancestors of today's animals. Ancestral sponges and jellyfish appear in the oldest layers, followed later by ancestors of worms, mollusks, and arthropods.

The full range of modern invertebrate animals, however, does not appear in the fossil record until the Cambrian period, marking the beginning of the Paleozoic era, about 541 million years ago. (The phrase "fossil record" is a shorthand reference to the entire collection of all fossil evidence that has been found to date.) These Cambrian fossils reveal an adaptive radiation (see Chapter 17) that had already yielded a diverse array of complex body plans. Almost all of the major groups of animals on Earth today were already present in the early Cambrian. The seemingly sudden appearance of so many different kinds of animals suggests that these groups actually arose earlier, but their early evolutionary history is not preserved in the fossil record.

Predation Favored the Evolution of Improved Mobility and Senses

The early diversification of animals was probably driven in part by the emergence of predatory lifestyles. For example, coevolution of predator and prey favored animals that were more mobile than their predecessors. Mobile predators gained an advantage from an ability to travel over wide areas in search of suitable prey; mobile prey benefited if they were able to make a speedy escape. The evolution of efficient movement was often associated with the evolution of greater sensory capabilities and more complex nervous systems. Senses for detecting touch, chemicals, and light became highly developed, along with nervous systems capable of handling the sensory information and directing appropriate behaviors.

By the Silurian period (444 million to 419 million years ago), life in Earth's seas included an array of anatomically complex animals, including armored trilobites, shelled ammonites, and the chambered nautilus (**FIG. 18-7**). The nautilus survives today in almost unchanged form in deep Pacific waters.

Skeletons Improved Mobility and Protection

In many Paleozoic animal species, mobility was enhanced in part by the origin of hard external body coverings known as **exoskeletons**. Exoskeletons improved mobility by providing hard surfaces to which muscles attached. These attachments made it possible for animals to use muscles to move their appendages to swim through the water or crawl over the seafloor. Exoskeletons also provided support for animals' bodies and protection from predators.

About 530 million years ago, one group of animals—the fishes—developed a new form of body support and muscle attachment: an internal skeleton. These early fishes were inconspicuous members of the ocean community, but by 400 million years ago, fishes were a diverse and prominent

(a) Silurian scene

(b) Trilobite

(c) Ammonite

(d) *Nautilus*

▲ **FIGURE 18-7 Diversity of ocean life during the Silurian period (a)** An artist's rendition of life characteristic of the oceans during the Silurian period, 444 million to 419 million years ago. Among the most common fossils from that time are **(b)** the trilobites and their predators, the nautiloids, and **(c)** the ammonites. **(d)** This living *Nautilus* is very similar in structure to the Silurian nautiloids, showing that a successful body plan may exist virtually unchanged for hundreds of millions of years.

group. By and large, fishes proved to be faster than the invertebrates, with more acute senses and larger brains. Eventually, they became the dominant predators of the open seas.

CHECK YOUR LEARNING

Can you . . .

- describe fossil evidence of the earliest multicellular organisms and the earliest animals?
- describe the advantages that fostered the origin of multicellularity?
- describe the adaptations associated with the later increase in animal diversity?

18.4 HOW DID LIFE INVADE THE LAND?

A compelling subplot in the long tale of life's history is the story of life's invasion of land. In moving to solid ground after more than 3 billion years of a strictly watery existence, organisms had many obstacles to overcome. Life in the sea provides buoyant support against gravity, but on land an organism must bear its weight against the crushing force of gravity. The sea provides ready access to life-sustaining water, but adequate water may not be easily available to a terrestrial organism. Sea-dwelling plants and animals can reproduce by means of mobile gametes that swim or drift to each other through the water. The sperm and eggs of land-dwellers, however, must be protected from drying out.

Despite the obstacles to life on land, the vast empty spaces of the Paleozoic landmass represented a tremendous evolutionary opportunity. The potential rewards of terrestrial life were especially great for plants. Water strongly absorbs light, so even in the clearest water, photosynthesis is at best possible only within a few hundred meters of the surface, and usually only at much shallower depths. Out of the water, the dazzling brightness of the sun permits rapid photosynthesis. Furthermore, terrestrial soils are rich storehouses of nutrients, whereas seawater tends to be low in nutrients, particularly nitrogen and phosphorus. Finally, the Paleozoic sea swarmed with plant-eating animals, but the land was devoid of animal life. Thus, the plants that first colonized the land would have had ample sunlight, abundant nutrients, and no predators.

Some Plants Became Adapted to Life on Dry Land

In moist soils at the water's edge, a few small green algae began to grow, taking advantage of the sunlight and nutrients. These algae didn't have large bodies to support them against the force of gravity, and, living right in the film of water on the soil, they could easily obtain water. About 475 million years ago, some of these algae gave rise to the first multicellular land plants. Initially simple and low-growing, land plants eventually evolved solutions to two of the main difficulties of plant life on land: obtaining and conserving water and staying upright despite gravity and winds. New adaptations that helped obtain and conserve water included water-resistant coatings on aboveground parts that reduced water loss by evaporation, rootlike structures that delved into the soil to absorb water and minerals, and specialized tissues (called vascular tissues) that contained tubes to conduct water from roots to leaves. Extra-thick walls surrounding certain cells enabled stems to stand erect, and the rootlike structures helped anchor erect plant bodies firmly to the soil.

Early Land Plants Retained Swimming Sperm and Required Water to Reproduce

Reproduction out of water presented challenges. Plants produce sperm and eggs, as animals do, and these gametes must meet to produce the next generation. The first land plants had swimming sperm, presumably much like those of today's mosses and ferns. Consequently, the earliest plants were restricted to swamps and marshes or to areas with abundant rainfall, where the ground would occasionally be covered with water. Here, the sperm and eggs could be released into the water, and sperm could swim to reach an egg. Later plants with swimming sperm prospered during periods in which the climate was warm and moist. For example, the Carboniferous period (359 million to 299 million years ago) was characterized by vast forests of giant tree ferns, club mosses, and horsetails (**FIG. 18-8**).

Seed Plants Encased Sperm In Pollen Grains

Meanwhile, some plants inhabiting drier regions had evolved a means of reproduction that no longer depended on water. The eggs of these plants were retained on the parent plant, and the sperm were encased in drought-resistant pollen grains that were carried by the wind from plant to plant. When the pollen grains landed near an egg, they released sperm cells directly into living tissue, eliminating the need for a surface film of water. The fertilized egg remained on the parent plant, where it developed inside a seed, which provided protection and nutrients for the embryo.

The earliest seed-bearing plants appeared in the late Devonian period (375 million years ago) and produced their seeds along branches, without any specialized structures to hold them. By the middle of the Carboniferous period, however, a new kind of seed-bearing plant had arisen. These plants, called **conifers**, protected their developing seeds inside cones. Conifers, which are wind-pollinated and do not depend on water for reproduction, flourished and spread during the Permian period (299 to 252 million years ago), when mountains rose, swamps drained, and the climate became much drier. The conifers' good fortune, however, was not shared by the tree ferns and giant club mosses, which, with their swimming sperm, largely went extinct.

Flowering Plants Enticed Animals to Carry Pollen

About 140 million years ago, during the Cretaceous period, the flowering plants appeared, having evolved from a group of conifer-like plants. Many flowering plants are pollinated by animals, especially insects, and this mode of pollination seems to have conferred an evolutionary advantage. Flower pollination by animals can be far more efficient than pollination by wind. Wind-pollinated plants must produce an enormous amount of pollen because the vast majority of pollen grains fail to reach their target. Today, flowering plants dominate the land, except in cold northern regions, where conifers still prevail. In some cases, flowering plants have re-evolved wind pollination, most likely in response to a past or ongoing reduction in the availability of animal pollinators.

Some Animals Became Adapted to Life on Dry Land

After land plants evolved, providing potential food sources for other organisms, animals emerged from the sea. The

◀ **FIGURE 18-8 The swamp forest of the Carboniferous period** Many of the treelike plants in this artist's reconstruction are extinct relatives of today's club mosses and horsetails.

THINK CRITICALLY Why are today's ferns, horsetails, and club mosses so small in comparison to their giant ancestors?

earliest evidence of land animals comes from fossils that are about 430 million years old. The first animals to move onto land were **arthropods** (the group that today includes insects, spiders, scorpions, centipedes, and crabs). Why arthropods? The answer seems to be that they already possessed certain structures that, purely by chance, were suited to life on land. Foremost among these structures was an exoskeleton, such as the shell of a lobster or crab. Exoskeletons are both waterproof and strong enough to support a small animal against the force of gravity.

For millions of years, arthropods had the land and its plants to themselves, and for tens of millions of years more, they were the dominant land animals. Dragonflies with a wingspan of 28 inches (70 centimeters) flew among the Carboniferous tree ferns, while millipedes 6.5 feet (2 meters) long munched their way across the swampy forest floor. Eventually, however, the arthropods' splendid isolation came to an end.

Amphibians Evolved from Lobefin Fishes

About 400 million years ago, a group of Devonian fishes called the lobefins appeared, probably in fresh water. **Lobefins** had two important features that would later enable their descendants to colonize land: (1) stout, fleshy fins with which they crawled about on the bottoms of shallow, quiet waters, and (2) an outpouching of the digestive tract that could be filled with air, like a primitive lung. One group of lobefins inhabited very shallow ponds and streams, which shrank during droughts and often became oxygen poor. By taking air into their lungs, these lobefins could still obtain oxygen. Some began to use their fins to crawl from pond to pond in search of prey or water, as some fish do today (**FIG. 18-9**).

▲ **FIGURE 18-9 A fish that walks on land** Some modern fishes, such as this mudskipper, walk on land. As did the ancient lobefin fishes that gave rise to amphibians, mudskippers use their strong pectoral fins to move across dry areas in their swampy habitats.

THINK CRITICALLY Does the mudskipper's ability to walk on land constitute evidence that lobefin fishes were the ancestors of amphibians?

The benefits of feeding on land and moving from pool to pool favored the evolution of a group of animals that could stay out of water for longer periods and that could move about more effectively on land. With improvements in lungs and legs, **amphibians** evolved from lobefins, first appearing in the fossil record about 370 million years ago. To an amphibian, the Carboniferous swamp forests were a kind of paradise: no predators to speak of, abundant prey, and a warm, moist climate. As had the insects and millipedes, some amphibians evolved gigantic size, including salamanders more than 10 feet (3 meters) long.

Despite their success, early amphibians were not fully adapted to life on land. Their lungs were simple sacs without very much surface area, so they had to obtain some of their oxygen through their skin. Therefore, their skin had to be kept moist, a requirement that restricted them to swampy habitats. Further, amphibian sperm and eggs could not survive in dry surroundings and had to be deposited in water. So, although amphibians could move about on land, they could not stray too far from the water's edge. Along with the tree ferns and club mosses, amphibians declined when the climate turned dry at the beginning of the Permian period about 299 million years ago.

Reptiles Evolved from Amphibians

As the conifers were evolving on the fringes of the swamp forests, a group of amphibians was also evolving adaptations to drier conditions. These amphibians ultimately gave rise to the **reptiles**, which had three major adaptations to life on land. First, reptiles evolved shelled, waterproof eggs that enclosed a supply of food and water for the developing embryo. Thus, reptiles could lay eggs on land and avoid the dangerous swamps full of fish and amphibian predators. Second, ancestral reptiles evolved scaly, water-resistant skin that reduced the loss of body water to the dry air. Finally, reptiles evolved improved lungs that were able to provide the entire oxygen supply of an active animal. As the climate dried during the Permian period, reptiles became the dominant land vertebrates, relegating amphibians to the swampy backwaters where most remain today.

A few tens of millions of years later, the climate became wetter again. This period saw the evolution of some very large reptiles, in particular, the dinosaurs (**FIG. 18-10**). The variety of dinosaur forms was enormous—large and small, fleet-footed and ponderous, predators and plant-eaters. Dinosaurs were among the most successful animals ever, if we consider persistence as a measure of success. They flourished for more than 100 million years, until about 66 million years ago, when the last dinosaurs went extinct. No one is certain why they died out, but the aftereffects of a gigantic meteorite's impact with Earth seem to have been the final blow (as discussed in Section 18.5).

FIGURE 18-10 A reconstruction of a Cretaceous forest By the Cretaceous period, flowering plants dominated terrestrial vegetation. Dinosaurs, such as the predatory pack of 6-foot-long velociraptors shown here, were the preeminent land animals. Although small by dinosaur standards, *Velociraptors* were formidable predators with great running speed, sharp teeth, and deadly, sickle-like claws on their hind feet.

Even during the age of dinosaurs, many reptiles remained quite small. One major difficulty faced by small reptiles is maintaining a high body temperature. A warm body is advantageous for an active animal, because warmer nerves and muscles work more efficiently. But a warm body loses heat to the environment unless the air is also warm. Heat loss is an especially big problem for small animals, which have a larger surface area per unit of volume than do larger animals. Many species of small reptiles have slow metabolisms and cope with the heat loss problem by confining activity to times when the air is sufficiently warm. One group of reptiles, however, followed a different evolutionary pathway. Members of this group, the birds, evolved insulation, in the form of feathers. (Birds were formerly placed in their own taxonomic group, separate from reptiles.)

In ancestral birds, feathers, which are modified scales, helped retain body heat. Consequently, these animals could be active in cool habitats and during the night, when their scaly relatives became sluggish. Later, some ancestral birds evolved longer, stronger feathers on their forelimbs, perhaps under selection for better ability to glide from trees or to jump after insect prey. Ultimately, feathers evolved into structures capable of supporting powered flight. Fully developed, flight-capable feathers are present in 150-million-year-old fossils, so the earlier insulating structures that eventually developed into flight feathers must have been present well before that time.

Reptiles Gave Rise to Mammals

Unlike the egg-laying reptiles, **mammals** evolved live birth and the ability to feed their young with secretions of the mammary (milk-producing) glands. Ancestral mammals also developed hair, which provided insulation.

CASE STUDY CONTINUED

Ancient DNA Has Stories to Tell

Can ancient DNA reveal the secrets of dinosaur evolutionary history? Sadly, no. DNA decays far too quickly to be present in fossils as old as dinosaur fossils are. But all is not lost; the paleontologist Mary Schweitzer and her colleagues have discovered, in some exceptionally well preserved dinosaur fossils, what appear to be preserved soft tissues, such as blood, bone marrow, and skin. These discoveries were initially met with great skepticism that soft tissues could be preserved for so long, but as additional evidence has accumulated, an increasing number of paleontologists have accepted the discoveries. Researchers have extracted proteins such as hemoglobin, keratin, and collagen from the fossil tissue, and the amino acid sequences of these proteins may reveal previously unknown information about the evolution of dinosaurs. Nonetheless, evolution's historians must, for the most part, rely on more traditional methods. What have such methods revealed about the dinosaurs' successors as Earth's dominant large animals, the mammals?

Because soft tissues like the uterus and mammary glands do not generally fossilize, we may never know when these structures first appeared or what their intermediate forms looked like. Hair, however, is occasionally preserved in fossils. The oldest known hair was fossilized about 160 million years ago, so mammals have presumably had hair for at least that long.

The earliest mammals arose more than 200 million years ago. Early mammals thus coexisted with the dinosaurs. They were mostly small creatures. The largest known mammal from the dinosaur era was about the size of a modern raccoon, but most early mammal species were far smaller. When the dinosaurs went extinct, however, mammals colonized newly empty habitats, prospered, and diversified into the array of forms that we see today.

CHECK YOUR LEARNING

Can you . . .

- describe the transitions and innovations associated with the origin and evolution of the major groups of land plants and vertebrates?
- describe the advantages gained by the first plants and animals to colonize land?

CASE STUDY CONTINUED

Ancient DNA Has Stories to Tell

Although it may never be possible to recover DNA from dinosaurs, ancient DNA of more recent vintage can help us understand more about the physiology and behavior of extinct animals. For example, researchers have extracted DNA from 43,000-year-old wooly mammoths that were preserved in the permafrost of Siberia. (Cold climates are especially favorable for preserving ancient DNA.) The investigators were able to sequence some of the DNA, including the genes that produced hemoglobin (a protein that transports oxygen in the blood). The researchers then inserted the mammoth hemoglobin genes into bacteria. The bacteria produced hemoglobin molecules just like those that circulated in the mammoth's blood when it was alive.

Unlike hemoglobin from modern elephants, mammoth hemoglobin releases oxygen readily not only at core body temperature, but also at temperatures near freezing. Thus, though a modern elephant must keep its legs warm in order to provide oxygen to its leg muscles, a mammoth's legs could get very cold and still function, an adaptation that helped the animals survive in ice-age Siberia.

In the end, mammoths became extinct, as do all species, eventually. What was responsible for history's largest waves of extinction?

18.5 WHAT ROLE HAS EXTINCTION PLAYED IN THE HISTORY OF LIFE?

If there is a lesson in the great tale of life's history, it is that nothing lasts forever. The story of life can be read as a long series of evolutionary dynasties, with each new dominant group rising, ruling the land or the seas for a time, and, inevitably, falling into decline and extinction. Dinosaurs are the most famous of these fallen dynasties, but the list of extinct groups known only from fossils is impressively long. Despite the inevitability of extinction, however, the overall trend has been for species to arise at a faster rate than they disappear, so the number of species on Earth has tended to increase over time.

Have You Ever Wondered ... If Extinct Species Can Be Revived by Cloning?

Scientists have cloned a number of animal species, including mice, dogs, cats, horses, and cows. Could the technology of cloning be used to bring back extinct species? In principle, yes, provided that perfectly preserved DNA of the extinct species is available. Such DNA could be transferred to an egg from a closely related, living species, and the egg implanted in a surrogate mother of that species.

For example, researchers have suggested that it might be possible to clone a woolly mammoth, using an elephant surrogate mother and DNA extracted from 20,000-year-old mammoths found frozen beneath the Siberian tundra. Most scientists, however, believe that any DNA recovered from a fossil mammoth would be far too degraded for use in cloning, and synthesizing an entire mammoth genome (its sequence is now almost fully known) is beyond the capabilities of current technology. The odds of success might be greater for another proposed project, which would use DNA from a preserved museum specimen to revive the Tasmanian tiger, an Australian mammal that has been extinct for only 70 years. If cloning recently extinct species proves to be possible, do you think it would be a good idea?

Evolutionary History Has Been Marked by Periodic Mass Extinctions

Over much of life's history, the origin and disappearance of species have proceeded in a steady, relentless manner. This slow and steady turnover of species, however, has been interrupted by episodes of **mass extinction**. These mass extinctions are characterized by the relatively sudden disappearance of a wide variety of species over a large part of Earth. The most dramatic episode of all, which occurred 252 million years ago, at the end of the Permian period, wiped out more than 90% of the world's species in only 60,000 years. Life came perilously close to disappearing altogether.

Climate Change Contributed to Mass Extinctions

Mass extinctions have had a profound impact on the course of life's history. What could have caused such dramatic changes in the fortunes of so many species? Many evolutionary biologists believe that changes in climate must have played an important role. When the climate changes, as it has done many times over the course of Earth's history, organisms that are adapted for survival in one climate may be unable to survive in a drastically different climate. In particular, at times when warm climates gave way to drier, colder climates with more variable temperatures, species may have gone extinct after failing to adapt to the harsh new conditions.

One cause of climate change is the shifting positions of continents. These movements are often called *continental drift*. Continental drift is caused by **plate tectonics**, in which the Earth's surface, including the continents and the seafloor, is divided into plates that rest atop a viscous but

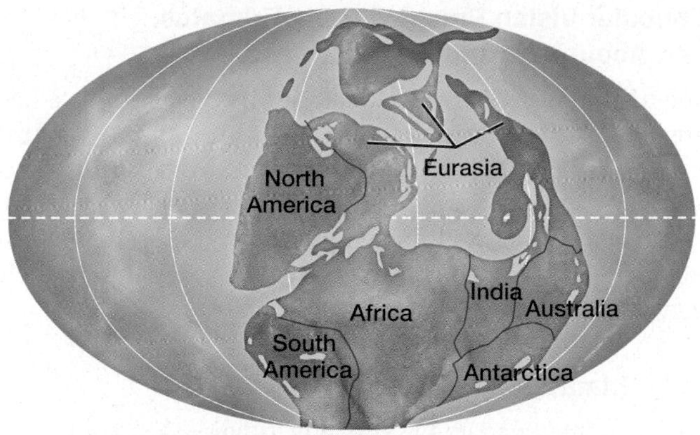

(a) 340 million years ago

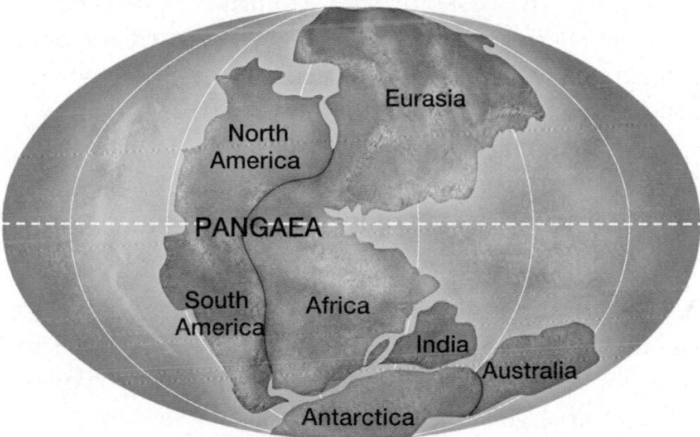

(b) 225 million years ago

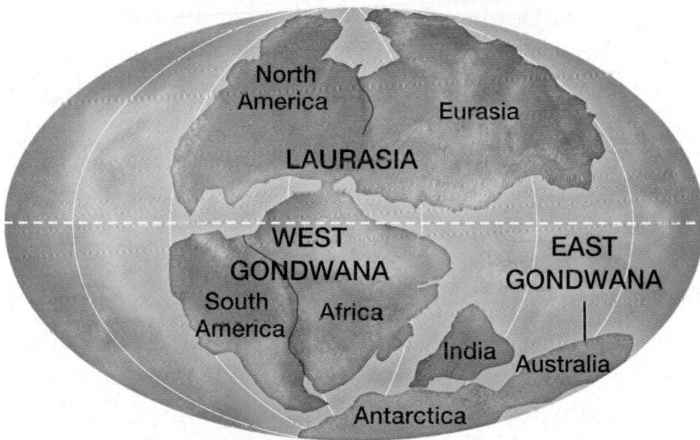

(c) 135 million years ago

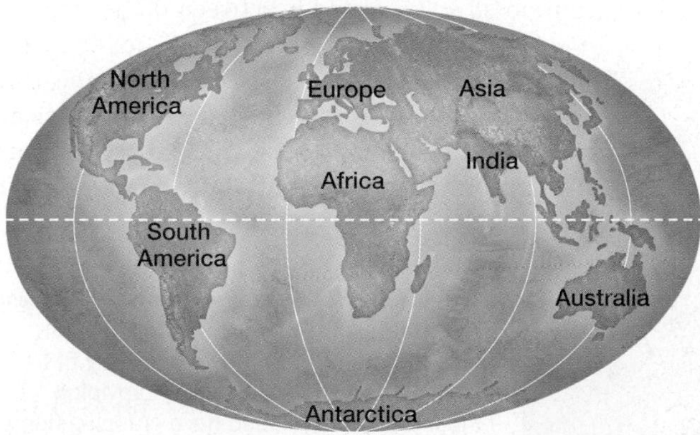

(d) Present

fluid layer and move slowly about. As the plates wander, their positions may change in latitude (**FIG. 18-11**). For example, 340 million years ago, much of North America was located at or near the equator, an area characterized by consistently warm and wet tropical weather. But as time passed, plate tectonics carried the continent up into temperate and arctic regions. As a result, the once tropical climate was replaced by a regime of seasonal changes, cooler temperatures, and less rainfall. Plate tectonics continues today; the Atlantic Ocean, for example, widens by a few centimeters each year.

Catastrophic Events May Have Caused the Biggest Mass Extinctions

Geological data indicate that most mass extinction events coincided with periods of climatic change. But more sudden events may also have played a role. For example, catastrophic geological events, such as massive volcanic eruptions, could rapidly kill many organisms. Geologists have found evidence of an immense eruption that began just prior to the end of the Permian, and many suspect that the volcanic activity may have been a cause of the subsequent mass extinction.

The search for the causes of mass extinctions took a fascinating turn in the early 1980s, when Luis and Walter Alvarez proposed that the extinction event of 66 million years ago, which wiped out the dinosaurs and many other species, was caused by the impact of a huge meteorite. The Alvarezes' idea was met with great skepticism when it was first introduced, but geological research since that time has generated a great deal of evidence that a massive impact did indeed occur 66 million years ago. In fact, researchers have identified the Chicxulub crater, a 100-mile wide crater buried beneath the Yucatan Peninsula of Mexico, as the impact site of a giant meteorite—6 miles (10 kilometers) in diameter—that collided with Earth around the time that dinosaurs disappeared.

Could this immense meteorite strike have caused the mass extinction that coincided with it? No one knows for sure, but scientists suggest that such a massive impact would have thrown so much debris into the atmosphere that the entire planet would have been plunged into darkness for a period of years. With little light reaching the planet, temperatures would have dropped precipitously and the photosynthetic capture of energy (on which all life ultimately depends) would have declined drastically. The worldwide "impact winter" would have spelled doom for the dinosaurs and a host of other species.

◀ **FIGURE 18-11 Continental drift from plate tectonics** The continents are passengers on plates moving on Earth's surface as a result of plate tectonics. **(a)** About 340 million years ago, much of what is now North America was positioned at the equator. **(b)** All the plates eventually fused together into one gigantic landmass, which geologists call Pangaea. **(c)** Gradually Pangaea broke up into Laurasia and Gondwanaland, which itself eventually broke up into West and East Gondwana. **(d)** Further plate motion eventually resulted in the current positions of the modern-day continents.

CHECK YOUR LEARNING

Can you . . .

- explain how extinction has affected the course of evolutionary history?
- describe the likely causes of mass extinctions in general and of the one that occurred 66 million years ago in particular?

18.6 HOW DID HUMANS EVOLVE?

Scientists are intensely interested in the origin and evolution of humans. The outline of human evolution that we present in this section represents an interpretation that is widely shared among paleontologists. However, fossil evidence of human evolution is comparatively scarce and therefore open to a variety of interpretations. Thus, some paleontologists would disagree with some aspects of the scenario we present.

Humans Inherited Some Early Primate Adaptations for Life in Trees

Humans belong to the mammal group known as **primates**, which also includes lemurs, monkeys, and apes. The oldest primate fossils are 55 million years old, but because primate fossils are relatively rare compared with those of many other animals, the first primates probably arose considerably earlier but left no fossil record. Early primates were adapted for life in the trees, and many modern primates retain the tree-dwelling lifestyle of their ancestors (**FIG. 18-12**). The common heritage of humans and other primates is reflected in a set of physical characteristics that was present in the earliest primates and that persists in many modern primates, including humans.

(a) Tarsier

(b) Lemur

(c) Macaque

◀ FIGURE 18-12 Representative primates The **(a)** tarsier, **(b)** lemur, and **(c)** lion-tail macaque monkey all have a relatively flat face, with forward-looking eyes providing binocular vision. All also have grasping hands. These features, retained from the earliest primates, are shared by humans.

Binocular Vision Provided Early Primates with Accurate Depth Perception

One of the earliest primate adaptations seems to have been large, forward-facing eyes (see Fig. 18-12). Jumping from branch to branch is risky business unless an animal can accurately judge where the next branch is located. Accurate depth perception was made possible by binocular vision, provided by forward-facing eyes with overlapping fields of view.

Early Primates Had Grasping Hands

Early primates had long, grasping fingers that could wrap around and hold onto tree limbs. This adaptation to tree dwelling was the basis for later evolution of human hands that could perform both a *precision grip* (used for delicate maneuvers such as picking up small objects and sewing) and a *power grip* (used for powerful actions, such as thrusting with a spear or swinging a hammer).

A Large Brain Facilitated Hand–Eye Coordination and Complex Social Interactions

Primates have brains that are larger, relative to their body size, than the brains of almost all other animals. No one really knows for certain which environmental factors favored the evolution of large brains. It seems reasonable, however, that controlling and coordinating rapid locomotion through trees, the dexterous movements of the hands in manipulating objects, and binocular vision would be facilitated by increased brain power. Most primates also have fairly complex social systems, which require relatively high intelligence. If sociality promoted increased survival and reproduction, the benefits to individuals of successful social interaction might have favored the evolution of a larger brain.

The Oldest Hominin Fossils Are from Africa

Based on analysis of human mutation rates and DNA sequences from modern chimpanzees, gorillas, and humans, researchers have estimated that the **hominin** line (humans and their fossil relatives) diverged from the ape lineage at least 7 million years ago. The fossil record is in accord with this estimate, as the oldest hominin fossil so far found is between 6 and 7 million years old (**FIG. 18-13**). This fossil species, *Sahelanthropus tchadensis*, was unearthed in the African country Chad and is clearly a hominin because it shares several anatomical features with later members of the

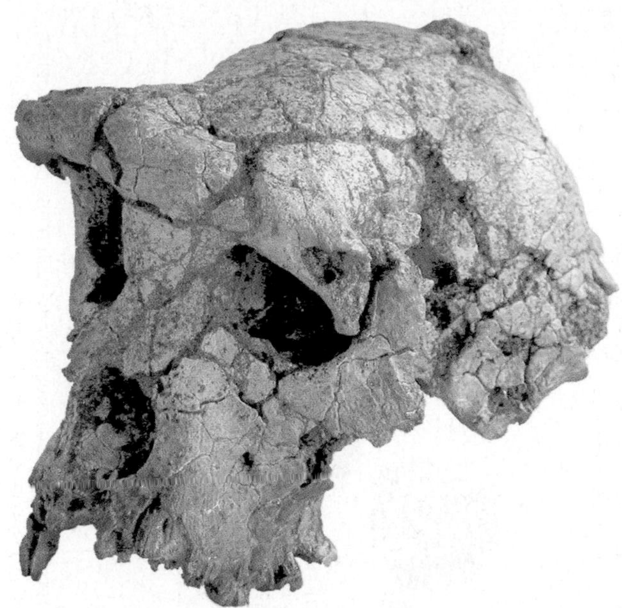

▲ **FIGURE 18-13 The earliest hominin** This nearly complete skull of *Sahelanthropus tchadensis,* which is more than 6 million years old, is the oldest hominin fossil yet found.

group. But because this oldest known member of our family also exhibits other features that are more characteristic of apes, it may represent a point on our family tree that is close to the split between apes and hominins.

Two other hominin species—*Orrorin tugenensis* and *Ardipithecus ramidus*—are known from African fossils appearing in rocks that are between 4 million and 6 million years old. Most of our knowledge of these species is based on fossil finds that include only small portions of skeletons. But one specimen, a fairly complete 4.4-million-year-old *Ardipithecus* skeleton, has revealed some intriguing features of this early hominin. The structure of its legs, feet, hands, and pelvis suggests that *Ardipithecus* could walk upright, though it probably also climbed trees in its forest habitat. Its canine teeth were small, like those of modern humans and unlike the large, fang-like canines of today's apes.

A more extensive record of early hominin evolution begins about 4 million years ago. That date marks the beginning of the fossil record of the genus *Australopithecus* (**FIG. 18-14**), a group of African hominin species with brains larger than those of their forebears but still much smaller than those of modern humans.

Early Hominins Could Stand and Walk Upright

It is possible that even the earliest hominins walked upright. The discoverers of *Sahelanthropus* and *Orrorin* argue that the leg and foot bones of these earliest hominins have characteristics that indicate bipedal locomotion, but this conclusion will remain speculative until more complete skeletons of these species are found. However, the *Ardipithecus* skeleton shows that hominins capable of upright posture had arisen by 4.4 million years ago, and the earliest australopithecines (as the various species of *Australopithecus* and the related genus *Paranthropus* are collectively known) had knee joints that allowed them to straighten their legs fully, permitting efficient bipedal (upright, two-legged) locomotion. Footprints almost 4 million years old, discovered in Tanzania by anthropologist Mary Leakey, show that the earliest australopithecines could, and at least sometimes did, walk upright.

The reasons for the evolution of bipedal locomotion among the early hominins remain poorly understood. Perhaps hominins that could stand upright gained an advantage in gathering or carrying food. Whatever its cause, the early evolution of upright posture was extremely important in the evolutionary history of hominins, because if the hands were no longer needed for walking, they were free to serve other functions. Later hominins were thus able to carry weapons, manipulate tools, and eventually achieve the cultural revolutions produced by modern *Homo sapiens*.

Several Species of *Australopithecus* Emerged in Africa

The oldest australopithecine species, represented by fossilized teeth, skull fragments, and arm bones, was unearthed near an ancient lake bed in Kenya from sediments that were dated as being between 3.9 million and 4.1 million years old. The species was named *Australopithecus anamensis* by its discoverers. The second most ancient australopithecine, called *Australopithecus afarensis*, was discovered in the Afar region of Ethiopia. Fossil remains of this species as old as 3.9 million years have been unearthed. The *A. afarensis* line apparently gave rise to at least two distinct forms: smaller species such as *A. africanus* and *A. sediba*, and larger, more robust species such as *Paranthropus robustus* and *P. boisei* that had bigger molar teeth and heavier jaws, suggesting that their diet included hard foods such as nuts. All of the australopithecine species had gone extinct by 1.2 million years ago. Before disappearing, however, one of these species gave rise to a new branch of the hominin family tree, the genus *Homo* (see Fig. 18-14).

The Genus *Homo* Diverged from the Australopithecines 2.5 Million Years Ago

Hominins that are sufficiently similar to modern humans to be placed in the genus *Homo* first appear in African fossils that are about 2.5 million years old. Among the earliest African *Homo* fossils are *H. habilis* (see Fig. 18-14), a species whose body and brain were larger than those of the australopithecines but that retained the apelike long arms and short legs of their australopithecine ancestors. In contrast, the skeletal anatomy of *H. ergaster*, a species whose fossils first appear 2 million years ago, has limb proportions more like those of modern humans. This species is believed by many paleontologists to be on the evolutionary branch that led ultimately to our own species, *H. sapiens*. In this view, *H. ergaster* was the common ancestor of two distinct

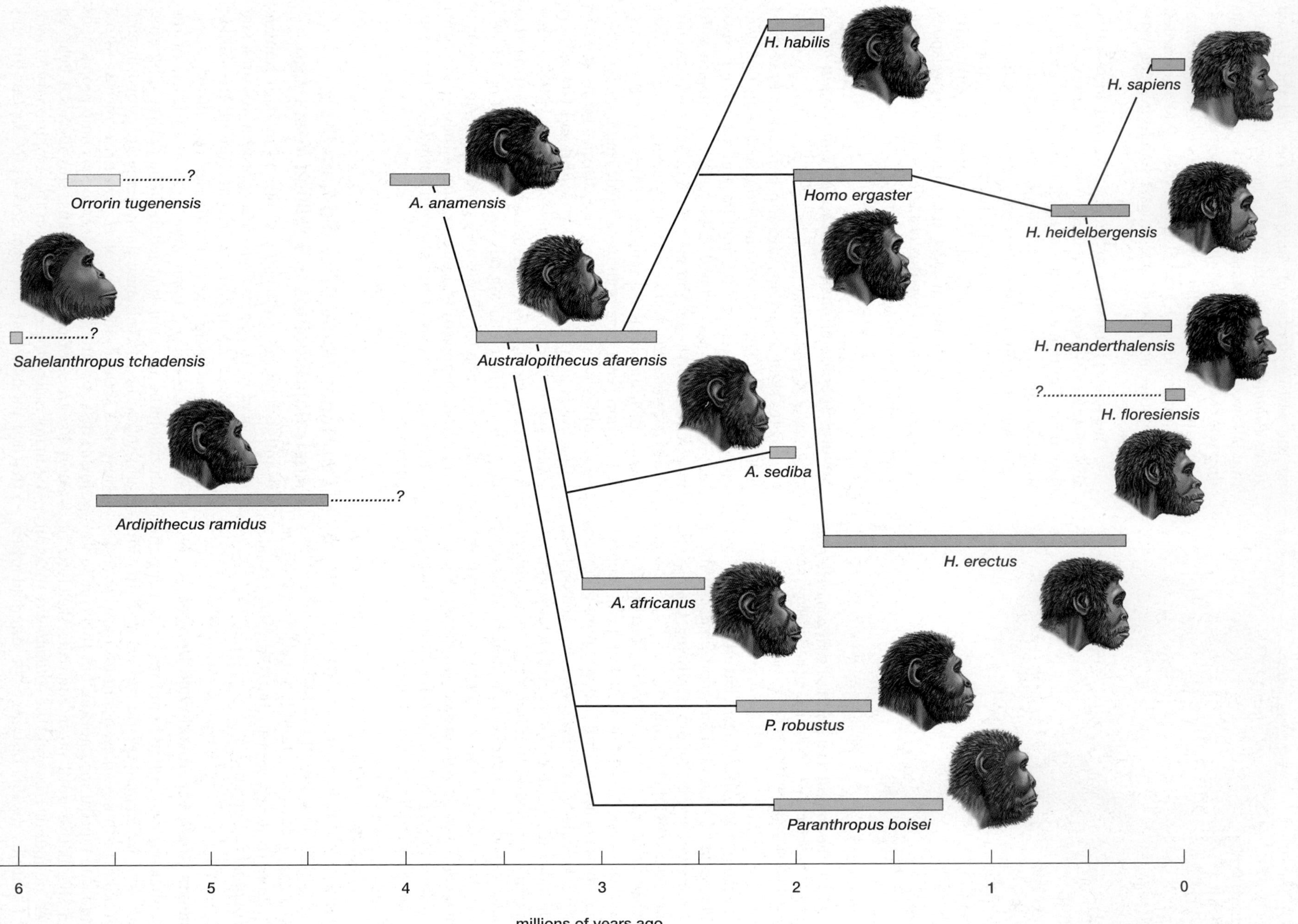

▲ **FIGURE 18-14 A possible evolutionary tree for humans** This hypothetical family tree shows facial reconstructions of representative specimens. Although many paleontologists consider this to be the most likely human family tree, there are several alternative interpretations of the known hominin fossils. Fossils of the earliest hominins are scarce and fragmentary, so the relationship of these species to later hominins remains unknown.

branches of hominins. The first branch led to *H. erectus*, which was the first hominin species to leave Africa. The second branch from *H. ergaster* ultimately led to *H. heidelbergensis*, some of which migrated to Europe and gave rise to the Neanderthals, *H. neanderthalensis*. Meanwhile, back in Africa, another branch split off from the *H. heidelbergensis* lineage. This branch became *H. sapiens*—modern humans.

The Evolution of *Homo* Was Accompanied by Advances in Tool Technology

Hominin evolution is closely tied to the development of tools, a hallmark of hominin behavior. The oldest tools discovered so far were found in 2.5-million-year-old East African rocks, concurrent with the early emergence of the genus *Homo*. Early *Homo*, whose molar teeth (the rearmost teeth in the jaw) were much smaller than those of the australopithecines, might first have used stone tools to break and crush tough foods that were hard to chew. Hominins constructed their earliest tools by striking one rock with another to chip off fragments. During the next several hundred thousand years, toolmaking techniques in Africa gradually became more advanced. By 1.7 million years ago, tools had become more sophisticated. Flakes were chipped symmetrically from both sides of a rock to form double-edged tools ranging from hand axes, used for cutting and chopping, to points, probably used on spears (**FIG. 18-15a, b**). *Homo ergaster* and other bearers of these weapons presumably ate meat, probably acquired both from hunting and from scavenging for the remains of prey killed by other predators. Double-edged tools were carried to Europe at least 600,000 years ago by migrating populations of *H. heidelbergensis*, and the Neanderthal descendants of these emigrants took stone tool construction to new heights of skill and delicacy (**FIG. 18-15c**).

(a) *Homo habilis*

(b) *Homo ergaster*

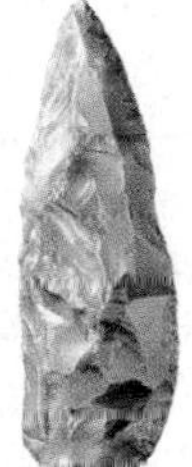

(c) *Homo neanderthalensis*

▲ **FIGURE 18-15 Representative hominin tools (a)** *Homo habilis* produced only fairly crude chopping tools called hand axes, usually unchipped on one end to hold in the hand. **(b)** *Homo ergaster* manufactured much finer tools. The tools were typically sharp all the way around the stone; at least some of these blades were probably tied to spears rather than held in the hand. **(c)** Neanderthal tools were works of art, with extremely sharp edges made by flaking off tiny bits of stone. In comparing these weapons, note the progressive increase in the number of flakes taken off the blades and the corresponding decrease in flake size. Smaller, more numerous flakes produce a sharper blade and suggest more insight into toolmaking and finer control of hand movements.

Neanderthals Had Large Brains and Excellent Tools

Neanderthals first appeared in the European fossil record about 150,000 years ago. By about 70,000 years ago, they had spread throughout Europe and western Asia. By 30,000 years ago, however, the species was extinct.

Contrary to the popular image of a hulking "caveman," Neanderthals were quite similar to modern humans in many ways. They walked fully erect, were dexterous enough to manufacture finely crafted stone tools, and had brains that, on average, were slightly larger than those of modern humans. Many European Neanderthal fossils show heavy brow ridges and a broad, flat skull, but others, particularly from areas around the eastern shores of the Mediterranean Sea, are somewhat more physically similar to *H. sapiens*.

Despite the physical and technological similarities between *H. neanderthalensis* and *H. sapiens*, there is no solid archaeological evidence that Neanderthals ever developed an advanced culture that included such characteristically human endeavors as art, music, and rituals. In general, the available evidence of the Neanderthal way of life is limited and open to different interpretations, and anthropologists are engaged in a sometimes heated debate about how advanced Neanderthal culture became.

Neanderthals and *Homo sapiens* May Have Interbred

Our understanding of the evolutionary relationship between *H. sapiens* and *H. neanderthalensis* has improved dramatically in recent years, thanks to evidence from ancient DNA. Researchers have sequenced the entire Neanderthal genome from DNA that was extracted from 38,000-year-old bones found in a cave in Croatia. Based on comparison of the

Neanderthal genome to whole-genome sequences of modern humans, researchers have deduced that the evolutionary branch leading to Neanderthals diverged from the ancestral human line at least 400,000 years ago, thousands of years before the emergence of modern *H. sapiens*. However, the sequence comparison also revealed that up to 4% of a modern non–African human's DNA is similar to distinctively Neanderthal sequences. This finding suggests that our ancestors interbred with Neanderthals, probably about 60,000 years ago. Because modern Africans do not carry the Neanderthal sequences but all other people do, interbreeding with Neanderthals must have occurred after *H. sapiens* had left Africa but before modern humans spread around the world.

Two Other *Homo* Species Survived Until Relatively Recently

Scientists' ability to obtain DNA from ancient bones has also revealed a previously unknown hominin. The new hominin's existence was discovered by sequencing DNA extracted from a single finger bone found in deposits laid down between 30,000 and 50,000 years ago in Denisova Cave in Siberia. Analysis of the sequence showed that the bone came from a hominin that is evolutionarily distinct from both *H. neanderthalensis* and *H. sapiens*. Though the Denisovan hominin is so far known only by its DNA, one bone, and two teeth, paleoanthropologists suspect that it's only a matter of time until skeletons turn up.

Skeletons did turn up beneath the floor of a cave on the Indonesian island of Flores, where researchers discovered 18,000-year-old bones that they at first believed to be the fossil skeleton of a human child. Closer examination of the skeleton, however, revealed that it belonged to a fully grown adult, but one that was no more than 3 feet tall. The researchers gave this creature the nickname "Hobbit." Unlike today's small humans, such as pygmies or pituitary dwarves, Hobbit had a very small brain, smaller even than the brain of a typical chimpanzee (**FIG. 18-16**). Furthermore, the shapes and arrangements of the bones in Hobbit's wrist, shoulder, and other parts of its skeleton were unlike those of anatomically modern humans. On the basis of these findings, researchers concluded that Hobbit is not simply a small *H. sapiens* but represents a different species, now named *Homo floresiensis*. Thus, it appears that modern humans at one time shared the planet (or at least some parts of it) not only with Neanderthals, but also with Denisovans and *H. floresiensis*.

Modern Humans Emerged Less Than 300,000 Years Ago

The fossil record shows that anatomically modern humans appeared in Africa at least 160,000 years ago and possibly as long as 300,000 years ago. The location of these fossils suggests that *Homo sapiens* originated in Africa, but most of our knowledge about our own early history comes from European and Middle Eastern fossils of *H. sapiens*, collectively known as Cro-Magnons (after the district in France in which their remains were first discovered). Cro-Magnons appeared about 90,000 years ago. They had domed heads, smooth brows, and prominent chins (just like us). Their tools were precision instruments similar to the stone tools that were still used in a few cultures as recently as the 1960s.

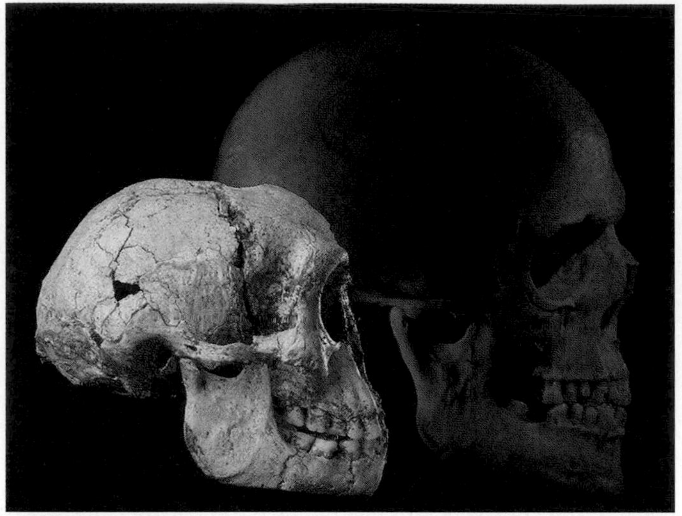

▲ **FIGURE 18-16 The hominin known as "Hobbit"** The skull of *Homo floresiensis,* a recently discovered diminutive human relative, is dwarfed by the skull of a modern *Homo sapiens*.

Behaviorally, Cro-Magnons seem to have been similar to, but more sophisticated than, Neanderthals. Artifacts from 30,000-year-old Cro-Magnon archaeological sites include elegant bone flutes, graceful carved ivory sculptures, and evidence of elaborate burial ceremonies (**FIG. 18-17**). Perhaps

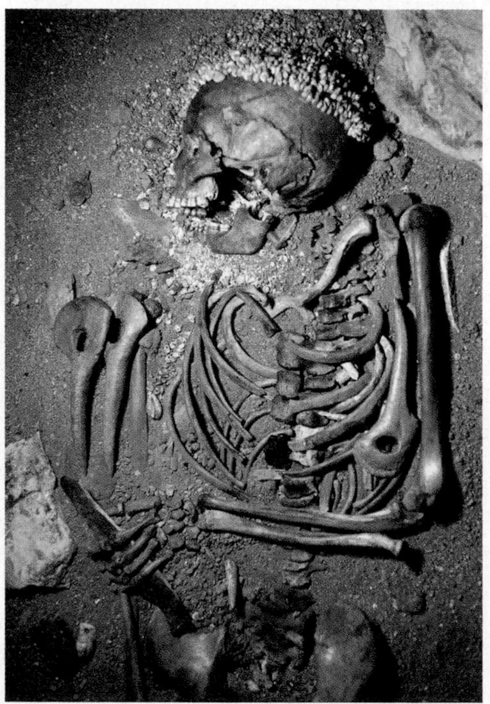

▲ **FIGURE 18-17 Paleolithic burial** This 24,000-year-old grave shows evidence that Cro-Magnon people ritualistically buried their dead. The body was covered with a dye known as red ocher and then buried with a headdress made of snail shells and a flint tool in its hand.

▲ **FIGURE 18-18 The sophistication of Cro-Magnon people** Cave paintings by Cro-Magnons have been remarkably preserved by the relatively constant underground conditions of the Chauvet-Pont-d'Arc cave in France.

the most remarkable accomplishment of Cro-Magnons is the magnificent art left in caves in places such as Altamira in Spain and Lascaux and Chauvet in France (**FIG. 18-18**). The oldest cave paintings so far found are more than 30,000 years old, and even these make use of sophisticated artistic techniques. No one knows exactly why these paintings were made, but they attest to minds as capable as our own.

Cro-Magnons and Neanderthals Lived Side by Side

Cro-Magnons coexisted with Neanderthals in Europe and the Middle East for perhaps as many as 50,000 years before the Neanderthals disappeared. The genetic analyses described earlier show that Cro-Magnons interbred with Neanderthals, so some researchers hypothesize that Neanderthals were essentially absorbed into the human genetic mainstream. Other scientists disagree, noting that the DNA evidence reveals only relatively limited interbreeding, and suggest that the later-arriving Cro-Magnons simply overran and displaced the less-well-adapted Neanderthals.

Several Waves of Hominins Emigrated from Africa

The human family tree is rooted in Africa, but hominins found their way out of Africa on numerous occasions. For example, *H. erectus* reached tropical Asia almost 2 million years ago and apparently thrived there, eventually spreading across the continent (**FIG. 18-19a**). Similarly, *H. heidelbergensis* made it to Europe at least 780,000 years ago. It is increasingly clear that genus *Homo* made repeated long-distance emigrations. What is less clear is how all this wandering is related to the origin of modern *H. sapiens*. According to the "African replacement" hypothesis (the basis of the scenario outlined earlier), *H. sapiens* emerged in Africa and dispersed less than 120,000 years ago, spreading into the Near East, Europe, and Asia and replacing all other hominins. But some paleoanthropologists believe that populations of *H. sapiens* evolved simultaneously in many regions from the already widespread populations of *H. erectus*. According to this "multiregional origin" hypothesis, continued migrations and interbreeding among *H. erectus* populations in different regions of the world maintained them as a single species as they gradually evolved into *H. sapiens* (**FIG. 18-19b**). Although an increasing number of studies of modern human DNA support the African replacement model, both hypotheses are consistent with the fossil record. Therefore, the question remains unsettled.

CASE STUDY | CONTINUED

Ancient DNA Has Stories to Tell

We might be able to more easily distinguish between the African replacement and multiregional origin hypotheses (and to answer a host of other unanswered questions about the origin and early evolution of *H. sapiens*) if we had access to DNA sequences from the earliest representatives of our genus. Is it possible that researchers will one day extract useable DNA from, say, early *H. erectus*? Perhaps, but the odds of success are not great. *H. erectus* fossils are up to 1.8 million years old, but the oldest ancient genome so far obtained is from a 700,000-year old fossil horse. What's more, the fossil horse was found in northern Canada, where the cold climate is excellent for preserving DNA. *H. erectus*, however, inhabited warmer regions where DNA degrades more quickly. Nonetheless, some evolutionary biologists hold out hope that useable DNA might be found in early *H. erectus* bones that fossilized in an environment conducive to preservation, perhaps in a deep cave or in oxygen-depleted underwater sediments. Could DNA survive for more than a million years under the right conditions? Current evidence says no, but then it wasn't too long ago that recovering Neanderthal DNA seemed like an impossible dream.

The Evolutionary Origin of Large Brains May Be Related to Meat Consumption and Cooking

The main physical features that distinguish us from our closest relatives, the apes, are our upright posture and large, highly developed brains. As described earlier, upright posture arose very early in hominin evolution, and hominins walked

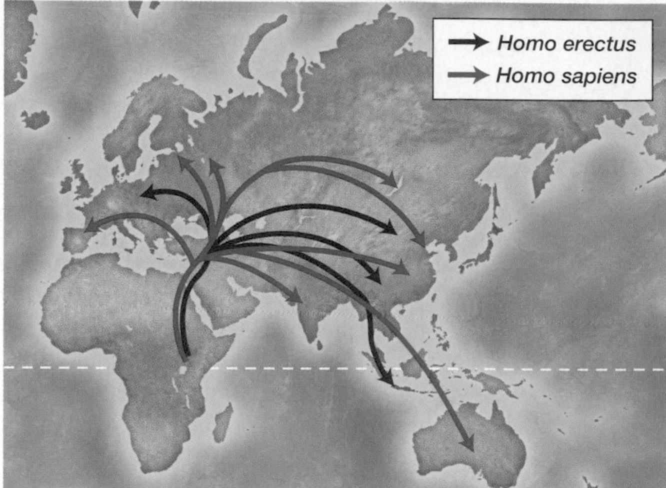

(a) African replacement hypothesis

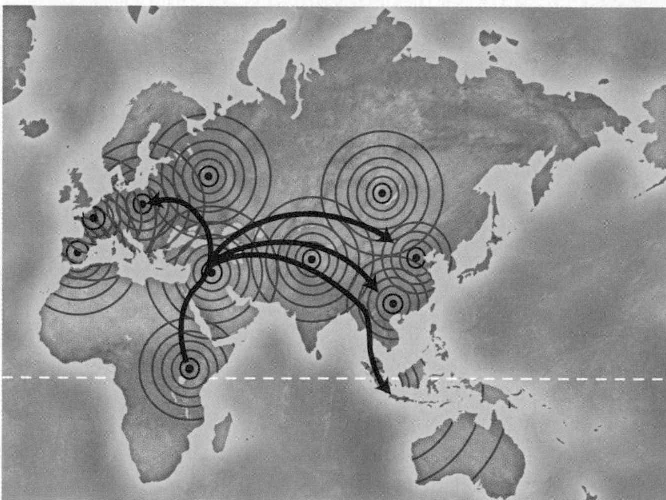

(b) Multiregional hypothesis

▲ **FIGURE 18-19 Competing hypotheses for the evolution of *Homo sapiens* (a)** The "African replacement" hypothesis suggests that *H. sapiens* evolved in Africa and then migrated throughout the Near East, Europe, and Asia, displacing the other hominin species that were present in those regions. **(b)** The "multiregional" hypothesis suggests that populations of *H. sapiens* evolved in many regions simultaneously from the already widespread populations of *H. erectus*.

THINK CRITICALLY Paleontologists recently discovered fossil hominins with features characteristic of modern humans in 160,000-year-old sediments in Africa. Which hypothesis does this new evidence support?

upright for several million years before large-brained *Homo* species arose. What circumstances might have caused the evolution of increased brain size? Many explanations have been proposed, but little direct evidence is available; hypotheses about the evolutionary origins of large brains are necessarily speculative.

One proposed explanation for the origin of large brains suggests that they evolved in response to increasingly complex social interactions. In particular, fossil evidence suggests that, beginning about 2 million years ago, hominin social life began to include a new type of activity—the cooperative hunting of large game. The resulting access to significant amounts of meat must have fostered a need to develop methods for distributing this valuable, limited resource among group members. Some anthropologists hypothesize that the individuals best able to manage this social interaction would have been more successful at gaining a large share of meat and using their share advantageously. Perhaps this social management was best accomplished by individuals with larger, more powerful brains, and natural selection therefore favored such individuals. Observations of chimpanzee societies have shown that the distribution of group-hunted meat often involves intricate social interactions in which meat is used to form alliances, repay favors, gain access to sexual partners, placate rivals, and so on. Perhaps the mental skill required to plan, assess, and remember such interactions was the driving force behind the evolution of our large, clever brains.

Whatever the nature of the advantages that favored individuals with larger brains, such brains could not have evolved without some mechanism to provide the large amount of energy necessary to grow and maintain a large volume of brain tissue. Some researchers speculate that cooking was the breakthrough that freed up the required extra energy. Cooked food is more digestible than raw food and requires far less chewing, so cooked food provides more nutrients with less effort expended. Thus, cooking by early hominins might have removed the limit that had previously restricted brain size. However, larger brains first arose in *H. erectus* at least 2 million years ago, and the earliest direct archaeological evidence of controlled fires is only 1 million years old. Proponents of the cooking hypothesis suggest that cooking actually did arise 2 million years ago, and the lack of evidence of cooking fires that old is simply due to the incompleteness of the hominin fossil record.

Sophisticated Culture Arose Relatively Recently

Even after the evolution of comparatively large brains in species such as *H. erectus*, more than a million years passed before the origin of modern humans and their extremely large brains. And even after the first appearance of modern *H. sapiens*, more than 100,000 years passed before the appearance of any archaeological evidence of the distinctively human characteristics that were made possible by a large brain: language, abstract thought, and advanced culture. The evolutionary origin of these human traits is another unresolved question, in part because direct evidence of the transition to advanced culture may never be found. Early humans capable of language and symbolic thought would not necessarily have created artifacts that indicated these capabilities. We can uncover some clues by studying our ape relatives, which possess less-complex versions of many human behaviors and mental processes. Their behavior might resemble that of ancestral hominins. Nonetheless, the late, seemingly rapid origin of advanced human culture remains a puzzle.

Biological Evolution Continues in Humans

Until recently, most evolutionary biologists agreed that the evolution of human bodies by natural selection had slowed or halted after we began to live in advanced societies. Today, however, our ever-growing ability to rapidly sequence DNA has made it possible for researchers to analyze sequences of a large and growing number of human genomes, and these analyses have led to a surprising conclusion: People have evolved rapidly since the advent of music, art, language, and the other hallmarks of advanced culture, and we continue to evolve today. Many of our genes show the telltale signs of evolution by natural selection in recent millennia. In many cases, the exact functions of these genes remain unknown, but researchers have determined the functions of some recent evolutionary changes. For example, the alleles required to digest milk as adults have arisen and become fixed in some populations within the past 4,000 years.

Culture Also Evolves

Human evolution in recent millennia has also included a great deal of cultural evolution, the evolution of information and behaviors that are transmitted from generation to generation by learning. Our recent evolutionary success, for example, was engendered not so much by new physical adaptations as by a series of cultural and technological revolutions. The first such revolution was the development of tools, which began with the early hominins. Tools increased the efficiency with which food and shelter could be acquired and thus increased the number of individuals that could survive within a given ecosystem. About 10,000 years ago, human culture underwent a second revolution as people discovered how to grow crops and domesticate animals. This agricultural revolution dramatically increased the amount of food that could be extracted from the environment, and the human population surged, increasing from about 5 million at the dawn of agriculture to around 750 million by 1750. The subsequent Industrial Revolution gave rise to the modern economy and its attendant improvements in public health. Longer lives and lower infant mortality led to truly explosive population growth, and today Earth's population is more than 7 billion and still growing.

Human cultural evolution and the accompanying increases in human population have had profound effects on the continuing biological evolution of other life-forms. Our agile hands and minds have transformed much of Earth's terrestrial and aquatic habitats. Humans have become the most powerful agent of natural selection. In the words of the late evolutionary biologist Stephen Jay Gould, "We have become, by the power of a glorious evolutionary accident called intelligence, the stewards of life's continuity on Earth. We did not ask for this role, but we cannot renounce it. We may not be suited for it, but here we are."

CHECK YOUR LEARNING

Can you . . .

- describe the evolutionary history of humans and the factors that may have fostered humans' distinctive adaptations?
- name and describe some characteristics of the hominin species that played key roles in humans' evolutionary history?
- describe the key features of the most recent phase of human evolution?

CASE STUDY REVISITED

Ancient DNA Has Stories to Tell

The unexpected discovery that humans interbred with Neanderthals was a triumph for the experts who developed the techniques for extracting, isolating, and sequencing ancient DNA. But perhaps the most stunning revelation made possible by ancient DNA was the discovery of the Denisovans, a hominin species whose existence would still be unknown if not for analysis of its ancient DNA. The fossil fragments from which the DNA was extracted were too few, too small, and too nondescript to have even been recognized as belonging to a previously unknown species. A newfound ability to identify new extinct species on the basis of DNA alone raises the intriguing possibility of future discovery of other previously unsuspected species, hominin and otherwise.

Like Neanderthals, Denisovans left a genetic trace in modern humans. One example is the Denisovan gene variant that helps Tibetans live at high altitude. Additionally, the people native to New Guinea and other Pacific islands carry a substantial number of Denisovan sequences. Almost 5% of the genome of these people is of Denisovan origin. This finding suggests that Denisovans interbred with the ancestors of Pacific Islanders, either in mainland Asia before the islands were first colonized by people, or later, if Denisovans were somehow able to get to multiple islands.

CONSIDER THIS Knowledge of the genomes of ancient hominins might help us better understand not only the evolutionary history of hominins, but also the traits that differ between us and our relatives—the traits that make us human. Our understanding of the functions of different genes is growing rapidly, so detailed comparisons of our genes to those present in, say, Neanderthals and Denisovans are very revealing. What do you think are the functions most likely to be related to the genetic differences between us and our hominin relatives?

CHAPTER REVIEW

Go to **Mastering Biology** to access the Pearson eText, vocabulary review, practice quizzes, activities, videos, current events, and more.

*Answers to **Think Critically** and **Thinking Through the Concepts** questions can be found in the **Answers** section at the back of the book.*

Summary of Key Concepts

18.1 How Did Life Begin?

Before life arose, energy from lightning, ultraviolet light, and heat formed organic molecules from water and the components of primordial Earth's atmosphere. The organic molecules formed probably included nucleic acids, amino acids, and lipids. By chance, some molecules of RNA may have had enzymatic properties, catalyzing the assembly of copies of themselves from nucleotides in Earth's waters. Protein-lipid vesicles enclosing these ribozymes may have formed protocells, the forerunners of life.

18.2 What Were the Earliest Organisms Like?

The oldest fossils, about 3.4 billion years old, are of prokaryotic cells that fed by absorbing organic molecules from their environment. Because there was no free oxygen in the atmosphere, their energy metabolism must have been anaerobic. As the cells multiplied, they depleted the organic molecules that had been formed by prebiotic synthesis. Some cells developed the ability to synthesize their own food molecules by using simple inorganic molecules and the energy of sunlight. These earliest photosynthetic cells were probably ancestors of today's cyanobacteria.

Photosynthesis releases oxygen as a by-product, and by about 2.4 billion years ago significant amounts of free oxygen had accumulated in the atmosphere. Aerobic metabolism, which generates more cellular energy than does anaerobic metabolism, probably arose about this time.

Eukaryotic cells had evolved by about 1.7 billion years ago. The first eukaryotic cells probably arose as symbiotic associations between predatory prokaryotic cells and other bacteria. Mitochondria may have evolved from aerobic bacteria engulfed by predatory cells. Chloroplasts may have evolved from photosynthetic bacteria by a similar process.

18.3 What Were the Earliest Multicellular Organisms Like?

Multicellular organisms evolved from eukaryotic cells and first appeared in the seas about 1.2 billion years ago. Multicellularity offers several advantages, including greater size. In plants, increased size offered some protection from predation. Specialization of cells allowed plants to anchor themselves in the nutrient-rich, well-lit waters near the shore. For animals, multicellularity allowed more efficient predation and more effective escape from predators. These in turn provided environmental pressures for faster locomotion, improved senses, and greater intelligence. Diverse animal forms appear in the fossil record beginning about 600 million years ago; fish were the predominant marine animals by about 400 million years ago.

18.4 How Did Life Invade the Land?

The first land organisms were probably algae. The first multicellular land plants appeared about 475 million years ago. Life on land required special adaptations for support of the body, reproduction, and the acquisition, distribution, and retention of water, but the land also offered abundant sunlight and freedom from aquatic herbivores. Soon after land plants evolved, arthropods invaded the land.

The earliest land vertebrates evolved from lobefin fishes, which had leglike fins and a primitive lung. A group of lobefins evolved into the amphibians about 370 million years ago. Reptiles evolved from amphibians, with several further adaptations for life on land. One reptile group, the birds, evolved feathers that provided insulation and facilitated flight. Mammals, whose bodies are insulated by hair, descended from a reptile group.

18.5 What Role Has Extinction Played in the History of Life?

The history of life has been characterized by constant turnover of species as some go extinct and are replaced by new ones. Mass extinctions, in which large numbers of species disappear within a relatively short time, have occurred periodically. Mass extinctions were probably caused by some combination of climate change and catastrophic events, such as volcanic eruptions and meteorite impacts.

18.6 How Did Humans Evolve?

One group of mammals evolved into the tree-dwelling primates. Some primates descended from the trees, and these were the ancestors of apes and humans. The oldest known hominin fossils are between 6 million and 7 million years old and were found in Africa. The australopithecines arose in Africa about 4 million years ago. These hominins walked erect, had larger brains than did their forebears, and fashioned primitive tools. One group of australopithecines gave rise to a line of hominins in the genus *Homo*. *Homo* arose in Africa, but populations of several *Homo* species migrated from Africa and spread to other geographic areas. In the last of these migrations, *Homo sapiens*, characterized by a large brain and advanced tool technology, dispersed from Africa to Asia and Europe.

Thinking Through the Concepts

Bloom's: Remembering, Understanding

Multiple Choice

1. Almost all of the oxygen gas in today's atmosphere is present as a result of
   a. outgassing from volcanoes.
   b. aerobic respiration.
   c. endosymbiosis.
   d. photosynthesis.

2. Extinction
   a. generally does not occur except during unpredictable mass extinctions.
   b. ordinarily occurs at a relatively slow but steady rate.
   c. has eliminated species at a faster rate than they have been formed.
   d. has not played a major role in the history of life.
3. In the endosymbiotic origin of the mitochondrion, the host cell benefited from the engulfed cell's ability to ___________; in the endosymbiotic origin of the chloroplast, the host cell benefited from the engulfed cell's ability to ___________.
   a. detoxify waste material; produce toxins that deterred predators
   b. use aerobic respiration to produce ATP; use photosynthesis to produce sugar
   c. use photosynthesis to generate oxygen; provide a shield against UV radiation
   d. move rapidly using flagella; form a cell nucleus
4. Which of the following does *not* list evolutionary events in chronological order (oldest to most recent)?
   a. first algae, first animals, first land plants
   b. first mammals, first reptiles, first hominins
   c. first amphibians, first dinosaurs, first flowering plants
   d. first photosynthesis, first exoskeletons, first hair
5. Which of the following lists hominin traits in the order in which they evolved?
   a. upright posture, large brain, symbolic culture
   b. large brain, upright posture, symbolic culture
   c. large brain, stone tools, symbolic culture
   d. upright posture, symbolic culture, stone tools

Fill-in-the-Blank

1. Because there was no oxygen in the earliest atmosphere, the first cells must have derived energy by ___________ metabolism of organic molecules. Oxygen was introduced into the atmosphere when some microbes developed the ability to ___________ and released oxygen as a by-product. Oxygen was ___________ to many of the earliest cells, but some cells evolved the ability to use oxygen in ___________ respiration, which provided far more ___________.
2. The molecule ___________ became a candidate for the first self-replicating information-carrying molecule when Thomas Cech and Sidney Altman discovered that some of these molecules can act as ___________, which they called ___________.
3. Complex cells that contain a nucleus and other organelles are called ___________ cells. A compelling explanation for the origin of these complex cells is the ___________ hypothesis. One observation that supports this hypothesis is that mitochondria have their own ___________.
4. The sperm of early land plants had to reach the egg by ___________, limiting them to ___________ environments. An important adaptation of plants to dry land was the evolution of ___________, which enclosed sperm in a drought-resistant coat.
5. Early plants that protected their seeds within cones are called ___________. These relied on ___________ to carry their pollen. Later, some plants evolved ___________, which attracted animals, particularly ___________ that carried their pollen. Animal pollination is much more ___________ than wind pollination.
6. The first animals to live on land were ___________ because their external skeletons, also called ___________, supported the animals' weight, while protecting their bodies from ___________.
7. Amphibians gave rise to ___________, which had three important adaptations to life on dry land: shelled, waterproof ___________; scaly, water-resistant ___________; and more efficient ___________.

Review Questions

1. What is the evidence that life might have originated from nonliving matter on early Earth?
2. How did the origin of photosynthesis affect subsequent evolution of life on Earth?
3. Explain the endosymbiont hypothesis for the origin of chloroplasts and mitochondria.
4. Name two advantages of multicellularity for plants and two for animals.
5. What advantages and disadvantages would terrestrial existence have had for the first plants to invade the land? For the first land animals?
6. Outline the major adaptations that emerged during the evolution of vertebrates from fish to amphibians to reptiles to birds and mammals. Explain how these adaptations increased the fitness of the various groups for life on land.
7. Outline the evolution of humans from early primates. Include in your discussion such features as binocular vision, grasping hands, bipedal locomotion, toolmaking, and brain expansion.

Applying the Concepts

Bloom's: Applying, Analyzing, Evaluating

1. Extinctions have occurred throughout the history of life on Earth. Why should we care if humans are causing a mass extinction event now?
2. In biological terms, what do you think was the most significant event in the history of life? Explain your answer.

19

Systematics: Seeking Order Amid Diversity

CASE STUDY

Origin of a Killer

ONE OF THE WORLD'S most frightening diseases is also one of its most mysterious. Acquired immune deficiency syndrome (AIDS) appeared seemingly out of nowhere, and when it was first recognized in the early 1980s, no one knew what caused it or where it came from. Scientists raced to solve the mystery and, within a few years, identified the infectious agent that causes AIDS: human immunodeficiency virus (HIV). Once HIV had been identified, researchers turned their attention to the question of its origin.

Finding the origin of HIV required an evolutionary approach. To ask "Where did HIV come from?" is really to ask "What kind of virus was the ancestor of HIV?" To answer this question, researchers began by identifying the closest relatives of HIV; when a biologist concludes that two viruses are closely related, it means that they share a recent common ancestor from which both evolved. Thus, comparing HIV with its closest relatives allowed researchers to infer the characteristics of their common ancestor.

The researchers who explored the ancestry of HIV discovered that its closest relatives are found not among other viruses that infect humans, but among those that infect monkeys and apes. In fact, the latest research on HIV's evolutionary history has concluded that the closest relative of HIV-1 (the type of HIV that is most responsible for the worldwide AIDS epidemic) is a virus strain that infects a chimpanzee subspecies that inhabits a limited range in West Africa. Therefore, the ancestor of the virus that we now know as HIV-1 did not evolve from a preexisting human virus. Instead, a chimpanzee virus must have acquired mutations that allowed it to infect humans and cause a deadly disease.

In this chapter, we examine *systematics*, the branch of biology that helped solve the mystery of HIV's origin. What methods did the researchers use to discover the relatives of HIV? What other kinds of questions can be answered with these methods?

Biologists studying the evolutionary history of type 1 human immunodeficiency virus (HIV-1) discovered that the virus, which causes AIDS, probably originated in chimpanzees.

AT A GLANCE

19.1 HOW ARE ORGANISMS NAMED AND CLASSIFIED?

To study and discuss organisms, biologists must name them. The branch of biology that is concerned with naming and classifying organisms is known as **taxonomy**. (A *taxon*—plural, *taxa*—is a named species or a named group of species). The basis of modern taxonomy was established by the Swedish naturalist Carl von Linné (1707–1778), who called himself Carolus Linnaeus, a Latinized version of his name. One of Linnaeus's most enduring achievements was the introduction of the two-part scientific name.

Each Species Has a Unique, Two-Part Name

The **scientific name** of an organism is a two-part Latin name that designates its genus and species. A **genus** is a group that includes a number of very closely related species; each **species** within a genus includes populations of organisms that can potentially interbreed under natural conditions. For example, the genus *Sialia* (bluebirds) includes three species: the eastern bluebird (*Sialia sialis*), the western bluebird (*Sialia mexicana*), and the mountain bluebird (*Sialia currucoides*) (**FIG. 19-1**). Although the three species are similar, bluebirds normally breed only with members of their own species.

In a scientific name, the genus name is presented first, followed by the species name. By convention, scientific names are always underlined or *italicized*. The first letter of the genus name is always capitalized, and the first letter of the species name is always lowercase. The species name is never used alone but is always paired with its genus name.

Each two-part scientific name is unique, so referring to an organism by its scientific name rules out any chance of ambiguity or confusion. For example, the bird *Gavia immer* is commonly known as the common loon in North America, as the northern diver in Great Britain, and by still other names in non-English-speaking countries. But the Latin scientific name *Gavia immer* is recognized by biologists worldwide, overcoming language barriers and allowing precise communication.

Modern Classification Emphasizes Patterns of Evolutionary Descent

In addition to naming species, biologists also classify them. Prior to the 1859 publication of Darwin's *On the Origin of Species*, classification served mainly to facilitate the study and discussion of organisms, much as a library's online catalog facilitates our ability to find a book. But after Darwin demonstrated that all organisms are linked by common ancestry, biologists began to recognize that classification ought to reflect and describe the pattern of evolutionary relatedness among organisms. Today, the process of classification focuses almost exclusively on reconstructing **phylogeny**, or evolutionary history. The science of reconstructing phylogeny is known as **systematics**. Systematists communicate their hypotheses about phylogeny by constructing evolutionary trees (see Fig. 17-11).

Systematists Identify Features That Reveal Evolutionary Relationships

As systematists seek to reconstruct the tree of life, they must do so without much direct knowledge of evolutionary history. Because systematists can't see into the past, they must infer it as best they can on the basis of similarities among living organisms. Not all similarities are useful for constructing phylogenetic trees, however. Some observed similarities stem from convergent evolution (see Chapter 15) in organisms that are not closely related, and such similarities are not useful for

(a) Eastern bluebird

(b) Western bluebird

(c) Mountain bluebird

FIGURE 19-1 Three species of bluebird Despite their obvious similarity, these three species of bluebird—**(a)** the eastern bluebird (*Sialia sialis*), **(b)** the western bluebird (*Sialia mexicana*), and **(c)** the mountain bluebird (*Sialia currucoides*)—evolve independently because they do not interbreed.

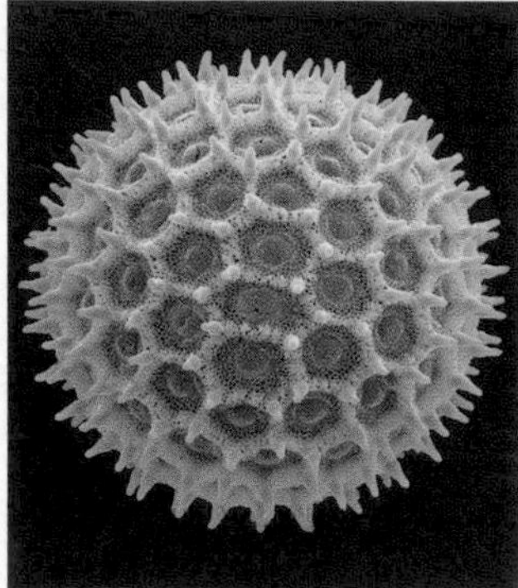

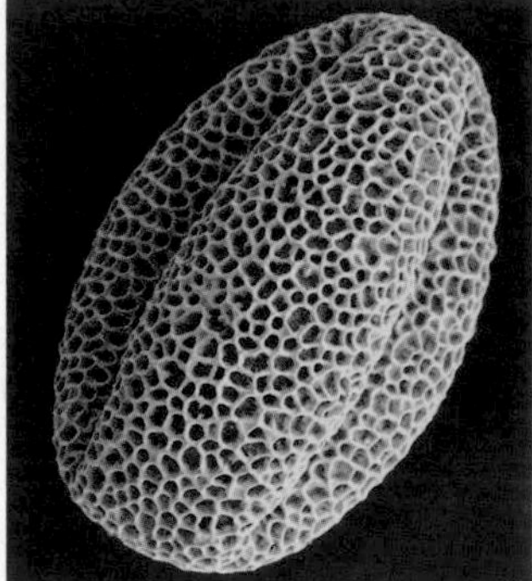

Morning glory pollen **Bitter melon pollen**

▲ **FIGURE 19-2 Microscopic structures may be used to classify organisms** The shape and surface features of pollen grains are among the finely detailed structures that can be useful in classification. Such structures can reveal similarities and differences between species that are not apparent in larger and more easily visible structures.

inferring evolutionary history. Instead, systematists use similarities that exist because two kinds of organisms both inherited a characteristic from a common ancestor. In the search for these informative similarities, biologists look at many kinds of characteristics.

Historically, the most important and useful distinguishing characteristics have been anatomical. Systematists look carefully at similarities in both external body structure and internal structures, such as skeletons and muscles. For example, homologous structures such as the finger bones of dolphins, bats, seals, and humans provide evidence of a common ancestor (see Fig. 15-8). To detect relationships among more closely related species, biologists may use microscopes to discern finer details, such as the external structure of the pollen grains of a flowering plant (**FIG. 19-2**).

Modern Systematics Relies on Molecular Similarities to Reconstruct Phylogeny

Recent advances in the techniques of molecular genetics have revolutionized studies of evolutionary relationships by allowing scientists to determine genetic similarities among organisms. Today's systematists rely mainly on the nucleotide sequences of DNA (that is, organisms' genotypes) to investigate relatedness among different types of organisms.

The logic underlying such molecular systematics is straightforward. It is based on the observation that when a single species divides into two species, the gene pool of each resulting species begins to accumulate mutations. The particular mutations present in each species' gene pool, however, will differ because the species are now evolving separately, with no gene flow between them. As time passes, more and more genetic differences accumulate. So a systematist who has obtained DNA sequences from representatives of both species can compare the two species' nucleotide sequences. Fewer differences indicate more closely related organisms (species with a relatively recent common ancestor).

In some cases, similarity of DNA sequences will be reflected in the structure of chromosomes. For example, both the DNA sequences and the chromosomes of chimpanzees and humans are extremely similar, showing that these two species shared a common ancestor in the not too distant past (**FIG. 19-3**).

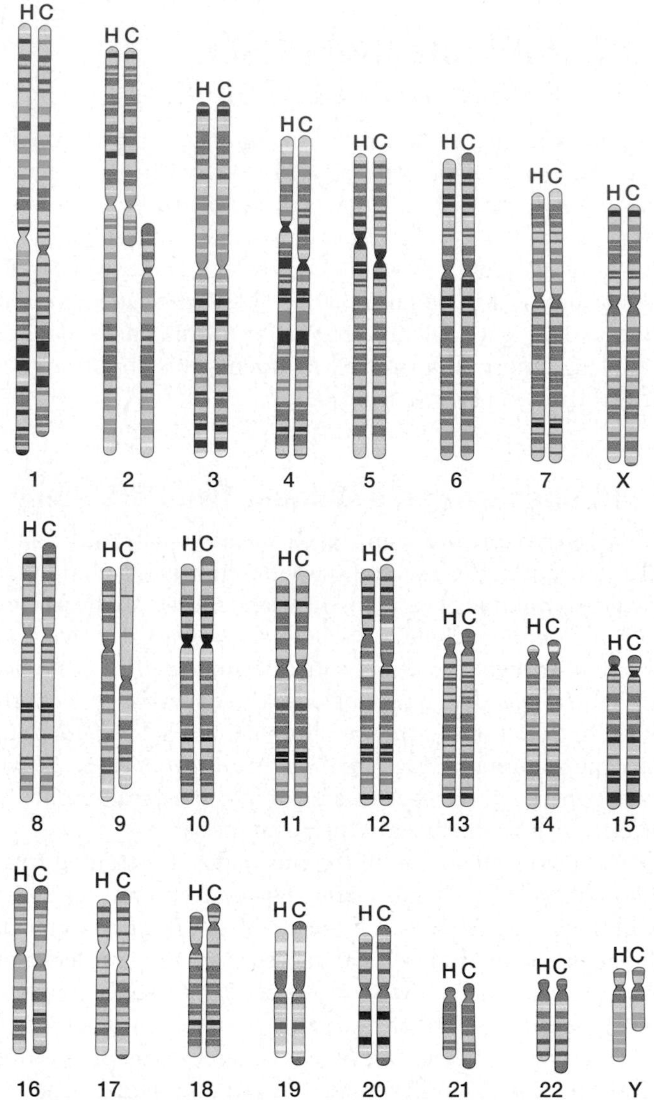

▲ **FIGURE 19-3 Human and chimpanzee chromosomes are similar** Chromosomes from different species can be compared by means of banding patterns that are revealed by staining. The comparison illustrated here, between human chromosomes (left member of each pair; H) and chimpanzee chromosomes (C), reveals that the two species are genetically very similar. The numbering system shown is that used for human chromosomes; note that human chromosome 2 corresponds to a combination of two chimp chromosomes. Data from Yunis, J. J., et al. 1980. *Science* 208:1145–1148.

THINK CRITICALLY Analysis of human chromosome 2 revealed that it contains both a functional centromere and the remnants of a second one. What does this finding suggest about the evolutionary origin of chromosome 2?

CASE STUDY CONTINUED

Origin of a Killer

Analysis of nucleotide sequences was absolutely required to construct the phylogeny of viruses that revealed the ancestor of HIV. The closely related viruses included in the phylogeny are all but indistinguishable on the basis of appearance and structure; the differences between them are revealed only by their nucleotide sequences. Thus, scientific sleuthing about the origin of HIV would have been impossible before the modern era of routine DNA sequencing.

DNA analysis has also shown that HIV is a member of a group called the lentiviruses. How do taxa get their names? Can the names tell us anything about the evolutionary histories of taxa?

Systematists Name Groups of Related Species

Although systematists communicate their findings about phylogeny mainly by presenting evolutionary trees, they also name groups of species. In keeping with their emphasis on reconstructing evolutionary history, systematists give formal names only to groups that include all the organisms descended from a common ancestor. Such groups are known as **clades**. If you examine an evolutionary tree, you will see that clades can be arranged in a hierarchy, with smaller clades nested within larger ones (**FIG. 19-4**).

Ranks Add Information to the Clade Name

When systematists name a clade, the name itself does not convey much information about the clade. For example, the name does not reveal much about the clade's size or breadth. Is the clade a relatively large, broadly inclusive one, such as the one that includes all mammals? Or is it a smaller, narrower clade, perhaps one that includes only the three species of zebra? The clade's size and breadth would be obvious if we could view its evolutionary tree, but if we don't have access to the tree, we will need clues other than the clade's name in order to understand its scope.

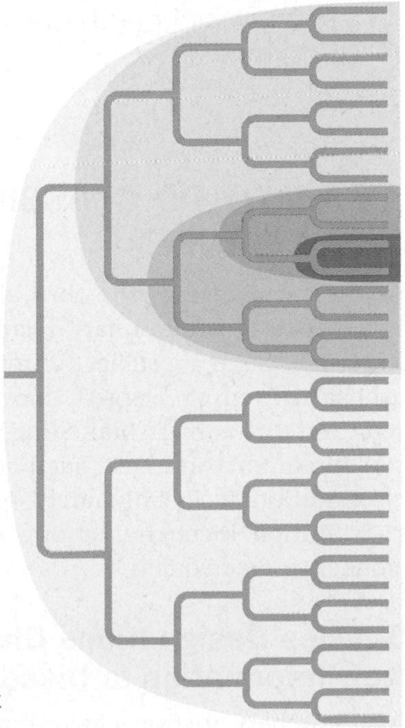

One possible way to signal the relative size and inclusiveness of named clades is to place them into categories called *taxonomic ranks*. This approach has a long history; Linnaeus placed each species into a series of ranked categories on the basis of its resemblance to other species. The Linnaean classification system eventually came to include eight major ranks: *domain, kingdom, phylum, class, order, family, genus,* and *species*. These ranks form a nested hierarchy in which each level includes all of the other levels below it; each domain contains a number of kingdoms; each kingdom contains a number of phyla; each phylum includes a number of classes; each class includes a number of orders; and so on. As we move down the hierarchy, smaller and smaller groups are included. Thus, if you knew that a certain named clade was a phylum, and a second named clade was a genus within the phylum, you would understand that the second clade was a subset of the first one.

FIGURE 19-4
Clades form a nested hierarchy Any group that includes all the descendants of a common ancestor is a clade. Some of the clades represented on this evolutionary tree are shaded in different colors. Note that smaller clades nest within larger clades.

Use of Taxonomic Ranks Is Declining

Although the use of taxonomic ranks has a long history, today's systematists have de-emphasized the Linnaean ranking system. Historically assigned ranks may misrepresent evolutionary history as it is currently understood, and implementing a scientifically sound revision of the ranking system would present some difficult technical challenges. These days, many systematists do not assign taxonomic ranks to the clades they name and instead concentrate on using data to construct accurate evolutionary trees, rather than on subjective evaluations of whether a given clade should be called a kingdom, a phylum, a class, an order, or a family. As a result, use of Linnaean taxonomic ranks is declining.

In the chapters of this text that describe the diversity of life (Chapters 20–25), our use of taxonomic ranks will vary according to the practices of the biologists who study the different organisms we discuss. In most chapters, we will follow the emerging convention of avoiding ranks, instead using the term "taxonomic group" (as a synonym for clade) to describe a collection of related species. In some chapters, however, we will make selective use of a few Linnaean ranks. For example, we will follow the tradition of using "kingdom" to refer to the three clades that contain, respectively, all animals, all plants, and all fungi. Similarly, in the two chapters about animals, we will designate certain clades as phyla, in keeping with tradition in animal systematics. And we will refer to the three broadest, most inclusive of life's clades as **domains**.

CHECK YOUR LEARNING

Can you . . .

- explain why scientific names are necessary?
- describe the type of similarities that systematists use to reconstruct phylogeny?
- describe the system of Linnaean taxonomic ranks?

19.2 WHAT ARE THE DOMAINS OF LIFE?

If we picture the common ancestor of all living things as the trunk at the very base of the tree of life, we might ask: Which clades arose from the earliest branching of the trunk?

By the 1970s, most systematists had concluded from the evidence then available that early splits in the tree of life divided all species into five kingdoms. The five-kingdom system placed all prokaryotic organisms into a single kingdom and divided the eukaryotes into four kingdoms. Among the eukaryotes, the five-kingdom system recognized three kingdoms of multicellular organisms (plants, animals, and fungi) and placed all of the remaining, mostly single-celled, eukaryotes in a single kingdom.

As new data accumulated and understanding of phylogeny grew, however, scientific assessment of life's fundamental categories was gradually revised. A key element of this revision stemmed from the pioneering work of microbiologist Carl Woese, who showed that biologists had overlooked a key event in the early history of life, one that demanded a new and more evolutionarily accurate classification.

Woese and other biologists interested in the evolutionary history of microorganisms studied the biochemistry of prokaryotic organisms. The researchers, by studying nucleotide sequences of the genes that encode the RNA that is found in organisms' ribosomes, discovered that prokaryotes fall into two large groups, each with its own distinctive version of ribosomal RNA. Woese dubbed these two groups the Bacteria and the Archaea (**FIG. 19-5**). The very large number of differences between the ribosomal RNA sequences of Bacteria and Archaea indicate that their common ancestor may have lived more than 3 billion years ago.

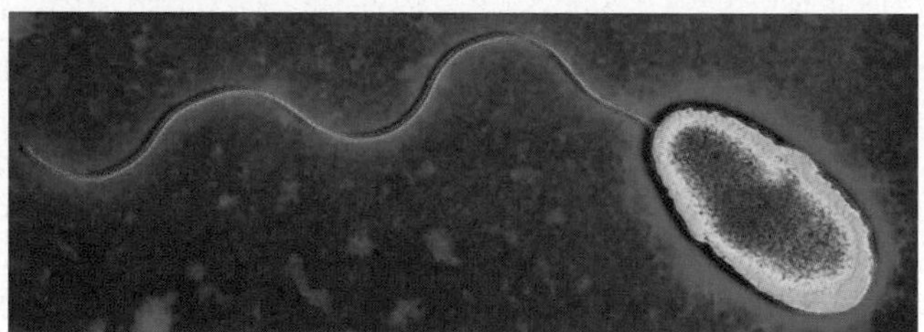

(a) A bacterium

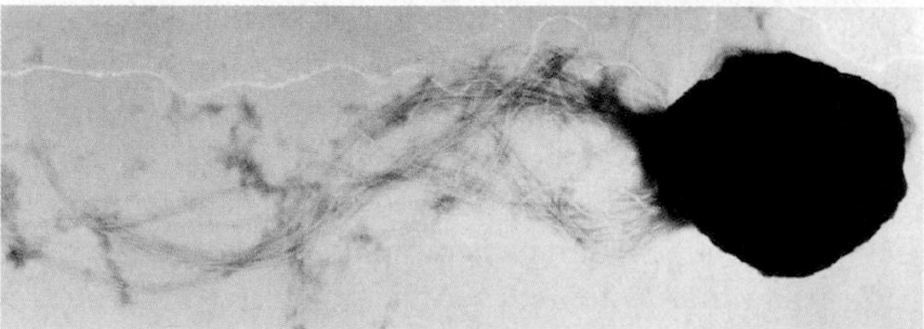

(b) An archaean

▲ **FIGURE 19-5 Two domains of prokaryotic organisms** Although similar in appearance, **(a)** *Pseudomonas aeruginosa*, in the domain Bacteria, and **(b)** *Methanococcus jannaschii*, in the domain Archaea, are less closely related than a mushroom and an elephant.

THINK CRITICALLY Given that bacteria and archaea are only very distantly related, why are they often similar in appearance?

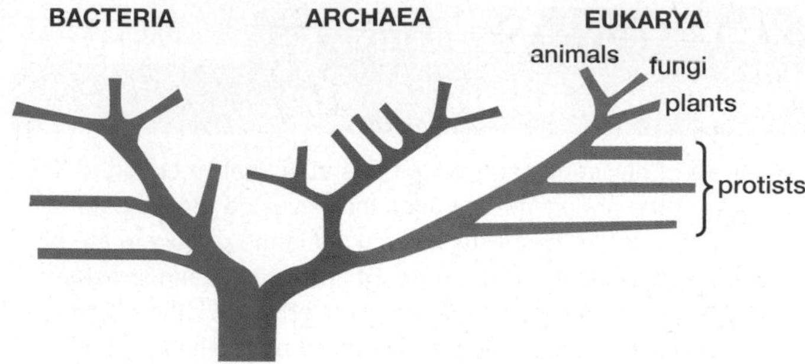

▲ **FIGURE 19-6 The tree of life** The three domains of life represent the three main "branches" on the tree of life. The term "protist" refers to the many eukaryotes that are not plants, animals, or fungi.

Despite superficial similarities in their appearance under the microscope, Bacteria and Archaea differ at the most fundamental molecular level; for example, they differ dramatically in the chemical composition of their cell walls and in the structure of their RNA polymerase molecules.

Bacteria and Archaea are no more closely related to one another than either one is to any eukaryote. The tree of life split into three parts very early in the history of life, long before the appearance of plants, animals, and fungi. This early split is reflected in a modern classification scheme that divides life into three domains: **Bacteria**, **Archaea**, and **Eukarya** (**FIG. 19-6**). Eukarya includes all organisms with eukaryotic cells, including plants, fungi, animals, and an array of clades of mostly singled-celled organisms collectively known as *protists*. **FIGURE 19-7** shows the evolutionary relationships among some members of the domain Eukarya.

CHECK YOUR LEARNING

Can you . . .

- name and briefly describe the three domains of life?
- explain how scientists discovered that prokaryotes fall into two domains?

19.3 WHY DO CLASSIFICATIONS CHANGE?

As the emergence of the three-domain system shows, the hypotheses of evolutionary relationships on which classification is based are subject to revision as new data become available. Even the largest, most inclusive clades—which represent the earliest branchings of the tree of life—must sometimes be rearranged. Such changes at the top levels of classification occur only rarely, but at the other end of the classification hierarchy, among species designations, revisions are more frequent.

Species Designations Change When New Information Is Discovered

As researchers uncover new information, systematists regularly propose changes in species-level classifications. For

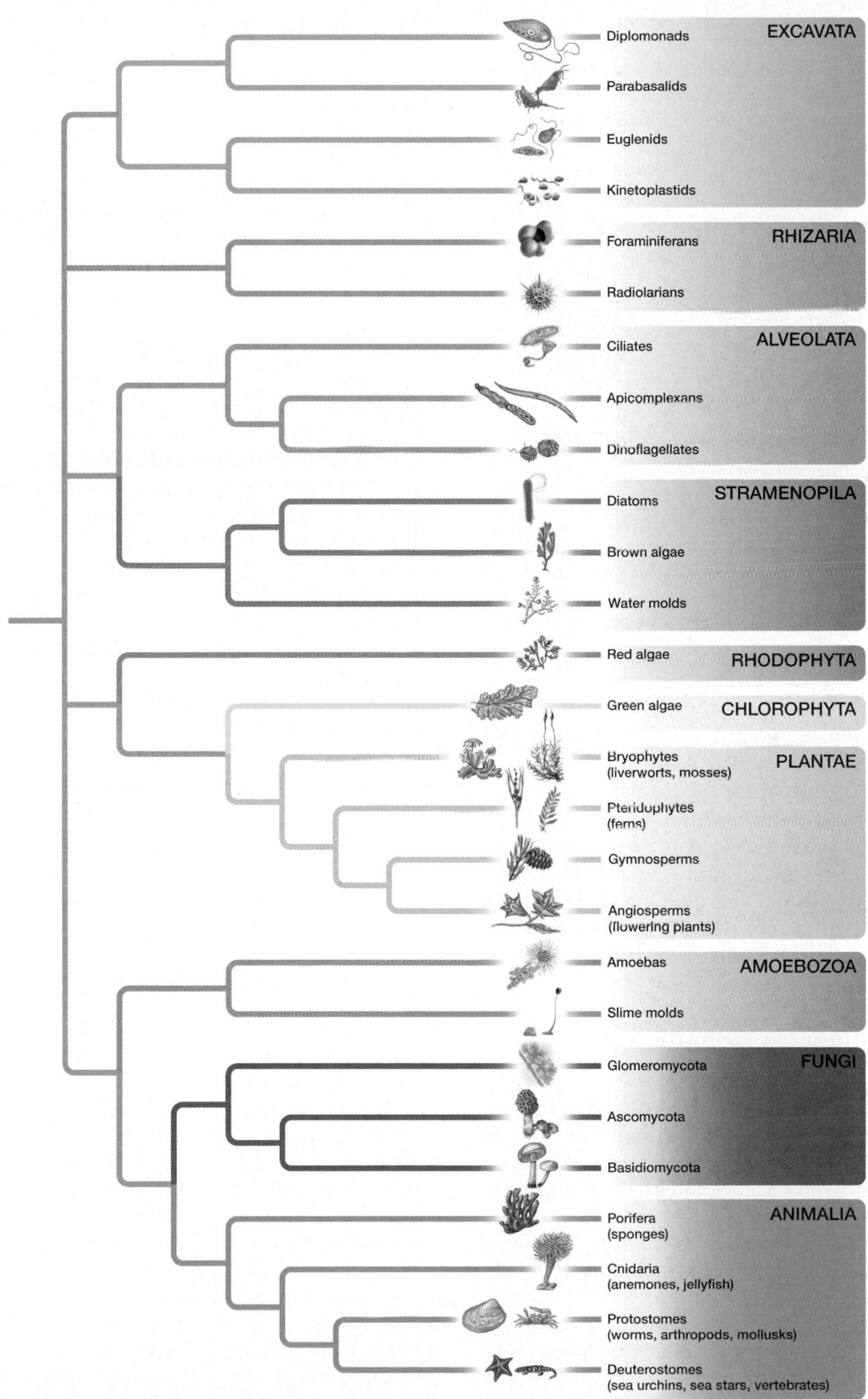

▲ **FIGURE 19-7 An evolutionary tree of eukaryotes** Some of the major evolutionary lineages within the domain Eukarya are shown.

Have You Ever Wondered ... When People Started Wearing Clothes?

Clothing decays pretty rapidly, so archeological discoveries can't answer the question of when our naked ancestors first started wearing clothes. But you might be surprised to learn that systematists can provide an answer. You might be even more surprised to learn that the answer comes from systematists who study lice. Most species of these parasitic insects live on only one host species, and human lice are no exception. The lice that can infest people include head lice that can survive only on the scalp, and body lice that can survive only in clothing (they leave the clothes temporarily to feed on the body, but can't thrive there for long). Systematists comparing the DNA of a number of louse species found that human head lice and body lice are close relatives. And, using a method known as the molecular clock, they estimated that the split between the two occurred between 80,000 and 170,000 years ago, most likely at the earlier end of that range. Because the common ancestor of the two kinds of lice could not have moved from the head to the body until there were clothes to live in, people must have been wearing clothes at least 80,000 years ago, probably even earlier. This timing suggests that we were already clothed when we left Africa and began to spread across Earth.

Body louse

example, until recently, systematists recognized two species of elephant, the African elephant and the Indian elephant. Now, however, we recognize three elephant species; the former African elephant is now divided into two species, the savanna elephant and the forest elephant. Why the change? Genetic analysis of elephants in Africa revealed that there is little gene flow between forest-dwelling and savanna-dwelling elephants. It turns out that the two groups of elephants are no more genetically similar than lions are to tigers.

The Biological Species Definition Can Be Difficult or Impossible to Apply

In some cases, systematists find themselves unable to say with certainty where one species ends and another begins. As discussed earlier (see Chapter 17), asexually reproducing organisms pose a particular challenge to systematists, because the criterion of interbreeding (the basis of the biological species definition that we have used in this text) cannot be used to distinguish among species. The irrelevance of this criterion in studies of asexual organisms leaves plenty of room for investigators to disagree about which asexual populations constitute a species, especially when comparing groups with similar phenotypes.

The difficulty of applying the biological species definition to asexual organisms applies to a significant portion of Earth's organisms. Most bacteria, archaea, and protists, for example, reproduce asexually most of the time. Some systematists argue that we need a more universally applicable definition of species, one that won't exclude asexual organisms and that doesn't depend on the criterion of reproductive isolation.

A number of alternative species definitions have been proposed, but none has been sufficiently compelling to displace completely the biological species definition. One alternative definition, however, has been gaining adherents in recent years. The *phylogenetic species concept* defines a species as the smallest possible group whose members descended from a common ancestor and share defining characteristics that distinguish them from other groups. In other words, if we draw an evolutionary tree that describes the pattern of ancestry among a collection of organisms, each distinctive branch on the tree constitutes a separate species. As you might suspect, rigorous application of the phylogenetic species concept would vastly increase the number of different species recognized by systematists.

Proponents and opponents of the phylogenetic species concept are currently engaged in a vigorous debate about the merits of this definition of species. Perhaps one day the phylogenetic species concept will replace the biological species concept as the "textbook definition" of species. In the meantime, classifications will continue to be debated and revised as systematists learn more about evolutionary relationships, particularly with the application of techniques used in molecular genetics.

CHECK YOUR LEARNING

Can you . . .

- explain why phylogenetic classifications sometimes change?
- describe the limitations of the biological species concept and explain how it differs from the phylogenetic species concept?

19.4 HOW MANY SPECIES EXIST?

The challenge of reconstructing the evolutionary history of Earth's species is complicated by the fact that most species remain undiscovered and undescribed. Scientists do not know exactly how many species share our world. Each year, around 15,000 new species are named, most of them insects, many from tropical rain forests. The total number of named species is currently about 1.6 million. However, scientists agree that many more species exist. A recent, well-received study estimated that 8.7 million species are present on Earth, and other studies have produced even higher estimates.

The number and variety of Earth's species constitute its **biodiversity**. Of all the species that have been identified thus far, about 5% are prokaryotes and protists. An additional 20% or so are plants and fungi, and the rest are animals. This distribution has little to do with the actual diversity of these organisms and a lot to do with the size of the organisms, how easy they are to classify, how accessible they are, and the number of scientists studying them. Historically, systematists have chiefly focused on large or conspicuous organisms in temperate regions, but biodiversity is greatest among small, inconspicuous organisms in the Tropics. In addition to the overlooked species on land and in shallow waters, an entire "continent" of species lies largely unexplored on the deep-sea floor. From the relatively limited samples available, scientists estimate that hundreds of thousands of unknown species may reside there.

Although about 10,000 species of prokaryotes have been described and named, most prokaryotic diversity remains undiscovered. Consider a study by Norwegian scientists, who analyzed DNA to count the number of different bacterial species present in a small sample of forest soil. To distinguish among species, the researchers arbitrarily defined bacterial DNA as coming from separate species if it differed by at least 30% from that of any other bacterial DNA in the sample. Using this criterion, they reported more than 4,000 species of bacteria in their soil sample and an equal number of species in a sample of shallow marine sediment.

Our ignorance of the full extent of life's diversity adds a new dimension to the tragedy of the destruction of tropical rain forests. Although these forests cover only about 6% of Earth's land area, they are believed to be home to two-thirds of the world's existing species, most of which have never been studied or named. Because these forests are being destroyed so rapidly, Earth is losing many species that we will never even know existed! Consider this: In 1990, a new species of primate, the black-faced lion tamarin, was discovered in a small patch of dense rain forest on an island just off the east coast of Brazil (**FIG. 19-8**). Had the patch of forest been cut before this squirrel-sized monkey was discovered, its existence would have remained undocumented. At current rates of deforestation, most of the tropical rain forests, with their undescribed wealth of life, will be gone by the end of the century.

CHECK YOUR LEARNING

Can you . . .

- explain why the number of described species is much lower than the actual number?
- explain why it is difficult to accurately estimate how many species are on Earth?

▶ FIGURE 19-8 The black-faced lion tamarin Researchers estimate that no more than 400 individuals remain in the wild; captive breeding may be the black-faced lion tamarin's only hope for survival.

CASE STUDY REVISITED

Origin of a Killer

What evidence has persuaded evolutionary biologists that HIV originated in apes and monkeys? To understand the evolutionary thinking behind this conclusion, examine the evolutionary tree shown in **FIGURE 19-9**. This tree illustrates the phylogeny of HIV and its close relatives, the simian immunodeficiency viruses (SIVs), as revealed by a comparison of RNA sequences among different viruses.

Notice the positions on the tree of the four human viruses (two strains of HIV-1 and two of HIV-2; a strain is a genetically distinct subgroup of a particular type of virus). The branch leading to strain 1 of HIV-1 is directly adjacent to the branch leading to strain 1 of chimpanzee SIV. These adjacent branches indicate that strain 1 of HIV-1 is more closely related to a chimpanzee virus than to strain 2 of HIV-1. Similarly, strain 1 of HIV-2 is more closely related to pig-tailed macaque SIV than to strain 2 of HIV-2.

(continued on next page)

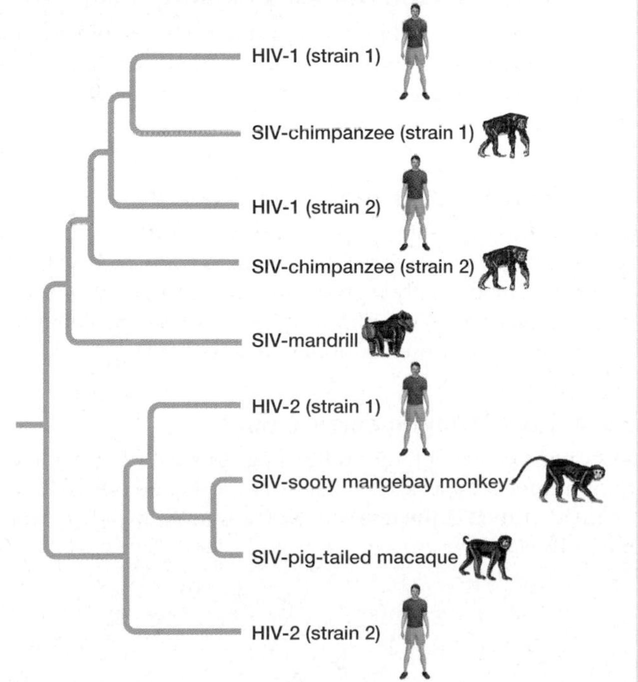

▶ FIGURE 19-9 Evolutionary analysis helps reveal the origin of HIV In this phylogeny of some immunodeficiency viruses, the viruses with human hosts do not cluster together. This lack of congruence between the evolutionary histories of the viruses and their host species suggests that the viruses must have jumped between host species.

Both HIV-1 and HIV-2 are more closely related to ape or monkey viruses than to one another.

The only way for the evolutionary history shown in the tree to have emerged is if viruses jumped between host species. If HIV had evolved strictly within human hosts, the human viruses would be each other's closest relatives. Because the human viruses do not cluster together on the phylogenetic tree, we can infer that cross-species infection occurred, probably on multiple occasions. The most likely means of transmission is human consumption of monkeys (HIV-2) and chimpanzees (HIV-1). Recently, a strain of SIV that is especially closely related to HIV-1 was found in members of a population of chimpanzees that inhabit forests in the southeastern corner of the Central African country of Cameroon. It is likely that viruses from this chimpanzee population made the chimpanzee-to-human jump that started the HIV epidemic.

CONSIDER THIS Can understanding the evolutionary origin of HIV help researchers devise better ways to treat AIDS and control its spread? More generally, how can evolutionary thinking help advance medical research?

CHAPTER REVIEW

Go to **Mastering Biology** to access the Pearson eText, vocabulary review, practice quizzes, activities, videos, current events, and more.

Answers to ***Think Critically*** *and* ***Thinking Through the Concepts*** *questions can be found in the* ***Answers*** *section at the back of the book.*

Summary of Key Concepts

19.1 How Are Organisms Named and Classified?

The scientific name of an organism is composed of its genus name and species name. Systematists use anatomical and molecular similarities among organisms to reconstruct the evolutionary relationships among species, and depict the results of their reconstructions in tree diagrams. On the basis of these evolutionary trees, systematists name clades (groups that include the species descended from a common ancestor). Clades nest within larger clades to form a nested hierarchy of categories. In Linnaean classification, the different clades in a hierarchy are assigned taxonomic ranks. The eight major ranks, in order of decreasing inclusiveness, are domain, kingdom, phylum, class, order, family, genus, and species.

19.2 What Are the Domains of Life?

The three domains of life, each representing one of three main branches of the tree of life, are Bacteria, Archaea, and Eukarya. Plants, fungi, and animals are among the clades within the domain Eukarya.

19.3 Why Do Classifications Change?

Classifications are subject to revision as new information is discovered. Species boundaries may be hard to define, particularly in the case of asexually reproducing species. However, systematics is essential for precise communication and contributes to our understanding of the evolutionary history of life.

19.4 How Many Species Exist?

Although only about 1.6 million species have been named, the total number of species is 8.7 million or higher. New species are being identified at the rate of 15,000 annually, mostly in tropical rain forests.

Thinking Through the Concepts

Bloom's: Remembering, Understanding

Multiple Choice

1. To say that species A is more closely related to species B than to species C is to say that A and B
   a. have a more ancient common ancestor than do A and C.
   b. have a more recent common ancestor than do A and C.
   c. are more similar in appearance than are A and C.
   d. live in the same area, but C does not live there.
2. To be informative for reconstructing the phylogeny of a group of taxa, a characteristic must be
   a. present in all the taxa due to its presence in a common ancestor.
   b. present in all the taxa due to convergent evolution.
   c. an anatomical feature.
   d. a DNA sequence.
3. Which of the following illustrates the correct way to present the scientific name of humans?
   a. *Homo Sapiens*
   b. Homo sapiens
   c. *homo sapiens*
   d. *Homo sapiens*
4. In modern systematics, classifications are expected to
   a. group organisms that share similar appearance.
   b. include Linnaean taxonomic ranks.
   c. reflect evolutionary history.
   d. never change once established.
5. Which of the following includes all the domains that consist of prokaryotic organisms?
   a. Bacteria and Fungi
   b. Archaea and Eukarya
   c. Bacteria and Archaea
   d. Bacteria and Eukarya

Fill-in-the-Blank

1. The science of naming and classifying organisms is called ____________. The related science of reconstructing and depicting evolutionary history is called ____________. A group consisting of all organisms descended from a particular common ancestor is a(n) ____________.

2. A scientific name consists of a(n) ____________ name followed by a(n) ____________ name. Both parts of a scientific name are in ____________ (a language). The first letter of the first word in a scientific name is always ____________, and both parts of the name are printed in ____________ letters.
3. In Linnaean classification, the eight major taxonomic ranks, in descending order of inclusiveness, are ____________, ____________, ____________, ____________, ____________, ____________, ____________, and ____________. The three domains of life are ____________, ____________, and ____________.
4. Systematists determine the evolutionary relationships among species mainly on the basis of similarities in ____________ and ____________.
5. The biological species definition is difficult to apply to organisms that ____________. An alternative species definition, known as the ____________, does not have this limitation.
6. The number of named species is about ____________, but the actual number of species on Earth is estimated to be about ____________ or higher.

Review Questions

1. What contributions did Linnaeus and Darwin make to modern taxonomy?
2. What features would you study to determine whether a dolphin is more closely related to a fish or to a bear?
3. What techniques might you use to determine whether the extinct cave bear is more closely related to a grizzly bear or to a black bear?
4. Only a small fraction of the total number of species on Earth has been scientifically described. Why?
5. In England, "daddy longlegs" refers to a long-legged fly, but the same name refers to a spider-like animal in the United States. How do scientists attempt to avoid such confusion?
6. Why are species designations of asexually reproducing organisms more likely to differ among different systematists than are the species designations of sexually reproducing organisms?

Applying the Concepts

Bloom's: Applying, Analyzing, Evaluating

1. The pressures created by human population growth and economic expansion place storehouses of biological diversity such as the Tropics in peril. The seriousness of the situation is clear when we consider that probably only 1 out of every 20 tropical species is known to science. What arguments can you make for preserving biological diversity in poor and developing countries, such as those in many areas of the Tropics? Does such preservation require that these countries sacrifice economic development? Suggest some solutions to the conflict between the growing demand for resources and the importance of conserving biodiversity.
2. During major floods, only the topmost branches of submerged trees may be visible above the water. If you were asked to sketch the branches below the surface of the water solely on the basis of the positions of the exposed tips, you would be attempting a reconstruction somewhat similar to the "family tree" by which systematists link various organisms according to their common ancestors (analogous to branching points). What sources of error do both exercises share? What advantages do modern systematists have?
3. Consider the following list of groups: (1) protists, (2) fungi, (3) seedless plants (ferns, mosses, and liverworts), (4) prokaryotes (bacteria and archaea), and (5) animals. Using Figures 19-6, 19-7, 22-4, 23-5, and 24-1 for reference, identify the monophyletic groups on the list.

20 The Diversity of Prokaryotes and Viruses

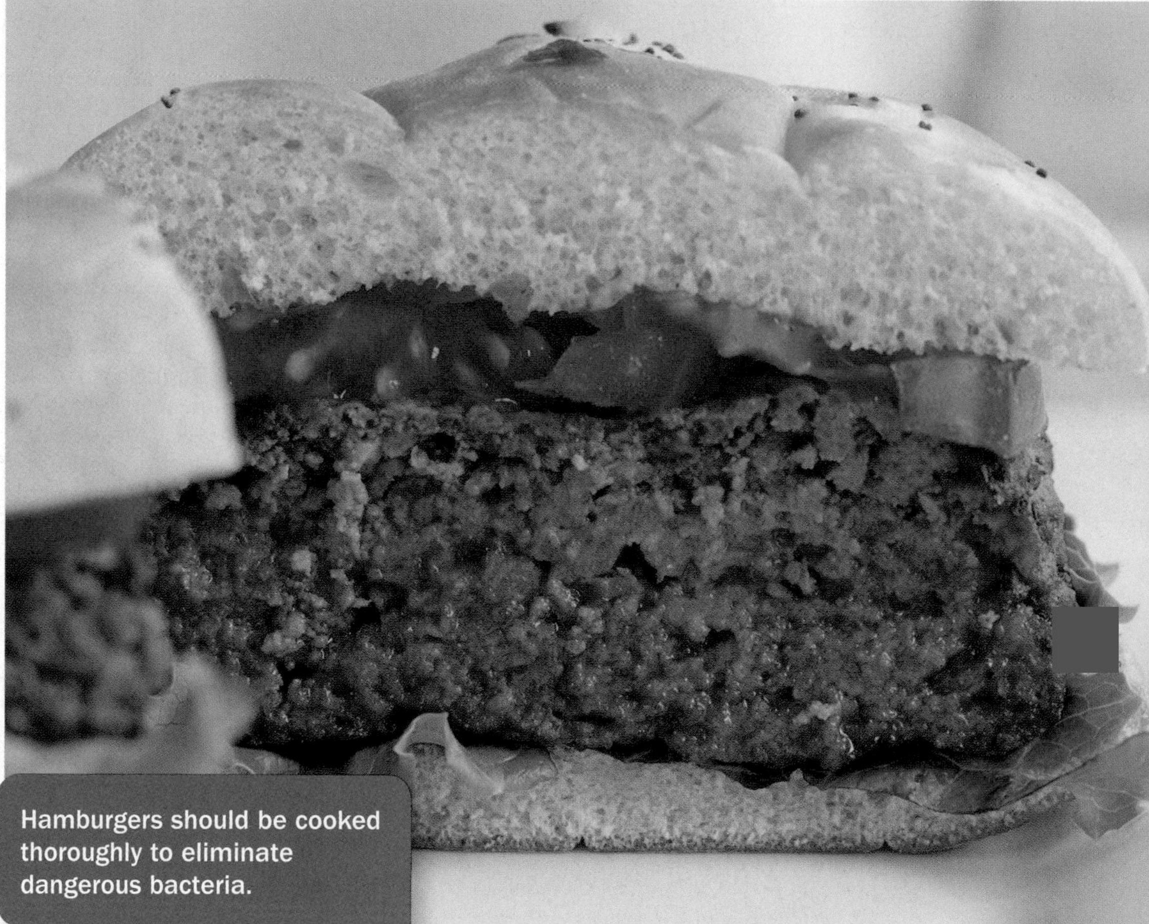

Hamburgers should be cooked thoroughly to eliminate dangerous bacteria.

CASE STUDY

Unwelcome Dinner Guests

ONE EVENING during his time as a student at the University of Michigan, Andrew Lekas ate a burrito at a favorite restaurant near campus. A few hours later, he began to feel sick, with symptoms that included vomiting, diarrhea, headache, and weakness. The illness ultimately became severe enough to require a hospital stay and more than a week in bed. What sickened Lekas? The lettuce in his burrito was contaminated.

As bad as Lekas's experience was, it could have been much worse. Other victims of foodborne illness have suffered more serious consequences. For example, Stephanie Smith, a former dance instructor from Minnesota, is paralyzed from the waist down as a result of the severe illness (hemolytic uremic syndrome) she developed after eating a tainted hamburger.

Regrettably, the ill effects of consuming contaminated food are all too common. The Centers for Disease Control and Prevention estimates that U.S. residents experience an astonishing 48 million cases of foodborne illness each year. Some of these cases are severe. For example, in 2011, at least 30 deaths were caused by a single source of cantaloupes, and in Germany, sprouts from a single supplier caused more than 850 cases of hemolytic uremic syndrome and 53 deaths. Overall, consumption of contaminated food results in about 125,000 hospitalizations and 3,000 deaths in the United States each year.

What is it that contaminates food and causes so much illness? Bacteria. The nutrients in the food you consume during meals and snacks can also provide sustenance for a wide variety of disease-causing bacteria. Some of these invisible diners may accompany your lunch to your digestive tract and take up residence there, causing unpleasant symptoms or, in many cases, serious illness.

Devising effective strategies for protecting our food supply against bacterial contamination depends in part on how well we understand the biology of bacteria. What do scientists know about bacteria and their fellow prokaryotes, the archaea?

AT A GLANCE

20.1 WHICH ORGANISMS ARE MEMBERS OF THE DOMAINS ARCHAEA AND BACTERIA?

Earth's first organisms were prokaryotes, single-celled organisms that lacked organelles such as the nucleus, chloroplasts, and mitochondria. (See Chapter 4 for a comparison of prokaryotic and eukaryotic cells.) For the first 1.5 billion years or more of life's history, all life was prokaryotic. Even today, prokaryotes are extraordinarily abundant. A drop of seawater contains hundreds of thousands of prokaryotic organisms, and a spoonful of soil contains billions. The average human body is home to trillions of prokaryotes, which live on the skin, in the mouth, and in the stomach and intestines. In terms of abundance, prokaryotes are Earth's predominant form of life.

Prokaryotes are usually very small, ranging from about 0.2 to 10 micrometers in diameter. In comparison, the diameters of eukaryotic cells range from about 10 to 100 micrometers. About 250,000 average-sized prokaryotes could congregate on the period at the end of this sentence, though a few species are larger. The largest known bacterium (*Thiomargarita namibiensis*) can be up to 700 micrometers in diameter, as big as the tip of a ballpoint pen and visible to the naked eye.

The cell walls produced by prokaryotic cells give characteristic shapes to different types of prokaryotes. The most common shapes are spherical, rod-shaped, and corkscrew-shaped (**FIG. 20-1**).

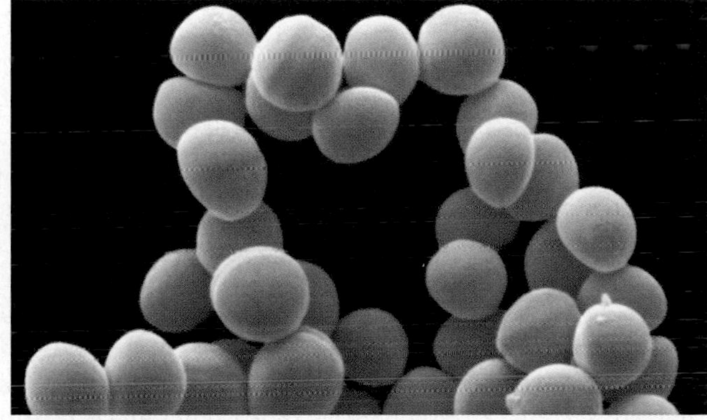

(a) Spherical

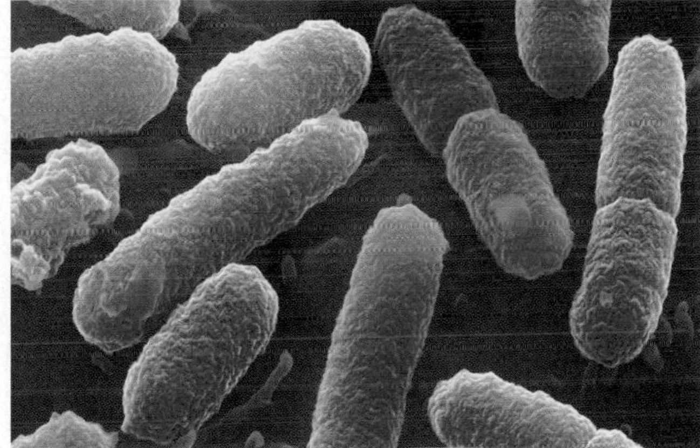

(b) Rod-shaped

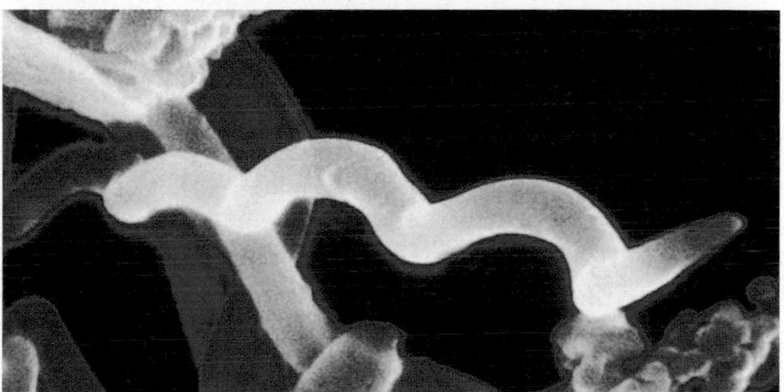

(c) Corkscrew-shaped

▲ **FIGURE 20-1 Three common prokaryote shapes (a)** Spherical bacteria of the genus *Staphylococcus*, **(b)** rod-shaped bacteria of the genus *Escherichia*, and **(c)** corkscrew-shaped bacteria of the genus *Borrelia*.

Bacteria and Archaea Are Fundamentally Different

Two of life's three domains, **Bacteria** and **Archaea**, consist entirely of prokaryotes. Bacteria and archaea are superficially similar in appearance under the microscope, but they have striking structural and biochemical differences that reveal the ancient evolutionary separation between the two groups (**TABLE 20-1**). For example, the cell walls of bacteria are strengthened by molecules of *peptidoglycan*, a polysaccharide that also incorporates some amino acids. Peptidoglycan is unique to bacteria, and the cell walls of archaea do not contain it. Bacteria and archaea also differ in the structure and composition of their plasma membranes, ribosomes, and enzymes involved in RNA synthesis, as well as in the mechanics of basic processes such as transcribing the instructions encoded in DNA and synthesizing proteins.

TABLE 20-1 Differences Between Organisms in the Domains Archaea and Bacteria

	Archaea	Bacteria
Peptidoglycan in cell wall	Absent	Present
Membrane lipid structure	Branched hydrocarbons	Unbranched hydrocarbons
Histone proteins associated with DNA	Present	Absent
Introns	Present in some genes	Absent
RNA polymerase	Several types	One type
Amino acid that initiates protein synthesis	Methionine	Formylmethionine

(For a refresher on the functions of histones, introns, and RNA polymerase, see Chapter 13.)

Classification Within the Prokaryotic Domains Is Based on DNA Sequences

The sharp differences between archaea and bacteria make distinguishing the two domains a straightforward matter, but classification within each domain is challenging. Historically, the main challenge arose because prokaryotes are very small and structurally simple; they do not exhibit the huge array of anatomical differences that can be used to infer the evolutionary history of plants, animals, and other eukaryotes. Consequently, prokaryotes have historically been classified on the basis of such features as shape, means of locomotion, pigments, nutrient requirements, the appearance of colonies (clusters of individuals that descended from a single cell), and staining properties. For example, the Gram stain, a staining technique, distinguishes two types of cell wall construction in bacteria. Based on the results of the stain, many bacteria can be classified as *gram-positive* or *gram-negative*.

In recent years, DNA sequence comparisons have revealed that many of the observable similarities and differences used in traditional prokaryotic classification do not accurately reflect evolutionary history. Today, prokaryote classification is based almost entirely on DNA sequence data. Systematists have used such data to determine the extent to which the traditional groupings within Bacteria and Archaea represent *clades* (groups of species united by descent from a common ancestor) and to identify additional clades. DNA sequence data have also allowed systematists to include in their classifications species that could not previously be confidently classified, because they could not be cultured in the lab and therefore could not be readily observed. Prokaryote classifications now even include some species that are so far known only from DNA sequences found during exploratory censuses in which all the DNA present in a microbial community is sampled and sequenced.

At present, systematists recognize about 30 major named groups within Bacteria and five within Archaea. Examples of some of the larger clades in Bacteria include Cyanobacteria, which includes many ecologically important photosynthetic species; Firmicutes, which includes the species that convert milk to yogurt and the species that causes tetanus; and Proteobacteria, which includes many of the species that cause food poisoning. The largest clades in Archaea are Crenarchaeota, which includes some species that live in super-hot environments, and Euryarchaeota, which includes some species that live in our intestines and produce methane (the main component of natural gas).

Determining the Evolutionary History of Prokaryotes Is Difficult

You may have noticed that, unlike the chapters in this text devoted to eukaryotic organisms, this chapter does not present an evolutionary tree for prokaryotes. No tree is shown because there is very little agreement among systematists with regard to the evolutionary branching patterns within Bacteria and Archaea. Although DNA comparisons have revealed clades within the two domains, they have not yielded a consensus view of the evolutionary relationships between those clades. Why haven't the usual methods of systematics yielded a widely accepted tree? One likely reason is that in prokaryotes, processes such as conjugation (described later in this chapter) make it possible for genes to move from one species to another. If a prokaryote species acquired some of its genes from other species, then different parts of its genome will have different evolutionary histories. Past (and ongoing) genetic mixing among even distantly related prokaryotes has resulted in a very complex evolutionary history that is extremely difficult to reconstruct.

CHECK YOUR LEARNING

Can you . . .

- describe some differences between bacteria and archaea?
- describe the typical sizes and shapes of prokaryotes?

20.2 HOW DO PROKARYOTES SURVIVE AND REPRODUCE?

The abundance of prokaryotes is due in large part to adaptations that allow members of the two prokaryotic domains to inhabit and exploit a wide range of environments. In this section, we discuss some of the traits that help prokaryotes survive and thrive.

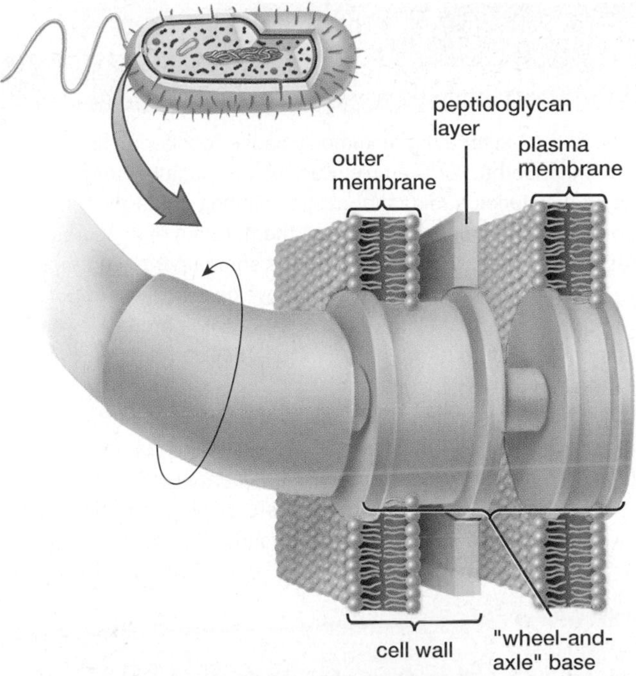

(a) The structure of the bacterial flagellum

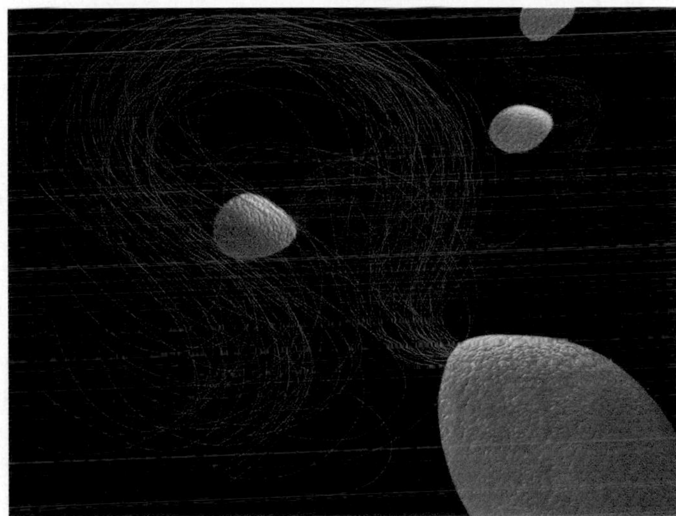

(b) A flagellated archaean

▲ **FIGURE 20-2 The prokaryote flagellum (a)** In bacteria, a unique "wheel-and-axle" arrangement anchors the flagellum within the cell wall and plasma membrane, enabling the flagellum to rotate rapidly. This diagram represents the flagellum of a gram-negative bacterium; the flagella of gram-positive bacteria lack the outermost "wheels." **(b)** A depiction of the archaean *Pyrococcus furiosus*, showing multiple flagella at one end of the cell.

Some Prokaryotes Are Motile

Many bacteria and archaea adhere to a surface or drift passively in liquid surroundings, but some are motile—they can move about. Many of these motile prokaryotes have **flagella** (singular, flagellum), hair-like extensions that can rotate rapidly to propel the organism through its liquid environment (**FIG. 20-2**). Prokaryotic flagella may appear singly at one end of a cell, in pairs (one at each end of the cell), as a tuft at one end of the cell, or scattered over the entire cell surface. The use of flagella to move allows prokaryotes to disperse into new habitats, migrate toward nutrients, and leave unfavorable environments.

The structure of prokaryote flagella is different from the structure of eukaryotic flagella (see also Chapter 4). In bacterial flagella, a unique wheel-like structure embedded in the bacterial membrane and cell wall allows the flagellum to rotate. Archaeal flagella are thinner than bacterial flagella and are constructed of different proteins. The structure of the archaeal flagellum, however, is not yet as well understood as that of the bacterial flagellum.

Many Bacteria Form Protective Films on Surfaces

Many prokaryotes continually synthesize signaling molecules and secrete them to the surrounding environment. If a large number of prokaryotes gathers in one place, the signaling molecules become concentrated enough that the molecules begin to move across the plasma membranes of neighboring cells and combine with receptors inside the cells. The activated receptors trigger cellular processes that would not otherwise be activated. Thus, the behavior of prokaryotes may change when population density grows sufficiently high, a process known as *quorum sensing*.

Among the most common changes induced by quorum sensing is formation of biofilms. In a **biofilm**, one or more species of prokaryote aggregate to form a community that is typically surrounded by sticky protective slime. The slime, composed of polysaccharide or protein, is secreted by the prokaryotes and both protects them and helps them adhere to surfaces. One familiar biofilm is dental plaque, which is formed by the bacteria that inhabit the mouth (**FIG. 20-3**).

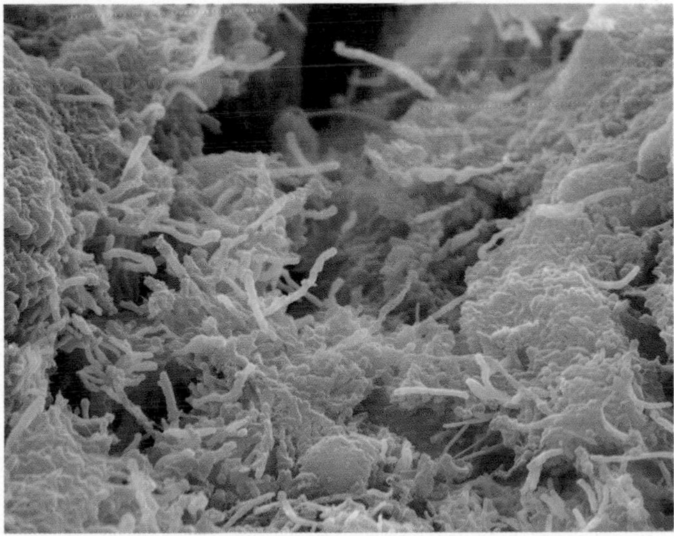

▲ **FIGURE 20-3 The cause of tooth decay** Bacteria in the human mouth form a slimy biofilm that helps them cling to tooth enamel and protects them from threats in the environment. In this micrograph, individual bacteria (colored green and yellow) are visible, embedded in the brown biofilm. The bacteria-laden biofilm can cause tooth decay.

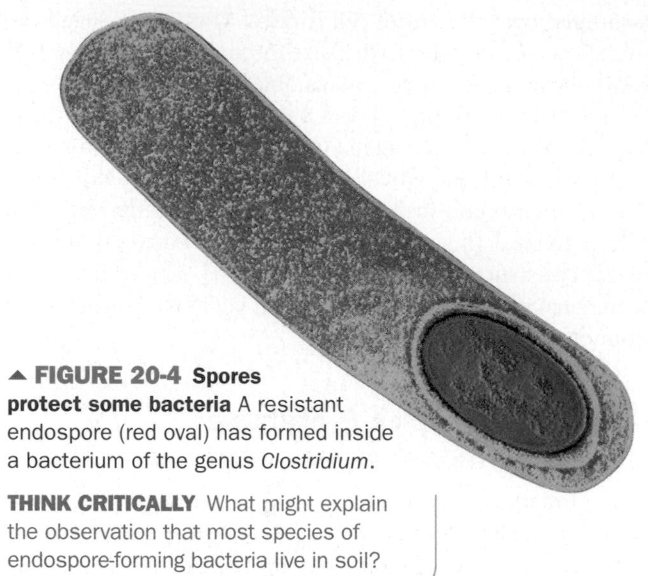

▲ **FIGURE 20-4 Spores protect some bacteria** A resistant endospore (red oval) has formed inside a bacterium of the genus *Clostridium*.

THINK CRITICALLY What might explain the observation that most species of endospore-forming bacteria live in soil?

CASE STUDY CONTINUED

Unwelcome Dinner Guests

A few of the bacteria that commonly cause foodborne illness form endospores. For example, *Bacillus cereus*, a bacterial species that includes strains that cause vomiting or diarrhea in people unlucky enough to consume them, forms endospores that are widespread in soil and dust. If some spores find their way into warm, moist food, they can develop and give rise to a thriving population of bacteria. *B. cereus* spores are somewhat resistant to heat, but fortunately for us, they can be destroyed by thorough cooking.

Although it is unsurprising that prokaryotes might thrive amidst the energy-rich substances that make up human food, our foodstuffs are hardly the only environment in which prokaryotes flourish. What are some of the more extreme environments in which bacteria and archaea can be found?

The protection afforded by biofilms helps defend the embedded prokaryotes against a variety of attacks, including those launched by antibiotics and disinfectants. As a result, biofilms can be very difficult to eradicate. Many infections of the human body take the form of biofilms, including those responsible for tooth decay, gum disease, and ear infections. Biofilms also cause many of the hospital-acquired infections that affect 725,000 Americans each year and kill 75,000 of them. Such biofilms can form in wounds and surgical incisions, as well as on implanted medical devices such as catheters, pacemakers, and artificial hips and knees.

Protective Endospores Allow Some Bacteria to Withstand Adverse Conditions

When environmental conditions become inhospitable, many rod-shaped bacteria form protective structures called **endospores**. An endospore, which forms inside a bacterium, consists of the bacterium's genetic material and a few enzymes encased within a thick protective coat (**FIG. 20-4**). After an endospore forms, the bacterial cell that contains it breaks open, and the spore is released to the environment. Metabolic activity ceases until the spore encounters favorable conditions, at which time metabolism resumes and the spore develops into an active bacterium.

Endospores are resistant even to extreme environmental conditions. Some can withstand boiling for an hour or more. Endospores are also able to survive for extraordinarily long periods. In the most astonishing example of such longevity, scientists discovered endospores that had been sealed inside rock for 250 million years. After being carefully extracted from their rocky tomb, the spores were incubated in test tubes. Amazingly, live bacteria developed from the ancient spores, which were older than the oldest dinosaur fossils.

Prokaryotes Are Specialized for Specific Habitats

Prokaryotes occupy virtually every habitat, including those where extreme conditions keep out other forms of life. For example, some bacteria thrive in near-boiling environments, such as the hot springs of Yellowstone National Park (**FIG. 20-5**). Many archaea live in even hotter environments, including deep-sea vents, where superheated water is spewed through cracks in Earth's crust at temperatures

▲ **FIGURE 20-5 Some prokaryotes thrive in extreme conditions** Hot springs harbor bacteria and archaea that are both heat and mineral tolerant. Several prokaryote species paint these hot springs in Yellowstone National Park with vivid colors; each species is confined to a specific area determined by temperature range.

THINK CRITICALLY Some of the enzymes that have important uses in molecular biology procedures are extracted from prokaryotes that live in hot springs. Can you guess why?

Have You Ever Wondered ... What Causes Bad Breath?

Unpleasant breath odors are caused mainly by prokaryotes that live in the mouth. The warm, moist human mouth cavity hosts a diverse microbial community that includes more than 2,000 prokaryote species. Many of these species acquire energy and nutrients by breaking down mucus, food particles, and dead cells. The by-products of this breakdown can include foul-smelling gases, some of which are also emitted by feces or decaying bodies.

The highest concentration of bad-breath prokaryotes is found at the base of the tongue. This location may be especially hospitable to microbes owing to the accumulation of mucus that drains down into the back of the throat from the nose. So, if you gargle with antiseptic mouthwash to control bad breath, thrust your tongue forward so the mouthwash can reach the base of your tongue.

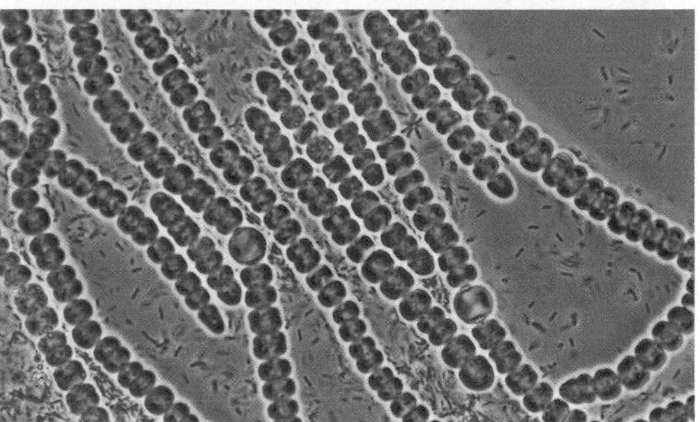

▲ **FIGURE 20-6 Photosynthetic prokaryotes** Photosynthetic cyanobacteria make up the filaments shown in this micrograph.

of up to 230°F (110°C). Prokaryotes can also survive at the extremely high pressures found on the sea floor and deep beneath Earth's surface, and in very cold environments, such as in Antarctic sea ice.

Extreme chemical conditions do not prevent colonization by prokaryotes, either. Thriving colonies of bacteria and archaea live in the Dead Sea, for example, where a salt concentration seven times that of the oceans precludes all other life, and in waters that are as acidic as vinegar or as alkaline as household ammonia. Given their ability to survive such extreme environments, it is not surprising that prokaryote communities also reside in a full range of more moderate habitats, including in and on the human body.

No single species of prokaryote, however, is as versatile as these examples may suggest. In fact, most prokaryotes are specialists. One species of archaea that inhabits deep-sea vents, for example, grows optimally at 223°F (106°C) and stops growing altogether at temperatures below 194°F (90°C). A bacteria species that lives deep underground has never been found less than 1.2 miles beneath the Earth's surface. Bacteria that live on the human body are also specialized; different species colonize the skin, the mouth, the respiratory tract, the large intestine, and the urogenital tract. (To learn more about how the bacteria in our bodies might affect our health, see "Health Watch: Is Your Body's Ecosystem Healthy?" on page 323.)

Prokaryotes Have Diverse Metabolisms

Prokaryotes are able to colonize diverse habitats in part because they have evolved diverse methods of acquiring energy and nutrients from the environment. For example, unlike eukaryotes, many prokaryotes are **anaerobes**; their metabolisms do not require oxygen. Their ability to inhabit oxygen-free environments allows prokaryotes to exploit habitats in which eukaryotes could not survive. Some anaerobes, such as many of the archaea found in hot springs and the bacterium that causes tetanus, are actually poisoned by oxygen. Others are opportunists, engaging in anaerobic respiration when oxygen is lacking and switching to aerobic respiration (a more efficient process; see Chapter 8) when oxygen becomes available. Many prokaryotes are strictly aerobic and require oxygen at all times.

Whether aerobic or anaerobic, different prokaryote species acquire energy from an amazing array of sources. Some species of bacteria use photosynthesis to capture energy directly from sunlight (**FIG. 20-6**). Like green plants, photosynthetic bacteria possess chlorophyll. Most species produce oxygen as a by-product of photosynthesis; the extremely abundant marine photosynthetic bacterium *Prochlorococcus* produces about 20% of all the oxygen gas in the atmosphere. However, not all photosynthetic bacteria produce oxygen. Some, known as the sulfur bacteria, use hydrogen sulfide (H_2S) instead of water (H_2O) in photosynthesis, releasing sulfur instead of oxygen. No photosynthetic archaea are known.

Nonphotosynthetic prokaryotes extract energy from a wide assortment of substances. Prokaryotes subsist not only on the sugars, carbohydrates, fats, and proteins that we usually think of as foods, but also on compounds that are inedible or even poisonous to humans, including petroleum, methane, and solvents such as benzene and toluene. Some prokaryotes can even metabolize inorganic molecules, including hydrogen, sulfur, ammonia, and iron. This ability is one of the reasons that prokaryotes can live in habitats in which sunlight is absent and organic molecules scarce. For example, one bacterial species that lives deep underground metabolizes hydrogen that is produced when radiation from radioactive uranium breaks apart water molecules.

Prokaryotes Reproduce by Fission

Most prokaryotes reproduce asexually by **prokaryotic fission** (also called binary fission), a form of cell division that is much simpler than mitotic cell division (see

Health WATCH Is Your Body's Ecosystem Healthy?

Microbiologists like to say, "You are born 100% human, but you die 50% microbial." A human fetus in the womb is more or less sterile, but as it passes down the birth canal to be born, it picks up some of the bacteria that live there, and more are transferred from the first hands to touch the new baby. From that point on, microbes accumulate steadily in and on the body, acquired from the environment, from food, and from other humans. By the time a child is 3 years old, its body is home to an entire ecosystem of microorganisms. This "microbiome" includes hundreds of different species, living in huge numbers in the nose and mouth, on the scalp, in the urogenital tract, in the gut, and on almost every skin surface. All told, the microbiome of a typical person includes about 30 trillion microbial cells, one microbial cell for every human one.

It is becoming increasingly apparent to physicians and researchers that the inhabitants of the microbiome are not mere hitchhikers, but instead play an important role in human health. Although some of the benefits provided by our microbial partners are well understood, such as the production by gut bacteria of nutrients essential to humans, most of the details about how the microbiome contributes to our health are more mysterious. However, much evidence suggests that its contribution is important. For example, researchers have shown that in infants with necrotizing enterocolitis, an often-fatal bowel disease, the composition of the community of microbes in the gut is different from that present in healthy babies. Similarly aberrant digestive tract microbiomes have been found in adults with bowel disorders, and even in people with disorders not directly related to digestion, such as diabetes and autoimmune diseases. Such findings have led to the hypothesis that, because a person's microbiome, like a coral reef or a tropical rain forest, is a diverse ecosystem characterized by complex interactions among species, disruptions of this ecosystem compromise its ability to perform functions essential to human health. However, scientists are not yet able to say for sure if the disrupted microbiomes of diseased individuals are a cause or a consequence of disease.

Confirmation of the microbiome's possible role in maintaining health depends on improved knowledge of its composition and characteristics. For that reason, a major ongoing research effort is directed at identifying and characterizing all of the microbial species that compose the microbiome, and at identifying exactly how microbiomes differ among different people. The job is challenging, because the traditional way of identifying a prokaryotic species—by growing it in a culture dish in the lab—is not effective for the many species that cannot be easily cultured. Scientists are now focused on the alternative approach of taking a sample of a whole microbiome and sequencing all of the DNA present in it. Some of the initial results of this approach are astonishing. For example, samples taken from 18 different locations on the skin and in the airways, mouths, vaginas, and digestive tracts of 242 people yielded a total of about 10,000 microbial species and 8 million different genes, about 350 times as many genes as are present in the human genome. Further study of these microbial genes may reveal their functions and how microorganisms that make up the microbiome contribute to human health and disease.

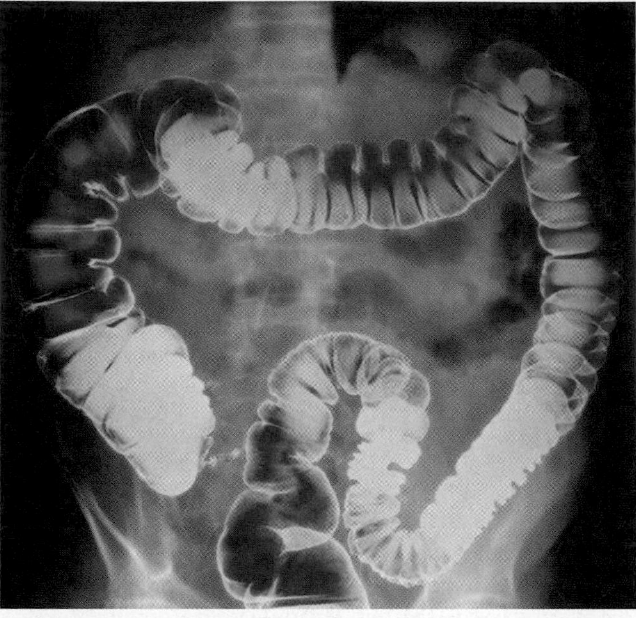

▲ **FIGURE E20-1 Transplanted feces could cure diseased bowels** By restoring a healthy microbial ecosystem in the recipient, fecal transplants can treat infected bowels and perhaps diabetes, obesity, and other disorders as well.

In the meantime, some physicians are pushing ahead with treatments designed to restore ecological health to the microbiomes of sick people. For example, a few doctors have begun using fecal transplants to treat patients with severe bowel infections (**FIG. E20-1**). In this treatment, a small amount of feces from a healthy donor is transplanted into the bowel of the sick person, with the hope that the transplanted microbial community will become established and spread, displacing the harmful microbes responsible for the infection. The physicians using this treatment report very high success rates, much higher than those typical of treatment with antibiotics, which have been shown to devastate the gut microbiome. These reports have encouraged funding agencies in some countries to overcome the yuck factor and fund large-scale clinical trials of the fecal transplant treatment.

THINK CRITICALLY As part of a study on the relationship between inflammatory bowel disease and the microbiome, researchers analyzed samples from the digestive tracts of 20 sets of identical twins in which one twin had the disease and the other did not. The researchers were able to determine the microbial diversity in each sample. Do you expect that samples from the healthy group will differ from samples from the ill group? If so, in what way? Why? Why did the researchers use twins in this study?

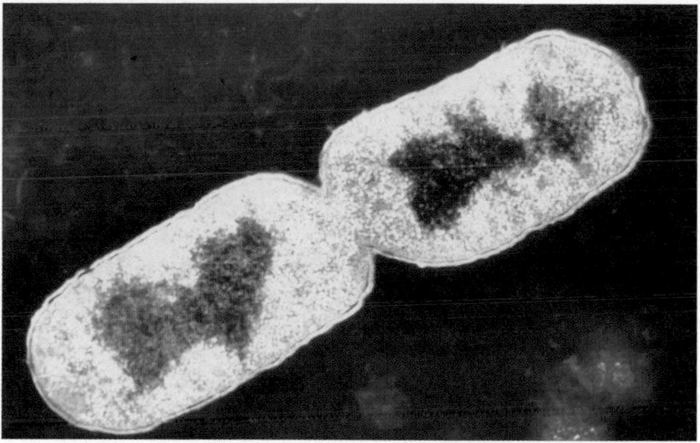

▲ **FIGURE 20-7 Reproduction in prokaryotes** Prokaryotic cells reproduce by prokaryotic fission. In this color-enhanced electron micrograph, an *Escherichia coli*, a normal inhabitant of the human intestine, is dividing.

THINK CRITICALLY What is the main advantage of prokaryotic fission, compared to sexual reproduction?

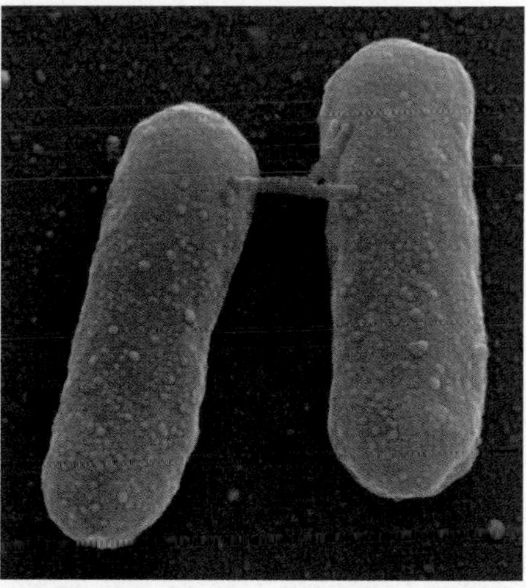

▲ **FIGURE 20-8 Conjugation: prokaryotic "mating"** During conjugation, one prokaryote acts as a donor, transferring DNA to the recipient. In this micrograph, two *Escherichia coli* are connected by a sex pilus. The sex pilus will retract, drawing the recipient bacterium to the donor bacterium.

Chapter 9 for a detailed description of prokaryotic fission). Prokaryotic fission produces genetically identical copies of the original cell (**FIG. 20-7**). Under ideal conditions, some prokaryotic cells can divide about once every 20 minutes, potentially giving rise to sextillions (10^{21}) of offspring in a single day.

Rapid reproduction allows bacterial populations to evolve quickly. Recall that many mutations, the source of genetic variability, are the result of mistakes in DNA replication during cell division. Thus, the rapid, repeated cell division of prokaryotes provides ample opportunity for new mutations to arise and also allows mutations that enhance survival to spread quickly.

Prokaryotes May Exchange Genetic Material Without Reproducing

Although prokaryotic reproduction is generally asexual and does not involve genetic recombination, some bacteria and archaea nonetheless exchange genetic material. In these species, DNA is transferred from a donor to a recipient in a process called **conjugation**. The plasma membranes of two conjugating prokaryotes fuse temporarily to form a cytoplasmic bridge across which DNA travels. In bacteria, donor cells may use specialized extensions called *sex pili* that attach to a recipient cell, drawing it closer to allow conjugation (**FIG. 20-8**). Conjugation produces new genetic combinations that may allow the resulting bacteria to survive under a greater variety of conditions. In some cases, genetic material may be exchanged even between individuals of different species. Much of the DNA transferred during conjugation is contained within a structure called a **plasmid**, a small, circular DNA molecule that is separate from the single prokaryote chromosome.

CHECK YOUR LEARNING

Can you . . .

- describe the range of environments inhabited by prokaryotes and the variety of methods by which they acquire energy?
- describe adaptations that help protect prokaryotes from environmental threats?
- explain how prokaryotes reproduce and exchange genetic material?

20.3 HOW DO PROKARYOTES AFFECT HUMANS AND OTHER ORGANISMS?

Although they are largely invisible to us, prokaryotes play a crucial role in life on Earth. Plants and animals (including humans) are utterly dependent on prokaryotes. Prokaryotes help plants and animals obtain vital nutrients and help break down and recycle wastes and dead organisms. We could not survive without prokaryotes, but their impact on us is not always beneficial. Some of humanity's most deadly diseases stem from microbes.

Prokaryotes Play Important Roles in Animal Nutrition

Many eukaryotic organisms depend on close associations with prokaryotes. For example, most animals that eat leaves—including cattle, rabbits, koalas, and deer—can't themselves digest cellulose, the principal component of plant cell walls. Instead, these animals depend on certain bacteria that have the ability to break down cellulose. These bacteria live in the animals' digestive tracts, where they

(a) Nodules on roots
(b) Nitrogen-fixing bacteria within nodules

▲ **FIGURE 20-9 Nitrogen-fixing bacteria in root nodules (a)** Special chambers called nodules on the roots of a legume provide a protected environment for nitrogen-fixing bacteria. **(b)** This scanning electron micrograph shows the nitrogen-fixing bacteria inside cells within the nodules.

THINK CRITICALLY If all of Earth's nitrogen-fixing prokaryotes were to die suddenly, what would happen to the concentration of nitrogen gas in the atmosphere?

liberate nutrients from plant tissue that the animals cannot digest themselves. Without such bacteria, leaf-eating animals could not survive.

Prokaryotes also have important effects on human nutrition. Many foods, including cheese, yogurt, and sauerkraut, are produced by the action of bacteria. Prokaryotes also inhabit your intestines. There they feed on undigested food, and some synthesize nutrients such as vitamin K and vitamin B_{12}, which the human body absorbs. In fact, good health and nutrition are highly dependent on the prokaryotes in our digestive system; we can't thrive without them.

Prokaryotes Capture the Nitrogen Needed by Plants

Humans could not live without plants, and plants are entirely dependent on bacteria. In particular, plants are unable to capture nitrogen from that element's most abundant reservoir: the atmosphere. Plants need nitrogen to grow. To acquire it, they depend on **nitrogen-fixing bacteria**, which live both in soil and in specialized nodules, which are small, rounded lumps on the roots of legumes (a group of plants that includes alfalfa, soybeans, lupines, and clover; **FIG. 20-9**). The nitrogen-fixing bacteria capture nitrogen gas (N_2) from air trapped in the soil and combine it with hydrogen to produce ammonium (NH_4^+), a nitrogen-containing nutrient that plants can use directly.

Prokaryotes Are Nature's Recyclers

Prokaryotes play a crucial role in recycling waste. Many prokaryotes obtain energy by breaking down complex organic molecules (molecules that contain carbon and hydrogen). Such prokaryotes find a plentiful source of organic molecules in the waste products and dead bodies of plants and animals. By consuming and thereby decomposing these wastes, prokaryotes prevent wastes from accumulating in the environment. In addition, decomposition by prokaryotes releases the nutrients contained in wastes. Once released, the nutrients become available for reuse by other living organisms.

Prokaryotes perform their recycling service wherever organic matter is found. They are important decomposers in lakes and rivers, in the oceans, and in the soil and groundwater of forests, grasslands, deserts, and other terrestrial environments. The recycling of nutrients by prokaryotes and other decomposers provides the raw materials needed for continued life on Earth.

Prokaryotes Can Clean Up Pollution

Many of the pollutants that are produced as by-products of human activity are organic compounds. As such, these pollutants can potentially serve as food for archaea and bacteria. Nearly anything that human beings can synthesize—including detergents, many toxic pesticides, and harmful industrial chemicals such as benzene and toluene—can be broken down by some prokaryote.

Even oil can be broken down by prokaryotes. Soon after the tanker *Exxon Valdez* spilled 11 million gallons of crude oil into Prince William Sound, Alaska, in 1989, researchers sprayed oil-soaked beaches with a fertilizer that encouraged the growth of natural populations of oil-eating bacteria. Within days, the oil deposits on these beaches were noticeably reduced in comparison with unsprayed areas. However, oil-eating bacteria are having a much slower effect on the 200 million gallons of oil released into the Gulf of Mexico during the *Deep Water Horizon* well blowout in 2010. The oil from the blowout was released deep underwater where the temperature is cold and prokaryote metabolism is slow. In addition, it is not practical to encourage bacterial growth by fertilizing the vast area over which the oil has spread.

The practice of manipulating conditions to stimulate breakdown of pollutants by living organisms is known as **bioremediation**. Improved methods of bioremediation could dramatically increase our ability to clean up toxic waste sites and polluted groundwater. A great deal of current research is therefore devoted to identifying prokaryote species that are especially effective for bioremediation and discovering practical methods for manipulating these organisms to improve their usefulness.

Some Bacteria Pose a Threat to Human Health

Despite the benefits some bacteria provide, the feeding habits of certain bacteria threaten our health and well-being. These **pathogenic** (disease-producing) bacteria synthesize toxic substances that cause disease symptoms. So far, no pathogenic archaea have been identified.

Some Anaerobic Bacteria Produce Dangerous Poisons

Some bacteria produce toxins that attack the nervous system. One such toxin is produced by *Clostridium tetani*, the bacterium that causes tetanus, a sometimes fatal disease whose symptoms include painful, uncontrolled contraction of muscles throughout the body. *C. tetani* are anaerobic bacteria that survive as spores until introduced into a favorable, oxygen-free environment. A deep puncture wound may allow tetanus bacteria to penetrate a human body and reach a place where they will be protected from contact with oxygen. As they multiply, the bacteria release their toxin into the body's bloodstream.

Another anaerobic *Clostridium* species that produces a dangerous neurotoxin is *Clostridium botulinum*. *C. botulinum* occurs naturally in soil, but may also thrive in a sealed container of canned food that has been incompletely sterilized. Foodborne *C. botulinum* is dangerous because botulinum toxin is among the most toxic substances known; a single gram is enough to kill 15 million people.

Humans Battle Bacterial Diseases Old and New

Bacterial diseases have had a significant impact on human history. Perhaps the most infamous example is bubonic plague, or "Black Death," which killed 100 million people during the mid-fourteenth century. In many parts of the world, one-third or more of the population died. Plague is caused by the highly infectious bacterium *Yersinia pestis*, which is spread by fleas that feed on infected rats and then move to human hosts. Although bubonic plague has not reemerged as a large-scale epidemic, about 2,000 to 3,000 people worldwide are diagnosed with the disease each year.

Some bacterial pathogens seem to emerge suddenly. Lyme disease, for example, was unknown until 1975 but is now diagnosed in about 30,000 people per year in the United States. This disease, named after the town of Old Lyme, Connecticut, where it was first described, is caused by the corkscrew-shaped bacterium *Borrelia burgdorferi* (see Fig. 20-1c). The bacterium is carried by deer ticks, which transmit it to the humans they bite. At first, the symptoms—chills, fever, and body aches—resemble those of flu. If untreated, weeks or months later the victim may experience rashes, bouts of arthritis, and in some cases abnormalities of the heart and nervous system.

Perhaps the most frustrating pathogens are those that come back to haunt us long after we believed that we had them under control. Two sexually transmitted bacterial diseases, gonorrhea and syphilis, have reached epidemic proportions around the globe. Cholera, a water-transmitted bacterial disease that flourishes when raw sewage contaminates drinking water or fishing areas, is under control in developed countries but remains a major killer in poorer parts of the world.

CHECK YOUR LEARNING

Can you . . .

- explain how prokaryotes affect animal and plant nutrition?
- explain prokaryotes' role in nutrient recycling?
- describe how prokaryotes help clean up pollution?
- describe some of the pathogenic bacteria that threaten human health?

CASE STUDY **CONTINUED**

Unwelcome Dinner Guests

Many of the bacteria responsible for foodborne illnesses do their damage by producing toxins. For example, different populations of the bacterial species *Escherichia coli* may differ genetically, and some genetic differences can transform this normally harmless inhabitant of the human digestive system into a toxin-producing pathogen. If one of these toxic strains, such as the ones designated O157:H7 and O104:H4, finds its way into a human digestive system, the bacteria attach firmly to the wall of the intestine and begin to release a toxin called shiga. Shiga toxin causes intestinal bleeding that results in painful cramping and bloody diarrhea. The toxin can damage other organs as well; victims of O157:H7 and O104:H4 often develop hemolytic uremic syndrome, a dangerous condition characterized by kidney failure and loss of red blood cells.

Bacteria are responsible for most foodborne illnesses. But besides bacteria, which other infectious agents can wreak havoc on the human body?

20.4 WHAT ARE VIRUSES, VIROIDS, AND PRIONS?

Although this chapter is devoted primarily to an overview of the prokaryotic domains, we also discuss here *viruses*, *viroids*, and *prions*, which are not organisms but have significant effects on organisms.

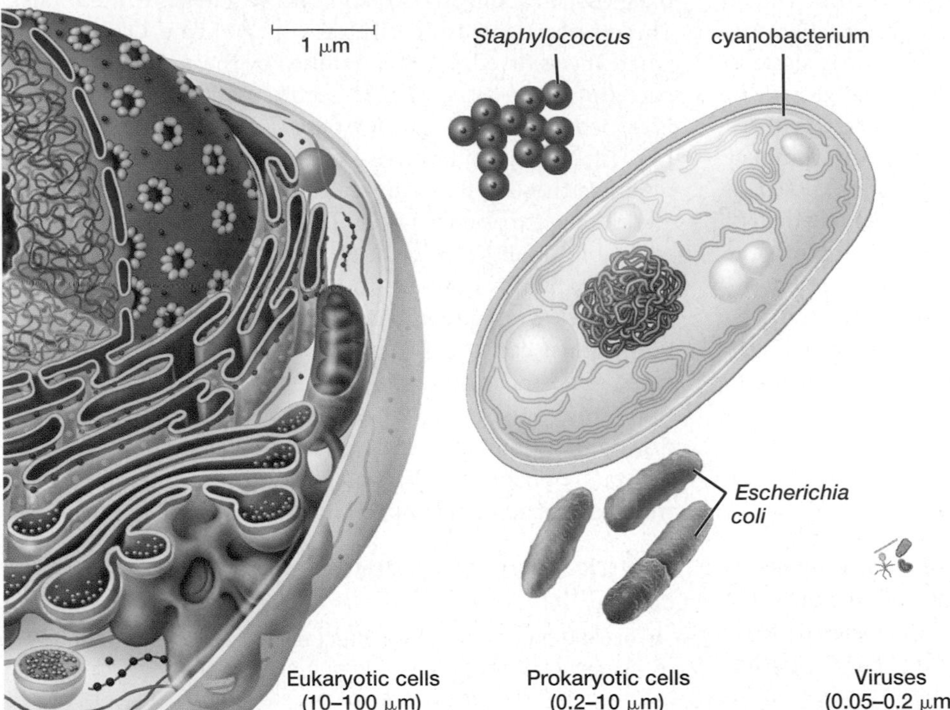

◀ **FIGURE 20-10 The sizes of microorganisms** The relative sizes of eukaryotic cells, prokaryotic cells, and viruses (1 μm = 1/1,000 millimeter).

Viruses Are Nonliving Particles

Although **viruses** are generally found in close association with living organisms, most biologists do not consider viruses to be alive because they lack many of the traits that characterize life. For example, they are not cells—nor are they composed of cells. Further, they cannot, on their own, accomplish the basic tasks that living cells perform. Viruses have no ribosomes on which to make proteins, no cytoplasm, no ability to synthesize organic molecules, and no capacity to extract and use the energy stored in such molecules. They possess no membranes of their own and cannot grow or reproduce on their own. The simplicity of viruses seems to place them outside the realm of living things. They can, however, evolve.

A Virus Consists of a Molecule of DNA or RNA Surrounded by a Protein Coat

Viruses are tiny; most are much smaller than even the smallest prokaryotic cell (**FIG. 20-10**). Virus particles are so small (0.05 to 0.2 micrometer in diameter) that they can be seen only under the enormous magnification of an electron microscope. Under such magnification, one can see that viruses assume a great variety of shapes (**FIG. 20-11**).

Viruses consist of two major parts: a molecule of hereditary material and a coat of protein surrounding the molecule. Depending on the type of virus, the hereditary molecule may be either DNA or RNA and may be single-stranded or

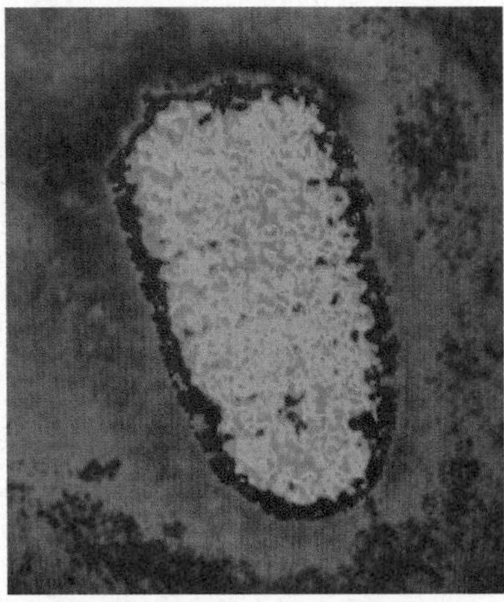
(a) Rabies virus

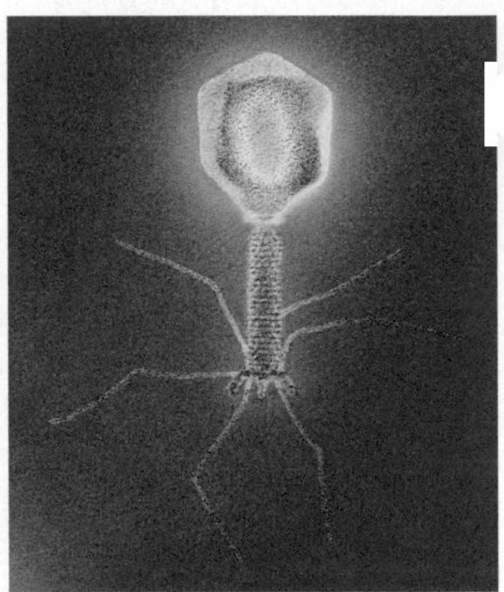
(b) Bacteriophage

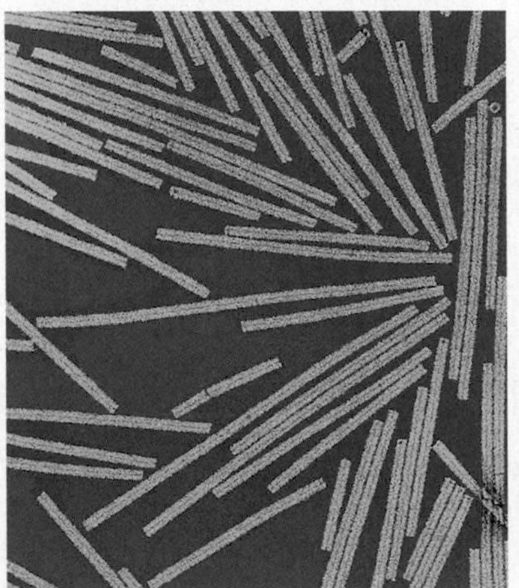
(c) Tobacco mosaic viruses

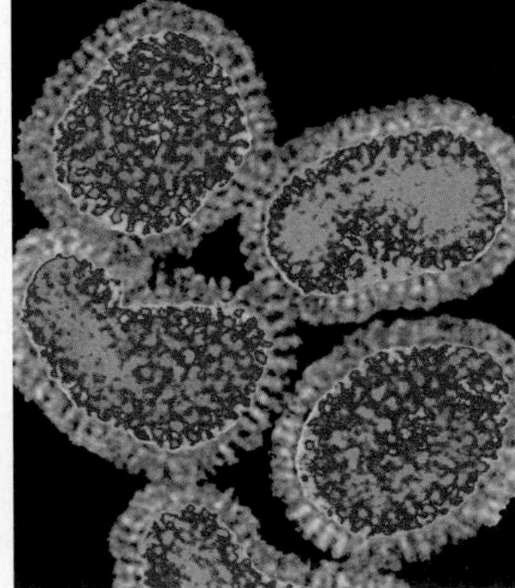
(d) Influenza viruses

▶ **FIGURE 20-11 Viruses come in a variety of shapes** Viral shape is determined by the nature of the virus's protein coat.

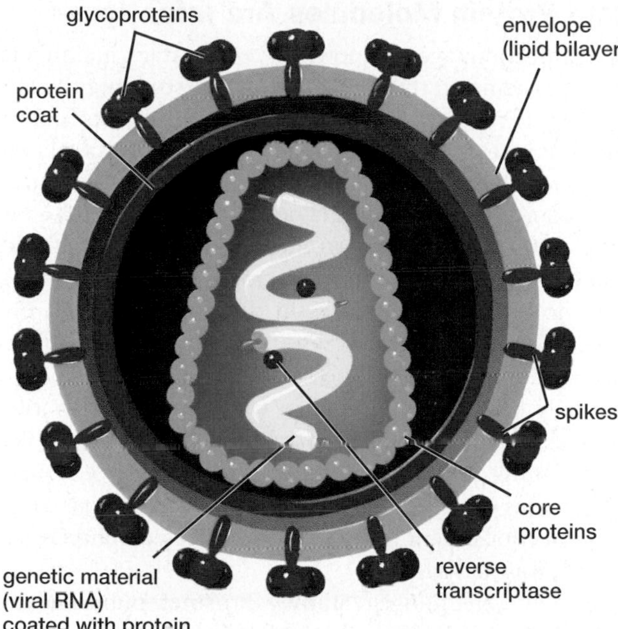

▲ **FIGURE 20-12 Viral structure and replication** A cross-section of HIV, the virus that causes AIDS. This virus is among those that have an outer envelope formed from the host cell's plasma membrane. Spikes made of glycoprotein (protein and carbohydrate) project from the envelope and help the virus attach to its host cell. Inside, a protein coat surrounds genetic material and molecules of reverse transcriptase, an enzyme that catalyzes the transcription of DNA from the viral RNA template after the virus enters a host cell.

THINK CRITICALLY Why are viruses unable to replicate outside of a host cell?

double-stranded, linear or circular. The protein coat may be surrounded by an envelope formed from the plasma membrane of the host cell (**FIG. 20-12**).

Viruses Require a Host to Reproduce

A virus can reproduce only inside a **host** cell—the cell that the virus infects. Viral reproduction begins when a virus penetrates a host cell. After the virus enters the host cell, the viral genetic material takes command. The hijacked host cell then uses the instructions encoded in the viral genes to produce the components of new viruses. The pieces are rapidly assembled, and an army of new viruses bursts forth to invade and conquer neighboring cells.

Viruses Are Host Specific

Each type of virus is specialized to attack a specific host cell. As far as we know, no organism in any of life's three domains is immune to all viruses. The viruses that infect archaea or bacteria are often called **bacteriophages**, sometimes shortened to *phages* (**FIG. 20-13**). Phages may soon become important in treating diseases caused by bacteria because many disease-causing bacteria have become

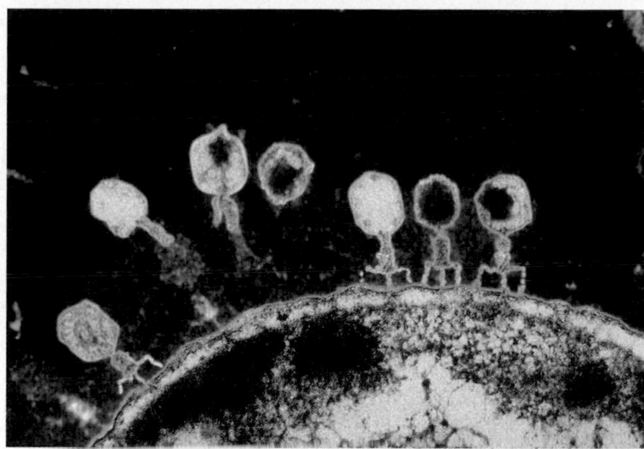

▲ **FIGURE 20-13 Some viruses infect bacteria** In this electron micrograph, bacteriophages are seen attacking a bacterium. They have injected their genetic material inside, leaving their protein coats clinging to the bacterial cell wall.

THINK CRITICALLY Biotechnologists often use viruses to transfer genes from the cells of one species to the cells of another. Which properties of viruses make them useful for this purpose?

increasingly resistant to antibiotics. (Bacteria may evolve resistance to phages, but phages remain effective because they evolve faster.) Treatments based on phages could also take advantage of the viruses' specificity, attacking only the targeted bacteria and not the harmless or beneficial bacteria in the body.

In multicellular organisms such as plants and animals, different viruses specialize in attacking particular cell types. Viruses responsible for the common cold, for example, attack the membranes of the respiratory tract, and rabies viruses attack nerve cells. One type of herpes virus specializes in the mucous membranes of the mouth and lips, causing cold sores; a second type produces similar sores on or near the genitals. Herpes viruses take up permanent residence in the body, erupting periodically (typically during times of stress) as infectious sores. The devastating disease AIDS (acquired immune deficiency syndrome), which cripples the body's immune system, is caused by a virus that attacks a specific type of white blood cell that controls the body's immune response. Viruses also cause some types of cancer, such as T-cell leukemia (a cancer of the white blood cells), liver cancer, and cervical cancer.

Viral Infections Are Difficult to Treat

Because viruses depend on the cellular machinery of their hosts, the illnesses they cause are difficult to treat. The antibiotics that are often effective against bacterial infections are useless against viruses, and antiviral agents may destroy host cells as well as viruses. Despite the difficulty of attacking viruses as they "hide" within cells, a number of antiviral drugs have been developed. Many of these drugs destroy or block the function of enzymes that the targeted virus requires for replication.

Unfortunately, the benefits of most antiviral drugs are limited because many viruses quickly evolve resistance to the drugs. Mutation rates can be very high in viruses, in part because many viruses lack mechanisms for correcting errors that occur during replication of genetic material. Thus, when a population of viruses is under attack by an antiviral drug, a mutation will often arise that confers resistance to the drug. The resistant viruses prosper and replicate in great numbers, eventually spreading to new human hosts. Ultimately, resistant viruses predominate, and a formerly helpful antiviral drug is rendered ineffective.

CASE STUDY CONTINUED

Unwelcome Dinner Guests

A few foodborne illnesses are caused by viruses. For example, the virus that causes hepatitis A is often transmitted in food, usually when the food has been handled by an infected person who has been lax about hand washing. Some people infected with hepatitis A do not exhibit symptoms, but many experience flu-like symptoms accompanied by jaundice (yellowish skin). Although additional complications can arise, most victims recover within a few months. Hepatitis A is one of the few foodborne diseases for which a vaccination exists.

Some Plant Diseases Are Caused by Infectious Agents Simpler Than Viruses

Viroids are infectious particles that lack a protein coat and consist of nothing more than short, circular strands of RNA (**FIG. 20-14**). Despite their simplicity, viroids are able to enter the nucleus of a host cell and direct the synthesis of new viroids. About a dozen crop diseases, including cucumber pale fruit disease, avocado sunblotch, and potato spindle tuber disease, are caused by viroids. No viroid is known to infect animals.

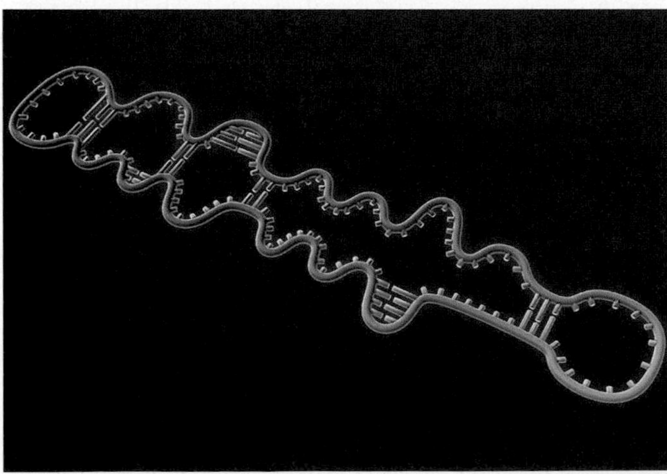

▲ **FIGURE 20-14 Viroids are circular strands of infectious RNA**

Some Protein Molecules Are Infectious

Simple infectious agents known as *prions* attack mammalian nervous systems. In the 1950s, physicians studying the Fore, a primitive tribe in New Guinea, were puzzled to observe numerous cases of a fatal degenerative disease of the nervous system, which the Fore called kuru. The symptoms of kuru—loss of coordination, dementia, and ultimately death—were similar to those of the rare but more widespread Creutzfeldt-Jakob disease in humans and of scrapie and bovine spongiform encephalopathy diseases in domestic livestock. Each of these diseases typically results in brain tissue that is spongy—riddled with holes. The researchers in New Guinea eventually determined that kuru was transmitted by ritual cannibalism; members of the Fore tribe honored their dead by consuming their brains. This practice has since stopped, and kuru has virtually disappeared. Clearly, kuru was caused by an infectious agent transmitted by infected brain tissue—but what was that agent?

In 1982, neurologist Stanley Prusiner published evidence that scrapie (and, by extension, kuru, Creutzfeldt-Jakob disease, and a number of other, similar afflictions) is caused by an infectious agent that consists only of protein. This idea seemed preposterous at the time, because most scientists believed that infectious agents must contain genetic material such as DNA or RNA to replicate. But Prusiner and his colleagues were able to isolate the infectious agent and demonstrate that it contained no nucleic acids. The researchers called these infectious protein particles **prions** (**FIG. 20-15**).

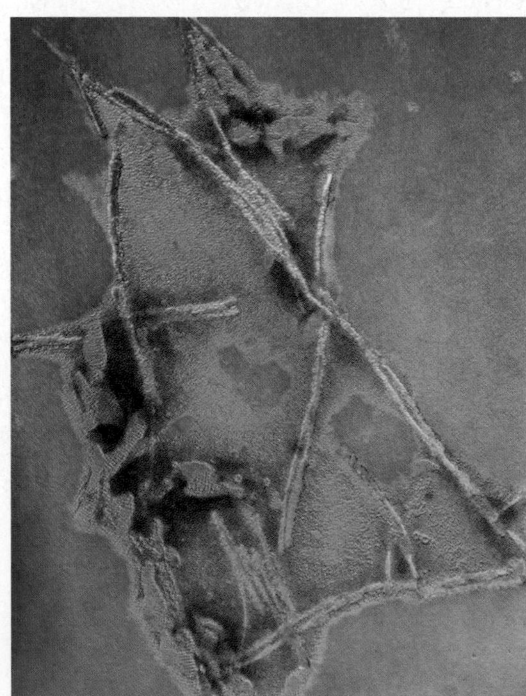

▲ **FIGURE 20-15 Prions: puzzling proteins** A section from the brain of a cow infected with bovine spongiform encephalopathy contains fibrous clusters of prion proteins.

How can a protein replicate itself and be infectious? Research over the decades since Prusiner's discovery has shown that prions are misfolded versions of a common protein called PrP. PrP is found in the membranes of neurons and is required for normal neuron function. Sometimes, copies of the PrP molecule become folded into the wrong shape and are thus transformed into infectious prions. If misfolded prion protein molecules are introduced into a healthy mammal, they can induce other, normal copies of the PrP molecule to become transformed into prions, which in turn induce still other conversions of normal PrP to the prion version. Eventually, this chain reaction leads to a concentration of prions high enough to cause nerve cell damage and degeneration. Why would a slight alteration to a normally benign protein turn it into a dangerous cell killer? No one knows.

CHECK YOUR LEARNING

Can you . . .

- describe the structure and characteristics of viruses, viroids, and prions?
- describe the effects that viruses, viroids, and prions have on host organisms?
- explain why viral infections are difficult to treat?

CASE STUDY REVISITED

Unwelcome Dinner Guests

How do harmful bacteria get into our food? Many foodborne illnesses result from consumption of contaminated beef. The intestinal tracts of about a third of the cattle in the United States carry bacteria that are harmful to humans, and these bacteria can be transmitted to humans when a meatpacker accidentally grinds some gut contents into hamburger. Similarly, chicken feces may splash onto eggs, setting the stage for harmful bacteria to enter the eggs through tiny cracks or when the consumer breaks the egg and its contents contact the shell. Produce such as lettuce, spinach, tomatoes, and melons can also become contaminated if farm fields are exposed to animal feces, which can be deposited by deer or wandering domestic animals or carried from nearby ranches and feedlots in dust or runoff. The warm, moist environments in which sprouts are grown provide excellent growing conditions for any harmful bacteria that may have been present on the seeds from which the sprouts were produced.

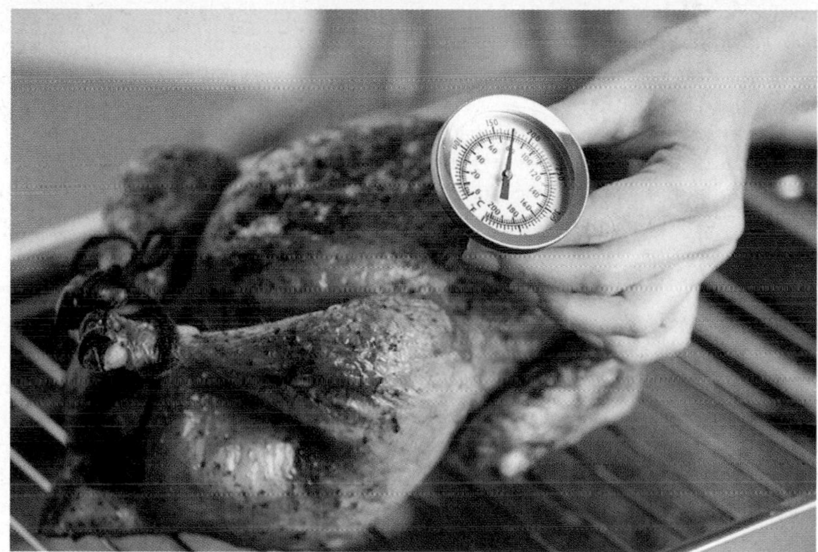

▲ **FIGURE 20-16 Use a meat thermometer to make sure your meal is safe**

How can you protect yourself from the bacteria that share our food supply? It's easy: Clean, cook, and chill. Cleaning helps prevent the spread of pathogens. Wash your hands before preparing food, and wash all utensils and cutting boards after preparing each item. Thorough cooking is the best way to ensure that any bacteria present in food are killed. Meats, in particular, must be thoroughly cooked; food safety experts recommend using a meat thermometer (**FIG. 20-16**) to ensure that the thickest part of cooked poultry has reached 165°F. The safe temperature for cuts of beef, veal, or lamb is 145°F; for pork or ground beef, 160°F. The color of cooked meat can be an unreliable indicator of safety, but when a meat thermometer is unavailable, try to avoid eating meat that is still pink inside, especially ground beef. Fish should be cooked until it is opaque and flakes easily with a fork; cook eggs until both white and yolk are firm. Finally, keep stored food cold. Pathogens multiply most rapidly at temperatures between 40° and 140°F. So get your groceries home from the store and into the refrigerator or freezer as quickly as possible. Don't leave cooked leftovers unrefrigerated for more than 2 hours. Thaw frozen foods in the refrigerator or the microwave, not at room temperature. A little bit of attention to food safety can save you from unwelcome guests in your food.

CONSIDER THIS Consumer groups contend that we can improve food safety by giving government agencies additional funding and greater authority to inspect food processing plants and order recalls of contaminated food. Opponents of such steps argue that we need not empower government agencies because the best protection against food contamination is informed consumers, who will stop buying products from companies that have produced unsafe foods. Would you support or oppose additional government oversight of food safety?

CHAPTER REVIEW

Go to **Mastering Biology** to access the Pearson eText, vocabulary review, practice quizzes, activities, videos, current events, and more.

*Answers to **Think Critically** and **Thinking Through the Concepts** questions can be found in the **Answers** section at the back of the book.*

Summary of Key Concepts

20.1 Which Organisms Are Members of the Domains Archaea and Bacteria?

Archaea and bacteria are unicellular and prokaryotic. Although archaea and bacteria are morphologically similar, they are not closely related and differ in several fundamental features, including cell wall composition, ribosomal RNA sequence, and membrane lipid structure. A cell wall determines the characteristic shapes of prokaryotes: spherical, rod-shaped, or corkscrew-shaped.

20.2 How Do Prokaryotes Survive and Reproduce?

Certain types of bacteria can move about using flagella; others form spores that disperse widely and withstand inhospitable environmental conditions. Bacteria and archaea have colonized nearly every habitat on Earth, including hot, acidic, very salty, and anaerobic environments.

Prokaryotes obtain energy in a variety of ways. Some rely on photosynthesis; others break down inorganic or organic molecules to obtain energy. Many are anaerobic, able to obtain energy when oxygen is not available. Prokaryotes reproduce by prokaryotic fission and may exchange genetic material by conjugation, in which DNA is transferred from a donor to a recipient.

20.3 How Do Prokaryotes Affect Humans and Other Organisms?

Some bacteria are pathogenic, causing disorders such as pneumonia, tetanus, botulism, and the sexually transmitted infections gonorrhea and syphilis. Most prokaryotes, however, are harmless to humans and play important roles in natural ecosystems. Some live in the digestive tracts of animals that eat leaves, where the prokaryotes break down cellulose. Nitrogen-fixing bacteria enrich the soil and aid in plant growth. Many other bacteria live off the dead bodies and wastes of other organisms, liberating nutrients for reuse.

20.4 What Are Viruses, Viroids, and Prions?

Viruses are parasites consisting of a protein coat that surrounds genetic material. They are noncellular and unable to move, grow, or reproduce outside a living cell. They invade cells of a specific host and use the host cell's energy, enzymes, and ribosomes to produce more virus particles, which are liberated when the cell ruptures. Many viruses are pathogenic to humans, including those causing colds and flu, herpes, AIDS, and certain forms of cancer.

Viroids are short strands of RNA that can invade a host cell's nucleus and direct the synthesis of new viroids. To date, viroids are known to cause only certain diseases of plants.

Prions have been implicated in diseases of the nervous system, such as kuru, Creutzfeldt-Jakob disease, scrapie, and bovine spongiform encephalopathy. Prions lack genetic material and are composed solely of mutated prion protein.

Thinking Through the Concepts

Bloom's: Remembering, Understanding

Multiple Choice

1. The name of the process by which DNA is transferred from one prokaryote to another via a cytoplasmic bridge is
   a. conjugation.
   b. prokaryotic fission.
   c. meiosis.
   d. bioremediation.
2. A community of prokaryotes surrounded by slime and adhering to a surface is called a(n)
   a. plasmid.
   b. flagellum.
   c. endospore.
   d. biofilm.
3. Which of the following statements about archaea is *not* true?
   a. Some archaea can survive and thrive at temperatures above the boiling point of water.
   b. The cell walls of archaea contain peptidoglycan.
   c. Archaea are not closely related to bacteria.
   d. Archaea are not known to cause any disease of plants or animals.
4. Viruses
   a. are usually photosynthetic.
   b. consist of a single cell.
   c. consist of DNA or RNA and a protein coat.
   d. consist of ATP and a lipid coat.
5. Applying fertilizer near an oil spill to increase the population of oil-consuming bacteria is an example of
   a. bioremediation.
   b. conjugation.
   c. genetic engineering.
   d. lateral gene transfer.

Fill-in-the-Blank

1. ____________ have peptidoglycan in their ____________, but ____________ do not.
2. Prokaryotic cells are ____________ (larger/smaller) than eukaryotic cells. The most common shapes of prokaryotes are ____________, ____________, and ____________.
3. Many prokaryotes use ____________ to move about. Some prokaryotes secrete slime that protects them when they aggregate in communities called ____________. Other prokaryotes can survive long periods and extreme conditions by producing protective structures called ____________.
4. ____________ bacteria inhabit environments that lack oxygen. ____________ bacteria capture energy from sunlight.
5. Prokaryotes reproduce by ____________ and may sometimes exchange genetic material through the process of ____________.
6. The plant nutrient ammonium is produced by ____________ bacteria in the soil and in nodules. Prokaryotes that live

in the digestive tracts of cows and rabbits break down ____________ in the leaves that those mammals eat.

7. Cholera, gonorrhea, and pneumonia are some of the diseases caused by pathogenic ____________. Harmful strains of *E. coli* can be transmitted to humans by consumption of ____________, ____________, or ____________.
8. A virus consists of a molecule of ____________ or ____________ surrounded by a(n) ____________ coat. A virus cannot reproduce unless it enters a(n) ____________ cell. A virus that infects bacteria is known as a(n) ____________.

Review Questions

1. Describe some of the ways in which prokaryotes obtain energy and nutrients.
2. What are nitrogen-fixing bacteria, and what role do they play in ecosystems?
3. Describe some of the extreme environments in which prokaryotes are found. What parts of the human body are inhabited by prokaryotes?
4. What is an endospore? What is its function?
5. What is conjugation? What role do plasmids play in conjugation?
6. Why are prokaryotes especially useful in bioremediation?
7. Describe the structure of a typical virus. How do viruses replicate?
8. Describe some examples of how prokaryotes are helpful to humans and some examples of how they are harmful to humans.
9. How do archaea and bacteria differ? How do prokaryotes and viruses differ?

Applying the Concepts

Bloom's: Applying, Analyzing, Evaluating

1. In some developing countries, antibiotics can be purchased without a prescription. Why do you think this is done? What biological consequences would you predict?
2. Before the discovery of prions, many (perhaps most) biologists would have agreed with the statement "It is a fact that no infectious organism or particle can exist that lacks nucleic acid (such as DNA or RNA)." What lessons do prions teach us about nature, science, and scientific inquiry? (You may wish to review Chapter 1 to help answer this question.)

21 The Diversity of Protists

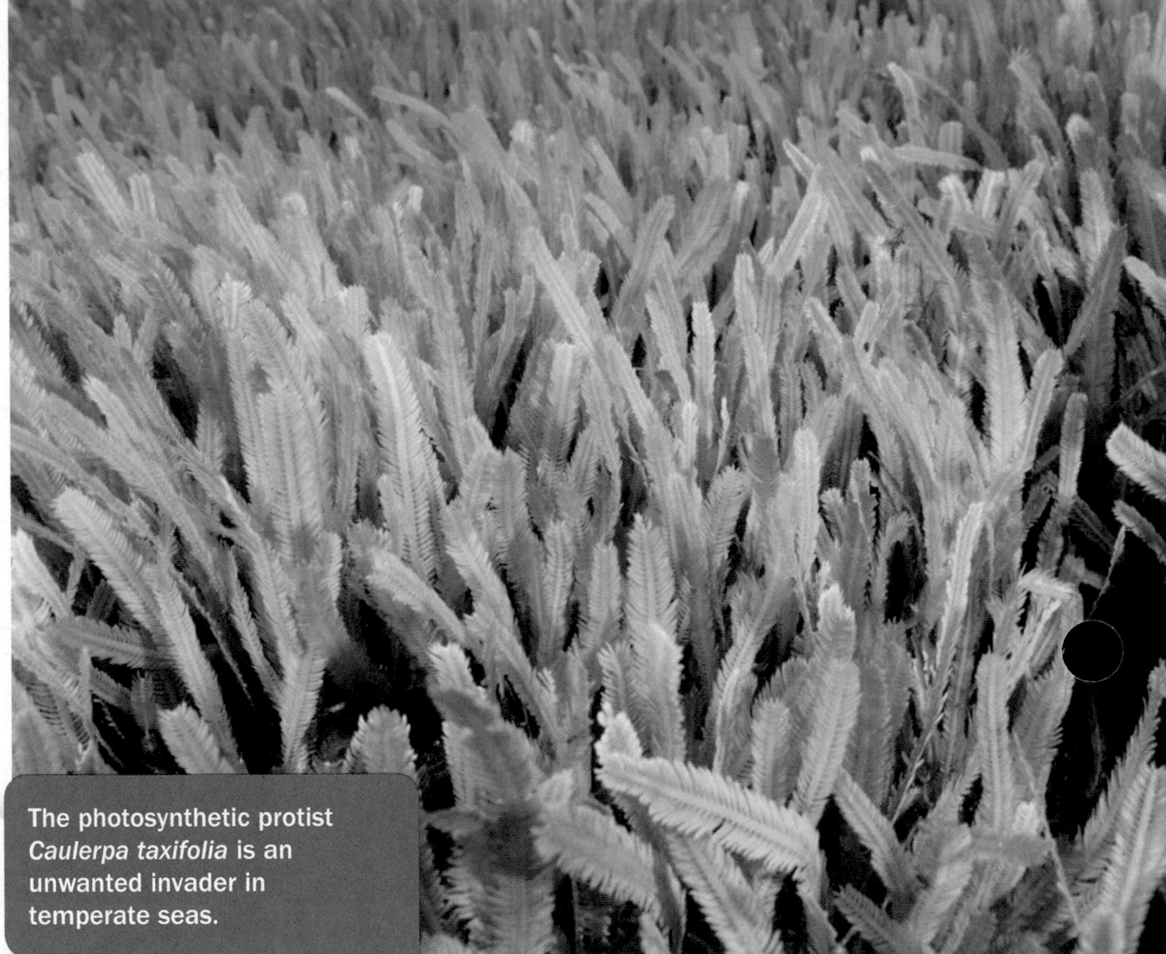

The photosynthetic protist *Caulerpa taxifolia* is an unwanted invader in temperate seas.

CASE STUDY

Green Monster

IN CALIFORNIA, IT IS A CRIME to possess, transport, or sell *Caulerpa*. Is *Caulerpa* an illicit drug or some kind of weapon? No—it is a small green seaweed. Why, then, do California's lawmakers want to ban it from their state?

The story of *Caulerpa*'s rise to public enemy status begins in the early 1980s at the Wilhelma Zoo in Stuttgart, Germany. There, the keepers of a saltwater aquarium found that the tropical seaweed *Caulerpa taxifolia* was an attractive companion and background for the tropical fish on display. Even better, years of captive breeding at the zoo had yielded a strain of the seaweed that was well suited to life in an aquarium. The new strain was particularly hardy and could survive in waters considerably cooler than the tropical waters in which wild *Caulerpa* is found. The zoo staff was happy to send cuttings to other institutions that wished to use it in aquarium displays.

One institution that received a cutting was the Oceanographic Museum of Monaco, located on the shore of the Mediterranean Sea. In 1984, a visiting marine biologist discovered a small patch of *Caulerpa* growing in the waters just below the museum. Presumably, someone cleaning an aquarium had dumped the water into the sea and thereby inadvertently introduced *Caulerpa* to the Mediterranean.

By 1989, the *Caulerpa* patch had grown to cover a few acres. It formed a continuous mat that seemed to exclude most of the other organisms that normally inhabit the Mediterranean Sea floor. The local herbivores, such as sea urchins and fish, did not feed on *Caulerpa*.

It soon became apparent that *Caulerpa* spreads rapidly, is not controlled by predation, and displaces native species. By the mid-1990s, biologists were alarmed to find *Caulerpa* all along the Mediterranean coast from Spain to Italy. Today, this invasive species grows in extensive beds throughout the Mediterranean and covers an ever-expanding area on the seafloor.

Despite the threat it poses to ecosystems, *Caulerpa* is a fascinating organism. Green algae such as *Caulerpa* are part of a diverse group of organisms informally known as protists. What other kinds of organisms are protists? How and where do protists live?

AT A GLANCE

21.1 WHAT ARE PROTISTS?

Two of life's domains, Bacteria and Archaea, contain only prokaryotes. The third domain, Eukarya, includes all eukaryotic organisms. The most conspicuous Eukarya are plants, fungi, and animals (covered in Chapters 22 through 25). The remaining eukaryotes constitute a diverse array of organisms collectively known as **protists**. Protists do not form a clade—a group consisting of all the descendants of a particular common ancestor—so systematists do not use the term "protist" as a formal group name. Instead, protist is a term of convenience that refers to any eukaryotc that is not a plant, animal, or fungus.

Most protists are single celled and are invisible to us as we go about our daily lives. If we could somehow shrink to their microscopic scale, we would be impressed with their beautiful forms, their varied lifestyles, their astonishingly diverse modes of reproduction, and the structural and physiological complexity that is possible within the limits of a single cell.

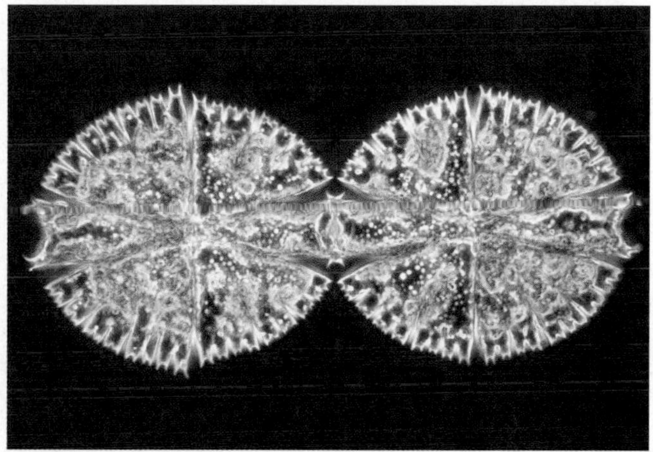

(a) Reproducing by cell division

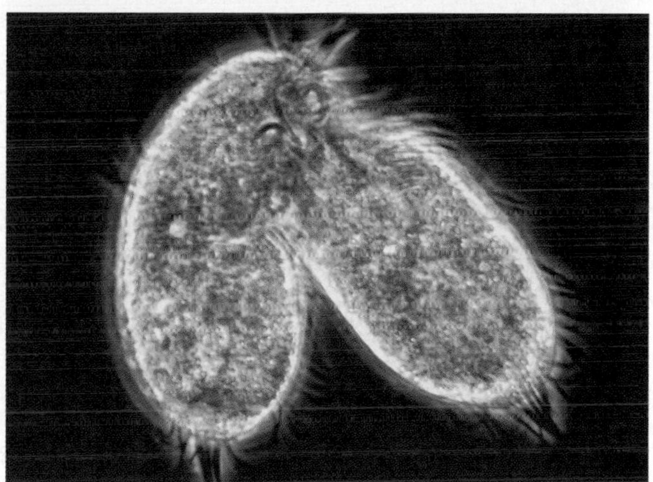

(b) Exchanging genetic material

▲ **FIGURE 21-1 Protistan reproduction and gene exchange**
(a) *Micrasterias*, a green alga, reproduces asexually by cell division. **(b)** Two *Euplotes* ciliates exchange genetic material.

THINK CRITICALLY What do biologists mean when they say that sex and reproduction are uncoupled in most protists?

Protists Use Diverse Modes of Reproduction

Most protists reproduce asexually; an individual divides by mitotic cell division to yield two individuals that are genetically identical to the parent cell (**FIG. 21-1a**). Many protists, however, are also capable of sexual reproduction, in which two individuals contribute genetic material to an offspring that is genetically different from either parent. Nonreproductive processes that combine the genetic material of different individuals are also common among protists (**FIG. 21-1b**).

In many protist species that are capable of sexual reproduction, most reproduction is nonetheless asexual. Sexual reproduction occurs only infrequently, at a particular time of year or under certain circumstances, such as a crowded environment or a shortage of food. The details of sexual reproduction and the resulting life cycles vary tremendously among different types of protists.

Protists Use Diverse Modes of Nutrition

Three major modes of nutrition are represented among protists. Protists may ingest their food, absorb nutrients from their surroundings, or capture solar energy by photosynthesis. Protists that ingest their food are predators. Predatory single-celled protists may have flexible cell membranes that can change shape to surround and engulf food such as bacteria or another protist. Protists that feed in this manner typically use finger-like extensions called **pseudopods** to engulf prey. Other predatory protists create tiny currents that sweep food particles into mouth-like openings in the cell. Whatever the means by which food is ingested, once it is inside the protist cell, it is typically packaged into a membrane-bound **food vacuole** for digestion.

Protists that absorb nutrients directly from the surrounding environment may be free living or may live inside the bodies of other organisms. The free-living types live in soil and other environments that contain dead organic matter, where they act as decomposers. Most absorptive feeders, however, live inside other organisms. In most cases, these protists are parasites whose feeding activity harms the host species.

Photosynthetic protists are abundant in oceans, lakes, and ponds. Most float suspended in the water, but some live inside the tissues of other organisms, such as corals or clams. These associations appear to be mutually beneficial; some of the solar energy captured by the photosynthetic protists is used by the host organism, which provides shelter and protection for the protists.

Protist photosynthesis takes place in chloroplasts (see Chapter 7). Chloroplasts are the descendants of ancient photosynthetic bacteria that took up residence inside a larger cell in a process known as *endosymbiosis* (see Chapter 18). In addition to the original instance of endosymbiosis that created the first protist chloroplast, there have been several later occurrences of *secondary endosymbiosis* in which a nonphotosynthetic protist engulfed a photosynthetic, chloroplast-containing protist. Ultimately, most components of the engulfed species disappeared, leaving only a chloroplast surrounded by four membranes. Two of these membranes are from the original, bacteria-derived chloroplast; one is from the engulfed protist; and one is from the food vacuole that originally contained the engulfed protist. Multiple occurrences of secondary endosymbiosis account for the presence of photosynthetic species in a number of different, unrelated protist groups.

Protists Affect Humans and Other Organisms

Protists have important impacts, both positive and negative, on human lives. The primary positive impact actually benefits all living things and stems from the ecological role of photosynthetic marine protists. Just as plants do on land, photosynthetic protists in the oceans capture solar energy and make it available to the other organisms in the ecosystem. Thus, the marine ecosystems on which humans depend for food in turn depend on protists. Further, in the process of using photosynthesis to capture energy, the protists release oxygen gas that helps replenish the oxygen removed from the atmosphere by respiration (recall from Chapter 8 that cellular respiration consumes oxygen).

On the negative side of the ledger, many human diseases are caused by parasitic protists. The diseases caused by protists include some of humanity's most prevalent ailments and some of its deadliest afflictions. Protists also cause a number of plant diseases, some of which damage crops that are important to humans.

CHECK YOUR LEARNING

Can you . . .

- define protist and describe the various ways in which protists acquire nutrients and reproduce?
- describe a scenario for the evolutionary origin of protist chloroplasts?
- describe the major effects of protists on people and other organisms?

21.2 WHAT ARE THE MAJOR GROUPS OF PROTISTS?

Taking advantage of the advent of fast, inexpensive DNA sequencing, systematists have used genetic comparisons to gain a better understanding of protist clades and the evolutionary relationships among them. Some of these clades are listed in **TABLE 21-1**, along with the key characteristics of their members. An evolutionary tree that includes the major protist clades is depicted in Figure 19-7.

Past classifications of protists grouped species according to their mode of nutrition, but the old categories are now obsolete because they do not accurately reflect our current understanding of phylogeny. Nonetheless, biologists still use terminology that refers to groups of protists that share particular characteristics but are not necessarily related. For example, photosynthetic protists are collectively known as **algae** (singular, alga), and single-celled, nonphotosynthetic protists are collectively known as **protozoa** (singular, protozoan).

In the following sections, we explore a sample of protist diversity.

Excavates Lack Mitochondria

Excavates are named for a feeding groove that gives the appearance of having been "excavated" from the surface of the cell. Most excavates are anaerobes (can live and grow without oxygen). Many lack mitochondria, but do possess other organelles that are probably evolutionarily derived from mitochondria. It is therefore likely that the excavates' ancestors did possess mitochondria, but these organelles were later lost in some excavate groups. The excavates include the diplomonads, parabasalids, euglenids, and kinetoplastids.

Diplomonads Have Two Nuclei

The single cells of **diplomonads** have two nuclei and move about by means of multiple flagella. A parasitic diplomonad, *Giardia* (**FIG. 21-2**), poses a health problem worldwide. *Cysts*

▲ **FIGURE 21-2** ***Giardia*: The curse of campers** A diplomonad (genus *Giardia*) that may infect water—causing gastrointestinal disorders for the people who drink it—is shown here in a human small intestine.

TABLE 21-1 The Major Groups of Protists

Group	Subgroup	Locomotion	Nutrition	Representative Features	Representative Genus
Excavates	Diplomonads	Swim with flagella	Heterotrophic (i.e., consume other organisms)	Lack mitochondria; inhabit soil or water, or may be parasitic; unicellular	*Giardia* (intestinal parasite of mammals)
	Parabasalids	Swim with flagella	Heterotrophic	Lack mitochondria; parasites or mutualistic symbionts; unicellular	*Trichomonas* (causes the sexually transmitted infection trichomoniasis)
	Euglenids	Swim with one flagellum	Photosynthetic	Have an eyespot; all freshwater	*Euglena* (common pond-dweller)
	Kinetoplastids	Swim with flagella	Heterotrophic	Inhabit soil or water, or may be parasitic; unicellular	*Trypanosoma* (causes African sleeping sickness)
Stramenopiles (chromists)	Water molds	Swim with flagella (gametes)	Heterotrophic	Filamentous	*Plasmopara* (causes downy mildew)
	Diatoms	Glide along surfaces	Photosynthetic	Have silica shells; most marine; unicellular	*Navicula* (glides toward light)
	Brown algae	Nonmotile	Photosynthetic	Seaweeds of temperate oceans; multicellular	*Macrocystis* (forms kelp forests)
Alveolates	Dinoflagellates	Swim with two flagella	Photosynthetic	Many bioluminescent; often have cellulose walls; unicellular	*Gonyaulax* (causes red tide)
	Apicomplexans	Nonmotile	Heterotrophic	All parasitic; form infectious spores; unicellular	*Plasmodium* (causes malaria)
	Ciliates	Swim with cilia	Heterotrophic	Include the most complex single cells; unicellular	*Paramecium* (fast-moving pond-dweller)
Rhizarians	Foraminiferans	Extend thin pseudopods	Heterotrophic	Have calcium carbonate shells; unicellular	*Globigerina* (shells cover large areas of ocean floor)
	Radiolarians	Extend thin pseudopods	Heterotrophic	Have silica shells; unicellular	*Actinomma* (found in oceans worldwide)
Amoebozoans	Amoebas	Extend thick pseudopods	Heterotrophic	Have no shells; unicellular	*Amoeba* (common pond-dweller)
	Acellular slime molds	Slug-like mass oozes over surfaces	Heterotrophic	Form multinucleate plasmodium	*Physarum* (forms a large, bright orange mass)
	Cellular slime molds	Amoeboid cells extend pseudopods; slug-like mass crawls over surfaces	Heterotrophic	Form pseudoplasmodium with individual amoeboid cells	*Dictyostelium* (often used in laboratory studies)
Red algae		Nonmotile	Photosynthetic	Some deposit calcium carbonate; mostly marine; most multicellular	*Porphyra* (used to make sushi wrappers)
Chlorophyte algae		Swim with flagella (some species)	Photosynthetic	Close relatives of clade that includes land plants; unicellular and multicellular	*Ulva* (sea lettuce)

(tough structures that enclose the organism during one phase of its life cycle) of this diplomonad are released in the feces of infected humans, dogs, or other animals; a single gram of feces may contain 300 million cysts. The excreted cysts may enter streams and lakes, municipal water supplies, and even swimming pools and hot tubs. If a mammal drinks infected water, the cysts develop into the adult form in the small intestine. In humans, infections can cause severe diarrhea, dehydration, nausea, vomiting, and cramps. Fortunately, these infections can be cured with drugs, and deaths from *Giardia* infections are uncommon.

Parabasalids Include Mutualists

Parabasalids are anaerobic, flagellated protists named for the presence in their cells of a distinctive structure called the parabasal body, which consists of densely packed Golgi vesicles (see Chapter 4). All known parabasalids live inside animals. For example, this group includes several species that inhabit the digestive systems of some species of wood-eating termites (**FIG. 21-3**). The termites cannot themselves digest the cellulose in wood, but the parabasalids can. Thus, the insect and the protist are in a mutually beneficial relationship. The termite delivers food to the parabasalids in its gut; as the parabasalids digest the food, some of the nutrients released become available for use by the termite.

Some parabasalids are parasitic. For example. the parabasalid *Trichomonas vaginalis* infects the mucous layers of the urinary and reproductive tracts in people, causing the sexually transmitted disease trichomoniasis. Trichomoniasis affects about 3.7 million people in the United States each year.

Euglenids Lack a Rigid Covering and Swim by Means of Flagella

Euglenids are single-celled protists that live mostly in fresh water and are named after the group's best-known representative, *Euglena*, a complex single cell that moves about by whipping its flagellum through water (**FIG. 21-4**). Euglenids lack a rigid outer covering, so some can move by wriggling as well as by whipping their flagella. Many euglenids are photosynthetic, but some species instead absorb or engulf food. Some euglenids possess simple light-sensing organelles consisting of a photoreceptor, called an *eyespot*, and an adjacent patch of pigment. The pigment shades the photoreceptor only when light strikes from certain directions, enabling the organism to determine the direction of the light source. Using information from the eyespot, the flagellum propels the protist toward light levels appropriate for photosynthesis.

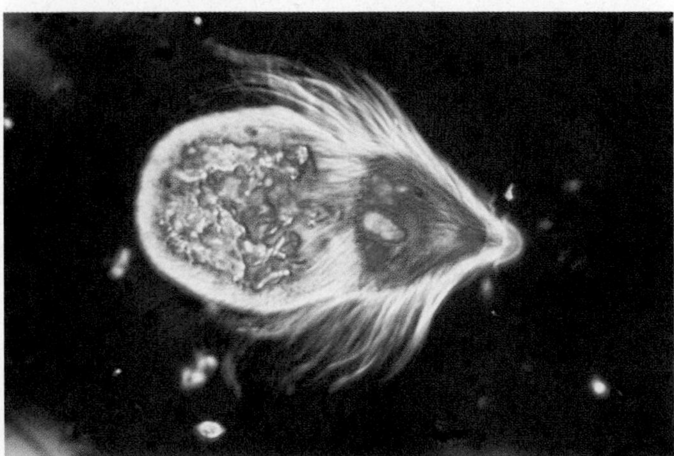

▲ **FIGURE 21-3 Parabasalids inhabit termite digestive tracts** The parabasalid *Trichonympha* lives in termite guts, where it digests cellulose in the woody plant material that termites consume.

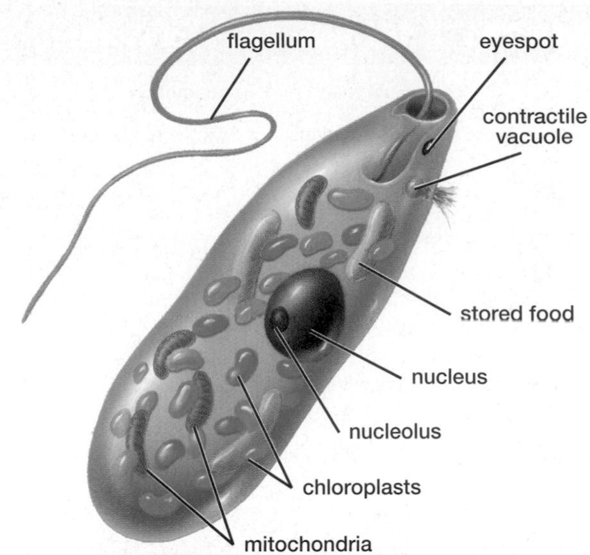

▲ **FIGURE 21-4** ***Euglena*, a representative euglenid** *Euglena*'s elaborate single cell is packed with green chloroplasts, which will disappear if the protist is kept in darkness.

Some Kinetoplastids Cause Human Diseases

The DNA in the mitochondria of **kinetoplastids** is arranged in complex assemblies called kinetoplasts, in which many copies of the circular mitochondrial genome are interlinked to form distinctive disk-shaped structures. Most kinetoplastids possess at least one flagellum, which may propel the organism, sense the environment, or ensnare food. Some kinetoplastids are free living, inhabiting soil and water; others live inside other organisms in a relationship that may be mutually beneficial or parasitic. A dangerous parasitic kinetoplastid in the genus *Trypanosoma* is responsible for African sleeping sickness, a potentially fatal disease (**FIG. 21-5**). Like many parasites, this organism has a complex life cycle, part of which is spent in the tsetse fly. While feeding on the blood of a mammal, an infected fly can transmit saliva containing the trypanosome to the mammal. The parasite then develops in the new host (which may be a human) and enters the bloodstream. The trypanosome may then be ingested by another tsetse fly that bites the host, thus beginning a new cycle of infection. For information on another disease-causing kinetoplastid, see "Health Watch: Neglected Protist Infections" on page 339.

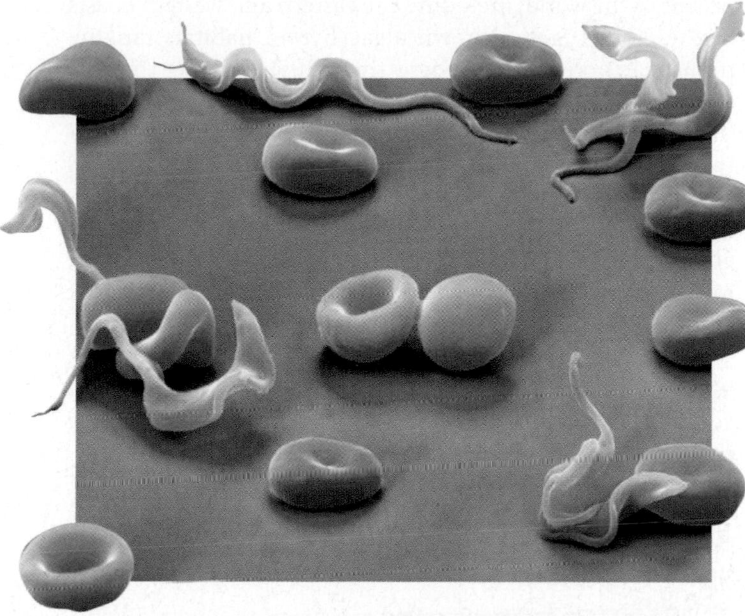

▲ **FIGURE 21-5 A disease-causing kinetoplastid** This photomicrograph shows human blood that is heavily infested with the corkscrew-shaped parasitic kinetoplastid *Trypanosoma*, which causes African sleeping sickness.

Stramenopiles Have Distinctive Flagella

All **stramenopiles** have fine, hair-like projections on their flagella (though in many stramenopiles, flagella are present only at certain stages of the life cycle). Despite their shared evolutionary history, stramenopiles display a wide range of forms. Some are photosynthetic and some are not; most are single celled, but some are multicellular. Three major stramenopile groups are the water molds, the diatoms, and the brown algae.

Water Molds Have Had Important Impacts on Humans

The **water molds** form a small group of protists, many of which form long filaments that aggregate into cottony tufts. These tufts are superficially similar to structures produced by some fungi, but this resemblance is due to convergent evolution, not shared ancestry. Many water molds are decomposers that live in water and damp soil. Some species have profound economic impacts on humans. For example, a water mold causes a disease of grapes known as *downy mildew*. Its inadvertent introduction into France from the United States in the late 1870s nearly destroyed the French wine industry. A water mold is also responsible for *late blight*, a disease of potatoes. When this protist was accidentally introduced into Ireland in about 1845, it destroyed nearly the entire potato crop, causing a devastating famine during which as many as 1 million people in Ireland starved and many more emigrated to the United States to escape the famine.

Diatoms Are Encased Within Glassy Walls

The **diatoms**, photosynthetic stramenopiles found in both fresh and salt water, produce protective cell walls that contain *silica* (glass) (**FIG. 21-6**). Accumulations of diatoms' glassy shells over millions of years have produced fossil deposits of "diatomaceous earth" that may be hundreds of meters thick. This slightly abrasive substance is widely used in products such as toothpaste and metal polish.

Diatoms form part of the **phytoplankton**, the single-celled photosynthesizers that float passively in the upper layers of Earth's lakes and oceans. Phytoplankton play an immensely important ecological role. Marine phytoplankton account for about 50% of all photosynthetic activity on Earth, absorbing carbon dioxide, recharging the atmosphere with oxygen, and supporting the complex web of aquatic life.

Brown Algae Are Multicellular

Though most photosynthetic protists, such as diatoms, are single celled, some form multicellular structures that are commonly known as seaweeds. Although some seaweeds seem to resemble plants, they lack many of the distinctive features of plants. For example, no seaweed has true roots.

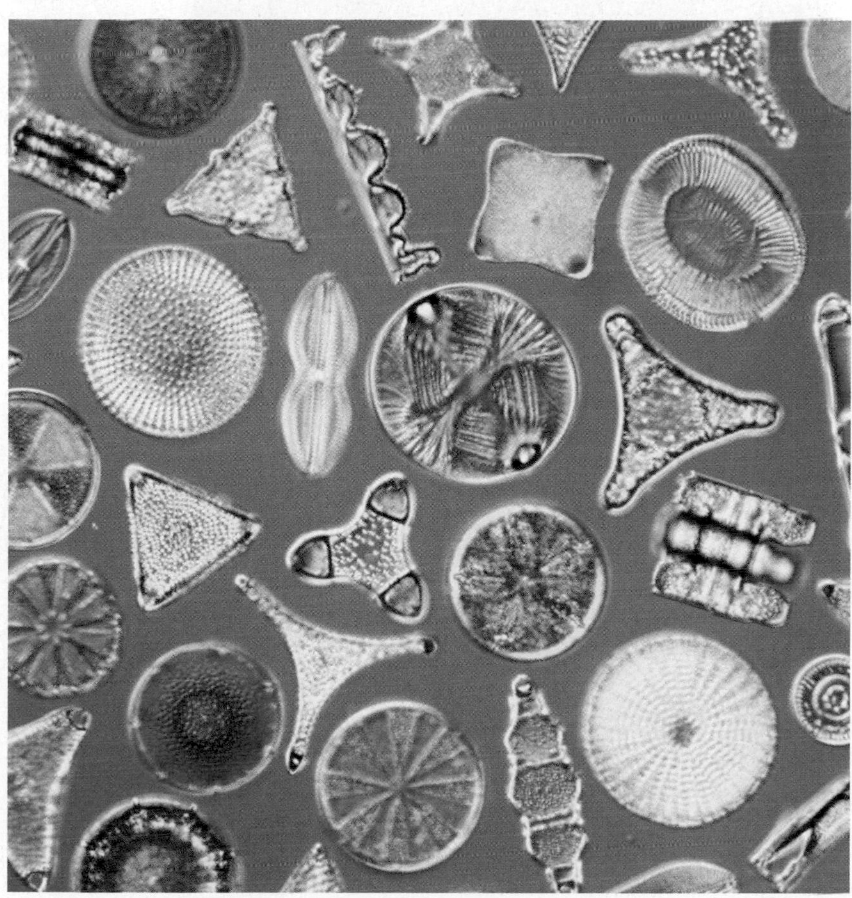

▶ **FIGURE 21-6 Some representative diatoms** This photomicrograph illustrates the intricate, microscopic beauty and variety of the glassy walls of diatoms.

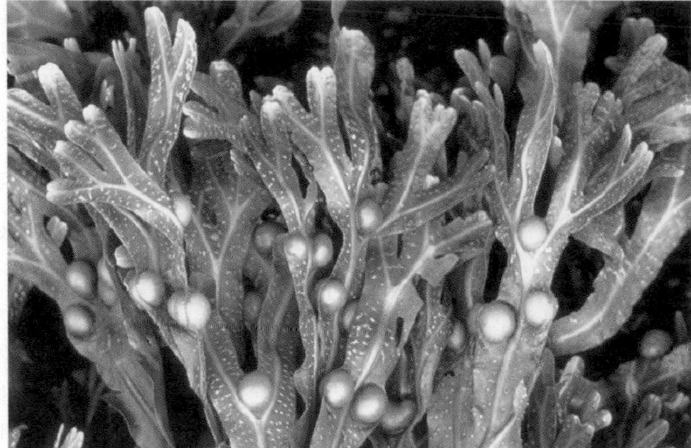

(a) *Fucus*

(b) Kelp forest

▲ **FIGURE 21-7 Brown algae are multicellular protists**
(a) *Fucus*, a genus found near shores, is shown here exposed at low tide. Notice the gas-filled floats, which provide buoyancy in water. **(b)** The giant kelp *Macrocystis* forms underwater forests off southern California.

The stramenopiles include one group of seaweeds, the brown algae, which are named for the brownish-yellow pigments that (in combination with green chlorophyll) increase the seaweed's light-gathering ability. Almost all brown algae are marine. The group includes the dominant seaweed species that dwell along rocky shores in the temperate (cooler) oceans of the world, including the eastern and western coasts of the United States. Brown algae live in habitats ranging from nearshore waters, where they cling to rocks that are exposed at low tide, to far offshore. Several species use gas-filled floats to support their bodies (**FIG. 21-7a**). Some of the giant kelp found along the Pacific coast reach heights of 175 feet (53 meters) and may grow more than 6 inches (15 centimeters) in a single day. With their dense growth and towering height, kelp form undersea forests that provide food, shelter, and breeding areas for a large variety of marine animals (**FIG. 21-7b**).

Brown algae also contribute to familiar consumer goods. Sodium alginate, extracted from brown algae such as giant kelp, forms a gel that is used to thicken and stabilize ice cream, paint, shaving cream, and many other products.

Alveolates Include Parasites, Predators, and Phytoplankton

The **alveolates** are single-celled organisms that have distinctive, small cavities beneath the surface of their cells. Some alveolates are photosynthetic, some are parasitic, and some are predatory. The major alveolate groups are the dinoflagellates, apicomplexans, and ciliates.

Dinoflagellates Swim by Means of Two Whip-Like Flagella

Though most **dinoflagellates** are photosynthetic, there are also some nonphotosynthetic species. Dinoflagellates (from the Latin for "whirling whips") are named for the two whip-like flagella that propel them (**FIG. 21-8**). One flagellum encircles the cell, and the second projects behind it. Some dinoflagellates are enclosed only by a plasma membrane; others have cellulose walls that resemble armor plates. Although some species live in fresh water, dinoflagellates are especially abundant in the ocean, where they are an important component of the phytoplankton and a food source for larger organisms. Many dinoflagellates are bioluminescent, producing a brilliant blue-green light when disturbed by motion in the water.

Nutrient-rich warm water may bring on a dinoflagellate population explosion. Dinoflagellates can become so numerous that the water is dyed by

◀ **FIGURE 21-8 A dinoflagellate**
This dinoflagellate has two flagella: a longer one that extends from a slot, and a shorter one that lies in a groove that encircles the cell.

Health WATCH Neglected Protist Infections

A number of protist species can live in or on the human body, and some of these species are parasites that cause infectious diseases. The best known and most feared of these diseases is malaria, which infects millions each year around the globe and kills hundreds of thousands of those infected. As a result, a great deal of money and effort are devoted to understanding and combating the *Plasmodium* species that cause the disease. Other protist infections, however, receive much less attention from researchers, physicians, and public health officials, even though they harm many people. Two of these under-the-radar diseases—Chagas disease and toxoplasmosis—have been included on the Centers for Disease Control's list of "neglected parasitic infections" because they affect large numbers of people in the United States.

Chagas disease is caused by the kinetoplastid *Trypanosoma cruzi*, which is transmitted from person to person mainly by blood-sucking insects known as triatomine bugs (sometimes called "kissing bugs"). However, it can also be transmitted from mother to child or by blood transfusions (blood donations in the United States are now screened for *T. cruzi*). Infections initially cause few or no symptoms, so the infection often goes undiagnosed. But years after infection, many victims develop heart diseases including abnormal rhythms and heart failure that can lead to sudden death. Because the bugs that transmit the disease are most common in Mexico, Central America, and South America, most Chagas disease cases occur in those regions. However, an estimated 300,000 victims of the disease live in the United States; many of them probably became infected elsewhere before moving to the United States. Chagas infections can be cured with drugs during the weeks immediately following infection, but the disease is much more difficult to treat once it reaches the chronic, heart-damaging stage.

Toxoplasmosis is caused by the apicomplexan parasite *Toxoplasma gondii* (**FIG. E21-1**). The parasite's primary host is cats, which shed *Toxoplasma* in their feces, from which they are transmitted to rodents that are later consumed by cats, completing the parasite's life cycle. However, humans can become infected if they come into contact with cat feces (for example, while gardening or cleaning a litter box) or consume contaminated food. Sources of contaminated food include unwashed produce and undercooked meat from animals that had themselves ingested *Toxoplasma*. Initial infection usually results in, at worst, mild symptoms such as fever, but infections acquired during pregnancy are very dangerous to the baby, which may develop severe illness leading to epilepsy, blindness, and developmental disorders. In addition, even infected adults who are initially asymptomatic may be at risk, as the infection persists for life and can ultimately lead to reduced immune function and increased susceptibility to other infectious diseases.

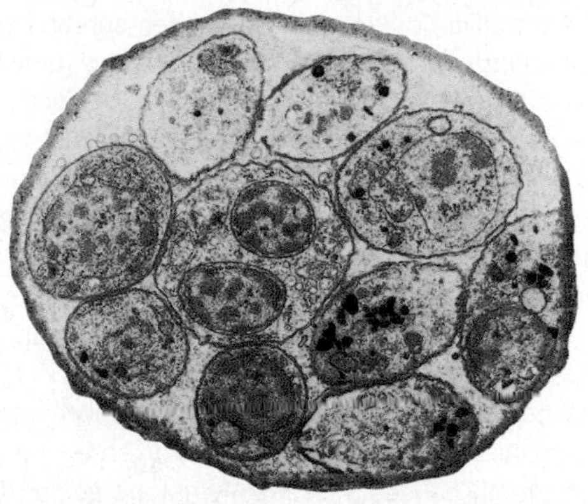

▲ **FIGURE E21-1** *Toxoplasma gondii* cyst The cyst contains several parasites, which may multiply until thousands are present.

One of *Toxoplasma*'s most fascinating adaptations is its ability to affect the behavior of infected mice and rats. The parasite invades the rodent's brain and induces neurological changes that make the animal less fearful and more reckless. That is, *Toxoplasma* changes rodent psychology in a way that makes the infected animal more likely to be eaten by a cat, thereby allowing the parasite to complete its life cycle and reproduce. Could *Toxoplasma* be affecting the behavior of the many millions of people it currently infects? Very preliminary evidence suggests that the parasite does indeed influence aspects of personality and may even trigger severe mental illness in susceptible individuals.

THINK CRITICALLY Imagine that you are a physician consulting with a patient who has recently become pregnant for the first time, and you are trying to decide whether to recommend blood tests for Chagas disease or toxoplasmosis. Before making your decision, what would you want to know about the patient's background and habits? Explain your answer.

the color of their bodies (**FIG. 21-9**). The water may turn yellow, pink, orange, or brown, but the most common result is a reddish tint, so dinoflagellate blooms are commonly called "red tides." During red tides, fish often die by the thousands, suffocated by clogged gills or by the oxygen depletion that results from the decay of billions of dinoflagellates. But dinoflagellate explosions can benefit oysters, mussels, and clams, which have a feast, filtering millions of the protists from the water and consuming them. In the process, however, the mollusks' bodies accumulate concentrations of a nerve toxin produced by the dinoflagellates. Dolphins, seals, sea otters, and humans who eat affected mollusks may be stricken with potentially lethal paralytic shellfish poisoning.

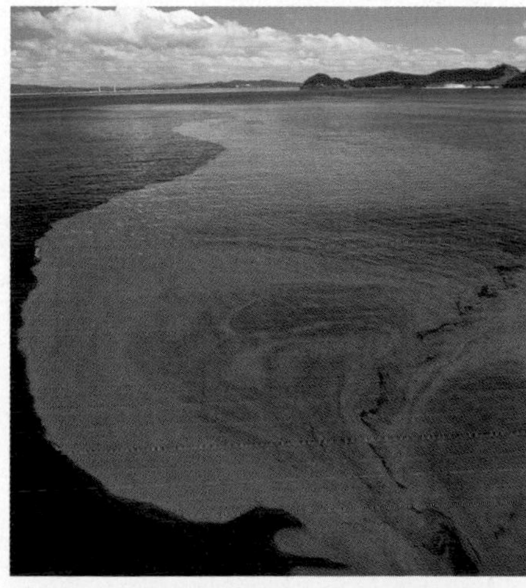

◀ **FIGURE 21-9** **A red tide** The explosive reproductive rate of certain dinoflagellates under the right environmental conditions can produce concentrations so great that their microscopic bodies dye the seawater.

CASE STUDY CONTINUED

Green Monster

Just as the invasive *Caulerpa* seaweed often spreads uncontrollably when introduced to environments free of its normal predators and parasites, populations of toxin-producing dinoflagellates may grow explosively when released into new waters. Red tides have become increasingly common in recent years. One reason for this increased incidence is that dinoflagellate species that can cause red tides have been inadvertently spread around the world by humans. The dinoflagellates travel mainly in seawater that is pumped into the ballast tanks of cargo ships and then discharged at distant ports.

Sometimes, a protist released into a new environment has a damaging impact not because it overwhelms an ecosystem but because it directly causes a disease. What are some examples of such an introduced disease among the alveolates?

Apicomplexans Are Parasitic and Have No Means of Locomotion

All **apicomplexans** are parasitic, living inside the bodies and sometimes inside the individual cells of their hosts. They form infectious spores—resistant structures transmitted from one host to another through food, water, or the bite of an infected insect. As adults, apicomplexans have no means of locomotion. Many have complex life cycles, a common feature of parasites. A well-known example is the malaria parasite *Plasmodium* (**FIG. 21-10**). Parts of its life cycle are spent in the body of a female *Anopheles* mosquito. The mosquito is not harmed by the presence of *Plasmodium* and may eventually bite a human and pass the protist to the unfortunate victim. The protist develops in the victim's liver, then enters the blood, where it reproduces rapidly in red blood cells. When the blood cells rupture, they release large quantities of spores, which cause the recurrent fever of malaria. Uninfected mosquitoes may acquire the parasite by feeding on the blood of a malaria victim, spreading the parasite when they bite another person.

Plasmodium species infect many kinds of animals. *Plasmodium relictum*, for example, infects birds and causes avian malaria. It was introduced to Hawaii, which had previously been free of avian malaria, in the blood of exotic bird species intentionally released on the islands. Native Hawaiian birds, which had

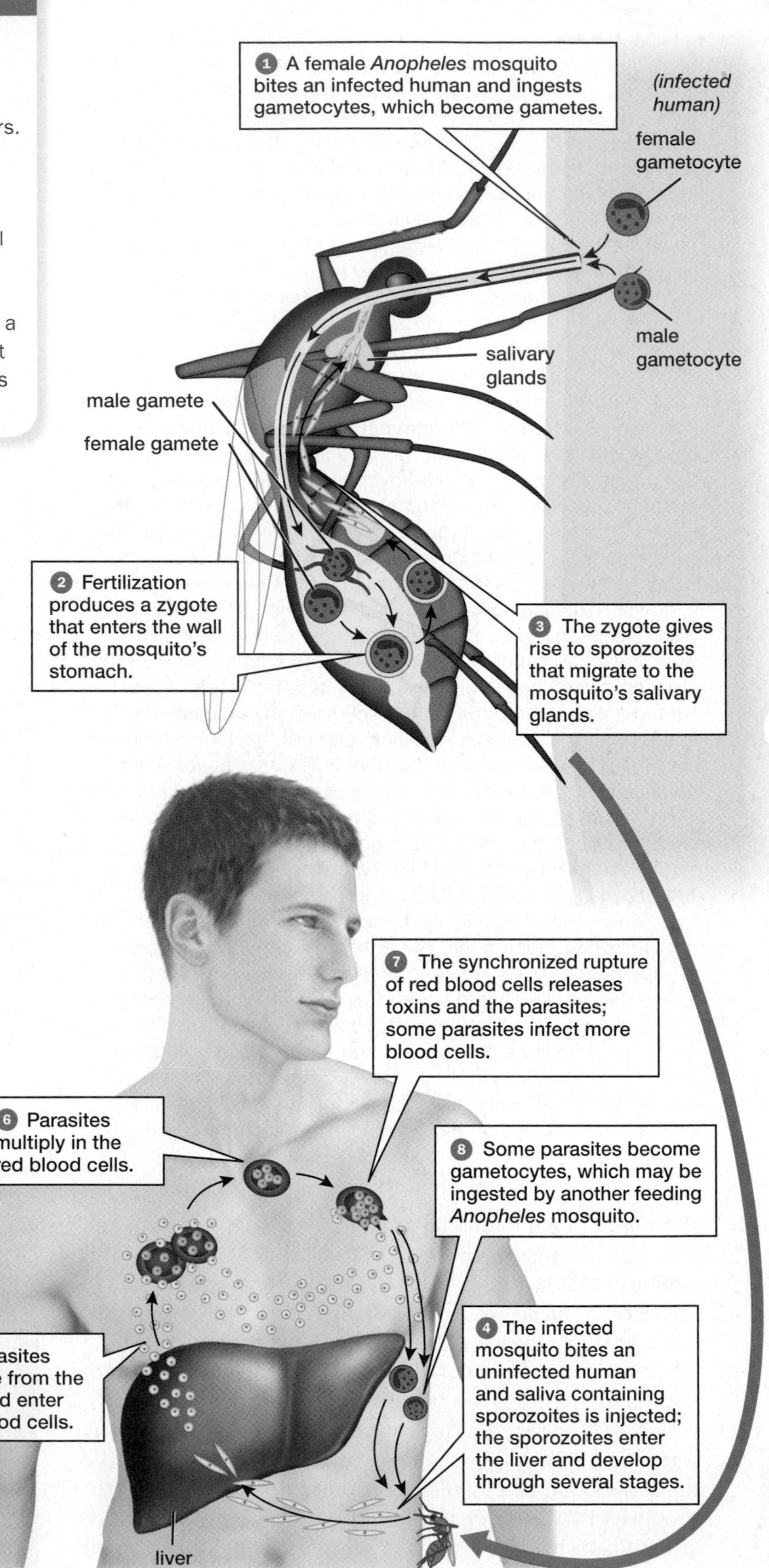

1 A female *Anopheles* mosquito bites an infected human and ingests gametocytes, which become gametes.
(infected human)
female gametocyte
male gametocyte
salivary glands
male gamete
female gamete
2 Fertilization produces a zygote that enters the wall of the mosquito's stomach.
3 The zygote gives rise to sporozoites that migrate to the mosquito's salivary glands.
7 The synchronized rupture of red blood cells releases toxins and the parasites; some parasites infect more blood cells.
6 Parasites multiply in the red blood cells.
8 Some parasites become gametocytes, which may be ingested by another feeding *Anopheles* mosquito.
5 Parasites emerge from the liver and enter red blood cells.
4 The infected mosquito bites an uninfected human and saliva containing sporozoites is injected; the sporozoites enter the liver and develop through several stages.
liver

FIGURE 21-10 The life cycle of the malaria parasite

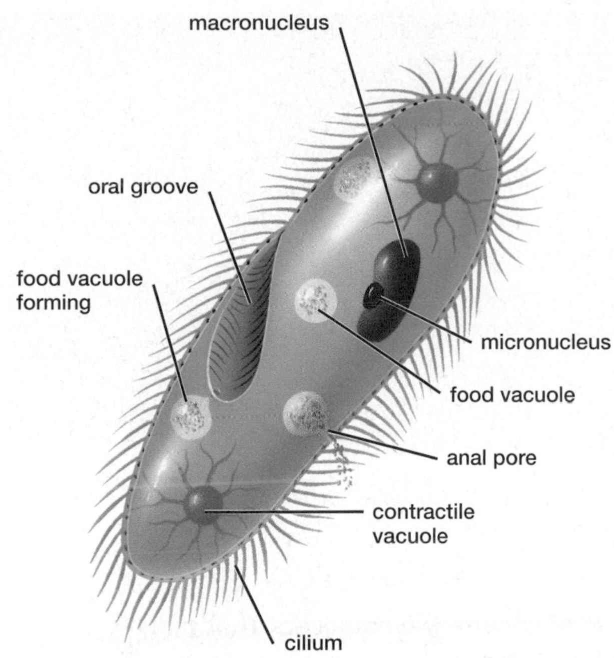

▲ **FIGURE 21-11 The complexity of ciliates** The ciliate *Paramecium* illustrates some important ciliate organelles. The oral groove acts as a mouth, food vacuoles—miniature digestive systems—form at its apex, and waste is expelled by exocytosis through an anal pore. Contractile vacuoles regulate water balance.

not evolved immune defenses against malaria, nonetheless mostly remained uninfected at first. In the 1920s, however, the mosquito species that transmits avian malaria was accidentally introduced to the island. Malaria then spread rapidly through native bird populations and was a major contributor to the extinction of many native species.

Ciliates Are the Most Complex of the Alveolates

Ciliates, which inhabit fresh and salt water, represent the peak of unicellular complexity. They possess many specialized organelles, including cilia, the short hair-like outgrowths for which they are named. In the well-known freshwater genus *Paramecium*, rows of cilia cover the organism's entire body surface (**FIG. 21-11**). The coordinated beating of the cilia propels the cell through the water at a rate of 1 millimeter (about 10 body lengths) per second—a protistan speed record. Although only a single cell, *Paramecium* responds to its environment as if it had a well-developed nervous system. Confronted with a noxious chemical or a physical barrier, the cell immediately backs up by reversing the beating of its cilia and then proceeds in a new direction. Some ciliates, such as *Didinium*, are accomplished predators (**FIG. 21-12**).

▲ **FIGURE 21-12 A microscopic predator** In this scanning electron micrograph, the predatory ciliate *Didinium* attacks a *Paramecium*. Notice that the cilia of *Didinium* are confined to two bands, whereas *Paramecium* has cilia over its entire body. Ultimately, the predator will engulf and consume its prey. This microscopic drama could take place on a pinpoint with room to spare.

Rhizarians Have Thin Pseudopods

Protists in a number of different groups possess flexible plasma membranes that they can extend in any direction to form finger-like projections called pseudopods, which they use in locomotion and for engulfing food. The pseudopods of **rhizarians** are thin and thread-like. In many species in this group, the pseudopods extend through hard shells. Rhizarians include the foraminiferans and the radiolarians.

Foraminiferans Have Chalky Shells

The **foraminiferans** are primarily marine protists that produce beautiful shells. Their shells are constructed mostly of calcium carbonate (chalk; **FIG. 21-13**). These elaborate shells are pierced by myriad openings through which pseudopods extend. The chalky shells of dead foraminiferans,

▲ **FIGURE 21-13 The chalky shell of a foraminiferan** In a living foraminiferan, thin pseudopods would extend out through the openings in the shell, to sense the environment and capture food.

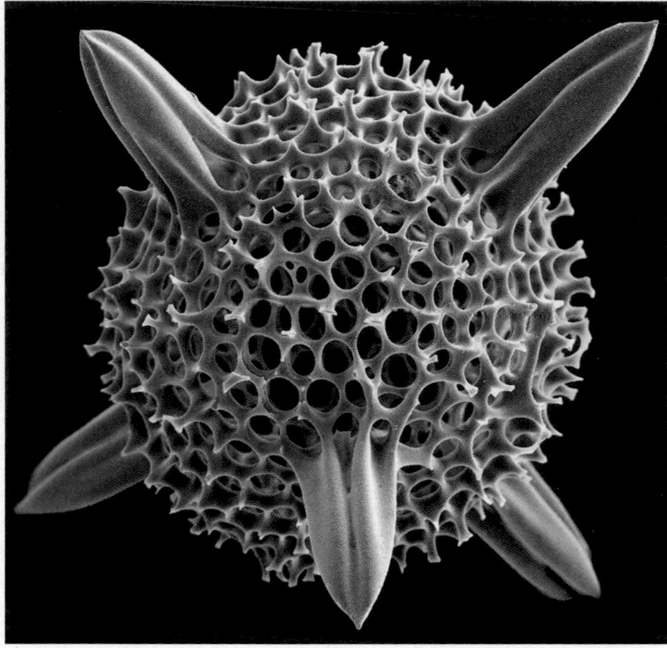

▲ **FIGURE 21-14 A radiolarian** Only the delicate, glassy shell is shown, so the pseudopods present in the living organism are not evident.

sinking to the ocean bottom and accumulating over millions of years, have resulted in immense deposits of limestone, such as those that form the famous White Cliffs of Dover, England.

Radiolarians Have Glassy Shells

Like foraminiferans, **radiolarians** have thin pseudopods that extend through hard shells. The shells of radiolarians, however, are made of glass-like silica (**FIG. 21-14**). The beauty of these microscopic glassy shells has long impressed architects, artists, and scientists. The prominent nineteenth-century biologist Ernst Haeckel wrote that "every morning I am newly amazed at the inexhaustible richness of these tiny and delicate structures," and said that his early study of radiolarians inspired him to pursue a career in science.

Amoebozoans Have Pseudopods and No Shells

Amoebozoans move by extending finger-shaped pseudopods, which may also be used for feeding. Amoebozoans generally do not have shells. The major groups of amoebozoans are the amoebas and the slime molds.

Amoebas Have Thick Pseudopods

Amoebas are common in freshwater lakes and ponds (**FIG. 21-15**). Many amoebas are predators that stalk and engulf prey, but some species are parasites. One parasitic form causes amoebic dysentery, a disease that is prevalent in warm climates. The dysentery-causing amoeba multiplies in the intestinal wall, triggering severe diarrhea.

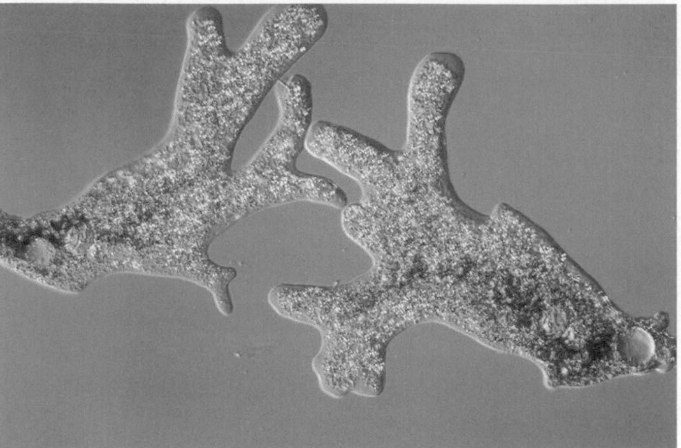

▲ **FIGURE 21-15 Amoebas** Amoebas are active predators that move through water to engulf food with thick, blunt pseudopods.

Slime Molds Are Decomposers That Inhabit the Forest Floor

The physical form of *slime molds* seems to blur the boundary between a colony of separate individuals and a single, multicellular individual. The life cycle of the slime mold consists of two phases: a mobile feeding stage and a stationary reproductive stage called a *fruiting body*. There are two main types of slime molds: acellular and cellular.

Acellular Slime Molds Form a Multinucleate Mass of Cytoplasm Called a Plasmodium The **acellular slime molds**, also known as *plasmodial* slime molds, consist of a mass of cytoplasm that may spread thinly over an area of several square yards. Although the mass contains thousands of diploid nuclei, the nuclei are not confined in separate cells surrounded by plasma membranes. This structure, called a **plasmodium**, explains why these protists are described as "acellular" (without cells). The plasmodium oozes through decaying leaves and rotting logs, engulfing food such as bacteria and particles of organic material. The mass may be bright yellow or orange (**FIG. 21-16a**). Dry conditions or starvation stimulate the plasmodium to form fruiting bodies, on which haploid spores are produced (**FIG. 21-16b**). The spores are dispersed and germinate to produce mobile haploid cells. Two such cells may meet and fuse, forming a diploid zygote that gives rise to a new plasmodium.

Cellular Slime Molds Live as Independent Cells but Aggregate into a Pseudoplasmodium When Food Is Scarce The **cellular slime molds**, also known as *social amoebas*, live in soil as independent haploid cells that move and feed by producing pseudopods. In the best-studied genus, *Dictyostelium*, individual cells release a chemical signal when food becomes scarce. This signal attracts nearby cells into a dense aggregation that forms a slug-like mass called a **pseudoplasmodium** ("false plasmodium") because, unlike a true plasmodium, it consists of individual cells (**FIG. 21-17**).

(a) Plasmodium
(b) Fruiting bodies

▲ **FIGURE 21-16 An acellular slime mold (a)** A plasmodium oozes over a stone on the damp forest floor. **(b)** When food becomes scarce, the mass differentiates into fruiting bodies in which spores are formed.

A pseudoplasmodium can be viewed as a colony of individuals because the cells that compose it are not all genetically identical. In some ways, however, a pseudoplasmodium is more like a multicellular organism, because its cells differentiate into different cell types, with different types serving different functions. A pseudoplasmodium moves about in slug-like fashion, migrating toward an aboveground spot suitable for spore dispersal, where its cells differentiate to convert the structure to a fruiting body. Haploid spores formed within the fruiting body are dispersed by wind and germinate directly into new single-celled individuals. In some circumstances, two independent cells may fuse to form a diploid zygote, which develops into a larger cyst that ultimately releases haploid spores.

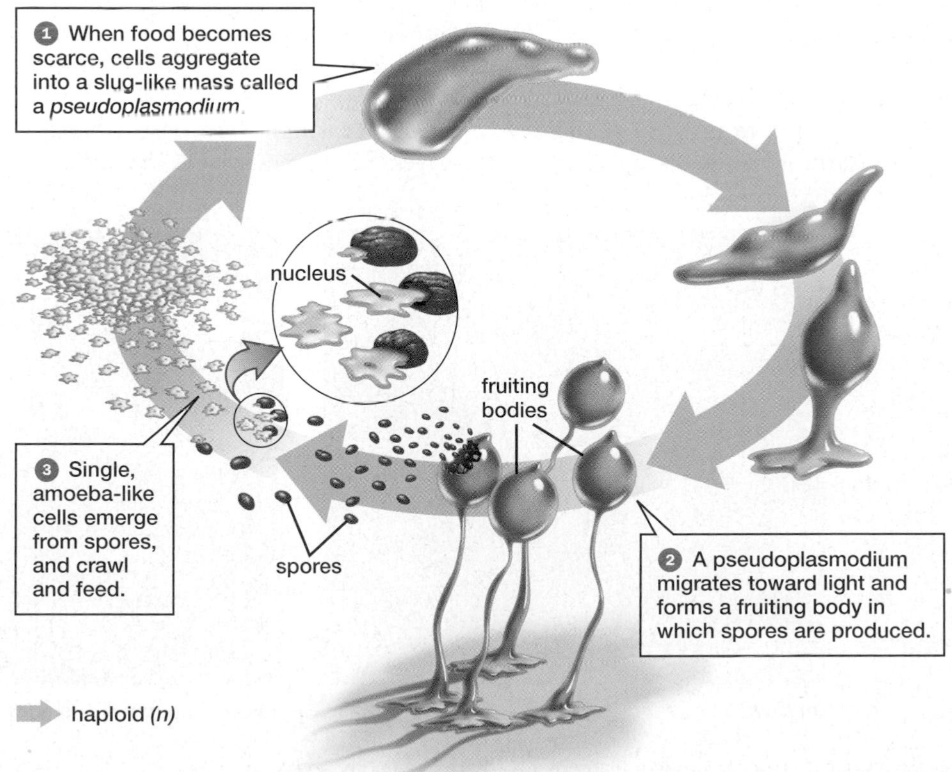
1 When food becomes scarce, cells aggregate into a slug-like mass called a *pseudoplasmodium*.
nucleus
fruiting bodies
3 Single, amoeba-like cells emerge from spores, and crawl and feed.
spores
2 A pseudoplasmodium migrates toward light and forms a fruiting body in which spores are produced.
haploid (*n*)

▲ **FIGURE 21-17 The life cycle of a cellular slime mold**

▲ **FIGURE 21-18 Red algae** Red coralline algae from the Mediterranean Sea. Coralline algae, which deposit calcium carbonate within their bodies, contribute to coral reefs in tropical waters.

Red Algae Contain Red Photosynthetic Pigments

The red algae are multicellular, photosynthetic seaweeds (**FIG. 21-18**). These protists range in color from bright red to nearly black; they derive their hue from red pigments that mask their green chlorophyll. Red algae are found almost exclusively in marine environments. They dominate in deep, clear tropical waters, where their red pigments absorb deeply penetrating blue-green light and transfer this light energy to chlorophyll, where it is used in photosynthesis. The solar energy that red algae capture helps support nonphotosynthetic organisms in marine ecosystems.

Red algae contain gelatinous substances with commercial uses. One of these substances is carrageenan, a combination of polysaccharides extracted from the cell walls of various species of red algae. Carrageenan melts at a relatively low temperature and, after cooling, forms a gel that remains stable at room temperature. These properties have proved useful to food processors, and carrageenan is widely used as a thickener and stabilizer in commercially produced foods including ice cream, yogurt, chocolate milk, soymilk, jellies, soups, salad dressings, and lunchmeats.

Have You Ever Wondered ... What Sushi Wrappers Are Made of?

If you like sushi, you've probably eaten sushi roll, in which rice and other foods are surrounded by a tasty, blackish-green wrapper. The wrapper is made from the dried bodies of a multicellular protist, the red alga *Porphyra*. *Porphyra*, also known as nori, is grown commercially, often in large coastal "farms" where the seaweed grows attached to vast nets that extend down from the ocean's surface. After harvest, the seaweeds are shredded, pulped, pressed into sheets, and dried, in a process very similar to papermaking.

Chlorophytes Are Green Algae

The **chlorophytes** are a clade of green algae that includes both multicellular and unicellular species. Most chlorophyte algae live in freshwater ponds and lakes, but some live in the oceans. Some, such as *Oedogonium*, form thin filaments from long chains of cells (**FIG. 21-19a**). Other chlorophyte species form colonies containing clusters of cells that are somewhat interdependent and represent a structure intermediate between unicellular and multicellular forms. These colonies range from a few cells to a few thousand cells, as in *Volvox*. Most chlorophytes are small, but some marine species are large. For example, *Ulva*, or sea lettuce, is similar in size to the leaves of its namesake (**FIG. 21-19b**).

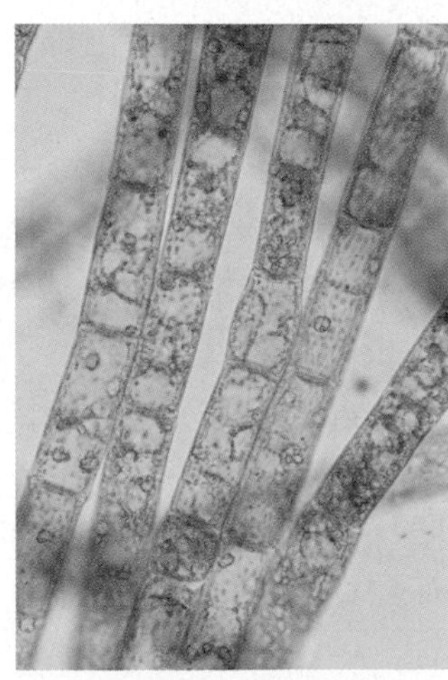

(a) *Oedogonium*

(b) *Ulva*

(c) Growing green algae for biofuel

▲ **FIGURE 21-19 Chlorophytes (a)** *Oedogonium* is a filamentous green alga composed of strands only one cell thick. **(b)** *Ulva* is a multicellular green alga that assumes a leaflike shape. **(c)** Biofuels produced from algae grown at a facility like the one pictured may one day fill a significant portion of our energy needs, if technical obstacles can be overcome.

Some chlorophyte species are currently under intensive cultivation by companies that hope to use them for commercial production of biofuels (**FIG. 21-19c**). Fuels based on chlorophyte algae could in principle replace dwindling fossil fuels with a renewable fuel whose production and use release less carbon dioxide into the atmosphere. However, efforts to develop an efficient, economically viable process for converting algae to fuel have not yet been successful. Some scientists have argued that bioengineering holds the key to success and have begun research aimed at engineering chlorophyte genomes to produce a novel organism capable of producing fuel efficiently under industrial conditions.

CHECK YOUR LEARNING

Can you . . .

- list the major protist taxonomic groups and the key characteristics of each group?
- describe some examples of how members of each group affect humans?

CASE STUDY REVISITED

Green Monster

Caulerpa taxifolia, the invasive seaweed that threatens to overrun the Mediterranean, is a chlorophyte. This species and other members of its genus have very unusual bodies. Outwardly, they appear plantlike, with rootlike structures that attach to the seafloor and with other structures that look like stems and leaves, rising to a height of several inches. Despite its seeming similarity to a plant, however, a *Caulerpa* body consists of a single, extremely large cell. The entire body is surrounded by a single, continuous cell membrane. The interior consists of cytoplasm that contains many nuclei but is not subdivided. That a single cell can take such a complex shape is extraordinary.

A potential problem with *Caulerpa*'s single-celled organization might arise when its body is damaged, perhaps by wave action or when a predator takes a bite out of it. When the cell membrane is breached, all of the organism's cytoplasm could potentially leak out, an event that would be fatal. But *Caulerpa* has evolved a defense against this potential calamity. Shortly after the cell membrane breaks, it is quickly filled with a "wound plug" that closes the gap. After the plug is established, the cell begins to grow and regenerates any lost portion.

This ability to regenerate is a key component of the ability of *Caulerpa taxifolia* to spread rapidly in new environments. If part of a *Caulerpa* breaks off and drifts to a new location, the fragment can regenerate a whole new body. The regenerated individual becomes the founder of a new, quickly growing colony—and these quickly growing colonies might appear anywhere in the world. Authorities in many countries worry that the aquarium strain of *Caulerpa* could invade their coastal waters, unwittingly transported by ships from the Mediterranean or released by careless aquarists. In fact, invasive *Caulerpa* is no longer restricted to the Mediterranean. It now thrives in at least 13 locations in Australia. It also appeared at two locations in California, but authorities there were able to eradicate it after 7 years of intensive, expensive effort. Australia has not been so fortunate, and *Caulerpa taxifolia* continues to spread there.

CONSIDER THIS Is it important to stop the spread of *Caulerpa*? Governments invest substantial resources to combat introduced species and prevent their populations from increasing and dispersing. Is this a wise use of funds? Can you think of some arguments against spending time and money for this purpose?

CHAPTER REVIEW

Go to **Mastering Biology** to access the Pearson eText, vocabulary review, practice quizzes, activities, videos, current events, and more.

*Answers to **Think Critically** and **Thinking Through the Concepts** questions can be found in the **Answers** section at the back of the book.*

Summary of Key Concepts

21.1 What Are Protists?

"Protist" is a term of convenience that refers to any eukaryote that is not a plant, animal, or fungus. Most protists are single, highly complex eukaryotic cells, but some form colonies, and some are multicellular. Protists exhibit diverse modes of nutrition, reproduction, and locomotion. Photosynthetic protists form much of the phytoplankton, which plays a key ecological role. Some protists cause human diseases; others are crop pests.

21.2 What Are the Major Groups of Protists?

Protist groups include excavates (diplomonads, parabasalids, euglenids, and kinetoplastids), stramenopiles (water molds, diatoms, and brown algae), alveolates (dinoflagellates, apicomplexans, and ciliates), rhizarians (foraminiferans and radiolarians), amoebozoans (amoebas and slime molds), red algae, and chlorophyte algae.

Thinking Through the Concepts

Bloom's: Remembering, Understanding

Multiple Choice

1. Which of the following statements about protists is false?
   a. Some protists are photosynthetic.
   b. All protists are eukaryotes.
   c. Although protists are diverse, they form a single clade.
   d. Protists include both unicellular and multicellular species.
2. The harmful protist blooms known as "red tides" are produced by
   a. apicomplexans.
   b. dinoflagellates.
   c. red algae.
   d. foraminiferans.
3. The organism that causes malaria belongs to which of the following groups?
   a. alveolates
   b. slime molds
   c. ciliates
   d. apicomplexans
4. The finger-like extensions of the cell membrane that some protists use for feeding or locomotion are called
   a. pseudopods.
   b. cilia.
   c. flagella.
   d. food vacuoles.
5. A person who develops severe diarrhea after drinking untreated water on a camping trip is likely to have been infected by
   a. *Plasmodium*, an apicomplexan.
   b. *Ulva*, a chlorophyte.
   c. *Paramecium*, a ciliate.
   d. *Giardia*, a diplomonad.

Fill-in-the-Blank

1. Protists that absorb nutrients from their surroundings may act as ____________ of dead organic matter or as harmful ____________ of larger living organisms.
2. Photosynthetic protists are collectively known as ____________; nonphotosynthetic, single-celled protists are collectively known as ____________.
3. Protist chloroplasts surrounded by four-layer membranes arose evolutionarily through ____________, in which an ancestral nonphotosynthetic protist engulfed but did not digest a(n) ____________.
4. The disease-causing protist that causes malaria is a member of the ____________ group, and the protist that causes sleeping sickness is a member of the ____________ group.
5. The plant diseases downy mildew and late blight are caused by protists in the ____________ group. Slime molds are members of the ____________ group.
6. Protists that make up a large proportion of Earth's phytoplankton include ____________ and ____________. The protist group containing the species most likely to one day be cultivated for biofuel production is ____________.

Review Questions

1. List the major differences between prokaryotes and protists.
2. What is secondary endosymbiosis?
3. What is the importance of dinoflagellates in marine ecosystems? What can happen to marine ecosystems when certain dinoflagellate species reproduce rapidly?
4. What is the major ecological role played by single-celled algae?
5. Which protist group consists entirely of parasitic forms?
6. Which protist groups include seaweeds?
7. Which protist groups include species that use pseudopods?

Applying the Concepts

Bloom's: Applying, Analyzing, Evaluating

1. The internal structure of many protists is much more complex than that of cells of multicellular organisms. Does this mean that the protist is engaged in more complex activities than the multicellular organism is? If not, why are protistan cells more complicated?
2. What are some important benefits and services provided by protists to Earth's other organisms?

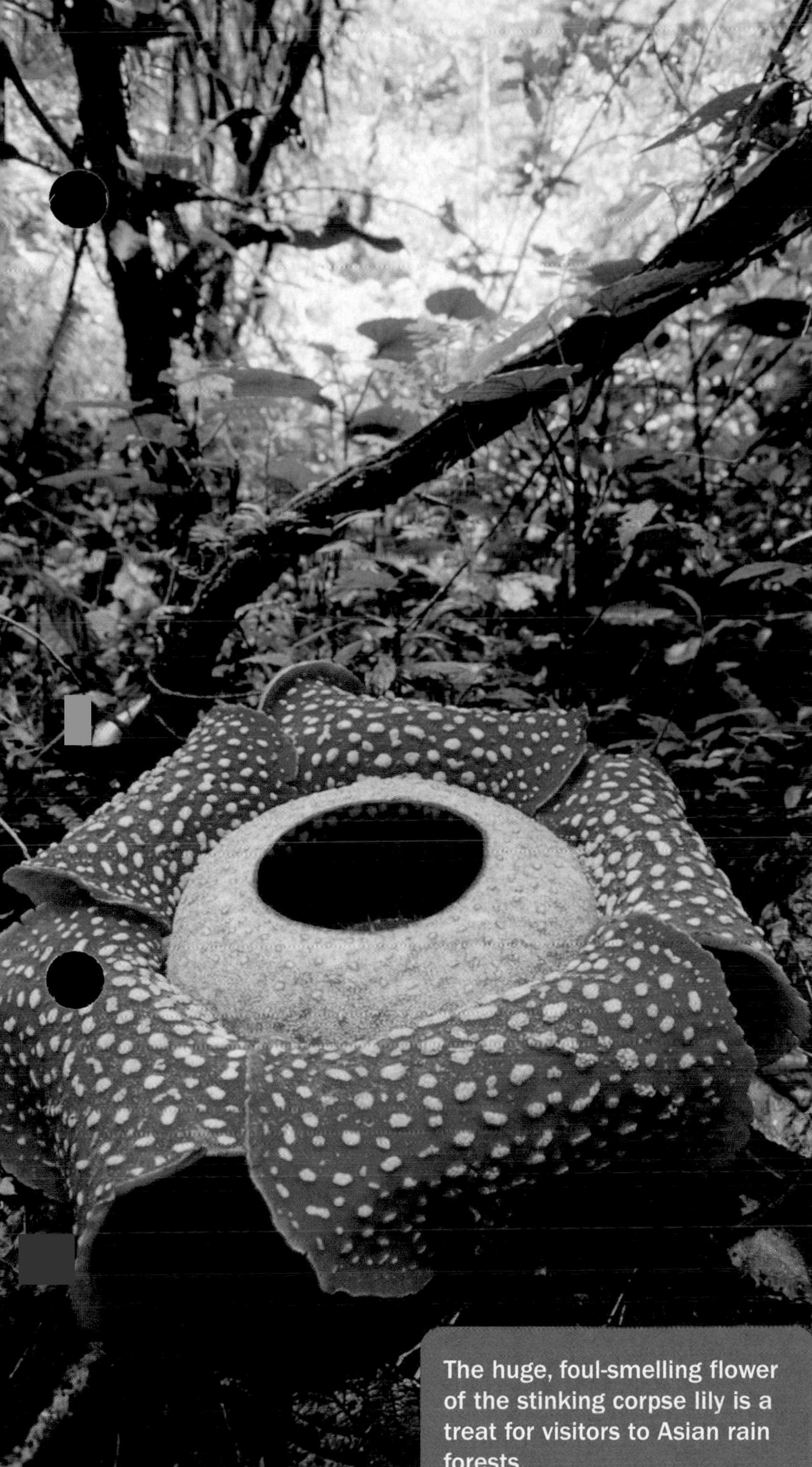

The huge, foul-smelling flower of the stinking corpse lily is a treat for visitors to Asian rain forests.

22 The Diversity of Plants

CASE STUDY

Queen of the Parasites

THE FLOWER OF THE STINKING CORPSE LILY makes a strong impression. For one thing, it's huge; a single flower may be 3 feet across. It also has a rather strange appearance, consisting largely of fleshy lobes that are almost fungus-like. But as its name implies, the thing that makes a stinking corpse lily almost impossible to ignore is its aroma, which has been described as "a penetrating smell more repulsive than any buffalo carcass in an advanced stage of decomposition."

Unlike most plants, a stinking corpse lily has no visible leaves, roots, or stems. In fact, it is a parasite, and its body is completely embedded in the tissue of its host, a vine in the grape family. Because it has no leaves, the stinking corpse lily cannot produce any food of its own, but instead draws all of its nutrition from its host. The parasite becomes visible outside the body of its host only when one of its cabbage-shaped flower buds pushes through the surface of the host's stem and its gigantic, stinking flower opens for a week or so before shriveling and falling off. If a male and a female flower happen to be open simultaneously and close together, the female flower may be fertilized and produce seeds. A seed that is dispersed in the droppings of an animal that has consumed it, and that happens to land on a stem of the host species, may germinate and penetrate a new host.

When you think of plants, you might first think of their most obvious feature: green leaves that capture solar energy by photosynthesis. It may seem odd, then, that this chapter about plants begins with a peculiar plant that does not photosynthesize. Oddities such as the stinking corpse lily, however, serve as reminders that evolution does not always follow a predictable pathway, and that even an adaptation as seemingly valuable as the ability to live on sunlight can be lost. What other interesting characteristics have appeared over the evolutionary history of plants?

AT A GLANCE

22.1 WHAT ARE THE KEY FEATURES OF PLANTS?

What distinguishes plants from other organisms? Most plants exhibit three characteristic traits: photosynthesis, multicellular embryos, and alternation of generations, as explained below. Each of these traits also occurs in some other kinds of organisms, but only plants combine all three.

Plants Are Photosynthetic

Perhaps the most noticeable feature of nearly all plants is their green color. The color comes from the presence of chlorophyll in many plant tissues. Chlorophyll plays a crucial role in photosynthesis, the process by which plants use energy from sunlight to convert water and carbon dioxide to sugar (see Chapter 7). Chlorophyll and photosynthesis, however, are not unique to plants; they are also present in many types of protists and prokaryotes.

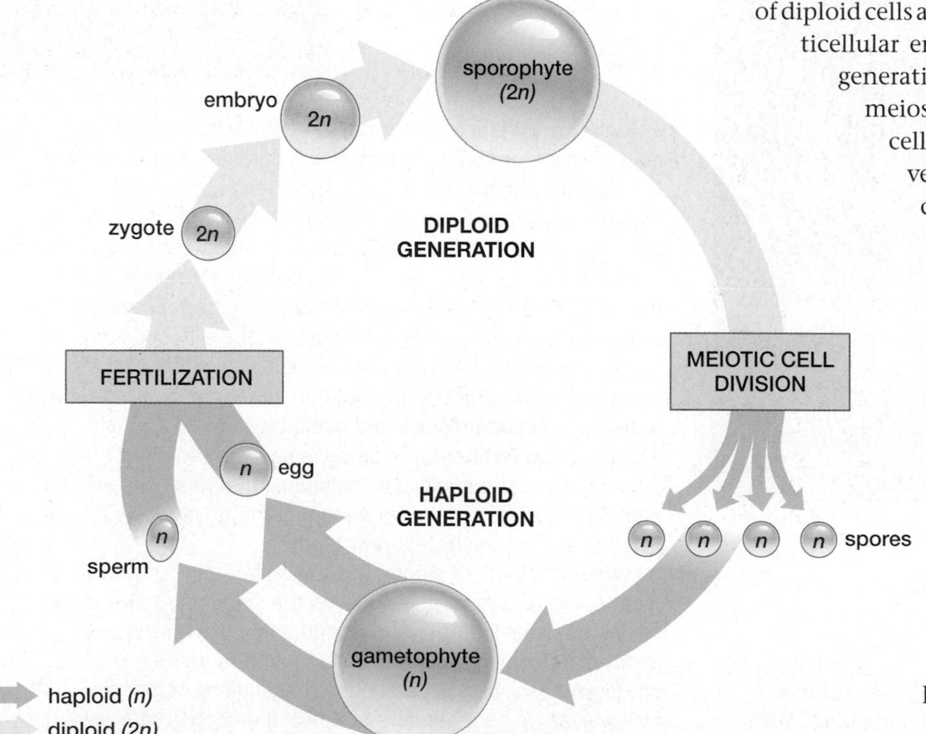

▲ **FIGURE 22-1 Alternation of generations in plants** As shown in this generalized depiction of a plant life cycle, a diploid sporophyte generation produces haploid spores through meiotic cell division. The spores develop into a haploid gametophyte generation that produces haploid gametes by mitotic cell division. The fusion of these gametes results in a diploid zygote that develops into the sporophyte plant.

Plants Have Multicellular, Dependent Embryos

Plants are distinguished from other photosynthetic organisms by their characteristic embryos. A plant embryo is multicellular and is attached to and dependent on its parent. As it grows and develops, the embryo receives nutrients from the tissues of the parent plant. Multicellular, dependent embryos are not found among photosynthetic protists.

Plants Have Alternating Multicellular Haploid and Diploid Generations

Plant reproduction is characterized by a type of life cycle called **alternation of generations** (**FIG. 22-1**). In organisms with alternation of generations, separate multicellular diploid and haploid generations alternate with one another. (Recall that a diploid organism has paired chromosomes; a haploid organism has unpaired chromosomes.) In the diploid (2*n*) generation, the body consists of diploid cells and is known as the **sporophyte**. The multicellular embryo is part of the diploid sporophyte generation. Certain cells of sporophytes undergo meiosis to produce haploid (*n*) reproductive cells called *spores*. The haploid spores develop into multicellular, haploid bodies called **gametophytes**.

A gametophyte ultimately produces male and female haploid gametes (sperm and eggs) by mitosis. Gametes, like spores, are reproductive cells but, unlike spores, an individual gamete by itself cannot develop into a new individual. Instead, two gametes of opposite sexes must meet and fuse to form a new diploid individual. In plants, gametes produced by gametophytes fuse to form a diploid zygote (a fertilized egg), which develops into a diploid embryo. The embryo develops into a mature sporophyte, and the cycle begins again.

CHECK YOUR LEARNING

CAN YOU . . .

- describe the features that distinguish plants from other kinds of organisms?

22.2 HOW HAVE PLANTS EVOLVED?

Plants form a clade within a larger clade that also includes several groups of green algae collectively known as *charophytes*. (A clade is a group consisting of all the descendants of a particular common ancestor.) The charophyte algae include plants' closest living relatives, the stoneworts (**FIG. 22-2**). The close evolutionary link between stoneworts and plants was revealed by DNA comparisons and is reflected in additional similarities between plants and charophytes. For example, plants and charophytes both store food as starch and have cell walls made of cellulose, and both use the same types of chlorophyll in photosynthesis (*chlorophylls a* and *b*).

The Ancestors of Plants Lived in Water

The ancestors of plants were photosynthetic protists, perhaps similar to stoneworts. Like modern stoneworts, the protists that gave rise to plants presumably lacked true roots, stems, leaves, and complex reproductive structures such as flowers or cones, features that appeared only later in the evolutionary history of plants. In addition, the ancestors of plants were confined to watery habitats.

For these ancestors of plants, life in water had many advantages. For example, in water, a body is bathed in a nutrient-rich solution, is supported by buoyancy, and is not likely to dry out. In addition, life in water facilitates reproduction, because gametes and zygotes can be carried by water currents or propelled by flagella.

Early Plants Invaded Land

Despite the benefits of aquatic environments, early plants invaded habitats on land. Today, most plants live on land. The move to land brought its own advantages, including access to sunlight unimpeded by water that might block its rays, access to nutrients contained in surface rocks, and freedom from predators. However, the move to land also imposed some challenges; plants could no longer rely on watery surroundings to provide support, moisture, nutrients, and transportation for gametes and zygotes. As a result, life on land has favored the evolution in plants of traits that help meet these environmental challenges.

Plant Bodies Evolved to Resist Gravity and Drying

Some of the key adaptations to life on land arose early in plant evolution and are now found in virtually all land plants (**FIG. 22-3**). These early adaptations include:

- Roots or rootlike structures that anchor the plant and absorb water and nutrients from the soil.
- A waxy **cuticle** that covers the surfaces of leaves and stems and that limits the evaporation of water.
- Pores called **stomata** (singular, stoma) in the leaves and stems that open to allow gas exchange but close when water is scarce, reducing the amount of water lost to evaporation.

Other key adaptations occurred somewhat later in the transition to terrestrial life and are now widespread but not universal among plants (most nonvascular plants, described later, lack these traits):

- Conducting tissues called **xylem** and **phloem** that transport water and dissolved substances. Xylem conducts water and minerals upward from the roots; phloem

▲ **FIGURE 22-2** ***Chara*, a stonewort** The green algae known as stoneworts are plants' closest living relatives.

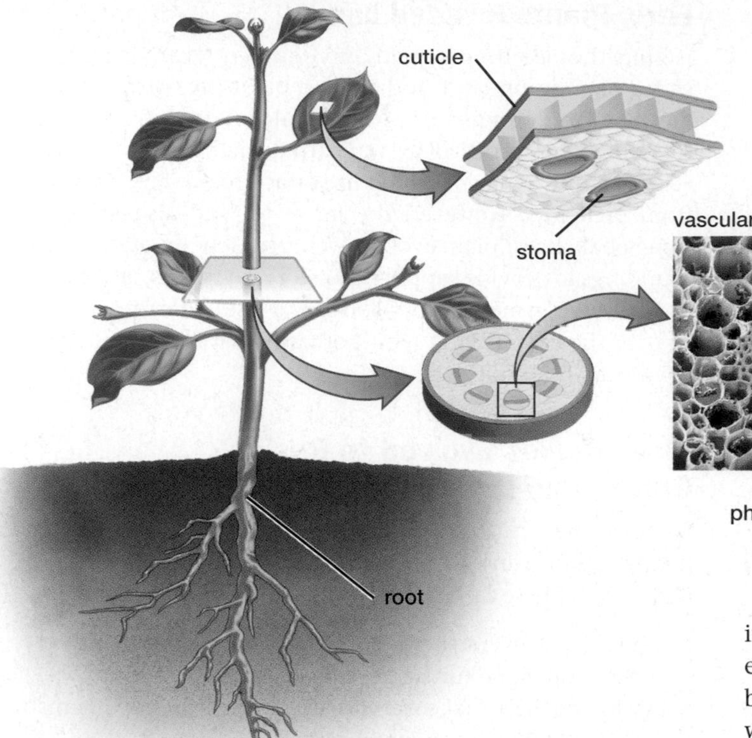

FIGURE 22-3 Early adaptations for life on land Adaptations for life on land include roots that anchor the plant, a waxy surface cuticle that reduces evaporation, stomata that can be closed to conserve water, and (in vascular plants) lignin-impregnated xylem and phloem tissues that transport water and nutrients and help support the plant body.

conducts the products of photosynthesis to different parts of the plant body.

- The stiffening substance **lignin**, a rigid polymer that impregnates the cells of the conducting tissues and supports the plant body against the force of gravity.

Plants Evolved Sex Cells That Disperse Without Water and Protection for Their Embryos

The most widespread groups of plants, collectively known as seed plants, are characterized by sex cells that do not rely on water for dispersal and by especially well-protected and well-provisioned embryos. The key adaptations of these plant groups are pollen, seeds, and, in the flowering plants, flowers and fruits.

Early seed plants gained an advantage over their competitors by producing dry, microscopic pollen grains that allowed wind, instead of water, to carry the male gametes. Early seed plants also produced seeds, which provided protection and nourishment for developing embryos and the potential for more effective dispersal. Later came the evolution of flowers, which enticed animal pollinators that were able to deliver pollen more precisely than did wind. Fruits also attracted animals, which consumed the fruit and dispersed its seeds in their feces.

More Recently Evolved Plants Have Smaller Gametophytes

The evolutionary history of plants has been marked by a tendency for the sporophyte generation to become increasingly prominent and for the longevity and size of the gametophyte generation to shrink. Thus, the earliest plants are believed to have been similar to today's nonvascular plants, which have a sporophyte that is smaller than the gametophyte and remains attached to it. In contrast, plants that originated somewhat later, such as ferns and the other seedless vascular plants, feature a life cycle in which the sporophyte is dominant, and the gametophyte is a much smaller, independent plant. Finally, in the most recently evolved group of plants, the seed plants, gametophytes are microscopic and barely recognizable as an alternate generation. These tiny gametophytes, however, still produce the eggs and sperm that unite to form the zygote that develops into the diploid sporophyte.

CHECK YOUR LEARNING

Can you . . .

- describe the probable ancestor of plants?
- identify the closest living relatives of plants and explain their similarities to and differences from plants?
- describe the adaptations that equip plants for life on land?

CASE STUDY CONTINUED

Queen of the Parasites

The stinking corpse lily, with its huge, 3-foot-wide flowers, apparently evolved from an ancestor with tiny flowers. A recent analysis of DNA sequences revealed that the plant group most closely related to the group that includes the stinking corpse lily is the spurges, plants with mostly tiny flowers. The analysis also showed that the common ancestor of spurges and corpse lilies probably had flowers that were about 1/80th the size of modern stinking corpse lily flowers.

In stark contrast to the stinking corpse lily and its relatives, many plants have no flowers at all. What are these flowerless plants like?

22.3 WHAT ARE THE MAJOR GROUPS OF PLANTS?

Two major groups of land plants arose from ancient algal ancestors (**FIG. 22-4** and **TABLE 22-1**). Members of one group, the **nonvascular plants** (also called *bryophytes*), require a moist environment to reproduce and thus straddle the boundary between aquatic and terrestrial life. The other group, the **vascular plants** (also called *tracheophytes*), has colonized drier habitats.

Nonvascular Plants Lack Conducting Structures

Nonvascular plants retain some characteristics of their algal ancestors. Their gametes are dispersed by water, and they lack true roots, leaves, and stems. They do possess rootlike anchoring structures called *rhizoids* that bring water and nutrients into the plant body, but nonvascular plants lack well-developed structures for conducting water and nutrients. They must instead rely on slow diffusion or poorly developed conducting tissues to distribute water and other nutrients. As a result, their body size is limited. Size is also limited by the absence of the stiffening agent lignin in their bodies. Without lignin, nonvascular plants cannot grow upward very far. Most nonvascular plants are less than 1 inch (2.5 centimeters) tall.

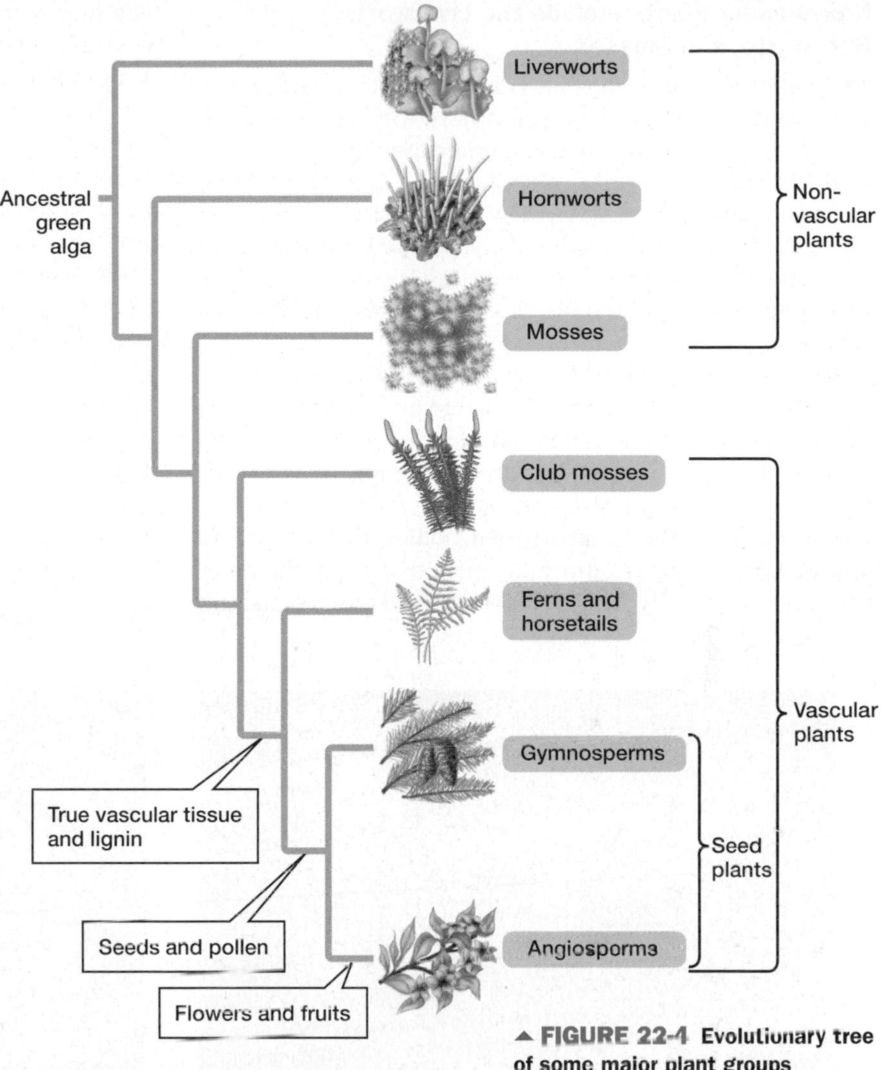

▲ **FIGURE 22-4 Evolutionary tree of some major plant groups**

TABLE 22-1 Features of the Major Plant Groups

Group	Subgroup	Relationship of Sporophyte and Gametophyte	Transfer of Reproductive Cells	Early Embryonic Development	Dispersal	Water and Nutrient Transport Structures
Nonvascular plants	Liverworts Hornworts Mosses	The gametophyte is dominant—the sporophyte develops from a zygote retained on a gametophyte	Motile sperm swim to a stationary egg retained on a gametophyte	Occurs within the archegonium of a gametophyte	Haploid spores are carried by wind	Absent
Vascular plants	Club mosses Horsetails and ferns	The sporophyte is dominant—it develops from a zygote retained on a gametophyte	Motile sperm swim to a stationary egg retained on a gametophyte	Occurs within the archegonium of a gametophyte	Haploid spores are carried by wind	Present
	Gymnosperms	The sporophyte is dominant—the microscopic gametophyte develops within a sporophyte	Wind-dispersed pollen carries sperm to a stationary egg in a cone	Occurs within a protective seed containing a food supply	Seeds containing a diploid sporophyte embryo are dispersed by wind or animals	Present
	Angiosperms	The sporophyte is dominant—the microscopic gametophyte develops within a sporophyte	Pollen, dispersed by wind or animals, carries sperm to a stationary egg within a flower	Occurs within a protective seed containing a food supply; the seed is encased within fruit	Fruit, carrying seeds, is dispersed by animals, wind, or water	Present

Nonvascular Plants Include the Liverworts, Hornworts, and Mosses

The nonvascular plants include three groups: liverworts, hornworts, and mosses. Liverworts and hornworts are named for their shapes. The gametophytes of certain liverwort species have a lobed form reminiscent of the shape of a liver (**FIG. 22-5a**). Hornwort sporophytes generally have a spiky, somewhat hornlike shape (**FIG. 22-5b**). Liverworts and hornworts are most abundant in areas where moisture is plentiful, such as in swamps and bogs, in moist forests, and near the banks of streams and ponds.

Mosses are the most diverse and abundant of the nonvascular plants (**FIG. 22-5c**). Like liverworts and hornworts, mosses are most likely to be found in moist habitats. Some mosses, however, have a waterproof covering that reduces water loss. Many of these mosses are also able to survive the loss of much of the water in their bodies; they dehydrate and become dormant during dry periods but absorb water and resume growth when moisture returns. Such mosses can survive in deserts, on bare rock, and in far northern and southern latitudes where humidity is low and liquid water is scarce for much of the year.

Mosses of the genus *Sphagnum* are especially widespread, living in moist habitats in northern regions around the world. In many of these wet northern habitats, *Sphagnum* is the most abundant plant, forming extensive mats (**FIG. 22-5d**). Because decomposition is slow in cold climates and because *Sphagnum* contains compounds that inhibit bacterial growth, dead *Sphagnum* may decay very slowly. As a result, partially decayed moss tissue may accumulate in deposits that can, over thousands of years, become hundreds of feet thick. These deposits are known as peat. Peat has long been harvested for use as fuel, a practice that continues today in Ireland, Finland, Russia, and other northern countries. Now, however, peat is more often harvested for use in horticulture. Dried peat can absorb many times its own weight in water, making it useful as a soil conditioner and as a packing material for transporting live plants.

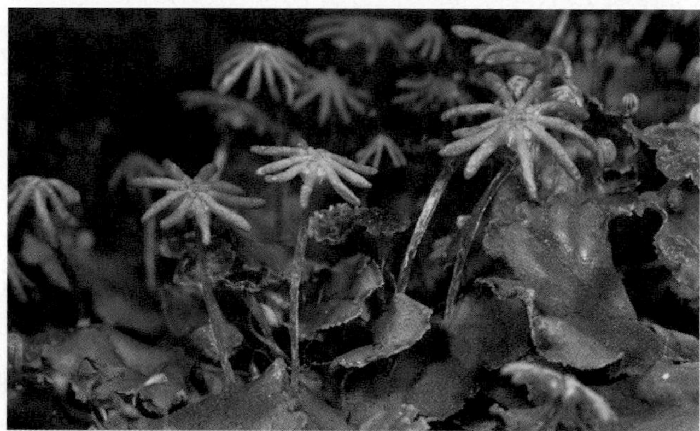

(a) Liverwort

(b) Hornwort

(c) Moss

(d) *Sphagnum* bog

▲ **FIGURE 22-5 Nonvascular plants** These plants are less than a half-inch (about 1 centimeter) in height. **(a)** Liverworts grow in moist, shaded areas. The palmlike structures on the female plants shown here hold eggs. Male plants produce sperm that swim through a film of water to reach and fertilize the eggs. **(b)** The hornlike sporophytes of hornworts grow upward from the gametophyte body. **(c)** Moss plants, showing the stalks that carry spore-bearing capsules. **(d)** Mats of *Sphagnum* moss cover moist bogs in northern regions.

THINK CRITICALLY Why are all nonvascular plants short?

The Reproductive Structures of Nonvascular Plants Are Protected

Nonvascular plants require moisture to reproduce, but they have evolved some traits that facilitate reproduction on land. For example, the reproductive structures of nonvascular plants are enclosed, which prevents the gametes from drying out (**FIG. 22-6**). There are two types of reproductive structures in which gametes are produced by mitotic cell division: **archegonia** (singular, archegonium), in which eggs develop, and **antheridia** (singular, antheridium), where sperm are formed ❶. In some nonvascular plant species, both archegonia and antheridia are located on the same plant; in other species, each individual plant is either male or female.

In all nonvascular plants, the sperm must swim to the egg through a film of water ❷, so nonvascular plants that live in drier areas can reproduce only when it rains. After fertilization, the zygote is retained in the archegonium, where the embryo grows and matures into a small diploid sporophyte that remains attached to the parent gametophyte plant ❸. At maturity, the sporophyte produces reproductive capsules. Within each capsule, haploid spores are produced by meiotic cell division ❹. When the capsule is opened, spores are released and dispersed by the wind ❺. If a spore lands in a suitable environment, it may develop into another haploid gametophyte plant ❻.

Vascular Plants Have Conducting Cells That Also Provide Support

Vascular plants are distinguished by the presence of xylem and phloem, specialized tissues consisting of tube-shaped conducting cells (see Fig. 22-3). These cells are impregnated

▼ **FIGURE 22-6 Life cycle of a moss** The photo shows moss plants; the short, leafy green plants are haploid gametophytes; the reddish brown stalks are diploid sporophytes.

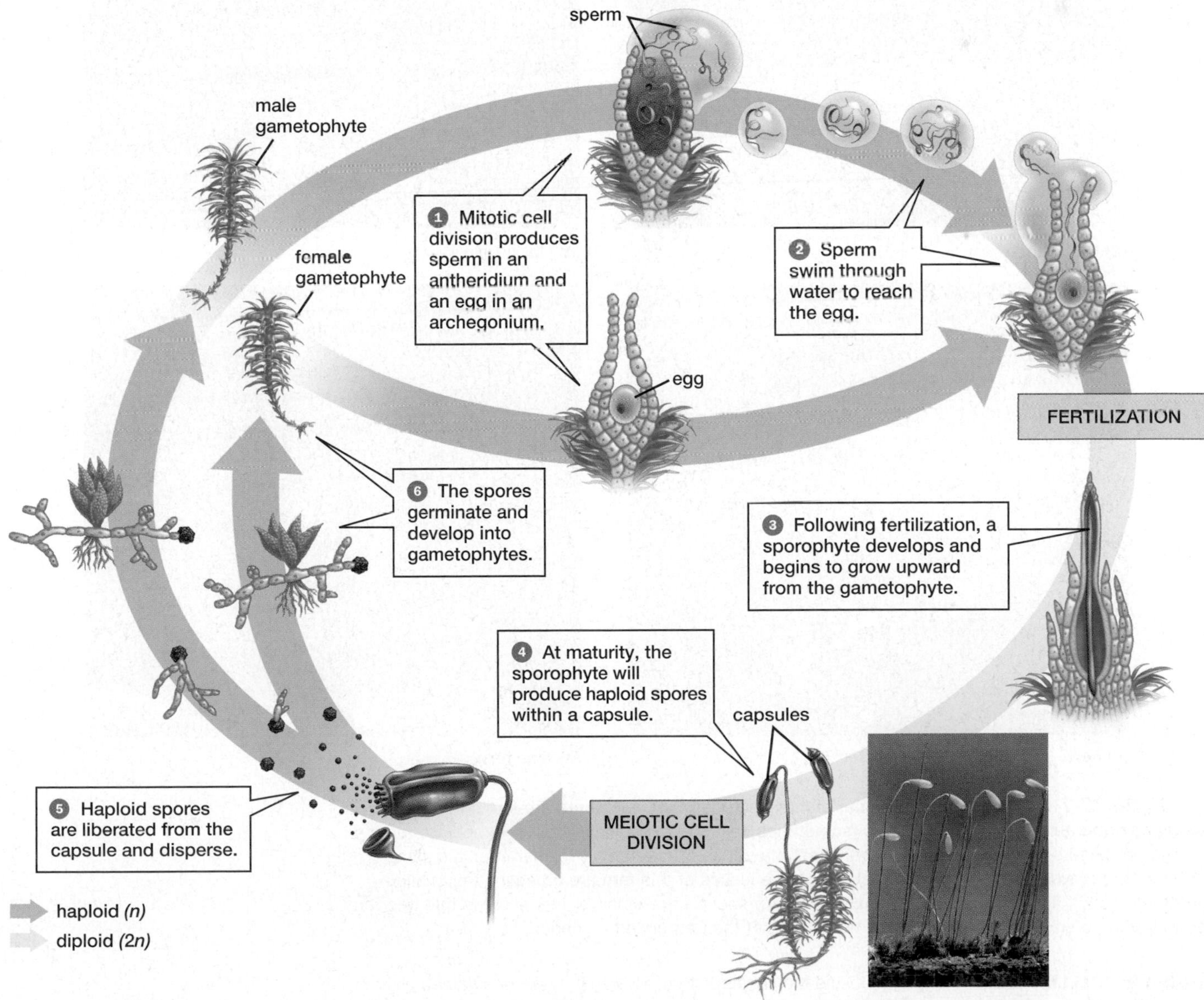

sperm
male gametophyte
female gametophyte
❶ Mitotic cell division produces sperm in an antheridium and an egg in an archegonium.
egg
❷ Sperm swim through water to reach the egg.
FERTILIZATION
❸ Following fertilization, a sporophyte develops and begins to grow upward from the gametophyte.
❻ The spores germinate and develop into gametophytes.
❹ At maturity, the sporophyte will produce haploid spores within a capsule.
capsules
MEIOTIC CELL DIVISION
❺ Haploid spores are liberated from the capsule and disperse.
haploid (n)
diploid (2n)

with the stiffening substance lignin and serve both supportive and conducting functions. They allow vascular plants to grow taller than nonvascular plants, both because of the extra support provided by lignin and because the conducting cells allow water and nutrients absorbed by the roots to move to the upper portions of the plant. Another difference between vascular plants and nonvascular plants is that in vascular plants, the diploid sporophyte is the larger, more conspicuous generation; in nonvascular plants, the haploid gametophyte is more evident. The vascular plants can be divided into two groups: the seedless vascular plants and the seed plants.

The Seedless Vascular Plants Include the Club Mosses, Horsetails, and Ferns

Like nonvascular plants, seedless vascular plants (**FIG. 22-7**) have swimming sperm and require water for reproduction. As their name implies, they propagate by spores rather than seeds. Present-day seedless vascular plants—the club mosses, horsetails, and ferns—are much smaller than their ancestors, which dominated the landscape in the Carboniferous period (359 million to 299 million years ago; see Fig. 18-8). Despite their historically diminished size, seedless vascular plants today are widespread in temperate and tropical regions worldwide.

(a) Club moss

(b) Horsetail

(c) Fern

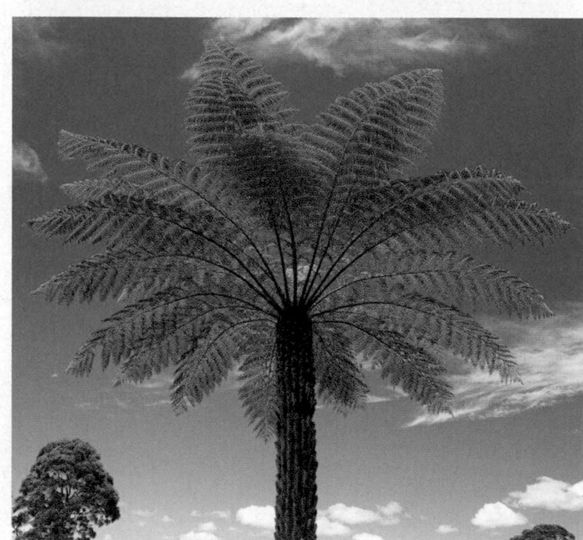

(d) Tree fern

▲ **FIGURE 22-7 Some seedless vascular plants** Seedless vascular plants are found in moist woodland habitats. **(a)** The club mosses (sometimes called ground pines) grow in temperate forests. **(b)** The horsetail extends long, narrow branches in a series of rosettes at regular intervals along the stem. Its leaves are insignificant scales. **(c)** The leaves of this fern are emerging from coiled structures called fiddleheads. **(d)** Although most fern species are small, some, such as this tree fern, retain the large size that was common among ferns of the Carboniferous period.

THINK CRITICALLY In each of these photos, is the pictured structure a sporophyte or a gametophyte?

Club Mosses and Horsetails Are Seedless Plants with Tiny, Scalelike Leaves

The club mosses, which despite their common name are not actually mosses, are now limited to representatives a few inches in height (**FIG. 22-7a**). Their leaves are small and scalelike, resembling the leaflike structures of mosses. Club mosses of the genus *Lycopodium*, commonly known as ground pine, form a beautiful ground cover in some temperate coniferous and deciduous forests. Because club mosses are evergreen, they are especially conspicuous on the forest floor in winter.

Modern horsetails belong to a single genus, *Equisetum*, that contains only 15 species, most less than 3 feet tall (**FIG. 22-7b**). The bushy branches of some species lend them the common name horsetails; the leaves are reduced to tiny scales on the branches. They are also called scouring rushes because all species of *Equisetum* have large amounts of silica (glass) in their outer layer of cells, giving them an abrasive texture. Early European settlers of North America used horsetails to scour pots and floors.

Ferns Are Broad-Leaved and Diverse

The ferns, with 12,000 species, are the most diverse of the seedless vascular plants (**FIG. 22-7c**). In the tropics, tree ferns still reach heights reminiscent of their ancestors from the Carboniferous period (**FIG. 22-7d**). Ferns are the only seedless vascular plants that have broad leaves.

In fern reproduction (**FIG. 22-8**), gametes are produced by mitotic cell division in archegonia and antheridia on the tiny fern gametophyte ❶. Sperm are released into water and swim to reach an egg in an archegonium ❷. If fertilization occurs, the resulting zygote develops into a sporophyte plant, which grows upward from its parent, the gametophyte ❸. On a mature sporophyte fern plant, which is much larger than the gametophyte, haploid spores are produced in structures called *sporangia* that form on special leaves of the sporophyte ❹. The sporangia open to release the spores, which are dispersed by the wind ❺. If a spore lands in a spot with suitable conditions, it germinates and develops into a gametophyte plant ❻.

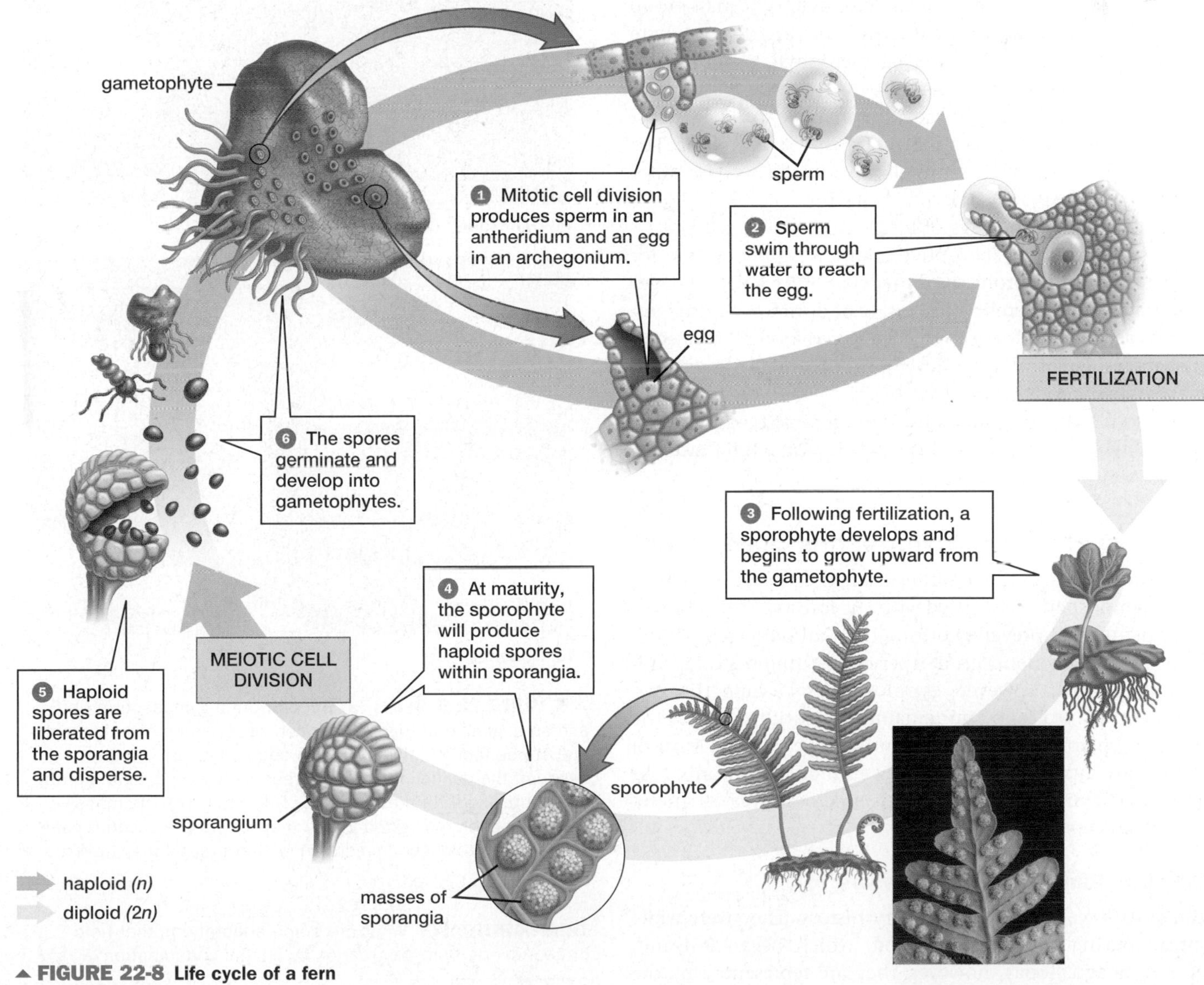

gametophyte
❶ Mitotic cell division produces sperm in an antheridium and an egg in an archegonium.
sperm
❷ Sperm swim through water to reach the egg.
egg
FERTILIZATION
❸ Following fertilization, a sporophyte develops and begins to grow upward from the gametophyte.
❻ The spores germinate and develop into gametophytes.
❹ At maturity, the sporophyte will produce haploid spores within sporangia.
MEIOTIC CELL DIVISION
❺ Haploid spores are liberated from the sporangia and disperse.
sporophyte
sporangium
masses of sporangia
haploid (n)
diploid (2n)

▲ **FIGURE 22-8 Life cycle of a fern**

The windborne spores of ferns make them especially effective at colonizing locations that lack abundant plant life. For example, just 2 years after a massive volcanic eruption that destroyed most life on the island of Krakatoa in 1883, visitors reported that ferns blanketed the previously denuded landscape. Similarly, fern abundance increased dramatically following the catastrophic asteroid impact that caused the extinction of dinosaurs and many other species about 66 million years ago. Spores of fossil ferns are extremely abundant in 66-million-year-old rocks at many locations around the world; these "spore spikes" are interpreted as evidence that massive fires followed the asteroid impact, burning up most vegetation and creating an opening for widespread colonization by ferns.

The Seed Plants Are Aided by Two Important Adaptations: Pollen and Seeds

The seed plants are distinguished from nonvascular plants and seedless vascular plants by their production of pollen and seeds. In seed plants, gametophytes (which produce the sex cells) are tiny. The female gametophyte is a small group of haploid cells that remains within the larger sporophyte and produces the egg. The male gametophyte is the **pollen** grain. Pollen grains are dispersed by wind or by animal pollinators such as bees. In this way, sperm move through the air to fertilize egg cells. This airborne transport means that the distribution of seed plants is not limited by the need for water through which sperm can swim to the egg.

Analogous to the eggs of birds and reptiles, **seeds** consist of an embryonic sporophyte plant, a supply of food for the embryo, and a protective outer coat (**FIG. 22-9**). The *seed coat* maintains the embryo in a state of dormancy until conditions are suitable for growth. The stored food helps sustain the emerging plant until it develops roots and leaves and can make its own food by photosynthesis.

Seed plants are grouped into two general types: gymnosperms, which lack flowers, and angiosperms, the flowering plants.

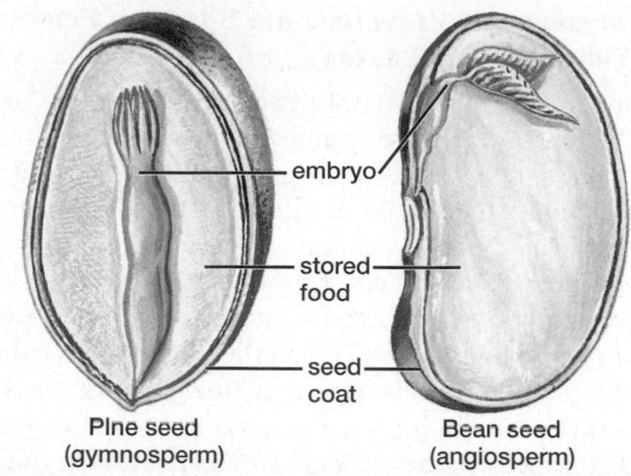

(a) Seeds

(b) Dandelion

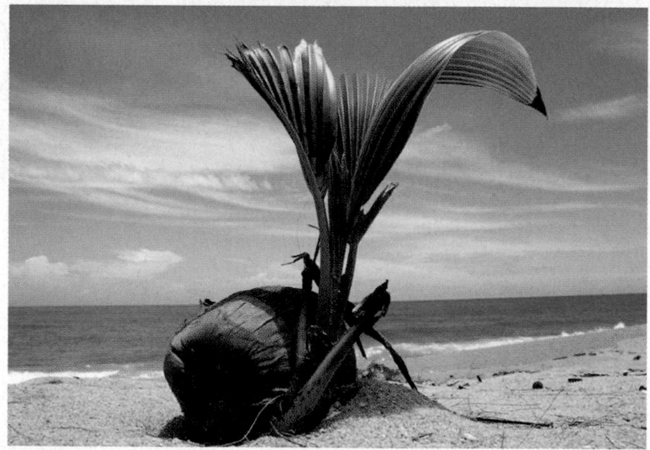
(c) Coconut

▲ **FIGURE 22-9 Seeds (a)** Seeds from a gymnosperm (left) and an angiosperm (right). Both consist of an embryonic plant and stored food confined within a seed coat. **(b)** The tiny seeds of the dandelion are dispersed by the wind, held aloft by parachute-like tufts that are part of the fruit. **(c)** The massive, armored seeds (protected inside the fruit) of the coconut palm can survive prolonged immersion in seawater as they traverse oceans.

THINK CRITICALLY What are some adaptations that help protect seeds from destruction by animal consumption?

Gymnosperms Are Nonflowering Seed Plants

Gymnosperms evolved earlier than the flowering plants. Early gymnosperms coexisted with the forests of seedless vascular plants that prevailed during the Carboniferous period. During the subsequent Permian period (299 million to 252 million years ago), however, gymnosperms became the predominant plant group and remained so until the rise of the flowering plants more than 100 million years later. Most of these early gymnosperms are now extinct. Today, only four groups of gymnosperms survive: ginkgoes, cycads, gnetophytes, and conifers.

Only One Ginkgo Species Survives

Ginkgoes have a long evolutionary history. They were widespread during the Jurassic period, which began 201 million years ago. Today, however, they are represented by the single species *Ginkgo biloba*, the maidenhair tree (**FIG. 22-10a**). Ginkgo trees are either male or female; female trees bear foul-smelling, fleshy seeds the size of cherries. Because they are more resistant to pollution than are most other trees, ginkgoes (usually the male trees) have been extensively planted in U.S. cities.

Cycads Are Restricted to Warm Climates

Like ginkgoes, cycads were diverse and abundant in the Jurassic period but have since dwindled. Today approximately 160 species survive, most of which dwell in tropical or subtropical climates. Cycads have large, finely divided leaves and bear a superficial resemblance to palms or large ferns (**FIG. 22-10b**). Most cycads are about 3 feet (1 meter) in height, although some species can reach 65 feet (20 meters).

The tissues of cycads contain potent toxins. Despite the presence of these toxins, people in some parts of the world use cycad seeds, stems, and roots for food. Careful preparation and processing removes the toxins before the plants are consumed. Nonetheless, cycad toxins are the suspected cause of neurological problems that occur in societies, such as the Chamorro people of the Mariana Islands, that use cycads for food. Cycad toxins can also harm grazing livestock.

About half of all cycad species are classified as threatened or endangered. The main threats to cycads are habitat destruction, competition from introduced species, and harvesting for the horticultural trade. A large specimen of a rare cycad highly prized by collectors can sell for thousands of dollars. Because cycads grow slowly, recovery of endangered populations is uncertain.

Gnetophytes Include the Odd *Welwitschia*

The gnetophytes include about 70 species of shrubs, vines, and small trees. Leaves of gnetophyte species in the genus *Ephedra* contain alkaloid compounds that act in humans as stimulants and appetite suppressants. For this reason, *Ephedra* was once widely used as an energy booster and weight-loss aid. However, following reports of sudden deaths of *Ephedra* users and publication of several studies linking *Ephedra* consumption to increased risk of heart problems, the U.S. Food and Drug Administration banned the sale of products containing this gnetophyte.

(a) Gingko

(b) Cycad

(c) Gnetophyte

(d) Conifer

▲ **FIGURE 22-10 Gymnosperms (a)** The ginkgo, or maidenhair tree, is widely cultivated as a shade or ornamental tree. **(b)** Common in the age of dinosaurs, cycads are now limited to about 160 species. **(c)** The leaves of the gnetophyte *Welwitschia* can live 1,000 years. **(d)** The needle-shaped leaves of conifers are protected by a waxy surface layer.

The gnetophyte *Welwitschia mirabilis* is among Earth's most distinctive plants (**FIG. 22-10c**). Found only in the extremely dry deserts of southwest Africa, *Welwitschia* has a deep taproot that can extend as far as 100 feet (30 meters) down into the soil. Above the surface, the plant has a fibrous stem. Only two leaves ever grow from the stem. The leaves are never shed and remain on the plant for its entire life, which can be very long. A typical life span is about 1,000 years, and the oldest *Welwitschia* are more than 2,000 years old. The strap-like leaves continue to grow for that entire period, spreading over the ground. The older portions of the leaves, whipped by the wind for centuries, may shred or split, giving the plant its characteristic gnarled and tattered appearance.

Conifers Are Adapted to Cool Climates

Though other gymnosperm groups like gingkoes and cycads are drastically reduced from their former prominence, the **conifers** still dominate large areas of our planet. Conifers, whose 500 species include pines, firs, spruce, hemlocks, and cypresses, are most abundant in the far north and at high elevations, places where winters are long and conditions are dry (because water in the soil remains frozen and unavailable during winter).

Conifers are adapted to these dry, cold conditions in three ways. First, most conifers retain green leaves throughout the year, enabling these plants to continue photosynthesizing and growing slowly during times when most other plants become dormant. For this reason, conifers are often called evergreens. Second, conifer leaves are thin needles covered with a thick, waterproof surface that minimizes evaporation (**FIG. 22-10d**). Finally, conifers produce an "antifreeze" in their sap that enables them to continue transporting nutrients in below-freezing temperatures and also gives them their fragrant piney scent.

Reproduction is similar in all conifers, so let's examine the reproductive cycle of a pine tree (**FIG. 22-11**). The tree itself is the diploid sporophyte, and it produces both male and female cones ❶. Male cones are relatively small (typically about $^3/_4$ inch long), delicate structures consisting of scales in which pollen (the male gametophyte) develops. Each female cone consists of a series of woody scales arranged in a spiral around a central axis. At the base of each scale are two **ovules** (unfertilized seeds), within which diploid spore-forming cells arise.

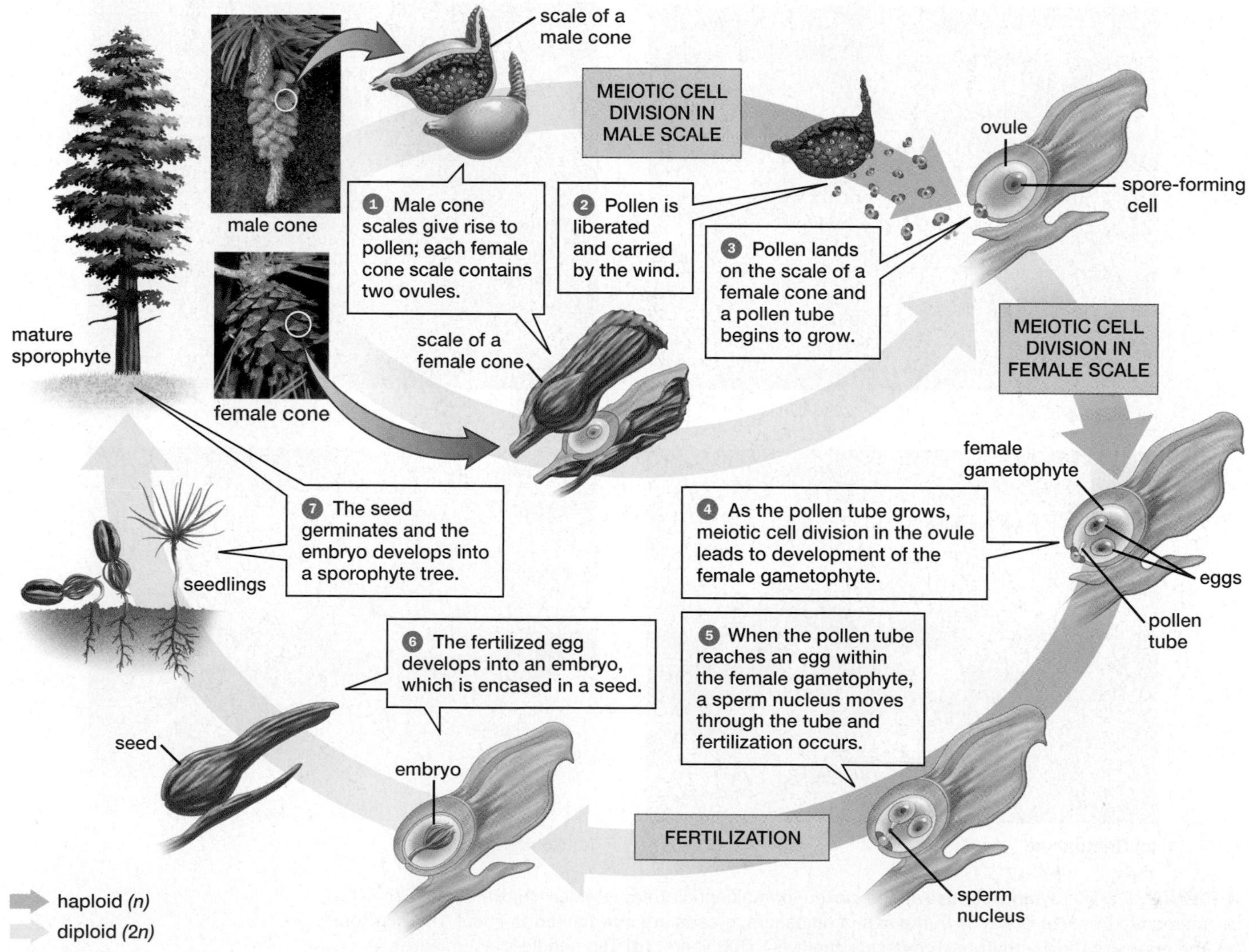

▲ **FIGURE 22-11 Life cycle of the pine**

Male cones release pollen during the reproductive season and then disintegrate ❷. The amount of pollen released is immense; inevitably, some pollen grains land by chance on female cone scales ❸. The pollen grain then sends out a pollen tube that slowly burrows into an ovule. As the pollen tube grows, the diploid spore-forming cell in the ovule undergoes meiosis to produce haploid spores, one of which gives rise to a haploid female gametophyte, within which egg cells develop ❹. After nearly 14 months, the pollen tube finally reaches the egg cell and releases the sperm that fertilize it ❺. The resulting zygote becomes enclosed in a seed as it develops into an embryo—a tiny embryonic sporophyte plant ❻. The seed is liberated when the cone matures and its scales separate. If it lands in a suitable patch of soil, it may germinate and grow into a sporophyte tree ❼.

Angiosperms Are Flowering Seed Plants

Flowering plants, or **angiosperms**, have been Earth's predominant plants for more than 100 million years. The group is incredibly diverse, with more than 230,000 species. Angiosperms range in size from the diminutive duckweed (**FIG. 22-12a**) to the towering eucalyptus tree (**FIG. 22-12b**). From desert cactus to tropical orchids to grasses to parasitic stinking corpse lilies, angiosperms dominate the plant kingdom. Their enormous success is due in part to three major adaptations: flowers, fruits, and broad leaves.

(a) Duckweed

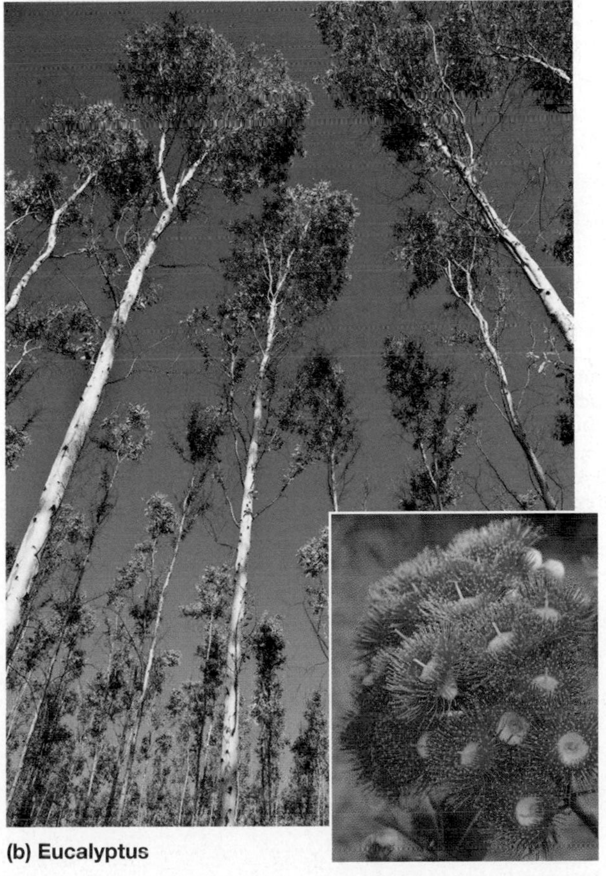

(b) Eucalyptus

(c) Grass

(d) Butterfly weed

FIGURE 22-12 Angiosperms (a) The smallest angiosperm is the duckweed, found floating on ponds. These specimens are about 1/8 inch (3 millimeters) in diameter. **(b)** The largest angiosperms are eucalyptus trees, which can reach 325 feet (100 meters) in height. **(c)** Grasses (and many trees) have inconspicuous flowers and rely on wind for pollination. More conspicuous flowers, such as those on **(d)** this butterfly weed and on a eucalyptus tree (**b**, inset), entice insects and other animals that carry pollen between individual plants.

THINK CRITICALLY What are the advantages and disadvantages of wind pollination? What are the advantages and disadvantages of pollination by animals? Why do both types of pollination persist among the angiosperms?

Flowers Attract Pollinators

Flowers, the structures in which both male and female gametes are formed, probably evolved when gymnosperm ancestors formed an association with animals (most likely insects) that carried their pollen from plant to plant. According to this scenario, the relationship between these ancient gymnosperms and their animal pollinators was so advantageous that natural selection favored the evolution of showy flowers that advertised the presence of pollen to insects and other animals (**FIGS. 22-12b, d**). The animals benefited by eating some of the protein-rich pollen, whereas the plant benefited from the animals' unwitting transportation of pollen from plant to plant. With this animal assistance, many flowering plants no longer needed to produce prodigious quantities of pollen and send it flying on the fickle winds to ensure fertilization. But there are nonetheless many wind-pollinated angiosperms (**FIG. 22-12c**). These probably evolved from animal-pollinated ancestors when environmental changes resulted in the decline or extinction of the affected species' pollinators.

In the angiosperm life cycle, flowers develop on the dominant sporophyte plant (**FIG. 22-13**). In the flower, female gametophytes develop from ovules within a structure called the *ovary*; male gametophytes (pollen) are formed inside a structure called the *anther* ❶. During the reproductive season, pollen is released from the anthers and carried away on the wind or by animal pollinators ❷. If a pollen grain lands on a *stigma*, a sticky pollen-catching structure of the flower, a pollen tube begins to grow from the pollen grain ❸. The tube bores through the stigma and extends toward the female gametophyte, within which an egg cell has developed. Fertilization occurs when the pollen tube reaches the egg cell and releases sperm cells ❹. The resulting zygote develops into an embryo enclosed in a seed formed from the ovule ❺. After it is dispersed, the seed may germinate and give rise to a sporophyte plant ❻.

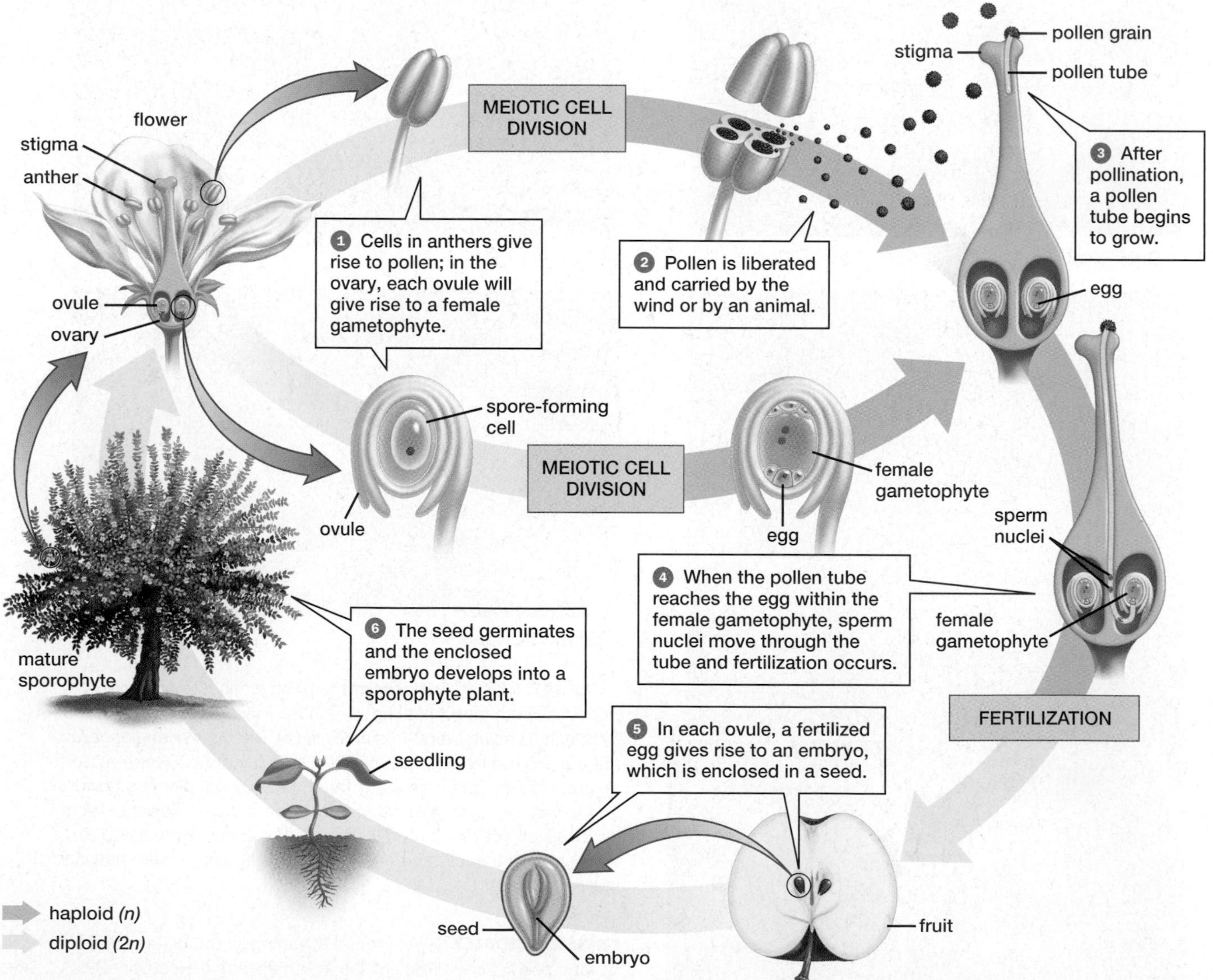

▲ **FIGURE 22-13 Life cycle of a flowering plant**

Fruits Encourage Seed Dispersal

The ovary surrounding the seeds of an angiosperm matures into a **fruit**, the second adaptation that has contributed to the success of angiosperms. Just as flowers encourage animals to transport pollen, so do many fruits entice animals to disperse seeds. If an animal eats a fruit, many of the enclosed seeds may pass through the animal's digestive tract unharmed, perhaps to fall at a suitable location for germination. Not all fruits, however, depend on edibility for dispersal. Dog owners are well aware, for example, that some fruits (called burrs) disperse by clinging to animal fur. Other fruits, such as those of maples, form wings that carry the seed through the air. The variety of dispersal mechanisms made possible by fruits has helped the angiosperms invade nearly all terrestrial habitats.

Broad Leaves Capture More Sunlight

The third feature that gives angiosperms an advantage in warmer, wetter climates is broad leaves. Broad leaves provide an advantage by collecting more sunlight for photosynthesis than the slender needles of conifers can. However, in regions with seasonal variations in growing conditions, many trees and shrubs drop their leaves during periods when water is in short supply, because being leafless reduces evaporative water loss. In temperate climates, such periods occur during the fall and winter. In the tropics and subtropics, species that inhabit areas where periods of drought are common may drop their leaves to conserve water during the dry season.

Broad leaves have costs as well as benefits. In particular, broad, tender leaves are much more appealing to herbivores than are the tough, waxy needles of conifers. As a result, angiosperms have evolved a range of defenses against mammalian and insect herbivores. These adaptations include physical defenses such as thorns, spines, and tough leaves. The evolutionary struggle for survival has also led to a host of chemical defenses—compounds that make plant tissue poisonous or distasteful to potential predators. Many of the compounds responsible for chemical defense have properties that humans have exploited for medicinal and culinary uses. Medicines such as aspirin and codeine, stimulants such as nicotine and caffeine, and spices such as mustard and pepper are all derived from angiosperm plants.

Have You Ever Wondered ...

Which Plants Provide Us with the Most Food?

Although thousands of plant species have edible parts, people exploit only a small proportion of them for food. In fact, the vast majority of the plant-derived food consumed by humans comes from only 20 species. The fruits (grains) of just three grass species—corn (maize), wheat, and rice—provide about half the calories consumed by people worldwide. The average person in the United States consumes about 200 pounds of these grains each year. In terms of annual production, the big three are followed, in order, by soybeans, barley, sorghum, millet, and peanuts. Looking further down the list, the world's most abundantly produced foods that are *not* grains or legumes are potatoes and cassava (a root that is a staple in parts of Africa and South America).

CHECK YOUR LEARNING

Can you . . .

- explain how vascular plants and nonvascular plants differ?
- describe the major plant taxonomic groups and representative members of each group?
- describe the key steps in the life cycles of mosses, ferns, gymnosperms, and flowering plants?

CASE STUDY CONTINUED

Queen of the Parasites

Why does the flower of a stinking corpse lily smell like rotting meat? Though the smell is utterly revolting to humans, it is attractive to blowflies and other insects that normally feed on and lay their eggs in decaying flesh. When such insects visit a male stinking corpse lily, they may carry away pollen that can fertilize a nearby female flower.

In many angiosperm species, flowers contain nectar that provides food for animal pollinators. But no such nectar reward awaits a fly that enters the flower of a stinking corpse lily. Instead, a fly attracted by the flower's stench searches in vain for putrefying meat, its movement guided toward the flower's cache of sticky pollen by grooves and hairs inside the flower. Eventually, the fly departs, coated in pollen. In essence, the plant tricks the fly into providing a service for no reward. Thus, the stinking corpse lily is a master exploiter: It takes advantage of both the host vines that provide its food and the flies that facilitate its reproduction.

The stinking corpse lily harms species it interacts with, but many plants benefit other species. How do other species benefit?

22.4 HOW DO PLANTS AFFECT OTHER ORGANISMS?

As plants survive, grow, and reproduce, they alter and influence Earth's landscape and atmosphere in ways that are tremendously beneficial to the rest of the planet's inhabitants, including humans. Humans also reap additional benefits by actively exploiting plants.

Plants Play a Crucial Ecological Role

The complex ecosystems that host terrestrial life could not be maintained without the help of plants. Plants make vital contributions to the food, air, soil, and water that sustain life on land.

Green Lifesaver

Many of the drugs that physicians use to treat diseases contain substances that were originally discovered in plants. In most cases, once a pharmaceutically useful plant component has been identified and isolated, researchers devise a method to synthesize the drug without using any actual plants. An important exception, at least until recently, is the antimalarial drug artemisinin. Artemisinin is an extremely important medicine, a key component of the best available treatment for the millions of people infected with *Plasmodium*, the protist parasite that causes malaria. The parasite has evolved resistance to most of the other drugs used to treat the disease.

Artemisinin is found in the sweet wormwood plant, *Artemisia annua* (**FIG. E22-1**). Its effectiveness as an antimalarial was first discovered by Youyou Tu and her colleagues, who tested hundreds of different herbs used in traditional chinese medicine before discovering that an extract of wormwood is an effective treatment for malaria. The researchers eventually identified artemisinin as the molecule responsible for wormwood's antimalarial capability. In 2015, Tu received the Nobel Prize for this accomplishment.

▲ **FIGURE E22-1 Sweet wormwood, ready for harvest**

Until recently, the only way to produce artemisinin was to extract it from wormwood plants. Now, however, researchers working on the cutting edge of synthetic biology have discovered how to insert *Artemisia* genes into yeast cells, and how to prompt the yeasts to excrete artemisinin while housed in industrial-scale fermentation vats. The synthetic drug can be produced in weeks, rather than the 14–18 months it takes to grow and process wormwood plants. This change is expected to greatly increase the availability of artemisinin, though at the expense of the mostly poor farmers who have been growing the wormwood formerly needed to make the drug.

Although the discovery of a high-tech way to ramp up artemisinin production is good news for malaria victims, it comes at a time when some *Plasmodium* parasites are developing resistance to the drug. Will the next great antimalarial come from a plant? The big pharmaceutical companies have largely abandoned their efforts to systematically screen Earth's plant diversity for new drugs, but researchers working in the African country of Mali might be on to something. In early trials, they have found that an extract of the poppy species *Argemone mexicana* is quite effective against malaria.

THINK CRITICALLY In the initial trials with *Argemone mexicana* in Mali, researchers observed patients treated with different herbal medicines by a traditional healer. They performed a blood test on each patient to determine which ones had malaria and tracked each patient to determine if and how quickly he or she recovered. If you were in charge of the next follow-up study, how would you design it? Would you use the traditional healer's preparations of the plant? Would you have the healer determine dosage and frequency of treatment?

Plants Capture Energy That Other Organisms Use

Plants provide food, directly or indirectly, for all of the animals, fungi, and nonphotosynthetic microbes on land. Plants use photosynthesis to capture solar energy, and they convert part of the captured energy into leaves, shoots, seeds, and fruits that are eaten by other organisms. Many of these consumers of plant tissue are themselves eaten by still other organisms. Plants are the main providers of energy and nutrients to terrestrial ecosystems, and life on land depends on plants' ability to manufacture food from sunlight.

Plants Help Maintain the Atmosphere

In addition to providing food, plants produce oxygen gas as a by-product of photosynthesis. By doing so, they continually replenish oxygen in the atmosphere. Without plants' contribution, atmospheric oxygen would be rapidly depleted by the oxygen-consuming respiration of Earth's multitude of organisms. Plant photosynthesis also removes carbon dioxide from the atmosphere, converting it to compounds such as starch and cellulose, which are stored in plant bodies. Without plants, atmospheric carbon dioxide would soar to levels that would be fatal for almost all organisms.

Plants Build and Protect Soil

Plants also help create and maintain soil. When a plant dies, its stems, leaves, and roots become food for fungi, prokaryotes, and other decomposers. Decomposition breaks the plant tissue into tiny particles of organic matter that become part of the soil. Organic matter improves the ability of soil to hold water and nutrients, thereby making the soil more fertile and better able to support the growth of living plants. The roots of plants help stabilize soil and reduce erosion by wind and water.

Plants Help Keep Ecosystems Moist

Plants take up water from the soil and retain some of it in their tissues. By doing so, plants slow the rate at which water escapes from terrestrial ecosystems and increase the amount of water available to meet the needs of the ecosystems' inhabitants. By reducing the amount of water runoff, plants also reduce the chances of destructive flooding. Thus, floods can be more frequent in areas in which forests, grasslands, or marshes have been destroyed by human activities.

Plants Provide Humans with Necessities and Luxuries

It would be difficult to exaggerate the degree to which people depend on plants. Neither our explosive population growth nor our rapid technological advance would have been possible without plants.

Plants Provide Shelter, Fuel, and Medicine

Plants provide wood that is used to construct housing for much of Earth's human population. In addition, wood has historically been the main fuel for warming dwellings and cooking, and it remains so in many parts of the world. Coal, another important fuel, is composed of the remains of ancient plants that have been transformed by geological processes.

Plants have also supplied many of the medicines on which modern health care depends. Important drugs that were originally found in and extracted from plants include the painkillers aspirin, codeine, and morphine, the heart medication digoxin, the cancer treatments Taxol and vinblastine, and many more (see "Health Watch: Green Lifesaver").

In addition to harvesting useful material from wild plants, humans have domesticated a host of useful plant species. Through generations of selective breeding, people have modified the seeds, stems, roots, flowers, and fruits of favored plant species to provide themselves with food and fiber. It is difficult to imagine life without corn, rice, potatoes, apples, tomatoes, cooking oil, cotton, and the myriad other staples that domestic plants provide.

Plants Provide Pleasure

Though we appreciate the practical value of wheat and wood, our most emotionally powerful connections with plants are purely sensual. We delight in the beauty and fragrance of flowers and present them to others as symbols of our most sublime and inexpressible emotions. Many of us spend hours of leisure time tending gardens and lawns, for no reward other than the pleasure and satisfaction we derive from observing the fruits of our labor. In our homes, we reserve space for our houseplant companions. We line our streets with trees and seek refuge from the stress of daily life in parks with abundant plant life. Clearly, plants help fill our emotional needs.

CHECK YOUR LEARNING

Can you . . .

- describe some of the effects that plants have on other organisms, including humans?

CASE STUDY REVISITED

Queen of the Parasites

The 17 or so parasitic plant species of the genus *Rafflesia*, which includes the stinking corpse lily, are found in the moist forests of Southeast Asia, a habitat that is disappearing rapidly as forests are cleared for agriculture and development. The geographic range of the stinking corpse lily is limited to the dwindling forests of the Malaysian peninsula and the Indonesian islands of Borneo and Sumatra; the species is rare and endangered. The government of Indonesia has established parks and reserves in an effort to help protect the stinking corpse lily, but—as is often the case in developing countries—a forest that is protected on paper may still be vulnerable in reality.

Perhaps the best hope for the continued survival of the largest *Rafflesia* is the growing realization among the rural residents of Sumatra and Borneo that the spectacular, putrid-smelling flowers of the stinking corpse lily might lure interested tourists to their countries. Under an innovative conservation program that seeks to take advantage of this potential for ecotourism, people who live in the vicinity of the stinking corpse lily can become caretakers of the plants. These assigned caretakers watch over the plants and, in return, may charge a small fee to curious visitors. Local inhabitants have been given an economic incentive to protect this rare parasitic plant.

THINK CRITICALLY Perhaps surprisingly, a parasitic lifestyle is not terribly rare among plants. More than 4,400 plant species are parasites, and systematists estimate that parasitism has evolved at least 12 different times over the evolutionary history of plants. Given the obvious benefits of photosynthesis, why has parasitism (which is often accompanied by loss of photosynthetic capability) evolved repeatedly in photosynthetic plants?

CHAPTER REVIEW

Go to **Mastering Biology** to access the Pearson eText, vocabulary review, practice quizzes, activities, videos, current events, and more.

Answers to ***Think Critically*** *and* ***Thinking Through the Concepts*** *questions can be found in the* ***Answers*** *section at the back of the book.*

Summary of Key Concepts

22.1 What Are the Key Features of Plants?

Plants are multicellular organisms that exhibit alternation of generations, in which a haploid gametophyte generation alternates with a diploid sporophyte generation. Most plants are photosynthetic. Unlike their green algae relatives, plants have multicellular, dependent embryos.

22.2 How Have Plants Evolved?

Photosynthetic protists, probably aquatic green algae, gave rise to the first plants. Ancestral plants were probably similar to modern multicellular algae such as stoneworts, which are plants' closest living relatives.

Early plants invaded terrestrial habitats, and modern plants exhibit a number of key adaptations for terrestrial existence: rootlike structures for anchorage and for absorption of water and nutrients; a waxy cuticle that slows the loss of water through evaporation; stomata that can open, allowing gas exchange, or close, preventing water loss; the conducting tissues xylem and phloem that transport water and nutrients throughout the plant; and a stiffening substance, called lignin, that impregnates the conducting cells and helps support the plant body.

Plant reproductive structures suitable for life on land include a smaller male gametophyte (pollen) that allows wind to replace water in carrying sperm to eggs; seeds that nourish, protect, and help disperse developing embryos; flowers that attract animals, which carry pollen more precisely and efficiently than wind; and fruits that entice animals to disperse seeds.

There has been a general evolutionary trend toward a reduction in size of the haploid gametophyte, which is dominant in nonvascular plants but microscopic in seed plants.

22.3 What Are the Major Groups of Plants?

Two major groups of plants, nonvascular plants and vascular plants, arose from their ancient algal ancestors. Nonvascular plants, including the hornworts, liverworts, and mosses, are small, simple land plants that lack conducting cells and mostly live in moist habitats. Nonvascular plant reproduction requires water through which the sperm swim to the egg.

In vascular plants, a system of conducting cells—stiffened by lignin—conducts water and nutrients absorbed by the roots into the upper portions of the plant and supports the plant body. Thanks to this support system, seedless vascular plants, including the club mosses, horsetails, and ferns, can grow larger than nonvascular plants. The sperm of seedless vascular plants must swim to the egg for sexual reproduction to occur.

Gymnosperms and angiosperms are vascular plants with two major additional adaptive features for life on dry land: pollen and seeds. Gymnosperms, which include ginkgoes, cycads, gnetophytes, and conifers, were the first fully terrestrial plants to evolve.

Angiosperms, the flowering plants, dominate much of the land today. In addition to pollen and seeds, angiosperms also produce flowers and fruits.

22.4 How Do Plants Affect Other Organisms?

Plants play a key ecological role, capturing energy through photosynthesis for use by inhabitants of terrestrial ecosystems, replenishing atmospheric oxygen, sequestering carbon dioxide, creating and stabilizing soils, and slowing the loss of water from ecosystems. Plants are also used by humans to provide food, fuel, building materials, medicines, and aesthetic pleasure.

Thinking Through the Concepts

Bloom's: Remembering, Understanding

Multiple Choice

1. In an alternation of generations life cycle, spores develop into ____________ that produce gametes that fuse to give rise to ____________.
   a. haploid gametophytes; diploid sporophytes
   b. diploid gametophytes; haploid sporophytes
   c. haploid sporophytes; diploid gametophytes
   d. diploid sporophytes; haploid gametophytes
2. Which of the following are *not* nonvascular plants?
   a. mosses
   b. liverworts
   c. ferns
   d. hornworts
3. Which of the following structures is present in angiosperms but not in gymnosperms?
   a. cones
   b. fruits
   c. seeds
   d. xylem
4. In which of the following is the gametophyte stage larger and more prominent than the sporophyte stage?
   a. nonvascular plants
   b. seed plants
   c. angiosperms
   d. gymnosperms
5. Which of the following is *not* an adaption that helps plants resist gravity and/or dry conditions on land?
   a. xylem
   b. lignin
   c. cuticle
   d. photosynthesis

Fill-in-the-Blank

1. Scientists hypothesize that the ancestors of plants were ____________. There are two major types of plants; those that lack conducting cells are called ____________ and those with conducting cells are called ____________. All plants produce multicellular ____________ and exhibit a complex life cycle called ____________.

2. Plant adaptations to life on land include a(n) ____________, which reduces evaporation of water, and ____________, which open to allow gas exchange but close when ____________ is scarce. In addition, the bodies of vascular plants gain increased support from ____________ and ____________ impregnated with the polymer ____________; these structures also help ____________ and ____________ move within the plant body.
3. Seedless vascular plants must reproduce when conditions are wet because their sperm must ____________. Two adaptations that allow seed plants to reproduce more efficiently on dry land are ____________ and ____________. The seed plants fall into two major categories: the nonflowering ____________ and the flowering ____________. Flowers were favored by natural selection because they ____________. Fruits were favored by natural selection because they ____________.
4. Three groups of nonvascular plants are ____________, ____________, and ____________. Three groups of seedless vascular plants are ____________, ____________, and ____________. Today, the most diverse group of plants is the ____________.

Review Questions

1. What is meant by "alternation of generations"? What two generations are involved? How does each reproduce?
2. Explain the evolutionary changes in plant reproduction that adapted plants to increasingly dry environments.
3. Describe evolutionary trends in the life cycles of plants. Emphasize the relative sizes of the gametophyte and sporophyte.
4. From which algal group did green plants probably arise? Explain the evidence that supports this hypothesis.
5. List the structural adaptations necessary for the invasion of dry land by plants. Which of these adaptations are possessed by nonvascular plants? By ferns? By gymnosperms and angiosperms?
6. The number of species of flowering plants is greater than the number of species in the rest of the plant kingdom combined. What feature(s) are responsible for the enormous success of angiosperms? Explain why.
7. List the adaptations of gymnosperms that have helped them become the dominant trees in dry, cold climates.
8. What is a pollen grain? What role has it played in helping plants colonize dry land?
9. The majority of all plants are seed plants. What is the advantage of a seed? How do plants that lack seeds meet the needs served by seeds?

Applying the Concepts

Bloom's: Applying, Analyzing, Evaluating

1. Prior to the development of synthetic drugs, more than 80% of all medicines were of plant origin. Even today, indigenous tribes in remote Amazonian rain forests can provide a plant product to treat virtually any ailment. Herbal medicine is also widely and successfully practiced in China. Most of these drugs are unknown to the Western world. But the forests from which much of this plant material is obtained are being converted to agriculture. We are in danger of losing many of these potential drugs before they can be evaluated by Western medicine. What steps can you suggest to preserve these natural resources while also allowing nations to direct their own economic development?
2. Only a few hundred of the more than 200,000 species of plants have been domesticated for human use. One example is the almond. The domestic almond is nutritious and harmless, but its wild precursor can cause cyanide poisoning. The oak makes potentially nutritious seeds (acorns) that contain very bitter-tasting tannins. If we could breed the tannin out of acorns, they might become a delicacy. Why do you suppose we have failed to domesticate oaks?

23 The Diversity of Fungi

These honey mushrooms are part of the visible portion of the largest organism on Earth.

CASE STUDY

Humongous Fungus

WHAT IS THE LARGEST organism on Earth? A reasonable guess might be the world's largest animal, the blue whale, which can be 100 feet long and weigh 400,000 pounds. But the blue whale is dwarfed by the General Sherman tree, a giant sequoia that is 275 feet high and whose weight is estimated at 6,200 *tons*. Even these two behemoths, however, can't match the real record-holder, the fungus *Armillaria solidipes*, also known as the honey mushroom.

The largest known *Armillaria* is a specimen in Oregon that spreads over almost 2,400 acres and weighs more than 7,500 tons. Despite its huge size, no one has actually seen the monster fungus, because it is largely underground. Its only aboveground parts are brown mushrooms that sprout occasionally from its gigantic body. Just beneath the surface, however, the fungus spreads through the soil by means of long, string-like structures that extend until they encounter the tree roots on which *Armillaria* subsists.

How can researchers be sure that the Oregon fungus is truly one single individual and not many intertwined individuals? The strongest evidence is genetic. Researchers gathered *Armillaria* tissue samples from throughout the area thought to be inhabited by a single individual and compared DNA extracted from the samples. All were genetically identical, demonstrating that they came from the same individual.

The lives of fungi typically take place outside of our view, but they play a fascinating role in human affairs. How do fungi affect us and other organisms? How and where do they live, grow, and reproduce?

AT A GLANCE

23.1 WHAT ARE THE KEY FEATURES OF FUNGI?

When you think of a fungus, you probably picture a mushroom. Most fungi, however, do not produce mushrooms. And even in those that do, the mushrooms are just temporary reproductive structures. The main body is typically concealed beneath the soil or inside a piece of decaying wood. So, to fully appreciate fungi, we must look beyond the conspicuous structures we encounter on the forest floor, at the edges of our lawns, or on top of a pizza. A closer look at fungi reveals a group of eukaryotic, mostly multicellular organisms that play a key role in the web of life and whose lifestyle differs in fascinating ways from that of plants or animals.

Fungal Bodies Consist of Slender Threads

The body of almost every fungus is a **mycelium** (plural, mycelia; **FIG. 23-1a**), which is an interwoven mass of one-cell-thick, thread-like filaments called **hyphae** (singular, hypha; **FIGS. 23-1b, c**). In some species, hyphae consist of single elongated cells with numerous nuclei; in other species, hyphae are subdivided—by partitions called **septa** (singular, septum)—into many cells, each containing one or more nuclei. Pores in the septa allow cytoplasm to stream between cells, distributing nutrients. Like plant cells, fungal cells are surrounded by cell walls. Unlike plant cells, however, fungal cell walls are strengthened by *chitin*, the same substance found in the hard outer surface (exoskeleton) of insects, crabs, and their relatives.

Most fungi cannot move. They compensate for this lack of mobility with hyphae that can grow rapidly in any direction within a suitable environment. In this way, the fungal mycelium can quickly spread into aging bread or cheese, beneath the bark of decaying logs, or into the soil. Periodically, the hyphae differentiate into reproductive structures that project above the surface. These structures, including mushrooms, puffballs, and the powdery molds on spoiled food, represent only a fraction of the complete fungal body, but are typically the only part of the fungus that we can easily see.

Fungi Obtain Their Nutrients from Other Organisms

Like animals, fungi survive by breaking down nutrients stored in the bodies or wastes of other organisms. Some fungi digest

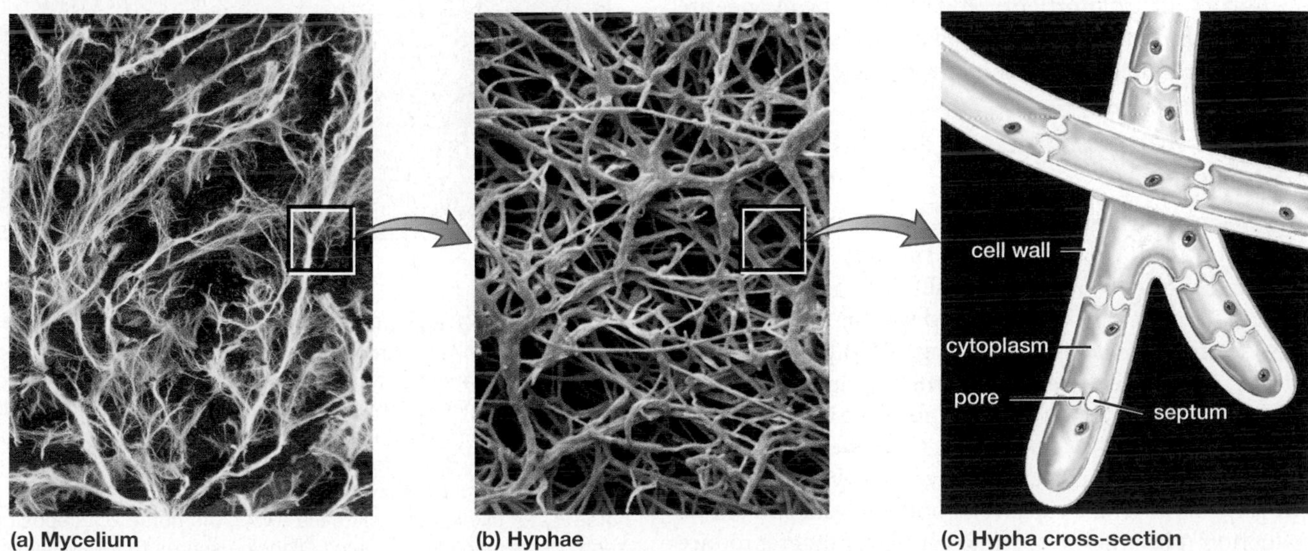

▲ **FIGURE 23-1 The filamentous body of a fungus (a)** A fungal mycelium spreads over decaying vegetation. The mycelium is composed of **(b)** a tangle of microscopic hyphae, only one cell thick, portrayed in cross-section **(c)** to show their internal organization.

THINK CRITICALLY Which features of a fungus's body structure are adaptations related to its method of acquiring nutrients?

▲ **FIGURE 23-2 Nemesis of nematodes** The fungus *Arthrobotrys*, also known as the nematode (roundworm) strangler, traps its prey in a noose-like modified hypha. When a nematode wanders into the noose, its presence stimulates the noose cells to swell with water. In a fraction of a second, the noose constricts, trapping the worm. Fungal hyphae then penetrate and feast on their prey.

the bodies of dead organisms. Others are parasitic, feeding on living organisms and causing disease. Some live in close, mutually beneficial relationships with other organisms that provide food. There are even a few predatory fungi, which attack tiny worms in soil (**FIG. 23-2**).

Unlike most animals, fungi do not ingest food. Instead, they secrete enzymes that digest complex molecules outside their bodies, breaking down the molecules into smaller subunits that can be absorbed. Fungal hyphae can penetrate deeply into a source of nutrients, and because the hyphae are only one cell thick, each cell in a fungal body is in position to absorb nutrients directly from the surrounding environment.

Fungi Can Reproduce Both Asexually and Sexually

Fungi develop from **spores**—haploid cells that can give rise to a new individual. Fungal spores are tiny and extraordinarily mobile, even though most—but not all—lack a means of self-propulsion. They are distributed far and wide as hitchhikers on the outside of animal bodies, as passengers inside the digestive systems of animals that have eaten them, or as airborne drifters, cast aloft by chance or shot into the atmosphere by elaborate reproductive structures (**FIG. 23-3**). Spores are often produced in great numbers; a single giant puffball may contain 5 trillion spores.

In general, fungi are capable of both asexual and sexual reproduction (**FIG. 23-4**). For the most part, fungi reproduce asexually under stable conditions, but reproduce sexually under conditions of environmental change or stress.

Asexual Reproduction Produces Haploid Spores by Mitosis

The mycelia and spores of fungi are haploid. A haploid mycelium produces haploid asexual spores by mitosis. If an asexual spore is deposited in a favorable location, it will begin mitotic divisions and develop into a new mycelium. This simple reproductive cycle results in the rapid production of a genetically identical clone of the original mycelium.

(a) Earthstar

(b) *Pilobolus*

▲ **FIGURE 23-3 Some fungi can eject spores (a)** A ripe earthstar mushroom, struck by a drop of water, releases a cloud of spores that will be dispersed by air currents. **(b)** The delicate, translucent reproductive structures of *Pilobolus*, which inhabits horse manure, literally blow their tops when ripe, dispersing the black, spore-containing caps up to 3 feet away. Spores that adhere to grass remain there until consumed by a grazing herbivore, perhaps a horse. Later (likely some distance away), the horse will deposit a fresh pile of manure containing *Pilobolus* spores that have passed unharmed through its digestive tract.

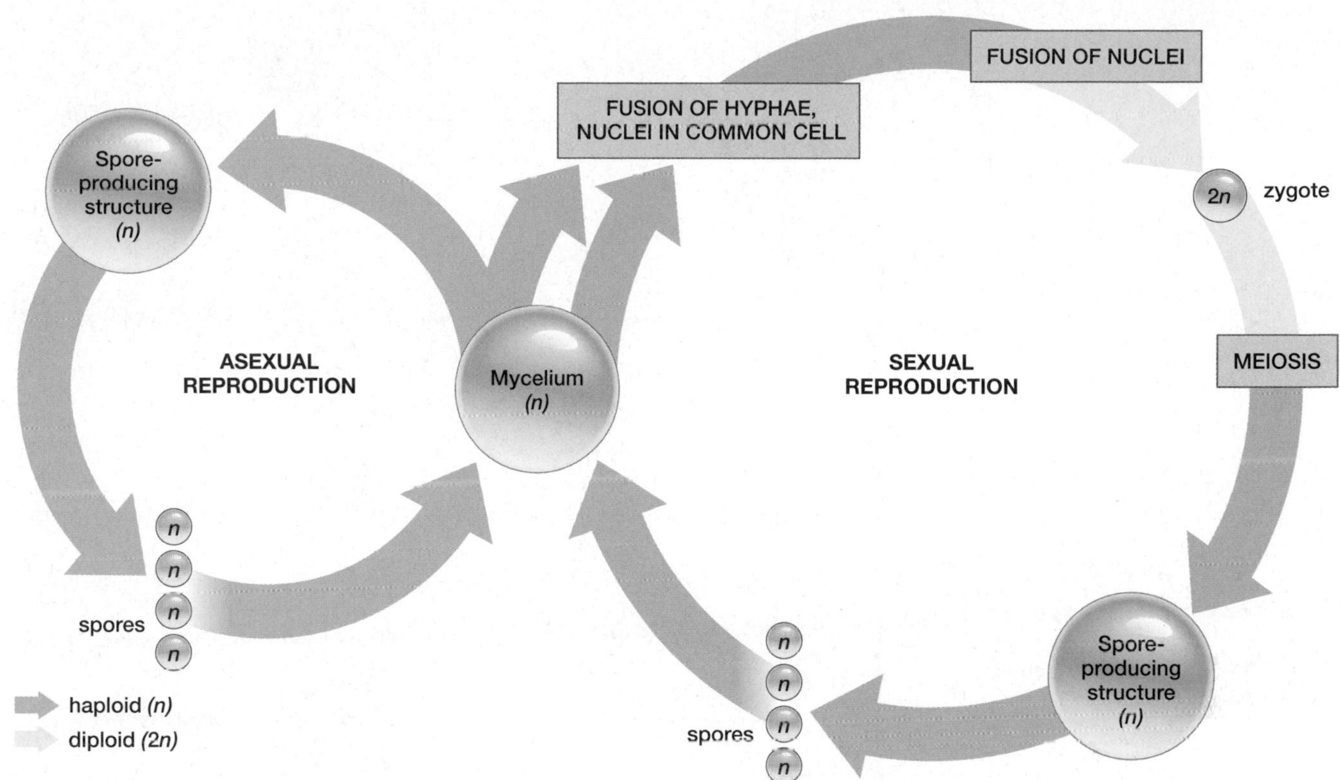

▲ **FIGURE 23-4 Generalized life cycle of fungi** In asexual reproduction, haploid hyphae in a mycelium give rise to structures that produce haploid spores by mitotic cell division. In sexual reproduction, haploid hyphae of different, compatible mating types fuse, resulting in cells that contain nuclei from both parents. These nuclei subsequently fuse, generating a diploid zygote that undergoes meiosis to yield haploid spores.

Sexual Reproduction Produces Haploid Spores by Meiosis

Diploid structures form only during a brief period of the sexual portion of the fungal life cycle. Sexual reproduction begins when a hypha of one mycelium comes into contact with a hypha from a second mycelium that is of a different, but compatible, mating type. (The different mating types of fungi are comparable to the different sexes of animals, except that in fungi there are often more than two mating types.) If conditions are suitable, the two hyphae may fuse, so that nuclei from the two different hyphae share a common cell. The merger of hyphae is followed by fusion of haploid nuclei, one from each of the two mating types, to form a diploid zygote. The zygote then undergoes meiosis to form haploid sexual spores. These spores are dispersed, germinate, and divide by mitosis to form new haploid mycelia. Unlike the cloned offspring produced by asexual spores, these sexually produced fungal bodies are genetically distinct from either parent.

CHECK YOUR LEARNING

Can you . . .

- describe the structure of a typical fungus?
- explain how fungi obtain energy and nutrients and how they reproduce?

23.2 WHAT ARE THE MAJOR GROUPS OF FUNGI?

Nearly 140,000 species of fungi have been described, but this number represents only a fraction of the true diversity of these organisms. Many new species are discovered and described each year, and mycologists (scientists who study fungi) estimate that the number of undiscovered species of fungi is at least 1.5 million. Fungi are classified into six main taxonomic groups: Chytridiomycota (chytrids), Neocallimastigomycota (rumen fungi), Blastocladiomycota (blastoclades), Glomeromycota (glomeromycetes), Basidiomycota (basidiomycetes), and Ascomycota (ascomycetes). Some fungi, however, are not members of any of these six groups. Most of these unclassified fungal species were historically placed in the taxonomic group Zygomycota (zygomycetes), but recent analysis of DNA sequences has revealed that zygomycetes do not constitute a clade and so cannot form a named taxonomic group. (A clade is a group consisting of all the descendants of a particular common ancestor.) Systematists are gathering additional data on the evolutionary history of the species previously classified as zygomycetes, with the goal of classifying them into new named groups.

Mass sequencing of DNA extracted from soil and water samples has also revealed a host of new fungal species, most of them so far known only from their DNA sequences. Many of

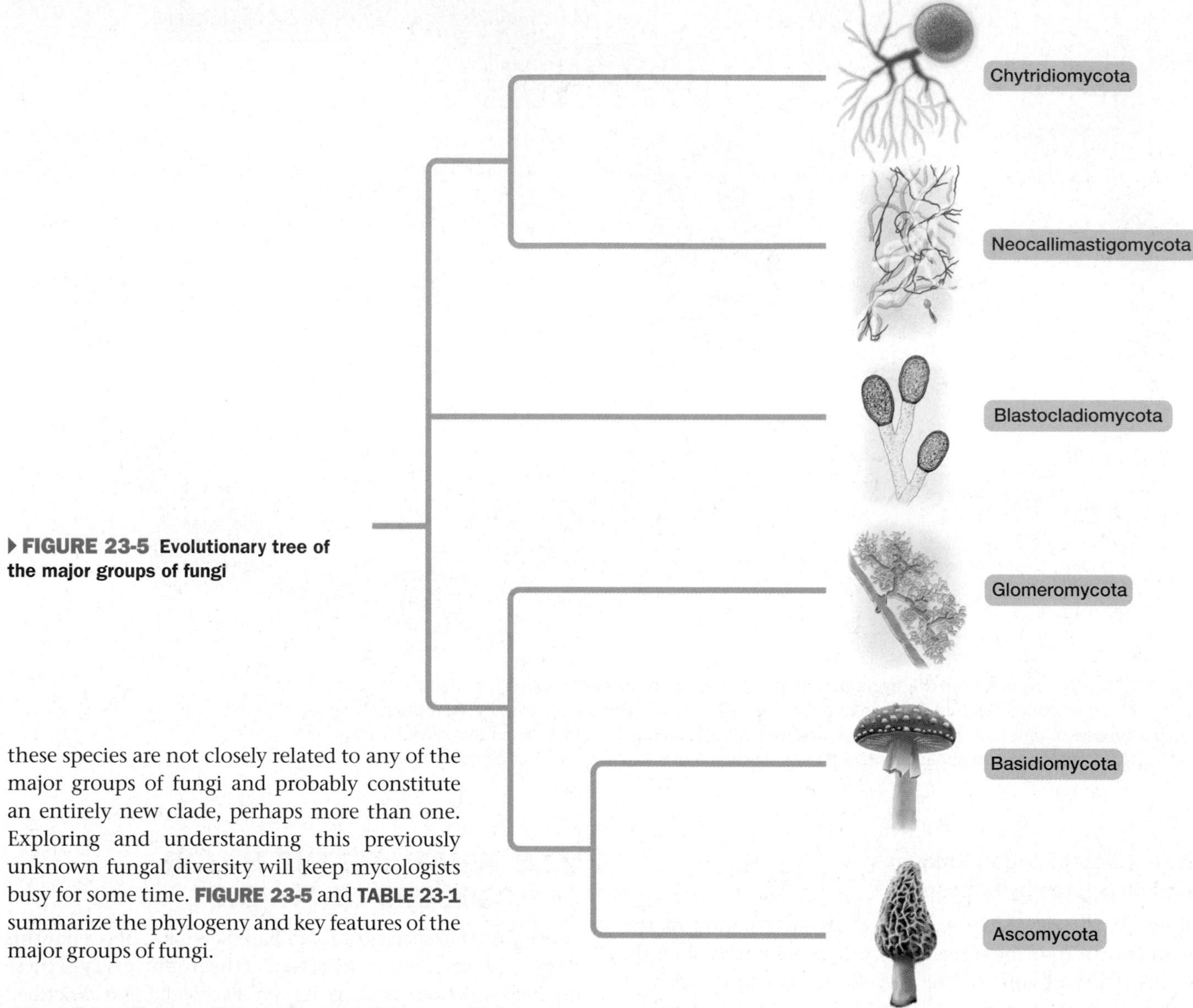

▶ **FIGURE 23-5 Evolutionary tree of the major groups of fungi**

these species are not closely related to any of the major groups of fungi and probably constitute an entirely new clade, perhaps more than one. Exploring and understanding this previously unknown fungal diversity will keep mycologists busy for some time. **FIGURE 23-5** and **TABLE 23-1** summarize the phylogeny and key features of the major groups of fungi.

Chytrids, Rumen Fungi, and Blastoclades Produce Swimming Spores

The members of three taxonomic groups of fungi—the chytrids, rumen fungi, and blastoclades—are distinguished by their swimming spores, which require water for dispersal. Many members of these groups live in water, and even those that live on land require a film of water for reproduction. The spores propel themselves through the water by means of one or more flagella.

Chytrids Are Mostly Aquatic

Most **chytrids** live in fresh water, but a few species are marine. Chytrid spores have a single flagellum on one end. The oldest known fossil fungi are chytrids that are found in rocks more than 600 million years old. Ancestral fungi may well have been similar to today's aquatic and marine chytrids, so fungi probably originated in a watery environment before colonizing land.

Most chytrid species feed on dead aquatic plants or other debris in watery environments, but some species are parasites of plants or animals. Parasitic chytrids are a major cause of the current worldwide die-off of frogs and salamanders, which threatens many species and has already caused the extinction of several. (For more on the decline of frogs, see "Earth Watch: Frogs in Peril" in Chapter 25.)

Rumen Fungi Live in Animal Digestive Tracts

The **rumen fungi** are anaerobic (they do not require oxygen) and reside mainly in the digestive tracts of plant-eating animals such as cows, sheep, kangaroos, elephants, and iguanas. These animals are not able to digest cellulose (a major component of plant tissue) themselves but instead rely on symbiotic organisms that inhabit their guts. The rumen fungi are among these organisms; they produce enzymes that digest cellulose, and the resulting breakdown product nourishes both the fungi and their animal hosts. The spores of most rumen fungi have multiple flagella, which may form a tuft at one end of the spore.

TABLE 23-1 The Major Taxonomic Groups of Fungi

Common Name (Latin Name)	Reproductive Structures	Cellular Characteristics	Economic and Health Impacts	Representative Genera
Chytrids (Chytridiomycota)	Form haploid or diploid flagellated spores	Septa are absent	Contribute to the decline of frog populations	*Batrachochytrium* (frog pathogen)
Rumen fungi (Neocallimastigomycota)	Form haploid or diploid flagellated spores	Septa are absent	Help enable cattle, horses, sheep to subsist on plants	*Neocallimastix* (lives in herbivore digestive systems)
Blastoclades (Blastocladiomycota)	Form haploid or diploid flagellated spores	Septa are absent	Cause brown spot disease of corn, crown wart disease of alfalfa	*Allomyces* (aquatic decomposer)
Glomeromycetes (Glomeromycota)	Form haploid asexual spores, often in clusters	Septa are absent	Form mycorrhizae (mutualistic, symbiotic associations with plant roots)	*Glomus* (widespread mycorrhizal partner)
Basidiomycetes (Basidiomycota)	Sexual reproduction involves formation of haploid basidiospores on club-shaped basidia	Septa are present	Cause smuts and rusts on crops; include some edible mushrooms	*Amanita* (poisonous mushroom); *Polyporus* (shelf fungus)
Ascomycetes (Ascomycota)	Form haploid sexual ascospores in saclike ascus	Septa are present	Cause molds on fruit; can damage textiles; cause Dutch elm disease and chestnut blight; include yeasts and morels	*Saccharomyces* (yeast); *Ophiostoma* (causes Dutch elm disease)
"Zygomycetes" (not a formally designated taxonomic group)	Form diploid sexual zygospores	Septa are absent	Cause soft fruit rot and black bread mold	*Rhizopus* (causes black bread mold); *Pilobolus* (dung fungus)

Blastoclades Have a Nuclear Cap

Blastoclades (**FIG. 23-6**) are distinguished by some characteristic features, such as a distinctive structure called the *nuclear cap* that is found near the nucleus of blastoclade spores. The nuclear cap consists of ribosomes. Blastoclades live in fresh water or in soil, and some are parasites of plants or aquatic invertebrates such as water fleas or mosquito larvae. Their spores have a single flagellum.

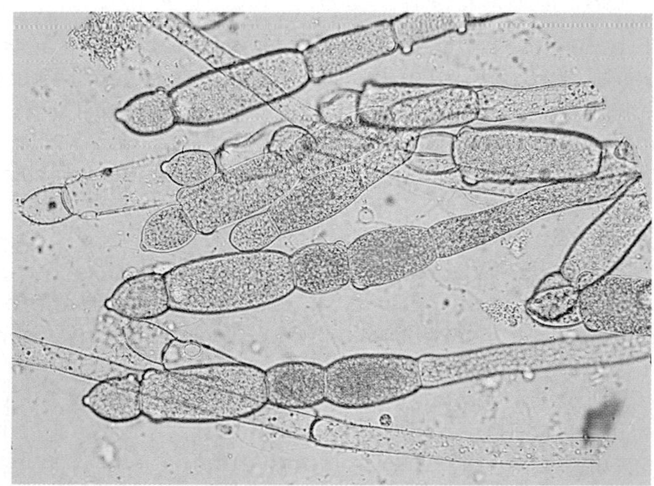

▲ **FIGURE 23-6 Blastoclade filaments** These filaments of the blastoclade fungus *Allomyces* are in the midst of sexual reproduction. The orange structures visible on many of the filaments will release male gametes; the clear swollen structures will release female gametes. Blastoclade gametes are flagellated, and these swimming reproductive structures aid dispersal of members of this mostly aquatic group.

▲ **FIGURE 23-7 Glomeromycete in a plant cell** Glomeromycete hyphae penetrate the cells of plants with which the fungus forms mutually beneficial associations. In this photo, a "trunk" hypha (red arrow) has branched off from the larger hypha at the bottom and divided into many smaller "fine branch" hyphae inside of a plant root cell.

Glomeromycetes Associate with Plant Roots

Almost all **glomeromycetes** live in intimate contact with the roots of plants. In fact, the hyphae of glomeromycetes actually penetrate the cells of the roots and form microscopic branching structures inside the cells (**FIG. 23-7**). This invasion of the plant's cells does not appear to harm the plant. On the contrary, glomeromycetes provide benefits to the plants they inhabit. This type of beneficial association between fungi and plant roots is known as a *mycorrhiza* and is described in more detail later in this chapter.

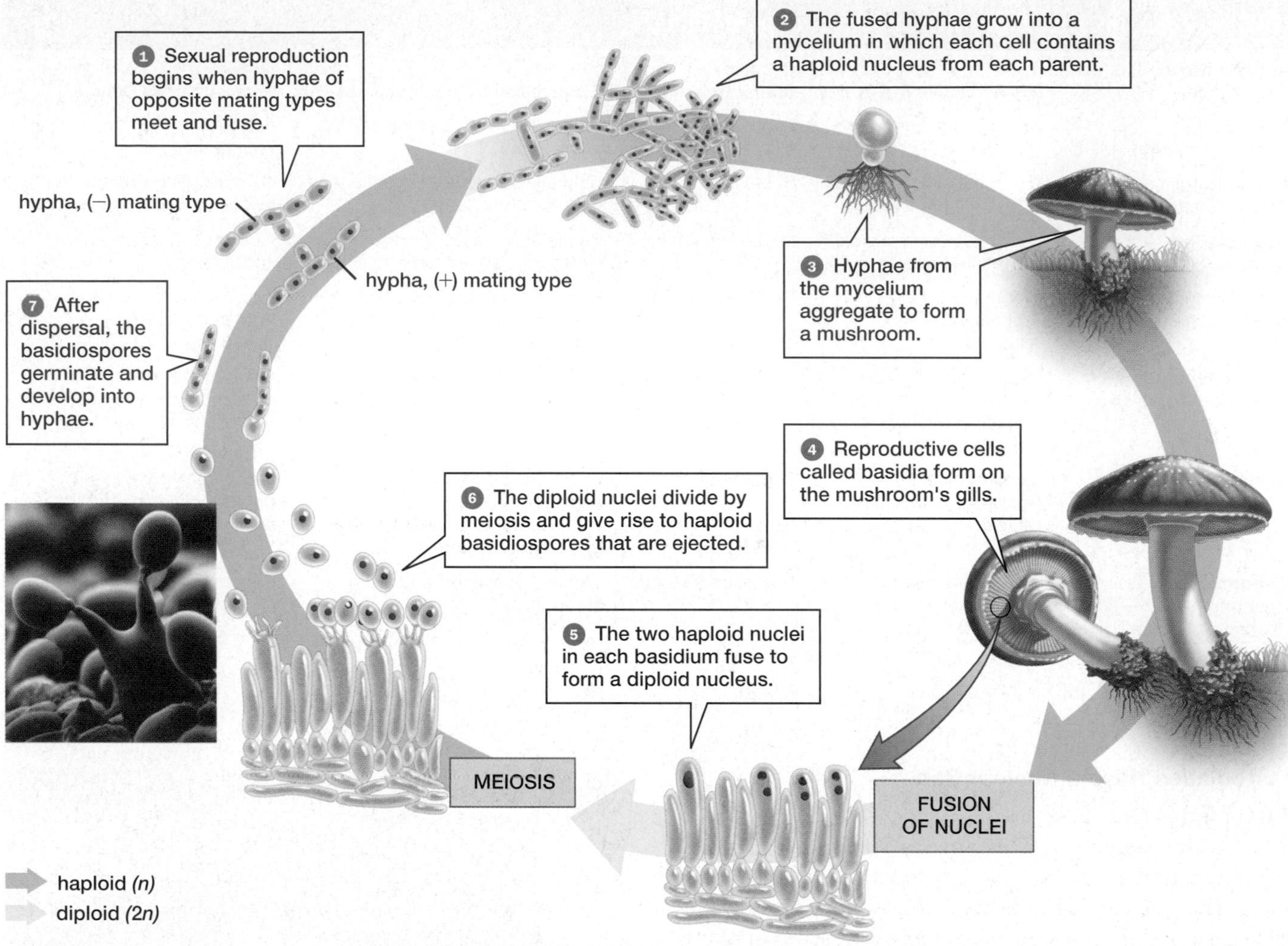

1 Sexual reproduction begins when hyphae of opposite mating types meet and fuse.
2 The fused hyphae grow into a mycelium in which each cell contains a haploid nucleus from each parent.
hypha, (–) mating type
hypha, (+) mating type
3 Hyphae from the mycelium aggregate to form a mushroom.
7 After dispersal, the basidiospores germinate and develop into hyphae.
4 Reproductive cells called basidia form on the mushroom's gills.
6 The diploid nuclei divide by meiosis and give rise to haploid basidiospores that are ejected.
5 The two haploid nuclei in each basidium fuse to form a diploid nucleus.
MEIOSIS
FUSION OF NUCLEI
haploid (n)
diploid (2n)

▲ **FIGURE 23-8 The life cycle of a typical basidiomycete** The photo shows two basidiospores attached to a basidium.

Glomeromycete reproduction is not fully understood; sexual reproduction by a member of the group is yet to be observed. During asexual reproduction, glomeromycetes produce clusters of spores by mitotic cell division. The spores form at the tips of hyphae that typically remain outside the host plant cell. When the spores germinate, hyphae grow into the surrounding soil, but the new fungus survives only if its germinating hyphae reach a plant root.

Basidiomycetes Produce Club-Shaped Reproductive Cells

Basidiomycetes typically reproduce sexually (**FIG. 23-8**). Hyphae of different mating types (designated "+" and "–") fuse 1 to form hyphae in which each cell contains two nuclei, one from each parent 2. These hyphae grow into an underground mycelium that, in response to appropriate environmental conditions, gives rise to an aboveground fruiting body that consists of densely aggregated hyphae 3. Some of the hyphae in the fruiting body develop into club-shaped reproductive cells called **basidia** (singular, basidium), which contain two haploid nuclei 4. In each basidium, the two nuclei fuse to yield a diploid nucleus 5. The diploid nucleus divides by meiosis and gives rise to four haploid *basidiospores* 6. If a basidiospore falls on fertile ground, it may germinate and form haploid hyphae 7.

Basidiomycete fruiting bodies are familiar to most of us as mushrooms, puffballs, shelf fungi, and stinkhorns (**FIG. 23-9**). In many basidiomycetes, basidia are produced in leaflike gills on the undersides of mushrooms. In puffballs, the basidia are enclosed within the fruiting body. Basidiospores are released by the billions through openings in the tops of puffballs or from the gills of mushrooms and are dispersed by wind and water. In many cases, spores give rise to hyphae that grow outward from the original spore in a roughly circular pattern as the older hyphae in the center die. The subterranean body periodically sends up numerous mushrooms, which emerge in a ring-like pattern called a fairy ring (**FIG. 23-10**).

(a) Puffball

(b) Shelf fungus

(c) Stinkhorn

▲ **FIGURE 23-9 Diverse basidiomycetes (a)** The giant puffball *Lycoperdon giganteum* may produce up to 5 trillion spores. **(b)** Shelf fungi, some the size of dessert plates, are conspicuous on trees. **(c)** The spores of stinkhorns are carried on the outside of a slimy cap that smells terrible to humans, but appeals to flies. The flies lay their eggs on the stinkhorn and inadvertently disperse the spores that stick to their bodies.

CASE STUDY CONTINUED

Humongous Fungus

Because the underground bodies of basidiomycetes such as *Armillaria* grow at a relatively steady rate, the age of a mycellum can be estimated by measuring the area over which its above-ground reproductive structures spread. On the basis of such measurements, it is apparent that basidiomycetes can live for hundreds of years. Some are even older than that. For example, the researchers who discovered the gigantic *Armillaria* in Oregon estimate that it took at least 2,400 years to grow to its current size. Do the life cycles and habitats of other fungal groups permit them to grow as large and old as some basidiomycetes?

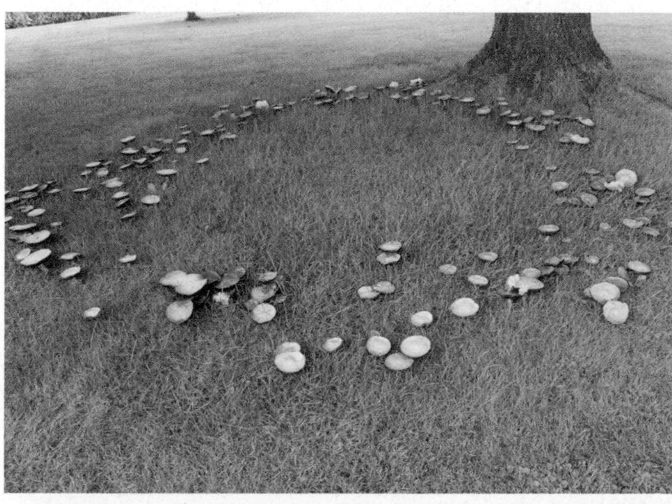
▲ **FIGURE 23-10 A mushroom fairy ring** Mushrooms emerge in a fairy ring from an underground fungal mycelium, growing outward from a central point where a single spore germinated, perhaps centuries ago.

Ascomycetes Form Spores in a Saclike Case

The **ascomycetes**, or **sac fungi**, reproduce both asexually and sexually (**FIG. 23-11**). In asexual reproduction, spores are produced at the tips of specialized hyphae and, after dispersal, develop into new hyphae ❶. During sexual reproduction, spores are produced by a more complex sequence of events that begins when hyphae of different mating types (+ and −) come into contact ❷. The two hyphae form reproductive structures that become linked by a connecting bridge. Haploid nuclei move across the bridge from the (−) reproductive structure to the (+) one, so that the (+) structure contains multiple nuclei from both parents ❸. The (+) structures that now contain the pooled nuclei develop into hyphae that are incorporated into a fruiting body ❹. At the tips of some of these hyphae, a saclike case called an **ascus** (plural, asci) forms ❺. At this stage, each ascus contains two haploid nuclei. These nuclei fuse to yield a single diploid nucleus ❻, which then divides by meiosis to yield four haploid nuclei ❼. These four nuclei divide by mitosis and develop into eight haploid spores known as *ascospores* ❽. Eventually, the ascus ruptures, liberating its ascospores. If the spores land in an appropriate location, they may germinate and develop into hyphae ❾.

Some ascomycetes live in decaying forest vegetation and form either beautiful cup-shaped reproductive structures (**FIG. 23-12a**) or corrugated, mushroom-like fruiting bodies called *morels* (**FIG. 23-12b**). The ascomycetes also include the species that produces penicillin, the first antibiotic, as well as many of the colorful molds that attack stored food and destroy fruit and grain crops and other plants. Ascomycetes may also harm animals (for an example, see "Earth Watch: Killer in the Caves" on page 378). The unicellular fungi known as yeasts are also ascomycetes. (When yeasts reproduce sexually, their single cell becomes the ascus.)

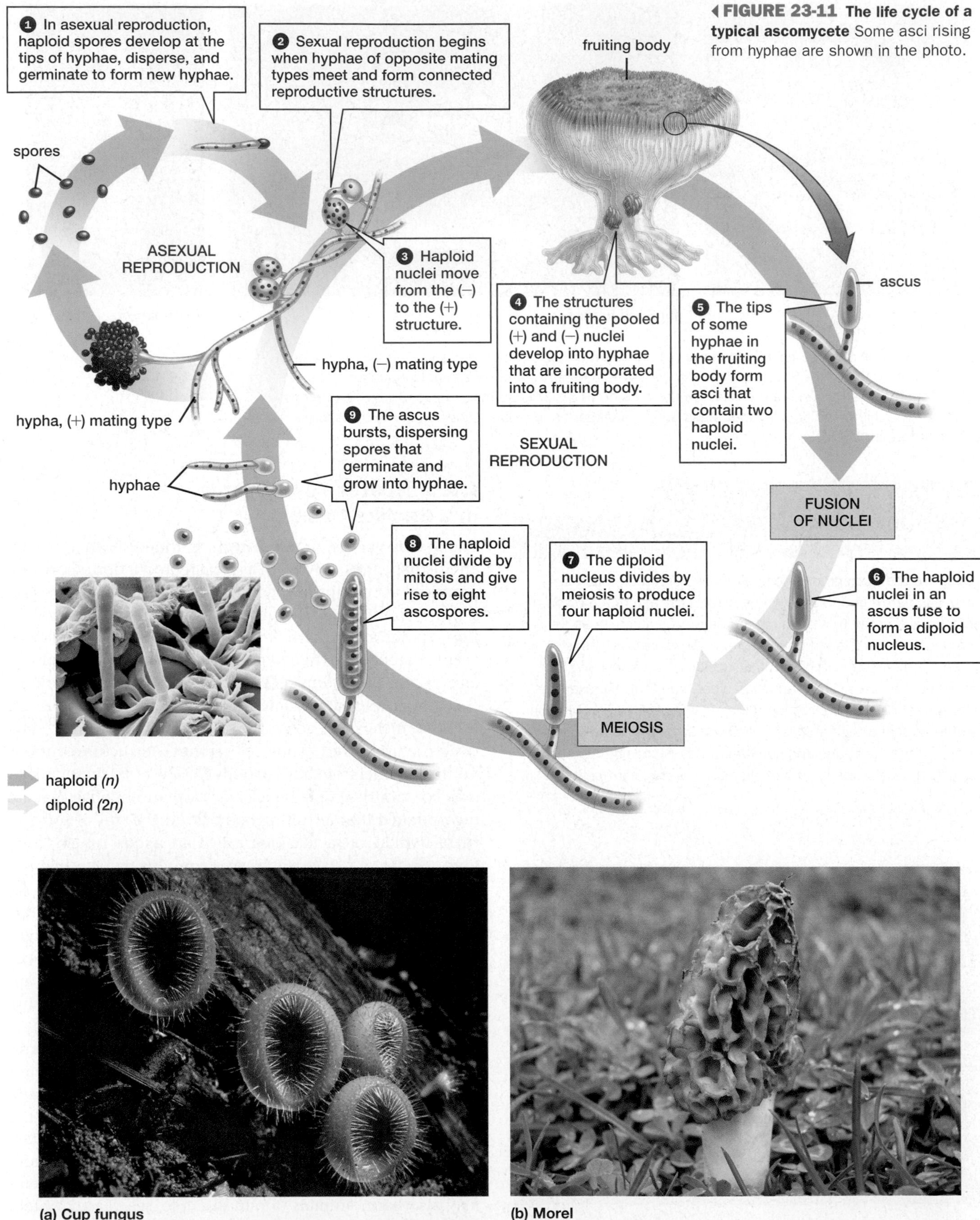

1 In asexual reproduction, haploid spores develop at the tips of hyphae, disperse, and germinate to form new hyphae.
2 Sexual reproduction begins when hyphae of opposite mating types meet and form connected reproductive structures.
fruiting body
spores
ASEXUAL REPRODUCTION
3 Haploid nuclei move from the (–) to the (+) structure.
4 The structures containing the pooled (+) and (–) nuclei develop into hyphae that are incorporated into a fruiting body.
5 The tips of some hyphae in the fruiting body form asci that contain two haploid nuclei.
ascus
hypha, (–) mating type
hypha, (+) mating type
9 The ascus bursts, dispersing spores that germinate and grow into hyphae.
SEXUAL REPRODUCTION
hyphae
FUSION OF NUCLEI
8 The haploid nuclei divide by mitosis and give rise to eight ascospores.
7 The diploid nucleus divides by meiosis to produce four haploid nuclei.
6 The haploid nuclei in an ascus fuse to form a diploid nucleus.
MEIOSIS
haploid (n)
diploid (2n)

◀ **FIGURE 23-11 The life cycle of a typical ascomycete** Some asci rising from hyphae are shown in the photo.

(a) Cup fungus

(b) Morel

▲ **FIGURE 23-12 Diverse ascomycetes (a)** The cup-shaped fruiting body of the scarlet cup fungus. **(b)** The morel, an edible delicacy. (Consult an expert before sampling any wild fungus—some are deadly!)

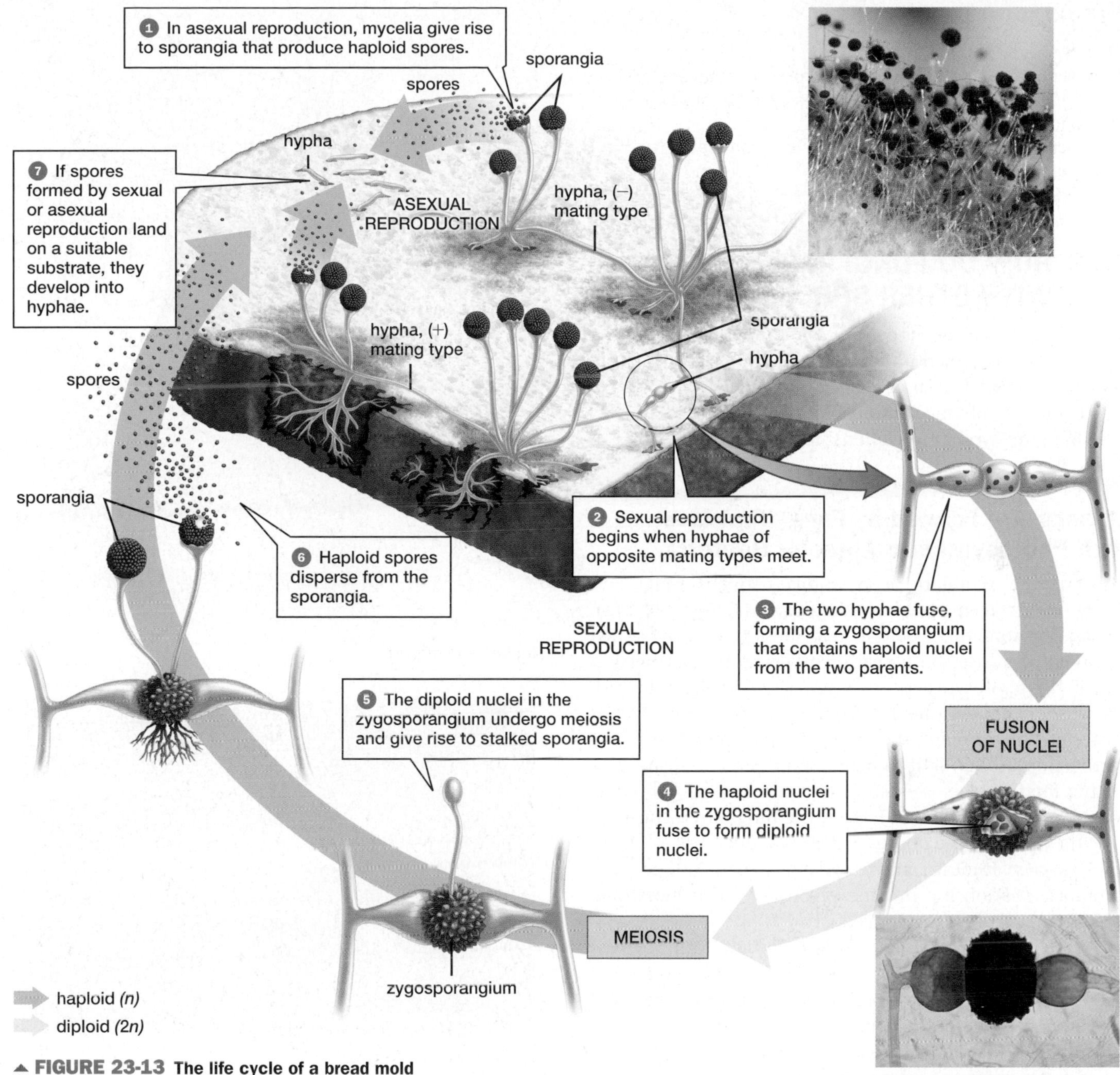

1 In asexual reproduction, mycelia give rise to sporangia that produce haploid spores.
sporangia
spores
hypha
7 If spores formed by sexual or asexual reproduction land on a suitable substrate, they develop into hyphae.
ASEXUAL REPRODUCTION
hypha, (−) mating type
hypha, (+) mating type
sporangia
hypha
spores
sporangia
6 Haploid spores disperse from the sporangia.
2 Sexual reproduction begins when hyphae of opposite mating types meet.
SEXUAL REPRODUCTION
3 The two hyphae fuse, forming a zygosporangium that contains haploid nuclei from the two parents.
FUSION OF NUCLEI
5 The diploid nuclei in the zygosporangium undergo meiosis and give rise to stalked sporangia.
4 The haploid nuclei in the zygosporangium fuse to form diploid nuclei.
MEIOSIS
zygosporangium
haploid (n)
diploid (2n)

▲ **FIGURE 23-13 The life cycle of a bread mold**

Bread Molds Are Among the Fungi That Can Reproduce by Forming Diploid Spores

Many of the species formerly assigned to the zygomycetes live in soil or on decaying plant or animal material. These species include those belonging to the genus *Rhizopus*, which cause the familiar annoyances of soft fruit rot and black bread mold. In the bread mold, reproduction may be asexual or sexual (**FIG. 23-13**). Asexual reproduction is initiated by the formation of haploid spores in black spore cases called **sporangia** (singular, sporangium) ❶. These spores disperse through the air and, if they land on a suitable substrate (such as a piece of bread), germinate to form new haploid hyphae.

If two hyphae of different mating types (+ and −) come into contact, sexual reproduction may ensue ❷. The two hyphae fuse to form a *zygosporangium* that contains multiple haploid nuclei from the two parents ❸. As the zygosporangium develops, it becomes tough and resistant, and it can remain dormant for long periods until environmental conditions are favorable for growth. Inside the zygosporangium, the haploid nuclei fuse to produce diploid nuclei ❹. When conditions are favorable, the diploid nuclei undergo meiosis and give rise to stalked sporangia ❺. The sporangia produce haploid spores that disperse ❻, germinate, and develop into new haploid hyphae ❼.

CHECK YOUR LEARNING

Can you . . .

- describe the six main taxonomic groups of fungi and explain why some fungal species are not assigned to any of these groups?
- describe the life cycles of a typical basidiomycete, ascomycete, and bread mold?

23.3 HOW DO FUNGI INTERACT WITH OTHER SPECIES?

Many fungi live in direct contact with another species for a prolonged period. Such intimate, long-term relationships are known as *symbiotic* relationships. In many cases, the fungal member of a symbiotic relationship is parasitic and harms its host. But some symbiotic relationships are mutually beneficial.

Lichens Are Formed by Fungi That Live with Photosynthetic Algae or Bacteria

Lichens are symbiotic associations between fungi and single-celled green algae or cyanobacteria (**FIG. 23-14a**). Lichens are sometimes described as fungi that have learned to garden, because the fungal members of the partnership "tend" the photosynthetic algal or bacterial partner by providing shelter and protection from harsh conditions. In this protected environment, the photosynthetic members of the partnership use sunlight to manufacture simple sugars, producing food for themselves but also some excess food that is consumed by the fungi. In fact, the fungi often consume the lion's share of the photosynthetic product (up to 90% in some species), leading some researchers to conclude that the symbiotic relationship in lichens is really much more one-sided than it is usually portrayed.

Thousands of different fungal species form lichens (**FIGS. 23-14b, c**). In most cases, one ascomycete species and one basidiomycete species combine with an algal or bacterial species. Together, these organisms form a unit so tough and self-sufficient that lichens are among the first living things to colonize newly formed volcanic islands, because many lichens can grow on bare rock. Brightly colored lichens also invade other inhospitable habitats ranging from deserts to the Arctic. Understandably, lichens in extreme environments grow very slowly; arctic colonies, for example, may expand as slowly as

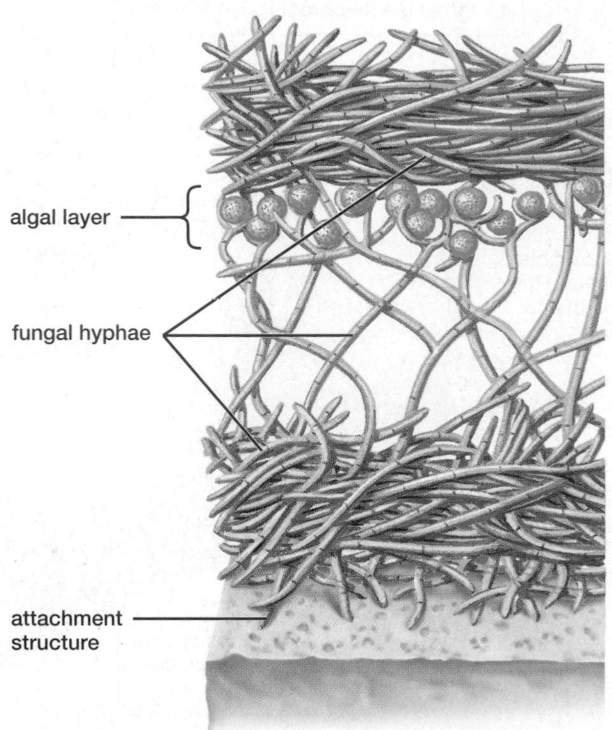

(a) Lichen structure

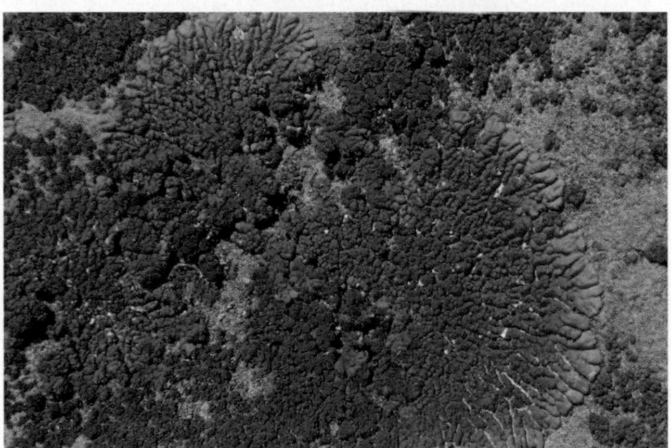

(b) Encrusting lichen

(c) Leafy lichen

FIGURE 23-14 The lichen: a symbiotic partnership (a) Most lichens have a layered structure bounded on the top and bottom by an outer layer formed from fungal hyphae. The fungal hyphae emerge from the lower layer, forming attachments that anchor the lichen to a surface, such as a rock or a tree. An algal layer in which the alga and fungus grow in close association lies beneath the upper layer of hyphae. **(b)** A colorful encrusting lichen, growing on dry rock, illustrates the tough independence of this symbiotic combination of fungus and algae. Pigments produced by the fungal partner are responsible for the bright orange color. **(c)** A leafy lichen grows on a rock.

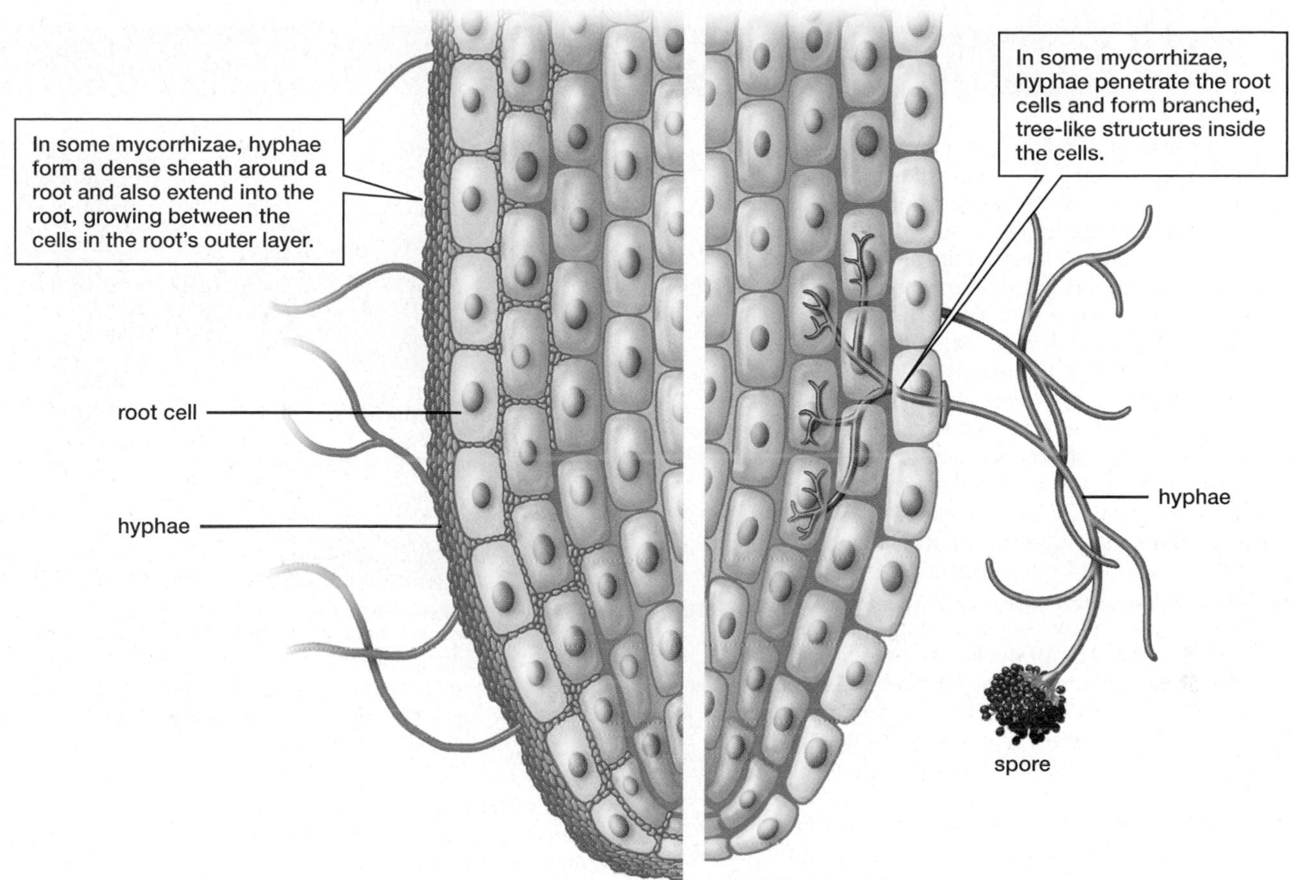

In some mycorrhizae, hyphae form a dense sheath around a root and also extend into the root, growing between the cells in the root's outer layer.
In some mycorrhizae, hyphae penetrate the root cells and form branched, tree-like structures inside the cells.
root cell
hyphae
hyphae
spore

▲ **FIGURE 23-15 Mycorrhizae enhance plant growth** Mycorrhizal associations between fungi and plant roots fall into two general types, shown in the left and right parts of this illustration (see also FIG. 23-7).

THINK CRITICALLY Fossil evidence suggests an important link between mycorrhizae and the successful invasion of land by plants. Why might mycorrhizae have been important in the colonization of terrestrial habitats by plants?

1 to 2 inches per 1,000 years. Despite their slow growth, lichens can persist for long periods of time; some arctic lichens are more than 4,000 years old.

Mycorrhizae Are Associations Between Plant Roots and Fungi

Mycorrhizae (singular, mycorrhiza) are important symbiotic associations between fungi and plant roots. More than 5,000 species of mycorrhizal fungi grow in intimate association with vascular plant roots, including those of most tree species. The hyphae of mycorrhizal fungi surround and invade roots (**FIG. 23-15**).

The association between plants and mycorrhizal fungi benefits both the fungi and their plant partners. The mycorrhizal fungi receive energy-rich sugar molecules that are produced photosynthetically by plants and passed from their roots to the fungi. In return, the fungi absorb mineral nutrients from the soil, passing some of them directly into the root cells. Mycorrhizal fungi also absorb water and pass it to the plant—an advantage for plants in dry, sandy soils.

The partnership between mycorrhizae and plants makes a crucial contribution to the health of Earth's plants. Plants without mycorrhizal fungi tend to be smaller and less vigorous than plants with mycorrhizal partners. Thus, the presence of mycorrhizae increases the overall productivity of Earth's plant communities, enhancing their ability to support the animals and other organisms that depend on them.

Endophytes Are Fungi That Live Inside Plant Stems and Leaves

The intimate association between fungi and plants is not limited to root mycorrhizae. Fungi have also been found living inside the aboveground tissues of virtually every plant species that has been tested for their presence. Some of these *endophytes* (organisms that live inside plants) are parasites that cause plant diseases, but many, perhaps most, are beneficial to the host plant. The best-studied examples of beneficial fungal endophytes are the ascomycete species that live inside the leaf cells of many species of grass. These fungi produce substances that are distasteful or toxic to insects

Killer in the Caves

The American chestnut and American elm have vanished from North American landscapes as a result of fungal diseases imported from other continents, and bats are now threatened with the same fate. An ascomycete species that arrived in the United States from Europe causes white-nose syndrome, an infectious disease that has killed an estimated 7 million bats of 11 species since its discovery in 2006. The disease is named for the white fungal growth that appears on the bare skin, especially the nose, of affected bats (**FIG. E23-1**).

The bat species that are most vulnerable to white-nose syndrome are those that hibernate during the winter. The fungus that causes the disease is adapted to cold temperatures; infections occur mainly during the winter and spread through groups of bats hibernating close together in caves and mines. White-nose syndrome has spread to bats throughout the eastern half of North America, and in recent years to locations farther west as well.

The disease has greatly reduced populations of the affected species. This abrupt decline of bats has serious implications for ecosystem health. All of the species that have been devastated by white-nose syndrome are insect-eaters that play an important role in regulating insect populations, helping ensure that plant communities are not overwhelmed by plant-eating insects. A recent study estimated that insect consumption in eastern North America has declined by 2,300 tons per year as a result of the onset of white-nose syndrome. This ecological loss has economic effects as well; insect consumption by bats saves U.S. farmers billions of dollars by reducing the amount of pesticide they must apply. Unfortunately, it is far from certain that we will continue to enjoy the ecological benefits of bats; there is so far no treatment or cure for white-nose syndrome.

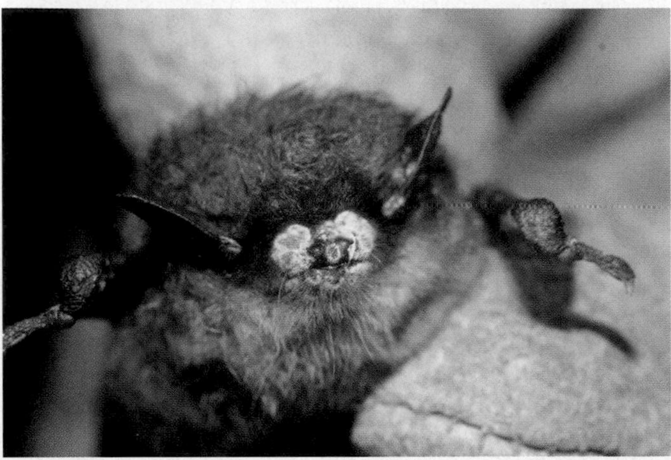

▲ **FIGURE E23-1 White nose syndrome** The fungus that causes the disease infects the skin of bats' faces, wings, and ears, breaking down skin and connective tissues. In addition, irritation from the infection causes bats to become active when they should be hibernating, which can lead to starvation.

THINK CRITICALLY The ascomycete that causes white-nose syndrome can remain viable on a cave floor for at least 5 years without infecting a bat host. As a result, even a cave in which all resident bats have become infected and died contains infectious spores that can potentially be carried to other caves. Given this circumstance, what steps could governments take to slow the spread of white-nose syndrome to previously unaffected areas?

and grazing mammals and thus help protect the grass plants from those predators.

The antipredator protection provided by fungal endophytes is sufficiently effective that agricultural scientists are working hard to discover a way to grow grasses free of endophytes and therefore more palatable to grazing animals. Horses, cows, and other agriculturally important grazers tend to avoid eating grasses that contain endophytes. When the only available food is endophyte-containing grass, animals that eat it experience poor health and slow growth.

Some Fungi Are Important Decomposers

Some fungi, acting as mycorrhizae and endophytes, play a major role in the growth and preservation of plant tissue. Other fungi, however, play a similarly major role in its destruction by acting as decomposers. Many fungal species can digest lignin or cellulose, the molecules that make up wood; some species can digest both molecules. Thus, when a tree or other woody plant dies, fungi can completely decompose its remains.

Fungal decomposition is not limited to wood; fungi digest dead organisms of all kinds. The fungi that are *saprophytes* (feeding on dead organisms) return the dead tissues' component substances to the ecosystems from which they came. The extracellular digestive activities of saprophytic fungi liberate nutrients that can be used by plants. If fungi and prokaryotes were suddenly to disappear, the consequences would be disastrous. Nutrients would remain locked in the bodies of dead plants and animals, the recycling of nutrients would grind to a halt, soil fertility would rapidly decline, and waste and organic debris would accumulate. In short, ecosystems would collapse.

CHECK YOUR LEARNING

Can you . . .

- describe and explain the significance of some symbiotic associations involving fungi, including lichens, mycorrhizae, and endophytes?
- explain how fungi help recycle nutrients?

23.4 HOW DO FUNGI AFFECT HUMANS?

Most people give little thought to fungi, but they affect our lives in more ways than you might imagine.

Fungi Attack Plants That Are Important to People

Fungi cause the majority of plant diseases, and some of the plants that they infect are important to humans. For example, fungal pathogens have a devastating effect on the world's food supply. Especially damaging are the basidiomycete plant pests descriptively called *rusts* and *smuts*, which cause billions of dollars' worth of damage to crops each year (**FIG. 23-16**). For example, Ug99, an especially virulent strain of the wheat disease called black stem rust, is currently a major threat to the world's supply of wheat. (Wheat is one of the world's most important crops in terms of total calories contributed.) None of the wheat varieties that feed much of the world's population is resistant to Ug99, whose spores can travel long distances by wind. As a result, wheat crops have been decimated over a large and growing area of Africa, Central Asia, and the Middle East. If Ug99 spreads to the massive wheat crops of China and India, serious food shortages might occur.

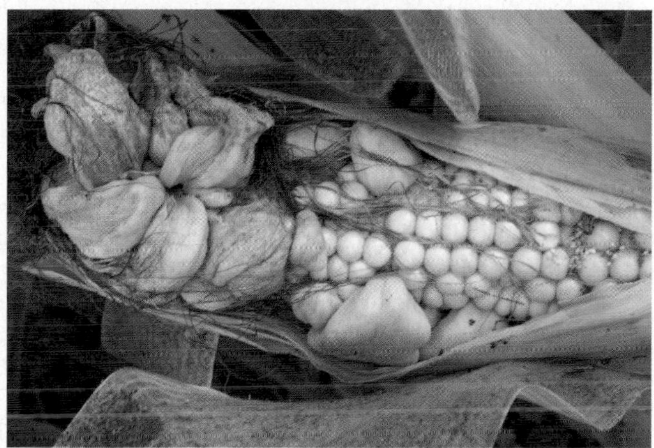

(a) Corn smut

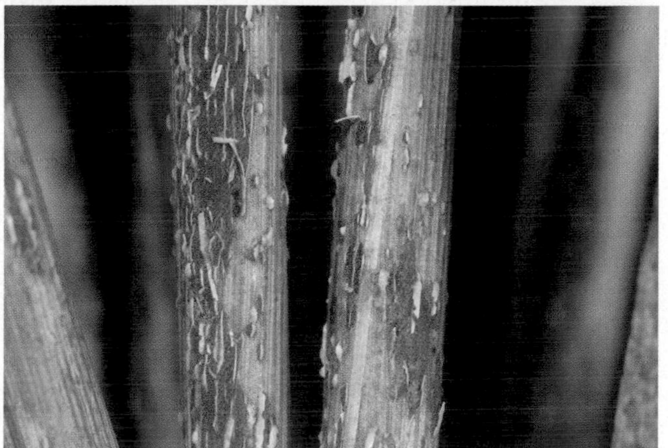

(b) Black stem rust

▲ **FIGURE 23-16 Smuts and rusts (a)** Corn smut is a basidiomycete pathogen that destroys millions of dollars' worth of corn each year. Even a pest like corn smut has its admirers, though. In Mexico this fungus is known as *huitlacoche* and is considered a great delicacy. **(b)** Another basidiomycete, black stem rust, currently threatens millions of acres of wheat in Africa, Central Asia, and the Middle East.

▲ **FIGURE 23-17 Pest-killing fungus** Fungi such as *Cordyceps* are used by farmers to control insect pests. The white spikes erupting from the body of this moth are *Cordyceps* fruiting bodies.

Fungal diseases also affect the appearance of our landscape. The American elm and the American chestnut—two tree species that were once prominent in many of America's parks, yards, and forests—were destroyed on a massive scale by the ascomycetes that cause Dutch elm disease and chestnut blight. Today, few people can recall the graceful forms of large elms and chestnuts, which are now almost entirely absent from the landscape. There may be hope for a return of the chestnut, however. Researchers have engineered trees that are genetically modified to contain a gene that confers resistance to chestnut blight and have demonstrated that the modified trees pass the resistance gene to their offspring. The researchers are seeking regulatory approval to plant the transgenic trees in the wild.

The fungal impact on agriculture and forestry is not entirely negative. Fungal parasites that attack insects and other arthropod pests can be an important ally in pest control (**FIG. 23-17**). Farmers who wish to reduce their dependence on toxic and expensive chemical pesticides are increasingly turning to biological methods of pest control, including the application of "fungal pesticides." Fungal pathogens are currently used to control termites, rice weevils, tent caterpillars, aphids, citrus mites, and other pests.

CASE STUDY CONTINUED

Humongous Fungus

The *Armillaria* fungus species that grew to massive size in Oregon harms trees in the forests it inhabits. As the fungus feeds on roots, it causes "root rot" that weakens or kills trees. This root rot provides aboveground evidence of *Armillaria*'s existence; the giant Oregon specimen was first identified by examining aerial photos to find forested areas with many dead trees. Can people, as well plants, be victims of fungal attack?

Fungi Cause Human Diseases

The fungi include parasitic species that attack humans directly. Some of the most familiar fungal diseases are those caused by ascomycetes such as the yeast *Candida albicans*, which causes vaginal infections, and those that attack the skin, resulting in athlete's foot, jock itch, and ringworm. These diseases, though unpleasant, are not life threatening and can usually be treated with antifungal ointments.

Fungi can infect the lungs if victims inhale spores of disease-causing fungal species such as the ascomycetes that cause valley fever and histoplasmosis. In the United States, valley fever is found mainly in the southwest, whereas histoplasmosis is more prevalent in central and eastern regions. Like other fungal infections, these diseases can, if promptly diagnosed, be controlled with antifungal drugs. If untreated, however, they can develop into serious, systemic infections.

Serious disease can also result from inhaling spores of *Cryptococcus gattii*, a basidiomycete that is normally found in the tropics but has in recent years appeared in the Pacific Northwest. In all environments, the fungus has proved to be dangerous; the death rate of those infected ranges from 13% to 33%, depending on location. Fortunately, infections are so far relatively rare: about 30 cases per year in the United States, for example, compared to more than 20,000 annually for valley fever. However, the area affected by *C. gattii* has grown steadily, so the incidence of infections is likely to increase.

In addition to causing human diseases, fungi can also help combat them. Biologists have discovered that certain fungi attack and kill the mosquito species that transmit malaria (**FIG. 23-18**). Plans are under way to enlist these fungi in the fight against malaria, one of the world's deadliest diseases.

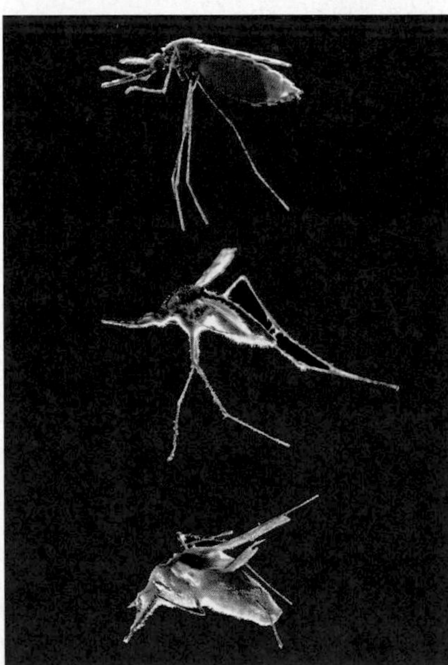

▲ **FIGURE 23-18 Disease-fighting fungus** A healthy malaria-carrying mosquito (top) infected by *Beauveria* is transformed to a fungus-encrusted corpse in less than 2 weeks.

Fungi Can Produce Toxins

In addition to their role as agents of infectious disease, some fungi produce toxins that are dangerous to humans. Of particular concern are toxins produced by fungi growing on grains and other foodstuffs that have been stored in moist conditions. For example, molds of the genus *Aspergillus* produce highly toxic, carcinogenic compounds known as aflatoxins. Some foods, such as peanuts, seem especially susceptible to attack by *Aspergillus*. Since aflatoxins were discovered in the 1960s, food growers and processors have developed methods for reducing the growth of *Aspergillus* in stored crops, so aflatoxins have been largely eliminated from the food supply in developed countries.

One infamous toxin-producing fungus is the ascomycete *Claviceps purpurea*, which infects rye plants and causes a disease known as ergot. This fungus produces several toxins, which can affect humans if infected rye is ground into flour and consumed. This happened frequently in northern Europe in the Middle Ages, with devastating effects. At that time, ergot poisoning was typically fatal, and victims experienced terrible symptoms before dying. One ergot toxin constricts blood vessels and reduces blood flow. The effect can be so extreme that gangrene develops and limbs shrivel and fall off. Other ergot toxins cause symptoms that include a burning sensation, vomiting, convulsive twitching, and vivid hallucinations. Today, new agricultural techniques have effectively eliminated ergot poisoning, but the hallucinogenic drug LSD, which is derived from a component of the ergot toxins, remains as a legacy of this disease.

Some of the deadliest poisons known to humankind are found in mushrooms. Especially noted for their poisons are certain species in the genus *Amanita*, which have suggestive common names such as death cap and destroying angel (**FIG. 23-19**). These names are apt, because even a single bite of one of these mushrooms can be lethal. Damage from *Amanita* toxins is most severe in the liver, where the toxins tend to accumulate. Often, a victim of *Amanita* poisoning

▲ **FIGURE 23-19 The destroying angel** Mushrooms produced by the basidiomycete *Amanita virosa* can be deadly.

can be saved only by undergoing a liver transplant. Each year, a number of small children, inexperienced collectors, and unlucky guests at gourmet dinners make unexpected trips to the hospital after eating poisonous wild mushrooms. So if you decide to collect some wild mushrooms to eat, be careful. Protect your health by inviting an expert to join your mushroom-hunting expeditions.

Many Antibiotics Are Derived from Fungi

Fungi also have positive impacts on human health. The modern era of lifesaving antibiotic medicines began with the discovery of penicillin, which is produced by an ascomycete mold (**FIG. 23-20**; also see Fig. 1-13). Penicillin is still used, along with other fungi-derived antibiotics such as oleandomycin and cephalosporin, to combat bacterial diseases. Other important drugs are also derived from fungi, including cyclosporin, which is used to suppress the immune response after an organ transplant so that the body is less likely to reject the transplanted organs.

Fungi Make Important Contributions to Gastronomy

Fungi make important contributions to the human diet. We consume some fungi directly, including wild and cultivated basidiomycete mushrooms and ascomycetes such as morels and the rare and prized truffle. The role of fungi in cuisine also has less-visible manifestations. For example, some of the world's most famous cheeses, including Roquefort, Camembert, Stilton, and Gorgonzola, gain their distinctive flavors from ascomycete molds that grow on them as they ripen. Perhaps the most important and pervasive fungal contributors to our food supply, however, are the single-celled ascomycetes (and a few species of basidiomycetes) known as yeasts.

▲ **FIGURE 23-20** ***Penicillium*** *Penicillium* growing on an orange. Reproductive structures, which coat the fruit's surface, are visible, while hyphae, beneath, draw nourishment from inside. The antibiotic penicillin was first isolated from this fungus.

THINK CRITICALLY Why do some fungi produce antibiotic chemicals? How and where would you search for new antibiotics produced by fungi?

Have You Ever Wondered ... Why Truffles Are So Expensive?

Although many fungi are prized as food, none is as avidly sought as the truffle. Truffles are the underground fruiting bodies of ascomycetes that form mycorrhizal associations with the roots of oak trees. The finest Italian truffles may sell for as much as $3,400 per pound, and unusually large specimens can fetch spectacular prices. A 3.3-pound Italian white truffle once sold at auction for $330,000!

Why such high prices? Truffles develop underground, and it takes some work to find one. In fact, humans can't do it alone and need help from other species. Some animals, especially pigs, are attracted to the aroma of a mature truffle. If a pig follows the smell to a truffle, it will dig the fungus up and devour it. That's why, traditionally, truffle collectors have used muzzled pigs to hunt their quarry.

Today, trained dogs are the most common assistants to truffle hunters. Dogs are necessary even on the farms where much of today's truffle crop is laboriously grown. The difficulty of cultivating and harvesting truffles accounts in part for their high price. And no one has figured out how to cultivate the prized white truffle. The only way to acquire one is to follow the nose of a truffle-hunting dog or pig.

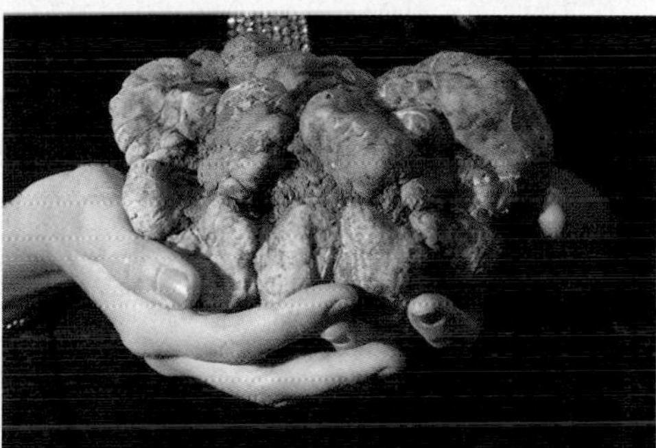

A prized Italian truffle

Wine and Beer Are Made Using Yeasts

The discovery that yeasts could be harnessed to enliven our culinary experience is surely a key event in human history. Among the many foods and beverages that depend on yeasts for their production are bread, wine, and beer, which have been consumed so widely for so long that it is difficult to imagine a world without them. All derive their special qualities from fermentation by yeasts. Fermentation occurs when yeasts extract energy from sugar and, as by-products of the metabolic process, emit carbon dioxide and ethyl alcohol.

As yeasts consume the fruit sugars in grape juice, the sugars are converted to alcohol, and wine is the result. Eventually, the increasing concentration of alcohol kills the yeasts, ending fermentation. If the yeasts in fermenting wine die before they have consumed all the available grape sugar, the wine will be sweet; if the yeasts have exhausted the supply of sugar, the wine will be dry.

Beer is brewed from grain (usually barley), but yeasts cannot effectively consume the carbohydrates in grain. For the yeasts to do their work, the barley grains must have sprouted (recall that grains are actually seeds). Germination converts the grains' carbohydrates to sugar, so the sprouted barley provides an excellent food source for the yeasts. As with wine, fermentation converts sugars to alcohol, but beer brewers capture the carbon dioxide by-product as well, giving the beer its characteristic bubbly carbonation.

Yeasts Make Bread Rise

In bread making, carbon dioxide is the crucial fermentation product. The yeasts added to bread dough do produce alcohol as well as carbon dioxide, but the alcohol evaporates during baking. In contrast, the carbon dioxide is trapped in the dough, where it forms the bubbles that give bread its light, airy texture (and saves us from a life of eating sandwiches made with crackers).

So the next time you're enjoying a slice of French bread with Camembert cheese and a nice glass of chardonnay, or a slice of pizza and a cold bottle of your favorite brew, you might want to quietly give thanks to the yeasts. Our diets would certainly be a lot duller without the help we get from fungal assistants.

CHECK YOUR LEARNING

Can you . . .

- explain how fungi affect agriculture?
- describe examples of how fungi affect human health?
- describe the role of fungi in the production of cheese, wine, beer, and bread?

CASE STUDY REVISITED

Humongous Fungus

Why do *Armillaria* fungi grow so large? Their size is due in part to their ability to form structures called rhizomorphs, which consist of hyphae bundled together inside a protective rind. The rhizomorphs can extend long distances through nutrient-poor areas to reach new sources of food. The *Armillaria* fungus can thus grow beyond the boundaries of a particular food-rich area.

Another factor that may contribute to the gigantic size of the Oregon *Armillaria* is the climate in which it was found. In the dry climate of eastern Oregon, fungal fruiting bodies form only rarely, so the colossal *Armillaria* rarely produces spores. In the absence of spores that might grow into new individuals, the existing individual faces little competition for resources and is free to grow and fill an increasingly large area.

The discovery of the Oregon specimen is merely the latest chapter in a long-running, good-natured "fungus war" that began with the discovery of the first humongous fungus, a 37-acre *Armillaria gallica* growing in Michigan. Since that initial landmark discovery, research groups in Michigan, Oregon, and elsewhere have engaged in a friendly competition to find the largest fungus. Will the current record ever be topped? Stay tuned.

CONSIDER THIS Because the entire Oregon Armillaria grew from a single spore, all of its cells are genetically identical. However, it is unlikely that any substances are transported through the entire 2,400-acre mycelium. And there is no continuous skin or bark or membrane that covers the entire mycelium and separates it from the environment as a unit. Is the fungus's genetic unity sufficient evidence for it to be considered a single individual, or do you need additional evidence? Do you think that the claim of "world's largest organism" is valid?

CHAPTER REVIEW

Go to **Mastering Biology** to access the Pearson eText, vocabulary review, practice quizzes, activities, videos, current events, and more.

*Answers to **Think Critically and Thinking Through the Concepts** questions can be found in the **Answers** section at the back of the book.*

Summary of Key Concepts

23.1 What Are the Key Features of Fungi?

Fungal bodies generally consist of filamentous hyphae, which are either multicellular or multinucleated and form large, intertwined networks called *mycelia.* Fungal nuclei are generally haploid. A cell wall of chitin surrounds fungal cells. All fungi feed by secreting digestive enzymes outside their bodies and absorbing the liberated nutrients.

Fungal reproduction is varied and complex. Asexual reproduction can occur through mitotic production of haploid spores. Sexual reproduction occurs when compatible haploid nuclei fuse to form a diploid zygote, which undergoes meiosis to form haploid spores. Both asexual and sexual spores produce haploid mycelia through mitosis.

23.2 What Are the Major Groups of Fungi?

The major taxonomic groups of fungi are Chytridiomycota (chytrids), Neocallimastigomycota (rumen fungi), Blastocladiomycota (blastoclades), Glomeromycota (glomeromycetes), Basidiomycota (basidiomycetes), and Ascomycota (ascomycetes).

23.3 How Do Fungi Interact with Other Species?

A lichen is a symbiotic association between a fungus and algae or cyanobacteria. The fungal partner provides shelter for the algae or cyanobacteria, which convert sunlight to energy-rich molecules that nourish the fungus. Mycorrhizae are associations between fungi and the roots of most vascular plants. The fungus derives photosynthetic nutrients from the plant roots and, in return, carries water and nutrients into the root from the surrounding soil. Endophytes are fungi that grow inside the leaves or stems of plants and that may help protect the plants that harbor them. Saprophytic fungi are extremely important decomposers in ecosystems. Their filamentous bodies penetrate soil and decaying organic material, liberating nutrients through extracellular digestion.

23.4 How Do Fungi Affect Humans?

Many plant diseases are caused by parasitic fungi. Some parasitic fungi can help control insect crop pests. Others can cause human diseases, including ringworm, athlete's foot, and common vaginal infections. Some fungi produce toxins that can harm humans. Nonetheless, fungi add variety to the human food supply, and fermentation by fungi helps make wine, beer, and bread. Fungi are also the source of many antibiotics.

Thinking Through the Concepts

Bloom's: Remembering, Understanding

Multiple Choice

1. Fungi have cell walls composed primarily of
   a. lignin.
   b. cellulose.
   c. chitin.
   d. glucose.
2. Which of the following diseases is *not* caused by a fungus?
   a. valley fever
   b. Dutch elm disease
   c. histoplasmosis
   d. malaria
3. A symbiotic association of plant roots and fungi is known as a
   a. lichen.
   b. mycorrhiza.
   c. sporangium.
   d. chytrid.
4. The alcohol in beer and wine is a by-product of __________ by__________.
   a. infection; *Candida albicans*
   b. respiration; rumen fungi
   c. fermentation; yeast
   d. decomposition; saprophytes
5. Which is an ecologically important function of fungi?
   a. Photosynthesis
   b. Decomposition
   c. Alcohol production
   d. Antibiotic production

Fill-in-the-Blank

1. The portions of a fungus that are visible to the naked eye are often structures specialized for __________. These structures release tiny __________, which are dispersed to produce new fungi.
2. The fungal body is a(n) __________ and is composed of microscopic threads called __________ that may be subdivided into many cells by __________. The cell walls of fungi are strengthened by __________.
3. In fungi, asexual spores are produced by __________ cell division and have __________ set(s) of chromosomes. Sexual spores are produced by __________ cell division in a(n) __________ and have __________ set(s) of chromosomes.
4. Fill in the blanks in the following sentences with the common names of fungal taxonomic groups. Almost all __________ live in intimate association with plant roots. Flagellated, swimming spores are produced by __________. Mushrooms and puffballs are reproductive structures of __________.
5. __________ are symbiotic associations of fungi and green algae. __________ are symbiotic, mutually beneficial associations of fungi and plant roots. Some fungi are __________ that live inside the aboveground tissues of plants.
6. Fungi are the only decomposers that can digest __________. The fungi that cause breads to rise and wines to ferment are __________. Human ailments caused by fungi include __________ and __________.

Review Questions

1. Describe the structure of the fungal body. How do fungal cells differ from most plant and animal cells?
2. What portion of the fungal body is represented by mushrooms, puffballs, and similar structures? Why are these structures elevated above the ground?
3. What two plant diseases, caused by parasitic fungi, have had an enormous impact on forests in the United States? In which taxonomic group are these fungi found?
4. List some fungi that attack crops. To which taxonomic group do they belong?
5. Describe asexual reproduction in fungi.
6. List the major taxonomic groups of fungi, describe some key features of each group, and give an example of a fungus in each group.
7. Describe how a fairy ring of mushrooms is produced. Why is its diameter related to its age?
8. Describe two symbiotic relationships between a fungus and another organism. In each case, explain how each partner in these associations is affected.

Applying the Concepts

Bloom's: Applying, Analyzing, Evaluating

1. Dutch elm disease in the United States is caused by an *exotic*—that is, an organism (in this case, a fungus) introduced from another part of the world. What damage has this introduction done? What other fungal pests fall into this category? Why are parasitic fungi particularly likely to be transported out of their natural habitat? What can governments do to limit this importation?
2. What ecological consequences would occur if humans, using a new and deadly fungicide, destroyed all fungi on Earth?

24 Animal Diversity I: Invertebrates

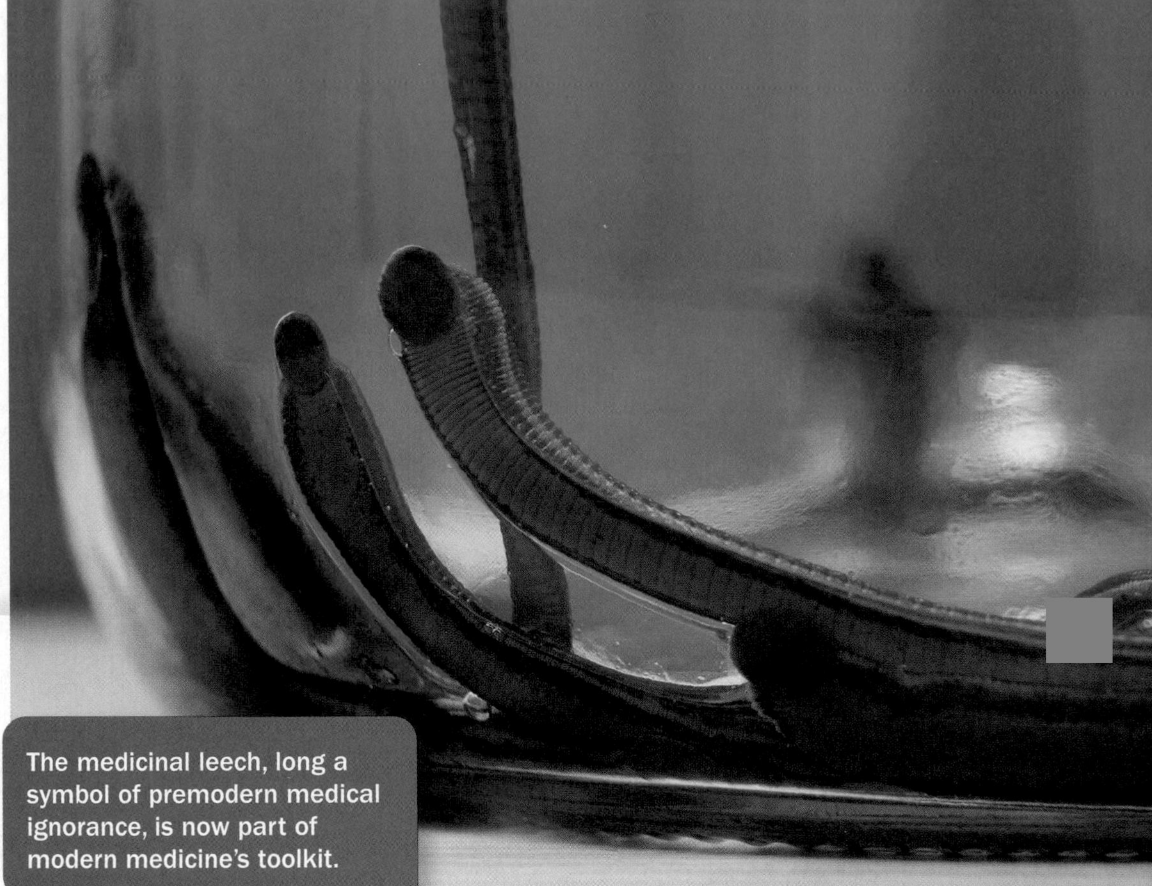

The medicinal leech, long a symbol of premodern medical ignorance, is now part of modern medicine's toolkit.

CASE STUDY

Physicians' Assistants

MODERN MEDICINE is a high-tech enterprise, dependent on multimillion-dollar machines, sophisticated medical devices, and drugs developed via cutting-edge chemistry and biotechnology. But despite the prevalence of advanced technology in medicine, physicians also get low-tech assistance from an unlikely source: *invertebrates* (animals without a backbone). Consider, for example, the medicinal leech. For more than 2,000 years, healers have enlisted these parasitic segmented worms for treatment of a wide range of illnesses. For much of human medical history, treatment with leeches was based on the hope that the creatures would suck out the "tainted" blood that was believed to be the primary cause of disease. As the actual causes of disease were discovered, however, medical use of leeches declined. By the beginning of the twentieth century, leeches no longer had a place in the toolkit of modern medicine and had become a symbol of the ignorance of an earlier age. Today, however, medicinal leeches are making a surprising comeback.

Currently, leeches are used to treat a surgical complication known as venous insufficiency. This complication is especially common in reconstructive surgery, such as surgery to reattach a severed finger or repair a disfigured face. In such cases, surgeons are often unable to reconnect all of the veins that would normally carry blood away from tissues. Eventually, new veins will grow, but in the meantime blood may accumulate in the repaired tissue. Unless the excess blood is removed, it will coagulate, causing clots that can deprive the tissue of the oxygen and nutrients it needs to live. Fortunately, leeches can help. Applied to the affected area, the leeches get right to work, making small, painless incisions and sucking blood into their stomachs. To aid them in their blood-removal task, the leeches' saliva contains a mixture of chemicals that dilate blood vessels and prevent blood from clotting. Although the chemical brew in the saliva is an adaptation that helps leeches consume blood more efficiently, it also helps the patient by promoting blood flow in the damaged tissue. In this way, leeches provide a painless, effective treatment for venous insufficiency.

Although relatively few invertebrate animals have medical uses, invertebrates account for a large share of Earth's known biodiversity. What have scientists learned about these diverse and abundant organisms?

AT A GLANCE

24.1 WHAT ARE THE KEY FEATURES OF ANIMALS?

It is difficult to devise a concise definition of the term "animal." No single feature uniquely defines animals, so the group is defined by a list of characteristics. None of these characteristics is unique to animals, but together they distinguish animals from members of other taxonomic groups:

- Animals are eukaryotes.
- Animals are multicellular.
- Animal cells lack a cell wall.
- Animals obtain their energy by consuming other organisms.
- Animals typically reproduce sexually.
- Animals are motile (able to move about) during some stage of their lives.
- Most animals are able to respond rapidly to external stimuli.

CHECK YOUR LEARNING

Can you . . .

- list the characteristics that collectively distinguish animals from other kinds of organisms?

24.2 WHICH ANATOMICAL FEATURES MARK BRANCH POINTS ON THE ANIMAL EVOLUTIONARY TREE?

By the Cambrian period, which began 541 million years ago, most of the animal phyla that currently populate Earth were already present. Unfortunately, the Precambrian fossil record is very sparse and does not reveal much about the early evolutionary history of animals. Therefore, systematists have looked to anatomy, embryological development, and DNA sequences for clues about animal history. These investigations have shown that certain features mark major branching points on the animal evolutionary tree and represent milestones in the evolution of the different body plans of modern animals (**FIG. 24-1**). In the following sections, we will explore these evolutionary milestones and their legacies in the bodies of modern animals.

Lack of Tissues Separates Sponges from All Other Animals

One of the earliest major innovations in animal evolution was the appearance of **tissues**—groups of similar cells integrated into a functional unit, such as muscle tissue or nerve tissue. Today, the bodies of almost all animals include tissues; the only animals that have retained the ancestral lack of tissues are the sponges. In sponges, individual cells often have specialized functions, but they act more or less independently and are not organized into tissues. This unique feature of sponges suggests that the split between sponges and the evolutionary branch leading to all other animal phyla must have occurred very early in the history of animals.

Animals with Tissues Exhibit Either Radial or Bilateral Symmetry

The first appearance of tissues coincided with the first appearance of body symmetry; all animals with true tissues also have symmetrical bodies. An animal is said to be symmetrical if it can be bisected along at least one plane such that the resulting halves are mirror images of one another.

The symmetrical, tissue-bearing animals can be divided into two groups, one containing animals that exhibit **radial symmetry** (**FIG. 24-2a**) and one with animals that exhibit **bilateral symmetry** (**FIG. 24-2b**). In radial symmetry, any plane through a central axis divides the object into roughly equal halves. In contrast, a bilaterally symmetrical animal can be divided into roughly mirror-image halves only along one particular plane through the central axis.

The difference between radially and bilaterally symmetrical animals reflects another major branching point in the animal evolutionary tree. This split separated the ancestors of the radially symmetrical cnidarians (sea jellies, corals, and anemones) and ctenophores (comb jellies) from the ancestors of the remaining animal phyla, all of which are bilaterally symmetrical.

Radially Symmetrical Animals Have Two Embryonic Tissue Layers; Bilaterally Symmetrical Animals Have Three

The distinction between radial and bilateral symmetry in animals is closely tied to a corresponding difference in the number of tissue layers, called *germ layers*, that arise during embryonic development. Embryos of animals with radial symmetry have two germ layers: an inner layer of **endoderm** (which gives rise to the tissues that line the gut cavity) and an outer layer of **ectoderm** (which gives rise to the tissues that cover the outside of the body.) Embryos of bilaterally symmetrical animals add a third germ layer, **mesoderm**, which lies between the endoderm and the ectoderm.

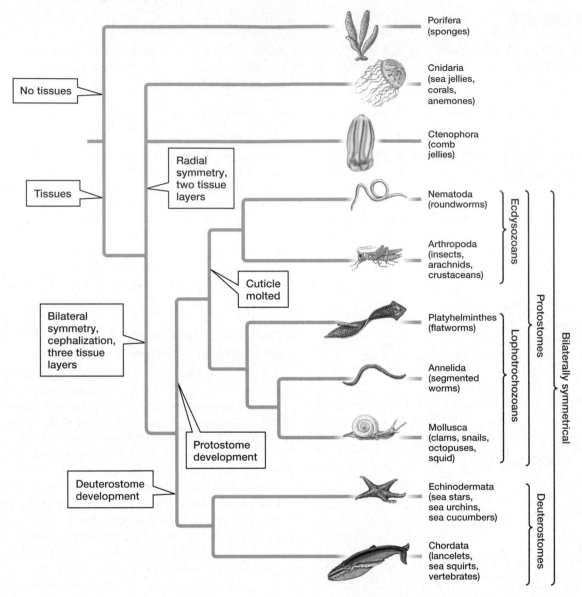

No tissues
Tissues
Radial symmetry, two tissue layers
Bilateral symmetry, cephalization, three tissue layers
Cuticle molted
Protostome development
Deuterostome development

Porifera (sponges)
Cnidaria (sea jellies, corals, anemones)
Ctenophora (comb jellies)
Nematoda (roundworms)
Arthropoda (insects, arachnids, crustaceans)
Platyhelminthes (flatworms)
Annelida (segmented worms)
Mollusca (clams, snails, octopuses, squid)
Echinodermata (sea stars, sea urchins, sea cucumbers)
Chordata (lancelets, sea squirts, vertebrates)

Ecdysozoans
Lophotrochozoans
Protostomes
Deuterostomes
Bilaterally symmetrical

▲ **FIGURE 24-1 An evolutionary tree of some major animal phyla**

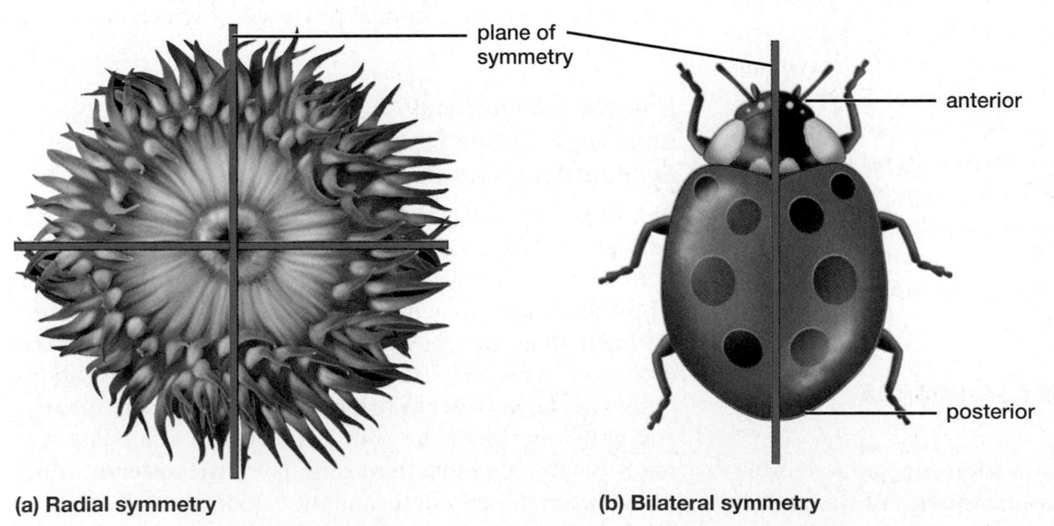

◀ **FIGURE 24-2 Body symmetry and cephalization** **(a)** Animals with radial symmetry, such as this sea anemone, lack a well-defined head. Any plane that passes through the central axis divides the body into mirror-image halves. **(b)** Animals with bilateral symmetry, such as this beetle, have an anterior head end and a posterior tail end. The body can be split into two mirror-image halves only along a particular plane that runs down the midline.

In bilaterally symmetrical animals, endoderm differentiates to form the tissues that line respiratory surfaces, the gut, and most hollow organs. (An **organ** is a discrete structure, such as a heart or stomach, in which two or more tissue types work together to perform functions.) Ectoderm forms nerve tissue and the tissues on the outer surface of the body. Mesoderm forms muscle and, when present, the circulatory and skeletal systems.

The connection between symmetry type and number of germ layers helps us make sense of the potentially puzzling case of the echinoderms (sea stars, sea urchins, and sea cucumbers). Adult echinoderms are radially symmetrical, yet our evolutionary tree places them squarely within the bilaterally symmetrical group. Why? Echinoderms have three germ layers, as well as several other characteristics (some described later) that unite them with the bilaterally symmetrical animals. So the immediate ancestors of echinoderms must have been bilaterally symmetrical, and the group subsequently evolved radial symmetry (a case of convergent evolution). To this day, larval echinoderms retain bilateral symmetry.

Bilaterally Symmetrical Animals Have Heads

Radially symmetrical animals tend either to be *sessile* (fixed to one spot, like sea anemones) or to drift around on currents (like sea jellies). Because such animals do not actively propel themselves in a particular direction, all parts of their bodies are more or less equally likely to encounter food. In contrast, most bilaterally symmetrical animals are *motile* (move under their own power), and resources such as food are most likely to be encountered by the part of the animal that is closest to the direction of movement. The evolution of bilateral symmetry was therefore accompanied by **cephalization**, the concentration of sensory organs and a brain in a defined head region. Cephalization produces an *anterior* (head) end, where sensory cells, sensory organs, clusters of nerve cells, and organs for ingesting food are concentrated. The other end of a cephalized animal is designated *posterior* (see Fig. 24-2b) and may feature a tail.

Most Bilateral Animals Have Body Cavities

The members of many bilateral animal phyla have a fluid-filled cavity between the digestive tube (or gut, where food is digested and absorbed) and the outer body wall. In an animal with a body cavity, the gut and body wall are separated by a fluid-filled space, creating a "tube-within-a-tube" body plan. No radially symmetrical animal has a body cavity, so it is likely that this feature arose sometime after the split between radially and bilaterally symmetrical animals.

A fluid-filled body cavity can serve a variety of functions. In many invertebrate animals it acts as a kind of skeleton, providing support for the body and a framework against which muscles can act. In other animals, internal organs are suspended within the fluid-filled cavity, which serves as a protective buffer between the organs and the outside world.

Body Cavity Structure Varies Among Phyla

The most widespread type of body cavity is a **coelom**, a fluid-filled cavity that is completely lined with a thin layer of tissue that develops from mesoderm (**FIG. 24-3a**). Phyla whose members have a coelom are called *coelomates*. The annelids (segmented worms), arthropods (insects, spiders, crustaceans), mollusks (clams and snails), echinoderms, and chordates (which include humans) are coelomates.

Some animals have a body cavity that is *not* completely surrounded by mesoderm-derived tissue. This type of cavity is known as a **pseudocoelom**. Phyla whose members have a pseudocoelom are collectively known as *pseudocoelomates* (**FIG. 24-3b**). The roundworms (nematodes) are the largest pseudocoelomate group.

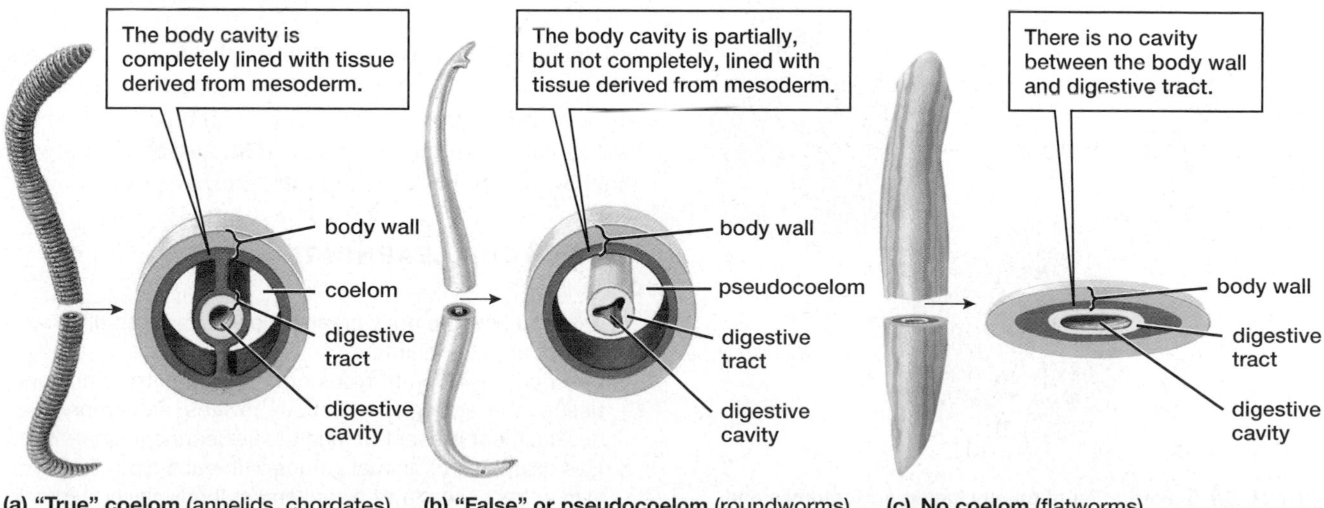
The body cavity is completely lined with tissue derived from mesoderm.
body wall
coelom
digestive tract
digestive cavity
The body cavity is partially, but not completely, lined with tissue derived from mesoderm.
body wall
pseudocoelom
digestive tract
digestive cavity
There is no cavity between the body wall and digestive tract.
body wall
digestive tract
digestive cavity
(a) "True" coelom (annelids, chordates)
(b) "False" or pseudocoelom (roundworms)
(c) No coelom (flatworms)

▲ **FIGURE 24-3 Body cavities (a)** Annelids have a true coelom. **(b)** Roundworms are pseudocoelomates. **(c)** Flatworms have no cavity between the body wall and digestive tract. (Tissues shown in blue are derived from ectoderm, those in red from mesoderm, and those in yellow from endoderm.)

Some phyla of bilateral animals have no body cavity and are known as *acoelomates*. For example, flatworms have no cavity between their gut and body wall; instead, the space is filled with solid tissue (**FIG. 24-3c**).

Bilateral Organisms Develop in One of Two Ways

Among the bilateral animal phyla, embryological development follows a variety of pathways. These diverse developmental pathways, however, can be grouped into two categories, known as **protostome** development and **deuterostome** development (**FIG. 24-4**). In protostome development, the coelom (when present) forms within the space between the body wall and the digestive cavity. In deuterostome development, the coelom forms as an outgrowth of the digestive cavity. The two types of development also differ in the pattern of cell division immediately after fertilization and in the method by which the mouth and anus are formed. Protostomes and deuterostomes represent distinct evolutionary branches within the bilateral animals. Annelids, arthropods, flatworms, roundworms, and mollusks exhibit protostome development; echinoderms and chordates are deuterostomes.

Protostomes Include Two Distinct Evolutionary Lines

The protostome animal phyla fall into two groups, which correspond to two different lineages that diverged early in the evolutionary history of protostomes. One group, the *ecdysozoans*, includes phyla such as the arthropods and roundworms, whose members have bodies covered by an outer layer that is periodically shed. The other group is known as the *lophotrochozoans* and includes phyla whose members have a special feeding structure called a lophophore (**FIG. 24-5a**), as well as phyla whose members pass through a particular type of developmental stage called a trochophore larva (**FIG. 24-5b**). The flatworms, annelids, and mollusks are lophotrochozoan phyla.

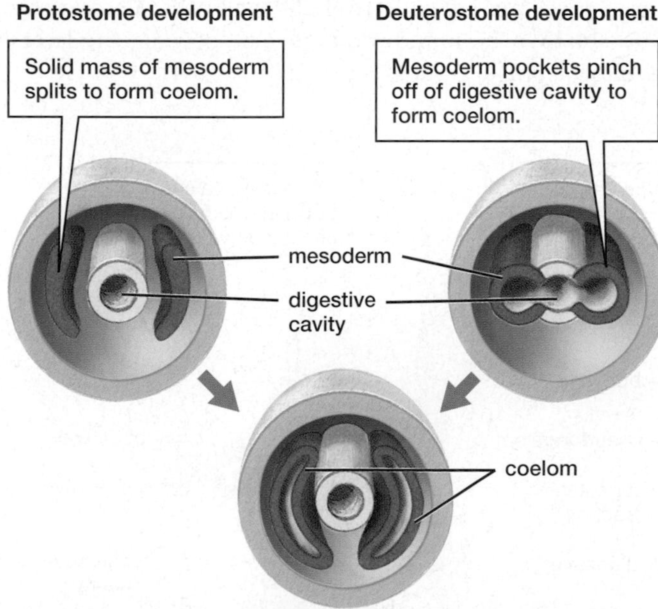
Protostome development
Deuterostome development
Solid mass of mesoderm splits to form coelom.
Mesoderm pockets pinch off of digestive cavity to form coelom.
mesoderm
digestive cavity
coelom

▲ **FIGURE 24-4 Formation of the coelom in protostomes and deuterostomes** The difference depicted here is one of several between protostome and deuterostome embryonic development. (Tissues shown in blue are derived from ectoderm, those in red from mesoderm, and those in yellow from endoderm.)

(a) Lophophore

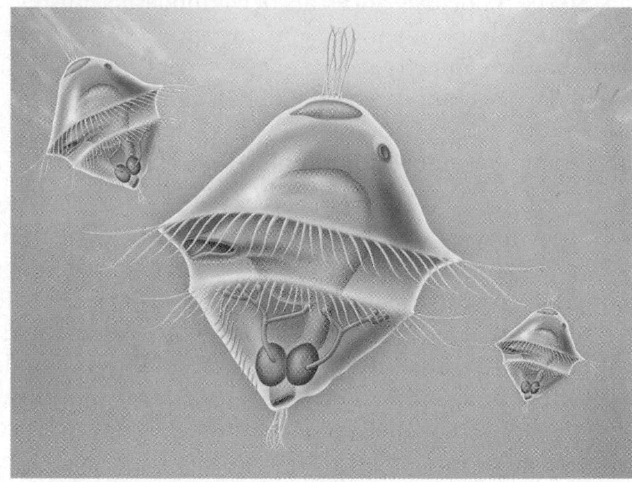

(b) Trochophore larva

▲ **FIGURE 24-5 Lophotrochozoan characteristics** Members of the lophotrochozoan phyla, which include flatworms, segmented worms, and mollusks, exhibit either **(a)** a feeding structure known as a lophophore or **(b)** a distinctive swimming larval form called a trochophore.

CHECK YOUR LEARNING

Can you . . .

- describe how the body organization of sponges differs from that of all other animals?
- describe the different types of body symmetry, embryonic tissue layer arrangements, body cavities, and embryonic development present among tissue-bearing animals?
- give examples of animal groups with each type of body symmetry, tissue layer arrangement, body cavity, and embryonic development?
- name and describe the two main lineages within the protostome animals?

24.3 WHAT ARE THE MAJOR ANIMAL PHYLA?

For convenience, biologists often place animals in one of two major categories: **vertebrates**, those with a backbone (or vertebral column), and **invertebrates**, those lacking a backbone. The vertebrates—fish, amphibians, reptiles, and mammals (see Chapter 25)—are perhaps the most conspicuous animals from a human point of view, but less than 3% of all known animal species are vertebrates. The vast majority of animals are invertebrates. Biologists recognize about 32 phyla of animals; some key invertebrate phyla are summarized in **TABLE 24-1**.

Sponges Are Simple, Sessile Animals

Sponges (Porifera) are found in most aquatic environments. Most of Earth's 8,800 known sponge species inhabit ocean waters, but some live in freshwater habitats such as lakes and rivers. Adult sponges live attached to rocks or other underwater surfaces. They are generally sessile, though researchers have demonstrated that some species are able to move about (very slowly—a few millimeters per day). Sponges come in a variety of shapes and sizes. Some species have a well-defined shape, but others grow free-form over underwater rocks (**FIG. 24-6**). The largest sponges can grow to more than 3 feet (1 meter) in height.

Most sponges are **hermaphroditic**—they possess both male and female sexual organs. Sponges usually reproduce sexually, a process that begins when sperm are released into the water. Sperm may enter the body of another sponge and be transported to eggs, which are retained in the sponge's body. Fertilized eggs develop inside the adult into active larvae that escape through the openings in the sponge body. Water currents disperse the larvae to new areas, where they settle and develop into adult sponges.

Sponges Lack Tissues

Sponge bodies do not contain tissues; in some ways, a sponge resembles a colony of single-celled organisms. The colony-like properties of sponges were revealed in an experiment performed by embryologist H. V. Wilson in 1907. Wilson mashed a sponge through a piece of silk, thereby breaking it apart into single cells and cell clusters. He then placed these tiny bits of sponge into seawater and waited for 3 weeks. By the end of the experiment, the cells had reaggregated into a functional sponge, demonstrating that individual sponge cells had been able to survive and move about independently.

A sponge's body is perforated by numerous tiny pores, through which water enters, and by fewer, large openings, through which water is expelled (**FIG. 24-7**). Within the sponge, water travels through canals. As water passes through the sponge, oxygen is extracted, microorganisms are filtered out and taken into individual cells where they are digested, and wastes are released.

Sponge Cells Are Specialized for Different Functions

Sponges have three major cell types, each with a specialized role (see Fig. 24-7). Flattened *epithelial cells* cover the animal's

(a) Encrusting sponge

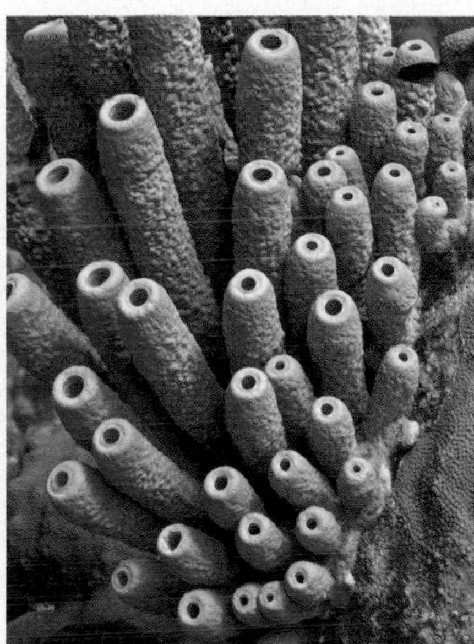

(b) Tubular sponge

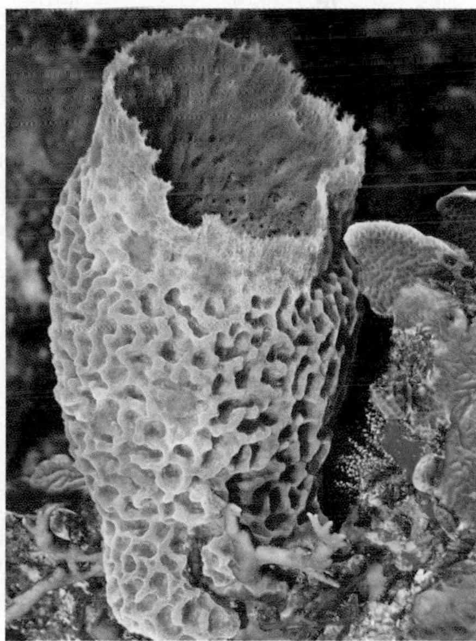

(c) Vase-shaped sponge

▲ **FIGURE 24-6 The diversity of sponges** Sponges come in a wide variety of sizes, shapes, and colors. Some, such as **(a)** this encrusting sponge, grow in a free-form pattern over undersea rocks. Others may be **(b)** tubular or **(c)** vase shaped.

THINK CRITICALLY Sponges are often described as the most "primitive" of animals. How can such a primitive organism have become so diverse and abundant?

TABLE 24-1 Comparison of the Major Animal Phyla

Common Name (Phylum)		Sponges (Porifera)	Sea Jellies, Corals, Anemones (Cnidaria)	Flatworms (Platyhelminthes)
Body plan	**Level of organization**	Cellular—lack tissues and organs	Tissue—lack organs	Organ system
	Germ layers	Absent	Two	Three
	Symmetry	Absent	Radial	Bilateral
	Cephalization	Absent	Absent	Present
	Body cavity	Absent	Absent	Absent
	Segmentation	Absent	Absent	Absent
Internal systems	**Digestive system**	Intracellular	Gastrovascular cavity; some intracellular	Gastrovascular cavity
	Circulatory system	Absent	Absent	Absent
	Respiratory system	Absent	Absent	Absent
	Excretory system (fluid regulation)	Absent	Absent	Canals with ciliated cells
	Nervous system	Absent	Nerve net	Head ganglia with longitudinal nerve cords
	Reproduction	Sexual; asexual (budding)	Sexual; asexual (budding)	Sexual (some hermaphroditic); asexual (body splits)
	Support	Endoskeleton of spicules	Hydrostatic skeleton	Hydrostatic skeleton
	Number of known species	8,800	10,000	10,000

outer body surfaces. Some epithelial cells are modified into *pore cells*, which surround the pores in the body wall, controlling their size and thereby regulating the entry of water. *Collar cells* bear flagella that extend into the inner cavity. The beating flagella maintain a flow of water through the sponge. The collars that surround the flagella act as fine sieves, filtering out microorganisms that are then ingested by the cell. Some of the food is passed to the *amoeboid cells*. These cells roam freely between the epithelial and collar cells, digesting and distributing nutrients, producing reproductive cells, and secreting small skeletal projections called *spicules*. Spicules, which may be composed of calcium carbonate (chalk), silica (glass), or protein, form an internal skeleton that provides support for the sponge's body (see Fig. 24-7).

Cnidarians Are Well-Armed Predators

The *cnidarians* (Cnidaria) include sea jellies (also known as jellyfish), corals, sea anemones, and hydrozoans (**FIG. 24-8**). The roughly 10,000 known species of cnidarians are confined to watery habitats; most are marine. Most species are small, from a few millimeters to a few inches in diameter, but the largest sea jellies can be 8 feet across and have tentacles 150 feet long. All cnidarians are predators.

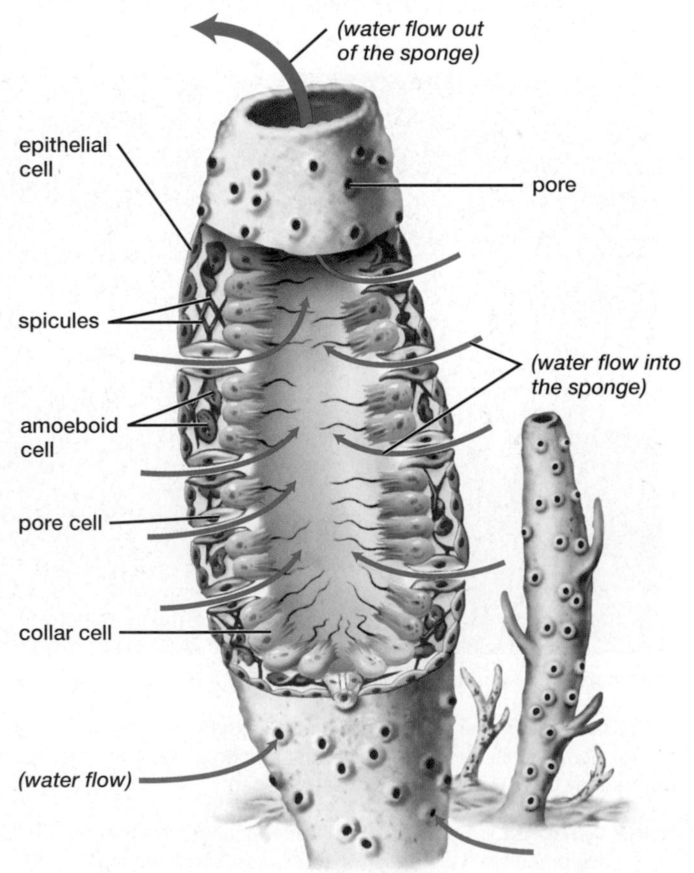

FIGURE 24-7 The body plan of sponges Water enters through numerous tiny pores in the sponge body and exits through a larger opening. Microscopic food particles are filtered from the water.

Segmented Worms (Annelida)	Clams, Snails, Octopuses, Squid (Mollusca)	Insects, Arachnids, Crustaceans (Arthropoda)	Roundworms (Nematoda)	Sea Stars, Sea Urchins, Sea Cucumbers (Echinodermata)
Organ system	Organ system	Organ system	Organ system	Organ system
Three	Three	Three	Three	Three
Bilateral	Bilateral	Bilateral	Bilateral	Bilateral larvae, radial adults
Present	Present	Present	Present	Absent
Coelom	Coelom	Coelom	Pseudocoel	Coelom
Present	Absent	Present	Absent	Absent
Separate mouth and anus	Separate mouth and anus	Separate mouth and anus	Separate mouth and anus	Separate mouth and anus (usually)
Closed	Open	Open	Absent	Absent
Absent	Gills, lungs	Tracheae, gills, or lungs	Absent	Tube feet, skin gills
Nephridia	Nephridia	Excretory glands resembling nephridia	Excretory gland	Absent
Head ganglia with paired ventral cords; ganglia in each segment	Well-developed brain in some cephalopods; several paired ganglia, most in the head; nerve network in the body wall	Head ganglia with paired ventral nerve cords; ganglia in segments, some fused	Head ganglia with dorsal and ventral nerve cords	Head ganglia absent; nerve ring and radial nerves; nerve network in the skin
Sexual (some hermaphroditic)	Sexual (some hermaphroditic)	Usually sexual	Sexual (some hermaphroditic)	Sexual (some hermaphroditic); asexual by regeneration (rare)
Hydrostatic skeleton	Hydrostatic skeleton	Exoskeleton	Hydrostatic skeleton	Endoskeleton of plates beneath outer skin
13,000	50,000	1,000,000	12,000	6,800

(a) Anemone

(b) Sea jelly

(c) Coral

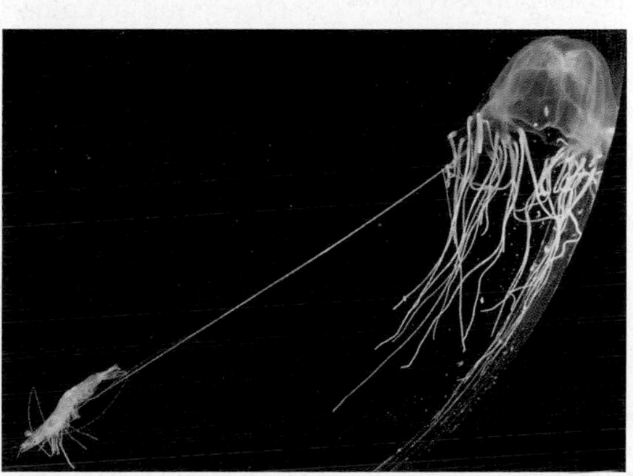

(d) Sea wasp

FIGURE 24-8 Cnidarian diversity **(a)** A red-spotted anemone spreads its tentacles to capture prey. **(b)** A sea jelly drifts in the ocean, its tentacles hanging down. **(c)** A close-up of coral reveals the extended tentacles of the polyps. **(d)** The sea wasp is a type of sea jelly whose stinging cells contain one of the most toxic of all known venoms. The venom quickly kills prey, such as the shrimp shown here, that brush against a sea wasp's tentacles.

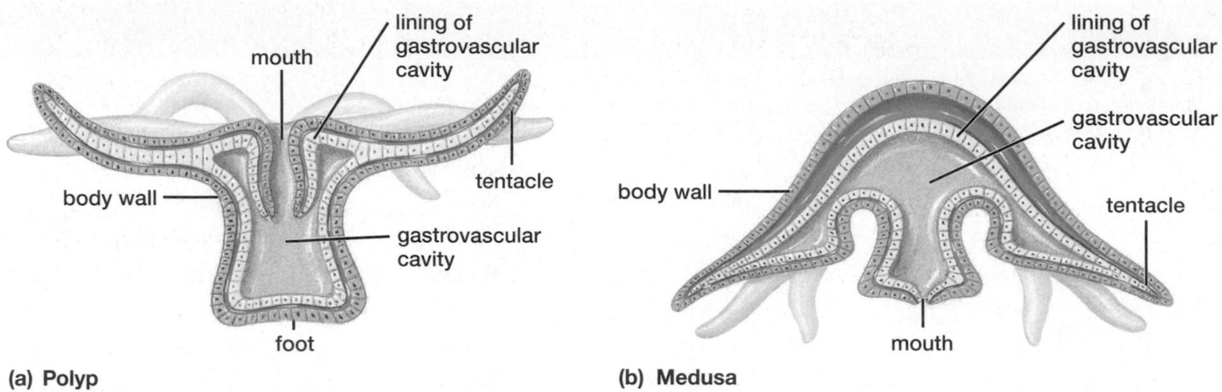

▲ **FIGURE 24-9 Polyp and medusa (a)** The polyp form is seen in sea anemones and the individual polyps within a coral. **(b)** The medusa form, seen in the sea jelly, resembles an inverted polyp. (Tissues shown in blue are derived from ectoderm, those in yellow from endoderm.)

Cnidarians Have Tissues and Two Body Types

The cells of cnidarians are organized into distinct tissues, including contractile tissue that acts like muscle. Cnidarian nerve cells are organized into tissue called a *nerve net,* which branches through the body and stimulates the contractile tissue, allowing the body to move and capture prey. However, most cnidarians lack organs and have no brain.

Cnidarians come in a variety of forms, all of which are variations on two basic body plans: the *polyp* (**FIG. 24-9a**) and the *medusa* (**FIG. 24-9b**). The tubular polyp is adapted to a life spent quietly attached to rocks; it has tentacles that reach upward to grasp and immobilize prey. The medusa floats in the water and is carried by currents, its bell-shaped body trailing tentacles like multiple fishing lines.

Many cnidarian life cycles include both polyp and medusa stages, though some species live only as polyps and others only as medusae. Both polyps and medusae develop from just two germ layers—the interior endoderm and the exterior ectoderm; between those layers is a jelly-like substance. Polyps and medusae are radially symmetrical, with body parts arranged in a circle around the mouth and digestive cavity (see Fig. 24-2a).

Reproduction varies considerably among different types of cnidarians, but one pattern is fairly common in species with both polyp and medusa stages. In such species, polyps typically reproduce by asexual **budding** in which the polyp produces miniature versions of itself that drop off and assume an independent existence. Under certain conditions, however, budding may instead give rise to medusae. After a medusa grows to maturity, it may release gametes (sperm or eggs) into the water. If a sperm and egg meet, they may fuse to form a zygote that develops into a free-swimming, ciliated larva. The larva may eventually settle on a hard surface, where it develops into a polyp.

Cnidarians Have Stinging Cells

Cnidarian tentacles are armed with *cnidocytes,* cells containing structures that, when touched, explosively inject poisonous or sticky filaments into prey (**FIG. 24-10**). These stinging cells, found only in cnidarians, are used to capture prey. Cnidarians do not actively hunt. Instead, they wait for their victims to blunder, by chance, into the grasp of their enveloping tentacles. Stung and firmly grasped, the prey is forced through an expandable mouth into a digestive sac, the *gastrovascular cavity* (see Fig. 24-9). Digestive enzymes secreted into this cavity break down some of the food, and further digestion occurs within the cells lining the cavity. Because the gastrovascular cavity has only a single opening, undigested material is expelled through the mouth when digestion is completed. Although this two-way traffic prevents continuous feeding, it is adequate to support the low energy demands of these animals.

The venom of some cnidarians can cause painful stings in humans; the stings of a few sea jelly species can even be life threatening. The most deadly of these species is the sea wasp, *Chironex fleckeri* (see Fig. 24-8d), which is found in the waters off northern Australia and Southeast Asia and can grow to about 12 inches (30 centimeters) in diameter. The amount of venom in a single sea wasp could kill up to 60 people, and the victim of a serious sting may die within minutes.

Many Corals Secrete Hard Skeletons

One group of cnidarians, the corals, is of particular ecological importance (see Fig. 24-8c). In many coral species, polyps form colonies, and each member of the colony secretes a hard skeleton of calcium carbonate. The skeletons persist long after the organisms die, serving as a base to which other individuals attach themselves. The cycle continues until, after thousands of years, massive coral reefs are formed. Many of these reefs are threatened by climate change, as we describe in "Earth Watch: When Reefs Get Too Warm" on page 396.

Coral reefs are found in both cold and warm oceans. Cold-water reefs form in deep waters and, though widely distributed, have only recently attracted the attention of researchers and are not yet well studied. The more familiar

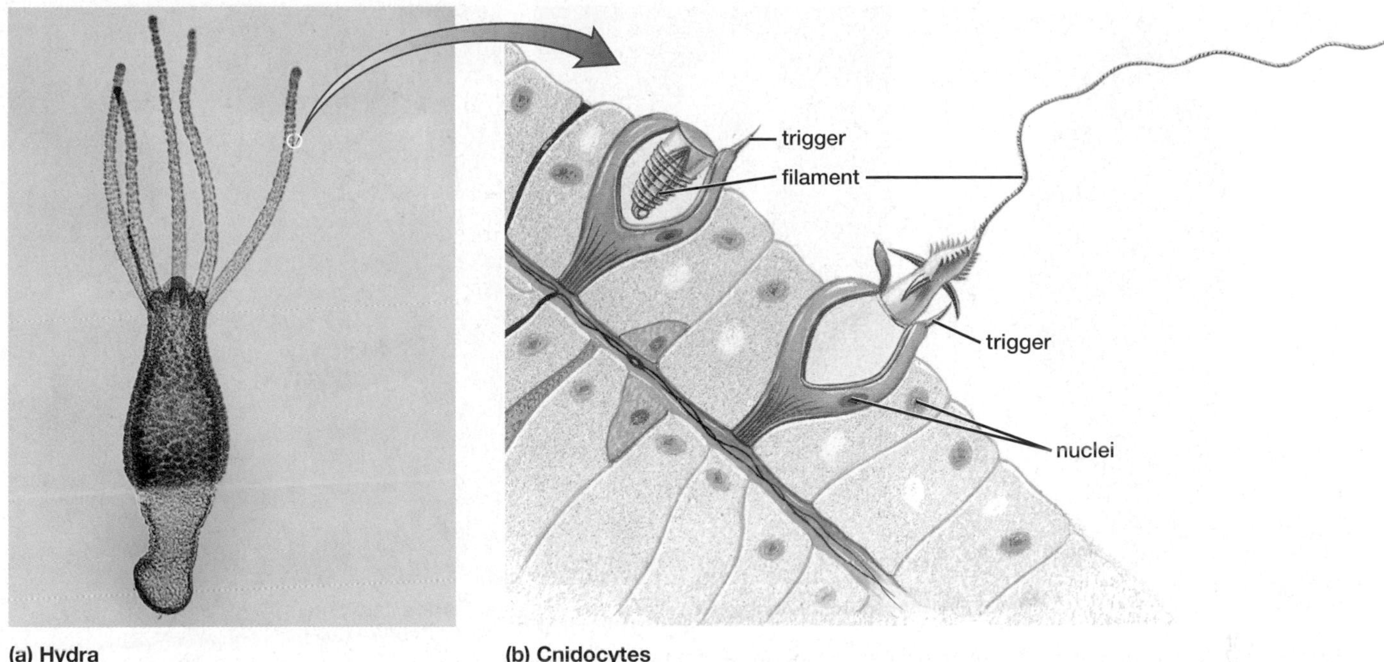

▲ **FIGURE 24-10 Cnidarian weaponry: the cnidocyte (a)** Cnidarian tentacles, such as those of the hydra, contain cnidocyte cells. **(b)** At the slightest touch to its trigger, a structure within a cnidocyte cell violently expels a poisoned filament, which may penetrate prey.

warm-water coral reefs are restricted to clear, shallow waters in the tropics. Here, coral reefs form undersea habitats that are the basis of an ecosystem of stunning diversity and unparalleled beauty.

Comb Jellies Use Cilia to Move

The roughly 150 known species of radially symmetrical *comb jellies* (Ctenophora) are superficially similar in appearance to some cnidarians, but form a distinct evolutionary lineage. Most comb jellies are less than an inch (2.5 cm) in diameter, but a few species can grow to more than 3 feet (1 m) across. Comb jellies move by means of cilia, which are arranged in eight rows known as combs. Although most comb jellies are colorless and transparent or translucent, light scattered by the beating cilia of the combs can appear as eight ever-changing rainbows of color on the comb jelly's body (**FIG. 24-11**).

All comb jellies are carnivores. Most species inhabit coastal or oceanic waters and eat tiny invertebrate animals (including, in some cases, other, smaller comb jellies), which are captured with sticky tentacles. Almost all comb jellies are hermaphroditic. Each individual releases both sperm and eggs to the surrounding water. Fertilized eggs float freely in the water until they hatch into larvae that gradually develop into adult comb jellies.

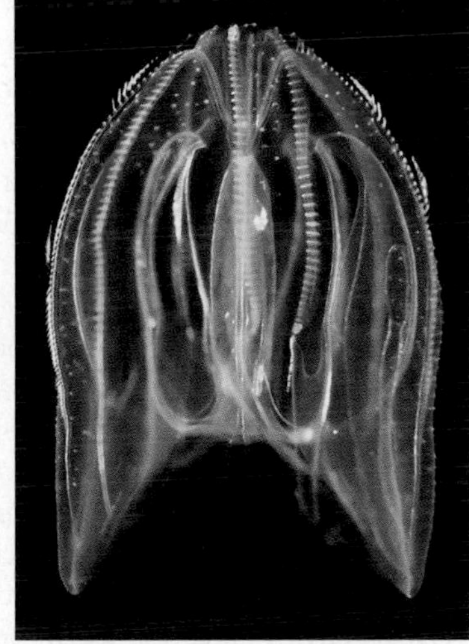

▲ **FIGURE 24-11 A comb jelly** This comb jelly's body is unpigmented, but rows of cilia refract light to produce iridescent colors.

Flatworms May Be Parasitic or Free Living

The *flatworms* (Platyhelminthes) are aptly named; they have a flat, ribbon-like shape. They are bilaterally symmetrical (see Fig. 24-2b). Many of the approximately 10,000 described flatworm species are parasites (**FIG. 24-12a**). (**Parasites** are organisms that live in or on the body of another organism, called a *host*, which is harmed as a result of the relationship.) Nonparasitic, free-living flatworms inhabit freshwater, marine, and moist terrestrial habitats. They tend to be small and inconspicuous (**FIG. 24-12b**), but some are brightly colored, spectacularly patterned residents of tropical coral reefs (**FIG. 24-12c**).

All flatworm species can reproduce sexually; most are hermaphroditic. This trait allows a flatworm to reproduce through self-fertilization, a great advantage to a parasitic worm that may be the only one of its kind present in its host. Some flatworms can also reproduce asexually. For example, free-living species may reproduce by cinching themselves around the middle until they separate into two halves, each of which regenerates its missing parts.

Flatworms Have Organs but Lack Respiratory and Circulatory Systems

Unlike cnidarians, flatworms have organs. For example, most free-living flatworms have sense organs, including eyespots (see Fig. 24-12b) that detect light and dark, and cells that respond to

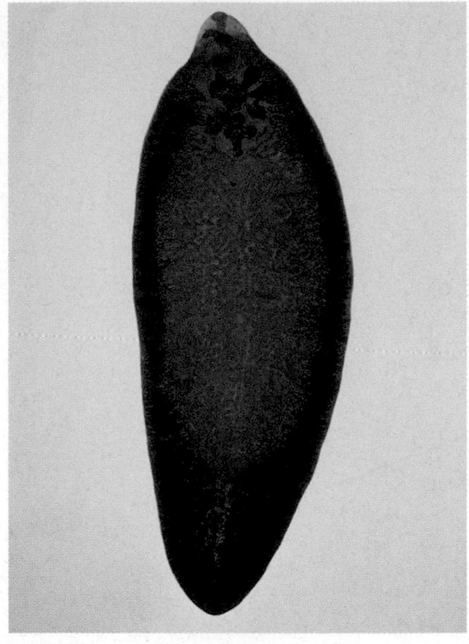

(a) Fluke

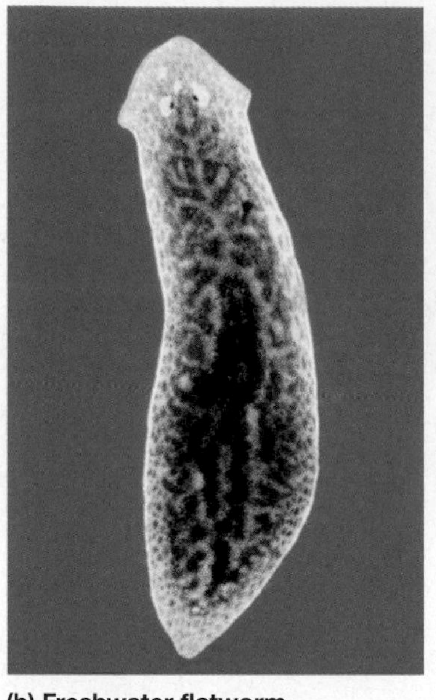

(b) Freshwater flatworm

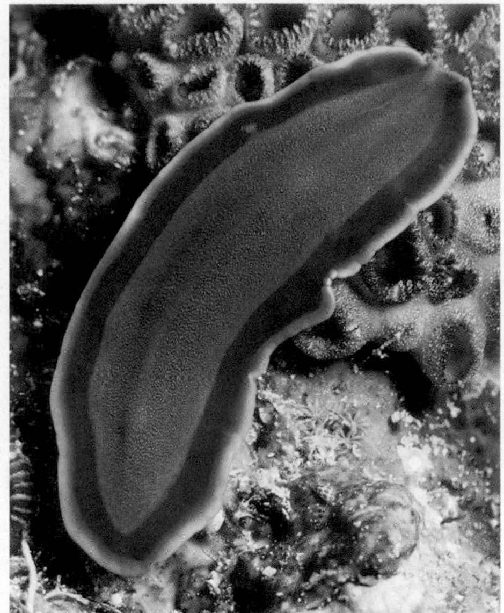

(c) Marine flatworm

▲ **FIGURE 24-12 Flatworm diversity (a)** This fluke is an example of a parasitic flatworm. **(b)** Eyespots are clearly visible in the head of this free-living, freshwater flatworm. **(c)** Many of the flatworms that inhabit tropical coral reefs are brightly colored.

chemical and tactile stimuli. To process information, a flatworm has clusters of nerve cells called **ganglia** (singular, ganglion) in its head, forming a simple brain. Paired neural structures called **nerve cords** extend down the body and conduct nervous signals to and from the ganglia.

Flatworms lack respiratory and circulatory systems. Oxygen and carbon dioxide are exchanged by direct diffusion between body cells and the environment. This mode of respiration is possible because the small size and flat shape of flatworm bodies ensure that no body cell is very far from the surrounding environment. In the absence of a circulatory system, nutrients must move directly from the digestive tract to body cells. The digestive cavity has a branching structure that reaches all parts of the body and allows digested nutrients to diffuse into nearby cells. As in cnidarians, the digestive cavity has only one opening to the environment, so wastes pass out through the mouth.

Some Flatworms Are Harmful to Humans

Some parasitic flatworms can infect humans. For example, tapeworms can infect people who eat undercooked beef, pork, or fish that has been infected by the worms. Tapeworm larvae form encapsulated resting structures, called *cysts*, in the muscles of the cows, pigs, or fish. The cysts hatch in the human digestive tract, where the young tapeworms attach themselves to the lining of the intestine. There they may grow to a length of more than 20 feet (7 meters), absorbing digested nutrients directly through their outer surface and eventually releasing packets of eggs that are shed in the host's feces. If a pig, cow, or fish eats food contaminated with infected human feces, the eggs hatch in the animal's digestive tract, releasing larvae that burrow into its muscles and form cysts, thereby continuing the infective cycle (**FIG. 24-13**).

Another group of parasitic flatworms includes the *flukes* (see Fig. 24-12a). Of these, the most devastating are liver flukes (common in Asia) and blood flukes, such as those of the genus *Schistosoma*, which cause the disease schistosomiasis. Like most parasites, flukes have a complex life cycle that includes an intermediate host (a snail, in the case of *Schistosoma*). Prevalent in Africa and parts of South America, schistosomiasis affects an estimated 200 million people worldwide. Its symptoms include diarrhea, anemia, and sometimes brain damage.

Annelids Are Segmented Worms

The bodies of *annelids* (Annelida) are divided into a series of similar repeating segments. Externally, this **segmentation** appears as a series of ring-like depressions on the surface. Internally, most of the segments contain identical copies of nerves, excretory structures, and muscles.

Sexual reproduction is common among annelids. Some species are hermaphroditic; others have separate sexes. Fertilization may be external or internal. External fertilization, in which sperm and eggs are released into the surrounding environment, is found mainly in species that live in water. In internal fertilization, two individuals copulate and sperm are transferred directly from one to the other. In hermaphroditic species, sperm transfer may be mutual, with each partner both donating and receiving sperm. In addition, some annelids can reproduce asexually, typically by a process in which the body breaks into two pieces, each of which regenerates the missing part.

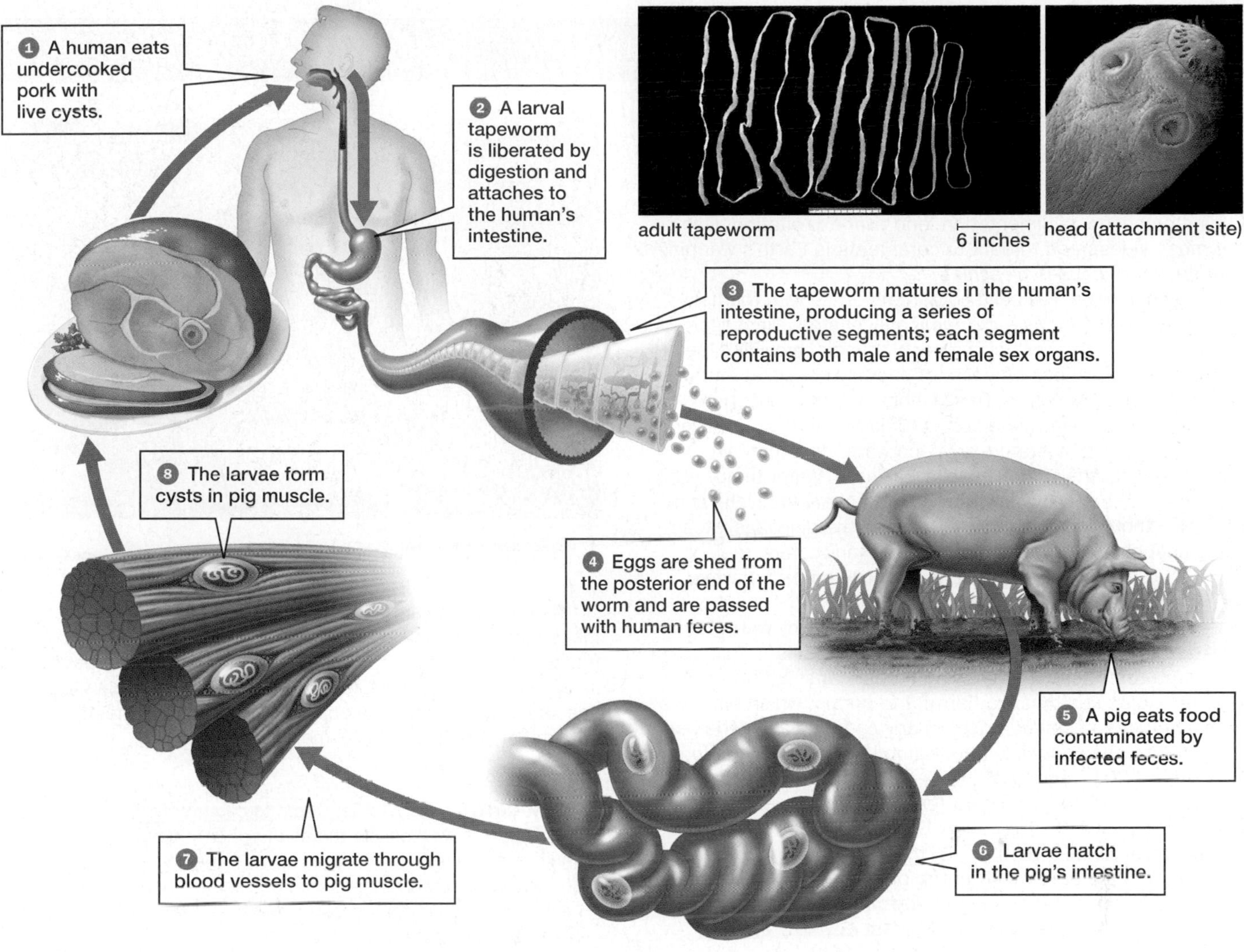

1 A human eats undercooked pork with live cysts.
2 A larval tapeworm is liberated by digestion and attaches to the human's intestine.
adult tapeworm
6 inches
head (attachment site)
3 The tapeworm matures in the human's intestine, producing a series of reproductive segments; each segment contains both male and female sex organs.
4 Eggs are shed from the posterior end of the worm and are passed with human feces.
5 A pig eats food contaminated by infected feces.
6 Larvae hatch in the pig's intestine.
7 The larvae migrate through blood vessels to pig muscle.
8 The larvae form cysts in pig muscle.

▲ **FIGURE 24-13 The life cycle of the human pork tapeworm**

THINK CRITICALLY What advantages favored the evolution of tapeworms' long, flat shape?

Annelids Are Coelomates and Have Organ Systems

Annelids have a fluid-filled true coelom between the body wall and the digestive tract (see Fig. 24-3a). The incompressible fluid in the coelom of many annelids is confined by the partitions between the segments and serves as a **hydrostatic skeleton**, a supportive framework against which muscles can act. A hydrostatic skeleton allows earthworms to burrow through soil.

Annelids have well-developed organ systems (groups of organs that act in a coordinated manner). For example, annelids have a **closed circulatory system** that distributes gases and nutrients throughout the body. In closed circulatory systems (including yours), blood remains confined to the heart and blood vessels. In the earthworm, for example, blood with oxygen-carrying hemoglobin is pumped through well-developed vessels by five pairs of "hearts" (**FIG. 24-14**). These hearts are actually short segments of specialized blood vessels that contract rhythmically. The blood is filtered and wastes are removed by excretory organs called *nephridia* (singular, nephridium) and excreted to the environment through small pores. Nephridia resemble the individual tubules of the vertebrate kidney. The annelid nervous system consists of a simple brain in the head and a series of paired segmental ganglia joined by a pair of ventral nerve cords that pass along the length of the body. The annelid digestive system includes a tubular gut that runs from the mouth to the anus. This kind of digestive tract, with two openings and a one-way digestive path, is much more efficient than the single-opening digestive systems of cnidarians and flatworms. Digestion in annelids occurs in a series of compartments in the digestive tract, each specialized for a different phase of food processing.

When Reefs Get Too Warm

The coral reefs that form in shallow tropical oceans are among Earth's most diverse, beautiful, and economically important ecosystems. Unfortunately, coral reef ecosystems are declining worldwide, threatened by pollution, destructive fishing practices, sedimentation, and over-exploitation. One the most widespread threats to coral reefs is Earth's warming climate, which causes *bleaching*.

The cnidarians that build coral reefs do so with the help of photosynthetic protists called dinoflagellates. Dinoflagellates live inside the cells of the coral animals, providing the animals with nutritious products of photosynthesis in return for shelter and protection. This symbiotic relationship benefits both the dinoflagellates and the coral animals. However, when the water temperature gets too warm, the coral animals may eject their dinoflagellate partners. Water temperature as little as 1°C higher than the normal annual highs can result in dinoflagellate ejection. Because the dinoflagellates are responsible for the bright colors of many corals, reefs from which dinoflagellates have been expelled have a white, bleached appearance (**FIG. E24-1**).

Scientists do not yet understand exactly why warm temperatures cause coral bleaching; one hypothesis is that warm temperatures stress the dinoflagellates, causing them to release molecules that are harmful to corals, which eject the dinoflagellates in self-defense. In any case, bleaching is bad for corals; if they remain without dinoflagellates for too long, they die. In 2016 and 2017, large-scale bleaching episodes occurred in back-to-back years for the first time, killing corals across the planet. In Australia's Great Barrier Reef—Earth's largest coral reef—a third of all corals have perished, with 70% dead in the hardest-hit areas of the reef. In some other parts of the world, the damage is even worse, with a coral death rate higher than 80% in the most severely affected regions.

▲ **FIGURE E24-1 Bleached coral**

As the planet warms due to human-caused climate change, mass coral bleaching events are expected to become more frequent (**FIG. E24-2**). The widespread coral deaths that will likely result are bad news for the thousands of reef-dwelling species (and millions of people) that directly or indirectly depend on healthy corals for their survival.

THINK CRITICALLY Another threat to coral reefs is sediment that washes into the sea from farm field runoff and other sources of soil erosion. Why might such sediment be harmful to corals? In your answer, consider what you have learned about the symbiotic relationship between dinoflagellates and corals.

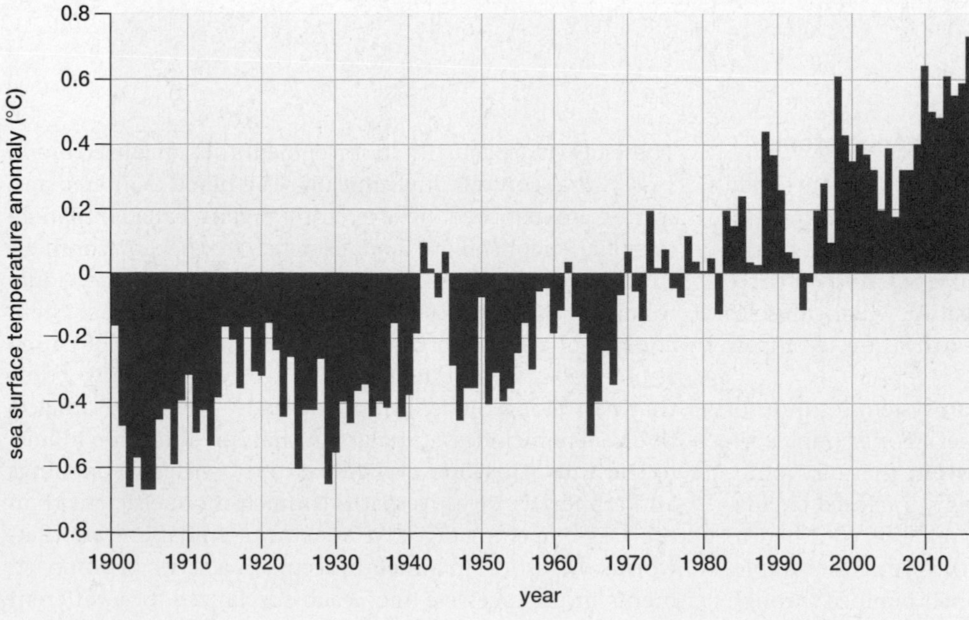

◀ **FIGURE E24-2 Sea surface temperatures at the Great Barrier Reef, 1900–2016** Temperatures warm enough to cause coral bleaching have become increasingly frequent, as illustrated by these measurements from the waters around Earth's largest reef system, the Great Barrier Reef of Australia. The vertical axis represents "temperature anomaly," which measures the amount by which the average temperature in a given year differs from the long-term average.

▶ **FIGURE 24-14 An annelid, the earthworm** This diagram shows an enlargement of segments, many of which are repeating similar units separated by partitions.

THINK CRITICALLY What advantage does a digestive system with two openings have relative to digestive systems with only a single opening (like that of the flatworms)?

Annelids Include Oligochaetes, Polychaetes, and Leeches

The 13,000 known species of annelids fall into three main subgroups, called the *oligochaetes*, the *polychaetes*, and the *leeches*. The oligochaetes include the familiar earthworm and its relatives. Charles Darwin devoted substantial time to the study of earthworms and was especially impressed by their role in improving soil fertility. More than a million earthworms may live in an acre of land, tunneling through soil, consuming and excreting soil particles and organic matter. These actions help ensure that air and water can move easily through the soil and that organic matter is continually mixed into it, creating conditions that are favorable for plant growth. The impact of earthworms, however, can also be negative. In some areas of North America, non-native, invasive earthworms disrupt the normal structure of forest soils, harming native forest plants.

Polychaetes live primarily in the ocean. In some polychaetes, most body segments have paired fleshy paddles that are used in locomotion. Other polychaetes live in tubes from which they project feathery gills that both exchange gases and filter the water for microscopic food (**FIGS. 24-15a, b**).

Leeches (**FIG. 24-15c**) live in freshwater or moist terrestrial habitats. Many leeches prey on smaller invertebrates; some suck the blood of larger animals.

(a) Polychaete gills

(b) Deep-sea polychaete

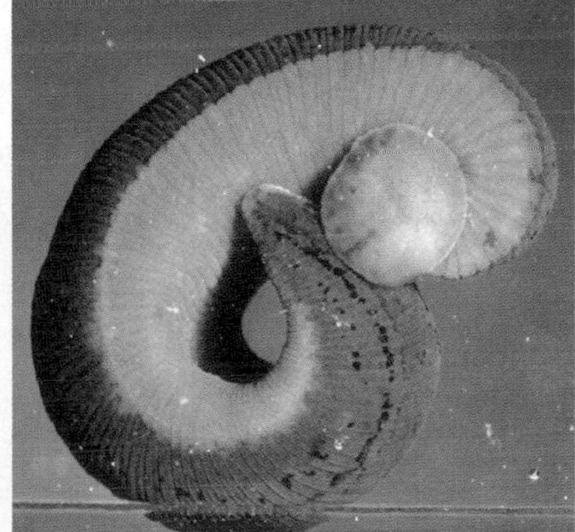

(c) Leech

▲ **FIGURE 24-15 Diverse annelids (a)** The brightly colored gills of a polychaete annelid. The rest of the worm's body is hidden inside a tube embedded in the coral that is visible in the background. **(b)** This polychaete lives near deep-sea vents where the water temperature may reach 175°F (80°C). **(c)** This leech, a freshwater annelid, shows numerous segments. The sucker encircles its mouth, allowing it to attach to its prey.

THINK CRITICALLY Why does pouring salt on a leech harm it?

CASE STUDY CONTINUED

Physicians' Assistants

The medicinal leech has been formally approved as a "medical device" by the U.S. Food and Drug Administration (FDA), mainly on the basis of its effectiveness as an aid to post-surgical healing. However, leeches may have additional medical uses. Especially promising is their potential for treatment of pain. Leech saliva contains pain-killing chemicals that help leeches evade detection when they puncture a victim's skin, so physicians hope that leeches can help reduce patients' pain. Several clinical trials have found that leeches provide substantial relief to people with osteoarthritis, a painful disorder of the joints. Anecdotal evidence suggests that leeches may help with other kinds of pain as well.

In addition to leeches, other invertebrate animals have medical potential, including insects and roundworms. You'll learn more about the biology of these animals later in this section.

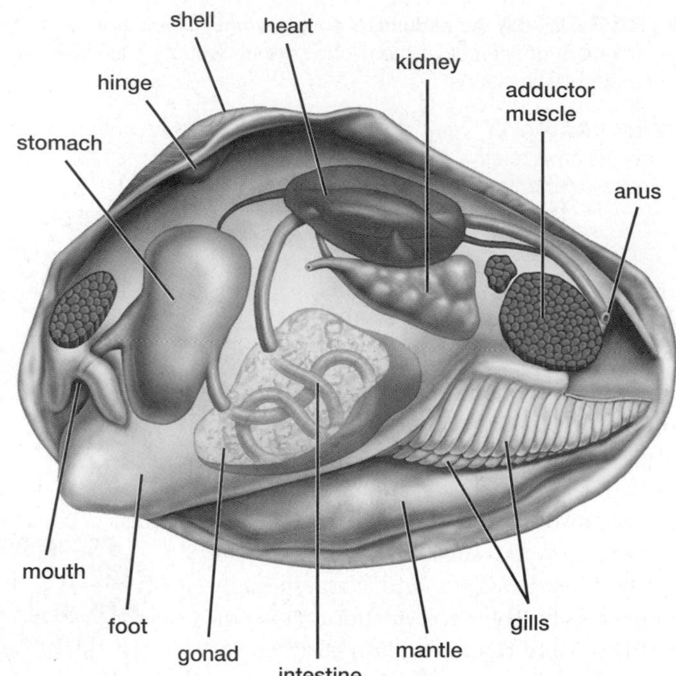

▲ **FIGURE 24-16 A bivalve mollusk** The body plan of a clam, showing the mantle, foot, gills, shell, and other features that are seen in most (but not all) mollusk species. The clam uses its adductor muscles to close its shell.

Most Mollusks Have Shells

If you have ever enjoyed a bowl of clam chowder, a plate of oysters on the half shell, or some sautéed scallops, you are indebted to *mollusks* (Mollusca). Mollusks are very diverse; in terms of number of known species, the 50,000 mollusks are second (albeit a distant second) only to the arthropods. These diverse species exhibit a range of lifestyles, from sessile forms such as mussels that spend their adult lives in one spot, filtering microorganisms from the water, to active, voracious predators of the ocean depths, such as the giant squid. Mollusks, with the exception of some snails and slugs, are aquatic. Although most mollusks are protected by hard shells of calcium carbonate, some lack shells.

The circulatory systems of most mollusks include a feature not seen in annelids: the **hemocoel**, or blood cavity. Blood empties into the hemocoel, where it bathes the internal organs directly. This arrangement is known as an **open circulatory system**. Mollusks also have a *mantle*, an extension of the body wall that forms a chamber for the gills and, in shelled species, also secretes the shell (**FIG. 24-16**). The mollusk nervous system, like that of annelids, consists of ganglia connected by nerves, but in many mollusks more of the ganglia are concentrated in the brain. Reproduction is sexual; some species have separate sexes, and others are hermaphroditic.

Among the many subgroups of mollusks, we will discuss three in more detail: gastropods, bivalves, and cephalopods.

Gastropods Are One-Footed Crawlers

Snails and slugs—collectively known as *gastropods*—crawl on a muscular *foot*, and many are protected by shells that vary widely in form and color (**FIG. 24-17a**). Not all gastropods are shelled, however. Sea slugs, for example, lack shells, but their brilliant colors warn potential predators that they are poisonous or foul tasting (**FIG. 24-17b**).

(a) Snail

(b) Sea slug

▲ **FIGURE 24-17 The diversity of gastropod mollusks (a)** A Florida tree snail displays a brightly striped shell and eyes at the tip of stalks that retract instantly if touched. **(b)** The brilliant colors of many sea slugs warn potential predators that they are distasteful.

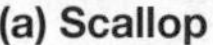

(a) Scallop

(b) Mussels

FIGURE 24-18 The diversity of bivalve mollusks **(a)** This scallop parts its hinged shells to allow the intake of water from which food will be filtered. The blue spots visible along the mantle just inside the upper and lower shells are simple eyes. **(b)** Mussels attach to rocks in dense aggregations exposed at low tide. White barnacles (which are arthropods) are attached to the mussel shells and surrounding rock.

Gastropods feed with a *radula*, a flexible ribbon of tissue studded with spines that is used to scrape algae from rocks or to grasp larger plants or prey. For respiration, most gastropods use gills. In shelled species, gills are enclosed in a cavity beneath the shell. Gases can also diffuse readily through the skin of most gastropods. The few gastropod species that live in terrestrial habitats (including garden snails and slugs) use a simple lung for breathing.

Bivalves Are Filter Feeders

Included among the *bivalves* are scallops, oysters, mussels, and clams (**FIG. 24-18**). Bivalves possess two shells connected by a flexible hinge. A strong muscle clamps the shells closed in response to danger; this muscle is what you are served when you order scallops in a restaurant.

Clams use a muscular foot for burrowing in sand or mud. In mussels, which live attached to rocks, the foot is smaller and helps secrete threads that anchor the animal to the rocks. Scallops lack a foot and move by a sort of jet propulsion achieved by flapping their shells together.

Bivalves are filter-feeders, using their gills as both respiratory and feeding structures. Water circulates over the gills, which are covered with a thin layer of mucus that traps microscopic food particles. Beating cilia on the gills sweep food to the mouth.

Cephalopods Are Marine Predators

The *cephalopods* include octopuses, nautiluses, cuttlefish, and squids (**FIG. 24-19**). The largest invertebrates, the giant squid and the colossal squid, belong to this group (see "Doing Science: Searching for a Sea Monster" on page 400). All cephalopods are carnivores, and all are marine. In these mollusks, the foot has evolved into tentacles with well-developed sensory abilities for detecting prey. Prey are grasped by suction disks on the tentacles and may be immobilized by a paralyzing venom in the saliva before being torn apart by beaklike jaws.

Cephalopods move rapidly by jet propulsion, which is generated by the forceful expulsion of water from the mantle

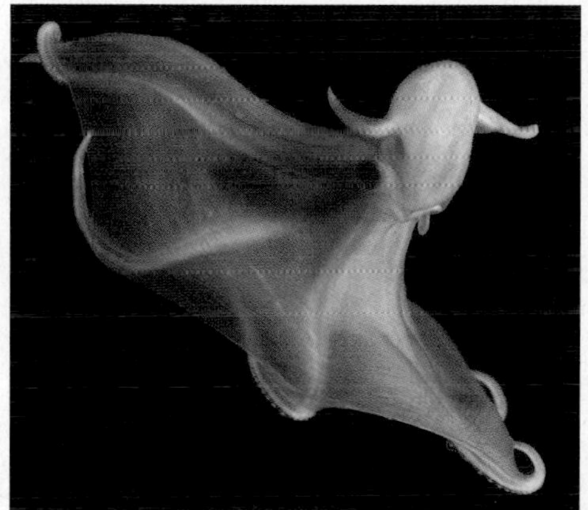

(a) Octopus

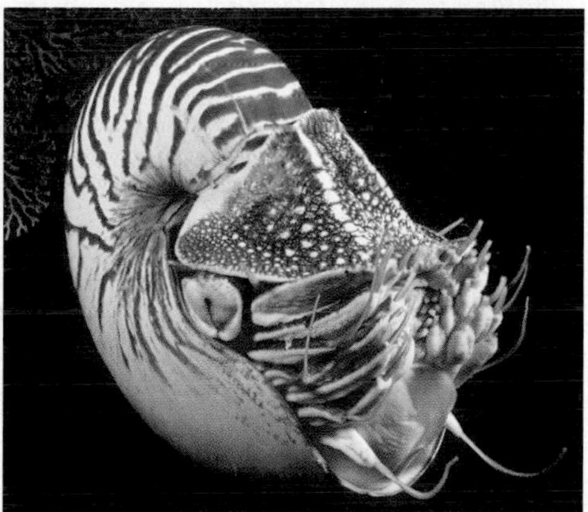

(b) Nautilus

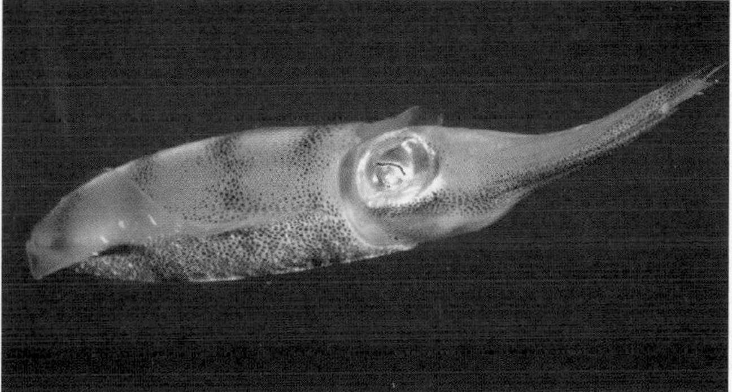

(c) Squid

FIGURE 24-19 The diversity of cephalopod mollusks (a) In emergencies, an octopus can jet backward by vigorously contracting its mantle. Octopuses and squid can emit clouds of dark purple ink to confuse pursuing predators. **(b)** The chambered nautilus secretes a shell with internal, gas-filled chambers that provide buoyancy. Note the well-developed eyes and the tentacles used to capture prey. **(c)** A squid can move by contracting its mantle to generate jet propulsion, which pushes the animal backward through the water.

DOING Science Searching for a Sea Monster

The giant squid, one of the world's largest invertebrate animals, is also one of the most mysterious. The squid reaches lengths of 40 feet (13 meters) or more, with eyes as large as a human head. The squid's 10 tentacles are covered with powerful suckers. The suckers contain sharp hooks used to grasp prey, which is then pulled toward the squid's mouth, where a heavily muscled beak tears the food. Clearly, the giant squid is one of the most imposing organisms on Earth. However, we know almost nothing about its habits and lifestyle because almost all observations of giant squid are of dead animals washed ashore or hauled up in fishing nets.

It was not until 2004 that a living giant squid was observed. That year, a team placed a camera on a baited fishing line that was dragged at a depth of 3,000 feet. This effort yielded a few photos of a giant squid that attacked the bait. Two years later, a giant squid was hooked on a fishing line and pulled to the surface (**FIG. E24-3**). Then, in 2012, Tsunemi Kubodera, Steve O'Shea, and Edith Widder mounted a joint expedition to see if their combined efforts might yield better observations.

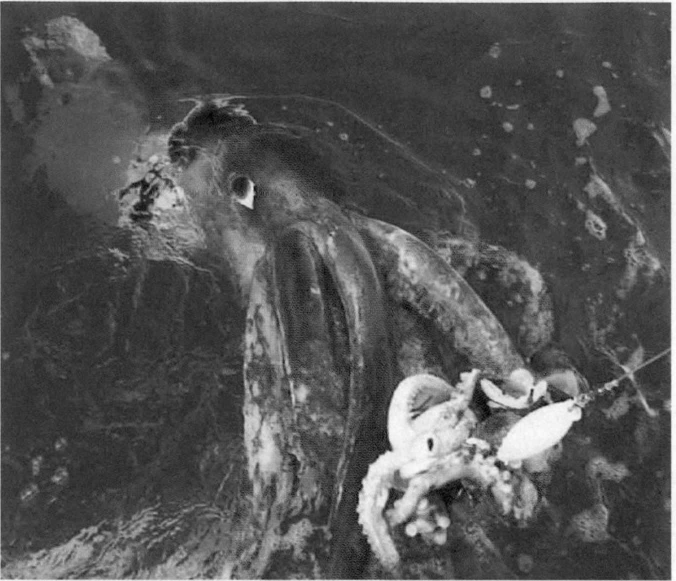

▲ **FIGURE E24-3 Hooked Giant** One of the first glimpses of a live giant squid occurred in 2006, when researchers hooked a large specimen in deep Pacific waters and raised it to the surface.

What Question Was Asked?

Although the three researchers shared the goal of observing giant squids in their natural, deep-ocean habitat, each one had a different hypothesis about how giant squids hunt, and therefore about the best way to attract one. O'Shea hypothesized that giant squids hunt by smell and would be attracted to substances secreted by their prey. Kubodera had a similar hypothesis, but thought that whole prey animals would be a better attractant. Widder, an expert on bioluminescence (the light produced naturally by many marine organisms), hypothesized that because giant squids eat small predators that in turn eat sea jellies, giant squids would be attracted to the flashing bioluminescent alarm signal that certain deep-water sea jellies produce when under attack.

How Was Evidence Gathered?

Each researcher tried a different approach to attracting the animals. O'Shea descended in an illuminated submarine and released a mixture of chemicals extracted from dead squids. He hoped that the chemicals would draw a giant squid into the circle of bright light around his submarine. Kubobera also used a submarine and bait, but his bait was a small squid, and his submarine was illuminated only by dim red light (undetectable by most deep-sea animals) and drifted quietly, with the engines off. Widder also wished to minimize noise and bright light, so she devised a lure that emitted pulses of blue light similar to those of jellyfish and tethered this "imitation jelly" beside a camera rig that illuminated the surroundings with red light. The whole contraption (**FIG. E24-4**) was tethered to a buoy and suspended 2,300 feet below the surface.

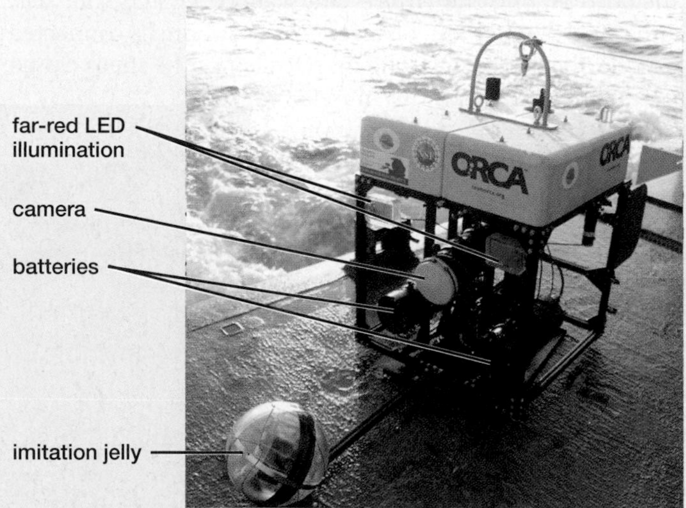

▲ **FIGURE E24-4 Search tools** This imitation jelly lure and camera rig were used to capture the first-ever live video of a giant squid in its deep-ocean habitat.

What Was Learned?

O'Shea's chemical lures attracted many squids to his submarine, but none were giant squids. Perhaps the bright lights kept them away, or the smell of prey alone was not enough to attract them. Fortunately, however, the lures used by the other two researchers did attract giant squids. Widder's imitation jelly rewarded the researchers (after many hours of waiting) with the first-ever video of a giant squid hunting in the deep ocean, its tentacles spread as its mouth moved toward the camera. And Kubodura's live-prey lure, after many long excursions with no results, finally attracted a giant squid to the submarine. After eagerly observing the animal for a few minutes, Kubodera took a big gamble by turning on some bright spotlights. To his great relief, the giant squid did not immediately flee, and the elated researcher captured high-definition color video of one of Earth's strangest and most magnificent creatures in its deep-ocean habitat.

THINK CRITICALLY The giant squids that were attracted to the imitation jelly pictured in Figure E24-4 seemed to be attacking the camera rig beside the lure, rather than the lure itself. How does this observation support Edith Widder's hypothesis about giant squid foraging?

cavity. Octopuses may also travel along the seafloor by using their tentacles like multiple undulating legs. The rapid movements and active lifestyles of cephalopods are made possible in part by their closed circulatory systems. Cephalopods are the only mollusks with a closed circulatory system, which transports oxygen and nutrients more efficiently than does an open circulatory system.

Cephalopods have highly developed brains and sensory systems. The cephalopod eye rivals our own in complexity. The cephalopod brain, especially that of the octopus, is exceptionally large and complex. It endows the octopus with highly developed capabilities to learn and remember. In the laboratory, octopuses can rapidly learn to solve a maze, associate symbols with food, or open a screw-cap jar to obtain food. In the wild, some octopuses use tools. For example, veined octopuses clean mud from buried coconut shells, stack them up for transport, and turn them into a protective shelter.

▲ **FIGURE 24-21 The exoskeleton must be molted periodically**
A cicada emerges from its outgrown exoskeleton (left).

Arthropods Are the Most Diverse and Abundant Animals

In terms of both number of individuals and number of species, no other animal phylum comes close to the *arthropods* (Arthropoda), which include insects, arachnids, myriapods, and crustaceans. About 1 million arthropod species have been discovered, and scientists estimate that millions more remain undescribed.

Arthropods Have Appendages and an External Skeleton

All arthropods have paired, jointed appendages and an **exoskeleton**, an external skeleton that encloses the arthropod body like a suit of armor. Secreted by the *epidermis* (the outer layer of skin), the exoskeleton is composed chiefly of protein and a polysaccharide called *chitin* (see Fig. 3-11). The exoskeleton provides rigid attachment sites for muscles, but also becomes thin and flexible at joints, thereby allowing the appendages to move. This combination of rigidity and flexibility makes possible the flight of the bumblebee and the intricate, delicate manipulations of the spider as it weaves its web (**FIG. 24-20**). The exoskeleton also provides protection against predators. In addition, it contributed enormously to the arthropod invasion of land by providing a watertight covering for delicate, moist tissues such as those used for gas exchange.

Despite its benefits, the arthropod exoskeleton also poses some problems. For example, because it cannot expand as the animal grows, the exoskeleton must periodically be shed, or **molted**, and replaced with a larger one (**FIG. 24-21**). Molting uses energy and leaves the animal temporarily vulnerable to predators until the new skeleton hardens.

Arthropods Have Specialized Segments and Adaptations for Active Lifestyles

Arthropods are segmented, but their segments tend to be few and specialized for different functions such as sensing the environment, feeding, and movement (**FIG. 24-22**). For example, in insects, sensory and feeding structures are concentrated on the front segment, known as the *head*, and

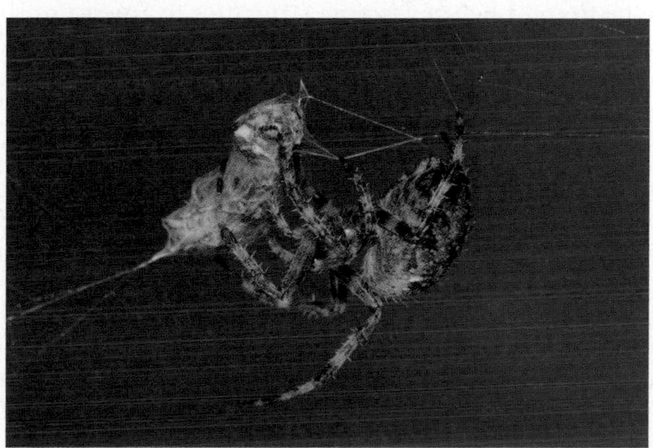

▲ **FIGURE 24-20 The exoskeleton allows precise movements**
A garden orb spider begins to wrap a captured wasp in silk. Such dexterous manipulations are made possible by the exoskeleton and jointed appendages characteristic of arthropods.

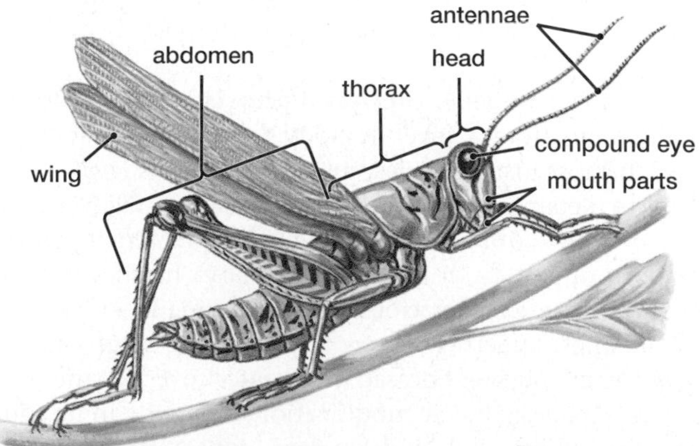

▲ **FIGURE 24-22 Segments are fused and specialized in insects**
Insects, such as this grasshopper, show fusion and specialization of body segments into a distinct head, thorax, and abdomen. Segments are visible on the abdomen.

digestive structures are largely confined to the *abdomen*, which is the rear segment. Between the head and the abdomen is the *thorax*, the segment to which structures used in locomotion, such as wings and legs, are attached.

Efficient gas exchange is required to supply adequate oxygen to the muscles, which allows the rapid flying, swimming, or running displayed by many arthropods. In aquatic arthropods such as crustaceans, gas exchange is accomplished by gills. In terrestrial arthropods, gas exchange is performed either by lungs (in arachnids) or by *tracheae* (singular, trachea), a network of narrow, branching respiratory tubes that open to the surrounding environment and that penetrate all parts of the body. Most arthropods have open circulatory systems, like those of mollusks, in which blood directly bathes the organs in a hemocoel.

Most arthropods possess well-developed sensory and nervous systems. Arthropod sensory systems often include **compound eyes**, which have multiple light detectors (**FIG. 24-23**), and acute chemical and tactile senses. The arthropod nervous system consists of a brain composed of fused ganglia and a series of additional ganglia along the length of the body that are linked by a ventral nerve cord. This well-developed nervous system, combined with sophisticated sensory abilities, has permitted the evolution of complex behaviors in many arthropods.

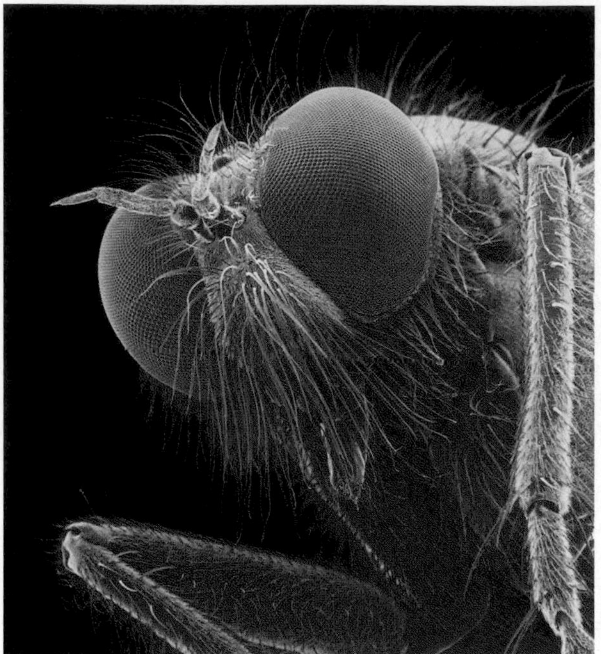

▲ **FIGURE 24-23 Arthropods possess compound eyes** This scanning electron micrograph shows the compound eye of a horse fly. Compound eyes consist of an array of similar light-gathering and sensing elements whose orientation gives the arthropod a wide field of view. Insects have reasonably good image-forming ability and good color discrimination.

Insects Are the Only Flying Invertebrates

The number of known *insect* species is about 850,000, roughly three times the total number of known species in all other groups of animals combined (**FIG. 24-24**). Insects have a single pair of antennae and three pairs of legs, usually supplemented by two pairs of wings. Insects' capacity for flight distinguishes them from all other invertebrates and has contributed to their enormous success. As anyone who has pursued a fly can testify, flight helps in escaping from predators. It also allows insects to find widely dispersed food. Swarms of locusts, for example, can travel 200 miles a day in search of food; researchers tracked one swarm on a journey that totaled almost 3,000 miles. Flight requires rapid and efficient gas exchange, which insects accomplish by means of tracheae.

All insects undergo **metamorphosis**, a change during development from a juvenile body form to an adult body form. In most insect species, metamorphosis is *complete*. In complete metamorphosis, the immature stage, called a **larva**, is worm shaped (for example, the maggot of a blowfly or the caterpillar of a moth or butterfly). The larva hatches from an egg, grows by eating voraciously and shedding its exoskeleton several times, and then forms a nonfeeding stage called a **pupa** (plural, pupae). Encased in an outer covering, the pupa body undergoes a drastic modification, emerging in its adult form. The adults mate and lay eggs, continuing the cycle. Metamorphosis may include a change in diet as well as in shape, thereby eliminating competition for food between adults and juveniles. Some insects, such as grasshoppers and crickets, undergo a more gradual metamorphosis (called *incomplete* metamorphosis), hatching as young that bear some resemblance to the adult, then gradually acquiring more adult features as they grow and molt.

Biologists classify the amazing diversity of insects into several dozen groups. We will describe three of the largest ones here.

Butterflies and Moths The butterflies and moths make up what is perhaps the most conspicuous and best-studied group of insects. The brightly colored, often iridescent wing patterns of many butterfly and moth species arise from pigments and light-refracting structures in the scales that cover the wings of all members of this group. (The scales are the powdery substance that rubs off onto your hand when you handle a butterfly or moth.) Butterflies fly mainly during the day and moths fly at night, though there are exceptions to this general rule, such as the hummingbird-like hawk moths that are often seen feeding by day in flower gardens.

The evolution of butterflies and moths has been closely tied to the evolution of flowering plants. Butterflies and moths feed almost exclusively on flowering plants, both as caterpillars and as adults. Many species of flowering plants depend in turn on butterflies and moths for pollination.

Bees, Ants, and Wasps The bees, ants, and wasps are known to many people by their painful stings. Many species in this group are equipped with a stinger that extends from

(a) Aphid

(b) Ant

(c) Beetle flying

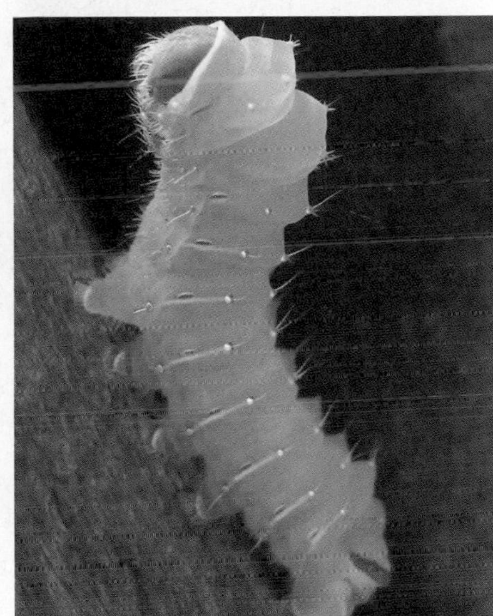
(d) Moth larva

FIGURE 24-24 The diversity of insects **(a)** Aphids suck sugar-rich juice from plants. **(b)** The bullet ant can inflict an extremely painful sting. **(c)** A June beetle displays its two pairs of wings as it comes in for a landing. The outer wings protect the abdomen and the inner wings, which are relatively thin and fragile. **(d)** Caterpillars are larval forms of moths or butterflies. This caterpillar larva of a silk moth can produce clicking sounds with its mouthparts. The sounds may serve to warn predators that the caterpillar is distasteful.

THINK CRITICALLY Explain how wings might have influenced the diversity and abundance of insects.

the abdomen and can be used to inject venom into the victim. The venom may be extremely toxic, but fortunately for human stinging victims, each insect carries only a tiny amount. Nonetheless, the amount is often sufficient to cause considerable pain. Only females have stingers, which are used by many stinging species to help defend a nest from attack by a potential predator. Defense, however, is not the only use for stingers. Many wasps, for example, act as parasites of other insects when reproducing and may sting a host insect to paralyze it. The parasitic wasp then lays an egg inside the paralyzed body of the host, typically the caterpillar of a moth or butterfly, which becomes food for the wasp larva after it hatches.

The social behavior of some ant and bee species is extraordinarily intricate. They may form huge colonies with complex organization in which individuals specialize in particular tasks such as foraging, defense, reproduction, or rearing larvae. The organization and division of labor in these insect societies require sophisticated communication and learning. Social insects accomplish remarkable tasks. For example, honeybees manufacture and store food (honey), and some ant species "farm" by cultivating fungi in underground chambers or "milking" aphids by inducing them to secrete a nutritious liquid.

Beetles Roughly one-third of all known insect species are beetles. Beetles exhibit a huge variety of shapes, sizes, and lifestyles. All beetles, however, have hard, protective coverings over their wings. Many destructive agricultural pests are beetles, such as the Colorado potato beetle, the grain weevil, and the Japanese beetle. However, others, such as lady beetles, are predators that help control insect pests.

One of this group's most impressive adaptations is found in the bombardier beetle. This species defends itself against ants and other enemies by emitting a toxic spray from a nozzle-like structure at the end of its abdomen. The beetle is able to precisely aim the spray, which emerges with explosive force and at temperatures higher than 200°F (93°C). To avoid harming itself, the beetle produces this hot, toxic brew only when needed, by combining two otherwise harmless substances.

Most Arachnids Are Carnivores

The *arachnids* include spiders, mites, ticks, and scorpions (**FIG. 24-25**). All arachnids have eight walking legs, and most are carnivorous. Many subsist on a liquid diet of blood or predigested prey. For example, spiders, the most numerous arachnids, first immobilize their prey with a paralyzing venom. They then inject digestive enzymes into the helpless victim (typically an insect) and suck in the resulting soup. Arachnids breathe using tracheae, lungs, or both.

In contrast to the compound eyes of insects and crustaceans, arachnids have simple eyes, each with a single lens. Most spiders have eight eyes placed in such a way as to give them a panoramic view of predators and prey. The eyes are sensitive to movement, and in some spider species—especially those that hunt actively and have no webs—the eyes are thought to form images. Most spider perception,

(a) Spider

(b) Scorpion

(c) Ticks

▲ **FIGURE 24-25 The diversity of arachnids (a)** The tarantula is one of the largest spiders but is relatively harmless. **(b)** Scorpions, found in warm climates such as the deserts of the southwestern United States, paralyze their prey with venom from a stinger at the tip of the abdomen. A few species can harm humans. **(c)** In a tick that has not yet fed (left), the exoskeleton is flexible and folded. This allows the animal to become grotesquely bloated while feeding on blood (right).

however, is not through their eyes but through sensory hairs found over much of the body. Some of a spider's hairs are touch sensitive and help the animal perceive prey, mates, and surroundings. Other hairs are sensitive to chemicals and function as organs of smell and taste. Hairs also respond to vibrations in the air, ground, or web, allowing spiders to detect nearby movement by predators, prey, or other spiders.

Among the distinctive features of spiders is their production of protein threads known as silk. Spiders manufacture silk in special glands in their abdomens and use it to perform a variety of functions, such as building webs that capture prey, wrapping up and immobilizing captured prey (see Fig. 24-20), constructing protective shelters for themselves, making cocoons to surround their eggs, and making "draglines" that connect a spider to a web or other surface and support its weight if it drops from its perch. Spider silk is an amazingly light, strong, and elastic fiber. It can be stronger than a steel wire of the same size, yet is as elastic as rubber. Human engineers have long sought to develop a fiber with this combination of strength and elasticity. Despite careful study of the structure of spider silk, no comparable human-made substance has been successfully manufactured.

Myriapods Have Many Legs

The *myriapods* include the centipedes and millipedes, whose most prominent feature is an abundance of legs (**FIG. 24-26**). Most millipede species have between 100 and 300 legs; the species with the largest number of legs can have up to 750.

Have You Ever Wondered ...

Why Spiders Don't Stick to Their Own Webs?

If you have ever accidentally walked into a spider web, then you know that the strands of the web can be very sticky indeed. And the small insects one sees trapped in spider webs seem to be quite firmly snared. So how does the spider that built the web scuttle so easily across it without getting stuck? One key factor is that the web is not entirely sticky. The web's scaffolding is made from non-sticky strands, with sticky ones restricted to the capture area. So a moving spider can often simply stay on the non-sticky strands and avoid the sticky ones. But the spider's distinctive claws also help. These claws, in conjunction with specialized hairs on the tips of its legs, allow a spider leg to grip a single web strand and then release it. A spider moving in this fashion, grasping strands with only the very tips of its legs, will have only a tiny surface area in contact with the web at any one time. Thus, even if the spider grips a sticky strand with one of its legs, it can easily pull loose, just as you can if you step on a wad of chewing gum. In contrast, an unsuspecting fly that crashes into the web will contact a number of strands with many parts of its body simultaneously and be stuck fast.

(a) Centipede

(b) Millipede

▲ **FIGURE 24-26 The diversity of myriapods (a)** Centipedes and **(b)** millipedes are common nocturnal arthropods. Each segment of a centipede's body holds one pair of legs, while each millipede segment has two pairs.

Centipedes are not quite as leggy; most have around 70 legs. Both centipedes and millipedes have one pair of antennae. The legs and antennae of centipedes are longer and more delicate than those of millipedes. Most myriapods have very simple eyes that detect light and dark but do not form images. In some species, the number of eyes can be high—up to 200—but other species lack eyes altogether. Myriapods respire by means of tracheae.

Myriapods inhabit terrestrial environments exclusively, living mostly in soil or leaf litter or under logs and rocks. Centipedes are generally carnivorous, capturing prey (mostly other arthropods) with their frontmost legs, which are modified into sharp claws that inject venom into prey. Bites from large centipedes can be painful to humans. In contrast, most millipedes are not predators but instead feed on decaying vegetation and other debris. When attacked, many millipedes defend themselves by secreting a foul-smelling, distasteful liquid.

Most Crustaceans Are Aquatic

The *crustaceans*, including crabs, crayfish, lobsters, shrimp, and barnacles, are the only arthropods that live primarily in the water (**FIG. 24-27**). Crustaceans range in size from microscopic

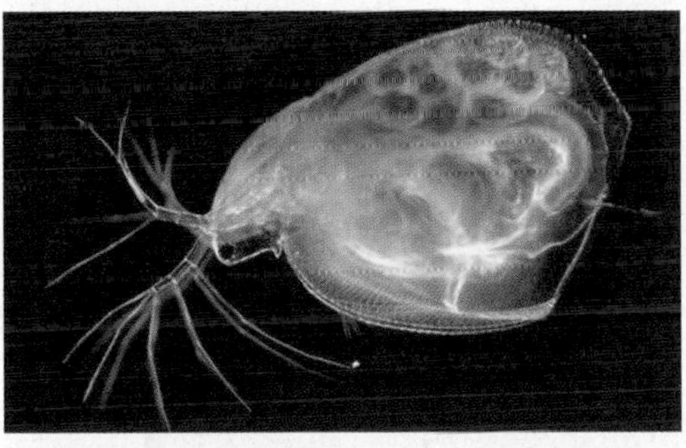

(a) Water flea

(b) Sowbug

(c) Hermit crab

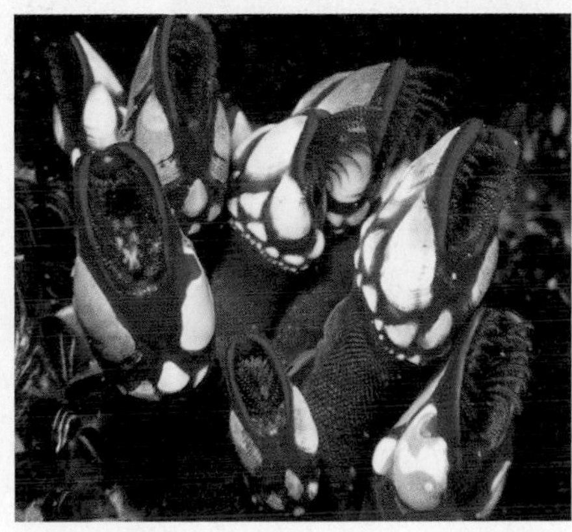

(d) Barnacles

◀ **FIGURE 24-27 The diversity of crustaceans (a)** The microscopic water flea is common in freshwater ponds. Notice the eggs developing within the body. **(b)** The sowbug, found in dark, moist places such as under rocks, leaves, and decaying logs, is one of the few crustaceans to invade the land successfully. **(c)** This hermit crab protects its soft abdomen by inhabiting an abandoned snail shell. **(d)** The gooseneck barnacle uses a tough, flexible stalk to anchor itself to rocks, boats, or even animals such as whales. Other types of barnacles attach with shells that resemble miniature volcanoes (see Fig. 24-18b). Early naturalists thought barnacles were mollusks until they observed barnacles' jointed legs (seen here extending into the water).

species that live in the spaces between grains of sand to the largest of all arthropods, the Japanese spider crab, whose legs span nearly 12 feet (4 meters). Crustaceans have two pairs of sensory antennae, but the rest of their appendages are highly variable in form and number, depending on the habitat and lifestyle of the species. Most crustaceans have compound eyes similar to those of insects, and nearly all respire using gills.

Crustaceans are an important food source for larger animals. For example, small crustaceans called krill are abundant in the Southern Ocean and are the main food of whales, seals, seabirds, and other animals. People eat a lot of crustaceans, too. Shrimp, for example, are the most consumed seafood in the United States by far. Today, most of the shrimp we eat are farm raised, mainly in coastal areas of Asia and South America. Unfortunately, widespread shrimp farming has had adverse ecological consequences, mainly because large areas of ecologically important mangrove forests have been cleared for shrimp farming.

Roundworms Are Abundant and Mostly Tiny

Although you may be blissfully unaware of their presence, *roundworms* (Nematoda) are nearly everywhere. Roundworms, also called *nematodes,* have colonized nearly every habitat on Earth, and they play an important role in breaking down organic matter. They are extremely numerous; a single rotting apple may contain 100,000 roundworms. Billions thrive in each acre of topsoil. In addition, almost every plant and animal species hosts several parasitic nematode species.

In addition to being abundant and ubiquitous, roundworms are diverse. Although only about 12,000 roundworm species have been named, there may be as many as 500,000. Most, such as the one shown in **FIGURE 24-28**, are microscopic, but some parasitic forms reach a meter in length.

Roundworms Are Pseudocoelomates with a Simplified Body Plan

Roundworms have a rather simple body plan, featuring a tubular gut and a fluid-filled pseudocoelom that surrounds the organs and forms a hydrostatic skeleton (see Fig. 24-3b). A tough, flexible, nonliving cuticle encloses and protects the thin, elongated body and is periodically molted. Sensory organs in the roundworm head transmit information to a simple brain composed of a nerve ring.

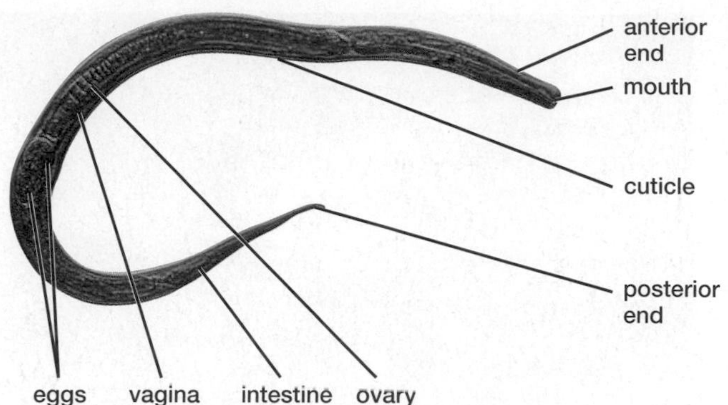

▲ **FIGURE 24-28 A freshwater nematode** In this micrograph, eggs can be seen inside a female freshwater nematode.

Nematodes lack circulatory and respiratory systems. Because most nematodes are extremely thin and have low energy requirements, diffusion suffices for gas exchange and distribution of nutrients. Most nematodes reproduce sexually, and the sexes are separate.

A Few Roundworm Species Are Harmful to Humans

Fifty roundworm species are known to infect humans. Most such worms are relatively harmless, but there are important exceptions. For example, hookworm larvae (found in soil in some tropical regions) can bore into human feet, enter the bloodstream, and travel to the intestine, where they cause continuous bleeding. Another dangerous roundworm parasite, *Trichinella*, causes the disease trichinosis. *Trichinella* worms can infect people who eat undercooked infected pork, which can contain up to 15,000 larval cysts per gram (**FIG. 24-29a**). The cysts hatch in the human digestive tract and invade blood vessels and muscles, causing bleeding and muscle damage.

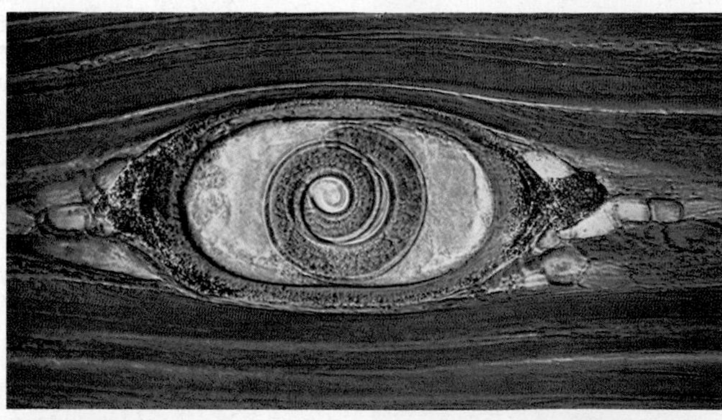

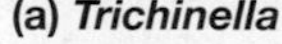

(a) *Trichinella*

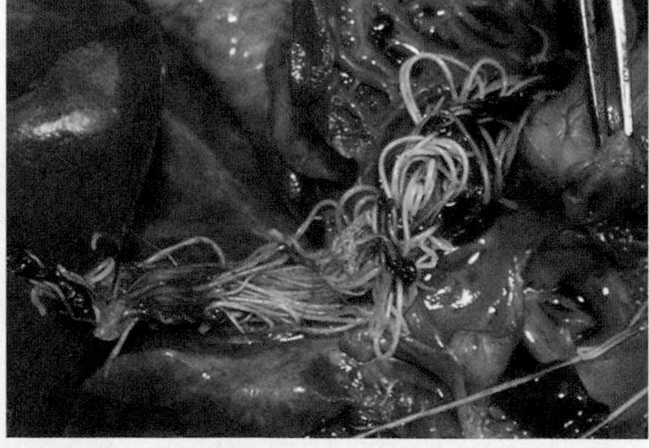

(b) Heartworms

▲ **FIGURE 24-29 Some parasitic nematodes (a)** Encysted larva of the *Trichinella* worm in the muscle tissue of a pig, where it may live for up to 20 years. **(b)** Adult heartworms in the heart of a dog. The juveniles are released into the bloodstream, where they may be ingested by mosquitoes and passed to another dog by the bite of an infected mosquito.

Parasitic roundworms can also endanger domestic animals. Dogs, for example, are susceptible to heartworm (**FIG. 24-29b**), which is transmitted by mosquitoes. In the southern United States, and increasingly in other parts of the country, heartworm poses a threat to the health of unprotected pets.

CASE STUDY CONTINUED

Physicians' Assistants

Leeches are not the only parasites that have been used to treat health problems. Some people have used hookworms and other parasitic roundworms to treat autoimmune diseases or severe allergies, in which the immune system inappropriately attacks the body's own tissues. Parasitic roundworms have evolved the ability to survive longer in their host by suppressing the host's immune system, and this immune suppression seems to ease the symptoms of people with immune system disorders who have infected themselves with parasitic worms. Unlike medical treatment with leeches, however, parasitic roundworms do not have the FDA's stamp of approval. Worm infections, after all, can make a person sick, and immune suppression associated with worm infections can increase susceptibility to other infectious diseases. Despite the risks, however, patients desperate for relief from debilitating, chronic disorders may infect themselves. Clinical researchers are working hard to discover methods to make treatment with parasitic worms safe and effective.

Echinoderms Have a Calcium Carbonate Skeleton

Echinoderms (Echinodermata) are found only in marine environments, and their common names tend to evoke their saltwater habitats: sand dollars, sea urchins, sea stars (or starfish), sea cucumbers, and sea lilies (**FIG. 24-30**). The name "echinoderm" ("hedgehog skin") stems from the bumps or spines that extend from the skin of most echinoderms. These spines are especially well developed in sea urchins and much reduced in sea stars and sea cucumbers. Echinoderm bumps and spines are actually extensions of an **endoskeleton** (internal skeleton) composed of plates of calcium carbonate that lie beneath the outer skin.

Echinoderms Are Bilaterally Symmetrical as Larvae and Radially Symmetrical as Adults

Echinoderms exhibit deuterostome development and are linked by common ancestry with the other deuterostome phyla, including the chordates (described in Chapter 25). Deuterostomes form a group of branches on the larger evolutionary tree of bilaterally symmetrical animals, but in echinoderms, only embryos and free-swimming larvae are bilaterally symmetrical. An adult echinoderm, in contrast, is radially symmetrical and lacks a head. This absence of cephalization is consistent with the sluggish lifestyle of echinoderms. Most echinoderms move very slowly as they feed on algae or small particles sifted from sand or water. Some echinoderms, however, are predators. Sea stars, for example, slowly pursue even slower-moving prey, such as snails or clams.

Echinoderms Have a Water-Vascular System

Echinoderms move on numerous tiny *tube feet*, delicate cylindrical projections that extend from the lower surface of the body and terminate in a suction cup. Tube feet are part of a unique echinoderm feature, the *water-vascular system*, which functions in locomotion, respiration, and food capture (**FIG. 24-31**). Seawater enters through an opening (the *sieve plate*) on the animal's upper surface and is conducted through a circular central canal, from which branch a number of radial canals. These canals conduct water to the tube feet, each of which is controlled by a muscular squeeze bulb known as an *ampulla*. Contraction of the bulb forces water into the tube foot, causing it to extend. The

(a) Sea cucumber

(b) Sea urchin

(c) Sea star

▲ **FIGURE 24-30 The diversity of echinoderms (a)** A sea cucumber feeds on debris in the sand. **(b)** A sea urchin's spines are actually projections of the internal skeleton. **(c)** Most sea stars have five arms, but some species have 20 or more.

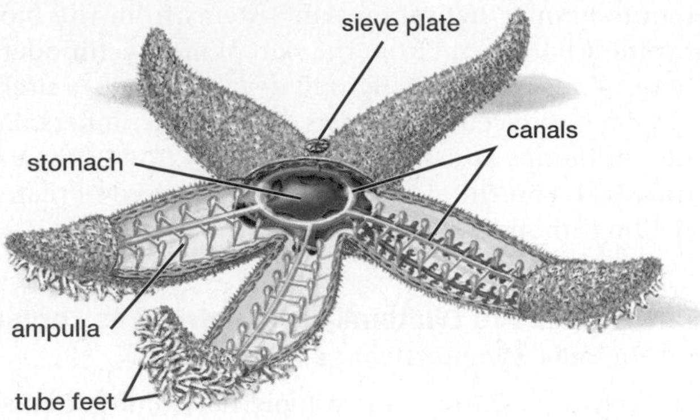

(a) Sea star body plan

(b) Sea star consuming a mussel

▲ **FIGURE 24-31 The water-vascular system of echinoderms (a)** Changing pressure inside the seawater-filled water-vascular system extends or retracts the tube feet. **(b)** The sea star often feeds on mollusks such as this mussel. A feeding sea star attaches numerous tube feet to the mussel's shells, exerting a relentless pull that slightly separates the shells. Then, the sea star turns the delicate tissue of its stomach inside out, extending it through its centrally located ventral mouth. The stomach can fit through an opening in the bivalve shells that is as narrow as 1 millimeter wide. Once pushed between the shells, the stomach tissue secretes digestive enzymes that weaken the mollusk, causing it to open further. Partially digested food is transported to the upper portion of the stomach, where digestion is completed.

suction cup may be pressed against the seafloor or a food object, to which it adheres tightly until its internal pressure is released.

Some Echinoderm Organ Systems Are Simplified

Echinoderms have a relatively simple nervous system with no distinct brain. Movements are loosely coordinated by a system consisting of a nerve ring that encircles the esophagus, radial nerves to the rest of the body, and a nerve network through the epidermis. In sea stars, simple receptors for light and chemicals are concentrated on the arm tips, and sensory cells are scattered over the skin. In some brittle star species, light receptors are associated with tiny lenses, smaller than the width of a human hair, that gather light and focus it on receptors. These "microlenses" are formed from crystals of calcite (calcium carbonate) and their optical quality is excellent, far superior to that of any human-created lens of comparable size. Researchers hypothesize that each of the thousands of lenses on a brittle star forms a tiny image and that the animal integrates the resulting information to detect changes in its surroundings, such as the approach of a predator.

Echinoderms lack a circulatory system, although movement of the fluid in their well-developed coelom serves this function. Gas exchange occurs through the tube feet and, in some forms, through numerous tiny "skin gills" that project through the epidermis. Most species have separate sexes and reproduce by shedding sperm and eggs into the water, where fertilization occurs.

Many echinoderms are able to regenerate lost body parts, and these regenerative powers are especially potent in sea stars. In fact, a single arm of a sea star is capable of developing into a whole animal, provided that part of the central body is attached to it. Before this ability was widely appreciated, mussel fishermen often tried to rid mussel beds of predatory sea stars by hacking them into pieces and throwing the pieces back. Needless to say, the strategy backfired.

Some Chordates Are Invertebrates

The *chordates* (Chordata) include the vertebrate animals and also a few groups of invertebrates, such as the sea squirts and the lancelets. We will discuss these invertebrate chordates and their vertebrate relatives in Chapter 25.

CHECK YOUR LEARNING

Can you . . .

- describe the basic body plans of sponges, cnidarians, comb jellies, flatworms, annelids, mollusks, arthropods, roundworms, and echinoderms?
- list some member organisms in each of these groups and describe the organisms' nervous and sensory systems and methods of circulation, gas exchange, digestion and reproduction?
- give examples of the effects invertebrate animals have on humans?

CASE STUDY REVISITED

Physicians' Assistants

Another invertebrate animal that has been recruited for medical duty is the blowfly or, more precisely, blowfly larvae, commonly known as maggots (**FIG. 24-32**). Blowfly maggots have proved to be effective at ridding wounds and ulcers of dead and dying tissue. If such tissue is not removed, it can interfere with healing or lead to infection. Traditionally, dead tissue in wounds is removed by a physician wielding a scalpel, but maggots offer an increasingly common alternative treatment. In this treatment, a bandage containing day-old, sterile maggots is applied to the wound. The maggots consume dead or dying tissue, secreting digestive enzymes that do not harm healthy skin or bone. After a few days, the maggots have grown to the size of rice kernels and are removed. The treatment is repeated until the wound is clean.

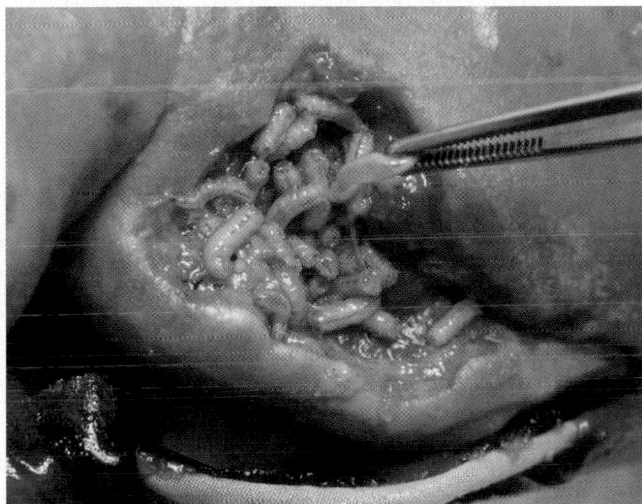

▲ **FIGURE 24-32 Blowfly maggots can clean wounds**

For maggot therapy, not just any maggot will do. In particular, the maggots must be sterile, which rules out getting them from rotting meat or restaurant garbage bins. Instead, physicians order flies from government-approved medical supply companies. Some companies provide sterile maggots in opaque bags that can be slipped over wounds, so that squeamish patients and physicians need not see the maggots at work in the wound. But other suppliers believe that bagging limits the maggots' effectiveness, and offer only loose, "free-range" maggots.

CONSIDER THIS Medical treatment with invertebrate animals is typically much less expensive than drugs, surgery, and other options. Nonetheless, very little research funding is directed to promising but challenging medical uses of animals, such as treatment of autoimmune diseases with parasitic roundworms. Would it be a good idea to increase research on such treatments, or are they too risky?

CHAPTER REVIEW

Go to **Mastering Biology** to access the Pearson eText, vocabulary review, practice quizzes, activities, videos, current events, and more.

*Answers to **Think Critically** and **Thinking Through the Concepts** questions can be found in the **Answers** section at the back of the book.*

Summary of Key Concepts

24.1 What Are the Key Features of Animals?

Animals are eukaryotic, multicellular, sexually reproducing organisms that acquire energy by consuming other organisms. Most animals can perceive and react rapidly to environmental stimuli and are motile at some stage in their lives. Their cells lack cell walls.

24.2 Which Anatomical Features Mark Branch Points on the Animal Evolutionary Tree?

The earliest animals had no tissues, a feature retained by modern sponges. All other modern animals have tissues. Animals with tissues can be divided into radially symmetrical and bilaterally symmetrical groups. During embryonic development, radially symmetrical animals have two germ layers; bilaterally symmetrical animals have three. Bilaterally symmetrical animals also tend to have sense organs and clusters of neurons concentrated in the head, a process called cephalization. Bilateral phyla can be divided into two main groups, one of which undergoes protostome development, the other of which undergoes deuterostome development. Protostome phyla can in turn be divided into ecdysozoans and lophotrochozoans. Some phyla of bilaterally symmetrical animals lack body cavities, but most have either pseudocoeloms or true coeloms.

24.3 What Are the Major Animal Phyla?

The bodies of sponges (Porifera) come in a variety of shapes and are generally sessile. Sponges lack tissues but have three different types of specialized cells. Digestion occurs exclusively within the individual cells.

The sea jellies, corals, anemones, and hydrozoans (Cnidaria) have tissues. A simple network of nerve cells directs the activity of contractile cells, allowing loosely coordinated movements. Digestion is extracellular, occurring in a central gastrovascular cavity with a single opening. Cnidarians exhibit radial symmetry, an adaptation to both the free-floating lifestyle of the medusa and the sedentary existence of the polyp.

Comb jellies (Ctenophora) are superficially similar to sea jellies but form a separate taxonomic group. Comb jellies are small, radially symmetrical carnivores that move using eight rows of cilia.

Flatworms (Platyhelminthes) have a distinct head with sensory organs and a simple brain. A system of canals that form a network throughout the body aids in excretion. They lack a body cavity.

The segmented worms (Annelida) are the most complex worms, with a well-developed closed circulatory system and excretory organs. The segmented worms have a compartmentalized digestive system, which processes food in a sequence. Annelids also have a true coelom, a fluid-filled space between the body wall and the internal organs.

The clams, snails, octopuses, and squid (Mollusca) lack a skeleton; some forms protect the soft, moist, muscular body with a single shell (many gastropods and a few cephalopods) or a pair of hinged shells (the bivalves). Most mollusks live in aquatic or moist terrestrial environments and have an open circulatory system, where blood directly bathes internal organs in a hemocoel. The octopus has the most complex brain and the best-developed learning capacity of any invertebrate.

Arthropods (Arthropoda), the insects, arachnids, millipedes and centipedes, and crustaceans, are the most diverse and abundant animals on Earth. Jointed appendages and well-developed nervous systems make possible complex, finely coordinated behavior. The exoskeleton (which conserves water and provides support) and specialized respiratory structures (which remain moist and protected) enable the insects and arachnids to inhabit dry land. The diversification of insects has been enhanced by their ability to fly. Crustaceans, which include the largest arthropods, are restricted to moist, usually aquatic habitats and respire using gills.

The pseudocoelomate roundworms (Nematoda) possess a separate mouth and anus and a cuticle layer that is molted.

The sea stars, sea urchins, and sea cucumbers (Echinodermata) are an exclusively marine group. Echinoderm larvae are bilaterally symmetrical; however, the adults show radial symmetry. This, in addition to a primitive nervous system that lacks a definite brain, adapts them to a relatively sedentary existence. Echinoderm bodies are supported by an endoskeleton that sends projections through the skin. The water-vascular system, which functions in locomotion, feeding, and respiration, is a unique echinoderm feature.

The chordates (Chordata) include two invertebrate groups, the lancelets and sea squirts, as well as the vertebrates.

Thinking Through the Concepts

Bloom's: Remembering, Understanding

Multiple Choice

1. Which of the following groups contains radially symmetrical organisms?
 a. arthropods
 b. cnidarians
 c. mollusks
 d. roundworms

2. To acquire nutrients, sponges
 a. filter microorganisms from water brought in through pores.
 b. use stinging cells to capture small prey.
 c. parasitize larger animals.
 d. suck nutrients from prey after injecting enzymes that dissolve prey tissues.

3. A coelom is
 a. a body cavity that is partially lined with tissue derived from mesoderm.
 b. a body cavity that is completely lined with tissue derived from mesoderm.
 c. found in sponges and cnidarians.
 d. never found in bilaterally symmetrical organisms.

4. Which of the following is *not* an advantage of an arthropod exoskeleton?
 a. It provides protection from predators.
 b. It provides attachment sites for muscles.
 c. It improves sensory perception.
 d. It protects against water loss on land.

5. Which of the following groups does *not* include organisms in which food enters and digestive wastes exit through the same opening?
 a. annelids
 b. flatworms
 c. comb jellies
 d. cnidarians

Fill-in-the-Blank

1. Animals obtain energy by ____________; they generally reproduce ____________, and their cells lack ____________.

2. Bilaterally symmetrical animals have ____________ embryonic tissue layers, known as ____________, ____________, and ____________. Radially symmetrical animals have ____________ tissue layers; they lack the ____________ layer.

3. Animals that have an anterior and posterior end are said to be ____________. The anterior end of such animals often contains structures used for ____________ and ____________. Coelomate animals have a body ____________ that is completely lined with tissue derived from ____________.

4. Lophotrochozoans and ecdysozoans (molting animals) are two large clades of animals that exhibit ____________ development. The other major type of development found in bilateral animals is known as ____________ development, and is present in animals in the clades known as ____________ and ____________.

5. Animals that lack a backbone are described as ____________; those that possess a backbone are ____________. The vast majority of all animals fall into which of these two groups? ____________ The only animals that lack tissues are ____________, whose bodies resemble a colony of ____________. Sea anemones and corals are ____________. Earthworms and leeches are ____________.

6. Three major groups within the mollusks are the two-shelled clams and scallops called ____________; the foot-crawling snails and slugs called ____________; and the tentacled squid and octopuses called ____________. Members of the largest animal phylum are called ____________. Three important groups within this phylum are the six-legged, often flying ____________; the eight-legged spiders and mites called ____________; and the mostly aquatic ____________.

7. Phyla that contain animals with segmented bodies include ____________ and ____________. In a(n) ____________ system, blood is confined to blood vessels. In a(n) ____________ system, blood bathes internal organs within a cavity called the ____________.
8. For each of the following distinctive structures, name the animal group in which it is found:
mantle: ____________
cnidocyte: ____________
water-vascular system: ____________
paired, jointed appendages: ____________

Review Questions

1. List the characteristics that, taken together, distinguish animals from other kinds of organisms.
2. List the distinguishing characteristics of each phylum discussed in this chapter, and give an example of a member of each phylum.
3. Briefly describe each of the following adaptations, and explain its adaptive significance: bilateral symmetry, radial symmetry, cephalization, closed circulatory system, coelom, segmentation.
4. Describe and compare the respiratory systems of the four major arthropod groups.
5. Describe the advantages and disadvantages of the arthropod exoskeleton.
6. In which of the three major mollusk groups is each of the following characteristics found?
   a. two hinged shells
   b. a radula
   c. tentacles
   d. some sessile members
   e. the best-developed brains
   f. numerous eyes
7. Give three functions of the water-vascular system of echinoderms.
8. List some invertebrate animals that can harm humans, and, for each animal that you list, name its taxonomic group.

Applying the Concepts

Bloom's: Applying, Analyzing, Evaluating

1. Insects are the largest group of animals on Earth. Insect diversity is greatest in the tropics, where habitat destruction and species extinction are occurring at an alarming rate. What biological, economic, and ethical arguments can you advance to persuade people and governments to preserve this biological diversity?
2. Discuss and defend the attributes you would use to define biological success among animals. Are humans a biological success by these standards? Why?

25 Animal Diversity II: Vertebrates

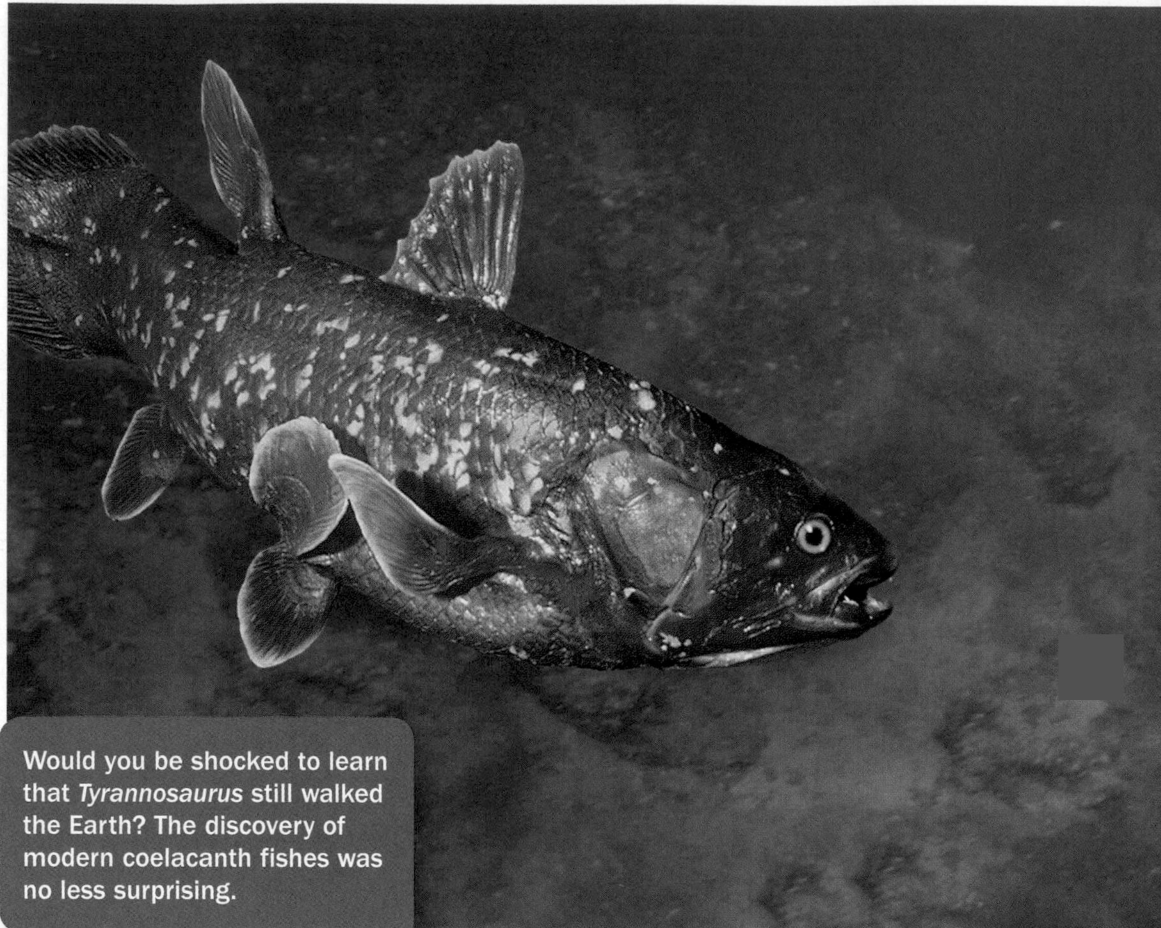

Would you be shocked to learn that *Tyrannosaurus* still walked the Earth? The discovery of modern coelacanth fishes was no less surprising.

CASE STUDY

Fish Story

ON DECEMBER 22, 1938, Marjorie Courtenay-Latimer received a phone call that would lead to one of the most spectacular discoveries in biological history. The call was from a local fisherman whom Courtenay-Latimer, the curator of a small museum in South Africa, had asked to collect some fish specimens for the museum. His boat had returned from its most recent voyage and was waiting at the town dock. Dutifully, Courtenay-Latimer went to the boat and began sorting through the fish that were strewn across the deck. Later, she wrote, "I noticed a blue fin sticking up from beneath the pile. I uncovered the specimen, and, behold, there appeared the most beautiful fish I had ever seen." In addition to its beauty, the fish had some odd features, including fins that were stumpy and lobed, unlike the fins of any other living species.

Courtenay-Latimer did not recognize the strange fish, but she knew it was unusual. She tried to find a place to refrigerate it, but in her small town she was unable to find a cold storage facility willing to store a fish. In the end, she was able to save only the skin. Undaunted, she made some drawings of the fish and used them to attempt an identification. To her amazement, the creature did not resemble any species known to inhabit the waters off South Africa, but did seem similar to members of a family of fishes known as coelacanths. The only problem was that coelacanths were known only from fossils. As far as anyone knew, coelacanths had been extinct for 80 million years!

Perplexed, Courtenay-Latimer sent her drawings to J. L. B. Smith, a fish expert at Rhodes University. Smith was astounded when he saw the sketch, later writing that "a bomb seemed to burst in my brain." Although bitterly disappointed that the specimen's bones and internal organs had been lost, Smith arranged to view the preserved skin. Ultimately, he confirmed the astonishing news that coelacanths still swam in Earth's waters.

Although coelacanths remained hidden from science until the twentieth century, they share a phylum with frogs, dogs, snakes, and many other familiar animals, including humans. What do we know about these animals?

AT A GLANCE

25.1 What Are the Key Features of Chordates?

25.2 Which Animals Are Chordates?

25.3 What Are the Major Groups of Vertebrates?

25.1 WHAT ARE THE KEY FEATURES OF CHORDATES?

Humans are members of a taxonomic group known as the chordates (Chordata). Chordates (**FIG. 25-1**) include not only bony animals like birds and apes, but also the barrel-shaped tunicates and small fishlike creatures called lancelets. What characteristics do we share with these animals?

All Chordates Share Four Distinctive Structures

All chordates have deuterostome development (which is also characteristic of echinoderms; see Chapter 24) and are further united by four features they all possess at some stage of their lives: a dorsal nerve chord, a notochord, pharyngeal gill slits, and a post-anal tail.

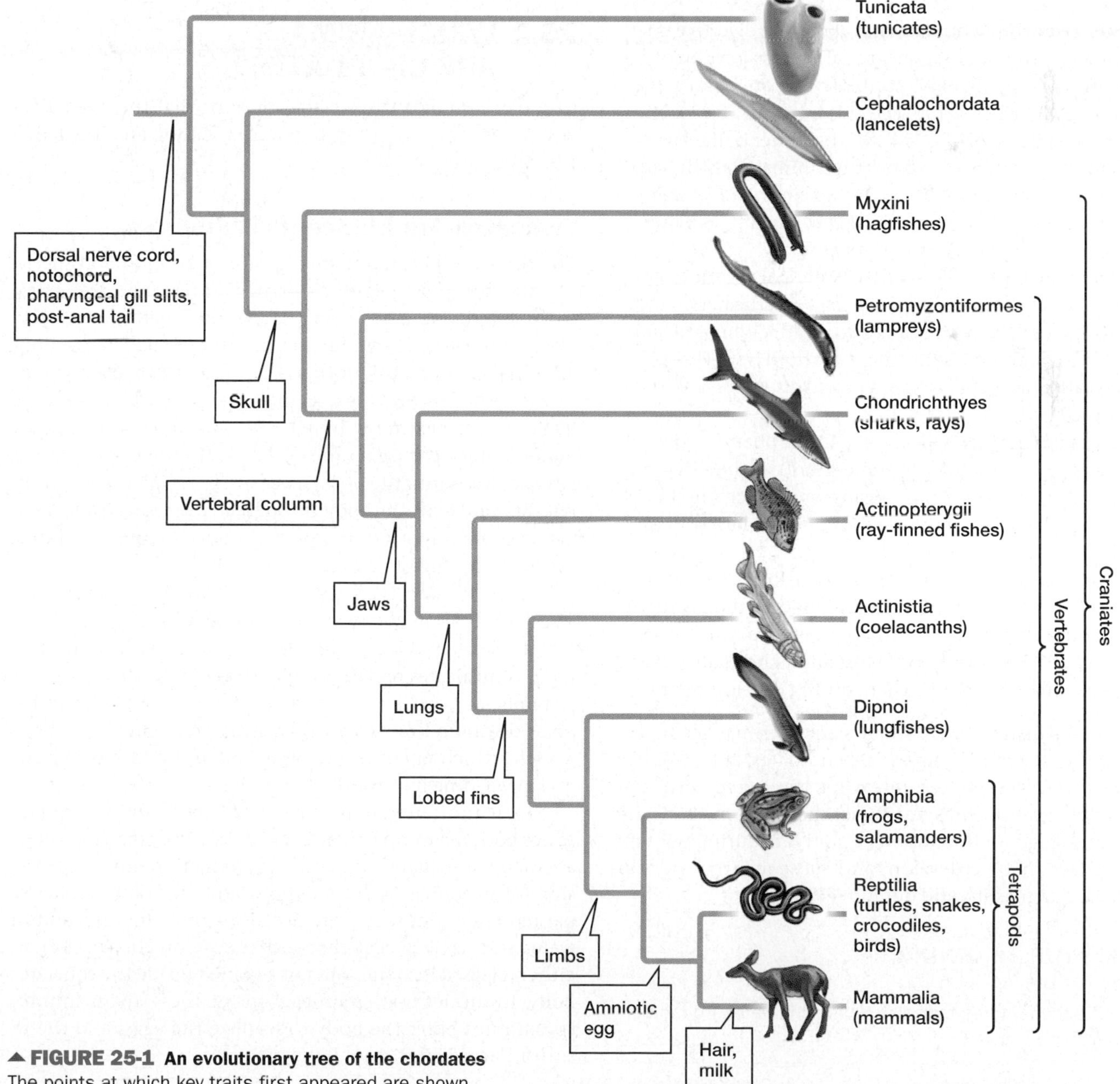

▲ **FIGURE 25-1 An evolutionary tree of the chordates**
The points at which key traits first appeared are shown.

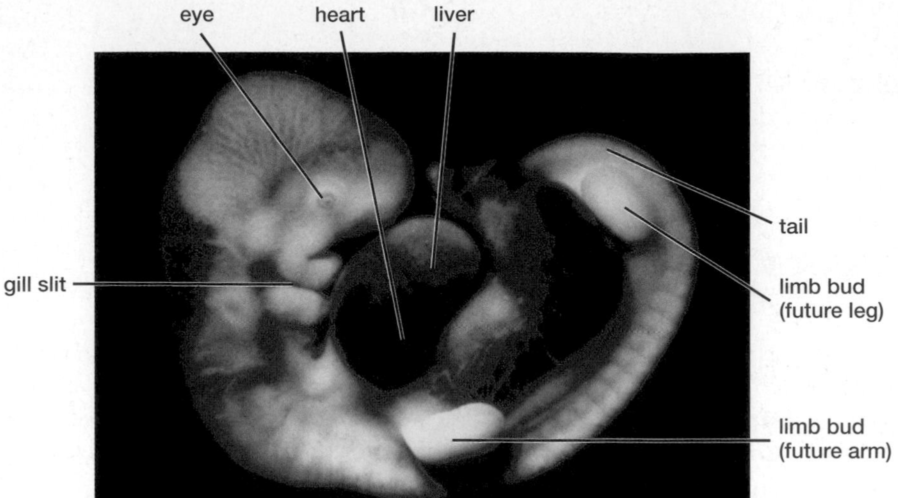

FIGURE 25-2 Chordate features in the human embryo This 5-week-old human embryo is about 1 centimeter long and clearly shows a tail and external gill slits (more properly called grooves, since they do not penetrate the body wall). Although the tail will disappear completely, the gill grooves contribute to the formation of the lower jaw.

- The **nerve cord** of chordates lies above the digestive tract, running lengthwise along the dorsal (upper) portion of the body. In contrast, the nerve cords of other animals lie in a ventral position, below the digestive tract (see Fig. 24-14). A chordate's nerve cord is hollow—its center is filled with fluid, unlike the nerve cords of other animals, which have solid nerve tissue throughout. During embryonic development in chordates, the nerve cord develops a thickening at its anterior end that becomes the brain.
- The **notochord** is a stiff rod that extends along the length of the body, between the digestive tract and the nerve cord. It provides support for the body and an attachment site for muscles. In many chordates, the notochord is present only during early stages of development and disappears as a skeleton develops.
- **Pharyngeal gill slits** are located in the pharynx (the cavity behind the mouth). In some chordates the slits form functional openings for gills (organs for gas exchange in water); in others they appear only as grooves during an early stage of development.
- The **post-anal tail** is a posterior extension of the chordate body that extends past the anus and contains muscle tissue and the rearmost portion of the nerve cord. Other animals lack this kind of tail. Most adult chordates have tails, but some species lose them during development.

This list of distinctive chordate structures may seem puzzling because, although humans are chordates, at first glance we seem to lack every feature except the nerve cord. But evolutionary relationships are sometimes seen most clearly during early stages of development, and it is during our embryonic phase that we develop, and subsequently lose, our notochord, our gill slits, and our tail (**FIG. 25-2**).

CHECK YOUR LEARNING

Can you . . .

- describe the features that distinguish chordates from other animals?

25.2 WHICH ANIMALS ARE CHORDATES?

Chordates include three clades (groups that include all the descendants of a common ancestor): the tunicates, the lancelets, and the craniates.

Tunicates Are Marine Invertebrates

The tunicates (Tunicata) are a group of about 2,300 species of marine invertebrate chordates. Tunicates are small, with lengths ranging from a few millimeters to 1 foot (30 centimeters). The group includes immobile, filter-feeding, vase-shaped animals known as sea squirts (**FIG. 25-3**). Much of a sea squirt's body is occupied by its pharynx, which is like a basket perforated by gill slits and lined with mucus. Water enters the sea squirt's body through an *incurrent siphon*, passes into the pharynx at its top, moves through the gill slits, and exits the body through an *excurrent siphon*. Food particles are trapped in the basket's mucous lining and then moved to the digestive tract.

Adult sea squirts are sessile—they live firmly attached to a surface. Sea squirt larvae, however, swim actively and possess the four chordate features (see Fig. 25-3, left). Some other types of tunicates remain mobile throughout their lives. For example, barrel-shaped tunicates known as salps live in the open ocean and move by contracting an encircling band of muscle, which forces a jet of water out of the back of the animal, propelling it forward.

Most tunicates are *hermaphroditic* (each individual possesses both male and female sex organs). They may reproduce asexually or sexually. In asexual reproduction, miniature versions of an adult grow from its body and then drop off. In sexual reproduction, sperm are broadcast into the surrounding water and fertilize eggs that (depending on the species) are either released into the water or retained inside the tunicate's body. In species that retain eggs inside the body, swimming sperm must enter the body to fertilize the eggs, and the resulting larvae must swim out.

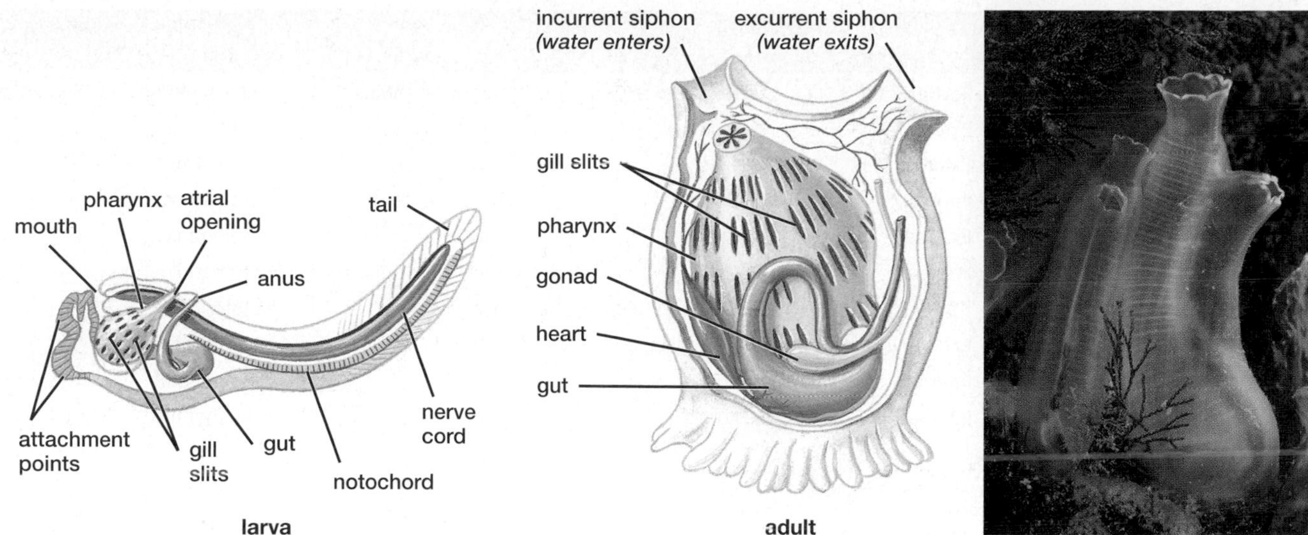

▲ **FIGURE 25-3 Sea squirt** The larva (left) of a sea squirt (a type of tunicate) exhibits all the diagnostic features of chordates. The adult sea squirt (middle) has lost its tail and notochord and has assumed a sedentary life (right).

Lancelets Live Mostly Buried In Sand

The 30 or so species of lancelets (Cephalochordata) form another group of invertebrate chordates. Lancelets are small (2 inches, or about 5 centimeters, long), fishlike animals that retain all four chordate features as adults (**FIG. 25-4**). An adult lancelet spends most of its time half-buried in the sandy sea bottom, with only the anterior end of its body exposed. The motion of cilia in the pharynx draws seawater into the lancelet's mouth. As the water passes through the pharyngeal gill slits, a film of mucus filters tiny food particles from the water. The captured food particles are transported to the lancelet's digestive tract.

Lancelets have separate sexes and always reproduce sexually. At particular times of year, most of the males and females in an area simultaneously release gametes (eggs and sperm) into the surrounding water. Fertilized eggs develop into microscopic larvae, which swim slowly and drift about during several weeks of continuing growth and development before dropping to the seabed and completing their transformation to the adult form.

Craniates Have a Skull

The **craniates** include all chordates that have a skull that encloses the brain. The skull may be composed of bone or cartilage, a tissue that resembles bone but is less brittle and more flexible. The earliest known craniates, whose fossils were found in 530-million-year-old rocks, resembled lancelets but had skulls and eyes. The mouths of the earliest craniates lacked jaws.

▼ **FIGURE 25-4 Lancelet** A lancelet, a fishlike invertebrate chordate. The adult organism exhibits all the chordate features.

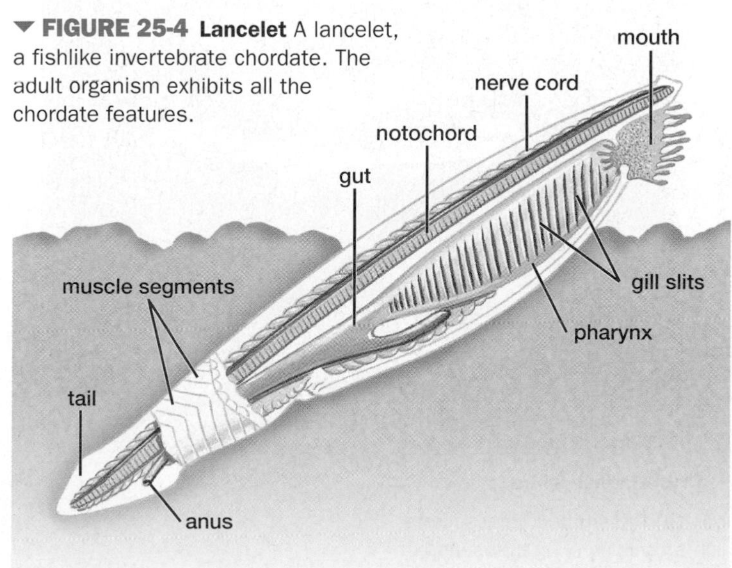

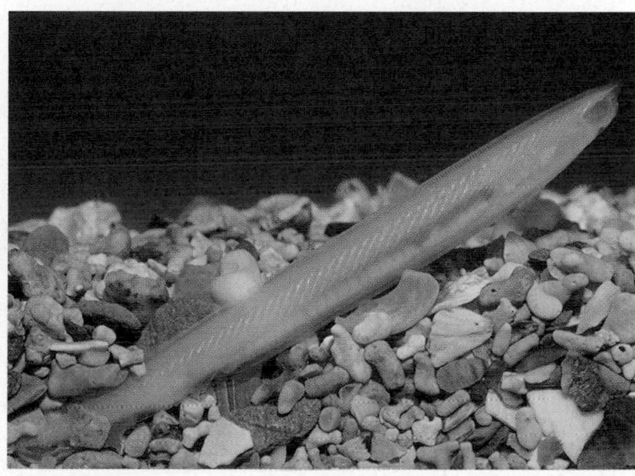

TABLE 25-1 Comparison of Craniate Groups

Group	Fertilization	Respiration	Heart Chambers	Body Temperature Regulation
Hagfishes (Myxini)	External	Gills	Two	Ectothermic
Lampreys (Petromyzontiformes)	External	Gills	Two	Ectothermic
Cartilaginous fishes (Chondrichthyes)	Internal	Gills	Two	Ectothermic
Ray-finned fishes (Actinopterygii)	External[1]	Gills	Two	Ectothermic
Coelacanths (Actinistia)	Internal	Gills	Two	Ectothermic
Lungfishes (Dipnoi)	External	Gills and lungs	Two	Ectothermic
Amphibians (Amphibia)	External or internal[2]	Skin, gills, and lungs	Three	Ectothermic
Reptiles (Reptilia)	Internal	Lungs	Three[3]	Ectothermic[4]
Mammals (Mammalia)	Internal	Lungs	Four	Endothermic

[1]A relatively small number of ray-finned fish have internal fertilization.
[2]External in most frogs and toads; internal in caecilians and most salamanders.
[3]Except for birds and crocodilians, which have four chambers.
[4]Except for birds, which are endothermic.

Today, craniates include two subgroups: the hagfishes and the **vertebrates**, which are animals in which the embryonic notochord is replaced during development by a backbone, or **vertebral column**, composed of bone or cartilage. **TABLE 25-1** summarizes some characteristics of the craniate groups described in the rest of this chapter.

Hagfishes Are Slimy Residents of the Ocean Floor

As did ancestral craniates, hagfishes (Myxini) lack jaws. Instead, they use a tongue-like, tooth-bearing structure to grind and tear food. A hagfish body is stiffened by a notochord, but its skeleton is limited to a few small cartilaginous elements, one of which forms a rudimentary skull. Because hagfishes lack skeletal elements that surround the nerve cord to form a vertebral column, most systematists do not consider them to be vertebrates, although they are the vertebrates' closest relatives.

The 75 or so species of hagfishes are exclusively marine (**FIG. 25-5a**). They respire using gills, have a two-chambered heart, and are ectothermic—that is, their body temperature depends on the temperature of their external environment. (Gills, two-chambered hearts, and ectothermy are also found in all vertebrate fishes.) Hagfishes live near the ocean floor, often burrowing in the mud, and feed primarily on worms. They will, however, eagerly attack dead and dying fish, using their teeth to burrow into a fish's body and consume its soft internal organs.

Hagfishes secrete slime as a defense against predators (**FIG. 25-5b**). When attacked by a predatory fish such as a shark, a hagfish quickly secretes a massive quantity of slime, which fills the mouth and gills of the would-be attacker, causing it to flee to avoid suffocation. A hagfish removes slime from its own body by twisting its body into a knot, which it slides forward over its head, scraping off the slime. Hagfish

(a) Hagfish

(b) Hagfish slime

FIGURE 25-5 Hagfishes (a) Hagfishes forage near the ocean floor. **(b)** If disturbed, they secrete copious slime. This scene shows the aftermath of an accident in which a truck carrying thousands of live hagfishes overturned, spilling the animals onto the roadway and causing an outpouring of slime and a multicar crash.

slime contains a mixture of mucus and protein threads that are extremely long, highly elastic, and very strong. Researchers are currently investigating the structure of the protein threads in hope of developing useful materials, such as fabrics that mimic the slime's strength and elasticity.

Vertebrates Have a Backbone

The bony or cartilaginous vertebral column of a vertebrate supports its body, provides attachment sites for muscles, and protects the delicate nerve cord and brain. It is also part of a living internal skeleton that can grow and repair itself.

The early history of vertebrates was characterized by an array of strange, now-extinct jawless fishes, many of which were protected by bony armor plates. About 425 million years ago, jawless fishes gave rise to a group of fish that possessed an important new structure: jaws. Jaws allowed fish to grasp, tear, or crush their food, permitting them to exploit a much wider range of food sources than could jawless fish. Today, most (but not all) vertebrates have jaws.

Vertebrates have other adaptations that have contributed to their successful invasion of most habitats. One such adaptation is paired appendages. These first appeared as fins in fish and served as stabilizers for swimming. Over millions of years, some fins were modified by natural selection into legs that allowed animals to crawl onto dry land, and later into wings that allowed some to take to the air. Another adaptation that has contributed to the success of vertebrates is an increase in the size and complexity of their brains and sensory structures, which allow vertebrates to perceive their environment in detail and to respond to it in a great variety of ways.

CHECK YOUR LEARNING

Can you . . .

- name and describe the chordates that are not craniates?
- name and describe the craniates that are not vertebrates?
- describe the key adaptations of vertebrates?

CASE STUDY **CONTINUED**

Fish Story

In the years since Courtenay-Latimer's discovery that coelacanths are not extinct, scientists have had the opportunity to investigate the creature's anatomy. The coelacanth body has some unusual features. For example, adult coelacanths retain a notochord, the body-stiffening rod that most other vertebrates lose during embryonic development. In addition, a coelacanth's brain is very small relative to its body size. The brain of a 90-pound (40-kilogram) coelacanth weighs only 1 or 2 grams (less than a tenth of an ounce). The tiny brain occupies less than 2% of the space in the cranial cavity; the rest is filled with fat.

Their anatomical oddities aside, coelacanths are vertebrates. What features distinguish them from other types of vertebrates? More generally, how do major groups of vertebrates differ?

25.3 WHAT ARE THE MAJOR GROUPS OF VERTEBRATES?

Vertebrates include lampreys, cartilaginous fishes, ray-finned fishes, coelacanths, lungfishes, amphibians, reptiles, and mammals.

Some Lampreys Parasitize Fish

Like hagfishes, the roughly 50 species of lampreys (Petromyzontiformes) are jawless. A lamprey is recognizable by the large, rounded sucker that surrounds its mouth and by the single nostril on the top of its head. The nerve cord of a lamprey is protected by segments of cartilage, so lampreys are considered to be true vertebrates. They live in both fresh and salt water, but the marine forms must return to fresh water to spawn. Lampreys migrate up shallow streams to spawn; eggs are deposited and fertilized in depressions that groups of lampreys excavate in the streambed. The adults die a short time after spawning. After the young hatch, they spend several years in the stream as larvae, eating algae, before maturing and moving downstream to their adult habitat in an ocean, lake, or river.

Adult lampreys of some species are parasitic. A parasitic lamprey uses its tooth-lined mouth to attach itself to a larger fish (**FIG. 25-6**). Using rasping teeth on its tongue, the lamprey excavates a hole in the host's body wall, through which it sucks blood and body fluids. Beginning in the 1920s, parasitic lampreys spread into the Great Lakes. There, in the absence of effective predators, they have multiplied prodigiously and greatly reduced commercial fish populations. Vigorous measures to control the lamprey population have allowed some recovery of fish populations in the Great Lakes.

Cartilaginous Fishes Are Marine Predators

The cartilaginous fishes (Chondrichthyes) include about 1,200 marine species, among them the sharks, skates, and rays

▲ **FIGURE 25-6 Lampreys** Some adult lampreys are parasitic, and use sucker-like mouths lined with rasping teeth to attach to fish.

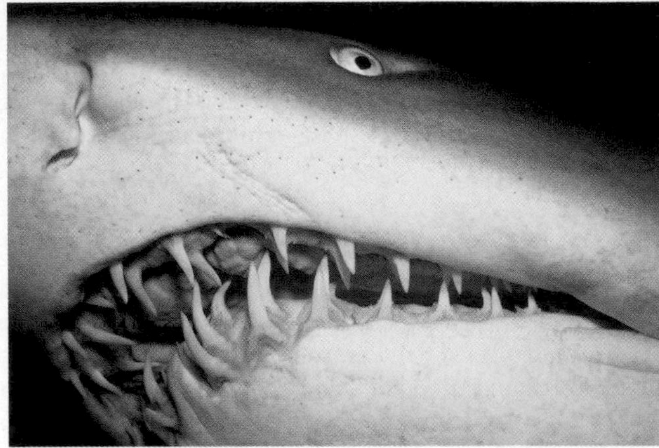

(a) Shark

(b) Ray

▲ **FIGURE 25-7 Cartilaginous fishes (a)** A shark displaying several rows of teeth. As the frontmost teeth are lost, they are replaced by the new ones behind them. **(b)** The tropical blue-spotted stingray swims by graceful undulations of lateral extensions of its body. Both sharks and rays lack a swim bladder and tend to sink toward the bottom when they stop swimming.

(**FIG. 25-7**). Unlike hagfishes and lampreys (but like all other vertebrates), cartilaginous fishes have jaws. They are graceful predators whose skeleton is formed entirely of cartilage. Their bodies are protected by leathery skin roughened by tiny scales. Although some must swim to circulate water through their gills, most can pump water across their gills. In contrast to the external fertilization that characterizes reproduction in almost all other fish, cartilaginous fish have internal fertilization, in which a male deposits sperm directly into a female's reproductive tract. Some cartilaginous fishes are very large. A whale shark, for example, can grow to more than 45 feet (14 meters) in length, and a manta ray may be more than 20 feet (6 meters) wide.

Although some sharks feed by filtering plankton (tiny animals and protists) from the water, most are predators of larger prey such as other fishes, marine mammals, sea turtles, crabs, or squid. Many sharks attack their prey with strong jaws that contain several rows of razor-sharp teeth; the back rows move forward as the front teeth are lost to age and use.

Have You Ever Wondered ... How Often Sharks Attack People?

Most sharks avoid humans, but large sharks of some species can be dangerous to swimmers and divers. However, shark attacks on people are rare. In the United States, dog attacks kill 30 times more people than sharks do. A U.S. resident is 75 times more likely to die from a lightning strike than from a shark attack, and a beachgoer is far more likely to drown than to be bitten by a shark. Nonetheless, unprovoked shark attacks do occur. During 2017, for example, there were 88 documented attacks in the world, 5 of them fatal. About 60% of attacks were on people who were surfing. To reduce the (already very small) risk of a shark attack, experts recommend several precautions. For example, stay in a group while in the water, because most shark attacks are on lone individuals. Stay out of the water at night, dawn, or dusk, when sharks are most active. Refrain from entering the water when bleeding from an open wound, because sharks can detect blood in the water. And avoid areas that are being actively fished, because sharks are attracted to baitfish.

Skates and rays are mostly bottom dwellers with flattened bodies, wing-shaped fins, and thin tails. Rays are generally larger than skates, but the most notable difference between the two groups is that rays give birth to live young, whereas skates lay eggs. Most skates and rays eat invertebrates. Some ray species defend themselves with a spine near their tail that can inflict dangerous wounds, and others produce a powerful electrical shock that can stun their prey. The largest ray species, the giant oceanic manta ray, can grow to 23 feet (7 meters) across and can weigh up to 3,000 pounds (1,360 kilograms).

Ray-Finned Fishes Are the Most Diverse Vertebrates

The vertebrate diversity crown belongs to the ray-finned fishes (Actinopterygii). About 32,000 species have been identified, and scientists estimate that perhaps twice this number exist, with many undiscovered species inhabiting deep waters and remote areas. Ray-finned fishes are found in nearly every watery habitat, both freshwater and marine.

Ray-finned fishes are distinguished by the structure of their fins, which consist of webs of skin supported by bony spines. In addition, ray-finned fishes have skeletons made of bone, a trait they share with the lobe-finned fishes and limbed vertebrates discussed later in this chapter. The skin of ray-finned fishes is covered with interlocking scales that provide protection while allowing for flexibility. Most ray-finned fishes have a swim bladder, a sort of internal balloon that allows a fish to float effortlessly at any level in the water. The swim bladder evolved from lungs, which were present (along with gills) in the ancestors of modern ray-finned fishes.

The ray-finned fishes include not only a large number of species but also a huge variety of different forms and lifestyles (**FIG. 25-8**). These range from snakelike eels to flattened flounders; from sluggish bottom feeders to speedy,

▼ **FIGURE 25-8 The diversity of ray-finned fishes** Ray-finned fishes have colonized nearly every aquatic habitat. **(a)** This female deep-sea anglerfish attracts prey with a living lure that projects just above her mouth. The fish is ghostly white; at the 6,000-foot (1,800-meter) depth where anglers live, no light penetrates and thus colors are unnecessary. Male deep-sea anglerfish are extremely small and remain permanently attached to the female, always available to fertilize her eggs. Two parasitic males can be seen attached to this female. **(b)** This tropical green moray eel lives in rocky crevices. The small fish (a banded cleaner goby) on its lower jaw eats parasites that cling to the moray's skin. **(c)** A sea horse may anchor itself with its prehensile tail (adapted for grasping) while feeding on small crustaceans.

THINK CRITICALLY With regard to water regulation (maintaining the proper amount of water in the body), how does the challenge faced by a freshwater fish differ from that faced by a saltwater fish?

(a) Anglerfish

(b) Moray eel

(c) Sea horse

streamlined predators; from brightly colored reef dwellers to translucent, luminescent deep-sea dwellers; from the massive 3,000-pound (1,350-kilogram) mola to the tiny stout infantfish, which weighs in at about 0.00003 ounce (1 milligram).

Ray-finned fishes are an extremely important source of food for humans. Unfortunately, however, the growing human population's appetite for ray-finned fishes, combined with increasingly effective high-tech methods for finding and catching them, has had a devastating impact on fish populations. Populations of almost all economically important ray-finned fish species have declined drastically. If overfishing continues, fish stocks are likely to collapse.

Coelacanths and Lungfishes Have Lobed Fins

Although almost all fish with bony skeletons belong to the ray-finned group, some bony fishes are coelacanths (Actinistia) or lungfishes (Dipnoi). Coelacanths are described in this chapter's case study. The six species of lungfishes are found in freshwater habitats in Africa, South America, and Australia (**FIG. 25-9**). Lungfishes have both gills and lungs. They tend to live in stagnant waters that may be low in oxygen, and their lungs allow them to supplement their supply of oxygen by breathing air. Lungfishes of several species are able to survive even if the pools they inhabit dry up completely. These fish burrow into mud and seal themselves in mucus-lined chambers. There, they breathe through their lungs and their metabolic rate declines drastically. When the rains return and the pools refill, the lungfishes leave their burrows and resume their underwater way of life.

Lungfishes and coelacanths are sometimes called *lobefins* because members of both groups have fleshy fins that contain rod-shaped bones surrounded by a thick layer of muscle. This trait is indicative of the groups' shared ancestry, though the two lineages have been evolving separately for hundreds of millions of years.

In addition to the coelacanths and lungfishes, several other lineages of lobefins arose early in the evolutionary history of jawed fish. Members of one of these other lineages evolved modified fleshy fins that, in an emergency, could be used as legs, allowing the fish to drag itself from a drying puddle to a deeper pool. This lineage left descendants that survive today. These survivors are the **tetrapods** (from the Greek for "four feet"), which instead of fins have limbs that can support their weight on land. Tetrapods also have digits (fingers or toes) on the ends of their limbs. The tetrapods include amphibians, reptiles, and mammals.

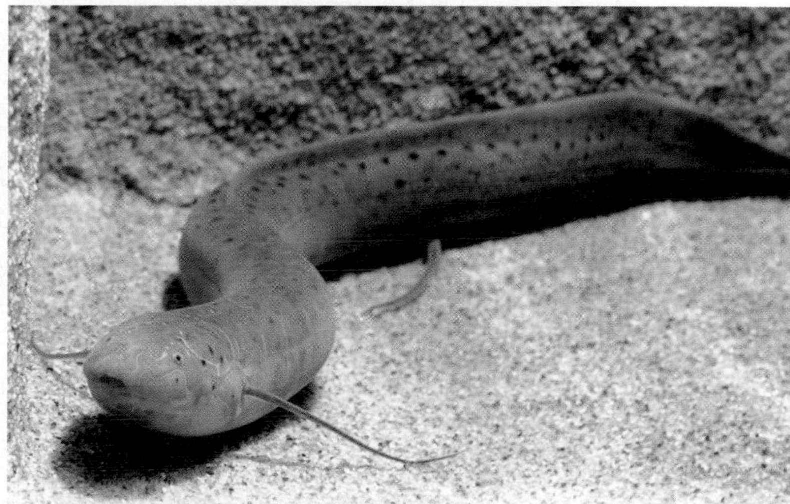

▲ **FIGURE 25-9 Lungfishes are lobe-finned fish** Among the fishes, lungfishes are the group most closely related to land-dwelling vertebrates.

CASE STUDY CONTINUED

Fish Story

In recent years, advances in DNA sequencing technology have greatly increased the number of species whose genomes have been sequenced. In 2013, the African coelacanth joined the list. Comparing the coelacanth sequence with that of cartilaginous fishes revealed that the coelacanth genome has changed very slowly since the two groups diverged. Thus, coelacanth genes, like coelacanth bodies, are today much the same as they were in the group's heyday 300 million years ago.

Amphibians Live a Double Life

The first tetrapods to invade land were amphibians. Today, the 6,500 species of amphibians (Amphibia) straddle the boundary between aquatic and terrestrial existence (**FIG. 25-10**). The limbs of amphibians show varying degrees of adaptation to movement on land, from the belly-dragging crawl of salamanders to the long leaps of frogs. A three-chambered heart (in contrast to the two-chambered heart of fishes) circulates blood more efficiently, and lungs replace gills in most adult forms. Amphibian lungs, however, are relatively inefficient and must be supplemented by the skin, which serves as an additional respiratory organ. This respiratory function requires that the skin remain moist, a constraint that greatly restricts the range of amphibian habitats on land.

Many amphibians are also tied to moist habitats by their breeding behavior, which requires water. For example, as in most fishes, fertilization in frogs and toads is generally external and takes place in water, where the sperm can swim to the eggs. The eggs must remain moist, because they are protected only by a jelly-like coating that leaves them vulnerable to water loss by evaporation. Different amphibian species keep

(a) Tadpole

(b) Frog

(c) Salamander

(d) Caecilian

▲ **FIGURE 25-10 "Amphibian" means "double life"** The double life of amphibians is illustrated by the bullfrog's transition from **(a)** a completely aquatic larval tadpole to **(b)** an adult leading a semi-terrestrial life. **(c)** The red salamander is restricted to moist habitats in the eastern United States. **(d)** Caecilians are legless, mostly burrowing amphibians.

THINK CRITICALLY What advantages might amphibians gain from their "double life"?

their eggs moist in different ways, but many species simply lay their eggs in water. In some amphibian species, fertilized eggs develop into aquatic larvae such as the tadpoles of some frogs and toads. These aquatic larvae undergo a dramatic transformation into semiterrestrial adults, a metamorphosis that gives the amphibians their name, which means "double life." Their double life and thin, permeable skin have made amphibians particularly vulnerable to pollutants and environmental degradation, as described in "Earth Watch: Frogs in Peril" on page 422.

Frogs and Toads Are Adapted for Jumping

The frogs and toads, with 5,700 species, are the most diverse group of amphibians. Adult frogs and toads move about by hopping and leaping, and their bodies are well adapted for this mode of locomotion, with hind legs that are long relative to their body size (much longer than their forelegs). The names "frog" and "toad" do not describe distinct evolutionary groups, but are instead used informally to distinguish two combinations of characteristics that are common among members of this branch of the amphibians. In general, frogs have smooth, moist skin, live in or near water, and have long hind limbs suitable for leaping; toads have bumpy, drier skin, live on land, and have shorter hind limbs suitable for hopping. During mating season, male frogs and toads utter loud calls, distinctive to each species, that advertise their location and readiness to mate.

Many frogs and toads (and other amphibians) contain toxic substances that make them distasteful to predators. In a few species, such as the golden poison dart frog of South America, the protective chemical is extremely toxic. The toxin from a single golden poison dart frog could kill several adult humans.

Most Salamanders Have Tails

Most salamanders have a lizard-like body: slender, with four legs of roughly equal size and a long tail. Some salamanders, however, have only very small legs; these species may have an eel-like appearance. Most of the roughly 580 species of salamanders live on land, often in moist, protected places, such as beneath rocks or logs on a forest floor. But members of some species are fully aquatic and spend their entire lives in the water. Even land-dwelling species generally move to ponds or streams to breed. In almost all salamander species, eggs hatch into aquatic larvae that use external gills to breathe. In some species, the larvae do not metamorphose, but instead retain the larval form throughout life.

Alone among vertebrates, salamanders can regenerate lost limbs. This ability has attracted the attention of researchers interested in regenerative medicine, which seeks treatments that would enable human bodies to repair or regenerate damaged tissues and organs. The researchers hope that learning how salamanders regenerate limbs will lead to effective treatments for humans.

Caecilians Are Limbless, Burrowing Amphibians

The caecilians form a small (175 species) group of legless amphibians that live in tropical regions. At first glance, a caecilian's appearance is reminiscent of an earthworm, though the larger species, which can be up to 5 feet (1.5 meters) long, might be mistaken for a snake. Most caecilians are burrowing animals that live underground, though a few species are aquatic. Caecilians eyes are very small and often covered by skin. As a result, caecilian vision is probably limited to detecting light.

Reptiles Are Adapted for Life on Land

The reptiles (Reptilia) include lizards, snakes, alligators, crocodiles, turtles, and birds (**FIG. 25-11**). Reptiles evolved from an amphibian ancestor about 250 million years ago.

(a) Alligator

(b) Snake

(c) Tortoise

▲ **FIGURE 25-11 The diversity of reptiles (other than birds) (a)** The outward appearance of the American alligator, found in swampy areas of the South, is almost identical to that of 150-million-year-old fossil alligators. **(b)** This scarlet king snake has a color pattern very similar to that of the poisonous coral snake, which potential predators avoid. This mimicry helps the harmless king snake elude predation. **(c)** The tortoises (a type of turtle) of the Galápagos Islands, Ecuador, may live to be more than 100 years old.

Frogs in Peril

During the past three decades, herpetologists (biologists who study reptiles and amphibians) from around the world have documented an alarming decline in amphibian populations. Thousands of species of frogs, toads, and salamanders are dramatically decreasing in number, and many have gone extinct.

This is a worldwide phenomenon; population crashes have been reported from every part of the globe. Of nearly 100 species of harlequin frog known from Central and South America, only 10 can still be found. In South Africa, the only remaining population of Rose's ghost frog has shrunk dramatically and the species is now critically endangered. The southern corroboree frog of Australia was once abundant, but there are now fewer than 50 of them in the wild (**FIG. E25-1**).

In the United States, a recent study showed that populations of virtually all frog and toad species are shrinking (**FIG. E25-2**). Endangered species are disappearing most quickly, but more common species are declining as well. The declines are occurring even in protected areas, such as national parks.

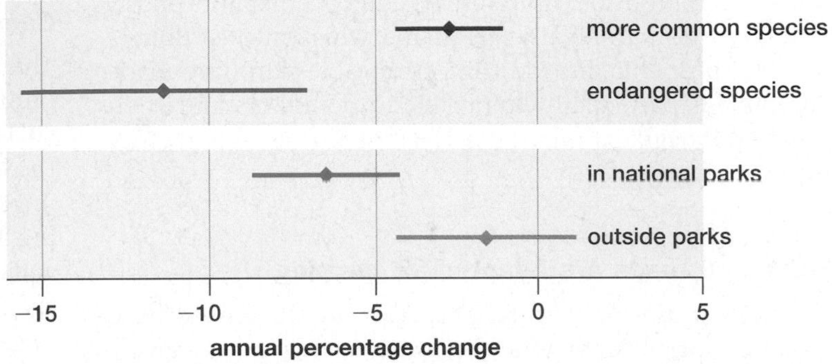

▲ **FIGURE E25-2 Shrinking populations** This graph shows estimated annual percentage population change of amphibians in the United States between 2001 and 2011, based on repeated counts at many locations across the country. The top part of the graph compares endangered species to more common species, and the bottom part compares populations in national parks to populations outside of parks. The horizontal lines passing through each data point show the margin of error for the estimated rates of change.

The causes of the worldwide decline in amphibian diversity are not fully understood, but researchers have discovered that frogs and toads in many places are succumbing to infection by a pathogenic fungus. The fungus has been found in the skin of dead and dying amphibians of hundreds of different species at locations on every continent (except Antarctica, which lacks amphibians). Presence of the fungus has coincided with frog and toad die-offs, and most herpetologists agree that the fungus is causing the deaths.

It seems unlikely, however, that the fungus alone is responsible for the worldwide decline of amphibians. Many herpetologists believe that the fungal epidemic would not have arisen if the frogs and toads had not first been weakened by other stressors. What are the other possible causes of stress? All of the most likely causes stem from human modification of the biosphere—the portion of Earth that sustains life.

Habitat destruction, especially the draining of wetlands that are hospitable to amphibian life, is one major cause of the decline. Amphibians are also vulnerable to toxic substances in the environment because amphibian bodies are protected only by a thin, permeable skin that pollutants can easily penetrate. For example, researchers found that frogs exposed to trace amounts of atrazine, a widely used herbicide that washes from farm fields into streams and lakes and is found in virtually all fresh water in the United States, suffer severe damage to their reproductive tissues.

Many scientists believe that the troubles of amphibians signal an overall deterioration of Earth's ability to support life. According to this line of reasoning, the highly sensitive amphibians are providing an early warning of environmental degradation that will eventually affect more resistant organisms as well.

▲ **FIGURE E25-1 Amphibians in danger** The corroboree frog is rapidly declining in its native Australia.

THINK CRITICALLY Consider the graph shown in Figure E25-2. Imagine that the population of one particular endangered species has declined at the same rate each year, and that the rate of decline was equal to the estimated average rate for endangered species. If the initial population of this species included 1,000 individuals, what would its population size be 10 years later? Draw a graph showing how the species' population changed over 10 years, and then extend the graph to show its projected population after 50 years. What assumptions underlie your projection?

▲ **FIGURE 25-12 The amniotic egg** A crocodile struggles free of its egg. The amniotic egg encapsulates the developing embryo in a fluid-filled membrane (the amnion), ensuring that development occurs in a watery environment, even if the egg is far from water.

Reptiles Haves Scales and Shelled Eggs

Most reptiles live on land. A number of adaptations make reptiles' life on land possible, three of which are especially notable: (1) Reptiles evolved a tough, scaly skin that reduces water loss and protects the body. (2) Reptiles evolved internal fertilization, in which the male deposits sperm within the female's body, eliminating the need to breed in water. (3) Reptiles evolved a shelled **amniotic egg**. The shell prevents the egg from drying out on land, and an internal membrane, the **amnion**, encloses the embryo in the watery environment that all developing animals require (**FIG. 25-12**).

In addition to these features, reptiles have more efficient lungs than do amphibians and do not use their skin as a respiratory organ. Reptile circulatory systems include a three-chambered or (in birds, alligators, and crocodiles) four-chambered heart that segregates oxygenated and deoxygenated blood more effectively than do amphibian hearts.

Lizards and Snakes Share a Common Evolutionary Heritage

Lizards and snakes together form a distinct lineage containing about 9,400 species. The common ancestor of snakes and lizards had limbs, which are retained by most lizards but have been lost in snakes. The limbed ancestry of snakes is revealed by remnants of hind limb bones found in some snake species.

Most lizards are small predators that eat insects or other small invertebrates, but a few lizard species are quite large. The Komodo dragon, for example, can reach 10 feet (3 meters) in length and weigh more than 200 pounds (90 kilograms). These giant lizards live in Indonesia and have powerful jaws and inch-long teeth that enable them to prey on large animals including deer, goats, and pigs. The Komodo dragon, however, does not rely on its teeth alone to kill its prey. It also produces a potent venom that flows from a gland in its jaw into the wound of a bitten victim. If an animal bitten by a Komodo dragon is not immediately killed, the venom ensures that it will likely die soon after the attack. The lizard simply waits patiently until its wounded, poisoned prey dies.

Most snakes are active, predatory carnivores and have a variety of adaptations that help them acquire food. For example, many snakes have special sense organs that help track prey by detecting small temperature differences between a prey's body and its surroundings. Some snake species immobilize prey with venom that is delivered through hollow teeth. Snakes also have a distinctive jaw joint that allows the jaws to distend so that the snake can swallow prey much larger than its head. A snake's ribs are not attached to a breastbone (which snakes lack), so the ribs are easily pushed outward to accommodate passage of a large prey item down the body. Following one of its infrequent but large meals, a snake's body gears up to digest the food. The snake's heart, liver, kidneys, and intestine grow rapidly, almost doubling in size, and its metabolic rate increases dramatically, as if it were a sprinting racehorse rather than its motionless self. When digestion is complete, the snake's organs and metabolism return to their pre-meal state.

Alligators and Crocodiles Are Adapted for Life in Water

Crocodilians, as the 25 species of alligators and crocodiles are collectively known, are found in coastal and inland waters of the warmer regions of Earth. They are well adapted to an aquatic lifestyle, with eyes and nostrils located high on their heads so that they are able to remain submerged for long periods with only the uppermost portion of the head above the water's surface. Crocodilians have strong jaws and conical teeth that they use to crush and kill the fish, birds, mammals, turtles, and amphibians that they eat.

Parental care is extensive in crocodilians, which bury their eggs in mud nests. Parents guard the nest until the young hatch and then carry their newly hatched young in their mouths, moving them to safety in the water. Young crocodilians may remain with their mother for several years.

Turtles Have Protective Shells

The 325 species of turtles occupy a variety of habitats, including deserts, streams and ponds, and the ocean. These diverse habitats have fostered a variety of adaptations, but all turtles are protected by a hard, boxlike shell that is fused to the vertebrae, ribs, and collarbone. Turtles have no teeth but have instead evolved a horny beak. The beak is used to eat a variety of foods; some turtles are carnivores, some are herbivores, and some are scavengers. The largest turtle, the leatherback, is an ocean dweller that can grow to more than 6 feet (2 meters) in length and feeds largely on sea jellies. Leatherbacks and other marine turtles must return to land to breed and often undertake extraordinary long-distance migrations to reach the beaches on which they bury their eggs.

(a) Hummingbird

(b) Toucan

(c) Ostrich

▲ **FIGURE 25-13 The diversity of birds (a)** The delicate hummingbird beats its wings about 60 times per second and weighs about 0.15 ounce (4 grams). **(b)** Toucans are fruit-eaters that inhabit the forests of Central and South America. **(c)** The ostrich, the largest of all birds, weighs more than 300 pounds (135 kilograms); its eggs weigh more than 3 pounds (1,500 grams).

THINK CRITICALLY Although the ancestor of all birds could fly, some bird species—such as the ostrich—cannot. Why do you suppose flightlessness has evolved repeatedly among birds?

Birds Are Feathered Reptiles

One very distinctive group of reptiles is the birds (**FIG. 25-13**). Although the 10,300 species of birds have traditionally been classified as a group separate from reptiles, biologists have shown that birds are really a subset of the reptiles. Birds descended from dinosaur ancestors, and birds first appear in the fossil record roughly 150 million years ago. Modern birds are distinguished from other reptiles by feathers, which are essentially a highly specialized version of reptilian scales. Birds retain scales on their legs—evidence of the ancestry they share with the rest of the reptiles.

Bird anatomy and physiology are dominated by adaptations that help the animals fly. In particular, most birds are exceptionally light for their size. Lightweight bones reduce the weight of the bird skeleton, and many bones present in other reptiles have been lost in the course of evolution. Bird reproductive organs shrink considerably during nonbreeding periods, and female birds possess only a single ovary, further minimizing weight. Feathers serve as lightweight extensions of the wings and the tail, providing the lift and control required for flight; feathers also provide lightweight protection and insulation for the body.

Birds are also able to maintain body temperatures high enough to allow their muscles and metabolic processes to operate at peak efficiency, regardless of the temperature of the external environment. This physiological ability to maintain an internal temperature that is usually higher than that of the surrounding environment is characteristic of both birds and mammals, which are therefore sometimes described as warm blooded or endothermic. In contrast, the body temperature of ectothermic (cold-blooded) animals—invertebrates, fish, amphibians, and reptiles other than birds—varies with the temperature of their environment.

Endothermic animals such as birds have a high metabolic rate, which requires efficient oxygenation of tissues. Therefore, birds possess circulatory and respiratory adaptations that help meet the need for efficiency. A bird's four-chambered heart prevents mixing of oxygenated and deoxygenated blood. The respiratory system of birds is supplemented by air sacs that provide a continuous supply of oxygenated air to the lungs, even while the bird exhales.

Mammals Provide Milk to Their Offspring

One branch of the tetrapod evolutionary tree gave rise to a group that evolved hair and diverged to form the mammals (Mammalia). Mammals are named for the milk-producing **mammary glands** used by all female mammals to feed their young. The mammals first appeared approximately 250 million years ago but did not diversify and become prominent until after the dinosaurs went extinct roughly 66 million years ago. In most mammals, hair protects and insulates the warm body. Like birds and crocodilians, mammals have four-chambered hearts that increase the amount of oxygen delivered to the tissues. The 4,900 species of mammals include three main lineages: monotremes, marsupials, and placental mammals.

Monotremes Are Egg-Laying Mammals

Unlike other mammals, **monotremes** lay eggs rather than giving birth to live young. This group includes only five species: the platypus and four species of spiny anteaters, also known as echidnas (**FIG. 25-14**). Monotremes are found only in New Guinea (the four echidna species) and Australia (the platypus and one of the echidna species).

Echidnas are terrestrial and eat insects or earthworms that they dig out of the ground. Platypuses forage for food in the water, diving below the surface to capture small

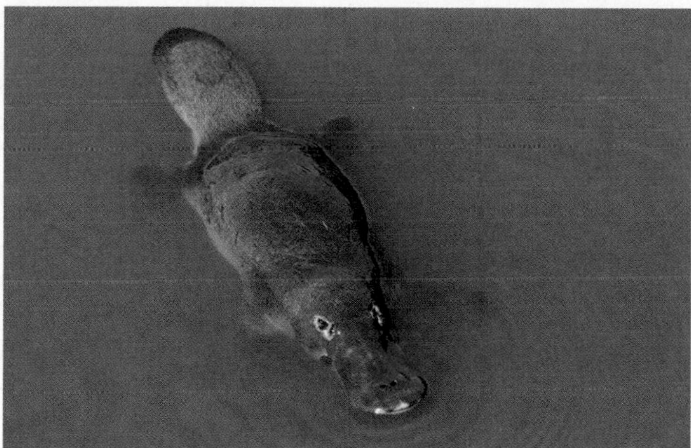

(a) Platypus

(b) Spiny anteater

▲ **FIGURE 25-14 Monotremes (a)** Monotremes, such as this platypus, lay leathery eggs resembling those of reptiles. Platypuses live in burrows that they dig in the banks of rivers, lakes, or streams. **(b)** The short limbs and heavy claws of spiny anteaters (also known as echidnas) help them unearth insects and earthworms to eat. The stiff spines that cover a spiny anteater's body are modified hairs.

vertebrate and invertebrate animals. Platypus bodies are well adapted for this aquatic lifestyle, with a streamlined shape, webbed feet, a broad tail, and a fleshy bill.

Monotreme eggs have leathery shells and are incubated for 10 to 12 days by the mother. Echidnas have a special pouch for incubating eggs, but a female platypus incubates her eggs by holding them between her tail and belly. Newly hatched monotremes are tiny and helpless and feed on milk secreted by the mother. Monotremes, however, lack nipples. Milk from the mammary glands oozes through ducts on the mother's abdomen and soaks the hair around the ducts. The young then suck the milk from the hair.

Marsupial Diversity Reaches Its Peak in Australia

In all mammals except the monotremes, embryos develop in the uterus, a hollow, muscular organ in the female reproductive tract. The lining of the uterus combines with membranes derived from the embryo to form the **placenta**, a structure that allows gases, nutrients, and wastes to be exchanged between the circulatory systems of the mother and embryo.

In **marsupials** (**FIG. 25-15**), embryos develop in the uterus for only a short period. Marsupial young are born at a very immature stage of development. Immediately after birth, a marsupial crawls to a nipple, firmly grasps it, and, nourished by milk, completes its development. In most but

(a) Wallaby

(b) Wombat

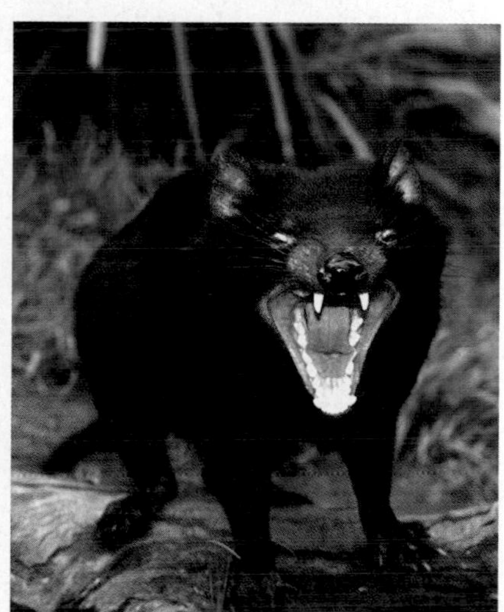

(c) Tasmanian devil

▲ **FIGURE 25-15 Marsupials (a)** Marsupials, such as the wallaby, give birth to extremely immature young who develop within the mother's protective pouch. **(b)** The wombat is a burrowing marsupial whose pouch opens toward the rear of its body to prevent dirt and debris from entering the pouch during tunnel digging. One of the wombat's predators is **(c)** the Tasmanian devil.

not all marsupial species, this postbirth development takes place in a protective pouch.

The majority of the 330 species of marsupials are found in Australia, where marsupials such as kangaroos have come to be seen as emblematic of the island continent. Kangaroos are the largest and most conspicuous of Australia's marsupials; the largest species, the red kangaroo, may be 7 feet tall (about 2 meters) and can make 30-foot (9-meter) leaps when moving at top speed. Though kangaroos are perhaps the most familiar marsupials, the group encompasses species with a range of sizes, shapes, and lifestyles, including koalas, wombats, and the Tasmanian devil. Only one marsupial species, the Virginia opossum, is native to North America.

The Tasmanian devil, a carnivorous predator the size of a small dog, is among the marsupial species at risk of extinction. Tasmanian devil populations were decimated by hunting until the species was protected by law and began to recover in the 1940s. Today, however, the recovery is threatened by a new form of cancer that appeared suddenly around 1996. Unlike most cancers, the one that afflicts Tasmanian devils is transmissible—it spreads from animal to animal. Tumors grow on the faces of affected animals. Tasmanian devils often bite each other on the face during fighting or sex, so tumor cells may enter a bite wound. The resulting facial tumors usually kill infected animals within a few months. Researchers estimate that the population of Tasmanian devils has decreased by 60% to 80% since the cancer epidemic began.

Placental Mammals Inhabit Land, Air, and Sea

Most mammal species are **placental mammals** (**FIG. 25-16**), so named because their placentas are far more complex than those of marsupials. Compared to marsupials, placental mammals retain their young in the uterus for a much longer period, so that offspring are more developed at birth.

The largest groups of placental mammals, in terms of number of species, are the rodents and the bats. Rodents account for almost 40% of all mammal species. Most rodent species are rats or mice, but the group also includes squirrels, hamsters, guinea pigs, porcupines, beavers, woodchucks, chipmunks, and voles. The largest rodent, the capybara, is found in South America and can weigh up to 110 pounds (50 kilograms).

About 20% of mammal species are bats, the only mammals to have evolved wings and powered flight. Bats are nocturnal and spend the daylight hours roosting in caves, rock

(a) Capybara

(b) Bat

(c) Whale

(d) Cheetah

(e) Orangutan

◀ FIGURE 25-16 The diversity of placental mammals (a) The South American capybara is the world's largest rodent. It stands 2 feet tall and can weigh well over 100 pounds. **(b)** A bat, the only type of mammal capable of true flight, navigates at night by using a kind of sonar. Large ears help the animal detect echoes as its high-pitched cries bounce off nearby objects. **(c)** A humpback whale may migrate 15,000 or more miles each year. **(d)** Mammals are named after the mammary glands with which females nurse their young, as illustrated by this mother cheetah. **(e)** Orangutans are gentle, intelligent apes that occupy swamp forests in limited areas of the Tropics but are endangered by hunting and habitat destruction.

crevices, or trees. Most bat species have evolved adaptations for feeding on a particular kind of food. Some bats eat fruit; others feed on nectar from night-blooming flowers. Most bats are predators, including species that hunt frogs, fish, or even other bats. A few species (vampire bats) subsist entirely on blood that they lap up from incisions they make in the skin of sleeping mammals or birds. Most predatory bats, however, feed on flying insects, which they detect by echolocation. To echolocate, a bat emits short pulses of high-pitched sound (too high for humans to hear). The sounds bounce off objects in the surrounding environment to produce echoes, which the bat hears and uses to identify and locate insect prey.

Although the majority of placental mammal species are rodents or bats, the other placental mammals are diverse in form and include many species that loom large in the human imagination. For example, many people are fascinated by the sometimes human-like social behavior of our closest relatives, the chimpanzees, gorillas, and other great apes. Some of us are awed by the grace and power of large carnivores such as lions, cheetahs, tigers, and wolves. And others are fascinated by the 70 species of whales, placental mammals that evolved from terrestrial ancestors and recolonized the ocean. The largest whale species, the blue whale, can grow to more than 100 feet long (more than 30 meters) and is the largest animal known to have existed in the history of Earth.

CHECK YOUR LEARNING

Can you . . .

- describe the key features of lampreys, cartilaginous fishes, ray-finned fishes, coelacanths, lungfishes, amphibians, reptiles, and mammals?
- name and describe the main subgroups included within each of these groups?

CASE STUDY REVISITED

Fish Story

After Marjorie Courtenay-Latimer's discovery of the coelacanth, J. L. B. Smith dedicated himself to searching for more coelacanth specimens in the waters off South Africa. He didn't find one until 1952, when fishermen from the Comoro Islands, having seen leaflets that offered a reward for a coelacanth, contacted Smith with the news that they had one in their possession. Smith immediately booked a flight to the Comoros, and reportedly wept for joy upon holding the 88-pound coelacanth awaiting him.

In the years since Smith's trip, about 200 additional coelacanths have been caught by fishermen, mostly in waters around the Comoros but also around nearby Madagascar and off the coasts of Mozambique and South Africa. Scientists thought that the fish's range was restricted to this relatively small area in the western Indian Ocean, and it was a surprise when a few specimens were discovered in Indonesia, more than 6,000 miles away. DNA tests showed that these Indonesian coelacanths were members of a second species.

The known populations of coelacanths are small, consisting of a few hundred individuals, and appear to be declining. Part of this decline is due to fishing, though coelacanths are mostly caught accidentally by fishermen searching for more commercially desirable species. Conservation efforts in South Africa and the Comoros thus focus largely on introducing fishing methods that will reduce the chances of accidentally snaring a coelacanth.

CONSIDER THIS Is it worth spending money to try to protect coelacanths from being accidentally killed by fishermen, or should scarce resources be instead devoted to preserving more ecologically important species and habitats?

CHAPTER REVIEW

Go to **Mastering Biology** to access the Pearson eText, vocabulary review, practice quizzes, activities, videos, current events, and more.

*Answers to **Think Critically** and **Thinking Through the Concepts** questions can be found in the **Answers** section at the back of the book.*

Summary of Key Concepts

25.1 What Are the Key Features of Chordates?

At some stage in their development, all chordates possess a notochord; a dorsal, hollow nerve cord; pharyngeal gill slits; and a post-anal tail.

25.2 Which Animals Are Chordates?

The chordates include three taxonomic groups: the tunicates, the lancelets, and the craniates. Tunicates are invertebrate filter-feeders that include the sessile sea squirts and the motile salps. The lancelets are also invertebrate filter-feeders and live partially buried in sandy seafloors. Craniates include all animals with skulls: the hagfishes (jawless, eel-shaped craniates that lack a backbone) and the vertebrates.

25.3 What Are the Major Groups of Vertebrates?

Lampreys are jawless vertebrates; the best-known lamprey species are parasites of fish. Cartilaginous fishes have skeletons made entirely of cartilage and bodies protected by leathery skin. They breathe with gills and reproduce using internal fertilization. Ray-finned fishes have bony skeletons and fins consisting of webs of skin supported by bony spines. Their skin in protected by interlocking scales, and they breathe with gills.

Coelacanths and lungfishes are collectively known as lobefins because of their fleshy, bone-containing fins. Coelacanths breathe with gills; lungfishes have both gills and lungs and can survive out of the water during the dry season.

Most amphibians have simple lungs for breathing air. Most are confined to relatively damp terrestrial habitats because of their need to keep their skin moist for respiration, their need for water to facilitate external fertilization, and their aquatic larvae.

Reptiles—with their well-developed lungs, dry skin covered with relatively waterproof scales, internal fertilization, and an amniotic egg with its own water supply—are well adapted to the driest terrestrial habitats. One group of reptiles, the birds, has additional adaptations, such as an elevated body temperature, that allow the muscles to respond rapidly regardless of the temperature of the environment. The bird body is adapted for flight, with feathers, lightweight bones, and efficient circulatory and respiratory systems.

Mammals have insulating hair and nourish their young with milk. Except for monotreme mammals, they give birth to live young.

Thinking Through the Concepts

Bloom's: Remembering, Understanding

Multiple Choice

1. The two groups of mammals with the largest number of species are
   a. marsupials and monotremes.
   b. carnivores and whales.
   c. bats and rodents.
   d. apes and lampreys.

2. More than half of all vertebrate species are
   a. mammals.
   b. ray-finned fishes.
   c. amphibians.
   d. cartilaginous fishes.

3. Which of the following is *not* a characteristic common to all chordates?
   a. a notochord
   b. a dorsal, hollow nerve cord
   c. pharyngeal gill slits
   d. a vertebral column

4. Adult frogs and toads obtain oxygen through
   a. gills only.
   b. gills and lungs.
   c. lungs and skin.
   d. lungs only.

5. Hagfishes
   a. are predators of live fishes.
   b. feed on worms and dead and dying fishes.
   c. feed on algae attached to rocks.
   d. are filter-feeders.

Fill-in-the-Blank

1. In chordates, the nerve cord is ____________ and runs along the ____________ side of the body. During at least one stage of a chordate's life, it has a tail that extends past its ____________ and its body is stiffened by a(n) ____________ that runs along its length.

2. Animals that are chordates but not vertebrates include ____________, ____________, and ____________. Craniates are animals that have a(n) ____________. Animals that are craniates but not vertebrates include ____________.

3. Both gills and lungs are present in adult ____________. Sharks and rays have internal skeletons composed of ____________. The vertebrate group with the largest number of species is ____________. Lampreys have teeth but lack ____________.

4. Among tetrapod groups, hair is found in ____________; the skin is a respiratory organ in ____________; shelled, amniotic eggs are found in ____________; aquatic larvae with gills are found in ____________.

5. The only mammals that lay eggs are ____________. The only vertebrates that regenerate lost limbs are ____________. The only mammals with powered flight are ____________.

Review Questions

1. Briefly describe each of the following adaptations, and explain the adaptive significance of each: vertebral column, jaws, limbs, amniotic egg, feathers, placenta.

2. List the vertebrate groups that have each of the following.
   a. a skeleton of cartilage
   b. a two-chambered heart
   c. amniotic egg
   d. endothermy
   e. a four-chambered heart
   f. a placenta
   g. lungs supplemented by air sacs

3. List four distinguishing features of chordates.

4. Describe the ways in which amphibians are adapted to life on land. In what ways are amphibians still restricted to a watery or moist environment?

5. List the adaptations that distinguish reptiles from amphibians and help reptiles adapt to life in dry terrestrial environments.

6. List the adaptations of birds that contribute to their ability to fly.

7. How do mammals differ from birds, and what adaptations do they share?

Applying the Concepts

Bloom's: Applying, Analyzing, Evaluating

1. Are hagfishes vertebrates or invertebrates? On which characteristics did you base your answer? Is it important to be able to place them in one category or the other? Why?

2. Is the decline of amphibian populations of concern to humans? Why is it important to understand the causes of the decline?

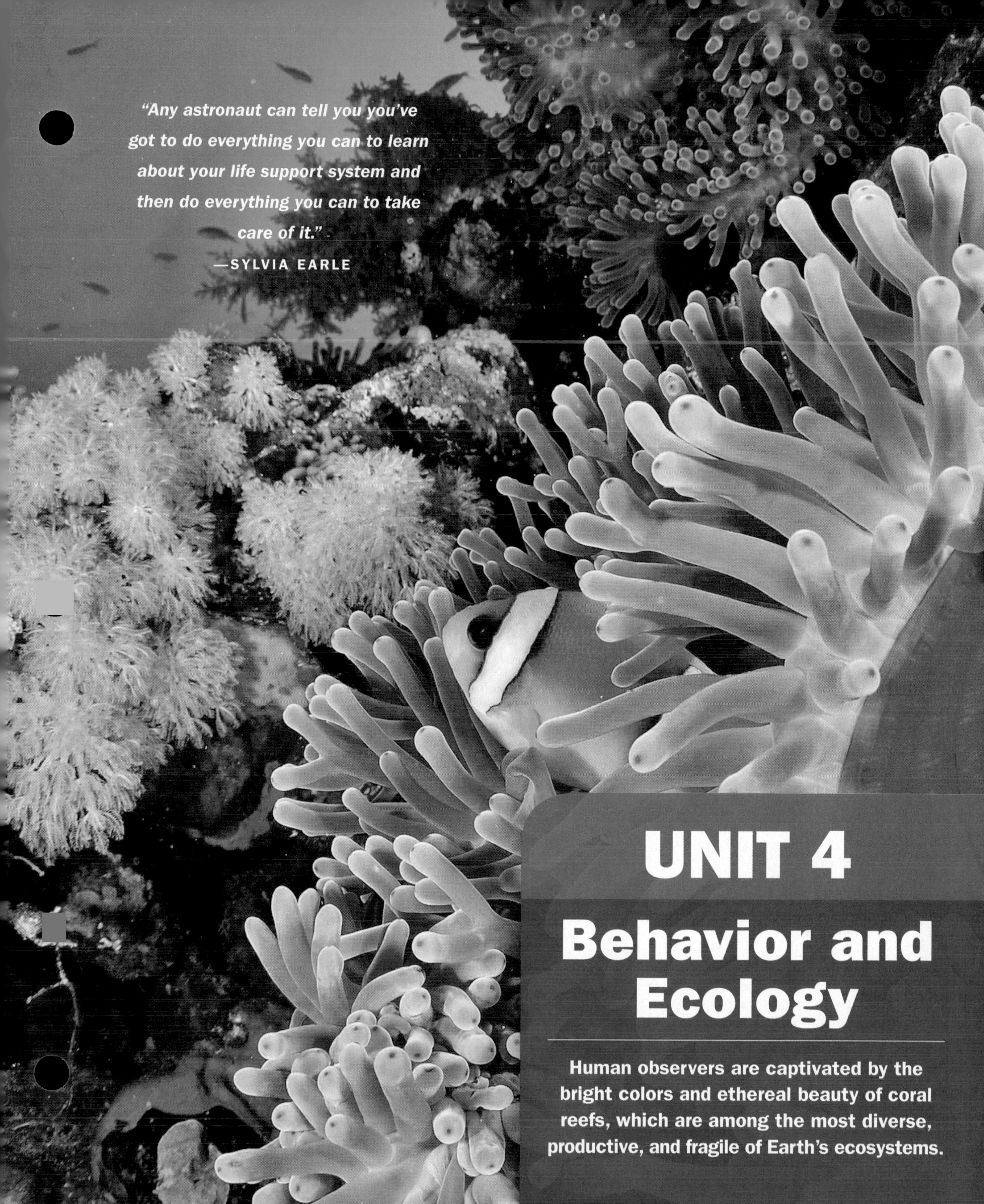

"Any astronaut can tell you you've got to do everything you can to learn about your life support system and then do everything you can to take care of it."

—SYLVIA EARLE

UNIT 4

Behavior and Ecology

Human observers are captivated by the bright colors and ethereal beauty of coral reefs, which are among the most diverse, productive, and fragile of Earth's ecosystems.

26 Animal Behavior

CASE STUDY

Sex and Symmetry

WHAT MAKES A MAN SEXY? According to a growing body of research, it's his symmetry. Sexual preference for symmetrical males was first documented in insects. For example, biologist Randy Thornhill found that symmetry accurately predicts the mating success of male Japanese scorpionflies. In Thornhill's experiments and observations, the most successful males were those whose left and right wings were equal or nearly equal in length. Males with one wing longer than the other were less likely to copulate; the greater the difference between the two wings, the lower the likelihood of mating success.

Male scorpionfly

Both this male scorpionfly and this male human are exceptionally attractive to potential mates. The secret to their sex appeal may be that both have highly symmetrical bodies.

Thornhill's work with scorpionflies led him to wonder if the effects of male symmetry also extend to humans. To test the hypothesis that female humans find symmetrical males more attractive, Thornhill and colleagues began by measuring symmetry in some young adult males. Each man's degree of symmetry was assessed by measurements of his ear length and the width of his foot, ankle, hand, wrist, elbow, and ear. From these measurements, the researchers derived an index that summarized the degree to which the size of these features differed between the right and left sides of the body.

The researchers next gathered a panel of heterosexual female observers who were unaware of the nature of the study and showed them photos of the faces of the measured males. As predicted by the researchers' hypothesis, men judged by the panel to be most attractive were also the most symmetrical. Apparently, a man's attractiveness to women is correlated with his body symmetry.

Why might females prefer symmetrical males? Consider this question as you read about animal behavior.

AT A GLANCE

26.1 HOW DOES BEHAVIOR ARISE?

Behavior is any observable activity of a living animal. For example, a moth flies toward a bright light, a honeybee flies toward a cup of sugar water, and a housefly flies toward a piece of rotting meat. Bluebirds sing, wolves howl, and frogs croak. Mountain sheep butt heads in ritual combat, chimpanzees groom one another, ants attack a termite that approaches an anthill. Humans dance, play sports, and wage wars. Even the most casual observer encounters many fascinating examples of animal behavior each day.

An animal's behavior is influenced by its genes and by its environment. All behavior develops out of an interaction between the two.

Genes Influence Behavior

Several lines of evidence indicate that variation in behavior is influenced by variation in genes. The evidence includes observation of innate behaviors, behavioral experiments, and genetic analysis.

Innate Behaviors Can Be Performed Without Prior Experience

One indication that genes influence the development of behavior comes from behaviors that are performed by newborn animals and that therefore appear to be inherited. Such **innate** behaviors occur in reasonably complete form the first time an animal encounters a particular stimulus. For instance, a gull chick pecks at its parent's bill very soon after hatching, which stimulates the parent to feed it. Innate behavior also occurs in the common cuckoo, a bird species in which females lay eggs in the nests of other bird species, to be raised by the unwitting adoptive parents. Soon after a cuckoo egg hatches, the cuckoo chick performs the innate behavior of shoving the nest owner's eggs (or baby birds) out of the nest (**FIG. 26-1**).

(a) A cuckoo chick ejects an egg

(b) A foster parent feeds a cuckoo

▲ **FIGURE 26-1 Innate behavior (a)** A cuckoo chick, just hours after it hatches evicts the eggs of its foster parents from the nest. **(b)** The parents, responding to the stimulus of the cuckoo chick's wide-gaping mouth, feed the chick for weeks after it hatches, unaware that it is not related to them.

THINK CRITICALLY The cuckoo chick benefits from its innate behavior, but the foster parent harms itself with its innate response to the cuckoo chick's begging. Why hasn't natural selection eliminated the foster parent's disadvantageous innate behavior?

The gull and the cuckoo benefit from their behaviors. The young gull acquires nutrition, and the young cuckoo eliminates its competitors for food. These examples illustrate a more general point about behavior: Much of it is adaptive and therefore may have evolved by natural selection.

Experiments Show That Behavior Can Be Inherited

Although the widespread occurrence of innate behaviors provides circumstantial evidence that behaviors can be inherited and are therefore influenced by genes, stronger evidence comes from experiments. One such experiment began when researchers noticed that fruit fly larvae exhibit two phenotypes with respect to feeding behavior. Some larvae are "rovers" and move about continually to search for food. Others are "sitters" and remain in more or less one place and eat whatever is there. When the researchers crossed adult rovers and sitters, all the offspring were rovers. When these first-generation rovers bred with one another, however, the resulting generation contained both rovers and sitters in a ratio of roughly 3:1. As you may recall from your study of inheritance in Chapter 11, the pattern observed in this experiment is the one expected for a trait controlled by a single gene with two alleles, one of which is dominant.

Another cross-breeding experiment, this time with blackcap warblers, showed that these birds have a genetically influenced tendency to migrate in a particular direction. Blackcap warblers breed in Europe and migrate to Africa, but populations from different areas travel by different routes. Blackcaps from western Europe travel in a southwesterly direction to reach Africa, whereas birds from eastern Europe travel to the southeast. If birds from the two populations are crossbred in captivity, however, the hybrid offspring try to migrate due south, which is intermediate between the directions of the two parents. This result suggests that genes influence migratory direction.

Geneticists Can Identify Particular Genes That Influence Behavior

Sometimes researchers can pinpoint the particular genes that influence behavior. To determine which genes affect a behavior, an investigator may select a candidate gene and then examine the effects of mutations that inactivate the gene. In some species, researchers may be able to engineer "knockout" animals in which the candidate gene is disabled. For example, mice in which the *V1aR* gene has been knocked out exhibit increased levels of risky behaviors, such as lingering in brightly lit, open areas (which normal mice avoid, presumably because risk of predation is higher in such places). *V1aR* codes for a protein that is a receptor for the hormone arginine vasopressin (AVP). When AVP binds to receptors in the brain, it influences behavior, and mice lacking the receptor fail to respond appropriately to dangerous circumstances.

Before researchers can knock out candidate genes, they must have some idea of which genes are likely to affect the behavior of interest. To identify likely candidates, researchers often use techniques that identify chromosomal locations at which genetic variation is correlated with variation in a behavior. They also use techniques that reveal which genes are expressed in particular tissues (especially the brain) when a particular behavior occurs. These methods often reveal that complex behaviors are influenced by many genes. For example, researchers have identified more than 800 different genes whose expression in a zebra finch brain changes each time the bird sings.

The Environment Influences Behavior

The development and expression of behavior can be influenced by variation in an animal's environment, including both the animal's physical environment and its experiences.

Behaviors Are Influenced by the Physical Environment

The environment in which an animal develops can affect its adult behavior. Consider, for example, the zebrafish. This species is found in diverse habitats, including fast-flowing streams in which the water contains ample oxygen and stagnant ponds in which oxygen levels are very low. In an experiment, genetically similar zebrafish were reared from eggs in two different conditions: oxygen-rich or oxygen-poor water. When the fish grew up, the experimenters measured their response to **aggression**, or antagonistic behavior, from another fish in both an oxygen-rich and an oxygen-poor environment. In the high-oxygen environment, the behavior of fish that were reared in a high-oxygen environment was much more aggressive. In the low-oxygen environment, however, the most aggressive fish were those reared in a low-oxygen environment. It appears that the environment in which a fish develops causes the fish to develop the ability to behave appropriately in the environment that it is most likely to encounter as an adult.

An animal's behavior can also be influenced by the living part of its environment, especially its interactions with other animals. For example, in prairie voles (a small rodent), pups raised by a single mother receive less licking and grooming than do pups raised by a mother and father together. In adulthood, the mating and parental behavior of voles varies, depending on their experience as pups. Those cared for by two parents mate sooner and provide more attentive parental care than do voles raised by a single parent.

Behaviors Are Influenced by the Experiential Environment

The capacity to make relatively permanent changes in behavior on the basis of experience is called **learning**. This deceptively simple definition encompasses a vast array of phenomena. A toad learns to avoid distasteful insects; a baby shrew learns which adult is its mother; a human learns to speak a language; a sparrow learns to use the stars for

(a) Prairie dog alarm call

(b) Habituated prairie dogs

FIGURE 26-2 Habituation (a) When a prairie dog detects the approach of a potential predator, it may produce a loud alarm call. However, if harmless intruders approach repeatedly, as when **(b)** human hikers routinely pass by, prairie dogs learn to stop responding with an alarm call.

navigation. Each of the many examples of animal learning represents the outcome of a unique evolutionary history, so learning is as diverse as animals themselves. Nonetheless, it can be useful to categorize types of learning, keeping in mind that the categories are only rough guides and that many examples of learning will not fit neatly into any category.

Habituation Is a Decline in Response to a Repeated Stimulus One common form of simple learning is **habituation**, defined as a decline in response to a repeated stimulus. The ability to habituate prevents an animal from wasting its energy and attention on irrelevant stimuli. For example, a sea anemone will retract its tentacles when first touched, but gradually stops retracting if touched repeatedly. Prairie dogs, which are ground-dwelling rodents, utter loud alarm calls and race to their burrows when potential predators approach their colony (**FIG. 26-2**). But prairie dogs habituate to repeated approaches by animals that prove to be nonthreatening, such as people on a well-used hiking path that passes near the colony, and stop responding with alarm calls.

The ability to habituate is generally adaptive. If a prairie dog ran to its burrow every time a harmless human passed by, the animal would waste a great deal of time and energy that could otherwise be spent on beneficial activities such as acquiring food. Habituation can also fine-tune an organism's innate responses to environmental stimuli. For example, newborn chicks instinctively crouch when any object moves over their heads, but birds that are a few weeks old crouch down when a hawk flies over but ignore harmless birds such as geese. The birds have habituated to things that pass overhead harmlessly and frequently, such as leaves, songbirds, and geese. Predators are much less common, and the novel shape of a hawk continues to elicit instinctive crouching. Thus, learning modified the innate response, making it more advantageous.

Researchers may intentionally induce habituation in animal subjects so they can be studied. For example, most of what we know about the social behavior of primates such as chimpanzees, gorillas, and baboons comes from studies of wild populations in which animals have been laboriously habituated to human presence. Only then can researchers approach closely enough to observe behavior (**FIG. 26-3**).

Imprinting Is Rapid Learning by Young Animals Learning often occurs within limits that help increase the chances that only the appropriate behavior is acquired. Such constraints on learning are perhaps most strikingly illustrated by **imprinting**, a form of learning in which an animal's nervous system is rigidly programmed to learn a certain thing only during a certain period of development. The information learned during this *sensitive period* is incorporated into behaviors that are not easily altered by later experience.

Imprinting is best known in birds such as geese, ducks, and chickens. These birds learn to follow the animal or object that they most frequently encounter during an early sensitive period. In nature, a mother bird is likely to be nearby during

FIGURE 26-3 Jane Goodall observes habituated chimpanzees Scientific investigation of animal behavior often relies on animals that have habituated to human presence.

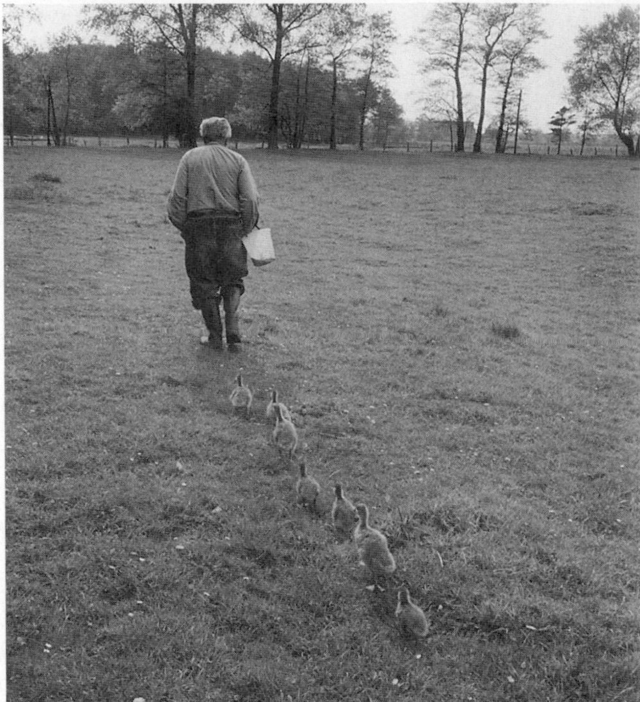

▲ **FIGURE 26-4 Konrad Lorenz and imprinting** Konrad Lorenz, a pioneer in the scientific study of animal behavior, is followed by goslings that imprinted on him shortly after they hatched. They follow him as they would their mother.

the sensitive period, so her offspring imprint on her. In the laboratory, however, these birds may imprint on a toy train or other moving object (**FIG. 26-4**).

Imprinting is a concern for conservationists who aim to preserve endangered species by rearing animals in captivity for release into the wild. Such programs often go to great pains to ensure that captive-reared animals do not imprint on their human caretakers (**FIG. 26-5a**), so that released animals will be attracted to others of their species and not to people. However, conservationists can also take advantage of imprinting to help ensure that captive-reared animals develop behaviors necessary for survival (**FIG. 26-5b**).

Conditioning Is a Learned Association Between a Stimulus and a Response Behaviors generally occur in response to a particular stimulus, and many animals can learn to associate a behavior with a different stimulus. For example, in a classic experiment conducted by Ivan Pavlov, dogs that normally salivated in response to the sight of food were trained to salivate in response to hearing a ringing bell. This kind of learning, in which an animal learns a new association between a stimulus and an innate response, is known as **classical conditioning**. For example, lemon damselfish perform innate predator avoidance behavior—hiding and scanning their surroundings—in response to a chemical alarm signal released by other damselfish. After experimenters repeatedly exposed young damselfish to the alarm signal mixed with the scents of several unfamiliar fish species, the damselfish performed the alarm response when exposed only to the scent of any of the previously unfamiliar species. As the damselfish experiment suggests, learning by classical conditioning can result in adaptive behavior, such as avoiding novel predators.

A more complex form of learning is **trial-and-error learning**, in which animals learn through experience to associate a behavior with a positive or negative outcome. Many animals are faced with naturally occurring rewards and punishments and learn to modify their behavior in response to them. For example, a hungry toad that captures a bee quickly learns to avoid future encounters with bees (**FIG. 26-6**). After only one experience with a stung tongue, a toad ignores bees and even other insects that resemble them.

Trial-and-error learning is sometimes known as **operant conditioning**, especially when the learning results from training in a laboratory setting. For example, an animal may learn to perform a behavior (such as pushing a lever or

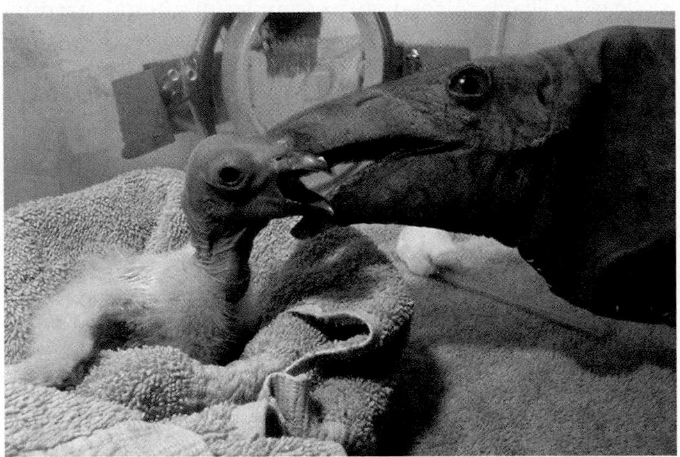

(a) Feeding a condor chick

(b) Leading a flock of cranes

▲ **FIGURE 26-5 Imprinting and endangered species (a)** To prevent young California condors from imprinting on humans, caretakers use a condor puppet to feed them. When the captive-reared birds are released to the wild, they will be drawn to other condors rather than to people. **(b)** Early in life, captive-reared whooping cranes are exposed to and imprint on ultralight aircraft. The young cranes later follow an ultralight to learn the route for their southward migration.

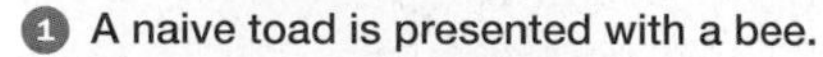

1 A naive toad is presented with a bee.

2 While trying to eat the bee, the toad is stung painfully on the tongue.

3 Presented with a harmless robber fly, which resembles a bee, the toad cringes.

4 The toad is presented with a dragonfly.

5 The toad immediately eats the dragonfly, demonstrating that the learned aversion is specific to bees.

▲ **FIGURE 26-6 Trial-and-error learning in a toad**

pecking a button) to receive a reward or to avoid punishment. This technique is most closely associated with the psychologist B. F. Skinner, who designed the "Skinner box," in which an animal is isolated and allowed to train itself. The box might contain a lever that, when pressed, ejects a food pellet. If the animal accidentally bumps the lever, a food reward appears. After a few such occurrences, the animal learns the connection between pressing the lever and receiving food and begins to press the lever repeatedly.

Operant conditioning has been used to train animals to perform tasks far more complex than pressing a lever. For example, Gambian giant pouched rats have been trained to sniff out buried land mines and, in return for a banana or peanut reward, scratch the ground vigorously when they find a mine (**FIG. 26-7**). Unexploded land mines pose a major threat to the safety of millions of people in countries around the world; more than 100 million mines remain buried where they were planted during past wars and forgotten. The rats are very good at detecting mines and are too light to detonate the ones they find. Gambian giant pouched rats can also detect the scent of the bacteria that cause tuberculosis (TB) and have been trained to distinguish TB-infected mucus from uninfected mucus. This ability may soon become the basis of a cheap and effective diagnostic test for TB.

Social Learning Is Learning from Other Animals In **social learning**, animals learn behaviors by watching or listening to others of their species. By observing and copying the behavior of others, animals may learn which foods to eat, where to find food or breeding sites, or how to avoid predators. Songbirds of many species acquire their songs by copying the songs of other birds. In animals that use tools, information about how to use them is usually gained by watching other animals use them. For example, some dolphins in Shark Bay, Australia, use sponges to help protect their snouts as

▲ **FIGURE 26-7 A Gambian giant pouched rat at work, detecting land mines**

they dig for prey in the stony seabed (**FIG. 26-8a**). Young dolphins learn this behavior by watching their mother perform it. Similarly, chimpanzees may use stones to crack open nuts or sticks to fish termites out of their mounds (**FIG. 26-8b**). These skills are transmitted from one chimpanzee to another by social learning.

Insight Is Problem Solving Without Trial and Error In certain situations, animals seem able to solve problems suddenly, without the benefit of prior experience. This kind of sudden problem solving is sometimes called **insight learning**, because it seems at least superficially similar to the process by which humans mentally manipulate concepts to arrive at a solution. We cannot, of course, know for sure if non-human animals experience similar mental states when they solve problems.

One species that seems to be good at solving problems is the New Caledonian crow. These birds not only use tools, but also manufacture them, shaping twigs and leaves into hooks that the birds use to extract insects from their hiding places. In a lab experiment, a New Caledonian crow was presented with a straight piece of wire and a bucket of meat that had been placed down a well, so the bird could not reach it. The crow quickly used its beak to bend the wire into a hook, which the bird used to lift the bucket up so it could eat the meat. New Caledonian crows also readily solved, on the first try, a multistep puzzle that required them to use a short stick to reach a long stick that was in turn used to reach a food reward. In experiments that tested crows' ability to gain access to food rewards floating on water at the bottom of a clear, vertical tube, researchers found that crows quickly reacted by dropping objects into the tube so that the water level rose, bringing the food to within the birds' reach (**FIG. 26-9**). What's more, the crows dropped solid objects that would sink rather than hollow objects that would float.

▲ **FIGURE 26-9 Insight learning** New Caledonian crows learn to solve fairly complex problems without prior training. Here, a crow obtains a reward by selecting the object that is most effective for raising the water level in a tube.

(a) Dolphin wearing a sponge tool

(b) Chimpanzees probing for termites

▲ **FIGURE 26-8 Social learning** Animals may learn helpful behaviors from other animals. By observing others, **(a)** bottlenose dolphins learn to use a sponge plucked from the seafloor as a snout protector, and **(b)** chimpanzees learn to use a twig to extract termites from a mound.

CHECK YOUR LEARNING

Can you . . .

- describe evidence that genes influence behavior?
- describe evidence that the physical and experiential environments influence behavior?
- explain habituation, imprinting, classical conditioning, trial-and-error learning, social learning, and insight learning?

26.2 HOW DO ANIMALS COMPETE FOR RESOURCES?

Resources such as food, space, and mates are scarce relative to the reproductive potential of populations, leading to a contest to survive and reproduce. The resulting competition underlies many of the most frequent types of interactions between animals.

▲ **FIGURE 26-10 Combat** Fighting can be dangerous for combatants, such as these male elephants.

Aggressive Behavior Helps Secure Resources

One of the most obvious manifestations of competition for resources is aggression between members of the same species. Aggressive behavior includes physical combat between rivals (**FIG. 26-10**). A fight, however, can injure its participants; even the victorious animal might not survive to pass on its genes. As a result, natural selection has favored the evolution of displays or rituals for resolving conflicts. Aggressive displays allow competitors to assess each other and determine a winner on the basis of size, strength, and motivation, rather than on the basis of wounds inflicted. Thanks to these displays, most aggressive encounters end without physical damage to the participants. We discuss signals of aggression in more detail in Section 26.5.

Dominance Hierarchies Help Manage Aggressive Interactions

Even when they do not cause injuries, aggressive interactions use a lot of energy and can disrupt other important tasks, such as finding food, watching for predators, or raising young. Thus, there are advantages to resolving conflicts with minimal aggression. When animals live in social groups in which individuals interact repeatedly, they may form a **dominance hierarchy**, in which each animal establishes a rank that determines its access to resources. Although aggressive encounters occur frequently while the dominance hierarchy is being established, once each animal learns its place in the hierarchy, disputes are infrequent, and the dominant individuals obtain the most access to the resources needed for reproduction, including food, space, and mates. For example, domestic chickens, after some initial squabbling, sort themselves into a reasonably stable "pecking order." Thereafter, all birds in the group defer to the dominant bird, all but the dominant bird give way to the second most dominant, and so on. In wolf packs, one member of each sex is the dominant, or "alpha," individual to whom all others of that sex are subordinate.

Animals May Defend Territories That Contain Resources

In many animal species, competition for resources takes the form of **territoriality**, the defense of an area where important resources are located. As with dominance hierarchies, territoriality tends to reduce aggression, because once a territory is established through aggressive interactions, relative peace prevails as boundaries are recognized and respected. One reason for this stability is that an animal is highly motivated to defend its territory and will often repel even larger, stronger animals that attempt to invade it.

Territories are most commonly defended by individual males, but may be defended by individual females, mated pairs, families, or larger social groups. The defended area may include places to mate, raise young, feed, or store food. Territories are as diverse as the animals defending them. For example, a territory can be a tree where woodpeckers store acorns (**FIG. 26-11**), a small depression in a lake floor used as a nesting site by a cichlid fish, a hole in the sand that is home to a crab, or a mouse carcass defended by a pair of burying beetles whose offspring are developing inside it.

Territoriality Occurs When Benefits Outweigh Costs

Acquiring and defending a territory require a costly investment of time and energy, but making the investment can increase an animal's reproductive success enough to offset the cost. For example, American redstarts are migratory songbirds that defend territories both on their breeding grounds in northern forests and in their wintering areas in the tropics. Birds that defend winter territories in lush, moist forests have a larger number of offspring than do birds whose winter territories are in less-productive habitats. Thus, in redstarts, resources acquired via territorial behavior in the nonbreeding season have an important effect on breeding success months

▲ **FIGURE 26-11 A feeding territory** Acorn woodpeckers live in communal groups that excavate acorn-sized holes in dead trees and stuff the holes with green acorns for dining during the lean winter months. The group defends the trees vigorously against other groups of acorn woodpeckers and against acorn-eating birds of other species, such as jays.

later, probably because the birds with good winter territories are in better condition and can migrate north more quickly to acquire the best breeding territories and get an earlier start on reproduction.

The costs and benefits of defending a territory may change as conditions change, so many animals are territorial only at certain times or only under certain conditions. For example, nonbreeding pied wagtail birds defend feeding territories on days when the invertebrates they feed on are relatively scarce, but abandon territoriality on days when prey animals are especially abundant. When the breeding season arrives, male wagtails defend territories continuously, regardless of the abundance of food. If all goes well for a male, a female will join him on the territory. Many animal species defend territories only while breeding, when it is critical to monopolize the resources required to produce offspring.

CHECK YOUR LEARNING

Can you . . .

- describe the function of aggressive behavior?
- explain why natural selection has favored the evolution of aggressive signals and displays, dominance hierarchies, and territoriality?

26.3 HOW DO ANIMALS BEHAVE WHEN THEY MATE?

Sexual selection (see Chapter 16) has fostered the evolution of behaviors that help animals compete for access to mates and that help them choose suitable mates. In most cases, males experience strong selection for traits that help them compete, and females experience selection for traits that help them make advantageous choices.

Males May Fight to Mate

The aggressive behaviors that we described in Section 26.2 often occur in the context of competition among males for opportunities to mate. For example, in the wasp *Sycoscapter australis*, males often must fight in order to mate. In this species, eggs develop within a fig, and newly hatched wasps emerge in the hollow space inside the fruit. Males hatch first and move about looking for emerging females to mate with. But when males encounter one another, they battle fiercely. The fights are intense; up to a quarter of combatants suffer fatal wounds. Winners earn the chance to mate.

In many species, males compete not for direct access to females, but for control of a territory that will attract females. In such species, males who successfully defend territories have the greatest chance of mating. Females usually prefer high-quality territories, which might have features such as large size, abundant food, and secure nesting areas. For example, male side-blotched lizards that defend territories containing many rocks are more successful in attracting mates than are males that defend territories with few rocks. Rockier territories provide more vantage points for detecting predators and a greater range of microclimates, which is important for "cold-blooded" animals like lizards that regulate body temperature by moving between warmer and cooler spots. Females that select males with the best territories increase their own reproductive success. When experimenters moved rocks from one male's territory to another, females strongly preferred to settle on the improved territories, showing that their preference was based on the territory rather than on the male defending it.

▲ **FIGURE 26-12 Nuptial gifts** A male dance fly copulates with a female that has accepted his gift of a dead insect.

Males May Provide Gifts to Mates

In some species, females are induced to mate by males that provide resources more directly, in the form of a meal. For example, female dance flies mate only with males that bring them a dead insect to eat (**FIG. 26-12**). Copulation occurs while the female eats the insect. If the insect is too small, copulation does not last long enough for sperm to be transferred, so males that present larger insects have greater reproductive success. Similar *nuptial gifts* are presented by males of a number of spider and insect species and may consist of nutritious secretions rather than prey items. Perhaps the ultimate in nutritious gifts is presented by the male red-backed spider. Following copulation, a male throws his own body into the much larger female's jaws, and she usually obliges by eating him. The reproductive cost of this sacrifice is probably small, as even the males that are not eaten after mating are very unlikely to survive long enough to mate with a second female. The tiny males are preyed upon by many predators, and among those that do not sacrifice themselves during mating, almost all are killed and eaten before they can reach a second female's web.

Competition Between Males Continues After Copulation

Among animal species, it is common for both males and females to copulate with multiple partners. When females have multiple partners, males may evolve behaviors that increase

▲ **FIGURE 26-13 Mate guarding** After mating, a male may continue to guard his mate to prevent her from mating with other males. A male swamp milkweed beetle guards his mate by riding around on her back.

the odds that a male's sperm, and not those of a competitor, fertilize the female's eggs. For example, a male may continue to guard a female after copulation to prevent her from mating with other males. In many species of birds, lizards, fish, and insects, males spend days or weeks closely following previous sexual partners, fighting off the advances of other males. In some cases, the male even rides around on the female (**FIG. 26-13**).

If a male copulates with a female and does not subsequently guard her, his reproductive success may be usurped by competitors. For example, in the dunnock, a small songbird in which a female may copulate with multiple males, a male ready to copulate first pecks the female's genital opening. In response, the female ejects any sperm from recent prior copulations, increasing the odds that the current male's sperm will be the ones that fertilize her eggs. Other males take an even more direct approach. Before a male black-winged damselfly copulates with a female that has landed in his territory, he uses the tip of his penis to scoop out any sperm that a previous sexual partner may have left in the female's sperm storage organ (**FIG. 26-14**).

Reproductive competition among males may continue even after conception. Lions live in social groups called *prides* that contain a few males that maintain exclusive reproductive access to a larger number of females. As ruling a pride is the only way for a male to reproduce, there is intense competition for the role; roaming males without a pride may attempt to overthrow a pride's current males and oust them from the group. If a takeover attempt is successful, the new males are likely to kill any cubs present in the pride. By eliminating the offspring of other males, the new males ensure that the females will reproduce again soon, this time investing their reproductive effort in the offspring of the new males. This kind of *infanticide* is widespread among animals.

Multiple Mating Behaviors May Coexist

In some species, a relatively small number of males can monopolize a large proportion of the opportunities for reproduction. For example, a few males in a population may be especially proficient at producing signals that females find attractive, or scarcity of resources may allow only a few males to secure a territory.

When some males are excluded from the optimal way of reproducing, alternative mating behaviors may arise. For example, in bluegill fish, two types of males, dubbed "parentals" and "cuckolders," coexist and exhibit distinctively different mating behavior (**FIG. 26-15**). Parental males are large, colorful, and territorial; they build and defend nests scooped out of the streambed. Females visit the nests and choose a territorial male to spawn with, laying eggs that mix with the male's sperm. Cuckolder males are very small "sneakers" that hide in vegetation near a nest and wait for an opportunity to deposit some sperm in the nest of a spawning pair without being noticed by the territory holder. When cuckolder males grow a bit larger, they may become female mimics, sporting drab colors that resemble those of females and behaving as a spawning female might. This subterfuge may allow a female-mimic male to slip between a spawning pair and add his sperm to the mix.

(a) Damselflies mating

(b) Damselfly penis

◀ **FIGURE 26-14 Sperm competition (a)** During copulation, a male damselfly (the upper fly in the photo) grasps the female behind her head with claspers on his rearmost segment. The male's penis is located on his underside just below his wings and transfers sperm to the genital opening on the female's rearmost segment, which she swings forward and upward toward the male's body. Before copulating, however, the male uses his **(b)** spiny penis to scrape out any sperm that a competing male may have deposited in the female's sperm storage organ.

(a) Territorial male

(b) Sneaker males lurk near a mating pair

(c) A female-mimic male approaches a mating pair

▲ **FIGURE 26-15 Alternative mating strategies** Different male mating behaviors coexist in populations of bluegills. **(a)** Some males are territorial, defending nests that attract females. **(b)** Other males are sneakers that do not defend territories but may be able to opportunistically fertilize eggs. **(c)** Female-mimic males look and act like females and may not be recognized as competitors by territorial males.

CHECK YOUR LEARNING

Can you . . .

- describe some types of behavior that have evolved as a result of competition for mates?
- give specific examples of these types of behavior?
- explain why multiple mating behaviors may coexist in a population?

26.4 HOW DO ANIMALS COMMUNICATE?

Animals frequently broadcast information. If this information evokes a response from other individuals, and if that response tends to benefit the sender and the receiver, then a communication channel can form. **Communication** is the production of a signal by one organism that causes another organism to change its behavior. These changes in behavior, on average, benefit both the signaler and receiver. However, the overall benefit may be reduced if the communication channel is exploited by other animals that harm the communicators.

Although animals of different species may communicate, as when a cat hisses at a dog, most animals communicate primarily with members of their own species. The ways in which animals communicate are astonishingly diverse and use all of the senses. In the following sections, we will look at communication by visual displays, sound, chemicals, and touch.

Visual Communication Is Most Effective over Short Distances

Animals with well-developed eyes use visual signals to communicate. Visual communication may involve passive signals in which the size, shape, or color of the signal conveys important information. For example, when female mandrills become sexually receptive, they develop a large, brightly colored swelling on their buttocks (**FIG. 26-16**). Alternatively, visual signals can be active, consisting of specific movements or postures that convey a message. For example, a wolf signals aggression by lowering its head, raising its hackles (fur on its neck and back), and exposing its fangs (**FIG. 26-17a**).

Using visual signals to communicate has both advantages and disadvantages. On the positive side, visual signals communicate instantaneously, and active visual signals can be rapidly changed to convey a variety of messages in a short period. For example, an initially aggressive wolf that encounters a more dominant individual can quickly shift to a submissive display, crouching with its rump lowered and its tail tucked (**FIG. 26-17b**). Visual communication is quiet and unlikely to alert distant predators, although the signaler does make itself conspicuous to those nearby. On the negative side, visual signals are generally ineffective in dense vegetation or in darkness and are limited to close-range communication.

Communication by Sound Is Effective over Longer Distances

Acoustic signals (messages broadcast by sound) overcome many of the shortcomings of visual displays. Like visual

▲ **FIGURE 26-16 A passive visual signal** The female mandrill's colorfully swollen buttocks serve as a passive visual signal that she is fertile and ready to mate.

(a) Aggressive display

(b) Submissive display

▲ **FIGURE 26-17 Active visual signals (a)** A wolf communicates aggressive intent by facing its opponent, lowering its head, erecting the fur along its backbone, and exposing its fangs. These signals can vary in intensity, communicating different levels of aggression. **(b)** A wolf communicates submission by lowering its rump and tucking its tail.

displays, acoustic signals reach receivers almost instantaneously. But unlike visual signals, acoustic signals can be transmitted through darkness, dense forests, and murky water. Acoustic signals can also be effective over longer distances than visual signals. For example, the low, rumbling calls of an African elephant can be heard by elephants several miles away, and the songs of humpback whales are audible for hundreds of miles. Even the small kangaroo rat produces a sound (by striking the desert floor with its hind feet) that is audible 150 feet (45 meters) away. The advantages of long-distance transmission, however, are offset by an important disadvantage: Predators and other unwanted receivers can also detect an acoustic signal from a distance and can use that signal to locate the signaler.

Like visual displays, acoustic signals can be varied to convey rapidly changing messages. An individual can convey different messages by varying the pattern, volume, or pitch of a sound. A wolf's vocal repertoire, for example, includes a variety of barks, howls, whimpers, and growls.

Some animals communicate with vibrations akin to the stimuli that humans perceive as sounds. For example, male water striders vibrate their legs, sending species-specific patterns of vibrations through the water to be detected by other water striders (**FIG. 26-18**). Caterpillars of some moth species communicate with other caterpillars by scraping or drumming on a leaf with a specialized structure, thereby sending vibrations through the surrounding vegetation.

Chemical Messages Persist Longer but Are Hard to Vary

Chemical substances that are produced by animals and that influence the behavior of other members of the species are called **pheromones**. Unlike visual or sound signals that may attract predators, pheromones are typically not detectable by other species. In addition, a pheromone can act as a kind of signpost, persisting over time and conveying a message long after the signaling animal has departed.

Pheromones can travel fairly far in air or water and so are effective for long-range communication. However, chemical communication requires animals to synthesize a different substance for each message. As a result, chemical signaling systems communicate fewer and simpler messages than do sight- or sound-based systems. In addition, pheromone signals cannot easily convey rapidly changing messages.

Humans have harnessed the power of pheromones to combat insect pests. The sex attractant pheromones of some agricultural pests, such as the Japanese beetle and the gypsy moth, have been successfully synthesized. These synthetic pheromones can be used to disrupt mating or to lure these insects into traps. Controlling pests with pheromones has major environmental advantages over conventional pesticides, which kill beneficial as well as harmful insects and foster the evolution of pesticide-resistant insects. In contrast,

▲ **FIGURE 26-18 Communication by vibration** The light-footed water strider relies on the surface tension of water to support its weight. By vibrating its legs, the water strider sends signals that radiate out over the surface of the water. These vibrations advertise the strider's species and sex to others nearby.

each pheromone is specific to a single species and does not promote the spread of resistance because insects resistant to the attraction of their own pheromones do not reproduce successfully.

CASE STUDY CONTINUED

Sex and Symmetry

Does symmetry have a scent? In one study, researchers measured the body symmetry of 80 men and then issued a clean T-shirt to each one. Each subject wore his shirt to bed for two consecutive nights. A panel of 82 women sniffed the shirts and rated their scents for "pleasantness" and "sexiness." Which shirts had the sexiest, most pleasant scents? The ones worn by the most symmetrical men. The researchers concluded that women can identify symmetrical men by their scent.

What other kinds of messages do animals send with pheromones and other signals? Find out in Section 26.5.

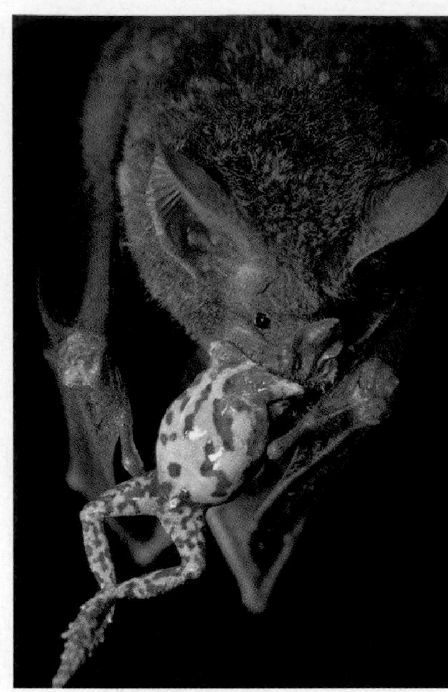

▲ **FIGURE 26-20 Illegitimate receivers** A tungara frog's loud calls may be intercepted by a predatory fringe-lipped bat.

Communication by Touch Requires Close Proximity

Communication by physical contact is, for obvious reasons, limited to signaling between animals that are very close together. As a result, it is most common in species that spend a great deal of time in social groups. However, even members of nonsocial species may come into close contact with other individuals during courtship or combat. So communication by touch may also occur in those contexts (**FIG. 26-19**).

Communication Channels May Be Exploited

Even though sending or receiving a signal is in general beneficial, communicators are sometimes harmed by organisms that exploit a communication channel. The exploiter may be an illegitimate receiver that intercepts a signal. For example, parasitic flies that lay their eggs in the bodies of field crickets find their hosts by moving toward the chirps that male crickets produce. Similarly, predatory fringe-lipped bats find tungara frogs to eat by homing in on the loud calls of male frogs (**FIG. 26-20**).

▲ **FIGURE 26-19 Communication by touch** Touch is important in sexual communication. These land snails engage in courtship behavior that will culminate in mating.

The exploiter of a communication channel may also be an illegitimate signaler. For example, as male fireflies in the genus *Photinus* fly about, they emit flashes in a distinctive pattern that identifies them as members of their species. If a receptive female on the ground sees the flashes, she may respond with flashes of her own, and the male flies down to mate with her. However, fireflies of the genus *Photuris* have evolved the ability to imitate the female *Photinus* flashing pattern. If a male *Photinus* approaches the deceptive signal, the larger *Photuris* firefly kills and eats him.

CHECK YOUR LEARNING

Can you . . .

- compare the advantages and disadvantages of visual, acoustic, and chemical signals?
- describe examples of visual, acoustic, chemical, and tactile (touch) signals?

26.5 WHAT DO ANIMALS COMMUNICATE ABOUT?

The information shared by communicating helps animals manage a range of interactions with other animals. Animals may communicate to help resolve conflicts. They may communicate about sex, about food, or about predators. Members of a social group may communicate to help coordinate activities.

(a) A male baboon

(b) Sarcastic fringeheads

▲ **FIGURE 26-21 Aggressive displays (a)** Threat display of the male baboon. Despite the potentially lethal fangs so prominently displayed, aggressive encounters between baboons rarely cause injury. **(b)** The aggressive display of many male fishes such as these sarcastic fringeheads, includes elevating the fins and flaring the gill covers, thus making the body appear larger.

Animals Communicate to Manage Aggression

As you learned in our earlier discussion of aggression, the costliness of direct combat has fostered the evolution of signals that communicate aggression. During aggressive displays, animals may exhibit weapons, such as claws and fangs (**FIG. 26-21a**), and often make themselves appear larger (**FIG. 26-21b**). Competitors often stand upright and erect their fur, feathers, ears, or fins. These visual signals are typically accompanied by vocal signals such as growls, croaks, roars, or chirps. Fighting tends to be a last resort when displays fail to resolve a dispute.

In addition to aggressive visual and vocal displays, many animal species engage in ritualized combat. Deadly weapons may clash harmlessly (**FIG. 26-22**) or may be displayed without contacting the opponent. The ritual thus allows contestants to assess the strength and the motivation of their rivals, and the loser retreats. Ritual contests often involve communication by touch, as when two male sandperch fish lock mouths in ritualized wrestling, or two male zebras begin a contest over access to mates by placing their heads on one another's shoulders.

Aggressive visual and vocal displays are commonly deployed in territorial defense. For example, a male roe deer warns intruders away from its territory by displaying its antlers and a male anole lizard does so by displaying the colorful dewlap that extends from his throat. A male sea lion defends a strip of beach by swimming up and down in front of it, calling continuously. A male cricket uses a special structure on its wings to produce chirps that advertise its ownership of its burrow. The often beautiful and elaborate vocalizations

▲ **FIGURE 26-22 Displays of strength** The oversized claws of fiddler crabs, which could severely injure another animal, grasp harmlessly. Eventually one crab, sensing greater vigor in his opponent, retreats unharmed.

of songbirds are also used in territory defense. For example, the husky trill of a male seaside sparrow warns other males to steer clear of his territory.

An animal that has a territory but cannot always be present may use pheromones to scent-mark the boundaries. For example, male sac-winged bats use secretions from their chin glands to mark the perimeter of the small territory that each male defends within a communal roost site that is occupied only during daytime. Scent-marking is also useful for territories that are too large for continuous monitoring of boundaries. A solitary tiger, for example, uses pheromones to mark its presence in a territory that may be as large as 385 square miles (1000 square kilometers).

▲ **FIGURE 26-23 Sexual vocalizations** The huge head of a male hammer-headed bat acts as a resonator that amplifies that male's loud courtship calls.

Mating Signals Encode Sex, Species, and Individual Quality

Before animals can successfully mate, they must identify one another as members of the same species, as members of the opposite sex, and as being sexually receptive. In many species, finding an appropriate potential partner is only the first step. Often the male must demonstrate his quality before the female will accept him as a mate. The need to fulfill all of these requirements has resulted in the evolution of a diverse and fascinating array of courtship signals.

Animals often use sounds to advertise their sex and species. Male grasshoppers and crickets advertise their sex and species with chirps, and a male fruit fly does so with a buzz he produces by vibrating one wing. Acoustic signals may also be used by potential mates to compare rival suitors. During mating season, male hammer-headed bats gather together and produce a chorus of loud, honking calls. The males have a very large head that helps the vocalizations resonate (**FIG. 26-23**). Females fly among the gathered males, listening to their calls before choosing a male to mate with.

Many species use visual displays for courting. Male fence lizards, for example, bob their heads in a species-specific rhythm, and females prefer the rhythm of their own species. The elaborate construction projects of the male gardener bowerbird and the scarlet throat of the male frigatebird serve as flashy advertisements of sex, species, and male quality (**FIG. 26-24**).

Pheromones can also play an important role in reproductive behavior. A sexually receptive female silk moth, for example, sits quietly and releases a chemical message that can be detected by males up to 3 miles (5 kilometers) away. The exquisitely sensitive and selective receptors on the antennae of the male silk moth respond to just a few molecules of the substance, allowing him to travel upwind along a concentration gradient to find the female (**FIG. 26-25a**). Water is an excellent medium for dispersing chemical signals, and fish commonly use a combination of pheromones and elaborate courtship movements to ensure the synchronous release of gametes. Mammals, with their highly developed sense of smell, often rely on pheromones released by the female during her fertile periods to attract males (**FIG. 26-25b**).

(a) A bowerbird bower

(b) A male frigatebird

◀ **FIGURE 26-24 Sexual displays** **(a)** During courtship, a male gardener bowerbird builds a bower out of twigs and decorates it with colorful items that he gathers. **(b)** A male frigatebird inflates his scarlet throat pouch to attract passing females.

THINK CRITICALLY The male bowerbird provides no protection, food, or other resources to his mate or offspring. Why, then, do females carefully compare the bowers of different males before choosing a mate?

(a) Antennae detect pheromones

(b) Noses detect pheromones

▲ **FIGURE 26-25 Sexual pheromones (a)** Male moths find females not by sight but by following airborne pheromones released by females. These odors are sensed by receptors on the male's huge antennae, whose enormous surface area maximizes the chances of detecting the female scent. **(b)** When dogs meet, they typically sniff each other near the base of the tail. Scent glands there broadcast information about the bearer's sex and interest in mating.

THINK CRITICALLY Female dogs use a pheromone to signal readiness to mate, but female mandrills (see Fig. 26-16) signal mating readiness with a visual signal. What differences would you predict between the two species' methods of searching for food?

CASE STUDY CONTINUED

Sex and Symmetry

If females are attracted to symmetrical males, we might expect females to assess the symmetry of male visual signals that function in mate attraction. For example, a male house finch has a patch of bright red feathers on the crown of his head, and researchers have shown that males whose crown patches are brightly colored are more likely to attract a mate than are males with dull patches. But males with patches that are both colorful *and* highly symmetrical have the highest mating success of all.

Animals Warn One Another About Predators

Animals may communicate about threats, especially threats from predators. When aphids are attacked by predators, they secrete an alarm pheromone that causes receivers to stop feeding and move away from the signal. If a Belding's ground squirrel spots an approaching predator, it produces an alarm call that sends nearby squirrels scrambling for safety. In some cases, varying alarm signals can convey different messages. For example, vervet monkeys produce different calls in response to threats from each of their major predators: snakes, leopards, and eagles. The response of listening monkeys to each of these calls is appropriate to the particular predator. The "bark" that warns of a leopard or other four-legged carnivore causes monkeys on the ground to take to trees and those in trees to climb higher. The "rraup" call, which advertises the presence of an eagle or other hunting bird, causes monkeys on the ground to look upward and take cover, whereas monkeys already in trees drop to the shelter of lower, denser branches. The "chutter" call that indicates the presence of a snake causes the monkeys to stand up and search the ground for the predator.

Animals Share Information About Food

Animals that live in groups may share information about food. For example, foraging termites that discover food lay a trail of pheromones from the food to the nest, and other termites follow the trail (**FIG. 26-26**). Honeybees share

▲ **FIGURE 26-26 Communication about food** A trail of pheromones, secreted by termites from their own colony, orients foraging termites toward a source of food.

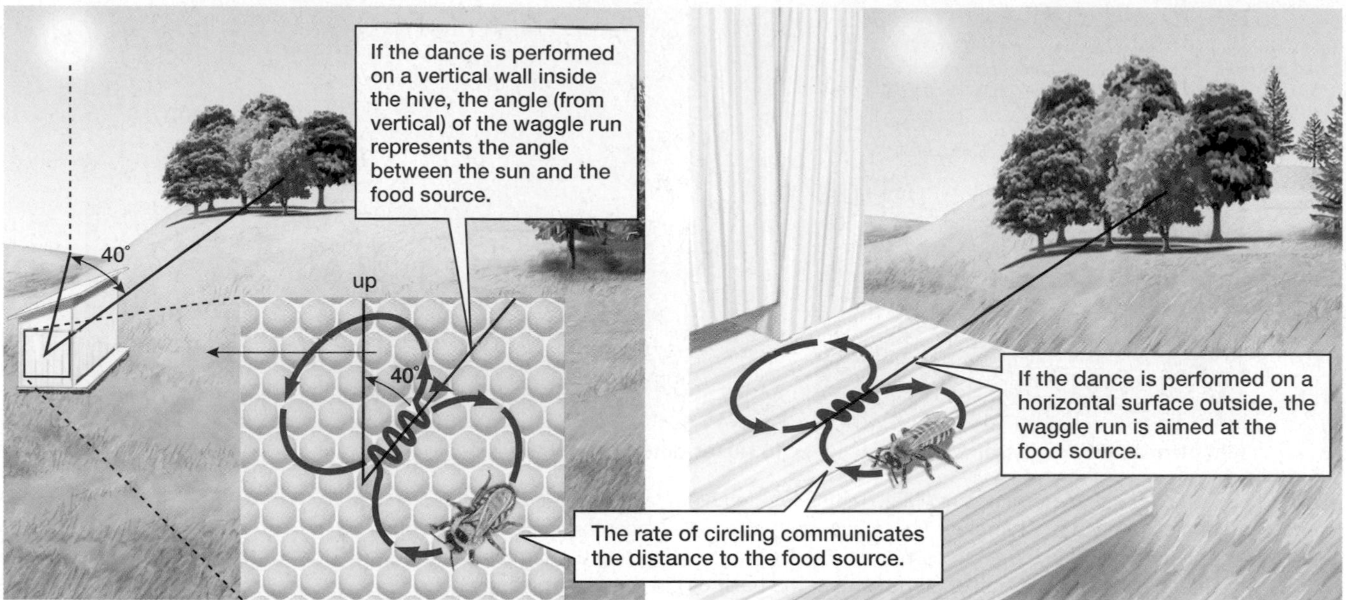

▲ **FIGURE 26-27 Bee language: the waggle dance** A forager, returning from a rich source of nectar, performs a waggle dance that communicates the distance and direction of the food source as other foragers crowd around her, touching her with their antennae. The bee moves in a straight line while shaking her abdomen back and forth ("waggling") and buzzing her wings. She repeats this dance over and over in the same location, circling back in alternating directions.

information about the location of food by means of a **waggle dance** (**FIG. 26-27**) that is usually performed in the darkness inside a hive. Other bees crowd around the dancer, and a message is transmitted by vibrations from the dancer's moving wings and body.

Communication Aids Social Bonding

Social bonds between animals may be established or reinforced through communication by touch. For example, the return of a familiar individual to its group is facilitated by touch, as when an elephant puts its trunk in a new arrival's mouth, two reunited coyotes touch noses, or an ant touches a returning colony member with its antennae. Social signaling through touch is especially apparent in primates, which have many gestures—including kissing, nuzzling, patting, petting, and grooming—that facilitate social cohesion (**FIG. 26-28**).

▲ **FIGURE 26-28 Grooming** An adult olive baboon grooms a juvenile. Grooming both reinforces social relationships and removes debris and parasites from the fur.

CHECK YOUR LEARNING

Can you . . .

- describe how signals function in aggression, courtship, and mating?
- describe how animals share information about predators and food?

26.6 WHY DO ANIMALS PLAY?

Many animals play. Pygmy hippopotamuses push one another, shake and toss their heads, splash in the water, and pirouette on their hind legs. Otters delight in elaborate acrobatics. Bottlenose dolphins balance fish on their snouts, throw objects, and carry them in their mouths while swimming. Baby vampire bats chase, wrestle, and slap each other with their wings. Even octopuses have been seen playing a game: pushing objects away from themselves and into a current, then waiting for the objects to drift back, only to push them back into the current to start the cycle over again.

Although play behavior seems easy to recognize, it is challenging to formulate a precise definition of play. A widely used definition describes **play** as behavior that seems to lack any immediate function, and that often includes modified versions of behaviors used in other contexts.

(a) Chimpanzees

(b) Polar bears

(c) Red foxes

FIGURE 26-29 Young animals at play

Animals Play Alone or with Other Animals

Play can be solitary, as when a single animal manipulates an object, such as a cat with a ball of yarn, or when a macaque monkey makes and plays with a snowball. Play can also be social. Often, young animals of the same species play together, but parents may join them. Social play typically includes chasing, fleeing, wrestling, kicking, and gentle biting (**FIG. 26-29**).

Young animals play more frequently than do adults. Play typically borrows movements from other behaviors (attacking, fleeing, stalking, and so on) and uses considerable energy. Also, play is potentially dangerous. Many young humans and other animals are injured, and some are killed, during play. In addition, play can distract an animal from the presence of danger while making it conspicuous to predators. So why do animals play?

Play Aids Behavioral Development

It is likely that play has survival value and that natural selection has favored those individuals who engage in playful activities. One of the best explanations for the survival value of play is the *practice hypothesis*. It suggests that play allows young animals to gain experience in behaviors that they will use as adults. By performing these acts repeatedly in play, an animal practices skills that will later be important in hunting, fleeing, and social interactions.

Play is most intense early in life when the brain develops and crucial neural connections form. Species with large brains tend to be more playful than species with small brains. Because larger brains are generally linked to greater learning ability, this relationship supports the hypothesis that adult skills are learned during juvenile play. Watch children roughhousing or playing tag, and you will see how play might foster strength and coordination and develop skills that might have helped our hunting ancestors survive.

CHECK YOUR LEARNING

Can you . . .

- describe the characteristics of animal play and some examples of it?
- describe a hypothesis about the function of play?

26.7 WHAT KINDS OF SOCIETIES DO ANIMALS FORM?

Sociality, the tendency to associate with others and form groups, is a widespread feature of animal life. Most animals interact at least a little with other members of their species. Many spend the bulk of their lives in the company of others, and a few species have developed complex, highly structured societies.

Group Living Has Advantages and Disadvantages

Living in a group has both costs and benefits, and a species will not evolve social behavior unless the benefits of doing so outweigh the costs. Benefits to social animals include the following:

- Increased abilities to detect, repel, and confuse predators.
- Increased hunting efficiency or ability to find food.
- Increased potential for division of labor within the group.
- Increased likelihood of finding mates.

On the negative side, social animals may encounter:

- Increased competition for limited resources.
- Increased risk of infection from contagious diseases.
- Increased risk that offspring will be killed by other members of the group.
- Increased risk of being spotted by predators.
- Increased competition for mates.

Sociality Varies Among Species

The degree to which animals of the same species cooperate varies from one species to the next. Some types of animals, such as the mountain lion, are basically solitary; interactions between adults consist of brief aggressive encounters and mating. Other animals form loose social groups, such as schools of fish, flocks of birds, and herds of musk oxen (**FIG. 26-30**). Still others form more structured social groups that may include more complex relationships among members. For example, members of a baboon troop often form alliances; if a baboon is threatened by another in the troop, the other members of its alliance may come to its defense. Similarly, male bottlenose dolphins often form alliances of two or three animals that cooperate to monopolize and defend reproductive females. Different dolphin alliances may band together to form super-alliances that engage in contests with other super-alliances to "steal" females.

▲ **FIGURE 26-30 Cooperation in loosely organized social groups** A herd of musk oxen functions as a unit when threatened by predators such as wolves. Males form a circle, horns pointed outward, around the females and young.

▲ **FIGURE 26-31 Cooperation in a complex society** Naked mole rats are highly social rodents. The individuals in a colony belong to different castes.

In some species, social behavior includes division of labor. Consider the spider *Anelosimus studiosus*. These spiders live in groups of up to 50 that build large communal webs that capture larger prey than a solitary spider could trap on its own. Group members cooperate to rear all of the group's offspring. In this spider society, different members perform different roles. Some watch over eggs and regurgitate food to feed young spiders. Others maintain the web, capture prey, and defend the colony.

Division of labor is more extreme in the rigidly structured societies of many bees, ants, and termites, and of naked mole rats (**FIG. 26-31**). In these societies, individuals are born into one of several castes. For example, in naked mole rats, which live in large underground colonies in southern Africa, the colony is dominated by the *queen*, the only reproducing female, to whom all other members are subordinate *workers*. Some workers clean the tunnels and gather food. Others dig new tunnels or defend the colony against predators and members of other colonies.

A honeybee hive also contains a single queen. Her main function is to produce eggs (up to 1,000 per day). Male bees, called *drones*, serve merely as mates for the queen and die as soon as their mating duties are complete. The hive is run by sterile female workers. The workers' tasks include carrying food to the queen and to developing larvae, producing hexagonal cells of wax in which the queen deposits her eggs, cleaning the hive and removing the dead, protecting the hive against intruders, and foraging for food by gathering pollen and nectar for the hive.

Reciprocity or Relatedness May Foster the Evolution of Cooperation

In the animal societies we have described, individuals sometimes behave in ways that help others but seem to place the helper at an immediate disadvantage. A baboon helps defend another, putting itself at risk of injury. A social spider spends time and energy capturing insects for other spiders to eat. Worker bees spend their lives doing hard labor but do not reproduce. There are many other examples: Vampire bats may regurgitate part of a blood meal to feed another bat that did not find food; ground squirrels may risk their own safety to warn the rest of their group about an approaching predator; young, mature Florida scrub jays, instead of breeding, may remain at their parents' nest and help them raise subsequent broods.

At first glance, such cooperative behavior appears to be **altruism**—behavior that decreases the reproductive success of one individual to benefit another. But true altruism presents an evolutionary puzzle. If individuals perform self-sacrificing deeds that reduce their survival and reproduction, why aren't the alleles that contribute to this behavior eliminated from the gene pool?

One possibility is that cooperative behaviors that reduce an individual's fitness over the short term actually increase it over the longer term. For example, helping another individual now may pay benefits later when the favor is returned. Such *reciprocity* is most likely to be the basis of cooperation in groups that remain together for an extended period of time and in which individuals recognize and remember one another. Cooperation may also be favored when other members of the group are close relatives of the cooperating individual. Because close relatives share alleles, the altruistic individual may promote the survival of its own alleles through behaviors that maximize the survival of its close relatives. This concept is called **kin selection**. Kin selection is believed to underlie cooperation in many animal societies, perhaps including those of social insects like honeybees, whose distinctive system of sex determination produces female workers that are very genetically similar to one another.

CHECK YOUR LEARNING

Can you . . .

- list the advantages and disadvantages of living in a group and describe the different degrees of sociality that occur among mammals?
- describe some hypotheses for the evolution of cooperation in animal societies?

26.8 CAN BIOLOGY EXPLAIN HUMAN BEHAVIOR?

The behaviors of humans, like those of all other animals, have an evolutionary history. Thus, the methods and concepts that help us understand and explain the behavior of other animals can help us understand and explain human behavior as well.

The Behavior of Newborn Infants Has a Large Innate Component

Because newborn infants have not had time to learn, we can assume that much of their behavior is innate. The rhythmic

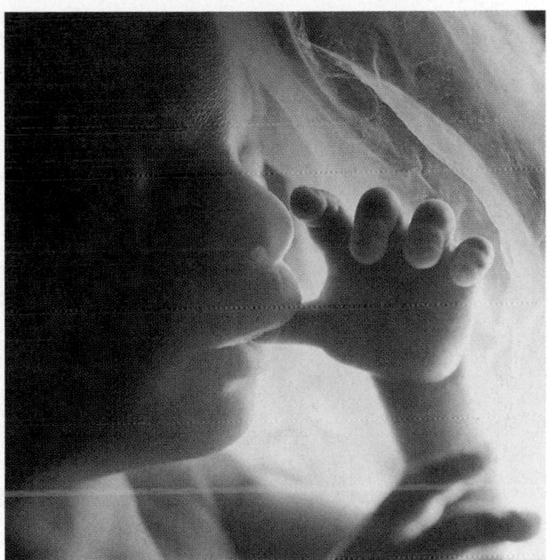

▲ **FIGURE 26-32 A human instinct** Thumb sucking is a difficult habit to discourage in young children because sucking on appropriately sized objects is an instinctive, food-seeking behavior. This fetus sucks its thumb at about 4 months of development.

movement of an infant's head in search of its mother's breast is an innate behavior that is expressed in the first days after birth. Sucking, which can be observed even in a human fetus, is also innate (**FIG. 26-32**). Other behaviors seen in newborns include grasping with the hands and feet and making walking movements when the body is held upright and supported.

Another example is smiling, which can occur soon after birth. Initially, smiling can be induced by almost any object looming over the newborn. This initial innate response, however, is soon modified by experience. Infants up to 2 months old will smile in response to a stimulus consisting of two eye-sized spots, which at that stage of development is a more potent stimulus for smiling than is an accurate representation of a human face. But as the child's development continues, learning and further development of the nervous system interact to limit the response to more correct representations of a face.

Newborns prefer their mothers' voices to other female voices. Even in their first 3 days of life, infants can be conditioned to produce certain rhythms of sucking when their mother's voice is used as reinforcement (**FIG. 26-33**). The infant's ability to learn his or her mother's voice and respond positively to it within days of birth has strong parallels to imprinting and may help initiate bonding with the mother.

Young Humans Acquire Language Easily

One of the most important insights from studies of animal learning is that animals tend to have an inborn tendency for specific types of learning that are important to their species' mode of life. In humans, one such inborn tendency is for the acquisition of language. Young children are able to acquire language rapidly and nearly effortlessly; they typically acquire a vocabulary of 28,000 words before the age of 8. Research suggests that we are born with a brain that is already primed for this early facility with language. For example, a human fetus

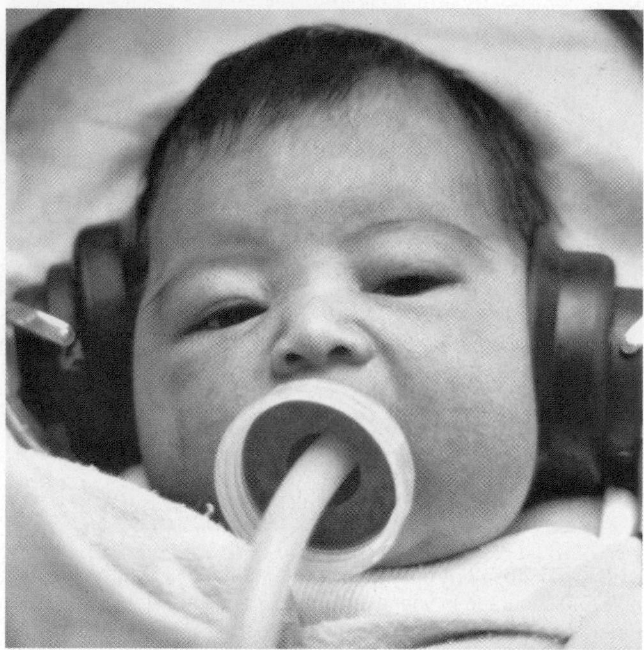

▲ **FIGURE 26-33 Newborns prefer their mother's voice** Using a nipple connected to a computer that plays audio tapes, researcher William Fifer demonstrated that newborns can be conditioned to suck at specific rates in order to listen to their own mothers' voices through headphones. For example, if the infant sucks faster than normal, her mother's voice is played; if she sucks more slowly, another woman's voice is played. Researchers found that infants easily learned and were willing to work hard at this task just to listen to their own mothers' voices, presumably because they had become used to her voice in the womb.

begins responding to sounds during the third trimester of pregnancy, and researchers have demonstrated that infants are able to distinguish among consonant sounds by 6 weeks after birth. In one experiment, infants sucked on a pacifier that contained a force transducer to record the sucking rate. The infants were conditioned to suck at a higher rate in response to playback of adult voices making various consonant sounds. When one sound (such as "ba") was presented repeatedly, the infants became habituated and decreased their sucking rate. But when a new sound (such as "pa") was presented, sucking rate increased, revealing that the infants perceived the new sound as different.

Behaviors Shared by Diverse Cultures May Be Innate

Another way to study the innate bases of human behavior is to compare simple acts performed by people from diverse cultures. This comparative approach has revealed several gestures that seem to form a universal, and therefore probably innate, human signaling system. Such gestures include facial expressions for pleasure, rage, and disdain, and greeting movements such as an upraised hand or the "eye flash" (in which the eyes are widely opened and the eyebrows rapidly elevated). The evolution of the neural pathways underlying these gestures presumably depended on the advantages that accrued to both senders and receivers from sharing information about the emotional state and intentions of the sender. A species-wide method of communication was perhaps especially important before the advent of language and later remained useful during encounters between people who shared no common language.

Certain complex social behaviors are widespread among diverse cultures. For example, the incest taboo (avoidance of mating with close relatives) seems to be universal across human cultures (and even across many species of non-human primates). It seems unlikely, however, that a shared belief could be encoded in our genes. Some biologists have suggested that the taboo is instead a cultural expression of an evolved, adaptive behavior. According to this hypothesis, close contact among family members early in life suppresses sexual desire toward these relatives, and this response arose because of the negative consequences of inbreeding (such as a higher incidence of genetic diseases and disorders). The hypothesis does not require us to assume an innate social belief, but rather proposes that we inherit a learning program that causes us to undergo a kind of imprinting early in life.

Humans May Respond to Pheromones

Although the main channels of human communication are through the eyes and ears, humans also seem to respond to chemical messages. In one experiment, for example, researchers asked nine female volunteers to wear cotton gauze in their armpits for 8 hours each day during their menstrual cycles. The gauze was then disinfected with alcohol and swabbed above the upper lips of another set of 20 female subjects (who reported that they could detect no odors other than alcohol on the gauze). The subjects were exposed to the gauze in this way each day for 2 months, with half the group sniffing armpit secretions from women in the early (preovulation) part of the menstrual cycle, while the other half was exposed to armpit secretions from later in the cycle (postovulation). Women exposed to early-cycle secretions had shorter-than-usual menstrual cycles, and women exposed to late-cycle secretions

Have You Ever Wondered ...

Which Is the World's Loudest Animal?

Sounds louder than about 120 decibels are painful to human ears, but some animals make sounds that are far louder than that. The songs of blue whales, for example, can reach 188 decibels, louder than the noise you would hear if you stood next to a jet airplane at takeoff. But the loudest known sound from an animal is produced by a much smaller organism, the pistol shrimp. The pistol shrimp doesn't sing, but it does use a specially modified claw to shoot a high-speed jet of water that stuns its prey. An air bubble forms behind the jet, and when the bubble collapses under the surrounding water pressure, the implosion produces a sound that can reach 200 decibels, much louder than a gun firing right beside your ear.

had delayed menstruation. It appears that women release different pheromones, with different effects on receivers, at different points in the menstrual cycle.

Other research suggests that people may also be able to detect chemical indicators of fear or stress. For example, researchers performed an experiment in which they collected sweat samples from 144 people. Half the people were sampled during a first-time skydive with a 1-minute freefall; the other half (the controls) were sampled during a bout of exercise. Test subjects then smelled one of the two types of samples while undergoing brain imaging. A brain region called the amygdala, which is associated with strong emotions such as fear and rage, was active in subjects who smelled the "stress sweat" from the skydivers, but not in the subjects who smelled the exercise sweat. A chemical present in the sweat of emotionally stressed people apparently triggered a similar emotion, or at least a similar brain response, in people exposed to the chemical.

Although the experiments described above offer strong evidence for the existence of human pheromones, little else is known about chemical communication in humans. The actual molecules that caused the effects documented by these and other experiments remain unknown. Receptors for chemical messages have not yet been found in humans, and we don't know if "menstrual pheromones" and "stress pheromones" are examples of an important communication system or merely isolated cases of a vestigial ability. Despite the hopeful advertisements for "sex attraction pheromones" on late-night television, chemical communication in humans is a scientific mystery awaiting a solution.

Biological Investigation of Human Behavior Is Controversial

The study of evolved, inherited human behavior is controversial, especially among nonscientists, because it challenges the long-held belief that environment is the most important determinant of human behavior. As discussed earlier in this chapter, we now recognize that all behavior has some genetic basis and that complex behavior in non-human animals typically combines elements of both innate and learned behaviors. Thus, it seems certain that our own behavior is influenced by both our evolutionary history and our cultural heritage. The debate over the relative importance of heredity and environment in determining human behavior continues and is unlikely ever to be fully resolved. The scientific study of human behavior will always be hampered because we can neither view ourselves with detached objectivity nor experiment with people as if they were laboratory rats. Despite these limitations, there is much to be learned about the interaction of learning and innate tendencies in humans.

CHECK YOUR LEARNING

Can you . . .

- describe evidence that some human behaviors and learning predispositions are innate?
- describe evidence that human behavior is influenced by pheromones?
- explain why biological investigation of human behavior is controversial?

CASE STUDY REVISITED

Sex and Symmetry

In the experiment described at the beginning of this chapter, women found males with the most symmetrical bodies to have the most attractive faces. But how did the women know which males were most symmetrical? After all, the researchers' measurement of male symmetry was based on small differences in the sizes of body parts that the female judges did not even see during the test.

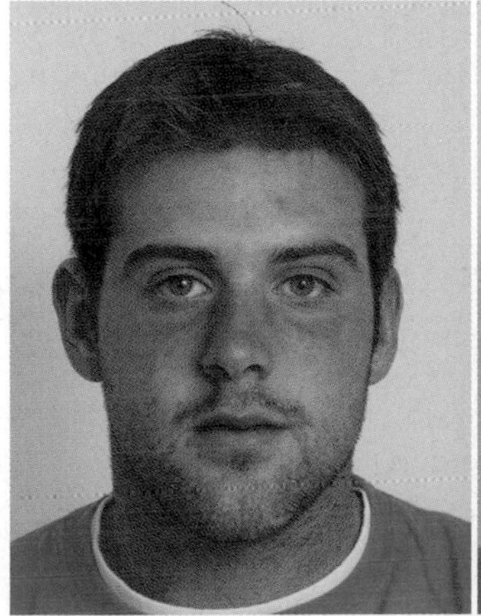

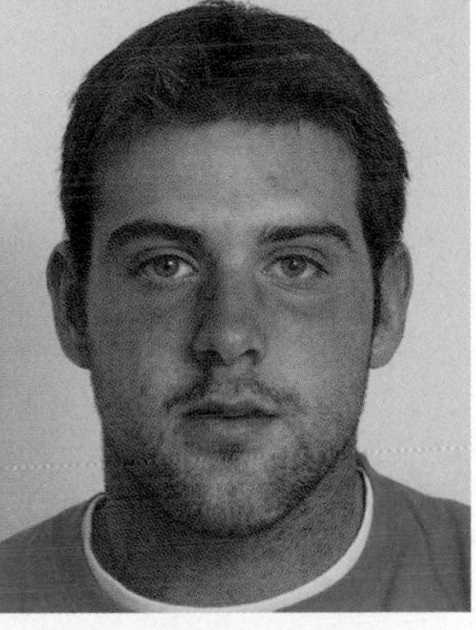

▲ **FIGURE 26-34 Faces of varying symmetry** Researchers used sophisticated software to modify facial symmetry. The face at left is unaltered; the one on the right has been modified to be more symmetrical.

Perhaps male body symmetry is reflected in facial symmetry, and females prefer symmetrical faces. To test this hypothesis, a group of researchers used computers to alter photos of male faces, either increasing or decreasing their symmetry (**FIG. 26-34**). Then heterosexual female observers rated each face for attractiveness. The observers had a strong preference for more symmetrical faces.

Why would females prefer to mate with symmetrical males? The most likely explanation is that symmetry indicates good physical condition. Disruptions of normal embryological development can cause bodies to be asymmetrical, so a highly symmetrical body indicates healthy, normal development. Females that mate with individuals whose health and vitality are announced by their symmetrical bodies are likely to have offspring that are similarly healthy and vital.

CONSIDER THIS Is our perception of human beauty determined by cultural standards, or is it part of our biological makeup, the product of our evolutionary heritage? What evidence would persuade you that beauty is a biological phenomenon? That it is a cultural one?

CHAPTER REVIEW

Go to **Mastering Biology** to access the Pearson eText, vocabulary review, practice quizzes, activities, videos, current events, and more.

*Answers to **Think Critically** and **Thinking Through the Concepts** questions can be found in the **Answers** section at the back of the book.*

Summary of Key Concepts

26.1 How Does Behavior Arise?

All animal behavior is influenced by both genetic and environmental factors. Biologists distinguish between innate behaviors, whose development is not highly dependent on external factors, and learned behaviors, which require more extensive experiences with environmental stimuli in order to develop. Innate behaviors can be performed properly the first time an animal encounters the appropriate stimulus, whereas learned behavior changes in response to the animal's social and physical environment.

26.2 How Do Animals Compete for Resources?

Although many competitive interactions are resolved through aggression, serious injuries are rare. Most aggressive encounters are settled by displays that communicate the motivation, size, and strength of the combatants. Some species establish dominance hierarchies that minimize aggression and determine access to resources. On the basis of initial aggressive encounters, each animal acquires a status in which it defers to more dominant individuals and, in turn, dominates subordinates. Territoriality, a behavior in which animals defend areas where important resources are located, also minimizes aggressive encounters.

26.3 How Do Animals Behave When They Mate?

Sexual selection has favored traits that help males gain access to mates and traits that help females choose beneficial mates. Males may fight to obtain mates, offer food gifts to females to induce them to mate, or defend territories that attract females. Females prefer to mate with males that win contests, offer high-quality gifts, or defend high-quality territories. Males may also compete by guarding their mates or by killing the offspring of other males. Multiple alternative mating behaviors may coexist in a species.

26.4 How Do Animals Communicate?

Communication allows animals of the same species to interact effectively in their quest for mates, food, shelter, and other resources. Animals communicate through visual signals, sound, chemicals (pheromones), and touch. Visual communication is quiet and can convey rapidly changing information. Visual signals are active (body movements) or passive (body shape and color). Sound communication can also convey rapidly changing information, and it is effective when vision is impossible. Pheromones can be detected after the sender has departed, conveying simple messages over time. Animals that associate in close proximity may communicate by touch. On average, signalers and receivers benefit from communication, but communication channels are sometimes exploited by animals that harm the communicators.

26.5 What Do Animals Communicate About?

Animals communicate to manage aggressive interactions and reduce the occurrence of potentially dangerous fighting. Mating signals help animals recognize the species, sex, sexual receptivity, and quality of potential mates. Alarm signals warn other animals of predators. Members of a social group may use signals to share information about the location of food. Physical contact reinforces social bonds and is a part of many premating rituals.

26.6 Why Do Animals Play?

Animals of many species engage in seemingly wasteful (and sometimes dangerous) play behavior. Play behavior in young animals has been favored by natural selection, probably because it provides opportunities to practice and perfect behaviors that will later be crucial for survival and reproduction.

26.7 What Kinds of Societies Do Animals Form?

Social living has both advantages and disadvantages, and species vary in the degree to which their members cooperate. Some species form cooperative societies. The most rigid and highly organized are those of the social insects such as the honeybee, in which the members fill rigidly defined roles throughout life.

26.8 Can Biology Explain Human Behavior?

Researchers are increasingly investigating the extent to which human behavior is influenced by evolved, genetically inherited factors. This emerging field is controversial. Because we cannot freely experiment on humans, and because learning plays a major role in nearly all human behavior, investigators must rely on studies of newborn infants and comparative cultural studies. Evidence is mounting that our genetic heritage plays a role in personality, intelligence, simple universal gestures, our responses to certain stimuli, and our tendency to learn specific things such as language at particular stages of development.

Thinking Through the Concepts

Bloom's: Remembering, Understanding

Multiple Choice

1. When a male lizard defends a certain area, he is exhibiting
   a. insight learning.
   b. kin selection.
   c. territoriality.
   d. altruism.
2. The benefits to an individual of living in a social group include
   a. increased protection against predators.
   b. decreased competition for food.
   c. less competition for mates.
   d. lower incidence of infectious disease.
3. Very early in their lives, young chicks crouch in fear when a leaf flutters overhead. Later they learn to disregard it. This kind of learning is called
   a. imprinting.
   b. habituation.
   c. operant conditioning.
   d. insight learning.

4. Young sea turtles head for the ocean immediately after they hatch. This behavior is most likely
   a. innate.
   b. learned through trial and error.
   c. classically conditioned.
   d. the result of habituation.

5. When animals engage in ____________, they often perform displays that make them look as large and dangerous as possible.
   a. courtship
   b. altruism
   c. kin selection
   d. aggression

Fill-in-the-Blank

1. In general, animal behaviors arise from an interaction between the animal's ____________ and its ____________. Some behaviors are performed correctly the first time an animal encounters the proper ____________. Such behaviors are described as ____________.
2. Play is almost certainly an adaptive behavior because it uses considerable ____________, and it can distract the animal from watching for ____________. The most likely explanation for why animals play is the "____________ hypothesis," which states that play teaches the young animal ____________ that will be useful as a(n) ____________. This hypothesis is supported by the observation that animals that have larger ____________ are more likely to play.
3. One of the simplest forms of learning is ____________, defined as a decline in response to a(n) ____________, harmless stimulus. A different type of learning in which an animal's nervous system is rigidly programmed to learn a certain behavior during a certain period in its life is called ____________. The time frame during which such learning occurs is called the ____________.
4. Animals often deal with competition for resources through ____________ behavior. Such conflicts are often resolved through ____________, which allow the competing animals to assess each other without ____________ each other. Animals that resolve conflicts this way often display their ____________ and make their bodies appear ____________.
5. The defense of an area where important resources are located is called ____________. Examples of important resources that may be defended include places to ____________, ____________, ____________, and ____________. Such resources are most commonly defended by which sex? ____________ Are these spaces usually defended against members of the same species or against members of different species? ____________
6. After each form of communication, list a major advantage in the first blank and a major disadvantage in the second blank. Pheromones: ____________; ____________. Visual displays: ____________; ____________.

Review Questions

1. Explain why animals play. Include the features of play in your answer.
2. List four senses through which animals communicate, and give one example of each form of communication. For each sense listed, present both advantages and disadvantages of that form of communication.
3. A bird will ignore a squirrel in its territory, but will act aggressively toward a member of its own species. Explain why.
4. Why are most aggressive encounters among members of the same species relatively harmless?
5. Discuss the advantages and disadvantages of group living.
6. What kinds of evidence might indicate that a particular human behavior is innate?

Applying the Concepts

Bloom's: Applying, Analyzing, Evaluating

1. Male mosquitoes orient toward the high-pitched whine of the female, and female mosquitoes, the only sex that sucks blood, are attracted to the warmth, humidity, and carbon dioxide exuded by their prey. Using this information, design a mosquito trap or killer that exploits a mosquito's innate behaviors. Then design one for moths.
2. Describe and give an example of a dominance hierarchy. What role does it play in social behavior? Give a human parallel, and describe its role in human society. Are the two roles similar? Why or why not? Repeat this exercise for territorial behavior in humans and in another animal.
3. You are the manager of an airport. Planes are being endangered by large numbers of flying birds, which can be sucked into and disable the engines. Without harming the birds, what might you do to discourage them from nesting and flying near the airport and its planes?

27 Population Growth and Regulation

Northern elephant seals were hunted almost to extinction in the 1800s. Today, they are a tourist attraction on the central California coast.

CASE STUDY

The Return of the Elephant Seals

For sheer drama in a natural spectacle, it's hard to top a battle between two elephant seal bulls. These bulls, which can be up to 15 feet long and weigh more than 5,000 pounds, fight pitched battles during the breeding season to determine which male will mate with the females gathered nearby. Elephant seal battles are often observed by crowds of curious human observers, who gather at many locations in coastal California to watch this fascinating marine mammal.

Although it is pretty easy to find an elephant seal these days, it was just about impossible to find one in the late 1800s. Because seal blubber can be rendered into oil that at the time was in high demand to fuel lamps and lubricate machine parts, elephant seals were hunted to near-extinction. They became easy targets when they crowded onto beaches to breed. Beginning in early winter, thousands of bulls and an even larger number of females would emerge from the Pacific onto beaches, mostly on islands off the coast of southern California and Baja California, Mexico. Some beaches were so packed with seals that hunters had to wait for the seals to get out of the way before they could land their boats.

Elephant seals were slaughtered by the tens of thousands, and it took only a few decades of hunting to eliminate almost all of them. Between 1884 and 1892, not a single elephant seal was sighted anywhere. Because elephant seals had become so rare, museum expeditions searched year after year, looking for specimens for their collections. In 1892, a Smithsonian expedition found nine seals on Isla de Guadalupe, about 150 miles off the coast of Baja California (and killed seven as specimens). Researchers estimate that as few as 10 to 20 seals remained alive in the late 1800s. These few seals bred on isolated beaches, and the population slowly increased. Still, an intensive search of Isla de Guadalupe in 1922 found only 264 seals. The Mexican government then banned the killing or capture of elephant seals and even put soldiers on Isla de Guadalupe to enforce the ban. The United States banned elephant seal hunting soon thereafter.

Once hunting was banned, the elephant seal population grew, and today there are more than 200,000 elephant seals, breeding on both islands and mainland beaches. How fast can populations grow? How long can populations keep growing? What are the factors that limit population growth?

AT A GLANCE

27.1 WHAT IS A POPULATION AND HOW DOES POPULATION SIZE CHANGE?

All the organisms in an area, along with their nonliving environment, form an **ecosystem**. The study of how these organisms interact with one another and their environment is known as **ecology**. We begin our exploration of ecology with an overview of populations. A **population** consists of all the members of a particular species that live in a specific area.

Changes in Population Size Result from Natural Increase and Net Migration

The size of some populations remains fairly constant over time. In other populations, the number of individuals changes cyclically or in response to changes in the environment. For example, if members of a species invade new territory, such as an island, their population may grow rapidly for a while, but then it usually stabilizes or plummets.

Population size changes as a result of births, deaths, and net migration. The contribution of births and deaths is summarized as the **natural increase** of a population, which is the difference between births and deaths. The natural increase is positive if births exceed deaths, or negative if deaths exceed births. Net migration is the difference between **immigration** (migration into the population) and **emigration** (migration out). A population grows when the sum of natural increase and net migration is positive and shrinks when this sum is negative. The equation for the change in population size within a given time span is:

change in population size =

natural increase (births − deaths)

+

net migration (immigration − migration)

Migration can significantly affect some natural populations. For simplicity, however, in this chapter we will ignore migration and consider only birth and death rates in our calculations of population growth. Also, although our discussion of population growth uses the term "birth," the concepts and conclusions we describe apply to all species, regardless of their mode of reproduction.

CASE STUDY — CONTINUED

The Return of the Elephant Seals

In 1960, about 14,000 elephant seals—more than 90% of the total population at the time—used the beaches of Isla de Guadalupe. As Guadalupe became increasingly crowded, many seals migrated to new territory in California. Immigration contributed greatly to the growth of elephant seal populations in California through the 1990s. But why did the Mexican beaches become crowded? And why are California beaches becoming crowded with elephant seals today?

Birth Rate, Death Rate, and Initial Population Size Affect Population Growth

Growing populations add individuals in proportion to the population's size, much like a bank account accumulates interest. Large bank accounts earn more money from interest than small ones do, even if the interest rate is the same, and large growing populations add more individuals that small ones do, even if the growth rate is the same.

The **growth rate (*r*)** of a population is the change in population size per individual per unit time. (In our examples, we will use 1 year as the unit of time). A population's growth rate equals its **birth rate (*b*)**, the births per individual per unit time, minus its **death rate (*d*)**, the deaths per individual per unit time:

$$\underset{\text{(birth rate)}}{b} - \underset{\text{(death rate)}}{d} = \underset{\text{(growth rate)}}{r}$$

If the birth rate exceeds the death rate, the population growth rate will be positive and population size will increase. If the death rate exceeds the birth rate, the growth rate will be negative and population size will decrease. Birth and death rates are expressed as decimal fractions or percentages; for example, if half the population dies each year, then the death rate is 0.5 per individual per year, or 50%.

Let's calculate the growth rate of a herd of 100 adult white-tailed deer, with 50 males and 50 females. Adult female white-tails typically give birth to twins each year, so the population produces 100 fawns (50 does × 2 fawns per doe = 100 fawns). The birth rate, therefore, is 1.0 birth per individual per year.

We'll simplify the biology of white-tailed deer and assume that female fawns mature and produce their own fawns the very next year. Predators and hunting kill many fawns and some adults, perhaps 70 each year, for a death rate of 0.7. The annual growth rate per individual, therefore, is 0.3:

1.0	−	0.7	=	0.3
(birth rate)		(death rate)		(growth rate)

To calculate population growth per unit time (*G*), we multiply the growth rate (*r*) by the population size (*N*) at the beginning of the time interval:

G	=	r	×	N
(population growth per unit time)		(growth rate)		(population size)

The growth (*G*) of the deer herd during the first year is $0.3 \times 100 = 30$. At a constant growth rate (*r*), *G* and *N* both increase with each successive time interval. In our example, the herd begins its second year with 130 deer (the new value of *N*). If *r* remains the same, then the herd adds 39 deer (the new value of *G*) during the second year ($0.3 \times 130 = 39$), for a total population of 169. In the third year, the herd adds 51 deer ($0.3 \times 169 = 51$), and so on.

Real populations, of course, never follow this equation exactly. Rather, the equation only provides estimates of population sizes over time. Environmental conditions never remain the same for very long, so the birth and death rates of real populations fluctuate from year to year.

A Constant Positive Growth Rate Results in Exponential Growth

As the deer example shows, a constant positive growth rate adds ever-increasing numbers to a population over time; this pattern is called **exponential growth**. A graph of change over time in the size of an exponentially growing population shows a characteristic shape called a **J-curve** (**FIG 27-1**). Note that exponential growth does not require a high growth rate. As you can see in Figure 27-1, exponential growth occurs at any constant positive rate of growth.

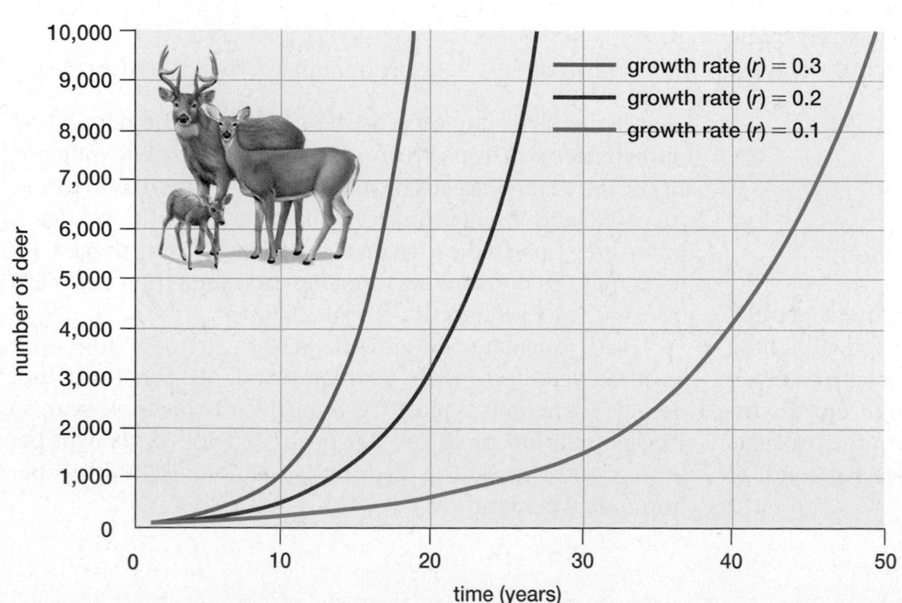

FIGURE 27-1 Exponential growth The graph shows the growth of three hypothetical populations of deer, each starting with 100 adults with the same birth rate (1.0 per individual per year), but differing in death rate (0.7 to 0.9 per individual per year), producing three different growth rates: 0.3, 0.2, and 0.1 per individual per year. Any positive rate of growth produces an exponential (J-shaped) growth curve, but higher growth rates lead to larger populations much more quickly.

Have You Ever Wondered ... How Many Children One Woman Can Bear?

Among women in modern times, Michelle Duggar of Arkansas probably came close to demonstrating the biotic potential of humans. Between 1988 and 2009, she gave birth to 19 children (including two sets of twins), all of whom remained healthy as of 2018. Duggar also had two miscarriages. Had all of her pregnancies resulted in live births, she would have borne about one child a year, for a little over 20 years. To see the population growth that this biotic potential could produce, let's imagine a woman bearing one child each year from age 20 to age 40. Let's assume that half her children are female and that, when the female children grow up, they bear children at this same reproductive rate. If we then factor in reasonable estimates for death rates at various ages, we find that in 100 years, this one woman could have 2.5 million descendants, and in 200 years, she could have almost 3 trillion.

The Biotic Potential Is the Maximum Rate at Which a Population Can Grow

A healthy organism can produce enough offspring during its lifetime to replace itself many times over, and the same is true of a population of organisms. The maximum possible rate at which a population can increase is known as its **biotic potential**. Estimation of a species' biotic potential assumes ideal conditions (unlimited resources, no predators, no diseases) that allow the maximum possible birth rate and the minimum possible death rate.

Factors that influence biotic potential include:

- The age of first reproduction
- The frequency of reproduction
- The average number of offspring produced each time the organism reproduces
- The organism's reproductive life span
- The death rate under ideal conditions

A species with a low biotic potential, such as elephants (females may reproduce only once every 5 years), may take many years to reach the same population size that a species with a very high biotic potential, such as houseflies (which reproduce every few weeks), could reach in just a few months. Regardless of a species' biotic potential, however, unchecked exponential growth would eventually result in a staggeringly large population size.

CHECK YOUR LEARNING

Can you . . .

- explain how migration and natural increase cause population sizes to change?
- describe how the growth rate interacts with population size to determine population growth per unit time?
- define *biotic potential* and list the factors that influence it?

CASE STUDY CONTINUED

The Return of the Elephant Seals

Elephant seals breed harem-style, with a single large bull typically monopolizing and mating with 30 to 100 females. Females reach sexual maturity at 3 to 4 years of age and produce one pup each year. Males reach sexual maturity at 5 to 7 years of age, but few are powerful enough to acquire a harem until 7 to 10 years of age. Further, although the maximum life span is about 20 years for females and 15 for males, only about 1 in 5 survives beyond 5 years of age. These factors combine to produce a biotic potential of about 12% per year. But elephant seal populations virtually never reproduce at their biotic potential. Why not?

27.2 HOW IS POPULATION GROWTH REGULATED?

In 1859, Charles Darwin wrote, "There is no exception to the rule that every organic being naturally increases at so high a rate that, if not destroyed, the Earth would soon be covered by the progeny of a single pair." But Earth is not, of course, completely blanketed with individuals of any species, so it is apparent that populations do not grow indefinitely. Populations fail to reach their biotic potential due to **environmental resistance**, which consists of all the living and nonliving factors that limit population growth. Environmental resistance includes interactions among organisms, such as predation and competition, and events such as freezing weather, extreme heat, storms, fires, floods, and droughts.

Exponential Growth in Natural Populations Is Always Temporary

Environmental resistance ensures that no natural population experiences exponential growth for very long.

Exponential Growth May Occur Temporarily in Populations with Boom-and-Bust Cycles

Temporary exponential growth occurs in populations that undergo regular cycles in which rapid population growth is followed by a sudden, massive die-off. These **boom-and-bust cycles** occur in a variety of organisms. Many short-lived, rapidly reproducing species, such as microbes and insects, have seasonal population cycles that are linked to changes in rainfall, temperature, or nutrient availability (**FIG. 27-2**). Boom-and-bust cycles can impact human health, as described in "Earth Watch: Boom-and-Bust Cycles Can Be Bad News" on page 458.

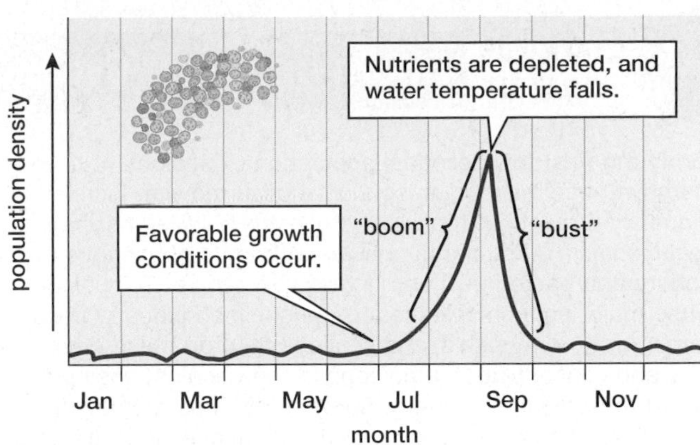

▲ **FIGURE 27-2 A boom-and-bust population cycle in photosynthetic bacteria** The density of a hypothetical population of photosynthetic bacteria in a lake. These microorganisms persist at a low level until early July. Then conditions become favorable for growth, and exponential growth occurs until early September, when the population plummets.

In temperate climates, insect populations grow rapidly during the spring and summer, then crash when the freezing weather of winter arrives. Houseflies, for example, mature within 2 weeks of hatching, so up to seven generations can occur during spring and summer. Each female fly lays about 120 eggs, so even though many flies die, the fly population becomes quite large by the end of the summer. But hard frosts in autumn kill virtually all of the adults, and the fly population crashes. Despite this "bust," some eggs survive the winter to start the cycle over again the following spring.

More complex factors can produce boom-and-bust patterns with longer cycles (**FIG. 27-3**). A lemming population,

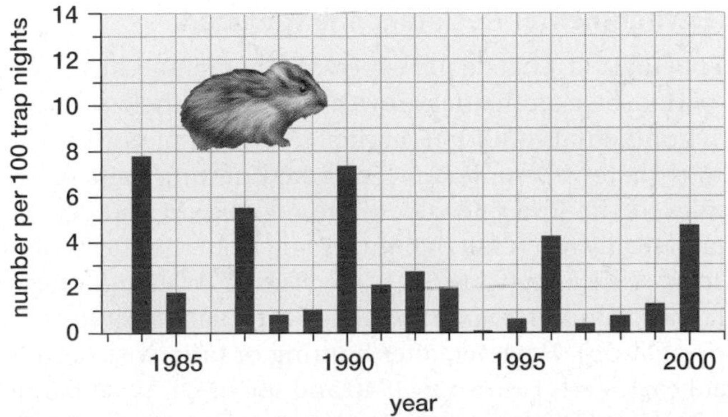

▲ **FIGURE 27-3 Boom-and-bust population cycles in a lemming population in the Canadian Arctic** This population follows a roughly 3- to 4-year cycle of boom and bust. The data are based on live trapping.

THINK CRITICALLY What factors might make these population data somewhat erratic?

Boom-and-Bust Cycles Can Be Bad News

During the past few decades, population explosions of toxic photosynthetic microorganisms have occurred with increasing frequency throughout the world. Collectively called *harmful algal blooms (HABs)*, these population booms of toxin-producing protists and bacteria kill fish and sicken people. They also cause major economic losses to the shellfish industry because clams, oysters, mussels, and scallops feed on these organisms and concentrate the poisons in their bodies, posing a hazard to human consumers. Some of the most damaging poisons are neurotoxins produced by dinoflagellates (protists) such as *Karenia brevis* (**FIG. E27-1**), which can reach densities of 20 million per liter of water.

◀ **FIGURE E27-1 One cause of harmful algal blooms** The dinoflagellate *Karenia brevis* (seen in this artificially colored SEM) causes harmful algal blooms in coastal waters of Florida and in the Gulf of Mexico. These blooms are sometimes called red tides, although the actual colors vary.

Many of the microorganisms that produce harmful algal blooms are common residents of lakes and coastal waters. What causes these species to bloom and boom? Although the precise causes are complex, warm water temperatures and abundant nutrients, such as phosphorus and nitrogen, are always required. Runoff of these nutrients from human agricultural activities has increased the frequency and intensity of HABs throughout the world. Global climate change may also contribute to the problem, because warmer waters foster more rapid growth of these microorganisms and extend their growing season. The booming populations "bust" when the enormous numbers of cells deplete the nutrients in the water. Falling water temperatures in the autumn and winter further decrease the cells' rate of reproduction.

THINK CRITICALLY Even algal blooms of species that do not produce toxins can be so harmful that most other organisms in the water die. If no toxins are present, can you explain why an algal bloom might be deadly?

for example, may grow rapidly until the lemmings overgraze their fragile arctic tundra habitat. At this point, the death rate may increase suddenly, as a result of lack of food, increased populations of predators, and social stress caused by crowding. Even more deaths result when many lemmings emigrate to escape crowding; the migrating lemmings become easy targets for predators or drown as they encounter bodies of water and attempt to swim across. Eventually, the reduced lemming population causes a decline in predator numbers and allows recovery of the plants on which the lemmings feed. These responses set the stage for the next round of explosive growth in the lemming population.

Exponential Growth May Occur Temporarily if Environmental Resistance Is Reduced

In populations that do not experience boom-and-bust cycles, a period of exponential growth may nevertheless occur if environmental conditions improve. For example, by the early 1960s there were only a few hundred nesting pairs of bald eagles in the lower 48 states. The population had declined dramatically as a result of hunting and from the effects of the widely used insecticide DDT (see "Health Watch: Biological Magnification of Toxic Substances" in Chapter 29 for more information). However, after hunting or otherwise harming bald eagles was banned in 1940 and use of DDT was banned in the United States in 1972, bald eagles began to recover. The bald eagle population grew exponentially between the early 1960s and 2014 (**FIG. 27-4**).

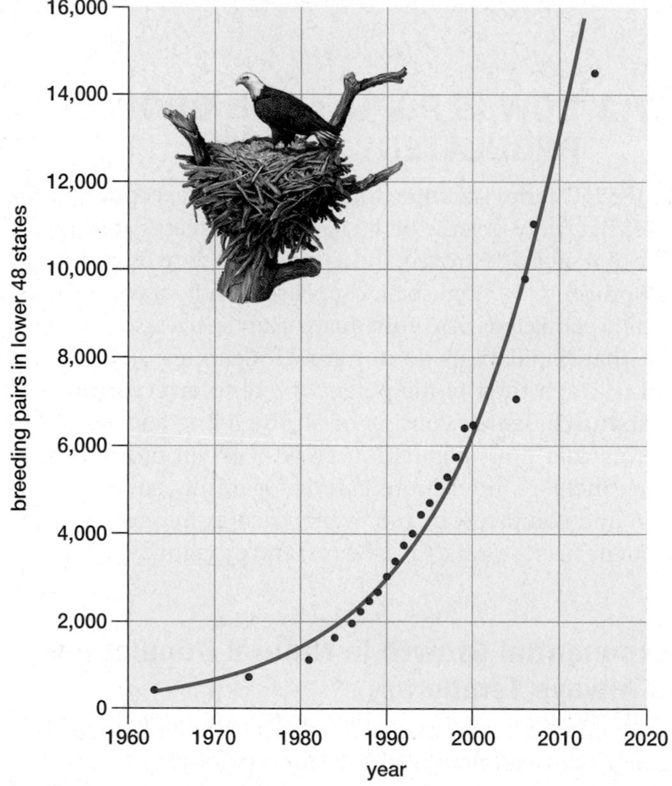

▲ **FIGURE 27-4 Exponential growth of the bald eagle population in the lower 48 United States** Protection from hunting and banning DDT allowed the eagle population in the lower 48 states to grow from fewer than 500 nesting pairs in the early 1960s to more than 14,500 pairs by 2014. The smooth line is an exponential J-curve fit to the data points. Data from the U.S. Fish and Wildlife Service and the Center for Biological Diversity.

Exponential growth may also occur when individuals invade a new habitat with favorable conditions and little competition. For example, exponential growth is often

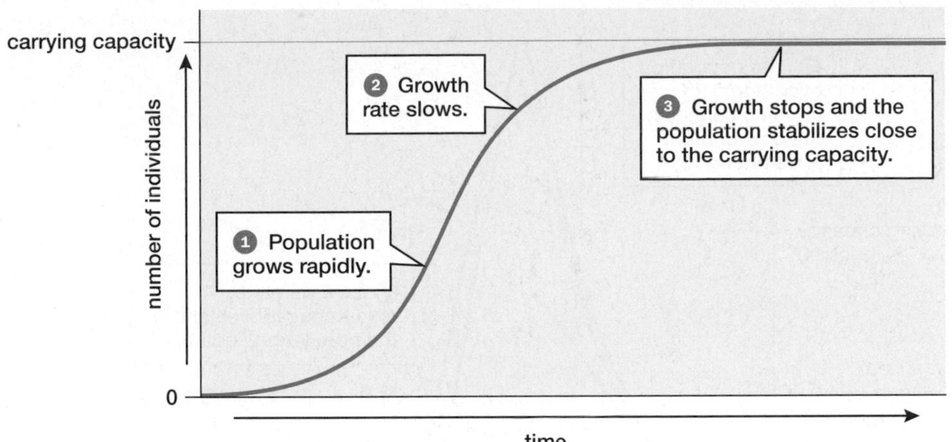

◀ **FIGURE 27-5 Logistic population growth** During logistic growth, a population begins growing exponentially, but the growth rate slows as the population encounters increasing density-dependent environmental resistance. Population growth finally ceases at or near carrying capacity (*K*). The result is a curve shaped like a "lazy S."

observed in **invasive species**, which are species with a high biotic potential that have been introduced (deliberately or accidentally) into ecosystems in which they did not evolve and where they encounter little environmental resistance. For example, many habitats in Europe and the United States are dominated by Japanese knotweed, an Asian species originally introduced as a garden plant. The knotweed population has grown explosively, outcompeting native plant species. Similarly, in the 1980s, people accidentally introduced Asian zebra mussels into Lake St. Clair, located in North America's Great Lakes region, when a ship arriving from Asia discharged ballast water that contained zebra mussel larvae. The introduced population grew explosively, because zebra mussels have a high biotic potential—a female can produce 100,000 eggs per year—and because the mussels suffered very little predation in their new environment. The exploding population spread to lakes and rivers across much of North America. Zebra mussels attach themselves in huge numbers to nearly any underwater surface, including piers, pipes, machinery, and boat hulls. The mussels also cover and suffocate other shellfish species, threatening some species with extinction. But even zebra mussels cannot increase their population forever. Why not?

Environmental Resistance Limits Population Growth Through Density-Dependent and Density-Independent Mechanisms

In all populations, environmental resistance eventually halts exponential growth.

Logistic Growth Occurs When New Populations Stabilize as a Result of Environmental Resistance

The maximum population size that can be sustained indefinitely without damage to an ecosystem is called the ecosystem's **carrying capacity (*K*)**. In most cases, carrying capacity is determined primarily by the availability of nutrients, energy, and space. A population that grows until it reaches its ecosystem's carrying capacity and then stabilizes is said to experience **logistic population growth**. A graph of logistic population growth shows a characteristic shape called an **S-curve** (**FIG. 27-5**). In nature, logistic population growth can occur when a species moves into a new habitat (**FIG. 27-6**).

A population may become larger than carrying capacity for a short time. However, a population above carrying capacity is living on resources that cannot regenerate as fast as they are being used up. Therefore, a small overshoot above *K* is likely to be followed by a decrease in population, to a level below the original carrying capacity, until the resources recover and the original carrying capacity is restored. Repeated episodes of small overshoots and undershoots result in a population size that fluctuates around the carrying capacity (the green line in **FIG. 27-7**).

If a population far exceeds the carrying capacity of its environment, the consequences are more severe. In this case, the excess demands placed on the ecosystem are likely to destroy essential resources that may be unable to recover. This loss of resources can severely and permanently reduce *K*, causing the population to decline to a fraction of its former size. After the decline, the population may stabilize (the blue line

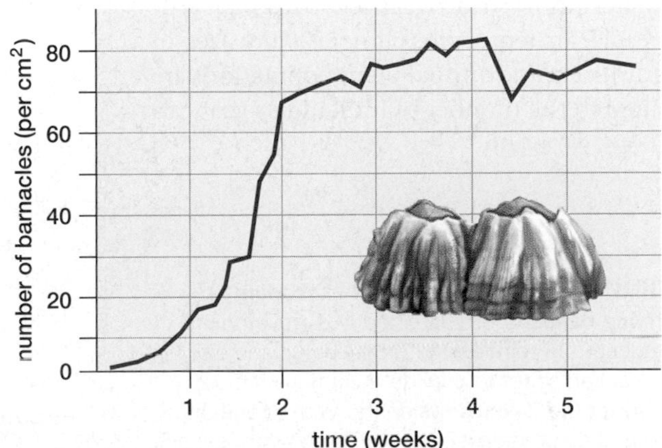

▲ **FIGURE 27-6 A logistic curve in nature** Barnacles are crustaceans whose larvae are carried in ocean currents to rocky seashores where they settle, attach permanently to rock, and grow into the shelled adult form. On a bare rock, the number of settling barnacles produces a logistic growth curve as competition for space limits their population density.

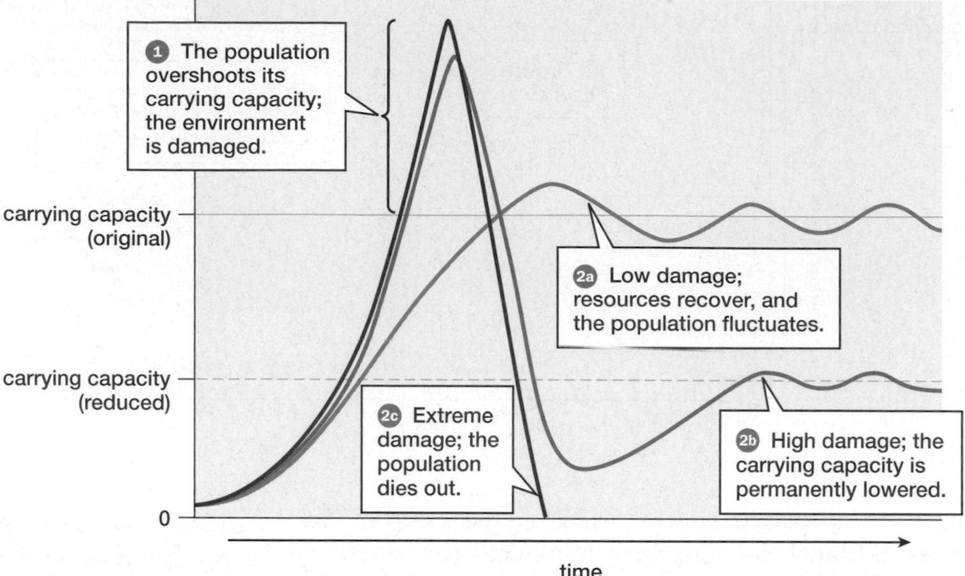

1 The population overshoots its carrying capacity; the environment is damaged.
carrying capacity (original)
carrying capacity (reduced)
2a Low damage; resources recover, and the population fluctuates.
2c Extreme damage; the population dies out.
2b High damage; the carrying capacity is permanently lowered.
0
time

▶ **FIGURE 27-7 Consequences of exceeding carrying capacity** Populations can overshoot carrying capacity (*K*), but only for a limited time. Three possible results are illustrated.

in Fig. 27-7) or disappear entirely (the red line in Fig. 27-7). For example, when reindeer were introduced onto St. Paul, an island off the coast of Alaska that lacks large predators, the reindeer population increased rapidly, seriously overgrazing the lichens that were their principal food source. Starvation then caused the reindeer population to plummet, as shown in **FIGURE 27-8**.

Density-Independent Factors Limit Populations Regardless of Density

A number of factors prevent populations from permanently exceeding their ecosystem's carrying capacity. Some of these limit population size regardless of the population's density (number of individuals per unit area). The most important of these **density-independent** factors are weather (short-term atmospheric conditions such as temperature, rainfall, and wind) and climate (long-term weather patterns). For example, in temperate climates, the sizes of many insect and annual plant populations are limited to the number of individuals that can be produced before the first hard freeze. Such populations typically do not reach carrying capacity before winter sets in. Hurricanes, floods, droughts, and fire can also have profound effects on local populations, regardless of their density.

Density-Dependent Factors Have Larger Effects as Population Density Increases

The effects of **density-dependent** factors, in contrast, increase as population density increases. For example, interactions such as predation, parasitism, and competition limit population growth more strongly as population density increases.

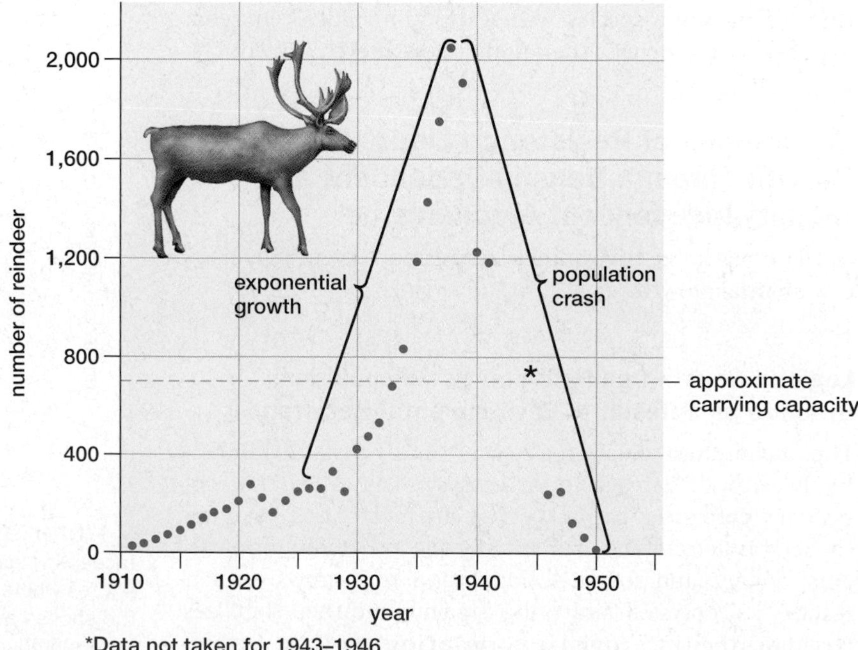

▶ **FIGURE 27-8 The effects of exceeding carrying capacity** In 1911, the U.S. government introduced 25 reindeer onto St. Paul Island off the Alaskan coast to provide a continuing meat supply for the island's residents. With abundant food and no predators, the herd grew exponentially (initial J-curve). By 1938, the herd had increased to 2,046 reindeer, roughly three times the island's estimated carrying capacity. The lichens on which the reindeer fed during the winter were seriously overgrazed and unable to recover, so the reindeer starved. By 1950, only eight reindeer remained.

Predators Exert Density-Dependent Controls on Populations

Predators are organisms that eat other organisms, called their **prey**. We will use a broad definition of predators, to include both *carnivores* (animals that eat other animals) and *herbivores* (animals that eat plants). Often, prey are killed directly and eaten (**FIG. 27-9a**), but not always. When deer browse on the buds of bushes and young trees, for example, or when gypsy moth larvae feed on the leaves of oaks, the plants are harmed but usually not killed.

Predator populations often grow as their prey becomes more abundant, making predators even more effective as control agents. For predators such as the arctic fox and snowy owl, which rely heavily on lemmings for food, the number of offspring they can produce is determined by the abundance of prey. Snowy owls (**FIG. 27-9b**) hatch up to 12 chicks when lemmings are abundant, but they may not reproduce at all in years when the lemming population has crashed.

(a) Predators often kill weakened prey

(b) Predator populations often increase when prey are abundant

▲ **FIGURE 27-9 Predators help control prey populations** **(a)** A pack of gray wolves has brought down an elk that may have been weakened by age or parasites. **(b)** The snowy owl produces more chicks when prey (such as lemmings) are abundant.

Parasites Spread More Rapidly in Dense Populations

A **parasite** is an organism that lives in or on a larger organism, called its **host**, feeding on the host and harming it. Parasites include tapeworms that live in the intestines of mammals, lice that cling to a host's skin or hair, and disease-causing microorganisms. Parasites have density-dependent effects on host populations. Because most parasites cannot easily travel long distances, they spread more readily among hosts with dense populations. For example, plant diseases spread rapidly through densely planted crops, and childhood diseases spread swiftly through schools and day-care centers. Parasites influence population size by killing their hosts outright or by weakening them and making them more susceptible to death from other causes, such as harsh weather or predators.

Competition for Resources Helps Control Populations

The resources that organisms need to survive are limited. When multiple individuals attempt to use the same limited resources, the result is **competition**—interactions that harm all of the interacting individuals. Competition limits population size in a density-dependent manner, because competition is more intense in denser populations. There are two major forms of competition: **interspecific competition** (competition among individuals of different species) and **intraspecific competition** (competition among individuals of the same species). Because members of the same species have nearly identical requirements for nutrients, shelter, breeding sites, and other resources, intraspecific competition is especially intense and acts as an important density-dependent mechanism of population control.

When population density increases and competition becomes intense, some types of animals—for example, lemmings and locusts—react by emigrating. Emigrating swarms of locusts periodically plague parts of Africa, consuming most of the vegetation in their path (**FIG. 27-10**).

▲ **FIGURE 27-10 Emigration** In response to overcrowding and lack of food, locusts emigrate in swarms, devouring nearly all vegetation (and even feeding on each other) as they go.

THINK CRITICALLY What benefits does mass emigration give to animals such as locusts or lemmings? Are there parallels to human emigrations?

Density-Independent and Density-Dependent Factors Interact

The size of a population at any given time is the result of complex interactions between density-independent and density-dependent forms of environmental resistance. For example, a population of pines weakened by drought (a density-independent factor) may more readily fall victim to predation by the pine bark beetle (density-dependent). Likewise, a caribou population weakened by competition that leads to hunger (density-dependent) and attack by parasites (density-dependent) is more likely to crash in an exceptionally cold winter (density-independent).

Human activities increasingly impose both density-independent and density-dependent limitations on natural populations. For example, when people cut down a rain forest, most of the organisms inhabiting the forest are exterminated regardless of their initial population density; the immediate effect of this human action is density-independent. However, some of the organisms that lived in the former forest may move to a nearby undisturbed forest, increasing the population density there. As a result, increased competition, predation, and parasitism will exert density-dependent controls on these now-larger populations.

CHECK YOUR LEARNING

Can you . . .

- describe exponential growth, the conditions under which it occurs, and the shape of the growth curve it produces?
- explain the stages of the logistic growth curve and describe the possible outcomes when populations overshoot carrying capacity?
- describe the two major types of environmental resistance and provide examples of each?

CASE STUDY **CONTINUED**

The Return of the Elephant Seals

Elephant seal populations experience both density-dependent and density-independent environmental resistance. Density-dependent factors include predation by great white sharks and orcas, potential food scarcity, and deaths during the breeding season, when males battle for access to mates. These battles not only injure and kill some males but stress the females and sometimes kill pups that get crushed beneath tons of charging bull. The denser the seal population on a beach, the more battles and deaths occur. Density-independent factors include bad weather, especially during El Niño years, which often bring strong storms to California. Storms erode the beaches, and high tides and waves may submerge beaches and wash away pups. In severe El Niño years, as many as half the pups in a colony may be lost. Overall, however, elephant seals have secure breeding sites, abundant food, and few predators, and their populations are stable or increasing. Does this mean that their future is secure? Find out in "Case Study Revisited" at the end of the chapter.

27.3 HOW DO LIFE HISTORIES DIFFER AMONG SPECIES?

The **life history** of an organism includes its age at first reproduction, the number of offspring it produces, and the amount of energy and resources devoted to each offspring. Life histories vary tremendously among species. Species representing the opposite extremes of life history are often called *r*-selected species and *K*-selected species, respectively.

Life Histories Reflect Trade-Offs Between Number of Offspring and Offspring Survival

At one extreme, ***r*-selected species** have evolved characteristics that favor rapid reproduction. They mature rapidly, have a short life span, produce a large number of small offspring, and provide little or no parental care (**FIG. 27-11a**). The vast majority of offspring die young. It is common for *r*-selected species to reproduce just a few times, sometimes only once, in their lives. Boom-and-bust species often have *r*-selected life histories.

At the other extreme, ***K*-selected species** have a long lifespan, produce small numbers of fairly large offspring, and either provide significant parental care (as large mammals do) or package significant nutrients along with the embryo (as do large-seeded trees such as oaks and walnuts). This parental investment ensures that a fairly high percentage of offspring survives to maturity (**FIG. 27-11b**). Many *K*-selected species produce offspring one or more times a year for several, sometimes many, years.

Most species do not fit neatly into these two extreme categories, but have a combination of traits from the two extremes. For example, giant clams are slow to mature (5 to 10 years), and spawn (release eggs and sperm) many times in a typical life span of scores of years. However, they produce hundreds of millions of eggs at a time, have very few nutrients packaged in their eggs, and provide no parental care. As a result, perhaps only 1 egg in a billion successfully matures into an adult clam. Similarly, redwood trees typically mature at 5 to 15 years of age and reproduce every year for 500 to 1,000 years, but the vast majority of their hundreds of thousands of tiny seeds never sprout.

A Species' Life History Predicts Survival Rates over Time

Species differ tremendously in their chances of dying at any given phase of their life cycle. These differences are recorded in **survivorship tables**, which track same-age groups of organisms throughout their lives, recording how many survive each year (or some other time period) (**FIG. 27-12a**). When the proportion of the population that survives is plotted against the species' life span, a **survivorship curve** is produced. The three principal types of survivorship curves are late-loss, constant-loss, and early-loss, named for the part of the life span during which most deaths occur (**FIG. 27-12b**).

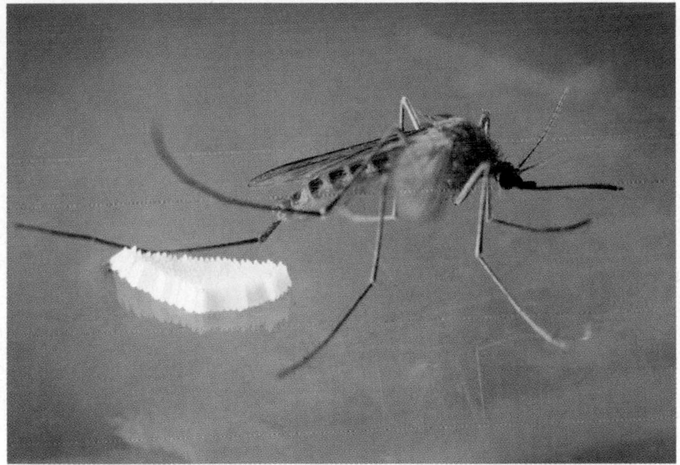

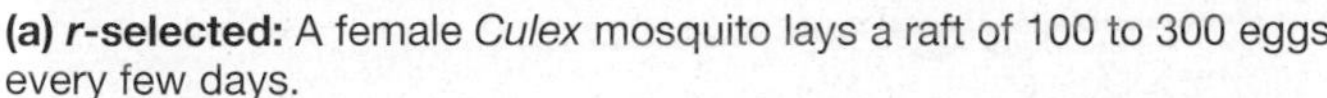

(a) *r*-selected: A female *Culex* mosquito lays a raft of 100 to 300 eggs every few days.

(b) *K*-selected: An African bush elephant and its calf.

▲ **FIGURE 27-11 Typical *r*-selected and *K*-selected species (a)** Most mosquitoes of the genus *Culex* have short life spans, produce numerous small eggs, and provide no parental care. Most offspring are eaten in the larval stage by predators. **(b)** African bush elephants have long life spans, produce single offspring after a 21-month-long pregnancy, and nurse their young for several years. Although a few infants fall prey to lions, elephants have no significant natural predators. Unless they fall victim to human poachers, elephants have a very high chance of surviving until old age.

Late-loss populations have relatively low juvenile death rates, and most individuals survive to old age. The resulting survivorship curve is convex. Late-loss survivorship curves are characteristic of humans and other large and long-lived animals such as elephants and mountain sheep. These species produce relatively few offspring, which are protected and nourished by their parents during early life.

Constant-loss populations consist of individuals that have an equal chance of dying at any time during their life span, producing a straight-line survivorship curve. This pattern is seen in some species of bird and turtles, and in laboratory populations of organisms that reproduce asexually, such as hydra and bacteria. *K*-selected species usually have late-loss or constant-loss survivorship curves.

Age	Number of survivors
0	100,000
10	99,224
20	98,910
30	98,011
40	96,798
50	94,295
60	88,770
70	78,069
80	57,188
90	23,619
100	1,968
110	0

(a) A survivorship table

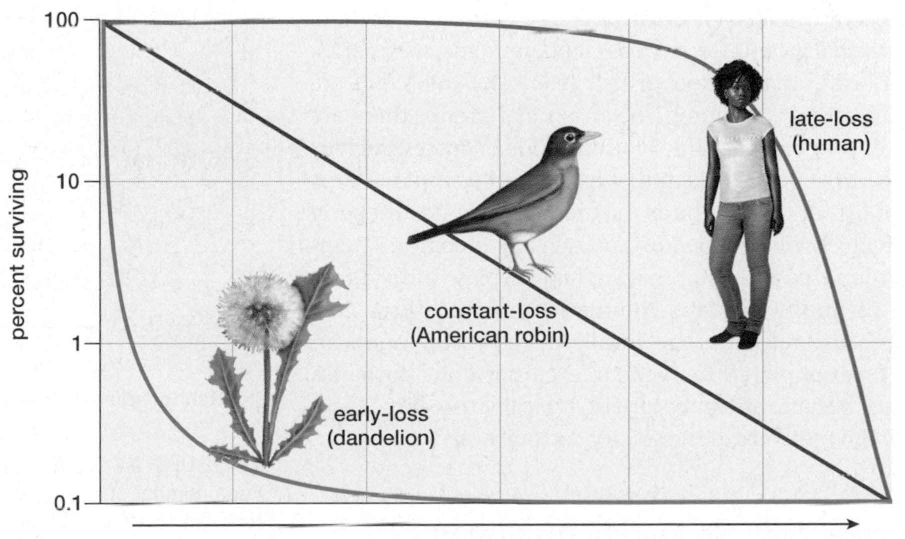

(b) Survivorship curves

▲ **FIGURE 27-12 Survivorship (a)** A survivorship table for the U.S. population in 2010, showing how many people are expected to remain alive at increasing ages, for each 100,000 people born. Plotting these data produces a curve similar to the blue curve in part (b). **(b)** Three types of survivorship curves are shown; note the logarithmic *y*-axis. Data in part (a) from Arias 2014. *National Vital Statistics Reports* 63(7): 1–63.

Early-loss populations have a high death rate early in life, so the survivorship curve is concave. Although the death rate is very high among the young, those individuals that reach adulthood have a reasonable chance of surviving to old age. Early-loss survivorship curves are characteristic of organisms that produce large numbers of offspring but give them little or no care after they hatch or germinate. Most invertebrates, many fish and amphibians, and most plants exhibit early-loss survivorship curves. Most *r*-selected species have early-loss survivorship curves.

CHECK YOUR LEARNING

Can you . . .

- describe the variables that affect a species' life history?
- compare the two most extreme types of life history?
- describe and graph the three principal types of survivorship curves?

27.4 HOW ARE ORGANISMS IN POPULATIONS DISTRIBUTED?

The pattern of spacing between individuals in a population differs among species and, within a species, may vary over time. For example, the spatial distribution of an animal population may differ between the breeding season and the rest of the year. Ecologists recognize three major types of spatial distribution: clumped, uniform, and random (**FIG. 27-13**).

Individuals May Clump Together in Groups

The members of many populations live clumped together in groups. Examples of such **clumped distributions** include family or social groupings, such as elephant herds, wolf packs, flocks of birds, and schools of fish (**FIG. 27-13a**). What are the advantages of clumping? In an animal group, there are many eyes that can search for localized food sources, such as a fruit-bearing tree or a lake full of fish. Membership in a large school of fish or flock of birds may reduce the odds that any one individual will be killed by an attacking predator. A temporary group that forms for mating may increase individuals' chances of finding a mate. Alternatively, populations may cluster because resources are localized. In North American prairies, for example, cottonwood trees cluster along the banks of streams, because streambanks are typically the only places with enough moisture in the soil for cottonwoods to grow.

Individuals May Be Evenly Dispersed

Populations with a **uniform distribution** maintain a relatively constant distance between individuals. Uniform distributions often result from competition. For example, mature desert creosote bushes are often spaced very evenly due to competition among their root systems, which occupy a roughly circular area around each plant. The roots so efficiently absorb water and other nutrients from the desert soil that they prevent the survival of nearby plants. A uniform distribution may also occur among animals, such as songbirds, that defend territories, thereby spacing themselves fairly evenly throughout suitable habitat. Sometimes a uniform distribution is simply a way to avoid harassment by neighbors. For example, some seabirds nest in colonies in which nests are evenly spaced, just out of pecking reach of one another (**FIG. 27-13b**).

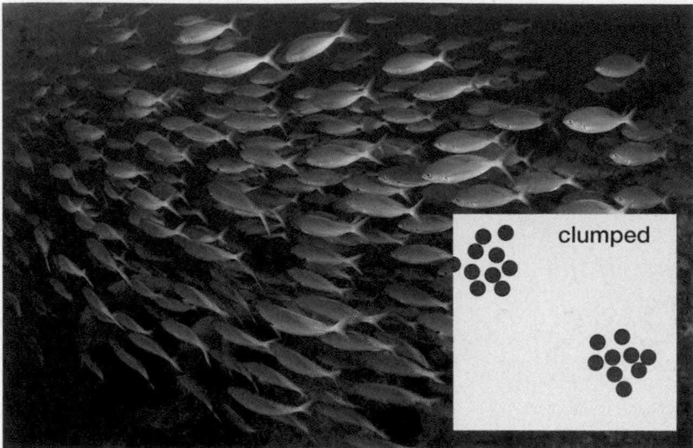

(a) Clumped distribution

(b) Uniform distribution

(c) Random distribution

▲ **FIGURE 27-13 Spatial distribution in populations (a)** The large number of individuals in a school of fish may confuse predators. **(b)** Nests in a breeding colony of gannets are arranged in a uniform pattern, with each nest just out of pecking range of its neighbors. **(c)** Dandelions and other plants with wind-borne seeds often have random population distributions because the seeds germinate wherever the wind deposits them.

Individuals May Be Distributed at Random

In a population with **random distribution**, the distance between individuals varies unpredictably. Individuals of species with random distributions do not consistently cluster together nor spread apart, and they do not form social groupings. The resources they need must be more or less equally available throughout the area they inhabit, and those resources must be fairly abundant, minimizing competition. The dandelions in a lawn are often randomly distributed, as they sprout from windblown seeds that land in random locations in well-fertilized soil (**FIG. 27-13c**).

CHECK YOUR LEARNING

Can you . . .

- describe the three types of spatial distribution and describe potential causes of each type of distribution?

27.5 HOW IS THE HUMAN POPULATION CHANGING?

Earth's human population has been growing rapidly for centuries, and the effects of our presence can be detected on virtually every part of the planet's surface. Unfortunately, the rapid growth of our population may now threaten us and the biosphere on which we depend.

The Human Population Has Grown Exponentially

Compare the graph of human population growth in **FIGURE 27-14a** with the exponential growth curves in Figures 27-1 and 27-2: Each has the J-curve characteristic of exponential growth. Initially, the human population grew slowly, and did not reach 1 million until about 10,000 B.C. By the early 1800s—almost 12,000 years later—the population reached 1 billion. But the second billion was added in about 125 years, the third billion in 33 years, and the fourth billion in just 14 years. Our population reached 7.5 billion in 2018. The human population has increased ‘so fast that about 6% of all the people who have *ever* lived on Earth are alive today.

Although the human population continues to grow rapidly, it may no longer be growing exponentially (**FIG. 27-14b**). According to the U.S. Census Bureau, the world's annual growth rate declined from 2.2% in 1963 to 1% in 2018. Our numbers now grow by about 76 million each year, down from a peak of about 88 million in 1990. Are humans starting to enter the final bend of an S-shaped logistic growth curve (see Fig. 27-5) that will eventually lead to a stable population? Only time will tell. Nevertheless, by 2050, the human population is expected to exceed 9 billion, and still be growing, although more slowly than at present. How has humanity managed to sustain such prolonged, rapid population

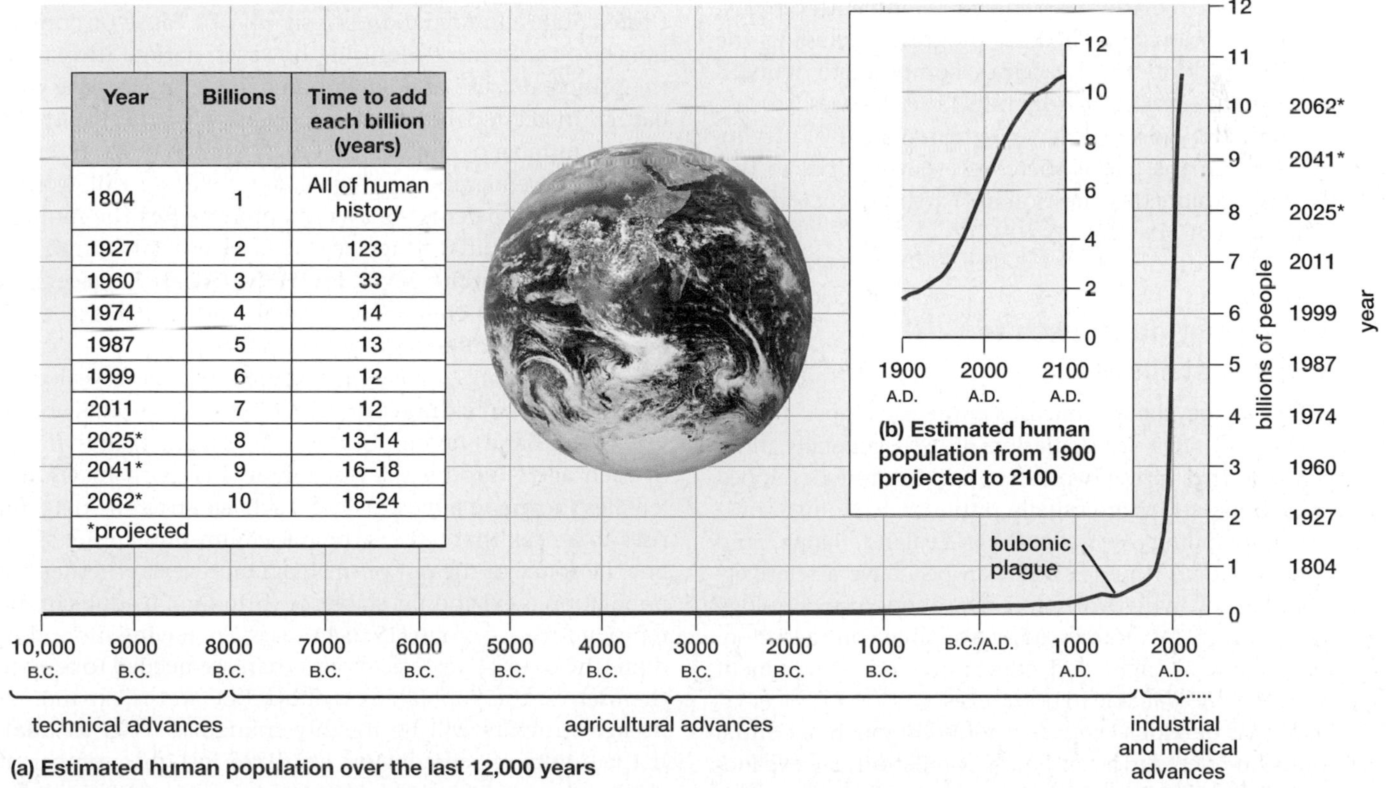

Year	Billions	Time to add each billion (years)
1804	1	All of human history
1927	2	123
1960	3	33
1974	4	14
1987	5	13
1999	6	12
2011	7	12
2025*	8	13–14
2041*	9	16–18
2062*	10	18–24
*projected		

▲ **FIGURE 27-14 Human population growth (a)** The human population from the Stone Age to the present has shown continued exponential growth as various advances overcame environmental resistance. Note the time intervals over which additional billions were added. **(b)** An expanded x-axis for the years 1900 to present, and projected to 2100, indicates that human population growth is slowing down.
Photo courtesy of NASA.

growth? Why hasn't environmental resistance put an end to our tremendous population growth?

People Have Increased Earth's Capacity to Support Our Population

Like all populations, humans have encountered environmental resistance, but unlike other populations, we have responded to environmental resistance by devising ways to overcome it. We have increased Earth's carrying capacity for people by virtue of a series of advances in which products of human culture conquered aspects of environmental resistance. The earliest of these advances occurred as early humans learned to construct tools and weapons, build shelters, fashion protective clothing, and control fire. Tools and weapons allowed people to hunt more effectively and obtain additional high-quality food, shelter and clothing allowed people to inhabit new areas, and controlled fire allowed people to cook, thereby expanding our food sources and improving our overall nutrition.

Carrying capacity for humans was further increased when people began cultivating domesticated crops in many parts of the world about 8,000 to 10,000 years ago. Agriculture provided people with a larger, more dependable food supply. An increased food supply resulted in a longer life span and more childbearing years. However, a high death rate from disease restrained population growth for thousands of years until industrial and medical advances permitted a population explosion, beginning in the mid-eighteenth century. Improved sanitation and medical progress, including the discovery of antibiotics and vaccines, combined to dramatically decrease the death rate from infectious diseases.

Technological advances have increased Earth's carrying capacity for humans, but is there an eventual limit to this capacity? We explore this question in "Earth Watch: Have We Exceeded Earth's Carrying Capacity?"

World Population Growth Is Unevenly Distributed

Countries can be characterized as more developed or less developed, and rates of population growth generally differ between countries in the two categories. In more-developed countries, populations are usually stable or declining. These countries—including Australia, New Zealand, Japan, and countries in North America and Europe—have a relatively high standard of living, with access to modern technology and medical care, including readily available contraception. Average income is high, and education and employment opportunities are available to both sexes.

In the less-developed countries found in much of Central and South America, Africa, and Asia, populations are expanding, sometimes rapidly. In these countries—home to more than 80% of humanity—the average person lacks the advantages that benefit most people in more-developed countries. How do differences in development affect population growth?

Birth Rate Declines as a Country Develops

In many more-developed countries, the population explosion fostered by the declining death rates that resulted from industrial and medical advances was eventually followed by a decline in birth rate. As a country industrializes, birth rate tends to fall as more people move from small farms to cities (where children are less important as a source of labor), contraceptives became more readily available, and employment opportunities for women increase. Today, the populations of most more-developed countries are stable, or even decreasing, with low rates of both births and deaths.

In less-developed countries, medical advances have decreased death rates, but birth rates remain relatively high. In many of these nations, population growth is driven by social and economic factors that promote large families. For example, families may depend on the labor or income of young children who work on farms or in factories, and on adult children who provide financial security for aging parents. Also, even people who do want to limit their family size often lack access to contraceptives. For example, in the West African nation of Nigeria, less than 20% of couples use modern contraceptive methods.

The Age Structure of a Population Predicts Its Future Growth

Populations change size due to a combination of natural increase and migration. In some countries, such as the United States, immigration is a significant cause of population growth. In most countries, however, natural increase is the principal cause of population changes. In the long run, natural increase depends on a population's *fertility rate*, the average number of children that each woman bears. If immigration and emigration are equal, a population will eventually stabilize if parents, on average, produce just the number of children required to replace themselves. This number is called **replacement level fertility (RLF)**. Replacement level fertility is 2.1 children per woman (rather than exactly 2) because not all children survive to maturity.

A population's capacity for future growth is revealed by an **age structure diagram**, which plots age groups on the vertical axis and the numbers or percentages of individuals in each age group on the horizontal axis, with males and females shown on opposite sides. Each age structure diagram rises to a peak that reflects the maximum human life span, but the shape of the rest of the diagram reveals whether the population is expanding, stable, or shrinking. If adults in the reproductive age group (15 to 44 years) are having more children (the 0- to 14-year age group) than are needed to replace themselves, the population is above RLF and is expanding. Its age structure will be roughly triangular (**FIG. 27-15a**). If the adults of reproductive age have just the number of children needed to replace themselves, the population is at RLF. A population that has been at RLF for many years will have an age structure diagram with relatively straight sides (**FIG. 27-15b**). In a shrinking population, the reproducing

Have We Exceeded Earth's Carrying Capacity?

In Côte d'Ivoire, a country in western Africa, the government is waging a battle to protect some of its rapidly dwindling tropical rain forest from illegal hunters, farmers, and loggers. Officials destroy the shelters of the squatters, who immediately return and rebuild. One such squatter, Sep Djekoule, explained, "I have ten children and we must eat. The forest is where I can provide for my family, and everybody has that right." His words exemplify the conflict between population growth and wise management of Earth's finite resources.

How many people can Earth sustain? The Global Footprint Network, an international group of scientists and other professionals, is attempting to calculate humanity's *ecological footprint*. This measurement compares human demand for resources to Earth's capacity to supply these resources in a sustainable manner. "Sustainable" means that the resources can be renewed indefinitely and that the ability of the biosphere to supply them is not diminished over time. Is humanity living on the "interest" produced by our global endowment, or are we eating into the "principal"? The Global Footprint Network concluded that, in 2014 (the most recent year for which complete data are available), humanity consumed more than 170% of the resources that were sustainably available. In other words, to avoid damaging Earth's resources (thus reducing Earth's carrying capacity), our population in 2014 would have required more than 1.7 Earths. We have added hundreds of millions of people since then, so our footprint is even larger now. Because we have used our technological prowess to overcome environmental resistance, our collective ecological footprint now dwarfs Earth's sustainable resource base, reducing Earth's future capacity to support us.

The human population now uses almost 40% of Earth's productive land for crops and livestock. Nonetheless, the United Nations estimates that more than 800 million people are undernourished, including an estimated 25% of the population of sub-Saharan Africa. The quest for more farmland drives people to clear-cut forests, often in places where the soil is poorly suited for agriculture. Demand for wood also causes large areas to be deforested annually. Deforestation causes runoff of much-needed fresh water, erosion of topsoil, and pollution of rivers and oceans. These harmful changes contribute to an overall reduction in the ability of the land and water to support not only crops and livestock (**FIG. E27-2**), but also fish and other wild animals that people harvest for food.

We are depleting many resources more quickly than they can be replenished. For example, the United Nations estimates that about 60% of commercial ocean fish populations are being harvested at their maximum sustainable yield, and another 30% are being overfished. Fresh water is also being depleted. In parts of India, China, Africa, and the United States, underground water stores are being depleted to irrigate cropland far faster than they are being refilled by rain and snow. Because irrigated land supplies about 40% of human food crops, water shortages can rapidly lead to food shortages.

Our present population, at its present level of technology, is clearly "overgrazing" the biosphere. Current trends are not sustainable. As the 6.2 billion people in less-developed countries strive to raise their standard of living, and the billion-plus people in more-developed countries continue to increase theirs, the damage to Earth's ecosystems accelerates. Earth's resources are not sufficient to provide all of Earth's inhabitants with the high standard of living currently enjoyed by people in more-developed countries. Supporting the world population sustainably at the average standard of living in the United States would require about 4.5 Earths. Technology can help us improve agricultural efficiency, conserve energy and water, reduce pollutants, and recycle far more of what we use. In the long run, however, it is extremely unlikely that technological innovation can compensate for continued population growth.

Hope for the future lies in recognizing the consequences of human population growth and responding by taking measures to reduce our population. Significant action will be required to prevent further damage to the biosphere and to ensure its ability to support people and the other precious and irreplaceable forms of life on Earth.

▲ **FIGURE E27-2 Overgrazing can lead to the loss of productive land** Human activities, including overgrazing, deforestation, and poor agricultural practices, reduce the productivity of the land.

CONSIDER THIS Much of the easily visible ecological damage caused by humans occurs in less-developed countries: overgrazed pastures, devastated rain forests, and critically endangered species such as rhinos and orangutans. Some of this damage results from economic demand from more-developed countries for imported products such as palm oil, gold, mahogany, or teak. Do you think developed countries should ban or limit imports of ecologically damaging goods from less-developed countries? Justify your answer by listing the pros and cons of such restrictions.

adults have fewer children than are required to replace themselves, causing the age structure diagram to narrow at the base (**FIG. 27-15c**). Age structures for the populations of developed and developing countries for 2018, with projections for 2050, are shown in **FIGURE 27-16**.

In Most Nations, Population Is Growing

Even if rapidly growing countries were to slow birth rates enough to achieve RLF, their populations would continue to increase for decades. Why? When the number of children exceeds the number of reproducing adults, this creates momentum for future growth, as these children mature and enter their reproductive years. For example, China reached RLF in the early 1990s, but at the time about 28% of its population was under age 15. Because a percentage below 20% is required to stabilize population size, China has since grown by almost 200 million people. China will probably have a stable population soon, however, because now only 17% of its people are younger than 15. In contrast, children make up about 40% of the population of Africa, so its population will continue to increase rapidly.

In general, population growth is highest in the less-developed countries that can least afford it. As more and more people compete for the same limited resources, it becomes increasing difficult to reduce poverty. Poverty diverts children away from schools and into activities that help support their families. The resulting lack of education, along with a lack of access to contraceptives, contributes to continued high birth rates. So the population increases, which tends to keep people poor, and the cycle continues. Of the 7.5 billion people on Earth in 2018, about 6.2 billion resided in less-developed countries. Although fertility rates in some less-developed countries, such as Brazil, have declined because of social changes and increased access to contraceptives, the fertility rate in most less-developed countries remains above RLF. Therefore, the human population will continue to grow, in less-developed countries and the world as a whole, for many years. **TABLE 27-1** shows current growth rates for various world regions.

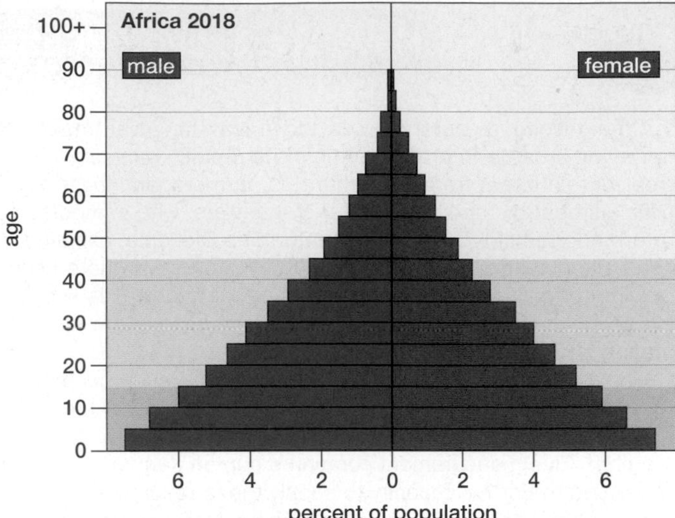

(a) Africa: A rapidly growing population

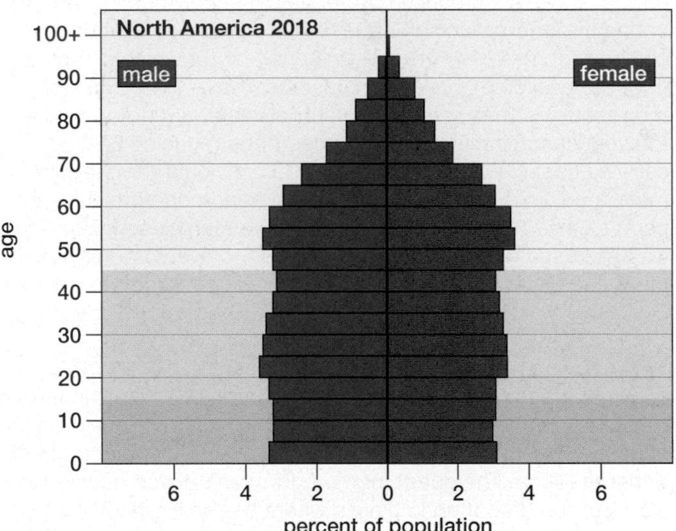

(b) North America: A slowly growing population

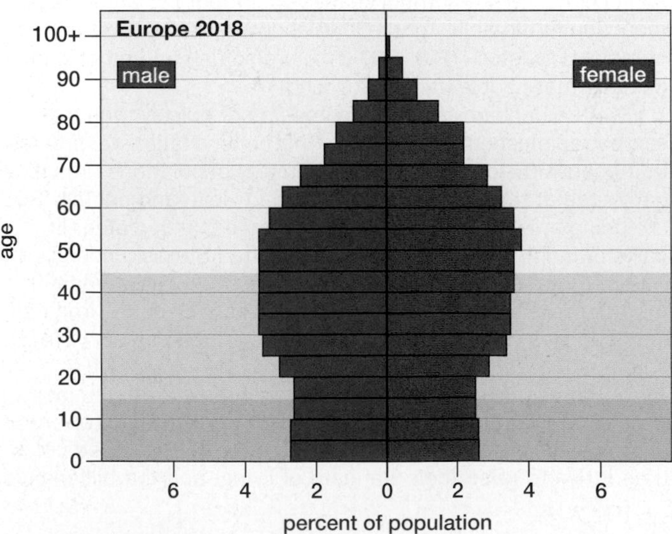

(c) Europe: A slowly declining population

▶ **FIGURE 27-15 Age structure diagrams (a)** The age structure for Africa illustrates a rapidly growing population that is projected to almost double by 2050. **(b)** North America represents a more slowly growing population that is still projected to add more than 100 million people by 2050. **(c)** Europe's age structure is of a slowly shrinking population, projected to decline by more than 35 million by 2050. Background colors from bottom to top indicate age groups: prereproductive children (0 to 14 years), reproductive age adults (15 to 44 years), and post-reproductive adults (45 to 100 years). Data from the U.S. Census Bureau, International Data Base.

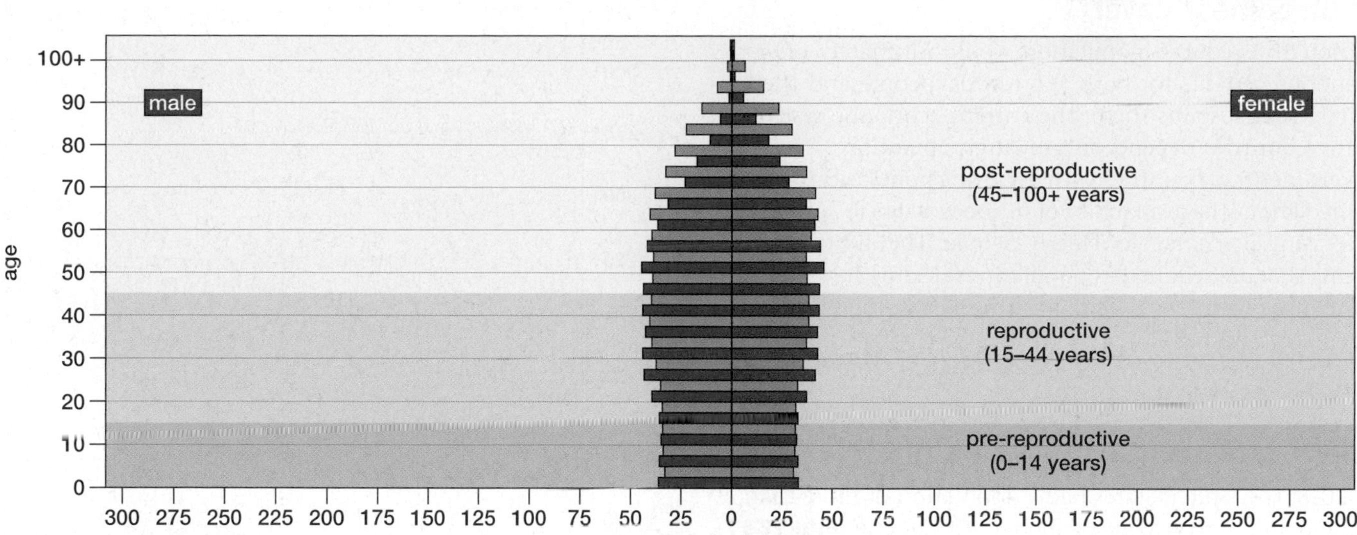

(a) More-developed countries

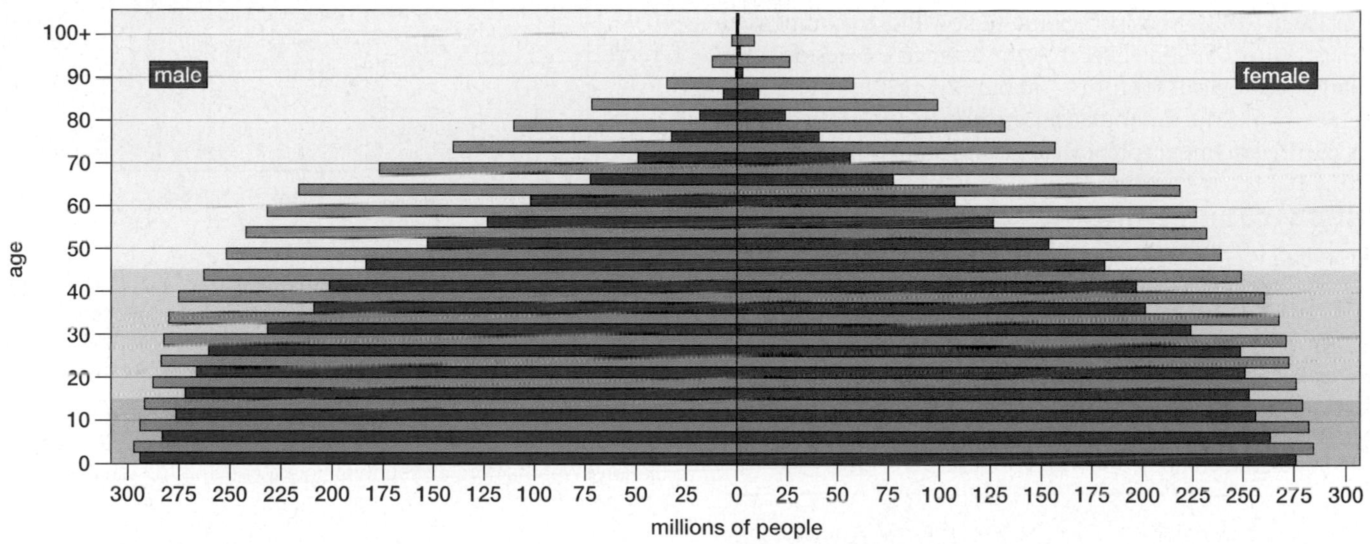

(b) Less-developed countries

▲ **FIGURE 27-16 Age structure diagrams of more-developed and less-developed countries** Note in part (b) that the predicted difference in the proportion of children compared to the proportion of adults in less-developed countries is smaller in 2050 than in 2018, as these populations approach RLF. However, the large numbers of young people in less-developed countries who will be entering their childbearing years will cause continued population growth. Data from the U.S. Census Bureau, International Data Base.

THINK CRITICALLY How does a fertility rate above RLF produce positive feedback (in which a change creates a situation that amplifies itself) for population growth?

TABLE 27-1 Average Population Statistics by World Region: 2018

Region	Fertility Rate	Rate of Natural Increase (%)
World	2.4	1.0
All less-developed countries	2.5	1.2
All more-developed countries	1.7	0.1
Africa	4.2	2.4
Latin America/Caribbean	2.0	1.0
Asia, excluding China	2.5	1.4
China	1.6	0.4
Europe	1.6	−0.1
North America	1.9	0.4

Data from the U.S. Census Bureau International Data Base.

Fertility in Some Nations Is Below Replacement Level

Although a reduced population would ultimately offer tremendous benefits for both the world's people and the biosphere that sustains them, the current economic systems of many countries depend on growing populations. Thus, the governments of countries with declining populations are concerned about the availability of future workers to support the increasing percentage of elderly people. The difficult adjustments that are required as populations decline have motivated many governments—such as those in Europe, where the population is shrinking by 0.1% per year—to adopt policies that encourage more childbearing and continued growth.

The U.S. Population Is Growing Rapidly

In 2018, the United States had a population of more than 29 million and a growth rate of about 0.8% per year (**FIG. 27-17**). The United States is one of the fastest-growing developed countries in the world. Continued immigration, which accounts for more than 30% of the population increase, will ensure growth for the indefinite future, unless the U.S. fertility rate (1.9 in 2018) drops sufficiently below RLF to compensate for the influx of people. Because the average U.S. resident has a large ecological footprint—about 2.5 times the global average—continued population growth in the United States has significant impact on both local and global environments.

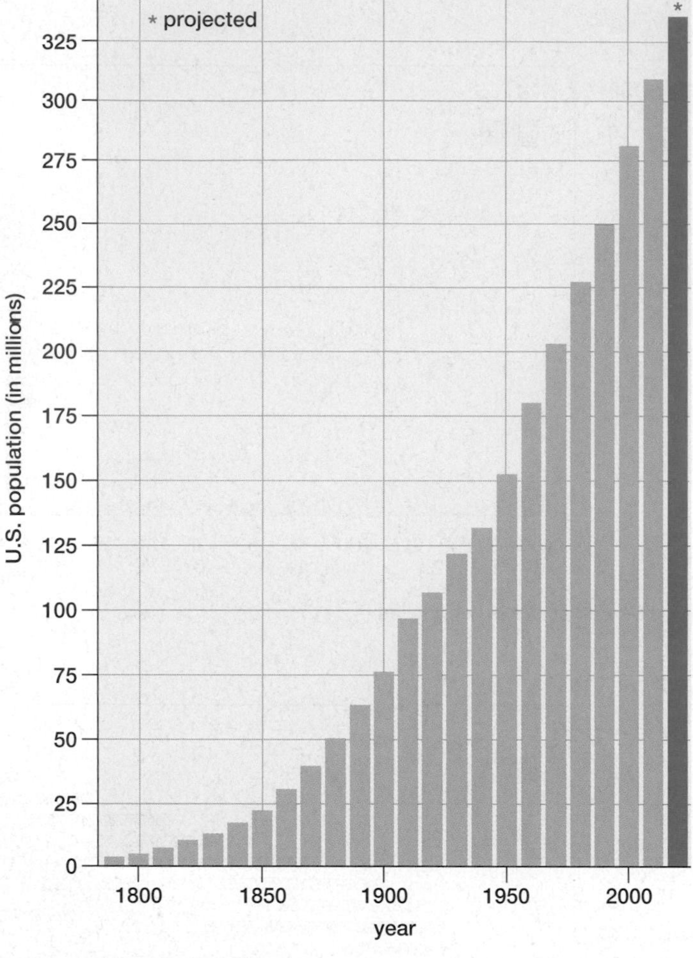

▲ **FIGURE 27-17 U.S. population growth** Since 1790, U.S. population growth has produced a J-shaped curve typical of exponential growth, with some slight slowing in recent decades. The U.S. Census Bureau predicts that the U.S. population will reach 334 million by 2020. Data from the U.S. Census Bureau.

THINK CRITICALLY At what stage of the S-curve is the U.S. population? What factors do you think might cause it to stabilize?

CHECK YOUR LEARNING

Can you . . .

- describe the advances that have allowed exponential growth of the human population?
- explain why rapid population growth continues today?
- explain why birth rates eventually decline in more-developed countries?
- sketch age structure diagrams and describe how their shape predicts future changes in population size?

CASE STUDY REVISITED

The Return of the Elephant Seals

The rapid recovery of elephant seal populations, from near-extinction in 1892 to about 200,000 animals today, is a triumph of wildlife conservation. Protective measures allowed exponential growth of a small population with abundant food, safe breeding sites, and few predators.

However, all modern elephant seals are the offspring of no more than 20 ancestors who lived in the 1890s. Therefore, all of today's elephant seals may have descended from just one bull and a few females. Passing through such a "population bottleneck" (see Chapter 16) reduces the genetic diversity of future generations, because only the alleles that were present in the minuscule ancestral population were available to pass on to the entire present-day population. Indeed, when molecular biologists analyzed the genomes of modern elephant seals, they found that today's elephant seals are almost genetically identical to one another. Analysis of specimens collected before 1892 revealed that pre-bottleneck elephant seals were much more genetically diverse. It may take hundreds or even thousands of years for mutations to replenish allele diversity in elephant seals.

Does this lack of genetic diversity threaten the prospects of the elephant seal population? For the immediate future, probably not, although elephant seals do have more genetic abnormalities, such as cleft palate and defects in the heart or brain, than harbor seals and California sea lions do. A high incidence of genetic defects is fairly common in inbred populations with low genetic diversity.

THINK CRITICALLY Aside from an increased incidence of genetic defects, what other risk to a population might be associated with low genetic diversity?

CHAPTER REVIEW

Go to **Mastering Biology** to access the Pearson eText, vocabulary review, practice quizzes, activities, videos, current events, and more.

*Answers to **Think Critically** and **Thinking Through the Concepts** questions can be found in the **Answers** section at the back of the book.*

Summary of Key Concepts

27.1 What Is a Population and How Does Population Size Change?

Populations change size through births and deaths, which result in natural increase, and through immigration and emigration, which produce net migration. Ignoring migration, the population growth rate (r) is its birth rate (b) minus its death rate (d). Population growth (G), the increase during a given time interval, equals the growth rate (r) multiplied by the population size (N). All organisms have the biotic potential to more than replace themselves over their lifetimes, resulting in population growth. A constant growth rate produces exponential growth in population size.

27.2 How Is Population Growth Regulated?

Carrying capacity (K) is the maximum size at which a population may be sustained indefinitely by an ecosystem. K is determined by limited resources such as space, nutrients, and energy. Environmental resistance generally maintains populations at or below the carrying capacity. Above K, populations deplete their resource base, leading to (1) the population stabilizing near K; (2) reduction of K and a permanently reduced population; or (3) the population being eliminated from the area. Population growth is restrained by density-independent forms of environmental resistance (principally weather and climate) and density-dependent forms of resistance (competition, predation, and parasitism).

27.3 How Do Life Histories Differ Among Species?

Whereas r-selected species tend to mature early, produce large numbers of small offspring, and provide little or no parental care, K-selected species tend to mature slowly and provide substantial parental care to small numbers of offspring. The life histories of most species fall between these two extremes. Late-loss survivorship curves are characteristic of K-selected species. Species with constant-loss curves have an equal chance of dying at any age. Early-loss curves are typical of r-selected species.

27.4 How Are Organisms in Populations Distributed?

Clumped distribution may occur for social reasons or around limited resources. Uniform distribution is often the result of territorial spacing. Random distribution occurs when individuals do not interact socially and when resources are abundant and evenly distributed.

27.5 How Is the Human Population Changing?

The human population has exhibited exponential growth for centuries, the result of a combination of high birth rates and technological, agricultural, industrial, and medical advances that have reduced the effects of environmental resistance and increased Earth's carrying capacity for humans. Age structure diagrams depict numbers of males and females in groups of increasing age. Expanding populations, mostly in less-developed countries, have triangular age structures with a broad base. More-developed countries typically have stable populations with relatively straight-sided age structures, or they have shrinking populations with age structures that are narrow at the base. Most people live in less-developed countries with growing populations and birth rates greater than replacement level fertility (RLF). Even if birth rates decline to RLF or below, the large number of children born in earlier years will eventually mature into reproducing adults, ensuring continued population growth.

Thinking Through the Concepts

Bloom's: Remembering, Understanding

Multiple Choice

1. Density-independent environmental resistance includes
  a. predation.
  b. floods.
  c. parasitism.
  d. competition.

2. Exponential growth often occurs when
  a. a population nears carrying capacity.
  b. a population exceeds carrying capacity.
  c. organisms invade a new habitat.
  d. the birth rate exceeds the death rate for a single generation.

3. Which of the following did *not* contribute to increasing Earth's carrying capacity for humans?
  a. birth control
  b. agriculture
  c. medical advances
  d. improved sanitation

4. A population is likely to increase in size if
  a. there are more post-reproductive adults than juveniles.
  b. parasites increase in abundance.
  c. its population pyramid is narrower at the base than in the middle.
  d. its reproductive rate is greater than replacement level fertility.

5. Boom-and-bust populations
  a. do not experience exponential growth.
  b. do not experience environmental resistance.
  c. may be affected by seasonal changes.
  d. level off at the carrying capacity of their environment.

Fill-in-the-Blank

1. Graphs that plot how the numbers of same-aged individuals change over time are called ________. The specific type of curve that applies to a dandelion that releases 300 seeds, most of which never germinate, is called ________. The curve for humans is an example of ________.

2. The type of growth that occurs in a population that grows by a constant percentage per year is ________. Does this form of growth add the same number of individuals each year? ________ What shape of curve is generated if this type of growth is graphed? ________ Can this type of growth be sustained indefinitely? ________

3. The maximum population size that can be sustained indefinitely without damaging the environment is called the ________. A growth curve in which a population first grows exponentially and then levels off at (or below) this maximum sustainable size is called a(n) ________ curve, or a(n) ________ curve.
4. The type of spatial distribution likely to occur when resources are localized is ________. The type of spatial distribution that results when pairs of animals defend breeding territories is ________.
5. A population grows whenever the number of ________ plus ________ exceeds the number of ________ plus ________. The growth rate of a population increases when the frequency of reproduction ________.

Review Questions

1. Define *biotic potential*, list the factors that influence it, and explain why natural selection may favor a high biotic potential.
2. Write and describe the meaning of the equation for population growth using the variables *G*, *r*, and *N*.
3. Draw, name, and describe the properties of a growth curve of a population without environmental resistance.
4. Define *environmental resistance* and distinguish between density-independent and density-dependent forms of environmental resistance. Describe three examples of each.
5. What is logistic population growth? What is *K*?
6. Describe three different possible consequences of exceeding carrying capacity. Sketch these scenarios on a graph. Explain your answer.
7. Distinguish between populations showing concave and convex survivorship curves. What are their typical life histories?
8. Explain why environmental resistance has not prevented exponential human population growth since prehistoric times; provide examples. Can this continue for much longer? Explain why or why not.
9. Draw the general shape of age structure diagrams characteristic of (a) expanding, (b) stable, and (c) shrinking populations. Label both axes. Explain why you can predict the next several decades of growth by the current age structure of a population.

Applying the Concepts

Bloom's: Applying, Analyzing, Evaluating

1. Research a developing country (such as Nigeria, Afghanistan, or Uganda) with rapid population growth; find out what factors sustain that growth, and explain why. Assess the likelihood that the fertility rate of this country will drop in the near future.
2. The U.S. Endangered Species Act (ESA) seeks to increase population sizes of endangered species to the point at which each species is self-sustaining and no longer needs special protection. Research two species that have been listed as endangered: one that has recovered to the extent that it has been "delisted" and one that has remained on the list for more than 20 years. Based on what you have learned in this chapter, what characteristics of the two species and their ecosystems have significantly contributed to their differing fates?

28 Community Interactions

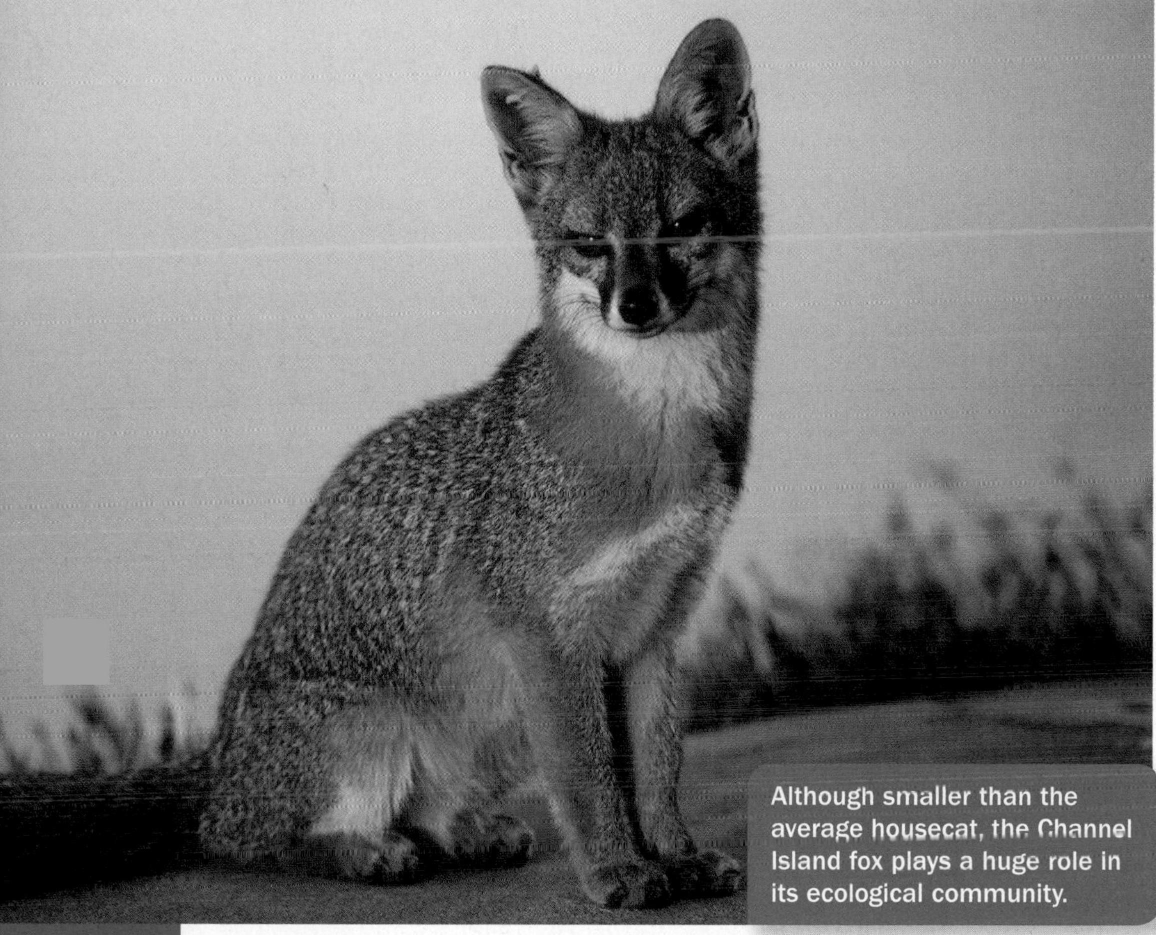

Although smaller than the average housecat, the Channel Island fox plays a huge role in its ecological community.

CASE STUDY

The Fox's Tale

THE ISLAND FOX, which weighs in at about 5 pounds, is one of the smallest members of the dog family. Nonetheless, it is the largest terrestrial predator on the Channel Islands, eight islands scattered off the coast of southern California. For thousands of years, island foxes flourished, eating mice, insects, fruit, and seeds. The foxes shared the islands with bald eagles, which are large enough to prey on the tiny foxes, but which fed primarily on the abundant fish in the surrounding ocean waters and posed no threat to the foxes. In the late 1900s, however, catastrophe struck the foxes, whose population crashed from more than 6,000 in the mid-1900s to just a few hundred by 2000. What happened?

The Channel Islands are home to a unique collection of species, some of which, like the island fox, are found nowhere else on Earth. But starting in the 1800s, the tapestry of life on the islands began to unravel. Settlers in search of farmland cleared large areas of native vegetation and brought in cattle, pigs, and sheep. Some of the pigs and sheep escaped, colonizing the remaining wild areas of the islands, where they consumed or uprooted native plants. Many of the native plants were replaced by invasive plant species brought to the islands, usually unintentionally, by people. People also inadvertently brought disease-causing bacteria and viruses that had not previously been present on the islands; these pathogens infected and killed native species with no prior exposure or immunity to the diseases.

In the mid-1900s, most of the bald eagles were wiped out by the insecticide DDT. The cumulative effects of these changes devastated the islands' native flora and fauna, particularly the foxes, which almost became extinct. Some islands were down to their last 15 or 20 foxes, while populations of mice, skunks, and non-native weeds increased.

These changes illustrate the interconnectedness of the species on the islands. Why would the addition of sheep and pigs, or the loss of bald eagles, harm the foxes? How might the loss of foxes affect other island species? Is there any way of restoring connections that have been torn apart?

AT A GLANCE

28.1 HOW DO SPECIES IN COMMUNITIES INTERACT?

An ecological **community** consists of all the populations of multiple species living in a defined area and interacting with one another. Interactions between the species in a community may be classified into four major categories: interspecific competition, consumer–prey interactions, mutualism, and commensalism (**TABLE 28-1**).

- In **interspecific competition** (*interspecific* means "between species"), two or more species that use the same resources interact in a way that is harmful to all of the species involved. Competitors are harmed because competitive interactions reduce the participants' access to resources.
- In **consumer–prey interactions**, one species (the consumer) uses another species (the prey) for food. As you might expect, consumers benefit from such interactions, whereas the prey are harmed. Consumer–prey interactions include predation and parasitism.
- In **mutualism**, two species interact in a way that benefits both species. Therefore, mutualism may make it possible for species to thrive together where neither could survive alone.
- In **commensalism**, two species interact in a way that benefits one species while the other is relatively unaffected.

Interactions between species are typically intermittent, as when a lion chases down a zebra or when a deer walks along nipping buds off bushes. Some interactions, however, are prolonged and intimate, so that members of the two species are almost always found together. Such intimate, ongoing associations are examples of **symbiosis**, and the participating species are often called *symbionts*. A symbiotic relationship may be beneficial to both participating species (mutualism), beneficial to one species and neutral to the other (commensalism), or beneficial to one species and harmful to the other (parasitism).

TABLE 28-1 Interactions Between Species

Type of Interaction	Effect on Species A	Effect on Species B
Interspecific competition between A and B	Harms	Harms
Consumer–prey interactions (A is the consumer, B is the prey)	Benefits	Harms
Mutualism between A and B	Benefits	Benefits
Commensalism between A and B	Benefits	No effect

Because interactions between species may help determine which members of each species survive and reproduce, such interactions can influence the evolution of the species involved. For example, predation favors evolution of prey with defenses against predators, and prey defenses favor the evolution of predators with traits that help them overcome the defenses. This process, by which interacting species act as agents of natural selection on one another, is called **coevolution**.

CHECK YOUR LEARNING

Can you . . .

- define *community*, *symbiosis*, and *coevolution*?
- name the three major types of community interactions and describe their effects on the species involved?

28.2 HOW DOES INTERSPECIFIC COMPETITION AFFECT COMMUNITIES?

Competition within a species is especially intense, because all members of a species require the same resources. Thus, it stands to reason that the intensity of interspecific competition depends on how similar the requirements of the competing species are. Ecologists express this idea by saying that the degree of competition depends on the degree to which the *ecological niches* of the competing species overlap.

Each Species Has a Unique Place in Its Ecosystem

Each species occupies a unique **ecological niche** that encompasses all aspects of its way of life. A species' niche includes the particular type of habitat in which it lives, the environmental factors (such as moisture levels and temperature) necessary for its survival and reproduction, and the nutrients it requires and the methods by which it obtains them. For an animal species, we can add to that list the types of food it eats, the time of day at which it forages for food, and the type of shelter it needs. A plant species' niche might include the range of soil pH it can tolerate, the amount of sunlight it needs, and the soil nutrients it requires. A species' predators, prey, symbionts, and competitors are also considered to be elements of its niche. In sum, a species' ecological niche is the combination of all factors that define how a species "makes a living" within its ecosystem.

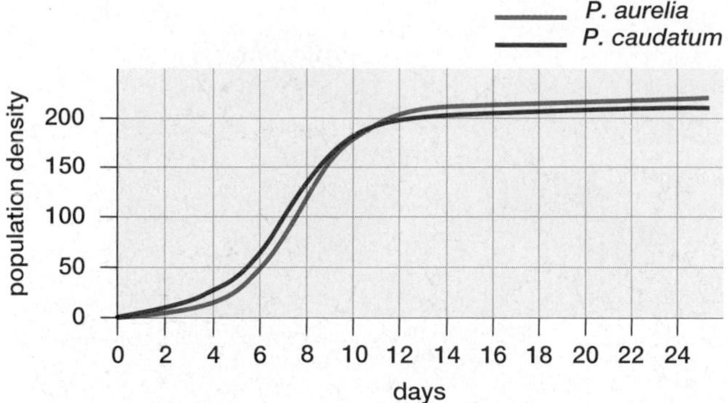

(a) Grown in separate flasks

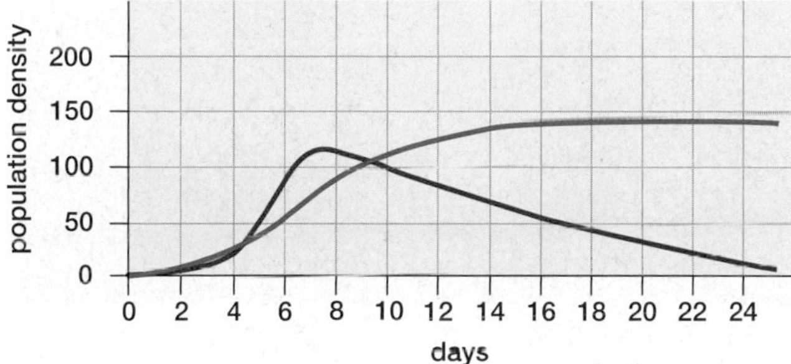

(b) Grown in the same flask

▲ **FIGURE 28-1 Interspecific competition in the lab (a)** Raised separately with a constant food supply, both *Paramecium aurelia* and *P. caudatum* show the S-curve typical of a population that initially grows rapidly and then stabilizes. **(b)** Raised together and forced to occupy the same niche, *P. aurelia* always eventually outcompetes *P. caudatum* and causes that population to die off. Data from Gause, G. F. 1934. *The Struggle for Existence.* Baltimore: Williams & Wilkins.

The Ecological Niches of Coexisting Species Never Overlap Completely

Although different species may share many aspects of their ecological niches, interspecific competition prevents any two species from occupying exactly the same niche. This concept, commonly called the **competitive exclusion principle**, was formulated in 1934 by the Russian microbiologist G. F. Gause.

Gause grew two species of the protist *Paramecium* (*P. aurelia* and *P. caudatum*) in laboratory cultures. Grown separately, with bacteria as their only food, both species thrived (**FIG. 28-1a**). However, when Gause placed the two species together, they could not coexist: *P. aurelia* grew more rapidly and always eliminated *P. caudatum* (**FIG. 28-1b**). Gause concluded that two species with exactly the same niche cannot coexist indefinitely.

Can species coexist if their niches differ slightly? To find out, Gause repeated the experiment, but this time pairing *P. caudatum* with a different species, *P. bursaria*. *P. caudatum* feeds on bacteria suspended in the culture medium, but *P. bursaria* feeds mostly on bacteria that settle to the bottom of the cultures. These two species of *Paramecium* were able to coexist indefinitely because they fed in different places and thus occupied slightly different niches.

Evolution in Response to Competition May Reduce Niche Overlap

Competitive exclusion may cause coexisting species with similar requirements to evolve so that each occupies a smaller niche than it would if it were by itself. Because natural selection favors individuals with fewer competitors, competing species tend to evolve physical and behavioral characteristics that reduce their competitive interactions. This phenomenon, called **resource partitioning**, is an outcome of the coevolution of species with extensive niche overlap.

Ecologist Robert MacArthur documented resource partitioning by carefully observing five species of North American warblers under natural conditions. All five species nest and hunt for insects in spruce trees. Although their niches overlap, MacArthur found that the different species search for food in different portions of the tree, employ different hunting tactics, and nest at slightly different times (**FIG. 28-2**). The five species of warblers have evolved behaviors that reduce the overlap of their niches, thereby reducing interspecific competition.

Another famous example of resource partitioning was discovered by Charles Darwin when he studied related

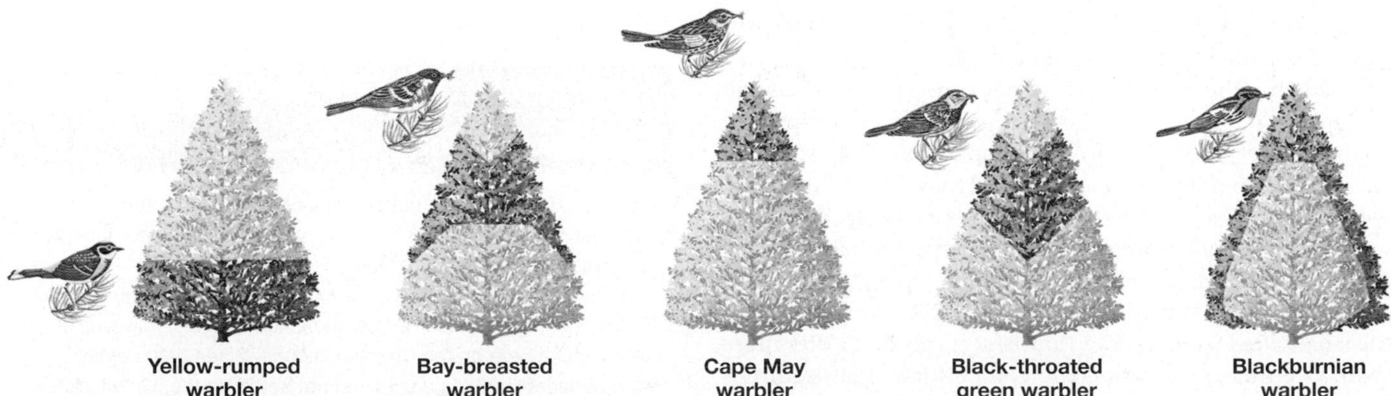

▲ **FIGURE 28-2 Resource partitioning** Each of these five insect-eating species of North American warblers searches for food in a slightly different part of a spruce tree. This reduces niche overlap and competition. Adapted from MacArthur, R. H. 1958. *Ecology* 39:599–619.

(a) Eurasian red squirrel

(b) Eastern gray squirrel

▲ **FIGURE 28-3 Competition limits population size and distribution (a)** Native red squirrels have been pushed out of most of England by competition from **(b)** gray squirrels imported from the United States.

species of finches on the Galápagos Islands. Different finch species that share the same island evolved different bill sizes and shapes and different feeding behaviors that reduce competition among them (see Fig.15-5).

Interspecific Competition May Influence the Size and Distribution of Populations

Although natural selection tends to reduce niche overlap, different species may occupy niches similar enough to generate interspecific competition that restricts population size and the distribution of species. Consider, for example, the two squirrel species in Great Britain. One species, the Eurasian red squirrel, *Sciurus vulgaris*, is the only squirrel native to Britain (**FIG. 28-3a**). In the late 1800s, however, rich landowners in Britain imported Eastern gray squirrels, *Sciurus carolinensis*, from the United States as "ornaments" for their manors (**FIG. 28-3b**). The two squirrel species competed for the same resources, and, at least in the deciduous forests of southern England, gray squirrels won the competition. Gray squirrels are more efficient at eating acorns, and they raid red squirrels' seed caches. As a result, red squirrels are practically extinct in southern England. However, red squirrels do much better in coniferous forests, probably because the larger gray squirrels don't get enough nutrition from small conifer seeds, whereas the smaller red squirrels can flourish on this diet. After more than a century of interspecific competition, gray squirrels dominate in southern England, the two species coexist in the mixed deciduous/conifer forests of northern England, and red squirrels still predominate in the coniferous forests of northern Scotland.

Gray squirrels are not the only species that humans have introduced to new habitats. In "Earth Watch: Invasive Species Disrupt Community Interactions", we describe what sometimes happens when people transport organisms into ecological communities in which the organisms did not evolve.

CHECK YOUR LEARNING

Can you . . .

- explain the *ecological niche* concept?
- explain how competitive exclusion leads to resource partitioning?
- explain how interspecific competition can affect the population size and geographic distribution of species?

CASE STUDY CONTINUED

The Fox's Tale

Channel Island foxes do not experience significant interspecific competition. Although foxes do compete with skunks for insects and mice, the competition seems to be very one-sided—the foxes suppress skunk populations, but skunks do not have much effect on fox populations. Bald eagles and golden eagles, however, do seem to compete on the Channel Islands. When bald eagles were abundant, golden eagles did not settle on the islands. After bald eagles were eliminated by DDT, golden eagles colonized the islands. Around the same time, the fox population plummeted. Did the switch from bald to golden eagles cause the decline in foxes?

Earth WATCH Invasive Species Disrupt Community Interactions

A non-native species that is introduced into an ecosystem in which it did not evolve often encounters few competitors, predators, or parasites in its new environment. As a result, its population may grow explosively, outcompete or prey on native species, and damage the ecosystem in ways that harm human health, the environment, or the local economy. Ecologists call these harmful invaders *invasive species*. An introduced species will not necessarily become invasive; the ones that do typically reproduce rapidly, disperse widely, and thrive under a wide range of environmental conditions.

English house sparrows were introduced into the United States on several occasions, starting in the 1850s, to control caterpillars feeding on shade trees. In 1890, European starlings were released into Central Park in New York City by a group attempting to introduce all the birds mentioned in the works of Shakespeare. Both bird species have spread throughout the continental United States. Their success has reduced populations of native songbirds, such as bluebirds and purple martins, with which they compete for nesting sites. Burmese pythons, probably released by pet owners, have become significant threats to natural communities in the Everglades of Florida (**FIG. E28-1a**). These huge predatory snakes have caused precipitous declines in populations of raccoons, opossums, deer, and rabbits.

Invasive plants also threaten natural communities. In the 1930s and 1940s, kudzu, a Japanese vine, was planted extensively in the southern United States to control soil erosion. Today, kudzu is a major pest, covering trees and engulfing abandoned buildings (**FIG. E28-1b**). Both water hyacinth and purple loosestrife were introduced into the United States as ornamental plants. Water hyacinth now clogs waterways in southern states, slowing boat traffic and displacing natural vegetation. Purple loosestrife aggressively invades wetlands, where it outcompetes native plants and reduces food and habitat for native animals.

Invasive species rank second only to habitat destruction in pushing endangered species toward extinction. Governments and land managers have sometimes tried to control invasive species by importing the species' natural predators or parasites (called *biocontrols*). However, biocontrols sometimes have unpredicted and even disastrous effects on native wildlife. The cane toad, for example, was introduced into Australia in the 1930s to control non-native beetles that threatened the sugarcane crop. Unfortunately, the toads have proven to be worse than the beetles; about 200 million toads now occupy northeastern Australia, outcompeting native frogs (**FIG. E28-1c**).

Despite the risks of biocontrols, there are often few realistic alternatives. Biologists now carefully screen proposed biocontrols to make sure they are likely to attack only the intended invasive species, and there have been successful introductions of biocontrols. For example, Eurasian beetles, released by the millions each year, are now among the most effective methods of controlling purple loosestrife in North America.

CONSIDER THIS Many invasive species were deliberately imported as pets or landscape plants and entered natural communities accidentally (escaped pets), deliberately (by owners who could no longer care for their exotic pets), or through natural spread (landscape plants). But probably the vast majority of non-native plants and animals never cause a major problem. Would you support laws to restrict the importation of non-native plants and animals? Why or why not?

(a) Burmese python

(b) Kudzu

(c) Cane toad

▲ **FIGURE E28-1 Invasive species (a)** Huge Burmese pythons introduced to the Florida Everglades can eat adult deer and even small alligators. **(b)** Kudzu, a Japanese vine imported to the southern United States, can rapidly cover entire trees and small buildings. **(c)** The cane toad (native to central and South America) was imported to Australia, where it outcompetes native toads and frogs and preys on many indigenous animals, such as this pygmy possum. Poisonous secretions on its skin protect the cane toad from predators.

(a) Eagle owl

(b) Koala

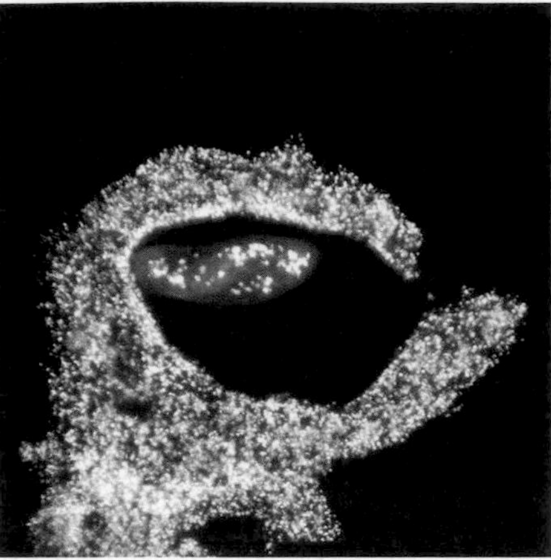

(c) Amoeba

▲ **FIGURE 28-4 Forms of predation (a)** An eagle owl feasts on a mouse. **(b)** The preferred food of koalas is eucalyptus leaves. **(c)** Amoebas are microscopic protists that eat bacteria and smaller protists.

28.3 HOW DO CONSUMER–PREY INTERACTIONS SHAPE EVOLUTIONARY ADAPTATIONS?

Consumer–prey interactions fall into two major categories: predation and parasitism. A **predator** is a free-living organism that eats other organisms. Predators include both *carnivores* (animals that eat other animals; **FIG. 28-4a**) and *herbivores* (animals that eat plants; **FIG. 28-4b**). Microorganisms that consume other microorganisms are also predators (**FIG. 28-4c**). Predators are generally larger and less abundant than their prey. Carnivorous predators typically kill their prey, whereas herbivorous predators usually do not immediately kill the plants they feed on.

A **parasite** usually lives symbiotically with its prey (called its **host**), dwelling on or inside the host for all or most of the parasite's life cycle. Parasites include tapeworms, fleas, and many disease-causing protists, fungi, bacteria, and viruses. Only a few vertebrates are parasites; the lamprey, which attaches itself to a host fish and sucks its blood, is one example (see Fig. 25-6). Most parasites are smaller and much more abundant than their hosts. Parasites harm or weaken their host, but often do not immediately kill it.

Predators and Prey Coevolve

To survive and reproduce, predators must feed and prey must avoid becoming food. Therefore, predator and prey exert intense natural selection on one another. The result is an ongoing cycle of adaptation and counter-adaptation, as predators evolve to become better at eating prey, and prey evolve to become more difficult to eat. Such coevolution has produced the keen eyesight of the hawk and owl, and the camouflaging colors of their mouse and ground squirrel prey. It has produced grasses with tough silica substances in their blades that make them difficult to chew, and the long, hard, grinding teeth of grazing animals, such as horses.

Both Structures and Behaviors Can Coevolve

The adaptations of echolocating bats and their moth prey (**FIG. 28-5**) provide excellent examples of how both body structures and behaviors are molded by coevolution. Echolocating bats are nighttime hunters that navigate and locate insect prey by emitting pulses of sound so high pitched that people can't hear them. The sounds bounce off nearby objects, and the bats use the echoes to produce a sonar "image" of their surroundings, allowing them to navigate around objects and detect moths and other prey.

In response to this prey-detection method, some moths have evolved ears that are particularly sensitive to the sound frequencies used by echolocating bats. When these moths hear a bat, they take evasive action, flying erratically or dropping to

▲ **FIGURE 28-5 Coevolution between bats and moths** A long-eared bat uses a sophisticated echolocation system to hunt moths, which in turn have evolved specialized sound detectors and behaviors to avoid capture.

▲ **FIGURE 28-6 Chemical warfare** A monarch caterpillar feeds on milkweed that contains a powerful toxin.

THINK CRITICALLY Why might the caterpillar be colored with conspicuous stripes?

the ground. In response to the moth's adaptation, some bat species have evolved the ability to change the pitch of their sound pulses away from the moth's sensitivity range. Some moth species have even evolved a way to interfere with the bats' echolocation by producing their own high-frequency clicks. In response, a bat may turn off its own sound pulses temporarily and zero in on the moth by following the moth's clicks.

Predators and Prey May Engage in Chemical Warfare

Coevolution may give rise to a kind of "chemical warfare" between predators and prey. Many plants synthesize toxic and distasteful chemicals that deter predators. As plants evolved these defensive toxins, many herbivorous insects evolved increasingly effective ways to detoxify or even use these substances. As a result, nearly every toxic plant is eaten by at least one type of insect. For example, monarch butterfly caterpillars eat nothing but milkweeds, which are toxic (**FIG. 28-6**). The caterpillars not only tolerate the milkweed poison but also store it in their tissues as a defense against their own predators.

Other defensive chemicals include the clouds of ink that certain mollusks (including squid and octopuses) emit when a predator attacks. These chemical "smoke screens" confuse predators and mask the prey's escape. Another dramatic example of chemical defense is seen in the bombardier beetle. In response to the bite of an ant or other threat of attack, the beetle releases secretions from defensive glands into a chamber in its abdomen. There, enzymes catalyze an explosive chemical reaction that shoots a toxic, boiling-hot spray onto the attacker.

Toxins can be used to attack as well as to defend. The venom of spiders and some snakes, such as rattlesnakes and cobras, both paralyzes their prey and deters predators.

Camouflage Conceals Both Predators and Prey

Have you ever heard the saying that the best hiding place may be in plain sight? Both predators and prey have evolved colors, patterns, and shapes that resemble their surroundings (**FIG. 28-7**). Such disguises, called **camouflage**, render plants and animals inconspicuous, even when they are in full view.

Have You Ever Wondered ...

Why Rattlesnakes Rattle?

Rattlesnakes are venomous, feared predators. So why bother rattling? Actually, most of the time they don't rattle. Camouflaged, silent hunters, rattlers sneak up on mice, rats, and small birds and strike without warning. However, even though rattlers have lethal venom, they are preyed upon by hawks, eagles, coyotes, badgers, and a few other animals. If a rattler detects a potential predator, it will quietly slink away if it can. But when confronted by a predator before it can escape, the snake will coil up and rattle, both startling and warning its adversary. In places where they are frequently hunted by humans, rattlers don't seem to rattle as much as they used to. Because rattling makes rattlesnakes more easily detectable by human hunters (and therefore more likely to be killed), natural selection may be favoring the evolution of silent rattlers.

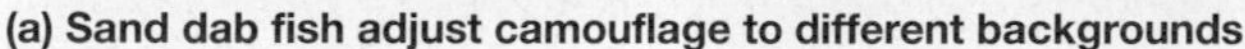

(a) Sand dab fish adjust camouflage to different backgrounds **(b) A camouflaged horned lizard**

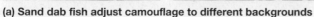

▲ **FIGURE 28-7 Camouflage by blending in (a)** Sand dabs are flat, bottom-dwelling ocean fish whose mottled colors closely resemble the sand on which they rest. Both their colors and patterns can be modified by nerve signals to better blend with their backgrounds. **(b)** This horned lizard helps protect itself from predation by snakes and hawks by resembling its surroundings of leaf litter.

THINK CRITICALLY How might predators evolve to detect camouflaged prey?

(a) Caterpillar of viceroy butterfly

(b) Leafy sea dragon

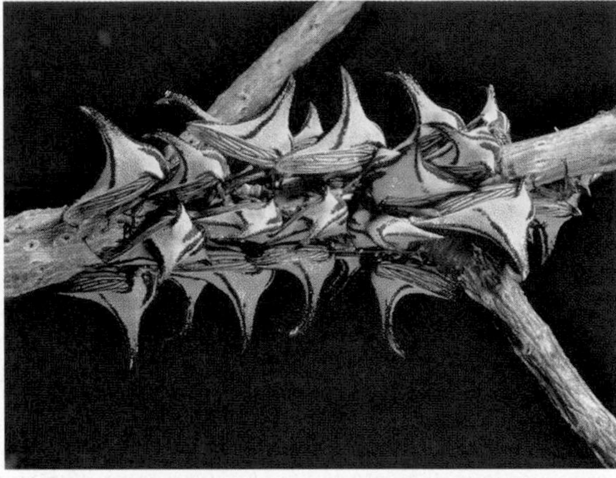

(c) Thorn treehoppers

(d) Living rock succulents

▲ **FIGURE 28-8 Camouflage by resembling specific objects (a)** A Viceroy butterfly caterpillar, whose color and shape resemble a bird dropping, sits motionless on a leaf. **(b)** The leafy sea dragon (a fish related to the seahorse) has evolved extensions of its body that mimic the algae in which it hides. **(c)** Thorn treehopper insects avoid detection by resembling thorns on a branch. **(d)** These South African succulents are appropriately called "living rocks."

THINK CRITICALLY How might such camouflage have evolved?

(a) A camouflaged snow leopard

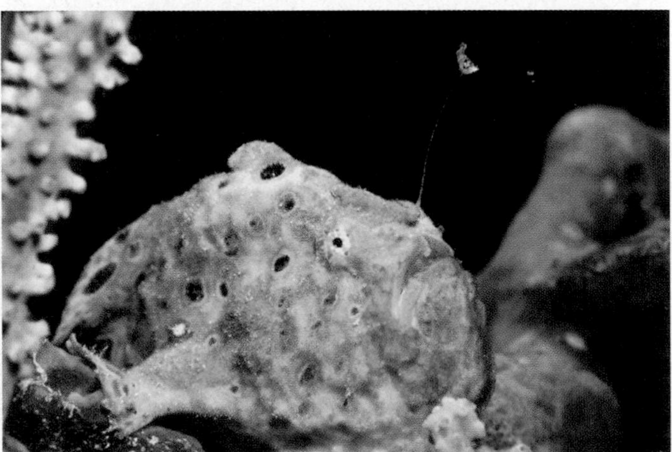

(b) A camouflaged frogfish

▲ **FIGURE 28-9 Camouflage assists predators (a)** As it waits for prey—such as gazelles, wild sheep, and deer—a snow leopard in the mountains of Mongolia avoids detection with camouflage coloration. **(b)** A camouflaged frogfish waits in ambush, its lumpy, yellow body matching the sponge-encrusted rock on which it rests.

Some prey animals closely resemble objects that carnivorous predators do not eat, such as leaves, twigs, seaweed, thorns, or even bird droppings (**FIGS. 28-8a–c**). Camouflaged animals tend to remain motionless; a crawling bird dropping would ruin the disguise. Whereas many camouflaged animals resemble parts of plants, a few succulent desert plants look like small rocks, which hides them from animals seeking the water stored in the plants' bodies (**FIG. 28-8d**).

Predators that ambush prey are also aided by camouflage. For example, the spotted snow leopard is almost invisible on a mountainside as it watches for prey (**FIG. 28-9a**). A frogfish, lurking motionless on the ocean floor awaiting smaller fish to prey upon, closely resembles sponges and algae-covered rocks (**FIG. 28-9b**).

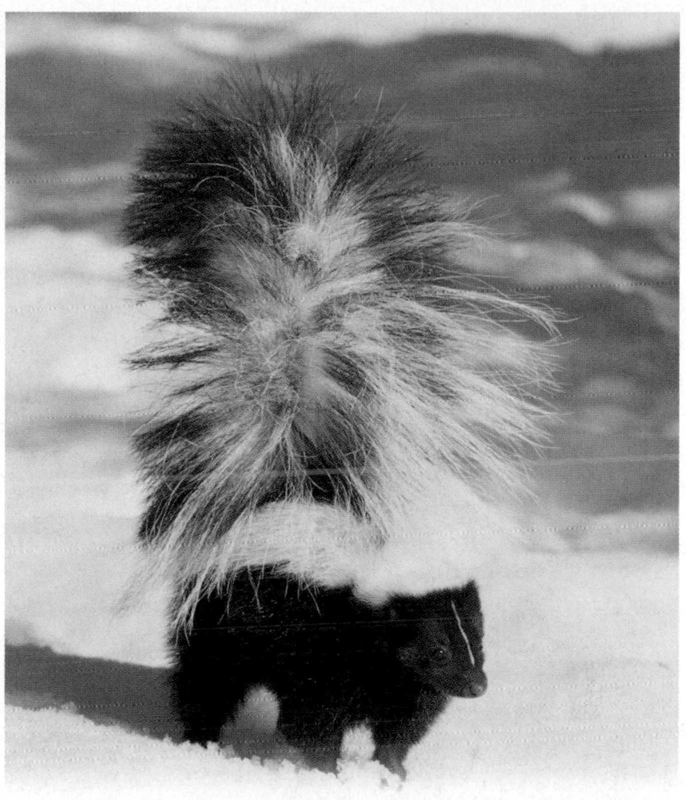

▲ **FIGURE 28-10 Warning coloration** The vivid stripe and the tail display behavior of the skunk advertise its ability to make any attacker miserable.

Conspicuous Colors Can Protect Toxic Prey

Some prey animals have evolved a very different strategy for visual defense: bright **warning coloration**. These animals may taste bad or make the predator sick (as monarch butterflies and their caterpillars do; see Fig. 28-6), inflict a venomous sting or bite (as bees and coral snakes do), or produce stinking chemicals when bothered (**FIG. 28-10**). The eye-catching colors seem to declare "Attack at your own risk!" A predator encountering such a prey for the first time might attack it, or even kill and eat it, but would avoid similar prey in the future, thus sparing others of the same species.

Some Prey Gain Protection Through Mimicry

In **mimicry**, members of one species have evolved to resemble members of another species. Mimicry evolves because mimics gain an advantage in predator–prey interactions. This benefit can arise in a number of different ways.

In *Müllerian mimicry*, two or more poisonous or distasteful species benefit by sharing a similar warning-color pattern. For example, both monarch butterflies and viceroy butterflies store plant toxins in their bodies, which deters birds from preying on them. The wing patterns of monarchs and viceroys are strikingly similar (**FIG. 28-11**), so birds that become ill from consuming one species are likely to avoid the other as well.

In *Batesian mimicry*, harmless animals evolve to resemble venomous or distasteful ones. For example, a harmless moth avoids predation by resembling a wasp (**FIG. 28-12**).

(a) Monarch (distasteful)

(b) Viceroy (distasteful)

▲ **FIGURE 28-11 Müllerian mimicry** Nearly identical warning coloration protects both **(a)** the distasteful monarch and **(b)** the equally distasteful viceroy butterfly.

(a) Hornet (venomous)

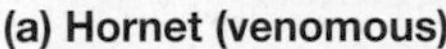

(b) Moth (nonvenomous)

▶ **FIGURE 28-12 Batesian mimicry (a)** A stinging hornet is mimicked by **(b)** a stingless moth.

Startle coloration is a form of mimicry in which animals have evolved color patterns that closely resemble the eyes of a much larger, possibly dangerous animal (**FIG. 28-13**). If a predator gets close, the prey suddenly reveals its eyespots, startling the predator and sometimes allowing the prey to escape. Some predators have evolved **aggressive mimicry**, in which they entice prey by resembling something attractive to the prey. For example, the orchid mantis is a predatory insect that resembles a flower (**FIG. 28-14**); its prey are insects that are attracted to what they perceive to be a source of nutritious nectar.

(a) False-eyed frog

(b) Peacock moth

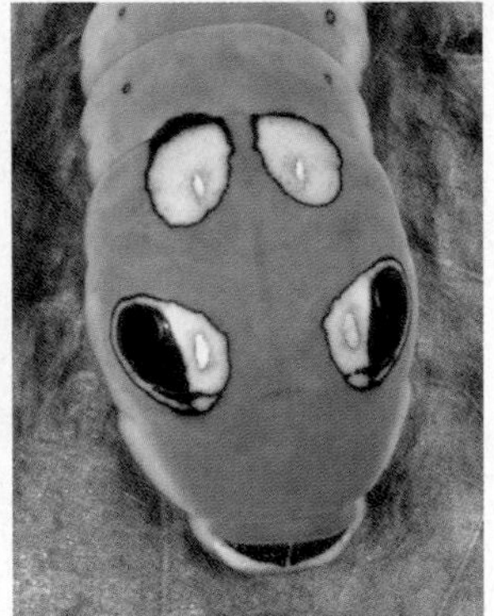

(c) Swallowtail caterpillar

▲ **FIGURE 28-13 Startle coloration (a)** When threatened, the false-eyed frog of South America raises its rump, which resembles the eyes of a large predator. **(b)** The peacock moth from Trinidad is well camouflaged, but should a predator approach, it suddenly opens its wings to reveal spots resembling large eyes. **(c)** Predators of this caterpillar larva of the Eastern tiger swallowtail butterfly are deterred by its resemblance to a snake.

▶ **FIGURE 28-14 Aggressive mimicry** An orchid mantis resembles a flower; the mantis eats the insects that approach in search of a nectar meal.

Parasites and Hosts Coevolve

Natural selection favors parasites that are better at invading hosts and hosts that can resist parasitic invasion, resulting in coevolution of parasites and hosts. Perhaps the best-studied example of parasite–host coevolution is the cellular warfare between infectious microorganisms and their mammal hosts. Mammals defend themselves against these parasites by secreting antimicrobial molecules; producing cells that engulf and destroy microbes or secrete antibodies that help to kill microbes; and even killing their own body cells that have been infected by viruses. In response, microbes have evolved coatings that prevent the immune system from recognizing them; an ability to cluster together in *biofilms* that help to keep antimicrobial secretions from reaching them; and sometimes even the capacity to invade and destroy parts of the host's defense system. We further explore parasite–host coevolution in "Health Watch: Parasitism, Coevolution, and Coexistence" on page 484.

CHECK YOUR LEARNING

Can you . . .

- compare how predators and parasites affect their prey and explain how these two types of consumer–prey interaction differ?
- describe why predators and prey coevolve and why parasites and hosts coevolve?
- describe some examples of chemical warfare between predators and prey?
- describe and provide examples of camouflage, warning coloration, Müllerian mimicry, Batesian mimicry, startle coloration, and aggressive mimicry?
- describe some outcomes of coevolution between parasites and hosts?

CASE STUDY CONTINUED

The Fox's Tale

Predation played a major role in the decline of Channel Island foxes. The loss of bald eagles opened up the islands to invasion by golden eagles. Meanwhile, pigs escaped from farms on the islands and multiplied, bearing litters year-round and providing a steady supply of piglets for the golden eagles to eat. Although piglets were the main course, foxes were tasty snacks. For thousands of years, island foxes had no predators. As a result, the foxes have not evolved strong defenses against predators. Unlike their mainland relatives, they hunt during the day and spend a lot of time in the open—traits that make them easy targets for golden eagles. As we will see, the loss of foxes reverberated throughout the islands' communities.

28.4 HOW DO MUTUALISMS BENEFIT DIFFERENT SPECIES?

Many interactions between species are mutualistic, beneficial to all participants. Lichens, for example, are symbiotic mutualistic associations between algae and fungi, in which the resulting structure appears to be a single organism (**FIG. 28-15a**). In many lichens, the body of one fungal species provides support and protection for (1) a second fungal species that produces a chemical defense against bacteria and (2) photosynthetic algae that provide sugar molecules for the fungi.

Many animals benefit from symbiotic mutualistic associations with microorganisms that inhabit their digestive systems. For example, bacteria and protists live in the guts of animals—such as cows, horses, rabbits, and termites—that consume cellulose-containing plant tissues such as leaves or wood. The microorganisms break down the cellulose, making its component sugar molecules available both to themselves and to the animals that harbor them. In human intestines, some species of mutualistic bacteria synthesize vitamins, such as vitamin K, which we absorb and use.

(a) Lichen

(b) Clownfish

▲ **FIGURE 28-15 Mutualism (a)** These lichens growing on a branch are formed by a mutualistic association between an alga and two types of fungus. Lichens take a variety of forms; both a flat lichen and a more upright species are visible here. **(b)** The clownfish is coated with a protective coat of mucus, allowing it to snuggle unharmed among the stinging tentacles of an anemone.

Plants may also benefit from mutualistic symbiosis with microorganisms. For example, plants and certain fungus species may form associations called *mycorrhizae* in which the fungal bodies penetrate and become entwined with plant roots. The fungi acquire sugars that the plants produce by photosynthesis, and the plants acquire mineral nutrients that the fungi extract from the soil.

In other symbiotic mutualisms, one partner does not live inside the other. For example, the clownfish of the southern Pacific Ocean shelters amid the venomous tentacles of certain species of anemones (**FIG. 28-15b**). The anemone provides the clownfish with protection from predators. In return, the clownfish drives away other fish that eat anemones and cleans dirt and debris from its host.

Many mutualistic relationships are not intimate and prolonged, so they are not symbiotic. Consider the relationship between plants and the animals that pollinate them, including bees, moths, hummingbirds, and even some bats. The animals help fertilize the plants' eggs by carrying plant sperm (found in pollen grains) from one plant to another, and they

Parasitism, Coevolution, and Coexistence

In most parasite species, many individual parasites infect a single host, resulting in competition that favors the parasites best able to exploit the host and reproduce rapidly. Thus, natural selection favors rapidly reproducing parasites. Because rapidly reproducing parasites tend to be especially harmful to the host, natural selection in essence favors the evolution of ever more-harmful parasites. At the same time, however, weakening and killing the host could endanger the parasite, because a dead host can no longer nourish the parasite. Thus, natural selection also favors parasites that do not have a rapid fatal effect on the host. Overall, then, parasites experience both selection to become more deadly and selection to become less deadly.

Which type of selection has the strongest effect on the evolution of a given parasite species? It depends in part on how the parasite is transmitted to new hosts. In particular, if transmission to a new host requires direct host-to-host contact, then selection will favor parasites whose hosts remain at least healthy enough to move about and come into contact with other potential hosts. But if a parasite has a way to be transported to a new host from even a very sick or dead host (such as by a tick or a mosquito), then exploiting the host's nutrients and reproducing as fast as possible can be a winning evolutionary strategy.

The bacterium that causes epidemic typhus is a good example of a parasite that thrives by being deadly. This bacterium is transmitted from person to person by lice (**FIG. E28-2**). People often die within a few weeks of being infected with typhus, yet the bacteria continue to thrive. How? Lice that feast on infected people leave after their hosts develop high fevers. The departing lice seek out other, healthy people as new hosts, bringing typhus along. A louse may thereby transmit typhus to a new host. Other lice may then bite the newly infected person, ingesting the typhus bacterium and starting the cycle over again. Rapid bacterial reproduction at the expense of its human host's health helps, rather than harms, the typhus bacterium.

Another disease that has evolved to become especially lethal is cholera. Cholera causes severe diarrhea. People with the disease quickly become too sick to move, but their feces, which contain cholera bacteria, may enter the water supply, especially in areas with poorly developed sanitation systems. Once in the water supply, the bacteria may be ingested by new hosts. Because the bacteria that cause cholera can be transmitted from even a very sick host, they have evolved to reproduce rapidly regardless of harm to the host.

Sometimes humans coevolve with their parasites, resulting in an uneasy truce. An example of such a parasite is *Helicobacter pylori*, a bacterium that colonizes the human stomach and often causes ulcers, sometimes even cancer. More than 3 billion people have *Helicobacter* in their stomachs. Strains of *Helicobacter* that have coevolved with particular human groups seem to cause little harm to the population they evolved with but may be deadly to outsiders. For example, Tumaco and Tuquerres, two villages in Colombia separated by only 125 miles, have a 25-fold difference in the incidence of stomach cancer, most of which is caused by *Helicobacter*. Researchers discovered that Tumaco, with a very low rate of stomach cancer, is mostly populated by the descendants of freed African slaves whose stomachs contain *Helicobacter* strains that originated in Africa. In contrast, Tuquerres, which has an extremely high rate of stomach cancer, is populated mostly by Native Americans whose *Helicobacter* strains are European, probably acquired originally from Spanish conquistadors. The researchers hypothesize that coevolution of African people with African *Helicobacter* has reduced their cancer risk, while the nearby Native Americans struggle to cope with introduced European *Helicobacter*.

▲ **FIGURE E28-2 A human body louse feeds on the blood of its host** Infected lice leave behind feces packed with the bacteria that cause epidemic typhus.

Humans and *Helicobacter* may even be coevolving into mutualism. In the human stomach, *Helicobacter* gets lots of nutrition. People infected with *Helicobacter* seem to benefit by having a lower incidence of acid reflux disease, as well as lower rates of asthma, allergies, and chronic inflammation. These benefits are likely a result of *Helicobacter* dampening its host's immune response.

THINK CRITICALLY Malaria is a common, often lethal, disease of tropical countries, with about 200 million cases worldwide, resulting in more than 600,000 deaths. Malaria is transmitted to people when they are bitten by infected mosquitoes. In many parts of the Tropics, the dominant malaria-carrying mosquitoes feed mostly indoors, at night. Therefore, an effective tool for reducing malaria infections is an insecticide-treated bednet, which protects people while they sleep. How might the mosquitoes evolve in response to widespread use of bednets? Do you think that bednets can permanently reduce transmission of malaria by mosquitoes?

benefit themselves by sipping nectar and sometimes eating pollen. The relationship is mutualistic, even though the partners spend little time together.

CHECK YOUR LEARNING

Can you . . .

- describe some mutualistic interactions and how they benefit the participants?
- explain why some mutualisms are not considered symbiotic?

28.5 HOW DO KEYSTONE SPECIES INFLUENCE COMMUNITY STRUCTURE?

In some communities, a particular species, called a **keystone species**, plays a major role in determining community structure—a role that is out of proportion to the species' abundance in the community. If the keystone species is removed, community interactions are significantly altered and the relative abundance of other species often changes dramatically.

In the African savanna, the bush elephant is a keystone species. By grazing on small trees and bushes (**FIG. 28-16a**), elephants prevent the encroachment of forests and help maintain the grassland community, along with its diverse population of grazing mammals and their predators.

Although elephants alter plant communities directly, large carnivores, such as wolves and cougars (**FIG. 28-16b**), may also have a major impact on plant communities. By keeping populations of deer and elk in check, wolves and cougars can help maintain the health of forests and stream banks that would otherwise be overgrazed. Because this vegetation provides food, nesting sites, and shelter for many species of smaller animals, the entire community structure depends on the presence of the predators.

Identifying a keystone species can be difficult. Many have been recognized only after their loss has had dramatic, unforeseen consequences. For example, the near-extinction of northern sea otters (**FIG. 28-16c**) along the coast of Alaska resulted in the destruction of kelp forest communities. (Long blades of kelp seaweed, anchored to the seafloor and floating upward, play a role similar to that of trees in a terrestrial forest, providing food, shelter, and protection to a diverse community of fish, mollusks, and crustaceans.) Beginning in the mid-1700s, fur hunters killed hundreds of thousands of sea otters, and by the early 1900s, the otters were nearly extinct. Fortunately, sea otter hunting was banned in 1911, and by the 1970s, the population had rebounded. The recovery, however, was uneven. In some areas, otter populations were large, but other areas supported only small populations or no sea otters at all. Ecologists found that areas with many sea otters also had extensive kelp forests, but areas without otters tended to lack kelp forests. The link between otters and kelp was sea urchins. These spiny animals eat underwater plants and algae, including kelp. They are a favorite food of sea otters, so urchin populations are small where otters are common. But in areas with few or no otters, sea urchins are abundant and consume kelp faster than it can grow. Thus, in the mid-1990s, when otter numbers plummeted once again due to increased predation by killer whales, the number of sea urchins again exploded, and they again rapidly deforested the seafloor. Clearly, sea otters are a keystone species in Alaskan coastal waters.

CASE STUDY CONTINUED
The Fox's Tale

Foxes are a keystone species on the Channel Islands. Predation by foxes controls populations of deer mice and ground-nesting birds. Foxes control skunk populations by competing with them for food. Foxes also disperse the seeds of many island plants by eating fruits and expelling the undigested seeds in their feces, often a considerable distance from the parent plant. When fox populations plummeted while pig populations skyrocketed, community interactions changed dramatically, and ecological communities on many parts of the islands became barely recognizable shadows of their former selves. Can such disturbed communities ever recover?

(a) African elephant

(b) Cougar

(c) Northern sea otter

▲ **FIGURE 28-16 Keystone species (a)** The bush elephant is a keystone species on the African savanna. **(b)** The cougar, found in isolated habitats throughout North and South America, helps to control herbivores such as deer. **(c)** A northern sea otter rests in a kelp bed while cradling a sea urchin, one of its favorite foods.

CHECK YOUR LEARNING

Can you . . .

- explain the concept of a keystone species?
- provide some examples of the effects that removing keystone species can have on their communities?

28.6 HOW DO SPECIES INTERACTIONS CHANGE COMMUNITY STRUCTURE OVER TIME?

Ecological communities emerge in stages over a long period of time. These stages constitute **succession**: a gradual change in a community and its nonliving environment in which groups of species replace one another in a reasonably predictable sequence. The precise changes during succession are as diverse as the environments in which it occurs, but most succession follows the same overall pattern. It begins with **pioneers**, hardy invaders that alter the environment in ways that favor the species that will eventually displace them, and progresses in stages to a diverse and relatively stable **climax community**.

Our discussion of succession will focus on plant communities, which dominate the landscape and provide both food and habitat for animals, fungi, and microorganisms.

There Are Two Major Forms of Succession: Primary and Secondary

During **primary succession**, a community gradually forms in a place where there are no remnants of a prior community and often no life at all. The disturbance that sets the stage for primary succession may be a glacier that scours the landscape down to bare rock or a volcano that produces a new island in the ocean or deposits a layer of lava on land that hardens into rock (**FIG. 28-17a**). Building a community from scratch

(a) Kilauea, in Hawaii (primary succession)

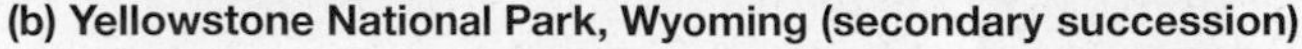

(b) Yellowstone National Park, Wyoming (secondary succession)

▲ **FIGURE 28-17 Succession in progress (a)** Primary succession. (Left) The Hawaiian volcano Kilauea has erupted repeatedly since 1983, sending rivers of lava over the surrounding countryside. (Right) A pioneer fern takes root in a crack in hardened lava. **(b)** Secondary succession. (Left) In the summer of 1988, extensive fires swept through the forests of Yellowstone National Park in Wyoming. (Right) Trees and flowering plants are thriving in the sunlight, and wildlife populations are increasing as secondary succession occurs.

THINK CRITICALLY People have suppressed fires for decades. How might fire suppression affect forest ecosystems and succession?

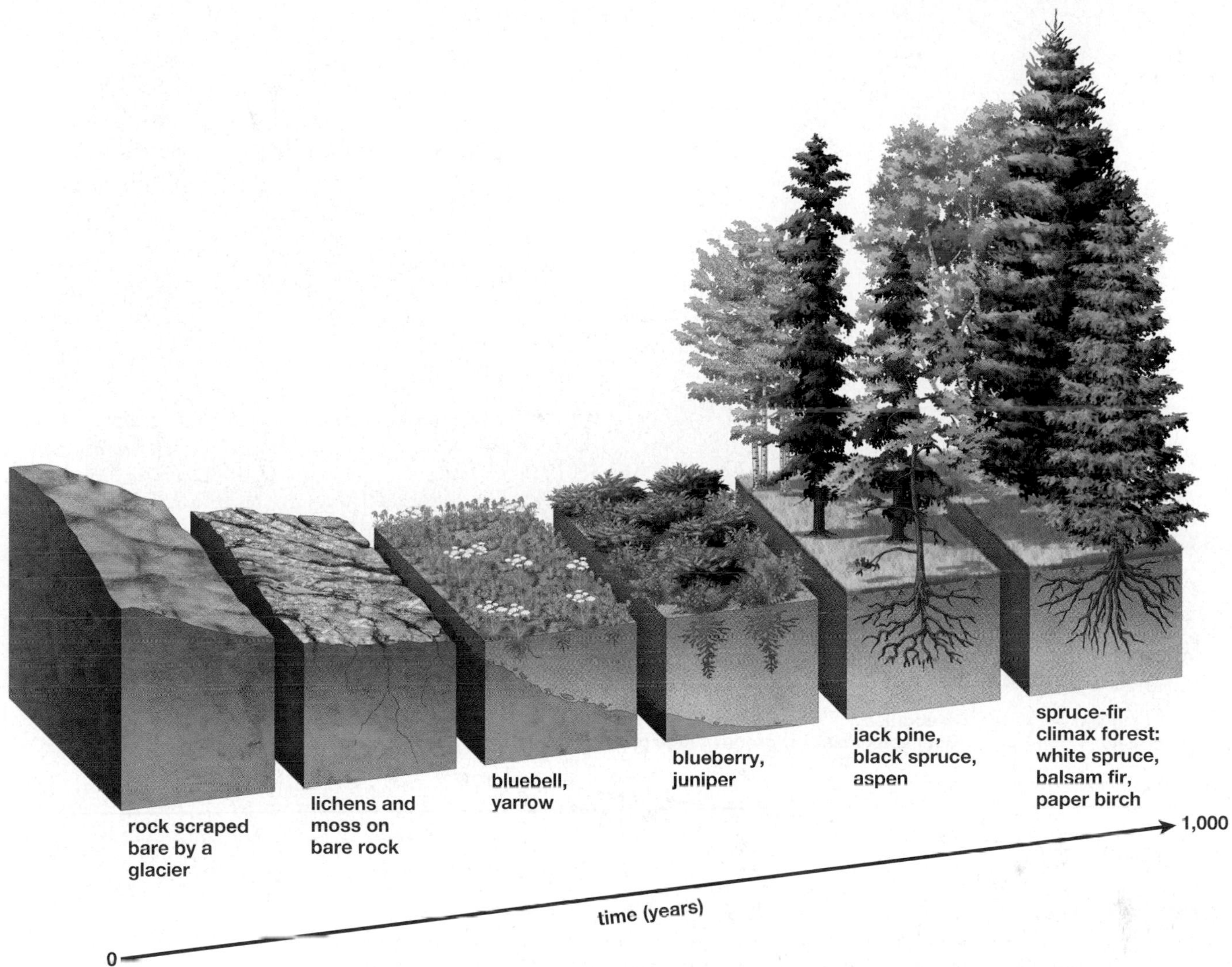

▲ **FIGURE 28-18 Primary succession** Primary succession on bare rock exposed as glaciers retreated from Isle Royale in Lake Superior in upper Michigan. Notice that the soil deepens over time, gradually burying the bedrock and allowing trees to take root.

through primary succession typically requires thousands or even tens of thousands of years.

During **secondary succession**, a new community develops after an existing ecosystem is disturbed in a way that leaves significant remnants of the previous community behind. For example, an abandoned farm typically has fertile soil and seeds from both crops and weeds. A clear-cut forest usually has soil, shrubs, and small trees remaining. Even a forest fire leaves behind the ingredients for secondary succession, such as residues of burnt trees that are high in plant nutrients, and seeds that can withstand fire or may even require it in order to sprout. Thus, fires may promote rapid regeneration of forests and other communities (**FIG. 28-17b**).

Primary Succession May Begin on Bare Rock

FIGURE 28-18 illustrates primary succession on Isle Royale, Michigan, an island in northern Lake Superior that was scraped down to bare rock by glaciers that retreated roughly 10,000 years ago. Bare rock is weathered by cycles of freezing and thawing that crack rocks and erode their surface layers, producing small particles. Rainwater dissolves some of the minerals from the rock particles, making the minerals available to pioneers.

Pioneering lichens attach to the weathered rock, obtaining energy through photosynthesis and secreting an acid that frees up additional mineral nutrients by dissolving rock. As the lichens spread over the rock surface, drought-tolerant mosses begin growing in cracks. Fortified by nutrients liberated by the lichens, the mosses form a dense mat that traps dust, tiny rock particles, and small bits of organic debris. The mosses eventually cover and kill many of the lichens that made their growth possible.

The mat of mosses acts like a sponge, absorbing and trapping moisture. As mosses die and decompose, their bodies add nutrients to a thin layer of new soil. Within the moss mat, seeds of larger plants, such as bluebell and yarrow, germinate. When these plants die, their decomposing bodies further thicken the layer of soil.

As woody shrubs such as blueberry and juniper take advantage of the deeper soil, the mosses and remaining

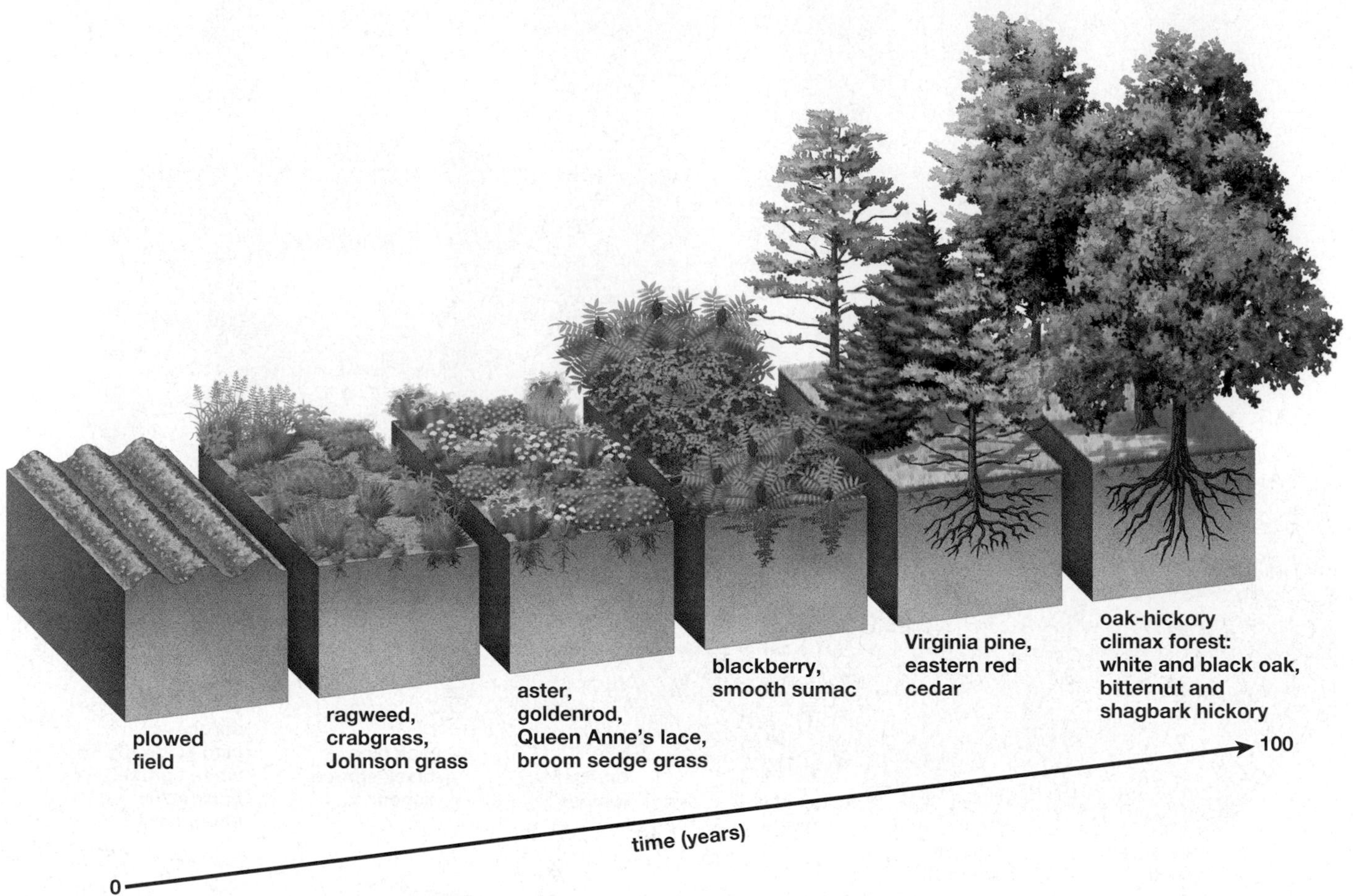

▲ **FIGURE 28-19 Secondary succession** Secondary succession as it might occur on a plowed, abandoned farm field in North Carolina, in the southeastern United States. Notice that a thick layer of soil is present from the beginning, which greatly speeds up the process compared to primary succession.

THINK CRITICALLY If a disturbed ecosystem is simply left alone, which environmental conditions tend to favor natural succession, and which conditions may slow succession or even prevent the ecosystem from recovering at all?

lichens may be shaded out or buried by decaying leaves and vegetation. Eventually, trees such as jack pine, black spruce, and aspen take root in the deeper crevices, and the sun-loving shrubs are shaded out. Within the forest, shade-tolerant seedlings of taller or faster-growing trees thrive, including balsam fir, paper birch, and white spruce. In time, these trees tower over and replace the earlier, shade-intolerant trees. After hundreds of years, a forest thrives on what was once bare rock.

An Abandoned Farm Undergoes Secondary Succession

FIGURE 28-19 illustrates secondary succession on an abandoned farm in the southeastern United States. The pioneer species are sun-loving, fast-growing plants such as ragweed, crabgrass, and Johnson grass. Such species generally produce large numbers of easily dispersed seeds that help them colonize open spaces. However, they don't compete well against longer-lived species that grow larger over the years and shade them out, so after a few years, plants such as asters, goldenrod, Queen Anne's lace, and perennial grasses take over, followed by woody shrubs such as blackberry and smooth sumac. About 20 years after a field is abandoned, the site becomes an evergreen forest dominated by pines. However, the new forest alters conditions in ways that favor its successors. The shade of the pine forest inhibits the growth of its own seedlings while favoring the growth of hardwood trees, whose seedlings are shade tolerant and can grow beneath the pines. After about 70 years, slow-growing hardwoods, such as oak and hickory trees, begin to replace the aging pines. Roughly a century after the field was abandoned, the former farm is covered by relatively stable forest dominated by oak and hickory.

Succession Also Occurs in Ponds and Lakes

Succession in freshwater ponds and lakes occurs as soil and rock particles, washed in from the shore or carried in by streams entering the pond, settle to the bottom. This sediment gradually makes the pond more and more shallow. Eventually, moisture-loving plants take root along the shore. As more sediment fills the pond, the shoreline slowly moves toward the center of the pond. The soil of the original shoreline becomes drier, and plants from the surrounding land take over (**FIG. 28-20**). Eventually, no open water remains, and the pond has been converted to a grassy meadow. If the meadow is in a forest, trees encroach around the meadow's edges, eventually converting the meadow to forest.

Succession Culminates in a Climax Community

The end point of succession is a reasonably stable climax community, which perpetuates itself if it is not disturbed by external forces (such as fire, parasites, invasive species, or human activities). The populations within a climax community have ecological niches that allow them to coexist without supplanting one another. In general, climax communities have more species and more community interactions than do communities in early stages of succession. The plant species that dominate climax communities generally live longer and tend to be larger than pioneer species, particularly in climax forests.

The makeup of a climax community is determined by geological and climatic factors—including temperature, rainfall, elevation, latitude, type of rock, and exposure to sun and wind—that differ dramatically in different areas. For example, if you drive through Colorado or Wyoming, you will see pine-spruce forests in the mountains, tundra on the mountain summits, and sagebrush-dominated communities in the western valleys.

Some Communities Are Maintained in Subclimax Stages

In some circumstances, succession does not reach the climax stage and a community remains in an earlier, **subclimax** stage. Frequent disturbances can maintain a community in subclimax stages for very long periods of time. The tallgrass prairie that once covered northern Missouri and most of Illinois was a subclimax stage of an ecosystem whose climax community is deciduous forest. The subclimax prairie was maintained by periodic fires, some set by lightning and others deliberately set by Native Americans to provide grazing land for bison. Today, people use controlled burns to maintain some tallgrass prairie preserves.

▼ **FIGURE 28-20 Succession in a freshwater pond** This pond is rapidly being converted to dry land. Although there is a little open water in the center of the pond, water lilies dominate the pond, showing that it is only a few feet deep in most places. Shrubs and small trees are beginning to grow around the drier edges of the pond.

Farms, gardens, and lawns are subclimax communities maintained by frequent, intentional disturbance. For example, cultivated fields of grains are maintained in an early stage of succession dominated by grasses such as wheat, corn, and rice. To maintain this early stage of succession, farmers may plow the soil, plant seeds, and use herbicides to prevent the grass crop from being displaced by competitors such as weeds, wildflowers, and woody shrubs.

CHECK YOUR LEARNING

Can you . . .

- explain the process of succession and its general stages?
- describe primary succession and secondary succession?
- explain what a climax community is and what a subclimax community is?

CASE STUDY **REVISITED**

The Fox's Tale

In the 1990s, the ecological health of the Channel Islands was in steep decline. Landscapes were scarred, native plants were replaced by invasive weeds, and the island fox population was decimated. Could the damaged communities of the Channel Islands be restored? Betting that the answer was yes, the National Park Service coordinated an ambitious restoration effort, focusing first on the largest island, Santa Cruz.

Restoration ecologists began by removing all cattle from Santa Cruz Island, and then turned their attention to the more than 20,000 feral sheep that also roamed the island. By the late 1990s, all the sheep had been removed. Skilled hunters then tracked the feral pigs that plagued the island, and the last pig was eliminated in 2007. Finally, the island's golden eagles, which had colonized Santa Cruz and had become major predators of the foxes, were all trapped and relocated to the mainland. A few bald eagles were successfully reintroduced, in hope that they would keep golden eagles from returning.

Free of predators, the island fox population has rebounded—Santa Cruz alone is now home to about 1,300 to 1,500 foxes, and the combined fox population on the islands is more than 6,000. Now that the pigs are gone, many native plants are making a comeback. Acorns, a favorite food of pigs, once again sprout into oak seedlings. Campaigns to control invasive weeds have helped the recovery of some of the Channel Islands' endangered plants, although several had already become extinct. The Channel Islands will probably always bear signs of past disturbance, but future generations of visitors may see nearly intact, native climax communities once again.

THINK CRITICALLY On another group of islands, the Gulf Islands off the coast of British Columbia, raccoons thrive along the shoreline. When ecologists deployed loudspeakers there, and broadcast the sound of barking dogs for several hours per day over the course of a month, they found that in the intertidal zone, populations of crabs, worms, snails, and small fish increased significantly. Propose a hypothesis to explain why.

CHAPTER REVIEW

Go to **Mastering Biology** to access the Pearson eText, vocabulary review, practice quizzes, activities, videos, current events, and more.

*Answers to **Think Critically and Thinking Through the Concepts** questions can be found in the **Answers** section at the back of the book.*

Summary of Key Concepts

28.1 How Do Species in Communities Interact?

Ecological communities consist of all the interacting populations of different species within an ecosystem. Community interactions include interspecific competition, in which both participating species are harmed; consumer–prey interactions, in which one species benefits at the expense of another; and mutualism, in which both participating species benefit. Populations of interacting species act as agents of natural selection on one another, resulting in coevolution.

28.2 How Does Interspecific Competition Affect Communities?

The ecological niche includes all aspects of a species' habitat and interactions with its living and nonliving environments. Each species occupies a unique ecological niche. Interspecific competition occurs whenever the niches of two species within a community overlap. When two species with the same niche are forced (under laboratory conditions) to occupy the same ecological niche, one species always outcompetes the other. Species in natural communities have evolved in ways that avoid extensive niche overlap, with behavioral and physical adaptations that allow resource partitioning. Interspecific competition limits both the population size and the distribution of competing species.

28.3 How Do Consumer–Prey Interactions Shape Evolutionary Adaptations?

Predators and parasites are consumers of other organisms. In general, predators are free-living organisms that are less abundant than their prey, while parasites live in or on their hosts for a significant part of their life cycle and are much more abundant than their prey. Both types of consumers and their prey act as strong agents of natural selection on one another. Predators and prey have evolved a variety of toxic chemicals for attack and defense. Plants have evolved defenses ranging from poisons to overall toughness. These defenses, in turn, have selected for predators that can detoxify poisons and grind down tough tissues. Many prey animals have evolved protective colorations that render them either inconspicuous (camouflage) or startling (startle coloration) to their predators. Some prey are poisonous, distasteful, or venomous and exhibit warning coloration by which they are readily recognized and avoided by predators. Some harmless species have evolved to resemble distasteful organisms. Parasite–host coevolution includes the mammalian immune response against disease microbes and the many ways that these microbes have evolved to avoid detection or destruction by the immune response.

28.4 How Do Mutualisms Benefit Different Species?

Mutualism benefits two or more interacting species. Some mutualistic interactions are symbiotic, such as those between cows and their cellulose-digesting microorganisms. Other mutualistic interactions are more temporary, such as those that occur between plants and the animals that pollinate them.

28.5 How Do Keystone Species Influence Community Structure?

Keystone species have a greater influence on community structure than can be predicted by their abundance. If a keystone species is removed from a community, the structure of the community is significantly altered.

28.6 How Do Species Interactions Change Community Structure over Time?

Succession is a change in a community and its nonliving environment over time. During succession, plants alter the environment in ways that favor certain competitors, thus producing a somewhat predictable progression of dominant species. Primary succession, which may take thousands of years, occurs where no remnant of a previous community exists, such as on bare rock. Secondary succession occurs much more rapidly because it builds on the results of primary succession or on the remains of a disrupted community, such as an abandoned field or the remnants of a forest after a fire. Uninterrupted succession ends with a climax community, which tends to be self-perpetuating unless acted on by outside forces. Some ecosystems, including tallgrass prairie and farm fields, are maintained in relatively early, subclimax stages of succession by periodic disruptions.

Thinking Through the Concepts

Bloom's: Remembering, Understanding

Multiple Choice

1. The orderly progress of communities, starting from bare rock with no soil or traces of a previous community, is called
   a. primary succession.
   b. secondary succession.
   c. interspecific succession.
   d. niche succession.
2. Which of the following statements is *not* true of an ecological niche?
   a. A niche includes the physical environment, such as climate and water availability.
   b. A niche includes predators and parasites.
   c. Niche overlap may help to drive evolution.
   d. Two species cannot have overlapping niches.
3. A keystone species
   a. is always a carnivore.
   b. has a minor influence on community structure.
   c. has a major influence on community structure out of proportion to its abundance.
   d. has a major influence on community structure only when present in large numbers.
4. Which of the following statements regarding invasive species is true?
   a. All introduced species are invasive.
   b. Almost all invasive species are plants.
   c. Invasive species have no predators or parasites in their new ecosystem.
   d. Invasive species multiply rapidly and may displace native species in a community.
5. Which of the following statements about predators and parasites is *not* true?
   a. Predators and parasites harm their prey.
   b. Predators live on or in their hosts for a significant part of their life cycle.
   c. Predators and parasites coevolve with their prey.
   d. Parasites may kill their prey.

Fill-in-the-Blank

1. Species that interact serve as agents of ____________ on one another. This results in ____________, the process by which two species evolve adaptations in response to their interspecific interactions.
2. Predators may be meat-eaters, called ____________, or plant-eaters, called ____________. Both predators and their prey may blend into their surroundings by using ____________.
3. The concept that no two species with identical niches can coexist indefinitely is called the ____________ principle.
4. Fill in the types of coloration or mimicry: used by prey to signal that it is distasteful: ____________; used by a moth with large eyespots on its wings: ____________; mimicry of a poisonous animal by a nonpoisonous animal: ____________; mimicry used by a predator to attract its prey: ____________.
5. Fill in the appropriate type of community interaction: bacteria, living in the human gut, that synthesize vitamin K: ____________; bacteria that cause illness: ____________; a deer eating grass: ____________; a bee pollinating a flower: ____________.
6. A somewhat predictable change in community structure over time is ____________. This process takes two forms. Which of these forms would start with bare rock? ____________ Which would occur after a forest fire? ____________ A relatively stable community that is the end product of this process is called a ____________ community. A mowed lawn in suburbia is an example of a ____________ community.

Review Questions

1. Define an *ecological community*, and describe the three major categories of community interactions, including the benefits and harms to the interacting species.
2. Explain how resource partitioning is a logical outcome of the competitive exclusion principle.
3. Describe examples of coevolution between consumers and their prey.
4. Define *succession*. Which type of succession would occur at the site of a clear-cut forest (where all trees have been logged) and why?
5. Provide examples of two climax and two subclimax communities. How do they differ?
6. What is an invasive species? Why are they destructive? What general adaptations do invasive species possess?
7. What is a keystone species? How can a keystone species within a community be identified?

Applying the Concepts

Bloom's: Applying, Analyzing, Evaluating

1. An ecologist visiting an island finds two species of birds, one of which has a slightly larger bill than the other. Interpret this finding with respect to the ecological niche concept and the competitive exclusion principle.
2. Why is succession hard to study? Choose a specific location, and describe how you would study succession there.
3. An ecologist observes a similar, brightly-colored wing pattern in two species of butterfly that occupy the same habitat, and she performs an experiment that shows that the similarity is due to mimicry. Describe an additional experiment she might perform to test if the mimicry is Müllerian or Batesian, using butterfly-eating birds that have not previously encountered the two species. What is the expected outcome for each hypothesis?

29

Energy Flow and Nutrient Cycling in Ecosystems

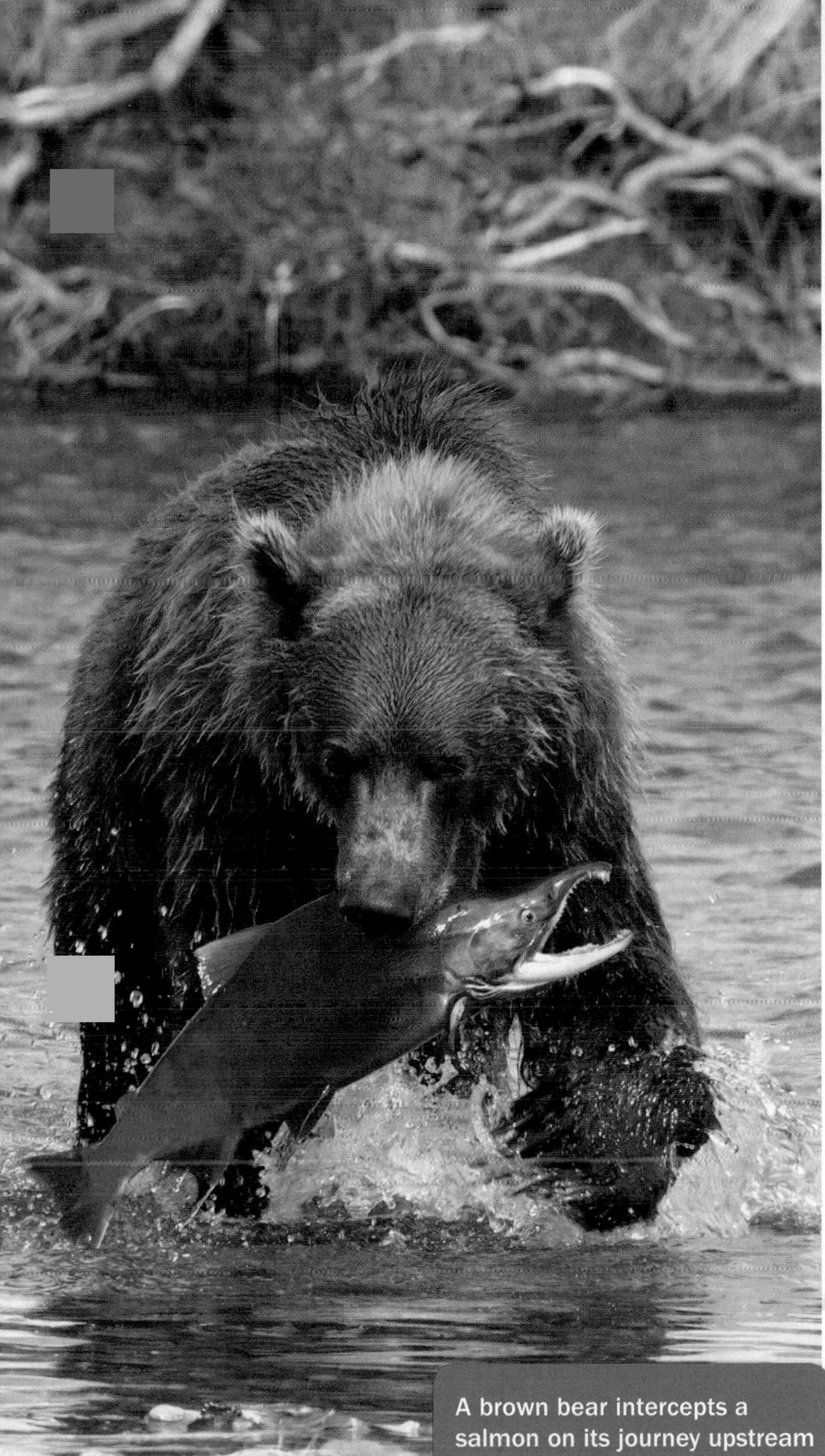

A brown bear intercepts a salmon on its journey upstream to spawn.

CASE STUDY

Dying Fish Feed an Ecosystem

SOCKEYE SALMON in Alaska have a remarkable life cycle that takes them from a swiftly flowing stream to the open ocean and back again. After emerging from eggs laid on the streambed, juvenile salmon remain in fresh water for 1 to 3 years, often in a nearby lake. Then the young fish begin the physiological changes that will prepare them for life in the ocean. They head downstream to estuaries, where fresh water and salt water mix. In the estuary, the salmon complete their transformation and then head out to sea.

In the ocean, the young salmon grow rapidly, feeding on small fish and crustaceans. A few years later, when they reach sexual maturity, an instinctive drive compels them to return to their home streams to spawn. If they are lucky enough to escape the jaws of brown bears and the talons of eagles on their journey, the salmon reach their spawning site, carrying a precious payload of sperm and eggs. The salmon die soon after spawning.

The fishes' trip back to their birthplace is remarkable in another way: Nutrients almost always move downstream, washed from the land into the ocean. But salmon, filled with muscle and fat acquired from feeding in the ocean, bring ocean nutrients back to the land. They bring energy upstream, too. This upstream movement of nutrients and energy supports the extraordinarily rich terrestrial ecosystems of the southern coast of Alaska. Where do nutrients and energy originate? How do living organisms acquire and transfer nutrients and energy?

AT A GLANCE

29.1 HOW DO NUTRIENTS AND ENERGY MOVE THROUGH ECOSYSTEMS?

An **ecosystem** consists of all the communities in a defined area, along with their nonliving environment. Thus, an ecosystem includes both a living **biotic** component and a nonliving **abiotic** component. Both nutrients and energy move through ecosystems, but nutrients cycle within and between ecosystems, whereas energy flows through ecosystems and is not recycled (**FIG. 29-1**).

Nutrients are atoms and molecules that organisms obtain from their environment. Although these chemicals may be transported around the planet or converted to different molecular forms, they do not leave Earth and its atmosphere, nor is their supply increased (except for small amounts in meteorites that reach Earth). The nutrient atoms now on Earth are the same ones that have been sustaining life for about 3.5 billion years. Your body contains oxygen, carbon, hydrogen, and nitrogen atoms that were once part of a dinosaur or a woolly mammoth.

Unlike nutrients, energy takes a one-way journey through ecosystems. Solar energy is captured by photosynthetic bacteria, algae, and plants and then moves from organism to organism. However, because the chemical reactions that occur in organisms are inefficient (see Chapter 6), most of the captured energy that flows through an ecosystem is ultimately converted to heat and lost to the environment. Therefore, life on Earth requires a continuous input of energy.

CHECK YOUR LEARNING

Can you . . .

- explain why nutrients cycle within and between ecosystems, whereas energy flows through ecosystems?

29.2 HOW DOES ENERGY FLOW THROUGH ECOSYSTEMS?

Thermonuclear reactions in the sun transform matter into enormous quantities of energy, much of it in the form of electromagnetic radiation, including heat (infrared light), visible light, and ultraviolet light. Because the sun releases energy in all directions and is 93 million miles from Earth, only a tiny fraction of the sun's energy (about 45 billionths of a percent) reaches our planet. Much of this fraction is absorbed or reflected by Earth's clouds and atmosphere. Of the portion that reaches Earth's surface as visible light, a small percentage is used in photosynthesis.

Energy Enters Ecosystems Through Photosynthesis

Organisms that can manufacture their own food are called **producers,** or **autotrophs** (Greek, meaning "self-feeders"). Almost all of Earth's producers are photosynthetic organisms that capture energy in sunlight and use it to join nutrient atoms into biological molecules, thereby storing some of the sun's energy in the chemical bonds of these molecules (see Chapter 7). However, in a few ecosystems, such as the deep-sea ecosystems that flourish near hot-water-spewing vents in the ocean floor, the producers are prokaryotes that produce food using energy stored in the chemical bonds of inorganic compounds such as hydrogen sulfide.

Energy Passes from One Trophic Level to the Next

As producers manufacture food for themselves, they also directly or indirectly produce food for nearly all other organisms as well. Organisms that cannot produce their own food, called **consumers** or **heterotrophs**, must acquire energy from the bodies of other organisms. Energy flows through communities from producers through several levels of consumers. Each category of organism through which energy passes is called a **trophic level**. Producers form the first trophic level; they generally produce their own food by converting the energy in sunlight to chemical energy. **Primary consumers** feed directly on producers. These **herbivores** (from Latin words meaning "plant-eaters") include animals such as grasshoppers, mice, and zebras, and form the second trophic level. **Carnivores** ("meat-eaters"), such as spiders, hawks, and salmon, are higher-level consumers. Carnivores that prey on herbivores act as **secondary consumers** and form the third trophic level. The carnivores that eat other carnivores act as **tertiary consumers** and occupy the fourth trophic level. Some ecosystems, particularly in the oceans, may include even higher trophic levels.

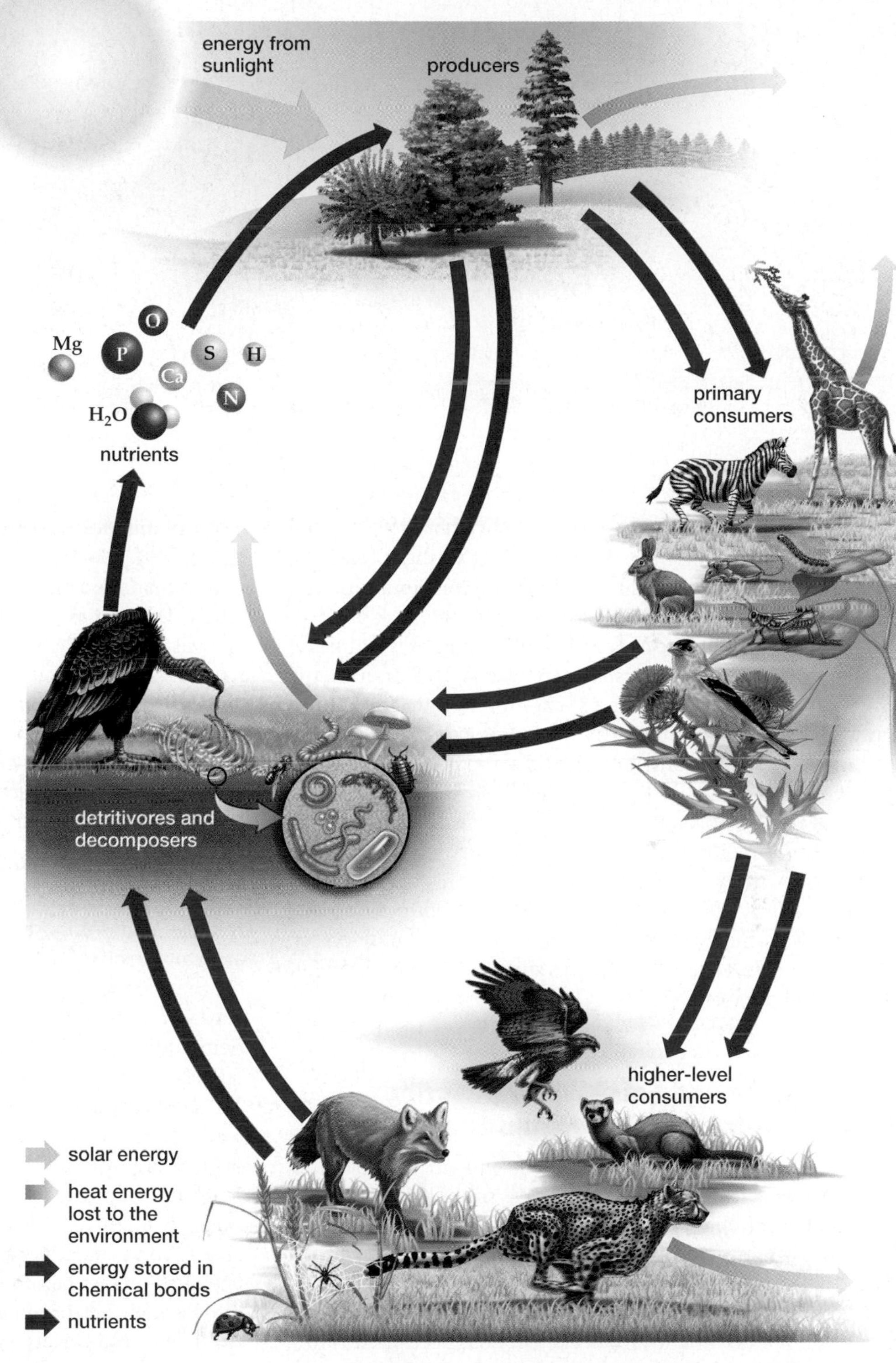

FIGURE 29-1 Energy flow, nutrient cycling, and feeding relationships in ecosystems The energy of sunlight (yellow arrow) enters an ecosystem during photosynthesis by organisms called producers, which store some of the energy in the biological molecules of their bodies. This biologically available energy (red arrows) is then passed to nonphotosynthetic organisms called consumers. Energy in wastes and dead bodies supports detritivores and decomposers. Every organism loses some energy as heat (orange arrows), so useful energy gradually becomes unavailable to living organisms. Therefore, ecosystems require a continuous input of energy. In contrast, nutrients (purple arrows) are recycled.

Net Primary Production Is a Measure of the Energy Stored in Producers

Because all of the energy used by consumers at all trophic levels originally comes from producers, the amount of life that an ecosystem can support is determined by the amount of energy captured by its producers. The amount of energy that photosynthetic organisms in a given area store in their bodies over a given period of time is called **net primary production**. Net primary production is often expressed in terms of the dry weight of biological material, or **biomass**, because it's easier to measure organisms' mass than to measure the energy they contain, and biomass is generally a good approximation of the energy stored in organisms' bodies. Thus, net primary production is usually given as grams of biomass per square meter per year.

FIGURE 29-2 Net primary production in ecosystems The average net primary production of some terrestrial and aquatic ecosystems is shown, measured in grams of biological material produced per square meter per year.

THINK CRITICALLY What factors contribute to the differences in primary production among ecosystems?

The net primary production of an ecosystem is influenced by many factors, including the amount of sunlight reaching the producers, the availability of water and nutrients, and the temperature (**FIG. 29-2**). In deserts, for example, lack of water limits production. In the open ocean, lack of light is a limiting factor in deep waters. In most surface waters, lack of nutrients limits production. In ecosystems where all resources are abundant, such as tropical rain forests, production is high.

An ecosystem's contribution to Earth's total production is determined both by the ecosystem's production per unit area and by the portion of Earth that the ecosystem covers. The oceans generally have low net primary production, so even though they cover about 70% of Earth's surface, they contribute only about 25% of Earth's total production. This is about the same overall contribution as tropical rain forests, which cover less than 5% of Earth's surface, but have high net primary production.

Food Chains and Food Webs Describe Feeding Relationships Within Communities

In illustrations of feeding relationships in an ecosystem, it is common to identify a representative of each trophic level such that each representative species eats another on the level below it. This linear feeding relationship is called a **food chain** (**FIG. 29-3**). Different types of ecosystems may support radically different food chains. In land-based (terrestrial) ecosystems, plants are the dominant producers; they support plant-eating insects, reptiles, birds, and mammals, each of which may be preyed on by other animals (**FIG. 29-3a**). In most aquatic food chains, the dominant producers are microscopic photosynthetic protists and bacteria, collectively called **phytoplankton**; they support a diverse group of consumers called **zooplankton,** which consist mainly of protists and small shrimp-like crustaceans. These primary consumers are eaten primarily by fish, which in turn are eaten by larger fish (**FIG. 29-3b**).

Animals in natural communities seldom fit neatly into simple food chains. Instead, the actual feeding relationships in a community form a **food web** of many interconnecting food chains (**FIG. 29-4**). Some animals, such as raccoons, bears, rats, and humans, are **omnivores** ("everything-eaters") that at different times act as primary, secondary, and occasionally tertiary consumers. A raccoon, for instance, is a primary consumer when it eats fruits and nuts, a secondary consumer when it eats grasshoppers, and a tertiary consumer when it eats frogs or carnivorous fish. A carnivorous plant such as the Venus flytrap can tangle the food web further by acting as both a photosynthesizing producer and an insect-trapping secondary consumer.

Detritivores and Decomposers Recycle Nutrients

Among the most important strands in a food web are the detritivores and decomposers. **Detritivores** ("debris-eaters") are an army of mostly small and often unnoticed organisms that feed on the wastes, dead bodies, or discarded parts (such as fallen leaves or molted skin) of other organisms. Detritivores include nematode worms, earthworms, millipedes, dung beetles, the larvae of some flies, and a few large vertebrates such as vultures.

Decomposers are primarily fungi and bacteria. They feed mostly on the same material as detritivores, but they do not ingest chunks of organic matter, as detritivores do. Instead, they secrete digestive enzymes and break down organic material outside their bodies. The decomposers absorb some of the resulting nutrient molecules, but many of the nutrients remain in the environment.

Because detritivores and decomposers recycle nutrients from all trophic levels back to the soil, water, and air, they are absolutely essential to life on Earth. They reduce the bodies and wastes of other organisms to simple molecules, such as carbon dioxide, ammonia, and minerals. Without detritivores and decomposers, ecosystems would gradually be buried by accumulated wastes and dead bodies, whose nutrients would be unavailable to other living organisms, including producers. If the producers died from a lack of nutrients, they would no longer capture energy from sunlight and store it in chemical bonds, and organisms at higher trophic levels, including people, would die as well.

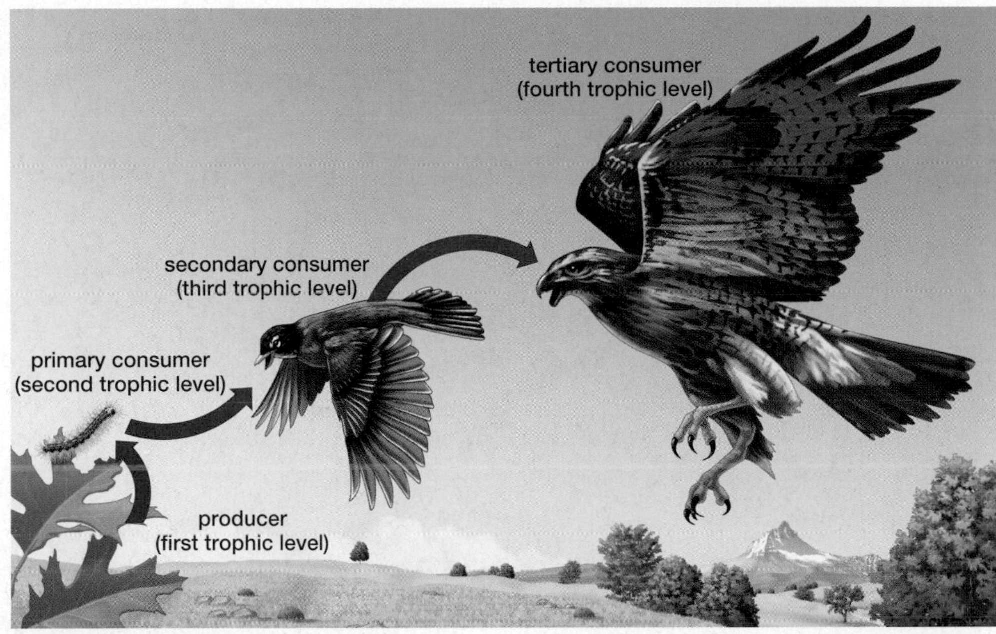

(a) A simple terrestrial food chain

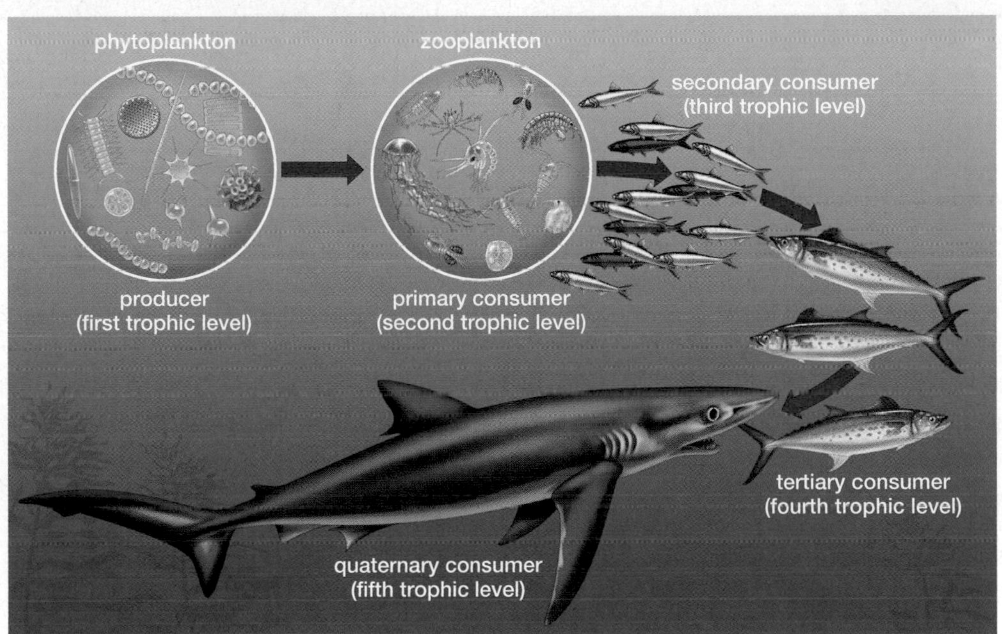

(b) A simple marine food chain

▲ **FIGURE 29-3 Food chains on land and sea**

Energy Transfer Between Trophic Levels Is Inefficient

A fundamental principle of thermodynamics is that energy use is never completely efficient (see Chapter 6). For example, as your car burns gasoline, about 80% of the energy released is immediately lost as heat and cannot be used to do the work of moving the car. The same inefficiency also applies in living systems. Waste heat is produced by all the biochemical reactions that keep cells alive. For example, splitting the chemical bonds of ATP to power muscle contraction releases heat; that is why shivering or walking briskly on a cold day warms your body.

Energy transfer from one trophic level to the next is also inefficient. When a grasshopper (a primary consumer) eats grass (a producer), only a portion of the solar energy originally trapped by the grass is available to the insect. Some of the original energy was used by the plant for growth and maintenance, and some was converted into the chemical bonds of molecules such as cellulose, which the grasshopper cannot digest. Even more energy was lost as heat during these processes. Therefore, only a fraction of the energy captured by the producers of the first trophic level can be used by organisms in the second trophic level.

▲ **FIGURE 29-4 A simplified grassland food web** The animals pictured in the foreground include a vulture (a detritivore), a bull snake, a ground squirrel, a burrowing owl, a badger, a mouse, and a shrew (which looks like a mouse but is carnivorous). In the middle distance you'll see a grouse, a meadowlark, a grasshopper, and a jackrabbit. In the distance, look for pronghorn antelope, a hawk, pheasants, a wolf, and bison.

Energy is lost in each transfer to a higher trophic level. If the grasshopper is eaten by a robin (the third trophic level), the bird will not obtain all the energy that the insect acquired from the plants. Some of the energy will have been used up to power hopping, flying, and eating. Some energy will be found in the grasshopper's indigestible exoskeleton. Most of the energy will have been lost as heat. Similarly, most of the energy in a robin's body will be unavailable to a hawk that may consume it.

Energy Pyramids Illustrate Energy Transfer Between Trophic Levels

In general, about 90% of the energy stored at a trophic level is lost in the transfer to the next level. Thus, the amount of energy stored in primary consumers is only about 10% of the energy stored in producers, and the energy stored in secondary consumers is roughly 10% of the energy stored in primary consumers. This inefficient energy transfer between trophic levels is called the "10% law." An **energy pyramid** illustrates the energy relationships between trophic levels—it is widest at the base and narrows progressively in higher trophic levels (**FIG. 29-5**).

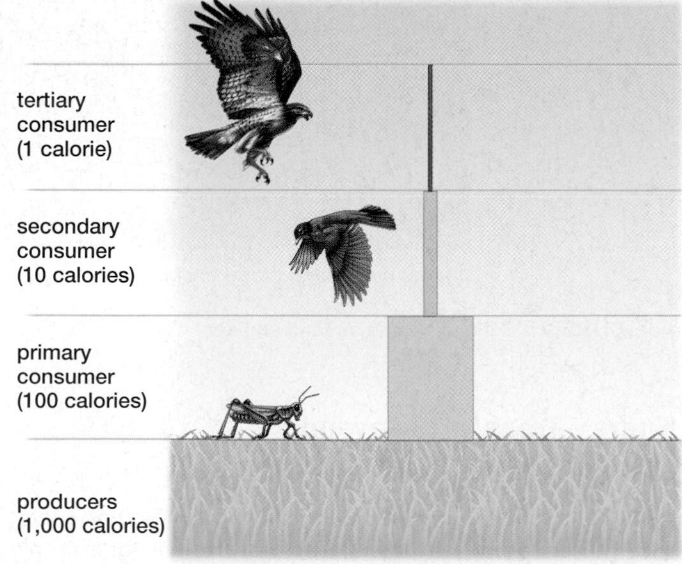

▲ **FIGURE 29-5 An energy pyramid for a grassland ecosystem** The width of each rectangle is proportional to the energy stored at that trophic level. Representative organisms for the first four trophic levels in a North American grassland ecosystem illustrated here are grass, a grasshopper, a robin, and a red-tailed hawk.

Because the amount of energy available decreases as the trophic level increases, primary producers are the most abundant organisms in an ecosystem, and top carnivores the rarest. An unfortunate side effect of less energy and lower abundance at higher tropic levels is that persistent toxic chemicals produced by human activities become concentrated in the bodies of carnivores. This **biological magnification** can lead to debilitating, even fatal, effects, as we explore in "Health Watch: Biological Magnification of Toxic Substances" on page 500.

CHECK YOUR LEARNING

Can you . . .

- name the trophic levels in a community and give examples of organisms found in each trophic level?
- describe how energy flows through an ecosystem?
- explain why detritivores and decomposers are essential to ecosystem function?
- explain how the inefficiency of energy transfer between trophic levels determines the relative abundance of organisms in different trophic levels?

CASE STUDY CONTINUED

Dying Fish Feed an Ecosystem

In the sockeye salmon's ecosystem, phytoplankton (producers) are eaten by zooplankton (primary consumers) that are in turn eaten by small fish (secondary consumers). So when a sockeye salmon eats a smaller fish, the salmon is a tertiary consumer on the fourth trophic level. And when an Alaskan brown bear eats a salmon, the bear is feeding on the fifth trophic level. In keeping with the 10% law, a food chain containing a single 1,000-pound bear will also contain 10,000 pounds of salmon, 100,000 pounds of smaller fish, a million pounds of zooplankton, and 10 million pounds of phytoplankton.

Organisms contain not only energy, but also many types of nutrients. How are these nutrients recycled and used by future generations of living things?

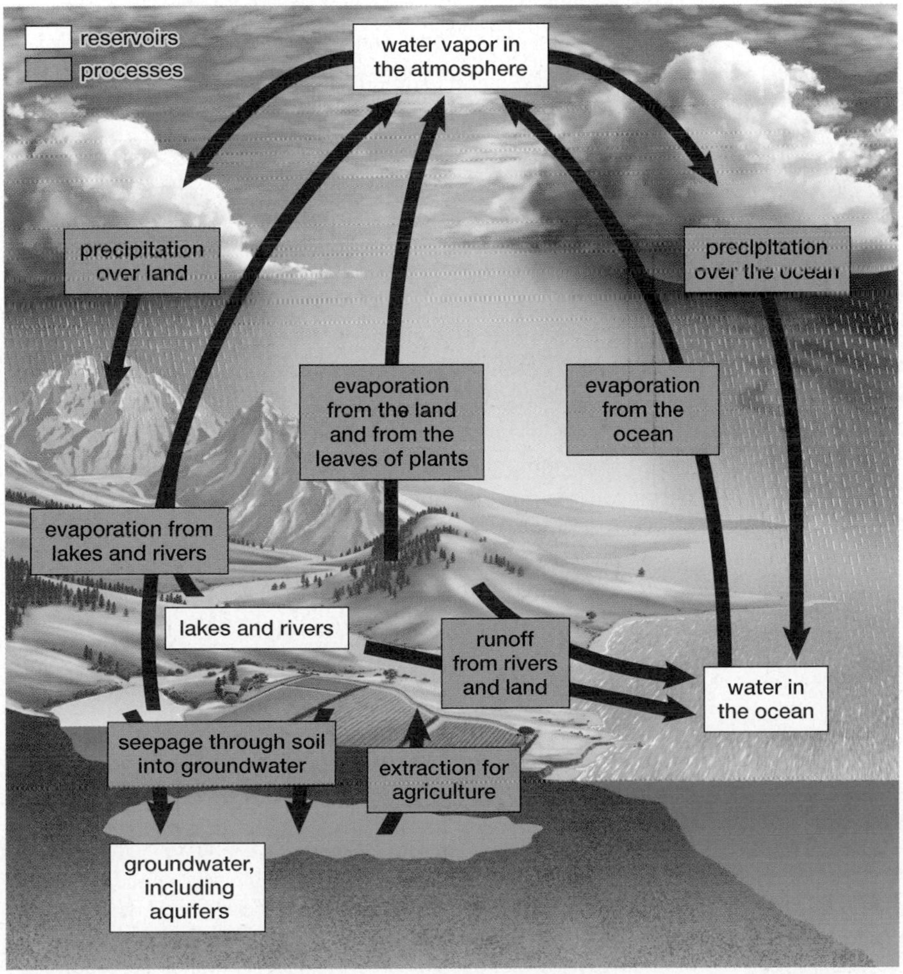

▲ **FIGURE 29-6 The water cycle**

29.3 HOW DO NUTRIENTS CYCLE WITHIN AND AMONG ECOSYSTEMS?

Earth's fixed supply of nutrients moves through ecosystems via **nutrient cycles**, also called *biogeochemical cycles*. These cycles are the pathways that nutrients take as they move from the nonliving environment through living communities and then back to the environment. As nutrients cycle, they may accumulate in one portion of the cycle, called a **reservoir**, and remain there for a long time.

Organisms require a wide variety of nutrients; here we describe the cycles of four nutrients that are used in large quantities: water, carbon, nitrogen, and phosphorus.

The Major Reservoirs for Water Are the Oceans

The **water cycle** (**FIG. 29-6**) is the pathway by which water travels from its major reservoir—the oceans—through the atmosphere, to smaller reservoirs in freshwater lakes, rivers, and groundwater, and then back again to the oceans. The oceans contain more than 97% of Earth's water. Another 2% of the total water is trapped in ice, leaving only 1% as fresh water.

Health WATCH Biological Magnification of Toxic Substances

Eating fish is supposed to be good for you, right? Not always. The U.S. Food and Drug Administration (FDA) advises that pregnant and breastfeeding women should avoid eating swordfish, shark, and king mackerel and should eat no more than one can of albacore (white) tuna per week. These fish species often contain a high concentration of mercury, which can damage the developing brains of fetuses and infants. Where does the mercury come from, and why do certain fish contain so much of it?

Most mercury in the environment comes from two sources: mining gold and burning coal. Many small-scale gold mining operations, especially in less-developed countries, use mercury to separate gold from crude ore. During the separation process, mercury is often released into groundwater and the atmosphere. Many power plants around the world are powered by coal, which contains traces of mercury. When coal is burned, the mercury is vaporized and may travel thousands of miles on the winds. Every part of Earth experiences mercury contamination.

Except in very localized areas, environmental concentrations of mercury are extremely low. Mercury becomes a health problem mainly through *biological magnification*, the process by which toxic substances become concentrated in animals occupying high trophic levels. Most substances that undergo biological magnification share two properties. First, they are stored for extended periods in living tissue, particularly in fatty tissue. Second, they are not easily *biodegradable*—neither animals nor decomposers can readily break these substances down into harmless materials. Mercury has both properties; it is stored in fat and muscle and, because it is an element, it cannot be broken down.

Biological magnification occurs because energy transfer between trophic levels is inefficient. Thus, even if producers absorb and store only low levels of toxic chemicals, an herbivore eats producers in large quantities and therefore absorbs and stores higher concentrations of the toxins. This sequence—eat, absorb, and store—continues up the trophic levels. Sharks, swordfish, tuna, and king mackerel—the species the FDA warns against—are long-lived, predatory fish that feed at the top of long food chains. Thus, they are more likely to accumulate high concentrations of mercury. In ecological terms, the FDA's advice is that people should avoid eating fish at a high trophic level.

Organic compounds, including many pesticides, may also build up in animals that occupy high trophic levels. Biological magnification first came to public attention in the 1950s and 1960s, when wildlife biologists witnessed an alarming decline in populations of several fish-eating birds such as cormorants, ospreys, brown pelicans, and bald eagles. They were being poisoned by DDT, a pesticide that had been sprayed on many aquatic ecosystems to control insects. Although the water in these ecosystems contained only low concentration of DDT, fish-eating birds contained DDT in concentrations a million times greater than the concentration in the water (**FIG. E29-1**). As a result, the birds laid thin-shelled eggs that broke under the parents' weight during incubation. Fortunately, DDT was banned in the United States in 1973, and populations of fish-eating birds have recovered significantly since then. About 180 countries worldwide have agreed to ban or restrict the production and use of DDT and a dozen other "persistent organic pollutants." And in 2011, the U.S. Environmental Protection Agency (EPA) issued rules that require power plants to reduce emissions of many toxic substances, including mercury, arsenic, nickel, and selenium. Unfortunately, however, the EPA has ordered a review of its toxics emissions rules for power plants, with the intention of relaxing the rules. So it has become increasingly possible that recent progress in reducing the perils of biomagnified toxins will be halted or reversed.

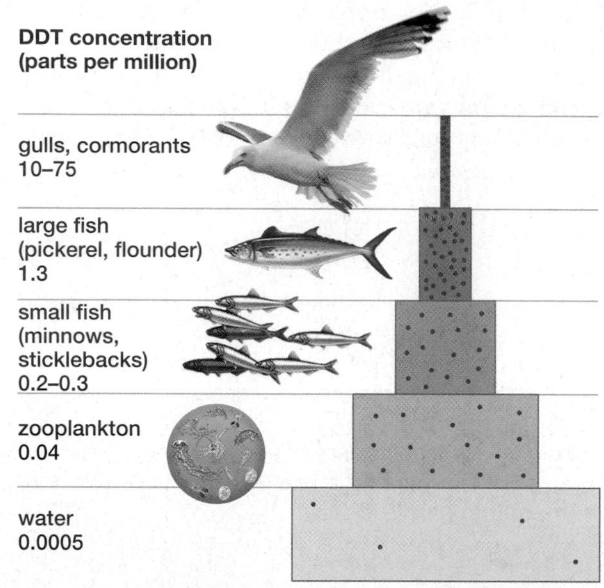

▲ **FIGURE E29-1 Biological magnification of DDT** During the time when DDT was widely used as an insecticide, concentrations of DDT in a marsh on Long Island, New York, increased about a million-fold with increasing trophic levels, from extremely low levels in the water to toxic levels in predatory birds. Based on Woodwell et al., *Science*, 1967.

THINK CRITICALLY As you're having lunch with your friend, she confides in you that she's planning to become pregnant this year. You notice that she is eating a tuna sandwich, and you remember reading that pregnant women should limit tuna intake because of mercury concerns. When you mention this to her, she asks if she should eat fish at all. Some research reveals that catfish, herring, haddock, and trout are among the fish species that usually contain low levels of mercury. Cod, halibut, and snapper contain moderate amounts of mercury. Explain to your friend why different fish species contain different levels of mercury.

The water cycle is driven by the sun's heat energy, which evaporates water. When water vapor condenses in the atmosphere, it falls back to Earth as rain or snow. Because the oceans cover about 70% of Earth's surface, most evaporation occurs from them and most precipitation falls back onto them. Of the water that falls on land, some evaporates from soil, lakes, and streams, and some runs downhill in rivers back to the oceans. Some is absorbed by the roots of plants; much of this water is returned to the atmosphere by evaporation from plant leaves. Some water seeps down through the soil to underground reservoirs, called **aquifers**. A tiny fraction of Earth's water is stored in the bodies of living organisms.

The Major Reservoirs of Carbon Are the Atmosphere and Oceans

Carbon is a key component of virtually every biological molecule. In the **carbon cycle** (**FIG. 29-7**), carbon moves from reservoirs in the atmosphere and oceans, through producers, consumers, detritivores, and decomposers, and then back to the reservoirs. Carbon enters the living part of an ecosystem when producers capture carbon dioxide (CO_2) during photosynthesis and incorporate the carbon atoms in organic molecules. On land, plants and other photosynthetic organisms acquire CO_2 from the atmosphere. In aquatic environments, phytoplankton and aquatic plants take up CO_2 dissolved in water.

Some of the carbon taken up by producers is returned as CO_2 to the atmosphere and to oceans and other bodies of water during cellular respiration, and the rest is incorporated into the producers' bodies. When primary consumers eat the producers, they acquire this stored carbon. The primary consumers and organisms in higher trophic levels release CO_2 during respiration, excrete carbon compounds in their feces, and store the rest of the carbon in their bodies. Both producers and consumers eventually die, and their bodies are broken down by detritivores and decomposers, whose cellular respiration releases CO_2.

The complementary processes of uptake by photosynthesis and release by cellular respiration continually recycle carbon from the abiotic to the biotic portions of ecosystems and back again. Some carbon, however, cycles much more slowly. Much of Earth's carbon is bound up in limestone, formed from calcium carbonate ($CaCO_3$) deposited on the ocean floor in the shells of prehistoric phytoplankton. The movement of carbon from this source to the atmosphere and back again takes millions of years. **Fossil fuels**, which include coal, oil, and natural gas, are also long-term reservoirs for carbon. These substances were produced from the remains of prehistoric organisms buried deep underground and subjected to high temperature and pressure.

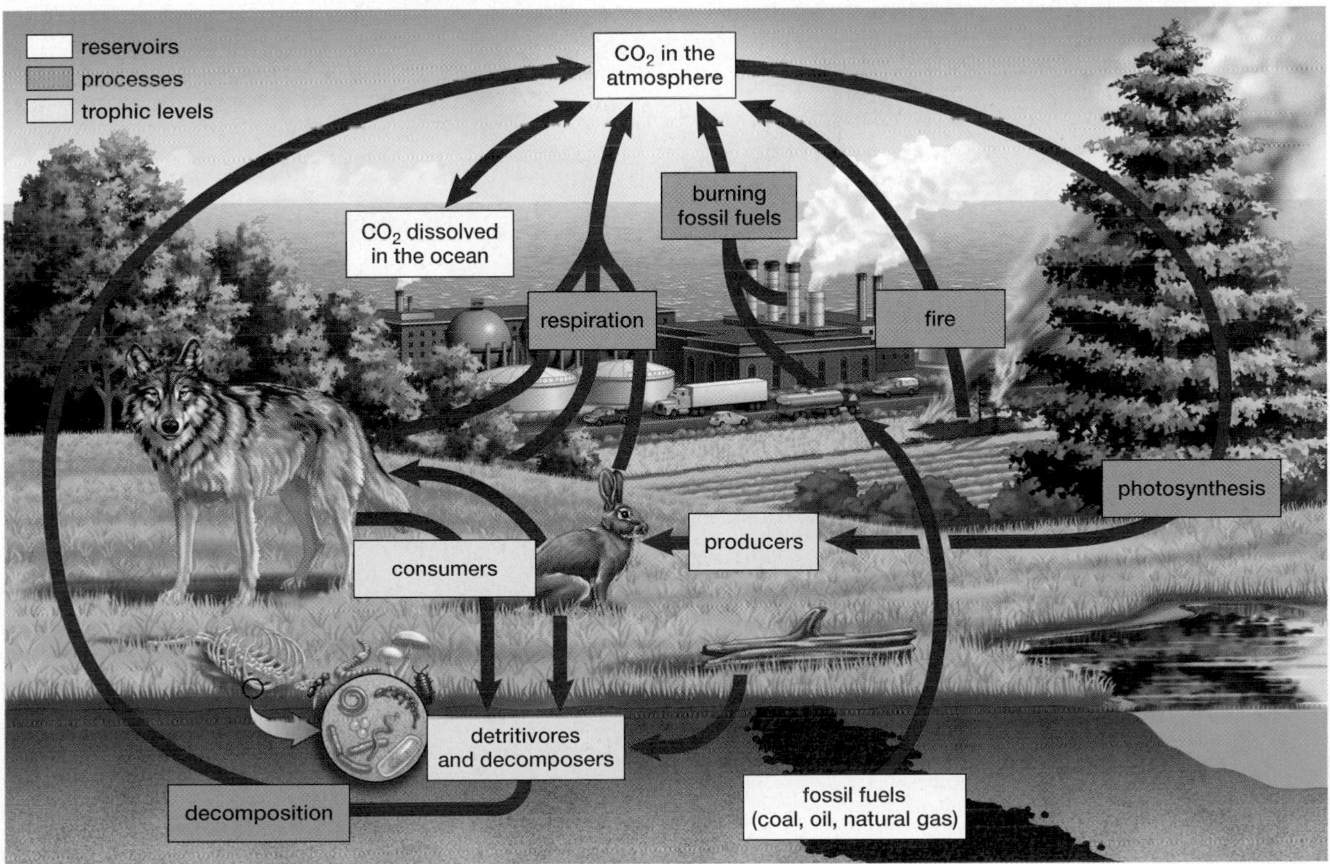

▲ **FIGURE 29-7 The carbon cycle**

The Major Reservoir of Nitrogen Is the Atmosphere

Nitrogen is a crucial component of proteins, many vitamins, nucleotides (such as ATP), and nucleic acids (such as DNA). In the **nitrogen cycle** (**FIG. 29-8**), nitrogen moves from its primary reservoir in the atmosphere to much smaller reservoirs of ammonia and nitrate in soil and water, through producers, consumers, detritivores, and decomposers, and back to its reservoirs.

The atmosphere consists of about 78% nitrogen gas (N_2), but most organisms cannot use nitrogen in this form. Plants and phytoplankton, however, can use ammonia (NH_3) and nitrate (NO_3^-). In a process called **nitrogen fixation**, certain bacteria in soil and water convert N_2 into ammonia, and other bacteria convert ammonia to nitrate. The bacteria do not use all the ammonia and nitrate they produce, and the excess is excreted into the surrounding soil or water. A small amount of nitrate is also produced during electrical storms, when the energy of lightning combines nitrogen and oxygen (O_2) gases to form nitrogen oxide molecules. These molecules dissolve in rain and fall to Earth's surface, where they are eventually converted to nitrate.

Producers absorb ammonia and nitrate and incorporate the nitrogen in biological molecules such as proteins. The nitrogen passes through successively higher trophic levels as primary consumers eat the producers and are themselves eaten. At each trophic level, bodies and wastes are broken down by decomposers, which liberate ammonia back into the soil and water, where bacteria convert it to nitrate. The nitrogen cycle is completed by **denitrifying bacteria**—residents of wet soil, swamps, and estuaries that break down nitrate, releasing nitrogen gas back into the atmosphere.

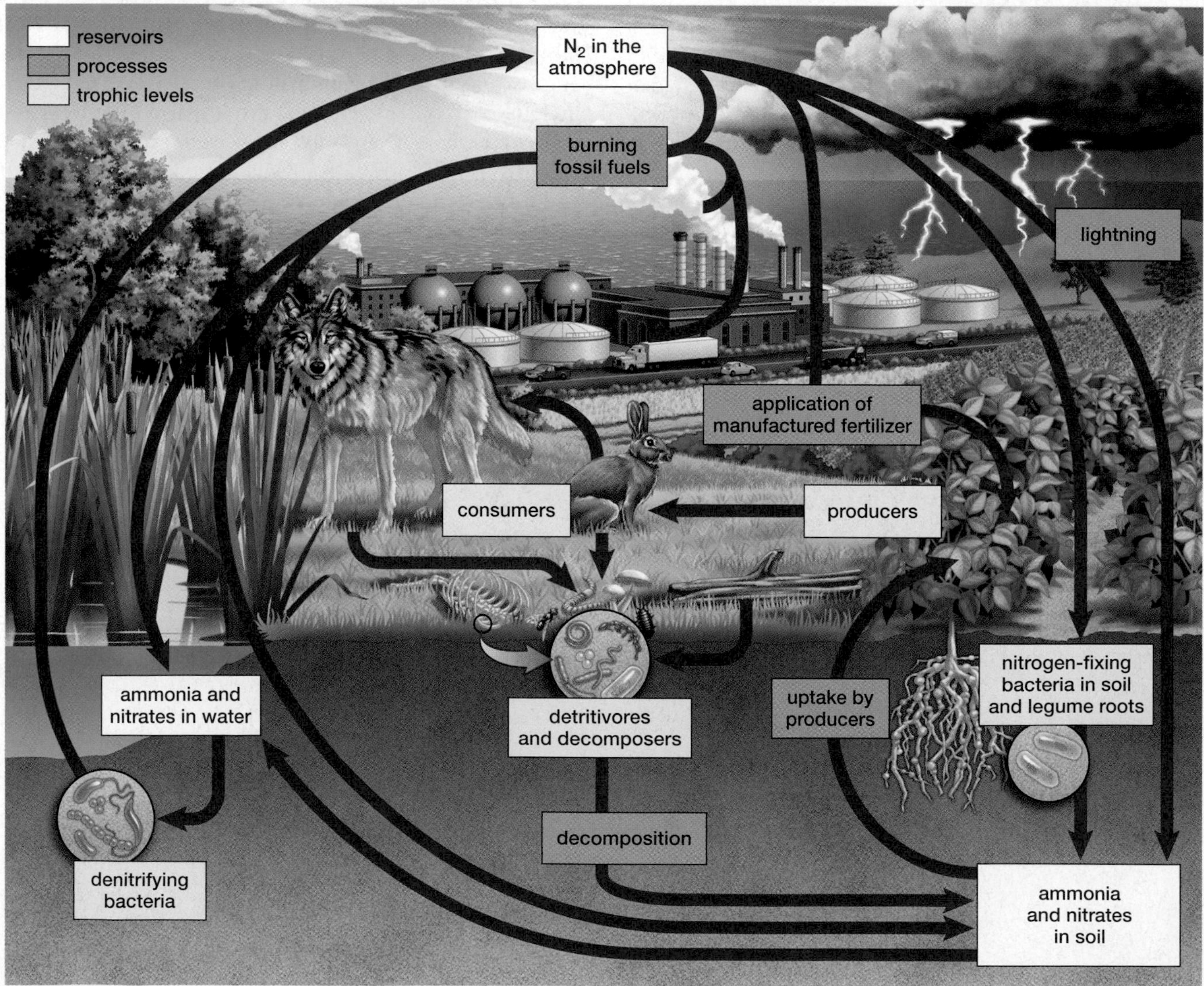

▲ **FIGURE 29-8 The nitrogen cycle**

THINK CRITICALLY What incentives cause humans to capture nitrogen from the air and pump it into the nitrogen cycle? What are some consequences of human influences on the nitrogen cycle?

People significantly manipulate the nitrogen cycle, both deliberately and unintentionally. Fertilizer factories combine N_2 from the atmosphere with hydrogen generated from natural gas, producing ammonia, which is then often converted to nitrate or urea (an organic nitrogen compound) for fertilizer to be used on farms, gardens, and lawns. In addition, the heat produced by burning fossil fuels combines atmospheric N_2 and O_2, generating nitrogen oxides that form nitrates. These human activities now dominate the nitrogen cycle.

CASE STUDY CONTINUED

Dying Fish Feed an Ecosystem

Nitrogen-fixing bacteria in ocean waters take up dissolved atmospheric nitrogen and produce ammonia, most of which is taken up by photosynthetic plankton. From these phytoplankton, nitrogen travels up the trophic levels, with some entering the bodies of salmon. Migrating salmon then bring nitrogen from the ocean to the terrestrial ecosystems of the Alaskan coast.

The Major Reservoir of Phosphorus Is in Rock

Phosphorus is found in nucleic acids and the phospholipids of cell membranes, and in phosphate ions, which are a major component of vertebrate teeth and bones. In the **phosphorus cycle** (**FIG. 29-9**), phosphorus moves from its primary reservoir in rocks to much smaller reservoirs in soil and water, through living organisms, and then back to its reservoirs. The phosphorus cycle does not include the atmosphere, because almost all phosphorus in the environment occurs as phosphate (PO_4^{3-}), which has no gaseous forms.

After phosphate-rich rocks are exposed by geological processes such as uplift, some of the phosphate is dissolved by rain and flowing water. The dissolved phosphate is carried into soil, lakes, and oceans, forming smaller reservoirs of phosphorus that are available to ecological communities. Dissolved phosphate is absorbed by producers, which incorporate it into biological molecules. From producers, phosphate passes to consumers at higher trophic levels. At each level, organisms excrete excess phosphate into their surroundings. Ultimately, detritivores and decomposers return the rest of the phosphate to the soil and water, where it may again be absorbed by producers or become bound to sediments and eventually re-formed into rock. Like the nitrogen cycle, the phosphorus cycle is now dominated by human activities. To produce fertilizer, we extract phosphate from rocks at a far higher rate than natural processes did in prehistoric times.

CHECK YOUR LEARNING

Can you . . .

- explain why nutrients cycle within and among ecosystems?
- summarize the water, nitrogen, carbon, and phosphorus cycles?

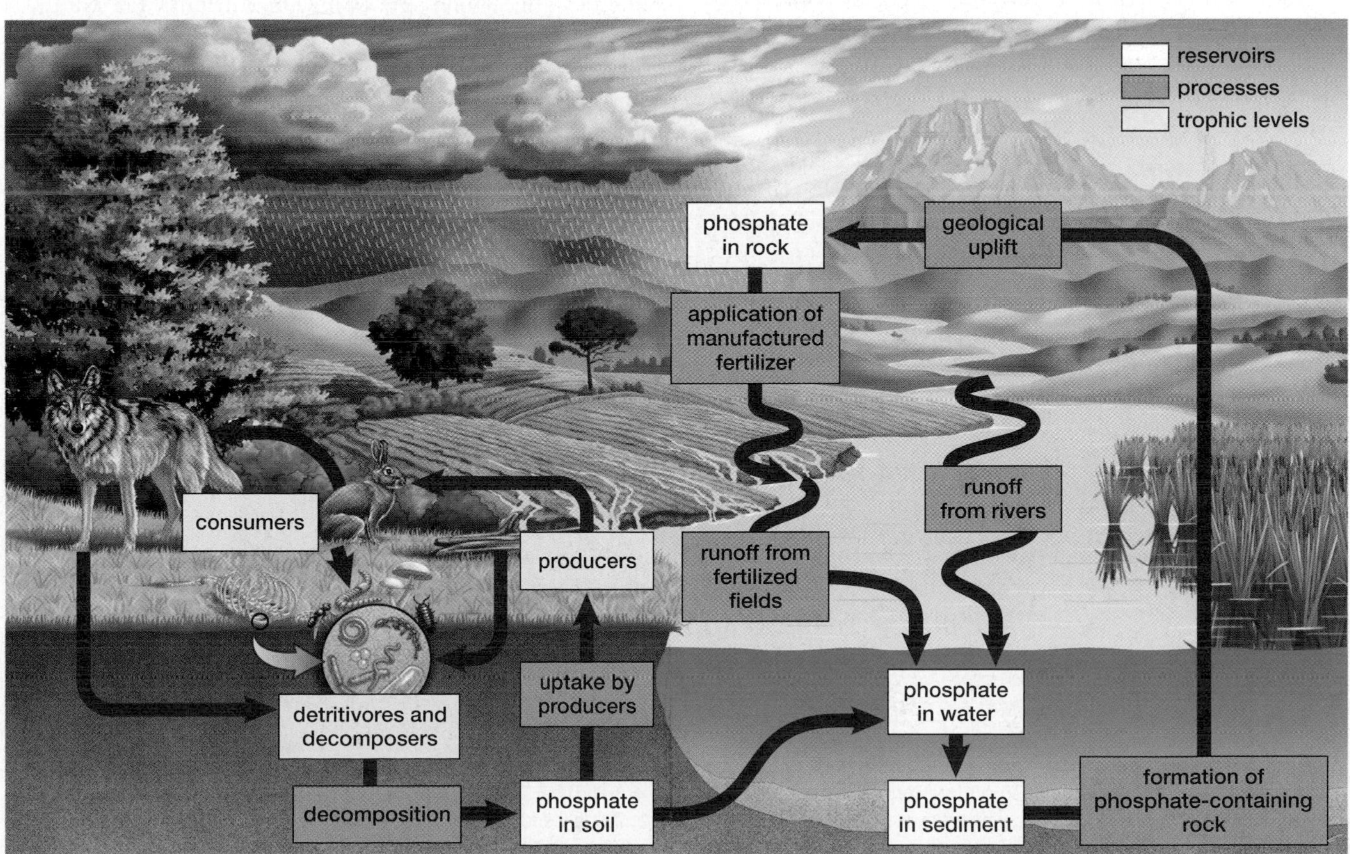

▲ **FIGURE 29-9 The phosphorus cycle**

29.4 WHAT HAPPENS WHEN HUMANS DISRUPT NUTRIENT CYCLES?

Ancient peoples, with small populations and limited technology, had relatively little impact on nutrient cycles. However, as the human population grew and new technologies were developed, people began to significantly alter many nutrient cycles. Today, human use of fossil fuels and chemical fertilizers has disrupted the global nutrient cycles of nitrogen, phosphorus, sulfur, and carbon.

Overloading the Nitrogen and Phosphorus Cycles Damages Aquatic Ecosystems

Each year, fertilizers containing about 45 million tons of phosphate and 115 million tons of nitrogen (as ammonium, nitrate, and urea) are applied to farm fields. When water from rainfall or irrigation washes over the land, it dissolves and carries away some of the nitrogen and phosphate. As the water drains into lakes, rivers, and the oceans, these nutrients can stimulate explosive growth of phytoplankton populations. As the phytoplankton die, their bodies sink into deeper water, where they provide a feast for decomposer bacteria. Cellular respiration by decomposers uses up most of the oxygen in the water. Deprived of oxygen, aquatic invertebrates and fish either leave the area or die, creating what is often called a dead zone. A huge dead zone appears each summer in the Gulf of Mexico off the coast of Louisiana. Spring runoff washes huge amounts of nitrates and phosphates into streams that flow into the Mississippi River, which carries the nutrients to the Gulf of Mexico. There, they cause an algal population explosion (**FIG. 29-10**) that soon produces an enormous dead zone. In 2017, the Gulf of Mexico dead zone covered almost 9,000 square miles.

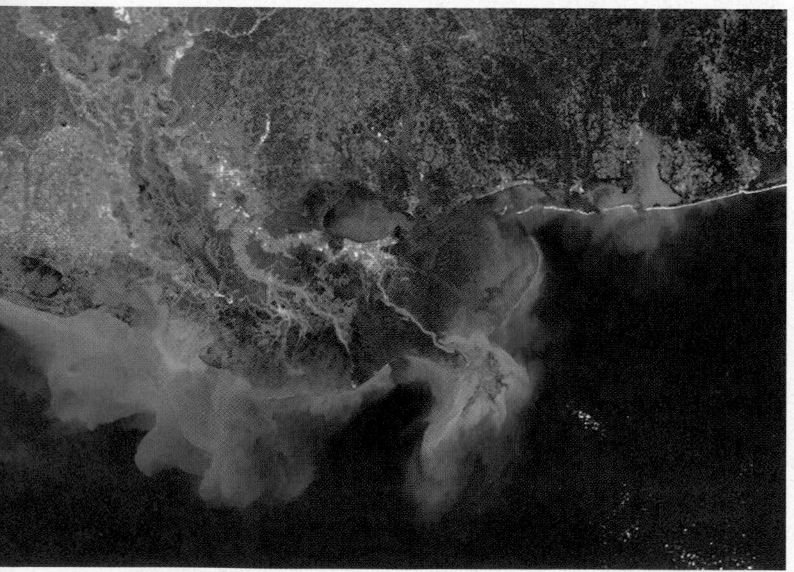

▲ **FIGURE 29-10 Harmful algal blooms** Nutrients from fertilizer, especially nitrate and phosphate, wash off farmland in the Midwest and travel down the Mississippi River, which flows from upper left to the center of the image, where it ends in the Mississippi Delta. When these nutrients enter the Gulf, they cause an explosive growth of algae, visible in this satellite photo as hazy green swirls near the coast.

Overloading the Sulfur and Nitrogen Cycles Causes Acid Deposition

Burning fossil fuels for industry and transportation releases huge amounts of nitrogen oxides and sulfur oxides into the atmosphere. Although nitrogen oxides are produced naturally by fires and lightning, and sulfur oxides are produced naturally by volcanoes, hot springs, and decomposers, most nitrogen and sulfur oxides entering the atmosphere are byproducts of human activity. When combined with water vapor in the atmosphere, these molecules are converted into nitric acid and sulfuric acid. Days later and often hundreds of miles away, these acids fall to Earth in rain or snow. This **acid deposition** was first recognized in New Hampshire, where a sample of rain collected in 1963 had a pH of 3.7—about the same as orange juice, and far more acidic than unpolluted rain, which usually has a pH between 5 and 6.

Acid deposition damages forests, can render lakes lifeless, and even eats away at buildings and statues (**FIG. 29-11**). Acid deposition is most damaging to ecosystems where the soils have little buffering capacity to neutralize acids, such as much of New England, the mid-Atlantic states, the upper Midwest, Western mountains, and Florida. Upstate New York and New England are doubly vulnerable, because the prevailing westerly winds that sweep across North America carry sulfates and nitrates from coal-burning power plants in the Midwest directly over these states.

Plants growing in acidified soil often become weak and vulnerable to infection and damage by insects. Calcium and magnesium, which are essential nutrients for plants, are leached out of the soil by acid precipitation. Acid rain also directly damages the needles of conifers such as spruce and fir. About half of the red spruce and one-third of the sugar maples in the Green Mountains of Vermont have been killed over the past 40 years due to acid deposition (**FIG. 29-12**).

▲ **FIGURE 29-11 Acid deposition is corrosive** These identical building decorations in Brooklyn, New York, show the effects of acid deposition. On the left, the decoration has been restored to its original state; on the right, an unrestored decoration has eroded almost completely away.

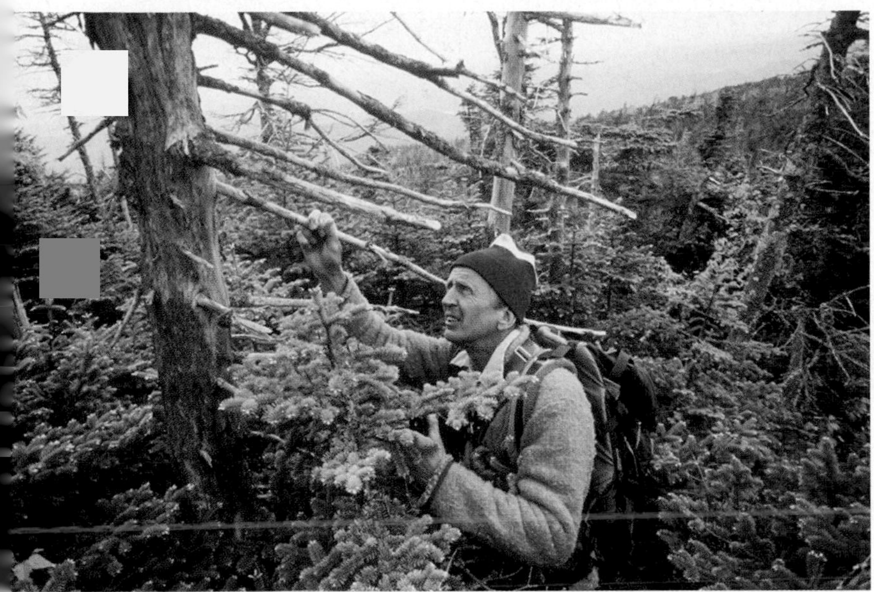

▲ **FIGURE 29-12 Acid deposition can destroy forests** On Camel's Hump in Vermont, virtually all of the older, mature trees are dead and bare. Since the 1990s, reduced sulfur and nitrogen emissions from power plants have decreased acid rain in New England, and young trees are beginning to recolonize the forest.

Since 1990, government regulations have resulted in substantial reductions in emissions of both sulfur dioxide and nitrogen oxides from U.S. power plants—sulfur dioxide emissions are down about 40%, and nitrogen oxide levels have been reduced by more than 50%. Air quality has improved and rain has become less acidic, although large areas of the northeastern United States still receive rain with a pH below 5.0.

Interfering with the Carbon Cycle Is Warming the Earth

The reservoir of carbon dioxide in the atmosphere not only provides producers with the starting material for photosynthesis, but also significantly affects Earth's climate. To understand why, let's begin with the fate of sunlight entering Earth's atmosphere (**FIG. 29-13**). Some of the energy from sunlight ❶ is reflected back into space by the atmosphere and by Earth's surface, especially by areas covered with snow or ice ❷. About half of the sunlight strikes relatively dark surfaces (land, vegetation, and open water) and is converted into heat ❸ that is radiated into the atmosphere ❹. Most of this heat continues on into space ❺, but water vapor, CO_2, and several other **greenhouse gases** trap some of the heat in the

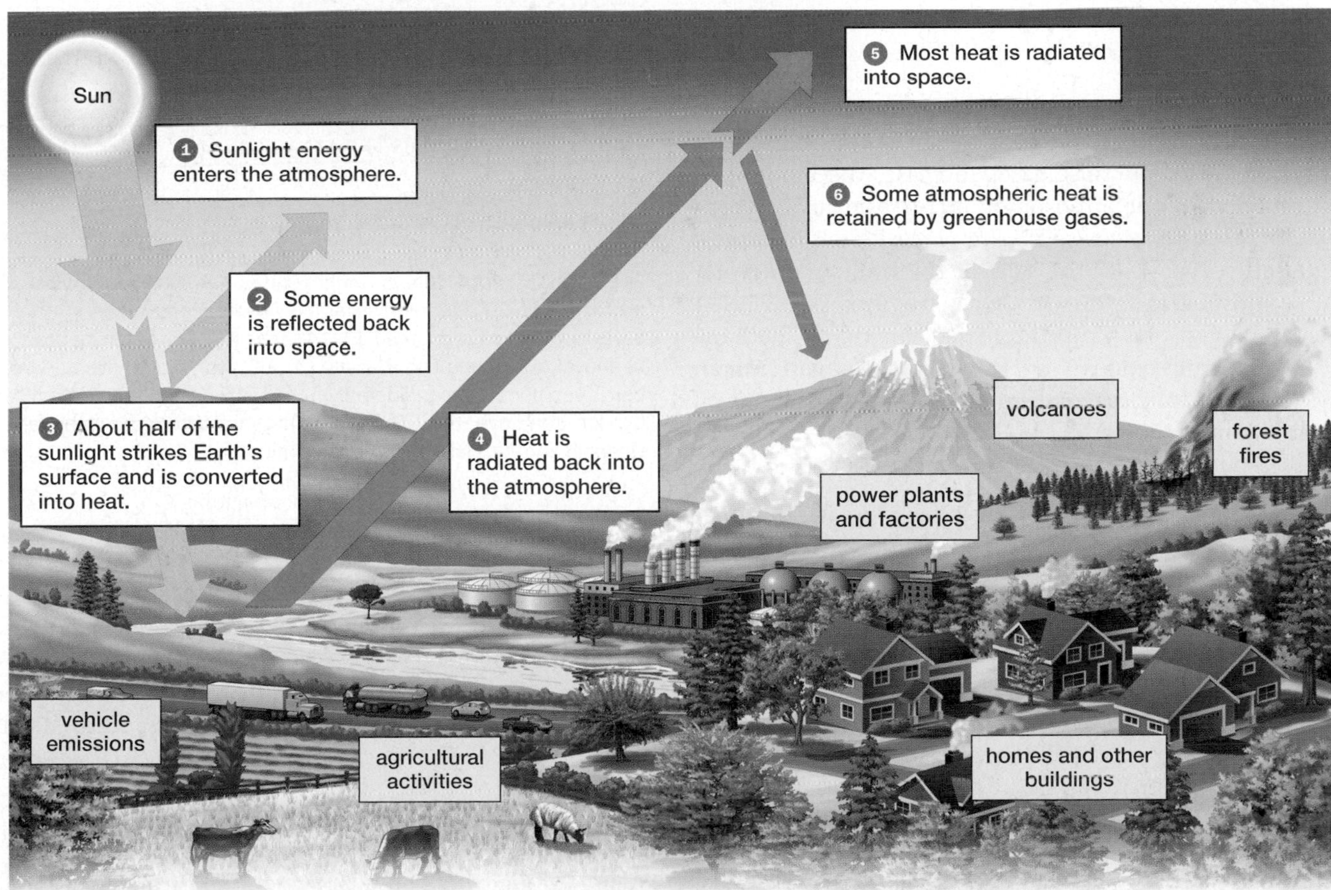
Sun
❶ Sunlight energy enters the atmosphere.
❷ Some energy is reflected back into space.
❸ About half of the sunlight strikes Earth's surface and is converted into heat.
❹ Heat is radiated back into the atmosphere.
❺ Most heat is radiated into space.
❻ Some atmospheric heat is retained by greenhouse gases.
volcanoes
forest fires
power plants and factories
vehicle emissions
agricultural activities
homes and other buildings

▲ **FIGURE 29-13 The greenhouse effect** Incoming sunlight warms Earth's surface and is radiated back to the atmosphere. Greenhouse gases, released by natural processes and substantially boosted by human activities (both shown in yellow rectangles), absorb increasing amounts of this heat, raising global temperatures.

Have You Ever Wondered ...

How Big Your Carbon Footprint Is?

Many human activities result in emissions of greenhouse gases and therefore have an effect on Earth's climate. One way of assessing the impact of an action is to measure its *carbon footprint*, the quantity of greenhouse gases emitted as a result of the action. Each of us has a personal carbon footprint that can help give us a sense of how our individual actions contribute to the greenhouse effect. For example, burning a gallon of gasoline releases 19.6 pounds (8.9 kg) of CO_2 into the air. So, if your car gets 20 miles to the gallon, each mile that you drive will add about a pound of CO_2 to the atmosphere.

To estimate your carbon footprint, you can use one of the online calculators provided by organizations such as the Nature Conservancy and the U.S. Environmental Protection Agency. These calculators evaluate how your daily choices affect carbon emissions by asking questions such as: What kind of foods do you eat? How fuel efficient is your car? How well is your home insulated, and what temperature is your thermostat set to?

atmosphere (6). This natural process is called the **greenhouse effect**, and it keeps our atmosphere relatively warm. Without the natural greenhouse effect, the average temperature of Earth's surface would be far below freezing, and Earth would probably be lifeless.

The release of greenhouse gases by human activities increases the greenhouse effect. When atmospheric concentrations of greenhouse gases increase, the amount of heat retained by the atmosphere increases, causing Earth to warm. Since people began burning fossil fuels in earnest about 150 years ago, at the start of the Industrial Revolution, atmospheric levels of CO_2 have increased. Other greenhouse gases have also increased, including methane (CH_4) and nitrous oxide (N_2O). However, the largest human contribution to the greenhouse effect, by far, results from our emissions of CO_2.

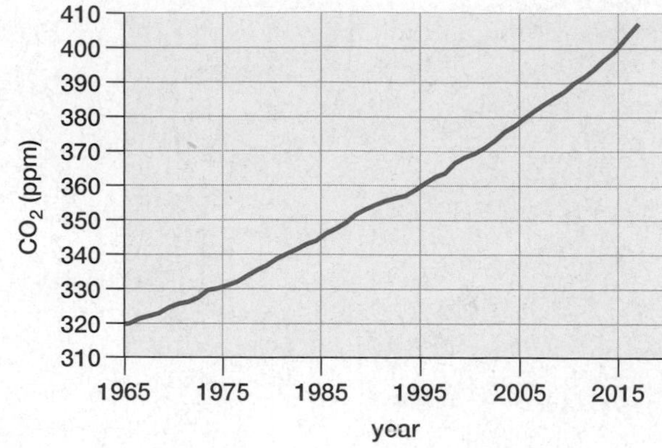

(a) Atmospheric CO_2

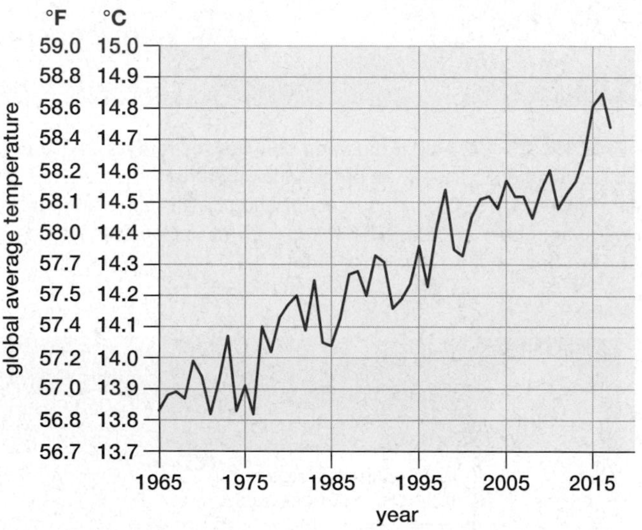

(b) Global surface temperature

▲ **FIGURE 29-14 Global temperatures rise in tandem with atmospheric CO_2 (a)** Yearly average CO_2 concentrations in parts per million, measured 11,141 feet (3,396 meters) above sea level, on Mauna Loa, Hawaii. **(b)** Although global average temperatures fluctuate considerably from year to year, there is a clear upward trend over time. Data for both graphs from the National Oceanic and Atmospheric Administration.

THINK CRITICALLY If people stopped emitting CO_2 next year, do you think that global temperature would begin to decline immediately? Why or why not?

Deforestation and Burning Fossil Fuels Increase Atmospheric Carbon Dioxide

Human activities release about 35 to 40 billion tons of CO_2 into the atmosphere each year. About 80% to 85% of this CO_2 comes from burning fossil fuels. Most of the rest is the result of **deforestation**; when trees are cut down and burned, the carbon stored in them returns to the atmosphere. Deforestation is occurring principally in the Tropics, where tens of millions of acres of rain forests are cleared each year for crops and cattle grazing.

About half of the CO_2 released by human activities is absorbed by the oceans and terrestrial plants. The rest remains in the atmosphere. Since 1850, the CO_2 content of the atmosphere has increased by over 40%—from 280 parts per million (ppm) to about 405 ppm in 2017—and is growing by more than 2 ppm annually (**FIG. 29-14a**). Atmospheric CO_2 is now higher than at any time in the past 800,000 years.

Carbon Dioxide Released by Humans Has Changed Earth's Climate

A large and growing body of evidence indicates that human release of CO_2 and other greenhouse gases has amplified the natural greenhouse effect, altering the global climate. Air

(a) Muir Glacier, 1941

(b) Muir Glacier, 2004

▲ **FIGURE 29-15 Glaciers are melting** Photos taken from the same vantage point in **(a)** 1941 and **(b)** 2004 document the retreat of the Muir Glacier in Glacier Bay National Park, Alaska.

temperatures at Earth's surface, recorded at thousands of sites on land and sea, show that Earth has warmed by about 2°F (1.1°C) since the late 1800s, including an increase of 1.6°F (0.9°C) just since the 1970s (**FIG. 29-14b**). All but 1 of the 17 warmest years on record have occurred since 2000.

The overall impact of increased greenhouse gases is usually called **climate change**, because greenhouse gases and the resulting global warming have many effects on Earth's climate. Our warming climate has already caused widespread changes. For example, glaciers are retreating worldwide: About 90% of the world's mountain glaciers are shrinking, and the trend seems to be accelerating (**FIG. 29-15**). Glacier National Park, Montana, named for its spectacular abundance of glaciers, had 150 glaciers in 1910; now, only 25 remain—and these are significantly smaller than they were in the recent past. During the past 30 years, the Arctic ice cap has become almost 50% thinner and 35% smaller in area. As the ice cap shrinks, the open water that appears in its place absorbs more light than the highly reflective ice, warming the water and further accelerating ice melt. On the continent of Antarctica, although some ice sheets are shrinking while others are expanding, the total volume of ice has dramatically decreased in recent years (see "Earth Watch: Monitoring Earth's Health" on page 508). The oceans are warming, which causes their water to expand and occupy more volume. This expansion, coupled with water flowing into the oceans from melting glaciers and ice sheets, is causing sea levels to rise.

Warming Will Continue and Extreme Weather Will Increase

What does the future hold? The Intergovernmental Panel on Climate Change, a consortium of climate scientists from 130 nations who work together to address climate change, predicts that even under the best-case scenario in which a concerted worldwide effort is made to reduce greenhouse gas emissions, the average global temperature will rise by another 1.3°F (0.7°C) by the year 2100. Without major reductions in emissions, global temperatures might rise as much as 5.8°F (3.2°C). These changes in climate will be difficult to stop, let alone reverse, as we explore in "Earth Watch: Climate Intervention—A Solution to Climate Change?" on page 509.

A warming climate is very likely to increase the frequency and intensity of extreme weather events. Climate scientists predict that a warming atmosphere will cause more severe storms, including stronger hurricanes; greater amounts of rain or snow in single storms; and more frequent, severe, and prolonged droughts.

Continued Climate Change Will Disrupt Ecosystems and Endanger Many Species

Even under the most optimistic scenarios, climate change will have profound effects on ecosystems. Already, many organisms have been affected by climate change. For example, thousands of species worldwide have shifted their ranges toward the poles, with the average species moving about 12.5 miles (20 kilometers) per decade in response to a warming planet. As climate change continues, some species will be more able to adjust than others, perhaps because they are more mobile (such as some birds) or because they can more readily disperse (such as plants that produce lightweight, wind-borne seeds). However, other species will not be able to move rapidly enough and will become rare or even go extinct.

Species on mountains or in the Arctic and Antarctic may have nowhere to go. For example, the loss of summer sea ice is bad news for polar bears and other marine mammals that rely on ice floes as nurseries for their young and as staging platforms for hunting fish or seals. As summer ice diminishes, walruses and polar bears must move onto land to give birth, putting the adults farther away from their prime hunting grounds. As walruses crowd

Earth WATCH Monitoring Earth's Health

How do we know that carbon dioxide in the atmosphere is increasing? Or that Earth is getting warmer, oceans are acidifying, glaciers are retreating, and Arctic sea ice is decreasing? The scientists who keep tabs on Earth's condition use a mix of long-term monitoring, ingenious methods, and cutting-edge technology.

Relatively recent changes in atmospheric CO_2 are measured at hundreds of stations in dozens of countries. The longest series of direct atmospheric measurements comes from the Mauna Loa Observatory in Hawaii, which has been tracking CO_2 levels continuously since 1958. However, past levels of CO_2, going back hundreds of thousands of years, are estimated by analyzing gas bubbles trapped in ancient Antarctic ice.

In some places, people have been keeping accurate records of temperature since well over a century ago. Today, air temperatures are recorded daily at about 1,500 locations, on both land and sea, and used to compute a global average. To estimate temperatures in the distant past, researchers rely on phenomena that vary with temperature and leave a long-lasting record. For example, the isotopes of oxygen in air vary with temperature, so measuring the isotopes in bubbles trapped in ancient ice can reveal the air temperature at the time the bubble formed.

Many measurements of Earth's environment rely on remote-sensing devices carried on satellites. Some of these measurements are relatively simple; for example, forest cover can be measured from satellite photos. Other measurements require more sophisticated sensors. The extent of Arctic sea ice is measured by satellites that detect microwave radiation emitted from Earth's surface. Ice emits more microwave radiation than liquid water does, so the satellites can easily distinguish the two. Satellite data show that the extent of Arctic sea ice has declined about 13% per decade since 1979 (**FIG. E29-2**). Many other features of Earth, from sulfur dioxide emitted by power plants to chlorophyll in the oceans, have distinctive "wavelength signatures" that satellite sensors can detect (**FIG. E29-3**).

Perhaps the most amazing measurements come from NASA's GRACE satellites—the Gravity Recovery and Climate Experiment. GRACE measures the strength of gravity, which varies among different parts of Earth because denser, heavier material exerts more gravitational force than does lighter, less-dense material. Water and ice are heavy, so large volumes of ice on land increase local gravity, allowing GRACE to measure the amount of ice present in an area. GRACE has found that ice sheets in Antarctica and Greenland have declined dramatically over the past decade. GRACE can even measure the amount of water underground. For instance, GRACE has shown that drought and groundwater pumping for agriculture have greatly depleted the aquifers underlying California's Central Valley (**FIG. E29-4**).

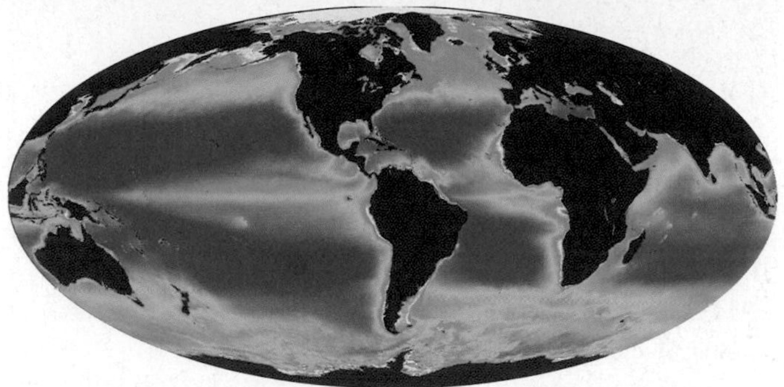

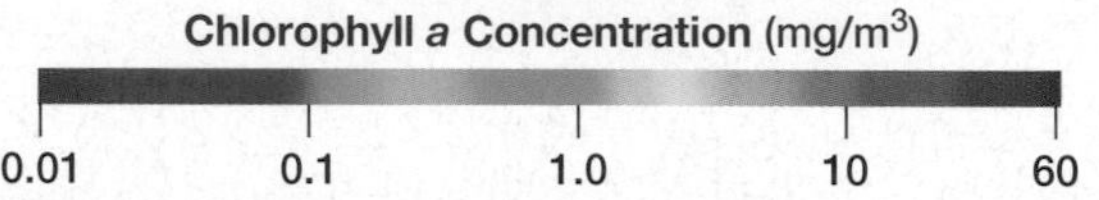

▲ **FIGURE E29-3 Ocean chlorophyll** Satellite measurements of chlorophyll show which areas of the ocean have the greatest amount of phytoplankton.

THINK CRITICALLY Use a ruler to estimate trend lines for the data in Figures 29-14 and E29-2. What do the trend lines predict about the future of atmospheric CO_2 concentrations, global temperatures, and Arctic sea ice? If these trends persist, when will the Arctic become ice-free in late summer? When will CO_2 concentrations double from preindustrial levels? Is it reasonable to extrapolate these trend lines into the future? Why or why not?

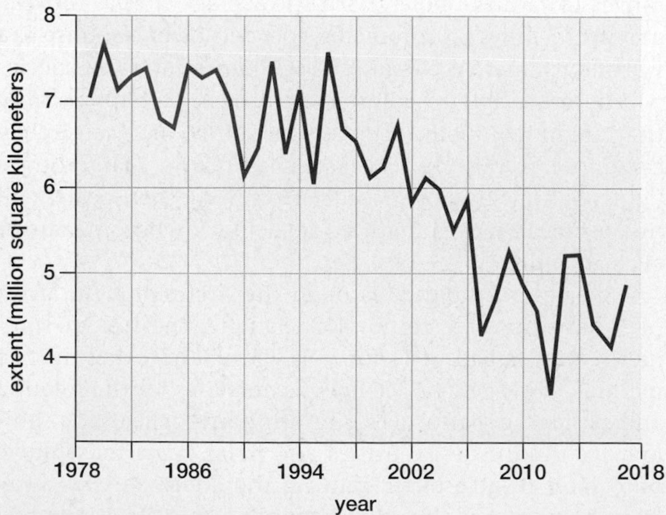

▲ **FIGURE E29-2 Changes in Arctic sea ice** Satellite measurements of Arctic sea ice began in 1979. By 2017, the area covered by ice at the end of the summer (September) had declined by more than a third.

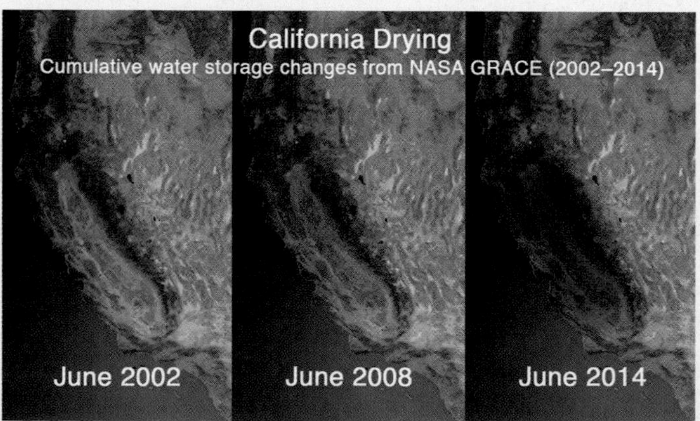

▲ **FIGURE E29-4 Changes in gravity show depletion of water in California's aquifers** Underground aquifers in California's Central Valley are losing about 4 trillion gallons of water each year. The transition from green to red in these false-color images shows water lost between 2002 and 2014.

Climate Intervention—A Solution to Climate Change?

What can we do to slow climate change? The obvious solution is to reduce our emissions of CO_2 and other greenhouse gases. But that isn't as easy as it sounds. Modern societies depend on the energy of fossil fuels and cannot just stop burning them overnight. As a result, some scientists and policymakers advocate *climate intervention*—altering fundamental characteristics of Earth to reduce greenhouse gases or counteract their effects until humanity makes the transition to carbon-neutral, renewable energy sources. Intervention proposals take two main approaches: shading the planet and removing CO_2 from the atmosphere.

Shading the Planet

Some molecules, such as certain sulfur compounds, can reflect sunlight back into space, cooling the planet. Whenever a massive volcano erupts, it spews millions of tons of sulfur dioxide miles high into the atmosphere. In 1991, when Mt. Pinatubo erupted in the Philippines (**FIG. E29-5**), it blasted about 20 million tons of sulfur dioxide as high as 22 miles skyward. For the next couple of years, global temperatures were slightly cooler. Some researchers have suggested that governments could similarly slow global warming by sending planes loaded with sulfur compounds high into the atmosphere, where they would release the sulfur. Other potential methods of shading the planet include spewing reflective metal particles from jet exhaust or modifying Earth's cloud cover.

▲ **FIGURE E29-5 Sulfur emissions both cool and pollute** The eruption of Mt. Pinatubo injected millions of tons of sulfur dioxide miles into the atmosphere, cooling the planet for a few years.

Capturing and Storing CO_2

One ambitious idea for capturing and storing carbon envisions thousands of towers, distributed planetwide, that would suck in air and remove the CO_2, which would then be concentrated and either injected underground or converted to solid forms (similar to limestone) that could be stored on land. Although this approach is technically feasible, it would be enormously expensive. It would be much less expensive to use similar methods to remove CO_2 at major sources such as power plants, rather than from the atmosphere as a whole. Carbon capture technology is currently in use at 20 or so power plants worldwide.

Another potential method of capturing CO_2 is enhanced weathering. Weathering is a natural process that slowly breaks down rocks. The chemical reactions of weathering consume CO_2 and produce bicarbonate ions (HCO_3^-). Ultimately, this bicarbonate washes into the ocean, where it is converted to carbon-containing minerals that sink to the bottom, where they remain indefinitely. Thus, natural weathering removes CO_2 from the atmosphere, but the process is slow. In enhanced weathering, rocks would be ground into a fine powder that would be applied to agricultural land on a vast scale. Powdered rock has a much larger surface area for weathering than a comparable weight of natural rock, so enhanced weathering would remove CO_2 from the atmosphere far more rapidly than natural weathering does.

A biological approach to removing CO_2 is to fertilize the oceans to encourage the growth of carbon-absorbing organisms. In many parts of the open ocean, phytoplankton growth is limited by the amount of available iron, an essential nutrient. Researchers have proposed spreading powdered iron on open ocean waters to trigger phytoplankton blooms. The phytoplankton would then take up CO_2 during photosynthesis and store some of the carbon in their bodies. After the phytoplankton died, they would sink, carrying the carbon to the ocean depths, where it would remain for many years.

Would Climate Intervention Work? Is It Worth the Risks?

Many people question the feasibility and desirability of climate intervention. As a panel of experts commissioned by the U.S. National Academy of Sciences put it, "Climate intervention is no substitute for reductions in carbon dioxide emissions." Why not? (1) Shading the planet with atmospheric sulfur would have undesirable side effects, because sulfur compounds cause respiratory damage in people, damage the ozone layer, and generate acid deposition that injures ecosystems; (2) removing CO_2 from the atmosphere by mechanical means would be extremely expensive; and (3) although small-scale tests have been promising, no one really knows if ocean fertilization or enhanced weathering would remove enough CO_2 to help very much, or what effects it might have on ecosystems. Still, the panel recommends studying both planetary shading and carbon storage. We must act to address the threat posed by climate change, and it would be best if our actions were informed by careful research.

CONSIDER THIS Some people argue against spending time and money on research and development of climate intervention strategies. In this view, investment in climate intervention comes at the unacceptable cost of reducing research and development of technologies and strategies aimed at reducing carbon emissions. In your opinion, should we invest exclusively in climate intervention, exclusively in reducing carbon emissions, or in some combination of the two? Think of some pros and cons for each option, and explain why you chose your preferred option.

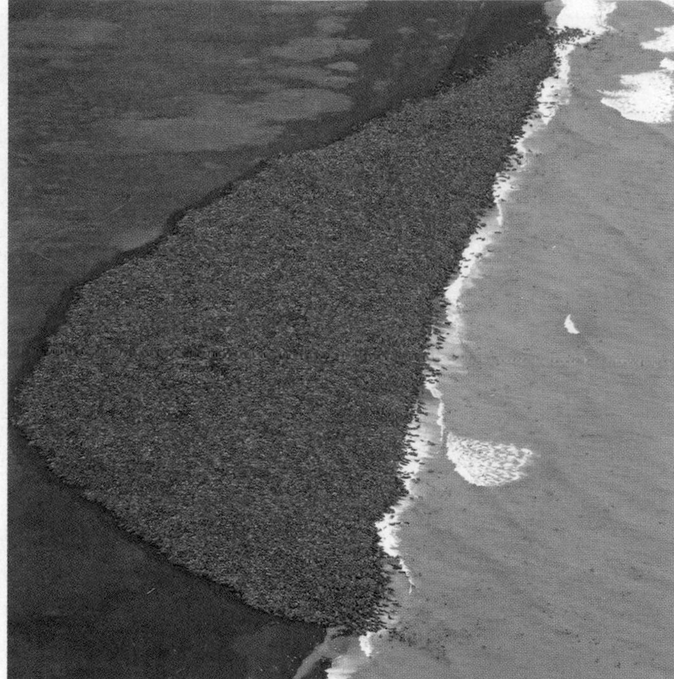

▲ **FIGURE 29-16 Walrus pack the beach at Point Lay, Alaska** Walrus are normally dispersed on floating sea ice in the summer. As the Arctic warms and sea ice disappears during the summer, they now congregate on a few beaches.

together onto small beaches (**FIG. 29-16**), instead of being spread out over sea ice, they inadvertently trample their own offspring to death. Complete loss of Arctic sea ice during the summer, which climate models predict could occur by mid-century, may cause the extinction of polar bears in the wild.

Some of the movement of species may have direct impacts on human health. For example, many diseases that are transmitted by mosquitoes are currently restricted to tropical or subtropical regions. Like other species, mosquitoes will spread poleward as temperatures warm, bringing with them microorganisms that cause diseases such as malaria, dengue fever, yellow fever, and Rift Valley fever. At the same time, though, global warming may cause some parts of the Tropics to become so hot and dry that mosquitoes will no longer thrive there, thus reducing mosquito-borne diseases in these regions. Climate change will affect the prevalence of diseases, but it is difficult to predict exactly how.

In addition to its effects on climate, increased CO_2 also makes the oceans more acidic. The increased acidity disturbs many natural processes. For example, acidification disrupts the process by which many marine animals, such as snails and corals, make their shells and skeletons. As a result, these organisms now grow more slowly and have weaker shells and skeletons. Such effects are likely to worsen as the oceans become increasingly acidic, with especially devastating consequences for coral reef ecosystems.

CHECK YOUR LEARNING

Can you . . .

- give examples of how human activities have disrupted nutrient cycles?
- describe how human interference with nutrient cycles causes acid deposition, damages aquatic ecosystems, enhances the greenhouse effect, and causes climate change?
- describe some of the evidence that Earth is warming and some of the impacts of climate change on Earth's ecosystems?

CASE STUDY REVISITED

Dying Fish Feed an Ecosystem

The sockeye salmon's return to an Alaskan stream is an unforgettable spectacle (**FIG. 29-17**). Even after the fish have run the gauntlet of brown bears and bald eagles, hundreds remain, their brilliant red bodies writhing in water so shallow that it barely covers them. The salmon have reached their destination, and it's time to spawn. Each breeding female digs a depression in the gravel where she releases her eggs; a male then showers them with sperm. But sperm and eggs aren't the only cargo the salmon carry upstream. Sockeye salmon acquire about 95% of their body mass during their years in the ocean, so as the fish move upstream, they carry large quantities of energy and nutrients from the ocean. These nutrients become available to terrestrial predators and scavengers, such as the brown bears that snatch the salmon from rushing waters and the minks that gorge on half-eaten salmon carcasses left behind by the bears.

▲ **FIGURE 29-17 Spawning sockeye salmon**

Of the nutrients that salmon transport from the ocean to the land, nitrogen is especially important. Ecologists have discovered that more than half of the nitrogen near some streams originated in the ocean and was brought upstream in the bodies of salmon. Salmon-derived nitrogen is important to Alaska's trees, such as Sitka spruce, which can grow three times faster near salmon streams than near streams without salmon. The nitrogen is also important to the next generation of salmon. Decaying salmon carcasses fertilize the water, stimulating the growth of phytoplankton and consequently the zooplankton upon which newly hatched salmon feed, so adult salmon, in death, indirectly feed their young.

CONSIDER THIS Dams, river pollution, and overfishing have depleted many salmon populations, causing salmon species to be listed as endangered or threatened under the Endangered Species Act. Some people argue that because these species are also raised commercially in fish farms, the depletion of wild populations doesn't really matter. Based on what you have learned in this chapter, do you think that wild populations of salmon should be protected and efforts made to increase their numbers?

CHAPTER REVIEW

Go to **Mastering Biology** to access the Pearson eText, vocabulary review, practice quizzes, activities, videos, current events, and more.

*Answers to **Think Critically** and **Thinking Through the Concepts** questions can be found in the **Answers** section at the back of the book.*

Summary of Key Concepts

29.1 How Do Nutrients and Energy Move Through Ecosystems?

Ecosystems are sustained by a continuous input of energy from sunlight and the recycling of nutrients. Energy from sunlight enters the biotic portion of ecosystems through photosynthesis and then flows through the ecosystem. Nutrients are obtained by organisms from their living and nonliving environments and are recycled within and among ecosystems.

29.2 How Does Energy Flow Through Ecosystems?

The energy of sunlight is captured by photosynthetic organisms (producers), the first trophic level in an ecosystem. Herbivores (plant-eaters, also called primary consumers) form the second trophic level. Carnivores (meat-eaters) are secondary consumers when they prey on herbivores and tertiary or higher-level consumers when they eat other carnivores. Omnivores occupy multiple trophic levels. Detritivores and decomposers feed on dead bodies and wastes. Decomposers (mostly bacteria and fungi) liberate nutrients as simple molecules that re-enter nutrient cycles.

The higher the trophic level, the less energy is available to sustain it. In general, only about 10% of the energy captured by organisms at one trophic level is available to organisms in the next higher level.

29.3 How Do Nutrients Cycle Within and Among Ecosystems?

A nutrient cycle depicts the movement of a particular nutrient from its reservoir, usually in the abiotic portion of an ecosystem, through the biotic portion, and back to its reservoir.

In the water cycle, the major reservoir of water is the oceans. Solar energy evaporates water, which returns to Earth as precipitation. Water flows into lakes and underground aquifers and is carried by rivers to the oceans.

In the carbon cycle, the short-term reservoirs are CO_2 in the oceans and the atmosphere. Carbon enters producers via photosynthesis. From producers, carbon is passed through food webs and released to the atmosphere as CO_2 during cellular respiration. Burning fossil fuels also releases CO_2 into the atmosphere.

In the nitrogen cycle, the major reservoir is N_2 in the atmosphere. N_2 is captured by nitrogen-fixing bacteria, which produce ammonia. Other bacteria convert ammonia to nitrate. Plants obtain nitrogen from nitrates and ammonia. Nitrogen passes from producers to consumers and is returned to the environment through excretion and the activities of detritivores and decomposers. N_2 is returned to the atmosphere by denitrifying bacteria.

In the phosphorus cycle, the principal reservoir consists of phosphate in rocks. Phosphate dissolves in water, is absorbed by photosynthetic organisms, and is passed through food webs. Some phosphate is excreted, and the rest is returned to the soil and water by decomposers. Some is carried to the oceans, where it may be deposited in marine sediments.

29.4 What Happens When Humans Disrupt Nutrient Cycles?

Human activities often produce and release more nutrients than nutrient cycles can efficiently process. Fertilizer use in agriculture has disrupted the nitrogen and phosphorus cycles of many aquatic ecosystems. By burning fossil fuels, humans have overloaded the natural cycles for sulfur, nitrogen, and carbon. In the atmosphere, sulfur dioxide and nitrogen oxide are converted to sulfuric acid and nitric acid, which fall to Earth in rain and snow as acid deposition, with harmful effects on lakes and forests.

Burning fossil fuels has substantially increased atmospheric carbon dioxide. Climate scientists have concluded that increased CO_2 causes increased global temperatures. Increased temperatures cause climate change that is manifested in more extreme weather, melting glaciers, thinning Arctic sea ice, warming and acidifying oceans, rising sea levels, and changing distributions and seasonal activities of wildlife.

Thinking Through the Concepts

Bloom's: Remembering, Understanding

Multiple Choice

1. Compared to the top predators in an ecosystem, producers are far more abundant and contain far more total biomass because
   a. a top carnivore requires less energy than a producer.
   b. as energy is transferred from one trophic level to the next, some is lost to the environment.
   c. as nutrient molecules are transferred from one trophic level to the next, some are destroyed.
   d. biological magnification of toxic substances kills many top predators.
2. Which of the following is *not* a major reservoir in the carbon cycle?
   a. consumers
   b. the atmosphere
   c. the oceans
   d. fossil fuels
3. Denitrifying bacteria
   a. convert ammonia to nitrate.
   b. convert nitrate to ammonia.
   c. convert N_2 to ammonia.
   d. convert nitrate to N_2.
4. Net primary production per unit area is likely to be highest in which of the following ecosystems?
   a. grasslands
   b. deserts
   c. temperate deciduous forests
   d. tropical rain forests
5. The effect of CO_2 emissions from burning fossil fuels is likely to be
   a. nothing; photosynthesis by plants will remove all the added CO_2 from the atmosphere.
   b. nothing; the carbon cycle will simply run faster.
   c. planetary warming due to an increased greenhouse effect.
   d. planetary cooling due to a reduced greenhouse effect.

Fill-in-the-Blank

1. Nearly all life gets its energy from ____________, which is captured by the process of ____________. In contrast, ____________ are constantly recycled during processes called ____________.
2. Photosynthetic organisms are called either ____________ or ____________. The energy that they store and make available to other organisms is called ____________.
3. Feeding levels within ecosystems are also called ____________. An illustration of these levels with only one organism at each level is called a(n) ____________. Feeding relationships are most accurately depicted as ____________.
4. In general, only about ____________% of the energy available in one trophic level is captured by the level above it.
5. Photosynthetic organisms make up the first trophic level. Organisms in higher trophic levels are collectively called ____________ or ____________. Photosynthetic organisms are consumed by organisms collectively called ____________ or ____________. Animals that feed on other animals are called ____________ or ____________. Organisms that feed on wastes and dead bodies are called ____________ and ____________.
6. During the nitrogen cycle, nitrogen gas is captured from its atmospheric reservoir by ____________ in the soil and is then returned to this reservoir by ____________. The two forms of nitrogen that are used by plants are ____________ and ____________.
7. Two relatively short-term reservoirs for carbon are the ____________ and ____________. Carbon in these reservoirs is in the form of ____________. Two long-term reservoirs for carbon are ____________ and ____________.

Review Questions

1. What makes the movement of energy through ecosystems fundamentally different from the movement of nutrients?
2. What is a producer? What trophic level does it occupy, and what is its importance in ecosystems?
3. Define *net primary production*. Would you predict higher productivity in a farm pond or an alpine lake? Explain your answer.
4. Name the first three trophic levels. Among the consumers, which are most abundant? Use the "10% law" to explain why you would predict that there will be a greater biomass of plants than herbivores in an ecosystem.
5. How do food chains and food webs differ? Which is the more accurate representation of feeding relationships in ecosystems?
6. Define *detritivore* and *decomposer* and explain their importance in ecosystems.
7. Trace the movement of carbon from one of its reservoirs through the biotic community and back to the reservoir. How have human activities altered the carbon cycle, and what are the implications for Earth's future climate?
8. Explain how nitrogen gets from the atmosphere into a plant's body.
9. Trace a pathway of a phosphorus molecule from a phosphate-rich rock into a carnivore. What makes the phosphorus cycle fundamentally different from the carbon and nitrogen cycles?

Applying the Concepts

Bloom's: Applying, Analyzing, Evaluating

1. Humans typically feed on several trophic levels. Discuss how the inefficiency of energy transfer between trophic levels would affect how many humans on the planet could be fed under two scenarios: all people ate a typical American diet including fish and red meat, or all people ate a strictly vegetarian diet. Discuss also the environmental impact of each diet.
2. Discuss the contribution of human population growth to (a) acid precipitation and (b) global climate change.

30 Earth's Diverse Ecosystems

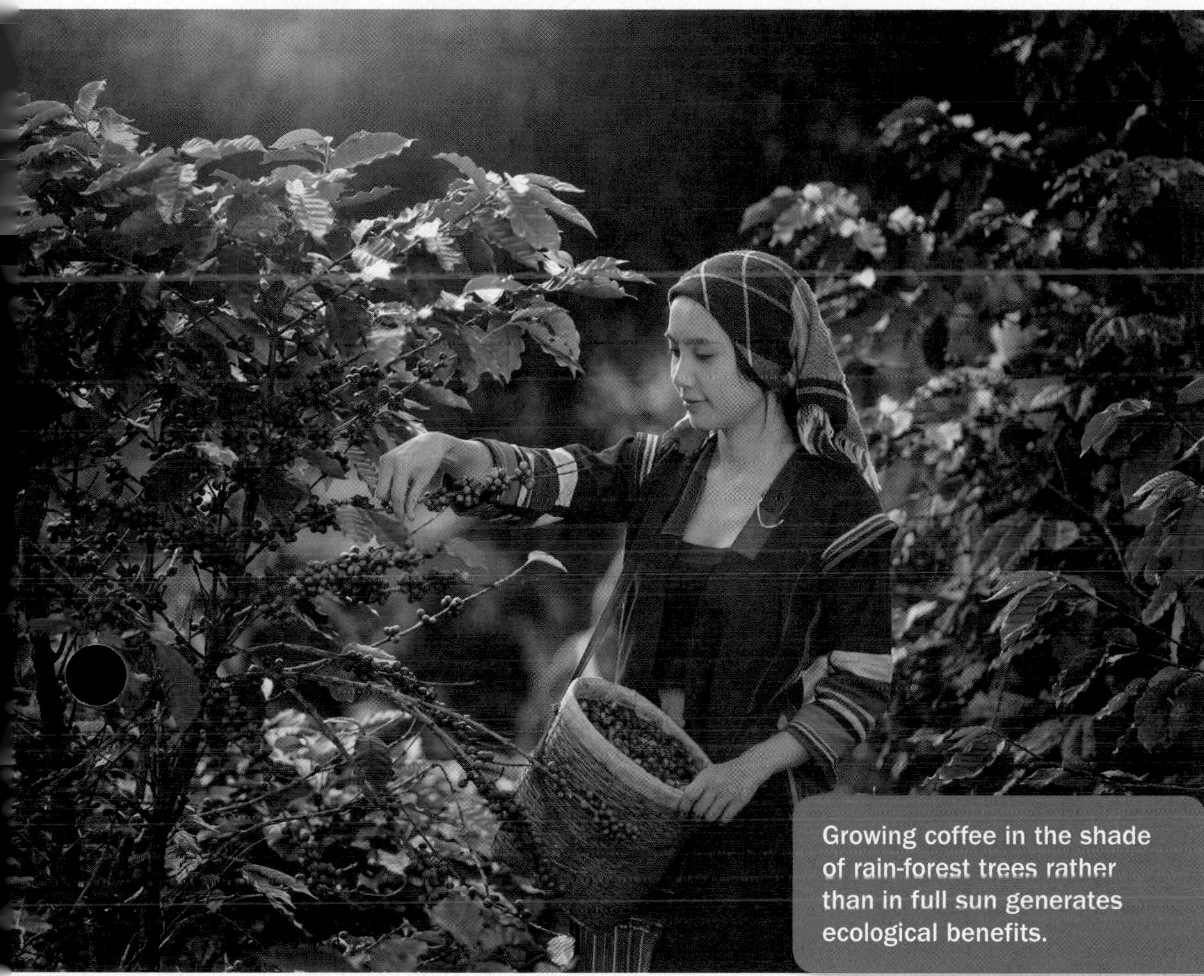

Growing coffee in the shade of rain-forest trees rather than in full sun generates ecological benefits.

CASE STUDY

Can Coffee Save Songbirds?

EVERY YEAR, the average person in the United States drinks about 340 cups of coffee. If you're a coffee drinker, you might be able to help preserve threatened species by choosing to brew beans grown in the shade.

Coffee "beans" are the seeds of two species of coffee plants, native to upland forests in the African country of Ethiopia. Coffee is now widely cultivated in tropical regions of South and Central America, Africa, and Southeast Asia. The original varieties grew in the shade of taller trees—in fact, full sunlight killed the plants, especially seedlings. In Central and South America, coffee was traditionally cultivated as an understory plant in the rain forest. Coffee plantations provided a multilevel, diverse habitat that supported monkeys, frogs, flowers, and hundreds of species of birds. The forest vegetation absorbed water and protected the soil from erosion. The shade discouraged weed growth, and detritivores and decomposers recycled fallen leaves into plant nutrients.

In the 1960s and 1970s, however, breeders developed new varieties of coffee plants that thrived in full sun and yielded more seeds. As world demand for coffee increases, more and more coffee plants are grown in full-sun plantations. What is lost when rain forests are cut down to make room for full-sun plantations? Or when we alter, or even destroy, natural communities in other ways, such as draining wetlands for agriculture or housing? To answer these questions, we must first understand the properties of the communities that make up life on Earth.

AT A GLANCE

30.1 WHAT DETERMINES THE DISTRIBUTION OF LIFE ON EARTH?

From Arctic tundra to tropical rain forests to open oceans, different places on Earth differ enormously in the identity and abundance of the species present. This variability arises because the resources required for life—energy, nutrients, liquid water, and suitable temperatures—are unevenly distributed.

Because life in most ecosystems depends on energy captured by photosynthetic producers (see Chapter 29), the geographic distribution of photosynthetic species largely determines the distribution of other species. The distribution of photosynthesizers is, in turn, determined by variation among locations in the abundance and availability of the essential resources.

In aquatic ecosystems, liquid water is almost always available, so the amount and type of life present in a location is generally determined by temperature and availability of nutrients and sunlight energy. Sunlight sufficient for photosynthesis is available only in the top 650 feet (200 meters) of a body of water, and light penetrates even less deeply if the water is at all cloudy. Even where light is available, photosynthesis may be limited because surface waters usually contain low levels of many nutrients. Temperature may also influence which organisms are present, because many aquatic species thrive only within a fairly narrow temperature range.

In terrestrial ecosystems, sunlight energy and most nutrients are relatively plentiful, so the amount and type of life present in a location is determined largely by temperature and availability of water. Terrestrial plants require liquid water in the soil, at least during part of the year. Generally speaking, areas with more precipitation have moister soil, but soil moisture is also strongly influenced by temperature. High temperatures evaporate water from the soil, and prolonged freezing temperatures turn soil water to ice, making it unavailable to plants.

The weather patterns, including patterns of temperature and precipitation, that prevail for years or centuries in a particular region make up its **climate**. Therefore, a region's climate is a good predictor of the kinds of plant species that will be present in its terrestrial ecosystems (**FIG. 30-1**). In the next section, we explore some of the factors that affect climate.

CHECK YOUR LEARNING

Can you . . .

- name the four main resources required for life on Earth?
- explain which of these requirements are most important in determining the distribution of life in aquatic ecosystems *versus* terrestrial ecosystems?
- define *climate*?

30.2 WHICH FACTORS INFLUENCE EARTH'S CLIMATE?

Weather and climate are driven by a great thermonuclear engine: the sun. When solar energy enters the atmosphere, some is reflected back into space. Most of the rest is absorbed by the atmosphere or by Earth's surface, thereby heating the planet. Most of sunlight's high-energy ultraviolet (UV) radiation, which can damage DNA and other biological molecules, is absorbed by an **ozone layer** in the middle atmosphere (see "Earth Watch: Plugging the Ozone Hole" on page 517).

The climate of different locations on Earth varies tremendously. Climate variation arises from physical properties of the planet that cause uneven heating of its surface. The factors that cause uneven heating include Earth's curvature, its tilted axis, and its orbit around the sun. Uneven heating causes surface temperatures to vary geographically and, in conjunction with Earth's rotation on its axis, generates air and ocean currents that greatly influence climate.

Earth's Curvature and Tilt Determine the Angle at Which Sunlight Strikes the Surface

The average temperature of a region depends on the amount of sunlight that reaches the surface, which in turn depends on latitude (**FIG. 30-2a**). Latitude is a measure of a location's distance north or south of the equator, expressed in degrees. The equator is defined as 0° latitude, and the poles

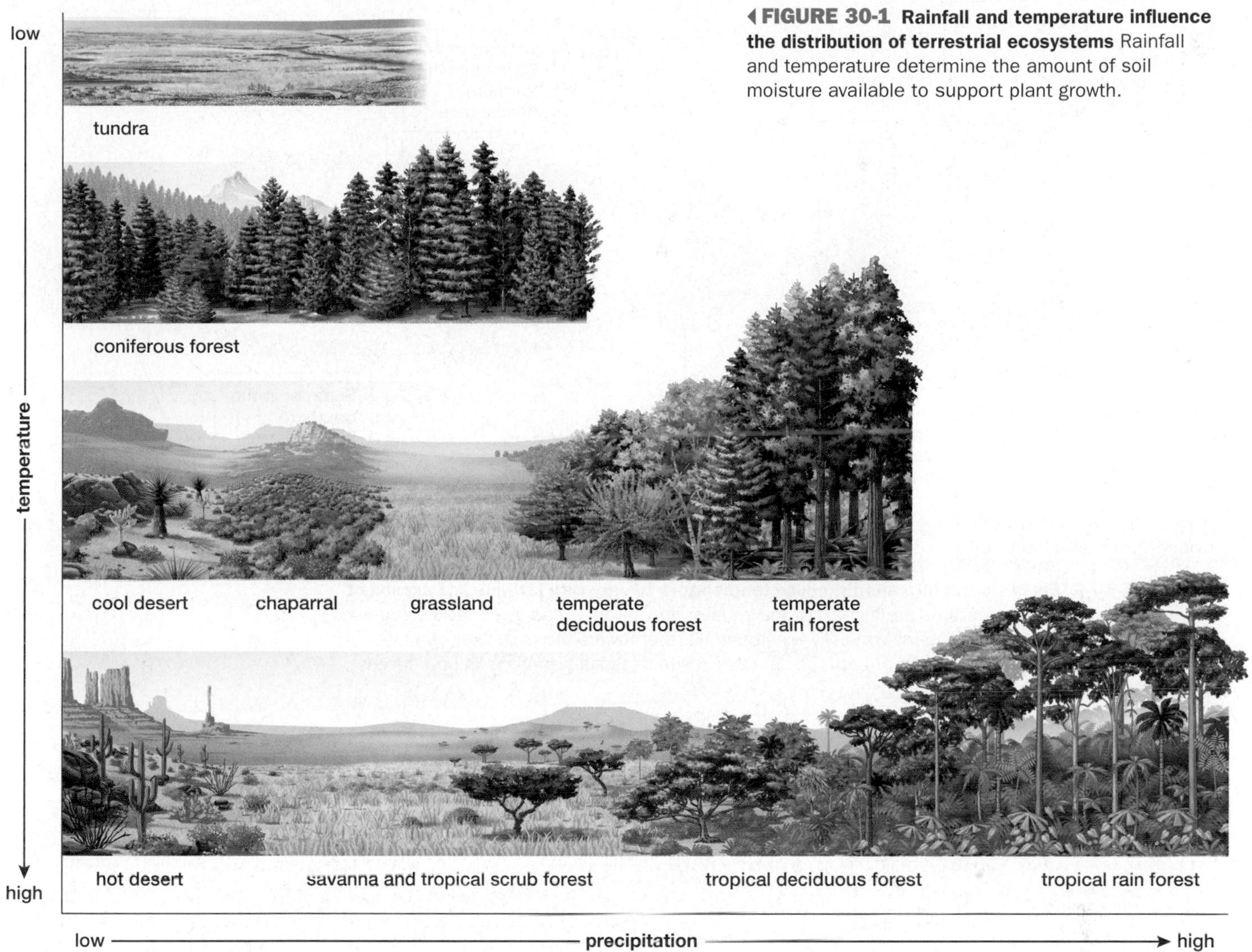

◀ **FIGURE 30-1 Rainfall and temperature influence the distribution of terrestrial ecosystems** Rainfall and temperature determine the amount of soil moisture available to support plant growth.

are at 90° north and south latitudes. At the equator, sunlight hits Earth's surface relatively directly (perpendicular to the surface), so the average temperature there is high. At higher latitudes, sunlight strikes Earth's surface at a greater slant, spreading the same amount of sunlight over a larger area and producing lower average temperatures.

Earth is tilted on its axis, so the angle of the sunlight at higher latitudes changes as the planet makes its yearly trip around the sun (**FIG. 30-2b**). This change causes pronounced seasons. When Earth's position in its orbit causes the Northern Hemisphere to be tilted toward the sun, this hemisphere receives relatively direct sunlight and experiences summer (Fig. 30-2b, left). Simultaneously, the Southern Hemisphere is tilted away from the sun, receives slanted sunlight, and experiences winter. Six months later, conditions are reversed; it is summer in the Southern Hemisphere and winter in the Northern Hemisphere (Fig. 30-2b, right). Because sunlight hits the equator fairly directly throughout the year, the Tropics remain warm year-round.

Air Currents Produce Large-Scale Climatic Zones That Differ in Temperature and Precipitation

Because the angle at which sunlight strikes Earth's surface differs at different latitudes, Earth has different climatic zones (**FIG. 30-3**). Annual warming by sunlight is greatest at the equator ❶, and the heat there evaporates water from Earth's surface, especially from the oceans. Thus, equatorial air is warm and contains a lot of moisture. Warm air is less dense than cool air, so this warm, moist air rises. As it rises, the air cools, and water condenses out, falling as rain ❷. This abundant rainfall, combined with the direct rays of sunlight the equator receives, results in a warm, wet climate in which rain forests flourish.

After the moisture has fallen from the rising equatorial air, cooler, drier air remains. The continuing upward flow of air from the equator pushes this cool, dry air to the north and south. By the time the air reaches about 30° N and 30° S

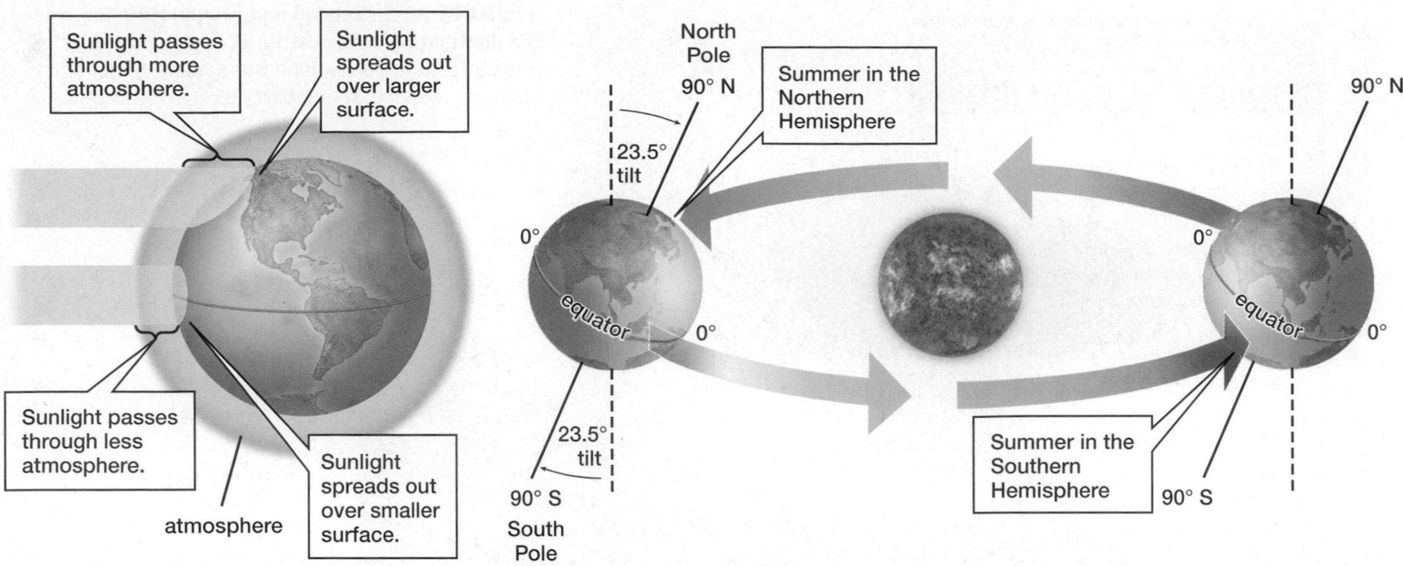

(a) The intensity of sunlight hitting Earth's surface varies with latitude.

(b) The intensity of sunlight hitting different latitudes on Earth's surface varies with the seasons.

▲ **FIGURE 30-2 Earth's curvature and tilt cause temperature to vary with latitude and season of the year (a)** Average temperatures are highest at the equator and lowest at the poles. At the equator, sunlight strikes Earth's surface nearly vertically year-round. Further toward the poles, sunlight hits Earth's surface at a slant. **(b)** The tilt of Earth on its axis causes seasonal variations in how directly sunlight strikes different latitudes.

THINK CRITICALLY The Arctic tern is a migratory seabird that experiences more hours of sunlight per year than any other species. Based on the information shown in part (b) of this figure, describe the migration pattern that allows the Arctic tern to experience so much daylight.

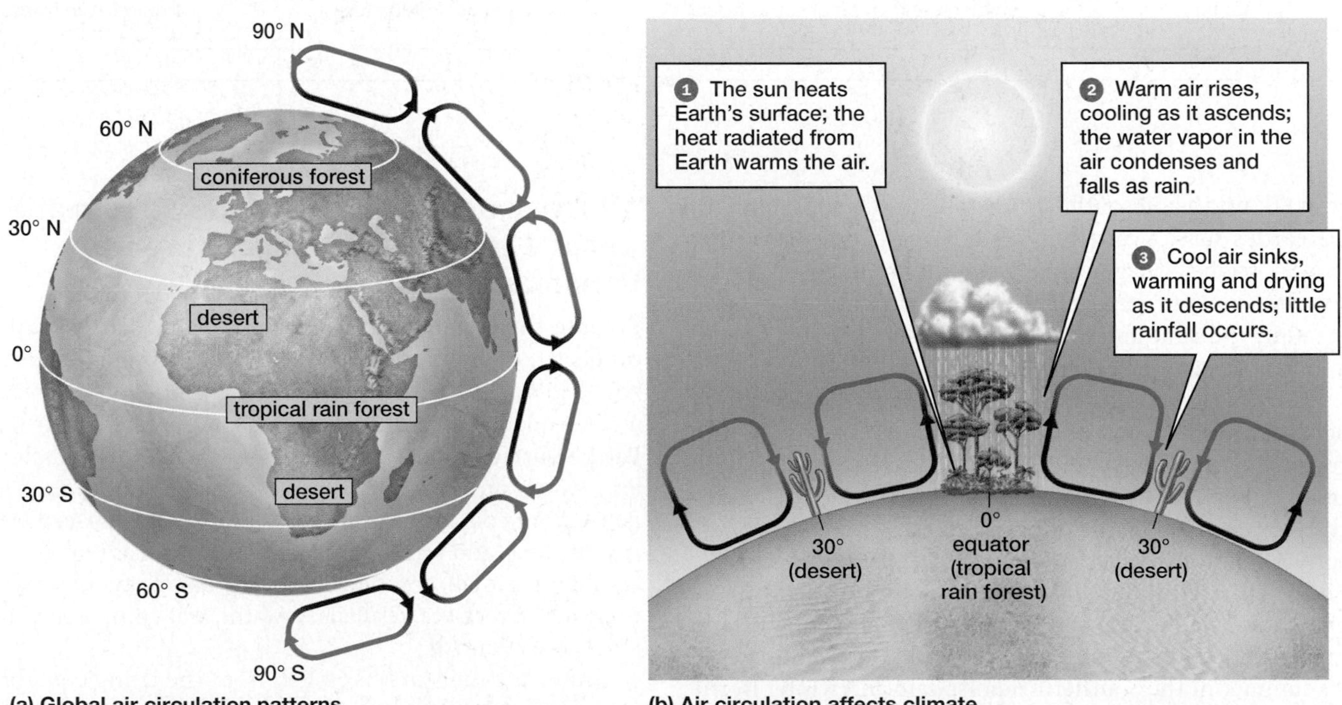

(a) Global air circulation patterns

(b) Air circulation affects climate

▲ **FIGURE 30-3 Air currents and climatic zones (a)** Warm air (red) rises at about 0° and 60° latitudes, and cool air (blue) falls at about 30° and 90° latitudes. **(b)** Air circulation patterns produce broad climatic zones.

Earth WATCH Plugging the Ozone Hole

Ultraviolet (UV) light is so energetic that it can damage biological molecules. It causes sunburn, premature aging of the skin, and skin cancer. Fortunately, more than 97% of the UV radiation that reaches Earth is filtered out by an ozone-enriched region of the upper atmosphere called the ozone layer, which begins about 6 miles (10 kilometers) above Earth's surface and extends up to about 30 miles (50 kilometers). In the ozone layer, UV light strikes molecules of ozone (O_3) and oxygen gas (O_2), causing reactions that both break down and regenerate ozone. In the process, the UV radiation is converted to heat, and the overall level of ozone remains reasonably constant—or it did before humans intervened.

In 1985, British atmospheric scientists published the startling news that springtime levels of ozone in the upper atmosphere over Antarctica had declined by more than 30% since 1979. By the mid-1990s, the **ozone hole** over Antarctica had worsened, with springtime ozone only about 50% of its original levels (**FIG. E30-1**). As a result of the diminished ozone layer, more UV light reaches Earth's surface, reducing photosynthesis by phytoplankton—the most important producers in marine ecosystems and the basis of food webs that support penguins, seals, and whales. Depletion of the ozone layer is most severe over Antarctica but is also seen over most of the world.

The ozone hole is caused primarily by human production and release of chlorofluorocarbons (CFCs). These chemicals were once widely used in the production of foam plastic, as coolants in refrigerators and air conditioners, as aerosol spray propellants, and as cleansers for electronic parts. CFCs are very stable and were considered safe. Their stability, however, is a major problem because they remain chemically unchanged as they slowly rise into the upper atmosphere. There, UV light causes them to break down and release chlorine atoms, which in turn catalyze the breakdown of ozone.

Fortunately, major steps have been taken toward plugging the ozone hole. The 1987 Montreal Protocol set limits and established phase-out periods for several ozone-depleting chemicals. In a remarkable worldwide effort, 197 countries have signed the treaty. Since 2000, the ozone layer has begun to show signs of recovery. However, because CFCs persist in the atmosphere for many years, full recovery is not expected until 2050, possibly even later.

THINK CRITICALLY UV light damages some of the molecules involved in photosynthesis, not only in phytoplankton, but also in many terrestrial plants. If the Montreal Protocol had not been implemented and ozone-depleting chemicals continued to be released in ever-increasing amounts, what effects on global climate would you expect? Explain your answer.

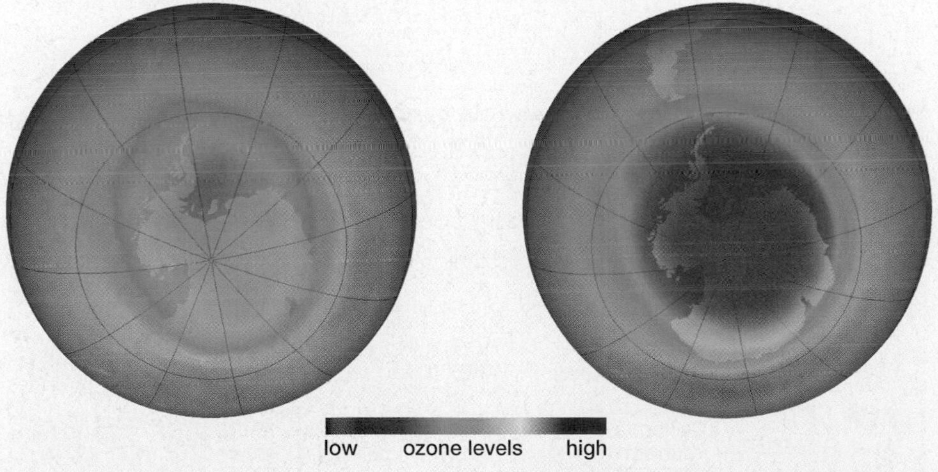

(a) Antarctic ozone hole, September 1979

(b) Antarctic ozone hole, September 2017

FIGURE E30-1 The ozone hole over Antarctica Satellite images show ozone concentrations above the Antarctic in September 1979, before significant ozone depletion occurred, and in September 2017. Low concentrations of ozone are shown in blue and purple. There are fluctuations from year to year, but the area of the hole (about 9.5 million square miles; 25 million square kilometers) and its ozone concentration (about 50% of the concentration before depletion began) have mostly stabilized since the mid-1990s. Images courtesy of NASA.

latitudes, it has cooled enough to sink. As it sinks, the air is warmed by heat radiated from Earth's surface. This descending mass of warm, dry air produces little rainfall ❸. As a result, the major deserts of the world are found near these latitudes.

After reaching the desert surface, the warm air flows to the north and south, some moving back toward the equator and some moving toward the poles. As the air moving poleward flows near the surface, its warmth evaporates water, so the air gradually becomes moister. This moderately warm, moist air rises at about 60° N and 60° S. The air cools as it rises, so water precipitates out as rain or snow, resulting in climatic conditions that favor the growth of temperate forests. The high-altitude dry air that remains flows to the poles and sinks once more, producing very little precipitation. Because they receive highly slanted sunlight in summer and no sunlight at all in the winter, the poles are also extremely cold.

These patterns of air circulation suggest that climate zones should occur in bands, with an area's climate determined by its latitude. Earth's actual climate zones are roughly consistent with this prediction, but there are some departures from the expected pattern. The deviations are largely caused by the effects of winds, ocean currents, continents, and mountains.

Terrestrial Climates Are Affected by Prevailing Winds and Ocean Currents

As air flows north and south from 30° N and 30° S latitudes, its movement is modified by the effects of Earth's rotation. This interaction determines the average direction of the wind. On average, wind blows from east to west between 30° N and 30° S latitudes, and from west to east between those latitudes and the poles. These prevailing winds produce ocean currents, as friction between winds and the ocean surface causes water to move. If there were no continents, then ocean currents would flow around the globe, east to west near the equator, and west to east in northern and southern regions. However, the continents interrupt the currents, breaking them into roughly circular patterns called **gyres**, which circulate clockwise in the Northern Hemisphere and counterclockwise in the Southern Hemisphere (**FIG. 30-4**). Gyres move water over vast distances, influencing the temperature and nutrient content of ocean waters worldwide.

Ocean gyres may affect coastal climates. Some gyres carry warm water from the Tropics to coastal regions located relatively far from the equator, creating warmer, moister climates than would be expected at these latitudes. For example, the Gulf Stream, carrying warm water from the Caribbean up the coast of North America and across the Atlantic (see **Fig. 30-4**), is responsible for the mild, infamously moist climate of the British Isles. Other ocean currents, such as the California current, move cold water from near the poles down toward the equator, causing cooler climates than would be expected in regions adjacent to these currents.

Terrestrial Climates Are Affected by Proximity to the Ocean

Earth's huge continental landmasses strongly affect terrestrial climate. Because water both heats and cools more slowly than land or air does, the coasts of continents experience less extreme temperatures than do interior regions at similar latitudes. For example, the average high temperatures in San Francisco, on the coast of California, range from 58°F (14°C) in winter to 71°F (22°C) in summer. But in St. Louis, Missouri, about 1,700 miles east of San Francisco and 600 miles from the nearest ocean, the range of temperatures is much wider; average high temperature is about 38°F (3°C) in winter and 90°F (32°C) in summer.

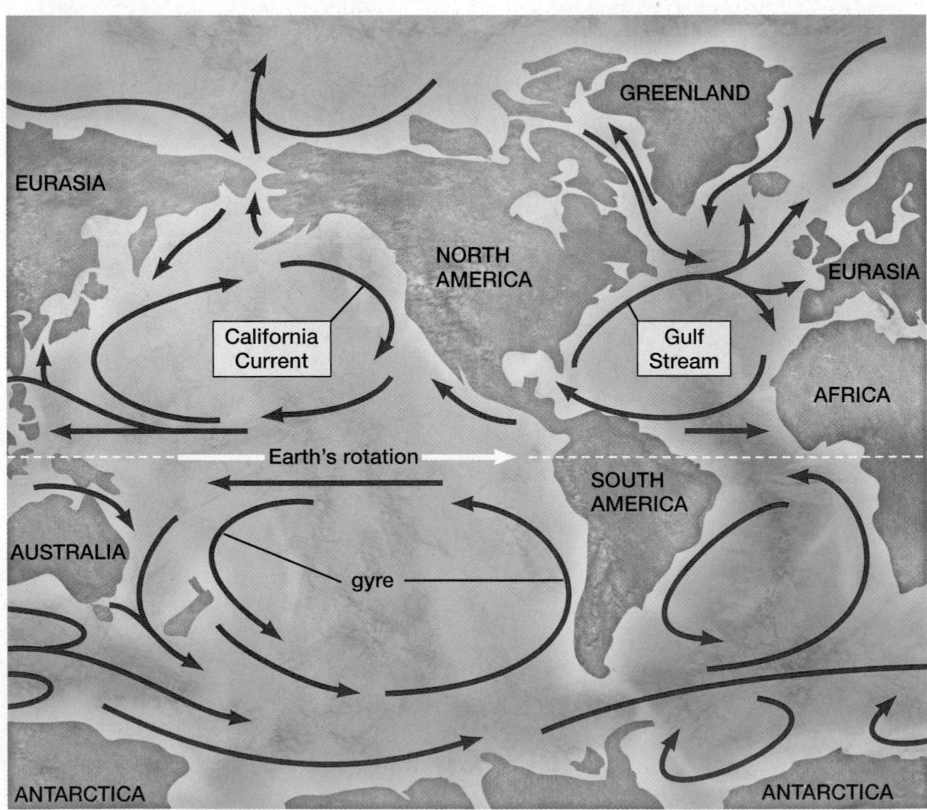

FIGURE 30-4 Ocean circulation patterns Gyres flow clockwise in the Northern Hemisphere and counterclockwise in the Southern Hemisphere. Some ocean currents, such as the Gulf Stream, carry warm water from the Tropics toward the poles. Others, such as the California Current, carry cold water from polar regions toward the equator.

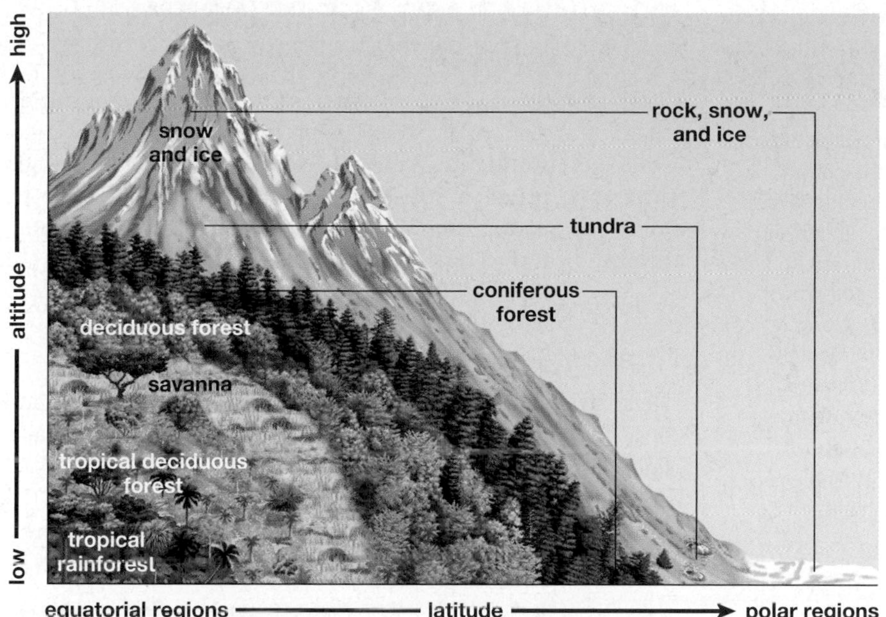

FIGURE 30-5 The similar effects of altitude and latitude on the distribution of the biomes Climbing a mountain in the Northern Hemisphere is like heading north; in both cases, increasingly cool temperatures produce a similar series of biomes.

Mountains Complicate Climate Patterns

Variation in elevation within continents also significantly affects climate. As elevation increases, air becomes thinner and cooler. Air temperature drops approximately 3.5°F (2°C) for every 1,000 feet (305 meters) in elevation, so increasing elevation and increasing latitude have similar effects on climate (**FIG. 30-5**). Even near the equator, lofty mountains, such as Mount Kilimanjaro in Tanzania (19,341 feet) and Chimborazo in Ecuador (20,565 feet), may be snowcapped much of the year.

Mountains also influence patterns of precipitation. When water-laden air meets a mountain and is forced to rise, it cools. Because cooling reduces the air's ability to hold water, the water condenses and falls as rain or snow on the windward side of the mountain. The air warms again as it travels down the far side of the mountain, so it absorbs water from the land, creating a localized dry area called a **rain shadow** (**FIG. 30-6**). For example, the Sierra Nevada range of California wrings moisture from westerly winds blowing off the Pacific Ocean. On the western side of the mountains, heavy winter snows provide moisture for forests of pine, fir, and massive sequoias. In contrast, in the rain shadow on the eastern side of the Sierra Nevada, the Great Basin Desert, the Owens Valley, and the northern Mojave Desert receive only 5 to 7 inches of rain a year and support mostly cacti and drought-resistant bushes.

CHECK YOUR LEARNING

Can you . . .

- explain how Earth's curvature, tilt on its axis, and orbit around the sun affect climate?
- explain how temperature and precipitation interact to determine soil moisture?
- describe how winds, ocean currents, continents, and mountains affect climate?

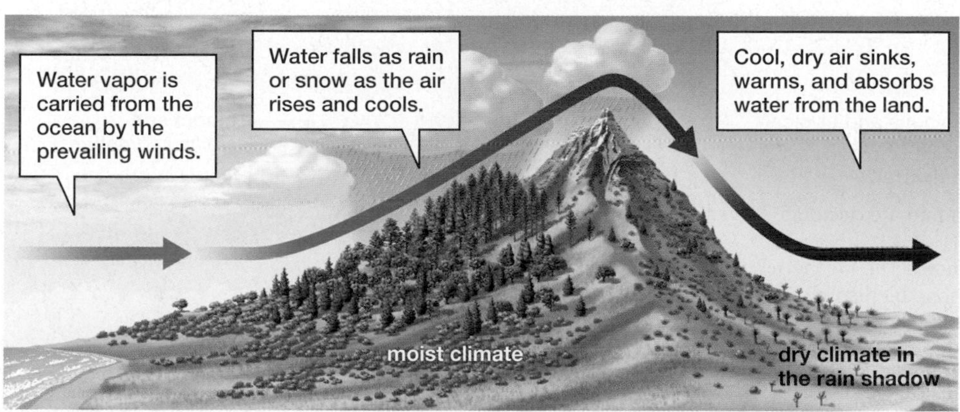

FIGURE 30-6 Mountains create rain shadows

CASE STUDY CONTINUED

Can Coffee Save Songbirds?

In recent years, coffee production in Central America has been threatened by a fungus that infects the leaves of the coffee plant and causes a disease called coffee rust. Infected leaves produce a brown powder that looks like rust. If enough leaves are infected, the entire plant may die. Coffee rust thrives at warm temperatures; at lower temperatures, the fungus grows slowly or not at all. Most Central American coffee is grown in cool climates at altitudes above 4,000 feet, so until recently coffee rust was not a significant problem. However, climate change has brought warmer temperatures and wetter weather, and the rust has proliferated, even in high-altitude farms. Coffee rust is especially prevalent in full-sun plantations. Could growing coffee in a more natural rain forest habitat help modern farmers to defeat coffee rust?

30.3 WHAT ARE THE PRINCIPAL TERRESTRIAL BIOMES?

Earth supports a variety of different terrestrial communities, each of which may extend over large areas. These large-scale communities are called **biomes** and are often named after their principal types of vegetation (**FIG. 30-7**). In the following sections, we discuss the major terrestrial biomes, beginning at the equator and working our way poleward.

Tropical Rain Forests

Near the equator, the average temperature is between 77° and 86°F (25° and 30°C), with little variation during the year. Rainfall ranges from 100 to 160 inches (250 to 400 centimeters) annually. These evenly warm, moist conditions create the most productive biome on Earth, the **tropical rain forest**, which is dominated by broadleaf evergreen trees (**FIG. 30-8**).

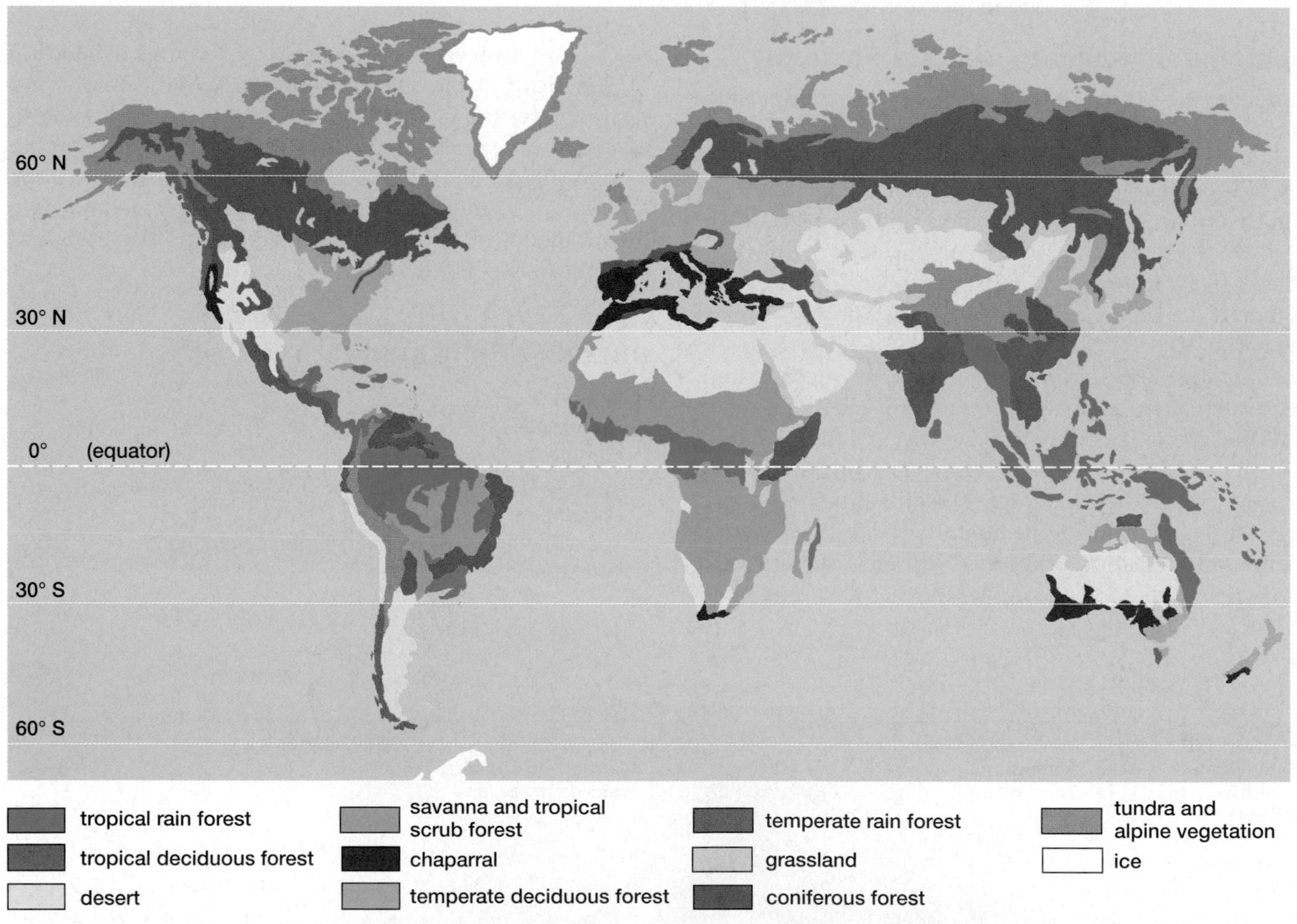

▲ **FIGURE 30-7 The world's terrestrial biomes** The northernmost parts of the Northern Hemisphere contain tundra and coniferous forests, but these biomes are rare in the Southern hemisphere, because there is little land between 45° S and the Antarctic continent. The deserts of North America, the Sahara, Saudi Arabia, Southern Africa, and Australia are located around 30° N and 30° S latitudes. Tropical rain forests are found near the equator.

Extensive rain forests are found in Central and South America, Africa, and Southeast Asia.

Rain forests have the highest **biodiversity**, or total number of species, of any biome on Earth. Although rain forests cover less than 5% of Earth's total land area, ecologists estimate that they contain half of the world's biodiversity. For example, in a 3-square-mile tract of rain forest in Peru (about 8 square kilometers), scientists counted more than 1,300 butterfly species and 600 bird species. For comparison, the entire continental United States is home to only about 600 butterfly species and 800 bird species.

Tropical rain forests typically have several layers of vegetation. The tallest trees may be more than 200 feet (60 meters) high, towering above the rest of the forest. Below these giants is a fairly continuous canopy of treetops at about 90 to 120 feet (30 to 40 meters). Another layer of shorter trees typically stands below the canopy. Woody vines grow up the trees. Collectively, these plants capture most of the sunlight. Only about 2% of the sunlight reaches the forest floor, where the plants often have enormous, dark-green leaves, an adaptation that allows them to carry out photosynthesis in dim light.

Because there is little sunlight and therefore little edible plant material close to the ground, most rain forest animals—including birds, monkeys, and insects—inhabit the trees. Competition for the nutrients that do reach the ground is intense among both plants and animals. For example, when a monkey defecates high up in the canopy, hundreds of dung beetles converge on the droppings within minutes after the waste hits the ground. Plants absorb nutrients almost as soon as soil decomposers release them from wastes or dead plants and animals. This rapid recycling means that almost all the nutrients in a rain forest are stored in the vegetation, leaving the soil relatively infertile.

Human Impacts Because of infertile soil and heavy rains, agriculture in rain forests is risky and often destructive. If the trees are cut and carried away for lumber, few nutrients

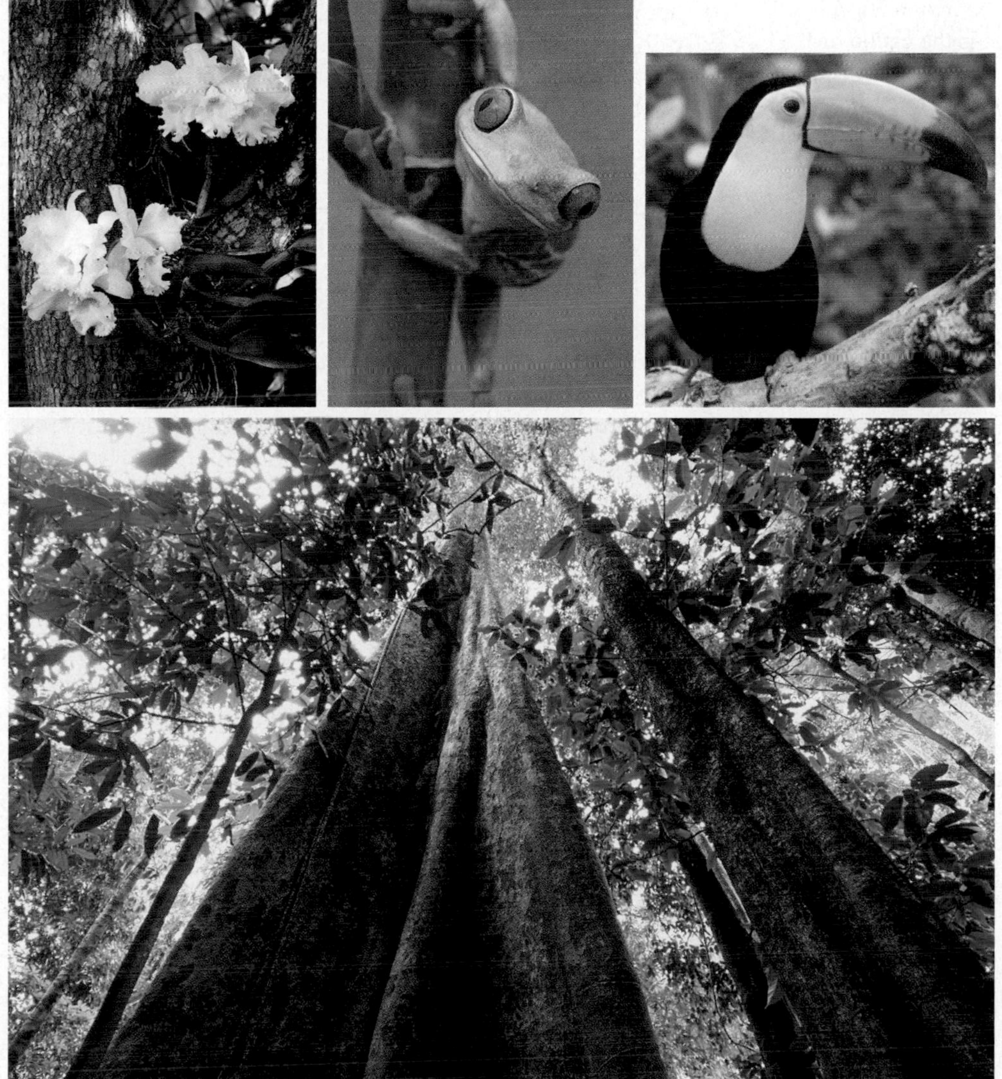

FIGURE 30-8 The tropical rain-forest biome Towering trees reach for the light in the dense tropical rain forest. Amid their branches dwells the most diverse assortment of life on Earth, including (from left to right) tree-dwelling orchids, red-eyed tree frogs, and fruit-eating toucans.

THINK CRITICALLY How does a biome with such poor soil support the highest plant productivity and the greatest animal diversity on Earth?

remain to support crops. If the trees are burned, they release their nutrients into the soil, but the heavy year-round rainfall quickly dissolves the nutrients and carries them away, depleting the soil after only a few seasons of cultivation.

Nevertheless, rain forests are being felled for lumber or burned for ranching or farming at an alarming rate. Satellite images indicate that about 20 to 30 million acres of tropical rain forest are lost each year—an area the size of a football field every 1 to 1.5 seconds. About half of the world's rain forests have now been lost. This loss intensifies the greenhouse effect and accelerates climate change, because about 10% of the CO_2 released into the atmosphere by human activities comes from cutting and burning tropical rain forests.

CASE STUDY CONTINUED

Can Coffee Save Songbirds?

When coffee is grown in the shade of tall trees in a rain forest environment, coffee rust fungus often causes relatively little damage. In shady environments, but not in full sun, a second fungus, called white halo, attacks the coffee rust fungus without harming the coffee plants. In addition, coffee plants growing slowly in shade are generally less stressed than plants growing rapidly in full sun, and are therefore less susceptible to coffee rust and other diseases. Insect pests are also less common in the shade than in full-sun plantations, mainly because birds are much more abundant in shady plantations. The birds eat harmful insects, such as coffee berry borers.

Although the ever-increasing demand for coffee means that growers are likely to convert more rain forest to plantations, shady plantations containing a diverse assortment of forest trees can help preserve many of the benefits of an intact rain forest. Can similar strategies that combine production and conservation be applied to other biomes?

Tropical Deciduous Forests

Slightly farther from the equator, annual rainfall is still high, but there are pronounced wet and dry seasons. In these areas, which include much of India as well as parts of Southeast Asia, South America, and Central America, **tropical deciduous forests** grow. During the dry season, trees in these forests cannot get enough water from the soil to compensate for evaporation from their leaves. Many shed their leaves during the dry season ("deciduous" literally means "falling off"), minimizing water loss.

Human Impacts Human activities impact tropical deciduous forests in much the same ways that they affect tropical rain forests. Logging, burning to clear land for agriculture, and cutting for firewood all contribute to deforestation of tropical deciduous forests. Fortunately, many tropical deciduous trees "stump-sprout" after logging. Therefore, if disturbances are not too severe and not too frequent, tropical deciduous forests often recover fairly quickly, with nearly the same species as were present before the disturbances occurred.

Tropical Scrub Forests and Savannas

In tropical areas with rainfall too low to support dense forests, the drier conditions produce **tropical scrub forests** dominated by deciduous trees that are shorter and more widely spaced than the trees in tropical deciduous forests. Between the scattered trees, sunlight penetrates to ground level, which allows grass to grow. In even drier regions, grasses become the dominant vegetation, with only scattered trees present; this biome is the **savanna** (**FIG. 30-9**).

Rainfall in tropical scrub forests and savannas ranges from about 12 to 40 inches (30 to 100 centimeters) per year, almost all of it falling during a rainy season lasting 3 or 4 months. When the dry season arrives, rain might not fall for months, and the soil becomes hard, dry, and dusty. Grasses are well adapted to this type of climate, growing very rapidly during the rainy season and dying back to drought-resistant roots during the dry season. Only a few specialized trees, such as the thorny acacia and the water-storing baobab, can survive the dry seasons.

The African savanna supports the most diverse array of large mammals on Earth. These include herbivores such as antelope, wildebeest, rhinos, elephants, and giraffes, and carnivores such as lions, leopards, hyenas, and wild dogs.

Human Impacts Africa's rapidly expanding human population threatens the wildlife of the savanna there. The abundant grasses that make the savanna a suitable habitat for so much wildlife also make it suitable for grazing domestic cattle. Fences erected to contain cattle disrupt the migration of herds of wild herbivores as they search for food and water. In addition, black market sales, usually in Asia, of products derived from rare African animals may be the death knell for some species. Black market demand for rhino horns has already driven the black rhinoceros to the brink of extinction; only a few thousand remain. Poaching for ivory endangers African elephants; more than 20,000 elephants are killed each year by poachers, a number that far exceeds the elephants' reproductive rate.

Deserts

Even drought-resistant grasses need at least 10 to 20 inches (25 to 50 centimeters) of rain a year, depending on the temperature and the precipitation's seasonal distribution. In areas with less than 10 inches of annual rainfall, **deserts** are found. Although we tend to think of deserts as hot, they are defined by lack of precipitation rather than by temperature.

In the Gobi Desert of Asia, for example, average temperatures are below freezing for half the year (though the summers are very hot). Desert biomes are found on every continent, typically at around 30° N and 30° S latitudes and in the rain shadows of mountain ranges.

Deserts vary in just how dry they are. At one extreme are the Atacama Desert in Chile and parts of the Sahara Desert in Africa, where it almost never rains and no vegetation grows (**FIG. 30-10a**). More commonly, deserts are characterized by widely spaced vegetation and large areas of bare ground (**FIG. 30-10b**).

Only highly specialized plants can grow in deserts. Cacti (found mostly in the Western Hemisphere; **FIG. 30-11a**) and euphorbs (found mostly in the Eastern Hemisphere; **FIG. 30-11b**) have shallow, spreading roots that rapidly absorb rainwater before it has a chance to evaporate. Their thick stems store water when it is available. Spines protect the plants from herbivores that would otherwise eat the stems for both nutrition and water. Evaporation is minimized because the leaves, if any, are very small; most photosynthesis occurs in the green, fleshy stems. A heavy wax coating on the stems further reduces water loss.

FIGURE 30-9 The African savanna Giraffes feed on savanna trees and share this biome with (from left to right) lions, rare black rhinos, and zebras.

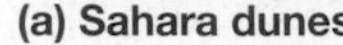

(a) Sahara dunes

(b) Utah desert

▶ **FIGURE 30-10 The desert biome** **(a)** Under the most extreme conditions of heat and drought, deserts can be almost devoid of life, such as these sand dunes of the Sahara Desert in Africa. **(b)** Throughout much of Utah and Nevada, the Great Basin Desert presents a landscape of widely spaced shrubs, such as sagebrush and greasewood.

(a) Cactus

(b) Euphorb

▲ **FIGURE 30-11 Environmental demands shape physical characteristics** Evolution in response to similar desert conditions has resulted in the bodies of **(a)** cacti and **(b)** euphorbs being nearly identical in shape, although they are not closely related to one another.

Some deserts have a very brief rainy season, in which a whole year's rain falls in just a few weeks. Annual wildflowers take advantage of the brief period of moisture to sprout from seed, grow, flower, and produce seeds of their own in a month or two (**FIG. 30-12**).

Desert animals are also adapted to survive heat and drought. Few animals are active during the hot summer days; many take refuge from the heat in underground burrows that stay relatively cool and moist. In North American deserts, nocturnal (night-active) animals include jackrabbits, bats, kangaroo rats, and burrowing owls (**FIG. 30-13**). Reptiles such as snakes, turtles, and lizards adjust their activity cycles depending on the temperature. In summer, they may be active only around dawn and dusk. Kangaroo rats and many other small desert animals survive without ever drinking. They obtain water from their food and as a by-product of cellular respiration. Larger animals, such as desert bighorn sheep, depend on permanent water holes during the driest times of the year.

Human Impacts Desert ecosystems are fragile. Desert soil is stabilized by strands of bacteria that intertwine among sand grains. Driving motorized vehicles on the desert destroys this crucial bacterial network, causing soil erosion and reducing the nutrients available to the desert's slow-growing plants. Regeneration is extremely slow: In the Mojave Desert of California, tread marks left by tanks during World War II are still visible today. Desert soil may require hundreds of years to fully recover from heavy vehicle use.

Human activities also contribute to **desertification**, the process by which relatively dry regions are converted to desert as a result of drought coupled with misuse of the land. When people overharvest bushes and trees for firewood, graze too many livestock, and deplete both surface and groundwater to grow crops, the native vegetation becomes extremely vulnerable to drought. Loss of vegetation in turn

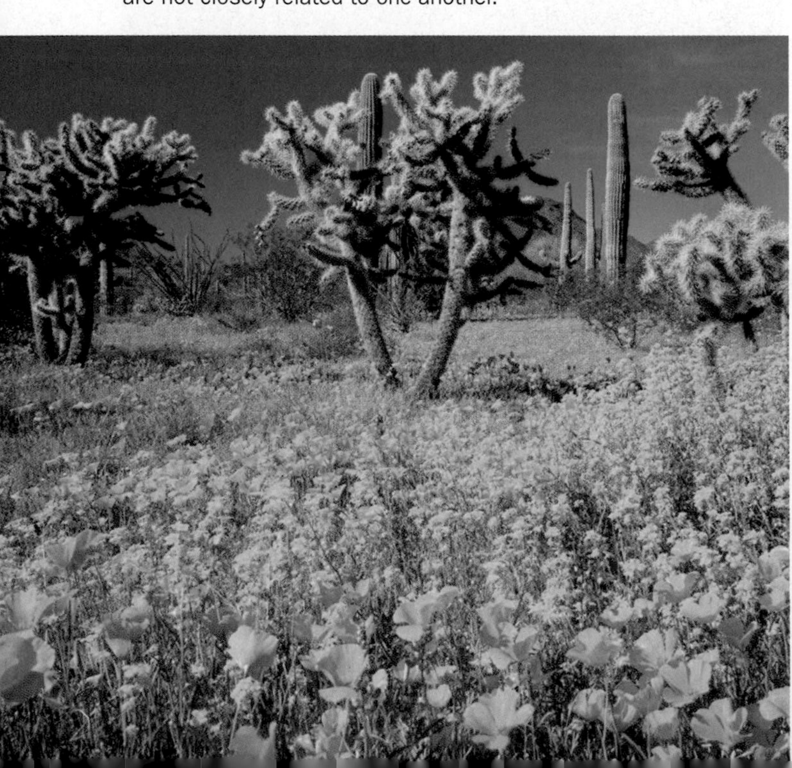

◀ **FIGURE 30-12 Desert wildflowers** After a relatively wet spring, this Arizona desert is carpeted with wildflowers. Through much of the year—and sometimes for several years—annual wildflower seeds lie dormant, waiting for adequate rains to fall.

(a) A kangaroo rat

(b) A burrowing owl

▲ **FIGURE 30-13 Desert dwellers (a)** Kangaroo rats and **(b)** burrowing owls spend the hottest part of the day in burrows, emerging at night to feed.

allows the soil to erode, further decreasing the land's productivity. Desertification has severely impacted the Sahel region in Africa just south of the Sahara Desert (**FIG. 30-14**). To combat desertification, 11 African countries plan to build a "Great Green Wall" of trees and bushes, about 9 miles (15 kilometers) wide, clear across the continent, in an effort to revegetate the degraded environment and stop desertification in the Sahel. Senegal, on the Atlantic coast, has planted 50,000 acres of drought-resistant trees, including acacia trees that provide food for livestock and gum arabic that can be sold overseas as a food additive. In Niger, preventing overgrazing and planting grasses and bushes has reduced erosion and helped forests to grow back naturally.

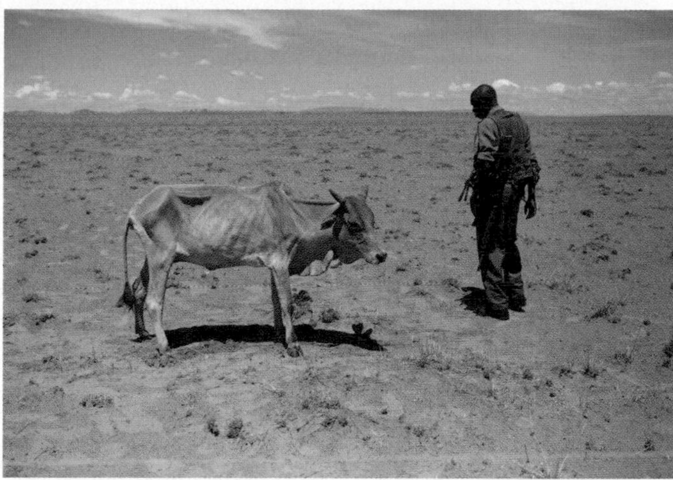

▲ **FIGURE 30-14 Desertification in the Sahel** A rapidly growing human population, coupled with drought and poor land use, has reduced the ability of many dry regions to support life.

Chaparral

Many coastal regions that border deserts, such as those in southern California and much of the Mediterranean region, support a distinctive type of vegetation called **chaparral** (**FIG. 30-15**). Annual rainfall in these areas is up to 30 inches (about 75 centimeters), nearly all of which falls during cool, wet winters. Summers are hot and dry. Chaparral plants consist mainly of drought-resistant shrubs and small trees. Their leaves are usually small and are often coated with tiny hairs or waxy layers that reduce evaporation during the dry summer months. Chaparral is adapted to withstand frequent fires; many chaparral shrubs can regrow from their roots after fires. Others have seeds that are stimulated to germinate by the high heat of fire or by chemicals found in smoke.

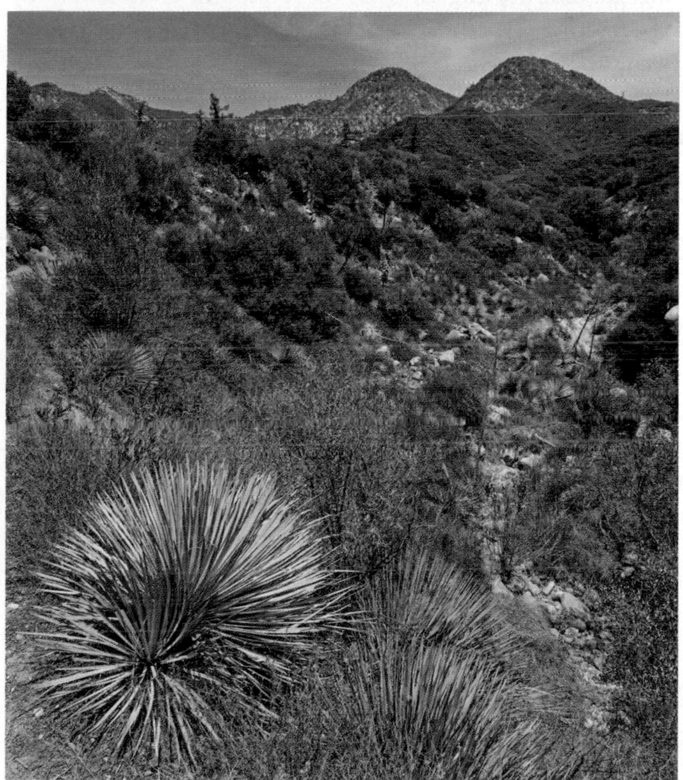

▲ **FIGURE 30-15 The chaparral biome** Limited to warm, dry coastal regions and maintained by fires caused by lightning, chaparral is characterized by drought-resistant shrubs and small trees, such as the ones seen here in the foothills of the San Gabriel mountains in Southern California.

Human Impacts People enjoy living in warm, dry climates adjacent to oceans, so development for housing is a major threat to chaparral biomes. In more rugged terrain, especially in southern Europe, chaparral has been cleared for grazing, olive groves, and other types of agriculture.

Grasslands

In the interior regions of some continents, such as North America and Eurasia, **grassland**, or *prairie,* biomes predominate (**FIG. 30-16**). Grassland biomes typically have hot summers and cold winters and receive 15 to 30 inches of rain annually. In general, these biomes have a continuous cover of grass and virtually no trees, except along rivers. In tallgrass prairie—in North America, originally found from Texas to southern Canada—grasses reach up to 6 feet in height. Areas further west in North America, which receive less rainfall, support midgrass and shortgrass prairies. In these grasslands, prairie dogs and ground squirrels provide food for eagles, foxes, coyotes, and bobcats. Pronghorns browse in western grasslands, and bison thrive in preserves.

Why do grasslands lack trees? Water and fire are the crucial factors in the competition between grasses and trees. The hot, dry summers and frequent droughts of the midgrass and shortgrass prairies can be tolerated by grasses but are fatal to trees. In tallgrass prairies, which receive more precipitation, forests are the climax ecosystems. Historically, however, tree growth was suppressed by a combination of occasional severe drought and frequent fires caused by lightning or set by Native Americans. Although fire kills trees, the root systems of grasses survive.

Human Impacts Grasses growing and decomposing for thousands of years produced the most fertile soil in the world. In the early nineteenth century, North American grasslands supported an estimated 60 million bison. Today, the Midwestern U.S. grasslands have been largely converted to farm and range land, and cattle have replaced bison.

Prairie dog colonies and the eagles and ferrets that hunted them have become a rare sight as their habitat shrinks. In some regions, overgrazing has destroyed the native grasses, allowing woody sagebrush to flourish (**FIG. 30-17**). Undisturbed grasslands are now largely confined to protected areas. Tallgrass prairie is one of the most endangered ecosystems in the world. Only about 1% of its original area remains, in tiny remnants restored by planting native species and maintained by controlled burning.

Temperate Deciduous Forests

At their eastern edge, the North American grasslands merge into the **temperate deciduous forest** biome (**FIG. 30-18**). Temperate deciduous forests are also found in much of Europe and eastern Asia. Annual precipitation in this biome is higher than in grasslands (30 to 60 inches, or 75 to 150 centimeters). The soil retains enough moisture for trees to grow, shading out most grasses.

The temperate deciduous forest biome has cold winters, often with long periods of below-freezing weather, when liquid water is not available. To conserve water when it is in short supply, deciduous trees drop their leaves in

▶ **FIGURE 30-16 Shortgrass prairie** Shortgrass prairie is characterized by low-growing grasses. In addition to many wildflowers, life in shortgrass prairies includes (from left to right) bison (in nature preserves), prairie dogs, and pronghorns.

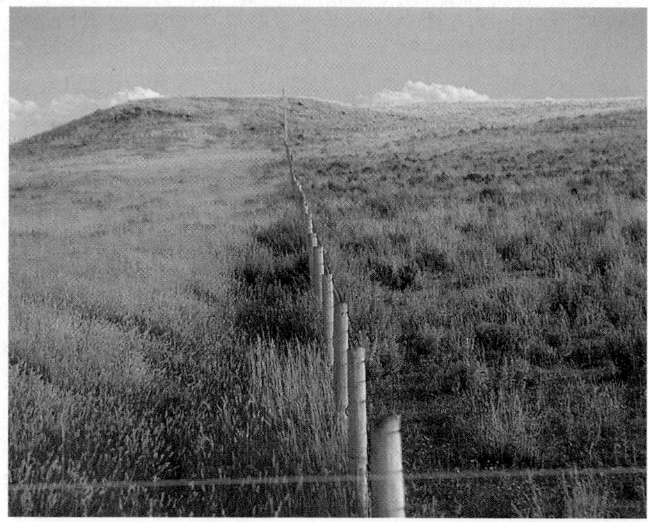

▲ **FIGURE 30-17 Sagebrush desert or shortgrass prairie?** Biomes are influenced by human activities as well as by temperature, rainfall, and soil. The shortgrass prairie field on the right has been overgrazed by cattle, causing the grasses to be replaced by sagebrush. In the ungrazed area on the left, sagebrush cannot outcompete native grasses.

the fall and remain dormant through the winter. They produce leaves again in the spring, when liquid water becomes available.

Decaying leaf litter on the forest floor provides food and suitable habitat for bacteria, earthworms, fungi, beetles, salamanders, and small plants. A variety of mammals—including mice, shrews, squirrels, raccoons, deer, bears, coyotes, and beavers—dwell in deciduous forests. Birds characteristic of deciduous forests include chickadees, grouse, and warblers.

Human Impacts Large predatory mammals such as black bears, wolves, bobcats, and mountain lions were formerly abundant in the eastern United States, but hunting and habitat loss have severely reduced their numbers. Consequently, in many areas deer populations have skyrocketed due to a lack of predators. Clearing for lumber, agriculture, and housing has dramatically reduced deciduous forests in the United States from their original extent, and virgin (uncut) deciduous forests are now almost nonexistent. Over the last century, however, deciduous forest cover has increased due to regrowth on abandoned farms and formerly logged land.

◀ **FIGURE 30-18 The temperate deciduous forest biome** Temperate deciduous forests of the eastern United States are inhabited by (from top to bottom) white-tailed deer and birds such as this blue jay; in spring, a profusion of woodland wildflowers (such as these hepaticas) blooms briefly before the trees produce leaves that shade the forest floor.

THINK CRITICALLY Both tropical deciduous and temperate deciduous forests are dominated by trees that drop their leaves for part of the year. Explain how dropping leaves is an effective adaptation in these two very different biomes.

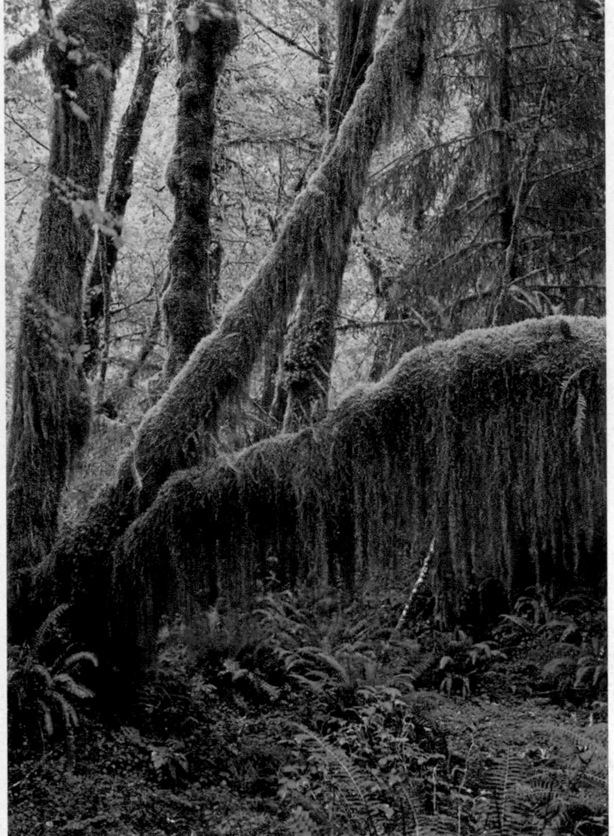

▲ **FIGURE 30-19 The temperate rain-forest biome** The Hoh River temperate rain forest in Olympic National Park receives about 12 feet of rain annually. Ferns, mosses, and wildflowers grow in the pale green light of the forest floor. Denizens of this rain forest include (from top to bottom) ferns, such as this lady fern; elk; and flowering foxglove.

Temperate Rain Forests

On the Pacific coast of the United States and Canada, from northern California to southeast Alaska, lies a **temperate rain forest** (**FIG. 30-19**). Temperate rain forests are also located along the southeastern coast of Australia, the southwestern coast of New Zealand, and parts of Chile and Argentina. In North America, these forests typically receive more than 55 inches (140 centimeters) of rain annually—and as much as 12 feet per year in some areas. The nearby ocean keeps the temperature moderate.

Most of the trees in a temperate rain forest are huge conifers, such as spruce, Douglas fir, and hemlock, commonly 250 to 300 feet tall. The forest floor and tree trunks are typically covered with mosses and ferns. Fungi thrive in the moisture and enrich the soil. As in tropical rain forests, so little light reaches the forest floor that tree seedlings usually cannot become established. Whenever one of the forest giants falls, however, it opens up a patch of light, and new seedlings quickly sprout, often right atop the fallen log.

Human Impacts Tall, straight trees are extremely valuable for lumber, and consequently many temperate rain forests have been logged. In the mild, wet climate, the forests regrow quickly, providing a renewable supply of lumber. However, some animals, such as the spotted owl, dwell mainly in old-growth forests that are hundreds of years old and do not thrive in second-growth forests. Fortunately, some pristine temperate rain forest is preserved in national parks, including Olympic in Washington and Glacier Bay in Alaska.

Northern Coniferous Forests

North of the grasslands and temperate forests are areas of **northern coniferous forest** (also called the *taiga*; **FIG. 30-20**). Northern coniferous forests, which make up the largest terrestrial biome on Earth, stretch across Scandinavia, Siberia, central Alaska, Canada, and parts of the northern United States. Similar forests occur in many mountain ranges, including the Cascades, the Sierra Nevada, and the Rocky Mountains.

◀ **FIGURE 30-20 The northern coniferous forest biome** The small needles and conical shape of conifers allow them to shed heavy snows. (Upper left) A Canada lynx chases a snowshoe hare. (Upper right) A male spruce grouse performs its mating display.

▲ **FIGURE 30-21 Clear-cutting** Coniferous forests are vulnerable to clear-cutting, as seen in this forest in Alberta, Canada. Clear-cutting is a relatively simple and inexpensive means of logging compared to selective harvesting of trees, but its environmental costs are high. Erosion diminishes the fertility of the soil, slowing new growth. Further, the dense stands of similarly aged trees that typically regrow are more susceptible to fires and parasites than a natural stand of trees of various ages would be.

Conditions in northern coniferous forests are much harsher than in temperate deciduous forests, with long, cold winters and short growing seasons. About 16 to 40 inches (40 to 100 centimeters) of precipitation fall annually, much of it as snow. The conical shape and narrow, stiff needles of evergreen conifers allow them to shed snow efficiently. The waxy coating on the needles minimizes water loss during the long winters, when water remains frozen. By retaining their leaves during the winter, evergreen conifers conserve the energy that deciduous trees must expend to grow new leaves in the spring. Therefore, when spring arrives, conifers can begin photosynthesis immediately. Large mammals—including black bears, moose, deer, and wolves—still roam the northern coniferous forest, as do wolverines, lynxes, foxes, bobcats, and snowshoe hares. These forests also serve as breeding grounds for many migratory bird species.

Human Impacts Clear-cutting for papermaking and lumber has leveled huge expanses of northern coniferous forest in both Canada and the U.S. Pacific Northwest (**FIG. 30-21**). Additional northern coniferous forest has been cleared to allow mining and extraction of natural gas and oil. Nevertheless, much of Earth's coniferous forest remains intact.

Tundra

The most northerly biome is the arctic **tundra**, a vast treeless region bordering the Arctic Ocean (**FIG. 30-22**). Conditions in the tundra are severe. Winter temperatures are often −40°F (−55°C) or lower, with howling winds. Precipitation averages 10 inches (25 centimeters) or less each year, making this region a freezing desert. Even during the summer, frosts are frequent, and the growing season may last only a few weeks. Similar climates and tundra vegetation are found at high elevations on mountains worldwide.

The cold climate of the arctic tundra results in **permafrost**, a permanently frozen layer of soil. Soil above the permafrost thaws each summer, often to a depth of 2 feet (60 centimeters) or more. When the summer thaws arrive, the underlying permafrost limits the ability of soil to absorb the water from melting snow and ice, and so the tundra becomes a marsh.

◀ **FIGURE 30-22 The tundra biome** Life on the tundra is seen here in Denali National Park, Alaska, turning color in autumn. (From left to right) Perennial plants such as this frost-covered bearberry grow low to the ground, avoiding the chilling tundra wind. Tundra animals, such as arctic fox and caribou, can regulate blood flow in their legs, keeping them just warm enough to prevent frostbite while preserving precious body heat for the brain and other vital organs.

Because of the tundra's extreme cold, brief growing season, and permafrost (which limits the depth of roots), trees do not grow there. Nevertheless, the ground is carpeted with small perennial flowers, dwarf willows, and large lichens called "reindeer moss," a favorite food of caribou. The summer marshes also provide superb mosquito habitat. The mosquitoes and other insects feed about 100 different species of birds, most of which are not year-round residents but migrate here to nest and raise their young during the brief summer feast. The tundra vegetation also supports arctic hares and lemmings (small rodents) that are eaten by wolves, owls, and arctic foxes.

Human Impacts The tundra is among the most fragile of all terrestrial biomes because of its short growing season. A willow 4 inches (10 centimeters) high may be 50 years old. Alpine tundra is easily damaged by off-road vehicles and even hikers. These impacts of civilization are mostly localized around oil-drilling sites, pipelines, mines, and scattered military bases, but another significant threat to the tundra—climate change—affects much larger areas. As temperatures rise, shrubs and trees are replacing tundra along its southern margins; climate models suggest that more than a third of Earth's tundra may be lost by the end of this century.

CHECK YOUR LEARNING

Can you . . .

- describe the principal terrestrial biomes and discuss how temperature and precipitation interact to determine their characteristic plant life?
- describe human impacts on terrestrial biomes?

30.4 WHAT ARE THE PRINCIPAL AQUATIC BIOMES?

In aquatic ecosystems, the assortment of species present is determined largely by the availability of nutrients and sunlight. Because these factors vary among aquatic environments, Earth's waters support a variety of biomes. In the following sections, we discuss the major aquatic biomes.

Freshwater Lakes

Freshwater lakes form when natural depressions fill with water from groundwater seepage, streams, and runoff from rain or melting snow. Large lakes in temperate climates have distinct zones of life: the littoral zone, the limnetic zone, and the profundal zone (**FIG. 30-23**).

Life Zones Are Determined by Light and Nutrients

Near the shore is the shallow **littoral zone**, which receives abundant sunlight and nutrients. Plants in the littoral zone include cattails, bullrushes, and water lilies, anchored in the bottom near the shore, and fully submerged plants that flourish in slightly deeper waters. Littoral waters are home to photosynthetic protists and bacteria collectively called **phytoplankton**, as well as **zooplankton**—tiny crustaceans that feed on phytoplankton. A great diversity of animal life is also found in the littoral zone. Littoral vertebrates

▼ **FIGURE 30-23 Lake life zones** A typical large lake has three life zones: a nearshore littoral zone with rooted plants, an open-water limnetic zone, and a deep, dark profundal zone.

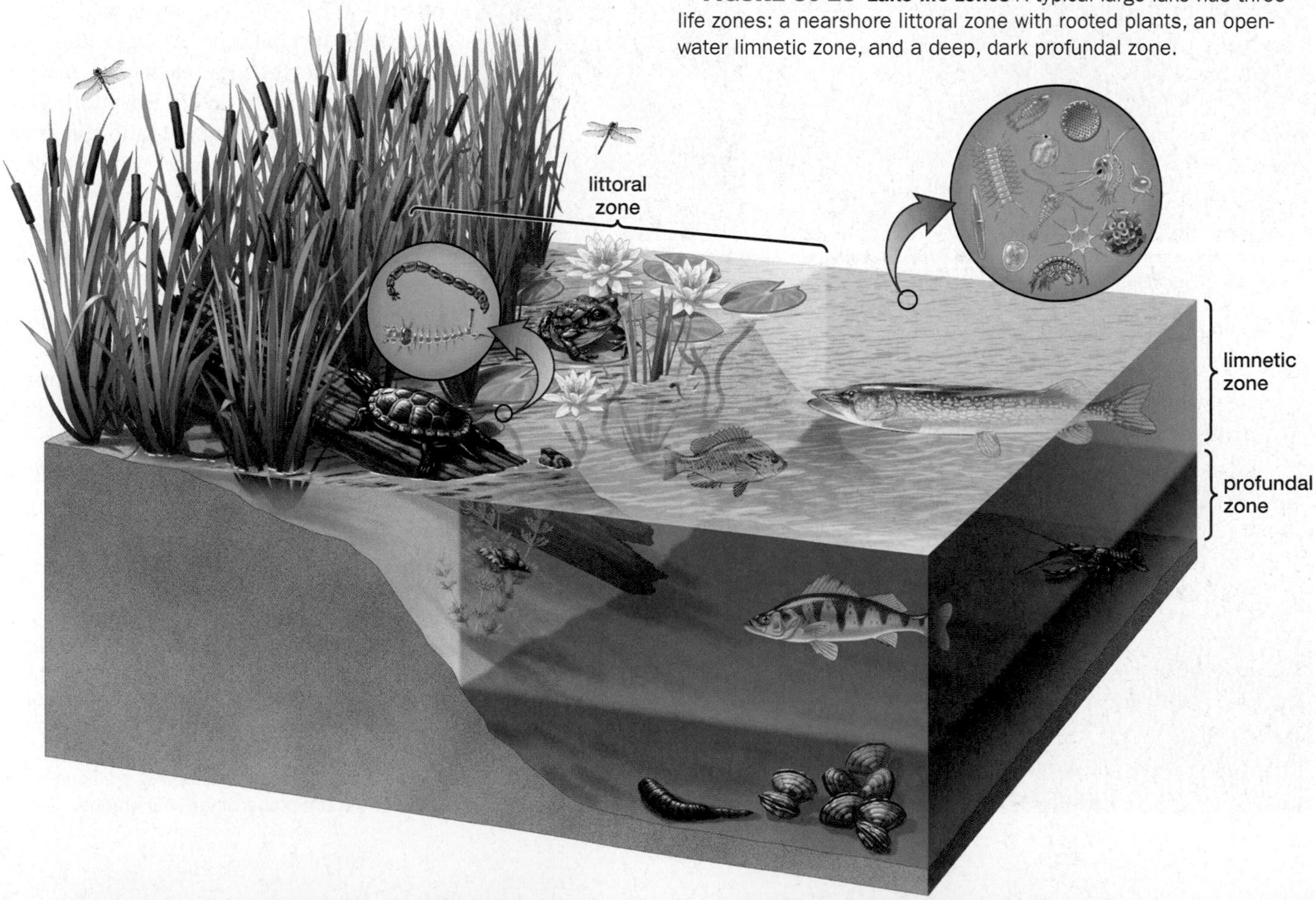

Have You Ever Wondered ... If People Can Re-Create Ancient Biomes?

Russian scientists Sergey and Nikita Zimov plan to convert Siberian tundra to grassland, restoring a biome that was present in the area prior to the arrival of humans. Until about 10,000 years ago, Siberia was home to mammoths, bison, woolly rhinos, and other large herbivores. The Zimovs hypothesize that grazing and trampling by these herbivores destroyed mosses, shrubs, and small trees—the vegetation that currently dominate the tundra—but that grasses thrived. So, when prehistoric people wiped out most of Siberia's large animals about 10,000 years ago, the grasslands disappeared. The Zimovs want to re-create the "mammoth steppe"—a vast grassland supporting herds of herbivores and the carnivores that prey on them. Of course, there aren't any mammoths or woolly rhinos anymore, but the Zimovs have introduced other large mammals to a large protected area in Siberia called Pleistocene Park. Of the introduced species—which include Yakutian horses, musk ox, European bison, and elk—the horses seem to be especially important to restoring the grasslands. Wherever there are enough horses, grasslands are returning. The "re wilding" of Pleistocene Park is well under way.

Yakutian horses thrive in Pleistocene Park

include frogs, snakes, turtles, and fish such as pike, bluegill, and perch; invertebrates include insect larvae, snails, flatworms, and crustaceans such as crayfish.

Further from shore, the water deepens, and light sufficient for photosynthesis does not penetrate to the lake bottom. Thus, plants cannot survive on the lake bottom in this open-water area, which is divided into two zones: a limnetic zone closer to the surface and a profundal zone below. In the **limnetic zone**, enough light penetrates to support photosynthesis by phytoplankton, but in the **profundal zone**, light is too weak for photosynthesis to occur (see Fig. 30-23). Plankton and fish dominate in the limnetic zone. Organisms that live in the profundal zone are nourished by organic matter that drifts down from the littoral and limnetic zones and by sediment washed in from the land. Inhabitants of the profundal zone include catfish, which mainly feed on the bottom, and detritivores and decomposers such as crayfish, worms, clams, leeches, and bacteria.

Freshwater Lakes Are Classified According to Their Nutrient Content

Freshwater lakes may be described on the basis of their nutrient content as *oligotrophic* (Greek, "poorly fed"), *eutrophic* ("well fed"), or *mesotrophic* (between these two extremes, or "middle fed"). Here, we describe the characteristics of oligotrophic and eutrophic lakes.

Oligotrophic lakes contain few nutrients and support relatively little life. Many oligotrophic lakes were formed by glaciers that scraped depressions in bare rock and are now fed by mountain streams and snowmelt. Because there is little sediment or microscopic life to cloud the water, oligotrophic lakes are clear, and light penetrates deeply. Fish that require well-oxygenated water, such as trout, thrive in oligotrophic lakes.

Eutrophic lakes receive relatively large inputs of sediments, organic material, and inorganic nutrients (such as phosphates and nitrates) from their surroundings, allowing them to support dense plant communities (**FIG. 30-24**). They are murky from suspended sediment and dense phytoplankton populations, so the limnetic zone is shallow. The dead bodies of limnetic zone inhabitants sink into the profundal zone, where they are consumed by decomposer organisms. The metabolic activities of these decomposers use up oxygen, so the profundal zone of eutrophic lakes is often very low in oxygen and supports little other life.

Human Impacts Nutrients carried into lakes from farms, feedlots, sewage, and even fertilized suburban lawns sometimes cause massive algal blooms. These blooms are followed by die-offs and decomposition that deplete the water of oxygen and kill most of the fish. Harmful algal blooms could be reduced by more effective sewage treatment, reduced fertilizer use, and proper location and operation of feedlots.

▲ **FIGURE 30-24 A eutrophic lake** Rich in dissolved nutrients carried from the land, eutrophic lakes support dense growths of algae, phytoplankton, and both floating and rooted plants.

Streams and Rivers

Streams originate in a *source region*, which is often in the mountains (**FIG. 30-25**). In the mountain source region, runoff from rain and melting snow cascades over rock. Little sediment reaches the streams, phytoplankton is sparse, and the water is clear and cold. Algae grow on rocks in the streambed, where insect larvae find food and shelter. Mountain streams are often turbulent and well oxygenated, good conditions for trout that feed on insect larvae and smaller fish.

At lower elevations, in the *transition zone*, small streams merge, forming wider, slower-moving streams and small rivers. The water warms slightly, and more sediment is carried in by tributaries, providing nutrients that allow aquatic plants, algae, and phytoplankton to proliferate. Fish such as bass, bluegills, and yellow perch (all of which require less oxygen than trout do) are found in such waterways.

As the land becomes lower and flatter, the river warms, widens, and slows, meandering back and forth. The water becomes murky with dense populations of phytoplankton. Decomposer bacteria deplete the oxygen in deeper water, but carp and catfish can still thrive despite the low oxygen levels. When precipitation or snowmelt is high, the river may flood the surrounding flat land, called a *floodplain*, depositing sediment over the adjoining terrestrial ecosystem. Ultimately. the river drains into a lake or an estuary, the area where a river meets the ocean (described below).

Human Impacts In the United States, salmon populations have been greatly reduced by hydroelectric dams, water diversion for agriculture, erosion from logging operations, and overfishing. Rivers are sometimes deepened and straightened to facilitate boat traffic, to prevent flooding, and to allow farming along their banks. In a river that has been altered in this way, water flows more rapidly, increasing erosion of the riverbed and banks. In addition, where natural flooding has been prevented, floodplain soil no longer receives the nutrients formerly deposited by floodwaters.

Freshwater Wetlands

Freshwater **wetlands**, which include marshes, swamps, and bogs, are regions where the soil is covered or saturated with water. Wetlands support dense growths of algae and phytoplankton, as well as both floating and rooted plants, including cattails, marsh grasses, and water-tolerant trees, such as bald cypress. Wetlands provide breeding grounds, food, and shelter for a great variety of birds (cranes, grebes, herons, kingfishers, and ducks), mammals (beavers, muskrats, and otters), freshwater fish, and invertebrates such as crayfish and dragonflies.

Freshwater wetlands are among the most productive ecosystems. Many occur around the margins of lakes or in the floodplains of rivers. Wetlands act as giant sponges, absorbing water and then gradually releasing it into rivers, making wetlands important safeguards against flooding and erosion. Wetlands also serve as natural water filters and purifiers. As water flows slowly through wetlands, suspended particles fall to the bottom. Wetland plants and phytoplankton absorb nutrients such as nitrates and phosphates that have washed from the land. Bacteria break down many organic pollutants, rendering them harmless.

Human Impacts About half of the freshwater wetlands in the United States (outside of Alaska) have been lost as a result of being drained and filled for agriculture, housing, and commercial uses. Destruction of wetlands makes nearby water more susceptible to pollutants, reduces wildlife habitat, and may increase the severity of floods. Fortunately, local, state, and federal agencies have cooperated to protect existing wetlands and restore some that have been degraded. These actions have combined to slow wetland loss in the United States.

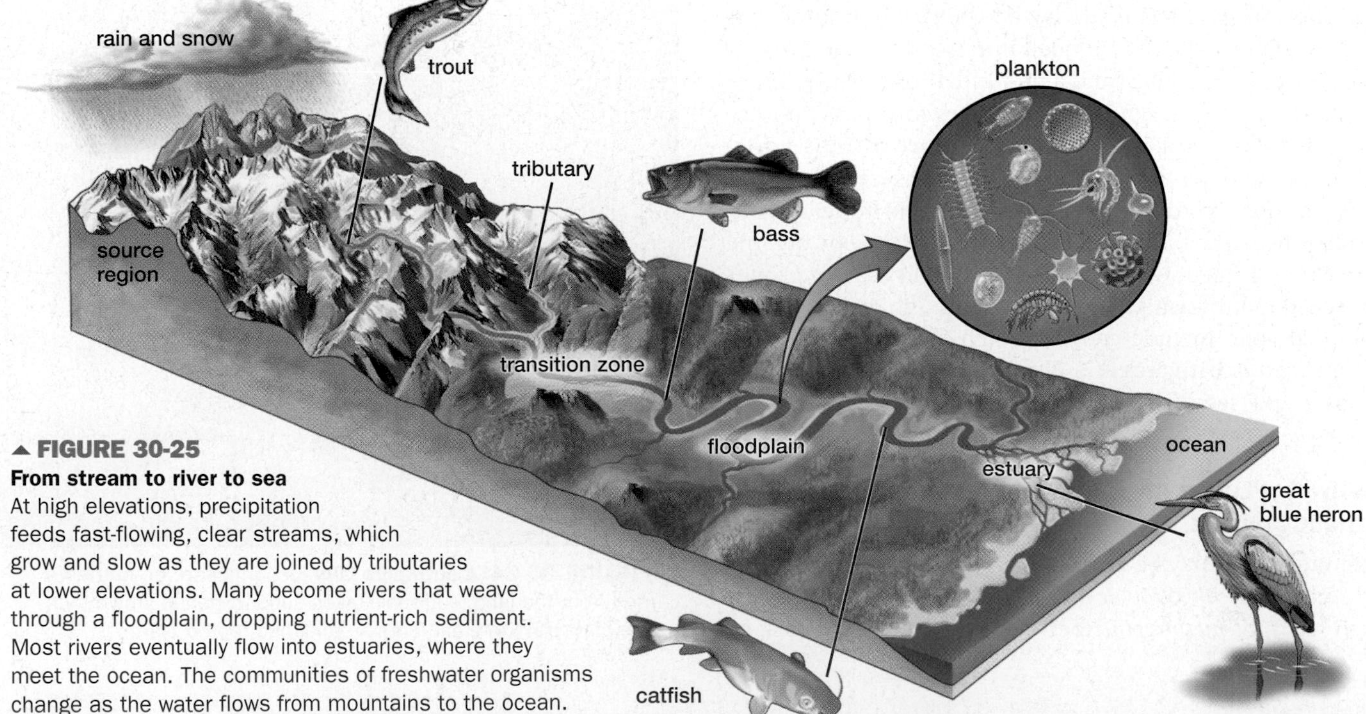

▲ **FIGURE 30-25**
From stream to river to sea
At high elevations, precipitation feeds fast-flowing, clear streams, which grow and slow as they are joined by tributaries at lower elevations. Many become rivers that weave through a floodplain, dropping nutrient-rich sediment. Most rivers eventually flow into estuaries, where they meet the ocean. The communities of freshwater organisms change as the water flows from mountains to the ocean.

Marine Biomes

The oceans can be divided into life zones characterized by the amount of light they receive and their proximity to the shore (**FIG. 30-26**). The **photic zone** consists of relatively shallow waters (to a depth of about 650 feet, or 200 meters) where the light is strong enough to support photosynthesis. Below the photic zone lies the **aphotic zone**, which extends to the ocean floor, with a maximum depth of about 36,000 feet (11,000 meters) in the Marianas Trench in the Pacific Ocean. Light in the aphotic zone is inadequate for photosynthesis. Therefore, nearly all the energy to support life must be extracted from the excrement and bodies of organisms that sink down from the photic zone above.

Shallow Water Marine Biomes

As in freshwater lakes, the major concentrations of life in the oceans are found in shallow waters where both nutrients and light are abundant. Such locations include estuaries, the intertidal zone, and kelp forests and coral reefs.

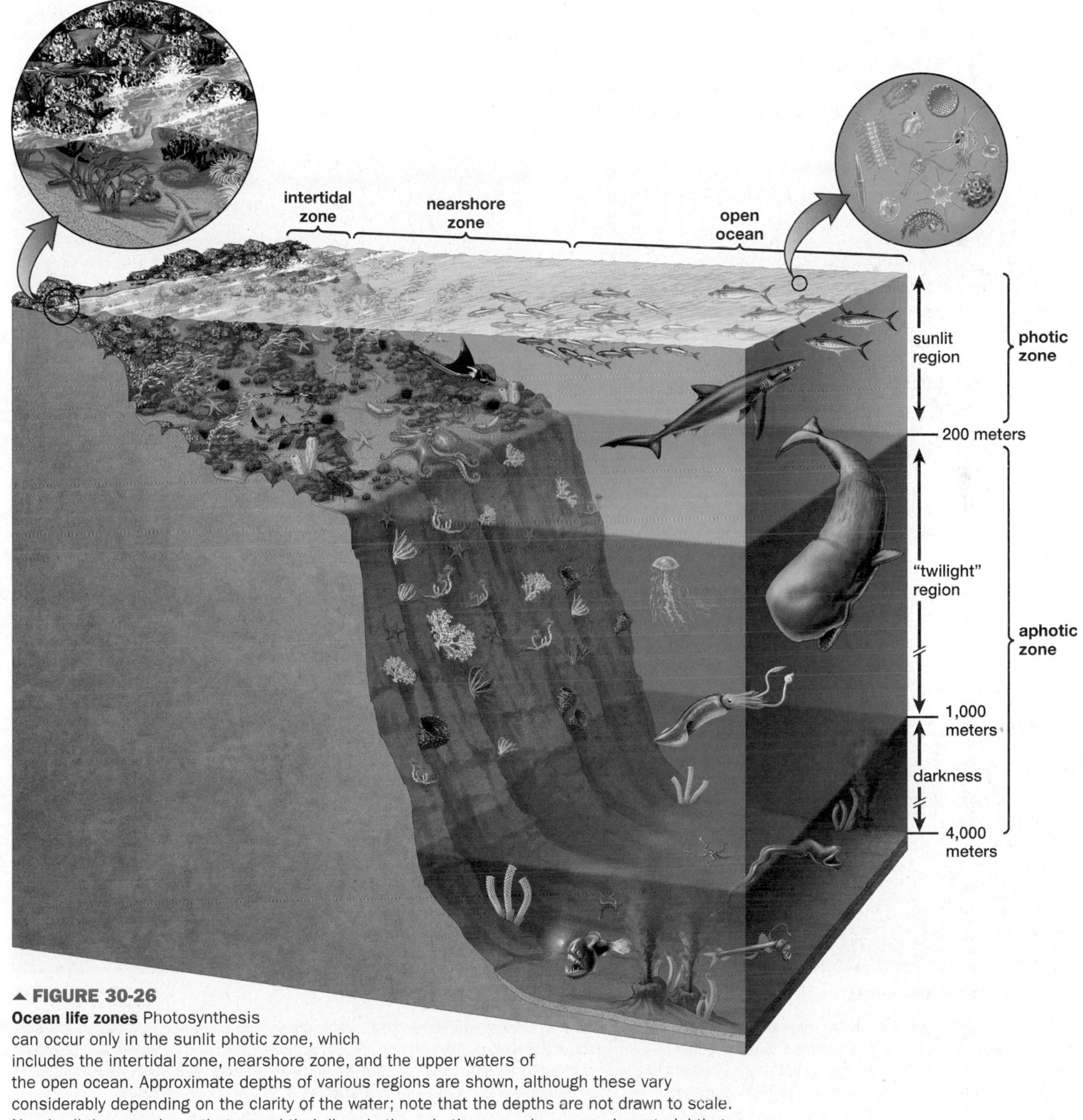

▲ **FIGURE 30-26**
Ocean life zones Photosynthesis can occur only in the sunlit photic zone, which includes the intertidal zone, nearshore zone, and the upper waters of the open ocean. Approximate depths of various regions are shown, although these vary considerably depending on the clarity of the water; note that the depths are not drawn to scale. Nearly all the organisms that spend their lives in the aphotic zone rely on organic material that drifts down from the photic zone above.

Estuaries An **estuary** is an area of brackish water where fresh water from one or more rivers mixes with seawater (**FIG. 30-27a**). Estuaries support enormous biological productivity and diversity. Many commercially important animal species, including shrimp, oysters, clams, crabs, and a variety of fish, spend part of their lives in estuaries. Dozens of species of birds, including ducks, swans, and shorebirds, feed and nest in estuaries.

Intertidal Zone The **intertidal zone**, found where the land meets the ocean, is alternately covered and exposed by the tides. The organisms that live in the intertidal zone must be able to survive both submerged in seawater and exposed to the air. During heavy rains, organisms in tide pools and mudflats may also experience significantly diluted seawater. On rocky shores, barnacles (crustaceans) and mussels (mollusks) filter plankton from the water at high tide and close their shells at low tide to resist drying. At high tide, sea stars pry open mussels to eat, sea urchins feast on algae coating the rocks, and anemones spread their tentacles to catch passing crustaceans and small fish (**FIG. 30-27b**). On sandy shores and mudflats, intertidal life is typically less diverse, but includes organisms such as sand crabs and burrowing worms.

Nearshore Zone Beyond the intertidal zone, the **nearshore zone** extends seaward from the low-tide line; water depth increases as the seafloor slopes downward. Ecosystems in the nearshore zone include kelp forests and coral reefs.

In cool nearshore waters, enormous brown algae called kelp grow as tall as 160 feet (50 meters), often occurring in dense stands called **kelp forests** (**FIG. 30-27c**). Kelp forests provide food and shelter for an amazing variety of animals, including annelid worms, sea anemones, sea urchins, snails, sea stars, lobsters, crabs, fish, seals, and otters.

(a) An estuary

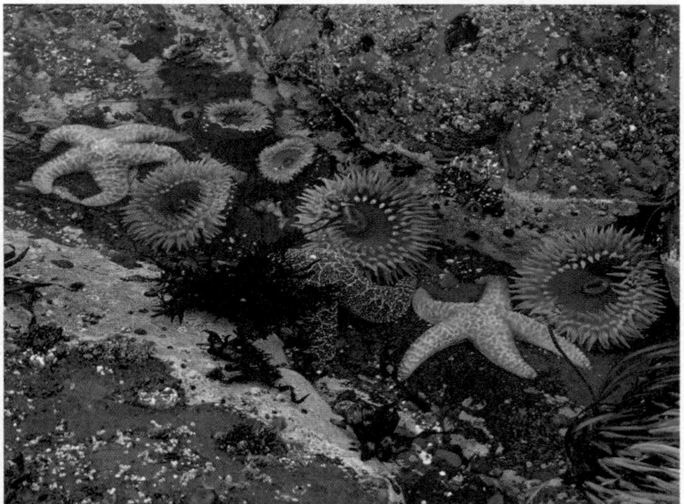

(b) A tide pool

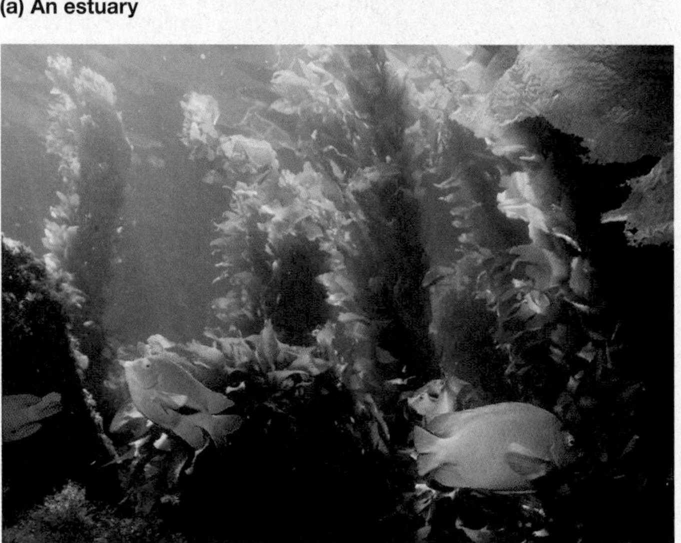

(c) An underwater kelp forest

(d) A tropical coral reef

▲ **FIGURE 30-27 Shallow water marine biomes (a)** Life flourishes in estuaries, where fresh river water mixes with seawater. Salt marsh grasses provide shelter for fish and invertebrates that are eaten by egrets (shown here) and many other birds. **(b)** Although pounded by waves and baked by the sun, tide pools in the intertidal zone harbor a brilliant diversity of invertebrates. **(c)** Kelp forests are home to a stunning array of invertebrates and fish, such as these bright orange Garibaldi damselfish. **(d)** Coral reefs provide habitat for many species of fish and invertebrates.

THINK CRITICALLY Why do estuaries and other coastal ecosystems have higher productivity than the open ocean?

In warm, tropical nearshore waters, the calcium carbonate skeletons of corals (relatives of anemones and sea jellies) accumulate over hundreds or thousands of years, building **coral reefs** (**FIG. 30-27d**). Large reefs are found in the Pacific and Indian Oceans, the Caribbean, and the Gulf of Mexico as far north as southern Florida. Coral reefs provide shelter and food for a diverse community of algae, fish, and invertebrates such as shrimp, sponges, and octopuses. The reefs are home to more than 90,000 known species, with possibly a million yet to be discovered.

Most reef-building corals harbor single-celled photosynthetic protists, called dinoflagellates, in their bodies. The relationship is mutually beneficial. The dinoflagellates provide the corals with food produced by photosynthesis while benefitting from high nutrient and carbon dioxide levels within the corals. Because their resident dinoflagellates require sunlight for photosynthesis, reef-building corals can thrive only within the photic zone, usually at depths of less than 130 feet (40 meters).

Human Impacts Human population growth is increasing the conflict between preserving coastal ecosystems as wildlife habitat and developing these areas for energy extraction, harbors, and marinas. Estuaries are threatened by runoff from farming operations, which often provide a glut of nutrients from fertilizer and livestock excrement. As in eutrophic lakes, these excess nutrients stimulate a chain of events that depletes the water of oxygen, killing both fish and invertebrates. Coral reefs are also threatened by runoff from land, which diminishes the water's clarity, harming the coral's photosynthetic partners and hindering coral growth. Mollusks, turtles, fish, crustaceans, and the corals themselves are often harvested from reefs faster than they can reproduce. Overexploitation of herbivorous fish and invertebrates often leads to an explosion of algae that smother the reefs.

The biggest threat to coral reefs comes from global warming and ocean acidification caused by increased CO_2 in the atmosphere. When waters become too warm, corals expel their colorful photosynthetic dinoflagellates and appear to be bleached. The dinoflagellates return if the water cools, but when water temperatures remain too high for too long, the corals may starve. Increased CO_2 is also slowly acidifying the oceans, reducing the ability of corals to build their skeletons of calcium carbonate.

The Open Ocean

The nearshore zone ends, and the **open ocean** begins, where the water is deep enough that wave action no longer affects the bottom. In the open ocean, the bottom is too deep to allow plants to anchor and still receive enough light to grow. Therefore, most life in the open ocean depends on photosynthesis by phytoplankton drifting in the photic zone. Phytoplankton are consumed by zooplankton, which in turn are eaten by larger invertebrates, small fish, and even some marine mammals, such as humpback and blue whales (**FIG. 30-28**).

FIGURE 30-28 The open ocean The open ocean supports fairly abundant life in the photic zone, including whales, such as these humpbacks feeding on krill. The animals in open ocean, including whales, krill (lower left), and sea jellies (lower middle), all ultimately depend on phytoplankton, the photic zone's photosynthetic producers (lower right).

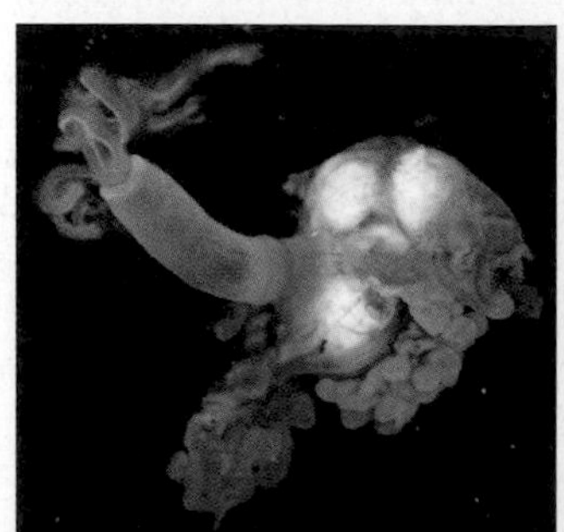

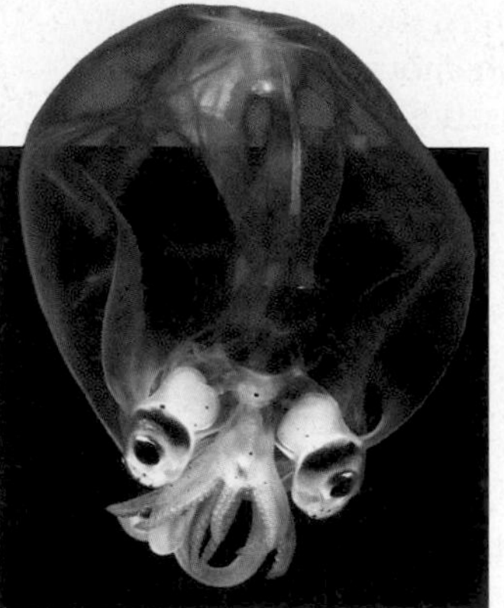

◀ **FIGURE 30-29 Denizens of the deep** The skeleton of a whale (bottom) provides a nutrient bonanza on the ocean floor. A zombie worm (upper left) inserts its rootlike lower body into the bones of the decomposing whale carcass, much of which is covered by a colorful mat of bacteria. Other denizens of the deep include viperfish (upper middle), whose huge jaws and sharp teeth allow it to grasp and swallow its prey whole, and nearly transparent squid with short tentacles below bulging eyes (upper right).

Even in the photic zone, the amount of life in the open ocean varies tremendously from place to place, largely due to differences in nutrient availability. Nutrients are provided by two major sources: runoff from the land and upwelling from the ocean depths. **Upwelling** brings cold, nutrient-laden water from the ocean depths to the surface. Major areas of upwelling occur around Antarctica and along western coastlines, including those of California, Peru, and West Africa. Nutrient-rich waters that support a large phytoplankton community are greenish and relatively murky. In contrast, the blue clarity of many tropical waters is due to a lack of nutrients, which limits the concentration of phytoplankton in the water.

Human Impacts Two major threats to the open ocean are pollution and overfishing. For example, plastic refuse, blown off the land or deliberately dumped at sea, is often mistaken for food by sea turtles, gulls, porpoises, seals, and whales. Animals that consume this refuse may die from clogged digestive tracts. Oil from oil-tanker spills, runoff from improper oil disposal on land, and leakage from offshore oil wells contaminates the open ocean.

Many fish populations are harvested unsustainably as a result of increased demand for fish and highly efficient fishing technologies. Populations of cod, bluefin tuna, haddock, mackerel, and many other fish have declined dramatically because of overfishing. The United Nations Food and Agriculture Organization estimates that 30% of marine fish stocks are being fished at unsustainable levels.

Efforts are now being made to prevent overfishing. Many countries have established quotas on fish whose populations are threatened. Fishing restrictions have succeeded in rebuilding the stocks of fish such as the Acadian redfish and Atlantic swordfish. Marine reserves, where fishing is prohibited, are increasingly being established throughout the world, causing substantial improvements in the diversity, number, and size of marine animals. Nearby areas also benefit because the reserves act as nurseries, helping to restore populations outside the reserves.

The Ocean Floor

Because the amount of light in the aphotic zone is insufficient for photosynthesis, most of the food on the ocean floor comes from the excrement and dead bodies that drift down from above. Nevertheless, life is found on the ocean floor in amazing quantity and variety, including worms, sea cucumbers, sea stars, mollusks, squid, and bizarre fish (**FIG. 30-29**). Little is known of the behavior and ecology of these creatures, which almost never survive being brought to the surface.

Entire communities feed on the dead bodies of whales, each of which contains an average of 40 tons of food. When a whale carcass reaches the ocean floor, fish, crabs, worms, and snails swarm over it, extracting nutrients from its flesh and bones. Bone-eating "zombie worms" tunnel into the bone and absorb nutrients. Anaerobic bacteria complete the breakdown of bone, and the bacteria themselves provide food for clams, worms, mussels, and crustaceans.

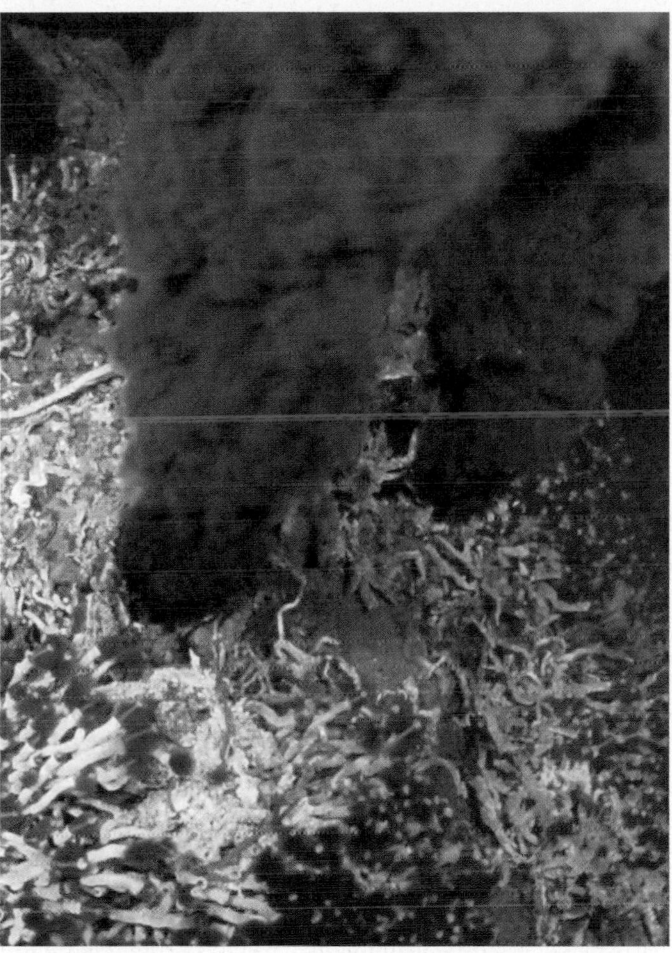

◀ **FIGURE 30-30 Hydrothermal vent communities** Hydrothermal vents spew superheated water rich in minerals that provide both energy and nutrients to the vent community. Giant red tube worms may reach 9 feet (nearly 3 meters) in length and live up to 250 years.

Hydrothermal vents on the ocean floor emit superheated water containing sulfides and other minerals, and are surrounded by **hydrothermal vent communities** (**FIG. 30-30**). These communities include pink fish, blind white crabs, enormous mussels, white clams, sea anemones, giant tube worms, and a species of snail sporting iron-laden plates of armor.

In this unique, completely dark ecosystem, sulfur bacteria serve as the producers. Instead of photosynthesis, sulfur bacteria manufacture organic molecules from carbon dioxide, harvesting energy from a source that is deadly to most other forms of life—hydrogen sulfide, which is discharged from the vents. Many vent animals consume the sulfur bacteria directly; others, such as the giant tube worm, harbor the bacteria within their bodies and live off the by-products of bacterial metabolism. The tube worms derive their red color from a unique form of hemoglobin that transports hydrogen sulfide to its symbiotic bacteria.

CHECK YOUR LEARNING

Can you . . .

- describe the principal freshwater and marine biomes?
- explain how water depth and proximity to the shore help to determine the nature and abundance of life in each biome?
- describe some effects humans have on aquatic biomes?

CASE STUDY | REVISITED

Can Coffee Save Songbirds?

World demand for coffee will probably continue to soar. Can coffee farmers meet this demand, enjoy a reasonable income, and simultaneously help to preserve rain forests? That largely depends on how the coffee plants are grown.

Full-sun plantations, in which the original vegetation has been completely removed and replaced with a monoculture of coffee plants, often have high production, but require substantial inputs of fertilizers and pesticides and provide little biodiversity. Shady plantations typically provide higher-quality coffee and more biodiversity. However, some shady plantations are better than others. For example, most coffee from Mexico is grown in shady plantations, but often under only a sparse cover of a few species of trees, which does not provide a sufficiently diverse habitat for rain-forest birds and other species. In contrast, on plantations that use a growing method referred to as "rustic," the canopy remains mostly intact, because only low-growing plants are removed from the rain forest to make way for coffee plants. Rustic plantations that promote diversity and sustainable production can be certified by the Smithsonian Migratory Bird Center, the Rainforest Alliance, and similar organizations. To achieve certification, plantations must include a diversity of tree species. Some rustic plantations provide a home for more than 150 different species of birds.

Rustic plantations are good for biodiversity, but are they bad for farmers? Not if they're well managed. In many cases, the canopy trees serve as an additional source of food or income for the farmers, providing citrus fruit, bananas, guavas, and lumber. Higher-quality coffee, combined with certification by the Bird Center or Rainforest Alliance, usually mean the beans can be sold at higher prices, with more profit for the farmers.

CONSIDER THIS Some of the farmers who decide to grow shade coffee will do so not by converting sun plantations, but by converting undisturbed forest. Given this reality, why do you think conservation organizations devote effort to encouraging consumers to purchase certified rustic plantation coffee, instead of focusing all of their resources on preserving undisturbed habitat? Explain your answer. If you were in charge of directing research related to this topic, what information would you request to help inform your opinion?

CHAPTER REVIEW

Go to **Mastering Biology** to access the Pearson eText, vocabulary review, practice quizzes, activities, videos, current events, and more.

*Answers to **Think Critically** and **Thinking Through the Concepts** questions can be found in the **Answers** section at the back of the book.*

Summary of Key Concepts

30.1 What Determines the Distribution of Life on Earth?

The requirements for life on Earth include nutrients, energy, liquid water, and an appropriate temperature range. In aquatic ecosystems, liquid water is readily available; sunlight, nutrients, and temperature determine the distribution and abundance of life. On land, sunlight energy and nutrients are usually plentiful; the distribution of life is largely determined by soil moisture, which in turn is determined by precipitation and temperature. The requirements for life occur in specific patterns on Earth, resulting in characteristic large-scale communities called biomes.

30.2 Which Factors Influence Earth's Climate?

Because of Earth's curvature, at the equator the sun's rays are nearly vertical and pass through the least amount of atmosphere; toward the poles, the rays are more slanted and must penetrate more atmosphere. Thus, the equator is uniformly warm, whereas higher latitudes have lower overall temperatures. Earth's tilt on its axis causes seasonal variations in climate at northern and southern latitudes as Earth orbits the sun. Rising warm air and sinking cool air in regular patterns from north to south produce areas of low and high moisture. These patterns are modified by prevailing winds, the presence and features of continents, and proximity to oceans.

30.3 What Are the Principal Terrestrial Biomes?

The tropical rain-forest biome, located near the equator, is warm and wet year-round. Tropical rain forests are dominated by broadleaf evergreen trees and have the highest productivity and biodiversity on Earth. Slightly farther from the equator, wet seasons alternate with dry seasons producing tropical deciduous forests in which trees shed their leaves during winter. Scrub forests and savannas, with less rain than tropical deciduous forests and extended dry seasons, are characterized by widely spaced trees with grass growing beneath. In deserts, which receive less than 10 inches of rain annually, plants are often widely spaced and have adaptations to conserve water. Desert animals have behavioral and physiological adaptations that help them to avoid excessive heat and to conserve water. Chaparral, located near coastlines, receives slightly more precipitation than deserts but is dry enough that its vegetation consists of drought-resistant bushes and small trees. Grasslands occur in interior regions of continents and have largely been converted to agriculture. When intact, these biomes have a continuous grass cover and few trees, whose growth is prevented by relatively low precipitation, fires, and severe droughts. Temperate deciduous forests, whose broadleaf trees drop their leaves in winter, dominate the eastern half of the United States and are also found in Europe and eastern Asia. Moderate precipitation and lack of severe droughts allow the growth of deciduous trees, which shade the forest floor, preventing the growth of grasses. Temperate rain forests, dominated by conifers, occur in coastal regions with both high rainfall and moderate temperatures. The northern coniferous forest nearly encircles Earth below the arctic region. It is dominated by conifers whose small, waxy needles are adapted to conserve water and take advantage of the short growing season. The tundra is a frozen desert where permafrost prevents the growth of trees and the bushes remain stunted. Tundra is found both in the Arctic and on mountain peaks.

30.4 What Are the Principal Aquatic Biomes?

In freshwater lakes, the littoral zone receives both sunlight and nutrients and supports the most life. The limnetic zone is the well-lit region of open water where phytoplankton thrive. In the deep profundal zone of large lakes, light is inadequate for photosynthesis, and most energy is provided by detritus. Oligotrophic lakes are clear, are low in nutrients, and support sparse communities. Eutrophic lakes are rich in nutrients and support dense communities.

Streams begin at a source region, often in mountains, where water is provided by rain and snow. Source water is generally clear, high in oxygen, and low in nutrients. In the transition zone, streams join to form rivers that carry sediment from land and support a larger community. On their way to lakes or oceans, rivers enter relatively flat floodplains, where they deposit nutrients, take a meandering path, and spill over the land during floods.

Most life in the oceans is found in shallow water, where sunlight can penetrate, and is concentrated near the continents, particularly in areas of upwelling, where nutrients are most plentiful. Estuaries are highly productive areas where rivers meet the ocean. The intertidal zone, alternately covered and exposed by tides, harbors organisms that can withstand waves and exposure to air. Kelp forests grow in cool, nutrient-rich coastal areas and provide food and shelter for many animals. Coral reefs, formed by the skeletons of corals, are primarily found in shallow water in warm tropical seas. Coral reefs support an extremely diverse ecosystem. In the open ocean, most life is found in the photic zone, where light supports photosynthesis by phytoplankton. In the aphotic zone, life is supported by nutrients that drift down from the photic zone. The deep ocean floor lies within the aphotic zone. Whale carcasses provide a nutrient bonanza that supports a succession of unique communities. Specialized hydrothermal vent communities are supported by bacteria that harvest energy from inorganic molecules discharged from the vents.

Thinking Through the Concepts

Bloom's: Remembering, Understanding

Multiple Choice

1. Which of the following does *not* characterize the tropical rain-forest biome?
   a. warm temperatures year-round
   b. abundant rainfall
   c. nutrient-rich soil
   d. high biodiversity

2. The biome that is mostly covered by grass and scattered trees, with warm temperatures year-round and pronounced wet and dry seasons, is the
   a. tropical deciduous forest.
   b. savanna.
   c. desert.
   d. tropical scrub forest.
3. The part of a freshwater lake that typically contains the most abundant plant and animal life is the
   a. profundal zone.
   b. aphotic zone.
   c. limnetic zone.
   d. littoral zone.
4. Which of the following is true about ocean ecosystems?
   a. Surface waters typically have abundant sunlight but low levels of nutrients.
   b. Because there isn't enough sunlight for photosynthesis, deep waters typically have no life.
   c. Surface waters in the open ocean have the highest productivity.
   d. Life near hydrothermal vents relies on photosynthesis for energy.
5. As the global climate warms, which of the following changes in the distribution of biomes is likely to occur?
   a. spread of northern coniferous forests to lower elevations on mountains
   b. expansion of the tundra biome
   c. spread of northern coniferous forests farther north
   d. spread of northern coniferous forests farther south

Fill-in-the-Blank

1. The tilt of Earth on its axis produces ____________. Coastal climates are more moderate because of ____________. A dry region on the side of a mountain range that faces away from the direction of prevailing winds is called a(n) ____________.
2. Of the four major requirements of life, which are the most important in determining the nature and distribution of terrestrial biomes? ____________, ____________ Which are most important for aquatic ecosystems? ____________, ____________, ____________
3. The most biologically diverse terrestrial ecosystems are ____________. The most biologically diverse aquatic ecosystems are ____________.
4. The shallow portion of a large freshwater lake is called the ____________. Photosynthetic plankton are called ____________; nonphotosynthetic plankton are called ____________. The open-water portion of a lake is divided into two zones, the ____________ at top and the ____________ below. Lakes that are low in nutrients are described as ____________. Lakes high in nutrients are described as ____________. The most diverse freshwater ecosystems are ____________.
5. The primary producers of the open ocean are mainly ____________. Hydrothermal vent communities are supported by bacteria that obtain energy from the compound ____________.

Review Questions

1. Explain how air currents contribute to the formation of rain forests and large deserts.
2. What are large, roughly circular ocean currents called? What effect do they have on climate, and where is that effect strongest?
3. Explain why traveling up a mountain in the Northern Hemisphere takes you through biomes similar to those you would encounter by traveling north for a long distance.
4. Why is life in the tropical rain forest concentrated high above the ground?
5. List some adaptations of desert cactus plants and desert animals to heat and drought.
6. What is desertification?
7. How are trees of the northern coniferous forest adapted to a lack of water and a short growing season?
8. How do deciduous and coniferous forest biomes differ?
9. What environmental factor best explains why the natural biome is shortgrass prairie in eastern Colorado, tallgrass prairie in Illinois, and deciduous forest in Ohio? Explain.
10. Where is life in the oceans most abundant, and why?
11. Distinguish among the littoral, limnetic, and profundal zones of lakes in terms of their location and the communities they support.
12. Distinguish between oligotrophic and eutrophic lakes. Describe a natural scenario and a human-created scenario under which an oligotrophic lake might be converted to a eutrophic lake.
13. Compare the source region, transition zone, and floodplain of streams and rivers.
14. Distinguish between the photic and aphotic zones of the ocean. How do organisms in the photic zone obtain nutrients? How are nutrients obtained in the aphotic zone?

Applying the Concepts

Bloom's: Applying, Analyzing, Evaluating

1. Fairbanks, Alaska, the plains of eastern Montana, and Tucson, Arizona, all have about the same annual precipitation. Explain why these locations contain very different vegetation.
2. Using Figures 30-3 and 30-7 as starting points, explain why terrestrial biomes are not evenly distributed in bands of latitude across Earth's surface. Explain how your proposed mechanisms apply to two specific locations.

31 Conserving Earth's Biodiversity

Reintroducing wolves to Yellowstone National Park has helped restore an ecosystem.

CASE STUDY

The Wolves of Yellowstone

IN 1926, THE LAST TWO WOLVES in Yellowstone National Park were killed by park rangers. Why were wolves intentionally exterminated in a national park? Because the park's managers had been instructed to protect the elk in the park, and wolves eat elk. But once the wolves were gone, the elk population exploded, and the newly abundant elk devastated the park's vegetation. Within a few years of the wolves' disappearance, scientists found heavy damage to aspen, cottonwoods, and willows.

In 1987, the U.S. Fish and Wildlife Service, charged with restoring endangered species, proposed transplanting wolves from Canada to Yellowstone. After considerable debate, 31 wolves were released in Yellowstone in 1995 and 1996. The transplanted wolves survived, founding a population that in recent years has fluctuated in size, ranging from about 100 to 170 wolves. After wolves were reintroduced to the park, elk populations began to decline. The elk changed their behavior as well, spending more time scanning their surroundings for predators and less time feeding. Aspen saplings suffered less browsing and began to grow again, providing increased habitat for many other species, including wildflowers and songbirds. Wolf restoration has caused profound changes in the Yellowstone ecosystem.

Is a wide variety of species important to ecosystem function, or are just a few species the key players? If species diversity is essential, why? How do human activities endanger species and potentially put at risk the ecosystems upon which all life on Earth depends?

AT A GLANCE

31.1 WHAT IS CONSERVATION BIOLOGY?

Conservation biology is the scientific discipline devoted to understanding and preserving Earth's **biodiversity**, the sum total of Earth's variety of life. Biodiversity includes:

- *Genetic Diversity* Genetic diversity is the variety and relative frequency of alleles in the gene pool of a species. It allows species to adapt to changing environments.
- *Species Diversity* Species diversity is the variety and relative abundance of species in a community. It helps buffer communities against disturbances such as drought, climate change, and invasive species.
- *Ecosystem Diversity* Ecosystem diversity is the variety of ecosystems in the biosphere, including the different habitats, communities, and ecological processes they encompass.

CHECK YOUR LEARNING

Can you . . .

- describe the goals of conservation biology?
- explain the importance of the three levels of biodiversity that conservation biologists study and seek to protect?

31.2 WHY IS BIODIVERSITY IMPORTANT?

Most of us live in cities or suburbs, and we may spend months without glimpsing an ecosystem in its natural state. So why should we care about preserving biodiversity? Many people would say that species and ecosystems are worth preserving for their own sake. But even if you disagree, a very practical reason for preserving biodiversity is simple self-interest. Ecosystems, and the biodiversity that sustains them, are essential for human well-being.

Ecosystems Provide Services That Support Human Needs

Ecosystem services are the benefits that people obtain from ecosystems. These services fall into four interconnected categories: provisioning services, regulating services, cultural services, and supporting services.

Provisioning Services Supply Products Directly

Ecosystems provide a variety of natural resources directly to people. These provisioning services include food, materials, and sources of energy.

- *Food* Most of our food comes from farms (agricultural ecosystems), but people also eat a substantial amount of wild-caught food from natural ecosystems. For example, worldwide per-person consumption of wild fish and other seafood is about 20 pounds (9 kilograms) per year. In parts of Africa, Asia, and South America, wild land animals provide an important source of protein for an often poorly nourished population.
- *Raw Materials* Natural ecosystems provide wood for construction, furniture, and paper, and provide almost all of the fresh water used for agriculture, industry, and drinking.
- *Energy* Hydroelectric power provides electricity in almost every country with enough rainfall and suitable dam sites. In less-developed countries, rural residents often rely on wood for heating and cooking.

Regulating Services Control Ecosystem Processes

Regulating services affect the quality or abundance of resources that people obtain from ecosystems, and they help control weather and climate.

- *Water Purification* Natural ecosystems, including forests, grasslands, and wetlands, purify water by removing sediments and pollutants.
- *Erosion and Flood Control* Vegetation helps to hold soil in place and prevent erosion. Plant roots also increase the soil's capacity to hold water, reducing both erosion and flooding. Thus, deforestation can make flooding and mudslides more likely (**FIG. 31-1**).

▲ **FIGURE 31-1 Loss of flood control services** Catastrophic mudslides triggered by heavy rains in Sierra Leone were worsened by massive deforestation.

- ***Pollination*** Bees and other insects pollinate most plants, including important crops such as coffee, cocoa, and many fruits.
- ***Pest Control*** Animals such as bats, frogs, birds, and wasps feed on insects, many of which are agricultural pests or transmit human diseases such as malaria.
- ***Climate Regulation*** By providing shade, reducing temperatures, and serving as windbreaks, plant communities affect local climates. Forests dramatically influence the water cycle, as water evaporating from leaves returns to the atmosphere. In the Amazon rain forest, one-third to one-half of the rain is water that evaporated from leaves.
- ***Carbon Storage*** Plants remove CO_2 from the atmosphere during photosynthesis. Some of this CO_2 is stored in the plants, especially in the trunks and roots of trees, after photosynthesis incorporates it in organic molecules. Thus, forests slow down the increase in atmospheric CO_2 that causes climate change and acidifies the oceans.

Cultural Services Supply Nonmaterial Benefits Desired by People

People use natural ecosystems to increase their enjoyment of life.

- ***Recreation*** Many, perhaps most, people take pleasure in "returning to nature." In the United States, more than 430 million visitors flock to national parks and national forests each year. Hundreds of millions more go to wildlife refuges and state parks.
- ***Tourism*** Ecotourism, in which people travel to observe unique biological communities, is a rapidly growing recreational industry. Examples of ecotourism destinations include tropical coral reefs and rain forests, the Galápagos Islands, the African savanna, and even Antarctica (**FIG. 31-2**).
- ***Mental and Physical Health*** Scientific studies have found that being in, or even just looking at, natural environments hastens healing after surgery, reduces stress hormone levels, improves mental focus in children with ADHD, and boosts the immune system.

Have You Ever Wondered ...

Why Cutting Back on Meat Helps the Environment?

Most of the animals we raise for meat are fed grain, and it takes many pounds of grain to produce a pound of meat. So, when we eat meat rather than food from plants, we indirectly use a much larger amount of the resources used for agriculture—land, water, pesticides, fertilizers, and fossil fuels. Consumption of these resources reduces biodiversity and creates pollution, including greenhouse gases. Plus, clearing land for cattle is a major cause of deforestation, cattle release a tremendous amount of methane (a potent greenhouse gas), and runoff from industrial-scale livestock feeding operations pollutes waterways. Cutting back on meat is a great way to contribute to sustainability.

Eating less meat helps preserve ecosystems

Supporting Services Are the Foundation of Ecosystem Services

Ecosystem services that affect people only indirectly or take a very long time to affect human welfare are usually classed as supporting services.

- ***Photosynthesis*** Photosynthesis by plants, algae, and bacteria provides the oxygen needed for life on Earth and is the first step in providing food for almost all living organisms, including people.

(a) Scuba diving in a coral reef

(b) Spotting penguins in Antarctica

▲ **FIGURE 31-2 Ecotourism** Carefully managed ecotourism represents a sustainable use of natural ecosystems, generating revenue and providing an incentive to preserve wildlife habitat.

- *Genetic Resources* The wealth of genes found in wild plants may help protect our food supply. Most of our food is supplied by only 12 crop plants, including rice, wheat, and corn. Researchers have identified genes in wild relatives of these domesticated plants that might be transferred into crops to increase their productivity or provide greater resistance to disease, drought, or salty soil.
- *Soil Formation* Natural ecosystems on land slowly build soil. For example, the rich soils of the midwestern United States accumulated under natural grasslands over thousands of years. Farmers have converted these grasslands into one of the most productive agricultural regions in the world.
- *Nutrient Cycling* As we described in Chapter 29, nutrients cycle within and between ecosystems, often moving from reservoirs that are not available to living organisms to chemical forms that organisms can use. For example, nitrogen-fixing bacteria in soil convert atmospheric nitrogen into ammonia and nitrate, which plants can then use to synthesize proteins and nucleic acids.

Earth's Ecosystem Services Have Enormous Monetary Value

Historically, people have assumed that ecosystem services are free and unlimited. Therefore, the value of ecosystem services has seldom been taken into account when making decisions about land use, farming practices, power generation, and other economic activities. But ecosystem services are actually enormously economically valuable. An international team of ecologists, economists, and geographers has calculated that ecosystem services provide benefits to humanity worth between $125 trillion and $145 trillion per year, about twice the value of all goods and services produced everywhere in the world.

Biodiversity Supports Ecosystem Function

The value of ecosystem services tends to be greatest in areas with high biodiversity. Biodiversity enhances ecosystem function because more diverse communities tend to have higher productivity (see Chapter 29), and because they are often better able to withstand disturbances, such as drought, severe winters, pollution, or invasive species.

Why are some diverse ecosystems more resistant to stress and disturbance? One hypothesis is that in diverse communities, several species may have similar ecological roles, so that even if a species is lost due to disturbance, its ecological function will not be lost. For example, several species of bees in an ecosystem may pollinate flowers, and if a few of these species are lost, the remaining ones may increase their population size and still pollinate the flowers.

Species that are crucially important to ecosystem function are called **keystone species** (see Chapter 28). The term is inspired by the keystone that sits at the top of a stone arch and holds all the other pieces in place. Remove the keystone, and the whole arch collapses. Similarly, in a biological community, a keystone species is one whose role is much more important than would be predicted by the size of its population or by a superficial glance at its position in the food web. In the oceans, the great whales may be keystone species, as we explore in "Earth Watch: Whales—The Biggest Keystones of All?"

CHECK YOUR LEARNING

Can you . . .

- describe the major categories of ecosystem services and provide examples of each?
- explain why biodiversity helps to maintain functioning ecosystems?

CASE STUDY CONTINUED

The Wolves of Yellowstone

After wolves were exterminated from Yellowstone, there were still plenty of grizzly bears, coyotes, foxes, and mountain lions in the park. However, all of these predators put together could not fill the role of the wolf in the Yellowstone ecosystem. In Yellowstone, the wolf is a top predator and keystone species that helps to determine populations not only of its direct prey, but also of other species. In the early twentieth century, no one expected that exterminating wolves would wreak havoc on aspens and the many species that depend on them. Unfortunately, in many parts of the world, loss of biodiversity, including keystone predators, continues today. What are the principal causes?

31.3 WHAT ARE THE MAJOR THREATS TO BIODIVERSITY?

No species lasts forever. Over the course of evolutionary time, species arise, flourish for various periods of time, and go extinct. If all species are fated to eventual extinction, why should we worry about modern extinctions? Because the rate of extinction during modern times has become extraordinarily high.

Extinction Rates Have Risen Dramatically in Recent Years

The fossil record indicates that, in the absence of cataclysmic events, extinctions occur naturally at a very low rate. This *background extinction rate* ranges from about 0.1 to 1 extinction per million species per year. However, the fossil record also provides evidence of five major **mass extinctions**, during which many species were eradicated in a relatively short period of time (see Chapter 18). The most recent mass extinction happened roughly 66 million years ago, abruptly ending the age of dinosaurs. Sudden changes in the environment, such as might be caused by an enormous meteor impact or rapid climate change, are the most likely reasons for these mass extinctions.

Many biologists have concluded that we are now in the midst of a sixth mass extinction, this one caused by humans. Researchers estimate that the modern extinction rate is about 1,000 times higher than the background rate, perhaps as high as 1,000 extinctions per million species per year. Extinctions of birds and mammals are best documented, although these represent only about 0.1% of the world's species. Since the 1500s, we have lost about 1.7% of all mammal species and about 2% of all bird species, an extinction rate more than 100 times the background rate.

Whales—The Biggest Keystones of All?

The oceans are a lot emptier than they used to be. No one knows for sure how many whales originally roamed the seas, but researchers estimate that commercial whaling reduced many whale populations by 90%. Even now, decades after almost all commercial whaling ceased, most whale populations are much less than half their pre-whaling size. Because whales are at the top of the food chain, you would think that whale prey, from giant squid to shrimp-like krill, would be experiencing a population boom. But they're not. Populations of krill have not increased and may have even declined a little. How can that be? Even though krill are eaten by whales, they also depend on them. Marine ecologists hypothesize that krill populations rely on nutrients carried by whales from ocean depths to surface waters.

Krill eat photosynthetic phytoplankton, which can only live in well-lit waters near the surface. Surface waters, however, generally contain only low levels of the nutrients plankton require, such as iron and nitrogen. Nutrients are scarce at the surface because they tend to be concentrated in organisms that are denser than seawater and sink after death, carrying nutrients from the surface down to the depths. Ocean currents and winter storms bring some of these nutrients back up to the surface, but not all.

Enter the whales. Many whales feed at substantial depths, up to a half mile below the surface, but they return to the surface to breathe and to eliminate wastes (**FIG. E31-1**). Whales release huge plumes of buoyant feces and urine that contain nutrients from the depths. The feces and urine bring nitrogen to the surface in the form of ammonia and urea; the feces are also rich in iron. By bringing nutrients to sunlit surface water, the "whale pump" enhances the productivity of the oceans.

◀ **FIGURE E31-1 Whale feces fertilize the oceans** Whales, such as the sperm whale shown here, release huge plumes of semi-liquid feces that drift in surface waters, providing essential nutrients for photosynthetic algae.

Today, almost all whaling has ceased, and most whale populations are increasing. As these giant keystone predators recover, populations of krill and other species in whales' ocean ecosystems should start to recover, too.

THINK CRITICALLY How would you test the whale pump hypothesis? Assume that you can measure populations of whales, krill, and phytoplankton and the concentrations of iron, nitrogen, and other nutrients in ocean waters.

Each year, a Red List of at-risk species is published by the International Union for Conservation of Nature (IUCN), the world's largest conservation network. Species are described as vulnerable, endangered, or critically endangered, depending on how likely they are to become extinct in the near future. Species that fall into any of these categories are described as threatened. In 2018 the Red List contained 25,821 threatened species. The U.S. Fish and Wildlife Service lists more than 1,500 threatened species in the United States alone. Why are so many species in danger of extinction? The greatest threats to biodiversity are habitat destruction, overexploitation, invasive species, pollution, and global climate change.

Habitat Destruction Is the Most Serious Threat to Biodiversity

Habitat loss imperils more than 85% of all endangered mammals, birds, and amphibians. The most serious threat is the loss of tropical rain forests, home to about half of Earth's plant and animal species. Satellite images show that about 30,000 to 45,000 square miles of rain forest are lost each year, or the area of a football field every 1 to 1.5 seconds (**FIG. 31-3**). The primary cause of the destruction of tropical rain forests is converting the land to agriculture, to create either small subsistence farms or huge plantations and ranches that supply beef, soybeans, palm oil, sugarcane, and biofuels, mostly to developed countries (see "Earth Watch: Biofuels—Are Their Benefits Bogus?" in Chapter 7).

▲ **FIGURE 31-3 Habitat destruction** Loss of habitat due to human activities is the greatest single threat to biodiversity worldwide.

Even when a natural ecosystem is not destroyed, it may become broken into small pieces, separated by roads, farms,

DOING Science Detecting the Effects of Forest Fragmentation

How do scientists study the effects of habitat fragmentation? In most studies on this topic, researchers simply compare fragmented and intact habitats. Such observational comparisons, however, yield results that are less conclusive than findings from the "gold standard" of scientific evidence: controlled experiments. Although it is challenging to carry out an experiment on a scale large enough to test hypotheses about habitat fragmentation, a few research groups have managed to accomplish the feat. The longest-running example of such an experiment is the Biological Dynamics of Forest Fragments Project.

What Question Was Asked?

Tropical rain forest is among Earth's most diverse and productive biomes. Unfortunately, rain forests are being rapidly cleared and fragmented, so scientists are urgently interested in the biological consequences of habitat fragmentation in this biome. The Forest Fragments Project was designed to test for the effects of fragmentation on species, community structure, and ecosystem function, and to test whether any of these effects differ between fragments of different sizes.

How Was Evidence Gathered?

The project began in the early 1980s, in undisturbed Amazonian rain forest in central Brazil. Researchers measured a variety of ecological variables in three large areas (3,000 to 5,000 hectares each) that they knew were slated to be cleared for cattle ranches. The researchers persuaded the landowners to leave 11 forest fragments in place when the land was cleared—five fragments of 1 hectare, four of 10 hectares, and two of 100 hectares (**FIG. E31-2**). The forest surrounding the fragmented area remains intact, extensive, and undisturbed, and thus serves as a control. Researchers have measured ecological variables more or less continuously over the life of the experiment, continuing to the present day.

What Was Learned?

Over four decades, the experiment has produced a variety of results, and some general conclusions have emerged. Compared to control plots in intact forest, species diversity declined dramatically in fragments, with the largest declines in the smallest fragments. What's more, the decline continued over time, with additional species disappearing even decades after the initial fragmentation (**FIG. E31-3**).

Other measures of ecosystem health also declined in fragments. For example, total animal and plant biomass was reduced, the rate of nutrient cycling became much slower, soil nutrient concentrations declined, and the conditions became hotter and drier. As with the effects on biodiversity, all of these effects were greater in smaller fragments and continued to worsen over long periods.

▲ **FIGURE E31-2 Biological Dynamics of Forest Fragments Project** Shown here are two of the rain forest fragments created as part of the longest-running habitat fragmentation experiment.

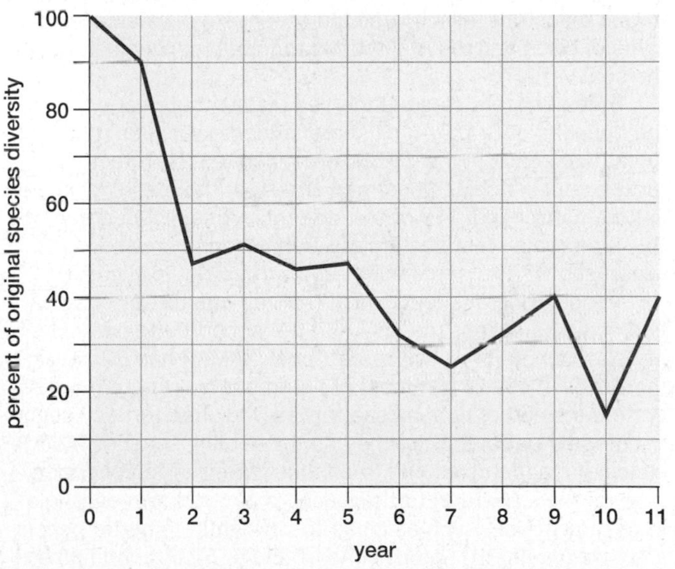

▲ **FIGURE E31-3 Biodiversity is lost in fragmented forests** This graph shows the loss of bird species in 1-hectare forest fragments during the first 11 years after experimental fragmentation.

THINK CRITICALLY Why do some species that eventually become extinct in small forest fragments persist for years before disappearing?

or housing developments. This **habitat fragmentation** is a serious threat to wildlife. For example, in small forest patches, reproductive success of many songbird species is unsustainably low. Big cats are also threatened by habitat fragmentation. For example, India has set aside 47 forest reserves intended to protect the endangered Bengal tiger, but many of the reserves have become islands in a sea of development, forcing the tigers into isolated patches that may be too small to support a **minimum viable population**, the smallest isolated population that can persist in spite of natural events such as disease and drought. For Bengal tigers, most wildlife experts think that a minimum viable population must include at least 50 females—more than are found in most of India's tiger reserves. For an example of how researchers study habitat fragmentation, see "Doing Science: Detecting the Effects of Forest Fragmentation."

Governments in many countries are taking action to preserve critical habitat. One of the world's largest protected

Earth Watch: Saving Sea Turtles

Six of the seven species of sea turtles are threatened with extinction. Their distinctive life history makes them especially vulnerable. Most sea turtles don't begin to breed until they are 20 to 50 years old. When females reach reproductive age, they must swim hundreds, even thousands, of miles to reach their nesting grounds, which are often on the same beaches where they hatched. The turtles drag themselves ashore, excavate a hole in the sand, deposit their eggs, and return to the sea (**FIG. E31-4a**). The eggs may be eaten by domestic dogs, foxes, wild pigs, raccoons, and a host of other predators. After about 2 months, baby turtles emerge from the surviving eggs and begin their difficult journey to the sea. Seabirds and crabs attack them as they crawl along the sand (**FIG. E31-4b**). Once in the water, the hatchlings may become a tasty morsel for fish.

As if these natural dangers weren't enough, female turtles and their eggs are easy prey for human poachers. Turtle meat and eggs are a delicacy in many cultures, turtle shells make beautiful jewelry, and turtle skin makes fancy leather. Turtles are also caught, both deliberately and accidentally, in fishing lines and nets. Hatchlings, which evolved to find the sea by crawling toward moonlight reflecting off the ocean, are disoriented by the lights of cities and resorts; they head toward danger instead of the sea.

These dangers plagued the sea turtles that nest on the beaches of Brazil—until some students vacationing on Rocas Atoll back in the 1970s were inspired to protect the turtles after watching fishermen slaughter them as they came ashore to lay eggs. Two of the students, José Albuquerque and Guy Marcovaldi, founded Projeto TAMAR (an abbreviation of *tartarugas marinhas*, meaning "sea turtles" in Portuguese).

Albuquerque and Marcovaldi realized sea turtle conservation could be successful only with the support and participation of fishermen and local villagers. Today, TAMAR has 22 bases on the Brazilian coast, and most of its employees are former fishermen. Instead of hunting sea turtles, they free turtles caught in nets and patrol the beaches during nesting season. TAMAR biologists tag females and trace their travels. The fishermen fend off the (now rare) turtle poachers and relocate eggs laid in risky locations (such as close to city lights) to better beach sites or to a nearby hatchery. As of 2018, TAMAR had helped more than 25 million hatchlings reach the sea.

(a) A green turtle excavating a nest

(b) A turtle hatchling heads for the sea

▲ **FIGURE E31-4 Endangered sea turtles (a)** A female green turtle scoops sand with powerful flippers, creating a cavity where she will bury about 100 eggs. **(b)** After incubating in the sand for about 2 months, the eggs hatch. Here a hatchling heads for the sea, where (if it survives) it will spend 20 to 50 years before reaching sexual maturity.

habitats is the Papahānaumokuākea Marine National Monument of Hawaii, which stretches across 375 million acres of Pacific Ocean, an area more than twice the size of Texas. The reserve encompasses the northwest Hawaiian Islands and is home to extensive coral reefs and more than 7,000 species of birds, fish, invertebrates, and marine mammals, including a number of endangered species. Some species depend on such huge reserves; for others, critical habitat may be a few patches of sandy beach. "Earth Watch: Saving Sea Turtles" on this page discusses an innovative sea turtle conservation program in Brazil, which not only preserves turtle nesting sites but also helps local communities to prosper.

CASE STUDY CONTINUED

The Wolves of Yellowstone

As you learned in Chapter 29, large predators at the top of food webs, such as wolves and grizzly bears, are always relatively rare, because of energy losses between trophic levels. Even a national park the size of Yellowstone—almost 3,500 square miles—often cannot sustain a minimum viable population of such animals over long time periods, primarily because of the loss of genetic diversity and the dangers of epidemic diseases. How can a minimum viable population be sustained for these animals? We will return to this crucial concern in Section 31.5.

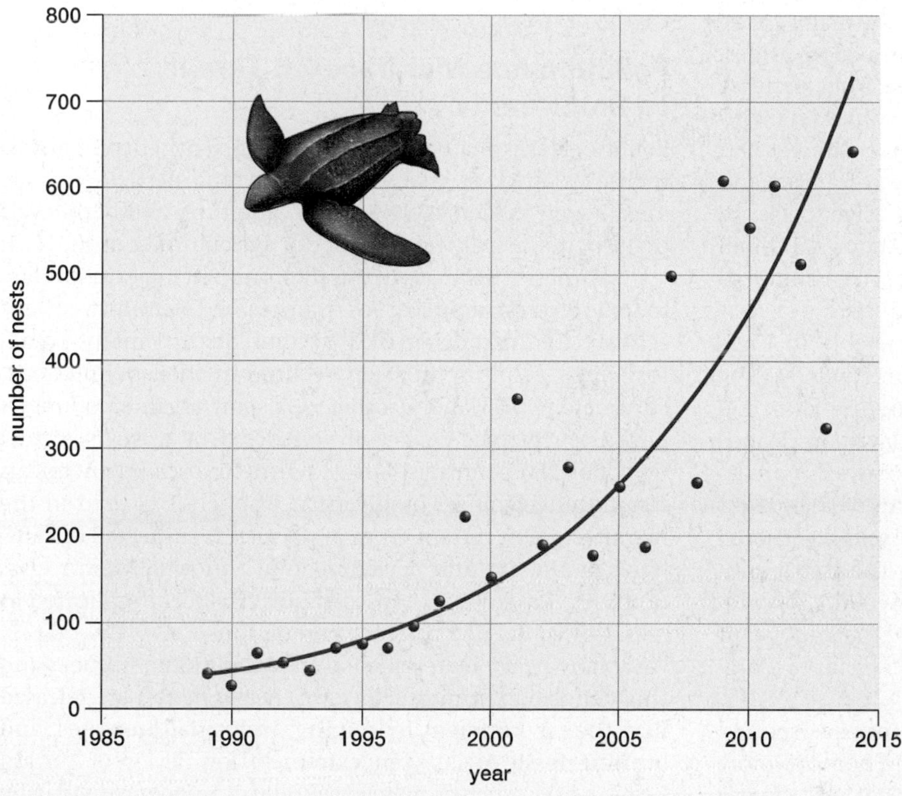

◀ **FIGURE E31-5 Atlantic leatherback sea turtle populations in Florida are growing** Atlantic leatherback sea turtle nests on a group of beaches in Florida were surveyed each year from 1989 to 2014. Because sea turtles are difficult to count at sea, nests are used as an indicator of population size. The population has been growing exponentially since the late 1980s (smooth curve).

TAMAR has been successful because the project organizers have engaged local communities as partners in turtle protection. Money flows into the local economies as ecotourists come to see baby turtles, visit turtle museums, and buy souvenirs made by local residents. TAMAR also sponsors communal gardens, day-care centers, and environmental education activities. Recognizing that the economic benefits derived from preserving turtles far outweigh the money that can be made by hunting them, local residents eagerly participate in turtle conservation. As IUCN sea turtle specialist Neca Marcovaldi put it, "Brazil's sea turtles are now worth more alive."

THINK CRITICALLY In 1970, Atlantic leatherback sea turtle population sizes were very small, and the U.S. Fish and Wildlife Service listed the species as endangered. Over the next several decades, steps were taken to protect turtle nests and prevent accidental killing of turtles at sea by fishing fleets. As a result, the population of leatherbacks nesting on beaches in Florida has been growing exponentially (**FIG. E31-5**). What factors probably contributed to this exponential growth? What factors are likely to eventually cause the population to stabilize?

Overexploitation Decimates Populations

Overexploitation is the hunting or harvesting of natural populations at a rate that exceeds their ability to replenish their numbers. Overexploitation of many species has increased as a growing demand for wild animals and plants has been coupled with technological advances that have increased our efficiency at harvesting them. For example, overharvesting is the single greatest threat to marine life, causing dramatic declines of many species, including invertebrates such as abalone, oyster, and corals, and fish such as cod, many sharks, haddock, tuna, and mackerel. The UN Food and Agriculture Organization estimates that about 30% of global fish populations are overexploited. Some overexploited populations, such as Atlantic bluefin tuna, are in danger of extinction.

Both poverty and wealth can contribute to overexploitation, particularly of endangered species. Rapidly growing populations in less-developed countries increase the demand for animal products, as hunger and poverty drive people to harvest all that can be sold or eaten, legally or illegally, without regard to its rarity. For example, illegal hunting of chimpanzees for meat is a major cause of declining chimp populations. Rich consumers often fuel the exploitation of endangered species by paying high prices for illegal products such as elephant-tusk ivory, rare orchids, and exotic birds. A single rhinoceros horn can sell for $200,000 or more in Asian markets. Although good data about black market activities are difficult to come by, the sale of endangered species, or products derived from them, is extremely lucrative, thought to total about $19 billion a year.

Invasive Species Displace Native Wildlife and Disrupt Community Interactions

Humans have transported a multitude of species around the world—everything from thistles to camels. In many cases, the introduced species cause no great harm. Sometimes, however, non-native species become invasive: They increase in number at the expense of native species, competing with them for food or habitat or preying on them directly (see Chapter 28). According to the Center for Invasive Species and Ecosystem Health, there are almost 2,900 invasive species in North America, mostly plants and insects. About half of all threatened species in the United States suffer from competition with or predation by invasive species.

Island ecosystems are particularly vulnerable to invasive species. For example, invasive species contributed to the extinction of many of the 1,000 or so species of plants and animals that have disappeared from the Hawaiian Islands since humans first arrived. Extinctions due to invasive predators and competitors began when pigs and rats were brought to Hawaii by the original Polynesian settlers, and accelerated in the nineteenth and twentieth centuries. The threat continues for most of Hawaii's remaining native wildlife. According to the U.S. Fish and Wildlife Service, more than 430 plant and animal species in Hawaii are endangered, by far the largest number in any state.

Lakes are also especially vulnerable to invasive species. For example, Lake Victoria in Africa was once home to more than 400 different species of cichlid fish that were found nowhere else on Earth (**FIG. 31-4a**). Enormous predatory Nile perch (**FIG. 31-4b**) and much smaller plankton-feeding tilapia were introduced into Lake Victoria in the mid-1900s. The combination of predation by Nile perch and competition from tilapia has helped cause a mass extinction of cichlids; only about 200 species remain.

Pollution Is a Multifaceted Threat to Biodiversity

Pollution takes many forms, including synthetic chemicals such as plasticizers, flame retardants, and pesticides; toxic metals such as mercury, lead, and cadmium; and high levels of nutrients, usually from sewage or agricultural runoff.

Synthetic chemicals in the environment may accumulate to toxic levels in animals. For example, in the mid-twentieth century the insecticide DDT accumulated in many predatory birds, resulting in steep declines in their populations. Fortunately, DDT and 11 other persistent organic pollutants have been banned or heavily restricted by a treaty signed by about 180 countries. Other synthetic organic chemicals also negatively affect biodiversity. Bisphenol A, used in the manufacture of certain types of plastics, is suspected of causing reproductive and developmental abnormalities. A class of insecticides called neonicotinoids has been implicated in massive declines in honeybee populations.

Many heavy metals are naturally bound up in rocks and thus rendered harmless. However, heavy metals are released into the environment by mining, industrial processes, and burning fossil fuels. Even extremely low levels of certain heavy metals, such as mercury and lead, are toxic to virtually all organisms.

Nutrients in excessive amounts can also become pollutants (see Chapter 29). For example, burning fossil fuels

(a) Blue rock hunter cichlid

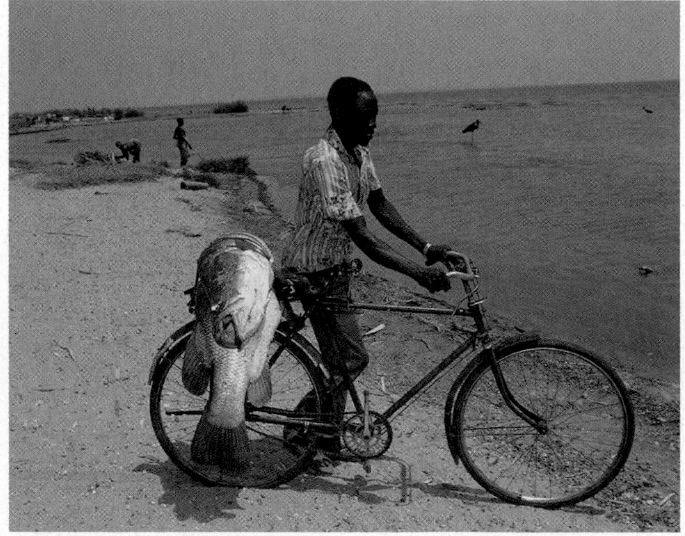

(b) Nile perch

▲ **FIGURE 31-4 Invasive species endanger native wildlife (a)** Lake Victoria was home to hundreds of species of stunningly colored cichlid fish, such as the blue rock hunter cichlid pictured here. **(b)** The Nile perch, introduced into Lake Victoria for fishermen, has proven to be a disaster for native fish.

THINK CRITICALLY Why might native species on islands be especially vulnerable to invasive species?

releases nitrogen and sulfur compounds, disrupting their natural biogeochemical cycles and causing acid precipitation that threatens forests and lakes. In addition, fertilizer runoff from farms and lawns often enters nearby waters and may cause harmful algal blooms.

Global Climate Change Is an Emerging Threat to Biodiversity

The rapid pace of human-induced climate change is challenging the ability of species to adapt. Scientists at the Convention on Biological Diversity, an international organization with more than 150 member countries, have concluded that warmer conditions have already contributed to some extinctions and are likely to cause many more. Although it is difficult to predict all the impacts of global climate change, they likely include the following:

- Deserts will likely become hotter and drier.
- Warmer conditions are forcing some species to retreat toward the poles or up mountains to stay within the climate zones in which they can survive and reproduce. Many species may be unable to retreat fast enough to stay within a suitable temperature range.
- Cool habitat will probably disappear completely from mountaintops. Animals that live at high altitudes, such as pikas in the Rocky Mountains (**FIG. 31-5a**), face shrinking habitat as the mountains warm. Populations of some species on isolated mountains have already vanished.
- Insect pests that were previously killed by frost or sustained freezes may spread and thrive. For example, populations of pine bark beetles, formerly limited by sustained cold weather in the winter, have reached epidemic levels in many areas of the northern and central Rocky Mountains (**FIG. 31-5b**).
- Parasites and insect-borne diseases are spreading closer to the poles. From lungworms in musk ox in Arctic Canada to disease-carrying mosquitoes infecting people in Sweden with tularemia, a warmer climate allows many pathogens and their carriers to move poleward.
- Coral reefs require warm water, but too much warming causes bleaching and coral death (**FIG. 31-5c**). The ocean acidification that accompanies climate change also reduces the ability of corals to build their external skeletons. Coral reefs have already suffered massive damage in the Seychelles Islands, American Samoa, Sri Lanka, the coasts of Tanzania and Kenya, and much of the Australian Great Barrier Reef.

CHECK YOUR LEARNING

Can you . . .

- define *mass extinction*?
- explain why biologists fear that a mass extinction is occurring as a result of human activities?
- describe how habitat destruction, overexploitation, invasive species, pollution, and global climate change threaten biodiversity?

(a) A pika gathers plants for the winter

(b) Pine bark beetles are killing pine trees

(c) Bleached corals (white) are usually dead or dying

▲ **FIGURE 31-5 Global climate change threatens biodiversity** **(a)** Pikas live at high altitudes in the Rocky Mountains. As the climate warms, suitable pika habitat may disappear from the tops of the mountains. **(b)** A forest infested with pine bark beetles often consists of a mosaic of uninfected trees (with green needles), newly killed trees (with reddish needles), and trees killed several years earlier (gray, without needles). **(c)** Corals usually contain photosynthetic algae that provide nourishment for the coral. When the water warms too much, corals eject their algae and become white; without the algae to help feed them, they often die.

31.4 WHY IS HABITAT PROTECTION NECESSARY TO PRESERVE BIODIVERSITY?

As we have seen, human activities pose many threats to biodiversity. To preserve biodiversity, we must counter all of these threats by, for example, reducing overexploitation, controlling invasive species, and curbing pollution, including greenhouse gas emissions. But perhaps the most essential component of the effort to preserve diversity is habitat protection, because without suitable natural habitat, species cannot survive. It is crucial that we set aside habitat in protected reserves and connect the reserves with wildlife corridors.

Core Reserves Preserve All Levels of Biodiversity

Core reserves are natural areas protected from most human uses except low-impact recreation. Ideally, a core reserve encompasses enough space to preserve ecosystems and all their biodiversity, even if natural disturbances such as storms, fires, and floods occur.

To establish effective core reserves, ecologists must estimate the smallest areas required to sustain minimum viable populations of the species that require the most space. The sizes of these *minimum critical areas* vary significantly among species and also depend on the availability of food, water, and shelter. In general, minimum critical areas are largest for large predators, especially in arid environments.

Wildlife Corridors Connect Habitats

An individual core reserve is seldom large enough to maintain biodiversity. This deficiency of core reserves can be offset by **wildlife corridors**, which are strips of protected land that link core reserves, allowing animals to move freely and safely between habitats that would otherwise be isolated. In India, for example, government and private groups are working to preserve forested corridors linking tiger reserves. Tigers now travel through the corridors to mate with other tigers in reserves as far as 230 miles away. In the increasingly fragmented Atlantic forest of Brazil, forested corridors connect reserves that are home to endangered species. On a smaller scale, wildlife tunnels or bridges may help wildlife cross roads that restrict movement within a habitat (**FIG. 31-6**).

Some Reserves Balance Preservation and Human Use

In some cases, especially in less-developed countries, it has proved difficult to establish core reserves in which human uses are prohibited. If governments lack the resources or ability to enforce protections, designated reserves may not actually be protected from exploitation that destroys habitat and damages wildlife. A possible solution to this dilemma is to develop reserves that allow low-impact uses by local inhabitants, who then have a vested interest in helping to protect the area.

Consider the Monarch Butterfly Biosphere Reserve in Mexico. Monarch butterflies have declined dramatically in recent years; the monarch population is only about 20% as large as it was 20 years ago. Each fall, the monarch butterflies in eastern North America migrate south to spend the winter in just a handful of forest groves in the mountains of central Mexico (**FIG. 31-7**). Without these wintering sites, the entire monarch population east of the Rocky Mountains would vanish.

To protect the butterflies, the Mexican government designated a 140,000-acre reserve that includes the overwintering sites. The land in the reserve, however, is

FIGURE 31-6 Wildlife bridges connect fragmented habitats Roads that pass through wildlife habitat can restrict movement that is needed for survival or breeding. Specially designed structures can make it possible for animals to cross roads.

THINK CRITICALLY What would be the likely effect of isolated small reserves on the genetic diversity of endangered species? How would genetic diversity be affected by connecting the small reserves with wildlife corridors?

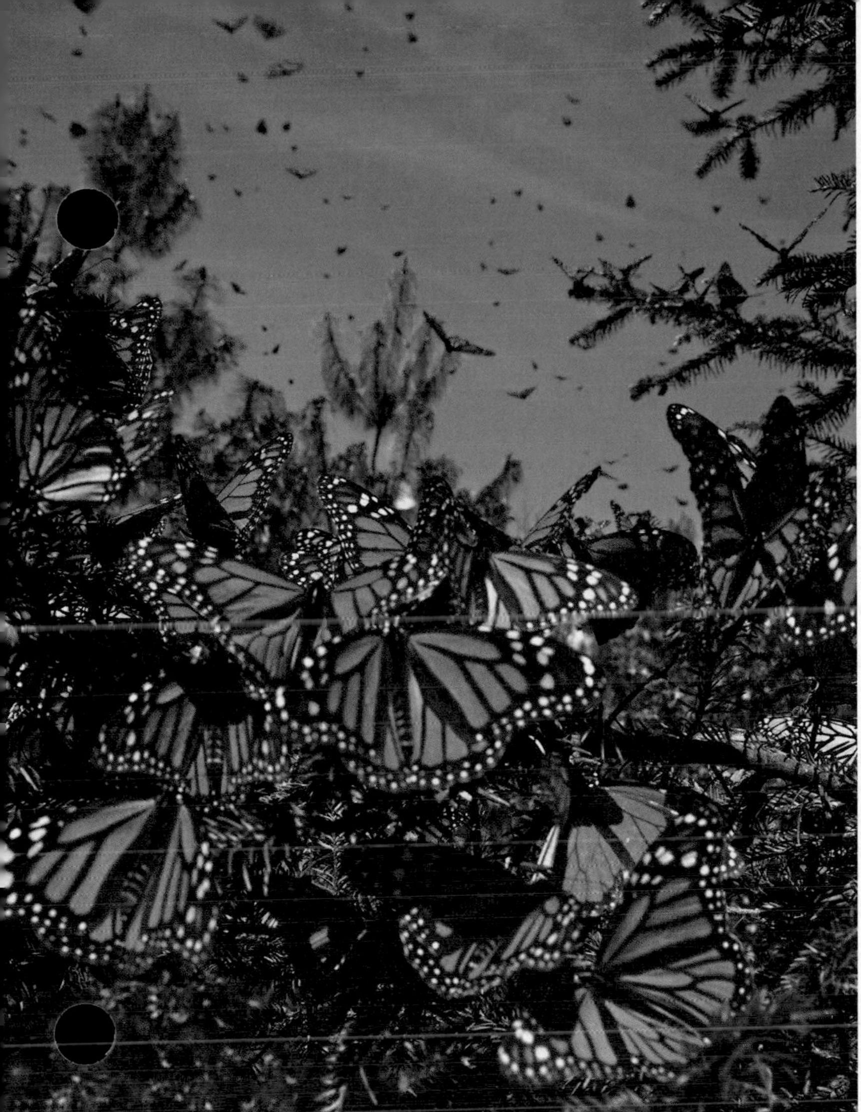

▲ **FIGURE 31-7 Preserving monarch butterflies** In the winter, so many monarchs roost in a handful of forest groves in the mountains of central Mexico that their weight bends the branches of the trees.

owned not by the Mexican government but by the local people, most of whom are low-income farmers. The stands of large trees that are essential for monarch survival have traditionally been an important economic resource for the farmers, providing firewood and lumber. How can the reserve meet the needs of both butterflies and farmers? One strategy is to provide funds and training that allow the local farmers to use some areas as tree farms, harvesting and replanting trees on a sustainable basis. If tree farming takes place in some spots, the farmers are more able to leave the old-growth groves alone.

Another source of income for the farmers is ecotourism. Tourists flock to the reserve each year to see the butterflies. If properly regulated, ecotourism can both preserve the forest and provide significant income opportunities for the local people, who serve as guides to the monarch groves and offer lodging and souvenirs for the tourists. For now, efforts to accommodate both people and wildlife in the monarch reserve have succeeded in preserving the monarch wintering grounds. However, many groves used by monarchs in and near the reserve are still being cleared for lumber and avocado plantations. Demand for avocados has skyrocketed in the United States, increasing pressure to clear more groves.

CHECK YOUR LEARNING

Can you . . .

- describe some strategies that can preserve natural ecosystems and their associated biodiversity?
- define the terms *core reserve* and *wildlife corridor* and explain the relationship between them?
- explain why it is sometimes better to allow human use of reserves rather than prohibiting it?

CASE STUDY CONTINUED

The Wolves of Yellowstone

For large predators such as wolves, grizzlies, and mountain lions, wildlife corridors are often required to provide enough habitat for a minimum viable population. Accordingly, the Yellowstone to Yukon Conservation Initiative has proposed a plan to provide corridors connecting habitat in the Rocky Mountains, from Yellowstone and Grand Teton National Parks in Wyoming all the way to the Yukon Territory in northwestern Canada. The proposed corridor would be nearly 2,000 miles long, and would include 500,000 square miles of habitat.

Wildlife corridors such as these must include private land interspersed between national parks and national forests. Can private landowners preserve wildlife habitat while still making a living and enjoying their land?

31.5 WHY IS SUSTAINABILITY ESSENTIAL FOR A HEALTHY FUTURE?

Natural ecosystems share features that allow them to persist and flourish. Healthy ecosystems include diverse communities, relatively stable populations that remain within the carrying capacity of the environment, efficient cycling of nutrients, and reliable sources of energy. Environments that have been modified by human development often lack these features. As a result, many human-modified ecosystems may not be sustainable in the long run. How can we meet our needs in ways that sustain the ecosystems on which we depend?

Sustainable Development Promotes Long-Term Ecological and Human Well-Being

Sustainable development is development that meets current human needs without compromising the ability of future generations to meet theirs. In *Caring for the Earth: A Strategy for Sustainable Living*, the IUCN described the goal of sustainable development as "improving the quality of human life while living within the carrying capacity of supporting ecosystems." Achieving this goal will require new approaches to many human endeavors. Here we will explore two aspects of sustainable development: renewable resources and sustainable agriculture.

Sustainable Development Relies on Renewable Resources

Development is sustainable only if we extract no more from ecosystems than the ecosystems can replenish. Thus, we must minimize our use of nonrenewable resources and use renewable resources in a manner that allows continued use, far into the future. For example, trees can be a renewable resource, but only if our consumption of wood and paper does not exceed the amount provided by sustainably managed forests. Worldwide, this balance has not been achieved, with consumption currently outpacing regrowth of forests. Although recycling and reuse can help bridge the current gap between consumption and forest production, trees cannot become a truly renewable resource until we reduce consumption.

Unlike trees and their products, fossil fuels are not renewable and cannot be recycled. Further, burning fossil fuels releases carbon dioxide, which is the principal cause of global climate change, with profound effects on humans and natural ecosystems alike. Therefore, a concerted effort to switch from fossil fuels to renewable energy sources, such as solar, wind, and geothermal power, is an essential part of sustainable development.

Sustainable Agriculture Preserves Productivity with Reduced Impact on Natural Ecosystems

Most of the land that is appropriated by people is used for agriculture, so the agricultural practices chosen by farmers have a huge impact on sustainability. To be sustainable, farms must not only produce enough food to feed humankind and ensure economic benefits to farmers, but must also maintain ecosystem services for future generations. Sustainable agriculture differs from traditional agriculture in many respects (**TABLE 31-1**). Here we explore in more detail two methods that increase sustainability: no-till farming and organic farming.

No-till Farming With conventional tillage, farmers prepare a field for planting by removing all residues of the prior year's crop and plowing the soil. In contrast, in **no-till farming**, farmers leave all crop residue in the field as mulch, and they do not plow. No-till farming is more sustainable than conventional tillage, because soil undisturbed by plowing experiences much less erosion and fertilizer runoff. No-till farming also helps to improve soil structure and increase the amount of organic matter in the soil. In addition, because no-till farming requires no plowing, it reduces use of nonrenewable fossil fuel.

Because plowing is a key method for controlling weeds, no-till farmers must rely on other methods of weed control. For example, many no-till farmers suppress weeds by planting a cover crop (such as winter wheat or clover) in the "off-season" between the growing periods for the cash crop (such as cotton or soybeans). In many circumstances, cover crops can suppress weeds so effectively that they reduce or eliminate the need for herbicides on the cash crop (**FIG. 31-8**). However, no-till farmers may use herbicides to kill the cover crop in the spring before planting the cash crop. Currently, no-till farming is used on about 35% of U.S. cropland.

Organic Farming Most farmers use synthetic herbicides, insecticides, and fertilizers, all of which degrade natural ecosystems. Organic farmers, however, do not use these chemicals, instead relying on natural predators to control pests and on soil microorganisms to provide fertilizing nutrients by decomposing animal and crop wastes. To control weeds, organic

TABLE 31-1 Agricultural Practices Affect Sustainability

	Unsustainable Agriculture	Sustainable Agriculture
Soil erosion	Allows soil to erode far faster than it can be replenished because the remains of crops are plowed under, leaving the soil exposed until new crops grow.	No-till agriculture greatly reduces soil erosion. Planting strips of trees as windbreaks reduces wind erosion.
Pest control	Uses large amounts of pesticides to control crop pests.	Trees and shrubs near fields provide habitat for insect-eating birds and predatory insects. Reducing insecticide use helps to protect birds and insect predators.
Fertilizer use	Uses large amounts of synthetic fertilizer.	Animal wastes are used as fertilizer. Legumes that replenish soil nitrogen (such as soybeans and alfalfa) are alternated with crops that deplete soil nitrogen (such as corn and wheat).
Water quality	Allows runoff from bare soil to contaminate water with pesticides and fertilizers. Allows excessive amounts of animal wastes to drain from feedlots.	Animal wastes are used to fertilize fields. Plant cover left by no-till agriculture reduces nutrient runoff.
Irrigation	May excessively irrigate crops, often using groundwater pumped from aquifers at a rate faster than the water is replenished by precipitation.	Modern irrigation technology reduces evaporation and delivers water only when and where it is needed. No-till agriculture reduces evaporation.
Crop diversity	Relies on a small number of high-profit crops, which encourages outbreaks of insects or plant diseases and leads to reliance on large quantities of pesticides.	Alternating crops and planting a wider variety of crops reduce the likelihood of major outbreaks of insects and diseases.
Fossil fuel use	Uses large amounts of nonrenewable fossil fuels to run farm equipment, produce fertilizer, and apply fertilizers and pesticides.	No-till agriculture reduces the need for plowing and fertilizing.

(a) Cotton seedlings emerge in a no-till field in North Carolina

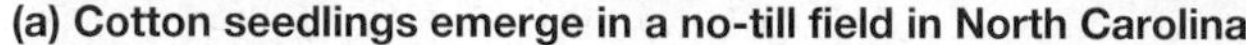

(b) The same field one month later

▲ **FIGURE 31-8 No-till agriculture (a)** A cover crop of wheat has been killed with an herbicide. Cotton seedlings thrive amid the dead wheat, which anchors soil and reduces evaporation. **(b)** Later in the season, the same field shows a healthy cotton crop mulched by the dead wheat.
Photos courtesy of Dr. George Naderman.

farmers may use cover crops and occasional plowing. And to reduce outbreaks of pests and diseases that attack a single type of plant, organic farmers often use crop rotation, planting different crops in the same field year to year. Pests may also be controlled by introducing predators that feed on the pest species.

Widespread adoption of organic farming has been slowed by its sometimes lower yields compared to conventional farming. However, organic food is usually sold at a higher price than nonorganic food, which may help organic farmers offset this disadvantage. Even farmers who do not choose to go fully organic can increase the sustainability of their farming by adopting some aspects of the organic method. Many projects, including the University of California's Sustainable Agriculture Research and Education Program, support research and educate farmers about the advantages of sustainable agriculture and how to practice it.

The Future of Earth Is in Your Hands

Can we manage our planet so that it provides a healthy, satisfying life for the current generation of people, while retaining biodiversity and the resources needed for future generations? Achieving this goal will require not just a sustainable approach to development, such as the use of renewable resources and adoption of sustainable agriculture methods, but a significant reduction in the human population growth rate.

Human Population Growth Is Unsustainable

The root cause of environmental degradation is simple: too many people using too many resources and generating too much waste. As the IUCN eloquently stated in *Who Will Care for the Earth?*, ". . . the central issue [is] how to bring human populations into balance with the natural ecosystems that sustain them."

In the long run, that balance cannot be achieved if the human population continues to grow. Given the lifestyle to which the vast majority of people on Earth aspire, it may not be possible to maintain the balance even with our current population, yet we add 75 to 80 million people each year. No matter how simple our diets, how efficient our housing, how low-impact our farming techniques, or how much we reuse and recycle, continued population growth will eventually overwhelm our best efforts.

The Global Footprint Network estimates that Earth's current human population uses resources and generates wastes at a rate that is about 1.7 times higher than the planet's long-term capacity to provide the resources and absorb the wastes. Although the average person's individual demands on Earth's ecosystems have not changed much over the past 50 years, our population has grown so rapidly that our cumulative demands have increased steadily, as has the gap between those demands and Earth's ability to meet them. As population continues to grow, the gap will widen. Thus, controlling population growth is essential if we wish to improve the quality of life for the 7.5 billion people currently on Earth, provide a similar quality of life for our descendants, and save Earth's biodiversity for future generations.

The Choices Are Yours

Achieving sustainable development will require action by governments, industry, and other institutions, but individual actions also play a role. As Canadian educator and philosopher Marshall McLuhan noted 50 years ago, "There are no passengers on Spaceship Earth. We are all crew." On the next page, we describe are some ways to make a difference.

Conserve Energy

- ***Heating and Cooling*** Set your thermostat to 68°F or lower in winter and 78°F or higher in summer. Reduce heating or cooling when you're away from home. When you purchase or remodel a home, consider energy-efficient features such as passive solar heating, good insulation, an attic fan, double-glazed windows (with "low-E" coating to reduce heat transfer), and tight weather stripping. If possible, purchase electricity that comes from renewable sources such as wind or solar power, rather than from fossil fuels.
- ***Lighting*** Buy energy-efficient compact fluorescent or LED bulbs, which use about a quarter to a sixth as much energy as an incandescent bulb with the same light output and last 8 to 30 times longer.
- ***Hot Water*** Take shorter showers and switch to low-flow shower heads. Wash only full loads in your washing machine and dishwasher; use cold water to wash clothes; don't prewash your dishes. Insulate and turn down the temperature on your water heater.
- ***Appliances*** When you choose a major appliance, look for the most energy-efficient models. Don't use your dryer in the summer—put up a clothesline.
- ***Transportation*** Choose the most fuel-efficient car that meets your needs. When possible, use public transit, carpool, walk, bicycle, or telecommute.

Conserve Materials

- ***Recycling*** Look into recycling options in your community and recycle everything that is accepted (**FIG. 31-9**). Don't forget to recycle your cell phone when you replace it. Purchase recycled paper products.
- ***Reuse*** Reuse anything possible, such as manila envelopes, file folders, and both sides of paper. Consider buying used furniture. Refill your water bottle. Reuse your grocery bags. Give away—rather than throw away—serviceable clothing, toys, and furniture. Make rags out of unusable old clothes and use them instead of disposable cleaning materials.
- ***Conserve Water*** If you live in a dry area, plant drought-resistant vegetation around your home to reduce water usage. Use a low-flow shower head to save both energy and water.
- ***Avoid Harmful Chemicals*** Limit your use of harsh cleaners, insecticides, and herbicides that may contaminate water and soil.

Support Sustainable Practices

- ***Food Choices*** When possible, buy locally grown produce that does not require long-distance shipping. Look for coffee with the "Bird-Friendly" or "Rainforest Alliance Certified" seal of approval, or other evidence of sustainable production such as the Starbucks "Coffee and Farmer Equity" program. Reduce your meat consumption. Consult Internet sites such as the Monterey Bay Aquarium Seafood Watch program at www.seafoodwatch.org to find out which fish at your local supermarket have been harvested sustainably.

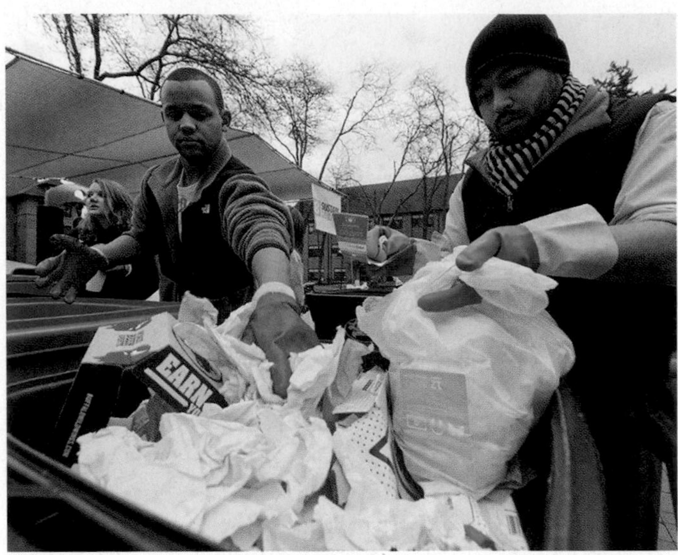

▲ **FIGURE 31-9 Recyclemania** Students at colleges across the United States and Canada raise awareness of recycling by participating in Recyclemania, a friendly competition to see which school can recycle the largest proportion of its trash. Recycling slows depletion of natural resources and reduces the air and water pollution associated with resource extraction.

Magnify Your Efforts

- ***Support Organized Conservation Efforts*** Join conservation groups and donate money to conservation efforts. Sign up for e-mail alerts that educate you about environmental legislation and make it easy for you to contact your government representatives and express your views.
- ***Volunteer*** Volunteer for local campus and community projects that improve the environment.
- ***Make Your Vote Count*** Investigate candidates' stands and voting records on environmental issues and consider this information when choosing which candidate to support.
- ***Educate*** Through your words and actions, share your concern for sustainability with your family, friends, and community. Write letters to the editor of your school or local newspaper, to local businesses, and to elected officials. Talk with your campus administrators about reducing energy use on your campus. Recruit other concerned people and lobby for change.
- ***Reduce Population Growth*** Consider the consequences of human population growth when you plan your family. Adoption, for example, allows people to raise children while simultaneously contributing to the welfare of humanity and the environment.

CHECK YOUR LEARNING

Can you . . .

- describe the principles of sustainable development?
- explain how agriculture and use of resources can become sustainable?
- explain how human population size affects sustainability?

CASE STUDY REVISITED

The Wolves of Yellowstone

Wolves affect other species at many trophic levels in the Yellowstone ecosystem. For example, berry-producing bushes thrive when wolves reduce the population of grazing elk. Grizzly bears benefit by feeding on the additional berries provided indirectly by wolves and by scavenging the carcasses of elk killed by wolves (**FIG. 31-10**). Coyotes are harmed by wolves, which chase coyotes and sometimes kill them. Pronghorns increase where wolves reduce coyotes, because coyotes are major predators of pronghorn fawns. Birds also benefit from the presence of wolves; crows and vultures scavenge on wolf-killed elk carcasses, and many songbirds are abundant in the aspen groves and willow thickets that flourish where wolves reduce elk populations.

Because ecosystems benefit so much from the presence of top predators, ecologists often propose that they be reintroduced to places where they have declined or disappeared. For example, advocates have proposed reintroducing wolves to Scotland, grizzly bears to the mountains of Washington state, and lynx to England. Despite the potential ecological benefits, however, proposed reintroductions are often controversial, mainly due to concerns about how top predators will coexist with people and their livestock. Restoring biodiversity while meeting the needs of local people is among the greatest challenges facing conservation biologists.

THINK CRITICALLY Before 1700, wolves were found across almost all of North America. Coyotes were found mostly west of the Mississippi, and white-tailed deer lived throughout the eastern half of the continent. When Europeans arrived, they exterminated the wolves. Based on what you have learned in this chapter and its Case Study, what do you think happened to the populations of coyotes and white-tailed deer between 1700 and now, in terms of their range and abundance? Why?

◀ FIGURE 31-10 A grizzly bear approaches a carcass as wolves attempt to defend their kill

CHAPTER REVIEW

Go to **Mastering Biology** to access the Pearson eText, vocabulary review, practice quizzes, activities, videos, current events, and more.

*Answers to **Think Critically** and **Thinking Through the Concepts** questions can be found in the **Answers** section at the back of the book.*

Summary of Key Concepts

31.1 What Is Conservation Biology?

Conservation biology is the scientific discipline devoted to understanding and preserving Earth's biodiversity, including diversity at the genetic, species, and ecosystem levels.

31.2 Why Is Biodiversity Important?

Biodiversity is necessary for healthy ecosystems, which are valuable for their own sake but also provide people with ecosystem services that have great monetary value. Ecosystems are a source of provisions such as food, fuel, building materials, and medicines. Other ecosystem services include forming soil, purifying water, controlling floods, moderating climate, and providing recreational opportunities.

31.3 What Are the Major Threats to Biodiversity?

Natural communities have a low background extinction rate. Many biologists have concluded that human activities are currently causing a mass extinction, increasing extinction rates perhaps 1,000-fold. Major threats to biodiversity include habitat destruction and fragmentation, overexploitation, invasive species, pollution, and global climate change.

31.4 Why Is Habitat Protection Necessary to Preserve Biodiversity?

Conservation efforts include conserving wild ecosystems by establishing core reserves connected by wildlife corridors, which helps to preserve functional communities and self-sustaining wildlife populations. Establishing reserves that allow some human use can help ensure that local people have a vested interest in habitat protection.

31.5 Why Is Sustainability Essential for a Healthy Future?

Sustainable development meets present needs without compromising the future. Such development requires that people maintain biodiversity, recycle raw materials, and rely on renewable resources. A shift to more sustainable agriculture using no-till or organic methods would help conserve soil and water, reduce pollution and energy use, and preserve biodiversity. Global human population growth is driving consumption of resources beyond nature's ability to replenish them; reducing the population growth rate is imperative to a sustainable future. Individual lifestyle and purchasing choices can also promote sustainability.

Thinking Through the Concepts

Bloom's: Remembering, Understanding

Multiple Choice

1. Which of the following is *not* a major threat to biodiversity?
   a. No-till agriculture
   b. Habitat destruction
   c. Invasive species
   d. Human-caused climate change

2. A species that plays an essential role in an ecosystem, usually out of proportion to its population size, is called a
   a. predator.
   b. keystone species.
   c. genetically diverse species.
   d. redundant species.

3. The modern rate of extinction is
   a. much higher than the background rate.
   b. the same as the background rate.
   c. somewhat higher than the background rate.
   d. having little effect on biodiversity.

4. Sustainable development does *not* include
   a. agriculture that reduces use of pesticides and disturbance of the soil.
   b. a shift from nonrenewable to renewable resources.
   c. reuse and recycling of materials.
   d. increased use of fossil fuels.

5. Which of the following is *not* true of a population of large predators in a habitat fragment?
   a. The population may disappear from the fragment.
   b. The population will probably grow due to immigration.
   c. The population will probably lose genetic diversity.
   d. The population may overeat its prey, causing a reduction in prey population.

Fill-in-the-Blank

1. Three levels of biodiversity are ____________, ____________, and ____________. If the population of a species becomes too small, it is likely to have lost much of its ____________ diversity.

2. Products or processes by which functioning ecosystems benefit humans are collectively called ____________. Four important categories of these benefits are ____________, ____________, ____________, and ____________.

3. The major threats to biodiversity include ____________, ____________, ____________, ____________, and ____________. For most endangered species, ____________ is probably the major threat.

4. The smallest population of a species that is likely to be able to survive in the long term is called the ____________. When suitable habitat for a given species is split up into areas that are too small to support a large enough population, this is called ____________. One way to maintain large enough populations is to set up core reserves of suitable habitat, connected by ____________.

5. A popular saying tells us that "We do not inherit the Earth from our ancestors, we borrow it from our children." If this principle guided our activities, we would practice ____________ development.

Review Questions

1. What are the three different levels of biodiversity, and why is each one important?

2. List the major categories of services that natural ecosystems provide, and give three examples of each.

3. What five specific threats to biodiversity are described in this chapter? Provide an example of each.

4. Why are efforts to protect monarch butterflies a good model for conservation and sustainable development?

Applying the Concepts

Bloom's: Applying, Analyzing, Evaluating

1. List some reasons that residents of affluent countries like the United States have an especially large negative effect on Earth's ecosystems. How could you reduce your personal impact?

2. Research and describe some examples of habitat destruction, pollution, and invasive species in the region around your home or campus. Predict how each of these might affect specific local populations of native animals and plants.

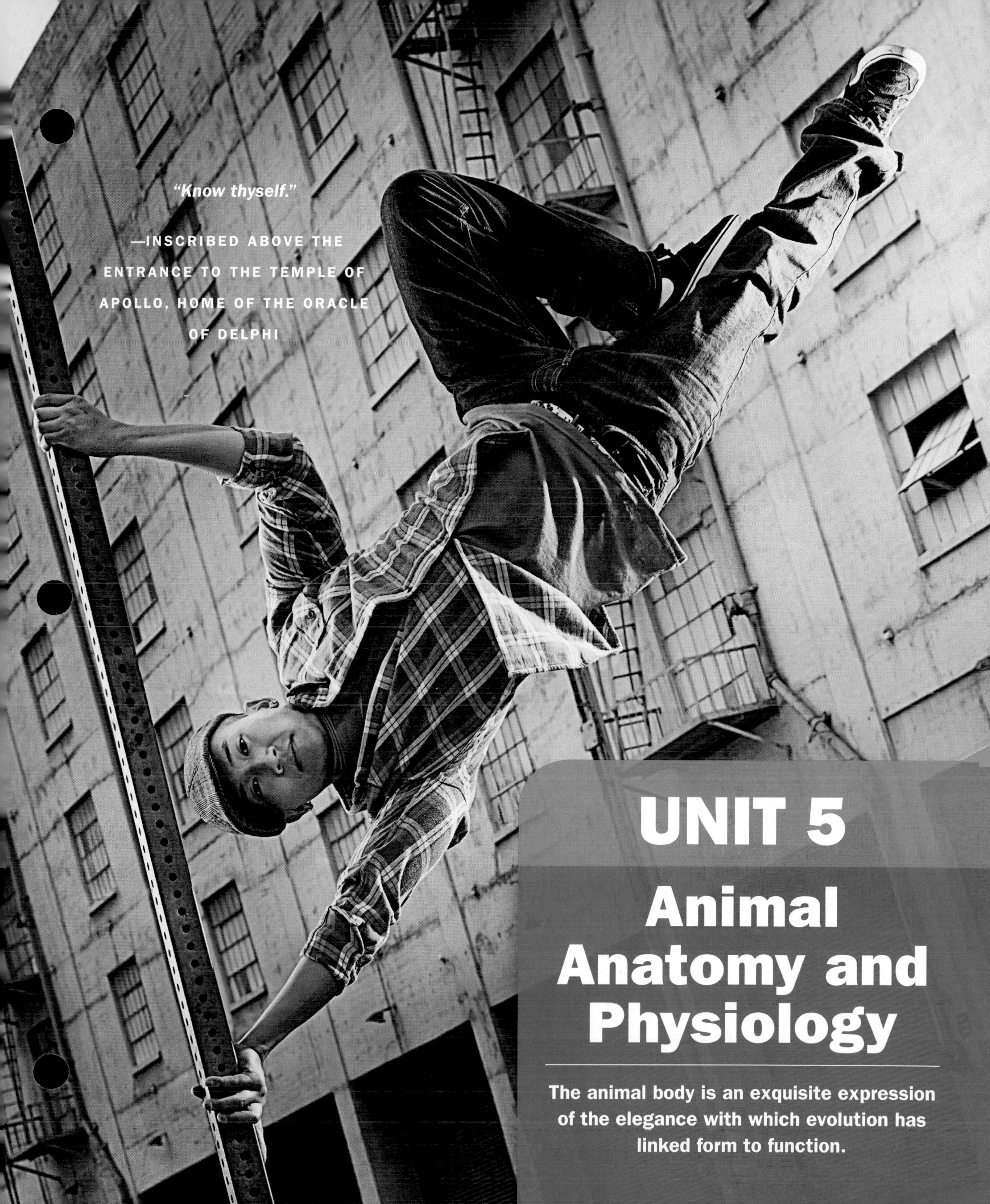

"Know thyself."

—INSCRIBED ABOVE THE ENTRANCE TO THE TEMPLE OF APOLLO, HOME OF THE ORACLE OF DELPHI

UNIT 5 Animal Anatomy and Physiology

The animal body is an exquisite expression of the elegance with which evolution has linked form to function.

32 Homeostasis and the Organization of the Animal Body

When the temperature soars, athletes can be in danger.

CASE STUDY

Overheated

"I'M HAVING THE RACE OF MY LIFE. I've just swum and biked harder than I ever thought I could.... My mental focus has reached a new level. Nothing is going to get in my way, not the 90° heat and 90% humidity, not the Cal girl in front of me, not the fire in my feet that is spreading up my legs." These were the thoughts of Kierann Smith, a medical student and member of the Stanford triathlon team, just before she collapsed at the USA Triathlon's Collegiate Nationals in Tuscaloosa, Alabama. At the time of her collapse, she was approaching the 5-mile mark on the third leg of the grueling competition, and her core body temperature had climbed to 106°F. Smith was rushed to the medical tent, where an ice water dousing and ice packs brought her temperature to just below 104°F. At this point, she was moved to make room for the many other elite college athletes who had been overcome by the heat on this April day in Alabama. Her body temperature eventually returned to normal.

Kierann Smith was relatively lucky. She survived, unlike wrestler Ben Richards, a 20-year-old sophomore at Darton State College in Georgia, who died 10 days after collapsing during a 5-mile training run. His temperature was 107°F when he arrived at the hospital. Marquese Meadow, a freshman attending Morgan State College in Maryland on a football scholarship, became disoriented after football practice and died in a nearby hospital. Autopsy results confirmed heat stroke as the cause of death.

What control mechanisms normally maintain human body temperature within narrow limits? Why must body temperature be closely regulated, and what happens to the body during heat stroke?

AT A GLANCE

32.1 HOMEOSTASIS: WHY AND HOW DO ANIMALS REGULATE THEIR INTERNAL ENVIRONMENT?

Whether you are sitting in your room, sweltering in a jungle, or shivering in a blizzard, most of the cells of your body—for instance, those in your heart, brain, and muscles—maintain an almost constant temperature. Further, whether you are standing in parched desert air or swimming in a pool, the ocean, or the Great Salt Lake, your cells are bathed in a liquid, called interstitial fluid, which has an almost constant composition.

The "internal constancy" of animal bodies was first described by French physiologist Claude Bernard in the mid-nineteenth century. In the 1920s, Walter Cannon coined the term **homeostasis** to describe the ability of an organism to maintain its internal environment within the narrow limits that allow optimal cell functioning. In addition to maintaining temperature and concentrations of water and salt in body fluids, homeostatic mechanisms regulate a host of other conditions, including glucose concentrations, pH (acid-base balance), hormone secretion, and concentrations of oxygen and carbon dioxide.

Although the word "homeostasis" (meaning "to stay the same") implies a static, unchanging state, the internal environment actually seethes with activity as the body continuously adjusts to maintain constancy amongst varying internal and external conditions. For example, exercise challenges the body's homeostatic mechanisms; more oxygen must be supplied to sustain the energy demands of working muscles, and the extra heat muscles produce must be dissipated. The body rises to the occasion with increased respiratory rate, increased activity of sweat glands, and greater blood flow to the skin. As you progress through this unit, you will find numerous examples of how interacting systems throughout the body help maintain homeostasis.

Homeostasis Allows Enzymes to Function

Why do cells require constant conditions in their surroundings? A big part of the reason boils down to proteins, particularly enzymes. Almost every biochemical reaction in a cell is catalyzed by a specific enzyme whose ability to function depends on its precise three-dimensional structure, maintained, in part, by hydrogen bonds (see Chapters 2 and 3). These crucial but vulnerable bonds can be disrupted by an environment that is too salty, too acidic, too basic, or too hot. Thus, the need to maintain hydrogen bonds and the protein function that depends on them helps explain why cells require surroundings that remain within a narrow range of salt concentration, pH, and temperature.

Temperature is particularly crucial to enzyme function for another reason as well: The rate at which enzymes catalyze reactions is temperature dependent. Low temperatures slow molecular motion, reducing both the number and the speed of interactions between enzymes and the molecules upon which they act. As temperature increases, these reactions speed up, sometimes to a harmful level.

CASE STUDY CONTINUED

Overheated

Both football player Marquese Meadow and wrestler Ben Richards had been hospitalized and treated for over a week before their deaths, and their body temperatures were brought down to normal within a few hours of suffering heat stroke. Why didn't they recover? The metabolic chaos from heat stroke damages vital organs, particularly the liver and kidneys. Muscle cells die and release substances into the bloodstream that can lead to kidney failure. Elevated body temperature also damages the liver, which is crucial for removing toxic substances from the blood. Kierann Smith wrote of her recovery from heat stroke: "A big blow was when I found out that I had sustained marked muscle damage and even some liver damage.... There is no quick fix to this, mentally or physically."

Human metabolism functions only within a narrow range of temperatures. Is this true for all animals?

Animals Differ in How They Regulate Body Temperature

Probably the best-studied mechanisms of homeostasis are methods for regulating body temperature, which differ dramatically among different animals. You may have heard mammals and birds described as "warm-blooded" and other reptiles, amphibians, fish, and invertebrates as "cold-blooded." However, these terms are often misleading. For example, the bodies of desert pupfish (**FIG. 32-1a**) may reach over 100°F (37.8°C) when the summer sun heats the springs and small ponds in which they live, so sometimes pupfish are quite warm-blooded. Hummingbirds have body temperatures as high as 105°F (40.5°C) while foraging during the day (**FIG. 32-1b**), but may cool down to 55°F (13°C) during the night.

(a) Desert pupfish

(b) Ruby-throated hummingbird

(c) Iguana

▲ **FIGURE 32-1 Warm-blooded or cold-blooded? (a)** Because "cold-blooded" fish such as this desert pupfish may be quite warm, and **(b)** "warm-blooded" animals like this hummingbird can become quite cool, biologists classify animals as endothermic or ectothermic, depending on their principal source of body warmth. **(c)** This iguana basking in the sun illustrates a behavioral mechanism that many ectotherms use to regulate body temperature.

To avoid confusion, scientists usually classify animals according to their major source of body warmth. **Ectotherms** (Greek for "outside heat") derive most of their heat from the environment. Reptiles (except for birds), amphibians, and most fish and invertebrates are ectotherms. In the simplest cases—an earthworm in its burrow or a fish in a stream—an ectotherm's body temperature will be the same as the temperature of its environment. **Endotherms** (Greek for "inside heat") produce most of their heat by metabolic reactions. Birds and mammals are the most common types of endotherms, although some other types of animals occasionally generate significant amounts of heat metabolically.

The body temperatures of most ectotherms vary quite a bit as the external temperature changes over the course of hours, days, or weeks. Many ectotherms, such as butterflies, bees, and most lizards, cool down and become inactive at night, thereby conserving energy, but warm up and become active during the day. At night, butterflies and bees become so cool that they cannot fly; lizards are often too sluggish to hunt or escape predators. At daybreak, bees shiver and butterflies beat their wings to generate metabolic heat, while lizards seek a warm, sunlit stone, providing the heat they need to resume their active lifestyles.

Although variable body temperatures are common in ectotherms, some can maintain quite stable body temperatures, through behavior or by occupying a very constant environment. For example, insects and lizards often warm up by basking in the sun (**FIG. 32-1c**) and cool off by moving to a shady spot. The pupfish mentioned earlier can tolerate water temperatures ranging from 36° to 113°F (2.2° to 45°C), but can breed only within a narrow range of temperatures. During breeding season, a pupfish regulates its temperature quite precisely by swimming to different areas of its pool as the water temperature changes. In the deep ocean, the temperature is so constant (typically from about 32° to 37.5°F, or 0° to 3°C) that ectothermic deep-sea fish experience little variation in body temperature throughout their lives.

Although a few endotherms, such as hummingbirds, allow their body temperatures to fall at night, most maintain a fairly constant body temperature that, depending on the species, falls between about 95°F and 106°F (about 35° to 41°C). There are both benefits and costs to keeping this warm. A major benefit is that a warm body usually can sense its environment better, respond more quickly, and move faster than a cold body. The major cost is the energy required to maintain a high body temperature.

How does an animal sense the conditions within its body and adjust them when necessary? The internal environment is maintained by mechanisms collectively known as feedback systems.

Feedback Systems Regulate Internal Conditions

There are two types of feedback systems: (1) negative feedback systems, which counteract the effects of changes in the internal environment and are principally responsible for maintaining homeostasis; and (2) positive feedback systems, which create cycles in which changes amplify themselves.

Negative Feedback Reverses the Effects of Changes

The most important mechanism governing homeostasis is **negative feedback**, in which a change causes responses that counteract the change. The overall result of negative feedback is to return the system to its original condition.

Negative feedback is a common feature of both living and nonliving systems. Negative feedback regulates almost every aspect of an organism's physiology, including body temperature; levels of glucose, hormones, water, salts, and oxygen in the blood; and even standing upright. Mechanical devices also often incorporate negative feedback, for example, to keep a constant temperature in your house, fill your toilet tank, or keep your car set on cruise control at a constant speed.

All negative feedback systems contain three principal components: a *sensor*, a *control center*, and an *effector*. The sensor detects the current condition, the control center compares that condition to a desired state called the set point, and the effector produces an output that restores the desired condition. Let's see how negative feedback systems work, first in the familiar example of heating your home and then in the control of body temperature.

In the negative feedback system that controls the temperature of your home on a cold day, the sensor is a thermometer, the control center is a thermostat, and the effector is a heater (**FIG. 32-2**). The thermometer detects the room temperature and sends that information to the thermostat, where the actual temperature is compared to the set point of the desired temperature. If the room temperature is below the set point, then the thermostat signals the heater to turn on and generate heat. The heater warms the room, restoring the temperature to the set point, which causes the thermostat to turn off the heater.

Negative feedback systems maintain many physiological parameters within narrow limits. The key homeostatic

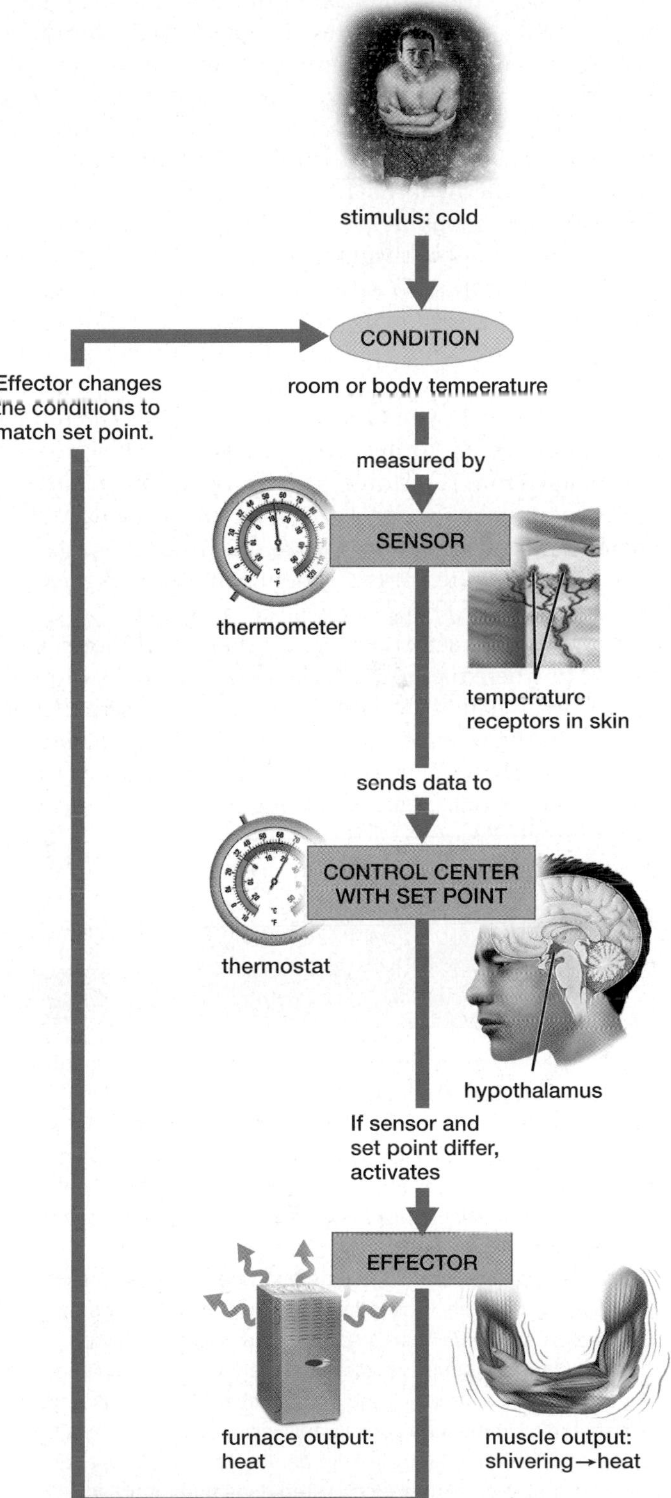

▲ **FIGURE 32-2 Negative feedback maintains homeostasis** In negative feedback, responses to a stimulus counteract the effects of the stimulus. Negative feedback regulates the temperature of your house as well as body temperature.

THINK CRITICALLY What would happen if a cold, shivering mammal ingested a poison that destroyed all of its body's nerve endings that detect heat?

Have You Ever Wondered ... Can You Drink Too Much Water?

The simple answer is yes, too much of almost anything, even water, can be harmful. Hydration has become a buzzword in the popular press, which sometimes encourages people to drink considerably more than they need. It seems logical that drinking extra water will help you sweat and stay cool, and some athletic coaches, wary of heat stroke among young football players, may encourage excessive water consumption. Deaths from over-hydration are rare, but they do occur. In one such tragedy, Georgia football player Zyrees Oliver, a 17-year-old high school senior, reportedly drank 2 gallons of water and 2 gallons of Gatorade during a football practice session, hoping it would relieve his muscle cramps. After returning home, he collapsed and later died after being helicoptered to a hospital. Oliver was killed by hyponatremia, which literally means "too little sodium," referring to the dilution of sodium ions (Na^+) in the blood by excess water. A study of Boston marathon participants found that 13% of those tested at the finish line had drunk so much water during the race that they had at least mild hyponatremia, and a few had dangerously low blood Na^+ levels.

The sodium ion is essential to physiological processes including neuronal signaling and muscle contraction, and elaborate homeostatic mechanisms have evolved that regulate Na^+ in body fluids within a very narrow range. Symptoms of hyponatremia include nausea and vomiting, muscle cramps, convulsions, unconsciousness, coma, and occasionally death.

regulator in vertebrate animals is a region buried deep in the brain called the *hypothalamus*. In birds and mammals, the hypothalamus coordinates nervous signals, hormone release, and behaviors to maintain body temperature despite large fluctuations in environmental temperature (see Fig. 32-2). Nerve endings in the skin and other parts of the body act as temperature sensors and transmit this information to the hypothalamus. If the body temperature falls below the set point (the normal body temperature), the hypothalamus activates mechanisms that generate and conserve heat, such as shivering (small, rapid muscular contractions) or moving to a warmer place. The blood vessels supplying nonvital areas of the body (such as the face, hands, feet, and skin) are constricted, reducing heat loss and diverting warm blood to the body core (including the brain, heart, and other internal organs). When normal body temperature is restored, the temperature sensors signal the hypothalamus, which switches off the actions that generate and conserve heat.

The temperature control system can also act to reduce body temperature. If body temperature exceeds the set point, the hypothalamus sends out signals that divert more blood to the skin where the heat can be radiated to the surrounding air. Sweat glands produce a watery secretion that cools the body as the water evaporates from the skin. The fatigue and discomfort caused by elevated body temperature usually stimulate behavioral changes, causing people to rest and seek shade or cooling water.

CASE STUDY CONTINUED

Overheated

During heat stroke, the negative feedback mechanisms that usually keep body temperature relatively constant malfunction. One form of feedback from overheating is a feeling of exhaustion, which normally would cause a person to rest. But a symptom of heat stroke is mental confusion, which disrupts this negative feedback and makes it easy for an athlete to ignore the body's warning signs.

A heat stroke victim may also stop sweating, as skyrocketing body temperature causes the temperature control systems of the brain to malfunction. Also, because enzyme activity increases with temperature, metabolic rate may rise and generate still more heat, causing the body temperature to soar. If not terminated by immediate cooling, this vicious cycle—an example of positive feedback—may lead to death. What is positive feedback? Is it always harmful? How often does it naturally occur?

Positive Feedback Enhances the Effects of Changes

In **positive feedback**, a change produces a response that amplifies that change. Imagine a thermostat gone haywire—you nudge up the temperature setting, but when the room reaches the set point, instead of turning off, the thermostat calls for more heat, and so on—a very uncomfortable situation.

Because its effects could spiral out of control, positive feedback is relatively rare in biological systems, and under normal physiological conditions, positive feedback events are halted eventually by negative feedback. An excellent example is the birth of babies in people and other mammals. The early contractions of labor push the baby's head against the cervix at the base of the uterus, causing the cervix to stretch and begin to open. Nerve cells in the cervix react to the stretching by signaling the hypothalamus, which responds by triggering the release of the hormone oxytocin. Oxytocin stimulates stronger uterine contractions, which cause the baby's head to stretch the cervix even further, stimulating more oxytocin release. This positive feedback cycle is eventually ended by negative feedback when the contractions cause birth, relieving pressure on the cervix. Another example of positive feedback is the let-down reflex, which releases milk during breastfeeding, a process that also involves oxytocin and the hypothalamus (described in Chapter 38). When the baby becomes full and stops suckling (negative feedback), the cycle stops.

Positive feedback can also occur in nonliving systems. In "Earth Watch: Positive Feedback in the Arctic," we explore how the shrinking Arctic ice cap is likely to cause a positive feedback cycle influencing Earth's climate.

CHECK YOUR LEARNING

Can you . . .

- define *homeostasis* and explain why organisms maintain homeostasis?
- explain the difference between ectotherms and endotherms, and give examples of each?
- define *negative feedback* and explain why it is crucial in achieving homeostasis?
- define *positive feedback* and explain why it is rare in living organisms?

32.2 HOW IS THE ANIMAL BODY ORGANIZED?

Animals maintain homeostasis by performing a multitude of tasks simultaneously. For example, we have seen that the hypothalamus, temperature receptors throughout the body, sweat glands, and blood vessels in the skin all work together to control human body temperature. This coordination of body functions relies on a hierarchy of structures:

cells → tissues → organs → organ systems

Cells are the fundamental units of all living organisms (see Chapters 1 and 4). In an animal body, a **tissue** is composed of dozens to billions of structurally similar cells that work together to carry out a particular task. Tissues are the building blocks of **organs**, which are discrete structures that perform complex functions. Examples of organs include the stomach, small intestine, kidneys, and urinary bladder. Organs, in turn, are organized into **organ systems**, groups of

Positive Feedback in the Arctic

In positive feedback, changing conditions cause responses that enhance the change. Climate scientists are convinced that global climate change has triggered positive feedback in the Arctic and that the result may be the planetary equivalent of hyperthermia.

The North Pole lies about 400 miles from the nearest land. It was formerly covered with ice 6 to 15 feet (2 to 5 meters) thick. However, as the planet heats up, the Arctic is warming about twice as fast as the global average. As a result, during the past 30 years the Arctic ice cap has become almost 50% thinner and 35% smaller in area (**FIG. E32-1**). The cap is currently shrinking by an average of about 10% per decade.

What does this have to do with positive feedback? The answer lies in how much sunlight is reflected by ice compared to open ocean waters. Sea ice typically reflects 50% to 70% of incoming sunlight back into space. Snow cover on the ice may reflect as much as 80% of sunlight. In contrast, seawater reflects only 5% to 10% of the sunlight hitting it, so nearly all of the solar energy is absorbed, warming the Arctic water. As the Arctic ice sheet melts, more water is exposed. The exposed water absorbs more sunlight than the ice did, and becomes warmer, which melts more ice, exposing still more water—and so on. This is a classic example of positive feedback, with unpredictable global consequences.

▲ **FIGURE E32-1 Arctic sea ice is disappearing** Polar bears depend on Arctic ice to hunt and rest, so the species' survival is threatened by a shrinking ice cap. Loss of ice due to a warming climate stimulates positive feedback that results in the loss of even more ice.

CONSIDER THIS By pumping carbon dioxide and other greenhouse gases into the atmosphere, we are increasing the temperature of our planet. How might negative feedback eventually slow human-induced warming?

organs that act in a coordinated manner. For example, the urinary system is an organ system consisting of the kidneys, ureters, bladder, and urethra; these organs work together to collect wastes from the bloodstream and to form, store, and finally release the urine from the body. An example of this hierarchy is illustrated in **FIGURE 32-3**.

Animal Tissues Are Composed of Similar Cells That Perform a Specific Function

A tissue consists of aggregations of cells that are similar in structure and functions. A tissue may also include extracellular components produced by its cells, as in blood, cartilage, and bone. Here we present a brief overview of the four major categories of animal tissues: epithelial tissue, connective tissue, muscle tissue, and nerve tissue.

Epithelial Tissue Covers the Body, Lines Its Cavities, and Forms Glands

Epithelial tissue forms both membranes and glands. Epithelial membranes cover both internal and external body surfaces, forming the epidermis of skin and coating the

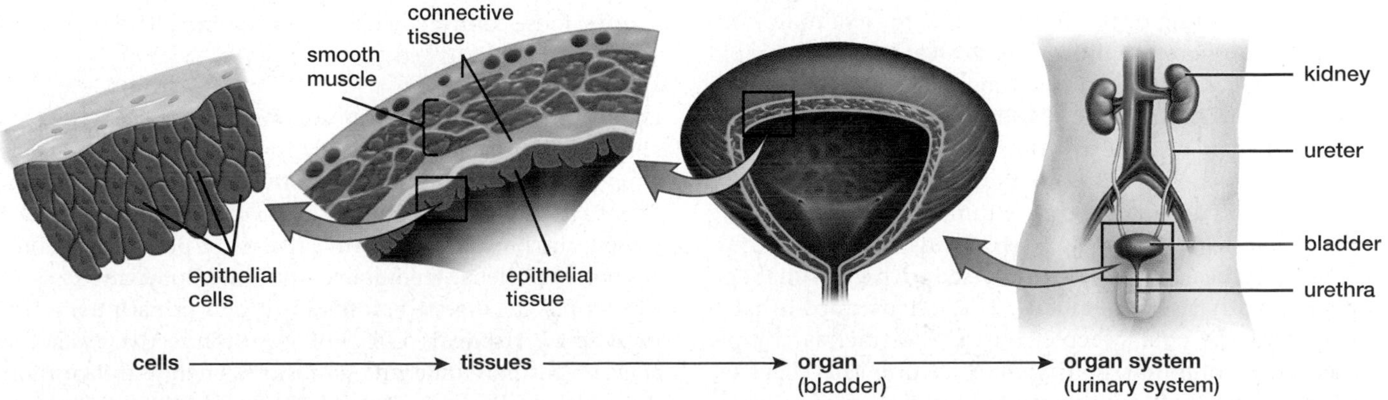

▲ **FIGURE 32-3 Cells, tissues, organs, and organ systems** The animal body is composed of cells, which make up tissues, which combine to form organs that work together as organ systems.

outer surfaces of internal organs. Epithelial membranes line ducts and hollow organs such as the uterus, bladder, lungs, heart and blood vessels, and the digestive tract. Epithelial membranes provide protection, facilitate gas exchange, and regulate the movement of nutrients and wastes across the walls of internal structures. In addition, epithelial membranes often contain cells that secrete substances, such as mucus.

Epithelial membranes have some general characteristics:

- Epithelial membranes have a free surface that usually faces a cavity inside the body or a much smaller space within a gland. They also cover the outside of the body.
- Epithelial membranes are anchored to and supported by an underlying *basement membrane*, a thin, noncellular layer composed primarily of fibrous proteins.
- Epithelial membranes are thin and lack blood vessels. Their cells acquire nutrients, exchange gases, and eliminate wastes by diffusion through their upper or lower surfaces.
- The cells of epithelial membranes can usually regenerate. Many, such as those lining the digestive tract, are exposed to harsh conditions. Consider your mouth: scalded by coffee and scraped by corn chips, its epithelium must replace itself continuously.
- The cells of epithelial membranes are firmly attached to one another. In the urinary bladder, for example, tight connections seal the spaces between epithelial cells, preventing urine from leaking into the body.

Epithelial membranes (**FIG. 32-4**) can be classified as *simple epithelium*, which is only one cell thick, or *stratified epithelium*, which contains more than one cell layer. Within each category, epithelial tissues are named according to cell shape, which may be *squamous* (flat and thin, looking a bit like fried eggs), *cuboidal* (cube-shaped), or *columnar* (elongated, like a column). These diverse tissues are also sometimes *ciliated* (bearing cilia on their upper surfaces; movements of the cilia may cause adjacent particles or fluids to move). Each type of epithelial tissue is specialized for its role in a particular organ.

Various types of **simple epithelium** line the digestive, urinary, reproductive, circulatory, and respiratory systems. The lining of the lung air sacs, for example, consists of squamous epithelium (**FIG. 32-4a**) that allows rapid diffusion of gases between the lungs and the bloodstream. Cuboidal epithelium (**FIG. 32-4b**) protects the inside of ducts that carry secretions from glands and also lines the kidney tubules, where it contributes to urine formation. Simple columnar ciliated epithelium, such as that lining the trachea (windpipe; **FIG. 32-4c**), consists of both short and tall cells bearing cilia, interspersed with glandular epithelial cells that secrete mucus. The mucus traps inhaled debris, and the cilia sweep it out of the trachea. Simple nonciliated columnar epithelial cells line the digestive tract; many are involved in either the secretion of digestive substances (such as those lining the stomach) or absorption of nutrients (intestinal lining).

Stratified epithelium is two or more cells thick, allowing it to withstand considerable wear and tear. Stratified epithelium is found in the esophagus, in the skin (**FIG. 32-4d**), and just inside body openings that are continuous with the skin, such as the mouth and anus.

Glands are clusters of cells that are specialized for secretion. They are classified into two categories: exocrine glands and endocrine glands. **Exocrine glands** secrete substances into a body cavity or onto the skin surface through a narrow tube or duct. Some exocrine glands consist of epithelial cells lining microscopic pits; examples are sweat and oil glands in skin (see Fig. 32-9) and several types of glands within the stomach lining. Larger exocrine glands contain secretory epithelial cells within a framework of connective tissue; these include milk glands, salivary glands, and others that release digestive secretions into the stomach and small intestine. **Endocrine glands**, such as the ovaries, testes, thyroid, and pituitary, do not have ducts. Secretory epithelial cells within these glands secrete hormones into the interstitial fluid, from which the hormones diffuse into nearby capillaries. **Hormones** are chemicals that are produced in small quantities and are transported in the bloodstream to distant parts of the body, where they regulate the activity of other cells (see Chapter 38).

Connective Tissues Have Diverse Structures and Functions

Connective tissue forms a diverse group of tissues that support and strengthen other tissues and help to bind the cells of other tissues together in coherent structures, such as skin or muscles. A general feature of connective tissue is that it has a large amount of extracellular matrix relative to cells. This extracellular matrix may be thin and watery (as in blood and lymph), gelatinous (as in the connective tissue layer that underlies epithelia), stretchy (as in the dermis of skin), tough and flexible (as in cartilage), or rigid (as in bone). Except in blood and lymph, the extracellular matrix contains protein fibers. The most abundant of these fibers is collagen, which confers strength, while other fibrous proteins provide support and elasticity.

Connective tissues can be grouped into three main categories: loose connective tissue, dense connective tissue, and specialized connective tissue.

Loose Connective Tissue Loose connective tissue is the most abundant form (**FIG. 32-5a**). This flexible tissue connects, supports, surrounds, and cushions other tissue types and forms a supple internal framework for organs such as the liver, spleen, and breast. Loose connective tissue also underlies and supports epithelial membranes that line body cavities such as those of the digestive, respiratory, and urinary tracts. Fat, or **adipose tissue**, is a form of loose connective tissue that acts as a cushion under the skin, stores energy, and cushions and insulates the body (**FIG. 32-5b**). In addition to providing insulation, some fat can generate heat (see "Health Watch: Can Some Fat Burn Calories?" on page 569).

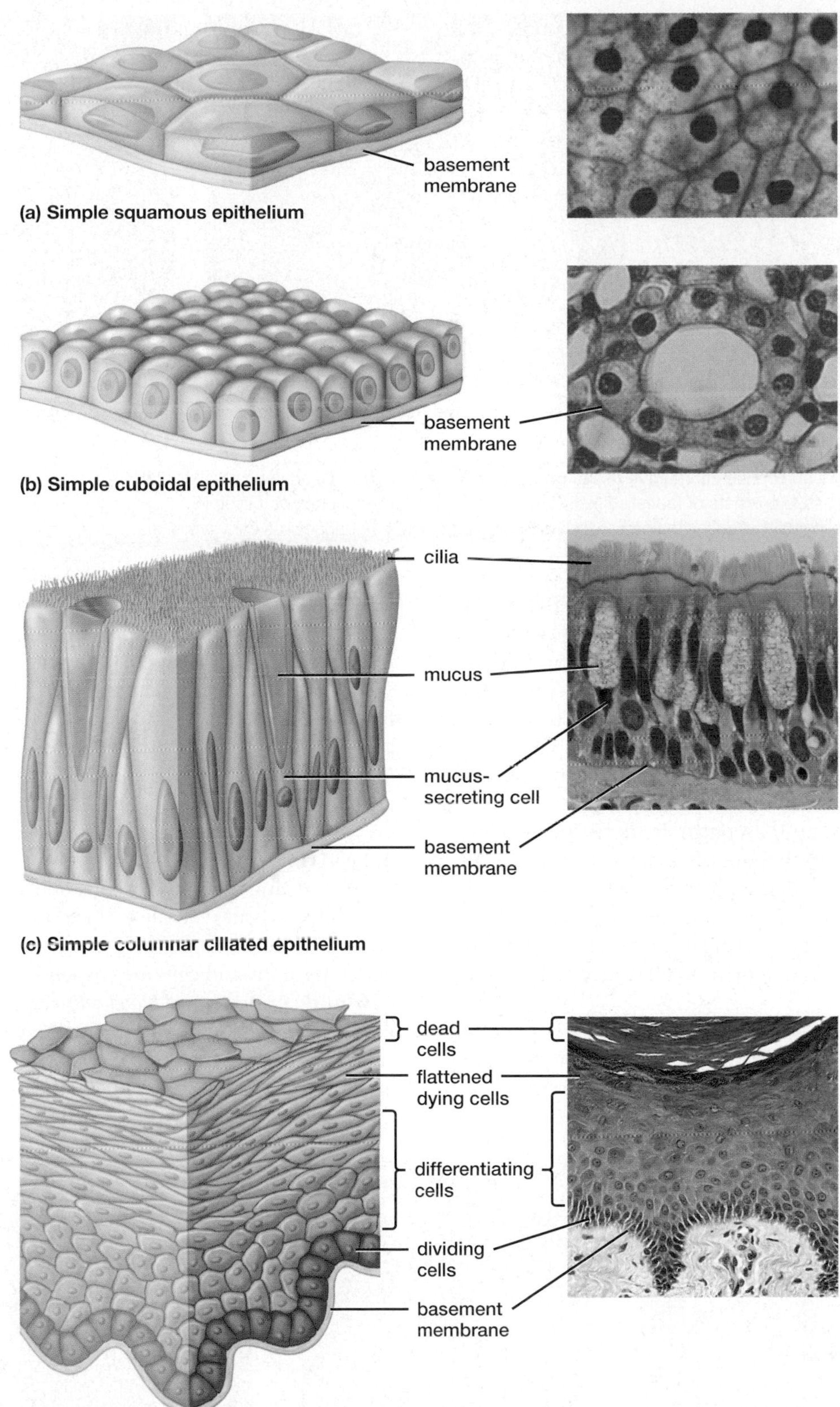

▲ **FIGURE 32-4 Epithelial tissue (a)** In the simple squamous epithelial tissue that lines the lungs, thin, flattened cells in a single layer allow rapid exchange of gases by diffusion. **(b)** Cuboidal epithelial cells line kidney tubules and the ducts that carry secretions from glands. **(c)** Simple columnar ciliated epithelium lining the trachea both secretes and sweeps away mucus. **(d)** Multilayered, stratified epithelial tissue such as the skin epidermis can withstand wear and tear. Dry, dead cells protect the underlying living layers.

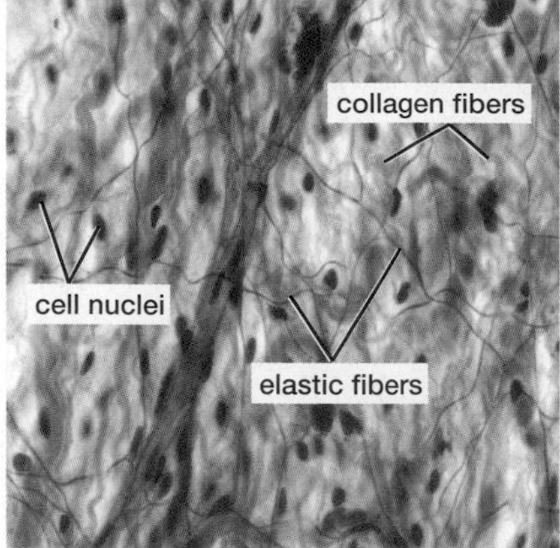

(a) Loose connective tissue underlying epithelial tissue

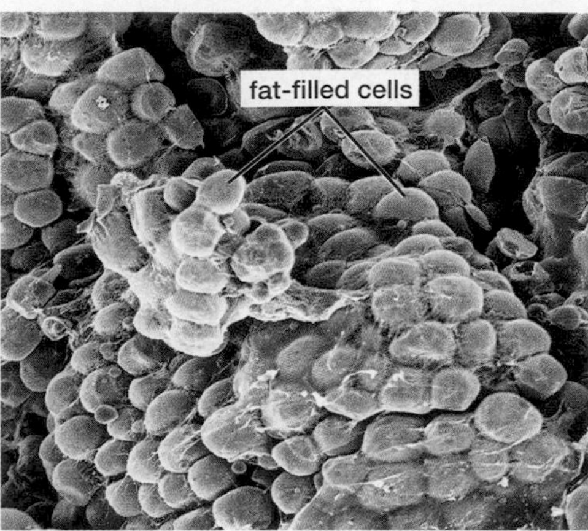

(b) Loose connective tissue forms adipose tissue

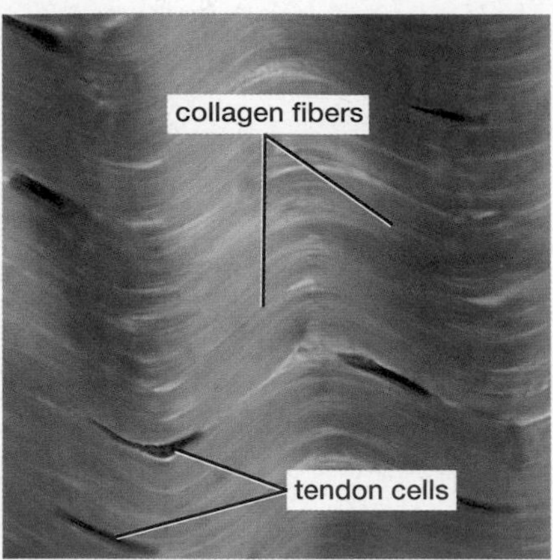

(c) Dense connective tissue of a tendon

▲ **FIGURE 32-5 Loose and dense connective tissue (a)** Loose connective tissue found beneath epithelial tissue. **(b)** Adipose (fat) tissue consists of fat cells filled with stored fat. **(c)** Dense connective tissue in a tendon shows the parallel alignment of collagen fibers, which are secreted by cells that appear as dark inclusions.

Dense Connective Tissue Most dense connective tissue is tightly packed with collagen fibers. In **tendons** (which connect muscles to bones) and **ligaments** (which connect bones to bones), collagen fibers are arranged parallel to one another (**FIG. 32-5c**). In other types of dense connective tissue, such as the dermis of the skin and the tough capsules that surround many internal organs and muscles, the collagen fibers form an irregular meshwork. Both arrangements provide differing degrees of flexibility and strength. Tendons and ligaments have tremendous strength, but only in the direction in which the collagen fibers are oriented (which is why twisting a knee can rupture a ligament). The irregular mesh of collagen in skin and muscle capsules resists tearing in all directions, but isn't as strong as the parallel arrangement in tendons and ligaments.

Specialized Connective Tissue This diverse group includes cartilage, bone, blood, and lymph. **Cartilage** consists of widely spaced cells embedded in a matrix rich in collagen and elastic fibers (**FIG. 32-6a**). Cartilage covers the ends of bones at joints, provides the supporting framework for the respiratory passages, supports the nose and ears, and forms shock-absorbing pads between the vertebrae. Cartilage is fairly flexible, but can snap if bent too far. **Bone** consists of a collagen matrix secreted by bone cells and hardened by deposits of calcium compounds. Bone is laid down in concentric circles around a central canal, which contains blood vessels and a nerve (**FIG. 32-6b**). (We discuss cartilage and bone in Chapter 41.)

Blood and lymph are considered specialized forms of connective tissues because they are composed mostly of an extracellular matrix (in these tissues, a watery liquid) in which proteins and cells are suspended. The cellular portion of **blood** consists of *red blood cells* (which transport oxygen), *white blood cells* (which fight infection), and cell fragments called *platelets*, which aid in blood clotting (**FIG. 32-6c**). These are all suspended in a liquid called *plasma*. **Lymph** consists

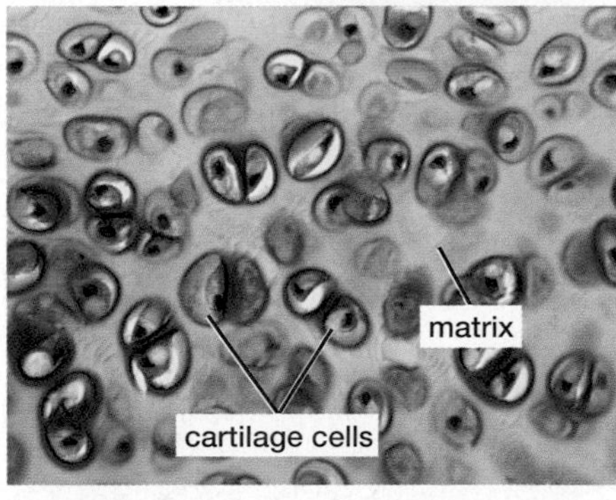

(a) Cartilage

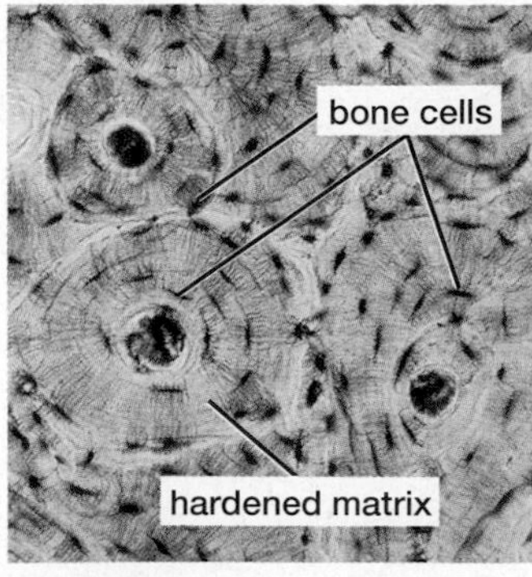

(b) Bone

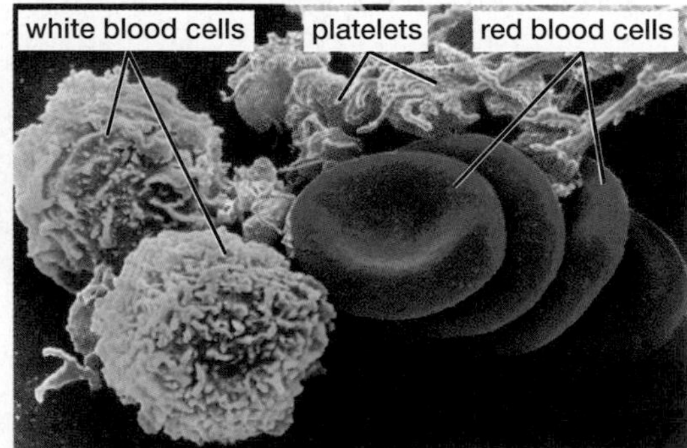

(c) Blood

▲ **FIGURE 32-6 Specialized connective tissue (a)** Cartilage consists of scattered cells within a firm collagenous matrix. **(b)** Bone cells appear as dark spots within the hardened collagenous matrix in concentric layers around a central canal. **(c)** The cellular components of blood are shown in this colorized SEM. The platelets are enmeshed in protein strands that help to form a clot. Blood cells are suspended in an extracellular matrix of fluid plasma.

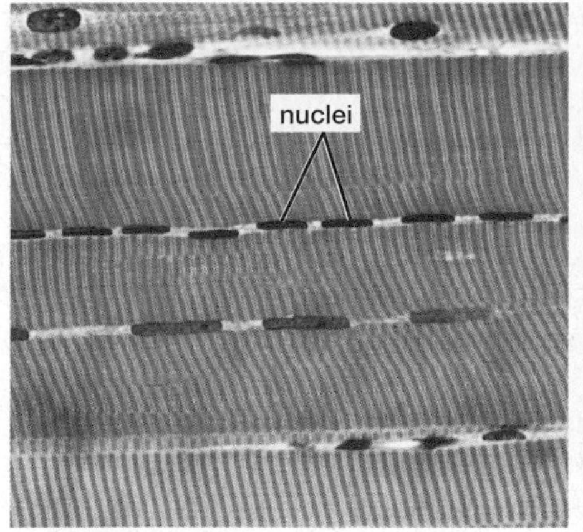

(a) Skeletal muscle

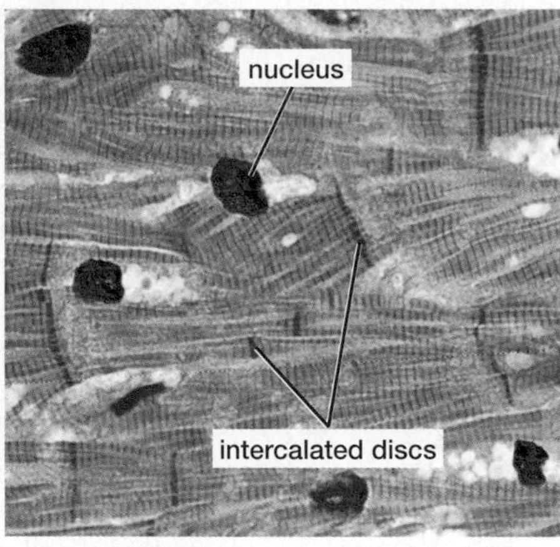

(b) Cardiac muscle

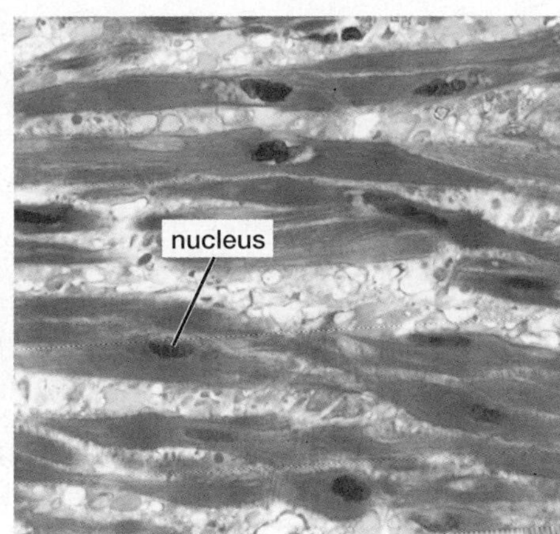

(c) Smooth muscle

▲ **FIGURE 32-7 Muscle tissue (a)** In skeletal muscle, the regular arrangement of fibrous proteins makes it appear striped, or "striated." These large cells each have many nuclei. **(b)** Cardiac muscle is also striated; intercalated discs are visible as darker bands between adjacent cardiac muscle cells. Each cell has a single nucleus. **(c)** Smooth muscle fibers are spindle-shaped cells in which the contractile proteins are not consistently aligned; hence, they appear smooth. Each cell has a single nucleus.

largely of liquid that has leaked out of blood capillaries (the smallest of the blood vessels), plus white blood cells. (Blood and lymph are covered in Chapter 33.)

Muscle Tissue Has the Ability to Contract

The long, thin cells of muscle tissue are packed with two types of fibrous proteins that slide past one another when stimulated, shortening (contracting) the muscle cell. These cellular contractions allow muscle tissue to do work. The cells relax passively when the stimulation stops. There are three types of muscle tissue: skeletal, cardiac, and smooth.

Skeletal muscle (FIG. 32-7a) is stimulated by the nervous system and is generally under voluntary, or conscious, control. As its name implies, its main function is to move the skeleton, as occurs when you walk or turn the pages of this text.

Cardiac muscle (FIG. 32-7b) is located only in the heart. Unlike skeletal muscle, it is spontaneously active, under involuntary (unconscious) control. Cardiac muscle cells are connected by **intercalated discs** that allow electrical signals to spread rapidly throughout the heart, causing coordinated cardiac muscle contraction. In both skeletal and cardiac muscle cells, orderly arrangements of fibrous proteins produce a striped appearance.

Smooth muscle (FIG. 32-7c), named because its cells do not appear striped, is found throughout the body, embedded in the walls of the digestive and respiratory tracts, uterus, bladder, larger blood vessels, skin, and in the iris of the eye. Smooth muscle produces slow, sustained contractions that are typically involuntary and may be stimulated by the nervous system, by stretching, or by hormones and certain other chemicals. (Muscles are covered in Chapter 41.)

Nerve Tissue Is Specialized to Produce and Conduct Electrical Signals

You owe your ability to sense and respond to the world to **nerve tissue**, which makes up the brain, the spinal cord, and the nerves in all parts of the body. Nerve tissue is composed of two types of cells: nerve cells, also called neurons, and glial cells.

Neurons (FIG. 32-8a) are specialized to generate electrical signals and to conduct these signals to other neurons, muscles, or glands. A typical nerve cell consists of *dendrites*,

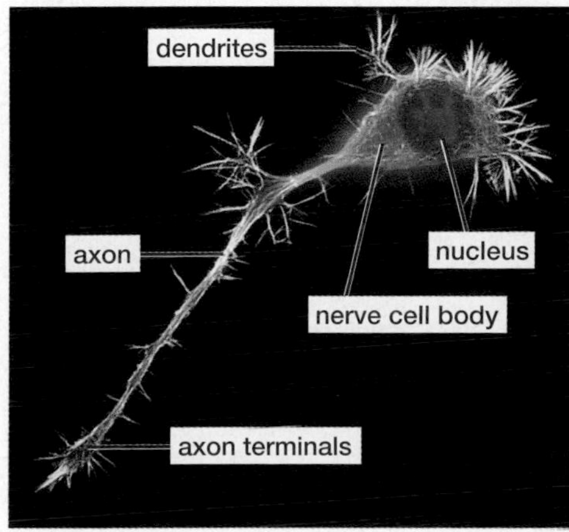

(a) Neuron

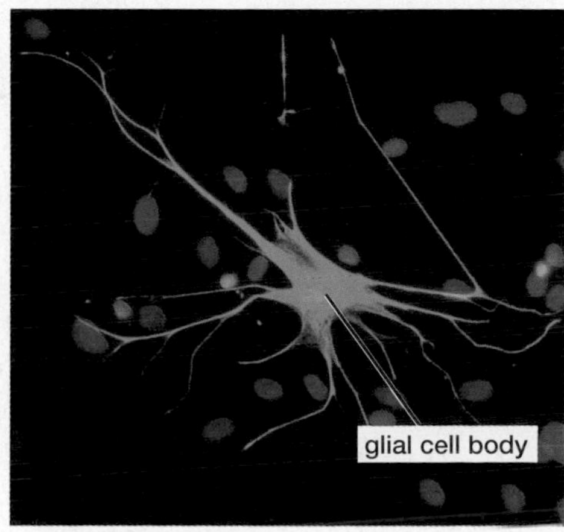

(b) Glial cell

◀ **FIGURE 32-8 Nerve tissue** Nerve tissue consists of neurons and glial cells. **(a)** Neurons are specialized to receive and transmit signals. The long strand emerging from the cell body is the axon, which sends signals to other cells; the short yellow spikes on the cell body are dendrites, which receive signals from other cells. **(b)** The brain has several different types of glial cells; this one, called an astrocyte, helps to nourish neurons and protect them from damage.

which receive signals from other neurons or from the environment; a *cell body*, which contains the nucleus and carries out most of the neuron's metabolism; an *axon*, which carries the neuron's electrical signal to a muscle, gland, or other neuron; and the *terminals* on the ends of an axon, which transmit information to other cells.

Glial cells (**FIG. 32-8b**) surround, support, electrically insulate, and protect neurons. They exert important effects on neurons by regulating the composition of the interstitial fluid. (We discuss nerve tissue in Chapter 39.)

Organs Include Two or More Interacting Tissue Types

Organs are formed from at least two tissue types that function together. Most organs, however, consist of all four tissue types, with different kinds and proportions of epithelial, connective, muscle, and nerve tissues. Most organs function as part of organ systems, structures that work together to carry out basic life functions and maintain homeostasis. In the following section, we describe the skin (part of the integumentary system) as a representative organ that includes all four tissue types.

The Skin Illustrates the Properties of Organs

The skin consists of an outer layer of epithelial tissue underlain by connective tissue. Coursing through the connective tissue are blood vessels, nerves, muscles, and glands (**FIG. 32-9**). Far more than a simple covering for the body, the skin is essential for maintaining homeostasis and, thus, life. Several processes within skin help regulate body temperature. Skin also provides an essential barrier against the evaporation of water and the entry of disease-causing microorganisms. Large-scale destruction of skin (such as occurs with extensive burns) can prove fatal.

The **epidermis**, or outer layer of the skin, is a stratified epithelial tissue (see Fig. 32-4d). Epithelial cells resting on the basement membrane divide continuously, producing daughter cells that become filled with keratin protein as they differentiate. Keratin makes the skin elastic, tough, and fairly waterproof. The differentiating cells are displaced upward as new cells form beneath them. Eventually, the skin cells die, forming the dry, protective, outer skin surface and then flaking off roughly 4 to 6 weeks after they were produced.

Immediately beneath the epidermis lies a layer of connective tissue, the **dermis**. Embedded in the dermis are a variety of structures composed of epithelial, muscle, and nervous

▼ **FIGURE 32-9 Skin is an organ** Mammalian skin contains epithelial, connective, muscle, and nervous tissue.

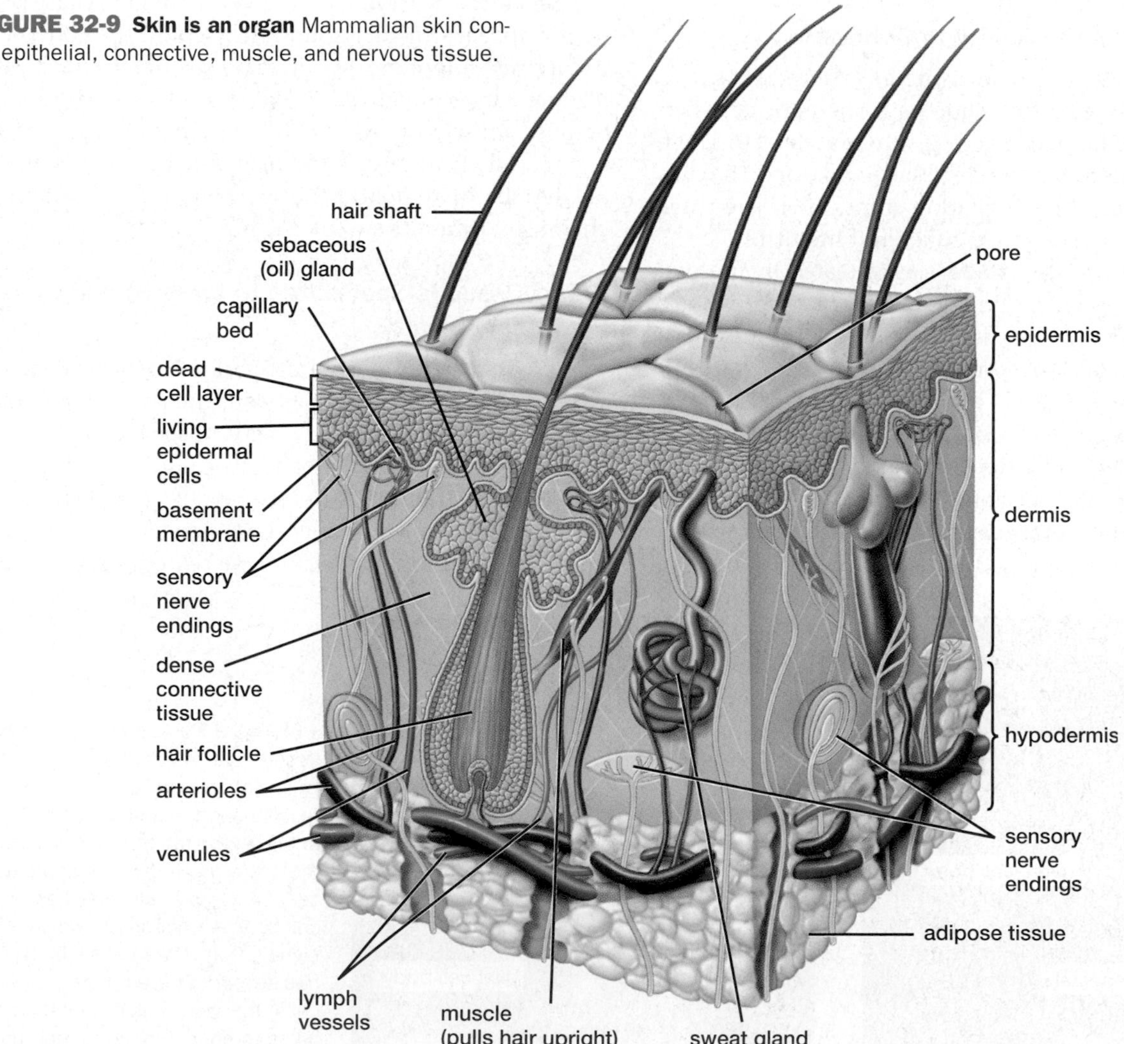

Can Some Fat Burn Calories?

Instead of shivering, human babies and some mammals use brown fat to keep warm. Unlike the white fat that many of us carry as excess weight, *brown fat* is packed with mitochondria and lipids. These mitochondria (which give brown fat its color) have a specialized membrane protein called UCP1 that causes them to "waste" energy by releasing it as heat instead of capturing it in ATP.

Although babies lose brown fat as they develop, researchers have recently discovered that human adults also have small amounts of heat-generating, mitochondria-rich fat cells interspersed with white fat above the collarbone. This newly discovered tissue, called *beige fat*, closely resembles white fat until it is activated, for example, by prolonged exposure to cold. It then starts burning glucose and triglycerides and generating heat. People with higher levels of beige fat are more likely to maintain a normal weight and healthy blood glucose and triglyceride levels. Smaller than average amounts of beige fat are associated with obesity, which in turn is linked to type 2 diabetes and high blood triglycerides, disorders that are reduced by exercise. Two recently identified hormones, irisin and Metrnl, help explain the link between beige fat and weight. Both hormones are released into the blood by muscle tissue during exercise and enhance beige fat activity. Medical researchers are excited by the possibility that these hormones may provide the basis for new drugs to treat type 2 diabetes and obesity, which are reaching epidemic proportions in modern society.

THINK CRITICALLY A substance called 2,4-DNP affects mitochondria much as UCP1 does. It was used as a weight-loss drug in the 1930s but was banned after causing many deaths. Imagine you're a physician and an overweight patient comes to you because she saw 2,4-DNP advertised on the Internet and wants to know if she should take it as a weight-loss aid. Explain the symptoms likely to occur if she uses the product and why it might lead to death.

tissues. Arterioles (small arteries) snake throughout the dermis, carrying blood pumped from the heart into a meshwork of capillaries that nourish both the dermal and epidermal tissue. The capillaries empty into venules (small veins) in the dermis. Loss of heat through the skin is regulated by neurons controlling the diameter of the arterioles. To conserve body heat, smooth muscles in the arterioles contract, reducing the diameter of the arterioles and restricting blood flow through them. To cool the body, these smooth muscles relax, allowing the arterioles to open up and flood the capillary beds with blood, thus releasing excess heat. Lymph vessels collect and carry off interstitial fluid within the dermis. Different sensory nerve endings that respond to temperature, touch, pressure, vibration, and pain are scattered throughout the dermis and epidermis and provide information to the nervous system.

Specialized epithelial cells dip down from the epidermis into the dermis, forming **hair follicles**. Cells in the base of the follicles divide rapidly, and their daughter cells fill with keratin. As new cells form in the base of the follicle, older, dying, keratin-filled cells are pushed upward to the surface of the skin, emerging as hairs.

The dermis also contains glands derived from epithelial tissue. Sweat glands produce watery secretions that cool the skin and excrete substances such as salts and urea. Sebaceous glands secrete an oily substance (sebum) that lubricates the epithelium.

Skin also contains muscle tissue. All skin has muscle cells within the walls of its arterioles. In addition, hairy skin has tiny muscles attached to the hair follicles that can cause the hairs of the skin to "stand on end" in response to signals from motor neurons. Most mammals can increase the height of their insulating fur in cold weather by erecting their hairs. This reaction is useless in people; we merely experience "goose bumps" when these muscles contract.

Finally, just below the dermis (and thus technically not part of the skin) is the hypodermis (Gk. *hypo*, below). This consists of adipose tissue interspersed with fibrous proteins produced by connective tissue cells. The fat helps to insulate the body, preserving heat and making it easier to maintain a constant body temperature in cool weather. Adipose tissue can also be broken down to provide energy.

Organ Systems Consist of Two or More Interacting Organs

Organ systems consist of two or more individual organs (in some cases, located in different parts of the body) that work together to perform a common function. The major organ systems of the vertebrate body and their representative organs and functions are summarized in **TABLE 32-1**.

CHECK YOUR LEARNING

Can you . . .

- explain the differences between tissues, organs, and organ systems and describe their relationships to one another?
- describe the four types of tissues?
- name and describe the major human organ systems?

TABLE 32-1 Major Vertebrate Organ Systems

Organ System	Major Structures	Physiological Role	Organ System	Major Structures	Physiological Role
Integumentary system	Skin, hair, nails, sensory receptors, various glands	Protects the underlying structures from damage; regulates body temperature; senses many features of the external environment	**Respiratory system**	Nose, pharynx, trachea, lungs (mammals, birds, reptiles, amphibians), gills (fish and some amphibians)	Provides a large area for gas exchange between the blood and the environment; allows oxygen acquisition and carbon dioxide elimination
Circulatory system	Heart, blood vessels, blood	Transports nutrients, gases, hormones, metabolic wastes; also assists in temperature control	**Lymphatic/ immune system**	Lymph, lymph nodes and vessels, white blood cells	Carries fat and excess fluids to the blood; destroys invading microbes
Digestive system	Mouth, esophagus, stomach, small and large intestines, glands producing digestive secretions	Supplies the body with nutrients that provide energy and materials for growth and maintenance	**Urinary system**	Kidneys, ureters, bladder, urethra	Maintains homeostatic conditions within the bloodstream; filters out cellular wastes, certain toxins, and excess water and nutrients
Nervous system	Brain, spinal cord, peripheral nerves	Controls physiological processes in conjunction with the endocrine system; senses the environment, directs behavior	**Endocrine system** (Male / Female)	A variety of hormone-secreting glands and organs, including the hypothalamus, pituitary, thyroid, pancreas, adrenals, ovaries, and testes	Controls physiological processes, typically in conjunction with the nervous system
Skeletal system	Bones, cartilage, tendons, ligaments	Provides support for the body, attachment sites for muscles, and protection for internal organs	**Muscular system**	Skeletal muscle	Moves the skeleton
				Smooth muscle	Controls movement of substances through hollow organs (digestive tract, large blood vessels)
				Cardiac muscle	Initiates and implements heart contractions
Male reproductive system	Testes, seminal vesicles, prostate gland, penis	Produces sperm and sex hormones, inseminates female	**Female reproductive system**	Ovaries, oviducts, uterus, vagina, mammary glands	Produces egg cells and sex hormones, nurtures developing offspring

CASE STUDY REVISITED

Overheated

Death from heat stroke is not limited to athletes. In fact, when the air temperature becomes extremely high, especially when coupled with high humidity, people can die of heat stroke while sitting in their living rooms. Although athletes typically receive more news coverage, death from heat stroke is far more likely to occur among the elderly during summer heat waves and among children. Farm workers, toiling under the hot sun, are also at considerable risk.

Children are especially vulnerable to heat stroke. In very young children, temperature homeostasis is not fully developed. Children produce more heat per pound of body weight than adults would under similar conditions, and they do not sweat as much as adults do. About 35 to 40 children in the United States die of heat stroke each year when left alone in cars. When it's over 90°F outside (32°C), the temperature in a closed car can skyrocket to 140°F (60°C) in less than an hour.

CONSIDER THIS Computer models of global climate change caused by increasing carbon dioxide in the atmosphere predict not only higher overall temperatures, but also more intense and frequent heat waves. During the devastating European heat wave in the summer of 2003, between 35,000 and 52,000 "excess deaths" occurred. Before taking specific actions to reduce carbon dioxide emissions, economists and politicians try to estimate the costs. Should excess deaths be included as a cost of inaction? If so, what value should we place on a life?

CHAPTER REVIEW

Go to **Mastering Biology** for practice quizzes, activities, eText, videos, current events, and more.

*Answers to **Think Critically** and **Thinking Through the Concepts** questions can be found in the **Answers** section at the back of the book.*

Summary of Key Concepts

32.1 Homeostasis: Why and How Do Animals Regulate Their Internal Environment?

Homeostasis refers to the dynamic equilibrium within the animal body in which internal conditions (including temperature, salt, oxygen, glucose, pH, and water levels) are maintained within a range in which proteins can function and energy can be made available. Animals differ in temperature regulation. Ectotherms derive most of their body warmth from the environment. Endotherms derive most of their heat from metabolic activities and tend to regulate their body temperatures within a narrow range.

Homeostatic conditions are maintained through negative feedback, in which a change triggers a response that counteracts the change and restores the original conditions. There are a few instances of positive feedback, in which a change initiates events that intensify the change, but these situations are all self-limiting.

32.2 How Is the Animal Body Organized?

The animal body is composed of organ systems, each consisting of two or more organs. Organs, in turn, are made up of tissues. A tissue is a group of cells and extracellular material that form a structural and functional unit specialized for a specific task. Animal tissues include epithelial, connective, muscle, and nerve tissue.

Epithelial tissue forms coverings over internal and external body surfaces and also gives rise to glands. Connective tissue usually contains considerable extracellular material in which cells and proteins are embedded and includes the dermis of the skin, bone, cartilage, tendons, ligaments, fat, and blood. Muscle tissue is specialized to produce movement by contraction. There are three types of muscle tissue: skeletal, cardiac, and smooth. Nerve tissue, composed of neurons and glial cells, generates and conducts electrical signals.

An organ contains at least two tissue types. In mammalian skin, the epidermis, an epithelial tissue, covers and protects the dermis beneath it. The dermis consists of connective tissue that contains blood and lymph vessels, sweat and sebaceous glands, hair follicles, muscles that erect the hairs, and a variety of sensory nerve endings.

Vertebrate organ systems include the integumentary, respiratory, circulatory, lymphatic/immune, digestive, urinary, nervous, endocrine, skeletal, muscular, and reproductive systems, summarized in Table 32-1.

Thinking Through the Concepts

Bloom's: Remembering, Understanding

Multiple Choice

1. Which of the following statements is false?
   a. Overheating to 108°F is likely to disrupt covalent bonds in enzymes.
   b. Overly acidic conditions can disrupt protein structure.
   c. Low body temperature will reduce the rate at which enzymes catalyze reactions.
   d. Considerable metabolic activity is required to maintain a relatively constant environment.
2. Which of the following statements is true?
   a. Ectotherms generate most of their body heat metabolically.
   b. Ectotherms include birds and mammals.
   c. Ectotherms experience variable body temperatures.
   d. Ectotherms are known as "warm-blooded" animals.

3. Negative feedback systems do *not*
   a. include thermostats that maintain room temperature.
   b. include a sensor, a control center, and an effector.
   c. control body temperature.
   d. regulate uterine contractions during labor.
4. Which of the following statements is *not* true of epithelial tissue?
   a. It is classified as loose, dense, and specialized.
   b. It is anchored to and supported by a basement membrane.
   c. It may secrete mucus.
   d. It lines the digestive tract.
5. Which of the following statements is true?
   a. Glands consist primarily of specialized epithelial tissue.
   b. The outer layer of all epithelial tissue consists of dead cells.
   c. Simple epithelial tissue may be several layers thick.
   d. Adipose tissue is a form of dense connective tissue.

Fill-in-the-Blank

1. The ability of the body to maintain its internal conditions within the narrow range required by cells to function is called ____________. The most important general mechanism for maintaining these conditions is ____________.
2. The four levels of organization of the animal body, from the smallest to the most inclusive, are ____________, ____________, ____________, and ____________.
3. Fill in the appropriate tissue type: Supports and strengthens other tissues: ____________; forms glands: ____________; includes the blood: ____________; includes the dermis of the skin: ____________; covers the body and lines its cavities: ____________; can shorten when stimulated: ____________; includes glial cells: ____________; includes adipose tissue: ____________.
4. Glands with ducts connecting them to the epithelium are called ____________ glands. Glands with no ducts are called ____________ glands; most of these secrete ____________ (a general term).
5. Fill in the appropriate type(s) of muscle: Contracts rhythmically and spontaneously: ____________; is controlled voluntarily: ____________; contains orderly arrangements of fibrous proteins, giving a striped appearance: ____________; is under involuntary control: ____________; is found in the walls of the digestive tract: ____________; moves the skeleton: ____________.

Review Questions

1. Define *homeostasis*, and explain how negative feedback helps to maintain it. Describe one example of homeostasis in the human body.
2. Define and compare *ectotherms* and *endotherms*. Provide an example of each. Are "cold-blooded" and "warm-blooded" accurate ways to describe them? Explain.
3. Explain positive feedback, and provide one physiological example. Explain why this type of feedback is relatively rare in physiological processes.
4. Explain what goes on in your body to restore temperature homeostasis when you become overheated by exercising on a hot, humid day.
5. Name and describe the structure of the three general types of epithelial cells, and describe the general structure and functions of epithelial tissue.
6. What property distinguishes connective tissue from all other tissue types? Describe the characteristics and functions of loose connective tissue, dense connective tissue, and specialized connective tissue.
7. Describe the skin as an organ. Include the various tissues that compose it, and briefly describe the role of each tissue.
8. Name the human organ systems, and briefly describe the components and functions of each.

Applying the Concepts

Bloom's: Applying, Analyzing, Evaluating

1. Why does life on land present particular difficulties in maintaining temperature homeostasis compared to life in water?
2. Third-degree burns are usually painless. Skin regenerates only from the edges of these wounds. Second-degree burns are often very painful. Skin regenerates from cells located at the burn edges, in hair follicles, and in sweat glands. First-degree burns are painful but heal rapidly from undamaged epidermal cells. Using this information, draw the depth of first-, second-, and third-degree burns on Figure 32-9.
3. Imagine you are a health care professional teaching a prenatal class for expecting parents. Create a design for a machine with sensors, electrical currents, motors, and so forth that would illustrate the feedback relationships involved in labor and childbirth in a way that a layperson could understand.

33 Circulation

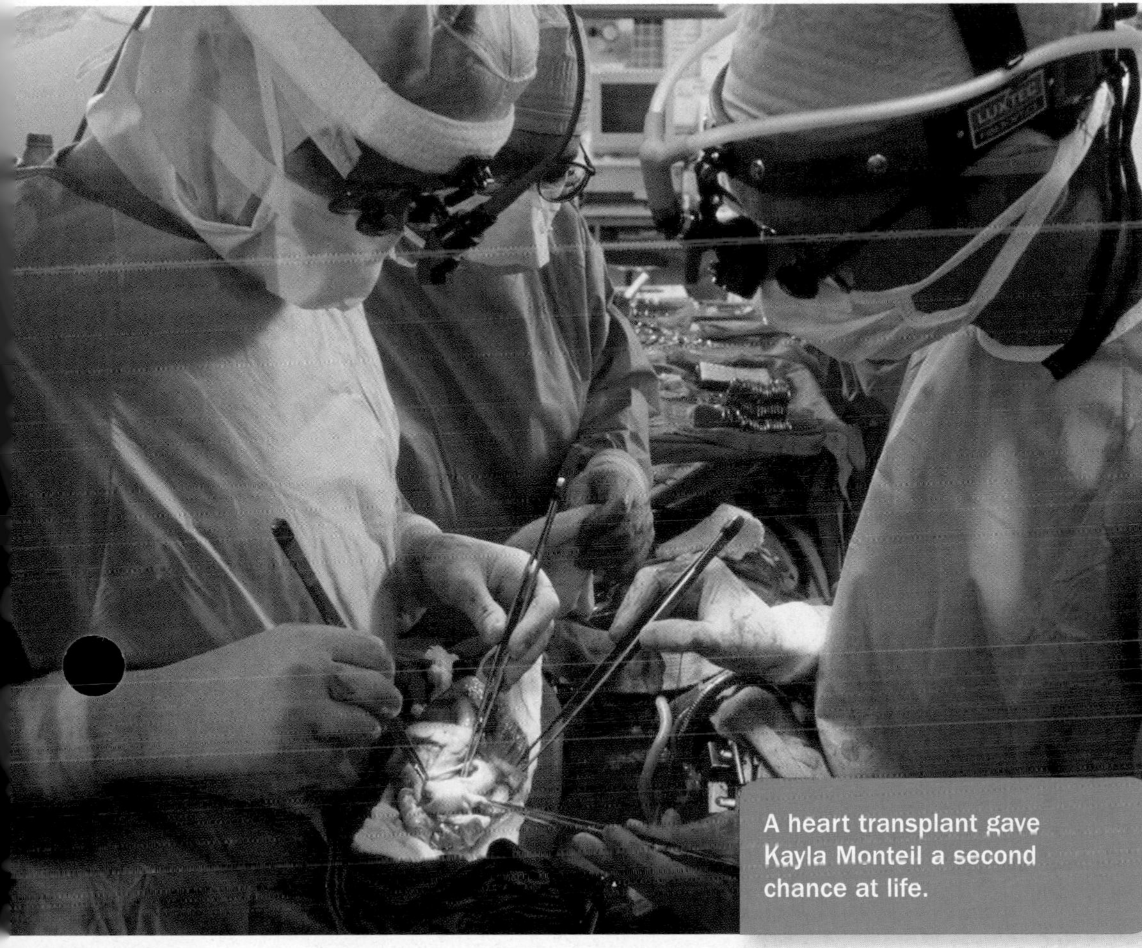

A heart transplant gave Kayla Monteil a second chance at life.

CASE STUDY

Living from Heart to Heart

AFTER 37 DAYS in the intensive care unit awaiting a heart and kidney transplant, 23-year-old Kayla Monteil was physically and emotionally drained. She had received her first donated heart when she was less than 2 years old. A congenital heart valve defect had caused her heart muscle to become stretched and thin as it struggled—and eventually failed—to supply enough blood to her body, a condition called dilated cardiomyopathy. At that time, Kayla's doctors considered 10 years to be an optimistic estimate for the life of her donated heart. But defying the odds, the transplanted organ served her through high school before problems arose. Now, after 22 years, the heart was failing and so were Kayla's kidneys, damaged irrevocably from decades of taking drugs to prevent her body from rejecting the donor heart. Because heroic medical interventions were keeping Kayla alive, she was at the top of the priority list for a heart and kidney transplant. Then, suddenly, the tragic death of a young man whose tissues were a good match for Kayla's gave her both these organs and another chance at life.

Dilated cardiomyopathy prevents the heart from contracting with sufficient force to circulate blood normally. In serious cases such as Kayla's, this condition leads to heart failure, in which a weakening of the heart prevents it from supplying enough oxygen to the body's tissues. Heart failure renders its victims exhausted after even minor exertion. Although older people are common victims, cardiomyopathy is a complex disorder with a variety of causes, so it can strike at any age. Some people, like Kayla, are born with a predisposition to cardiomyopathy. Cocaine, amphetamines, or excessive use of alcohol can also cause or contribute to this condition at a young age. High blood pressure is another major risk factor.

How do circulatory systems normally supply the body with oxygen and nutrients, and why does the heart sometimes fail to do its job adequately? How does high blood pressure strain the heart? How does heart failure differ from a heart attack?

AT A GLANCE

33.1 WHAT ARE THE MAJOR FEATURES AND FUNCTIONS OF CIRCULATORY SYSTEMS?

Billions of years ago, Earth's early cells were nurtured by the primordial sea. The sea provided nutrients, which diffused into the cells, and washed away wastes, which diffused out from the cells. But diffusion is only efficient over very short distances, and today, only microorganisms and some simple multicellular animals rely almost exclusively on diffusion to exchange nutrients and wastes with the environment. Sponges, for example, circulate seawater through pores in their bodies, bringing the environment close to each cell (see Figs. 34-1a and 35-8). In more complex animals, individual cells are farther from the outside world, and diffusion alone is inadequate to ensure that nutrients reach the cells and that the animals aren't poisoned by their own wastes. With the evolution of circulatory systems, a sort of "internal sea" arose that performs a function similar to that which the sea performed for the earliest cells.

All circulatory systems have three major components:

- A liquid, **blood**, that serves as a medium of transport for gases, nutrients, and cellular wastes.
- A pump, the **heart**, that keeps the blood circulating.
- A system of tubes, **blood vessels**, that consist of *arteries* that carry blood away from the heart, *veins* that carry blood toward the heart, and *capillaries* that link arteries and veins and exchange materials through their walls.

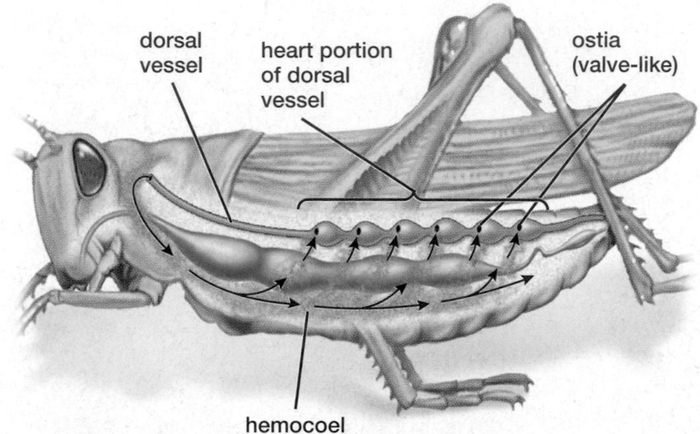

(a) Open circulatory system (grasshopper)

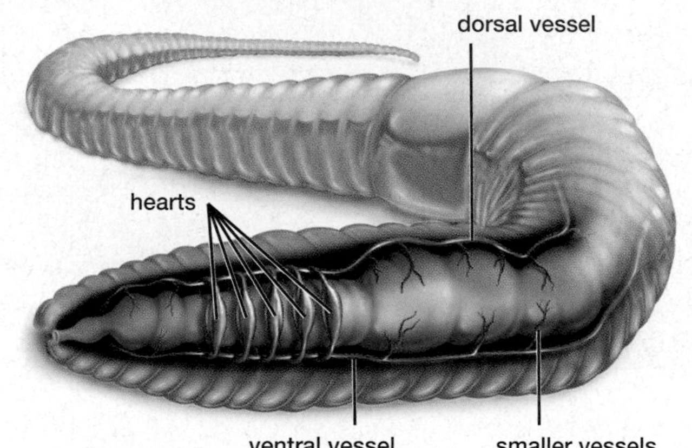

(b) Closed circulatory system (earthworm)

▲ **FIGURE 33-1 Open and closed circulatory systems (a)** In the open circulatory systems of most invertebrates, one or more hearts pump hemolymph through vessels into the hemocoel, where blood directly bathes the internal organs. The hemolymph of insects, such as this grasshopper, is transparent and sometimes pale green. **(b)** In a closed circulatory system, blood remains confined within the heart(s) and the blood vessels. In the earthworm, five contractile vessels serve as hearts that pump blood through major ventral and dorsal vessels, from which smaller interconnecting vessels branch. Earthworm blood contains red hemoglobin.

THINK CRITICALLY Why doesn't insect hemolymph need hemoglobin?

Two Types of Circulatory Systems Are Found in Animals

The circulatory systems of animals take two different forms. Both use hearts to circulate blood, but **open circulatory systems** bathe organs directly in blood, whereas **closed circulatory systems** confine the blood within blood vessels.

Open circulatory systems are present in most invertebrates that possess circulatory systems, including all arthropods (such as crustaceans, spiders, and insects) and most mollusks (such as snails and clams). An animal with an open circulatory system has one or more simple hearts, some blood vessels, and a series of interconnected chambers within the body, collectively called a **hemocoel** (**FIG. 33-1a**). Within the hemocoel, which occupies 20% to 40% of the body volume, tissues and internal organs are directly bathed in **hemolymph**, a fluid that functions both as blood and as the interstitial fluid that surrounds each cell. In insects, a single large dorsal blood vessel is modified into a series of contracting chambers in the abdomen, which form the insect heart. Each chamber contains a slit-like opening called an *ostium* (plural, ostia) that functions as a one-way valve. The ostia are forced shut as the heart chambers contract, propelling the hemolymph forward through the dorsal vessel. The hemolymph is pumped into hemocoel compartments of the head and then flows through the body into the abdominal portions of the hemocoel. As the heart chambers relax, they expand, drawing hemolymph back through the ostia and into the heart.

Animals with open circulatory systems expend less energy on circulating blood than do animals with closed systems, but open circulatory systems are less efficient at supplying oxygen and nutrients to tissues. Such systems are fully adequate for relatively sedentary animals, but how can an active, flying insect obtain enough oxygen with its open circulatory system? In fact, an insect's blood does not carry oxygen to its tissues. Insect evolution has outsourced the gas-exchange function to a system of gas-filled tubes (called tracheae; see Fig. 34-4) that have openings to the air and branch extensively throughout their tissues, carrying oxygen close to each cell.

In a closed circulatory system, blood pressure and flow rates are higher than is possible in an open system. A closed circulatory system can also adjust the amount of blood flowing through different vessels, directing blood to specific tissues as needed—for example, to muscles during exercise or to the digestive tract after a meal. Such systems are well adapted to an active lifestyle. Closed circulatory systems are present in all vertebrates (such as fishes, reptiles, and mammals) and in a few invertebrates, including very active mollusks (squid and octopuses) and, perhaps surprisingly, earthworms and many of their close relatives (**FIG. 33-1b**). Although earthworms have a sluggish reputation, they must perform feats of burrowing through dense soil where little oxygen is available, so a closed system is advantageous for them.

The Vertebrate Circulatory System Has Diverse Functions

The circulatory system supports all the other organ systems in the body. In vertebrates, the circulatory system performs the following functions:

- Transports oxygen from the gills or lungs to the rest of the body and transports carbon dioxide from the tissues to the gills or lungs.
- Distributes nutrients from the digestive system to all body cells.
- Transports toxic substances to the liver for detoxification and transports cellular wastes to the kidneys, where they are filtered from the blood and excreted.
- Distributes hormones from the glands and organs that produce them to the tissues upon which they act.
- Helps regulate body temperature by adjusting blood flow.
- Helps stop bleeding and heal wounds by producing blood clots.
- Protects the body from diseases by circulating white blood cells and antibodies.

In the following sections we examine the three components of the circulatory system: the heart, the blood, and the blood vessels, with emphasis on the human system. Finally, we describe the lymphatic system, which works closely with the circulatory system.

CHECK YOUR LEARNING

Can you . . .

- explain the key features shared by all circulatory systems?
- compare open and closed circulatory systems?
- describe the functions of the vertebrate circulatory system?

33.2 HOW DOES THE VERTEBRATE HEART WORK?

The vertebrate heart consists of muscular chambers capable of strong contractions. Chambers called **atria** (singular, atrium) collect blood. Contractions of the atria send blood into the **ventricles**, chambers whose contractions circulate blood through the lungs and to the rest of the body.

The Two-Chambered Heart of Fishes Was the First Vertebrate Heart to Evolve

Fish hearts consist of two main contractile chambers: a single atrium that empties into a single ventricle (**FIG. 33-2a**). Blood pumped from the ventricle passes first through the gills, where the blood picks up oxygen and releases carbon dioxide. The blood travels directly from the gills through the rest of the body, delivering oxygen to the tissues and picking up carbon dioxide. Blood from the body then returns to the single atrium. The bodies of fishes are supported by water, so their hearts do not need to pump blood against gravity. This allows the blood pressure of fishes to be lower than that of most terrestrial vertebrates. The pressure drops considerably as the blood travels through the microscopic gill capillaries and enters the blood vessels. Undulations of the tail and body during swimming help to drive blood back toward the heart.

Increasingly Complex and Efficient Hearts Evolved in Terrestrial Vertebrates

Over the course of evolution, vertebrates emerged from the sea. As fishes gave rise to amphibians (such as salamanders and frogs), a three-chambered heart evolved, with two atria and one ventricle (**FIG. 33-2b**). Reptiles evolved from amphibians, and reptiles such as snakes, turtles, and lizards (but not birds) also have three-chambered hearts.

An important circulatory adaptation in these terrestrial vertebrates is **double circulation**, which creates two separate circuits of blood. The **pulmonary circuit** ("pulmonary" refers to lungs) directs blood from the heart through the lungs, where carbon dioxide is exchanged for oxygen, and back to the heart. The **systemic circuit** carries blood between the heart and the rest of the body, where oxygen is exchanged for carbon dioxide. In the three-chambered heart, blood from the systemic circuit enters the right atrium, blood from the pulmonary circuit enters the left atrium, and both atria empty into the single ventricle. Amphibian and reptile hearts have internal features that direct most of the oxygen-poor blood into the right portion of the ventricle, where it is pumped to the lungs, and direct most of the oxygenated blood into the left portion of the ventricle, which pumps it to the rest of the body.

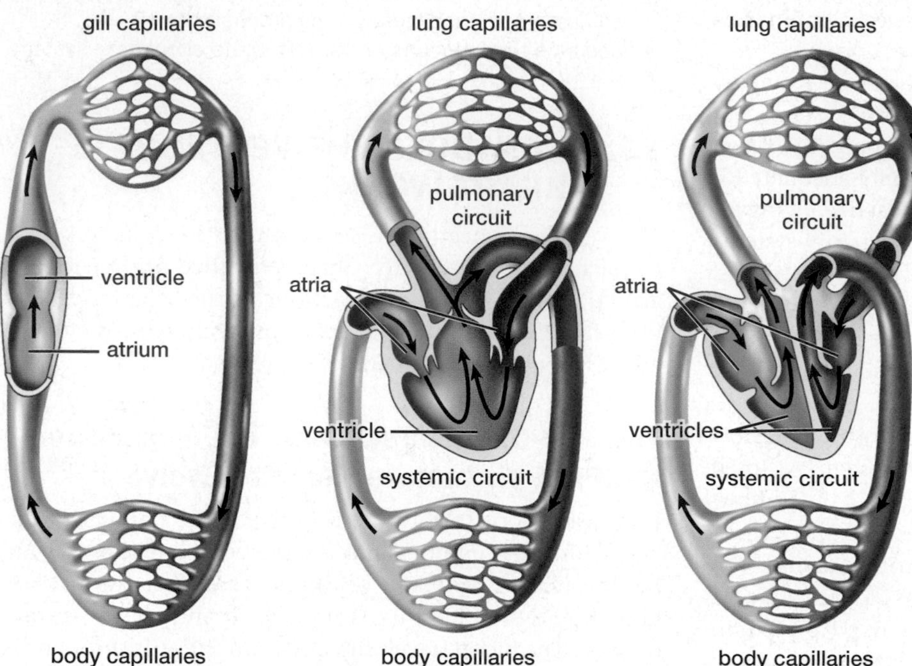

FIGURE 33-2 Evolution of the vertebrate heart (a) The earliest vertebrate heart to evolve was the two-chambered heart of fishes. **(b)** Amphibians and most reptiles have three-chambered hearts in which two atria empty into a single ventricle. **(c)** The hearts of crocodiles, birds, and mammals combine two separate pumps that maintain very different pressures and prevent mixing of oxygenated and oxygen-poor blood. (Because these hearts face the reader, left and right appear reversed. In art throughout this text, oxygen-rich blood is red and oxygen-poor blood is blue.)

Four-Chambered Hearts Consist of Two Separate Pumps

A few groups of reptiles, such as birds (now classified as reptiles) and crocodiles, as well as all mammals, have four-chambered hearts (**FIG. 33-2c**). The four-chambered heart—with its right atrium and right ventricle completely isolated from its left atrium and left ventricle—acts like two hearts beating as one (**FIG. 33-3**). The "right heart" deals with oxygen-poor blood. The right atrium receives oxygen-depleted blood from the body through the two largest **veins** (vessels that carry blood toward the heart), the superior vena cava and the inferior vena cava. After filling with blood, the right atrium contracts, forcing the blood into the right ventricle. Contraction of the right ventricle then sends the oxygen-poor blood to the lungs through the pulmonary **arteries** (vessels that carry blood away from the heart). The "left heart" deals with oxygenated blood. Oxygen-rich blood from the lungs enters the left atrium through the pulmonary veins and is then squeezed into the left ventricle. A strong contraction of the left ventricle, the heart's most muscular chamber, sends the oxygenated blood coursing out through the largest artery, the aorta, and then to the rest of the body.

Valves Maintain the Direction of Blood Flow

The directionality of blood flow is maintained by one-way valves (see Figs. 33-3 and 33-5). Pressure from one side opens the valves easily, but pressure from the other side forces them closed. **Atrioventricular valves** allow blood to flow from the atria into the ventricles, but prevent the blood from flowing back into the atria when the ventricles contract. **Semilunar valves** (Latin for "half-moon") allow blood to enter the pulmonary artery and the aorta when the ventricles contract, but prevent blood from returning as the ventricles relax.

CASE STUDY **CONTINUED**

Living from Heart to Heart

Kayla Monteil's first heart transplant, when she was only 18 months old, was necessitated by a defect in her left atrioventricular valve that prevented it from closing completely. When her left ventricle contracted, blood was forced back up into her left atrium, so less blood entered her aorta to be circulated throughout her body. Her tiny heart was forced to work harder to provide adequate oxygen-rich blood to her tissues. Also, in response to inadequate blood flow, Kayla's body retained fluid in her bloodstream to help maintain her blood pressure. But this extra blood volume overstretched her ventricles, weakening them. She was within a few months of death when a donor heart from a 2-year-old boy became available.

Kayla's heart muscles were abnormally stretched, preventing them from contracting forcibly and circulating her blood adequately. What does normal heart muscle look like, and how are its contractions coordinated to circulate blood effectively?

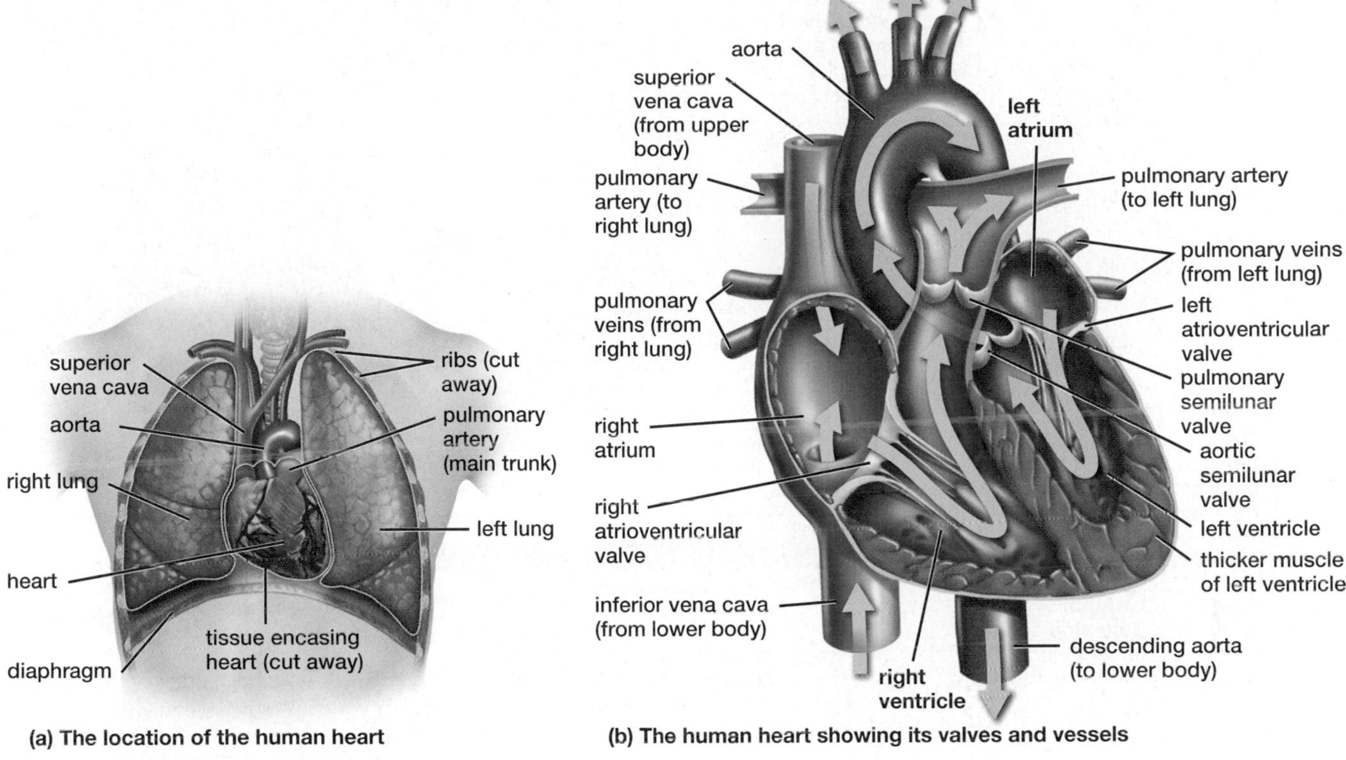

(a) The location of the human heart **(b) The human heart showing its valves and vessels**

▲ **FIGURE 33-3 The human heart (a)** The heart nestles between the lungs, sitting just above the diaphragm, a sheet of muscle between the chest cavity and the abdomen. **(b)** One-way semilunar valves separate the aorta from the left ventricle and the pulmonary artery from the right ventricle. Atrioventricular valves separate each atrium from its corresponding ventricle. Notice that the muscular wall of the left ventricle is thicker than the right because it must pump blood throughout the body.

Cardiac Muscle Is Present Only in the Heart

Most of the heart consists of a specialized type of muscle, **cardiac muscle**, found nowhere else in the body. Cardiac muscle cells are small, branched, and packed with an orderly array of protein strands that give them a striped appearance (**FIG. 33-4**). Cardiac muscle cells are linked to one another by **intercalated discs**, which appear as bands between the cells. Here, adjacent cell membranes are tightly attached to one another, which prevents the strong heart contractions from pulling the muscle cells apart. Intercalated discs also contain membrane-lined channels, which allow the electrical signals that trigger contractions to spread directly and rapidly from one muscle cell to adjacent ones. This rapid spreading causes the interconnected regions of cardiac muscle to contract almost synchronously, thus providing sufficient force to pump blood throughout the body.

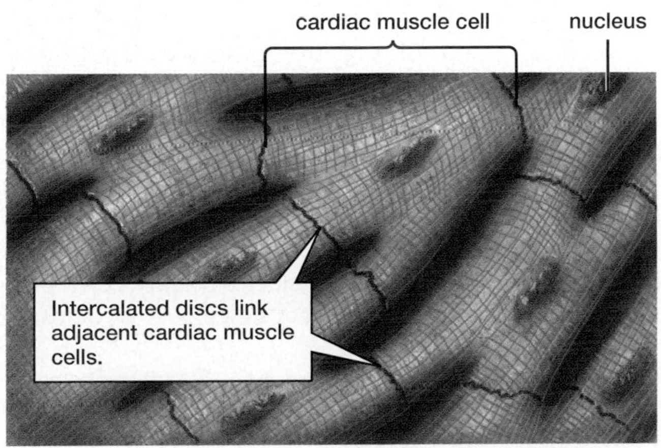

▲ **FIGURE 33-4 The structure of cardiac muscle**

THINK CRITICALLY If a muscle is exercised regularly, it increases in size. Why is the resting heart rate of a well-conditioned athlete slower than the heart rate of a less active person?

The Coordinated Contractions of Atria and Ventricles Produce the Cardiac Cycle

The human heart beats about 100,000 times each day. Each heartbeat is actually a series of coordinated events, called the

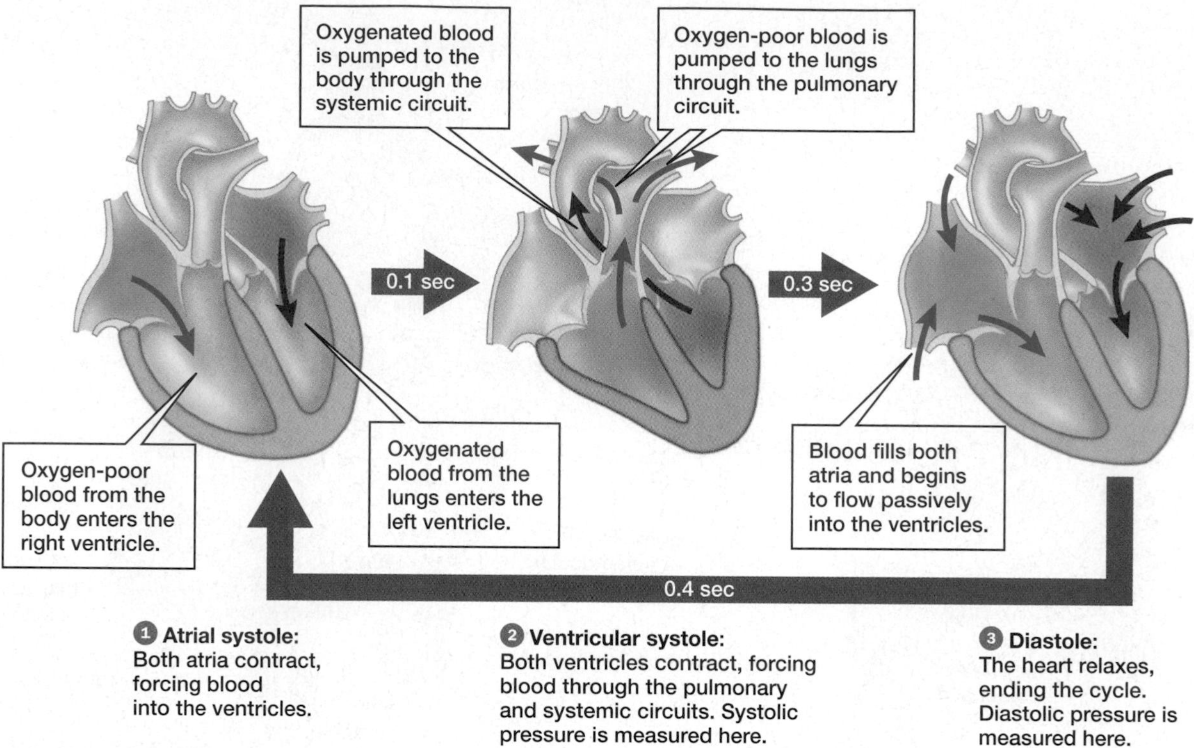

Oxygenated blood is pumped to the body through the systemic circuit.
Oxygen-poor blood is pumped to the lungs through the pulmonary circuit.
0.1 sec
0.3 sec
Oxygen-poor blood from the body enters the right ventricle.
Oxygenated blood from the lungs enters the left ventricle.
Blood fills both atria and begins to flow passively into the ventricles.
0.4 sec
1 Atrial systole: Both atria contract, forcing blood into the ventricles.
2 Ventricular systole: Both ventricles contract, forcing blood through the pulmonary and systemic circuits. Systolic pressure is measured here.
3 Diastole: The heart relaxes, ending the cycle. Diastolic pressure is measured here.

▲ **FIGURE 33-5 The cardiac cycle**

cardiac cycle (**FIG. 33-5**). During each cycle, the two atria first contract in synchrony, emptying their contents into the ventricles, a process called *atrial systole* ①. A fraction of a second later, during *ventricular systole,* the two ventricles contract simultaneously, forcing blood into arteries that exit the heart ②. Then, during *diastole,* both atria and both ventricles relax briefly and begin to fill with blood before the cardiac cycle repeats ③. In a typical resting person, the complete cycle occurs in just under 1 second, or about 70 times per minute. **Heart rate** is the number of cardiac cycles (heartbeats) per minute.

Blood pressure consists of two measurements (**FIG. 33-6**). **Systolic pressure** (the higher of the two) is generated in the arteries by the muscular left ventricle as it pumps blood through the systemic circuit. **Diastolic pressure** (the lower of the two) is the pressure in the arteries as the heart rests between contractions.

A blood pressure reading below 120/80 mmHg (millimeters of mercury) and above 90/60 is considered healthy. Lower blood pressure is generally not a problem unless it is accompanied by symptoms such as dizziness. A pressure of 140/90 mmHg or higher is defined as high blood pressure, or **hypertension**. This condition forces the heart to work harder to pump blood throughout the body, and the strain can weaken the heart, leading to heart failure. Some people have a genetic tendency toward hypertension, but it is also associated with smoking, obesity, lack of exercise, high alcohol consumption, stress, and aging.

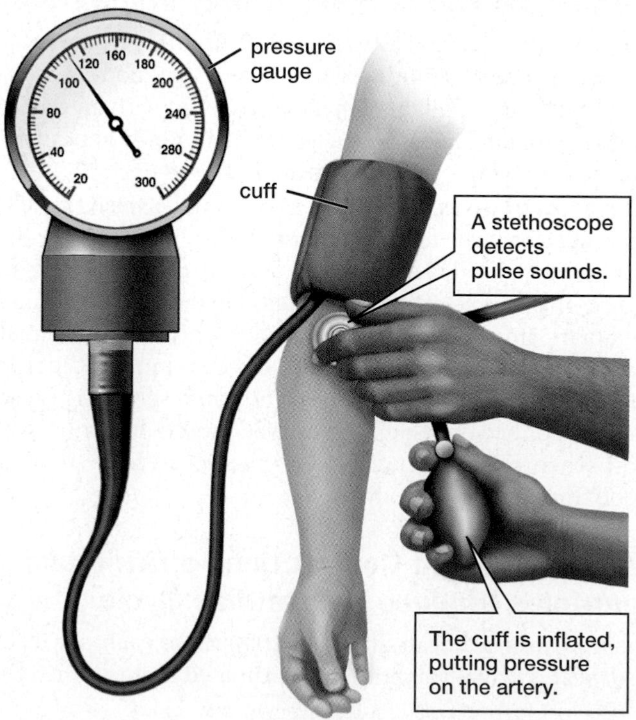

pressure gauge
cuff
A stethoscope detects pulse sounds.
The cuff is inflated, putting pressure on the artery.

▶ **FIGURE 33-6 Measuring blood pressure** The cuff is inflated until its pressure closes off the arm's main artery. The pressure is then gradually reduced until rhythmic blood sounds are first heard through the stethoscope. This is systolic pressure, when some blood is getting through the artery with each heartbeat and the pressure produced by the left ventricle has just overcome the cuff pressure. The cuff pressure is then further reduced until no pulse sounds are audible. This is diastolic pressure, when the blood pressure *between* ventricular contractions just barely overcomes the cuff pressure. The numbers are in millimeters of mercury, a standard measure of pressure also used in barometers.

Have You Ever Wondered ...

How a Giraffe's Heart Can Pump Blood Up to the Brain?

The long legs and 8-foot (2.5-meter) neck of a giraffe allow this amazing animal to browse for food high in trees, but these adaptations put enormous demands on its circulatory system. A giraffe's heart can meet these demands because it weighs about 22 pounds and is about 2 feet from top to bottom. If a giraffe weighed the same as an average human, its heart would be twice as large as the human's heart. The giraffe's enormous heart beats about 170 times per minute and generates blood pressure of about 280/140 mm Hg; both measurements are roughly double those of a human. These adaptations help the blood make the long uphill journey to the giraffe's brain.

Electrical Impulses Coordinate the Sequence of Heart Chamber Contractions

The contraction of the heart is initiated and coordinated by a *pacemaker,* a cluster of specialized heart muscle cells that produces spontaneous electrical signals at a regular rate. The heart's pacemaker is the **sinoatrial (SA) node**, located in the upper wall of the right atrium (**FIG. 33-7**). Electrical signals from the SA node pass freely and rapidly into the connecting cardiac muscle cells and then to muscle cells throughout the atria.

During the cardiac cycle, the atria contract first and empty their contents into the ventricles. This requires a slight delay between the atrial and ventricular contractions to allow the ventricles to be filled before they contract. How is this accomplished? First, the SA node initiates a wave of contraction ❶ that sweeps through the right and left atria, which contract in synchrony ❷. The signal then reaches a barrier of tissue between the atria and the ventricles that cannot conduct electrical signals. Here, the excitation is channeled through the **atrioventricular (AV) node**, a small mass of specialized muscle cells located in the floor of the right atrium ❸. The impulse is conducted slowly through the AV node, briefly delaying conduction into tracts that stimulate ventricular contraction. This delay gives the atria time to complete the transfer of blood into the ventricles before ventricular contraction begins.

From the AV node, the signal to contract spreads along specialized tracts of rapidly conducting muscle fibers. These tracts start with the thick cluster of fibers called the *atrioventricular bundle* (*AV bundle*), which sends branches to the lower portion of both ventricles ❹. Here, the AV bundle branches separate and give rise to *Purkinje fibers*, which transmit the electrical signal to the surrounding cardiac muscle cells, sending a wave of contraction within the ventricular walls from the base of

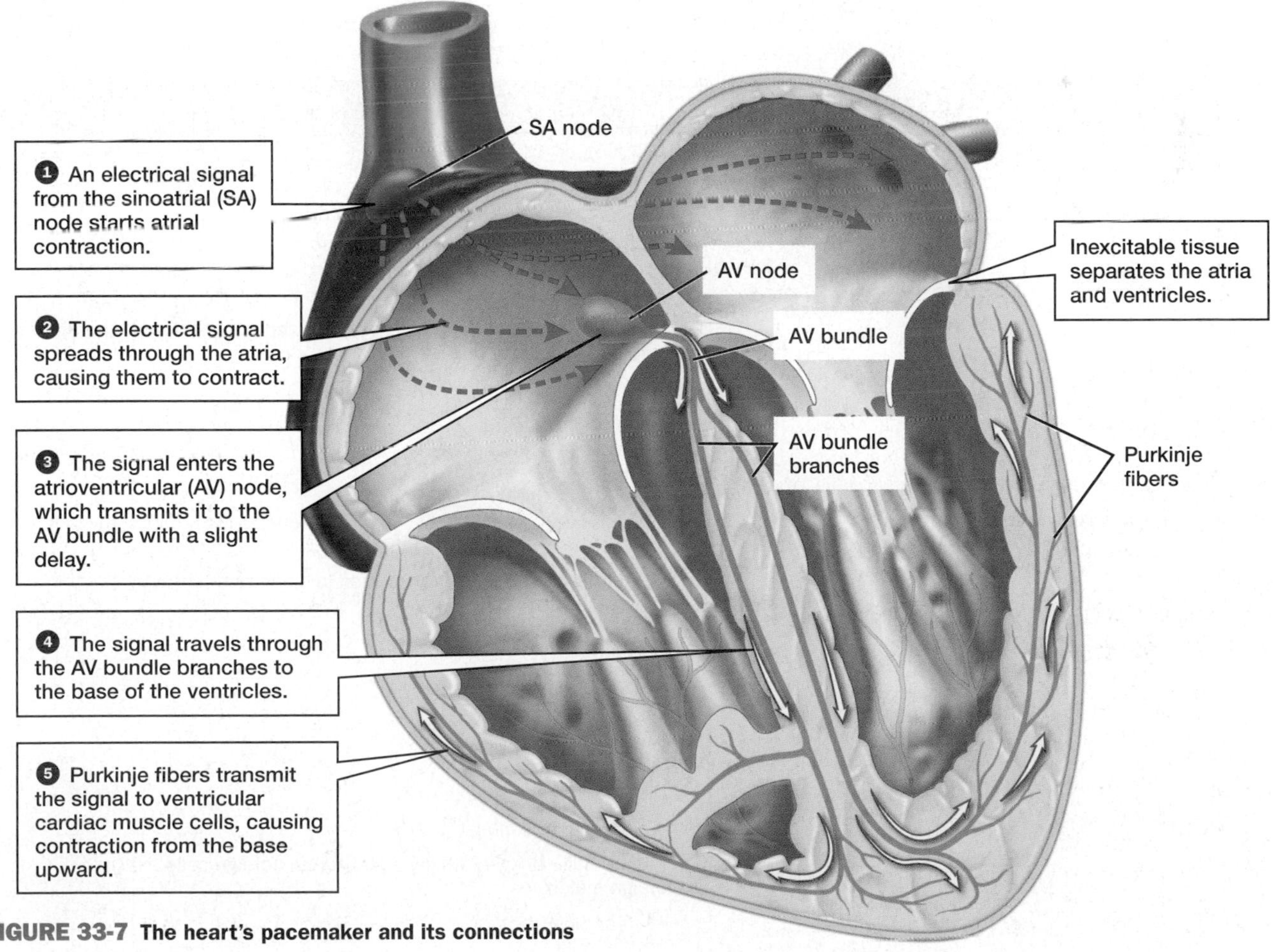

SA node
❶ An electrical signal from the sinoatrial (SA) node starts atrial contraction.
❷ The electrical signal spreads through the atria, causing them to contract.
❸ The signal enters the atrioventricular (AV) node, which transmits it to the AV bundle with a slight delay.
❹ The signal travels through the AV bundle branches to the base of the ventricles.
❺ Purkinje fibers transmit the signal to ventricular cardiac muscle cells, causing contraction from the base upward.
AV node
AV bundle
AV bundle branches
Inexcitable tissue separates the atria and ventricles.
Purkinje fibers

▲ **FIGURE 33-7 The heart's pacemaker and its connections**

the ventricles upward ❺. This ventricular contraction forces blood up into the pulmonary artery and the aorta.

A variety of disorders can interfere with the complex series of events that produce the normal cardiac cycle. When the pacemaker fails—or if other areas of the heart become more excitable and usurp the pacemaker's role—rapid, uncoordinated, weak contractions called **fibrillation** may occur. Fibrillation of the ventricles is soon fatal, because blood is not pumped by the quivering muscle. To treat this condition, a device called a defibrillator is used to apply a jolt of electricity to the heart, synchronizing the contraction of the cardiac muscle and (if successful) allowing the pacemaker to resume its normal coordinating function.

The Nervous System and Hormones Influence Heart Rate

Your heart rate is finely tuned to your body's activity level, whether you are running to class or basking in the sun. On its own, the SA node pacemaker would maintain a steady rhythm of about 100 beats per minute. However, nerve impulses and hormones significantly alter the heart rate. In a resting individual, the parasympathetic nervous system, which regulates body systems during periods of rest, slows the heart rate to roughly 70 beats per minute. When exercise or stress creates a demand for greater blood flow to the muscles, the sympathetic nervous system, which prepares the body for emergency action, accelerates the heart rate and increases the force of cardiac muscle contractions. The adrenal glands simultaneously release the hormone epinephrine (adrenaline), which reinforces these effects.

CHECK YOUR LEARNING

Can you . . .

- describe the three types of vertebrate hearts and the structure of cardiac muscle?
- trace the flow of blood through a four-chambered heart, naming the structures through which the blood passes and explaining the function of each?
- explain the cardiac cycle and how electrical impulses are conducted through the human heart?

33.3 WHAT IS BLOOD?

Blood, sometimes called the "river of life," has two major components: (1) a liquid called **plasma**, which constitutes 55% to 60% of the blood volume, and (2) a cell-based component consisting of *red blood cells, white blood cells*, and *platelets* suspended in the plasma (**FIG. 33-8**). The average person has roughly 5.3 quarts (about 5 liters) of blood, so if you donate a pint of blood, you are giving only about 10% of your blood (which your body will soon replenish). The components of blood are summarized in **TABLE 33-1**.

TABLE 33-1 Blood Components and Their Functions

Plasma Components (about 55% of blood volume)	Functions
Water	Dissolves other components; gives blood its fluidity
Major Proteins	
Albumin	Maintains blood osmotic strength; binds and transports some hormones and fatty acids
Globulins	Serve as antibodies that fight infection; bind and transport some hormones, ions, and other molecules
Fibrinogen	Gives rise to fibrin, which promotes clotting
Ions (sodium, potassium, calcium, magnesium, chloride, bicarbonate, hydrogen)	Maintain pH; allow neuronal activity; allow muscle contraction; facilitate enzyme activity
Nutrients (simple sugars, amino acids, lipids, vitamins, oxygen)	Provide materials for cellular metabolism
Wastes (urea, carbon dioxide, ammonia)	By-products of cellular metabolism that are transported in blood to sites of elimination
Hormones	Signaling molecules that are transported in blood to their target cells
Cell-Based Components (about 45% of blood volume)	**Functions**
Erythrocytes (5,000,000 per mm^3)	Transport oxygen
Leukocytes (5,000–10,000 per mm^3)	All fight infection and disease
Neutrophils	Engulf and destroy bacteria
Eosinophils	Kill parasites
Basophils	Produce inflammation
Lymphocytes	Mount an immune response
Monocytes	Mature into macrophages, which engulf debris, foreign cells, and foreign molecules
Platelets (250,000 per mm^3)	Essential for blood clotting

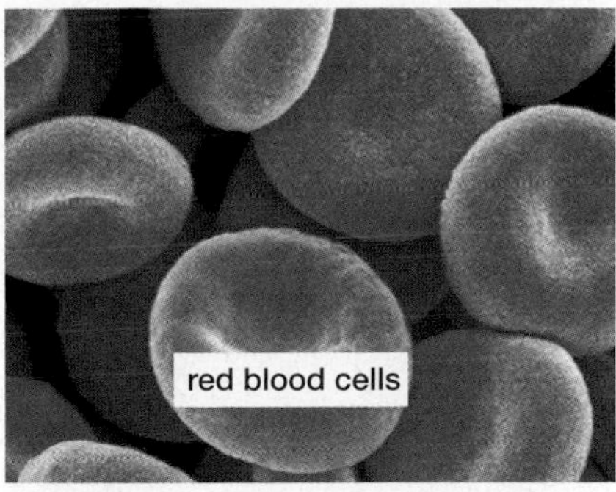

(a) Erythrocytes (red blood cells)

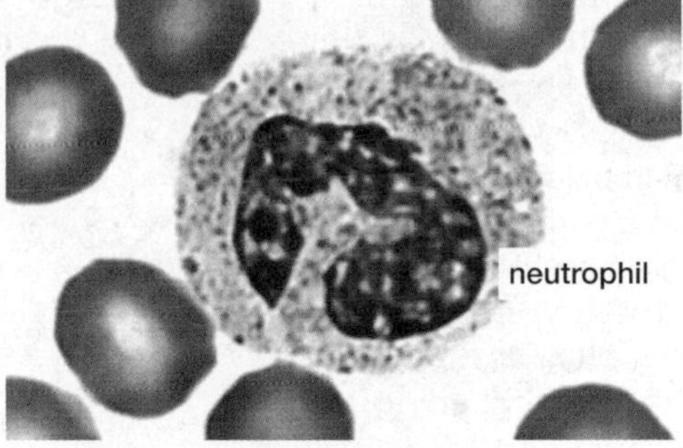

(b) Leukocyte (white blood cell)

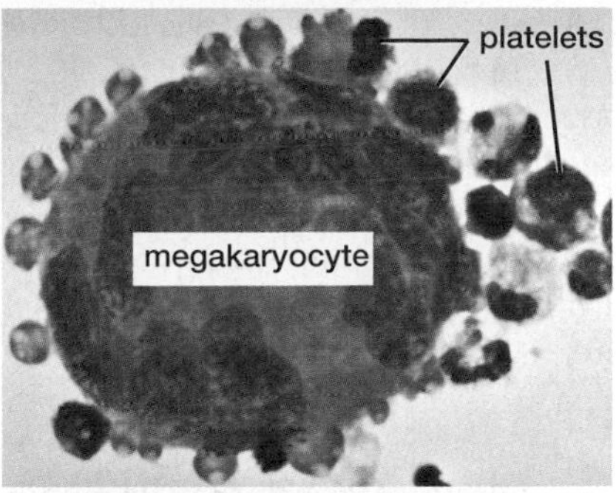

(c) Megakaryocyte forming platelets

▲ **FIGURE 33-8 Types of blood cells (a)** This scanning electron micrograph shows the biconcave shape of erythrocytes. **(b)** A leukocyte called a neutrophil is shown in this light micrograph (LM) surrounded by much smaller red blood cells. **(c)** Platelets are membrane-enclosed fragments of megakaryocytes seen in this LM.

THINK CRITICALLY Why does dietary iron deficiency cause anemia (too few erythrocytes)?

Plasma Is Primarily Water in Which Proteins, Salts, Nutrients, and Wastes Are Dissolved

Although plasma is about 90% water, this clear, pale-yellow fluid has more than 100 different types of molecules dissolved in it. The plasma transports proteins, hormones, nutrients, and cellular wastes. It also contains a variety of ions; some of these maintain blood pH, while others are crucial for the functioning of nerve and muscle cells.

Proteins make up the largest component of dissolved molecules by weight. The three most common plasma proteins are *albumin*, *globulins*, and *fibrinogen*. Albumin helps to maintain the blood's osmotic strength, thus preventing too much fluid from diffusing out of the plasma through capillary walls. Some globulins are antibodies that play an important role in the immune response (see Chapter 37). Fibrinogen is important in blood clotting, described later in this chapter.

The Cell-Based Components of Blood Are Formed in Bone Marrow

All three cell-based components of blood—red blood cells, white blood cells, and platelets—originate from cells that reside in **bone marrow**, a tissue within the cavities of bones (see Chapter 41). Of these three, only white blood cells have a full complement of organelles. Red blood cells of mammals lose their nuclei and mitochondria during development, and platelets are actually small fragments of cells.

Red Blood Cells Carry Oxygen from the Lungs to the Tissues

About 99% of all blood cells are red blood cells, also called **erythrocytes**, whose major function is to transport oxygen. A red blood cell is shaped like a ball of clay that has been squeezed between a thumb and forefinger (**FIG. 33-8a**). The red color of erythrocytes is produced by the large, iron-containing protein **hemoglobin** (**FIG. 33-9**), which transports almost all of the oxygen carried in the blood. Each hemoglobin molecule can bind and carry four molecules of oxygen, one on each heme group. Hemoglobin takes on a bright cherry-red color when bound to oxygen and becomes a deeper maroon-red color after it releases oxygen, appearing bluish in veins seen through the skin. Hemoglobin binds loosely to oxygen, picking it up in the capillaries of the lungs, where the oxygen concentration is high, and releasing it in other tissues of the body where the oxygen concentration is lower (see Chapter 34).

Carbon Monoxide Displaces Oxygen from Hemoglobin

Unfortunately, oxygen is not the only molecule that binds to hemoglobin. Carbon monoxide (CO) does so as well, and CO poisoning causes roughly 450 accidental deaths each year in the United States. Carbon monoxide is produced when fuel is not completely burned, and some is released by engines, furnaces, charcoal grills, and cigarettes. Like oxygen,

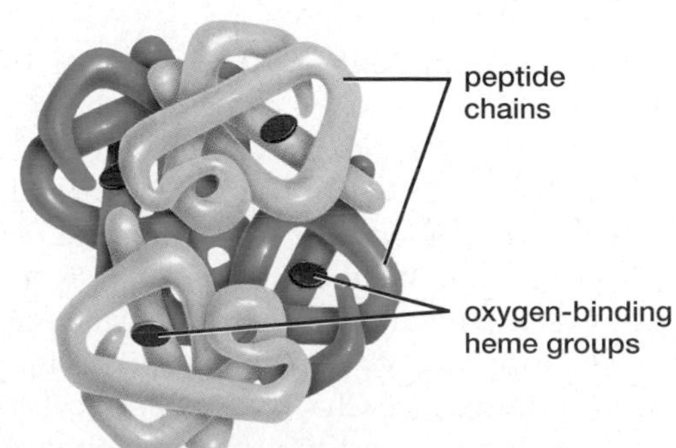

▲ **FIGURE 33-9 Hemoglobin** Each of four peptide chains (two each of two types) surrounds an iron-containing heme molecule (red disk) which binds oxygen.

it binds to the heme groups on hemoglobin, but it adheres more than 200 times as tightly. As a result, carbon monoxide remains bound to hemoglobin for several hours, preventing the hemoglobin from transporting oxygen and starving body tissues of the oxygen they require. Hemoglobin bound to either carbon monoxide or oxygen appears bright red, so although victims of asphyxiation generally have bluish lips and nail beds (because their hemoglobin is oxygen-poor), victims of CO poisoning maintain a healthy color.

Erythrocytes live about 4 months. Every second, more than 2 million red blood cells (about 200 billion daily) die and are replaced by new ones formed in the bone marrow. Dead erythrocytes are broken down in the spleen and liver. Iron from hemoglobin is returned to the bone marrow where it is reused to synthesize hemoglobin for use in new red blood cells. Although this recycling process is efficient, some iron is lost during bleeding from injury or from menstruation, and a small amount is excreted daily in feces, so some iron must be provided in the diet.

Negative Feedback Regulates Red Blood Cell Numbers

The number of erythrocytes determines how much oxygen the blood can carry. Red blood cell number is maintained by a negative feedback system that involves the hormone **erythropoietin**. Erythropoietin, produced by the kidneys and released into the blood in response to low oxygen levels, stimulates the bone marrow to increase production of red blood cells (**FIG. 33-10**). Low oxygen may be caused by a loss of blood, insufficient production of hemoglobin, high altitude (where less oxygen is available), or conditions that interfere with gas exchange in the lungs, such as lung disease and heart failure, which often causes fluid to accumulate in the lungs. When a healthy oxygen level is restored, erythropoietin production declines, and the rate of red blood cell production returns to normal. People with advanced kidney disease often suffer from *anemia*—an inadequate number of erythrocytes—because their failing kidneys do not produce enough erythropoietin.

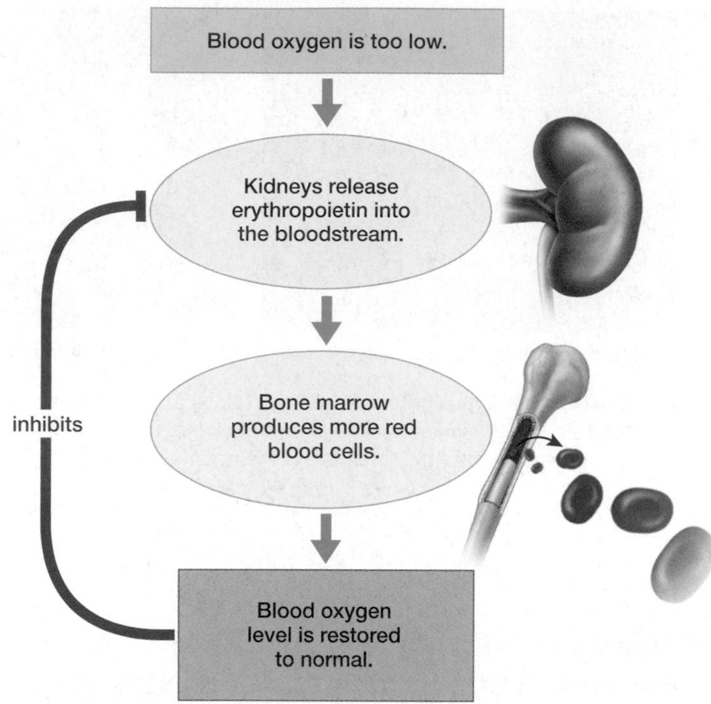

▲ **FIGURE 33-10 Red blood cell production is regulated by negative feedback**

THINK CRITICALLY Some endurance athletes cheat by injecting large doses of erythropoietin. How does this provide a competitive advantage?

Platelets Are Cell Fragments That Aid in Blood Clotting

Platelets are pieces of large cells called *megakaryocytes*. Megakaryocytes remain in the bone marrow, where they pinch off membrane-enclosed chunks of cytoplasm that become platelets (see Fig. 33-8c). The platelets, which survive for about 10 days, enter the blood and play a central role in blood clotting.

White Blood Cells Defend the Body Against Disease

White blood cells, called **leukocytes**, are larger than red blood cells, but they can crawl, change shape, and ooze through far narrower spaces than erythrocytes can, including through capillary walls. There are five types of leukocytes: *neutrophils*, *eosinophils*, *basophils*, *lymphocytes*, and *monocytes* (see Fig. 33-8b; see also Table 33-1). Their life spans range from hours to years, and altogether they make up less than 1% of the cellular portion of the blood. Leukocytes protect the body against disease (see Chapter 37). Monocytes, for example, enter tissues and transform into macrophages (literally, "big-eaters"). Macrophages engulf bacteria (**FIG. 33-11**) and cellular debris; in the spleen and liver they also break down dead red blood cells.

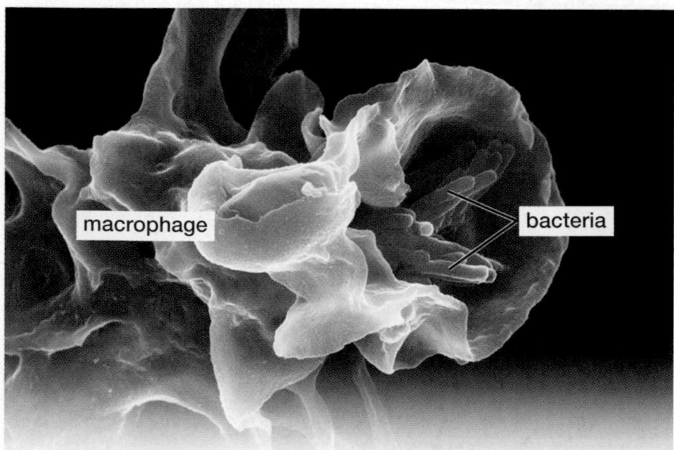

▲ **FIGURE 33-11 A white blood cell attacks bacteria** This macrophage has formed cytoplasmic extensions that are engulfing a cluster of rod-shaped tuberculosis bacteria.

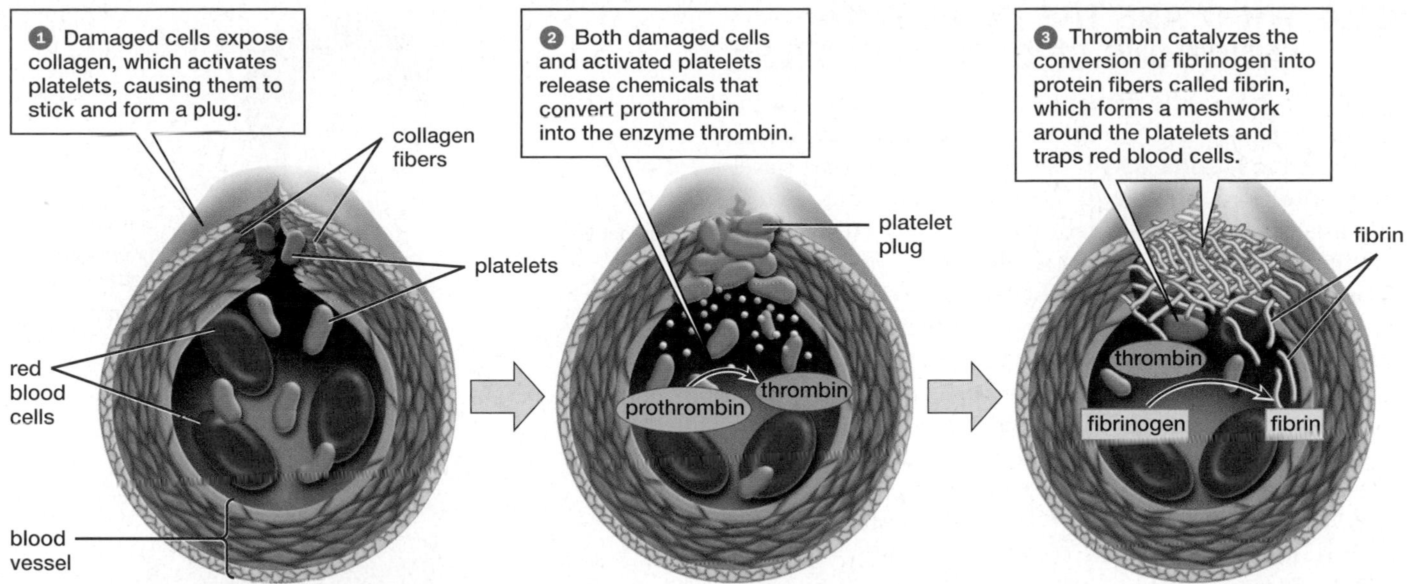

1 Damaged cells expose collagen, which activates platelets, causing them to stick and form a plug.
collagen fibers
platelets
red blood cells
blood vessel
2 Both damaged cells and activated platelets release chemicals that convert prothrombin into the enzyme thrombin.
platelet plug
prothrombin
thrombin
3 Thrombin catalyzes the conversion of fibrinogen into protein fibers called fibrin, which forms a meshwork around the platelets and traps red blood cells.
fibrin
thrombin
fibrinogen
fibrin

▲ **FIGURE 33-12 Blood clotting** Injured tissue and adhering platelets cause a series of biochemical reactions among blood proteins that lead to clot formation. A simplified sequence is shown here.

Blood Clotting Plugs Damaged Blood Vessels

Blood clotting is a complex process that protects animals from losing excessive amounts of blood, not only from trauma, but also from the minor wear and tear that occurs with normal activities. Clotting begins when blood comes in contact with injured tissue, for example, a break in a blood vessel wall (**FIG. 33-12**). Platelets adhere to collagen proteins exposed on the ruptured wall and form a *platelet plug* that partially blocks the opening ❶. Both the adhering platelets and the ruptured blood vessel cells initiate a cascade of complex reactions among *clotting factors*, which are mostly circulating plasma proteins. The end result of this series of reactions is to produce a substance that activates the plasma protein *prothrombin*, converting prothrombin (an inactive protein) into the active enzyme **thrombin** ❷. Thrombin cleaves the plasma protein *fibrinogen*, starting a series of reactions that form insoluble protein strands called **fibrin** ❸. Fibrin strands adhere to one another, forming a fibrous network around the aggregated platelets.

This web of fibrin traps more platelets and blood cells, primarily erythrocytes (**FIG. 33-13**), increasing the density of the clot. Platelets adhering to the fibrous mass send out sticky projections that grip one another. The cross-linked platelets contract in 30 to 60 minutes, squeezing the fibrin web and forcing fluid out. This creates a denser, stronger clot (you will see this on the skin as a scab). The contraction also pulls the damaged surfaces of the wound closer together, which promotes healing.

Blood clotting is crucial, not only to prevent excessive bleeding from wounds, but also to stop bleeding that occurs naturally as tiny blood vessels are damaged by everyday bumps, normal joint movements, and muscle contractions. People with *hemophilia*, a genetic disorder, lack a specific plasma protein required for blood clotting. In severe cases, internal bleeding can occur and continue destructively without an obvious cause. Fortunately, most cases of hemophilia can be controlled by injecting the missing clotting factor into the bloodstream.

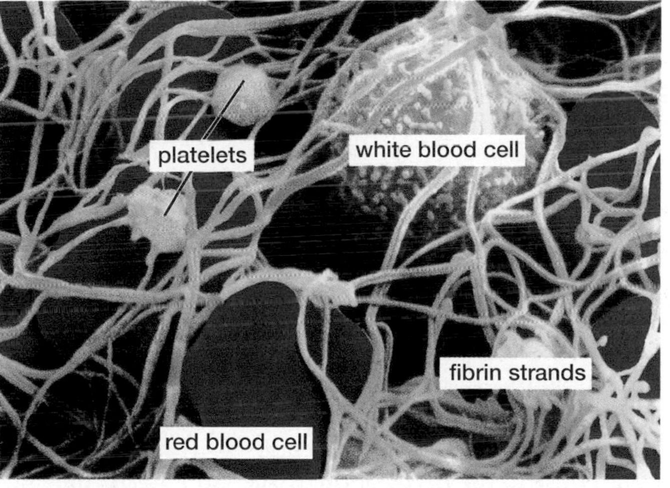

▲ **FIGURE 33-13 A blood clot** Thread-like fibrin protein strands produce a tangled, sticky mass that traps blood cells and eventually forms a clot.

CHECK YOUR LEARNING

Can you . . .

- describe each component of blood and explain its function?
- explain how the number of red blood cells in the body is regulated?
- explain the sequence of events during blood clotting?

33.4 WHAT ARE THE TYPES AND FUNCTIONS OF BLOOD VESSELS?

Blood circulates within a network of blood vessels; some of the major blood vessels of the human circulatory system are illustrated in **FIGURE 33-14**. Blood leaving the heart travels from arteries to arterioles to capillaries, then into venules, and finally to veins, which return it to the heart (**FIG. 33-15**). Capillary walls are made of a single layer of *endothelium* (a single layer of specialized epithelial cells; see Chapter 32). The larger vessels are lined with endothelium surrounded by a thin layer of connective tissue, and have two additional cell layers: a middle layer of smooth muscle cells and an outer layer of connective tissue (see Fig. 33-15). What happens when blood vessels rupture, are narrowed by deposits of cholesterol, or are blocked by clots, damming the river of life? We explore these questions in "Health Watch: Repairing Broken Hearts" on page 586.

Arteries and Arterioles Carry Blood Away from the Heart

Arteries carry blood away from the heart. Compared to veins, artery walls are thicker and far more elastic (see Fig. 33-15). With each surge of blood from the ventricles, the arteries expand slightly, like thick-walled balloons. As their elastic walls recoil between heartbeats, the arteries actually help pump the blood and keep it flowing steadily into the smaller vessels. Arteries branch into smaller diameter vessels called **arterioles**, which play a major role in determining how blood is distributed within the body.

Arterioles Control the Distribution of Blood Flow

Arterioles are microscopically narrow vessels (most are less than 300 micrometers, or 0.01 inches, in diameter) that carry blood to capillaries. Their muscular walls are influenced by nerves, hormones, and chemicals produced by nearby tissues. This allows arterioles to contract and relax in response to the needs of the tissues and organs they supply. In a suspense novel, you might read "Her face went pale as she gazed at the bloodstained floor." Skin becomes pale when the sympathetic nervous system and epinephrine from the adrenal glands stimulate contraction of the smooth muscle in the walls of the arterioles that supply the skin. The resulting constriction of the arterioles redirects blood away from the skin and into the muscles, where it may be needed for vigorous action.

In extremely cold weather, fingers and toes can become frostbitten because the sympathetic nervous system causes

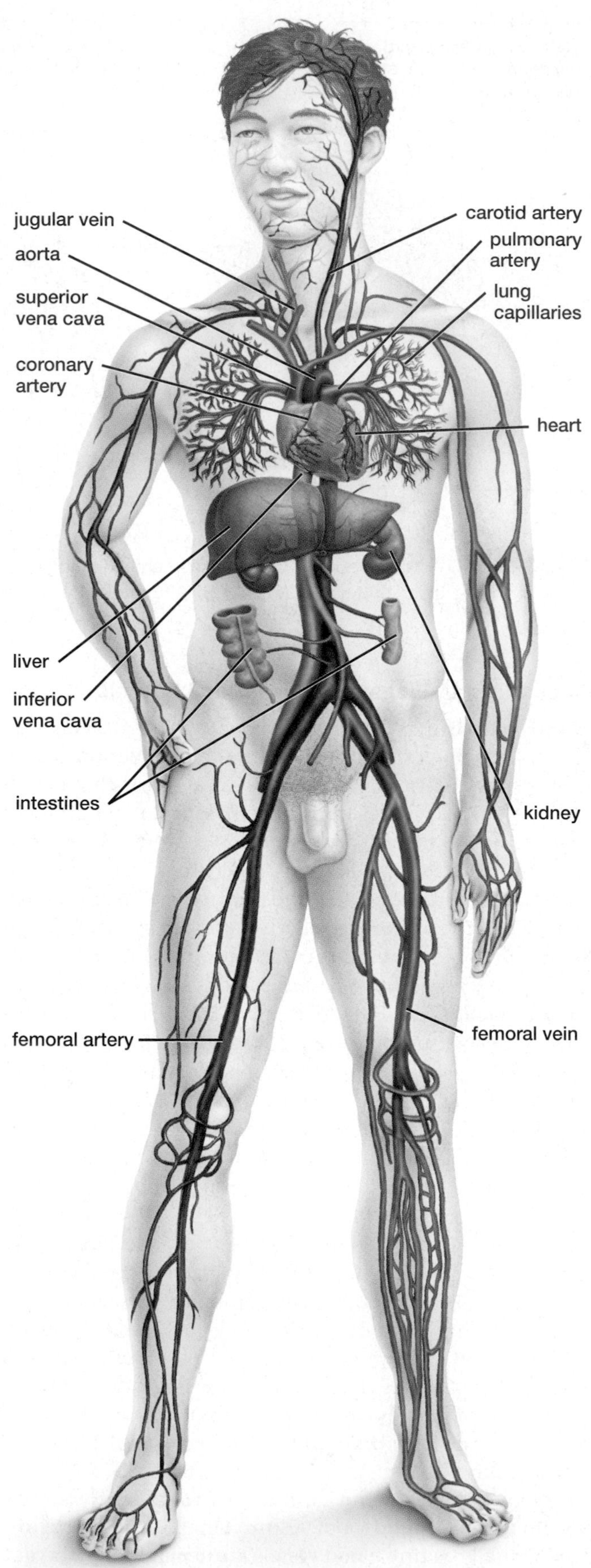

▶ **FIGURE 33-14 The human circulatory system** Arteries (emphasized on the person's right side) carry blood from the heart, and veins (emphasized on the person's left side) carry blood toward the heart. All arteries except the pulmonary artery carry oxygenated blood, and all veins except the pulmonary veins carry oxygen-poor blood. The microscopic lung capillaries are shown greatly enlarged.

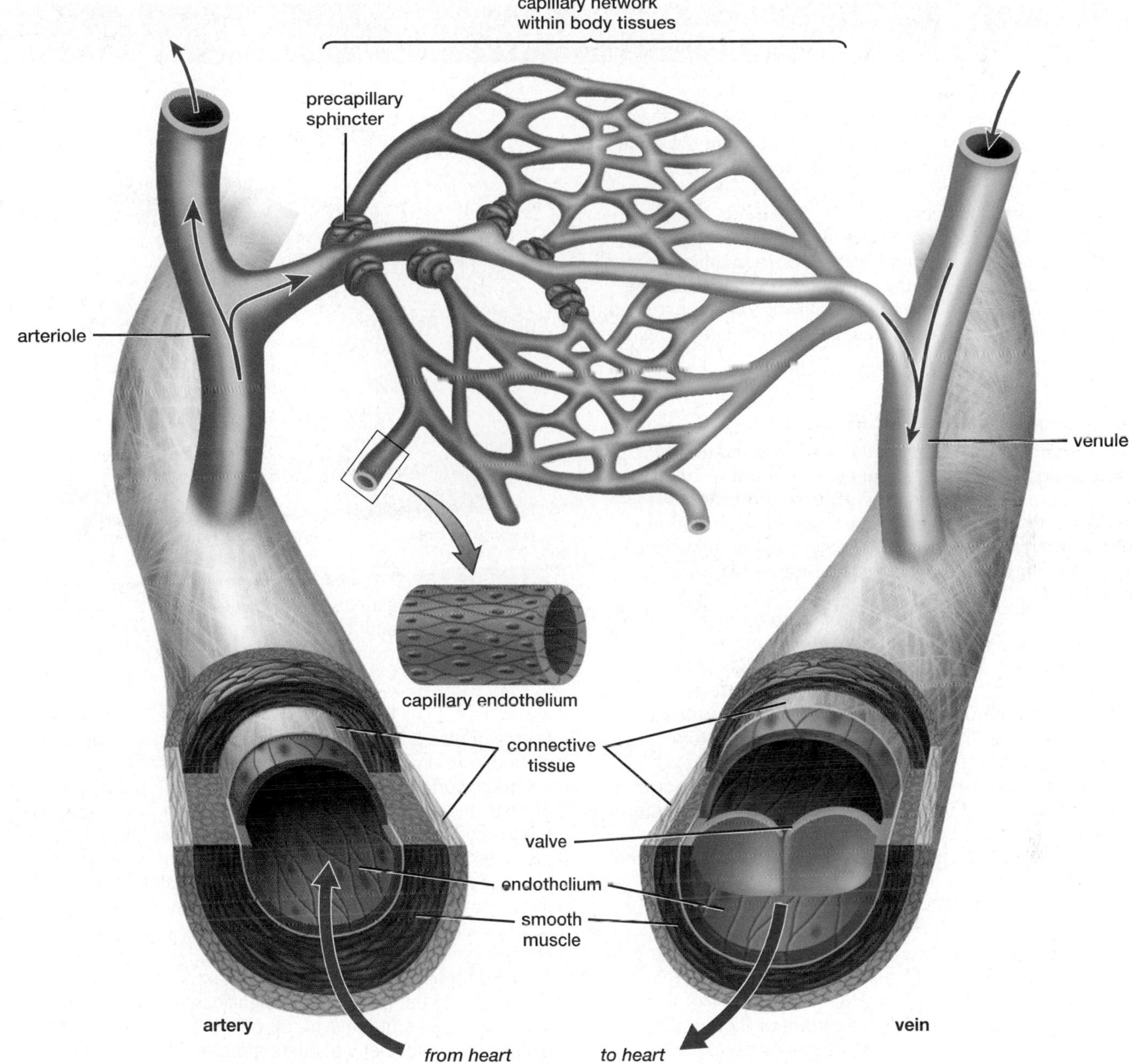

▲ **FIGURE 33-15 Structures and interconnections of blood vessels** Arteries and arterioles are more muscular than veins and venules. Oxygenated blood moves from arteries to arterioles to capillaries. Capillary walls are only one cell thick, allowing them to exchange gases and nutrients with their surroundings.

THINK CRITICALLY Could the vessels as they are color-coded here be part of the circulation to the lungs? Why or why not?

arterioles that supply blood to the extremities to constrict. This shunts blood to vital organs, such as the heart and brain, which cannot function properly if their temperature drops. By minimizing blood flow to heat-radiating extremities, the body conserves heat. On a hot summer day, in contrast, you become flushed as arterioles in the skin expand and deliver more blood to the skin capillaries. This enables the body to dissipate excess heat to the air outside, helping to reduce the body temperature.

Capillaries Allow Exchange of Nutrients and Wastes

Elaborate networks of far smaller **capillaries** receive blood from arterioles (see Fig. 33-15, middle). Capillaries, whose walls are only one cell thick, are the only vessels that allow individual body cells to exchange nutrients and wastes with the blood by diffusion. In contrast, larger vessels are adapted to conduct blood rather than exchange substances,

Repairing Broken Hearts

Cardiovascular disease (CVD; disorders of the heart and blood vessels) is the leading cause of death in the United States. According to the American Heart Association, CVD is responsible for about one out of every three deaths annually in the United States, a loss of about 800,000 lives, and no wonder. The heart must contract vigorously more than 2.5 billion times during an average lifetime without once stopping to rest, forcing blood through a lengthy network of vessels. Because the heart may weaken or the vessels may become constricted, blocked, or ruptured, the cardiovascular system is a prime candidate for malfunction.

Atherosclerosis Obstructs Arteries

Atherosclerosis (from Greek meaning "hard paste") is caused by deposits, called **plaques**, within the artery walls. These deposits cause the walls of the arteries to thicken and lose their elasticity. They tend to form in people with high levels of "bad" cholesterol and low levels of "good" cholesterol. What is the difference between the two forms?

Cholesterol is transported through the bloodstream in two types of packets called *low-density lipoprotein* (LDL; often called "bad" cholesterol) and *high-density lipoprotein* (HDL; often called "good" cholesterol). Both LDL and HDL consist of cholesterol surrounded by a shell of proteins and phospholipids that make it soluble in the watery plasma. Their cholesterol molecules are the same, but their predominant proteins differ, and LDL has less total protein in its shells than does HDL. LDL carries cholesterol from the liver to the body cells, including the cells of the artery walls, where it may contribute to plaques. HDL transports cholesterol to the liver, thus removing it from the bloodstream and reducing overall blood cholesterol levels.

Plaque formation is often initiated by minor damage to the endothelium lining an artery. The damaged endothelium attracts macrophages, which burrow beneath it and ingest large quantities of LDL cholesterol and other lipids (**FIG. E33-1**). The bloated bodies of these macrophages contribute to a growing fatty core of plaque. Meanwhile, smooth muscle cells from below the endothelium migrate into the core, absorb more fat and cholesterol, and add to the plaque. They also produce proteins that form a fibrous cap that covers the fatty core.

As the plaque grows, its fibrous cap may rupture, exposing clot-promoting factors within the plaque. A blood clot then forms, which further obstructs the artery. The clot may completely block the artery (see Fig. E33-1), or it may break free and be carried in the blood until it blocks a narrower part of the artery. Arterial clots are responsible for the most serious consequences of atherosclerosis: heart attacks and strokes.

A **heart attack** occurs if a *coronary artery*, which supplies blood to the heart muscle, is blocked. This deprives the hard-working cardiac muscle of blood and the oxygen it carries, leading to the death of some heart muscle cells. If a sufficiently large area of cardiac muscle dies, the heart stops. Although heart attacks are the major cause of death

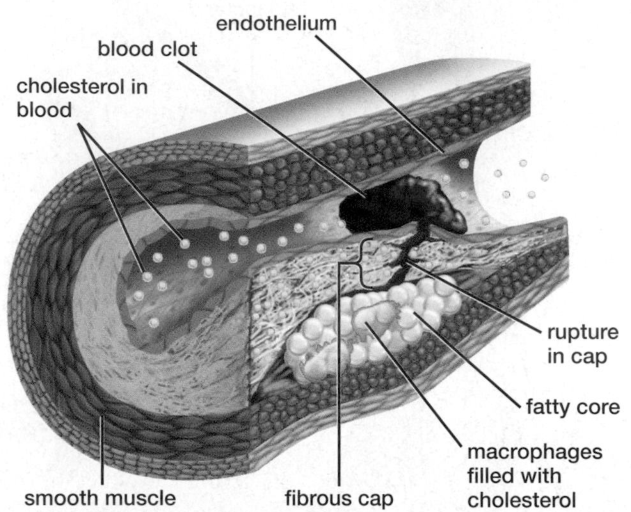

▲ **FIGURE E33-1 Plaque clogs an artery** Compare this artery to that shown in Figure 33-15. The yellow deposits inside this artery form a plaque. If the fibrous cap ruptures (as illustrated here), a blood clot forms, obstructing the artery.

from atherosclerosis, this disease also causes plaques and clots to form in arteries other than the coronary artery. A **stroke**, sometimes called a "brain attack," is the death of brain cells from lack of oxygen when their blood supply is interrupted by a clot or the rupture of a vessel. Depending on the extent and location of brain damage, strokes can be fatal, and stroke survivors may suffer neurological problems, including partial paralysis; difficulty remembering, communicating, or learning; mood swings; or personality changes.

When a heart attack or stroke strikes, rapid treatment can minimize the damage and significantly increase the victim's chances of survival. Blood clots can be dissolved by injecting a special clot-busting protein; this treatment works best if administered within a few hours after the attack occurs.

Treatment of Atherosclerosis

Atherosclerosis is promoted by high blood pressure, cigarette smoking, obesity, diabetes, lack of exercise, genetic predisposition, and high LDL cholesterol levels in blood. Treatment of atherosclerosis includes changes in diet and lifestyle, but if this fails, drugs may be prescribed to lower cholesterol. If a person has had a heart attack or suffers from **angina**, which is chest pain caused by insufficient blood flow to the heart, he may be a candidate for surgery to widen or bypass the obstructed artery.

Angioplasty refers to techniques that widen obstructed coronary arteries (**FIG. E33-2**). A physician threads a flexible tube through an artery in the upper leg or arm and guides it into the clogged artery. The tube may be equipped with a tiny drill bit, which shears off the plaque in microscopic pieces that are carried away in the blood (**FIG. E33-2a**), or it may have a small

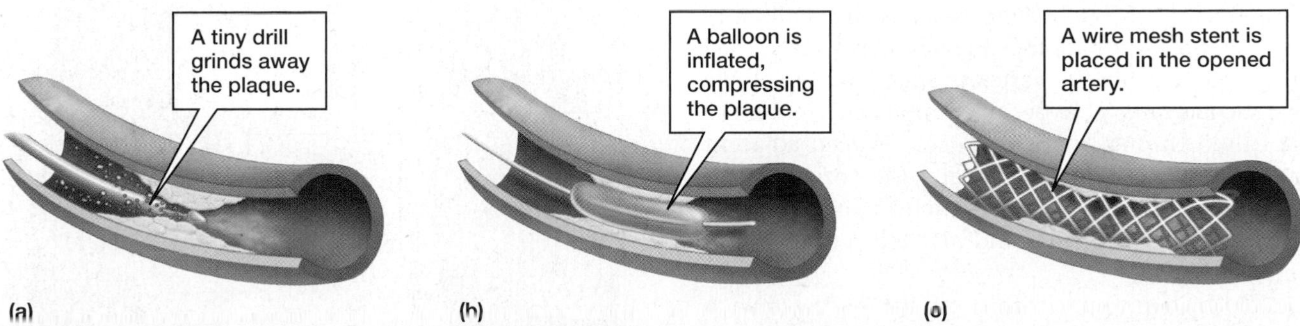

A tiny drill grinds away the plaque.
A balloon is inflated, compressing the plaque.
A wire mesh stent is placed in the opened artery.

▲ **FIGURE E33-2 Angioplasty unclogs arteries** A narrowed artery may be opened **(a)** by drilling out the plaque or **(b)** by inflating a tiny balloon inside. **(c)** Following angioplasty, a metal mesh stent is often inserted to maintain the opening.

balloon at its tip, which is inflated to compress the plaque (**FIG. E33-2b**). After the procedure, physicians may insert a wire mesh tube, called a stent, into the artery to help keep it open (**FIG. E33-2c**). In more severe cases, coronary bypass surgery may be performed. This operation bypasses obstructed coronary arteries with segments of artery from the patient's forearm or with segments of vein from the patient's leg (**FIG. E33-3**).

THINK CRITICALLY While Aletha was cramming for finals, her mother called to say that her father Bill, who had high blood pressure and was moderately obese and sedentary but otherwise healthy, had been rushed to the emergency room and then placed in intensive care. He had first complained of a severe headache, then over the next few hours his speech became slurred and one side of his body became numb and weak. What is a likely cause of Bill's symptoms? Explain your reasoning.

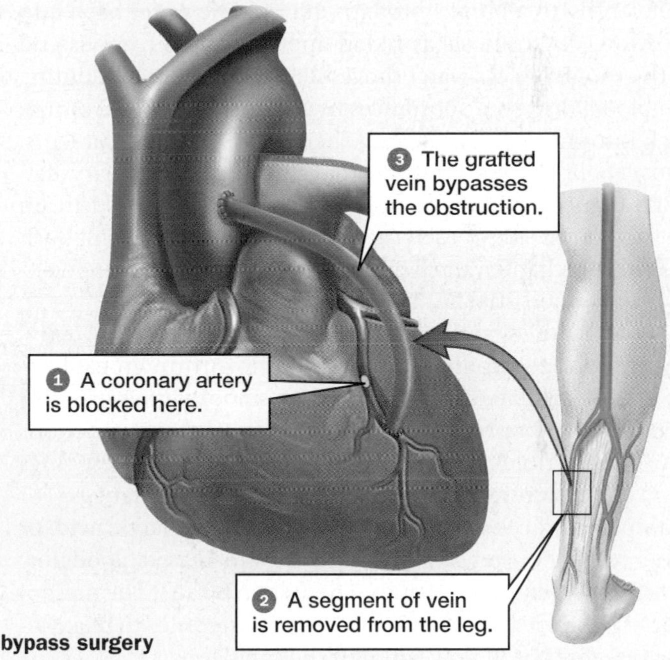

3 The grafted vein bypasses the obstruction.
1 A coronary artery is blocked here.
2 A segment of vein is removed from the leg.

▶ **FIGURE E33-3 Coronary bypass surgery**

and their multilayered walls are relatively impermeable. Capillaries are so narrow (about 10 micrometers in diameter) that red blood cells must pass through them single file (**FIG. 33-16**). In addition, capillaries are so numerous that most body cells are no more than 100 micrometers (0.004 inch; about four book pages thick) from a capillary, allowing diffusion to effectively exchange dissolved substances. A recent estimate puts the body's total cell number in the ballpark of 30 trillion. Supplying all of these cells requires an estimated 60,000 miles (about 96,500 kilometers) of capillaries—enough to encircle Earth more than twice. As blood moves through this narrow, lengthy capillary network, the rate of blood flow drops considerably, allowing more time for diffusion to occur.

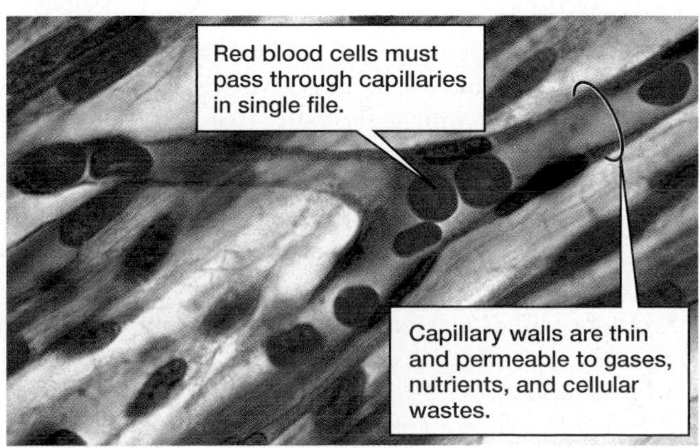

▲ **FIGURE 33-16 Red blood cells travel single file through a capillary**

Substances take various routes across the thin capillary walls. All must pass through the **interstitial fluid** that acts as an intermediary between body cells and capillary blood. Gases, water, lipid-soluble hormones, and fatty acids can diffuse directly through the capillary cell membranes into the interstitial fluid. Some small proteins are ferried across the endothelium in vesicles. Small, water-soluble nutrients, such as salts, glucose, and amino acids, enter the interstitial fluid through narrow spaces between adjacent capillary cells. White blood cells can also ooze through these crevices to engulf foreign particles. Large proteins such as albumin, erythrocytes, and platelets remain inside the capillaries.

Relatively high pressure within capillaries that branch directly from arterioles causes large quantities of fluid to leak out through the capillary walls. Pressure within the capillaries then drops as blood travels toward the venules (see Fig. 33-15). The high osmotic pressure of the blood inside the capillaries (due largely to albumin protein) draws some water back into the vessels by osmosis as blood approaches the venous end of the capillaries. As water moves into the capillaries, diluting the blood, dissolved substances in the interstitial fluid diffuse back into the capillaries along their concentration gradients. Thus, about 85% of the fluid that leaks out of capillary networks branching from arterioles is restored to the bloodstream on the venous side of each capillary network. As you will learn later in this chapter, the lymphatic system returns the remaining excess interstitial fluid to the blood.

You learned earlier that arterioles control the delivery of blood to capillaries, but blood flow through capillaries is also regulated by tiny rings of smooth muscle called **precapillary sphincters**, which surround the junctions between arterioles and capillaries (see Fig. 33-15). These rings open and close in response to local chemical changes. For example, the accumulation of carbon dioxide, lactic acid, or other cellular wastes signals the need for increased blood flow to bring oxygen to the tissues. These signals cause the precapillary sphincters and muscles in nearby arterioles to relax, allowing more blood flow through the capillaries.

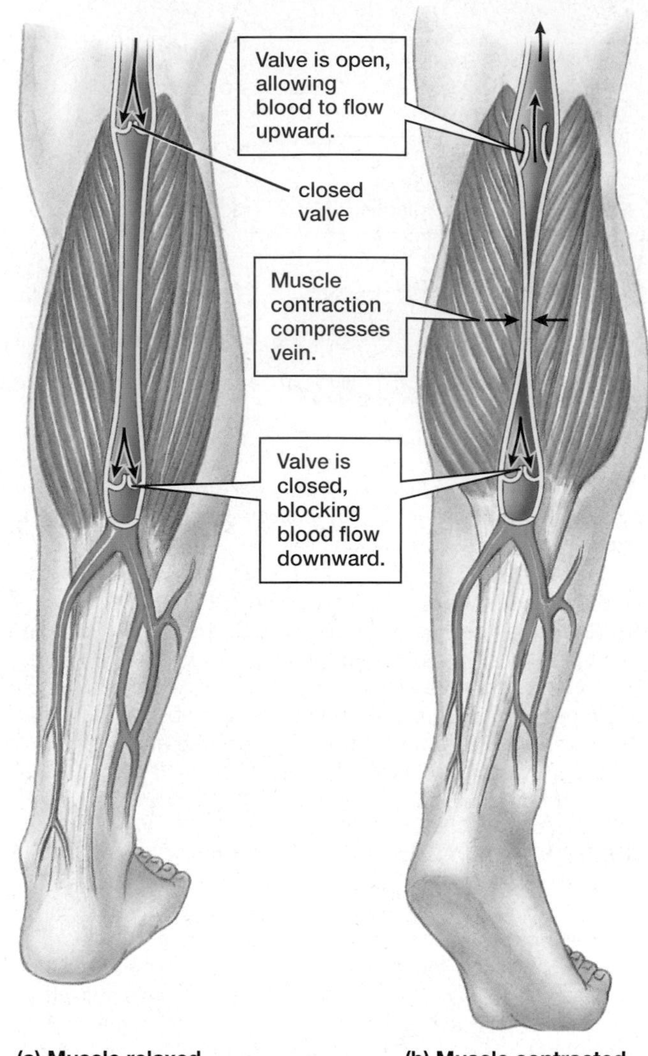
Valve is open, allowing blood to flow upward.
closed valve
Muscle contraction compresses vein.
Valve is closed, blocking blood flow downward.
(a) Muscle relaxed
(b) Muscle contracted

▲ **FIGURE 33-17 Valves direct blood flow in veins** A specific muscle in the calf is sometimes referred to as a "second heart" because of its role in forcing venous blood upward.

Veins and Venules Carry Blood Back to the Heart

After picking up carbon dioxide and other wastes from cells, capillary blood drains into larger vessels called **venules**, which empty into still larger veins (see Fig. 33-15, right). Veins provide a low-resistance pathway that conducts blood back toward the heart. The walls of veins are thinner and expand more readily than those of arteries, largely because they contain far less smooth muscle. The internal diameter of veins is also generally larger than that of arteries. When veins are compressed, as occurs when nearby skeletal muscles are contracted, one-way valves keep blood flowing toward the heart (**FIG. 33-17**).

When people sit or stand, venous blood pressure is too low to return all the blood to the heart from lower body parts such as the feet and legs without some help from skeletal muscles. The internal pressure changes caused by breathing, as well as the enlargement in the diameter of skeletal muscles as they contract during normal daily movements and exercise, squeezes the nearby veins, forcing blood through their one-way valves toward the heart (see Fig. 33-17). Prolonged sitting or standing still can cause swollen ankles because without muscle contractions to compress the veins, venous blood tends to pool in the lower legs. This pooling can lead to varicose veins, in which veins just below the skin become permanently swollen with blood because their valves have been stretched and weakened by pooling blood.

If blood pressure should fall—for instance, after extensive bleeding—veins can help restore it. In such cases, the sympathetic nervous system stimulates contraction of the smooth muscles in the walls of veins (and arteries). This decreases the internal volume of the vessels and raises blood pressure.

CHECK YOUR LEARNING

Can you . . .

- describe and compare the structures of arteries, arterioles, capillaries, venules, and veins?
- explain the functions of each of these types of blood vessels?
- describe how the distribution of blood flow is controlled?

33.5 HOW DOES THE LYMPHATIC SYSTEM WORK WITH THE CIRCULATORY SYSTEM?

The **lymphatic system** includes the lymphatic organs as well as an extensive system of lymphatic vessels, which eventually feed into the circulatory system (**FIG. 33-18**). This organ system performs the following functions:

- Returns excess interstitial fluid to the bloodstream.
- Transports fats from the small intestine to the bloodstream.
- Filters aged blood cells and other debris from the blood.
- Defends the body by exposing bacteria and viruses to lymphocytes and macrophages.

In the following sections, we emphasize the first three functions, in which the lymphatic system works intimately with the circulatory system (the role of the lymphatic system in defense of the body is covered in Chapter 37).

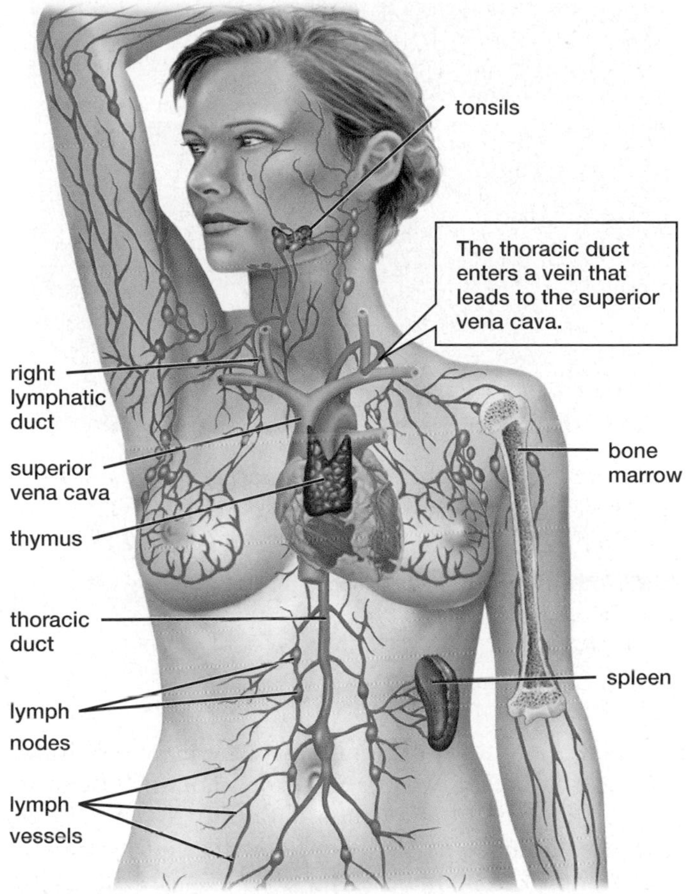
tonsils
The thoracic duct enters a vein that leads to the superior vena cava.
right lymphatic duct
superior vena cava
thymus
bone marrow
thoracic duct
spleen
lymph nodes
lymph vessels

▲ **FIGURE 33-18 The human lymphatic system**

Lymphatic Vessels Resemble the Capillaries and Veins of the Circulatory System

The smallest lymphatic vessels, called **lymphatic capillaries**, resemble blood capillaries in that they branch extensively throughout the body and their walls are only one cell thick. Lymphatic capillaries, however, are far more permeable than blood capillaries and are absent from bone and the central nervous system. Unlike blood capillaries, which form a continuous interconnected network, lymphatic capillaries "dead-end" in the interstitial fluid surrounding body cells (**FIG. 33-19**). Interstitial fluid flows into the lymphatic capillaries; once inside the lymphatic capillaries, this fluid is called **lymph**.

From the lymphatic capillaries, lymph is channeled into increasingly large lymphatic vessels, which resemble the veins of the circulatory system in that both have similar walls and both possess one-way valves that control the direction of fluid

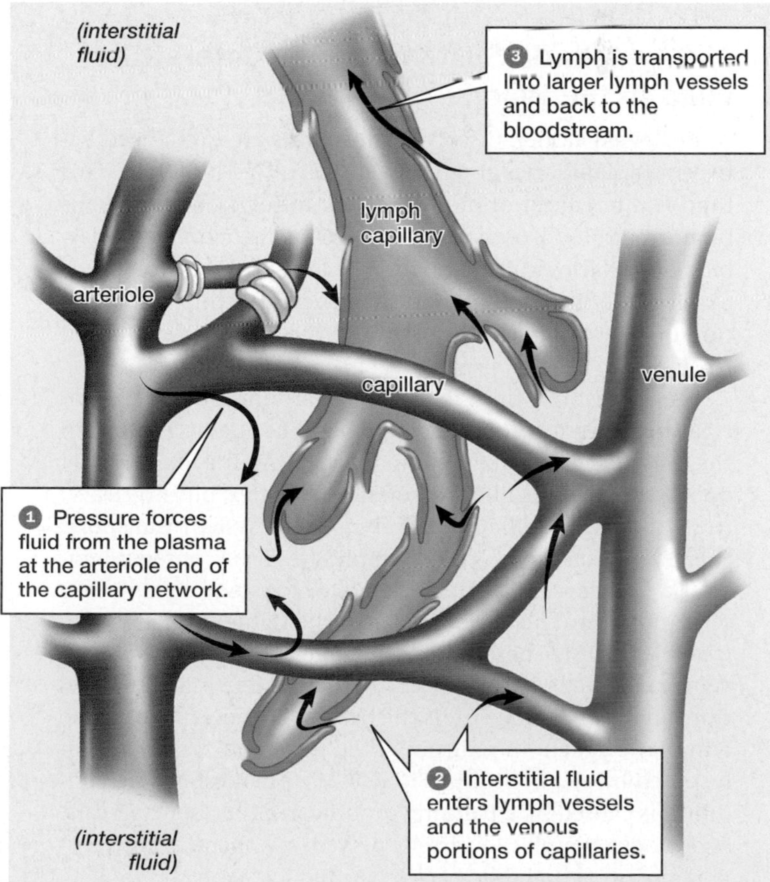
(interstitial fluid)
3 Lymph is transported into larger lymph vessels and back to the bloodstream.
lymph capillary
arteriole
venule
capillary
1 Pressure forces fluid from the plasma at the arteriole end of the capillary network.
2 Interstitial fluid enters lymph vessels and the venous portions of capillaries.
(interstitial fluid)

▶ **FIGURE 33-19 Lymph capillary structure** Lymph capillaries end in the body tissues. Here, pressure from the accumulation of interstitial fluid leaking from capillaries forces the fluid into the lymph capillaries as well as back into the venous side of the capillary network.

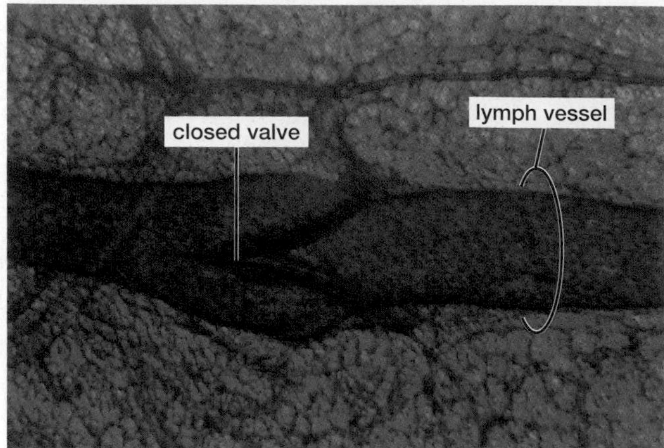

▲ **FIGURE 33-20 A valve in a lymph vessel** Lymphatic vessels (like blood veins) have internal one-way valves that direct the flow of lymph toward the large lymphatic ducts into which they empty.

THINK CRITICALLY Which direction is the lymph flowing in this vessel?

movement (**FIG. 33-20**). As the larger lymphatic vessels fill, stretching stimulates contractions of smooth muscles in their walls, pumping the lymph toward the heart. As with veins, further impetus for lymph flow through lymphatic vessels comes from internal pressure changes caused by breathing and the contraction of nearby skeletal muscles during exercise.

The Lymphatic System Returns Interstitial Fluid to the Blood

As described earlier, dissolved substances are exchanged between the capillaries and body cells via interstitial fluid. This fluid is filtered out of blood plasma through capillary walls by normal blood pressure. In an average person, each day, blood capillaries leak out about 3 or 4 more quarts (roughly 3 to 4 liters) of interstitial fluid than they reabsorb. One function of the lymphatic system is to return this excess fluid to the blood.

As interstitial fluid accumulates around body cells, increasing pressure forces it through flap-like openings between the cells of the lymphatic capillary walls. Acting like one-way doors, these valves allow substances to enter, but not leave, the lymphatic capillaries. The lymphatic system transports this interstitial fluid—called lymph after it has entered lymphatic vessels—back to the circulatory system. Lymph vessels empty into the thoracic duct or the right lymphatic duct (see Fig. 33-18). These ducts discharge the lymph into large veins near the base of the neck, which merge into the superior vena cava, which enters the heart. The importance of the lymphatic system in returning fluid to the bloodstream is illustrated by elephantiasis (**FIG. 33-21**). This disfiguring condition is caused by a parasitic roundworm that infects, scars, and blocks lymphatic vessels, preventing them from transporting interstitial fluid back to the bloodstream.

CASE STUDY CONTINUED

Living from Heart to Heart

A buildup of fluid in Kayla Monteil's abdomen was a sure sign that something was wrong with her transplanted heart. The excess fluid that Kayla's body was retaining in her blood caused more fluid to leak from her blood capillaries than her lymphatic capillaries could absorb. For about a year prior to surgery, fluid needed to be drained from Kayla's abdomen every 1 to 2 weeks. The lymphatic vessels usually absorb all the interstitial fluid that leaks from capillaries and return it to the bloodstream. What other functions does the lymphatic system perform?

The Lymphatic System Transports Fatty Acids from the Small Intestine to the Blood

After a fatty meal, fat-transporting particles may make up 1% of the lymph, giving it a milky white color. How does this occur? The small intestine is richly supplied with lymph capillaries called *lacteals* (see Fig. 35-16). After absorbing digested fats, intestinal cells release fat-transporting particles into the interstitial fluid. The particles are too large to diffuse into blood capillaries but can easily move into the lacteals. The lymphatic system then delivers the fatty particles to the blood, and the blood carries them to muscles, where they provide energy, and to fat tissue for storage.

Lymphatic Organs Filter Blood and House Cells of the Immune System

Organs of the lymphatic system (see Fig. 33-18) include bone marrow (where lymphocytes are produced) and the **thymus** (where some lymphocytes mature). The remaining lymphatic organs—tonsils, lymph nodes, and the spleen—house macrophages and lymphocytes that proliferate in readiness to

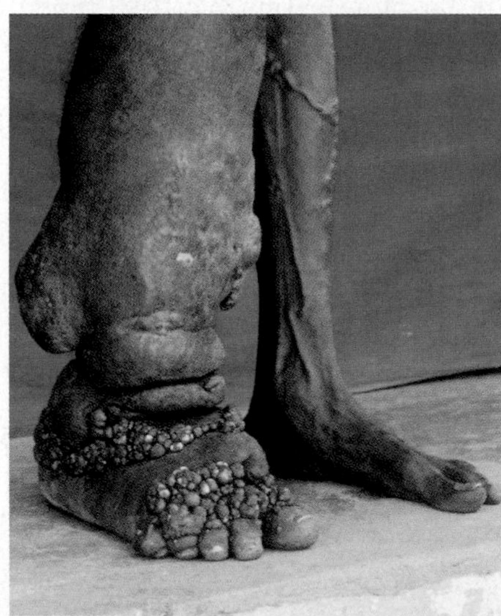

▲ **FIGURE 33-21 Elephantiasis results from blocked lymphatic vessels** When a parasitic worm scars and blocks lymphatic vessels, preventing fluid from returning to the bloodstream, the affected area can become massively swollen.

mount an immune response (described in Chapter 37). The **tonsils** are patches of lymphatic tissue that stand guard within the pharynx. Along the lymphatic vessels, hundreds of small swellings called **lymph nodes** filter the lymph past white blood cells. The **spleen**, a fist-sized lymphatic organ located between the stomach and diaphragm, filters blood past macrophages and lymphocytes much as the lymph nodes filter lymph. The spongy interior of the spleen also breaks down red blood cells and stores blood, which can be released if bleeding causes blood volume to fall.

CHECK YOUR LEARNING

Can you . . .

- explain how the lymphatic system returns interstitial fluid to the blood, and why this is important?
- describe the role of the lymphatic system in fat absorption from the intestines?
- list the organs of the lymphatic system and describe their role in filtering blood?

CASE STUDY REVISITED

Living from Heart to Heart

At the age of 23, Kayla Monteil was for the second time on the brink of dying while awaiting a heart transplant. A transplant is possible only if a compatible person dies and if grieving relatives agree to donate that person's heart. Each year, roughly 2,500 heart transplants are performed in the United States, but more than 4,000 individuals are left waiting for one.

Kayla had already survived an unexpectedly long time since her first transplant as a toddler. About 70% of donor heart recipients survive for 5 years, and about half remain alive at the 10-year mark, but only 16% survive 20 years. Survival declines over time mainly because, even though transplant recipients receive immune-suppressant drugs, the body continues a low-level immune attack on the donor organ, which is never a perfect match to the recipient unless the donor is an identical twin. In addition, the immune-suppressant drugs make the patient more susceptible to infectious diseases, cancer, and sometimes—as in Kayla's case—kidney damage.

Advances in bioengineering may one day dramatically reduce the need for donor hearts and immune-suppressant drugs. Research teams around the world are creating heart scaffolds consisting of extracellular matrix (see Chapter 4) from which all the original cells have been removed and replaced with stem cells from another organism. In preliminary research, human heart stem cells have been infused into a mouse heart scaffold, where they differentiated into endothelium and heart muscle cells and generated spontaneous contractions. A long-term goal is to create these "ghost heart" scaffolds from pig hearts, which are similar in size to a human heart (**FIG. 33-22**), and then seed them with cardiovascular stem cells from the transplant recipient to create a personalized heart. Will this be possible anytime soon? No. But with good fortune Kayla may benefit from this research if her second donor heart eventually fails.

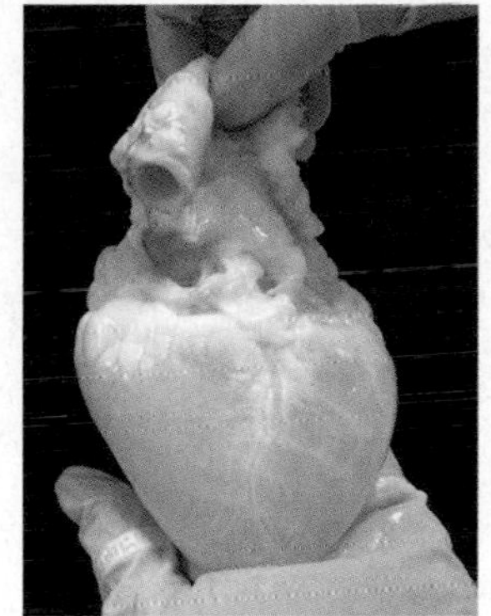

FIGURE 33-22 Heart scaffold The ghostly extracellular matrix of a pig heart may one day provide the scaffold for a personalized human heart.

CONSIDER THIS Kelly Muzzi, a 20-year-old student at Bennington College, died tragically when she fell through a plate glass window. Her devastated parents made the decision to donate her organs. Her heart saved a patient with heart failure, her liver and one kidney together saved another recipient, and her second kidney and pancreas saved a third. The parents' decision was reaffirmed when they later found Kelly's driver's license and saw that Kelly had designated herself as an organ donor. Are you a designated organ donor, or would you consider becoming one? What factors and personal experiences might influence a person's decision to become a donor?

CHAPTER REVIEW

Go to **Mastering Biology** to access the Pearson eText, vocabulary review, practice quizzes, activities, videos, current events, and more.

*Answers to **Think Critically** and **Thinking Through the Concepts** questions can be found in the **Answers** section at the back of the book.*

Summary of Key Concepts

33.1 What Are the Major Features and Functions of Circulatory Systems?

Circulatory systems transport blood rich in dissolved nutrients and oxygen close to each cell, releasing nutrients and absorbing wastes. All circulatory systems have three major parts: one or more hearts that pump blood, the blood itself, and a system of blood vessels. In open circulatory systems, found in most invertebrates, hemolymph is pumped by the heart into a hemocoel, where the hemolymph directly bathes the internal organs. A few invertebrates and all vertebrates have closed circulatory systems, in which blood is confined to the heart and blood vessels. Vertebrate circulatory systems transport gases, hormones, nutrients, and wastes; they help regulate body temperature; and they help defend the body against blood loss and disease.

33.2 How Does the Vertebrate Heart Work?

The vertebrate heart evolved from two chambers in fishes to three in amphibians and some reptiles to four in birds, crocodiles, and mammals. In the four-chambered heart, blood is pumped first to the lungs and then to the rest of the body. The heart also maintains complete separation of oxygenated and oxygen-poor blood. Oxygen-poor blood is collected from the body in the right atrium and then passed to the right ventricle, which pumps it to the lungs. Oxygenated blood from the lungs enters the left atrium, passes to the left ventricle, and is pumped to the rest of the body.

Cardiac muscle cells are small, branched, and connected by intercalated disks. The cardiac cycle consists of atrial contraction followed by ventricular contraction. The direction of blood flow is maintained by valves within the heart. The contractions of the heart are initiated by the sinoatrial node, the heart's pacemaker, and coordinated by the atrioventricular node, the AV bundle, and Purkinje fibers. The heart rate can be modified by the nervous system and by hormones such as epinephrine.

33.3 What Is Blood?

Blood is composed of fluid plasma and cell-derived components (Table 33-1). Plasma consists of water that contains proteins, hormones, nutrients, gases, and wastes. Red blood cells, or erythrocytes, are packed with hemoglobin, which carries oxygen. Their numbers are regulated by the hormone erythropoietin. There are five types of white blood cells, or leukocytes, which fight infection. Platelets, which are fragments of megakaryocytes, are important for blood clotting.

33.4 What Are the Types and Functions of Blood Vessels?

Blood leaving the heart travels through arteries, arterioles, capillaries, venules, veins, and then back to the heart. Each vessel is specialized for its role. Elastic, muscular arteries conduct blood from the heart, leading to smaller arterioles that empty into capillaries. The distribution of blood is regulated by the constriction and dilation of arterioles by the sympathetic nervous system and by local factors, such as the amount of carbon dioxide in the tissues. Local factors also regulate precapillary sphincters, which control blood flow to the capillaries. The microscopically narrow, thin-walled capillaries allow exchange of materials between the body cells and the blood. Venules and veins provide a path of low resistance back to the heart, with one-way valves that maintain the direction of blood flow.

33.5 How Does the Lymphatic System Work with the Circulatory System?

The human lymphatic system consists of the lymphatic vessels and the lymphatic organs, which are the bone marrow, thymus, tonsils, lymph nodes, and spleen. The lymphatic system removes excess interstitial fluid that leaks through blood capillary walls and delivers it back to the circulatory system. It transports fats to the bloodstream from the small intestine. It filters debris from the blood and fights infection by providing sites where lymphocytes can mature, proliferate, and mount an immune response.

Thinking Through the Concepts

Bloom's: Remembering, Understanding

Multiple Choice

1. Which of the following describes the layers of the walls of arteries and veins from inside to outside?
   a. endothelium, connective tissue, smooth muscle
   b. endothelium, smooth muscle, connective tissue
   c. smooth muscle, connective tissue, endothelium
   d. connective tissue, smooth muscle, endothelium
2. Which of the following is true?
   a. Arteriole diameter is controlled by nerves, hormones, and chemicals from nearby tissues.
   b. The sympathetic nervous system causes constriction of arterioles in hot conditions.
   c. Precapillary sphincters are controlled by the nervous system.
   d. In response to major blood loss, the veins and arteries increase in diameter.
3. Which of the following is *not* a function of the lymphatic system?
   a. transport of sugar from the small intestine to the bloodstream
   b. filtration of aged blood cells and other debris from the blood
   c. defense of the body by exposing disease organisms to white blood cells
   d. return of excess interstitial fluid to the blood
4. Which of the following accurately describes the flow of blood from lung capillaries to the aorta?
   a. lung capillaries, pulmonary vein, right atrium, right ventricle, aorta
   b. lung capillaries, pulmonary artery, left atrium, left ventricle, aorta
   c. lung capillaries, pulmonary vein, left atrium, left ventricle, aorta
   d. lung capillaries, vena cava, left atrium, left ventricle, aorta
5. Which of the following is true of blood pressure?
   a. If it is too high, it can contribute to atherosclerosis.
   b. It is recorded as diastolic over systolic pressure.
   c. Diastolic pressure is recorded during ventricular contractions.
   d. It is often measured by compressing a vein in the arm.

Fill-in-the-Blank

1. Animals with open circulatory systems utilize a fluid called ____________ that serves the function of both blood and ____________ fluid. This fluid is passed among chambers in the body that are collectively called the ____________.
2. Fill in the following with the appropriate heart chamber, including the side it is on. Receives blood from the body: ____________; has the thickest wall: ____________; contains the SA node: ____________; pumps blood to the lungs: ____________; pumps blood into the aorta: ____________.

3. The heart's pacemaker is called the (complete term) ____________. The pacemaker is composed of specialized ____________. The pacemaker first sends impulses that stimulate contraction of both ____________. The (complete term) ____________ introduces a delay in the transmission of pacemaker signals into the ventricles. Signals are conducted from this node directly into fibers called the ____________. If the pacemaker loses control of heart contractions, ineffective quivering of the heart muscle, called ____________, may occur.
4. Fill in the following with the appropriate type of blood vessel. Allow exchange of wastes and nutrients between blood and body cells: ____________; have the thickest walls: ____________; transport blood toward the heart: ____________; receive blood from the capillaries: ____________; deliver blood into capillaries: ____________; carry blood away from the heart: ____________.
5. During clot formation, damaged cells activate cell fragments called ____________. These cell fragments release chemicals that cause the enzyme ____________ to be produced. This enzyme catalyzes the conversion of a blood protein called ____________ into strands of ____________, which create a framework for the clot.
6. Fill in the following with the appropriate cell-based blood component, using the scientific term. Formed by megakaryocytes: ____________; there are five types of these: ____________; carry oxygen: ____________; include macrophages: ____________; lack a nucleus: ____________; fight infection and disease: ____________; contain hemoglobin: ____________.
7. Lymph is ____________ that has entered lymphatic vessels. The lymphatic system returns excess fluid that has leaked out of blood ____________ to the circulatory system. Another function of the lymphatic system is to transport ____________ from the small intestine. Lymphatic vessels have ____________ that keep lymph flowing in the proper direction. The lymphatic organ that filters blood is the ____________.

Review Questions

1. List the major structures of all circulatory systems and the specific functions of vertebrate circulatory systems.
2. Describe and compare the features of open and closed circulatory systems; include the animal groups in which they are found and some advantages and disadvantages of each.
3. Explain how two- and three-chambered vertebrate hearts work and in which animal groups each is found.
4. Explain the structure of cardiac muscle and how it helps produce coordinated contraction in each heart chamber.
5. List the three cell-based components of blood and describe the structure and principal functions of each.
6. Explain the sequence of events that causes the mammalian heart to beat, including the names of the structures involved.
7. Describe veins, capillaries, and arteries, noting their similarities and differences.
8. Distinguish among blood plasma, interstitial fluid, and lymph.
9. Describe the cardiac cycle, and relate the contractions of the atria and ventricles to the two readings taken during the measurement of blood pressure.
10. List the components of the lymphatic system, and describe three important functions of this system.
11. Explain how the sympathetic nervous system influences blood distribution and blood pressure.
12. Describe how the number of red blood cells is regulated by a negative feedback system.
13. In what way do veins and lymphatic vessels resemble one another? How are they different?

Applying the Concepts

Bloom's: Applying, Analyzing, Evaluating

1. In addition to completely separating oxygenated and oxygen-poor blood, what major advantage is there to the four-chambered heart with its two separate pumps?
2. Considering the prevalence of cardiovascular disease and the increasingly high costs of treating it, certain treatments may not be available to all who might need them. What factors would you take into account if you had to ration cardiovascular procedures, such as heart transplants or bypass surgery?

34 Respiration

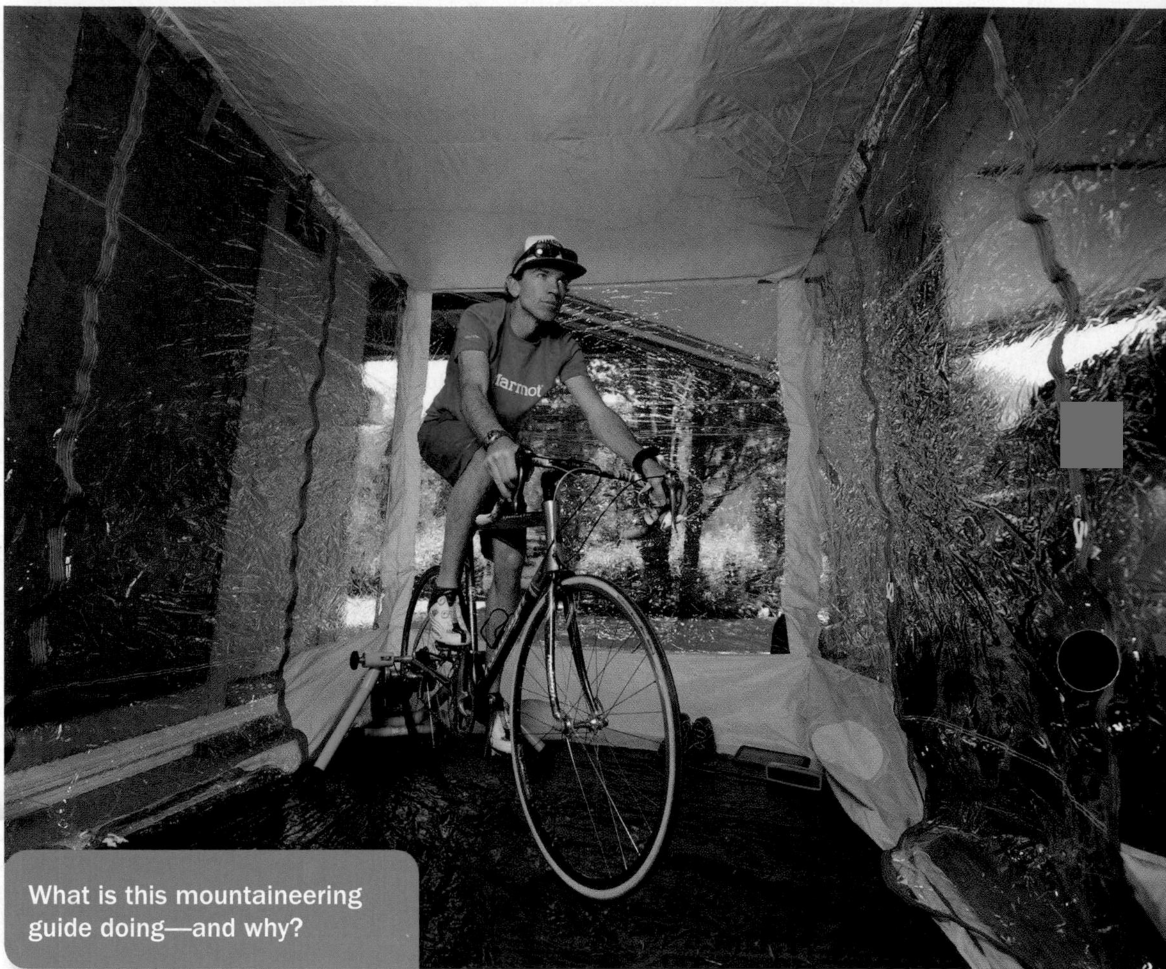
What is this mountaineering guide doing—and why?

CASE STUDY

Straining to Breathe—with High Stakes

DURING A TELEVISED INTERVIEW in January 2013, the seven-time Tour de France winner Lance Armstrong admitted to decades of deceit. The cycling world was rocked by news that he had illegally transfused his own blood and also taken synthetic erythropoietin (EPO) to pack in more red blood cells. Denying allegations of illegal blood doping in an earlier interview, Armstrong had stated, "I survived cancer; why would I take drugs? Why would I do that to myself?"

Why indeed? In elite sports, mere fractions of a second can separate winners from losers, and anything that increases oxygen delivery to hardworking muscles provides a competitive edge. Some athletes, such as Armstrong, cheat their way to the top with drugs or other illegal methods. Other professional endurance athletes attempt to legally gain this edge by living or training at high altitude, where atmospheric pressure is lower and each lungful of air contains fewer oxygen molecules. This stimulates a compensatory response in the athletes' bodies that allows their circulatory and respiratory systems to deliver oxygen to their muscles more efficiently. What about athletes who can't travel to high altitude to live or train? Lowlanders seeking high-altitude advantages can sleep or train in altitude tents (like the one shown in the photo) filled with air that has a reduced oxygen content.

As elite athletes strain to gain a split-second edge, millions of people struggle at times to get the minimum amount of oxygen they need to stay alive. People with sleep apnea, chronic bronchitis, emphysema, asthma, or cystic fibrosis cannot take normal breathing for granted. What is the sequence of events during breathing? How do respiratory disorders interfere with these processes? Consider these questions as we explore the respiratory system and then revisit the issue of exercise training at high altitudes.

AT A GLANCE

34.1 WHY EXCHANGE GASES AND WHAT ARE THE REQUIREMENTS FOR GAS EXCHANGE?

Late again! Sprinting up two flights of stairs to your classroom, your calves are burning. (Remembering Chapter 8, you think, "Ah ha! That's lactic acid building up—my muscle cells are using fermentation because they can't get enough oxygen for cellular respiration.") As you slip into your seat, quietly panting and feeling your heart pounding, the discomfort eases. Less exertion coupled with rapid breathing ensure that adequate oxygen (O_2) is now available. You are experiencing first-hand the connection between breathing and generating ATP through cellular respiration.

The Exchange of Gases Supports Cellular Respiration

Cellular respiration is the primary source of energy for all animals and for most other forms of life on Earth. As cellular respiration converts the energy in nutrients (such as sugar) into the ATP that supplies cellular energy, the process requires a steady supply of O_2 and generates carbon dioxide (CO_2) as a waste product. Organismal respiration, which we will simply call **respiration**, is the process by which organisms exchange gases with the environment, taking in oxygen and releasing carbon dioxide (CO_2), in support of cellular respiration. The rapid beating of your heart as you relax after your sprint reminds you that your circulatory system works in close harmony with your respiratory system to deliver O_2 and remove CO_2 from each cell of your body.

Gas Exchange Through Cells and Tissues Relies on Diffusion

All organisms that obtain energy through cellular respiration rely on the process of diffusion to acquire O_2 and to eliminate CO_2. Although animal respiratory structures are quite diverse, they all facilitate diffusion through three adaptations:

- Respiratory surfaces—even those in contact with air—remain moist, because cell membranes are always moist, and only gases dissolved in water can diffuse into or out of cells.
- Respiratory surfaces are very thin to minimize diffusion distances.
- Respiratory surfaces have a sufficiently large surface area to allow enough gas exchange by diffusion to meet the needs of the organism.

During diffusion, individual molecules move from areas where they are in higher concentration to areas of lower concentration (see Chapter 5), so gas exchange by diffusion requires concentration gradients of gases. To maintain these gradients, the air or water in which an animal lives must flow past the animal's respiratory surface, continuously supplying O_2 and carrying away CO_2. This movement is called **bulk flow** because the masses of molecules that form the flowing air or water move together "in bulk" through relatively large spaces.

CHECK YOUR LEARNING

Can you . . .

- explain how organismal respiration supports cellular respiration?
- describe the adaptations of respiratory surfaces for diffusion?
- explain why bulk flow is required for respiration?

34.2 HOW DO RESPIRATORY ADAPTATIONS MINIMIZE DIFFUSION DISTANCES?

Because movement of gases by diffusion is slow, diffusion can move dissolved gases over only very short distances. Evolution has provided a variety of ways to keep diffusion distances short—usually less than a millimeter—as described in the examples that follow.

Relatively Inactive Animals May Lack Specialized Respiratory Organs

For animals that are sluggish and don't require large amounts of ATP, diffusion alone may provide adequate oxygen. Sponges (**FIG. 34-1a**), Earth's simplest animals, live in aquatic—mostly marine—environments. Sponges use ciliated cells to create a current of water, which moves by bulk flow through pores in their bodies into a central chamber. The water then flows out through one or more larger openings. This circulation, aided by ocean currents,

(a) Sponge

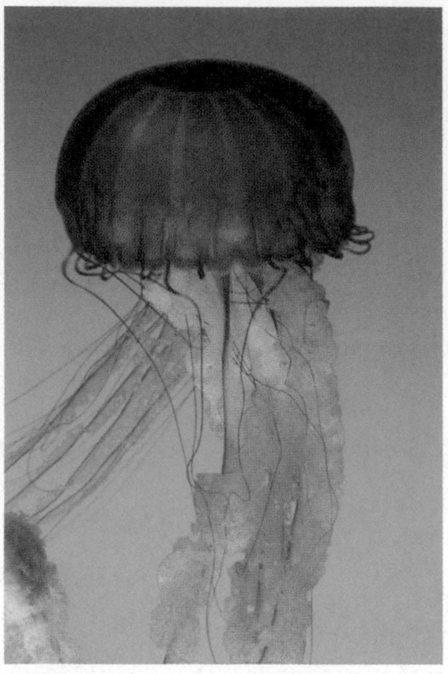

(b) Sea jelly

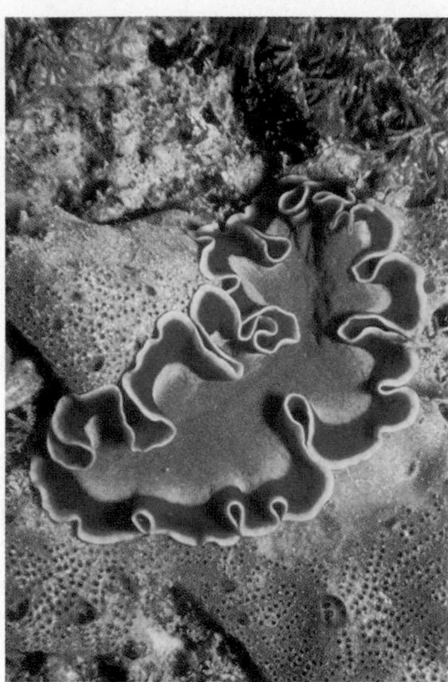

(c) Flatworm

▲ **FIGURE 34-1 Some animals lack specialized respiratory structures** Animals without respiratory systems have relatively low metabolic demands and large, moist body surfaces. **(a)** Flagellated cells draw currents of oxygen-rich water through numerous pores in the body of the sponge and expel oxygen-poor water through one or more larger openings. **(b)** Cells in the bell-shaped body of a sea jelly have a low metabolic rate; ocean currents and swimming movements provide adequate gas exchange. **(c)** The large outer surface of this marine flatworm exchanges gases with the water.

brings a continuous supply of water that is relatively high in O_2 and low in CO_2 within diffusing distance of each cell of the sponge. Sea jellies, corals, and anemones (cnidarians; **FIG. 34-1b**) have an extremely thin outer skin that allows gases to move readily by diffusion into and out of their outer cell layers as seawater flows over their bodies. Cnidarians use muscle cells to generate additional bulk flow of water into and out of a central chamber called the *gastrovascular cavity* (also used in digestion; see Fig. 35-9). The gastrovascular cavity brings water close enough to the internal cells that diffusion meets their needs for gas exchange. Flatworms also have a gastrovascular cavity; in some species it is used for gas exchange as well as for digestion. In all flatworms, gas exchange by diffusion occurs through an extensive gas-permeable skin surface. Their flattened shapes ensure that all their cells are close to the skin (**FIG. 34-1c**). Bulk flow of water over the bodies of these aquatic animals maintains the necessary diffusion gradients.

Some more complex animals, such as earthworms, combine a large skin surface for diffusion—produced by the worms' elongated shape—with a well-developed circulatory system (see Fig. 33-1b) that transports gases in the blood by bulk flow. As oxygen diffuses through the skin and into the blood, capillaries carry it away, thus maintaining the concentration gradient that promotes continued inward diffusion of O_2. Meanwhile, CO_2 diffuses from cells into the blood and is carried in capillaries to the skin, where it diffuses out. If an earthworm's skin dries out, it will suffocate because diffusion of gases into and out of cells requires a moist surface.

Respiratory Systems and Circulatory Systems Often Work Together to Facilitate Gas Exchange

Most relatively large, active animals have **respiratory systems** that consist of respiratory organs that work together to facilitate gas exchange between the animal and its environment. Respiratory systems often work closely with circulatory systems that carry gases to and from all body cells. Major organs of gas exchange include gills in many aquatic vertebrates and invertebrates, tracheae in insects, and lungs in terrestrial vertebrates. For animals with interacting circulatory and respiratory systems, gas exchange occurs in the following stages, illustrated for mammals in **FIGURE 34-2**:

1. Bulk flow carries the surrounding air or water, which is relatively high in O_2 and low in CO_2, past a respiratory surface. The flow is usually propelled by muscular movements.
2. O_2 diffuses from the air or water through the respiratory surface and into the capillaries of the circulatory system, while CO_2 diffuses out of the capillaries and through the respiratory surface into the air or water.
3. Bulk flow of blood transports gases between the respiratory system and the tissues. Blood is pumped throughout the body by the heart.
4. Diffusion transfers O_2 out of the capillaries into nearby tissues and transfers CO_2 from the tissues into capillaries.

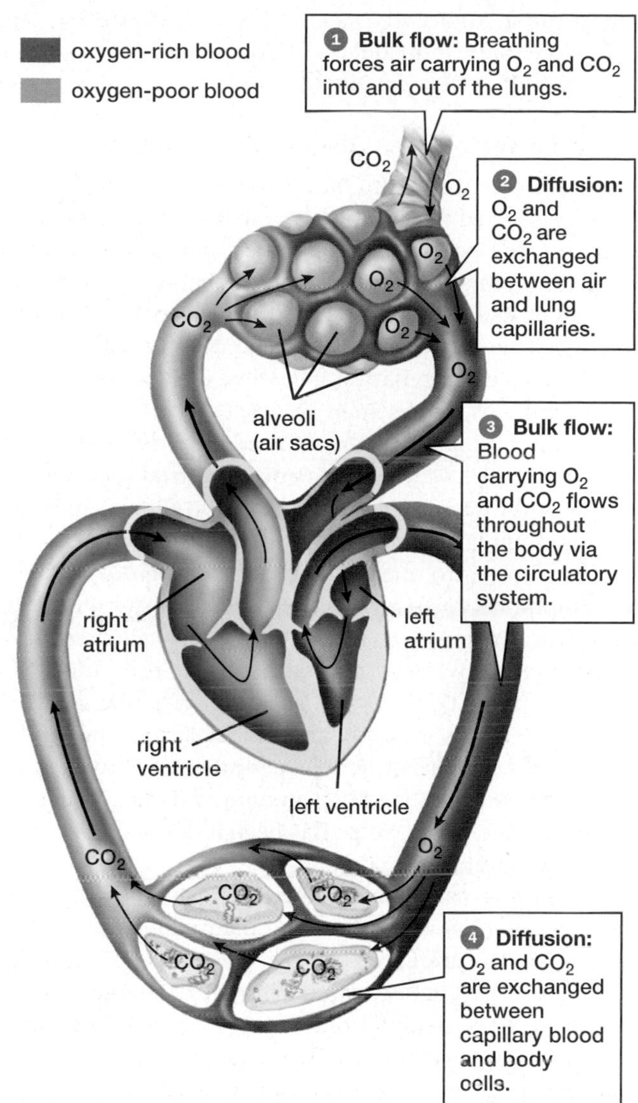

oxygen-rich blood
oxygen-poor blood
1 Bulk flow: Breathing forces air carrying O_2 and CO_2 into and out of the lungs.
CO_2
O_2
2 Diffusion: O_2 and CO_2 are exchanged between air and lung capillaries.
alveoli (air sacs)
3 Bulk flow: Blood carrying O_2 and CO_2 flows throughout the body via the circulatory system.
right atrium
left atrium
right ventricle
left ventricle
4 Diffusion: O_2 and CO_2 are exchanged between capillary blood and body cells.

▲ **FIGURE 34-2 An overview of gas exchange in mammals**

Gills Facilitate Gas Exchange in Aquatic Environments

Gills are the respiratory structures of many aquatic animals. The simplest type of gill, found in some amphibians (see Fig. 34-5a) and in nudibranch (literally, "naked gill") mollusks, consists of many thin projections of the body surface that protrude into the surrounding water (**FIG. 34-3**).

In general, gills are elaborately branched or folded, increasing their surface area. Just beneath their delicate outer membranes, gills have a dense profusion of capillaries that carry blood close to the gill surface, where gas exchange occurs. Most types of fish protect their delicate gills beneath a bony flap, the *operculum*. A fish creates a continuous water current through its mouth, over its gills, and out its operculum, using pumping movements of the operculum and mouth or by swimming rapidly with its mouth open.

Fish face a challenge in extracting enough O_2 from water to support their active lifestyles. There are only about 3% as many oxygen molecules in a given volume of fresh water as there are in the same volume of air (salt water contains even less). Because water is about 800 times as dense as air, pumping enough water over gills to obtain adequate oxygen uses far more energy than breathing air does. Fish have evolved a very efficient method for exchanging gases with water, using a process called **countercurrent exchange**.

▲ **FIGURE 34-3 External gills on a mollusk** The feathery projections from the back of this nudibranch mollusk are gills used for gas exchange.

Have You Ever Wondered ... If Sharks Really Need to Keep Swimming to Stay Alive?

You may have heard that sharks will suffocate if they stop swimming because swimming with their mouths open is the only way they can force water past their gills. In fact, many types of sharks spend most of their time feeding and resting on the seafloor and get plenty of O_2 by pumping water over their gill using muscles around their mouths. But others, beautifully streamlined, seem to be almost constantly in motion with open mouths, as seen in this great white shark. These sharks primarily or entirely use "ram ventilation"—forcing water over their gills as they swim. So the great white and closely related sharks, as well as a few other fishes such as tuna, will indeed suffocate if they stop swimming for long.

Although the human lung only extracts about 25% of the O_2 from the inhaled air, some fish can remove 80% of the O_2 from water flowing over their gills. Countercurrent exchange also allows efficient diffusion of CO_2 from capillary blood into the surrounding water.

Terrestrial Animals Have Internal Respiratory Structures

Gills are useless in air because they collapse and dry out, so as animals made the evolutionary transition to life on dry land, new respiratory adaptations arose. Respiratory structures evolved that protected, supported, and retained the moisture of the delicate membranes through which gas exchange occurs. These structures include tracheae in insects and lungs in nearly all terrestrial vertebrates and some scorpions, spiders, and snails.

Insects Respire Using Tracheae

The respiratory system of an insect consists of **tracheae** (singular, trachea), a system of branched, air-filled tubes. Tracheae conduct air into and out of the body through openings called **spiracles**, which are arrayed in rows along each side of the body surface (**FIG. 34-4**, left). Tracheae, which are reinforced with chitin (a principal component of the insect's external skeleton), penetrate the body tissues and branch into microscopic, open-ended tubes called *tracheoles* (Fig. 34-4, top right). Tracheoles deliver air close to each body cell, minimizing diffusion distances for O_2 and CO_2. Some large insects use pumping movements of their abdomens to increase air movement in and out of the tracheae.

In most animals, the respiratory system moves O_2 into the circulatory system, which distributes it to the body cells. However, the open circulatory systems of insects (see Fig. 33-1a) cannot move blood fast enough to supply sufficient oxygen to support strenuous activity, such as flying. Tracheae, by bringing O_2 close to the individual cells, compensate for the insect's relatively inefficient circulatory system.

Terrestrial Vertebrates Respire Using Lungs

Lungs are chambers containing moist respiratory surfaces that are protected within the body, where water loss is minimized and the body wall provides support and protection. The first vertebrate lung to evolve probably appeared in a freshwater fish and consisted of an outpocketing of the digestive tract. This simple lung supplemented the gills, helping the fish survive in stagnant water where O_2 is scarce.

Amphibians, represented primarily by frogs (which include toads) and salamanders, evolved from fish and straddle the boundary between aquatic and terrestrial life. Amphibians use gills during their aquatic larval (tadpole) stage, but most lose their gills and develop simple, saclike lungs as they metamorphose into more terrestrial adults (**FIGS. 34-5a–c**). Most amphibians also rely to a considerable extent on diffusion of gases through their skin, a process called *cutaneous respiration* (literally, "skin respiration"). The thin, moist skin that covers their bodies is rich in capillaries that allow gas exchange. A disadvantage of amphibian skin in the modern world is that it makes them very sensitive to environmental pollutants, which readily penetrate or damage their thin skins. Since the 1980s, there has been a dramatic decline in amphibian numbers worldwide.

In reptiles (snakes, lizards, turtles, crocodiles, and birds—now also classified as reptiles) and mammals, relatively waterproof skin is covered with scales (**FIG. 34-5d**), feathers, or fur. This covering reduces water loss in dry environments, but also prevents the skin from serving as a respiratory organ. Compensating for this loss, the lungs of reptiles and mammals have a far larger surface area for gas exchange than do those of amphibians.

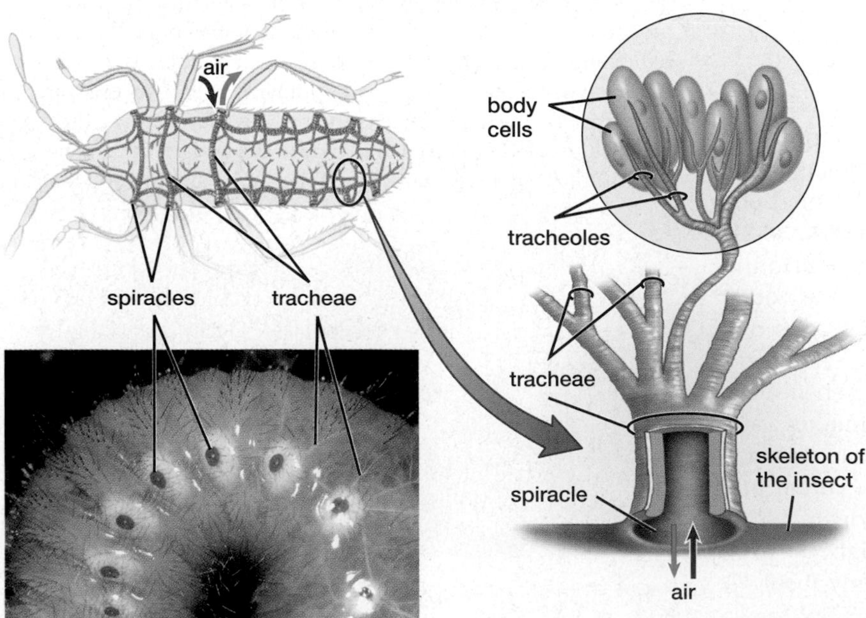

FIGURE 34-4 Insects breathe using tracheae The tracheae of insects branch intricately throughout the body; air moves in and out through spiracles in the body wall. Tracheoles branch from the tracheae and bring air within diffusing distance of each body cell.

THINK CRITICALLY How does the insect tracheal system compensate for insects' lack of a closed circulatory system?

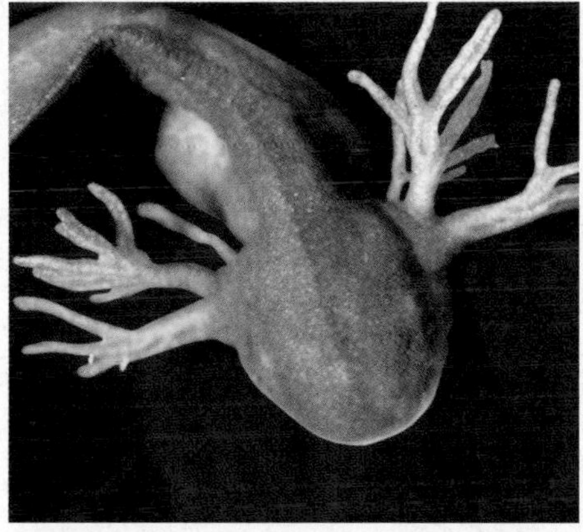

(a) Tadpole

(b) Adult bullfrog

(c) Axolotl

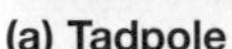

(d) Snake

◀ **FIGURE 34-5 Amphibians and reptiles have different respiratory adaptations** **(a)** The bullfrog, an amphibian, begins life as a fully aquatic tadpole with feathery external gills. **(b)** During metamorphosis into an air-breathing adult, the frog's gills are lost and replaced by simple saclike lungs. In both tadpole and adult, gas exchange also occurs through cutaneous respiration. **(c)** The axolotl salamander, endangered in its native habitat in Mexico, has rudimentary lungs but also retains its gills and aquatic lifestyle throughout life. **(d)** Scaled reptiles, such as this green mamba snake, cannot use cutaneous respiration; their lungs compensate with a larger surface area for gas exchange.

THINK CRITICALLY How do the respiratory adaptations of amphibians influence the range of habitats in which amphibians are found?

Bird respiratory systems differ substantially from those of other terrestrial vertebrates, including other reptiles. The bird lung has adaptations that allow exceptionally efficient gas exchange, providing adequate O_2 to support the enormous energy demands of flight. In the bird respiratory system, the lungs are rigid and connected to seven to nine inflatable air sacs, which serve as reservoirs for air. For clarity, **FIGURE 34-6** represents these as a single anterior (front) air sac and a posterior (rear) air sac.

The lungs of birds are filled with perforated stiff tubes called *parabronchi* barely visible to the naked eye. Parabronchi conduct air through surrounding capillary-rich tissue. The

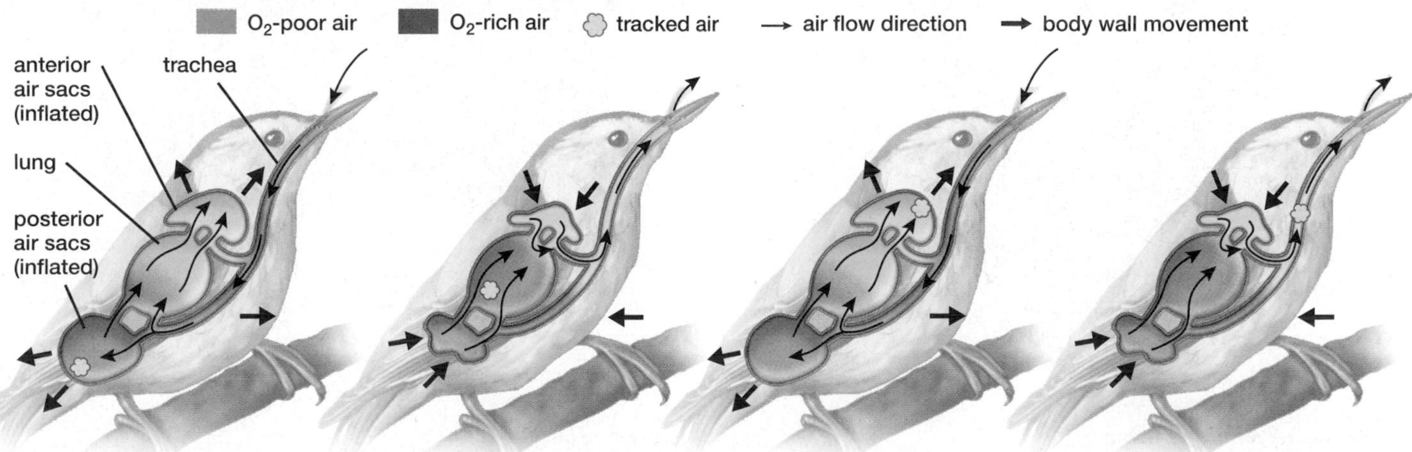
O_2-poor air
O_2-rich air
tracked air
air flow direction
body wall movement
anterior air sacs (inflated)
trachea
lung
posterior air sacs (inflated)
❶ Inhalation #1 inflates the air sacs, drawing O_2-rich air in through the trachea, filling the posterior sacs and lungs.
❷ Exhalation #1 compresses the air sacs, forcing O_2-rich air from the posterior sacs into the lungs.
❸ Inhalation #2 inflates the air sacs, drawing O_2-poor air from the lungs into the anterior air sacs.
❹ Exhalation #2 compresses the air sacs, expelling O_2-poor air from the anterior sacs and out through the trachea.

▲ **FIGURE 34-6 The bird respiratory system** Two breaths are required to move a given volume of air (tracked air) completely through the system. Bird lungs contain rigid, tubular parabronchi where gas exchange occurs. The flexible air storage sacs do not exchange gases.

tissue is riddled with interconnected microscopic air spaces where gas is exchanged with the capillary blood. Parabronchi allow air to flow completely through the lungs, from the posterior air sacs that supply fresh air to the anterior sacs that temporarily store air from the lungs before it is breathed out. These sacs cause bird lungs to receive fresh air during both inhalation and exhalation.

CHECK YOUR LEARNING

Can you . . .

- explain the four stages of gas exchange in animals with circulatory systems and lungs?
- explain the respiratory adaptations of fish?
- describe the general respiratory adaptations for life on land and the specific adaptations of insects, amphibians and reptiles including birds?

34.3 HOW IS AIR CONDUCTED THROUGH THE HUMAN RESPIRATORY SYSTEM?

The respiratory system in humans and other mammals can be divided into two parts: the **conducting portion** and the **gas-exchange portion**. The conducting portion consists of a series of passageways that carry air into and out of the gas-exchange portion within the lungs, where O_2 and carbon dioxide are exchanged with blood in lung capillaries.

The Conducting Portion of the Respiratory System Carries Air to the Lungs

Air enters through the nose or the mouth, passes through the nasal cavity or oral cavity into a chamber, the **pharynx** (which is shared with the digestive tract), and then travels through the **larynx**, where sounds are produced (**FIG. 34-7**).

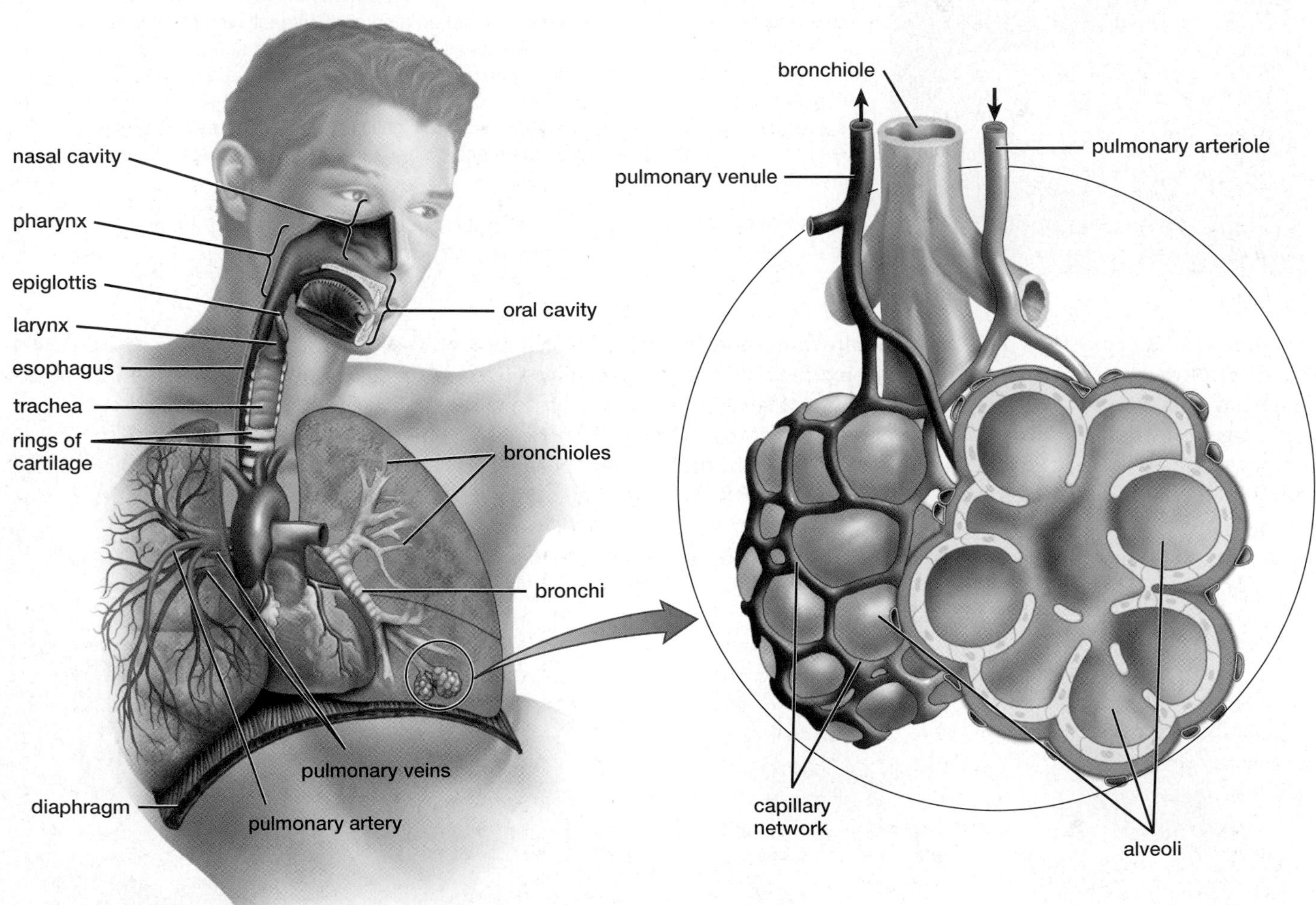

▲ **FIGURE 34-7 The human respiratory system (a)** The major structures of the human respiratory system are illustrated here. The pulmonary artery carries oxygen-poor blood (blue) to the lungs; the pulmonary vein carries oxygen-rich blood (red) back to the heart. **(b)** Close-up of alveoli (interiors shown in cutaway section), their arterioles, and their surrounding network of capillaries.

THINK CRITICALLY Why is oxygen-rich blood depicted as red and oxygen-poor blood as blue?

The opening to the larynx is guarded by the **epiglottis**, a flap of tissue supported by cartilage. During breathing, the epiglottis is tilted upward, allowing air to flow freely into the larynx. During swallowing, the epiglottis folds downward and covers the larynx, directing substances into the esophagus (see Fig. 35-15). If an individual attempts to inhale and swallow at the same time, this reflex may fail and food can become lodged in the larynx, blocking air from entering the lungs. What should you do if you see this happen? Use the lifesaving **Heimlich maneuver**, described in **FIGURE 34-8**.

Within the larynx, or "voice box," are the **vocal cords**, bands of elastic tissue controlled by muscles. Contraction of these muscles causes the vocal cords to partially obstruct air flow through the larynx. Exhaled air then causes the vocal cords to vibrate, producing sound. Stretching the vocal cords changes the sound pitch, which movements of the tongue and lips then articulate into speech or song.

Inhaled air travels past the larynx into the **trachea**, a flexible tube whose walls are reinforced with semicircular bands of stiff cartilage. Within the chest, the trachea splits into two large branches called **bronchi** (singular, bronchus), one leading to each lung. Inside the lung, each bronchus branches repeatedly into ever-smaller tubes. Finally, these divide into **bronchioles** about 1 millimeter in diameter. The walls of bronchi and bronchioles are encased in smooth muscle, which regulates their diameter. During activities that require extra oxygen, such as exercise, the smooth muscle relaxes, allowing more air to enter. The increasingly small bronchioles conduct air to **alveoli** (singular, alveolus), microscopic air sacs where gas exchange occurs (see Fig. 34-7).

During its passage through the conducting system, air is warmed and moistened. Much of the dust and bacteria it carries is trapped in mucus secreted by cells that line the respiratory passages. The mucus, with its trapped debris, is continuously swept upward toward the pharynx by cilia that line the bronchioles, bronchi, and trachea. Upon reaching the pharynx, the mucus is coughed up or swallowed. Smoking interferes with this cleansing process by paralyzing the cilia (see "Health Watch: Smoking—A Life and Breath Decision" on page 603).

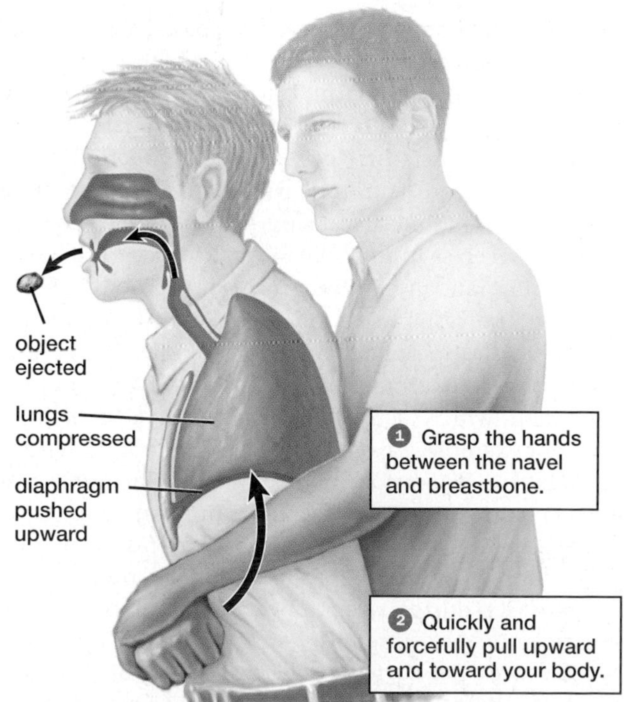

object ejected
lungs compressed
diaphragm pushed upward
1 Grasp the hands between the navel and breastbone.
2 Quickly and forcefully pull upward and toward your body.

▲ **FIGURE 34-8 The Heimlich maneuver can save lives** Pressing forcefully and suddenly upward and inward just below the diaphragm forces air from the lungs, which can expel food blocking the trachea. The maneuver can be repeated if necessary.

Air Is Inhaled Actively and Exhaled Passively

Breathing occurs as inhalation and exhalation cause bulk flow of air into and out of the lungs. Both the lungs and the chest cavity are elastic; they can be enlarged by stretching, but when the tension is relaxed, they recoil passively back to their original size. The lower boundary of the chest cavity is formed by a sheet of muscle, the **diaphragm**, which domes upward when relaxed. During **inhalation**, the intake of air into the lungs, the diaphragm is actively contracted, which pulls it downward. The rib or *intercostal* muscles are also contracted during inhalation, lifting the ribs upward and outward. Both of these muscular movements actively enlarge the chest cavity (**FIG. 34-9a**). As the chest cavity expands, the lungs inflate within it, because the lungs are sealed to the inner chest wall by fluid that fills the space between these structures. As the lungs expand, their increased volume reduces the pressure inside them, drawing air in. A puncture wound to the chest is dangerous in part because it can allow air to penetrate between the chest wall and the lungs, breaking the seal and preventing the lungs from inflating when the chest cavity expands.

Exhalation occurs spontaneously and passively when the muscles that cause inhalation relax. As the diaphragm relaxes, it domes upward; at the same time, the ribs move downward and inward. Both of these movements decrease the size of the chest cavity, forcing air out of the lungs (**FIG. 34-9b**). Additional air can be forced out by contracting the abdominal muscles. After exhalation, the lungs still contain some air, which helps prevent the thin alveoli from collapsing. A typical breath in an average-sized adult moves about a pint (roughly 500 milliliters) of air through the respiratory system. Because the air must also fill the conducting portion, only about three-quarters of the inhaled and exhaled air reaches the alveoli. During exercise, deeper breathing can move four times this volume of air during each breath.

Breathing Rate Is Controlled by the Respiratory Center of the Brain

Imagine having to think about every breath. Fortunately, breathing occurs rhythmically and automatically,

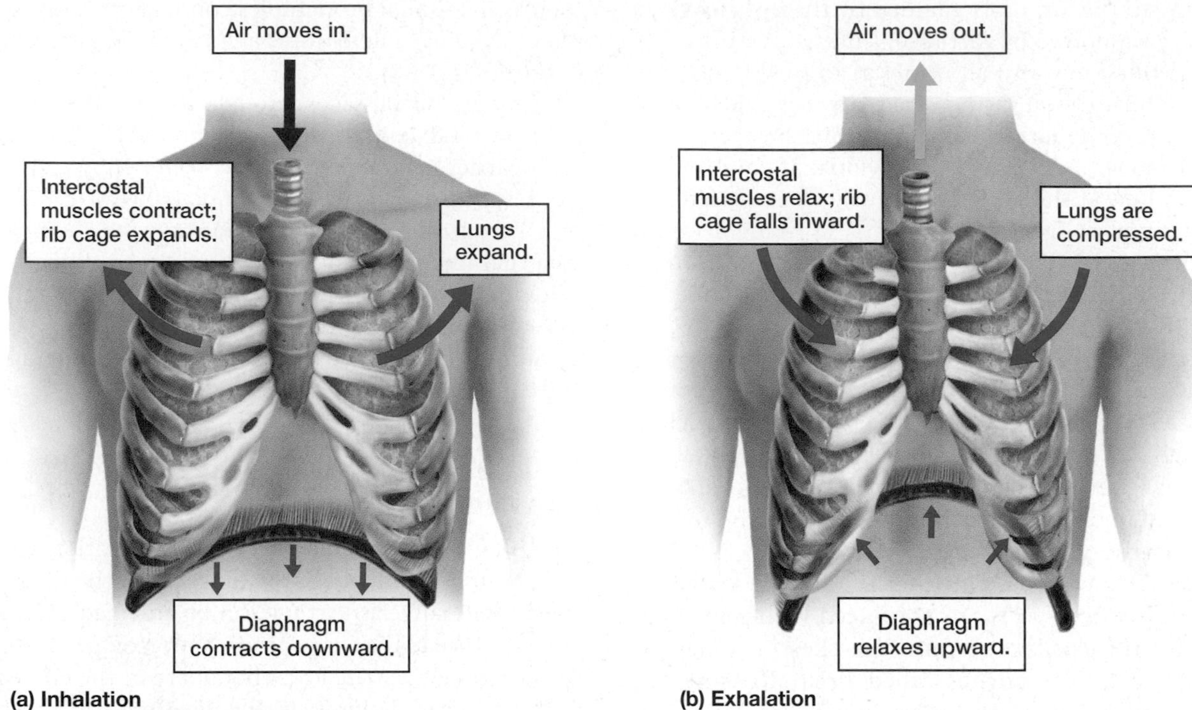

▲ **FIGURE 34-9 The mechanics of breathing (a)** During inhalation, rhythmic nerve impulses from the brain stimulate the diaphragm and the intercostal muscles to contract. This increases the size of the chest cavity, causing air to rush in. **(b)** During exhalation, all these muscles relax, forcing air out of the lungs.

THINK CRITICALLY How does contracting the diaphragm muscle and the intercostal muscles increase the volume of the chest cavity?

CASE STUDY **CONTINUED**

Straining to Breathe—with High Stakes

If you have sleep apnea, you never get a good night's sleep. Although you may not realize it, your sleep is briefly interrupted from 30 to more than 100 times each night when you temporarily stop breathing for 10 to 60 seconds ("apnea" means "without breath"). Your blood O_2 falls and CO_2 rises, causing you to partially awaken, often with a snort. What causes this? Obstructive sleep apnea, the most common form, occurs when muscles that support the tongue and soft tissues of the pharynx relax, causing these tissues to collapse and obstruct the airway. People who snore, are overweight, drink alcohol before bed, or smoke are all more susceptible to sleep apnea. The resulting daytime drowsiness, making it hard to concentrate and dangerous to drive, is bad enough. But the repeated O_2 deprivation and CO_2 buildup caused by sleep apnea puts your heart at risk as well, so this is a condition to take seriously. How does your body know that you have stopped breathing?

without conscious thought. But, unlike the heart muscle, the diaphragm and rib muscles used in breathing are not self-activating; each contraction causing inhalation is stimulated by impulses from nerve cells. These impulses originate in the **respiratory center**, which is located in the medulla, a portion of the brain just above the spinal cord (see Fig. 39-12). Nerve cells in the respiratory center generate cyclic bursts of electrical signals (action potentials) that cause contraction. During the intervals between bursts, the muscles relax, causing exhalation. The respiratory center receives input from several sources and adjusts the breathing rate and volume to meet the body's changing needs. The respiratory rate is primarily regulated by CO_2 receptors, also located in the medulla. These adjust the breathing rate to maintain a constant low level of CO_2 in the blood, which also ensures that O_2 levels remain adequate. For example, if you run to class, your muscles need extra ATP. This requires an increase in cellular respiration, which generates more CO_2. Receptors in the medulla detect the increased CO_2 and cause you to breathe faster and more deeply, eliminating more CO_2 and taking in more O_2. The buildup of CO_2 during sleep apnea will also activate them. These receptors are extremely sensitive; a 16% increase in CO_2 will cause a person to double the amount of air moving through the lungs. There are also O_2 and CO_2 receptors in the aorta and carotid arteries that stimulate the respiratory center. These receptors are primarily sensitive to increased CO_2 in the blood, but a major decline in blood O_2 will also activate them.

Health WATCH Smoking—A Life and Breath Decision

Smoking-related diseases kill nearly half a million people each year in the United States—one out of every five deaths. These preventable afflictions include lung and several other types of cancer, chronic obstructive pulmonary disease (COPD), and cardiovascular disease. Smoking accounts for nearly all lung cancer deaths and one-third of all cancer deaths in the United States. Despite decades of warning about smoking's dangers, about 18% of U.S. adults and 9% of high school students continue to smoke.

As smoke is inhaled, toxic substances paralyze the cilia lining the trachea and bronchi. Because cilia remove inhaled particles, smoking inhibits them just when they are needed most. The visible portion of cigarette smoke consists of microscopic carbon particles bearing dozens of cancer-causing substances. With the cilia incapacitated, these particles enter the lungs. Like a car air filter that's never been changed, the lungs of a heavy smoker are literally blackened by soot (**FIG. E34-1**). Cigarette smoke also impairs white blood cells that engulf foreign particles and bacteria, rendering the lungs more vulnerable to these invaders. The irritated respiratory system produces more mucus, which builds up and can obstruct the airways, producing the familiar "smoker's cough."

Long-time smokers are especially vulnerable to COPD, which usually refers to chronic bronchitis and emphysema occurring together. Incurable and worsening over time, COPD is the third leading cause of death in the United States, after cardiovascular disease and cancer. *Chronic bronchitis* is a permanent inflammation of the bronchi and bronchioles that decreases air flow to the alveoli. It causes swelling, a decrease in the number and activity of cilia, an increase in mucus production, and a persistent cough that attempts to clear the airways. The mucus provides a fertile breeding ground for bacteria that cause frequent respiratory infections. Emphysema occurs when toxic substances from cigarette smoke cause the alveolar walls and respiratory membranes to break down. As emphysema develops, magnified clusters of alveoli progress from looking like a pink sponge to resembling blackened Swiss cheese.

Smoking causes cancer because many of the chemicals found in tobacco smoke are mutagens—they cause mutations by damaging the DNA in cells. Other chemicals in smoke interfere with cells' DNA-repair mechanisms, further compounding the damage. Cells with many mutations are more likely to become cancerous. Because tobacco smoke weakens the body's immune system, it reduces the body's ability to attack and destroy cancerous cells before they proliferate.

Smoking also promotes *atherosclerosis*, thickening of the arterial walls by fatty deposits that can lead to blood clots that cause heart attacks and strokes (see Chapter 33). Carbon monoxide in cigarette smoke reduces the blood's oxygen-carrying capacity and increases the workload on the heart. COPD makes it harder for the circulatory system to extract enough oxygen from the air, further stressing the heart. As a result, smokers are two to four times as likely as nonsmokers to die of heart disease.

When pregnant women smoke, the carbon monoxide and nicotine contribute to reproductive problems by reducing oxygen delivery to the developing fetus. Smoking mothers are more likely to have miscarriages, and their babies are more likely to have lower birth weight and a higher incidence of certain birth defects and learning and behavioral problems.

(a) A normal lung

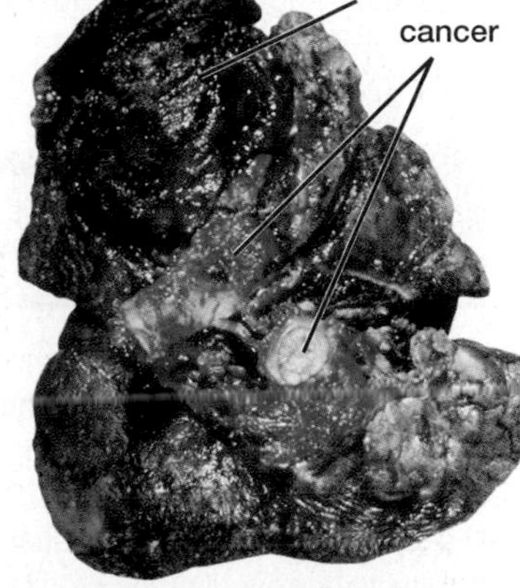

(b) A smoker's lung with cancer and emphysema

▲ **FIGURE E34-1 To breathe, or not to breathe?**

Passive smoking, or breathing secondhand smoke, poses health hazards to nonsmokers. Children whose parents smoke have decreased lung capacity and are at increased risk of pneumonia, ear infections, coughs, colds, asthma, and sudden infant death syndrome (SIDS). Nonsmoking adults exposed to smoke at home or at work are 20% to 30% more likely to die of cardiovascular disease, stroke, and lung cancer.

So why do people keep smoking? The nicotine in tobacco is as addictive as cocaine or heroin. Like other addictive drugs, nicotine activates the brain's reward center more intensely than natural stimuli do. The brain adapts by becoming less sensitive to these drugs, requiring larger quantities to experience the same rewarding effect. When a person tries to quit smoking, withdrawal symptoms may include nicotine craving, depression, anxiety, irritability, headaches, difficulty concentrating, and disturbed sleep. But a smoker who quits will begin to feel better almost immediately, as her blood CO drops, and her heart rate and blood pressure begin to return to normal. In a few weeks, her smoker's cough will diminish and blood circulation will improve. After several months, the ability of her lungs to exchange gases will increase, her sense of smell will improve, and food will taste better. People who quit at any age reduce their chances of dying of smoking-related diseases, but those who quit by age 30 are 90% less likely to die of these causes than if they had continued.

THINK CRITICALLY A 6-year-old child is brought to the school health clinic from the playground, wheezing and having difficulty breathing. The school nurse recognizes the girl, who has been in his clinic several times for coughs and severe colds. Concerned, the nurse invites the parents to a conference and notices that the mother is pregnant. What questions should he ask? What advice should he give?

CHECK YOUR LEARNING

Can you . . .

- describe the structures and functions of the conducting portion of the human respiratory system?
- explain the mechanics of breathing?
- explain how the brain controls the breathing rate?

CASE STUDY CONTINUED

Straining to Breathe—with High Stakes

If you have asthma, you are vividly aware of the tight-chested wheezing struggle to inhale and exhale during an asthma attack. What is happening? An attack occurs when the bronchioles become inflamed. Bronchial walls swell and often release more mucus than usual into the air passages. Smooth muscle surrounding the bronchioles constricts, further narrowing the airways. Triggers for attacks differ among individuals and include allergic reactions, inhaled dust or pollutants, and respiratory infections. Physical activity is a common trigger, but medication can allow asthmatics to participate in high-intensity sporting events; several Olympic medalists have asthma. No single cause of asthma has been identified, but it often begins in childhood and probably results from a combination of genetic tendencies and environmental factors. Parents who smoke, for example, greatly increase their children's likelihood of developing asthma. Although asthma is incurable, most cases can be well controlled with a combination of drugs and lifestyle precautions. If an attack occurs, a fast-acting inhaled "rescue medication" can relax the smooth muscles around the bronchioles. A severe untreated attack can lead to death by suffocation, so asthmatics must monitor and control their condition and carry rescue inhalers. Asthma can cause suffocation because it prevents gas exchange in the alveoli of the lungs. How do the alveoli provide life-sustaining O_2 to the bloodstream?

34.4 HOW DOES GAS EXCHANGE OCCUR IN THE HUMAN RESPIRATORY SYSTEM?

As air moves by bulk flow into and out of the lungs, gases enter and leave the blood by diffusion within the lungs (see Fig. 34-2). The human lung provides a moist, well-protected air cavity where exquisitely fragile membranes separate atmospheric air from the bloodstream.

Gas Exchange Occurs in the Alveoli

Inside each lung, a dense, tree-like branching system of bronchioles conducts air to the alveoli, which cluster around the end of each bronchiole like a bunch of grapes. In an average adult, the two lungs combined have approximately 300 million alveoli. These microscopic chambers (0.2 millimeter in diameter) give magnified lung tissue the appearance of a pink sponge. The inside of each alveolus is coated with a thin layer of watery fluid containing a surfactant (a detergent-like substance composed of proteins and lipids), which prevents the fragile alveolar surfaces from sticking together and collapsing when air is exhaled. Gases dissolve in this fluid as they pass in or out of the alveolus.

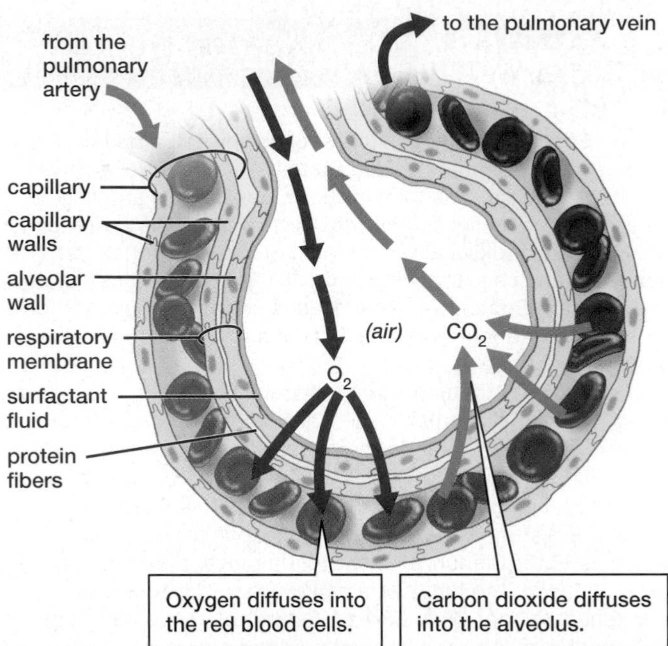

▲ **FIGURE 34-10 Gas exchange between alveoli and capillaries** The respiratory membrane consists of alveolar and capillary walls that are each only one cell thick. Gases diffuse through this membrane between the lungs and the circulatory system. The inner alveolus is coated in a slippery surfactant fluid that prevents the alveolar membranes from adhering to one another.

A network of capillaries covers most of the alveolar surface (see Fig. 34-7b). The walls of the alveoli consist of a single layer of epithelial cells. Gases diffuse through the **respiratory membrane**, which consists of the alveolar wall and the capillary wall, glued together by protein fibers (**FIG. 34-10**). The respiratory membrane creates a microscopically small diffusion distance (0.5 to 1 micrometer) between the air in the lungs and the blood in capillaries, which allows gases to be transferred efficiently.

The lungs provide an enormous moist surface for gas exchange. In an average adult, the total surface area of the alveoli is about 1,500 square feet (roughly 145 square meters), about 80 times the skin surface area of a human adult, and about one-third the area of a basketball court.

Oxygen and Carbon Dioxide Are Transported in Blood Using Different Mechanisms

Blood picks up O_2 from the air in the lungs and supplies it to the body tissues, simultaneously absorbing CO_2 from the tissues and releasing it into the lungs. These exchanges occur because the diffusion gradients favor them;

in the lungs, O_2 is high and CO_2 is low, whereas in body cells, CO_2 is high and O_2 is low (see Fig. 34-10). Nearly all (about 98%) of the O_2 carried by blood is bound to **hemoglobin** (**FIG. 34-11a**), a large protein that gives red blood cells their color (see Fig. 33-9). By removing O_2 from solution in the plasma, hemoglobin increases the gradient that causes O_2 to diffuse from the lungs into the plasma. Each hemoglobin molecule can bind to up to four O_2 molecules. This binding allows blood to carry far more O_2 than if the O_2 were simply dissolved in the plasma. As O_2 binds hemoglobin, the protein changes its shape, which alters its color. Oxygenated or "O_2-rich" blood is a bright cherry-red, and deoxygenated or "oxygen-poor" blood is maroon-red and appears bluish through the skin. For this reason, we depict oxygen-rich blood as red, and oxygen-poor blood as blue.

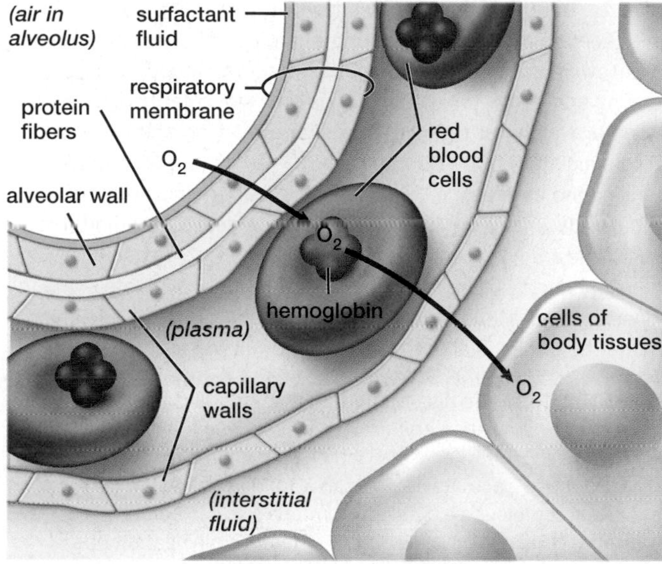

(a) O_2 transport from the lungs to the tissues

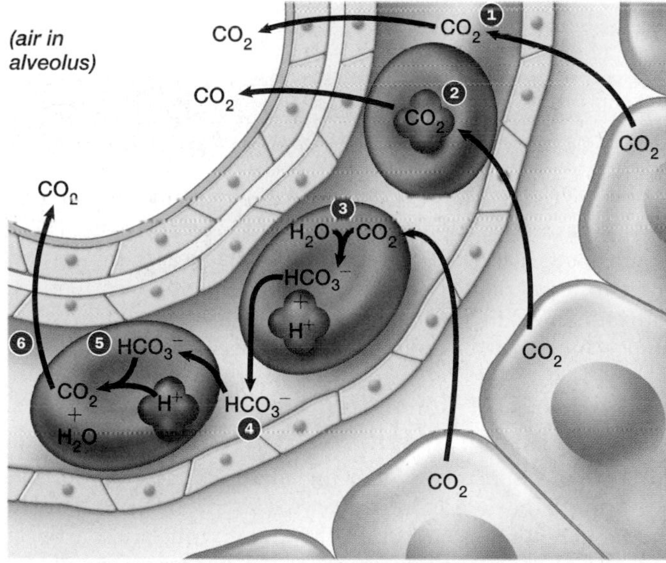

(b) CO_2 transport from the tissues to the lungs

▲ **FIGURE 34-11 Oxygen and carbon dioxide transport**
(a) The high O_2 content of the air in the alveoli favors diffusion of O_2 through the respiratory membrane into the alveolar capillaries. Here, O_2 loosely binds to hemoglobin and is transported to the body cells. At the body cells where O_2 is lower, O_2 diffuses off the hemoglobin, out of the capillaries, through the interstitial fluid, and into the cells. **(b)** CO_2 diffuses from body cells and into capillaries by one of three routes; see numbered steps in the main text.

THINK CRITICALLY Why is it important for the hydrogen ions (generated when bicarbonate is formed) to remain bound to hemoglobin?

Carbon dioxide produced by cellular respiration diffuses from cells through the interstitial fluid into nearby capillaries. The CO_2 is then carried in the bloodstream to the respiratory membranes of the alveoli. Blood transports CO_2 in three different ways, as shown in the numbered sequence in **FIGURE 34-11b**. Roughly 10% is dissolved in the plasma ❶, about 20% is bound to hemoglobin after it diffuses into red blood cells ❷, and most of it (about 70%) combines with water in red blood cells to form bicarbonate ions (HCO_3^-) ❸. This reaction is catalyzed by the enzyme *carbonic anhydrase*:

$$H_2O + CO_2 \xrightarrow[\text{(carbonic anhydrase)}]{} H^+ + HCO_3^-$$

Most of the H^+ that is liberated during this reaction remains bound to hemoglobin in the red blood cells, which helps prevent the blood plasma from becoming too acidic (a potentially fatal condition). The HCO_3^- then diffuses into the plasma ❹. These reactions keep the amount of CO_2 dissolved in plasma low, which increases the gradient for CO_2 to diffuse into the plasma from the body cells where CO_2 levels are relatively high.

The CO_2 gradient is reversed when the blood reaches the lung capillaries, where the dissolved CO_2 diffuses from the plasma into the alveoli. The reduction in plasma CO_2 favors the reverse reaction that regenerates CO_2 and H_2O from bicarbonate ❺:

$$H^+ + HCO_3^- \xrightarrow[\text{(carbonic anhydrase)}]{} H_2O + CO_2$$

The resulting CO_2 and H_2O diffuse from red blood cells into the plasma. The increased plasma CO_2 promotes diffusion of CO_2 through the respiratory membrane and into the air within the alveoli, which is expelled during exhalation ❻.

CHECK YOUR LEARNING

Can you . . .

- explain how the structure of the alveoli facilitates gas exchange?
- describe how O_2 and CO_2 are transported and exchanged between the lungs and body cells?

CASE STUDY REVISITED

Straining to Breathe—with High Stakes

Why does training at high altitude increase the amount of oxygen in an athlete's blood? It's because the air is thinner at altitude, and the body responds by increasing its red blood cell count. Although air at any altitude is about 21% O_2, at higher altitudes, each lungful of air contains fewer molecules. For example, at 7,500 feet, each inhalation brings in only 77% as much O_2 as it would at sea level. As a result, fewer O_2 molecules diffuse into alveolar capillaries, and blood oxygen is reduced. This reduction triggers production of erythropoietin (EPO), which stimulates production of more red blood cells. Blood hemoglobin levels increase, and the blood transports more O_2.

Athletes who train at high altitude compete well at high altitude, because their bodies adjust to permit extreme exertion with reduced oxygen. However, such training does not improve subsequent performance at lower altitudes, even though elevated hemoglobin levels persist for weeks after a return to sea level. Why doesn't high-altitude training—like EPO injections—improve low-altitude performance? One reason is that, even though athletes who train at altitude have elevated hemoglobin, muscles exercised at high altitude never receive as much oxygen as they do at sea level, so they are never pushed to their maximum potential. As a result, high-altitude training sessions do not allow athletes to reach or maintain their peak low-altitude condition.

Much better results in low-altitude competitions are achieved with a "live high–train low" regimen, in which athletes live at 7,000 to 8,000 feet and descend to much lower altitudes for workouts. But this method is inconvenient and expensive. Altitude tents (see the chapter opening photo) are an appealing alternative, allowing athletes to train intensely at low altitude, while sleeping and resting under conditions that simulate the lower oxygen availability at high altitude. At sea level, an altitude tent set to simulate 7,500 feet removes 23% of the oxygen without changing the air pressure. But does this really mimic living at high altitude, where overall air pressure is reduced? Do the tents work? Research on these questions has produced inconclusive results, in part because people are so variable. An athlete's response to an altitude tent depends on genes, training level, and whether he or she is able to sleep soundly in the tent.

CONSIDER THIS Training at altitude and altitude tents are legal ways of stimulating erythropoietin production and increasing red blood cell count to enhance the O_2-carrying capacity of the blood. Do these approaches differ, in a fundamental way, from transfusing red blood cells taken from one's own blood or injecting synthetic EPO? Also, how do the different training methods described here affect the fairness of sporting competitions?

CHAPTER REVIEW

Go to **Mastering Biology** to access the Pearson eText, vocabulary review, practice quizzes, activities, videos, current events, and more.

*Answers to **Think Critically** and **Thinking Through the Concepts** questions can be found in the **Answers** section at the back of the book.*

Summary of Key Concepts

34.1 Why Exchange Gases and What Are the Requirements for Gas Exchange?

The respiratory systems of animals support cellular respiration by supplying O_2, which allows ATP production during cellular respiration, and by carrying away the waste product CO_2. Gas exchange in all animals relies on both bulk flow and diffusion. The transfer of gases between the environment and body cells requires air or water carrying dissolved O_2 and CO_2 to move by bulk flow past the respiratory surface. Respiratory surfaces must be moist, thin, and sufficiently large to provide adequate O_2 for cellular respiration. The exchange of O_2 and CO_2 between the respiratory surface and cells of the body occurs by diffusion across respiratory membranes, bulk flow in the circulatory system, and then diffusion across cell membranes.

34.2 How Do Respiratory Adaptations Minimize Diffusion Distances?

In watery environments, animals with very small or flattened bodies and low metabolic demands may lack specialized respiratory organs and rely exclusively on diffusion through the body surface. Larger, more active animals have evolved specialized respiratory systems. Aquatic animals, such as fish and larval amphibians, often possess gills. On land, respiratory surfaces must be protected, supported, and kept moist internally. These requirements have led to the evolution of tracheae in insects and lungs in terrestrial vertebrates. Amphibians, non-bird reptiles, and birds have increasingly efficient respiratory systems that support increased demands for gas exchange.

34.3 How Is Air Conducted Through the Human Respiratory System?

The human respiratory system includes a conducting portion consisting of the nose, mouth, pharynx, larynx, trachea, bronchi, and bronchioles, which lead to the alveoli. Inhalation actively draws air into the lungs by contracting the diaphragm and the rib muscles, which expands the chest cavity. Exhalation occurs passively when the diaphragm and rib muscles relax, reducing the volume of the chest cavity and expelling the air. Respiration is controlled by the medulla's respiratory center, using signals from receptors that monitor CO_2 and O_2 levels in the blood.

34.4 How Does Gas Exchange Occur in the Human Respiratory System?

Bronchioles conduct air to microscopic air sacs called alveoli where gas exchange with capillary blood occurs. Gases diffuse through the respiratory membrane, which consists of a single layer each of alveolar and capillary cells. Blood in the dense capillary network surrounding the alveoli absorbs O_2 from the alveolar air and releases CO_2 into it.

Oxygen entering lung capillaries is picked up by hemoglobin within red blood cells. Blood transports the O_2 to the body tissues, where it diffuses out. Carbon dioxide diffuses from the tissues into the blood. Most is transported as bicarbonate ions, some is bound to hemoglobin, and a small amount is dissolved in blood plasma. Upon reaching the lung capillaries, CO_2 diffuses into the alveoli to be exhaled.

Thinking Through the Concepts

Bloom's: Remembering, Understanding

Multiple Choice

1. Which of the following is *not* true of respiratory systems?
  a. They transport carbon dioxide.
  b. They require moist membranes for gas exchange.
  c. They minimize the distances gases must diffuse.
  d. They all work closely with circulatory systems.

2. Countercurrent exchange of gases
  a. occurs in the tracheae of insects.
  b. relies on both bulk flow and diffusion.
  c. is important for gas exchange in flatworms.
  d. forces fish to keep swimming to get enough oxygen.

3. Which of the following statements is false?
  a. Gills may be used by mollusks.
  b. Tracheoles carry air to insect body cells.
  c. Parabronchi are found in the conducting portion of mammalian respiratory systems.
  d. The human trachea branches to form two bronchi.

4. Cutaneous respiration
  a. is important for many reptiles.
  b. causes the lungs of mammals to expand.
  c. helps protect animals from air pollution.
  d. cannot occur through dry skin.

5. Which of the following statements is true?
  a. Carbonic anhydrase is found in plasma.
  b. Hemoglobin changes color slightly when it binds oxygen.
  c. All blood vessels can exchange gases with their surrounding tissues.
  d. The respiratory center of the brain is most sensitive to changes in the oxygen level in blood.

Fill-in-the-Blank

1. Three types of organisms that lack specialized respiratory systems are __________, __________, and __________. These animals rely entirely on the process of __________ for gas exchange.

2. Air enters insect bodies through openings called __________. Insect __________ are tubes that conduct air close to each body cell; this is necessary because the __________ of insects cannot adequately distribute gases.

3. Which part of the conducting portion of the respiratory system is shared with the digestive tract? __________ What structure normally keeps food from entering the larynx? __________ After passing through the larynx, air travels in sequence through the __________, __________, and __________, after which it enters the gas-exchange portion of the respiratory system, which consists of the __________.

4. In the lungs, oxygen diffuses into blood vessels called __________. Most oxygen in the blood is carried by a protein called __________. Most carbon dioxide in the blood is transported in the form of __________ ions. Cells require oxygen and release carbon dioxide because they generate energy using the process of __________.

5. The respiratory membrane is found in the __________. This membrane consists of the wall of a(n) __________ and the wall of a(n) __________ joined by protein strands. The respiratory membrane is (how many?) __________ cell(s) thick. The inside of the respiratory membrane is coated with a watery fluid containing __________.

6. Air is inhaled when the __________ and the rib muscles contract, making the chest cavity __________. In contrast, exhalation is a(n) __________ process, caused by allowing these muscles to __________. Breathing is driven by neurons of the __________ located in the __________ of the brain. The respiratory rate is increased when receptors detect an excess of __________ in the blood.

Review Questions

1. Describe how fish gills work, including the basic concept of countercurrent exchange. Why is this process important to allow fish to extract oxygen from water?

2. How does the respiratory system of a frog change when it undergoes metamorphosis? Why are these changes necessary?

3. Describe the respiratory system of birds and how it allows oxygen-rich air to enter the lungs during both inhalation and exhalation.

4. Trace the route taken by air in the mammalian respiratory system, listing the structures through which air flows and the structure in which gas exchange occurs.

5. Explain some traits of animals in moist habitats that make specialized respiratory systems unnecessary.

6. How is breathing initiated? How are breathing rate and depth adjusted, and which blood gas is most tightly regulated?

7. What events occur during human inhalation? During exhalation? Which of these is always an active process?

8. Compare bulk flow and diffusion. Explain how bulk flow and diffusion interact to promote gas exchange between air and blood and between blood and tissues.

9. Compare CO_2 and O_2 transport in the blood. Include the source and destination of each and what physiological process makes gas exchange necessary.

10. Explain how the structure and arrangement of alveoli make them well suited for their role in gas exchange.

Applying the Concepts

Bloom's: Applying, Analyzing, Evaluating

1. Heart–lung transplants are performed in some cases in which both organs have been damaged by cigarette smoking. Given a scarcity of organ donors, what criteria would you use in selecting a recipient for such a transplant?

2. Nicotine is responsible for keeping smokers addicted. Discuss the advantages and disadvantages of both low-nicotine tobacco cigarettes and artificial "e-cigs" that deliver nicotine in appealing-scented water vapor, compared to smoking regular cigarettes.

3. Mary, a strong-willed 3-year-old, threatens to hold her breath until she dies if she doesn't get her way. Can she carry out her threat? Explain.

35 Nutrition and Digestion

Troian Bellisario speaks to the National Eating Disorder Association.

CASE STUDY

Dying to Be Thin

TROIAN BELLISARIO IS AN ACCOMPLISHED ACTRESS who is in recovery from an eating disorder. "There is a part of my brain that defies logic," she writes. "Once, it completely convinced me I should live off 300 Calories a day." During her first year at college, Bellisario's disorder left her emaciated and hospitalized. Untreated eating disorders can be fatal, but Bellisario survived and recovered. Today, healthy and successful, she is outspoken about her past struggles and is working to help others avoid the damage her body suffered. Young people are especially vulnerable; about 95% of eating disorders occur among people between 12 and 26 years of age.

Eating disorders include two especially serious conditions: anorexia and bulimia. Perhaps 1% of women will suffer from anorexia during their lives. People with anorexia experience an intense fear of gaining weight; although their bodies may become skeletal, they perceive themselves as fat. In response, they eat very little food and sometimes exercise compulsively, burning off essentially all their body fat and breaking down muscle tissue to supply their energy needs. Anorexia also disrupts digestive, reproductive, endocrine, and cardiac functions. About 2 in 10 anorexia sufferers will die prematurely of causes related to the disorder, including heart disease and suicide.

People with bulimia—some of whom are also anorexic—engage in binge eating, consuming enormous amounts of food in a short period of time. To purge the food from their bodies, they induce vomiting or overdose with laxatives, and they may also engage in excessive exercise.

Eating disorders occur roughly three times more frequently in women, but men also fall victim. One 17-year-old male with anorexia explained, "I would wake up a few minutes early, . . . then pump out about a hundred push-ups, do some crunches and then get in the shower, get dressed, come downstairs, hide [my breakfast], then flush it when I was going up to brush my teeth. . . . And then I'd pump out some more push-ups." Celebrities who have battled eating disorders include Russel Brand, Ashley Hamilton, Demi Lovato, Kesha, and Lady Gaga, as well as Isabella Caro, one of several fashion models who have died from complications of anorexia.

What happens when people starve themselves or, conversely, overeat? What nutrients do our bodies require, and how do our bodies extract nutrients from the food we eat?

AT A GLANCE

35.1 WHAT NUTRIENTS DO ANIMALS NEED?

Whether you eat cauliflower or a candy bar, your food contains important nutrients. **Nutrients** are substances obtained from food that organisms need to maintain health. Animal nutrients fall into six major categories: carbohydrates, lipids, proteins, minerals, vitamins, and water. These substances provide energy and raw materials to synthesize the molecules of life.

Energy from Food Powers Metabolic Activities

Cells rely on a continuous supply of energy to maintain their incredible complexity and wide range of activities. Deprived of energy, cells begin to die within minutes. The nutrients that provide energy are carbohydrates (sugars and starches), lipids (fats and oils), and proteins. In the average U.S. diet, carbohydrates supply about 50% of the calories, fats and oils about 35%, and proteins about 15%. The energy content of these nutrients is expressed in **Calories** (capital "C"), a unit that represents 1,000 calories (lowercase "c"). A **calorie** is the amount of energy required to raise the temperature of 1 gram of water by 1°C. The average human body at rest burns roughly 70 Calories per hour, but this value is influenced by several factors. For example, people differ in their resting **metabolic rate**, their resting rate of energy expenditure. Muscle tissue burns more calories than fat does, so a muscular individual, even at rest, consumes more calories than a person carrying the same weight in fat. Well-trained athletes can temporarily burn nearly 20 Calories per minute during vigorous exercise. See **FIGURE 35-1** for a rough idea of how certain activities burn calories.

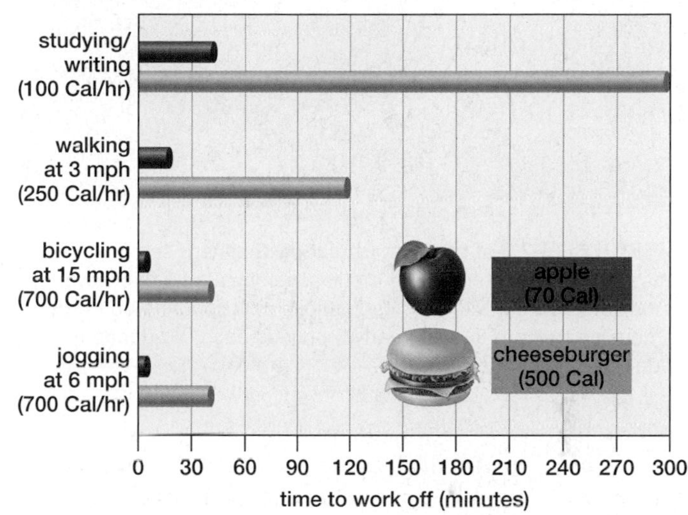

▲ **FIGURE 35-1 Consider Calories when eating** If you have mayo on your cheeseburger, you add another 100 Calories! You would need to walk for almost 30 more minutes to burn it off.

Carbohydrates Are a Source of Quick Energy

Carbohydrates include sugars such as glucose, from which cells derive most of their energy; sucrose (table sugar); and poly saccharides, long chains of sugar molecules (see Chapter 3). Cellulose, starch, and glycogen are polysaccharides composed of chains of glucose. Cellulose, the major structural component of plant cell walls, is the most abundant carbohydrate on the planet, but only a few types of organisms are able to digest it, as described later. Starch is the principal energy-storage material of plants and a major energy source for humans and many other animals.

Athletes sometimes "carbo-load" before competing by eating meals rich in starch, such as potatoes and pasta. This allows their bodies to produce a carbohydrate that is not found in food: glycogen. Glycogen is synthesized and stored in the liver and muscles of animal bodies to provide a source of quick energy. Although humans can accumulate hundreds of pounds of fat, most can store less than a pound of glycogen. During exercise such as running, one pound of glycogen can power about 18 miles of running.

Fats and Oils Are the Most Concentrated Energy Sources

Fats and oils are the most concentrated sources of energy, containing more than twice as many Calories per unit weight as carbohydrates or proteins do (about 9 Calories per gram for fats compared to about 4 Calories per gram for proteins and carbohydrates). When an animal's diet provides more energy than it expends, most of the excess energy is stored as body fat. In addition to its high caloric content, fat is hydrophobic, so it neither attracts water nor dissolves in water, as carbohydrates and proteins do. For this reason, fat deposits do not cause extra water to accumulate in the body, so fat stores more Calories for a given amount of weight than other molecules do. Minimizing weight allows an animal to move faster (important for escaping predators and hunting prey) and to use less energy when it moves (important when food supplies are limited).

In addition to storing energy, fat deposits may provide insulation. Fat, which conducts heat at only one-third the rate of other body tissues, is often stored in a layer directly beneath the skin. Birds and mammals that live in polar climates and in cold ocean waters—such as penguins, seals, whales,

▲ **FIGURE 35-2 Fat provides insulation** This harp seal pup can withstand the icy waters of the arctic seas because a thick layer of fat insulates it from the cold. Its mother's milk contains up to 48% fat, causing the pup to gain nearly 5 pounds each day during the 12-day nursing period.

and walruses—are particularly dependent on this insulating layer, which reduces the amount of energy they must expend to keep warm (**FIG. 35-2**).

An Evolved Tendency to Store Fat Can Lead to Obesity When Food Is Abundant

Because humans evolved under conditions in which the food supply was unpredictable, we have inherited a strong tendency to eat when food is available. People in some modern societies, however, now have virtually unlimited access to high-calorie food. In this environment, our natural tendency to overeat can become a liability, and we often eat more than we need. Some of us must exert considerable willpower to avoid storing excessive amounts of fat. The **body mass index (BMI)** is a common tool for estimating a healthy weight, but is imprecise because it is based on height and weight only and does not account for differences in build or muscle mass. To calculate your BMI, divide your weight in pounds by your height in inches squared and multiply by 703—or use one of the many online calculators. For people with average amounts of muscle, a BMI between 18.5 and 24.9 is considered healthy. About 34% of all U.S. adults are overweight (BMI between 25 and 29.9) and an additional 35% are obese (BMI of 30 or more).

Essential Nutrients Provide the Raw Materials for Health

Our cells can synthesize carbohydrates and most of the other molecules our bodies require, but they cannot synthesize certain raw materials, called **essential nutrients**, which must be supplied in the diet.

Essential nutrients differ for different animals. For example, vitamin C (ascorbic acid) is an essential nutrient for people, but not for most other animals because they can synthesize it. Essential nutrients for humans include certain fatty acids and amino acids, a variety of minerals and vitamins, and water.

Have You Ever Wondered ...

If Pears Are Healthier Than Apples?

You have probably seen articles warning of the danger of being apple-shaped. It's true that where you store your body fat is at least as important as how much fat you are carrying. "Apples" are said to store visceral belly fat, which is far less healthy than storing fat just beneath the skin of the hips and buttocks, as "pears" do. What's the difference? Visceral fat surrounds organs inside the abdominal cavity and releases a variety of cell-signaling proteins that normally balance one another to promote a healthy metabolism. But in obese individuals, too much visceral fat leads to changes in its cellular makeup, causing imbalances in its secretions. This imbalance causes metabolic changes that lead to an increased risk of cardiovascular disorders, type 2 diabetes, and several types of cancer. The tendency to store visceral fat is influenced by genes and also increases with age. A round, firm, "pot belly" is a tell-tale sign of visceral fat, but this can be hidden by rolls of subcutaneous ("beneath the skin") fat—the squishy kind you can grab by the handful. Subcutaneous fat is less metabolically active and has a far lower health risk.

Back to apples and pears: The ratio of waist to hip circumference is often used as an indicator of visceral fat storage. An apple has a ratio greater than 0.85 for females and 0.90 for males; the pear's ratio is lower. But if you have small hips and lots of subcutaneous fat around your waist, your ratio could classify you as an unhealthy apple. Or you might have lots of visceral fat but be classified as a healthier pear if you happen to have large hips. So forget the fruit. In general, a large waist (>35 inches for females, >40 inches for males) is a pretty reliable indicator of visceral fat and its associated health risks. Fortunately, when you lose weight, particularly through aerobic exercise, visceral fat tends to be the first to go.

Certain Fatty Acids Are Essential in the Human Diet

Fats and oils are more than just a source of energy—some provide **essential fatty acids**. Essential fatty acids serve as raw materials to synthesize molecules involved in a wide range of physiological activities: They help us to absorb fat-soluble vitamins (described later) and are important in cell division, fetal development, and the immune response. Important sources of essential fatty acids include fish oils, canola oil, soybean oil, flaxseed, and walnuts.

Amino Acids Form the Building Blocks of Protein

Proteins form muscle, connective tissue, nails, and hair. They also act as enzymes, receptors on cell membranes, and antibodies. Of the 20 different amino acids used in proteins, the human body cannot synthesize 9 (adults) or 10 (infants). These **essential amino acids** must be obtained from protein-rich foods such as meat, milk, eggs, nuts, beans, and soybeans. Because many plant proteins are deficient in some of the essential amino acids, vegetarians must consume a variety of plants (for example, legumes, grains, and corn) whose proteins collectively provide all of them. Protein deficiency can cause a debilitating

TABLE 35-1 Important Minerals for Humans

Mineral	Dietary Sources	Important Roles in the Body	Deficiency Symptoms
Calcium	Milk, cheese, leafy vegetables	Helps in bone and tooth formation and maintenance; aids in blood clotting; contributes to nerve impulse transmission and muscle contraction	Stunted growth, rickets, osteoporosis
Phosphorus	Milk, cheese, meat, poultry, grains	Helps maintain pH of body fluids; contributes to bone and tooth formation; component of ATP and of phospholipids in cell membranes	Muscular weakness, weakening of bone
Potassium	Meats, milk, fruits	Helps maintain pH and osmotic strength of body fluids; important in nervous system activity	Nausea, muscular weakness, paralysis
Chlorine	Table salt	Helps maintain pH and osmotic strength of body fluids; component of HCl produced by gastric glands; important in nervous system activity	Muscle cramps, apathy, reduced appetite
Sodium	Table salt	Helps maintain pH and osmotic strength of body fluids; important in nervous system activity	Muscle cramps, nausea
Magnesium	Whole grains, leafy vegetables, dairy products, legumes, nuts	Helps to activate many enzymes	Tremors, muscle spasms, weakness, irregular heartbeat, hypertension
Iron	Meats, legumes, nuts, whole grains, leafy vegetables	Component of hemoglobin and many enzymes	Iron-deficiency anemia (weakness, reduced resistance to infection)
Fluorine	Fluoridated water, seafood	Component of teeth and bones	Increased tooth decay; may increase risk of osteoporosis
Zinc	Seafood, meat, cereals, nuts, legumes	Constituent of several enzymes; component of proteins required for normal growth, smell, and taste	Retarded growth, learning impairment, depressed immunity
Iodine	Iodized salt, seafood, dairy products, many vegetables	Component of thyroid hormones	Goiter (enlarged thyroid gland)
Chromium	Meats, whole grains	Helps maintain normal blood glucose levels	Elevated insulin in blood; increased risk of type 2 diabetes

and potentially fatal condition called kwashiorkor (**FIG. 35-3**), which occurs most frequently in poverty-stricken countries.

Minerals Are Elements Required by the Body

Minerals are elements that play many crucial roles in animal nutrition. Minerals such as calcium, magnesium, and phosphorus are major constituents of bones and teeth. Sodium, calcium, and potassium are essential for muscle contraction and the conduction of nerve impulses. Iron is a central component of blood hemoglobin, and iodine is found in thyroid hormones. We also require trace amounts of several other minerals listed in **TABLE 35-1**. Because no organism can manufacture elements, all essential minerals must be obtained from food or drinking water.

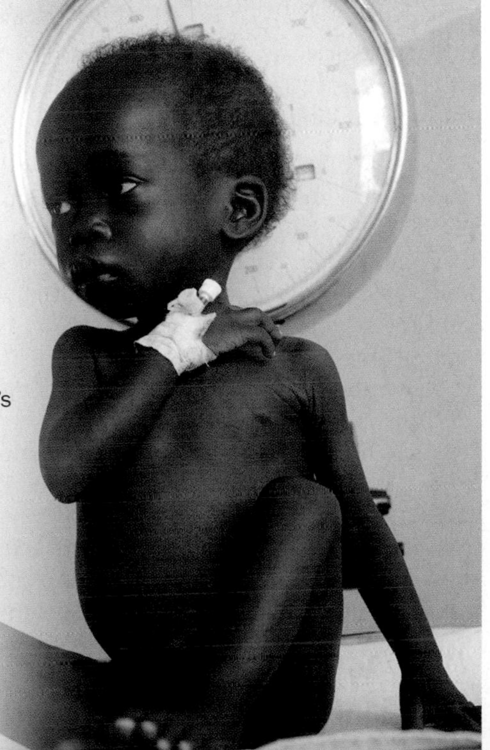

FIGURE 35-3 Kwashiorkor is caused by protein deficiency This young victim's muscles are atrophied by lack of protein. His pot belly is caused by slack abdominal muscles and fluid accumulation because low levels of blood protein decrease the blood's osmotic strength, causing more fluid to leak out of blood capillaries.

Vitamins Play Many Roles in Metabolism

"Take your vitamins!" is a familiar refrain in many households with children. **Vitamins** are a diverse group of organic molecules that animals cannot synthesize, but that are necessary for cell function, growth, and development. The vitamins essential in human nutrition are listed in **TABLE 35-2**. Human vitamins are grouped into two categories: water soluble and fat soluble.

Water-Soluble Vitamins Water-soluble vitamins include vitamin C as well as the eight compounds that make up the B-vitamin complex. Because these substances dissolve in the watery blood plasma and are filtered out by the kidneys, they are not stored in the body in appreciable amounts.

Most water-soluble vitamins act as *coenzymes*; that is, they work in conjunction with enzymes to promote chemical reactions that supply energy or synthesize biological molecules. Because each vitamin participates in several metabolic processes, a deficiency of a single vitamin can have wide-ranging effects (see Table 35-2). For example, deficiency of the B vitamin niacin causes the swollen tongue and skin

TABLE 35-2 Important Vitamins for Humans

Vitamin	Sources	Functions in Body	Deficiency Symptoms
Water Soluble			
Vitamin C (ascorbic acid)	Citrus and other fruits, tomatoes, leafy vegetables	Helps maintain cartilage, bone, and dentin (hard tissue of teeth); helps in collagen synthesis	Scurvy (degeneration of skin, teeth, gums, and blood vessels, and epithelial hemorrhages)
B Complex			
Vitamin B_1 (thiamin)	Meat, cereals, peas and soybeans, fish	Coenzyme in glucose metabolism	Beriberi (muscle weakness, peripheral nerve changes, heart failure)
Vitamin B_2 (riboflavin)	Shellfish, dairy products, eggs	Component of coenzymes involved in energy metabolism	Reddened lips, cracks at the corners of the mouth, light sensitivity, blurred vision
Vitamin B_3 (niacin)	Meats, fish, leafy vegetables	Component of coenzymes involved in energy metabolism	Pellagra (skin and gastrointestinal lesions and nervous and mental disorders)
Vitamin B_5 (pantothenic acid)	Meats, whole grains, legumes, eggs; product of intestinal bacteria	Component of coenzyme A, with a role in cellular respiration	Fatigue, sleep disturbances, impaired coordination
Vitamin B_6 (pyridoxine)	Meats, whole grains, tomatoes, potatoes	Coenzyme in amino acid metabolism	Irritability, convulsions, skin disorders, increased risk of heart disease
Biotin	Legumes, vegetables, meats; product of intestinal bacteria	Coenzyme in amino acid metabolism and cellular respiration	Fatigue, depression, nausea, skin disorders, muscular pain
Folic acid (folate)	Meats, green and leafy vegetables, whole grains, eggs; product of intestinal bacteria	Coenzyme in nucleic and amino acid metabolism	Anemia, gastrointestinal disturbances, diarrhea, nervous system defects in fetus and low birth weight
Vitamin B_{12}	Meats, eggs, dairy products	Coenzyme in nucleic acid metabolism and other metabolic pathways	Anemia, neurological disturbances
Fat Soluble			
Vitamin A	Green, yellow, orange, and red vegetables; liver; fortified dairy products	Component of visual pigment; helps maintain skin and other epithelial cells; promotes normal development of teeth and bones	Night blindness, permanent blindness, increased susceptibility to infections
Vitamin D	Tuna, salmon, eggs, fortified dairy products and cereals; synthesized by skin in sunlight	Promotes bone growth and mineralization; increases calcium absorption; may improve immune function	Rickets (bone deformities), skeletal deterioration
Vitamin E (tocopherol)	Nuts, whole grains, leafy vegetables, vegetable oils	Antioxidant; may reduce cellular damage from free radicals	Neurological damage
Vitamin K	Leafy vegetables; product of intestinal bacteria	Important in blood clotting	Bleeding, internal hemorrhages

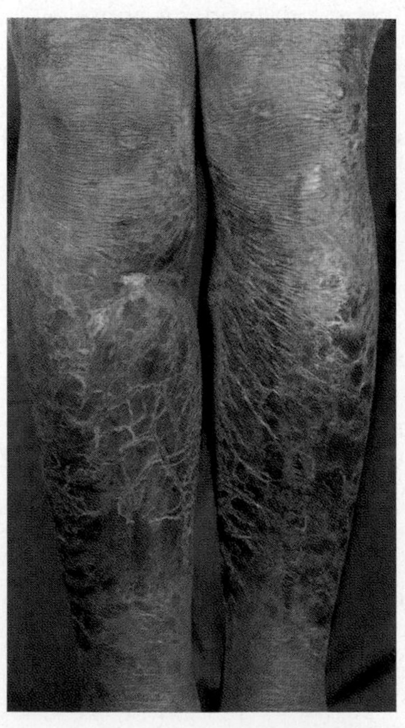

FIGURE 35-4 Pellagra is caused by niacin deficiency
Symptoms of pellagra include the scaly lesions visible on this individual's legs.

lesions of pellagra (**FIG. 35-4**), as well as some nervous system disorders. Folic acid, another B vitamin, is required to synthesize thymine, a component of DNA; folic acid deficiency impairs cell division throughout the body. It is particularly important for pregnant women to get enough folic acid to supply the rapidly growing fetus. People only obtain vitamin B_{12} by eating meat and dairy products or foods that are supplemented with vitamin B_{12}. Although deficiencies are more common than overdoses, a variety of side effects can occur if large excesses of B vitamins are consumed as supplements.

Fat-Soluble Vitamins The fat-soluble vitamins A, D, E, and K have a variety of functions (see Table 35-2). Vitamin A is used to synthesize the light-capturing molecules in the retina of the eye. Vitamin D is required for normal bone formation; a deficiency can lead to rickets (**FIG. 35-5**). It also contributes to proper functioning of the immune system. Sunlight stimulates vitamin D synthesis in the skin; however, people who get little exposure to sunlight may not synthesize enough. Researchers have discovered that many adult women in the United States, particularly those with dark skin (which reduces the penetration of sunlight), have inadequate levels of vitamin D. Breast-fed children born to vitamin D–deficient mothers are at particular risk for rickets. Vitamin E is an antioxidant, neutralizing free radicals that form in the body (see Chapter 2). Vitamin K helps to regulate blood clotting. Fat-soluble vitamins may accumulate in the body fat over time;

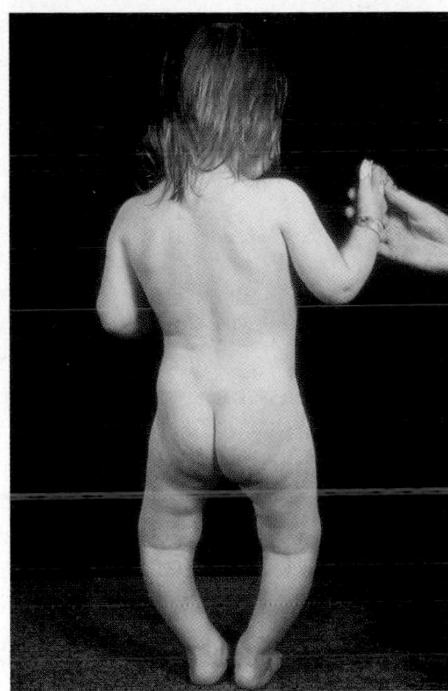

▲ **FIGURE 35-5 Rickets is caused by vitamin D deficiency** This child's bones have not absorbed adequate calcium, making them soft and unable to bear weight without bending. The resulting bowed legs may be permanent.

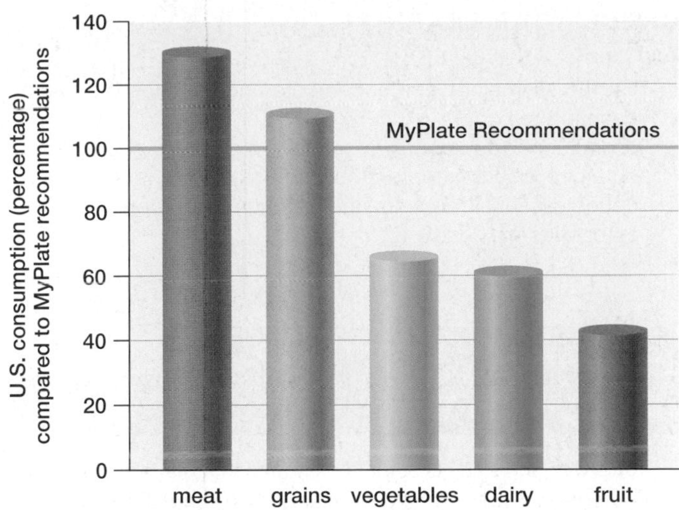

▲ **FIGURE 35-6 Average U.S. diets compared to MyPlate recommendations** The "grains" group includes all food made from grains (including breads, pasta, tortillas, corn chips, breakfast cereals, and crackers). Meat includes poultry, beef, pork, and fish. Source: Calculated by Economic Research Services RS/USDA based on data from various sources. Note: Rice is not included in grains.

THINK CRITICALLY If the MyPlate recommendations for vegetables were doubled, how would the height of the vegetable bar be changed?

as a result, some can reach toxic levels if taken in megadoses as supplements.

The Human Body Is About Sixty Percent Water

A person can survive far longer without food than without water, which makes up roughly 60% of total body weight. All metabolic reactions occur in a watery solution, and water participates directly in hydrolysis reactions (see Chapter 3) that break down proteins, carbohydrates, and fats into simpler molecules. Water is the principal component of saliva, blood, lymph, interstitial fluid, and the cytosol within each cell. By sweating, people use the evaporation of water to keep from overheating. Urine, which is mostly water, is necessary to eliminate cellular waste products from the body (see Chapter 36).

Many People Choose an Unbalanced Diet

The overwhelming variety of tempting and convenient processed foods in a typical U.S. supermarket or convenience store makes it easy to make poor nutritional choices. Guidelines for healthy nutrition are available on a Web site called "Choose My Plate," published by the U.S. Department of Agriculture. However, the average American diet differs considerably from these guidelines (**FIG. 35-6**). To help people make better food choices, commercially packaged foods are required to provide complete information about Calorie, fiber, fat, sugar, and vitamin content (**FIG. 35-7**). In addition, chain restaurants and vending machines in the United States are now required to provide customers with the Calorie content of their offerings. Many fast-food chains are now including healthy choices on their menus.

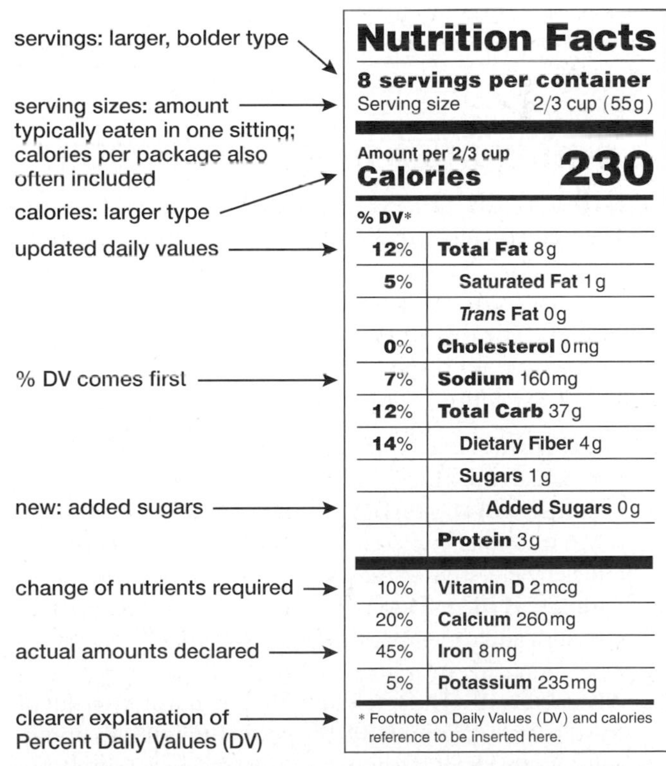

▲ **FIGURE 35-7 Revised food labeling** The U.S. FDA will soon require updated labels like this on nearly all packaged food and drinks. The labels will reflect updated dietary guidelines and actual typical serving sizes, and will be easier to read and interpret. To learn more about the updated labels, visit the "Changes to the Nutrition Facts Label" page at the FDA Web site.

CHECK YOUR LEARNING

Can you . . .

- list the six major types of nutrients and describe the role of each in the body?
- explain how food energy is measured and how energy is obtained, used, and stored by the body?
- define *essential nutrients* and describe five different categories of these?

CASE STUDY CONTINUED

Dying to Be Thin

People with anorexia consume so few nutrients that they literally starve, sometimes to death. Malnutrition can damage every organ in the body, including the liver, kidneys, and heart. As fat stores are eliminated, the body fuels itself with protein from muscle tissue, including cardiac muscle, causing the heart to weaken. Bones may become weak and brittle due to deficiencies in calcium and phosphorus and from hormonal imbalances that reduce bone formation. Anorexia sufferers are at risk of sudden death because of an imbalance of electrolytes. These minerals—including potassium, chloride, sodium, and calcium—form ions in the blood whose concentrations must be precisely regulated. Fatal arrhythmias (irregular heartbeats that do not circulate blood adequately) may result when electrolyte homeostasis is disrupted.

Because people with anorexia consume so little food, muscular activity within the digestive tract slows, and the stomach shrinks. How does digestion proceed in healthy individuals?

35.2 HOW DOES DIGESTION OCCUR?

Digestion is the process that physically grinds up food and then chemically breaks it down. The animal **digestive system** consists of a series of compartments in which food is processed, as well as organs that produce secretions that aid in digestion. The digestive tracts of animals are diverse, having evolved to meet the challenges posed by a variety of diets. All digestive systems, however, must accomplish five tasks:

1. **Ingestion** Food is brought into the digestive tract through an opening, usually called a **mouth**.
2. **Mechanical digestion** Food is physically broken down into smaller pieces with a larger surface area for attack by digestive enzymes.
3. **Chemical digestion** Particles of food are exposed to enzymes and other digestive secretions that break down large molecules into smaller subunits.
4. **Absorption** The small subunits are transported out of the digestive tract and into the body through cells lining the digestive tract. Although the digestive tract surrounds and acts on food, nutrients do not actually enter the body until they are absorbed.
5. **Elimination** Indigestible materials are expelled from the body.

In the following sections, we explore a few of the diverse mechanisms by which animal digestive systems accomplish these functions.

In Sponges, Digestion Occurs Within Single Cells

Sponges are the only animals that lack a digestive system and rely exclusively on **intracellular digestion**, in which all digestion occurs within individual cells (**FIG. 35-8**). As you might suspect, sponges only consume microscopic particles, primarily bacteria. Sponges circulate seawater through pores in their bodies ①. Specialized collar cells inside the sponge filter food from the water ② and ingest them using the process of phagocytosis ("cell eating"; see Chapter 5). Once ingested by a cell, the food is enclosed in a temporary sac called a **food vacuole** ③. The vacuole fuses with a **lysosome**, a membrane-enclosed packet of digestive enzymes within the cell ④. Food is broken down within the vacuole into smaller molecules that can be absorbed into the cell cytoplasm. Undigested remnants are expelled by exocytosis ⑤ and exit the sponge through an opening in its body wall ⑥.

The Simplest Digestive System Is a Chamber with One Opening

All animals except sponges have evolved a chamber within the body where chunks of food are broken down by enzymes that act outside the cells, a process called **extracellular digestion**. One of the simplest of these chambers is found in cnidarians, such as sea anemones, hydra, and sea jellies. The cnidarian digestive chamber is called a **gastrovascular cavity** and has a single opening through which food is ingested and wastes are ejected (**FIG. 35-9**). The cnidarian's stinging tentacles capture smaller animals and move them to the mouth and into the gastrovascular cavity ①. Gland cells lining the cavity secrete enzymes that begin digesting the prey ②. Nutritive cells lining the cavity then absorb the nutrients and also engulf food particles by phagocytosis. Further digestion is intracellular, within food vacuoles in the nutritive cells ③. Because undigested wastes are expelled back through the mouth, only one meal can be processed at a time.

Most Animals Have Tubular Digestive Systems with Specialized Compartments

A saclike digestive system is unsuitable for animals that must eat frequently. Most animals, including invertebrates such as mollusks, arthropods, echinoderms, and earthworms, as well as all vertebrates, have digestive systems that are basically one-way tubes that begin with a mouth and end with an anus. Specialized regions within the tube process the food in an orderly sequence, taking it in through the mouth, physically grinding it up, enzymatically breaking it down, absorbing the nutrients, and, finally, expelling the wastes through the anus.

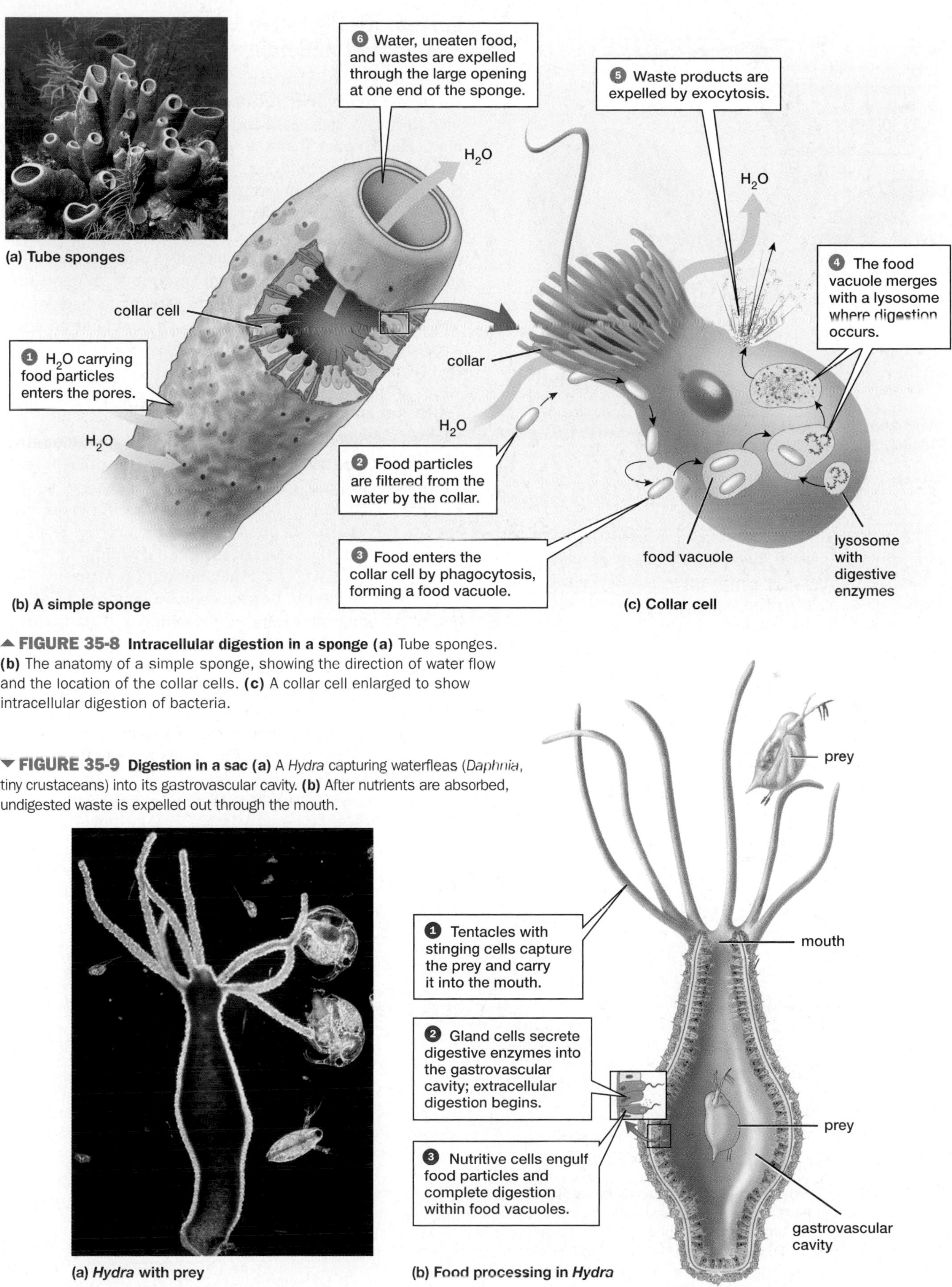
(a) Tube sponges
collar cell
1 H_2O carrying food particles enters the pores.
H_2O
6 Water, uneaten food, and wastes are expelled through the large opening at one end of the sponge.
H_2O
5 Waste products are expelled by exocytosis.
H_2O
4 The food vacuole merges with a lysosome where digestion occurs.
collar
H_2O
2 Food particles are filtered from the water by the collar.
3 Food enters the collar cell by phagocytosis, forming a food vacuole.
food vacuole
lysosome with digestive enzymes
(b) A simple sponge
(c) Collar cell
prey
mouth
1 Tentacles with stinging cells capture the prey and carry it into the mouth.
2 Gland cells secrete digestive enzymes into the gastrovascular cavity; extracellular digestion begins.
prey
3 Nutritive cells engulf food particles and complete digestion within food vacuoles.
gastrovascular cavity
(a) *Hydra* with prey
(b) Food processing in *Hydra*

▲ **FIGURE 35-8 Intracellular digestion in a sponge (a)** Tube sponges. **(b)** The anatomy of a simple sponge, showing the direction of water flow and the location of the collar cells. **(c)** A collar cell enlarged to show intracellular digestion of bacteria.

▼ **FIGURE 35-9 Digestion in a sac (a)** A *Hydra* capturing waterfleas (*Daphnia*, tiny crustaceans) into its gastrovascular cavity. **(b)** After nutrients are absorbed, undigested waste is expelled out through the mouth.

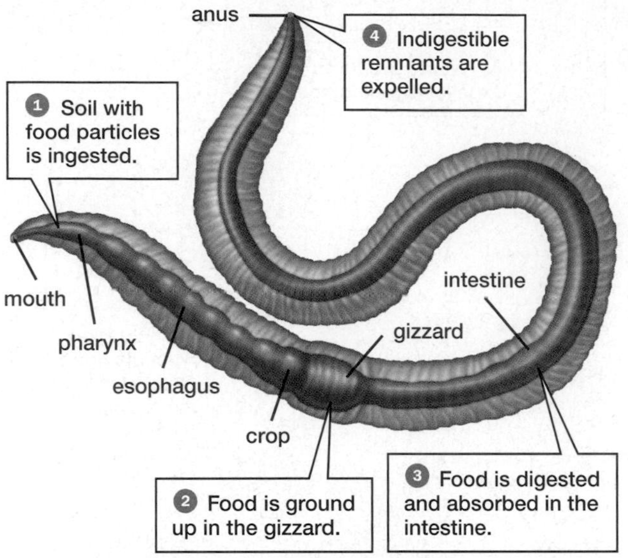

▲ **FIGURE 35-10 A tubular digestive system**

The earthworm provides a good example (**FIG. 35-10**). As it burrows, the worm ingests soil and bits of plant material that pass through the **esophagus**, a muscular tube leading from the mouth to the *crop*, an expandable sac where food is stored ❶. The material is then gradually released into the *gizzard*, where ingested sand grains and muscular contractions grind it into smaller particles ❷. In the intestine, enzymes attack the food particles, and the resulting small molecules are absorbed into the worm's body ❸. The remaining soil containing undigested organic material is expelled through the anus ❹.

Vertebrate Digestive Systems Are Specialized According to Their Diets

Different types of vertebrate animals have radically different diets. **Carnivores** (L. *carne*, flesh, and *vorare*, to devour), such as wolves, cats, seals, and predatory birds, eat other animals. **Herbivores** (L. *herba*, plant), which eat only plants, include seed-eating birds; grazing animals, such as deer, camels, and cows; and many rodents, such as mice. **Omnivores** (L. *omnis*, all), such as humans, bears, and raccoons, consume and are adapted to digest both animal and plant sources of food. Specialized digestive systems allow animals with different diets to extract the maximum amount of nutrients from their foods. The major organs of the vertebrate digestive system are the mouth, esophagus, stomach, small intestine, and large intestine—but some vertebrates have additional specialized chambers.

Teeth Accommodate Different Diets

Teeth are adapted to diet (**FIG. 35-11**). The varied, omnivorous diet of humans has selected for our distinctive set of teeth. We have thin, flat incisors for shearing food. Because we don't catch prey with our mouths, our canines are small and not very sharp. Our premolars and molars have relatively large, irregular surfaces for crushing and grinding.

If you have a dog (a cat may not put up with this), look carefully in its mouth. Carnivores have very small incisors, but greatly enlarged canines for stabbing and tearing flesh. Their molars and premolars have sharp edges for shearing through tendon and bone. None of their teeth are adapted for grinding or chewing; they tend to swallow their food in chunks.

Herbivores, such as horses, rabbits, deer, and cows, generally lack canines. Horses and rabbits have well-developed

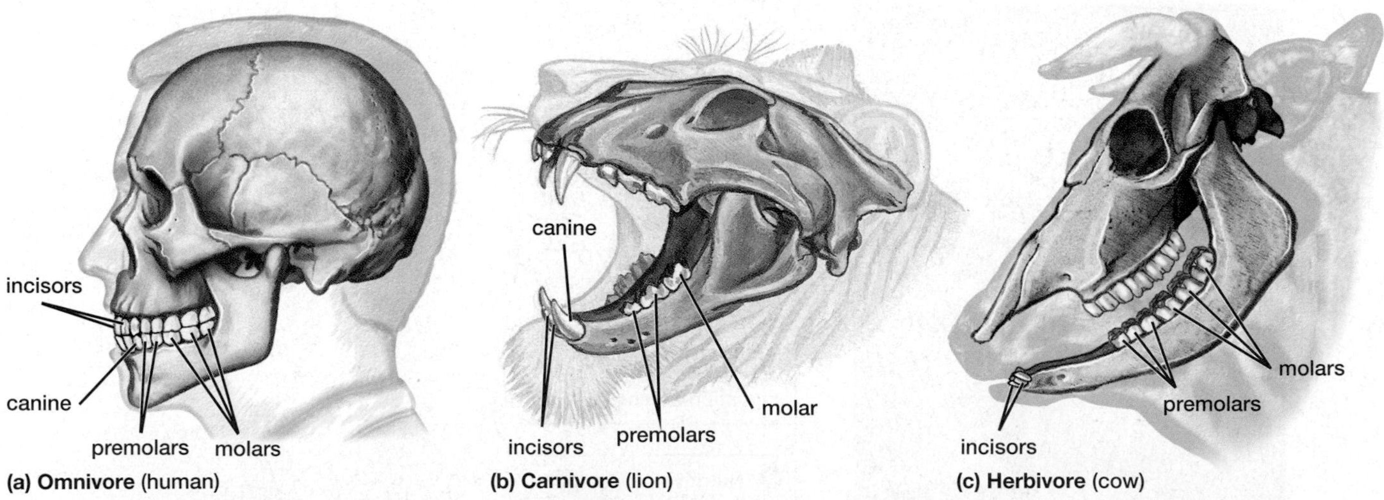

▲ **FIGURE 35-11 Teeth have evolved to suit different diets (a)** Humans have cutting incisors, reduced canines, and flattened premolars and molars for grinding up plant and animal food. **(b)** Carnivores have large canines for grasping and killing prey, reduced incisors, and premolars and molars adapted for cutting rather than grinding. **(c)** Herbivores usually lack canine teeth and have flattened premolars and molars for grinding up tough plant fibers.

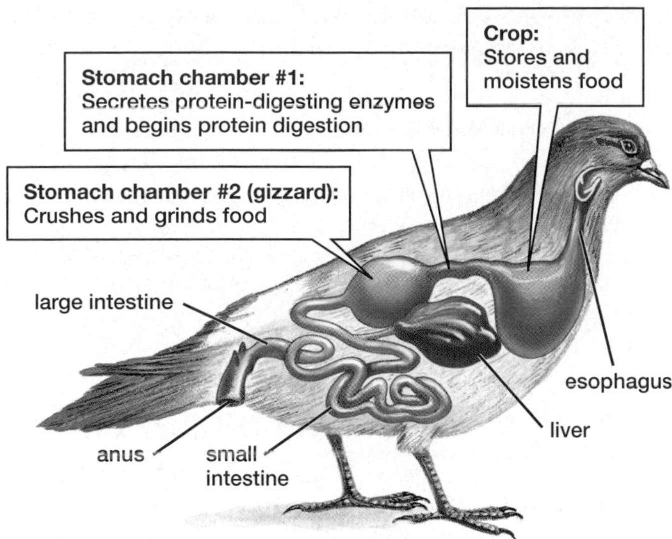
Crop:
Stores and moistens food
Stomach chamber #1:
Secretes protein-digesting enzymes and begins protein digestion
Stomach chamber #2 (gizzard):
Crushes and grinds food
large intestine
esophagus
liver
anus
small intestine

▲ **FIGURE 35-12 Bird digestive adaptations**

upper and lower incisors for snipping plants. In **ruminant** herbivores such as cows, who regurgitate and re-chew their partially digested food, the upper incisors are replaced by a tough, leathery *dental pad*. Ruminant herbivores pull vegetation into their mouths with large flexible tongues and bite it off between their lower incisors and dental pads. The premolars and molars of all herbivores have large flat surfaces for grinding up tough plant material.

Birds' Stomachs Grind Food

Birds lack teeth and swallow their food whole. In many birds, food is stored in an expandable crop (**FIG. 35-12**). The food then passes gradually into two stomach chambers. The tube-like first chamber secretes protein-digesting enzymes that begin protein breakdown, while the second, the gizzard, is a thick-walled, muscular, grinding chamber lined with ridges or plates made of the protein keratin (which also forms the bird's beak). The gizzard crushes and grinds food using muscular contractions. Many birds swallow sand or small stones that lodge in the gizzard and aid in the grinding process. From the gizzard, pulverized food particles are released into the small intestine, where they are further digested and their nutrients are absorbed.

Specialized Stomach Chambers Allow Ruminants to Digest Cellulose

The cellulose surrounding each plant cell is potentially one of the most abundant food energy sources on Earth; nevertheless, if humans were restricted to a cow's diet of grass, we would soon starve. Although cellulose, like starch, consists of long chains of glucose molecules, it resists the attack of vertebrate digestive enzymes because of the way bonds link its glucose molecules (see Chapter 3). Only certain microorganisms and a few invertebrates have enzymes that break down cellulose.

Most herbivorous vertebrates, including rabbits and horses and ruminants such as cows, sheep, goats, deer, elk, and camels, obtain energy from cellulose only because of their **microbiomes**. These specialized colonies of microorganisms in chambers within their digestive tracts help them digest plant material, with bacteria playing the largest role in cellulose breakdown.

Ruminants have multiple stomach chambers (**FIG. 35-13**). Digestion begins in the *rumen*. A cow rumen can hold nearly 40 gallons (about 150 liters), and it houses most of the microorganisms that break down cellulose and other carbohydrates. The microorganisms ferment the resulting sugars for energy. In the process, they release small organic molecules that supply at least half of the cow's energy needs; the cow absorbs most of these molecules through the rumen wall.

After partial digestion in the rumen, the semi-digested plant material enters the reticulum, where it forms masses called cud. The cud is regurgitated, chewed, and then swallowed back into the rumen. (Ruminant animals can often be seen placidly ruminating, or chewing their cud.) The extra chewing exposes more of the cellulose and cell contents to the rumen's microorganisms, which digest it further.

Gradually, the partially digested plant material and many microorganisms are released into the *omasum*, where water, salts, and the remaining small organic molecules released by the microorganisms are absorbed. The food then enters the

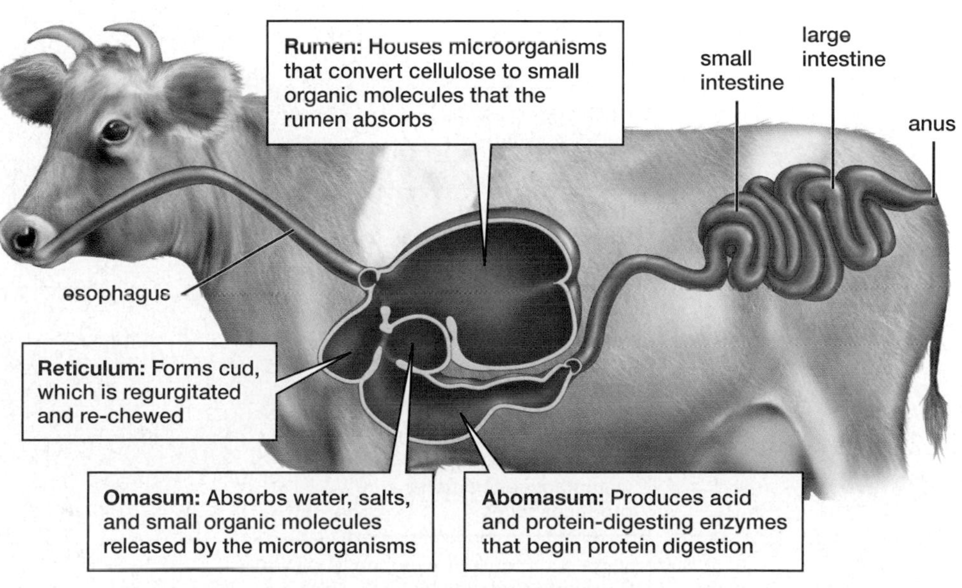
Rumen: Houses microorganisms that convert cellulose to small organic molecules that the rumen absorbs
small intestine
large intestine
anus
esophagus
Reticulum: Forms cud, which is regurgitated and re-chewed
Omasum: Absorbs water, salts, and small organic molecules released by the microorganisms
Abomasum: Produces acid and protein-digesting enzymes that begin protein digestion

▶ **FIGURE 35-13 Ruminants have a multichambered stomach**

abomasum, where acid and protein-digesting enzymes are secreted and protein digestion begins. Here, the cow digests not only plant proteins, but also the microorganisms that accompany the partially digested food from the rumen. The cow then absorbs most nutrients through the walls of its small intestine.

Small Intestine Length Is Correlated with Diet

Most chemical digestion and absorption of nutrients in vertebrates occur in the small intestine. Herbivores have relatively long small intestines, which provide more opportunity to extract nutrients from their diet of relatively hard-to-digest plants. Carnivores have shorter small intestines than herbivores because proteins are relatively easy to digest, and protein digestion begins in the stomach.

The link between diet and small intestine length is evident during frog development. A juvenile tadpole is an algae-eating herbivore with an elongated small intestine. As it metamorphoses into a carnivorous (usually insect-eating) adult frog, its small intestine shortens by about two-thirds.

CHECK YOUR LEARNING

Can you . . .

- list and describe the five basic tasks of all digestive systems?
- compare the various ways that invertebrates digest food?
- explain how vertebrate digestive systems are specialized for different diets?

35.3 HOW DO HUMANS DIGEST FOOD?

The human digestive system (**FIG. 35-14**) is adapted for processing the wide variety of foods in our omnivorous diet. Each compartment has particular digestive secretions associated with it, as summarized in **TABLE 35-3**.

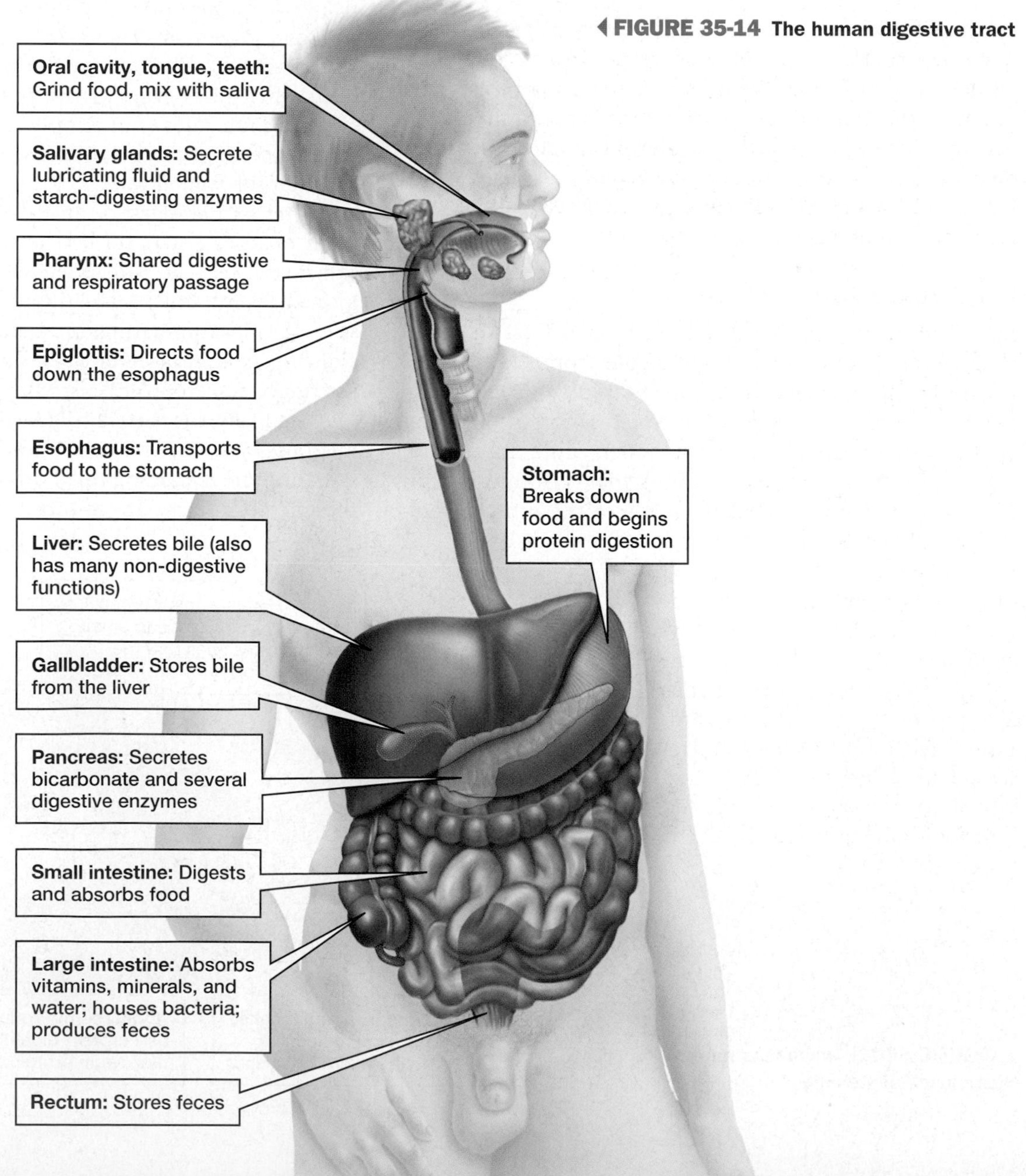
Oral cavity, tongue, teeth: Grind food, mix with saliva
Salivary glands: Secrete lubricating fluid and starch-digesting enzymes
Pharynx: Shared digestive and respiratory passage
Epiglottis: Directs food down the esophagus
Esophagus: Transports food to the stomach
Liver: Secretes bile (also has many non-digestive functions)
Gallbladder: Stores bile from the liver
Pancreas: Secretes bicarbonate and several digestive enzymes
Small intestine: Digests and absorbs food
Large intestine: Absorbs vitamins, minerals, and water; houses bacteria; produces feces
Rectum: Stores feces
Stomach: Breaks down food and begins protein digestion

FIGURE 35-14 The human digestive tract

TABLE 35-3 Digestive Secretions in Humans

Site of Action	Secreted Substance	Source of Secretion	Role in Digestion
Mouth	Salivary amylase	Salivary glands	Breaks down starch into disaccharide sugars
	Mucus, water		Lubricates and dissolves food
Stomach	Pepsin	Gastric glands in stomach	Breaks down proteins into peptides
	Hydrochloric acid		Allows pepsin to work; kills some bacteria; aids in mineral absorption
	Mucus	Cells lining the stomach	Protects the stomach from digesting itself
Small intestine	Bile	Liver	Emulsifies lipids
	Proteases	Pancreas	Break down peptides into shorter peptides and amino acids
	Lipase		Breaks lipids into fatty acids and glycerol
	Sodium bicarbonate		Neutralizes acidic chyme from the stomach
	Pancreatic amylase		Breaks starch into disaccharides
	Peptidase	Epithelial cells of small intestine	Breaks small peptides into amino acids
	Disaccharidases		Split disaccharides into monosaccharides
	Mucus		Protects the intestine from digestive secretions

Digestion Begins in the Mouth

As you take a bite of food, your mouth waters and you begin chewing; these activities begin the mechanical and chemical breakdown of food. While the teeth begin mechanical digestion by pulverizing food, the first phase of chemical digestion occurs as three pairs of salivary glands pour out saliva, which is over 99% water. Human salivary glands produce up to 1.5 quarts (about 1.5 liters) of saliva during waking hours (almost none at night); while eating, the average secretion rate increases by a factor of five. Saliva has many functions. Water and mucus in saliva lubricate food to ease swallowing. Antibacterial agents in saliva help to guard against infection. Saliva also contains the digestive enzyme **amylase**, which begins breaking down starches into disaccharides (sugars with two glucose subunits; see Chapter 3). The water in saliva dissolves some molecules, such as acids and sugars, exposing them to taste receptor cells, located in clusters called *taste buds*, on the tongue. Taste buds help to identify the type and the quality of the food.

The muscular tongue manipulates chewed food into a soft mass called a *bolus* and presses it back into the **pharynx**, a cavity that connects the mouth with the esophagus (**FIG. 35-15a**). The pharynx is shared with the respiratory system, which conducts air from the nose and mouth into the larynx and trachea. This arrangement occasionally causes problems, as anyone who has ever choked on a piece of food well knows (see Fig. 34-8). Normally, however, the **epiglottis**, a flap of tissue containing flexible cartilage, blocks the respiratory passage during swallowing and directs food into the esophagus (**FIG. 35-15b**).

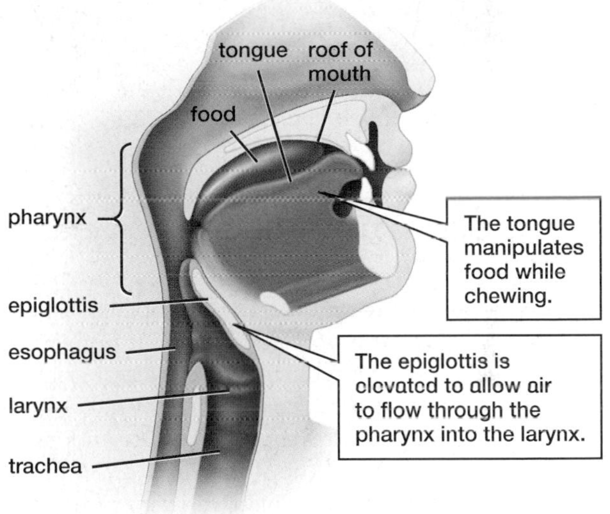
tongue
roof of mouth
food
pharynx
epiglottis
esophagus
larynx
trachea
The tongue manipulates food while chewing.
The epiglottis is elevated to allow air to flow through the pharynx into the larynx.
(a) Before swallowing

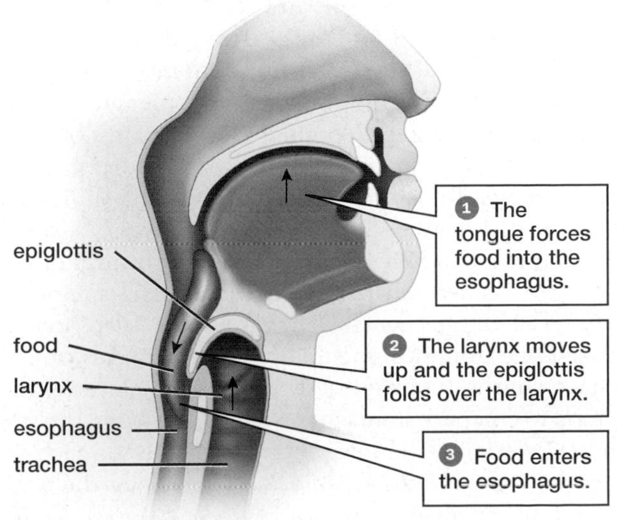
epiglottis
food
larynx
esophagus
trachea
1 The tongue forces food into the esophagus.
2 The larynx moves up and the epiglottis folds over the larynx.
3 Food enters the esophagus.
(b) During swallowing

FIGURE 35-15 The challenge of swallowing (a) Both the esophagus and the larynx emerge from the pharynx. **(b)** During swallowing, the larynx moves upward and the epiglottis folds down over it, directing food down the esophagus.

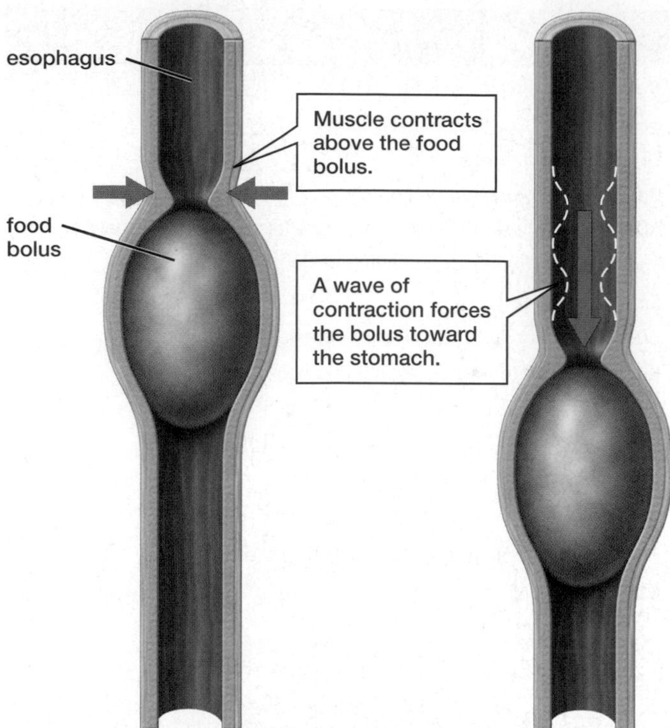

esophagus
Muscle contracts above the food bolus.
food bolus
A wave of contraction forces the bolus toward the stomach.

▲ **FIGURE 35-16** **Peristalsis in the esophagus**

The Esophagus Conducts Food to the Stomach, Where Digestion Continues

Swallowing forces food into the esophagus, a muscular tube that propels food from the mouth to the stomach. Mucus secreted by cells lining the esophagus protects it from abrasion and also lubricates the food during its passage. Muscles surrounding the esophagus produce waves of contraction called **peristalsis** that begin just above the bolus and progress along the esophagus toward the stomach (**FIG. 35-16**). Peristalsis occurs throughout the digestive tract, pushing food material through the esophagus, stomach, intestines, and, finally, out through the anus. Peristalsis is so effective that a person can actually swallow liquids when upside down.

The **stomach** in humans and other vertebrates is a muscular sac with a folded inner lining that allows it to expand so that we can eat rather large, infrequent meals (**FIG. 35-17**). Some carnivores take this ability to an extreme. A lion, for instance, may consume 40 pounds (18 kilograms) of meat at one meal and then spend the next few days digesting it. In adult humans, the stomach can comfortably hold about 2 pounds, or 1 quart (1 liter), although this varies with body size. Food is retained in the stomach by two rings of circular muscles, called **sphincter muscles**. The sphincter at the top, called the *lower esophageal sphincter*, keeps food and stomach acid from sloshing up into the esophagus while the stomach churns. It opens briefly just after swallowing, allowing food to enter the stomach. A second sphincter, the *pyloric sphincter*, separates the lower portion of the stomach from the upper small intestine. This muscle regulates the passage of food into the small intestine.

The stomach has four main functions. First, the muscular stomach walls produce churning contractions that break up chunks of food into smaller pieces that are more readily attacked by digestive enzymes. Second, the stomach begins protein breakdown using secretions from the gastric glands (see Table 35-3). Third, the stomach's gastric glands secrete hormones that regulate digestive activity. Fourth, the stomach stores and gradually releases partially digested food into the small intestine at a suitable rate to allow the small intestine to completely digest the food and absorb its nutrients.

The **gastric glands** are clusters of specialized epithelial cells that line millions of microscopic pits in the stomach lining. Gastric gland secretions include mucus, hydrochloric acid (HCl), and the protein pepsinogen. The hydrochloric acid gives the stomach fluid a very acidic pH of 1 to 3, about the same as lemon juice. This destroys many microbes, such as bacteria and viruses, that are inevitably swallowed along with food. Pepsinogen is the inactive form of *pepsin*, a type of **protease**—a protein-digesting enzyme that breaks proteins into shorter chains of amino acids called peptides. The stomach's acidity converts pepsinogen into pepsin (which works best in this acidic environment). Why not secrete pepsin in the first place? Gastric glands secrete inactive pepsinogen because pepsin would digest the very cells that synthesize it. Mucus, secreted by gastric gland cells and by epithelial cells throughout the stomach, coats the stomach lining and serves as a barrier to self-digestion. The protection, however, is not perfect, so cells of the stomach epithelium are replaced every few days. The digestive substances produced by the stomach can cause this organ to self-digest if its protective mucous barriers are breached. Indeed, this is what happens when a person develops ulcers, as described in "Doing Science: Identifying the Cause of Ulcers."

Food in the stomach is gradually converted to a thick, acidic liquid called **chyme**, which consists of digestive secretions and partially digested food. Peristaltic waves (about three per minute) then propel the chyme toward the small intestine, forcing about a teaspoon of chyme through the pyloric sphincter with each wave. Small chunks of food cannot

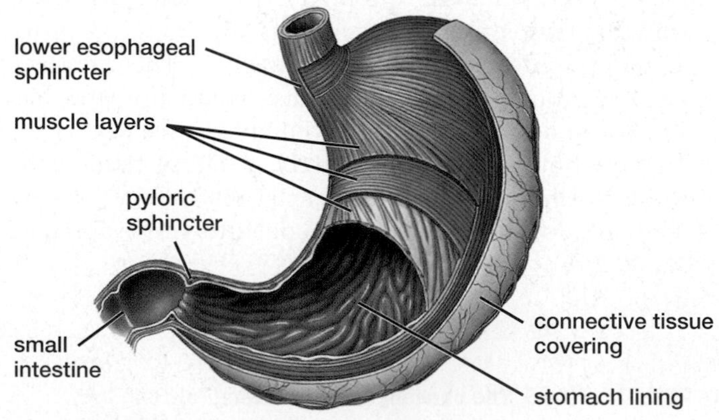

▲ **FIGURE 35-17** **The human stomach**

DOING SCIENCE Identifying the Cause of Ulcers

An open sore in the lining of the stomach or the adjacent duodenum of the small intestine is called an ulcer (**FIG. E35-1**). Ulcer victims can experience burning pain, nausea, and, in severe cases, bleeding. Before the 1990s, doctors believed that most ulcers were caused mainly by overproduction of stomach acid. Consequently, they treated their patients with antacids, a bland diet, and stress-reduction programs. Unfortunately, the ulcers commonly recurred when the treatment stopped. Researchers began to wonder whether something other than stomach acid is actually the root cause of ulcers.

What Question Was Asked?

In the 1980s, pathologist J. Robin Warren noticed that samples of inflamed stomach tissue were commonly infected with a spiral-shaped bacterium. He discussed his work with Barry Marshall, then a trainee in internal medicine, and they proposed that the bacterium, later named *Helicobacter pylori*, caused stomach inflammation and ulcers. The medical community was skeptical of this hypothesis because this bacterial species is also found in the stomachs of many people without ulcers.

How Was Evidence Gathered?

To test the hypothesis that *Helicobacter* causes ulcers, Warren and Marshall followed a protocol that researchers often use to find disease-causing microbes. First, confirm the presence of the bacterium in all animals infected with the disease. Second, grow the bacterium in culture. Third, infect experimental animals with the cultured bacterium and demonstrate that they develop the disease. Fourth, isolate and culture the identical type of bacterium from the diseased animals.

Marshall and Warren tried to grow *H. pylori* from the stomachs of ulcer patients by incubating the bacterium at body temperature in culture dishes on a nutrient medium, with no success. However, as is common in research, chance provided an opportunity for scientific insight. Lab technicians had been discarding cultures after 2 days if no growth was visible, but when a technician accidentally left discarded dishes in an incubator over a holiday break, the resulting 5-day-old cultures contained colonies of the slow-growing bacteria.

Next, after unsuccessfully attempting to infect piglets with the bacterium, a frustrated Marshall resorted to a dangerous approach: He experimented on himself. After undergoing an exam with an endoscope (a tiny camera threaded down his esophagus) that showed his stomach free of inflammation, Marshall swallowed a culture of about a billion *H. pylori* cells.

What Was Learned?

During the week after he infected himself, Marshall began feeling ill, and samples of his stomach lining showed it to be damaged, thinned, and heavily infected with *H. pylori.* (Antibiotics relieved Marshall's symptoms, presumably by killing the bacteria in his stomach.)

Marshall's experiment on himself was dangerous, had a sample size of just one, and was unlikely to be repeated by others. But it bolstered his hypothesis and set the stage for further research that eventually confirmed the hypothesis. Scientists now know that *H. pylori* colonize the protective mucus that coats the lining of the stomach and duodenum. In people who develop ulcers from *H. pylori,* the ulcers weaken the tissue and increase acid production, making the lining of the stomach and duodenum more susceptible to attack by stomach acid and protein-digesting enzymes. The body's immune response to the infection further contributes to the tissue destruction. In the United States, *H. pylori* causes about 90% of ulcers; most will be cured by a 2-week treatment with antibiotics. For their findings, based on careful observations, chance, and the scientific method, Warren and Marshall were awarded the Nobel Prize in Physiology or Medicine in 2005.

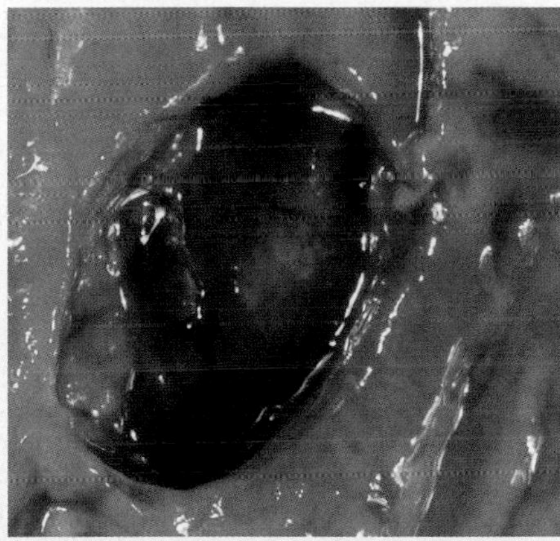

▲ **FIGURE E35-1 An ulcer viewed through an endoscope**

THINK CRITICALLY Why can antacids relieve ulcers?

pass through the sphincter and remain in the stomach for further breakdown. Depending on the amount and type of food eaten, it takes about 4 to 6 hours to empty the stomach after a meal. Churning movements of an empty stomach cause rumblings and hunger pangs.

Only a few substances, including alcohol and certain drugs, can enter the bloodstream through the stomach wall. Food in the stomach slows alcohol absorption, so following the advice "never drink on an empty stomach" will help reduce alcohol's intoxicating effects.

Most Digestion and Nutrient Absorption Occur in the Small Intestine

The **small intestine** is a long, muscular tube that receives chyme from the stomach, completes the chemical digestion of food molecules in the chyme, and absorbs the resulting nutrient molecules into the body. The small intestine consists of three segments: the duodenum, the jejunum, and the ileum. The short *duodenum* receives chyme from the stomach, receives digestive secretions from the gallbladder and

CASE STUDY CONTINUED

Dying to Be Thin

Stomach acid can be very destructive to the digestive tracts of people with bulimia, many of whom vomit several times daily. The strong acid in the stomach contents dissolves the protective enamel of teeth, making them extremely prone to decay. Stomach acid also damages tissues of the gums, throat, and esophagus. In extreme cases, the explosive pressure of repeated vomiting can tear or rupture the esophagus—a true medical emergency. Frequent vomiting weakens the stomach lining, allowing acid to attack the stomach wall, which may produce ulcers, and can even cause the stomach to rupture, a situation that can be fatal.

During normal digestion, food moves from the stomach to the small intestine—how is it processed there?

the pancreas, secretes hormones that help control digestion, and begins the absorption of nutrients. In the *jejunum* and the *ileum* (which empties into the large intestine), digestion and nearly all nutrient absorption are completed. Like the stomach, the small intestine is protected from digesting itself by mucus secreted by epithelial cells.

A Variety of Digestive Substances Are Found in the Small Intestine

After the stomach releases chyme into the small intestine, chemical digestion is accomplished with the aid of other digestive secretions from the liver, the pancreas, and cells lining the small intestine itself (**FIG. 35-18**).

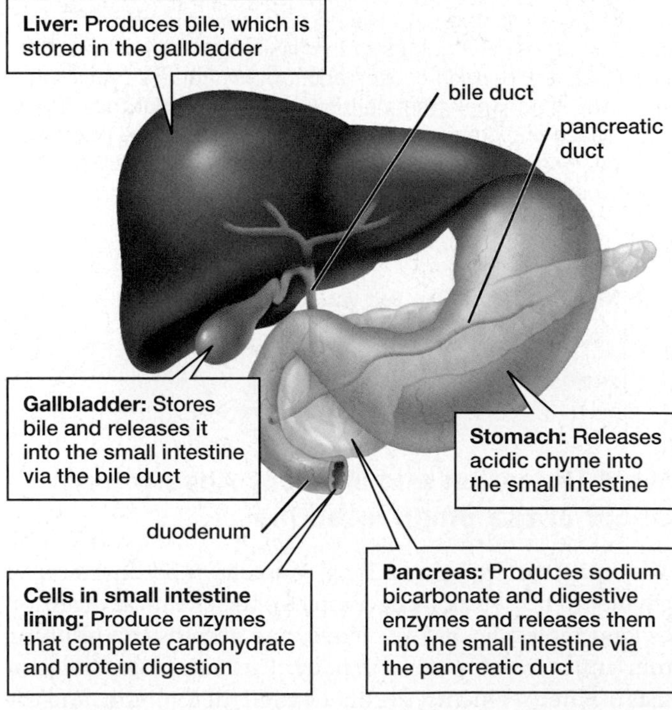
Liver: Produces bile, which is stored in the gallbladder
bile duct
pancreatic duct
Gallbladder: Stores bile and releases it into the small intestine via the bile duct
Stomach: Releases acidic chyme into the small intestine
duodenum
Cells in small intestine lining: Produce enzymes that complete carbohydrate and protein digestion
Pancreas: Produces sodium bicarbonate and digestive enzymes and releases them into the small intestine via the pancreatic duct

▲ **FIGURE 35-18 Sources of digestive secretions in the small intestine**

The Liver Produces Bile The **liver** is perhaps the most versatile organ in the body. This master of multitasking stores fats and carbohydrates for energy, regulates blood glucose levels, synthesizes blood proteins, stores iron and certain vitamins, converts toxic ammonia (released when amino acids are broken down) into urea, and detoxifies harmful substances such as nicotine and alcohol.

The liver also makes **bile**, a greenish liquid consisting primarily of water and bile salts, that breaks up lumps of fat. *Bile salts*, synthesized from cholesterol, have a hydrophilic end that is attracted to water and a hydrophobic end that interacts with fats. They disperse lipids into microscopic particles in the watery chyme, much as dish detergent disperses fat from a frying pan. The particles expose a large surface area to attack by **lipases**, lipid-digesting enzymes.

Bile synthesized in the liver is stored and concentrated in the **gallbladder**. Hormonal signals from the duodenum in response to the entry of chyme cause the gallbladder to contract and expel bile through the *bile duct*, which empties into the duodenum (see Fig. 35-18).

The Pancreas Supplies Pancreatic Juice The **pancreas** lies in the loop between the stomach and small intestine (see Fig. 35-18). In addition to releasing hormones that help regulate blood sugar (see Chapter 38), the pancreas produces a digestive secretion called **pancreatic juice**. In response to hormones secreted by the duodenum as chyme enters it, the pancreas sends pancreatic juice into the duodenum through the *pancreatic duct*. Pancreatic juice is a mixture of water, sodium bicarbonate (which neutralizes the acidic chyme), and several digestive enzymes, including pancreatic amylase, lipase, and proteases (see Table 35-3). Pancreatic digestive enzymes work best in the slightly alkaline (basic) environment created by sodium bicarbonate in the pancreatic juice. Pancreatic amylase breaks down carbohydrates, lipase attacks lipids, and proteases split proteins and peptides, whose breakdown began in the stomach.

Enzymes in the Intestinal Wall Complete Digestion The epithelial cells of the small intestine contain enzymes such as *peptidases*, which break down peptides into amino acids. They also contain *disaccharidases* that split disaccharide sugars into monosaccharides. Lactase, for example, breaks down lactose (milk sugar) into glucose and galactose (see Chapter 6). As they are formed, these subunits are absorbed into the epithelial cells.

Most Absorption Occurs in the Small Intestine

In addition to being the principal site of digestion, the small intestine is also the major site of nutrient absorption into the body.

The Intestinal Lining Provides a Huge Surface Area for Absorption In a living human adult, the small intestine is about 1 inch (2.5 centimeters) in diameter and averages 10 feet (3 meters) in length; in a cadaver, the length can double due to loss of muscle tone. In addition to being quite long, the small intestine has numerous folds and projections, giving it an internal surface area that is roughly 600 times that of a smooth tube of the same length (**FIG. 35-19**). Tiny protrusions called

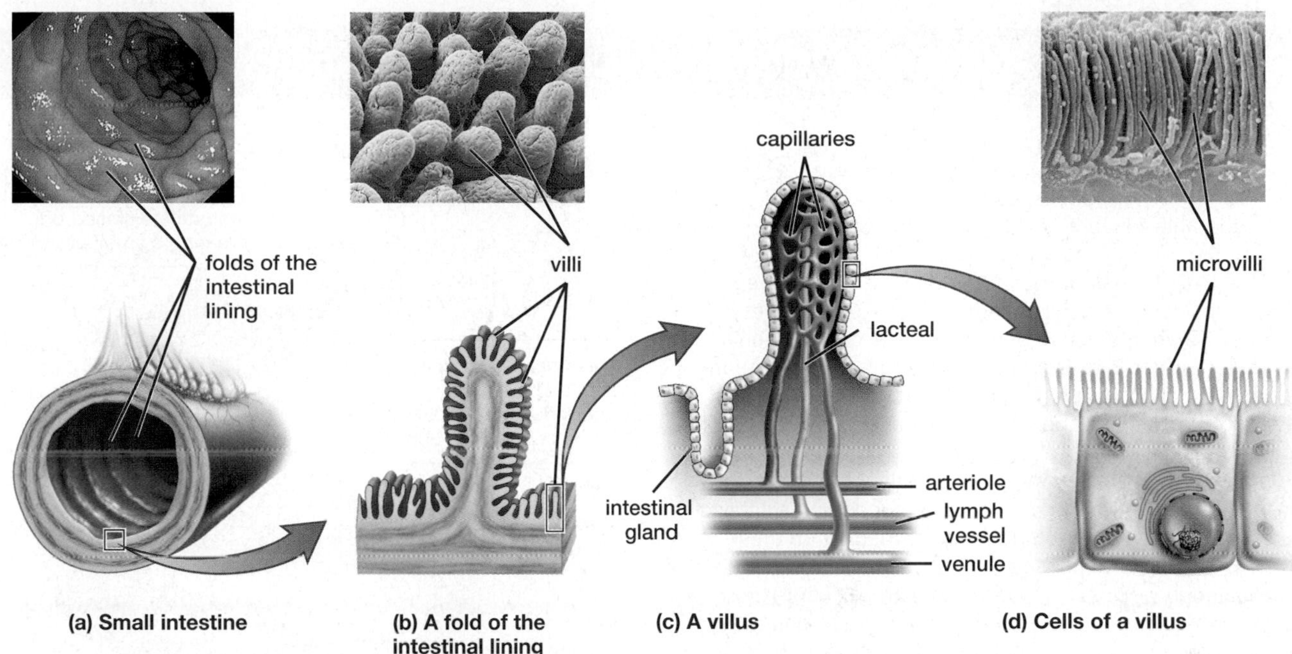

▲ **FIGURE 35-19 The structure of the small intestine (a)** Visible folds in the intestinal lining are carpeted with **(b)** tiny projections called villi. **(c)** Each villus contains a network of capillaries and a central lymph capillary called a lacteal. Most digested nutrients enter the capillaries, but fats enter the lacteal. **(d)** Intestinal epithelial cells bear microvilli.

THINK CRITICALLY What might the anatomy of the digestive system be like if the internal folds, villi, and microvilli of the small intestine had not evolved?

villi (singular, villus; from Latin, meaning "hair") cover the entire folded surface of the intestinal wall. Villi, which are about 1/25th of an inch (about 1 millimeter) long, make the intestinal lining appear velvety to the naked eye. Finally, the plasma membranes of the epithelial cells are folded into **microvilli** that are studded with digestive enzymes. Taken together, the specializations of the lining of the adult small intestine give it a surface area of about 2,700 square feet (about 250 square meters), almost the size of a doubles tennis court.

Unsynchronized contractions of the circular muscles of the small intestine, called *segmentation movements*, slosh the chyme back and forth, homogenizing the mix of food and digestive juices and bringing nutrients into contact with the enormous absorptive surface of the small intestine. When absorption is complete, coordinated peristaltic waves conduct the leftovers into the large intestine.

Nutrients Are Absorbed Through Various Pathways Nutrients absorbed by the small intestine include water, monosaccharides, amino acids and short peptides, fatty acids, vitamins, and minerals. Some nutrients enter the cells lining the small intestine by diffusion and others by active transport. Water follows by osmosis. The water and most other nutrients then enter the network of blood capillaries located within each of the villi.

Fatty acids released by the digestion of fats and oils take a distinctive pathway. Clustered together with cholesterol, they diffuse directly through intestinal epithelial cell membranes. Inside the epithelial cells, these substances are assembled and coated with proteins to form particles called **chylomicrons**, which are then released into the interstitial fluid. The chylomicrons are too large to enter blood capillaries and instead diffuse through the porous wall of the **lacteal**, a lymph capillary that ends blindly within each villus (see Fig. 35-19c). Chylomicrons are then transported in lymph by the lymphatic system, which eventually empties into a large vein near the heart (see Fig. 33-18). Excess fat can accumulate to health-threatening levels, and some obese people choose surgery to lose weight, as described in "Health Watch: Overcoming Obesity: A Complex Challenge" on page 624.

Water Is Absorbed and Feces Are Formed in the Large Intestine

The **large intestine** in a living adult human is about 2.5 inches (6.5 centimeters) in diameter and about 5 feet (1.5 meters) long, wider but far shorter than the small intestine (see Fig. 35-14). Most of the large intestine is called the **colon**; its final 6-inch chamber is the **rectum**. The leftovers of digestion and absorption in the small intestine—indigestible fiber, small amounts of unabsorbed nutrients, and water—flow into the large intestine. The large intestine is home to a flourishing population of bacteria. Some of the bacteria in the large intestine earn their keep by synthesizing the B vitamins B_1, B_2, B_{12}, and folic acid as well as vitamin K (a typical human diet would be deficient in vitamin K without them).

Overcoming Obesity: A Complex Challenge

Patrick Deuel's doctor gave him an ultimatum: Lose weight or die. At 1,072 pounds, Deuel needed a wall to be removed to be able to leave his bedroom. Although Deuel's case is extreme, obesity is a growing epidemic. Obese individuals have a higher risk of liver disease, gallstones, sleep apnea (interruptions of breathing during sleep), type 2 diabetes (see Chapter 38), some cancers, arthritis, and cardiovascular disease (see Chapter 33). Even recognizing the dangers, many people repeatedly fail in their weight-loss attempts. Why is it so difficult to lose weight, and why do people differ so much in their ability to control their weight?

The maintenance of body weight is an enormously complex homeostatic process involving multiple hormones and signaling molecules. Individuals differ both in their production of these molecules and in their responses to them. Many genes have been implicated in weight regulation, and complex interactions among different alleles of these genes contribute to enormous differences in the propensity of individuals to gain weight. One weight-regulating gene is *FTO*. One out of six people is homozygous for a particular allele of the *FTO* gene. These individuals (*AA*) eat more than average and are far more likely to be obese. Immediately after a meal, when they should be satiated, *AA* individuals maintain higher levels of the hunger-stimulating hormone ghrelin and respond more positively to images of high-calorie food than do individuals homozygous for a different *FTO* allele. In short, *AA* individuals tend to be chronically hungrier than average. Findings such as these help explain why many people fail to lose weight or fail to maintain hard-earned weight loss.

The difficulty that dangerously obese people face in trying to diet their way to better health has spurred development of surgical fixes that act directly on the digestive tract. Despite the risks involved, doctors may recommend surgery in extreme cases in which remaining obese would pose even greater health risks than the surgery. Patrick Deuel underwent *gastric bypass surgery*, the most commonly performed weight loss surgery in the United States. This procedure staples off most of the stomach, leaving only a small pouch which is connected directly to the small intestine below the duodenum (**FIG. E35-2**). As a result, only very small amounts of solid food can be consumed at a sitting, and the absorptive area of the small intestine is reduced. The surgery may reduce ghrelin levels and can cause dramatic improvements in type 2 diabetes. Many have benefited from this surgery; although Deuel remains morbidly obese, he has reduced his body weight by about half.

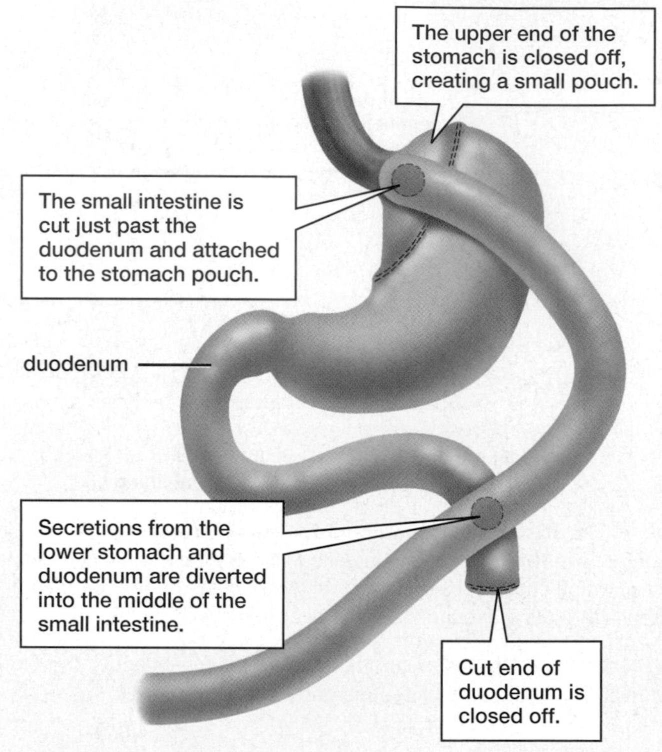

The upper end of the stomach is closed off, creating a small pouch.
The small intestine is cut just past the duodenum and attached to the stomach pouch.
duodenum
Secretions from the lower stomach and duodenum are diverted into the middle of the small intestine.
Cut end of duodenum is closed off.

▲ **FIGURE E35-2 Gastric bypass surgery**

An entirely new approach to weight loss, called VBLOC therapy, was approved by the U.S. FDA in 2014. An implanted pacemaker-like device uses electrical signals to interrupt communication between the brain and stomach along the vagus nerve, reducing sensations of hunger.

THINK CRITICALLY A physical exam reveals that a 5'10", 45-year-old, 410-pound man has type 2 diabetes, hypertension, and arthritic knees. The patient has a large and smoothly distended abdomen as well as rolls of fat. What would you recommend, and why?

The Large Intestine Produces Feces

Epithelial cells of the large intestine secrete lubricating mucus and absorb the newly synthesized vitamins, leftover salts, and much of the water, compacting the remaining material into semisolid **feces**. Feces consist primarily of residual water, indigestible fiber, mucus, some unabsorbed nutrients, and a host of bacteria, which make up about one-third of fecal dry weight. Feces are transported by peristaltic movements until they reach the rectum, the final chamber of the large intestine. Expansion of this chamber stimulates the urge to defecate. The anal opening is controlled by two sphincter muscles—an inner one that is under involuntary control and an outer muscle that can be consciously controlled. Thus, although defecation is a reflex, it comes under voluntary control in early childhood—to the delight of parents.

The Large Intestine Hosts an Extensive Bacterial Ecosystem

The trillions of cells that make up your body are complemented by a roughly equal number of bacterial cells on and in your body, with the largest bacterial population residing in your large intestine. This assemblage of microorganisms, called the microbiome, differs based on your sex, genes, environment, and diet. The diverse intestinal microbiome helps control immune responses, prevents some disease-causing bacteria from establishing themselves, and releases both cancer-fighting and cancer-promoting substances from food. In addition, these bacteria convert some indigestible food fiber into small fatty acids that contribute 4% to 10% of your dietary calories.

A rapidly expanding area of research seeks to unravel the influence of microbial gut residents on obesity. In general, obese people harbor a lower diversity of gut bacteria than is found in lean subjects—but is this a cause or an effect of obesity? It is possible that different proportions of bacterial types in individuals may affect how much energy different people extract from the same diet. There is some evidence that experimental mice lacking gut microbiomes will gain fat if they are colonized with gut microbiomes from obese, but not from lean, mice. Although some tantalizing headlines suggest that modifying the gut microbiome might be a magic bullet for weight loss, the field is still in its infancy, and the complex individuality of people makes this unlikely to ever become a miracle cure—although it may someday help with weight control.

Digestion Is Controlled by the Nervous System and Hormones

As you begin eating a meal, your body is coordinating a complex series of events that will convert the meal into nutrients circulating in your blood. The secretions and muscular activity of the digestive tract are coordinated by both nerve signals and hormones.

Food Triggers Nervous System Responses

Especially if you are hungry, the sight, smell, taste, and even the thought of food stimulates the hypothalamus, which controls many responses of the nervous system that help maintain homeostasis (see Chapter 32). The hypothalamus stimulates nervous pathways that prepare the digestive system to process food. For example, salivation increases, and the stomach produces more acid and protective mucus. As food moves through the digestive tract, its bulk activates stretch receptors. The stretch receptors stimulate local nervous reflexes that cause peristalsis and segmentation movements.

Hormones Help Regulate Digestive Activity

Hormones secreted by the digestive system enter the bloodstream and circulate through the body, acting on specific receptors within the digestive tract or brain. Like most hormones, digestive hormones are regulated by negative feedback. For example, nutrients in chyme, particularly amino acids and peptides from protein digestion, stimulate gastric glands in the stomach lining to release the hormone **gastrin** into the bloodstream. Gastrin travels back to the stomach and stimulates further acid secretion, which promotes protein digestion. When the chyme becomes acidic, this inhibits gastrin secretion, reducing acid production.

Secretin and **cholecystokinin** are digestive hormones released into the bloodstream by cells of the duodenum in response to the acidity of chyme and the nutrients—particularly peptides and fats—in chyme. These two hormones help regulate the chemical environment within the small intestine and the rate at which the chyme enters the small intestine, promoting optimal digestion and absorption of nutrients. Together, secretin and cholecystokinin reduce stomach acid production, preventing excess acidity. They also slow peristaltic stomach contractions, reducing the rate at which chyme is forced into the small intestine; this allows more time for digestion and absorption to occur. They increase bile production by the liver and bile release from the gallbladder. Finally, secretin and cholecystokinin increase the production and release of pancreatic juice into the small intestine.

Hormones Regulate Hunger

Scientists are discovering an ever-increasing array of hormones related to fat storage and hunger; some of these act on the brain, and many contribute in complex ways to a person's weight. Two important appetite-regulating hormones are leptin and ghrelin.

Leptin, a peptide secreted by fat cells, helps to regulate fat storage in mammals (**FIG. 35-20**). Leptin secretion increases when fat stores increase and decreases when fat stores fall below an optimum level. Acting on the hypothalamus of the brain, increased leptin decreases hunger and increases metabolic rate (which would cause loss of stored

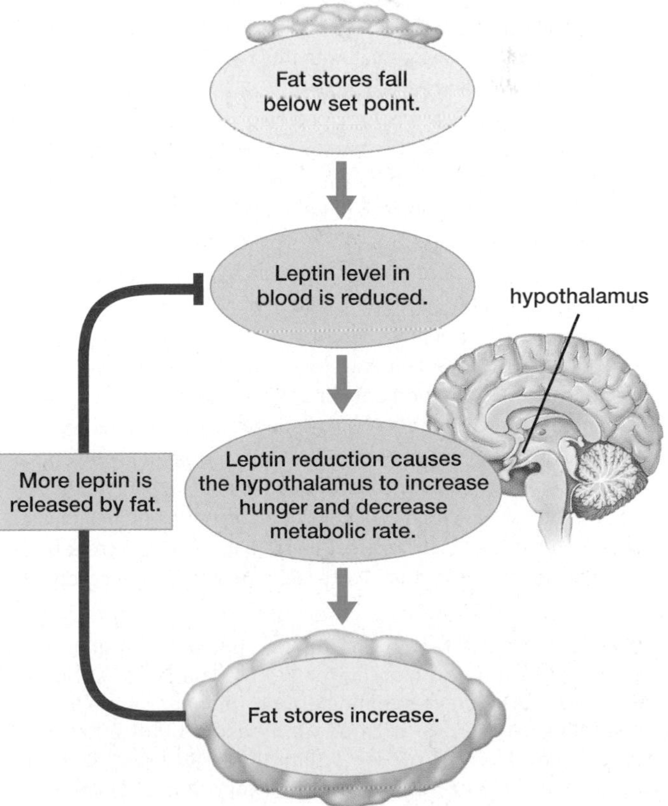

▲ **FIGURE 35-20 Leptin helps maintain fat stores**

fat), and decreased leptin increases hunger and decreases metabolic rate (which would cause increased fat storage). Unfortunately for a modern dieter, the weight-loss effects of increased leptin are minor, whereas the weight-gain effects of decreased leptin are much greater, which makes evolutionary sense. Few prehistoric people had easy access to surplus food, so storing fat when food was available was highly advantageous, whereas even small amounts of weight loss could cause increased susceptibility to cold, lack of endurance, and (in women) inability to adequately nourish a developing fetus or nursing infant.

Ghrelin is a peptide secreted by gastric gland cells in the stomach lining when the stomach is empty. An increase in circulating ghrelin occurs prior to mealtime, triggering release of substances that act on the hypothalamus to stimulate hunger. Ghrelin release diminishes after a meal, reducing hunger. In this way, ghrelin exerts short-term control over food intake. Like essentially all hormones, ghrelin has multiple roles in the body. Not surprisingly, pharmaceutical researchers are seeking safe ways to block ghrelin's ability to stimulate hunger while retaining its beneficial roles.

CHECK YOUR LEARNING

Can you . . .

- trace the path of food through the human digestive tract, describing the structures and functions of each digestive compartment?
- explain how the small intestine absorbs nutrients and how its structure facilitates nutrient absorption?
- describe feces and how they are produced and eliminated?
- explain how the nervous system and hormones regulate digestion and food intake?

CASE STUDY REVISITED

Dying to Be Thin

The causes of eating disorders are complex and poorly understood. Genes play a role; people with a close relative who has an eating disorder are about five times more likely to develop such a disorder themselves. Psychological problems (such as anxiety and depression), a history of emotional or sexual trauma, and personality traits (such as perfectionism, low self-esteem, and a high need for achievement) seem to predispose people to eating disorders. Cases often begin during adolescence, when bodies and brains are undergoing rapid changes. Magazines, TV, and Internet ads bombard susceptible young people with the message that being thin is a route to beauty and acceptance.

Isabelle Caro was among those who fell victim to an eating disorder (**FIG. 35-21**). Caro, a French model who died at age 28, had suffered from anorexia since the age of 13. She is best known for her efforts to help raise awareness and help other victims of this disorder by posing nude for billboards in Italy featuring the words: "No—Anorexia." The year before Caro went public with her battle, the weight of her 5-foot, 4-inch body had dipped to 55 pounds. Determined to fight her disorder, she struggled to make herself eat. Sadly, her efforts came too late for her ravaged body, and she died 2 years later.

Anorexia and bulimia are difficult to overcome. Victims are given nutritional therapy to help them recover from malnutrition. Psychotherapy is usually necessary, and antidepressant drugs may be helpful. Because many victims hide or deny their problems, and because treatment is expensive, the majority of sufferers are inadequately treated; fewer than half experience complete recovery.

CONSIDER THIS Media and fashion houses that glamorize slimness have been blamed for the prevalence of eating disorders. Why do you think extreme thinness is made so appealing? Are there appropriate measures that a free society can take to reverse or limit this message? Do you support the idea of a minimum BMI for models?

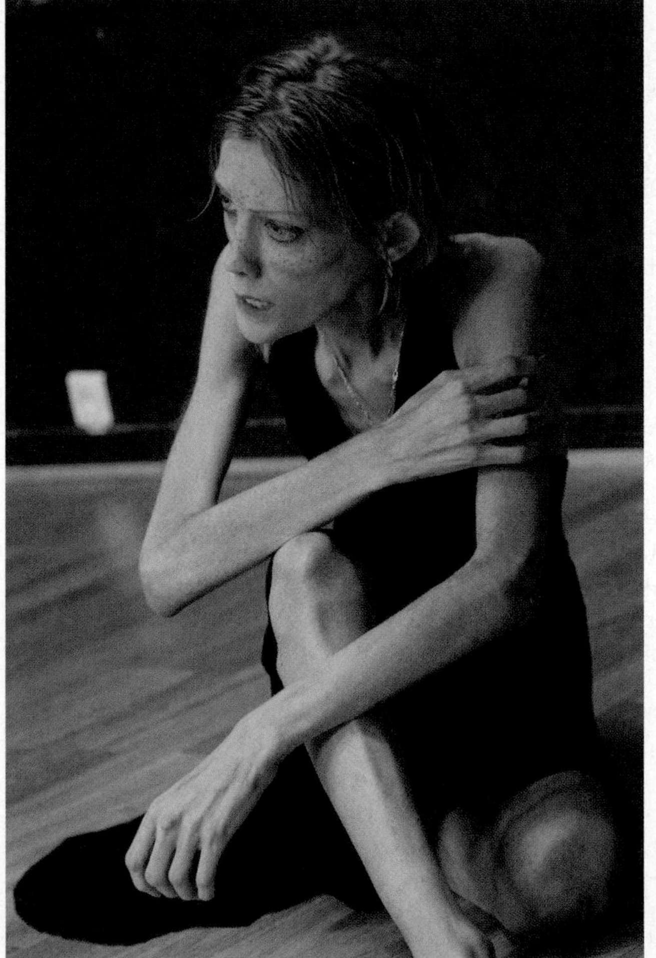

▲ **FIGURE 35-21 Anorexia** The model Isabelle Caro 2 years before her death.

CHAPTER REVIEW

Go to **MasteringBiology** for practice quizzes, activities, eText, videos, current events, and more.

Answers to ***Think Critically*** *and* ***Thinking Through the Concepts*** *questions can be found in the* ***Answers*** *section at the back of the book.*

Summary of Key Concepts

35.1 What Nutrients Do Animals Need?

All animals require nutrients that provide energy; for humans, these are primarily carbohydrates and lipids, with a small percent derived from protein. Food energy is measured in Calories. Excess energy from food is stored in body fat, the most concentrated energy source. Each type of animal has specific require ments for essential nutrients that it cannot synthesize, but that are required for cellular function. For humans, these include essential fatty acids, essential amino acids, minerals, vitamins, and water.

35.2 How Does Digestion Occur?

Digestion is the mechanical and chemical breakdown of food that converts complex molecules into simpler molecules that can be absorbed and used by the organism. In sponges, digestion is entirely intracellular. Digestive systems must accomplish five tasks: ingestion of food, mechanical digestion, chemical digestion, absorption of nutrients, and elimination of wastes. The simplest digestive system is the saclike gastrovascular cavity in organisms such as *Hydra*. Most digestive systems consist of a one-way tube along which specialized compartments process food in an orderly sequence. Specialized digestive systems allow different animals to utilize a wide variety of foods.

35.3 How Do Humans Digest Food?

In humans, digestion begins in the mouth, which begins both the physical breakdown of food (chewing) and its chemical breakdown (saliva). Swallowing followed by peristalsis in the esophagus directs food to the stomach, which continues mechanical breakdown and begins protein digestion. The resulting chyme is released into the small intestine, where secretions from the pancreas and liver, and cells of the intestinal epithelium, complete the chemical breakdown of proteins, fats, and carbohydrates. The simple nutrient molecules enter the epithelial cells and are released into blood or (for lipids) lymph capillaries from which they enter the bloodstream. The colon of the large intestine absorbs water, salts, and vitamins manufactured by intestinal bacteria. The remaining feces are conducted to the rectum; distention of the rectum triggers the urge to defecate. Digestive secretions are summarized in Table 35-3.

Digestion is regulated by the interaction of the nervous system and hormones. The hypothalamus of the brain helps maintain homeostasis in part by regulating food intake, as it responds to the hunger-regulating hormones ghrelin (produced by stomach cells) and leptin (secreted by adipose tissue).

Thinking Through the Concepts

Bloom's: Remembering, Understanding

Multiple Choice

1. Which of the following is false?
   a. Carbohydrates are the major energy storage molecule of plants.
   b. Carbohydrates are the major source of rapidly available energy stores in people.
   c. Carbohydrates include cellulose.
   d. Carbohydrates are hydrophobic and don't retain water in the body.
2. Which of the following is true?
   a. Most fat-soluble vitamins serve as coenzymes.
   b. Water-soluble vitamins are vitamin C and the B complex vitamins.
   c. Water soluble vitamins accumulate in adipose tissue.
   d. Most vitamins can be synthesized by the body.
3. Which of the following occurs in the small intestine?
   a. the first stage of starch breakdown
   b. the initial breakdown of proteins into peptides
   c. fat absorption
   d. vitamins production by bacteria
4. Which of the following is true?
   a. The human stomach maintains a very high pH.
   b. Gastric glands in the stomach wall secrete pepsinogen.
   c. Bile is released into the stomach from the liver.
   d. Many different nutrients are absorbed through the stomach.
5. Which of the following is false?
   a. Segmentation movements conduct food along the digestive tract.
   b. Vitamin K is produced by bacteria in the large intestine.
   c. Expansion of the rectum helps stimulate defecation.
   d. Bacteria contribute about one-third of the dry weight of feces.

Fill-in-the-Blank

1. Two general roles for nutrients are to provide __________ and __________. Nutrients required in small amounts that often assist in enzyme function are called __________. Nutrients that are elements are called __________. Nutrients that cannot be synthesized by the body are called __________ nutrients.
2. Sponges rely exclusively on __________ digestion. Cnidarians such as *Hydra* digest food in a(n) __________. Earthworms have a(n) __________ digestive system. Ruminants can break down __________ only because of microorganisms in their stomachs.
3. The enzyme called __________ is present in saliva and begins breaking down __________. Digestion of __________ begins in the stomach. Stomach acid converts the inactive substance __________ into the active enzyme __________. Nearly all fat digestion occurs in the __________.

4. The five major processes carried out by the digestive systems are ____________, ____________, ____________, ____________, and ____________.
5. In humans, a cavity called the ____________ is shared by the ____________ and ____________ systems. A flap called the ____________ prevents food from entering the trachea during the act of ____________. Food is pushed through the digestive system by muscular contractions called ____________. Circular muscles that control movement into and out of organs such as the stomach are called ____________.
6. Fats are dispersed by a secretion called ____________. This secretion is produced by the ____________ and is stored in the ____________. Products of fat digestion are formed into chylomicrons inside intestinal ____________ cells. The chylomicrons are released inside the villus and diffuse through the wall of the ____________, which is part of the ____________ system.
7. Fill in the hormones. Release is stimulated by amino acids and peptides in the stomach: ____________; increased secretion causes hunger: ____________; released by fat cells: ____________; reduce stomach acid production: ____________ and ____________; produced by cells in stomach gastric glands: ____________ and ____________; stimulate pancreatic juice secretion: ____________ and ____________.

Review Questions

1. List the six general types of nutrients. Which two provide the most energy for humans? Which of these stores the most energy in the body, and why?
2. Give an example of a vertebrate and an invertebrate that both use gizzards. Describe the general structure and function of the gizzard and how it works.
3. Vertebrates can be grouped into three categories based on their diets; list and briefly define these categories, giving one example of each. Which group has the shortest small intestine, and why?
4. Why is the stomach both muscular and expandable?
5. List and describe the functions of the secretions of the gastric glands.
6. List the digestive substances secreted into the small intestine, and describe the origin and function of each.
7. Name and describe the muscular movements of the human digestive system, where they occur, and their functions.
8. Vitamin C is an essential vitamin for humans but not for dogs. Certain amino acids are essential for humans but not for plants. Explain.
9. Name four structural or functional adaptations of the human small intestine that contribute to effective digestion and absorption.
10. Describe the processing of a piece of steak, starting in the mouth and ending with the absorption of individual amino acids.
11. Explain the composition of feces and how the large intestine forms and expels them.
12. Describe how and where the three important digestive system hormones described in this chapter coordinate digestion.

Applying the Concepts

Bloom's: Applying, Analyzing, Evaluating

1. Wanting to lose weight, Jason decides to use artificial sweetener in the cookies he's baking. Since the sweetener is hundreds of times as sweet as sugar, the amount he uses to replace the 2 cups of sugar in his cookie recipe is only 1 teaspoon. His friend smirks and uses sugar to make the same recipe. The other ingredients are the same: 1.5 cups of butter and 3 cups of flour. If Jason and his friend's cookies are each the same size, how do Jason's cookies compare in calories? Explain.
2. When leptin was discovered, researchers were disappointed to discover that excess leptin in the blood does not reduce hunger or food intake. Suggest what caused their initial hope, and suggest an evolutionary explanation for leptin's role in fat regulation.
3. Discuss how modern processed foods are contributing to an epidemic of obesity. In a typical day, what stimuli do you encounter that tend to make you (or your friends) overeat?

36 The Urinary System

Barbara Asofsky hugs Good Samaritan Anthony DeGiulio, whose kidney restored her health and started a chain of kidney donations.

CASE STUDY

Paying It Forward

Anthony DeGiulio had a dream—he wanted to save someone's life. At first, it was a nebulous ambition, but it started to take shape as he watched a segment of a TV show that highlighted living kidney donation. DeGiulio, who hadn't realized that a living person could donate a kidney, immediately saw this as a way to achieve his ambition. He called New York-Presbyterian Hospital and initiated a series of events that gave new life not just to one person, but to four. The "domino chain" of events was possible because three people in the area needed kidneys, and each had a family member who was eager to donate but was not able to do so because of incompatible tissue types.

Barbara Asofsky, a nursery school teacher, had known for 5 years that she would need a kidney transplant. When DeGiulio's tissue type was found to be a good match for Asofsky, her husband, Douglas, was happy to donate the kidney he had hoped to give to his wife—but couldn't because of tissue incompatibility—to a stranger instead. The fortunate stranger was Alina Binder, a student at Brooklyn College. Alina's father, Michael, was a good match for Andrew Novak, a telecommunications technician. Finally, Andrew's sister, Laura Nicholson, donated her kidney to Luther Johnson, a hotel kitchen steward.

Do kidney transplant recipients lead completely normal lives? Why are kidneys so important? What is their function, and how do they work? Do all animals have kidneys?

AT A GLANCE

36.1 WHAT ARE THE MAJOR FUNCTIONS OF URINARY SYSTEMS?

Urinary systems are organ systems that excrete cellular wastes and help maintain homeostasis. To accomplish these functions, the body produces **urine**, a watery fluid that contains a variety of substances that have been removed from the blood or (in many invertebrates) from the interstitial fluid that bathes all cells. Urine contains waste products from proteins, various ions and other water-soluble nutrients in excess of the body's needs, and certain foreign substances (such as drugs or their metabolic by-products). After urine is produced, it is excreted.

Urinary Systems Excrete Cellular Wastes

Excretion is a general term that encompasses the elimination of any form of waste from the body. Urinary systems excrete cellular wastes, primarily the *nitrogenous wastes* ammonia, urea, and uric acid. To a large extent, an animal's lifestyle determines the type of nitrogenous waste it excretes (**TABLE 36-1**). Nitrogenous wastes are primarily formed by the degradation of excess amino acids that have been liberated by protein digestion (nucleic acids also make a small contribution). Amino acid degradation starts with the removal of the amino group ($—NH_2$), producing the simplest—but most toxic—nitrogenous waste: **ammonia** (NH_3). Ammonia is the primary nitrogenous waste of aquatic organisms, including many invertebrates, nearly all bony fish, and amphibian tadpoles. These animals are able to release ammonia continually into their watery environments, often through their skin or gills.

Terrestrial vertebrates collect and store their urine. Most terrestrial vertebrates generate ammonia in their livers, where it is immediately converted to far less toxic **urea**. Marine sharks and rays (cartilaginous fish) produce urea, which they excrete and also store in high concentrations in their tissues, as described later.

In the bird liver, ammonia is transformed through a complex series of reactions into **uric acid**, a water-insoluble substance that forms harmless crystals. Birds void uric acid along with feces through a common opening. You've undoubtedly seen this white paste (accompanied by dark feces) decorating the heads of outdoor statues or the windshield of your car. Because uric acid requires almost no water to produce and store, this adaptation helps keep birds lightweight, which facilitates flight. Embryonic birds digest egg albumin protein, forming uric acid, which remains confined within a membrane inside the egg. This isolation of wastes protects the embryos from stewing in toxic urea or ammonia. Insects and terrestrial snails conserve water by excreting uric acid in a nearly dry state.

Some non-bird reptiles such as crocodiles and alligators excrete ammonia; others excrete urea or uric acid depending on their habitat and sometimes their developmental stage. Some animals occupy different environments at different stages

TABLE 36-1 Nitrogenous Wastes

Nitrogenous waste	Chemical formula	Advantages	Disadvantages	Conditions that favor excreting this form of waste	General groups of organisms that excrete this waste product
Ammonia	H–N(–H)–H	Takes almost no energy to produce	Highly toxic	Aquatic	Bony fish; crocodiles and alligators; fully aquatic amphibians; most aquatic invertebrates
Urea	NH_2–C(=O)–NH_2	Less toxic and requires less water to excrete than ammonia	Somewhat toxic; requires more energy than ammonia to produce	Terrestrial, with adequate moisture; marine environments	All mammals; semiaquatic and terrestrial amphibians; cartilaginous fish; some reptiles
Uric acid	[structural formula: purine ring with O, HN, N, H, =O]	Nontoxic and insoluble in water; requires almost no water to excrete	Requires more energy than ammonia and urea to produce	Dry terrestrial environments; within shelled eggs	Birds, snakes, and many other terrestrial reptiles; insects; land snails

in their life cycles, and they may change their predominant nitrogenous wastes accordingly. For example, frog tadpoles (which are fully aquatic) primarily excrete ammonia, but more terrestrial adult frogs excrete most nitrogenous waste as urea.

Urinary Systems Help to Maintain Homeostasis

The second function of the urinary system is to help maintain **homeostasis**, the relatively constant internal environment that cells need in order to function properly (see Chapter 32). Urinary systems play a crucial role in homeostasis by adjusting the water content, pH, and concentrations of a variety of ions and small organic molecules in body fluids.

CHECK YOUR LEARNING

Can you . . .

- describe the two major functions of urinary systems?
- define *homeostasis* and describe how urinary systems help to maintain it?
- describe how different nitrogenous wastes support different lifestyles?

36.2 WHAT ARE SOME EXAMPLES OF INVERTEBRATE URINARY SYSTEMS?

Sponges (see Fig. 35-8) and cnidarians (such as *Hydra*; see Fig. 35-9) lack urinary systems. For these simple animals, diffusion and active transport through cell membranes into the surrounding water are adequate to excrete cellular wastes and maintain homeostasis. The urinary systems of more complex invertebrates remove wastes and maintain homeostasis by regulating water balance and the composition of body fluids.

Protonephridia Filter Interstitial Fluid in Flatworms

The simplest urinary systems are the **protonephridia** of flatworms, which consist of tubules that branch throughout the body within the interstitial fluid that surrounds the flatworm's cells and tissues (**FIG. 36-1a**). Each of the many branches is capped by a ciliated "flame cell" packed with beating cilia that resemble a flickering flame. Flame cells produce a current that draws interstitial fluid into the tubules through slit-like openings. Most dissolved substances are reabsorbed as the fluid flows through the tubule, and watery urine is excreted through the excretory pores. Protonephridia are most highly developed in freshwater planarian flatworms. In these residents of streams and lakes, protonephridia primarily maintain osmotic homeostasis by removing the excess water that continually diffuses into their bodies. The large body surface of flatworms allows most cellular wastes to diffuse directly into the environment without passing through the urinary system.

Malpighian Tubules Produce Urine from the Hemolymph of Insects

Insects have open circulatory systems, in which *hemolymph* (a fluid that serves as both blood and interstitial fluid) fills the *hemocoel* (the body cavity) and bathes the internal tissues and organs directly. Insect urinary systems consist of **Malpighian tubules**, small tubes that extend outward from the intestine and end blindly within the hemolymph (**FIG. 36-1b**). Wastes and dissolved nutrients move from the

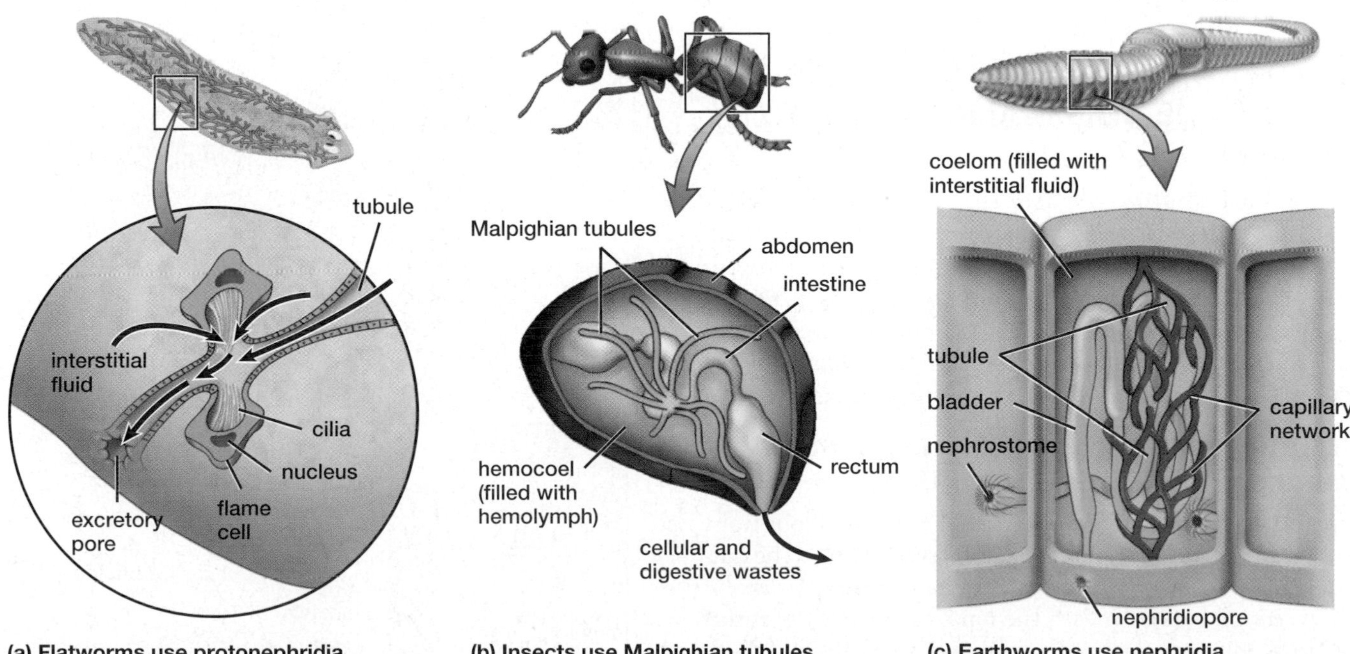

▲ **FIGURE 36-1 Some invertebrate urinary systems (a)** Tubules of the protonephridia of freshwater flatworms contain ciliated flame cells that propel urine to the excretory pores. **(b)** Malpighian tubules of insects produce concentrated urine (mostly uric acid), which is excreted with the feces. **(c)** Most segments of the earthworm have paired nephridia that produce urine from interstitial fluid.

hemolymph into the Malpighian tubules both by diffusion and by active transport, and water follows by osmosis. The urine flows through the Malpighian tubules into the intestine, where active transport returns important dissolved substances to the insect's hemolymph, and water follows by osmosis. The remaining urine, primarily uric acid, is excreted from the intestine along with feces.

Nephridia Produce Urine from Interstitial Fluid in Annelid Worms and Mollusks

In some invertebrates, such as earthworms (and other annelids) and mollusks, the urinary system consists of tubular structures called **nephridia** (singular, nephridium). In an earthworm, most segments contain a pair of nephridia. The nephridia lie within the **coelom**, the body cavity that encloses the internal organs and is filled with interstitial fluid into which wastes and nutrients from the blood diffuse. Each nephridium begins with a funnel-like opening, the *nephrostome*, ringed with cilia that conduct interstitial fluid into a narrow, twisted tubule surrounded by a network of capillaries (**FIG. 36-1c**). As the fluid traverses the tubule, salts and other nutrients are reabsorbed back into the capillary blood, leaving wastes and excess water behind. The resulting urine is collected in a bladder and then excreted through an opening called the *nephridiopore* in the body wall of the adjacent segment. As you study the nephrons of vertebrates, notice their similarities to nephridia.

CHECK YOUR LEARNING

Can you . . .

- describe and compare the urinary systems of freshwater flatworms, insects, and earthworms?

36.3 WHAT ARE THE STRUCTURES OF THE MAMMALIAN URINARY SYSTEM?

The mammalian urinary system consists of the paired kidneys and ureters and a single bladder and urethra. These organs filter small nutrient and waste molecules and ions out of the blood and then help maintain homeostasis by returning essential ions and nutrients to the blood, while collecting and excreting excess substances and cellular wastes. In the sections that follow, we focus on the human urinary system.

Structures of the Human Urinary System Produce, Store, and Excrete Urine

Human **kidneys** are fist-sized organs located at about waist level on either side of the spinal column (**FIG. 36-2**). The outermost layer of each kidney is the **renal cortex** (L. *renalis*, kidney; *cortex*, bark). Beneath the renal cortex lies the **renal medulla** ("kidney marrow"), which allows the kidney to produce concentrated urine, thus conserving water. The renal medulla surrounds a branched, funnel-like chamber called the **renal pelvis** ("kidney bucket"), which collects urine and conducts it into the ureter (**FIG. 36-3**). The **ureter** is a narrow, muscular tube that contracts rhythmically to

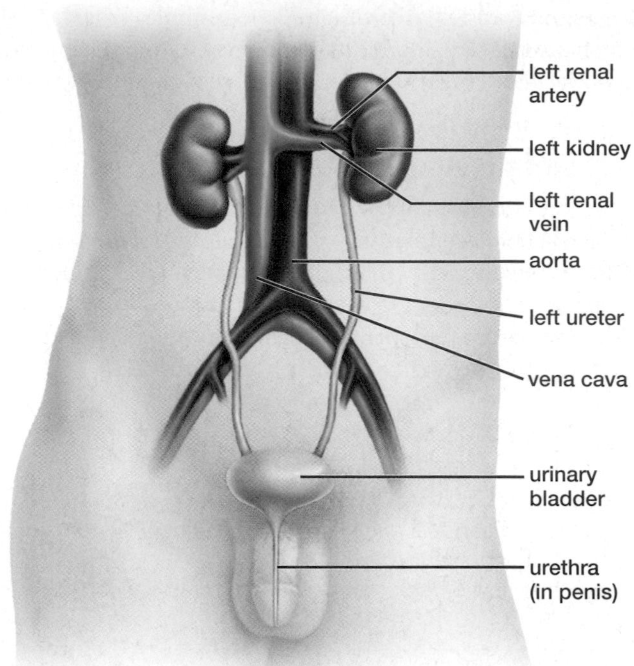

▲ **FIGURE 36-2 The human urinary system and its blood supply**

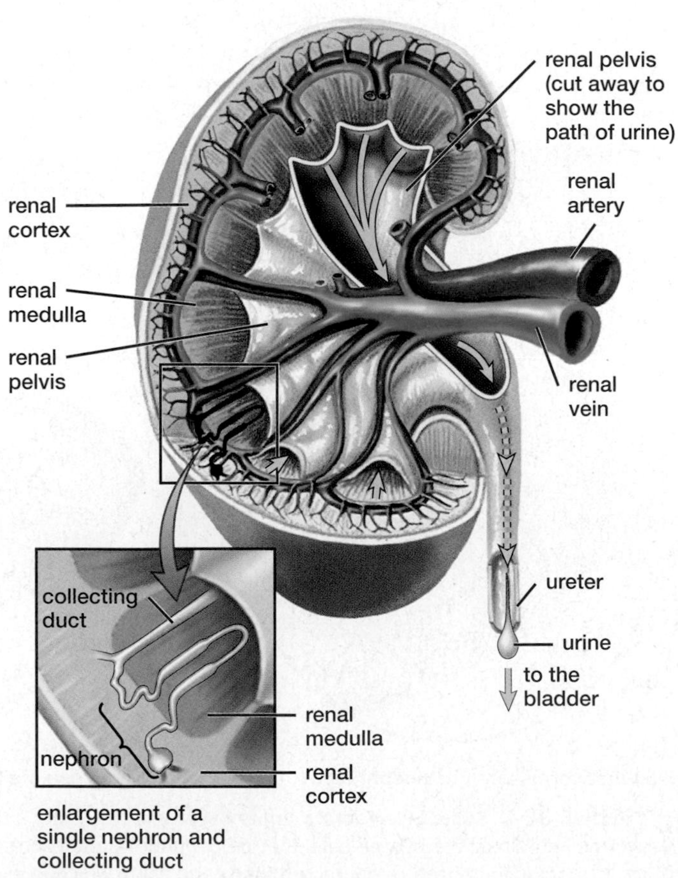

▲ **FIGURE 36-3 The structure and blood supply of the human kidney** Yellow arrows show the path of urine flow.

propel the urine from the kidney to the **bladder**, a hollow, muscular chamber that collects and stores urine.

The average adult bladder can hold about a pint (500 milliliters) of urine, but the desire to urinate is triggered by smaller amounts. As accumulating urine expands the bladder wall, the pressure eventually activates stretch receptors that trigger reflexive contractions. Urine is retained in the bladder by two circular sphincter muscles. The internal sphincter, located at the junction of the bladder and the urethra, opens automatically during these contractions. The external sphincter, slightly below the internal sphincter, is under voluntary control, allowing the brain to suppress urination unless the bladder becomes overly full. Urine exits the body through the **urethra**, a single narrow tube about 1.5 inches long in women and about 8 inches long in men (because it extends through the prostate gland and penis).

CASE STUDY CONTINUED

Paying It Forward

Each year in the United States, roughly 6,000 kidneys are transplanted from living donors like Anthony DeGiulio in domino donation chains. How can kidney donors and kidney recipients each survive with only one kidney? The immense filtering capacity of the kidneys is the key; half of their usual capacity is fully adequate. How do kidneys maintain homeostasis by filtering the blood and regulating its composition?

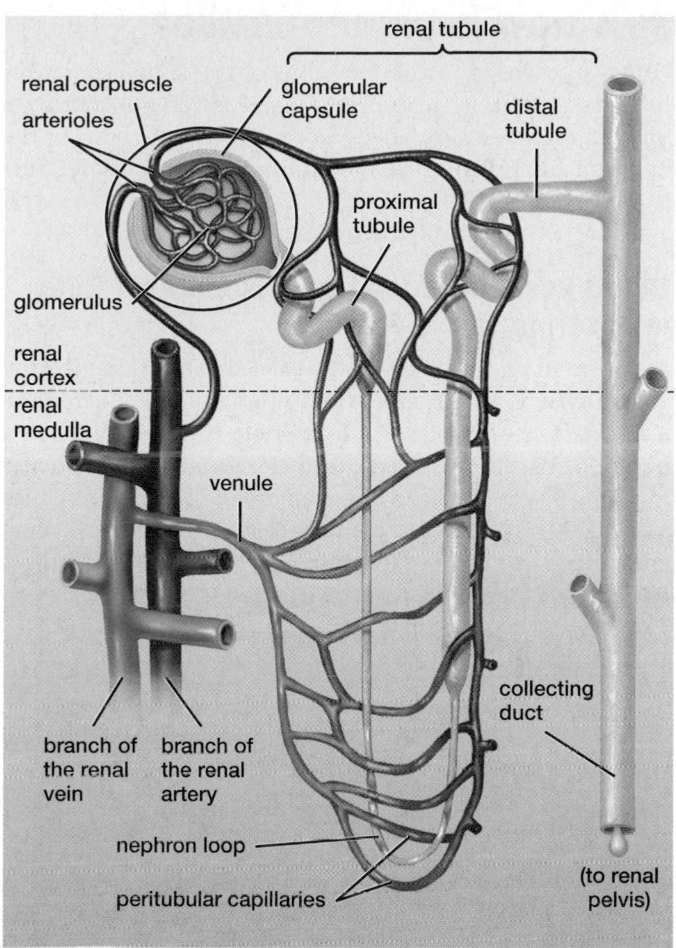

▲ **FIGURE 36-4 An individual nephron and its blood supply** The dashed line marks the boundary between the renal cortex and renal medulla.

Nephrons in the Kidneys Filter Blood and Produce Urine

Your entire blood volume passes through your kidneys about 60 times daily, allowing them to fine-tune its composition. Each kidney contains roughly 1 million microscopic urine-forming units called **nephrons** (**FIG. 36-4**). Nephrons are packed together in the renal cortex, with a thin extension of each nephron extending into the renal medulla (see Fig. 36-3, inset).

Each nephron has two major parts: the renal corpuscle and the renal tubule. The role of the **renal corpuscle** is to pressure-filter the blood and collect the resulting fluid, called **filtrate**. The renal corpuscle consists of two parts: the glomerulus and the glomerular capsule. The **glomerulus** is a knot of exceptionally porous capillaries that allow water and small molecules dissolved in the blood plasma to ooze out as blood flows through them. The surrounding cup-shaped **glomerular capsule** captures this blood filtrate (see Fig. 36-4). The filtrate then enters the **renal tubule**, which conducts the filtrate as it is converted to urine. The renal tubule consists of three parts. The first portion is the **proximal tubule**, which returns water and most essential molecules and ions to the blood. The filtrate then enters the second portion of the tubule, the **nephron loop** (also called the **loop of Henle**). In most nephrons, the nephron loop extends from the renal cortex into the uppermost portion of the renal medulla, but some extend much more deeply, as described later. The main function of the nephron loop is to produce and maintain a high concentration of salt ions (Na^+ and Cl^-) in the interstitial fluid of the renal medulla. This high interstitial concentration of solutes helps the kidney produce concentrated urine and maintain water in the blood, as described later. The filtrate is finally converted into urine in the **distal tubule**, where more substances are removed from and secreted into the blood.

The distal tubule empties urine into a **collecting duct**, a larger tube adjacent to the nephron. There are thousands of collecting ducts within the kidney; each receives urine from many nephrons. Collecting ducts conduct urine from the renal cortex, through the renal medulla, and into the renal pelvis (see Fig. 36-4). As urine within the collecting ducts flows through the renal medulla, additional water may be reclaimed into the blood to maintain homeostasis, as described later.

CHECK YOUR LEARNING

Can you . . .

- list and describe the structures of the human urinary system?
- diagram and describe the structures within the kidney?
- describe the blood supply to each kidney?
- draw and label a nephron?

36.4 HOW IS URINE FORMED?

Urine is produced in the nephrons of the kidneys by three processes: filtration, reabsorption, and secretion. As urine is formed, dissolved substances move between the parts of the nephron and the interstitial fluid that surrounds these structures. The interstitial fluid in turn exchanges substances with a nearby network of microscopic capillaries.

Blood Vessels Support the Nephron's Role in Filtering the Blood

As shown in Figure 36-4, blood is carried to the kidney by the **renal artery**, which gives rise to thousands of microscopically narrow arterioles. Each arteriole supplies blood to a nephron. Within the renal corpuscle, the arteriole branches to form the capillaries of the glomerulus. These empty into an outgoing arteriole (in contrast to most capillaries, which empty into venules; see Chapter 33). The outgoing arteriole gives rise to **peritubular capillaries** (*peritubular*, around the tubule) which form a network surrounding the renal tubule. The peritubular capillaries conduct the blood into a venule that joins the **renal vein** (see Fig. 36-4).

Filtration Removes Small Molecules and Ions from the Blood

Filtration, the first step in urine formation, occurs when fluid is forced by blood pressure through the walls of the nephron's glomerular capillaries (**FIG. 36-5** ①). Two factors facilitate glomerular filtration. First, the glomerular capillaries are far more porous than most other capillaries, and second, the arterioles that collect blood from the capillaries are narrower than the arterioles that supply them, creating an unusually high pressure within the glomerular capillaries. As a result, about 20% of the blood's fluid, along with its small dissolved molecules, is forced out through the glomerular capillary walls. Blood cells and plasma proteins, which are too large to penetrate the

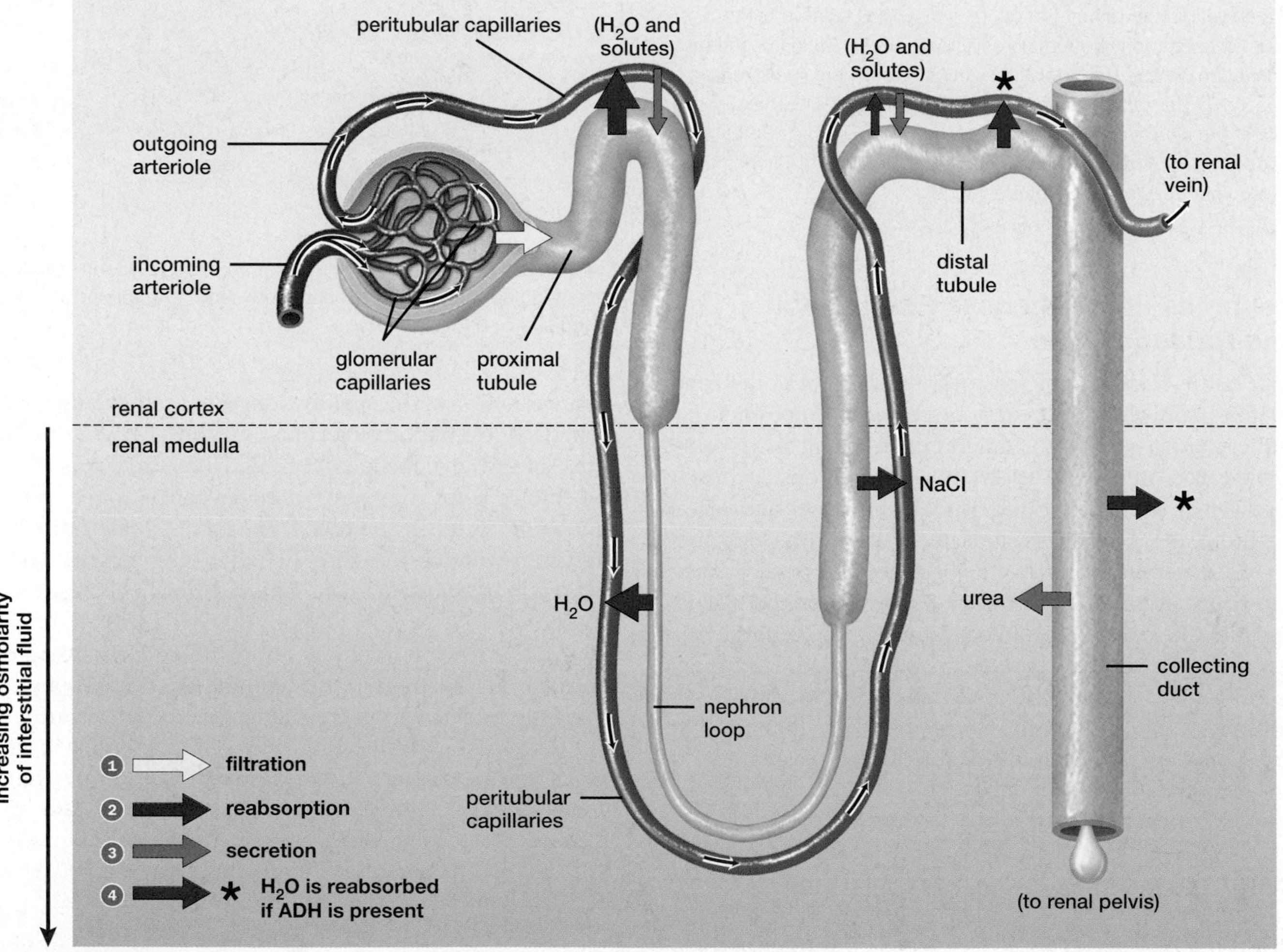

▲ **FIGURE 36-5 Urine formation and concentration** The major functions of each part of the nephron and the collecting duct are shown in this simplified diagram. The numbered steps correspond to the colored arrows in the diagram. The dashed line marks the boundary between the renal cortex and the renal medulla.

capillary walls, remain in the blood. The filtrate (essentially plasma minus its large proteins) is collected by the surrounding glomerular capsule, which conducts it into the proximal tubule. Urea makes up about 40% of the solutes in the glomerular filtrate.

Reabsorption Returns Important Substances to the Blood

Reabsorption returns to the blood nearly all the water, ions (Na^+, Cl^-, K^+, Ca^{2+}, HCO_3^-), and organic nutrients such as vitamins, glucose, and amino acids that were previously removed during filtration ❷. The ions Na^+, Cl^-, K^+, and Ca^{2+} are critical for nerve and muscle function, and Na^+ levels in blood exert a major influence on blood volume and pressure. The bicarbonate ion (HCO_3^-) is crucial for maintaining the constant pH required for metabolic reactions.

The reabsorbed molecules move by diffusion or active transport through the walls of the renal tubule and into the peritubular capillaries, which return them to the bloodstream. Most reabsorption takes place in the proximal tubule, but reabsorption also occurs in the nephron loop and the distal tubule. In the proximal tubule, reabsorption is generally not under hormonal control, but in the distal tubule, it is under the control of hormones that help maintain homeostasis. Thus, the distal tubule fine-tunes blood composition by regulating the reabsorption of water and ions to maintain homeostasis based on the changing needs of the body. The fluid that has travelled through the nephron becomes urine as it leaves the distal tubule.

Secretion Actively Transports Substances into the Renal Tubule for Excretion

Secretion, which occurs mainly through active transport, moves wastes and excess ions from the blood into the renal tubule ❸. Secreted substances include excess K^+ and H^+, small quantities of ammonia, some medicinal and recreational drugs (including penicillin, aspirin, morphine, nicotine, and cocaine), and the breakdown products of these drugs. Certain food additives and pesticides are secreted as well. Secretion occurs primarily in the proximal tubule, but some also occurs in the distal tubule during the final stages of urine formation. As with reabsorption, secretion by the distal tubule is regulated by circulating hormones to maintain homeostasis.

CHECK YOUR LEARNING

Can you . . .

- describe the blood supply of the nephron and how it supports the nephron's function?
- explain the three stages of urine formation, including the nutrients and wastes that are involved in each stage of the process?
- describe where in the nephron each stage occurs?

36.5 HOW DO VERTEBRATE URINARY SYSTEMS HELP MAINTAIN HOMEOSTASIS?

To accomplish their functions of excreting cellular wastes and maintaining homeostasis, the kidneys filter enormous quantities of blood. Nearly one-quarter of the volume of blood pumped by each heartbeat travels through the kidneys. This high volume of flow allows the kidneys to maintain the composition of the blood within narrow limits. Should both kidneys fail, death would be rapid without the treatments described in "Health Watch: When the Kidneys Collapse."

Vertebrate urinary systems help maintain homeostasis in several ways, including:

- Regulating small organic nutrients and ions within the blood and interstitial fluid (described earlier).
- Regulating the water and ion content of the blood to maintain the proper blood osmolarity.
- Maintaining the proper pH of the blood by regulating H^+ and HCO_3^- concentrations.
- Secreting substances that help regulate blood pressure and blood oxygen content.

The Kidneys Regulate the Water and Ion Content of the Blood

The kidneys receive about 5 cups (40 ounces or 1,200 milliliters) of blood every minute, and from this they remove about half a cup (about 125 milliliters) of water and solutes from the blood. If the kidneys were unable to return any fluid to the blood, we would produce roughly 45 gallons of urine a day—and need to drink almost constantly to replace the lost water. Instead, the urinary system restores nearly all of the water filtered out through the glomeruli, and we typically excrete only about 1.5 quarts (about 1,500 milliliters) of urine daily.

Water Balance Is Essential for Homeostasis—and Life

An important function of the kidney is osmoregulation. **Osmoregulation** is the process of maintaining blood **osmolarity**—the concentration of ions and other solutes in the blood plasma—within very strict limits. If a person consumes excess water faster than the kidneys can excrete it, the surplus water in the blood will move by osmosis into the interstitial fluid and then into cells, causing them to swell. Swelling in brain cells causes headaches, nausea and vomiting, seizures, coma, and sometimes death. In contrast, if a person becomes dehydrated (if water is unavailable or illness causes prolonged diarrhea and vomiting), blood volume decreases and blood osmolarity increases. If the kidneys cannot conserve enough water, dehydration can cause low blood pressure, dizziness, and confusion. In extreme cases, loss of water in brain cells can lead to coma and death.

When the Kidneys Collapse

When a person's kidneys fail, wastes accumulate in the blood, and imbalances in ion concentrations occur. Each year in the United States, about 90,000 people die of kidney failure, also called end-stage renal disease (ESRD). The most common causes of ESRD are diabetes and high blood pressure, which damage the glomerular capillaries, but the kidneys can also be compromised by infection or overdoses of some painkilling medicines.

Unfortunately, although more than 100,000 people await kidney transplants in the United States, only about 17,000 kidneys become available annually. People awaiting kidney transplants are kept alive using *hemodialysis*, a treatment in which wastes and excess water are filtered from the blood by an elaborate machine (**FIG. E36-1**). During hemodialysis, the patient's blood is pumped through a machine containing tubes made of a membrane suspended in dialyzing fluid. Like glomerular capillaries, the membrane has pores that are too small to permit the passage of blood cells and large proteins but large enough to pass smaller molecules. Dialyzing fluid has normal blood levels of salts and nutrients. Thus, only molecules or ions whose concentrations are higher than normal in the patient's blood (such as urea and H^+) diffuse into the dialyzing fluid.

Although people may remain on hemodialysis for many years, the treatment is far from ideal. Whereas healthy kidneys work nonstop, hemodialysis treatments are intermittent. Patients must monitor their diets carefully between sessions to avoid dangerous ion imbalances and limit their fluid intake to minimize the accumulation of water in their blood. A typical hemodialysis session takes at least 4 hours and is usually done three times a week. The 400,000 people in the United States who rely on hemodialysis (some have conditions that make them ineligible for transplants) can survive for many years, but their lives are disrupted, blood composition fluctuates, and toxic substances accumulate between sessions. Encouragingly, small home dialysis machines are now available (see Fig. E36-1, right). After thorough training, qualified patients can use them for a few hours five to seven times a week, more closely mimicking a functioning kidney. These units are portable and offer patients considerably more freedom and control over their condition.

A bioartificial kidney, which would be a major advance over hemodialysis and human kidney donation, is under development. The device, about the size of a coffee cup, uses a cartridge filled with tiny hollow tubes lined with kidney tubule cells. Blood enters the cartridge, where it is filtered through a membrane that (like the glomerulus) removes much of the water and dissolved substances, leaving blood cells and proteins behind. The filtrate then passes through compartments lined with renal cells, which process the filtrate much like the renal tubule processes the blood filtrate. Preliminary trials in animals have been encouraging.

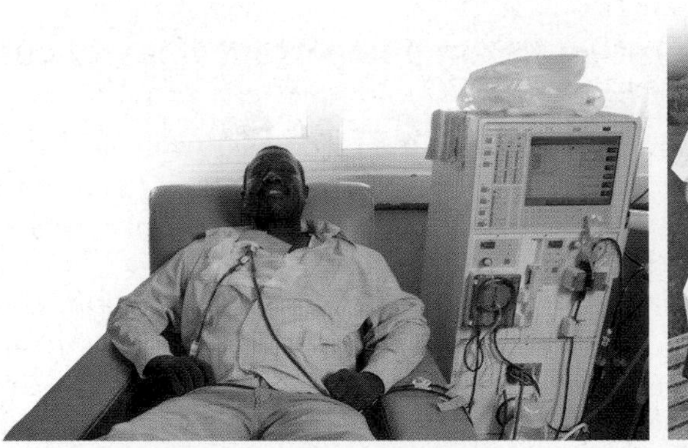
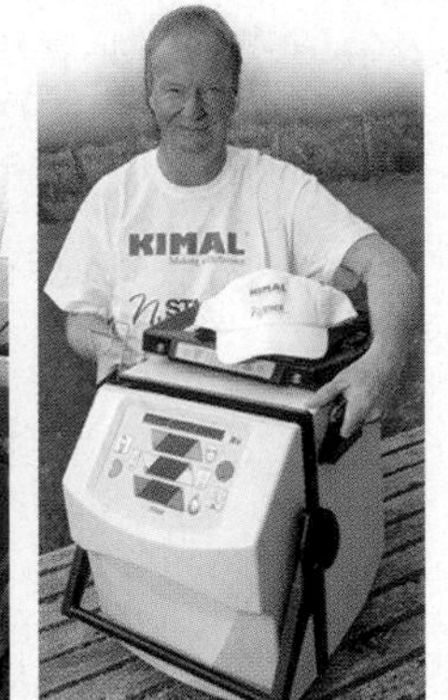

▲ **FIGURE E36-1 Hemodialysis** A patient receives hemodialysis at a hospital (left). A portable dialysis machine (right) accompanied another patient on a 6-day canoe trip.

Meanwhile, other research groups are hoping to grow functional kidneys in the lab, by seeding stem cells onto a scaffold consisting of a kidney with all its cells removed, leaving only the extracellular matrix. To prepare the scaffold, cells are removed from a cadaver kidney by injecting a detergent solution into the renal artery. In early trials using rat kidneys, scaffolds were successfully repopulated with blood vessel stem cells and kidney cells from newborn rats. Stimulated by growth factors within the extracellular matrix, these cells developed into renal capillaries and nephrons. The lab-grown kidney even produced some urine when implanted into a rat, although not as efficiently as a real kidney. The future holds promise for people with ESRD.

THINK CRITICALLY In a patient with kidney failure, urea in the blood increases rapidly. What would happen to blood urea in a patient whose kidneys were functioning normally but whose liver failed? Why?

Concentration of Urine Occurs in the Distal Tubule and Collecting Duct

When the filtrate enters the distal tubule, about 80% of its water has already been reabsorbed in the proximal tubule and nephron loop, but the filtrate is still considerably more dilute than the surrounding interstitial fluid in the renal cortex. From this point on, additional reabsorption of water is precisely regulated to maintain the blood's osmolarity within narrow limits. If fluid intake has been high, more water will be left behind in the filtrate, and watery urine will be produced until the normal blood volume is restored. If fluid intake has been low, concentrated urine will be produced. The urine becomes concentrated because the distal tubule and collecting duct become more permeable to water, which leaves the urine by osmosis and returns to the blood (see Fig. 36-5 ④).

The collecting duct is able to return water to the blood, thereby producing concentrated urine and preventing dehydration, because the concentration of solutes in the surrounding interstitial fluid is elevated. The solute concentration is especially high in the renal medulla, where salt enters the interstitial fluid from the nephron loop and some urea diffuses into the fluid from the collecting duct (see Fig. 36-5). Thus, as the urine in the collecting duct is carried from the renal cortex to the renal medulla, it passes through increasingly concentrated interstitial fluid. Because the collecting duct is water permeable, the difference in osmolarity between the urine and the interstitial fluid causes water to leave the urine by osmosis.

Antidiuretic Hormone Controls the Water Permeability of the Distal Tubule and Collecting Duct

The amount of water reabsorbed into the blood from the distal tubule and collecting duct depends on the number of water channel proteins called **aquaporins** in the cell membranes of those structures. The membranes of the proximal tubule and the descending portion of the nephron loop have large numbers of aquaporins at all times, so these membranes remain highly permeable to water. In contrast, aquaporins in the membranes of the distal tubule and collecting duct are adjusted as the body's needs change. Aquaporin numbers are controlled by **antidiuretic hormone (ADH)**. "Diuretic" means "increasing urine production," so an "*anti* diuretic" decreases urine production.

ADH is secreted by the posterior pituitary gland and carried in the bloodstream, and it causes cells of the distal tubule and collecting duct to insert aquaporin proteins into their membranes. In the absence of ADH, these membranes are nearly impermeable to water. As ADH levels rise, more aquaporins are inserted and the permeability of the membranes to water increases. In this way, ADH levels control the osmolarity of the blood by controlling the amount of water in urine.

How are ADH levels controlled? Receptors in the hypothalamus monitor blood osmolarity (**FIG. 36-6**). For example, when your body becomes dehydrated—as might occur if you were dripping sweat while exercising in the hot sun—your blood osmolarity rises. When blood osmolarity exceeds an optimal level, the hypothalamus stimulates the pituitary gland to release ADH. In response to ADH, cells of the distal tubule and collecting duct insert additional aquaporins into their membranes. As urine flows through the distal tubule and collecting duct, the more concentrated interstitial fluid draws water out by osmosis. The water enters peritubular capillaries and is restored to the bloodstream. The urine in the collecting duct can become as concentrated as the surrounding interstitial fluid; in humans, its osmolarity can reach four times that of the blood.

But reducing water loss will not, by itself, restore blood to its normal osmolarity. To replace lost water, receptors in the hypothalamus simultaneously activate a thirst center (also located in the hypothalamus) that stimulates the desire to drink and restore water to the blood.

Like most homeostatic conditions, blood osmolarity is controlled by negative feedback. When the osmolarity returns to normal, receptors in the hypothalamus signal the pituitary to reduce ADH release down to a baseline level. If you drink too much water, ADH secretion will be temporarily blocked, causing you to excrete large amounts of very dilute urine (about one-third the osmolarity of the blood). When normal blood osmolarity is restored, ADH secretion is increased to baseline levels.

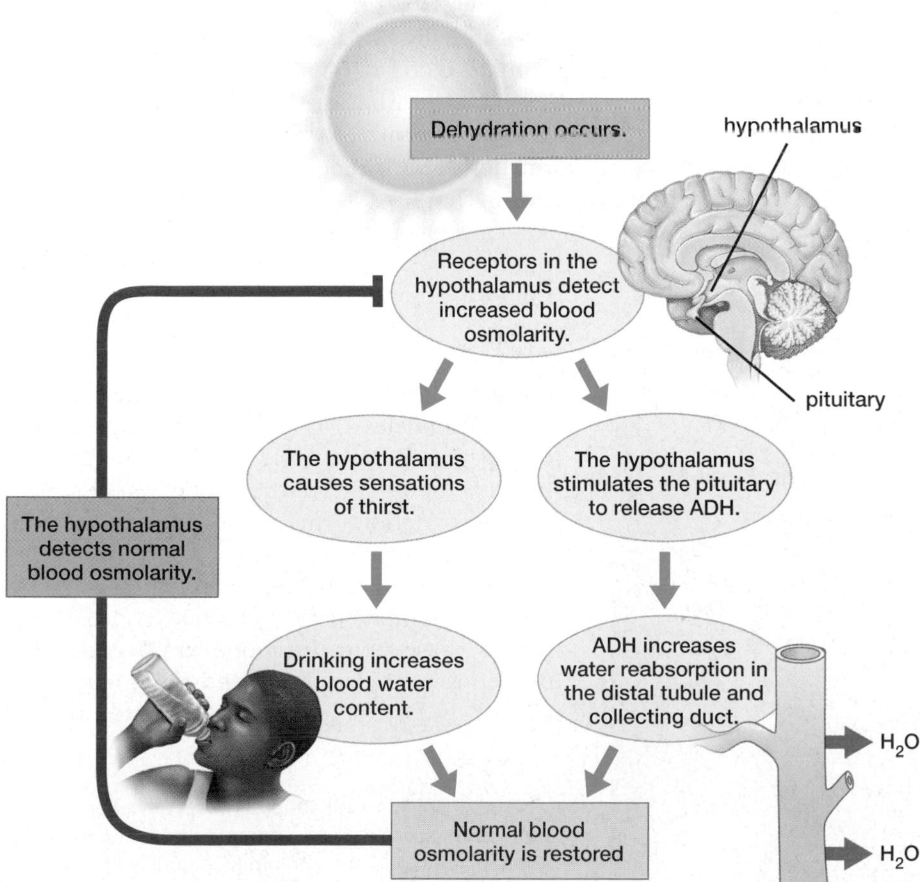

FIGURE 36-6 Dehydration stimulates ADH release and water retention

THINK CRITICALLY Describe the feedback process that would occur if you drank far more water than your body needed.

Have You Ever Wondered ...

Why Alcohol Makes You Pee a Lot?

One of alcohol's many effects on people who consume it is to inhibit ADH release. With ADH levels very low, the body produces urine that is very dilute and watery. As a result, after a bout of drinking, you may excrete more water than you drank. So, ironically, having too much to drink can actually dehydrate you. Dehydration makes you thirsty, which may lead to additional alcohol consumption and further suppression of ADH. Dehydration also contributes to the misery of the hangover you may experience the morning after.

Mammalian Nephrons Are Adapted to the Availability of Water

In the human kidney, about 20% of nephrons have long (about 1.25 inches, or 3 centimeters) loops that extend deep into the renal medulla (see Fig. 36-3, inset). These long nephron loops play a major role in maintaining the high salt concentration in the interstitial fluid of the medulla.

Mammals adapted to dry climates (**FIG. 36-7**) generally have kidneys containing a much higher percentage of nephrons with long nephron loops. These long loops produce a higher salt concentration in the interstitial fluid of the medulla, allowing more water to be reclaimed from the urine as it travels through the collecting ducts. The masters of urine concentration are desert rodents such as kangaroo rats (see Fig. 30-13a), which can produce urine with 14 times the osmolarity of their blood. Not surprisingly, all the nephrons in kangaroo rat kidneys have very long loops. With their extraordinary ability to conserve water, kangaroo rats do not need to drink; they rely entirely on water contained in their food and on metabolic reactions that produce water.

In contrast, mammals adapted to habitats with abundant fresh water typically have far more short nephron loops. For example, beavers, which live along streams, have only short-looped nephrons and can only concentrate their urine to about twice their blood osmolarity.

The Kidneys Help Maintain Blood pH

The pH of human blood is maintained within the extremely narrow range of 7.38 to 7.42. This slightly basic pH is crucial to a multitude of cellular metabolic processes, including the functioning of enzymes and neurons. However, many normal cellular metabolic activities—such as breaking down proteins and fats, fermenting lactic acid in muscles, and synthesizing ATP—produce an excess of H^+, which makes solutions acidic. The most important defense against pH change is provided by buffers in the blood, particularly HCO_3^-, which can combine with H^+ to form H_2CO_3 by the reaction $H^+ + HCO_3^- \rightarrow H_2CO_3$.

The kidneys help maintain blood pH by reabsorbing HCO_3^- into the blood and by secreting excess H^+ into the renal tubule. If the blood becomes too acidic, the nephrons will increase H^+ secretion and increase HCO_3^- reabsorption. If the blood becomes too basic, the nephrons will decrease H^+ secretion and reduce HCO_3^- reabsorption. These processes occur in both the proximal and distal tubules of the nephrons.

The Kidneys Help Regulate Blood Pressure and Oxygen Levels

The kidneys release substances that help regulate blood pressure and maintain blood oxygen levels. When blood pressure falls, as can occur with excessive blood loss, the kidneys release the enzyme **renin** into the bloodstream. Renin catalyzes the formation of the hormone **angiotensin** from a protein circulating in the blood. This hormone helps combat low blood pressure in three major ways: (1) Angiotensin stimulates release of the hormone **aldosterone** (from the adrenal cortex; see Chapter 38), which causes the proximal tubules of the nephrons to reabsorb more Na^+ into the blood. More water follows by osmosis, increasing the blood volume. (2) Angiotensin stimulates ADH release, causing more water to be reclaimed from the urine as it passes through the distal tubule and collecting duct. (3) Angiotensin causes arterioles throughout the body to constrict, increasing blood pressure.

The kidneys also help maintain blood oxygen at levels that support the body's needs. The kidneys release the hormone **erythropoietin** in response to reduced oxygen, which may occur if blood is lost, if lung disease reduces oxygen

FIGURE 36-7 A well-adapted desert dweller Camels, which are native to deserts, have extremely long nephron loops and can produce urine with over nine times their blood osmolarity.

uptake, or at high altitudes, where each lungful of air supplies less oxygen (see Chapter 34). Erythropoietin stimulates the bone marrow to produce more oxygen-transporting red blood cells.

CASE STUDY CONTINUED

Paying It Forward

Kidney failure almost always prevents the kidneys from producing adequate erythropoietin, which stimulates red blood cell production. This results in *anemia* (too few red blood cells to supply adequate oxygen to the tissues) so that victims feel chronically tired and short of breath. Fortunately, anemia sufferers can be given human erythropoietin produced by genetically engineered cells grown in culture. But a functioning kidney will respond to the changing needs of the body, producing the proper amounts of this hormone at the appropriate times—a definite advantage for transplant recipients. How are living kidney donations accomplished? We revisit this at the end of the chapter.

Fish Face Homeostatic Challenges in Their Aquatic Environments

Osmoregulation is especially challenging for many animals that live in water, because they are constantly immersed in a solution that has either a lower (hypotonic) or a higher (hypertonic) osmolarity than their body fluids. Such animals have evolved mechanisms that maintain a homeostatic balance of water and salt within their bodies.

Freshwater fish, which live in a hypotonic environment, maintain a plasma osmolarity far above that of their freshwater surroundings. As these fish circulate water over their gills to exchange gases, some water continuously leaks into their bodies by osmosis, and salt diffuses out (**FIG. 36-8a**). Freshwater fish acquire salt from their food, and also through their gills, which use active transport to pump salt into their bodies from the surrounding water. Freshwater fish never drink (although they take in some water as they feed). Their kidneys retain salt and excrete large quantities of extremely dilute urine.

Saltwater fish live in a hypertonic environment; seawater has a solute concentration two to three times that of their body fluids. As a result, water continuously leaves their tissues by osmosis, and salt continuously diffuses in. Most saltwater fish drink to restore their lost water and excrete the excess salt by active transport through their gills (**FIG. 36-8b**). Fish nephrons completely lack nephron loops, so fish cannot produce urine that is more concentrated than their blood. Instead, to conserve water, the kidneys of most saltwater fish excrete only very small quantities of urine, which contains some salts that their gills did not eliminate.

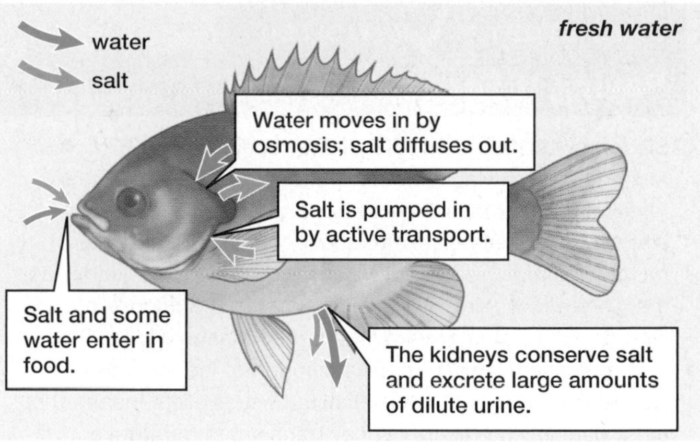
water
salt
fresh water
Water moves in by osmosis; salt diffuses out.
Salt is pumped in by active transport.
Salt and some water enter in food.
The kidneys conserve salt and excrete large amounts of dilute urine.

(a) Freshwater fish

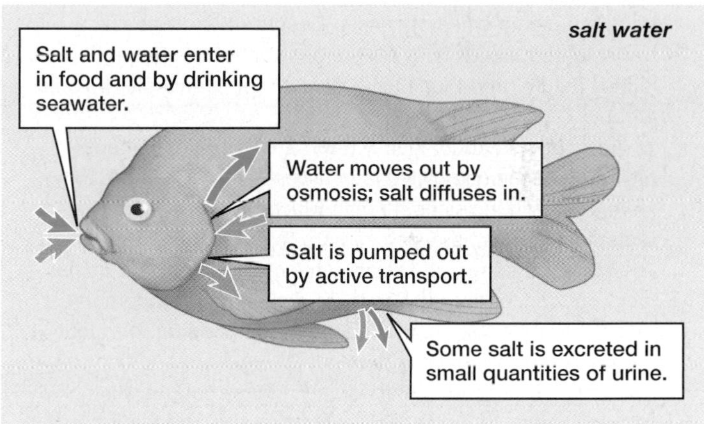
salt water
Salt and water enter in food and by drinking seawater.
Water moves out by osmosis; salt diffuses in.
Salt is pumped out by active transport.
Some salt is excreted in small quantities of urine.

(b) Saltwater fish

▲ **FIGURE 36-8 Osmoregulation in fish (a)** Freshwater fish must contend with large amounts of water entering their bodies. **(b)** Saltwater fish must conserve water, which constantly diffuses out of their bodies into the more salty surrounding seawater.

THINK CRITICALLY What osmoregulatory problems would occur if a freshwater trout were placed in the ocean, and why?

Cartilaginous fish such as sharks and rays have evolved a different solution for conserving water. These saltwater fish excrete urea instead of ammonia. They also store urea in their tissues at a concentration so high that it would kill most other vertebrates. The stored urea gives the tissues in sharks and rays approximately the same osmolarity as the surrounding seawater, so they avoid losing water by osmosis.

CHECK YOUR LEARNING

Can you . . .

- explain how the nephron and collecting duct help control blood osmolarity?
- explain the role of ADH in water reabsorption?
- explain how kidneys help control blood pH, blood pressure, and blood oxygen content?
- describe how mammalian kidneys are adapted to wet and dry environments?
- explain how and why the urinary systems of freshwater and saltwater fish differ?

CASE STUDY REVISITED

Paying It Forward

Since the 1950s, when living kidney donation was first recognized as a viable alternative to cadaver organ donors, family and friends have come forward to offer a kidney to a victim of kidney failure. To reduce the chance that the recipient's immune system will attack the donated kidney as if it were an invading microbe or parasite, the donor's blood type and several important glycoproteins should match those of the recipient. But, with the exception of identical twins, no two people have perfectly matching tissues. This means that most people who receive kidney transplants must take immune-suppressing drugs for the remainder of their lives. These drugs make transplant recipients vulnerable to infections and some types of cancer. Despite this drawback, a transplanted kidney is by far the best option for patients with kidney failure, and those lucky enough to receive one benefit greatly.

To remove a donor kidney (**FIG. 36-9**), surgeons generally use a technique called *laparoscopic* surgery, where they make half-inch incisions through which they insert surgical tools, including a tiny video camera to guide the operation. The kidney is extracted through an incision about 2½ inches long, put on ice, and rushed to its recipient. The operation takes 3 to 4 hours; donors remain in the hospital for about 3 days and return to work in about 3 weeks. In addition to the risks associated with major surgery, kidney donors will lack a back-up kidney in the unlikely event that their remaining kidney fails. But a recent study of deaths among 80,000 kidney donors during a 15-year period found no greater mortality among this group (once they had recovered from their surgery) than among non-donors.

Domino donations are almost always started spontaneously by someone inspired to make a difference. Since 2008, when Anthony DeGiulio's donated kidney started a chain that saved four lives (**FIG. 36-10**), such domino donation chains have become longer and more frequent. For example, during a 4-month period, 17 hospitals in 11 states from California to New Jersey matched 30 people—who might otherwise have died—with kidneys from 30 donors they had never met. This heroic enterprise was started by Good Samaritan Rick Ruzzamenti, who got the idea from the desk clerk at his yoga studio, who had mentioned to him that she had donated a kidney to a friend. "People think it's so odd that I'm donating a kidney," he told the transplant coordinator at his hospital, but "I think it's so odd that they think it's so odd. . . . It causes a shift in the world."

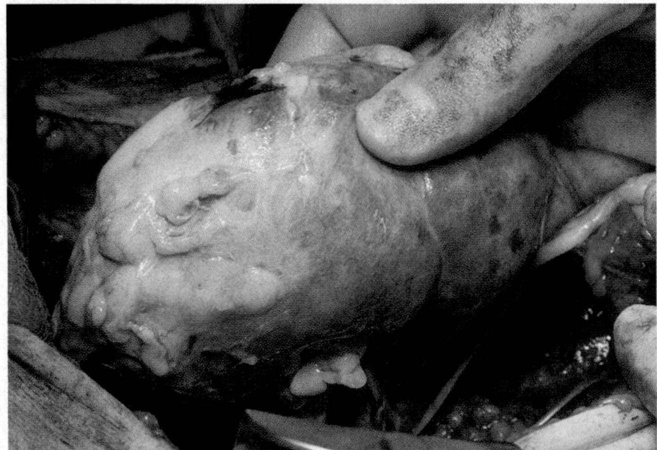

▲ **FIGURE 36-9 Surgeons transplant a kidney**

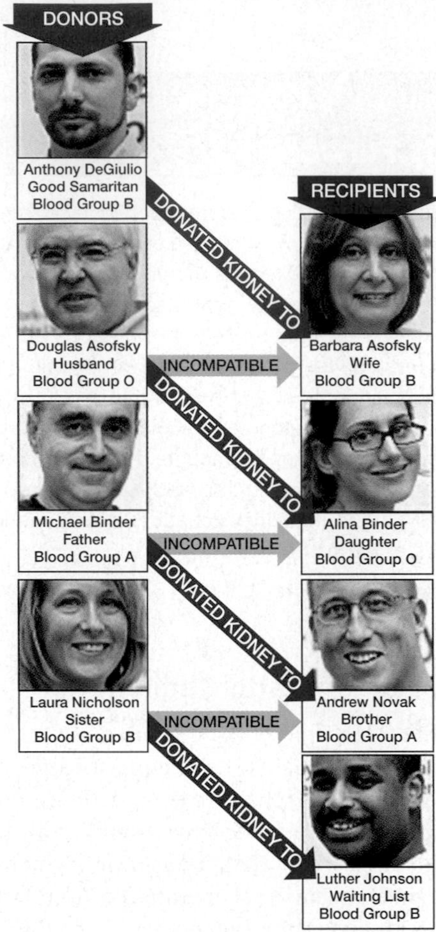

▲ **FIGURE 36-10 Domino donations** Kidneys from compatible strangers saved the lives of these four recipients.

The more than 100,000 eligible individuals awaiting a kidney transplant ardently hope that domino donation chains continue to be forged and to lengthen.

CONSIDER THIS Would you donate a kidney to a friend or family member whose kidneys were failing? Would you consider donating a kidney to a stranger? Explain your reasoning.

CHAPTER REVIEW

Go to **Mastering Biology** to access the Pearson eText, vocabulary review, practice quizzes, activities, videos, current events, and more.

*Answers to **Think Critically** and **Thinking Through the Concepts** questions can be found in the **Answers** section at the back of the book.*

Summary of Key Concepts

36.1 What Are the Major Functions of Urinary Systems?

Urinary systems produce and eliminate urine, which contains waste products of cellular metabolism (particularly nitrogenous wastes; see Table 36-1), ions, foreign substances, and some nutrients that have been ingested in excess of the body's needs. Urinary systems also maintain homeostasis by adjusting the water content, pH, and the concentrations of ions in body fluids.

36.2 What Are Some Examples of Invertebrate Urinary Systems?

The flatworm's simple urinary system consists of protonephridia, a network of branching tubules that collect wastes and excess water. Flame cells create a current that forces the urine out of the body through urinary pores. Insects use Malpighian tubules that process hemolymph within the hemocoel of their open circulatory systems. Malpighian tubules release urine into the intestine for elimination. Earthworms use nephridia to process the interstitial fluid within the body cavity. After nutrients are reabsorbed, the urine is released through nephridiopores.

36.3 What Are the Structures of the Mammalian Urinary System?

The mammalian urinary system consists of paired kidneys and ureters, a bladder, and a urethra. Kidneys produce urine, which is conducted by the ureters to the bladder, where it is stored. Distention of the bladder wall triggers urination, during which urine passes out of the body through the urethra. Kidneys have an outer renal cortex packed with filtering units called nephrons, whose nephron loops extend into the underlying renal medulla. The innermost kidney chamber, the renal pelvis, funnels urine into the ureter. Blood enters each kidney through a renal artery and exits through a renal vein. A nephron consists of the renal corpuscle, composed of the glomerulus and glomerular capsule, and the renal tubule, consisting of the proximal tubule, the nephron loop, and the distal tubule. The distal tubule empties into the collecting duct, which carries urine through the renal medulla and releases it into the renal pelvis.

36.4 How Is Urine Formed?

Urine formation in the nephron begins in the renal corpuscle, where blood is filtered through the glomerulus. The resulting filtrate of water and small dissolved molecules is collected in the glomerular capsule and enters the renal tubule. Reabsorption throughout the renal tubule restores most water, nutrients, and ions to the blood, leaving wastes such as urea behind. Secretion in the proximal and distal tubule actively transports remaining wastes and excess ions from the blood into the filtrate. Peritubular capillaries surrounding the nephron allow exchange of substances between the blood and the filtrate via the interstitial fluid. When the filtrate exits the nephron, it has become urine.

36.5 How Do Vertebrate Urinary Systems Help Maintain Homeostasis?

Vertebrate urinary systems regulate the water and ion content of the blood, maintain the proper blood pH, retain nutrients, and secrete substances that help regulate blood pressure and blood oxygen content. Water is returned to blood from the filtrate by reabsorption in the renal tubule and collecting duct. As urine flows through the collecting duct to the renal pelvis, it passes through the interstitial fluid in the renal medulla. This interstitial fluid becomes increasingly concentrated in salt (from the nephron loop) and urea (from the collecting duct).

Dehydration increases blood osmolarity, which causes the posterior pituitary to secrete antidiuretic hormone (ADH) into the blood. ADH causes aquaporins to be inserted into the distal tubule and collecting duct membranes, increasing their water permeability. High water permeability allows water to be reabsorbed into the blood from the distal tubule and from the collecting duct as it passes through the increasing concentration of salt and urea in the renal medulla.

The kidneys help regulate blood pressure by secreting the enzyme renin when blood pressure falls. Renin catalyzes the formation of angiotensin from a blood protein. Angiotensin increases Na^+ reabsorption, stimulates ADH release, and constricts arterioles, elevating blood pressure. Erythropoietin, released from the kidneys when the oxygen content of the blood is low, stimulates bone marrow to produce red blood cells.

Mammals that live where water is scarce have long nephron loops and produce concentrated urine; where water is abundant, mammals tend to have short nephron loops and produce dilute urine. Freshwater fish generate large quantities of dilute urine and actively transport salt in through their gill tissues. Saltwater fish drink seawater, actively transport salt out of their gill tissues, and produce very little urine.

Thinking Through the Concepts

Bloom's: Remembering, Understanding

Multiple Choice

1. Which of the following is correct?
   a. Freshwater organisms are likely to excrete ammonia.
   b. Ammonia is mostly excreted in urine.
   c. Ammonia is the most complex of the three major nitrogenous wastes.
   d. Humans convert ammonia to urea in the kidney.

2. Which of the following matches is correct?
   a. nephrons—earthworms
   b. Malpighian tubules—insects
   c. protonephridia—cnidarians
   d. nephridia—flatworms

3. Long nephron loops
   a. are common in animals with access to abundant fresh water.
   b. allow urine to become concentrated within the collecting duct.
   c. are the most common type of nephron loop in humans.
   d. extend deep into the renal cortex.

4. Which of the following is true?
   a. Collecting ducts empty into the renal medulla.
   b. Aquaporins are inserted into the proximal tubule when ADH is secreted.
   c. Ureters conduct urine out of the bladder.
   d. Blood is pressure-filtered through the glomerulus.
5. Which is the correct sequence and direction of blood flow?
   a. renal vein → renal artery → arterioles → peritubular capillaries
   b. venule → renal vein → peritubular capillaries → arterioles
   c. arteriole → glomerulus → arteriole → peritubular capillaries
   d. renal vein → venule → peritubular capillaries → arteriole

Fill-in-the-Blank

1. Fill in the primary nitrogenous waste excreted by the following animals: fish: ____________; birds: ____________; insects: ____________; snakes: ____________; camels: ____________.
2. The human kidney consists of an outer layer called the ____________, an underlying layer called the ____________, and a funnel-like, central chamber called the ____________. Urine is funneled from the central chamber into the ____________, which leads to a storage organ, the ____________, which empties through the ____________.
3. List the parts of the nephron through which the filtrate passes in their correct order. ____________, ____________, ____________, ____________, ____________.
4. Fill in with the appropriate term: Most reabsorption occurs here: ____________; final concentration of urine occurs here: ____________; creates a salt concentration gradient in the interstitial fluid: ____________; location of the concentration gradient within the kidney: ____________.
5. Fill in the following substances: produced from ammonia and excreted in urine of mammals: ____________; secreted into the tubule when blood pH is too low: ____________; actively transported out of the ascending portion of the nephron loop: ____________; leaves the tubule by osmosis: ____________.
6. The kidney helps maintain an internal balance called ____________. To help regulate blood pressure, the kidney secretes the enzyme ____________, which converts a substance circulating in the bloodstream into the hormone ____________. To control oxygen levels, the kidneys secrete the hormone ____________, which acts on the bone marrow to increase production of ____________.
7. If blood osmolarity increases, receptors in the ____________ detect this and signal the ____________ gland to increase release of ____________. This hormone acts on the walls of the ____________ and the ____________, causing them to insert more water pores called ____________ into their membranes and, thus, cause the urine to become more ____________.

Review Questions

1. Explain the two major functions of urinary systems, and list the processes that accomplish these functions.
2. Trace an ammonia molecule through the mammalian body, starting in the bloodstream and ending outside the body.
3. What is the function of the nephron loop? The collecting duct? Antidiuretic hormone?
4. Discuss the differences in function of the two major capillary beds in the kidneys: the glomerular capillaries and the peritubular capillaries.
5. Describe the processes of filtration, reabsorption, and secretion, including the primary site in the nephron where each occurs.
6. Describe the role of the kidneys as organs of homeostasis.
7. Briefly describe and compare the urinary systems of flatworms, insects, and earthworms.
8. Explain and contrast osmoregulation in freshwater and saltwater fish.

Applying the Concepts

Bloom's: Applying, Analyzing, Evaluating

1. In his poem "The Rime of the Ancient Mariner," Samuel Taylor Coleridge wrote: "Water, water, everywhere, nor any drop to drink." Seawater has more than four times the osmolarity of blood. Why can't a person avoid dying of thirst by drinking seawater?
2. Some people advocate drinking one's own urine, claiming significant health benefits. What problems do you see with this "alternative therapy"? Are there conditions under which this might be advisable?

37 Defenses Against Disease

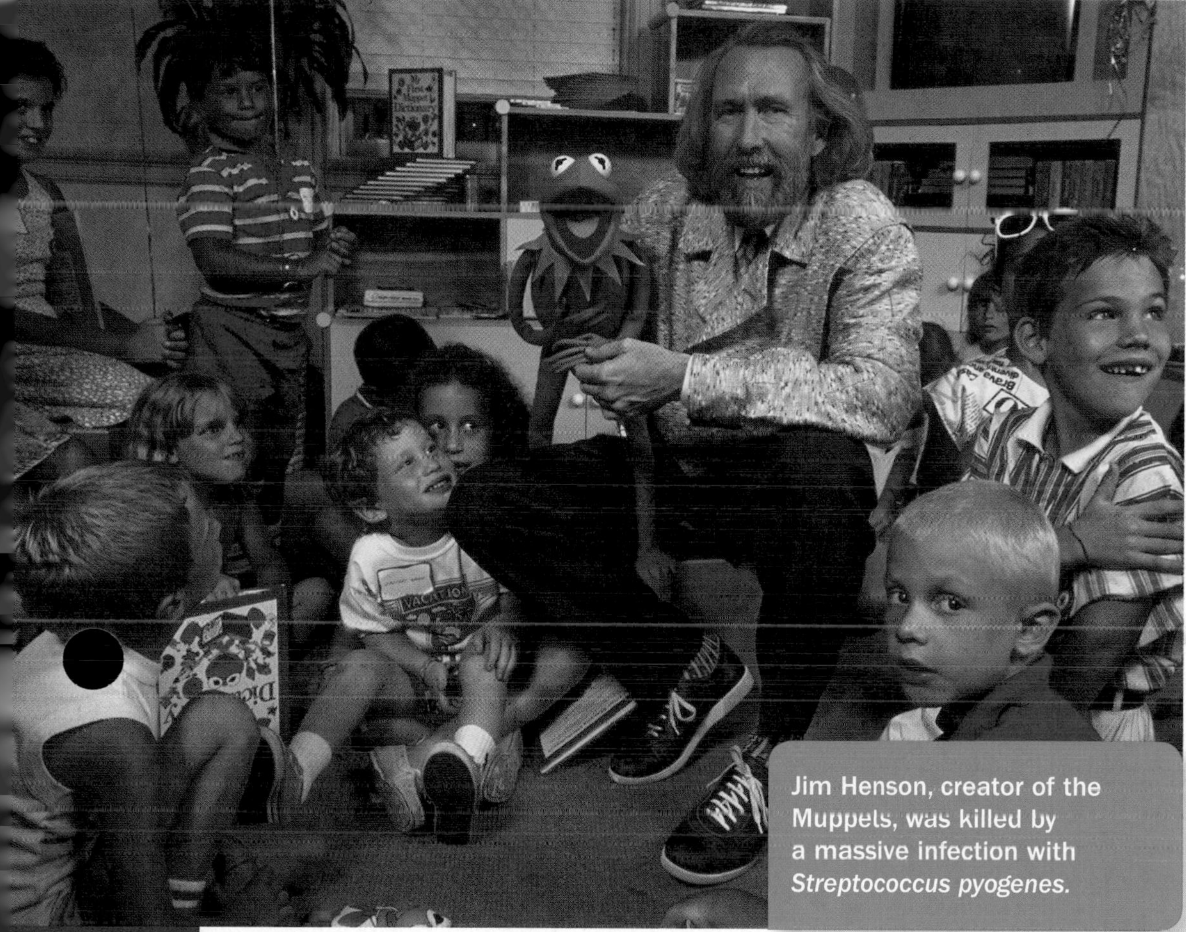

Jim Henson, creator of the Muppets, was killed by a massive infection with *Streptococcus pyogenes*.

CASE STUDY

Flesh-Eating Bacteria

ON MAY 4, 1990, JIM HENSON, creator of the Muppets and the original voice of Kermit the Frog, felt unusually tired and had a sore throat. By May 13, he felt considerably worse, but his physician didn't find any signs of pneumonia or other serious illness. However, on May 15, Henson began coughing up blood. Three hours later, at New York–Presbyterian Hospital, he couldn't breathe and was placed on a respirator. Despite the best care and multiple antibiotics, Henson died just 20 hours later.

Henson was killed by *Streptococcus pyogenes*. Chances are, you've been infected by this type of bacterium—it causes strep throat. However, some strains of *S. pyogenes* are nastier than others, and some people are more susceptible than average. Henson died of streptococcal toxic shock syndrome when *S. pyogenes* invaded his lungs, causing massive inflammation and multiple organ failure.

Sometimes, if *S. pyogenes* or certain other species of bacteria enter a cut or open sore, they may rapidly destroy huge amounts of tissue, leading to the common name "flesh-eating bacteria" (the medical term is necrotizing soft tissue infection or necrotizing fasciitis). Although this sounds overly dramatic, it's pretty accurate. Consider Aimee Copeland, a graduate student at the University of West Georgia. She was zip-lining over the Little Tallapoosa River in Georgia when the line broke, cutting her left leg and dumping her into the river. The Little Tallapoosa, like thousands of other streams in the United States, contains *Aeromonas hydrophila*, a bacterium that usually infects fish and frogs. *A. hydrophila* is frequently swallowed by swimmers, who usually don't get sick at all or might develop a mild fever, cramps, and diarrhea. In Copeland's case, however, the bacteria entered deep into her cut leg. Physicians closed the wound, but the bacteria inside began to destroy tissue in multiple parts of her body. Three days later, doctors had to perform extensive amputations to cut out the bacteria and save her life.

Despite occasional devastating failures such as these, the body's defenses against infection by bacteria and other microbes are usually highly effective. As you learn about these defenses, consider a few questions: How does your body distinguish "self" from "invader"? How does the immune system kill invaders? How can microbes such as *S. pyogenes* and *A. hydrophila* evade the immune system and cause deadly diseases?

AT A GLANCE

37.1 HOW DOES THE BODY DEFEND ITSELF AGAINST DISEASE?

The environment teems with **microbes**, which include microscopic living organisms such as bacteria, protists, and fungi, as well as viruses, which are usually not considered to be alive. Most microbes, even those that live in animal bodies, are harmless, and some are beneficial. For example, cattle do not produce enzymes that break down cellulose; without cellulose-metabolizing bacteria in their digestive tracts, cattle would starve in a pasture full of grass.

Some microbes, however, are **pathogens**, a term derived from Greek words meaning "to produce disease." As pathogens reproduce and spread to new victims, they harm their hosts. Some pathogens, such as cold viruses, are mostly an inconvenience. Other pathogens are far more dangerous. Cholera bacteria, for example, cause diarrhea that may be so overwhelming that the victim dies of dehydration. In places with inadequate sanitation, diarrhea may allow the bacteria to enter the water supply and spread to more people.

Most microbial diseases, such as cholera, measles, tetanus, and chicken pox, have infected people for hundreds or even thousands of years. In recent decades, new pathogens, and more deadly strains of familiar pathogens, have emerged. Many are viruses that you've heard about in the news: HIV, Ebola, West Nile, Zika, swine flu, and avian flu. We are also endangered by deadly, sometimes new, strains of bacteria. For example, *Staphylococcus aureus* bacteria ("staph") often occur (usually harmlessly) on the skin and in the nasal passages, but some mutated varieties cause toxic shock syndrome or prolonged infections if they penetrate through the skin or mucous membranes. And, as described in the case study, some strains of *Streptococcus pyogenes* and *Aeromonas hydrophila* can destroy much of a person's body in just a few hours.

Given the diversity of disease-causing organisms, you might wonder, "Why don't we get sick more often?" Over evolutionary time, animals and their pathogens have engaged in constant warfare. As animals evolved sophisticated defense systems, pathogens evolved more effective ways of overcoming those defenses, which in turn favored the evolution of still more powerful defenses that resist most attacks by microbes.

Vertebrate Animals Have Three Major Lines of Defense

Vertebrates have evolved three major forms of protection against disease (**FIG. 37-1**):

- *Nonspecific External Barriers* These barriers prevent most disease-causing microbes from entering the body. They are primarily anatomical structures, such as skin and

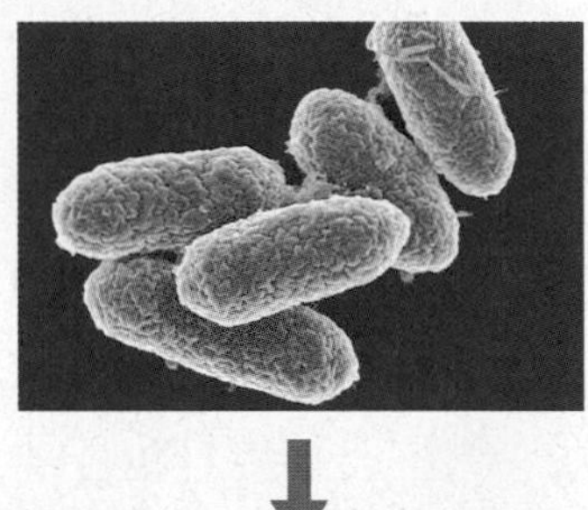

Nonspecific External Barriers
skin, mucous membranes

If these barriers are penetrated, the body responds with

Innate Immune Response
phagocytic and natural killer cells, inflammation, fever

If the innate immune response is insufficient, the body responds with

Adaptive Immune Response
humoral immunity, cell-mediated immunity

▲ **FIGURE 37-1 Levels of defense against infection**

TABLE 37-1 Characteristics of Innate and Adaptive Immune Responses to Invasion

	Specificity of response to invasion	Speed of response to the first invasion by a given microbe	Magnitude of the response to a second invasion by the same microbe	Distribution in the animal kingdom
Innate Immune Response	Nonspecific	Immediate	Identical to the first response	Invertebrates and vertebrates
Adaptive Immune Response	Specific for particular microbes	Delay of several days to 2 weeks	Enormously enhanced response; usually no disease symptoms occur (the animal has become immune to that specific microbe)	Vertebrates only

cilia, and secretions, such as tears, saliva, and mucus. The barriers cover the external surfaces of the body and line body cavities that open to the outside world, including the respiratory, digestive, and urogenital tracts.

- *Nonspecific Internal Defenses* If the external barriers are breached, the **innate immune response** swings into action. Some white blood cells engulf foreign particles or destroy infected cells. Chemicals released by damaged body cells and proteins released by white blood cells trigger inflammation and fever. Like the external barriers, the innate immune response operates regardless of the exact nature of the invader.
- *Specific Internal Defenses* The final line of defense is the **adaptive immune response**, in which immune cells selectively destroy the specific invading toxin or microbe and then "remember" the invader, allowing a faster response if it reappears in the future.

The major similarities and differences between innate immunity and adaptive immunity are summarized in **TABLE 37-1**. The internal defenses involve a large number of different cell types, briefly described in **TABLE 37-2**. Most of these cells and their roles in defending the body will be unfamiliar to you now, but Table 37-2 will be a useful guide as you read the rest of the chapter.

Invertebrate Animals Possess Nonspecific Lines of Defense

Invertebrates are protected only by nonspecific external barriers and nonspecific internal defenses. Different invertebrates possess an enormous range of external barriers, ranging from exoskeletons to slimy secretions. Internally, invertebrates have white blood cells that attack pathogens and secrete proteins that neutralize invaders or their toxins.

In the remainder of this chapter, we focus on antimicrobial defenses in mammals, especially humans.

CHECK YOUR LEARNING

Can you . . .

- compare and contrast the terms "microbe" and "pathogen"?
- describe the defenses that are characteristic of vertebrates and those that are characteristic of invertebrates?

TABLE 37-2 The Body's Cellular Arsenal Against Disease

Type of Cell	Function
Neutrophils	White blood cells that engulf invading microbes
Dendritic cells	White blood cells that engulf invading microbes and present antigens to lymphocytes
Macrophages	White blood cells that engulf invading microbes and present antigens to lymphocytes
Natural killer cells	White blood cells that destroy infected or cancerous cells
Mast cells	Connective tissue cells that release histamine; important in the inflammatory response
B cells	Lymphocytes (a type of white blood cell) that produce antibodies
Memory B cells	Offspring of B cells that provide future immunity against invasion by the same antigen
Plasma cells	Offspring of B cells that secrete antibodies into the bloodstream
T cells	Lymphocytes (a type of white blood cell) that regulate the immune response or kill infected cells or cancerous cells
Cytotoxic T cells	T cells that destroy infected body cells or cancerous cells
Helper T cells	T cells that stimulate immune responses by both B cells and cytotoxic T cells
Memory T cells	Offspring of cytotoxic or helper T cells that provide future immunity against invasion by the same antigen
Regulatory T cells	T cells that suppress immune attack against the body's own cells and molecules; important in preventing autoimmune diseases

37.2 HOW DO NONSPECIFIC DEFENSES FUNCTION?

The ideal defenses against disease are barriers, such as the skin and mucous membranes, that prevent invaders from entering the body in the first place. If these barriers are breached, the body also has several nonspecific internal defenses that can kill a wide variety of invading microbes.

The Skin and Mucous Membranes Form Nonspecific External Barriers to Invasion

The first line of defense consists of the surfaces with direct exposure to the environment: the skin and the mucous membranes of the digestive, respiratory, and urogenital tracts.

The Skin, Its Secretions, and Harmless Microbes Reduce Invasion by Pathogens

The outer surface of human skin consists of dry, dead cells filled with proteins similar to those in hair and nails. The skin is protected by secretions from sweat glands and sebaceous (oil) glands. These secretions contain natural antibiotics, such as lactic acid, that inhibit the growth of many bacteria and fungi. In addition, the skin is host to several microbial ecosystems, including oily sites (scalp, face, and trunk), moist sites (the crook of your elbow, between your toes), and dry sites (most of the rest of your body). In these ecosystems, harmless microbes often produce secretions that inhibit the growth of pathogenic microbes. Nevertheless, skin almost always carries populations of potentially harmful bacteria such as *Streptococcus* and *Staphylococcus*. However, if you practice reasonable personal hygiene, their populations are usually fairly low, and they seldom penetrate through unbroken skin into the tissues below.

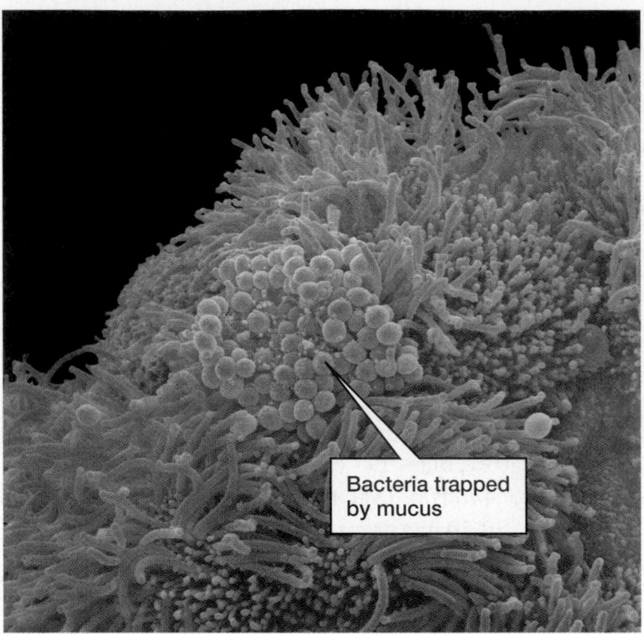

▲ **FIGURE 37-2 The protective function of mucus** Mucus traps microbes and debris in the respiratory tract. In this colored micrograph, bacteria are caught in mucus atop cilia. The cilia then sweep both the mucus and bacteria out of the body.

Mucus, Antibacterial Proteins, and Ciliary Action Defend the Mucous Membranes Against Microbes

Mucous membranes secrete mucus that traps microbes entering the nose or mouth (**FIG. 37-2**). Further, mucus contains antibacterial proteins, including *lysozyme,* which kills bacteria by digesting their cell walls, and *defensin,* which makes holes in bacterial plasma membranes. Finally, cilia on the membranes sweep up the mucus, microbes and all, until it is either coughed or sneezed out of the body or is swallowed.

If microbes are swallowed, they enter the stomach, where they encounter protein-digesting enzymes and extreme acidity, both of which are often lethal. Further along, the intestines contain bacteria that are harmless to people but that secrete substances that destroy invading bacteria or fungi. In females, acidic secretions and mucus help protect the vagina. Finally, fluids released by the body, including tears, urine, diarrhea, and vomit, help to expel invaders.

The Innate Immune Response Nonspecifically Combats Invading Microbes

Pathogens that penetrate the external barriers, for example, through a cut in the skin, encounter three types of nonspecific innate immune responses: (1) protection by white blood cells, including phagocytes and natural killer cells, (2) the inflammatory response, and (3) fever.

Phagocytes and Natural Killer Cells Destroy Invading Microbes

The body has a standing army of white blood cells, or **leukocytes**, many of which are specialized to attack and destroy invading cells. One brigade, collectively called **phagocytes**, ingests invaders by *phagocytosis* (see Chapter 5). Three important types of phagocytes are **macrophages** (literally, "big eaters"), **neutrophils**, and **dendritic cells**. These cells travel within the bloodstream, ooze out through capillary walls, and patrol the body's tissues (**FIG. 37-3a**). There, they consume bacteria and foreign substances that have penetrated the skin or mucous membranes (**FIG. 37-3b**). Dendritic cells and macrophages also play a critical role in the adaptive immune response, described later.

Another type of leukocyte, called the **natural killer cell**, does the job of providing nonspecific defense against harmful viruses. Viral pathogens enter body cells and use the cells' own metabolism to manufacture more viruses (see Fig. 14-2). Natural killer cells destroy virus-infected cells, often

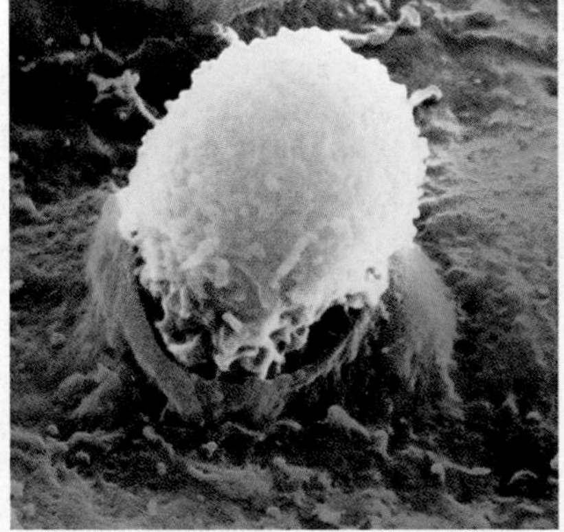

(a) A macrophage leaves a capillary and enters a wound

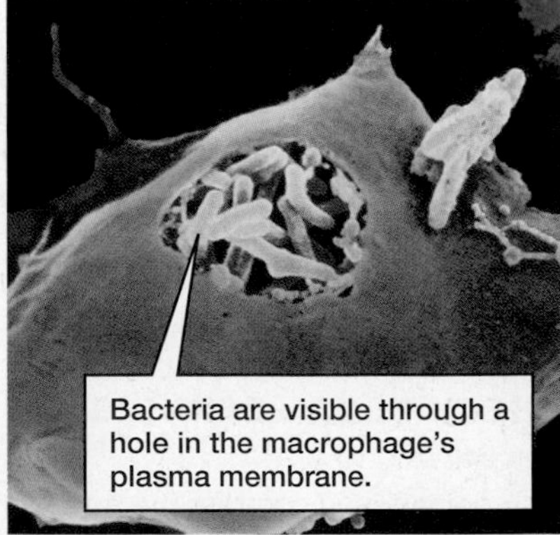

Bacteria are visible through a hole in the macrophage's plasma membrane.

(b) A macrophage stuffed with bacteria that it has ingested

▲ **FIGURE 37-3 The attack of the macrophages**

killing infected cells before the viruses have had enough time to reproduce and spread to other cells. Thus, natural killer cells can stop viral infections before they do very much damage to the body as a whole. (For examples of harmful viruses, see "Health Watch: Emerging Deadly Viruses" on page 651.) Natural killer cells are not phagocytic; instead, they release proteins that bore holes in the membranes of infected cells and secrete enzymes through the holes. The attacked cells soon die.

How do natural killer cells distinguish virus-infected cells from healthy cells? The surfaces of normal body cells display a set of proteins, collectively called the **major histocompatibility complex (MHC)**. MHC proteins differ between species and between individuals within a species. Therefore, MHC proteins are unique to each person (except identical twins, who have the same MHC proteins), and identify the body's own cells as "self." Natural killer cells patrol the body, killing any "non-self" cells that they encounter, while sparing self cells. Virus-infected cells often lack some MHC proteins, so they look like non-self to natural killer cells.

Most cancerous cells also have missing or altered MHC proteins. Therefore, they are identified as non-self by natural killer cells and are destroyed in the same way that virus-infected cells are.

The Inflammatory Response Attracts Phagocytes to Injured or Infected Tissue

A wound, with its combination of tissue damage and invading microbes, provokes an **inflammatory response**, which recruits phagocytes to the site of injury and walls off the injured area, isolating the infected tissue from the rest of the body. The inflammatory response causes injured tissues to become warm, red, swollen, and painful (*inflame* literally means "to set on fire").

A typical inflammatory response begins with an injury that damages cells and allows bacteria to enter the wounded area (**FIG. 37-4** ❶). Damaged cells release chemicals that cause certain cells in the connective tissue, called **mast cells**, to release **histamine** and other chemicals ❷, ❸. Histamine relaxes the smooth muscle surrounding arterioles, thereby increasing blood flow, and makes capillary walls leaky ❹. Extra blood flowing through leaky capillaries drives fluid from the blood into the wounded area. Thus, histamine accounts for the redness, warmth, and swelling of the inflammatory response. Other chemicals released by wounded cells and mast cells, and some substances produced by invading microbes, attract macrophages and other types of leukocytes. These cells squeeze out through the leaky capillary walls and ingest bacteria, dirt, and cellular debris ❺. In some cases, inflammation results in an accumulation of pus, a thick whitish mixture of dead bacteria, debris, and living and dead leukocytes.

Other chemicals released by injured cells initiate blood clotting (see Chapter 33). Clots plug up damaged blood vessels, reducing blood loss and preventing microbes from entering the bloodstream. Swelling and some of the chemicals released by the injured tissue cause pain, which usually leads to protective behaviors that reduce the likelihood of further injury.

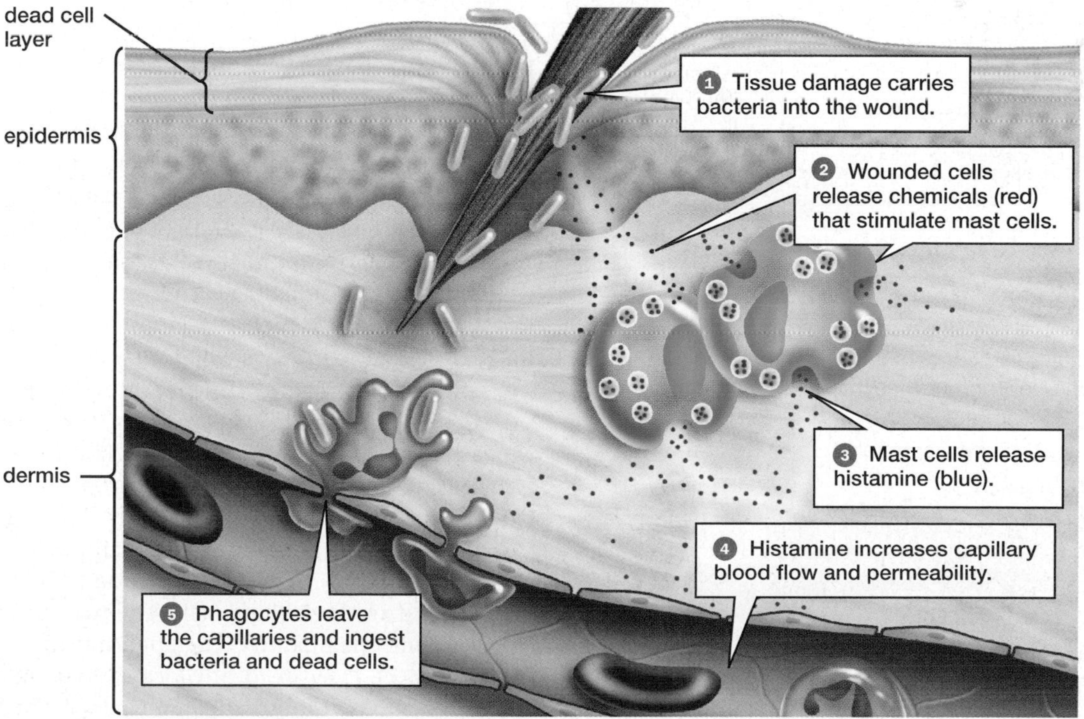

dead cell layer
epidermis
dermis
❶ Tissue damage carries bacteria into the wound.
❷ Wounded cells release chemicals (red) that stimulate mast cells.
❸ Mast cells release histamine (blue).
❹ Histamine increases capillary blood flow and permeability.
❺ Phagocytes leave the capillaries and ingest bacteria and dead cells.

▲ **FIGURE 37-4 The inflammatory response**

THINK CRITICALLY The inflammatory response causes fluid to leak from capillaries, so tissues swell up. Under what circumstances might this be dangerous, even life-threatening?

Fever Combats Large-Scale Infections

If invaders survive these defenses and cause a serious infection, they may trigger a **fever**, which both slows down microbial reproduction and enhances the body's own fighting abilities. The onset of fever is controlled by the hypothalamus, the part of the brain housing temperature-sensing nerve cells that serve as the body's thermostat. In humans, the thermostat is set at about 97° to 99°F (36° to 37°C). Certain types of bacteria, as well as the phagocytic cells that respond to an infection, produce chemicals called *pyrogens* (literally, "fire-producers"). Pyrogens travel in the bloodstream to the hypothalamus and raise the thermostat's set point. The body responds with increased fat metabolism (which generates more heat), constriction of blood vessels supplying the skin (which reduces heat loss through the skin), and behaviors such as shivering (which produce heat). Pyrogens also cause other cells to reduce the concentration of iron in the blood.

How does fever combat infection? An elevated body temperature increases the activity of phagocytic white blood cells while simultaneously slowing reproduction in some types of bacteria. The iron deficiency accompanying a fever also hampers bacterial multiplication. Meanwhile, the high temperature causes the cells of the adaptive immune system to multiply more rapidly, hastening the onset of an effective adaptive immune response. In addition, fever stimulates cells infected by viruses to produce a protein called interferon, which travels to other cells and increases their resistance to viral attack. Interferon also stimulates the natural killer cells that destroy virus-infected body cells.

On the other hand, high fevers are uncomfortable and may be dangerous: Extremely high fevers may cause brain damage, although this is rare. Because fevers have the potential to be either beneficial or harmful, you should consult your physician about whether you should take fever-reducing medicines such as aspirin, acetaminophen, or ibuprofen.

CHECK YOUR LEARNING

Can you . . .

- describe the external barriers to infection, including how they function and why they are nonspecific?
- name the major components of the innate immune response and describe how each of them combats invasion by microbes?

CASE STUDY CONTINUED

Flesh-Eating Bacteria

If phagocytes kill most species of bacteria that enter mucous membranes or a wound, how did infection kill Jim Henson and destroy Aimee Copeland's tissues? Both *S. pyogenes* and deadly strains of *A. hydrophila* are surrounded by polysaccharide capsules that phagocytic cells often do not recognize as foreign. Consequently, the phagocytes do not attack these bacteria. If such nonspecific defenses fail, can the adaptive immune response step in?

37.3 WHAT ARE THE KEY COMPONENTS OF SPECIFIC INTERNAL DEFENSES?

If the nonspecific defenses are breached, the body mounts a highly specific and coordinated response directed against the particular species that has successfully colonized the body. The essential features of this adaptive immune response were recognized more than 2,000 years ago by the Greek historian Thucydides. He observed that sometimes a person would contract a disease, recover, and never catch that particular disease again—the person had become immune. With rare exceptions, immunity to one disease confers no protection against other diseases. Thus, the adaptive immune response attacks one specific type of microbe, overcomes it, and provides future protection against that microbe but no others.

The adaptive immune response is produced by interactions among several types of leukocytes, including macrophages, dendritic cells, and lymphocytes, which collectively make up the **adaptive immune system**. Some cells, such as macrophages and dendritic cells, play a role in both the innate and adaptive immune responses, whereas **lymphocytes** are specialized white blood cells that are unique to the adaptive immune response. There are two types of lymphocytes: **B cells** and **T cells**. Both B cells and T cells arise from stem cells in bone marrow. Some of the resulting daughter cells complete their development in bone marrow, becoming B (for *bone*) cells. Others leave the marrow, travel through the circulatory system, and enter the thymus, where they develop into T (for *thymus*) cells.

As befits a system that must patrol the entire body for invading microbes, the adaptive immune system is distributed throughout the body, with concentrations of cells in specific locations, including bone marrow, vessels of the lymphatic system, lymph nodes, thymus, spleen, and patches of specialized connective tissue, such as the tonsils (**FIG. 37-5**).

The human body contains approximately 500 **lymph nodes** scattered along the lymph vessels. Lymph nodes contain masses of macrophages and lymphocytes lining narrow passages through which the lymph flows. When you have a disease that causes "swollen glands," these are actually lymph nodes swollen with leukocytes, bacteria, debris from dead cells, and fluid. The **thymus** is located beneath the breastbone slightly above the heart. It is large in infants and young children but starts to shrink after puberty. The **spleen** is a fist-sized organ located on the left side of the abdominal cavity, between the stomach and diaphragm. The spleen filters blood, exposing it to white blood cells that destroy microbes and aged red blood cells. The **tonsils** are located in a ring around the pharynx (the uppermost part of the throat; see Fig. 34-7). They are ideally located to sample microbes entering the body through the mouth. Macrophages and other leukocytes in the tonsils directly destroy many invading microbes and often trigger an adaptive immune response. Immune cells are also widely, if thinly, distributed in mucous membranes, especially those lining the airways, digestive tract, and vagina.

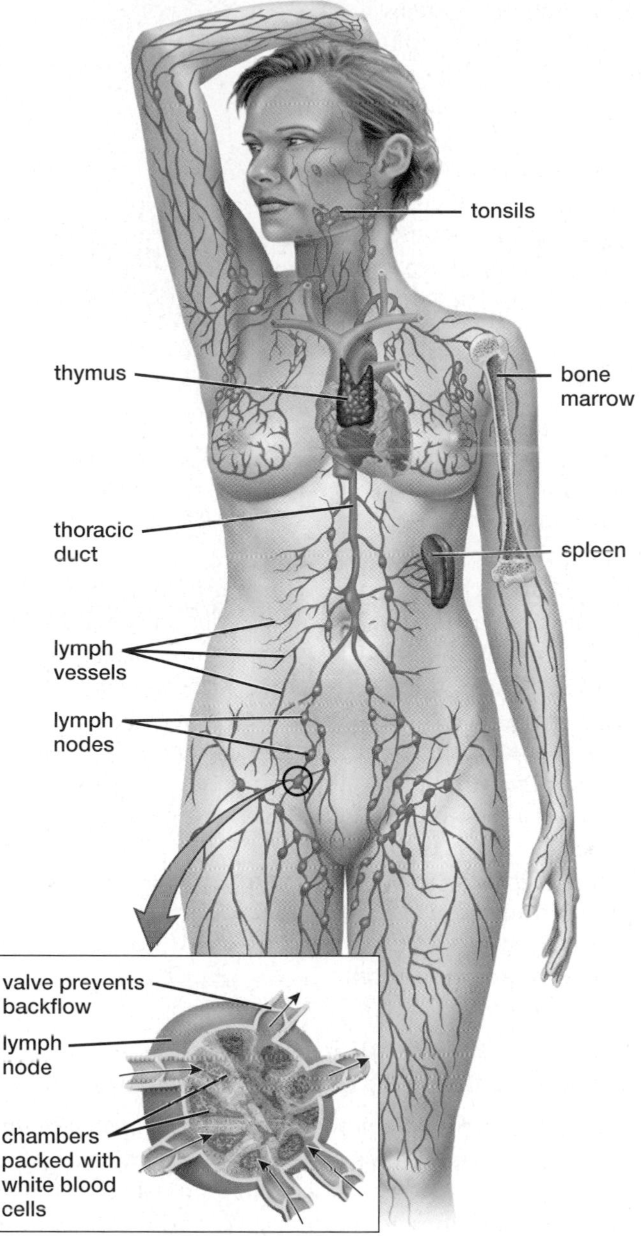

▲ **FIGURE 37-5 The lymphatic system contains much of the immune system** The cells of the immune system are formed in the bone marrow and mature either there or in the thymus. Most mature immune cells reside in the lymph nodes or spleen. As lymph travels through the lymph nodes, most microbial invaders are attacked and killed.

Many different proteins are involved in the adaptive immune response. For example, there are dozens of different **cytokines**, produced by a wide variety of cells, including macrophages and lymphocytes. Cytokines are used for communication between cells. Their functions include stimulating cell division in lymphocytes during the immune response, stimulating the inflammatory response, and enhancing the resistance of cells to viral infection. **Antibodies**, which are proteins produced by B cells and their offspring, help the immune system to recognize invading microbes and destroy them. **Complement** proteins are mostly synthesized by the liver and circulate in the blood plasma. They assist the immune system in killing invading microbes. Although antibodies are exclusively part of the adaptive immune response, some cytokines and complement proteins are involved in both the innate and adaptive responses.

All adaptive immune responses include the same three steps. First, lymphocytes recognize an invading microbe, identifying it as non-self; second, they launch an attack; and third, they retain a memory of the invader that allows them to ward off future infections by the same type of microbe.

CHECK YOUR LEARNING

Can you . . .

- describe the adaptive immune response and explain how it differs from the innate immune response?
- list the major components of the adaptive immune system and describe how each component functions?

37.4 HOW DOES THE ADAPTIVE IMMUNE SYSTEM RECOGNIZE INVADERS?

To understand how the adaptive immune system recognizes invaders and initiates a response, we must answer three related questions: (1) How do lymphocytes recognize foreign cells and molecules? (2) How can lymphocytes produce specific responses to so many different types of cells and molecules? (3) How do they avoid mistaking the body's own cells and molecules for invaders?

The Adaptive Immune System Recognizes Invaders' Complex Molecules

Large, complex molecules—usually proteins, polysaccharides, or glycoproteins—that can trigger an adaptive immune response are called **antigens**, which is short for *anti*body-*gen*erating molecules. Antigens are often located on the surfaces of invading bacteria, fungi, or other microbes. Viral antigens may become incorporated into the plasma membranes of the body cells that they infect. Dendritic cells and macrophages that engulf viruses or bacteria also "display" antigens from the microbes on their plasma membranes. Other antigens, such as toxins released by bacteria, may be dissolved in blood plasma, lymph, or interstitial fluid.

However, antigens are not confined to invading microbes and their toxins. Any large, complex organic molecule, including molecules of the cells of your own body, may be an antigen and provoke an adaptive immune response. So, if you donate a kidney to another person, antigens on your kidney can trigger an immune response in the recipient. Without both careful tissue matching and drugs that suppress the immune system, the recipient's adaptive immune response will attack the transplant.

Antibodies and T-Cell Receptors Trigger Immune Responses to Antigens

Lymphocytes produce two types of proteins that bind to specific antigens and trigger an immune response to them: antibodies and T-cell receptors.

Antibodies are manufactured only by B cells and their offspring. They are Y-shaped proteins composed of two pairs of peptide chains: one pair of identical large (heavy) chains and one pair of identical small (light) chains (**FIG. 37-6**). Both heavy and light chains consist of a **variable region**, which differs among antibodies, and a **constant region**, which is the same in all antibodies of a given category (see below).

The light and heavy chains combine to form the two functional parts of an antibody: the "arms" and the "stem" of the Y. The tips of the arms contain the variable regions, which form sites that bind antigens (see Fig. 37-6). Each binding site has a particular size, shape, and electrical charge, which allow only certain antigen molecules to fit in and bind. The sites are so specific that each antibody can bind only a few, very similar types of antigens. Every B cell produces its own unique antibody, with variable regions that differ from those of the antibodies produced by all other B cells.

The stem of the Y is composed of the constant region of the heavy chains. The stem determines where the antibody is located (for example, on the surface of a B cell or secreted into breast milk) and what role it plays in the immune response. There are five major categories of antibodies, given the abbreviations IgM, IgD, IgG, IgA, and IgE. These differ in location and function. IgM and IgD antibodies are attached to the surfaces of B cells. When an IgM or IgD antibody on the surface of a B cell binds an antigen, this triggers multiplication of the B cell. IgG antibodies are the most common category found in blood, lymph, and interstitial fluid; these are the major antibodies that help to destroy invading microbes. IgG antibodies also cross the placenta and defend the developing fetus against disease. IgA antibodies are secreted onto the surfaces of the digestive and respiratory tracts and into saliva and breast milk; they help to combat invaders entering via the mouth and nasal passages and provide temporary defenses for nursing infants whose immune systems have not yet fully developed. IgE antibodies are responsible for allergic reactions.

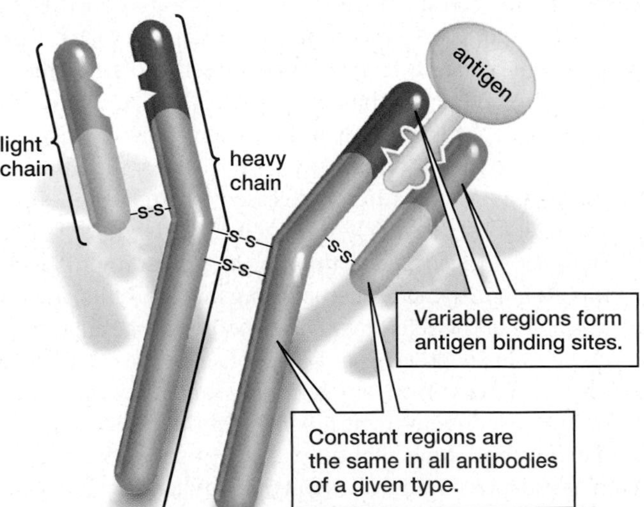

▲ **FIGURE 37-6 Antibody structure** Antibodies are Y-shaped proteins composed of two pairs of peptide chains (light chain and heavy chains). Constant regions form the stem of the Y. The variable regions on the two chains form a specific binding site at the end of each arm of the Y. Different antibodies have different variable regions, forming unique binding sites.

Antibodies serve two functions in the adaptive immune response: (1) recognizing foreign antigens and triggering the response against the invaders and (2) helping to destroy invading cells or molecules that bear the antigens (described in Section 37.5). In an antibody's recognition role, the stem of an antibody anchors it in the plasma membrane of the B cell that produced it. The antibody's two arms stick out from the B cell, sampling the blood and lymph for antigen molecules (**FIG. 37-7**). When an arm of the antibody encounters an antigen with a compatible chemical structure, it binds to it. Antigen–antibody binding stimulates the B cell to divide.

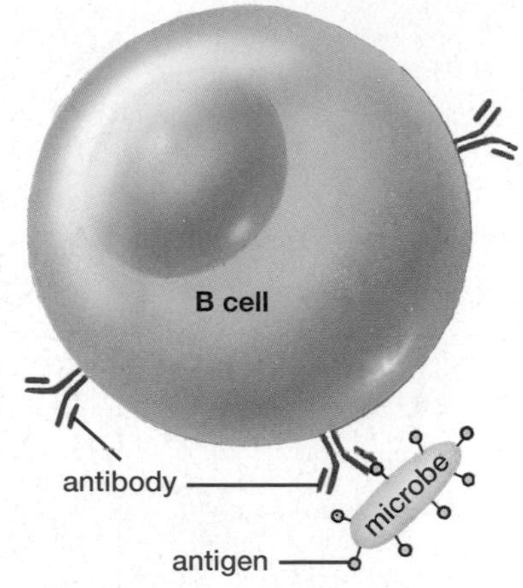

▲ **FIGURE 37-7 Antibodies on the surface of B cells bind to antigens on invading microbes**

T cells have a different type of protein, called the **T-cell receptor**, that recognizes and binds antigens. Every T cell produces T-cell receptors that differ from those of all other T cells. Variable regions, which form the antigen binding sites, protrude from the cell surface. When an antigen binds to a T-cell receptor, the T cell becomes activated in ways that depend on the type of T cell.

The Adaptive Immune System Can Recognize Millions of Different Antigens

During your lifetime, your body will be challenged by a multitude of invaders. Your classmates may sneeze cold and flu viruses into the air you breathe. Your food may contain bacteria or molds. A mosquito carrying West Nile virus may bite you. Fortunately, the adaptive immune system recognizes and responds to millions of potentially harmful antigens. How can it accomplish this remarkable feat?

B and T cells cannot anticipate the future to design and build antibodies and T-cell receptors with variable regions that are just right for binding antigens on the next pathogen that happens to invade your body. Rather, B and T cells randomly synthesize millions of different antibodies and T-cell receptors. At any given time, the human body contains perhaps 100 million different antibodies. The number of different T-cell receptors is even higher. This vast array of antibodies and receptors is simply there, waiting. Think of clothing hanging on racks in a department store. The clothing wasn't designed to your specific body measurements, but given enough garments to choose from, you can probably find something that fits fairly well. Similarly, although antibodies and T-cell receptors are not synthesized to match any particular antigen, an invading antigen will almost always encounter at least a few antibodies and T-cell receptors that will bind it.

Emerging Deadly Viruses

Viral diseases have plagued humanity for millennia—Egyptian mummies from at least 3,000 years ago, including Pharaoh Ramses V, have what appear to be smallpox scars on their faces. But in recent decades, deadly viruses seem to be emerging with frightening frequency, including HIV, West Nile, Zika, avian flu, swine flu, and Ebola.

Ebola

The first cases of Ebola virus disease occurred in equatorial Africa in 1976, infecting 602 people, 70% of whom died. There were sporadic small outbreaks in Africa in succeeding decades. Then, in 2014, Ebola exploded into an epidemic that infected tens of thousands, with about a 40% fatality rate.

The Ebola virus is so deadly because it both evades the immune system and damages vital organs. Normally, when a virus invades the body, the innate immune response steps in: Virus-infected cells release interferon, which increases the resistance of other, not-yet-infected cells. Ebola, however, reduces the ability of infected cells to produce interferon and the ability of healthy cells to respond appropriately. Ebola also sabotages the adaptive immune response. In most viral diseases, virus-infected cells display antigens to macrophages and dendritic cells, which start up the adaptive response. But Ebola infects macrophages and dendritic cells; instead of kick-starting the adaptive response, these cells are commandeered into making more Ebola viruses (**FIG. E37-1a**). This flood of viruses infects B and T cells and kills them, further reducing the adaptive response.

Meanwhile, the rapidly increasing numbers of viruses infect almost all the cells of the body. Cells that form the lining of blood vessels die, so the blood vessels leak. To make things worse, Ebola infection also gradually shuts down blood clotting. The result is catastrophic failure of the vascular system: As the cells lining the blood vessels die in ever-greater numbers, the vessels develop bigger and bigger leaks that cannot be plugged with clots. Blood vessels become so leaky that the victim bleeds both internally and externally.

Eventually, the immune system does respond to the infection, but the response is often the death blow for the weakened body: a "cytokine storm." Immune cells release so many cytokines that they cause an overwhelming, body-wide inflammatory response. Severe vomiting and diarrhea also occur. And at this time, the blood and other body fluids are full of Ebola viruses. Contact with any of these body fluids can be a death sentence for relatives, caregivers, and medical personnel.

With intensive medical care, many people do recover from Ebola. Their blood, which contains anti-Ebola antibodies, has sometimes been used as a source of effective treatment for newly infected victims.

Avian and Swine Flu

Every winter, a wave of seasonal human flu sweeps across the world. The flu virus invades cells of the respiratory tract, multiplying inside the cells and then killing them as new generations of viruses emerge. Hundreds of thousands of the elderly, newborn, or infirm die, because the flu worsens preexisting diseases or because their immune systems are weak. Most of the millions of healthy adults who catch the flu each year suffer respiratory distress, fever, and muscle aches, but they survive because their adaptive immune systems eventually overcome the virus. Their immune systems also produce memory cells that lie in wait for the next flu season.

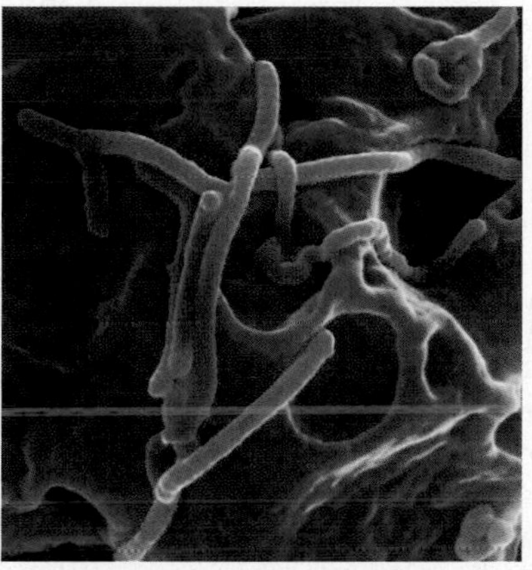
(a) Ebola virus

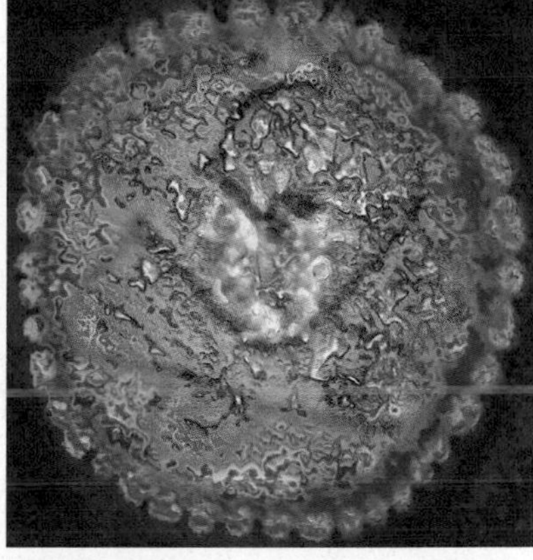
(b) Avian flu virus

▲ **FIGURE E37-1 Lethal viruses (a)** Ebola viruses (red) emerging from an infected cell. **(b)** The "beads" on the outside of the avian flu virus are proteins that allow the virus to enter and leave its host cells.

Flu viruses have a high mutation rate, causing the virus to change from year to year. Further, in some years, fundamentally new strains of flu arise when flu viruses that normally infect other animals mutate and become infectious to people or when genes from human and bird or pig viruses combine in a single virus. Modern avian flu (**FIG. E37-1b**) first appeared in 1997 in Hong Kong. Health authorities concluded that the flu came from infected chickens, so the government ordered every chicken in Hong Kong—about 1.5 million birds—to be killed. Nevertheless, avian flu spread across the world, with deaths reported in countries as far apart as Turkey, Indonesia, and Egypt. About 60% of the infected people died.

Fortunately, today's avian flu isn't very efficient at passing from birds to people and seems almost totally incapable of passing from person to person. Public health officials keep a watchful eye, however, in case the virus acquires mutations that might make person-to-person transmission more likely.

The H1N1 swine flu virus is a combination of genes from human, avian, and pig viruses. Some strains of swine flu can be extremely infectious and lethal: In 1918, swine flu infected perhaps 500 million people worldwide, killing more than 50 million of them. The modern H1N1 swine flu virus is much less deadly, with a fatality rate below 0.1%. Nevertheless, in the 2015–2016 flu season, the U.S. Centers for Disease Control and Prevention estimated that H1N1 flu infected about 25 million Americans, causing more than 11,000 deaths.

THINK CRITICALLY Makayla, a 19-year-old college student, visits the student health center at the beginning of the school year and asks the physician why she needs a flu vaccine this year, considering she just had the flu last winter. Describe the advice the physician should give her, including why Makayla could catch the flu again this year and why she should get the flu vaccine.

The Adaptive Immune System Distinguishes Self from Non-Self

As we described earlier, the cells of your body manufacture thousands of complex molecules that can stimulate immune responses in other people's bodies, causing, for example, transplant rejection. Why don't your body's antigens arouse your own immune system? Two important mechanisms normally prevent self-immunity: the continuous presence of the body's own antigens during immune cell development and modulation of the immune response by regulatory T cells.

When B and T cells first form in the bone marrow, they can bind antigens, but they are not yet able to trigger an immune response. Instead, if these immature cells bind antigen, they undergo *apoptosis*, or programmed cell death, in which they essentially commit cellular suicide. Obviously, immature immune cells are constantly exposed to the body's own antigens, so almost all B and T cells that happen to produce "self-reactive" antibodies or T-cell receptors are eliminated before they mature. B and T cells that produce antibodies or T-cell receptors that can only bind foreign antigens usually do not encounter those antigens while they are still developing. Therefore, these "foreign-reactive" cells survive, mature, and can then trigger an immune response if the appropriate foreign antigens later invade the body.

Not all self-reactive B and T cells are eliminated from the body during immune cell development. Any surviving self-reactive cells are suppressed by a type of T cell called a **regulatory T cell**. Regulatory T cells prevent self-reactive lymphocytes from attacking the body's own cells.

CHECK YOUR LEARNING

Can you . . .

- explain the mechanisms by which the adaptive immune system recognizes invading microbes?
- describe the structure of antibodies and how antibodies form specific binding sites for antigens?
- explain why the immune system does not attack the body's own molecules and cells?

37.5 HOW DOES THE ADAPTIVE IMMUNE SYSTEM ATTACK INVADERS?

The benefit to having millions of unique B and T cells is that almost any invader will bear antigens that can bind to a few antibodies and T-cell receptors and stimulate an adaptive immune response. The drawback is that there will be only a few immune cells that can recognize any given invader, and a handful of cells isn't enough to kill the invaders immediately. To get a good immune response, the responding cells must first multiply and differentiate, a process that takes 1 or 2 weeks. In the meantime, you may become ill and may even die, if the development of the immune response loses its race with the multiplying microbes and the damage they cause to your body.

When faced with invading microbes, the adaptive immune system simultaneously launches two types of attack: humoral immunity and cell-mediated immunity.

Humoral Immunity Is Produced by Antibodies Dissolved in the Blood

Humoral immunity is provided by B cells and the antibodies they secrete into the bloodstream (**FIG. 37-8**). Each of the millions of different B cells in your body bears its own unique type of antibody on its surface. When a microbe enters the body, the antibodies on only a few of these B cells can bind to antigens on the invader ❶. Antigen–antibody binding causes these B cells, but no others, to multiply rapidly ❷. This process is called **clonal selection**, because the antigens "select" which B cells will multiply, and the resulting daughter cells are clones—cells that are genetically identical to the selected B cells ❸.

The daughter cells differentiate into two cell types: *memory B cells* and **plasma cells** ❹. Memory B cells do not release antibodies, but they play an important role in future immunity to the invader that stimulated their production, as we will see shortly. Plasma cells become enlarged and packed with rough endoplasmic reticulum, which synthesizes huge quantities of antibodies. These antibodies are released into the bloodstream ❺ (hence the name "humoral" immunity; to the ancient Greeks, blood was one of the four "humors," or body fluids). The secreted antibodies have the same antigen binding site that was found in the antibodies located on the surface of the original parent B cell.

Humoral Antibodies Have Multiple Modes of Action Against Invaders

Antibodies in the blood combat invading molecules or microbes in three principal ways. First, the circulating antibodies may bind to a foreign molecule, virus, or cell, and render it harmless, a process called neutralization. For example, if the active site of a toxic enzyme in snake venom is covered with antibodies, it cannot harm your body (**FIG. 37-9a**). Many viruses gain entry into your body's cells when a protein on the virus binds to a molecule on the surface of a cell. If antibodies cover up this viral protein, the neutralized virus cannot enter a cell.

Second, antibodies may coat the surface of invading molecules, viruses, or cells, and make it easier for macrophages and other phagocytes to destroy them (**FIG. 37-9b**). Remember, the variable regions on the "arms" of an antibody bind to antigens on invaders, so the constant regions that make up the "stems" of humoral antibodies stick out into the blood or interstitial fluid. Macrophages bind to the antibody stems, engulf the antibody-coated invaders, and digest them.

Third, when antibodies bind to antigens on the surface of a microbe, the antibodies interact with complement proteins that are always present in the blood. Some complement proteins punch holes in the plasma membrane of the microbe, killing it. Other complement proteins make it easier for phagocytes to ingest the invaders.

Humoral Immunity Fights Invaders That Are Outside Cells

Antibodies are large proteins that cannot readily cross plasma membranes, so they usually do not enter cells. Therefore, a humoral response is effective against microbes or toxins only when they are outside cells, generally in the blood, lymph, or interstitial fluid. Bacteria, bacterial toxins, and some fungi and protists are usually vulnerable to the humoral immune response. Viruses are exposed to the humoral response when they are outside of the body's cells—for example, when they are spreading from cell to cell in the interstitial fluid—but are safe from antibody attack when they are inside a cell. Fighting viral infections, therefore, requires help from the cell-mediated immune response.

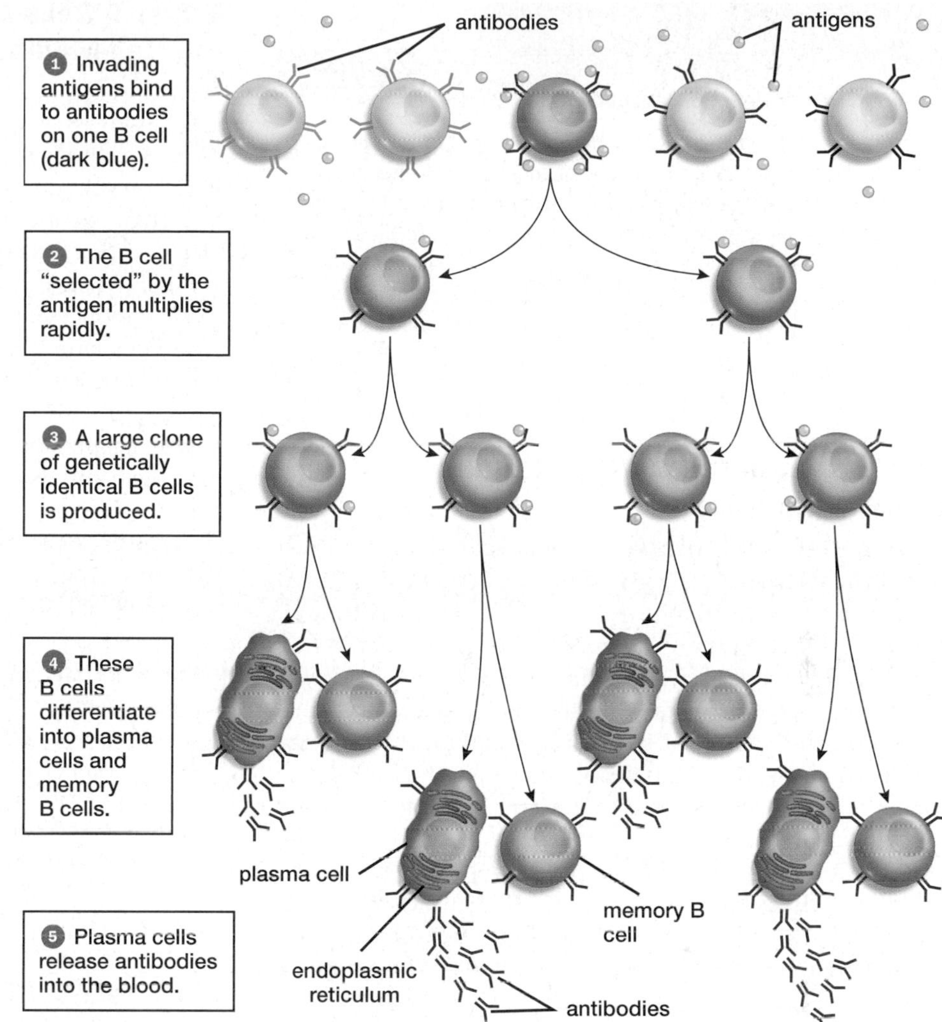
antibodies
antigens
1 Invading antigens bind to antibodies on one B cell (dark blue).
2 The B cell "selected" by the antigen multiplies rapidly.
3 A large clone of genetically identical B cells is produced.
4 These B cells differentiate into plasma cells and memory B cells.
plasma cell
memory B cell
5 Plasma cells release antibodies into the blood.
endoplasmic reticulum
antibodies

▲ **FIGURE 37-8 Clonal selection of B cells by invading antigens**

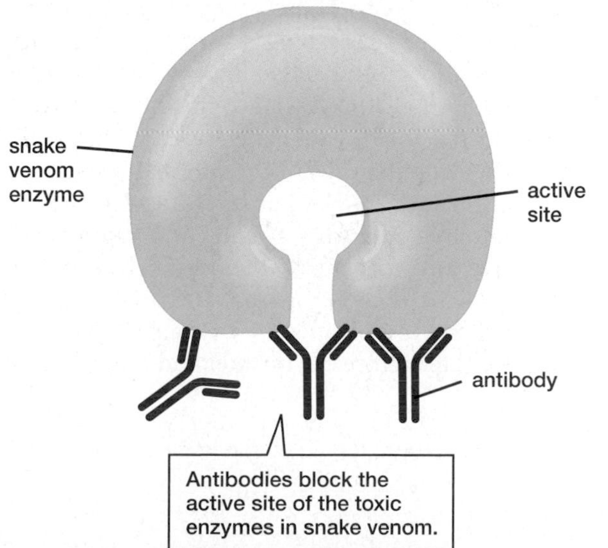
snake venom enzyme
active site
antibody
Antibodies block the active site of the toxic enzymes in snake venom.

(a) Antibodies neutralize toxic molecules by covering up their active sites

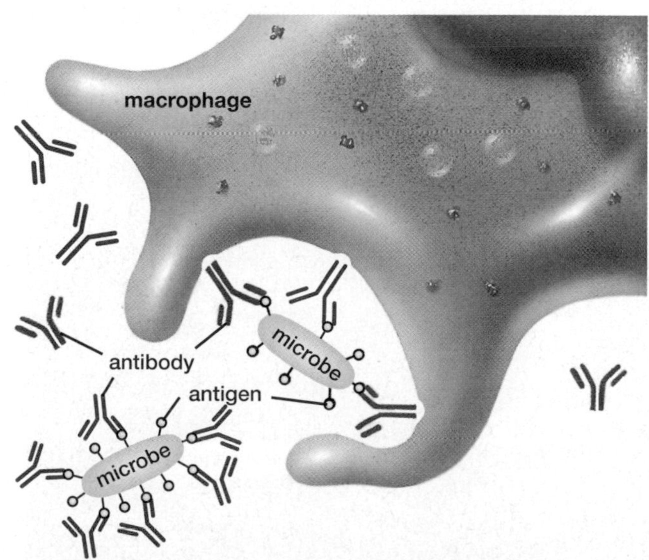
macrophage
antibody
antigen
microbe
microbe

(b) Antibodies bind to antigens on a microbe and promote phagocytosis by macrophages

▲ **FIGURE 37-9 Antibody action**

CASE STUDY CONTINUED

Flesh-Eating Bacteria

S. pyogenes and some strains of *A. hydrophila* produce proteins on their surfaces that inhibit the binding of complement proteins, thus protecting the bacteria from complement-stimulated phagocytosis. Fortunately, antibodies secreted by plasma cells bind to these bacterial proteins and stimulate phagocytosis by macrophages. But the very antibodies that help to destroy *S. pyogenes* sometimes turn against the victim's own body, as we describe in Section 37.8.

Cell-Mediated Immunity Is Produced by Cytotoxic T Cells

Cell-mediated immunity is produced by a type of T cell called the **cytotoxic T cell**, which attacks virus-infected body cells and cells that have become cancerous. Although the process is complex, in essence, it works like this: When a cell is infected by a virus, some pieces of viral proteins are brought to the surface of the infected cell and displayed on the outside of its plasma membrane. Cytotoxic T cells, each bearing its own unique type of T-cell receptor, drift about, occasionally bumping into the displayed viral antigens. When a cytotoxic T cell with an appropriate matching T-cell receptor binds to a viral antigen, the cytotoxic T cell secretes proteins onto the surface of the infected cell. These proteins form pores in the infected cell's plasma membrane. Enzymes, also secreted by the cytotoxic T cell, pass through the pores, killing the infected cell. If the infected cell is killed before the virus has finished multiplying, then no new viruses are produced, and the viral infection cannot spread to other cells. Cancer cells may be killed by the same mechanism, because they often display unusual proteins on their surfaces that cytotoxic T cells recognize as foreign (**FIG. 37-10**).

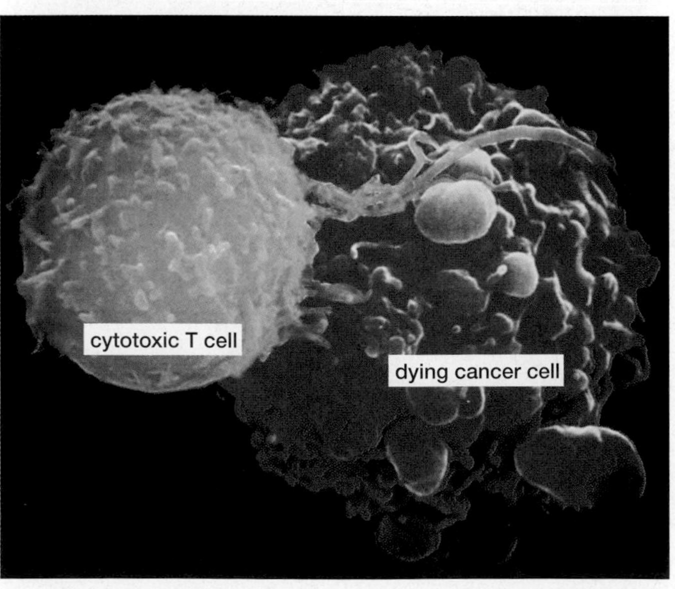

FIGURE 37-10 Cell-mediated immunity in action This scanning electron micrograph captures a cytotoxic T cell in the act of attacking a cancer cell.

Helper T Cells Enhance Both Humoral and Cell-Mediated Immune Responses

B cells and cytotoxic T cells cannot fight microbial invasions by themselves; they require the assistance of **helper T cells**. Helper T cells bear T-cell receptors that bind to antigens displayed on the surfaces of dendritic cells or macrophages that have engulfed and digested invading microbes. Only helper T cells bearing matching T-cell receptors can bind to any particular antigen. When its receptor binds an antigen, a helper T cell multiplies rapidly. Its daughter cells differentiate and release cytokines that stimulate cell division and differentiation in both B cells and cytotoxic T cells. In fact, B cells and cytotoxic T cells usually make a significant contribution to defense against disease only if they simultaneously bind to an antigen and receive stimulation by cytokines from helper T cells.

FIGURE 37-11 compares the humoral immune response with the cell-mediated immune response and shows the role of helper T cells in both.

CHECK YOUR LEARNING

Can you . . .

- describe the processes by which the adaptive immune system works, including humoral immunity and cell-mediated immunity?
- explain why helper T cells are so important in the adaptive immune response?

37.6 HOW DOES THE ADAPTIVE IMMUNE SYSTEM REMEMBER ITS PAST VICTORIES?

After recovering from a disease, you remain immune to the particular microbe that caused it for many years, perhaps a lifetime. How does the adaptive immune system accomplish this feat? Although the plasma cells and cytotoxic T cells that conquered the disease generally live only a few days, some of the daughter cells of the original B cells, cytotoxic T cells, and helper T cells that responded to the infection differentiate into **memory B cells** and **memory T cells** that survive for many years. Memory cells bear the same antibodies or T-cell receptors that the original parent cells did. If the body is reinvaded by the same type of microbe, memory cells recognize the invader and mount an immune

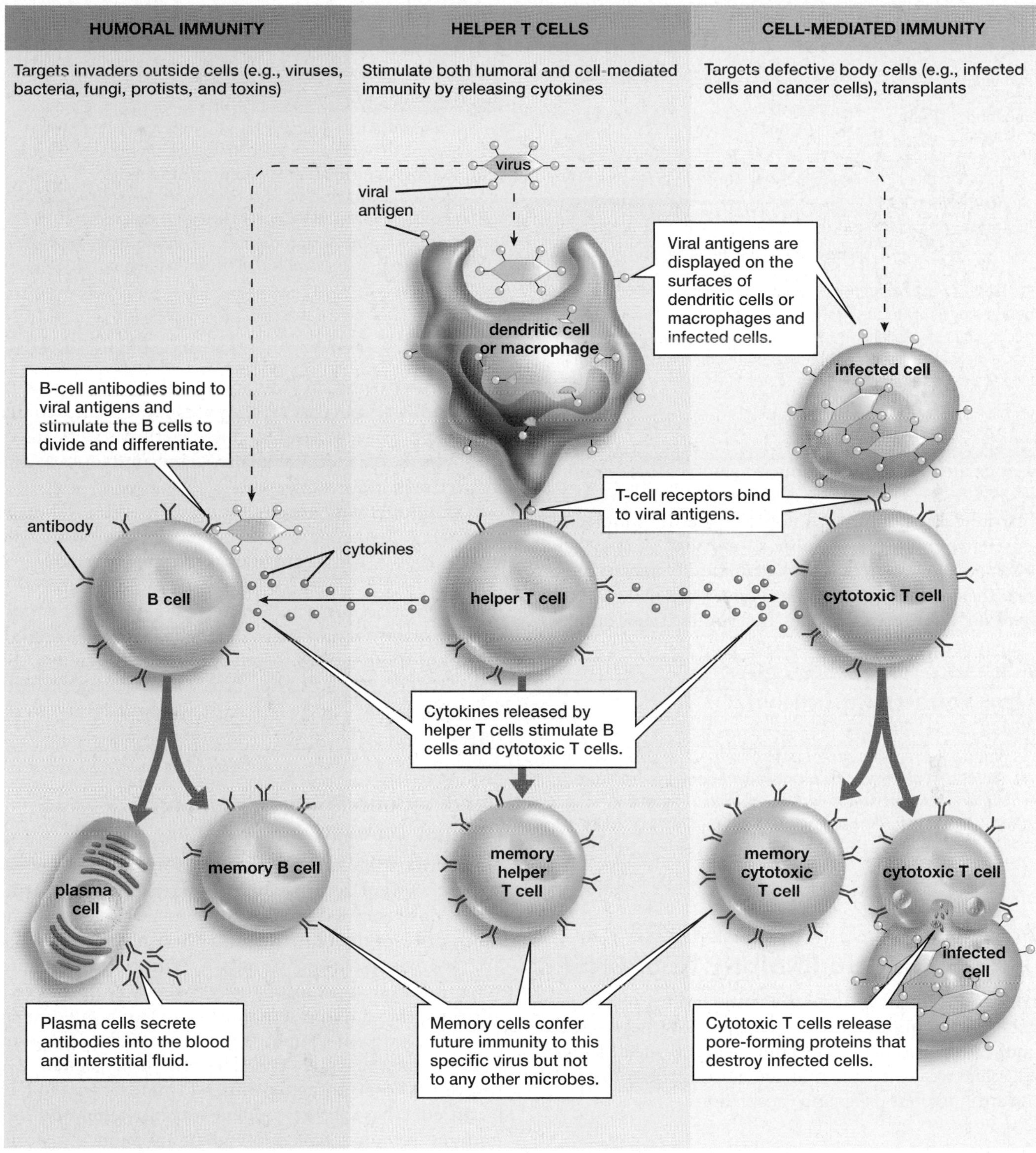

HUMORAL IMMUNITY
HELPER T CELLS
CELL-MEDIATED IMMUNITY
Targets invaders outside cells (e.g., viruses, bacteria, fungi, protists, and toxins)
Stimulate both humoral and cell-mediated immunity by releasing cytokines
Targets defective body cells (e.g., infected cells and cancer cells), transplants
virus
viral antigen
Viral antigens are displayed on the surfaces of dendritic cells or macrophages and infected cells.
dendritic cell or macrophage
infected cell
B-cell antibodies bind to viral antigens and stimulate the B cells to divide and differentiate.
T-cell receptors bind to viral antigens.
antibody
cytokines
B cell
helper T cell
cytotoxic T cell
Cytokines released by helper T cells stimulate B cells and cytotoxic T cells.
plasma cell
memory B cell
memory helper T cell
memory cytotoxic T cell
cytotoxic T cell
infected cell
Plasma cells secrete antibodies into the blood and interstitial fluid.
Memory cells confer future immunity to this specific virus but not to any other microbes.
Cytotoxic T cells release pore-forming proteins that destroy infected cells.

▲ **FIGURE 37-11 A summary of humoral and cell-mediated immune responses**

response. Memory B cells rapidly produce a clone of plasma cells, secreting antibodies that combat this second invasion. Memory T cells produce clones of helper T cells and cytotoxic T cells that are also specific for the "remembered" invader. There are far more memory cells than original B, cytotoxic T, or helper T cells that responded to the first infection. Further, memory cells respond faster than the original parent cells could. Therefore, memory cells usually produce an immune response to a second infection that is so fast and so large that the body fends off the attack before

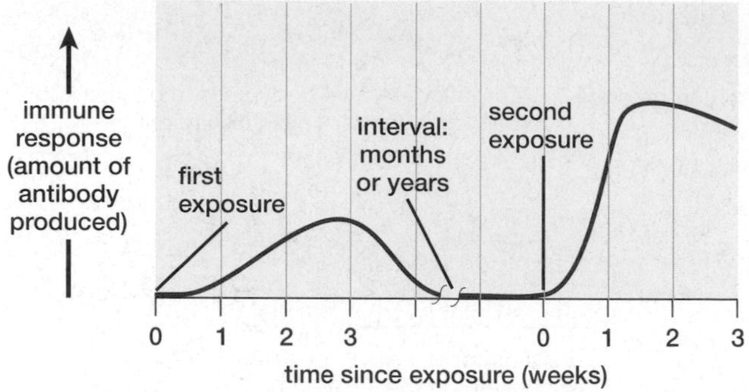

▲ **FIGURE 37-12 Acquired immunity** The immune system responds sluggishly to the first exposure to a disease-causing organism as B and T cells are selected and multiply. A second exposure activates memory cells formed during the first response, making the second response both faster and larger.

you suffer any disease symptoms: You have become immune (**FIG. 37-12**).

Acquired immunity confers long-lasting protection against many diseases—smallpox, measles, mumps, and chicken pox, for example. Acquired immunity may fail, however, if the disease organisms evolve rapidly. Rapid evolution may result in adaptive changes that produce new antigens that memory cells do not recognize.

CHECK YOUR LEARNING

Can you . . .

- explain why the immune response to the first exposure to a microbe is relatively slow and why responses to future infections by the same microbe are much faster?
- describe the role of memory cells in acquired immunity?

37.7 HOW DOES MEDICAL CARE ASSIST THE IMMUNE RESPONSE?

For most of human history, the battle against disease was fought by the immune response alone. Now, however, the immune response has a powerful assistant: medical treatment. Here we describe two of the most important medical tools: antimicrobial drugs and vaccinations.

Antimicrobial Drugs Kill Microbes or Slow Down Microbial Reproduction

Antibiotics are chemicals that help to combat infection by killing or interfering with the multiplication of bacteria, fungi, or protists. Although antibiotics usually do not destroy every disease-causing microbe in the body, they may kill enough to give the immune system time to finish the job. However, antibiotics are potent agents of natural selection, favoring the survival and reproduction of microbes that can withstand their effects (see the Case Study "Evolution of a Menace" in Chapter 16). Mutant microbes that are resistant to an antibiotic will pass on the genes for resistance to their offspring. The result: Resistant microbes thrive, while susceptible microbes die off. Eventually, many antibiotics become ineffective in treating diseases.

Antibiotics are not useful against viruses, because they target metabolic processes that viruses do not possess. However, **antiviral drugs** are now available that target different stages of the viral cycle of infection, including attachment to a host cell, replication of viral parts, assembly of new viruses within the host cell, and the release of these viruses to infect more of the body's cells (see Chapter 20). Antiviral drugs are available to treat HIV, herpes (cold and genital sores), hepatitis B and C, and flu viruses.

Have You Ever Wondered . . .

Why You Get Colds So Often?

If the immune system is so good at remembering past assaults and developing long-lasting immunity, why do many people get several colds every year? Because there are more than 200 different viruses that cause cold symptoms, and immunity to one usually doesn't confer immunity to the others. Cold viruses also mutate rapidly, and some of the mutations allow an "old" virus to look "new" to the immune system. Thus, a large number of constantly mutating cold viruses keeps one step ahead of the human immune system.

Vaccinations Produce Immunity Against Disease

A **vaccine** stimulates an immune response by exposing a person to antigens produced by a pathogen. Vaccines often consist of weakened or killed microbes (that cannot cause disease) or some of the pathogen's antigens, usually synthesized using genetic engineering techniques. When the body is exposed to a weakened pathogen or its antigens, the immune system produces an army of memory cells that confer immunity against living, dangerous microbes of the same type. Like immunity acquired by recovering from an actual illness, immunity stimulated by an effective vaccine produces such a rapid and large immune response to a subsequent invasion by the living pathogen that you don't experience any symptoms at all. Learn more about vaccines and their origin in "Doing Science: Discovering How to Prevent Infectious Diseases."

CHECK YOUR LEARNING

Can you . . .

- explain how antibiotics, antiviral drugs, and vaccines supplement the human immune response?
- describe how antibiotic resistance arises?

DOING Science Discovering How to Prevent Infectious Diseases

It has been known for thousands of years that people gain lifelong immunity after recovering from certain diseases. Today, vaccination allows us to obtain such immunity without ever being sick. The story of vaccination unfolded over centuries, beginning with informal attempts to replicate the natural process of disease-caused immunity, and eventually moving to formal scientific investigation.

What Question Was Asked?

The first application of vaccines focused on smallpox, so named because of the pus-filled blisters that disfigure its victims. Historically, the smallpox virus killed about 30% of its victims. Some outbreaks were very lethal, others less so. Early attempts to protect people from smallpox took place over 1,000 years ago, in China. Pioneering practitioners dipped a sharp knife in pus from a person with a mild case of the disease and used it to cut the skin of uninfected people. Although some of the inoculated people died, most developed relatively minor symptoms and resisted later exposure to severe forms of smallpox. By the 1700s, investigators intrigued by such folk remedies began to ask whether the potentially beneficial effects of inoculation could be more rigorously tested. The most fruitful of these tests followed up on a key observation: It appeared that dairy farmers and milkmaids, who often contracted a disease called cowpox, seldom caught smallpox.

How Was Evidence Gathered?

In 1796, Edward Jenner, an English biologist and surgeon, inoculated an 8-year-old boy, James Phipps, with fluid from cowpox blisters on the hand of a milkmaid. A few months later, Jenner tested Phipps for immunity by infecting him with smallpox. Jenner subsequently repeated the procedure several times with other children, and published his findings.

What Was Learned?

James Phipps remained healthy after being infected with smallpox, and so did the other inoculated children. Jenner had demonstrated that inoculation with cowpox provided protection against smallpox, and his method was rapidly adopted in Europe and eventually worldwide. We now know that the inoculation worked because cowpox and smallpox viruses have some extremely similar antigens, so the immune response to cowpox also protects against smallpox.

Despite the success of inoculation against smallpox, nearly a century passed before the technique was applied to other infectious diseases. In the late 1800s, the French microbiologist Louis Pasteur, who was among the first to recognize the role of microbes in causing disease, discovered that inoculation with weakened cholera bacteria provided protection against that disease. He coined the term "vaccine" (L., *vacca*, meaning "cow"), in recognition of Jenner's pioneering work. Pasteur later applied the technique to anthrax in sheep and to rabies in humans.

How effective are vaccines? For the measles vaccine, see for yourself in **FIGURE E37-2**. Vaccines have eradicated smallpox worldwide, except for a few vials of virus kept in labs in the United States and Russia for research purposes. Vaccines have also nearly eliminated polio; fewer than 30 cases occurred in 2017, confined to a handful of countries in Asia and Africa. Vaccines are not only medically effective, they're cost-effective, too. Vaccines reduce direct medical costs in the United States by more than $13 billion each year and overall costs to society by about $69 billion annually.

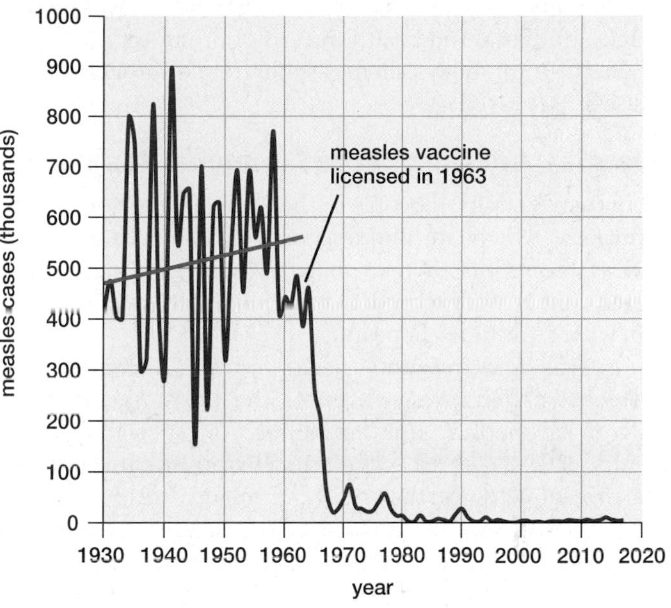

▲ **FIGURE E37-2 Measles in the United States, 1930–2017**
This graph shows the number of cases of measles diagnosed and reported to medical authorities each year. Most cases of measles occur in children; many parents do not take children with measles to the doctor, so the actual number of cases may be much higher than the numbers shown here. Data from the U.S. Census Bureau.

Despite decades of clinical practice and scientific evidence, some people question the safety of vaccines. Probably the most notorious example was an article published in 1998 in *The Lancet* by Andrew Wakefield and coauthors. Wakefield claimed that the measles-mumps-rubella (MMR) vaccine was linked to an increase in bowel disorders and autism. However, reviews of MMR vaccine safety by numerous scientific and medical societies, including the U.S. Centers for Disease Control and Prevention (CDC) and the National Academy of Sciences, found no association between the MMR vaccine and autism. An investigation by the British General Medical Council concluded that Wakefield's paper was fraudulent, and it was retracted by *The Lancet*. Nevertheless, fears of vaccination persist. As thousands of parents decline to vaccinate their children, and travelers from abroad bring the measles virus home with them, the incidence of measles in the United States rose from a few dozen cases in the early 2000s to 667 cases in 2014. The number of cases has declined somewhat since then, but outbreaks still regularly occur in communities with low vaccination rates.

THINK CRITICALLY The straight blue line in Figure E37-2 shows the trend (linear regression) of the number of cases from 1930 through 1963. Although measles occurred in cycles every few years, the number of measles cases was generally increasing over that time period. Do you think the increase means that measles was getting more infectious, or might there be another reason? Suggest a reason why the number of measles cases didn't immediately decline in 1964, although the vaccine was licensed in 1963. What additional data would you need to test your hypotheses?

37.8 WHAT HAPPENS WHEN THE IMMUNE SYSTEM MALFUNCTIONS?

Occasionally, the immune system launches inappropriate attacks, undermining health instead of promoting it. In addition, the immune system may suffer from disorders that decrease its effectiveness.

Allergies Are Misdirected Immune Responses

More than 50 million people in the United States suffer from **allergies**, which are immune reactions to harmless substances. Common allergies include those to pollen, mold spores, bee or wasp venoms, and foods such as milk, eggs, fish, wheat, tree nuts, or peanuts.

An allergic reaction begins when allergy-causing antigens, called *allergens*, enter the body and bind to "allergy antibodies" (IgE antibodies) on special types of B cells (**FIG. 37-13** ❶). These B cells proliferate, producing plasma cells ❷ that pour out allergy antibodies into the bloodstream. The antibodies attach to mast cells ❸, mostly in the respiratory and digestive tracts. If allergens later bind to these attached antibodies ❹, they cause the mast cells to release histamine, which causes leaky capillaries and other symptoms of inflammation ❺. In the respiratory tract, histamine also increases the secretion of mucus. Thus, airborne substances such as pollen grains, which typically enter the nose and throat, may trigger the runny nose, sneezing, and congestion typical of "hay fever." Food allergies usually cause intestinal cramps and diarrhea. In some cases, such as severe allergic reactions to peanuts or bee stings, the inflammatory response in the airways is so strong that the airways may completely close, causing suffocation.

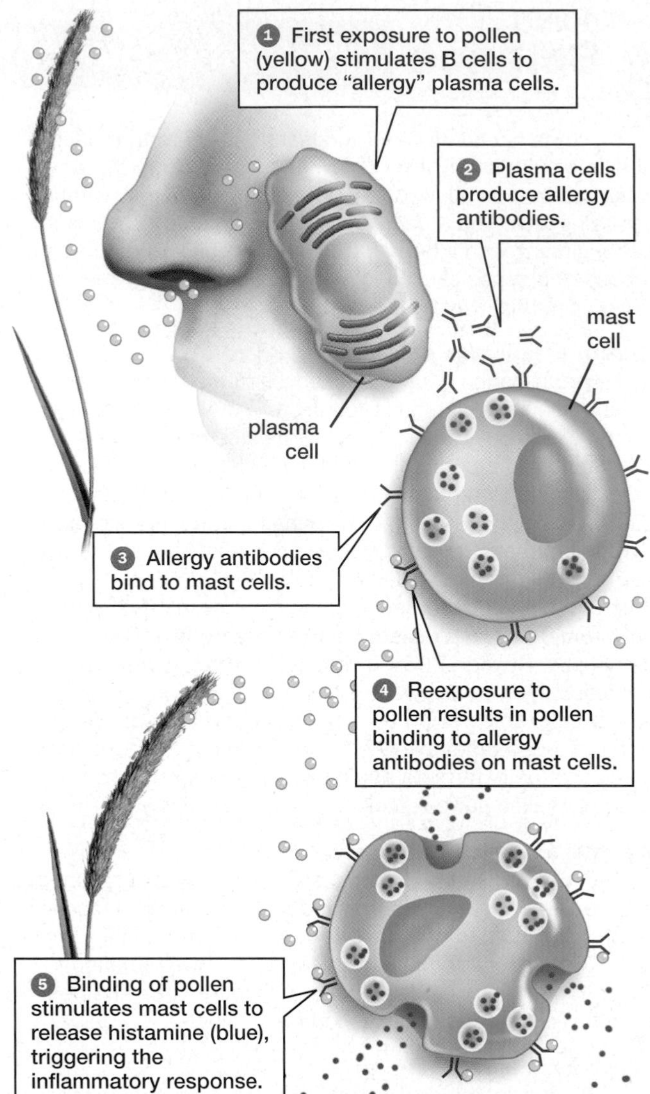

❶ First exposure to pollen (yellow) stimulates B cells to produce "allergy" plasma cells.
❷ Plasma cells produce allergy antibodies.
mast cell
plasma cell
❸ Allergy antibodies bind to mast cells.
❹ Reexposure to pollen results in pollen binding to allergy antibodies on mast cells.
❺ Binding of pollen stimulates mast cells to release histamine (blue), triggering the inflammatory response.

▲ **FIGURE 37-13 An allergic reaction to pollen**

THINK CRITICALLY What might be the evolutionary advantage of allergic reactions? (*Hint:* Are there harmful substances or organisms that might provoke allergic reactions?)

An Autoimmune Disease Is an Immune Response Against the Body's Own Molecules

Our immune systems rarely mistake our own cells for invaders. Occasionally, however, something goes awry, and "anti-self" antibodies are produced. The result is an **autoimmune disease**, in which the immune system attacks a component of one's own body. Some types of anemia, for example, are caused by antibodies that destroy a person's red blood cells. Many cases of type 1 diabetes begin when the immune system attacks the insulin-secreting cells of the pancreas. Other autoimmune diseases include rheumatoid arthritis (which affects the joints), myasthenia gravis (skeletal muscle), multiple sclerosis (central nervous system), and systemic lupus (almost any part of the body).

Unfortunately, there are no cures for autoimmune diseases. For some, replacement therapy can alleviate the symptoms—for instance, by giving insulin to diabetics. Some autoimmune disorders can be reduced with drugs that suppress the immune response. Immune suppression, however, also reduces responses to the assaults of disease microbes, so this therapy has major drawbacks. In the future, it might be possible to stimulate the activity of specific types of regulatory T cells that can damp down autoimmune responses; however, treatments based on regulatory T cells appear to be many years away from the doctor's office.

CASE STUDY **CONTINUED**

Flesh-Eating Bacteria

Certain proteins on the surface of *S. pyogenes* are very similar to proteins on cells of the heart and kidney. As a result, the same antibodies that help to destroy *S. pyogenes* may also attack these organs: About 1% to 3% of people with untreated strep throat develop an autoimmune disease called rheumatic fever, in which these antibodies damage the heart muscle.

Immune Deficiency Diseases Occur When the Body Cannot Mount an Effective Immune Response

There are two very different types of disorders in which the immune system cannot combat routine infections. About 1 in 500 people has some type of genetic immune deficiency, such as severe combined immune deficiency. Much more frequent are acquired immune deficiencies that arise during a person's lifetime. Acquired immune deficiencies may arise from several causes, including radiation, chemotherapy, immune suppressive drugs, diabetes, leukemia, and pathogenic microbes.

Severe Combined Immune Deficiency Is an Inherited Disorder

About 1 in 60,000 children is born with **severe combined immune deficiency (SCID)**, which occurs when any of several genetic defects cause few or no immune cells to be formed. A child with SCID may survive for a few months after birth, protected by IgG and IgA antibodies acquired from the mother during pregnancy or in her milk. Once these antibodies are lost, however, common infections can prove fatal because the child cannot generate an effective immune response. One form of therapy is to transplant bone marrow (from which immune cells arise) from a healthy donor into the child. Bone marrow transplants sometimes result in enough immune cell production to confer normal immune responses. Gene therapy, an experimental treatment in which genetic engineering is used to insert functional genes into a child's own cells, has been used to create a working immune system in a few dozen children with SCID (see Chapter 14).

AIDS Is an Acquired Immune Deficiency Disease

The most common immune deficiency disease is **acquired immune deficiency syndrome (AIDS)**. AIDS is caused by **human immunodeficiency viruses (HIV)**. These viruses undermine the immune system by infecting and destroying helper T cells, which stimulate both cell-mediated and humoral immune responses (see Fig. 37-11). The United Nations Program on HIV/AIDS estimates that in 2017, about 940,000 people worldwide died of AIDS and nearly 1.8 million more became infected with HIV, bringing the total infected population to 37 million. About 35 million people have died of AIDS since the first official diagnosis in 1981.

HIV enters a helper T cell and hijacks the cell's metabolic machinery, forcing it to make more viruses, which then emerge, taking an outer coating of T-cell membrane with them (**FIG. 37-14**). Early in the infection, as the immune system fights the virus, the victim may develop a fever, rash, muscle aches, headaches, and enlarged lymph nodes. After several months, the rate of viral replication slows. Enough helper T cells remain that infected individuals are able to resist disease, and they generally feel quite well. This condition often persists for several years. However, helper T-cell numbers continue to decline, and eventually the immune response becomes too weak to overcome routine infections. At this point the person is considered to have AIDS. The life expectancy for untreated AIDS victims is 1 to 2 years.

▲ **FIGURE 37-14 HIV causes AIDS** In this electron micrograph, human immunodeficiency viruses are seen emerging from a helper T cell, acquiring an outer envelope of plasma membrane (green) in the process. The plasma membrane coating will help them infect new cells.

Several drugs can slow down the replication of HIV and thereby slow the progress of AIDS. Combinations of drugs targeting different stages of viral replication have been particularly effective, and a complete AIDS treatment has now been combined into a once-a-day pill. HIV-positive people who receive the best medical care might now live a normal life span, although nobody knows for sure, because the most effective drug combinations have been available for only about 20 years.

Because HIV cannot survive for long outside the body, it can be transmitted only through direct contact with an infected person's broken skin, mucous membranes, or virus-laden body fluids, including blood, semen, vaginal secretions, and breast milk. HIV infection can be spread by sexual activity, by sharing a needle with an intravenous drug user, or through a blood transfusion (this is now rare in developed countries because all donated blood is screened for HIV). A woman infected with HIV can transmit the virus to her child during pregnancy, childbirth, or breast-feeding.

The best solution would be a vaccine against HIV. Developing an effective vaccine is a major challenge, partly because HIV has an incredibly high mutation rate. Therefore, there are many different strains of HIV in the world. In fact, single infected individuals may harbor multiple strains of HIV resulting from mutations that occurred within their bodies after they were first infected. Despite years of effort, as of 2018, there are still no effective vaccines against HIV.

CHECK YOUR LEARNING

Can you . . .

- describe allergies, autoimmune diseases, and immune deficiency diseases, including how each occurs?

37.9 HOW DOES THE IMMUNE SYSTEM COMBAT CANCER?

Cancer, the uncontrolled replication of the body's own cells, is one of the most dreaded diseases in the world, and with good reason. About 600,000 people in the United States will die of cancer this year, a fatality rate second only to heart disease. Worldwide, cancer will kill more than 8 million people. Cancers may be triggered by many causes, including environmental factors (for example, UV radiation or smoking), faulty genes, mistakes during cell division, and viruses. All of these triggers produce cancer by sabotaging the mechanisms that normally control the multiplication of the body's cells (see Chapter 9).

The Immune System Recognizes Most Cancerous Cells as Foreign

Cancer cells form in our bodies every day. Fortunately, the immune system destroys nearly all of them before they have a chance to spread. How are cancer cells weeded out? The surfaces of most cancer cells either lack some of the usual proteins found on normal body cells (such as certain MHC proteins) or bear proteins that are unique to cancers. Natural killer cells and cytotoxic T cells recognize these differences as markers of non-self cells and destroy the cancer cells (see Fig. 37-10). However, some cancer cells evade detection because they do not bear antigens that allow the immune system to recognize them as foreign.

Vaccines May Prevent Some Types of Cancer

Viruses cause some cancers of the liver, mouth, throat, and genitals, some types of leukemia and lymphoma, and probably all cases of cervical cancer. In the United States, preventive vaccines have been approved that protect against the virus that causes hepatitis B, which triggers some cases of liver cancer, and against the types of human papillomaviruses (HPV) that are most often spread by sexual contact and cause most cases of cervical cancer.

Medical Treatments for Cancer Depend on Selectively Killing Cancerous Cells

Medical care for most cancers still relies on early detection and the traditional treatments of surgery, radiation, and chemotherapy. Surgically removing the tumor is the first step in treating many cancers, but it can be difficult to remove every bit of cancerous tissue. Tumors can be bombarded with radiation, which can destroy even microscopic clusters of cancer cells by disrupting their DNA, thus preventing cell division. Unfortunately, neither surgery nor radiation is effective against cancer that has spread throughout the body. Chemotherapy is commonly used to supplement surgery or radiation or to combat cancers that cannot be treated any other way. Chemotherapy drugs attack the machinery of cell division, and so they are somewhat selective for cancer cells, which divide more frequently than normal cells do. Unfortunately, chemotherapy also kills some healthy, dividing cells. Damage to rapidly dividing cells in patients' hair follicles and intestinal lining produces the well-known side effects of hair loss, nausea, and vomiting.

More selective cancer treatments are in the works. A number of promising treatments stimulate or enhance the patient's immune response. For example, in one approach, T cells are extracted from a patient and genetically engineered to produce receptors that bind to antigens found only on cancer cells (**FIG. 37-15**). The engineered T cells are then

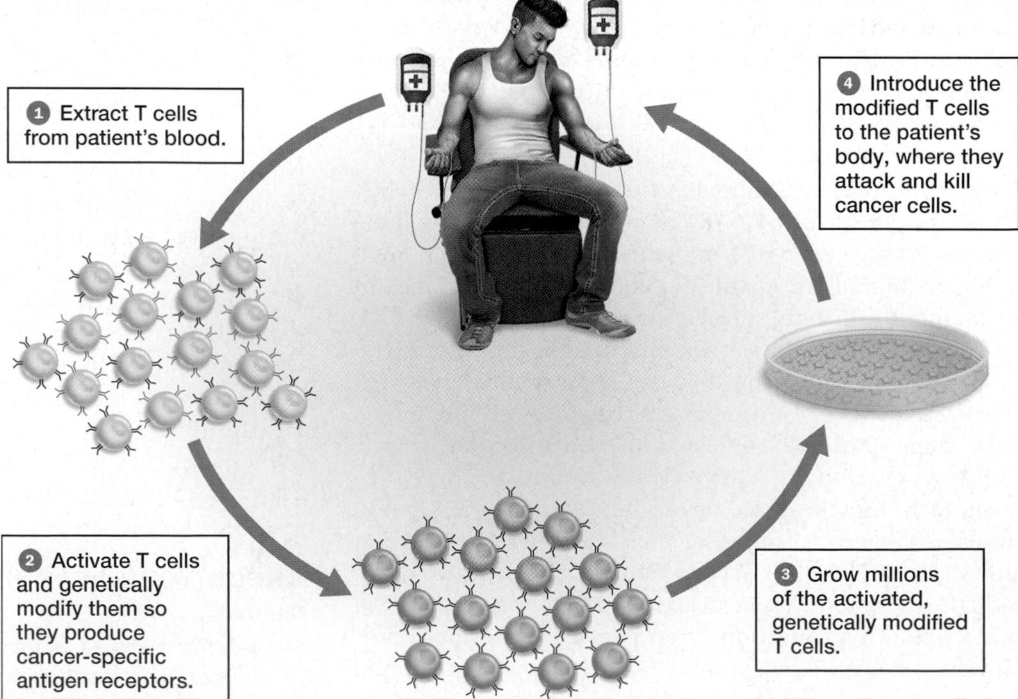

1 Extract T cells from patient's blood.
2 Activate T cells and genetically modify them so they produce cancer-specific antigen receptors.
3 Grow millions of the activated, genetically modified T cells.
4 Introduce the modified T cells to the patient's body, where they attack and kill cancer cells.

FIGURE 37-15
Immunotherapy for cancer
The treatment method shown here is called CAR-T, because the receptors on the surface of the genetically engineered T cells are known as ***c**himeric **a**ntigen **r**eceptors*.

reintroduced to the patient. Treatments using engineered T cells are already in use, and more are in clinical trials.

Another approach uses drugs that act as "checkpoint inhibitors." The immune system normally includes checkpoint proteins that put the brakes on T-cell proliferation, providing negative feedback that prevents an immune response from growing out of control. Many cancers resist attack by inducing this checkpoint function. Checkpoint-inhibitor drugs block or inactivate checkpoint proteins, permitting a much stronger immune attack on tumor cells. Unfortunately, a turbocharged immune response may also attack healthy tissues, but researchers are developing treatments that minimize this side effect.

A third approach is development of viruses that selectively bind to and enter cancerous cells. The viruses replicate inside cancer cells, causing the cells to burst open and release more viruses, which continue the killing process. The many fragments of the broken cancer cells also stimulate a stronger immune response than the intact cells did. Numerous clinical trials of these viruses are under way.

CHECK YOUR LEARNING

Can you . . .

- explain how the immune system attacks cancerous cells, even though they are your own body's cells?
- describe current medical treatments against cancer?

CASE STUDY REVISITED

Flesh-Eating Bacteria

Flesh-eating bacteria are masters of both disguise and attack. Infections occur when *S. pyogenes* or *A. hydrophila* enter a wound and hide from the immune system behind their capsules. Inside the body, they release toxins and enzymes that kill both ordinary body cells and phagocytic cells that might otherwise destroy them. Other secretions dissolve blood clots and the molecules that attach cells to one another, thereby allowing the bacteria to enter the bloodstream, spread rapidly throughout the body, and attack new cells. Some strains of both bacteria are also resistant to multiple types of antibiotics.

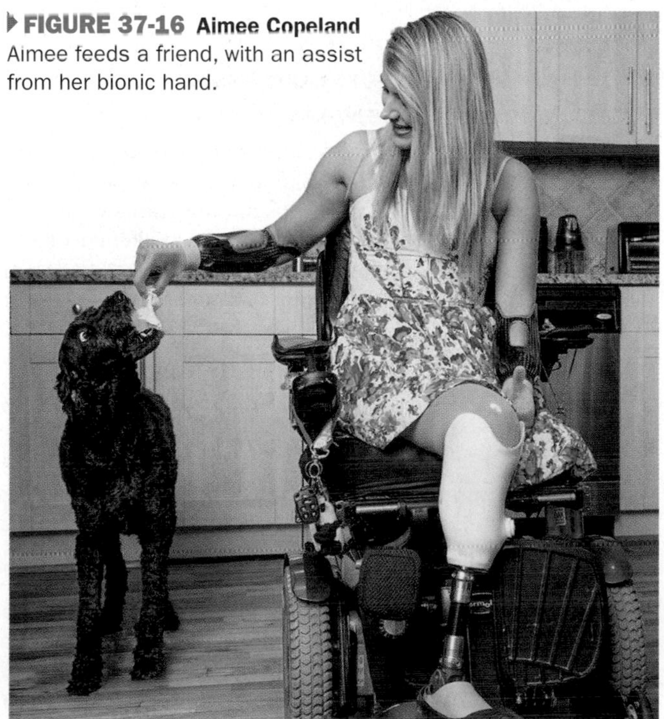

FIGURE 37-16 Aimee Copeland Aimee feeds a friend, with an assist from her bionic hand.

As in Aimee Copeland's case, sometimes the patient's life can be saved only by drastic surgery to cut out the infected tissue. Copeland lost her left leg, right foot, and both hands. Fortunately, Copeland has now received prostheses for her amputated limbs, including bionic hands that allow her to perform remarkably precise movements, such as picking up small objects and using a knife to cut food (**FIG. 37-16**).

Finally, *S. pyogenes* sometimes uses the immune system to kill its host. *S. pyogenes* can produce proteins that are "superantigens," meaning that they tremendously overactivate the immune system—causing a response 10,000 to 100,000 times stronger than the response to infection by most other bacteria. This enormous immune response releases huge quantities of cytokines, causing the overwhelming inflammation of streptococcal toxic shock syndrome. Remember, inflammation causes leaky capillaries with increased blood flow, so affected tissues swell up with extra fluid. If massive inflammation occurs in the lungs, then the victim cannot breathe and may die. Inflammation caused by superantigens is probably what killed Jim Henson.

CONSIDER THIS Flesh-eating infections are quite rare—perhaps 500 to 1,000 cases occur each year in the United States. However, 25% to 70% of the victims die. A physician who suspects a flesh-eating infection begins antibiotic treatment immediately.

Antibiotic treatment for anyone with an infected cut, even if there is no evidence of flesh-eating bacteria, would certainly reduce suffering, disfigurement, and mortality in those patients. On the other hand, antibiotics also kill friendly bacteria, especially in the digestive tract, that are beneficial. And antibiotics are potent agents of natural selection, favoring the evolution of antibiotic-resistant bacteria. How do you think that the medical community should balance the benefits of antibiotic treatment for individuals against the overall risk to society of increased antibiotic resistance among bacteria?

CHAPTER REVIEW

Go to **Mastering Biology to access** the Pearson eText, vocabulary review, practice quizzes, activities, videos, current events, and more.

Answers to ***Think Critically*** *and* ***Thinking Through the Concepts questions*** *can be found in the* ***Answers*** *section at the back of the book.*

Summary of Key Concepts

37.1 How Does the Body Defend Itself Against Disease?

Nonspecific external barriers, including the skin and mucous membranes, prevent disease-causing organisms from entering the body. Nonspecific internal defenses, collectively called the innate immune response, consist of white blood cells, inflammation, and fever. These defenses destroy microbes, toxins, and both virus-infected body cells and cancerous cells. The adaptive immune response selectively destroys a specific toxin or microbe and "remembers" the invader, allowing a faster response if the invader reappears.

37.2 How Do Nonspecific Defenses Function?

The skin and its secretions physically block the entry of microbes into the body and inhibit their growth. The mucous membranes of the respiratory and digestive tracts secrete antibiotic substances and mucus that traps microbes. If microbes do enter the body, they are engulfed by phagocytes. Natural killer cells secrete proteins that kill infected or cancerous cells. Injuries stimulate the inflammatory response, in which chemicals are released that attract phagocytes, increase blood flow, and make capillaries leaky. Blood clots wall off the injury site. Fever is caused by chemicals produced by bacteria or released by white blood cells in response to infection. High temperatures inhibit bacterial growth and accelerate the immune response.

37.3 What Are the Key Components of Specific Internal Defenses?

Most of the cells of the adaptive immune system are located in bone marrow, thymus, spleen, and lymphatic vessels and nodes. These cells include macrophages, dendritic cells, and two types of lymphocytes: B cells and T cells. The cells secrete cytokine proteins that allow communication between cells. B cells also secrete antibodies that combat infection.

37.4 How Does the Adaptive Immune System Recognize Invaders?

Cells of the adaptive immune system recognize large, complex molecules, called antigens, produced by invading microbes and cancerous cells. Antibodies (on B cells) and T-cell receptors (on T cells) have specific sites that bind to one or a few types of antigens and trigger immune responses. The body contains millions of different antibodies and T-cell receptors, so any antigen entering the body is likely to bind to some antibodies and T-cell receptors.

Both foreign invaders and the body's own cells have antigens that can potentially bind antibodies and T-cell receptors. However, immature immune cells die if they bind antigen. Because the body's own proteins are continuously present during immune cell differentiation, self-reactive, immature immune cells are destroyed. Therefore, only foreign antigens normally provoke an immune response.

37.5 How Does the Adaptive Immune System Attack Invaders?

Only those B and T cells that are activated by antigen binding multiply and produce a specific immune response to an invading microbe, a process called clonal selection. B cells give rise to plasma cells, which secrete antibodies into the bloodstream, causing humoral immunity. Antibodies destroy microbes or their toxins while they are outside of body cells. Cytotoxic T cells destroy cancer cells, virus-infected cells, and some microbes, causing cell-mediated immunity. Helper T cells stimulate both the humoral and cell-mediated immune responses.

37.6 How Does the Adaptive Immune System Remember Its Past Victories?

Some progeny cells of both B and T cells are long-lived memory cells. If the same antigen reappears in the bloodstream, these memory cells are immediately activated, dividing rapidly and causing an immune response that is much faster and more effective than the original response.

37.7 How Does Medical Care Assist the Immune Response?

Antimicrobial drugs kill microbes or slow down their reproduction, allowing the body's defenses more time to respond and exterminate the invaders. Vaccinations are injections of antigens from disease organisms, in some cases the weakened or dead microbes themselves. An immune response is evoked by the antigens, providing memory and a rapid response should a real infection occur later.

37.8 What Happens When the Immune System Malfunctions?

Allergies are immune responses to normally harmless foreign substances. Certain B cells treat these as antigens and produce IgE antibodies that attach to mast cells. When antigens bind to the IgE antibodies attached to mast cells, the cells release histamine, causing a local inflammatory response. Autoimmune diseases arise when the immune system mistakes the body's own cells for foreign invaders and destroys them. Immune deficiency diseases occur when the immune system cannot respond strongly enough to ward off usually minor diseases. Immune deficiency diseases may be innate, such as severe combined immune deficiency (SCID), or acquired during a person's lifetime, such as acquired immune deficiency syndrome (AIDS).

37.9 How Does the Immune System Combat Cancer?

Cancerous cells are often recognized as non-self by the immune system and destroyed by natural killer cells and cytotoxic T cells. Cancer may be triggered by genetic factors, environmental factors, mistakes during cell division, or viruses. Vaccines can help to prevent certain virus-caused cancers. Other vaccines are under development to cure cancer once it occurs.

Thinking Through the Concepts

Bloom's: Remembering, Understanding

Multiple Choice

1. An individual B cell
   a. produces antibodies with sites that can bind many different antigens.
   b. produces antibodies with sites that can bind one or a few antigens.
   c. is capable of responding to many unrelated pathogens.
   d. kills body cells infected with viruses.
2. Which of the following cell types is important in nonspecific responses to infection?
   a. natural killer cell
   b. cytotoxic T cell
   c. B cell
   d. regulatory T cell
3. Molecules that label your cells as "self" are
   a. antibodies.
   b. T-cell receptors.
   c. major histocompatibility complex proteins.
   d. interferon.
4. Cells that secrete antibodies into the bloodstream are
   a. daughter cells of helper T cells.
   b. macrophages.
   c. plasma cells.
   d. mast cells.
5. The immune system usually does not attack your body's own cells because
   a. your body does not produce antigens.
   b. although your body produces antigens, you do not produce immune cells that can bind those antigens.
   c. immune cells that might respond to your body's own antigens are usually killed during their development.
   d. antibodies that bind to your body's own antigens are harmless.

Fill-in-the-Blank

1. External defenses against microbial invasion include the ____________ and the mucous membranes that line the ____________, ____________, and ____________ tracts.
2. Nonspecific internal defenses against disease include ____________, which engulf and digest microbes; ____________, which destroy cells that have been infected by viruses; the ____________, provoked by injury; and ____________, an elevation of body temperature that slows microbial reproduction and enhances the body's defenses.
3. The adaptive immune response is stimulated when the body is invaded by complex proteins or polysaccharides collectively called ____________. These molecules bind to one of two types of protein receptors of the immune system: ____________ or ____________.
4. An antibody consists of four protein chains, two ____________ chains and two ____________ chains. Each is composed of a(n) ____________ region and a(n) ____________ region. The ____________ regions form the binding site for antigens.
5. ____________ immunity is provided by B cells and their daughter cells, called ____________, which secrete antibodies into the blood plasma. ____________ immunity is provided by T cells. ____________ T cells kill body cells that have been infected by viruses. ____________ T cells produce cytokines that stimulate immune responses in both B cells and T cells. Protection against future invasions by microbes bearing the same antigens is provided by ____________ cells of both B and T types.
6. A(n) ____________ occurs when the immune system produces a response to a harmless substance such as pollen. In a(n) ____________, the immune system cannot mount an effective response, even to dangerous infections; these may be inborn or acquired. When the immune system attacks a person's own body, this is called a(n) ____________.

Review Questions

1. What are the human body's three lines of defense against invading microbes? Which are nonspecific (that is, act against all types of invaders) and which are specific (act only against a particular type of invader)?
2. How do natural killer cells and cytotoxic T cells destroy their targets?
3. Describe humoral immunity and cell-mediated immunity. Include in your answer the types of immune cells involved in each, the location of antibodies and receptors that bind foreign antigens, the mechanisms by which invading cells are destroyed, and the role of helper T cells in facilitating both responses.
4. Diagram the structure of an antibody. What parts bind to antigens? Why does each antibody bind only to one or a few specific antigens?
5. How does the immune system combat cancer?

6. Describe some cancer treatments that work by stimulating or modifying the natural immune response.
7. How does the body distinguish "self" from "non-self"? Why is it important that the body makes this distinction?
8. What are memory cells? How do they contribute to long-lasting immunity to specific diseases?
9. What is a vaccination? How does it confer immunity to a disease?
10. How does the inflammatory response help the body resist disease? What symptoms does it cause?
11. Describe the process that leads to an allergic reaction.
12. Distinguish between autoimmune diseases and immune deficiency diseases, and give one example of each.
13. Describe the causes and eventual outcome of AIDS. How is HIV spread?

Applying the Concepts

Bloom's: Applying, Analyzing, Evaluating

1. Why is it essential that antibodies and T-cell receptors bind only large molecules (such as proteins) and not small molecules (such as amino acids)?
2. Patients who have had an organ transplant often receive the drug cyclosporine. This drug inhibits the production of a cytokine that stimulates helper T cells to proliferate. How does cyclosporine prevent rejection of transplanted organs? Some patients who received cyclosporine following transplants many years ago are now developing various kinds of cancers. Propose a hypothesis to explain this phenomenon.

38 Chemical Control of the Animal Body: The Endocrine System

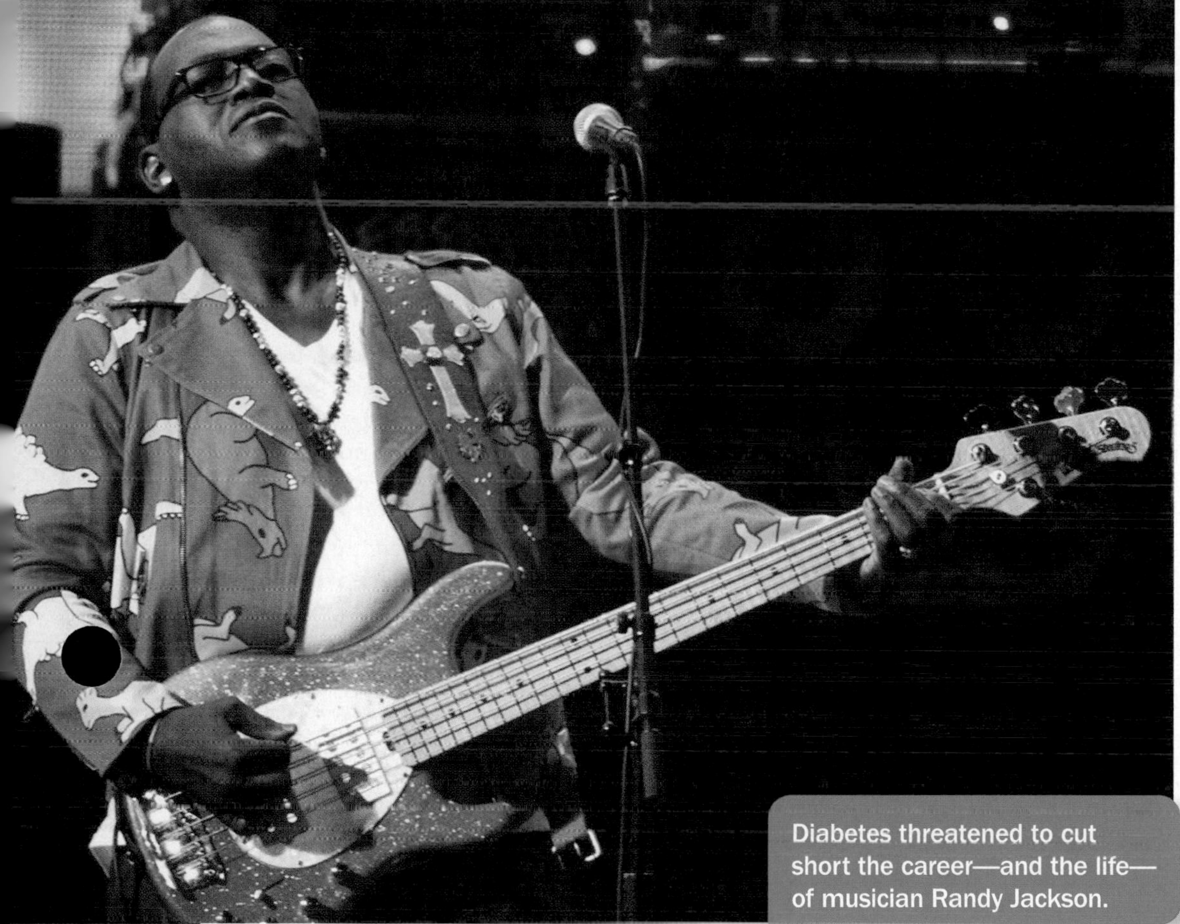

Diabetes threatened to cut short the career—and the life—of musician Randy Jackson.

CASE STUDY

Insulin Resistance

RANDY JACKSON, RENOWNED BASS GUITARIST, music producer, and judge on *American Idol,* should have been on top of the world as the hit show entered its second season. Except he wasn't. At 46 years old, he weighed over 300 pounds and suffered from raging thirst and copious urination. When his physician ordered a blood test, the result was shocking: a glucose level five times higher than that of a healthy person.

Blood glucose is regulated by insulin, a hormone produced by the pancreas. In most people, after eating a meal, the pancreas releases insulin, which stimulates cells of muscle, fat, and other tissues to take up glucose from the blood. These cells then either use the glucose for energy or convert it into a storage carbohydrate called glycogen. But in Randy Jackson and about 30 million other Americans with **diabetes mellitus**, glucose metabolism goes awry.

In diabetes mellitus, either the pancreas produces insufficient amounts of insulin (the cause of type 1 diabetes) or the body's cells fail to respond to insulin, a condition called *insulin resistance* (the principal cause of type 2 diabetes). In both type 1 and type 2 diabetes, glucose is not efficiently taken up by muscles and fat, but remains in the blood. The result: blood glucose skyrockets and fluctuates wildly with food intake, especially intake of sugary foods. Jackson has type 2 diabetes: His high blood glucose levels result from insulin resistance. Fortunately, as we will see, type 2 diabetes can often be managed very successfully by weight control, exercise, or medication.

People with diabetes can benefit from treatment only if they are aware that they have the condition. The Centers for Disease Control estimates that, in the United States, almost a quarter of people with diabetes are undiagnosed, mainly because they have not visited a physician for a blood glucose test.

Insulin is one of the many hormones that affect cells throughout the body. How do hormones alter cellular activity? How are the levels of hormones produced by the endocrine glands regulated to maintain homeostasis in the body?

AT A GLANCE

38.1 HOW DO ANIMAL CELLS COMMUNICATE?

The cells of an animal's body must communicate with one another to ensure the proper functioning of the whole organism. Should leg muscles just hold up a standing body or should they make the legs run? Should blood flow be directed to the intestines for digestion or to the muscles for movement? These and hundreds of other messages are sent throughout the body every minute. Methods of communication among cells fall into four broad categories: direct, synaptic, paracrine, and endocrine (**TABLE 38-1**).

In direct communication, gap junctions link the cytoplasm of adjacent cells, allowing ions and electrical signals to flow between them (see Chapter 5). Direct communication occurs in many tissues, including the heart (see Chapter 33) and brain. Direct communication is very fast but requires the cells to be in intimate contact with one another.

In the other three types of communication—synaptic, paracrine, and endocrine—the "sending" cells release chemical messengers. In synaptic communication, the messengers are chemicals known as **neurotransmitters**, which are released at specialized junctions called synapses. In paracrine and endocrine communication, the messengers are given a variety of names. A **hormone** is a chemical messenger that is secreted by a cell and carried by the interstitial fluid or bloodstream to other cells. In paracrine communication, cells release **local hormones** that diffuse through the interstitial fluid and affect nearby cells ("para" means "alongside" in Greek). Local hormones may be produced by small glands, clusters of cells, or even single cells. In endocrine communication, cells, usually in glands, release **endocrine hormones** that travel throughout the body in the bloodstream. Endocrine hormones may influence cells either close to the gland or in distant parts of the body.

Messenger molecules move to "receiving" cells and alter the cells' physiology by binding to **receptors**, which are specialized proteins located either on the surface of or inside the receiving cells. When a messenger binds to a receptor, the receiving cell responds in a way that is determined by the messenger, the receptor, the type of cell, and

TABLE 38-1 How Cells Communicate

Communication	Chemical Messengers	Mechanism of Transmission	Examples
Direct	Ions, small molecules	Direct movement through gap junctions linking the cytoplasm of adjacent cells	Ions flowing between cardiac muscle cells
Synaptic	Neurotransmitters	Diffusion from a neuron across a narrow space (synaptic cleft) to an adjacent cell	Acetylcholine, dopamine
Paracrine	Local hormones	Diffusion through the interstitial fluid to nearby cells	Prostaglandins, histamine
Endocrine	Endocrine hormones	Carried in the bloodstream to nearby or distant cells	Insulin, estrogen, growth hormone

the receiving cell's metabolic state. Every cell has dozens of receptor proteins, each capable of binding a specific chemical messenger and stimulating a particular response. Cells with receptors that bind a messenger and respond to it are **target cells** for that messenger. Cells without the appropriate receptors cannot respond to the messenger and are not target cells.

Synaptic Communication Is Used in the Nervous System

In **synaptic communication**, signals may travel, in just a fraction of a second, from a nerve cell to other nearby nerve cells, to muscles in the farthest reaches of the body, or throughout the brain. In this type of communication, electrical activity in a nerve cell stimulates the release of neurotransmitters that cross a tiny space between the end of the nerve cell and its target. The neurotransmitters bind to receptors on the surface of the target cell, stimulating a response. The response in the target cells may be very brief, such as a reflex, or long lasting, such as learning. (We will explore the nervous system and synaptic communication in detail in Chapter 39.)

Paracrine Communication Acts Locally

In **paracrine communication**, the sending and receiving cells are very close to one another, so communication tends to be very fast. The local hormones used in paracrine communication have mostly short-range actions, either because they are degraded soon after release or because nearby cells take them up out of the interstitial fluid so fast that they cannot get very far from the cells that secrete them.

Local hormones include histamine, which is released as part of the allergic and inflammatory responses, and many of the cytokines by which cells of the immune system communicate with one another (see Chapter 37). Nitric oxide, a gas produced by cells lining blood vessels, can act as a local hormone. It diffuses rapidly into muscle cells surrounding the vessels, causing them to relax. This causes the blood vessel to expand, which increases blood flow.

Prostaglandins are an important group of local hormones. Prostaglandins are modified fatty acids secreted by cells throughout the body. They have diverse effects, depending on the type of prostaglandin and the target cell. For example, during childbirth, prostaglandins both cause the cervix to dilate and help to stimulate uterine contractions. Prostaglandins also contribute to pain and inflammation (such as occurs in arthritic joints). Drugs such as aspirin, acetaminophen (Tylenol), and ibuprofen provide relief from these symptoms by inhibiting the enzymes that synthesize prostaglandins.

Endocrine Communication Uses the Circulatory System to Carry Hormones to Target Cells Throughout the Body

Endocrine communication begins with the secretion of hormones by **endocrine glands** (**FIG. 38-1**). An endocrine gland may be a well-defined mass of cells whose principal function is hormone secretion, as is the case with the thyroid and pituitary glands. Other endocrine glands consist of clusters of cells, or even scattered individual cells, embedded in organs that have multiple functions, such as the pancreas, ovary, or testes. In all cases, the secretory cells of an endocrine gland are embedded within a network of capillaries. The cells secrete their hormones into the interstitial fluid surrounding the capillaries ①. The hormones diffuse into the capillaries and are carried in the blood throughout the body ②.

Although an endocrine hormone may reach nearly all of the body's cells ③, only target cells, with receptors capable of binding that specific hormone, can respond ④. The hormone *oxytocin*, for example, stimulates the contraction of uterine muscles during childbirth because uterine muscle cells have receptors that bind oxytocin. However, oxytocin does not cause most of the other muscles of the body to contract, because their cells do not have the necessary receptors ⑤.

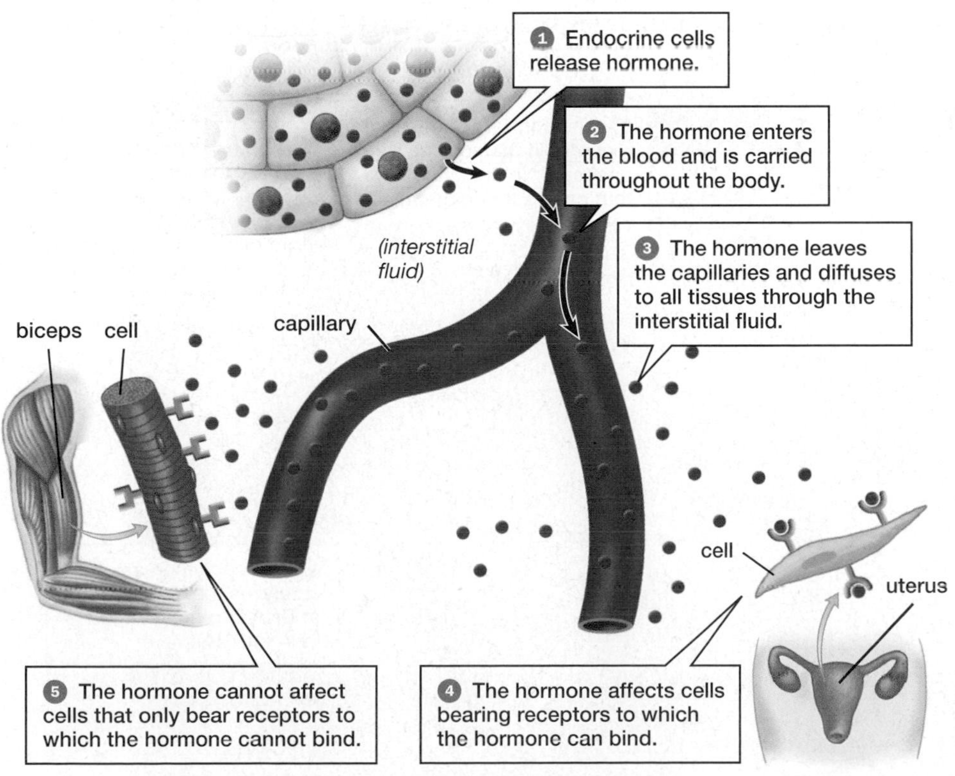

① Endocrine cells release hormone.
② The hormone enters the blood and is carried throughout the body.
③ The hormone leaves the capillaries and diffuses to all tissues through the interstitial fluid.
(interstitial fluid)
capillary
biceps
cell
cell
uterus
⑤ The hormone cannot affect cells that only bear receptors to which the hormone cannot bind.
④ The hormone affects cells bearing receptors to which the hormone can bind.

▲ **FIGURE 38-1 Hormone release, distribution, and reception**

CHECK YOUR LEARNING

Can you . . .

- describe the four methods of communication between cells in an animal body?
- explain why some cells respond to chemical messengers and others do not?
- describe the pathway by which endocrine hormones move from secreting cells to target cells, often in distant parts of the body?

38.2 HOW DO ENDOCRINE HORMONES PRODUCE THEIR EFFECTS?

There are three classes of vertebrate endocrine hormones: **steroid hormones**, which are synthesized from cholesterol; **peptide hormones**, which are chains of amino acids; and **amino acid derived hormones**, which are composed of one or two modified amino acids. These hormones bind to receptors located on the surface of, or inside, a target cell. Hormone–receptor binding then causes one or both of two major effects: (1) regulating gene transcription, thereby changing the amounts or types of proteins that the cell synthesizes, and (2) stimulating changes in the metabolism of the cell, usually by activating or inhibiting enzymes.

Steroid Hormones Usually Bind to Receptors Inside Target Cells

Steroid hormones are lipid soluble, so they can diffuse through plasma membranes. Although they sometimes bind to receptors on the surface of a target cell, most steroid hormones bind to intracellular receptors. Steroid hormone action begins with the hormone diffusing through the plasma membrane into the cytoplasm of a cell (**FIG. 38-2** ❶). Steroid hormones diffuse into every cell they encounter, but only cells with appropriate receptors can respond to a particular hormone. Once inside a target cell, steroid hormones attach to receptors located either in the cytoplasm or in the nucleus, forming a hormone–receptor complex ❷. If the receptors are in the cytoplasm, the hormone–receptor complex moves into the nucleus. The complex then binds to specific genes ❸ and stimulates the transcription of messenger RNA ❹. The messenger RNA travels to the cytoplasm and directs protein synthesis ❺. The steroid hormone testosterone, for example, increases the synthesis of proteins involved in the development of the testes, muscles, beard, pubic hair, and many other parts of the body.

Although thyroid hormone (thyroxine) is an amino acid derived hormone, not a steroid, it has a similar mechanism of action. Thyroxine is transported across the membrane by carrier proteins. Once inside cells, thyroxine binds to intracellular receptors and activates transcription of specific genes.

Peptide Hormones and Amino Acid Derived Hormones Usually Bind to Receptors on the Surfaces of Target Cells

Most peptide hormones and amino acid derived hormones are soluble in water but not in lipids, so these hormones cannot diffuse through the phospholipid bilayer of the plasma membrane. Therefore, peptide hormones and amino acid derived hormones bind to receptors on the surface of a target cell (**FIG. 38-3** ❶). Hormone–receptor binding activates an enzyme that synthesizes a molecule, called a **second messenger**, inside the cell ❷. There are many different

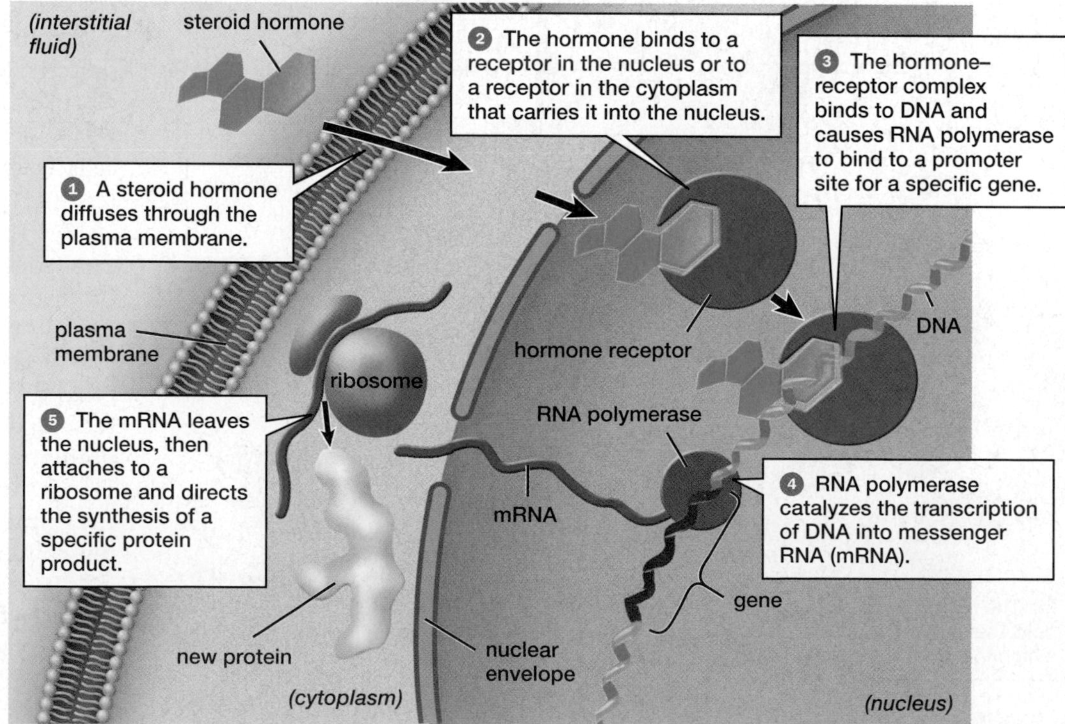

▶ **FIGURE 38-2 Steroid hormone action on target cells** Steroid hormones often stimulate target cells by binding to intracellular receptors, which creates a hormone–receptor complex that activates gene transcription and ultimately results in the synthesis of new, or increased amounts of, specific proteins.

second messenger molecules, including **cyclic adenosine monophosphate (cyclic AMP**; see Chapter 3), modified lipids, calcium ions, and nitric oxide gas. The second messenger transfers the signal from the first messenger—the hormone—to other molecules within the cell, often activating specific intracellular enzymes ❸, which then initiate a chain of biochemical reactions ❹.

These intracellular reactions vary depending on the hormone, the second messenger, and the target cell. For example, epinephrine (an amino acid derived hormone also called adrenaline) prepares the body to deal with emergency situations. It stimulates the synthesis of cyclic AMP in both heart muscle cells and liver cells. Cyclic AMP acts differently, however, in the two cell types. Cyclic AMP causes heart muscle cells to contract more strongly, which increases blood flow. In liver cells, cyclic AMP activates enzymes that break down glycogen (a starch-like polysaccharide) to glucose. The glucose is released into the bloodstream, providing the entire body with a source of quick energy.

Hormone Release Is Regulated by Feedback Mechanisms

The release of most hormones is controlled by **negative feedback**, in which a change causes responses that counteract the change and restore the system to its original condition (see Chapter 32). For example, suppose you have jogged a few miles on a hot, sunny day and have lost a quart of water through perspiration. In response to the loss of water from your bloodstream, your pituitary gland releases antidiuretic hormone (ADH), which causes your kidneys to reabsorb water and therefore to produce only small amounts of very

CASE STUDY CONTINUED
Insulin Resistance

Insulin, the hormone whose regulation is disrupted in people with diabetes mellitus, is a large, water-soluble molecule. It binds to receptors on the outside surface of muscle, fat, and several other cell types, activating proteins that stimulate or inhibit enzymes inside the cell. One of the activated enzymes starts a cascade of reactions inside the cell, with the result that glucose-transporting proteins are moved to the plasma membrane. Once in the plasma membrane, these proteins facilitate the diffusion of glucose into the cell.

Insulin is essential to life, but it can have a dark side, too. Excessive insulin, particularly if combined with fasting, can stimulate muscle and fat cells to take up too much glucose from the blood. In extreme cases, blood glucose can drop so low that a person goes into insulin shock. Deprived of glucose, the brain shuts down, and the victim may faint, go into a coma, and die. But insulin shock never happens in a properly functioning human body. Why not? How does the body regulate insulin release and glucose metabolism, preventing insulin shock? Or regulate the release of other hormones, so you don't grow so tall that you can't stand up, or become so frightened by a loud noise that your heart fails?

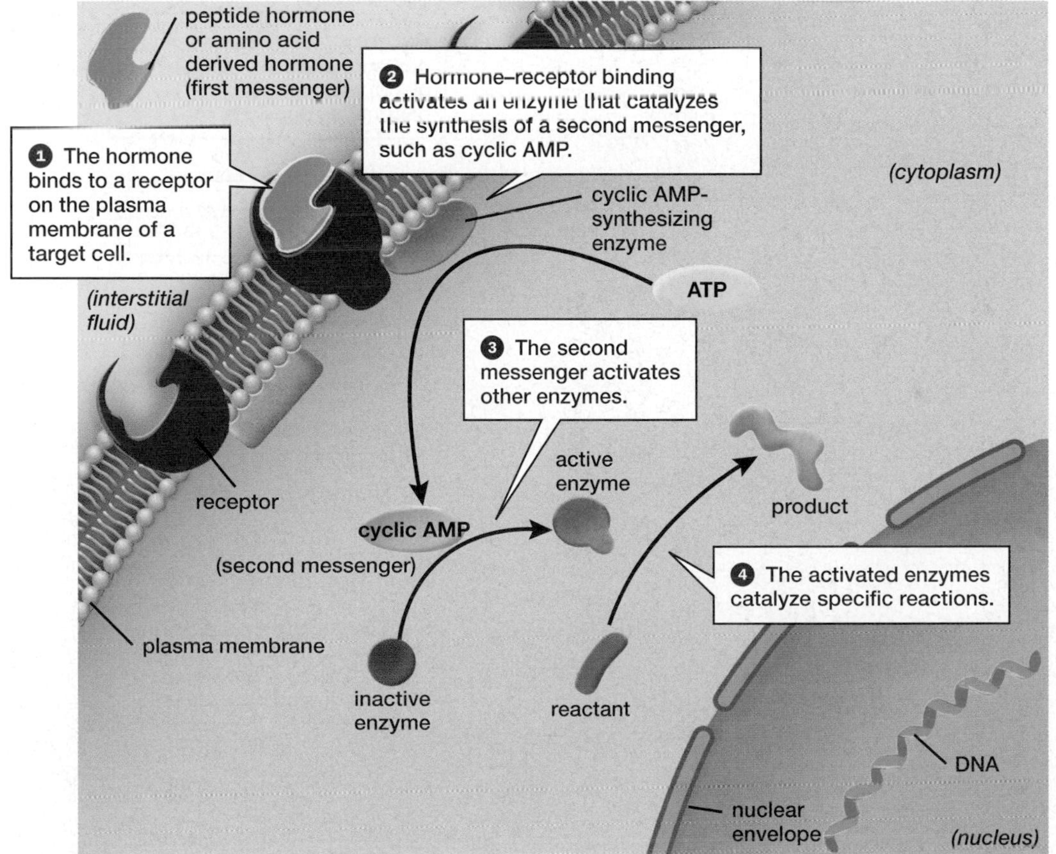
peptide hormone or amino acid derived hormone (first messenger)
❷ Hormone–receptor binding activates an enzyme that catalyzes the synthesis of a second messenger, such as cyclic AMP.
❶ The hormone binds to a receptor on the plasma membrane of a target cell.
(cytoplasm)
cyclic AMP-synthesizing enzyme
ATP
(interstitial fluid)
❸ The second messenger activates other enzymes.
active enzyme
product
receptor
cyclic AMP
(second messenger)
❹ The activated enzymes catalyze specific reactions.
plasma membrane
inactive enzyme
reactant
DNA
nuclear envelope
(nucleus)

◀FIGURE 38-3 Actions of peptide hormones and amino acid derived hormones on target cells Peptide hormones and amino acid derived hormones usually stimulate target cells by binding to receptors on the plasma membrane, which causes the cell to synthesize a second messenger molecule that sets off a cascade of intracellular biochemical reactions.

concentrated urine (see Chapter 36). But let's say that you arrive home and drink 2 quarts of water, twice as much as you lost in sweat. If your body retained this extra water, it might raise your blood pressure and damage your heart. Negative feedback, however, acts to restore the original condition. When enough water has entered your blood to bring its volume back to normal, ADH secretion is turned off. Your kidneys start producing watery urine, ridding your body of the extra quart of water. Look for other examples of negative feedback throughout this chapter, including how negative feedback controls blood sugar.

In a few cases, hormone release is temporarily controlled by **positive feedback**, in which a change produces a response that enhances the change. For example, contractions of the uterus early in childbirth push the baby's head against the cervix (a ring of connective tissue between the uterus and the vagina), which causes the cervix to stretch. Stretching the cervix sends nervous signals to the mother's brain, triggering the release of oxytocin. Oxytocin stimulates further contractions of the uterine muscles, pushing the baby harder against the cervix, which stretches further, causing still more oxytocin to be released. However, positive feedback cannot continue indefinitely. In the case of childbirth, the positive feedback between oxytocin release and uterine contractions ends when the infant is born. After delivery, the cervix is no longer stretched, so oxytocin release stops.

CHECK YOUR LEARNING

Can you . . .

- name the three classes of endocrine hormones?
- describe how hormones affect target cells?
- explain the processes of negative and positive feedback and provide an example of how each is used in regulating hormone release?

38.3 WHAT ARE THE STRUCTURES AND FUNCTIONS OF THE MAMMALIAN ENDOCRINE SYSTEM?

The mammalian **endocrine system** consists of the endocrine glands and the hormones they produce (**FIG. 38-4** and **TABLE 38-2**). In the following sections, we will focus on the hypothalamus–pituitary complex, the thyroid gland, the pancreas, the sex organs, and the adrenal glands.

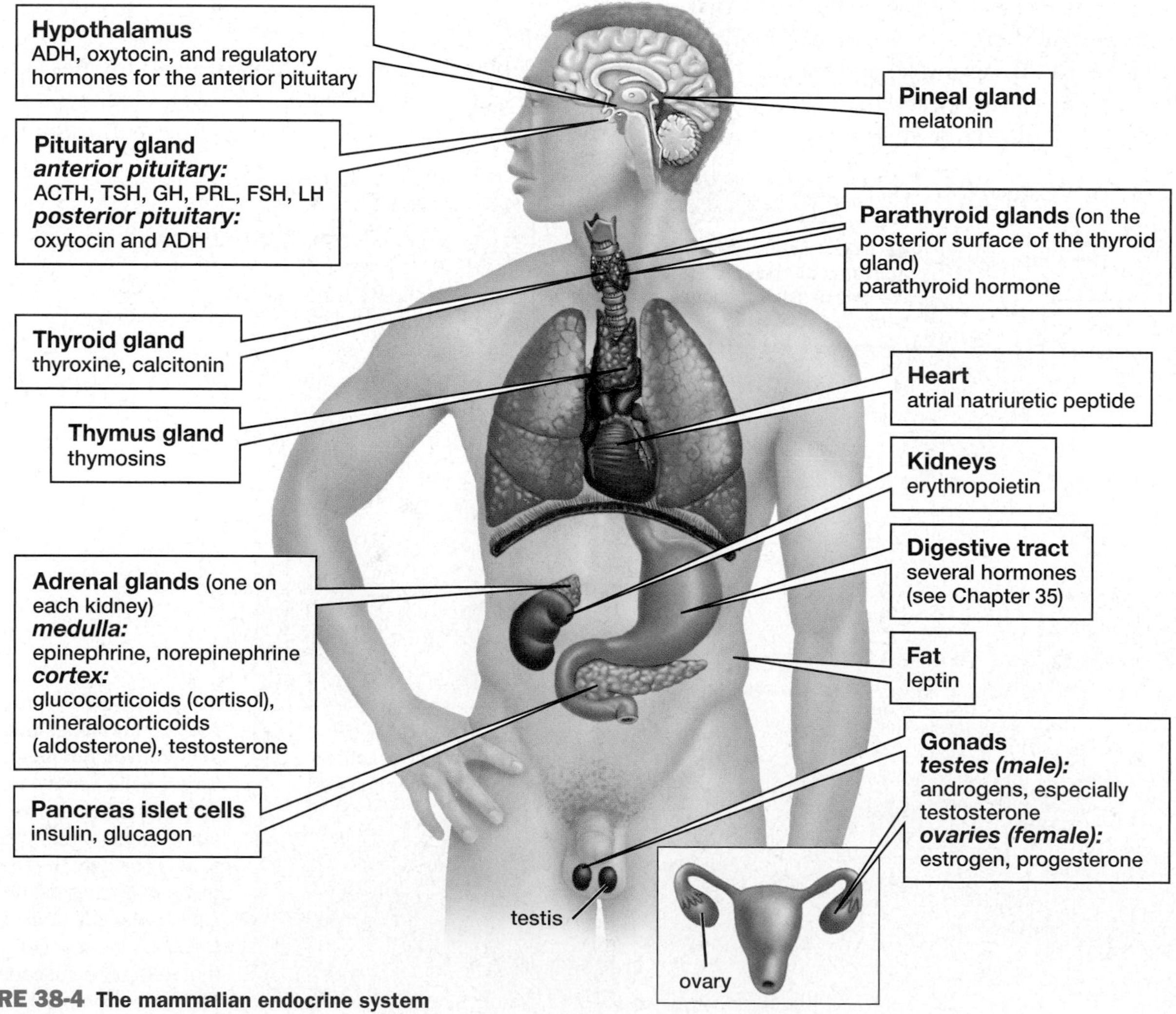

▲ **FIGURE 38-4 The mammalian endocrine system**

TABLE 38-2 The Mammalian Endocrine System

Endocrine Gland	Hormone	Type of Chemical	Principal Function
Hypothalamus (to anterior pituitary)	Releasing and inhibiting hormones	Peptide	Releasing hormones stimulate the release of hormones from the anterior pituitary; inhibiting hormones inhibit the release of hormones from the anterior pituitary
Anterior pituitary	Follicle-stimulating hormone (FSH)	Peptide	Females: stimulates the growth of follicles in the ovary, secretion of estrogen, and perhaps ovulation Males: stimulates sperm development
	Luteinizing hormone (LH)	Peptide	Females: stimulates ovulation, growth of the corpus luteum, and the secretion of estrogen and progesterone Males: stimulates the secretion of testosterone
	Thyroid-stimulating hormone (TSH)	Peptide	Stimulates the thyroid to release thyroxine
	Adrenocorticotropic hormone (ACTH)	Peptide	Stimulates the adrenal cortex to release hormones, especially glucocorticoids such as cortisol
	Prolactin (PRL)	Peptide	Stimulates milk synthesis and secretion by the mammary glands
	Growth hormone (GH)	Peptide	Stimulates growth, protein synthesis, and fat metabolism; inhibits sugar metabolism
Hypothalamus (via posterior pituitary)	Antidiuretic hormone (ADH)	Peptide	Promotes reabsorption of water from the kidneys; constricts arterioles
	Oxytocin	Peptide	Females: stimulates contraction of uterine muscles during childbirth, milk ejection, and maternal behaviors Males: may facilitate ejaculation of sperm
Thyroid	Thyroxine	Amino acid derivative	Increases the metabolic rate of most body cells; increases body temperature; regulates growth and development
	Calcitonin	Peptide	Inhibits the release of calcium from bones; decreases the blood calcium concentration
Parathyroid	Parathyroid hormone	Peptide	Increases blood calcium by stimulating calcium release from bones, absorption by the intestines, and reabsorption by the kidneys
Pancreas	Insulin	Peptide	Decreases blood glucose by increasing uptake of glucose into cells and conversion of glucose to glycogen, especially in the liver; regulates fat metabolism
	Glucagon	Peptide	Converts glycogen to glucose, raising blood glucose levels
Testes[1]	Testosterone	Steroid	Stimulates the development of genitalia and male secondary sexual characteristics; stimulates sperm development
Ovaries[1]	Estrogen	Steroid	Causes the development of female secondary sexual characteristics and the maturation of eggs; promotes the development of the uterine lining
	Progesterone	Steroid	Stimulates the development of the uterine lining and the formation of the placenta
Adrenal cortex	Glucocorticoids (e.g., cortisol)	Steroid	Increase blood sugar; regulate sugar, lipid, and fat metabolism; have anti-inflammatory effects
	Mineralocorticoids (e.g., aldosterone)	Steroid	Increase reabsorption of salt in the kidney
	Testosterone	Steroid	Causes masculinization of body features, growth
Adrenal medulla	Epinephrine (adrenaline) and norepinephrine (noradrenaline)	Amino acid derivatives	Increase levels of sugar and fatty acids in the blood; increase metabolic rate; increase the rate and force of contractions of the heart; constrict some blood vessels
Pineal gland	Melatonin	Amino acid derivative	Regulates seasonal reproductive cycles and sleep–wake cycles; may regulate onset of puberty
Thymus	Thymosin	Peptide	Stimulates maturation of T cells of the immune system
Kidney[2]	Erythropoietin	Peptide	Stimulates red blood cell synthesis in bone marrow
Digestive tract[3]	Secretin, gastrin, ghrelin, cholecystokinin, and others	Peptide	Control secretion of mucus, enzymes, and salts in the digestive tract; regulate peristalsis; regulate appetite
Fat cells	Leptin	Peptide	Regulates appetite; stimulates immune function; promotes blood vessel growth; required for the onset of puberty
Heart	Atrial natriuretic peptide (ANP)	Peptide	Increases salt and water excretion by the kidneys; lowers blood pressure

[1]See Chapters 42 and 43.
[2]See Chapter 36.
[3]See Chapter 35.

Hormones of the Hypothalamus and Pituitary Gland Regulate Many Functions Throughout the Body

The **hypothalamus** is a part of the brain that contains clusters of specialized nerve cells called **neurosecretory cells**, which synthesize peptide hormones, store them, and release them when stimulated (**FIG. 38-5**, top). Some hormones produced by the hypothalamus are released into the general circulatory system and produce effects throughout the body. Other hypothalamic hormones are produced in minuscule amounts and control the release of hormones produced in the **pituitary gland**, a pea-sized gland connected to the hypothalamus (**FIG. 38-5**, bottom). The pituitary consists of two distinct parts: the **anterior pituitary** and the **posterior pituitary**.

The Anterior Pituitary Produces and Releases Multiple Hormones

Four of the hormones released by the anterior pituitary regulate hormone production in other endocrine glands. **Follicle-stimulating hormone (FSH)** and **luteinizing hormone (LH)** stimulate the production of sperm and testosterone in the testes of males and the production of eggs, estrogen, and progesterone in the ovaries of females. (We will discuss the roles of FSH and LH in reproduction in Chapter 42.) **Thyroid-stimulating hormone (TSH)** stimulates the thyroid gland to release its hormones, and **adrenocorticotropic hormone** (**ACTH**; "hormone that stimulates the adrenal cortex") causes the release of the hormone *cortisol* from the adrenal cortex. We will examine the effects of thyroid and adrenal cortical hormones later in this chapter.

The remaining hormones of the anterior pituitary do not act on other endocrine glands. **Prolactin (PRL)**, in conjunction with other hormones, stimulates the development of milk-producing mammary glands in the breasts during pregnancy. **Growth hormone (GH)** acts on nearly all the body's cells by increasing protein synthesis, promoting the use of fats for energy, and regulating carbohydrate metabolism. During childhood, growth hormone stimulates bone growth. Much of the normal variation in human height is due to differences in the secretion of, or responses to, growth hormone. Too little growth hormone—or defective receptors for it—causes some cases of dwarfism; too much can cause gigantism.

A major advance in the treatment of pituitary dwarfism occurred in 1981 when molecular biologists successfully inserted the gene for human growth hormone into bacteria, which then churned out large quantities of the hormone. Previously, the main commercial source of growth hormone was human cadavers, from which tiny amounts were extracted at great cost. Thanks to the new, cheaper source, children with underactive pituitary glands, who would previously have been extremely short, can now achieve normal height.

Growth hormone is one of several hormones that are sometimes used by athletes to increase strength, speed, or endurance. Because hormone supplements can be dangerous and may give athletes a competitive advantage, the International Olympic Committee and professional sports leagues ban most hormone supplements, as we discuss in "Health Watch: Performance-Enhancing Drugs—Fool's Gold?" on page 677.

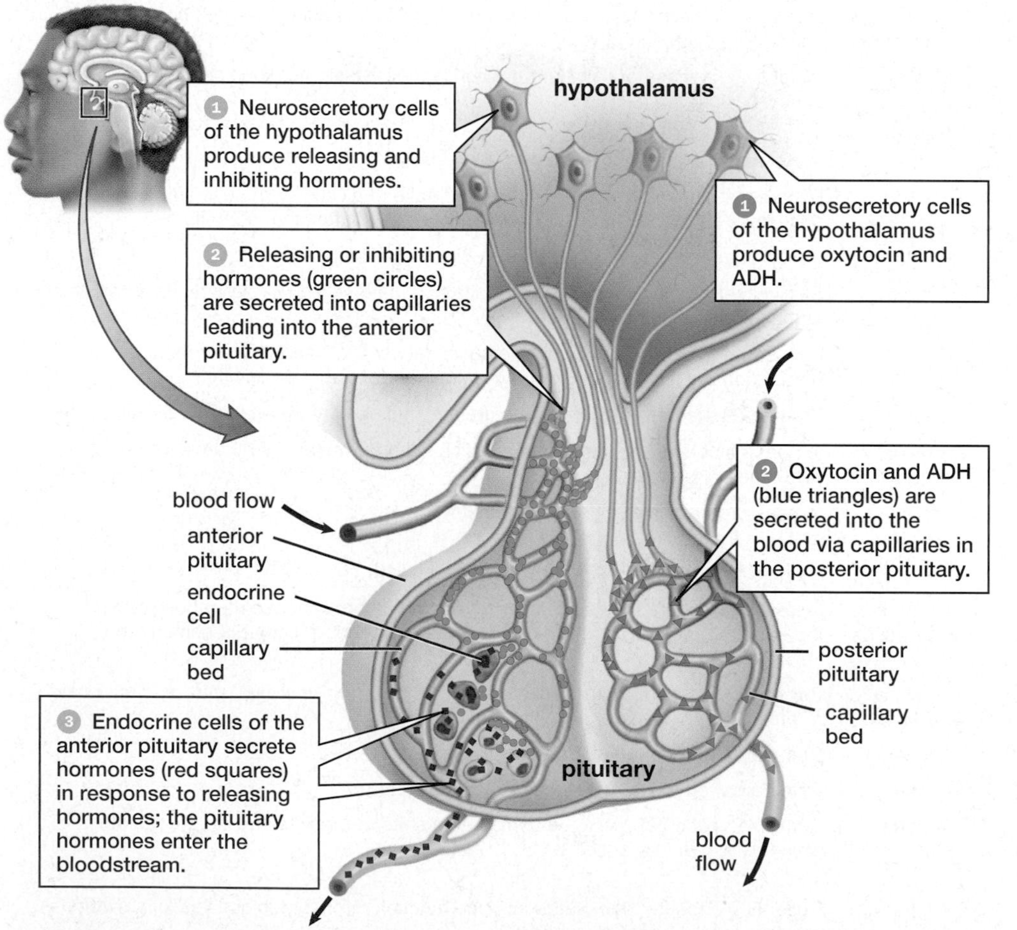

▲ **FIGURE 38-5 The hypothalamus–pituitary system** The left side of the diagram (green circled numbers) shows the relationship between the hypothalamus and the anterior pituitary, and the right side (blue circled numbers) shows the relationship between the hypothalamus and the posterior pituitary. Releasing hormones are shown as green circles, hormones from the anterior pituitary as red squares, and hormones from the hypothalamus/posterior pituitary as blue triangles.

Hypothalamic Hormones Control the Release of Hormones in the Anterior Pituitary Neurosecretory cells of the hypothalamus produce six hormones that regulate the release of the anterior pituitary hormones that we have just described (Fig. 38-5, ①). These hypothalamic regulatory hormones are called **releasing hormones** or **inhibiting hormones**, depending on whether they stimulate or inhibit the release of a particular hormone in the anterior pituitary.

The neurosecretory cells grow thin fibers, called axons, that end on a capillary bed in the stalk connecting the hypothalamus to the anterior pituitary. There, the axons secrete releasing or inhibiting hormones into the capillary bed ②. The hormones travel a short distance through blood vessels to a second capillary bed that surrounds the endocrine cells of the anterior pituitary. There, the releasing and inhibiting hormones diffuse out of the capillaries and bind to receptors on the surfaces of the endocrine cells, regulating the release of their hormones ③.

CASE STUDY CONTINUED

Insulin Resistance

In the 17th century, diabetes mellitus was called the "pissing evil." In diabetes mellitus, high blood glucose levels cause excessive amounts of glucose to enter the kidneys, overwhelming the kidneys' capacity for glucose reabsorption. Consequently, glucose remains in the urine (*mellitus* means "sweet"). Water enters the glucose-laden urine by osmosis, producing a high volume of urine. A few thousand people in the United States are afflicted with *diabetes insipidus*, in which the urine is extremely watery (*insipidus* means "without taste") because the posterior pituitary doesn't produce adequate amounts of ADH. People with either form of diabetes drink a lot, as Randy Jackson did, to replace the water lost in their urine. Drinking water seems like a pretty straightforward solution to excessive urination, but as we will see, diabetes mellitus can cause other, very serious health effects, including heart disease.

The Posterior Pituitary Releases Hormones Synthesized by Cells in the Hypothalamus

The hypothalamus contains other neurosecretory cells that synthesize either oxytocin or antidiuretic hormone (ADH) ①. The axons of these neurosecretory cells extend down into the posterior pituitary. The axons end in a capillary bed in the posterior pituitary into which they release hormones that are then carried by the bloodstream to the rest of the body ②.

As we described earlier, **oxytocin** causes contractions of the muscles of the uterus during childbirth. It also triggers the milk let-down reflex in nursing mothers by causing muscle tissue within the mammary glands of the breasts to contract in response to stimulation by the suckling infant. This contraction ejects milk from the saclike milk glands into the nipples (**FIG. 38-6**).

Oxytocin also acts directly in the brain, causing behavioral effects. In rats, for example, injecting oxytocin into the brain causes virgin females to exhibit maternal behaviors, such as building a nest and retrieving pups that have strayed. In humans (both men and women), oxytocin may play a role in emotions, including trust and both romantic and maternal love (see the case study "How Do I Love Thee?" in Chapter 39).

Antidiuretic hormone (ADH; literally, "a hormone that reduces urination") helps prevent dehydration by causing the kidneys to absorb water and return it to the bloodstream. However, wastes must still be excreted from the body, so with less water to dilute salts and waste products, the urine is highly concentrated and often deep yellow.

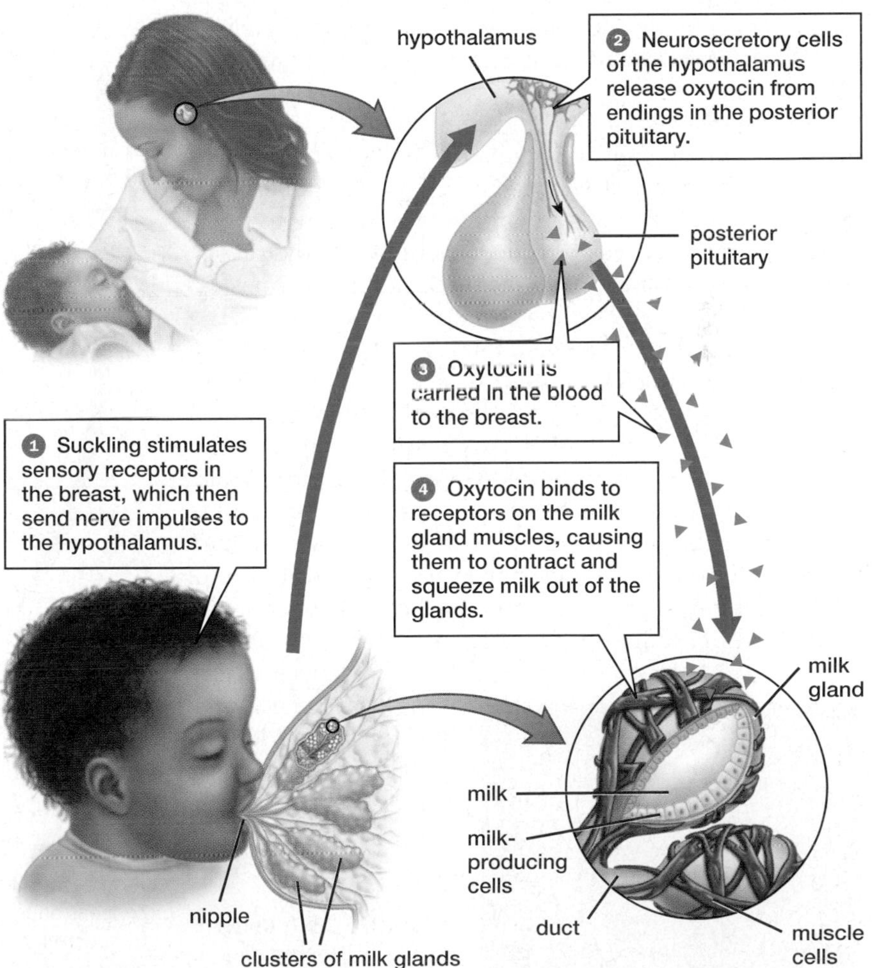

hypothalamus
② Neurosecretory cells of the hypothalamus release oxytocin from endings in the posterior pituitary.
posterior pituitary
③ Oxytocin is carried in the blood to the breast.
① Suckling stimulates sensory receptors in the breast, which then send nerve impulses to the hypothalamus.
④ Oxytocin binds to receptors on the milk gland muscles, causing them to contract and squeeze milk out of the glands.
milk gland
milk
milk-producing cells
duct
muscle cells
nipple
clusters of milk glands

▲ **FIGURE 38-6 Hormones and breast-feeding** Interactions between an infant and its mother regulate the control of milk let-down by oxytocin during breast-feeding. The cycle begins with the infant's suckling and continues until the infant is full and stops suckling. When the nipple is no longer stimulated, oxytocin release stops, the muscles relax, and milk flow ceases.

The Thyroid and Parathyroid Glands Influence Metabolism and Calcium Levels

The **thyroid gland** lies at the front of the neck, nestled just below the larynx (see Fig. 38-4). The thyroid produces two hormones: thyroxine and calcitonin. The **parathyroid gland** consists of two pairs of small clusters of endocrine cells, one pair on each side of the back of the thyroid gland. These cells release parathyroid hormone.

Thyroxine Influences Energy Metabolism

Thyroxine, or thyroid hormone, is an iodine-containing amino acid derivative. By stimulating the synthesis of enzymes that break down glucose and provide energy, and by directly stimulating the mitochondria, thyroxine increases metabolic rate in most of the cells of the body. More than 10 million Americans suffer from hypothyroidism, in which the thyroid does not produce enough thyroxine. People with hypothyroidism feel mentally and physically sluggish. They may lose appetite but still gain weight, and they become less tolerant of cold, because the body generates less heat when its metabolic rate is low. Hyperthyroidism is much less common; the resulting excess thyroxine leads to restlessness and irritability, increased appetite, and intolerance to heat.

In juvenile vertebrates, including humans, thyroxine helps regulate growth by stimulating both metabolic rate and nervous system development. In people, undersecretion of thyroid hormone early in life can cause cretinism, a condition characterized by reduced mental and physical development. Fortunately, early diagnosis and thyroxine supplementation can reverse this condition.

One common, but preventable, cause of hypothyroidism is an iodine-deficient diet. A lack of iodine reduces production of thyroxine, and the body compensates by increasing the number of thyroxine-producing cells in the thyroid. If the hypothyroidism is severe, the thyroid becomes enlarged, a condition called **goiter** (**FIG. 38-7**). Worldwide, hundreds of millions of people have iodine-deficient diets. Iodine deficiency in pregnant women and young children is the leading preventable cause of mental retardation. Iodized salt is a simple, cheap solution to iodine deficiency; it costs less than $1.50 per ton to add iodine to table salt.

◀ FIGURE 38-7 Goiter An iodine-deficient diet often causes enlargement of the thyroid gland. Goiter is all too common in less-developed countries where people lack iodized salt in their diets.

Thyroxine Release Is Controlled by the Hypothalamus and Anterior Pituitary Levels of thyroxine in the bloodstream are controlled by negative feedback (**FIG. 38-8**). Thyroid-stimulating hormone–releasing hormone (TSH-releasing

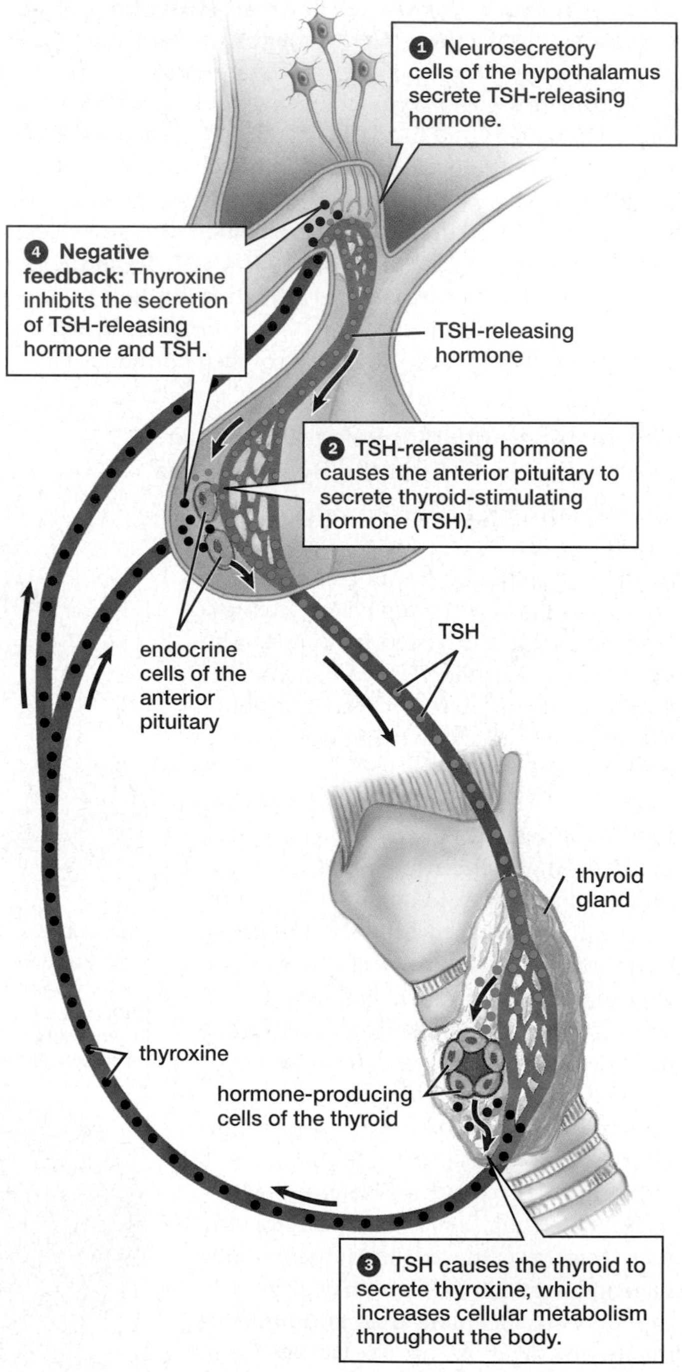

1 Neurosecretory cells of the hypothalamus secrete TSH-releasing hormone.
4 Negative feedback: Thyroxine inhibits the secretion of TSH-releasing hormone and TSH.
TSH-releasing hormone
2 TSH-releasing hormone causes the anterior pituitary to secrete thyroid-stimulating hormone (TSH).
TSH
endocrine cells of the anterior pituitary
thyroid gland
thyroxine
hormone-producing cells of the thyroid
3 TSH causes the thyroid to secrete thyroxine, which increases cellular metabolism throughout the body.

▲ FIGURE 38-8 Negative feedback in thyroid gland function The concentration of thyroxine in the bloodstream (black dots) regulates the secretion of TSH-releasing hormone (blue dots) and TSH (green dots) by negative feedback.

THINK CRITICALLY A common test of thyroid gland function is to measure the amount of thyroid-stimulating hormone circulating in the blood. What would you hypothesize is wrong in a person who has an abnormally high level of TSH?

hormone) produced by neurosecretory cells in the hypothalamus ❶ travels to the anterior pituitary and causes the release of TSH ❷. TSH travels in the bloodstream to the thyroid and stimulates the release of thyroxine ❸. Adequate levels of thyroxine circulating in the bloodstream inhibit the secretion of both TSH-releasing hormone from the hypothalamus and TSH from the anterior pituitary, thus inhibiting further release of thyroxine from the thyroid ❹.

Thyroxine Has Varied Effects in Different Vertebrates In amphibians and many fish, thyroxine stimulates metamorphosis from a larval to an adult body form. A frog hatches out of its egg as an aquatic tadpole, which looks a bit like a fat-headed fish, with gills, a large finned tail, and no legs. A newly hatched tadpole has low levels of thyroxine. A few weeks later, its thyroxine level increases, causing it to sprout legs and resorb its tail. After another several weeks, the transformation into a small frog is complete. In lampreys and salmon, thyroxine helps trigger the bodily transformations that allow the young fish, which are born in fresh water, to thrive in seawater, where they grow up and mature. Thyroxine also regulates the seasonal molting of many terrestrial vertebrates. From snakes to birds to your family dog, surges of thyroxine stimulate the shedding of skin, feathers, or hair.

Parathyroid Hormone and Calcitonin Regulate Calcium Metabolism

The proper concentration of calcium is essential to nerve and muscle function. **Parathyroid hormone** from the parathyroid gland and **calcitonin** from the thyroid work together to maintain nearly constant calcium levels in the blood and body fluids.

The skeleton serves as a "bank" into which calcium can be deposited or withdrawn as necessary. If blood calcium levels drop, parathyroid hormone causes the bones to release calcium. It also causes the kidneys to reabsorb more calcium during urine production and to return the calcium to the blood. The increased blood calcium then inhibits further release of parathyroid hormone in a negative feedback loop. If blood calcium gets too high, the thyroid releases calcitonin, which inhibits the release of calcium from bone. In humans, the effects of calcitonin appear to be minor compared to those of parathyroid hormone.

The Pancreas Has Both Digestive and Endocrine Functions

The **pancreas** has multiple functions. In its role in digestion, it produces bicarbonate and several enzymes that are released into the small intestine, where they promote the breakdown of food (see Chapter 35). The endocrine portion of the pancreas consists of clusters of **islet cells**. Each islet cell produces one of two peptide hormones: **insulin** or **glucagon**.

Insulin and Glucagon Control Glucose Levels in the Blood

Insulin and glucagon have opposing effects on carbohydrate and fat metabolism: Insulin reduces blood glucose, whereas glucagon increases it. Together, the two hormones help keep blood glucose nearly constant (**FIG. 38-9**). When

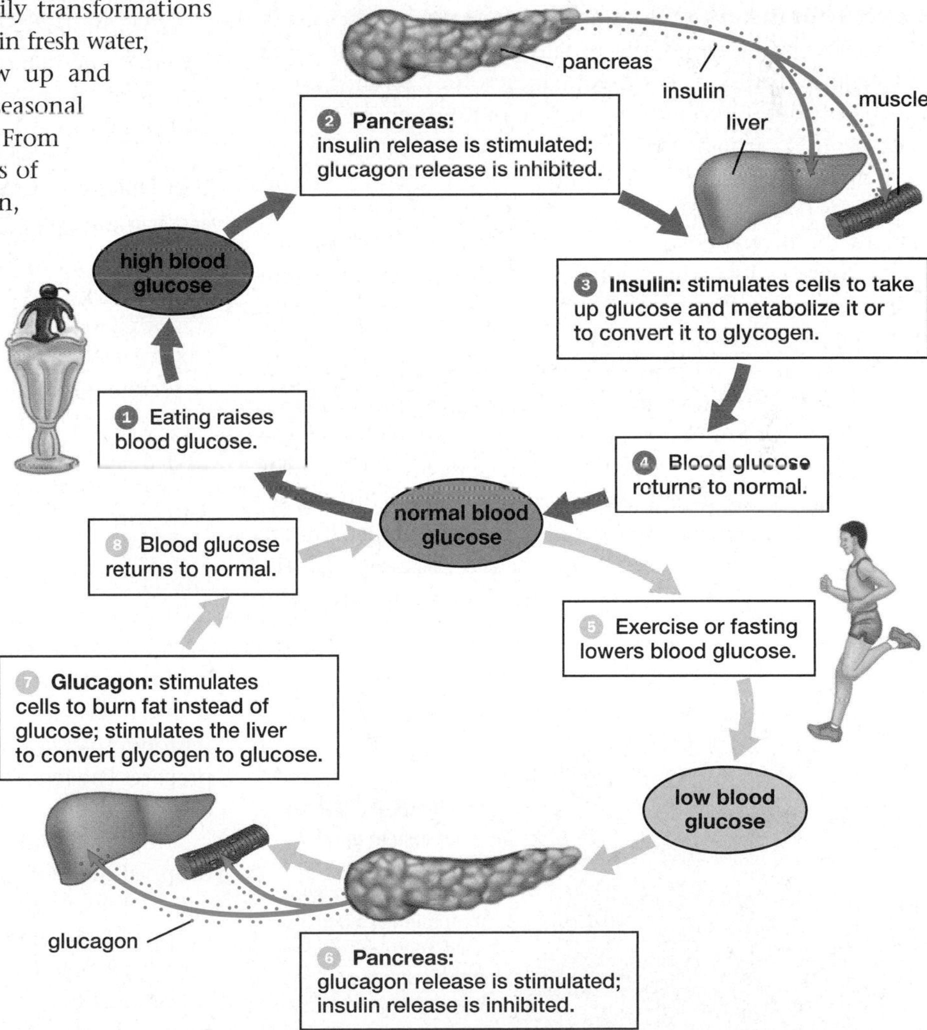
pancreas
insulin
liver
muscle
❷ Pancreas:
insulin release is stimulated;
glucagon release is inhibited.
high blood glucose
❸ Insulin: stimulates cells to take up glucose and metabolize it or to convert it to glycogen.
❶ Eating raises blood glucose.
❹ Blood glucose returns to normal.
normal blood glucose
❽ Blood glucose returns to normal.
❺ Exercise or fasting lowers blood glucose.
❼ Glucagon: stimulates cells to burn fat instead of glucose; stimulates the liver to convert glycogen to glucose.
low blood glucose
glucagon
❻ Pancreas:
glucagon release is stimulated;
insulin release is inhibited.

▲ **FIGURE 38-9 The pancreas controls blood glucose levels** The actions of insulin and glucagon form a two-part negative feedback loop that keeps blood glucose concentrations from varying too much, despite episodes of eating, fasting, and exercise throughout the day.

THINK CRITICALLY How would blood glucose be affected in a person who was born with a mutation that prevented glucagon receptors from binding glucagon?

blood glucose rises (for example, after you have eaten), the pancreas releases more insulin and less glucagon. Insulin causes many body cells to take up glucose from the blood and either metabolize it for energy or convert it to glycogen, thus returning blood glucose concentrations back to normal.

If blood glucose levels drop (for example, if you've skipped breakfast or run a 10-kilometer race), insulin secretion is inhibited and glucagon secretion is stimulated. Glucagon activates an enzyme in the liver that breaks down glycogen to glucose, which is then released into the blood. Glucagon also promotes fat breakdown. The resulting fatty acids can be metabolized to produce energy, sparing glucose. These actions increase blood glucose and return it to normal.

Diabetes Results from Defective Insulin Control

About 5% of people with diabetes have type 1, an autoimmune disease in which the immune system destroys the insulin-secreting cells in the pancreas (see Chapter 37). Insulin replacement therapy profoundly improves the health of people with type 1 diabetes, but usually requires frequent blood testing and insulin injections and doesn't fully mimic natural control of energy metabolism. Insulin pumps—about the size of a cell phone, usually worn attached to a belt—can deliver insulin virtually continuously, eliminating the need for insulin injections (**FIG. 38-10**). An experimental pump system, often called a "bionic pancreas," measures blood glucose in real time and automatically delivers appropriate doses of either insulin or glucagon accordingly. In 2014, researchers reported coaxing stem cells to differentiate into insulin-producing cells that release insulin in response to physiological levels of glucose, perhaps the first step toward stem cell therapy for type 1 diabetes.

The other 95% of diabetics have type 2 diabetes. In the early stages of the disorder, they produce adequate amounts of insulin, but, like Randy Jackson, their bodies are insulin resistant. The vast majority of people with type 2 diabetes are overweight, which probably causes insulin resistance: Fat tissue releases certain types of fatty acids that enter muscle and fat cells and make them less responsive to insulin. As a result, glucose transporters are not moved to the plasma membrane, so glucose is not transported efficiently into the cells, but remains in the blood. Insulin resistance also causes liver cells to break down glycogen to glucose and release the glucose into the blood. To make things even worse, insulin resistance causes fat cells to break down fats to fatty acids, which are released into the bloodstream. Then these fatty acids enter muscle and fat cells, continuing a positive feedback loop that further increases insulin resistance.

◀ FIGURE 38-10 An insulin pump Some insulin pumps continuously monitor the patient's glucose levels, thereby reducing the frequency of the "fingersticks" needed to obtain blood samples. Having real-time glucose levels also allows the patient to rapidly correct insulin levels after exercise (which uses up glucose) or a high-carbohydrate meal (which raises blood glucose).

The Sex Organs Produce Both Gametes and Sex Hormones

The **testes** in males and the **ovaries** in females are important endocrine organs. The testes secrete several steroid hormones, collectively called **androgens**; the most important of these is **testosterone**. The ovaries secrete two types of steroid hormones: **estrogen** and **progesterone**. Sex hormones influence development in both sexes and affect brain function and behavior throughout life. (The roles of sex hormones in sperm and egg production, the menstrual cycle, and pregnancy are discussed in Chapter 42.)

Sex Hormone Levels Increase During Puberty

Sex hormones play a key role in puberty, the phase of life during which the reproductive system becomes functional. Puberty begins when, for reasons not fully understood, the hypothalamus starts to secrete increasing amounts of releasing hormones, which in turn stimulate the anterior pituitary to secrete luteinizing hormone (LH) and follicle-stimulating hormone (FSH). LH and FSH stimulate cells in the testes or ovaries to produce higher levels of sex hormones.

The resulting increase in sex hormones affects target cells throughout the body. Both sexes develop pubic and underarm hair. Testosterone, secreted by the testes, stimulates the development of male secondary sexual characteristics, including body and facial hair, a deep voice, and relatively large muscles. Testosterone also promotes sperm cell production. In females, estrogen secreted by the ovaries stimulates breast development and maturation of the female reproductive system, including egg production. Progesterone prepares the female reproductive tract to receive and nourish a fertilized egg.

In developed countries, the average age of puberty has dropped markedly since the 1800s; for example, the average age at which girls have their first menstrual period has dropped from about 17 in the mid-1800s to about 12 today. Boys typically become sexually mature at age 13 or 14. Many researchers believe that better health and nutrition account for much of the reduction in age at puberty. In girls, the trend toward earlier puberty is probably influenced by earlier accumulation of stored fat; natural selection may have favored puberty that began when the body accumulated enough fat to sustain a successful pregnancy. Other factors, such as environmental pollutants and changes in social interactions, might also play a role. Studies in a wide variety of animals, as well as a few studies in humans, have revealed

Health WATCH Performance-Enhancing Drugs—Fool's Gold?

Many athletes, at levels from high school to the Olympics and professional leagues, take hormones—natural or synthetic—to become stronger or faster or to increase endurance. Some even admit taking performance-enhancing drugs (PEDs), but usually not until after they're caught. Top athletes who have been penalized for taking PEDs include former Tour de France champion Lance Armstrong (**FIG. E38-1**, top), former world and Olympic champion sprinter Marion Jones (Fig. E38-1, bottom), and New York Yankees slugger Alex Rodriguez. Let's look at the actions of three common PEDs.

Anabolic Steroids

Anabolic steroids include testosterone and a variety of synthetic steroids that act similarly, but with stronger effects. Athletes often take anabolic steroids because they bind to testosterone receptors and stimulate muscle development. Many synthetic steroids activate these receptors at lower concentrations than testosterone can. Further, if you inject testosterone into your body, it will be completely metabolized within a few hours, whereas some synthetic steroids remain in your system for weeks. Therefore, synthetic anabolic steroids often have more powerful, longer-lasting effects than natural testosterone does.

Taking steroids may cause unwanted side effects, some serious. In both men and women, anabolic steroids may suppress the immune system, increase blood pressure, decrease HDL ("good") cholesterol, and alter mood. In men, the anterior pituitary, tricked by anabolic steroid injections, releases smaller amounts of hormones than are required for testes development and sperm production, so the testes often shrink and sperm count drops. Men also produce enzymes that convert testosterone and some synthetic steroids into estrogen, which may cause partial breast development. In fact, some male steroid abusers take tamoxifen, an estrogen antagonist, to block these estrogenic effects. In women, anabolic steroids promote male-like body changes, including deepening of the voice, increased facial hair, and even pattern baldness. Anabolic steroids also interfere with egg development and ovulation.

Erythropoietin

Erythropoietin stimulates the bone marrow to produce more red blood cells, thereby increasing the delivery of oxygen to the muscles (see Chapter 33). A drug-free method of achieving the same effect is to draw blood from an athlete a few weeks before a major event, separate out and store the red blood cells, allow the body to replace the depleted red blood cells naturally, and then reinject the stored blood to boost red blood cell counts. EPO and blood transfusions probably both improve endurance. However, the resulting high red blood cell count can thicken the blood, causing clots and leading to strokes.

Growth Hormone

Growth hormone not only helps children to grow taller, but also increases muscle tissue, strengthens bones, and reduces fat. Growth hormone has become a drug of choice for cheating athletes, even though there haven't been any rigorous studies proving that it actually enhances performance. Possible side effects of growth hormone include joint and muscle pain, heart disease, high cholesterol, high blood pressure, and diabetes.

Are bulging muscles worth the risks? Some athletes say they would willingly risk long-term damage to their bodies to win Olympic gold. And the payoff in professional sports can be astronomical—Alex Rodriguez, although suspended for the 2014 baseball season for prior PED abuse, had a guaranteed salary of $20 million a year for the 2015–2017 baseball seasons. For the more typical man or woman, whose principal benefits are admiring glances from potential dating partners and competitors, well. . . . Before you decide, be sure to find reliable information from authoritative sources such as the National Institutes of Health or the Mayo Clinic—not from body-building Web sites or online drug retailers.

▲ **FIGURE E38-1 From superhero to loser** (Top) In 2012, Lance Armstrong was stripped of his seven Tour de France titles. The U.S. Anti-Doping Agency called Armstrong's cheating "the most sophisticated, professionalized and successful doping program that sport has ever seen." Armstrong has admitted to using erythropoietin, blood transfusions, testosterone, growth hormone, and cortisone. (Bottom) Marion Jones was forced to return five Olympic medals because of her steroid abuse.

THINK CRITICALLY Baseball player Juan M. reports to training camp 20 pounds heavier than at the end of last season, and it looks to be all muscle. What changes in Juan's physical exam and urine and blood analyses might suggest that he has been taking PEDs? Which PEDs would you suspect?

Have You Ever Wondered ...

Why You Often Get Sick When You're Stressed?

If you get sick during final exam week, stress-induced cortisol release may be responsible. In addition to its effects on glucose metabolism, cortisol also inhibits the immune response. Why should the body ever suppress its immune response? Although finals are not life-threatening, many other stresses, such as searching for scarce food or encountering predators, can be. Because activating the immune system takes a lot of energy and makes people and other mammals feel sluggish, suppressing the immune response in favor of dealing with immediate stresses has been favored by natural selection.

that pollutants from agricultural and industrial activities can disrupt hormone signaling. These "endocrine disruptors" frequently mimic or block the actions of sex hormones and consequently might alter sexual development, as we describe in "Earth Watch: Endocrine Deception."

The Adrenal Glands Secrete Hormones That Regulate Metabolism and Responses to Stress

The **adrenal glands** (Latin for "on the kidney") consist of two very different parts: the adrenal cortex and the adrenal medulla.

The Adrenal Cortex Produces Steroid Hormones

The outer layer of each adrenal gland is the **adrenal cortex** (*cortex* is the Latin word for the bark of a tree). The cortex secretes three types of steroid hormones: glucocorticoids, mineralocorticoids, and small amounts of testosterone. As their names imply, **glucocorticoids** help control glucose metabolism, while **mineralocorticoids** regulate salt metabolism.

Glucocorticoid release is stimulated by adrenocorticotropic hormone (ACTH) from the anterior pituitary, which in turn is stimulated by a releasing hormone produced by the hypothalamus. Glucocorticoids are released in response to stimuli such as stress, trauma, or temperature extremes. **Cortisol** is by far the most abundant glucocorticoid. Cortisol increases blood glucose levels by stimulating glucose production, inhibiting the uptake of glucose by muscle cells, and promoting the use of fats for energy.

Mineralocorticoid hormones regulate the mineral (salt) content of the blood. The most important mineralocorticoid is **aldosterone**, which helps to control sodium concentrations. Sodium ions are the most abundant positive ions in blood and interstitial fluid. The sodium ion concentration gradient across plasma membranes (high in the interstitial fluid, low in the cytoplasm) is crucial to many cellular events, including the production of electrical signals by nerve cells. If blood sodium falls, the adrenal cortex releases aldosterone, which causes the kidneys and sweat glands to retain sodium and return it to the blood. Meanwhile, eating helps to replenish the body's sodium stores, because almost all foods contain sodium. When the combination of dietary sodium intake and aldosterone-induced sodium conservation returns blood sodium concentrations back to normal levels, aldosterone secretion is shut off—another example of negative feedback.

In both women and men, the adrenal cortex also produces small amounts of the male sex hormone testosterone. Tumors of the adrenal cortex can lead to excessive testosterone release, causing masculinization of women.

The Adrenal Medulla Produces Amino Acid Derived Hormones

The **adrenal medulla** is located in the center of each adrenal gland. It produces two hormones in response to stress or exercise—**epinephrine** and a small quantity of **norepinephrine** (also called adrenaline and noradrenaline, respectively). These hormones prepare the body for emergency action. They increase heart and respiratory rates, increase blood pressure, cause blood glucose levels to rise, and direct blood flow away from the digestive tract and toward the brain and muscles. They also cause the air passages to the lungs to expand, allowing larger volumes of air to enter and leave the lungs. For this reason, epinephrine is often administered to people whose airways become constricted, such as during a severe allergic reaction or an asthma attack.

Hormones Are Also Produced by the Pineal Gland, Thymus, Kidneys, Digestive Tract, Fat Cells, and Heart

The **pineal gland** is located between the two hemispheres of the brain (see Fig. 38-4). The pineal produces the hormone **melatonin**, an amino acid derivative. Melatonin is secreted in a daily cycle, very little during the day and much more during the night. In mammals, the cycle is driven by light detected by the eyes. In some vertebrates with thin, translucent skulls, such as frogs, the pineal itself contains photoreceptive cells. In these animals, the pineal directly responds to day length. By responding to day lengths characteristic of different seasons, the pineal helps to regulate the reproductive cycles of many animals.

Despite years of research, the function of the pineal gland and melatonin in humans is still not well understood. Melatonin secretion may influence sleep–wake cycles. Darkness increases melatonin production and bright light inhibits it, and there is some evidence that secretion of melatonin at night promotes sleep. Consequently, melatonin is sometimes used as a sleeping aid or to overcome jet lag, although most research suggests that the effects are fairly small.

The **thymus** is located in the chest cavity behind the breastbone (see Fig. 38-4). The thymus produces the hormone **thymosin**, which stimulates the development of specialized white blood cells (T cells) that play crucial roles in the immune response (see Chapter 37). The thymus is large in children but, under the influence of sex hormones, gradually decreases in size after puberty. As a result, the elderly produce fewer new T cells than younger people do and, hence, are more susceptible to new diseases.

The kidneys release **erythropoietin** when the oxygen content of the blood is low. Erythropoietin stimulates the

Earth WATCH Endocrine Deception

Some synthetic organic compounds that enter the environment mimic or block the actions of certain hormones, most commonly estrogen, testosterone, or thyroxine. These **endocrine disruptors** include some pesticides, herbicides, ingredients in plastics, flame retardants, detergents, sunscreens, and polychlorinated biphenyls (formerly used in sealants and paints and as insulating fluids in power transformers). Probably the most potent endocrine disruptor in the environment is ethinylestradiol, a synthetic estrogen commonly found in birth control pills.

In the most common form of endocrine disruption, a synthetic chemical enters cells and binds to estrogen receptors. Endocrine disruptors exert a wide variety of harmful effects, including feminization of males, masculinization of females, reproductive cancers, malformed sex organs, altered blood hormone levels, and reduced fertility. Feminization of males due to estrogenic endocrine disruptors can be extreme; in some cases, eggs develop in a feminized male's testes.

In one of the best known cases, agricultural runoff and a pesticide spill near Lake Apopka in Florida polluted the lake's water with large quantities of several endocrine disruptors, including the pesticide DDT and its breakdown products. Wildlife biologists noted an alarming decline in the alligator population of the lake. Many eggs never hatched. Males had high estrogen, low testosterone, small penises, and abnormal testes. Females typically had exceptionally high estrogen levels and abnormal ovaries.

Although alarming, the Lake Apopka results were the result of massive exposure to endocrine disruptors. Do smaller doses, more likely to be found in fairly clean water, also have harmful effects? Indeed they may. In 2003 and 2004, researchers from the University of Colorado sampled a common minnow in Boulder Creek, which runs through Boulder, Colorado (**FIG. E38-2**). They found that upstream of the city's sewage outfall, about half the fish were males and half were females. Downstream of the outfall, more than 80% of the fish were females. In lab studies, male fish exposed to wastewater from the treatment plant were rapidly feminized. In 2007, the city of Boulder upgraded its sewage treatment plant, converting to a system that uses microbes to more thoroughly digest organic molecules, including endocrine disruptors. It worked—a follow-up study in 2011 found that the wastewater no longer feminized fish.

Are endocrine disruptors harmful to people? Some, such as polychlorinated biphenyls, certainly are. Others probably are harmful at high enough concentrations. The debates among toxicologists, reproductive biologists, industry officials, and government regulators focus on several questions: What are the minimum concentrations that affect people? Do those concentrations actually occur in people? Might mixtures of several endocrine disruptors, all at very low concentrations, add up to harmful effects? How could we definitively prove either safety or harm, given that human experiments cannot be performed?

▲ **FIGURE E38-2 Sampling water in Boulder Creek** Although the creek's water is now fairly clean, it formerly contained chemicals that disrupted the reproductive systems of fish.

Some known endocrine disruptors, such as DDT and polychlorinated biphenyls, have been banned in most countries. For suspected endocrine disruptors, the situation is more unsettled. In 2008, Canada banned the use of a chemical called bisphenol A (BPA) in plastic baby bottles; the European Union followed suit in 2011. In 2012, in response to consumer concerns, manufacturers in the United States stopped using BPA in baby bottles and sippy cups. Although the U.S. Food and Drug Administration regards BPA as safe at the levels typically found in humans, the National Toxicology Program of the National Institutes of Health finds that BPA merits "some concern for effects on the brain, behavior, and prostate gland in fetuses, infants, and children at current human exposures." More research is clearly needed. In 2014, the U.S. Environmental Protection Agency began the Endocrine Disruptor Screening Program, examining over 100 chemicals for possible endocrine effects.

THINK CRITICALLY Endocrine disruptors have most frequently been associated with reproductive effects in fish and amphibians. Why might these animals be more susceptible than birds or mammals?

bone marrow to increase red blood cell production (see Chapter 33). The kidneys also produce an enzyme called *renin* in response to low blood pressure, for example, after profuse bleeding from a wound. Renin catalyzes the production of the hormone **angiotensin** from proteins in the blood. Angiotensin raises blood pressure by constricting arterioles. It also stimulates the release of aldosterone by the adrenal cortex, causing the kidneys to return sodium to the blood. The resulting high salt concentration attracts and retains water, increasing blood volume and pressure.

The stomach and small intestine produce numerous peptide hormones, including gastrin, ghrelin, secretin, and cholecystokinin (see Chapter 35). These hormones regulate several aspects of nutrition, such as the secretion of digestive enzymes, movement of food within the stomach and small intestine, and the sensations of hunger and fullness.

Fat cells release the peptide hormone **leptin**: The more fat the body has stored, the more leptin is released. Mutant mice that cannot produce leptin eat a lot, have low metabolic rates, and become obese (**FIG. 38-11**). They also develop many of the symptoms of type 2 diabetes, including high blood glucose. Injecting them with leptin causes them to lose weight and helps to restore normal blood glucose levels.

Working together with a number of other hormones, including thyroxine and insulin, ghrelin and leptin control appetite, metabolic rate, and fat storage. Ghrelin is released when the stomach is empty. Ghrelin stimulates hunger and eating, which tends to increase body weight and fat storage. Leptin is released when fat stores are high. Leptin decreases hunger and increases metabolic rate, which tends to decrease body weight and fat storage. Therefore, you might think that giving leptin to overweight people would be a powerful weight-loss aid. Unfortunately, the appetite-reducing effects of high leptin are minor compared to the appetite-stimulating effects of low leptin. In fact, many obese people have high levels of leptin but seem to be insensitive to it.

The heart releases a hormone, **atrial natriuretic peptide (ANP)**, that helps regulate blood volume and pressure. If the blood volume is too high—for example, if you drink too much water—the atria become overfilled, which stretches their walls and stimulates the release of ANP. ANP inhibits the release of ADH and aldosterone and increases the excretion of sodium. By reducing reabsorption of water and salt by the kidneys, ANP helps to lower blood volume.

▲ **FIGURE 38-11 Leptin helps regulate body fat** The mouse on the left has a mutation that stops production of the hormone leptin.

CHECK YOUR LEARNING

Can you . . .

- name the endocrine glands described in this section, identify the hormones each one produces, and give an example of an effect of each hormone?
- explain how negative feedback regulates the secretion and actions of thyroxine, oxytocin, and insulin?

CASE STUDY REVISITED

Insulin Resistance

Type 2 diabetes is a growing problem in the United States. It increases the incidence of cardiovascular disease, which has roughly doubled over the last 20 years. Most people with type 2 diabetes are overweight before they develop diabetes, and they already have high blood levels of triglycerides and cholesterol (especially low-density lipoprotein, or LDL, the "bad" cholesterol). Once people develop type 2 diabetes, insulin resistance interferes with lipid metabolism, further increasing triglycerides and LDL in the blood. Some of this fat and cholesterol settles in the blood vessels, forming plaques.

Plaques narrow the diameter of arteries, increasing resistance to blood flow, which causes hypertension. Hypertension itself, by damaging blood vessels, can also promote plaque formation. Thus diabetes can set off a positive feedback cycle: increased plaque formation → hypertension → more plaque formation → even higher hypertension. All too often, only a heart attack or stroke ends this positive feedback, by ending life. About 65% of people with diabetes die of cardiovascular disease or strokes.

Weight loss and exercise are the usual first-line treatments for type 2 diabetes. Reducing body fat decreases blood concentrations of fatty acids that can cause insulin resistance. Even a single bout of moderate exercise greatly increases insulin sensitivity and glucose uptake in muscles. What's more, insulin sensitivity remains elevated for two or three days after exercise, so exercising two or three times a week can make a big difference. Weight training increases muscle mass, which further enhances glucose uptake from the blood. If exercise and weight loss fail, a drug called metformin is usually prescribed. Metformin reduces glucose production by the liver and increases insulin responsiveness in many cells.

Fortunately, Randy Jackson's physician diagnosed his type 2 diabetes in time. Thanks to a better diet, gastric bypass surgery (see Chapter 35), and exercise, Jackson lost over 100 pounds. He became a spokesman for the American Heart Association's "Heart of Diabetes" outreach program, trying to help others to recognize the early symptoms of diabetes and take action.

THINK CRITICALLY Diabetes can be diagnosed with a glucose tolerance test. For several hours before the test, the patient eats nothing and drinks only water. Then, blood glucose level is measured. Next, the patient consumes a sugary drink, after which blood glucose is measured periodically over several hours. How would test results differ between a person who has diabetes and a person who does not? Why?

CHAPTER REVIEW

Go to **Mastering Biology** to access the Pearson eText, vocabulary review, practice quizzes, activities, videos, current events, and more.

*Answers to **Think Critically** and **Thinking Through the Concepts** questions can be found in the **Answers** section at the back of the book.*

Summary of Key Concepts

38.1 How Do Animal Cells Communicate?

In multicellular organisms, communication among cells occurs through gap junctions directly linking cells, by diffusion of chemicals to nearby cells, or by transport of chemicals within the bloodstream (endocrine hormones). A chemical messenger acts only on target cells bearing receptors that can bind the messenger molecule and trigger a response in the cell. Vertebrate endocrine hormones are produced by glands embedded in capillary beds. The hormones are secreted into the interstitial fluid, diffuse into the capillaries, and are transported in the bloodstream to other parts of the body, where they bind to receptors on target cells and exert their effects.

38.2 How Do Endocrine Hormones Produce Their Effects?

Vertebrate endocrine hormones fall into one of three classes: peptides, amino acid derivatives, and steroids. Most hormones act on their target cells in one of two ways: (1) Steroid hormones usually diffuse through the plasma membrane of a target cell and bind to receptors inside the cell. The hormone–receptor complex changes the transcription of specific genes. Thyroxine is also transported across the plasma membrane into a cell, where it binds to receptors and changes gene transcription. (2) Peptide hormones and amino acid derived hormones bind to receptors on the surface of a target cell and cause the synthesis of intracellular second messengers, such as cyclic AMP, which then alter the cell's metabolism or change the rate of transcription of specific genes, or both.

Hormone action is usually regulated by negative feedback, a process in which a hormone causes changes that inhibit further secretion of that hormone. In rare instances, such as childbirth, hormone release may be temporarily controlled by positive feedback.

38.3 What Are the Structures and Functions of the Mammalian Endocrine System?

The major endocrine glands of the human body are the hypothalamus–pituitary complex, the thyroid and parathyroid glands, the pancreas, the sex organs, and the adrenal glands. The hormones released by these glands and their actions are summarized in Table 38-2. Other structures that produce hormones include the pineal gland, thymus, kidneys, digestive tract, fat cells, and heart.

Thinking Through the Concepts

Bloom's: Remembering, Understanding

Multiple Choice

1. Neurosecretory cells in the hypothalamus
   a. release growth hormone and prolactin.
   b. release hormones that control hormone secretion by cells in the posterior pituitary.
   c. produce hormones that regulate water reabsorption in the kidney.
   d. produce hormones that stimulate glucose uptake by muscle cells.
2. The response of a target cell to a hormone depends on
   a. the receptor in the target cell that binds the hormone.
   b. which second messengers are activated in the cell.
   c. the type of cell.
   d. all of these factors.
3. Steroid hormones include
   a. testosterone.
   b. growth hormone.
   c. melatonin.
   d. testosterone, growth hormone, and melatonin.
4. Peptide hormones
   a. typically bind to receptors in the cytoplasm of a target cell.
   b. typically bind to receptors on the surface of a target cell.
   c. always activate gene transcription in their target cells.
   d. never activate gene transcription in their target cells.
5. In negative feedback,
   a. a change causes responses that counteract the change.
   b. a response to a change damages the body.
   c. a change produces a response that enhances the change.
   d. a response to a change permanently alters the body.

Fill-in-the-Blank

1. Endocrine hormones are molecules released by cells that are parts of the ____________. These cells are embedded in capillary beds, so the hormones enter the bloodstream and move throughout the body. Only specific cells of the body, called ____________, can respond to any given hormone, because only these cells bear proteins, called ____________, that can bind the hormone.
2. Most endocrine hormones fall into three chemical classes: ____________, ____________, and ____________. ____________ and ____________ are mostly water soluble and bind to receptors on the surfaces of cells. These typically stimulate the synthesis of intracellular molecules called ____________, which activate enzymes and change the cell's metabolism. ____________ are lipid soluble and bind to receptors in the cytoplasm or nucleus. The hormone–receptor complex typically binds to DNA and causes ____________.
3. A part of the brain called the ____________ controls the activity of the pituitary gland. Specialized nerve cells in this brain area, called ____________, release the hormones ____________ or ____________ from the endings of their axons in the posterior pituitary.
4. The major hormones produced by the anterior pituitary gland are (in any order): ____________, ____________, ____________, ____________, ____________, and ____________.

5. The pancreas releases the hormone ____________ when blood glucose levels become too high; it causes many cells of the body to take up glucose. When the pancreas produces too little of this hormone, or body cells cannot respond to it, a disorder called ____________ results. ____________ is released when blood glucose levels become too low; it causes the liver to break down the starch-like storage molecule ____________ and release glucose into the blood.
6. The male sex organs, called the ____________, release the sex hormone ____________. The female sex organs, called the ____________, release two hormones, ____________ and ____________.
7. The adrenal cortex releases three major types of steroid hormones: ____________, ____________, and ____________. The adrenal medulla releases the amino acid derived hormones ____________ and ____________.

Review Questions

1. Which chemical class of hormones usually attaches to membrane receptors on target cells? What cellular events usually follow?
2. Which chemical class of hormones usually binds to receptors inside target cells? What cellular events usually follow?
3. What are the major endocrine glands in the human body, and where are they located?
4. Describe the structure and function of the hypothalamus–pituitary complex. Describe how releasing hormones regulate the secretion of hormones by cells of the anterior pituitary. Name the hormones of the anterior pituitary, and give one function of each.
5. Describe how the hormones of the pancreas act together to regulate the concentration of glucose in the blood.
6. Compare the adrenal cortex and adrenal medulla by answering the following questions: Where are they located within the adrenal gland? Which hormones do they produce? Which organs do their hormones target?

Applying the Concepts

Bloom's: Applying, Analyzing, Evaluating

1. A student researcher decides to perform an experiment on the effect of the thyroid gland on frog metamorphosis. She sets up three aquaria with tadpoles. She adds thyroxine to the water of one, the drug thiouracil to a second, and nothing to the third. Thiouracil destroys thyroxine. Assuming that the student uses appropriate concentrations, predict what will happen.
2. Suggest a hypothesis about the endocrine system to explain why many birds lay their eggs in the spring. Why do poultry farmers who produce eggs often keep the lights on at night in their chicken coops?

39 The Nervous System

Is love "a fire sparkling in lovers' eyes ... a madness most discreet," or just the right mixture of chemicals in lovers' brains?

CASE STUDY

How Do I Love Thee?

"But, soft! What light through yonder window breaks?

It is the east, and Juliet is the sun."

—*Romeo and Juliet,*
Act II, scene II

IN SHAKESPEARE'S *Romeo and Juliet,* two teenage lovers defy their families, risk their fortunes, jeopardize their futures, and ultimately give up their lives, all for love. Shakespeare's play provides a dramatic portrayal of the power of romantic love, but other forms of love can be equally powerful. A mother's love for her child is deep, and people have willingly died for love of friends, God, or country. Are these types of love related? What happens in the brain when two lovers meet or a mother cradles her infant? No one knows for sure—at least not in people. But when it comes to a small rodent called the prairie vole, scientists know quite a bit about love—or at least about pair bonding and sex.

If Juliet had been a prairie vole, her first encounter with Romeo would have released a flood of oxytocin, the same hormone that causes uterine contractions during childbirth. Oxytocin would have bound to receptors in a few small areas of her brain, causing nerve cells to release dopamine, often called the reward chemical. She would have felt wonderful. What's more, she would have linked that euphoric feeling to Romeo. In a Romeo vole, some of the molecules and brain regions would have differed, but the end result would have been similar: a flood of dopamine, giving him the ultimate high and making him believe that he could attain that feeling again only with Juliet. So, the two voles would mate for life. They would build a nest together and raise their young.

How does a vole, or Romeo for that matter, perceive that an object standing in front of him is a potential mate, and not food or a predator? How do animals respond to stimuli with appropriate behaviors such as courting, nesting, eating, or fleeing? The answers lie in the nervous system.

AT A GLANCE

39.1 WHAT ARE THE STRUCTURES AND FUNCTIONS OF NERVE CELLS?

The nervous system contains two principal cell types: **neurons**, often called nerve cells, and **glia**. Neurons are the cells that carry out the principal jobs of all nervous systems: They receive, process, and transmit information and control movements of the body. Glia assist neuronal function by providing nutrients, regulating the composition of the interstitial fluid that bathes the neurons, protecting against infection, helping to repair damage, fine-tuning communication among neurons, and speeding up the movement of electrical signals within neurons. Some glia also guide nerve cells to their proper places in the brain during development. Although the nervous system could not function without glia, in this chapter we will focus on the structure and function of neurons.

The Functions of a Neuron Are Localized in Separate Parts of the Cell

A neuron is a highly specialized cell. A typical neuron includes four major structures: dendrites, a cell body, an axon, and synaptic terminals (**FIG. 39-1**). These four structures perform the four main functions of a neuron:

1. Receive information from the internal or external environment or from other neurons.
2. Process this information, often along with information from other sources, and produce an electrical signal.
3. Conduct the electrical signal, sometimes for a considerable distance, to a junction where the neuron meets another cell.
4. Transmit information to other cells, such as other neurons or the cells of muscles or glands.

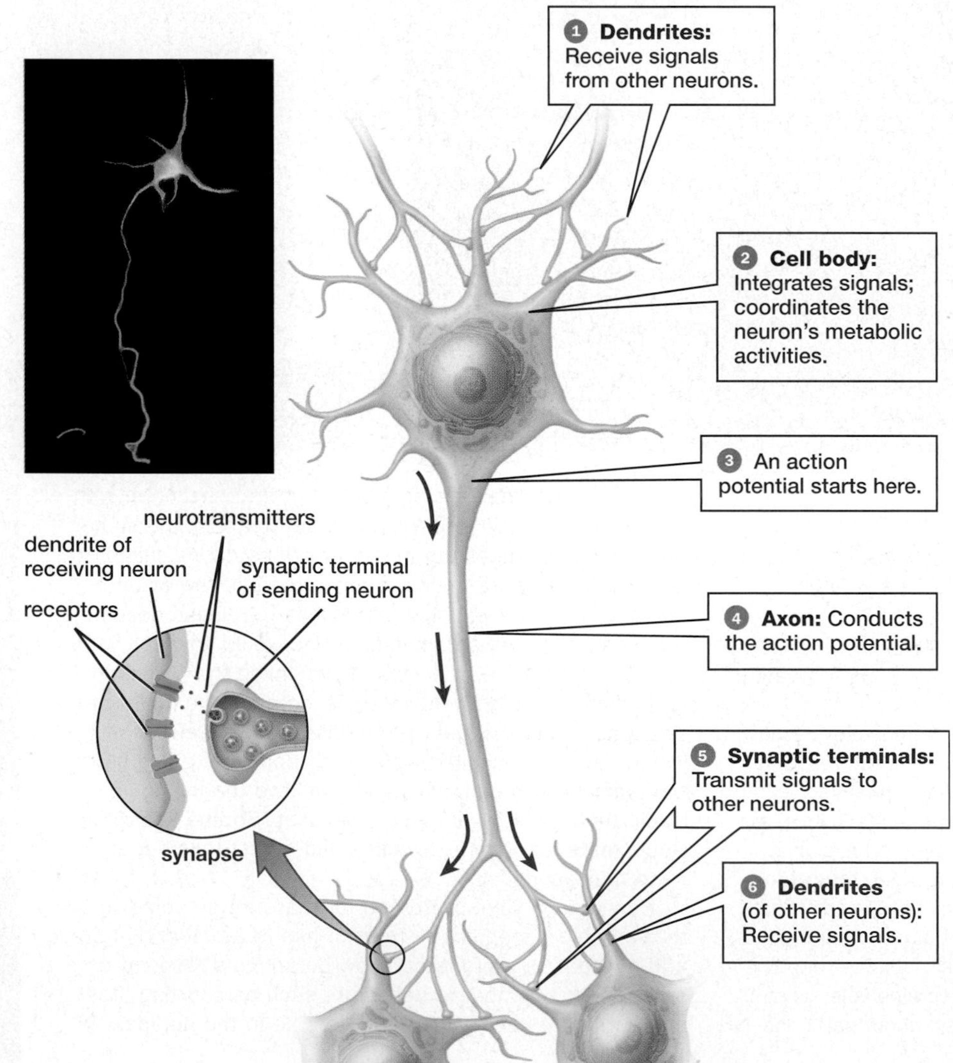

1 Dendrites: Receive signals from other neurons.
2 Cell body: Integrates signals; coordinates the neuron's metabolic activities.
3 An action potential starts here.
4 Axon: Conducts the action potential.
5 Synaptic terminals: Transmit signals to other neurons.
6 Dendrites (of other neurons): Receive signals.
neurotransmitters
dendrite of receiving neuron
synaptic terminal of sending neuron
receptors
synapse

FIGURE 39-1 A neuron, its specialized parts, and their functions The red arrows indicate action potentials moving from the cell body down the axon to the synaptic terminals.

TABLE 39-1 Some Important Neurotransmitters

Neurotransmitter	Location in the Nervous System	Some Major Functions
Acetylcholine	Motor neuron-to-muscle synapses; autonomic nervous system, many areas of the brain	Activates skeletal muscles; activates target organs of the parasympathetic nervous system
Dopamine	Midbrain	Important in emotion, rewards, and the control of movement
Norepinephrine (noradrenaline)	Sympathetic nervous system	Activates target organs of the sympathetic nervous system
Serotonin	Midbrain, pons, and medulla	Influences mood and sleep
Glutamate	Many areas of the brain and spinal cord	Major excitatory neurotransmitter in the CNS
Glycine	Spinal cord	Major inhibitory neurotransmitter in the spinal cord
GABA (gamma aminobutyric acid)	Many areas of the brain and spinal cord	Major inhibitory neurotransmitter in the brain
Endorphins	Many areas of the brain and spinal cord	Influence mood, reduce pain sensations
Nitric oxide	Many areas of the brain	Important in forming memories

Dendrites Receive Information

Dendrites, branched tendrils protruding from the cell body, perform the "receive information" function ❶. Dendrites provide a large surface area for receiving signals, either from the environment or from other neurons. Dendrites of sensory neurons produce electrical signals in response to specific stimuli from the internal environment, such as body temperature or blood pH, or from the external environment, such as touch, odor, heat, or cold. Dendrites of neurons in the brain and spinal cord usually respond to chemicals, called **neurotransmitters**, which are released by other neurons. **TABLE 39-1** lists important neurotransmitters and some of their functions.

The Cell Body Processes Signals from the Dendrites

Electrical signals travel down the dendrites and converge on the neuron's **cell body**, which performs the "process information" function ❷. The cell body adds up the electrical signals that it receives from the dendrites. As we will see, some of these signals are positive and some are negative. If their sum is sufficiently positive, the neuron will produce a large, rapid electrical signal called an **action potential** ❸. The neuron's cell body also contains organelles, such as the nucleus, mitochondria, endoplasmic reticulum, and Golgi apparatus, and performs typical cellular activities, such as synthesizing complex molecules and coordinating the cell's metabolism.

The Axon Conducts Action Potentials Long Distances

In a typical neuron, a long, thin strand called an **axon** extends outward from the cell body. The axon performs the "conduct signals" function by conducting action potentials from the cell body to the axon's end ❹, which contacts other cells. Some neurons have axons that stretch from your spinal cord to your toes, a distance of about 3 feet (approximately a meter), making these neurons the longest cells in the body. Axons are bundled together, much like wires in an electrical cable, to form **nerves**. In vertebrates, nerves emerge from the brain and spinal cord and extend to all regions of the body.

At Synapses, Signals Are Transmitted from One Cell to Another

A neuron transmits information to another cell at a site called a **synapse** (see the enlarged drawing of a synapse in Fig. 39-1). A typical synapse consists of the **synaptic terminal** ❺, which is a swelling at the end of an axon of the "sending" neuron; a "receiving" cell (usually a muscle cell, a gland cell, or, most frequently, another neuron ❻); and a small gap separating the two. Most synaptic terminals contain neurotransmitters that are released in response to an action potential reaching the terminal. The neurotransmitters diffuse across the gap, bind to receptors on the plasma membrane of the receiving neuron, and stimulate a response in this cell. Therefore, at a synapse, the output of the first cell becomes the input to the second cell.

CHECK YOUR LEARNING

Can you . . .

- describe the structure of a typical neuron?
- explain the functions of each part of a neuron?

39.2 HOW DO NEURONS PRODUCE AND TRANSMIT INFORMATION?

As a general rule, information is carried *within* a neuron by electrical signals, and information is transmitted *between* neurons by neurotransmitters released from one neuron and received by a second neuron.

Information Within a Neuron Is Carried by Electrical Signals

An inactive neuron maintains a constant electrical voltage, or potential, across its plasma membrane, similar to the

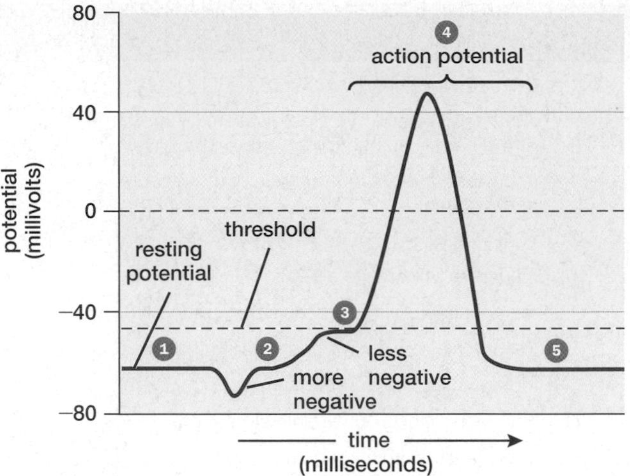

▲ **FIGURE 39-2 Electrical events in a neuron**

voltage across the poles of a battery (**FIG. 39-2** ❶). This voltage, called the **resting potential**, is always negative inside the cell and ranges from about −40 to −90 millivolts (mV; thousandths of a volt).

If a neuron is stimulated, the voltage inside it may become either more negative or less negative ❷. If the potential becomes sufficiently less negative, it reaches a level called **threshold** ❸ and triggers an action potential ❹. During an action potential, the neuron's voltage rises rapidly to about +50 mV. An action potential lasts a few milliseconds (thousandths of a second) before the cell's negative resting potential is restored ❺.

Action potentials are conducted from a neuron's cell body to the axon's synaptic terminals. The plasma membranes of axons are specialized to perform this conducting function. Unlike electrical voltages conducted in metal wires, which become smaller the farther they travel, action potentials are conducted along the axon from cell body to axon terminal with no change in voltage. Action potentials do not weaken with distance traveled.

Myelin Speeds Up the Conduction of Action Potentials

The speed at which an action potential travels varies tremendously among axons. In general, the thicker the axon, the faster the action potential travels. However, for an axon of any given thickness, conduction speed is greatly increased if the axon is surrounded by a fatty insulation called **myelin** (**FIG. 39-3**). Myelin is formed by glial cells that wrap themselves around the axon. Each myelin wrapping covers about 0.2 to 2 millimeters of axon, leaving short segments of naked axon, called nodes, in between. In an axon without a myelin covering, action potentials travel continuously but fairly slowly, typically about 3 to 6 feet (1 to 2 meters) per second. In contrast, action potentials in myelinated axons "jump" rapidly from node to node. In some myelinated axons, action potentials move as fast as 330 feet (100 meters) per second.

At Synapses, Neurons Use Chemicals to Communicate with One Another

Think of an action potential as a packet of information moving down an axon. Once the information reaches the synaptic terminal, it must be transmitted to another cell, either another neuron or a cell in a muscle or gland. At *electrical synapses*, electrical activity can pass directly from neuron to neuron through junctions that connect the cells. Although electrical synapses occur in many places in the nervous system, neurons far more frequently use chemicals to transmit information to other cells. We will confine our discussion to these *chemical synapses*.

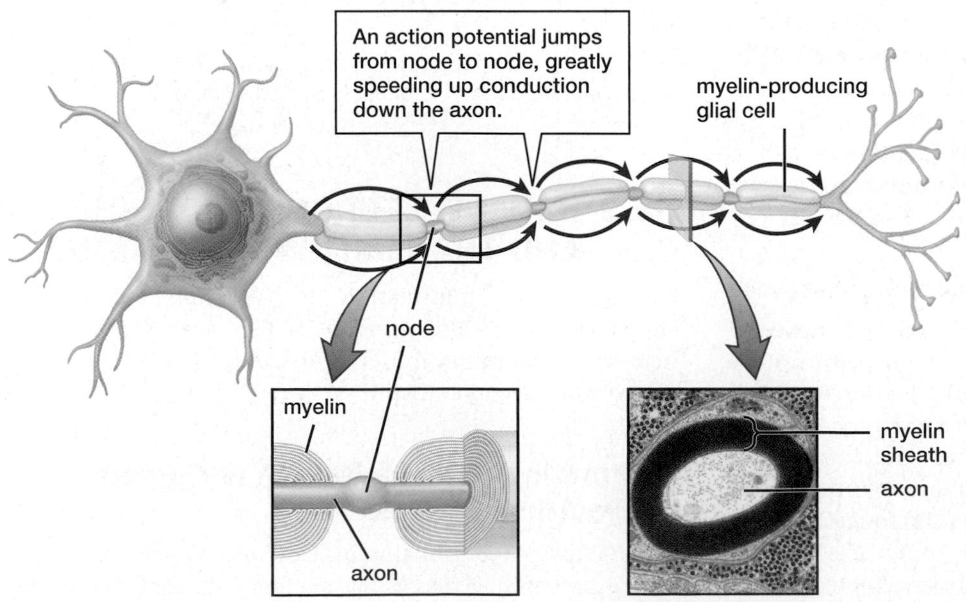

◀ **FIGURE 39-3 A myelinated axon** Many vertebrate axons are covered with insulating myelin. Action potentials occur only at the nodes between each myelin wrapping, seeming to "jump" from node to node (red arrows), with almost no time spent traveling beneath the myelin.

In ordinary English, the word "transmit" means "to send something," and that is exactly what happens at a synapse (**FIG. 39-4**). Two neurons do not actually touch at a synapse: A tiny gap, the **synaptic cleft**, separates the sending **presynaptic neuron** from the receiving **postsynaptic neuron**. The presynaptic neuron sends neurotransmitter molecules across the gap to the postsynaptic neuron.

Communication between neurons begins with an action potential in a presynaptic neuron, usually beginning near the cell body ❶ and traveling down the axon until it reaches a synaptic terminal ❷. The synaptic terminal contains scores of tiny sacs, called vesicles, each full of neurotransmitter molecules. When the action potential invades the synaptic terminal, the inside of the terminal becomes positively charged, triggering a cascade of changes that causes some of the vesicles to release neurotransmitters into the synaptic cleft ❸. The neurotransmitter molecules diffuse across the cleft and bind to receptor proteins in the plasma membrane of the postsynaptic neuron ❹.

Synapses Produce Inhibitory or Excitatory Postsynaptic Potentials

At most synapses, the receptor proteins on the postsynaptic neuron control ion channels that span the neuron's plasma membrane. When neurotransmitter molecules bind to these receptor proteins, the channels open. Depending on the type of channels associated with the receptor, to which a neurotransmitter binds, ions such as sodium (Na^+), potassium (K^+), calcium (Ca^{2+}), or chloride (Cl^-) may move through the channels ❺, causing a small, brief change in voltage, called a **postsynaptic potential (PSP)**.

PSPs differ from action potentials in three major ways. First, PSPs vary in voltage, depending on many factors, such as the amount and type of transmitter released by the presynaptic neuron. Second, PSPs decrease with distance, so a PSP produced in a dendrite will have a smaller voltage by the time it reaches the cell body. Third, PSPs can make a neuron either more negative or less negative. If the postsynaptic neuron becomes more negative (the downward deflection in

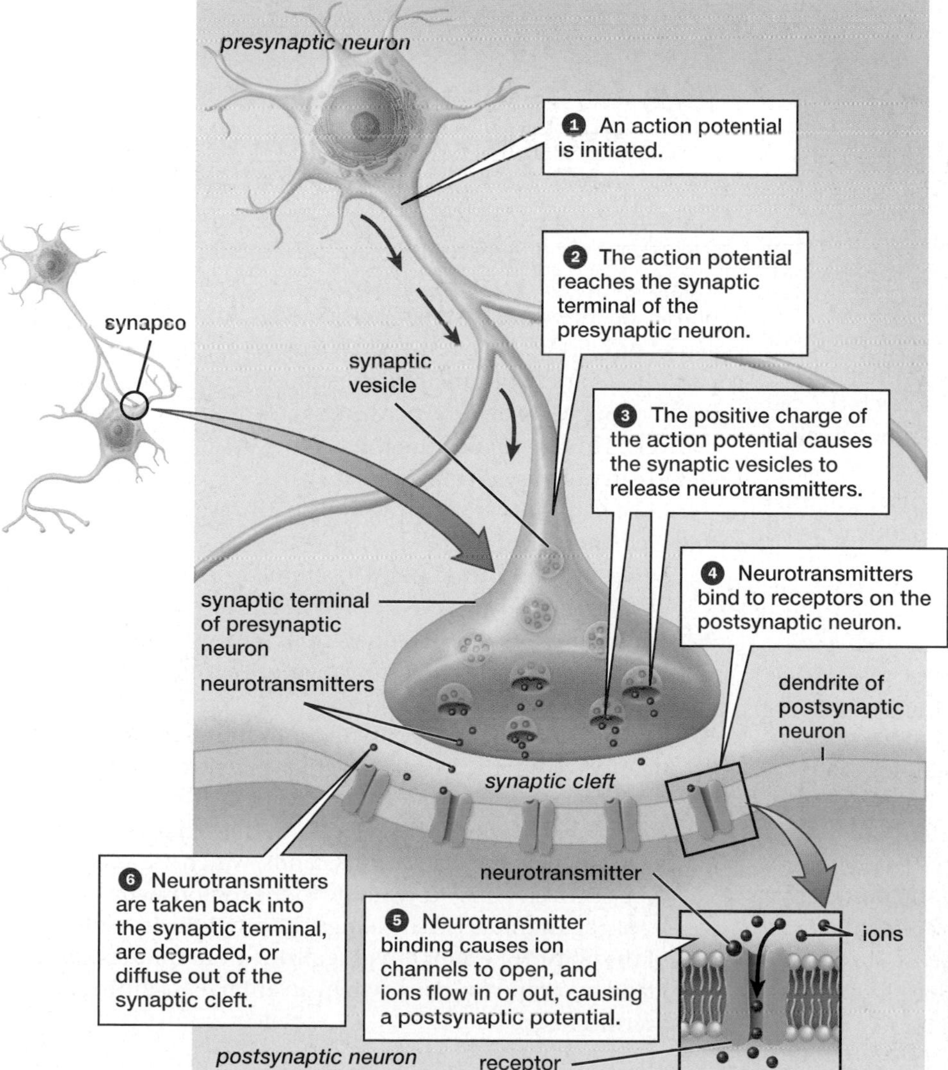
presynaptic neuron
❶ An action potential is initiated.
❷ The action potential reaches the synaptic terminal of the presynaptic neuron.
synapse
synaptic vesicle
❸ The positive charge of the action potential causes the synaptic vesicles to release neurotransmitters.
❹ Neurotransmitters bind to receptors on the postsynaptic neuron.
synaptic terminal of presynaptic neuron
neurotransmitters
dendrite of postsynaptic neuron
synaptic cleft
neurotransmitter
❻ Neurotransmitters are taken back into the synaptic terminal, are degraded, or diffuse out of the synaptic cleft.
❺ Neurotransmitter binding causes ion channels to open, and ions flow in or out, causing a postsynaptic potential.
ions
postsynaptic neuron
receptor

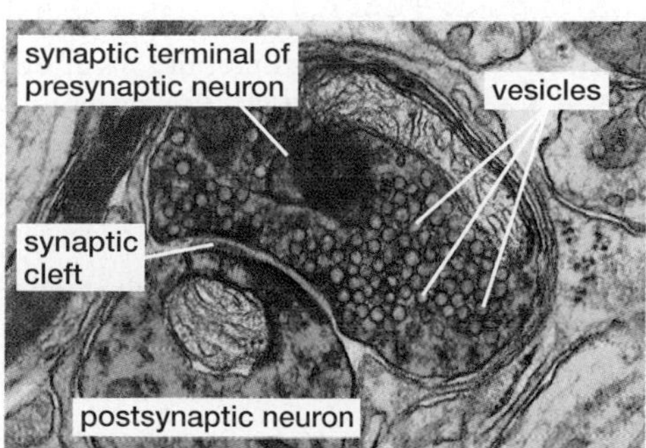
synaptic terminal of presynaptic neuron
vesicles
synaptic cleft
postsynaptic neuron

◀ **FIGURE 39-4 The structure and function of a synapse** The electron micrograph shows a synapse between two neurons in the brain.

THINK CRITICALLY Imagine an experiment in which the neurons pictured here are bathed in a solution containing a nerve poison. The presynaptic neuron is stimulated and produces an action potential, but this does not result in a PSP in the postsynaptic neuron. When the experimenter adds some neurotransmitter to the synapse, the postsynaptic neuron still produces no PSP. How does the poison act to disrupt nerve function?

Fig. 39-2 ①), its resting potential moves farther away from threshold. This change in voltage is called an **inhibitory postsynaptic potential (IPSP)** because it inhibits the postsynaptic neuron, making it less likely to fire an action potential. If the postsynaptic neuron becomes less negative (the upward deflection in Fig. 39-2 ②), its resting potential moves closer to threshold. Consequently, this voltage change is called an **excitatory postsynaptic potential (EPSP)** because it excites the postsynaptic neuron, making it more likely to fire an action potential.

In general, whether a PSP is inhibitory or excitatory depends on the type of ion channel that is associated with the receptor proteins on the postsynaptic neuron. If the channels are permeable to K^+, then when the channels open in response to neurotransmitter binding, K^+ diffuses out of the postsynaptic neuron, making the cell more negative. Making the cell more negative inhibits the production of action potentials in the postsynaptic cell, so the resulting PSP is inhibitory. If the channels linked to the receptors are permeable to Na^+, then Na^+ diffuses down its concentration gradient into the postsynaptic neuron. Na^+ entry makes the inside of the cell less negative. This voltage change constitutes an excitatory PSP, because if the postsynaptic neuron becomes sufficiently less negative, it will reach threshold and produce an action potential.

Integration of Postsynaptic Potentials Determines the Activity of a Neuron

The dendrites of a single neuron may produce EPSPs and IPSPs in response to transmitters received from the synaptic terminals of hundreds or even thousands of presynaptic neurons. Most of these PSPs are small, rapidly fading signals, but they travel far enough to reach the cell body of the postsynaptic neuron. The voltages of all the PSPs that reach the cell body at about the same time are added up, a process called **integration.** If the excitatory and inhibitory postsynaptic potentials, when added together, raise the electrical potential inside the neuron above threshold, the postsynaptic cell will produce an action potential.

Neurotransmitter Action Is Usually Brief

Consider what would happen if transmitters from a presynaptic neuron stimulated a postsynaptic cell, and the postsynaptic cell never stopped responding. You might, for example, contract your biceps muscle, flex your arm, and have it stay flexed forever! Fortunately, the nervous system has several ways to end neurotransmitter action (see Fig. 39-4 ⑥). Many neurotransmitters are transported back into the presynaptic neuron. Some neurotransmitters—notably acetylcholine, the transmitter that stimulates skeletal muscle cells—are rapidly broken down by enzymes in the synaptic cleft. Neurotransmitters also diffuse out of the synaptic cleft.

CHECK YOUR LEARNING

Can you . . .

- describe resting and action potentials?
- explain how an action potential in a presynaptic neuron causes a response in a postsynaptic neuron?
- explain the difference between inhibitory postsynaptic potentials and excitatory postsynaptic potentials?

CASE STUDY CONTINUED

How Do I Love Thee?

Social and emotional responses differ among people, and some of these differences stem from variation in neurotransmitter receptors. For example, people with social phobias, who avoid many social situations for fear of embarrassment, have relatively few dopamine receptors in reward areas of the brain; people who are good at social interactions have far more. How do such receptors and the electrical and chemical signals associated with them allow animals to perceive their environments and generate behaviors?

39.3 HOW DOES THE NERVOUS SYSTEM PROCESS INFORMATION AND CONTROL BEHAVIOR?

The nervous system performs marvelous feats of computation, stores prodigious amounts of information, and directs a wide variety of complex behaviors. The remarkable capabilities of the nervous system arise largely from three interacting properties. First, individual neurons are specialized to perform a variety of functions. Second, the nervous system includes huge, yet orderly, networks of connections between neurons. Third, outputs from the nervous system direct muscles to carry out a wide range of behaviors.

Most behaviors are controlled by pathways composed of four elements:

1. **Sensory neurons** respond to stimuli from inside or outside the body.
2. **Interneurons** receive signals from sensory neurons, hormones, neurons that store memories, and many other sources.
3. **Motor neurons** receive instructions from sensory neurons or interneurons and activate muscles or glands.
4. **Effectors**, usually muscles, perform the behavior directed by the nervous system. Glands, which are another type of effector, may contribute to the responses by releasing hormones that change the physiological state of the body. For example, epinephrine from the adrenal gland boosts performance when an animal is running away from a predator.

These four elements, when properly connected, carry out the basic operations required of any nervous system:

1. Determine the type of stimulus (sensory neurons).
2. Determine and signal the intensity of a stimulus (sensory neurons and interneurons).
3. Integrate information from many sources (interneurons).
4. Direct appropriate behaviors (interneurons, motor neurons, and effectors).

The Nature of a Stimulus Is Encoded by Sensory Neurons and Their Connections to Specific Parts of the Brain

The senses inform the brain about the properties of the environment, both outside the body (such as images, sounds, or odors) and inside the body (such as the body temperature or the concentration of salts in the blood). Information gathered by the senses is converted to action potentials, either directly in sensory neurons, such as in touch receptors in the skin, or in interneurons, such as in the retina. Sensory information is then sent to the brain in the form of these action potentials. Given that all action potentials are fundamentally the same, how can the brain determine the nature of a stimulus?

Sensory neurons are specialized to respond to specific stimuli. For example, some sensory neurons respond to light, but not temperature, touch, or chemicals; others respond to specific chemicals but not to touch or light, or even to other chemicals (see Chapter 40). The nervous system encodes the type of stimulus by two processes: first, which sensory neurons respond to the stimulus and second, which parts of the brain are activated when those sensory neurons are stimulated. For example, light stimulates photoreceptors in your retina, which produce electrical signals that cause action potentials to fire in interneurons whose axons make up the optic nerves. The brain interprets all action potentials in optic nerve axons as the sensation of light. This was demonstrated long ago, when a German physiologist sat in a dark room and poked himself in the eye. The ensuing minor damage to his retina caused neurons to produce action potentials that traveled in his optic nerve to his brain. (As they say on TV, do *not* try this yourself!) The result? He "saw stars" because his brain interpreted action potentials in his optic nerve as light. Thus, you distinguish hot from cold temperatures, or the bitterness of coffee from the sweetness of sugar, because these diverse stimuli activate different sensory neurons that connect, sometimes by way of interneurons, to different areas of your brain.

The Intensity of a Stimulus Is Encoded by the Frequency of Action Potentials

Because all action potentials are roughly the same size and duration, no information about the strength, or intensity, of a stimulus (for example, how hard an object pushes on your skin) can be encoded in a single action potential. Intensity is coded in two ways (**FIG. 39-5**). First, intensity can be signaled by the frequency of action potentials in a single neuron. The more intense the stimulus, the faster the neuron fires action potentials. Second, many neurons may respond to the same type of stimulus. Stronger stimuli excite more of these neurons, whereas weaker stimuli excite fewer. These two ways of coding intensity work together. For example, a gentle touch may cause a single touch receptor in your skin to fire action potentials very slowly (**FIG. 39-5a**), whereas a hard poke may cause several touch receptors to fire, some very rapidly (**FIG. 39-5b**).

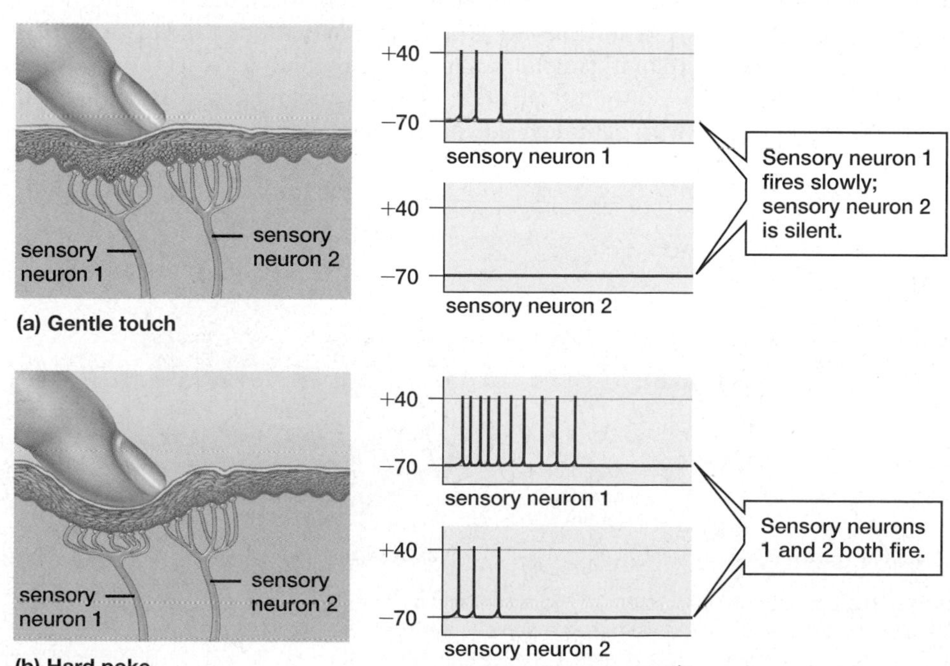

◀ FIGURE 39-5 Signaling stimulus intensity The intensity of a stimulus is signaled by the rate at which individual sensory neurons produce action potentials and by the number of sensory neurons activated. **(a)** A gentle touch activates only the closest sensory neuron, which fires action potentials at a low rate. **(b)** A hard poke activates multiple sensory neurons, causing the closest one to fire rapidly and more distant ones to fire more slowly.

THINK CRITICALLY How do you think skin areas that are especially sensitive to touch differ from less sensitive areas?

The Nervous System Processes Information from Many Sources

An animal's brain is continuously bombarded by sensory stimuli from both inside and outside the body. Interneurons in the brain evaluate these stimuli, determine which ones are important, and direct an appropriate response. A mouse, for example, might simultaneously receive visual inputs from sunflower seeds on a feeder, hunger stimuli from its digestive tract, and auditory inputs from a nearby cat creeping through the grass. Interneurons in various parts of the mouse's brain integrate these inputs and direct appropriate behaviors—in this case, probably to run away, no matter how hungry it may be.

The Nervous System Produces Outputs to Effectors

Motor neurons in the brain, the spinal cord, or the sympathetic and parasympathetic nervous systems (described later) stimulate activity in effectors. The same principles of connectivity and intensity coding that we described for sensory inputs are used for the brain's outputs to effectors. Which effectors are activated is determined by their connections with the brain or spinal cord. For example, different motor neurons activate your biceps muscles and the muscles in your face. How strongly a muscle contracts is determined by how many motor neurons connect to it and how fast those motor neurons fire action potentials.

Behaviors Are Controlled by Networks of Neurons in the Nervous System

Simple behaviors, such as reflexes (see Section 39.5), may be controlled by activity in as few as two or three neurons—a sensory neuron, a motor neuron, and perhaps an interneuron in between—ultimately stimulating a single muscle. More complex behaviors are organized by interconnected neural pathways, in which several types of sensory input (along with memories, hormones, and other factors) converge on a set of interneurons. The interneurons integrate PSPs from these multiple sources and stimulate motor neurons that direct activity in the appropriate muscles and glands. Millions of neurons, mostly in the brain, may be required to perform complex actions such as playing the piano.

CHECK YOUR LEARNING

Can you . . .

- name the four components of a neural pathway and explain how these components carry out the functions of a nervous system?
- explain how the brain interprets the type and intensity of a sensory stimulus?
- explain how the brain determines which muscles to contract and how strongly to contract them?

39.4 HOW ARE NERVOUS SYSTEMS ORGANIZED?

All animals have one of two basic types of nervous systems: a diffuse nervous system, such as that of cnidarians (*Hydra*, jellyfish, anemones, and their relatives; **FIG. 39-6a**), or a centralized nervous system, found in more complex organisms. Nervous system architecture is highly correlated with an animal's body plan and lifestyle.

Cnidarians are radially symmetrical (see Chapter 24). Because they have no "front end," natural selection has not favored concentrating the senses in one place. For example, a *Hydra* sits anchored to a rock at the bottom of a pond, so prey or predators are equally likely to come from any direction. A cnidarian nervous system is composed of a network of neurons, often called a **nerve net**, woven more or less equally throughout the animal's tissues, which produces roughly equal ability to detect and respond to stimuli in all directions. Here and there we find a cluster of neurons, called a **ganglion** (plural, ganglia), but nothing resembling a real brain.

Almost all other animals are bilaterally symmetrical, with definite head and tail ends. Because the head is usually

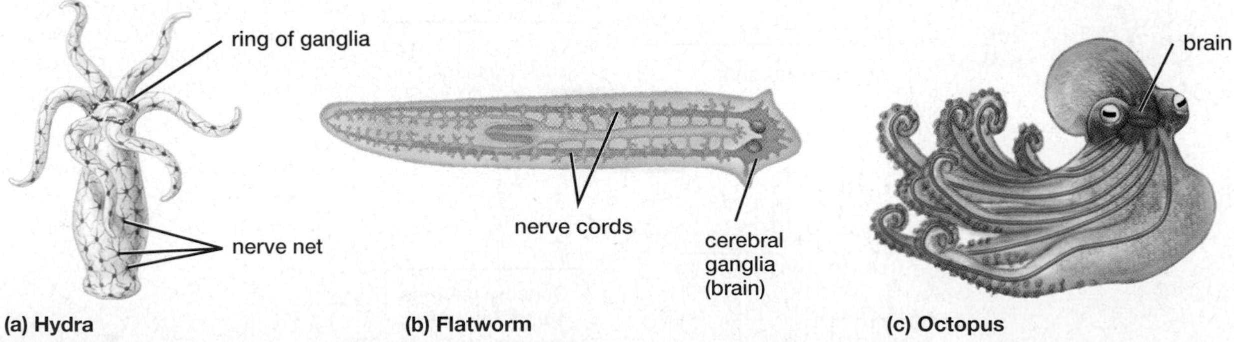

▲ **FIGURE 39-6 Nervous system organization (a)** The diffuse nervous system of *Hydra* contains a few clusters of neurons at the bases of tentacles, but no brain. **(b)** The flatworm has a nervous system that is less diffuse, with a cluster of ganglia in the head. **(c)** An octopus has a large, complex brain and learning capabilities rivaling those of some mammals.

the first part of the body to encounter food, danger, and potential mates, it is advantageous to have sense organs concentrated there. Sizable ganglia evolved that integrate the information gathered by the senses and direct appropriate actions. Over evolutionary time, the major sense organs became localized in the head, and the ganglia became centralized into a brain. This trend is clearly seen in invertebrates (**FIGS. 39-6b, c**), but reaches its peak in vertebrates, in which nearly all the cell bodies of the nervous system reside in the brain or spinal cord.

CHECK YOUR LEARNING

Can you . . .

- describe the anatomy of diffuse and centralized nervous systems and provide examples of animals with each type?

39.5 WHAT ARE THE STRUCTURES AND FUNCTIONS OF THE HUMAN NERVOUS SYSTEM?

The nervous systems of all mammals, including humans, can be divided into two parts: central and peripheral. Each of these has further subdivisions (**FIG. 39-7**). The **central nervous system (CNS)** consists of the **brain** and **spinal cord.** The **peripheral nervous system (PNS)** consists of neurons and axons that lie outside the CNS.

The Peripheral Nervous System Links the Central Nervous System with the Rest of the Body

The cell bodies of some sensory neurons, including many that detect touch, pressure, temperature, stretch, and pain, are in the PNS, often in clusters called dorsal root ganglia, located alongside the spinal cord (see Fig. 39-9). The cell bodies of neurons that control involuntary movements and activities, such as dilation or constriction of the pupils and airways, secretion of saliva, and sexual arousal, are also mostly, but not exclusively, found in the PNS.

The peripheral nerves consist of three major categories of axons: (1) axons of sensory neurons that carry information to the CNS, (2) axons of motor neurons that innervate the skeletal muscles, thereby controlling voluntary movements, and (3) axons of neurons that control involuntary movements and activities.

Motor activities of the PNS are controlled by either the somatic nervous system or the autonomic nervous system.

The Somatic Nervous System Controls Voluntary Movement

Motor neurons of the **somatic nervous system** form synapses with skeletal muscles and control voluntary movement. As you take notes or lift a coffee cup, your somatic nervous system is in charge. The cell bodies of most somatic motor neurons are located in the spinal cord (part of the CNS); their axons (part of the PNS) extend out to the muscles they control.

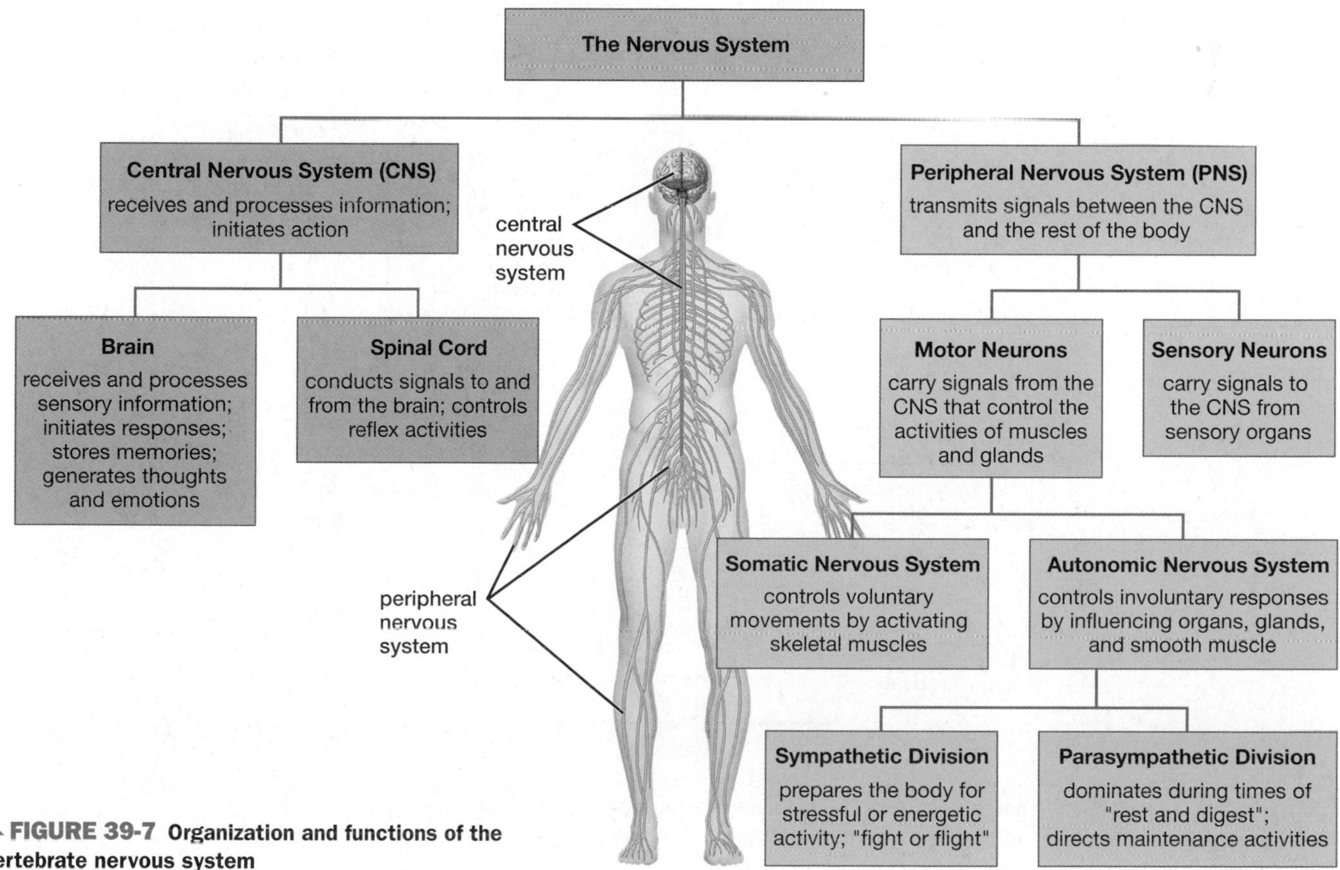

▲ **FIGURE 39-7 Organization and functions of the vertebrate nervous system**

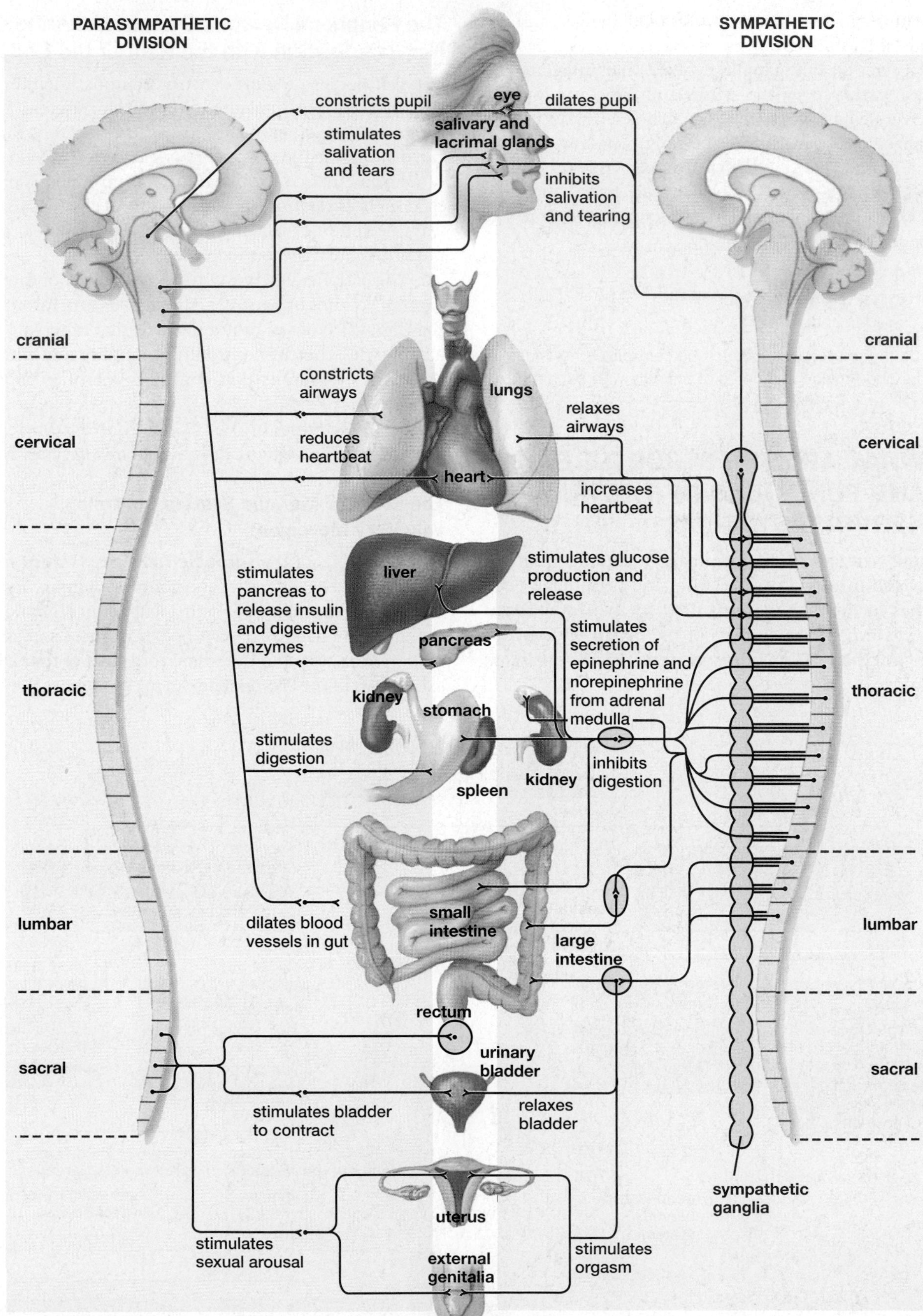

▲ **FIGURE 39-8 The autonomic nervous system** The autonomic nervous system has two divisions, sympathetic and parasympathetic, which innervate many of the same organs but generally produce opposite effects.

The Autonomic Nervous System Controls Involuntary Actions

Neurons of the **autonomic nervous system** innervate many parts of the body, including the heart, smooth muscles in the respiratory tract and blood vessels, and many glands. These neurons produce mostly involuntary actions. The autonomic nervous system consists of two parts: the **sympathetic division** and the **parasympathetic division** (**FIG. 39-8**). In general, these two divisions innervate the same organs, but produce opposite effects.

The neurons of the sympathetic division release the neurotransmitter norepinephrine (also called noradrenaline) onto their target organs, preparing the body for stressful or energetic activity, such as fighting, escaping, or taking an exam. During such "fight-or-flight" activities, the sympathetic nervous system reduces activity in the digestive tract, redirecting some of its blood supply to the muscles of the arms and legs. Heart rate accelerates. The pupils of the eyes open wider, admitting more light, and the air passages in the lungs expand, letting in more air.

The neurons of the parasympathetic division release the neurotransmitter acetylcholine onto their target organs. The parasympathetic division dominates during maintenance activities that can be carried out at leisure, often called "rest and digest." Under parasympathetic control, the digestive tract becomes active. Heart rate slows and the air passages in the lungs constrict, because the body requires less blood flow and less oxygen.

The Central Nervous System Consists of the Spinal Cord and Brain

The spinal cord and brain make up the central nervous system (CNS). The CNS receives and processes sensory information, generates thoughts and emotions, and directs responses. The CNS consists primarily of interneurons—probably about 100 billion of them.

The brain and spinal cord are protected from physical damage in three ways. The first line of defense is a bony armor, consisting of the skull, which surrounds the brain, and a chain of vertebrae that protect the spinal cord. The second defense consists of three layers of connective tissues, called meninges, that lie beneath these bones and surround the CNS (see Fig. 39-12a). The third defense, lying between the layers of the meninges, is the cerebrospinal fluid, a clear liquid similar to blood plasma. Cerebrospinal fluid cushions the brain and spinal cord and nourishes the CNS.

The brain is protected from damaging chemicals in the bloodstream because the cells that make up the walls of brain capillaries are sealed tightly together, making brain capillaries far less permeable than those in the rest of the body. This **blood–brain barrier** selectively transports needed materials into the brain, while keeping out harmful bacteria and many dangerous substances, such as toxins.

The Spinal Cord Controls Many Reflexes and Conducts Information to and from the Brain

The spinal cord (**FIG. 39-9**) extends from the base of the brain to the lower back. Nerves carrying axons of sensory neurons arise from the dorsal (back) part of the spinal cord, and nerves carrying axons of motor neurons arise from the ventral (front) part. These nerves come together to form the spinal nerves that innervate most of the body. Because the nerves resemble the roots of a tree merging into a single trunk, they are called the dorsal and ventral roots of the spinal nerves. Swellings on each dorsal root, called the **dorsal root ganglia**, contain the cell bodies of sensory neurons.

In the center of the spinal cord is a butterfly-shaped area of **gray matter**. ("Gray matter" is really a misnomer, because most nervous tissue is pinkish-tan. It turns gray when it is preserved.) Gray matter in the spinal cord consists of the cell bodies of motor neurons that control voluntary muscles and the autonomic nervous system, plus interneurons that communicate with the brain and other parts of the spinal

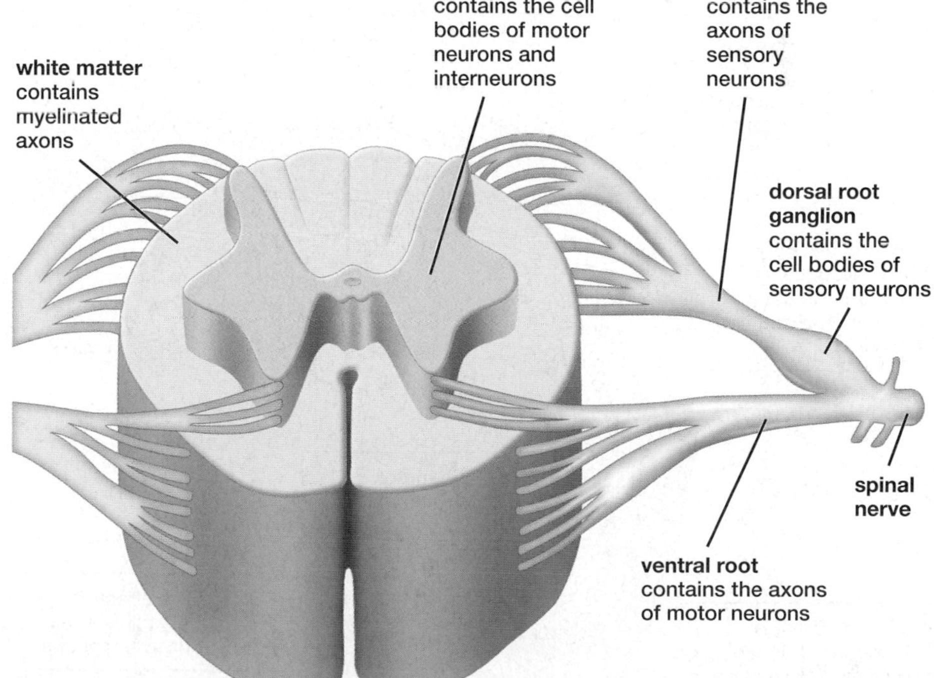
gray matter contains the cell bodies of motor neurons and interneurons
dorsal root contains the axons of sensory neurons
white matter contains myelinated axons
dorsal root ganglion contains the cell bodies of sensory neurons
spinal nerve
ventral root contains the axons of motor neurons

▲ **FIGURE 39-9 The spinal cord** In cross-section, the spinal cord has an outer region of myelinated axons (white matter) that travel to and from the brain, and an inner, butterfly-shaped region of dendrites and the cell bodies of interneurons and motor neurons (gray matter). The cell bodies of the sensory neurons are outside the spinal cord in the dorsal root ganglion and thus are part of the peripheral nervous system.

cord. The gray matter is surrounded by **white matter**, containing myelin-coated axons that extend up and down the spinal cord—the fatty myelin covering the axons is white. Some axons carry sensory signals from internal organs, muscles, and the skin up to the brain. Other axons extend downward from the brain, carrying signals that regulate the activity of motor neurons.

If the spinal cord is severed, sensory input from below the cut cannot reach the brain, and motor output from the brain cannot reach motor neurons located below the cut. Therefore, body parts that are innervated by motor and sensory neurons located below the injury are paralyzed and have no sensation, even though the motor and sensory neurons, the spinal nerves, and the muscles remain intact.

The Neurons That Control Many Reflexes Reside in the Spinal Cord and Peripheral Nervous System

The simplest type of behavior is the **reflex**, a largely involuntary movement of a body part in response to a stimulus. In vertebrates, many reflexes are produced by neurons in the spinal cord and the peripheral nervous system.

Let's examine the pain-withdrawal reflex (**FIG. 39-10**). If you bring your hand too close to a flame, the resulting tissue damage activates a pain sensory neuron ❶. Action potentials in the axons of these neurons travel up a spinal nerve and enter the spinal cord through a dorsal root ❷. Within the gray matter of the cord, the pain sensory neuron stimulates an interneuron, which stimulates a motor neuron ❸. Action potentials in the axon of the motor neuron leave the spinal cord through a ventral root and travel in a spinal nerve to a muscle. Synaptic terminals of the axon stimulate the muscle ❹, causing it to contract and withdraw your hand away from the flame ❺.

Many spinal cord interneurons also have axons that extend up to the brain ❻. Action potentials in these axons inform the brain about burnt hands and may trigger more complex behaviors, such as shrieks and learning about the dangers of open flames. The brain in turn sends action potentials down axons that connect to interneurons and motor neurons in the spinal cord. These signals from the brain can modify spinal reflexes. With enough training or motivation, you can suppress the pain-withdrawal reflex: To rescue a child from a burning crib, you could reach into the flames.

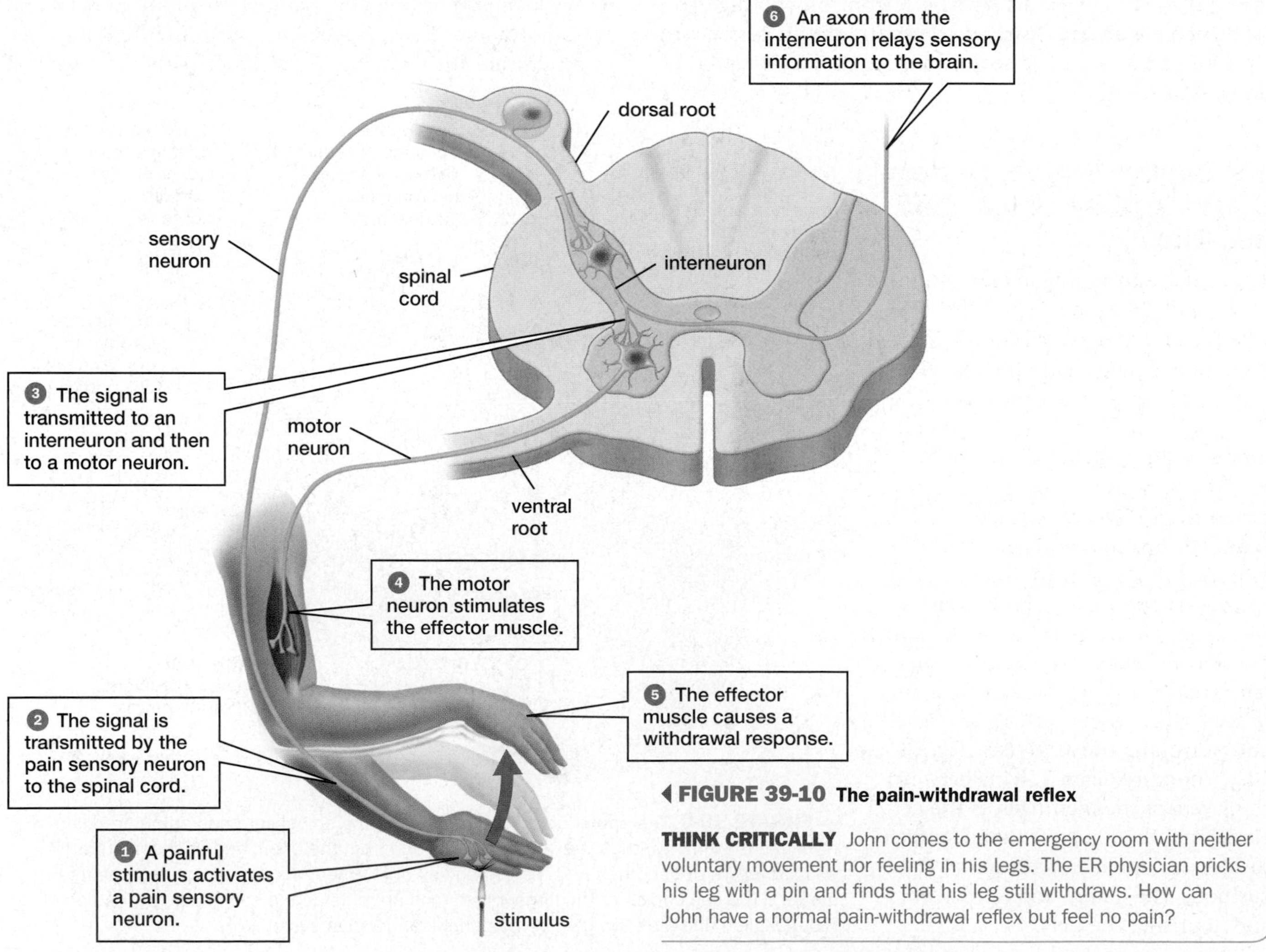

❻ An axon from the interneuron relays sensory information to the brain.
dorsal root
sensory neuron
spinal cord
interneuron
❸ The signal is transmitted to an interneuron and then to a motor neuron.
motor neuron
ventral root
❹ The motor neuron stimulates the effector muscle.
❺ The effector muscle causes a withdrawal response.
❷ The signal is transmitted by the pain sensory neuron to the spinal cord.
❶ A painful stimulus activates a pain sensory neuron.
stimulus

◀ **FIGURE 39-10 The pain-withdrawal reflex**

THINK CRITICALLY John comes to the emergency room with neither voluntary movement nor feeling in his legs. The ER physician pricks his leg with a pin and finds that his leg still withdraws. How can John have a normal pain-withdrawal reflex but feel no pain?

Some Complex Actions Are Coordinated Within the Spinal Cord

The wiring for some fairly complex activities, such as walking, resides in the spinal cord. The advantage of this arrangement is probably an increase in speed and coordination, because messages do not have to travel all the way up to the brain and back down again merely to swing a leg forward. The brain's role in these behaviors is to initiate, guide, and modify the firing of spinal motor neurons, based on conscious decisions (Where are you going? How fast should you walk?). To maintain balance, the brain also uses sensory input from the muscles to regulate motor neuron activity and adjust the way the muscles move.

The Brain Consists of Many Parts That Perform Specific Functions

All vertebrate brains consist of three major parts: the **hindbrain**, **midbrain**, and **forebrain** (**FIG. 39-11a**). Scientists hypothesize that, in early vertebrates, these three anatomical divisions were also functional divisions: The hindbrain governed automatic behaviors such as breathing and heart rate, the midbrain controlled vision, and the forebrain dealt largely with the sense of smell. In nonmammalian vertebrates, all three divisions remain prominent (**FIGS. 39-11b, c**). However, in mammals—particularly in humans—the midbrain has shrunk, while the forebrain has greatly expanded (**FIGS. 39-11d, e**). The major structures of the human brain are shown in **FIGURE 39-12**.

The Hindbrain Consists of the Medulla, Pons, and Cerebellum

In humans and other mammals, the hindbrain is composed of the medulla, the pons, and the cerebellum (**FIG. 39-12a**). The **medulla** resembles an enlarged extension of the spinal cord, with neuron cell bodies at its center, surrounded by a layer of myelin-covered axons. The medulla and the **pons**, located just above the medulla, control automatic functions such as breathing, heart rate, blood pressure, and swallowing. Other clusters of neurons in the pons influence bladder control, balance, posture, and sleep.

The **cerebellum** is crucial for coordinating movements. It receives information both from command centers in the forebrain that control movement and from position sensors in muscles and joints. By comparing information from these two sources, the cerebellum guides smooth, accurate motions and body position. The cerebellum is also involved in *motor learning.* As you learn to write or play the guitar, your forebrain directs each separate movement. After you have become skilled, your forebrain still determines what behavior to perform (for example, what words to write), but your cerebellum becomes largely responsible for ensuring that the actions are carried out appropriately. Not surprisingly, the cerebellum is especially large in flying animals, such as bats and birds (see Fig. 39-11c), which perform intricate aerial maneuvers while navigating around obstacles and capturing prey.

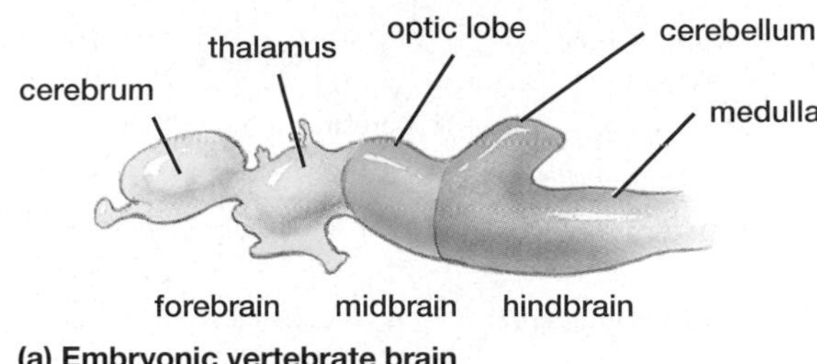

(a) Embryonic vertebrate brain

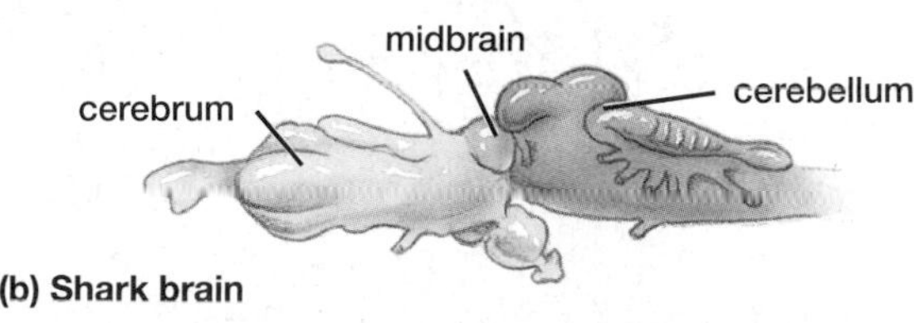

(b) Shark brain

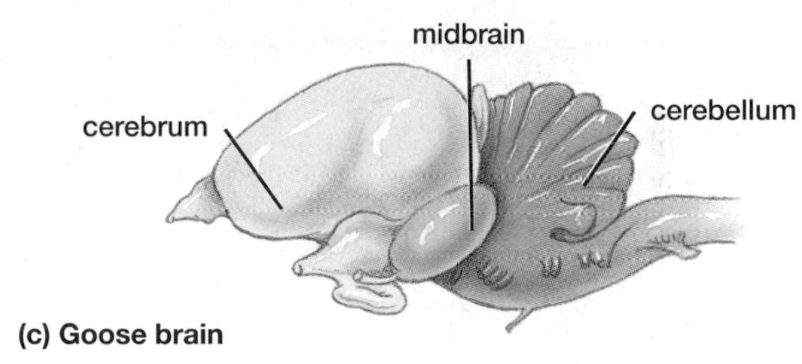

(c) Goose brain

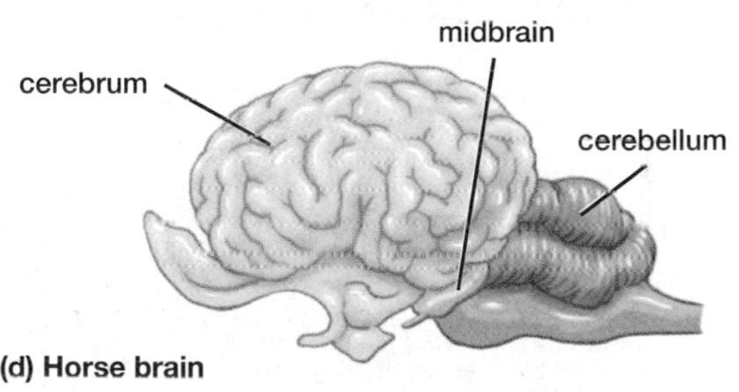

(d) Horse brain

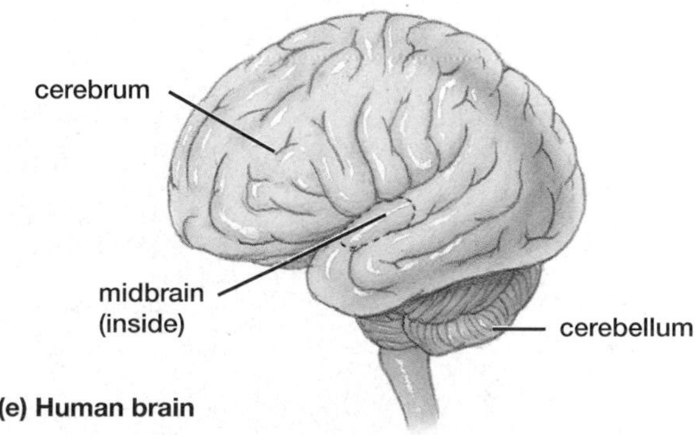

(e) Human brain

▲ **FIGURE 39-11 A comparison of vertebrate brains (a)** The brains of vertebrate embryos consist of three regions: the forebrain, midbrain, and hindbrain. **(b)** The brain of an adult shark maintains this basic organization, although the midbrain is smaller. **(c)** In the goose, the midbrain remains small, whereas both the cerebrum and cerebellum are enlarged. **(d, e)** In mammals, especially humans, the cerebrum is very large compared to the other brain regions.

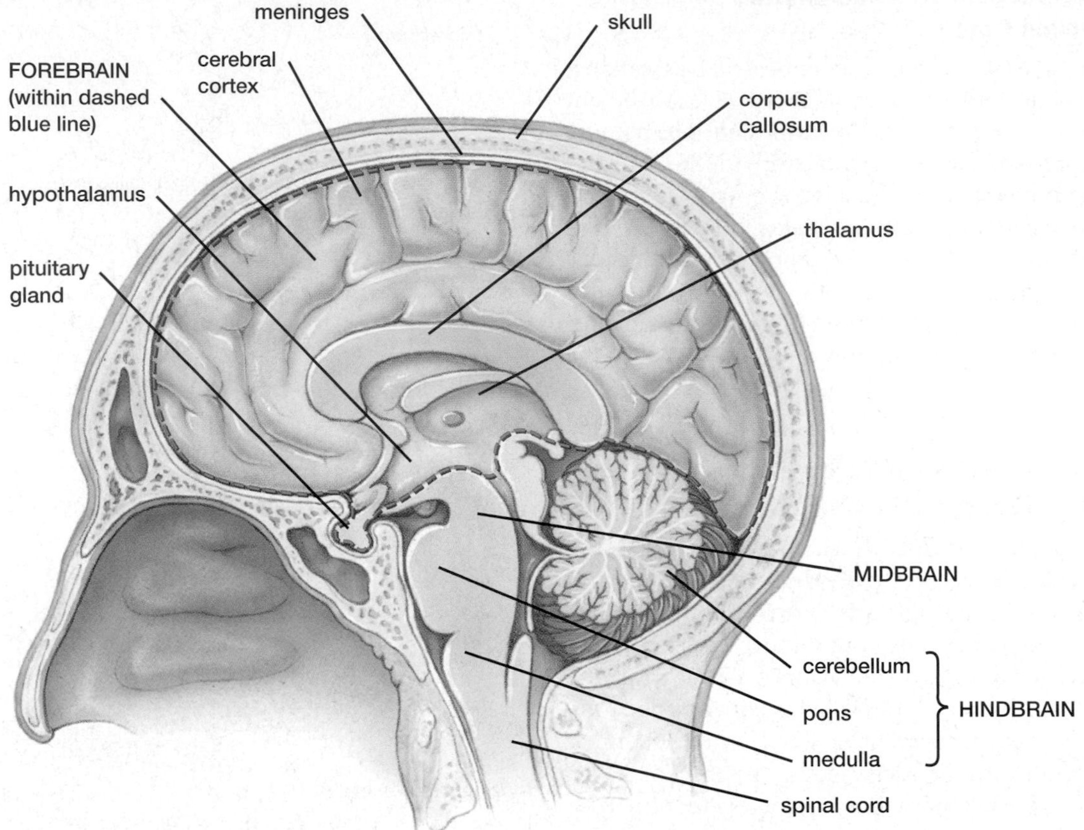

(a) A lateral section of the human brain

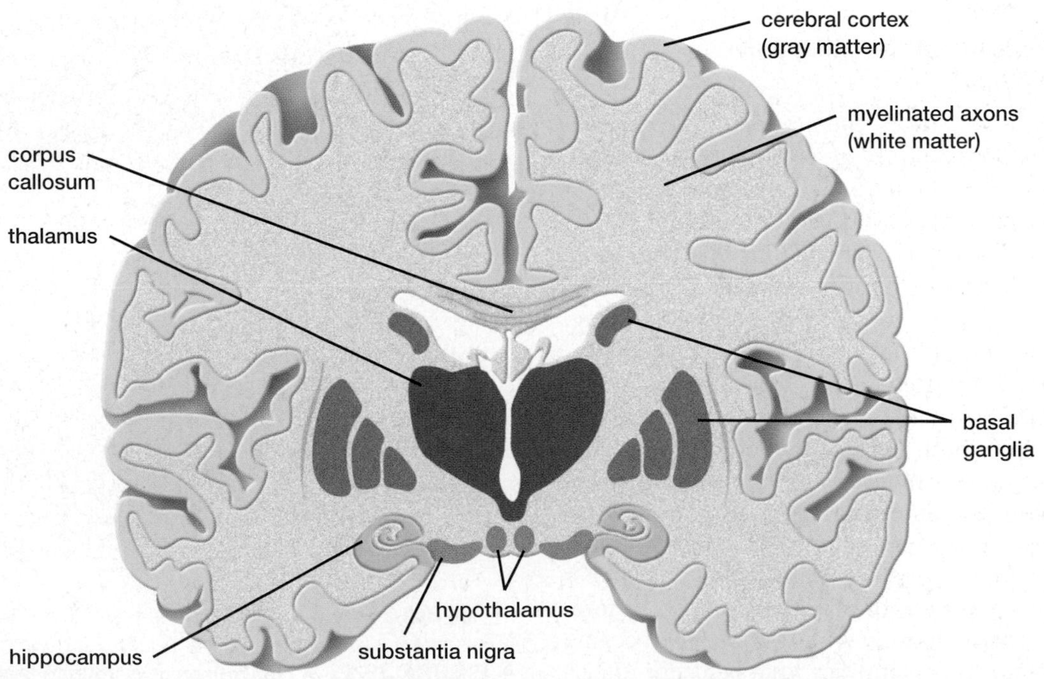

(b) A cross-section of the brain

▲ **FIGURE 39-12 The human brain** Structures of the human brain, seen as if the brain were sliced **(a)** through the midline between the cerebral hemispheres, showing the hindbrain, midbrain, and forebrain, and **(b)** from ear to ear, showing mostly the cerebrum. Not all brain structures are visible in these sections.

The Midbrain Contains Clusters of Neurons That Contribute to Movement, Arousal, and Emotion

The midbrain is quite small in humans (see Figs. 39-11e and 39-12a). It contains an auditory relay center and clusters of neurons that control reflex movements of the eyes. For example, if you're sitting in class and someone races through the door, the resulting sight and sound activate the visual and auditory centers in your midbrain, which direct your gaze to the unexpected arrival and track his movement through the room. The midbrain also contains neurons that produce the transmitter dopamine. One of these clusters of neurons, called the *substantia nigra* (**FIG. 39-12b**), helps to control movement, as we will describe in our discussion of the forebrain, below. Another cluster of dopamine neurons is an essential part of the "reward circuit" that is responsible for pleasurable sensations and, unfortunately, addiction, as we explore in "Health Watch: Drugs, Neurotransmitters, and Addiction" on page 698.

The midbrain also contains a portion of the **reticular formation**, which plays a role in sleep and wakefulness, emotion, and some movements and reflexes. The reticular formation consists of interconnected groups of neurons in the medulla, pons, and midbrain, many of which send axons to the forebrain. These neurons receive input from virtually every sense, from every part of the body, and from many areas of the brain as well. The reticular formation filters sensory inputs before they reach conscious regions of the brain, allowing you to read and concentrate in the presence of a variety of distracting stimuli, such as music or the smell of coffee. The fact that a mother wakens upon hearing the faint cry of her infant but sleeps through loud traffic noise outside her window testifies to the effectiveness of the reticular formation in screening sensory inputs.

The Forebrain Includes the Thalamus, Hypothalamus, and Cerebrum

The **thalamus** channels sensory information from all parts of the body to the cerebral cortex (see Figs. 39-12a, b). In fact, input from all of the senses, except olfaction, passes through the thalamus on its way to the cerebral cortex. Signals traveling from the spinal cord, cerebellum, medulla, pons, and reticular formation also pass through the thalamus.

The **hypothalamus** (literally, "under the thalamus") contains many clusters of neurons. Some release hormones into the blood or control the release of hormones from the pituitary gland (see Chapter 38). Other parts of the hypothalamus direct the activities of the autonomic nervous system. The hypothalamus helps to maintain homeostasis by influencing body temperature, food intake, water balance, heart rate, blood pressure, the menstrual cycle, and circadian rhythms.

The **cerebrum** consists of two **cerebral hemispheres**. Each hemisphere is composed of an outer **cerebral cortex**; bundles of axons, some that interconnect the two hemispheres and others that link the hemispheres with the midbrain and hindbrain; and several clusters of neurons beneath the cortex near the thalamus, which we describe next (see Figs. 39-12a, b).

CASE STUDY CONTINUED

How Do I Love Thee?

Although far from completely understood, the emotion of love in humans evokes changes in brain electrical and chemical activity that are similar to the changes that occur during pair bonding in prairie voles. When people fall in love, oxytocin stimulates dopamine release by neurons in the midbrain that send axons to several parts of the forebrain. In the forebrain, dopamine contributes to emotional attachment, lack of fear and critical judgment of the loved one, and fond memories of times spent together. Which parts of the forebrain allow us to experience these emotions and memories?

Structures in the Interior of the Cerebrum The paired **hippocampi** (singular, hippocampus), nestled at the base of the cerebrum (see Fig. 39-12b), are crucial for the formation of long-term memory and are thus required for learning, as discussed later in this chapter. The hippocampi are particularly important for "place learning" in most, perhaps all, vertebrates. For example, blue jays, Steller's jays, and nutcrackers store seeds for the winter and must remember where their caches are. These birds have larger hippocampi than most birds do. In humans, London taxi drivers, who must memorize that city's maze of more than 25,000 streets, have larger-than-average hippocampi; repeated, intensive place learning causes their hippocampi to become larger.

The **basal ganglia** consist of structures deep within the cerebrum, as well as the substantia nigra in the midbrain (see Fig. 39-12b). These structures are important in the overall control of movement. The motor part of the cerebral cortex directs specific movements, such as which motor neurons to fire to activate just the right muscles to pick up a pen. The basal ganglia are essential to the decision to initiate a particular movement and suppress other movements, for example, to pick up the pen but not simultaneously attempt to comb your hair with the same hand. This function is most clearly illustrated by two major disorders of the basal ganglia. In Parkinson's disease, the substantia nigra degenerates, and affected people have a hard time starting a movement—the "freezing" symptom. In Huntington's disease, basal ganglia in the cerebrum degenerate, and affected people make involuntary, undirected movements.

The Cerebral Cortex The cerebral cortex is the thin outer layer of each cerebral hemisphere, in which billions of neurons are packed in a highly organized way into a sheet just a few millimeters thick. The cortex is folded into *convolutions*,

Drugs, Neurotransmitters, and Addiction

Chances are you know someone who is addicted. How can substances such as cocaine, alcohol, and nicotine so profoundly alter people's lives? The answer lies in how these drugs affect neurotransmitter action and how the nervous system adjusts to those effects.

Many addictive drugs, including cocaine, meth (methamphetamine), and ecstasy (MDMA), target synapses in the brain's reward circuitry that use the transmitters dopamine or serotonin. Normally, after releasing a transmitter, the presynaptic neurons in these synapses rapidly pump most of the transmitter back in, thus limiting its effects. Cocaine, meth, and ecstasy block dopamine or serotonin pumps, or both, so the transmitter concentration increases around the postsynaptic receptors, enhancing synaptic transmission and increasing pleasurable feelings.

Cocaine and meth cause addiction mainly by blocking dopamine pumps. Because dopamine makes people feel good, cocaine and meth are highly rewarding, making users want to repeat the experience. Meanwhile, the brain, in response to excessive dopamine stimulation, reduces the number of dopamine receptors in the synapses and increases the number of dopamine pumps (**FIG. E39-1**). Therefore, when not taking the drugs, the user feels "down," because his reward circuitry has less dopamine available (the increased number of pumps rapidly remove dopamine from the synaptic cleft) and fewer receptors to respond to what little dopamine remains. Over time, he needs more and more cocaine or meth just to feel OK, let alone to get high: He has become an addict.

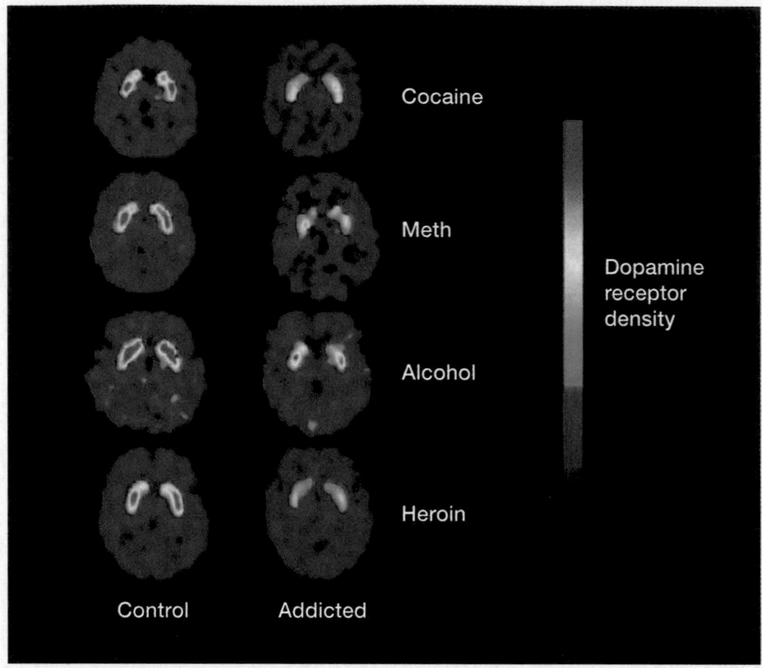

▲ **FIGURE E39-1 Addiction changes the brain** Repeated exposure to many drugs of abuse, including cocaine, meth, alcohol, and heroin, causes a reduction in the number of dopamine receptors in reward areas of the brain, leading to addiction. In these PET scans, red and yellow represent the highest concentration of dopamine receptors; green, blue, and black represent decreasing concentrations.

Ecstasy blocks serotonin pumps, causing a temporary but massive increase in serotonin, which in turn causes increased release of oxytocin. Users report feelings of pleasure, increased energy, heightened sensory awareness, and improved rapport with other people. Ecstasy users may incur long-term damage to serotonin-producing neurons and may suffer deficits in learning and memory. Ecstasy may also damage dopamine-producing neurons. Because ecstasy is usually taken in combination with other recreational drugs, it is difficult to assess its potential to cause addiction. However, although ecstasy seems to be less addictive than cocaine, meth, or heroin, ecstasy users often need higher doses over time to achieve the same reward and suffer withdrawal if deprived of the drug, which are two prominent hallmarks of addiction.

Alcohol stimulates receptors for the neurotransmitter gamma aminobutyric acid (GABA), thereby enhancing inhibitory neuronal signals, and blocks receptors for glutamate, reducing excitatory signals. Together, these changes produce alcohol's well-known relaxing effects. However, when a person drinks frequently, the brain compensates by decreasing GABA receptors and increasing glutamate receptors. Without alcohol, an alcoholic feels jittery and nervous—in short, overstimulated. Alcohol also initially increases dopamine neurotransmission, so the drinker feels good. Chronic alcohol consumption, however, both suppresses dopamine release and reduces the number of dopamine receptors (see Fig. E39-1). These effects combine to produce addiction—an alcoholic needs alcohol just to feel normal, and more and more of it to feel good.

Nicotine in cigarette smoke stimulates receptors that normally respond to acetylcholine. Overstimulation of these receptors activates other neurons that increase dopamine release, thereby contributing to smoking's pleasurable and addictive properties.

To overcome addiction, drug users must undergo the misery of a nervous system deprived of a drug to which it has adjusted. Although transmitter release and receptor concentrations eventually return to normal, drug cravings often recur periodically, suggesting that the brains of addicts have been permanently altered, in poorly understood but important ways.

THINK CRITICALLY A start-up drug company hopes to make its fortune with a drug that blocks dopamine receptors, advertising a rapid cure for addiction. However, most patients drop out of early clinical trials. If the drug company hired you to fix this problem, how would you explain to the company's directors that a high drop-out rate is inevitable?

which are wrinkled ridges that increase its surface area to over 2 square yards—about the area of a twin bed! Neurons in the cortex receive sensory information, process it, direct voluntary movements, create memories, and allow us to be creative and even envision the future. The cortexes in the two hemispheres communicate with each other through a large band of axons, the **corpus callosum** (see Figs. 38-12a, b).

Each hemisphere of the cerebral cortex is divided into four anatomical regions: the frontal, parietal, occipital, and temporal lobes (**FIG. 39-13**). The cortex can also be divided into functional regions. Primary sensory areas are regions of the parietal, temporal, and occipital lobes that receive input from senses such as the ears, eyes, temperature receptors in the skin, and so on. In the primary sensory areas, these inputs are converted into subjective impressions, such as sound, light, heat, or cold. Nearby association areas interpret the stimuli. Sounds, for example, may be interpreted as speech, music, dogs barking, and so forth. Primary motor areas in the frontal lobe command movements by stimulating motor neurons in the spinal cord that activate muscles, allowing you to walk to class or play a video game. An adjacent area in the frontal lobe, called the premotor area, receives inputs from sensory association areas and other parts of the cortex and directs the motor area to produce movements. The prefrontal cortex, the part of the frontal lobe situated directly behind the bones of the forehead, is involved in complex brain functions such as short-term memory, decision making, planning for the future, and social behaviors such as predicting the consequences of actions and controlling aggression.

Neurons in the motor areas of each hemisphere of the cerebral cortex send axons down through the hindbrain into the spinal cord. The axons cross the midline to the opposite side of the spinal cord, where they stimulate motor neurons that innervate the skeletal muscles (see Figs. 39-9 and 39-10). Similarly, sensory pathways from most of the body enter the spinal cord and cross over to the opposite side on their way to the two hemispheres. Because both sensory and motor pathways cross the midline on their way to and from the cerebral cortex, the left hemisphere of the cortex receives sensation from the right side of the body and controls movement of the right side of the body. The right hemisphere senses and controls the left side of the body.

Damage to the cortex from trauma, stroke, or a tumor results in specific deficits, such as problems with speech, difficulty reading, or the inability to sense or move specific parts of the body, depending on where the damage occurs. Historically, the capabilities of patients with brain damage provided important insights into the localization of brain functions. Now, neuroscientists have sophisticated, often noninvasive, methods of examining brain function, as we describe in "Doing Science: Neuroimaging: Observing the Brain in Action" on page 700.

The Limbic System Contributes to Emotions, Memories, and Maintaining Homeostasis

The **limbic system** is a diverse group of forebrain structures, located in a ring between the thalamus and cerebral cortex. It includes the hypothalamus, hippocampus, and amygdala, as well as nearby regions of the cerebral cortex (**FIG. 39-14**). These structures help to generate emotions such as fear, rage, and sexual desire. For example, electrical stimulation of clusters of neurons in the **amygdala** produces sensations of pleasure, fear, or sexual arousal. Damage to the amygdala early in life

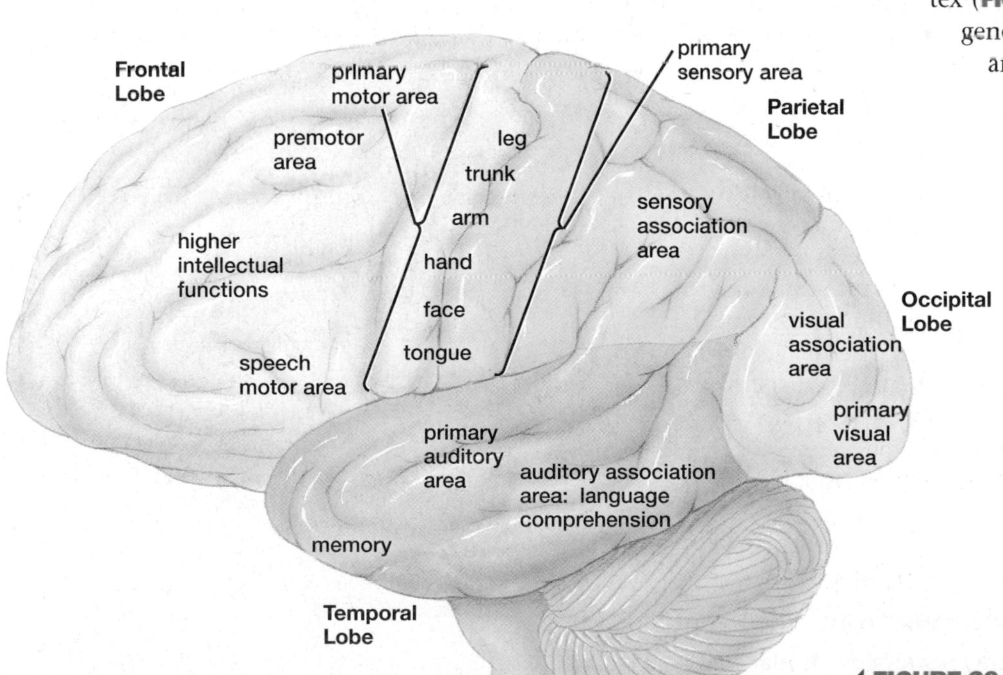

FIGURE 39-13 The cerebral cortex Different regions of the human left cerebral cortex are specialized to perform specific functions. A map of the right cerebral cortex would be similar, except that speech and language areas would not be as well developed.

DOING Science Neuroimaging: Observing the Brain in Action

Until about 30 years ago, the only way to study the functions of different parts of the brain was to observe changes in the behaviors and abilities of people who suffered traumatic brain injuries. Consider the case of Phineas Gage, who in 1848 suffered severe damage to his brain's frontal lobes when a 13-pound steel rod was blasted through his skull during a construction accident (**FIG. E39-2**). Amazingly, Gage somehow survived, but his personality changed radically.

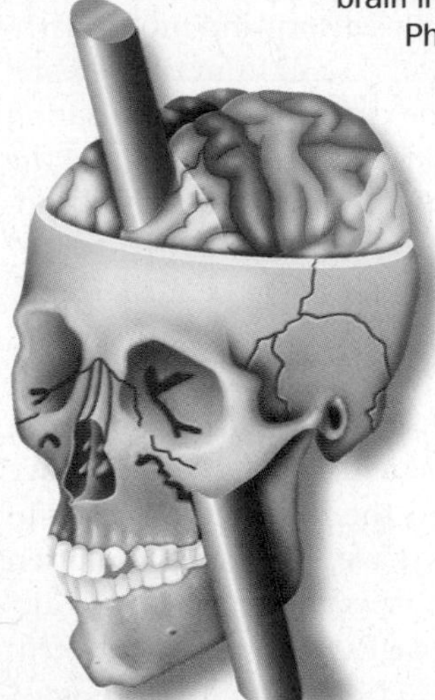

▲ **FIGURE E39-2 A revealing accident** Studies of the skull of Phineas Gage have enabled scientists to re-create the path taken by the steel rod that was blown through his head.

Today, neuroscientists have powerful, noninvasive techniques to visualize activity in the intact brain, including positron emission tomography (PET) and functional magnetic resonance imaging (fMRI). Both rely on the fact that regions of the brain that are most active need the most energy and, therefore, use more glucose and attract a greater flow of oxygenated blood than less-active regions do.

What Question Was Asked?

Researchers have used PET and fMRI to test hundreds of hypotheses about brain function. Here we will review two examples that illustrate the kinds of questions that these techniques can help answer. In one study, investigators used PET to help answer the question of whether language usage and language processing occur in different parts of the brain. A second study used fMRI to help answer the question of whether the pattern of brain activity during facial recognition differs predictably between different faces.

How Was Evidence Gathered?

In the PET scan study of language processing, scientists injected subjects with a radioactive form of glucose and then monitored levels of radioactivity (which reflect differences in glucose metabolism and hence in brain activity) as the subjects completed different tasks. The tasks were to listen to a spoken word, silently

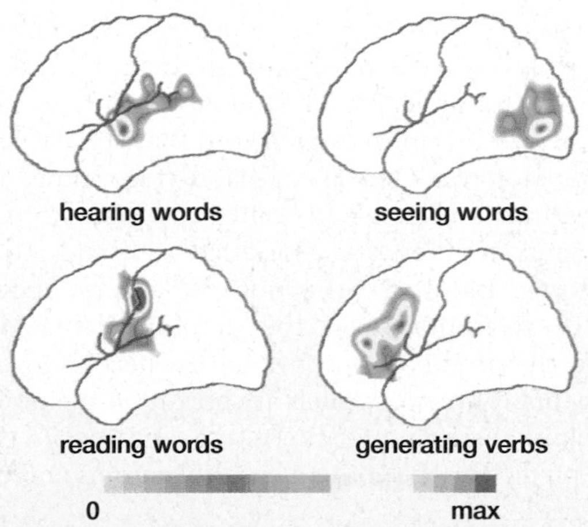

▲ **FIGURE E39-3 Localization of language tasks** PET scans reveal the different cortical regions involved in language-related tasks. The scale ranges from white (lowest brain activity) to red (highest).

eliminates the ability both to feel fear and to recognize fearful facial expressions in other people. Several areas in the limbic system are innervated by dopamine-containing neurons from the midbrain, forming part of the reward circuit responsible for pleasure, love, and addiction. The limbic system, especially the hypothalamus, is also responsible for generating drives, such as hunger and thirst, that are crucial to maintaining homeostasis of the body. Finally, several parts of the limbic system are required for various types of memory, including memories of places (hippocampus) and fearful situations (amygdala).

The Left and Right Sides of the Brain Are Specialized for Different Functions

Although the two cerebral hemispheres are very similar in appearance, this symmetry does not extend to function. We

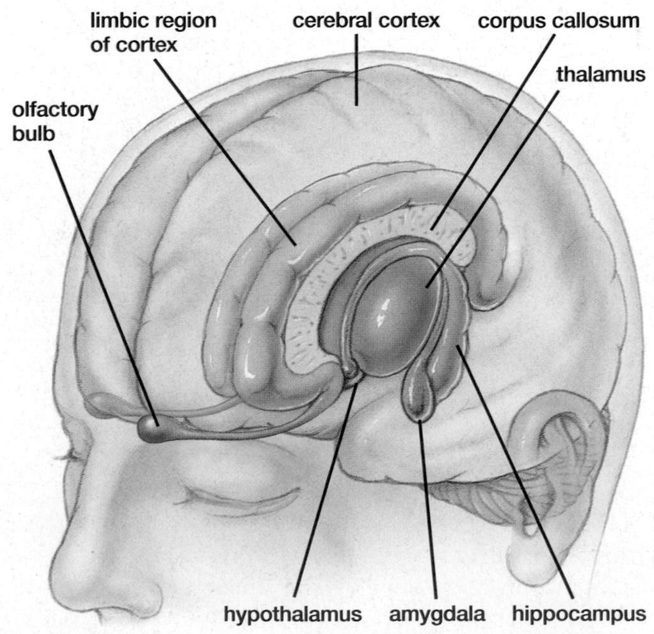

▶ **FIGURE 39-14 The limbic system**

read a printed word, read a printed word aloud, and generate a verb related to a printed noun (for example, if the word was "cake," the subject might say "eat" or "bake").

In the facial recognition study, researchers first collected fMRI data while subjects viewed images of 300 different faces. Functional MRI works by detecting differences in the way oxygenated blood and deoxygenated blood respond to a powerful magnetic field, so it does not use radioactivity and has become the favored tool for localizing brain function. After the subjects had viewed the faces, the investigators searched the fMRI data for patterns of brain activity that were associated with viewing particular facial features. Finally, the subjects returned for a session in which their brain activity was monitored as they viewed new, previously unseen faces.

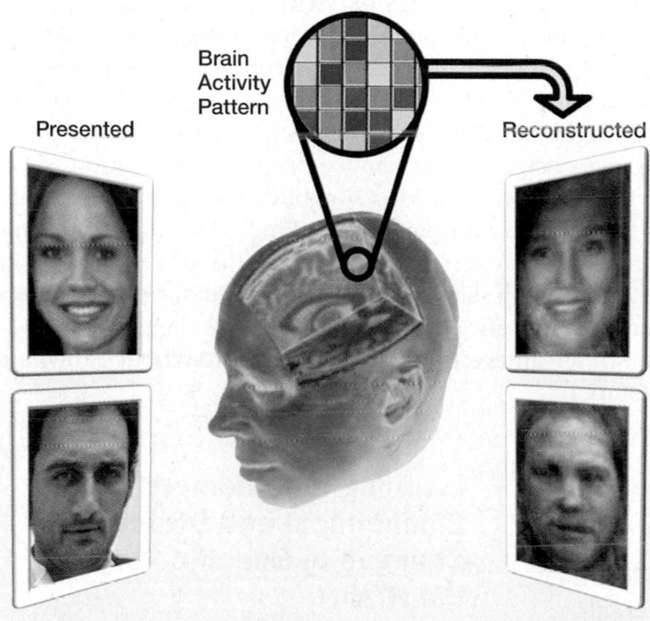

▲ **FIGURE E39-4 Functional MRI reconstructs faces from brain scans** Computer algorithms can use fMRI data to translate photos of faces (left) into reasonable facsimiles (right).

What Was Learned?

In the PET scan language study, brain activity while performing different tasks was translated into colors on images of the brain. The results demonstrated that specific aspects of language processing occur in distinct areas of the cerebral cortex (**FIG. E39-3**). In the fMRI facial recognition study, when the subjects viewed new faces, a computer was able to use the previously established relationships between brain activity and facial features to construct a rough approximation of what the newly viewed faces looked like (**FIG. E39-4**). When asked to choose which of two faces the computer was trying to reconstruct, both naive observers and computer algorithms picked out the correct face about 60% to 70% of the time. Thus, in a limited sense, fMRI was able to read people's minds.

CONSIDER THIS At least two commercial companies are trying to develop fMRI into a lie detector. Conventional lie detectors (polygraphs) measure breathing and pulse rates, blood pressure, and sweating, on the principle that these measures will increase with the stress induced by lying. The U.S. National Research Council found that polygraph tests are about 80% to 90% accurate: much better than chance, but far from perfect. Many people can teach themselves to remain calm while lying and fool lie detectors. However, a person has little control over the brain's activity, which is very likely to differ if someone is telling the truth or telling a lie. Assuming that fMRI can be applied to lie detection, what level of accuracy do you think should be required before fMRI is allowed as evidence in the courtroom? If that level of accuracy were achieved, would you support compulsory fMRI lie detection for criminal defendants? Why or why not?

have already seen that the two hemispheres receive sensory information from, and control movements of, opposite sides of the body. The two hemispheres also differ significantly in the types of higher-level functions they control, playing different roles in the control of language, thought, emotion, and other mental processes.

Beginning in the 1950s, Roger Sperry, Michael Gazzaniga, and other researchers studied people whose hemispheres had been separated when surgeons cut the corpus callosum to prevent the spread of epilepsy from one hemisphere to the other. Severing the corpus callosum also prevents the two hemispheres from communicating with one another. However, the surgery does not affect input to the hemispheres from the optic nerves, which follow a pathway that causes the left half of each visual field to be "seen" by the right hemisphere and the right half to be seen by the left hemisphere (**FIG. 39-15**). Under everyday conditions, even in someone with a severed corpus callosum, rapid eye movements inform both hemispheres about both the left and right visual fields. To investigate the individual responses of each hemisphere to visual inputs, the researchers designed an ingenious device that projects different images onto the left and right visual fields and thus sends different signals to each hemisphere.

When the neuroscientists projected an image of a nude figure onto only the left visual field, the subjects would

Have You Ever Wondered ...

How Con Artists Fool Their Victims?

One of the foundations of human society is trust in the good intentions of other people. Oxytocin quiets the amygdala, allowing us to trust rather than fear. It also helps us to feel good when we do nice things for other people. Unfortunately, con artists take advantage of this. They seem helpful, kind, and sometimes needy. They pretend to offer their victim a reward in exchange for trust, thereby activating the victim's oxytocin trust system.

smile or blush but claim to have seen nothing. The same figure projected onto the right visual field was readily described verbally. These experiments and other more complex ones in people with intact brains revealed that the human brain is *lateralized*: The left hemisphere dominates in performing some functions, and the right hemisphere dominates in performing other functions. In most people, the left hemisphere is dominant in speech, reading, writing, language comprehension, and rote mathematics (like adding or multiplying numbers). The right hemisphere is superior to the left in musical skills, artistic ability, recognizing faces, spatial visualization, and the ability to recognize and express emotions.

However, the left–right dichotomy is not complete. For example, the popular media often say that the left hemisphere is dominant for mathematics. However, although the left hemisphere is usually dominant in rote math skills, the right hemisphere tends to be dominant in tasks such as estimating quantities, for example, guessing how many marbles are in a jar. Further, despite frequent self-improvement advice to "use the creative (right) side of your brain," a simple left-brain-logical, right-brain-creative duality is incorrect.

If necessary, the hemispheres can retool to assume new functions. A person who has suffered a stroke that damaged the left hemisphere typically shows symptoms such as loss of speaking ability. These deficits can often be partially overcome through training, even though the left hemisphere itself has not recovered. This observation suggests that the right hemisphere has some language capabilities that can be further developed, if necessary.

Most Vertebrates Have Lateralized Brains

People aren't the only animals with lateralized brains. Many fish, reptiles, and amphibians preferentially snatch prey on their right side, using their left hemisphere to control the movement. About 80% of humpback whales also feed on their right side; interestingly, this is not too different from the percentage of right-handed people, which various studies put at 70% to 95%. Most songbirds control singing—more or less the bird equivalent of speaking—with their left hemisphere. Why are vertebrate brains lateralized? Experiments in chicks strongly suggest that lateralized brains process information more efficiently than symmetrical brains do, providing an advantage that has persisted throughout vertebrate evolution.

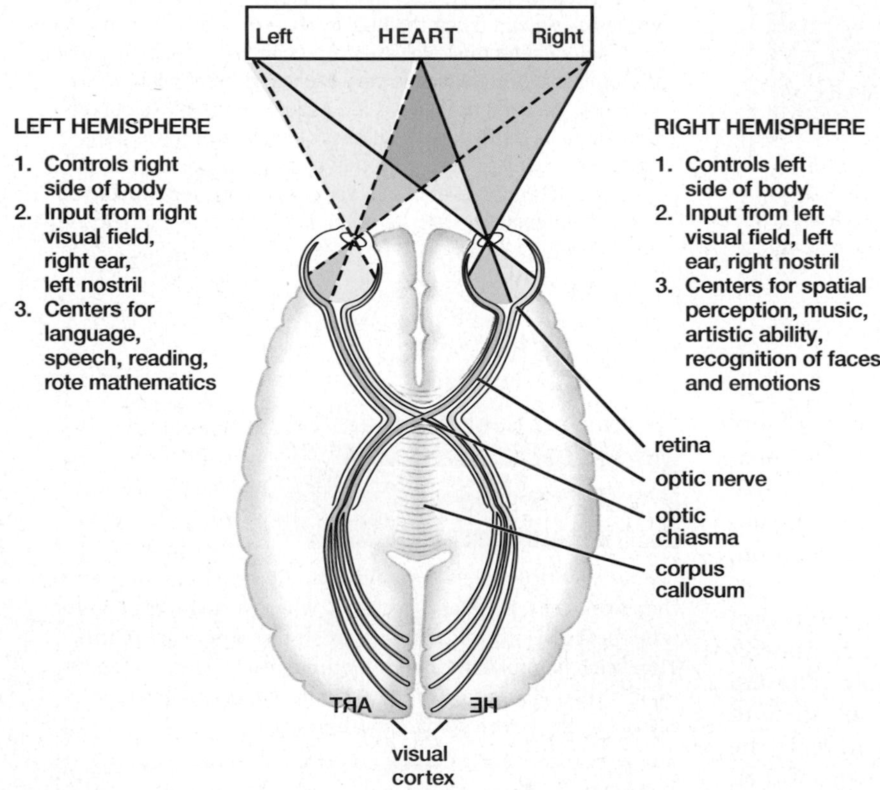

▲ **FIGURE 39-15 Specialization of the cerebral hemispheres** Each half of the retina of each eye "sees" the opposite visual field. The axons from the half-retinas that see the left visual field send information to the right hemisphere, and vice versa. Therefore, with a quick glance at the word "heart" (before you had a chance to move your eyes), the right hemisphere would perceive "he" and the left hemisphere would perceive "art." In addition to receiving inputs from different parts of the visual field, the two hemispheres typically control the opposite sides of the body and are specialized for a variety of functions.

Learning and Memory Involve Biochemical and Structural Changes in Specific Parts of the Brain

Most neurobiologists and psychologists agree that learning has two phases: **short-term memory** and **long-term memory.** For example, when you read a recipe for chocolate chip cookies, you can probably remember "2¼ cups of flour" long enough to measure out the flour and put it in the bowl, but you probably won't remember the correct amount months later. This is short-term memory, which typically lasts for minutes or less and has a very limited capacity to store information. However, if you make the same cookies twice a week for several months, you probably won't need the recipe anymore—it has been stored in long-term memory. A professional chef may have scores of recipes, and information about hundreds of ingredients, stored in long-term memory, which seems to have no practical size limitations.

The frontal and parietal lobes of the cerebral cortex, the hippocampus, and some of the basal ganglia deep in the cerebrum are important sites of short-term memory. Most short-term memory probably requires the repeated activity of a particular neural circuit in the brain. As long as the circuit is active, the memory stays. In other cases, short-term memory may rely on short-lived biochemical changes that temporarily strengthen synapses between specific neurons.

In contrast, long-term memory seems to be structural. It may require the formation of new, long-lasting synaptic connections between neurons or long-term strengthening of existing synapses (for example, by increasing neurotransmitter release or increasing the number of receptors for the neurotransmitter). For many memories, including memories of places, facts, and specific events, converting short-term memory into long-term memory involves the hippocampus, which is believed to process new memories and both store them in long-term memory and transfer them to the cerebral cortex for even longer-term storage. Although long-term memory probably resides in many areas of the cerebrum, some research suggests that the temporal and frontal lobes are particularly important. The cerebellum and basal ganglia are crucial for learning and storing habits and physical skills (motor learning).

CHECK YOUR LEARNING

Can you . . .

- distinguish between the central and peripheral nervous systems and between the somatic and autonomic nervous systems?
- describe the autonomic nervous system and its divisions and provide some examples of activities controlled by each division?
- label diagrams of the human brain and spinal cord, naming the principal structures and describing their functions?
- describe the different functions that are usually controlled by the left and right halves of the human cerebral hemispheres?
- distinguish between short-term and long-term memory?

CASE STUDY REVISITED

How Do I Love Thee?

What happens in the human brain when we fall in love? Although people aren't just big prairie voles, people and prairie voles show striking similarities in both brain function and hormones during emotional encounters. For example, oxytocin levels increase in women and men during sex, just as they do in prairie voles. Oxytocin also reduces stress and inhibits the amygdala, reducing the fear of others, which is probably an important prerequisite for forming long-lasting emotional bonds with another person. Magnetic resonance imaging has shown that parts of the human brain that contain oxytocin and dopamine, including reward areas, are strongly activated when people are shown pictures of their lovers, but not as much when they are shown images of equally attractive people to whom they feel no emotional bond. Thus in humans, as in voles, oxytocin probably plays an important role in attraction and commitment. Given the prevalence of one-night stands and adultery, oxytocin obviously doesn't guarantee monogamy, but it seems to help.

Prairie vole

What about different kinds of love? Brain scans reveal that some of the same brain areas are activated when subjects see either photos of their lovers or photos of their children. Other areas are activated by one or the other, but not both. And some brain areas, particularly those involved in critical decision making and social judgments, are turned off when subjects see photos of their lovers or children.

CONSIDER THIS When sprayed into the nose, which provides direct access to olfactory neurons and the brain, oxytocin enhances trust, even between complete strangers. You can buy oxytocin online, in the form of a cologne, perfume, or nasal spray. The claim is that these oxytocin products will make other people trust you more, thus aiding both business and sexual conquests. Assuming that wearing oxytocin cologne produces a high enough concentration of oxytocin in the air to affect anyone's brain, would you use it? Why or why not?

CHAPTER REVIEW

Go to **Mastering Biology** to access the Pearson eText, vocabulary review, practice quizzes, activities, videos, current events, and more.

*Answers to **Think Critically and Thinking Through the Concepts questions** can be found in the Answers section at the back of the book.*

Summary of Key Concepts

39.1 What Are the Structures and Functions of Nerve Cells?

A neuron has four major functions, which are reflected in its structure: (1) dendrites receive information from the environment or from other neurons; (2) the cell body adds together electrical signals from synapses on the dendrites and on the cell body itself, with the resulting electrical potential determining whether or not the neuron produces an action potential; (3) the axon conducts the action potential to its synaptic terminal; and (4) synaptic terminals release neurotransmitters that transmit the signal to other neurons, muscles, or glands.

39.2 How Do Neurons Produce and Transmit Information?

An unstimulated neuron maintains a negative resting potential inside the cell. Signals received from other neurons are small, rapidly fading changes in potential called postsynaptic potentials.

Inhibitory and excitatory postsynaptic potentials (IPSPs and EPSPs) make the neuron less likely or more likely, respectively, to produce an action potential. If postsynaptic potentials, added together within the cell body, bring the neuron to threshold, an action potential will be triggered. An action potential is a wave of positive charge that travels, undiminished in magnitude, along an axon to its synaptic terminals.

A synapse consists of the synaptic terminal of the presynaptic neuron, a specialized region of the postsynaptic neuron, and the synaptic cleft between the neurons. Neurotransmitters from the presynaptic neuron, released in response to an action potential, diffuse across the synaptic cleft, bind to receptors in the postsynaptic cell's plasma membrane, and produce either an EPSP or an IPSP.

39.3 How Does the Nervous System Process Information and Control Behavior?

Neural pathways usually have four elements: (1) sensory neurons, (2) interneurons, (3) motor neurons, and (4) effectors (muscles or glands). These elements carry out four operations: (1) determining the type of stimulus, (2) determining and signaling the intensity of the stimulus, (3) integrating information from many sources, and (4) directing appropriate behaviors.

39.4 How Are Nervous Systems Organized?

Nervous systems may be diffuse (distributed throughout the body, usually in radially symmetrical animals) or centralized (with most of the senses and nervous system in the head, in bilaterally symmetrical animals). Complex nervous systems are centralized.

39.5 What Are the Structures and Functions of the Human Nervous System?

The nervous system of humans and other mammals consists of the central nervous system and the peripheral nervous system. The peripheral nervous system is divided into sensory and motor portions. The motor portion consists of the somatic nervous system (which controls voluntary movement) and the autonomic nervous system (which directs involuntary responses). The autonomic nervous system is further subdivided into the sympathetic and parasympathetic divisions.

The central nervous system consists of the brain and spinal cord. The spinal cord contains neurons that control voluntary muscles and the autonomic nervous system; neurons that communicate with the brain and other parts of the spinal cord; axons leading to and from the brain; and neural pathways for reflexes and certain simple behaviors.

The brain consists of three parts: the hindbrain, midbrain, and forebrain. The hindbrain consists of the medulla and pons, which control involuntary functions, and the cerebellum, which coordinates complex motor activities. In humans, the small midbrain contains clusters of neurons that help to control movement, arousal, and emotion. The midbrain and hindbrain contain the reticular formation, which is a filter and relay for sensory stimuli. The forebrain includes the thalamus, a sensory relay station that shuttles information to and from conscious centers in the forebrain; the hypothalamus, which is responsible for maintaining homeostasis and directs much of the activity of the autonomic nervous system; and the cerebrum, the center for information processing, memory, and initiation of voluntary actions. The cerebral cortex includes primary sensory and motor areas, together with association areas that analyze sensory information and plan movements. A group of forebrain structures called the limbic system contributes to maintaining homeostasis, learning and storing memories, and perceiving and expressing emotions.

The cerebral hemispheres are specialized. In general, the left hemisphere controls the right side of the body and dominates speech, reading, writing, language comprehension, and rote mathematical ability. The right hemisphere controls the left side of the body and specializes in recognizing faces and spatial relationships, producing artistic and musical abilities, and recognizing and expressing emotions.

Memory has two stages: Short-term memory is electrical or chemical, whereas long-term memory probably involves structural changes that increase the effectiveness or number of synapses. The hippocampus is an important site for the transfer of information from short-term into long-term memory.

Thinking Through the Concepts

Bloom's: Remembering, Understanding

Multiple Choice

1. A change in the voltage across a neuron's plasma membrane that makes the neuron less likely to fire an action potential is
   a. a change that makes the resting potential less negative.
   b. an excitatory postsynaptic potential.
   c. an inhibitory postsynaptic potential.
   d. a change that lowers the threshold.
2. Neurotransmitters are typically released by a
   a. presynaptic neuron.
   b. postsynaptic neuron.
   c. muscle cell.
   d. gland cell.
3. Automatic bodily functions such as breathing and swallowing are controlled by the
   a. pons and medulla.
   b. thalamus.
   c. midbrain.
   d. cerebral cortex.
4. Jane suffers a stroke and can no longer speak or comprehend language. The stroke most likely damaged her
   a. left cerebral hemisphere.
   b. right cerebral hemisphere.
   c. medulla.
   d. cerebellum.
5. Place learning requires the ____________, whereas motor learning usually requires the ____________.
   a. midbrain; cerebellum
   b. hypothalamus; occipital lobes
   c. cerebellum; hippocampus
   d. hippocampus; cerebellum

Fill-in-the-Blank

1. An individual nerve cell, also called a(n) ____________, includes structures specialized to perform different functions. The "input" end of a nerve cell, called a(n) ____________, receives information from the environment or from other nerve cells. The ____________ contains the nucleus and other typical organelles of a eukaryotic cell.

Electrical signals are sent down the ____________, a long, thin strand that leads to the ____________, where the nerve cell sends out its signal to other cells.

2. When they are not being stimulated, neurons have an electrical charge across their membranes called the resting potential. This potential is ____________ charged inside. When a neuron receives a sufficiently large stimulus, and reaches a potential called the ____________, it produces an action potential. This causes the neuron to become ____________ charged inside.
3. When an action potential reaches a synaptic terminal, it causes the release of a chemical called a(n) ____________. This chemical binds to protein ____________ on the postsynaptic cell, causing a change in potential. If the postsynaptic cell becomes less negative, this change in potential is called a(n) ____________. If the postsynaptic cell becomes more negative, it is called a(n) ____________.
4. The ____________ is part of the peripheral nervous system that innervates skeletal muscles. Smooth muscles and glands are innervated by the ____________, which has two divisions: the ____________, active during "fight or flight," and the ____________, active during "rest and digest."
5. The human hindbrain consists of three parts: the ____________, the ____________, and the ____________. One of these, the ____________, is important is coordinating complex movements such as typing.
6. The cerebral cortex consists of four lobes: the ____________, ____________, ____________, and ____________. The visual centers are located in the ____________ lobe. The primary motor areas are in the ____________ lobe.

Review Questions

1. Diagram a synapse. How are signals transmitted from one neuron to another at a synapse?
2. How does the brain perceive the intensity of a stimulus? The type of stimulus?
3. What are the four elements of a neuronal pathway, beginning with a sensory neuron and ending with a muscle? Describe how these elements function in the human pain-withdrawal reflex.
4. Draw a cross-section of the spinal cord. What types of neurons are located in the spinal cord? Explain why severing the spinal cord paralyzes the body below where the cut occurs.
5. Describe the functions of the following parts of the human brain: medulla, cerebellum, reticular formation, thalamus, amygdala, and cerebral cortex.
6. What structure connects the two cerebral hemispheres? Describe the usual functions of each hemisphere.
7. Explain the differences between short-term memory and long-term memory.

Applying the Concepts

Bloom's: Applying, Analyzing, Evaluating

1. If the axons of human spinal cord neurons were unmyelinated, would you expect the spinal cord to be larger or smaller? Would you move faster or slower? Explain your answer.
2. What is the adaptive value of reflexes? Why couldn't all behaviors be controlled by reflexes?

40

The Senses

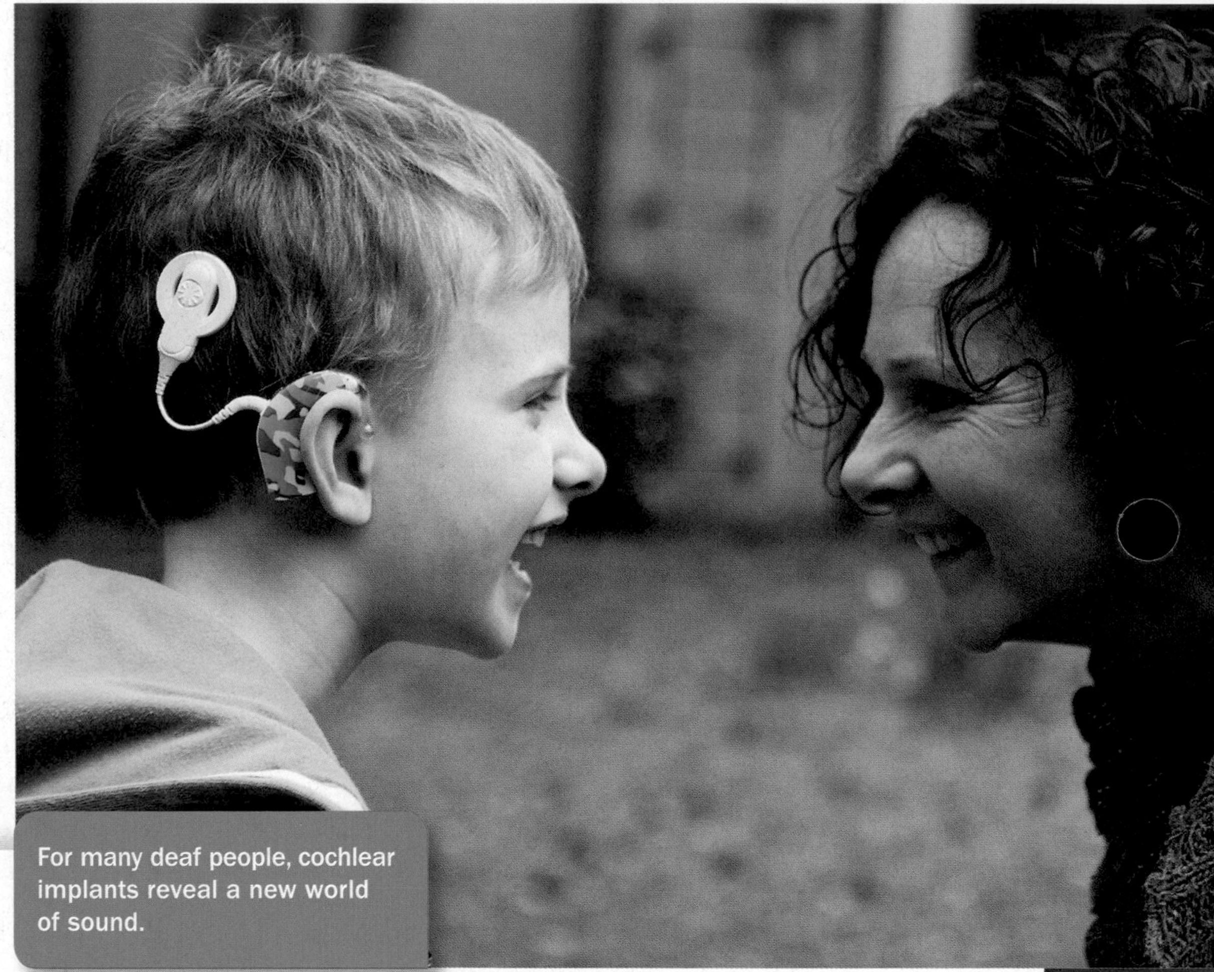

For many deaf people, cochlear implants reveal a new world of sound.

CASE STUDY

Bionic Ears

SAMANTHA BRILLING DOWNTON, like the boy shown here, was born with impaired hearing, although she was not totally deaf. She wore hearing aids for many years, but they didn't provide normal hearing. Sounds that most of us take for granted, like music or conversation with friends, were missing for Samantha. Today, however, Samantha can hear almost normally, thanks to a cochlear implant. As you will learn in this chapter, the cochlea is the part of the ear that converts sound to electrical signals, which then travel in the auditory nerve to the brain. A cochlear implant is a microprocessor-controlled device that partially replaces cochlear function in a hearing-impaired person by converting sound to electrical pulses that stimulate the auditory nerve.

Although Samantha's cochlear implant does not make her hearing perfect, she can now enjoy music, carry on a conversation in a noisy room, and know when someone out of sight is calling to her. In fact, Samantha can hear so well that she has held jobs that require accurate verbal communication, including serving as a receptionist at her college alumni center, volunteering at a hospital, and now working as a marketing coordinator at the Auditory-Verbal Center of Atlanta.

How do your ears produce the sensation of sound? How can electrical impulses, whether generated biologically or bionically, be understood by the brain as speech, bird calls, or music? And what about your other senses—how do you perceive the odor of pine trees, the colors of blooming flowers, the silky feel of a baby's skin, or the "burn" of hot peppers?

AT A GLANCE

40.1 HOW DO ANIMALS SENSE THEIR ENVIRONMENT?

You've probably heard that animals have five senses—touch, hearing, vision, smell, and taste. However, almost all animals also sense temperature, gravity, and many other external stimuli. Some animals can detect magnetic or electrical fields and use this information to move around in murky water, to migrate, or to find prey. Animals also sense their internal environment, including conditions such as oxygen levels, blood pH, body temperature, and how full their bladders are.

All sensory perception begins with a **receptor**—a molecule, cell, or multicellular structure that produces a response when it is acted on by a stimulus. A **sensory receptor** is a specialized cell (often a neuron) that produces an electrical signal in response to an environmental stimulus—that is, it translates an environmental stimulus into the language of the nervous system. Sensory receptors can be grouped into five major categories, according to the stimuli to which they respond (**TABLE 40-1**).

Many sensory receptors are nerve cells with specialized dendrites, often called free nerve endings, that are located in the skin, the digestive and respiratory tracts, the bladder, and many other body parts. The plasma membranes of these dendrites contain receptor proteins that respond to stimuli such as heat, cold, or touch. Other sensory receptors are housed in a **sense organ**—a structure that includes both the sensory receptors and accessory structures that play essential roles in detecting specific stimuli. The most familiar sense organs are the eyes and ears, which contain accessory structures such as the lens and the eardrum. However, sensory receptors for vibration, pressure, odors, and tastes are also located within sense organs. Perhaps the most striking example of the importance of the accessory structures for sensory reception occurs in the mammalian inner ear. As we will describe in Sections 40.4 and 40.5, although the sensory receptor cells in the middle ear are all virtually identical, the accessory structures that enclose them determine whether the cells respond to the pull of gravity, motion of the head, or music.

The Senses Inform the Brain About the Nature and Intensity of Environmental Stimuli

For sensory information about the environment to be useful, the brain must determine the nature of the stimulus—light or sound, for example—and the strength, or **intensity**, of the stimulus—bright or dim light, loud or soft sounds, and so on. All sensory receptors produce electrical signals in response to environmental stimuli, and all sensory information ultimately reaches the brain as action potentials traveling in axons that connect the sense organ to the brain (see Chapter 39). Given that all action potentials are fundamentally the same, how can the brain recognize the nature and intensity of a stimulus?

Every sensory receptor cell contains receptor molecules that respond to some stimuli but not to others. Further, each sensory receptor is linked to a specific set

TABLE 40-1 Principal Categories of Vertebrate Sensory Receptors

Category of Receptor	Stimuli	Sensory Cell Type	Location
Thermoreceptor	Heat, cold	Free nerve ending	Skin, brain
Mechanoreceptor	Vibration produced by sound waves, motion, or gravity	Hair cell	Inner ear
	Vibration, pressure, touch	Free nerve endings and endings surrounded by accessory structures	Skin
	Stretch	Specialized nerve endings in muscles or joints	Muscles, tendons
Photoreceptor	Light	Rod, cone	Retina of the eye
Chemoreceptor	Odor (airborne molecules)	Olfactory receptor	Nasal cavity
	Taste (waterborne molecules)	Taste receptor	Tongue and oral cavity
Pain receptor	Chemicals released by tissue injury; extreme heat or cold; excessive stretch; acid	Free nerve ending	Widespread in the body

of axons that connect to particular locations in the brain or spinal cord. Electrical activity in these locations is then interpreted as a specific form of sensory perception. In mammals, for example, neurons that detect odors send axons to a part of the brain called the olfactory bulb. The activity of specific sets of neurons in the olfactory bulb results in the perception of particular odors, such as coffee or roses.

The linkage of stimulus type to sensory receptor to axons that lead to a particular brain region provides the first principle of sensory perception: The nature of a stimulus is encoded by which neurons in the brain are activated.

When a sensory receptor cell is stimulated, it produces an electrical signal called a **receptor potential** (**FIG. 40-1**). Unlike action potentials, which are always the same size (see Chapter 39), receptor potentials vary in size with the intensity of a stimulus—the stronger the stimulus, the larger the receptor potential (**FIG. 40-2**). Some sensory receptor cells, such as those that detect touch or temperature of the skin, are neurons with axons that connect to the central nervous system (CNS). In these cells, a receptor potential may cross threshold and trigger action potentials. A small receptor potential may barely reach threshold and produce only a few action potentials, whereas a large receptor potential goes far above threshold, causing a high frequency of action potentials.

Other sensory receptor cells, such as those in the inner ear, do not have axons. Most of these receptor cells form synapses with neurons that have axons connecting to the CNS (see Fig. 39-4 for a description of the anatomy and function of a synapse). A receptor potential in this type of sensory receptor cell causes neurotransmitters to be released onto a neuron, ultimately stimulating action potentials that move down the neuron's axon to the CNS. In such cases, the stronger the stimulus, the larger the receptor potential, which causes the release of larger amounts of transmitter onto the neurons, producing a higher frequency of action potentials traveling to the CNS.

The linkage of stimulus intensity to receptor potential size to action potential frequency provides the second principle of sensory perception: The intensity of a stimulus is encoded by the frequency of action potentials reaching the brain.

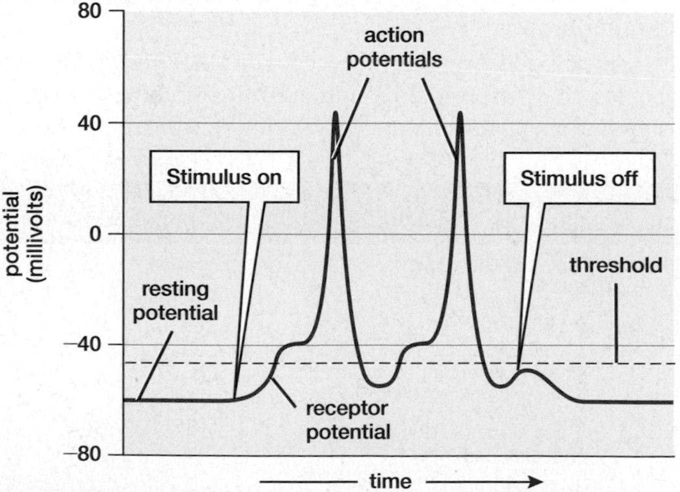

▲ **FIGURE 40-1 Converting an environmental stimulus to action potentials** In most sensory receptor neurons, an environmental stimulus generates a receptor potential that makes the resting potential less negative. If the receptor potential is large enough, it reaches threshold and triggers action potentials.

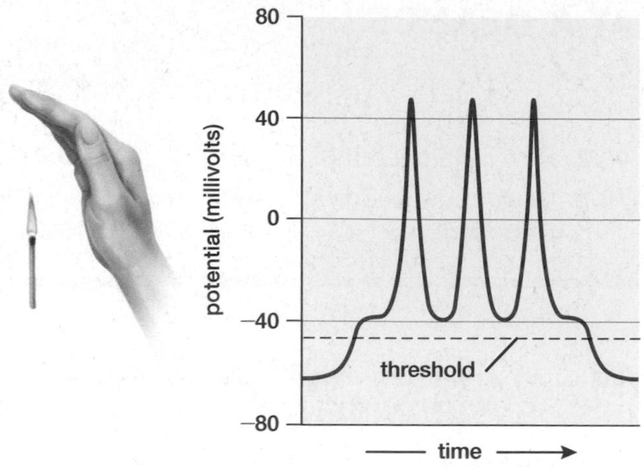

(a) Weak stimulus

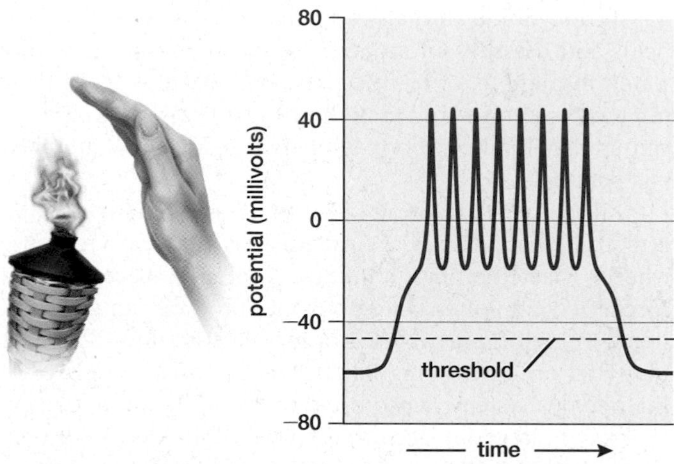

(b) Strong stimulus

▲ **FIGURE 40-2 Stimulus intensity is encoded by the frequency of action potentials (a)** Weak stimuli cause small receptor potentials that barely reach threshold (dashed line) and produce only a few action potentials. **(b)** Strong stimuli cause large receptor potentials that reach far above threshold, producing many action potentials.

CHECK YOUR LEARNING

Can you . . .

- list and describe the five major types of sensory receptors and give an example of each?
- describe how the nervous system codes for the nature of a stimulus and the intensity of a stimulus?

CASE STUDY CONTINUED

Bionic Ears

Cochlear implants work because the brain interprets action potentials in axons of the auditory nerve as sound, regardless of the actual stimulus triggering them. This principle was discovered by Alessandro Volta, who invented the battery in 1800. He stuck a metal rod in his ear and connected the rod to a battery. He felt a jolt and heard a sound like boiling water. Like Volta's rod, a modern cochlear implant stimulates action potentials in axons of the auditory nerve, although with much more sensitivity and precision. We will see how the ear normally detects sound in Section 40.4.

40.2 HOW IS TEMPERATURE SENSED?

Thermoreceptors respond to heat or cold. External temperatures that are typically not harmful to the body are primarily sensed by free nerve endings in the skin and oral cavity. (Damaging extremes of temperature activate pain receptors, described in Section 40.8.) Generally, both heat and cold receptors fire action potentials spontaneously at usual skin temperatures of about 77° to 91°F (about 25° to 33°C; at typical room temperatures, the skin is cooler than the core of the body). Cold receptors fire more rapidly at temperatures below 77°F, whereas warm receptors fire more rapidly at temperatures above 91°F. These receptors can also be activated by certain chemicals. Menthol, for example, stimulates cold receptors, while camphor and clove oil stimulate heat receptors, which explains their use in ointments commonly given names such as deep heat or icy hot.

Thermoreceptors in the brain detect core body temperature and activate homeostatic responses to maintain an appropriate body temperature (see Chapter 32).

CHECK YOUR LEARNING

Can you . . .

- define the term thermoreceptor and describe how thermoreceptors respond to changing temperatures?

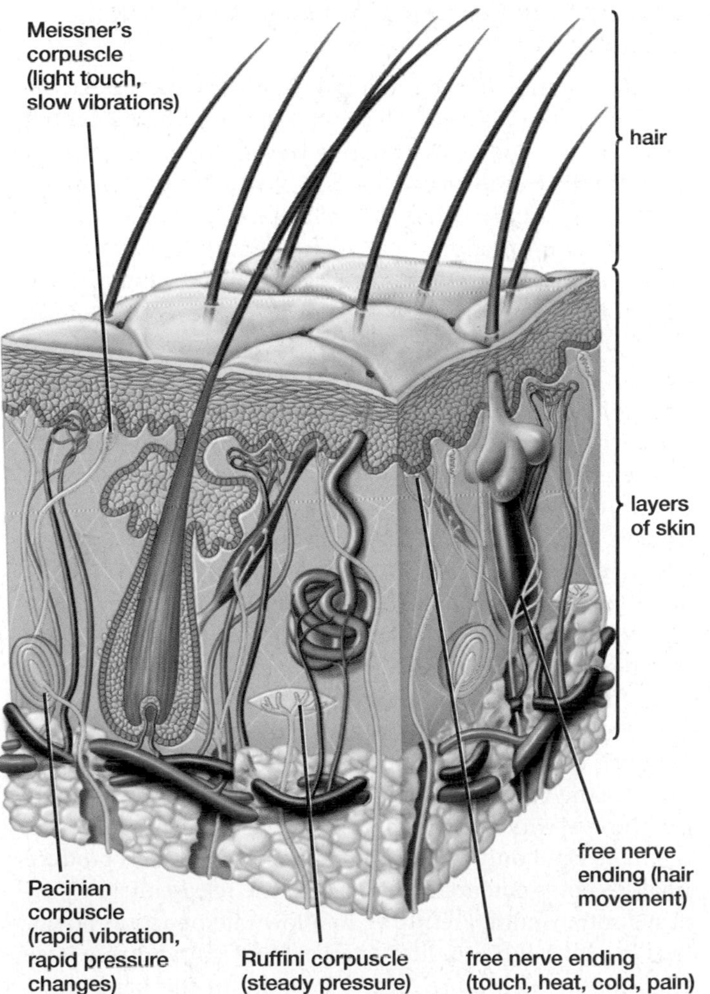

FIGURE 40-3 Receptors in the human skin The diversity of receptors in the skin allows us to perceive mechanical stimuli such as touch, pressure, and vibration, as well as other sensations such as pain, heat, and cold.

40.3 HOW ARE MECHANICAL STIMULI DETECTED?

Mechanoreceptors, found throughout the human body, respond to physical deformation such as stretching, dimpling, or bending various parts of the body. Mechanoreceptors produce receptor potentials when their membranes are stretched or dented. The body contains many types of mechanoreceptors, including receptors in the skin that respond to touch, vibration, or pressure; stretch receptors in many internal organs, including the intestines, stomach, urinary bladder, and muscles; and receptors in the inner ear that respond to sound, gravity, or movement (see Sections 40.4 and 40.5).

In the skin, free nerve endings of some mechanoreceptors produce sensations of touch, itching, or tickling (**FIG. 40-3**). The endings of other mechanoreceptors are enclosed in accessory structures—for example, Pacinian corpuscles, which respond to changes in pressure such as rapid vibrations or a sharp poke; Meissner's corpuscles, which respond to light touch or slow vibrations; and Ruffini corpuscles, which respond to steady pressure. The density of mechanoreceptors in the skin varies tremendously over the surface of the body. For example, each square inch of fingertip has hundreds of touch receptors, but on the back, there may be less than one per square inch.

Mechanoreceptors in many hollow organs, such as the stomach and urinary bladder, signal fullness by responding to stretch. Mechanoreceptors in the joints and muscles, also responding primarily to stretch, let us know whether the joints are straight or bent and how much force is being applied. Therefore, you don't usually have to look at your hands, arms, legs, or feet to tell where they are or consciously think about what they're doing. Imagine how annoying it would be if you had to watch your fork on its way to your mouth to avoid stabbing yourself in the face!

Mechanoreceptors are important to all animals. Spiders, for example, use mechanoreceptors in their legs to detect vibrations of their webs. A spider can sense whether the vibrations are from small fluttering objects that might be good to eat, like flies or moths; from larger, possibly predatory, animals; or from potential mates. Fish have mechanoreceptors in organs called the lateral lines. These mechanoreceptors detect movement and vibration of the water around them, allowing the fish to detect prey, avoid objects, and orient with respect to other fish in a school.

CHECK YOUR LEARNING

Can you . . .

- describe the types of stimuli that are detected by mechanoreceptors?
- give some examples of mechanoreceptors in your body and their functions?

40.4 HOW IS SOUND DETECTED?

The mammalian ear performs several different functions: perceiving sounds, determining the direction of gravity, and detecting the orientation and movement of the head. In this section, we will describe the role of the ear in sound perception.

The Ear Converts Sound Waves into Electrical Signals

Sound is produced by vibrating objects—drums, vocal cords, or the speaker in your cell phone. Our ears convert the resulting sound waves into electrical signals that our brains interpret as sound, including its pitch and loudness.

The ears of humans and other mammals consist of three parts: the outer, middle, and inner ear (**FIG. 40-4a**). The **outer ear** consists of the pinna and the auditory canal. The **pinna**, a flap of skin-covered cartilage attached to the surface of the head, collects sound waves. Humans and other large animals determine the location of a sound's source by differences in when the sound arrives at the two ears and in how loud the sound is in each ear. The shape of the pinna and, in many animals, the ability to swivel it around further help to locate the source of a sound.

The **auditory canal** conducts sound waves from the pinna to the **middle ear**, which consists of the **tympanic membrane**, or eardrum; three tiny bones called the **hammer** (malleus), **anvil** (incus), and **stirrup** (stapes); and the **auditory tube** (Eustachian tube). The auditory tube connects the middle ear to the pharynx and equalizes air pressure across the eardrum, between the middle ear and the outside atmosphere. The auditory tube may become swollen shut if you have a cold or an ear infection. If this happens, air pressure changes (such as those experienced during takeoff and landing in an airplane) cause the pressure on one side of the eardrum to be higher than on the other side. The resulting pressure difference pushes painfully on the eardrum.

Sound waves traveling through the auditory canal vibrate the tympanic membrane, which in turn vibrates the hammer, anvil, and stirrup. These small bones transmit the vibrations to the **inner ear**. The hollow bones of the inner ear form a spiral-shaped structure called the **cochlea** (Latin for "snail"). The cochlea contains two fluid-filled compartments, which are most easily visualized if we mentally unroll the cochlea (**FIG. 40-4b**): (1) a U-shaped tube (colored blue in Fig. 40-4b) enclosing (2) a central tube (colored gray).

The stirrup bone transmits sound waves to the fluid in the U-shaped tube by vibrating the **oval window**, a flexible membrane covering the opening at the beginning of the tube. A second membrane, the **round window**, covers an opening at the far end of the tube. When the crest of a sound wave pushes the oval window inward, the fluid moves around the tip of the cochlea (black arrows in Fig. 40-4b) and pushes the round window outward; when the trough of a sound wave pulls the oval window outward, the round window flexes inward. Thus the fluid in the cochlea shifts back and forth as the stirrup bone vibrates the oval window.

Vibrations Are Converted into Electrical Signals in the Cochlea

In a cross-section of the cochlea (**FIG. 40-4c**), we can see the two arms of the U-shaped tube (blue) surrounding the central compartment (gray). The floor of the central compartment is the **basilar membrane**, on top of which sit mechanoreceptors called **hair cells**. Hair cells have cell bodies topped by hair-like projections that resemble stiff cilia. Some of these hairs are embedded in a gelatinous structure called the **tectorial membrane** (**FIGS. 40-4c, d**). As the fluid in the U-shaped tube moves back and forth in synchrony with incoming sound waves, it moves the basilar membrane relative to the tectorial membrane. Movement of the membranes bends the hairs, causing receptor potentials in the hair cells. The receptor potentials cause the hair cells to release neurotransmitters onto neurons whose axons form the **auditory nerve**. These axons produce action potentials that travel to auditory centers in the brain.

How do we perceive loudness (the magnitude of sound vibrations) and pitch (the musical note, or the frequency of sound vibrations)? Remember that the intensity of a stimulus is encoded by the rate of action potentials traveling to the brain, and the type of stimulus is encoded by which nerve cells fire action potentials. Soft sounds cause small vibrations of the tympanic membrane, the bones of the middle ear, the oval window, and the basilar membrane. In response, the hairs bend only a little. Therefore, the hair cells produce small receptor potentials that trigger the release of a tiny bit of neurotransmitter, resulting in a low rate of action potentials in axons of the auditory nerve. Loud sounds cause large vibrations, which cause greater bending of the hairs and a

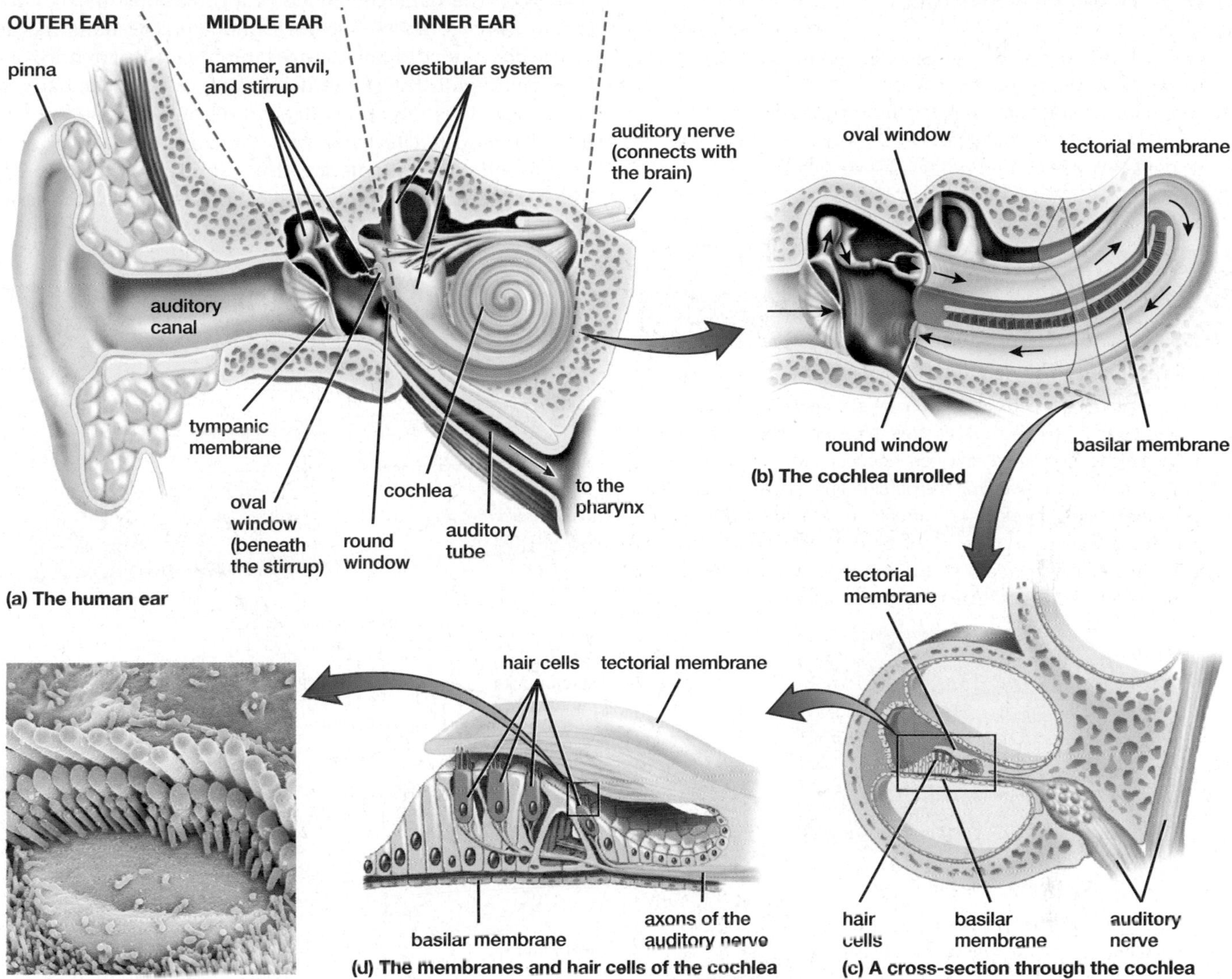

▲ **FIGURE 40-4 The human ear (a)** Overall anatomy of the ear. **(b)** If we could unroll the cochlea, we would see that it consists of a U-shaped, fluid-filled compartment, with a central compartment nestled between the arms of the U. **(c)** The hair cells sit atop the basilar membrane in the central compartment of the cochlea. **(d)** The hairs of hair cells span the gap between the basilar and tectorial membranes. Sound vibrations move the membranes relative to one another, bending the hairs and producing a receptor potential in the hair cells. The hair cells then release neurotransmitters that stimulate action potentials in the axons of the auditory nerve.

larger receptor potential, producing a high rate of action potentials in the auditory nerve. Very loud sounds damage the hair cells, resulting in hearing loss, a fate suffered by some rock musicians and their fans. Many sounds in our modern environment, including those coming from jets, trains, lawn mowers, and earbuds plugged into phones, have the potential to cause hearing loss.

The perception of pitch is a little more complex. The basilar membrane is narrow and stiff at the end near the oval window, wider and more flexible near the tip of the cochlea. This progressive change in width and stiffness causes each portion of the membrane to vibrate most strongly when stimulated by a particular frequency of sound: high notes near the oval window and low notes near the tip of the cochlea. The brain interprets signals originating in hair cells near the oval window as high-pitched sound; signals from hair cells located progressively closer to the tip of the cochlea are interpreted as progressively lower in pitch. Young people with undamaged cochleas can hear sounds from about 20 vibrations per second (very low bass) to about 20,000 vibrations per second (very, very high treble) and distinguish about 1,400 different pitches.

Complex processing of pitch and loudness by the brain enables us to comprehend language and appreciate music. If our surroundings become too noisy, however, it can be difficult to carry on a conversation or figure out what song is playing. Humans aren't the only animals to suffer from noise pollution, as we explore in "Earth Watch: Say Again? Ocean Noise Pollution Interferes with Whale Communication."

CHECK YOUR LEARNING

Can you . . .

- describe the parts of the human ear and explain how sound waves move through the ear?
- describe the structures and mechanisms by which sound waves are converted to electrical activity in the inner ear?
- explain how pitch and loudness are encoded?

CASE STUDY CONTINUED

Bionic Ears

A cochlear implant functions on the same principles by which a functioning inner ear perceives loudness and pitch. The implant contains 16 to 22 platinum electrodes threaded through the cochlea. Each electrode can be activated independently of the others. A sound-receiving unit, worn on the outside of the head, picks up sounds and sends tiny electrical currents to the appropriate electrodes in the cochlea. The currents stimulate action potentials in axons of the auditory nerve. The louder the sound, the stronger the current and the faster the auditory nerve axons fire.

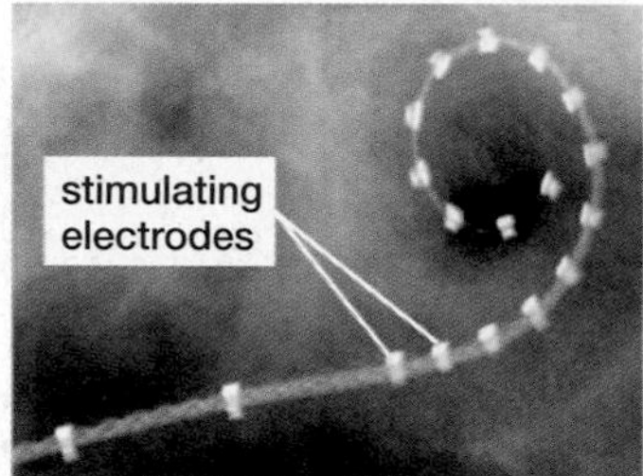

A cochlear implant

Which electrodes are activated depends on pitch: For low notes, electrodes near the tip of the cochlea are turned on. For progressively higher notes, currents are passed through electrodes progressively closer to the oval window. The electrical currents stimulate action potentials in roughly the same axons that would have been stimulated by hair cells in these locations in a functioning cochlea. As we describe in the Case Study Revisited, even a small array of electrodes provides the brain with useful information about sound.

40.5 HOW ARE GRAVITY AND MOVEMENT DETECTED?

The inner ear not only detects sound; it also contains structures, collectively called the **vestibular apparatus**, that detect gravity and the orientation and movement of the head. The vestibular apparatus is a fluid-filled tube embedded in the bones of the skull, consisting of the vestibule (a small chamber at the entrance to the apparatus) and the semicircular canals (**FIG. 40-5**).

The vestibule contains the **utricle** and **saccule**, which detect the direction of gravity and the orientation of the head. The utricle and saccule each contain a cluster of hair cells, with the hairs embedded in a gelatinous matrix containing tiny stones of calcium carbonate. The hairs in the utricle are vertical, whereas those in the saccule are horizontal. Gravity pulls the stones downward, causing the hairs to bend in various directions depending on the angle of the head. People can detect about a half-degree tilt.

Beyond the vestibule are three **semicircular canals**, which detect head movement. Each semicircular canal consists of a fluid-filled tube with a bulge at one end, called an ampulla.

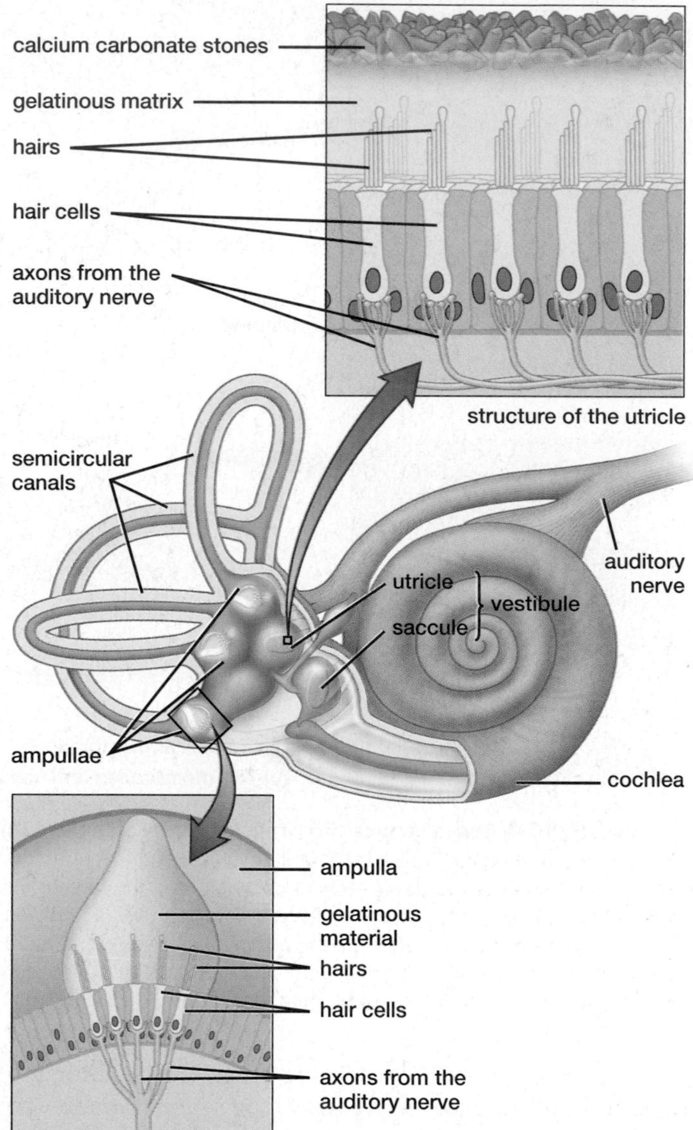

▲ **FIGURE 40-5 The vestibular apparatus detects gravity and the orientation and movement of the head** (Top) The hairs of hair cells in the utricle and saccule bend under the weight of calcium carbonate stones, providing information about the direction of gravity and the orientation of the head. (Bottom) The hairs of hair cells in the ampullae of the semicircular canals bend when head movement causes the fluid in the canals to slosh around.

THINK CRITICALLY The vestibular apparatus is located completely inside the head. If you close your eyes, how can you tell if the rest of your body is tilted with respect to gravity?

Say Again? Ocean Noise Pollution Interferes with Whale Communication

"Whatever man does or produces, noise seems to be an unavoidable by-product. Perhaps he can, as he now tends to believe, do anything. But he cannot do it quietly."

—Joseph Wood Krutch
Grand Canyon: Today and All Its Yesterdays (1957)

Have you ever had trouble following a conversation in a crowded restaurant because of all the background noise? That's what whales face in today's oceans.

Although whale sounds vary tremendously, they fall into two main categories: sounds used for echolocation and sounds used for communication. Most toothed whales, including orcas (killer whales), bottlenose dolphins, and sperm whales, locate prey by echolocation. Like many bats, they emit high-frequency clicks and squeaks, which are often far too high-pitched to be audible to to humans. The sound waves bounce off potential prey, and the whales use the returning echoes to determine the prey's distance, direction, and speed. Both toothed and baleen whales (such as blue, humpback, and fin whales) communicate with one another using lower-frequency sounds. Low-frequency sounds travel much farther in water than high-frequency sounds do. Some baleen whale species can communicate over vast distances—at least hundreds of miles—using very low-frequency sound. Whales probably communicate with one another to advertise a rich food source, to coordinate attacks on prey, or to find mates (**FIG. E40-1**).

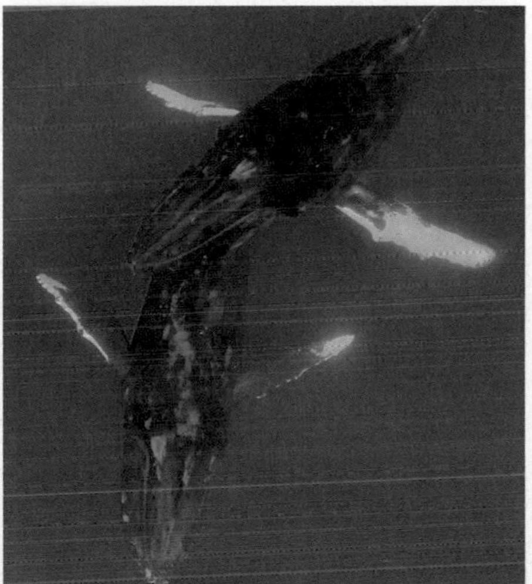

▲ **FIGURE E40-1 Mating humpback whales** If oceanic noise pollution continues to increase, disrupted whale communication might make it more difficult for whales to find mates.

Whale inner ears are quite similar to those of other mammals, but natural selection favoring adaptations for perceiving sound underwater has led to the evolution of very different structures in the outer and middle ears. Whales do not have external pinnas, and their auditory canals are usually plugged up with wax. Instead of traveling through the auditory canal, sound is transmitted through fat-filled regions of the lower jaw directly to a thin bone called the tympanic plate, which vibrates the middle ear bones that transfer the sound waves to the oval window and inner ear. Baleen whales, which typically communicate by producing extremely low-frequency sound, have very broad, thin basilar membranes, suitable for vibrating in response to, and therefore detecting, low-frequency sound. Toothed whales, which use extremely high-frequency sound to detect prey with echolocation, have thick basilar membranes, with extra stiffening provided by bony supports.

Unfortunately for whales, humans have greatly increased the ocean's background noise. The enormous propellers of large ships produce prolonged loud sounds with frequencies that are similar to the communication calls of baleen whales; this noise has probably increased by a factor of 15 to 20 since motors replaced sails on ocean-going ships. Researchers believe that the distance over which blue whales can hear their calls has decreased from 1,000 miles in the 1800s to only a couple of hundred miles today. The frequency of naval sonar overlaps extensively with that of the echolocation signals of toothed whales. When blue whales hear the sound produced by sonar, they stop feeding, stop calling, and swim away from the source of the sonar. Researchers hypothesize that blue whales become alarmed because they mistake sonar for the echolocation sounds of orcas, which prey on blue whales.

Blasts of extremely loud sound may directly damage whales. For example, the U.S. Navy is testing low-frequency active sonar, at incredibly loud levels, to detect distant submarines. There is some evidence that whales swimming near the source of the sonar signal suffer damage to their internal organs, including their ears. Whales that have been injured in this way are more likely to beach themselves as a result.

THINK CRITICALLY Whales aren't the only animals affected by human sounds. What animals might be impacted by noisy environments on land? What cascading effects might altered animal communication have on ecological communities?

Hair cells sit inside each ampulla, with their hairs embedded in a gelatinous capsule (but without the stones found in the utricle and saccule). Acceleration of the head—for example, if you lurch sideways in a roller coaster—pushes the fluid against the capsule, bending the hairs. The three semicircular canals are arranged perpendicularly to each other, similar to the two walls and the floor in the corner of a room, allowing you to detect head movement in any direction.

As in the cochlea, axons from the auditory nerve innervate the hair cells of the vestibular apparatus. When the hairs bend, the hair cells release neurotransmitter onto the endings of these axons, triggering action potentials that travel in the axons to balance and motion centers in the brain.

CHECK YOUR LEARNING

Can you . . .

- describe the structures that are used to detect gravity and the orientation and movement of the head?
- explain how a single type of sensory receptor, the hair cell, can respond to sound, gravity, or movement?

40.6 HOW IS LIGHT PERCEIVED?

The vast majority of animals can detect light. Some, such as flatworms and jellyfish, can distinguish light from dark, but their eyespots, as their light-sensing organs are called, cannot form an image. Most arthropods, some mollusks such as octopus and squid, and almost all vertebrates have eyes that form images of the world around them.

In all animals, vision begins with cells called **photoreceptors**. These cells contain **photopigments**, receptor molecules that change shape when they absorb light. This shape change sets off chemical reactions inside the photoreceptors that result in receptor potentials.

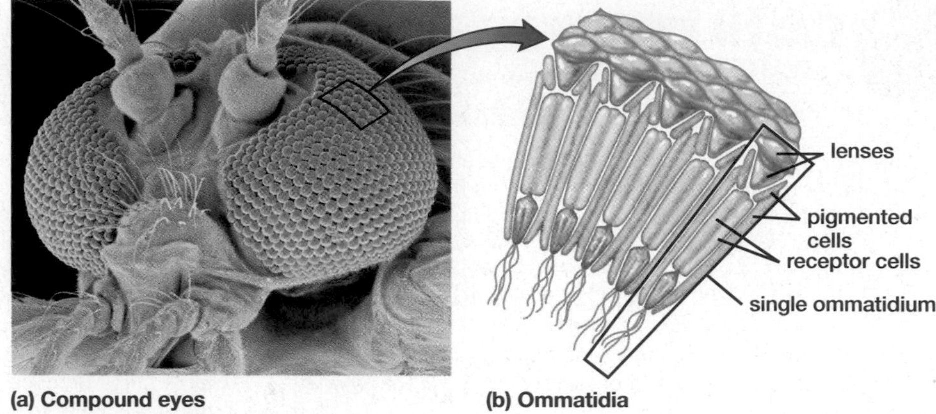

▲ **FIGURE 40-6 Compound eyes (a)** A scanning electron micrograph of the head of a mosquito, showing a compound eye on each side of the head. **(b)** Each eye is made up of numerous ommatidia. Within each ommatidium are several receptor cells, capped by a lens. Pigmented cells surrounding each ommatidium prevent the passage of light to adjacent receptors.

The Compound Eyes of Arthropods Produce a Pixelated Image

Many arthropods, including insects and crustaceans, have **compound eyes**, which consist of an array of light-sensitive subunits called **ommatidia** (singular, ommatidium; **FIG. 40-6**). Each ommatidium functions as an individual light detector, like the pixels in a digital camera. However, even the best arthropod eyes—those of dragonflies, for example—contain only about 30,000 ommatidia. A human eye, in contrast, contains more than 100 million photoreceptors. By human standards, therefore, the image formed by a compound eye is very pixelated, like an extremely low-resolution digital image. However, compound eyes are excellent at detecting movement, as light and shadow flicker across adjacent ommatidia, which is an advantage in avoiding predators and in hunting. Many insects can also see ultraviolet light; UV patterns in many flowers guide bees to sources of nectar and pollen (see Chapter 45).

The Mammalian Eye Collects and Focuses Light and Converts Light into Electrical Signals

A mammalian eye is structured somewhat like a camera (**FIG. 40-7**). The eye consists of two major modules: (1) accessory structures that hold the eye in a fairly fixed shape, control the amount of light that enters, and focus the light rays (comparable to the body and lens of a camera), and (2) the retina, which contains the photoreceptors (comparable to the image sensors in a digital camera).

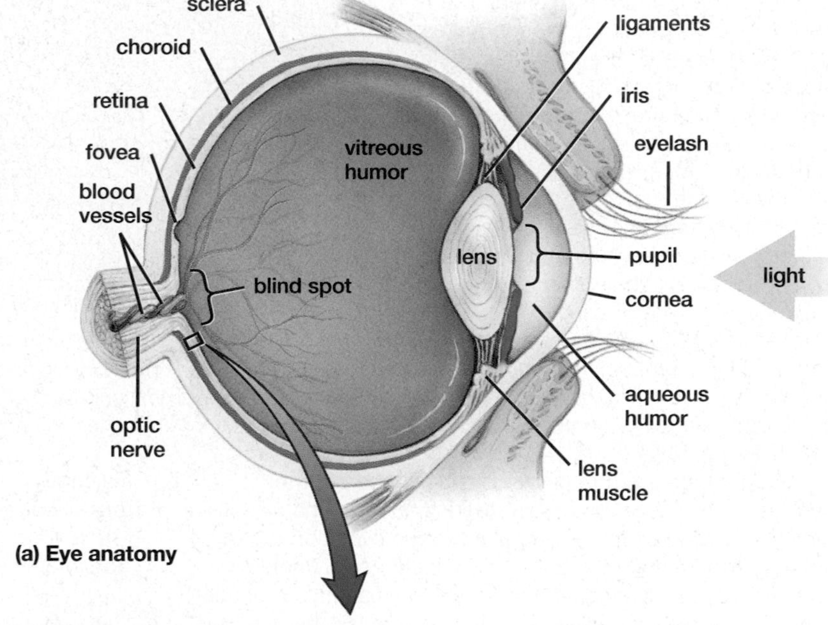

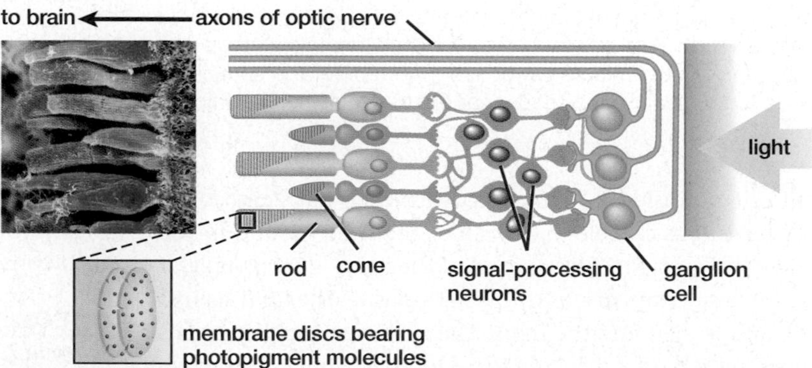

▲ **FIGURE 40-7 The human eye (a)** The anatomy of the human eye. **(b)** The retina contains photoreceptors (rods and cones), signal-processing neurons, and ganglion cells. In the scanning electron micrograph, rods are colored green and cones are colored blue.

The eyeball is surrounded by the **sclera**, a tough connective tissue layer that is visible as the white of the eye and is continuous with the transparent **cornea** at the front. Light enters the eye through the cornea. The light then travels through a chamber filled with a watery fluid called **aqueous humor**, which provides nourishment for the cells of both the lens and the cornea. The light continues through the **pupil**, a circular opening in the center of the colored **iris**. Contraction and relaxation of the muscles of the iris regulate the size of the pupil and, hence, the amount of light entering the rest of the eye. Light next encounters the **lens**, a structure composed of transparent proteins and shaped like a flattened sphere. The lens is suspended behind the pupil by a ring of smooth muscle. Behind the lens is a large chamber filled with **vitreous humor**, a clear jelly-like substance that helps maintain the shape of the eyeball.

After passing through the vitreous humor, light hits the **retina**. Here, light energy is converted into action potentials that are conducted to the brain. Behind the retina is the **choroid**. The choroid's rich blood supply helps nourish the cells of the retina. In people, the choroid is darkly pigmented. It absorbs stray light, preventing the light from bouncing around inside the eyeball and interfering with sharp vision. In nocturnal animals, the choroid is often reflective rather than dark. By reflecting light back through the retina, a mirror-like choroid gives the photoreceptors a second chance to capture scarce photons of light they may have missed the first time through. Although light reflected back and forth inside the eyeball degrades the image, at night, it's better to have blurry vision than no vision at all.

The Lens Focuses Light on the Retina

The lens focuses incoming light on a small area of the retina called the **fovea**. Although focusing begins at the cornea, whose rounded contour bends light rays, the lens is responsible for final, sharp focusing. The shape of the lens is adjusted by its encircling muscle. When viewed from the side, the lens is either rounded, to focus on nearby objects, or flattened, to focus on distant objects (**FIG. 40-8a**).

If your eyeball is too long or your cornea is too rounded, you will be **nearsighted**—light from distant objects will focus in front of the retina, so you will not see them clearly. If your eyeball is too short or your cornea is too flat, you will be **farsighted**—light from nearby objects will focus behind the retina. These conditions can be corrected by contact lenses or eyeglasses with lenses of the appropriate shape (**FIGS. 40-8b, c**). Both nearsightedness and farsightedness can also be corrected with laser surgery that reshapes the cornea. As people age, the lens stiffens. As a result, the lens can no longer round up enough to focus on nearby objects. By their mid-40s, most people require reading glasses for close work.

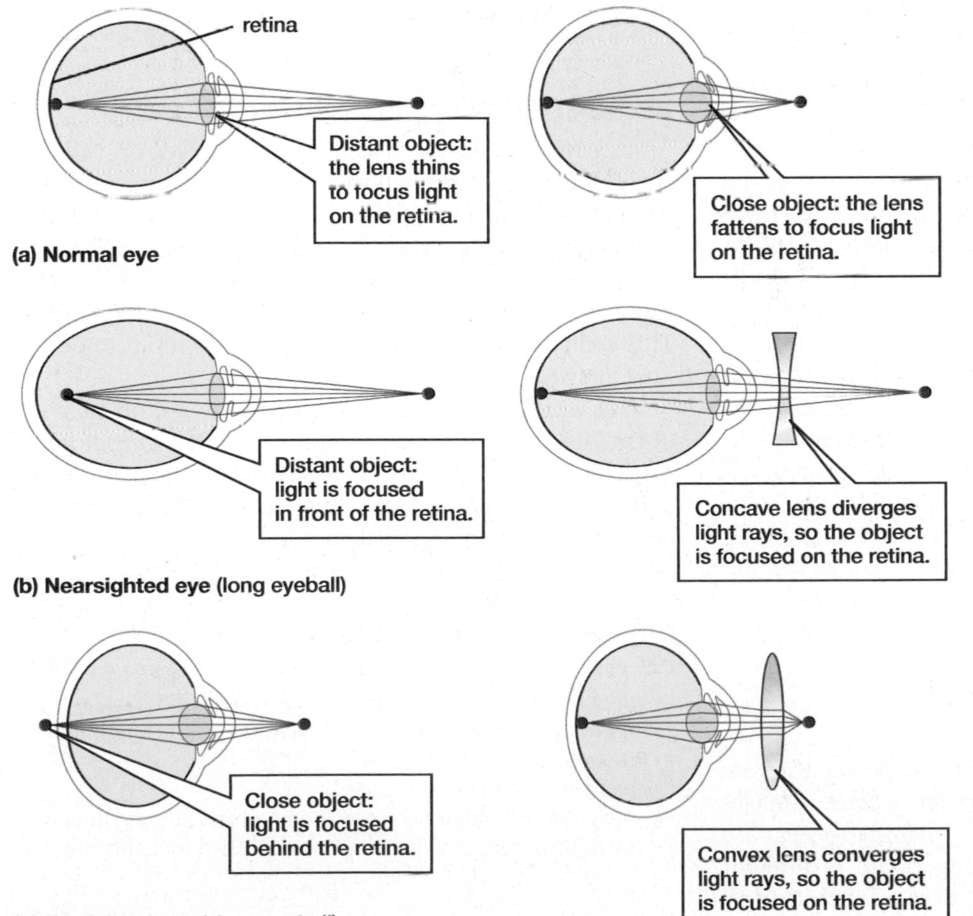
retina
Distant object: the lens thins to focus light on the retina.
Close object: the lens fattens to focus light on the retina.
(a) Normal eye
Distant object: light is focused in front of the retina.
Concave lens diverges light rays, so the object is focused on the retina.
(b) Nearsighted eye (long eyeball)
Close object: light is focused behind the retina.
Convex lens converges light rays, so the object is focused on the retina.
(c) Farsighted eye (short eyeball)

FIGURE 40-8 Focusing in the human eye (a) The lens changes shape to focus on objects at different distances. **(b)** Nearsightedness is corrected by eyeglasses with concave lenses. **(c)** Farsightedness is corrected by eyeglasses with convex lenses.

THINK CRITICALLY Assume that you are an ophthalmologist. One of your patients has been plagued by nearsightedness his whole life and is fed up with glasses and contacts. You suggest laser surgery on his corneas as a possible solution. The patient asks, "How will reshaping my corneas fix my nearsightedness? Aren't the lenses what focus light on the retina?" Respond to his question.

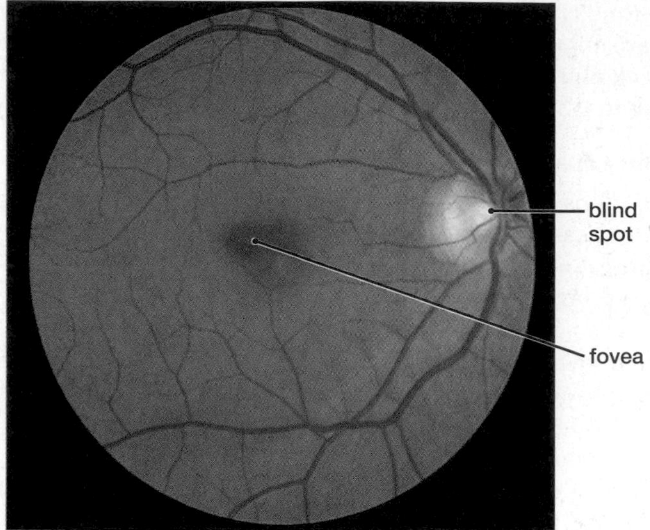

▲ **FIGURE 40-9 The human retina** A portion of the human retina, photographed through the cornea and lens of a living person. The blind spot and fovea are visible. Blood vessels supply oxygen and nutrients; note that the vessels are dense over the blind spot (where they won't interfere with vision) and scarce near the fovea.

THINK CRITICALLY Despite the presence of the blind spot, you do not consciously experience a "hole" in your vision. Why not?

The Retina Detects Light and Produces Action Potentials in the Optic Nerve

The photoreceptors, called **rods** and **cones** because of their shapes, are located at the rear of the retina (see Fig. 40-7b). Both rods and cones contain membranes packed with photopigment molecules. Photoreception begins when the photopigments absorb light, which triggers chemical reactions that produce receptor potentials in the photoreceptor cells.

Between the photoreceptors and the incoming light lie several layers of neurons that process signals from the photoreceptors. These neurons enhance our ability to detect edges, movement, and changes in light intensity. The cells nearest the vitreous humor (at the front of the retina) are the **ganglion cells**, whose axons make up the **optic nerve**. Ganglion cells convert signals from photoreceptors and the intervening neurons into action potentials. To reach the brain, ganglion cell axons pass through the retina at a location called the **blind spot** (**FIG. 40-9**). This area lacks photoreceptors, so images focused there cannot be seen.

Rods and Cones Differ in Distribution and Light Sensitivity

Although cones are located throughout the retina, they are concentrated in the fovea, where the lens focuses images most sharply (see Figs. 40-7 and 40-9). Human eyes have three varieties of cones, each containing a slightly different photopigment. Each type of photopigment is most strongly stimulated by a particular wavelength of light, corresponding roughly to red, green, or blue. The brain distinguishes color according to the relative intensity of stimulation of different cones. For example, the sensation of blue is produced by wavelengths of light that mostly stimulate blue cones; yellow is produced by roughly equal stimulation of red and green cones, with much less stimulation of blue cones. Most humans can distinguish tens of thousands of different colors. However, about 7% of men have difficulty distinguishing red from green, because they possess an allele on their single X chromosome that codes for a defective red or green photopigment (see Fig. 11-18a). Although often called "color-blind," these men are more accurately described as "color-deficient." True color blindness, in which a person perceives the world only in shades of gray, is extremely rare.

Rods are most abundant outside the fovea. Rods are longer than cones and thus contain more photopigment (see Fig. 40-7b), so they are more sensitive to light than cones are. Therefore, rods are largely responsible for vision in dim light. All rods contain identical photopigments, so rods do not provide color vision. In dim light, the world appears in shades of gray.

Not all vertebrates have both rods and cones. Those that are active almost entirely during the day (certain lizards, for example) may have all-cone retinas, whereas many nocturnal animals (such as rats and ferrets) and those dwelling in dimly lit habitats (such as deep-sea fishes) have retinas that contain mostly or only rods.

Binocular Vision Allows Depth Perception

Among mammals, the placement of the eyes on the head differs with the lifestyle of the animal. Predators and omnivores usually have both eyes facing forward (**FIG. 40-10a**), producing slightly different but extensively overlapping visual fields. This **binocular vision** allows depth perception, the accurate judgment of the distance of an object. This ability is obviously important to a cat about to pounce on a mouse.

In contrast, most herbivores have one eye on each side of the head (**FIG. 40-10b**), with little overlap in their visual fields. Some depth perception is sacrificed in favor of a nearly 360-degree field of view, allowing prey animals to spot a predator approaching from any direction.

CHECK YOUR LEARNING

Can you . . .

- describe the structures of the human eye and explain the pathway taken by light from outside the eye to the photoreceptors?
- explain how color is encoded?
- describe the differences between normal, nearsighted, and farsighted eyes and explain how defective focusing can be corrected by artificial lenses?

(a) Binocular vision

(b) Almost 360° vision

▲ **FIGURE 40-10 Eye position differs in predators and prey (a)** Most predatory mammals, such as this lynx, have eyes in front of the head; both eyes can be focused on a target, providing binocular vision. **(b)** Most prey animals, such as rabbits, have eyes at the sides, which allows them to scan for predators.

THINK CRITICALLY Why do some herbivorous or fruit-eating mammals, such as monkeys and fruit bats, have both eyes in front?

40.7 HOW ARE CHEMICALS SENSED?

Chemoreceptors respond to chemicals in the internal or external environments. Chemoreceptors in some large blood vessels and in the hypothalamus of the brain monitor levels of crucial molecules in the blood such as oxygen and glucose. Terrestrial vertebrates have two senses that respond to chemicals from outside the body: Olfaction, the sense of smell, allows animals to detect airborne molecules; gustation, the sense of taste, allows animals to detect chemicals dissolved in water or saliva in the mouth.

Olfactory Receptors Detect Airborne Chemicals

Receptor cells for olfaction are neurons located in a patch of mucus-covered tissue in the upper portion of the nasal cavity (**FIG. 40-11**). Olfactory receptor neurons bear long dendrites that protrude into the nasal cavity and are embedded in the mucus. Odorous molecules in the air diffuse through the mucus and bind to receptor proteins on the dendrites. Olfactory receptor neurons send axons to the olfactory bulb of the brain.

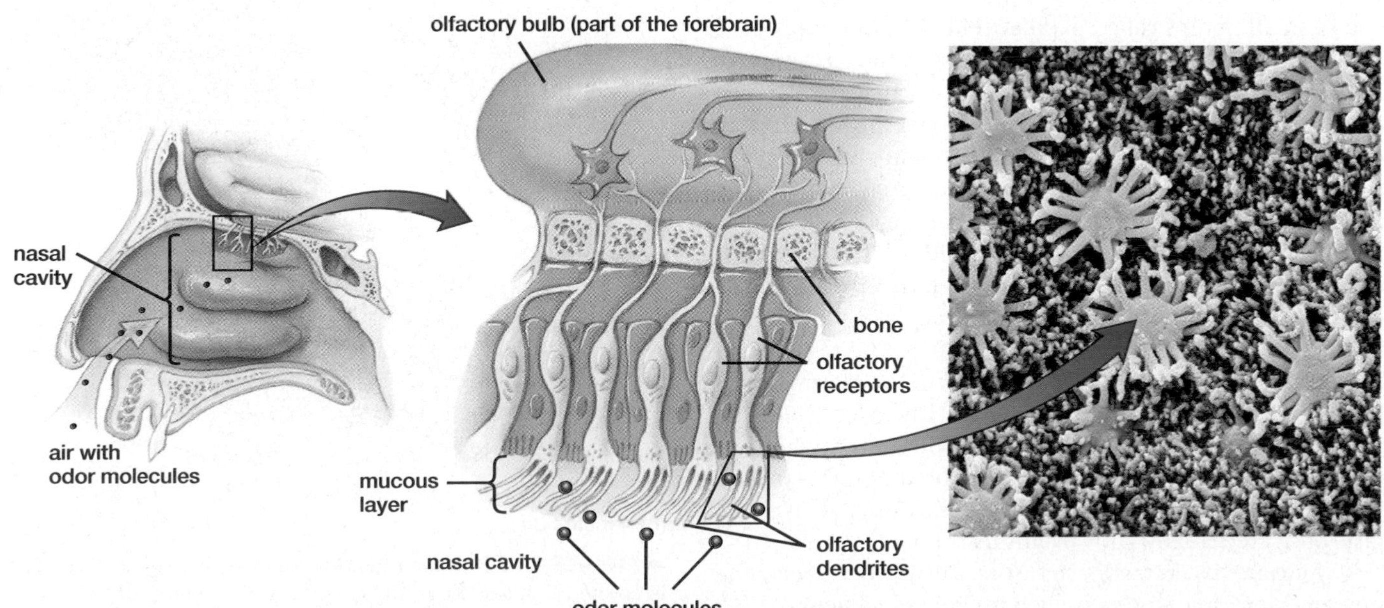

▲ **FIGURE 40-11 Olfaction** Human olfactory receptors are neurons bearing microscopic hair-like projections that protrude into the nasal cavity. The projections are embedded in a layer of mucus in which odor molecules dissolve before contacting the receptors.

Humans have about 5 to 10 million olfactory receptor neurons and about 400 different types of olfactory receptor proteins. Each olfactory neuron bears many copies of a single type of receptor protein. Each receptor protein is specialized to bind a particular type of odor molecule and cause the olfactory neuron to produce a receptor potential. If the receptor potential is large enough, it exceeds threshold, producing action potentials that travel along the neuron's axon to the olfactory bulb.

Most odors are complex mixtures of molecules that stimulate several different receptor proteins, so our perception of odors arises as the brain interprets signals from many different olfactory receptor neurons. A recent study found that people can probably distinguish more than a trillion odors, composed of mixtures of different types of odor molecules (for example, "an orange bouquet with a hint of menthol and just a touch of dead fish"). Interestingly, most people do not express the full range of receptor proteins, which explains why some people are quite insensitive to certain odors.

Many other animals detect odors better than we do, because they have more olfactory receptor neurons, more types of receptor proteins, or both. Dogs, for example, have a couple of hundred million olfactory neurons—possibly over a billion in some breeds, such as bloodhounds. They also have about 800 different types of olfactory receptor proteins. Dogs can detect certain odors, such as some found in human sweat, about 10,000 to 100,000 times better than people can.

Taste Receptors Detect Chemicals Dissolved in Liquids

The human tongue bears about 5,000 **taste buds**, embedded in small bumps called papillae (**FIG. 40-12a**). Taste buds are also found in the back of the mouth, in the airways, and in the intestines. On the tongue, each taste bud consists of a cluster of cells in a small pit in a papilla, opening into the oral cavity through a taste pore. A taste bud contains 50 to 150 cells of several types: supporting cells, stem cells, and taste receptor cells (**FIG. 40-12b**). Supporting cells regulate the composition of the interstitial fluid and help the receptor cells to function properly. The stem cells produce replacement receptor and support cells, following normal wear and tear or a close encounter with scalding-hot coffee. The taste receptor cells bear microvilli (thin projections of the plasma membrane) that protrude into the taste pore. Dissolved chemicals enter the pore and contact these microvilli. Although it was once thought that taste buds for specific tastes are concentrated on specific areas of the tongue, the different types of taste buds are actually fairly evenly distributed.

There are five known tastes: sour, salty, sweet, bitter, and umami (a Japanese word loosely translated as "delicious"). Recent research suggests that fatty acids may elicit a sixth taste sensation. Each taste receptor cell responds to only a single taste. Sour and salty sensations are caused by hydrogen ions or sodium ions, respectively, entering certain taste receptor cells through channels in the plasma membranes of their microvilli. Sweet, bitter, and umami sensations are caused by specific organic molecules binding to receptor proteins on the surface of the microvilli of other taste receptor cells. Umami receptor cells respond to the amino acid glutamate. High concentrations of glutamate found in foods such as meat, fish, and cheese stimulate the umami receptor, producing a sensation sometimes described as "savory." Monosodium glutamate (MSG) has long been used as a seasoning because it enhances the flavor of meat, fish, and vegetable dishes.

We perceive a great variety of tastes in two ways. First, a food may stimulate two or more receptor types, making the substance taste "sweet and sour," for example. Second, foods usually release molecules into the air inside the mouth. These odorous molecules bind to olfactory receptors, which contribute an odor component to the basic flavor.

To prove that what we call taste is often mostly olfaction, try holding your nose and closing your eyes while you eat different flavors of jelly beans. The flavors—from cherry to

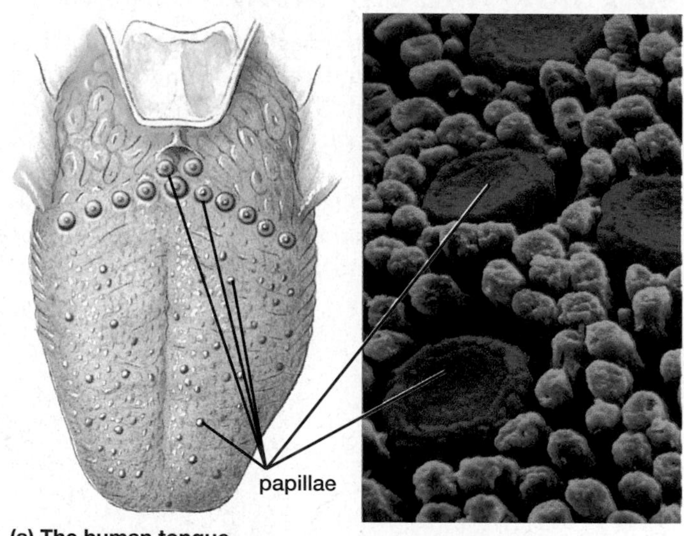

(a) The human tongue

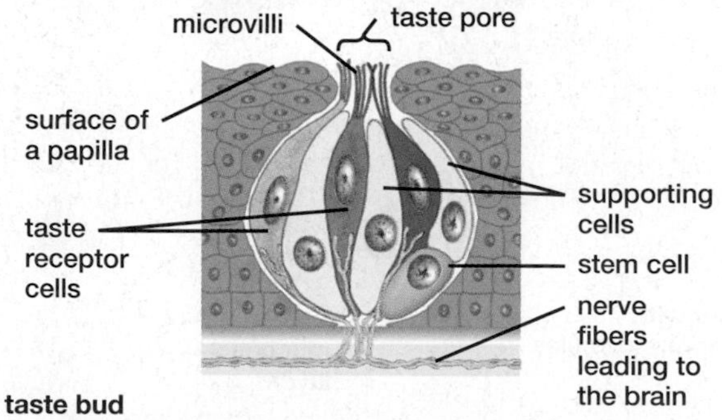

(b) A taste bud

▲ **FIGURE 40-12 Human taste receptors (a)** The human tongue is covered with papillae in which taste buds are embedded. **(b)** Each taste bud contains taste receptor cells of several types, supporting cells, and stem cells. Microvilli of taste receptor cells, bearing protein receptor molecules on their plasma membranes, protrude into the taste pore.

buttered popcorn—will be indistinguishably sweet. Likewise, when you have a bad cold, otherwise tasty foods seem bland.

CHECK YOUR LEARNING

Can you . . .

- define the term *chemoreceptor* and explain the difference between olfaction and taste?
- explain how different, complex odors and tastes are perceived?

40.8 HOW IS PAIN PERCEIVED?

If you are burned or cut, or spill acid on your skin, you will experience pain. Pain is a subjective feeling arising in the brain, produced by the stimulation of **pain receptors** (also called nociceptors), found in most parts of the body. Pain perception is crucially important to well-being and survival, teaching humans and other animals to avoid behaviors and objects that may damage the body. A few people are born without the ability to perceive pain. They are highly accident prone, suffering many bruises, cuts, and even broken bones.

There are just about as many types of pain receptors as there are ways to harm your body. For example, some pain receptors respond to high temperatures, with a typical threshold of about 109°F (43°C). Others respond to low temperatures, below about 59°F (15°C). These temperatures may not seem to be very hot or very cold, but remember, these are not air temperatures, but the temperature of the skin where the pain receptors are located: Cold-sensitive pain receptors reach a temperature as low as 59°F only if the external temperature is cold enough to overwhelm the warming effect of blood flowing just below the surface of the skin.

Other pain receptors respond to excessive stretching, such as might be caused by bloating in the intestine (gas pains). Many pain receptors are activated by chemical stimuli, such as acid or chemicals released during injury, or during the inflammatory response to injury. For example, if you are cut or bruised, damaged cells release their contents, including enzymes that convert certain blood proteins to bradykinin, a chemical that activates pain receptors. Chemicals released during the inflammatory response, including histamine and prostaglandins, increase the sensitivity of pain receptors, making otherwise mild discomfort much more painful. Certain individual pain receptors can respond to several different damaging stimuli, including excessive heat, acid, or certain chemicals.

CHECK YOUR LEARNING

Can you . . .

- name the types of stimuli that are perceived as painful?
- explain why it is useful to have pain receptors?

Have You Ever Wondered . . . Why Chili Peppers Taste Hot?

The "hot" ingredient in chili peppers is a chemical called capsaicin. The heat of chili peppers depends on how much capsaicin they contain, ranging from very mild pimentos to jalapenos (about 20 times hotter) to habaneros (50 times hotter than jalapenos) to the excruciatingly painful Carolina Reaper (7 to 15 times hotter than habaneros). Capsaicin activates receptor proteins on many pain receptor cells, including some in the mouth, that are also activated by high temperatures. Recall that what type of stimulus we perceive depends on which sensory cells fire action potentials. Because individual pain receptors respond to both damaging heat and capsaicin, the brain interprets both of these stimuli as burning pain.

CASE STUDY REVISITED

Bionic Ears

As amazing as cochlear implants are, they are very unsophisticated compared to a functioning biological cochlea. A cochlea contains about 3,500 hair cells that are primarily involved in detecting sound and stimulating axons in the auditory nerve. By comparison, a cochlear implant, with 16 to 22 electrodes more-or-less mimicking the hair cells, might seem hopelessly crude, but the brain can learn to do wondrous things with very little information. Most people with cochlear implants learn to understand spoken language quite well and can appreciate music.

There may always be limitations to what cochlear implants can do. For example, the physics of electricity dictates that the electrodes in the cochlea must be relatively far apart, which sets an upper limit on how many electrodes there can be, which in turn restricts how well a user can hear music and perhaps detect emotional overtones in speech. But in Samantha Downton's words, her cochlear implant ushered her "into a new world." Samantha is paying it forward at the Auditory-Verbal Center of Atlanta, a nonprofit organization dedicated to helping hearing-impaired people discover the world of sound.

THINK CRITICALLY Retinal implants have been developed to provide limited vision for people who have become blind due to retinitis pigmentosa or macular degeneration, in which the photoreceptors degenerate but many of the ganglion cells survive. Compare the structure of the retina and the cochlea. Based on what you know about cochlear implants, what would be the general guiding principles for designing a retinal implant?

CHAPTER REVIEW

Go to **Mastering Biology** to access the Pearson eText, vocabulary review, practice quizzes, activities, videos, current events, and more.

*Answers to **Think Critically and Thinking Through the Concepts** questions can be found in the **Answers** section at the back of the book.*

Summary of Key Concepts

40.1 How Do Animals Sense Their Environment?

Sensory receptors convert a stimulus from the internal or external environment to an electrical signal called a receptor potential. Sensory receptors are categorized according to the stimulus to which they respond. Many sensory receptors are contained within sense organs that help the receptors to respond to a specific stimulus. Either directly or indirectly, receptor potentials result in action potentials in specific axons that connect to the brain. The type of stimulus perceived is determined by which neurons in the brain are activated. The intensity of a stimulus is encoded by the frequency of action potentials reaching the brain.

40.2 How Is Temperature Sensed?

Thermoreceptors respond to either cold or warmth by increasing their rate of firing. Most thermoreceptors are free nerve endings in either the skin or brain.

40.3 How Are Mechanical Stimuli Detected?

Mechanoreceptors detect stimuli such as touch, vibration, pressure, stretch, or sound. Some mechanoreceptors, including those for touch and the sensation of itching, are free nerve endings. Other mechanoreceptors are surrounded by accessory structures that regulate which stimulus, such as pressure, sound, or gravity, is detected.

40.4 How Is Sound Detected?

In the vertebrate ear, air vibrates the tympanic membrane, which transmits vibrations to the bones of the middle ear and then to the oval window of the cochlea in the inner ear. Within the cochlea, vibrations bend hairs of mechanoreceptors called hair cells, producing receptor potentials that cause the hair cells to release neurotransmitters onto the endings of axons of the auditory nerve, triggering action potentials that travel to auditory centers in the brain. The pitch of sound is encoded by which hair cells are stimulated by a given frequency of sound vibrations. The loudness of sound is encoded by the frequency of action potentials in the axons of the auditory nerve.

40.5 How Are Gravity and Movement Detected?

The vestibular apparatus of the inner ear consists of the utricle and saccule in the vestibule, which detect the direction of gravity and the orientation of the head, and the semicircular canals, which detect movement of the head.

40.6 How Is Light Perceived?

In the vertebrate eye, light enters the cornea and passes through the pupil to the lens, which focuses an image on the retina. The retina contains two types of photoreceptors, rods and cones, which produce receptor potentials in response to light. These signals are processed through neurons in the retina and are translated into action potentials in the ganglion cells, whose axons form the optic nerve. Rods are more abundant and more light-sensitive than cones, providing black-and-white vision in dim light. Cones, which are concentrated in the fovea, provide color vision.

40.7 How Are Chemicals Sensed?

Terrestrial vertebrates detect chemicals in the external environment either by smell (olfaction) or by taste (gustation). Each olfactory or taste receptor cell type responds to only one or a few specific types of molecules, allowing discrimination among tastes and odors. The olfactory neurons of vertebrates are located in the nasal cavity. Taste receptors are primarily located in taste buds on the tongue.

40.8 How Is Pain Perceived?

Pain receptors respond to damaging stimuli such as cuts, burns, acid, or extremely high or low temperatures. Some pain receptors also respond to chemicals that are produced by the body during tissue damage.

Thinking Through the Concepts

Bloom's: Remembering, Understanding

Multiple Choice

1. Sensory receptors that respond to movement, gravity, or sound are
   a. chemoreceptors.
   b. mechanoreceptors.
   c. photoreceptors.
   d. thermoreceptors.

2. The brain's perception of the nature of a sensory stimulus is determined by
   a. the frequency of action potentials in a sensory receptor.
   b. the specific part of the brain that is activated.
   c. the size of the receptor potential in a sensory receptor.
   d. the size of action potentials traveling to the brain.

3. Sensory receptors that respond to oxygen, pH, and glucose are
   a. mechanoreceptors.
   b. photoreceptors.
   c. thermoreceptors.
   d. chemoreceptors.

4. A large receptor potential in a sensory receptor cell
   a. will usually be caused by a mild stimulus.
   b. will usually produce a low frequency of action potentials in axons leading from the sense organ to the brain.
   c. will usually stimulate a high frequency of action potentials in axons leading from the sense organ to the brain.
   d. is unrelated to the brain's perception of the intensity of a stimulus.

5. The cells that send axons from the mammalian eye to the brain are
   a. ommatidia.
   b. rods.
   c. cones.
   d. ganglion cells.

Fill-in-the-Blank

1. Sensory receptors respond to an appropriate stimulus with an electrical signal called a(n) ____________. Larger stimuli cause these signals to be larger than the signals from small stimuli. Ultimately, the intensity of a stimulus is conveyed to the brain encoded by the ____________ of action potentials in axons connected to specific sensory areas of the brain.
2. The mammalian ear consists of three parts: the outer, middle, and inner ear. The flap of skin-covered cartilage on the outside of the head is the ____________, which collects sound waves and funnels them down the auditory canal to a flexible membrane called the ____________. This connects to three small bones, the ____________, ____________, and ____________. The final bone vibrates the ____________, the beginning of the cochlea. Within the cochlea, the vibrations move the cilia of specialized mechanoreceptors called ____________.
3. Light enters the human eye through the ____________ and ____________ and then passes through the pupil, which is a hole in the ____________. Light then passes through the ____________ and ____________ before finally striking the retina. Both the ____________ and ____________ are involved in focusing light.
4. The retina of the human eye contains two types of photoreceptors, called ____________ and ____________. ____________ are larger, are more sensitive to light, and provide vision in dim light. ____________ are mostly located in the central region of the retina, called the ____________. These photoreceptors are smaller and less sensitive to light, but provide color vision.
5. In humans, the five principal kinds of taste sensations are ____________, ____________, ____________, ____________, and ____________. Molecules that leave the food and enter the air inside the mouth are detected by the sense of ____________, which plays a major role in the brain's perception of taste.

Review Questions

1. How do the senses encode the intensity of a stimulus? The type of stimulus?
2. What are the names of the specific receptors used for taste, vision, hearing, smell, and touch?
3. Why are we apparently able to distinguish hundreds of different flavors if we have only five types of taste receptors? How are we able to distinguish so many different odors?
4. Describe the structure and function of the various parts of the human ear by tracing the route of a sound wave from the air outside the ear to action potentials in the auditory nerve.
5. How does the structure of the inner ear allow for the perception of pitch? Of loudness?
6. Diagram the overall structure of the human eye. Label the cornea, iris, lens, sclera, retina, and choroid. Describe the function of each labeled structure.
7. How does the eye's lens change shape to allow focusing of distant objects? What defects make focusing on distant objects impossible, and what is this condition called? What type of lens can be used to correct it, and how does the lens do so?
8. Compare and contrast rods and cones.
9. List the types of stimuli that activate pain receptors. What properties do these stimuli have in common?

Applying the Concepts

Bloom's: Applying, Analyzing, Evaluating

1. We don't merely identify odors. We also label them good or bad, fragrant or disgusting. What do you think might be the evolutionary advantage of emotional responses to odors? Do you think that all animals have the same emotional responses to odors that humans do?
2. Many people like to eat spicy foods, but most other mammals don't. Historically, the general trend was that the hotter the climate, the spicier the foods a society traditionally favored. Using these facts as a starting point, what might be the selective advantage for plants that can manufacture spicy chemicals? Why might it be useful for humans to tolerate, and even enjoy, spicy food?

41 Action and Support: The Muscles and Skeleton

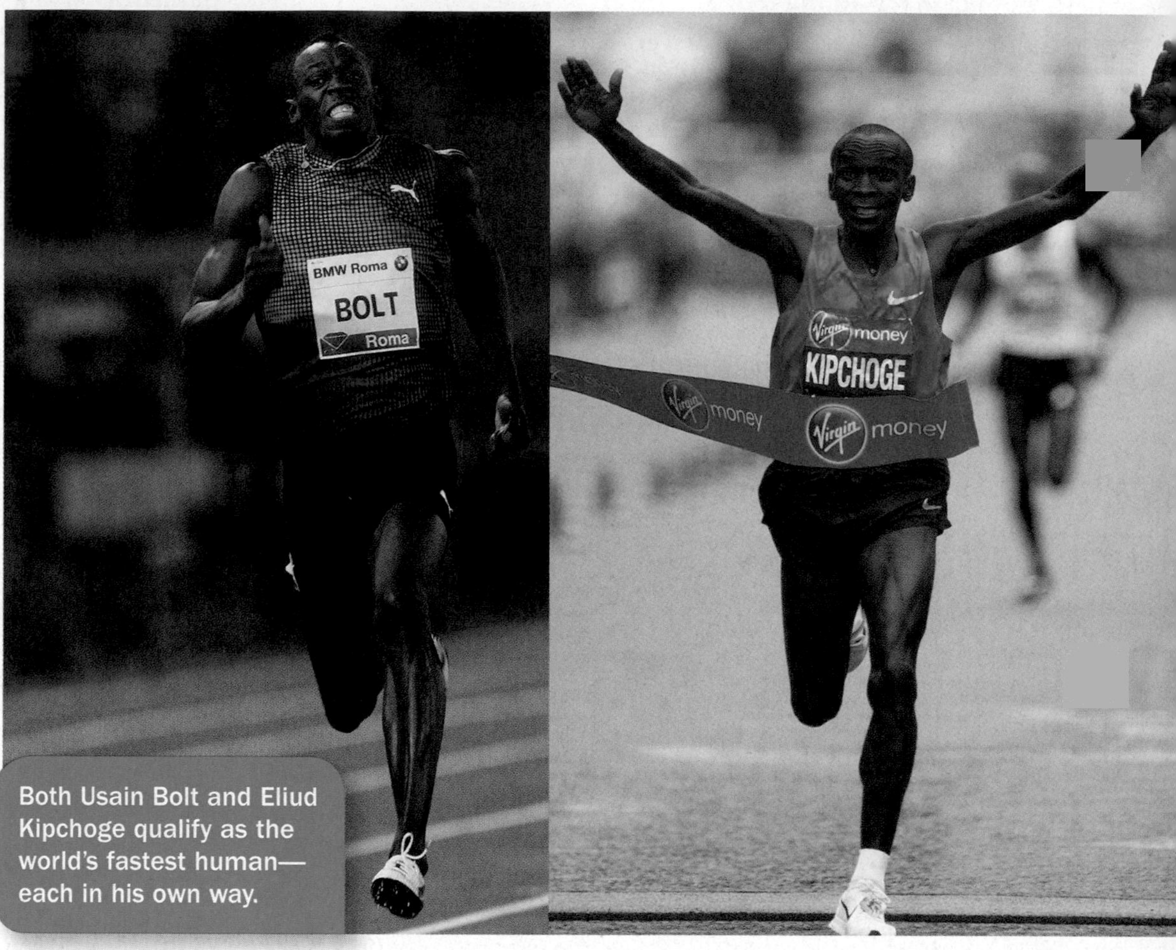

Both Usain Bolt and Eliud Kipchoge qualify as the world's fastest human—each in his own way.

CASE STUDY

Legs of Gold

AFTER USAIN "LIGHTNING" BOLT (left) blasted out of the starting blocks at the 100-meter dash in the 2009 World Championships in Berlin, his leg muscles propelled him to the finish line in a mere 9.58 seconds, a world record. His pace was a stunning 2.58 minutes per mile. By some measures Bolt is the world's fastest human, but fastest over what distance? Could he have completed a mile at this rate? No—in fact, no one has run a mile faster than 3.71 minutes.

In 2018, Kenya's Eliud Kipchoge set a new marathon world record: 2 hours, 1 minute, and 39 seconds. True, Kipchoge's average speed was "only" about 17.3 seconds per 100 meters (4.64-minute miles, or 12.9 mph), but he kept this pace up for 26.2 miles.

Could Bolt beat Kipchoge in a marathon? No way. And Kipchoge couldn't even qualify for the Olympic 100-meter dash. Why not? Compare their bodies. Bolt has bigger muscles than Kipchoge, but, as you will learn in this chapter, the muscles of world-class sprinters and marathoners differ at the cellular level as well. Bolt's larger leg muscles are essential for powering off the blocks and driving for 100 meters, but Kipchoge's are superior over the long distance of a marathon.

As you read this chapter, consider the muscles of Bolt and Kipchoge, and also your own muscles. With enough effort, could you become a world-class sprinter or marathoner? How do Bolt's and Kipchoge's muscles differ—not only in size, but on a cellular level as well? Are their skeletons also likely to differ? How do muscles and bones work together to support and move the body?

AT A GLANCE

41.1 HOW DO MUSCLES CONTRACT?

Almost all animals have **muscle**, a tissue composed of cells that are capable of contracting and thereby moving the parts of the body. Even animals that lack muscle tissue, such as sponges, have cells that can contract. What's more, these cells contract using the same types of proteins, interacting in basically the same way, as do human muscle cells. In this chapter, we focus on vertebrate muscles, but the basic principles of muscle function are similar across most of the animal kingdom; the ability to move and the fundamental cellular mechanisms that produce movement are extremely ancient.

The three types of vertebrate muscle—skeletal, cardiac, and smooth—differ somewhat in function, appearance, and control. We begin by describing the structure and function of skeletal muscle, and we describe cardiac and smooth muscle later in the chapter.

Vertebrate Skeletal Muscles Have Highly Organized, Repeating Structures

Skeletal muscle, so named because it moves the skeleton, is also known as *striated muscle* (meaning "striped") because its cells appear striped under a microscope. Most skeletal muscles are attached to the skeleton by tough, fibrous **tendons**, which consist of collagen strands bundled into increasingly large groups by connective tissue sheaths. Nearly all skeletal muscle is under voluntary, or conscious, control by the nervous system. Skeletal muscles can produce contractions ranging from quick twitches (blinking your eye) to powerful, sustained tension (carrying an armload of textbooks). An individual skeletal muscle consists of repeated components nested within one another (**FIG. 41-1**). Let's start at the outside of a muscle and work our way in.

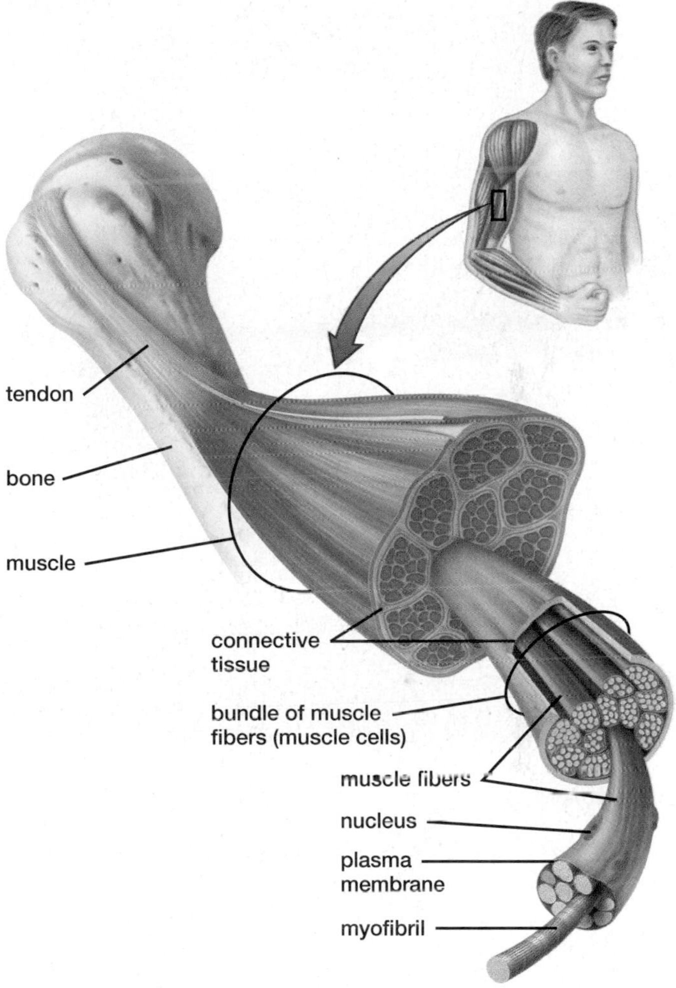

▲ **FIGURE 41-1 Skeletal muscle structure** A muscle is surrounded by connective tissue and is usually attached to bones by tendons. Muscle cells (fibers) are bundled within the muscle; each fiber is packed with myofibrils.

Skeletal muscles are encased in connective tissue sheaths, which merge into their attaching tendons. Inside a muscle's outer sheath, individual muscle cells, called **muscle fibers**, are grouped into bundles that are also encased by connective tissue. Blood vessels and nerves pass through the muscle in the spaces between the bundles. Each individual muscle fiber also has its own thin connective tissue wrapping. These multiple connective tissue coverings, each joined to the others, allow the muscle to contract as a unit and provide the strength that keeps the muscle from bursting apart during contraction.

Muscle fibers range from about 10 to 100 micrometers in diameter. Some extend the entire length of a muscle, which can be up to 2 feet (60 cm) in the longest muscle of the body (the sartorius, located in the leg). Each skeletal muscle fiber contains many nuclei, located just beneath the cell's plasma membrane. The largest fibers have several thousand nuclei to direct the synthesis of the enzymes and structural proteins that these very active cells require.

Individual muscle fibers contain many parallel cylindrical **myofibrils** (**FIG. 41-2**; also see Fig. 41-1). Each myofibril is surrounded by a specialized type of endoplasmic reticulum called the **sarcoplasmic reticulum (SR**; **FIG. 41-2a**), consisting of flattened, membrane-enclosed compartments filled with fluid. The fluid contains a high concentration of calcium ions (Ca^{2+}) that play a crucial role in muscle contraction (described later). The plasma membrane that surrounds each muscle fiber forms tiny tubes, called **T tubules**, that tunnel deep into the muscle fiber at regular intervals. T tubules encircle the myofibrils, running between, and closely attached to, segments of the SR (Fig. 41-2a).

Each myofibril, in turn, consists of repeating subunits called **sarcomeres**, aligned end to end along the length of the myofibril (**FIG. 41-2b**) and connected to one another by complexes of protein called **Z discs**. With its cylindrical sarcomere subunits, each myofibril looks a bit like thousands of miniature soup cans glued together end to end. Within each sarcomere lies a precise arrangement of thin and thick protein filaments (**FIG. 41-2c**). Each **thin filament** is anchored to a Z disc at one end; the opposite end remains free. Suspended between the thin filaments are **thick filaments**. The regular arrangement of thin and thick filaments within each myofibril gives a skeletal muscle fiber its striped appearance. The myofibrils are composed primarily of two proteins, **actin** in thin filaments and **myosin** in thick filaments, which interact with one another to contract the muscle fiber (**FIG. 41-2d**). Thin filaments are formed from two strings of roughly spherical actin proteins wound about each other like two pearl necklaces twisted together. Thin filaments also contain two smaller accessory proteins called *troponin* and *tropomyosin*, which regulate the interaction between thick and thin filaments.

Thick filaments are formed from bundles of myosin proteins. Each myosin protein is shaped somewhat like a hockey stick, with a head attached at an angle to a long shaft (Fig. 41-2d). The **myosin head** is hinged to the shaft and can swivel back and forth. Within each thick filament, myosin proteins are bundled together with their shafts in the middle of the bundle and their heads protruding out from opposite ends.

Muscle Fibers Contract Through Interactions Between Thin and Thick Filaments

The molecular architecture of thin and thick filaments allows them both to grip and to slide past one another. Thick and thin filaments can grip one another because each actin sphere has a site that can bind to a myosin head. In a resting muscle, these sites are covered by tropomyosin. When a neuron signals a muscle to contract, this causes the tropomyosin to move off the binding sites. **FIGURE 41-3** shows the sequence of events during a cycle of myosin head movements that cause contraction (the energy transformations that drive these events are described in the next section). When the

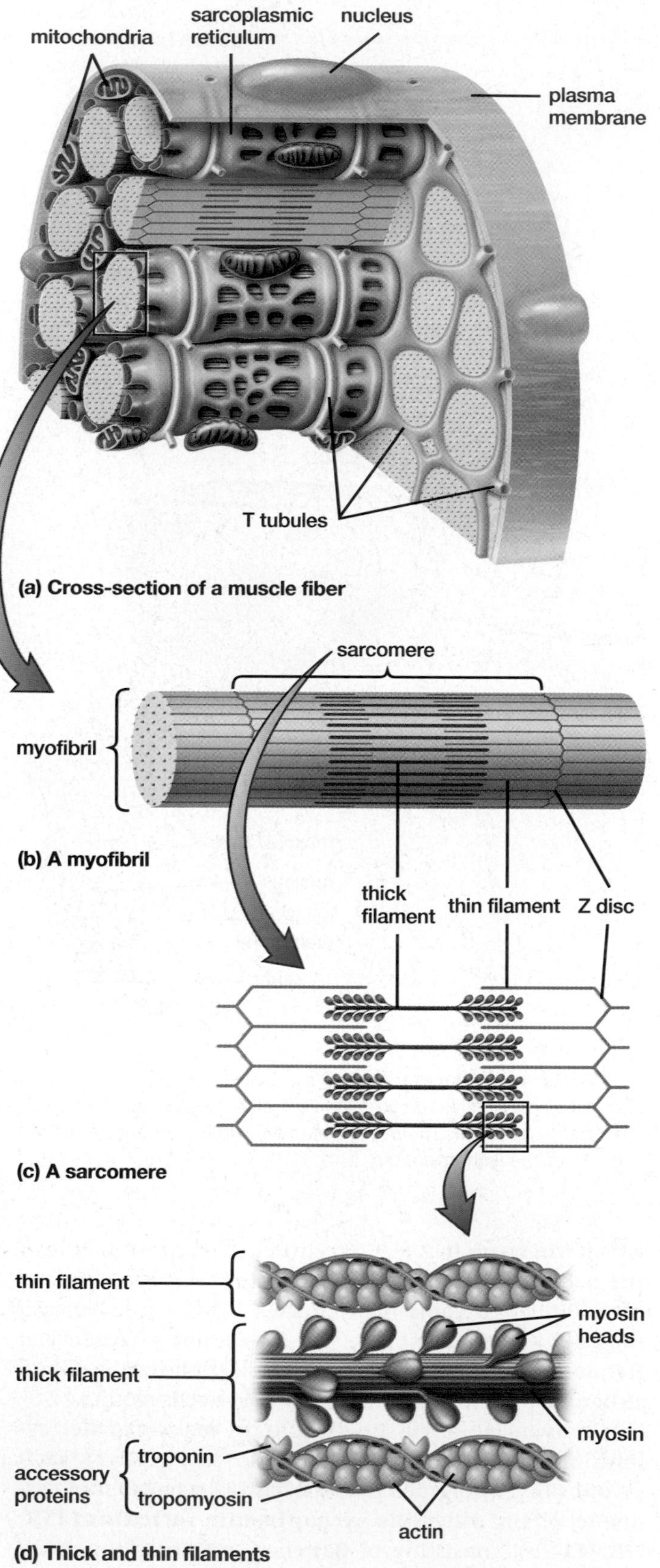

FIGURE 41-2 A skeletal muscle fiber (a) Each muscle fiber is surrounded by plasma membrane that extends into the fiber via T tubules. Sarcoplasmic reticulum surrounds each myofibril within the fiber. **(b)** Each myofibril consists of a series of sarcomeres, attached end to end by protein Z discs. **(c)** Within each sarcomere are alternating thin filaments, composed of actin, troponin, and tropomyosin, and thick filaments, composed of myosin. **(d)** Details of thick and thin filament structure are shown here.

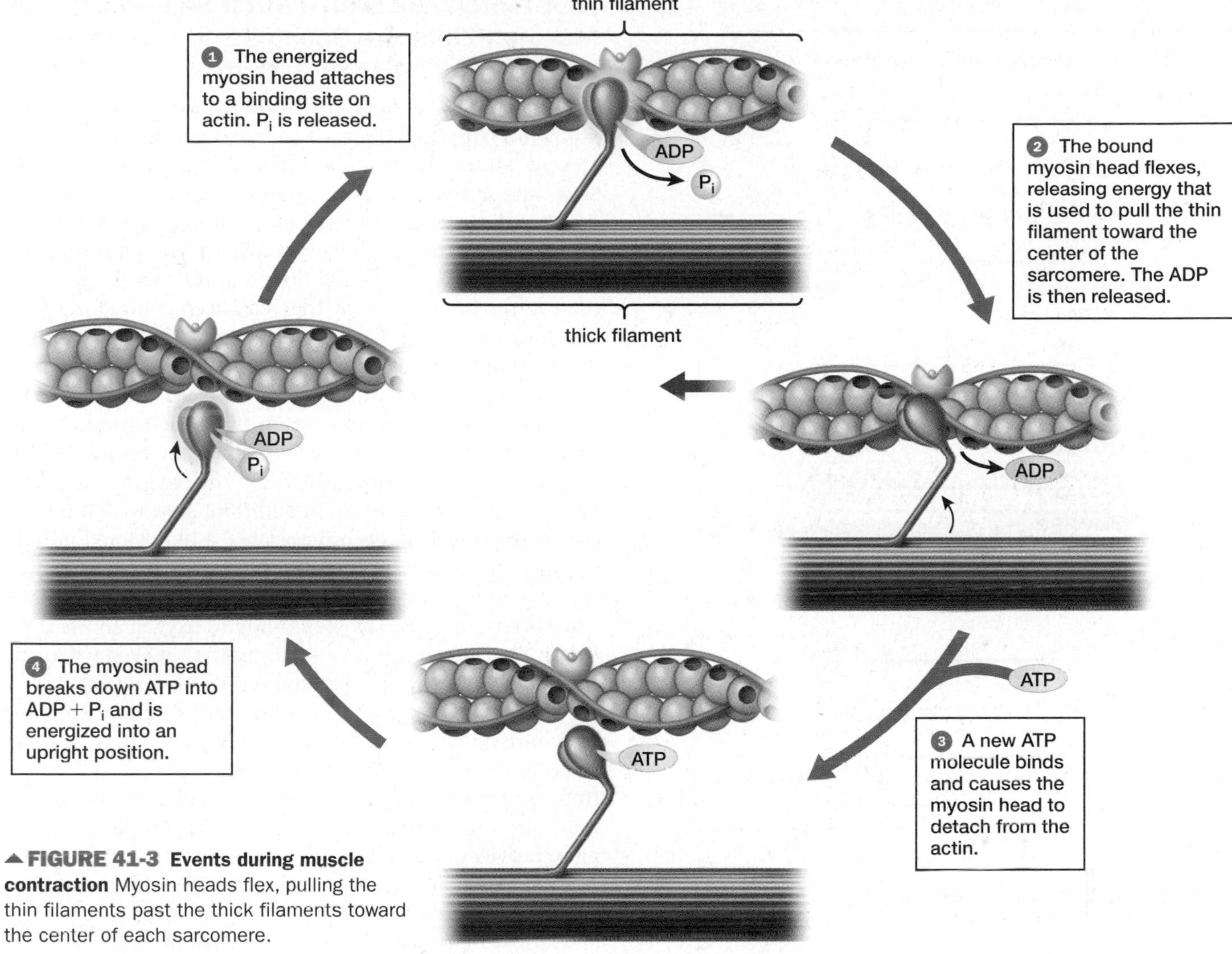
thin filament
1 The energized myosin head attaches to a binding site on actin. P_i is released.
ADP
P_i
2 The bound myosin head flexes, releasing energy that is used to pull the thin filament toward the center of the sarcomere. The ADP is then released.
thick filament
ADP
P_i
ADP
4 The myosin head breaks down ATP into ADP + P_i and is energized into an upright position.
ATP
ATP
3 A new ATP molecule binds and causes the myosin head to detach from the actin.

▲ **FIGURE 41-3 Events during muscle contraction** Myosin heads flex, pulling the thin filaments past the thick filaments toward the center of each sarcomere.

binding sites on the actin molecules are exposed, extended myosin heads immediately attach to them ❶, temporarily linking the thick and thin filaments. Binding stimulates the myosin heads to relax into a bent position ❷, pulling the thin filaments a short distance along the thick filament toward the middle of the sarcomere. The myosin heads then release the thin filament ❸, re-extend ❹, and reattach to binding sites farther along the thin filament in the direction of the Z disc to which the thin filament is attached. They continue to repeat the sequence, much like a sailor hauling in a long anchor line hand over hand, a little at a time, by grasping the line, pulling it in, releasing it, then grasping it further along. As the thin filaments slide past the thick filaments, the sarcomeres shorten, which shortens the myofibrils, causing the entire muscle fiber to contract. The cycle repeats as long as the binding sites on the actin are exposed, or until the muscle fiber is maximally contracted. The coordinated contraction of many muscle fibers causes the entire muscle to contract and generate movement.

This contraction process, called the *sliding filament mechanism* (**FIG. 41-4**), shortens all the sarcomeres along the muscle fiber simultaneously. Notice that in a fully contracted fiber the thick filaments of myosin run into the proteins of the Z discs. In contrast, a fully extended muscle has very little overlap between thick and thin filaments.

Muscle Contraction Uses ATP Energy

Figure 41-3 also shows how ATP provides the energy for muscle contraction. For clarity, we will start at position ❸ in the cycle, where ATP binds to a relaxed myosin head, causing it to release the actin filament. In position ❹, the myosin head catalyzes the breakdown of ATP into ADP + P_i (which remain bound to the myosin) and uses the energy released to cock itself into an upright position (imagine this as stretching the rubber band of a slingshot). In position ❶, this energized myosin head attaches to a binding site on actin and releases P_i. Continuing around to position ❷, in its "power stroke," the myosin head expends its stored energy by pulling the actin filament toward the center of the sarcomere (like releasing the band of the slingshot to shoot its pebble). The relaxed myosin head

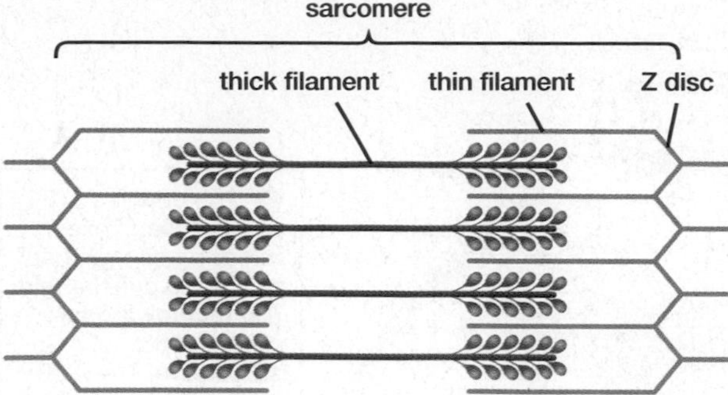

(a) Relaxed muscle

(b) Fully contracted muscle

(c) Fully extended muscle

▲ **FIGURE 41-4 The sliding filament mechanism of muscle contraction**

THINK CRITICALLY Why is it so difficult to hold a heavy weight in fully extended arms; in other words, why does a highly stretched muscle generate very little force?

releases its ADP and is ready to bind another ATP, which will cause it to detach from the binding site and begin a new cycle by attaching to a binding site farther along the actin toward the nearest Z disc. This cycle must repeat many times to cause a noticeable muscle contraction. The cycling involves all the myosin heads in the stimulated muscle cell, and continues as long as the binding sites on actin are exposed or until the muscle is fully contracted (see Fig. 41-4b).

A skeletal muscle's reserves of ATP are used up after only a few seconds of high-intensity exercise. For brief, high-intensity exertion, muscle cells can generate a little more ATP using glycolysis, a process that does not require oxygen, but is also not very efficient (see Chapter 8). For prolonged and/or low-intensity exercise, muscle cells produce ATP using cellular respiration. This requires a continuous supply of oxygen to be delivered to the muscle cells by the capillaries of the cardiovascular system.

Fast-Twitch and Slow-Twitch Skeletal Muscle Fibers Are Specialized for Different Types of Activity

Skeletal muscle fibers come in two basic types, *slow-twitch* and *fast-twitch* (although there are subtypes within each of these groups). Most muscles contain some of each type (**FIG. 41-5**). Slow- and fast-twitch fibers have different forms of myosin, which pull the thin filaments along either relatively slowly (slow twitch) or more rapidly (fast twitch). There are other differences as well. Slow-twitch fibers contract with less power than fast-twitch fibers, but they can keep contracting for a very long time. How? Slow-twitch muscle fibers resist fatigue due to features that allow cellular respiration to provide them with an abundant, continuous source of ATP to power contraction. For example, slow-twitch fibers are surrounded by capillaries that deliver oxygen (O_2), and they contain a large number of mitochondria, which use the oxygen to produce ATP by cellular respiration. In addition, slow-twitch muscle cells contain high levels of *myoglobin*, a red-colored protein similar to hemoglobin. Myoglobin stores O_2, binding it when the O_2 concentration is high and releasing it when the O_2 concentration drops. Myoglobin-bound oxygen allows slow-twitch fibers to continue cellular respiration even when the amount of O_2 in the blood is temporarily inadequate.

Fast-twitch fibers, on the other hand, contain far more myofibrils, giving them a larger diameter and allowing them to contract with greater force than do slow-twitch fibers. However, fast-twitch fibers have a smaller blood supply, much less myoglobin, and fewer mitochondria. They rely far more heavily on glycolysis to generate ATP. Glycolysis generates ATP much faster than does cellular respiration, but is far less efficient, so fast-twitch fibers deplete their energy stores and fatigue more rapidly than slow-twitch fibers do.

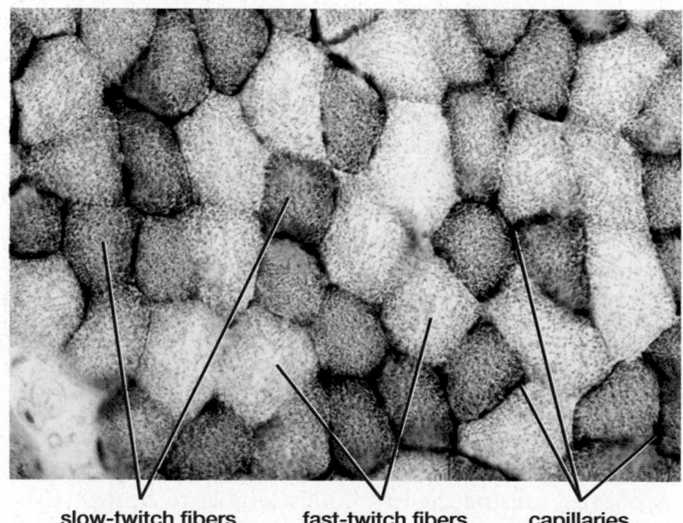

▲ **FIGURE 41-5 Muscle cross-section** Mitochondria are stained blue and capillaries black.

THINK CRITICALLY Why are the capillaries clustered around the fibers with the most mitochondria?

CASE STUDY CONTINUED

Legs of Gold

The legs of champion sprinters like Usain Bolt have about 80% fast-twitch fibers capable of the rapid, explosive contractions that are so essential to blasting off the starting blocks. The legs of world-class marathoners like Eliud Kipchoge, on the other hand, have about 80% slow-twitch fibers, which are less explosive but can rapidly contract again and again, each leg stepping about 11,000 times to complete a marathon. Both athletes probably have roughly the same number of muscle fibers in their legs, but Bolt's muscles are larger because he has mostly thick, fast-twitch fibers, while Kipchoge's muscles have a predominance of thin, slow-twitch fibers. Are there likely to be differences in Bolt's and Kipchoge's tendons and bones as well? You'll find out in Section 41.3.

The Nervous System Controls the Contraction of Skeletal Muscles

Skeletal muscle contraction is controlled by the nervous system and is mostly voluntary (with the exception of certain reflexes such as the knee-jerk reflex). How does the nervous system initiate skeletal muscle contraction?

Motor Neurons Excite Skeletal Muscle Fibers at Neuromuscular Junctions

Skeletal muscle fibers are activated by *motor neurons*, which (like motors) cause movement. The cell bodies of most motor neurons are in the spinal cord; their axons exit the cord in spinal nerves and contact muscle fibers at specialized synapses called **neuromuscular junctions** (**FIG. 41-6**). At a neuromuscular junction, an action potential (AP) in the motor neuron releases the neurotransmitter *acetylcholine* onto the muscle fiber. The acetylcholine produces a huge excitatory postsynaptic potential in the muscle fiber, causing the muscle fiber to generate an AP, much like a neuron does ❶.

Recall that the plasma membrane of a muscle fiber sends T tubules into the fiber alongside the sarcoplasmic reticulum surrounding each myofibril. The AP in the muscle travels down the T tubules to the SR ❷, where it causes Ca^{2+} to be released from the SR into the cytosol surrounding the thin and thick filaments of the myofibril ❸. The presence of Ca^{2+} is what allows contraction. How? The Ca^{2+}

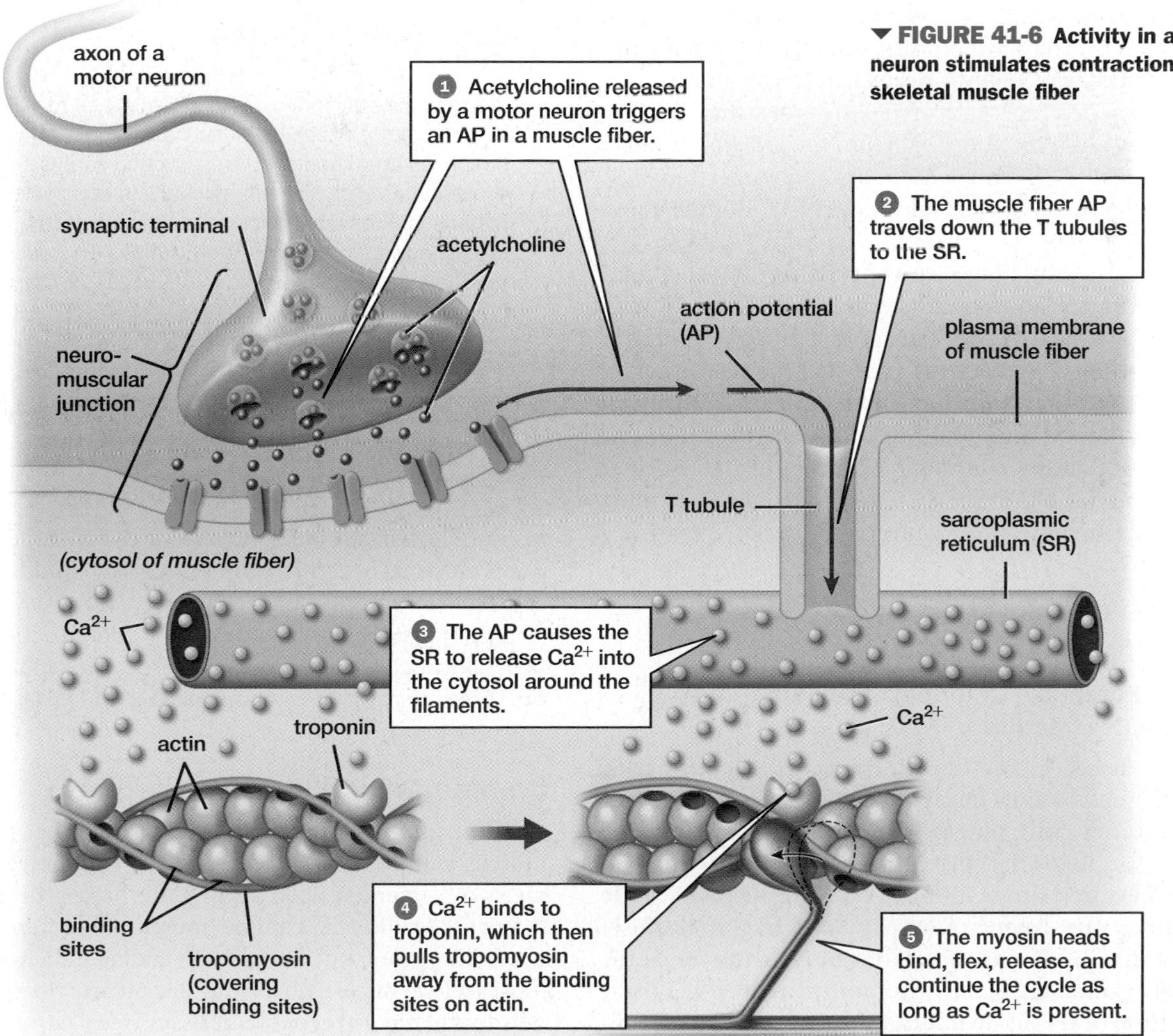
axon of a motor neuron
❶ Acetylcholine released by a motor neuron triggers an AP in a muscle fiber.
synaptic terminal
acetylcholine
neuro-muscular junction
action potential (AP)
❷ The muscle fiber AP travels down the T tubules to the SR.
plasma membrane of muscle fiber
T tubule
sarcoplasmic reticulum (SR)
(cytosol of muscle fiber)
Ca2+
❸ The AP causes the SR to release Ca2+ into the cytosol around the filaments.
Ca2+
troponin
actin
binding sites
tropomyosin (covering binding sites)
❹ Ca2+ binds to troponin, which then pulls tropomyosin away from the binding sites on actin.
❺ The myosin heads bind, flex, release, and continue the cycle as long as Ca2+ is present.

▼ **FIGURE 41-6 Activity in a motor neuron stimulates contraction of a skeletal muscle fiber**

Have You Ever Wondered ...

How White and Dark Meat Differ?

White meat or dark? It seems almost everyone has a preference. Dark leg meat has a richer flavor and more fat that makes it moister. But some people prefer the milder, leaner white meat of the breast. The differences between dark and white meat are due to their predominant type of muscle fiber. Wild chickens and turkeys are ground-dwelling and usually fly only to escape a fox or coyote or to reach a safe perch for the night. You can probably predict that their breast meat is packed with fast-twitch muscle fibers—good for a brief frantic flutter (as well as our eating pleasure). Given the opportunity, these birds will spend much of the day walking about hunting for food, and their leg muscles are adapted for this prolonged, low-intensity effort with primarily slow-twitch fibers. Blood-carrying capillaries surrounding slow-twitch fibers and oxygen-storing myoglobin within them provide more iron, a darker color, and more intense flavor to the legs of these birds. But what about migratory ducks and geese that fly long distances? As you might predict, their breast muscles consist of dark meat.

binds to troponin, causing it to change shape and pull the tropomyosin off the actin binding sites ❹. With tropomyosin out of the way, myosin heads can bind to actin. This triggers the repeating cycle of flexing, releasing, and extending that pulls the thin filaments toward the center of each sarcomere ❺. A single AP in a muscle fiber causes all of its sarcomeres to shorten simultaneously, slightly shortening the fiber.

What makes the fiber stop contracting? When the AP ends (in just a few thousandths of a second), the SR stops releasing Ca^{2+}, and active transport proteins in the SR membrane pump Ca^{2+} back into the SR. This causes Ca^{2+} to diffuse off the troponin, which then reverts to its resting shape and allows tropomyosin to slide back over the actin binding sites, halting contraction. Pumping Ca^{2+} released by a single AP back into the SR takes long enough to allow hundreds of myosin head movement cycles, creating a visible twitch.

The Size of Motor Units and the Frequency of Action Potentials Determine the Force of Muscle Contraction

The force, distance, and duration of muscle contraction are determined by how many fibers in a muscle contract, how much they contract, and for how long.

A motor neuron typically synapses with multiple muscle fibers in a single muscle, forming a **motor unit**. Motor units vary tremendously in size. In muscles used for fine control, such as those that move the eyes or fingers, motor units are small; the motor neuron may synapse on just a few muscle fibers, allowing delicate, precise movements. In muscles used for large-scale movements, such as those of the thigh and buttocks, motor units are large; the neuron may synapse on dozens or even hundreds of muscle fibers, allowing powerful—but less precise—movements (imagine trying to write by moving your thigh).

The nervous system controls the force of contraction of a given muscle by varying both the number of motor units (and thus the number of muscle fibers) stimulated and the rate of AP production in each fiber. Because each motor neuron synapses on multiple muscle fibers in a given muscle, and because the muscle fibers are attached to one another and to the muscle's tendons, a single AP in a single motor neuron will cause a small twitch of the entire muscle. If the motor neuron fires multiple APs in rapid succession, the twitches from each AP overlap to produce a larger contraction. Several motor neurons, each stimulating multiple fibers in the same muscle, will produce a still greater and more powerful shortening of the muscle. Rapid firing of all of the motor neurons that innervate all of the fibers in the muscle will cause a maximal and forceful contraction. The contraction will be sustained as long as the motor neurons continue to produce APs.

CHECK YOUR LEARNING

Can you . . .

- describe the structure of vertebrate skeletal muscle, from a whole muscle through individual muscle cells and subcellular structures down to the proteins involved in muscle contraction?
- explain the sliding filament mechanism of muscle contraction?
- distinguish between fast-twitch and slow-twitch muscles?
- explain how the nervous system causes contraction of skeletal muscles and how it controls the strength of contraction?

41.2 HOW DO CARDIAC AND SMOOTH MUSCLES DIFFER FROM SKELETAL MUSCLE?

All muscle cells are built on the same fundamental principles: Filaments of actin and myosin attach and slide past one another, causing the cells to contract. However, cardiac and smooth muscles differ structurally and functionally from skeletal muscles and from each other. The characteristics of the three muscle types are summarized in **TABLE 41-1**.

Cardiac Muscle Powers the Heart

Cardiac muscle is found only in the heart. Like skeletal muscle, cardiac muscle is striated due to its regular arrangement of sarcomeres containing alternating thick and thin filaments. However, cardiac muscle fibers are branched, contain only one nucleus per cell, and are much shorter than most skeletal muscle fibers. Cardiac muscle fibers connect to one another by **intercalated discs** (see Chapter 33), which allow the action potentials that stimulate contraction to spread rapidly among the fibers. Intercalated discs also bind cardiac muscle fibers tightly to one another, preventing the force of contraction from pulling the heart's fibers apart.

Because cardiac muscles must contract roughly 70 times each minute for your whole life, each cell requires a large, uninterrupted supply of O_2 for producing ATP. To support their demand for oxygen, cardiac muscle fibers are surrounded by an extensive network of blood vessels and contain oxygen-storing myoglobin. They also house enormous numbers of mitochondria; these ATP-producing organelles make up roughly 35% of the mass of cardiac muscle.

Unlike skeletal muscle fibers, cardiac muscle contractions are involuntary. Cardiac muscle fibers contract spontaneously, without input from the nervous system, although both the rate and force of contraction are influenced by the nervous system and by hormones (see Chapter 33). The ability to contract spontaneously is particularly well developed in the specialized cardiac muscle fibers clustered in the heart's natural pacemaker. Action potentials from the pacemaker spread rapidly through the intercalated discs that interconnect cardiac muscle fibers, coordinating the contraction of the heart's chambers.

Smooth Muscle Produces Slow, Involuntary Contractions

Smooth muscle surrounds blood vessels and most hollow organs, including the uterus, bladder, and digestive tract. The cells of smooth muscle are not striated because their thin and thick filaments are scattered (see Table 41-1). Like cardiac muscle fibers, smooth muscle fibers each contain a single nucleus and communicate with adjacent smooth muscle fibers, causing the cells to contract in synchrony. These contractions may be slow and sustained, such as the constriction of arteries that elevates blood pressure during times of stress (see Chapter 33) or slow and wavelike, such as the waves of contraction that move food through the digestive tract (see Chapter 35). Contraction of smooth muscles can be stimulated

TABLE 41-1 Properties of the Three Muscle Types

	Type of Muscle		
Property	**Skeletal**	**Cardiac**	**Smooth**
Muscle appearance	Striated	Striated	Non-striated
Cell shape	Tapered at both ends	Branched	Tapered at both ends
Number of nuclei	Many per cell	One per cell	One per cell
Speed of contraction	Slow to rapid	Intermediate	Slow
Contraction stimulus	Nervous system	Spontaneous	Nervous system, hormones, spontaneous, stretch
Function	Moves the skeleton	Pumps blood	Controls movement of substances through tubes and hollow organs
Under voluntary control?	Yes	No	No
Art (left) and micrograph (right)	nuclei	nucleus; intercalated discs	nucleus

by stretch, hormones, signals from the autonomic nervous system, or combinations of these stimuli. Smooth muscle contraction is generally not under voluntary control.

CHECK YOUR LEARNING

Can you . . .

- describe the similarities and differences between cardiac and smooth muscle, including their appearance, where they are found, what stimulates them to contract, and what kinds of movements they usually cause?
- compare cardiac and smooth muscle to skeletal muscle?

41.3 HOW DO MUSCLES AND SKELETONS WORK TOGETHER TO PROVIDE MOVEMENT?

Despite enormous differences in body form and structure, nearly every animal—whether earthworm, crab, horse, or human—moves using the same fundamental mechanism: Contracting muscles exert forces on a structure that supports the body, called a **skeleton**, causing the body to change shape.

The Actions of Antagonistic Muscles on Skeletons Move Animal Bodies

Coordinated movement of an animal's body is produced by alternating contractions of **antagonistic muscles**—pairs of muscles with opposing actions that compress or pull on its skeleton. Across the animal kingdom, there are three different types of skeleton: hydrostatic skeletons, exoskeletons, and endoskeletons.

Worms, cnidarians (sea jellies, anemones, and their relatives), and many mollusks (slugs, octopuses, and their relatives) have a **hydrostatic skeleton**, which is basically a sac or tube filled with liquid. "Hydrostatic" means "to stand with water," something like a water-filled balloon. A water balloon's volume is constant because water cannot be compressed, but its shape can be changed by squeezing it. Similarly, muscular squeezing changes the shape of animals with hydrostatic skeletons.

To squeeze its hydrostatic skeleton, the animal uses two sets of antagonistic muscles in its surrounding body wall (**FIG. 41-7a**). One muscle set is circular; contracting these

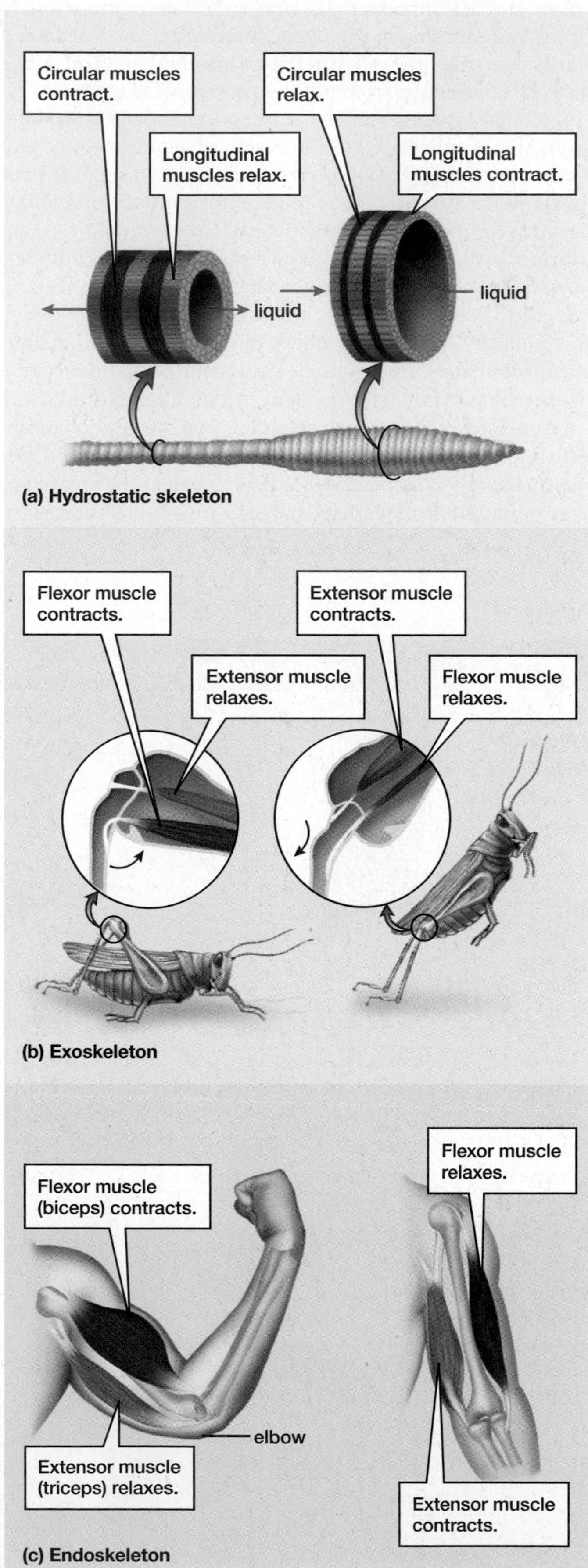

Circular muscles contract.
Circular muscles relax.
Longitudinal muscles relax.
Longitudinal muscles contract.
liquid
liquid
(a) Hydrostatic skeleton
Flexor muscle contracts.
Extensor muscle contracts.
Extensor muscle relaxes.
Flexor muscle relaxes.
(b) Exoskeleton
Flexor muscle (biceps) contracts.
Flexor muscle relaxes.
elbow
Extensor muscle (triceps) relaxes.
Extensor muscle contracts.
(c) Endoskeleton

▶ FIGURE 41-7 Antagonistic muscles move hydrostatic skeletons, exoskeletons, and endoskeletons (a) A hydrostatic skeleton is essentially a liquid-filled tube enclosed within walls containing antagonistic circular and longitudinal muscles. Circular muscle contraction makes the tube long and thin (left); longitudinal muscle contraction makes the tube short and thick (right). **(b)** Antagonistic flexor and extensor muscles attach to the inner surfaces of an exoskeleton on opposite sides of a flexible joint. **(c)** Antagonistic flexor and extensor muscles attach to an endoskeleton on opposite sides of the outer surfaces of joints.

makes the body longer and thinner. The opposing set is longitudinal; contracting these shortens and fattens the body. To burrow through soil, an earthworm anchors its body using tiny bristles in its skin, then contracts circular muscles in its head segments, causing them to elongate forward. Then it shortens and expands the head segments by contracting its longitudinal muscles, pulling the rest of its body forward. To move efficiently, the worm alternately contracts the longitudinal and circular muscles in fairly short sections of its body in a wavelike pattern known as peristalsis. (Peristaltic waves also move food through one-way digestive tracts; see Chapter 35).

The bodies of arthropods (such as spiders, crustaceans, and insects) are encased by rigid **exoskeletons** (literally, "outside skeletons"; **FIG. 41-7b**). Movement of an exoskeleton typically occurs only at joints where thin, flexible tissue joins stiff sections of exoskeleton. In arthropods, joints are located in the legs, mouthparts, antennae, bases of the wings, and between body segments. Antagonistic muscles attach to the inside of the exoskeleton across each joint. Contraction of a **flexor** muscle bends a joint; contraction of an **extensor** muscle straightens a joint. Alternating contraction of the antagonistic muscles moves joints back and forth, allowing the animal to walk, fly, or eat. Although an exoskeleton provides an effective support system for the muscles that move the body, it comes with a major problem: It cannot significantly expand to accommodate growth. Therefore, an arthropod must periodically molt its exoskeleton so that it can grow (**FIG. 41-8**). Molting exposes a partially formed soft exoskeleton that renders the animal vulnerable to predators, including people (soft-shelled crabs are eaten at this stage). If the animal survives after molting, it will then expand its body with air or water and deposit minerals in its shell to harden it.

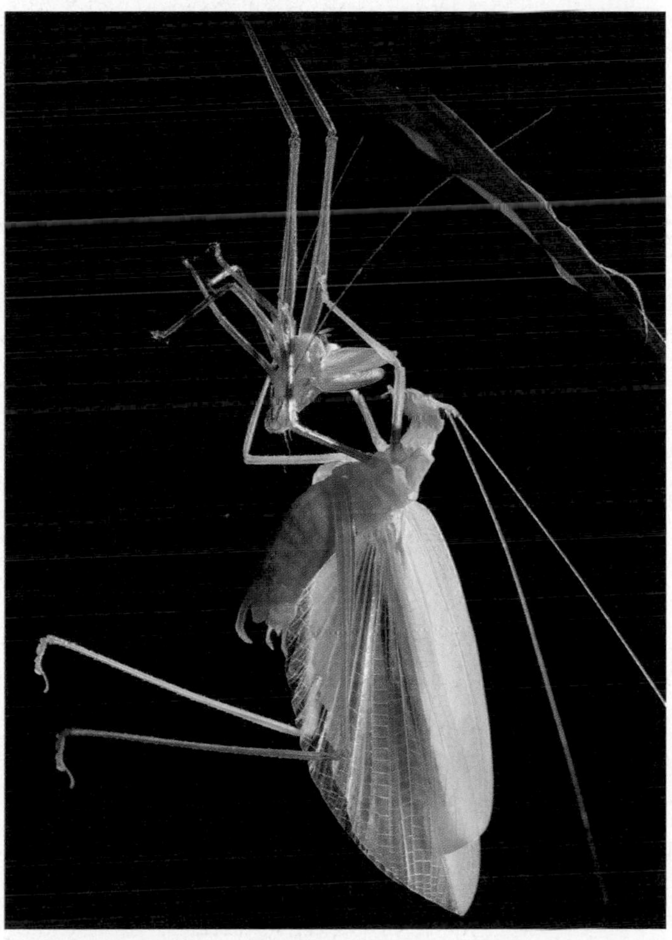

▲ **FIGURE 41-8 A katydid molts** Here a bright green, newly molted katydid has just abandoned its old exoskeleton.

THINK CRITICALLY Why are thick, armor-like exoskeletons found mostly in water-dwelling animals, whereas land-dwelling insects and spiders tend to have thinner exoskeletons?

Endoskeletons ("internal skeletons") are rigid structures found inside the bodies of echinoderms (sea stars and their relatives) and chordates (vertebrates and their relatives). In vertebrates, movement occurs primarily at **joints**, where two parts of the skeleton are firmly but flexibly attached to one another. Antagonistic muscles, such as the biceps (a flexor) and the triceps (an extensor), attach on opposite sides of the outside of a joint (in this case, the elbow; **FIG. 41-7c**). Antagonistic muscles move joints back and forth, or rotate them in one direction or the other.

Why do all animals use pairs of antagonistic muscles to move their skeletons? Because muscles can only produce force by contracting; they cannot actively lengthen. A contracted muscle will lengthen only if it is pulled on by other forces, such as antagonistic muscles or gravity. For example, if you've bent your elbow by contracting your biceps, you can only lengthen the biceps and straighten your arm by relaxing it and letting gravity do the job, or by contracting your antagonistic triceps muscle.

The Vertebrate Endoskeleton Serves Multiple Functions

The endoskeleton of humans and other vertebrates provides a wide range of functions:

- The skeleton allows locomotion. Different types of vertebrates have evolved skeletons that permit them to crawl, walk, run, jump, swim, fly, or perform various combinations of these actions.
- The skeleton provides a rigid framework that supports the body and protects its internal organs. The brain and spinal cord are almost completely enclosed within the skull and vertebral column, the rib cage protects the lungs and the heart, and the pelvic girdle supports and partially protects abdominal organs.
- In mammals, the bones of the middle ear are essential to hearing; these bones transmit sound vibrations between the eardrum and the cochlea (see Chapter 40).
- Red bone marrow produces red blood cells, white blood cells, and platelets (see Chapter 33).
- Bones store calcium and phosphorus, absorbing and releasing these minerals as needed.

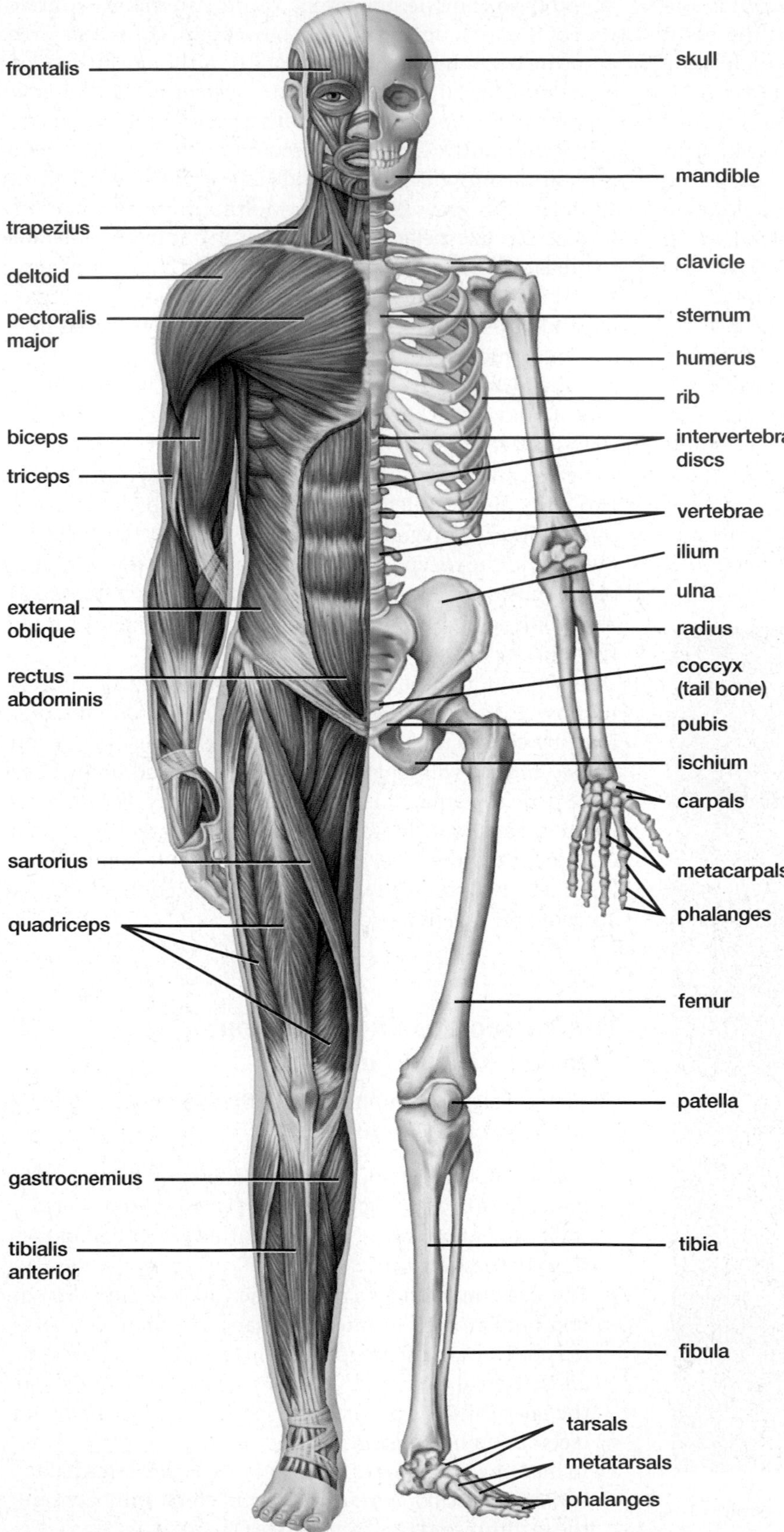

▲ **FIGURE 41-9 The human muscular and skeletal systems function together** Some of the major skeletal muscles (of about 640 total) and bones in the human body are illustrated. The 206 bones of the skeleton are grouped into the axial skeleton (blue) and the appendicular skeleton (beige).

The vertebrate skeleton consists of two parts (**FIG. 41-9**). The **axial skeleton** includes the bones of the head, vertebral column, and rib cage. The **appendicular skeleton** includes the pectoral and pelvic girdles, and the appendages attached to them: the forelimbs (in humans, the arms and hands) and hind limbs (in humans, the legs and feet). The pectoral girdle, which consists of the clavicle and scapula in humans, links the arms to the axial skeleton and provides attachment sites for muscles of the trunk and arms. Hip bones form the pelvic girdle, which links the legs to the axial skeleton, helps protect the abdominal organs, and forms attachment sites for muscles of the trunk and legs.

The Vertebrate Skeleton Is Composed of Cartilage, Ligaments, and Bones

The skeleton is composed of three types of connective tissue: cartilage, ligaments, and bone (see Chapter 32). All consist of living cells embedded in an extracellular matrix of a protein called *collagen,* with various amounts of other substances included in the matrix. In **cartilage**, the extracellular matrix contains large amounts of glycoproteins and often includes elastic fibers composed of stretchy protein. The extracellular matrix of **ligaments** (which connect bones to one another at joints) consists principally of slightly wavy collagen fibers, arranged parallel to one another. In the **bone** extracellular matrix, strands of collagen form a scaffold for crystals of bone mineral composed primarily of calcium and phosphate, which make the bone hard and rigid.

Cartilage Provides Flexible Support and Connections

Cartilage plays many roles in the vertebrate skeleton. In some fishes, such as sharks and rays, the entire skeleton is composed of cartilage. During the embryonic development of other vertebrates, the skeleton (except for the skull and collarbone) first forms from cartilage. In humans, the cartilaginous skeleton begins to be replaced by bone about 8 weeks after conception, and by 12 weeks, both bone and remaining cartilage are clearly visible (**FIG. 41-10**).

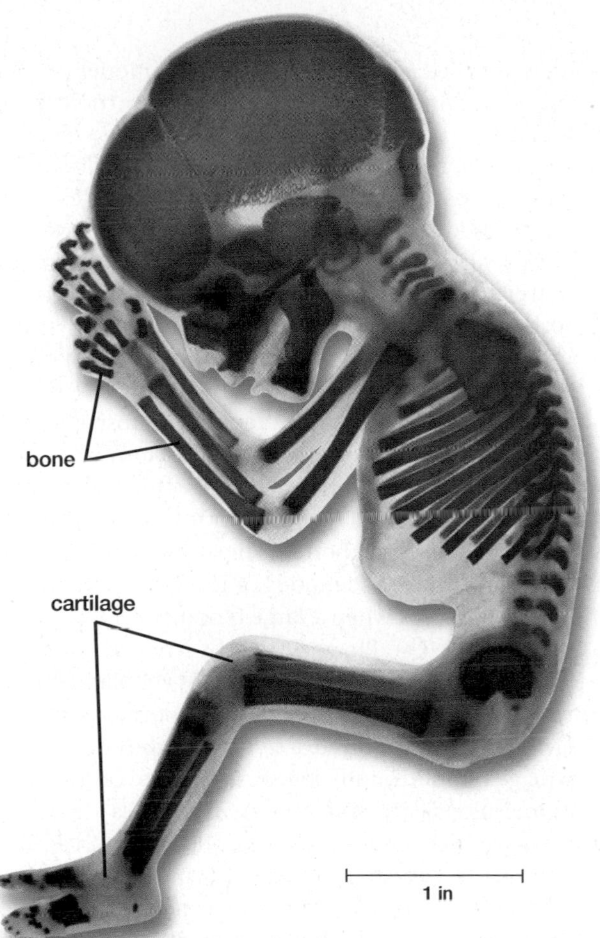

▲ **FIGURE 41-10 Bone replaces cartilage during development** In this 16-week-old human fetus, bone is stained magenta. The clear areas at the wrists, knees, ankles, elbows, and breastbone show cartilage that will later be replaced by bone.

Cartilage also covers the ends of bones at joints (**FIG. 41-11**), supports the flexible portions of the nose and external ears, and provides the framework for the larynx, trachea, and bronchi of the respiratory system. In addition, cartilage forms the tough, shock-absorbing intervertebral discs between the vertebrae of the backbone (see Fig. 41-9).

The living cells of cartilage are called **chondrocytes**. These cells secrete the glycoproteins and collagen that make up most of the extracellular matrix of cartilage (**FIG. 41-11a**). No blood vessels penetrate cartilage. To exchange wastes and nutrients, chondrocytes rely on diffusion of materials through the collagen matrix. As you might predict, cartilage cells have a very low metabolic rate, and damaged cartilage repairs itself very slowly, if at all.

Ligaments Connect Bone to Bone in Joints

The bones of most movable joints are attached to one another by strong, flexible ligaments. The parallel orientation of collagen in ligaments gives them tremendous strength against pulling forces, but twisting forces or impacts perpendicular to the collagen fibers can rupture a ligament, as sometimes happens to a knee in sports such as football, basketball, or skiing.

Bone Provides a Strong, Rigid Framework for the Body

A bone generally includes spongy bone surrounded by a hard outer shell of compact bone. **Spongy bone** (**FIG. 41-11b**) consists of an open network of bony fibers. It is porous, lightweight, and rich in blood vessels. **Compact bone** is dense and strong and provides an attachment site for muscles. Compact bone consists of subunits called *osteons*, each formed of concentric layers of bone surrounding a central canal containing a tiny nerve, artery, and vein (**FIG. 41-11c**).

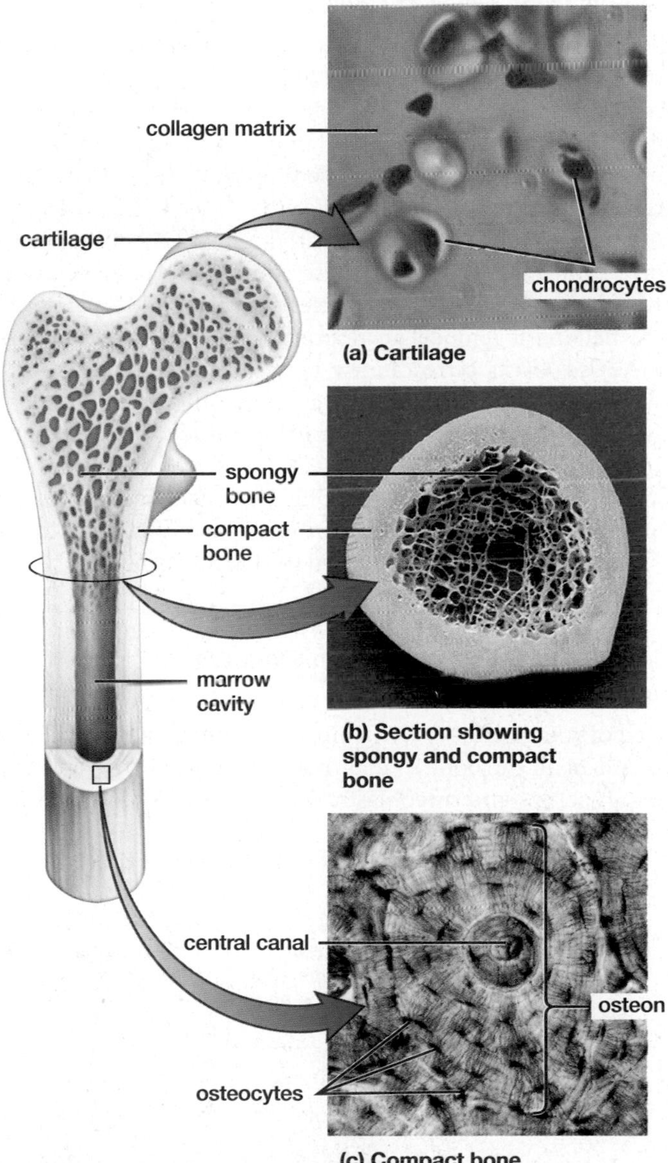

▲ **FIGURE 41-11 Cartilage and bone (a)** In cartilage, the chondrocytes secrete a surrounding extracellular matrix of collagen. **(b)** Spongy bone fills compact bone shown here near the end of a long bone. **(c)** Compact bone is composed of osteons with embedded osteocytes. Their central canals each contain a nerve, artery, and vein.

THINK CRITICALLY Why isn't the entire skeleton made of strong compact bone?

The shafts of long bones of the arms and legs consist of compact bone surrounding a canal filled with bone marrow. Red bone marrow, where blood cells form, fills the spongy bone and the long bone marrow cavities of infants and children up to about age seven. Then these marrow cavities fill with fatty yellow marrow, and blood cells form primarily at the ends of the femur and humerus, and in spongy flat bones of the sternum and pelvis (ilium, ischium, and pubis).

Bone Is Formed by Interactions Among Three Types of Cells

There are three types of bone cells: **osteoblasts** (bone-forming cells), **osteocytes** (mature bone cells), and **osteoclasts** (bone-dissolving cells that secrete an acidic mixture of enzymes that degrade bone matrix and dissolve its minerals). During embryonic development, osteoblasts produce bone within the cartilage skeleton by secreting a collagen matrix that becomes infiltrated with minerals. New osteoblasts form through cell division; these continue to produce bone throughout life, allowing the skeleton to grow and repair itself and helping it remodel in response to stresses placed upon it. As osteoblasts generate new bone, many become trapped within the hardened matrix and mature into osteocytes. Osteocytes are essential to bone health because they coordinate the activity of osteoclasts and osteoblasts, which are constantly breaking down and replacing bone. This renewal helps maintain bone strength and also maintains blood calcium homeostasis by exchanging Ca^{2+} between the blood and bones.

Bone Remodeling Allows the Skeleton to Adapt and Repair Itself

Each of your bones is a dynamic, living organ. Every year, 5% to 10% of all the bone in your body is removed and replaced by osteoclasts and osteoblasts. This process, coordinated by osteocytes, is called *bone remodeling*. Bone remodeling allows the skeleton to alter its shape in response to the demands placed on it.

Early in life, the activity of osteoblasts outpaces that of osteoclasts, so bones become larger and thicker as a child grows. As the body ages, however, the balance of power shifts to favor osteoclasts, and bones become more fragile, particularly if weight-bearing exercise decreases. Although both sexes lose bone mass with age, the loss is typically greater in women, as we explore in "Health Watch: Osteoporosis—When Bones Become Brittle."

Bone remodeling constantly repairs microscopic breaks that occur with everyday activities, but the ultimate in bone remodeling occurs after a fracture. A fracture ruptures the thin layer of connective tissue, rich in capillaries and osteoblasts, that surrounds the bone. Typically, a physician aligns and immobilizes the broken ends of the bone with a cast or splint. Healing begins when a large blood clot surrounds the break (**FIG. 41-12** ①). Phagocytic cells from the blood and osteoclasts from the damaged bone ingest cellular debris and dissolve bone fragments. Osteoblasts, in conjunction with cartilage-forming cells, secrete a *callus*, a porous mass of bone and cartilage that surrounds the break ②. The callus replaces the original blood clot and temporarily holds the ends of the break together. Osteoclasts, osteoblasts, and capillaries invade the callus. Osteoclasts break down cartilage while osteoblasts add new bone ③. Finally, osteoclasts remove excess bone, mostly restoring the bone's original shape, although often leaving a slight thickening ④. It takes about 6 weeks for a fracture to heal completely.

Bones, Tendons, and Ligaments Remodel in Response to Exercise

You know that exercise, especially high-intensity exercise like weightlifting, increases muscle size and strength. But did you know that exercise also affects tendons, ligaments,

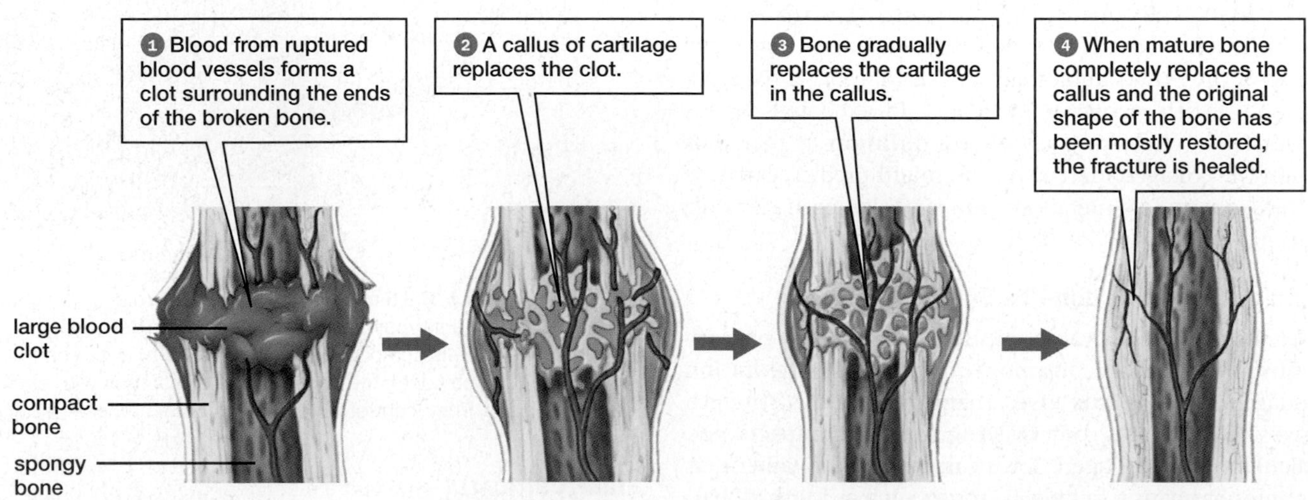

① Blood from ruptured blood vessels forms a clot surrounding the ends of the broken bone.
② A callus of cartilage replaces the clot.
③ Bone gradually replaces the cartilage in the callus.
④ When mature bone completely replaces the callus and the original shape of the bone has been mostly restored, the fracture is healed.
large blood clot
compact bone
spongy bone

▲ **FIGURE 41-12 Bone repair**

Health WATCH Osteoporosis—When Bones Become Brittle

Bone density in people peaks between ages 25 and 35. In middle age, the activity of osteoclasts starts to exceed that of osteoblasts, and bone density begins a slow decline. In the United States, about 10% of women and 2% of men over age 50 have **osteoporosis** (literally, "porous bones"; **FIG. E41-1b**). People with osteoporosis who fall are more likely to break a bone, and in severe cases, everyday activities such as opening a window can cause a minor fracture. The vertebrae of individuals with severe osteoporosis may become partially crushed, causing a hunched back (**FIG. E41-1c**).

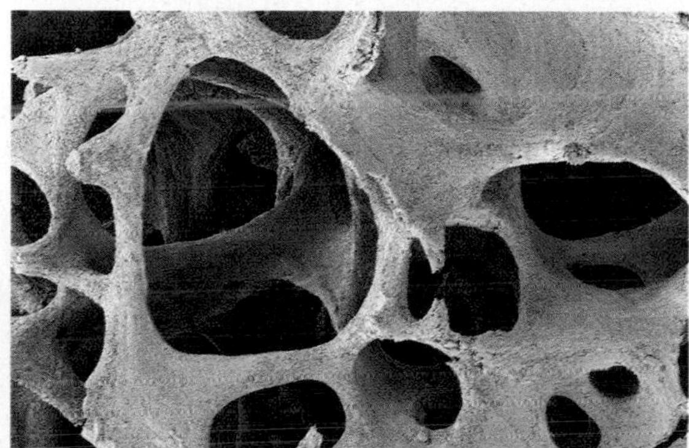

(a) Normal bone

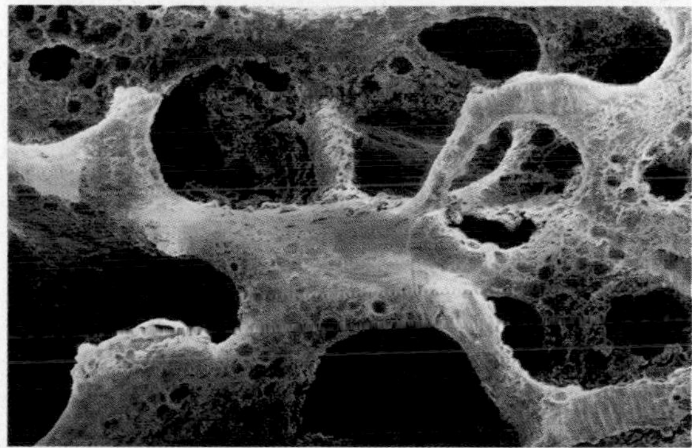

(b) Osteoporotic bone

(c) Effects of osteoporosis

▲ **FIGURE E41-1 Osteoporosis** A cross-section of normal spongy bone **(a)** compared with spongy bone with osteoporosis **(b)**. People with osteoporosis may acquire a hunched back when the spongy bone that fills vertebrae becomes partially crushed **(c)**.

Women have a much higher incidence of osteoporosis than men, in part because women's bones are smaller and less dense than men's to begin with. Further, although the high estrogen levels of premenopausal women stimulate osteoblasts and help maintain bone density, estrogen production drops dramatically after menopause (around age 50), causing a rapid decline in bone density. In men, testosterone stimulates osteoblast activity. Although testosterone levels decline with age, they usually do not drop as fast or as far as women's estrogen levels do after menopause. In both sexes, alcoholism and smoking also contribute to bone loss.

Can osteoporosis be prevented? Calcium, of course, is a major component of bone. In addition, bones thrive on moderate stress. Therefore, regular exercise throughout life, combined with a diet containing adequate calcium and vitamin D (which is required for calcium absorption from the intestine), helps to ensure that bone mass is as high as possible before age-related losses begin, and slows the rate of loss with aging. Older people tend to be less active, resulting in loss of bone minerals, but low-impact, weight-bearing exercise such as walking or dancing can reduce bone loss and sometimes even increase bone mass. People with osteoporosis, in consultation with their physicians, may choose to take drugs that inhibit osteoclast activity to help maintain their bone density.

THINK CRITICALLY A woman on a highly competitive college gymnastics team limps into your medical office with a swollen lower left leg, having experienced sudden severe pain upon landing after a dismount from the high bar. She is extremely thin and cannot recall having menstruated during the past year. What would you look for in a blood test, and what would you expect to see on an X-ray of her leg?

and bones? Weight-bearing exercise increases the activity of tendon- and ligament-building cells, causing them to produce more of the collagen proteins that make up the bulk of tendons and ligaments. Exercise that stresses bones also stimulates bone-building osteoblasts. As a result, the bone remodels, somewhat as it does when a broken bone heals.

Tendons and bones thicken and strengthen in proportion to the stresses placed upon them. For example, tibia (lower leg bones) of sprinters are stronger and thicker than those of distance runners because the explosive movements of sprinting apply greater forces to these bones. Professional tennis players and baseball players have significantly thicker and stronger bones in their racket and throwing arms, respectively (**FIG. 41-13**). Archaeologists can even tell which skeletons found in medieval battlefields are the remains of English longbowmen—their skeletons typically have very broad shoulders and thick left arms, the result of endless hours of shooting bows with pull strengths well over 100 pounds.

If weight-bearing exercise and other stresses increase bone mass, what happens if bones are freed from stress? The weightlessness experienced by astronauts poses a potentially serious threat to the skeleton. Studies of astronauts who have spent months aboard space stations show that space travelers lose 1% to 2% of bone mass each month, particularly in the (normally) weight-bearing bones of the hip, legs, and lower spine. Extensive research to develop exercise, dietary,

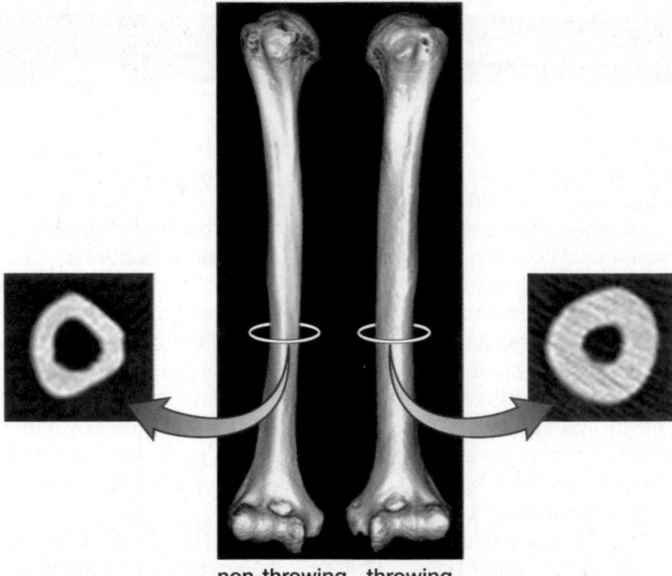

▲ **FIGURE 41-13 Stress causes bones to thicken** The throwing and non-throwing upper arm (humerus) bones of professional baseball players show significant differences in the thickness of the compact bone to which muscles attach.

and drug regimens to help prevent bone loss in space may eventually benefit earthbound individuals who are confined to bed, suffer from injuries that prevent them from engaging in healthy, bone-stressing activities, or are simply experiencing bone loss due to aging.

Antagonistic Muscles Move Joints in the Vertebrate Skeleton

As noted earlier, almost all animal movement is produced by pairs of antagonistic muscles working on a skeleton. When one member of a pair of antagonistic muscles contracts, it moves a bone around a joint and simultaneously stretches the relaxed opposing muscle. Depending on the configuration of the joint, antagonistic muscles allow for a range of motions, including moving bones back and forth or side to side, or rotating them. Here, we focus on the arrangement and movement of muscles around the movable joints of the human skeleton (**FIG. 41-14**). Note, however, that not all joints are movable; for example, immobile joints called sutures join the bones of the skull.

In movable joints, the portion of each bone that forms the joint is coated with a layer of cartilage. The smooth, resilient cartilage allows the bone surfaces to slide past one another with little friction. Joints are held together by ligaments, which are strong, flexible, and somewhat stretchy to allow the joint to move. Muscles are attached to bones by tendons. The tendon at one end of a muscle, called the *origin*, is fixed to a bone that remains stationary relative to the joint. The tendon at the other end of the muscle, the *insertion*, is attached to the movable bone on the far side of the joint.

Hinge joints are located in the ankle, elbows, knees, and fingers. Like a hinged door, these joints move in one plane (**FIG. 41-15a**). When the flexor muscle contracts, it bends the joint; when the extensor muscle contracts, it straightens the joint. In Figure 41-14, for example, contraction of the biceps femoris (the flexor) bends the leg at the knee, while contraction of the quadriceps (the extensor) straightens it. Alternating contractions of flexor and extensor muscles cause the lower leg bones to swing back and forth at the knee joint.

In **ball-and-socket joints**, such as those of the hip and shoulder, the round end of one bone fits into a hollow depression in another, allowing the joint to rotate (**FIG. 41-15b**). Ball-and-socket joints allow movement in several directions—compare

CASE STUDY CONTINUED

Legs of Gold

As Bolt's and Kipchoge's athletic feats attest, muscles can contract with amazing force. But sometimes they contract more than tendons can withstand. Achilles tendon injuries have sidelined a number of basketball stars such as Kobe Bryant, as well as baseball star Zach Britton and football star Eric Berry. Because tendons have a very limited blood supply, recovery from a serious tear (tendon rupture) can take a full year, and elite athletes may never regain their former levels of performance. The Achilles tendon—the largest and strongest tendon in the body—attaches the calf muscles to the heel bone and is vulnerable to sudden muscle contractions, such as those that occur during running or jumping. This tendon flexes a hinge joint in the ankle. How do hinge joints work?

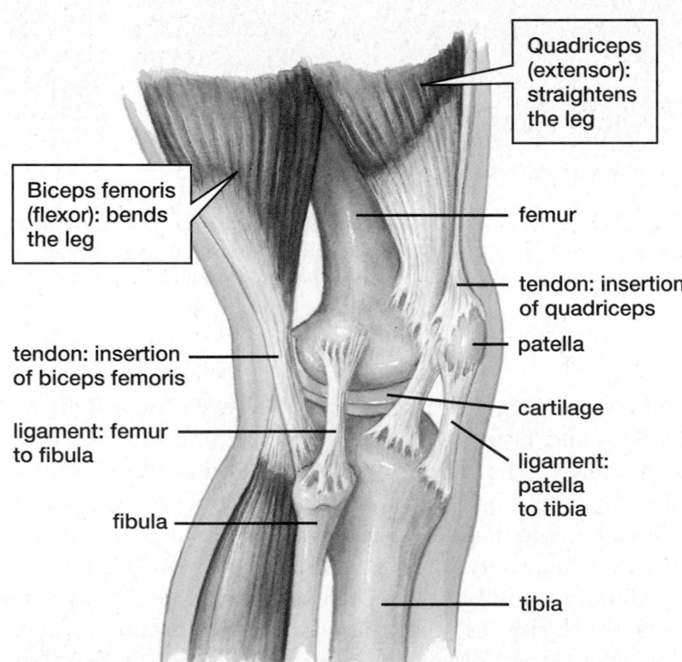

▲ **FIGURE 41-14 The human knee** The human knee, showing antagonistic muscles (the biceps femoris and the quadriceps), tendons, and ligaments. The complexity of this joint, coupled with the extreme stresses placed on it, makes the knee very susceptible to injury.

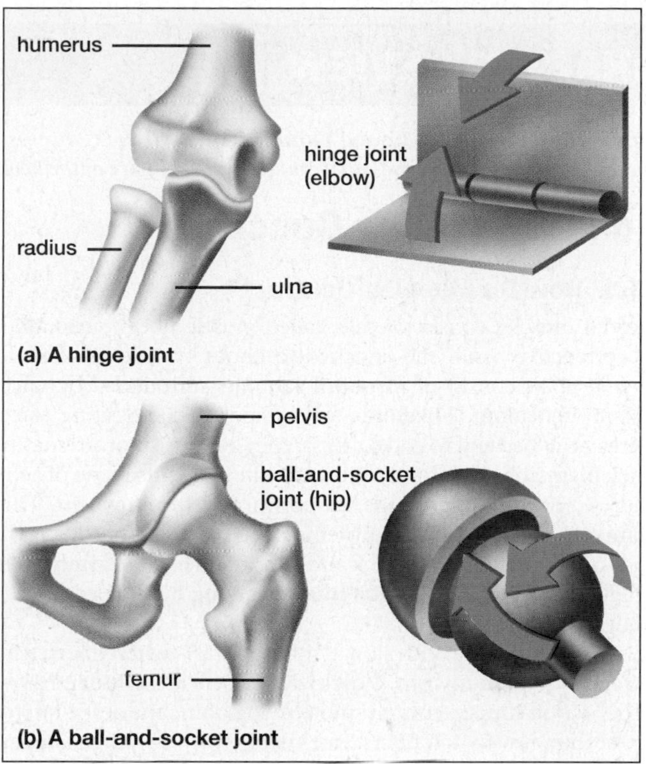

▶ **FIGURE 41-15 Hinge and ball-and-socket joints (a)** The human elbow is a hinge joint. **(b)** The human hip can rotate because it has a ball-and-socket joint. The rounded end of the femur (the ball) fits into a cuplike depression (the socket) in the pelvic bone.

the wide range of movement of your shoulder with the limited bending of your knee. The range of motion in ball-and-socket joints is made possible by at least two pairs of antagonistic muscles, attached at angles to each other, so the joints can rotate.

CHECK YOUR LEARNING

Can you . . .

- describe hydrostatic skeletons, exoskeletons, and endoskeletons and explain how antagonistic muscles act on each of these skeletons to move an animal's body?
- explain the functions of vertebrate skeletons?
- list and describe the functions of the different types of connective tissue that make up a vertebrate skeleton, both in adulthood and during embryonic development?
- explain how a bone fracture is repaired and how bone remodels in response to mechanical stresses?
- explain how hinge joints and ball-and-socket joints work?

CASE STUDY REVISITED

Legs of Gold

Without an overwhelming number of bulky fast-twitch muscle fibers in his legs, Usain Bolt certainly could not win gold in the sprints. Likewise, most top marathoners have legs like Eliud Kipchoge's, dominated by thinner slow-twitch fibers. In the average person, the two forms are roughly equal in number, with some variability in the proportions of fast-twitch and slow-twitch fibers, based partly on genetics. People with an unusually high proportion of slow-twitch fibers are likely to excel in, and therefore favor, endurance sports, whereas those with more fast-twitch fibers will generally find that sports requiring bursts of speed or strength are more rewarding. Researchers have found little evidence that training causes conversion between slow and fast-twitch types. Elite athletes really are different from birth (**FIG. 41-16**).

Around 240 chromosomal genes and 18 mitochondrial genes contribute in some way to physical fitness. No one knows which alleles of each of these genes are carried by any given elite athlete, but Bolt and Kipchoge undoubtedly have a superb set. No one can reach the Olympics—let alone bring home the gold—without first winning the genetic lottery.

CONSIDER THIS For a fee, a few companies will test young children to see which allele they carry for a gene that encodes an important protein in muscle cells. One allele is associated with fast-twitch muscles and ability in sports that require sprinting or strength. Presumably, kids homozygous for this allele should be steered into sprinting or football rather than distance running or soccer. What are some advantages and disadvantages of parents having this information?

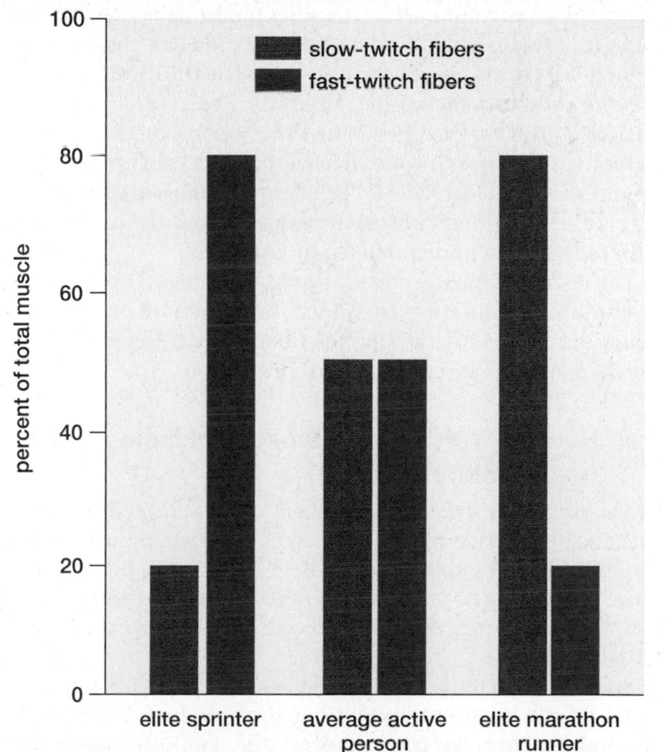

▲ **FIGURE 41-16 Muscle composition differs** Percentages of slow- and fast-twitch muscle fibers differ among elite sprinters, marathoners, and average active people.

CHAPTER REVIEW

Go to **Mastering Biology** to access the Pearson eText, vocabulary review, practice quizzes, activities, videos, current events, and more.

*Answers to **Think Critically** and **Thinking Through the Concepts** questions can be found in the **Answers** section at the back of the book.*

Summary of Key Concepts

41.1 How Do Muscles Contract?

Skeletal muscles consist of cells called muscle fibers, surrounded by connective tissue and attached to bones by tendons. Skeletal muscle fibers consist of myofibril subunits surrounded by sarcoplasmic reticulum. Myofibrils are composed of repeating sarcomeres, attached end to end. Each sarcomere consists of alternating thick filaments made of myosin and thin filaments made of actin and two accessory proteins, troponin and tropomyosin. Thin filaments are attached to proteins called Z discs that form the ends of sarcomeres. The thick and thin filaments are arranged in a regular pattern in a skeletal muscle, giving it a striped, or striated, appearance.

Muscles include both slow-twitch and fast-twitch fibers, with different types of myosin. Slow-twitch fibers are thinner, have a richer blood supply, contain more myoglobin, and resist fatigue better than fast-twitch fibers. Fast-twitch fibers are larger and are adapted for short-term bursts of effort.

A motor neuron innervates each muscle fiber at the neuromuscular junction. An action potential in the motor neuron causes it to release acetylcholine onto the muscle fiber, stimulating an excitatory postsynaptic potential that triggers an action potential in the fiber. The AP stimulates release of Ca^{2+} from the sarcoplasmic reticulum, which exposes binding sites on the thin filaments. Myosin heads bind to these sites and flex, detach, then reattach in a continuing cycle that causes the thin and thick filaments to slide past each other, shortening each sarcomere, and contracting the muscle fiber. When the action potential is over, calcium is actively transported back into the sarcoplasmic reticulum, ending contraction. The energy for muscle contraction comes from ATP, which may be produced by cellular respiration, or, for bursts of strenuous activity, by glycolysis.

The degree of muscle contraction is determined by the number of muscle fibers stimulated by a motor neuron and the frequency of action potentials in each fiber. Rapid firing in all of the fibers of a muscle causes maximum contraction.

41.2 How Do Cardiac and Smooth Muscles Differ from Skeletal Muscle?

Cardiac muscle is striated like skeletal muscle. Its cells contract rhythmically and spontaneously, synchronized by action potentials initiated by specialized muscle fibers in the heart's pacemaker. Cardiac muscle fibers are interconnected by intercalated discs, which conduct APs between cells, producing coordinated contraction.

Smooth muscles lack organized sarcomeres. They are connected to one another by junctions that conduct APs between cells and synchronize their contractions. Smooth muscle surrounds hollow organs such as the uterus, digestive tract, bladder, arteries, and veins, producing involuntary, slow, and sustained or rhythmic contractions.

41.3 How Do Muscles and Skeletons Work Together to Provide Movement?

Animals have hydrostatic skeletons, exoskeletons, or endoskeletons. Pairs of antagonistic muscles act on the skeleton to move the body.

The vertebrate endoskeleton provides support for the body, attachment sites for muscles, and protection for internal organs. Red blood cells, white blood cells, and platelets are produced in bone marrow. Bone acts as a storage site for calcium and phosphorus. The axial skeleton includes the skull, vertebral column, and rib cage. The appendicular skeleton consists of the pectoral and pelvic girdles and the bones of the arms, legs, hands, and feet.

The vertebrate skeleton consists of cartilage, ligaments, and bone, whose cells are embedded in a extracellular matrix of collagen and other substances. Cartilage, located at the ends of bones, forms pads in joints and supports the nose, ears, and respiratory passages. During embryological development, cartilage is the precursor of bone. Ligaments connect bones at movable joints. Bone is formed by osteoblasts, which secrete a collagen matrix that becomes hardened by minerals. A typical bone consists of an outer shell of compact bone, to which muscles are attached, and inner spongy bone with marrow. Some marrow produces blood cells. Bone remodeling occurs through the coordinated action of bone-dissolving osteoclasts and bone-forming osteoblasts.

In the vertebrate skeleton, movement occurs around joints. Tendons attach pairs of antagonistic muscles to the bones on either side of a joint. The contraction of one muscle stretches out its antagonistic muscle. In hinge joints, contraction of the flexor muscle bends the joint; contraction of its antagonistic extensor straightens it. At ball-and-socket joints, antagonistic muscles rotate one bone relative to another.

Thinking Through the Concepts

Bloom's: Remembering, Understanding

Multiple Choice

1. Which of the following lists the structures from largest to smallest?
   a. motor unit, muscle, myofibril, muscle fiber, thick and thin filaments
   b. muscle, muscle cell, myofibril, muscle fiber, actin and myosin
   c. muscle cell, thick and thin filaments, sarcomere, myofibril
   d. muscle, muscle fiber, myofibril, thick and thin filaments
2. Which of the following statements is true?
   a. Tendons contract using the sliding filament mechanism.
   b. Tendons are composed primarily of cartilage fibers.
   c. Tendons attach bones to bones at joints.
   d. Tendons attach muscles to bone.

3. Muscles
   a. can actively contract, but cannot actively lengthen.
   b. generally contain only slow-twitch or only fast-twitch fibers.
   c. require that all their motor units fire in order to contract.
   d. are each stimulated by a single motor neuron.
4. Exoskeletons
   a. are typical of vertebrates.
   b. are hydrostatic.
   c. must be shed periodically.
   d. are moved by muscles attached to the outside of joints.
5. Which of the following statements is false?
   a. Smooth muscle contractions are involuntary.
   b. Both skeletal and cardiac muscle are striated.
   c. Skeletal, smooth, and cardiac muscle each have one nucleus per fiber.
   d. Cardiac muscle cells are branched.

Fill-in-the-Blank

1. The three types of skeletal systems found in animals are ____________, ____________, and ____________. All three move the skeleton using pairs of ____________ muscles.
2. ____________ and ____________ muscles are striped because they have a regular alignment of sarcomeres. ____________ and ____________ muscles are usually not under conscious control. The cells of ____________ and ____________ muscles are interconnected to allow coordinated contraction.
3. A skeletal muscle cell is called a(n) ____________. It consists of bundles of cylindrical structures called ____________, which in turn consist of much smaller cylinders, the ____________, attached end to end at their ____________.
4. Muscle contraction results from interactions between thin filaments, composed mainly of the protein ____________, and thick filaments, composed mainly of the protein ____________. The "heads" of the thick filament protein grab on to the thin filament protein and flex. The energy of ____________ is used to extend the head, which stores that energy to power the flexing movement of the head, pulling the thin filament past the thick filament.
5. Motor neurons innervate muscle fibers at specialized synapses called ____________. The release of acetylcholine at these synapses ultimately triggers an action potential in the muscle fiber, which invades the interior of the fiber through ____________ and stimulates the release of calcium ions from the ____________.
6. Skeletons consist of three types of connective tissue: ____________, ____________, and ____________. All consist of cells embedded in a matrix of ____________ protein and other extracellular components, such as glycoproteins or calcium phosphate.
7. A(n) ____________ joint moves in two dimensions, while a(n) ____________ joint has significant motion in three dimensions. At a two-dimensional joint, the ____________ muscle bends the joint and the ____________ muscle straightens the joint.

Review Questions

1. Sketch a relaxed muscle fiber containing a myofibril, sarcomeres, and thick and thin filaments. How would a contracted muscle fiber look by comparison?
2. Describe the process of skeletal muscle contraction, beginning with an action potential in a motor neuron and ending with the relaxation of the muscle. Your answer should include the following words: *neuromuscular junction, T tubule, sarcoplasmic reticulum, calcium, thin filaments, binding sites, thick filaments, sarcomere, Z disc,* and *active transport.*
3. Explain the following two statements: Muscles can only actively contract; muscle fibers lengthen passively.
4. What are the three types of skeletons found in animals? For one of them, describe how the muscles are arranged around the skeleton and how contractions of the muscles result in movement of the skeleton.
5. Compare the structures of the following pairs: spongy and compact bone, smooth and skeletal muscle, and cartilage and bone.
6. Explain the functions of osteoblasts and osteoclasts.
7. Describe a hinge joint and how it is moved by antagonistic muscles.

Applying the Concepts

Bloom's: Applying, Analyzing, Evaluating

1. Discuss some of the problems that would result if the human heart were made of skeletal muscle instead of cardiac muscle.
2. Myasthenia gravis is caused by the abnormal production of antibodies that destroy acetylcholine receptors within the neuromuscular junction. Drugs such as neostigmine, which inhibits the enzyme that breaks down acetylcholine, are used to treat myasthenia gravis. Predict the symptoms of myasthenia gravis and explain how neostigmine could help alleviate them.

42 Animal Reproduction

Cronos, a white rhino calf produced by artificial insemination, poses with her mother at the Madrid Zoo.

CASE STUDY

To Breed a Rhino

RHINOS ARE MASSIVE, aggressive, and seemingly indestructible. Nevertheless, three of the five rhinoceros species are critically endangered. Why are rhinoceroses in danger of extinction? Their name—derived from a Greek word meaning "nose horn"—tells you why: They are being killed for their horns.

Some cultures place enormous value on rhino horn. In China, Vietnam, and some other East Asian nations, powdered rhino horn is believed to reduce fevers and cure rheumatism, gout, cancer, and many other disorders. In recent years, the black market price for rhino horn has soared to about $30,000 a pound, higher than the price of gold or cocaine. As a result, rhino poaching is rapidly increasing; in South Africa, for example, 122 rhinos were killed in 2009, 668 in 2012, and 1,028 in 2017.

To help stave off extinction of endangered rhinos, scientists at several zoos are breeding them in captivity. By mating carefully selected rhinos, they hope to preserve the rhinos' genetic diversity, which is crucial to a species' survival. Captive breeding sounds simple enough—put a male and female into the same enclosure and let nature take its course. Unfortunately, this often doesn't work with large, aggressive loners like rhinos, who sometimes prefer combat to copulation. Even if a male and female don't fight, they may not mate.

Artificial insemination avoids both the dangers of natural mating and a possible lack of sexual interest. However, both mating and artificial insemination result in pregnancy only if the female has an egg ready to fertilize. Therefore, scientists need to know when the female ovulates naturally or, alternatively, they must be able to induce her to ovulate at the right time.

How do scientists assisting rhinos to reproduce—or physicians collecting human eggs for *in vitro* fertilization—determine when ovulation occurs naturally or induce ovulation? How do rhinos and other animals produce gametes, and how do the reproductive systems of males and females differ?

AT A GLANCE

42.1 HOW DO ANIMALS REPRODUCE?

Animals reproduce either sexually or asexually. In **asexual reproduction**, a single animal produces offspring, usually through repeated mitotic cell divisions in some part of its body. The offspring are therefore genetically identical to the parent. In **sexual reproduction**, organs called **gonads** produce haploid sperm or eggs through meiotic cell division. A sperm and egg—usually from separate parents—fuse to produce a diploid fertilized egg, or **zygote**, which then undergoes repeated mitotic cell divisions to produce an offspring. Because the offspring receives genes from two parents, it is not genetically identical to either.

In Asexual Reproduction, an Organism Reproduces Without Mating

Asexual reproduction is efficient in effort (no need to search for mates, court members of the opposite sex, or battle rivals) and materials (no wasted gametes). There are several common methods of asexual reproduction among animals.

Budding Produces a Miniature Version of the Adult

Many sponges and cnidarians (such as corals and anemones) reproduce by **budding**. A miniature version of the adult—a bud—grows directly on the body of the adult (**FIG. 42-1**). When the bud has grown large enough, it breaks off and becomes independent, eventually growing into an adult.

Fragmentation Followed by Regeneration Can Produce a New Individual

Among animals that are capable of regeneration—regrowth of body parts—some can reproduce by **fragmentation**. These animals can split their bodies apart at specific locations. Each of the resulting pieces then regenerates the missing parts of a complete body. In addition, if accidents or predators break these animals apart, sufficiently large fragments can often regenerate whole new individuals. Reproduction by fragmentation occurs in a variety of animals, including flatworms, corals, sea jellies, and brittle stars.

During Parthenogenesis, Eggs Develop Without Fertilization

Females of some animal species can reproduce by a process known as **parthenogenesis**, in which egg cells develop into offspring without being fertilized. In some species, parthenogenetically produced offspring are haploid. For example, male honeybees develop from unfertilized eggs and are haploid; their diploid sisters develop from fertilized eggs.

Some parthenogenetic fish, amphibians, and reptiles produce diploid offspring by doubling the number of chromosomes in the eggs, either before or after meiosis. In most such species, all the resulting diploid offspring are female. In fact, some species of fish and lizards, such as the whiptail lizard of the southwestern United States and Mexico (**FIG. 42-2**), consist entirely of parthenogenetically reproducing females.

▲ **FIGURE 42-1 Budding** Some cnidarians, such as this anemone, reproduce asexually by budding. Cnidarians can also reproduce sexually.

▲ **FIGURE 42-2 Courting female whiptail lizards** Although all members of this species are female, they still engage in courtship rituals. These mating behaviors increase sex hormone concentrations, with the result that courting whiptails lay more eggs than isolated whiptails do.

Still other animals, such as some aphids, can reproduce either sexually or parthenogenetically, depending on environmental factors such as the season of the year or the availability of food (**FIG. 42-3**).

In Sexual Reproduction, an Organism Reproduces Through the Union of Sperm and Egg

Even though sexual reproduction is less efficient than asexual reproduction, most animals reproduce sexually. No one is certain why sexual reproduction evolved. The most likely explanation is that sexual reproduction was favored by natural selection because the resulting offspring are genetically different from one another. For example, some offspring may inherit new combinations of alleles that enable them to exploit novel habitats or foil parasites.

In most sexually reproducing animal species, an individual is either male or female, defined by the type of gamete that it produces. The male gonad, called the **testis** (plural, testes), produces small haploid **sperm**, which have almost no cytoplasm and hence no food reserves. Sperm swim by thrashing their tails. The female gonad, called the **ovary**, produces **eggs**, which are large, haploid cells containing food reserves. The food reserves provide nourishment for the **embryo**, the offspring in its early stages of development before birth or hatching. Eggs cannot swim. **Fertilization**, the union of sperm and egg, produces a diploid zygote.

In some sexually reproducing animals, such as earthworms and many snails, single individuals produce both sperm and eggs. Such individuals are called **hermaphrodites**. Most hermaphrodites still engage in sex, with partners exchanging

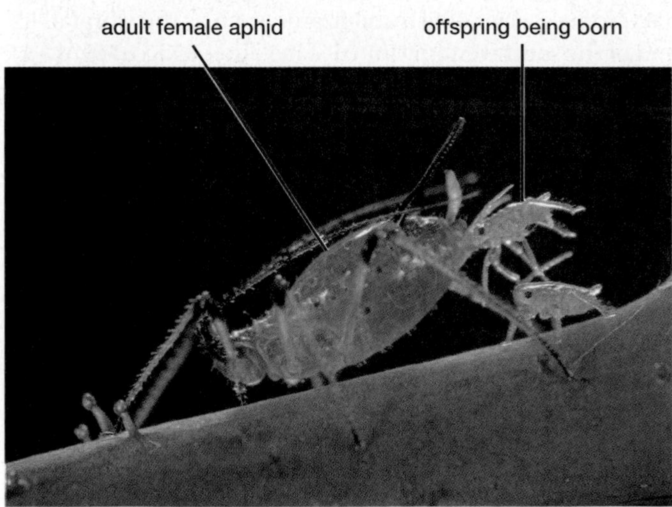

▲ **FIGURE 42-3 A female aphid gives birth** In spring and early summer, when food is abundant, aphid females reproduce parthenogenetically; in fact, the females are born pregnant! In fall, they reproduce sexually.

THINK CRITICALLY Why might natural selection favor aphids that alternate between asexual reproduction and sexual reproduction at different seasons of the year?

▲ **FIGURE 42-4 Hermaphroditic earthworms exchange sperm**

sperm (**FIG. 42-4**). Some hermaphrodites, however, can fertilize their own eggs. In many cases, hermaphroditic animals, including tapeworms and some types of snails, are relatively immobile and may find themselves isolated from other members of their species, so self-fertilization may be the only way they can reproduce.

For species with two separate sexes or with hermaphrodites that cannot self-fertilize, successful reproduction requires that sperm and eggs from different individuals be brought together for fertilization.

External Fertilization Occurs Outside the Parents' Bodies

In **external fertilization**, sperm and egg unite outside the bodies of the parents. Gametes are typically released into water, a process called **spawning**, and the sperm swim to reach the eggs. Because sperm and eggs are generally short-lived, spawning animals must synchronize their reproductive behaviors, both temporally (male and female spawn at the same time) and spatially (male and female spawn in the same place). Coordination may be achieved by using environmental cues, chemical signals, courtship behaviors, or a combination of factors.

Most spawning animals rely on environmental cues to some extent. Seasonal changes in day length often stimulate the physiological changes required for reproduction, which restricts breeding to certain times of year. However, the actual release of sperm and egg must be more precisely synchronized. For example, each spring, many corals of Australia's Great Barrier Reef orchestrate their spawning by the phase of the moon. Near sunset, a few days after the full moon, millions of corals release sperm and eggs into the water (**FIG. 42-5**). Many corals also release chemicals along with their gametes; when neighboring corals sense these chemicals, they also spawn, thereby providing even more precise synchronization and helping to ensure fertilization.

Many animals rely on mating behaviors to synchronize spawning. Most fish, for example, have courtship rituals, ensuring that they release their gametes in the same place and at the same time (**FIG. 42-6**). Frogs and toads mate in shallow

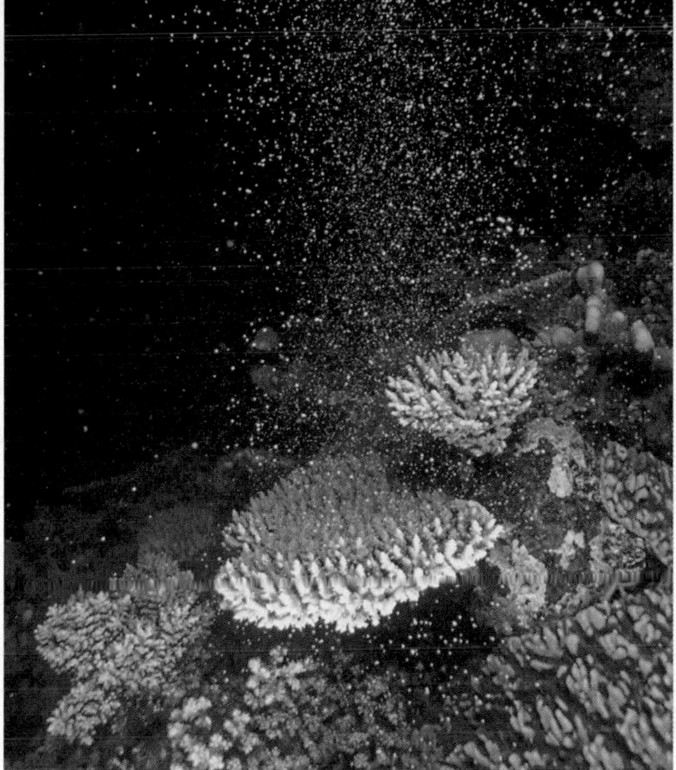

▲ **FIGURE 42-5 Environmental cues may synchronize spawning** In the Great Barrier Reef of Australia, thousands of corals spawn simultaneously, creating this "blizzard" effect.

water near the edges of ponds and lakes. The male mounts the female and prods the sides of her abdomen (**FIG. 42-7**). This stimulates her to extrude her eggs, while he deposits sperm onto them.

Internal Fertilization Occurs Within the Female's Body

During **internal fertilization**, sperm are placed within the female's moist reproductive tract, where her eggs are

▲ **FIGURE 42-6 Courtship rituals synchronize spawning** During spawning, male and female gourami fish perform a courtship dance that culminates with both partners releasing gametes simultaneously. The male will collect the fertilized eggs in his mouth and spit them into a floating nest of bubbles. The male guards the eggs and the newly hatched babies in the nest until they are able to swim and fend for themselves.

THINK CRITICALLY In addition to ensuring synchronized release of gametes, what other advantages do courtship rituals provide?

▲ **FIGURE 42-7 Splendid tree frogs mating** The smaller male clutches the female and stimulates her to release eggs. Although they spend most of their lives high up in the trees of Central American rain forests, the tree frogs descend to small pools of water to breed.

fertilized. Internal fertilization is an essential adaptation to terrestrial life because sperm quickly die if they dry out. Even in aquatic environments, internal fertilization may increase the likelihood of reproductive success because the sperm and eggs are confined in a small space rather than dispersed in a large volume of water.

Internal fertilization usually occurs by **copulation**, in which the male deposits sperm directly into the female's reproductive tract (**FIG. 42-8**). Some animals do not copulate but still have internal fertilization. In these species, including some salamanders, scorpions, and grasshoppers, males package their sperm in a container called a *spermatophore* (Greek for "sperm carrier"). Males and females usually engage in a mating display, after which the male may insert his spermatophore into the female's reproductive tract or drop the spermatophore on the ground for the female to pick up if she chooses. Once inside the female, the spermatophore releases sperm.

▲ **FIGURE 42-8 Internal fertilization allows reproduction on land** Ladybugs mate on a blade of grass.

In most animals, neither sperm nor eggs live very long. Therefore, **ovulation**, the release of a mature egg cell from the ovary of the female, usually must occur about the same time that sperm are deposited in the female's reproductive tract. Most mammals, for example, use courtship displays that synchronize mating with ovulation. Many copulate only when the female signals readiness to mate, which usually occurs about the same time as ovulation. In some animals, such as rabbits, the act of mating stimulates ovulation, so new, healthy sperm and eggs are almost guaranteed to meet.

CHECK YOUR LEARNING

Can you . . .

- explain the difference between sexual and asexual reproduction and describe some advantages of each?
- define *budding*, *fragmentation*, and *parthenogenesis* and provide some examples of animals that use each of these methods of reproduction?
- describe internal and external fertilization and provide examples of animals that use each of these methods of fertilization?

CASE STUDY CONTINUED

To Breed a Rhino

To attract a mate, rhinos urinate, raise their tails, vocalize, and bump each other on the head and genitals with their snouts. Sometimes this courtship gets violent enough to injure or even kill one or both animals, particularly if the animals are confined in a small area. However, if they do copulate, rhinos use structures and hormonal controls that are very similar to those used by humans. How do these structures and hormones function in the reproductive systems of mammals?

42.2 WHAT ARE THE STRUCTURES AND FUNCTIONS OF HUMAN REPRODUCTIVE SYSTEMS?

In people and other mammals, the sexes are separate, and eggs are fertilized internally following copulation. Although most mammals reproduce only during certain seasons of the year, men produce sperm more or less continuously, and women ovulate about once a month.

The Ability to Reproduce Begins at Puberty

Both the ability to reproduce and the development of secondary sexual characteristics begin with the onset of **puberty**, the process that results in sexual maturity. The timing of puberty in humans varies considerably, but usually occurs between the ages of 10 and 14 in girls and between 12 and 16 in boys. In girls, the first outward sign of puberty is usually breast enlargement; the first menstrual period typically occurs a couple of years later. Other secondary sexual characteristics that develop during female puberty include widening of the hips, growth of pubic and underarm hair, and often a moderate growth spurt. In boys, growth of the testes signals the beginning of puberty, followed by onset of fertility a year or two later. Boys also experience enlargement of the penis, muscular development, growth of facial, pubic, and underarm hair, deepening of the voice (caused by enlargement of the larynx), and a prolonged growth spurt. In both sexes, courtship behaviors start to appear at puberty.

In both sexes, puberty is triggered by brain maturation, causing the hypothalamus to release **gonadotropin-releasing hormone (GnRH)**, which stimulates the anterior pituitary to produce **luteinizing hormone (LH)** and **follicle-stimulating hormone (FSH)**. Although LH and FSH derive their names from their functions in females, they are equally essential in males. These hormones stimulate the testes to produce **testosterone** and the ovaries to produce **estrogen**. Testosterone and estrogen stimulate the bodily changes that occur during puberty and of course play major roles in the continuing physiological and behavioral events that occur during a person's reproductive life.

The Male Reproductive System Includes the Testes and Accessory Structures

The male reproductive system includes the testes, which produce testosterone and sperm; glands that secrete substances that activate and nourish sperm; and tubes that store sperm and conduct them out of the body (**FIG. 42-9** and **TABLE 42-1**). Testosterone and sperm are produced nearly continuously, beginning at puberty and continuing until death.

TABLE 42-1 The Human Male Reproductive Tract

Structure	Function
Testis (male gonad)	Produces sperm and testosterone
Epididymis and vas deferens (ducts)	Store sperm; conduct sperm from the testes to the urethra
Urethra (duct)	Conducts semen from the vas deferens and urine from the urinary bladder to the tip of the penis
Penis	Deposits sperm in the female reproductive tract
Seminal vesicles (glands)	Secrete alkaline fluid containing fructose and prostaglandins into the semen
Prostate gland	Secretes alkaline fluid containing nutrients and enzymes into the semen
Bulbourethral glands	Secrete alkaline fluid containing mucus into the semen

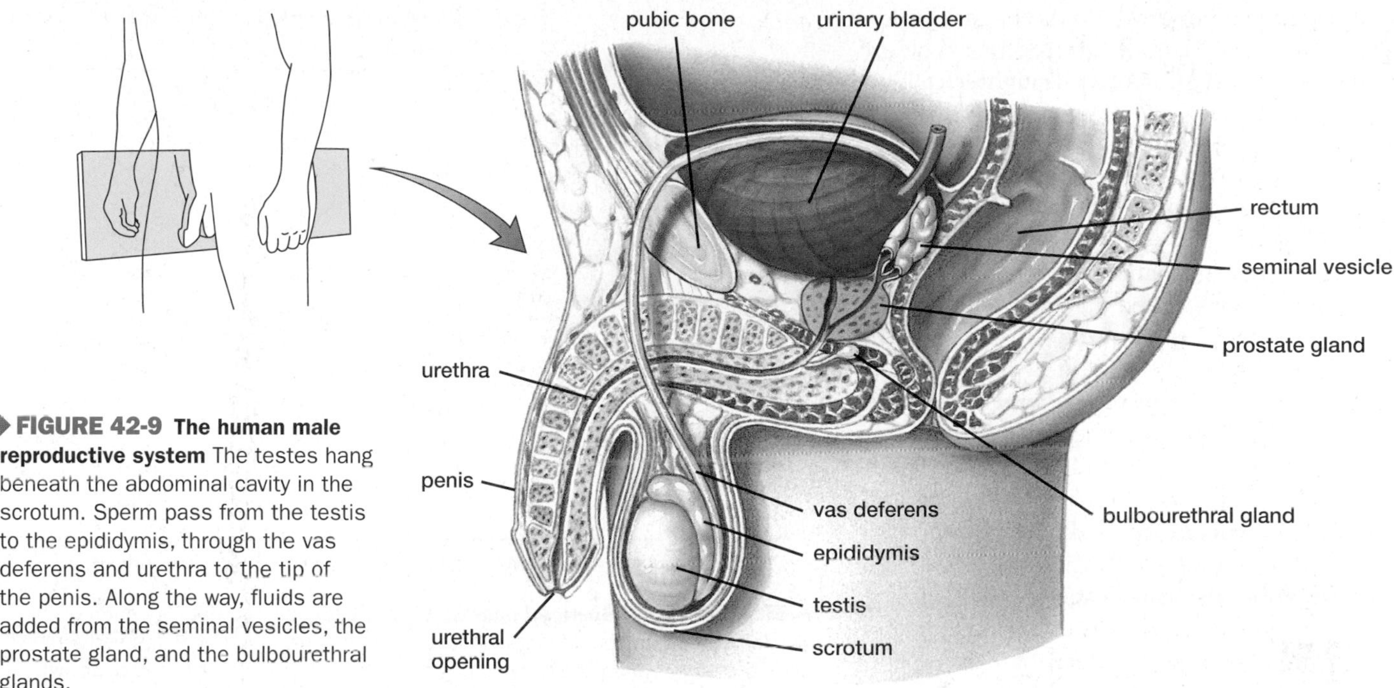

FIGURE 42-9 The human male reproductive system The testes hang beneath the abdominal cavity in the scrotum. Sperm pass from the testis to the epididymis, through the vas deferens and urethra to the tip of the penis. Along the way, fluids are added from the seminal vesicles, the prostate gland, and the bulbourethral glands.

Sperm Are Produced in the Testes

The testes are located in the **scrotum**, a pouch that hangs outside the main body cavity. This location keeps the testes about 1° to 6°F (0.5° to 3°C) cooler than the core of the body, depending on whether the man is standing (cooler) or sitting (warmer) and what type of clothing he is wearing. Cooler temperatures promote sperm development. Each testis is nearly filled with coiled, hollow **seminiferous tubules**, in which sperm are produced (**FIG. 42-10a**). **Interstitial cells**, which synthesize testosterone, are located in the spaces between the tubules (**FIG. 42-10b**).

Just inside the wall of each seminiferous tubule lie two major types of cells: (1) stem cells called **spermatogonia** or spermatogonial stem cells, which give rise to sperm, and (2) much larger **Sertoli cells**, which nourish the developing sperm and regulate their growth.

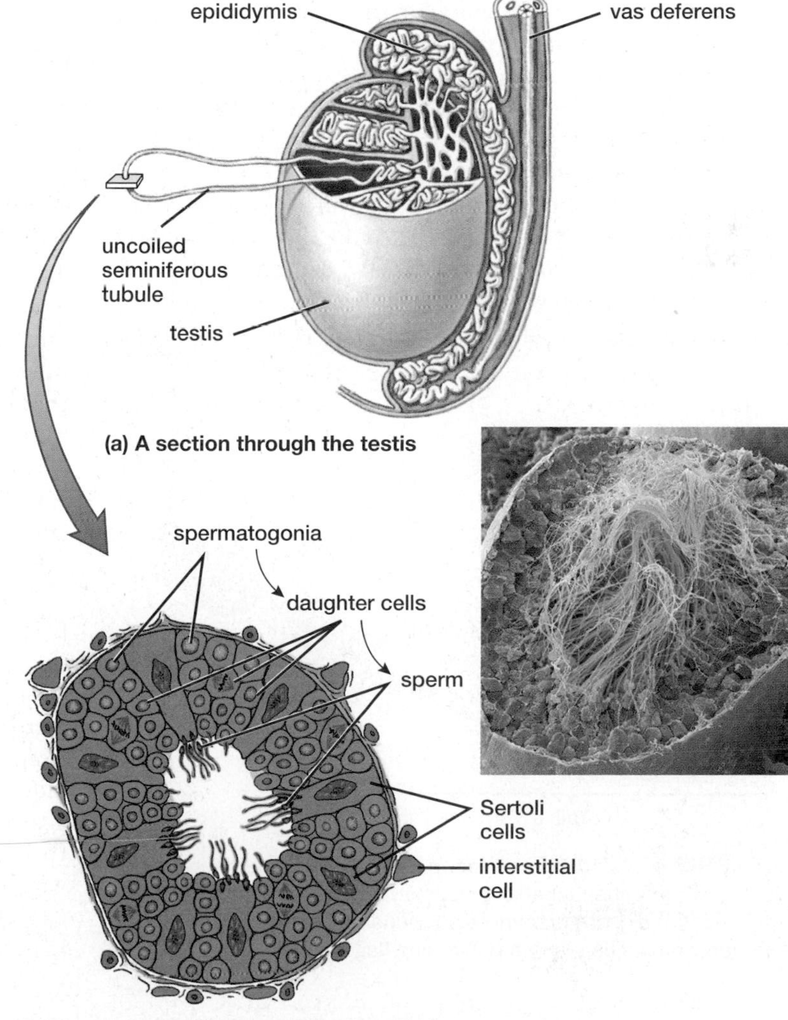

FIGURE 42-10 The anatomy of the testes (a) A section of the testis, showing the seminiferous tubules, epididymis, and vas deferens. **(b)** The walls of the seminiferous tubules are lined with Sertoli cells and spermatogonia. As the spermatogonia divide, the daughter cells move inward. Mature sperm are freed into the central cavity, which is continuous with the inside of the epididymis. Testosterone is produced by the interstitial cells.

Spermatogonia are diploid cells that undergo mitotic cell division, forming two types of daughter cells (**FIG. 42-11**). Daughter cells of one type remain spermatogonia, ensuring a steady supply throughout a man's life. Daughter cells of the other type become committed to **spermatogenesis**, the processes that produce sperm.

In a process stimulated by testosterone and FSH, a committed daughter cell passes through several rounds of mitotic cell division. Each final offspring cell differentiates into a large cell called a **primary spermatocyte**, which then undergoes meiotic cell division (see Chapter 10). At the end of meiosis I, each primary spermatocyte gives rise to two haploid **secondary spermatocytes**. Each secondary spermatocyte undergoes meiosis II, producing two **spermatids**. Thus, each diploid primary spermatocyte generates a total of four haploid spermatids. Spermatids differentiate into sperm without further cell division. Spermatogonia, spermatocytes, and spermatids are enfolded in the Sertoli cells. As spermatogenesis proceeds, the developing sperm migrate to the central cavity of the seminiferous tubule into which the mature sperm are released (see Fig. 42-10b).

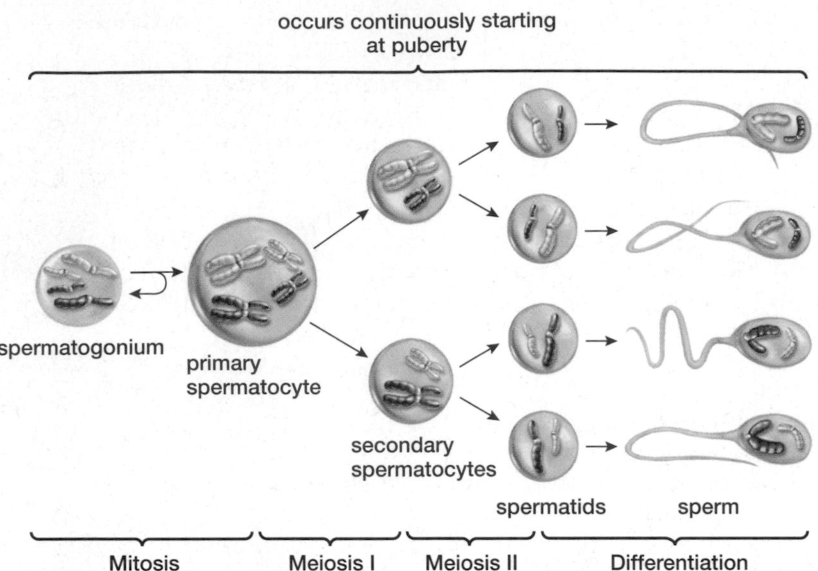

▲ **FIGURE 42-11 Spermatogenesis** A spermatogonium divides by mitotic cell division; one of its two daughter cells remains a spermatogonium (curved arrow), while the other enlarges and becomes a primary spermatocyte (straight arrow), which will undergo meiotic cell division followed by differentiation, producing haploid sperm. For clarity, only two pairs of chromosomes are shown; however, human cells have 23 pairs of chromosomes.

A human sperm (**FIG. 42-12**) is unlike any other cell of the body. It contains little cytoplasm, so the haploid nucleus nearly fills the sperm's head. Atop the nucleus lies a sac of enzymes called an **acrosome**. As we will see, the enzymes dissolve protective layers that surround the egg, enabling the sperm to penetrate through them. Behind the head is the midpiece, packed with mitochondria. The mitochondria provide the energy needed to move the tail, which is actually a long flagellum. Whip-like movements of the tail propel the sperm through the female reproductive tract.

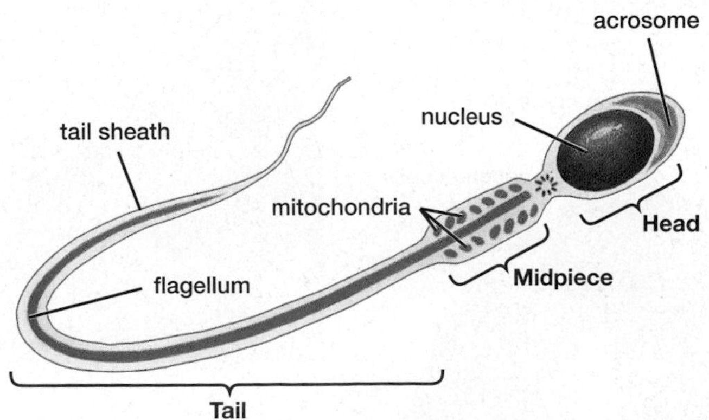

▲ **FIGURE 42-12 A human sperm cell** A mature sperm contains little more than a haploid nucleus, an acrosome (containing enzymes that digest the barriers surrounding the egg), mitochondria for energy production, and a tail (a long flagellum) for locomotion.

Accessory Structures Contribute to Semen and Conduct the Sperm Outside the Body

The seminiferous tubules merge to form the **epididymis**, a long folded tube (see Figs. 42-9 and 42-10a). In the epididymis, sperm are stored and continue to mature. Sperm then move from the epididymis into the **vas deferens**, a tube that carries sperm out of the scrotum. Most of the roughly hundred million sperm produced each day are stored in the vas deferens and epididymis. The vas deferens joins the **urethra**, which runs to the tip of the **penis**. The urethra conducts urine out of the body during urination and sperm out of the body during ejaculation.

Semen consists of about 5% sperm mixed with secretions from three types of glands that empty into the vas deferens or the urethra: the seminal vesicles, the prostate gland, and the bulbourethral glands (see Fig. 42-9 and Table 42-1). Paired **seminal vesicles** secrete about 60% of the semen. Liquid from the seminal vesicles is rich in fructose, a sugar that provides energy for the sperm. The liquid's slightly alkaline pH protects the sperm from the acidity of urine remaining in the man's urethra and from acidic secretions in a woman's vagina. Seminal vesicle fluid also contains prostaglandins (see Chapter 38), which stimulate uterine contractions that help to transport the sperm up the female reproductive tract.

The **prostate gland** produces an alkaline, nutrient-rich secretion that makes up about 30% of the semen. The prostate fluid includes enzymes that increase the fluidity of the semen after it is released into the vagina, allowing the sperm to swim more freely. Paired **bulbourethral glands** secrete a small amount of alkaline mucus into the urethra, generally before ejaculation, which helps to neutralize residual urine before sperm arrive and may provide some lubrication if the fluid leaks out of the penis. In some men, this pre-ejaculatory fluid contains sperm, although usually at a low level.

CASE STUDY CONTINUED

To Breed a Rhino

Mammalian sperm don't live very long after ejaculation. Therefore, for a rhino cow to become pregnant, the sperm must be placed in her vagina or uterus, whether by mating or by artificial insemination, at about the same time that ovulation occurs. The simplest way to ensure correct timing is to induce ovulation and have mating or artificial insemination the same day. How might ovulation be induced, in rhinos or in other mammals?

The Female Reproductive System Includes the Ovaries and Accessory Structures

The female reproductive system consists of the ovaries and structures that accept sperm, conduct the sperm to the egg, and nourish the developing embryo (**FIG. 42-13** and **TABLE 42-2**).

TABLE 42-2 The Human Female Reproductive Tract

Structure	Function
Ovary (female gonad)	Produces eggs, estrogen, and progesterone
Fimbriae (at the opening of the uterine tube)	Bear cilia that sweep the egg into the uterine tube
Uterine tube	Conducts the egg to the uterus; site of fertilization
Uterus	Muscular chamber where the embryo develops
Cervix	Nearly closes off the lower end of the uterus; supports the developing embryo during pregnancy
Vagina	Receptacle for semen; birth canal

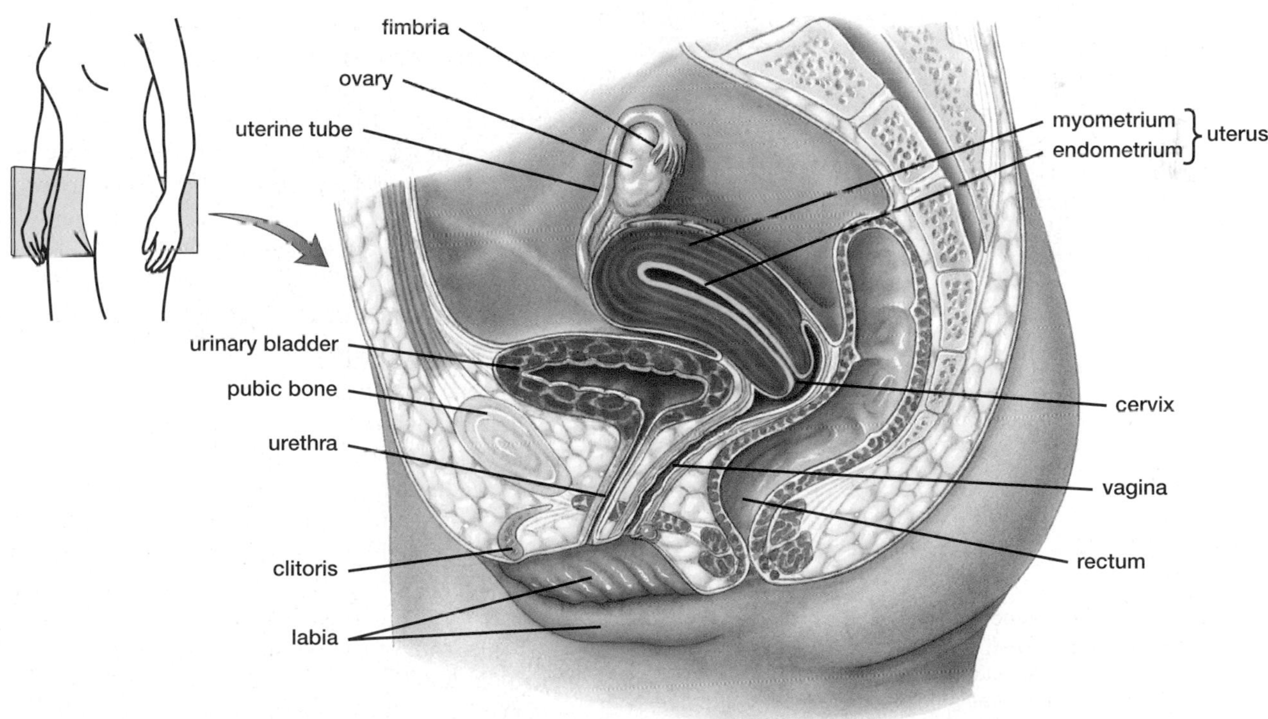

▲ **FIGURE 42-13 The human female reproductive system** Eggs are produced in the ovaries and enter the uterine tube. Sperm and egg usually meet in the uterine tube, where fertilization and very early development occur. The early embryo implants in the lining of the uterus, where development continues. The vagina receives sperm and serves as the birth canal.

Egg Development in the Ovaries Begins Before Birth

Oogenesis, the formation of egg cells, begins in the developing ovaries of a female embryo (**FIG. 42-14**). Oogenesis starts with the formation of diploid cells called **oogonia** (singular, oogonium) as early as the 6th week of embryonic development. From about the 9th through the 20th weeks, the oogonia enlarge and differentiate, becoming **primary oocytes**. By about the 20th week, all of the primary oocytes have begun meiotic cell division, but meiosis is suspended during prophase of meiosis I. None of the primary oocytes will resume meiotic cell division until puberty.

Therefore, a woman is born with her lifetime's supply of primary oocytes—about 1 to 2 million. Many of these die each day, but about 400,000 remain at puberty. A few oocytes resume meiotic cell division during each month of a woman's reproductive span, from puberty to menopause at about age 50.

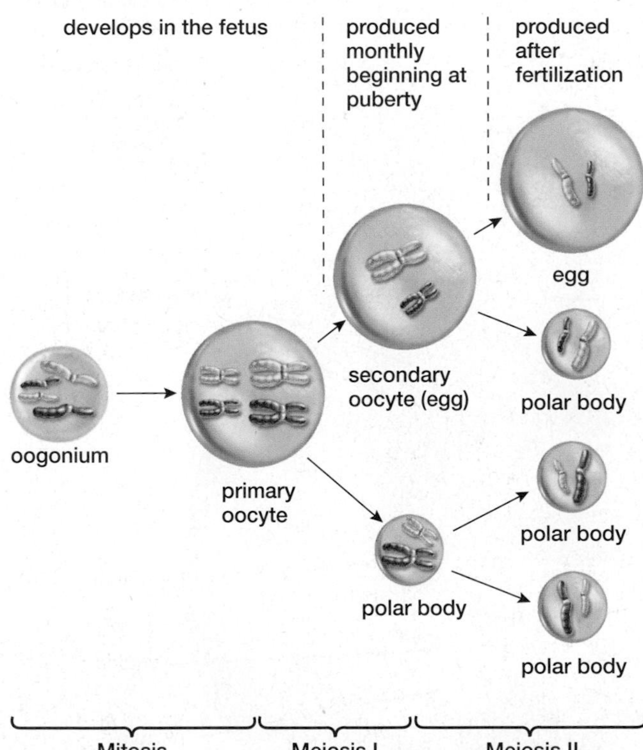

▲ **FIGURE 42-14 Oogenesis** An oogonium enlarges to form a primary oocyte. At meiosis I, almost all the cytoplasm is included in one daughter cell, the secondary oocyte. The other daughter cell is a small polar body that contains chromosomes but little cytoplasm. At meiosis II, almost all the cytoplasm of the secondary oocyte is included in the egg, and a second small polar body discards the remaining "extra" chromosomes. The first polar body may also undergo the second meiotic division. The polar bodies eventually degenerate. In humans, meiosis II does not occur unless a sperm penetrates the egg.

The Ovary Produces Eggs, Estrogen, and Progesterone During the Menstrual Cycle

In a mature ovary, each oocyte is surrounded by a layer of smaller cells. Together, the oocyte and these accessory cells make up a **follicle** (**FIG. 42-15** ①). Follicle development and ovulation are governed by interactions among hormones produced by the hypothalamus, anterior pituitary, and ovary during the **menstrual cycle** (from the Latin "mensis," meaning "month").

Roughly once a month, the hypothalamus secretes GnRH, which stimulates the anterior pituitary to release LH and FSH. FSH stimulates about a dozen follicles to begin developing ②. The small accessory cells multiply, providing nourishment for the developing oocyte. The follicle cells also release estrogen into the bloodstream. Usually, only one follicle completely matures during each menstrual cycle ③.

The maturing follicle secretes increasing amounts of estrogen, which stimulates a surge of LH that causes the primary oocyte to complete meiosis I, dividing into a single **secondary oocyte** and the first **polar body** (see Fig. 42-14). The polar body is a small cell, with very little cytoplasm; it cannot be fertilized by sperm. The surge of LH also causes ovulation, as the follicle erupts through the surface of the ovary, releasing its secondary oocyte ④. The secondary oocyte will not undergo meiosis II unless it is fertilized. For convenience, we will refer to the ovulated secondary oocyte as the egg.

Some of the accessory follicle cells leave the ovary with the egg, but most remain in the ovary ⑤, where they enlarge, forming a temporary gland called the **corpus luteum** ⑥. The corpus luteum secretes both estrogen and a second hormone called **progesterone**. The combination of estrogen and progesterone inhibits further release of GnRH, LH, and FSH, thereby preventing the development of any more follicles. If the egg is not fertilized, the corpus luteum will degenerate within a few days ⑦.

Accessory Structures Include the Uterine Tubes, Uterus, and Vagina

Each ovary sits near the open end of a **uterine tube** (also called the oviduct or Fallopian tube), which leads from the ovary to the uterus. The opening to the uterine tube is fringed with ciliated projections, called *fimbriae* (see Fig. 42-13). The cilia create a current that sweeps the ovulated egg into the uterine tube. There, the egg may encounter sperm and be fertilized. Cilia lining the uterine tube transport the fertilized egg down the tube into the **uterus**.

The wall of the uterus has two layers that correspond to its dual functions of nourishing the developing embryo and delivering a child. The inner lining, or **endometrium**, will form the mother's contribution to the **placenta**, the structure that transfers oxygen, carbon dioxide, nutrients, and wastes between mother and embryo (see Chapter 43). The outer muscular wall of the uterus, called the **myometrium**, contracts during childbirth, expelling the infant out of the uterus.

The lower end of the uterus is nearly closed off by the **cervix**, a ring of connective tissue that encircles a tiny opening. The cervix holds the developing baby in the uterus and then expands during labor, permitting passage of the child. Beyond the cervix is the **vagina**, which opens to the outside. The vagina serves both as the receptacle for the penis and sperm during intercourse and as the birth canal. The vaginal lining is acidic, which reduces the likelihood of infections.

Estrogen and Progesterone Levels Regulate the Development of the Endometrium

Developing follicles secrete estrogen, which stimulates the endometrium to become thicker and grow an extensive network of blood vessels and glands that secrete carbohydrates, lipids, and proteins. After ovulation, estrogen and progesterone released by the corpus luteum further stimulate the development of the endometrium. Thus, if an egg is fertilized, when the developing embryo reaches the uterus, it encounters a rich environment for growth. If fertilization does not occur, however, the corpus luteum disintegrates, estrogen and progesterone levels fall, and the enlarged endometrium disintegrates. The uterus then contracts (often causing menstrual cramps) and expels the excess endometrial tissue, a process called **menstruation**.

The Embryo Sustains Its Own Pregnancy

If fertilization does not occur, degeneration of the corpus luteum ends the current menstrual cycle and allows a new one to begin. However, if fertilization does occur, the embryo starts to secrete an LH-like hormone called **chorionic gonadotropin (CG)** shortly after implanting in the uterus. This hormone travels in the bloodstream to the ovary, where it functions to keep the corpus luteum alive. The corpus luteum continues to secrete estrogen and progesterone for the first few months of pregnancy. These hormones continue to stimulate the development of the endometrium, nourishing the embryo and sustaining the pregnancy. Some CG is excreted in the mother's urine, where it can be detected to confirm pregnancy.

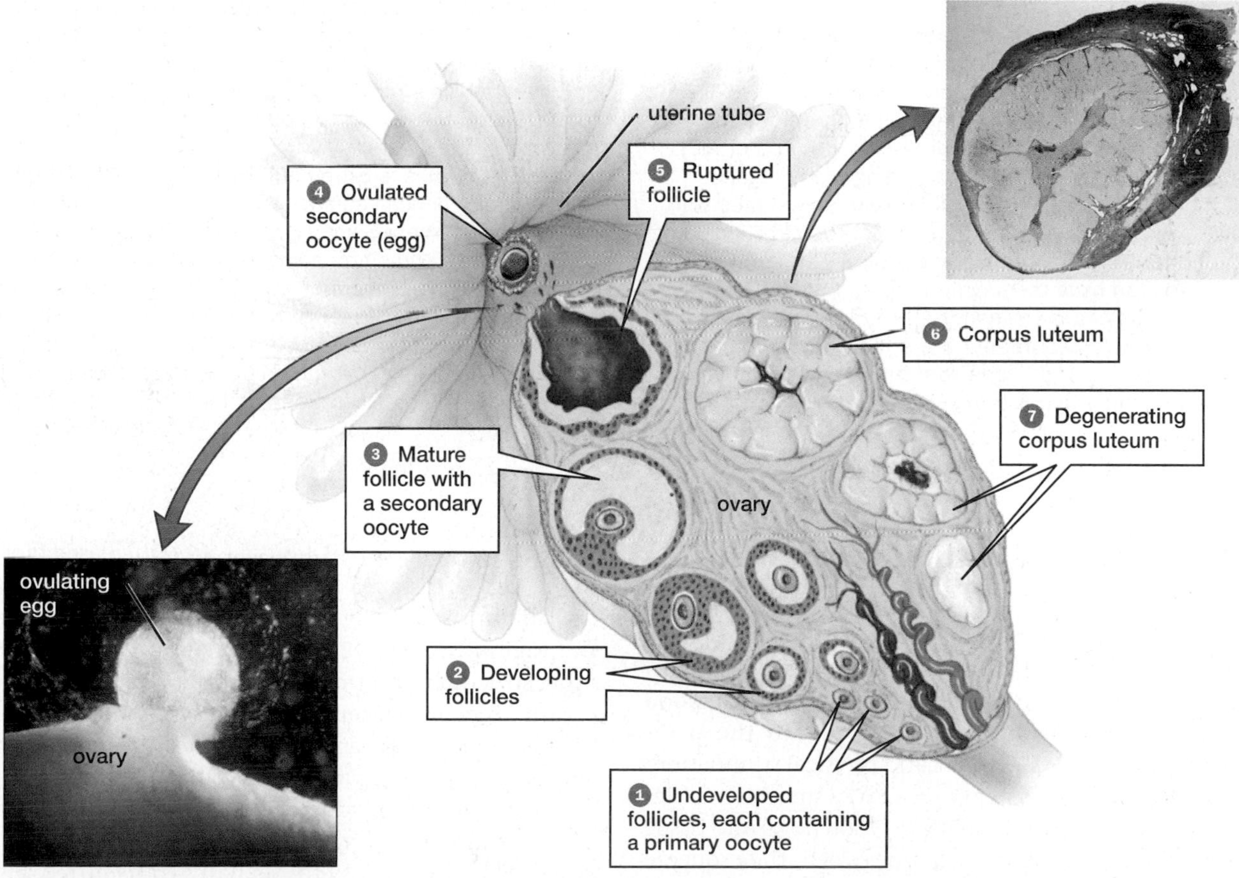

▲ **FIGURE 42-15 Follicle development** For convenience, this diagram shows all of the stages of follicle development throughout a complete menstrual cycle, going clockwise from lower right. In a real ovary, all the stages would not be present at the same time, and the follicle would not circle around the ovary as it develops.

Sexually Transmitted Diseases

Sexually transmitted diseases (STDs) are caused by bacteria, viruses, protists, or arthropods that infect the sexual organs and reproductive tract. As their name implies, they are transmitted primarily through sexual contact.

Bacterial Infections

The U.S. Centers for Disease Control and Prevention (CDC) estimates that there are about 800,000 new cases of **gonorrhea** each year in the United States. Gonorrhea bacteria penetrate membranes lining the urethra, anus, cervix, uterus, uterine tubes, and throat. Infected men may experience painful urination and discharge of pus from the penis. Female symptoms include vaginal discharge and painful urination, but the symptoms are usually less severe than they are in men. Gonorrhea infections may also cause a buildup of scar tissue in the uterine tubes, resulting in infertility. Many infected individuals experience few or no symptoms, so they may not realize they have the disease and seek treatment. Therefore, they may continue to infect their sexual partners. Although gonorrhea can be cured with the right antibiotics, many strains of gonorrhea have evolved resistance to common antibiotics. Gonorrhea bacteria may also attack the eyes of infants born to infected mothers.

About 55,000 new cases of **syphilis** occur in the United States each year. Syphilis bacteria penetrate the mucous membranes of the genitals, lips, anus, or breasts. Syphilis begins with a sore at the site of infection, sometimes followed by a rash, often on a totally different site on the body. If untreated, the bacteria spread, damaging many organs, including the skin, kidneys, heart, and brain, sometimes with fatal results. Syphilis is easily cured with antibiotics, including penicillin and tetracycline, but many people in the early stages of syphilis infection have mild symptoms and do not seek treatment. Syphilis can be transmitted to the embryo during pregnancy. Some infected infants are stillborn or die shortly after birth; others suffer damage to the skin, teeth, bones, liver, and central nervous system.

Chlamydia is the most common bacterial STD: The CDC estimates that there are about 2.9 million new infections in the United States each year. Chlamydia causes inflammation of the urethra in males and of the urethra and cervix in females. In many cases, there are no obvious symptoms, so the infection goes untreated and the bacteria may be passed to sexual partners. A chlamydia infection may block the uterine tubes, resulting in sterility. Chlamydia can cause eye inflammation in infants born to infected mothers and is a major cause of blindness in developing countries. Chlamydia is readily cured with antibiotics such as doxycycline or azithromycin.

Viral Infections

Acquired immune deficiency syndrome (AIDS) is caused by the human immunodeficiency virus (HIV; see Chapter 37). Worldwide, each year about 2 million people become infected with HIV and 1.2 million people die of AIDS. In the United States, HIV causes about 50,000 new infections and 13,000 deaths each year. As its name implies, AIDS weakens the immune system. When first infected with HIV, most people have either mild flu-like symptoms or no symptoms at all. Later, as the immune system continues to deteriorate, HIV causes a wide variety of symptoms, including a greatly increased susceptibility to other, often rare illnesses and to certain types of cancer. HIV is spread primarily by sexual activity, contaminated blood and needles, and from mother to newborn. There is no cure for AIDS, but drug combinations can keep the disease under control for many years.

Genital herpes is extremely common; the CDC estimates that 16% of American adults—more than 24 million people—have genital herpes. Genital herpes may cause painful blisters on the genitals and surrounding skin and is transmitted primarily when blisters are present. Even when the blisters have healed, herpes virus remains in the body, emerging unpredictably, possibly in response to stress. Antiviral drugs reduce the severity of outbreaks, but there is no cure. A pregnant woman with active symptoms can transmit the virus to her developing

During Copulation, Sperm Are Deposited in the Vagina

The male role in copulation begins with erection of the penis. Before erection, the penis is relaxed (flaccid) because smooth muscles surrounding the arterioles that supply it with blood are contracted, allowing little blood flow into the penis (**FIG. 42-16a**). Under psychological and physical stimulation, signals from the nervous system cause these smooth muscles to relax. The arterioles dilate and more blood flows into tissue spaces within the penis. As these tissues swell, they squeeze off the veins that drain the penis (**FIG. 42-16b**). Blood fills the penis, causing an erection.

After the penis is inserted into the vagina, movements further stimulate touch receptors on the penis, triggering ejaculation. Muscles encircling the epididymis, vas deferens, and urethra contract, forcing semen out through the penis and into the vagina. A typical ejaculation consists of about 2 to 5 milliliters of semen containing about 100 million to 400 million sperm.

In the female, sexual arousal causes increased blood flow to the vagina, to paired folds of tissue called the **labia** (singular, labium), and to the **clitoris**, a small structure just above the vagina (see Fig. 42-13). Stimulation of the clitoris and other structures near the entrance to the vagina may result in orgasm. Female orgasm is not necessary for fertilization, but contractions of the vagina and uterus during orgasm probably help to move sperm up toward the uterine tubes.

Intimate contact during copulation provides a favorable environment for transmitting disease organisms, as we describe in "Health Watch: Sexually Transmitted Diseases" on this page.

During Fertilization, the Sperm and Egg Nuclei Unite

During intercourse, the penis releases sperm into the vagina. The sperm move through the cervix, into the uterus, and finally enter the uterine tubes. Sperm, under ideal conditions,

embryo, in very rare cases causing mental or physical disability or stillbirth. Herpes can also be transmitted to babies during childbirth.

Human papillomavirus (HPV) (**FIG. E42-1a**) infects the majority of sexually active people at some time in their lives; at any given time, about 80 million Americans have HPV. Most experience no symptoms and never know that they have been infected. However, the virus sometimes causes warts on the labia, vagina, cervix, or anus in women and on the penis, scrotum, or groin in men. Certain strains of HPV cause most, and possibly all, cases of cervical cancer, which kills more than 4,000 women each year in the United States. Two vaccines are now available that prevent infections with the most common cancer-causing forms of HPV. The CDC recommends HPV vaccination for males as well as females, preferably before they become sexually active. The HPV vaccines cannot cure existing infections.

Protist and Arthropod Infections

Trichomoniasis is caused by a protist that colonizes the mucous membranes lining the urinary tract and genitals of both males and females. The CDC estimates that about 3.7 million Americans have trichomoniasis. Symptoms may include itching or burning sensations and, sometimes, a discharge from the penis or vagina. However, most people don't have any symptoms, so they aren't treated and may pass the disease to their sexual partners. Prolonged infections can cause sterility. Trichomoniasis can be cured with oral antibiotics such as metronidazole.

Pubic lice are tiny insects that live and lay their eggs in pubic hair (**FIG. E42-1b**). Their mouthparts are adapted for penetrating skin and sucking blood and body fluids, a process that causes severe itching. Although pubic lice probably do not directly spread disease, bacterial infections frequently occur when people scratch the itchy bites. About 3 million people in the U.S. contract pubic lice each year. Topical insecticides, such as permethrin, are used to kill the lice.

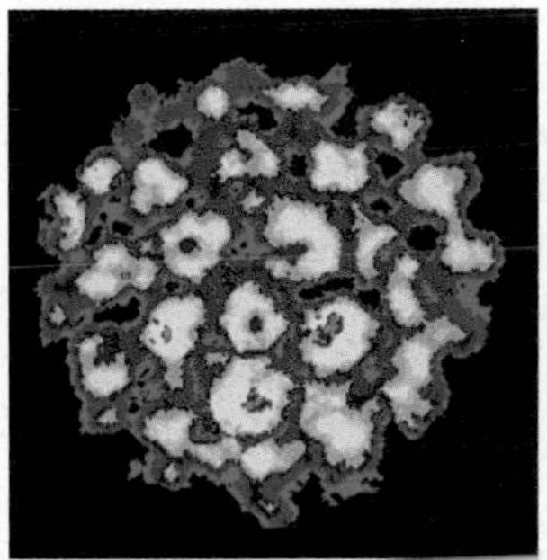

(a) The human papillomavirus

(b) Pubic lice

▲ **FIGURE E42-1 Agents of sexually transmitted diseases** **(a)** The DNA of some strains of the human papillomavirus becomes inserted into chromosomes of cells of the cervix. Proteins synthesized from the instructions in the viral DNA can promote uncontrolled cell division, causing cancer. **(b)** Pubic lice are called "crabs" because of their shape and the tiny claws on the ends of their legs, which they use to cling to pubic hair.

THINK CRITICALLY Envision yourself as a physician whose patients sometimes ask about birth control. Provide a contraceptive method for each of the following patients, and explain why it would be appropriate: couple A, a man and woman who have intercourse regularly but never want to have children; couple B, a man and woman who have intercourse regularly and want to have children someday; and person C, a single woman who occasionally has intercourse with single men and does not wish to become pregnant. Be sure to include the level of protection against STDs that each person is likely to need.

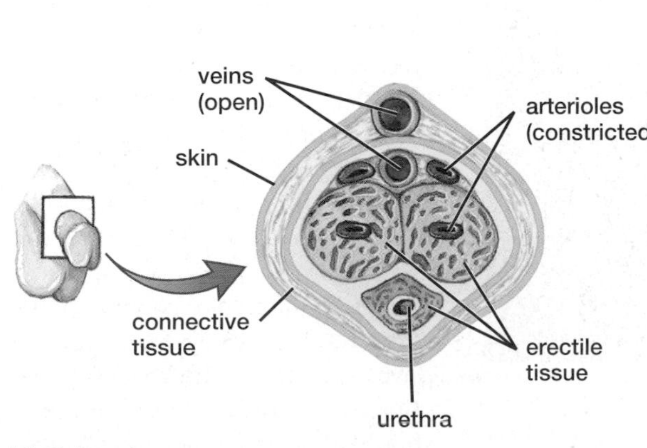

(a) Relaxed

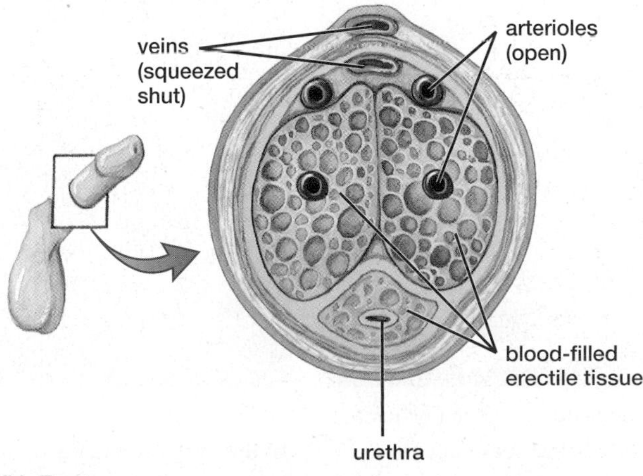

(b) Erect

▲ **FIGURE 42-16 Changes in blood flow in the penis cause erection** **(a)** Smooth muscles encircling the arterioles leading into the penis are usually contracted, limiting blood flow. **(b)** During sexual stimulation, these muscles relax, and blood flows into spaces within the penis. The swelling penis squeezes off the veins through which blood would leave the penis, increasing its blood pressure and causing it to become elongated and firm.

may live for 2 to (rarely) 4 days inside the female reproductive tract, and an unfertilized egg remains viable for a day or so. Therefore, if copulation occurs within a day or two of ovulation, the sperm may meet an egg in one of the uterine tubes.

When it leaves the ovary, the egg is surrounded by accessory follicle cells. These cells, now called the **corona radiata**, form a barrier between the sperm and the egg (**FIG. 42-17a**). A second barrier, the jelly-like **zona pellucida** ("clear area"), lies between the corona radiata and the egg. In the uterine tube, hundreds of sperm surround the corona radiata (**FIG. 42-17b**). Each sperm releases enzymes from its acrosome. These enzymes weaken both the corona radiata and the zona pellucida, allowing the sperm to wriggle through to the egg. If there aren't enough sperm, not enough enzymes are released, and none of the sperm will reach the egg. Perhaps 1 in 100,000 of the sperm deposited in the vagina reaches the uterine tube, and 1 in 20 of those encounters the egg, so only a few hundred join in attacking the barriers around the egg.

When the first sperm contacts the egg's surface, the plasma membranes of egg and sperm fuse, and the sperm's head enters the egg cytoplasm. Sperm entry triggers two critical changes in the egg: First, vesicles near the surface of the egg release chemicals into the zona pellucida that reinforce it and prevent additional sperm from entering. Second, the egg undergoes the second division of meiosis (see Fig. 42-14). Fertilization occurs as the haploid nuclei of sperm and egg fuse, forming the diploid nucleus of the zygote.

Defects in the male or female reproductive system can prevent fertilization. For example, a blocked uterine tube can prevent sperm from reaching the egg. A man with a low sperm count (fewer than 20 million sperm per milliliter of semen) may be unable to impregnate a woman through sexual intercourse because too few sperm reach the egg. Today, many couples who have difficulty conceiving a child seek help through artificial insemination or *in vitro* fertilization (IVF; see "Health Watch: High-Tech Reproduction").

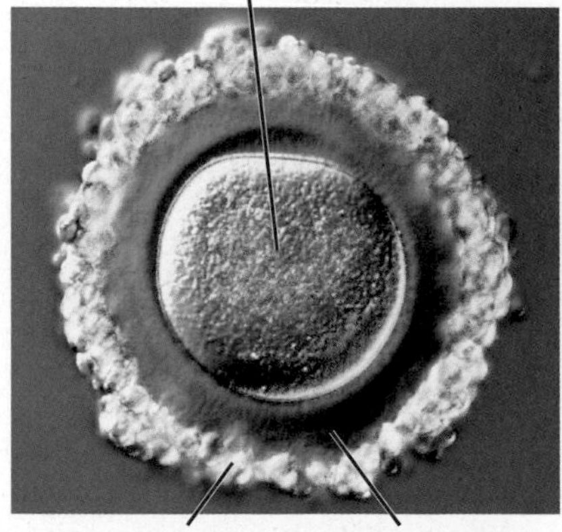

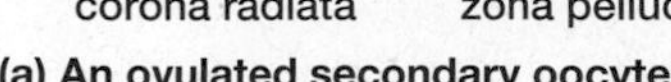

(a) An ovulated secondary oocyte

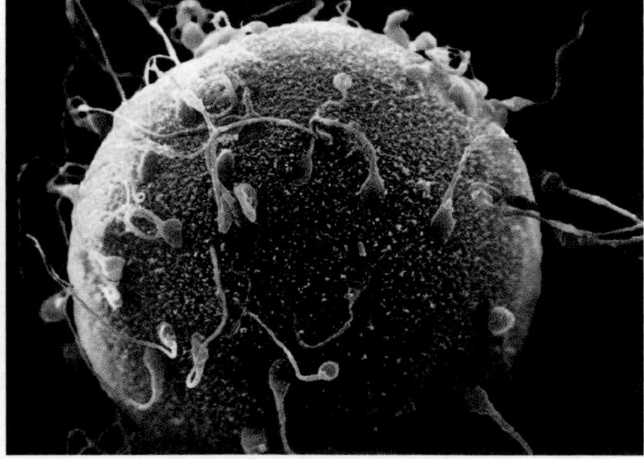

(b) Sperm surrounding an oocyte

▲ **FIGURE 42-17 The secondary oocyte and fertilization (a)** A human secondary oocyte (egg) shortly after ovulation. Sperm must digest their way through the corona radiata and the zona pellucida to reach the oocyte. **(b)** Sperm surround the oocyte, attacking the corona radiata and zona pellucida.

Have You Ever Wondered ...
How Porcupines Mate?

The 30,000 or so needle-sharp quills that protect a porcupine from predators seem like they might also prevent close contact between two porcupines. Nonetheless, porcupines do manage to mate. When the female is ready to mate (which happens only once a year), she emits a scent that attracts males. The arriving males compete for her attention with displays that include loud screams and a forceful spray of urine that soaks the female. After she chooses a mate from among her suitors, the female flattens down the quills on the rear potion of her body and raises her tail to expose its quill-free underside. These changes make it safe for the male to mount the female, and copulation proceeds.

CHECK YOUR LEARNING

Can you . . .

- describe the human male and female reproductive tracts, including the gonads and accessory structures, and the functions of each structure?
- explain how hormonal interactions control reproduction in men and women?
- describe spermatogenesis and oogenesis, the locations and timing of these two processes, the important features of sperm and eggs, and how fertilization occurs?

42.3 HOW CAN PEOPLE PREVENT PREGNANCY?

Many people want to engage in sex without risking pregnancy. Historically, limiting fertility has not been easy. In the past, women in some cultures have tried such inventive techniques as swallowing froth from the mouth of a camel or placing crocodile dung in the vagina. Most of these ancient remedies were not, of course, effective methods of **contraception** (preventing pregnancy). But even after effective methods were developed, their use was often limited by social or legal restrictions. In the United States, contraception did not become legal nationwide until 1965. Since the 1970s, however, use of several effective contraceptive techniques has beomce widespread. All forms of birth control have possible drawbacks. The choice of contraceptive method should always be made in consultation with a health professional.

High-Tech Reproduction

About 10% to 15% of couples have difficulty conceiving a child. When this difficulty stems from absent or irregular ovulation, fertility drugs may help promote conception. These drugs (which contain or cause the release of FSH and LH) stimulate ovulation. However, fertility drugs often cause several eggs to be released simultaneously, with the result that the rate of multiple births in the United States has nearly doubled since 1980. Multiple births are much riskier than single births, for both the mother and her children.

Worldwide, about 5 million people now alive were conceived by *in vitro* fertilization (IVF; literally, "fertilization in glass"). In most cases, the first step in IVF is administering fertility drugs that stimulate follicle development. When the follicles are about to ovulate, a surgeon inserts a long needle into each ripe follicle and sucks out its oocyte. The oocytes are placed in a dish with sperm. After fertilization, the zygotes divide. The resulting embryos may be cultured for about 3 days, usually reaching the eight-cell stage, or for 5 days, becoming more advanced *blastocysts* (see Chapter 43). The advantage of using 3-day-old embryos is that they spend less time in possibly stressful conditions in culture. However, if the IVF lab is proficient at embryo culture and routinely produces healthy blastocysts, then blastocysts are often preferred, because this is the stage at which an embryo would normally reach the uterus following natural fertilization, so implantation is more likely to be successful.

To transfer the embryos into the uterus, embryos are gently sucked into a tube and expelled into the uterus. For young women, usually only one or two embryos are transferred. In women over age 40, who often have difficulty carrying a baby to term, as many as four embryos may be transferred. Transplanting more than one embryo increases the success rate, but also increases the probability of multiple births. Embryos that are not immediately implanted can be frozen for later use.

Conceiving a child through *in vitro* fertilization can be quite expensive. In the United States, IVF costs about $10,000 to $15,000 per attempt. In young women, IVF has an average live birth rate of about 40%, so the average cost of a successful birth via IVF is $25,000 to $50,000. The live birth rate is much lower for women over 35 years of age, dropping to below 15% for women over 40. Therefore, the likely cost for a successful birth is much higher for older women.

Even men whose sperm are incapable of swimming or normal fertilization may be able to father children through intracytoplasmic sperm injection (ICSI). In ICSI, immature sperm cells are extracted from the testes and injected through a tiny, sharp pipette directly into an egg's cytoplasm (**FIG. E42-2**). The live birth rate for IVF using ICSI is about the same as for eggs fertilized by sperm in a dish.

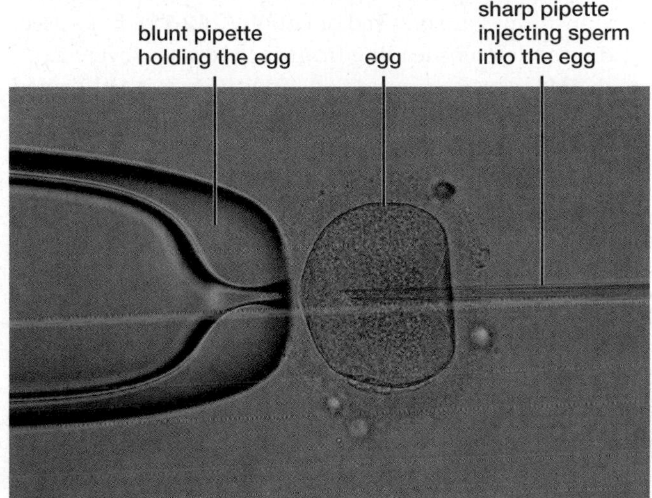

▲ **FIGURE E42-2 Intracytoplasmic sperm injection** An egg is held in place on the tip of a smooth glass pipette. A much smaller, sharp pipette injects a single sperm cell directly into the egg's cytoplasm.

Using sperm-sorting technology, parents can even change the odds of having a male or female child. This may be medically important if the parents are carriers of sex-linked disorders, but some parents are just seeking to balance the sex of their children. Sperm-sorting methodology, based on the difference in the amount of DNA in X-bearing sperm (more) versus Y-bearing sperm (less), provides 80% to 90% separation. The sorted sperm can be used either for artificial insemination or *in vitro* fertilization. Sperm sorting is now in clinical trials in the United States; in a few other countries, sperm sorting is already available in specialized IVF clinics. For absolute sex certainty during IVF, the sex of an embryo can be determined before implantation with 100% accuracy by carefully removing one of the cells and analyzing its karyotype.

Finally, it is now possible to produce children with *three* "parents": a nuclear father, a nuclear mother, and a mitochondrial mother. Why would anyone do that? About 1 in 2,000 to 5,000 children is born with defective mitochondria. In severe cases, the defects can cause nerve or muscle damage, blindness, or heart failure. In mammals, sperm do not provide any mitochondria to a fertilized egg—all the mitochondria were already present in the unfertilized oocyte. With three-parent IVF, a prospective mother who has defective mitochondria could still have healthy children. The nucleus would be removed from an unfertilized donor egg, obtained from a woman with normal mitochondria. Then, a nucleus from an embryo produced through IVF would be injected into the donor egg. The resulting cell would have nuclear DNA from the IVF mother and father and mitochondria from the egg donor. Because mitochondria contain some DNA, the child would have three genetic parents.

Producing three-parent children has not yet been approved in United States, on the grounds that researchers have have not yet demonstrated that the procedure will have no negative long-term effects on the health of the child. Some other countries, however, do not restrict the procedure, and a growing number of children with three biological parents can be found in Ukraine, Great Britain, and elsewhere.

CONSIDER THIS In 2017, scientists used CRISPR-Cas9 DNA editing to repair a defective gene in a human embryo. Should this be allowed? Is preventing disease and disability in the children sufficient justification for changing the genetic makeup of a human being? What about "designer" children whose DNA may be changed solely to produce an athlete or supermodel body type?

Sterilization Provides Permanent Contraception

Perhaps the most foolproof method of contraception is **sterilization**, in which the pathways through which sperm or eggs must travel are blocked or cut (**FIG. 42-18**). In a vasectomy, the vas deferens leading from each testis is severed and the ends tied, clamped, or sealed. The surgery is performed under a local anesthetic, usually requires no stitches, and has no known effects on health or sexual performance. Sperm are still produced, but they cannot leave the epididymis, where they die. Phagocytic white blood cells and the cells that make up the lining of the epididymis remove the debris.

A somewhat more complex operation, called a tubal ligation, renders a woman infertile by clamping or cutting her uterine tubes and tying or sealing off the cut ends. Ovulation still occurs, but sperm cannot travel to the egg, nor can the egg reach the uterus. An alternative procedure is a tubal implant. Tiny springs are guided through the vagina, cervix, and uterus and inserted into each uterine tube. The springs cause the uterine tubes to form scar tissue that blocks passage of both sperm and eggs. Tubal implants require no incisions and only local anesthesia.

If a sterilized woman or man wishes to reverse the operation, a surgeon can attempt to reconnect the uterine tubes or vas deferens. About 70% to 90% of young women are able to become pregnant after their uterine tubes have been reconnected by a skilled, experienced surgeon. Pregnancy rates vary, however, according to the age of the woman (older women have a lower success rate) and the method of the original tubal ligation (higher success rates occur if the uterine tubes were clamped rather than cut or cauterized). Tubal implants are not readily reversible.

After surgical reversal of a vasectomy, sperm reappear in the ejaculate in 70% to 98% of cases, depending mostly on the skill of the surgeon. However, the pregnancy rate of the men's sexual partners is only 30% to 75%. The longer the interval between vasectomy and reconnection, the lower the pregnancy rate. In many cases, a vasectomized man gradually develops an immune response against his sperm, so even if the vasectomy is successfully reversed, he may ejaculate damaged sperm that cannot reach or fertilize an egg.

The vas deferens is severed and its ends are sealed.

testis

scrotum

(a) Vasectomy

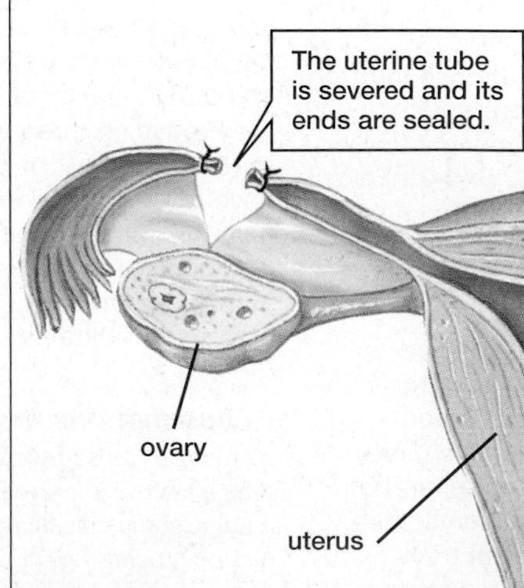
The uterine tube is severed and its ends are sealed.
ovary
uterus

(b) Tubal ligation

▲ **FIGURE 42-18 Sterilization provides permanent contraception**

Temporary Birth Control Methods Are Readily Reversible

Temporary methods of birth control prevent pregnancy in the immediate future, while leaving open the option of later pregnancies. Temporary birth control methods use one or more of three major mechanisms: (1) preventing ovulation, (2) preventing sperm and egg from meeting, and, less commonly, (3) preventing implantation in the uterus (**TABLE 42-3**). Note that birth control methods provide no protection against sexually transmitted diseases (STDs) unless they prevent physical contact between the penis and vagina. Complete abstinence, of course, prevents sperm and egg from encountering one another and also provides total protection against pregnancy and STDs.

Synthetic Hormones Prevent Pregnancy by Multiple Mechanisms

Combination birth control pills contain synthetic versions of estrogen and progesterone. As you learned earlier in this chapter, follicle development is stimulated by FSH, and ovulation is triggered by a midcycle surge of LH. The estrogen in birth control pills prevents FSH release, so follicles do not develop. Even if one were to develop, the progesterone in the pill would suppress the surge of LH needed for ovulation. Progesterone also thickens the cervical mucus, making it more difficult for sperm to move from the vagina into the uterus. Minipills, which contain synthetic varieties of progesterone and no estrogen, inhibit ovulation in about 60% to 97% of menstrual cycles, depending on the type of synthetic progesterone and the dose. Minipills probably work mostly by thickening the cervical mucus so that sperm cannot leave the vagina and reach an egg. Both combination pills and minipills also alter the uterine lining, so in the rare cases where fertilization occurs, the embryo may be unable to implant in the endometrium.

Contraceptive patches, rings, injections, and implants containing estrogen and progesterone are also available, as described in Table 42-3. These devices are usually effective for a few weeks to as long as a few years.

Emergency contraception, popularly known as "morning-after" pills, contain either a high dose of synthetic progesterone or a chemical that interacts with progesterone receptors. Both types of pills act primarily by preventing ovulation or by delaying ovulation long enough that any sperm in a woman's reproductive tract die before ovulation occurs.

TABLE 42-3 Methods of Temporary Contraception

Method	Description	Pregnancy Rate per Year[1]	Protection Against STDs
None	Frequent intercourse with no contraception	About 95% for women under 25 years old, declining to about 45% by age 40	None
Abstinence	No sexual activity	0%	Excellent
Birth control pill	Pill containing either synthetic estrogen and synthetic progesterone (combination pill) or progesterone only (minipill); taken daily	0.3% to 8%	None
Vaginal ring	Flexible plastic ring containing synthetic estrogen and progesterone, inserted into the vagina around the cervix; replaced every 4 weeks	0.3% to 8%	None
Contraceptive patch	Skin patch containing synthetic estrogen and progesterone; replaced weekly	0.3% to 8%[2]	None
Birth control injection	Injection of synthetic progesterone that blocks ovulation; repeated at 3-month intervals	0.3% to 3%	None
Contraceptive implant	Small plastic rod containing synthetic progesterone that blocks ovulation; replaced every 3 years	0.1%	None
Emergency contraception ("morning-after" pill)	Concentrated dose of the hormones in birth control pills, usually taken within 72 hours after unprotected intercourse	5% to 15% (less effective the later the pill is taken after intercourse)[3]	None
Condom (male)	Thin latex or polyurethane sheath placed over the penis just before intercourse, preventing sperm from entering the vagina; more effective when used with spermicide	2% to 15%	Good
Condom (female)	Lubricated polyurethane pouch inserted into the vagina just before intercourse, preventing sperm from entering the cervix; more effective when used with spermicide	5% to 21%	Good (probably about the same as a male condom)
Sponge	Domed disposable sponge containing spermicide, inserted in the vagina up to 24 hours before intercourse	9% to 20% (failure rate doubles for women who have given birth)	Poor
Diaphragm or cervical cap	Reusable, flexible, domed rubber-like barrier; spermicide is placed within the dome, and the diaphragm (larger) or cap (smaller) is fitted over the cervix just before intercourse	6% to 14% (failure rate for the cervical cap is higher for women who have given birth)	Poor
Spermicide	Sperm-killing foam is placed in the vagina just before intercourse, forming a chemical barrier to sperm	18% to 29%	Probably none
IUD (intrauterine device)	Small plastic device treated with hormones or copper and inserted through the cervix into the uterus; replaced every 5 to 10 years	0.2% to 0.9%	None
Fertility-awareness-based	Measuring the time since the last menstruation, or changes in body temperature and cervical mucus, to estimate the time of ovulation so that intercourse can be avoided during the fertile period	1% to 24% (rarely performed correctly)	None

[1]The percentage of women becoming pregnant per year. The low numbers represent the pregnancy rate with consistent, correct contraceptive use; the higher numbers represent the pregnancy rate with more typical use that is not always consistent or correct. It is likely that many women do not report incorrect usage, so the actual failure rates with correct use may be lower.

[2]Patches and birth control pills are about equally effective; however, the patch is more likely to be used properly. The patch is less effective in women weighing more than 200 pounds.

[3]The likely percentage of women who will become pregnant after a single episode of unprotected intercourse. Some emergency contraceptives are less effective in women over 165 pounds.

Barrier Methods Prevent Sperm from Reaching an Egg

Barrier methods include male and female condoms, cervical caps, diaphragms, vaginal sponges, and spermicides (sperm-killing chemicals). A female condom completely lines the vagina; a male condom covers the penis. Therefore, both types of condom keep sperm from contacting the vagina. Because the penis does not directly touch the vagina, both types also offer fairly good protection against the spread of STDs.

Other barrier methods provide little or no protection against STDs. The diaphragm and cervical cap cover the cervix, preventing sperm from leaving the vagina and entering the uterus. A vaginal sponge is soaked with spermicide, killing the sperm before they can leave the vagina. All barrier devices are more effective when combined with a spermicide, as additional protection in case the barriers are breached.

Intrauterine Devices Can Work for Several Years

There are two common forms of intrauterine device (IUD). A copper IUD has copper wire wound around a plastic "T." A hormonal IUD contains synthetic progesterone instead of the copper wrapping. The primary action of both types of IUD is to prevent fertilization by interfering with sperm motility or survival and by thickening the cervical mucus so sperm cannot enter the uterine tubes. Both also alter the uterine lining, which reduces the likelihood of implantation of an embryo, should fertilization occur. An IUD is inserted through the cervix into the uterus, where it typically remains in place for several years. IUDs are highly effective (less than 1% pregnancy rate) and do not require any further action to prevent pregnancy. Most women can become pregnant soon after removal of an IUD.

Other Contraceptive Methods Are Generally Less Reliable

In principle, abstaining from intercourse during the ovulatory period of the menstrual cycle can be very effective in preventing pregnancy. In practice, fertility-awareness-based contraceptive methods tend to be relatively unreliable because ovulation is difficult to predict precisely. In addition, sperm can survive for a few days in the female reproductive tract, so intercourse must be avoided for several days before ovulation. In the calendar method (formerly called the rhythm method), a woman keeps track of when she menstruates and calculates when her next ovulation should occur. Unfortunately, in many women the duration of the menstrual cycle varies somewhat from month to month. The calendar method can be improved somewhat by keeping track of body temperature, which rises slightly around the time of ovulation, and monitoring the quantity and texture of cervical mucus, which increases in volume and becomes wet and slippery just before ovulation.

Highly unreliable methods include withdrawal (removing the penis from the vagina before ejaculation) and douching (attempting to wash sperm out of the vagina before they have entered the uterus).

Male Birth Control Methods Are Under Development

You may have noticed that most birth control techniques are designed for use by women. There are probably three major reasons for this. First, the woman, not the man, becomes pregnant, bears the health risks of the pregnancy and childbirth, and usually plays a larger role in child care than men do. Second, it was relatively simple to design "use it and forget about it" birth control methods for women, including birth control pills and intrauterine devices, that do not interfere with sexual desire or performance. It has proven much more difficult to design similar methods for men. Third, opinion surveys find that a significant number of men, often more than 25%, claim they would never use hormonal birth control methods, even if they are assured that the hormones would not interfere with their sexual performance or other secondary sexual characteristics.

Nevertheless, researchers are working on male contraception. One of the options under investigation is to inject plugs into the vas deferens; the plugs block sperm from moving out of the testes to the urethra. Other options include drugs, usually administered by injection, that block the action of GnRH, FSH, or LH, thus preventing sperm from being produced. These drugs would need to be supplemented with testosterone injections, because testosterone production by the testes would be inhibited. Non-hormonal drugs that selectively alter sperm differentiation have shown success in animal trials, but these are probably years away from human use.

CHECK YOUR LEARNING

Can you . . .

- describe the principal methods of contraception, their mechanisms of action, and their effectiveness?

CASE STUDY REVISITED

To Breed a Rhino

Whether using artificial insemination or nature's way, rhino breeders are most successful when they can time sperm transfer to coincide with ovulation. The surest way to achieve this synchrony is to induce the female to ovulate when a male rhino or sperm sample is ready. As you have learned, ovulation in mammals is stimulated by a surge of LH and FSH, which in turn is stimulated by a surge in GnRH. Thus, ovulation in rhino cows can be induced by injecting them with a synthetic version of GnRH.

The next step depends on how the cow will be fertilized. For normal mating, the biologists wait until the female shows signs of mating interest. The male and female are then housed together and watched carefully for signs of mating or aggression. For artificial insemination, the biologists use ultrasound to check for the presence of ripe follicles. Semen collected from a bull rhino is then inserted into the cow's uterus. For IVF, the biologists suck ripe eggs out of the cow's ovary, again using ultrasound to guide their efforts, mix the eggs with sperm, and implant the resulting embryos into the cow's uterus.

These procedures have all proven successful. White, black, Indian, and Sumatran rhinos have been successfully bred using one or more of these techniques. With highly endangered species like the Sumatran, Javan, and black rhinos, preserving genetic diversity is extremely important, which means that every male and female should be given the chance to pass on their genes by reproducing with rhinos around the world. Artificial insemination using frozen sperm is much safer and less expensive than shipping rhinos between zoos. Artificial insemination can even help to preserve the genes of deceased bull rhinos. Tashi, the Indian rhino cow at the Buffalo Zoo, was artificially inseminated with sperm from Jimmy, a rhino bull who died at the Cincinnati Zoo about 10 years earlier. Jimmy's sperm had been stored all that time, frozen in liquid nitrogen. The Buffalo Zoo named the calf Monica, after Dr. Monica Stoops of the Cincinnati Zoo, who performed the artificial insemination.

Similar techniques have also been used with other endangered species. For example, the Smithsonian National Zoo in Washington, D.C., houses black-footed ferrets that were born a decade after their fathers had died. In addition to the Cincinnati Zoo, several other facilities, including the Audubon Nature Institute and the San Diego Zoo, maintain Frozen Zoos—liquid nitrogen storage tanks containing frozen sperm, tissue samples, and even embryos from endangered species.

CONSIDER THIS By rhino standards, southern white rhinos are abundant—there are about 20,000 of them. In contrast, there are about 60 Javan rhinos and 100 Sumatran rhinos in the world. How might white rhinos be used to increase reproduction of Javan and Sumatran rhinos? Describe advantages and obstacles to your proposals.

CHAPTER REVIEW

Go to **Mastering Biology** to access the Pearson eText, vocabulary review, practice quizzes, activities, videos, current events, and more.

*Answers to **Think Critically** and **Thinking Through the Concepts** questions can be found in the **Answers** section at the back of the book.*

Summary of Key Concepts

42.1 How Do Animals Reproduce?

Animals reproduce either sexually or asexually. Asexual reproduction produces offspring that are genetically identical to the parent. In sexual reproduction, haploid gametes, usually from two separate parents, unite and produce offspring that are genetically different from either parent.

During sexual reproduction, a male gamete (a small, motile sperm) fertilizes a female gamete (a large, nonmotile egg). Some animal species are hermaphroditic, producing both sperm and eggs, but most have separate sexes. Fertilization can occur outside the bodies of the animals (external fertilization) or inside the body of the female (internal fertilization). External fertilization must occur in water so that the sperm can swim to meet the egg. Internal fertilization generally occurs by copulation, in which the male deposits sperm directly into the female's reproductive tract.

42.2 What Are the Structures and Functions of Human Reproductive Systems?

The human male reproductive system consists of paired testes, which produce sperm and testosterone; accessory structures that conduct sperm out of the man's body into the female's reproductive system; and three sets of glands. The seminal vesicles, prostate gland, and bulbourethral glands secrete fluids that provide energy for the sperm, activate the sperm to swim, and provide the proper pH for sperm survival. Spermatogenesis and testosterone production are stimulated by FSH and LH. Spermatogenesis and testosterone production begin at puberty and continue throughout life.

The human female reproductive tract consists of paired ovaries, which produce eggs and the hormones estrogen and progesterone, and accessory structures that conduct sperm to the egg and receive and nourish the embryo during prenatal development. Oogenesis, hormone production, and development of the lining of the uterus repeat in a monthly menstrual cycle. The cycle is controlled by hormones from the hypothalamus (GnRH), anterior pituitary (FSH and LH), and ovaries (estrogen and progesterone). The production of estrogen and mature eggs begins at puberty and lasts until menopause.

During copulation, the male ejaculates semen into the female's vagina. The sperm swim through the vagina and uterus into the uterine tube, where fertilization usually takes place. The unfertilized egg is surrounded by two barriers, the corona radiata and the zona pellucida. Enzymes released from the acrosomes in the heads of sperm digest these layers, permitting sperm to reach the egg. Only one sperm enters the egg and fertilizes it.

42.3 How Can People Prevent Pregnancy?

Permanent contraception can be achieved by sterilization, usually by severing the vas deferens in males (vasectomy) or the uterine tubes in females (tubal ligation). Temporary contraception techniques include those that prevent ovulation by delivering estrogen and progesterone—for example, birth control pills, vaginal rings, contraceptive patches and implants, and hormone injections. Emergency contraceptives delay or prevent ovulation. Sperm may be prevented from reaching an egg by methods such as the diaphragm, cervical cap, vaginal sponge, and condom, usually accompanied by spermicide. Intrauterine devices block sperm movement and may prevent implantation of the early embryo. Fertility-awareness-based methods, which typically have a relatively high failure rate, require abstinence around the time of ovulation. Withdrawal and douching are unreliable.

Thinking Through the Concepts

Bloom's: Remembering, Understanding

Multiple Choice

1. In humans, sperm are produced in the
   a. spermatophore.
   b. seminiferous tubules.
   c. epididymis.
   d. urethra.
2. In humans, fertilization usually occurs in the
   a. fimbriae.
   b. vagina.
   c. uterus.
   d. uterine tubes.
3. The corpus luteum
   a. produces LH and FSH.
   b. produces estrogen and progesterone.
   c. nourishes the developing oocyte.
   d. forms the corona radiata.
4. In a menstrual cycle that does not result in pregnancy,
   a. LH stimulates follicle development early in the cycle.
   b. GnRH stimulates follicle development late in the cycle.
   c. ovulation typically occurs a day or two before menstruation.
   d. negative feedback by estrogen and progesterone inhibits the secretion of GnRH.
5. Sexually transmitted diseases
   a. are mostly caused by protists.
   b. can often, but not always, be cured with appropriate antibiotics.
   c. always produce symptoms that alert the victim of the disease.
   d. can easily be transmitted by kissing.

Fill-in-the-Blank

1. Reproduction by a single animal, without the need for sperm fertilizing an egg, is called ____________ reproduction. A(n) ____________ is a new individual that grows on the body of the adult and eventually breaks off to become independent. During ____________, an adult animal splits into two or more pieces, and each piece regenerates a complete organism.
2. In mammals, the male gonad is called the ____________. It produces both sperm and the sex hormone ____________. Within the male gonad, spermatogenesis occurs within the hollow, coiled structures called ____________.

3. A sperm consists of three regions, the head, midpiece, and tail. The head contains very little cytoplasm and consists mostly of the ____________, in which the chromosomes are found, and a sac of enzymes, the ____________. Organelles in the midpiece, the ____________, provide energy for movement of the tail.
4. Sperm are stored in the ____________ and ____________ until ejaculation, when the sperm, mixed with fluids from three glands, the ____________, ____________, and ____________, flow through the ____________ to the tip of the penis.
5. In mammals, the female gonad is the ____________. It produces eggs and two hormones, ____________ and ____________. Although usually referred to as an "egg," female mammals actually ovulate a cell called a ____________, which has completed only meiosis I. Meiosis I also produces a much smaller cell, the ____________, that serves mainly as a way to discard chromosomes. The egg is swept up by ciliated structures, called fimbriae, that form the entrance to the ____________. Fertilization usually occurs in this structure. The fertilized egg then implants in the ____________, where it will develop until birth.
6. Oocytes develop in a multicellular structure, the follicle. After ovulation, most of the follicle cells remain in the ovary, forming a temporary endocrine gland, the ____________. This gland disintegrates about 10 days after ovulation unless stimulated by a hormone, ____________, secreted by the developing embryo.

Review Questions

1. Describe the advantages and disadvantages of asexual reproduction and sexual reproduction. Provide three examples of animals that engage in each type of reproduction.
2. Compare the structures of the egg and sperm. What structural modifications do sperm have that facilitate movement, energy use, and gaining access to the egg?
3. What is the role of the corpus luteum in a menstrual cycle? In early pregnancy? What determines its survival after ovulation?
4. List the structures, in order, through which a sperm passes, starting with the seminiferous tubules of the testis and ending in the uterine tube of the female.
5. Name the three accessory glands of the male reproductive tract. What are the functions of the secretions they produce?
6. Describe the principal methods of contraception, including their methods of action, likely failure rates, and protection against STDs.

Applying the Concepts

Bloom's: Applying, Analyzing, Evaluating

1. After menopause, a woman's follicles no longer develop. Predict what happens to levels of FSH, LH, and estrogen in post-menopausal women. Explain your answer.

43 Animal Development

Salamanders, such as this axolotl, can regrow lost legs.

CASE STUDY

Rerunning the Program of Development

FLATWORMS DO IT. Sea stars do it. To a more limited but still very impressive extent, so do some insects, crabs, crayfish, salamanders, juvenile alligators, and lizards.

What can these animals do? They can regrow lost body parts. Flatworms and sea stars can even regenerate most of a lost body, starting with a fairly small part. Cut a flatworm in half, and the tail portion can grow a new head. Cut off the arms of some species of sea stars and each arm can regenerate a whole animal. Insects, crabs, and crayfish can regrow lost legs and antennae. Some salamanders can regenerate their legs and tails. Juvenile alligators and a few species of lizards can regrow their tails.

Regeneration is a lot more complicated than healing a wound. In wound healing, a few cell types proliferate and essentially stitch the edges of the wound together. In adult mammals, the job usually isn't done very well, and a scar remains as a permanent reminder of the injury. For some wounds, such as damage to the spinal cord, healing binds the broken parts together and helps reestablish the blood supply but leaves a scar that prevents the recovery of any significant function. In contrast, regeneration is like a rerun of the developmental program for the lost body part. When a salamander regenerates a lost leg, it must grow bone, tendons, blood vessels, nerves, muscle, and skin, all in the right places and all integrated properly with one another to restore leg function.

Human regeneration has been a dream of physicians for decades. If a human embryo can grow a leg during development, why can't an adult regrow a leg to replace one lost to amputation? To find out if regeneration in humans might be possible, biomedical researchers first need to understand how normal development is controlled.

AT A GLANCE

43.1 WHAT ARE THE PRINCIPLES OF ANIMAL DEVELOPMENT?

Development is the process by which a multicellular organism grows and increases in organization and complexity. Development begins with a fertilized egg and ends with a sexually mature adult. Three principal processes contribute to development. First, individual cells multiply. Second, some of their daughter cells **differentiate**, or become specialized in structure and function, for example, as nerve or muscle cells. Third, groups of differentiating cells move to different locations in the body and become organized into multicellular structures, such as a brain or a biceps muscle.

All of the cells of an individual animal's body (except gametes) are genetically identical to one another and to the fertilized egg from which they descended. How can genetically identical cells differentiate into remarkably different structures? As we will see, the answer is that specific sets of genes are turned on and off in different places in an animal's body, at specific times during an animal's life.

We begin with a brief survey of the diverse ways in which different animal species develop.

CHECK YOUR LEARNING

Can you . . .

- define *development* and describe how cell division and differentiation underlie animal development?

43.2 HOW DO DIRECT AND INDIRECT DEVELOPMENT DIFFER?

Animals undergo one of two types of development as they progress from newborn to adult: **direct development**, in which the newborn animal resembles the adult (**FIG. 43-1**), or **indirect development**, in which the newborn has a very different body structure than the adult (**FIG. 43-2**).

(a) Seahorses

(b) Snails

(c) Polar bears

▶ **FIGURE 43-1 Direct development** The offspring of animals with direct development closely resemble their parents from the moment of hatching or birth. **(a)** A male seahorse gives birth. The female seahorse deposits her eggs in his pouch, where they develop for a few weeks. Muscular contractions of the pouch then squirt out as many as 200 young. **(b)** Many land and freshwater snails hatch from small, yolk-rich eggs. **(c)** Mammalian mothers nourish their developing young within their bodies before birth and with milk from their mammary glands after birth.

(a) Caterpillar (larva)

(b) Butterfly (adult)

▲ **FIGURE 43-2 Indirect development** The larvae of animals with indirect development are very different from the adult form, in structure, behaviors, and ecological niches. **(a)** Caterpillars of butterflies, such as the blue morpho caterpillar shown here, feed on leaves, usually of a limited number of host plant species. **(b)** The principal food of most adult butterflies is flower nectar. Although blue morpho adults may drink nectar, they prefer fluids from fermenting fruits and even decomposing animal carcasses.

Many animal species undergo direct development, including some invertebrates, fish, and amphibians, and all mammals and reptiles (including birds). Compared to the typically very small offspring of indirectly developing species, the hatchlings or newborns of directly developing species are generally fairly large relative to their parents. Therefore, they need significant amounts of nourishment before emerging into the world. Two strategies have evolved to meet this food requirement. Birds and most other reptiles, and many fish, produce eggs that contain large amounts of a food reserve called **yolk**, which nourishes the embryo before it hatches. Mammals, some snakes, and a few fish have relatively little yolk in their eggs; instead, their embryos are nourished within the mother's body. Despite extensive development in the egg or in the mother's body, the young of many directly developing animals, such as those of birds and mammals, require additional care and feeding after birth. Because of these demands both before and after birth, few offspring are produced. However, the costly parental investment helps ensure that a high proportion of the offspring survives to adulthood.

Indirect development occurs in most amphibians and many invertebrates. In these species, females typically produce large numbers of eggs, each containing a small amount of yolk. The yolk nourishes the developing embryo until it hatches into a small, sexually immature feeding stage called a **larva** (plural, larvae). Depending on the species, larvae may eat algae (tadpoles of many species of frogs), plants (many insect larvae; see Fig. 43-2a), dead animals (many types of fly larvae), or feces (dung beetles). Some, such as harvester butterfly larvae, prey on other animals. Still other larvae, such as those of most aquatic invertebrates, filter protists from ponds, lakes, or marine environments. After feeding for a few weeks to several years, larvae undergo a revolution in body form, known as **metamorphosis**, and become sexually mature adults.

Most larvae not only look very different from the adults of the species, but also play different roles in their ecosystems. For instance, most adult butterflies sip nectar from flowers, whereas their caterpillar larvae typically eat leaves (see Fig. 43-2a). Most toads spend the majority of their adult life on land, eating insects, worms, and snails; their tadpole larvae are aquatic and usually feed on algae.

CHECK YOUR LEARNING

Can you . . .

- describe direct and indirect development and name some animal groups that have each type of development?

43.3 HOW DOES ANIMAL DEVELOPMENT PROCEED?

Most of the mechanisms of development are similar in all animals. Here, we will focus on vertebrate development. We will begin by describing development in amphibians such as frogs, newts, and salamanders. Amphibians have long been favored subjects for the study of development because they can be induced to breed at any time of year; they produce numerous large eggs and embryos; and the embryos develop in water, where they can be easily observed.

Cleavage of the Zygote Begins Development

Development begins when a fertilized egg, or **zygote**, undergoes a series of mitotic cell divisions collectively called

cleavage (**FIG. 43-3** ①). The zygote is a very large cell; a frog zygote, for example, may be a million times larger than an average cell in an adult frog. During cleavage, there is little or no cell growth between cell divisions, so as cleavage progresses, the available cytoplasm is split up into ever smaller cells, whose size gradually approaches that of cells in an adult. After a few cell divisions, a solid ball of cells, the **morula**, is formed. As cleavage continues, a cavity opens within the morula ②, and the cells become the outer covering of a hollow structure called the **blastula**.

The details of cleavage differ among species and are partly determined by the amount of yolk, because yolk hinders the division of the cytoplasm. In frogs, cells in the yolky portion of the zygote (the pale bottom of the zygote in Fig. 43-3) divide more slowly than cells in the almost yolk-free portion (the dark top of the zygote), so the morula has larger cells on the bottom than on the top. Eggs with extremely large yolks, such as a hen's egg, cannot divide all the way through; in such cases, cleavage produces a flat cluster of cells atop the yolk. Nevertheless, a hollow blastula (or its equivalent; see Section 43.5) is always produced, although in birds and other egg-laying reptiles it resembles a flattened bag rather than a ball.

Gastrulation Forms Three Tissue Layers

In amphibians and many other animals, the location of cells on the surface of the blastula forecasts their ultimate developmental fate in the adult. In Figure 43-3, we have colored these cells blue, yellow, or pink. These colors indicate the parts of the adult body that the cells are destined to produce.

The cells on the surface of the blastula move to their proper destinations during **gastrulation** ③ (literally, "producing the stomach"). Gastrulation begins when a dimple called the **blastopore** forms on one side of the blastula. The dimple enlarges, going deeper and deeper into the blastula and forming a cavity that will become the digestive tract.

The migrating cells eventually form three tissue layers in the embryo, which is now called a **gastrula** ④. The cells that move through the blastopore to line the future digestive tract (yellow) are called **endoderm** (meaning "inner skin"). Endoderm also forms the liver, pancreas, and the lining of the respiratory tract. The cells remaining on the outside of the developing gastrula (blue) are called **ectoderm** ("outer skin"). These cells mostly form surface structures, such as skin, hair, and nails. Ectoderm also forms the nervous system. Cells that migrate between the endoderm and ectoderm form the third layer (pink), called **mesoderm** ("middle skin"). Mesoderm forms structures that are generally located between the skin and the lining of the digestive tract, including muscles, the skeleton, and the circulatory system. **TABLE 43-1** lists the major structures produced from each layer of cells.

The Major Body Parts Develop During Organogenesis

Organogenesis is the development of the body's organs from the three embryonic layers. Organogenesis proceeds by

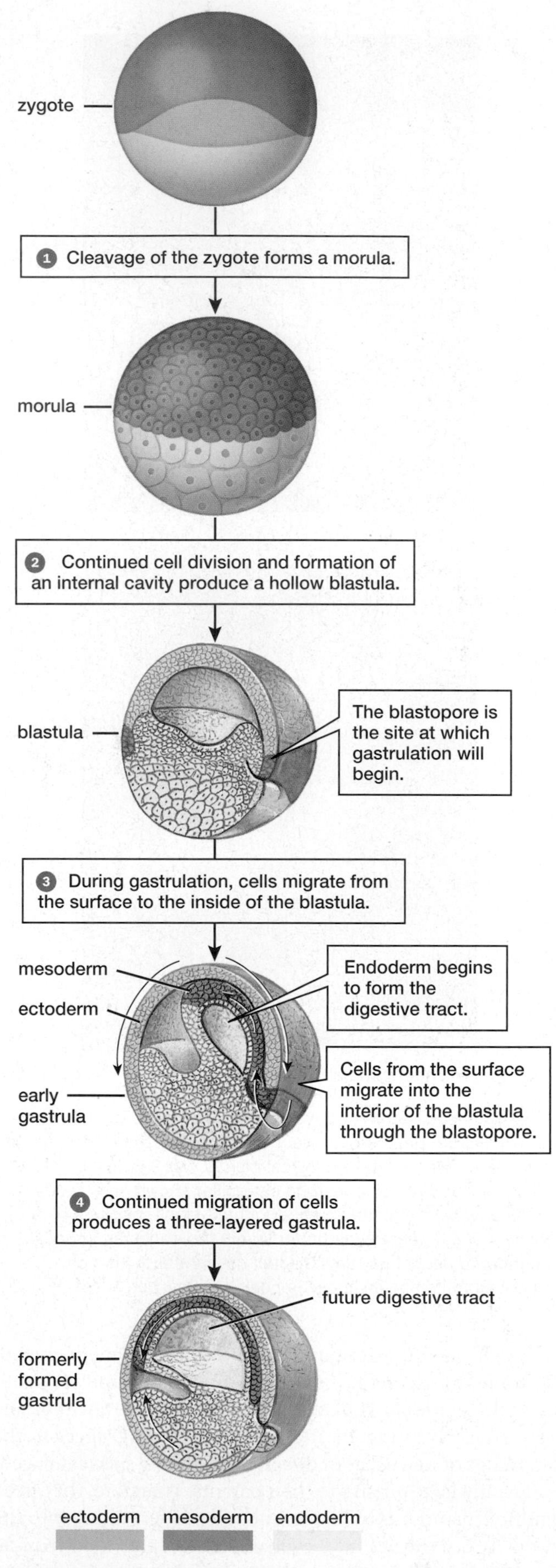
zygote
1 Cleavage of the zygote forms a morula.
morula
2 Continued cell division and formation of an internal cavity produce a hollow blastula.
blastula
The blastopore is the site at which gastrulation will begin.
3 During gastrulation, cells migrate from the surface to the inside of the blastula.
mesoderm
ectoderm
Endoderm begins to form the digestive tract.
early gastrula
Cells from the surface migrate into the interior of the blastula through the blastopore.
4 Continued migration of cells produces a three-layered gastrula.
future digestive tract
formerly formed gastrula
ectoderm
mesoderm
endoderm

▲ **FIGURE 43-3 From zygote to gastrula**

TABLE 43-1	Derivation of Adult Tissues from Embryonic Cell Layers
Embryonic Layer	**Adult Tissue**
Ectoderm	Epidermis of the skin; hair; lining of the mouth and nose; glands of the skin; nervous system
Mesoderm	Dermis of the skin; muscle, skeleton; circulatory system; gonads; kidneys; outer layers of the digestive and respiratory tracts
Endoderm	Lining of the digestive and respiratory tracts; liver; pancreas

two major processes. First, a series of "master" genes turns on and off in specific cells. Somewhat like the ignition switch in your car, which turns on the engine, headlights, seat-belt alarm, power steering, and so on, each master gene controls the activity of many individual genes involved in producing, say, an arm or a backbone. We will return to the role of master genes in development in Section 43.4.

Second, organogenesis prunes away superfluous cells, much as Michelangelo chiseled away "extra" marble, revealing his statue of David. In development, the sculpting of body parts often requires the death of excess cells. For example, early human embryos have tails and webbed fingers and toes. As cells in the tail and webbing die, the tail disappears and the fingers and toes become separate.

Development in Reptiles and Mammals Depends on Extraembryonic Membranes

All animal embryos develop in water. Not only does this ensure that the embryo does not dehydrate, but the water also supplies the embryo with oxygen and carries away its wastes. Keeping embryos moist is not a problem for fish, which live and reproduce in water, or for amphibians, which may spend their adult lives on land but lay their eggs in water. For terrestrial vertebrates, however, providing a watery environment for embryonic development is a major challenge.

Fully terrestrial vertebrate life was not possible until the evolution of the **amniotic egg**. This innovation first arose in reptiles and persists today in that group (including birds) and its descendants, the mammals. Inside an amniotic egg, the embryo develops in a watery environment, even if, as in reptiles, the egg is laid on land. The amniotic egg includes four **extraembryonic membranes**: the chorion, amnion, allantois, and yolk sac (**TABLE 43-2**). In reptiles, the

TABLE 43-2 Reptile and Mammal Embryonic Membranes

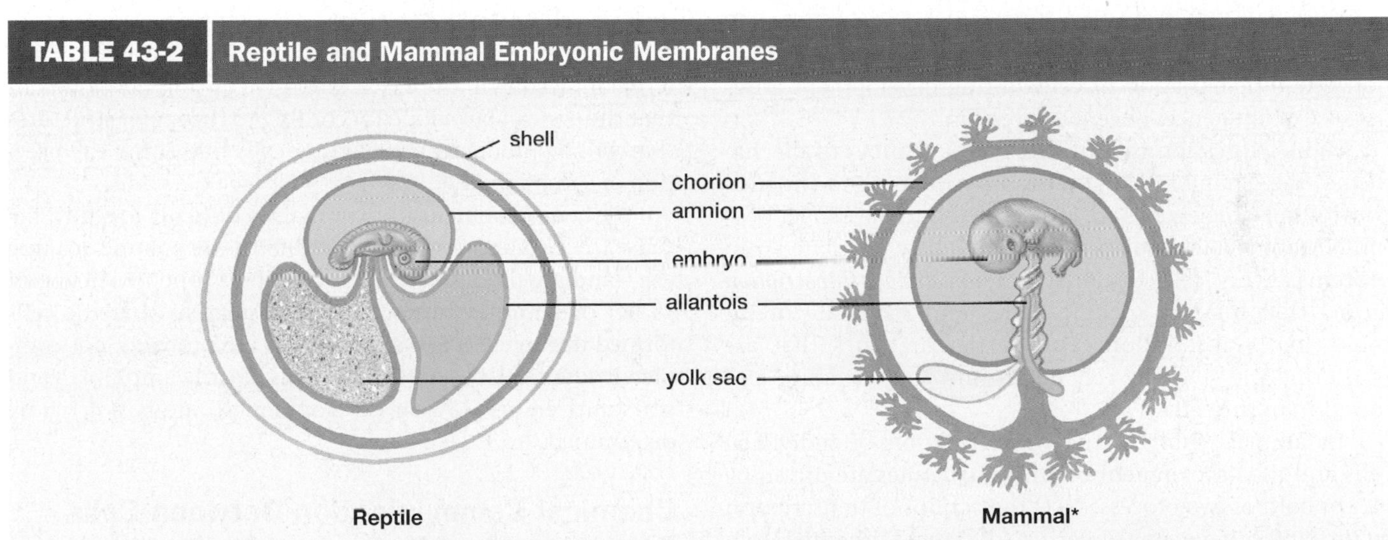

Membrane	Structure (Reptile)	Function (Reptile)	Structure (Mammal*)	Function (Mammal*)
Chorion	Membrane lining the inside of the shell	Acts as a respiratory surface; regulates the exchange of gases and water between the embryo and the air	Fetal contribution to the placenta	Provides for the exchange of gases, nutrients, and wastes between the embryo and the mother
Amnion	Sac surrounding the embryo	Encloses the embryo in fluid	Sac surrounding the embryo	Encloses the embryo in fluid
Allantois	Sac connected to the embryonic urinary tract; a capillary-rich membrane lining the inside of the chorion	Stores wastes (especially urine); acts as a respiratory surface	Membranous sac arising from the gut; varies in size	May store metabolic wastes; contributes to the umbilical cord blood vessels
Yolk sac	Membrane surrounding the yolk	Contains yolk as food; digests yolk and transfers its nutrients to the embryo	Small, membranous, fluid-filled sac	Helps absorb nutrients from the mother; forms blood cells; contributes to the umbilical cord

*Marsupial mammals, such as kangaroos and opossums, and monotremes, such as platypuses and echidnas, have the same four extraembryonic membranes. However, the placenta in marsupials is usually derived largely from the yolk sac rather than from the chorion and allantois. Monotremes lay eggs with extraembryonic membranes that are similar to those of reptiles.

chorion lines the egg's shell and allows for exchange of oxygen and carbon dioxide between the embryo and the air. The **amnion** encloses the embryo in its own "private pond." The **allantois** surrounds and isolates wastes. The **yolk sac** contains the yolk.

In mammals (except for platypuses and echidnas, which lay eggs), the embryo develops within the mother's body until birth. Nevertheless, all four extraembryonic membranes still persist, and in fact these membranes are essential for development, as described in Table 43-2.

CHECK YOUR LEARNING

Can you . . .

- describe early development in amphibians, including the processes of cleavage, blastula formation, and organogenesis?
- name the extraembryonic membranes present in reptiles and mammals and describe their functions?

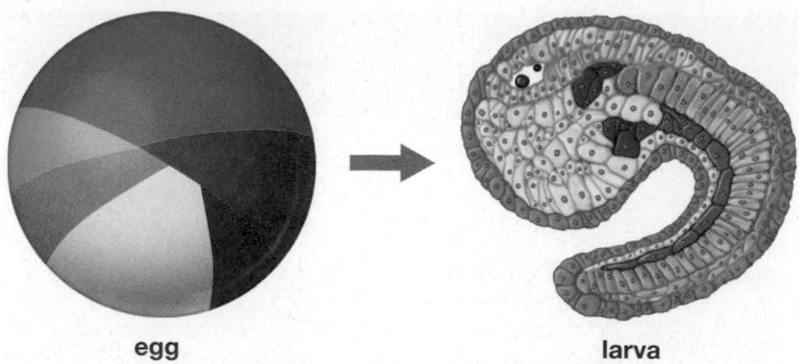

▲ **FIGURE 43-4 A "fate map" of the sea squirt egg** Gene-regulating substances in the cytoplasm of the egg of a sea squirt control early development. In this drawing, the coloring of the egg and larva shows which parts of the egg will give rise to which parts of the larva.

THINK CRITICALLY If development in humans were as thoroughly determined as it is in sea squirts, would identical twins be more or less common? Explain.

43.4 HOW IS DEVELOPMENT CONTROLLED?

In any animal species, zygotes and almost all the cells of both embryos and adults contain the same genes—all the genes needed to produce an entire adult animal. In any given cell, however, some genes are used, or expressed, while others are not. The differentiation of cells during development arises because of differences in gene expression.

Gene expression is controlled by a number of mechanisms (see Chapter 13). One important method involves controlling which genes are transcribed into messenger RNA (mRNA) molecules, which are subsequently translated into proteins. Every cell contains proteins called *transcription factors* that bind to specific genes and stimulate or inhibit their transcription. The particular genes that are transcribed in a given cell determine the structure and function of the cell.

In animal embryos, the differentiation of individual cells and the development of entire structures are driven by one or both of two processes: (1) the actions of transcription factors and other gene-regulating substances inherited from the mother in her egg, and (2) chemical communication between the cells of the embryo.

Maternal Molecules in the Egg May Direct Early Embryonic Differentiation

Virtually all the cytoplasm in a zygote is already present in the egg before it is fertilized; the sperm contributes little more than a nucleus (see Chapter 42). In most invertebrates and some vertebrates, specific protein molecules become localized in different places in the egg's cytoplasm. Some of these proteins are transcription factors that determine which genes are turned on and off, in some cases turning on master genes that control the activity of many other genes.

During the first few cleavage divisions, the zygote and its daughter cells divide at specific places and in specific orientations. As a result, each of these early embryonic cells receives different transcription factors from the egg. Therefore, different cells transcribe different genes, start differentiating into distinct cell types, and in many cases, ultimately give rise to specific adult structures. In some animals, the position of maternal molecules in the egg so strongly controls development that the egg can be mapped according to the major structures that will be produced by daughter cells inheriting each section of cytoplasm (**FIG. 43-4**).

Early development in mammals is difficult to study, because the eggs are very small—1/1,000th the volume of a frog egg—and are produced in small numbers (one to a dozen or so per ovulation). Current evidence suggests that the cells formed during cleavage in mammals are probably not completely identical, but, unlike the cells of early amphibian and sea squirt embryos, their developmental fate is not rigidly determined.

Chemical Communication Between Cells Regulates Most Embryonic Development

As development proceeds, cells differentiate in response to chemical messengers released by other, usually nearby cells, a process called **induction**. Although discovering cellular mechanisms of induction had to wait until the techniques of genetics and molecular biology were invented in the late twentieth century, the principles of induction were discovered over a century ago by researchers who carefully observed the effects of transplanting small clusters of cells between amphibian embryos.

In the early 1900s, embryologists transplanted bits of light-colored amphibian embryos (the donors) to various locations in dark-colored embryos (the hosts), and vice versa. In this way, they could use the color difference to determine whether a structure that developed in the host embryo

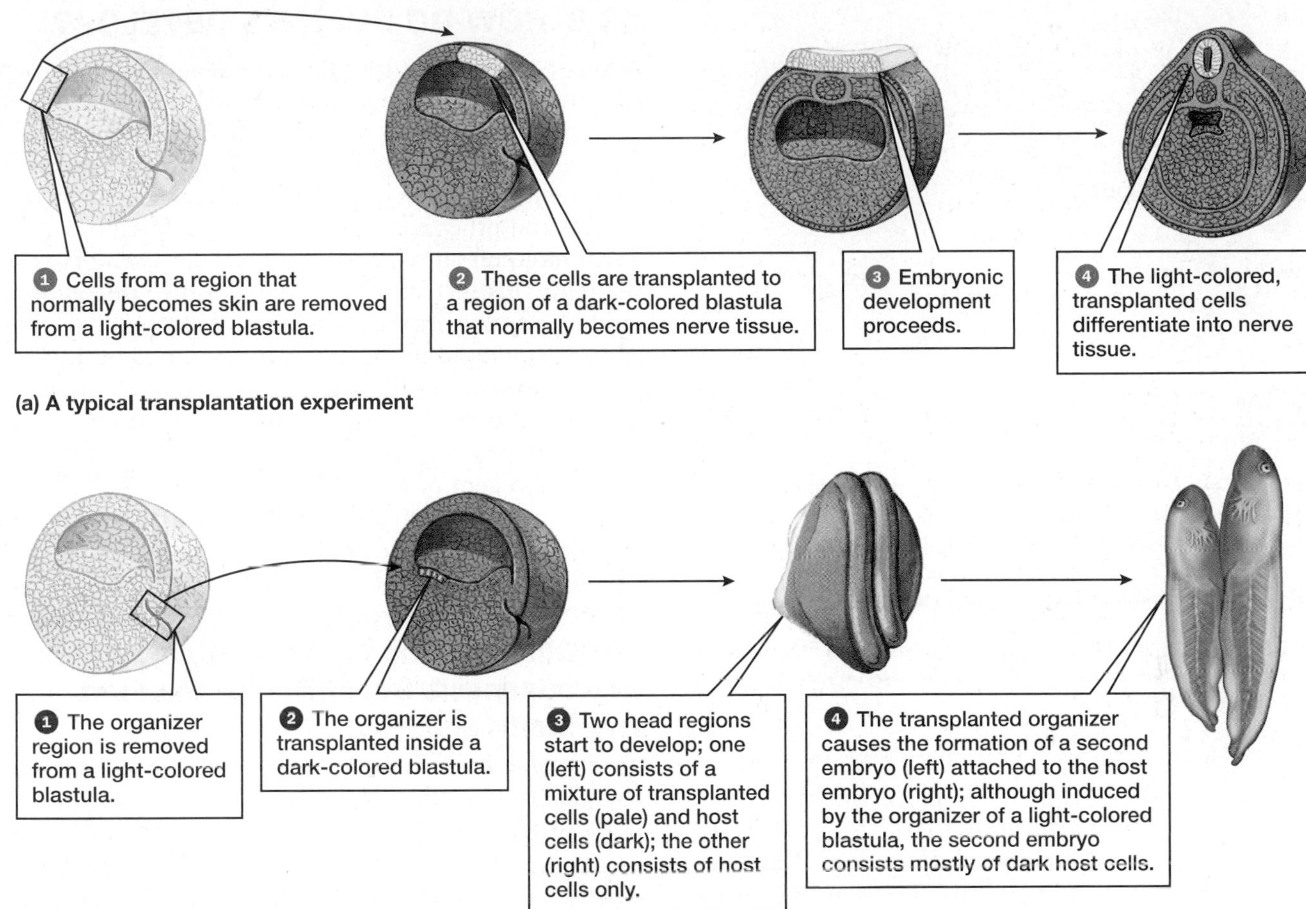

1 Cells from a region that normally becomes skin are removed from a light-colored blastula.
2 These cells are transplanted to a region of a dark-colored blastula that normally becomes nerve tissue.
3 Embryonic development proceeds.
4 The light-colored, transplanted cells differentiate into nerve tissue.
(a) A typical transplantation experiment
1 The organizer region is removed from a light-colored blastula.
2 The organizer is transplanted inside a dark-colored blastula.
3 Two head regions start to develop; one (left) consists of a mixture of transplanted cells (pale) and host cells (dark); the other (right) consists of host cells only.
4 The transplanted organizer causes the formation of a second embryo (left) attached to the host embryo (right); although induced by the organizer of a light-colored blastula, the second embryo consists mostly of dark host cells.
(b) Transplantation of the organizer region

▲ **FIGURE 43-5 Induction** During most of development, the fate of any given cell is strongly influenced by the cells that surround it. **(a)** When part of an amphibian blastula is transplanted into a second blastula, the surrounding cells usually induce the transplant to assume the characteristics of the region into which the transplant was placed. **(b)** However, if cells of the organizer, near the opening of the blastopore, are transplanted into another embryo, they induce the surrounding cells of the host to develop into most of the structures of a secondary embryo.

consisted of cells that originally came from the donor or from the host. They found that the fate of the transplanted cells was, in general, not predetermined. Instead, the cells of the host embryo induced the donor cells to assume the developmental fate of the area of the host into which they were transplanted (**FIG. 43-5a**).

However, not all transplants simply blended in, seamlessly becoming a part of the host embryo. In the 1920s, Hans Spemann and Hilde Mangold discovered that a specific cluster of cells located near the blastopore of an amphibian embryo, now called the *organizer*, determines whether nearby cells will become ectoderm or mesoderm and even where the head and nervous system will form. A transplant from the organizer region of a donor embryo induced cells of a host embryo to form parts of a second head and, occasionally, to become an almost complete secondary embryo (**FIG. 43-5b**).

We now know that cells of the organizer release proteins that interact with other messenger molecules to stimulate or inhibit the expression of specific master genes in nearby cells. These master genes often encode transcription factors that alter the transcription of many other genes. Which groups of genes are expressed determines the structures and functions of the cells. As these cells differentiate, they in turn release chemicals that alter the fate of still other cells, in a cascade that culminates in the development of the tissues and organs of the adult body.

Homeobox Genes Regulate the Development of Entire Segments of the Body

Homeobox genes, found in animals as diverse as fruit flies, frogs, and humans, comprise a particularly important set of master genes. Although their functions differ somewhat in different animals, homeobox genes generally code for transcription factors that affect the development of a particular region of the body. Homeobox genes were discovered in fruit flies, in which mutations of these genes can cause entire parts of the body to be duplicated, replaced, or omitted. For example, one mutant homeobox gene causes the development of an extra body segment, complete with an extra set of wings.

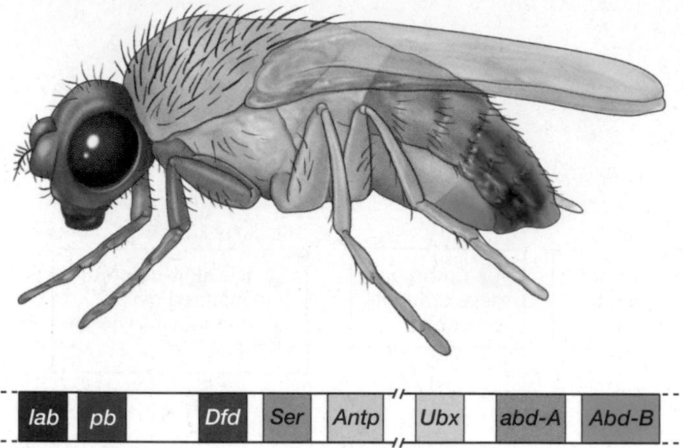

▲ **FIGURE 43-6 Homeobox genes regulate development of body segments** The sequence of homeobox genes on the chromosome corresponds to their role in the development of different body segments. In fruit flies, each homeobox gene is active in the body segment shown in the same color, in a head-to-tail order.

THINK CRITICALLY Snakes have ribs all the way from just behind the head almost to the tip of the tail. Snakes also lack legs. Propose a possible genetic mechanism for development of this body structure, based on homeobox genes.

Homeobox genes are arranged on the chromosomes in a head-to-tail order and are transcribed in cells in the specific locations in the body whose development they control. For example, "head" homeobox genes are transcribed in the head of the embryo, and "tail" homeobox genes are transcribed in the tail (**FIG. 43-6**).

CHECK YOUR LEARNING

Can you . . .

- explain the roles of gene-regulating substances in the egg cytoplasm, transcription factors, and induction between cells in development?
- describe the role of homeobox genes?

CASE STUDY **CONTINUED**

Rerunning the Program of Development

In salamanders and other amphibians, secreted messenger proteins alter the expression of master genes, including homeobox genes. The master genes in turn induce the development of forelegs with a humerus (the bone found in the upper arm of a person), radius, and ulna (lower arm), carpals (wrist), metacarpals (palm), and phalanges (fingers). Most of these same master genes become active during limb regeneration, as well. Does the development of a human limb, and the rest of the human body, use similar processes?

43.5 HOW DO HUMANS DEVELOP?

As mentioned earlier, specific placement of gene transcription factors in the egg does not seem to be the major mechanism of differentiation in mammalian development. Even in very early embryos, all the cells seem to be functionally equivalent. Induction is the principal mechanism by which human and other mammalian embryos develop, with different molecules driving various parts of the embryo to develop into distinct structures. Many of the same molecules produced by the organizer region of an amphibian embryo are also employed in mammalian development, including similar transcription factors inside cells and secreted proteins that alter the development of nearby cells. In addition, very similar molecules and intracellular pathways induce the development of the limbs of all tetrapods. These similarities increase biologists' confidence that research into amphibian development will provide insight into mammalian development.

Cell Differentiation, Gastrulation, and Organogenesis Occur During the First Two Months

Fertilization of a human egg usually takes place in the uterine tube. The resulting zygote undergoes a few cleavage divisions in the uterine tube, becoming a morula on its way to the uterus (**FIG. 43-7**). By about the fifth day after fertilization, the zygote has developed into a hollow ball of cells, known as a **blastocyst** (the mammalian version of a blastula; **FIG. 43-8a**). A human blastocyst consists of an outer layer of cells surrounding a cluster of cells called the **inner cell mass**. Beginning around the sixth to ninth day after fertilization, the outer cell layer attaches to, and then burrows into, the lining of the uterus (the endometrium), a process called **implantation** (**FIGS. 43-8a, b**). The outer cell layer of the blastocyst will become the chorion. The complex intermingling of the chorion and the endometrium forms the placenta, which we will describe shortly.

All the cells of the inner cell mass have the potential to develop into any type of tissue in the human body. This remarkable flexibility allows the inner cell mass to produce both the entire embryo and the amnion, allantois, and yolk sac. The inner cell mass is also the usual source of human embryonic stem cells, as described in "Health Watch: The Promise of Stem Cells" on page 768.

Although they still retain the potential to develop into any cell type, the cells of the inner cell mass of an early blastocyst start to differentiate through the process of induction. For example, the cells in contact with the outer cell layer usually produce the embryo proper, while the cells exposed to the blastocyst fluid produce extraembryonic membranes, especially the yolk sac.

Gastrulation Occurs After Implantation

During the second week of development, the inner cell mass grows and splits, forming two fluid-filled sacs that are

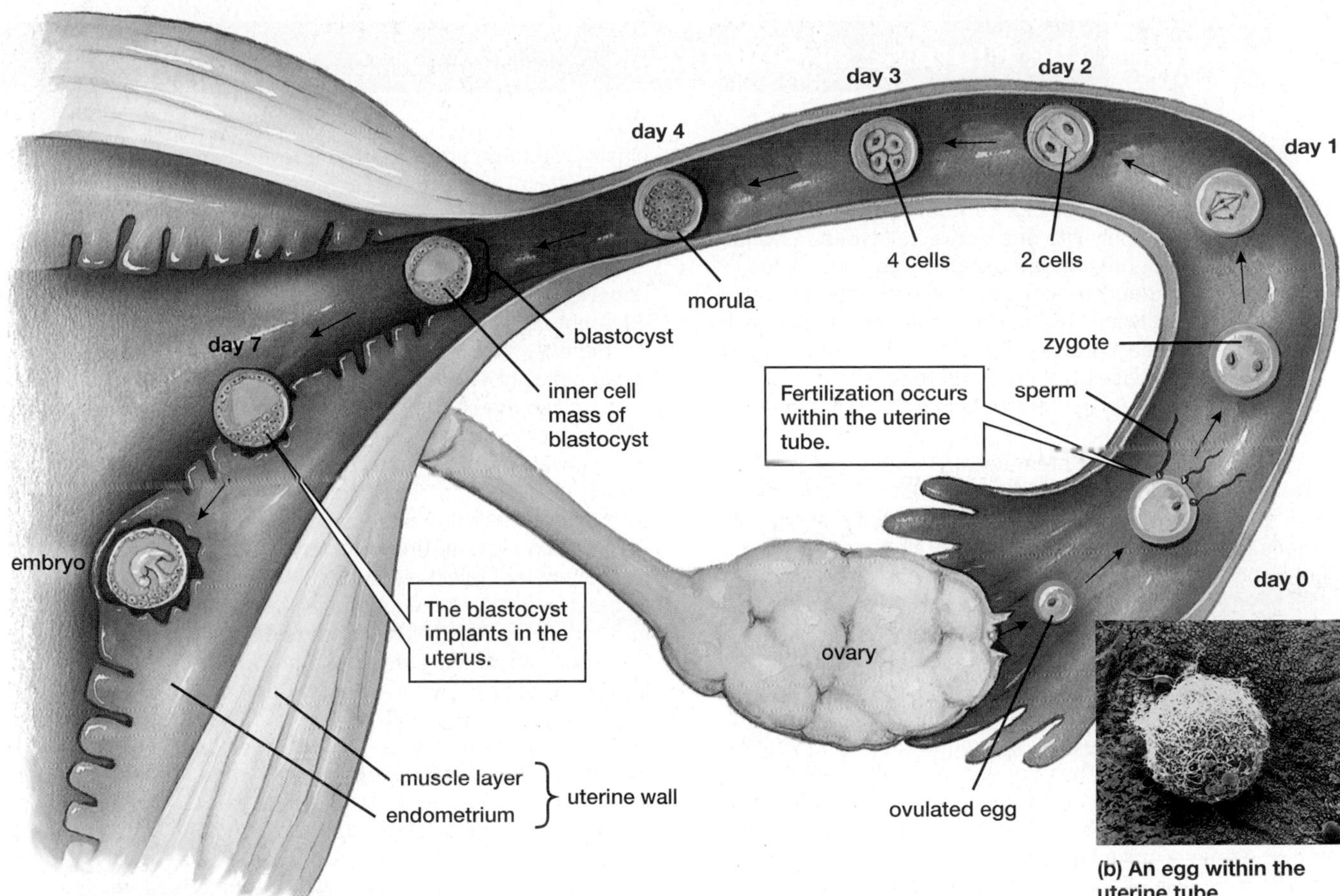

(b) An egg within the uterine tube

(a) The first week of development

▲ **FIGURE 43-7 The journey of the egg (a)** A human egg is fertilized in the uterine tube and slowly travels down to the uterus. Along the way, the zygote divides a few times, becoming first a morula and then a blastocyst. When the blastocyst reaches the uterine endometrium, it burrows in. **(b)** An egg, surrounded by sperm (yellow in this colored micrograph), is cradled in the uterine tube.

THINK CRITICALLY The corona radiata and zona pellucida surround a fertilized egg. As the egg develops into a blastocyst, what must happen before the blastocyst can implant in the uterus?

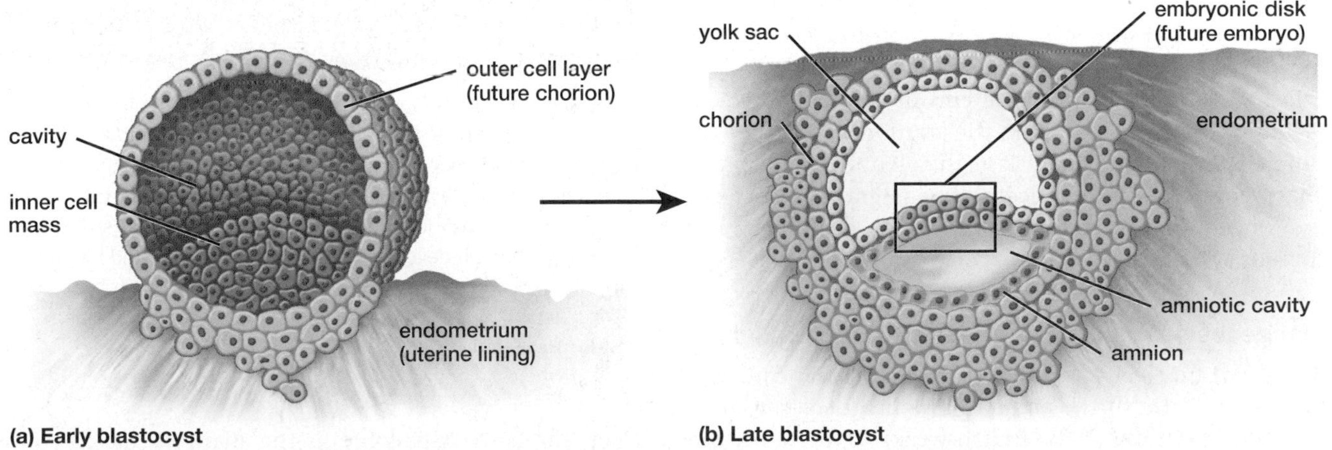

(a) Early blastocyst **(b) Late blastocyst**

▲ **FIGURE 43-8 A blastocyst implants (a)** As it burrows into the uterine lining, the outer cell layer of the blastocyst forms the chorion, the embryonic contribution to the future placenta. **(b)** A few days later, the blastocyst has completely submerged beneath the uterine lining. The inner cell mass begins to develop into the yolk sac, the amnion, and the embryonic disk.

The Promise of Stem Cells

Many cells in adult animals are permanently differentiated as a specific cell type, such as nerve or muscle cells, and cannot divide. Some adult cells can divide, but their daughter cells can differentiate only into one or two cell types. A **stem cell**, however, has not differentiated, can divide many times, and can give rise to daughter cells that differentiate into any of several different cell types. The medical potential of stem cells is vast—victims of heart attacks, strokes, spinal cord injuries, and degenerative diseases from arthritis to Parkinson's disease would benefit if their damaged tissues could be regenerated from stem cells.

There are three types of stem cells. The first type, **embryonic stem cells (ESCs)**, are usually derived from the inner cell mass of a blastocyst (see Fig. 43-8). In an intact embryo, the stem cells of the inner cell mass produce all the cell types of the entire body. ESCs grown in cell culture can also generate any cell type in the body, if they are exposed to the correct mixture of differentiation factors—mostly proteins either secreted by nearby cells or part of the extracellular matrix—that nudge them into one differentiation pathway or another (**FIG. E43-1**).

The second type of stem cell is present in small numbers in most parts of the body—including muscle, skin, bone marrow, fat, brain, and heart. These are usually called **adult stem cells (ASCs)**, although they are also present in children. ASCs can differentiate into only a few cell types, generally the cell types that normally make up the tissue from which they were taken. The third type of stem cell is produced by biotechnology, when genes or molecules that regulate gene transcription are inserted into adult (non-stem) cells, transforming them into **induced pluripotent stem cells (iPSCs)**. iPSCs can differentiate into any cell type.

Significant obstacles must be surmounted before stem cells can become effective therapies. First, embryos at the blastocyst stage are usually destroyed to obtain ESCs, which raises ethical concerns for some people, even though the embryos are typically "extras" that were created for *in vitro* fertilization and would eventually be discarded anyway. To alleviate such concerns, some researchers obtain ESCs by removing a single cell from the morula stage, which generally does not significantly damage the embryo.

Second, the immune system rejects stem cells that are not genetically identical to the cells of the recipient, or at least a very close match. Rejection is generally not a problem with ASCs, which can often be obtained from the patient, but it seems likely that ESCs may be rejected as foreign. Researchers initially expected that iPSCs taken from a patient and transplanted back into the same patient would not be rejected by the immune system. Unfortunately, it turns out that such iPSC transplants are rejected, at least in mice. Subsequent studies suggest that differentiating the iPSCs before transplantation may reduce or possibly eliminate rejection.

Third, researchers do not yet know for sure if ESCs and iPSCs are safe to use in people. Clinical studies are under way to evaluate ESC and/or iPSC safety in tissues as diverse as spinal cord and retina. Although designed primarily to evaluate safety, the studies also measure disease progression. In one study, 10 of 18 patients treated with ESCs for retinal degeneration showed improved vision, offering real hope for a cure.

ASCs are the only stem cells currently used in clinical practice. For example, ASCs in bone marrow, which can produce all of the types of both red and white blood cells in the body, have been used for decades to treat some types of leukemia, anemia, and immune deficiency syndromes. Nowadays, blood-forming ASCs are often "filtered out" of the donor's blood, rather than actually puncturing a bone and

separated by a double layer of cells called the **embryonic disk** (see Fig. 43-8b). One layer of cells is continuous with the yolk sac. The second layer of cells is continuous with the amnion.

Gastrulation begins near the end of the second week. Cells migrate in through a slit in the amnion side of the embryonic disk. This slit is the disk's equivalent of the amphibian blastopore. Once inside the disk, the migrating cells form mesoderm, endoderm, and the allantois. The cells remaining on the surface become ectoderm.

Organogenesis Begins During Weeks Three to Eight

During the third week of development, the spinal cord and brain begin to form in the embryo. The heart starts to beat around the beginning of the fourth week. At this time, the embryo, bathed in fluid contained within the amnion, bulges into the uterine cavity (**FIG. 43-9**). Meanwhile, the umbilical cord forms from the fusion of the *yolk stalk* and *body stalk*. The yolk stalk connects the yolk sac to the embryonic digestive tract. The body stalk contains the allantois, which contributes the blood vessels that will become the umbilical arteries and vein. The umbilical cord now connects the embryo to the placenta, which has formed from the merger of the chorion of the embryo and the lining of the uterus.

CASE STUDY CONTINUED

Rerunning the Program of Development

The activation of homeobox genes and other master genes endows a human embryo with the ability to develop all of the parts of the body, including, of course, arms and legs. Rerunning a similar program of development allows adult salamanders to regrow lost limbs. Might it be possible, some day, to restart human development at specific times and locations in the body and thus regenerate fingers, toes, or even entire arms and legs? We investigate this possibility in the Case Study Revisited.

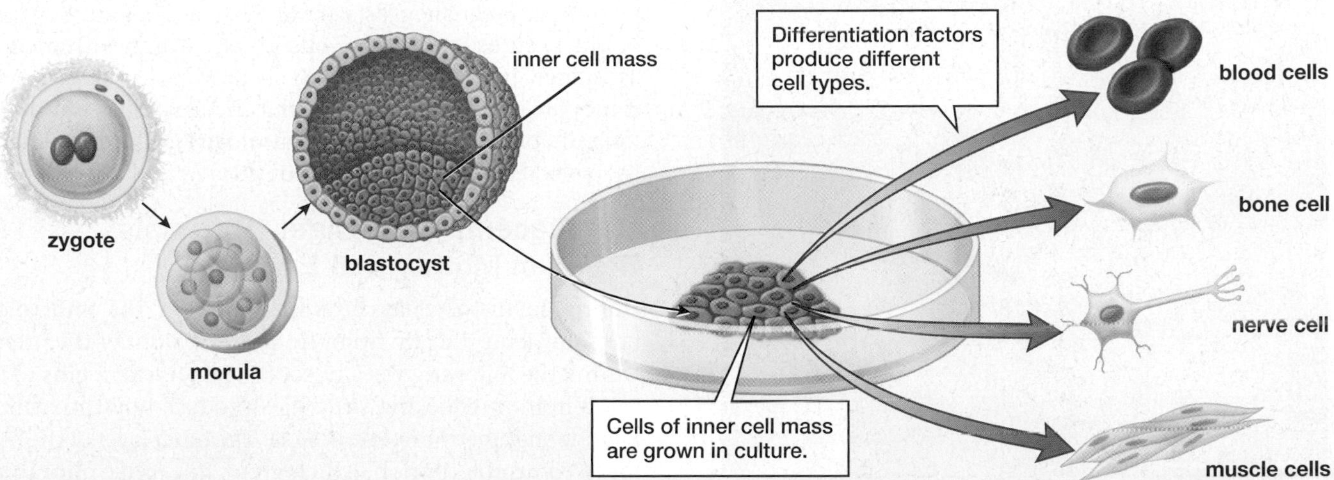

▲ **FIGURE E43-1 Culturing embryonic stem cells from the inner cell mass of a blastocyst**

withdrawing marrow. ASCs have also been used for more than a decade to treat horses and dogs with bone, ligament, or tendon injuries. Controlled studies show that stem cells injected directly into the injury site help horses with tendon and ligament damage to recover faster and heal more completely. A few people have already been injected with ASCs to treat tendon or ligament injuries. For example, professional baseball pitcher Bartolo Colón received ASCs for a shoulder injury that never fully healed after rotator cuff surgery (see the Case Study for Chapter 9). A year later, he was pitching for the New York Yankees. Of course, no one really knows if the ASCs were crucial to Colón's recovery or if he would have healed anyway. Therapies using adult stem cells to treat Crohn's disease, congestive heart failure, atherosclerosis of peripheral arteries, and several other disorders are now in clinical trials.

THINK CRITICALLY The amazing feats of top flight athletes sometimes require exertions that are beyond what their muscles, ligaments, and tendons can withstand. Surgical procedures—such as Tommy John elbow surgery for baseball pitchers or rotator cuff shoulder surgery for gymnasts—can often repair the damage, but not always. If you were a physician advising an athlete about treatments for an injured joint, would you recommend stem cell therapy? What are the possible risks?

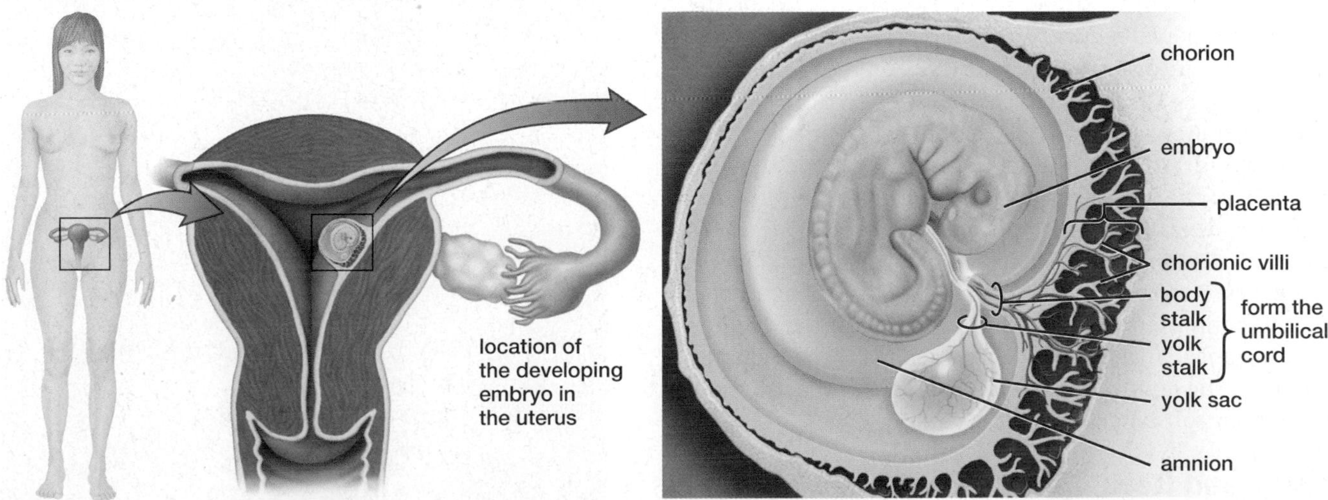

▲ **FIGURE 43-9 Human development during the fourth week** The embryo bulges into the uterus, and the placenta is restricted to one side. The body stalk and yolk stalk combine to form the umbilical cord, which exchanges wastes and nutrients between embryo and mother.

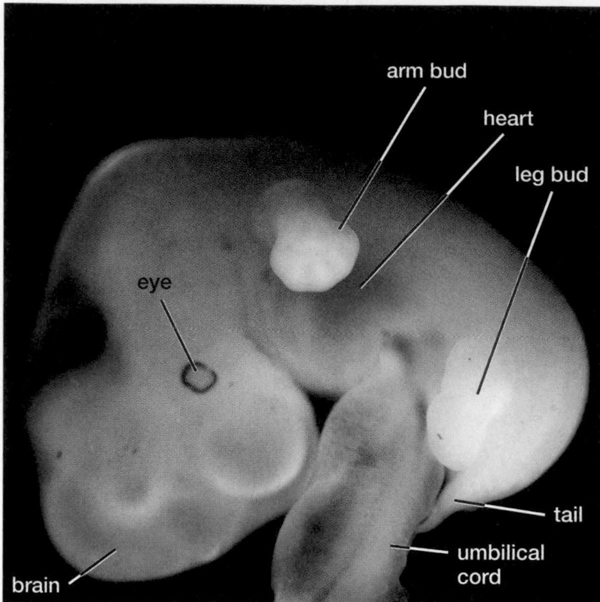

▲ **FIGURE 43-10 A 5-week-old human embryo** At the end of the fifth week, a human embryo is almost half head. A tail is clearly visible, evidence of our evolutionary relationship to other vertebrates.

During the fourth and fifth weeks (**FIG. 43-10**), the embryo develops a prominent tail and pharyngeal grooves (indentations behind the head that are homologous to a fish embryo's developing gills). These structures are reminders that we share evolutionary ancestry with other vertebrates that retain their tails and gills in adulthood. In humans, however, the tail and grooves disappear as development continues. By the seventh week, the embryo has rudimentary eyes and a rapidly developing brain, and the webbing between its fingers and toes is disappearing.

After Two Months, the Embryo Becomes a Fetus

As the second month draws to an end, nearly all of the major organs have begun to develop. The gonads appear and develop into testes or ovaries. Sex hormones are secreted—testosterone from testes or estrogen from ovaries. These hormones affect the development of many structures, including the reproductive organs, brain, and urinary tract. At the end of the second month, the embryo has taken on a generally human appearance and is now called a **fetus** (**FIG. 43-11**).

The first 2 months of pregnancy are a time of extremely rapid differentiation and growth for the embryo, and a time of considerable danger. Although vulnerable throughout development, the rapidly developing organs are especially sensitive to toxic substances during this time.

Growth and Development Continue During the Last Seven Months

Pregnancy usually continues for another 7 months. As the brain and spinal cord grow, they begin to generate behaviors. As early as the third month of pregnancy, the fetus can move, respond to stimuli, and even suck its thumb. The lungs, stomach, and intestine become functional. The kidneys also begin to function; in fact, fetal urine makes up most of the amniotic fluid during the last 6 months of pregnancy. Although a full-term pregnancy lasts about 38 weeks, a fetus 32 weeks or older can usually survive outside the womb with medical assistance. Infants born as early as 26 weeks usually survive if they are given intensive care, but they may suffer from lifelong disabilities, particularly neurological disorders. Infants born before 26 weeks often do not survive.

The Placenta Exchanges Materials Between Mother and Embryo

During the first few days after implantation, the embryo obtains nutrients directly from the endometrium of the uterus. During the following week or so, the **placenta** begins to develop from interlocking structures produced by the embryo and the endometrium (**FIG. 43-12**). The outer layer of the blastocyst forms the chorion, which grows finger-like **chorionic villi** that extend into the endometrium. Blood vessels of the umbilical cord connect the embryo's circulatory system with a dense network of capillaries in the villi. Meanwhile, some of the blood vessels of the endometrium erode away, producing pools of maternal blood that bathe the chorionic villi.

The embryo's blood and the mother's blood remain separated by the walls of the villi and their capillaries, so the two blood supplies do not actually mix to any great extent. The walls of the capillaries and chorionic villi restrict the movement of many substances, including most large proteins and cells. Many small molecules, on the other hand, readily move between the mother's blood and the embryo's blood. Oxygen diffuses from the mother to the embryo. Nutrients, many aided by active transport, also travel from the mother to the embryo. Carbon dioxide and other wastes, such as urea, diffuse from the embryo to the mother. Certain types of antibodies, even though they are quite large, are selectively

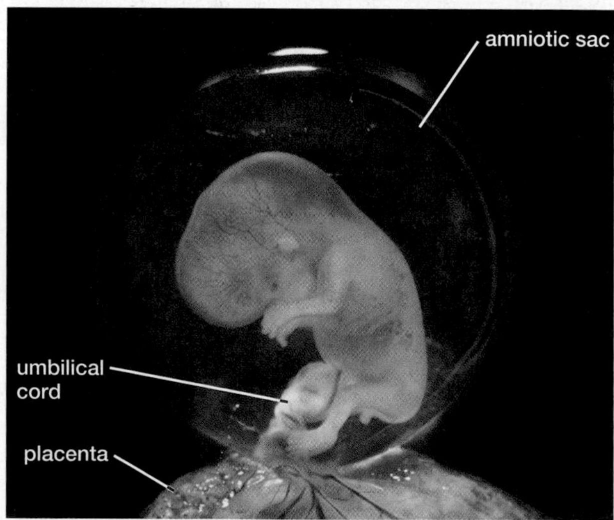

▲ **FIGURE 43-11 An 8-week-old human embryo** At the end of the eighth week, the embryo is clearly human in appearance and is now called a fetus. Most of the major organs of the adult body have begun to develop.

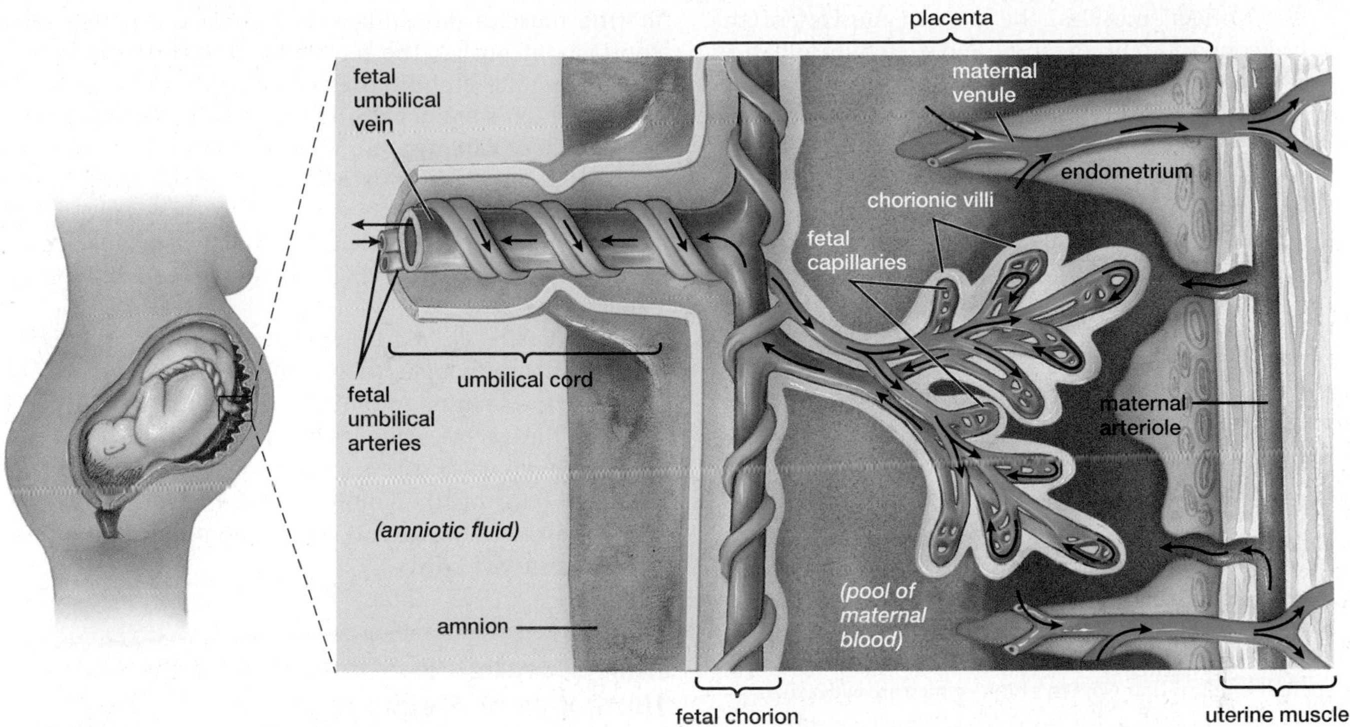

▲ **FIGURE 43-12 The placenta** The placenta allows exchange of wastes and nutrients between fetal capillaries and maternal blood pools, while keeping the fetal and maternal blood supplies separate. The umbilical arteries carry deoxygenated blood from the fetus to the placenta, and the umbilical vein carries oxygenated blood back to the fetus.

THINK CRITICALLY Marsupial mammals, such as kangaroos, have a much more rudimentary placenta. What would you predict about the degree of development of newborn marsupials?

transported across the placenta from mother to embryo, especially late in pregnancy, and play an important role in defending the newborn infant against disease.

Although the placenta isolates the fetus from many assaults, it does not provide complete protection: Some disease-causing organisms and many harmful chemicals can pass through the placenta, as described in "Health Watch: The Placenta—Barrier or Open Door?" on page 774.

Pregnancy Culminates in Labor and Delivery

During the last months of pregnancy, the fetus usually becomes positioned head downward in the uterus, with the crown of the skull resting against, and held up by, the cervix. Around the end of the ninth month, coordinated contractions of the uterus, called **labor**, result in delivery of the child (**FIG. 43-13**).

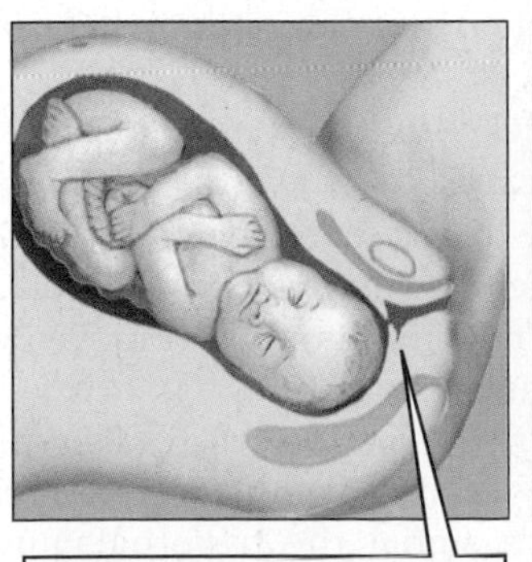

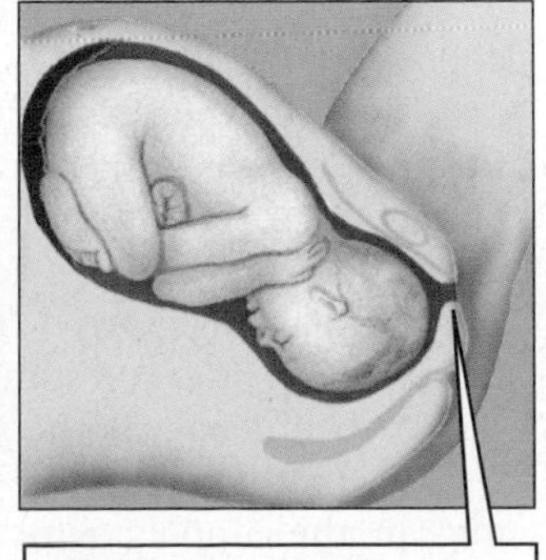

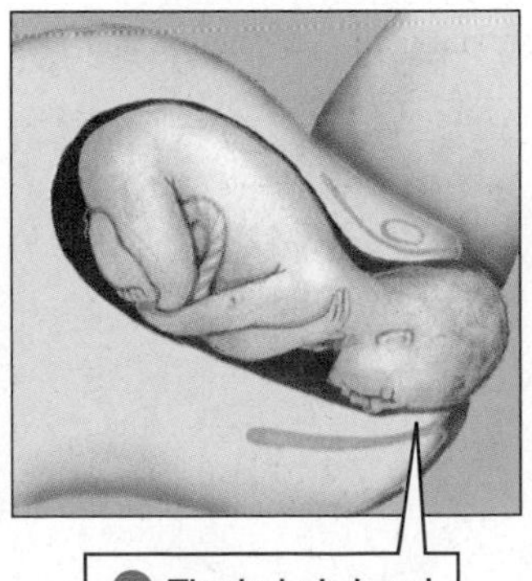

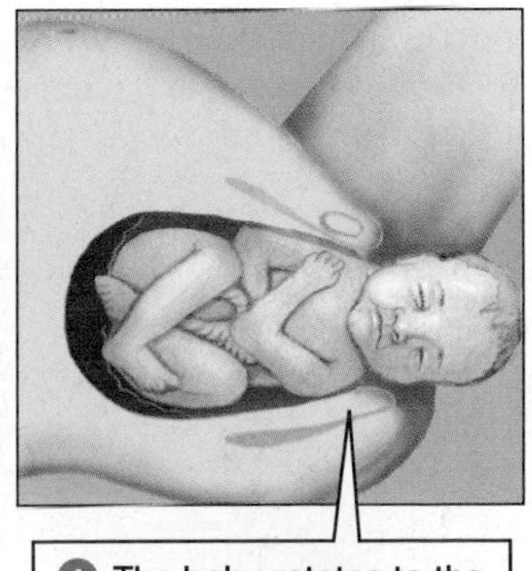

1 The baby orients head downward, facing the mother's side; the cervix begins to thin and expand in diameter (dilate).

2 The cervix dilates to 10 centimeters (almost 4 inches wide), and the baby's head enters the vagina; the baby rotates to face the mother's back.

3 The baby's head emerges.

4 The baby rotates to the side once again as the shoulders emerge.

▲ **FIGURE 43-13 Childbirth**

Unlike skeletal muscles, the smooth muscles of the uterus can contract spontaneously, and stretching enhances their tendency to contract. As the baby grows, it stretches the uterine muscles, which occasionally contract weeks before delivery (these are called Braxton Hicks contractions or false labor contractions).

Chemical signals from the uterus, placenta, and fetus all seem to be involved in triggering the onset of labor. The placenta responds to those signals by releasing prostaglandins and other messenger molecules that make the uterine muscles more likely to contract. As the uterus contracts, it pushes the fetus's head against the cervix, stretching it. Stretching the cervix sends nervous signals to the mother's brain, causing the release of the hormone oxytocin (see Chapter 38). Oxytocin stimulates contractions of the uterine muscles, pushing the baby harder against the cervix, which stretches further, causing still more oxytocin to be released. This positive feedback cycle continues until the cervix expands far enough for the baby to emerge.

Shortly after childbirth, the uterus resumes its contractions and shrinks remarkably. During these contractions, the placenta is sheared from the uterine wall and expelled through the vagina. Muscles surrounding the blood vessels in the umbilical cord contract and shut off blood flow. Although tying off the umbilical cord is standard practice, it is not usually necessary; if it were, non-human mammals would not survive birth.

Have You Ever Wondered ...

Why Childbirth Is So Difficult?

Compared to the birth of other mammals, childbirth is a lengthy and painful ordeal. Why? A human infant, especially its head, is very large compared to the size of a woman's body and pelvic opening. A newborn's head actually must be squashed a bit to fit through the pelvic opening. So why didn't women evolve to have either a bigger pelvis or smaller babies? The "obstetric dilemma" hypothesis maintains that women's hips—which are wider than the hips of a man of similar size, with a larger central opening—are about as wide as they can get and still provide efficient bipedal locomotion. A recent study, however, found that men and women are about equally efficient at walking despite differences in hip anatomy. On the other hand, factors such as speed or agility were not measured, so the "obstetric dilemma" hypothesis cannot be ruled out.

What about delivering younger, smaller infants, or at least infants with smaller heads? Some research suggests that a lot of prenatal brain growth is crucial to producing a large adult brain. In fact, some investigators hypothesize that human pregnancies last as long as possible and that childbirth occurs about the time that a pregnant woman's body would no longer be able to meet the demands of her infant's brain growth. In this scenario, women's hips evolved to be just wide enough for a large-headed infant to squeeze through, and efficient locomotion was not a selective force. Perhaps early in human evolution, poorer nutrition meant babies were born smaller, relative to the size of their mothers, so childbirth wasn't quite so traumatic. The debate continues, but the facts remain: In modern humans, childbirth usually lasts several hours, causes considerable pain to the mother and probably the infant, and often requires assistance. They don't call it "labor" for nothing!

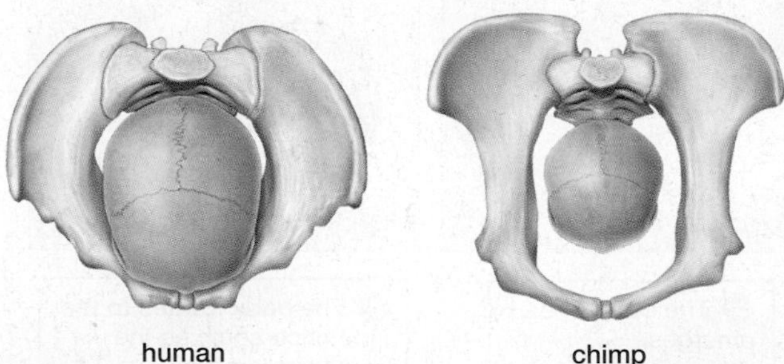

A tight fit A human newborn's head can barely pass through its mother's pelvis (left). In other primates, such as chimpanzees, the infant's head slips through easily (right).

Milk Secretion Is Stimulated by the Hormones of Pregnancy

The human female breast consists of milk-producing **mammary glands** embedded in fatty tissue and connected to the chest wall by ligaments (**FIG. 43-14**). The mammary glands consist of clusters of hollow, milk-producing glands, called alveoli, arranged in a circle around the nipple and extending back almost to the chest wall. An alveolus consists of a single layer of milk-secreting cells surrounding a hollow cavity, which connects to a milk duct leading to the nipple.

Providing an infant with breast milk occurs in three stages. The first stage is preparing the breasts for milk production. During pregnancy, large quantities of estrogen and progesterone are secreted by the placenta. Estrogen and progesterone stimulate the release of prolactin by the anterior pituitary. Together, these three hormones stimulate the mammary glands, causing them to enlarge and develop the capacity to secrete milk. However, high levels of estrogen and progesterone inhibit the actual secretion of milk before the child is born.

The second stage is **lactation**, the secretion of milk by the alveolar cells into the cavities of the alveoli. When the placenta is ejected from the uterus soon after childbirth, estrogen and progesterone levels plummet, while prolactin levels remain high. Freed from inhibition by estrogen and progesterone, prolactin stimulates lactation.

The third stage is the milk ejection reflex (also called "let-down"). The infant's suckling stimulates nerve endings in the nipples, which signal the hypothalamus to cause the pituitary gland to release a surge of oxytocin and prolactin. Oxytocin causes tiny muscles surrounding the mammary glands to contract, ejecting the milk out of the alveolar sacs into the ducts that lead to the nipples. The prolactin surge stimulates milk production for the next feeding.

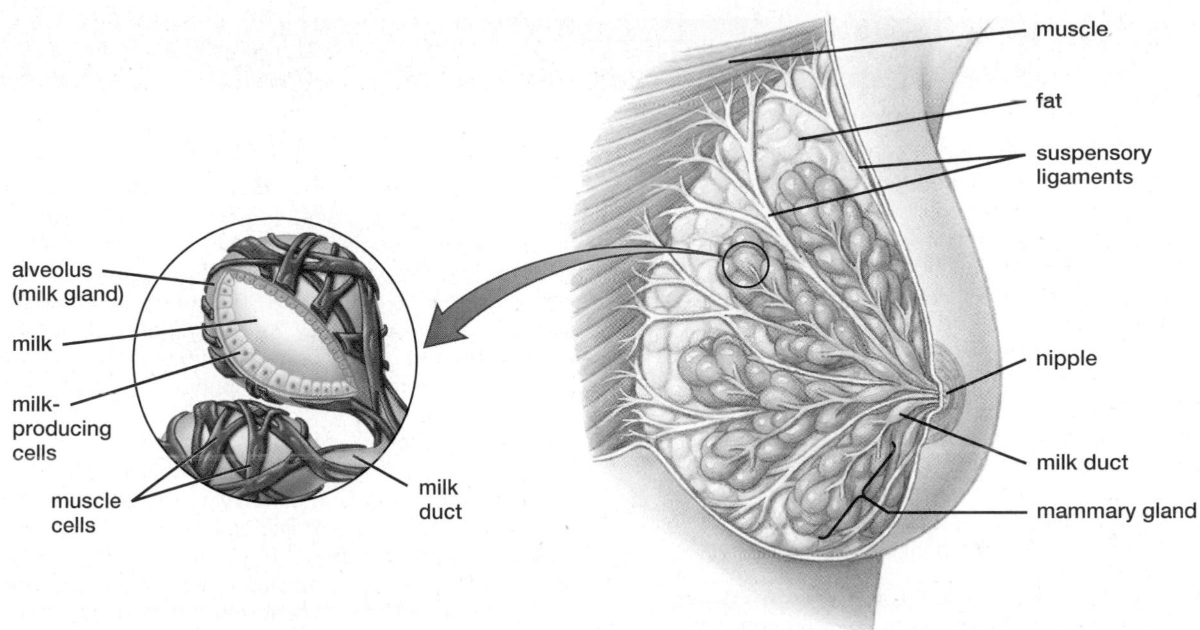

▲ **FIGURE 43-14 The structure of the human breast** During pregnancy, fatty tissue, milk-secreting glands, and milk ducts all increase in size.

During the first few days after birth, the mammary glands secrete a yellowish fluid called **colostrum**. Colostrum is high in protein and contains antibodies from the mother that help protect the infant against disease while its immune system develops. Colostrum is gradually replaced by mature milk, which is higher in fat and milk sugar (lactose) and lower in protein.

CHECK YOUR LEARNING

Can you . . .

- describe human embryonic development from fertilization to birth?
- describe the structure and function of the placenta?
- name the hormones involved in milk production and secretion and describe their roles in these processes?

43.6 IS AGING THE FINAL STAGE OF DEVELOPMENT?

Aging is the gradual accumulation of damage to essential biological molecules, particularly DNA, resulting in defects in cell functioning, declining health, and ultimately death. The damage results from errors in DNA replication, radiation from the sun or from radioactive rocks in Earth's crust, and chemicals in food, cigarettes, and industrial products. Young bodies can often repair such damage or compensate for it. As most animals age, however, their repair abilities diminish, and the body's tolerance for damage is exceeded. Muscle and bone mass are lost, skin elasticity decreases, reaction time slows, and senses such as vision and hearing become less acute. A less-robust immune response renders the aging individual more vulnerable to disease. Eventually, the animal dies.

Why do older animals fail to repair damage to their bodies? Is aging inevitable? Many animals seem to have *programmed aging*, in which genetically determined characteristics influence the timing and speed of aging. In some cases, aging is very rapid. In some mayfly species, adults live for only 30 minutes or so, just long enough to lay some eggs (females) or fertilize them (males). Probably the most extreme case of programmed, rapid aging among mammals occurs in marsupial mice (miniature relatives of kangaroos; **FIG. 43-15**). In some species of marsupial mice, males mature in just under a year. Then, in a brief mating frenzy, they burn out and die

▲ **FIGURE 43-15 Programmed aging** Following a week or two of intense mating activity, male marsupial mice suffer from infectious diseases, internal bleeding, and gangrene, and then they die.

The Placenta—Barrier or Open Door?

The fetal and maternal blood supplies are separated by the walls of the chorionic villi and fetal capillaries (see Fig. 43-12). Perhaps the most important protective layer is the outer covering of the villi, which consists of a remarkable structure: a thin, gigantic single cell containing billions of nuclei. In a full-term fetus, this cell may have a surface area of 10 to 15 square yards—about the size of the typical parking space for a car. Why a single cell? By eliminating the "seams" between cells, this single-cell layer isolates the fetus from many harmful substances that may be present in the mother's blood—from toxins and bacteria to the mother's white blood cells—much better than a multicellular membrane could. Nevertheless, many molecules, and even some infectious microbes, can still move from mother to fetus, either directly across this single-cell layer or through occasional damaged villi.

Microbes

Most microbes that cross the placenta must be able to penetrate through the plasma membrane of the single-cell layer on its maternal side, survive in the cytoplasm, and then cross through the plasma membrane on the fetal side. Most bacteria fail to cross through. Those that can penetrate through the cell layer include *Treponema pallidum*, which causes syphilis, some types of *Streptococcus* that cause blood infections, and *Listeria*, which is often found in contaminated food. Viral invaders that can cross the placenta include those that cause genital herpes, German measles (rubella), hepatitis B, Zika virus disease, and HIV.

Many of these microbes cause severe disorders in the fetus. For example, 30% to 40% of fetuses infected with syphilis are stillborn. *Listeria* can cause reduced fetal growth, premature delivery, and meningitis (a sometimes fatal infection of the membranes surrounding the brain). Zika virus can cause severe birth defects, including damaged eyesight and *microcephaly*, a condition in which the head and brain are significantly smaller than normal (**FIG. E43-2**). Infections with any of several other types of viruses may cause premature birth, mental retardation, limb and joint defects, and many other disorders.

Medicinal Drugs

Many medicinal drugs are lipid soluble and can easily move across plasma membranes (see Chapter 5), including those of the placenta. Some lipid-soluble drugs can harm the fetus. Thalidomide, for example, was commonly prescribed in Europe in the late 1950s and early 1960s to combat morning sickness during pregnancy. Its devastating effects on embryos were discovered only when many babies were born with missing or extremely abnormal limbs. In the late 1980s, pregnant women who took isotretinoin, a potent anti-acne medication, were found to have a high risk of miscarriage and premature birth. Their children were also at increased risk for congenital deformities. Preliminary research suggests that a number of commonly prescribed medicines, including the anti-depressant paroxetine (Paxil) and the cholesterol-reducing drug atorvastatin, are associated with an increased risk of birth defects and should generally not be used by pregnant women.

Recreational Drugs

Many recreational drugs can cross the placenta and harm a developing fetus. Mothers who use cocaine, for example, increase their risk of miscarriage, premature birth, and delivering infants with low birth weight. Use of methamphetamine carries similar risks. The children of drug users, especially users of opioid drugs such as heroin and fentanyl, may be born addicted to the drug, a condition called *neonatal abstinence syndrome* (NAS). Soon after birth, babies with NAS experience symptoms of withdrawal, which may include vomiting, diarrhea, and seizures.

Cigarette Smoke

Many women who smoke continue the habit during pregnancy, thus exposing their developing children to nicotine, carbon monoxide, and a host of carcinogens. Nicotine is also present in the vapor produced by e-cigarettes and other devices used for "vaping," so women who vape during pregnancy also put their children at risk. Mothers who smoke have a higher

just a week or two after putting all of their body's resources into big testes, lots of sperm, and copulation, even shutting down their immune systems and digesting their muscles for energy. In contrast, females often live for a couple of years.

At the other extreme, some animals are genetically programmed for *negligible senescence*, in which aging occurs extremely slowly, if at all. Ocean quahogs (a species of clam)

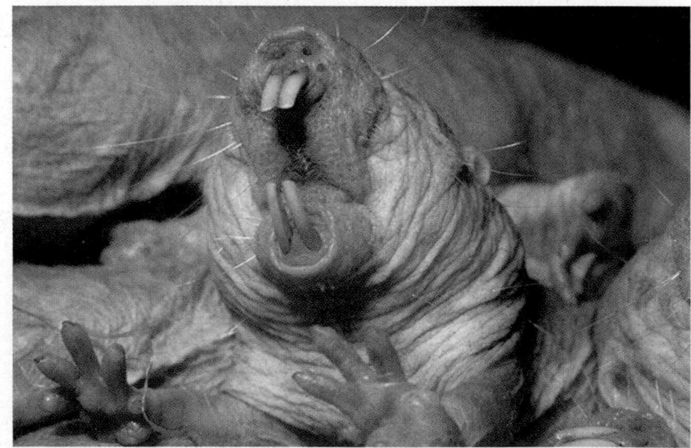

FIGURE 43-16 Negligible senescence Naked mole rats far outlive other rodents of similar size. They have much lower body temperatures, respiratory rates, and metabolisms than most other rodents, and virtually never develop cancer. These factors might contribute to their long life.

frequency of miscarriages and tend to give birth to underweight infants. Children born to smokers have a higher than normal incidence of behavioral and intellectual problems.

Alcohol

Alcohol is soluble in both water and lipids. It dissolves in the blood and then easily passes through the placenta: When a pregnant woman drinks, the blood of her unborn child has the same concentration of alcohol as her own blood does. Worse, the fetus cannot metabolize alcohol as rapidly as an adult, so alcohol's effects in the fetus are greater and longer lasting.

According to the U.S. Centers for Disease Control and Prevention, about 8% of U.S. women drink during pregnancy. Each year, they give birth to more than 100,000 infants with a variety of disabilities collectively called fetal alcohol spectrum disorders. Women who drink during pregnancy tend to have smaller babies who are more likely to grow up to be anxious, depressed, or aggressive.

If a pregnant woman drinks heavily on a regular basis (four or more drinks per day), or goes on alcoholic binges (five or more drinks at a time), her child has a significant risk of developing full-blown **fetal alcohol syndrome (FAS)**, including below-average intelligence, hyperactivity, irritability, and poor impulse control. Children with FAS may also have reduced growth and defects of the heart and other organs. In extreme cases, children afflicted with fetal alcohol syndrome may have small, improperly developed brains. The damage is irreversible and can be fatal.

Fetal alcohol syndrome is thought to be the most common cause of mental retardation in the United States, and it is 100% preventable. The U.S. Surgeon General advises pregnant women and those who are likely to become pregnant to avoid alcohol completely.

The Placenta Cannot Be Trusted to Protect the Embryo

A pregnant woman should assume that any drugs she takes will cross the placenta to her developing infant. Crucial stages of embryonic development can occur before a woman even realizes that she is pregnant, so any woman having sex without contraception should take the same precautions she would take if she were already pregnant.

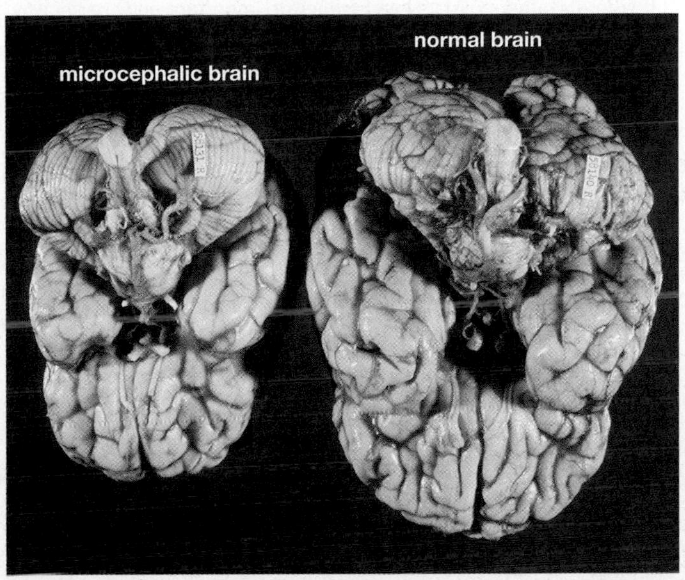

▲ **FIGURE E43-2 Fetal infections can cause birth defects** Viral and bacterial infections that are transmitted from mother to fetus can cause permanent harm to the fetus. Some pathogens, such as the Zika virus, can cause microcephaly, in which the head and brain are abnormally small.

THINK CRITICALLY So many substances can pass through the placenta. Why do you think that it is biologically useful to have the fetal and maternal bloodstreams separated by membranes, rather than have their blood vessels fuse into one combined circulatory system?

live for up to 400 years. Giant tortoises have been recorded to live 150 to 250 years; the rougheye rockfish can live at least 205 years. The naked mole rat (**FIG. 43-16**) can live for 30 years. Although that may not seem very long, similarly sized mice and dwarf hamsters typically live only 2 to 3 years, even under ideal conditions; for a rodent, naked mole rats have an extremely long life span.

For thousands of years, people have attempted to delay aging and live longer. Modern medical care can prevent or cure many diseases and can fix or replace damaged organs through procedures such as bypass surgery and transplants. Some dietary changes, particularly eating very small amounts of highly nutritious food (euphemistically called "caloric restriction"), can prolong life, at least in animal experiments. However, what seems to be the maximum human life span, about 130 years, has not changed. Can researchers learn the secrets of negligible senescence enjoyed by giant tortoises, rockfish, and naked mole rats? If so, can these findings be applied to humans? Perhaps you'll live long enough to find out!

CHECK YOUR LEARNING

Can you . . .

- define *aging* and describe the mechanisms thought to cause it?
- describe *programmed aging* and give an example of it?
- describe *negligible senescence* and give an example of it?

CASE STUDY REVISITED

Rerunning the Program of Development

If so many other animals can regenerate important body parts, why can't we? Well, humans *can* regenerate, a little. Young children, up to the age of 10 or so, can often regenerate a fingertip, nail and all, that has been accidentally cut off, as stem cells in the severed stump replace the lost structures. Similar results have been reported in a few adults with severed fingertips. But regeneration never occurs, even in children, if a finger is cut off far enough that no knuckles are left. No human, of course, has ever regenerated a lost arm or leg.

However, things aren't as hopeless as this might suggest. Researchers are trying to uncover the genetic programs that run during the development of specific body parts and find out how to turn on those programs in adult mammals. A tantalizing hint comes from mice. Unlike most mice, a strain called MRL can regrow damaged ears, including cartilage, skin, and hair. MRL mice can also partially regenerate amputated toes. MRL mice differ from other mice in many ways, including a reduced inflammatory response to injury, prolific regrowth of blood vessels after injury, and lower oxidative stress. MRL mice also differ in some of the molecules that control the cell cycle. Some of these characteristics probably underlie their regenerative abilities. Mice are a lot more similar to people than sea stars, salamanders, or lizards are, so studying the mechanisms of regeneration in MRL mice provides hope that, someday, we may discover how to regenerate damaged or lost human body parts.

THINK CRITICALLY MRL mice have lower levels of a protein called p21 than other mice do. The p21 protein is involved in the control of cell division, suppressing the division of cells with damaged DNA. "Knocking out" the *p21* gene allows otherwise normal mice to regenerate injured ears. If someone devised a way of selectively inactivating the p21 protein in humans, what do you think the effects might be, both for regeneration and for health?

CHAPTER REVIEW

Go to **Mastering Biology** to access the Pearson eText, vocabulary review, practice quizzes, activities, videos, current events, and more.

Answers to ***Think Critically*** *and* ***Thinking Through the Concepts*** *questions can be found in the* ***Answers*** *section at the back of the book.*

Summary of Key Concepts

43.1 What Are the Principles of Animal Development?

Development is the process by which an organism grows and increases in organization and complexity, using three mechanisms: (1) Individual cells multiply; (2) daughter cells differentiate; and (3) groups of cells move about and become organized into multicellular structures.

43.2 How Do Direct and Indirect Development Differ?

Animals undergo either direct or indirect development. In direct development, the newborn animal is sexually immature but otherwise resembles a small adult. Animals with direct development usually either produce large, yolk-filled eggs or nourish the developing embryo within the mother's body. In indirect development, eggs (usually with relatively little yolk) hatch into larvae, which undergo metamorphosis to become adults with notably different body forms.

43.3 How Does Animal Development Proceed?

Animal development occurs in several stages. *Cleavage and blastula formation*: A fertilized egg undergoes cell divisions with little intervening growth. Cleavage results in the formation of the morula, a solid ball of cells. A cavity then opens up within the morula, forming a hollow ball of cells called a blastula. *Gastrulation*: A dimple forms in the blastula, and cells migrate from the surface into the interior of the ball, eventually forming a three-layered gastrula. These three cell layers—ectoderm, mesoderm, and endoderm—give rise to all the adult tissues. *Organogenesis*: The cell layers of the gastrula form organs characteristic of the animal species (see Table 43-1).

In mammals and reptiles (including birds), extraembryonic membranes (the chorion, amnion, allantois, and yolk sac) encase the embryo in a fluid-filled space and regulate the exchange of nutrients and wastes between the embryo and its environment (see Table 43-2).

43.4 How Is Development Controlled?

All the cells of an animal body contain a full set of genetic information, yet different cells are specialized for particular functions. Cells differentiate by stimulating and repressing the transcription of specific genes, processes that are controlled by (1) differences in gene-regulating substances inherited from the mother in her egg and/or (2) chemical communication between the cells of the embryo, a process called induction. A particularly important group of genes, called homeobox genes, regulates the differentiation of body segments and their associated structures, such as wings or legs.

43.5 How Do Humans Develop?

A human egg is fertilized in the uterine tube. The resulting zygote develops into a blastocyst and implants in the endometrium. The outer wall of the blastocyst becomes the chorion and forms the embryonic contribution to the placenta; the inner cell mass develops into the embryo and the other three extraembryonic membranes. During the third and fourth weeks of gestation, the digestive tract begins to form, the heart begins to beat, and the rudiments of a nervous system appear. By the end of the second month, the major organs have begun to form, and the embryo—now called a fetus—appears human. In the next 7 months before birth, the fetus continues to grow; the lungs, stomach, intestine, kidneys, and nervous system enlarge, develop, and become functional.

Intermingling of the chorion of the embryo and the endometrium of the uterus forms the placenta, which provides the embryo with nutrients and oxygen from its mother, and carries away its wastes. The placenta also isolates the embryo from many, but not all, potentially toxic substances that may be present in its mother's circulatory system.

During pregnancy, mammary glands in the mother's breasts enlarge under the influence of estrogen, progesterone, and prolactin. After about 9 months, uterine contractions are triggered by a complex interplay of uterine stretch and the release of prostaglandin and oxytocin. As a result, the uterus expels the baby and then the placenta. After its birth, the infant begins suckling the breast and activates the release of prolactin, which stimulates milk production, and oxytocin, which triggers milk to flow to the nipple.

43.6 Is Aging the Final Stage of Development?

Aging is the gradual accumulation of cellular damage (particularly to DNA) over time that leads to loss of functionality and, eventually, to death. The hypothesis of programmed aging proposes that evolution has favored specific genetic processes that cause different timing and rates of aging in different animals. Some animal species have negligible senescence, in which aging occurs extremely slowly.

Thinking Through the Concepts

Bloom's: Remembering, Understanding

Multiple Choice

1. In a reptile egg, the extraembryonic membrane that encloses the embryo in a watery environment is the
   a. allantois.
   b. amnion.
   c. chorion.
   d. yolk sac.
2. Animals with indirect development
   a. develop within the bodies of their mothers.
   b. undergo metamorphosis from the larval stage to the adult stage.
   c. include most reptiles and mammals.
   d. typically have newborn offspring that are similar in appearance to the adults of the species.
3. During what stage of development do the three primary tissue layers first appear?
   a. fertilization
   b. gastrulation
   c. cleavage
   d. differentiation
4. Bone develops from
   a. ectoderm.
   b. colostrum.
   c. endoderm.
   d. mesoderm.
5. In vertebrate development, a zygote undergoes cleavage, forming a solid ball of cells called a
   a. morula.
   b. gastrula.
   c. blastula.
   d. blastocyst.
6. The inner cell mass of a human blastocyst
   a. gives rise to the chorion.
   b. gives rise to the placenta.
   c. gives rise to the embryo.
   d. consists of highly differentiated cells.
7. The placenta
   a. is formed by the chorion and the endometrium of the uterus.
   b. is required for reproduction in terrestrial vertebrates.
   c. allows the blood supply of the mother and embryo to mix together.
   d. is absorbed back into the body of the mother after delivery.
8. In humans, implantation in the uterus occurs at which stage of embryonic development?
   a. zygote
   b. morula
   c. blastocyst
   d. gastrulation

Fill-in-the-Blank

1. A fertilized egg, also called a ____________, is a very large cell. One of the first events of animal development is the division of this cell, a process called ____________, to produce a solid ball of much smaller cells, the ____________. Soon, a cavity opens in this ball of cells, producing a hollow ball called the ____________. Movements of cells in this hollow ball during gastrulation produce the three embryonic tissue layers, the ____________, ____________, and ____________.
2. Undifferentiated cells that can multiply and produce daughter cells of many different types are called ____________. Embryonic cells of this type can differentiate into any cell type of the body; they are derived from the ____________.
3. Genes that control the development of entire body segments and their accompanying limbs (if any) are called ____________. They usually code for proteins called ____________ that regulate the transcription of many other genes.
4. Milk production in the mammary glands is stimulated by a hormone from the pituitary gland called ____________. Another hormone, ____________, stimulates contraction of both the uterine muscles during childbirth and the muscles around the mammary glands that cause milk to be ejected from the glands.

Review Questions

1. Distinguish between direct and indirect development and give examples of animals in each category.
2. Describe the structure and function of the four extraembryonic membranes found in reptiles, including birds. Are these four present in placental mammals? In what ways are their roles similar in reptiles and mammals? How do they differ?

3. What is gastrulation? Describe gastrulation in amphibians.
4. Name two structures derived from each of the three embryonic tissue layers: endoderm, ectoderm, and mesoderm.
5. Describe the process of induction in animal development. Describe the experiment that first demonstrated induction.
6. In humans, where does fertilization occur, and what stages of development occur before the fertilized egg reaches the uterus?
7. Describe how the human blastocyst gives rise to the embryo and its extraembryonic membranes.
8. Explain how the structure of the placenta prevents mixing of fetal and maternal blood while allowing for the exchange of substances between the mother and the fetus.
9. How do changes in the breast prepare a mother to nurse her newborn? How do hormones influence these changes and stimulate milk production?
10. What is aging? What causes it?

Applying the Concepts

Bloom's: Applying, Analyzing, Evaluating

1. A researcher obtains two frog embryos at the gastrula stage of development. She carefully removes a cluster of cells from a location that she knows would normally become neural tube tissue and transplants it into the second gastrula in a location that would normally become skin. Does the second gastrula develop two neural tubes? Explain your answer.
2. In almost all animals that produce eggs with very little yolk, the blastula stage is roughly spherical. Mammalian eggs, however, contain virtually no yolk but have flat embryonic disks, which undergo gastrulation much like spherical blastulas do. Propose an explanation for this difference in shape.

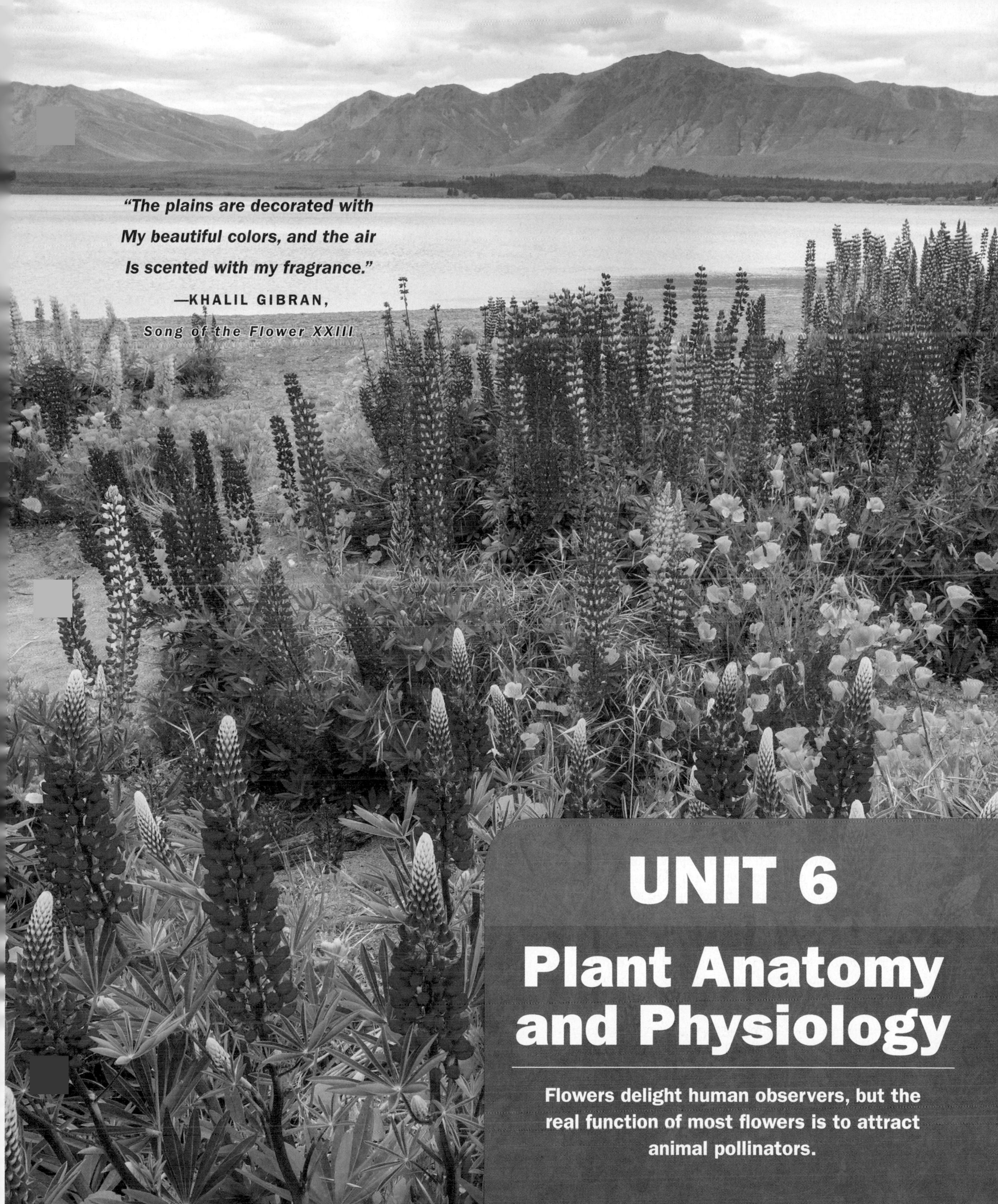

"The plains are decorated with My beautiful colors, and the air Is scented with my fragrance."

—KHALIL GIBRAN, *Song of the Flower XXIII*

UNIT 6 Plant Anatomy and Physiology

Flowers delight human observers, but the real function of most flowers is to attract animal pollinators.

44 Plant Anatomy and Nutrient Transport

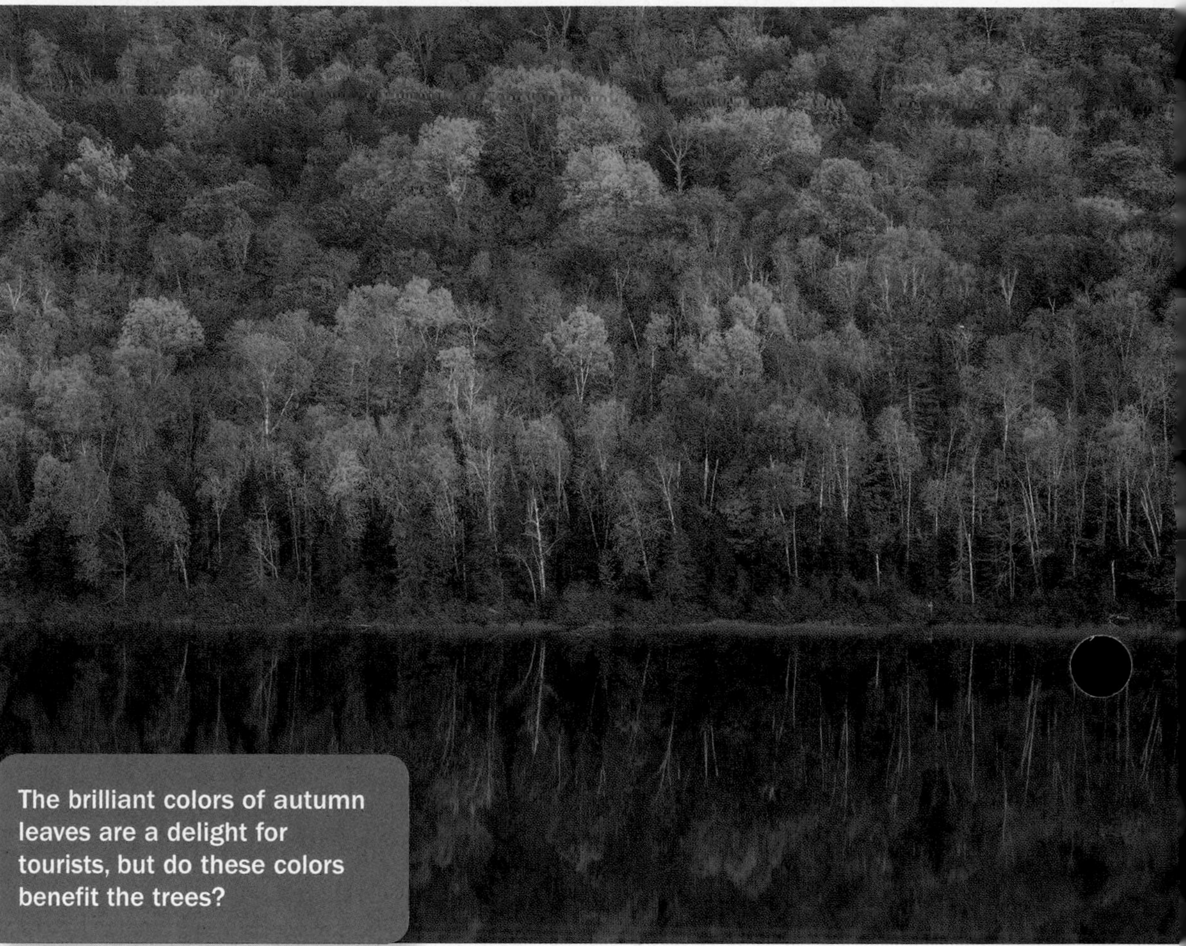

The brilliant colors of autumn leaves are a delight for tourists, but do these colors benefit the trees?

CASE STUDY

Autumn in Vermont

EACH AUTUMN, more than 3.5 million "leaf peepers" tour Vermont to admire the dazzling red, yellow, and orange trees. Stunning displays of fall color also attract visitors to the rest of New England and to upstate New York, southeastern Canada, and the northern Appalachian Mountains. The residents of these areas reap an abundant harvest of tourist dollars from the fall foliage. But what's in it for the trees? The colors didn't evolve to entertain visitors. The changing colors presumably help the trees in some way—but how?

During most of the growing season, leaves are green, colored by the chlorophyll molecules that are essential for photosynthesis. Other molecules, however, are responsible for the autumn colors of the leaves of deciduous trees and shrubs. For example, pigments known as carotenoids provide the yellow and orange colors of autumn leaves (as well as the bright colors of daffodils, marigolds, oranges, and carrots). Red colors in autumn leaves are produced by pigments called anthocyanins.

Carotenoids are present in green leaves, but anthocyanins are not. The carotenoids in green leaves assist photosynthesis by absorbing some colors of light that chlorophyll cannot, and they also protect chlorophyll molecules from damage by absorbing excess energy when sunlight is too bright. Anthocyanins play no role in green leaves because, unlike carotenoids, they are synthesized only in the fall.

Autumn colors prompt a host of questions about plants and their leaves. Why are the leaves of deciduous trees large and flat, and why are they shed in the autumn? Why do they turn color before they fall? Why are carotenoids visible only in the autumn? Why would natural selection favor plants that spend energy synthesizing anthocyanin in leaves that are about to drop off?

AT A GLANCE

44.1 How Are Plant Bodies Organized?

44.2 How Do Plants Grow?

44.3 What Are the Differentiated Tissues and Cell Types of Plants?

44.4 What Are the Structures and Functions of Leaves?

44.5 What Are the Structures and Functions of Stems?

44.6 What Are the Structures and Functions of Roots?

44.7 How Do Plants Acquire Nutrients?

44.8 How Do Plants Move Water and Minerals from Roots to Leaves?

44.9 How Do Plants Transport Sugars?

44.1 HOW ARE PLANT BODIES ORGANIZED?

Most people admire fields of wildflowers and groves of giant sequoias, but seldom stop to think about the adaptations that allow these plants to thrive. Plants cannot move to seek food or water, escape predators, avoid winter, or find a mate. Yet dwarf willows survive in the harsh tundra of Alaska, mangroves perch in salt water along coastlines throughout the Tropics, and cacti endure the searing heat of the Mojave Desert. Evolution has produced a wide variety of distinctly different types of plants (see Chapter 22). In this chapter, we will focus principally on the flowering plants, or angiosperms, the most widespread and diverse group of plants.

The bodies of flowering plants are composed of two major parts: the root system and the shoot system (**FIG. 44-1**). The **root system** consists of all the roots of a plant. **Roots** are branched portions of a plant body, usually embedded in soil, that typically carry out six major functions: (1) anchor the plant in the ground; (2) absorb water and minerals from the soil; (3) transport water, minerals, sugars, and hormones to and from the shoot; (4) store surplus sugars and starches; (5) produce hormones; and (6) interact with soil fungi and bacteria that help the plant to acquire nutrients.

The **shoot system** consists of stems, leaves, buds, flowers, and fruits and usually grows aboveground. The shoot system performs five major functions: (1) capture sunlight energy and synthesize sugars during photosynthesis; (2) transport materials to and from various parts of the plant;

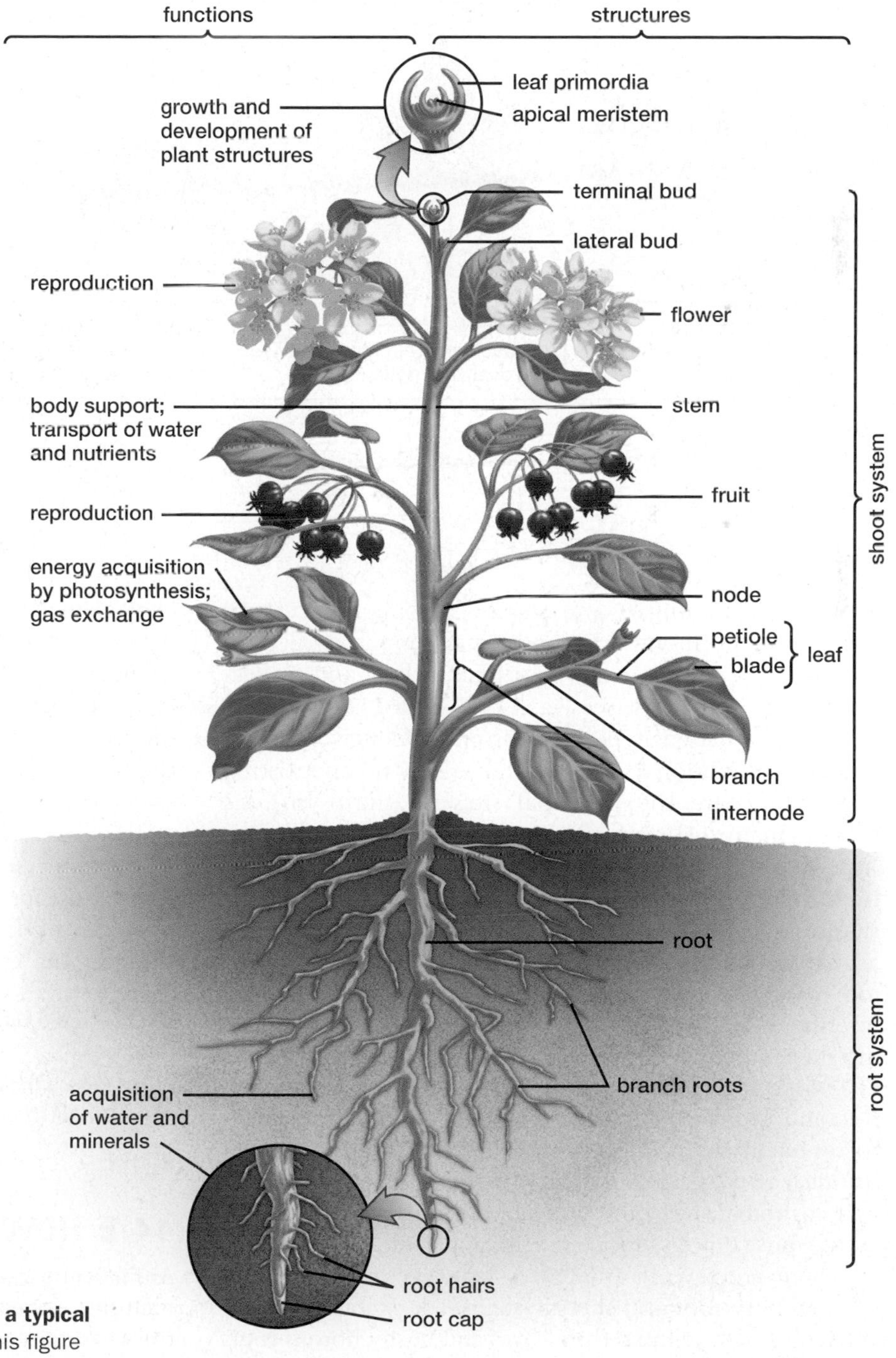

FIGURE 44-1 The structures and functions of a typical flowering plant It will be helpful to refer back to this figure as you read the rest of the chapter.

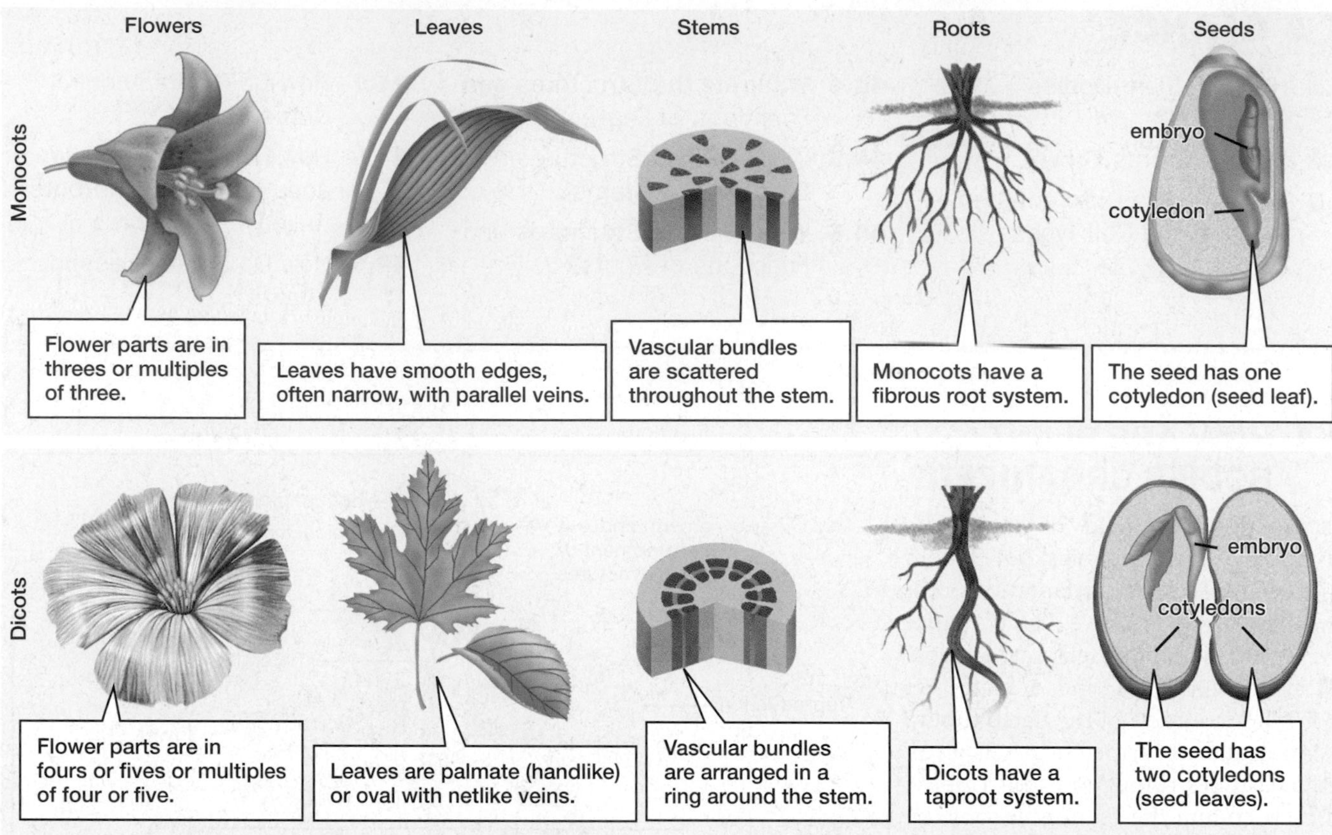

Flowers
Leaves
Stems
Roots
Seeds
Monocots
embryo
cotyledon
Flower parts are in threes or multiples of three.
Leaves have smooth edges, often narrow, with parallel veins.
Vascular bundles are scattered throughout the stem.
Monocots have a fibrous root system.
The seed has one cotyledon (seed leaf).
Dicots
embryo
cotyledons
Flower parts are in fours or fives or multiples of four or five.
Leaves are palmate (handlike) or oval with netlike veins.
Vascular bundles are arranged in a ring around the stem.
Dicots have a taproot system.
The seed has two cotyledons (seed leaves).

▲ **FIGURE 44-2 Characteristics of monocots and dicots**

(3) store surplus sugars and starches; (4) reproduce; and (5) produce hormones. The principal support structures of a shoot are **stems**, which bear buds, leaves, and (in season) flowers and fruits (see Fig. 44-1). A **bud** is an embryonic shoot. Different types of buds may produce branches, flowers, or additional growth at the top of an existing stem. Leaves are the principal sites of photosynthesis in most plants. Flowers are the plant's reproductive organs, producing seeds enclosed in fruits which protect the developing seeds and often help disperse them (see Chapter 45).

Earth supports an amazing diversity of flowering plants, from tender tulips to prickly cacti, from short grasses to towering oaks. All flowering plants, however, can be placed into one of two large groups, called monocots and dicots (**FIG. 44-2**). **Monocots** include lilies, daffodils, tulips, palm trees, and a wide variety of grasses, not only the familiar lawn grasses but also wheat, rice, corn, oats, and bamboo. **Dicots** include virtually all "broad-leafed" plants, such as deciduous trees and bushes, most vegetables, cacti, and many of the plants whose flowers grace our fields and gardens.

There are several differences between monocots and dicots in both root and shoot systems. For example, monocots typically have long, thin leaves and bushy fibrous roots (think of grasses), while dicots have broader leaves and thick taproots (as in dandelions or carrots). However, the characteristic that gives the groups their names is the number of *cotyledons*. Monocots have a single cotyledon (Gk. *mono,* one), and dicots have two (Gk. *di,* two). A cotyledon is the part of a plant embryo that absorbs and often stores food reserves in the seed and then transfers the food to the rest of the embryo when the seed sprouts. When you eat a peanut, each half of the "nut" consists almost entirely of a cotyledon, which has stored starches and fats that power the growth of a peanut seedling. Cotyledons are often called "seed leaves," because they sometimes become the first green, photosynthetic leaves during seed sprouting (described in Chapter 45).

CHECK YOUR LEARNING

Can you . . .

- list the principal functions of shoots and roots?
- distinguish between monocots and dicots?

44.2 HOW DO PLANTS GROW?

Animals and plants develop in dramatically different ways. At maturity, animals of a given species resemble each other in their basic size and shape. In contrast, most flowering plants grow throughout their lives, responding to a variety of environmental stimuli. As a result, many never reach a stable body size or precise shape, a property called **indeterminate growth**. Moreover, whereas young animals grow by increasing the size of all parts of their bodies, most plants grow longer or taller only at the tips of their branches and roots.

Plants are composed of two fundamentally different types of cells: meristem cells and differentiated cells. **Meristem cells**, like the stem cells of animals, are not specialized and are capable of mitotic cell division (see Chapter 9). Most of their daughter cells lose the ability to divide, transforming into **differentiated cells** with specialized structures and functions.

Plants grow as a result of the division of meristem cells found in two general locations in the plant body (**FIG. 44-3**). **Apical meristems** (L. *apex*, tip) are located at the tips of roots and shoots, and (in dicots) at the tips of branches as well. Growth produced by apical meristem cells is called **primary growth**, which includes an increase in the height or length of a shoot or root as well as the development of specialized parts of the plant, such as leaves and buds. Primary growth from apical meristems explains why a swing tied to a tree branch doesn't get any higher off the ground over the years—a tree grows taller as its main stem elongates at the tip and it grows wider as its branches extend from their tips. **Lateral meristems** (L. *lateralis*, side) are concentric cylinders of meristem cells that extend though the roots, stems, and branches of many dicot plants, but are absent in most monocots. Cell division in lateral meristems and the differentiation of the resulting daughter cells produce **secondary growth**, typically an increase in the diameter and strength of roots and shoots. Secondary growth occurs in woody plants, including deciduous trees, many shrubs, and most conifers such as pines and spruces. Some woody plants become very tall and thick and may live for hundreds or—in rare instances—thousands of years. Secondary growth is why the branch holding up a swing, the trunk of the tree bearing the branch, and the roots that support the entire tree all became thicker and stronger over the years.

Many plants, such as strawberries, grasses, and annual garden flowers, do not undergo secondary growth. As you might predict, most plants that lack secondary growth are

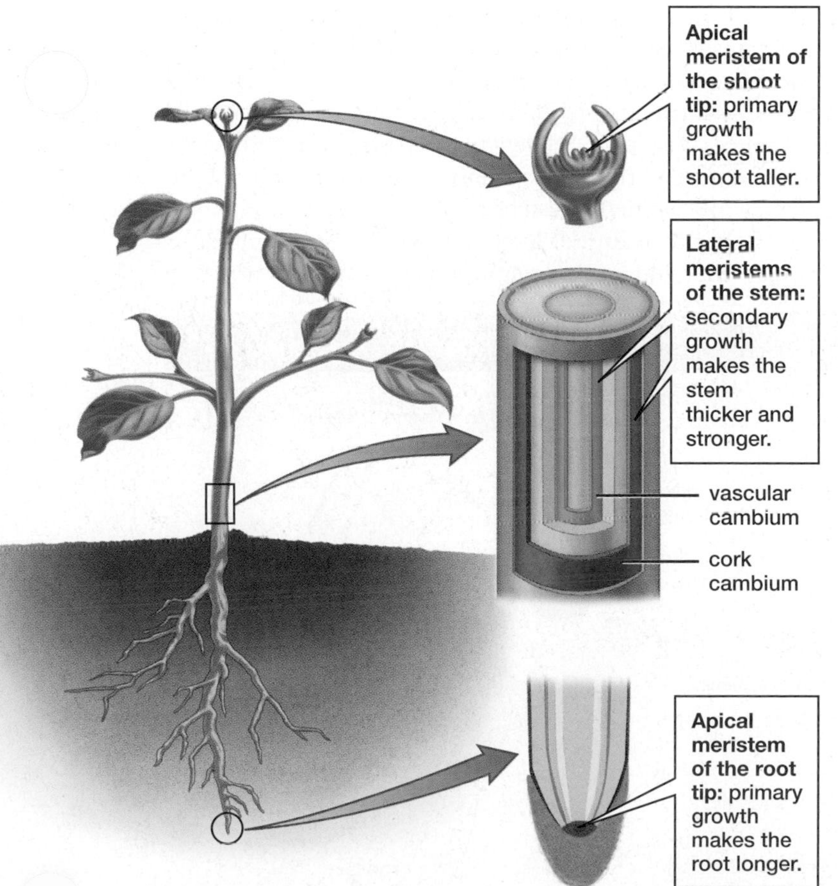

Apical meristem of the shoot tip: primary growth makes the shoot taller.
Lateral meristems of the stem: secondary growth makes the stem thicker and stronger.
vascular cambium
cork cambium
Apical meristem of the root tip: primary growth makes the root longer.

▲ **FIGURE 44-3 Meristems** A dicot has apical meristems at the tips of shoots, shoot branches, roots, and root branches. Lateral meristems occur as cylinders within the root and shoot in woody plants.

Have You Ever Wondered ...

How Trees Can Live So Long?

Some trees, such as giant sequoias and bristlecone pines, can live more than 1,000 years; the oldest known bristlecone pine, called Methuselah, is more than 4,800 years old! What allows this extreme longevity? (1) Essential parts are spread throughout the tree and not concentrated in one location like a brain or a heart. If a branch or part of a root dies, the rest of the tree lives on. (2) Because their meristems never stop producing new cells, trees always have some young roots and branches. (3) Many trees produce resins or other substances that prevent attack from bacteria or fungi. (4) Ancient trees often live in cold or dry environments, which limit the growth of bacteria and fungi and reduce undergrowth that might catch fire and burn down the trees. Barring accidents such as a lightning strike or major climate change, a bristlecone pine, which has all of these advantages, could potentially live for tens of thousands of years.

soft bodied, with short, flexible stems that survive for only one growing season.

CHECK YOUR LEARNING

Can you . . .

- explain the difference between meristem cells and differentiated cells?
- describe primary and secondary growth?

44.3 WHAT ARE THE DIFFERENTIATED TISSUES AND CELL TYPES OF PLANTS?

As apical meristem cells differentiate, they produce a variety of cell types. When one or more specialized types of cells work together to perform a specific function, such as conducting water and minerals, they form a **tissue**. Functional groups of more than one tissue are called **tissue systems**. The plant body is composed of three tissue systems: the ground tissue system, the dermal tissue system, and the vascular tissue system, each of which extends continuously through the roots, stems, and leaves (**TABLE 44-1**).

The Ground Tissue System Makes Up Most of the Young Plant Body

Most of the body of a young plant consists of cells of the **ground tissue system**. There are three tissue types within the ground tissue system, each composed of only a single type of cell: parenchyma, collenchyma, and sclerenchyma. These cell types can be distinguished by their cell walls and by whether they remain alive in the mature plant.

Parenchyma tissue consists of living parenchyma cells with thin cell walls and diverse functions (**FIG. 44-4a**). Parenchyma cells carry out most of the plant's metabolic activities, including photosynthesis, secretion of hormones, and food storage. Storage structures—including seeds (such as rice), fruits (such as apples), tubers (such as potatoes), and roots (such as carrots)—are packed with parenchyma cells that contain sugars and starches. Thin-walled parenchyma cells, inflated with water, also help support the bodies of many non-woody plants. Houseplants wilt if you forget to water them because their parenchyma cells deflate. Parenchyma cells are also found in the dermal and the vascular tissue systems.

Collenchyma tissue is composed of living collenchyma cells that are typically elongated, with irregularly thickened, but still flexible, cell walls (**FIG. 44-4b**). Collenchyma cells store nutrients. They also form a kind of flexible skeleton that helps to support the bodies of young and non-woody plants, providing them with supple strength against bending forces, such as wind. Collenchyma also stiffens thick leaf veins and the leaf stalks, or *petioles*, of all plants. For example, celery stalks (which are actually extremely thick petioles) are supported by ribs composed of bundles of collenchyma cells located just beneath the outer dermal cells; these bundles form the celery strings that may catch in your teeth.

Sclerenchyma tissue is composed of sclerenchyma cells with thickened cell walls (**FIG. 44-4c**). Like collenchyma, sclerenchyma cells support and strengthen the plant body. However, unlike collenchyma and parenchyma cells, sclerenchyma cells die after they differentiate, leaving behind their cell walls as a source of support. Sclerenchyma tissue forms nut shells and the outer covering of peach pits. Scattered throughout the parenchyma cells in the flesh of a pear, sclerenchyma cells give pears their gritty texture. Sclerenchyma cells also contribute to vascular tissue and form an important component of wood.

TABLE 44-1 Tissue Systems of Plants

Type	Tissues Within the Tissue System	Functions	Locations of the Tissue Systems
Ground tissue system	Parenchyma	Photosynthesizes (leaves and young stems); stores nutrients; supports non-woody plants; secretes hormones; contributes to the dermal and vascular tissue systems	
	Collenchyma	Stores nutrients; supports non-woody plant bodies and all leaf stalks	
	Sclerenchyma	Supports the plant body as strengthening fibers in both xylem and phloem; its cell walls are a major component of wood and nut shells.	
Dermal tissue system	Epidermis	Protects plant surfaces; regulates the movement of O_2, CO_2, and water between the plant and the air or soil	
	Periderm (secondary growth)	Thickens stems and roots with a protective cork layer	
Vascular tissue system	Xylem	Transports water and dissolved minerals from root to shoot	
	Phloem	Transports sugars and other organic molecules (including amino acids, proteins, and hormones) throughout the plant body	

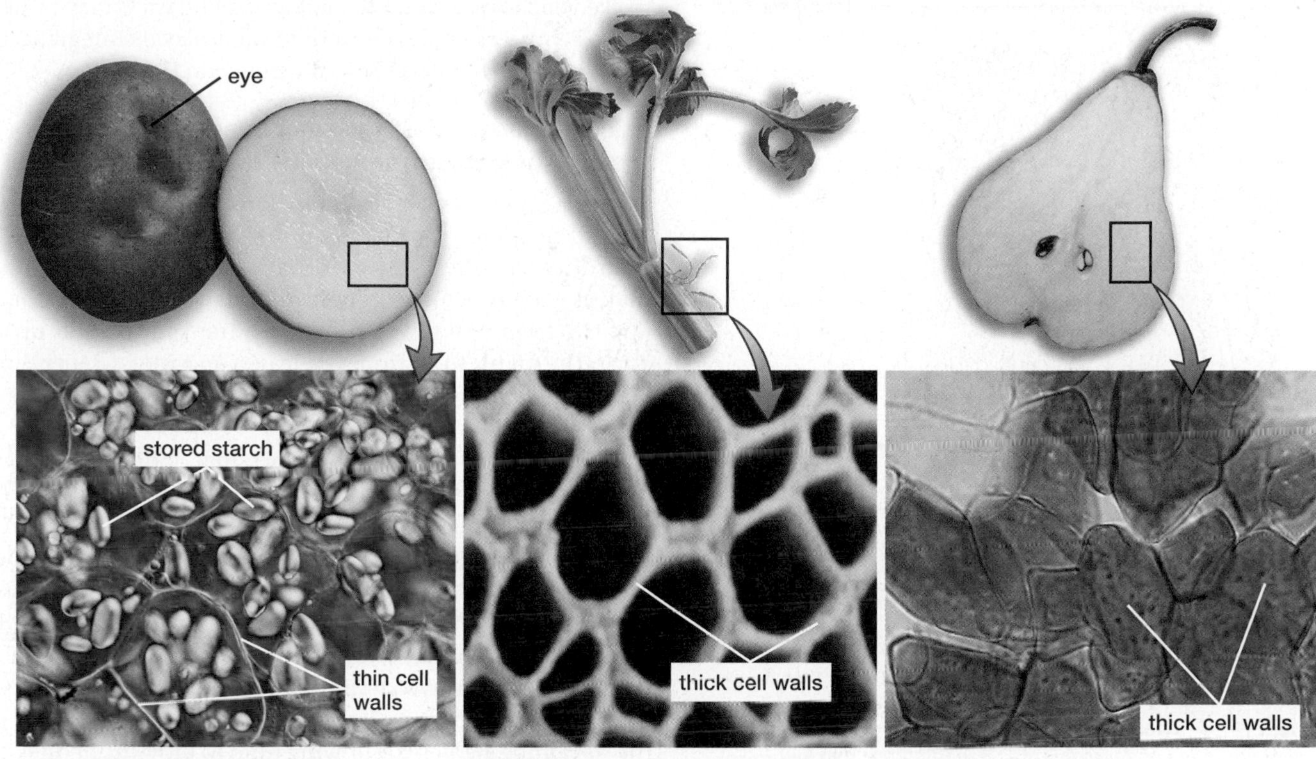

(a) Parenchyma cells in a potato (b) Collenchyma cells in a celery stalk (c) Sclerenchyma cells in a pear

▲ **FIGURE 44-4 Ground tissue (a)** Parenchyma cells are living, with thin cell walls. They perform many functions, including photosynthesis, hormone secretion, and storage. **(b)** Collenchyma cells are living and have thickened, but somewhat flexible, cell walls that help support the plant body. **(c)** Sclerenchyma cells have thick, hard walls and die after they differentiate.

The Dermal Tissue System Covers the Plant Body

The **dermal tissue system** (L. *derma,* skin) covers and protects the entire plant body. Cells of the epidermal tissue, collectively called the **epidermis**, form the outermost cell layer covering the leaves, stems, and roots of young plants and new growth in older plants (see Fig. 44-8). The epidermis is generally composed of flattened, tightly packed, thin-walled epidermal cells that lack chloroplasts. Some epidermal cells produce **trichomes**, projections that can be either outgrowths of a single cell or multicellular. Trichomes have a variety of shapes and functions. Some trichomes (such as root hairs) absorb water, others form a hairy coating on leaves, and some (such as cotton fibers; **FIG. 44-5**) emerge from the epidermis of seeds and aid in dispersal. The aboveground parts of a plant are covered with a waterproof, waxy **cuticle** secreted by the epidermal cells. The cuticle reduces the evaporation of water from the plant and helps protect it from pathogens. As described later, specialized guard cells regulate the movement of water vapor, O_2, and CO_2 across the epidermis in stems and leaves. In woody plants, a type of protective dermal tissue called **periderm** replaces epidermal tissue on roots and stems as they undergo secondary growth (see Section 44.5).

▲ **FIGURE 44-5 Cotton fibers are epidermal trichomes produced on seeds**

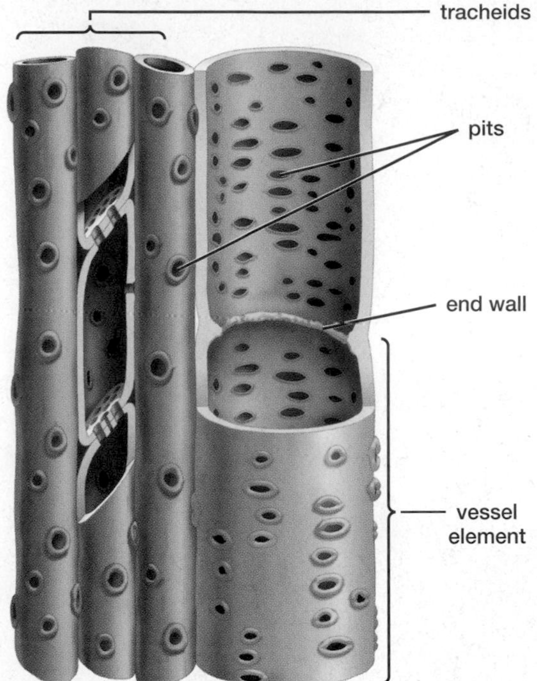

▲ **FIGURE 44-6 Xylem** Xylem contains two types of conducting cells: tracheids and vessel elements. Tracheids are thin with tapered ends; vessel elements are much larger in diameter and usually have blunt ends. Both the ends and sides of adjacent tracheids are connected by pits. Pits in their side walls also connect tracheids and vessel elements.

The Vascular Tissue System Transports Water and Nutrients

The **vascular tissue system** of plants conducts water and dissolved substances throughout the plant body. It consists of two conducting tissues: xylem and phloem. These tissues contain specialized conducting cells and often include sclerenchyma cells for added support and parenchyma cells that store various materials. Here we focus on the conducting cells that are unique to the vascular tissue.

Xylem Transports Water and Dissolved Minerals from the Roots to the Rest of the Plant

Xylem transports water and dissolved minerals from the roots to all parts of the shoot system. Xylem contains two types of conducting sclerenchyma cells: tracheids and vessel elements (**FIG. 44-6**). Both tracheids and vessel elements develop thick cell walls and then die, leaving behind hollow tubes of nonliving cell wall. These tubes are perforated by **pits**, which are thin, porous dimples in the cell walls. Pits allow water and minerals to pass among the conducting cells of the xylem, and also into adjacent cells of the plant body.

Tracheids are thin, elongated cells with tapered, overlapping ends connected by pits.

Vessel elements, which are larger in diameter than tracheids, form pipelines called **vessels**. Vessel elements are stacked end to end, and their adjoining end walls may be perforated with large holes, or the end walls may disintegrate entirely, leaving an open tube as illustrated in Figure 44-6.

Phloem Transports Dissolved Sugars and Other Organic Molecules Throughout the Plant Body

Phloem transports solutions containing organic molecules, including sugars, amino acids, and hormones, from the structures that synthesize them to the structures that use them. Unlike xylem, in which fluids take a one-way trip from root to shoot, phloem transports fluids up or down the plant, to or from the leaves and roots, depending on the metabolic state of various parts of the plant at any given time.

Phloem contains two cell types that cooperate in conducting solutions rich in sugar and other organic nutrients: sieve-tube elements and companion cells (**FIG. 44-7**). **Sieve-tube elements** are joined end to end to form pipes called sieve tubes. As sieve-tube elements mature, they lose their nuclei and most other organelles, leaving behind only a thin layer of cytoplasm lining the plasma membrane. The junction between two sieve-tube elements is called a **sieve plate**. Here, membrane-lined pores connect the insides of the sieve-tube elements, allowing fluid to move from one cell to the next.

How can sieve-tube elements maintain and repair their plasma membranes when they lack nuclei and most other organelles? Life support for sieve-tube elements is provided by adjacent **companion cells**, which are connected to sieve-tube elements by pores called **plasmodesmata** (singular,

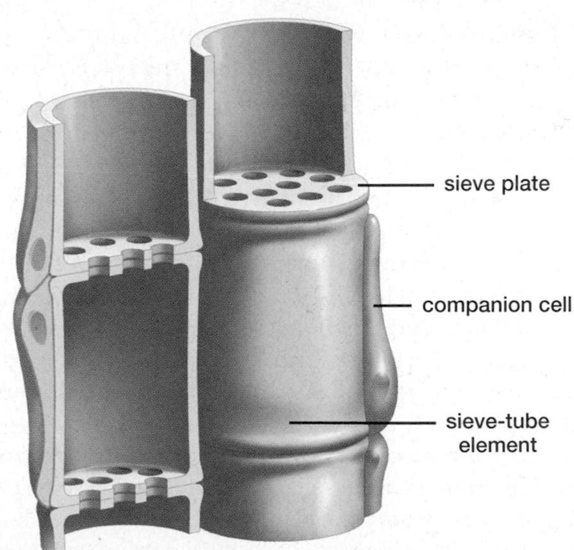

▲ **FIGURE 44-7 Phloem** The conducting system of phloem consists of sieve-tube elements, stacked end to end. Where they join at sieve plates, large membrane-lined pores allow fluid to move between them. Each sieve-tube element has a companion cell that nourishes it and regulates its function.

plasmodesma). Companion cells supply sieve-tube elements with proteins and ATP. They also help move the molecules transported by phloem into the sieve-tube elements.

CHECK YOUR LEARNING

Can you . . .

- name the three tissue systems in flowering plants and describe their locations in the plant body?
- describe the cell types in each tissue system and explain their functions?
- explain the difference between xylem and phloem, in both structure and function?

44.4 WHAT ARE THE STRUCTURES AND FUNCTIONS OF LEAVES?

A typical **leaf** consists of a large, flat portion, the **blade**, connected to the stem by a stalk called the **petiole**. Inside the blade and petiole, xylem and phloem connect the leaf to the vascular system of the rest of the plant (**FIG. 44-8**).

Leaves are the principal photosynthetic structures of most plants. Photosynthesis uses the energy of sunlight (captured by chlorophyll) to convert water (H_2O) and carbon dioxide (CO_2) into sugar, releasing oxygen (O_2) as a by-product (see Chapter 7). Water is absorbed from the soil by a plant's roots and transported to the leaves, and CO_2 enters leaves from the air. You might think that maximum photosynthesis would occur in a porous leaf (which would allow CO_2 to diffuse easily from the air into the leaf) with a large surface area (which would intercept the most sunlight). However, on a hot, dry day, a plant with large, porous leaves might lose more water through evaporation than it could replace from the soil, threatening the entire plant with death by dehydration. Therefore, the structure of leaves reflects an elegant compromise between the need to acquire sunlight and CO_2 and the conflicting need to conserve water. Most leaves have a large, relatively waterproof surface, with pores that can open to admit CO_2 and close to restrict water evaporation, as needed.

The Epidermis Regulates the Movement of Gases into and out of a Leaf

The leaf epidermis consists of a layer of nonphotosynthetic, transparent cells that secrete a waxy cuticle on their outer surfaces. The cuticle is nearly waterproof and reduces the evaporation of water from the leaf. Epidermal cells of many types of leaves bear trichomes, which serve a variety of functions; for example, trichomes may reduce the evaporation of water, reflect harmful ultraviolet radiation, or deter insect predators. The epidermis and cuticle are pierced by openings called **stomata** (singular, stoma) that regulate the diffusion of CO_2, O_2, and water vapor into and out of the leaf. Most stomata are in the shaded lower epidermis to minimize evaporation. A stoma consists of two sausage-shaped **guard cells** that enclose, and adjust the size of, the pore between them. Unlike other epidermal cells, guard cells contain chloroplasts and carry out photosynthesis.

Photosynthesis Occurs in Mesophyll Cells

Transparent epidermal cells allow sunlight to reach the **mesophyll** ("middle of the leaf"), which consists of loosely packed parenchyma cells containing chloroplasts. Many leaves contain two types of mesophyll cells—an upper layer of columnar *palisade cells* and a lower layer of irregularly shaped *spongy cells* (see Fig. 44-8). Mesophyll cells carry out most of the photosynthesis in a leaf. Air spaces between mesophyll cells allow CO_2 from the atmosphere to diffuse to each cell and O_2 produced during photosynthesis to diffuse away.

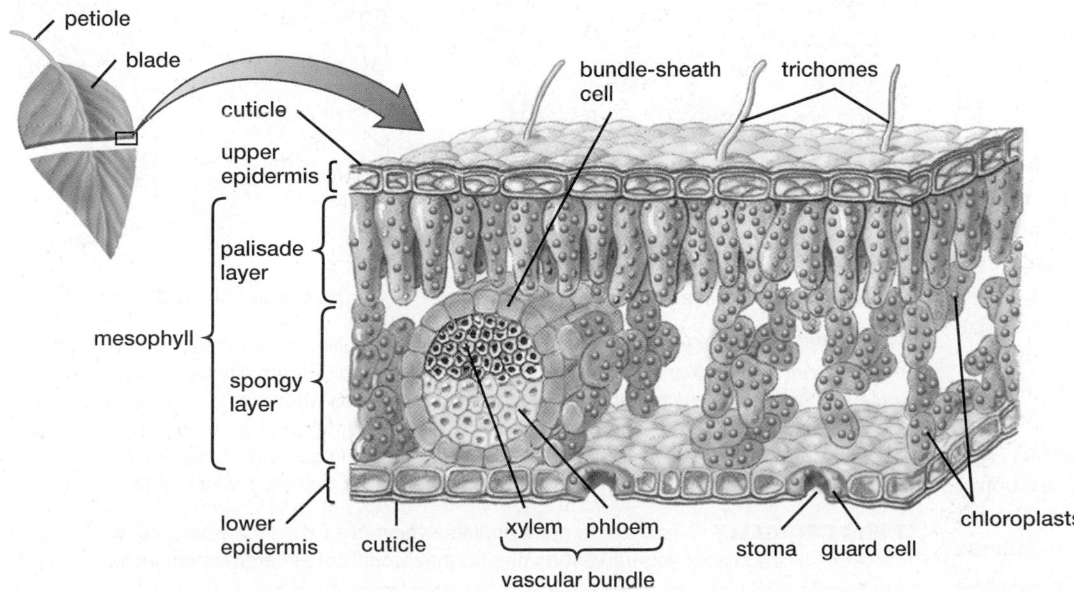

FIGURE 44-8 A typical dicot leaf A waterproof cuticle covers the outside surfaces of the epidermal cells. Epidermal cells of many plants produce trichomes. Epidermal cells are transparent, allowing sunlight to penetrate to the chloroplast-containing mesophyll cells beneath. Stomata within the lower epidermis allow gas exchange and help control water loss.

CASE STUDY CONTINUED

Autumn in Vermont

Summer leaves are green due to the large amount of chlorophyll in the chloroplasts that are packed within the leaves' mesophyll cells. The amount of chlorophyll in a leaf is about 50 times larger than the amount of carotenoids. But once fall arrives, carotenoids have their season of glory. Chlorophyll breaks down, revealing the striking orange and yellow colors of the carotenoids that were there all along. Before leaves fall off, they undergo a process that allows the plant to reclaim their nutrients. How do nutrients leave the leaf?

Veins Transport Water and Nutrients Throughout the Leaf

Vascular bundles (in leaves, these are also called **veins**) containing xylem and phloem conduct materials between the leaf and the rest of the plant body. Xylem delivers water and minerals to the mesophyll cells of the leaf. Phloem transports soluble organic molecules. For example, it carries sugar produced by mature leaves to other parts of the plant to provide energy or to store it for later use. As leaves die, the phloem transports recaptured nutrients, particularly nitrogen-containing amino acids, into seeds, stems, and roots for storage.

Many Plants Produce Specialized Leaves

Differences between different habitats in temperature and the availability of water and light have influenced natural selection's effect on leaves. Plants growing in the dim light that reaches the floor of a tropical rain forest often have very large leaves, an adaptation demanded by the low light level and permitted by the abundant water of the rain forest (**FIG. 44-9a**). The leaves of desert-dwelling cacti, in contrast, have been reduced to spines that provide almost no surface area for evaporation (**FIG. 44-9b**). Spines also protect the juicy cactus stem from being eaten by herbivores. The plump leaves of other drought-resistant succulents store water in their parenchyma cells and are covered with a thick cuticle that greatly reduces water evaporation (**FIG. 44-9c**).

Some plants have evolved leaves with unexpected structures and functions. Onions and daffodil bulbs are actually extremely short underground stems enveloped by thick, overlapping leaves, which store water, sugars, and other nutrients during the summer to support new growth the following spring (**FIG. 44-9d**). Carnivorous plants, such as Venus flytraps and sundews, have leaves that are modified into snares that trap and digest unwary insects (see the opening photo of Chapter 46). Pea plants produce thin, modified leaves called tendrils that coil around other plants, or the trellis in your garden, helping the plant to reach sunlight.

(a) Elephant ear leaves

(b) Cactus spines

(c) Succulent leaves

(d) Onion leaves

▲ **FIGURE 44-9 Specialized leaves (a)** Some plants that live on the rain-forest floor (such as this elephant ear) have enormous leaves that capture the limited light filtering through the taller trees. **(b)** Spines of desert cacti are nonphotosynthetic leaves whose minuscule surface area reduces evaporation. Sharp spines also protect the plant from browsing animals. **(c)** Succulent desert plants have fleshy leaves that store water from the infrequent rains. **(d)** An onion consists of a short central stem surrounded by thickened leaves that store water and food.

THINK CRITICALLY Onion plants are biennials, which live for only 2 years. As a gardener, when should you harvest onions so they would be as large as possible?

CHECK YOUR LEARNING

Can you . . .

- diagram an angiosperm leaf and describe the structure and function of each part?
- describe some specialized leaf adaptations and explain the function of each?

44.5 WHAT ARE THE STRUCTURES AND FUNCTIONS OF STEMS?

The stems of a plant support its leaves, lifting them up to the sunlight. Stems also transport water and dissolved minerals from the roots to the leaves and transport sugars produced in the leaves to the roots and other parts of the shoot, such as buds, flowers, and fruits. In most dicots, stems undergo primary growth during their first year and at the tips of their branches throughout life. In perennial dicots (which live for more than 2 years), older stems and branches undergo secondary growth.

Primary Growth Produces the Structures of a Young Stem

At the tip of a newly developing shoot sits a **terminal bud** consisting of apical meristem cells surrounded by developing leaves, or **leaf primordia** (singular, primordium) (**FIG. 44-10**). During primary growth, the meristem cells divide. Most of their daughter cells differentiate into the specialized cell types of leaves, buds, and the structures of the stem. The remaining daughter cells remain meristematic, providing for future growth of the shoot.

Below the terminal bud, the stem surface can be divided into two regions: nodes and internodes (**Fig. 44-10a**). A **node** is the site of attachment of the petiole of a leaf to the stem. When nodes first form from the apical meristem, they are close together. Then the stem between the nodes elongates, forming naked **internodes** (literally, "between nodes"). As most dicot shoots grow, small clusters of meristem cells, the **lateral buds**, are left behind at the nodes, usually in the angle between the leaf petiole and the stem surface. Under appropriate hormonal conditions, lateral buds sprout and grow into branches (see Chapter 46). As a branch grows, it duplicates the development of the stem, including new leaves and both terminal and lateral buds. During the reproductive season, usually spring or summer, meristem cells form **flower buds**, generally in the same locations in which a terminal or lateral bud would otherwise develop.

The apical meristem of a shoot also produces the internal structures of the stem (see Fig. 44-10); these are typically grouped into four tissue types: epidermis, cortex, pith, and vascular tissues. Monocots and dicots differ somewhat in their arrangement of vascular tissues; here we discuss only dicot stems.

The Epidermis of the Stem Reduces Water Loss While Allowing Carbon Dioxide to Enter

The epidermis of a young stem, like that of leaves, secretes a waxy cuticle that reduces water loss. As in leaves, the stem epidermis is perforated by stomata that regulate the movement of carbon dioxide, oxygen, and water vapor into and out of the stem.

The Cortex and Pith Support the Stem, Store Food, and May Photosynthesize

In dicots, the **cortex** lies between the epidermis and the vascular tissues, and the **pith** fills the central part of the stem, surrounded by the vascular tissues. Cortex and pith consist of parenchyma cells that fill most of the young stem. Cells of the cortex and pith perform three major functions:

- ***Support*** In very young stems, water filling cortex and pith cells pushes the cytoplasm up against the cell wall, causing turgor pressure (see Chapter 5). Turgor pressure stiffens the cells, much as air inflates a tire, and helps to hold the stem erect.
- ***Storage*** Parenchyma cells in both cortex and pith convert sugar into starch and store the starch as a food reserve.
- ***Photosynthesis*** In many young dicot stems, the outer cells of the cortex contain chloroplasts and carry out photosynthesis. In plants such as cacti, in which the leaves are reduced to spines, the cortex of the stem may be the only green, photosynthetic part of the plant.

Vascular Tissues in the Stem Transport Water, Dissolved Nutrients, and Hormones

Most young dicot stems contain bundles of vascular tissue, often roughly triangular in cross-section, that extend the full height of the stem. A dozen or more bundles are arranged in a ring within the stem just inside the epidermis (**FIG. 44-10b**, top). The pointed inner edge of each bundle consists of *primary xylem*, and the rounded outside edge consists of *primary phloem*, with *procambium* (that will later become vascular cambium) separating them (**FIG. 44-10c**, top). These structures are described as "primary" because they are produced from an apical meristem during primary growth. Xylem carries water and dissolved minerals from the roots up through the stem. Phloem transports solutions containing a variety of organic molecules up and down the stem in response to changing metabolic conditions, as described later.

Secondary Growth Produces Thicker, Stronger Stems

The stems of most conifers and perennial dicots undergo secondary growth, resulting from cell division in the lateral meristems of the vascular cambium and cork cambium (see Fig. 44-10). Such stems may survive for decades, centuries, and, in rare cases, even millennia, becoming thicker and stronger each year.

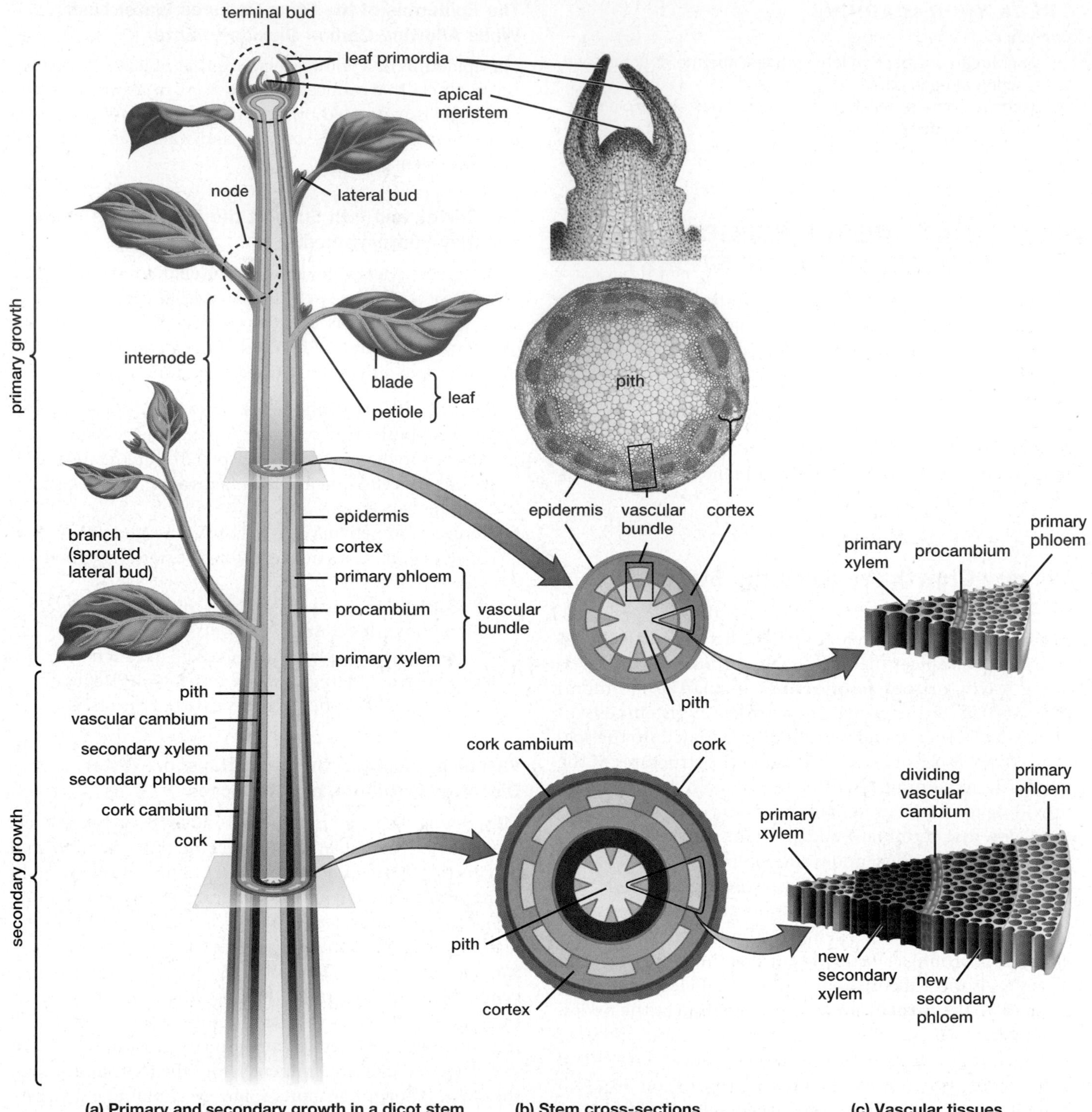

▲ **FIGURE 44-10 Primary and secondary growth in a dicot shoot** A dicot shoot after 2 years of growth, showing primary growth in the upper part of the shoot and secondary growth farther down. **(a)** A vertical slice of the stem, showing its internal structures. Cross-sections of **(b)** the entire stem and **(c)** vascular tissue after primary growth (top) and secondary growth (bottom). During secondary growth, the vascular cambium forms a complete ring around the stem, so the resulting secondary xylem and phloem also form complete rings, rather than the separate vascular bundles of primary xylem and phloem that form during primary growth.

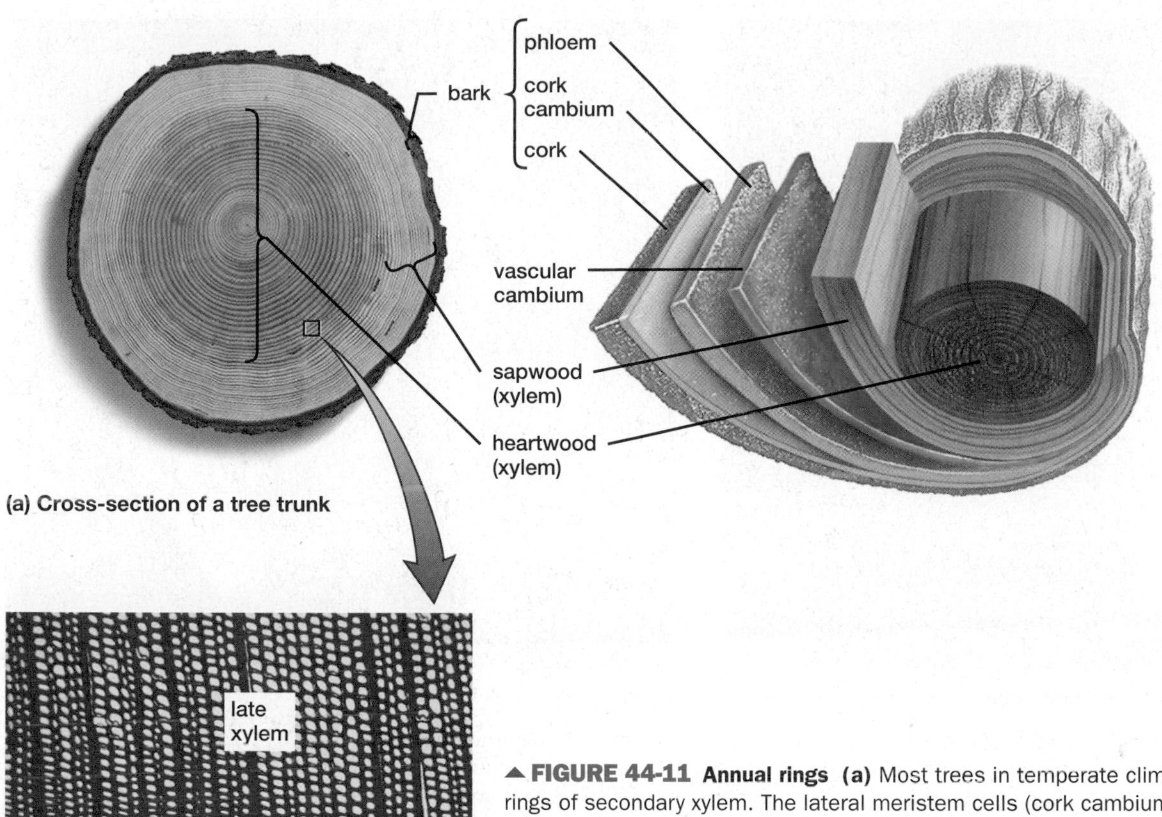

(a) Cross-section of a tree trunk

late
xylem

early
xylem

(b) An annual ring

▲ **FIGURE 44-11 Annual rings (a)** Most trees in temperate climates form annual rings of secondary xylem. The lateral meristem cells (cork cambium and vascular cambium) are responsible for secondary growth. Early wood, formed during spring and early summer, consists of large cells with thin cell walls and is pale. Late wood, usually formed in late summer, consists of smaller cells with thicker walls and is dark. **(b)** The junction between spring and summer wood in a single annual ring.

THINK CRITICALLY Would you expect a tree from a climate that was uniformly warm and wet all year to have growth rings? Explain your answer.

Vascular Cambium Produces Secondary Xylem and Secondary Phloem

The **vascular cambium** forms a cylinder of lateral meristem cells that lies between the primary xylem and primary phloem. Daughter cells of the vascular cambium produced toward the inside of the stem differentiate into secondary xylem; those produced toward the outside of the stem differentiate into secondary phloem (**FIG. 44-10c**, bottom).

The cells of secondary xylem, with their thick cell walls, form the wood in trees and woody shrubs. Young secondary xylem, called **sapwood**, transports water and minerals and is located just inside the vascular cambium (**FIG. 44-11**). Older secondary xylem, the **heartwood**, fills the central portion of older stems. Heartwood no longer carries water and minerals, but continues to provide support and strength to the stem.

Phloem cells are much weaker than xylem cells. Sieve-tube elements and companion cells are split apart as the stem enlarges and are also crushed between the hard xylem on the inside of the stem and the tough outer covering, called cork. Only a thin strip of recently formed phloem remains alive and functioning.

Cell division in the vascular cambium ceases during the cold of winter. In spring and summer, the cambium cells divide, forming new secondary xylem and phloem. In spring and early summer, when water is plentiful, xylem cells are large, with thin cell walls; in late summer, when water becomes scarce, xylem cells are smaller in diameter, with thicker cell walls. As a result, cross sections of tree trunks often show a pattern of **annual rings** of alternating pale regions (large cells with thin walls) and dark regions (small cells with thick walls), as shown in Figure 44-11. The tree's age can be determined by counting the annual rings. The widths of the rings also provide information about past climate, because warm, wet years produce larger cells and wider rings.

Cork Cambium Replaces the Epidermis with Cork Cells

As the stem increases in girth with new secondary xylem and phloem, the nondividing epidermal cells flake off. To protect the growing stem, some underlying cells produce a layer of lateral meristem cells called **cork cambium** (see Figs. 44-10 and 44-11). These cells divide, forming daughter **cork cells**, with tough, waterproof cell walls that protect the trunk both

(a) Cork protects this giant sequoia tree

(b) Harvesting cork from a cork oak

▲ **FIGURE 44-12 Cork (a)** The cork of a giant sequoia in the Sierra Nevada of California may be up to 3 feet thick (almost a meter). This massive, fire-resistant cork layer contributes to the sequoia's great longevity—sequoias in the Sierra Nevada commonly live to be more than 2,000 years old, and some are more than 3,000 years old. Blackened areas are from past fires. **(b)** A layer of cork is stripped from a cork oak. The cork will regrow and can be harvested again in about a decade.

from drying out and from physical damage. Cork cambium and its daughter cork cells make up the periderm.

Cork cells die as they mature, forming a protective layer that can be more than 2 feet thick in some tree species, such as the giant sequoia (**FIG. 44-12a**). As the trunk expands from year to year, the outermost layers of cork split apart or peel off, accommodating the growth. Corks used to plug wine bottles are made from the outermost layer of cork from cork oaks, carefully peeled off by harvesters so as not to harm the underlying cork cambium (**FIG. 44-12b**), which will regenerate the next crop of cork.

Bark includes all of the tissues outside the vascular cambium: phloem, cork cambium, and cork cells (see Fig. 44-11). Removing a strip of bark all the way around a tree, called girdling, kills the tree because it severs the phloem. Without phloem, sugars synthesized in the leaves cannot reach the roots. Deprived of energy, the roots can no longer take up minerals, and the tree dies.

Many Plants Produce Specialized Stems or Branches

Flowering plants have evolved a variety of stem specializations. For example, many stems are adapted for storage. The fat, bulging trunk of the baobab tree contains water-storing parenchyma cells, which allow it to thrive in climates with sporadic rainfall (**FIG. 44-13a**). Cactus stems both photosynthesize and store water. The common white potato is actually an underground stem, packed with parenchyma specialized to store starch. Each "eye" of a potato (see Fig. 44-4a) is a lateral bud, ready to sprout a branch when conditions become favorable. Strawberries send out horizontal stems called runners that spread over the soil and sprout new plants from their buds (**FIG. 44-13b**). Certain branches of grapes and Boston ivy form grasping tendrils that coil around trees and trellises or adhere to buildings, giving the plant better access to sunlight (**FIG. 44-13c**). Thorns are branch adaptations that protect certain plants, such as hawthorns and honey locust trees, from browsing by large herbivores (**FIG. 44-13d**).

CHECK YOUR LEARNING

Can you . . .

- diagram the internal and external structures of a dicot stem after primary growth and after secondary growth?
- explain the function of each major structure of the stem?
- describe some specializations of stems and the functions of each?

44.6 WHAT ARE THE STRUCTURES AND FUNCTIONS OF ROOTS?

Most dicots, such as carrots and dandelions, develop a **taproot system**, consisting of a central root with many

(b) Strawberry plants

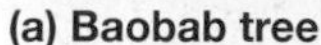

(a) Baobab tree

(c) Grape vine tendril

(d) Honey locust tree

▲ **FIGURE 44-13 Specialized stems and branches (a)** The enormously expanded water-storing trunk of the baobab tree allows it to thrive in a dry climate. **(b)** Strawberry plants can reproduce using runners, which are horizontal stems. Where a node of a runner touches the soil, its lateral bud may sprout into a complete plant. **(c)** Tendrils are specialized branches that allow grape vines to cling to trees or trellises. **(d)** Honey locusts protect themselves with branches modified into strong, sharp, often forked thorns.

smaller roots branching out from its sides (**FIG. 44-14a**). Most monocots, such as grasses and daffodils, produce a **fibrous root system** in which many roots of roughly equal size emerge from the base of the stem (**FIG. 44-14b**).

In young roots of both taproot and fibrous root systems, divisions of the apical meristem and differentiation of the resulting daughter cells give rise to four distinct regions. Daughter cells produced on the lower side of the apical meristem differentiate into the root cap. Daughter cells

(a) A taproot system

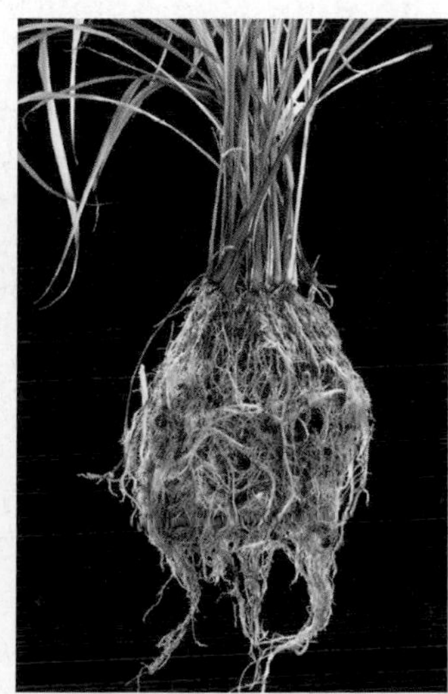

(b) A fibrous root system

▶ **FIGURE 44-14 Taproots and fibrous roots (a)** Dicots typically have a taproot system, consisting of a long central root with many smaller, secondary roots branching from it. **(b)** Monocots usually have a fibrous root system, with many roots of roughly equal size.

FIGURE 44-15 Primary growth in roots Primary growth in roots results from mitotic cell division in the apical meristem near the tip. The root is composed of the root cap, epidermis, cortex, and vascular cylinder.

produced on the upper side of the apical meristem differentiate into cells of the epidermis, cortex, and vascular cylinder (**FIG. 44-15**).

The Root Cap Shields the Apical Meristem

The **root cap** is located at the very tip of the root and protects the apical meristem, the source of the root's primary growth, from being scraped off as the root pushes down between the rocky particles of the soil. Root cap cells have thick cell walls and secrete a slimy lubricant that helps ease the root between soil particles. Nevertheless, root cap cells wear away and must be continuously replaced by new cells produced by the apical meristem.

The Epidermis of the Root Is Permeable to Water and Minerals

The root's outermost covering of cells is the epidermis, which is in contact with the soil and the water trapped among the soil particles. Unlike stem epidermis, which is covered by a waxy cuticle that reduces evaporation, root epidermis lacks a cuticle. Consequently, the walls of root epidermal cells are highly permeable to water and minerals. Further, many epidermal cells send trichomes called **root hairs** into the surrounding soil (**FIG. 44-16**). By expanding the root's surface

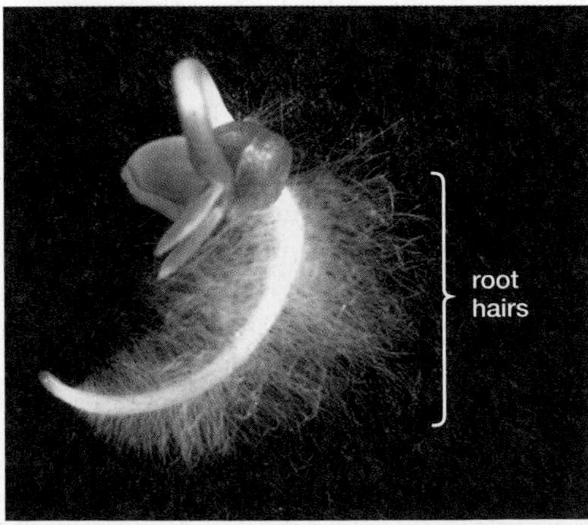

FIGURE 44-16 Root hairs Root hairs, shown here on a sprouting radish, greatly increase a root's surface area, enhancing the absorption of water and minerals from the soil.

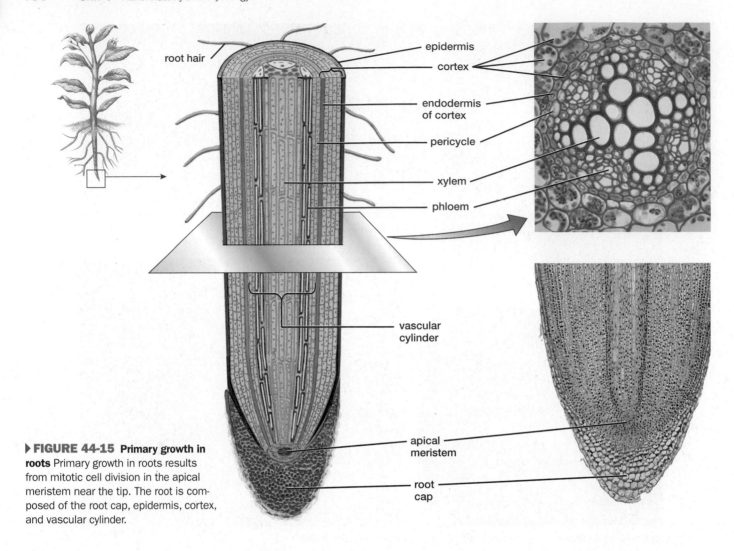

area, root hairs greatly increase the root's ability to absorb water and minerals from the soil.

The Cortex Stores Food and Controls Mineral Absorption into the Root

The cortex occupies most of the inside of a young root between the epidermis and the vascular cylinder (see Fig. 44-15). The cortex is mainly composed of large, loosely packed parenchyma cells with porous cell walls, but its innermost layer, the **endodermis**, consists of a single layer of smaller, tightly packed cells. Sugar produced in the shoot by photosynthesis is transported via phloem to cortex parenchyma cells. Roots of perennial plants convert sugar to starch, which is stored through the cold winter months. In the spring, root cells break down the stored starch into sugar, which is transported upward in the phloem to power new growth of the shoot system. The cortex is especially large in roots that are specialized for carbohydrate storage, such as those of sweet potatoes, beets, carrots, and radishes (**FIG. 44-17**).

Water and inorganic nutrients that enter the root travel through or between the cells of the cortex until they reach the endodermis. Here, the nutrient-laden water is forced to travel through the endodermal cells' plasma membranes, which control the entry of solutes, as described in Section 44.7.

The Vascular Cylinder Contains Conducting Tissues and Forms Branch Roots

The **vascular cylinder** contains the conducting tissues of xylem and phloem. The outermost layer of the root vascular cylinder is the **pericycle**, located just inside the endodermis of the cortex and outside the xylem and phloem (see Fig. 44-15). The pericycle is the source of new branches in roots. Under the influence of hormones, pericycle cells become meristematic and divide, forming the apical meristem of a **branch root** (**FIG. 44-18**). Branch root development is similar to that of primary roots except that the branch must break out through the cortex and epidermis of the primary root. It does so both by crushing the cells that lie in its path and by secreting enzymes that digest them. The vascular tissues of the branch root connect with the vascular tissues of the primary root.

▲ **FIGURE 44-17 Specialized roots** Dicot taproots modified for nutrient storage include (left to right) sweet potatoes, radishes, carrots, and beets.

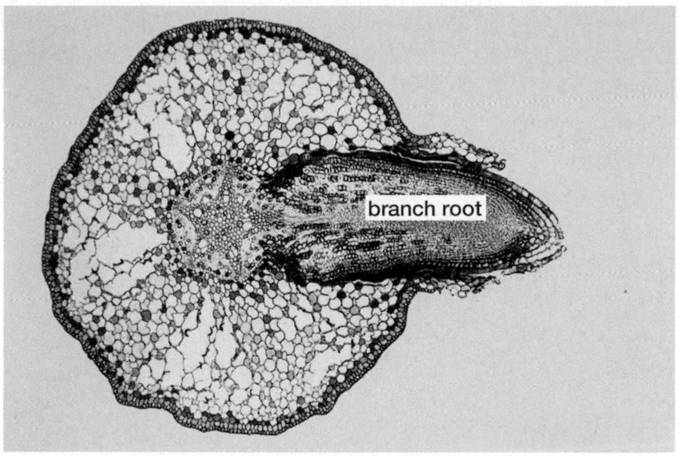

▲ **FIGURE 44-18 Branch roots** Branch roots emerge from the pericycle of a root. The center of this branch root is already differentiating into vascular tissue.

Roots May Undergo Secondary Growth

The roots of woody plants, including conifers and deciduous trees and shrubs, become thicker and stronger through secondary growth. Although there are some differences between secondary growth in stems and roots, the essentials are similar: Vascular cambium produces secondary xylem and phloem in the interior of the root, and cork cambium produces protective layers of cork cells on the outside of the root. Secondary growth in roots is woody and unable to absorb water or nutrients; it provides more secure anchorage and sometimes food storage for the plant.

CHECK YOUR LEARNING

Can you . . .

- diagram the structure of a dicot root after primary growth and explain the function of each major structure?
- describe how branch roots form?

44.7 HOW DO PLANTS ACQUIRE NUTRIENTS?

Nutrients are substances obtained from the environment that are required for the growth and survival of an organism. Plants need only inorganic nutrients, because, unlike

TABLE 44-2 Essential Nutrients Required by Plants

Element*	Major Source	Function
Macronutrients		
Carbon (C)	CO_2 in air	Component of all organic molecules
Oxygen (O)	O_2 in air and dissolved in soil water	Component of most organic molecules
Hydrogen (H)	Water in soil (as H_2O)	Component of all organic molecules
Nitrogen (N)	Dissolved in soil water [as nitrate (NO_3^-) and ammonium (NH_4^+)]	Component of proteins, nucleotides, and chlorophyll
Potassium (K)	Dissolved in soil water (K^+)	Helps control osmotic pressure; regulates stomata opening and closing
Calcium (Ca)	Dissolved in soil water (Ca^{2+})	Component of cell walls; involved in enzyme activation and the control of responses to environmental stimuli
Phosphorus (P)	Dissolved in soil water [as phosphate (PO_4^{3-})]	Component of ATP, nucleic acids, and phospholipids
Magnesium (Mg)	Dissolved in soil water (Mg^{2+})	Component of chlorophyll; activates many enzymes
Sulfur (S)	Dissolved in soil water [as sulfate (SO_4^{2-})]	Component of some amino acids and proteins; component of coenzyme A
Micronutrients		
Iron (Fe)	Dissolved in soil water (Fe^{2+})	Component of some enzymes; activates some enzymes; is required for chlorophyll synthesis
Chlorine (Cl)	Dissolved in soil water (Cl^-)	Helps maintain ionic balance across membranes; participates in splitting water during photosynthesis
Copper (Cu)	Dissolved in soil water (Cu^{2+})	Component of some enzymes; activates some enzymes
Manganese (Mn)	Dissolved in soil water (Mn^{2+})	Activates some enzymes; participates in splitting water during photosynthesis
Zinc (Zn)	Dissolved in soil water (Zn^{2+})	Component of some enzymes; activates some enzymes
Boron (B)	Dissolved in soil water [as $B(OH)_4^-$]	Found in cell walls
Molybdenum (Mo)	Dissolved in soil water [as $Mo(O_4)^{2-}$]	Component of some enzymes involved in nitrogen utilization

*Listed in approximate order of abundance in the plant body.

animals, plants can synthesize all of their own organic molecules. Plants require some nutrients, called macronutrients, in large quantities; collectively, these make up more than 99% of the dry weight of the plant body. Others, called micronutrients, are needed only in trace amounts (**TABLE 44-2**).

Plants obtain carbon from carbon dioxide in the air, oxygen from the air or dissolved in water, and hydrogen from water. These three elements—the atomic building blocks of carbohydrates such as cellulose, starch, and sugar—make up more than 95% of the dry weight of most plants. The other nutrients are obtained when plant roots absorb **minerals**, inorganic substances such as nitrate (NO_3^-), potassium (K^+), calcium (Ca^{+2}), or phosphate (PO_4^{3-}) from the soil.

Finally, a large part of the mass of the living portion of a plant is water. Water provides internal support for plant cells and dissolves and transports minerals, sugars, hormones, and other organic molecules throughout the plant body. For most plants, the primary source of water is the soil.

Roots Transport Minerals and Water from the Soil into the Xylem of the Vascular Cylinder

Roots absorb minerals from the soil and transport them to the shoot. Soil consists of rock particles, air, water, and organic matter. Although the rock particles and organic matter contain minerals, only those dissolved in the soil water can be taken up by roots. Dissolved minerals are transported from root to shoot in xylem; therefore, a root must conduct minerals from the soil water to the xylem in the root's vascular cylinder (**FIG. 44-19**). How does this happen?

Water Moves Through Roots via Two Major Pathways

A young root is made up of (1) living cells; (2) extracellular space, mostly filled with the porous, nonliving cell walls of root cells; and (3) the tracheids and vessel elements of xylem, which consist solely of cell walls (see Fig. 44-6). All soil water first enters the root through the cell walls of epidermal cells. This water then travels through extracellular pathways or intracellular pathways en route to the endodermis (Fig. 44-19).

The extracellular pathways ❶ (blue arrows) wander through cell walls and extracellular spaces that surround the epidermal and cortex cells. Cell walls, which consist primarily of a matrix of cellulose fibers, are porous and allow free movement of soil water and its dissolved minerals.

Intracellular pathways ❷ (red arrows) begin wherever water and minerals enter into cells through the cells' selective plasma membranes. The intracellular route most frequently

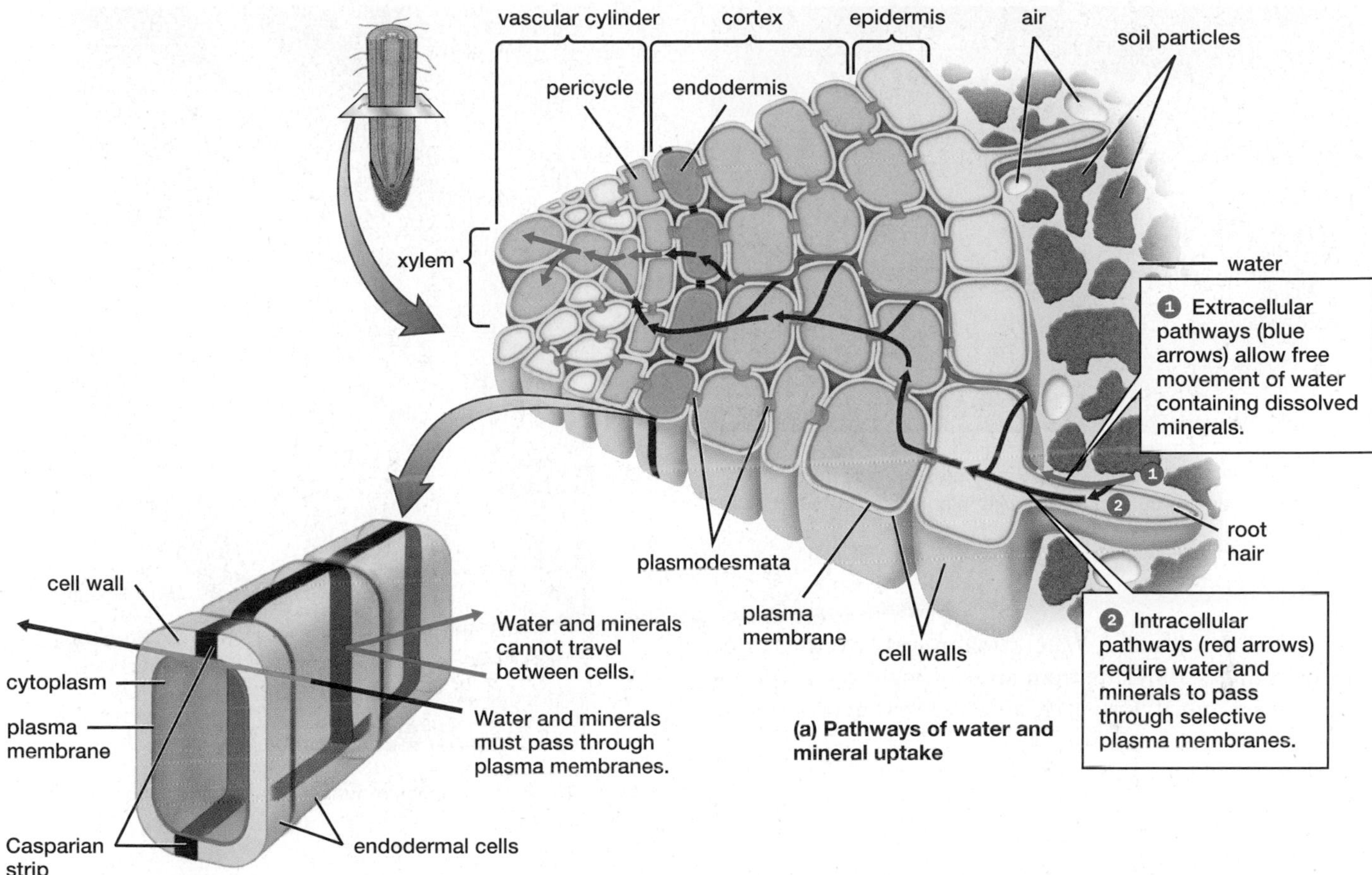
vascular cylinder
cortex
epidermis
air
soil particles
pericycle
endodermis
xylem
water
1 Extracellular pathways (blue arrows) allow free movement of water containing dissolved minerals.
root hair
plasmodesmata
plasma membrane
cell walls
2 Intracellular pathways (red arrows) require water and minerals to pass through selective plasma membranes.
(a) Pathways of water and mineral uptake
cell wall
cytoplasm
plasma membrane
Casparian strip
Water and minerals cannot travel between cells.
Water and minerals must pass through plasma membranes.
endodermal cells
(b) Casparian strip forces water to travel through endodermal cells.

▲ **FIGURE 44-19 Water and mineral uptake by roots (a)** Water with dissolved minerals within the extracellular pathway flows through porous cell walls and extracellular space (blue arrows). Water and minerals enter cell cytoplasm by different mechanisms, but then both flow through the intracellular pathway via plasmodesmata (red arrows). **(b)** The Casparian strip (blow-up) diverts all extracellular water and its solutes from the extracellular to the intracellular pathway at the endodermis.

begins at the root hairs of the epidermis, which provide an enormous surface of plasma membrane for water and mineral uptake. But water and selected nutrients may also enter the intracellular pathway at any point by moving from extracellular spaces into adjacent cortex cells. Both water and mineral nutrients then continue within the intracellular pathway because adjacent plant cells are connected by large pores called plasmodesmata. These pores, which are lined with plasma membrane and filled with cytosol, allow molecules to pass from the cytoplasm of one cell to the cytoplasm of neighboring cells.

The intracellular and extracellular pathways converge at the endodermis when all of the water and minerals are forced to take the intracellular path through endodermal cells to reach the vascular cylinder. This occurs at the waterproof **Casparian strip**, which encircles the endodermal cells and fills the extracellular space between them, much as mortar in a brick wall surrounds the edges of each brick. The Casparian strip forces the water to pass through endodermal cell membranes. The plasma membranes of the endodermal cells (as well other cortical and epidermal cells) act as gatekeepers, allowing nutrient minerals (such as phosphate ions) to enter the vascular cylinder, but blocking many toxic substances (such as aluminum ions) and some infectious microorganisms that would harm the plant. The Casparian strip prevents this filtered solution from diffusing back through the extracellular spaces of the cortex. After water and selected solutes enter the vascular cylinder, the nutrient solution flows upward through the pits and holes perforating the dead cell walls that make up the xylem.

Minerals Enter Root Cells by Active Transport, and Water Follows by Osmosis

The intracellular pathway requires that water and minerals move separately through the plasma membrane, which is freely permeable to water but which requires selective transporter proteins for various minerals. Soil water has a very low concentration of mineral nutrients compared to the cytosol of plant cells, so root cells use active transport to concentrate

▲ **FIGURE 44-20 Root pressure** Root pressure may force water out of the leaf tips of certain low-growing plants, such as this strawberry.

minerals from soil water. Water then moves into the cortex cells by osmosis, diffusing from areas of high water concentration (low mineral content) to areas of low water concentration (high mineral content).

In some plants, this osmotic entry of water following mineral uptake is so powerful that it actually pushes the solution up into the shoot, a phenomenon called **root pressure**. Occasionally, the effect of root pressure is visible as droplets are forced out of the veins at the tips of leaves in low-growing plants (**FIG. 44-20**).

However, in most plants, under most conditions, root pressure is not the major force moving water up through xylem. The flow of water and its dissolved minerals through the xylem from roots to the uppermost plant parts is driven by transpiration, the evaporation of water through stomata, which occurs primarily in the leaves.

Symbiotic Relationships Help Plants Acquire Nutrients

Minerals essential to plant growth, particularly nitrogen and phosphorus, are often too scarce in soil water to support plants. However, most plants have evolved mutually beneficial relationships with fungi that help them acquire these nutrients. In addition, certain plants have forged relationships with bacteria that can harvest nitrogen from the air.

Fungal Mycorrhizae Help Most Plants Acquire Minerals

The roots of roughly 90% of land plants form symbiotic relationships with fungi. The resulting complexes, called **mycorrhizae** (singular, mycorrhiza; "fungus root" in Greek) help the plant obtain scarce minerals from the soil, particularly phosphorus and nitrogen (**FIG. 44-21**). Microscopic fungal strands extend from the soil into the root, where they

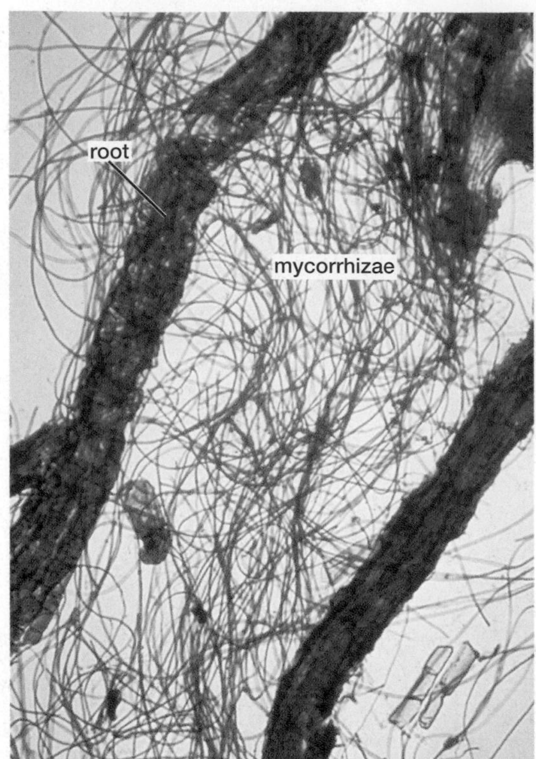

▲ **FIGURE 44-21 A mesh of fungal strands surrounds and penetrates a root**

intertwine among the root cells. The web of fungal filaments greatly increases the volume of soil from which minerals can be absorbed, compared to the volume in contact with the plant root alone. Minerals absorbed by the fungus are then transferred to the root. The fungus, in return, receives sugars from the plant; in fact, most mycorrhizal fungi could not survive without this energy source.

Nitrogen-Fixing Bacteria Help Plants Acquire Nitrogen

Amino acids, nucleic acids, and chlorophyll all contain nitrogen, so plants need large amounts of it. Although nitrogen gas (N_2) makes up about 78% of the atmosphere and readily diffuses into the air spaces in the soil, plants can absorb nitrogen from soil only in the form of ammonia (NH_3) or nitrate (NO_3^-). Some **nitrogen-fixing bacteria** in the soil combine atmospheric N_2 with hydrogen to make NH_3, a process called **nitrogen fixation**. The bacteria then use NH_3 to synthesize amino acids and nucleic acids. However, nitrogen fixation requires a lot of energy, using eight ATPs to make a single molecule of NH_3. Consequently, nitrogen-fixing bacteria don't manufacture a lot of extra NH_3.

Legumes and a few other plants enter into a mutually beneficial relationship with certain species of nitrogen-fixing bacteria. The bacteria enter the legume's root hairs and make their way into cortex cells. Both the bacteria and

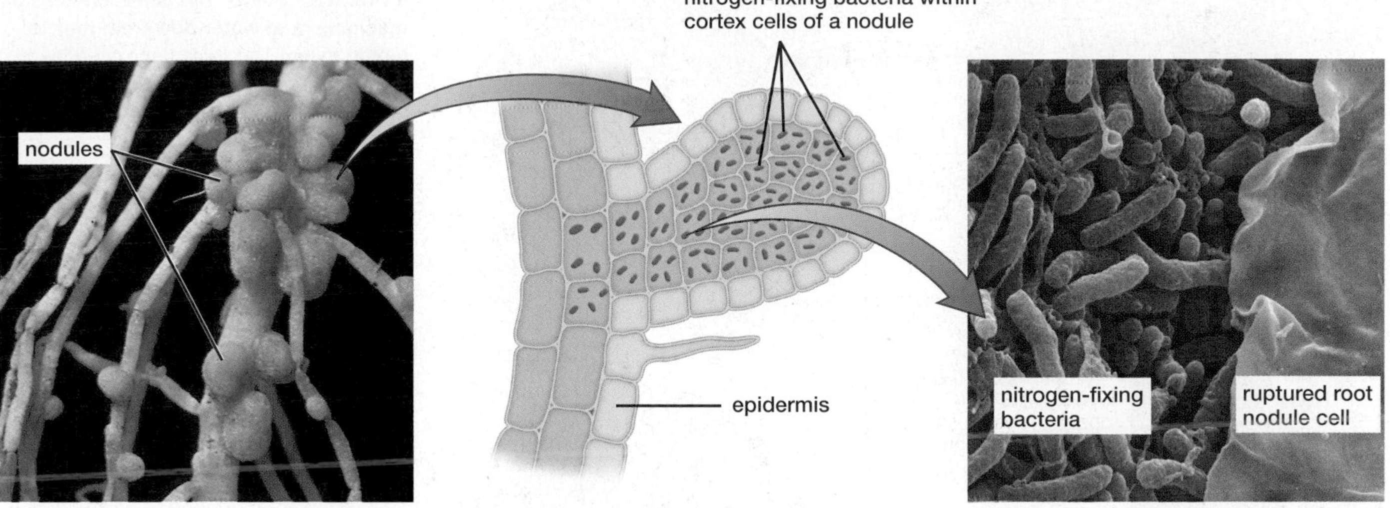

▲ **FIGURE 44-22 Root nodules in legumes are packed with nitrogen-fixing bacteria**

cortex cells multiply, forming a swelling, or **nodule**, composed of cortex cells full of bacteria (**FIG. 44-22**). The bacteria live off the root's food reserves, obtaining so much energy that they produce more NH_3 than they need. The surplus NH_3 diffuses into the host cortex cells, providing the plant with usable nitrogen. Most legumes also associate with mycorrhizae, thus benefitting from both bacterial and fungal partners.

CHECK YOUR LEARNING

Can you . . .

- explain how minerals and water are taken up by a root?
- explain the function of the Casparian strip in mineral and water uptake?
- describe how root pressure occurs?
- explain the importance of mycorrhizae and nitrogen-fixing bacteria in plant nutrition?

CASE STUDY CONTINUED

Autumn in Vermont

The brilliant colors of autumn last just a few weeks, and then the leaves fall. Why do deciduous trees drop their leaves? In temperate climates, the soil is frozen for long periods during the winter, preventing the trees from absorbing water from the soil. Without a continuous supply of water from the roots, transpiration from leaves would eventually cause the tree to die from lack of water and dissolved nutrients. By dropping their leaves and entering a dormant state, deciduous trees reduce water loss during winter when they would be unable to absorb replacement water from the frozen soil. After leaves reappear for the growing season, transpiration carries nutrients from the soil to the tips of branches. How?

44.8 HOW DO PLANTS MOVE WATER AND MINERALS FROM ROOTS TO LEAVES?

After entering the root xylem, water flows to the rest of the plant, and minerals dissolved in the water are passively carried along as the water moves upward. But in the tallest redwood trees, the topmost parts may be 380 feet (115 meters) from the roots. How do plants overcome the force of gravity and make water flow upward? In most plants, at least 90% of the water absorbed by the roots evaporates through the stomata of leaves. This evaporation, called **transpiration**, provides the force that pulls water upward through the plant body.

The Cohesion–Tension Mechanism Explains Water Movement in Xylem

The **cohesion–tension mechanism** explains how water is pulled up the xylem by transpiration from the leaves (**FIG. 44-23**). Hydrogen bonds among water molecules link them together (see Chapter 2). Just as individually weak cotton threads together create the strong fabric of your jeans, the network of individually weak hydrogen bonds in water produces a strong *cohesion*, the resistance of a substance to being pulled apart. Cohesion creates a chain of water in the xylem—extending the entire height of the plant—that is at least as strong as a steel wire of the same diameter. The tension component of the cohesion–tension mechanism is supplied by water evaporating from the leaves during transpiration. Water first evaporates from mesophyll cells into the air spaces within the leaf and then exits through the stomata into the atmosphere ❶. As water exits the mesophyll cells, their water concentration decreases. This causes water to move by osmosis from nearby xylem into the mesophyll cells to replace the water that evaporated.

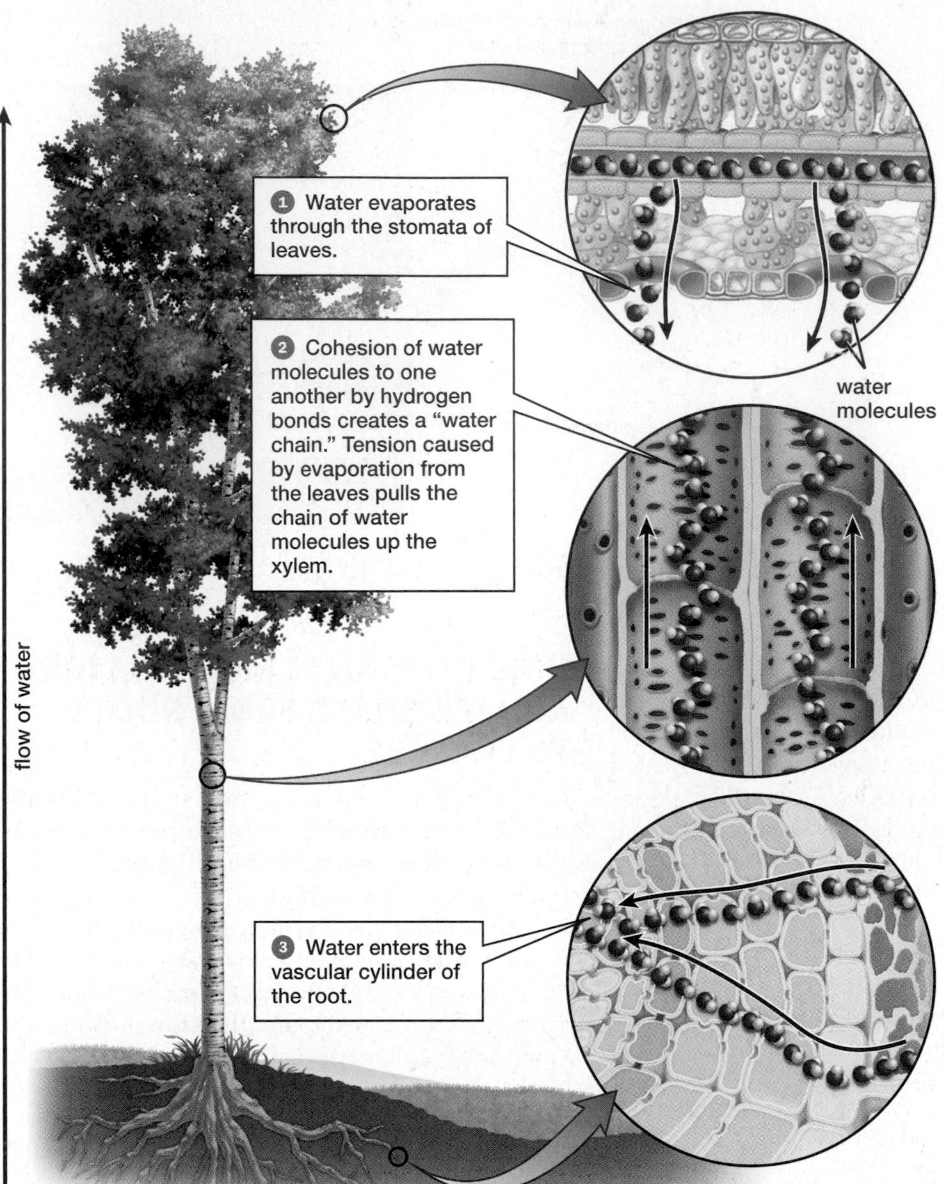

1 Water evaporates through the stomata of leaves.
2 Cohesion of water molecules to one another by hydrogen bonds creates a "water chain." Tension caused by evaporation from the leaves pulls the chain of water molecules up the xylem.
3 Water enters the vascular cylinder of the root.
water molecules
flow of water

FIGURE 44-23 The cohesion–tension mechanism of water flow from root to leaf in xylem The force generated by transpiration can drag water from soil to the tops of the tallest trees.

Cohesion and Tension Work Together to Move Water Up the Xylem

Water molecules leaving the xylem are linked by cohesion to other water molecules in the same xylem tube. Therefore, as water molecules leave the xylem to replace the water evaporating from mesophyll cells, they pull more water up the xylem ❷. The tension in xylem is strong enough to lift water more than 500 feet (150 meters), much higher than the height of any tree. The tension in the xylem extends all the way down to the roots, where water in the extracellular space of the vascular cylinder is pulled in through the porous pits in the walls of the vessel elements and tracheids of the xylem, replenishing the water molecules at the bottom of the chain ❸. The movement of water into the root xylem causes more water to enter the root from the soil. Only the aboveground parts of a plant, usually the shoot, can transpire, so water flow in xylem is unidirectional, upward from root to shoot.

A large maple tree may transpire about 250 gallons of water a day—almost a ton of water lifted up more than 50 feet, every day. What supplies the energy to transport so much water? Ultimately, the sun. Sunlight warms both the leaves and the air, powering the evaporation of water from the leaves. Now imagine a whole forest, with each tree releasing hundreds of gallons of water into the air through transpiration each day. This process can have a major effect on the local climate, as we describe in "Earth Watch: Forests Water Their Own Trees."

Earth WATCH Forests Water Their Own Trees

Tropical rain forests grow in the Amazon River basin because of the year-round warm temperatures and abundant rainfall. Where does the rain come from? A lot of it comes from moist air masses that develop near the equator (see Chapter 30). However, tropical rain forests are a significant source of their own high humidity and plentiful rainfall.

In the Amazon rain forest, an acre of soil supports hundreds of towering trees, each bearing hundreds of thousands of leaves. The combined surface area of the leaves dwarfs the surface area of the soil. Therefore, most of the water evaporating from the forest comes from the trees—as little as 2 or 3 gallons a day for small understory trees, but as much as 200 gallons a day for large trees of the canopy. This evaporation produces perpetually high humidity. Some of the transpired water falls again as rain, right there in the rain forest. In fact, up to half of the rainfall in the Amazon region is water transpired by the trees themselves.

When rain forests are clear-cut, the local climate changes dramatically. In an intact forest, water evaporating from the leaves cools the air. When trees are removed and transpiration is reduced, the region becomes not only drier, but also hotter. Rainfall decreases and temperature increases in uncut adjacent forests as well; the rain-forest trees there do not thrive under these conditions, so many begin to die. These uncut but damaged forests also transpire less water, making the local climate even drier and hotter, and the vicious cycle continues.

Rain-forest transpiration may also affect forests that are relatively distant from the rain forest. For example, the Monteverde Cloud Forest Reserve in Costa Rica depends on an almost constant shroud of fog for its survival. Much of the water in the fog comes from lowland forests, where transpiration pumps moisture into air that is carried up mountain slopes by winds. As the air cools, the moisture condenses into fog. However, logging has eliminated about 80% of Costa Rica's rain forests, so the air contains less moisture. Drier air requires colder temperatures for water to condense, which means that the fog forms higher up the mountains than in years past. As a result, the forest in the lower part of the mountains is changing, with different species of trees, birds, and lizards colonizing the slopes.

The habitats in which trees affect their own growth are not limited to tropical rain forests. On Tenerife, one of the Canary Islands off the northwest coast of Africa, rainfall averages less than 8 inches (20 centimeters) a year (a level typical of deserts), with average high temperatures between 68° and 88°F (20°and 31°C) year-round. Nevertheless, some of Tenerife's mountain slopes support pine forests. How is this possible? Of course, on the mountainsides it's a little cooler, with a little more rain, than on the coast. The major factor, however, is the forest itself. The trees intercept fog drifting across the island from the Atlantic Ocean. Water from the fog condenses on the trees and trickles down to the soil. More than half of the water in Tenerife's streams comes from condensing fog.

People have created cloud-trapping structures and are creating cloud-trapping forests as well. For example, just south of Lima, Peru, the mountain town of Bellavista is thriving in a region that gets about ½ inch (about 1.5 centimeters) of rainfall each year. Formerly reliant on expensive, trucked-in water, the town now harvests hundreds of gallons of water daily during the winter from fog carried in for free by winds off the Atlantic Ocean. The fog is trapped as condensation on fine mesh nets (**FIG. E44-1**), drips into pipes, and is collected in large storage containers. The clear water is used not only for drinking, gardening, and a small brewery, but is also transforming the landscape. Fog water has allowed residents to plant hundreds of evergreens whose fine leaves act as natural fog-catchers. With the help of plastic funnels that direct water down to their roots, the trees are able to water themselves and will likely create a self-sustaining forest.

▲ **FIGURE E44-1 Fog-catchers** Fog condensing on the fine mesh of these nets provides liquid water.

THINK CRITICALLY Under what environmental conditions can Bellavista continue to thrive? What changes could undermine the progress being made there?

Minerals Dissolved in Water Move Up the Xylem

The minerals absorbed by plant roots are dissolved in water, where they form charged ions, including NO_3^-, NH_4^+ K^+, Ca^{2+}, PO_4^{3-}, and Cl^-. Ions dissolve in water because either the negative or the positive parts of water molecules are electrically attracted to the ions (see Chapter 2). In xylem, this electrical attraction loosely holds the minerals to the water. Therefore, the minerals are carried along with the water from root to shoot.

Stomata Control the Rate of Transpiration

Transpiration powers the transport of water and minerals, but it is also a great threat to plants because it is by far the largest source of water loss—a loss that can be fatal in hot, dry weather. Water loss can be reduced by closing the stomata through which the water evaporates. Closing the stomata, however, also prevents CO_2 from entering the leaf—and CO_2 is required for photosynthesis. Therefore, a plant must regulate its stomata to achieve a balance between acquiring carbon dioxide and losing water. In most plants, stomata open during the day, when sunlight can power photosynthesis, and close during the night, conserving water. However, if excessive evaporation threatens the plant with dehydration, the stomata will close regardless of the time of day.

Guard Cells Regulate Opening and Closing of Stomata

A stoma consists of two guard cells that surround a central pore (**FIG. 44-24**). How do plants open and close their stomata? This is actually a two-part question: (1) mechanically, how is the size of the opening changed? and (2) physiologically, how do guard cells respond to stimuli, such as sunlight or dehydration, and adjust the size of the opening?

Guard cells adjust the size of the opening of a stoma by changing their volume and shape, which they accomplish by taking up or losing water. Guard cells have both an unusual shape and a specific arrangement of cellulose fibers in their cell walls (**FIG. 44-25a**). The two guard cells of a stoma are curved slightly outward and linked together at their ends, resembling two sausage links tied together at their tips. Cellulose fibers in the guard cell walls encircle the cells like dozens of tiny belts.

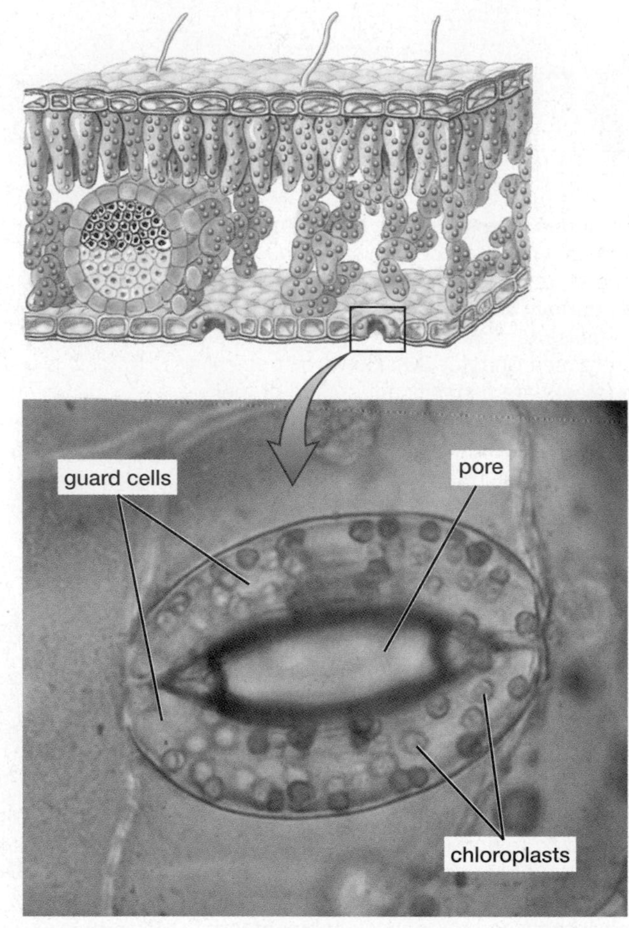
guard cells
pore
chloroplasts

▲ **FIGURE 44-24 A stoma in the leaf epidermis** Guard cells are packed with chloroplasts, but the surrounding epidermis is transparent and lacks chloroplasts.

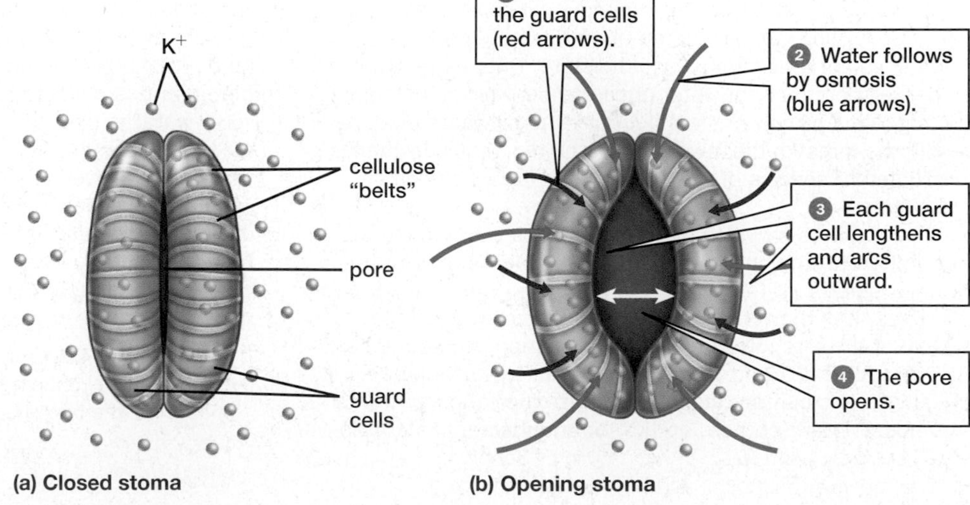
K^+
cellulose "belts"
pore
guard cells
(a) Closed stoma
1 K^+ enters the guard cells (red arrows).
2 Water follows by osmosis (blue arrows).
3 Each guard cell lengthens and arcs outward.
4 The pore opens.
(b) Opening stoma

▶ **FIGURE 44-25 How guard cells open a stoma (a)** The structure of a stoma, showing the central pore closed. The closed guard cells have a relatively low K^+ concentration (similar to other epidermal cells), low water content, and low volume. **(b)** When K^+ is transported into the guard cells, water follows by osmosis, increasing guard cell volume and forcing the pore open.

THINK CRITICALLY When the stomata close, how is photosynthesis affected? How is the movement of water into the roots affected?

A stoma opens when its guard cells take up water, increasing their volume. The cellulose belts prevent the cells from getting fatter, so they are forced to get longer instead. The curved shape of the guard cells and their attachment to one another at their ends mean that they can get longer only by curving outward, which opens the central pore. A stoma closes when its guard cells lose water and decrease in volume. The cells become shorter and less curved, closing the pore.

How does a guard cell change its water content? Water always moves across plasma membranes by osmosis, following a concentration gradient of dissolved solutes. Guard cells create osmotic gradients across their plasma membranes by adjusting the potassium ion (K^+) concentration in their cytoplasm in response to stimuli such as light and CO_2 (see below). Moving K^+ into the cell causes water to follow by osmosis, increasing the volume of the cells and opening the pore (**FIG. 44-25b**). Moving K^+ back out again causes water to leave by osmosis, shrinking the cells and closing the pore.

Three important stimuli control K^+ movement into and out of guard cells:

- ***Light*** Light is absorbed by pigments in the guard cells, triggering a series of reactions that causes K^+ to enter the cells. Water follows by osmosis, and the guard cells swell, opening the pore. In the dark, the reactions promoting K^+ entry stop, allowing K^+ to diffuse back out of the guard cells.
- ***Carbon Dioxide*** Recall that guard cells contain chloroplasts and can carry out photosynthesis. Carbon fixation during photosynthesis uses up CO_2, reducing its concentration inside the guard cells. Low CO_2 concentrations stimulate K^+ entry into the guard cells. At night, photosynthesis stops but cellular respiration continues, so CO_2 levels rise. K^+ transport into the guard cells stops. K^+ diffuses out, water follows by osmosis, and the guard cells close the pore.
- ***Water*** When a plant begins to dry out, it synthesizes a hormone called abscisic acid (see Chapter 46). Abscisic acid binds to receptors on the guard cells and shuts down K^+ entry, allowing K^+ to diffuse out. As water follows by osmosis, the stomata close. Dangerous dehydration is most likely to occur on sunny days. As you might expect, the effects of abscisic acid are powerful, overcoming the effects of light and CO_2, and causing the stomata to close.

CHECK YOUR LEARNING

Can you . . .

- describe how water moves in xylem?
- explain how minerals are transported in xylem?
- describe how potassium and water movements open and close stomata and describe the stimuli that trigger these movements?

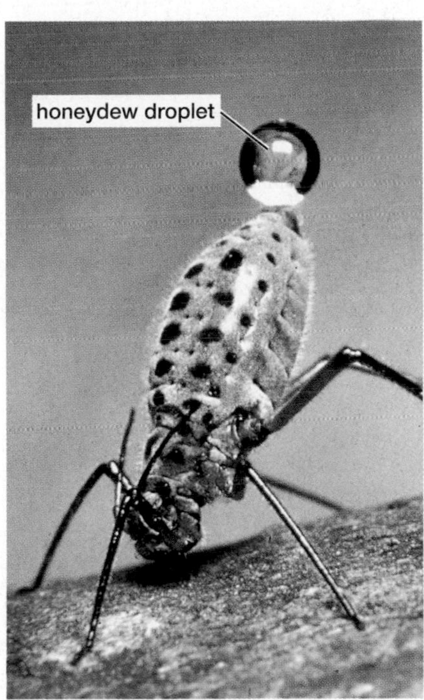

(a) An aphid sucks sap

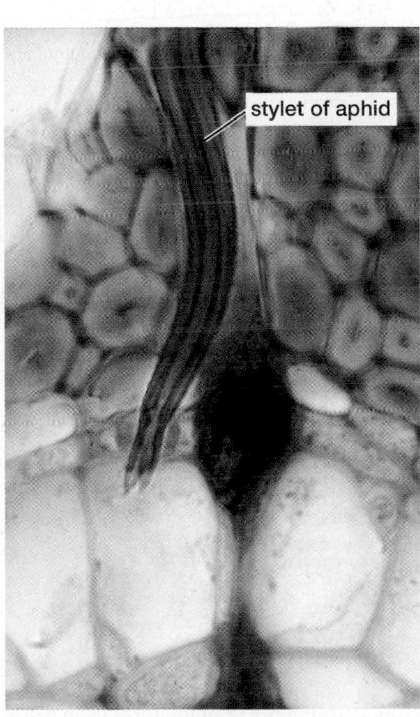

(b) A stylet penetrates into phloem

▲ **FIGURE 44-26 An aphid feeds on the sugary fluid in phloem sieve tubes** **(a)** Pressure in the sieve tube forces fluid out of the phloem and into the aphid's digestive tract. The aphid excretes excess fluid from its anus, as a sugar-rich "honeydew." This fluid is collected by certain species of ants that, in turn, defend the aphids from predators. **(b)** In this micrograph, the stylet of an aphid penetrates a sieve-tube element.

44.9 HOW DO PLANTS TRANSPORT SUGARS?

Sugars synthesized in the leaves must be moved to other parts of the plant, where they nourish nonphotosynthetic structures such as roots or flowers or are stored in cortex cells of roots and stems. Sugar transport is the function of phloem.

Botanists studying phloem sometimes employ unlikely lab assistants: aphids. An aphid is an insect that feeds by sucking fluids through a sharp, hollow tube called a stylet, similar to the more familiar stylet used by a mosquito to suck blood from its victims. An aphid, however, feeds on the fluid in phloem. The insect inserts its stylet through the epidermis and cortex of a young stem and into a sieve tube (**FIG. 44-26**). As we will see, phloem fluid is under pressure, which drives the fluid into the aphid's digestive tract; sometimes, the aphid inflates like a balloon.

After an aphid has penetrated a sieve tube, botanists can remove most of its body, leaving the stylet in place. Phloem fluid will flow out of the stylet for several hours. Chemical analysis shows that the phloem fluid consists of water containing about 10%

to 20% dissolved sugar (mostly sucrose), with lower amounts of other substances such as amino acids, proteins, and hormones. What drives the movement of this sugary solution in phloem?

The Pressure–Flow Mechanism Explains Sugar Movement in Phloem

The most widely accepted explanation for fluid transport in phloem is the **pressure–flow mechanism**, in which differences in water pressure drive the flow of fluid through the sieve tubes. These pressure differences are created indirectly by the production and use of sugar in different parts of the plant. Any part of a plant that synthesizes more sugar than it uses, such as a photosynthesizing leaf, is a sugar **source**. Any structure that uses more sugar than it produces is a sugar **sink**. Sinks include meristems and developing flowers and fruits. Sources and sinks can change with the seasons. For example, the roots of deciduous trees are sugar sinks during the summer, when they convert sucrose to starch for storage. The following spring, the roots become sugar sources as they convert the starch back to sucrose, which travels upward in the phloem and supplies energy for the development of new leaves, which are sinks. As the leaves mature and their photosynthetic capacities develop, they become sugar sources.

According to the pressure–flow mechanism, phloem sieve tubes carry fluid away from sugar sources and toward sugar sinks, as illustrated by the numbered steps in **FIGURE 44-27**. Sugar produced by a source cell (for example, in a photosynthesizing leaf) is actively transported into a phloem sieve tube ①. This raises the sugar concentration in the phloem fluid in that portion of the sieve tube. Water from nearby xylem follows the sugar into the sieve tube by osmosis ②. Because their rigid cell walls prevent sieve-tube cells from expanding, water entering the sieve tube increases the pressure of the fluid inside. The resulting water pressure at a source region drives the fluid through the phloem sieve tubes to regions of lower pressure ③. How is lower pressure created? Cells of a sugar sink (such as a fruit) actively transport sugar out of the phloem ④. Water follows by osmosis, producing lower water pressure in this part of the sieve tube. The phloem fluid thus moves from a source, where water pressure is high, to a sink, where water pressure is lower, carrying the dissolved sugar with it.

The pressure–flow mechanism explains why sieve-tube elements must have intact plasma membranes but very little cytoplasm. Intact plasma membranes, which confine the sugar solution within the sieve tube cells but are permeable to water, allow both accumulation of sugar and the resulting entry of water by osmosis. The lack of cytoplasm reduces resistance to the flow of fluid in the sieve tube.

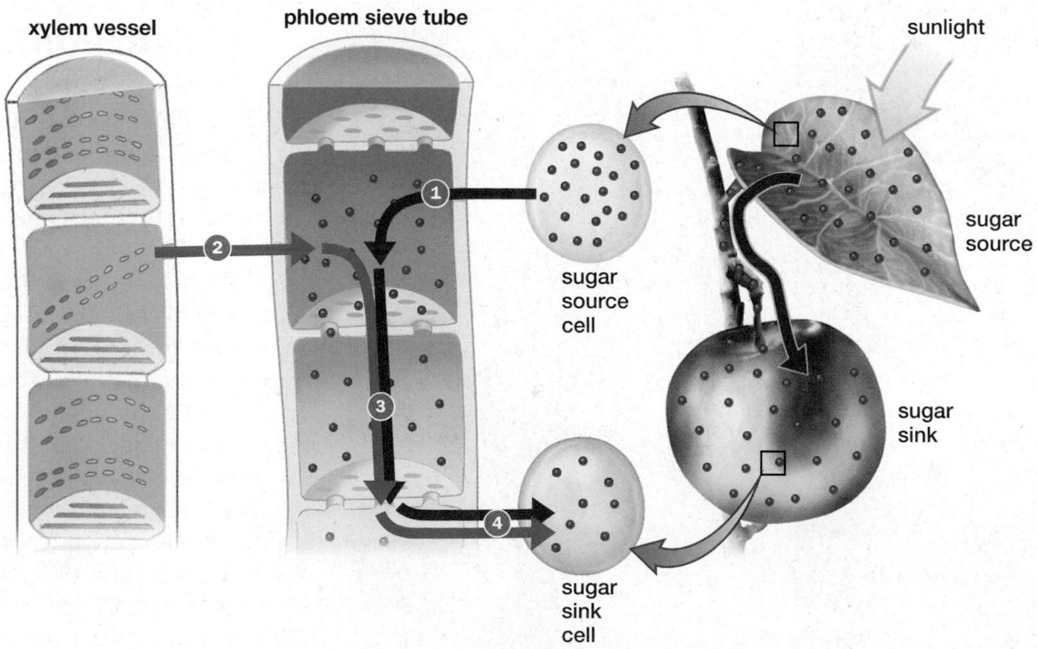

▲ **FIGURE 44-27 The pressure–flow mechanism of sugar transport in phloem** Differences in water pressure drive phloem fluid from the leaf (a sugar source) to the fruit (a sugar sink). Red and blue arrows indicate sugar and water movement, respectively. The blue gradient in the phloem sieve tube represents water pressure, which is higher in the source end of the tube and lower in the sink end of the same tube.

THINK CRITICALLY At what stage of growth would a leaf be a sugar sink? What would be its sugar source?

Note that sinks of sucrose may be either above or below sources of sucrose. For example, a photosynthesizing leaf will be below the apical meristem but above the roots. The pressure–flow mechanism explains how phloem fluid can move either up or down the plant, from sugar sources to sugar sinks.

CHECK YOUR LEARNING

Can you . . .

- explain how sugar solutions move in phloem and why they always move from sources to sinks of sugars?

CASE STUDY REVISITED

Autumn in Vermont

Of all the fall colors, the striking reds are the most intriguing. In most trees, yellow and orange carotenoids play essential roles in photosynthesis throughout the summer, but the red anthocyanins are synthesized only during the autumn, just a few weeks before the leaves are shed. What good are anthocyanins? They act as sunscreen for leaves.

At the onset of autumn, as temperatures cool but days remain fairly long and the sun is still very bright, a leaf's photosynthetic pathways become less efficient and the leaf is no longer able to use all of the light it absorbs. The excess light energy can harm the chloroplasts and reduce photosynthesis even more. Anthocyanins block ultraviolet and blue light and act as antioxidants, scavenging harmful free radicals that might be formed by UV light or by inefficient photosynthesis. Therefore, red, anthocyanin-containing leaves are better protected against intense light than are leaves without anthocyanin and hence can carry out photosynthesis more effectively. In some plants, red leaves dominate the outer parts of the plant, which receive the most sunlight, whereas shaded inner leaves become orange and yellow.

Why spend energy to protect leaves that are just about to fall off? To reclaim nutrients, especially nitrogen and phosphate, that a plant cannot afford to lose. In the autumn, complex organic molecules in leaves are broken down, and the resulting simpler molecules are transported to stworage cells in stems and roots, where they remain throughout the winter. Photosynthesis provides the energy needed for this breakdown and transport. Compared to normal plants, those with a mutation that prevents them from producing anthocyanin retrieve far less nitrogen from their leaves under conditions of intense light and low temperatures.

Newly synthesized red pigment

This indicates that anthocyanin indeed helps a plant to reclaim nutrients from its leaves.

CONSIDER THIS Sugar maples are a major asset to the New England states, producing both maple syrup and brilliant red fall colors that attract tourists. But sugar maples are heat sensitive, and climate change projections suggest that, by the end of the century, rising temperatures may shift the range of these trees northward and higher into the mountains. Already changes are being noted in the timing of fall colors, whose appearance is partly stimulated by falling temperatures. What are some implications of global warming for the future economy of the New England states?

CHAPTER REVIEW

Go to **Mastering Biology** to access the Pearson eText, vocabulary review, practice quizzes, activities, videos, current events, and more.

*Answers to **Think Critically** and **Thinking Through the Concepts** questions can be found in the **Answers** section at the back of the book.*

Summary of Key Concepts

44.1 How Are Plant Bodies Organized?

The body of a land plant consists of root and shoot systems. Roots are usually underground. Their functions include anchoring the plant in the soil; absorbing water and minerals; transporting water, minerals, sugars, and hormones; storing surplus sugar and starch; producing some hormones; and interacting with soil fungi and microorganisms that provide nutrients. Shoots are generally located aboveground and consist of stems, leaves, buds, and (in season) flowers and fruit. Shoot functions include photosynthesis, transport of materials, storage of sugar and starch, reproduction, and hormone production. The two principal groups of flowering plants are the monocots and dicots (see Fig. 44-2).

44.2 How Do Plants Grow?

Plant bodies are composed of meristem cells and differentiated cells. Meristem cells are undifferentiated and retain the capacity for mitotic cell division. Differentiated cells arise from divisions of meristem cells, become specialized for particular functions, and usually do not divide. Most meristem cells are located in apical meristems at the tips of roots and shoots and in lateral meristems in the shafts of roots, stems, and branches. Primary growth (growth in length or height and the differentiation of parts) results from the division and differentiation of cells derived from apical meristems. Secondary growth (growth in diameter and strength) results from the division and differentiation of cells derived from lateral meristems.

44.3 What Are the Differentiated Tissues and Cell Types of Plants?

Plant bodies consist of three tissue systems: ground, dermal, and vascular systems (see Table 44-1). The ground tissue system consists of several cell types, including parenchyma, collenchyma, and sclerenchyma cells. Most are involved in photosynthesis, support, or storage. Ground tissue makes up most of the body of a young plant during primary growth. The dermal tissue system forms the outer covering of the plant body. The dermal tissue system of leaves and of primary roots and stems is usually a single cell layer of epidermis. After secondary growth, dermal tissue is called periderm and consists of cork cells and the cork cambium that produces them. The vascular tissue system consists of xylem, which transports water and minerals from the roots to the shoots, and phloem, which transports water, sugars, and other organic molecules throughout the plant body.

44.4 What Are the Structures and Functions of Leaves?

A leaf is composed of a petiole and a blade. The petiole attaches the blade to a stem or branch. The blade consists of a transparent outer epidermis covered by a waterproof cuticle; mesophyll cells, which have chloroplasts and carry out photosynthesis; and vascular bundles of xylem and phloem, which carry water, minerals, and photosynthetic products to and from the leaf. The lower epidermis is perforated by adjustable pores called stomata that regulate the exchange of gases and water. Leaves may be specialized to store nutrients, catch prey, or to form spines and tendrils.

44.5 What Are the Structures and Functions of Stems?

The dicot stem produced by primary growth consists of (from outside in): the epidermis with its waterproof cuticle, the cortex (comprised of photosynthetic, supporting, and storage cells), vascular tissue (xylem and phloem), and central pith (whose cells also store nutrients and provide support). Lateral buds emerge from nodes located at intervals along the stem. Under the proper hormonal conditions, lateral buds sprout into branches. Secondary growth in stems results from cell divisions in the vascular cambium and cork cambium. Vascular cambium produces secondary xylem and secondary phloem, increasing the stem's diameter. Cork cambium produces waterproof cork cells that cover the outside of the stem. Stems may be specialized to store nutrients or to form thorns and tendrils.

44.6 What Are the Structures and Functions of Roots?

Primary growth in roots results in a structure consisting of an outer epidermis and an inner vascular cylinder of xylem and phloem, with cortex between the two. The apical meristem near the tip of a root is protected by a root cap. Cells of the root epidermis absorb water and minerals from the soil. Root hairs are projections of epidermal cells that increase the surface area for absorption. Most cortex cells store surplus sugars (usually in the form of starch) produced by photosynthesis. The innermost layer of cortex cells is the endodermis, which controls the movement of water and minerals from the soil into the vascular cylinder. Root branching is initiated by meristematic pericycle cells that surround the vascular cylinder. Secondary growth in roots is similar to secondary growth in stems.

44.7 How Do Plants Acquire Nutrients?

Plant nutrients are described in Table 44-2. Water and dissolved minerals may diffuse through cell walls and spaces of the epidermis and cortex until they are forced (by the Casparian strip) to move through the plasma membranes of endodermal cells. These membranes allow water and selected minerals into the vascular cylinder.

Most minerals are taken up from the soil water selectively by active transport into the root hairs. These minerals diffuse from cell to cell via plasmodesmata, ultimately to the endodermis, which transfers them to the extracellular space of the vascular cylinder. Water follows the high mineral concentration intracellularly, moving by osmosis across the plasma membranes of epidermal, cortex, and endodermal cells.

Most plants form root–fungus complexes called mycorrhizae that help absorb some soil nutrients. Legumes have evolved a cooperative relationship with nitrogen-fixing bacteria that invade legume roots. The plant provides the bacteria with sugars, and the bacteria use some of the energy in those sugars to convert atmospheric nitrogen to ammonia, which the plant then absorbs.

44.8 How Do Plants Move Water and Minerals from Roots to Leaves?

The cohesion–tension mechanism explains xylem function. Water molecules are electrically attracted to one another, forming hydrogen bonds between molecules. The resulting cohesion holds together the water within xylem tubes. As water molecules evaporate from the leaves during transpiration, the hydrogen bonds pull other water molecules up the xylem to replace them all the way from the root. Minerals move up the xylem dissolved in the water.

Stomata in the epidermal layers of leaves or young stems control the evaporation of water (transpiration). The opening of a stoma is regulated by the shape and volume of the guard cells that form the pore. Open stomata allow more rapid transpiration. In most plants, stomata open during the day (admitting carbon dioxide needed for photosynthesis); the stimuli that trigger opening of the stomata are light and a low carbon dioxide concentration in the guard cells. If the plant is losing too much water, a hormone called abscisic acid is released, causing the guard cells to close the stomata.

44.9 How Do Plants Transport Sugars?

The pressure-flow mechanism explains sugar transport in phloem. Parts of the plant that synthesize sugar (for example, leaves) export sugar into the sieve tubes. High sugar concentrations cause water to enter the sieve tubes by osmosis, increasing the local water pressure in the phloem. Other parts of the plant (for example, fruits) consume sugar, reducing the sugar concentration in the sieve tube and causing water to leave the tubes by osmosis, which reduces pressure. Water and dissolved sugar move in the sieve tubes from high pressure to low pressure.

Thinking Through the Concepts

Bloom's: Remembering, Understanding

Multiple Choice

1. Taproots
   a. are characteristic of dicots.
   b. act as sugar sources when leaves are fully developed.
   c. are filled with sclerenchyma cells.
   d. include white potatoes.
2. Guard cells
   a. are found primarily on the upper surface of leaves.
   b. open their stomata by shrinking.
   c. lack chloropolasts.
   d. close in response to abscisic acid.
3. Bark
   a. is a type of epidermal tissue.
   b. includes the vascular cambium.
   c. allows sugar synthesized in leaves to reach the roots.
   d. consists primarily of living cork cells.
4. Minerals dissolved in water
   a. enter root epidermal cells by active transport.
   b. take an extracellular pathway through roots to the vascular cylinders.
   c. are transported up through phloem to the leaves.
   d. enter root cortex cells by diffusion.
5. Transpiration
   a. occurs as a result of root pressure.
   b. occurs in the xylem.
   c. allows soil minerals to reach leaves.
   d. brings sugars down to the roots.
6. Which of the following tissues gives rise to secondary growth?
   a. Apical meristem
   b. Vascular cambium
   c. Terminal buds
   d. Parenchyma
7. Casparian strips are present in the ________ of the root.
   a. pericycle
   b. epidermis
   c. endodermis
   d. xylem
8. Which of the following is characteristic of dicots?
   a. Fibrous root system
   b. Flower parts in multiples of three
   c. Parallel leaf veins
   d. Vascular bundles arranged in a ring

Fill-in-the-Blank

1. Plants grow through division of ____________ cells and differentiation of the resulting daughter cells. These dividing cells reside in two locations: at the tip of a shoot or root, called the ____________, and in cylinders along the sides of roots and stems, called ____________. Which is responsible for primary growth? ____________ Which is responsible for secondary growth? ____________
2. The three tissue systems of a plant body are ____________, ____________, and ____________. Which covers the outside of the plant body? ____________ Which conducts water, minerals, and sugars within the plant body? ____________ Which stores starches? ____________
3. Water travels upward through plant roots and shoots within hollow tubes of ____________, which contains two types of conducting cells, ____________ and ____________. Water molecules within these tubes are interconnected by forces called ____________, which allow a chain of water molecules to be pulled up the plant, driven by the evaporation of water from the leaves, a process called ____________. The ____________ in the epidermis of a leaf control the rate of evaporation from the leaf.
4. ____________ transports sugars and other organic molecules from sources of sugar to sinks of sugar. The conducting cell type in this tissue is the ____________, which is supported by adjacent cells called ____________.
5. The very tip of a young root is protected by the cells of the ____________. The surface area of a young root is increased by projections from epidermal cells called ____________, which move minerals from the soil water into their cytoplasm by the process of ____________. Many young roots have a mutually beneficial relationship with fungi to help absorb some minerals; these root–fungus complexes are called ____________.

Review Questions

1. What are the main functions of roots, stems, and leaves?
2. Describe the locations and functions of the three tissue systems in land plants.
3. Distinguish between primary growth and secondary growth, and describe the cell types involved in each.
4. Distinguish between meristem cells and differentiated cells.
5. Diagram the internal structure of a dicot root after primary growth, labeling and describing the function of epidermis, cortex, endodermis, pericycle, xylem, and phloem.
6. Diagram the internal structure of a dicot stem after primary growth, labeling and describing the function of the epidermis, cortex, pith, xylem, and phloem. How does the structure change after secondary growth?
7. Describe xylem and phloem, including their cell types and the solutions they carry.
8. What types of cells form root hairs? What is the function of root hairs?
9. Diagram the structure of a leaf. What structures regulate water loss and CO_2 absorption by a leaf?
10. Describe the daily cycle of opening and closing of stomata, and their response to dehydration.

11. Describe how water and minerals are absorbed by a root.
12. Describe how water and minerals move in xylem. Why is water and mineral movement in xylem unidirectional?
13. Describe how sugars are moved in phloem, and explain why phloem fluid may move up or down the plant.
14. Describe the surface structures of a stem after primary growth. Where are buds located, and what are their functions?

Applying the Concepts

Bloom's: Applying, Analyzing, Evaluating

1. An important goal of molecular botanists is to insert the genes for nitrogen fixation into crop plants such as corn or wheat. Why would the insertion of such genes be useful? What changes in farming practices would this technique allow?
2. When a hurricane sweeps over low-lying coastal areas, it leaves behind a lot of salt water. Many land plants die soon thereafter. Why would salt water kill plants?
3. Grasses (monocots) form their primary meristem near the ground surface rather than at the tips of branches the way dicots do. How does this feature allow you to grow a lawn and mow it every week in the summer? What would happen if you had a dicot lawn and mowed it?

45 Plant Reproduction and Development

Amorphophallus titanum is truly breathtaking—in more ways than one! The flower can be up to 10 feet tall and smells like a rotting carcass.

CASE STUDY

Some Like It Hot—and Stinky!

MOST FLOWERS ARE PRETTY, delicate, fragrant, and fairly small. And flowers usually remain at the same temperature as their environment. But not all flowers obey these rules. Behold *Amorphophallus titanum,* known in its native Sumatra (an island of Indonesia) as bunga bangkai, the corpse flower. (People seldom translate its scientific name in print, but you can probably figure it out.)

The corpse flower gets its common name from the odor of its flowers, which can be up to 10 feet tall. The odor is variously described as resembling decomposing fish, decaying pumpkin, or just plain rotting carrion. The corpse flower also gets hot. It generates pulses of warm water vapor at temperatures as high as 97°F (36°C). In the Sumatran forest, the huge flower acts like a chimney, blasting its smelly steam high into the air and dispersing it throughout the vicinity.

Flesh flies and carrion beetles are attracted by the smell of rotting flesh. They swarm around animal carcasses and lay their eggs in them. When the eggs hatch, the larvae eat the decaying meat, then eventually pupate and hatch into another generation of flies and beetles. When a corpse flower emits "parfum de carcass," it attracts these scavengers, which inadvertently serve as pollinators.

Several other flowers are warm, or stinky, or both. The related dead-horse arum smells up Corsica and other islands in the northern Mediterranean. The stinking corpse lily shares Sumatran forests with the corpse flower. In South Africa, the starfish flower attracts flies with its five-armed, putrid flowers—and believe it or not, some people grow them as houseplants! Several relatives of the corpse flower, including philodendrons and skunk cabbages, bear heat-generating flowers whose aromas are not particularly offensive (although the leaves of skunk cabbages smell like skunk if they are damaged).

Why would a plant produce an enormous, hot, putrid flower? For that matter, why do plants produce flowers at all? And how does a plant benefit by deceiving beetles and flies into mistaking its flower for a mass of rotting flesh?

AT A GLANCE

45.1 HOW DO PLANTS REPRODUCE?

Many plants can reproduce either sexually or asexually. During asexual reproduction, mitotic cell division in an existing plant produces offspring that are genetically identical to the parent. For example, aspen trees can sprout new shoots from their roots. When the arching branches of blackberries or the spreading horizontal runners of strawberries (see Fig. 44-13b) touch the ground, they take root and produce new plants. The bulbs of tulips, daffodils, and amaryllis can reproduce by growing new, smaller bulbs.

Although biologists have developed a number of hypotheses to explain the origin of sexual reproduction, the causes of its evolution are not fully understood. However, the benefits of sexual reproduction must be large, because most eukaryotes, including the plants mentioned above, reproduce sexually at least some of the time.

The Plant Sexual Life Cycle Alternates Between Diploid and Haploid Stages

The sexual life cycle of plants is described as **alternation of generations** because it alternates between multicellular diploid and multicellular haploid generations that give rise to each other in a continuing cycle (**FIG. 45-1**). We begin our description of the cycle at the multicellular diploid stage, in which specialized reproductive cells undergo meiotic cell division, producing haploid cells called **spores** ❶. Because the diploid stage produces spores, it is named the **sporophyte**, or "spore-bearing plant." Why are these haploid cells called *spores* rather than *gametes*? Gametes (sperm or eggs) never divide, but two gametes may fuse to form a diploid fertilized egg, or **zygote**. In contrast, spores never fuse to form a diploid cell; instead, plant spores undergo mitotic cell division to produce a multicellular haploid stage called the **gametophyte** ("gamete-bearing plant") ❷. Specialized reproductive cells in the gametophyte differentiate into haploid gametes ❸. Two gametes then fuse to form a zygote ❹. The zygote grows by mitotic cell division, becoming the next multicellular diploid generation ❺.

Although alternation of generations is the sexual life cycle of all plants, the relative size, complexity, and life span of the sporophyte and gametophyte stages vary considerably among different types of plants. In mosses, for example, the gametophyte stage is an independent plant that dominates the life cycle (see Chapter 22). Sperm fertilize eggs that are retained in the gametophyte. The resulting zygotes develop into sporophytes that grow directly on the gametophyte and rely on the gametophyte for nourishment. A sporophyte is never an independent plant.

In ferns, both the haploid and diploid stages are free-living, independent plants, but the life cycle is dominated by the diploid sporophyte. Reproductive cells of the sporophyte—the fern commonly seen in moist, shady woods—undergo meiotic cell division, producing clusters of haploid spores, typically on the undersides of the fronds (**FIG. 45-2**). If a spore lands on moist soil, it germinates into a tiny haploid gametophyte. Cells in separate male and female structures on the gametophyte differentiate into sperm and eggs. A sperm

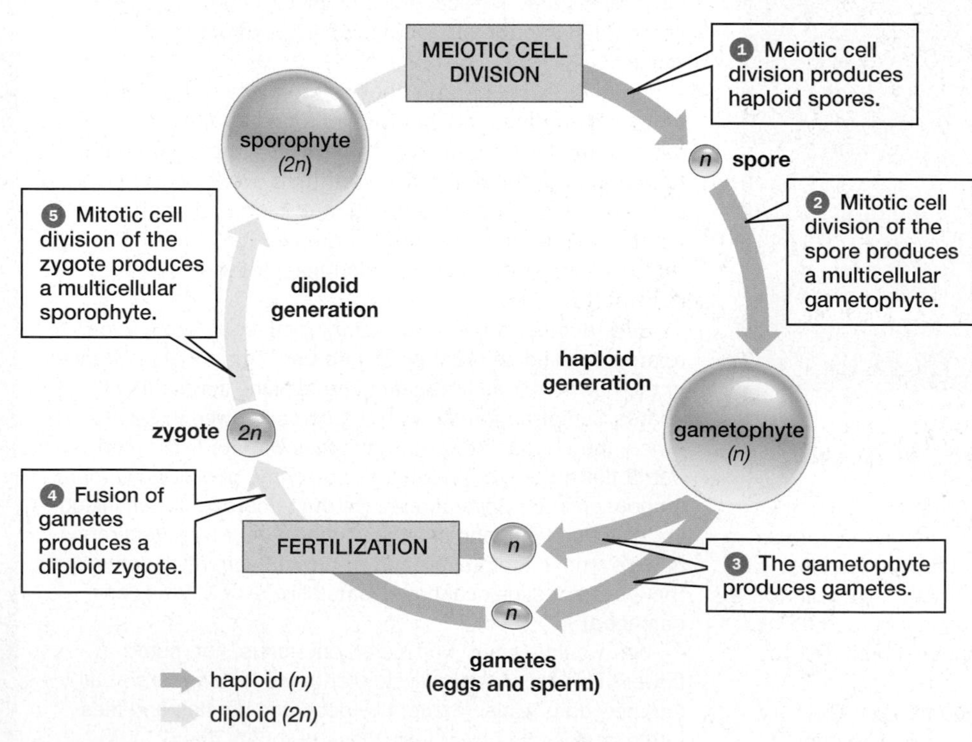

MEIOTIC CELL DIVISION
❶ Meiotic cell division produces haploid spores.
sporophyte (2n)
n spore
❷ Mitotic cell division of the spore produces a multicellular gametophyte.
❺ Mitotic cell division of the zygote produces a multicellular sporophyte.
diploid generation
haploid generation
zygote 2n
gametophyte (n)
❹ Fusion of gametes produces a diploid zygote.
FERTILIZATION
n
❸ The gametophyte produces gametes.
n
gametes (eggs and sperm)
haploid (n)
diploid (2n)

▲ **FIGURE 45-1 Alternation of generations**

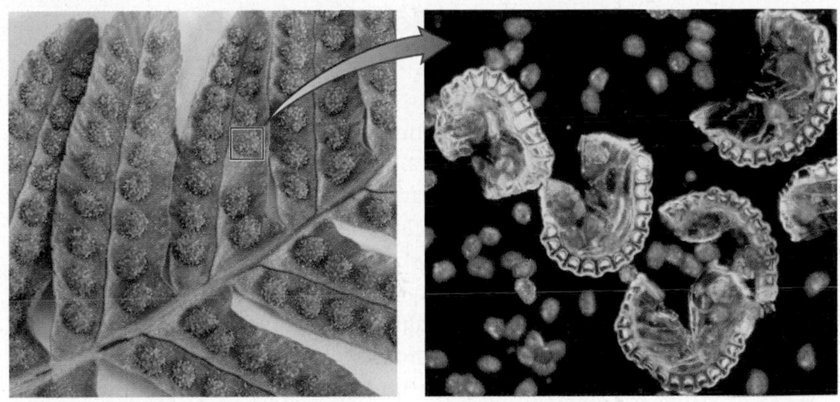

▲ **FIGURE 45-2 Spore production in ferns** In most ferns, clusters of reproductive cells on the undersides of the fronds of the sporophyte stage (left) produce spores that burst out of the capsules in which they are produced (right) and are dispersed primarily by the wind.

fertilizes an egg inside the female reproductive structure, creating a diploid zygote. Cell divisions of the zygote produce an embryonic sporophyte that begins growing atop the gametophyte, but soon the sporophyte develops its own roots and leaves.

In the seed plants, which include gymnosperms (conifers and their relatives) and angiosperms (flowering plants), the diploid sporophyte is dominant; the haploid gametophyte is never a free-living, independent plant.

Let's look at the sexual life cycle of flowering plants in more detail (**FIG. 45-3**). The multicellular sporophyte stage is the plant you see in gardens, orchards, forests, and fields. At the appropriate time of year, it produces flowers. Male and female reproductive structures in the flower produce specialized *mother cells* ❶ that undergo meiotic cell division to form haploid male and female spores ❷. The spores undergo mitotic cell division to produce haploid male and

mother cell
spores
MEIOTIC CELL DIVISION
flower
stigma
anther
nucleus
male gametophyte (pollen grain)
ovule
ovary
❶ In the flower, diploid mother cells develop in the reproductive structures: anthers (male) and ovaries (female).
❷ Meiotic cell division of mother cells in the sporophyte produces haploid spores.
❸ Mitotic cell division of the spores forms male gametophytes (pollen), which produce sperm, and female gametophytes, which produce eggs.
sperm
❹ Pollen carries the sperm to the female reproductive structure of a flower; sperm travel within a pollen tube to the female gametophyte.
❺ A sperm fertilizes an egg within the female gametophyte, producing a diploid zygote.
mother cell
pollen tube
MEIOTIC CELL DIVISION
ovule
mature sporophyte
spores
sperm nuclei
female gametophyte
❻ The zygote develops into an embryo, a seedling, and eventually, a new mature sporophyte.
egg
female gametophyte
FERTILIZATION
seedling
seed
haploid *(n)*
diploid *(2n)*
seed
embryo
fruit

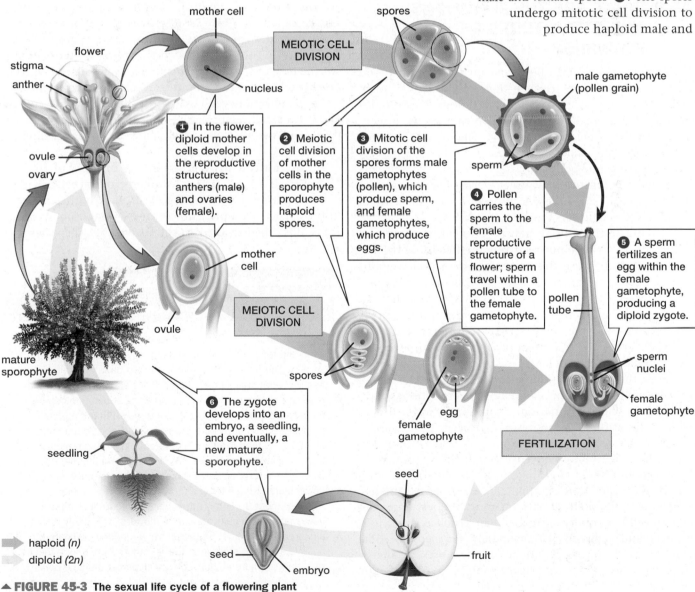

▲ **FIGURE 45-3 The sexual life cycle of a flowering plant**

female gametophytes ❸. The female gametophyte, which remains in the flower, usually consists of seven cells, one of which is the egg. The male gametophyte is the **pollen grain**. It is multicellular, but just barely; it consists of only three cells, two of which are sperm. Pollen grains are carried by wind or animals to the female reproductive structure of the flower ❹. One of the pollen's sperm fertilizes an egg, producing a diploid zygote ❺. The zygote undergoes repeated mitotic cell divisions to form an embryo and eventually a new adult sporophyte plant ❻.

CHECK YOUR LEARNING

Can you . . .

- describe the plant sexual life cycle?
- diagram the life cycles of ferns and flowering plants and explain the reproductive function of each stage?

45.2 WHAT ARE THE FUNCTIONS AND STRUCTURES OF FLOWERS?

Both angiosperms and the more ancient gymnosperms have seeds, and both package their sperm inside pollen grains. But only angiosperms have flowers. The flowers of most angiosperms entice animals—especially insects, birds, and bats—to carry pollen from one plant to another. In exchange for this pollen transport, pollinators generally receive food—some pollen or a sip of nectar. The mutually beneficial relationship between flowering plants and their pollinators has led to the evolution of a great diversity of colorful, scented flowers that attract animals.

Relying on the wind to transport pollen, as all gymnosperms do, is a viable but relatively inefficient method of reproduction, because the vast majority of pollen grains do not reach their targets. Relying on animals to carry pollen increases the efficiency of the transfer considerably. This increased efficiency probably provided the advantage that favored the evolutionary origin of flowers and the rise of angiosperms. Today, animal-pollinated angiosperms are the dominant plants on Earth.

Even though wind pollination is less efficient overall than animal pollination, some angiosperm species have reverted to wind pollination. Evolutionary reversion to wind pollination is hypothesized to occur when angiosperm species evolve in environments that contain too few animal pollinators to ensure reliable reproduction. Today, about 10% of angiosperm species rely on the wind for pollination. Wind-pollinated angiosperms include deciduous trees such as oaks, maples, birches, and poplars, as well as a variety of smaller plants such as sagebrush, grasses, and ragweed. Unfortunately for people, some wind-borne pollen ends up inside the noses of allergy sufferers, as we explore in "Health Watch: Are You Allergic to Pollen?"

Flowers Are the Reproductive Structures of Angiosperms

Flowers are the sexual reproductive structures of angiosperms, produced by the diploid sporophyte. A **complete flower**, such as that of a petunia, rose, or lily, consists of four sets of modified leaves: sepals, petals, stamens, and carpels. The **sepals** are located at the base of the flower. In dicots such as roses, strawberries, or apples, sepals are usually green and leaflike (**FIG. 45-4a**); in monocots such as daffodils, tulips, or amaryllis (**FIG. 45-4b**), sepals typically resemble the petals. In both dicots and monocots, sepals surround and protect the flower bud as the remaining three flower structures develop.

Just above the sepals are the **petals**, which are often brightly colored and fragrant, advertising the location of the flower to potential pollinators. The male reproductive structures, the **stamens**, are attached just above the petals. Each stamen usually consists of a slender **filament** bearing an **anther** that produces pollen. In the center of the flower are one or more female reproductive structures, called **carpels** (see Fig. 45-4a). A typical carpel is vase shaped, with a sticky **stigma** mounted atop an elongated **style**. The style connects the stigma with the **ovary**, at the base of the carpel. Inside the ovary are one or more **ovules**; a female gametophyte develops inside each ovule. After fertilization, each ovule will become a **seed**, consisting of a small, embryonic plant and stored food for the embryo. The ovary will develop into a **fruit** enclosing the seeds.

Incomplete flowers lack one or more of the four floral parts (sepals, petals, stamens, or carpels). For example, grass flowers (see Fig. 45-8) lack both petals and sepals. If an incomplete flower lacks either stamens or carpels, it is called an **imperfect flower**. Plant species with

CASE STUDY CONTINUED

Some Like It Hot—and Stinky!

A corpse flower is actually a mass of separate, tiny imperfect male and female flowers consisting only of stamens and carpels. The flowers are clustered at the base of a tall, fleshy central stalk and enclosed in a giant modified leaf that resembles a maroon vase. This amazing structure blooms for only two days. How does a plant ensure that the pollen it produces does not fertilize the female flowers on the same plant, but instead fertilizes a different plant? The female flowers open first and have wilted by the time the male flowers bloom. Therefore, pollen is transferred only to other corpse flower plants whose female flowers happen to be open.

Only about 35 botanical gardens in the entire United States have managed to cultivate corpse flowers, which only bloom about once a decade. So the plants are people-pollinated; botanists collect, store, and transfer the pollen among these rare plants. Where and how do angiosperms produce pollen?

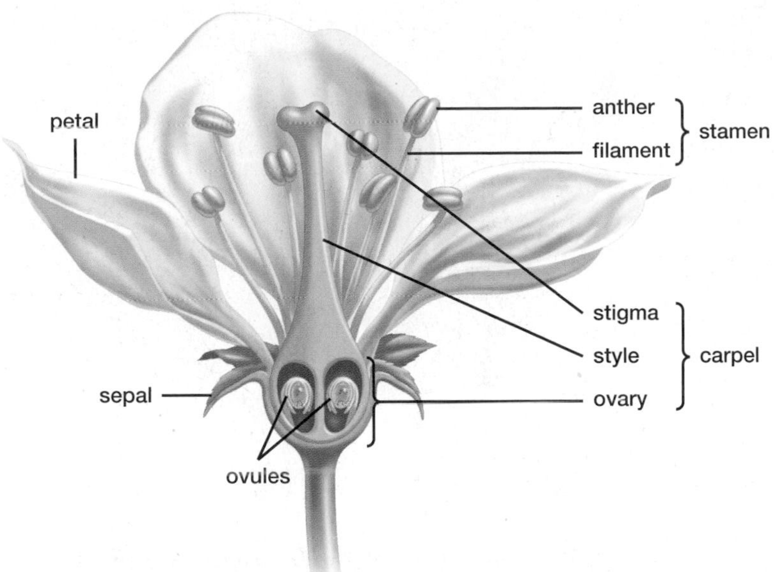

(a) A representative dicot flower

(b) A representative monocot flower (amaryllis)

▲ **FIGURE 45-4 A complete flower (a)** A complete flower has four parts: sepals, petals, stamens (the male reproductive structures), and at least one carpel (the female reproductive structure). This illustration shows a complete dicot flower. **(b)** The amaryllis is a complete monocot flower, with three sepals (virtually identical to the petals), three petals, six stamens, and three carpels (fused into a single structure). The anthers are well below the stigma, making self-pollination unlikely.

Are You Allergic to Pollen?

Wind pollination can succeed only if plants release huge quantities of pollen into the air. Unfortunately for allergy sufferers, people often inhale these microscopic male gametophytes. Proteins on the surfaces of pollen grains activate the immune systems of sensitive individuals causing itchy eyes, runny noses, coughing, and sneezing. If you are among these unlucky "hay fever" sufferers, your immune system is hypersensitive, creating the same symptoms that might occur if you were infected by a virus.

Individuals are often sensitive only to specific pollens, usually from the inconspicuous flowers of wind-pollinated plants. In temperate climates in springtime, tree pollen may be the culprit, whereas summer and fall allergies are often caused by grasses or (in North America) ragweed (**FIG. E45-1**, left). A single ragweed plant can release a billion pollen grains during its lifetime; collectively, ragweed plants release about a million tons of pollen in North America each year. Ragweed pollen has been collected 400 miles out to sea and 2 miles up in the atmosphere, so it is nearly impossible to avoid ragweed pollen completely.

Plants pollinated by bees and other animals rarely cause allergies because their pollen is sticky and produced in small amounts. Goldenrod, for example, blooms during the ragweed season and is often blamed for allergies that are actually caused by ragweed. Goldenrod's yellow blooms attract bee and butterfly pollinators, and most people can enjoy goldenrod in perfect comfort (**FIG. E45-1**, right).

▲ **FIGURE E45-1 Ragweed versus goldenrod** The inconspicuous flowers of wind-pollinated ragweed (left, with micrograph inset of its pollen) are a major cause of allergies, whereas the brightly colored insect-pollinated goldenrod flowers (right) are not.

THINK CRITICALLY Researchers have hypothesized that the amount of ragweed pollen in the air, and hence the severity of ragweed allergies, will increase as a result of global warming and/or the increased levels of atmospheric carbon dioxide that cause the warming. Explain the reasoning underlying this hypothesis, and describe an experiment that could help test the hypothesis.

imperfect flowers produce separate male and female flowers, sometimes on a single plant, as in zucchini (**FIG. 45-5**). Other species with imperfect flowers produce male and female flowers on separate plants, so only the female plants produce fruit. Female American holly trees, for example, yield decorative red fruit, so they are favored as ornamentals (although a few males must be nearby to pollinate them).

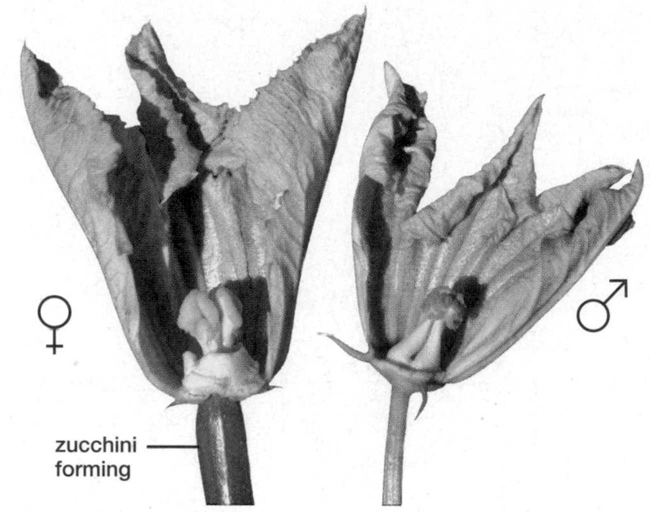

♀
♂
zucchini forming

▲ **FIGURE 45-5 Male and female imperfect flowers** Plants of the squash family, such as zucchinis, bear separate female (left) and male (right) flowers. Note the small zucchini (actually a fruit) forming from the ovary at the base of the female flower.

THINK CRITICALLY In species with separate male and female flowers on the same plant, why would natural selection favor individuals whose male and female flowers bloom at different times?

The Pollen Grain Is the Male Gametophyte

Male gametophytes develop within the anthers of a flower (**FIG. 45-6**). Each anther consists of four chambers called pollen sacs. Within each pollen sac, hundreds to thousands of diploid **microspore mother cells** develop ❶. Each microspore mother cell undergoes meiotic cell division (see Chapter 10) to produce four haploid **microspores** ❷. Each microspore then undergoes one mitotic cell division to produce an immature pollen grain. This is an immature male gametophyte consisting of two cells: a large **tube cell** occupying most of the pollen grain and a smaller **generative cell** that resides within the cytoplasm of the tube cell ❸. Mitotic cell division of the generative cell produces two haploid sperm cells ❹. As it matures, the pollen grain becomes surrounded by a

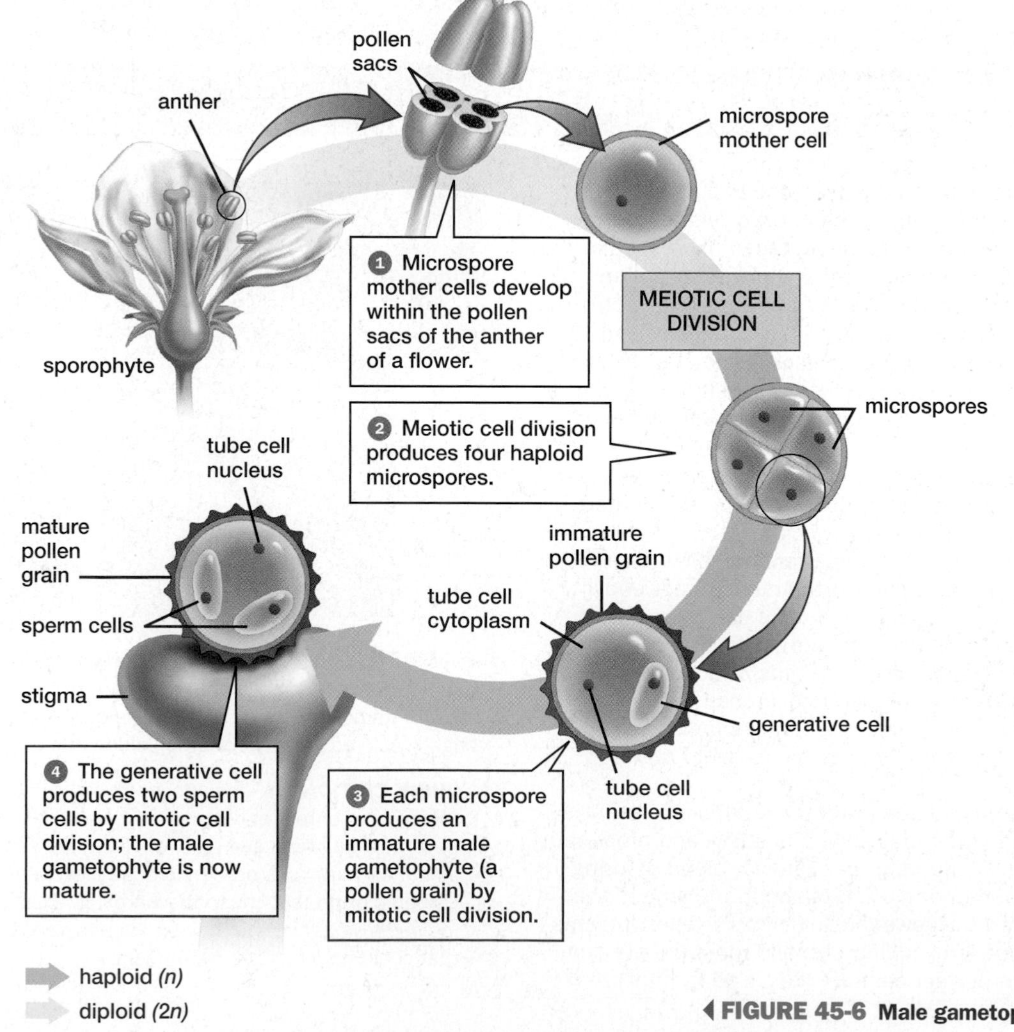

pollen sacs
anther
microspore mother cell
❶ Microspore mother cells develop within the pollen sacs of the anther of a flower.
MEIOTIC CELL DIVISION
sporophyte
microspores
❷ Meiotic cell division produces four haploid microspores.
tube cell nucleus
mature pollen grain
immature pollen grain
tube cell cytoplasm
sperm cells
stigma
generative cell
❹ The generative cell produces two sperm cells by mitotic cell division; the male gametophyte is now mature.
❸ Each microspore produces an immature male gametophyte (a pollen grain) by mitotic cell division.
tube cell nucleus
haploid (n)
diploid (2n)

◀ **FIGURE 45-6 Male gametophyte development**

▲ **FIGURE 45-7 Pollen grains** The tough outer coats of many pollen grains are elaborately sculptured in species-specific shapes and patterns, as shown in this colorized scanning electron micrograph.

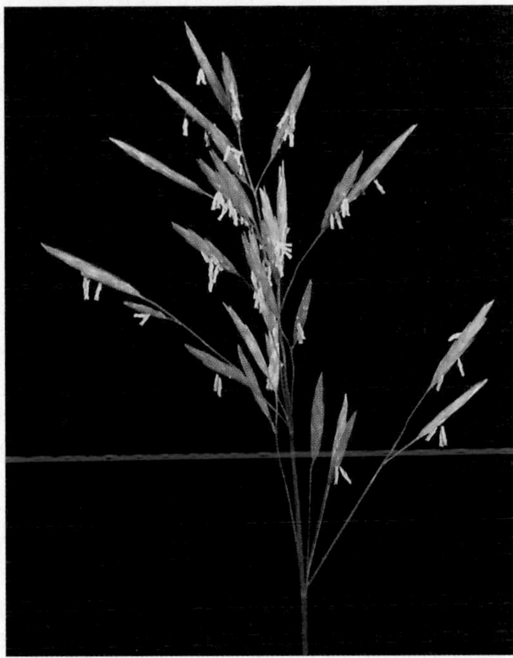

▲ **FIGURE 45-8 Wind-pollinated flowers** The flowers of grasses are wind pollinated, with anthers (dangling yellow structures) exposed to the wind.

tough, waterproof coat, often sculpted with an elaborate pattern of pits and protrusions characteristic of the plant species (**FIG. 45-7**). The coat protects the sperm during their journey to a sometimes distant female carpel.

When the pollen has matured, the pollen sacs of the anther split open. In animal-pollinated flowers, the pollen adheres weakly to the anther until a pollinator comes along and brushes or picks it off. In wind-pollinated flowers, such as those of grasses (**FIG. 45-8**) and oaks, the anthers usually protrude from small, often inconspicuous flowers. The slightest breeze carries off the pollen grains.

The Female Gametophyte Forms Within the Ovule

Within the ovary of a carpel, clusters of cells differentiate into ovules (**FIG. 45-9**). Depending on the plant species, an ovary may have as few as one ovule or as many as several dozen. Each young ovule consists of protective, multicellular, outer layers called **integuments**, which surround a single, diploid **megaspore mother cell** ❶. The megaspore mother cell undergoes meiotic cell division, producing four haploid **megaspores** ❷. Only one megaspore survives; the other three degenerate. The nucleus of the surviving megaspore undergoes three rounds of mitosis, producing eight haploid nuclei ❸. Plasma membranes and cell walls then form, dividing the cytoplasm into the seven (not eight) cells that make up the female gametophyte ❹. There are

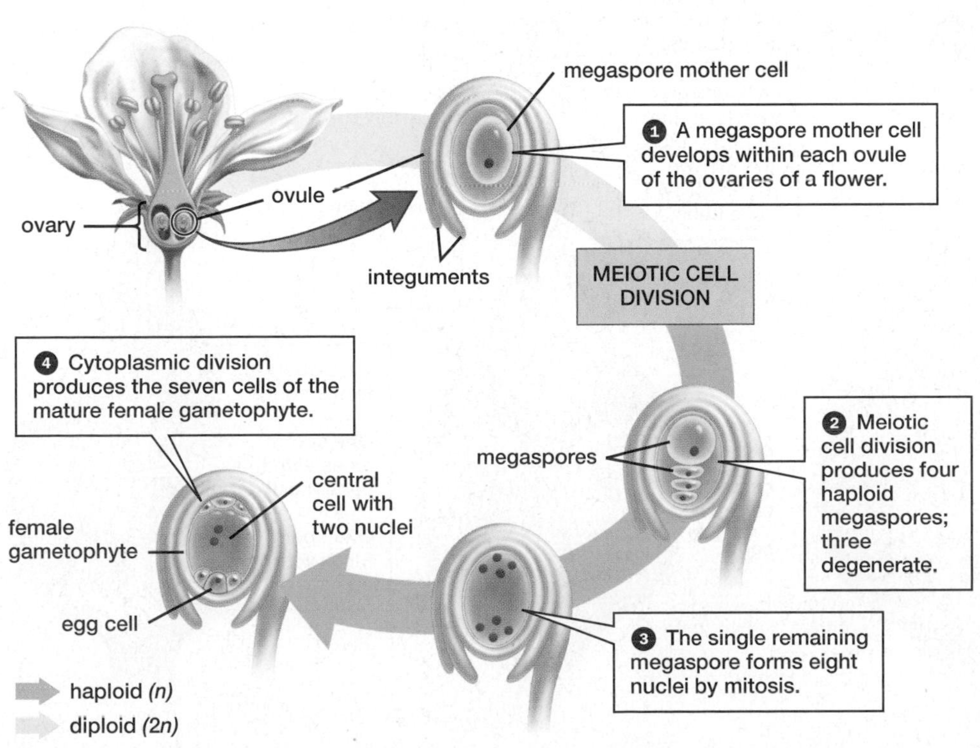
megaspore mother cell
❶ A megaspore mother cell develops within each ovule of the ovaries of a flower.
ovule
ovary
integuments
MEIOTIC CELL DIVISION
❹ Cytoplasmic division produces the seven cells of the mature female gametophyte.
megaspores
❷ Meiotic cell division produces four haploid megaspores; three degenerate.
central cell with two nuclei
female gametophyte
egg cell
❸ The single remaining megaspore forms eight nuclei by mitosis.
haploid (n)
diploid (2n)

◀ **FIGURE 45-9 Female gametophyte development**

three small cells at each end, each with a single nucleus, and one large central cell with two nuclei. The egg is one of the three cells at the lower end, located near an opening in the integuments of the ovule.

Pollination of the Flower Leads to Fertilization

Pollination is necessary for fertilization, but these are two distinct events, just as copulation and fertilization are separate events in mammals. **Pollination** occurs when a pollen grain lands on the stigma of a flower of the same plant species (**FIG. 45-10** ❶). The pollen grain absorbs water from the stigma and swells, splitting open the pollen coat. The tube cell, which makes up most of the pollen grain (see Fig. 45-6), elongates through the opening in the pollen coat and burrows through the style, producing a tube that will conduct sperm down the style to an ovule within the ovary ❷.

When the pollen tube reaches the opening in the integuments of an ovule and enters the female gametophyte, the tip of the tube ruptures, releasing the two sperm. In a process unique to flowering plants, called **double fertilization**, both sperm fuse with cells of the female gametophyte ❸. One sperm fertilizes the egg, producing a diploid zygote that will develop into an embryo and eventually into a new sporophyte. The second sperm enters the large central cell, where its nucleus fuses with the two nuclei already present, forming a triploid nucleus containing three sets of chromosomes. Through repeated mitotic cell divisions, the central cell will develop into the triploid **endosperm**, a food-storage tissue within the seed. The other five cells of the female gametophyte degenerate.

CHECK YOUR LEARNING

Can you . . .

- diagram the structure of a complete flower and explain the function of each part?
- describe the development of male and female gametophytes in flowering plants?
- describe the processes of pollination and double fertilization?

45.3 HOW DO FRUITS AND SEEDS DEVELOP?

After double fertilization, the female gametophyte and the surrounding integuments of the ovule develop into a seed surrounded by the ovary. The petals and stamens shrivel and fall away as the ovary ripens.

The Fruit Develops from the Ovary

When you eat a fruit, you are consuming the plant's ripened ovary, sometimes accompanied by other flower parts (**FIG. 45-11**). In a bell pepper, for example, the edible flesh develops from the wall of the ovary. Each of the seeds develops from an individual ovule within the ovary. As you will learn in Section 45.6, fruits are not always edible; some are hard, feathery, winged, spiked, adhesive, or even explosive. These various shapes, colors, and textures all serve the

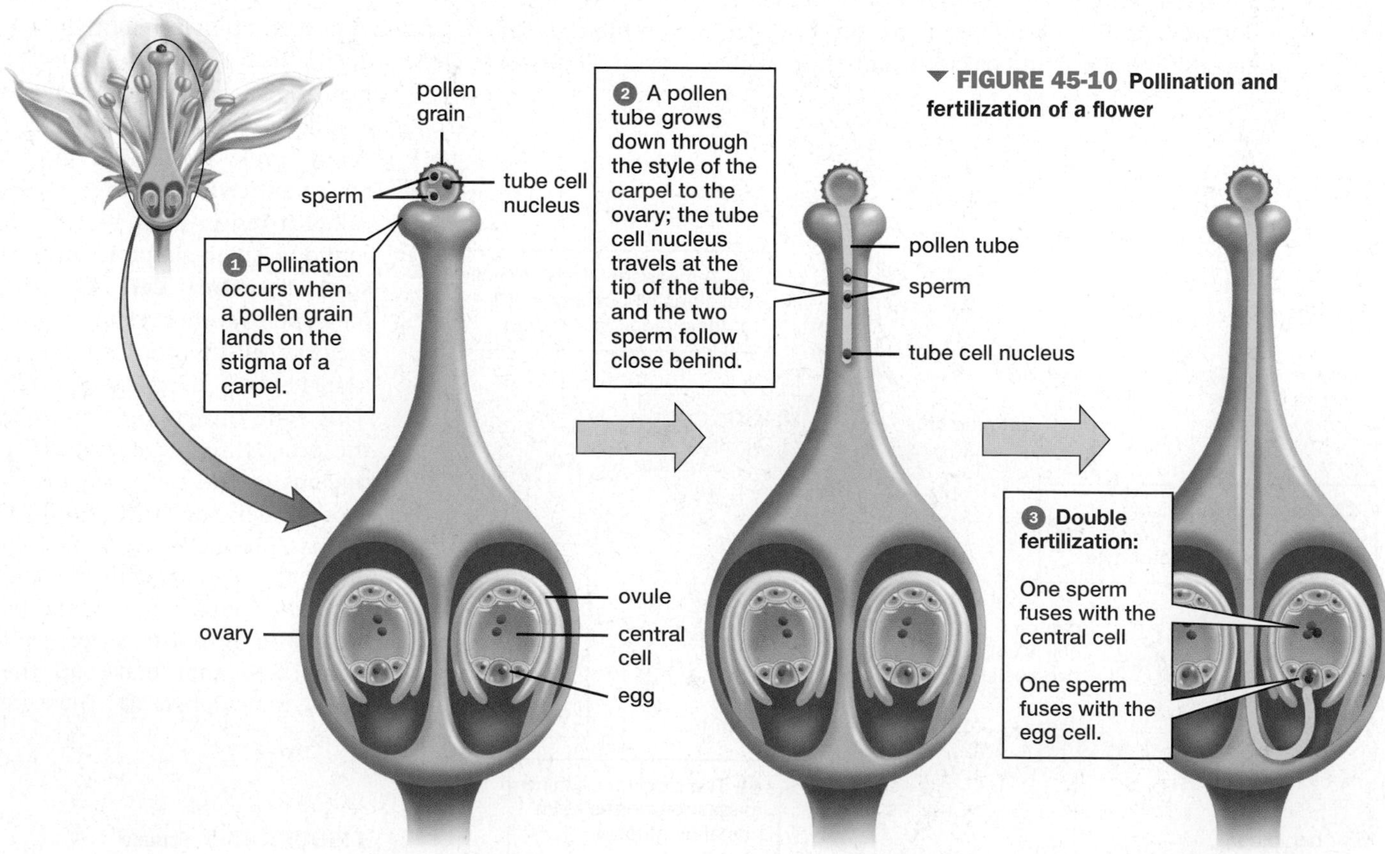
pollen grain
sperm
tube cell nucleus
❶ Pollination occurs when a pollen grain lands on the stigma of a carpel.
❷ A pollen tube grows down through the style of the carpel to the ovary; the tube cell nucleus travels at the tip of the tube, and the two sperm follow close behind.
pollen tube
sperm
tube cell nucleus
ovary
ovule
central cell
egg
❸ Double fertilization:
One sperm fuses with the central cell
One sperm fuses with the egg cell.

▼ **FIGURE 45-10 Pollination and fertilization of a flower**

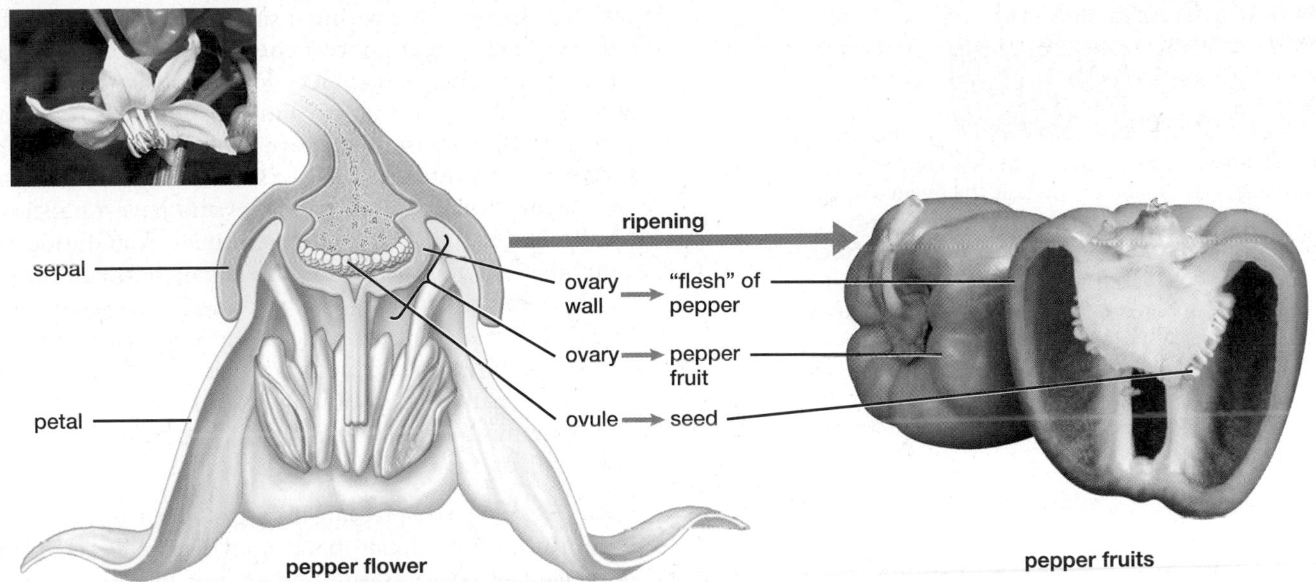

▲ **FIGURE 45-11 Development of fruit and seeds in a bell pepper** Fruits and seeds develop from flower parts. The ovary wall ripens into the fruit flesh, and its many ovules develop into seeds.

same function: They help disperse seeds away from the parent plant, in many cases by taking advantage of the mobility of animals.

The Seed Develops from the Ovule

Three distinct developmental processes transform an ovule into a seed (**FIG. 45-12a**). First, the integuments of the ovule thicken, harden, and become the **seed coat** that surrounds and protects the seed. Second, the triploid central cell divides rapidly. The resulting daughter cells absorb nutrients from the parent plant, forming a food-filled endosperm. Third, the zygote develops into an embryo.

As the seed matures, the embryo begins to differentiate into shoot and root (**FIGS. 45-12b, c**). The shoot portion consists of a short stem, often with one or two developing leaves, and either one or two **cotyledons**, or "seed leaves," that absorb food molecules from the endosperm and transfer them to other parts of the embryo. The seeds of monocots, as their name implies, have a single cotyledon ("mono" means "one"). In most monocots, including grasses, rice, corn, and wheat, the cotyledon absorbs some endosperm during development, but most of the endosperm remains in the mature seed until the seed sprouts. Wheat flour is ground-up endosperm; wheat germ is made from the embryo. A monocot embryo is enclosed in a pair of sheaths, one surrounding the developing root and a second, called the **coleoptile**, surrounding the developing shoot tip.

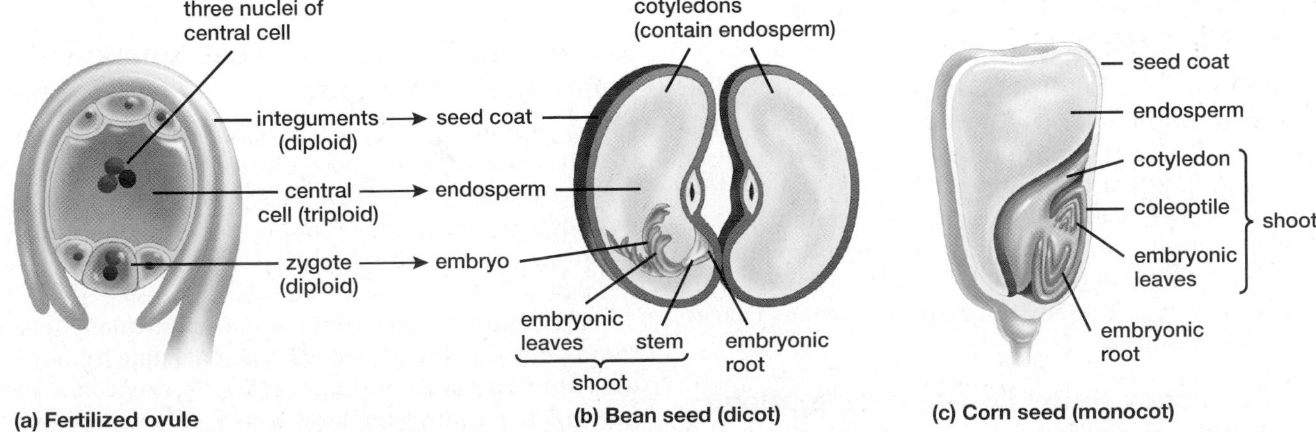

▲ **FIGURE 45-12 Seed development (a)** The embryo develops from the zygote, the endosperm develops from the triploid central cell, and the integuments of the ovule form the seed coat.
(b) The two cotyledons of dicot seeds usually absorb most of the endosperm before germination.
(c) Monocot seeds have a single cotyledon and usually retain most of their endosperm until germination.

Have You Ever Wondered ...

When a Fruit Is a Vegetable?

To a botanist, a fruit is the ripened, seed-containing ovary of an angiosperm, often including some other parts of the flower. Botanists define vegetables as edible plant parts that do not include the ripened ovary, including flowers (think broccoli), leaves (lettuce), roots (carrots), and stems (rhubarb). The culinary and legal worlds, however, beg to differ. Many botanical fruits are deemed vegetables by chefs and cookbooks: tomatoes, zucchinis, cucumbers, bell and chili peppers, and eggplants, to name just a few. The U.S. Supreme Court actually ruled on the status of the tomato in 1893, calling it a vegetable. Arkansas decisively straddles the fence: In 1987, the South Arkansas Vine Ripe Pink Tomato was named both the state fruit and the state vegetable!

Dicot seeds have two cotyledons ("di" means "two"). In the seeds of most dicots, including peas, beans, peanuts, and walnuts, the cotyledons absorb most of the endosperm during seed development, so the mature seed is virtually filled by the cotyledons. If you strip the thin seed coat from a bean or peanut, you will find that the inside splits easily into two halves; each is a cotyledon. The tiny white nub adhering to one of the cotyledons is the rest of the embryo.

CHECK YOUR LEARNING

Can you . . .

- explain how the parts of a flower develop into the parts of a fruit and its seeds?
- describe the differences between monocot and dicot seeds?

45.4 HOW DO SEEDS GERMINATE AND GROW?

Germination, often called sprouting, occurs when the embryonic plant within a seed grows, breaks out of the seed, and forms a seedling. In addition to nutrients, seeds need warmth and moisture to germinate, but even under ideal conditions, newly matured seeds may not germinate immediately. Instead, they often enter a period of **dormancy** during which they will not sprout. Dormant seeds are typically able to resist adverse environmental conditions such as freezing and drying.

Seed Dormancy Helps Ensure Germination at an Appropriate Time

Seed dormancy solves two problems. First, it prevents seeds from germinating within a moist fruit, where the sprout might be eaten by a fruit-eating animal or destroyed by mold growing in rotting fruit. Even if they survived, multiple seedlings germinating within a single fruit would grow in a dense cluster, competing with one another for nutrients and light. Second, environmental conditions that are suitable for germination (such as warmth and moisture) may sometimes be followed by conditions that would not allow the seedling to survive and mature. For example, seeds in temperate climates are usually produced in late summer and have a harsh winter ahead of them. Most do not germinate even during mild autumn weather. Instead, the seeds remain dormant through both autumn and winter, so that tender new sprouts do not freeze. Germination usually occurs the following spring.

The requirements to break dormancy are finely tuned to the plant's native environment and dispersal mechanisms. The three most common are drying, exposure to cold, and disruption of the seed coat.

- ***Drying*** Seeds that require drying often are dispersed by fruit-eating animals that cannot digest the seeds. The animals excrete the seeds in their feces. Exposed to the air, the seeds dry out. Later, when temperature and moisture levels are favorable, they germinate.
- ***Cold*** Seeds of many temperate and arctic plants will not germinate unless they are exposed to prolonged subfreezing temperatures, followed by sufficient warmth and moisture. Requiring a substantial cold spell keeps them from sprouting until the following spring.
- ***Seed Coat Disruption*** The seed coat may need to be weathered or partially digested by passing through an animal digestive tract before germination can occur. Some coats contain chemicals that inhibit germination. In deserts, for example, years may go by without enough water for a plant to complete its life cycle. The seed coats of many desert plants have water-soluble chemicals that inhibit germination, and only a hard rainfall (weathering) can wash away enough of the inhibitors to allow sprouting.

In warm, moist, tropical regions, where environmental conditions are suitable for germination throughout the year, seed dormancy is much less common.

During Germination, the Root Emerges First, Followed by the Shoot

During germination, the embryo absorbs water, which makes it swell and burst its seed coat. The root usually emerges first and grows rapidly, absorbing water and minerals from the soil (**FIG. 45-13**). Much of the water is transported to the shoot, whose cells elongate and push upward through the soil toward the light.

The nutrients for germination, particularly energy-storing molecules such as starch and oil, come from the endosperm of the seed. In dicot seeds, the cotyledons absorb most of the endosperm long before germination, so the cotyledons merely transfer these nutrients to the embryo as germination occurs. Monocot seeds, however, retain most of their endosperm until germination. During germination, the cotyledon digests the endosperm, transferring its nutrients to the rest of the growing embryo.

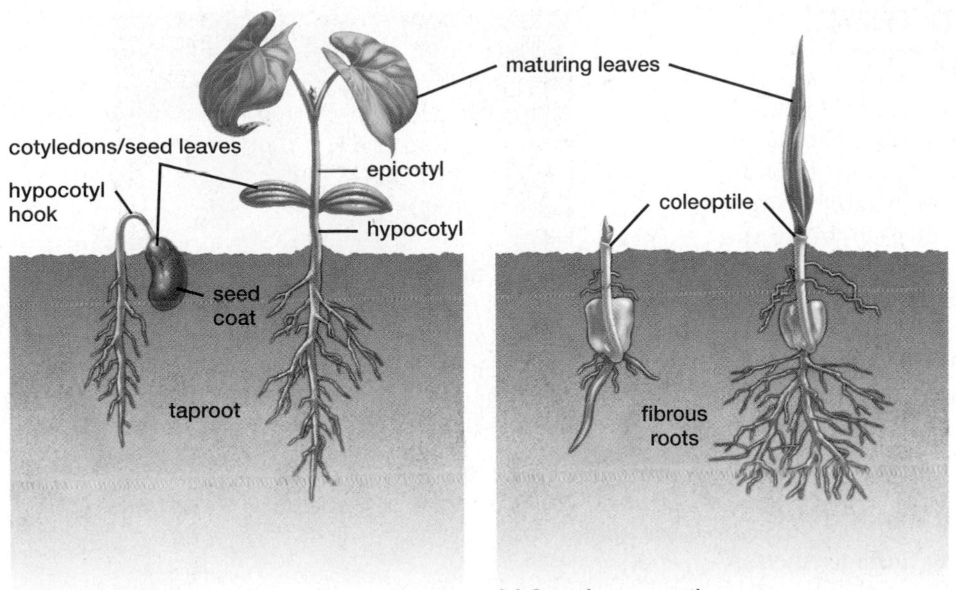

FIGURE 45-13 Seed germination First, the root grows rapidly, absorbing water and minerals. Using these resources, the shoot pushes upward through the soil. **(a)** In some dicots, such as the bean shown here, the hypocotyl bends, forming a hook that pushes through the soil first, protecting the downward-pointing shoot tip. In other dicots, such as the pea, the bend forms in the epicotyl. **(b)** In monocots such as corn, the shoot tip is protected within a tough coleoptile.

Most seeds germinate beneath the soil, so the delicate apical meristems of the embryo must be protected from sharp soil particles during germination. The apical meristem of a root tip is protected by a root cap throughout the life of the plant. A shoot, however, is underground only during germination, so its apical meristem only needs temporary protection. Dicots form protective hooks in their embryonic shoots. If the hook is above where the cotyledons attach to the stem, it is called an **epicotyl hook** (Gk. *epi*; above). If below this point, the structure is called a **hypocotyl hook** (Gk. *hypo*; below; see Fig. 45-13a). The hook, encased in thick-walled epidermal cells, forces its way up through the soil, clearing the path for the downward-pointing apical meristem and its delicate new leaves. In dicots with hypocotyl hooks, such as zucchini and beans, the elongating shoot carries the cotyledons out of the soil into the air. These cotyledons typically become green and photosynthetic (hence their name of "seed leaves"; **FIG. 45-14**) and transfer both previously stored food and newly synthesized sugars to the shoot before they wither away. In dicots with epicotyl hooks, the cotyledons remain below the ground, shriveling up as the embryo absorbs their stored food. In all dicots, the shoot straightens after it emerges, orienting its leaves toward the sunlight.

In monocots, such as corn, the coleoptile encloses the shoot tip like a glove around a finger (see Fig. 45-13b), protecting it and pushing aside soil particles as the tip grows. Once out in the air, the coleoptile degenerates, allowing the shoot to emerge. The cotyledon stays belowground in the remnants of the seed, absorbing the endosperm and transferring it to the shoot.

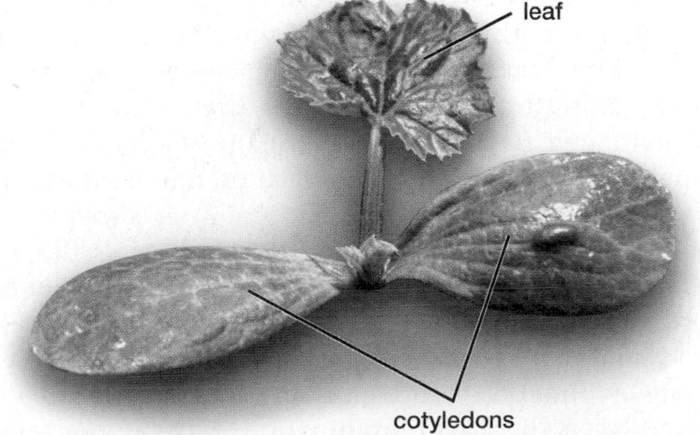

FIGURE 45-14 Cotyledons nourish the developing plant In some dicots, such as the zucchini shown here, the cotyledons emerge from the soil, expand, and photosynthesize. The crinkled leaves develop a bit later. Eventually, the cotyledons shrivel up.

CHECK YOUR LEARNING

Can you . . .

- explain why many seeds undergo dormancy before germinating?
- describe the process of germination in monocots and dicots?

CASE STUDY CONTINUED

Some Like It Hot—and Stinky!

The warmth of hot flowers attracts pollinators and helps broadcast the flowers' scent. Corpse flower stalks can reach 97°F, nearly as warm as our bodies. How do these flowers get so hot? *Amorphophallus* and other heat-producing flowers have evolved mechanisms that disconnect cellular respiration from ATP synthesis. In most cells, cellular respiration uses about 40% of the energy in glucose to synthesize ATP, with the rest given off as heat (see Chapter 8). Hot flowers, on the other hand, synthesize very little ATP; instead, almost all of the energy in glucose is released as heat. Besides the scent of carrion, what other features have flowers evolved that attract animal pollinators?

45.5 HOW DO PLANTS AND THEIR POLLINATORS INTERACT?

Plants and their pollinators have coevolved; that is, each has acted as an agent of natural selection on the other. Animal-pollinated flowers have evolved traits that attract useful pollinators and frustrate undesirable visitors that might eat nectar or pollen without pollinating the flower in return. Pollinators have evolved senses, behaviors, and body structures that help them locate and identify nutritious flowers and extract their nectar or pollen. Animal-pollinated flowers can be loosely grouped into three categories, on the basis of the benefits (real or perceived) that they offer to potential pollinators: food, sex, or a nursery.

Some Flowers Provide Food for Pollinators

Many flowers provide food for foraging animals such as bees, moths, butterflies, hummingbirds, and even a few mammals, such as bats and lemurs, as we explore in "Earth Watch: Pollinators, Seed Dispersers, and Ecosystem Tinkering." In return for food, the animals unwittingly distribute pollen from flower to flower. Most flying pollinators are able to locate flowers from a distance because the flower colors contrast with the mass of green leaves surrounding them. The colors providing the best contrast differ depending on the pollinator. For example, bee-pollinated flowers are usually white, blue, yellow, or orange, but not red, because bees do not perceive red as a distinct color. Bees can, however, see ultraviolet light (**FIG. 45-15**), so many bee-pollinated flowers have markings that reflect UV light, including central spots or lines pointing toward the center, almost like bull's-eyes. We can also thank the bees for most of the sweet-smelling flowers because "flowery" odors attract bees.

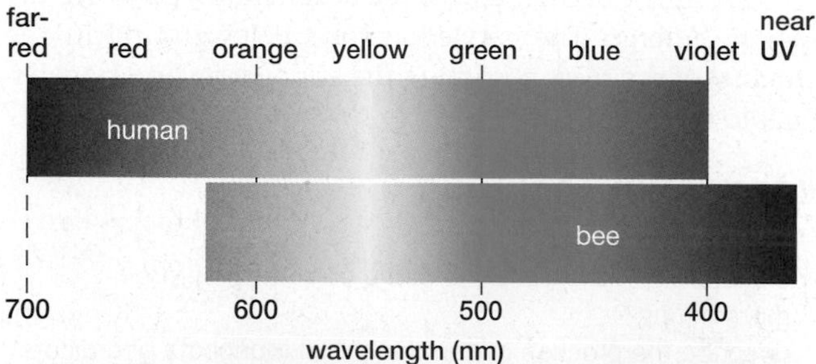

(a) A comparison of color vision in humans and bees

(b) Flower color patterns seen by humans and bees

▲ **FIGURE 45-15 Ultraviolet patterns guide bees to nectar**
(a) The spectra of color vision for humans and bees overlap, but are not identical. Humans (top) are sensitive to red, which bees (bottom) do not perceive; bees can see near-UV light, which is invisible to the human eye. **(b)** Many flowers photographed under visible light (left) and under UV light (right) show patterns that presumably direct bees to the nectar- and pollen-containing centers of the flowers.

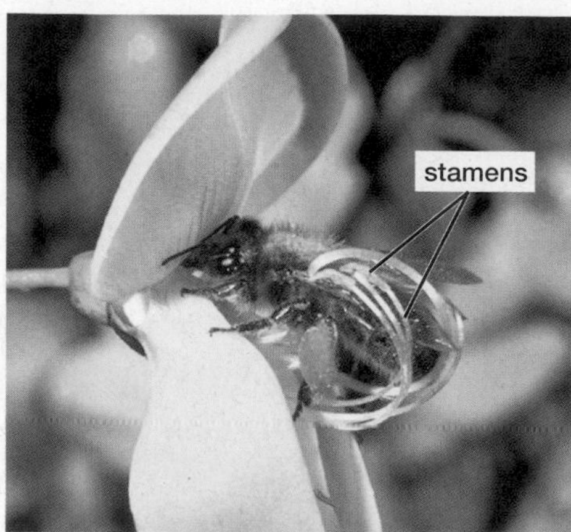

▲ **FIGURE 45-16 "Pollinating" a pollinator** Pollen-laden stamens on this scotch broom flower have popped up and are covering the bee's hairy back with pollen.

Bee-pollinated flowers also have structural adaptations that help ensure pollen transfer. In the Scotch broom flower, for example, nectar forms in a crevice between enclosing petals. In newly opened flowers, pollen-laden stamens lurk within the crevice. When a bee visits a young flower, the stamens emerge, brushing pollen onto its back as the bee's weight deflects the petals downward (**FIG. 45-16**). In older flowers, the carpel's style elongates, pushing the sticky stigma out through the crevice; therefore, when a pollen-coated bee delves for nectar, it leaves pollen behind on the stigma.

Many flowers adapted for moth and butterfly pollinators have nectar-containing tubes that accommodate the long tongues of these insects. Flowers pollinated by night-flying moths open only in the evening. Most are white, which makes them more visible in the dark. Some also give off strong, musky odors that attract moths. Bat-pollinated flowers are also usually white and open at night. Beetles and flies often feed on animal wastes or carrion, so flowers pollinated by these insects frequently smell like dung or rotting flesh. These flowers deceive their pollinators by smelling like a nutritious meal but offering no food.

Hummingbirds are one of the few important vertebrate pollinators, although some mammals also pollinate flowers. Because hummingbirds have a poor sense of smell, hummingbird-pollinated flowers seldom synthesize fragrant chemicals. However, they often produce more nectar than other flowers, because hummingbirds need more energy than insects do and will favor flowers

Earth WATCH Pollinators, Seed Dispersers, and Ecosystem Tinkering

The flowering plants that dominate most terrestrial ecosystems depend on mutually beneficial relationships with animals that pollinate their flowers and disperse their seeds. If pollinators or seed dispersers are severely reduced or eliminated, entire ecosystems may be endangered.

Consider Madagascar, an island off the African coast, where a unique and irreplaceable community of plants and pollinators has evolved. This is the only native habitat for primates called lemurs, which are probably the most important seed-dispersers on Madagascar. But Madagascar's rapidly growing human population has eliminated most of Madagascar's original forests, and lemurs are hunted for food. As a result, they are now the world's most endangered vertebrate, with 94% of the 103 lemur species threatened with extinction. Lemurs are important seed-dispersers for at least 40 species of trees, most of them unique to Madagascar. After swallowing the seeds intact, these primates excrete the seeds some distance from the parent tree. Passing through a lemur's digestive tract often makes the seeds more likely to sprout.

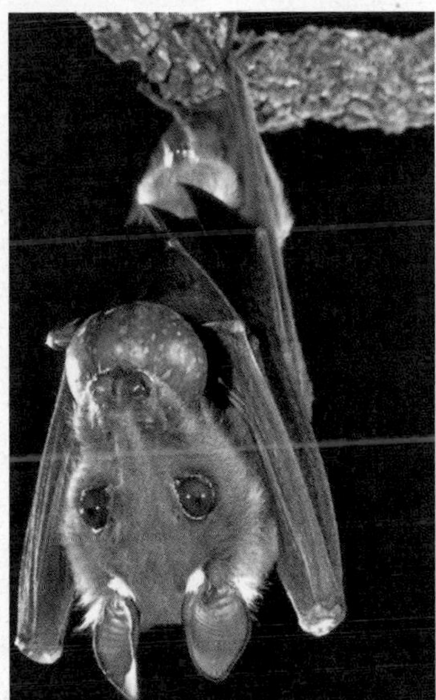

▲ **FIGURE E45-3 Fruit bats are important for seed dispersal**

The critically endangered black-and-white ruffed lemur (**FIG. E45-2a**) not only disperses seeds, but also is the world's largest pollinator; the travelers palm (also native only to Madagascar; **FIG. E45-2b**) probably depends on it. This 40-foot-tall plant produces large, tough, spiky flowers that must be ripped apart for pollen to escape or enter, and black-and-white ruffed lemurs seem to be the only animal on Madagascar with enough strength and dexterity to accomplish this. Seeking the plant's tasty nectar, they bury their snouts deep inside, incidentally collecting pollen on their noses and fur, to be deposited in the next travelers palm they encounter.

In many other tropical forests, monkeys and fruit-eating bats are important agents of seed dispersal (**FIG. E45-3**), and the trip through the digestive tracts of monkeys or bats helps seeds to germinate. Fruit bat populations in Africa are declining rapidly because they are considered a delicacy and also killed as pests. In Mexican rain forests, spider monkeys and howler monkeys disperse the seeds of dozens of tree species, but both are endangered due to habitat destruction and hunting for food.

Even large-scale farms are vulnerable to the loss of pollinators. For example, most cultivated fruits and nuts, and many vegetables, depend on pollination by introduced European honeybees, which pollinate an estimated $15 billion worth of crops annually in the United States. Since 2006, a phenomenon called colony collapse disorder has devastated bee colonies. One hypothesis is that pesticide exposure weakens the bees, making them more susceptible to a wide variety of viruses as well as parasitic mites. Can native bees replace honeybees? Not without major changes in our agricultural system. Honeybee colonies are carefully managed and often transported considerable distances to pollinate crops. Most native bees do not form transportable colonies and reproduce far more slowly than honeybees. Honeybees forage for much longer distances than most native bees, so they are better adapted to pollinate the enormous, single-crop fields of commercial farms. Further, the diverse hedgerows, natural meadows, and forest edges where the native bees thrive have often been replaced by these huge commercial fields.

Flowering plants, their pollinators, and their seed dispersers often form an intricate, interconnected web, with each supporting the others. As ecologist Aldo Leopold wrote in *A Sand County Almanac,* "To keep every cog and wheel is the first precaution of intelligent tinkering." When humans tinker with Earth's ecosystems, we must be careful not to lose critical parts.

THINK CRITICALLY If raising honeybees as pollinators becomes unprofitable because of colony collapse, what alternatives do farmers have?

(a) Black-and-white ruffed lemur

(b) Travelers palm showing spiky flowers

▲ **FIGURE E45-2 The black-and-white ruffed lemur and the tree that depends on it**

▲ **FIGURE 45-17 Hummingbirds are effective pollinators** The hibiscus flower's anthers are positioned so that they deposit pollen on the bird's head.

THINK CRITICALLY What advantage might a hibiscus gain by encouraging hummingbirds to approach, while discouraging bees (which cannot reach the nectar in these flowers)?

that provide it. Hummingbird-pollinated flowers may have a deep, tubular shape that accommodates the birds' long bills and tongues while preventing most insects from reaching the nectar (**FIG. 45-17**). In addition, these flowers are often red, which is attractive to hummingbirds, but not bees.

Some Flowers Are Mating Decoys

A few plants, most notably some orchids, take advantage of the mating drives of male wasps, bees, and flies. These orchid flowers mimic female wasps, bees, or flies in both shape (**FIG. 45-18**) and scent (the orchids release a sexual attractant similar to that produced by the female insect). When a male insect lands atop these faux females and attempts to copulate, a packet of pollen often becomes attached to the insect. When the amorous insects repeat their mating attempts on other orchids of the same species, they transfer pollen packets from one flower to another.

▲ **FIGURE 45-18 Sexual deception promotes pollination** This male wasp is trying to copulate with an orchid flower. The result is successful reproduction—for the orchid, but not the wasp!

Some Flowers Provide Nurseries for Pollinators

Perhaps the most elaborate relationships between plants and pollinators occur in a few cases in which insects fertilize a flower and then lay their eggs in the flower's ovary. This arrangement occurs between milkweeds and milkweed bugs, figs and fig wasps, and yuccas and yucca moths (**FIG. 45-19**). For example, when a female moth

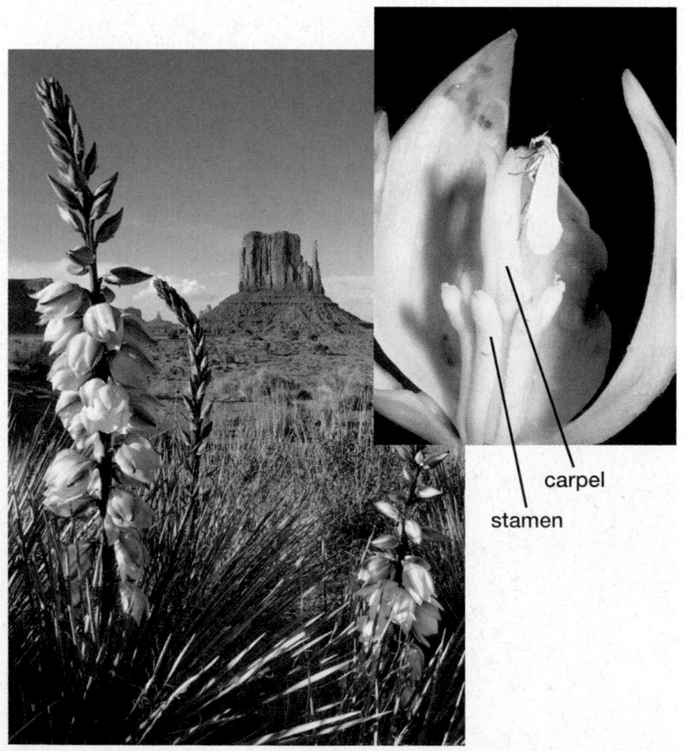

▲ **FIGURE 45-19 A mutually beneficial relationship** Yuccas bloom in Monument Valley, Arizona. (Inset) A yucca moth places pollen on the stigma of the carpel of a yucca flower.

visits a yucca flower, she collects pollen and rolls it into a compact ball. She carries the pollen ball to another yucca flower, drills a hole in the ovary wall, and lays her eggs inside the ovary. Then she smears the pollen ball on the stigma of the flower. By pollinating the yucca, the moth ensures that the plant will provide a supply of developing seeds for its caterpillar offspring to eat. Because the caterpillars eat only a fraction of the seeds, the yucca also reproduces successfully. The mutual adaptation of yucca and yucca moth is so complete that neither can reproduce without the other.

CHECK YOUR LEARNING

Can you . . .

- describe how the structures of flowers and their specific animal pollinators facilitate efficient pollination?

45.6 HOW DO FRUITS HELP TO DISPERSE SEEDS?

A plant benefits if its seeds are dispersed far enough away so that its offspring don't compete with it for light and nutrients. Dispersal also helps tender seedlings avoid browsing animals attracted to their parent plants. Finally, evolution has favored plant species that at least occasionally disperse their seeds to distant habitats, allowing them to extend their range so that these species persist even if their original location becomes uninhabitable.

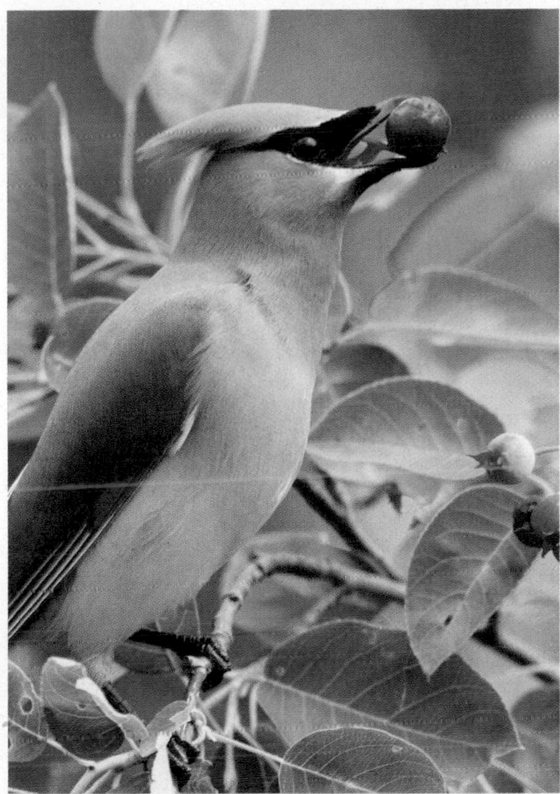

▲ **FIGURE 45-21 The colors of ripe fruits attract animals** These bright red serviceberries attract cedar waxwings in summer and autumn. Only ripe fruits containing mature seeds are sweet and brightly colored.

Clingy or Edible Fruits Are Dispersed by Animals

Clingy fruits, such as burdocks and sticktights, grasp animal fur (or human clothing) with prongs, hooks, spines, or adhesive hairs (**FIG. 45-20**). The parent plant holds its ripe fruit very loosely, so that even slight contact with fur pulls the fruit off the plant, leaving it stuck to the animal. Some of these fruits later fall off as the animal brushes against objects, grooms itself, or sheds its fur.

▲ **FIGURE 45-20 The cocklebur fruit uses hooked spines to hitch a ride on furry animals** This dog—or its owner—will eventually dislodge the cockleburs, often far away from the parent plant.

Unlike hitchhiking fruits, edible fruits have features that attract and benefit the animal disperser. The plant stores appealing flavors and nutritious sugars and starches in a fleshy fruit that surrounds the seeds, enticing hungry animals (**FIG. 45-21**), including people.

Some edible fruits, including peaches, plums, and avocados, contain large, hard seeds that animals usually do not eat, but drop somewhere along their way. Other fruits, such as blackberries, raspberries, strawberries, and tomatoes, have small seeds that animals swallow. These seeds are eventually excreted unharmed. Some have seed coats that must be scraped or weakened by passing through an animal's digestive tract before they will germinate. Besides being transported away from its parent plant, a seed that is swallowed and excreted benefits in another way: It ends up with its own supply of fertilizer! It is often important for the right type of animal to eat the fruit and seeds. For example, chili peppers have evolved a "hot" taste because it discourages mammals (whose digestive tracts destroy pepper seeds) from eating them. Birds, on the other hand, are apparently insensitive to the burning chemicals in chili seeds, and pepper seeds that have passed through a bird's digestive tract germinate at three times the rate of those that just fall to the ground. Next time you eat a fruit with seeds, think about how the seeds would be dispersed in nature.

CASE STUDY CONTINUED

Some Like It Hot—and Stinky!

The *Amorphophallus titanum* flower produces about 500 olive-sized, bright orange-red fruits that attract rain-forest birds called great hornbills. The hornbills eat the fruits and disperse the seeds. Both these unique plants and the hornbills that disperse them are threatened by poaching and by massive destruction of Indonesian rain forests. What are some other ways that fruits are adapted to disperse their seeds?

Explosive Fruits Shoot Out Seeds

A few plants develop explosive fruits that eject their seeds away from the parent plant. Himalayan balsam (**FIG. 45-22a**), an annual plant native to Asia, produces fruits that can shoot seeds up to 20 feet (about 6 meters). This explosive seed dispersal has helped the species spread rapidly when introduced to new habitats, and Himalayan balsam has become invasive in Europe and North America.

Lightweight Fruits May Be Carried by the Wind

Dandelions, milkweeds, elms, and maples produce lightweight fruits with large wind-catching surfaces. Each hairy tuft on a dandelion puff is a separate fruit bearing a single small seed that it can carry for miles if the winds cooperate. The single wing of the maple fruit, in contrast, causes its seed to spin like a propeller as it falls, usually taking it only a few yards from its parent tree (**FIG. 45-22b**).

Floating Fruits Allow Water Dispersal

Many fruits can float on water for a time and may be carried in streams or rivers, although this is usually not their principal method of dispersal. The coconut fruit, however, is a champion floater. Round, buoyant, and waterproof, the coconut drops from its parent palm, often on a sandy shore. It may germinate there, or it may be washed into the sea and float for weeks or months until it washes ashore on some distant isle (**FIG. 45-22c**). There, it may germinate, possibly establishing a new coconut colony where none previously existed.

CHECK YOUR LEARNING

Can you . . .

- explain why it is beneficial for seeds to disperse away from their parent plant?
- describe how fruit structures aid in seed dispersal?

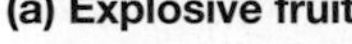
(a) Explosive fruit

(b) Wind-dispersed fruits

(c) Water-dispersed fruit

▲ **FIGURE 45-22 Seed dispersal via air and water (a)** As the fruit of a Himalayan balsam grows, tension builds in its walls. A mature fruit ruptures at the slightest touch, shooting out seeds at speeds up to 10 mph. **(b)** Maple fruits resemble miniature glider–helicopters, whirling away from the tree as they fall. **(c)** This sprouting coconut will soon wash ashore after a long journey at sea.

THINK CRITICALLY Although many fruits can float, why is water seldom their primary method of dispersal?

CASE STUDY REVISITED

Some Like It Hot—and Stinky!

Putrid-smelling flowers like the corpse flower, starfish flower, stinking corpse lily, and dead-horse arum are pollinated by carrion-loving beetles and flies. Chemical analysis shows that these species produce some of the same putrid chemicals that rotting carcasses do. Decaying carcasses also warm up, a by-product of the metabolism of bacteria decomposing the flesh. Heat production in these flowers may be another aspect of corpse mimicry; researchers have found that flies prefer warm, smelly flowers to cool, equally smelly ones.

The corpse flower seems to offer no reward in exchange for pollination. This is not always the case with warm flowers, however. Many species of philodendron, plants found in tropical rain forests and in many people's houses, produce hot flowers (up to 114°F for one species) with mildly pleasant aromas. The flowers of some species serve as warm, cozy orgy rooms for beetles, who crawl into the bloom in the early evening and spend the night mating (**FIG. 45-23**). The beetles use only half as much energy keeping warm in a philodendron bloom as they would in the cool night air outside. The beetles also feed on pollen and other parts of the flower. Philodendrons, therefore, give the beetles real rewards in exchange for their services as pollinators.

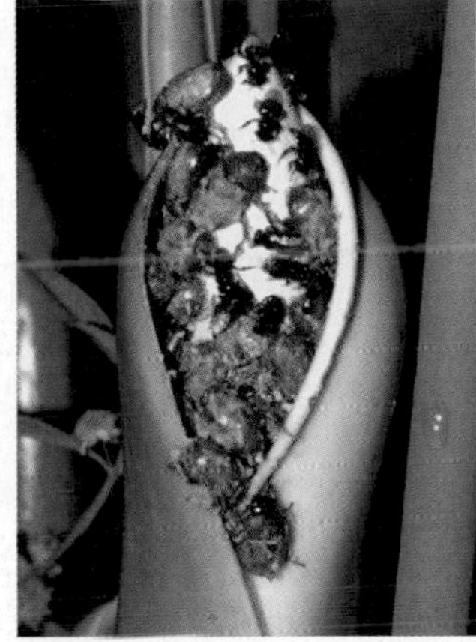

FIGURE 45-23 Hot spot Beetles congregate on a warm philodendron flower cluster that is encased in a petal-like modified leaf.

THINK CRITICALLY Heat-producing flowers are rare, and many are members of evolutionarily ancient groups. Some botanists hypothesize that warmth was an early adaptation that attracted beetle pollinators. Today, most plants do not have warm flowers, but supply their pollinators with a sip of nectar. Develop a hypothesis to explain why the more recently evolved "fast-food" flowers predominate today.

CHAPTER REVIEW

Go to **Mastering Biology** to access the Pearson eText, vocabulary review, practice quizzes, activities, videos, current events, and more.

*Answers to **Think Critically** and **Thinking Through the Concepts** questions can be found in the **Answers** section at the back of the book.*

Summary of Key Concepts

45.1 How Do Plants Reproduce?

The sexual life cycle of plants, called alternation of generations, includes both a diploid sporophyte stage and a haploid gametophyte stage. Meiotic cell division in the diploid sporophyte produces haploid spores, which undergo mitotic cell division to produce the haploid gametophyte. Reproductive cells of the gametophyte differentiate into sperm and eggs, which fuse to produce a diploid zygote. Mitotic cell division of the zygote gives rise to another sporophyte generation. In mosses, the sporophyte develops on the dominant gametophyte and never lives independently. In ferns, both the smaller gametophyte and dominant sporophyte are independent plants. In gymnosperms and angiosperms, the gametophytes are very small and never live independently of the dominant sporophyte.

45.2 What Are the Functions and Structures of Flowers?

Complete flowers consist of sepals, petals, stamens (male reproductive structures), and carpels (female reproductive structures). The sepals form the outer covering of the flower bud. Most petals are brightly colored, attracting pollinators. The stamen consists of a filament that bears an anther, in which pollen develops. The carpel consists of the ovary, in which female gametophytes develop, and a style that bears a sticky stigma to which pollen adheres during pollination.

The male gametophyte of flowering plants is the pollen grain. An immature pollen grain consists of a tube cell and a generative cell that will later divide to produce two sperm cells. The female gametophyte develops within the ovules of the ovary, producing eight nuclei enclosed in seven cells. One of these becomes the egg and another becomes a large central cell with two nuclei.

When a pollen grain lands on a stigma, its tube cell grows through the style to the female gametophyte. The two sperm cells travel down the style within the tube and enter the female gametophyte. One sperm fuses with the egg to form a diploid zygote. The other sperm fuses with the two nuclei of the central cell, producing a triploid cell that will give rise to the endosperm.

45.3 How Do Fruits and Seeds Develop?

The function of fruit is to disperse seeds. A fruit is a ripened ovary, often with contributions from other parts of the flower. Seeds develop from the ovules. The integuments of the ovules form the seed coat. Within the seed coat, a seed contains an embryo and variable amounts of food-storing endosperm.

45.4 How Do Seeds Germinate and Grow?

Seed germination requires warmth and moisture. Energy for germination comes from food stored in the endosperm, which is transferred to the embryo by the cotyledons. Seeds may remain dormant for some time after fruit ripening. To break dormancy and germinate, some seeds require drying, exposure to cold, or disruption of the seed coat. The root emerges first from the germinating seed, absorbing water and nutrients that are transported to the shoot. Dicot shoots form epicotyl or hypocotyl hooks that protect the delicate shoot tip. Monocot shoots are protected by a coleoptile that covers the shoot tip during germination.

45.5 How Do Plants and Their Pollinators Interact?

Plants and their animal pollinators act as agents of natural selection on one another. Flowers attract animals with scent, food such as nectar, and colors and shapes that make them visible and accessible to their pollinators. Some flowers deceive pollinators, attracting insects with food scents or by resembling a mate. Some plants and their pollinators, such as the yucca plant and yucca moth, depend completely on one another.

45.6 How Do Fruits Help to Disperse Seeds?

Clinging fruits adhere to animals. Edible fruits pass through animal digestive tracts, usually without harming the seeds. Explosive fruits shoot seeds away from the parent plant. Lightweight fruits are carried by the wind, while floating fruits are dispersed by water.

Thinking Through the Concepts

Bloom's: Remembering, Understanding

Multiple Choice

1. When a dicot seed sprouts,
   a. a coleoptile protects the emerging shoot.
   b. the shoot develops more rapidly than the root.
   c. the shoot absorbs nutrients directly from the endosperm.
   d. a root cap protects the emerging root.
2. Which of the following is false?
   a. A pollen grain is a male gametophyte.
   b. The tube cell nucleus fuses with the central cell.
   c. Microspores produce pollen by mitotic cell division.
   d. The male gametophyte produces two sperm cells.
3. The female gametophyte in angiosperms
   a. is an independent plant.
   b. has ten cells.
   c. is produced from a megaspore.
   d. contains a diploid egg cell.
4. Which of the following is true?
   a. Moth-pollinated flowers are usually white.
   b. Flowers always benefit their pollinators.
   c. Flowers pollinated by beetles generally smell sweet.
   d. Some flowers guide bees to their centers with patterns visible under infrared light.
5. Seed dormancy
   a. is common in the tropics.
   b. is often maintained by moisture.
   c. can be ended by partial digestion of the seed coat.
   d. is usually broken in the fall or winter.

Fill-in-the-Blank

1. The plant sexual life cycle is called ____________. The diploid generation is called the ____________. Reproductive cells in this stage of the cycle produce spores through ____________ (type of cell division). The spores germinate to produce the haploid generation, called the ____________.
2. In a flowering plant, the male gametophyte is the ____________. It is formed in the ____________ of a flower. Pollination occurs when pollen lands on the ____________ of a flower of the same plant species. The pollen grain grows a tube through the ____________ of the carpel to the ovary at the base of the carpel. The tube enters an ovule through an opening in the ____________ of the ovule.
3. Double fertilization in flowering plants occurs when one sperm cell fuses with an egg to form a diploid cell, the ____________, and a second sperm fuses with the two nuclei of the central cell of the female gametophyte. This triploid cell will divide mitotically to form a food-storage tissue called the ____________. This food is transferred to the developing plant embryo by structures called ____________.
4. The fruit of a flowering plant is formed from the ____________ of a flower, possibly with additional contributions from other flower parts. The seed forms from the ____________. The seed coat develops from the outer covering, or ____________, of this structure.
5. ____________ is the emergence of the embryo from a seed. Usually, the ____________ (structure of the embryo) emerges first. In monocots, the shoot is protected by a sheath called the ____________. In dicots, a(n) ____________ in the hypocotyl or epicotyl protects the apical meristem and developing leaves.

Review Questions

1. Diagram the general plant life cycle. Which stages are haploid and which are diploid? At which stage do gametes form?
2. Diagram a complete flower. Where are the male and female gametophytes formed?
3. Describe the development of the female gametophyte in a flowering plant. How does double fertilization occur?
4. What is a pollen grain, and how is it formed?
5. What are the parts of a seed, and how does each part contribute to the development of a seedling?
6. Describe the characteristics you would expect to find in flowers that are pollinated by the wind, beetles, bees, and hummingbirds, respectively.
7. What is the endosperm? From which cell of the female gametophyte is it derived? Compare the storage site of endosperm in monocot and dicot seeds.
8. Describe three mechanisms that break seed dormancy in different types of seeds. How are these mechanisms related to the typical environment of the plant?
9. How do monocot and dicot seedlings protect the delicate shoot tip during seed germination?
10. Describe three types of fruits and the mechanisms whereby these fruit structures help disperse their seeds.

Applying the Concepts

Bloom's: Applying, Analyzing, Evaluating

1. A friend gives you some seeds to grow in your yard. When you plant them, nothing happens. What might you try to get the seeds to germinate?
2. Charles Darwin once described a flower that produced nectar at the bottom of a tube nearly a foot deep (30 centimeters). He predicted that there must be a moth or other animal with a 30-centimeter-long "tongue" to match. He was right; it's a moth. Such specialization almost certainly means that this particular flower could be pollinated only by that specific moth. What are the advantages and disadvantages of such specialization?

46

Plant Responses to the Environment

A fly lands on a Venus flytrap, dangerously close to bumping the fine hairs (arrows) that trigger the trap to snap shut.

Predatory Plants

A foraging fly cruising over a bog finds itself attracted to the brightly colored leaves and flowery scent of a Venus flytrap. The fly lands on the plant to investigate if there is any nectar to eat, when suddenly the leaves snap shut, trapping the hapless insect. During the next week or so, the leaf excretes enzymes that break down the protein in the fly's body. The resulting nitrogen-containing molecules are absorbed by the plant. Soon afterward, the trap reopens and awaits its next victim. Why would a plant eat an insect?

CASE STUDY

In bogs, plants are often hungry—for nitrogen. They need this nutrient for the same reasons that we and the fly do—to synthesize ATP, nucleic acids, and a variety of proteins. But available nitrogen can be scarce in boggy soils, which tend to be acidic. Acidic conditions discourage the growth of nitrogen-fixing bacteria, which capture atmospheric nitrogen and trap it in ammonia (NH_3) that plants can absorb through their roots. The lack of nitrogen in bogs has selected for plants that can ingest and digest animals, whose bodies provide abundant nitrogen-containing protein. As a result, small-scale drama abounds in bogs.

What brought the fly into the trap's gaping maw? Scientists have identified three types of attractants produced by this predatory plant. From the air, flies can detect a flowery, fruity scent that results from a combination of dozens of volatile compounds released by the plant. In addition, the inner faces of the trapping leaves emit a blue fluorescence when hit by the ultraviolet component of sunlight. This wavelength, though invisible to us, is attractive to insects. Finally, small glands lining the edge of its trap secrete a sweet nectar. To insects, these deadly leaves resemble flowers.

Plants have neither a nervous system nor muscles. So how do predatory plants detect their prey and then move quickly enough to trap them? More generally, how do plants both sense and respond to environmental stimuli that include temperature, gravity, light, water availability, and even munching predators? Read on to learn more about the amazing abilities of plants.

AT A GLANCE

46.1 WHAT ARE SOME MAJOR PLANT HORMONES?

Plant hormones are chemicals secreted by specific cells and transported—often via xylem or phloem—to other parts of the plant body. Various plant hormones influence every aspect of the plant life cycle. Each plant hormone can elicit a variety of responses depending on the type of cell it stimulates, the stage of the plant life cycle, the concentration of the hormone, the presence and concentrations of other hormones, and the species of plant. In the sections that follow, we focus on six important types of plant hormones: auxin, gibberellin, cytokinin, ethylene, abscisic acid, and florigen (**TABLE 46-1**). Of these, only ethylene and abscisic acid are identical among different plant species; the other hormone names represent classes of hormones that are similar, but not identical, in chemical structure and function.

Auxin exerts a wide variety of effects. For example, in germinating seeds it promotes or inhibits elongation in different target cells. Auxin inhibits the sprouting of lateral buds, but stimulates formation of lateral root branches. Auxin also stimulates the differentiation of vascular tissues (xylem and phloem) in newly formed plant parts. In the mature plant, auxin stimulates fruit development, and it inhibits both fruits and leaves from falling prematurely. The primary site of auxin synthesis is the shoot apical meristem (the growing shoot tip). Some auxin is also synthesized in young leaves, as well as in developing seeds and plant embryos, where it influences development and germination.

Several forms of synthetic auxin allow growers to manipulate plants. An auxin called 2,4-D is widely used to kill dicot plants by disrupting the normal balance among auxin and other plant hormones. Synthetic auxin is also used commercially to promote root formation in plant cuttings, to stimulate fruit development, and to delay fruit fall. Auxin was the first plant hormone to be recognized, as described in "Doing Science: Discovering How Plants Grow Toward Light" on page 830.

Gibberellin promotes stem and root elongation by increasing both cell division and cell elongation. Gibberellin also stimulates seed germination, fruit production, and fruit development. This type of hormone is named after the fungus

TABLE 46-1 Some Major Plant Hormones and Their Functions

Hormone	Some Major Effects	Major Site(s) of Synthesis
Auxins	Promote cell elongation in shoots Inhibit growth of lateral buds (apical dominance) Promote root branching Control phototropism and gravitropism in shoots and roots Stimulate vascular tissue development Stimulate fruit development Delay senescence of leaves and fruit	Shoot apical meristem
Gibberellins	Stimulate stem elongation by promoting cell division and cell elongation Stimulate fruit development and seed germination	Shoot apical meristem Plant embryos Young leaves
Cytokinins	Stimulate cell division throughout the plant Stimulate lateral bud sprouting Inhibit formation of branch roots Delay senescence of leaves and flowers	Root apical meristem
Ethylene	Promotes growth of shorter, thicker stems in response to mechanical disturbance Stimulates ripening in some fruits Promotes senescence in leaves Promotes leaf and fruit drop	Throughout the plant, particularly during stress and aging
Abscisic acid	Causes stomata to close Inhibits stem growth and stimulates root growth in response to drought Maintains dormancy in buds and seeds	Throughout the plant
Florigens	Stimulate flowering in response to day length	Mature leaves

Gibberella fujikuroi, from which it was first isolated in the 1930s. The fungus causes "foolish seedling" disease in rice because it produces gibberellin and stimulates these plants to grow exceptionally tall and spindly. The effects of added gibberellin can be dramatic (**FIG. 46-1**). Gibberellin is produced in plant embryos, in the shoot apical meristem, and in young leaves.

Cytokinin participates in many aspects of plant development. It promotes cell division, which is required for the growth of all plant tissues. In roots, cytokinin inhibits the formation of branch roots; in shoots, cytokinin promotes the formation of branches by stimulating cell division in lateral bud meristems. Cytokinin causes nutrients to be transported into plant leaves, which stimulates chlorophyll production and delays aging. The major site of cytokinin synthesis is in the root apical meristem, although some is also generated in the shoot. Commercially, a form of cytokinin is sprayed on cut flowers to keep them fresh.

▲ **FIGURE 46-1 Gibberellin stimulates growth** The cabbage plant on the right was treated with gibberellin for 5 months, while the control plant on the left remained untreated.

Ethylene is an unusual plant hormone because it is a gas. Produced in most plant tissues, it is released in response to a wide range of environmental stimuli and has diverse effects on its target cells. Ethylene serves as a plant stress hormone synthesized by plant tissues to stimulate adaptive changes in response to wind, wounding, flooding, drought, and temperature extremes. Ethylene also causes leaves, flower petals, and fruit to drop off at appropriate times during the year and the plant life cycle. Ethylene is commercially important for its ability to cause certain fruits to ripen.

Like ethylene, **abscisic acid** is also a plant stress hormone that helps plants to withstand unfavorable environmental conditions. For example, abscisic acid causes stomata to close when water is scarce (see Chapter 44). It also promotes root growth and inhibits stem and leaf growth under dry conditions. Abscisic acid in seeds helps to maintain **dormancy**—a state in which growth ceases and metabolic process are greatly slowed—at times when germination would lead to death. Abscisic acid, synthesized in tissues throughout the plant body, was named based on the mistaken hypothesis that it caused leaf abscission (dropping), now known to be caused by ethylene.

Florigen is synthesized in leaves, where it stimulates flowering in response to environmental cues, particularly light. This hormone is a protein that travels in the phloem from leaves to the flower-producing apical meristem.

CHECK YOUR LEARNING

Can you . . .

- list six important types of plant hormones?
- describe the major site of synthesis for each hormone and examples of the hormone's important actions?

46.2 HOW DO HORMONES REGULATE PLANT LIFE CYCLES?

Hormones control nearly all aspects of the plant life cycle by altering gene expression, often in response to stimuli from the environment. In the sections that follow, we summarize some of the effects of the best-studied hormones on plant life cycles. As you read, notice how the same hormone exerts radically different responses depending on the tissue upon which it is acting and the stage of the plant's life cycle.

The Plant Life Cycle Begins with a Seed

The timing of seed germination is crucial to ensure that new seedlings will have adequate water and appropriate temperatures to thrive. Two hormones play major roles in seed germination: abscisic acid and gibberellin.

Abscisic Acid Maintains Seed Dormancy

Seed dormancy is maintained by abscisic acid. The seeds of many temperate plants require a period of cold temperatures to allow germination; this ensures that seedlings do

DOING **Science**

Discovering How Plants Grow Toward Light

Anyone who keeps houseplants knows that they bend toward sunlight streaming through a window. Have you ever wondered how this happens? Charles Darwin did.

Charles and Francis Darwin Determined the Source of the Signal

Charles Darwin was brilliant and insatiably curious. In 1880, Darwin and his son Francis investigated how plants grow toward light. In a series of experiments that exemplify the scientific method, the researchers manipulated the coleoptiles (protective sheaths that surround emerging monocot shoots) of grass. The Darwins illuminated the plants from different angles and observed that the bending occurred in a region just below the coleoptile tip, causing the tip to point toward the light source. When they covered the coleoptile tip with an opaque cap, the coleoptile didn't bend (**FIG. E46-1a**). A clear cap, in contrast, allowed the stem to bend. This bending occurred even if the bending region (below the coleoptile tip) was covered by an opaque sleeve (**FIG. E46-1b**). The Darwins concluded that the tip of the coleoptile perceives the direction of light, although bending occurs farther down. As routinely happens in scientific inquiry, the Darwins' conclusion led to a new hypothesis: The coleoptile tip transmits information about light direction down to the bending region via (in Darwin's words) "a matter which transmits its effects from one part of the plant to another." But what is this "matter"?

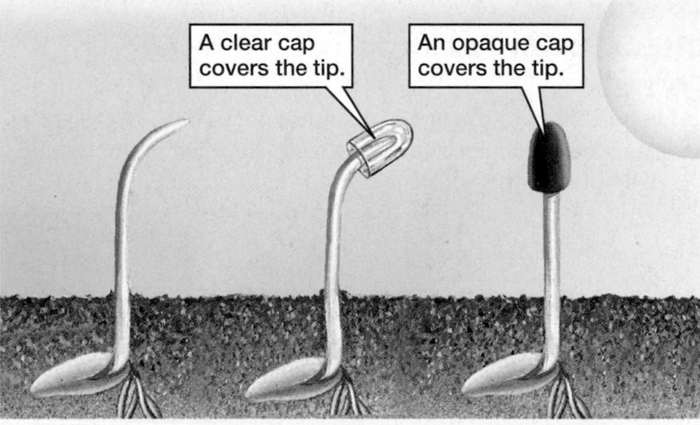

A clear cap covers the tip.
An opaque cap covers the tip.

(a)

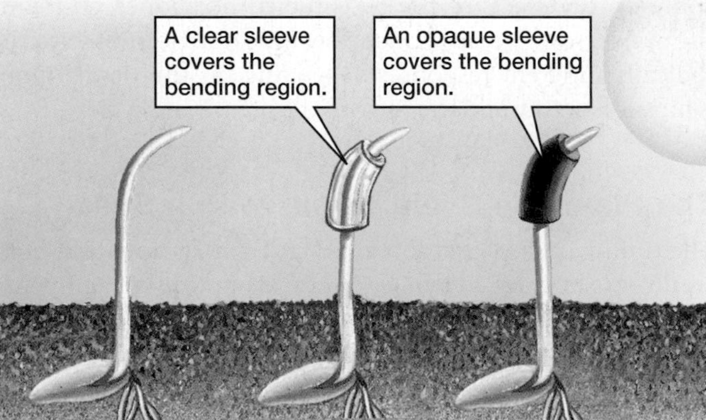

A clear sleeve covers the bending region.
An opaque sleeve covers the bending region.

(b)

▲ **FIGURE E46-1 The shoot tip senses light and sends a signal to the bending region**

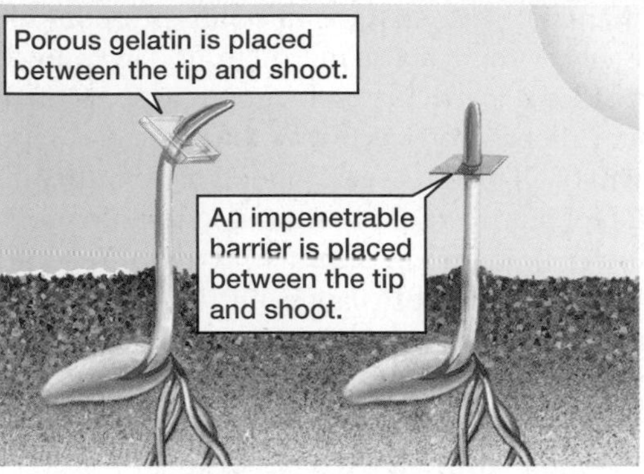

Porous gelatin is placed between the tip and shoot.
An impenetrable barrier is placed between the tip and shoot.

▲ **FIGURE E46-2 A chemical diffuses downward from the coleoptile**

Peter Boysen-Jensen Demonstrated That the Signal Is a Chemical

In 1913, Peter Boysen-Jensen of Denmark cut the tips off coleoptiles and found that the remaining stump neither elongated nor bent toward the light. If he replaced the tip and placed the patched-together coleoptile in the dark, it elongated straight up. In the light, it bent toward the light source. When he inserted a thin layer of gelatin that prevented direct contact but permitted substances to diffuse between the severed tip and the stump, he observed the same elongation and bending. In contrast, an impenetrable barrier eliminated these responses (**FIG. E46-2**). Boysen-Jensen concluded that a chemical produced by the tip diffused through the gelatin and transmitted the phototropic signal.

Frits Went Collected the Chemical and Named It Auxin

In 1926, Frits Went, working in the Netherlands, placed the tips of oat coleoptiles on agar, allowing the unidentified chemical to migrate into this gelatinous material (**FIG. E46-3**). He then cut up the agar and placed small pieces on the tops of coleoptile stumps in darkness. A piece of agar placed squarely atop the stump caused it to elongate straight up; all

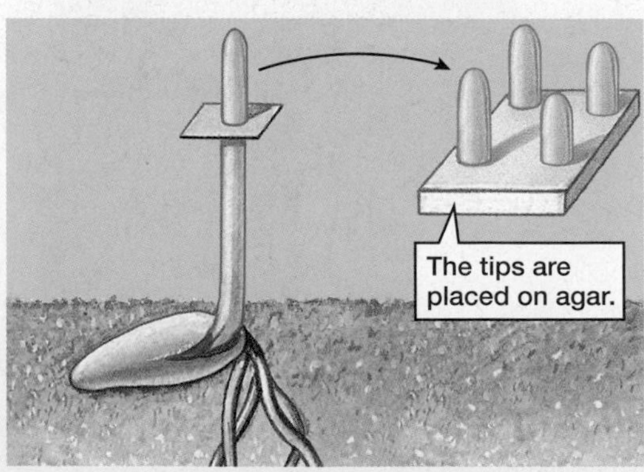

The tips are placed on agar.

▲ **FIGURE E46-3 The chemical accumulates in agar**

of the stump cells received equal amounts of the chemical and elongated at the same rate. When Went placed agar on only one side of the cut stump, the stump bent away from the side with agar (**FIG. E46-4**). Confirming the hypothesis that a diffusing chemical stimulated elongation, he named it "auxin," from the Greek word meaning "to increase."

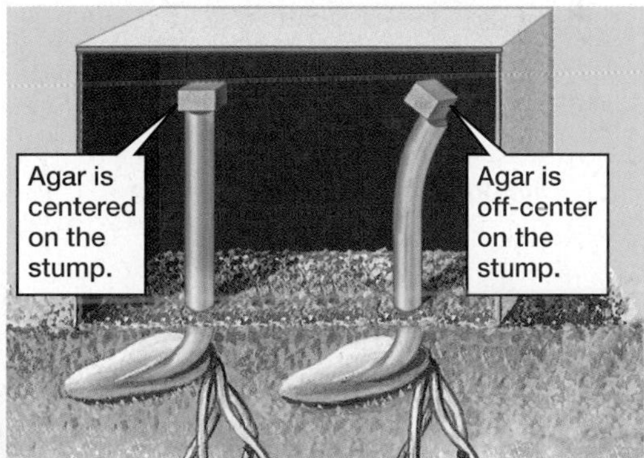

▲ **FIGURE E46-4 The chemical stimulates growth**

Kenneth Thimann Determined the Chemical Structure of Auxin

Working at Caltech in the early 1930s, Kenneth Thimann purified auxin and determined its chemical structure. He and other researchers then determined which features of the molecule were important for its action in plants. This knowledge led to the production of synthetic auxins, such as the herbicide 2,4-D, developed in 1946.

The discovery and characterization of auxin spanned two continents over a 50-year period. It relied on clear communication among scientists, and illustrates how each new scientific discovery leads to more questions. Research on auxin continues throughout the world today, steadily opening new lines of inquiry.

THINK CRITICALLY Explain how the Darwins' experiments using caps and sleeves on the coleoptile followed the scientific method (see Chapter 1) by providing an *observation*, *question*, *hypothesis*, *prediction*, *experiment*, and *conclusion* for each experiment. What were the control groups, and what did each group control for?

not emerge until winter is over. In such plants, the extended cold of winter gradually reduces the amount of abscisic acid within the seed, preparing the plant to germinate during the warming days of spring.

In deserts, where lack of water is the greatest threat to seedling survival, some plants produce seeds with high levels of abscisic acid in their coats. It requires a hard rain to wash the hormone away, allowing germination only when enough moisture is available for the plant to develop and complete its life cycle. Abundant winter rains may trigger germination in desert seeds that have remained dormant for years, briefly carpeting some North American deserts with wildflowers (**FIG. 46-2**). In contrast, certain plants—particularly those of grasslands, chaparral, and some types of forests—require fire to germinate, as described in "Earth Watch: Where There's Smoke, There's Germination" on page 833.

Gibberellin Stimulates Seed Germination

Germination occurs as levels of abscisic acid fall and the embryo produces more gibberellin. Gibberellin activates genes that code for enzymes that break down the stored starch reserves of the endosperm. This releases sugar that is used by the developing embryo to provide both energy and the carbon atoms needed to synthesize organic molecules.

▲ **FIGURE 46-2 A desert in bloom** After abundant rainfall, annual desert plants often germinate in large numbers, grow rapidly, and carpet the desert floor with blossoms, as shown here in Picacho Peak State Park, Arizona.

Auxin Controls the Orientation of the Sprouting Seedling

The growing embryo emerges from its seed coat, and the new seedling responds to light and gravity by directing its root downward and its shoot upward. These orienting movements are called **tropisms**, movements toward or away from specific environmental stimuli. Tropisms occur when a stimulus enhances or inhibits cell elongation without necessarily influencing the rate of cell division. Auxin is important for both **gravitropism** (a directional response to gravity) and

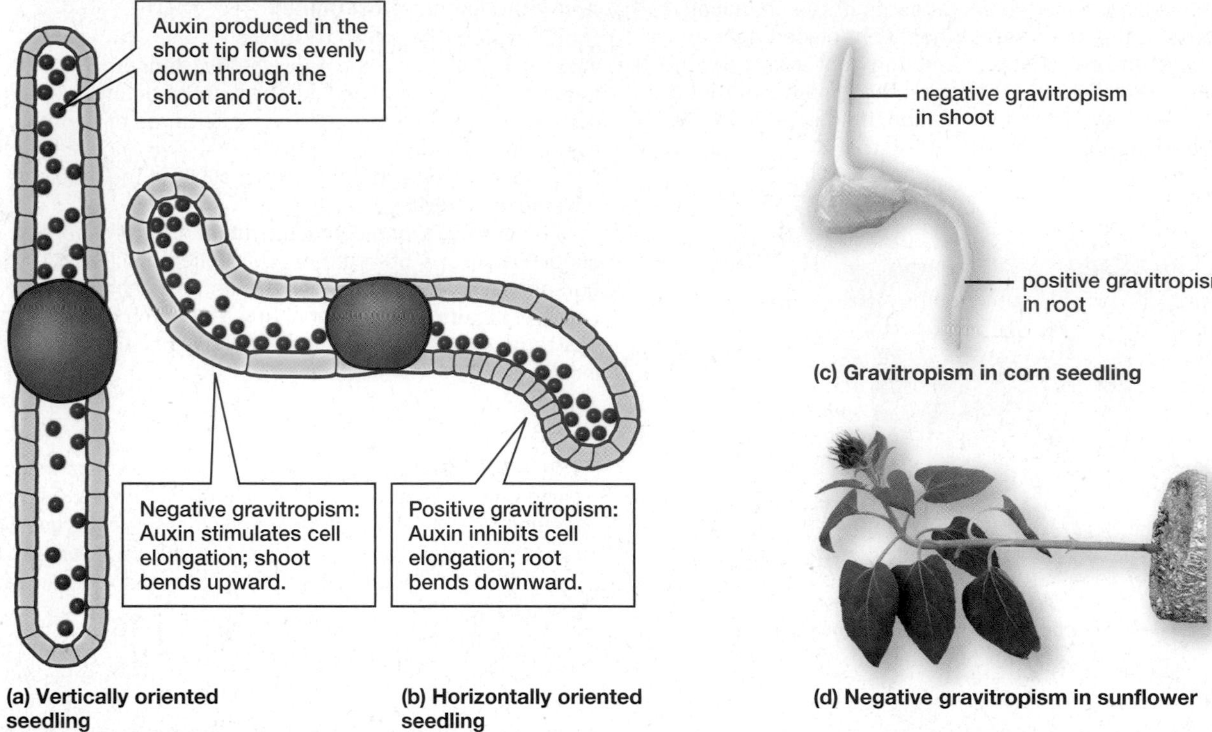
Auxin produced in the shoot tip flows evenly down through the shoot and root.
Negative gravitropism: Auxin stimulates cell elongation; shoot bends upward.
Positive gravitropism: Auxin inhibits cell elongation; root bends downward.
negative gravitropism in shoot
positive gravitropism in root
(c) Gravitropism in corn seedling
(d) Negative gravitropism in sunflower
(a) Vertically oriented seedling
(b) Horizontally oriented seedling

▲ **FIGURE 46-3 Auxin causes gravitropism in shoots and roots**

phototropism (a directional response to light) in shoots and roots.

The Shoot and Root Respond Oppositely to Gravity

Auxin is synthesized primarily in the shoot apical meristem, from which it travels down the shoot and into the root. If the seedling is oriented with the shoot upward and root downward, auxin is distributed evenly on either side of the stem or root, so they elongate straight up or down, respectively (**FIG. 46-3a**). But if the seedling is on its side, auxin accumulates on the lower side of both the shoot and root, inhibiting elongation in root cells but stimulating elongation in shoot cells. As the cells on the lower part of the root are inhibited, the root bends in the direction of gravity's pull, a response called *positive gravitropism* (**FIGS. 46-3b, c**). In contrast, increased auxin in the lower side of the shoot stimulates these shoot cells to elongate faster, causing the shoot to bend away from gravity, a response called *negative gravitropism* (**FIGS. 46-3b, d**).

How do plants sense gravity? Specialized starch-filled plastids called *statoliths* play a role in gravitropism. Statoliths are found in endodermal cells of the shoot and in the root cap. They are denser than the surrounding cytosol, so they settle into the lower part of a cell (**FIG. 46-4**). Auxin accumulates on the side of a cell with the greatest density of statoliths.

The Shoot and Root Respond Oppositely to Light

Auxin also causes phototropism, a directional response to light. Photoreceptor molecules in roots and shoots sense light and cause auxin to accumulate in the side away from the light. As in gravitropism, auxin causes shoot cells to elongate, but inhibits the normal elongation of root cells. When illuminated from the side, the shoot bends toward the light

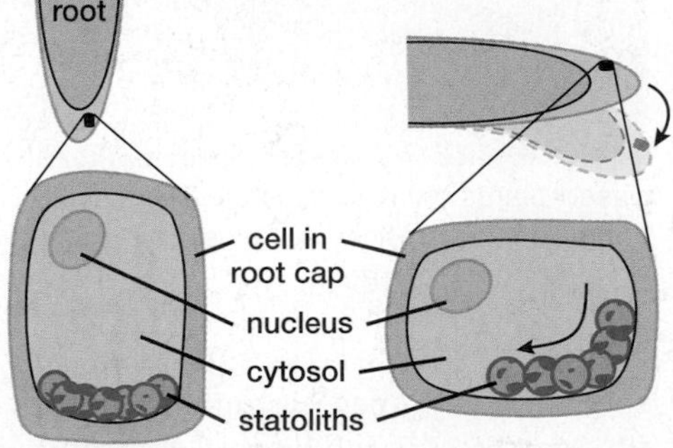
root
cell in root cap
nucleus
cytosol
statoliths

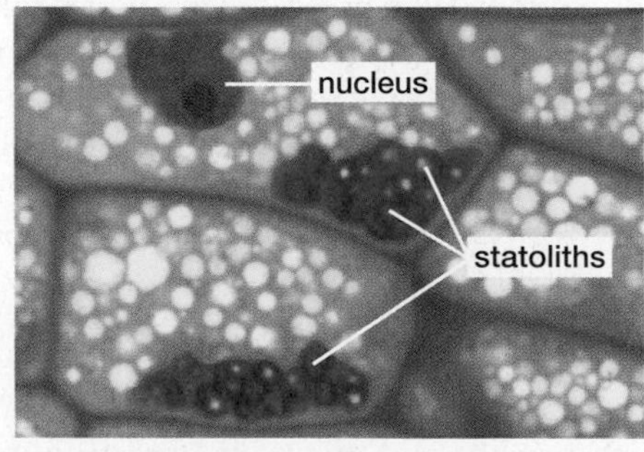
nucleus
statoliths

▲ **FIGURE 46-4 Statoliths respond to gravity** Statoliths fall to the lowest part of the root cap cells, where they stimulate elongation in the direction of gravity.

Earth WATCH Where There's Smoke, There's Germination

Wildfires are necessary to maintain ecosystems such as tallgrass prairie and coastal chaparral and to regenerate aging climax forests (see Chapter 30). Without human intervention, lightning-set fires naturally sweep through temperate forests every few decades, keeping them relatively open and providing a diversity of habitats. But people suppressed fires in the United States for most of the twentieth century. As a result, many forests are now overly dense, with less-healthy trees competing for space, water, nutrients, and light. They are also underlain with branches, leaves, and dead trees that have been accumulating for decades. When fire comes, it burns these crowded forests with an unnatural intensity, killing far more of the trees and destroying seeds and root systems that would otherwise survive. Far from being restored, the forest is devastated. Human suppression of forest fires has backfired.

Land plants have been evolving in the presence of fire throughout their evolutionary history. As a result, many plants are adapted to take advantage of the open space, light, and nutrients that become newly available in the aftermath of a fire. Many species of fire-adapted plants are stimulated to germinate by chemicals in smoke; for example, nitrogen dioxide gas in smoke triggers germination in a variety of chaparral species (**FIG. E46-5**). Australian researchers recently discovered a group of compounds (called "karrikins," from the Aboriginal word for "smoke") that are released by burning cellulose and trigger germination in several smoke-responsive species. Scientists are beginning to discover how smoke influences the plant hormones involved in germination. For example, wild tobacco is an important early colonizer of burned areas. Its seeds show a dramatic decrease in abscisic acid (which inhibits germination) coupled with an increase in sensitivity to gibberellin (which stimulates germination) shortly after exposure to smoke. Seed germination in response to chemicals in smoke ensures that fire-blackened landscapes speedily regain their carpet of green. Recognizing the value of fire and the importance of recurring fire to maintain forest health, forest managers now often allow forest fires to continue burning if they don't endanger human lives or property.

▲ **FIGURE E46-5 A chaparral plant emerges from a charred landscape**

CONSIDER THIS In many places, the U.S. Forest Service is deliberately setting controlled burns in national parks and national forests to reduce accumulated debris, kill invasive plants, and protect against far more devastating accidental wildfires. But controlled burns sometimes get out of control, and their smoke can impact local residents and visitors to scenic areas. How can foresters make decisions that balance the benefits and hazards of controlled burns? Think of an alternative to a controlled burn, and describe some advantages and disadvantages of this alternative approach.

(*positive phototropism*) and the root bends away from it (*negative phototropism*). If you've grown plants on a windowsill, you've seen positive phototropism in action (**FIG. 46-5**). Negative phototropism in roots has been more difficult to study because its effects tend to be overwhelmed by positive gravitropism in roots. To study root phototropism, scientists have grown plants in microgravity aboard space shuttles.

The Growing Plant Emerges and Reaches Upward

As a seed germinates, pressure of the root and shoot against the surrounding soil induces ethylene production. Ethylene causes both the root and shoot to slow their elongation and

▶ **FIGURE 46-5 Radish seedlings show positive phototropism**

THINK CRITICALLY Would you expect gravitropism or phototropism to dominate the response of these seedlings?

become thicker, stronger, and better able to force their way through the soil. In dicots, ethylene also causes the emerging shoot to form a hook, so that the tender new leaves point downward. The bend of the hook protects the leaves as the plant forces its way up through the soil (see Fig. 45-13a).

The apical meristems of the shoot and root divide rapidly, producing new cells that differentiate into stem and root tissues. Auxin produced in the apical meristem travels down the stem and stimulates gibberellin production in the internode regions. Gibberellin causes the internode regions to elongate through both cell division and cell elongation, helping to determine the ultimate height of the plant. In plants such as wheat and rice, shorter varieties have mutations that cause them to produce less gibberellin or be less sensitive to it. Selective breeding of such plants, which are sturdy and resist damage by wind and rain, has allowed substantial increases in production of these crops.

Vine plants have flexible stems that rely on nearby objects for support as they grow upward toward sunlight. Vines display **thigmotropism**, a directional movement or change in growth in response to touch. Most thigmotropic plants wrap specialized *tendrils* (modified leaves or stems) or the entire stem around supporting structures (**FIG. 46-6**). Twining occurs when cell elongation is inhibited on the side of the stem in contact with the object, while elongation continues on the opposite side. Electrical signals in response to touch seem to be involved in this plant behavior.

▲ **FIGURE 46-6 Thigmotropism** Cells on the side of the stem opposite an object elongate more rapidly than cells in contact with the object, causing the stem to twine.

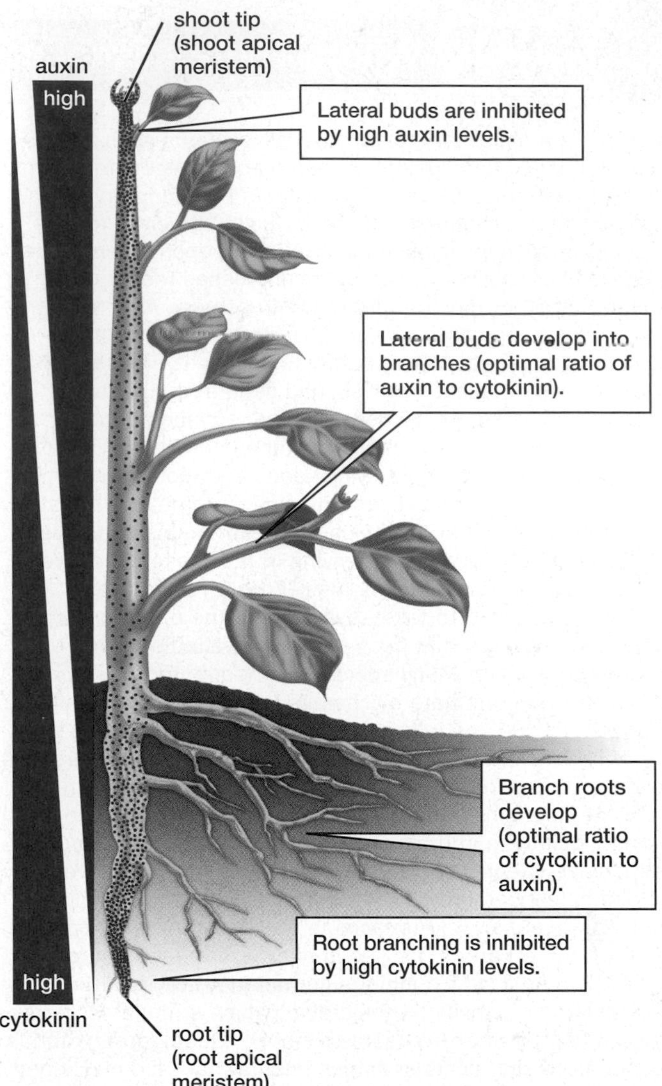
shoot tip
(shoot apical
meristem)
auxin
high
Lateral buds are inhibited
by high auxin levels.
Lateral buds develop into
branches (optimal ratio of
auxin to cytokinin).
Branch roots
develop
(optimal ratio
of cytokinin to
auxin).
Root branching is inhibited
by high cytokinin levels.
high
cytokinin
root tip
(root apical
meristem)

▲ **FIGURE 46-7 The role of auxin and cytokinin in lateral bud sprouting** Auxin (blue) and cytokinin (red) control the sprouting of lateral buds and the development and branching of lateral roots. Auxin from the shoot apical meristem moves downward; cytokinin from the root apical meristem moves upward.

THINK CRITICALLY If you removed a plant's shoot apical meristem and applied auxin to the cut surface, how would you expect the lateral buds to respond? Why?

Auxin and Cytokinin Control Stem and Root Branching

The size of the root and shoot systems of plants must be balanced so that as the shoot grows, the roots also grow enough to provide adequate anchorage, water, and nutrients to support the needs of the stem. Auxin produced by the shoot apical meristem and cytokinin produced by the root apical meristem promote this balance (**FIG. 46-7**).

Branching in Stems Is Inhibited by Auxin and Stimulated by Cytokinin

Auxin produced in the shoot apical meristem travels down the stem and inhibits lateral buds from developing into branches, a phenomenon called **apical dominance** (**FIG. 46-8**, left). Auxin levels decrease with distance from the shoot tip, so lower lateral buds are more likely to sprout and become branches. Pinching off the tip of a growing plant removes the auxin-producing apical meristem, allowing lateral buds to sprout and causing the plant to become bushier (**FIG. 46-8**, right).

Cytokinin produced in the root apical meristem stimulates the development of lateral stem buds into stem branches. Because the lower buds receive more cytokinin and less auxin, they are more likely to form branches than are buds near the apical meristem.

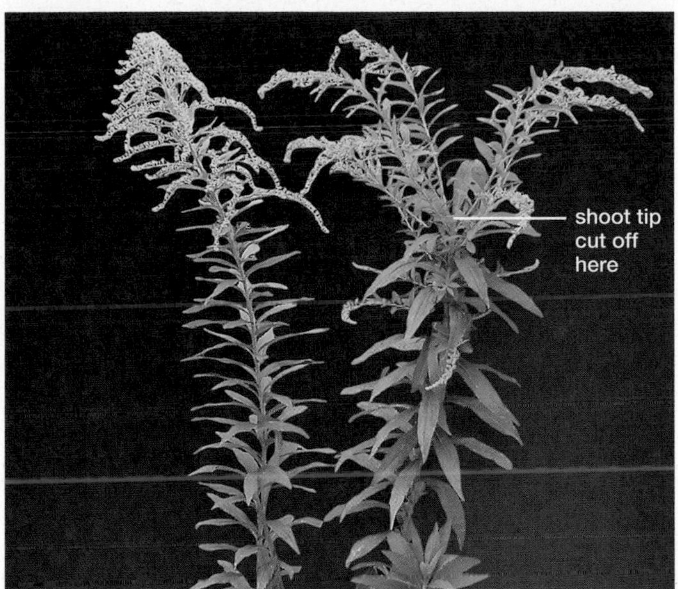

▲ **FIGURE 46-8 Apical dominance** The goldenrod plant on the left has produced a shoot whose meristem is suppressing side branches. The plant on the right has had its apical meristem cut off, allowing lateral buds to develop into branches.

Branching in Roots Is Stimulated by Auxin and Inhibited by Cytokinin

Auxin transported down from the shoot apical meristem stimulates root pericycle cells to divide and form branch roots. Synthetic auxin powder also has this effect; gardeners can produce a new plant by dipping the cut end of a stem into auxin powder, which stimulates the stem to develop roots. In contrast, the cytokinin produced in the root apical meristem inhibits root branching. Roots closer to the root tip receive more cytokinin and less auxin, so their branching is suppressed. Root branching is stimulated closer to the shoot, where it receives more auxin and less cytokinin.

Plants Use Differing Cues to Time Their Flowering

The timing of flowering and seed production must be finely tuned to the environment. In temperate climates, plants must flower early enough that their seeds mature before the killing frosts of late autumn. Depending on the species and how quickly its seeds develop, plants may flower at any time during the growing season from spring through early fall. Although day length is the only precise and consistent indicator of seasonal change, certain plants rely on other signals. Some, including corn, roses, tomatoes, and cucumbers, rely on cues such as temperature and the availability of water; these are described as **day-neutral plants**. Some day-neutral plants flower when they have reached the proper developmental stage after germination. In temperate climates, day-neutral plants that live more than 1 year often require the cold temperatures of winter to intervene between one flowering cycle and the next. Many tropical plants (mangos, for example) are day-neutral; the tropical climate supports reproduction year-round, and day length near the equator remains almost constant.

In many temperate plants, the onset of flowering is stimulated by seasonal changes in day length. Observers categorized such plants as long-day or short-day plants, but subsequent experiments showed that the plants actually respond to the duration of uninterrupted darkness. **Long-day plants** (short-night plants), including iris, lettuce, spinach, and hollyhocks, flower only when uninterrupted darkness is shorter than a specific length of time that varies with the species of plant. **Short-day plants** (long-night plants), such as cocklebums, chrysanthemums, asters, potatoes, and goldenrods, flower when uninterrupted darkness exceeds a species-specific duration (**FIG. 46-9**). For example, cocklebur, a short-day plant, flowers only if uninterrupted darkness lasts more than 8.5 hours, whereas spinach,

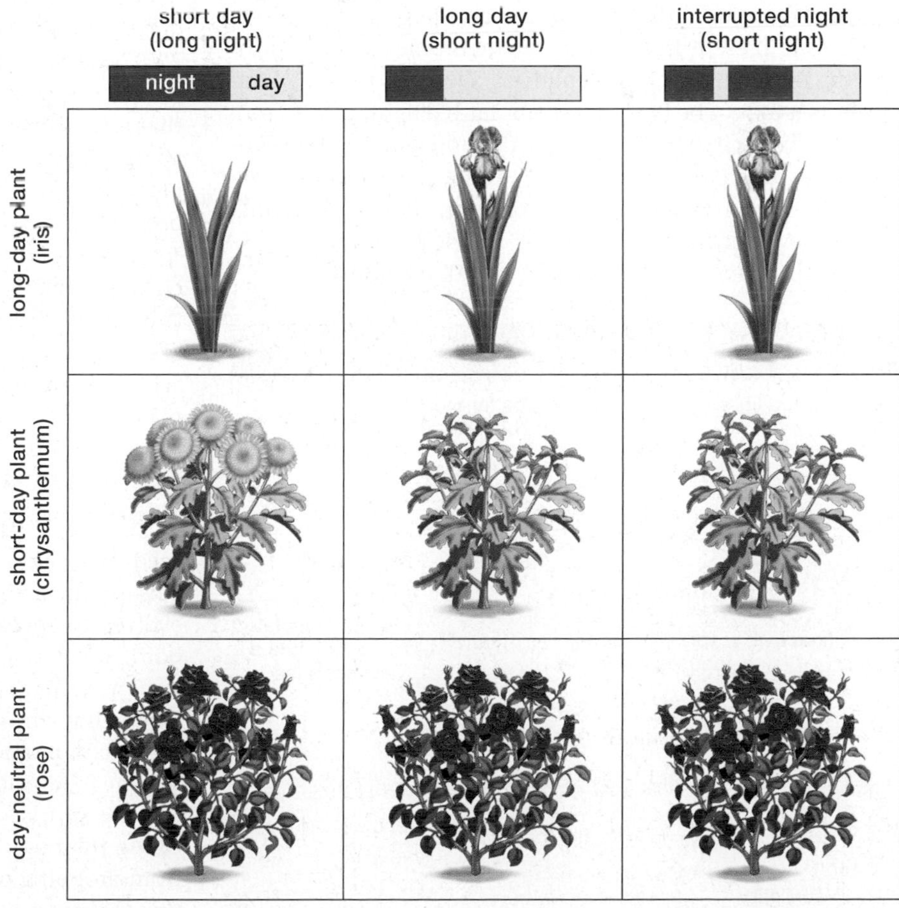

▲ **FIGURE 46-9 The effects of the duration of uninterrupted darkness on flowering**

a long-day plant, flowers only if uninterrupted darkness lasts 10 hours or less; both will flower with 9-hour nights. How do these plants sense the duration of continuous darkness and thus respond to seasonal changes in day length?

Photopigments Called Phytochromes Control Responses to Day Length

Organisms detect light using molecules called photopigments, proteins linked to pigment molecules that absorb specific wavelengths of light. Light energy causes the photopigment to alter its chemical configuration, initiating a series of biochemical changes that cause a response in the organism. The eyes of animals contain photopigments that allow vision, whereas the bodies of plants contain photopigments that capture light. Some plant photopigments allow photosynthesis and phototropisms; others, called **phytochromes** (literally, "plant colors"), detect day length.

Plants measure the duration of darkness using an internal **biological clock** that relies on complex biochemical reactions. If a plant in darkness is exposed to specific wavelengths of light, its phytochrome molecules will "reset" the clock, something like stopping and resetting a stopwatch. Thus, if an 8-hour night is interrupted after 4 hours with a few minutes of light, the plant will respond only to the uninterrupted 4 hours of darkness that occur after the interval of light.

How does this work? A phytochrome occurs in two similar forms—P_r and P_{fr}—that are converted from one to the other by specific wavelengths of light: red and far-red. Both wavelengths are present in sunlight. Phytochrome in the P_r form absorbs red light (r; a wavelength of about 660 nm), which converts it to P_{fr}. The P_{fr} form absorbs far-red light (fr; 730 nm), which converts it to P_r. The P_r is inactive and remains stable unless it is exposed to red light. The P_{fr} is the active form of phytochrome that triggers responses to light, such as flowering. It is, however, unstable, so that over a period of hours in darkness, P_{fr} spontaneously reverts back to P_r (**TABLE 46-2**).

TABLE 46-2 Light and Phytochrome Activity

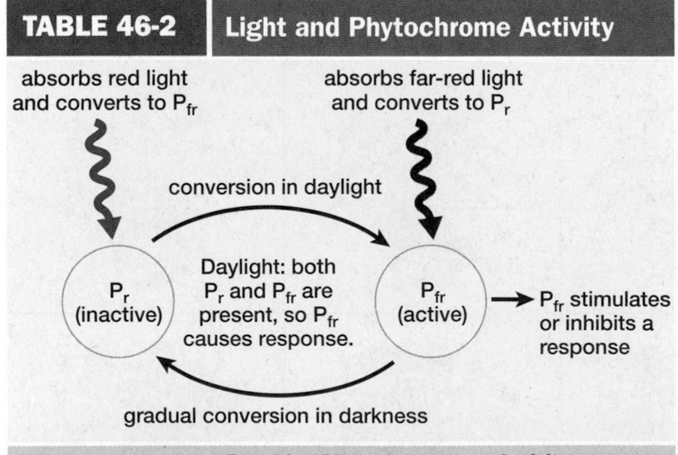

Light	Resulting Phytochrome	Activity
Red	P_{fr}	Active
Far-red	P_r	Inactive
White (sunlight)	$P_r + P_{fr}$	Active
Prolonged darkness	P_r	Inactive

Different types of plants respond in different ways to P_{fr}. For example, P_{fr} stimulates flowering in long-day plants, but inhibits flowering in short-day plants. In the short-day cocklebur, interrupting a 9-hour dark period with just a minute or two of red or white light (which includes red wavelengths) prevents flowering by converting P_r to P_{fr}.

Phytochromes Influence Many Plant Responses

Light is life for plants, which use photosynthesis to trap and store light energy as chemical energy, so it is not surprising that the light-responsive phytochromes influence many aspects of plant development. In addition to flowering, phytochromes help control seed germination, elongation of the shoot as it emerges under the soil, expansion of newly developed leaves, development of chloroplasts, and stem elongation that helps shaded plants reach the sunlight.

Florigen Stimulates Flowering in Response to Light Cues

In experiments that began in the 1930s, plant physiologists found that flowering in long-day or short-day plants can be induced by exposing just a single leaf to the appropriate schedule of light and darkness, even if the apical meristem that produces the flower receives no light. Such experiments supported the hypothesis that an unknown factor must be produced in leaves and transported to the apical meristem, where it promotes flowering. This hypothetical flowering hormone was dubbed "florigen" (derived from *flor*, "flower," and *gen*, "to produce"). But the florigen molecule itself was not identified until 2007, when researchers found a gene called *FT*, which is active in leaves but not in the apical meristem. *FT* codes for a protein that is transported from the leaves through the phloem to the apical meristem. The pathway was verified in several different plant species, including both monocots and dicots, and the FT protein was finally confirmed as the elusive florigen.

Since plants measure light-dark cycles using phytochromes and produce florigen in response to these cycles, this provides strong evidence that phytochromes control the expression of the *FT* gene that codes for florigen synthesis. Scientists are now working to unravel the complex mechanisms by which this occurs.

Hormones Coordinate the Development and Ripening of Fruits and Seeds

Seed maturation is closely coordinated with fruit ripening because most seeds are dispersed in fruits. Fruits that have evolved to appeal to animal seed dispersers are often green, hard, and bitter (or even poisonous) when they are unripe and the seeds are immature. Ripening is influenced by auxin, gibberellin, and (often) ethylene.

Auxin promotes the growth of the plant ovary, causing it to store food materials and to develop into a mature fruit. Synthetic auxin is sometimes sprayed on cultivated fruits to increase their size and sweetness and also to prevent the fruit from dropping prematurely. Gibberellin, which also contributes to fruit growth, is commercially applied to grapes, which grow larger and form looser clusters as a result (**FIG. 46-10**).

▲ **FIGURE 46-10 Commercial uses for plant hormones** The grapes on the left were sprayed with gibberellin, producing looser clusters of larger grapes. Those on the right have developed naturally, forming smaller grapes in tighter clusters.

THINK CRITICALLY Agricultural biotechnologists have developed genetically modified tomato plants in which ethylene production is blocked. Why might such a plant be valuable to tomato growers?

The effects of ethylene on ripening differ among different fruits. Some do not require ethylene for ripening, including strawberries, grapes, cherries, and citrus fruits. In other fruits such as bananas, apples, peaches, pears, avocados, tomatoes, and watermelons, ripening is accompanied by a burst of ethylene production.

Ethylene stimulates the production of enzymes that break down chlorophyll, revealing yellow, orange, or red pigments that were already present or are newly synthesized. Ethylene also initiates synthesis of enzymes that convert starches to sugars, sweetening the fruit, and enzymes that break down pectin (a constituent of cell walls), making the fruit softer. Brightly colored, sweet, soft fruits attract animals (**FIG. 46-11**), just as they appeal to people perusing the produce section of the supermarket.

▲ **FIGURE 46-11 Ripe fruit becomes attractive to animal seed dispersers** A desert tortoise is attracted to the ripe and tasty fruit of a prickly pear cactus.

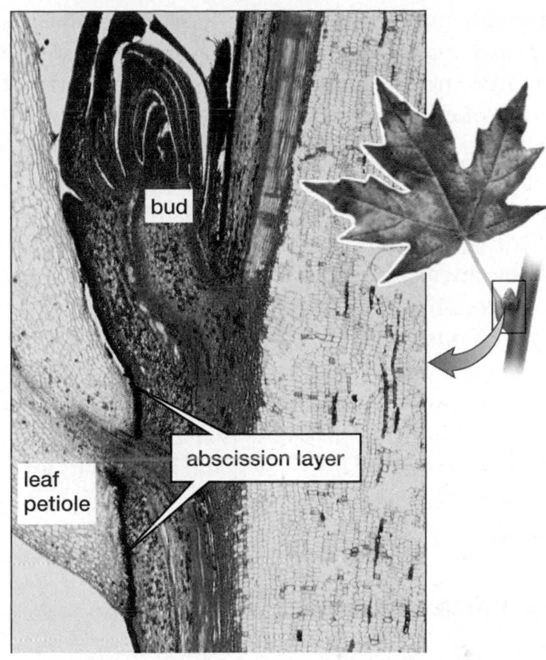

▲ **FIGURE 46-12 The abscission layer** This light micrograph of the base of a maple leaf clearly shows the dark abscission layer. A lateral bud is visible above the senescing leaf petiole.

The discovery of the role of ethylene in ripening has revolutionized modern fruit marketing. Bananas grown in Central America can be picked green and tough, then shipped to North American markets, where they are ripened with ethylene. Green tomatoes, which withstand transport better than ripe ones, may also be ripened at central warehouses using ethylene. But after purchasing ripe fruits, we want the ripening to stop, so our bananas don't become black and our tomatoes mushy as we store them. Special produce bags are now available that absorb ethylene released from the fruit, helping to keep it fresh.

Pectin breakdown in response to ethylene not only softens fruit, but also allows the ripening fruit to fall from the plant by weakening a layer of cells, called the **abscission layer**, located where the fruit stalk joins the stem (**FIG. 46-12**). Having dropped its fruit, the parent plant has completed its reproductive cycle. Annual plants die soon afterward, whereas perennials in temperate climates prepare for the coming winter.

Senescence and Dormancy Prepare the Plant for Winter

In autumn, under the influence of shortening days and falling temperatures, plants undergo **senescence**, a genetically programmed series of events in which certain parts of the plant (such as leaves) die. Ethylene production increases, and the production of auxin and cytokinin (which delay senescence) declines. During senescence, starch and chlorophyll in the leaf are broken down into simpler molecules that are transported to the stem and roots for winter storage.

Although many hormones influence senescence, ethylene is particularly important. As in ripening fruits, ethylene promotes the breakdown of chlorophyll and triggers the production of enzymes that weaken a leaf's abscission layer, located where the leaf stalk joins the stem, allowing the leaf to fall (see Fig. 46-12). Senescence may also occur in response to environmental stresses that cause a rapid increase in ethylene production. This increase occurs if the leaves are infected by microorganisms or exposed to temperature extremes or drought—as you may observe if you forget to water your houseplants.

In the autumn, new buds become tightly wrapped and dormant rather than developing into leaves or branches. Dormancy in buds, as in seeds, is maintained by abscisic acid. The plant's metabolism slows, and it enters its long winter sleep, awaiting signals of warmth, moisture, and longer spring days before awakening once again.

CHECK YOUR LEARNING

Can you . . .

- explain the roles of plant hormones in seed dormancy, seed germination, and directional growth of the emerging root and shoot?
- explain thigmotropism?
- describe how plants balance root and shoot growth?
- explain how some plants sense and respond to light and darkness?
- explain how hormones coordinate the maturation of seeds and fruit and control fruit ripening?
- describe how hormones control senescence?

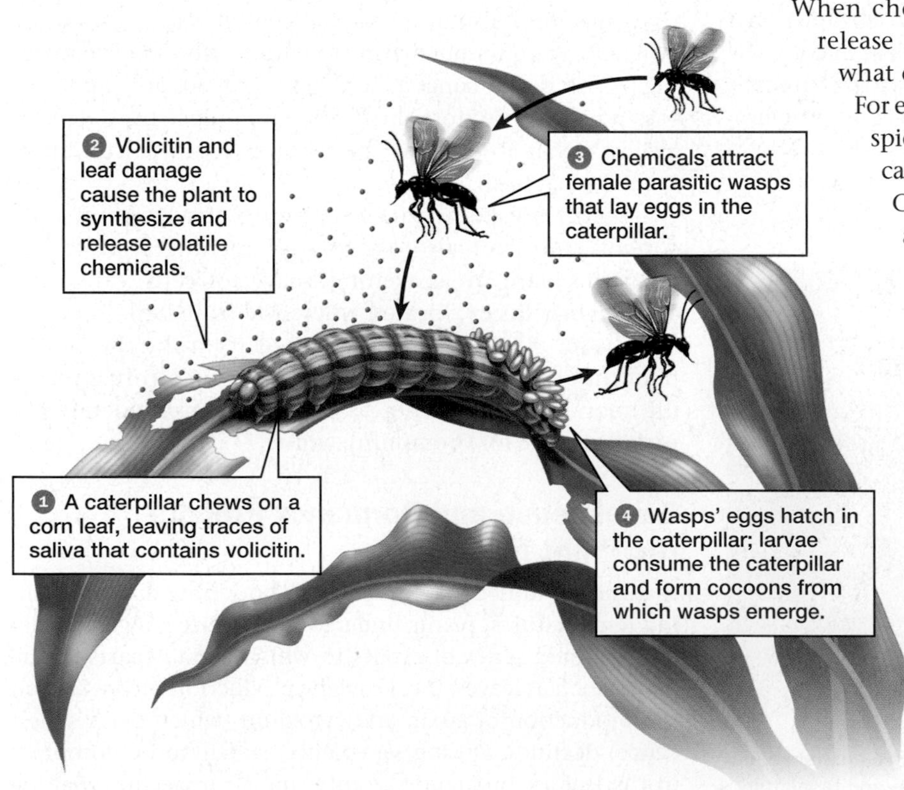

1 A caterpillar chews on a corn leaf, leaving traces of saliva that contains volicitin.
2 Volicitin and leaf damage cause the plant to synthesize and release volatile chemicals.
3 Chemicals attract female parasitic wasps that lay eggs in the caterpillar.
4 Wasps' eggs hatch in the caterpillar; larvae consume the caterpillar and form cocoons from which wasps emerge.

▲ **FIGURE 46-13 A chemical cry for help**

46.3 HOW DO PLANTS COMMUNICATE, DEFEND THEMSELVES, AND CAPTURE PREY?

Flowering plants have been coevolving with animals and microorganisms for well over 100 million years. In addition to communicating with their insect pollinators (see Chapter 45), plants have evolved surprisingly sophisticated behaviors under natural selection imposed by parasitism, predation, and limited nutrients.

CASE STUDY CONTINUED

Predatory Plants

Although insects are major plant predators, carnivorous plants turn the tables on them, enticing insects with sugary secretions and bright colors. Scientists have found that the color red, such as found on the inner faces of the Venus flytrap leaf, enhances the attractiveness of certain carnivorous plants to insects. Sundews attract their insect prey using red hairs and an appealing scent.

Very few plants "eat" insects as Venus flytraps and sundews do. Do most plants instead stand stoically and allow themselves to be chewed on by insects or invaded by pathogens—or can they defend themselves?

Plants May Summon Insect "Bodyguards" When Attacked

When chewed on by hungry insects, many plants release volatile chemicals into the air, producing what could be considered a chemical cry for help. For example, lima bean plants that are attacked by spider mites release a chemical that attracts a carnivorous mite that preys on the spider mites. Corn leaves attacked by caterpillars also release a chemical alarm signal (**FIG. 46-13**). The corn alarm signal is stimulated by a compound called volicitin in the saliva of the caterpillar; damage from other causes (such as hail) will not stimulate this response. The alarm signal attracts parasitic female wasps that lay their eggs in the caterpillar's body. Upon hatching, the wasp larvae consume the caterpillar from the inside, then form cocoons from which adult wasps emerge.

Wild tobacco plants munched on by hornworms (hawk moth caterpillars) respond by releasing chemicals that differ depending on the time of attack. During the day, when parasitic wasps are active and seeking hornworm caterpillars, the tobacco plants produce wasp-attracting chemicals. At night, when adult hawk moths are active, the tobacco plants release chemicals that deter the moths from laying eggs on them.

Attacked Plants May Defend Themselves

Leaf damage from insects causes many plants to produce a signaling molecule that travels through the plant body, where it stimulates responses that make the plant more distasteful, difficult to eat, or toxic. For example, when tobacco leaves are damaged, they produce more nicotine (a poison used commercially as an insecticide). Radish plants attacked by caterpillars produce a bitter-tasting chemical and grow more spiny hairs on their leaves. Seeds of these damaged plants produce seedlings with enhanced bitterness and spines, indicating that the damaged parents incorporate a chemical signal into their seeds that causes specific genes to be activated, triggering the development of defenses in their offspring.

Plants also have a very effective immune system. They produce salicylic acid (aspirin) as a hormone, and one of its many roles is to help plants defend themselves. Salicylic acid production is increased in response to attack by infectious microorganisms. It then stimulates both plasmodesmata (channels that interconnect cells) and stomata to close, limiting the ability of the pathogen to enter and spread within the plant. Salicylic acid also activates genes coding for proteins in metabolic pathways that help the plants resist their current infection and future infections as well.

Wounded Plants Warn Their Neighbors

If plants under siege can summon help, might neighboring plants also get the message? Evidence is accumulating that some healthy plants sense chemicals released by nearby members of their species that are infected by microbes such as viruses. Tobacco plants infected with a virus produce large amounts of salicylic acid, which helps boost their immune responses. Some of the salicylic acid produced by the infected plants is converted to methyl salicylate (wintergreen oil), a highly volatile compound. Airborne methyl salicylate released by infected plants activates a gene in healthy neighboring plants that helps them resist the virus. Communication among plants may also cross species lines. Wild tobacco plants exposed to damaged sagebrush leaves produce stronger and more rapid defenses when attacked by the tobacco hornworm than unexposed tobacco plants do.

Have You Ever Wondered …

What People Took for Pain Before Aspirin Was Invented?

Hippocrates, a Greek physician born in 460 B.C., wrote of a powder made from the bark and leaves of willow trees that relieved headache, pain, and fever. Native Americans independently discovered the pain-relieving power of willow bark, which they chewed or made into a tea. In the early 1800s, researchers isolated the active compound, salicylic acid—allowing chemists to synthesize aspirin (acetylsalicylic acid), first marketed by the Bayer Company in 1899. Salicylic acid, a plant hormone, relieves pain in people by inhibiting an enzyme involved in synthesizing prostaglandins (hormone-like substances that sensitize nerve endings to pain). Today, this molecule—"invented" by plants to protect themselves from attack by herbivores—is one of the most widely consumed human medications in the world. As one researcher observed, "Plants can't run away and they can't make noises. But they are wonderful chemists."

Sensitive Plants React to Touch

If you touch a sensitive plant such as a *Mimosa* (**FIG. 46-14**), the rows of leaflets along each side of the petiole immediately fold inward, and the petiole droops. No one is sure why

(a) Before the leaves are touched

(b) After the leaves are touched

▲ **FIGURE 46-14 A rapid response to touch (a)** A leaf of the sensitive plant (*Mimosa*) consists of an array of leaflets emerging from a central stalk, which is attached to the main stem by a short petiole. **(b)** Touching causes the leaflets to fold together.

the response evolved, but botanists hypothesize that this rapid movement may surprise and discourage leaf-eating insects. It is stimulated by electrical signals in *motor cells* at the base of each leaflet and where the petiole joins the stem. The signal allows potassium ions, normally maintained at a high internal concentration, to diffuse rapidly out of the motor cells. As water follows by osmosis, the motor cells shrink quickly, pulling the leaflets together and causing the petioles to droop.

Carnivorous Sundews and Bladderworts Respond Rapidly to Prey

In nitrogen-deficient bogs like the one where the Venus flytrap of the Case Study grows, other plants have also evolved to supplement their diet with insects. The sundew, for example, is named for the sweet gluey droplets—resembling dewdrops in sunlight—secreted from the tips of hairs projecting from its rounded leaves. The leaves attract insects that soon find themselves struggling helplessly in a sticky mass (**FIG. 46-15**). The vibrations from an insect's thrashing to escape cause an electric current in the hairs, stimulating them to curl around the insect. The struggle also stimulates the sundew to secrete a cocktail of digestive enzymes into the enveloping goo, where they rapidly break down the insect's body. The sundew's leaves then absorb the liberated nitrogen compounds.

Beneath the bog's surface, bladderworts dangle hundreds of water-filled chambers into the water. Each chamber is sealed by a watertight trapdoor whose lower edge is fringed with bristles. If a water flea (related to shrimp, but barely visible to us) bumps a bladder, it is sucked into the chamber within 1/60th of a second (**FIG. 46-16**). How? Cells lining the bladder actively transport ions out of bladder water and into the pond. Water follows the ions out of the bladder by osmosis, reducing pressure inside. The reduced pressure causes the bladder walls to be drawn inward under tension, setting the trap. If a tiny aquatic organism bumps into the bristles that surround the trapdoor, the bristles push the door inward and break the seal around it. In a split second, the bladder walls spring outward to their resting position, suddenly increasing the volume of the bladder and sucking in the prey. Enzymes secreted into the chamber gradually digest the organism, releasing nutrients that the bladderwort absorbs.

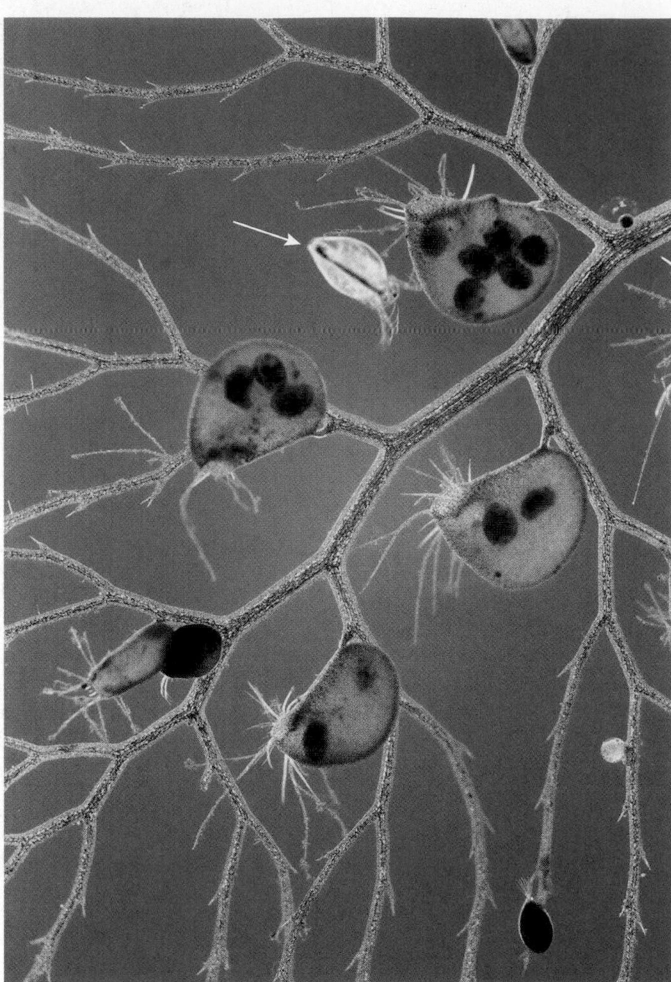

▲ **FIGURE 46-16 The bladderwort snares tiny aquatic organisms** Bladderworts are studded with prey-catching bladders. A tiny water flea (arrow) has settled on one of the bladders, unaware of the danger. If the water flea triggers the bladder, it will be sucked inside and digested.

▲ **FIGURE 46-15 A sundew with its insect prey** The leaf of this sundew uses sweet, red, sticky hairs to entice and capture a lacewing insect.

CHECK YOUR LEARNING

Can you . . .

- explain how plants that are attacked by predators recruit insects to help with defense?
- explain how plants that are attacked by predators and parasites defend themselves with chemicals and hormones?
- describe how wounded plants communicate with other plants?
- describe why and how sensitive plants, sundews, and bladderworts produce their rapid movements?

CASE STUDY REVISITED

Predatory Plants

Prey capture by Venus flytraps is an amazing process. First, the flytrap must determine whether the object invading its leaf is alive, so it doesn't waste energy closing on a piece of debris. Three to five tiny trigger hairs on each inside face of the leaf (see the chapter-opening photo) detect movement as the insect seeks nectar. The leaf will only close if the insect bumps into the hairs more than once in 30 seconds, triggering an action potential—literally a "hair-trigger" response!

But how do the leaves close quickly enough to trap a fly? The trap remains open when each half of the leaf maintains a gradient of solutes between its inner and outer cell layers, causing it to assume a slightly convex shape. The action potential triggered by a trapped insect opens aquaporins (water channels) that allow water to move rapidly by osmosis from the inner to the outer layers, causing the leaf to snap into a concave shape that envelops the insect. But a small gap remains between the two halves of the leaf. Researchers hypothesize that this allows very small insects to escape, but entraps insects large enough to be worth the time and energy to digest (**FIG. 46-17**). More struggling by the trapped prey causes the leaves to seal tightly shut. The inner leaf surfaces then secrete digestive enzymes, and for the following week or so this temporary stomach digests the insect and the plant feasts on the liberated nutrients.

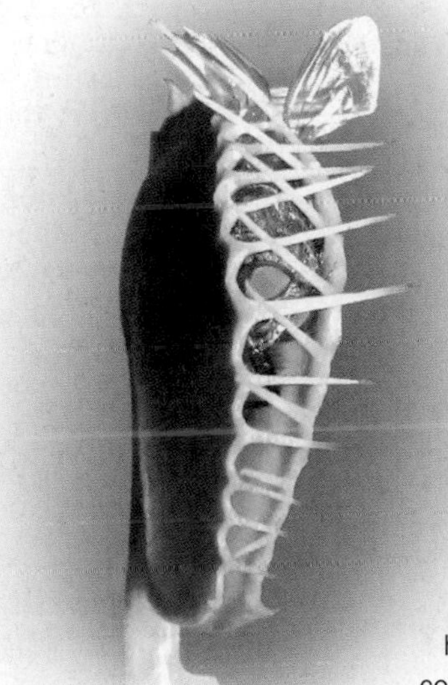

▲ **FIGURE 46-17 Success!**

In 1875, Charles Darwin described the Venus flytrap as "one of the most wonderful plants in the world," and in 2005 North Carolina designated the Venus flytrap as the official state carnivorous plant. Unfortunately, these plants are rare, growing wild only in certain wetlands of North and South Carolina. Poachers, often dressed in camouflage clothing, dig them from private lands as well as protected reserves to sell illegally. Development also threatens these helpless carnivores; about 70% of their original habitat has now been transformed into landscapes such as golf courses, parking lots, and suburban sprawl.

THINK CRITICALLY Many wetlands in the United States are threatened by runoff water from nearby farms, which may be heavily fertilized or rich in animal wastes. Carnivorous plants thrive in bogs partly because other species that can't snare nitrogen-rich food cannot compete with them. Explain why runoff from farms poses a threat to carnivorous plants in nearby wetlands.

CHAPTER REVIEW

Go to **Mastering Biology** to access the Pearson eText, vocabulary review, practice quizzes, activities, videos, current events, and more.

*Answers to **Think Critically** and **Thinking Through the Concepts** questions can be found in the **Answers** section at the back of the book.*

Summary of Key Concepts

46.1 What Are Some Major Plant Hormones?

Plants have evolved the ability to sense and react to environmental stimuli including touch, gravity, moisture, light, and day length. Plant hormones, secreted in response to changes in the environment, influence every stage of the plant life cycle. Six major types of plant hormones that are important in the flowering plant life cycle are auxins, gibberellins, cytokinins, ethylene, abscisic acid, and florigens (see Table 46-1).

46.2 How Do Hormones Regulate Plant Life Cycles?

Hormones regulate plant growth and development in response to environmental stimuli. Dormancy in seeds is enforced by abscisic acid. Falling levels of abscisic acid and rising levels of gibberellin trigger germination. As the seedling grows, auxin stimulates positive phototropism and negative gravitropism in the shoot and negative phototropism and positive gravitropism in the root. Statoliths help plant shoot and root tip cells detect gravity and influence auxin accumulation toward the pull of gravity. Gibberellin stimulates the stem to elongate. Ethylene slows elongation and promotes thickening of both roots and shoots as the plant pushes through soil.

Branching in stems and roots is controlled by auxin and cytokinin, which keep these two systems in balance. Auxin, synthesized in the shoot apical meristem, inhibits stem branching but stimulates root branching. Cytokinin, synthesized in the root apical meristem, inhibits root branching, but stimulates stem branching.

Plants detect day length using photopigments called phytochromes, which respond to specific wavelengths of light by transitioning between active (P_{fr}) and inactive (P_r) forms. Active phytochrome is involved in flowering in long- and short-day plants, seed germination, shoot elongation, leaf expansion, and chloroplast synthesis. In response to the appropriate day length and under the control of phytochromes, the flowering hormone florigen is produced in plant leaves and travels to the flower bud in the apical meristem, where it initiates flowering.

Fruit and growth and development are influenced by auxin and gibberellin, and, in some fruits, ethylene triggers ripening. During senescence, fruit and leaves fall in response to ethylene, which weakens their abscission layers. Buds become dormant as a result of high concentrations of abscisic acid.

46.3 How Do Plants Communicate, Defend Themselves, and Capture Prey?

Some plants under attack by insects release volatile chemicals that attract other insects that prey on or parasitize the plant predators. Damaged plants use internal chemical signals to strengthen their own defenses and, sometimes, to produce better-defended offspring. Chemicals released by injured or infected plants may also stimulate defenses in neighboring plants.

Sensitive plant leaves fold together when touch sensors in the leaves generate electrical signals that cause specialized cells to rapidly lose water. The sundew traps and digests insects in sticky droplets on projecting hairs. The aquatic bladderwort generates tension within its bladders. The tension is released when a bladder is touched by prey, expanding the bladder and sucking in water and prey.

Thinking Through the Concepts

Bloom's: Remembering, Understanding

Multiple Choice

1. Short-day plants
   a. flower in response to maximum duration of uninterrupted light.
   b. flower in response to a minimum duration of uninterrupted darkness.
   c. are likely to be common in the Tropics.
   d. are stimulated to flower by P_{fr}.
2. Which of the following is true?
   a. Ethylene stimulates abscission layers to form.
   b. Abscisic acid causes abscission layers to form.
   c. Ethylene is sometimes sprayed on flowers to keep them fresh.
   d. Abscisic acid stimulates seed germination.
3. Tropisms
   a. are primarily caused by cell division.
   b. allow bladderworts to catch water fleas.
   c. always cause bending toward a stimulus.
   d. are primarily caused by unequal cell elongation.
4. Phytochromes
   a. primarily respond to ultraviolet light.
   b. exist in two different stable forms.
   c. have the same effects in all flowering plants.
   d. have an active form called P_{fr}.
5. Which of the following is false?
   a. Thigmotropism is a directed response to touch.
   b. Auxin from apical meristems inhibits stem branching.
   c. Gibberellin stimulates elongation of stem internodes.
   d. Root branching is stimulated by cytokinin.

Fill-in-the-Blank

1. Auxin produced by the apical meristem of the shoot ____________ sprouting of lateral branches; this phenomenon is called ____________. In roots, auxin ____________ development of root branches.
2. The hormone ____________ was discovered by researchers investigating "foolish seedling" disease in rice. During stem elongation, this hormone stimulates both cell ____________ and cell ____________.
3. The hormone ____________ is a gas. During senescence, this hormone stimulates the formation of the ____________, which allows dead leaves and ripe fruits to fall.
4. The major site of cytokinin synthesis is the ____________. Cytokinin ____________ sprouting of lateral root branches and ____________ branching of ____________.
5. The hormone ____________ causes stomata to close when water is scarce. It also maintains ____________ in seeds during unfavorable environmental conditions.

Review Questions

1. Compare phototropism and gravitropism in roots and shoots. How are the two stimuli sensed? How do these responses differ in each plant part?
2. What is apical dominance? How do auxin and cytokinin interact in determining the growth of lateral buds?
3. What is a phytochrome? How do the two forms of phytochrome help control the plant life cycle?
4. Which hormones cause fruit development? Which hormone causes ripening in bananas?
5. What is senescence? Describe some changes that accompany leaf senescence in the fall. Which hormone causes abscission?
6. What is a major agricultural use of gibberellin? Of ethylene?
7. Describe one example of a plant chemical defense mechanism.
8. Describe how a sensitive plant closes its leaves. Why might this behavior have evolved?

Applying the Concepts

Bloom's: Applying, Analyzing, Evaluating

1. A student reporting on a project said that one of her seeds did not grow properly because it was planted upside down, so it got confused and tried to grow down. Should the teacher accept this explanation? Why or why not?
2. Would you expect a predominance of long-day, short-day, or day-neutral plants near the equator? Explain.

APPENDIX I Biological Vocabulary: Common Roots, Prefixes, and Suffixes

Biology has an extensive vocabulary often based on Greek or Latin rather than English words. Rather than memorizing every word as if it were part of a new, foreign language, you can figure out the meaning of many new terms if you learn a much smaller number of word roots, prefixes, and suffixes. We have provided common meanings in biology rather than literal translations from Greek or Latin. For each item in the list, the following information is given: meaning; part of word (prefix, suffix, or root); example from biology.

a–, an–: without, lack of (prefix); *abiotic*, without life

acro–: top, highest (prefix); *acrosome*, vesicle of enzymes at the tip of a sperm

ad–: to (prefix); *adhesion*, property of sticking to something else

allo–: other (prefix); *allopatric* (literally, "different fatherland"), restricted to different regions

amphi–: both, double, two (prefix); *amphibian*, a class of vertebrates that usually has two life stages (aquatic and terrestrial; e.g., a tadpole and an adult frog)

andro: man, male (root); *androgen*, a male hormone such as testosterone

ant–, anti–: against (prefix); *antibiotic* (literally "against life"), a substance that kills bacteria

antero–: front (prefix or root); *anterior*, toward the front of

apic–: top, highest (prefix); *apical meristem*, the cluster of dividing cells at the tip of a plant shoot or root

aqu–, aqua–: water (prefix or root); *aquifer*, an underground water source, usually rock saturated with water

arthr–: joint (prefix); *arthropod*, animals, such as spiders, crabs, and insects, with exoskeletons that include jointed legs

–ase: enzyme (suffix); *protease*, an enzyme that digests protein

auto–: self (prefix); *autotrophic*, self-feeder (e.g., photosynthetic)

bi–: two (prefix); *bipedal*, having two legs

bio–: life (prefix or root); *biology*, the study of life

blast: bud, precursor (root); *blastula*, embryonic stage of development, a hollow ball of cells

bronch: windpipe (root); *bronchus*, a branch of the trachea (windpipe) leading to a lung

carcin, –o: cancer (root); *carcinogenesis*, the process of producing a cancer

cardi, –a–, –o–: heart (root); *cardiac*, referring to the heart

carn–, –i–, –o–: flesh (prefix or root); *carnivore*, an animal that eats other animals

centi–: one hundredth (prefix); *centimeter*, a unit of length, 1 one-hundredth of a meter

cephal–, –i–, –o–: head (prefix or root); *cephalization*, the tendency for the nervous system to be located principally in the head

chloro–: green (prefix or root); *chlorophyll*, the green, light-absorbing pigment in plants

chondr–: cartilage (prefix); *Chondrichthyes*, class of vertebrates including sharks and rays, with a skeleton made of cartilage

chrom–: color (prefix or root); *chromosome*, a thread-like strand of DNA and protein in the nucleus of a cell (*Chromosome* literally means "colored body," because chromosomes absorb some of the colored dyes commonly used in microscopy.)

–cide: killer (suffix); *pesticide*, a chemical that kills "pests" (usually insects)

–clast: break down, broken (root or suffix); *osteoclast*, a cell that breaks down bone

co–: with or together with (prefix); *cohesion*, property of sticking together

coel–: hollow (prefix or root); *coelom*, the body cavity that separates the internal organs from the body wall

contra–: against (prefix); *contraception*, acting to prevent conception (pregnancy)

cortex: bark, outer layer (root); *cortex*, outer layer of kidney

crani–: skull (prefix or root); *cranium*, the skull

cuti–: skin (prefix or root); *cuticle*, the outermost covering of a leaf

–cyte, cyto–: cell (root or prefix); *cytokinin*, a plant hormone that promotes cell division

de–: from, out of, remove (prefix); *decomposer*, an organism that breaks down organic matter

dendr: tree-like, branching (root); *dendrite*, highly branched input structures of nerve cells

derm: skin, layer (root); *ectoderm*, the outer embryonic germ layer of cells

deutero–: second (prefix); *deuterostome* (literally, "second opening"), an animal in which the coelom is derived from the gut

di–: two (prefix); *dicot*, an angiosperm with two cotyledons in the seed

diplo–: both, double, two (prefix or root); *diploid*, having paired homologous chromosomes

dys–: difficult, painful (prefix); *dysfunction*, an inability to function properly

eco–: house, household (prefix); *ecology*, the study of the relationships between organisms and their environment

ecto–: outside (prefix); *ectoderm*, the outermost tissue layer of animal embryos

–elle: little, small (suffix); *organelle* (literally, "little organ"), a subcellular structure that performs a specific function

end–, endo–, ento–: inside, inner (prefix); *endocrine*, pertaining to a gland that secretes hormones inside the body

epi–: outside, outer (prefix); *epidermis*, outermost layer of skin

equi–: equal (prefix); *equidistant*, the same distance

erythro–: red (prefix); *erythrocyte*, red blood cell

eu–: true, good (prefix); *eukaryotic*, pertaining to a cell with a true nucleus

ex–, exo–: out of (prefix); *exocrine*, pertaining to a gland that secretes a substance (e.g., sweat) outside of the body

extra–: outside of (prefix); *extracellular*, outside of a cell

–fer: to bear, to carry (suffix); *conifer*, a tree that bears cones

gastr–: stomach (prefix or root); *gastric*, pertaining to the stomach

–gen–: to produce (prefix, root, or suffix); *antigen*, a substance that causes the body to produce antibodies

glyc–, glyco–: sweet (prefix); *glycogen*, a starch-like molecule composed of many glucose molecules bonded together

gyn, –o: female (prefix or root); *gynecology*, the study of the female reproductive tract

haplo–: single (prefix); *haploid*, having a single copy of each type of chromosome

hem–, hemato–: blood (prefix or root); *hemoglobin*, the molecule in red blood cells that carries oxygen

hemi–: half (prefix); *hemisphere*, one of the halves of the cerebrum

herb–, herbi–: grass (prefix or root); *herbivore*, an animal that eats plants

hetero–: other (prefix); *heterotrophic*, an organism that feeds on other organisms

hom–, homo–, homeo–: same (prefix); *homeostasis*, to maintain constant internal conditions in the face of changing external conditions

hydro–: water (usually prefix); *hydrophilic*, being attracted to water

hyper–: above, greater than (prefix); *hyperosmotic*, having a greater osmotic strength (usually higher solute concentration)

hypo–: below, less than (prefix); *hypodermic*, below the skin

inter–: between (prefix); *interneuron*, a neuron that receives input from one (or more) neuron(s) and sends output to another neuron (or many neurons)

intra–: within (prefix); *intracellular*, pertaining to an event or substance that occurs within a cell

iso–: equal (prefix); *isotonic*, pertaining to a solution that has the same osmotic strength as another solution

–itis: inflammation (suffix); *hepatitis*, an inflammation (or infection) of the liver

kin–, kinet–: moving (prefix or root); *cytokinesis*, the movements of a cell that divide the cell in half during cell division

lac–, lact–: milk (prefix or root); *lactose*, the principal sugar in mammalian milk

leuc–, leuco–, leuk–, leuko–: white (prefix); *leukocyte*, a white blood cell

lip–: fat (prefix or root); *lipid*, the chemical category to which fats, oils, and steroids belong

–logy: study of (suffix); *biology*, the study of life

lyso–, –lysis: loosening, split apart (prefix, root, or suffix); *lysis*, to break open a cell

macro–: large (prefix); *macrophage*, a large white blood cell that destroys invading foreign cells

medulla: marrow, middle substance (root); *medulla*, inner layer of kidney

mega–: large (prefix); *megaspore*, a large, haploid (female) spore formed by meiotic cell division in plants

–mere: segment, body section (suffix); *sarcomere*, the functional unit of a vertebrate skeletal muscle cell

meso–: middle (prefix); *mesophyll*, middle layers of cells in a leaf

meta–: change, after (prefix); *metamorphosis*, to change body form (e.g., developing from a larva to an adult)

micro–: small (prefix); *microscope*, a device that allows one to see small objects

milli–: one-thousandth (prefix); *millimeter*, a unit of measurement of length; 1 one-thousandth of a meter

mito–: thread (prefix); *mitosis*, cell division (in which chromosomes appear as thread-like bodies)

mono–: single (prefix); *monocot*, a type of angiosperm with one cotyledon in the seed

morph–: shape, form (prefix or root); *polymorphic*, having multiple forms

multi–: many (prefix); *multicellular*, pertaining to a body composed of more than one cell

myo–: muscle (prefix); *myofibril*, protein strands in muscle cells

neo–: new (prefix); *neonatal*, relating to or affecting a newborn child

neph–: kidney (prefix or root); *nephron*, functional unit of mammalian kidney

neur–, neuro–: nerve (prefix or root); *neuron*, a nerve cell

neutr–: of neither gender or type (usually root); *neutron*, an uncharged subatomic particle found in the nucleus of an atom

non–: not (prefix); *nondisjunction*, the failure of chromosomes to distribute themselves properly during cell division

oligo–: few (prefix); *oligomer*, a molecule made up of a few subunits (see also *poly–*)

omni–: all (prefix); *omnivore*, an animal that eats both plants and animals

oo–, ov–, ovo–: egg (prefix); *oocyte*, one of the stages of egg development

opsi–: sight (prefix or root); *opsin*, protein part of light-absorbing pigment in eye

opso–: tasty food (prefix or root); *opsonization*, process whereby antibodies and/or complement render bacteria easier for white blood cells to engulf

–osis: a condition, disease (suffix); *atherosclerosis*, a disease in which the artery walls become thickened and hardened

oss–, osteo–: bone (prefix or root); *osteoporosis*; a disease in which the bones become spongy and weak

para–: alongside (prefix); *parathyroid*, referring to a gland located next to the thyroid gland

pater, patr–: father (usually root); *paternal*, from or relating to a father

path–, –i–, –o–: disease (prefix or root); *pathology*, the study of disease and diseased tissue

–pathy: disease (suffix); *neuropathy*, a disease of the nervous system

peri–: around (prefix); *pericycle*, the outermost layer of cells in the vascular cylinder of a plant root

phago–: eat (prefix or root); *phagocyte*, a cell (e.g., some types of white blood cell) that eats other cells

–phil, philo–: to love (prefix or suffix); *hydrophilic* (literally, "water loving"), pertaining to a water-soluble molecule

–phob, phobo–: to fear (prefix or suffix); *hydrophobic* (literally, "water fearing"), pertaining to a water-insoluble molecule

photo–: light (prefix); *photosynthesis*, the manufacture of organic molecules using the energy of sunlight

–phyll: leaf (root or suffix); *chlorophyll*, the green, light-absorbing pigment in a leaf

–phyte: plant (root or suffix); *gametophyte* (literally, "gamete plant"), the gamete-producing stage of a plant's life cycle

plasmo, –plasm: formed substance (prefix, root, or suffix); *cytoplasm*, the material inside a cell

ploid: chromosomes (root); *diploid*, having paired chromosomes

pneumo–: lung (prefix or root); *pneumonia*, a disease of the lungs

-pod: foot (root or suffix); *gastropod* (literally, "stomach-foot"), a class of mollusks, principally snails, that crawl on their ventral surfaces

poly-: many (prefix); *polysaccharide*, a carbohydrate polymer composed of many sugar subunits (see also *oligo-*)

post-, postero-: behind (prefix); *posterior*, pertaining to the hind part

pre-, pro-: before, in front of (prefix); *premating isolating mechanism*, a mechanism that prevents gene flow between species, acting to prevent mating (e.g., having different courtship rituals or different mating seasons)

prim-: first (prefix); *primary cell wall*, the first cell wall laid down between plant cells during cell division

pro-: before (prefix); *prokaryotic*, pertaining to a cell without (that evolved before the evolution of) a nucleus

proto-: first (prefix); *protocell*, a hypothetical evolutionary ancestor to the first cell

pseudo-: false (prefix); *pseudopod* (literally, "false foot"), the extension of the plasma membrane by which some cells, such as *Amoeba*, move and capture prey

quad-, quat-: four (prefix); *quaternary structure*, the "fourth level" of protein structure, in which multiple peptide chains form a complex three-dimensional structure

ren: kidney (root); *adrenal*, gland attached to the mammalian kidney

retro-: backward (prefix); *retrovirus*, a virus that uses RNA as its genetic material; this RNA must be copied "backward" to DNA during infection of a cell by the virus

sarco-: muscle (prefix); *sarcoplasmic reticulum*, a calcium-storing, modified endoplasmic reticulum found in muscle cells

scler-: hard, tough (prefix); *sclerenchyma*, a type of plant cell with a very thick, hard cell wall

semi-: one-half (prefix); *semiconservative replication*, the mechanism of DNA replication, in which one strand of the original DNA double helix becomes incorporated into the new DNA double helix

-some, soma-, somato-: body (prefix or suffix); *somatic nervous system*, part of the peripheral nervous system that controls the skeletal muscles that move the body

sperm, sperma-, spermato-: seed (usually root); *gymnosperm*, a type of plant producing a seed not enclosed within a fruit

-stasis, stat-: stationary, standing still (suffix or prefix); *homeostasis*, the physiological process of maintaining constant internal conditions despite a changing external environment

stoma, -to-: mouth, opening (prefix or root); *stoma*, the adjustable pore in the surface of a leaf that allows carbon dioxide to enter the leaf

sub-: under, below (prefix); *subcutaneous*, beneath the skin

sym-: same (prefix); *sympatric* (literally "same father"), found in the same region

tel-, telo-: end (prefix); *telophase*, the last stage of mitosis and meiosis

test-: witness (prefix or root); *testis*, male reproductive organ (derived from the custom in ancient Rome that only males had standing in the eyes of the law; *testimony* has the same derivation)

therm-: heat (prefix or root); *thermoregulation*, the process of regulating body temperature

trans-: across (prefix); *transgenic*, having genes from another organism (usually another species); the genes have been moved "across" species

tri-: three (prefix); *triploid*, having three copies of each homologous chromosome

-trop-, tropic: change, turn, move (suffix); *phototropism*, the process by which plants orient toward the light

troph: food, nourishment (root); *autotrophic*, self-feeder (e.g., photosynthetic)

ultra-: beyond (prefix); *ultraviolet*, light of wavelengths beyond the violet

uni-: one (prefix); *unicellular*, referring to an organism composed of a single cell

vita: life (root); *vitamin*, a molecule required in the diet to sustain life

-vor: eat (usually root); *herbivore*, an animal that eats plants

zoo-, zoa-: animal (usually root); *zoology*, the study of animals

APPENDIX II Periodic Table of the Elements

atomic number (number of protons)

element (chemical symbol)

atomic mass (total mass of protons + neutrons + electrons)

1 H 1.008																	2 He 4.003
3 Li 6.941	4 Be 9.012											5 B 10.81	6 C 12.01	7 N 14.01	8 O 16.00	9 F 19.00	10 Ne 20.18
11 Na 22.99	12 Mg 24.31											13 Al 26.98	14 Si 28.09	15 P 30.97	16 S 32.07	17 Cl 35.45	18 Ar 39.95
19 K 39.10	20 Ca 40.08	21 Sc 44.96	22 Ti 47.87	23 V 50.94	24 Cr 52.00	25 Mn 54.94	26 Fe 55.85	27 Co 58.93	28 Ni 58.69	29 Cu 63.55	30 Zn 65.39	31 Ga 69.72	32 Ge 72.61	33 As 74.92	34 Se 78.96	35 Br 79.90	36 Kr 83.80
37 Rb 85.47	38 Sr 87.62	39 Y 88.91	40 Zr 91.22	41 Nb 92.91	42 Mo 95.94	43 Tc (98)	44 Ru 101.1	45 Rh 102.9	46 Pd 106.4	47 Ag 107.9	48 Cd 112.4	49 In 114.8	50 Sn 118.7	51 Sb 121.8	52 Te 127.6	53 I 126.9	54 Xe 131.3
55 Cs 132.9	56 Ba 137.3	57 *La 138.9	72 Hf 178.5	73 Ta 180.9	74 W 183.8	75 Re 186.2	76 Os 190.2	77 Ir 192.2	78 Pt 195.1	79 Au 197.0	80 Hg 200.6	81 Tl 204.4	82 Pb 207.2	83 Bi 209.0	84 Po (209)	85 At (210)	86 Rn (222)
87 Fr (223)	88 Ra (226)	89 †Ac (227)	104 Rf (261)	105 Db (262)	106 Sg (263)	107 Bh (264)	108 Hs (265)	109 Mt (268)	110 Ds (281)	111 Rg (280)	112 Cn (285)	113 Nh (286)	114 Fl (289)	115 Mc (290)	116 Lv (293)	117 Ts (294)	118 Og (294)

*Lanthanide series	58 Ce 140.1	59 Pr 140.9	60 Nd 144.2	61 Pm (145)	62 Sm 150.4	63 Eu 152.0	64 Gd 157.3	65 Tb 158.9	66 Dy 162.5	67 Ho 164.9	68 Er 167.3	69 Tm 168.9	70 Yb 173.0	71 Lu 175.0
†Actinide series	90 Th 232.0	91 Pa 231	92 U 238.0	93 Np (237)	94 Pu (244)	95 Am (243)	96 Cm (247)	97 Bk (247)	98 Cf (251)	99 Es (252)	100 Fm (257)	101 Md (258)	102 No (259)	103 Lr (262)

The periodic table of the elements was first devised by Russian chemist Dmitri Mendeleev. The atomic numbers of the elements (the numbers of protons in the nucleus) increase in normal reading order: left to right, top to bottom. The table is "periodic" because all the elements in a column possess similar chemical properties, and such similar elements therefore recur "periodically" in each row. For example, elements that usually form ions with a single positive charge—including H, Li, Na, K, and so on—occur as the first element in each row.

The gaps in the table are a consequence of the maximum numbers of electrons in the most reactive, usually outermost, electron shells of the atoms. It takes only two electrons to completely fill the first shell, so the first row of the table contains only two elements, H and He. It takes eight electrons to fill the second and third shells, so there are eight elements in the second and third rows. It takes 18 electrons to fill the fourth and fifth shells, so there are 18 elements in these rows.

The lanthanide series and actinide series of elements are usually placed below the main body of the table for convenience—it takes 32 electrons to completely fill the sixth and seventh shells, so the table would become extremely wide if all of these elements were included in the sixth and seventh rows.

The important elements found in living things are highlighted in color. The elements in pale red are the six most abundant elements in living things. The elements that form the five most abundant ions in living things are in purple. The trace elements important to life are shown as dark blue (more common) and lighter blue (less common). For radioactive elements, the atomic masses are given in parentheses, and represent the most common or the most stable isotope.

APPENDIX III Metric System Conversions

To Convert Metric Units:	Multiply by:	To Get English Equivalent:
Length		
Centimeters (cm)	0.394	Inches (in)
Meters (m)	3.281	Feet (ft)
Meters (m)	1.094	Yards (yd)
Kilometers (km)	0.621	Miles (mi)
Area		
Square centimeters (cm^2)	0.155	Square inches (in^2)
Square meters (m^2)	10.764	Square feet (ft^2)
Square meters (m^2)	1.196	Square yards (yd^2)
Square kilometers (km^2)	0.386	Square miles (mi^2)
Hectare (ha) (10,000 m^2)	2.471	Acres (a)
Volume		
Cubic centimeters (cm^3)	0.0610	Cubic inches (in^3)
Cubic meters (m^3)	35.315	Cubic feet (ft^3)
Cubic meters (m^3)	1.308	Cubic yards (yd^3)
Cubic kilometers (km^3)	0.240	Cubic miles (mi^3)
Liters (L)	1.057	Quarts (qt), U.S.
Liters (L)	0.264	Gallons (gal), U.S.
Mass		
Grams (g)	0.0353	Ounces (oz)
Kilograms (kg)	2.205	Pounds (lb)
Metric ton (tonne) (t)	1.102	Ton (tn), U.S.
Speed		
Meters/second (mps)	2.237	Miles/hour (mph)
Kilometers/hour (kmph)	0.621	Miles/hour (mph)

To Convert English Units:	Multiply by:	To Get Metric Equivalent:
Length		
Inches (in)	2.540	Centimeters (cm)
Feet (ft)	0.305	Meters (m)
Yards (yd)	0.914	Meters (m)
Miles (mi)	1.609	Kilometers (km)
Area		
Square inches (in^2)	6.452	Square centimeters (cm^2)
Square feet (ft^2)	0.0929	Square meters (m^2)
Square yards (yd^2)	0.836	Square meters (m^2)
Square miles (mi^2)	2.590	Square kilometers (km^2)
Acres (a)	0.405	Hectare (ha) (10,000 m^2)
Volume		
Cubic inches (in^3)	16.387	Cubic centimeters (cm^3)
Cubic feet (ft^3)	0.0283	Cubic meters (m^3)
Cubic yards (yd^3)	0.765	Cubic meters (m^3)
Cubic miles (mi^3)	4.168	Cubic kilometers (km^3)
Quarts (qt), U.S.	0.946	Liters (L)
Gallons (gal), U.S.	3.785	Liters (L)
Mass		
Ounces (oz)	28.350	Grams (g)
Pounds (lb)	0.454	Kilograms (kg)
Ton (tn), U.S.	0.907	Metric ton (tonne) (t)
Speed		
Miles/hour (mph)	0.447	Meters/second (mps)
Miles/hour (mph)	1.609	Kilometers/hour (kmph)

Metric Prefixes

Prefix		Meaning	
giga-	G	$10^9 =$	1,000,000,000
mega-	M	$10^6 =$	1,000,000
kilo-	k	$10^3 =$	1,000
hecto-	h	$10^2 =$	100
deka-	da	$10^1 =$	10
		$10^0 =$	1
deci-	d	$10^{-1} =$	0.1
centi-	c	$10^{-2} =$	0.01
milli-	m	$10^{-3} =$	0.001
micro-	μ	$10^{-6} =$	0.000001

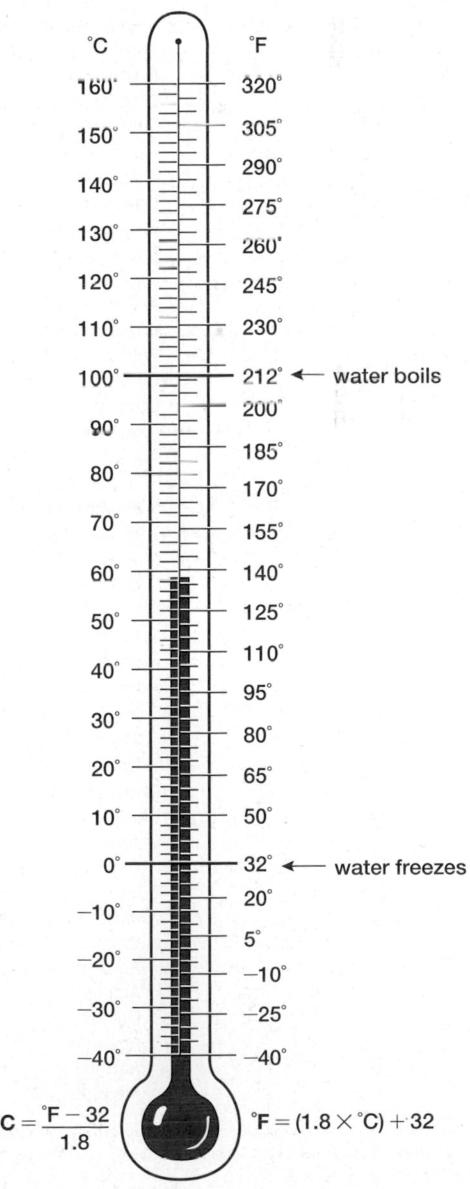

APPENDIX IV Classification of Major Groups of Eukaryotic Organisms*

Kingdom†	Phylum or Class	Common Name
Excavata	Parabasalia	parabasalids
	Diplomonadida	diplomonads
	Euglenida	euglenids
	Kinetoplastida	kinetoplastids
Stramenopila	Oomycota	water molds
	Phaeophyta	brown algae
	Bacillariophyta	diatoms
Alveolata	Apicomplexa	sporozoans
	Pyrrophyta	dinoflagellates
	Ciliophora	ciliates
Rhizaria	Foraminifera	foraminiferans
	Radiolaria	radiolarians
Amoebozoa	Tubulinea	amoebas
	Myxomycota	acellular slime molds
	Acrasiomycota	cellular slime molds
Rhodophyta		red algae
Chlorophyta		green algae
Plantae	Marchantiophyta	liverworts
	Anthocerotophyta	hornworts
	Bryophyta	mosses
	Lycopodiopsida	club mosses
	Polypodiopsida	ferns, horsetails
	Gymnospermae	cycads, ginkgos, gnetophytes, conifers
	Anthophyta	flowering plants
Fungi	Chytridiomycota	chytrids
	Neocallimastigomycota	rumen fungi
	Blastocladiomycota	blastoclades
	Glomeromycota	glomeromycetes
	Ascomycota	sac fungi
	Basidiomycota	club fungi
Animalia	Porifera	sponges
	Cnidaria	hydras, sea anemones, sea jellies, corals
	Ctenophora	comb jellies
	Platyhelminthes	flatworms
	Annelida	segmented worms
	Oligochaeta	earthworms
	Polychaeta	tube worms
	Hirudinea	leeches
	Mollusca	mollusks
	Gastropoda	snails
	Pelecypoda	mussels, clams
	Cephalopoda	squid, octopuses
	Nematoda	roundworms
	Arthropoda	arthropods
	Insecta	insects
	Arachnida	spiders, ticks
	Crustacea	crabs, lobsters
	Myriapoda	millipedes, centipedes
	Chordata	chordates
	Tunicata	tunicates
	Cephalochordata	lancelets
	Myxini	hagfishes
	Petromyzontiformes	lampreys
	Chondrichthyes	sharks, rays
	Actinopterygii	ray-finned fishes
	Actinistia	coelacanths
	Dipnoi	lungfishes
	Amphibia	amphibians (frogs, salamanders)
	Reptilia	reptiles (turtles, crocodiles, birds, snakes, lizards)
	Mammalia	mammals

*This table lists only those taxonomic categories described in the textbook.

†Although the major protist groups are not generally called "kingdoms," they are approximately the same taxonomic rank as the kingdoms Plantae, Fungi, and Animalia.

Glossary

abiotic (ā-bī-ah′-tik): nonliving; the abiotic portion of an ecosystem includes soil, rock, water, and the atmosphere.

abscisic acid (ab-sis′-ik): a plant hormone that generally inhibits the action of other hormones, enforcing dormancy in seeds and buds and causing the closing of stomata.

abscission layer: a layer of thin-walled cells, located at the base of the petiole of a leaf, that produces an enzyme that digests the cell walls holding the leaf to the stem, allowing the leaf to fall off.

absorption: the process by which nutrients enter the body through the cells lining the digestive tract.

accessory pigment: a colored molecule, other than chlorophyll *a*, that absorbs light energy and passes it to chlorophyll *a*.

acellular slime mold: a type of organism that forms a multinucleate structure that crawls in amoeboid fashion and ingests decaying organic matter; also called *plasmodial slime mold*. Acellular slime molds are members of the protist clade Amoebozoa.

acid: a substance that releases hydrogen ions (H^+) into solution; a solution with a pH less than 7.

acid deposition: the deposition of nitric or sulfuric acid, either in rain (acid rain) or in the form of dry particles, as a result of the production of nitrogen oxides or sulfur dioxide through burning, primarily of fossil fuels.

acidic: referring to a solution with an H^+ concentration exceeding that of OH^-; referring to a substance that releases H^+.

acquired immune deficiency syndrome (AIDS): an infectious disease caused by the human immunodeficiency virus (HIV); attacks and destroys T cells, thus weakening the immune system.

acrosome (ak′-rō-sōm): a vesicle, located at the tip of the head of an animal sperm, that contains enzymes needed to dissolve protective layers around the egg.

actin (ak′-tin): a major muscle protein whose interactions with myosin produce contraction; found in the thin filaments of the muscle fiber; see also *myosin*.

action potential: a rapid change from a negative to a positive electrical potential in a nerve cell. An action potential travels along an axon without a change in amplitude.

activation energy: in a chemical reaction, the energy needed to force the electron shells of reactants together, prior to the formation of products.

active site: the region of an enzyme molecule that binds substrates and performs the catalytic function of the enzyme.

active transport: the movement of materials across a membrane through the use of cellular energy, normally against a concentration gradient.

adaptation: a trait that increases the ability of an individual to survive and reproduce compared to individuals without the trait.

adaptive immune response: a response to invading toxins or microbes in which immune cells are activated by a specific invader, selectively destroy that invader, and then "remember" the invader, allowing a faster response if that type of invader reappears in the future; see also *innate immune response*.

adaptive immune system: a widely distributed system of organs (including the thymus, bone marrow, and lymph nodes), cells (including macrophages, dendritic cells, B cells, and T cells), and molecules (including cytokines and antibodies) that work together to combat microbial invasion of the body; the adaptive immune system responds to and destroys specific invading toxins or microbes; see also *innate immune response*.

adaptive radiation: the rise of many new species in a relatively short time; may occur when a single species invades different habitats and evolves in response to different environmental conditions in those habitats.

adenine (A): a nitrogenous base found in both DNA and RNA; abbreviated as A.

adenosine diphosphate (a-den′-ō-sēn dī-fos′-fāt; ADP): a molecule composed of the sugar ribose, the base adenine, and two phosphate groups; a component of ATP.

adenosine triphosphate (a-den′-ō-sēn trī fos′-fāt; ATP): a molecule composed of the sugar ribose, the base adenine, and three phosphate groups; the major energy carrier in cells. The last two phosphate groups are attached by "high-energy" bonds.

adhesion: the tendency of polar molecules (such as water) to adhere to polar surfaces (such as glass).

adhesive junctions: attachment structures that link cells to one another within tissues.

adipose tissue (a′-dipōs): tissue composed of fat cells.

adrenal cortex: the outer part of the adrenal gland, which secretes steroid hormones that regulate metabolism and salt balance.

adrenal gland: a mammalian endocrine gland, adjacent to the kidney; secretes hormones that function in water and salt regulation and in the stress response.

adrenal medulla: the inner part of the adrenal gland, which secretes epinephrine (adrenaline) and norepinephrine (noradrenaline) in the stress response.

adrenocorticotropic hormone (a-drēn′-ō-kor-tik-ō-trō′-pik; ACTH): a hormone, secreted by the anterior pituitary, that stimulates the release of hormones by the adrenal cortex, especially in response to stress.

adult stem cell (ASC): any stem cell not found in an early embryo; can divide and differentiate into any of several cell types, but usually not all the cell types of the body.

aerobic: using oxygen.

age structure diagram: a graph showing the distribution of males and females in a population according to age groups.

aggression: antagonistic behavior, normally among members of the same species, that often results from competition for resources.

aggressive mimicry (mim′ik-rē): the evolution of a predatory organism to resemble a harmless animal or a part of the environment, thus gaining access to prey.

aging: the gradual accumulation of damage to essential biological molecules, particularly DNA in both the nucleus and mitochondria, resulting in defects in cell functioning, declining health, and ultimately death.

albinism: a recessive hereditary condition caused by defective alleles of the genes that encode the enzymes required for the synthesis of melanin, the principal pigment in mammalian skin and hair; albinism results in white hair and pink skin.

alcoholic fermentation: a type of fermentation in which pyruvate is converted to ethanol (a type of alcohol) and carbon dioxide, using hydrogen ions and electrons from NADH; the primary function of alcoholic fermentation is to regenerate NAD^+ so that glycolysis can continue under anaerobic conditions.

aldosterone: a hormone, secreted by the adrenal cortex, that helps regulate ion concentration in the blood by stimulating the reabsorption of sodium by the kidneys and sweat glands.

alga (al′-ga; pl., algae, al′-jē): any photosynthetic protist.

allantois (al-an-tō′-is): one of the embryonic membranes of reptiles (including birds) and mammals; in reptiles, serves as a waste-storage organ; in mammals, forms most of the umbilical cord.

allele (al-ēl′): one of several alternative forms of a particular gene.

allele frequency: for any given gene, the relative proportion of each allele of that gene in a population.

allergy: an inflammatory response produced by the body in response to invasion by foreign materials, such as pollen, that are themselves harmless.

allopatric speciation (al-ō-pat′-rik): the process by which new species arise following physical separation of parts of a population (geographic isolation).

allosteric regulation: the process by which enzyme action is enhanced or inhibited by small organic molecules that act as regulators by binding to the enzyme at a regulatory site distinct from the active site and altering the shape and/or function of the active site.

alternation of generations: a life cycle, typical of plants, in which a diploid sporophyte (spore-producing) generation alternates with a haploid gametophyte (gamete-producing) generation.

altruism: a behavior that benefits other individuals while reducing the fitness of the individual that performs the behavior.

alveolate (al-vē′-ō-lāt): a member of the Alveolata, a large protist clade. The alveolates, which are characterized by a system of sacs beneath the cell membrane, include ciliates, dinoflagellates, and apicomplexans.

alveolus (al-vē′-ō-lus; pl., alveoli): a tiny air sac within the lungs, surrounded by capillaries, where gas exchange with the blood occurs.

amino acid: the individual subunit of which proteins are made, composed of a central carbon atom bonded to an amino group ($-NH_2$), a carboxyl group ($-COOH$), a hydrogen atom, and a variable group of atoms denoted by the letter *R*.

amino acid derived hormone: a hormone composed of one or two modified amino acids. Examples include epinephrine and thyroxine.

ammonia: NH_3; a highly toxic nitrogen-containing waste product of amino acid breakdown. In the mammalian liver, it is converted to urea.

amniocentesis (am-nē-ō-sen-tē′-sis): a procedure for sampling the amniotic fluid surrounding a fetus: A sterile needle is inserted through the abdominal wall, uterus, and amniotic sac of a pregnant woman, and 10 to 20 milliliters of amniotic fluid are withdrawn. Various tests may be performed on the fluid and the fetal cells suspended in it to provide information on the developmental and genetic state of the fetus.

amnion (am′-nē-on): one of the embryonic membranes of reptiles (including birds) and mammals; encloses a fluid-filled cavity that envelops the embryo.

amniotic egg (am-nē-ot′-ik): the egg of reptiles, including birds; contains a membrane, the amnion, that surrounds the embryo, enclosing it in a watery environment and allowing the egg to be laid on dry land.

amoeba: an amoebozoan protist that uses a characteristic streaming mode of locomotion by extending a cellular projection called a *pseudopod*. Also known as a *lobose amoeba*.

amoeboid cell: a protist or animal cell that moves by extending a cellular projection called a pseudopod.

amoebozoan: a member of the Amoebozoa, a protist clade. The amoebozoans, which generally lack shells and move by extending pseudopods, include the lobose amoebas and the slime molds.

amphibian: a member of the chordate clade Amphibia, which includes the frogs, toads, and salamanders, as well as the limbless caecilians.

amygdala (am-ig′-da-la): part of the forebrain of vertebrates that is involved in the production of appropriate behavioral responses to environmental stimuli.

amylase (am′-i-lās): an enzyme, found in saliva and pancreatic secretions, that catalyzes the breakdown of starch.

anaerobe (an-e-rōb): an organism that can live and grow in the absence of oxygen.

anaerobic: not using oxygen.

analogous structure: structures that have similar functions and superficially similar appearance but very different anatomies, such as the wings of insects and birds. The similarities are the result of similar environmental pressures rather than a common ancestry.

anaphase (an′-a-fāz): in mitosis, the stage in which the sister chromatids of each chromosome separate from one another and are moved to opposite poles of the cell; in meiosis I, the stage in which homologous chromosomes, consisting of two sister chromatids, are separated; in meiosis II, the stage in which the sister chromatids of each chromosome separate from one another and are moved to opposite poles of the cell.

androgen: a male sex hormone.

angina (an-jī′-nuh): chest pain associated with reduced blood flow to the heart muscle; caused by an obstruction of the coronary arteries.

angiosperm (an′-jē-ō-sperm): a flowering vascular plant.

angiotensin (an′-jē-ō-ten′-sun): a hormone that functions in water regulation in mammals by stimulating physiological changes that increase blood volume and blood pressure.

annual ring: a pattern of alternating light (early) and dark (late) xylem in woody stems and roots; formed as a result of the unequal availability of water in different seasons of the year, normally spring and summer.

antagonistic muscles: a pair of muscles, one of which contracts and in so doing extends the other, relaxed muscle; this arrangement allows movement of the skeleton at joints.

anterior pituitary: a lobe of the pituitary gland that produces prolactin, growth hormone, follicle-stimulating hormone, luteinizing hormone, adrenocorticotropic hormone, and thyroid-stimulating hormone.

anther (an′-ther): the uppermost part of the stamen, in which pollen develops.

antheridium (an-ther-id′-ē-um; pl., antheridia): a structure in which male sex cells are produced; found in nonvascular plants and certain seedless vascular plants.

antibiotic: chemicals that help to combat infection by destroying or slowing down the multiplication of bacteria, fungi, or protists.

antibody: a protein, produced by cells of the immune system, that combines with a specific antigen and normally facilitates the destruction of the antigen.

anticodon: a sequence of three bases in transfer RNA that is complementary to the three bases of a codon of messenger RNA.

antidiuretic hormone (an-tē-dī-ūr-et′-ik; ADH): a hormone produced by the hypothalamus and released into the bloodstream by the posterior pituitary when blood volume is low; increases the permeability of the distal tubule and the collecting duct to water, allowing more water to be reabsorbed into the bloodstream.

antigen: a complex molecule, normally a protein or polysaccharide, that stimulates the production of a specific antibody.

antioxidant: any molecule that reacts with free radicals, neutralizing their ability to damage biological molecules. Vitamins C and E are examples of dietary antioxidants.

antiviral drug: a medicine that interferes with one or more stages of the viral life cycle, including attachment to a host cell, replication of viral parts, assembly of viruses within a host cell, and release of viruses from a host cell.

anvil: the second of the small bones of the middle ear that link the tympanic membrane (eardrum) to the oval window of the cochlea; also called the *incus*.

aphotic zone: the region of the ocean below 200 meters where sunlight does not penetrate.

apical dominance: the phenomenon whereby a growing shoot tip inhibits the sprouting of lateral buds.

apical meristem (āp′-i-kul mer′-i-stem): the cluster of meristem cells at the tip of a shoot or root (or one of their branches).

apicomplexan (ā-pē-kom-pleks′-an): a member of the protist clade Apicomplexa, which includes mostly parasitic, single-celled eukaryotes such as *Plasmodium*, which causes malaria in humans. Apicomplexans are part of a larger group known as the alveolates.

appendicular skeleton (ap-pen-dik′-ū-lur): the portion of the skeleton consisting of the bones of the extremities and their attachments to the axial skeleton; the appendicular skeleton therefore consists of the pectoral and pelvic girdles and the arms, legs, hands, and feet.

aquaporin: a channel protein in the plasma membrane of a cell that is selectively permeable to water.

aqueous humor (ā′-kwē-us): the clear, watery fluid between the cornea and lens of the eye; nourishes the cornea and lens.

aquifer (ok′-wifer): an underground deposit of fresh water, often used as a source of water for irrigation.

archaea: prokaryotes that are members of the domain Archaea, one of the three domains of living organisms; only distantly related to members of the domain Bacteria.

Archaea: one of life's three domains; consists of prokaryotes that are only distantly related to members of the domain Bacteria.

archegonium (ar-ke-gō′-nē-um; pl., archegonia): a structure in which female sex cells are produced; found in nonvascular plants and certain seedless vascular plants.

arteriole (ar-tēr′-ē-ōl): a small artery that empties into capillaries; constriction of arterioles regulates blood flow to various parts of the body.

artery (ar′-tuh-rē): a vessel with muscular, elastic walls that conducts blood away from the heart.

arthropod: a member of the animal phylum Arthropoda, which includes the insects, spiders, ticks, mites, scorpions, crustaceans, millipedes, and centipedes.

artificial selection: a selective breeding procedure in which only those individuals with particular traits are chosen as breeders; used mainly to enhance desirable traits in domesticated plants and animals; may also be used in evolutionary biology experiments.

ascomycete: a member of the fungus clade Ascomycota, whose members form sexual spores in a saclike case known as an ascus.

ascus (as′-kus; pl., asci): a saclike case in which sexual spores are formed by members of the fungus clade Ascomycota.

asexual reproduction: reproduction that does not involve the fusion of haploid gametes.

atherosclerosis (ath′-er-ō-skler-ō′-sis): a disease characterized by the obstruction of arteries by cholesterol deposits and thickening of the arterial walls.

atom: the smallest unit of an element that retains the properties of the element.

atomic mass: the total mass of all the protons, neutrons, and electrons within an atom.

atomic nucleus: the central part of an atom that contains protons and neutrons.

atomic number: the number of protons in the nuclei of all atoms of a particular element.

ATP synthase: a channel protein in the thylakoid membranes of chloroplasts and the inner membrane of mitochondria that uses the energy of H^+ ions moving through the channel down their concentration gradient to produce ATP from ADP and inorganic phosphate.

atrial natriuretic peptide (ANP) (ā′-trē-ul nā-trē-ū-ret′-ik; ANP): a hormone, secreted by cells in the mammalian heart, that reduces blood volume by inhibiting the release of ADH and aldosterone.

atrioventricular (AV) node (ā′-trē-ō-ven-trik′-ū-lar nōd): a specialized mass of muscle at the base of the right atrium through which the electrical activity initiated in the sinoatrial node is transmitted to the ventricles.

atrioventricular valve: a heart valve that separates each atrium from each ventricle, preventing the backflow of blood into the atria during ventricular contraction.

atrium (ā′-trē-um; pl., atria): a chamber of the heart that receives venous blood and passes it to a ventricle.

auditory canal (aw′-di-tor-ē): a relatively large-diameter tube within the outer ear that conducts sound from the pinna to the tympanic membrane.

auditory nerve: the nerve leading from the mammalian cochlea to the brain; it carries information about sound.

auditory tube: a thin tube connecting the middle ear with the pharynx, which allows pressure to equilibrate between the middle ear and the outside air; also called the *Eustachian tube.*

autoimmune disease: a disorder in which the immune system attacks the body's own cells or molecules.

autonomic nervous system: the part of the peripheral nervous system of vertebrates that synapses on glands, internal organs, and smooth muscle and produces largely involuntary responses.

autosome (aw′-tō-sōm): a chromosome that occurs in homologous pairs in both males and females and that does not bear the genes determining sex.

autotroph (aw′-tō-trōf): literally, "self-feeder"; normally, a photosynthetic organism; a producer.

auxin (awk′-sin): a plant hormone that influences many plant functions, including phototropism, gravitropism, apical dominance, and root branching.

axial skeleton: the skeleton forming the body axis, including the skull, vertebral column, and rib cage.

axon: a long extension of a nerve cell, extending from the cell body to synaptic endings on other nerve cells or on muscles.

B cell: a type of lymphocyte that matures in the bone marrow and that participates in humoral immunity; gives rise to plasma cells, which secrete antibodies into the circulatory system, and to memory cells.

bacteria (sing., bacterium): prokaryotes that are members of the domain Bacteria, one of the three domains of living organisms; only distantly related to members of the domain Archaea.

Bacteria: one of life's three domains; consists of prokaryotes that are only distantly related to members of the domain Archaea.

bacteriophage (bak-tir′-ē-ō-fāj): a virus that specifically infects bacteria.

ball-and-socket joint: a joint in which the rounded end of one bone fits into a hollow depression in another, as in the hip; allows movement in several directions.

bark: the outer layer of a woody stem, consisting of phloem, cork cambium, and cork cells.

Barr body: a condensed, inactivated X chromosome in the cells of female mammals that have two X chromosomes.

basal body: a structure derived from a centriole that produces a cilium or flagellum and anchors this structure within the plasma membrane.

basal ganglion: a cluster of neurons in the interior of the cerebrum, plus the substantial nigra in the midbrain, that functions in the control of movement. Damage to or degeneration of one or more basal ganglia causes disorders such as Parkinson's disease and Huntington's disease.

base: (1) a substance capable of combining with and neutralizing H^+ ions in a solution; a solution with a pH greater than 7; (2) one of the nitrogen-containing, single- or double-ringed structures that distinguishes one nucleotide from another. In DNA, the bases are adenine, guanine, cytosine, and thymine.

basic: referring to a solution with an H^+ concentration less than that of OH^-; referring to a substance that combines with H^+.

basidiomycete: a member of the fungus clade Basidiomycota, which includes species that produce sexual spores in club-shaped cells known as basidia.

basidiospore (ba-sid′-ē-ō-spor): a sexual spore formed by members of the fungus clade Basidiomycota.

basidium (pl., basidia): a diploid cell, typically club-shaped, formed by members of the fungus clade Basidiomycota; produces basidiospores by meiosis.

basilar membrane (bas′-eh-lar): a membrane in the cochlea that bears hair cells that respond to the vibrations produced by sound.

behavior: any observable activity of a living animal.

behavioral isolation: reproductive isolation that arises when species do not interbreed because they have different courtship and mating rituals.

bilateral symmetry: a body plan in which only a single plane through the central axis will divide the body into mirror-image halves.

bile (bīl): a digestive secretion formed by the liver, stored and released from the gallbladder, and used to disperse fats in the small intestine.

binocular vision: the ability to see objects simultaneously through both eyes, providing greater depth perception and more accurate judgment of the size and distance of an object than can be achieved by vision with one eye alone.

binomial system: the method of naming organisms by genus and species, often called the scientific name, usually using Latin or Greek words or words derived from Latin or Greek.

biocapacity: an estimate of the sustainable resources and waste-absorbing capacity actually available on Earth. Biocapacity calculations are subject to change as new technologies change the way people use resources.

biodegradable: able to be broken down into harmless substances by decomposers.

biodiversity: the diversity of living organisms; measured as the variety of different species, the variety of different alleles in species' gene pools, or the variety of different communities and nonliving environments in an ecosystem or in the entire biosphere.

biofilm: a community of prokaryotes of one or more species, in which the prokaryotes secrete and are embedded in slime that adheres to a surface.

biogeochemical cycle: the pathways of a specific nutrient (such as carbon, nitrogen, phosphorus, or water) through the living and nonliving portions of an ecosystem; also called a *nutrient cycle*.

biological clock: a metabolic timekeeping mechanism found in most organisms, whereby the organism measures the approximate length of a day (24 hours) even without external environmental cues such as light and darkness.

biological magnification: the increasing accumulation of a toxic substance in progressively higher trophic levels.

biological molecules: all molecules produced by living things.

biology: the study of all aspects of life and living things.

biomass: the total weight of all living material within a defined area.

biome (bī′-ōm): a terrestrial ecosystem that occupies an extensive geographic area and is characterized by a specific type of plant community; for example, deserts.

bioremediation: the use of organisms to remove or detoxify toxic substances in the environment.

biosphere (bī′-ō-sfēr): all life on Earth and the nonliving portions of Earth that support life.

biotechnology: any industrial or commercial use or alteration of organisms, cells, or biological molecules to achieve specific practical goals.

biotic (bī-ah′-tik): living.

biotic potential: the maximum rate at which a population is able to increase, assuming ideal conditions that allow a maximum birth rate and minimum death rate.

birth rate (*b*): the number of births per individual in a specified unit of time, such as a year.

bladder: a hollow muscular organ that stores urine.

blade: the flat part of a leaf.

blastoclade: a member of the fungus clade Blastocladiomycota, whose members have swimming spores with a single flagellum and ribosomes arranged to form a nuclear cap.

blastocyst (blas′-tō-sist): an early stage of human embryonic development, consisting of a hollow ball of cells, enclosing a mass of cells attached to its inner surface, which becomes the embryo.

blastopore: the site at which a blastula indents to form a gastrula.

blastula (blas′-tū-luh): in animals, the embryonic stage attained at the end of cleavage, in which the embryo usually consists of a hollow ball with a wall that is one or several cell layers thick.

blind spot: the area of the retina at which the axons of the ganglion cells merge to form the optic nerve; because there are no photoreceptors in the blind spot, objects focused at the blind spot cannot be seen.

blood: a specialized connective tissue, consisting of a fluid (plasma) in which blood cells are suspended; carried within the circulatory system.

blood–brain barrier: relatively impermeable capillaries of the brain that protect the cells of the brain from potentially damaging chemicals that reach the bloodstream.

blood clotting: a complex process by which platelets, the protein fibrin, and red blood cells block an irregular surface in or on the body, such as a damaged blood vessel, sealing the wound.

blood vessel: any of several types of tubes that carry blood throughout the body.

body mass index (BMI): a number derived from an individual's weight and height that is used to estimate body fat. The formula is weight (in kg)/height^2 (in meters^2).

bone: a hard, mineralized connective tissue that is a major component of the vertebrate endoskeleton; provides support and sites for muscle attachment.

bone marrow: a soft, spongy tissue that fills the cavities of large bones and generates the cell-based components of blood.

boom-and-bust cycle: a population cycle characterized by rapid exponential growth followed by a sudden massive die-off; seen in seasonal species, such as many insects living in temperate climates, and in some populations of small rodents, such as lemmings.

brain: the part of the central nervous system of vertebrates that is enclosed within the skull.

branch root: a root that arises as a branch of a preexisting root; occurs through divisions of pericycle cells and subsequent differentiation of the daughter cells.

bronchiole (bron′-kē-ōl): a narrow tube, formed by repeated branching of the bronchi, that conducts air into the alveoli.

bronchus (bron′-kus; pl., bronchi): a tube that conducts air from the trachea to each lung.

bud: in plants, an embryonic shoot, often dormant until stimulated by specific combinations of hormones. In asexually reproducing animals, a miniature copy of an animal that develops on some part of the adult animal's body; usually eventually separates from the adult and assumes independent existence.

budding: asexual reproduction by the growth of a miniature copy, or bud, of the adult animal on the body of the parent. The bud breaks off to begin independent existence.

buffer: a compound that minimizes changes in pH by reversibly taking up or releasing H^+ ions.

bulbourethral gland (bul-bō-ū-rē′-thrul): in male mammals, a gland that secretes a basic, mucus-containing fluid that forms part of the semen.

bulk flow: the movement of many molecules of a gas or liquid in unison (in bulk, hence the name) from an area of higher pressure to an area of lower pressure.

bundle sheath cells: cells that surround the veins of plants; in C_4 (but not in C_3) plants, bundle sheath cells contain chloroplasts.

C_3 pathway: in photosynthesis, the cyclic series of reactions whereby carbon from carbon dioxide is fixed as phosphoglyceric acid, the simple sugar glyceraldehyde-3-phosphate is generated, and the carbon-capture molecule, RuBP, is regenerated. Also called the *Calvin cycle*.

C_3 plant: a plant that relies on the C_3 pathway to fix carbon.

C_4 pathway: the series of reactions in certain plants that fixes carbon dioxide into a four-carbon molecule, which is later broken down for use in the Calvin cycle of photosynthesis. This reduces wasteful photorespiration in hot, dry environments.

C_4 plant: a plant that relies on the C_4 pathway to fix carbon.

calcitonin (kal-si-tōn′-in): a hormone, secreted by the thyroid gland, that inhibits the release of calcium from bone.

calorie (kal′-ō-rē): the amount of energy required to raise the temperature of 1 gram of water by 1 degree Celsius.

Calorie: a unit used to measure the energy content of foods; it is the amount of energy required to raise the temperature of 1 liter of water 1 degree Celsius; also called a *kilocalorie*, equal to 1,000 calories.

Calvin cycle: in photosynthesis, the cyclic series of reactions whereby carbon from carbon dioxide is fixed as phosphoglyceric acid, the simple sugar glyceraldehyde-3-phosphate is generated, and the carbon-capture molecule, RuBP, is regenerated. Also called the *C_3 pathway*.

cambium (kam′-bē-um; pl., cambia): a lateral meristem, parallel to the long axis of roots and stems, that causes secondary growth of woody plant stems and roots. See also *cork cambium; vascular cambium*.

camouflage (cam′-a-flaj): coloration and/or shape that renders an organism inconspicuous in its environment.

cancer: a disease in which some of the body's cells escape from normal regulatory processes and divide without control.

capillary: the smallest type of blood vessel, connecting arterioles with venules; capillary walls, through which the exchange of nutrients and wastes occurs, are only one cell thick.

capillary action: the movement of water within narrow spaces resulting from its properties of adhesion and cohesion.

carbohydrate: a compound composed of carbon, hydrogen, and oxygen, with the approximate chemical formula $(CH_2O)_n$; includes sugars, starches, and cellulose.

carbon cycle: the biogeochemical cycle by which carbon moves from its reservoirs in the atmosphere and oceans through producers and into higher trophic levels, and then back to its reservoirs.

carbon fixation: the process by which carbon derived from carbon dioxide is captured in organic molecules during photosynthesis.

cardiac cycle (kar′-dē-ak): the alternation of contraction and relaxation of the heart chambers.

cardiac muscle (kar′-dē-ak): the specialized muscle of the heart; able to initiate its own contraction, independent of the nervous system.

carnivore (kar′-neh-vor): literally, "meat-eater"; a predatory organism that feeds on herbivores or on other carnivores; a secondary (or higher) consumer.

carotenoid (ka-rot′-en-oid): a red, orange, or yellow pigment, found in chloroplasts, that serves as an accessory light-gathering pigment in thylakoid photosystems.

carpel (kar′-pel): the female reproductive structure of a flower, composed of stigma, style, and ovary.

carrier: an individual who is heterozygous for a recessive condition; a carrier displays the dominant phenotype but can pass on the recessive allele to offspring.

carrier protein: a membrane protein that facilitates the diffusion of specific substances across the membrane. The molecule to be transported binds to the outer surface of the carrier protein; the protein then changes shape, allowing the molecule to move across the membrane.

carrying capacity (*K*): the maximum population size that an ecosystem can support for a long period of time without damaging the ecosystem; determined primarily by the availability of space, nutrients, water, and light.

cartilage (kar′-teh-lij): a form of connective tissue that forms portions of the skeleton; consists principally of cartilage cells and their major extracellular secretion, collagen protein.

Casparian strip (kas-par′-ē-un): a waxy, waterproof band, located in the cell walls between endodermal cells in a root, that prevents the movement of water and minerals into and out of the vascular cylinder through the extracellular space.

catalyst (kat′-uh-list): a substance that speeds up a chemical reaction without itself being permanently changed in the process; a catalyst lowers the activation energy of a reaction.

cell: the smallest unit of life, consisting, at a minimum, of an outer membrane that encloses a watery medium containing organic molecules, including genetic material composed of DNA.

cell body: the part of a nerve cell in which most of the common cellular organelles are located; typically a site of integration of inputs to the nerve cell.

cell cycle: the sequence of events in the life of a cell, from one cell division to the next.

cell division: splitting of one cell into two; the process of cellular reproduction.

cell-mediated immunity: an adaptive immune response in which foreign cells or substances are destroyed by contact with T cells.

cell plate: in plant cell division, a series of vesicles that fuse to form the new plasma membranes and cell wall separating the daughter cells.

cell theory: the scientific theory stating that every living organism is made up of one or more cells; cells are the functional units of all organisms; and all cells arise from preexisting cells.

cell wall: a nonliving, protective, and supportive layer secreted outside the plasma membrane of fungi, plants, and most bacteria and protists.

cellular respiration: the oxygen-requiring reactions, occurring in mitochondria, that break down the end products of glycolysis into carbon dioxide and water while capturing large amounts of energy as ATP.

cellular slime mold: a type of organism consisting of individual amoeboid cells that can aggregate to form a slug-like mass, which in turn forms a fruiting body. Cellular slime molds are members of the protist clade Amoebozoa.

cellulose: an insoluble carbohydrate composed of glucose subunits; forms the cell wall of plants.

central nervous system (CNS): in vertebrates, the brain and spinal cord.

central vacuole (vak′-ū-ōl): a large, fluid-filled vacuole occupying most of the volume of many plant cells; performs several functions, including maintaining turgor pressure.

centriole (sen′-trē-ōl): in animal cells, a short, barrel-shaped ring consisting of nine microtubule triplets; a pair of centrioles is found near the nucleus and may play a role in the organization of the spindle; centrioles also give rise to the basal bodies at the base of each cilium and flagellum that give rise to the microtubules of cilia and flagella.

centromere (sen′-trō-mēr): the region of a replicated chromosome at which the sister chromatids are held together until they separate during cell division.

cephalization (sef-ul-ī-zā′-shun): concentration of sensory organs and nervous tissue in the anterior (head) portion of the body.

cerebellum (ser-uh-bel′-um): the part of the hindbrain of vertebrates that is concerned with coordinating movements of the body.

cerebral cortex (ser-ē′-brul kor′-tex): a thin layer of neurons on the surface of the vertebrate cerebrum in which most neural processing and coordination of activity occurs.

cerebral hemisphere: one of two nearly symmetrical halves of the cerebrum, connected by a broad band of axons, the corpus callosum.

cerebrum (ser-ē′-brum): the part of the forebrain of vertebrates that is concerned with sensory processing, the direction of motor output, and the coordination of most of the body's activities; consists of two nearly symmetrical halves (the hemispheres) connected by a broad band of axons, the corpus callosum.

cervix (ser′-viks): a ring of connective tissue at the outer end of the uterus that leads into the vagina.

channel protein: a membrane protein that forms a channel or pore completely through the membrane and that is usually permeable to one or to a few water-soluble molecules, especially ions.

chaparral: a biome located in coastal regions, with very low annual rainfall; is characterized by shrubs and small trees.

checkpoint: a mechanism in the eukaryotic cell cycle by which protein complexes in the cell determine whether the cell has successfully completed a specific process that is essential to successful cell division, such as the accurate replication of chromosomes.

chemical bond: an attraction between two atoms or molecules that tends to hold them together. Types of bonds include covalent, ionic, and hydrogen.

chemical digestion: the process by which particles of food within the digestive tract are exposed to enzymes and other digestive fluids that break down large molecules into smaller subunits.

chemical energy: a form of potential energy that is stored in molecules and may be released during chemical reactions.

chemical reaction: a process that forms and breaks chemical bonds that hold atoms together in molecules.

chemiosmosis (ke-mē-oz-mō′-sis): a process of ATP generation in chloroplasts and mitochondria. The movement of electrons down an electron transport system is used to pump hydrogen ions across a membrane, thereby building up a concentration gradient of hydrogen ions; the hydrogen ions diffuse back across the membrane through the pores of ATP-synthesizing enzymes; the energy of their movement down their concentration gradient drives ATP synthesis.

chemoreceptor: a sensory receptor that responds to chemicals in either the internal or external environment.

chemosynthesis (kē-mō-sin′-the-sis): the process of oxidizing inorganic molecules, such as hydrogen sulfide, to obtain energy. Producers in hydrothermal vent communities, where light is absent, use chemosynthesis instead of photosynthesis.

chemosynthetic (kēm′-ō-sin-the′-tik): capable of oxidizing inorganic molecules to obtain energy.

chiasma (kī-as′-muh; pl., chiasmata): a point at which a chromatid of one chromosome crosses with a chromatid of the homologous chromosome during prophase I of meiosis; the site of exchange of chromosomal material between chromosomes.

chitin (kī′-tin): a compound found in the cell walls of fungi and the exoskeletons of insects and some other arthropods; composed of chains of nitrogen-containing, modified glucose molecules.

chlamydia (kla-mid′-ē-uh): a sexually transmitted disease, caused by a bacterium, that causes inflammation of the urethra in males and of the urethra and cervix in females.

chlorophyll (klor′-ō-fil): a pigment found in chloroplasts that captures light energy during photosynthesis; chlorophyll absorbs violet, blue, and red light but reflects green light.

chlorophyll *a* (klor′-ō-fil): the most abundant type of chlorophyll molecule in photosynthetic eukaryotic organisms and in cyanobacteria; chlorophyll *a* is found in the reaction centers of the photosystems.

chlorophyte (klor′-ō-fīt): a member of Chlorophyta, a protist clade. Chlorophytes are photosynthetic green algae found in marine, freshwater, and terrestrial environments.

chloroplast (klor′-ō-plast): the organelle in plants and plantlike protists that is the site of photosynthesis; is surrounded by a double membrane and contains an extensive internal membrane system that bears chlorophyll.

cholecystokinin (kō′-lē-sis-tō-ki′-nin): a digestive hormone produced by the small intestine that stimulates the release of pancreatic enzymes.

chondrocyte (kon′-drō-sīt): a living cell of cartilage. Together with their extracellular secretions of collagen, chondrocytes form cartilage.

chorion (kor′-ē-on): the outermost embryonic membrane in reptiles (including birds) and mammals. In reptiles, the chorion functions mostly in gas exchange; in mammals, it forms the embryonic part of the placenta.

chorionic gonadotropin (kor-ē-on′-ik gō-nādō-trō′-pin; CG): a hormone secreted by the chorion (one of the fetal membranes) that maintains the integrity of the corpus luteum during early pregnancy.

chorionic villus (kor-ē-on-ik; pl., chorionic villi): in mammalian embryos, a finger-like projection of the chorion that penetrates the uterine lining and forms the embryonic portion of the placenta.

chorionic villus sampling (kōr-ē-on′-ik; CVS): a procedure for sampling cells from the chorionic villi produced by a fetus: A tube is inserted into the uterus of a pregnant woman, and a small sample of villi is suctioned off for genetic and biochemical analyses.

choroid (kor′-oid): a darkly pigmented layer of tissue behind the retina; contains blood vessels and also pigment that absorbs stray light.

chromatid (krō′-ma-tid): one of the two identical strands of DNA and protein that forms a duplicated chromosome. The two sister chromatids of a duplicated chromosome are joined at the centromere.

chromatin (krō′-ma-tin): the complex of DNA and proteins that makes up eukaryotic chromosomes.

chromosome (krō′-mō-sōm): a DNA double helix and associated proteins that help to organize and regulate the use of the DNA.

chylomicron: a particle produced by cells of the small intestine; it consists of proteins, triglycerides, and cholesterol, and it transports the products of lipid digestion into the lymphatic system and ultimately into the circulatory system.

chyme (kīm): an acidic, soup-like mixture of partially digested food, water, and digestive secretions that is released from the stomach into the small intestine.

chytrid: a member of the fungus clade Chytridomycota, which includes species with flagellated swimming spores.

ciliate (sil′-ē-et): a member of a protist group characterized by cilia and a complex unicellular structure. Ciliates are part of a larger group known as the alveolates.

cilium (sil′-ē-um; pl., cilia): a short, hair-like, motile projection from the surface of certain eukaryotic cells that contains microtubules in a 9 + 2 arrangement. The movement of cilia may propel cells through a fluid medium or move fluids over a stationary surface layer of cells.

citric acid cycle: a cyclic series of reactions, occurring in the matrix of mitochondria, in which the acetyl groups from the pyruvic acids produced by glycolysis are broken down to CO_2, accompanied by the formation of ATP and electron carriers; also called the *Krebs cycle*.

clade: a group that includes all the organisms descended from a common ancestor, but no other organisms; a monophyletic group.

class: in Linnaean classification, the taxonomic rank composed of related orders. Closely related classes form a phylum.

classical conditioning: a type of learning in which an animal learns to associate an innate behavior with a novel stimulus, as when a dog is trained to salivate in response to the sound of a bell.

cleavage: the early cell divisions of embryos, in which little or no growth occurs between divisions; reduces the cell size and distributes gene-regulating substances to the newly formed cells.

climate: patterns of weather that prevail for long periods of time (from years to centuries) in a given region.

climate change: a long-lasting change in weather patterns, which may include significant changes in temperature, precipitation, the timing of seasons, and the frequency and severity of extreme weather events.

climax community: a diverse and relatively stable community that forms the endpoint of succession.

clitoris: an external structure of the female reproductive system that is composed of erectile tissue; a sensitive point of stimulation during the sexual response.

clonal selection: the mechanism by which the adaptive immune response gains specificity; an invading antigen elicits a response from only a few lymphocytes, which proliferate to form a clone of cells that attack only the specific antigen that stimulated their production.

clone: offspring that are produced by mitosis and are, therefore, genetically identical to each other.

cloning: the process of producing many identical copies of a gene; also the production of many genetically identical copies of an organism.

closed circulatory system: a type of circulatory system, found in certain worms and vertebrates, in which the blood is always confined within the heart and vessels.

clumped distribution: the distribution characteristic of populations in which individuals are clustered into groups; the groups may be social or based on the need for a localized resource.

cochlea (kahk′-lē-uh): a coiled, bony, fluid-filled tube found in the mammalian inner ear; contains mechanoreceptors (hair cells) that respond to the vibration of sound.

codominance: the relation between two alleles of a gene, such that both alleles are phenotypically expressed in heterozygous individuals.

codon: a sequence of three bases of messenger RNA that specifies a particular amino acid to be incorporated into a protein; certain codons also signal the beginning or end of protein synthesis.

coelom (sē′-lōm): in animals, a space or cavity, lined with tissue derived from mesoderm, that separates the body wall from the inner organs.

coenzyme: an organic molecule that is bound to certain enzymes and is required for the enzymes' proper functioning; typically, a nucleotide bound to a water-soluble vitamin.

coevolution: the evolution of adaptations in two species due to their extensive interactions with one another, such that each species acts as a major force of natural selection on the other.

cohesion: the tendency of the molecules of a substance to stick together.

cohesion–tension mechanism: a mechanism for the transport of water in xylem; water is pulled up the xylem tubes, powered by the force of evaporation of water from the leaves (producing tension) and held together by hydrogen bonds between nearby water molecules (cohesion).

coleoptile (kō-lē-op′-tīl): a sheath surrounding the shoot in monocot seedlings that protects the shoot from abrasion by soil particles during germination.

collecting duct: a tube within the kidney that collects urine from many nephrons and conducts it through the renal medulla into the renal pelvis. Urine may become concentrated in the collecting ducts if antidiuretic hormone (ADH) is present.

collenchyma tissue (kō-len′-ki-muh): a tissue formed from collenchyma cells which are often elongated, with thickened, flexible cell walls. This cell type is alive at maturity and helps support the plant body.

colon: the longest part of the large intestine; does not include the rectum.

colostrum (kō-los′-trum): a yellowish fluid, high in protein and containing antibodies, that is produced by the mammary glands before milk secretion begins.

commensalism: an interaction between species in which one species benefits and the other species is relatively unaffected.

communication: the act of producing a signal that causes a receiver, normally another animal of the same species, to change its behavior in a way that is, on average, beneficial to both signaler and receiver.

community: populations of different species that live in the same area and interact with one another.

compact bone: the hard and strong outer bone; composed of osteons. Compare with *spongy bone*.

companion cell: a cell adjacent to a sieve-tube element in phloem; involved in the control and nutrition of the sieve-tube element.

competition: interaction among individuals who attempt to utilize a resource (for example, food or space) that is limited relative to the demand for that resource.

competitive exclusion principle: the concept that no two species can simultaneously and continuously occupy the same ecological niche.

competitive inhibition: the process by which two or more molecules that are somewhat similar in structure compete for the active site of an enzyme.

complement: a group of blood-borne proteins that participate in the destruction of foreign cells, especially those to which antibodies have bound.

complementary base pair: in nucleic acids, bases that pair by hydrogen bonding. In DNA, adenine is complementary to thymine, and guanine is complementary to cytosine; in RNA, adenine is complementary to uracil, and guanine to cytosine.

complete flower: a flower that has all four floral parts (sepals, petals, stamens, and carpels).

compound eye: a type of eye, found in many arthropods, that is composed of numerous independent subunits called *ommatidia*. Each ommatidium contributes a piece of a mosaic-like image perceived by the animal.

concentration: the amount of solute (often in moles, a measurement that is proportional to the number of molecules) in a given volume of solvent.

concentration gradient: a difference in the concentration of a solute between different regions within a fluid or across a barrier such as a membrane.

conclusion: in the scientific method, a decision about the validity of a hypothesis, based on experiments or observations.

conducting portion: the portion of the respiratory system in lung-breathing vertebrates that carries air to the alveoli.

cone: a cone-shaped photoreceptor cell in the vertebrate retina; not as sensitive to light as are the rods. In humans, the three types of cones are most sensitive to different colors of light and provide color vision; see also *rod*.

conifer (kon′-eh-fer): a member of a group of nonflowering vascular plants whose members reproduce by means of seeds formed inside cones; retains its leaves throughout the year.

conjugation: in prokaryotes, the transfer of DNA from one cell to another via a temporary connection; in single-celled eukaryotes, the mutual exchange of genetic material between two temporarily joined cells.

connection protein: a protein in the plasma membrane of a cell that attaches to the cytoskeleton inside the cell, to other cells, or to the extracellular matrix.

connective tissue: a tissue type consisting of diverse tissues, including bone, cartilage, fat, and blood, that generally contain large amounts of extracellular material.

conservation biology: the application of knowledge from ecology and other areas of biology to understand and conserve biodiversity.

constant-loss population: a population characterized by a relatively constant death rate; constant-loss populations have a roughly linear survivorship curve.

constant region: the part of an antibody molecule that is similar in all antibodies of a given class.

consumer: an organism that eats other organisms; a heterotroph.

consumer–prey interaction: an interaction between species where one species (the consumer) uses another (the prey) as a food source.

contraception: the prevention of pregnancy.

contractile vacuole: a fluid-filled vacuole in certain protists that takes up water from the cytoplasm, contracts, and expels the water outside the cell through a pore in the plasma membrane.

control: that portion of an experiment in which all possible variables are held constant; in contrast to the "experimental" portion, in which a particular variable is altered.

convergent evolution: the independent evolution of similar structures among unrelated organisms as a result of similar environmental pressures; see also *analogous structures*.

convolution: a folding of the cerebral cortex of the vertebrate brain.

copulation: reproductive behavior in which the penis of the male is inserted into the body of the female, where it releases sperm.

coral reef: an ecosystem created by animals (reef-building corals) and plants in warm tropical waters.

core reserve: a natural area protected from most human uses that encompasses enough space to preserve most of the biodiversity of the ecosystems in that area.

cork cambium: a lateral meristem in woody roots and stems that gives rise to cork cells.

cork cell: a protective cell of the bark of woody stems and roots; at maturity, cork cells are dead, with thick, waterproof cell walls.

cornea (kor′-nē-uh): the clear outer covering of the eye, located in front of the pupil and iris; begins the focusing of light on the retina.

corona radiata (kuh-rō′-nuh rā-dē-a′-tuh): the layer of cells surrounding an egg after ovulation.

corpus callosum (kor′ pus kal-ō′-sum): the band of axons that connects the two cerebral hemispheres of vertebrates.

corpus luteum (kor′-pus loo′-tē-um): in the mammalian ovary, a structure that is derived from the follicle after ovulation and that secretes the hormones estrogen and progesterone.

cortex: the part of a primary root or stem located between the epidermis and the vascular cylinder.

cortisol (kor′-ti-sol): a steroid hormone released into the bloodstream by the adrenal cortex in response to stress. Cortisol helps the body cope with short-term stressors by raising blood glucose levels; it also inhibits the immune response.

cotyledon (kot-ul-ē′-don): a leaflike structure within a seed that absorbs food molecules from the endosperm and transfers them to the growing embryo; also called *seed leaf*.

countercurrent exchange: a mechanism for the transfer of some property, such as heat or a dissolved substance, from one fluid to another, generally without the two fluids actually mixing; in countercurrent exchange, the two fluids flow past one another in opposite directions, and they transfer heat or solute from the fluid with the higher temperature or higher solute concentration to the fluid with the lower temperature or lower solute concentration.

coupled reaction: a pair of reactions, one exergonic and one endergonic, that are linked together such that the energy produced by the exergonic reaction provides the energy needed to drive the endergonic reaction.

covalent bond (kō-vā′-lent): a chemical bond between atoms in which electrons are shared.

craniate: an animal that has a skull.

crassulacean acid metabolism (CAM): a biochemical pathway used by some plants in hot, dry climates, that increases the efficiency of carbon fixation during photosynthesis. Mesophyll cells capture carbon dioxide at night and use it to produce sugar during the day.

CRISPR-Cas9: a molecular tool for making precisely targeted changes to a cell's DNA. The tool combines an enzyme that cuts DNA with a customized RNA molecule that guides the enzyme to the targeted site in a genome.

critically endangered species: a species that faces an extreme risk of extinction in the wild in the immediate future.

cross-fertilization: the union of sperm and egg from two individuals of the same species.

crossing over: the exchange of corresponding segments of the chromatids of two homologous chromosomes during meiosis I; occurs at chiasmata.

cuticle (kū′-ti-kul): a waxy or fatty coating on the surfaces of the aboveground epidermal cells of many land plants; aids in the retention of water.

cyclic adenosine monophosphate (cyclic AMP): a cyclic nucleotide, formed within many target cells as a result of the reception of amino acid derived or peptide hormones, that causes metabolic changes in the cell.

cytokine (sī′-tō-kīn): any of several chemical messenger molecules released by cells that facilitate communication with other cells and transfer signals within and among the various systems of the body. Cytokines are important in cellular differentiation and the adaptive immune response.

cytokinesis (sī-tō-ki-nē′-sis): the division of the cytoplasm and organelles into two daughter cells during cell division; normally occurs during telophase of mitotic and meiotic cell division.

cytokinin (sī-tō-ki′-nin): a group of plant hormones that promotes cell division, fruit development, and the sprouting of lateral buds; also delays the senescence of plant parts, especially leaves.

cytoplasm (sī′-tō-plaz-um): all of the material contained within the plasma membrane of a cell, exclusive of the nucleus.

cytosine (C): a nitrogenous base found in both DNA and RNA; abbreviated as C.

cytoskeleton: a network of protein fibers in the cytoplasm that gives shape to a cell, holds and moves organelles, and is typically involved in cell movement.

cytosol: the fluid portion of the cytoplasm; the substance within the plasma membrane exclusive of the nucleus and organelles.

cytotoxic T cell: a type of T cell that, upon contacting foreign cells, directly destroys them.

daughter cell: one of the two cells formed by cell division.

day-neutral plant: a plant in which flowering occurs as soon as the plant has undergone sufficient growth and development, regardless of day length.

death rate (*d*): the number of deaths per individual in a specified unit of time, such as a year.

decomposer: an organism, usually a fungus or bacterium, that digests organic material by secreting digestive enzymes into the environment, in the process liberating nutrients into the environment.

deductive reasoning: the process of generating hypotheses about the results of a specific experiment or the nature of a specific observation.

deforestation: the excessive cutting of forests. In recent years, deforestation has occurred primarily in rain forests in the Tropics, to clear space for agriculture.

dehydration synthesis: a chemical reaction in which two molecules are joined by a covalent bond with the simultaneous removal of a hydrogen from one molecule and a hydroxyl group from the other, forming water; the reverse of hydrolysis.

deletion mutation: a mutation in which one or more pairs of nucleotides are removed from a gene.

demographic transition: a change in population dynamic in which a fairly stable population with both high birth rates and high death rates experiences rapid growth as death rates decline and then returns to a stable (although much larger) population as birth rates decline.

demography: the study of the changes in human numbers over time, grouped by world regions, age, sex, educational levels, and other variables.

denature: to disrupt the secondary and/or tertiary structure of a protein while leaving its amino acid sequence intact. Denatured proteins can no longer perform their biological functions.

denatured: having the secondary and/or tertiary structure of a protein disrupted, while leaving the amino acid sequence unchanged. Denatured proteins can no longer perform their biological functions.

dendrite (den′-drīt): a branched tendril that extends outward from the cell body of a neuron; specialized to respond to signals from the external environment or from other neurons.

dendritic cell (den-drit′-ick): a type of phagocytic leukocyte that presents antigen to T and B cells, thereby stimulating an adaptive immune response to an invading microbe.

denitrifying bacteria (dē-nī′-treh-fī-ing): bacteria that break down nitrates, releasing nitrogen gas to the atmosphere.

density-dependent: referring to any factor, such as predation, that limits population size to an increasing extent as the population density increases.

density-independent: referring to any factor, such as floods or fires, that limits a population's size regardless of its density.

deoxyribonucleic acid (dē-ox′-ē-rī-bō-noo-klā′-ik; DNA): a molecule composed of deoxyribose nucleotides; contains the genetic information of all living cells.

dermal tissue system: a plant tissue system that makes up the outer covering of the plant body.

dermis (dur′-mis): the layer of skin beneath the epidermis; composed of connective tissue and containing blood vessels, muscles, nerve endings, and glands.

desert: a biome in which less than 10 inches (25 centimeters) of rain fall each year; characterized by cacti, succulents, and widely spaced, drought-resistant bushes.

desertification: the process by which relatively dry, drought-prone regions are converted to desert as a result of drought and overuse of the land, for example, by overgrazing or cutting of trees.

desmosome (dez′-mō-sōm): a strong cell-to-cell junction that attaches adjacent cells to one another.

detritivore (de-trī′-ti-vor): one of a diverse group of organisms, ranging from worms to vultures, that eat the wastes and dead remains of other organisms.

deuterostome (doo′-ter-ō-stōm): an animal with a mode of embryonic development in which the coelom is derived from outpocketings of the gut; characteristic of echinoderms and chordates.

development: the process by which an organism proceeds from fertilized egg through adulthood to eventual death.

diabetes mellitus (di-uh-bē′-tēs mel-ī′-tus): a disease characterized by defects in the production, release, or reception of insulin; characterized by high blood glucose levels that fluctuate with sugar intake.

diaphragm (dī′-uh-fram): in the respiratory system, a dome-shaped muscle forming the floor of the chest cavity; when the diaphragm contracts, it flattens, enlarging the chest cavity and causing air to be drawn into the lungs.

diastolic pressure (dī′-uh-stal′-ik): the blood pressure measured during relaxation of the ventricles; the lower of the two blood pressure readings.

diatom (dī′-uh-tom): a member of a protist group that includes photosynthetic forms with two-part glassy outer coverings; important photosynthetic organisms in fresh water and salt water. Diatoms are part of a larger group known as the stramenopiles.

dicot (dī′-kaht): short for dicotyledon; a type of flowering plant characterized by embryos with two cotyledons, or seed leaves, that are usually modified for food storage.

differentiate: the process whereby a cell becomes specialized in structure and function.

differentiated cell: a mature cell specialized for a specific function; in plants, differentiated cells normally do not divide.

diffusion: the net movement of solute particles from a region of high solute concentration to a region of low solute concentration, driven by a concentration gradient; may occur within a fluid or across a barrier such as a membrane.

digestion: the process by which food is physically and chemically broken down into molecules that can be absorbed by cells.

digestive system: a group of organs responsible for ingesting food, digesting food into simple molecules that can be absorbed into the circulatory system, and expelling undigested wastes from the body.

dinoflagellate (dī-nō-fla′-jel-et): a member of a protist group that includes photosynthetic forms in which two flagella project through armor-like plates; abundant in oceans; can reproduce rapidly, causing "red tides." Dinoflagellates are part of a larger group known as the alveolates.

diploid (dip′-loid): referring to a cell with pairs of homologous chromosomes.

diplomonad: a member of a protist group characterized by two nuclei and multiple flagella. Diplomonads, which include disease-causing parasites such as *Giardia*, are part of a larger group known as the excavates.

direct development: a developmental pathway in which the offspring is born as a miniature version of the adult and does not radically change in body form as it grows and matures.

directional selection: a type of natural selection that favors one extreme of a range of phenotypes.

disaccharide (dī-sak′-uh-rīd): a carbohydrate formed by the covalent bonding of two monosaccharides.

disruptive selection: a type of natural selection that favors both extremes of a range of phenotypes.

dissolve: the process by which solvent molecules completely surround and disperse the individual atoms or molecules of another substance, the solute.

distal tubule: the last section of a mammalian nephron, following the nephron loop and emptying urine into a collecting duct; most secretion and a small amount of reabsorption occur in the distal tubule.

disturbance: any event that disrupts an ecosystem by altering its community, its abiotic structure, or both; disturbance precedes succession.

disulfide bond: the covalent bond formed between the sulfur atoms of two cysteines in a protein; typically causes the protein to fold by bringing otherwise distant parts of the protein close together.

DNA cloning: any of a variety of technologies that are used to produce multiple copies of a specific segment of DNA (usually a gene).

DNA helicase: an enzyme that helps unwind the DNA double helix during DNA replication.

DNA ligase: an enzyme that bonds the terminal sugar in one DNA strand to the terminal phosphate in a second DNA strand, creating a single strand with a continuous sugar-phosphate backbone.

DNA polymerase: an enzyme that bonds DNA nucleotides together into a continuous strand, using a preexisting DNA strand as a template.

DNA probe: a sequence of nucleotides that is complementary to the nucleotide sequence of a gene or other segment of DNA under study; used to locate the gene or DNA segment during gel electrophoresis or other methods of DNA analysis.

DNA profile: the pattern of short tandem repeats of specific DNA segments; using a standardized set of 13 short tandem repeats, DNA profiles identify individual people with great accuracy.

DNA replication: the copying of the double-stranded DNA molecule, producing two identical DNA double helices.

DNA sequencing: the process of determining the order of nucleotides in a DNA molecule.

domain: the broadest category for classifying organisms; organisms are classified into three domains: Bacteria, Archaea, and Eukarya.

dominance hierarchy: a social structure that arises when the animals in a social group establish individual ranks that determine access to resources; ranks are usually established through aggressive interactions.

dominant: an allele that can determine the phenotype of heterozygotes completely, such that they are indistinguishable from individuals homozygous for the allele; in the heterozygotes, the expression of the other (recessive) allele is completely masked.

dormancy: a state in which an organism does not grow or develop; usually marked by lowered metabolic activity and resistance to adverse environmental conditions.

dorsal root ganglion: a ganglion, located on the dorsal (sensory) branch of each spinal nerve, that contains the cell bodies of sensory neurons.

double circulation: the separation of circulatory routes between (1) the heart and lungs and (2) the heart and the rest of the body.

double fertilization: in flowering plants, the fusion of two sperm nuclei with the nuclei of two cells of the female gametophyte. One sperm nucleus fuses with the egg to form the zygote; the second sperm nucleus fuses with the two haploid nuclei of the central cell to form a triploid endosperm cell.

double helix (hē′-liks): the shape of the two-stranded DNA molecule; similar to a ladder twisted lengthwise into a corkscrew shape.

Down syndrome: a genetic disorder caused by the presence of three copies of chromosome 21; common characteristics include learning disabilities, distinctively shaped eyelids, a small mouth, heart defects, and low resistance to infectious diseases; also called *trisomy 21*.

duodenum: the first section of the small intestine, in which most food digestion occurs; receives chyme from the stomach, buffers and digestive enzymes from the pancreas, and bile from the liver and gallbladder.

duplicated chromosome: a eukaryotic chromosome following DNA replication; consists of two sister chromatids joined at the centromeres.

early-loss population: a population characterized by a high birth rate, a high death rate among juveniles, and lower death rates among adults; early loss populations have a concave survivorship curve.

ecdysone (ek-dī′-sōn): a steroid hormone that triggers molting in insects and other arthropods.

ecological economics: the branch of economics that attempts to determine the monetary value of ecosystem services and to compare the monetary value of natural ecosystems with the monetary value of human activities that may reduce the services that natural ecosystems provide.

ecological footprint: the area of productive land needed to produce the resources used and absorb the wastes (including carbon dioxide) generated by an individual person, or by an average person of a specific part of the world (for example, an individual country), or of the entire world, using current technologies.

ecological isolation: reproductive isolation that arises when species do not interbreed because they occupy different habitats.

ecological niche (nitch): the role of a particular species within an ecosystem, including all aspects of its interaction with the living and nonliving environments.

ecology (ē-kol′-uh-jē): the study of the interrelationships of organisms with each other and with their nonliving environment.

ecosystem (ē′-kō-sis-tem): all the organisms and their nonliving environment within a defined area.

ecosystem services: the processes through which natural ecosystems and their living communities sustain and fulfill human life. Ecosystem services include purifying air and water, replenishing oxygen, pollinating plants, reducing flooding, providing wildlife habitat, and many more.

ectoderm (ek′-tō-derm): the outermost embryonic tissue layer, which gives rise to structures such as hair, the epidermis of the skin, and the nervous system.

ectotherm: an animal that obtains most of its body warmth from its environment; body temperatures of ectotherms vary with the temperature of their surroundings.

effector (ē-fek′-tor): a part of the body (normally a muscle or gland) that carries out responses as directed by the nervous system.

egg: the haploid female gamete, usually large and nonmotile; contains food reserves for the developing embryo.

electromagnetic spectrum: the range of all possible wavelengths of electromagnetic radiation, from wavelengths longer than radio waves, to microwaves, infrared, visible light, ultraviolet, X-rays, and gamma rays.

electron: a subatomic particle, found in an electron shell outside the nucleus of an atom, that bears a unit of negative charge and very little mass.

electron carrier: a molecule that can reversibly gain or lose electrons. Electron carriers generally accept high-energy electrons produced during an exergonic reaction and donate the electrons to acceptor molecules that use the energy to drive endergonic reactions.

electron shell: a region in an atom within which electrons orbit; each shell corresponds to a fixed energy level at a given distance from the nucleus.

electron transport chain (ETC): a series of electron carrier molecules, found in the thylakoid membranes of chloroplasts and the inner membrane of mitochondria, that extract energy from electrons and generate ATP or other energetic molecules.

element: a substance that cannot be broken down, or converted, to a simpler substance by ordinary chemical reactions.

elimination: the expulsion of indigestible materials from the digestive tract, through the anus, and outside the body.

embryo (em′-brē-ō): in animals, the stages of development that begin with the fertilization of the egg cell and end with hatching or birth; in mammals, the early stages in which the developing animal does not yet resemble the adult of the species.

embryonic disk: in human embryonic development, the flat, two-layered group of cells that separates the amniotic cavity from the yolk sac; the cells of the embryonic disk produce most of the developing embryo.

embryonic stem cell (ESC): a cell derived from an early embryo that is capable of differentiating into any of the adult cell types.

emerging infectious disease: a previously unknown infectious disease (one caused by a microbe), or a previously known infectious disease whose frequency or severity has significantly increased in the past two decades.

emigration (em-uh-grā′-shun): migration of individuals out of an area.

endangered species: a species that faces a high risk of extinction in the wild in the near future.

endergonic (en-der-gon′-ik): pertaining to a chemical reaction that requires an input of energy to proceed; an "uphill" reaction.

endocrine communication: communication between cells of an animal in which certain cells, usually part of an endocrine gland, release hormones that travel through the bloodstream to other cells (often in distant parts of the body) and alter their activity.

endocrine disrupter: an environmental pollutant that interferes with endocrine function, often by disrupting the action of sex hormones.

endocrine gland: a ductless, hormone-producing gland consisting of cells that release their secretions into the interstitial fluid from which the secretions diffuse into nearby capillaries; most endocrine glands are composed of epithelial cells.

endocrine hormone: a molecule produced by the cells of endocrine glands and released into the circulatory system. An endocrine hormone causes changes in target cells that bear specific receptors for the hormone.

endocrine system: an animal's organ system for cell-to-cell communication; composed of hormones and the cells that secrete them.

endocytosis (en-dō-sī-tō′-sis): the process in which the plasma membrane engulfs extracellular material, forming membrane-bound sacs that enter the cytoplasm and thereby move material into the cell.

endoderm (en′-dō-derm): the innermost embryonic tissue layer, which gives rise to structures such as the lining of the digestive and respiratory tracts.

endodermis (en-dō-der′-mis): the innermost layer of small, close-fitting cells of the cortex of a root that form a ring around the vascular cylinder; see also *Casparian strip*.

endomembrane system: internal membranes that create loosely connected compartments within the eukaryotic cell. It includes the nuclear envelope, the endoplasmic reticulum, vesicles, the Golgi apparatus, and lysosomes.

endometrium (en-dō-mē′-trē-um): the nutritive inner lining of the uterus.

endoplasmic reticulum (en-dō-plaz′-mik re-tik′-ū-lum; ER): a system of membranous tubes and channels in eukaryotic cells; the site of most protein and lipid synthesis.

endoskeleton (en′-dō-skel′-uh-tun): a rigid internal skeleton with flexible joints that allow for movement.

endosperm: a triploid food storage tissue in the seeds of flowering plants that nourishes the developing plant embryo.

endospore: a protective resting structure of some rod-shaped bacteria that withstands unfavorable environmental conditions.

endosymbiont hypothesis: the hypothesis that certain organelles, especially chloroplasts and mitochondria, arose as mutually beneficial associations between the ancestors of eukaryotic cells and captured bacteria that lived within the cytoplasm of the pre-eukaryotic cell.

endotherm: an animal that obtains most of its body heat from metabolic activities; body temperatures of endotherms usually remain relatively constant within a fairly wide range of environmental temperatures.

energy: the capacity to do work.

energy-carrier molecule: high-energy molecules that are synthesized at the site of an exergonic reaction, where they capture one or two energized electrons and hydrogen ions. They include nicotinamide adenine dinucleotide (NADH) and flavin adenine dinucleotide ($FADH_2$).

energy pyramid: a graphical representation of the energy contained in succeeding trophic levels, with maximum energy at the base (primary producers) and steadily diminishing amounts at higher levels.

energy-requiring transport: the transfer of substances across a cell membrane using cellular energy; includes active transport, endocytosis, and exocytosis.

entropy (en′-trō-pē): a measure of the amount of randomness and disorder in a system.

environmental resistance: any factor that tends to counteract biotic potential, limiting population growth and the resulting population size.

enzyme (en′zīm): a biological catalyst, usually a protein, that speeds up the rate of specific biological reactions.

epicotyl hook (ep′-ē-kot-ul): in dicots, a hook in the embryonic shoot that is located above the attachment point of the cotyledons but below the tip of the shoot.

epidermal tissue: dermal tissue in plants that forms the epidermis; the outermost cell layer that covers leaves, young stems, and young roots.

epidermis (ep-uh-der′-mis): in animals, specialized stratified epithelial tissue that forms the outer layer of the skin; in plants, the outermost layer of cells of a leaf, young root, or young stem.

epididymis (e-pi-di′-di-mus): a series of tubes that connect with and receive sperm from the seminiferous tubules of the testis and empty into the vas deferens.

epigenetics: the study of the mechanisms by which cells and organisms change gene expression and function without changing the base sequence of their DNA; usually, epigenetic controls over DNA expression involve modification of DNA, modification of chromosomal proteins, or alteration of transcription or translation through the actions of noncoding RNA molecules.

epiglottis (ep-eh-glah′-tis): a flap of cartilage in the lower pharynx that covers the opening to the larynx during swallowing; directs food into the esophagus.

epinephrine (ep-i-nef′-rin): a hormone, secreted by the adrenal medulla, that is released in response to stress and that stimulates a variety of responses, including the release of glucose from the liver and an increase in heart rate.

epithelial tissue (eh-puh-thē′-lē-ul): a tissue type that forms membranes that cover the body surface and line body cavities, and that also gives rise to glands.

equilibrium population: a population in which allele frequencies and the distribution of genotypes do not change from generation to generation.

erythrocyte (eh-rith′-rō-sīt): a red blood cell, which contains the oxygen-binding protein hemoglobin and thus transports oxygen in the circulatory system.

erythropoietin (eh-rith′-rō-pō-ē′-tin): a hormone produced by the kidneys in response to oxygen deficiency; stimulates the production of red blood cells by the bone marrow.

esophagus: a muscular, tubular portion of the mammalian digestive tract located between the pharynx and the stomach; no digestion occurs in the esophagus.

essential amino acid: an amino acid that is a required nutrient; the body is unable to manufacture essential amino acids, so they must be supplied in the diet.

essential fatty acid: a fatty acid that is a required nutrient; the body is unable to manufacture essential fatty acids, so they must be supplied in the diet.

essential nutrient: any nutrient that cannot be synthesized by the body, including certain fatty acids and amino acids, vitamins, minerals, and water.

estrogen: a female sex hormone, produced by follicle cells of the ovary, that stimulates follicle development, oogenesis, the development of secondary sex characteristics, and growth of the uterine lining.

estuary (es′-choō-ār-ē): a wetland formed where a river meets the ocean; the salinity is quite variable, but lower than in seawater and higher than in fresh water.

ethylene: a plant hormone that promotes the ripening of some fruits and the dropping of leaves and fruit; promotes senescence of leaves.

euglenid (ū-gle′-nid): a member of a protist group characterized by one or more whip-like flagella, which are used for locomotion, and by a photoreceptor, which detects light. Euglenids are photosynthetic and are part of a larger group known as excavates.

Eukarya (ū-kar′-ē-a): one of life's three domains; consists of all eukaryotes (plants, animals, fungi, and protists).

eukaryote (ū-kar′-ē-ōt): an organism whose cells are eukaryotic; plants, animals, fungi, and protists are eukaryotes.

eukaryotic (ū-kar-ē-ot′-ik): referring to cells of organisms of the domain Eukarya (plants, animals, fungi, and protists). Eukaryotic cells have genetic material enclosed within a membrane-bound nucleus, and they contain other membrane-bound organelles.

eutrophic lake: a lake that receives sufficiently large inputs of sediments, organic material, and inorganic nutrients from its surroundings to support dense communities, especially of plants and phytoplankton; contains murky water with poor light penetration.

evolution: (1) the descent of modern organisms, with modification, from preexisting life-forms; (2) the theory that all organisms are related by common ancestry and have changed over time; (3) any change in the genetic makeup (the proportions of different genotypes) of a population from one generation to the next.

excavate: a member of the Excavata, a protist clade. The excavates, many of which lack mitochondria, include the diplomonads, parabasalids, kinetoplastids, and euglenids.

excitatory postsynaptic potential (EPSP): an electrical signal produced in a postsynaptic cell that makes the resting potential of the postsynaptic neuron less negative and, hence, makes the neuron more likely to produce an action potential.

excretion: the elimination of waste substances from the body; can occur from the digestive system, skin glands, urinary system, or lungs.

exergonic (ex-er-gon′-ik): pertaining to a chemical reaction that releases energy (either as heat or in the form of increased entropy); a "downhill" reaction.

exhalation: the act of releasing air from the lungs, which results from a relaxation of the respiratory muscles.

exocrine gland: a gland that releases its secretions into ducts that lead to the outside of the body or into a body cavity, such as the digestive or reproductive system; most exocrine glands are composed of epithelial cells.

exocytosis (ex-ō-sī-tō′-sis): the process in which intracellular material is enclosed within a membrane-bound sac that moves to the plasma membrane and fuses with it, releasing the material outside the cell.

exon: a segment of DNA in a eukaryotic gene that codes for amino acids in a protein; see also *intron*.

exoskeleton (ex′-ō-skel′-uh-tun): a rigid external skeleton that supports the body, protects the internal organs, and has flexible joints that allow for movement.

experiment: in the scientific method, the use of carefully controlled observations or manipulations to test the predictions generated by a hypothesis.

exponential growth: a continuously accelerating increase in population size; this type of growth generates a curve shaped like the letter "J."

extensor: a muscle that straightens (increases the angle of) a joint.

external fertilization: the union of sperm and egg outside the body of either parent.

extinction: the death of all members of a species.

extracellular digestion: the physical and chemical breakdown of food that occurs outside a cell, normally in a digestive cavity.

extracellular matrix (ECM): material secreted by and filling the spaces between cells. Animal cells secrete supporting and adhesive proteins embedded in a gel of polysaccharides linked by proteins; plant cells secrete a matrix of cellulose that forms cell walls.

extraembryonic membrane: in the embryonic development of reptiles (including birds) and mammals, one of the following membranes: the chorion (functions in gas exchange), amnion (provision of the watery environment needed for development), allantois (waste storage), or yolk sac (storage of the yolk).

facilitated diffusion: the diffusion of molecules across a membrane, assisted by protein pores or carriers embedded in the membrane.

family: in Linnaean classification, the taxonomic rank composed of related genera. Closely related families make up an order.

farsighted: the inability to focus on nearby objects, caused by the eyeball being slightly too short or the cornea too flat.

fat (molecular): a lipid composed of three saturated fatty acids covalently bonded to glycerol; fats are solid at room temperature.

fatty acid: an organic molecule composed of a long chain of carbon atoms, with a carboxylic acid ($-COOH$) group at one end; may be saturated (all single bonds between the carbon atoms) or unsaturated (one or more double bonds between the carbon atoms).

feces: semisolid waste material that remains in the intestine after absorption is complete and that is voided through the anus. Feces consist principally of indigestible wastes and bacteria.

feedback inhibition: in enzyme-mediated chemical reactions, the condition in which the product of a reaction inhibits one or more of the enzymes involved in synthesizing the product.

fermentation: anaerobic reactions that convert the pyruvic acid produced by glycolysis into lactic acid or alcohol and CO_2, using hydrogen ions and electrons from NADH; the primary function of fermentation is to regenerate NAD^+ so that glycolysis can continue under anaerobic conditions.

fertilization: the fusion of male and female haploid gametes, forming a zygote.

fetal alcohol syndrome (FAS): a cluster of symptoms, including mental retardation and physical abnormalities, that occur in infants born to mothers who consumed large amounts of alcoholic beverages during pregnancy.

fetus: the later stages of mammalian embryonic development (after the second month for humans), when the developing animal has come to resemble the adult of the species.

fever: an elevation in body temperature caused by chemicals (pyrogens) that are released by white blood cells in response to infection.

fibrillation (fi-bri-lā′-shun): the rapid, uncoordinated, and ineffective contraction of heart muscle cells.

fibrin (fī′-brin): a clotting protein formed in the blood in response to a wound; binds with other fibrin molecules and provides a matrix around which a blood clot forms.

fibrinogen (fī-brin′-ō-jen): the inactive form of the clotting protein fibrin. Fibrinogen is converted into fibrin by the enzyme thrombin, which is produced in response to injury.

fibrous root system: a root system, commonly found in monocots, that is characterized by many roots of approximately the same size arising from the base of the stem.

filament: in flowers, the stalk of a stamen, which bears an anther at its tip.

filtrate: the fluid produced by filtration; in the kidneys, the fluid produced by the filtration of blood through the glomerular capillaries.

filtration: within renal corpuscle in each nephron of a kidney, the process by which blood is pumped under pressure through permeable capillaries of the glomerulus, forcing out water and small solutes, including wastes and nutrients.

first law of thermodynamics: the principle of physics that states that within any isolated system, energy can be neither created nor destroyed, but can be converted from one form to another; also called the law of conservation of energy.

fitness: the reproductive success of an organism, relative to the average reproductive success in the population.

flagellum (fla-jel′-um; pl., flagella): a long, hair-like, motile extension of the plasma membrane; in eukaryotic cells, it contains microtubules arranged in a $9 + 2$ pattern. The movement of flagella propels some cells through fluids.

flavin adenine dinucleotide (FAD or $FADH_2$): an electron-carrier molecule produced in the mitochondrial matrix by the Krebs cycle; subsequently donates electrons to the electron transport chain.

flexor: a muscle that flexes (decreases the angle of) a joint.

florigen: one of a group of plant hormones that may stimulate or inhibit flowering in response to day length.

flower: the reproductive structure of an angiosperm plant.

flower bud: a cluster of meristem cells (a bud) that forms a flower.

fluid: any substance whose molecules can freely flow past one another; "fluid" can describe liquids, cell membranes, and gases.

fluid mosaic model: a model of cell membrane structure; according to this model, membranes are composed of a double layer of phospholipids in which various proteins are embedded. The phospholipid bilayer is a somewhat fluid matrix that allows the movement of proteins within it.

follicle: in the ovary of female mammals, the oocyte and its surrounding accessory cells.

follicle-stimulating hormone (FSH): a hormone, produced by the anterior pituitary, that stimulates spermatogenesis in males and the development of the follicle in females.

food chain: a linear feeding relationship in a community, using a single representative from each of the trophic levels.

food vacuole: a membranous sac, within a single cell, in which food is enclosed. Digestive enzymes are released into the vacuole, where intracellular digestion occurs.

food web: a representation of the complex feeding relationships within a community, including many organisms at various trophic levels, with many of the consumers occupying more than one level simultaneously.

foraminiferan (for-am-i-nif′-er-un): a member of a protist group characterized by pseudopods and elaborate calcium carbonate shells. Foraminiferans are generally aquatic (largely marine) and are part of a larger group known as rhizarians.

forebrain: during development, the anterior portion of the brain. In mammals, the forebrain differentiates into the thalamus, the limbic system, and the cerebrum. In humans, the cerebrum contains about half of all the neurons in the brain.

fossil: the remains of a dead organism, normally preserved in rock. Fossils may be petrified bones or wood shells; eggs; feces; impressions of body forms, such as feathers, skin, or leaves; or markings made by organisms, such as footprints.

fossil fuel: a fuel, such as coal, oil, and natural gas, derived from the remains of ancient organisms.

founder effect: the result of an event in which an isolated population is founded by a small number of individuals; may result in genetic drift if allele frequencies in the founder population are by chance different from those of the parent population.

fovea (fō′-vē-uh): in the vertebrate retina, the central region on which images are focused; contains closely packed cones.

fragmentation: a mechanism of asexual reproduction in which an animal splits its body apart and the resulting pieces regenerate the missing parts of a complete body.

free nucleotide: a nucleotide that has not been joined with other nucleotides to form a DNA or RNA strand.

free radical: a molecule containing an atom with an unpaired electron, which makes it highly unstable and reactive with nearby molecules. By removing an electron from the molecule it attacks, it creates a new free radical and begins a chain reaction that can lead to the destruction of biological molecules crucial to life.

fruit: in flowering plants, the ripened ovary (plus, in some cases, other parts of the flower), which contains the seeds.

functional group: one of several groups of atoms commonly found in an organic molecule, including hydrogen, hydroxyl, amino, carboxyl, and phosphate groups, that determine the characteristics and chemical reactivity of the molecule.

gallbladder: a small sac located next to the liver that stores and concentrates the bile secreted by the liver. Bile is released from the gallbladder to the small intestine through the bile duct.

gamete (gam′-ēt): a haploid sex cell, usually a sperm or an egg, formed in sexually reproducing organisms.

gametic incompatibility: a postmating reproductive isolating mechanism that arises when sperm from one species cannot fertilize eggs of another species.

gametophyte (ga-mēt′-ō-fīt): the multicellular haploid stage in the life cycle of plants.

ganglion (gang′-lē-un; pl., ganglia): a cluster of neurons.

ganglion cell: in the vertebrate retina, the nerve cells whose axons form the optic nerve.

gap junction: a type of cell-to-cell junction in animals in which channels connect the cytoplasm of adjacent cells.

gas-exchange portion: the portion of the respiratory system in lung-breathing vertebrates where gas is exchanged in the alveoli of the lungs.

gastric gland: one of numerous small glands in the stomach lining; contains cells that secrete mucus, hydrochloric acid, or pepsinogen (the inactive form of the protease pepsin).

gastrin: a hormone produced by the stomach that stimulates acid secretion in response to the presence of food.

gastrovascular cavity: a saclike chamber with digestive functions, found in some invertebrates such as cnidarians (sea jellies, anemones, and related animals); a single opening serves as both mouth and anus.

gastrula (gas′-troo-luh): in animal development, a three-layered embryo with ectoderm, mesoderm, and endoderm cell layers. The endoderm layer usually encloses the primitive gut.

gastrulation (gas-troo-lā′-shun): the process whereby a blastula develops into a gastrula, including the formation of endoderm, ectoderm, and mesoderm.

gel electrophoresis: a technique in which molecules (such as DNA fragments) are placed in wells in a thin sheet of gelatinous material and exposed to an electric field; the molecules migrate through the gel at a rate determined by certain characteristics, most commonly size.

gene: the unit of heredity; a segment of DNA located at a particular place on a chromosome that usually encodes the information for the amino acid sequence of a protein and, hence, a particular trait.

gene flow: the movement of alleles from one population to another owing to the movement of individual organisms or their gametes.

gene linkage: the tendency for genes located on the same chromosome to be inherited together.

gene pool: the total of all alleles of all genes in a population; for a single gene, the total of all the alleles of that gene that occur in a population.

gene therapy: the attempt to cure a disease by inserting, deleting, or altering a patient's genes.

generative cell: in flowering plants, one of the haploid cells of a pollen grain; undergoes mitotic cell division to form two sperm cells.

genetic code: the collection of codons of mRNA, each of which directs the incorporation of a particular amino acid into a protein during protein synthesis or causes protein synthesis to start or stop.

genetic drift: a change in the allele frequencies of a small population purely by chance.

genetic engineering: the modification of the genetic material of an organism, usually using recombinant DNA techniques.

genetic recombination: the generation of new combinations of alleles on homologous chromosomes due to the exchange of DNA during crossing over.

genetically modified organism (GMO): a plant or animal that contains DNA that has been modified or that has been obtained from another species.

genital herpes: a sexually transmitted disease, caused by a virus, that can cause painful blisters on the genitals and surrounding skin.

genome: the complete set of genetic material in an organism.

genomic imprinting: a form of epigenetic control by which a given gene is expressed in an offspring only if the gene has been inherited from a specific parent; the copy of the gene that was inherited from the other parent is usually not expressed.

genotype (jēn′-ō-tīp): the genetic composition of an organism; the actual alleles of each gene carried by the organism.

genus (jē′-nus): in Linnaean classification, the taxonomic rank composed of related species. Closely related genera make up a family.

geographic isolation: reproductive isolation that arises when species do not interbreed because a physical barrier separates them.

germination: the growth and development of a seed, spore, or pollen grain.

ghrelin (grel′-in): a peptide hormone secreted by the stomach that acts via the hypothalamus to stimulate hunger.

gibberellin (jib-er-el′-in): one of a group of plant hormones that stimulates seed germination, fruit development, flowering, and cell division and elongation in stems.

gill: in aquatic animals, a branched tissue richly supplied with capillaries around which water is circulated for gas exchange.

gland: a cluster of cells that are specialized to secrete substances such as sweat, mucus, enzymes, or hormones.

glia: cells of the nervous system that provide nutrients for neurons, regulate the composition of the interstitial fluid in the brain and spinal cord, modulate communication between neurons, and insulate axons, thereby speeding up the conduction of action potentials. Also called *glial cells.*

glomeromycete: a member of the fungus clade Glomeromycota, which includes species that form mycorrhizal associations with plant roots and that form bush-shaped branching structures inside plant cells.

glomerular capsule: the cup-shaped portion of the renal corpuscle that surrounds the glomerulus and captures blood filtrate.

glomerulus (glō-mer′-ū-lus): a dense network of thin-walled capillaries, located within the renal corpuscle of each nephron of the kidney, where blood pressure forces water and small solutes, including wastes and nutrients, through the capillary walls into the nephron.

glucagon (gloo′-ka-gon): a hormone, secreted by the pancreas, that increases blood sugar by stimulating the breakdown of glycogen (to glucose) in the liver.

glucocorticoid (gloo-kō-kor′-tik-oid): a class of hormones, released by the adrenal cortex in response to the presence of ACTH, that makes energy available to the body by stimulating the synthesis of glucose.

glucose: the most common monosaccharide, with the molecular formula $C_6H_{12}O_6$; most polysaccharides, including cellulose, starch, and glycogen, are made up of glucose subunits covalently bonded together.

glycerol (glis′-er-ol): a three-carbon alcohol to which fatty acids are covalently bonded to make fats and oils.

glycogen (glī′-kō-jen): a highly branched polymer of glucose that is stored by animals in the muscles and liver and metabolized as a source of energy.

glycolysis (glī-kol′-i-sis): reactions, carried out in the cytoplasm, that break down glucose into two molecules of pyruvic acid, producing two ATP molecules; does not require oxygen but can proceed when oxygen is present.

glycoprotein: a protein to which a carbohydrate is attached.

goiter: a swelling of the thyroid gland, caused by iodine deficiency, that affects the functioning of the thyroid gland and its hormones.

Golgi apparatus (gōl′-jē): a stack of membranous sacs, found in most eukaryotic cells, that is the site of processing and separation of membrane components and secretory materials.

gonad: an organ where reproductive cells are formed; in males, the testes, and in females, the ovaries.

gonadotropin-releasing hormone (gō-na-dō-trō′-pin; GnRH): a hormone produced by the neurosecretory cells of the hypothalamus, which stimulates cells in the anterior pituitary to release follicle-stimulating hormone (FSH) and luteinizing hormone (LH). GnRH is involved in the menstrual cycle and in spermatogenesis.

gonorrhea (gon-uh-rē′-uh): a sexually transmitted bacterial infection of the reproductive organs; if untreated, can result in sterility.

gradient: a difference in concentration, pressure, or electrical charge between two regions.

grana: stacks of thylakoid membranes that form disks that surround the thylakoid space within the chloroplast.

grassland: a biome, located in the centers of continents, that primarily supports grasses; also called a *prairie*.

gravitropism: growth with respect to the direction of gravity.

gray matter: the outer portion of the brain and inner region of the spinal cord; composed largely of neuron cell bodies, which give this area a gray color in preserved tissue.

greenhouse effect: the process in which certain gases such as carbon dioxide and methane trap sunlight energy in a planet's atmosphere as heat; the glass in a greenhouse causes a similar warming effect. The result, global warming that causes climate change, is being enhanced by the production of these gases by humans.

greenhouse gas: a gas, such as carbon dioxide or methane, that traps sunlight energy in a planet's atmosphere as heat; a gas that participates in the greenhouse effect.

ground tissue system: a plant tissue system, consisting of parenchyma, collenchyma, and sclerenchyma cells, that makes up the bulk of a leaf or young stem, excluding vascular or dermal tissues. Most ground tissue cells function in photosynthesis, support, or carbohydrate storage.

growth factor: small molecules, usually proteins or steroids, that bind to receptors on or in target cells and enhance their rate of cell division or differentiation.

growth hormone: a hormone, released by the anterior pituitary, that stimulates growth, especially of the skeleton.

growth rate (*r*): a measure of the change in population size per individual per unit of time.

guanine (G): a nitrogenous base found in both DNA and RNA; abbreviated as G.

guard cell: one of a pair of specialized epidermal cells that surrounds the central opening of a stoma in the epidermis of a leaf or young stem and regulates the size of the opening.

gymnosperm (jim′-nō-sperm): a nonflowering seed plant, such as a conifer, gnetophyte, cycad, or gingko.

gyre (jīr): a roughly circular pattern of ocean currents, formed because continents interrupt the flow of the current; rotates clockwise in the Northern Hemisphere and counterclockwise in the Southern Hemisphere.

habitat fragmentation: the process by which human development and activities produce patches of wildlife habitat that may not be large enough to sustain minimum viable populations.

habituation (heh-bich-oo-ā′-shun): a type of simple learning characterized by a decline in response to a repeated stimulus.

hair cell: a type of mechanoreceptor cell found in the inner ear that produces an electrical signal when stiff hair-like cilia projecting from the surface of the cell are bent. Hair cells in the cochlea respond to sound vibrations; those in the vestibular system respond to motion and gravity.

hair follicle (fol′-i-kul): a cluster of specialized epithelial cells located in the dermis of mammalian skin, which produces a hair.

hammer: the first of the small bones of the middle ear that link the tympanic membrane (eardrum) with the oval window of the cochlea; also called the *malleus*.

haploid (hap′ loid): referring to a cell that has only one member of each pair of homologous chromosomes.

Hardy–Weinberg principle: a mathematical model proposing that, under certain conditions, the allele frequencies and genotype frequencies in a sexually reproducing population will remain constant over generations.

heart: a muscular organ responsible for pumping blood within the circulatory system throughout the body.

heart attack: a severe reduction or blockage of blood flow through a coronary artery, depriving some of the heart muscle of its blood supply.

heart rate: the number of cardiac cycles (heartbeats) per minute.

heartwood: older secondary xylem that usually no longer conducts water or minerals, but that contributes to the strength of a tree trunk.

heat of fusion: the energy that must be removed from a compound to transform it from a liquid into a solid at its freezing temperature.

heat of vaporization: the energy that must be supplied to a compound to transform it from a liquid into a gas at its boiling temperature.

Heimlich maneuver: a method of dislodging food or other obstructions that have entered the airway.

helix (hē′-liks): a coiled, spring-like secondary structure of a protein.

helper T cell: a type of T cell that helps other immune cells act against antigens.

hemocoel (hē′-mō-sēl): a cavity within the bodies of certain invertebrates in which a fluid, called hemolymph, bathes tissues directly; part of an open circulatory system.

hemodialysis (hē-mō-dī-al′-luh-sis): a procedure that simulates kidney function in individuals with damaged or ineffective kidneys; blood is diverted from the body, artificially filtered, and returned to the body.

hemoglobin (hē′-mō-glō-bin): the iron-containing protein that gives red blood cells their color; binds to oxygen in the lungs and releases it in the tissues.

hemolymph: in animals with an open circulatory system, the fluid that is located within the hemocoel and that bathes all the body cells, therefore serving as both blood and interstitial fluid.

hemophilia: a recessive, sex-linked disease in which the blood fails to clot normally.

herbivore (erb′-i-vor): literally, "plant-eater"; an organism that feeds directly and exclusively on producers; a primary consumer.

hermaphrodite (her-maf′-ruh-dīt): an organism that possesses both male and female sexual organs. Some hermaphroditic animals can fertilize themselves; others must exchange sex cells with a mate.

hermaphroditic (her-maf′-ruh-dit′-ik): possessing both male and female sexual organs. Some hermaphroditic animals can fertilize themselves; others must exchange sex cells with a mate.

heterotroph (het′-er-ō-trōf′): literally, "other-feeder"; an organism that eats other organisms; a consumer.

heterozygous (het′-er-ō-zī′-gus): carrying two different alleles of a given gene; also called *hybrid*.

hindbrain: the posterior portion of the brain, containing the medulla, pons, and cerebellum.

hinge joint: a joint at which the bones fit together in a way that allows movement in only two dimensions, as at the elbow or knee.

hippocampus (hip-ō-kam′-pus): a part of the forebrain of vertebrates that is important in emotion and especially learning.

histamine (his′-ta-mēn): a substance released by certain cells in response to tissue damage and invasion of the body by foreign substances; promotes the dilation of arterioles and the leakiness of capillaries and triggers some of the events of the inflammatory response.

homeobox gene (hō′-mē-ō-boks): a sequence of DNA coding for a transcription factor protein that activates or inactivates many other genes that control the development of specific, major parts of the body.

homeostasis (hōm-ē-ō-stā′-sis): the maintenance of the relatively constant internal environment that is required for the optimal functioning of cells.

hominin: a human or a prehistoric relative of humans; the oldest known hominin is *Sahelanthropus*, whose fossils are more than 6 million years old.

homologous chromosome: a chromosome that is similar in appearance and genetic information to another chromosome with which it pairs during meiosis; also called *homologue*.

homologous structures: structures that may differ in function but that have similar anatomy, presumably because the organisms that possess them have descended from common ancestors.

homologue (hō′-mō-log): a chromosome that is similar in appearance and genetic information to another chromosome with which it pairs during meiosis; also called *homologous chromosome*.

homozygous (hō-mō-zī′-gus): carrying two copies of the same allele of a given gene; also called *true-breeding*.

hormone: a chemical that is secreted by one group of cells and transported to other cells, whose activity is influenced by reception of the hormone.

host: the prey organism on or in which a parasite lives; the host is harmed by the relationship.

human immunodeficiency virus (HIV): a pathogenic virus that causes acquired immune deficiency syndrome (AIDS) by attacking and destroying the immune system's helper T cells.

human papillomavirus (pap-il-lo′-ma; HPV): a virus that infects the reproductive organs, often causing genital warts; causes most, if not all, cases of cervical cancer.

humoral immunity: an immune response in which foreign substances are inactivated or destroyed by antibodies that circulate in the blood.

Huntington disease: an incurable genetic disorder, caused by a dominant allele, that produces progressive brain deterioration, resulting in the loss of motor coordination, flailing movements, personality disturbances, and eventual death.

hybrid: an organism that is the offspring of parents differing in at least one genetically determined characteristic; also used to refer to the offspring of parents of different species.

hybrid infertility: a postmating reproductive isolating mechanism that arises when hybrid offspring (offspring of parents of two different species) are sterile or have low fertility.

hybrid inviability: a postmating reproductive isolating mechanism that arises when hybrid offspring (offspring of parents of two different species) fail to survive.

hydrogen bond: the weak attraction between a hydrogen atom that bears a partial positive charge (due to polar covalent bonding with another atom) and another atom (oxygen, nitrogen, or fluorine) that bears a partial negative charge; hydrogen bonds may form between atoms of a single molecule or of different molecules.

hydrolysis (hī-drol′-i-sis): the chemical reaction that breaks a covalent bond by means of the addition of hydrogen to the atom on one side of the original bond and a hydroxyl group to the atom on the other side; the reverse of dehydration synthesis.

hydrophilic (hī-drō-fil′-ik): pertaining to molecules that dissolve readily in water, or to molecules that form hydrogen bonds with water; polar.

hydrophobic (hī-drō-fō′-bik): pertaining to molecules that do not dissolve in water or form hydrogen bonds with water; nonpolar.

hydrostatic skeleton (hī-drō-stat′-ik): in invertebrate animals, a body structure in which fluid-filled compartments provide support for the body and change shape when acted on by muscles, which alters the animal's body shape and position or causes the animal to move.

hydrothermal vent community: a community of unusual organisms, living in the deep ocean near hydrothermal vents, that depends on the chemosynthetic activities of sulfur bacteria.

hypertension: arterial blood pressure that is chronically elevated above the normal level.

hypertonic (hī-per-ton′-ik): referring to a solution that has a higher concentration of solute (and therefore a lower concentration of free water) than has the cytosol of a cell.

hypha (hī′-fuh; pl., hyphae): in fungi, a thread-like structure that consists of elongated cells, typically with many haploid nuclei; many hyphae make up the fungal body.

hypocotyl (hī′-pō-kot-ul) hook: in dicots, a hook in the embryonic shoot that is located below the attachment point of the cotyledons but above the root.

hypothalamus: (hī-pō-thal′-a-mus): a region of the forebrain that controls the secretory activity of the pituitary gland; synthesizes, stores, and releases certain peptide hormones; and directs autonomic nervous system responses.

hypothesis (hī-poth′-eh-sis): a proposed explanation for a phenomenon based on available evidence that leads to a prediction that can be tested.

hypotonic (hī-pō-ton′-ik): referring to a solution that has a lower concentration of solute (and therefore a higher concentration of free water) than has the cytosol of a cell.

immigration (im-uh-grā′-shun): migration of individuals into an area.

immune system: a system of cells, including macrophages, B cells, and T cells, and molecules, such as antibodies and cytokines, that work together to combat microbial invasion of the body.

imperfect flower: a flower that is missing either stamens or carpels.

implantation: the process whereby the early embryo embeds itself within the lining of the uterus.

imprinting: a type of learning in which an animal acquires a particular type of information during a specific sensitive phase of development.

incomplete dominance: a pattern of inheritance in which the heterozygous phenotype is intermediate between the two homozygous phenotypes.

incomplete flower: a flower that is missing one of the four floral parts (sepals, petals, stamens, or carpels).

indeterminate growth: growth in plants that continues to elongate and produce new leaves and branches throughout their lives.

indirect development: a developmental pathway in which an offspring goes through radical changes in body form as it matures.

induced pluripotent stem cell (ploo-rē-pō′-tent; iPSC): a type of stem cell produced from nonstem cells by the insertion or forced expression of a specific set of genes that cause the cells to become capable of unlimited cell division and to be able to be differentiated into many different cell types, possibly any cell type of the body.

induction: the process by which a group of cells causes other cells to differentiate into a specific tissue type.

inductive reasoning: the process of creating a generalization based on many specific observations that support the generalization, coupled with an absence of observations that contradict it.

inflammatory response: a nonspecific, local response to injury to the body, characterized by the phagocytosis of foreign substances and tissue debris by white blood cells and by the walling off of the injury site by the clotting of fluids that escape from nearby blood vessels.

ingestion: the movement of food into the digestive tract, usually through the mouth.

inhalation: the act of drawing air into the lungs by enlarging the chest cavity.

inheritance: the genetic transmission of characteristics from parent to offspring.

inhibiting hormone: a hormone, secreted by the neurosecretory cells of the hypothalamus, that inhibits the release of specific hormones from the anterior pituitary.

inhibitory postsynaptic potential (IPSP): an electrical signal produced in a postsynaptic cell that makes the resting potential more negative and, hence, makes the neuron less likely to fire an action potential.

innate (in-āt′): inborn; instinctive; an innate behavior is performed correctly the first time it is attempted.

innate immune response: nonspecific defenses against many different invading microbes; the innate response involves phagocytic white blood cells, natural killer cells, the inflammatory response, and fever.

inner cell mass: in human embryonic development, the cluster of cells, on the inside of the blastocyst, that will develop into the embryo.

inner ear: the innermost part of the mammalian ear; composed of the bony, fluid-filled tubes of the cochlea and the vestibular apparatus.

inorganic: describing any molecule that does not contain both carbon and hydrogen.

insertion: the site of attachment of a muscle to the relatively movable bone on one side of a joint.

insertion mutation: a mutation in which one or more pairs of nucleotides are inserted into a gene.

insight learning: a type of learning in which a problem is solved by understanding the relationships among the components of the problem rather than through trial and error.

insulin: a hormone, secreted by the pancreas, that lowers blood sugar by stimulating many cells to take up glucose and by stimulating the liver to convert glucose to glycogen.

integration: the process of adding up all the electrical signals in a neuron, including sensory inputs and postsynaptic potentials, to determine the output of the neuron (action potentials and/or synaptic transmission).

integument (in-teg′-ū-ment): in plants, the outer layers of cells of the ovule that surround the female gametophyte; develops into the seed coat.

intensity: the strength of stimulation or response.

intercalated disc: junctions connecting individual cardiac muscle cells that serve both to attach adjacent cells to one another and to allow electrical signals to pass between cells.

intermediate filament: part of the cytoskeleton of eukaryotic cells that is composed of several types of proteins and probably functions mainly for support.

intermembrane space: the fluid-filled space between the inner and outer membranes of a mitochondrion.

internal fertilization: the union of sperm and egg inside the body of the female.

interneuron: in a neural network, a nerve cell that is postsynaptic to a sensory neuron and presynaptic to a motor neuron. In actual circuits, there may be many interneurons between individual sensory and motor neurons.

internode: the part of a stem between two nodes.

interphase: the stage of the cell cycle between cell divisions in which chromosomes are duplicated and other cell functions occur, such as growth, movement, and acquisition of nutrients.

interspecific competition: competition among individuals of different species.

interstitial cell (in-ter-sti′-shul): in the vertebrate testis, a testosterone-producing cell located between the seminiferous tubules.

interstitial fluid: fluid that bathes the cells of the body; in mammals, interstitial fluid leaks from capillaries and is similar in composition to blood plasma, but lacking the large proteins found in plasma.

intertidal zone: an area of the ocean shore that is alternately covered by water during high tides and exposed to the air during low tides.

intracellular digestion: the chemical breakdown of food within single cells.

intraspecific competition: competition among individuals of the same species.

intrinsically disordered proteins: proteins or segments of proteins with no stable secondary or tertiary structure.

intron: a segment of DNA in a eukaryotic gene that does not code for amino acids in a protein; see also *exon*.

invasive species: organisms with a high biotic potential that are introduced (deliberately or accidentally) into ecosystems where they did not evolve and where they encounter little environmental resistance and tend to displace native species.

inversion: a mutation that occurs when a piece of DNA is cut out of a chromosome, turned around, and reinserted into the gap.

invertebrate (in-vert′-uh-bret): an animal that lacks a vertebral column.

ion (ī-on): a charged atom or molecule; an atom or molecule that either has an excess of electrons (and, hence, is negatively charged) or has lost electrons (and is positively charged).

ionic bond: a chemical bond formed by the electrical attraction between positively and negatively charged ions.

iris: the pigmented muscular tissue of the vertebrate eye that surrounds and controls the size of the pupil, through which light enters the eye.

islet cell: a cell in the endocrine portion of the pancreas that produces either insulin or glucagon.

isolated system: in thermodynamics, a hypothetical space where neither energy nor matter can enter or leave.

isolating mechanism: a morphological, physiological, behavioral, or ecological difference that prevents members of two species from interbreeding.

isotonic (ī-sō-ton′-ik): referring to a solution that has the same concentration of solute (and therefore the same concentration of free water) as has the cytosol of a cell.

isotope (ī′-suh-tōp): one of several forms of a single element, the nuclei of which contain the same number of protons but different numbers of neutrons.

Jacob syndrome: a set of characteristics typical of human males possessing one X and two Y chromosomes (XYY); most XYY males are phenotypically normal, but XYY males tend to be taller than average and to have a slightly increased risk of learning disabilities.

J-curve: the J-shaped growth curve of an exponentially growing population in which increasing numbers of individuals join the population during each succeeding time period.

joint: a flexible region between two rigid units of an exoskeleton or endoskeleton, allowing for movement between the units.

karyotype: a preparation showing the number, sizes, and shapes of all the chromosomes within a cell.

***K*-selected species:** species that typically live in a stable environment and often develop a population size that is close to the carrying capacity of that environment. *K*-selected species usually mature slowly, have a long life span, produce small numbers of fairly large offspring, and provide significant parental care or nutrients for the offspring, so that a large percentage of the offspring live to maturity.

kelp forest: a diverse ecosystem consisting of stands of tall brown algae and associated marine life. Kelp forests occur in oceans worldwide in nutrient-rich cool coastal waters.

keratin (ker′-uh-tin): a fibrous protein in hair, nails, and the epidermis of skin.

keystone species: a species whose influence on community structure is greater than its abundance would suggest.

kidney: one of a pair of organs of the excretory system that is located on either side of the spinal column; the kidney filters blood, removing wastes and regulating the composition and water content of the blood.

kin selection: a type of natural selection that favors traits that enhance the survival or reproduction of an individual's relatives, even if the traits reduce the fitness of the individuals bearing them.

kinetic energy: the energy of movement; includes light, heat, mechanical movement, and electricity.

kinetochore (ki-net′-ō-kor): a protein structure that forms at the centromere regions of chromosomes; attaches the chromosomes to the spindle.

kinetoplastid: a member of a protist group characterized by distinctively structured mitochondria. Kinetoplastids are mostly flagellated and include parasitic forms such as *Trypanosoma*, which causes sleeping sickness. Kinetoplastids are part of a larger group known as excavates.

kingdom: the second broadest taxonomic category, consisting of related phyla. Related kingdoms make up a domain.

Klinefelter syndrome: a set of characteristics typically found in individuals who have two X chromosomes and one Y chromosome; these individuals are phenotypically males but are usually unable to father children without the use of artificial reproductive technologies, such as in vitro fertilization. Men with Klinefelter syndrome may have several female-like traits, including broad hips and partial breast development.

Krebs cycle: a cyclic series of reactions, occurring in the matrix of mitochondria, in which the acetyl groups from the pyruvic acids produced by glycolysis are broken down to CO_2, accompanied by the formation of ATP and electron carriers; also called the citric acid cycle.

labium (pl., labia): one of a pair of folds of skin of the external structures of the mammalian female reproductive system.

labor: a series of contractions of the uterus that result in birth.

lactation: the secretion of milk from the mammary glands.

lacteal (lak′-tēl): a lymph capillary; found in each villus of the small intestine.

lactic acid fermentation: anaerobic reactions that convert the pyruvic acid produced by glycolysis into lactic acid, using hydrogen ions and electrons from NADH; the primary function of lactic acid fermentation is to regenerate NAD^+ so that glycolysis can continue under anaerobic conditions.

lactose (lak′-tōs): a disaccharide composed of glucose and galactose; found in mammalian milk.

lactose intolerance: the inability to digest lactose (milk sugar) because lactase, the enzyme that digests lactose, is not produced in sufficient amounts; symptoms include bloating, gas pains, and diarrhea.

lactose operon: in prokaryotes, the set of genes that encodes the proteins needed for lactose metabolism, including both the structural genes and a common promoter and operator that control transcription of the structural genes.

large intestine: the final section of the digestive tract; consists of the colon and the rectum, where feces are formed and stored.

larva (lar′-vuh; pl., larvae): an immature form of an animal that subsequently undergoes metamorphosis into its adult form; includes the caterpillars of moths and butterflies, the maggots of flies, and the tadpoles of frogs and toads.

larynx (lar′-inks): the portion of the air passage between the pharynx and the trachea; contains the vocal cords.

late-loss population: a population in which most individuals survive into adulthood; late-loss populations have a convex survivorship curve.

lateral bud: a cluster of meristem cells at the node of a stem; under appropriate conditions, it grows into a branch.

lateral meristem: a meristem tissue that forms cylinders parallel to the long axis of roots and stems; normally located between the primary xylem and primary phloem (vascular cambium) and just outside the phloem (cork cambium); also called *cambium*.

law of conservation of energy: the principle of physics that states that within any isolated system, energy can be neither created nor destroyed, but can be converted from one form to another; also called the first law of thermodynamics.

law of independent assortment: the independent inheritance of two or more traits, assuming that each trait is controlled by a single gene with no influence from gene(s) controlling the other trait; states that the alleles of each gene are distributed to the gametes independently of the alleles for other genes; this law is true only for genes located on different chromosomes or very far apart on a single chromosome.

law of segregation: the principle that each gamete receives only one of each parent's pair of alleles of each gene.

laws of thermodynamics: the physical laws that define the basic properties and behavior of energy.

leaf: an outgrowth of a stem, normally flattened and photosynthetic.

leaf primordium (pri-mor′-dē-um; pl., primordia): a cluster of dividing cells, surrounding a terminal or lateral bud, that develops into a leaf.

learning: the process by which behavior is modified in response to experience.

legume (leg′-ūm): a member of a family of plants characterized by root swellings in which nitrogen-fixing bacteria are housed; includes peas, soybeans, lupines, alfalfa, and clover.

lens: a clear object that bends light rays; in eyes, a flexible or movable structure used to focus light on the photoreceptor cells of the retina.

leptin: a peptide hormone released by fat cells that helps the body monitor its fat stores and regulate weight.

less developed country: a country that has not completed the demographic transition; usually has relatively low income and educational level for most of its population, with high birth rates and sometimes fairly high death rates, resulting in moderate to rapid population growth.

leukocyte (loo′-kō-sīt): any of the white blood cells circulating in the blood.

lichen (lī′-ken): a symbiotic association between an alga or cyanobacterium and a fungus, resulting in a composite organism.

life history: the characteristic survivorship and reproductive features of a species, particularly when and how often reproduction occurs, how many offspring are produced, how many resources are provided to each offspring, and what proportion of the offspring survive to maturity.

life table: a data table that groups organisms born at the same time and tracks them throughout their life span, recording how many continue to survive in each succeeding year (or other unit of time). Various parameters such as sex may be used in the groupings. Human life tables may include many other parameters (such as socioeconomic status) used by demographers.

ligament: a tough connective tissue band connecting two bones.

light reactions: the first stage of photosynthesis, in which the energy of light is captured in ATP and NADPH; occurs in thylakoids of chloroplasts.

lignin: a hard material that is embedded in the cell walls of vascular plants and that provides support in terrestrial species; an early and important adaptation to terrestrial life.

limbic system: a diverse group of brain structures, mostly in the lower forebrain, that includes the thalamus, hypothalamus, amygdala, hippocampus, and parts of the cerebrum and is involved in basic emotions, drives, behaviors, and learning.

limnetic zone: the part of a lake in which enough light penetrates to support photosynthesis.

linkage: the inheritance of certain genes as a group because they are parts of the same chromosome. Linked genes do not show independent assortment.

lipase (lī′-pās): an enzyme that catalyzes the breakdown of lipids such as fats into their component fatty acids and glycerol.

lipid (li′-pid): one of a number of organic molecules containing large nonpolar regions composed solely of carbon and hydrogen, which make lipids hydrophobic and insoluble in water; includes oils, fats, waxes, phospholipids, and steroids.

littoral zone: the part of a lake, usually close to the shore, in which the water is shallow and plants find abundant light, anchorage, and adequate nutrients.

liver: an organ with varied functions, including bile production, glycogen storage, and the detoxification of poisons.

lobefin: fish with fleshy fins that have well-developed bones and muscles. Lobefins include two living clades: the coelacanths and the lungfishes.

local hormone: a general term for messenger molecules produced by most cells and released into the cells' immediate vicinity. Local hormones, which include prostaglandins and cytokines, influence nearby cells bearing appropriate receptors.

locus (pl., loci): the physical location of a gene on a chromosome.

logistic population growth: population growth characterized by an early exponential growth phase, followed by slower growth as the population approaches its carrying capacity, and finally reaching a stable population at the carrying capacity of the environment; this type of growth generates a curve shaped like a stretched-out letter "S."

long-day plant: a plant that will flower only if the length of uninterrupted darkness is shorter than a species-specific critical period; also called a *short-night plant.*

long-term memory: the second phase of learning; a more-or-less permanent memory formed by a structural change in the brain, brought on by repetition.

lung: in terrestrial vertebrates, one of the pair of respiratory organs in which gas exchange occurs; consists of inflatable chambers within the chest cavity.

luteinizing hormone (loo′-tin-īz-ing; LH): a hormone produced by the anterior pituitary that stimulates testosterone production in males and the development of the follicle, ovulation, and the production of the corpus luteum in females.

lymph (limf): a pale fluid found within the lymphatic system; composed primarily of interstitial fluid and white blood cells.

lymph node: a small structure located on a lymph vessel, containing macrophages and lymphocytes (B and T cells). Macrophages filter the lymph by removing microbes; lymphocytes are the principal components of the adaptive immune response to infection.

lymphatic capillary: the smallest vessel of the lymphatic system. Lymphatic capillaries end blindly in interstitial fluid, which they take up and return to the bloodstream.

lymphatic system: a system consisting of lymph vessels, lymph capillaries, lymph nodes, and the thymus, spleen, tonsils, and bone marrow. The lymphatic system helps protect the body against infection, carries fats from the small intestine to blood vessels, and returns excess fluid and small proteins to the blood circulatory system.

lymphocyte (lim′-fō-sit): a type of white blood cell (natural killer cell, B cell, or T cell) that is important in either the innate or adaptive immune response.

lysosome (lī′-sō-sōm): a membrane-bound organelle containing intracellular digestive enzymes.

macronutrient: a nutrient required by an organism in relatively large quantities.

macrophage (mak′-rō-fāj): a type of white blood cell that engulfs microbes and destroys them by phagocytosis; also presents microbial antigens to T cells, helping stimulate the immune response.

major histocompatibility complex (MHC): a group of proteins, normally located on the surfaces of body cells, that identify the cell as "self"; also important in stimulating and regulating the immune response.

Malpighian tubule (mal-pig′-ē-un): the functional unit of the excretory system of insects, consisting of a small tube protruding from the intestine; wastes and nutrients move from the surrounding blood into the tubule, which drains into the intestine, where nutrients are reabsorbed back into the blood, while the wastes are excreted along with feces.

maltose (mal′-tōs): a disaccharide composed of two glucose molecules.

mammal: a member of the chordate clade Mammalia, which includes vertebrates with hair and mammary glands.

mammary gland (mam′-uh-rē): a milk-producing gland used by female mammals to nourish their young.

marsupial (mar-soo′-pē-ul): a member of the clade Marsupialia, which includes mammals whose young are born at an extremely immature stage and undergo further development in a pouch, where they remain attached to a mammary gland; kangaroos, opossums, and koalas are marsupials.

mass extinction: a relatively sudden extinction of many species, belonging to multiple major taxonomic groups, as a result of environmental change. The fossil record reveals five mass extinctions over geologic time.

mass number: the total number of protons and neutrons in the nucleus of an atom.

mast cell: a cell of the immune system that releases histamine and other molecules used in the body's response to trauma and that are a factor in allergic reactions.

matrix: the fluid contained within the inner membrane of the mitochondrion.

mechanical digestion: the process by which food in the digestive tract is physically broken down into smaller pieces.

mechanical incompatibility: a reproductive isolating mechanism that arises when differences in the reproductive structures of two species make the structures incompatible and prevent interbreeding.

mechanoreceptor: a sensory receptor that responds to mechanical stimuli such as the stretching, bending, or dimpling of a part of the body.

medulla (med-ū′-luh): the part of the hindbrain of vertebrates that controls automatic activities such as breathing, swallowing, heart rate, and blood pressure.

megakaryocyte (meg-a-kar′-ē-ō-sīt): a large cell type in the bone marrow, which pinches off pieces of itself that enter the circulation as platelets.

megaspore: a haploid cell formed by meiotic cell division from a diploid megaspore mother cell; through mitotic cell division and differentiation, it develops into the female gametophyte.

megaspore mother cell: a diploid cell, within the ovule of a flowering plant, that undergoes meiotic cell division to produce four haploid megaspores.

meiosis (mī-ō′-sis): in eukaryotic organisms, a type of nuclear division in which a diploid nucleus divides twice to form four haploid nuclei.

meiosis I: the first division of meiosis, which separates the pairs of homologous chromosomes and sends one homologue from each pair into each of two daughter nuclei, which are therefore haploid.

meiosis II: the second division of meiosis, which separates the chromatids into independent chromosomes and parcels one chromosome into each of two daughter nuclei.

meiotic cell division: meiosis followed by cytokinesis.

melatonin (mel-uh-tōn′-in): a hormone, secreted by the pineal gland, that is involved in the regulation of circadian cycles.

memory B cell: a type of white blood cell that is produced by clonal selection as a result of the binding of an antibody on a B cell to an antigen on an invading microorganism. Memory B cells persist in the bloodstream and provide future immunity to invaders bearing that antigen.

memory T cell: a type of white blood cell that is produced by clonal selection as a result of the binding of a receptor on a T cell to an antigen on an invading microorganism. Memory T cells persist in the bloodstream and provide future immunity to invaders bearing that antigen.

menstrual cycle: in human females, a roughly 28-day cycle during which hormonal interactions among the hypothalamus, pituitary gland, and ovary coordinate ovulation and the preparation of the uterus to receive and nourish a fertilized egg. If pregnancy does not occur, the uterine lining is shed during menstruation.

menstruation: in human females, the monthly discharge of uterine tissue and blood from the uterus.

meristem cell (mer′-i-stem): an undifferentiated cell that remains capable of cell division throughout the life of a plant.

mesoderm (mēz′-ō-derm): the middle embryonic tissue layer, lying between the endoderm and ectoderm, and normally the last to develop; gives rise to structures such as muscles, the skeleton, the circulatory system, and the kidneys.

mesophyll (mez′-ō-fil): loosely packed, usually photosynthetic cells located beneath the epidermis of a leaf.

messenger RNA (mRNA): a strand of RNA, complementary to the DNA of a gene, that conveys the genetic information in DNA to the ribosomes to be used during protein synthesis; sequences of three bases (codons) in mRNA that specify particular amino acids to be incorporated into a protein.

metabolic pathway: a sequence of chemical reactions within a cell in which the products of one reaction are the reactants for the next reaction.

metabolic rate: the speed at which cellular reactions that release energy occur.

metabolism: the sum of all chemical reactions that occur within a single cell or within all the cells of a multicellular organism.

metamorphosis (met-a-mor′-fō-sis): in animals with indirect development, a radical change in body form from larva to sexually mature adult, as seen in amphibians (e.g., tadpole to frog) and insects (e.g., caterpillar to butterfly).

metaphase (met′-a-fāz): in mitosis, the stage in which the chromosomes, attached to spindle fibers at kinetochores, are lined up along the equator of the cell; also the approximately comparable stages in meiosis I and meiosis II.

microbe: a microorganism.

microbiome: populations of microorganisms (particularly bacteria) that reside in and on the bodies of animals. The largest human microbiome resides in the large intestine.

microfilament: part of the cytoskeleton of eukaryotic cells that is composed of the proteins actin and (in some cases) myosin; functions in the movement of cell organelles, locomotion by extension of the plasma membrane, and sometimes contraction of entire cells.

micronutrient: a nutrient required by an organism in relatively small quantities.

microRNA: small molecules of RNA that interfere with the translation of specific genes.

microspore: a haploid cell formed by meiotic cell division from a microspore mother cell; through mitotic cell division and differentiation, it develops into the male gametophyte.

microspore mother cell: a diploid cell contained within an anther of a flowering plant; undergoes meiotic cell division to produce four haploid microspores.

microtubule: a hollow, cylindrical strand, found in eukaryotic cells, that is composed of the protein tubulin; part of the cytoskeleton used in the movement of organelles, cell growth, and the construction of cilia and flagella.

microvillus (mī-krō-vi′-lus; pl., microvilli): a microscopic projection of the plasma membrane, which increases the surface area of a cell.

midbrain: during development, the central portion of the brain; contains most of an important relay center, the reticular formation.

middle ear: the part of the mammalian ear composed of the tympanic membrane, the auditory (Eustachian) tube, and three bones (hammer, anvil, and stirrup) that transmit vibrations from the auditory canal to the oval window.

mimicry (mim′-ik-rē): the situation in which a species has evolved to resemble something else, typically another type of organism.

mineral: an inorganic substance, especially one in rocks or soil, or dissolved in water. In nutrition, minerals such as sodium, calcium, and potassium are essential nutrients that must be obtained from the diet.

mineralocorticoid: a type of steroid hormone produced by the adrenal cortex that regulates salt retention in the kidney, thereby regulating the salt concentration in the blood and interstitial fluid.

minimum viable population: the smallest isolated population that can persist indefinitely and survive likely natural events such as fires and floods.

mitochondrion (mī-tō-kon′-drē-un; pl., mitochondria): an organelle, bounded by two membranes, that is the site of the reactions of aerobic metabolism.

mitosis (mī-tō′-sis): a type of nuclear division, used by eukaryotic cells, in which one copy of each chromosome (already duplicated during interphase before mitosis) moves into each of two daughter nuclei; the daughter nuclei are therefore genetically identical to each other.

mitotic cell division: mitosis followed by cytokinesis.

molecule (mol′-e-kūl): a particle composed of one or more atoms held together by chemical bonds; the smallest particle of a compound that displays all the properties of that compound.

molt: to shed an external body covering, such as an exoskeleton, skin, feathers, or fur.

monocot: short for monocotyledon; a type of flowering plant characterized by embryos with one seed leaf, or cotyledon.

monomer (mo′-nō-mer): a small organic molecule, several of which may be bonded together to form a chain called a polymer.

monosaccharide (mo-nō-sak′-uh-rīd): the basic molecular unit of all carbohydrates, normally composed of a chain of carbon atoms bonded to hydrogen and hydroxyl groups.

monotreme: a member of the clade Monotremata, which includes mammals that lay eggs; platypuses and spiny anteaters are monotremes.

more developed country: a country that has completed the demographic transition; usually has relatively high income and educational level for most of its population, with low birth and death rates that result in low or even negative population growth.

morula (mor′-ū-luh): in animals, an embryonic stage during cleavage, when the embryo consists of a solid ball of cells.

motor neuron: a neuron that receives instructions from sensory neurons or interneurons and activates effector organs, such as muscles or glands.

motor unit: a single motor neuron and all the muscle fibers on which it forms synapses.

mouth: the opening through which food enters a tubular digestive system.

multicellular: many-celled; most members of the kingdoms Fungi, Plantae, and Animalia are multicellular, with intimate cooperation among cells.

multiple alleles: many alleles of a single gene, perhaps dozens or hundreds, as a result of mutations.

muscle: an animal tissue composed of cells that are capable of contracting and thereby moving specific parts of the body.

muscle fiber: an individual muscle cell.

muscle tissue: tissue composed of one of three types of contractile cells (smooth, skeletal, or cardiac).

muscular dystrophy: an inherited disorder, almost exclusively found in males, in which defective dystrophin proteins cause the skeletal muscles to degenerate.

mutation: a change in the base sequence of DNA in a gene; often used to refer to a genetic change that is significant enough to alter the appearance or function of the organism.

mutualism (mū′-choo-ul-iz-um): an interaction between species in which both participating species benefit.

mycelium (mī-sēl′-ē-um; pl., mycelia): the body of a fungus, consisting of a mass of hyphae.

mycorrhiza (mī-kō-rī′-zuh; pl., mycorrhizae): a symbiotic association between a fungus and the roots of a land plant that facilitates mineral extraction and absorption.

myelin (mī′-uh-lin): a wrapping of insulating membranes of specialized glial cells around the axon of a vertebrate nerve cell; increases the speed of conduction of action potentials.

myofibril (mī-ō-fī′-bril): a cylindrical subunit of a muscle cell, consisting of a series of sarcomeres, surrounded by sarcoplasmic reticulum.

myometrium (mī-ō-mē′-trē-um): the muscular outer layer of the uterus.

myosin (mī′-ō-sin): one of the major proteins of muscle, the interaction of which with the protein actin produces muscle contraction; found in the thick filaments of the muscle fiber; see also *actin*.

myosin head: the part of a myosin protein that binds to the actin subunits of a thin filament; flexion of the myosin head moves the thin filament toward the center of the sarcomere, causing muscle fiber contraction.

natural causality: the scientific principle that natural events occur as a result of preceding natural causes.

natural increase: the difference between births and deaths in a population. This number will be positive if the population is increasing and negative if it is decreasing.

natural killer cell: a type of white blood cell that destroys some virus-infected cells and cancerous cells on contact; part of the innate immune system's nonspecific internal defense against disease.

natural laws: basic principles derived from the study of nature that have never been disproven by scientific inquiry. Natural laws include the laws of gravity, the behavior of light, and the way atoms interact with one another.

natural selection: unequal survival and reproduction of organisms due to heritable differences in their phenotypes, with the result that better adapted phenotypes become more common in the population.

nearshore zone: the region of coastal water that is relatively shallow but constantly submerged and that can support large plants or seaweeds; includes bays and coastal wetlands.

nearsighted: the inability to focus on distant objects caused by an eyeball that is slightly too long or a cornea that is too curved.

negative feedback: a physiological process in which a change causes responses that tend to counteract the change and restore the original state. Negative feedback in physiological systems maintains homeostasis.

nephridium (nef-rid′-ē-um; pl., nephridia): an excretory organ found in earthworms, mollusks, and certain other invertebrates; somewhat resembles a single vertebrate nephron.

nephron (nef′-ron): the functional unit of the kidney, where blood is filtered and urine is formed.

nephron loop (loop of Henle; hen′-lē): a specialized portion of the tubule in birds and mammals that creates an osmotic concentration gradient in the fluid immediately surrounding it. This gradient allows the production of urine that is more osmotically concentrated than blood plasma.

nerve: a bundle of axons of nerve cells, bound together in a sheath.

nerve cord: a major nervous pathway consisting of a cord of nervous tissue extending lengthwise through the body, paired in many invertebrates and unpaired in chordates.

nerve net: a simple form of nervous system, consisting of a network of neurons that extends throughout the tissues of an organism such as a cnidarian.

nerve tissue: the tissue that makes up the brain, spinal cord, and nerves; consists of neurons and glial cells.

net primary production: the energy stored in the autotrophs of an ecosystem over a given time period.

neuromuscular junction: the synapse formed between a motor neuron and a muscle fiber.

neuron (noor′-on): a single nerve cell.

neurosecretory cell: a specialized nerve cell that synthesizes and releases hormones.

neurotransmitter: a chemical that is released by a nerve cell close to a second nerve cell, a muscle, or a gland cell and that influences the activity of the second cell.

neutral mutation: a mutation that does not detectably change the function of the encoded protein.

neutron: a subatomic particle that is found in the nuclei of atoms, bears no charge, and has a mass approximately equal to that of a proton.

neutrophil (nū′-trō-fil): a type of white blood cell that engulfs invading microbes and contributes to the nonspecific defenses of the body against disease.

nicotinamide adenine dinucleotide (NAD^+ or NADH): an electron carrier molecule produced in the cytoplasmic fluid by glycolysis and in the mitochondrial matrix by the Krebs cycle; subsequently donates electrons to the electron transport chain.

nicotinamide adenine dinucleotide phosphate ($NADP^+$ or NADPH): an electron carrier molecule used in photosynthesis to transfer high-energy electrons from the light reactions to the Calvin cycle.

nitrogen cycle: the biogeochemical cycle by which nitrogen moves from its primary reservoir of nitrogen gas in the atmosphere via nitrogen-fixing bacteria to reservoirs in soil and water, through producers and into higher trophic levels, and then back to its reservoirs.

nitrogen fixation: the process that combines atmospheric nitrogen with hydrogen to form ammonia (NH_3).

nitrogen-fixing bacterium: a bacterium that possesses the ability to remove nitrogen (N_2) from the atmosphere and combine it with hydrogen to produce ammonia (NH_3).

no-till farming: a method of growing crops that leaves the remains of harvested crops in place, with the next year's crops being planted directly in the remains of last year's crops without significant disturbance of the soil.

node: in plants, a region of a stem at which the petiole of a leaf is attached; usually, a lateral bud is also found at a node.

nodule: a swelling on the root of a legume or other plant that consists of cortex cells inhabited by nitrogen-fixing bacteria.

noncompetitive inhibition: the process by which an inhibitory molecule binds to a site on an enzyme that is distinct from the active site. As a result, the enzyme's active site is distorted, making it less able to catalyze the reaction involving its normal substrate.

nondisjunction: an error in meiosis in which chromosomes fail to segregate properly into the daughter cells.

nonpolar covalent bond: a covalent bond with equal sharing of electrons.

nonvascular plant: a plant that lacks lignin and well-developed conducting vessels. Nonvascular plants include mosses, hornworts, and liverworts.

norepinephrine (nor-ep-i-nef′-rin): a neurotransmitter, released by neurons of the sympathetic nervous system, that prepares the body to respond to stressful situations; also called *noradrenaline*.

northern coniferous forest: a biome with long, cold winters and only a few months of warm weather; dominated by evergreen coniferous trees; also called *taiga*.

notochord (nōt′-ō-kord): a stiff, but somewhat flexible, supportive rod that extends along the head-to-tail axis and is found in all members of the phylum Chordata at some stage of development.

nuclear envelope: the double-membrane system surrounding the nucleus of eukaryotic cells; the outer membrane is typically continuous with the endoplasmic reticulum.

nuclear pore complex: an array of proteins that line pores in the nuclear membrane and control which substances enter and leave the nucleus.

nucleic acid (noo-klā′-ik): an organic molecule composed of nucleotide subunits; the two common types of nucleic acids are ribonucleic acid (RNA) and deoxyribonucleic acid (DNA).

nucleoid (noo′-klē-oid): the location of the genetic material in prokaryotic cells; not membrane enclosed.

nucleolus (noo-klē′-ō-lus; pl., nucleoli): the region of the eukaryotic nucleus that is engaged in ribosome synthesis; consists of the genes encoding ribosomal RNA, newly synthesized ribosomal RNA, and ribosomal proteins.

nucleotide: a subunit of which nucleic acids are composed; a phosphate group bonded to a sugar (deoxyribose in DNA), which is in turn bonded to a nitrogen-containing base (adenine, guanine, cytosine, or thymine in DNA). Nucleotides are linked together, forming a strand of nucleic acid, by bonds between the phosphate of one nucleotide and the sugar of the next nucleotide.

nucleus (atomic): the central region of an atom, consisting of protons and neutrons.

nucleus (cellular): the membrane-bound organelle of eukaryotic cells that contains the cell's genetic material.

nutrient: a substance acquired from the environment and needed for the survival, growth, and development of an organism.

nutrient cycle: the pathways of a specific nutrient (such as carbon, nitrogen, phosphorus, or water) through the living and nonliving portions of an ecosystem; also called a *biogeochemical cycle*.

observation: in the scientific method, the recognition of and a statement about a specific phenomenon, usually leading to the formulation of a question about the phenomenon.

oil: a lipid composed of three fatty acids, some of which are unsaturated, covalently bonded to a molecule of glycerol; oils are liquid at room temperature.

oligotrophic lake: a lake that is very low in nutrients and hence supports little phytoplankton, plant, and algal life; contains clear water with deep light penetration.

ommatidium (ōm-ma-tid′-ē-um; pl., ommatidia): an individual light-sensitive subunit of a compound eye; consists of a lens and several receptor cells.

omnivore: an organism that consumes both plants and animals.

oogenesis (ō-ō-jen′-i-sis): the process by which egg cells are formed.

oogonium (ō-ō-gō′-nē-um; pl., oogonia): in female animals, a diploid cell that gives rise to a primary oocyte.

open circulatory system: a type of circulatory system found in some invertebrates, such as arthropods and most mollusks, that includes an open space (the hemocoel) in which blood directly bathes body tissues.

open ocean: that part of the ocean in which the water is so deep that wave action does not affect the bottom, even during strong storms.

operant conditioning: a laboratory training procedure in which an animal learns to perform a response (such as pressing a lever) through reward or punishment.

operator: a sequence of DNA nucleotides in a prokaryotic operon that binds regulatory proteins that control the ability of RNA polymerase to transcribe the structural genes of the operon.

operon (op′-er-on): in prokaryotes, a set of genes, often encoding the proteins needed for a complete metabolic pathway, including both the structural genes and a common promoter and operator that control transcription of the structural genes.

optic nerve: the nerve leading from the eye to the brain; it carries visual information.

order: in Linnaean classification, the taxonomic rank composed of related families. Related orders make up a class.

organ: a structure (such as the liver, kidney, or skin) composed of two or more distinct tissue types that function together.

organ system: two or more organs that work together to perform a specific function; for example, the digestive system.

organelle (or-guh-nel′): a membrane-enclosed structure found inside a eukaryotic cell that performs a specific function.

organic: describing a molecule that contains both carbon and hydrogen.

organic molecule: a molecule that contains both carbon and hydrogen.

organism (or′-guh-niz-um): an individual living thing.

organogenesis (or-gan-ō-jen′-uh-sis): the process by which the layers of the gastrula (endoderm, ectoderm, mesoderm) rearrange to form organs.

origin: the site of attachment of a muscle to the relatively stationary bone on one side of a joint.

osmolarity: a measure of the total number of dissolved solute particles in a solution.

osmoregulation: homeostatic maintenance of the water and salt content of the body within a limited range.

osmosis (oz-mō′-sis): the diffusion of water across a differentially permeable membrane, normally down a concentration gradient of free water molecules. Water moves into the solution that has a lower concentration of free water from a solution that has a higher concentration of free water.

osteoblast (os′-tē-ō-blast): a cell type that produces bone.

osteoclast (os′-tē-ō-klast): a cell type that dissolves bone.

osteocyte (os′-tē-ō-sīt): a mature bone cell.

osteoporosis (os′-tē-ō-por-ō′-sis): a condition in which bones become porous, weak, and easily fractured; most common in elderly women.

outer ear: the outermost part of the mammalian ear, including the external ear and auditory canal leading to the tympanic membrane.

oval window: the membrane-covered entrance to the cochlea.

ovary: (1) in animals, the gonad of females; (2) in flowering plants, a structure at the base of the carpel that contains one or more ovules and develops into the fruit.

overexploitation: hunting or harvesting natural populations at a rate that exceeds those populations' ability to replenish their numbers.

ovulation: the release of a secondary oocyte, ready to be fertilized, from the ovary.

ovule: a structure within the ovary of a flower, inside which the female gametophyte develops; after fertilization, it develops into the seed.

oxytocin (oks-ē-tō′-sin): a hormone, released by the posterior pituitary, that stimulates the contraction of uterine and mammary gland muscles.

ozone hole: a region of severe ozone loss in the stratosphere caused by ozone-depleting chemicals; maximum ozone loss occurs from September to early October over Antarctica.

ozone layer: the ozone-enriched layer of the upper atmosphere (stratosphere) that filters out much of the sun's ultraviolet radiation.

pacemaker: a cluster of specialized muscle cells in the upper right atrium of the heart that produce spontaneous electrical signals at a regular rate; the sinoatrial node.

pain receptor: a receptor cell that stimulates activity in the brain that is perceived as the sensation of pain; responds to very high or very low temperatures, mechanical damage (such as extreme stretching of tissue), and/or certain chemicals, such as potassium ions or bradykinin, that are produced as a result of tissue damage; also called *nociceptor*.

pancreas (pan′-krē-us): a combined exocrine and endocrine gland located in the abdominal cavity next to the stomach. The endocrine portion secretes the hormones insulin and glucagon, which regulate glucose concentrations in the blood. The exocrine portion secretes pancreatic juice (a mixture of water, enzymes, and sodium bicarbonate) into the small intestine; the enzymes digest fat, carbohydrate, and protein; the bicarbonate neutralizes acidic chyme entering the intestine from the stomach.

pancreatic juice: a mixture of water, sodium bicarbonate, and enzymes released by the pancreas into the small intestine.

parabasalid: a member of a protist group characterized by mutualistic or parasitic relationships with the animal species inside which they live. Parabasalids are part of a larger group known as the excavates.

paracrine communication: communication between cells of a multicellular organism in which certain cells release hormones that diffuse through the interstitial fluid to nearby cells and alter their activity.

parasite (par′-uh-sīt): an organism that lives in or on a larger organism (its host), harming the host but usually not killing it immediately.

parasympathetic division: the division of the autonomic nervous system that produces largely involuntary responses related to the maintenance of normal body functions, such as digestion; often called the *parasympathetic nervous system*.

parathyroid gland: one of four small endocrine glands, embedded in the surface of the thyroid gland, that produces parathyroid hormone, which (with calcitonin from the thyroid gland) regulates calcium ion concentration in the blood.

parathyroid hormone: a hormone released by the parathyroid gland that stimulates the release of calcium from bone.

parenchyma tissue (par-en′-ki-muh): A tissue formed from parenchyma cells that are alive at maturity, normally have thin cell walls, and carry out most of the metabolism of a plant. Most dividing meristem cells in a plant are parenchyma.

parthenogenesis (par-the-nō-jen′-uh-sis): an asexual specialization of sexual reproduction, in which a haploid egg undergoes development without fertilization.

passive transport: the movement of materials across a membrane down a gradient of concentration, pressure, or electrical charge without using cellular energy.

pathogen: an organism (or a toxin) capable of producing disease.

pathogenic (path-ō-jen′-ik): capable of producing disease; referring to an organism with such a capability (a pathogen).

pedigree: a diagram showing genetic relationships among a set of individuals, normally with respect to a specific genetic trait.

pelagic (puh-la′-jik): free-swimming or floating.

penis: an external structure of the male reproductive and urinary systems; serves to deposit sperm into the female reproductive system and deliver urine to the outside of the body.

peptide (pep′-tīd): a chain composed of two or more amino acids linked together by peptide bonds.

peptide bond: the covalent bond between the nitrogen of the amino group of one amino acid and the carbon of the carboxyl group of a second amino acid, joining the two amino acids together in a peptide or protein.

peptide hormone: a hormone consisting of a chain of amino acids; includes small proteins that function as hormones.

pericycle (per′-i-sī-kul): the outermost layer of cells of the vascular cylinder of a root.

periderm: the outer cell layers of roots and stems that have undergone secondary growth; consists primarily of cork cambium and cork cells.

periodic table: a chart first devised by Russian chemist Dmitri Mendeleev that includes all known elements and organizes them according to their atomic numbers in rows and their general chemical properties in columns (see Appendix II).

peripheral nervous system (PNS): in vertebrates, the part of the nervous system that connects the central nervous system to the rest of the body.

peristalsis: rhythmic coordinated contractions of the smooth muscles of the digestive tract that move substances through the digestive tract.

peritubular capillaries: a capillary network surrounding each kidney tubule that allows the exchange of substances between the blood and the tubule contents during reabsorption and secretion.

permafrost: a permanently frozen layer of soil, usually found in tundra of the Arctic or high mountains.

petal: part of a flower, typically brightly colored and fragrant, that attracts potential animal pollinators.

petiole (pet′-ē-ōl): the stalk that connects the blade of a leaf to the stem.

pH scale: a scale, with values from 0 to 14, used for measuring the relative acidity of a solution; at pH 7 a solution is neutral, pH 0 to 7 is acidic, and pH 7 to 14 is basic; each unit on the scale represents a 10-fold change in H^+ concentration.

phagocyte (fā′-gō-sīt): a type of immune system cell that destroys invading microbes by using phagocytosis to engulf and digest the microbes. Also called a *phagocytic cell*.

phagocytosis (fa-gō-sī-tō′-sis): a type of endocytosis in which extensions of a plasma membrane engulf extracellular particles, enclose them in a membrane-bound sac, and transport them into the interior of the cell.

pharyngeal gill slit (far-in′-jē-ul): one of a series of openings, located just posterior to the mouth, that connects the throat to the outside environment; present (as some stage of life) in all chordates.

pharynx (far′-inks): in vertebrates, a chamber that is located at the back of the mouth, shared by the digestive and respiratory systems; in some invertebrates, the portion of the digestive tube just posterior to the mouth.

phenotype (fēn′-ō-tīp): the physical characteristics of an organism; can be defined as outward appearance (such as flower color), as behavior, or in molecular terms (such as glycoproteins on red blood cells).

pheromone (fer′-uh-mōn): a chemical produced by an organism that alters the behavior or physiological state of another member of the same species.

phloem (flō′-um): a conducting tissue of vascular plants that transports a concentrated solution of sugars (primarily sucrose) and other organic molecules up and down the plant.

phospholipid (fos-fō-li′-pid): a lipid consisting of glycerol bonded to two fatty acids and one phosphate group, which bears another group of atoms, typically charged and containing nitrogen. A double layer of phospholipids is a component of all cellular membranes.

phospholipid bilayer: a double layer of phospholipids that forms the basis of all cellular membranes. The phospholipid heads, which are hydrophilic, face the watery interstitial fluid, a watery external environment, or the cytosol; the tails, which are hydrophobic, are buried in the middle of the bilayer.

phosphorus cycle (fos′-for-us): the biogeochemical cycle by which phosphorus moves from its primary reservoir—phosphate-rich rock—to reservoirs of phosphate in soil and water, through producers and into higher trophic levels, and then back to its reservoirs.

photic zone: the region of an ocean where light is strong enough to support photosynthesis.

photon (fō′-ton): the smallest unit of light energy.

photopigment (fō′-tō-pig-ment): a chemical substance in a photoreceptor cell that, when struck by light, changes shape and produces a response in the cell.

photoreceptor: a receptor cell that responds to light; in vertebrates, rods and cones.

photorespiration: a series of reactions in plants in which O_2 replaces CO_2 during the Calvin cycle, preventing carbon fixation; this wasteful process dominates when C_3 plants are forced to close their stomata to prevent water loss.

photosynthesis: the complete series of chemical reactions in which the energy of light is used to synthesize high-energy organic molecules, usually carbohydrates, from low-energy inorganic molecules, usually carbon dioxide and water.

photosystem: in thylakoid membranes, a cluster of chlorophyll, accessory pigment molecules, proteins, and other molecules that collectively capture light energy, transfer some of the energy to electrons, and transfer the energetic electrons to an adjacent electron transport chain.

phototropism: growth with respect to the direction of light.

phylogeny (fī-lah′-jen-ē): the evolutionary history of a group of species.

phylum (fī′-lum): in Linnaean classification, the taxonomic rank composed of related classes. Related phyla make up a kingdom.

phytochrome (fī′-tō-krōm): a light-sensitive plant pigment that mediates many plant responses to light, including flowering, stem elongation, and seed germination.

phytoplankton (fī′-tō-plank-ten): photosynthetic protists that are abundant in marine and freshwater environments.

pigment molecule: a light-absorbing, colored molecule, such as chlorophyll, carotenoid, or melanin molecules.

pilus (pl., pili): a hair-like protein structure that projects from the cell wall of many bacteria. Attachment pili help bacteria adhere to structures. Sex pili assist in the transfer of plasmids.

pineal gland (pī-nē′-al): a small gland within the brain that secretes melatonin; controls the seasonal reproductive cycles of some mammals.

pinna: a flap of skin-covered cartilage on the surface of the head that collects sound waves and funnels them to the auditory canal.

pinocytosis (pi-nō-sī-tō′-sis): the nonselective movement of extracellular fluids and their dissolved substances, enclosed within a vesicle formed from the plasma membrane, into a cell.

pioneer: an organism that is among the first to colonize an unoccupied habitat in the first stages of succession.

pit: an area in the cell walls between two plant cells where the two cells are separated only by a thin, porous cell wall.

pith: cells forming the center of a root or stem.

pituitary gland: an endocrine gland, located at the base of the brain, that produces several hormones, many of which influence the activity of other glands.

placenta (pluh-sen′-tuh): in mammals, a structure formed by a complex interweaving of the uterine lining and the embryonic membranes, especially the chorion; functions in gas, nutrient, and waste exchange between embryonic and maternal circulatory systems and also secretes the hormones estrogen and progesterone, which are essential to maintaining pregnancy.

placental (pluh-sen′-tul) mammal: a mammal possessing a complex placenta (that is, species that are not marsupials or monotremes).

plankton: microscopic organisms that live in marine or freshwater environments; includes phytoplankton and zooplankton.

plant hormone: a chemical produced by specific plant cells that influences the growth, development, or metabolic activity of other cells, typically some distance away in the plant body.

plaque (plak): a deposit of cholesterol and other fatty substances within the wall of an artery.

plasma: the fluid, noncellular portion of the blood.

plasma cell: an antibody-secreting descendant of a B cell.

plasma membrane: the outer membrane of a cell, composed of a bilayer of phospholipids in which proteins are embedded.

plasmid (plaz′-mid): a small, circular piece of DNA located in the cytoplasm of many bacteria; usually does not carry genes required for the normal functioning of the bacterium, but may carry genes, such as those for antibiotic resistance, that assist bacterial survival in certain environments.

plasmodesma (plaz-mō-dez′-muh; pl., plasmodesmata): a cell-to-cell junction in plants that connects the cytosol of adjacent cells.

plasmodium (plaz-mō′-dē-um): a slug-like mass of cytoplasm containing thousands of nuclei that are not confined within individual cells.

plastid (plas′-tid): in plant cells, an organelle bounded by two membranes that may be involved in photosynthesis (chloroplasts), pigment storage, or food storage.

plate tectonics: the theory that Earth's crust is divided into irregular plates that are converging, diverging, or slipping by one another; these motions cause continental drift, the movement of continents over Earth's surface.

platelet (plāt′-let): a cell fragment that is formed from megakaryocytes in bone marrow; platelets lack a nucleus; they circulate in the blood and play a role in blood clotting.

play: behavior that seems to lack any immediate function and that often includes modified versions of behaviors used in other contexts.

pleated sheet: a form of secondary structure exhibited by certain proteins, such as silk, in which many protein chains lie side by side, with hydrogen bonds holding adjacent chains together.

pleiotropy (ple′-ō-trō-pē): a situation in which a single gene influences more than one phenotypic characteristic.

polar body: in oogenesis, a small cell, containing a nucleus but virtually no cytoplasm, produced by both the first meiotic division (of the primary oocyte) and the second meiotic division (of the secondary oocyte).

polar covalent bond: a covalent bond with unequal sharing of electrons, such that one atom is relatively negative and the other is relatively positive.

pollen: the male gametophyte of a seed plant; also called a *pollen grain.*

pollen grain: the male gametophyte of a seed plant.

pollination: in flowering plants, when pollen grains land on the stigma of a flower of the same species; in conifers, when pollen grains land within the pollen chamber of a female cone of the same species.

polygenic inheritance: a pattern of inheritance in which the interactions of two or more functionally similar genes determine phenotype.

polymer (pah′-li-mer): a molecule composed of three or more (perhaps thousands) smaller subunits called monomers, which may be identical (for example, the glucose monomers of starch) or different (for example, the amino acids of a protein).

polymerase chain reaction (PCR): a method of producing virtually unlimited numbers of copies of a specific piece of DNA, starting with as little as one copy of the desired DNA.

polypeptide: a long chain of amino acids linked by peptide bonds. A protein consists of one or more polypeptides.

polyploid (pahl′-ē-ploid): having more than two copies of each homologous chromosome.

polysaccharide (pahl-ē-sak′-uh-rīd): a large carbohydrate molecule composed of branched or unbranched chains of repeating monosaccharide subunits, normally glucose or modified glucose molecules; includes starches, cellulose, and glycogen.

pons: a portion of the hindbrain, just above the medulla, that contains neurons that influence sleep and the rate and pattern of breathing.

population: all the members of a particular species within a defined area, found in the same time and place and actually or potentially interbreeding.

population bottleneck: the result of an event that causes a population to become extremely small; may cause genetic drift that results in changed allele frequencies and loss of genetic variability.

population cycle: regularly recurring, cyclical changes in population size.

post-anal tail: a tail that extends beyond the anus and contains muscle tissue and the most posterior part of the nerve cord; found in all chordates at some stage of development.

positive feedback: a physiological mechanism in which a change causes responses that tend to amplify the original change.

posterior pituitary: a lobe of the pituitary gland that is an outgrowth of the hypothalamus and that releases antidiuretic hormone and oxytocin.

postmating isolating mechanism: any structure, physiological function, or developmental abnormality that prevents organisms of two different species, once mating has occurred, from producing vigorous, fertile offspring.

postsynaptic neuron: at a synapse, the nerve cell that changes its metabolic activity or its electrical potential in response to a chemical (the neurotransmitter) released by another (presynaptic) cell.

postsynaptic potential (PSP): an electrical signal produced in a postsynaptic cell by transmission across the synapse; it may be excitatory (EPSP), making the cell more likely to produce an action potential, or inhibitory (IPSP), tending to inhibit the production of an action potential.

potential energy: "stored" energy including chemical energy (stored in molecules), elastic energy (such as stored in a spring), or gravitational energy (stored in the elevated position of an object).

prairie: a biome, located in the centers of continents, that primarily supports grasses; also called *grassland*.

precapillary sphincter (sfink′-ter): a ring of smooth muscle between an arteriole and a capillary that regulates the flow of blood into the capillary bed.

predation (pre-dā′-shun): the act of eating another living organism.

predator: an organism that eats other organisms.

prediction: in the scientific method, a statement describing an expected observation or the expected outcome of an experiment, assuming that a specific hypothesis is true.

premating isolating mechanism: any structure, physiological function, or behavior that prevents organisms of two different species from exchanging gametes.

pressure-flow mechanism: the process by which sugars are transported in phloem; the movement of sugars into a phloem sieve tube causes water to enter the tube by osmosis, while the movement of sugars out of another part of the same sieve tube causes water to leave by osmosis. The resulting pressure gradient causes the bulk movement of water and dissolved sugars from the end of the sieve tube into which sugar is transported (a source) toward the end of the sieve tube from which sugar is removed (a sink).

presynaptic neuron: a nerve cell that releases a chemical (the neurotransmitter) at a synapse, causing changes in the electrical activity or metabolism of another (postsynaptic) cell.

prey: organisms that are eaten, and often killed, by another organism (a predator).

primary consumer: an organism that feeds on producers; an herbivore.

primary electron acceptor: a molecule in the reaction center of each photosystem that accepts an electron from one of the two reaction center chlorophyll *a* molecules and transfers the electron to an adjacent electron transport chain.

primary growth: growth in length and development of the initial structures of plant roots and shoots; results from cell division of apical meristems and differentiation of the daughter cells.

primary oocyte (ō′-ō-sīt): a diploid cell, derived from the oogonium by growth and differentiation, that undergoes meiotic cell division, producing the egg.

primary spermatocyte (sper-ma′-tō-sīt): a diploid cell, derived from the spermatogonium by growth and differentiation, that undergoes meiotic cell division, producing four sperm.

primary structure: the amino acid sequence of a protein.

primary succession: succession that occurs in an environment, such as bare rock, in which no trace of a previous community is present.

primate: a member of the mammalian clade Primates, characterized by the presence of an opposable thumb, forward-facing eyes, and a well-developed cerebral cortex; includes lemurs, monkeys, apes, and humans.

prion (prē′-on): a protein that, in mutated form, acts as an infectious agent that causes certain neurodegenerative diseases, including kuru and scrapie.

producer: a photosynthetic organism; an autotroph.

product: an atom or molecule that is formed from reactants in a chemical reaction.

profundal zone: the part of a lake in which light is insufficient to support photosynthesis.

progesterone (prō-ge′-ster-ōn): a hormone, produced by the corpus luteum in the ovary, that promotes the development of the uterine lining in females.

prokaryote (prō-kar′-ē-ōt): an organism whose cells are prokaryotic (their genetic material is not enclosed in a membrane-bound nucleus and they lack other membrane-bound organelles); bacteria and archaea are prokaryotes.

prokaryotic (prō-kar-ē-ot′-ik): referring to cells of the domains Bacteria or Archaea. Prokaryotic cells have genetic material that is not enclosed in a membrane-bound nucleus; they also lack other membrane-bound organelles.

prokaryotic fission: the process by which a single bacterium divides in half, producing two identical offspring.

prolactin (PRL): a hormone, released by the anterior pituitary, that stimulates milk production in human females.

promoter: a specific sequence of DNA at the beginning of a gene, to which RNA polymerase binds and starts gene transcription.

prophase (prō′-fāz): the first stage of mitosis, in which the chromosomes first become visible in the light microscope as thickened, condensed threads and the spindle begins to form; as the spindle is completed, the nuclear envelope breaks apart, and the spindle microtubules invade the nuclear region and attach to the kinetochores of the chromosomes. Also, the first stage of meiosis: In meiosis I, the homologous chromosomes pair up, exchange parts at chiasmata, and attach to spindle microtubules; in meiosis II, the spindle re-forms and chromosomes attach to the microtubules.

prostaglandin (pro-stuh-glan′-din): a family of modified fatty acid hormones manufactured and secreted by many cells of the body.

prostate gland (pros′-tāt): a gland that produces part of the fluid component of semen; prostatic fluid is basic and contains a chemical that activates sperm movement.

protease (prō′-tē-ās): an enzyme that digests proteins.

protein: a polymer composed of amino acids joined by peptide bonds.

protist: a eukaryotic organism that is not a plant, animal, or fungus. The term encompasses a diverse array of organisms and does not represent a monophyletic group.

protocell: the hypothetical evolutionary precursor of living cells, consisting of a mixture of organic molecules within a membrane.

proton: a subatomic particle that is found in the nuclei of atoms; it bears a unit of positive charge and has a relatively large mass, roughly equal to the mass of a neutron.

protonephridium (prō-tō-nef-rid′-ē-um; pl., protonephridia): the functional unit of the excretory system of some invertebrates, such as flatworms; consists of a tubule that has an external opening to the outside of the body, but lacks an internal opening within the body. Fluid is filtered from the body cavity into the tubule by a hollow cell at the end of the tubule, such as a flame cell in flatworms, and released outside the body.

protostome (prō′-tō-stōm): an animal with a mode of embryonic development in which the coelom is derived from splits in the mesoderm; characteristic of arthropods, annelids, and mollusks.

protozoan (prō-tuh-zō′-an; pl., protozoa): a nonphotosynthetic, single-celled protist.

proximal tubule: the initial part of the renal tubule; in mammals, most reabsorption occurs here.

pseudocoelom (soo′-dō-sēl′-ōm): in animals, a "false coelom," that is, a space or cavity, partially but not fully lined with tissue derived from mesoderm, that separates the body wall from the inner organs; found in roundworms.

pseudoplasmodium (soo′-dō-plaz-mō′-dē-um): an aggregation of individual amoeboid cells that form a slug-like mass.

pseudopod (soo′-dō-pod): an extension of the plasma membrane by which certain cells, such as amoebas, locomote and engulf prey.

puberty: a stage of development characterized by sexual maturation, rapid growth, and the appearance of secondary sexual characteristics.

pubic lice: arthropod parasites that can infest humans; can be transmitted by sexual contact.

pulmonary circuit: the pathway of blood from the heart to the lungs and back to the heart.

Punnett square method: a method of predicting the genotypes and phenotypes of offspring in genetic crosses.

pupa (pl., pupae): a developmental stage in some insect species in which the organism stops moving and feeding and may be encased in a cocoon; occurs between the larval and the adult phases.

pupil: the adjustable opening in the center of the iris, through which light enters the eye.

quaternary structure (kwat′-er-nuh-rē): the complex three-dimensional structure of a protein consisting of more than one peptide chain.

question: in the scientific method, a statement that identifies a particular aspect of an observation that a scientist wishes to explain.

***r*-selected species:** species that typically live in rapidly changing, unpredictable environments and usually do not have population sizes that approach carrying capacity. *r*-selected species usually mature rapidly, have a short life span, produce a large number of small offspring, and provide little parental care, so most offspring die before reaching maturity.

radial symmetry: a body plan in which any plane along a central axis will divide the body into approximately mirror-image halves. Cnidarians and many adult echinoderms exhibit radial symmetry.

radioactive: pertaining to an atom with an unstable nucleus that spontaneously breaks apart or decays, with the emission of radiation.

radiolarian (rā-dē-ō-lar′-ē-un): a member of a protist group characterized by pseudopods and typically elaborate silica shells. Radiolarians are largely aquatic (mostly marine) and are part of a larger group known as rhizarians.

rain shadow: a local dry area, usually located on the downwind side of a mountain range that blocks the prevailing moisture-bearing winds.

random distribution: the distribution characteristic of populations in which the probability of finding an individual is equal in all parts of an area.

reabsorption: the process by which cells of the renal tubule of the nephron remove water and nutrients from the filtrate within the tubule and return those substances to the blood.

reactant: an atom or molecule that is used up in a chemical reaction to form a product.

reaction center: two chlorophyll *a* molecules and a primary electron acceptor complexed with proteins and located near the center of each photosystem within the thylakoid membrane. Light energy is passed to one of the chlorophylls, which donates an energized electron to the primary electron acceptor, which then passes the electron to an adjacent electron transport chain.

receptor: (1) a protein, located in a membrane or the cytoplasm of a cell, that binds to specific molecules (for example, a hormone or neurotransmitter), triggering a response in the cell, such as endocytosis, changes in metabolic rate, cell division, or electrical changes; (2) a cell that responds to an environmental stimulus (chemicals, sound, light, pH, and so on) by changing its electrical potential.

receptor-mediated endocytosis: the selective uptake of molecules from the interstitial fluid by binding to a receptor located at a coated pit on the plasma membrane and pinching off the coated pit into a vesicle that moves into the cytosol.

receptor potential: an electrical potential change in a receptor cell, produced in response to the reception of an environmental stimulus (chemicals, sound, light, heat, and so on). The size of the receptor potential is proportional to the intensity of the stimulus.

receptor protein: a protein, located in a membrane or the cytosol of a cell, that binds to specific molecules (for example, a hormone or neurotransmitter), triggering a response in the cell, such as endocytosis, changes in metabolic rate, cell division, or electrical changes.

recessive: an allele that is expressed only in homozygotes and is completely masked in heterozygotes.

recognition protein: a protein or glycoprotein protruding from the outside surface of a plasma membrane that identifies a cell as belonging to a particular species, to a specific individual of that species, and in many cases to one specific organ within the individual.

recombinant DNA: DNA that has been altered by the addition of DNA from a different organism, typically from a different species.

recombination: the formation of new combinations of the different alleles of each gene on a chromosome; the result of crossing over.

rectum: the terminal portion of the vertebrate digestive tube where feces are stored until they are eliminated.

reflex: a simple, stereotyped movement of part of the body that occurs automatically in response to a stimulus.

regeneration: the regrowth of a body part after loss or damage; also, asexual reproduction by means of the regrowth of an entire body from a fragment.

regulatory gene: in prokaryotes, a gene encoding a protein that binds to the operator of one or more operons, controlling the ability of RNA polymerase to transcribe the structural genes of the operon.

regulatory T cell: a type of T cell that suppresses the adaptive immune response, especially by self-reactive lymphocytes, and appears to be important in the prevention of autoimmune disorders.

releasing hormone: a hormone, secreted by the hypothalamus, that causes the release of specific hormones by the anterior pituitary.

renal artery: the artery carrying blood to each kidney.

renal corpuscle: the portion of a nephron in which blood filtrate is collected from the glomerulus by the cup-shaped glomerular capsule.

renal cortex: the outer layer of the kidney, in which the largest portion of each nephron is located, including renal corpuscle and the distal and proximal tubules.

renal medulla: the layer of the kidney just inside the renal cortex, in which loops of Henle produce a highly concentrated interstitial fluid, allowing the production of concentrated urine.

renal pelvis: the inner chamber of the kidney, in which urine from the collecting ducts accumulates before it enters the ureter.

renal tubule (toob′-ūl): the tubular portion of a nephron composed of the proximal tubule, the nephron loop, and the distal tubule. Urine is formed from the blood filtrate as it passes through the tubule.

renal vein: the vein carrying blood away from each kidney.

renin: in mammals, an enzyme that is released by the kidneys when blood pressure falls. Renin catalyzes the formation of angiotensin, which causes arterioles to constrict, thereby elevating blood pressure.

replacement level fertility (RLF): the average number of offspring per female that is required to maintain a stable population.

repressor protein: in prokaryotes, a protein encoded by a regulatory gene, which binds to the operator of an operon and prevents RNA polymerase from transcribing the structural genes.

reproductive isolation: the failure of organisms of one population to breed successfully with members of another; may be due to premating or postmating isolating mechanisms.

reptile: a member of the chordate group that includes the snakes, lizards, turtles, alligators, birds, and crocodiles.

reservoir: the major source and storage site of a nutrient in an ecosystem, normally in the abiotic portion.

resource partitioning: the coexistence of two species with similar requirements, each occupying a smaller niche than either would if it were by itself; a means of minimizing the species' competitive interactions.

respiration: in terrestrial vertebrates, the act of moving air into the lungs (inhalation) and out of the lungs (exhalation); during respiration, oxygen diffuses from the air in the lungs into the circulatory system, and carbon dioxide diffuses from the circulatory system into the air in the lungs.

respiratory center: a cluster of neurons, located in the medulla of the brain, that sends rhythmic bursts of nerve impulses to the respiratory muscles, resulting in breathing.

respiratory membrane: within the lungs, the fusion of the epithelial cells of the alveoli and the endothelial cells that form the walls of surrounding capillaries.

respiratory system: a group of organs that work together to facilitate gas exchange—the intake of oxygen and the removal of carbon dioxide—between an animal and its environment.

resting potential: an electrical potential, or voltage, in unstimulated nerve cells; the inside of the cell is always negative with respect to the outside.

restriction enzyme: an enzyme, usually isolated from bacteria, that cuts double-stranded DNA at a specific nucleotide sequence; the nucleotide sequence that is cut differs for different restriction enzymes.

restriction fragment: a piece of DNA that has been isolated by cleaving a larger piece of DNA with restriction enzymes.

restriction fragment length polymorphism (RFLP): a difference in the length of DNA fragments that were produced by cutting samples of DNA from different individuals of the same species with the same set of restriction enzymes; fragment length differences occur because of differences in nucleotide sequences, and hence in the ability of restriction enzymes to cut the DNA, among individuals of the same species.

reticular formation (reh-tik′-ū-lar): a diffuse network of neurons extending from the hindbrain, through the midbrain, and into the lower reaches of the forebrain; involved in filtering sensory input and regulating what information is relayed to conscious brain centers for further attention.

retina (ret′-in-uh): a multilayered sheet of nerve tissue at the rear of camera-type eyes, composed of photoreceptor cells plus associated nerve cells that refine the photoreceptor information and transmit it to the optic nerve.

rhizarian: a member of Rhizaria, a protist clade. Rhizarians, which use thin pseudopods to move and capture prey and which often have hard shells, include the foraminiferans and the radiolarians.

ribonucleic acid (rī-bō-noo-klā′-ik; RNA): a molecule composed of ribose nucleotides, each of which consists of a phosphate group, the sugar ribose, and one of the bases adenine, cytosine, guanine, or uracil; involved in converting the information in DNA into protein; also the genetic material of some viruses.

ribosomal RNA (rī-bō-sō′-mul; rRNA): a type of RNA that combines with proteins to form ribosomes.

ribosome (rī′-bō-sōm): a complex consisting of two subunits, each composed of ribosomal RNA and protein, found in the cytoplasm of cells or attached to the endoplasmic reticulum, that is the site of protein synthesis, during which the sequence of bases of messenger RNA is translated into the sequence of amino acids in a protein.

ribozyme: an RNA molecule that can catalyze certain chemical reactions, especially those involved in the synthesis and processing of RNA itself.

RNA polymerase: in RNA synthesis, an enzyme that catalyzes the bonding of free RNA nucleotides into a continuous strand, using RNA nucleotides that are complementary to those of the template strand of DNA.

rod: a rod-shaped photoreceptor cell in the vertebrate retina, sensitive to dim light but not involved in color vision; see also *cone*.

root: the part of a plant body, normally underground, that provides anchorage, absorbs water and dissolved nutrients and transports them to the stem, produces some hormones, and in some plants serves as a storage site for carbohydrates.

root cap: a cluster of cells at the tip of a growing root, derived from the apical meristem; protects the growing tip from damage as it burrows through the soil.

root hair: a fine projection from an epidermal cell of a young root that increases the absorptive surface area of the root.

root pressure: pressure within a root caused by the transport of minerals into the vascular cylinder, accompanied by the entry of water by osmosis.

root system: all the roots of a plant.

round window: a flexible membrane at the end of the cochlea opposite the oval window that allows the fluid in the cochlea to move in response to movements of the oval window.

rubisco: in the carbon fixation step of the Calvin cycle, the enzyme that catalyzes the reaction between ribulose bisphosphate (RuBP) and carbon dioxide, thereby fixing the carbon of carbon dioxide in an organic molecule; short for ribulose bisphosphate carboxylase.

rumen fungus: a member of the fungus clade Neocallimastigomycota, whose members have swimming spores with multiple flagella. Rumen fungi are anaerobic and most live in the digestive tracts of plant-eating animals.

ruminant (roo′-min-ant): an herbivorous animal with a digestive tract that includes multiple stomach chambers, one of which contains cellulose-digesting bacteria, and that regurgitates the contents ("cud") of the first chamber for additional chewing ("ruminating").

S-curve: the S-shaped growth curve produced by logistic population growth, usually describing a population of organisms introduced into a new area; consists of an initial period of exponential growth, followed by a decreasing growth rate, and finally, relative stability around a growth rate of zero.

sac fungus: a member of the fungus clade Ascomycota, whose members form spores in a saclike case called an ascus.

saccule: a patch of hair cells in the vestibule of the inner ear; bending of the hairs of the hair cells permits detection of the direction of gravity and the degree of tilt of the head.

sapwood: young secondary xylem that transports water and minerals in a tree trunk.

sarcomere (sark′-ō-mēr): the unit of contraction of a muscle fiber; a subunit of the myofibril, consisting of thick and thin filaments, bounded by Z discs.

sarcoplasmic reticulum (sark′-ō-plas′-mik re-tik′-ū-lum; SR): the specialized endoplasmic reticulum in muscle cells; forms interconnected hollow tubes. The sarcoplasmic reticulum stores calcium ions and releases them into the interior of the muscle cell, initiating contraction.

saturated: referring to a fatty acid with as many hydrogen atoms as possible bonded to the carbon backbone (therefore, a saturated fatty acid has no double bonds in its carbon backbone).

savanna: a biome that is dominated by grasses and supports scattered trees; typically has a rainy season during which most of the year's precipitation falls, followed by a dry season during which virtually no precipitation occurs.

science: the organized and systematic inquiry, through observation and experiment, into the origins, structure, and behavior of our living and nonliving surroundings.

scientific method: a rigorous procedure for making observations of specific phenomena and searching for the order underlying those phenomena.

scientific name: the two-part Latin name of a species; consists of the genus name followed by the species name.

scientific theory: an explanation of natural phenomena developed through extensive and reproducible observations; more general and reliable than a hypothesis.

scientific theory of evolution: the theory that modern organisms descended, with modification, from preexisting life-forms.

sclera: a tough, white connective tissue layer that covers the outside of the eyeball and forms the white of the eye.

sclerenchyma tissue (skler-en′-ki-muh): a tissue formed from sclerenchyma cells with thick, hardened cell walls. Sclerenchyma cells usually die as they mature and generally support or protect the plant body.

scrotum (skrō′-tum): in male mammals, the pouch of skin containing the testes.

second law of thermodynamics: the principle of physics that states that any change in an isolated system causes the quantity of concentrated, useful energy to decrease and the amount of randomness and disorder (entropy) to increase.

second messenger: an intracellular chemical, such as cyclic AMP, that is synthesized or released within a cell in response to the binding of a hormone or neurotransmitter (the first messenger) to receptors on the cell surface; brings about specific changes in the metabolism of the cell.

secondary consumer: an organism that feeds on primary consumers; a type of carnivore.

secondary growth: growth in the diameter and strength of a stem or root due to cell division in lateral meristems and differentiation of their daughter cells.

secondary oocyte (ō′-ō-sīt): a large haploid cell derived from the diploid primary oocyte by meiosis I.

secondary spermatocyte (sper-ma′-tō-sīt): a large haploid cell derived from the diploid primary spermatocyte by meiosis I.

secondary structure: a repeated, regular structure assumed by a protein chain, held together by hydrogen bonds; for example, a helix.

secondary succession: succession that occurs after an existing community is disturbed—for example, after a forest fire; secondary succession is much more rapid than primary succession.

secretin: a hormone produced by the small intestine that stimulates the production and release of digestive secretions by the pancreas and liver.

secretion: the process by which cells of the tubule of the nephron remove wastes from the blood, actively secreting those wastes into the tubule.

seed: the reproductive structure of a seed plant, protected by a seed coat; contains an embryonic plant and a supply of food for it.

seed coat: the thin, tough, and waterproof outermost covering of a seed, formed from the integuments of the ovule.

segmentation (seg-men-tā′-shun): an animal body plan in which the body is divided into repeated, typically similar units.

selectively permeable: the quality of a membrane that allows certain molecules or ions to move through it more readily than others.

self-fertilization: the union of sperm and egg from the same individual.

semen: the sperm-containing fluid produced by the male reproductive tract.

semicircular canal: in the inner ear, one of three fluid-filled tubes, each with a bulge at one end containing a patch of hair cells; movement of the head moves fluid in the canal and consequently bends the hairs of the hair cells.

semiconservative replication: the process of replication of the DNA double helix; the two DNA strands separate, and each is used as a template for the synthesis of a complementary DNA strand. Consequently, each daughter double helix consists of one parental strand and one new strand.

semilunar valve: a valve located between the right ventricle of the heart and the pulmonary artery, or between the left ventricle and the aorta; prevents the backflow of blood into the ventricles when they relax.

seminal vesicle: in male mammals, a gland that produces a basic, fructose-containing fluid that forms part of the semen.

seminiferous tubule (sem-i-ni′-fer-us): in the vertebrate testis, a series of tubes in which sperm are produced.

senescence: in plants, a specific aging process, typically including deterioration and the dropping of leaves and flowers.

sense organ: a multicellular structure that includes both sensory receptor cells and accessory structures that assist in the detection of a specific stimulus; examples include the eye and ear.

sensory neuron: a nerve cell that responds to a stimulus from the internal or external environment.

sensory receptor: a cell (typically, a neuron) specialized to respond to particular internal or external environmental stimuli by producing an electrical potential.

sepal (sē′-pul): one of the group of modified leaves that surrounds and protects a flower bud; in dicots, usually develops into a green, leaflike structure when the flower blooms; in monocots, usually similar to a petal.

septum (pl., septa): a partition that separates the fungal hypha into individual cells; pores in septa allow the transfer of materials between cells.

Sertoli cell: in the seminiferous tubule, a large cell that regulates spermatogenesis and nourishes the developing sperm.

severe combined immune deficiency (SCID): a disorder in which no immune cells, or very few, are formed; the immune system is incapable of responding properly to invading disease organisms, and the individual is very vulnerable to common infections.

sex chromosome: either of the pair of chromosomes that usually determines the sex of an organism; for example, the X and Y chromosomes in mammals.

sex-linked: referring to a pattern of inheritance characteristic of genes located on one type of sex chromosome (for example, X) and not found on the other type (for example, Y); in mammals, in almost all cases, the gene controlling the trait is on the X chromosome, so this pattern is often called X-linked. In X-linked inheritance, females show the dominant trait unless they are homozygous recessive, whereas males express whichever allele, dominant or recessive, is found on their single X chromosome.

sexual reproduction: a form of reproduction in which genetic material from two parent organisms is combined in the offspring; usually, two haploid gametes fuse to form a diploid zygote.

sexual selection: a type of natural selection that acts on traits involved in finding and acquiring mates.

sexually transmitted disease (STD): a disease that is passed from person to person by sexual contact; also known as *sexually transmitted infection (STI)*.

shoot system: all the parts of a vascular plant exclusive of the root; usually aboveground. Consists of stem, leaves, buds, and (in season) flowers and fruits; functions include photosynthesis, transport of materials, reproduction, and hormone synthesis.

short-day plant: a plant that will flower only if the length of uninterrupted darkness exceeds a species-specific critical period; also called a *long-night plant*.

short tandem repeat (STR): a DNA sequence consisting of a short sequence of nucleotides (usually two to five nucleotides in length) repeated multiple times, with all the repetitions side by side on a chromosome; variations in the number of repeats of a standardized set of 13 STRs produce DNA profiles used to identify people by their DNA.

short-term memory: a memory that lasts only a brief time, usually while a specific neuronal circuit is active or while a change in biochemical activity temporarily strengthens specific synapses.

sickle-cell anemia: a recessive disease caused by a single amino acid substitution in the hemoglobin molecule. Sickle-cell hemoglobin molecules tend to cluster together, distorting the shape of red blood cells and causing them to break and clog capillaries.

sieve plate: in plants, a structure between two adjacent sieve-tube elements in phloem, where holes formed in the cell walls interconnect the cytoplasm of the sieve-tube elements.

sieve-tube element: in phloem, one of the cells of a sieve tube.

simple diffusion: the diffusion of water, dissolved gases, or lipid-soluble molecules through the phospholipid bilayer of a cellular membrane.

simple epithelium: a type of epithelial tissue, one cell layer thick, that lines many hollow organs such as those of the respiratory, digestive, urinary, reproductive, and circulatory systems.

sink: in plants, any structure that uses up sugars or converts sugars to starch, and toward which phloem fluids will flow.

sinoatrial (SA) node (sī′-nō-āt′-rē-ul nōd): a small mass of specialized muscle in the wall of the right atrium; generates electrical signals rhythmically and spontaneously and serves as the heart's pacemaker.

skeletal muscle: the type of muscle that is attached to and moves the skeleton and is under the direct, normally voluntary, control of the nervous system; also called *striated muscle*.

skeleton: a supporting structure for the body, on which muscles act to change the body configuration; may be external or internal.

small intestine: the portion of the digestive tract, located between the stomach and large intestine, in which most digestion and absorption of nutrients occur.

smooth muscle: the type of muscle that surrounds hollow organs, such as the digestive tract, bladder, and blood vessels; not striped in appearance (hence the name "smooth") and normally not under voluntary control.

social learning: learning that is influenced by observation of, or interaction with, other animals, usually of the same species.

sodium-potassium (Na^+-K^+) pump: an active transport protein that uses the energy of ATP to transport Na^+ out of a cell and K^+ into a cell; produces and maintains the concentration gradients of these ions across the plasma membrane, such that the concentration of Na^+ is higher outside a cell than inside, and the concentration of K^+ is higher inside a cell than outside.

solute: a substance dissolved in a solvent.

solution: a solvent containing one or more dissolved substances (solutes).

solvent: a liquid capable of dissolving (uniformly dispersing) other substances in itself.

somatic nervous system: that portion of the peripheral nervous system that controls voluntary movement by activating skeletal muscles.

source: in plants, any structure that actively synthesizes sugar, and away from which phloem fluid will be transported.

spawning: a method of external fertilization in which male and female parents shed gametes into water, and sperm must swim through the water to reach the eggs.

speciation: the process of species formation, in which a single species splits into two or more species.

species (spē′-sēs): the basic unit of taxonomic classification, consisting of a population or group of populations that evolves separately from other populations. In sexually reproducing organisms, a species can be defined as a population or group of populations whose members interbreed freely with one another under natural conditions but do not interbreed with members of other populations.

specific heat: the amount of energy required to raise the temperature of 1 gram of a substance by 1°C.

sperm: the haploid male gamete, normally small, motile, and containing little cytoplasm.

spermatid: a haploid cell derived from the secondary spermatocyte by meiosis II; differentiates into the mature sperm.

spermatogenesis: the process by which sperm cells form.

spermatogonium (pl., spermatogonia): a diploid cell, lining the walls of the seminiferous tubules, that gives rise to a primary spermatocyte.

spermatophore: a package of sperm formed by the males of some invertebrate animals; the spermatophore can be inserted into the female reproductive tract, where it releases its sperm.

sphincter muscle: a circular ring of muscle surrounding a tubular structure, such as the esophagus, stomach, or intestine; contraction and relaxation of a sphincter muscle controls the movement of materials through the tube.

spinal cord: the part of the central nervous system of vertebrates that extends from the base of the brain to the hips and is protected by the bones of the vertebral column; contains the cell bodies of motor neurons that form synapses with skeletal muscles, the circuitry for some simple reflex behaviors, and axons that communicate with the brain.

spindle: an array of microtubules that moves the chromosomes to opposite poles of a cell during mitotic and meiotic cell division.

spindle microtubule: one of the microtubules organized in a spindle shape that separate chromosomes during meiotic and meiotic cell division.

spiracle (spi′-ruh-kul): an opening in the body wall of insects through which air enters the tracheae.

spleen: the largest organ of the lymphatic system, located in the abdominal cavity; contains macrophages that filter the blood by removing microbes and aged red blood cells, and lymphocytes (B and T cells) that reproduce during times of infection.

spongy bone: porous, lightweight bone tissue in the interior of bones; the location of bone marrow. Compare with *compact bone*.

spontaneous generation: the proposal that living organisms can arise from nonliving matter.

sporangium (spor-an′-jē-um; pl., sporangia): a structure in which spores are produced.

spore: (1) in plants and fungi, a haploid cell capable of developing into an adult without fusing with another cell (without fertilization); (2) in bacteria and some other organisms, a stage of the life cycle that is resistant to extreme environmental conditions.

sporophyte (spor′-ō-fīt): the multicellular diploid stage in the life cycle of a plant; produces haploid, asexual spores through meiosis.

stabilizing selection: a type of natural selection that favors the average phenotype in a population.

stamen (stā′-men): the male reproductive structure of a flower, consisting of a filament and an anther, in which pollen grains develop.

starch: a polysaccharide that is composed of branched or unbranched chains of glucose molecules; used by plants as a carbohydrate-storage molecule.

start codon: the first AUG codon in a messenger RNA molecule.

startle coloration: a form of mimicry in which a color pattern (in many cases resembling large eyes) can be displayed suddenly by a prey organism when approached by a predator.

stem: the portion of the plant body, normally located aboveground, that bears leaves and reproductive structures such as flowers and fruit.

stem cell: an undifferentiated cell that is capable of dividing and giving rise to one or more distinct types of differentiated cell(s).

sterilization: a generally permanent method of contraception in which the pathways through which the sperm (vas deferens) or egg (oviducts) must travel are interrupted; the most effective form of contraception.

steroid: a lipid consisting of four fused carbon rings, with various functional groups attached.

steroid hormone: a class of hormones whose chemical structure (four fused carbon rings with various functional groups) resembles cholesterol; steroids, which are lipids, are secreted by the ovaries and placenta, the testes, and the adrenal cortex.

stigma (stig′-muh): the pollen-capturing tip of a carpel.

stirrup: the third of the small bones of the middle ear that link the tympanic membrane (eardrum) with the oval window; the stirrup is directly connected to the oval window; also called the *stapes*.

stoma (stō′-muh; pl., stomata): an adjustable opening in the epidermis of a leaf or young stem, surrounded by a pair of guard cells, that regulates the diffusion of carbon dioxide and water into and out of the leaf or stem.

stomach: the muscular sac between the esophagus and small intestine where food is stored and mechanically broken down and in which protein digestion begins.

stop codon: a codon in messenger RNA that stops protein synthesis and causes the completed protein chain to be released from the ribosome.

stramenopile: a member of Stramenopila, a large protist clade. Stramenopiles, which are characterized by hair-like projections on their flagella, include the water molds, the diatoms, and the brown algae.

strand: a single polymer of nucleotides; DNA is composed of two strands wound about each other in a double helix; RNA is usually single-stranded.

stratified epithelium: a type of epithelial tissue composed of several cell layers, usually strong and waterproof; mostly found on the surface of the skin.

stroke: an interruption of blood flow to part of the brain caused by the rupture of an artery or the blocking of an artery by a blood clot. Loss of blood supply leads to rapid death of the area of the brain affected.

stroma (strō′-muh): the semifluid material inside chloroplasts in which the thylakoids are located; the site of the reactions of the Calvin cycle.

structural gene: in the prokaryotic operon, the genes that encode enzymes or other cellular proteins.

style: a stalk connecting the stigma of a carpel with the ovary at its base.

subclimax: a community in which succession is stopped before the climax community is reached; it is maintained by regular disturbance—for example, a tallgrass prairie maintained by periodic fires.

substitution mutation: a mutation in which a single base pair in DNA has been changed.

substrate: the atoms or molecules that are the reactants for an enzyme-catalyzed chemical reaction.

succession (suk-seh′-shun): a structural change in a community and its nonliving environment over time. During succession, species replace one another in a somewhat predictable manner until a stable, self-sustaining climax community is reached.

sucrose: a disaccharide composed of glucose and fructose.

sugar: a simple carbohydrate molecule, either a monosaccharide or a disaccharide.

sugar-phosphate backbone: a chain of sugars and phosphates in DNA and RNA; the sugar of one nucleotide bonds to the phosphate of the next nucleotide in a DNA or RNA strand. The bases in DNA or RNA are attached to the sugars of the backbone.

surface tension: the property of a liquid to resist penetration by objects at its interface with the air, due to cohesion between molecules of the liquid.

survivorship curve: the curve that results when the number of individuals of each age in a population is graphed against their age, usually expressed as a percentage of their maximum life span.

survivorship table: a data table that groups organisms born at the same time and tracks them throughout their life span, recording how many continue to survive in each succeeding year (or other unit of time). Various parameters such as gender may be used in the groupings. Human life tables may include many other parameters (such as socioeconomic status) used by demographers.

sustainable development: human activities that meet current needs for a reasonable quality of life without exceeding nature's limits and without compromising the ability of future generations to meet their needs.

symbiosis: a relationship between species in which the two or more species share a close, long-term physical association.

sympathetic division: the division of the autonomic nervous system that produces largely involuntary responses that prepare the body for stressful or highly energetic situations; often called the *sympathetic nervous system*.

sympatric speciation (sim-pat′-rik): the process by which new species arise in populations that are not physically divided; the genetic isolation required for sympatric speciation may be due to ecological isolation or chromosomal aberrations (such as polyploidy).

synapse (sin′-aps): the site of communication between nerve cells. At a synapse, one cell (presynaptic) releases a chemical (the neurotransmitter) that changes the electrical potential or metabolism of the second (postsynaptic) cell.

synaptic cleft: in a synapse, a small gap between the presynaptic and postsynaptic neurons.

synaptic communication: communication between cells of an animal in which nerve cells affect the activity of other cells by releasing chemicals called neurotransmitters across a small gap separating the ending of a nerve cell from a target cell (usually another nerve cell, a muscle cell, or an endocrine cell).

synaptic terminal: a swelling at the branched ending of an axon, where the axon forms a synapse.

syphilis (si′-ful-is): a sexually transmitted bacterial infection of the reproductive organs; if untreated, can damage the nervous and circulatory systems.

systematics: the branch of biology concerned with reconstructing phylogenies and with naming clades.

systemic circuit: the pathway of blood from the heart through all the parts of the body except the lungs and back to the heart.

systolic pressure (sis-tal′-ik): the blood pressure measured at the peak of contraction of the ventricles; the higher of the two blood pressure readings.

T cell: a type of lymphocyte that matures in the thymus. Some types of T cells recognize and destroy specific foreign cells or substances; other types of T cells regulate the activity of other cells of the immune system.

T-cell receptor: a protein receptor, located on the surface of a T cell, that binds a specific antigen and triggers the immune response of the T cell.

T tubule: a deep infolding of the plasma membrane of a muscle cell; conducts the action potential inside the cell.

taiga (tī′-guh): a biome with long, cold winters and only a few months of warm weather; dominated by evergreen coniferous trees; also called *northern coniferous forest*.

taproot system: a root system, commonly found in dicots, that consists of a long, thick main root and many smaller lateral roots that grow from the main root.

target cell: a cell on which a particular hormone exerts its effect.

taste bud: a cluster of taste receptor cells and supporting cells that is located in a small pit beneath the surface of the tongue and that communicates with the mouth through a small pore.

taxonomy (tax-on′-uh-mē): the branch of biology concerned with naming and classifying organisms.

tectorial membrane (tek-tor′-ē-ul): one of the membranes of the cochlea in which the hairs of the hair cells are embedded. In sound reception, movement of the basilar membrane relative to the tectorial membrane bends the hairs.

telomere (tēl′-e-mēr): the nucleotides at the end of a chromosome that protect the chromosome from damage during condensation, and prevent the end of one chromosome from attaching to the end of another chromosome.

telophase (tēl′-ō-fāz): in mitosis and both divisions of meiosis, the final stage, in which the spindle fibers usually disappear, nuclear envelopes re-form, and cytokinesis generally occurs. In mitosis and meiosis II, the chromosomes also relax from their condensed form.

temperate deciduous forest: a biome having cold winters and warm summers, with enough summer rainfall for trees to grow and shade out grasses; characterized by trees that drop their leaves in winter (deciduous trees), an adaptation that minimizes water loss when the soil is frozen.

temperate rain forest: a temperate biome with abundant liquid water year-round, dominated by conifers.

template strand: the strand of the DNA double helix from which RNA is transcribed.

temporal isolation: reproductive isolation that arises when species do not interbreed because they breed at different times.

tendon: a tough connective tissue band connecting a muscle to a bone.

terminal bud: meristem tissue and surrounding leaf primordia that are located at the tip of a plant shoot or a branch.

territoriality: the defense of an area in which important resources are located.

tertiary consumer (ter′-shē-er-ē): a carnivore that feeds on other carnivores (secondary consumers).

tertiary structure (ter′-shē-er-ē): the complex three-dimensional structure of a single peptide chain; held in place by disulfide bonds between cysteines.

test cross: a breeding experiment in which an individual showing the dominant phenotype is mated with an individual that is homozygous recessive for the same gene. The ratio of offspring with dominant versus recessive phenotypes can be used to determine the genotype of the phenotypically dominant individual.

testis (pl., testes): the gonad of male animals.

testosterone: in vertebrates, a hormone produced by the interstitial cells of the testis; stimulates spermatogenesis and the development of male secondary sex characteristics.

tetrapod: an organism descended from the first four-limbed vertebrate. Tetrapods include all extinct and living amphibians, reptiles (including birds), and mammals.

thalamus: the part of the forebrain that relays sensory information to many parts of the brain.

therapeutic cloning: the production of a clone for medical purposes. Typically, the nucleus from one of a patient's own cells would be inserted into an egg whose nucleus had been removed; the resulting cell would divide and produce embryonic stem cells that would be compatible with the patient's tissues and therefore would not be rejected by the patient's immune system.

thermoreceptor: a sensory receptor that responds to heat or cold.

thick filament: in the sarcomere, a bundle of myosin that interacts with thin filaments, producing muscle contraction.

thigmotropism: growth in response to touch.

thin filament: in the sarcomere, a protein strand that interacts with thick filaments, producing muscle contraction; composed primarily of actin, plus the accessory proteins troponin and tropomyosin.

threatened species: all species classified as critically endangered, endangered, or vulnerable.

threshold: the electrical potential at which an action potential is triggered; the threshold is usually about 10 to 20 mV less negative than the resting potential.

thrombin: an enzyme produced in the blood as a result of injury to a blood vessel; catalyzes the production of fibrin, a protein that assists in blood clot formation.

thylakoid (thī′-luh-koid): a disk-shaped, membranous sac found in chloroplasts, the membranes of which contain the photosystems, electron transport chains, and ATP-synthesizing enzymes used in the light reactions of photosynthesis.

thymine (T): a nitrogenous base found only in DNA; abbreviated as T.

thymosin (thī′-mō-sin): a hormone, secreted by the thymus, that stimulates the maturation of T lymphocytes of the immune system.

thymus (thī′-mus): an organ of the lymphatic system that is located in the upper chest in front of the heart and that secretes thymosin, which stimulates maturation of T lymphocytes of the immune system.

thyroid gland: an endocrine gland, located in front of the larynx in the neck, that secretes the hormones thyroxine (affecting metabolic rate) and calcitonin (regulating calcium ion concentration in the blood).

thyroid-stimulating hormone (TSH): a hormone, released by the anterior pituitary, that stimulates the thyroid gland to release hormones.

thyroxine (thī-rox′-in): a hormone, secreted by the thyroid gland, that stimulates and regulates metabolism.

tight junction: a type of cell-to-cell junction in animals that prevents the movement of materials through the spaces between cells.

tissue: a group of (normally similar) cells that together carry out a specific function; a tissue may also include extracellular material produced by its cells.

tissue system: a group of two or more tissues that together perform a specific function.

tonsil: a patch of lymphatic tissue, located at the entrance to the pharynx, that contains macrophages and lymphocytes; destroys many microbes entering the body through the mouth and stimulates an adaptive immune response to them.

trachea (trā′-kē-uh): in terrestrial vertebrates, a flexible tube, supported by rings of cartilage, that conducts air between the larynx and the bronchi.

tracheae (trā′-kē): the respiratory organ of insects, consisting of a set of air-filled tubes leading from openings called spiracles and branching extensively throughout the body.

tracheid (trā′-kē-id): an elongated cell type in xylem, with tapered ends that overlap the tapered ends of other tracheids, forming tubes that transport water and minerals. Pits in the cell walls of tracheids allow easy movement of water and minerals into and out of the cells, including from one tracheid to the next.

trans fat: a type of fat, produced during the process of hydrogenating oils, that may increase the risk of heart disease. The fatty acids of trans fats include an unusual configuration of double bonds that is not normally found in fats of biological origin.

transcription: the synthesis of an RNA molecule from a DNA template.

transcription factor: a protein that binds to DNA, thereby enhancing or suppressing transcription of a gene.

transfection: the process by which foreign DNA is inserted into a host cell; usually along with mechanisms to regulate the expression of the DNA in the host cell.

transfer RNA (tRNA): a type of RNA that binds to a specific amino acid, carries it to a ribosome, and positions it for incorporation into the growing protein chain during protein synthesis. A set of three bases in tRNA (the anticodon) is complementary to the set of three bases in mRNA (the codon) that codes for that specific amino acid in the genetic code.

transformation: a method of acquiring new genes, whereby DNA from one bacterium (normally released after the death of the bacterium) becomes incorporated into the DNA of another, living bacterium.

transgenic: referring to an animal or a plant that contains DNA derived from another species, usually inserted into the organism through genetic engineering.

translation: the process whereby the sequence of bases of messenger RNA is converted into the sequence of amino acids of a protein.

translocation: a mutation that occurs when a piece of DNA is removed from one chromosome and attached to another chromosome.

transpiration (trans-per-ā′-shun): the evaporation of water through the stomata, chiefly in leaves.

transport protein: a protein that regulates the movement of water-soluble molecules through the plasma membrane.

trial-and-error learning: a type of learning in which behavior is modified in response to the positive or negative consequences of an action.

trichomes: projections from the epidermal cells of some plants. Trichomes may absorb water, form a hairy coating on leaves, or aid in seed dispersal.

trichomoniasis (trik-ō-mō-nī′-uh-sis): a sexually transmitted disease, caused by the protist *Trichomonas*, that causes inflammation of the mucous membranes that line the urinary tract and genitals.

triglyceride (trī-glis′-er-īd): a lipid composed of three fatty acid molecules bonded to a single glycerol molecule.

trisomy 21: see *Down syndrome*.

trisomy X: a condition of females who have three X chromosomes instead of the normal two; most such women are phenotypically normal and are fertile.

trophic level: literally, "feeding level"; the categories of organisms in a community, and the position of an organism in a food chain, defined by the organism's source of energy; includes producers, primary consumers, secondary consumers, and so on.

tropical deciduous forest: a biome, warm year-round, with pronounced wet and dry seasons; characterized by trees that shed their leaves during the dry season (deciduous trees), an adaptation that minimizes water loss.

tropical rain forest: a biome with evenly warm, evenly moist conditions year-round, dominated by broadleaf evergreen trees; the most diverse biome.

tropical scrub forest: a biome, warm year-round, with pronounced wet and dry seasons (drier conditions than in tropical deciduous forests); characterized by short, deciduous, often thorn-bearing trees with grasses growing beneath them.

tropism: directional growth in response to an environmental stimulus.

true-breeding: pertaining to an individual all of whose offspring produced through self-fertilization are identical to the parental type. True-breeding individuals are homozygous for a given trait.

tube cell: the outermost cell of a pollen grain, containing the sperm. When the pollen grain germinates, the tube cell produces a tube penetrating through the tissues of the carpel, from the stigma, through the style, and to the opening of an ovule in the ovary.

tundra: a biome with severe weather conditions (extreme cold and wind, and little rainfall) that cannot support trees.

turgor pressure: pressure developed within a cell (especially the central vacuole of plant cells) as a result of osmotic water entry.

Turner syndrome: a set of characteristics typical of a woman with only one X chromosome; women with Turner syndrome are sterile, with a tendency to be very short and to lack typical female secondary sexual characteristics.

tympanic membrane (tim-pan′-ik): the eardrum; a membrane that stretches across the opening of the middle ear and transmits vibrations to the bones of the middle ear.

unicellular: single-celled; most members of the domains Bacteria and Archaea and the kingdom Protista are unicellular.

uniform distribution: the distribution characteristic of a population with a relatively regular spacing of individuals, commonly as a result of territorial behavior.

unsaturated: referring to a fatty acid with fewer than the maximum number of hydrogen atoms bonded to its carbon backbone (therefore, an unsaturated fatty acid has one or more double bonds in its carbon backbone).

upwelling: an upward flow that brings cold, nutrient-laden water from the ocean depths to the surface.

urea (ū-rē′-uh): a water-soluble, nitrogen-containing waste product of amino acid breakdown; one of the principal components of mammalian urine.

ureter (ū′-re-ter): a tube that conducts urine from a kidney to the urinary bladder.

urethra (ū-rē′-thruh): the tube leading from the urinary bladder to the outside of the body; in males, the urethra also receives sperm from the vas deferens and conducts both sperm and urine (at different times) to the tip of the penis.

uric acid: a water-insoluble, nitrogen-containing waste product of amino acid breakdown; excreted as a solid or semi-solid by insects and terrestrial reptiles, including birds and snakes.

urinary system: the organ system that produces, stores, and eliminates urine. The urinary system is critical for maintaining homeostatic conditions within the bloodstream. In mammals, it includes the kidneys, ureters, bladder, and urethra.

urine: the fluid produced and excreted by the urinary system, containing water and dissolved wastes, such as urea.

uterine tube: the tube leading from the ovary to the uterus, into which the secondary oocyte (egg cell) is released; also called the *oviduct*, or, in humans, the *Fallopian tube*.

uterus: in female mammals, the part of the reproductive tract that houses the embryo during pregnancy.

utricle: a patch of hair cells in the vestibule of the inner ear; bending of the hairs of the hair cells permits detection of the direction of gravity and the degree of tilt of the head.

vaccine: an injection into the body that contains antigens characteristic of a particular disease organism and that stimulates an immune response appropriate to that disease organism.

vagina: the passageway leading from the outside of a female mammal's body to the cervix of the uterus; serves as the receptacle for semen and as the birth canal.

variable: a factor in a scientific experiment that is deliberately manipulated in order to test a hypothesis.

variable region: the part of an antibody molecule that differs among antibodies; the ends of the variable regions of the light and heavy chains form the specific binding site for antigens.

vas deferens (vaz de′-fer-enz): the tube connecting the epididymis of the testis with the urethra.

vascular bundle: a strand of xylem and phloem in leaves and stems; in leaves, commonly called a *vein*.

vascular cambium: a lateral meristem that is located between the xylem and phloem of a woody root or stem and that gives rise to secondary xylem and phloem.

vascular cylinder: the centrally located conducting tissue of a young root; consists of primary xylem and phloem, surrounded by a layer of pericycle cells.

vascular plant (vas′-kū-lar): a plant that has conducting vessels for transporting liquids; also called a *tracheophyte*.

vascular tissue system: a plant tissue system consisting of xylem (which transports water and minerals from root to shoot) and phloem (which transports water and sugars throughout the plant).

vein: in vertebrates, a large-diameter, thin-walled vessel that carries blood from venules back to the heart; in plants, a vascular bundle in a leaf.

ventricle (ven′-tre-kul): the lower muscular chamber on each side of the heart that pumps blood out through the arteries. The right ventricle sends blood to the lungs; the left ventricle pumps blood to the rest of the body.

venule (ven′-ūl): a narrow vessel with thin walls that carries blood from capillaries to veins.

vertebral column (ver-tē′-brul): a column of serially arranged skeletal units (the vertebrae) that protect the nerve cord in vertebrates; the backbone.

vertebrate: an animal that has a vertebral column.

vesicle (ves′-i-kul): a small, temporary, membrane-bound sac within the cytoplasm.

vessel: in plants, a tube of xylem composed of vertically stacked vessel elements with heavily perforated or missing end walls, leaving a continuous, uninterrupted hollow cylinder.

vessel element: one of the cells of a xylem vessel; elongated, dead at maturity, with thick lateral cell walls for support but with end walls that are either heavily perforated or missing.

vestibular apparatus: part of the inner ear, consisting of the vestibule and the semicircular canals, involved in the detection of gravity, tilt of the head, and movement of the head.

vestigial structure (ves-tij′-ē-ul): a structure that is the evolutionary remnant of structure that performed a useful function in an ancestor, but is currently either useless or used in a different way.

villus (vi′-lus; pl., villi): a finger-like projection of the wall of the small intestine that increases its absorptive surface area.

viroid (vī′-roid): a particle of RNA that is capable of infecting a cell and of directing the production of more viroids; responsible for certain plant diseases.

virus (vī′-rus): a noncellular parasitic particle that consists of a protein coat surrounding genetic material; multiplies only within a cell of a living organism (the host).

vitamin: one of a group of diverse chemicals that must be present in trace amounts in the diet to maintain health; used by the body in conjunction with enzymes in a variety of metabolic reactions.

vitreous humor (vit′-rē-us): a clear, jelly-like substance that fills the large chamber of the eye between the lens and the retina; helps to maintain the shape of the eyeball.

vocal cords: a pair of bands of elastic tissue that extend across the opening of the larynx and produce sound when air is forced between them. Muscles alter the tension on the vocal cords and control the size and shape of the opening, which in turn determines whether sound is produced and what its pitch will be.

vulnerable species: a species that is likely to become endangered unless conditions that threaten its survival improve.

waggle dance: a symbolic form of communication used by honeybee foragers to communicate the location of a food source to their hive mates.

warning coloration: bright coloration that warns predators that the potential prey is distasteful or even poisonous.

water cycle: the biogeochemical cycle by which water travels from its major reservoir, the oceans, through the atmosphere to reservoirs in freshwater lakes, rivers, and groundwater, and back into the oceans. The water cycle is driven by solar energy. Nearly all water remains as water throughout the cycle (rather than being used in the synthesis of new molecules).

water mold: a member of a protist group that includes species with filamentous shapes that give them a superficially fungus-like appearance. Water molds, which include species that cause economically important plant diseases, are part of a larger group known as the stramenopiles.

wax: a lipid composed of fatty acids covalently bonded to long-chain alcohols.

weather: short-term fluctuations in temperature, humidity, cloud cover, wind, and precipitation in a region over periods of hours to days.

wetlands: a region (sometimes called a marsh, swamp, or bog) in which the soil is covered by, or saturated with, water for a significant part of the year.

white matter: the portion of the brain and spinal cord that consists largely of myelin-covered axons and that gives these areas a white appearance.

wildlife corridor: a strip of protected land linking larger areas. Wildlife corridors allow animals to move freely and safely between habitats that would otherwise be isolated by human activities.

work: energy transferred to an object, usually causing the object to move.

working memory: the first phase of learning; short-term memory that is electrical or biochemical in nature.

X chromosome: the female sex chromosome in mammals and some insects.

xylem (zī-lum): a conducting tissue of vascular plants that transports water and minerals from root to shoot.

Y chromosome: the male sex chromosome in mammals and some insects.

yolk: protein-rich or lipid-rich substances contained in eggs that provide food for the developing embryo.

yolk sac: one of the embryonic membranes of reptiles (including birds) and mammals. In reptiles, the yolk sac is a membrane surrounding the yolk in the egg; in mammals, it forms part of the umbilical cord and the digestive tract but does not contain yolk.

Z disc: a fibrous protein structure to which the thin filaments of skeletal muscle are attached; forms the boundary of a sarcomere.

zona pellucida (pel-oo′-si-duh): a clear, noncellular layer between the corona radiata and the egg.

zooplankton: nonphotosynthetic protists that are abundant in marine and freshwater environments.

zygomycete: a fungus species formerly placed in the now-defunct taxonomic group Zygomycota. Zygomycetes, which include the species that cause fruit rot and bread mold, do not constitute a true clade and are now distributed among several other taxonomic groups.

zygosporangium (zī′-gō-spor-an′-jee-um): a tough, resistant reproductive structure produced by some fungi, such as bread molds; encloses diploid nuclei that undergo meiosis and give rise to haploid spores.

zygote (zī′-gōt): in sexual reproduction, a diploid cell (the fertilized egg) formed by the fusion of two haploid gametes.

Answers to Selected Questions

Answers to all the Think Critically questions and Thinking Through the Concepts questions are included here.

Chapter 1

Think Critically

Figure 1-3 The bun is made from wheat, a photosynthetic organism that can capture sunlight directly. The meat came from a cow, which feeds on photosynthetic organisms and fuels its body with energy stored from photosynthesis.

Figure 1-9 Of all the animals, plant-eaters have access to the largest amount of energy because they feed on plants, which capture the energy directly from sunlight. The huge size of some herbivores is also an adaptation for reaching leaves on tall trees.

Figure 1-10 Research at the molecular level of organization would be best suited for understanding the mechanism of photosynthesis.

Doing Science: Controlled Experiments Provide Reliable Data

Other things (besides flies) that the gauze blocked from entering might have produced the maggots in the uncovered jar, or they could have come from eggs that were in some of the meat before it was placed in the open jar. An experiment to support the hypothesis that flies produce maggots would be to use Redi's experimental setup with meat in each of two gauze covered jars, but place flies in one jar and not the other.

Multiple Choice

1. b; **2.** c; **3.** a; **4.** d; **5.** b

Fill-in-the-Blank

1. stimuli; energy, materials; complex, organized; evolve
2. atom; cell; tissues; population; community; ecosystem
3. scientific theory; hypothesis; scientific method
4. evolution; natural selection
5. deoxyribonucleic acid, DNA; genes

Review Questions

1. Organisms acquire and use materials and energy; actively maintain organized complexity; perceive and respond to stimuli; grow; reproduce; and, collectively, have the capacity to evolve.
2. Living things must use energy continuously to maintain their organized complexity. The ultimate source of energy for life on Earth is sunlight.
3. Evolution is change over time as a result of descent with modification from a common ancestor. Evolution by natural selection is the inevitable outcome of: (1) differences among members of a population that arise as a result of mutations; (2) inheritance of these differences by offspring; and (3) increased survival and reproduction by individuals that inherit beneficial traits. As a result of this increased survival and reproduction, beneficial traits (and the genes that underlie them) become increasingly common in the population.
4. The domains are Archaea, Bacteria, and Eukarya.
5. Eukaryotic cells of the domain Eukarya are complex and contain a variety of organelles, many surrounded by membranes. They contain a nucleus that encloses the cell's DNA within a membrane. Prokaryotic cells are far simpler and generally much smaller than eukaryotic cells, and they lack organelles enclosed by membranes. The DNA of prokaryotic cells is not confined within a nucleus. The domains Bacteria and Archaea consist entirely of prokaryotic cells.
6. The principles are the following: first, that all events have natural causes; second, that our universe is, and always has been, governed by natural laws; and third, that scientific data exist independently of subjective value systems.
7. A hypothesis is a proposed explanation for a phenomenon, based on preliminary evidence. In contrast, a scientific theory is more like a natural law or a basic principle derived from the study of nature that has never been disproven by scientific inquiry. Scientists describe these fundamental principles as "theories" rather than "facts" because science must be conducted with an open mind; if compelling evidence arises that renders a scientific theory invalid, that theory must be modified or discarded.
8. Redi's controls were the same kind of meat and jar and the same length of time exposed to air at the same temperature and location. Andersson controlled for the season, weather, timing of the experiment, and general location of the observations (by performing the experiments simultaneously in the same area), as well as for handling the birds and modifying their tail feathers.
9. Scientific inquiry begins with an *observation* of a specific phenomenon, which leads to the *question* such as, "What caused this?" Then a *hypothesis*, which is a proposed explanation for the phenomenon, is formed based on existing evidence. The hypothesis leads to a *prediction*, which is the expected outcome of testing if the hypothesis is correct. The prediction is tested by carefully designed additional observations or carefully controlled manipulations called *experiments*. Experiments produce results that either support or refute the hypothesis, allowing the scientist to reach a *conclusion* about whether the hypothesis is valid or not.

Chapter 2

Think Critically

Figure 2-1 The mass number of H is 1; the mass number of He is 4.

Figure 2-2 Atoms with outer shells that are not full become stable by filling (or emptying) their outer shells.

Figure 2-3 Heat from the fire excites electrons into higher energy levels. When the electrons spontaneously revert to their original stable level, they give off light as well as heat.

Figure 2-9 No. The hydrophobic oil exerts no attraction for the ions in salt, so the salt would remain in solid crystals.

Figure 2-10 The water drop would spread out on clean glass because of adhesion to the glass and cohesion among the water molecules. On an oil-covered slide, the droplet would ball up due to hydrophobic interactions.

Figure 2-11 There is more empty space between the molecules in ice than between molecules of liquid water, so ice is less dense and floats.

Doing Science: Radioactive Revelations

Fluid-filled space occupies a far larger portion of the brain of the Alzheimer's patient, indicating a significant loss of neurons.

Health Watch: Free Radicals—Friends and Foes?

Eating dark chocolate by itself is unlikely to reverse high blood pressure. If Thomas added a 6-ounce chocolate bar (about 1,000 calories) to his regular daily diet, he could gain 50 pounds over 6 months, which would likely worsen his high blood pressure. Thomas should maintain a healthy weight, gradually increase his exercise, and eat lots of fruits and vegetables. In addition, it would be okay for him to add a far smaller amount of dark chocolate daily.

Multiple Choice

1. d; **2.** a; **3.** d; **4.** b; **5.** d

Fill-in-the-Blank

1. protons, neutrons; electrons, electron shells
2. ion; positive; ionic

3. isotopes; elements; radioactive
4. inert; reactive; share
5. polar; hydrogen; cohesion

Review Questions

1. There are eight neutrons in oxygen, zero in hydrogen, and seven in nitrogen.
2. An atom is the smallest particle of an element that still retains the properties of that element. A molecule consists of two or more atoms chemically bonded together. Protons are positively charged subatomic particles found in the atomic nucleus. Neutrons are uncharged subatomic particles found in the atomic nucleus. Electrons are negatively charged subatomic particles found in electron shells around the atomic nucleus.
3. Covalent bonds involve the sharing of a pair of electrons between two atoms. Ions form when one or more electrons are transferred between atoms, and ionic bonds result from electrical charge attraction between ions.
4. Polar covalent bonds involve unequal sharing of electrons between two atoms. This produces slight opposite charges on different parts (poles) of the molecule, with one pole being slightly positive and the other slightly negative. Hydrogen bonds form as a result of the attraction between these small opposite charges on polar molecules. For example, water molecules form hydrogen bonds with one another between their slightly positive hydrogen poles and slightly negative oxygen poles.
5. Because water molecules are linked by hydrogen bonds, the molecules can move faster only if the hydrogen bonds are broken more often and more rapidly. Breaking hydrogen bonds uses up a considerable amount of heat energy, which is then not available to raise the water's temperature. This property of water is described as a high specific heat.
6. Because water is polar, the negative poles of water molecules are attracted to the sodium ions and the positive poles to the chloride ions of salt. This attraction to water breaks the ionic bonds that hold salt crystals together. As water molecules surround each ion in salt, the salt dissolves.
7. The pH scale is a range of values from 0 to 14 that defines the concentration of H^+ in a solution and thus how acidic or basic the solution is. The pH number is the negative log of the H^+ concentration. An acid is a substance that gives off H^+ in solution. A base is a substance that combines with H^+ in solution. Buffers are molecules that maintain a solution at a nearly constant pH by accepting or releasing H^+ in response to small changes in H^+ concentration. Buffers are important in organisms because biological molecules function only within a narrow pH range, and buffers help maintain body fluids within this range.

Chapter 3

Think Critically

Figure 3-1 Hydrogen cyanide is a polar molecule, because the nitrogen atom exerts a much stronger pull on carbon's electrons than does the hydrogen atom.

Figure 3-8 During hydrolysis of sucrose, water is split; a hydrogen from water is added to the oxygen from glucose (that formerly linked the two subunits), and the remaining OH from water is added to the carbon (that formerly bonded to oxygen) of the fructose subunit.

Figure 3-14 Other amino acids with hydrophobic functional groups are glycine, alanine, valine, isoleucine, methionine, tryptophan, and proline.

Figure 3-16 Heat energy can break hydrogen bonds and disrupt a protein's three-dimensional structure. This prevents the protein from carrying out its usual function(s).

Figure 3-22 Triglycerides are broken apart by hydrolysis reactions.

Figure 3-27 As lipids, steroids are soluble in the phospholipid-based cell membranes and can cross them to act inside the cell.

Health Watch: Cholesterol, Trans Fats, and Your Heart

The doctor should do the following:

- Do a blood test for high LDL cholesterol; this could cause partial artery blockage that would deprive the heart of adequate oxygen during exercise and cause chest pain.
- Recommend that the patient lose weight.
- Ask about the patient's exercise regimen (if any), and recommend that the patient try a variety of exercise classes to find one she enjoys.
- Ask if her diet includes a lot of saturated fat; if so, strongly recommend she switch to oil.
- Ask if she smokes, which contributes to heart disease; if so, encourage cessation.
- Suggest additional tests that are used to evaluate arterial health, and explain how obesity damages the heart (use reputable Web sites to research this information).

Multiple Choice

1. d; **2.** c; **3.** a; **4.** d; **5.** a

Fill-in-the-Blank

1. monomer, polymers; polysaccharides; hydrolysis; (any three) cellulose, starch, glycogen, chitin
2. hydrogen bond, hydrogen and disulfide bonds, hydrogen bond, peptide bond
3. dehydration, water; amino acids; primary; helix, pleated sheet; denatured
4. ribose sugar, base, phosphate; adenosine triphosphate (ATP); adenine, guanine, cytosine, thymine; deoxyribonucleic acid (DNA), ribonucleic acid (RNA); phosphate
5. oil, wax, fat, steroids, cholesterol, phospholipid

Review Questions

1. "Organic" describes molecules with a carbon backbone that usually also contain hydrogen and oxygen.
2. Carbohydrates—glucose, sucrose, starch, glycogen, cellulose, chitin
   Lipids—oil, fat, waxes, cholesterol
   Proteins—keratin, silk, hemoglobin
   Nucleic acids—DNA, RNA
3. ATP serves as a short-term energy storage molecule; cyclic AMP is an intracellular messenger. The nucleotides adenine, guanine, cytosine and thymine are subunit monomers used in assembling the nucleic acids DNA and RNA.
4. Fats and oils both consist of a glycerol molecule with each of its carbons bonded to a fatty acid "tail." They differ in that oils have fewer hydrogen atoms bonded to the carbons in their fatty acid tails, causing double bonds between the carbons, which cause the tails to kink. This keeps the molecules from packing closely together, so oil remains liquid at room temperature. Fats are saturated with hydrogen atoms, their tails are straight, and they pack closely, so fat remains solid at room temperature.
5. Dehydration synthesis forms polymers by removing a water molecule from two monomers that then become bonded together. A hydroxyl group is removed from one monomer and a hydrogen atom from the other. For example, dehydration synthesis links glucose and fructose to form sucrose. Hydrolysis breaks polymers into monomer subunits by adding water to split the bonded monomers apart. For example, sucrose is hydrolyzed into glucose and fructose.
6. Proteins are produced by dehydration synthesis of amino acids to form polypeptides linked by peptide bonds. Primary structure is the sequence of amino acids; secondary structure involves bending the polypeptide into helices or pleated sheets, stabilized by hydrogen bonds between amino acids; tertiary structure is the folding of a polypeptide into a complex, three-dimensional structure involving hydrogen and sometimes disulfide bonds between amino acids in different parts of the molecule; and quaternary structure is the

association of more than one polypeptide into a closely packed arrangement, usually maintained by hydrogen bonds, disulfide bonds, and/or attractions between oppositely charged polar portions of amino acids.

7. Cellulose is found in plant cell walls. Chitin is found in the exoskeletons of animals and in many fungal cell walls. They are similar in that they are composed of glucose subunits; however, in chitin, a hydroxyl group in the glucose subunits is replaced by a nitrogen-containing functional group.

Chapter 4

Think Critically

Figure 4-2 A light micrograph provides overall views of cellular structures in relationship to one another, and it also allows you to see a living cell in motion, without distortion from preservation techniques. In an SEM of an intact cell or organelle, you can observe its overall three-dimensional shape and surface structures. A TEM provides a more diagrammatic view of cells and organelles, with a clearer image of their internal organization.

Figure 4-10 Condensation allows the chromosomes to become organized and separated from one another so that a complete copy of genetic information can be distributed to each of the daughter cells that results from cell division.

Figure 4-18 The fundamentally similar composition of membranes allows them to merge with one another. This allows molecules in vesicles to be transferred from one membrane-enclosed structure (such as endoplasmic reticulum) to another (such as the Golgi apparatus). The material can be exported from the cell when the vesicle merges with the plasma membrane.

Figure 4-19 If lysosomal enzymes were active at a pH found in the cytoplasm, the enzymes would break down the membranes of the endoplasmic reticulum, Golgi, and the vesicles exposed to the enzymes.

Figure 4-21 Fluid would not move upward because the beating of the flagella would direct fluid straight out from the cell membranes. Mucus and trapped particles would accumulate in the trachea.

Earth Watch: Would You Like Fries with Your Cultured Cow Cells?

Scanning the graph reveals that China's per-person consumption reached half of the U.S. level sometime between 2003 and 2013. During that interval, China's consumption rose from about 105 pounds/year to about 135 pounds/year (increasing by about 3 pounds/year), and U.S. consumption fell from about 270 pounds/year to about 250 pounds/year (decreasing by about 2 pounds/year). Using these data to estimate each country's consumption in each year, we can determine that China reached half the level of consumption of the United States in 2011. Annual worldwide consumption of meat increased from about 235 billion pounds in 1973 (3.92 billion people × 60 pounds/person) to about 972 billion pounds in 2013 (7.2 billion people × 135 pounds/person). Annual worldwide consumption of meat increased from about 235 billion pounds in 1973 (3.92 billion people × 60 pounds/person) to about 972 billion pounds in 2013 (7.2 billion people × 135 pounds/person).

Multiple Choice

1. a; **2.** b; **3.** b; **4.** d; **5.** b

Fill-in-the-Blank

1. phospholipids, proteins; phospholipid, protein
2. microfilaments, intermediate filaments, microtubules; microtubules; microtubules; microfilaments; intermediate filaments
3. ribosomes, endoplasmic reticulum, nucleolus, Golgi apparatus, cell wall, messenger RNA
4. rough endoplasmic reticulum; vesicles, Golgi apparatus; carbohydrate; plasma membrane
5. mitochondria, chloroplasts, extracellular matrix, nucleoid, cilia, cytoplasm
6. mitochondria, chloroplasts; double, ATP, size
7. cell walls; nucleoid; plasmids; sex pili

Review Questions

1. (1) Every organism is made up of one or more cells. (2) The smallest organisms are single cells, and cells are the functional units of multicellular organisms. (3) All cells arise from preexisting cells.
2. Animal and plant cells share the following cytoplasmic structures: cytoskeleton, endoplasmic reticulum, Golgi apparatus, vesicles, vacuoles, mitochondria, and ribosomes. Lysosomes and centrioles are absent or rare in plant cells. Chloroplasts and the central vacuole are absent or rare in animal cells.
3. Microfilaments: smallest, twisted double strands, important in muscle contraction and cell division; intermediate filaments: midsize, internal scaffolding, help maintain cell shape; microtubules: largest, guide chromosomes during cell division, major component of cilia and flagella, transport organelles through the cytosol
4. The nucleus houses the DNA in eukaryotic organisms. It is surrounded by the nuclear envelope, which consists of two membranes containing numerous pores that allow various molecules to pass between the nucleus and the cytoplasm. Chromatin consists of chromosomes in an uncondensed state. A chromosome is a DNA molecule with associated proteins. DNA carries genes whose nucleotide sequence codes for the production of proteins and RNA. The nucleolus, a dark-staining region of the nucleus, is the site of ribosome synthesis. It consists of rRNA, proteins, ribosomes in various stages of synthesis, and the portions of chromosomes carrying genes that code for rRNA.
5. Mitochondria extract energy from food molecules and store it in the bonds of ATP; chloroplasts are the sites of photosynthesis. Their double membranes could have originated from their original host cell (outer membrane) and its prokaryotic guest (inner membrane). The organelles also have their own DNA, their own ribosomes, the ability to make ATP, and a size similar to that of prokaryotic cells, all supporting the endosymbiont hypothesis.
6. Ribosomes are sites for protein synthesis. They are embedded in the outer nuclear membrane of the nuclear envelope and the rough endoplasmic reticulum. They are also found free in the cytoplasm or attached to strands of mRNA as polyribosomes. Ribosomes are found in both eukaryotic and prokaryotic cells, although prokaryotic ribosomes are smaller and have different RNA and proteins.
7. The endoplasmic reticulum forms membrane enclosed channels in the cytoplasm. Smooth ER lacks ribosomes but contains enzymes. Some smooth ER enzymes synthesize steroid hormones; smooth ER in liver cells detoxifies drugs and metabolic wastes; smooth ER in muscle cells stores calcium ions. Rough ER is the major site of protein synthesis. Proteins synthesized by the ribosomes attached to the rough ER enter the channels of the endoplasmic reticulum, where they are chemically modified and folded and transported through the ER to pockets that pinch off as vesicles. Vesicles transport the proteins to the Golgi apparatus. The Golgi apparatus is a series of specialized flattened membranous sacs. The Golgi sorts, chemically alters, and packages proteins produced by the rough ER. These leave the Golgi in vesicles and are transported to other parts of the cell or to the plasma membrane for export.
8. Lysosomes are membrane-enclosed vesicles derived from the Golgi complex that contain digestive enzymes that are used to break down food particles and defective organelles.
9. Refer to Figure 4-8 for diagrams of cilia and flagella. Cilia (short and numerous) move in a way that resembles the oars of a rowboat. In multicellular organisms, cilia move fluid past a surface; in some protists they allow the cell to swim. Flagella (longer and fewer) propel sperm and some protists through fluid like a propeller on a motorboat.
10. Ribosomes, cell wall, flagella

Chapter 5

Think Critically

Figure 5-6 The distilled water made the solution hypotonic to the blood cells, causing enough water to flow in to burst their fragile cell membranes.

Figure 5-7 The rigid cell wall of plant cells counteracts the pressure exerted by water entering by osmosis, so although the cell would stiffen from internal water pressure, it would not burst.

Figure 5-8 No. Actively transporting water across a membrane against its concentration gradient would waste energy because the water would simply diffuse by osmosis back through the membrane.

Figure 5-12 Exocytosis uses cellular energy, whereas diffusion occurs passively. During exocytosis, materials are expelled without passing directly through the plasma membrane, allowing the cell to eliminate materials that are too large to pass through membranes.

Health Watch: Membrane Fluidity, Phospholipids, and Fumbling Fingers

It is adaptive to feel enhanced pain when parts of the body are in danger of being damaged by the cold, because this will stimulate urgent behavior to warm up the affected body region.

Doing Science: Discovering Aquaporins

The control eggs (without aquaporins) would have shrunk very slightly. Because they swelled slightly in distilled water, it is clear that they are somewhat permeable to water, and they would lose some water by osmosis in a hypertonic solution. The eggs with aquaporins inserted into their plasma membranes would have shrunk considerably and would be much smaller than the control eggs. Because water can flow either way through aquaporins, it would have flowed out into the hypertonic environment by osmosis.

Case Study Revisited: Vicious Venoms

Venom phospholipases are injected to injure and help immobilize the prey, whereas phospholipases in the digestive tract break down membrane phospholipids into nutrients that can be absorbed.

Multiple Choice

1. c; **2.** b; **3.** b; **4.** c; **5.** a

Fill-in-the-Blank

1. phospholipids; receptor, recognition, enzymes, attachment, transport
2. selectively permeable; diffusion; osmosis; aquaporins; active transport
3. channel, carrier; simple, lipids
4. simple diffusion, simple diffusion, facilitated diffusion, facilitated diffusion
5. exocytosis; yes; vesicles

Review Questions

1. See Figure 5-1 for a diagram of the plasma membrane. The plasma membrane can be described as a fluid mosaic, referring to the fluidity of the phospholipid bilayer and the mosaic of different proteins it contains. The two types of molecules are phospholipids and proteins. The function of the phospholipid bilayer is to isolate the cell from its surroundings. The proteins of the membrane allow selective interactions with the environment.
2. The phospholipid molecules would form a droplet with all their tails in the center and their heads facing the water. This is because the heads are polar (hydrophilic) and attracted to polar water molecules, whereas the tails are hydrophobic and not attracted to water.
3. The five categories of proteins found in the plasma membrane are (1) enzymes, which promote chemical reactions without being changed themselves; (2) recognition proteins, which serve to identify the cell; (3) transport proteins, which are used to regulate the movement of water-soluble molecules or ions through the plasma membrane; (4) receptor proteins, which respond to molecules in the extracellular fluid, such as hormones; and (5) attachment proteins, which anchor the plasma membrane to the cytoskeleton inside, to the extracellular matrix outside, or to adjacent cells.
4. Diffusion is the net movement of molecules in a fluid from regions of high concentration to regions of low concentration. Osmosis is the diffusion of water though selectively permeable membranes. Plant leaves are supported by the turgor pressure produced when plant cell vacuoles take in water by osmosis. Turgor pressure "inflates" plant cells, supporting them.
5. The terms *hypotonic, hypertonic,* and *isotonic* describe relative concentrations of solutes when one solution is compared to another solution separated by a membrane that is selectively permeable to water. A hypotonic solution has a lower solute content, a hypertonic solution has a higher solute content, and an isotonic solution has the same solute content as the compared solution. Water will move by osmosis through the membrane from a hypotonic solution into a hypertonic solution. If an animal cell is placed into a solution hypotonic to the cell's cytoplasm, water will enter the cell by osmosis and may eventually cause the cell to burst. A hypertonic solution causes water to leave a cell and will cause an animal cell to shrivel. An isotonic solution results in no net loss or gain of water by a cell, because both the cytoplasm and the extracellular solution contain the same concentration of solutes.
6. Simple diffusion is the diffusion of water, dissolved gases, or lipid-soluble molecules through the phospholipid bilayer of a membrane. Facilitated diffusion is the diffusion of molecules through a membrane, assisted by membrane transport proteins. Active transport is the movement of small molecules and ions through membrane proteins, using cellular energy, usually in the form of ATP. Pinocytosis is the introduction of extracellular fluid into a cell by a small piece of the plasma membrane that dimples inward and pinches off, forming a vesicle. Receptor-mediated endocytosis moves specific molecules into a cell when the particular molecule binds to receptor proteins on the plasma membrane, and the membrane is pinched off, forming a sac that contains the imported molecule. Phagocytosis is the movement of large food particles or whole organisms into a cell by extensions of the plasma membrane called pseudopods that encircle an extracellular particle and form a vesicle. In exocytosis, vesicles fuse with the plasma membrane, expelling the contents outside the cell.
7. Aquaporin allows facilitated diffusion of water. Peter Agre and associates inserted an unknown membrane protein from red blood cells into frog eggs (which are not very permeable to water) and found that the frog eggs with the protein took up water rapidly and burst in hypotonic solutions, whereas those without the protein were unchanged.
8. Glucose will diffuse from compartment A into compartment B, and water will follow by osmosis. The result will be greater (but equal) concentrations of glucose in both compartments and the same amount of water in both compartments.

Chapter 6

Think Critically

Figure 6-1 Yes, but this would be a boring roller coaster because each successive peak would need to be much lower than the previous one. When a car plunges down from a peak, potential energy is converted to kinetic energy, but some energy is lost as heat during the conversion. So as the car starts up the next peak, the amount of kinetic energy available to power the climb is lower than the amount of energy that was available at the top of the prior peak.

Figure 6-4 Glucose breakdown is exergonic; photosynthesis is endergonic.

Figure 6-7 Both parts of the coupled reaction occur with a loss of usable energy in the form of heat.

Figure 6-8 No, only exergonic reactions can occur spontaneously after the activation energy is surmounted.

Health Watch: Lack of an Enzyme Leads to Lactose Intolerance

He would suspect lactose intolerance, which would be even more likely if the child's birth parents were of East Asian, West African, or Native American origin. If tests confirmed his suspicions, he might suggest that the child reduce lactose-containing foods to a level she can tolerate and that the parents try giving her special formulations of dairy products that have lactase added to them or lactase pills that can be taken prior to a meal.

Case Study Revisited: Energy Unleashed

The plot line of enzyme activity will start at zero at a temperature near water's freezing point (0°C), rise relatively gradually to a peak at or near body temperature (37°C), and then decline, falling sharply to zero at some temperature not too far past the peak (perhaps 50°). The temperature of peak activity is near body temperature because most human enzymes have evolved to function optimally under normal conditions in the body. At lower temperatures, enzyme and substrate molecules move more slowly and therefore less frequently collide with sufficient energy to react; the contact rate will increase steadily with increased temperature. At temperatures above the optimal temperature, the proportion of denatured enzyme molecules increases, and when temperature reaches a critically high point, the proportion of denatured molecules approaches 100% and activity quickly plummets to zero.

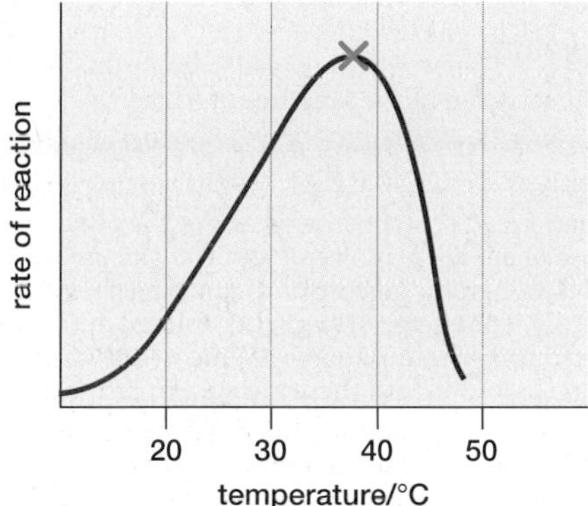

Multiple Choice

1. b; **2.** a; **3.** d; **4.** b; **5.** a

Fill-in-the-Blank

1. created, destroyed; kinetic, potential
2. less; entropy
3. exergonic; endergonic; exergonic; endergonic
4. adenosine triphosphate; adenosine diphosphate, phosphate; energy
5. proteins; catalysts, activation energy; active site
6. inhibiting; competitive

Review Questions

1. Life does not violate the second law of thermodynamics because Earth is not a closed system but instead is receiving a constant influx of energy from the sun, which is captured by organisms and used to maintain their organized complexity. The sun releases heat and undergoes a continuous increase in entropy.
2. Potential energy is energy that is stored, for example, in chemical bonds, in the elastic energy of a stretched spring, or as gravitational energy stored in water behind a dam or a car at the top of a roller coaster. Kinetic energy is the energy of movement, including radiant energy (waves of light or X-rays), heat energy (the motion of molecules), and any motion of larger objects, such as the plummeting roller-coaster car or running marathoners. Kinetic energy can be converted to potential energy (gravitational, for example) by moving an object up a hill. Potential energy can be converted to kinetic energy, as when an object falls or when ATP is used to power muscle contraction. Heat is always released when energy is converted from one form to another.
3. Metabolism is the sum total of all chemical reactions in a cell. Reactions are coupled so that one reaction releases energy (exergonic) to drive another reaction (endergonic). Energy-carrier molecules, particularly ATP, transfer chemical energy from exergonic to endergonic reactions. In every conversion, some energy is lost as heat.
4. Activation energy is an initial input of energy required to start a chemical reaction. Catalysts reduce the amount of activation energy required to start a reaction, which results in the reaction proceeding at a much faster rate.
5. Glucose breakdown in cells occurs in the small, enzyme-catalyzed steps of a metabolic pathway. Energy is released in small increments, and some is stored in chemical bonds such as in ATP. Heat is released in small amounts. When set on fire, glucose releases large quantities of heat energy at a rapid rate, and none is captured in chemical bonds. If the glucose is burned completely, the end products of both the metabolic pathway and fire are the same: carbon dioxide and water. Activation energy in the metabolic pathway is supplied by body heat, which is adequate because enzymes greatly reduce the need for activation energy. The heat of the match flame overcomes the activation energy when sugar is burned.
6. In competitive inhibition, a substance that is not the enzyme's normal substrate binds to the active site of the enzyme, competing with the substrate for the active site. In noncompetitive inhibition, a molecule binds to a noncompetitive inhibitor site on the enzyme that is distinct from the active site. As a result, the enzyme's active site is distorted, making it less able to catalyze the reaction.
7. Enzymes are nearly always proteins with complex three-dimensional shapes that bind to specific substrates. Enzymes promote a single reaction involving one or two specific molecules. Enzyme activity is regulated in three ways: (1) A cell may regulate how much of an enzyme it contains by turning genes on or off; (2) a cell may synthesize an enzyme in an inactive form that is activated only when necessary; and (3) a cell can temporarily activate or inhibit enzymes through allosteric regulation.

Chapter 7

Think Critically

Figure 7-5 **(a)** If only red light hit the leaf, much less light energy would be gathered from chlorophyll *b* and essentially none from the carotenoids, so photosynthesis and its resulting O_2 production would decline significantly. **(b)** No oxygen would be generated if only infrared light was present because there are no infrared-capturing pigments in chloroplasts. **(c)** Shining green light on the leaf would substantially reduce O_2 production because neither chlorophyll *a* nor *b* would absorb it. Most green light is reflected from the leaf.

Figure 7-9 As shown in the figure, capturing CO_2 to convert RuBP to G3P does not consume ATP and NADPH, but converting G3P to RuBP does. During the day, the light reactions provide a continual supply of ATP and NADPH, so production of G3P and regeneration of RuBP both proceed continually, and both molecules are present at steady (and roughly equal) levels. But at night, the light reactions cannot continue, and ATP and NADPH are not produced. Thus, regeneration of RuBP stops, and the amount present in the leaf declines steadily. Production of G3P continues, however, and since it is not being converted to RuBP, the amount present increases steadily. So, measurements made in the middle of the night would find the level of RuBP to be lower than during the day and the level of G3P higher.

Earth Watch: Biofuels—Are Their Benefits Bogus?

The raw material for cellulosic ethanol can be things like corn stalks and grass clippings that are currently often treated as waste, so ethanol production would not necessarily require diverting food crops or converting natural habitats to new cropland, as current ethanol production does. Also, even if land were devoted to a crop (such as corn) specifically for ethanol, the entire plant (i.e., including stems and leaves) could be used instead of just the high-sugar parts (e.g., corn kernels), so less cropland per unit of ethanol would be required.

Multiple Choice

1. a; **2.** b; **3.** a; **4.** b; **5.** d

Fill-in-the-Blank

1. stomata, oxygen (O_2), carbon dioxide (CO_2); chloroplasts, mesophyll
2. red, blue, violet; green; carotenoids; photosystems, thylakoid
3. reaction center; primary electron acceptor molecule; electron transport chain; hydrogen ions (H^+); chemiosmosis
4. water (H_2O), carbon dioxide (CO_2); Calvin cycle; carbon fixation
5. ATP, NADPH, Calvin; RuBP (or ribulose bisphosphate); G3P (glyceraldehyde-3-phosphate), glucose

Review Questions

1. Most forms of life would die soon after phosynthesis ceased. This is because maintaining life requires a constant expenditure of energy, and nearly all forms of life rely on solar energy trapped by photosynthesis.
2. The photosynthetic equation is $6\ CO_2 + 6\ H_2O + \text{light} \rightarrow C_6H_{12}O_6 + 6\ O_2$. Light energy drives the use of carbon dioxide and water to produce sugar and oxygen.
3. Refer to Figure 7-3 for a diagram of a typical leaf. The flattened shape allows maximum absorbance of light. Upper and lower leaf surfaces are protected by a layer of transparent cells of the epidermis, covered by the transparent, waxy, waterproof cuticle. A leaf obtains the CO_2 necessary for photosynthesis from the air, through adjustable pores in the epidermis called stomata. Inside the leaf, mesophyll cells contain most of the leaf's chloroplasts where photosynthesis occurs. Vascular bundles form veins that supply water and minerals to the cells and carry the sugars produced during photosynthesis to other parts of the plant.
4. Refer to Figures 7-3e and 7-8 for diagrams of a chloroplast. The thylakoid membranes create enclosed thylakoid spaces where H^+ is concentrated. The membranes contain precise arrays of photosystems, energy-transfer molecules (electron transport chains), and H^+ channels coupled to ATP-synthesizing enzymes. The membrane component molecules and the thylakoid spaces allow light capture and ATP synthesis by chemiosmosis. The stroma region inside the chloroplast but outside the thylakoids contains the enzymes that allow the reactions of the Calvin cycle to occur, producing the energy-storing sugar end products of photosynthesis.
5. Light strikes the light-harvesting complex of photosystem II, and the energy is absorbed by chlorophyll or carotenoids. The energy is funneled to the reaction center chlorophylls, where it is used to energize electrons that are captured by the primary electron acceptor molecule. The electrons then travel through an electron transport chain (ETC) within the thylakoid membranes. Here, the energy is released in small stages. Some of this energy is used to produce ATP indirectly by chemiosmosis. During chemiosmosis, energy (released from the light-energized electron as it moves through the ETC) is used to pump H^+ into the thylakoid space, creating an H^+ gradient. These ions flow out of the space into the stroma through a protein channel linked to an ATP-synthesizing enzyme (ATP synthase), which harnesses the energy of the hydrogen ion flow to generate ATP.
6. The reactions of the Calvin cycle occur in the stroma of the chloroplast, where overall, atmospheric CO_2 is fixed to produce the simple sugar G3P. This is accomplished using chemical energy in ATP and NADPH harnessed during the light reactions. The molecule RuBP is regenerated.

Chapter 8

Think Critically

Figure 8-3 Glycolysis produces a net of two NADH and two ATP molecules.

Figure 8-6 Without oxygen, electrons would be unable to continue entering the electron transport chain, and no further ATP would be produced.

Figure 8-11 NAD^+ would become unavailable for further glycolysis or cellular respiration.

Health Watch: How Can You Get Fat by Eating Sugar?

Hypothesis: Colin is eating many more carbohydrates than he did previously to stay satisfied. Questions: What is he eating instead of fat? Has he decreased his average daily caloric intake? Has he increased his exercise?

Case Study Revisited: Did the Dinosaurs Die from Lack of Sunlight?

Jeremy does not need to worry about a child inheriting his disorder, because mitochondrial mutations are only inherited from the mother.

Multiple Choice

1.d; **2.** a; **3.** b; **4.** d; **5.** b

Fill-in-the-Blank

1. glycolysis, cellular respiration; cytosol, mitochondria; aerobic
2. anaerobic; glycolysis, two; fermentation, NAD^+
3. ethanol (alcohol), carbon dioxide; lactic acid; lactate fermentation
4. matrix, intermembrane space, concentration gradient; chemiosmosis; ATP synthase
5. Krebs; acetyl CoA; two; NADH, $FADH_2$

Review Questions

1. Glucose breakdown in the presence of oxygen:
$C_6H_{12}O_6 \rightarrow 6\ CO_2 + 6\ H_2O + \text{ATP (chemical energy)}$
Photosynthesis: $6\ CO_2 + 6\ H_2O + \text{light energy} \rightarrow C_6H_{12}O_6 + 6\ O_2$
Each equation is essentially the reverse of the other. Photosynthesis captures solar energy in molecules such as glucose, synthesized from CO_2 and H_2O. Aerobic glucose breakdown releases CO_2 and H_2O from glucose. The energy of sunlight captured during photosynthesis is released during the breakdown of glucose and captured in ATP. Heat energy is lost in both these metabolic pathways.
2. See Figure 8-4. The inner mitochondrial membrane encloses the matrix, where enzymes for the Krebs cycle are localized. Embedded in this membrane are the electron transport chain molecules that allow transfer of high-energy electrons from the Krebs cycle and glycolysis and controlled energy release that is used to pump H^+ into the intermembrane space. This space encloses a high concentration of H^+. The inner membrane also contains H^+ channels linked to ATP synthase that capture energy in ATP as H^+ flows through the channels down its concentration gradient.
3. In *glycolysis*, a six-carbon glucose is split into two three-carbon pyruvate molecules, storing energy in ATP and the high-energy electron carrier NADH. Under aerobic conditions, *cellular respiration* breaks down the pyruvate, capturing its energy in ATP and the high-energy electron carriers NADH and $FADH_2$. *Chemiosmosis* generates ATP using energy from a hydrogen ion gradient, which in turn was created by the electron transport chain using energy released by NADH and $FADH_2$. Under anaerobic conditions, *fermentation* follows glycolysis and regenerates NAD^+ from NADH, which means NAD^+ is available to allow glycolysis to continue. *NAD^+* and *FAD* serve as electron acceptors, forming NADH and $FADH_2$. NADH carries high-energy electrons released during glycolysis, and both NADH and $FADH_2$ carry high-energy electrons from stage 1 of cellular respiration to the electron transport chain.
4. See Figure 8-3. During the energy investment stage, glucose is rearranged and activated (by adding one phosphate group and some of the energy from each of two ATPs), producing fructose bisphosphate. During the energy-harvesting stage, fructose bisphosphate is split into two G3P molecules, each retaining one phosphate. Each G3P is then converted to pyruvate, and energy is captured in two ATPs and two NADHs (one from each G3P). Glycolysis occurs in the cytosol.
5. Glycolysis produces two pyruvate molecules as end products. During stage 1 of cellular respiration, one carbon from pyruvate is released as CO_2 during the formation of acetyl CoA. Then during the Krebs cycle, acetyl CoA is broken down, releasing two more molecules of CO_2. Most of the energy captured during the Krebs cycle is in the high-energy electron carriers NADH and $FADH_2$.
6. The inner membrane of a mitochondrion has many copies of an electron transport chain (ETC), consisting of a series of electron transfer molecules. The high-energy electron carriers formed during glycolysis and the Krebs cycle deposit their electrons into the ETC.

As the electrons pass through the ETC, energy is released in stages and is used to pump hydrogen ions (H^+) across the inner membrane from the matrix to the intermembrane space. This creates an H^+ gradient. During the process of chemiosmosis, H^+ flows down its concentration gradient through the channel of the enzyme ATP synthase, which captures some of this energy to drive ATP synthesis.

7. Oxygen is necessary for cellular respiration because oxygen acts as the final electron acceptor for the electron transport chain (ETC). Without oxygen, the transfer of high-energy electrons through the ETC would stop, the hydrogen ion gradient would dissipate, and ATP formation would cease.
8. Chloroplasts and mitochondria are both large, membrane-enclosed organelles with additional inner membranes where important energy-related reactions occur. Both chloroplasts and mitochondria have ETCs embedded in their inner membranes; the ETC pumps H^+ into compartments formed by inner membranes (chloroplasts: interior of thylakoid; mitochondria: intermembrane compartment). Both chloroplasts and mitochondria use the H^+ gradient to generate ATP by chemiosmosis.

Chapter 9

Think Critically

Figure 9-8 If the sister chromatids of one replicated chromosome failed to separate, then one daughter cell would not receive any copy of that chromosome, whereas the other daughter cell would receive both copies.

Health Watch: Cancer—Running the Stop Signs at Cell Cycle Checkpoints

For a malignant tumor, the pathology lab would report mutated oncogenes, either causing excessive production of growth factors or making the cells of the tumor more likely to divide when growth factors are present (such as mutations in cyclin genes), and mutated tumor suppressor genes (such as p53), rendering the cell likely to divide even if it has damaged DNA.

Case Study Revisited: Body, Heal Thyself

Mutations that disrupt control of the cell cycle can lead to cancer, so the discovery of mutations in cells descended from the introduced stem cells could indicate a risk that the treatment might cause cancer. Such a risk is not tolerable in a clinical trial.

Multiple Choice

1. c; **2.** b; **3.** a; **4.** b; **5.** d

Fill-in-the-Blank

1. DNA (deoxyribonucleic acid)
2. prokaryotic fission
3. mitotic, differentiation; Stem
4. growth factors; Checkpoints; oncogenes, tumor suppressor genes
5. prophase, metaphase, anaphase, telophase; cytokinesis, telophase
6. kinetochores; polar

Review Questions

1. Refer to Figure 9-6 for a diagram of the eukaryotic cell cycle. The eukaryotic cell cycle consists of two major phases, interphase and cell division. During G_1 of interphase, the cell grows and may differentiate (possibly entering the G_1 stage for life). If the correct signals are received by the cell, it enters the S phase, in which the DNA of the chromosomes is replicated. The cell then enters the G_2 phase, where it may grow some more and prepares for cell division (usually mitotic cell division). During mitotic cell division, one copy of each chromosome and approximately half the cytoplasm are parceled out into each of the daughter cells that are produced from the parent cell.
2. *Mitosis* is division of the nucleus. *Cytokinesis* is division of the cytoplasm.
3. Refer to Figure 9-8 for a diagram of the stages of mitosis. During normal mitosis, the two sister chromatids of each replicated chromosome become attached to spindle microtubules leading to opposite poles of the cell. Therefore, during anaphase, when the chromatids separate and become independent chromosomes, one copy of each chromosome moves to each pole of the cell, forming two clusters of chromosomes, each containing one copy of all the chromosomes of the original parent cell.
4. A *centromere* is a specialized region of a chromosome to which the spindle fibers attach during cell division; the sister chromatids of a replicated chromosome are attached at the centromere. A *telomere* is a protective "cap" on the ends of chromosomes that is important for maintaining the integrity of the chromosome during replication. A *kinetochore* is a structure located at the centromere that becomes attached to spindle microtubules during mitosis. A *chromatid* is a DNA double helix formed by DNA replication; the replicated chromosome consists of two identical sister chromatids, attached to one another at the centromere. The *spindle* is an array of microtubules that moves the chromosomes (kinetochore microtubules) and pulls the poles apart (polar microtubules) during cell division.
5. Cytokinesis divides the cytoplasm into approximately equal halves at the end of cell division. In animal cells, the plasma membrane is pinched along the equator by a ring of microfilaments, separating the original cell into two daughter cells. In plant cells, new plasma membranes form along the equator by fusion of vesicles produced by the Golgi complex. The contents of the fused vesicles become the new cell wall between the two daughter cells.
6. The cell cycle is controlled by the interactions between many proteins, including cyclins, cyclin-dependent protein kinases (Cdks), growth factors, and a variety of other proteins whose activity is regulated by the integrity of DNA, successful completion of DNA replication, attachment of all of the chromosomes to the spindle, and correct positioning of chromosomes during metaphase. During a "normal" cell division, growth factors bind to receptors that start a chain of events resulting in the production of cyclins that bind to cyclin-dependent protein kinases. The kinases stimulate the cell to replicate its DNA and proceed through the cell cycle. The cell cycle must be regulated so that the body develops properly and repairs damaged or dying cells at the appropriate times and rates. When cells divide without regulation, they may form life-threatening cancers.
7. See Figure 9-4 for a diagram of the prokaryotic cell cycle, with a description of the major events.

Chapter 10

Think Critically

Figure 10-5 Draw the chromosomes, and follow them through meiosis. If one pair of homologues failed to separate at anaphase I, one of the resulting daughter cells (and the gametes produced from it) would have both homologues, and the other daughter cell (and the gametes produced from it) would not have any copies of that homologue. Assuming that these defective gametes fused with normal gametes, then the offspring would have either three copies or only one copy of that homologue.

Figure 10-6 Draw the chromosomes and follow them through meiosis, *including the chromosomes of each homologous pair that did not cross over at all.* Don't forget that the homologues are randomly separated during meiosis I. Therefore, some gametes would receive both of the incorrectly crossed-over chromosomes, but many would receive only one incorrectly crossed-over chromosome. When gametes were produced, some would contain a chromosome that was missing one of its own segments but contained a segment from another, nonhomologous chromosome. Assuming that these defective gametes fused with normal gametes, then the offspring would receive only one copy of some genes and three copies of other genes. For those genes that were removed from the incorrectly crossed-over chromosome, many offspring would receive those genes only from the other parent's gamete. For those genes that were added to the incorrectly crossed-over chromosome, many offspring would receive one copy of the genes added to the incorrect chromosome, a second copy of those same genes from a chromatid that never crossed over, and a third copy from the other parent's gamete.

Case Study Revisited: Diversity Runs in the Family

If skin cancer killed large numbers of pale-skinned people before they reproduced, it would select against pale skin. But skin cancer is mainly a disease of older, post-reproductive people, so it is not likely that it actually has selected against pale-skinned individuals (i.e., skin cancer does not have a big effect on reproductive success).

Multiple Choice

1.b; **2.** d; **3.** a; **4.** a; **5.** b

Fill-in-the-Blank

1. four; gametes *or* sperm and eggs
2. prophase, chiasmata; crossing over
3. anaphase, meiosis I; anaphase, meiosis II
4. shuffling of homologues, crossing over, union of gametes
5. Turner; do not, cannot; Klinefelter; XXY

Review Questions

1. Refer to Figure 10-5 for a diagram of meiosis. Homologous chromosomes separate during anaphase I, forming haploid daughter cells.
2. During prophase I, proteins bind the maternal and paternal homologues together so that they align precisely along their entire length. Enzymes then cut through the DNA of the paired homologues and graft the cut ends together, often joining part of one of the chromatids of the maternal homologue to part of one of the chromatids of the paternal homologue, and vice versa. The binding proteins and enzymes then depart, leaving chiasmata where the maternal and paternal chromosomes have exchanged parts. In addition to being the sites of exchange of DNA, chiasmata also hold the chromosomes together as they attach to the spindle microtubules during meiosis I.
3. Similarities between mitosis and meiosis: Both require DNA replication prior to division, and both involve the same sequence of steps using many of the same cellular components. Differences between mitosis and meiosis: Mitosis results in the formation of two daughter cells with the same chromosome complement as the original parent cell; therefore, the daughter cells are genetically identical to each other and to the parent cell. Meiosis separates homologous chromosomes, resulting in four haploid daughter cells. Because of crossing over and the random assortment of the paternal and maternal homologues into the daughter cells, meiosis produces genetically variable cells.
4. Meiosis provides for genetic variability in two ways. First, crossing over mixes segments of homologous pairs, creating new combinations of genes. Second, homologous pairs align independently of other pairs in metaphase I, resulting in daughter cells that contain different combinations of chromosomes than in the parent cell. If we ignore crossing over, an animal with a haploid number of 2 can produce $2^2 = 4$ genetically different gametes; with a haploid number of 5, the animal can produce $2^5 = 32$ genetically different gametes.
5. *Nondisjunction* is an error in meiosis that causes some cells to have too few or too many copies of some chromosomes. Common syndromes resulting from nondisjunction include Turner syndrome, trisomy X, Klinefelter syndrome, Jacob syndrome, and trisomy 21 or Down syndrome. See the text for descriptions of these conditions.

Chapter 11

Think Critically

Figure 11-8 Half of the gametes produced by a *Pp* plant will have the *P* allele, and half will have the *p* allele. All of the gametes produced by a *pp* plant will have the *p* allele. Therefore, half of the offspring of a *Pp* × *pp* cross will be *Pp* (purple) and half will be *pp* (white), whereas all of the offspring of a *PP* × *pp* cross will be *Pp* (purple). See Figure 11-9.

Figure 11-11 A plant grown from a wrinkled, green seed has the genotype *ssyy*. A plant grown from a smooth, yellow seed could be *SSYY*, *SsYY*, *SSYy*, or *SsYy*. Set up four Punnett squares to see if the smooth yellow plant's genotype can be revealed by a test cross.

Figure 11-12 Chromosomes, not individual genes, assort independently during meiosis. Therefore, if the genes for seed color and seed shape were on the same chromosome, they would tend to be inherited together and would not assort independently.

Figure 11-13 A palomino has the genotype C_1C_2. The only way to ensure a palomino foal is to breed a cremello (C_2C_2) with a chestnut (C_1C_1).

Figure 11-18 All daughters will inherit an X_N (normal) allele from their X_NY father, so no daughter will be color-deficient. All sons will inherit an X_n (deficiency) allele from their X_nX_n mother, so all sons will be color-deficient.

Figure 11-21 One of Victoria and Albert's sons, Leopold, had hemophilia. To be male, Leopold must have inherited Albert's Y chromosome. The X chromosome, not the Y chromosome, bears the gene for blood clotting, so Leopold must have inherited the hemophilia allele from his mother, Victoria.

Health Watch: The Genetics of Muscular Dystrophy

If the woman is a carrier, she is heterozygous for the defective dystrophin allele. Statistically, half of her eggs will receive the defective allele. Therefore, if her next child is a son, he has a 50% chance of having muscular dystrophy. If her next child is a daughter, she will not be affected (the father almost certainly does not have the defective allele, so daughters will receive at least one normal allele from him). Assuming that both of her parents were phenotypically normal, the woman inherited the defective allele from her mother, who would be heterozygous for the defective allele. Therefore, her sisters each have a 50% chance of being a carrier.

Multiple Choice

1.a; **2.** d; **3.** c; **4.** d; **5.** a

Fill-in-the-Blank

1. genotype, phenotype; heterozygous
2. independently; as a group; linked
3. XY, XX; sperm
4. sex-linked
5. incomplete dominance; codominance; polygenic inheritance

Review Questions

1. A *gene* is a segment of DNA coding for a specific protein. An *allele* is any of the alternative forms of the same gene. *Dominant* is a term that describes alleles that are expressed when a single copy is present. *Recessive* is a term that describes alleles that are masked in the presence of a dominant allele and are expressed only in the absence of a dominant allele (usually, when the organism has the recessive allele on both homologous chromosomes). *True-breeding* refers to organisms in which all the offspring produced by self-fertilization are genetically identical to the parent. *Homozygous* means having a pair of alleles that are the same for a particular gene. *Heterozygous* means having two different alleles for a particular gene. *Cross-fertilization* is the fertilization of an egg produced by one organism by a sperm produced by a different organism. *Self-fertilization* is fertilization of an egg by a sperm produced by the same organism.
2. Independent assortment occurs during meiosis I as pairs of homologous chromosomes are randomly separated during anaphase. Therefore, individual genes assort independently only if they are on different chromosomes. Genes on the same chromosome tend to be moved as a unit into the same daughter cell and tend to be inherited together. Crossing over may separate alleles of linked genes during meiosis.
3. *Polygenic inheritance* occurs when two or more genes influence a particular trait. These traits often involve genes that show incomplete dominance. The result is a gradient of different phenotypes depending on the numbers of particular alleles present. Let us assume that two parents both have "medium" skin color, because they are heterozygous at several gene loci that together

influence skin color. Because of independent assortment of alleles during meiosis, some of the gametes produced by these parents may, by chance, have all "pale" alleles of skin-color genes, while other gametes may have all "dark" alleles. If a "pale allele" sperm fertilizes a "pale allele" egg, the resulting offspring would have pale skin. If a "dark allele" sperm fertilizes a "dark allele" egg, the resulting offspring would have dark skin.

4. Sex linkage refers to genes that are found on one type of sex chromosome (for example, the X chromosome in mammals) but not on the other sex chromosome. In mammals, the X chromosome contains many more genes than the Y chromosome does, and most of the X chromosome genes do not have any corresponding gene on the Y chromosome. Therefore, males are most likely to exhibit recessive sex-linked traits, because they possess only one copy of the X chromosome and their Y chromosome cannot have dominant alleles of corresponding genes that might be able to mask recessive alleles present on the X chromosome. Females, on the other hand, might have a dominant allele of any given gene on one X chromosome, which would mask the expression of a recessive allele of that gene on the other X chromosome.

5. An organism's phenotype is its physical characteristics (for example, outward appearance, behaviors, or biochemical composition). An organism's genotype is the actual combination of alleles in the organism's genome. Knowing the phenotype of an organism does not always allow you to determine the genotype. For example, an organism with the dominant phenotype for a particular trait might be homozygous or heterozygous for the gene that controls that trait. Examining the offspring of a test cross (mating the phenotypically dominant individual with a homozygous recessive individual) allows the genotype of the phenotypically dominant individual to be determined.

6. The parental individuals showing the trait must be heterozygous. If they were homozygous, then all of their offspring would inherit a dominant allele and show the trait. However, only some of their offspring show the trait. Similarly, any given offspring must also be heterozygous, because they could be homozygous only by inheriting a dominant allele from both parents, which would mean that both of their parents would have to show the trait. However, in no case do both parents show the trait.

Genetics Problems

1. a. A red bull (R_1R_1) is mated to a white cow (R_2R_2). The bull will produce all R_1 sperm; the cow will produce all R_2 eggs. All the offspring will be R_1R_2 and will have roan hair (codominance).

b. A roan bull (R_1R_2) is mated to a white cow (R_2R_2). The bull produces half R_1 and half R_2 sperm; the cow produces R_2 eggs. Using the Punnett square method:

sperm \ eggs	R_2
R_1	R_1R_2
R_2	R_2R_2

Using probabilities:

Sperm	Egg	Offspring
$\frac{1}{2}R_1$	R_2	$\frac{1}{2}R_1R_2$
$\frac{1}{2}R_2$	R_2	$\frac{1}{2}R_2R_2$

The predicted offspring will be $\frac{1}{2}R_1R_2$ (roan) and $\frac{1}{2}R_2R_2$ (white).

2. a. *TtGg* × *TtGg*. This is a "standard" cross for differences in two traits. Both parents produce *TG, Tg, tG,* and *tg* gametes. The expected proportions of offspring are $\frac{9}{16}$ tall, green; $\frac{3}{16}$ tall, yellow; $\frac{3}{16}$ short, green; $\frac{1}{16}$ short yellow.

b. *TtGg* × *TTGG*. In this cross, the heterozygous parent produces *TG, Tg, tG,* and *tg* gametes. However, the homozygous dominant parent can produce only *TG* gametes. Therefore, all offspring will receive at least one *T* allele for tallness and one *G* allele for green pods, and thus all the offspring will be tall with green pods.

c. *TtGg* × *Ttgg*. The second parent will produce two types of gametes, *Tg* and *tg*. Using a Punnett square:

sperm \ eggs	*Tg*	*tg*
TG	*TTGg*	*TtGg*
Tg	*TTgg*	*Ttgg*
tG	*TtGg*	*ttGg*
tg	*Ttgg*	*ttgg*

The expected proportions of offspring are $\frac{3}{8}$ tall, green; $\frac{3}{8}$ tall, yellow; $\frac{1}{8}$ short, green; $\frac{1}{8}$ short, yellow.

3. If the genes are on separate chromosomes—that is, assort independently—then this would be a typical two-trait cross with expected offspring of all four types (about $\frac{9}{16}$ round, smooth; $\frac{3}{16}$ round, fuzzy; $\frac{3}{16}$ long, smooth; and $\frac{1}{16}$ long, fuzzy). However, only the parental combinations show up in the F_2 offspring, indicating that the genes are on the same chromosome.

4. The genes are on the same chromosome and are quite close together. On rare occasions, crossing over occurs between the two genes, producing recombination of the alleles.

5. a. *BBMM* (brown) × *BbMm* (brown). The first parent can produce only *BM* gametes, so all offspring will receive at least one dominant allele for each gene. Therefore, all offspring will have brown hair.

b. *BbMm* (brown) × *BbMm* (brown). Both parents can produce four types of gametes: *BM, Bm, bM,* and *bm*. Filling in the Punnett square:

sperm \ eggs	*BM*	*Bm*	*bM*	*bm*
BM	*BBMM*	*BBMm*	*BbMM*	*BbMm*
Bm	*BBMm*	*BBmm*	*BbMm*	*Bbmm*
bM	*BbMM*	*BbMm*	*bbMM*	*bbMm*
bm	*BbMm*	*Bbmm*	*bbMm*	*bbmm*

All *mm* offspring are unpigmented, so we get the expected proportions $\frac{9}{16}$ brown-haired, $\frac{3}{16}$ blond-haired, $\frac{4}{16}$ unpigmented.

c. *BbMm* (brown) × *bbmm* (brown):

sperm \ eggs	*bm*
BM	*Bb Mm*
Bm	*Bb mm*
bM	*bb Mm*
bm	*bb mm*

The expected proportions of offspring are $\frac{1}{4}$ brown-haired, $\frac{1}{4}$ blond-haired, $\frac{1}{2}$ unpigmented.

6. A man with normal color vision is *CY* (remember, the Y chromosome does not have the gene for color vision). His color-deficient wife is *cc*. Their expected offspring will be:

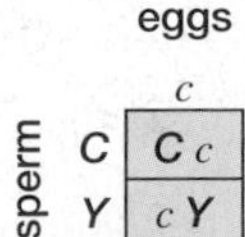

We therefore expect that all the daughters will have normal color vision and all the sons will be color-deficient.

7. The husband should win his case. All his daughters must receive one X chromosome, with the *C* allele, from him and therefore should have normal color vision. If his wife gives birth to a color-deficient daughter, her husband cannot be the father (unless there was a new mutation for color deficiency in his sperm line, which is very unlikely).

Chapter 12

Think Critically

Figure 12-4 It takes more energy to break apart a C–G base pair because these bases are held together by three hydrogen bonds, compared with the two hydrogen bonds that bind A to T.

Doing Science: Discovering the Hereditary Molecule

The experiment would not have succeeded because eukaryotic (and prokaryotic) chromosomes contain both DNA and protein. The original experiment worked because the phage genetic material that entered the bacterial cells consisted of DNA only, and the phage coat that consisted of protein only remained outside the cell. If the phage had injected a chromosome containing both DNA and protein, both labeled protein and labeled DNA would have entered the bacterial cells, and it would have been impossible to determine which was the phage's genetic material.

Case Study Revisited: Muscles, Mutations, and Myostatin

Fast running is probably highly adaptive in an animal that hunts speedy prey, so heterozygotes would leave more offspring than homozygotes for either allele (normal or defective). Selection favoring heterozygotes would result in the defective allele being preserved in the population, though at a lower level than would be the case if homozygotes for the defective allele ran the fastest. (Note: Homozygotes for the defective allele are probably the strongest phenotype, so your answer might be somewhat different if you considered the relative benefits of being strong and being fast.)

Multiple Choice

1. b; **2.** a; **3.** b; **4.** d; **5.** b

Fill-in-the-Blank

1. nucleotides; sugar (deoxyribose), phosphate, base
2. phosphate, sugar; double helix
3. thymine, cytosine; complementary
4. semiconservative
5. DNA polymerases; free nucleotides
6. mutations; substitution mutation

Review Questions

1. Bacteriophages infect bacteria by injecting their genetic material into a host bacterium, where the genetic material is used to make new phages. Phages consist of only DNA and protein. DNA contains phosphorus but not sulfur. Proteins contain sulfur but not phosphorus. Hershey and Chase labeled bacteriophages with either radioactive phosphorus (to label DNA) or radioactive sulfur (to label proteins). They found that radioactive phosphorus from labeled DNA enters bacteria from phages, but radioactive sulfur from labeled protein does not. After the protein coats have been shaken off, the infected bacteria still produce new phages, so the injected DNA must be the genetic material of phages.
2. Refer to Figure 12-1 for a diagram of a nucleotide. The four DNA nucleotides all have the same phosphate group and sugar (deoxyribose) but vary in the type of nitrogenous base they contain (adenine, guanine, thymine, or cytosine).
3. The DNA of a chromosome is composed of two strands of nucleotides wound around each other in a double helix. In each strand, the sugar of one nucleotide is covalently bonded to the phosphate of the next nucleotide in the strand, forming the backbones on each side of the double helix. The bases from each strand pair up in the middle of the helix and are held together by hydrogen bonds. Adenine and thymine are complementary; cytosine and guanine are complementary. Adenine is held to thymine by two hydrogen bonds, while cytosine is held to guanine by three hydrogen bonds.
4. Information is encoded in the sequence of the four bases (A, T, C, and G). Like letters in an alphabet, different sequences of bases spell out the information needed to synthesize different proteins.
5. General description: The two DNA strands of the double helix unwind and partially separate. DNA polymerase enzymes move along each strand, linking up free nucleotides into new DNA strands. The sequence of nucleotides in each newly formed strand is complementary to the sequence on the parent strand. As a result, two double helices are synthesized, each consisting of one parental DNA strand plus one newly synthesized complementary strand that is an exact copy of the other parental strand. This type of replication is called semiconservative replication.
6. Mutations arise because of mistakes during DNA replication and damage to DNA from radiation (such as ultraviolet light) or environmental chemicals. Types of mutations include (1) substitutions, in which a single nucleotide is changed; (2) insertions, in which one or more nucleotides are inserted into the DNA; (3) deletions, in which one or more nucleotides are removed from the DNA; (4) inversions, in which segments of DNA are removed from a chromosome, turned around and reinserted into the gap; and (5) translocations, in which a segment of DNA is moved from one chromosome to another.

Chapter 13

Think Critically

Figure 13-4 Cells produce far larger amounts of some proteins than others. Obvious examples include cells that produce antibodies or protein hormones, which they secrete into the bloodstream in vast quantities, affecting functioning throughout the body. If a cell needs to produce more of certain proteins, it will probably synthesize more mRNA that will be translated into that protein.

Figure 13-7 Grouped in codons, the original mRNA sequence visible here is AUG GGA GUU. Changing all G to U would produce the sequence AUU UUA UUU. Refer to the genetic code (Table 13-3). First, AUG encodes methionine, while AUU encodes isoleucine, so the first G → U change would substitute isoleucine for methionine in the protein. Second, GGA encode glycine, while UUA encodes leucine. Therefore, the second G → U change would substitute leucine for glycine. The final codon in the illustration, GUU, encodes valine, while UUU encodes phenylalanine. Overall, changing all G to U would change the peptide from met-gly-val to ile-leu-phe.

Health Watch: The Strange World of Epigenetics

Generally, methyl groups attached to DNA reduce transcription, and acetyl groups attached to histone proteins increase transcription. Therefore, you would expect that people with type 2 diabetes will probably have methylated DNA in the insulin gene and/or in its promoter. To increase insulin production, you could try to remove methyl groups on the insulin gene and/or its promoter or to add acetyl groups to histone proteins in the vicinity of the insulin gene.

Health Watch: Androgen Insensitivity Syndrome

To help explain the syndrome to the parents, you might designate the normal X chromosome X_A and the mutated one X_a. The 16-year-old daughter who tested XY must have inherited the mutated androgen receptor gene (X_a) from her mother. The father cannot have androgen insensitivity; otherwise, he would be phenotypically female. The mother can bear children and therefore must be heterozygous for androgen insensitivity: X_AX_a. All the phenotypically male children (X_AY) in the family inherited the normal X allele from mom and Y from dad. The daughters, however, may be either X_AX_A, X_AX_a, or X_aY, with equal probability. In other words, any daughter has a 33% chance of being X_aY and androgen insensitive. A Punnett square could be used to illustrate the probabilities.

Multiple Choice

1. b; **2.** a; **3.** a; **4.** c; **5.** d

Fill-in-the-Blank

1. transcription; translation; ribosome
2. messenger RNA, transfer RNA, ribosomal RNA; noncoding RNA
3. three; codon; anticodon
4. RNA polymerase; template; promoter; termination signal
5. start, stop; transfer RNA; peptide
6. substitution; Insertion; Deletion

Review Questions

1. RNA is different from DNA in three respects: (1) RNA is usually single-stranded; (2) RNA has a different type of sugar in its backbone—ribose instead of deoxyribose; and (3) the base thymine in DNA is replaced by uracil in RNA.
2. There are three types of RNA used in protein synthesis: (1) messenger RNA (mRNA) carries a copy of genetic information (to make a protein) from the nucleus into the cytoplasm; (2) transfer RNA (tRNA) carries amino acids to the site of protein synthesis in the cytoplasm (the ribosome); (3) ribosomal RNA (rRNA) is a structural component of the ribosomes that "reads" the code of nucleotides of mRNA to synthesize an amino acid sequence using the amino acids carried by the tRNAs.
3. The *genetic code* is the set of three-base sequences (codons in messenger RNA) corresponding to the amino acids that will be incorporated into a protein. A *codon* is a specific three-base sequence in mRNA that specifies a particular amino acid, or the start or termination of protein synthesis. An *anticodon* is a triplet of tRNA nucleotides that is complementary to one of the mRNA codons. The relationship between bases in DNA, codons, and anticodons is based on complementary base pairing: mRNA codons are complementary to three-base sequences in DNA, and tRNA anticodons are complementary to mRNA codons.
4. All RNA is formed from a gene through a process called transcription. Transcription is carried out by an enzyme called RNA polymerase. RNA polymerase binds to the template strand of DNA at a site called the promoter, which is a short sequence of DNA nucleotides that precedes the nucleotide sequence of the gene itself. The RNA polymerase causes the DNA to unwind. The RNA polymerase then travels along a DNA strand (the template strand), synthesizing a complementary mRNA molecule using free RNA nucleotides present in the nucleus. RNA polymerase continues transcription until it reaches a termination sequence of DNA nucleotides at the end of the gene. At this point, the RNA molecule is released, and the RNA polymerase enzyme leaves the DNA.

   In eukaryotic cells, messenger RNA is synthesized as a long transcript from the template strand, including both the introns and exons of a gene. Nucleotides are added at the beginning and the end of the RNA transcript, forming a cap and a tail. Enzymes in the nucleus then cut apart the RNA at the junctions between introns and exons, discarding the introns and splicing together the exons. The resulting RNA strand is the final mRNA.
5. Refer to Figure 13-7 for a diagram of protein synthesis. In translation, a protein is synthesized when a ribosome attaches to and reads the codons of an mRNA molecule. A preinitiation complex composed of a small ribosomal subunit, a "start" (methionine) tRNA, and several proteins binds to an mRNA molecule. The complex scans the mRNA until it finds the start codon, AUG. The large ribosomal subunit then binds to the complex. The completed ribosome moves along the mRNA message one codon at a time, while complementary tRNA molecules carrying specific amino acids position themselves alongside one another according to the nucleotide sequence of the mRNA. Peptide bonds are formed between the amino acids as the polypeptide is synthesized. This process continues until the ribosome reaches a stop codon on the mRNA sequence that causes the ribosome to release the mRNA and the newly synthesized polypeptide.
6. In transcription, complementary base pairing is used to copy the nucleotide sequence of a gene in DNA into a sequence of mRNA nucleotides. In translation, complementary base pairing is used to match each mRNA codon of nucleotides to a complementary anticodon of a tRNA molecule carrying a specific amino acid.
7. Mechanisms of gene regulation include regulation of the rate of transcription, synthesis of different mRNAs by alternative splicing of the original RNA transcript, differences in the "lifetime" of different mRNA molecules, regulation of the rate of translation of mRNA into protein, requirement for protein modification before it becomes "active," and regulation of the rate of breakdown of proteins in the cell.
8. A *mutation* is a change in the nucleotide sequence of a DNA molecule. A substitution mutation that produces a codon that codes for the same amino acid will have no effect on the amino acid composition of the protein or on protein function. A substitution that changes a codon to one that codes for a chemically very similar amino acid (for example, still about the same size, hydrophilic) will probably not affect protein function very much. A substitution that changes a codon to one that codes for a chemically very different amino acid will probably interfere with protein function. A substitution that changes an amino-acid-coding codon for a stop codon will create a short protein that is likely to be completely nonfunctional.

Chapter 14

Think Critically

Figure 14-8 Each person normally has two copies of each STR gene, one on each of a pair of homologous chromosomes. A person may be homozygous (two copies of the same allele) or heterozygous (one copy each of two alleles) for each STR. The bands on the gel represent individual alleles of an STR gene. Therefore, a single person can have one band (if homozygous) or two bands (if heterozygous). For a homozygote, the DNA from both (identical) alleles will end up in the same place on the gel, and that (single) band will have twice as much DNA as each of the two bands of DNA from a heterozygote. The more DNA, the brighter the band.

Figure 14-9 The genetic material of bacteriophages is DNA. Each bacterial restriction enzyme cuts DNA at a specific nucleotide sequence. A given bacterium is likely to have evolved restriction enzymes that cut DNA at sequences found in bacteriophages but not in its own chromosome.

Earth Watch: What's Really in That Sushi?

Animals leave feces and hair (caught on thorns, for example) in their habitat. Barcoding hair (if the hair samples included bits of the follicles, which contain live cells with DNA) could reveal what species of animals are in the rain forest. Barcoding feces (which contain cells from both predators and their prey) could reveal what species of predators are in the forest and what species of prey they eat.

Health Watch: Prenatal Genetic Screening

If we have DNA samples from the mother and at least one potential father, we can produce DNA profiles similar to those used in forensics. The fetus must share an allele at each locus with the mother; the other fetal allele at each locus can be compared with the alleles of the potential father(s). DNA profiles can definitively rule out or confirm a potential father with a high degree of confidence.

Multiple Choice

1. a; **2.** c; **3.** c; **4.** a; **5.** b

Fill-in-the-Blank

1. Genetically modified organisms
2. Transformation; plasmids
3. PCR (polymerase chain reaction)
4. short tandem repeats (STRs); size (differently sized alleles are of different lengths, because they have different numbers of repeats); DNA profile
5. electrophoresis *or* gel electrophoresis; base pairing

Review Questions

1. Two natural forms of genetic recombination include bacterial transformation and viral infection. Bacterial transformation is the acquisition of DNA by bacteria; this DNA is usually derived from other bacteria, possibly of different species, when those other bacteria die. Viral infection may transfer DNA from one organism (of the same or different species) to another; when viruses infect a host, daughter viruses acquire some of the host's DNA and then infect a second host. Both natural genetic recombination and recombinant DNA technology involve exchange of DNA between organisms, sometimes between species, but they are different in that natural recombination is relatively random and its usefulness is driven by natural selection, whereas recombinant DNA technology is deliberate and directed toward a specified use.
2. A plasmid is a small, circular DNA molecule found in the cytoplasm of bacteria. Living bacteria exchange plasmids, or bacteria may acquire plasmids when a bacterium dies and its cellular contents are released.
3. A restriction enzyme breaks a DNA molecule at specific short nucleotide sequences. Some restriction enzymes make a "staggered" cut across the DNA, leaving complementary single-stranded ends protruding from the cut DNA. Restriction enzymes can be used to insert human DNA into a plasmid by cutting both human DNA and the plasmid with the same restriction enzyme and then mixing them together. The single-stranded ends base-pair with one another, sometimes temporarily holding pieces of human DNA to pieces of plasmid DNA. After the cut ends of the DNA are joined together, some plasmids have a fragment of human DNA permanently spliced into them.
4. The polymerase chain reaction amplifies specific segments of DNA. Small pieces of DNA, called primers, are synthesized. One primer is complementary to the beginning of one DNA strand of the segment to be copied, and the other primer is complementary to the beginning of the other strand of the segment. The DNA sample, the primers, free nucleotides, and a special DNA polymerase that is active at very high temperatures are mixed together in a small test tube. The tube is heated to 90°C to separate the DNA strands, then cooled to 50°C to allow the primers and DNA polymerase to bind to the separated DNA strands. The tube is then heated to 72°C, allowing the DNA polymerase to copy the DNA. Repeated heating and cooling cycles copy the DNA, doubling the amount of copied DNA with each cycle.
5. CRISPR-Cas9 is a molecular tool for editing genomes. A CRISPR-Cas9 tool consists of a Cas9 enzyme molecule attached to RNA that includes a synthesized guide sequence complementary to a target DNA sequence. When the tool is introduced to a cell, the guide RNA finds its DNA match in the host genome, and the attached Cas9 enzyme cuts both strands of the DNA at that location. If a segment of DNA has been introduced to the cell along with the CRISPR-Cas9 tool, the cell's DNA repair mechanism will usually insert that segment at the site of the cut. Otherwise, the repair mechanism will attempt to reconnect the broken strands, but in so doing it will usually add a few random base pairs, thereby disabling the target gene.
6. A short tandem repeat (STR) is a short sequence of nucleotides in DNA (usually two to five nucleotides long), repeated many times side by side (tandem). Different people have the same sets of STRs, but differ in the number of repeats. By analyzing the numbers of repeats in a standard set of STRs, forensic scientists can identify any individual person's DNA with a probability of misidentification of less than one in a trillion.
7. Gel electrophoresis separates DNA pieces according to size. DNA samples are loaded at one end of the gel, and then an electric field is applied, negative at the end with the samples and positive at the other end, so that current moves through the gel. Because DNA is negatively charged, it moves through the gel toward the positive end. Smaller pieces of DNA move more rapidly than larger pieces, so gel electrophoresis separates the DNA pieces by size—the smaller the piece, the farther it moves through the gel in any given period of time.
8. Genetic engineering in agriculture usually consists of adding genes from other species to important crop plants or livestock. These added genes may produce crop plants that resist insects (for example, by expressing a protein from *Bacillus thuringiensis* that damages insects' digestive tracts), resist herbicides (so that herbicides kill weeds in the field without harming the crops), or resist disease. Other genetic manipulations may produce plants that are more nutritious or that produce medicinally useful proteins such as antibodies. In animals, added genes may increase the rate of growth, may cause the animal to produce more muscle (meat) and less fat, or cause the animal to produce novel useful substances such as spider silk or antibodies.
9. Genetic engineering in medicine may be used to diagnose and treat diseases. Restriction enzymes may be used to diagnose certain diseases, such as sickle-cell anemia, because certain restriction enzymes cut the normal allele but not the disease-causing mutant allele (or vice versa). DNA sequencing or binding of normal and mutant alleles to DNA probes may also be used to diagnose diseases such as cystic fibrosis. Many medical treatments, including insulin for diabetics, are made with recombinant DNA techniques. There are a few successful examples of using recombinant DNA technologies to cure inherited diseases. In severe combined immune deficiency, children are born with a defective allele that prevents the normal development of the immune system. After transfecting their bone marrow cells with disabled viruses containing the functional allele, some of these children develop normally functioning immune systems.
10. In amniocentesis, amniotic fluid is removed when the fetus is 15 weeks or older. This fluid contains some fetal cells, which are cultured and analyzed for particular genetic disorders, such as trisomy 21 (Down syndrome). The amniotic fluid may also contain molecules that indicate health problems with the fetus, such as spina bifida. In chorionic villus sampling, a few villi are removed from the chorion, a structure that is produced by the fetus and becomes part of the placenta. The cells of these villi are analyzed for genetic disorders. Chorionic villus sampling can be performed as early as the eighth week of pregnancy, and the sample contains far more fetal cells than can be obtained by amniocentesis. A disadvantage of chorionic villus sampling when compared with amniocentesis is that chorion cells are more likely to have chromosomal abnormalities, even in healthy fetuses. There is also a slightly higher risk of miscarriage or damage to the fetus. Finally, chorionic villus sampling does not obtain amniotic fluid, and so certain types of fetal defects cannot be detected. In maternal blood sampling, fetal DNA and cells are extracted from a sample of the mother's blood. The fetal cells may be karyotyped and the fetal DNA sequenced or otherwise analyzed to detect genetic disorders.

Chapter 15

Think Critically

Figure 15-6 No. Mutations, the ultimate source of the variation on which natural selection acts, occur in all organisms, including those that reproduce asexually.

Figure 15-7 No. Evolution can include changes in traits that are not revealed in morphology (the physical form of an organism), such as physiological systems and metabolic pathways. More generally, evolution in the sense of changes in a species' genetic makeup is inevitable; genetic evolution is not necessarily reflected in morphological change.

Figure 15-10 Analogous. Peacock tails consist of feathers; dog tails do not (they consist of bone, muscle, and skin). If the two structures were homologous, they would both consist of bone, muscle, and skin or both consist of feathers.

Doing Science: Charles Darwin and the Mockingbirds

The species of mockingbird that descended from the birds that originally colonized one of the islands arose as birds gradually dispersed, moving from island to island away from the site of the original colonization.

Earth Watch: People Promote High-Speed Evolution

Natural selection favoring pesticide resistance is absent in the pesticide-free zones, so among the insects that live there, the frequency of alleles that confer resistance remains very low. If insects hatched in these areas later interbreed with resistant insects hatched in the pesticide-used zones, gene flow between the nonresistant and resistant subpopulations will result in a lower frequency of resistance alleles in the pesticide-used zones, thus slowing the evolution of resistance.

Case Study Revisited: What Good Are Wisdom Teeth and Ostrich Wings?

Natural selection does not favor "improvement" in any absolute sense; it favors traits that are advantageous under current conditions. Thus, loss of a tail over evolutionary time would be adaptive under circumstances in which the cost of producing a tail outweighs the benefits of having one (as would have been the case in humans after the origin of upright posture). In any case, the value of vestigial traits as evidence of evolution stems from their homology with functional traits in other species, not from their status as adaptations.

Multiple Choice

1. c; **2.** d; **3.** c; **4.** c; **5.** b

Fill-in-the-Blank

1. wing, arm; analogous, convergent; vestigial
2. common ancestor; amino acids, ATP
3. catastrophism; uniformitarianism; old
4. evolution; mutations, DNA
5. natural selection; artificial selection
6. many traits are inherited; Gregor Mendel

Review Questions

1. Although natural selection determines an individual's likelihood of survival and reproduction, the changes in an individual during its lifetime do not constitute evolution. Evolution is a change in the genetic makeup of a population over time.
2. Catastrophism hypothesized that successive catastrophes produced the layers of rock, killing species and fossilizing them in the rock layers. Uniformitarianism hypothesized that ordinary natural processes that occurred repeatedly over long periods of time produced the layers of rock.
3. According to the hypothesis of inheritance of acquired characteristics, the bodies of living organisms can be modified through use or disuse of parts, and these modifications can be inherited by their offspring. This theory is invalid because characteristics such as these are not inherited.
4. Natural selection is the process that selects those individuals whose traits best adapt them to a particular environment. For example, faster barracudas obtain food more easily and therefore are more likely to survive and reproduce successfully than are slower barracudas.
5. Natural populations of all organisms have the potential to increase rapidly, because organisms can produce far more offspring than are required merely to replace the parents. Nevertheless, the sizes of most natural populations and the resources available to maintain them remain relatively constant over time; therefore, there is competition for survival and reproduction. In each generation, many individuals must die young, fail to reproduce, produce few offspring, or produce less-fit offspring that fail to survive and reproduce in their turn. Individual members of a population differ from one another in their ability to obtain resources, withstand environmental extremes, escape predators, and so on. On average, the most well adapted individuals in one generation will usually leave the most offspring. At least some of the variation in adaptive traits among individuals is due to genetic differences that may be passed on from parent to offspring. Over many generations, differential, or unequal, reproduction among individuals with different genetic makeup changes the overall genetic composition of the population.
6. Convergent evolution is evolution in unrelated organisms that leads to similar structures, given similar environmental demands. An example is the wings of flies and birds.
7. The presence in all organisms of DNA and other biological molecules is evidence of the shared ancestry of all living things.
8. Human activities alter the environment such that some traits that were previously disadvantageous become advantageous. Human-caused environmental change has favored evolution of pesticide resistance in insects, herbicide resistance in plants, antibiotic resistance in bacteria, and small adult size in fish.

Chapter 16

Think Critically

Figure 16-3 The surviving colonies would be in different places on each treated plate, and there would be a different number of surviving colonies on each plate (because the antibiotic-caused mutations would arise unpredictably, depending on which bacteria happened to interact with the antibiotic such that mutations were caused). Another possibility is that all of the colonies would survive (if the antibiotic always caused mutations in every colony).

Figure 16-4 For a locus with two alleles, one dominant and one recessive, there are two possible phenotypes. A mating between a heterozygote and a homozygote-recessive yields offspring with a 50:50 ratio of the two phenotypes.

	B	***b***
b	*Bb* (black)	*bb* (brown)
b	*Bb* (black)	*bb* (brown)

Figure 16-5 With a higher initial frequency, the frequency of allele B starts out closer to the top of the *y*-axis, so the random fluctuations of genetic drift are more likely to take the frequency line to the top of the plot (fixation of the allele, frequency of 100%) and less likely to take it to the bottom of the plot (loss of the allele, frequency of zero). In Figure 16-5a (large population), the lines for different simulations would fluctuate around 0.7 instead of around 0.5, and one or a few lines might reach 1.0 before 100 generations. In Figure 16-5b (medium-sized population), the number of lines reaching 1.0 before 100 generations would likely be larger than the number of lines reaching zero, and it's possible that no lines would reach zero. In Figure 16-5c (very small population), all simulations would still reach either 1.0 or zero in a few generations. It's likely that more simulations would go to 1.0 than to zero; the odds of this happening would increase if the number of simulations were increased.

Figure 16-6 Mutations inevitably and continually add variability to a population, and after the population becomes larger, the counteracting, diversity-reducing effects of drift decrease. The net result is an increase in genetic diversity.

Figure 16-10 It would be greater for males. A female's reproductive success is limited by her maximum litter size, but a male's potential reproductive success is limited only by the number of available females. When, as in bighorn sheep, males battle for access to females, the most successful males can impregnate many females, while unsuccessful males may not fertilize any females at all. Thus, the difference between the most and least successful male can be very large. In contrast, even the most successful female can have only one litter of offspring per breeding season, which is not that many more offspring than a female who fails to reproduce.

Figure 16-12 There is always a limit to directional selection. As a trait becomes more extreme, eventually the cost of increasing it further outweighs the benefits (for example, the cost of obtaining extra food may outweigh the benefit of larger size), or it may be physically impossible for the trait to become more extreme (for example, a limb's length may be limited by the maximum length that a bone can attain without breaking under its own weight).

Earth Watch: The Perils of Shrinking Gene Pools

One way to counter loss of genetic diversity due to genetic drift in small populations is to foster gene flow between populations. Because genetic drift is a random process, for many genes, different alleles will be lost (and therefore different alleles will be present) in different small populations. Thus, gene flow can introduce new alleles to an isolated population and increase its genetic diversity. Such gene flow is promoted by habitat corridors, which allow individuals, spores, or seeds to move between habitat patches.

Health Watch: Cancer and Darwinian Medicine

If you sequenced the genotypes in the original and new tumors of the relapsed patients, mutations that were present in both the original and new tumors could be the cause of the drug resistance. It is likely that these mutations would not be present in the original tumor of the patient who did not suffer a relapse. Natural selection by the chemotherapy drugs would allow tumors that happened to initially contain a few cells with resistance mutations to persist, and it would eliminate tumors that contained no such cells.

Case Study Revisited: Evolution of a Menace

Many species of bacteria and fungi that live in soil secrete antibiotic chemicals to help them compete for access to space and food. As a result of natural selection imposed by these poisons in their environment, some soil bacteria have evolved antibiotic resistance. Because the alleles that provide resistance may have negative side effects on the fitness of their possessors, they tend to be rare in environments that contain few antibiotic producers.

Multiple Choice

1. c; **2.** b; **3.** b; **4.** d; **5.** c

Fill-in-the-Blank

1. Hardy–Weinberg principle, equilibrium, allele; no
2. alleles; nucleotides; mutations; homozygous, heterozygous
3. genotype, phenotype; phenotype
4. genetic drift; small; founder effect, population bottleneck; founder effect
5. species; natural selection, coevolution; adaptations
6. reproduction; environment

Review Questions

1. A gene pool is all of the genes that are present in a population. Allelic frequencies are determined by counting all of the different alleles for a trait in a population and determining their relative proportions.
2. An *equilibrium population* is one in which the allele frequencies and the distribution of genotypes remain constant in succeeding generations. This can occur only if (1) there are no mutations, (2) there is no gene flow between populations, (3) the population is very large, (4) all mating is random, and (5) there is no natural selection.
3. Random changes in allele frequencies occur in all populations, but the effects of such changes are greater in small populations. Genetic drift can cause evolution.
4. This would not prove that natural selection was occurring in the population because other factors could be influencing the changes in allele frequencies.
5. This experiment demonstrates that mutation occurs spontaneously and not in response to specific selective pressures.
6. Natural selection can affect a population through directional selection, stabilizing selection, and disruptive selection. Stabilizing selection is most likely to occur in stable environments, and directional selection is most likely to occur in rapidly changing environments.
7. Sexual selection favors traits that improve success at mating. Sexual selection often seems to work in opposition to other forms of natural selection, for example, by favoring traits that reduce survival.

Chapter 17

Think Critically

Figure 17-1 The key question is whether the gray-furred and black-furred squirrels interbreed freely. Tests would involve careful observation of the squirrels in areas where both types occur to check for mixed mating and perhaps genetic comparisons to determine the degree of gene flow between the two types. If hybrid matings are observed, it would be important to determine if hybrid offspring are viable and fertile.

Figure 17-8 Possibilities include continental drift; climate changes (especially glacial advances) that cause habitat fragmentation; formation of islands by volcanic activity or rising sea level; movements of organisms to existing islands (including "islands" of isolated habitats such as lakes, mountaintops, and deep-ocean vents); and formation of barriers to movement (e.g., new mountain ranges, deserts, rivers). These processes are indeed sufficiently common and widespread to account for a multitude of speciation events over the history of life.

Figure 17-10 The key question is whether the two populations inhabiting the two species of trees (apple and hawthorn) interbreed. Tests might involve careful observation of flies under natural conditions, lab experiments in which captive flies of the two types are provided with opportunities to interbreed, or genetic comparisons to determine the degree of gene flow between the two types of flies.

Figure 17-11 It is likely that the island species arose though allopatric speciation, following the movement of part of the mainland population to the island. (Isolation by dispersal; divergence likely due at least in part to a founder effect.)

Figure 17-12 Small groups of organisms can colonize islands and become genetically isolated from the mainland population of their species. If such isolated island populations ultimately become separate species, the new species will (at least initially) be endemic to the island on which speciation occurred. Populations of species endemic to islands, especially small islands, are likely to be small. Species with small populations are at a higher risk of extinction.

Figure 17-13 Natural selection cannot look forward and ensure that the only traits that evolve are those that ensure survival of the species as a whole. Instead, natural selection ensures only the preservation of traits that help individuals survive and reproduce more successfully than individuals lacking the trait. So if, in a particular species, highly specialized individuals survive and reproduce better than less-specialized individuals, the specialized phenotype will eventually dominate, even if it ultimately puts the species at greater risk of extinction.

Doing Science: Seeking the Secrets of the Sea

The map reveals areas in which multiple species are concentrated for extended time periods. It would be helpful to protect these "hotspots."

Case Study Revisited: Discovering Diversity

The negative effects of very small population size on species will act more quickly than any potential new speciation could occur. Very small populations are at high risk of extinction, in part due to the negative effects of inbreeding and loss of genetic diversity though genetic drift (see Chapter 16). Even if new species were to arise in isolated populations, these populations would be small and at risk of extinction.

Multiple Choice

1. a; **2.** b; **3.** a; **4.** a; **5.** c

Fill-in-the-Blank

1. populations, independently; reproductive isolation; asexually
2. behavioral isolation, hybrid inviability, temporal isolation, gametic incompatibility, mechanical incompatibility
3. genetically isolated, diverge; allopatric speciation; genetic drift, natural selection
4. adaptive radiation; habitat (or environment)
5. small, specialized; habitat destruction

Review Questions

1. *Species* are all the populations of organisms that are capable of interbreeding under natural conditions and that are reproductively isolated from members of other populations. *Speciation* is the formation of new species through evolution. *Allopatric speciation* is speciation that occurs when two populations are geographically separated from each other. This may occur if the two isolated populations diverge genetically due to different selective pressures or genetic drift and are no longer able to interbreed. *Sympatric speciation* is speciation that occurs when two populations share the same

geographic area. Speciation may occur if environmental change results in two different habitats that lead to genetic divergence of organisms living in each, resulting in reproductive isolation.

2. If we use a strict reading of the biological species concept as our species definition, then naturally hybridizing populations are not technically separate species. But if hybridization is sufficiently rare that there is little gene flow between the two populations, most biologists would say that the two populations represent separate species.
3. Convincing data to support the conclusion that the two forms of fruit flies are separate species would be evidence that individuals of the two types do not mate under natural conditions or that matings between the two types do not produce viable offspring.
4. Apply colchicine to the developing flowers and pollinate them. Pollen and ovules should contain diploid sperm and eggs that will fuse in fertilization to produce a polyploid embryo.
5. Premating isolating mechanisms include geographic isolation (a physical barrier separates species), ecological isolation (species occupy different habitats), temporal isolation (species breed at different times), behavioral isolation (species have different mating behaviors), and mechanical incompatibility (physical differences prevent mating between species). Postmating isolating mechanisms include gametic incompatibility (sperm and eggs of different species cannot fuse), hybrid inviability (hybrid offspring do not survive), and hybrid infertility (hybrid offspring are sterile).

Chapter 18

Think Critically

Figure 18-2 The presence of oxygen would prevent the accumulation of organic compounds by quickly oxidizing them or their precursors. All of the successful abiotic synthesis experiments used oxygen-free "atmospheres."

Figure 18-5 The bacterial sequence would be most similar to that of the plant mitochondrion, because (as the descendant of the immediate ancestor of the mitochondrion) the bacterium shares with the mitochondrion a more recent common ancestor than it does with the chloroplast or the nucleus.

Figure 18-8 Today's ferns, horsetails, and club mosses are small most likely because of competition with seed plants, which had not yet arisen during the period when ferns and club mosses reached large sizes. After seed plants arose, competition from them eventually eliminated other types of plants from many ecological niches, presumably including those niches that favored evolution of large size.

Figure 18-9 No. The mudskipper merely demonstrates the plausibility of a hypothetical intermediate step in the proposed scenario for the origin of land-dwelling tetrapods. But the existence of a modern example similar in form to the hypothetical intermediate form does not provide information about the actual identity of that intermediate form.

Figure 18-19 The African replacement hypothesis. These fossils are the oldest modern humans found so far, and their presence in Africa suggests that modern humans were present in Africa before they were present anywhere else, which, if true, would mean that they originated in Africa.

Doing Science: Discovering the Age of a Fossil

The rock is 713 million years old. (A 1:1 ratio means that one-half of the original uranium-235 is left, so it has reached its half-life.)

Multiple Choice

1. d; **2.** b; **3.** b; **4.** b; **5.** a

Fill-in-the-Blank

1. anaerobic; photosynthesize; poisonous (toxic, harmful), aerobic, energy
2. RNA (ribonucleic acid), enzymes (catalysts), ribozymes
3. eukaryotic; endosymbiotic; DNA (deoxyribonucleic acid)
4. swimming, moist (wet); pollen
5. conifers; wind; flowers, insects; efficient
6. arthropods, exoskeletons, drying out
7. reptiles, eggs, skin, lungs

Review Questions

1. Evidence for prebiotic evolution is that organic molecules can be synthesized and can polymerize spontaneously under prebiotic conditions, and membrane-like structures can form spontaneously under prebiotic conditions.
2. The origin of water-based photosynthesis introduced significant amounts of oxygen gas to the atmosphere for the first time, which set the stage for the evolution of aerobic metabolism and multicellularity.
3. The endosymbiont hypothesis states that cells acquired the precursors of mitochondria and chloroplasts by engulfing certain types of bacteria. These cells and the bacteria trapped inside them gradually entered into a symbiotic relationship. Mitochondria may have evolved from engulfed aerobic bacteria. Chloroplasts may have evolved from engulfed photosynthetic cyanobacteria.
4. In plants, multicellularity conferred the advantage of being larger, which made it more difficult for unicellular predators to swallow them. Additionally, multicellularity conferred the ability of anchorage on plants. In animals, multicellularity conferred the ability to consume larger prey and also provided for greater locomotion, allowing animals to efficiently escape their predators.
5. The advantages of terrestrial existence for the first plants included abundant light, nutrients, and few predators. Disadvantages included limited water, a need to support their body weight, and problems in reproduction without an aquatic environment for sperm to swim in. The advantages of terrestrial existence for the first land animals included an abundant food source in newly evolved land plants and few predators. Disadvantages included having to develop an ability to breathe, to support body weight, and to survive with limited water.
6. Amphibians evolved both lungs and legs, which better adapted them to a terrestrial environment. However, their skin needed to be kept moist, and they deposited their sperm in water. Reptiles evolved internal fertilization, waterproof eggs, and waterproof skin, which allowed them to distance themselves from dependency on bodies of water. Birds evolved feathers as a means of insulation, which allowed them to inhabit cooler environments. Mammals evolved body hair for the same reason.
7. Early primates had grasping hands, which allowed them to perform powerful and precise manipulations. Early primates also evolved binocular vision, which provided accurate depth perception for moving from tree to tree. A large brain facilitated hand–eye coordination as well as complex social interactions. Australopithecines evolved bipedalism, which freed their hands from the task of walking. Later brain expansion allowed them to fashion and manipulate tools.

Chapter 19

Think Critically

Figure 19-3 The finding suggests that chromosome 2 arose from the fusion of two separate chromosomes, each of which contained a centromere. The two ancestral chromosomes must have been present in the common ancestor of chimpanzees and humans, and chimpanzees retained the separate chromosomes.

Figure 19-5 Over their long evolutionary histories, both bacteria and archaea retained a single-celled, prokaryotic structure, the complexity of which is limited by a lack of organelles. This simple structure limits the variety of forms that such an organism can take and still survive. The limited array of possible options made the evolution of similar forms in the two domains likely.

Multiple Choice

1. b; **2.** a; **3.** d; **4.** c; **5.** c

Fill-in-the-Blank

1. taxonomy; systematics; clade
2. genus, species; Latin; capitalized, italic
3. domain, kingdom, phylum, class, order, family, genus, species; Archaea, Bacteria, Eukarya
4. anatomy, DNA sequence
5. reproduce asexually; phylogenetic species concept
6. 1.6 million, 8.7 million

Review Questions

1. Linnaeus laid the groundwork for the modern classification system, using the resemblance of organisms to classify them into hierarchical categories. He developed binomial nomenclature, identifying each organism by a scientific name consisting of the genus and species. Darwin added to the significance of the taxonomic categories established by Linnaeus by indicating evolutionary relatedness.
2. DNA sequences; skeleton, muscle, and organ anatomy; and physiological systems.
3. Compare fossil skeletons (or ancient DNA if available) of cave bears to modern skeletons (or DNA) of grizzly bears and black bears.
4. There are many inconspicuous species living in areas that have not been investigated very thoroughly. Less-investigated areas, such as the Tropics, have high biodiversity. Less-investigated groups, such as fungi and bacteria, are very diverse.
5. Use of universal scientific names prevents the confusion of using regional or common names.
6. The biological species concept does not apply to asexually reproducing organisms, so systematists cannot designate species by testing for reproductive isolation.

Chapter 20

Think Critically

Figure 20-4 Protective structures like endospores are most likely to evolve in environments in which protection is especially advantageous. Compared to other environments inhabited by bacteria, soils are especially vulnerable to drying out, which can be fatal to unprotected bacteria. Bacteria that could resist long dry periods would gain an evolutionary advantage.

Figure 20-5 Enzymes from bacteria that live in hot environments are active at high temperatures (temperatures that usually denature enzymes in organisms from more temperate environments). This ability to function at high temperatures makes the enzymes useful in test tube reactions (such as the polymerase chain reaction) that are run at high temperatures.

Figure 20-7 The main advantage is efficiency. In prokaryotic fission, every individual produces new individuals. In sexual reproduction, only some individuals (e.g., females) produce offspring. So the average individual of a species produces twice as many offspring by fission as it would by sexual reproduction.

Figure 20-9 The concentration of nitrogen gas would increase, because the major process for removing atmospheric nitrogen would end, while the processes that add nitrogen gas to the atmosphere would continue.

Figure 20-12 Viruses lack ribosomes and the rest of the "machinery" required to manufacture proteins.

Figure 20-13 Viruses replicate by integrating their genetic material into the host cell's genome. Thus, if biotechnologists can insert foreign genetic material into a virus, the virus will naturally tend to transfer the foreign genes to the cells they infect.

Health Watch: Is Your Body's Ecosystem Healthy?

Given the increasing evidence that the composition of the microbiome is associated with the health of the digestive system, we might expect to find that the microbiomes of the healthy twins in the sample differ from those of the sick twins. Because a microbiome is an ecosystem, and healthier ecosystems typically have greater species diversity, the microbiomes of the healthy twins might well be more diverse than those of the sick twins. The researchers compared sets of twins to help rule out the possibility that the differences between the healthy and sick twins in a pair were due to genetic differences between them (identical twins are genetically identical).

Multiple Choice

1. a; **2.** d; **3.** b; **4.** c; **5.** a

Fill-in-the-Blank

1. Bacteria, cell walls, archaea
2. smaller; spherical, rod-shaped, corkscrew-shaped
3. flagella; biofilms; endospores
4. Anaerobic; Photosynthetic
5. prokaryotic fission, conjugation
6. nitrogen-fixing; cellulose
7. bacteria; meat, eggs, produce
8. DNA, RNA, protein; host; bacteriophage

Review Questions

1. Bacteria acquire energy through photosynthesis or by the breakdown of organic or inorganic molecules. Bacteria acquire nutrients through the breakdown of organic or inorganic molecules and by nitrogen fixation.
2. Nitrogen-fixing bacteria inhabit the roots of legume plants and capture nitrogen gas, releasing ammonia, which is important in plant nutrition and ultimately our own nutrition. This process is critical to the nitrogen cycle.
3. Prokaryotes are found in extremely hot environments such as boiling hot springs and deep-sea hydrothermal vents; in extremely cold environments such as Antarctic ice; in extremely salty environments such as the Dead Sea; and in extremely acidic or alkaline waters. In and on humans, prokaryotes live in the mouth, nose, urogenital tract, digestive system, and most parts of the skin.
4. An endospore is a protective resting structure of many rod-shaped bacteria that allows them to survive extremely unfavorable conditions for long periods of time.
5. Conjugation is the transfer of genetic material from one prokaryote to another. In most cases, the transferred material consists of small, circular pieces of DNA known as plasmids.
6. Prokaryotic species can extract energy from many different types of molecules, including those that are environmental pollutants and that cannot be broken down by eukaryotes.
7. A virus consists of a nucleic acid molecule (DNA or RNA) surrounded by a protein coat. Viruses replicate inside host cells using the replication machinery of the host cell.
8. Helpful: Act as decomposers; bioremediation of toxic pollutants; act as beneficial gut residents; fix nitrogen for crops; help domestic animals digest cellulose. Harmful: Produce toxins; cause food to rot or spoil; cause diseases.
9. Bacterial cell walls contain peptidoglycan, archaeal cell walls do not; the structure of bacterial ribosomal RNA differs from that of archaeal ribosomal RNA; bacteria and archaea use different versions of the enzymes that synthesize RNA (RNA polymerases). Viruses are much smaller than prokaryotes; the genetic material of prokaryotes is always double-stranded DNA, virus genetic material may be single- or double-stranded DNA or single- or double-stranded RNA; viruses cannot replicate outside of a host cell, prokaryotes can replicate independently; prokaryotes have ribosomes, viruses do not.

Chapter 21

Think Critically

Figure 21-1 Sex is the process that combines the genomes of two different individuals. In plants and animals, this mixing of genomes occurs only during reproduction. But in many protists (and prokaryotes), genome mixing may occur through conjugation and other processes that take place independently of reproduction (which in many cases occurs by mitotic cell division).

Health Watch: Neglected Protist Infection

Testing for Chagas disease would probably be most appropriate for women who have spent time in areas where triatomine bugs are present or whose parents have lived in such areas. Toxoplasmosis is not so clearly tied to a particular risk factor, so more widespread testing during pregnancy might be warranted. However, women with cats in their households may be at particular risk.

Multiple Choice

1.c; **2.** b; **3.** d; **4.** a; **5.** d

Fill-in-the-Blank

1. decomposers; parasites
2. algae; protozoa
3. secondary endosymbiosis; photosynthetic protist
4. apicomplexan (or alveolate); kinetoplastid
5. water mold (or oomycete or stramenopile); amoebozoan
6. dinoflagellates, diatoms; chlorophytes

Review Questions

1. Prokaryotes lack most organelles, whereas protists, as single-celled eukaryotes, contain most membrane-bound organelles.
2. Secondary endosymbiosis is the evolutionary process by which previously nonphotosynthetic protist species acquired chloroplasts by engulfing and retaining photosynthetic protists.
3. Dinoflagellates are an important food source for larger organisms. Upon fast reproduction of dinoflagellates, a "red tide" may occur, causing the death of thousands of fish due to oxygen depletion.
4. Unicellular algae account for nearly 50% of the photosynthetic activity on Earth.
5. apicomplexans
6. red algae, brown algae (stramenopiles), and chlorophytes
7. rhizarians and amoebozoans

Chapter 22

Think Critically

Figure 22-5 Bryophytes lack lignin (which provides stiffness and support) and xylem and phloem (conducting tissues that transport materials to distant parts of the body). Xylem, phloem, and stiff stems seem to be required to achieve heights greater than a few inches.

Figure 22-7 All of the pictured structures are sporophytes. In ferns, horsetails, and club mosses, the gametophyte is small and inconspicuous.

Figure 22-9 The most common adaptations are hard, protective shells and incorporation of toxic and/or distasteful chemicals.

Figure 22-12 Angiosperms

Type of Pollination	Advantages	Disadvantages
Wind	Not dependent on presence of animals; no investment in nectar or showy flowers; pollen can disperse over large distances	Larger investment in pollen because most fails to reach an egg; higher chance of failure to fertilize any egg
Animal	Each pollen grain has much greater chance of reaching suitable egg	Depends on presence of animals; must invest in nectar and showy flowers

Both types of pollination persist in angiosperms because the cost–benefit balance, and therefore the most adaptive pollination system, differs depending on the ecological circumstances of a species.

Health Watch: Green Lifesaver

Although the initial study took advantage of the traditional healer's experience in devising effective methods of preparing and administering herbal remedies, the design of a follow-up study should probably include methods to standardize preparations and treatments so as to test the effects of different preparations and different dosages. It would also be a good idea to introduce control groups, double-blind testing procedures, and standardized methods for assessing patients' condition.

Case Study Revisited: Queen of the Parasites

Photosynthesis is a very useful adaptation, but it comes with costs, such as the energy expenditure required to access and acquire the nutrients needed to produce the molecules and structures used in photosynthesis. In environments in which the necessary nutrients are scarce or competition for them is especially intense, a nonphotosynthetic plant might gain an advantage, provided it evolved an alternative means of acquiring energy, such as by stealing it from other photosynthetic plants.

Multiple Choice

1.; a; **2.** c; **3.** b; **4.** a; **5.** d

Fill-in-the-Blank

1. green algae; nonvascular plants (bryophytes), vascular plants (tracheophytes); embryos, alternation of generations
2. cuticle, stomata, water; xylem and phloem, lignin; water, nutrients
3. swim to the egg; seeds, pollen; gymnosperms, angiosperms; attract pollinators; facilitate seed dispersal
4. hornworts, liverworts, mosses; club mosses, horsetails, ferns; angiosperms

Review Questions

1. Plants produce alternating diploid and haploid generations. These two generations are called the sporophyte and gametophyte generations, respectively. The sporophyte plant produces haploid spores by meiosis, and the haploid plant produces haploid gametes by mitosis that fuse to form a diploid zygote, which begins the next sporophyte generation.
2. Male gametes evolved into pollen grains that allowed them to be carried on air currents; the evolution of seeds protected the developing embryo; the evolution of fruits improved seed dispersal; and the evolution of flowers improved the efficiency of pollination.
3. As evolution in the plant kingdom progressed, there was a shift from a small sporophyte that was dependent on a dominant gametophyte to a dominant sporophyte containing a small gametophyte generation within it.
4. Green plants probably arose from green algae, because the closest living relatives of plants are a type of green alga (the stoneworts) and because green algae use the same type of chlorophyll and accessory pigments as do land plants, have cell walls made of cellulose, and store food in starch. The ancestral green algae lived in fresh water, where they faced environmental challenges that resulted in adaptations that helped them cope with the challenges of dry land.
5. Structural adaptations to dry land included the development of roots or rootlike structures that anchor the plant and absorb water and nutrients from the soil (present in nonvascular plants, ferns, gymnosperms, angiosperms); conducting tissues that transport water and minerals up from the roots and move photosynthetic products from the leaves throughout the rest of the plant (present in ferns, gymnosperms, angiosperms); lignin-impregnated cell walls to support the plant body (present in ferns, gymnosperms, angiosperms); a waxy cuticle to limit evaporation of water (present in nonvascular plants, ferns, gymnosperms, angiosperms); and stomata to allow gas exchange and limit water loss when water is scarce (present in nonvascular plants, ferns, gymnosperms, angiosperms).
6. Success of angiosperms can be attributed to flowers, fruit, and broad leaves. Flowers attract pollinators, fruit assists in seed dispersal, and broad leaves allow for greater rates of photosynthesis.

7. Retaining green leaves throughout the year (evergreens) allows gymnosperms to continue photosynthesizing and grow slowly during times in which most other plants become dormant. Conifer needles are thin and covered with a thick cuticle that minimizes evaporation. Conifers produce an "antifreeze" in their sap that enables them to continue transporting nutrients in below-freezing temperatures.
8. Pollen grains are male gametophytes containing sperm. Pollen grains protect the sperm and produce a pollen tube that delivers the sperm to the ovule, bypassing the requirement of water for fertilization, thus allowing for colonization of dry land.
9. Seeds provide waterproof protection and nourishment for the developing embryos. Some seed coat modifications aid in seed dispersal. Plants that lack seeds produce spores.

Chapter 23

Think Critically

Figure 23-1 Its filamentous shape helps the fungal body penetrate and extend into its food sources and also maximizes the ratio of surface area to interior volume (which maximizes the area available for absorbing nutrients). The extreme thinness of the filaments ensures that no cell is very far from the surface at which nutrients are absorbed.

Figure 23-15 Compared to plants that lack mycorrhizae, plants that have them are much more effective at absorbing water and nutrients from the soil. Therefore, when early plants were first spreading over Earth's land, individuals whose roots were associated with fungi would have gained an advantage over those that lacked such associations.

Figure 23-20 In nature, bacteria compete with fungi for access to food and living space. The antibiotic chemicals produced by fungi serve as a defense against competition from bacteria. New antibiotics are most likely to be found in environments in which fungi and bacteria coexist, such as soil. One might begin a search by testing extracts of various candidate fungi to see if they kill bacterial cultures of different types.

Earth Watch: Killer in the Caves

The animal most likely to carry white-nose syndrome from cave to cave is people. Spelunkers (people who explore caves) can carry fungal spores between caves on their boots and equipment. So one way to slow the spread of the disease would be to ban people from exploring caves (and in fact such bans are currently active in many places).

Multiple Choice

1. c; **2.** d; **3.** b; **4.** c; **5.**b

Fill-in-the-Blank

1. reproduction; spores
2. mycelium, hyphae, septa; chitin
3. mitotic, one; meiotic, zygote, one
4. glomeromycetes; chytrids, rumen fungi, and blastoclades; basidiomycetes
5. Lichens; Mycorrhizae; endophytes
6. wood (cellulose and lignin); yeasts; athlete's foot, jock itch, vaginal infections, ringworm, histoplasmosis, valley fever (many possible answers)

Review Questions

1. The structure of the fungal body is a mycelium, which is an interwoven mass of thread-like filaments one cell thick called hyphae. Fungal cells differ from plant cells in that fungal cells are not photosynthetic and fungal cell walls contain chitin. Fungal cells are different from animal cells in that fungal body cells are usually haploid and produce asexual spores.
2. Reproductive structures are represented by mushrooms, puffballs, and similar structures; they are elevated above the ground for spore dispersal.
3. Dutch elm disease and chestnut blight are caused by ascomycetes.
4. Fungi that attack crops include molds on fruit (Ascomycota) and smuts and rusts (Basidiomycota).
5. Many fungi reproduce asexually through spore production. Haploid hyphae produce haploid spores by mitotic cell division. The spores germinate to produce new haploid hyphae.
6. The major groups of fungi are Chytridiomycota (chytrids), Neocallimastigomycota (rumen fungi), Blastocladiomycota (blastoclades), Glomeromycota (glomeromycetes), Basidiomycota (basidiomycetes), and Ascomycota (ascomycetes). Refer to Table 23-1 for descriptions of the characteristics of each group and examples.
7. A mushroom basidiospore germinates and produces haploid hyphae of two different mating types. When the two types meet, cells fuse and form an underground mycelium. The hyphae grow outward in a circular pattern and periodically send up numerous mushrooms that appear in a ring-like pattern. The diameter is related to the age of the fungus because the mycelium continues to grow outward as the innermost hyphae die off.
8. Mycorrhizae are fungi associated with plant roots. The fungi absorb nutrients and water that are then passed on to the plant. The fungi derive photosynthetic products from the plant roots. Lichens consist of a combination of fungi and single-celled green algae or cyanobacteria. Excess sugar produced by the bacteria provides nutrition for the fungi; the fungal body provides moisture and protection for the photosynthetic algae or bacteria.

Chapter 24

Think Critically

Figure 24-6 Sponges are "primitive" only in the sense that their lineage arose early in the evolutionary history of animals and their body plan is comparatively simple. But early origin and simplicity do not determine effectiveness, and the sponge body plan and way of life are clearly suitable for excellent survival and reproduction in many habitats.

Figure 24-13 Parasitic tapeworms have no gut and absorb nutrients across their body surfaces. Their ribbon-like shape maximizes surface area for absorption and allows the worm body to extend through the greatest possible area of the host's body (to be in contact with as many nutrients as possible).

Figure 24-14 Two openings allow one-way travel of food through the gut, which allows continuous feeding. One-way movement allows more efficient digestion than two-way movement; digestive waste from which all nutrients have been extracted can be excreted quickly without the need for reverse travel back along the gut, and food can be processed more quickly.

Figure 24-15 Water travels easily through the moist epidermis of a leech. When a high concentration of salt is dissolved in the moisture on the outside of a leech's body, water moves rapidly out of the leech's body by osmosis, dehydrating and ultimately killing the animal.

Figure 24-24 The mobility provided by flight may have allowed ancestral insects to more easily exploit new food sources, habitats, and geographic areas. This ability to disperse would have promoted formation of new species and increased population sizes.

Earth Watch: When Reefs Get Too Warm

Sediments can negatively affect reef-building corals in two ways. (1) Suspended sediments make the water murky, so less light passes through, reducing the light available to fuel photosynthesis in corals' dinoflagellate symbionts. (2) Sediments can settle directly on the corals, forcing them to spend a lot of energy keeping their surfaces clean, or even burying them so completely that they suffocate.

Doing Science: Searching for a Sea Monster

Widder hypothesized that a giant squid's preferred food is not jellyfish, but the small (compared to a giant squid) predators that eat jellyfish. The observation that a giant squid attacked an object near the imitation-jelly lure rather than the lure itself is consistent with the hypothesis's prediction that giant squids will attack animals near jellyfish, rather than jellyfish themselves.

Multiple Choice

1. b; **2.** a; **3.** b; **4.** c; **5.** a

Fill-in-the-Blank

1. consuming other organisms, sexually, cell walls
2. three, endoderm, mesoderm, ectoderm; two, mesoderm
3. cephalized; sensing the environment, ingesting food; cavity, mesoderm
4. protostome; deuterostome, echinoderms, chordates
5. invertebrates, vertebrates; invertebrates; sponges, single-celled organisms; cnidarians; annelids
6. bivalves, gastropods, cephalopods; arthropods; insects, arachnids, crustaceans
7. annelids, arthropods (chordates is also correct); closed circulatory; open circulatory, hemocoel
8. mollusks, cnidarians, echinoderms, arthropods

Review Questions

1. Animals are eukaryotic, multicellular, and motile; obtain energy by consuming other organisms; reproduce sexually; and respond rapidly to stimuli. Their cells lack cell walls.
2. Refer to Table 24-1.
3. In bilateral symmetry only one plane through a central axis divides the animal into equal halves. This is an adaptation for active movement.

   In radial symmetry any plane through a central axis divides the animal into roughly equal halves. Having a body facing all directions at once is advantageous.

   Cephalization is the concentration of sensory organs and a brain in a defined head region. It is important for the evolution of complex behavior.

   In a closed circulatory system, blood remains confined to the heart and blood vessels. It allows for more efficient distribution of gases and nutrients throughout the body than does an open circulatory system.

   The coelom is the fluid-filled body cavity. This cavity freed the gut from the constraints imposed by direct attachment to the body wall and created a space in which new organ systems could develop.

   Segmentation is repeated similar body parts. It allows the body to increase in size with a minimum of new genetic information.
4. Gills are thin, external respiratory membranes used in aquatic arthropods (the crustaceans). Tracheae are a network of narrow, branching respiratory tubes found in insects and myriapods. Arachnid respiratory structures include both tracheae and lungs.
5. An exoskeleton protects against predators, allows for increased agility, and provides a watertight covering. However, an exoskeleton cannot expand as an animal grows, and in the process of molting, the animal becomes vulnerable until the new exoskeleton hardens.
6. **a.** two hinged shells—bivalves
   b. a radula—gastropods
   c. tentacles—cephalopods
   d. some sessile members—bivalves
   e. the best-developed brains—cephalopods
   f. numerous eyes—cephalopods
7. The water-vascular system functions in locomotion, respiration, and food capture.
8. Sea wasp (cnidarian); tapeworm (flatworm); fluke, e.g., *Schistosoma* (flatworm); bees and wasps (arthropods); hookworm (roundworm); *Trichinella* (roundworm)

Chapter 25

Think Critically

Figure 25-8 A freshwater fish's body is immersed in a hypotonic solution, so water tends to continuously enter the body by osmosis. The physiological challenge is to get rid of all this excess water. For a saltwater fish, the challenge is reversed. The surrounding solution is hypertonic, so water tends to leave the body. The physiological challenge is to retain sufficient water.

Figure 25-10 One advantage is that adults and juveniles occupy different habitats and therefore do not compete with one another for resources (the niche occupied by an individual over its lifetime is broadened).

Figure 25-13 Flight is a very expensive trait (consumes a lot of energy, requires many special structures). In circumstances in which the benefits of flight are low, such as in habitats without predators, natural selection may favor individuals that forgo an investment in flight, and flightlessness can arise.

Earth Watch: Frogs in Peril

On the graph in the figure, the annual rate of decrease for endangered species appears to be about 12%. Assuming a decrease of 12% per year, the population after 10 years will be about 278. (Remember that each year's starting population differs from the starting population from the prior year.) A graph of population size extrapolated to 50 years shows that the population will shrink to nearly zero, assuming a constant rate of decrease (that is, exponential decrease).

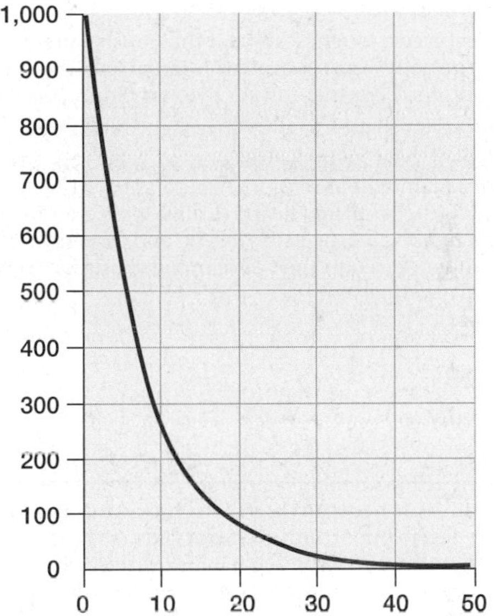

Multiple Choice

1. c; **2.** b; **3.** d; **4.** c; **5.** b

Fill-in-the-Blank

1. hollow, dorsal; anus, notochord
2. tunicates, lancelets, hagfishes; skull; hagfishes
3. lungfishes; cartilage; ray-finned fishes; jaws
4. mammals, amphibians, reptiles, amphibians
5. monotremes; salamanders; bats

Review Questions

1. The vertebral column is a bony column that encloses the spinal cord; it provides protection and support that allows increased body size and faster locomotion. A jaw is a bone or cartilage structure that forms the opening to the mouth; jaws allow more efficient feeding and exploitation of a wide range of foods. Limbs are jointed appendages; they allow varied forms of locomotion. Modifications of limbs enabled vertebrates to colonize land. An amniotic egg is an egg in which the embryo is surrounded by protective membranes; an amniotic egg allows embryos to develop in moist conditions even on land. Feathers are modified scales found in birds; they are used for insulation and in flight. The placenta is a structure, found in mammals, that is derived from a combination of embryonic membrane and the lining of the uterus; the placenta allows the exchange of nutrients, gases, and wastes between an embryo and its mother.
2. **a.** A skeleton of cartilage is found in cartilaginous fishes and lampreys.

b. A two-chambered heart is found in ray-finned fishes, cartilaginous fishes, coelacanths, lungfishes, lampreys, and hagfishes.
c. An amniotic egg is found in reptiles and mammals.
d. Endothermy is found in birds and mammals.
e. A four-chambered heart is found in crocodilians, birds, and mammals.
f. A placenta is found in mammals (except monotremes).
g. Lungs supplemented by air sacs are found in birds.

3. Four distinguishing features of the chordates are (1) a dorsal, hollow nerve cord; (2) a notochord; (3) pharyngeal gill slits; and (4) a post-anal tail.
4. Amphibians are adapted to dry land in that they possess lungs; however, they require their skin to be kept moist because if their moist skin dries out, it becomes impermeable to oxygen, and the animal suffocates. Amphibians are tied to moist habitats by their breeding behavior, which requires water to keep eggs moist and allow sperm to swim to eggs.
5. Reptiles evolved a tough, scaly skin that resists water loss and protects the body; reptiles evolved internal fertilization and a shelled amniotic egg that can survive away from water.
6. Adaptations for flight include wings, lightweight bones, reduced reproductive organs, and feathers.
7. Like birds, mammals are endothermic and have high metabolic rates and four-chambered hearts. Unlike birds, mammals have evolved a remarkable diversity of form and possess mammary glands, and most retain their developing young in the uterus (placental mammals).

Chapter 26

Think Critically

Figure 26-1 One possibility is that the variation necessary for selection has never arisen. (If no members of the species by chance gain the ability to discriminate between their own chicks and cuckoo chicks, then selection has no opportunity to favor the novel behavior.) Another possibility is that the cost of the behavior is relatively low. (If parasitism by cuckoos is rare, a parent that feeds any begging chick in its nest is, on average, much more likely to benefit than suffer.)

Figure 26-24 If females do indeed gain any benefits from their mates, those benefits would necessarily be genetic (as the male provides no material benefits). Male fitness may vary, and to the extent that this fitness can be passed to offspring, females would benefit by choosing the most-fit males. If a male's fitness is reflected in his ability to build and decorate a bower, females would benefit by preferring to mate with males that build especially good bowers.

Figure 26-25 Canines forage mainly by smell; mandrills are mostly visual foragers. Modes of sexual signaling are affected not only by the nature of the information to be encoded, but also by the sensory biases and sensitivities of the species involved. Communication systems may evolve to take advantage of traits that originally evolved for other functions.

Multiple Choice

1. c; **2.** a; **3.** b; **4.** a; **5.** d

Fill-in-the-Blank

1. genes, environment; stimulus; innate
2. energy, predators; practice, skills; adult; brains
3. habituation, repeated; imprinting; sensitive period
4. aggressive; displays, injuring (wounding, damaging); weapons (such as fangs, claws), larger
5. territoriality (territorial behavior); mate, raise young, feed, store food; male; same species
6. For pheromones, any of the following advantages are correct: long lasting, use little energy to produce, are species specific, don't attract predators, may convey messages over long distances. Any of the following disadvantages are correct: cannot convey changing information, convey fewer different types of information. For visual displays, any of the following advantages are correct: instantaneous, can change rapidly, many different messages can be sent, quiet. Any of the following disadvantages are correct: make animals conspicuous to predators, generally ineffective in darkness or dense vegetation, require that recipient be in visual range (close by).

Review Questions

1. Play has survival value, as young animals gain experience in a variety of behaviors that they will use as adults in hunting, fleeing, or social interactions. (1) Play seems to lack any clear goal. (2) Play is abandoned in favor of escaping from danger, feeding, or courtship. (3) Young animals play more frequently than adults. (4) Play often involves movements borrowed from other behaviors. (5) Play uses considerable energy. (6) Play is potentially dangerous.
2. Animals communicate through the senses of sight, sound, smell, and touch. Female mandrills communicate sexual receptivity by a brightly colored swelling of their buttocks. Monkeys use different calls to warn others of different types of predators. Insects use pheromones that are sensed through smell as sex attractants. Primates use touch to establish social bonds. Visual communication is instantaneous and conveys a great amount of information, but it can be used only over relatively short distances and is ineffective in the dark. Auditory communication is also instantaneous and can also be conveyed in the dark and in dense forests where visual communication is difficult. Making sounds alerts predators to your presence, however, and consumes a lot of energy. Chemical communication can be conveyed over long distances and takes little energy to produce. However, an individual can produce only a limited number of different chemical messages, and chemical messages cannot be quickly changed. Touch is important in developing social bonds and is often required for sexual activity, but can only be used between animals that are very close together.
3. Territories are usually defended against members of the same species who compete most directly for the resources being protected.
4. Natural selection favors the evolution of symbolic displays or rituals for resolving conflicts. Aggressive displays allow competitors to assess each other and acknowledge a winner on the basis of size, strength, and motivation rather than on the wounds they inflict.
5. Advantages of group living include (1) increased ability to detect, repel, and confuse predators; (2) increased hunting efficiency or increased ability to spot localized food resources; (3) potential for division of labor; (4) conservation of energy; and (5) increased likelihood of finding mates. Disadvantages of group living include (1) increased competition within the group for limited resources; (2) increased risk of infection from contagious diseases; (3) increased risk that offspring will be killed by other members of the group; and (4) increased risk of being spotted by predators.
6. Behaviors that are present in newborns or that are widely shared among diverse cultures may be innate.

Chapter 27

Think Critically

Figure 27-3 Many variables interact in complex ways to produce real population cycles. Weather, for example, affects the lemmings' food supply and thus their ability to survive and reproduce. Predation of lemmings is influenced by both the number of predators and the availability of other prey, which in turn is influenced by multiple environmental variables.

Figure 27-10 Emigration relieves population pressure in an overpopulated area, spreading the migrating animals into new habitats that may have more resources. Human emigration within and between countries is often driven by the desire or need for more resources, although social factors—such as wars and religious or racial persecution—also fuel human emigration. (This subjective question can lead to discussion of the extent to which overpopulation may drive human emigration.)

Figure 27-16 When fertility exceeds RLF, there are more children than parents. As the additional children mature and become parents themselves, this larger generation produces still more children, and so on, in a positive feedback cycle.

Figure 27-17 U.S. population growth resembles the rapidly rising "exponential" phase of the S-curve. Stabilization will require some combination of reduction in immigration rates and birth rates. An increase in death rates is less likely, but cannot be ruled out entirely.

Earth Watch: Boom-and-Bust Cycles Can Be Bad News

Explosive population growth of non-toxic photosynthetic microorganisms can produce a population so large that respiration by its living members, and by the bacteria that decompose its dead members, uses up almost all of the oxygen in the water. Without oxygen, most nearby aquatic organisms cannot survive.

Case Study Revisited: The Return of the Elephant Seals

Ordinarily, if there is a sudden environmental change, such as the appearance of an infectious disease, some members of a population will, by chance, carry alleles that help them survive and thrive in the changed environment. But in a population with low genetic diversity, there are few alleles, and the chance that a newly favorable one will happen to be present is small. Thus, populations with low genetic diversity may not be able to adapt to changing conditions and are therefore at risk. The threat from new infectious diseases is especially concerning.

Multiple Choice

1. b; **2.** c; **3.** a; **4.** d; **5.** c

Fill-in-the-Blank

1. survivorship curves, early-loss, late-loss
2. exponential; no; J-curve; no
3. carrying capacity; logistic, S-
4. clumped; uniform
5. births, immigrants, deaths, emigrants; increases

Review Questions

1. *Biotic potential* is the maximum rate of population growth, assuming ideal conditions allowing a maximum birth rate and a minimum death rate. Factors that influence biotic potential include the age at which the organism first reproduces, the frequency of reproduction, the average number of offspring produced each time the organism reproduces, the length of the organism's reproductive life span, and the death rate of individuals under ideal conditions. Natural selection may favor a high biotic potential because many species evolved under conditions where (because of environmental resistance) many offspring did not live to reproduce.
2. G (population growth) $= r$ (growth rate) $\times N$ (population size at the beginning of the time under consideration). This equation tells us that the change in population size depends on its rate of growth and its size. Because N will increase during each calculated time increment, this causes G to increase with each increment as well. If r remains constant, this equation will produce exponential growth.
3. These conditions produce exponential growth (see Figs. 27-1 and 27-4), during which the population adds increasing numbers of individuals during each successive time increment.
4. *Environmental resistance* consists of limits on population growth that are set by the living and nonliving environment. Density-dependent forms of environmental resistance strengthen as the population density increases. Density-dependent forms include competition and consumer–prey interactions such as predation and parasitism. Predation is density dependent because the more prey there are, the more predator–prey encounters there will be. Additionally, predators tend to prey on organisms that are most abundant. Parasitism is density dependent because most parasites have limited mobility and spread more readily among hosts at higher population densities. Competition both within a species and between species for resources such as food, water, and living and breeding spaces is also density dependent. Density-independent forms of environmental resistance (weather, climate, fire, drought, hurricanes, floods, habitat destruction) reduce or limit population size regardless of population density.
5. Logistic growth refers to the growth of a population in a region with finite resources, such as food and space. As the population increases, it encounters increasingly significant density-dependent environmental resistance, causing its growth rate to slow and then stop. The population size (N) stabilizes around carrying capacity (K), producing an S-curve, as is illustrated in Figure 27-5.
6. Exceeding carrying capacity damages an ecosystem, reducing its ability to support a population. Population growth can be exponential until critical resources within the ecosystem are damaged or eliminated, and then population size declines. The decline may be temporary until the ecosystem recovers, or the damage may permanently reduce K. In extreme cases, the population might disappear entirely. Refer to Figure 27-7.
7. A concave survivorship curve is characteristic of r-selected species that produce large numbers of young and provide them with few resources and little or no parental care, producing high mortality among the young. Those few that survive to adulthood have a reasonable chance of living until old age. A convex survivorship curve is characteristic of K-selected species that have small numbers of offspring that are provided with resources and protection. Many of these young survive to fairly old age. Death rates are low among the young and increase rapidly in older age groups.
8. Since prehistoric times, people have circumvented environmental resistance using brainpower and dexterity. We increased our space because clothing and shelter allowed us to spread worldwide (now we can also live atop one another in skyscrapers). Food has been increased by better hunting and fishing tools and by agriculture. The impact of disease has been reduced by medical advances including vaccines and antibiotics. This cannot continue indefinitely because Earth's resources are finite, so at some point, even with continuing technological advances, a growing population would indeed hit an absolute carrying capacity limit.
9. Figure 27-15 shows the three age structures. Over time, children reach reproductive age and have their own families. If there are more children than current parents, there will be more women having babies in the future. If the numbers of parents and children are equal, the population will be stable, and if there are fewer children than parents, as those children reach reproductive age, there will be fewer women having babies.

Chapter 28

Think Critically

Figure 28-6 Monarch caterpillars store toxic chemicals from the milkweeds they eat, so they would likely kill or at least sicken birds or other predators that might eat them. However, the caterpillar would still be dead. The bold stripes of monarch caterpillars probably evolved as a warning to predators, advertising that they are toxic. When a predator has eaten and been sickened by one monarch caterpillar, it would probably not want to eat another. The evolution of bright colors would make it easier for predators to learn to avoid monarch caterpillars. This would enhance survival of the monarchs.

Figure 28-7 Many predators hunt by detecting odors, sounds, or even electric fields, in addition to, or instead of, detecting their prey visually.

Figure 28-8 Ancestors of these organisms that happened to slightly resemble their surroundings would be a bit less likely to be eaten, so they would be slightly more likely to reproduce and leave offspring. Chance mutations that increased their camouflage would enhance their survival and reproduction even more, eventually resulting in the modern, superb camouflage.

Figure 28-17 Fire has been a natural part of the forest environment for many thousands of years. Some forest plants might directly depend on fire for reproduction (for example, opening cones to release seeds) or might benefit from the loss of trees, which lets sunlight hit the ground and reduces competition for water and nutrients. Fire might maintain some forests in permanent, or recurring, subclimax stages.

Figure 28-19 If reasonably deep soil remains after the disturbance, along with at least a few surviving plants to provide seeds, then succession may proceed. But if an ecosystem is severely degraded—for example, by overgrazing that almost completely eliminates native vegetation—then it would be difficult for succession to restore the original community. This is especially true on islands, which may be separated by many miles from sources of native seeds. If invasive species are introduced simultaneously with overgrazing, then the bare soil would provide an opening for seeds of invasive plants to sprout and possibly take over permanently.

Health Watch: Parasitism, Coevolution, and Coexistence

Natural selection imposed by bednets might result in the evolution of mosquitoes that are active earlier in the day. Mosquitoes of a given species have some genetic variability, including variation in genes that influence their preferred time and place of feeding. For simplicity, let's assume that the bednets are 100% effective, so that people are never bitten while sleeping beneath the net. A mosquito that flew about, especially inside houses, earlier in the day would have the opportunity to feed on people before they went to bed beneath a net. These "early" mosquitoes would reproduce more often than the rest of the mosquito population, passing their genes for early feeding on to their offspring. Eventually, one would expect that early mosquitoes would dominate the population, rendering bednets ineffective.

Case Study Revisited: The Fox's Tale

One hypothesis is that the sounds of dogs, which might attack or kill raccoons, made the islands' raccoons more fearful, reducing the amount of time they spent foraging for food on the shoreline. As a result, populations of their prey species in the intertidal zone increased. Community interactions were altered by changes in the raccoons' behavior.

Multiple Choice

1. a; **2.** d; **3.** c; **4.** d; **5.** b

Fill-in-the-Blank

1. natural selection; coevolution
2. carnivores, herbivores; camouflage
3. competitive exclusion
4. warning coloration, startle coloration, Batesian mimicry, aggressive mimicry
5. mutualism (or symbiotic mutualism), parasitism, predation, mutualism
6. succession; primary succession; secondary succession; climax; subclimax

Review Questions

1. An *ecological community* consists of all the interacting populations within an ecosystem. Community interactions that form the basis of a community are interspecific competition, consumer–prey interactions, and mutualism. Competition is an interaction in which individuals of the same or different species attempt to use the same, limited resources. Both species are harmed. In consumer–prey interactions, one species (the consumer) feeds on another species (the prey). The consumer benefits, and its prey is harmed. Consumer–prey interactions include predation and parasitism. Predation is an interaction between species in which one organism eats another; generally, but not always, predators are free-living organisms that are less abundant than their prey. Parasitism is an interaction between species in which one organism lives in or on its host, feeding on the host; generally, parasites are smaller and more abundant than their hosts. Typically, a parasite harms the host but often does not immediately kill it. Mutualism is an interaction between species in which both benefit.
2. If the niches of different species completely overlap, one will become extinct unless the species evolve in ways that reduce niche overlap. This can occur by resource partitioning in which each species specializes on particular resources, reducing competition for the same resources.
3. There are many examples of consumer–prey coevolution, including (1) faster, stronger carnivores and faster, more alert prey; (2) distasteful or toxic prey and consumers able to detoxify the distasteful or toxic chemicals; (3) grasses with tough silica in their leaves and herbivores with longer, stronger teeth; and (4) immune responses to disease-causing bacteria and protective coatings on bacteria that disguise them from immune cells.
4. *Succession* is a process that occurs after an ecological disturbance. It consists of a somewhat predictable series of changes in the community inhabiting an ecosystem over time. Succession occurs as organisms alter their environment in ways that favor the establishment and survival of other species that eventually outcompete the original colonizers. A clear-cut forest would experience secondary succession because a previous ecosystem had already been established by primary succession. Tree seedlings, shrubs, grasses, and other plants as well as soil would remain to hasten the process toward a climax community.
5. Subclimax communities: a suburban lawn, tallgrass prairies. Climax communities: rain forests, deciduous forests, coniferous forests, tundra (see the text for more examples). Climax and subclimax communities differ in that climax communities are relatively stable, whereas subclimax communities usually take active maintenance or recurring natural events (such as fires) to prevent succession from continuing to the natural climax communities.
6. An invasive species is a species introduced into an ecosystem where it did not evolve. Invasive species may harm human health, the environment, or the economy of a region. They are destructive because they lack natural controls on their population growth, often having no effective local predators or competitors. Generally, invasive species reproduce rapidly, disperse widely, are tolerant of a wide range of environmental conditions, and (if animals) eat a variety of foods.
7. A keystone species is a species that plays a major role in determining community structure, out of proportion to its abundance in the community. It is often difficult to identify a keystone species, unless it is removed, resulting in community interactions that are significantly altered and dramatic changes in the relative abundance of other species.

Chapter 29

Think Critically

Figure 29-2 On land, high productivity is supported by optimal temperatures for plant growth, a long growing season, and plenty of moisture, such as is found in rain forests. Lack of water limits desert productivity. Lack of nutrients limits the productivity of the open ocean, even in well-lit surface waters.

Figure 29-8 Humanity's need to grow crops to feed our growing population has led to the fixing of nitrogen for fertilizer using industrial processes. Nitrogen oxides are also generated when fossil fuels are burned in power plants, vehicles, and factories, and when forests are burned. Consequences include the overfertilization of lakes and rivers and the creation of dead zones in coastal waters that receive excessive nutrient runoff from land. Another important consequence is acid deposition, in which nitrogen oxides formed by combustion produce nitric acid in the atmosphere; this acid is then deposited on land.

Figure 29-14 Global temperatures would not begin to decline immediately. CO_2 stays in the atmosphere for years, providing a long lag time before its contribution to the greenhouse effect would decline significantly. Further, there are other greenhouse gases, such as methane, produced by human activities.

Health Watch: Biological Magnification of Toxic Substances

Assuming that the different fish species live in water with similar mercury concentrations, there are two major factors that contribute to the amount of mercury in fish: (1) Their average trophic level: the lower the level, the less mercury. (2) Their age when they are eaten: on average, older fish have eaten more prey and have had more chance to accumulate mercury. For example, catfish tend to have low levels of mercury because they feed at a low trophic level (algae, aquatic plants, insects, and crayfish) and the ones sent to supermarkets and restaurants tend to be young (less than 2 years old). In contrast, albacore tuna tend to have high levels of mercury because they feed at a high trophic level (they are top predators) and are generally eaten when older (2 to 4 years old).

Earth Watch: Monitoring Earth's Health

It isn't a trivial task to "eyeball" a trend line (scientists use computer programs). Further, not all trends are linear: the yearly increase in atmospheric CO_2 concentrations, for example, has been getting larger over time. Nevertheless, if you draw a straight line through the data for Arctic sea ice and, say, the last 25 years for CO_2 concentrations, you will find that the data predict the Arctic to become ice-free sometime in the 2050s, and CO_2 concentrations to have doubled by about 2070—both probably within your lifetime. Actually, if major changes in human activities do not occur, these effects may happen sooner. On the other hand, if people use more renewable energy and less fossil fuel energy, then CO_2 levels may eventually stabilize or even drop, slowing down or reversing the loss of ice.

Multiple Choice

1. b; **2.** a; **3.** d; **4.** d; **5.** c

Fill-in-the-Blank

1. sunlight, photosynthesis; nutrients, nutrient cycles
2. autotrophs, producers; net primary production
3. trophic levels; food chain; food webs
4. 10
5. heterotrophs, consumers; herbivores, primary consumers; carnivores, secondary consumers; detritivores, decomposers
6. nitrogen-fixing bacteria, denitrifying bacteria; ammonia, nitrate
7. atmosphere, oceans; CO_2 (carbon dioxide); limestone, fossil fuels

Review Questions

1. Energy enters ecosystems from sunlight and flows through the trophic levels. As it is used and transferred from trophic level to trophic level, some energy is lost as heat and cannot be reused. Therefore, ecosystems require a continuous input of energy. In contrast, nutrients cycle as they are taken up from their abiotic reservoirs, passed through trophic levels, and moved back to their reservoirs, from which they can be reused.
2. A producer manufactures its own food through photosynthesis. Autotrophs occupy the first trophic level and are almost always photosynthetic organisms. Autotrophs are important because they capture energy and make it available to the rest of the organisms in the ecosystem.
3. *Net primary production* is the energy that photosynthetic organisms capture and store in a given area over a given period of time. A farm pond would have a higher productivity because it has far more nutrients from nearby fields and animal wastes than would an alpine lake. Also, there would be a longer, warmer growing season in the farm pond.
4. The first three trophic levels are (1) producers, (2) primary consumers or herbivores, and (3) secondary consumers or carnivores. Primary consumers are much more abundant than secondary consumers. Because herbivores acquire their energy from plants, and because only a fraction of the energy contained in the plants is available to the herbivores for use in building their own biomass, plants usually have a higher biomass than their consumers do. The 10% law states that the average energy transfer from one trophic level to the next is about 10% efficient (although ecosystems actually vary considerably in the efficiency of energy transfer from trophic level to trophic level).
5. A food chain describes a linear feeding relationship among organisms in an ecosystem, considering a single type of animal as a representative of its trophic level. A food web depicts with far greater accuracy the complex and interconnected feeding relationships within an ecosystem.
6. *Detritivores* are organisms that eat wastes and dead bodies. They include some insects and crustaceans, earthworms, centipedes, and vultures. *Decomposers* are fungi and bacteria that secrete enzymes outside their bodies, where they digest the wastes and dead bodies of other organisms. The decomposers absorb some of the resulting nutrients; the rest remain in the environment. Both groups are important to ecosystems because they release the nutrients trapped in dead bodies and wastes, allowing them to be reused by producers.
7. Carbon is found in its atmospheric reservoir as carbon dioxide (CO_2). CO_2 moves into producers during photosynthesis. The carbon passes from producers through consumers. At each stage, some carbon is returned to its reservoir during cellular respiration, and detritivores and decomposers return the rest during their cellular respiration. Humans, by burning forests and fossil fuels, are adding CO_2 to the atmosphere. CO_2 is a greenhouse gas that traps heat in Earth's atmosphere, causing warmer temperatures and climate change. Earth's future climate will be warmer, with greater weather extremes. These climate changes will probably have major impacts on the biosphere.
8. Nitrogen-fixing bacteria—in soil and water, as well as in the roots of legume plants—combine nitrogen gas in the atmosphere with hydrogen atoms to form ammonia (NH_3). Other bacteria can convert ammonia into nitrate. Plants take up ammonia or nitrate and use the nitrogen to form proteins, nucleic acids, and other biological molecules.
9. Phosphate dissolved from phosphate-rich rocks or from fertilizers enters plants, where its phosphorus becomes incorporated into biological molecules such as nucleic acids and phospholipids. These molecules pass through the trophic levels via consumption of plants by herbivores, and then of herbivores by carnivores. The phosphorus cycle differs from that of carbon and nitrogen because it lacks an atmospheric reservoir.

Chapter 30

Think Critically

Figure 30-2 Earth's longest days of the year are in the Arctic during the northern summer and in the Antarctic during the northern winter (which is the southern summer). Arctic terns spend May through August in the far north, September and October migrating slowly southward, November through February in the far south, and March and April migrating back north to start the cycle again.

Figure 30-8 Nutrients are abundant in tropical rain forests, but they are not stored in the soil. The optimal temperature and moisture of tropical climates allow plants to make such efficient use of nutrients that nearly all nutrients are stored in plant bodies, and to a lesser extent, in the bodies of the animals they support. These growing conditions support a vast array of plants, and these, in turn, provide a wealth of habitats and food sources for diverse animals.

Figure 30-18 Tropical deciduous forests have a dry season in which the soil has little moisture. Temperate deciduous forests have freezing winter weather, in which soil moisture becomes frozen and unavailable to the trees. In both cases, dropping the leaves reduces the trees' water loss from evaporation at a time of year when the water cannot be replaced by absorbing water from the soil.

Figure 30-27 Coastal ecosystems have an abundance of the two limiting factors for life in water: nutrients and light to support photosynthetic organisms. Both upwelling from ocean depths and runoff from the land can provide nutrients, depending on the location of the ecosystem. The shallow water in these areas allows adequate light to penetrate to support rooted plants and/or anchored algae, which in turn provide food and shelter for a wealth of marine life.

Earth Watch: Plugging the Ozone Hole

Ozone depletion would lead to higher UV radiation, which would reduce photosynthesis. Less photosynthesis means that less CO_2 would be removed from the atmosphere. More CO_2 remaining in the atmosphere would increase the greenhouse effect, leading to more global warming and other changes in climate. Increased CO_2 in the atmosphere would also cause more CO_2 to dissolve in the oceans, increasing ocean acidification.

Multiple Choice

1. c; **2.** b; **3.** d; **4.** a; **5.** c

Fill-in-the-Blank

1. seasons; ocean currents *or* the presence of the ocean; rain shadow
2. suitable temperatures, availability of liquid water; light, nutrients, suitable temperature
3. tropical rain forests; coral reefs
4. littoral zone; phytoplankton, zooplankton; limnetic zone, profundal zone; oligotrophic; eutrophic; wetlands
5. phytoplankton; hydrogen sulfide

Review Questions

1. Direct rays of the sun warm Earth's surface at the equator, which warms the air and evaporates water. The warm air rises and cools. As

the air cools, water condenses and falls as rain. The resulting warm, moist climate provides the conditions necessary for the growth of rain forests. The remaining cooler, drier air flows north and south from the equator, where it sinks and is warmed by heat radiated from Earth. By the time it reaches Earth, it is both warm and dry, producing deserts at 30° north and south of the equator.

2. Ocean circulation patterns are called *gyres*. Gyres distribute heat from the equator to northern and southern coastal areas and return cold water from polar regions back down toward the equator. These effects are strongest in coastal regions.

3. In both cases, the average temperature gradually falls, which selects for cold-adapted plants and animals.

4. Nutrients are concentrated in the vegetation. Life is concentrated in the trees because light is blocked by the dense forest canopy of leaves, so relatively few plants—and thus few food sources—grow close to the ground.

5. Cacti have large, shallow root systems; spines that discourage predation; and thickened stems that store water. Some animals remain in shade or in burrows during the summer days, and some are able to derive water exclusively from the food they eat.

6. Desertification is the spread of deserts brought on by drought and misuse of the land in areas that typically receive little rain.

7. Conifers have waxy needles that reduce water loss, which is particularly important in winter when the soil is frozen and water is unavailable. They are evergreen, which allows them to immediately begin photosynthesis in spring, taking maximum advantage of the short growing season.

8. Rainfall is higher, temperatures are warmer, and the growing season is longer in deciduous biomes. In deciduous forests, trees shed their leaves when water is limited, whereas the evergreens in coniferous forests retain their leaves for the entire year. Retaining their leaves allows evergreen conifers to conserve the energy that deciduous trees must expend in growing new leaves each year.

9. Increasing precipitation with less frequent droughts as one goes east from Colorado to Ohio explains the differences among the three biomes.

10. Ocean diversity is highest in coastal waters, where light can penetrate to support producers. There is also a steady flow of nutrients from the land and, in some regions, upwelling from the seafloor.

11. The littoral zone is near the shore and supports the most diverse community because it has abundant light and nutrients. Littoral communities include rooted plants, plankton, algae, and vertebrates including fish, turtles, snakes, and frogs. The limnetic zone is the open water to a depth where there is sufficient light to support photosynthesis by algae and phytoplankton, which support a community of zooplankton and fish. The profundal zone lies below the limnetic zone, reaching to the bottom of large lakes. Light in this zone is insufficient to support photosynthesis, and life is therefore limited to decomposers and detritus feeders, including some worms and bacteria.

12. Oligotrophic lakes are low in nutrients and clear, due to low sediment levels and few phytoplankton. Eutrophic lakes are high in nutrients and support dense communities, and their waters are murky and greenish due to sediment and dense phytoplankton populations. Lakes progress toward a eutrophic state naturally as sediment from runoff carried by streams and rainwater accumulates over time. A human-created scenario involves the addition of nutrient-rich pollutants, such as sewage or fertilizers, and manure from farm runoff into waterways.

13. Stream sources are often in mountains, where water from rain and melting snow is cold, fast moving, well oxygenated, and low in nutrients. In the transition zone, tributaries join, the stream becomes larger, water warms, nutrient levels rise, and oxygen levels diminish somewhat. Within the floodplain, the river moves slowly, is even warmer, usually has a lower oxygen concentration, deposits considerable sediment, and takes a winding path to a lake or ocean.

14. The photic zone is the upper layer of ocean water, where light is strong enough to support photosynthesis. The aphotic zone lies below the photic zone. In the aphotic zone, light is too dim to support photosynthesis, and the only energy comes from the excrement and bodies of organisms that sink or swim down (or from hydrogen sulfide for vent communities). Photosynthetic organisms in the photic zone use nutrients dissolved in the water, those provided by runoff from land, or those brought up into the photic zone by upwelling. Photic zone consumers eat phytoplankton or other producers.

Chapter 31

Think Critically

Figure 31-4 Because an island is inhabited only by species whose ancestors happened to find their way to the island, the number of species present is typically smaller than in mainland areas. One possible result is that predation and competition are less intense on islands, so many island species have not evolved strong defenses against predation or effective adaptations for competition. If this hypothesis is correct, island species may be unable to coexist with newly introduced predators or competitors.

Figure 31-6 A small reserve will have small populations of many organisms, especially large, predatory animals. Small populations typically lose genetic diversity through genetic drift (see Chapter 16). As genetically diverse organisms migrate through corridors, they help to reduce genetic drift and maintain genetic diversity (by repopulating isolated small reserves, for example).

Earth Watch: Whales—The Biggest Keystones of All?

A good test would be to compare otherwise similar areas with differing numbers of whales. [Perhaps even better would be to assess an area (or several areas) at different times, before and after decimated whale populations recovered.] If the whale pump hypothesis is correct, then iron, nitrogen, and perhaps other nutrients in surface waters should be higher in areas or during times with larger populations of whales, which would in turn increase populations of phytoplankton and perhaps krill.

Doing Science: Detecting the Effects of Forest Fragmentation

Some longer-lived species, such as trees and large mammals, may be unable to successfully reproduce in forest fragments, but the individuals present before fragmentation may live for some years after fragmentation. Also, some species may still be able to reproduce in fragments, but less successfully than in intact habitat; their populations may persist until the declining growth rate causes them to fall below the minimum viable population size.

Earth Watch: Saving Sea Turtles

With a low turtle population, food supplies were probably abundant. Listing leatherbacks as endangered and enforcing other protection measures meant that they were much less likely to be killed in fishing nets. When turtle nesting beaches in Florida were protected, people no longer disturbed the nests and eggs were no longer collected. Abundant food, less predation by fishing, and protected nesting sites greatly increased turtle survival and reproduction, allowing exponential population growth (see Chapter 27). However, exponential growth cannot continue indefinitely. Density-dependent factors usually slow down population growth in long-lived species such as sea turtles. These factors might include increasing natural predation, for example, by gulls and raccoons learning that the beaches with large numbers of turtle nests are easy sources of food during turtle nesting season. As turtle populations increase, food supplies in the ocean might also become depleted.

Case Study Revisited: The Wolves of Yellowstone

Wolves harm coyotes, so the disappearance of wolves would be expected to result in increased coyote populations and the expansion of their range. (Coyotes are now found throughout the United States, including on farms and in cities.) Wolves prey on deer, so the loss of wolves would also be expected to increase white-tailed deer populations. (Deer populations have boomed in the eastern United States, assisted also by a decline in hunting by humans.)

Multiple Choice

1. a; **2.** b; **3.** a; **4.** d; **5.** b

Fill-in-the-Blank

1. genetic, species, ecosystem; genetic
2. ecosystem services; provisioning services, regulating services, cultural services, supporting services
3. habitat destruction, overexploitation, invasive species, pollution, global warming; habitat destruction
4. minimum viable population; habitat fragmentation; wildlife corridors
5. sustainable

Review Questions

1. Genetic diversity is the variety and relative frequencies of different alleles in the gene pool of a species. Genetic diversity is crucial for a species to adapt to changing environments. Species diversity is the variety of different species that make up natural communities. The integrity and sometimes even the survival of a community depend on its species composition. Ecosystem diversity is the variety of combinations of communities and their nonliving environments. The nonliving environment is modified by the activities of its community. Diversity of species in the community is often required for sustaining ecosystem services.
2. *Provisioning services*: food, raw materials, energy, medicines. *Regulating services*: water purification, pollination, pest control, erosion and flood control, climate regulation, carbon storage. *Cultural services*: recreation, tourism, mental and physical health. *Supporting services*: photosynthesis, genetic resources, soil formation, nutrient cycling.
3. Threats to biodiversity include *habitat destruction*, such as cutting tropical rain forests; *overexploitation*, such as overfishing of sharks and Pacific bluefin tuna; *invasive species* such as Nile perch and tilapia, which threaten cichlids in Lake Victoria in Africa; *pollution*, such as with pesticides, heavy metals, and excessive nutrients; and *global climate change*, which results from excess production of greenhouse gases, particularly carbon dioxide, and may cause deserts to become hotter, species to move poleward to find favorable climates, insect pests to move poleward, bringing diseases with them, and damage to coral reefs from overly warm water and ocean acidification.
4. In Mexico, the projects enlist the aid and support of local people who ultimately determine the fate of the forests upon which the overwintering monarchs depend. The projects seek to provide a higher, more sustainable income for the residents by improving agriculture, managing the forests both for monarchs and for revenue, and attracting ecotourists, thereby demonstrating that monarch preservation and sustainable development are compatible goals. It remains to be seen whether projects to provide increased monarch summer habitat in the United States and Canada will succeed.

Chapter 32

Think Critically

Figure 32-2 If the mammal's heat-sensing nerve endings became nonfunctional, the nervous system would not send a signal to the hypothalamus when the body reached the set-point temperature. Consequently, the hypothalamus would send continuous "turn on" signals to the body's heat-generating and heat-retention mechanisms, which would continue to increase body temperature indefinitely; this would end in death.

Health Watch: Can Some Fat Burn Calories?

2,4-DNP ingestion will affect mitochondria throughout the body, leading to increased metabolic rate and hyperthermia with its associated risks.

Multiple Choice

1. a; **2.** c; **3.** d; **4.** a; **5.** a

Fill-in-the-Blank

1. homeostasis; negative feedback
2. cells, tissues, organs, organ systems
3. connective; epithelial; connective; connective; epithelial; muscle; nerve; connective
4. exocrine; endocrine; hormones
5. cardiac; skeletal; cardiac and skeletal; cardiac and smooth; smooth; skeletal

Review Questions

1. *Homeostasis* is the ability of an organism to maintain its internal environment within the narrow range of conditions necessary for optimal cell functioning in the face of a changing external environment. A change in the internal environment is often regulated by negative feedback, in which the change triggers a response that tends to counteract the change and to restore and maintain the original conditions. For example, the maintenance of body temperature in humans is regulated by negative feedback. The set-point temperature is established by neurons in the hypothalamus. Other neurons located around the body act as sensors for body temperature and transmit this information to the hypothalamus. When body temperature drops, the hypothalamus activates mechanisms such as shivering to raise the body temperature. When normal body temperature is restored, the hypothalamus switches off these mechanisms.
2. *Ectotherms* derive much of their body warmth from the environment, while endotherms derive much of their warmth from metabolic reactions. Ectotherms include most fish, amphibians, reptiles (except for birds), and invertebrates. *Endotherms* include mammals and birds. Generally, endotherms have a higher metabolic rate and maintain higher body temperatures at more constant levels. Cold- and warm blooded are misleading terms because ectotherms may become very warm (such as the desert pupfish) and endotherms may become quite cool (for example, hummingbirds on very cool nights; hibernating mammals).
3. Positive feedback is a change that initiates a response that intensifies the original change. An example is during the process of childbirth. The early contractions of labor force the baby's head against the cervix, creating pressure that causes the cervix to dilate. Stretch-receptive neurons in the cervix respond to this expansion by signaling the hypothalamus, which responds by triggering the release of oxytocin, which causes further uterine contractions. Stronger contractions create more pressure on the cervix, which triggers the release of more oxytocin. This cycle of positive feedback is terminated by the birth of the baby. Another example of positive feedback is the let-down reflex, which releases milk during breastfeeding. Milk let-down is triggered by suckling, which causes nerves in the breast to stimulate the hypothalamus to release oxytocin, which stimulates milk let-down. Milk stimulates the baby to suckle more vigorously, releasing more oxytocin, until the baby becomes full. Positive feedback is rare because its effects could easily spiral out of control away from the set point.
4. As your body temperature begins to exceed its set point, blood is directed to the skin, radiating heat. Behavioral changes may be required to maintain normal body temperature. The sensation of being too hot may cause you to stop exercising, seek shade, or drink cool water.
5. Epithelial cells may be squamous (flattened), cuboidal (cube-shaped), or columnar (elongated like columns). Epithelial tissue is composed of cells in continuous sheets, underlain by a basement membrane of protein fibers. This tissue lines hollow organs, surrounds the body, and also forms glands.
6. The distinguishing property of connective tissue is that a large proportion of the tissue consists of a fluid matrix containing a high concentration of proteins, with connective tissue cells embedded in the matrix. Loose connective tissue connects, supports, surrounds, and cushions other tissue types and forms a flexible internal framework for organs. It includes adipose tissue (fat), which cushions, insulates, and provides energy reserves. Fibrous connective tissue contains collagen fibers densely packed in an orderly parallel arrangement. It includes tendons (which connect bones to muscles) and ligaments (which connect bones to bones). Specialized connective tissues include cartilage, bone, blood, and lymph. Cartilage covers the ends of bones at joints, provides the supporting framework for the respiratory passages, supports the

ears and nose, and forms the shock-absorbing pads between the vertebrae; bone forms the skeleton that protects and supports the body; blood carries dissolved oxygen and nutrients to cells and removes carbon dioxide and cellular wastes; lymph returns fluid that has leaked out of the capillaries back into the blood.

7. The skin, like any organ, contains several types of tissues. The epidermis is specialized epithelial tissue that is covered by a dry protective layer of dead cells packed with keratin protein that keeps the skin airtight and waterproof. Under the epidermis lies the dermis, a layer of loose connective tissue consisting of loosely packed cells permeated with arterioles that nourish both the epidermis and dermis. Temperature regulation through the skin is accomplished by neurons (nerve tissue) that control the degree of dilation of the arterioles. The dermis also contains glands derived from epithelial tissue, including the sweat glands and sebaceous glands. Smooth muscles (muscle tissue) are attached to the hair follicles and are used to erect the hairs to assist in heat retention. Smooth muscle also surrounds the arterioles and controls their diameter by contracting and relaxing. The hypodermis below the skin consists of connective tissue that pads the skin and insulates the body.
8. Human organ systems include the integumentary, respiratory, circulatory, lymphatic/immune, digestive, urinary, nervous, endocrine, skeletal, muscular, and reproductive systems. Their components and functions are summarized in Table 32-1.

Chapter 33

Think Critically

Figure 33-1 Insect circulatory systems are not important in gas exchange; in insects this function is assumed by the tracheae.

Figure 33-4 Because of an exercise-induced increase in muscle size, the hearts of well-conditioned athletes are larger than those of sedentary people, so more blood is pumped with each heartbeat. Since the body's resting demand for oxygen remains relatively unchanged, the demand can be met with fewer heartbeats per minute.

Figure 33-8 Iron is a key component of hemoglobin, which is necessary for building red blood cells capable of transporting oxygen. Consumption of iron in the diet is necessary to replace iron that is excreted with body waste because red blood cells are continually broken down after they die, and the iron recycling in the body is not 100% efficient.

Figure 33-10 The hormone erythropoietin stimulates production of additional red blood cells. These extra cells increase the blood's capacity to carry oxygen to muscles, thereby increasing the amount of time that the muscles can work without depleting their oxygen supply and becoming fatigued.

Figure 33-15 No, because arteries carry oxygen-poor blood (blue) to the lungs, where capillary blood picks up oxygen, and veins carry oxygenated blood (red) away.

Figure 33-20 From left to right

Health Watch: Repairing Broken Hearts

Bill is likely to have had a stroke (brain attack) because obesity, hypertension, and lack of exercise are associated with atherosclerosis, which can cause clots to form in arteries. These clots can block blood flow to the heart or brain. Bill's symptoms suggest damage to one side of his brain, rather than a heart attack.

Multiple Choice

1. b; **2.** a; **3.** a; **4.** c; **5.** a

Fill-in-the-Blank

1. hemolymph, interstitial; hemocoel
2. right atrium, left ventricle, right atrium, right ventricle, left ventricle
3. sinoatrial node; cardiac muscle cells; atria; atrioventricular node; atrioventricular (AV) bundle; fibrillations
4. capillaries, arteries, veins, venules, arterioles, arteries
5. platelets; thrombin; fibrinogen, fibrin
6. platelets, leukocytes, erythrocytes, leukocytes, erythrocytes, leukocytes, erythrocytes
7. interstitial fluid; capillaries; fats; one-way valves; spleen

Review Questions

1. All circulatory systems have a heart, blood or hemolymph, and blood vessels. *Functions of the vertebrate circulatory system:* Transports oxygen from the lungs or gills to the tissues and transports carbon dioxide from the tissues to the lungs or gills; distributes nutrients from the digestive system to all body cells; transports toxic substances to the liver for detoxification and delivers cellular wastes to the kidneys for excretion; distributes hormones from the glands and organs that produce them to the tissues upon which they act; helps to regulate body temperature by adjusting blood flow; helps to heal wounds and prevent bleeding by creating blood clots; protects the body from diseases by circulating white blood cells and antibodies.
2. Open circulatory systems are present in invertebrates, including all arthropods and most mollusks. Closed circulatory systems are present in all vertebrates and in a few invertebrates, including active mollusks (squid and octopuses) and earthworms. Open circulatory systems have one or more simple hearts, blood vessels, and a hemocoel where tissues and internal organs are bathed in hemolymph. A closed circulatory system confines the blood to the heart and to blood vessels, which carry blood to cells throughout the body.

   Animals with open circulatory systems expend less energy on circulation than do animals with closed systems. Open circulatory systems maintain relatively low blood pressures and are less efficient at supplying oxygen and nutrients to tissues than are closed systems. A closed circulatory system is better able to control and direct blood to specific tissues as needed.
3. The two-chambered hearts of fishes have a single atrium that empties into a single ventricle. Blood pumped from the ventricle passes first through the gills, where the blood picks up oxygen and releases carbon dioxide. The blood travels directly from the gills through the rest of the body, delivering oxygen to the tissues and picking up carbon dioxide. Blood from the body then returns to the single atrium. In amphibians and most reptiles, the heart has two atria and a single ventricle. Oxygen-poor blood from the body is delivered into the right atrium, while blood from the lungs enters the left atrium. Both atria empty into the single ventricle. Amphibian and reptile hearts have various structures that help ensure that most oxygen-poor blood remains in the right portion of the ventricle and is pumped into vessels that lead to the lungs, while most of the oxygenated blood remains in the left portion of the ventricle and is pumped to the rest of the body.
4. Cardiac muscle cells are small and branched. They communicate directly with one another through gap junctions in their adjacent membranes. These connecting pores allow electrical signals that cause contraction to pass freely and rapidly between heart cells, coordinating the contraction within each chamber.
5. Red blood cells are biconcave and lack a nucleus, they are red with hemoglobin, and they function to transport oxygen and a small amount of carbon dioxide. The five types of leukocytes—*neutrophils*, *eosinophils*, *basophils*, *lymphocytes*, and *monocytes*—are large, mobile, and able to ooze through openings. All help defend the body against disease. Platelets are cell fragments that bud off from megakaryocytes and aid in blood clotting.
6. The contraction of the heart is initiated by a pacemaker, the sinoatrial (SA) node, in the right atrium that produces spontaneous electrical signals at a regular rate. Signals from the SA node cause the atria to contract in synchrony. The impulse reaches a barrier of inexcitable tissue separating the atria from the ventricles, where it is channeled through the atrioventricular (AV) node, located in the floor of the right atrium. The impulse is delayed at the AV node, postponing the ventricular contraction for about 0.1 second after contraction of the atria. From the AV node, the signal is carried into the AV bundle and AV bundle branches and from there into Purkinje fibers at the base of both ventricles. The Purkinje fibers transmit the signal to cardiac muscle at the base of the ventricles, and it sweeps upward, causing the ventricles to contract from the bottom up.
7. Veins carry blood traveling to the heart. Vein walls are thinner and more expandable than the walls of arteries, and veins have

valves to maintain the direction of blood flow, whereas arteries do not. Both veins and arteries contain a layer of smooth muscle, but veins have less muscle than arteries do. Arteries are large blood vessels that carry blood away from the heart. Veins and arteries have the same three layers: outer connective tissue, middle smooth muscle, and inner endothelium. Capillaries are microscopic blood vessels where exchange of gases and nutrients occurs between the blood and body cells. Capillary walls consist of a single layer of endothelial cells.

8. Blood plasma is primarily water in which large proteins, salts, nutrients, and wastes are dissolved. Interstitial fluid is blood plasma that has leaked into the spaces surrounding the capillaries and tissues. It consists of water and dissolved nutrients, hormones, gases, wastes, and small proteins from the blood. Large plasma proteins are unable to leave the capillaries due to their size. Exchange of materials between capillary blood and nearby cells occurs through the interstitial fluid. Lymph is interstitial fluid after it has entered the lymphatic system through lymph capillaries, to be returned to the circulatory system.

9. The cardiac cycle is the alternating contraction and relaxation of the heart chambers. The two atria contract in synchrony, emptying their contents into the ventricles. A fraction of a second later, the two ventricles contract simultaneously, forcing blood into the arteries leaving the heart. Both chambers relax briefly before the cycle is repeated. The period of ventricular contraction is called systole. The rest of the cycle, including relaxation of all the chambers followed by contraction of the atria, is called diastole. When blood pressure is measured, the higher of the two readings is the systolic pressure, which is measured during ventricular contraction, and the lower reading is the diastolic pressure, measured between ventricular contractions.

10. The lymphatic system consists of lymph, lymph vessels, and lymphatic organs, which are the bone marrow, thymus, lymph nodes, and spleen. The lymphatic system (1) returns excess fluid and its dissolved proteins and other substances to the blood, (2) transports fats from the intestines to the blood, and (3) assists in defending the body against infectious disease.

11. The sympathetic nervous system stimulates smooth muscles in both vein and artery walls, increasing blood pressure. It also causes arterioles to constrict in specific regions, directing blood flow to places that need it most, depending on the situation.

12. The number of red blood cells is maintained at an adequate level through negative feedback involving the hormone erythropoietin. Erythropoietin is produced by the kidneys in response to oxygen deficiency. This hormone stimulates rapid production of new red blood cells by the bone marrow. When adequate oxygen levels are restored, erythropoietin production is inhibited, and the rate of red blood cell production returns to normal.

13. Both blood veins and lymphatic vessels have similar walls and one-way valves to prevent backflow. In both, fluid movement is promoted by nearby muscle contractions.

Chapter 34

Think Critically

Figure 34-4 The insect tracheole system replaces the capillaries of a closed circulatory system by bringing air for gas exchange close to every cell in the body.

Figure 34-5 The respiratory systems of amphibians are adapted to moist environments, so they are restricted to habitats that are reasonably moist. Most amphibian larvae develop in water, which limits them to regions where standing water is available.

Figure 34-7 When oxygen binds hemoglobin, the protein changes color, becoming cherry-red. When oxygen leaves hemoglobin, it becomes maroon-red in color and appears bluish through the skin.

Figure 34-9 In its relaxed state, the diaphragm muscle is expanded and domes upward, reducing the volume of the chest cavity. Contracting it causes it to become smaller and flatten downward, enlarging the chest cavity.

Figure 34-11 It is important for hemoglobin to be able to bind the hydrogen ions released when bicarbonate is formed because if the hydrogen ions were released into the blood plasma, they would make it acidic (lower pH), which could lead to death.

Health Watch: Smoking—A Life and Breath Decision

The nurse should ask the parents if either of them smoke. He should advise them of the health effects of secondhand smoke on their daughter, whose coughs, colds, and likely asthma are made worse by smoke. If the mother smokes, he should strongly caution her and provide her with educational material about the dangers of smoking to her developing child.

Multiple Choice

1. d; **2.** b; **3.** c; **4.** d; **5.** b

Fill-in-the-Blank

1. sponges, sea jellies, flatworms; diffusion

2. spiracles; tracheae, circulatory systems

3. the pharynx; the epiglottis; trachea, bronchi, bronchioles, alveoli

4. capillaries; hemoglobin; bicarbonate; cellular respiration

5. alveoli; alveolus, capillary; two; surfactant

6. diaphragm, larger; passive, relax; respiratory center, medulla; carbon dioxide

Review Questions

1. Fish gills have delicate membranes with a large surface area. They are often protected by a bony flap, the operculum. Fish create a continuous current over their gills by pumping water into their mouths and ejecting it out through the opercular opening. Just beneath their membranes, gills have a dense profusion of capillaries that pick up carbon dioxide and release oxygen. Within the gill, water and blood flow in opposite directions, maintaining a relatively constant concentration gradient, which can extract most of the oxygen from the water in a process called countercurrent exchange. Countercurrent exchange, which is extremely efficient, is important because water contains only a small percentage of the oxygen contained in air.

2. Frogs use gills during their aquatic larval (tadpole) stage because they must extract oxygen from water. They lose their gills and develop simple, saclike lungs as they metamorphose into their more terrestrial, air-breathing adult form.

3. Bird lungs are rigid and filled with tiny, perforated tubes called parabronchi that allow air to flow completely through the lungs. Parabronchi are surrounded by tissue containing microscopic air spaces surrounded by a dense capillary network that allows gas exchange. Birds have seven to nine inflatable air sacs that serve as reservoirs for air, some of which are posterior (behind) and some anterior to (in front of) the lungs. Anterior sacs temporarily store O_2-poor air from the lungs, while posterior sacs store fresh air and provide it to the lungs during exhalation. The organization of air sacs and lungs allows a one-way flow of fresh, oxygen-rich air through the lungs, from posterior to anterior, as the bird both inhales and exhales. This efficient process helps supply enough oxygen for the demands of flight.

4. Air enters through the nose or mouth; then passes through a chamber, the pharynx, which is shared with the digestive system; and then travels through the larynx. Inhaled air continues past the larynx into the trachea, which splits into two large branches called bronchi, one leading to each lung. Inside the lung, each bronchus branches repeatedly into smaller tubes called bronchioles. These finally lead to the microscopic alveoli, tiny air sacs where gas exchange occurs.

5. Some animal characteristics that make a specialized respiratory system unnecessary include (1) a size and body shape that allows for adequate gas exchange, as in small, elongated nematode worms and thin and flattened flatworms; (2) low energy demands in some cells of thicker-bodied animals, such as the jellyfish; (3) an environment close to the body cells, making possible gas exchange directly between the environment (water) and the body cells, as in sponges; (4) low energy demands, coupled with a large skin surface through which diffusion occurs, or a well-developed circulatory system, as in the earthworm.

6. Respiratory movements occur rhythmically and automatically. Nerve impulses that originate in the respiratory center, located in the medulla just above the spinal cord, cause the contraction of diaphragm and rib muscles that results in inhalation. The respiratory center receives input from several sources. One of the most important is from receptor neurons in the medulla that monitor the concentration of carbon dioxide in the blood. The respiratory rate is regulated to keep this level relatively constant. If blood oxygen levels fall drastically, oxygen receptors in the aorta and carotid arteries stimulate the respiratory center to increase the breathing rate. These controls are adaptive because they cause the respiratory system to adjust its activity to meet the changing needs of the tissues of the body.

7. Inhalation is always an active process, accomplished by enlarging the chest cavity. The diaphragm muscles contract, drawing the diaphragm downward, and the rib muscles contract, lifting the ribs up and outward. When the chest cavity expands, the lungs expand with it, because there is no air between them and the inner wall of the chest. As the lungs expand, their increased volume draws air into the lungs. Exhalation occurs automatically when the muscles causing inhalation are relaxed. As the chest cavity decreases in size, air is forced out of the lungs.
8. During bulk flow, fluids or gases move in "bulk" through relatively large spaces, from areas of higher pressure to areas of lower pressure. In diffusion, molecules move individually from areas of higher concentration to areas of lower concentration. In respiration: (1) Air or water containing oxygen is moved across a respiratory surface by bulk flow, often facilitated by muscular breathing movements. (2) Oxygen and carbon dioxide are exchanged through the respiratory membranes by diffusion. Diffusion across the respiratory membranes moves oxygen from the alveolar air into the capillaries of the circulatory system and moves carbon dioxide from alveolar capillaries into the air within the alveoli. (3) Gases are transported between the respiratory system and the tissues by bulk flow of blood as it is pumped throughout the body by the heart. (4) Gases are exchanged between tissues and the circulatory system by diffusion. At the tissues, oxygen diffuses out of the capillaries and carbon dioxide diffuses into them. Diffusion is driven by concentration gradients.
9. Nearly all of the oxygen from the lungs that diffuses into the blood is carried by hemoglobin. Oxygen then diffuses into body cells from the blood. About 70% of the carbon dioxide released from cells into the blood reacts with water in the presence of the enzyme carbonic anhydrase to form bicarbonate ion (HCO_3^-), which is carried in the blood plasma and released into the lungs by diffusion. About 20% of the carbon dioxide is returned to the lungs bound to hemoglobin, and about 10% is dissolved directly in the plasma. Gas exchange supports cellular respiration, which uses oxygen and releases carbon dioxide while producing ATP.
10. Each lung contains about 1.5 million alveoli. The thin-walled alveoli provide an enormous surface area for diffusion. The alveoli are enmeshed in capillaries. Because both the alveolar walls and adjacent capillary walls are only one cell thick, the air is extremely close to the blood in the capillaries. The lung cells remain moist because they are enclosed by the body, and the inside of each alveolus is coated with a watery fluid. Gases must dissolve in this water before they can diffuse through the respiratory membranes.

Chapter 35

Think Critically

Figure 35-6 The height of the vegetable bar would be halved.

Figure 35-19 The required absorptive surface area would have to be provided by some other adaptations, such as a much longer small intestine and a correspondingly larger abdominal cavity to hold it. Or the food might need to spend a much longer time in the intestine, being sloshed back and forth to expose nutrients to the smooth absorptive surface.

Doing Science: Identifying the Cause of Ulcers

Antacids relieve ulcers by counteracting the stomach acid that attacks the lining of the stomach and upper small intestine.

Health Watch: Overcoming Obesity: A Complex Challenge

The patient should first attempt weight loss with a change in eating habits and knee-friendly exercise. Failing success with that approach, this patient might be a candidate for gastric bypass surgery, which would help both his diabetes and weight, some of which is abdominal fat.

Multiple Choice

1. d; **2.** b; **3.** c; **4.** b; **5.** a

Fill-in-the-Blank

1. energy, raw materials; vitamins; minerals; essential
2. intracellular; gastrovascular cavity; tubular; cellulose
3. amylase, starch; protein; pepsinogen, pepsin; small intestine
4. ingestion, mechanical digestion, chemical digestion, absorption, elimination
5. pharynx, digestive, respiratory; epiglottis, swallowing; peristalsis; sphincters
6. bile; liver, gallbladder; epithelial; lacteal, lymphatic
7. gastrin; ghrelin; leptin; secretin, cholecystokinin; gastrin, ghrelin; secretin, cholecystokinin

Review Questions

1. There are six categories of nutrients: carbohydrates, lipids, proteins, minerals, vitamins, and water. Most energy is derived from carbohydrates (starches and sugars) and lipids (fats and oils). Most energy is stored in the body as fat, because it contains twice as much energy per gram as proteins or carbohydrates and because it does not attract extra water; thus, fat stores energy while adding the least weight. Minimizing weight helped our ancestors move quickly to catch prey and avoid predators.
2. Both birds and earthworms have gizzards. Gizzards use muscular contractions and small sharp stones (sand in an earthworm) to physically digest (grind up) food.
3. Vertebrates are classified as herbivores (such as horses, cows, deer, camels, and mice), which eat mainly plants; carnivores (such as cats, seals, wolves, and predatory birds), which mainly eat other animals; and omnivores (such as bears, raccoons, and humans), which eat both plants and other animals. Carnivores have the shortest intestines, because considerable protein digestion is accomplished in the stomach. Animal protein is not surrounded by cell walls and is easily accessed by the proteases in the small intestine, so it requires less time and muscular activity to digest.
4. The muscular walls of the stomach generate the churning action that is an important part of the digestive process. An expandable stomach enables us to consume large amounts of food when food is available. This capability is presumably a holdover from our earlier evolutionary history, when our food supply was not as predictable as it is for most of us today. The expanded stomach serves to store food until it can be processed by the rest of the digestive system.
5. Secretions of the stomach gastric glands are hydrochloric acid, pepsinogen, mucus, gastrin, and ghrelin. Hydrochloric acid provides the acidic environment needed to convert pepsinogen into pepsin. Pepsinogen is an inactive form of pepsin, a protease enzyme involved in the breakdown of proteins. Mucus is secreted to coat the stomach lining, providing a barrier to self-digestion. Gastrin is a hormone that increases stomach acid secretion and peristalsis. Ghrelin is a hormone that acts on the hypothalamus to trigger hunger.
6. Bile is produced in the liver, concentrated in the gallbladder, and released into the small intestine. Bile assists in the digestion of lipids by acting as a detergent to disperse fat particles into microscopic particles. Pancreatic juice is secreted by the pancreas into the small intestine to neutralize acidic chyme and to digest carbohydrates, lipids, and proteins.
7. Peristalsis, a sequential squeezing of circular muscles surrounding the digestive tract, pushes food through the digestive tract. Peristalsis begins in the esophagus and occurs throughout the digestive tract. In the stomach, muscular churning contractions assist in the breakdown of food particles, and peristalsis pushes chyme into the small intestine. Segmentation movements in the small intestine slosh the chyme back and forth, exposing the particles to digestive secretions and exposing the released nutrients to a large surface for absorption.
8. An essential nutrient is defined as one that is needed, but not synthesized, by the body. Dogs can synthesize ascorbic acid (which we call vitamin C), so it is not a vitamin for them, because vitamins are essential nutrients. Because humans can't synthesize ascorbic acid, it is a vitamin for us. Plants can synthesize all the amino acids that they need, but humans cannot and must obtain essential amino acids from plants or other animals.

9. The wall of the small intestine has numerous folds that increase its absorptive surface. Covering the folded wall are tiny, finger-like projections called villi, and individual villi bear a fringe of microscopic microvilli, both of which greatly enhance the absorptive surface area of the intestine. Finally, the small intestine is extremely long (about 10 feet), which greatly increases its total absorptive surface.
10. Chewing reduces the meat into small chunks. The acidity of the stomach converts pepsinogen (the inactive form of the protease pepsin) into pepsin, which begins protein digestion. Churning movements expose more surface area to the pepsin. In the stomach, proteins are digested into peptides that enter the small intestine. In the small intestine, the peptides in chyme are broken down into very short peptides and amino acids. This is accomplished first by proteases secreted by the pancreas into the duodenum and just prior to absorption by enzymes on the microvilli of intestinal epithelial cells. As the amino acids are liberated from the peptides, they are absorbed into the epithelial cells.
11. The large intestine absorbs water and salts from the chyme. Bacteria in the large intestine feed on nutrients remaining in the chyme, releasing some vitamins (notably vitamin K), which are absorbed by the large intestine. The large intestine forms feces from the water, mucus, indigestible fiber, some undigested nutrients, and bacteria and then propels the feces by peristaltic contractions into its final segment, the rectum. Distention of the rectum triggers the urge to defecate, eliminating the feces through the anus.
12. In the stomach, peptides from protein digestion trigger secretion of gastrin from the gastric glands. Gastrin is a hormone that increases stomach acid secretion and peristalsis to aid in protein digestion. As chyme containing peptides and fats enters the upper small intestine (duodenum), it triggers the release of secretin and cholecystokinin, which slows stomach peristalsis and acid secretion, hence controlling stomach acid production and slowing the movement of chyme into the small intestine. This allows more time for digestion and absorption to occur. Both secretin and cholecystokinin increase production and release of bile into the small intestine to disperse fats and pancreatic juice to neutralize the chyme and digest carbohydrates, fats, and proteins.

Chapter 36

Think Critically

Figure 36-6 If you consumed excess water, a drop in blood osmolarity would occur and would be detected by receptors in the hypothalamus. The hypothalamus would signal the posterior pituitary, causing it to reduce its release of ADH. Less ADH would result in fewer aquaporins in the membranes of the distal tubule and collecting duct, making them almost impermeable to water. This would result in production of large amounts of watery urine until the normal blood osmolarity was restored.

Figure 36-8 Freshwater fish are adapted to excrete large amounts of water. Placed in ocean water, a trout would likely lose too much water, become dehydrated, and die.

Health Watch: When the Kidneys Collapse

In a patient with liver failure, the level of urea in the blood would decrease, because the liver is the site at which nitrogenous wastes are converted to urea.

Multiple Choice

1. a; **2.** b; **3.** b; **4.** d; **5.** c

Fill-in-the-Blank

1. ammonia, uric acid, uric acid, uric acid, urea
2. renal cortex, renal medulla, renal pelvis; ureter, bladder, urethra
3. glomerulus, glomerular capsule, proximal tubule, nephron loop, distal tubule
4. proximal tubule, collecting duct, nephron loop, renal medulla
5. urea, H^+ (hydrogen ions), NaCl (salt); water
6. homeostasis; renin, angiotensin; erythropoietin, red blood cells
7. hypothalamus, posterior pituitary, antidiuretic hormone (ADH); distal tubule, collecting duct, aquaporins, concentrated

Review Questions

1. Urinary systems eliminate cellular wastes and toxic substances, using filtration to remove them from the blood and then tubular secretion to secrete remaining wastes into the nephron. Urinary systems also play a crucial role in homeostasis by regulating the chemical composition of the blood. This is accomplished by nonselective filtration, which removes most dissolved substances, followed by reabsorption of water, important nutrients, and the ions required to maintain homeostatic conditions in the blood.
2. Ammonia is released when cells break down amino acids. The ammonia diffuses into the bloodstream, which transports it to the liver. Here, the ammonia is converted into urea. Urea is transported in the blood to the kidney through the renal artery. In the kidney, the urea is filtered out through the glomerular capillaries into the nephrons. Urea travels through the nephron into the collecting duct. The urine flows through chambers in the renal pelvis and leaves the kidneys through the ureter, which carries it to the bladder. Urine containing urea exits the body through the urethra.
3. The *nephron loop* creates an osmotic gradient of salt (NaCl) within the renal medulla that reaches its highest concentration near the lower region of the loop. The *collecting duct* contributes urea to the high solute concentration in the renal medulla. The collecting duct carries urine from nephrons through this renal medulla solute concentration gradient. Depending on the permeability of the collecting duct to water, as urine flows through the collecting duct, it either becomes more concentrated or remains dilute. *Antidiuretic hormone (ADH)* is released continuously by the posterior pituitary and controls water permeability of the distal tubule and collecting duct. When receptors in the hypothalamus detect increased blood osmolarity (which signals the need to conserve water by producing concentrated urine), they trigger an increase in ADH release. ADH stimulates cells of the distal tubule and collecting duct to insert more aquaporins into their membranes, increasing their permeability to water and causing more water to leave the collecting duct as it passes through the renal medulla. When blood osmolarity falls, less ADH is released, and permeability of the distal tubule and collecting duct decreases, causing more dilute urine to be excreted.
4. The highly permeable capillaries of the glomerulus filter blood as it travels through them under pressure. Water and small dissolved molecules exit through the capillary walls, which retain proteins and blood cells. The filtered blood travels through peritubular capillaries, where it picks up water and nutrients that are removed from the renal tubule by reabsorption. The capillaries surrounding the tubule also contribute additional wastes from the blood, which enter the tubule during secretion.
5. *Filtration* is the process by which blood is forced through the glomerular capillaries, driving water and small dissolved molecules and ions (but not cells or proteins) into the glomerular capsule. *Reabsorption* occurs primarily in the proximal tubule but also in the distal tubule. During reabsorption, ions and nutrients (such as Ca^{2+}, Na^+, Cl^-, HCO_3^-, amino acids, and glucose) diffuse or are actively transported out of the tubule and into the peritubular capillaries. Water follows by osmosis. *Secretion*, which occurs primarily in the proximal tubule but also in the distal tubule, actively transports some drugs and excess K^+, H^+, and NH_3 into the tubule.
6. The kidneys regulate the water content of the blood by inserting aquaporins into distal tubule and collecting duct membranes in response to the release of ADH by the posterior pituitary. The kidneys regulate dissolved substances in the blood, maintaining appropriate levels of ions such as calcium, potassium, sodium, and chloride. The kidneys help regulate blood pH by controlling bicarbonate and hydrogen ion levels. In response to low blood pressure, kidneys release the hormone renin, which catalyzes formation of the hormone angiotensin. Angiotensin increases blood pressure by reabsorbing Na^+, which causes water to follow by osmosis and blood volume to increase; stimulates ADH release, which causes more water to be reabsorbed from the distal tubule and collecting duct; and causes the constriction of arterioles. By releasing erythropoietin, the kidneys stimulate red blood cell production and increase the blood's oxygen-carrying capacity if blood oxygen levels fall below normal.
7. The flatworm uses a simple urinary system of protonephridia consisting of two networks of tubules branching through the body. Flame cells circulate body fluid through the tubules. Excess water and some wastes

are excreted through numerous urinary pores. The insect urinary system consists of Malpighian tubules, which filter wastes and salts from the hemocoel and release them into the digestive tract, where some are reabsorbed and others are excreted as waste. Earthworms use nephridia, which somewhat resemble vertebrate nephrons and are found in each of the earthworm's segments. Interstitial fluid is drawn into a ciliated opening, the nephrostome, nutrients and water are reabsorbed, and wastes are excreted through nephridiopores.

These systems are similar because all of them regulate water content and remove wastes from body fluids. Interstitial fluid is filtered by the protonephridia of flatworms, the Malpighian tubules of insects, and the nephridia of earthworms. Although they differ in complexity, all three systems use tubules with selectively permeable walls to transport fluid and wastes, and all three help to maintain homeostasis.

8. Most freshwater fish experience a continuous influx of water and efflux of salt. They counteract this by excreting large amounts of dilute urine and actively transporting salt in through their gills. Saltwater fish experience a steady loss of water and influx of salt, which they counteract by excreting very little urine and actively pumping salt out through their gills.

Chapter 37

Think Critically

Figure 37-4 If swelling occurs in the respiratory passages or lungs, the inflammatory response could cause suffocation.

Figure 37-13 The same antibodies and bodily reactions that cause allergies also occur in response to some parasites, such as hookworms that invade the digestive tract. Strong allergic reactions, such as vomiting or diarrhea, can help to rid the body of these parasites.

Doing Science: Discovering How to Prevent Infectious Diseases

The graph in Figure E37-2 shows the total number of measles cases that were reported to health authorities each year. The U.S. population increased over time, so there is a larger population base each year. To determine the infectivity of measles, it would be better to graph the number of cases per unit population (for example, per 100,000 people; compare the graph below with Fig. E37-2). Although the measles vaccine was licensed in 1963, it was not mandatory. It took a couple of years before childhood measles vaccination became nearly universal in the United States.

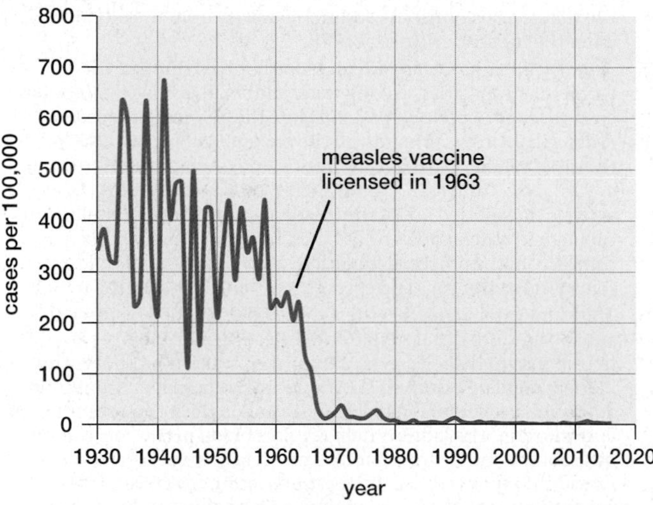

Health Watch: Emerging Deadly Viruses

Flu viruses mutate frequently, producing somewhat different viruses each year. Recombination between human, pig, and bird viruses can cause additional changes. Therefore, this year's flu virus will not be exactly the same as last year's virus, so the immunity produced as a result of last year's infection will not provide full protection against this year's flu. The vaccines are changed each year in an attempt to provide immunity against the flu viruses that are thought most likely to be common that year.

Multiple Choice

1. b; **2.** a; **3.** c; **4.** c; **5.** c

Fill-in-the-Blank

1. skin, digestive, respiratory, urogenital
2. phagocytes, natural killer cells, inflammatory response, fever
3. antigens; antibodies, T cell receptors
4. light, heavy; variable, constant; variable
5. Humoral, plasma cells; Cell-mediated; Cytotoxic; Helper; memory
6. allergy; immune deficiency; autoimmune disease

Review Questions

1. The body's three lines of defense against invading microbes are (1) nonspecific external barriers that keep microbes out of the body; (2) the innate immune response (nonspecific internal defenses) that combat many invading microbes without regard to their specific types; and (3) the adaptive immune response, which directs its assault against specific microbes.
2. Both natural killer cells and cytotoxic T cells secrete enzymes and pore-forming proteins onto target cells. The pore-forming proteins create large holes on the target cell's plasma membrane that allow entry of the enzymes, which destroy cellular components.
3. Humoral immunity is produced by antibodies in the blood. B cells that bind to an antigen divide rapidly, producing many more B cells; some of them become memory B cells and others become plasma cells. Plasma cells release antibodies into the blood or interstitial fluid. These antibodies specifically bind to antigens. Antibodies may cover up active sites on toxic molecules, or they may bind to microbes. Antibodies attached to microbes facilitate phagocytosis, or they may enhance binding by complement proteins, which may directly kill the microbe or further facilitate phagocytosis. Cell-mediated immunity is the primary defense against the body's own cells when they become cancerous or are infected with a virus. Cytotoxic T cells bind to antigens on the surface of the infected cell and release pore-forming proteins and enzymes that destroy the infected cell. Helper T cells stimulate both humoral and cell-mediated immunity. Helper T cells have receptors that bind to specific antigens displayed on the surfaces of dendritic cells or macrophages. The helper T cells then release cytokines that stimulate the activation and multiplication of B cells and cytotoxic T cells.
4. See Figure 37-6 for the structure of an antibody. The antigen binding site consists of the variable regions of the heavy and light chains of each arm of the antibody. Antibodies bind to specific antigens because the binding sites of the antibodies have specific sizes, shapes, and electrical charges that interact only with specific antigens.
5. Cancerous cells often lack surface proteins, such as MHC proteins, that are found on normal cells or have new, different proteins that are not usually found on normal cells. Natural killer cells and cytotoxic T cells recognize these protein differences and kill most cancerous cells.
6. Some of a cancer patient's immune cells, especially T cells, may be removed from the body, genetically engineered to produce receptors that bind to particular cancer cell antigens, and then reintroduced to the patient's body, where they attack cancer cells. Alternatively, patients may receive checkpoint-inhibitor drugs that block the action of proteins that inhibit the immune response, allowing a stronger immune attack on cancer cells.
7. Immature immune system cells that recognize and bind to antigens are destroyed. Because immature immune cells are almost certain to encounter "self antigens" during their development, any immature immune cells that bind self antigens will die, with the result that only immune cells that do not respond to "self" are retained and mature into functional B and T cells.
8. Memory cells are a subset of the offspring of activated B cells, cytotoxic T cells, and helper T cells, which survive for many years. Memory cells respond to the same antigens that the original parent B or T cells did, but there are far more memory cells, and each one responds much faster than the original B or T cell could. Therefore, during a second exposure to an antigen, they produce a rapid, large immune response, destroying the invaders so fast that usually there are no symptoms of disease.

9. Vaccination introduces weakened or dead microbes, or sometimes antigens that are normally found on specific microbes, into the body to stimulate the development of memory cells, conferring immunity against subsequent exposure to living, dangerous microbes.
10. The inflammatory response is generally a response to tissue damage, such as occurs in a wound in the skin. Chemicals released from damaged cells cause mast cells to release histamine and other chemicals, which increase blood flow, cause capillaries to become leaky, and thereby increase movement of fluid from the capillaries into the injured area. As a result, the injured tissues become warm, red, swollen, and painful. The inflammatory response has several functions: It attracts phagocytic cells to the area so they can engulf pathogens and debris, promotes blood clotting, and causes pain that stimulates protective behaviors.
11. An allergic reaction occurs when an allergen binds to antibodies on B cells, which proliferate and produce more antibodies. The antibodies bind to mast cells, which respond by producing histamines. The histamines trigger inflammation that results in allergy symptoms.
12. An autoimmune disease is an immune response against some of the body's own molecules and cells. Examples are some types of anemia and type 1 diabetes. An immune deficiency disease results from the inability to mount an effective immune response to infection. Examples include severe combined immune deficiency (SCID) and acquired immune deficiency syndrome (AIDS).
13. HIV viruses cause AIDS by infecting and destroying helper T cells. AIDS does not directly kill its victims, but as the helper T-cell population declines, the patient with AIDS becomes increasingly susceptible to other diseases, eventually dying of infection. AIDS is spread by direct contact with mucous membranes or broken skin or with body fluids such as semen, blood, vaginal secretions, and breast milk.

Chapter 38

Think Critically

Figure 38-8 There are several possible hypotheses, including (1) a tumor in the anterior pituitary caused multiplication of the TSH-secreting cells, (2) a tumor in the hypothalamus caused multiplication of the cells that secrete TSH-releasing hormone; and (3) an iodine-deficient diet caused the thyroid to produce too little thyroxine so that the negative feedback system regulating TSH release was shut off.

Figure 38-9 If the receptors could not bind glucagon, then target cells, such as those in the liver, would not be able to respond to glucagon. When blood glucose levels were too low, liver cells would not convert glycogen to glucose, so blood glucose levels would stay low.

Earth Watch: Endocrine Deception

Fish and amphibians are immersed in water, so if endocrine disruptors are present in a stream or lake, these animals will be continually exposed to them. Amphibians, with thin, moist, permeable skin, would be expected to take up more disruptors from the water than fish, birds, or mammals would. Finally, because fish and amphibians are endothermic ("cold-blooded"), their body temperatures would usually be considerably lower than the temperatures of birds or mammals. Lower body temperatures usually mean slower metabolism, so fish and amphibians probably break down endocrine disruptors in their bodies more slowly than birds or mammals would.

Health Watch: Performance-Enhancing Drugs—Fool's Gold?

Anabolic steroids can definitely cause increased muscle mass; although the evidence is weak, many people think that growth hormone can also increase muscle mass. In his preseason physical, Juan might show a decreased percentage of fat (if he has increased muscle mass, then the percent of fat would go down). If he has taken steroids, the exam might also show increased blood pressure, reduced natural testosterone, and possibly smaller testes and enlarged breast tissue. Increased levels of steroids and/or growth hormone in his urine or blood would be definitive proof of PED abuse.

Case Study Revisited: Insulin Resistance

The initial blood glucose concentration will be higher in a patient with untreated diabetes than in a patient without diabetes, because the mechanisms for controlling blood glucose do not function effectively in diabetics (insufficient insulin and/or insufficient sensitivity to insulin). In a non-diabetic, blood glucose will initially rise after consumption of a sugary drink but will begin to decline after a short period as the excess glucose stimulates release of insulin. Within a few hours, glucose levels will return to the initial baseline. In a diabetic, the initial increase in glucose level will not be counteracted by the action of insulin, so the level will remain high, decreasing only slightly and slowly as body cells use some of the glucose for cellular respiration.

Multiple Choice

1. c; **2.** d; **3.** a; **4.** b; **5.** a

Fill-in-the-Blank

1. endocrine system; target cells, receptors
2. peptide, amino acid derived, steroid; Peptide hormones, amino acid derived hormones; second messengers; Steroid hormones; changes in gene transcription (or synthesis of messenger RNA)
3. hypothalamus; neurosecretory cells, oxytocin, antidiuretic hormone
4. adrenocorticotropic hormone, follicle-stimulating hormone, luteinizing hormone, growth hormone, prolactin, thyroid-stimulating hormone
5. insulin; diabetes mellitus; Glucagon, glycogen
6. testes, testosterone; ovaries, estrogen, progesterone
7. glucocorticoids, mineralocorticoids, testosterone; norepinephrine, epinephrine

Review Questions

1. Most peptide and amino acid derived hormones bind to receptors on the plasma membrane of target cells. Hormone–receptor binding stimulates enzymes that synthesize second messenger molecules, such as cyclic AMP, inside the cell. The second messengers stimulate biochemical reactions, usually by stimulating the action of other enzymes. The cell's response may also include changes in gene transcription and protein synthesis.
2. Steroid hormones usually diffuse through the plasma membrane (of all cells) and attach to intracellular receptors that are present in the cytoplasm or nucleus of target cells. The hormone–receptor complex binds to specific regions of DNA, facilitating the binding of RNA polymerase to the promoters of specific genes, which stimulates transcription of those genes into messenger RNA. The mRNA is then translated into protein.
3. See Figure 38-4.
4. The hypothalamus contains clusters of specialized nerve cells, called neurosecretory cells. The pituitary is a pea-sized gland attached to the hypothalamus by a stalk. The pituitary gland consists of two distinct lobes called the posterior pituitary and the anterior pituitary. The hypothalamus controls release of hormones from both parts. The posterior pituitary secretes two neurosecretory hormones that are actually synthesized in the hypothalamus: antidiuretic hormone, which causes water to be reabsorbed from the urine and retained in the body, and oxytocin, which triggers milk secretion from the saclike milk glands into the nipples in nursing mothers.

   Releasing hormones and inhibiting hormones are synthesized in nerve cells in the hypothalamus, secreted into a capillary bed in the hypothalamus, and transported a short distance through blood vessels to a second capillary bed that surrounds the endocrine cells of the anterior pituitary. There, these hormones diffuse out of the capillaries and influence pituitary hormone secretion: Releasing hormones stimulate secretion from the anterior pituitary, and inhibiting hormones suppress secretion. The anterior pituitary produces six peptide hormones. Follicle-stimulating hormone and luteinizing hormone stimulate the production of sperm and testosterone in males and eggs and estrogen in females. Thyroid-stimulating hormone stimulates the thyroid gland to release its hormones. Adrenocorticotropic hormone causes release of hormones from the adrenal cortex. Prolactin stimulates the development of the mammary glands during pregnancy. Growth hormone regulates body growth and metabolism.

5. When blood glucose rises, insulin is released. Insulin causes many body cells to take up glucose and either metabolize it for energy or convert it to fat or glycogen for storage. When blood glucose levels drop, insulin release is inhibited and glucagon is released. Glucagon activates an enzyme in the liver that breaks down glycogen, releasing glucose into the blood. When glucose levels rise, glucagon release is inhibited. If blood glucose rises too far, insulin is released again.
6. The adrenal medulla is located in the center of each adrenal gland. It produces epinephrine and norepinephrine in response to stress or exercise. These hormones increase the heart and respiratory rates, cause blood glucose levels to rise, and direct blood flow away from the digestive tract and toward the brain and muscles. The outer layer of the adrenal gland forms the adrenal cortex, which secretes glucocorticoids that help control glucose metabolism, promote the use of fats for energy, and suppress the immune response and inflammation, and mineralocorticoids (such as aldosterone), which regulate the sodium content of the blood by stimulating the kidneys and sweat glands to retain sodium. The adrenal cortex also secretes small amounts of testosterone (in both sexes).

Chapter 39

Think Critically

Figure 39-4 Based on the evidence, the toxin must act on the postsynaptic part of the synapse by blocking neurotransmitter receptors. If the toxin acted by blocking release of neurotransmitters from the presynaptic neuron, then adding neurotransmitter to the synapse would have generated a postsynaptic action potential. If the toxin acted by preventing sodium channels from opening, then it would not have been possible to stimulate an action potential in the presynaptic neuron.

Figure 39-5 Sensitive areas have a higher density of touch-sensitive sensory neurons. This would allow two types of increased sensitivity: (1) the higher density of receptors would mean that touching any particular place on the sensitive skin area would be more likely to stimulate at least one touch receptor, and (2) a higher density of receptors would also mean that a touch of a particular force would stimulate a larger number of neurons. The touch thus would generate a larger number of action potentials in the sensitive area and would be perceived as a stronger stimulus.

Figure 39-10 John probably has a severed spinal cord. Cutting the spinal cord cuts axons that travel to and from the brain, thereby preventing the sensation of pain from being relayed to the brain and preventing the brain from stimulating motor neurons that activate muscles voluntarily. However, the cut cord often does not disrupt the withdrawal reflex circuit, which requires only transmission within a small portion of the spinal cord.

Health Watch: Drugs, Neurotransmitters, and Addiction

Both addictive drugs and emotional love and friendship cause the release of dopamine in brain regions that promote feelings of happiness, euphoria, and rapport with other people. A drug that combats drug addiction by reducing or eliminating satisfying relationships with other people is doomed to fail in the marketplace.

Multiple Choice

1. c; **2.** a; **3.** a; **4.** a; **5.** d

Fill-in-the-Blank

1. neuron; dendrite; cell body; axon, synaptic terminal
2. negatively; threshold; positively
3. neurotransmitter; receptors; excitatory postsynaptic potential; inhibitory postsynaptic potential
4. somatic nervous system; autonomic nervous system, sympathetic, parasympathetic
5. cerebellum, pons, medulla; cerebellum
6. frontal, parietal, temporal, occipital; occipital; frontal

Review Questions

1. Refer to Figure 39-4 for a diagram of a synapse. Signals are transmitted from one neuron to another by release of neurotransmitter molecules from the presynaptic neuron into the synaptic cleft between the two neurons. These neurotransmitters diffuse across the gap and bind to receptors on the postsynaptic neuron's dendrite. Activation of the receptors may cause a change in the resting potential of the postsynaptic neuron—either an excitatory (less negative than the resting potential) or inhibitory (more negative) postsynaptic potential.
2. The intensity of a stimulus is signaled by the frequency of action potentials in a single neuron and by the number of neurons firing in response to the same stimulus at the same time. The type of stimulus is determined by the area of the brain that is wired to the neurons that are firing: Action potentials conducted to specific areas of the brain are perceived as specific sensory types (e.g., action potentials in axons of the optic nerve travel to visual areas of the brain and are always perceived as light).
3. The four elements are the sensory neuron, interneuron, motor neuron, and effector (the muscle). The pain-withdrawal reflex involves one of each type of cell. A painful stimulus is received by a sensory neuron, which transmits the signal to an interneuron in the spinal cord, which signals the motor neuron to stimulate the muscle to contract, which causes withdrawal from the painful stimulus.
4. Refer to Figure 38-9 for a diagram of the spinal cord. The cell bodies of motor neurons and interneurons, the central axons and synaptic terminals of sensory neurons, and the axons and synaptic terminals of some neurons whose cell bodies are in the brain are all located in the spinal cord. Severing the spinal cord causes paralysis of the body below the cut because axons from the brain are cut and hence cannot stimulate spinal cord interneurons or motor neurons that are located below the cut.
5. The medulla controls several automatic functions, such as swallowing, breathing, heart rate, and blood pressure. The cerebellum coordinates body movements and is important in learning motor skills such as writing or typing. The reticular formation is an important relay center, which plays a role in sleep and wakefulness, emotion, and some movements and reflexes. The thalamus carries sensory information from most of the senses (but not olfaction) to the limbic system and the cerebrum. The amygdala produces sensations of pleasure, fear, and sexual arousal. The cerebral cortex processes sensory information, stores short-term and long-term memories, and directs voluntary movements.
6. The corpus callosum connects the two cerebral hemispheres. The left hemisphere almost always controls the right side of the body, and the right hemisphere controls the left side. In most (but not all) people, the left hemisphere is principally responsible for speech, reading, writing, language comprehension, and rote mathematical ability. The right side of the brain is usually responsible for musical skills, artistic ability, recognizing faces, spatial visualization, and the ability to recognize and express emotions.
7. Long-term memory is information that is retained permanently, probably caused by the formation of new, permanent synaptic connections between certain neurons or strengthening of weak connections. Working memory is short-term memory that appears to be electrical in nature or may involve temporary biochemical changes within neurons in a circuit.

Chapter 40

Think Critically

Figure 40-5 The joints and muscles in your neck and the rest of your body have stretch and position receptors. If your body is tilted but you keep your head vertical, position receptors in your neck will be activated. Your brain will compare information from the vestibular apparatus in your head and position receptors in your neck and calculate the amount of body tilt.

Figure 40-8 First, you should explain to the patient that the cornea actually does most of the focusing in the human eye. Then, you should explain that, for nearsightedness, laser surgery flattens the cornea (makes the cornea less convex), which causes the light rays to converge less before they reach the lens, so that the lens can now produce a sharp image on the retina. If the patient had been farsighted, the surgery would reshape the cornea so that it had a more rounded, convex shape that causes the light rays to converge more before they reach the lens.

Figure 40-9 Two reasons: (1) the blind spots in your two eyes do not see the same place in your visual field, so one of the eyes always perceives objects that would be in the "hole" in the visual field of the other eye. (2) Your eyes constantly move around, so no part of the visual field is in the blind spot of either eye for very long. The brain "fills in" the missing visual information with what the eye saw just a fraction of a second before.

Figure 40-10 Distance perception is useful for many behaviors, not just predation. Fruit-eating bats fly around during the daytime, and binocular vision is useful for avoiding obstacles such as branches. Monkeys often jump from branch to branch and need to know how far it is to the next branch.

Earth Watch: Say Again? Ocean Noise Pollution Interferes with Whale Communication

Many familiar terrestrial animals, particularly songbirds, communicate extensively by sound and live in noisy human environments. If the reproductive success of some of these animals is reduced by noise pollution, their predators might experience a lack of food, populations of their prey might increase, and populations of species with which they have mutualistic relationships might decline.

Case Study Revisited: Bionic Ears

In both the retina and cochlea, the receptor cells do not send axons to the brain. In the retina, axons of the ganglion cells make up the optic nerve and send information to visual centers in the brain. To provide vision for someone whose rods and cones have degenerated, a retinal implant system would need to have (1) an array of miniaturized light detectors, (2) a way of stimulating the ganglion cells in a pattern that replicates the detector array, and (3) possibly some sort of power supply and lenses to focus light, depending on the design. As of 2015, one retinal implant has been approved for clinical use; several others are under development. The designs of retinal implants still in the research stage are fascinating and well worth an Internet search.

Multiple Choice

1. b; **2.** b; **3.** d; **4.** c; **5.** d

Fill-in-the-Blank

1. receptor potential; frequency
2. pinna, eardrum (tympanic membrane); hammer (malleus), anvil (incus), stirrup (stapes); oval window; hair cells
3. cornea, aqueous humor, iris; lens, vitreous humor; cornea, lens
4. rods, cones; Rods; Cones, fovea
5. sweet, salty, sour, bitter, umami; olfaction *or* smell

Review Questions

1. The intensity of a stimulus is signaled by the frequency of action potentials in axons leading from sensory receptors to the brain (and subsequently in neurons in the brain). The type of stimulus is determined by the area of the brain that is connected to the axons that are firing: Action potentials conducted to specific areas of the brain are perceived as specific sensory types (e.g., action potentials in axons of the optic nerve travel to visual areas of the brain and are always perceived as light).
2. The senses of taste and smell are provided by chemoreceptors in the tongue and nasal epithelium, respectively. Photoreceptors are the receptors for light. Mechanoreceptors are the receptors for touch and hearing.
3. We can distinguish hundreds of different flavors because (1) a given flavor may stimulate two or more different taste receptor types to different degrees (for example, sweet and sour), and (2) most foods give off odor molecules, which diffuse to the olfactory receptors and add an odor component to the basic taste. We perceive a great variety of odors because (1) we have about 400 types of odor receptor proteins embedded in the olfactory dendrites, each specialized to bind a particular type of odor molecule, and (2) complex odors stimulate two or more different types of odor receptors.
4. Sound waves enter the auditory canal and vibrate the tympanic membrane. These vibrations are transmitted through the three bones of the middle ear: the hammer, anvil, and stirrup. These bones transmit the vibrations to a smaller membrane called the oval window, which covers the opening to the cochlea, where sound vibrations travel through fluid. The fluid in the cochlea is compartmentalized by several membranes, including the basilar and tectorial membranes. The actual sound receptors, the hair cells, are attached to the basilar membrane; their hairs are attached to the tectorial membrane. Vibrations of the fluid in the cochlea move these membranes and cause the hairs to bend, setting up receptor potentials in the hair cells, which in turn release neurotransmitters onto neurons whose axons form the auditory nerve. Action potentials in the auditory nerve travel to the auditory centers in the brain.
5. The perception of the pitch (musical note) of sound occurs because different frequencies of sound vibrate the basilar membrane in different locations and thus stimulate hair cells preferentially in these locations. These hair cells in turn stimulate different axons in the auditory nerve, which connect with different auditory neurons in the brain. Sound intensity is perceived by the amplitude of vibrations of the tympanic membrane, the bones of the middle ear, the oval window, and the basilar membrane, which finally cause different amounts of bending of the hairs of the hair cells. Soft sounds cause small vibrations, which bend hairs slightly, producing small receptor potentials and a low frequency of action potentials in the auditory nerve axons. Loud sounds produce large vibrations, causing great bending of the cochlear hairs, producing larger receptor potentials and a high frequency of action potentials in the auditory nerve axons.
6. Refer to Figure 40-7 for a diagram of the human eye. The cornea is the transparent covering of the front of the eye that aids in bending incoming light for focusing on the retina. The iris is a muscular tissue that controls the size of the opening of the pupil, thereby regulating the amount of light entering the eye. The lens is composed of transparent proteins, and its flexible shape is responsible for bending light and for final fine focusing of light on the retina. The sclera is tough connective tissue surrounding the outer portion of the eye. The retina is a multilayered sheet of photoreceptors and neurons where light is converted into electrical impulses that are transmitted to the brain. The choroid is a darkly pigmented tissue located behind the retina, whose blood supply helps nourish the cells of the retina. Its dark pigment also absorbs stray light, whose reflection inside the eyeball would interfere with clear vision.
7. A circular muscle surrounding the lens changes its shape. To focus on a distant object, the lens is flattened. A relatively flat lens does not bend light much, which allows light from distant objects to be focused on the retina. Nearsightedness does not allow the focusing of distant objects, because the eyeball is too long or the cornea is too curved and so the light is focused in front of the retina. A concave lens is used to diverge light rays so that they can be properly focused on the retina.
8. Rods and cones are both photoreceptors located in the retina. Both use the same general molecules and mechanisms to absorb and respond to light, although their photopigments differ. In humans, rods are far more numerous than cones and dominate in the periphery of the retina. Rods have longer stacks of pigment-bearing membrane than cones do and are therefore more sensitive to light. Rods are responsible for vision in dim light, but rods cannot provide color vision. Cones are concentrated in the fovea and respond to different wavelengths of light, giving us color vision.
9. Pain receptors may be activated by high or low temperatures, excessive stretching, chemicals such as acids, chemicals released by damaged cells, and chemicals released during the inflammatory response. All of these stimuli result from, or may cause, damage to the body.

Chapter 41

Think Critically

Figure 41-4 There is very little overlap between thick and thin filaments in a fully stretched muscle, so very little force can be generated to pull the filaments toward the center of the sarcomere.

Figure 41-5 Fibers with the most mitochondria are slow twitch. Slow-twitch fibers rely primarily on cellular respiration, which requires oxygen that must be supplied continuously by capillary blood.

Figure 41-8 Thick exoskeletons are heavy; their weight impedes movement on land, but water supports most of this weight.

Figure 41-11 Compact bone is much heavier than spongy bone, so it would require much more energy for an animal to stand or move. Spongy bone is necessary to allow space for bone marrow where blood cells are formed.

Health Watch: Osteoporosis—When Bones Become Brittle

Low estrogen due to overtraining may have caused cessation of menstruation and also low bone density, similar to that experienced by postmenopausal women. She also may be suffering from an eating disorder brought on by an attempt to control her weight. This could result in inadequate calcium intake, further contributing to brittle bones. A leg X-ray is likely to show a fracture of the tibia.

Multiple Choice

1. d; **2.** d; **3.** a; **4.** c; **5.** c

Fill-in-the-Blank

1. hydrostatic skeleton, endoskeleton, exoskeleton; antagonistic
2. skeletal, cardiac; cardiac, smooth; cardiac, smooth
3. muscle fiber; myofibrils, sarcomeres, Z lines
4. actin, myosin; ATP
5. neuromuscular junctions; T tubules, sarcoplasmic reticulum
6. cartilage, bone, ligaments; collagen
7. hinge, ball-and-socket; flexor, extensor

Review Questions

1. Refer to Figures 41-2 and 41-3. A contracted muscle fiber will exhibit shortened sarcomeres, and the filaments will overlap more.
2. Motor neurons send axons from the spinal cord to the muscles, where they form synapses called neuromuscular junctions on muscle fibers. The neurotransmitter acetylcholine released from the motor neuron binds receptors on the muscle fiber membrane, causing an action potential in the fiber. The action potential travels down the T tubules into the interior of the muscle. This AP causes the sarcoplasmic reticulum to release calcium ions into the myofibrils. The Ca^{2+} binds to troponin proteins on the thin filament, altering their shape and causing them to pull other tropomyosin off the binding sites on the actin. This allows the myosin heads of the thick filaments to attach to the actin, initiating contraction. The sarcomeres shorten as thin and thick filaments slide past each other and the thick filaments draw closer to the Z disc. As soon as the action potential fades away, active transport proteins in the sarcoplasmic reticulum pump calcium back inside.
3. Contraction of muscles requires signals from the nervous system and occurs as described in the answer to Review Question 2 when calcium and ATP are available. There is no mechanism for the sliding filaments to actively slide "backward," but at rest the thick and thin filaments are detached from one another. This allows muscles to extend when they are stretched by other forces, such as the weight of a limb or contractions of opposing muscles.
4. See Figure 41-7. Hydrostatic skeletons support soft-bodied animals including worms, cnidarians, and many mollusks. A hydrostatic skeleton is basically a sac or tube filled with liquid and surrounded by both circular and longitudinal muscles. Contractions of the muscles force the fluid to change shape, which changes the shape of the body. Exoskeletons encase the bodies of arthropods. They are rigid, except at joints, where they are thin and flexible to allow for movement. Antagonistic muscles are attached inside around the joints, causing them to bend or straighten. Endoskeletons are internal and found in echinoderms and chordates. In vertebrates, to move the skeleton, antagonistic skeletal muscles are attached to bones and span joints. Contraction of these muscles extends or bends the joint.
5. Compact bone forms the outer shell of bone. It is dense and strong and provides an attachment site for muscle. Spongy bone is found in the interior of compact bone. It is lightweight, rich in blood vessels, and highly porous. Smooth muscle is not striped in appearance like skeletal or cardiac muscle. It surrounds the walls of the digestive tract and large blood vessels and produces slow, sustained, involuntary contractions. Striated (striped) skeletal muscle is responsible for voluntary movements. Skeletal muscle contractions range from quick twitches to powerful, sustained tension. Cartilage is primarily composed of collagen protein, with various other proteins or glycoproteins. It is somewhat flexible. Cartilage covers the ends of bone joints, supports flexible regions, and provides a framework for the respiratory system. Cartilage forms tough pads that act as shock absorbers on the knees and between vertebrae. Bone, the most rigid connective tissue, is composed of collagen fibers hardened by minerals with large amounts of calcium and phosphate.
6. Osteoblasts are bone-forming cells that gradually become osteocytes. Osteocytes may secrete substances that control the continuous remodeling of bone. Osteoclasts are bone-dissolving cells that dissolve cartilage early in development as bone replaces it, and they break down bone so it may be remodeled in response to changing stresses.
7. In hinge joints, one bone moves while the other remains fixed. A hinge joint is movable in only two dimensions. Pairs of muscles, called antagonistic muscles, lie in roughly the same plane as the joint. One end of each muscle, called the origin, is fixed to a relatively immovable bone on one side of the joint; the other end, the insertion, is attached to a mobile bone on the far side of the joint. When the flexor muscle contracts, it bends the joint. When the extensor muscle contracts, it straightens the joint.

Chapter 42

Think Critically

Figure 42-3 Asexual reproduction produces offspring that are genetically identical to the mother aphid. This is an efficient way for her to maximize the number of her genes in the population, but provides no significant opportunity for any of her offspring to have new genotypes that might enhance survival or reproduction. Because the weather is favorable and food is so abundant in the spring and summer, natural selection is not very strong, so this is probably not a major difficulty. When autumn arrives, weather and food become more problematic. Sexual reproduction produces genetically variable offspring, some of which may be better adapted to the conditions and be favored by natural selection.

Figure 42-6 Courtship rituals help ensure that animals mate with other individuals of their own species. Courtship rituals are reproductive isolating mechanisms that help prevent potentially disadvantageous hybrid mating. The courtship activities in some species stimulate ovulation.

Health Watch: Sexually Transmitted Diseases

Couple A: If they have frequent intercourse and are fairly sure that they do not want to have children, then sterilization of one of the partners would be foolproof and eventually (after recovering from surgery) a very convenient method. However, if the partners later separate, one or both may discover that they want to have children with a new partner. Sterilization can often be reversed, but the success rate is not 100%. If the partners are completely monogamous, then STDs may not be a concern, but even rare infidelity could expose them to STDs. **Couple B:** Sterilization is not a suitable method for people who want to have children eventually. The methods of IUD, birth control pills, patches, rings, injections, or implants would provide highly effective contraception and are usually readily reversible for later childbearing. These methods, however, do not provide protection against STDs, which they could be exposed to via infidelity. **Person C:** This woman would presumably be having intercourse with multiple men, who in turn may have had intercourse with other women. STD protection, therefore, is a priority. She should wear a vaginal condom or insist that her male partners wear male condoms. Condoms provide protection against pregnancy, but are not nearly as effective as an IUD or birth control pills, patches, rings, injections, or implants. Her best approach would be to combine condoms with one of these other, more effective methods of preventing pregnancy.

Multiple Choice

1. b; **2.** d; **3.** b; **4.** d; **5.** b

Fill-in-the-Blank

1. asexual; bud; fragmentation
2. testis; testosterone (or androgen); seminiferous tubules
3. nucleus, acrosome; mitochondria
4. epididymis, vas deferens, seminal vesicles, prostate gland, bulbourethral gland, urethra
5. ovary; estrogen, progesterone; secondary oocyte; polar body; oviduct (or uterine tube); uterus
6. corpus luteum; chorionic gonadotropin

Review Questions

1. Asexual reproduction is more efficient because it does not require a mate, and there is no wasted production of eggs and sperm. Asexual reproduction produces offspring that are genetically identical to the parent, which may be an advantage if the parent is well adapted to an environment, but lack of genetic diversity is a serious disadvantage in changing environments. Many sponges, cnidarians, flatworms, brittle stars, and some fish and reptiles can reproduce asexually. Sexual reproduction allows for greater genetic diversity because it combines the genetic material of two organisms. Greater genetic diversity allows better survival and reproduction of some members of a species under a variety of environmental conditions. Disadvantages of sexual reproduction include finding and competing for a suitable mate and the fact that some of the genetically variable offspring may be less well adapted to their environment. Birds, mammals, most reptiles and amphibians, and many invertebrates, including many insects, mollusks, annelids, cnidarians, and many others reproduce sexually.
2. The egg is a large, nonmotile cell that includes nutrients required for the early development of the embryo. The mammalian egg is surrounded by a jelly-like barrier, the zona pellucida, and an outer layer of follicle cells, called the corona radiata, that form barriers between egg and sperm. A sperm is a very small, motile cell with virtually no stored nutrients. It possesses a flagellum to propel it toward the egg and mitochondria just ahead of the flagellum to provide energy for its movement. The head of the sperm consists almost entirely of the nucleus and the acrosome. The acrosome contains enzymes that can dissolve a path through the protective layers surrounding the egg.
3. The corpus luteum secretes estrogen and progesterone, which prevent the development of additional follicles after ovulation. The estrogen and progesterone also stimulate growth of the endometrial lining to prepare for embryo implantation. The corpus luteum disintegrates if fertilization does not occur; if fertilization does occur, the embryo secretes chorionic gonadotropin, which maintains the corpus luteum during early pregnancy. In this case, the corpus luteum continues to secrete estrogen and progesterone, maintaining the endometrium and sustaining the pregnancy.
4. Sperm travel from the seminiferous tubules to the epididymis and then to the vas deferens, where they are stored. During ejaculation, muscles around the epididymis, vas deferens, and urethra contract, squeezing sperm out of the epididymis and vas deferens, through the urethra, and out of the penis. The sperm are deposited into the vagina, enter the uterus via the opening in the cervix, travel through the uterus, and enter the uterine tube, where fertilization usually takes place.
5. The seminal vesicles secrete fluids containing fructose that nourishes the sperm and prostaglandins that cause uterine contractions to help transport sperm into the female reproductive tract. Alkaline pH of seminal fluid helps neutralize the acid environment of the vagina, providing conditions in which sperm can survive. The prostate gland secretes a nutrient-rich, alkaline fluid that includes enzymes that increase semen fluidity, allowing sperm to swim more freely. The bulbourethral glands secrete alkaline mucus that neutralizes any remaining urine in the urethra.
6. There are both permanent and temporary methods of contraception that preserve the ability to have intercourse while preventing pregnancy. Sterilization by vasectomy (in men) or tubal ligation (in women) provides permanent contraception, although either operation can often be reversed. Temporary methods include (1) preventing ovulation by administering synthetic versions of progesterone, often combined with estrogen, in birth control pills, patches, injections, implants, and cervical rings; (2) preventing sperm and egg from meeting in the female reproductive tract, including abstinence and barrier methods, such as IUDs, condoms, cervical caps, diaphragms, vaginal sponges, and spermicides; and (3) preventing implantation of an embryo in the uterus, which is a secondary action of IUDs. Failure rates tend to be lowest for birth control pills, patches, implants, cervical rings, and IUDs. Effective protection against STDs requires a method that prevents the penis from directly contacting the vagina. For details, see Table 42-3.

Chapter 43

Think Critically

Figure 43-4 Twins would probably never occur because even the first cell division of cleavage would produce daughter cells with different developmental fates, with the result that each of the daughter cells would be able to give rise to some, but not all, of the structures of the human adult.

Figure 43-6 *Ribs*: Homeobox genes are usually active in a head-to-tail sequence, such that gene *A* might be active in the head region, gene *B* would be active in the torso, gene *C* would be active in the pelvic region, and so on. Let's assume that gene *B* causes the production of ribs. In snakes, gene *B* might be active all the way from the end of the (very short) neck to the tail segments. *Lack of legs*: A specific homeobox gene would normally be responsible for triggering the production of the entire leg (or at least most of it). A severe mutation in this particular homeobox gene might prevent leg formation.

Figure 43-7 The ovulated egg is surrounded by barriers (the zona pellucida and the corona radiata; see Chapter 42). However, it is the outer layer of the blastocyst that attaches to, and implants in, the uterus. Therefore, the blastocyst must emerge from these barriers before implantation.

Figure 43-12 Because there is only a very small placenta for exchange of nutrients, wastes, and gases, the embryo is retained in the uterus for a much shorter time. Offspring are born in a much less developed state. Development continues outside the uterus (typically inside an external pouch).

Health Watch: The Placenta—Barrier or Open Door?

The embryo is genetically different from the mother. If the bloodstreams were truly combined into one, many blood cells from the embryo would enter the mother's circulation, prompting an immune response in her body. The mother's immune response would destroy the embryo.

Health Watch: The Promise of Stem Cells

Stem cells are capable of dividing indefinitely; typically, their daughter cells differentiate into specialized cells such as those that produce bones, tendons, or ligaments. Stem cell therapy is based on the premise that the cells would divide, and their daughter cells would differentiate appropriately, replacing those lost to injury. Successful stem cell therapies in horses and dogs provide a great deal of hope that similar successes might occur in humans, although there is generally little clinical evidence. The patient should be told that stem cell therapy is experimental and may not heal the injury. In addition, assuming that stem cell therapy repaired the injured joint, after healing it would be essential that stem cell division slow down or stop. If, however, the stem cells continue to divide rapidly, they might form a tumor (some people have developed tumors after stem cell therapy). Adult stem cells do not typically cause cancers in the tissues in which they normally reside, so they would seem to be least likely to cause cancer when injected into the appropriate site in a patient. Assuming that the proper molecular signals could be provided in cell culture, partially differentiating ESCs or iPSCs might minimize their cancer risk. Producing "next-generation" iPSCs by using protein messengers that alter gene transcription, rather than inserting genes into the recipient cells, might minimize the risk from iPSCs.

Case Study Revisited: Rerunning the Program of Development

If human tissue responses to injury are similar to those in mice, then blocking p21 activity might promote regeneration. On the other hand, because p21 suppresses division in cells with damaged DNA, blocking p21 activity might also promote cancer. Interestingly, knockout mice lacking p21 do not show increased rates of cancer, but in people, reduced levels of p21 promote the formation or increase the severity of some cancers, including cancers of the colon and breast.

Multiple Choice

1. b; **2.** b; **3.** b; **4.** d; **5.** a; **6.** c; **7.** a; **8.** c

Fill-in-the-Blank

1. zygote; cleavage, morula; blastula *or* blastocyst; ectoderm, mesoderm, endoderm
2. stem cells; inner cell mass
3. homeobox genes; transcription factors
4. prolactin; oxytocin

Review Questions

1. In direct development, the newborn animal closely resembles a small, but sexually immature, adult. Usually, animals that undergo direct development either have eggs with large amounts of yolk or nourish the embryo inside the body of the mother before birth. In indirect development, an egg, usually containing relatively little yolk, hatches into a sexually immature feeding stage called a larva. The larva has a body structure that differs greatly from that of the adult. During metamorphosis, the larva undergoes a radical change in structure to become a sexually mature adult. All mammals and reptiles undergo direct development, as do some snails and fish. Examples of animals with indirect development include butterflies and moths, most frogs and toads, and many marine invertebrates such as corals and sea urchins.
2. See Table 43-2 for the structures and functions of the amnion, chorion, allantois, and yolk sac in reptiles and birds and in mammals. Similarities between reptiles and mammals: In both reptiles and mammals, the chorion is the site of gas exchange for the embryo; the amnion surrounds the embryo and encloses it in fluid; the allantois stores wastes; and the yolk sac absorbs nutrients for the embryo. Differences between reptiles and mammals: In reptiles, gas exchange across the chorion exchanges is between the embryo and the air, whereas in mammals the exchange is between the embryo and the mother. In mammals but not in reptiles, wastes and nutrients are exchanged across the chorion. In reptiles but not in mammals, the allantois acts as a respiratory surface. In reptiles but not in mammals, the yolk sac contains yolk (food). The allantois and yolk sac contribute to the umbilical cord in mammals but not in reptiles (which don't have an umbilical cord).
3. Gastrulation is the process whereby a hollow blastula is transformed into a three-layered embryo containing ectoderm, mesoderm, and endoderm. In amphibians, cells from the outside of the blastula migrate through a site called the blastopore to the inside of the blastula, forming the endoderm of the future digestive tract. The cells remaining on the outside of the blastula will form ectoderm. Further migration of cells to a location between the ectoderm and the endoderm forms the mesoderm.
4. See Table 43-1.
5. Induction is the process whereby chemical signals produced by certain cells influence the developmental fate of other, usually nearby, cells. These chemicals activate transcription of specific genes in the recipient cells, causing them to differentiate into specific cell types. The first demonstration of induction came from an experiment in which cells were transplanted to a frog blastula from the organizer region of a different frog blastula, which ultimately resulted in development of a double embryo.
6. Fertilization occurs in the uterine tube. The zygote then undergoes cleavage to become a morula, which then continues development into a blastocyst as it moves toward the uterus. The blastocyst implants in the endometrium of the uterus.
7. The blastocyst consists of an outer layer of cells that will become the chorion and an inner cell mass that will become the embryo and the other three extraembryonic membranes. During implantation, the outer layer attaches to, and then moves into, the endometrium of the uterus. As it develops into the chorion, the outer layer of cells sends chorionic villi into the endometrium, beginning the formation of the placenta. The inner cell mass splits into two cavities separated by a double layer of cells called the embryonic disk. The inner cavity (closest to the endometrium) is the amniotic cavity and is surrounded by the amnion. The outer cavity is the yolk sac. The two layers of the embryonic disk are continuous with the amnion and yolk sac, respectively. A slit forms in the layer on the "amnion side" of the disk. Cells migrate through this slit, forming mesoderm and endoderm. Some of the migrating cells form the allantois, which will become part of the umbilical cord.
8. The chorionic villi, containing a dense network of fetal capillaries, are bathed in pools of maternal blood in the endometrium. Many small molecules including oxygen, carbon dioxide, and urea diffuse across the membranes of the villi and capillaries. Nutrients are moved from mother to fetus either by diffusion or by active transport. The membranes act as barriers to the passage of many large molecules and most cells. However, lipid-soluble molecules easily cross the placenta. Some infectious microbes can also cross the placenta.
9. During pregnancy, estrogen, progesterone, and prolactin stimulate the mammary glands to grow, branch, and develop the capacity to secrete milk. Progesterone and estrogen, however, inhibit the actual secretion of milk. After birth, progesterone levels fall, and prolactin stimulates milk secretion.
10. Aging is the final stage of development, in which an animal declines in vigor, physical condition, and ability resist disease. Aging is caused by the accumulation of damage to biomolecules, especially DNA.

Chapter 44

Think Critically

Figure 44-9 Harvest the onion at the end of the first growing season, when it has stored the maximum nutrients for use by the shoot during the second year. These nutrients would be used up by the end of the second season.

Figure 44-11 In a climate that was truly uniformly warm and wet year-round, the vascular cambium would divide at the same rate year-round, and the daughter secondary xylem cells would grow the same way year-round. All of the secondary xylem cells would have the same size and cell wall thickness, so they would all be the same color. There would be no annual rings.

Figure 44-25 CO_2 is required for photosynthesis. When stomata close, photosynthesis in the leaf slows and eventually stops because closed stomata prevent CO_2 in the air from entering the leaf and reaching the mesophyll cells where photosynthesis occurs. Closed stomata also greatly reduce transpiration from the leaf. Because transpiration provides the driving force for the upward movement of water in xylem, which in turn provides the major driving force for water entry into the roots, closing the stomata would slow down water entry into the roots.

Figure 44-27 A developing leaf would be a sink for sugars to fuel its growth. Possible sources would be fully developed photosynthesizing leaves and sugars or starches stored in the stem and roots.

Earth Watch: Forests Water Their Own Trees

Fog can only supply a limited amount of water and only during the winter. Population growth in Bellavista could easily outstrip the water supply. Only strict limitations on population and water consumption will allow the town to thrive on fog. Climate change could also reduce the influx of fog.

Multiple Choice

1. a; **2.** d; **3.** c; **4.** a; **5.** c; **6.** b; **7.** c; **8.** d

Fill-in-the-Blank

1. meristem; apical meristem, lateral meristem; apical meristem; lateral meristem
2. ground, dermal, vascular; dermal; vascular; ground
3. xylem, tracheids, vessel elements; hydrogen bonds, transpiration; stomata *or* guard cells

4. Phloem; sieve-tube element, companion cells
5. root cap; root hairs *or* trichomes, active transport; mycorrhizae

Review Questions

1. Root functions include anchorage, absorption and transport of water and minerals, synthesis and transport of some hormones, and storage of sugars. Stems support the leaves; transport water, minerals, and organic molecules such as sugars, amino acids, and hormones; and store sugars and starches. Some stems photosynthesize. Leaves function in photosynthesis, which captures energy, and in transpiration, which drives water and nutrient uptake from the roots.
2. The ground tissue system forms much of the internal substance of the plant body, functioning in photosynthesis, hormone secretion, support, and food storage. The dermal tissue system covers the plant body, protecting it and regulating the evaporation of water. Vascular tissues transport water and dissolved substances. See Table 44-1 for tissue system locations within the plant body.
3. Primary growth refers to root and shoot elongation and the production of specialized internal and external parts, produced by cell division of apical meristems at the tips of stems and roots. Secondary growth is produced by cell division of the lateral meristems, consisting of vascular cambium between the xylem and phloem, and the cork cambium inside the bark. Secondary growth results in an increase in the diameter and strength of stems, branches, and roots as the plant ages.
4. Meristem cells are unspecialized, are capable of cell division, and are responsible for growth of the plant. Differentiated cells have become specialized in structure and function and are not usually capable of cell division.
5. See Figure 44-15 for a diagram of a dicot root. The epidermal cells absorb water and minerals from the soil. Most of the cortex cells store starches and may also take up water and minerals. The endodermis of the cortex regulates water and mineral entry into the vascular cylinder and prevents mineral loss from the vascular cylinder. The vascular cylinder includes the outer pericycle, which produces root branches, and inner xylem and phloem. Xylem carries water and dissolved minerals from the root into the upper parts of the plant. Phloem transports sugar and other organic molecules from sources of sugars (locations that produce more sugar than they consume) to sinks of sugar (locations that consume more sugar than they produce).
6. See Figure 44-10 for a diagram of a dicot stem. The epidermis protects the stem and regulates water evaporation. The outer cortex cells may photosynthesize and store food and water. Cells of the pith store food and water. Xylem carries water and dissolved minerals from the root into the upper parts of the plant. Phloem transports sugar and some other organic molecules from sources of sugars (locations that produce more sugar than they consume) to sinks of sugar (locations that consume more sugar than they produce). During secondary growth, the vascular cambium, located between the primary xylem and primary phloem, produces secondary xylem to the inside of the stem and secondary phloem to the outside of the stem. This increases the thickness and strength of the stem. Some epidermal, cortex, or phloem cells develop into cork cambium, which produces tough cork cells that cover the outside of the stem.
7. Xylem consists of tracheids and larger vessel elements that both die at maturity, leaving behind rigid, porous, interconnected cell walls that conduct water with dissolved minerals up the plant from the roots. Phloem consists of living sieve tubes that lack most organelles and of companion cells that maintain the sieve tubes. Phloem sieve tubes conduct sugars, amino acids, and hormones throughout the plant body from sources of sugar to sinks of sugar.
8. Epidermal cells of a young root form root hairs, which are a type of trichome. Root hairs greatly increase the surface area of the young root and function to absorb minerals and water from the soil.
9. See Figure 44-8 for a diagram of the structure of a typical dicot leaf. A leaf usually has a large surface area, covered in transparent epidermal cells that produce a waterproof outer layer of cuticle. Stomata in the lower epidermis regulate the evaporation of water out of the leaf and the entry of CO_2 into the leaf. Most of the middle of the leaf consists of photosynthetic mesophyll cells. The veins of a leaf contain xylem and phloem, which are continuous with the xylem and phloem of the stem. Xylem transports water and minerals into the leaf. Phloem transports sugars dissolved in water out of the leaf.
10. Stomata generally open during daylight to allow CO_2 entry for photosynthesis and close at night to conserve water. Stomatal opening and closing are determined by light, CO_2 levels, and water availability. During the day, light and low CO_2 levels (because photosynthesis uses up CO_2) both trigger entry of K^+ into the guard cells. Water follows by osmosis, causing the guard cells to swell and the stomata to open. At night, darkness and increased CO_2 levels (when photosynthesis stops but cellular respiration, which produces CO_2, continues) cause K^+ to leave the guard cells, causing the stomata to close. Dehydration triggers synthesis of a hormone (abscisic acid) that reduces transport of K^+ into the guard cells, causing the stomata to close, thereby conserving water.
11. In the extracellular pathway, soil water, containing dissolved minerals, bathes the root and penetrates into the extracellular spaces of the root up to the Casparian strip, which blocks further inward movement through extracellular space. In the intracellular pathway, minerals are actively transported into the root hairs, epidermal cells, cortex cells, and endodermal cells of the outer part of a young root. Minerals move from cell to cell through plasmodesmata, going inward from root hairs through the epidermal cells, cortex cells, and endodermal cells, and into the pericycle cells. The minerals are transported out of the endodermal and pericycle cells into the extracellular spaces in the vascular cylinder and then diffuse into the tracheids and vessel elements of xylem through the porous pits in their cell walls. This mineral movement results in a higher concentration of minerals in the root cells, in the extracellular space of the vascular cylinder, and in the xylem cells of the vascular cylinder than in the soil water. Water therefore moves by osmosis across the plasma membranes of root hairs, epidermal cells, cortex cells, and endodermal cells into the cells and eventually into the xylem in the vascular cylinder.
12. Cohesion caused by hydrogen bonds among water molecules causes them to stick together, forming a "water chain" within the xylem. Tension is applied to the chain of water by transpiration, as solar energy causes water to evaporate through leaf stomata. This pulls the water molecules behind them up into the leaf. The tension is transmitted all the way down to the xylem of the roots. Because only shoots (in most plants, principally leaves) usually transpire, tension always pulls the water chain upward, from root to shoot.
13. Water pressure drives phloem fluid movement from regions of the plant that produce sugar (sources) to regions that absorb sugar (sinks). Source cells (such as photosynthesizing leaves) actively transport sugar into phloem sieve tubes, and water follows by osmosis, increasing the pressure within the sieve tubes near the source. Sugar is actively transported out of phloem sieve tubes and into sink cells; water follows by osmosis, creating a region of lower pressure near the sink. The phloem fluid is thus driven by differences in pressure from sugar-producing regions to sugar-absorbing regions. Different parts of the plant may be sources or sinks at different stages of plant growth or different times of year. For example, a newly developing leaf consumes sugar for growth (a sink), but a mature leaf produces sugar during photosynthesis (a source). Root cortex cells may be a sink in summer, when they are converting sugars to starches for storage, and a source in spring, when they convert the starches back to sugars, which need to be transported up to the developing leaves. Therefore, phloem may transport sugars up or down the plant, but always from source to sink.
14. During primary growth, the surface of a stem consists of nodes and internodes. A node is the location at which the petiole of a leaf is attached to the stem; an internode is the section of stem between nodes. A terminal bud is found at the top of the stem. It consists of an apical meristem surrounded by leaf primordia. The apical meristem of the terminal bud is responsible for growth in length and the development of the major tissues of the stem (primary growth). A lateral bud is often located in the angle between the petiole and the stem; a lateral bud may develop into a branch. A flower bud may be in the same position as a terminal or lateral bud, but develops into a flower rather than elongating the shoot or developing into a branch.

Chapter 45

Think Critically

Figure 45-5 Separate bloom times reduce the chances of self-pollination and hence self-fertilization, which would result in inbreeding. Plants that do not inbreed produce a greater variety of offspring, adapted to a wider range of environments. Inbred individuals are more likely to have homozygous recessive alleles that are deleterious.

Figure 45-17 A flower whose nectar is inaccessible to insects is more likely to have nectar available for a visiting hummingbird; it is thus more likely that hummingbirds will learn to repeatedly visit this type of flower, ensuring that the flowers will be pollinated and reproduce.

Figure 45-22 Many plants do not live near flowing water, but nearly all can use wind or animals. Also, flowing water would always carry the seeds downhill and would not necessarily carry them to suitable habitats.

Earth Watch: Pollinators, Seed Dispersers, and Ecosystem Tinkering

There are hundreds of native pollinators, but they are far more difficult to raise than honeybees. Relying on native pollinators would require more natural habitat for the native pollinators to be set aside around farm fields. The fields would need to be smaller than today's large commercial fields so the bees could reach the center of the field. This approach to farming would also provide natural habitat for birds, which would help control pest insects, and create natural windbreaks, which would reduce soil erosion from wind.

Health Watch: Are You Allergic to Pollen?

The hypothesis supposes that production of ragweed pollen will increase if higher levels of CO_2 cause ragweed plants to photosynthesize at a higher rate and thereby grow faster and/or if higher temperatures increase the growth rate of ragweed plants or increase the length of the growing system. These changes would result in larger, more robust plants that produce more pollen. Elements of the hypothesis could be tested with experiments designed to determine how temperature and CO_2 concentration affect ragweed growth and pollen production (a variety of experimental designs could accomplish this).

Case Study Revisited: Some Like It Hot—and Stinky!

A reasonable hypothesis is that flowers that produce small amounts of nectar and no heat will attract pollinators that sip the nectar (acquiring pollen in the process) and quickly move to the next flower. This will guarantee a high rate of pollination. Pollinators are likely to spend a much longer time in heat-producing flowers (such as the philodendron and dead-horse arum), making pollination less efficient.

Multiple Choice

1. d; **2.** b; **3.** c; **4.** a; **5.** c

Fill-in-the-Blank

1. alternation of generations; sporophyte; meiotic cell division; gametophyte
2. pollen grain; anthers; stigma; style; integuments
3. zygote; endosperm; cotyledons
4. ovary; ovule; integuments
5. Germination; root; coleoptile; bend or hook

Review Questions

1. Refer to Figure 45-1 for a generalized diagram of the plant life cycle. The sporophyte is the diploid stage, and the gametophyte is the haploid stage. Gametes are formed by mitotic cell division in the gametophyte stage. Spores are formed by meiotic cell division in the sporophyte stage.
2. Refer to Figure 45-4 for a diagram of a complete flower. Male gametophytes are formed in the anther, and female gametophytes are formed in the ovule within the ovary.
3. Refer to Figure 45-9. The female gametophyte develops through meiotic cell division of a megaspore mother cell. Three of four haploid megaspores degenerate; the nucleus of the remaining megaspore divides by mitosis to produce eight haploid nuclei. Cytokinesis then parcels these nuclei into a total of seven cells. One cell with one haploid nucleus is the egg. A large central cell contains two haploid nuclei. Double fertilization involves fertilization of the haploid egg by one sperm and the central cell by a second sperm, both from the same pollen grain. Both nuclei of the central cell fuse with a sperm nucleus, forming a triploid nucleus in this cell.
4. Refer to Figure 45-6. The pollen grain is the male gametophyte. It is formed by meiotic cell division of a microspore mother cell in an anther. The resulting four haploid microspores divide by mitotic cell division to produce the male gametophyte, consisting of a tube cell containing a generative cell within its cytoplasm. The generative cell divides to become two sperm cells. Meanwhile, the pollen grain develops a tough outer coat.
5. Seed parts include the seed coat, endosperm, and the embryo. The seed coat is a protective covering that, in some seeds, ensures dormancy until conditions are favorable for germination. The seed coat develops from the integuments of the ovule. Endosperm is a source of food for the developing embryo; it develops from the central cell. The embryo, which develops from the zygote, is a developing sporophyte, consisting of an embryonic root and shoot and one or two cotyledons.
6. Wind-pollinated flowers have anthers that hang outside of the flowers to allow the pollen to be carried away by a breeze. They usually do not have colorful petals or produce scents (which would serve to attract animal pollinators). Beetle-pollinated flowers often smell like dung or rotting flesh because beetles usually feed on animal material. Bee-pollinated flowers produce sweet scents, are colorful, and may have patterns that show in UV light, directing the bees to the center where nectar is stored. The flower shapes also often ensure that pollen is brushed against the insect. Hummingbird-pollinated flowers have deep, nectar-filled tubes and are often colored red (a color that bees do not perceive).
7. Refer to Figure 45-12. Endosperm is a food source contained in a seed. A sperm fuses with the central cell of a female gametophyte. The sperm nucleus and the two nuclei of the central cell fuse, making a triploid nucleus. Endosperm is generally most abundant in mature monocot seeds. In monocots, most endosperm is stored within the seed coat outside the embryo with very little stored in the cotyledon. In dicots, the endosperm is stored within the cotyledons.
8. Mechanisms of breaking dormancy include (1) drying, (2) cold, and (3) disruption of the seed coat. A requirement for drying prevents seed from germinating while it is still in the fruit. Requiring a cold period forces the seed to overwinter in a dormant state. Disruption of the seed coat, specifically, washing out chemicals that inhibit germination, is required in many desert plants. A requirement for hard rains to wash out inhibitory chemicals ensures that there will be enough soil moisture for the seed to germinate, grow, flower, and set seed in its turn.
9. Monocots protect the growing shoot tip with a tough sheath called the coleoptile that surrounds the shoot tip. Dicot shoots form a hook in their hypocotyl or epicotyl. The bend of the hook, encased in epidermal cells with tough cell walls, leads the way through the soil, clearing the way for the downward-pointing apical meristem with its delicate new leaves.
10. Most fleshy fruits entice animals to eat them. The undigested seeds are excreted unharmed, often in a distant location. Adhesive or pronged fruits attach to animals and are carried some distance before falling from the animal's fur or being removed during grooming or shedding. Lightweight fruits with wind-catching surfaces are carried by air currents. Explosive fruits shoot their seeds some distance from the parent plant. Water-dispersed fruits are buoyant and watertight and have substantial food stores, allowing them to take root on nutrient-poor sandy shores.

Chapter 46

Think Critically

Figure 46-5 Phototropism would dominate in the shoot, which requires light to survive.

Figure 46-7 The lateral buds would be inhibited, just as if the meristem were present to release auxin.

Figure 46-10 With no ethylene present, tomato ripening—and the rotting that follows—would be delayed. Tomatoes would be resistant to bruising during shipping. Shipping and storing could be done at room temperature, saving energy and greatly reducing waste. They could be ripened just before marketing by exposing them to ethylene.

Doing Science: Discovering How Plants Grow Toward Light

The scientific method applies to the initial experiment as follows: *observation*: a region below the coleoptile tip bends toward light; *question*: What causes this region to bend?; *hypothesis*: the tip senses the light and causes bending of the region below it; *prediction*: if light is blocked from the tip, no bending will occur; *experiment*: block light by covering only the tip with an opaque cap; *conclusion*: blocking light from the tip prevents bending, supporting the hypothesis. There are two appropriate controls for this experiment: first, uncovered coleoptiles (to control for light hitting the coleoptile), and second, clear caps (to control for the physical presence of a cap). The scientific method applies to the next experiment as follows: *observation (from previous experiment)*: an opaque cap prevents bending toward light; *question*: does the bending region (as well as the tip) need to be exposed to light for bending to occur?; *hypothesis*: only the tip needs to be exposed for bending to occur; *prediction*: if light is blocked from the bending region, bending will still occur; *experiment*: block light by covering only the bending region with an opaque sleeve; *conclusion*: blocking light from the bending region allows bending, supporting the hypothesis. There are two appropriate controls for this experiment: first, uncovered coleoptiles (to control for light hitting the bending region), and second, clear sleeves (to control for the physical presence of a sleeve).

Case Study Revisited: Predatory Plants

If bogs become nitrogen-enriched from fertilizer runoff, a variety of other plants would thrive there as well. These plants would likely outcompete the carnivorous plants, which expend considerable energy capturing and digesting insects.

Multiple Choice

1. b; **2.** a; **3.** d; **4.** d; **5.** d

Fill-in-the-Blank

1. inhibits, apical dominance; stimulates
2. gibberellin; division, elongation
3. ethylene; abscission layer
4. root apical meristem; inhibits, stimulates, stems
5. abscisic acid; dormancy

Review Questions

1. In both roots and shoots, light is sensed by photoreceptor molecules that cause auxin to accumulate on the side of the plant away from light. Auxin stimulates cell elongation in shoots but inhibits it in roots, causing opposite bending in response to light. In both shoots and roots, gravity is sensed by statoliths, which stimulate auxin accumulation on the downward side. The opposite effects of auxin on shoots and roots causes shoots to bend away from the force of gravity and roots to bend toward it.
2. Apical dominance is the process by which auxin production by the apical meristem reduces sprouting and growth of lateral buds, acting most strongly near the apical meristem where it is most concentrated. In contrast, root branches are stimulated by auxin as it is transported through the stem and down into the root. The effects of auxin are opposed by cytokinin, produced by the root apical meristem. As it travels upward through the root, cytokinin inhibits root branching, acting most strongly near the root apical meristem where it is most concentrated. The constantly changing ratios of auxin and cytokinin from the shoot tip to the root tip influence the shape of the maturing plant, as shown in Figure 46-7.
3. Phytochrome is a plant pigment that occurs in an active form (P_{fr}) and an inactive form (P_r). These change from one form to the other under the influence of light. P_r absorbs red light, producing P_{fr}; P_{fr} absorbs far-red light, producing P_r. In sunlight, both red and far-red wavelengths are present, and active P_{fr} controls the plant response to light. In the dark, P_{fr} breaks down spontaneously to P_r. Plants use this phytochrome system to detect night length. Responses mediated by phytochrome include flowering, seed germination, elongation of the emerging shoot, leaf expansion, the development of chloroplasts, and stem elongation that helps shaded plants reach the sunlight.
4. Fruit development is triggered by auxin and gibberellin. Banana ripening is stimulated by ethylene.
5. Senescence is a genetically programmed aging process during which starch and chlorophyll in leaves are broken down and their component molecules are stored in the stems and roots of the plant. The leaf abscission layer is also weakened, allowing the leaves to fall. Auxin and cytokinin levels decline, and ethylene levels increase. Ethylene stimulates leaf abscission.
6. Gibberellin is used to increase grape size and create larger, looser grape clusters. Ethylene is used to speed the ripening of fruit, such as bananas and tomatoes, after they are shipped to market.
7. In response to attack by caterpillars, corn plants release a mixture of volatile chemicals that attract a parasitic wasp, which then attacks the caterpillars. When spider mites attack lima bean plants, the beans release a chemical that attracts another type of mite that preys on the spider mite. Wild tobacco plants attacked by hawk moth larvae also release parasitic wasp-attracting chemicals. Tobacco plants infected with a plant virus produce large quantities of salicylic acid, which activates an immune response in the plants and helps them to fight off the viral attack. The plant also releases methyl salicylate ("wintergreen"), which is absorbed by nearby plants and enhances their immune defenses.
8. Sensitive plants transmit electrical signals that travel from the touched leaf to the petiole. The signal causes specialized "motor cells" (located at the base of each leaflet and in the petiole where it joins the stem) to increase their permeability to potassium ions. The ions flow out of the motor cells along their concentration gradients, causing water to follow by osmosis. As the cells shrink from water loss, both the leaflets and the petioles rapidly droop. The rapidly drooping leaf may startle an insect alighting on it to feed.

Credits

Unit Openers: Unit 1: Biophoto Associates/Science Source; Unit 2: Ariadne Van Zandbergen/Alamy Stock Photo; Unit 3: blickwinkel/Alamy Stock Photo; Unit 4: Mint Images Limited/Alamy Stock Photo; Unit 5: Carlos Villoch - MagicSea .com/Alamy Stock Photo; Unit 6: pitinan/123rf

Chapter 1 opener: Philippe Psaila/Science Source; 1-1: Biophoto Associates/Science Source 1-2: MELBA PHOTO AGENCY/Alamy Stock Photo; 1-4: Rido/Shutterstock; 1-5: Terry Audesirk; 1-6a: Dr. Tony Brain/Science Source; 1-6b: Terry Audesirk; 1-6c: Eric Baccega/AGE Fotostock; 1-7: NASA; 1-8: Penny Tweedie/Alamy Stock Photo; 1-9: Jose Maria Farfagl WENN Photos/ Newscom; 1-12: Roy Kaltschmidt/Lawrence Berkeley National Laboratory; Have You Ever Wondered: Susan Schmitz/Shutterstock; 1-13: John Durham/ Science Source

Chapter 2 opener: The Asahi Shimbun/Getty Images; E2-1: National Institutes of Health; E2-2: Science Source/Science Source; 2-4: Piotr Sosnowski; E2-3: Juergen Lenhardt/Getty Images; E2-4: Terry Audesirk; E2-4 inset: Valentyn Volkov/Shutterstock; 2-8a: Basil/Shutterstock; 2-8b: JanMiko/Getty Images; 2-8c: Terry Audesirk; 2-10: Renn Sminkey/Pearson Education, Inc.; 2-12: Westend61 GmbH/Alamy Stock Photo

Chapter 3 opener: Terry Audesirk; 3-9a: Jeremy Burgess/Science Source; 3-10a: Mike Norton/Shutterstock; 3-10b: Marilyn Schaller/Science Source; 3-10c: Biophoto Associates/Science Source; 3-11: Terry Audesirk; 3-12a: image BROKER/Alamy Stock Photo; 3-12b: BGSmith/Shutterstock; 3-12c: Becky Sheridan/Shutterstock; 3-18: Fotokostic/Shutterstock; Have You Ever Wondered top: VERSUSstudio/Fotolia, bottom: Valua Vitaly/Fotolia; 3-23a: Sylvie Bouchard/Fotolia; 3-23b: Terry Audesirk; E3-1: GJLP/Science Source; 3-24: Terry Audesirk; 3-25: Lotus studio/Shutterstock

Chapter 4 opener: Sam Ogden/Science Source; 4-1a left: Biophoto Associates/Science Source; 4-1a right: Dr. Cecil H. Fox/Science Source; 4-1b: The Print Collector/Alamy Stock Photo; 4-1c: M. I. Walker/Science Source; 4-2 left: Monty Rakusen/Getty Images; 4-2 top center: Stacie Kirsch/Electron Microscopy Sciences; 4-2 top right: CNRI/Science Source; 4-2 bottom center: SPL/Science Source; 4-2 bottom right: Professors P. M. Motta, S. Makabe & T. Naguro /Science Source; 4-5b: CNRI/Science Source; 4-5c: Centers for Disease Control and Prevention/Science Source; 4-5d: Biology Pics/Science Source; 4-5e: Dr. Kari Lounatmaa/Science Sources; Case Study: Steve Gschmeissner/Science Source; 4-9b: Dr. Elena Kiseleva/Science Source; 4-10: Melba/AGE Fotostock; 4-11: Dr. Elena Kiseleva/Science Source; 4-12: TKeith R. Porter/Science Source; 4-13: Omikron/Science Source; 4-14: Biophoto Associates/Science Source; 4-15b: Jennifer Waters/Science Source; E4-1: Toby Melville/Reuters; 4-16b top: MedImage/Science Source; 4-16b bottom: Don W. Fawcett/Science Source; 4-17: Biophoto Associates/Science Source; 4-20b: Walter Dawn/Science Source; 4-21a: Don W. Fawcett/Science Source; 4-21b: PHANIE AGENCY/Science Source; 4-21c: David M. Phillips/ Science Source

Chapter 5 opener: Kevin Wells Photography/Shutterstock; E5-2: Ted Kinsman/ Science Source; E5-3: Chrispeels, Maarten J. and Peter Agre. "Aquaporins: water channel proteins of plant and animal cells" Trends in Biochemical Sciences. Volume 19, Issue 10, October 1994, Pages 421–425. Elsevier; 5-6a: David M. Phillips/Science Source; 5-6b: Dr. Amar/Phanie/AGE Fotostock; 5-6c: David M. Phillips/Science Source; 5-7: Nigel Cattlin/Science Source; 5-9b: Don W. Fawcett/Science Source; 5-10: Republished with permission of The Company Biologists Ltd, Perry, M. M., & Gilbert, A. B. (1979). Yolk transport in the ovarian follicle of the hen (Gallus domesticus): lipoprotein-like particles at the periphery of the oocyte in the rapid growth phase. Journal of cell science, 39(1), 257–272; 5-11b: Eric V. Grave/Science Source; 5-11c: SPL/Science Source; 5-12: Linda Hufnagel; 5-14: Justin Schwartz

Chapter 6 opener: Lucas Jackson/Reuters; 6-1: David Wall/Alamy Stock Photo; 6-3: Zoonar RF/Getty Images; 6-5: Terry Audesirk; E6-1: Michael Brooke/Alamy Stock Photo; Have You Ever Wondered: Seonyeong Kwak/Michael Strano, Strano Research Group MIT; E6-2: Ashley Corbin-Teich/Getty Images; 6-13: Terry Audesirk

Chapter 7 opener: Roger Harris/Science Photo Library/Getty Images; 7-2: Dr. Jeremy Burgess/Science Source; 7-3a: Haveseen/Shutterstock; 7-3c: John Durham/Science Source; 7-3d: Omikron/Science Source; 7-6: Easyshutter/ Shutterstock; E7-2: Rich Carey/Shutterstock; E7-2 inset: Eric Gevaert/ Shutterstock

Chapter 8 opener: Classic Image/Alamy Stock Photo; 8-1b: Marek MIs/Science Photo Library Ltd.; 8-4b: CNRI/Science Source; 8-8: Mara Zemgaliete/Fotolia; E8-1 left: Tassii/E+/Getty Images; E8-1 right: William Leaman/Alamy Stock Photo; 8-10: Simon Balson/Alamy Stock Photo; 8-12: Rostislav_Sedlacek/ Shutterstock; 8-13: Rui Vieira/EMPPL PA Wire/AP Images

Chapter 9 opener: Jerry Holt/ZUMA Press/Newscom; 9-2b: Mark Rainey; 9-3a: Hazel Appleton/Health Protection Agency Centre for Infections/Science Source; 9-3b: M. I. Walker/Science Source; 9-3c: Biophoto Associates/Science Source; 9-3d: Terry Audesirk; 9-5: Andrew Syred/Science Source; 9-8a-g: Jennifer Waters/Science Source; 9-8h: Thomas Deerinck, NCMIR/Science Source; E9-1: Du Cane Medical Imaging Ltd./Science Source

Chapter 10 opener: Mikael Buck Photography Limited; 10-1: Biophoto Associates/Science Source; 10-10a: CNRI/SCIENCE PHOTO LIBRARY; 10-10b: philidor/Fotolia

Chapter 11 opener: Ronald C. Modra/Sports Illustrated/Getty Images; 11-2: Science Source/Science Source; 11-10: Martin Shields/Alamy Stock Photo; 11-14: Sarah Leen/National Geographic Stock; 11-15: David Hosking/Alamy Stock Photo; Have You Ever Wondered: Liliya Kulianionak/Shutterstock; 11-18: Terry Audesirk; 11-20: Science History Images/Alamy Stock Photo; 11-21: Mary Evans Picture Library/Science Source; E11-1: Vetpathologist/ Shutterstock; 11-22: David J. Phillip/AP Images

Chapter 12 opener: David Bagnall/Alamy Stock Photo; 12-2a: Omikron/ Science Source; 12-2b: NLM/Science Source; 12-3: A. Barrington Brown/ Science Source; 12-9: Bruce Stotesbury/Times Colonist

Chapter 13 opener: Luke Martineau; 13-4: Oscar L. Miller, Jr.; E13-1: AIS-DSD Support Group; E13-2: Randy Jirtle; 13-10: dpa picture alliance/Alamy Stock Photo

Chapter 14 opener: P. Kevin Morley/Richmond Times Dispatch/AP Images; 14-4: Kavram/Shutterstock; E14-1: RPMG srl/Getty Images; E14-2 top to bottom: Daniel Prudek/Shutterstock, Jan S./Shutterstock, Studiotouch/ Shutterstock, Stubblefield Photography/Shutterstock; 14-11: Science Photo Library/Alamy Stock Photo; 14-12: Angus Catchot; p. 216: Yodsawaj Suriyasirisin/123RF; 14-13: Tim Flach/Getty Images; 14-14: Kathy Hutchins/ Shutterstock; E14-4: Lada/Science Source; 14-16: International Rice Research Institute; 14-18: Carl Lynn/Richmond Times Dispatch/AP Images

Chapter 15 opener: Tony Heald/Nature Picture Library; 15-4a: Scott Orr/iStock/ Getty Images; 15-4b: Woudloper; 15-4c: Smithsonian Institution Libraries; p. 234: FineArt/Alamy Stock Photo; E15-1 top left: Ingo Schulz/imageBROKER/ AGE Fotostock; E15-1 top right: Stefan Huwiler/imageBROKER/AGE Fotostock; E15-1 bottom left: ARCO/Therin-Weise/AGE Fotostock; E15-1 bottom right: Ross Hoddinott/Nature Picture Library; E15-2: LETZ/SIPA/Newscom; 15-6: James Carmichael/NHPA//Science Source; 15-10: Stephen Dalton/Science Source; 15-11: TT Nyhetsbyrån AB; 15-13a: John Pitcher/Getty Images; 15-13b: GK Hart/Vikki Hart/Getty Images; 15-14: Robin M. Tinghitella; E15-3: Reinhard, H./Arco Images GmbH/Alamy Stock Photo; 15-15: VEM/Science Source

Chapter 16 opener: NIAID/Science Source; 16-6b: Fogstock/AGE Fotostock; E16-1: Rebecca Jackrel/Alamy Stock Photo; 16-7: The Alan Mason Chesney Medical Archives; 16-8: Johann Schumacher/Getty Images; 16-9: James Steinberg/Science Source; 16-10: Warren Metcalf/Shutterstock; 16-11: zjk/ Fotolia; 16-13: Photo by Thomas B. Smith/UCLA Center for Tropical Research

Chapter 17 opener: CB2/ZOB/Chris Austin/WENN.com/Newscom; 17-1: Jeremiah Easter/Tetrachromat Design; 17-2a: Rick and Nora Bowers/Alamy Stock Photo; 17-2b: Tim Zurowski/AGE Fotostock; 17-3a: Thomas & Pat Leeson/ Science Source; 17-3b: Thomas & Pat Leeson/Science Source; 17-4: Tim Laman/ National Geographic Creative; 17-5: Alex Mustard/naturepl.com/Nature Picture Library/Alamy Stock Photo; 17-6: Phil Savoie/Nature Picture Library; 17-7: Boaz Rottem/Alamy Stock Photo; E17-1: Gary Cranitch/Newscom; 17-9a: WWF/ Newscom; 17-9b: Mark Gurney/Smithsonian/Sipa USA/Newscom; 17-12a: Goss Images/Alamy Stock Photo; 17-12b: A. C. Medeiros/Carr Botanical Consultation; 17-12c: Jack Jeffrey/Photo Resource Hawaii/Alamy Stock Photo; 17-12d: A. C. Medeiros/Carr Botanical Consultation; 17-13: Steve Apps/Alamy Stock Photo; E17-3: Clark Palinggi/Thomson Reuters (Markets) LLC

Chapter 18 opener: Keren Su/Getty Images; 18-4: Mark Garlick/Science Source; 18-6: Michael Plewka; 18-7a: Chase Studio, OMNH; 18-7b: Peter Halasz; 18-7c: Grauy/Getty Images; 18-7d: Douglas Faulkner/Science Source/

Getty Images; 18-8: Richard Bizley/Science Source; 18-9: Terry Whittaker/Science Source; 18-10: Chris Butler/Science Source; 18-12a: Tom McHugh/Science Source; 18-12b: Kevin Schafer/Alamy Stock Photo; 18-12c: TUNS/Arco Images GmbH TUNS/Arco Images GmbH/Glow Images; 18-13: Michel Brunet; 18-15a: Javier Trueba/MSF/Science Source; 18-15b left: Pascal Goetgheluck/Science Source; 18-15b right: CM Dixon/Heritage Image/AGE Fotostock; 18-15c left: Dirk Wiersma/Science Source; 18-15c center: Mary Evans/Mary Evans Picture Library Ltd/AGE Fotostock; 18-15c right: Martin Land/Science Source; 18-16: Ira Block/National Geographic Stock; 18-17: David Frayer, Dept. of Anthropology, University of Kansas; 18-18: Stéphane Marc/PHOTOPQR/LE DAUPHINE/Newscom

Chapter 19 opener: Kristin Mosher/DanitaDelimont.com/Alamy Stock Photo; 19-1a: Steve Byland/Shutterstock; 19-1b: Nnehring/IStock/Getty Images; 19-1c: Steve Maslowski/Science Source; 19-2a: Ami Images/Science Source; 19-2b: Scimat/Science Source; 19-5a: Kwangshin Kim/Science Source; 19-5b: W. Jack Jones/Springer Publishing; Have You Ever Wondered: Ph. D. Frank Collins/Centers for Disease Control and Prevention (CDC); 19-8: Margraf, Celso

Chapter 20 opener: Rupp Tina/Bon Appetit/Alamy Stock Photo; 20-1a: Mediscan/Alamy Stock Photo; 20-1b: Science Source; 20-1c: Scott Camzine/Science Source; 20-2b: Power & Syred/Science Source; 20-3: Eye of Science/Science Source; 20-4: Dr. Kari Lounatmaa/Science Source; 20-5: Galyna Andrushko/Fotolia; 20-6: Michael Abbey/Science Source; E20-1: CNRI/Science Source; 20-7: CNRI/Science Source; 20-8: Eye of Science/Science Source; 20-9a: Nigel Cattlin/Science Source; 20-9b: Steve Gschmeissner/Science Source; 20-11a: Eye of Science/Science Source; 20-11b: Department of Microbiology, Biozentrum, University of Basel/Science Source; 20-11c-d: Dr. Linda M. Stannard/University of Cape Town/Science Source; 20-13: Eye of Science/Science Source; 20-14: Evan Oto/Science Source; 20-15: EM Unit, VLA/Science Source; 20-16: Africa Studio/Shutterstock

Chapter 21 opener: Alexis Rosenfeld/Science Source; 21-1a: M. I. Walker/Science Source; 21-1b: Eric V. Grave/Science Source; 21-2: Professors P. M. Motta & F. M. Magliocca/Science Source; 21-3: Eric V. Grave/Science Source; 21-5: Eye of Science/Science Source; 21-6: Astrid & Hanns-Frieder Michler/Science Source; 21-7a: David Hosking/D. P. Wilson/ScienceSource; 21-7b: Mark Conlin/Alamy Stock Photo; 21-8: David M. Phillips/Science Source; E21-1: Science Source/Science Source; 21-9: Pete Atkinson/Science Source; 21-12: Eye of Science/Science Source; 21-13: Andrew Syred/Science Source; 21-14: Manfred Kage/Science Source; 21-15: M. I. Walker/Science Source; 21-16a: Patrick W. Grace/Science Source; 21-16b: Ray Simons/Science Source; 21-18: Francis Abbott/Nature Picture Library; 21-19a: Ray Simons/Science Source; 21-19b: Nick Upton/Nature Picture Library; 21-19c: Pascal Goetgheluck/Science Source

Chapter 22 opener: A & J Visage/Alamy Stock Photo; 22-2: Andre Seale/AGE Fotostock; 22-2 inset: Emilie Nat/Dutch Information Centre on Charophytes; 22-3: Andrew Syred/Science Source; 22-5a: Adrian Davies/Nature Picture Library; 22-5b: BerndH; 22-5c: Anest/Shutterstock; 22-5d: Dr. Morley Read/Shutterstock; 22-6: JOSEPH SZALAPSKI/Alamy Stock Photo; 22-7a: Miika Silfverberg; 22-7b: imageBROKER/Alamy Stock Photo; 22-7c: GYRO PHOTOGRAPHY/Getty Images; 22-7d: David Wall/Alamy Stock Photo; 22-8: Ed Reschke/Getty Images; 22-9b: Brian Jackson/Getty Images; 22-9c: Khoroshunova Olga/Shutterstock; 22-10a: Terry Audesirk; 22-10b: Melvyn Longhurst/Alamy Stock Photo; 22-10c: Michael Schwab/Getty Images; 22-10d: Alamy Stock Photo; 22-11 top: Ivan Ulianovsky/Shutterstock; 22-11 bottom: Henk Bentlage/Shutterstock; 22-12a: Jin-liang Lin/Dreamstime; 22-12b: Yuttasak Jannarong/AGE Fotostock; 22-12b inset: Zoonar/S Caston/AGE Fotostock; 22-12c: Terry Audesirk; 22-12d: Larry West/Science Source; E22-1: ImageMore/AGE Fotostock

Chapter 23 opener: Philippe Clement/Nature Picture Library; 23-1a: blickwinkel/Alamy Stock Photo; 23-1b: Dr. Tony Brain/Science Source; 23-2: Biophoto Associates/Science Source; 23-3a: Jeff Lepore/Science Source; 23-3b: Darlyne A. Murawski/Getty Images; 23-6: Thomas J. Volk; 23-7: Mark Brundrett; 23-8: S. Lowry/Univ Ulster/The Image Bank/Getty Images; 23-9a: Fertnig/iStock/Getty Images; 23-9b: blickwinkel/Laule/Alamy Stock Photo; 23-9c: blickwinkel/Alamy Stock Photo; 23-10: Stan Rohrer/Alamy Stock Photo; 23-11: Andrew Syred/Science Source; 23-12a: Fletcher & Baylis/Science Source; 23-12b: Nedim Jukic/Dreamstime; 23-13 top: Gregory G. Dimijian/Science Source; 23-13 bottom: Ed Reschke; 23-14b: Terry Audesirk; 23-14c: Robin Chittenden/Frank Lane Picture Agency; E23-1: Ryan von Linden/New York Department of Environmental Conservation; 23-16a: Inga Spence/Science Source; 23-16b: Nigel Cattlin/Alamy Stock Photo; 23-17: Francesco Tomasinelli/Science Source; 23-18: Hugh Sturrock; 23-19: Matt Meadows/Photolibrary/Getty Images; 23-20: Terry Audesirk; Have You Ever Wondered: Tony Gentile/Reuters

Chapter 24 opener: Bildagentur Rm/AGE Fotostock; 24-5a: Gerd Guenther/Science Source; 24-6a: Masa Ushioda/Image Quest Marine; 24-6b: John Anderson/Alamy Stock Photo; 24-6c: Durden Images/Shutterstock; 24-8a: Gregory Ochocki/Science Source; 24-8b: Oxford Scientific/Getty Images; 24-8c: Constantinos Petrinos/Nature Picture Library; 24-8d: David Doubilet/Getty Images; 24-10a: Omikron/Science Source; 24-11: WaterFrame/Alamy Stock Photo; 24-12a: Biophoto Associates/Science Source; 24-12b: M I (Spike) Walker/Alamy Stock Photo; 24-12c: CK Ma/Shutterstock; 24-13 left: Andrew Syred/Science Source; 24-13 right: CDC/Science Source; E24-1: Ethan Daniels/Shutterstock; 24-15a: Tim Rock/Waterframe/AGE Fotostock; 24-15b: Peter Batson/Image Quest Marine; 24-15c: James H. Robinson/Science Source; 24-17a: Ray Coleman/Science Source; 24-17b: Juuyoh Tanaka; 24-18a: Stephen Frink/Photodisc/Getty Images; 24-18b: Kuttelvasrova Stuchelova/Shutterstock; 24-19a: David Shale/Nature Picture Library; 24-19b: David Fleetham/Alamy Stock Photo; 24-19c: Kjell Sandved/Science Source; E24-3: Dr. Tsunemi Kubodera, National Science Museum of Japan; E24-4: Edith Widder/Ocean Research and Conservation Association Inc.; 24-20: Malcolm Schuyl/Frank Lane Picture; 24-21: Brian1442; 24-23: Susumu Nishinaga/Science Source; 24-24a: J. Meul/Arco/AGE Fotostock; 24-24b: Hugh Lansdown/Shutterstock; 24-24c: Stephen Dalton/Science Source; 24-24d: Dean Evangelista/Shutterstock; 24-25a: Audrey Snider-Bell/Shutterstock; 24-25b: Picture Bank/Alamy Stock Photo; 24-25c: Terry Audesirk; 24-26a: Tom McHugh/Science Source; 24-26b: D Assmann/AGE Fotostock; 24-27a: Tom Branch/Science Source; 24-27b: George Grall/National Geographic/Getty Images; 24-27c: Beverly Speed/Dreamstime; 24-27d: Age Fotostock/SuperStock; 24-28: The Natural History Museum/Alamy Stock Photo; 24-29a: Jessica Wilson/Science Source; 24-29b: Nnehring/E+/Getty Images; 24-30a: Terry Audesirk; 24-30b: Andrew Martinez/Science Source; 24-30c: aquapix/Shutterstock; 24-31b: Michael Male/Science Source; 24-32: Volker Steger/Science Source

Chapter 25 opener: Hoberman Collection/Universal Images Group Editorial/Getty Images; 25-2: JOHN GIANNICCHI/Science Source; 25-3: Frank Hecker/Alamy Stock Photo; 25-4: Heather Angel/Natural Visions/Alamy Stock Photo; 25-5a: Tom McHugh/Science Source; 25-5b: CPEN1/ZOB/Cover Images/Newscom; 25-6: F Hecker/Blickwinkel/AGE Fotostock; 25-6 inset: Gena Melendrez/Shutterstock; 25-7a: Jeff Rotman/Nature Picture Library; 25-7b: DAVID HALL/Science Source; 25-8a: Peter David/The Image Bank/Getty Images; 25-8b: MICHAEL NEUMANN/Science Source; 25-8c: Reinhard Dirscherl/AGE Fotostock; 25-9: Tom McHugh/Science Source; 25-10a: Photononstop/SuperStock; 25-10b: ivkuzmin/iStock/Getty Images; 25-10c: A. Cosmos Blank/Science Source; 25-10d: Dante Fenolio/Science Source; 25-11a: hakoar/Fotolia; 25-11b: Joseph T. & Suzanne L. Collins/Science Source; 25-11c: McPHOTO/AGE Fotostock; E25-1: Stanley Breeden/Getty Images; 25-12: Victoria Stone & Mark Deeble/Oxford Scientific/Getty Images; 25-13a: OBXbchcmbr/E+/Getty Images; 25-13b: David & Sheila Glatz/Getty Images; 25-13c: Ainars Aunins/Alamy Stock Photo; 25-14a: Michael Mocanachie/Papilio/Alamy Stock Photo; 25-14b: Craig Ingram/Alamy Stock Photo; 25-15a: Mark Newman/SuperStock; 25-15b: Dave Watts/Alamy Stock Photo; 25-15c: Jean-Louis Klein & Marie-Luce Hubert/Science Source; 25-16a: aaprophoto/iStock/Getty Images; 25-16b: Jonathan Watts/Science Source; 25-16c: Masa Ushioda/AGE Fotostock; 25-16d: Brehm, H/Arco Images GmbH/Alamy Stock Photo; 25-16e: Samuel Maglione/Science Source

Chapter 26 opener left to right: Andy Sands/Nature Picture Library, Pascal Le Segretain-Staff/Getty Images; 26-1a: blickwinkel/Lammers/Alamy Stock Photo; 26-1b: Thomas Hinsche/imageBroker/Alamy Photo Stock; 26-2a: Tom/Pat Leeson/Mary Evans Picture Library Ltd/AGE Fotostock; 26-2b: Brian Miller/Photoshot License Limited; 26-3: Michael Nichols/National Geographic/Getty Images; 26-4: Thomas D. McAvoy/The LIFE Images Collection/Getty Images; 26-5a: STR New/Reuters; 26-5b: Mark J. Barrett/Alamy Stock Photo; 26-7: Howard Burditt/Reuters; 26-8a: Yann Hubert/Science Source; 26-8b: Steve Bloom Images/Alamy Stock Photo; 26-9: Sarah Jelbert; 26-10: Barbara von Hoffmann/Alamy Stock Photo; 26-11: William Leaman/Alamy Stock Photo; 26-12: Nick Upton/Nature Picture Library; 26-13: Bev Wigney; 26-14a: Chinahbzyg/Shutterstock; 26-14b: Andrew Syred/Science Source; 26-16: Bernard Walton/Nature Picture Library; 26-17a: Mark Hamblin/AGE Fotostock; 26-17b: Duncan Usher/Alamy Stock Photo; 26-18: H. Eisenbeiss/Frank Lane Picture Agency Limited; 26-19: Ken Preston-Mafham/PREMAPHOTOS/Nature Picture Library; 26-20: Danita Delimont & Christian Ziegler/Alamy Stock Photo; 26-21a: Arco Images GmbH/G. Lacz/Alamy Stock Photo; 26-21b: Richard Herrmann/SeaPics.com; 26-22: Rick Price/Stockbyte/Getty Images; 26-23: Jakob Fahr; 26-24a: Tim Laman/Nature Picture Library; 26-24b: Wayne Lynch/AGE Fotostock; 26-25a: Rolf Nussbaumer Photography/Alamy Stock Photo; 26-25b: Terry Audesirk; 26-26: Stuart Wilson/Science Source; 26-28: Creatas/AGE Fotostock; 26-29a: J Sohns/C Sohns/AGE Fotostock; 26-29b: Eric Baccega/AGE Fotostock; 26-29c: georgesanker.com/Alamy Stock Photo; 26-30: Fred Bruemmer/Photolibrary/Getty Images; 26-31: H Schmidbauer/AGE Fotostock; 26-32: Lennart Nilsson/Scanpix Sweden AB; 26-33: William P. Fifer; 26-34: Benedict Jones

Chapter 27 opener: Marc Moritsch/National Geographic Creative/Alamy Stock Photo; E27-1: Florida Fish and Wildlife Conservation Commission; 27-9a: Tom McHugh/Science Source; 27-9b: Winfried Wisniewski/age fotostock/Getty Images; 27-10: Francois Hellio & Nicolas Van Ingen/Science Source; 27-11a: Douglas Allen/E+/Getty Images; 27-11b: Gavriel Jecan/age fotostock/Alamy Stock Photo; 27-13a: Jurgen Freund/Nature Picture Library; 27-13b: Schimmelpfennig/Premium/AGE Fotostock America Inc.; 27-13c: Virunja/Shutterstock; 27-14: NASA; E27-2: David Keith Jones/Images of Africa Photobank/Alamy Stock Photo

Chapter 28 opener: B Moose Peterson/AGE Fotostock; 28-3a: U. Schanz/Juniors Bildarchiv GmbH/Alamy Stock Photo; 28-3b: Michael Steciuk/The Photolibrary Wales/Alamy Stock Photo; E28-1a: Dan Callister/Alamy Stock Photo; E28 1b: Photoshot License Limited; E28-1c: Auscape/UIG/AGE Fotostock; 28-4a: Arco Images GmbH/Alamy Stock Photo; 28-4b: Kamonrat/Shutterstock; 28-4c: Oxford Scientific/Getty Images; 28-5: Stephen Dalton/Science Source; 28-6: Cathy Keifer/Shutterstock; 28-7a: Terry Audesirk; 28-7b: Daniel L. Geiger/SNAP/Alamy Stock Photo; 28-8a: Clarence Holmes Wildlife/Alamy Stock Photo; 28-8b: Dr. Paul Zahl/Science Source; 28-8c-d: Science Source; 28-9a: Eric Dragesco/Nature Picture Library; 28-9b: Science Source; 28-10: Corbis/SuperStock; 28-11a: Millard Sharp/Science Source; 28-11b: Mark Cassino Photography/SuperStock; 28-12: Michael Boppré; 28-13a: Jany Sauvanet/Science Source; 28-13b: WILDLIFE GmbH/Alamy Stock Photo; 28-13c: Science Source; 28-14: Roger Meerts/Shutterstock; 28-15a: Custom Life Science Images/Alamy Stock Photo; 28-15b: Carol Buchanan/Dreamstime; E28-2: Centers for Disease Control and Prevention (CDC) 28-16a: tish1/Shutterstock; 28-16b: Jorn Friederich/imageBROKER/Alamy Stock Photo; 28-16c: Tom Mangelsen/Nature Picture Library; 28-17a left: Greg Vaughn/Alamy Stock Photo; 28-17a right: Steffen Foerster/Shutterstock; 28-17b left: Phil Farnes/Science Source; 28-17b right: Jim Zipp/Science Source; 28-20: Michael P. Gadomski/Science Source

Chapter 29 opener: alexmgn/Alamy Stock Photo; 29-10: NASA; 29-11: Monica Schroeder/Science Source; 29-12: Peter M. Miller/Science Source; 29-15: USGS Information Services; E29-4: NASA; E29-5: ARLAN NAEG/Stringer/Getty Images; 29-16: NOAA; 29-17: Accent Alaska.com/Alamy Stock Photo

Chapter 30 opener: Saravutpics/Shutterstock; E30-1: NASA; 30-8 top left: Ricardo Arnaldo/Alamy Stock Photo; 30-8 top center: Will E. Davis/Shutterstock; 30-8 top right: Schafer & Hill/Photolibrary/Getty Images; 30-8 bottom: David Noton Photography/Alamy Stock Photo; 30-9 top left: EcoPrint/Shutterstock; 30-9 top center: Federico Rizzato/Fotolia; 30-9 top right: Dirk Freder/E+/Getty Images; 30-9 bottom: iStockphoto/Getty Images; 30-10a: Vladimir Wrangel/Shutterstock; 30-10b: Louise A Heusinkveld/Alamy Stock Photo; 30-11a: Terry Audesirk; 30-11b: Tom McHugh/Science Source; 30-12: George H. H. Huey/Alamy Stock Photo; 30-13a: Tom McHugh/Science Source; 30-13b: Craig K. Lorenz/Science Source; 30-14: Siegfried Modola/Reuters; 30-15: Design Pics/Jack Goldfarb/Getty Images; 30-16 top: Bill Brooks/Alamy Stock Photo; 30-16 bottom left: Olivier Le Queinic/Dreamstime LLC; 30-16 bottom center: Eric Gevaert/Shutterstock; 30-16 bottom right: Alan Scheer/Shutterstock; 30-17: Terry Audesirk; 30-18 top left: Paul Tessier/Getty Images; 30-18 center left: fbosse/iStockphoto/Getty Images; 30-18 bottom left: blakisu/iStockphoto/Getty Images; 30-18 right: Alekander Bolbot/Shutterstock; 30-19 top left: iStock/Getty Images; 30-19 center left: Mark Wallace/Alamy Stock Photo; 30-19 bottom left: Haykai/Getty Images; 30-19 right: Robert Marien/Alamy Stock Photo; 30-20 top left: Tom & Pat Leeson/Science Source; 30-20 top right: Eric Baccega/age fotostock/Alamy Stock Photo; 30-20 bottom: Top-Pics Tbk/Alamy Stock Photo; 30-21: Bill Brooks/Alamy Stock Photo; 30-22 top left: Larry Trupp/Almay Stock Photo; 30-22 top center: DmitryND/E+/Getty Images; 30-22 top right: Science Source; 30-22 bottom: Lisa Dearing/Alamy Stock Photo; Have You Ever Wondered: Nastenok/Shutterstock; 30-24: Michael P. Gadomski/Science Source; 30-27a: Vince Bucci/Getty Images; 30-27b: Universal Images Group/SuperStock; 30-27c: NHPA/SuperStock; 30-27d: Vlad61/Shutterstock; 30-28 top, bottom left: WILDLIFE GmbH/Alamy Stock Photo; 30-28 bottom center: Michael Durham/Minden Pictures/National Geographic Creative; 30-28 bottom right: Jan Hinsch/Science Source; 30-29 top left: Greg Rouse; 30-29 top center: Kelvin Aitken/Image Quest Marine; 30-29 top right: Peter Batson/Image Quest Marine; 30-29 bottom: Craig R. Smith; 30-30: NOAA PMEL EOI

Chapter 31 opener: JudiLen/Getty Images; 31-1: Michael Duff/Getty Images; Have You Ever Wondered: Tim Scrivener/Alamy Stock Photo; 31-2a: Shutterstock; 31-2b: Robert Harding World Imagery; E31-1: Franco Banfi/AGE Fotostock; 31-3: Rich Carey/Shutterstock; E31-2: Richard Bierregaard; E31-4a: age fotostock/SuperStock; E31-4b: Doug Perrine/Nature Picture Library; 31-4a: Mark Smith/Science Source; 31-4b: Mike Goldwater/Alamy Stock Photo; 31-5a: Ed Reschke/Getty Images; 31-5b: Tom Nevesely/AGE Fotostock; 31-5c: Reinhard Dirscherl/Alamy Stock Photo; 31-6: Hans Blossey/Alamy Stock Photo; 31-7: Richard Ellis/Alamy Stock Photo; 31-8: Dr. George Naderman; 31-9: John Froschauer; 31-10: Martin Grosnick/AGE Fotostock

Chapter 32 opener: Paul Harding/Action Images/ZUMAPRESS.com; 32-1a: Tom McHugh/Science Source; 32-1b: Kwan Tse/Shutterstock; 32-1c: Ivan Cholakov/Shutterstock; E32-1: Russell Millner/Alamy Stock Photo; 32-4a: Ray Simons/Science Source; 32-4b: Biophoto Associates/Science Source; 32-4c: Ed Reschke/Oxford Scientific/Getty Images; 32-4d: Jose Luis Calvo/Shutterstock; 32-5a: Pearson Education, Inc.; 32-5b: Steve Gschmeissner/Science Source; 32-5c: Ed Reschke/Photolibrary/Getty Images; 32-6a-b: Ed Reschke/Photolibrary/Getty Images; 32-6c: Dr. Yorgos Nikas/Science Source; 32-7a: Innerspace Imaging/Science Source; 32-7b: Ed Reschke/Photolibrary/Getty Images; 32-7c: G W Willis/Getty Images; 32-8a: Dr. Torsten Wittmann/Science Source; 32-8b: Riccardo Cassiani-Ingoni/Science Source

Chapter 33 opener: Mark Harmet/Getty Images; 33-8a: Susumu Nishinaga/Science Source; 33-8b: Michael Ross/Science Source; 33-8c: M. D. Kenneth Kaushansky; 33-11: SPL/Science Source; 33-13: CNRI/Science Source; 33-16: Science Source; 33-20: ERIC GRAVE/Science Source; 33-21: Copyright World Health Organization (WHO), 2015. All Rights Reserved; 33-22: Texas Heart Institute www.texasheart.org

Chapter 34 opener: Josh Edelson/AFP/Getty Images; 34-1a: ANDREW MARTINEZ/Science Source; 34-1b: Terry Audesirk; 34-1c: Leslie Newman/Andrew Flowers/Science Source; 34-3: Mary Jonilonis/AGE Fotostock; Have You Ever Wondered: TsuneoMP/Shutterstock; 34-4: Valerie Giles/Science Source; 34-5a: Sinclair Stammers/Science Source; 34-5b: Design Pics/SuperStock; 34-5c: Paul Starosta/Getty Images; 34-5d: Lynn M. Stone/Nature Picture Library; E34-1a: Matt Meadows/Getty Images; E34-1b: Clinical Photography, Central Manchester University Hospitals NHS Foundation Trust, UK/Science Source

Chapter 35 opener: Craig Barritt/Getty Images; 35-2: Juergen Sohns/Alamy Stock Photo; 35-3: Mauro Fermariello/Science Source; 35-4: Dr. M.A. Ansary/Science Source; 35-5: BIOPHOTO ASSOC/Science Source; 35-7: U. S. Food and Drug Administration; 35-8a: Dennis Sabo/Shutterstock; 35-9a: Tierbild Okapia/Science Source; E35-1: Dr. E. Walker/Science Source; 35-19a: Gastrolab/Science Source; 35-19b: Science Photo Library/AGE Fotostock; 35-19d: Steve Gschmeissner/Science Source; 35-21: Lydie/SIPA/Newscom

Chapter 36 opener: Susan Watts/NY Daily News Archive via Getty Images; E36-1 left: Medicimage Ltd/AGE Fotostock; E36-1 right: Cascade News; 36-7: Pichugin Dmitry/Shutterstock; 36-9: Life in View/Science Source; 36-10: New York Daily News/Getty Images

Chapter 37 opener: Ted Thai/Getty Images; 37-1: Dr. Gary Gaugler/Science Source; 37-2: Juergen Berger/Science Source; 37-3: Science Source; E37-1a: National Institute of Allergy and Infectious Diseases (NIAID); E37-1b: James Cavallini/Science Source; 37-10: Science Source; 37-14: Science Source; 37-16: Courtesy of Aimee Copeland

Chapter 38 opener: Ethan Miller/Getty Images; 38-7: Alison Wright/Science Source; 38-10: Hilleborg/Fotolia; E38-1a: PIERRE ANDRIEU/AFP/Getty Images; E38-1b: Peter J. Thompson/UPI Photo Service/Newscom; E38-2: Jennifer A. Beck, U. S. Geological Survey; 38-11: Jackson Laboratory

Chapter 39 opener: Walter McBride/Contributor/Getty Images; 39-1: Dieter Brandner & Ginger Withers; 39-3: Steve Gschmeissner/Science Source; 39-4: Thomas Deerinck/NCMIR/Science Source; E39-1: Science Source; E39-4: Alan Cowen Case Study Revisted: Terry Audesirk

Chapter 40 opener: Life in View/Science Source; 40-4: Dr. David Furness, Keele University/Science Source; Case Study Continued: Advanced Bionics Corporation; E40-1: Masa Ushida/Stephen Frink Collection/Alamy Stock Photo; 40-6a: Steve Gschmeissner/Science Source; 40-7b: Omikron/Science Source; 40-9: Terry Audesirk; 40-10a: Gillmar/Shutterstock; 40-10b: Larsek/Shutterstock; 40-11: Steve Gschmeissner/Science Source; 40-12a: Omikron/Science Source

Chapter 41 Opener left to right: Paolo Bruno/Getty Images, Malcolm Park editorial/Alamy Stock Photo; 41-5: Biophoto Associates/Science Source; Table 41-1 left: Eric Grave/Science Source; Table 41-1 center: Ed Reschke/Getty Images; Table 41-1 right: Pearson Education Inc.; 41-8: fiftymm99/Getty Images; 41-10: SPL/Science Source; 41-11a: Wayne Schaefer; 41-11b: Ed Reschke/Getty Images; 41-11c: Michael Abbey/Science Source; E41-1a: Steve Gschmeissner/Science Source; E41-1b: Professor Pietro M. Motta/Science Source; E41-1c: Nordic Photos/Superstock; 41-13: Warden, Stuart J., Sara M. Mantila Roosaa, Mariana E. Kershc, Andrea L. Hurda, Glenn S. Fleisigd, Marcus G. Pandyc, and Robyn K. Fuchs. Physical activity when young provides lifelong benefits to cortical bone size and strength in men. vol. 111 no. 14. pp 5337–5342, doi: 10.1073/pnas.1321605111

Chapter 42 opener: Victor R. Caivano/AP images; 42-1: Stuart Westmorland/AGE Fotostock; 42-2: John Cancalosi/Getty Images; 42-3: blickwinkel/Alamy Stock Photo; 42-4: Robin Chittenden/Frank Lane Picture Agency; 42-5: Jurgen Freund/Nature Picture Library; 42-6: H. Schmidbauer/mauritius images GmbH/Alamy Stock Photo; 42-7: Nature's Images/Science Source; 42-8: iStock-photo/Getty Images; 42-10b: Susumu Nishinaga/Science Source; 42-15 left: C. Edelmann/Science Source; 42-15 right: Biophoto Associates/Science Source; E42-1a: Dr. Linda M. Stannard, University of Cape Town/Science Source; E42-1b: Eye of Science/Science Source; 42-17a: Scanpix Sweden AB; 42-17b: David M. Phillips/Science Source; E42-2: CC Studio/Science Source

Chapter 43 opener: Jane Burton/Nature Picture Library; 43-1a: Dr. Paul A. Zahl/Science Source; 43-1b: BSIP/UIG/Getty Images; 43-1c: Outdoorsman/Dreamstime LLC; 43-2a: Survivalphotos/Alamy Stock Photo; 43-2b: Tomas Kvidera/Shutterstock; 43-7b: Dr. Yorgos Nikas/Science Source; 43-10: Lennart Nilsson/TT Nyhetsbyran; 43-11: CNRI/Science Source; 43-15: B. G. Thomson/Science Source; 43-16: Ron Austing/Frank Lane Picture Agency; E43-2: Martin M. Rotker/Getty Images

Chapter 44 opener: Matt Anderson/Getty Images; Have You Ever Wondered: Terry Donnelly/Alamy Stock Photo; 44-4 top: Terry Audesirk; 44-4a: G. Guenther/blickwinkel/AGEFotostock; 44-4b: Biophoto Associates/Science Source; 44-4c: Garry DeLong/Science Source; 44-5: Deposit Photos/Glow Images; 44-9a: Emmanuel LATTES/Alamy Stock Photo; 44-9b-d: Terry Audesirk; 44-10 top: Graham Kent/Benjamin Cummings/Pearson Education, Inc.; 44-10 bottom: Ed Reschke/Getty Images; 44-11a: Alamy Stock Photo; 44-11b: Biology Pics/Science Source; 44-12a: Neale Clark/Alamy Stock Photo; 44-12b: Michael Newton/Alamy Stock Photo; 44-13a: Bernard Gagnon; 44-13b: Kenneth W. Fink/Science Source; 44-13c: Terry Audesirk; 44-13d: KentZilla; 44-14a: Lynwood M. Chace/Science Source; 44-14b: Stefan Diller/Science Source; 44-15 top: Ed Reschke/Getty Images; 44-15 bottom: M. I. Walker/Science Source; 44-16: Terry Audesirk; 44-17: Terry Audesirk; 44-18: Omikron/Science Source; 44-20: Mbogdan; 44-21: Biophoto Associates/Science Source; 44-22 left: Hugh Spencer/Science Source; 44-22 right: Power & Syred/Science Source; E44-1: Diane Cook/Len Jenshel/National Geographic Stock; 44-24: Ray Simon/Science Source; 44-26: Harvard Forest Archives, Harvard University; Case Study Revisted: Brian Scantlebury/Alamy Stock Photo

Chapter 45 opener: Karen Bleier/Getty Images; 45-2 left: Tom Joslyn/AGE Fotostock; 45-2 right: Andrew Syred/Science Source; 45-4b: Terry Audesirk; E45-1 left to right: Mediscan/Alamy Stock Photo, WILDLIFE GmbH/Alamy Stock Photo, Yevgeni Kacnelson/Alamy Stock Photo; 45-5: Terry Audesirk; 45-7: Ami Images/Science Source; 45-8: Terry Audesirk; 45-11 left: David Rhodes; 45-11 right: Terry Audesirk; 45-14: Terry Audesirk; 45-15b: Ted Kinsman/Science Source; 45-16: Nuridsany et Perennou/Science Source E45-2a: John Warburton Lee/SuperStock; E45-2b: joel zatz/Alamy Stock Photo; E45-3: Merlin D. Tuttle/Science Source; 45-17: Anthony Mercieca/Science Source; 45-18: Veronica Carter/Alamy Stock Photo; 45-19: Philippe Clement/Nature Picture Library; 45-19 inset: Gilbert S. Grant/Science Source; 45-20: Scott Camazine/Science Source; 45-21: Doris Dumrauf/Alamy Stock Photo; Case Study Continued left to right: Les Gibbon/Alamy Stock Photo, J & C Sohns/AGE Fotostock; 45-22a: Dirk V. Mal-linckrodt/Alamy Stock Photo; 45-22b: i4images premium/Alamy Stock Photo; 45-22c: Suzanne Long/Alamy Stock Photo; 45-23: Marc Gibernau

Chapter 46 opener: Cathy Keifer/Shutterstock; 46-1: Eiichi Tanimoto; 46-2: Anton Foltin/Shutterstock; 46-3c: Martin Shields/Alamy Stock Photo; 46-3d: Terry Audesirk; 46-4: Michael L. Evans; E46-5: David McNew/Getty Images; 46-5: Terry Audesirk; 46-6: Terry Audesirk; 46-8: Terry Audesirk; 46-10: Regents of the University of California; 46-11: John Cancalosi/age fotostock/Getty Images; 46-12: Biophoto Associates/Science Source; 46-14: Ed Reschke/Getty Images; 46-15: Reinhard Dirscherl/Alamy Stock Photo; 46-16: Hermann Eisenbeiss/Science Source; 46-17: Terry Audesirk

Index

Note: Citations followed by *b* refer to material in boxes; *f* refers to figures or illustrations; and *t* refers to tables.

C

F

H

J

K

L

O

P

U

V